A

TOPOGRAPHICAL DICTIONARY

OF

ENGLAND,

COMPRISING THE

SEVERAL COUNTIES, CITIES, BOROUGHS, CORPORATE AND MARKET TOWNS,

PARISHES, CHAPELRIES, AND TOWNSHIPS,

AND THE ISLANDS OF GUERNSEY, JERSEY, AND MAN,

WITH

HISTORICAL AND STATISTICAL DESCRIPTIONS;

ILLUSTRATED BY

MAPS OF THE DIFFERENT COUNTIES AND ISLANDS;

A Map of England,

SHEWING THE PRINCIPAL TOWNS, ROADS, RAILWAYS, NAVIGABLE RIVERS, AND CANALS;

AND A

PLAN OF LONDON AND ITS ENVIRONS;

AND EMBELLISHED WITH

ENGRAVINGS OF THE ARMS OF THE CITIES, BISHOPRICS, UNIVERSITIES, COLLEGES, CORPORATE TOWNS, AND BOROUGHS; AND OF THE SEALS OF THE SEVERAL MUNICIPAL CORPORATIONS.

BY SAMUEL LEWIS.

FOURTH EDITION.

IN FOUR VOLUMES.

VOL. IV.

LONDON:

PUBLISHED BY S. LEWIS & Co., 87, ALDERSGATE STREET.

MDCCCXL.

LONDON:
GILBERT AND RIVINGTON, PRINTERS,
ST. JOHN'S SQUARE.

A

TOPOGRAPHICAL DICTIONARY

OF

ENGLAND.

SABRIDGEWORTH, county of Hertford. — See SAWBRIDGEWORTH.

SACOMB (*St. Mary*), a parish, in the hundred of Broadwater, union and county of Hertford, 4 miles (N. by W.) from Ware; containing 360 inhabitants. The living is a rectory, valued in the king's books at £10. 3. 4.; patron, Samuel Smyth, Esq. The tithes have been commuted for a rent-charge of £335. 15., subject to the payment of rates, which on the average have amounted to £41.; the glebe comprises 11 acres, valued at £16. 10. per annum. The church is situated upon an eminence on the north side of the Ware and Wotton road; it has a tower on the south side of the nave, formerly embattled, and in the chancel two stone stalls, and a piscina under trefoil arches, with memorials of the Rolt family. A school is partly supported by Abel Smith, Esq. There is a bequest of £5 per annum by the Rev. John Meriton, late rector, for apprenticing one poor child.

SADBERGE, a chapelry, in the parish of Haughton-le-Skerne, union of Darlington, South-Western Division of Stockton ward, Southern Division of the county palatine of Durham, 4¼ miles (E. N. E.) from Darlington; containing 403 inhabitants. This was formerly a place of great importance, and the capital of a district, or county, of the same name, having its gaol, sheriff, coroner, and other civil officers. The Stockton and Darlington railway passes in the vicinity. The chapel is dedicated to St. Andrew, and contains 190 free sittings, the Incorporated Society having granted £120 in aid of the expense. Around the base of the hill upon which it stands are traces of an ancient intrenchment. A school on the National plan is partly supported by subscription and a small endowment.

SADDINGTON (*St. Helen*), a parish, in the union of Market-Harborough, hundred of Gartree, Southern Division of the county of Leicester, 6 miles (N. W. by W.) from Market-Harborough; containing 268 inhabitants. The living is a rectory, valued in the king's books at £19. 2. 6.; present net income, £280: it is in the patronage of the Crown. A day and Sunday school is supported by subscription. The Union canal passes through a tunnel partly in this parish and partly in that of Kibworth-Harcourt.

SADDLEWOOD, a tything, in the parish of Hawkesbury, Upper Division of the hundred of Grumbalds-Ash, Western Division of the county of Gloucester, 5 miles (W. S. W.) from Tetbury: the population is included in the return for Killcott.

SADDLEWORTH, a chapelry, in the parish of Rochdale, Upper Division of the wapentake of Agbrigg, West Riding of the county of York, 12 miles (S. W. by W.) from Huddersfield; containing 15,986 inhabitants. The Huddersfield canal passes through the chapelry, and the manufacture of woollen and cotton goods is carried on to a very great extent; the number of looms employed in the former exceeding 3500, and in the latter 400: there are more than 100 mills on the river Tame and its tributary streams. A few coal mines are worked, and excellent freestone abounds within the chapelry. The living is a perpetual curacy; patron, Vicar of Rochdale. The chapel, dedicated to St. Chad, has been enlarged, and 410 free sittings provided, the Incorporated Society having granted £400 in aid of the expense. There are five places of worship for dissenters. A free school was founded in 1729, by Ralph Hawkyard, who endowed it with £280; and, in augmentation of the master's salary, John Walker, in 1755, bequeathed £200. There are some interesting natural curiosities at Greenfield, consisting of huge caverns, rocks, and a stupendous rocking-stone, with many Druidical remains. Castle Shaw is said to have been a fortress of the Britons; round beads, similar to those contained in the barrows on Salisbury plain, and a brazen celt, having been discovered near it.

SAFFRON-WALDEN, county of Essex. — See WALDEN, SAFFRON.

SAHAM-TONEY (*St. George*), a parish, in the union of Swaffham, hundred of Wayland, Western Division of the county of Norfolk, 1¾ mile (N. W.) from Watton; containing 1060 inhabitants. The living is a rectory, valued in the king's books at £21. 19. 4½.; present net income, £796; patrons, Warden and Fellows of New College, Oxford. The tower of the church, erected about 1480, has upon it a representation of St. George and the Dragon, carved in stone. There is a place of worship for Wesleyan Methodists; also a National school for girls.

SAIGHTON, a township, in the chapelry of BUERTON, parish of ST. OSWALD, CHESTER, union of GREAT BOUGHTON, Lower Division of the hundred of BROXTON, Southern Division of the county of CHESTER, 4½ miles (S. E.) from Chester; containing 303 inhabitants. A school is partly supported by private charity.

SAINTBURY (*ST. NICHOLAS*), a parish, in the union of EVESHAM, Upper Division of the hundred of KIFTSGATE, Eastern Division of the county of GLOUCESTER, 2¼ miles (W.) from Chipping-Campden; containing 123 inhabitants. The living is a rectory, valued in the king's books at £19. 9. 4½.; present net income, £415; patron, James Roberts West, Esq., of Alscot Park. The church has undergone various alterations, but there is still a Norman door remaining. A Sunday school is supported by J. R. West, Esq. Castle Bank, an ancient camp in this parish, is ascribed to the Danes, and supposed to have been dependent upon a larger one upon the summit of the same hill, in the adjoining parish of Willersey.

SALCOMBE, a chapelry, in the parish of MALBOROUGH, union of KINGSBRIDGE, hundred of STANBOROUGH, Stanborough and Coleridge, and Southern, Divisions of the county of DEVON, 5 miles (S.) from Kingsbridge: the population is returned with the parish. This place, which, from the mild temperature of its climate and the salubrity of the air, is called the Montpelier of England, is much visited by tourists for the picturesque beauty of its scenery, and recommended by medical men as a place of resort for consumptive patients. The village, of which the population within the last 35 years has increased threefold, and which has become a place of very considerable trade, is pleasantly situated on the western side of the entrance to Kingsbridge harbour, for which it is one of the principal stations; the houses in general are well built and of pleasing appearance, and the neighbourhood contains several handsome villas and marine residences. There are 50 vessels belonging to this place, of which 30 are schooners of the first class, and are employed principally in the fruit and coasting trade. Passage vessels sail every day to Plymouth. Ship-building is carried on to some extent, there being four yards for that purpose. A peculiar kind of beer, called white ale, is brewed here: a pleasure fair is annually held at Whitsuntide. The chapel, which was originally erected prior to the year 1401, was rebuilt in 1801 by subscription, and is served by a curate appointed by the Vicar of West Alvington. There are places of worship for Baptists and Wesleyan Methodists. Some remains of an ancient castle, of which few particulars have been recorded, may be traced.

SALCOMBE, REGIS (*ST. PETER AND ST. MARY*), a parish, in the union of HONITON, hundred of EAST BUDLEIGH, Woodbury and Southern Divisions of the county of DEVON, 2 miles (E. N. E.) from Sidmouth; containing 448 inhabitants. It is bounded on the south by the English Channel, and was anciently held in royal demesne. The living is a discharged vicarage, in the patronage of the Dean and Chapter of Exeter (the appropriators), valued in the king's books at £14. 12. 8.; present net income, £143. There was formerly a chapel, dedicated to St. Clement and St. Mary Magdalene. Gypsum and chalk for lime are obtained here.

SALCOTT (*ST. MARY*), a parish, in the union of LEXDEN and WINSTREE, hundred of WINSTREE, Northern Division of the county of ESSEX, 8½ miles (S. S. W.) from Colchester; containing 154 inhabitants. This parish, called also Salcot Verley, from one of the proprietors of the manor, is bounded on the south by a creek, which separates it from the parish of Great Wigborough, and comprises about 2000 acres of marshy ground. Verley Hall, the mansion-house of the ancient manor, is near the church; and the village is pleasantly situated on the south bank of the Verley Channel, which is navigable to this place. The living is a rectory; net income, £112; it is in the patronage of Mrs. Clive. The church is a small ancient edifice.

SALDEN, a hamlet, in the parish of MURSLEY, union of WINSLOW, hundred of COTTESLOE, county of BUCKINGHAM, 4½ miles (E. N. E.) from Winslow: the population is returned with the parish.

SALE, a township, in the union of ALTRINCHAM, parish of ASHTON-UPON-MERSEY, though locally in that of Great Budworth, hundred of BUCKLOW, Northern Division of the county of CHESTER, 2 miles from Ashton-upon-Mersey; containing 1104 inhabitants. The Duke of Bridgewater's canal passes through the township.

SALEBY (*ST. MARGARET*), a parish, in the union of LOUTH, Wold Division of the hundred of CALCEWORTH, parts of LINDSEY, county of LINCOLN, 1½ mile (N. by E.) from Alford; containing 220 inhabitants. The living is a vicarage, valued in the king's books at £4; present net income, £227; patrons, Trustees of Alford free grammar school. A school is endowed with £6 per annum, the bequest of the Rev. T. Falkner, in 1762, and is further entirely supported by the vicar, who also supports two Sunday schools.

SALEHURST (*ST. MARY*), a parish, in the union of TICEHURST, hundred of HENHURST, rape of HASTINGS, Eastern Division of the county of SUSSEX, 6 miles (N.) from Battle; containing 2204 inhabitants. The living is a vicarage, valued in the king's books at £14; present net income, £503; patron, H. Winchester, Esq.; impropriator, Sir Sotherton B. P. Micklethwait, Bart. The church exhibits portions in the early and later styles of English architecture. A school is supported partly by an endowment, in 1829, by Geo. Munn, Esq., of £16. 17. 8. per annum, and partly by subscription. Here is a National school.

SALESBURY, a chapelry, in the parish, union, and Lower Division of the hundred, of BLACKBURN, Northern Division of the county palatine of LANCASTER, 4½ miles (N.) from Blackburn; containing 433 inhabitants. The living is a perpetual curacy; net income, £118; patron, Lord de Tabley, who supports a school for children of both sexes.

SALFORD (*ST. MARY*), a parish, in the union of WOBURN, hundred of MANSHEAD, county of BEDFORD, 4½ miles (N. by W.) from Woburn; containing 340 inhabitants. The living is a discharged vicarage, united in 1750 to the rectory of Holcutt, and valued in the king's books at £7. 16. 3. There is a place of worship for Wesleyan Methodists; and two Sunday schools are supported by Miss Eliza Deane.

SALFORD, county palatine of LANCASTER.—See MANCHESTER.

SALFORD (*St. Mary*), a parish, in the union of CHIPPING-NORTON, hundred of CHADLINGTON, county of OXFORD, 2 miles (W. N. W.) from Chipping-Norton; containing 341 inhabitants. The living is a rectory, valued in the king's books at £9. 11. 3.; present net income, £251; patron, Nash Skillicorne, Esq. There are some stone quarries, and the remains of two ancient crosses in the parish.

SALFORD (*St. Matthew*), a parish, in the union of ALCESTER, Stratford Division of the hundred of BARLICHWAY, Southern Division of the county of WARWICK, 5¾ miles (S. by W.) from Alcester; containing 899 inhabitants. The living is a vicarage, valued in the king's books at £9; present net income, £111; patron and impropriator, Sir Grey Skipwith, Bart. William Perkins, in 1656, gave £232 for the support of a free grammar school; the annual income, now upwards of £40, is applied to instructing, in English, the children of parishioners; the classics were formerly taught, but they have of late years been discontinued. Another school is partly supported by the vicar, and partly by small payments from the children; the mistress is allowed a salary of £15 annually. There are also two Sunday schools, supported by voluntary contributions, one in connection with the Established Church, the other appertaining to Dissenters of the Baptist denomination. An ancient mansion, the property of Mr. Berkeley, is now occupied as a nunnery, the society consisting of an abbess, sixteen professed nuns, and a school for young ladies, noviciates. The river Avon and its tributary stream, the Arrow, run through the parish.

SALHOUSE (*All Saints*), a parish, in the union of ST. FAITH, hundred of TAVERHAM, Eastern Division of the county of NORFOLK, 4¼ miles (S. E. by S.) from Coltishall; containing 539 inhabitants. The living is a discharged vicarage, united to that of Wroxham. There is a place of worship for Baptists.

SALING, LITTLE, or BARDFIELD, county of ESSEX.—See BARDFIELD-SALING.

SALING, GREAT (*St. James*), a parish, in the union of BRAINTREE, hundred of HINCKFORD, Northern Division of the county of ESSEX, 5 miles (N. W. by W.) from Braintree; containing 367 inhabitants. This parish, which formerly included also the parish of Little Saling, is about two miles in length and nearly of equal breadth; it is intersected by a rivulet which rises in the parish of Great Bardfield, and falls into the river Blackwater. The soil is various, but generally fertile, and the lands are in a state of profitable cultivation. Saling Hall is a handsome mansion, with extensive gardens and pleasure grounds; and Saling Grove, also situated near the village green, is a handsome residence. About half a mile from the church is a mansion called Parks. The village is pleasantly situated on a green of triangular form, comprising about five acres. The living is a discharged vicarage, valued in the king's books at £7; present net income, £148; patron, incumbent, and impropriator, Rev. Bartlet Goodrich. The church, a small ancient edifice with a tower, supposed to have been erected in the reign of Henry II., contains monuments to the Yeldham, Goodrich, and Sheddon families. A National school is supported by subscription.

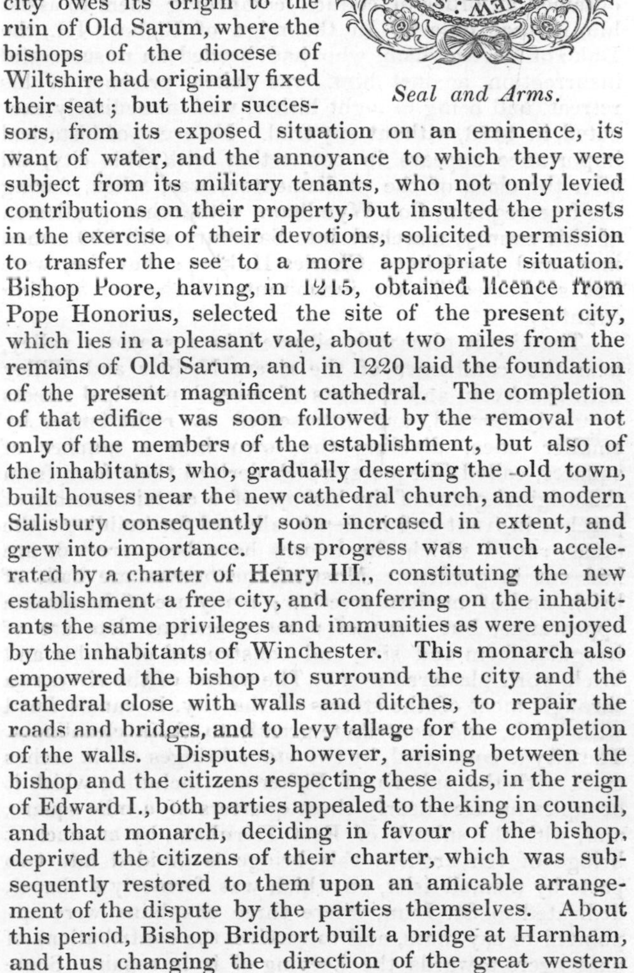

Seal and Arms.

SALISBURY, a city, having separate civil jurisdiction, locally in the hundred of Underditch, Southern Division of the county of WILTS, 82 miles (S. W. by W.) from London; containing 9876 inhabitants. This city owes its origin to the ruin of Old Sarum, where the bishops of the diocese of Wiltshire had originally fixed their seat; but their successors, from its exposed situation on an eminence, its want of water, and the annoyance to which they were subject from its military tenants, who not only levied contributions on their property, but insulted the priests in the exercise of their devotions, solicited permission to transfer the see to a more appropriate situation. Bishop Poore, having, in 1215, obtained licence from Pope Honorius, selected the site of the present city, which lies in a pleasant vale, about two miles from the remains of Old Sarum, and in 1220 laid the foundation of the present magnificent cathedral. The completion of that edifice was soon followed by the removal not only of the members of the establishment, but also of the inhabitants, who, gradually deserting the old town, built houses near the new cathedral church, and modern Salisbury consequently soon increased in extent, and grew into importance. Its progress was much accelerated by a charter of Henry III., constituting the new establishment a free city, and conferring on the inhabitants the same privileges and immunities as were enjoyed by the inhabitants of Winchester. This monarch also empowered the bishop to surround the city and the cathedral close with walls and ditches, to repair the roads and bridges, and to levy tallage for the completion of the walls. Disputes, however, arising between the bishop and the citizens respecting these aids, in the reign of Edward I., both parties appealed to the king in council, and that monarch, deciding in favour of the bishop, deprived the citizens of their charter, which was subsequently restored to them upon an amicable arrangement of the dispute by the parties themselves. About this period, Bishop Bridport built a bridge at Harnham, and thus changing the direction of the great western road, which formerly passed through Old Sarum, that place was completely deserted, and Salisbury became one of the most flourishing cities in the kingdom. Edward I. assembled a parliament here, to deliberate upon measures for recovering the province of Gascoigne, which had been seized upon by Philip of France; at this parliament none of the clergy assisted, the king having suspended them from the exercise of their secular functions for having refused him aid. In the reign of Edward III., a second parliament, for inquiring into the state of the kingdom, was held at Salisbury, at which Mortimer, Earl of March, and his partisans, were attended by their followers in arms: the Earls of Kent, Norfolk, and Lancaster, who, on being summoned to attend this parliament, were prohibited by Mortimer from appearing with an armed force, finding, on their arrival, that his own partisans were armed, retreated for the purpose of assembling their retainers, and, returning with an army, were about to take vengeance on Mortimer, when the quarrel was compromised through the intervention of the clergy. From the time of Edward I., the bishops and the citizens appear to have lived in a state of mutual harmony, till the reign of Richard II., when the prelate requiring the corporation to concur with him in his efforts to suppress the meetings of the Lollards, who assembled here in great numbers, the latter refused, and the bishop, appealing to the king, obtained an order in council compelling them to assist him in that object. In the reign of Richard III., the Duke of Buckingham, who had headed an unsuccessful insurrection against him, was taken prisoner in his retreat, and being brought hither, was immediately executed, in 1483, without any trial. No event of historical importance appears in connection with this city, till after the close of the parliamentary war, when, during the interregnum, Col. Wyndham, with other gentlemen of this county, marched into Salisbury with 200 armed men, and proclaimed Charles II. king; but they were not supported by the inhabitants of the surrounding country.

The city is pleasantly situated in a spacious valley, near the confluence of the rivers Nadder and Willey with the Avon, and consists of several principal streets regularly formed, and intersected at right angles by smaller ones, dividing the town into a number of squares, called Chequers, which derived their form from the original grant of the bishops of a certain number of perches in front and in depth allotted for building; the areas, round which the houses have been erected, are laid out in gardens. Most of the houses are built of brick, and of modern erection; many are of handsome appearance, but some of a more ancient date are of irregular form and size, and constructed of timber and brick-work plastered over. The waters of the rivers run through most of the streets of the city, in canals lined with brick, and contribute greatly to their cleanliness. The city is connected, by two stone bridges of six arches each, with the suburb of Fisherton, including which it occupies an area nearly three quarters of a mile square, and, with the suburb of East Harnham, by an ancient bridge of ten arches: this bridge is divided into two parts by a small islet, on which was formerly a chapel, dedicated to St. John, where three chaplains were appointed to say mass, and to receive the contributions of passengers towards the keeping of it in repair. Some improvement has lately taken place in paving and lighting the town, wich is amply supplied with water. The Salisbury and Wiltshire library and news-room was established in 1819, and is supported by a proprietary and by annual subscriptions; the library contains more than 2000 volumes, in the various departments of literature, and annexed to it is a small museum. A small neat theatre is opened for some months in the winter; and assemblies and concerts are held occasionally. Races take place annually in August, and are in general well attended. Salisbury was formerly celebrated for its manufactures of flannels, druggets, and the cloths called Salisbury Whites; but these branches of trade are now almost extinct, and what remains is confined to a very inconsiderable number of persons: the town is still noted for its manufacture of the more select articles of cutlery of superior quality, the sale of which is very limited; and a silk-factory, employing about 120 persons, has been lately established. The Salisbury canal, joining with the Andover canal near Romsey, was originally intended to be continued westward to Bath and Bristol, connecting the Bristol and English Channels, but the design was abandoned, barges at present proceeding no further in this direction than Romsey, though it has a communication with Southampton and the English Channel. The market days are Tuesday and Saturday; the former for corn, of which there is an abundant supply, and the latter for cheese and all kinds of provisions; there is also a large cattle market every alternate Tuesday. The fairs, which are falling into disuse, are on the Tuesday after January 6th, for cattle; Tuesday after the 25th of March, for cloth; Whit-Monday and Tuesday, for horses and pedlery; and October 20th, for butter and cheese. The poultry cross, which appears to have been built in the reign of Edward III., and of which only the lower part is remaining, is situated without the south-west corner of the market-place, an extensive quadrilateral area, well arranged for the general uses of the market.

The first charter granted to the city was by Henry III., in the eleventh year of his reign; this was subsequently confirmed and altered by several succeeding sovereigns, but the government, previously to the passing of the Municipal Reform Act, was wholly regulated by the charters bestowed by James I., Charles I. and II., and Anne. Under these the corporation consisted of a mayor, high steward, recorder, deputy recorder, twenty-four aldermen, and thirty assistants, aided by a town-clerk, two chamberlains, a clerk of the peace, three serjeants-at-mace, &c. The justices of the peace were the mayor, late mayor, recorder and his deputy, and ten aldermen who had passed the chair. The government is now vested in a mayor, six aldermen, and eighteen councillors, by the act of the 5th and 6th of William IV., cap. 76, for an abstract of which see the Appendix, No. I. The municipal boundaries are by this act made co-extensive with those for parliamentary purposes, and the city is divided into three wards, instead of four as before. The mayor and late mayor are justices of the peace by virtue of their office, and five other magistrates are appointed by the crown. The police force consists of a superintendent and eleven constables. The city exercised the elective franchise in the 23rd of Edward I., since which time it has regularly returned two members to parliament. The right of election was formerly vested in the mayor and corporation: the borough has been extended, and now comprises an area of 601 acres, the limits of which are minutely described in the Appendix, No. II.: the old borough contained only 166 acres. The present number of voters registered under the Reform Act is 670, of whom 34 are freemen: the mayor is the returning officer. The recorder holds quarterly courts of session. On the part of the bishop are a bailiff and deputy-bailiff, who hold a court of record, for the recovery of debts to any amount, the jurisdiction of which extends over the city and the cathedral close, but no process has issued from it for the last ten years; they also hold a court leet for the bishop, as lord of the manor. The bailiff and deputy-bailiff are both appointed by the bishop, and, by charter, are chosen for the same offices on behalf of the corporation: their powers resemble those of the sheriff of the county, but the clerk of the peace for the borough is by the statute of William IV. made summoning officer of jurors. The bishop is not entitled to a place in the common chapter, which consists of the dean and the residentiary canons only; but in an extraordinary, or what is commonly called a general, chapter, which is composed of all the members, the bishop is entitled to a seat, as prebendary of Pottern. The spring assize and the Lent quarter session for the county are regularly held here, and petty sessions are held every Monday. Salisbury is the place of election for the southern division of the shire. The council-house, having been destroyed by fire, was rebuilt in 1795, under the provisions of an act of parliament, at the expense of the late Earl of Radnor, on the site of the ancient guildhall; it is a substantial and handsome building of white brick, ornamented with rustic quoins and cornices of stone, and consists of two wings, connected by a central vestibule. The right wing is occupied entirely by the council-chamber and other offices, in which the business of the corporation is transacted, and the public entertainments are held. The left wing comprises the court-rooms, in which are held the sessions for the city, the assize and session for the county, and the bishop's court; above is a grand jury room and other offices. The building at first erected, being found too small for holding the assize, was, in 1819, enlarged by public subscription. The county gaol and bridewell, a substantial and spacious edifice, erected in 1818, at the western extremity of Fisherton-Anger, at an expense of about £30,000, comprises ten wards, ten day-rooms, and ten airing-yards, for the classification of prisoners.

The seat of the diocese was originally established at Wilton, in this county, about the beginning of the tenth century, where it remained under the superintendence of eleven successive bishops, of whom Hermannus, the last, having been appointed to the see of Sherborne, annexed that bishoprick to Wilton, and founded, for the united sees, a cathedral church at Old Sarum, which was afterwards completed by Osmund, who accompanied William the Conqueror into England, and was by that monarch appointed bishop. The see remained at Old Sarum till the year 1217, when Richard le Poore transferred the episcopal chair to Salisbury in 1220, where it has ever since remained. Under the provisions of the act of the 6th and 7th of William IV., cap. 77, a considerable alteration will be made in the territorial extent of this diocese: the deaneries of Cricklade and Malmesbury will be transferred to the diocese of Gloucester and Bristol; and the county of Berks, to the diocese of Oxford; and the county of Dorset will be added to this diocese. The establishment at present consists of a bishop, dean, precentor, chancellor, treasurer, six canons residentiary, who are also prebendaries, three archdeacons (for Berks, Sarum, and Wilts), subdean, succentor, 38 prebendaries, four priest vicars, six singing men, eight choristers, organist, and other officers. The cathedral, dedicated to the Blessed Virgin Mary, begun by Richard le Poore in 1220, and completed in 1258, is one of the most magnificent and interesting ecclesiastical edifices in the kingdom. It is in the form of a double cross, with a highly enriched tower rising from the intersection of the nave and larger transepts, and surmounted by a lofty spire rising to the height of 400 feet from the pavement, being the highest in England: the whole building, with the exception only of the upper part of the tower, and the spire, which are of later date, are in the purest style of early English architecture. The west front is divided into five compartments by buttresses ornamented with canopied niches filled with statues: between the two central buttresses is the principal entrance through a richly moulded arch of spacious dimensions, with a smaller on each side: above the entrance is a large and beautiful window, and at the angles of the front are square embattled towers beautifully enriched, crowned with angular pinnacles and surmounted by spires. The north front is of considerable beauty, and the end fronts of the transepts, projecting boldly from the sides of the main building, and displaying, in successive series of arches, a pleasing variety of composition, corresponding with the general style, are a fine relief to the exterior. The interior, of which the perspective is impressively striking, is exquisitely beautiful, from the loftiness of its elevation and the delicacy and lightness of its structure; the nave is separated from the aisles by a handsome range of ten clustered columns and finely pointed arches; the roof, which is plainly groined, is 84 feet high, and the space above the columns is occupied by a triforium of elegant design, and a range of clerestory windows of three lights, of which the central is higher than the rest, and which is continued round the whole extent of the building; the larger transepts, of the same character with the nave, consist of three arches of similar arrangement, and the smaller of two arches. The choir, separated from the nave by a screen of modern workmanship supporting the organ, which was the gift of his Majesty George III., consists of seven arches, and, by the removal of the altar-screen, has been connected with the Lady chapel, of which the roof, being lower than that of the choir, in a great degree destroys the effect: the bishop's throne, the pulpit, and the prebendal stalls, are of finely-executed tabernacle-work, and harmonise with the prevailing character of the building: the floor is of black and white marble; the east window is embellished with a painting of the Resurrection, by Eginton, from a design by Sir Joshua Reynolds; the choir is also ornamented with a painting of the Elevation of the Brazen Serpent in the Wilderness, from a design by Mortimer, executed by Pearson, the gift of the late Earl of Radnor; and many of the other windows are painted in Mosaic. The cathedral was repaired, under the superintendence of Mr. Wyatt, at an expense of £26,000. The chapels in the transepts have been removed, and their principal ornaments have been distributed in various parts of the building. In the nave, choir, and transepts, are numerous monuments to the bishops of the see, among which are those of Bishops Joceline and Roger, the latter perhaps the earliest specimen of monumental sculpture extant; also of a chorister bishop, one of the children of the choir, who died while personating the character of a bishop, according to custom, during the festival of St. Nicholas; exclusively of several to the Earls of Salisbury, and the neighbouring nobility and gentry. The cloisters

Arms of the Bishoprick.

are the largest and most magnificent of any in the kingdom, and the cathedral close has some entrance gateways of ancient character and of beautiful design. The chapter-house, of an octagonal form, of which the roof is supported by one central clustered column, is a beautiful building lighted by lofty windows; the frieze is ornamented with subjects from the sacred writings in bas relief, which are in tolerable preservation. The episcopal palace is the work of different times, and combines various styles of architecture; a considerable portion was added by the late Dr. Shute Barrington: it contains portraits of nearly all the modern prelates of the see.

The city comprises the parishes of *St. Edmund*, *St. Martin*, and *St. Thomas*, and the extra-parochial district of the Cathedral Close. The living of St. Edmund's is a perpetual curacy; net income, £176; patron, the Bishop. The church, formerly collegiate, is a handsome structure, in the later style of English architecture, with a tower, which, having fallen down in 1653, was rebuilt in a style of appropriate character: the interior is neatly arranged, but the chancel has been modernized: at the east end is a beautiful painted window of the Ascension, by Eginton, the gift of the late Samuel Whitchurch, Esq. The living of St. Martin's is a vicarage, valued in the king's books at £11. 3. 1½.; present net income, £188; patrons and appropriators, Dean and Chapter of Salisbury. The church is a spacious structure, combining different styles of architecture, with a tower surmounted by a spire; the chancel is in the early, and other parts in the decorated and later, styles of English architecture. The living of St. Thomas' is a perpetual curacy; net income, £118; patrons and appropriators, the Dean and Chapter. A Lectureship was founded in the year 1617 by Christopher Eyre, which has an endowment of £20 per annum. The church is a spacious and handsome structure in the later style of English architecture, with a tower on the south side of the south aisle; it has been enlarged, and 310 free sittings provided, the Incorporated Society having granted £200 in aid of the expense; the nave is lighted with a handsome range of clerestory windows; the chancel and other parts are specimens of considerable merit: among the monuments is one supposed to be that of the Duke of Buckingham, who was executed here, in the reign of Richard III. There are two places of worship for Independents, and one each for Baptists, Wesleyan Methodists, Unitarians, and Roman Catholics. The grammar school, in the Close, is for the education of eight boys: among the scholars educated in it was Addison, the poet and essayist, who was a native of Milston, near Amesbury, in this county, of which parish his father was rector. The city grammar school was founded by Queen Elizabeth, in 1540, for the education of the sons of citizens, and is under the control of the mayor and commonalty: it is endowed with £15. 9. 1. crown rents, and £10. 12. 2., previously appropriated to the schools at Trowbridge and Bradford, in lieu of which it was founded. The Lectureship of St. Thomas' is appropriated to the mastership, in addition to the above rents; and an usher, who has no salary, receives £1. 1. from each pupil, for reading, writing, and arithmetic, in addition to classical instruction, with residence, and permission to take pay scholars. A school, in which eight orphan females are educated, was founded by the Rev. John Talman, in 1755, who left a house in High-street, St. Thomas', for a mistress to teach reading and needle-work for the term of three years, with liberty to take pay scholars. Another school also is endowed by the late Charles Godolphin, Esq., for the maintenance and education of eight orphans, daughters of poor gentlemen, with whom the mistress is allowed £280 per annum, and £30 annually for house rent. A school for the choristers of the cathedral is supported by the bishop; and a National central and two Sunday National schools are supported by subscription: in the former, which is connected with the Parent Society, 208 boys and 135 girls are instructed.

The infirmary, a spacious and commodious brick building, near Fisherton bridge, owes its origin to Lord Feversham, who bequeathed £500 to the first institution of the kind which should be established in the county, and is further liberally supported by contributions. The College of Matrons was founded in 1683, for the maintenance of the widows of ten poor clergymen, by Seth Ward, bishop of the diocese, who endowed it with property producing £200 per annum; the income has been much augmented by the increased value of the original property; a bequest by William Benson Earle, in 1794; and some subsequent donations: the buildings are within the Close, and the establishment is under the direction of the Bishop, and the Dean and Chapter, who elect the matrons, to each of whom a handsome pension is allowed. Bishop Richard le Poore founded, near Harnham bridge, an hospital for a master, eight aged men, and four women, which was completed by his successor, Bishop Bingham: it is occupied by a master, six aged men, and six women. Trinity Hospital was founded in 1379 by Agnes Boltenham, and augmented in 1397 by John Chandler; it was placed upon its present foundation by charter of James I., and the endowment has since received several additions from W. Chiffinch, W. Moulton, and others. Among other similar establishments are Bricket's hospital, in Exeter-street, founded in 1534, for six aged men or women; Eyre's hospital in Winchester-street, founded in 1617, for six men and their wives; Blechynden's hospital, in Green Croft-street, founded in 1683, for six aged widows; Taylor's hospital, in Bedwin-street, founded in 1698, and the endowment subsequently augmented by Mathew Best and Francis Swanton, for six aged men; and Frowd's hospital, in Rolleston-street, founded in 1750, for six aged men and six women. There are also several other almshouses for the residence of poor people, of which the principal are three in St. Ann's-street, the bequest of Robert Sutton; six in Culver-street, supposed to be the donation of Bishop Poore; twenty-six in Bedwin-street, the gift of Madame Menks; and thirteen in Castle-street, presented to the corporation by William Hussey, Esq., in 1794, for the use of the poor: these last were subsequently endowed by the will of the donor, and are occupied by men and their wives. There are various charitable bequests for apprenticing poor children, and for distribution among the poor. The regulation of the affairs of the poor is under a local act, which embraces the three parishes: the Close is in the union of Alderbury.

A college was founded here by Egidius de Bridport, in 1260, in which many of the students who had retired from Oxford, in consequence of their quarrel with Otho, the Pope's legate, in 1238, afterwards continued their studies: and there were formerly remains of a monastery of Grey friars, founded by the Bishop of Salisbury, in the reign of Henry III., on a site of ground given by that monarch; of a convent of Black friars, to which Edward I., if not the founder, was at least a considerable benefactor; and of the hospital of St. Michael, and the college of St. Erith. The neighbourhood abounds with flints found in the alluvial soil, and with strata of chalk; these flints, both the nodular from the chalk, and the fractured found in the gravel, yield a variety of organic remains of the *spongia* and *alcyonia genera*; several valuable collections have been formed in the neighbourhood, and it has been reckoned that there are not less than twelve distinct species of that submarine substance. Among the eminent natives of this city were Walter Winterton, Cardinal of St. Sabric; William Herman, author of several works in prose and verse; John Thornborough, Bishop of Worcester; George Coryate, author of "The Crudities;" Michael Muschant, an able civilian and poet: Sir Toby Matthews, a celebrated Jesuit and politician; Dr. Thomas Bennet, a noted divine and writer; Thomas Chubb and John Eden, distinguished controversial writers; John Greenhill, a celebrated portrait-painter; William and Henry Lawes, musicians and composers; Dr. Harris, an eminent historian and biographer; James Harris, author of "Hermes;" and John Tobin, author of "The Honeymoon," and other dramatic works. Salisbury gives the title of Marquess to the family of Cecil.

SALKELD-GATE, county of CUMBERLAND.—See PLUMPTON-WALL.

SALKELD, GREAT (*St. Cuthbert*), a parish, in the union of PENRITH, LEATH ward, Eastern Division of the county of CUMBERLAND, 3 miles (S. by W.) from Kirk-Oswald; containing 445 inhabitants. The living is a rectory, valued in the king's books at £22. 10. 10.; present net income, £345; patron, Bishop of Carlisle. The tower of the church, which appears to have contained four rooms above each other, was formerly resorted to as a place of security: under it is a dungeon. There are places of worship for Presbyterians and Primitive Methodists. This parish is intersected by the river Eden, which is crossed by a bridge of singular construction, having elliptical, semicircular, and pointed arches; it was partly built with the materials of an old bridge taken down about 50 years ago: the remains of a pier belonging to a still more ancient structure, demolished by a great flood in 1360, are still visible in the stream of the Eden. In the neighbourhood are vestiges of an ancient encampment, the ramparts of which are twelve feet high; and on the common is a chalybeate spring. Among the eminent natives were, Dr. George Benson, a nonconformist divine and biblical critic, born in 1699; the late Lord Ellenborough, Lord Chief Justice of the King's Bench; and Rowland Wetheral, the celebrated mathematician and astronomer, born in the middle of the last century.

SALKELD, LITTLE, a township, in the parish of ADDINGHAM, union of PENRITH, LEATH ward, Eastern Division of the county of CUMBERLAND, $3\frac{1}{2}$ miles (S.) from Kirk-Oswald; containing 105 inhabitants. Here was anciently a chapel.

SALL (*St. Peter and St. Paul*), a parish, in the union of AYLSHAM, hundred of EYNSFORD, Eastern Division of the county of NORFOLK, $1\frac{3}{4}$ mile (N. N. E.) from Reepham; containing 298 inhabitants. The living is a rectory, valued in the king's books at £12. 19. 7.; patrons, Master and Fellows of Pembroke Hall, Cambridge. The tithes have been commuted for a rent-charge of £560, subject to the payment of rates, which on the average have amounted to £145; the glebe comprises 30 acres, valued at £60 per annum. The church is a stately cruciform structure, principally in the later style of English architecture, with a western tower of elegant proportions.

SALMONBY (*St. Margaret*), a parish, in the union of HORNCASTLE, hundred of HILL, parts of LINDSEY, county of LINCOLN, $5\frac{1}{4}$ miles (E. N. E.) from Horncastle; containing 90 inhabitants. The living is a discharged rectory, valued in the king's books at £5. 10. $2\frac{1}{2}$.; present net income, £308; patron and incumbent, Rev. W. Bowerbank.

SALPERTON (*All Saints*), a parish, in the union of NORTHLEACH, hundred of BRADLEY, Eastern Division of the county of GLOUCESTER, 5 miles (N. W. by N.) from Northleach; containing 216 inhabitants. The living is a perpetual curacy; net income, £95; patron, John Browne, Esq.

SALSEY FOREST, an extra-parochial liberty, in the hundred of CLELEY, Southern Division of the county of NORTHAMPTON, $6\frac{1}{2}$ miles (S. S. E.) from Northampton: the population is returned with Hartwell.

SALT, a joint township with Enson, in the parish of ST. MARY and ST. CHAD, STAFFORD, union of STAFFORD, Southern Division of the hundred of PIREHILL, Northern Division of the county of STAFFORD, 4 miles (N. E. by N.) from Stafford; containing 533 inhabitants. There is a place of worship for Wesleyan Methodists; also a school supported by Earl Talbot, for clothing and educating 40 poor children.

SALTASH, or ESSAY, a market-town and chapelry, formerly a representative borough, in the parish of ST. STEPHEN, and union of ST. GERMANS, having separate jurisdiction, though locally in the Southern Division of the hundred of East, Eastern Division of the county of CORNWALL, 4 miles (N. N. W.) from Plymouth (the post town), 21 miles (S. S. E.) from Launceston, and 220 (W. S. W.) from London; containing 1637 inhabitants. This is a place of considerable antiquity, the county assizes having been held here so early as 1393. In the civil commotions between Charles I. and the parliament, its local importance was evinced in the repeated contests for possession by both the conflicting parties, which terminated in its final abandonment by the royalists in 1646; during this collision the town was fortified. It is pleasantly situated on a steep rocky elevation rising from the western bank of the river Tamar, and consists principally of three narrow streets irregularly formed; the houses in general are of ancient appearance. The market is on Saturday; and fairs are held on Candlemas-day, and

Seal and Arms.

July 25th, for cattle; there are likewise four quarterly cattle shows. The inhabitants are, for the greater part, fishermen, or persons connected with the docks of Devonport. The first charter of incorporation was granted in the reign of Henry III., confirmed by Richard II., and renewed with additional privileges by Charles II., in 1682, under the provisions of which the municipal body consists of a mayor and six aldermen, styled "the council of the borough," with an indefinite number of free burgesses, assisted by a recorder and other officers. The mayor is chosen annually from among the aldermen, by the aldermen and free burgesses; the aldermen out of the free burgesses, and the town-clerk by the aldermen. The property of the oyster-fishery to the mouth of the Tamar, except between Candlemas and Easter, with river dues for anchorage, buoyage, and salvage, and a right of ferry, is vested in the corporation, and their coroner sits upon all bodies found drowned in the river. Holding the manor under the duchy of Cornwall, they are empowered by the charter to hold a court of admiralty for the borough, and liberty of the river Tamar. The borough first returned members to parliament in the reign of Edward VI., but was disfranchised by the act of the 2nd of William IV., cap. 45: the elective franchise was formerly in the freeholders possessing burgage tenements, about 70 in number: the mayor was the returning officer. A court of record, established by charter of the 35th of George III., for the recovery of debts to any amount, is held every week, at which the mayor and aldermen, or any two of them, preside. Sessions for the division are held quarterly in the guildhall: the assizes have not been held here for many years. The living is a perpetual curacy; net income, £45: it is in the patronage of the Mayor and Corporation, but the advowson must be sold under the Municipal Reform Act; appropriators, Dean and Canons of Windsor. The chapel, dedicated to St. Nicholas, is an ancient structure with a fine massive tower: in the interior is a magnificent monument to the memory of three brothers named Drew, who were drowned. There are places of worship for Baptists and Wesleyan Methodists; also a grammar school endowed with £6. 7. 6. per annum, and a National school for boys.

SALTBY (*St. Peter*), a parish, in the union of Melton-Mowbray, hundred of Framland, Northern Division of the county of Leicester, 8¼ miles (N. E.) from Melton-Mowbray; containing 263 inhabitants. The living is a discharged vicarage, consolidated with that of Sproxton, and valued in the king's books at £7.

SALTER, an extra-parochial district, in Allerdale ward above Derwent, Western Division of the county of Cumberland, 8 miles (E. by S.) from Whitehaven; containing, with Eskat, 42 inhabitants.

SALTERFORTH, a township, in the parish of Barnoldwick, union of Skipton, Eastern Division of the wapentake of Staincliffe and Ewcross, West Riding of the county of York, 8¼ miles (S. W. by W.) from Skipton; containing 725 inhabitants.

SALTERSFORD, a chapelry, in the parish of Prestbury, hundred of Macclesfield, Northern Division of the county of Chester, 6 miles (E. N. E.) from Macclesfield: the population is returned with the parish. The living is a perpetual curacy; net income, £47; patron, Vicar of Prestbury. Saltersford gives the inferior title of Baron to the family of Stanhope, Earls of Courtown.

SALTFLEET-HAVEN, a hamlet (formerly a market-town), in the parish of Skidbrook, union of Louth, Marsh Division of the hundred of Louth-Eske, parts of Lindsey, county of Lincoln, 38 miles (N. E. by E.) from Lincoln: the population is returned with the parish. This was a town of some importance about half a century ago, but it is now decayed: the old town is said to have been destroyed by an inundation of the sea. A fair is held on October 3rd, and is celebrated for the show of good foals. Here is a very fine bed of oysters. There is a place of worship for Wesleyan Methodists.

SALTFLEETBY (*All Saints*), a parish, in the union of Louth, Marsh Division of the hundred of Louth-Eske, parts of Lindsey, county of Lincoln, 10¾ miles (E. by N.) from Louth; containing 180 inhabitants. The living is a rectory, valued in the king's books at £12. 19. 4½.; present net income, £317; patrons, President and Fellows of Magdalene College, Oxford.

SALTFLEETBY (*St. Clement's*), a parish, in the union of Louth, Marsh Division of the hundred of Louth-Eske, parts of Lindsey, county of Lincoln, 10½ miles (E. N. E.) from Louth; containing 110 inhabitants. The living is a discharged rectory, valued in the king's books at £7. 0. 1.; present net income, £210; patron, Earl Brownlow.

SALTFLEETBY (*St. Peter's*), a parish, in the union of Louth, Marsh Division of the hundred of Louth-Eske, parts of Lindsey, county of Lincoln, 8½ miles (E. by N.) from Louth; containing 200 inhabitants. The living is a rectory, valued in the king's books at £5; present net income, £229; patrons, Provost and Fellows of Oriel College, Oxford.

SALTFORD (*St. Mary*), a parish, in the union and hundred of Keynsham, Eastern Division of the county of Somerset, 5¼ miles (W. N. W.) from Bath; containing 380 inhabitants. This parish is bounded on the east and north by the river Avon, on the banks of which there are extensive brass works, and is intersected by the line of the Great Western railway. The living is a discharged rectory, valued in the king's books at £10. 5. 10.; patron, Duke of Buckingham. The tithes have been commuted for a rent-charge of £185, subject to the payment of rates; the glebe comprises 13 acres, valued at £20 per annum. There is a National school.

SALT-HILL, a village, partly in the parish of Farnham-Royal, hundred of Burnham, and partly in the parish of Upton, hundred of Stoke, county of Buckingham, 2 miles (N.) from Eton: the population is returned with the parishes. This place, which is situated on the road to Bath, is distinguished by two very large and elegant inns, and is further noted as being connected with the triennial ceremony of the Eton scholars, termed the Montem, the procession repairing hither to a tumulus on the south side of the road, which probably acquired the name Salt-hill from the money collected by the boys being called "Salt-Money." The Great Western railway passes near the village.

SALTHOUSE (*St. Nicholas*), a parish, in the union of Erpingham, hundred of Holt, Western Division of the county of Norfolk, 2¼ miles (E.) from Clay; containing 262 inhabitants. The living is a discharged rectory, annexed to that of Kelling, and valued in the king's books at £20. The tithes have been commuted for a rent-charge of £219. 11., subject to the payment of rates, which on the average have amounted to £26; the glebe comprises one acre, valued at £1. 10. per annum.

SALTMARSH, a township, in the parish and union of HOWDEN, wapentake of HOWDENSHIRE, East Riding of the county of YORK, 4½ miles (S. E.) from Howden; containing 191 inhabitants.

SALTON (*St. John of Beverley*), a parish, in the union of MALTON, liberty of ST. PETER'S, East Riding, though locally in the wapentake of Ryedale, North Riding, of the county of YORK; containing 355 inhabitants, of which number, 156 are in the township of Salton, 6¾ miles (W. S. W.) from Pickering. The living is a discharged vicarage, valued in the king's books at £4. 10. 10.; present net income, £90; patron and impropriator, G. W. Dowker, Esq.

SALTWICK, a township, in the Northern Division of the parish of STANNINGTON, Western Division of CASTLE ward, Southern Division of the county of NORTHUMBERLAND, 5 miles (S. S. W.) from Morpeth: the population is returned with the parish.

SALTWOOD (*St. Peter and St. Paul*), a parish, in the union of ELHAM, hundred of HAYNE, lathe of SHEPWAY, Eastern Division of the county of KENT, ¾ of a mile (N. by W.) from Hythe; containing 534 inhabitants. This place was distinguished at an early period for its castle, which is said to have been first built by the son of Hengist, the Saxon, in 448, and, in the reign of John, to have become one of the palaces of the Archbishops of Canterbury: the remains, which are so considerable as to create some idea of its former magnificence, are situated on an eminence, commanding a fine view of the sea, which it is supposed formerly came up to this place, near which an anchor was dug up some years since. The living is a rectory, with that of Hythe annexed, in the patronage of the Archbishop of Canterbury, valued in the king's books at £34; present net income, £784. The church is principally in the decorated style of English architecture.

SALWARPE (*St. Michael*), a parish, in the union of DROITWICH, Upper Division of the hundred of HALFSHIRE, Droitwich and Eastern Divisions of the county of WORCESTER, 2⅙ miles (W. S. W.) from Droitwich; containing 475 inhabitants. The living is a rectory, valued in the king's books at £14. 14. 7.; present net income, £520; patron and incumbent, Rev. J. V. Vashon. The church exhibits portions in the Norman, and in the decorated and later styles of English architecture. The Droitwich canal passes through the parish. A parish school was erected a few years since by subscription, on the site of one in a ruinous state; the master occupies the house rent free, and receives £20 annually, the bequest of Talbot Barker, Esq., for instructing 20 poor children. An old mansion, erected in the time of Henry VIII., is supposed to occupy the site of an ancient religious house. Richard Beauchamp, the celebrated Earl of Warwick, was born here in 1351.

SAMBOURN, a hamlet, in the parish of COUGHTON, union of ALCESTER, Alcester Division of the hundred of BARLICHWAY, Southern Division of the county of WARWICK, 3¾ miles (N. W. by N.) from Alcester; containing 694 inhabitants, a few of whom are employed in making needles.

SAMLESBURY, a chapelry, in the parish and Lower Division of the hundred of BLACKBURN, union of PRESTON, Northern Division of the county palatine of LANCASTER, 4 miles (E. by N.) from Preston; containing 1948 inhabitants. The living is a perpetual curacy; net income, £110; patron, Vicar of Blackburn. The chapel is dedicated to St. Leonard. There is a place of worship for Roman Catholics. A school is partly supported by an endowment of £7 per annum; and another for the children of Roman Catholics is partly supported by the priest.

SAMPFORD-ARUNDEL (*Holy Cross*), a parish, in the union of WELLINGTON, hundred of MILVERTON, Western Division of the county of SOMERSET, 2¾ miles (S. W.) from Wellington; containing 427 inhabitants. The living is a discharged vicarage, valued in the king's books at £6. 3. 1⅐.; present net income, £146; patron and impropriator, Rev. Charles Sweet. The church has been enlarged, and 58 free sittings provided, the Incorporated Society having granted £25 in aid of the expense. A school is partly supported by subscription.

SAMPFORD-BRETT (*St. George*), a parish, in the union of WILLITON, hundred of WILLITON and FREEMANNERS, Western Division of the county of SOMERSET, 15 miles (N. W.) from Taunton; containing 197 inhabitants. This parish is supposed to have derived the adjunct to its name from the family of the De Bretts, to whom it anciently belonged, and to a member of whom the effigy of a cross-legged knight among the monuments in the church is traditionally assigned. The mail-coach road from Taunton to Minehead passes through the parish; and at Aller are the remains of an ancient manor-house, supposed to have been the residence of one of the Wyndham family, to whose memory is an ancient marble monument in the church, with an elegant inscription in Latin. The living is a rectory, valued in the king's books at £7. 19. 7.; present net income, £358; patron, Rev. Charles Tripp, D.D. The church, in addition to the monuments above named, contains several to the memory of deceased rectors of the parish.

SAMPFORD-COURTENAY (*St. Andrew*), a parish, in the union of OAKHAMPTON, hundred of BLACK TORRINGTON, Black Torrington and Shebbear, and Northern, Divisions of the county of DEVON, 5¼ miles (N. E. by N.) from Oakhampton; containing 1217 inhabitants. This parish is situated in a district abounding with beautiful and picturesque scenery, though of strikingly bold character. Near Sticklepath, which in the reign of Henry V. was a parish of itself, and where there is still a chapel, in which divine service is occasionally performed, a copper mine was opened some years since, but the produce was insufficient to remunerate the adventurers, and it has consequently been discontinued. A serious commotion broke out here, in 1549, in consequence of some alteration in the church service. The living is a rectory, valued in the king's books at £47. 12. 1.; present net income, £510; patrons, Provost and Fellows of King's College, Cambridge. At Brightley, in this parish, a monastery of Cistercians was founded in 1136, by Richard Fitz-

Baldwin de Brioniis, Baron of Oakhampton, which was afterwards removed to Ford; but the ruins of a chapel, supposed to have belonged to it, are still remaining.

SAMPFORD, GREAT (*St. Michael*), a parish, in the union of SAFFRON-WALDEN, hundred of FRESHWELL, Northern Division of the county of ESSEX, $3\frac{3}{4}$ miles (N. E. by N.) from Thaxted; containing 800 inhabitants. This parish is about three miles in length, and nearly of equal breadth: the surface is pleasingly undulated, and the scenery enriched with ornamental timber. The soil is luxuriantly fertile, and along the borders of the Freshwell rivulet are some fine tracts of meadow and pasture land. The village is pleasantly situated, and contains several good houses: the straw-plat manufacture, which is of recent introduction, affords employment to several of the inhabitants; and a fair was formerly held on Whit-Monday, but it has almost fallen into disuse. The living is a vicarage, with that of Hempstead annexed, valued in the king's books at £18; present net income, £255; patron, Sir William Eustace, K.C.H.; appropriators, Dean and Chapter of Canterbury. The church, situated on an eminence, is a handsome structure in the decorated English style, with a square embattled tower strengthened with buttresses; the interior is rich in architectural details, and in the chancel are some stone stalls of beautiful design, elaborately ornamented. On the south side of the chancel is an ancient chantry chapel, now used as a vestry-room; and there are several interesting monuments and ancient inscriptions. There is a place of worship for Baptists; and a National school is supported by subscription. Mrs. Catherine Riley, in 1820, bequeathed £200, which has been vested in Bank annuities; the dividends are distributed among the poor of Great and Little Sampford.

SAMPFORD, LITTLE (*St. Mary*), a parish, in the union of SAFFRON-WALDEN, hundred of FRESHWELL, Northern Division of the county of ESSEX, $3\frac{3}{4}$ miles (N. E. by E.) from Thaxted; containing 423 inhabitants. This parish is about four miles in length and three in breadth, and is watered by a stream, which in its course forms the river Pant. The living is a rectory, valued in the king's books at £11; present net income, £494; patrons, Warden and Fellows of New College, Oxford. The church is a plain edifice of stone, with a lofty tower surmounted by a spire, and contains several ancient and interesting monuments. A National school is supported by subscription. Near the manor-house of Friers are the foundations of an ancient chapel of the Knights Hospitallers, from whom, as its possessors, the manor derived its name.

SAMPFORD-PEVERELL (*St. John the Baptist*), a parish, in the union of TIVERTON, hundred of HALBERTON, Collumpton and Northern Divisions of the county of DEVON, 5 miles (E. by N.) from Tiverton; containing 787 inhabitants. This place is distinguished as having been the residence of Margaret, Countess of Richmond, and mother of Henry VII.; the manor-house in which she resided, and which subsequently belonged to Sir Amias Poulett, who had the custody of Mary, Queen of Scots, at the time of her execution, was a castellated building, erected in 1337, and taken down in 1775. The woollen trade was formerly carried on extensively at this place, which is said to have been anciently a borough. The parish abounds with excellent limestone, and there are several kilns for burning it, for the supply of the neighbourhood. The Grand Western canal, and the Bristol and Exeter railway, pass through the parish. The living is a rectory, valued in the king's books at £23. 8. $11\frac{1}{2}$.; present net income, £270; patron, Rev. Edward Pidsley; impropriators, Trustees of the late Rev. S. Pidsley. The church, of which one aisle is said to have been built by the Countess of Richmond, contains some interesting monuments, one of which has the effigy of an armed knight, and another, inscribed to the Poulett family, bears the date 1502: near the altar are two recesses, in which are a piscina and water drain. From the churchyard a fine view is obtained of the surrounding country, and of Sidmouth gap, about 25 miles distant. There are two schools, in which 20 children are paid for by subscription.

SAMPFORD-SPINEY, a parish, in the union of TAVISTOCK, hundred of ROBOROUGH, Tavistock and Southern Divisions of the county of DEVON, $4\frac{1}{4}$ miles (E. by S.) from Tavistock; containing 366 inhabitants. This parish is situated on the verge of Dartmoor, and is intersected by the Plymouth railway. Here is a mine where cobalt and silver have been found. The living is a perpetual curacy, with that of Shaugh annexed; net income of Shaugh, £107, and of Sampford-Spiney, £54; patrons and appropriators, Dean and Canons of Windsor. The church contains 30 free sittings, the Incorporated Society having granted £15 in aid of the expense.

SAMPSON'S, ST., a parish, in the union of ST. AUSTELL, Eastern Division of the hundred of POWDER and of the county of CORNWALL, 4 miles (S. S. E.) from Lostwithiel; containing 314 inhabitants. The living is a perpetual curacy; net income, £53; patron and impropriator, W. Rashleigh, Esq. The river Fowey is navigable on the east of this parish. Here was anciently a castle of the Earls of Salisbury, the site of which is called Castle-Dore.

SANCREED (*St. Creed*), a parish, in the union of PENZANCE, Western Division of the hundred of PENWITH and of the county of CORNWALL, 4 miles (W. by S.) from Penzance; containing 1069 inhabitants. The living is a vicarage, valued in the king's books at £8; present net income, £265; patrons and appropriators, Dean and Chapter of Exeter. Besides the church, here were three ancient chapels, of which there are still some remains. There are places of worship for Baptists, Bryanites, and Wesleyan Methodists; and a National school is supported by subscription. In the churchyard there is a fine ancient cross: there are also two rude upright stones at Drift, and another rising from a barrow in Boswen's Croft. In the tenement of Bodinnar is a curious monument, consisting of two flat walls; the outermost forming two circles, the larger 55 feet in diameter, and enclosing another of smaller dimensions, between the walls of which is a ditch four feet wide. Near the village of Brahan are some remains of an ancient castle called Caerbran: at Nenkridge is a cave, and there is also a holy well in the parish.

SANCTON (*All Saints*), a parish, in the union of POCKLINGTON, Hunsley-Beacon Division of the wapentake of HARTHILL, East Riding of the county of YORK;

containing 462 inhabitants, of which number, 377 are in the township of Sancton with Houghton, 2½ miles (S. E.) from Market-Weighton. The living is a discharged vicarage, valued in the king's books at £6. 1. 10½.; present net income, £49; patron and impropriator, J. Broadley, Esq. There is a place of worship for Wesleyan Methodists, also a Roman Catholic chapel at Houghton. A free school is endowed with a rent-charge of £20, which is paid for teaching about 30 children; another school for Roman Catholics is supported by the Hon. Charles Langdale.

SANDALL, GREAT (*St. Helen*), a parish, in the union of WAKEFIELD, Lower Division of the wapentake of AGBRIGG, West Riding of the county of YORK; containing 2878 inhabitants, of which number, 1075 are in the township of Great Sandall, 2 miles (S. by E.) from Wakefield. A castle built here about 1320, by John Plantagenet, the last Earl of Warren, for his favourite mistress, Maude, the wife of Thomas Earl of Lancaster, was occupied by Edward Baliol in the reign of Edward III., during the preparations for placing him on the Scottish throne. It subsequently became the residence of Richard Plantagenet, Duke of York, and lastly of the Duke of Gloucester, afterwards Richard III. During the great civil war it was held for Charles I., till surrendered in 1645, and in 1646 it was completely demolished, insomuch that there are now only a few very inconsiderable fragments. The living is a discharged vicarage, valued in the king's books at £13. 7. 8.; present net income, £157: it is in the patronage of the Crown; impropriator, S. W. Pilkington, Esq. Richard Taylor, in 1686, bequeathed certain houses in Wakefield, producing an annual income of £18, of which sum, £10 is paid for teaching eight poor children, and £6 to two widows in almshouses founded by the same individual. There are two other almshouses for poor women, founded by George Grice, and rebuilt in 1823.

SANDALL, KIRK, or LITTLE (*St. Oswald*), a parish, in the union of DONCASTER, Southern Division of the wapentake of STRAFFORTH and TICKHILL, West Riding of the county of YORK, 4¼ miles (N. E. by N.) from Doncaster; containing 200 inhabitants. The living is a rectory, valued in the king's books at £9. 0. 2½.; present net income, £393: it is in the patronage of the Crown. The church is a cruciform structure, with a tower rising from the intersection: it contains a curious monument to the memory of John Rokeby, a native of this place, and Archbishop of Dublin, who directed his body to be buried here, and his heart and bowels at Halifax. A free school was founded in 1626, by Robert Wood, who endowed it with 30 acres of land and two houses, for which the master educates about 30 children.

SANDALL, LONG, a joint township with Wheatley, in the parish, union, and soke of DONCASTER, West Riding of the county of YORK, 3½ miles (N. E. by N.) from Doncaster; containing, with Wheatley, 323 inhabitants.

SANDBACH (*St. Mary*), a market-town and parish, partly in the union of NORTHWICH and hundred of NANTWICH, but chiefly in the union of CONGLETON and hundred of NORTHWICH, Southern Division of the county of CHESTER; containing 7214 inhabitants, of which number, 3710 are in the town of Sandbach, 26 miles (E. by S.) from Chester, and 162 (N. W.) from London. This town occupies a pleasant eminence near the small river Wheelock, which falls into the Dane, and in the midst of a fertile tract, commanding, from certain points, extensive views of a rich landscape, embracing the Vale Royal, the hills of Staffordshire and Derbyshire, and the distant mountains of Wales. The worsted trade formerly prevailed, but has been superseded by the throwing and manufacture of silk, by which the town has considerably advanced in importance and prosperity within the last 20 years: the malt trade, which was formerly carried on, has also declined: here are some brine springs. The Grand Trunk canal passes through the parish. A charter for a market was obtained in the seventeenth century; it is held on Thursday; and fairs are on Easter Tuesday and Wednesday, the first Thursday after September 11th, and a statute and pleasure fair on December 27th, for cattle and wearing apparel: in the market-place are some ancient crosses, which were repaired in 1816. A court is held occasionally by the lord of the manor, and two constables are appointed at the petty sessions of the county magistrates. This town has been made a polling-place for the south division of the shire.

The living is a vicarage, valued in the king's books at £15. 10. 2½.; present net income, £1000; patron and incumbent, Rev. John Armistead; impropriator, Lord Crewe. The church is principally in the later English style. A new church has been recently erected at Wheelock, about a mile from the mother church. There are places of worship for Independents, and Primitive and Wesleyan Methodists. Some small benefactions have been made at different periods for the instruction of poor children, the principal of which are, a school-house, erected in 1694, at the expense of Francis Wells; and a bequest of £200, for teaching three boys, and preparing them for the University; the school is under the direction of Trustees, and affords instruction to 20 children in the rudiments of an English education. Here is also a Sunday National school. There are some very extensive charities originating in benefactions to the amount of £420, which was, in 1790, laid out in the purchase of land in the neighbourhood of Burslem, in the county of Stafford, under which some valuable strata of coal have been discovered; the income arising from this land, on an average, is about £1200 per annum.

SANDERINGHAM, (*St. Mary*), a parish, in the union and hundred of FREEBRIDGE-LYNN, Western Division of the county of NORFOLK, 3¼ miles (N. E.) from Castle-Rising; containing 81 inhabitants. The living is a discharged rectory, with that of Babingley annexed, valued in the king's books at £5. 6. 8.; patron, H. H. Henley, Esq. The tithes have been commuted for a rent-charge of £90, subject to the payment of rates, which on the average have amounted to £9; the glebe comprises 15 acres, valued at £10 per annum.

SANDERSTEAD (*All Saints*), a parish, in the union of CROYDON, First Division of the hundred of WALLINGTON, Eastern Division of the county of SURREY, 3 miles (S. S. E.) from Croydon; containing 242 inhabitants. This parish, which is pleasantly situated, is intersected by the line of the London and Brighton railway, which passes near Purley House. A collegiate establishment was founded here in the reign of Henry VI., by Reginald Lord Cobham, the revenue of which at the dissolution was valued at £79. 15. 10. The living is a rectory, valued in the king's books at £7; present net income, £352; patron, A. D. Wigsell, Esq. The church, which is picturesquely situated, is an ancient edifice, with a low tower surmounted with a spire, and contains some interesting details; the chancel was nearly rebuilt in 1832, at the expense of the Rev. J. Courtney, the present incumbent; among numerous interesting monuments is an altar-tomb of black marble, with the recumbent figure of a lady dressed in a winding sheet, exquisitely sculptured in white marble, erected to the memory of Maria Hawtry. A parochial school is supported by subscription; and the parish has the privilege of sending one aged person to the almshouse at Warlingham, founded and endowed by H. Atwood in 1675, and also partakes to a small extent of Henry Smith's charity. Purley House was the residence of John Horne Took, author of a treatise on English grammar, called from that circumstance "the Diversions of Purley"

SANDFORD, a chapelry, in the parish of St. Helen's Abingdon, union of Abingdon, hundred of Hormer, county of Berks, 3 miles (N. W. by N.) from Abingdon; containing 114 inhabitants.

SANDFORD, a joint township with Woodley, in the parish of Sonning, union of Wokingham, hundred of Sonning, county of Berks, $5\frac{1}{2}$ miles (E.) from Reading; containing, with Woodley, 759 inhabitants.

SANDFORD (*St. Swithin*), a parish, in the union and hundred of Crediton, Crediton and Northern Divisions of the county of Devon, 2 miles (N. by W.) from Crediton; containing 2011 inhabitants. The living is a perpetual curacy; net income, £205; the patronage and impropriation belong to the Governors of the Crediton Charity. The church was formerly a chapel of ease to that of Crediton. Sir Humphrey P. Davie, Bart., erected, in 1825, a commodious building for a National school, which was at that time incorporated with a school founded in 1677 by Sir John Davie, Bart., and endowed by him with a rent-charge of £16, for teaching 20 children, and clothing 10 of them every alternate year: the school fund is augmented by an annual allowance of £12. 7. from the Governors of Crediton Charity, and other contributions chiefly from the family of the founder, by which means a master is allowed £45 and a mistress £25 per annum; the school is open to all poor children of the parish, of whom about 150 boys and 100 girls receive gratuitous instruction; all are partly clothed by Sir H. P. Davie annually at Christmas. There are also two other charity schools with small endowments by the Rev. Robert Ham and Mary Lock.

SANDFORD (*St. Andrew*), a parish, in the hundred of Bullingdon, county of Oxford, 3 miles (S. S. E.) from Oxford; containing 229 inhabitants. This parish is situated on the east bank of the river Isis, and on the turnpike road from Oxford through Henley to London. It is a favourite place of resort for aquatic excursions of the collegians. On the banks of the Isis is a paper-mill, affording employment to fifteen men and thirty women. The living is a donative, in the patronage of the Duke of Marlborough; net income, £15. The church was built in the 12th century, and is in various styles of architecture, with a circular font at the west end. Some years ago a stone was dug up, which had been used as a common flag stone: on the reverse was found a very rich carving representing the Assumption, surrounded by a wreath of angels; it is now erected over the north end of the chancel, near the altar. There was anciently a preceptory of Knights Templars, the preceptor and brethren of which had the management of most of the estates belonging to that order in the neighbouring counties.

SANDFORD (*St. Martin*), a parish, in the union of Woodstock, hundred of Wootton, county of Oxford, $3\frac{3}{4}$ miles (E. N. E.) from Neat-Enstone; containing 534 inhabitants. The living is a discharged vicarage, valued in the king's books at £7. 0. 5.; present net income, £180; joint patrons, Duke of Marlborough and the Rev. Edward Marshall; impropriators, various proprietors of land. Fifteen children are educated for six guineas a year, the bequest of Henry Meads, in 1750; and 29 girls are paid for by a lady, who also supports an infants' school.

SANDFORD, a township, in the parish of Prees, Whitchurch Division of the hundred of North Bradford, Northern Division of the county of Salop, $5\frac{1}{2}$ miles (N. E.) from Wem; containing, with the hamlets of Darleston, Fauls, and Mickley, 487 inhabitants.

SANDFORD, a hamlet, in the parish of Warcop, East ward, county of Westmorland, $4\frac{1}{4}$ miles (W. N. W.) from Brough: the population is returned with the parish. There are several ancient intrenchments and tumuli in the vicinity: the largest of the latter was opened in 1766, and found to contain calcined human bones and some military weapons; near them was formerly a circle of stones, about 50 yards in diameter.

SANDFORD-ORCAS (*St. Nicholas*), a parish, in the union of Sherborne, hundred of Horethorne, Eastern Division of the county of Somerset, 6 miles (N. E.) from Yeovil; containing 353 inhabitants. The living is a rectory, valued in the king's books at £11. 9. $9\frac{1}{2}$.; patron, John Hutchins, Esq. The tithes have been commuted for a rent-charge of £260, subject to the payment of rates, which on the average have amounted to £12; the glebe comprises 47 acres, valued at £61. 15. per annum. There is a place of worship for Wesleyan Methodists; also a National school.

SANDGATE, a chapelry, partly within the liberty of the town of Folkestone, and partly in the parish of Cheriton, union of Elham, hundred of Folkestone, lathe of Shepway, Eastern Division of the county of Kent, $1\frac{1}{4}$ mile (W. by S.) from Folkestone, and 8 miles (W.) from Dovor. The name of this village is a contraction of *Sandygate*, and is derived from its situation in one of those openings from the sea between the hills, formerly called *gates*, and the sandy nature of the soil on which it stands: it emerged from obscurity and insignificance about 50 years since, when two yards were established here for ship-building, and six 28-gun frigates, of about 800 tons each, were built. A castle, similar to those at Deal and Walmer, was erected by Henry VIII., in 1539, on the site, as it is supposed, of a more ancient one which stood here in the reign of Richard II., and was formerly an object of much curiosity, but has undergone considerable alterations of late years, the large circular tower forming the centre having been converted into a martello tower: it is within the jurisdiction of the lord warden of the cinque-ports. During the late war with France there was a summer camp on Shorncliff, a hill at the north side of the village; where also, about 30 years since, some extensive barracks were erected. At the bottom of the hill commences the New Military canal, cut about 20 years since in a zig-zag course, which extends along the coast, passes Hythe, where it crosses the Romney road, and, following the course of the hills for 23 miles, terminates at Cliff End, in Sussex; it was originally about 20 yards wide, and six in depth. The situation of the village is in the highest degree salubrious and pleasant: it lies along the shore, with hills rising immediately behind it, consists principally of irregularly built houses, forming one long street, and possesses machines and every requisite for hot and cold bathing, with a circulating library and reading-rooms. It is rising into estimation as a watering-place: there are some elegant villas. The line of the South-Eastern railway passes within a short distance. A fair for toys is held on July 23rd. The living is a perpetual curacy; net income, £192; patron, Hon. J. D. Bligh, at whose death it will be in the gift of the perpetual curate of Folkestone. The late Earl of Darnley erected a neat chapel in 1822, which has been since enlarged, and 117 free sittings provided, the Incorporated Society having granted £200 in aid of the expense. There is a place of worship for Wesleyan Methodists; also a National school supported by subscription. On the summit of a hill in this neighbourhood is an ancient camp, of elliptic form, comprising nearly two acres: the north and west sides are defended by a triple ditch, the south by a single one very steep, and the east by a double one; its formation is attributed to King Ethelbert.

SANDHOE, a township, in the parish of St. John Lee, union of Hexham, Southern Division of Tindale ward and of the county of Northumberland, $2\frac{1}{2}$ miles (E. N. E.) from Hexham; containing 240 inhabitants. This township contains some fine mansions, commanding prospects of a richly diversified country; and near the gardens of Beaufront is a Roman Catholic chapel, now in disuse. A small school is partly supported by two gentlemen.

SANDHOLME, a joint township with Stockhill, in the parish of St. John, liberties of the borough of Beverley, East Riding of the county of York: the population is returned with Stockhill. There is a place of worship for Wesleyan Methodists.

SANDHURST (*St. Michael*), a parish, in the union of Easthampstead, hundred of Sonning, county of Berks, $5\frac{1}{4}$ miles (S. by E.) from Wokingham; containing 672 inhabitants. The living is a perpetual curacy; net income, £72; patron and appropriator, Dean of Salisbury. John Moseley, in 1773, bequeathed a trifling annuity for teaching six children; and there is a National school. In this parish is the Royal Military College, for the scientific instruction of cadets intended for the army, and of officers already possessing military commissions. The two branches of this national institution were first temporarily placed at High Wycombe, in 1799, and removed to Great Marlow, in 1802, by their founder, His Royal Highness the late Duke of York, on a plan furnished by Major-General J. G. Le Marchant, who fell gallantly fighting at the battle of Salamanca. In 1812, the establishment at Marlow was removed to the present magnificent structure, which had been erected at the national expense, and where, since the year 1820, both branches of the institution have been concentrated. The senior department is a school for the staff, where officers of all ranks already in the service are admitted to study: the junior department is appropriated to the professional education of young gentlemen intended for the cavalry and infantry. Since its foundation the college has afforded instruction to above 3000 young gentlemen for the service, besides qualifying above 450 other officers for the staff. Its affairs are under the control of a board of commissioners, under the presidency of the Commander in Chief, consisting of the Secretary at War, the Master-General of the Ordnance, and the principal general officers on the home staff of the army. The institution, however, is immediately governed by a general, having under him a colonel as lieutenant-governor, with other officers. The instruction, both of the senior students and the gentlemen cadets, is conducted under the superintendence of the military authorities of the college, by professors and masters in the various branches of study, of which the chief are mathematics, practical astronomy, the theory of fortification and actual construction of field works, military drawing and surveying, the principal modern languages, the Latin classics, and general history: the young gentlemen are also regularly instructed in military exercises and riding.

The college stands in the midst of extensive and picturesque grounds, with a fine sheet of water in front, and surrounded by many thriving and beautiful plantations. The edifice, which has a fine Doric portico of eight columns, is of a simple and majestic character: it is calculated for the reception of 400 gentlemen cadets, and 30 students of the senior department; the length of the main building being 434 feet, and that of the whole principal façade not less than 900. The house of the governor stands detached in its own grounds; that of the lieutenant-governor closes the western extremity of the front range; and the quarters of the officers of the establishment form, with the main building, a square in its rear; while the masters' houses, at the distance of about a quarter of a mile in front, are built on a terrace overlooking the high western road. A well-situated observatory, and a spacious riding-house, 110 feet by 50, are among the detached buildings; and the principal edifice, besides the halls of study, the dining-halls, and dormitories of the gentlemen cadets, and servants' offices, contains a handsome octagonal room, in which the public examinations are held, and a very neat and chastely decorated chapel.

SANDHURST (*St. Lawrence*), a parish, in the Upper Division of the hundred of Dudstone and King's Barton, union, and Eastern Division of the county, of Gloucester, 3 miles (N.) from Gloucester; containing 434 inhabitants. The navigable river Severn bounds the parish on the west and north: the sandy soil of that part of the parish being anciently overgrown with wood, probably gave name to the place. The living is a discharged vicarage; net income, £209; patron and appropriator, Bishop of Gloucester and Bristol. Here is a Sunday National school.

SANDHURST (*St. Nicholas*), a parish, in the union of Cranbrooke, hundred of Selbrittenden, Lower Division of the lathe of Scray, Western Division of the county of KENT, 7 miles (S. W. by W.) from Tenterden; containing 1307 inhabitants. The living is a rectory, valued in the king's books at £20; patron, Archbishop of Canterbury. The tithes have been commuted for a rent-charge of £880, subject to the payment of rates, which on the average have amounted to £230; the glebe comprises 9 acres, valued at £11. 5. per annum. The church is principally in the later style of English architecture. There are places of worship for Baptists and Wesleyan Methodists; also a National school, and two other schools, partly supported by subscription. The river Rother separates this extensive parish from the county of Sussex. A fair for cattle and pedlery is held on May 25th.

SAND-HUTTON, county of YORK.—See HUTTON, SAND.

SANDIACRE (*St. Giles*), a parish, in the union of SHARDLOW, hundred of MORLESTON and LITCHURCH, Southern Division of the county of DERBY, 9½ miles (E.) from Derby; containing 758 inhabitants. The living is a perpetual curacy, in the patronage of the Prebendary of Sandiacre in the Cathedral Church of Lichfield, the appropriator; net income, £95. The church exhibits an admixture of the various styles of English architecture, from the Norman downwards, though the decorated predominates; there are some slight remains of stained glass in the windows, and in the chancel three elegant stone stalls. There is a place of worship for Wesleyan Methodists. The Erewash and Derby canals form a junction near the village, at which a market and a fair were formerly held.

SANDON (*St. Andrew*), a parish, in the union and hundred of CHELMSFORD, Southern Division of the county of ESSEX, 2 miles (W. by S.) from Danbury; containing 525 inhabitants. This parish, which is partly skirted by the Chelmer and Blackwater navigation, takes its name from the quality of the elevated ground on which it is situated: the soil is sandy in the higher parts, and in the lower, chiefly a stiff wet loam on a substratum of clay. The living is a rectory, valued in the king's books at £13. 16. 8.; present net income, £607; patrons, President and Fellows of Queen's College, Cambridge. The church is a small edifice with a tower; and near it is the rectory-house, a neat residence. A National school is supported by subscription. The learned Dr. Brian Walton, author of the Polyglot Bible, was rector of this parish.

SANDON (*All Saints*), a parish, in the union of BUNTINGFORD, hundred of ODSEY, county of HERTFORD, 4¾ miles (N. W. by N.) from Buntingford; containing 716 inhabitants. The living is a vicarage, valued in the king's books at £9; present net income, £227; patron and appropriator, Dean of St. Paul's, London. Here is a National school.

SANDON (*All Saints*), a parish, in the union of STONE, Southern Division of the hundred of PIREHILL, Northern Division of the county of STAFFORD, 4½ miles (N. N. E.) from Stafford; containing 558 inhabitants. The living is a vicarage, endowed with a portion of the rectorial tithes, and valued in the king's books at £7. 10.; patron, and impropriator of the remainder of the rectorial tithes, Earl of Harrowby. The impropriate tithes have been commuted for a rent-charge of £366, and the vicarial for £356, subject to the payment of rates, which on the latter, on the average, have amounted to £15; the glebe comprises 8 acres, valued at £14 per annum. The church, situated in the middle of the park of Sandon Hall, is a neat edifice, containing an elegant monument to the memory of the well-known genealogist and antiquary, Sampson, the last of the Erdeswickes, formerly proprietors of the manor, who was born here, and died in 1603; the site of their ancient mansion, encompassed by a moat, is still distinguishable. There is a place of worship for Methodists. The Grand Trunk canal passes through the parish, in a line parallel with the Trent. In the grounds of Sandon Hall, a noble stone structure belonging to the Earl of Harrowby, is an obelisk, erected to the memory of the Rt. Hon. William Pitt, which bears date 1806, and is encircled with iron palisades. The stone with which this mansion and pillar are built was obtained in a quarry on the spot. Near the north-east end of the park is Percival's Cave, an arched recess, with seats, excavated in the red sandstone cliff, and so called in memory of that distinguished statesman, who was shot in the lobby of the House of Commons. An elegant National school has been erected in the park by its noble owner, and is supported at the sole expense of him and his Countess. Sandon confers the inferior title of Viscount on the family of Ryder, Earls of Harrowby.

SANDON-FEE, a tything, in the parish of HUNGERFORD, hundred of KINTBURY-EAGLE, county of BERKS, 1¾ mile (S. W. by S.) from Hungerford; containing 674 inhabitants.

SANDRIDGE (*St. Leonard*), a parish, in the union of ST. ALBANS, hundred of CASHIO, or liberty of ST. ALBANS, county of HERTFORD, 2¾ miles (N. E.) from St. Albans; containing 810 inhabitants. The living is a discharged vicarage, valued in the king's books at £8; present net income, £200; patron and impropriator, Earl Spencer. A National school was erected by subscription in 1824, on ground given by the late Earl Spencer, and is supported by voluntary contributions.

SANDWICH, a cinque-port, borough, and market-town, having separate jurisdiction, in the union of EASTRY, locally in the hundred of Eastry, lathe of St. Augustine, Eastern Division of the county of KENT, 39 miles (E.) from Maidstone, and 68 (E. by S.) from London; containing 3136 inhabitants. This place, which appears to have risen into reputation upon the decline of the *Portus Rutupensis*, derived its Saxon name *Sondwic*, signifying a town on the sands, from its situation on a point of land which had been gained from the sea, on its retiring from that ancient port. It is by most antiquaries supposed to have been also the *Lunden-wic*, noticed in the Saxon Chronicle as the principal place of resort for merchants trading with the port of London, and to have been at a very early period a place of considerable importance. In 851, Athelstan defeated a large party of the Danes, who had landed on this part of the coast, and destroyed nine of their ships; soon after which an army of those invaders landed from 350 ships, and plundered this town and Canterbury; and in 993, Anlaf, another Danish

Arms.

chieftain, arrived with a fleet of 90 sail, and laid waste the town. In 1011, a Danish fleet having landed at Sandwich, ravaged the coast of Kent and Sussex, besieged Canterbury, massacred the inhabitants, and set fire to that city. In 1014, Canute, on leaving England, touched at this port, and sent on shore his English hostages, whom he had barbarously mutilated: after being established on the throne of England, he granted the port of Sandwich, and all its revenues, to Christ Church, Canterbury, for the support of the monks, and partly rebuilt the town, which from this time began to flourish, and subsequently attained such eminence as to be made one of the principal cinque-ports of the kingdom by Edward the Confessor, who resided here for some time, and, in 1052, fitted out a fleet to oppose Earl Godwin and his sons, who in the same year entered this harbour, whence they sailed for London. In the Norman survey Sandwich is described as a borough, held by the Archbishop of Canterbury, and as a fort rendering to the king the same services as Dovor, yielding then a rent of £50, and 40,000 herrings for the monks' food. In the reign of Henry III., the French having effected a landing, burnt the town, which, from the opulence of the inhabitants, was soon rebuilt in a more substantial manner, and obtained from that monarch the grant of a weekly market and other privileges. Edward I. fixed the staple of wool here for a short time; and in the same reign, the monks of Canterbury, in exchange for other lands in Kent, surrendered to the king all their rights and customs in the town, with the exception only of their houses and quays, a free passage across the ferry, and the privilege of buying and selling in the market free of toll, which reservations were afterwards abandoned in exchange for lands in Essex, in the reign of Edward III. At this time Sandwich contributed to the armament destined for the invasion of France 22 ships and 504 mariners, and was the general place of rendezvous for the fleets of Edward, who usually embarked here on his several expeditions against that country. Richard II., in the seventh year of his reign, issued an order for enclosing and fortifying the town, which, from its naval importance, had become a principal object of attack with the French, who, preparing to invade England, had constructed a wall of wood, 3000 paces in length and 20 feet in height, with towers at short intervals, to protect their troops from the English archers, which it was their intention to fix upon the coast after they had effected a landing; parts of this wall, being found on board of two large ships which were taken in the following year, were used in strengthening the fortifications of the town. In 1416, Henry V., while waiting to embark for Calais, took up his residence in the monastery of the Carmelite friars. In the 16th and 35th of Henry VI., the French plundered the greater part of the town, which, however, in the reign of Edward IV., was in a very prosperous state, its trade having greatly increased. In 1456, the French made another attempt on the town; and in the following year, Marshal De Bréze landed a force of 5000 men, and, after a sanguinary battle, succeeded in obtaining possession of the town, which they plundered, and, after setting it on fire, returned to their ships and escaped: it was soon afterwards pillaged by the Earl of Warwick, in his insurrection against the king. To guard against similar assaults, Edward IV. fortified the town with a wall strengthened with bastions, and surrounded it with a fosse, appropriating £100 per annum of the custom-house dues towards its restoration, which, together with the advantages of its haven, soon enabled it to regain its former prosperity; and its trade so much increased, that the net amount of the customs was £16,000 per annum, and there were 95 vessels belonging to the port, furnishing employment to 1500 seamen. The harbour soon after this began to decay, from the quantity of light sand which was washed into it by the sea; and this detriment was further increased by the sinking of a large vessel at its mouth. In 1493, a mole was constructed, and many attempts were made, during the reigns of Henry VIII. and Elizabeth, to remove the obstructions and improve the harbour, but they were not attended with success; and so much had the trade declined in consequence, that in the eighth year of the reign of Elizabeth, there were only 62 seamen belonging to the port. The persecutions on account of religious tenets in the Netherlands drove away many artisans, who, with their families, sought an asylum in England; and Elizabeth encouraged the refugees, of whom not less than 400 were settled here by letters patent, dated at Greenwich in 1561, to whom she granted two weekly markets for the sale of their manufactures. They introduced the weaving of silk, and the manufacture of baizes and flannels, bringing them in a short time to a great degree of perfection; and, by their industry and good conduct, they soon became a flourishing and opulent community. Among them were some gardeners, who, finding the ground favourable for the production of esculent plants, employed themselves in their cultivation, to the great benefit of the landholders; they also introduced the cultivation of flax, teasel, and canary seed, which, shortly after their introduction, were propagated with success in every part of the Isle of Thanet. The settlement of the Flemings tended greatly to compensate for the decay of the harbour, and sustained the prosperity and importance of the town. Elizabeth paid it a visit in 1573, and was hospitably entertained by the corporation for three days; and, in 1670, Queen Catherine, with a large retinue, was entertained by the mayor. In the reign of James I., the trade of the port had revived in some degree, and the amount of the customs was £3000 per annum: the descendants of the Flemish refugees had laid aside their original employment, and were intermingled with the rest of the inhabitants in the general occupations of the town.

The town is situated on the navigable river Stour, about two miles from its influx with the sea, near the commencement of the Roman Watling-street, and is surrounded on all sides by a considerable extent of low ground; the houses, many of which are of very ancient appearance, are irregularly built: the streets are narrow, though some improvements have been effected under the provisions of an act passed in 1787, and it has been recently lighted with gas: the inhabitants are amply supplied with water from the river Stour, and from a small stream which rises near the village of Eastry, the water of which is conveyed to the town by a cut, nearly three miles in length, called the Delf, which was constructed under letters patent granted in the 13th of Edward I. Considerable portions of the walls are still remaining, and till the year 1784 five of the ancient gates were entire, the only one now standing being

Fisher's gate, a plain ancient structure, facing the quay. A bridge of two arches, in the centre of which is a swing-bridge, to admit vessels to pass without lowering their masts, connects the town with Stonar and the Isle of Thanet. The port extends from the North Foreland, in a north-easterly direction, to eleven fathoms of water, six miles distant from the shore, and in a southerly direction to the head of the Goodwin sands, along which it continues for five miles to Sandown castle, in a south-westerly direction up the haven, and thence in a southerly direction to the quay at the mouth of the Gestling; including within its jurisdiction as a cinque-port the ports of Ramsgate and Deal, the parish of Walmer, the town of Fordwich, and the ville of Sarre. The foreign trade is principally with Norway, Sweden, and the Baltic, for timber and iron; and the home trade with London, Wales, Scotland, and the north, in which corn, coal, flour, seeds, hops, malt, fruit, &c., are shipped. The custom-house is a branch of that of Ramsgate, though lately it was the superior. There are several large establishments of fellmongery and wool stapling, extensive breweries, malt-houses, and tan-yards; and the manufacture of coarse towelling and sackcloth is also carried on. The harbour is at present so choked up with sand that only vessels of very small burden can enter it with safety. The importance of forming a great national harbour at this place has from an early period attracted the attention of government; in the reign of Edward VI., the work was actually begun, and a considerable excavation made for that purpose, of which part is still remaining; but the undertaking was relinquished. In 1559, and in 1736, other attempts were also made and abandoned; and in 1825, a magnificent plan was laid down by the late Sir John Rennie, and an act of parliament passed to make the river navigable to Canterbury from the sea, and to form a capacious basin, docks, and haven, opening into the Downs; but though large sums were subscribed by the townspeople and others in the neighbourhood, this object of local and national importance was not carried into effect. A central Kent railway, having its terminus at this harbour, is now in contemplation, and it is not improbable that the long projected improvement of the harbour may still be revived in connection with the formation of the railway, and this very desirable object be ultimately accomplished. The market day is Wednesday, for corn, with which it is abundantly supplied: a large cattle market is held every alternate Monday, and an annual fair on Dec. 4th, which generally continues a week.

Old Corporation Seal, now disused.

Obverse. *Reverse.*

By a succession of charters, of which the last was granted by Charles II., in the 36th year of his reign, the government was vested in a mayor, high steward or recorder, twelve jurats, and twenty-four common-councilmen, assisted by a town-clerk and other officers: the mayor, recorder, and jurats, were justices of the peace within the town and liberties, and additional cinque-port magistrates were appointed by act of parliament, for the several members of the port. Since the passing of the Municipal Reform Act (for an abstract of which see the Appendix, No. I.), the corporation consists of a mayor, four aldermen, and twelve councillors, of whom the mayor and late mayor are justices of the peace, with others appointed by commission from the crown; and the usual cinque-port magistrates are appointed as before for the several members of the port.

Corporation Seal.

Among the privileges enjoyed by Sandwich, as a cinque-port, is that of sending three barons to assist in supporting the canopy over the king at coronations; and when a queen consort is crowned, six are sent, who enjoy the favour of dining at the coronation feast, at a table placed on the right of their Majesties. This borough first exercised the elective franchise in the 42nd of Edward III., since which time it has regularly returned two members to parliament, who are styled barons. The right of election was formerly vested in the mayor, jurats, and freemen, resident and non-resident, not receiving alms, in number about 900; but the non-resident electors, except within seven miles, have been disfranchised, and the privilege has been extended to the £10 householders of an enlarged district, including Deal and Walmer, which has been made to constitute the borough for parliamentary purposes, comprising by estimation an area of 2867 acres, the limits of which are minutely described in the Appendix, No. II.: the old borough contained only 684 acres. The present number of registered electors is 982, of whom 437 are freemen: the mayor is the returning officer. Since 1437, until the passing of the Municipal Reform Act, the town consisted of twelve wards, each of which was under the jurisdiction of a jurat, who appointed a constable and deputy constable. The old corporation had the power of inflicting capital punishment, which formerly was by drowning, and a document of the date of 1315, is still extant, in which a complaint is preferred against the prior of Christchurch, "for that he had diverted the course of a certain stream, called the Gestling, so that the felons could not be executed for want of water." The recorder holds quarterly courts of session for the trial of all offences within the town and liberties; their jurisdiction extends also to the town of Ramsgate, the ville of Sarre, and the parish of Walmer. He also holds a court of record every three weeks, for debts to any amount. A court of requests, for debts not exceeding £5, is held every second Tuesday in the month, by commissioners under an act passed in the 47th of George III., of which the jurisdiction extends over the town and port of Sandwich, the ville of Sarre, Stonar, Minster, Monkton, Walmer, St. Nicholas-at-Wade, Ash, Eastry, Wingham, Staple, Chillenden, Nonington, Goodnestone, Woodnesborough, Eythorne, Word otherwise Worth, Elmstone, Preston next Wingham

Ickham, Wickhambreux, Waldershare, Barfreston, Shepherds-well (otherwise Sibertswould), Womenswould, Barham, Patricksbourne, Bishopsbourne, Beaksbourne, Littlebourne, Stodmarsh, and Stourmouth. The commissioners of land and assessed taxes hold their sittings in the town. The guildhall, usually called the court-hall, was erected in 1579, and contains, on the basement story, the several rooms for holding the courts, and on the first story, the council-chamber, and offices in which the public business of the corporation and liberties is transacted; in the upper story are kept the ancient cucking-stool and wooden mortar, for the punishment of scolds, and arms for the train bands. All municipal elections, decrees, and ordinances, were till lately made by the whole corporate body in a general assembly, held not less than twice in the year at the guildhall, and convened, according to ancient usage, by the sound of a brass horn. The old borough gaol and house of correction, a small and inconvenient building, having been found inadequate for the classification of prisoners, a larger and more appropriate edifice was erected in 1831, at an expense of £6000: it consists of a centre and two wings, and contains a governor's house and a chapel, four day-rooms, four spacious airing-yards, a treadwheel, and a solitary cell, and is adapted to the reception of 52 prisoners.

The town comprises the parishes of *St. Clement*, *St. Mary the Virgin*, and *St. Peter the Apostle*, and the extra-parochial liberty of the hospital of St. Bartholomew. The living of St. Clement's is a vicarage, endowed with the rectorial tithes, and valued in the king's books at £13. 16. 10½.; present net income, £310; patron, Archdeacon of Canterbury. The church is an ancient and spacious structure, combining various styles of architecture, with a massive central tower of Norman character, enriched with several series of arches of very fine composition: the interior has portions in the early and later styles of English architecture, and contains several monuments, and an octagonal font. The living of St. Mary's is a discharged vicarage, endowed with the rectorial tithes, and valued in the king's books at £8. 1.; present net income, £117; patron, Archdeacon of Canterbury. The church, an ancient building, consists of a nave, north aisle, and chancel, in which are some interesting remains of the early style. The living of St. Peter's is a discharged rectory, valued in the king's books at £8; present net income, £144: it was in the alternate patronage of the Crown, and of the Mayor and Corporation, who have recently sold their right. The south aisle of this church was destroyed by the fall of the steeple, in 1661, but the latter was rebuilt with the materials of the former as high as the nave, and finished with bricks made from the mud in the harbour. There was formerly a chapel, dedicated to St. James, the cemetery of which is still used as a burial-place. There are places of worship for Calvinists, Independents, and Wesleyan Methodists.

The free grammar school was founded by subscription among the inhabitants, in the reign of Elizabeth, and in 1563 endowed with lands for its support by Sir Roger Manwood, then recorder of the borough, and subsequently Chief Baron of the Exchequer, who appointed the mayor and corporation governors: the revenue is about £43. 16. per annum, which, subject to a deduction for the repairs of the buildings, is paid to the master, who resides in the school-house. Mrs. Joan Trapps, of London, in 1568, founded four scholarships in Lincoln College, Oxford, of which two are in the appointment of the Governors of this school, and two in that of the Rector and Fellows of the college, without any distinction of place: the scholarships were afterwards augmented with £3 per annum, by Mrs. Joyce Frankland. Sir Roger Manwood, in 1581, founded four scholarships in Caius College, Cambridge, in the alternate nomination of the Governors of this school, and the Master and Fellows of the college. A charity school, established by subscription in 1711, and principally supported by the same means, is under the direction of the mayor and three trustees chosen from each parish: the rents for its support amount to £25 per annum, besides some bequests made in 1811 and 1817: 90 boys and 60 girls are instructed in this establishment, on the National system. St. Thomas' hospital was founded about the year 1392, by Mr. Thomas Ellis, a wealthy draper of this town, who endowed it for eight aged men and four women, each of whom, from the improved state of the funds, receives an allowance of £25 per annum. St. Bartholomew's was founded prior to the year 1244, when Sir Henry de Sandwich made a considerable addition to its original endowment; sixteen aged men and women receive £36 per annum each: the gross annual income is £766. 3. 6. The buildings occupy a spacious triangular area, and afford healthy and pleasant dwellings for the inmates: the site is extra-parochial. There is a small neat chapel attached to the charity; the chaplain is now appointed by the charity trustees of the borough. St. John's hospital, supposed to have been founded about the year 1287, has been taken down, and six small houses have been erected on its site, for the reception of six aged men and women, who were formerly appointed by the mayor, and receive each an annual sum of £20: the gross annual income is £139. 10. 1. Sir John Manwood, Chief Baron of the Exchequer, and author of the "Forest Laws;" and Mr. Richard Knolles, master of the grammar school, and author of the "History of the Turkish Empire," were natives of this place. Sandwich gives the title of Earl to the family of Montagu.

SANDWITH, a township, in the parish of St. Bees, union of Whitehaven, Allerdale ward above Derwent, Western Division of the county of Cumberland, 2¼ miles (S. by W.) from Whitehaven; containing 320 inhabitants. It extends to St. Bees' Head, where there is a lighthouse; and upon the cliffs adjacent grows an abundance of samphire. Here is a National school.

SANDY (*St. Swithin*), a parish, in the union of Biggleswade, partly in the hundred of Wixamtree, but chiefly in that of Biggleswade, county of Bedford, 3¾ miles (N. by W.) from Biggleswade; containing 1617 inhabitants. The living is a rectory, valued in the king's books at £32. 2. 11.; present net income, £769; patron, F. Pym, Esq. Three schools are supported by subscription. Owing to the sandy nature of the soil, cucumbers are cultivated in the open air in such abundance that Covent-Garden market, London, is almost wholly supplied with that vegetable from this place: carrots and other vegetables are also grown in abundance. Galley-hill is the site of the ancient Roman station *Salinæ*, which commanded another at Chesterfield, a piece of ground, still so called, near the village, through which passed the great road from Baldock in Herts, across this county, into Cambridgeshire. The ramparts, which enclose an area of 30 acres, are surrounded by a deep fosse, and in the centre is a mount, probably thrown up for the prætorium. At some distance, on the other side of the valley, are the remains of Cæsar's camp. Several Roman urns, coins, and fragments of a beautiful red pottery, have been discovered at Chesterfield; the latter, which was ornamented with figures, has been deemed the ancient Samian ware.

SANKEY, GREAT, a chapelry, in the parish of PRESCOT, union of WARRINGTON, hundred of WEST DERBY, Southern Division of the county palatine of LANCASTER, $2\frac{3}{4}$ miles (W.) from Warrington; containing 563 inhabitants. The living is a perpetual curacy; net income, £103; patron, Lord Lilford; impropriators, Provost and Fellows of King's College, Cambridge. The first canal navigation in modern times originated here, in 1755.

SANTON, a joint township with Melthwaite, in the parish of IRTON, union of BOOTLE, ALLERDALE ward above Derwent, Western Division of the county of CUMBERLAND, $4\frac{1}{2}$ miles (N. N. E.) from Ravenglass: the population is returned with the parish. This place is supposed to derive its name from those flying sands which abounded in the vicinity and laid waste most of the adjoining lands. A Roman pottery was discovered in these sands, with numerous fragments of urns, and at the bottom of one of the furnaces a large cross of brass, on which probably the urns were placed for baking or drying them. Several Roman coins have also been found here; and opposite to the village are the remains of a Roman road, on the east of which are the foundations of an Augustine priory, said to have been founded by King Stephen, but of which nothing further has been recorded. Near the road are several sand-hills, resembling barrows.

SANTON (*ST. HELEN*), a parish, in the union of THETFORD, hundred of GRIMSHOE, Western Division the county of NORFOLK, 4 miles (N. W. by N.) from Thetford; containing 18 inhabitants. The living is a discharged rectory, in the patronage of the Mayor and Corporation of Thetford; net income, £18.

SAPCOTE (*ALL SAINTS*), a parish, in the union of HINCKLEY, hundred of SPARKENHOE, Southern Division of the county of LEICESTER, $4\frac{1}{4}$ miles (E. by S.) from Hinckley; containing 871 inhabitants. The living is a rectory, valued in the king's books at £10. 11. $10\frac{1}{2}$.; present net income, £485; patron, Thomas Frewen Turner, Esq. There is a place of worship for Wesleyan Methodists. The river Soar runs through the parish, in which upwards of 200 frames are employed in the manufacture of hosiery. A house of industry, also a common mill for grinding corn, were built by subscription in 1806, the expense of each amounting to £1300; and in the same year handsome bathing rooms were erected at an expense of £600, defrayed by J. F. Turner, Esq., over a spring called Golden Well, the water of which is serviceable in nervous, consumptive, scorbutic, and scrofulous disorders. A school for the instruction of poor children was endowed by the Rev. Mr. Burroughs, formerly rector of this parish, and is further supported by T. F. Turner, Esq. There are vestiges of a mount and moat of an ancient castle, which once occupied the site of the family mansion of the Bassetts. In a field called Black Piece a curious tesselated pavement was discovered in 1770.

SAPEY-PRITCHARD (*ST. BARTHOLOMEW*), a parish, in the union of BROMYARD, Upper Division of the hundred of DODDINGTREE, Hundred-House and Western Divisions of the county of WORCESTER, $5\frac{3}{4}$ miles (N. E. by N.) from Bromyard; containing 250 inhabitants. The living is a discharged rectory, valued in the king's books at £4. 4. 2.; present net income, £220; patron, Francis Rufford, Esq.

SAPEY, UPPER (*ST. MICHAEL*), a parish, in the union of BROMYARD, hundred of BROXASH, county of HEREFORD, $6\frac{1}{2}$ miles (N. N. E.) from Bromyard; containing 357 inhabitants. The living is a rectory, valued in the king's books at £9. 5. $7\frac{1}{2}$.; patron, Sir. T. Winnington, Bart. The tithes have been commuted for a rent-charge of £287. 10., subject to the payment of rates, which on the average have amounted to £16; the glebe comprises 41 acres, valued at £42 per annum. A charity school, founded by Mr. Addingbroke, is endowed with £10 per annum. In the neighbourhood are the remains of a single intrenched Roman camp.

SAPISTON (*ST. ANDREW*), a parish, in the union of THETFORD, hundred of BLACKBOURN, Western Division of the county of SUFFOLK, $3\frac{1}{4}$ miles (N. by W.) from Ixworth; containing 234 inhabitants. The living is a perpetual curacy; net income, £78; patron, Duke of Grafton. The church contains some remains of Norman architecture.

SAPLEY, an extra-parochial district, in the hundred of HURSTINGSTONE, county of HUNTINGDON, 2 miles (E.) from Huntingdon: the population is included in the return for the parish of King's-Ripton.

SAPPERTON, a township, in the parish of CHURCH-BROUGHTON, hundred of APPLETREE, Southern Division of the county of DERBY, 12 miles (W.) from Derby: the population is returned with the parish.

SAPPERTON (*ST. KENELM*), a parish, in the union of CIRENCESTER, hundred of BISLEY, Eastern Division of the county of GLOUCESTER; containing 453 inhabitants, of which number, 275 are in the tything of Sapperton, $5\frac{1}{4}$ miles (W. N. W.) from Cirencester. The living is a rectory, valued in the king's books at £17; present net income, £367; patron, Earl Bathurst. Two schools are supported by endowment. The Thames and Severn canal, in its course through this parish, is conducted by a tunnel, 4180 feet long, underneath Hagley wood. At Frampton, two urns filled with dénarii and copper coins were discovered, in 1759, by a waggon passing over and breaking them, near which spot are vestiges of an ancient camp, and south-east of it there was a beacon.

SAPPERTON (*ST. NICHOLAS*), a parish, in the union and soke of GRANTHAM, parts of KESTEVEN, county of LINCOLN, 7 miles (E. by S.) from Grantham; containing 62 inhabitants. The living is a discharged rectory, valued in the king's books at £5. 9. $9\frac{1}{2}$.; present net income, £190; patron, Sir W. E. Welby, Bart.

SAREDON, GREAT and LITTLE, a township, in the parish of SHARESHILL, union of PENKRIDGE, Eastern Division of the hundred of CUTTLESTONE, Southern Division of the county of STAFFORD, 7 miles (N. N. E.) from Wolverhampton; containing 246 inhabitants.

SARK, or SERK, a small island, lying about 6 miles eastward of Guernsey, being one of its dependencies, and under its immediate jurisdiction, and containing 543 inhabitants. This island is supposed to be mentioned in the Itinerary of Antoninus under the name *Sarnica.* At one part, called the *Coupée,* it is nearly divided into two portions, connected only by a high and narrow ridge not many yards wide. It was early noted for the ancient convent of St. Maglorius, a British Christian, who, fleeing with many others from the persecutions of the Pagan Saxons into Armorica, was made Bishop of Dol, and first planted Christianity in these islands, about the year 565. Queen Elizabeth granted it in fee-farm, by letters patent under the great seal, dated in 1565, to Hilary de Carteret, Esq., by the twentieth part of a knight's fee. The surface of Sark is a table land, rising a little towards the west, but having no declivity to the sea at any part, except a trifling descent at the northern extremity. The surrounding cliffs, from 200 to 300 feet in height, are so very abrupt on the western side, that the largest ship may approach very near them without danger; but the eastern shore is beset with rocks running far out into the sea. The rocky scenery is very grand and picturesque; that of the *Port du Moulin* in particular, the descent to which is through a narrow pass, is uncommonly wild and romantic. Such is the natural strength of the island, that although there are five landing-places, yet, except at what is called the *Creux,* where a tunnel was cut through the rock in 1588, by one of the De Carterets, scarcely any entrance is to be found without the difficulty of climbing. The landing-place nearest to Guernsey is that of *Havre Gossetin,* which is formed between the land and the little *Isle des Marchands* on the western side. The high ridge, or isthmus, which joins the main island to the smaller portion of it, called Petit Sark, is about 100 yards long, with a precipice immediately overhanging the sea on the eastern side; the passage on the western being in some places only three or four feet wide, and over broken rocks of terrific appearance. To the south of Petit Sark is an isolated rock, called *Etat,* much resembling in shape the Mew-stone at Plymouth; and on the coast is a funnel, 200 feet deep, and 100 feet in diameter at the surface, called *Creux Terrible,* similar in appearance to the Buller of Buchan, or Tol Pedn, Penwith, near which is a spring of water, of which the specific gravity is one-eighth less than that of any other water found in the island. There are also numerous picturesque caverns excavated in the cliffs along the sea shore. The sky is usually serene, and the air remarkably salubrious; and the soil, which is extremely fertile, affords every necessary article of produce for the inhabitants, particularly apples, from which excellent cider is made, also turnips, parsnips, potatoes, and other vegetables, together with most kinds of grain. The grass is very sweet, and the mutton fine. Milk and butter are produced in sufficient quantities for the consumption of the inhabitants. Rabbits are also very abundant, and sea fish in great variety. The only branch of manufacture is the knitting of stockings, gloves, and waistcoats, called Guernsey jackets, which affords employment to many of the inhabitants: these are exported to Bristol and some other western ports of England, and various articles of domestic consumption brought back in return. The island, with the exception of the land held by the seignor, is divided into forty copyhold tenements, which are held under him on payment of a moderate rent. The inhabitants are principally employed in agriculture and in fishing, and dredging for oysters for the London market. A feudal court is held three times in the year, for the purpose of enacting by-laws for regulating the affairs of the island, which are in force when carried by a majority of the 40 tenants, and confirmed by the consent of the seignor. The executive power is vested in a seneschal, who has cognizance of civil cases, and from whose decision an appeal lies to the Royal Court at Guernsey. The chapel, dedicated to *St. Peter,* was erected in 1820, and consecrated by the Bishop of Winchester in 1829; it contains 170 free sittings, the Incorporated Society having granted £400 in aid of the expense. The monastery founded by St. Maglorius, was existing in the reign of Edward III., but it has long since gone to decay. In 1719, an earthen pot, bound with an iron hoop, was discovered, containing eighteen Gallic coins of silver gilt, which were engraved by Vertue in 1725.

SARNESFIELD (*All Saints*), a parish, in the union of Weobley, hundred of Wolphy, county of Hereford, 2½ miles (W. by S.) from Weobley; containing 98 inhabitants. The living is a discharged rectory, valued in the king's books at £5. 6. 8.; present net income, £203; patron, Thomas Monnington, Esq. Here is a place of worship for Roman Catholics.

SARRE, a ville, in the union of the Isle of Thanet, cinque-port liberty of Sandwich, though locally in the hundred of Ringslow, or Isle of Thanet, lathe of St. Augustine, Eastern Division of the county of Kent, 8¾ miles (N. E.) from Canterbury; containing 200 inhabitants. This place derives its name from an ancient ford at low water, which led from the Isle of Thanet to the main land, and which, previously to the landing of the Saxons, formed a communication with Chislet on the opposite coast. It was formerly a separate parish, in old documents designated "*St. Giles at Serre,*" but is now united with the parish of St. Nicholas, Sandwich, though it still maintains its own poor. The road from Canterbury to Ramsgate and Margate passes through this place, which formerly carried on a considerable trade; but on the failure of the river Wantsune, it declined, and the inhabitants removing to other places, the church fell into disuse, and afterwards into decay. In Archbishop Parker's visitation, in 1561, the living is returned as "Vicaria Sarre Dissoluta."

SARRATT (*Holy Cross*), a parish, in the union of Watford, hundred of Cashio, or liberty of St. Albans, county of Hertford, 3½ miles (N. W. by N.) from Rickmansworth; containing 452 inhabitants. The village is situated on a ridge of land forming the western boundary of a vale watered by a small river, commonly called the Sarret stream. The living is a vicarage, endowed with the rectorial tithes, and valued in the king's books at £9; present net income, £243; patron, J. A. Gordon, Esq. The church is a cruciform structure, with a square tower, and is built of a mixture of brick, stone, and flints: it contains a piscina beneath an embattled cornice ornamented with roses, and two stone seats. Three schools are chiefly supported by private charity. John Baldwin, in 1700, founded six almshouses; and Henry Day, in 1816, founded two others; they are occupied rent-free by poor families, who receive clothing from a bequest of £500 by Ralph Day, in 1828.

SARSDEN, a parish, in the union of Chipping-

NORTON, hundred of CHADLINGTON, county of OXFORD, $3\frac{3}{4}$ miles (S. W. by S.) from Chipping-Norton; containing 154 inhabitants. The living is a rectory, valued in the king's books at £8. 18. $1\frac{1}{2}$.; present net income, £262; patron, J. H. Langston, Esq. This place is supposed to have been the scene of a battle, in 1016, in which Canute was defeated by Edmund Ironside. The mansion of J. H. Langston, Esq., contains some ancient remains of what is supposed to have been a religious house. Ann Walker, in 1705, gave £600, now producing an annual income of £52. 10., for which 24 girls are educated.

SARSON, a tything, in the parish of AMPORT, hundred of ANDOVER, Andover and Northern Divisions of the county of SOUTHAMPTON: the population is returned with the parish.

SARUM, OLD, formerly a representative borough, in the parish of STRATFORD-UNDER-THE-CASTLE, union of ALDERBURY, hundred of UNDERDITCH, Southern Division of the county of WILTS, $1\frac{1}{2}$ mile (N.) from Salisbury: the population is returned with the parish. This place was originally a British settlement of some importance prior to the time of the Romans, who, on their establishment in the island, fixed here their station *Sorbiodunum*, situated on the *Via Iceniana*, or Ikeneld-street. By the Saxons, who, under their leader Kenric, son of Cerdic, second king of Wessex, took this town from the Britons in 552, it was called *Searesbyrig*, from the dryness of its situation, and continued to be a residence of the West Saxon kings till the union of the Heptarchal provinces under Egbert, after which time it still continued to be a royal castle. Alfred issued an order to the sheriff of Wiltshire to fortify this place with a trench and palisades, which order is given by Ledwiche, in his "*Antiquitates Sarisburienses*," and the remains of the present fortifications are evidently of Saxon character. In 960, Edgar convoked a witenagemot, or great council of the state here, the especial object of which was to deliberate upon the best mode of defending the northern counties against the incursions of the Danes, by whom this part of the kingdom was particularly infested. In 1003, Sweyn, King of Denmark, having landed on the western coast, to retaliate for the massacre of his countrymen in the reign of Ethelred, pillaged the town and burnt the castle. Soon after the Norman Conquest, pursuant to a decree of a synod held in St. Paul's cathedral, in 1076, for removing episcopal sees from obscure villages into fortified cities, the seat of the bishoprick of Wiltshire was, by Bishop Herman, removed from Sherborne (which had fallen into decay) to this place, where he laid the foundation of a cathedral church, which was completed by his successor, Bishop Osmund, in 1092. On the completion of the Norman survey, in 1086, William the Conqueror summoned all the bishops, abbots, barons, and knights of the kingdom, to attend him at Sarum, and do homage for the lands which they held by feudal tenure. In 1095, or 1096, William Rufus assembled a great council here, in which William, Count of Eu, was impeached of high treason against the king, in conspiring to raise Stephen, Earl of Albemarle, to the throne. Henry I. held his court here several months during the year 1100, where he received Archbishop Anselm, on his arrival in England, whom he required to do homage and swear fealty to him, and to accept from his hands the investiture of his see. This demand gave rise to a dispute between the king and the pope, which was at length compromised, the pope allowing the prelates to do homage to the king, and reserving to himself alone the right of investiture, which was the first attempt to establish papal supremacy in the island. This monarch again fixed his residence here in 1106, and in 1116 assembled the prelates and barons of the realm, to swear allegiance and do homage to his son William, as his successor on the English throne, previously to his embarkation for Normandy, on his return from which place that prince was unfortunately drowned. In the reign of Stephen, Bishop Roger held the castle for the king; and soon after the instalment of his successor, Joceline, in 1142, the partisans of the Empress Matilda took possession of the town, which, in the course of the contest, was alternately occupied by both parties. On the accession of Henry II., in 1154, the castle was found to be in a dismantled state, and a considerable sum was expended in putting it into repair. From the time of Stephen, disputes had arisen between the castellans and the clergy, which became so violent that, in the reign of Richard I., Herbert, then bishop, induced by these annoyances, and other inconveniences attending the situation of his church, among which was the difficulty of obtaining water, which could not be accomplished, except by permission of the governor, obtained licence from the king to remove the see, and to erect a new church in the valley, at the distance of nearly two miles from the castle. This design was carried into execution by his successor, who, having obtained a special indulgence from the pope, laid the foundation of the present cathedral of New Sarum, or Salisbury, to which place the episcopal chair was transferred. From that period the town of Old Sarum began to decay, and was gradually deserted by its inhabitants, who established themselves in the more immediate vicinity of the new cathedral church. Of the old town there is scarcely a single vestige, except a few fragments of foundation walls of some of the houses, on the declivity of an eminence rising from the western side of a valley, and forming the extremity of a ridge which extends towards the east. The vast ditches and ramparts of the ancient city, and the site of the castle, may be traced, and, while they constitute the only visible remains of this once flourishing city, are among the most interesting objects of antiquarian research: there were houses remaining in the time of Henry VIII., and service was performed in the old chapel of the cathedral until nearly the same period. Old Sarum was a borough by prescription, and first exercised the elective franchise in the 23rd of Edward I., but made no other return till the 34th of Edward III., from which time it continued to send two members to parliament until the 2nd of William IV., when it was disfranchised. The right of election was vested in the burgage freeholders, in number seven; the bailiff was the returning officer. A temporary house was erected under a large tree for holding the election, but this tree was blown down during the hurricane which occurred in August 1833. John of Salisbury, one of the most eminent scholars of his time, and celebrated as an historian and biographer, was born at Old Sarum, in the early part of the twelfth century.

SATLEY, a chapelry, in the parish and union of LANCHESTER, Western Division of CHESTER ward,

Northern Division of the county palatine of DURHAM, 5 miles (N. E. by N.) from Wolsingham; containing 112 inhabitants. The living is a perpetual curacy; net income, £68; patron, Perpetual Curate of Lanchester. There is a National school, supported by subscription.

SATTERLEIGH (*St. Peter*), a parish, in the union and hundred of SOUTH MOLTON, South Molton and Northern Divisions of the county of DEVON, $3\frac{3}{4}$ miles (S. W. by W.) from South Molton; containing 58 inhabitants. The living is a discharged rectory, valued in the king's books at £4. 0. $7\frac{1}{2}$.; present net income, £113; patron, James Gould, Esq.

SATTERTHWAITE, a chapelry, in the parish of HAWKSHEAD, union of ULVERSTONE, hundred of LONSDALE, north of the sands, Northern Division of the county palatine of LANCASTER, 4 miles (S. by W.) from Hawkshead; containing 403 inhabitants. This place, which comprises the three portions of Satterthwaite, Grisedale, and Graithwaite, is still overspread with coppice wood, from the abundance of which the smelting of iron-ore was formerly carried on to a considerable extent. The principal manufacture is that of bobbin, for which an extensive mill has been recently erected at Cunsey. Graithwaite Hall, an ancient mansion, is the residence of M. Sandys, Esq., whose ancestor founded and endowed the school at Hawkshead. The living is a perpetual curacy; net income, £71; patron, Incumbent of Hawkshead. The chapel was, in 1837, repaired and enlarged by 50 additional sittings, by the exertions of the Rev. H. Baines, the present minister.

SATTERTHWAITE, a chapelry, in the parish and union of ULVERSTONE, hundred of LONSDALE, north of the sands, Northern Division of the county palatine of LANCASTER, $7\frac{1}{2}$ miles (N. by W.) from Ulverstone; containing 163 inhabitants. Here are quarries of slate for roofing houses.

SAUGHALL, GREAT, a township, in the parish of SHOTWICK, union of GREAT BOUGHTON, Higher Division of the hundred of WIRRALL, Southern Division of the county of CHESTER, 4 miles (N. W. by W.) from Chester; containing 367 inhabitants.

SAUGHALL, LITTLE, a township, in the parish of SHOTWICK, union of GREAT BOUGHTON, Higher Division of the hundred of WIRRALL, Southern Division of the county of CHESTER, $3\frac{1}{4}$ miles (N. W. by W.) from Chester; containing 40 inhabitants.

SAUGHALL-MASSEY, a township, in the parish of BIDSTONE, union, and Lower Division of the hundred, of WIRRALL, Southern Division of the county of CHESTER, 9 miles (N. N. W.) from Great Neston; containing 143 inhabitants.

SAUL (*St. James*), a parish, in the union of WHEATENHURST, Upper Division of the hundred of WHITSTONE, Eastern Division of the county of GLOUCESTER, 9 miles (N. W. by W.) from Stroud; containing 443 inhabitants. The living is a perpetual curacy; net income, £125; patron, Vicar of Standish; appropriator, Bishop of Gloucester and Bristol. The appropriate tithes have been commuted for a rent-charge of £115, and the incumbent's for £42, subject to the payment of rates, which on the former have, on the average, amounted to £11; the glebe comprises three acres. The church has been enlarged, and contains 100 free sittings, the Incorporated Society having granted £50 in aid of the expense. There is a place of worship for Wesleyan Methodists. The navigable river Severn runs on the northern side of the parish, through which passes the Stroudwater, and the Gloucester and Berkeley, canals. Here is an ancient house, formerly belonging to the Earl of Leicester, surrounded by a moat.

SAUNDBY (*St. Martin*), a parish, in the union of GAINSBOROUGH, North-Clay Division of the wapentake of BASSETLAW, Northern Division of the county of NOTTINGHAM, $2\frac{1}{2}$ miles (S. W. by W.) from Gainsborough; containing 104 inhabitants. The living is a rectory, valued in the king's books at £14. 8. $6\frac{1}{2}$.; patron, Viscount Midleton. The tithes have been commuted for a rent-charge of £325. 15., subject to the payment of rates, which on the average have amounted to £25; the glebe comprises two acres, valued at £7 per annum. The church is in the later style of English architecture, with portions of an earlier date.

SAUNDERTON (*St. Mary*), a parish, in the union of WYCOMBE, hundred of DESBOROUGH, county of BUCKINGHAM, $1\frac{1}{2}$ mile (S. W.) from Princes-Risborough; containing 231 inhabitants. The living is a rectory, valued in the king's books at £13. 19. 7.; present net income, £377; patrons, President and Fellows of Magdalene College, Oxford. This place formerly constituted two parishes, but, coming into the possession of one individual, they were united in the year 1457, and the church dedicated to St. Nicholas was suffered to go to ruin.

SAUSTHORPE (*St. Andrew*), a parish, in the union of SPILSBY, hundred of HILL, parts of LINDSEY, county of LINCOLN, 3 miles (N. N. W.) from Spilsby; containing 200 inhabitants. The living is a discharged rectory, valued in the king's books at £6. 3. $6\frac{1}{2}$.; present net income, £216; patron, F. Swan, Esq. Here is a National school.

SAVERNAKE-PARK, or NORTH SIDE, an extra-parochial district, in the hundred of SELKLEY, Marlborough and Ramsbury, and Northern, Divisions of the county of WILTS, $1\frac{1}{2}$ mile (S. E. by E.) from Marlborough; containing 110 inhabitants.

SAVERNAKE-FOREST, or SOUTH SIDE, an extra-parochial district, in the hundred of KINWARDSTONE, Marlborough and Ramsbury, and Northern, Divisions of the county of WILTS, 2 miles (S.) from Marlborough.

SAWBRIDGEWORTH (*St. Michael*), a parish, in the union of BISHOP-STORTFORD, hundred of BRAUGHIN, county of HERTFORD, $11\frac{1}{2}$ miles (E. by N.) from Hertford; containing 2231 inhabitants. The living is a vicarage, valued in the king's books at £17; patron, Bishop of London; appropriators, Dean and Chapter of Westminster. The appropriate tithes have been commuted for a rent-charge of £1479. 12. 6., and the vicarial for £391, subject to the payment of rates, which on the average have respectively amounted to £182 and £60; the appropriate glebe comprises 128 acres, valued at £192. 15. per annum. There are places of worship for Independents and Wesleyan Methodists; and a National school has been established.

SAWDON, a township, in the parish of BROMPTON, union of SCARBOROUGH, PICKERING lythe, North Riding of the county of YORK, 9 miles (W. S. W.) from Scarborough; containing 146 inhabitants. There is a place of worship for Wesleyan Methodists.

SAWLEY (*All Saints*), a parish, in the union of

SHARDLOW, hundred of MORLESTON and LITCHURCH, Southern Division of the county of DERBY; containing 3750 inhabitants, of which number, 1009 are in the township of Sawley, 4 miles (N. by W.) from Kegworth. The living is a vicarage, with the perpetual curacy of Wilne annexed, in the patronage of the Prebendary of Sawley in the Cathedral Church of Lichfield, the appropriator; net income, £266. There is a place of worship for Wesleyan Methodists; and a day school and a Sunday National school are partly supported by subscription. The rivers Trent, Derwent, and Erewash run through the parish, which is also intersected by the Derby and Erewash canals. Harrington bridge, across the Trent, was completed in 1790. This place had anciently a market and a fair; the market, having fallen into disuse, was revived about 1760, and was again discontinued before 1770; the market-house still remains.

SAWLEY, a chapelry, in the parish and liberty of RIPON, West Riding of the county of YORK, $5\frac{1}{2}$ miles (S. W. by W.) from Ripon; containing 499 inhabitants. The living is a perpetual curacy; net income, £66; patrons, Dean and Chapter of Ripon. The chapel is dedicated to St. Michael. The Wesleyan Methodists have a place of worship here. Six poor children are taught for the interest arising from £100, the gift of Ralph Lowther, in 1770.

SAWLEY, an extra-parochial liberty, in the union of CLITHEROE, Western Division of the wapentake of STAINCLIFFE and EWCROSS, West Riding of the county of YORK, 4 miles (N. E.) from Clitheroe; containing with Tosside, 588 inhabitants. A Cistercian abbey, in honour of the Blessed Virgin, was founded here in 1146, by William de Percy, the revenue of which, at the dissolution, was estimated at £221. 15. 8. The ruined gate-house has been converted into a cottage, and the greater part of the nave and transept, with the foundations of the choir and chapter-house, still remain. A school is partly supported by Earl De Grey.

SAWSTON (*St. Mary*) a parish, in the union of LINTON, hundred of WHITTLESFORD, county of CAMBRIDGE, $5\frac{1}{4}$ miles (W. N. W.) from Linton; containing 771 inhabitants. This place is situated on the high road from London to Cambridge. An extensive paper-mill affords employment to about 60 persons, and a manufacture of parchment to about 20. The ancient manor house was built with materials given to the Huddleston family by Queen Mary, who spent some time at that mansion. The living is a vicarage, valued in the king's books at £13. 10. $2\frac{1}{2}$.; present net income, £118; patrons and impropriators, John Gosling and R. Huddleston, Esqrs. There is a place of worship for Independents. Four almshouses, built in 1819, from the proceeds of a bequest by John Huntingdon, in 1554, are inhabited rent-free by aged men and women.

SAWTRY (*All Saints*), a parish, in the hundred of NORMAN-CROSS, union and county of HUNTINGDON, $3\frac{1}{2}$ miles (S.) from Stilton; containing 510 inhabitants. The living is a rectory, valued in the king's books at £8. 15. $7\frac{1}{2}$.; present net income, £189; patrons, Duke of Devonshire and M. M. Middleton, Esq. The church, which is ancient, has three seats, or stalls, on the north side of the chancel. There is a place of worship for Wesleyan Methodists. A monastery, in honour of the Blessed Virgin, was founded in 1146, by Simon Earl of Northampton, who brought hither from the abbey of Wardon, or Sartis, in Bedfordshire, a convent of Cistercian monks, whose revenue at the dissolution was estimated at £199. 11. 8.

SAWTRY (*St. Andrew*), a parish, in the hundred of NORMAN-CROSS, union and county of HUNTINGDON, $3\frac{1}{2}$ miles (S. by E.) from Stilton; containing 320 inhabitants. The living is a rectory, valued in the king's books at £8. 1. $0\frac{1}{2}$.; present net income, £169; patron, A. A. Bletchington, Esq.

SAWTRY (*St. Judith*), an extra-parochial liberty (formerly a parish), in the hundred of NORMAN-CROSS, union and county of HUNTINGDON, 4 miles (S. by E.) from Stilton; containing 227 inhabitants. Here, it is said, was formerly a cell to the abbey of Ramsey: the church has long been demolished.

SAXBY (*St. Peter*), a parish, in the union of MELTON-MOWBRAY, hundred of FRAMLAND, Northern Division of the county of LEICESTER, $4\frac{1}{2}$ miles (E. by N.) from Melton-Mowbray; containing 206 inhabitants. The living is a discharged rectory, with the vicarage of Stapleford consolidated, valued in the king's books at £5; present net income, £168; patron, Earl of Harborough. The Melton-Mowbray and Oakham canal passes through the parish. The neighbourhood presents the appearance of having been the scene of some sanguinary contest; the skeletons of horses and men, earthen urns, which are supposed to have contained the hearts of the slain, bridle bits, fibulæ, &c., and weapons in use before the introduction of fire arms, having been discovered three feet below the soil, and immediately upon the surface of the gravel, in digging for which the workmen invariably found heaps of pebbles laid upon the bodies.

SAXBY (*St. Helen*), a parish, in the Eastern Division of the wapentake of ASLACOE, parts of LINDSEY, union and county of LINCOLN, $7\frac{1}{4}$ miles (W. by S.) from Market-Rasen; containing 124 inhabitants. This parish is bounded on the west by the ancient Roman road from Lincoln to the Humber, at the distance of two miles from the village, and by two inconsiderable streams which, uniting at the north-eastern extremity, form the river Ancholme, which becomes navigable at Bishopbridge, four miles distant. The living is a vicarage, with the rectory of Firsby united, valued in the king's books at £7. 4. 1.; present net income, £46; patron and impropriator, Earl of Scarborough. The church, a neat edifice in the Grecian style of architecture, is the place of interment for the family of the Earl of Scarborough. Stone is found in the parish, but is used only for common buildings, and for repairing the roads.

SAXBY (*All Saints*), a parish, in the union of GLANDFORD-BRIDGE, Northern Division of the wapentake of YARBOROUGH, parts of LINDSEY, county of LINCOLN, 5 miles (S. W.) from Barton-upon-Humber; containing 260 inhabitants. The living is a rectory, valued in the king's books at £12. 18. $6\frac{1}{2}$.; patron, J. W. Barton, Esq. The tithes have been commuted for a rent-charge of £432. 17. 6., subject to the payment of rates, which on the average have amounted to £22; the glebe comprises 8 acres, valued at £14 per annum.

SAXELBY (*St. Peter*), a parish, in the union of MELTON-MOWBRAY, hundred of EAST GOSCOTE, Northern Division of the county of LEICESTER, 4 miles (W. N. W.) from Melton-Mowbray; containing 120 inhabitants. The living is a rectory, valued in the king's books at £9; present net income, £215; patron, Earl of Aylesford. The late Mr. Houghton left £7 per annum for the education of poor children.

SAXELBY (*St. Botolph*), a parish, in the wapentake of Lawress, parts of Lindsey, union and county of Lincoln, 6½ miles (N. W. by W.) from Lincoln; containing 719 inhabitants. The living is a discharged vicarage, valued in the king's books at £10; present net income, £167; patron, Bishop of Lincoln; impropriator, Lord Monson. There is a place of worship for Wesleyan Methodists.

SAXHAM, GREAT (*St. Andrew*), a parish, in the union and hundred of Thingoe, Western Division of the county of Suffolk, 5 miles (W. by S.) from Bury-St. Edmund's; containing 260 inhabitants. The living is a rectory, valued in the king's books at £11. 13. 11½.; present net income, £330; patron, William Mills, Esq., who has a seat here.

SAXHAM, LITTLE (*St. Nicholas*), a parish, in the union and hundred of Thingoe, Western Division of the county of Suffolk, 3¾ miles (W.) from Bury-St. Edmund's; containing 198 inhabitants. The living is a rectory, valued in the king's books at £8. 11. 5½.; present net income, £300; patron, Marquess of Bristol. The tower of the church, which is round, is remarkable for the elegance of the design; this and the south door are the chief Norman features remaining. A school is chiefly supported by the Marchioness of Bristol.

SAXLINGHAM (*St. Margaret*), a parish, in the union of Walsingham, hundred of Holt, Western Division of the county of Norfolk, 3¾ miles (W. by N.) from Holt; containing 153 inhabitants. The living is a rectory, with that of Sharrington annexed, valued in the king's books at £12. 17. 3½.; present net income, £589; patron, Sir R. P. Jodrell, Bart. The church contains a large and costly monument in the form of an Egyptian pyramid, ornamented with numerous hieroglyphics, erected by Sir Christopher Heydon to the memory of his lady, who died in 1593. There is a place of worship for Wesleyan Methodists.

SAXLINGHAM-NETHERGATE (*St. Mary*), a parish, in the union and hundred of Henstead, Eastern Division of the county of Norfork, 3½ miles (N. E.) from Stratton-St. Mary; containing 666 inhabitants. The living is a discharged rectory, with that of Saxlingham-Thorpe united, valued in the king's books at £13. 11. 8.; present net income, £699; patron, Rev. J. H. Steward. The church contains monuments to the Rev. John Baron, Dean of Norwich, and the Rev. J. Gooch, Archdeacon of Sudbury, and his lady. There is a place of worship for Baptists. About 50 persons in this parish are employed in weaving.

SAXLINGHAM-THORPE (*St. Mary*), a parish, in the union and hundered of Henstead, Eastern Division of the county of Norfolk, 3¼ miles (N. E. by E.) from Stratton-St. Mary; containing 161 inhabitants. The living is a discharged rectory, united to that of Saxlingham-Nethergate, and valued in the king's books at £6. 13. 4. The church has fallen into ruins.

SAXMUNDHAM (*St. John the Baptist*), a market-town and parish, in the union and hundred of Plomesgate, Eastern Division of the county of Suffolk, 20 miles (N. E. by N.) from Ipswich, and 89 (N. E.) from London; containing 1048 inhabitants. This town, supposed to be of Saxon origin, is situated in a valley, on the high road to London, and near a small stream which flows on the eastern side into the Ore: it consists chiefly of one street running north and south, of modern and newly-fronted houses, and is of neat and respectable appearance: the inhabitants are plentifully supplied with water from springs: there is an assembly-room, in which balls and concerts are occasionally held. The only branch of business is that in malt. The market is on Thursday, and is noted for corn, which is shipped in large quantities from Snape and Iken wharfs for London. Fairs are on Whit-Tuesday, and on the first Thursday in October, for toys, &c. The town has been made a polling-place for the eastern division of the county. The living is a discharged rectory, valued in the king's books at £8. 15. 10.; present net income, £275; patron, William Long, Esq. The church is a neat edifice embosomed in trees, standing a little southward of the town; a gallery has been erected and 90 free sittings provided, the Incorporated Society having granted £75 in aid of the expense: it contains several monuments to the family of Long, who have their seat at Hurt's hall, in this parish. There is a place of worship for Independents; also a National school. A chantry was founded here by Robert Swan, about 1308.

SAXONDALE, a township, in the parish of Shelford, union, and Southern Division of the wapentake, of Bingham, Southern Division of the county of Nottingham, 8 miles (E.) from Nottingham; containing 116 inhabitants.

SAXTEAD (*All Saints*), a parish, in the union and hundred of Hoxne, Eastern Division of the county of Suffolk, 2¼ miles (N. W.) from Framlingham; containing 505 inhabitants. The living is annexed to the rectory of Framlingham. The tithes have been commuted for a rent-charge of £340, subject to the payment of rates, which on the average have amounted to £107. The steeple of the church fell down in 1806, and has not been rebuilt.

SAXTHORPE (*St. Andrew*), a parish, in the union of Aylsham, hundred of South Erpingham, Eastern Division of the county of Norfolk, 5¼ miles (N. W. by W.) from Aylsham; containing 362 inhabitants. The living is a discharged vicarage, valued in the king's books at £4. 13. 4.; present net income, £139; patrons and impropriators, Master and Fellows of Pembroke Hall, Cambridge.

SAXTON (*All Saints*), a parish, in the Upper Division of the wapentake of Barkstone-Ash, West Riding of the county of York; containing 522 inhabitants, of which number, 407 are in the township of Saxton with Scarthingwell, 4¼ miles (S. by W.) from Tadcaster. The living is a perpetual curacy; net income, £83; patron and impropriator, J. R. O. Gascoigne, Esq., who partly supports a school. Lords Dacre and Westmorland, with a vast number of the slain in the sanguinary battle of Towton, fought on March 29th, 1461, between the houses of York and Lancaster, were interred here.

SCACKLETON, a township, in the parish of Hovingham, union of Malton, wapentake of Bulmer, North Riding of the county of York, 10 miles (W. by N.) from New-Malton; containing 164 inhabitants. A school is supported by William Garforth, Esq.

SCAFTWORTH, a township, in the parish of Everton, union of East-Retford, liberty of Southwell and Scrooby, though locally in the wapentake of Bassetlaw, Northern Division of the county of NOTTINGHAM, 1¼ mile (S. E. by E.) from Bawtry; containing 78 inhabitants.

SCAGGLETHORPE, a township, in the parish of SETTRINGTON, union of MALTON, wapentake of BUCKROSE, East Riding of the county of YORK, 3 miles (E. by N.) from New Malton; containing 252 inhabitants. There is a place of worship for Primitive Methodists; and two schools are supported by the lord of the manor and the rector.

SCALBY, a township, in the parish of BLACKTOFT, union of HOWDEN, wapentake of HOWDENSHIRE, East Riding of the county of YORK, 6¼ miles (E. by N.) from Howden; containing 127 inhabitants.

SCALBY (*St. Lawrence*), a parish, in the union of SCARBOROUGH, PICKERING lythe, North Riding of the county of YORK; containing 1676 inhabitants, of which number, 583 are in the township of Scalby, 3¼ miles (N. W. by W.) from Scarborough. The living is a discharged vicarage, valued in the king's books at £6. 13. 4.; present net income, £302; patrons and appropriators, Dean and Chapter of Norwich. There is a chapel of ease at Cloughton. Four schools are supported by subscription. At Scalby Mill, in this parish, are tea-gardens, resorted to by the company that visit Scarborough during the season for bathing.

SCALDWELL (*St. Peter and St. Paul*), a parish, in the union of BRIXWORTH, hundred of ORLINGBURY, Northern Division of the county of NORTHAMPTON, 8½ miles (N. by E.) from Northampton; containing 387 inhabitants. The living is a rectory, valued in the king's books at £14. 0. 10.; present net income, £357; patron, Duke of Buccleuch. A school is supported by endowment.

SCALEBY (*All Saints*), a parish, in the union of LONGTOWN, ESKDALE ward, Eastern Division of the county of CUMBERLAND; containing 560 inhabitants, of which number, 212 are in the township of East Scaleby, 6½ miles, and 348 in that of West Scaleby, 5½ miles, (N. E. by N.) from Carlisle. The living is a discharged rectory, valued in the king's books at £7. 12. 1.; present net income, £107; patron, Bishop of Carlisle. The church was repaired in 1827. Richard Tilliol, called Richard the Rider, received a grant of this territory from Henry I., and built a castle upon it with materials brought from the Picts wall. In the early part of the parliamentary war, Scaleby castle was garrisoned for Charles I.; in 1645 it surrendered to the parliamentarians; in 1648 it again fell into the hands of the royalists, but was soon after re-captured and kept for the parliament. It is an interesting monument of antiquity; the more ancient portion is in ruins, but a portion has been rebuilt and is inhabited. There is a trifling sum for the support of a school left by Joseph Jackson, in 1773. The late Rev. William Gilpin, author of the lives of the Reformers, Forest Scenery, &c., was born in the castle in 1724.

SCALES, a joint township with Bromfield and Crookdake, in the parish of BROMFIELD, ALLERDALE ward, below Derwent, Western Division of the county of CUMBERLAND, 5½ miles (S. W. by W.) from Wigton: the population is returned with Blomfield.

SCALES, a joint township with Newton, in the parish of KIRKHAM, union of the FYLDE, hundred of AMOUNDERNESS, Northern Division of the county palatine of LANCASTER, 2¼ miles (S. E.) from Kirkham: the population is returned with Newton.

SCALFORD (*St. Egelwin the Martyr*), a parish, in the union of MELTON-MOWBRAY, hundred of FRAMLAND, Northern Division of the county of LEICESTER, 4 miles (N. by E.) from Melton-Mowbray; containing 467 inhabitants. The living is a vicarage, valued in the king's books at £8. 1. 10½.; present net income, £255; patron and impropriator, Duke of Rutland. A school is partly supported from the proceeds of land for the use of the poor, amounting to £13. 16. per annum.

SCAMBLESBY, a parish, in the union of HORNCASTLE, Northern Division of the wapentake of GARTREE, parts of LINDSEY, county of LINCOLN, 6¾ miles (N. by E.) from Horncastle; containing 413 inhabitants. The living is a perpetual curacy; net income, £71; patron and impropriator, Lord Yarborough.

SCAMMONDEN, a chapelry, in the parish and union of HUDDERSFIELD, Upper Division of the wapentake of AGBRIGG, West Riding of the county of YORK, 7½ miles (W.) from Huddersfield; containing 912 inhabitants. The living is a perpetual curacy; net income, £186; patron, Vicar of Huddersfield; impropriator, Sir J. Ramsden, Bart.

SCAMPSTON, a chapelry, in the parish of RILLINGTON, union of MALTON, wapentake of BUCKROSE, East Riding of the county of YORK, 5¼ miles (N. E. by E.) from New Malton; containing 231 inhabitants. The living is a perpetual curacy; net income, £59; patron, Vicar of Rillington.

SCAMPTON (*St. John the Baptist*), a parish, in the wapentake of LAWRESS, parts of LINDSEY, union and county of LINCOLN, 5¾ miles (N. N. W.) from Lincoln; containing 242 inhabitants. The living is a discharged rectory, valued in the king's books at £8. 16. 8.; present net income, £82; patron, Sir George Cayley, Bart. Two schools are partly supported by the rector and curate.

SCANSBY, a joint township with Pickburn, in the parish of BRODSWORTH, union of DONCASTER, Northern Division of the wapentake of STRAFFORTH and TICKHILL, West Riding of the county of YORK: the population is returned with the parish.

SCARBOROUGH (*St. Mary*), a borough, market-town, and parish, having separate jurisdiction, and the head of a union, locally in Pickering lythe, North Riding of the county of YORK, 39 miles (N. E.) from York, and 216 (N.) from London; containing 8760 inhabitants. The origin of this town has not been satisfactorily ascertained: it is supposed to have derived its name from the Saxon *Scear*, a rock, and *Burgh*, a fortified place. The earliest authentic record of it is a charter of Henry II., conferring certain privileges on the inhabitants; and in the reign of Henry III., a charter was granted for making a new pier at *Scardeburgh*, as it was then called. Prior to the construction of the pier, the town began to rise into importance, and was defended by walls and a fosse, of which some vestiges may still be traced. In the reign of

Corporation Seal.

Stephen, a castle had been erected by William le Gros, Earl of Albemarle and Holdernesse, which that nobleman was compelled to surrender to Henry II., who made considerable additions to it. In this castle Piers Gaveston took refuge from the attacks of the confederate barons, and for a considerable time maintained it against their assaults, till a scarcity of provisions obliged him to surrender. In this reign the town was burnt by the Scottish forces, who, headed by Robert Bruce, their king, made an irruption into England. Robert Aske, the leader of the insurrection called the Pilgrimage of Grace, made an unsuccessful attempt to gain possession of the castle in 1536; and during Wyat's rebellion in 1553, it was surprised and taken by a party headed by Thomas, second son of Lord Stafford, who had disguised themselves as peasants: but it was soon retaken by the Earl of Westmorland, and Stafford and three of his accomplices being made prisoners, were sent to London and executed for high treason.

During the civil war in the reign of Charles I., the parliamentarian forces, commanded by Sir John Meldrum, besieged the castle, which held out under its brave governor, Sir Hugh Cholmley, for more than twelve months; and after the death of their leader, who fell in the assault, the command devolved upon Sir Matthew Boynton, to whom, after the exhaustion of its military stores, the fortress was surrendered in 1645, upon honourable terms. Colonel Boynton, who succeeded Sir Matthew in the command of the castle, having declared for the king, it came again into the possession of the royalists; but it was finally surrendered to the parliament in 1648, and soon afterwards dismantled. George Fox, founder of the Society of Friends, was confined in it in 1665. During the rebellion in 1745, the castle was put into a state of temporary repair; and since that time batteries have been erected for the protection of the town and harbour, and within the enclosure are barracks for the accommodation of 120 men. This once formidable fortress comprised within the boundary walls an area of more than nineteen acres, and occupied the summit of an eminence 300 feet above the level of the sea, which surrounds it on all sides except on the west, by which it is connected with the town, and on the north, east, and south is a vast range of perpendicular rocks; the entrance is through an arched gateway, on the summit of a narrow isthmus, flanked by bastions, and formerly defended by two drawbridges within the gates, and a deep fosse. The principal parts now remaining are the keep, a square tower, the walls of which are 12 feet thick, and some portions of the semicircular towers which defended the ramparts, now falling rapidly to decay; some slight remains of the chapel are still discernible within the walls: the castle and its precincts are extra-parochial.

The town is beautifully and romantically situated in the recess of a fine open bay, on the coast of the North Sea, and consists of several spacious streets of handsome well-built houses, rising in successive tiers from the shore in the form of an amphitheatre: the beach, of firm and smooth sand, slopes gradually towards the sea, and affords at all times safe and commodious sea-bathing, for which the town is celebrated. On the cliffs are many new and handsome houses for private residence, and numerous lodging-houses have been erected for the accommodation of visiters, who repair hither, either for the convenience of sea-bathing, for which the water of the bay, unimpaired in its quality by the influx of any stream of fresh water, is peculiarly favourable; or for the benefit of the mineral springs, the efficacy of which, in numerous diseases, has been for more than two centuries in the highest repute. These springs, which are saline chalybeates, varying in the proportions of their several ingredients, were for some time lost by the sinking of a large mass of the cliff in 1737, but were recovered after a diligent search. The principal are the south and the north wells, situated at the base of the cliff, south of the town, near the sea-shore, where a handsome promenade-room, 75 feet in length, was erected in 1839 for the accommodation of visiters. The water of the south well contains 98 ounces, and that of the north well 100 ounces, of carbonic acid gas in a gallon; the former is purgative, and the latter tonic. A fine terrace, 100 feet above the level of the sands, forms a pleasant marine promenade, and leads to a handsome iron bridge of four arches on stone pillars, connecting the dissevered cliffs, in the chasm between which runs the stream called Millbeck. This bridge affords facility of access to the spas, and was erected in 1827: it is 414 feet in length and 75 in height, and constitutes one of the principal ornaments of the town. Adjacent to the bridge is the museum, an elegant circular building with a dome, erected by subscription in 1829, for the investigation and illustration of the natural history of the district, and already displaying many valuable specimens: it is supported by a Scientific Institution, or Philosophical Society. There are five separate bathing establishments, where warm sea-water baths may be obtained at any time: three of them are situated on the cliff, and the others near the pier; they are under the superintendence of medical practitioners residing in the town. There is also a general sea-bathing infirmary, supported by subscription, for the use of poor invalids, who, on the plan of the infirmary at Margate, are boarded and lodged upon very moderate terms, and, during their residence in it, have the gratuitous use of the waters. The theatre is open during the season; and assemblies are held occasionally in a handsome suite of rooms elegantly fitted up for that purpose. The environs are beautifully diversified with hill and dale, and include much picturesque and romantic scenery; Olivers' mount, about a mile from the town, approached by a gradual ascent, forms a magnificent natural terrace, 500 feet above the level of the sea, commanding an interesting view of the Castle hill, with its venerable ruins, the town, the harbour, and the piers on one side, and an extensive view of the ocean on the other. The rides are pleasant; and the salubrity of the air, and the numerous objects of interest with which the neighbourhood abounds, have rendered Scarborough a favourite place of fashionable resort. The town is supplied with fresh water by pipes from the hill, two miles distant, and likewise by a reservoir, capable of containing 4000 hogsheads.

The port is a member of the port of Hull, and its limits extend from the most easterly part of Flamborough Head, in a direction northward, to Peasholme Beck, including all the sea-coast to fourteen fathoms of water, at low water mark. The foreign trade is principally with France, Holland, and the Baltic, from which places it imports wine, brandy, geneva, timber, deals, hemp, flax, and iron: it carries on also a considerable coasting trade in corn, butter, bacon, and salt-fish, with Newcastle, Sunderland, and other places on the coast, and with the port of London for groceries. There are 173 ships belonging to the port, averaging a burden of 164 tons. The amount of duties paid at the custom-house, for the year ended Jan. 5th, 1837, was £2139. The harbour, though confined at the entrance, is easy of access, and safe and commodious within: it is protected by two piers, of which the one was considerably enlarged by act of parliament obtained in the 5th of George II.; it is 1200 feet in length, and 42 feet broad at the extremity, and in the intermediate line varies from thirteen to eighteen feet in breadth. This pier having been found insufficient to prevent the accumulation of sand in the harbour, an act was obtained for the construction of a new pier, of which the breadth at the foundation is 60 feet, and at the curvature, where it is most subject to the action of the waves, 63 feet; it is 40 feet high, 42 feet in breadth at the top, and 1200 feet in length, and was designed by Smeaton, the celebrated engineer. To defray the expense of this undertaking, a duty of one halfpenny per chaldron was granted on all coal brought from Newcastle, with other duties on shipping frequenting the port. The custom-house, a neat building on the sand side, is under the superintendence of the usual officers; and several steam-packets touch at this port every week, on their passage between London and Edinburgh. The fishery was formerly carried on to a considerable extent, and was a source of great profit to the town, but has of late greatly declined. There are some establishments for ship-building, and several manufactories for cordage. The market days are Thursday and Saturday, the former for corn; the fish market is held on the sands near the harbour: the fairs are on Holy Thursday and Nov. 23d, chiefly for cattle. Scarborough is a borough by prescription; the first charter was granted by Henry II. in 1181, and was subsequently confirmed and extended by various sovereigns; the government was vested in two bailiffs, who were justices of the peace, two coroners, four chamberlains, 36 capital burgesses or common councilmen, a recorder, town-clerk, and subordinate officers. The corporation now consists of a mayor, six aldermen, and 18 councillors, by the act of the 5th and 6th of William IV., cap. 76, for an abstract of which see the Appendix, No. I. The borough is divided into two wards, and the municipal boundaries are co-extensive with those for parliamentary purposes. The mayor and late mayor are justices of the peace, with five others appointed by commission. The borough first exercised the elective franchise in the 23rd of Edward I., since which time it has regularly returned two members to parliament. The right of election was formerly vested in the bailiffs and corporation; but, by the act of the 2nd of William IV., cap. 45, it has been extended to the £10 householders within the borough; the number of voters now registered is 558, of whom 26 are freemen; the mayor is the returning officer. The corporation hold quarterly courts of session, for all offences not capital; a court of pleas, for the recovery of debts to any amount; and manorial courts. Petty sessions are held every Wednesday; and the town is a polling-place for the North Riding. The town-hall is a spacious and commodious building, in which the several courts are held, and the public business of the corporation is transacted. The borough gaol and the house of correction, distinct buildings, are but ill adapted for the classification of prisoners; the former, chiefly for debtors, contains four rooms, and the latter only three, without airing-yards.

The living is a discharged vicarage, valued in the king's books at £13. 6. 8.; present net income, £243; patron, Lord Hotham; the impropriation belonged to the Corporation, who allowed the tithes to be redeemed by individual proprietors, so that the parish is now nearly tithe-free. The church anciently belonged to the Cistercian monastery, and was formerly a spacious and magnificent cruciform structure, with three noble towers: it sustained considerable damage during the sieges of the castle, in the time of the parliamentary war, and retains but few portions of its ancient character: the present steeple stands at the eastern end. Christ-church, a handsome edifice, in the later style of English architecture, with a square embattled tower crowned with pinnacles, was erected in 1828, at an expense of £5000, by grant from the parliamentary commissioners, exclusively of a local subscription of £3000, and the stone, which was the gift of Sir John V. B. Johnstone, Bart. The incumbency is a perpetual curacy; net income, about £200; patron, Vicar of St. Mary's. A chapel has been lately erected, containing 440 sittings, 320 of which are free, the Incorporated Society having granted £300 in aid of the expense. There are places of worship for Baptists, the Society of Friends, Independents, and Wesleyan Methodists, and a Roman Catholic chapel. The grammar school is of very remote origin; in 1648, the corporation ordered a part of St. Mary's church to be fitted up and appropriated to its use, the expense of which was defrayed by the sale of the Charnel Chapel, or old school-house; the income, arising from donations of land and money, is about £12 per annum: gratuitous instructiou is afforded to four scholars only. The Amicable Society, now consisting of about 300 members, was established in 1729, under the patronage of Robert North, Esq., for clothing and educating the poor children of the town, which is, to a considerable extent, accomplished by weekly contributions of its members, and general subscription. A National school is supported by subscription, and there are Sunday schools in connection with the established church and the dissenting congregations. The Seamen's hospital was erected in 1752, by the ship-owners of the town, for the maintenance and support of aged seamen, their widows, and children: it is supported by a contribution of sixpence per month from the owner of every vessel belonging to the port, for each person on board during the time the vessel is at sea, or in actual service, and is under the superintendence of a president and trustees, annually elected: the income of this hospital, arising from donations, is about £200 per annum. The Trinity-house, originally founded by subscription for similar purposes, in 1602, was rebuilt in 1832. Wilson's Mariner's Asylum, a beautiful range of buildings in the Elizabethan style, for the reception of 14 married persons, was erected and endowed in the lifetime of the donor, in 1837. Taylor's free dwellings for the poor were erected in 1810. St. Thomas' hospital was founded by the corporation, for aged and infirm persons; the buildings are low and of ancient appearance. There are also several friendly societies, and various charitable bequests for distribution among the poor. The poor-law union of Scarborough comprises 33 parishes or places, under the care of 35 guardians, and, according to the census of 1831, contains a population of 17,920. To the north of St. Sepulchre's-street is the site of a Franciscan convent, supposed to have been founded about the 29th of Henry III.; and among other monastic establishments anciently existing here were, a monastery of Dominicans, founded in the reign of Edward II. by Adam Say, Knt., or by Henry Percy, Earl of Northumberland; and a house of Carmelite friars, founded by Edward II., in 1319. Scarborough gives the title of Earl to the family of Lumley.

SCARCLIFF (*St. Leonard*), a parish, in the union of MANSFIELD, hundred of SCARSDALE, Northern Division of the county of DERBY, 6 miles (N. N. W.) from Mansfield; containing 524 inhabitants. The living is a discharged vicarage, valued in the king's books at £5; present net income, £68; patron, Duke of Devonshire; impropriator, Earl Bathurst. The church contains a monument of the eleventh century, representing a lady in robes, with a coronet on her head, and an infant on her left arm: the inscription, in Lombardic capitals, has become illegible from time and mutilation. Ten poor children are instructed for an annuity of £6 bequeathed by Kithe Vaughan, in 1813, and another of 10*s*. by Elizabeth Saxton, in 1815.

SCARCROFT, a township, in the parish of THORNER, Lower Division of the wapentake of SKYRACK, West Riding of the county of YORK, 6¼ miles (S. S. W.) from Wetherby; containing 168 inhabitants.

SCARGILL, a township, in the parish of BARNINGHAM, union of TEESDALE, wapentake of GILLING-WEST, North Riding of the county of YORK, 3½ miles (S. W. by W.) from Greta-Bridge; containing 119 inhabitants.

SCARISBRICK, a township, in the parish and union of ORMSKIRK, hundred of WEST DERBY, Southern Division of the county palatine of LANCASTER, 2 miles (N. W.) from Ormskirk; containing 1783 inhabitants. Here is a Roman Catholic chapel.

SCARLE, NORTH (*All Saints*), a parish, in the union of NEWARK, Lower Division of the wapentake of BOOTHBY-GRAFFO, parts of KESTEVEN, county of LINCOLN, 10 miles (W. S. W.) from Lincoln; containing 479 inhabitants. The living is a discharged rectory, valued in the king's books at £4. 17. 3½.; present net income, £252: it is in the patronage of the Crown. There is a place of worship for Wesleyan Methodists; also a Sunday National school. The river Trent forms a boundary of this parish.

SCARLE, SOUTH (*St. Helen*), a parish, in the union, and Northern Division of the wapentake, of NEWARK, Southern Division of the county of NOTTINGHAM, 7½ miles (N. E. by N.) from Newark; containing 479 inhabitants. The living is a discharged vicarage, with the perpetual curacy of Girton annexed, valued in the king's books at £5. 2. 5.; present net income, £168; patron, Prebendary of South Scarle in the Cathedral Church of Lincoln; impropriators, G. Hutton, Esq., and others. The impropriate tithes have been commuted for a rent-charge of £190. 11. 9., and the vicarial for £54. 18. 6., subject to the payment of rates, which on the average have respectively amounted to £24 and £6; the impropriate glebe comprises 112 acres, valued at £125 per annum.

SCARNING (*St. Peter and St. Paul*), a parish, in the union of MITFORD and LAUNDITCH, hundred of LAUNDITCH, Western Division of the county of NORFOLK, 2 miles (W. S. W.) from East Dereham; containing 603 inhabitants. The living is a discharged vicarage, endowed with a moiety of the rectorial tithes, and valued in the king's books at £9. 19.; present net income, £389; patron, E. Lombe, Esq.; impropriator of the remainder of the rectorial tithes, E. Long, Esq. A free school was founded here and endowed by William Secker, in 1604.

SCARRINGTON, a parish, in the union, and Northern Division of the wapentake, of BINGHAM, Southern Division of the county of NOTTINGHAM, 12½ miles (E. by N.) from Nottingham; containing 188 inhabitants. The living is annexed, with that of Thoroton, to the vicarage of Orston. There is a place of worship for Wesleyan Methodists.

SCARTHINGWELL, a joint township with Saxton, in the parish of SAXTON, Upper Division of the wapentake of BARKSTONE-ASH, West Riding of the county of YORK, 4¾ miles (S. by E.) from Tadcaster: the population is returned with Saxton.

SCARTHO (*St. Giles*), a parish, in the union of CAISTOR, wapentake of BRADLEY-HAVERSTOE, parts of LINDSEY, county of LINCOLN, 2¼ miles (S.) from Great Grimsby; containing 147 inhabitants. The living is a discharged rectory, valued in the king's books at £8. 10. 10.; present net income, £231; patrons, Principal and Fellows of Jesus' College, Oxford. A day and Sunday school is partly supported by subscription.

SCATHWAITERIGG-HAY, a joint township with Hutton-i'-th'-Hay, in the parish, union, and ward of KENDAL, county of WESTMORLAND, 2 miles (N. E.) from Kendal; containing 380 inhabitants.

SCATTERGATE, a township, in the parish of APPLEBY-ST. LAWRENCE, EAST ward, county of WESTMORLAND; containing 179 inhabitants. It adjoins the town of Appleby on the south, and within the township are the venerable remains of Appleby Castle. Dr. Waugh, Bishop of Carlisle, was a native of this place.

SCAWBY (*St. Hibald*), a parish, in the union of GLANDFORD-BRIDGE, Eastern Division of the wapentake of MANLEY, parts of LINDSEY, county of LINCOLN, 2½ miles (W. S. W.) from Glandford-Bridge; containing 942 inhabitants. This place, which is of considerable antiquity, has from a remote period belonged to the family of Nelthorpe, of whom Sir John, who died in London, in 1669, was the first baronet, and whose descendant, Sir John Nelthorpe, Bart., is now lord of the manor. The living is a discharged vicarage, valued in the king's books at £7; present net income, £170; patron and impropriator, Sir John Nelthorpe, Bart. The church contains a good monument to Richard Nelthorpe, Gent., who died in 1640, and another to Sir John, the first baronet. There are places of worship for Calvinists, and Primitive and Wesleyan Methodists. A free school was founded in 1705, by Sir Henry Nelthorpe, who endowed it with land now producing £30 per annum. At Weston, a hamlet in this parish, are evident remains of a Roman station. In the garden of Henry Grantham, Esq., are two tesselated pavements, one about 16 feet square, and the other 12 feet long and 8 feet wide, communicating by a narrow passage; the latter appears to have been used as a dressing-room, and at the south end is a semicircular bath. Several coins of Constantine have been found here; and at a farm-house, about 300 yards distant, are vestiges of a fortified camp, on the site of which a religious house appears to have been erected, in the walls of which some heads sculptured in stone are still visible. Here have been found a crucifix of jet, an ear-drop of stone, which are now in the possession of Mr. Grantham, and several rings which fell to pieces on being exposed to the air. Mr. Grantham, at a short distance from his house, discovered a stone vault, in which were a small earthen vase, some human bones, a plated bracelet, a few rings, part of an old stone coffin, and other relics; the same gentleman has also two ancient British battle axes, fragments of pottery, and other relics found in different parts of his premises. There is a mineral spring in the parish, but not used for medicinal purposes.

SCAWTON (*St. Mary*), a parish, in the union of Helmsley-Blackmoor, wapentake of Ryedale, North Riding of the county of York, 5 miles (W.) from Helmsley; containing 148 inhabitants. The living is a discharged rectory, valued in the king's books at £2. 19. 2.; present net income, £145; patron, William Worsley, Esq. The church is in the early style of English architecture.

SCHOLES, a joint township with Morwick, in the parish of Barwick-in-Elmett, Lower Division of the wapentake of Skyrack, West Riding of the county of York, 7 miles (E. N. E.) from Leeds: the population is returned with the parish. There is a place of worship for Wesleyan Methodists.

SCILLY ISLANDS. These Islands, of which there are 17, varying in extent from 1640 acres to 10, besides 22 smaller islets, and numerous naked rocks, form a cluster lying off the south-west coast, and annexed to the Western Division of the county of Cornwall, about 17 leagues due west from the Lizard point, and 10 leagues nearly west by south from the Land's End. By the Greeks they were called *Hesperides* and *Cassiterides*; by the Romans, *Sellinæ* and *Siluræ Insulæ*: their present name of Scilly, anciently written Sully, or Sulley, appears to be British, and they are reported to take it from a small island, containing only one acre, which is called Scilly Island. Except what relates to their trading intercourse with the Phœnicians and the Romans, and the circumstance of their having been occasionally appropriated by the latter as a place of banishment for state criminals, the first mention we find of them in history is in the tenth century, when they were subdued by King Athelstan. From this period there is no record of any remarkable historical event respecting them until the reign of Charles I., when they became of considerable importance as a military post. In 1645, they afforded a temporary asylum to Prince Charles and his friends, Lords Hopton and Capel. In 1649, Sir John Grenville being governor of the Scilly Islands, fortified and held them for Charles II. The parliament, finding their trading vessels much annoyed by his frigates, fitted out an expedition for the reduction of the islands, under the command of Admiral Blake and Sir George Ascue, who first took possession of those of Trescaw and Bryer, and threw up fortifications for the purpose of attacking Sir John Grenville, at St. Mary's. The Dutch Admiral, Van Tromp, is said to have made insidious, but ineffectual, proposals to the governor to take the islands under his protection. Resistance being found vain, they were delivered up to the parliament, in the beginning of June of the same year, this having been one of the last rallying points for the royalists: the garrison consisted of 800 soldiers, with numerous commissioned officers.

The total surface of the islands is about 4700 acres, and the number of inhabitants 2465. The extent of St. Mary's Island, the largest, including the garrison, which is joined to it by an isthmus, is 1640 acres, and its population is 1311. The principal town, called Hugh-town, or Heugh town, was much damaged by inundation during the great storm in 1744. The pier was finished in 1750, at the expense of Lord Godolphin; vessels of 150 tons' burden may ride here in safety. Near this town are the ruins of an old fortress, with a mount, and the remains of several block-houses and batteries, supposed to have been constructed in the civil war. About two furlongs to the eastward is a bay, called Pomellin, or Porthmellin, where a fine white sand, composed of chrystals and talc, much esteemed as a writing sand, and for other purposes, is procured in abundance. About a mile from Hugh-town is the Church-town, consisting of a few houses and the church, in the chancel of which are interred Sir John Narborough, Bart., son of the celebrated admiral of that name; Henry Trelawney, son of a Bishop of Winchester; and Captain Edmund Loades, of the Association man of war, all of whom shared the fate of Rear-Admiral Sir Cloudesley Shovel, who was lost on the Gilston rock, October 22nd, 1707. Two furlongs further, bordering on the sea, is Old-town, formerly the principal town of the island. On a promontory, called the Giant's Castle, are traces of an ancient fortress, supposed to be of remote origin. On the west side of the island are St. Mary's garrison, with the barracks and several batteries, and Star castle: the latter was built by Sir Francis Godolphin, in 1593. The island next in magnitude is Trescaw, anciently called Iniscaw, and St. Nicholas, which contains 880 acres, and 470 inhabitants. In it are some remains of the conventual church of St. Nicholas, the ruins of Old-castle, and Oliver's Battery. Old-castle, which appears to have been built in or about the reign of Henry VIII., is spoken of by Leland as "a little pile, or fortress:" it appears to have been afterwards enlarged, as its ruins shew it to have been a considerable building. Oliver's castle, as it is called, from its having been built by the parliamentarians, was repaired in 1740; but is described by Borlase, in 1756, as being then already much decayed. St. Martin's island, though next in size to St. Mary's and Trescaw, containing 720 acres, was uninhabited until the reign of Charles II.: it now contains 230 inhabitants. Mr. Ekins, in 1683, built a tower on this island, as a land-mark, 20 feet high, with a spire upon it of the same height. On St. Agnes' island, which contains 289 inhabitants, is a lighthouse. Bryer, or Brehar, contains 330 acres, and 128 inhabitants.

The principal employment and trade of the islanders consist in fishing and making kelp: about 100 boats are used for fishing, piloting &c.: the quantity of kelp annually made varies from 100 to 200 tons. Tin is found in several of the islands, and in some, lead and copper; but no mines are now worked. Others produce grain, chiefly barley, peas, and oats, with a small portion of wheat; a few acres are sown with the pillis, or naked oat: potatoes are cultivated in great qnantities in St. Mary's. Cattle are fed on most of them, and though not very abundant, are sometimes sold to masters of vessels. Samphire, for pickling, is collected in abundance in the isle of Trescaw. The tamarisk and *lavatera arborea* grow plentifully in that of St. Mary.

The property and temporal jurisdiction of these islands were anciently attached to the earldom, as they now are to the duchy, of Cornwall, excepting those of St. Nicholas (now Trescaw), St. Sampson (which contains 37 inhabitants), St. Elid, St. Teon, and Nullo, and some lands in other islands, which were given, in or before the reign of Edward the Confessor, to certain monks, or hermits, who had their abode in the island of St. Nicholas, and were subsequently granted by Henry I. to the abbot of Tavistock. The present lessee of the whole is the Duke of Leeds, the representative of the Godolphin family, to whom they appear to have been first leased in the 13th of Elizabeth. The government of them appears to have been vested, at least since the Reformation, uniformly in the proprietors, except in the instances of Sir John Grenville and Joseph Hunkin, Esq., during the interregnum, and Major Bennet, previously to the year 1733. Before the Reformation, it appears that the proprietor kept the peace of the islands, with the assistance of twelve armed men; and that there were frequent feuds between them and the king's coroner, who came hither to hold assizes for the trial of prisoners accused of greater offences. It is most probable that all minor offences were cognizable, as they now are, by a court delegated by the lord proprietor, whose authority for exercising the civil jurisdiction is derived from a patent of the 10th of King William. The lord proprietor appoints a court, or council of twelve, consisting of some of the principal inhabitants, among whom are generally the military commandant, steward, chaplain, and commissary of musters. Vacancies are supplied by election; but the whole may be dissolved, and a fresh appointment made, at any time, by the lord proprietor. After the death of a lord proprietor, a new council is necessarily appointed. The court generally sits monthly, for the trial of plaints, suits, &c., between the islanders, excepting such causes as affect life and limb, and such as are cognizable by the court of Admiralty. Persons charged with transportable offences, such as receiving stolen goods, &c., are tried here; but the punishment is only fines, or whipping, and sometimes imprisonment. Those accused of murder, burglaries, &c., are conveyed before the nearest Cornish magistrate, and sent to be tried at the assizes for the county of Cornwall.

These islands are under the spiritual jurisdiction of the Bishop of Exeter, and form part of the archdeaconry of Cornwall. In early times the abbot of Tavistock held the tithes of the whole, and certain lands, by the title of finding two monks to reside there, and to provide for the spiritual wants of the inhabitants. Since the Reformation the tithes have been vested in the lord proprietor, who is patron of the donative, and pays the minister an optional salary. Until of late years the minister of St. Mary's was the only clergyman in the islands, officiating constantly at St. Mary's, where a register of baptisms and marriages was kept for all the islands; at Trescaw, only on the Sunday after Easter; and at St. Martin's on Trinity Sunday: the chapels of the other islands were served by laymen, or, as they were called, island clerks, usually fishermen. The Society for Promoting Christian Knowledge now employs two missionaries, who officiate at what are called the Off-islands. There are chapels at Trescaw, St. Martin's, St. Agnes', Bryer, and St. Sampson's, for the most part built by the Godolphin family, since the Reformation; that of Bryer, about 1746: that of St. Agnes' was built at the expense of the Society for Promoting Christian Knowledge, which also gave £400 towards erecting a house for the missionary, at Trescaw; it contains 130 free sittings, the Incorporated Society having granted £75 in aid of the expense. St. Martin's chapel contains 100 free sittings, the Incorporated Society having granted £200 for that purpose. The Wesleyan Methodists have four places of worship in these islands. The Earl of Godolphin, in 1747, established a school for instructing twelve boys. The Rev. Richard Corbet Hartshorne, rector of Broseley in Shropshire, about the year 1753, gave the sum of £250 towards the support either of a minister or schoolmaster at Trescaw, under the direction of the Society for Promoting Christian Knowledge. The only considerable benefaction which the Society has received towards the religious instruction of the islanders, since that time, is the sum of £500, given by Charles Etty, Esq. About £300 per annum is expended by the society, on the missions and schools, chiefly out of their general funds. On St. Helen's island, now uninhabited, are the ruins of houses, and of an ancient chapel.

SCOLE, otherwise OSMONDISTON (*St. Andrew*), a parish, in the union of DEPWADE, hundred of DISS, Eastern Division of the county of NORFOLK, 19½ miles (S. S. W.) from Norwich; containing 617 inhabitants. The living is a discharged rectory, valued in the king's books at £9; present net income, £205; patron, Sir E. Kerrison, Bart. The village is a great thoroughfare on the high road from Ipswich to Norwich and Yarmouth, and contains a very good inn, built in the seventeenth century by a merchant of London, at the expense of £1500.

SCOPWICK (*Holy Cross*), a parish, in the union of SLEAFORD, Second Division of the wapentake of LANGOE, parts of KESTEVEN, county of LINCOLN, 8¾ miles (N.) from Sleaford; containing 278 inhabitants. The living is a discharged vicarage, valued in the king's books at £8; present net income, £185; patron and appropriator, Bishop of Lincoln.

SCORBROUGH (*St. Leonard*), a parish, in the union of BEVERLEY, Bainton-Beacon Division of the wapentake of HARTHILL, East Riding of the county of YORK, 4½ miles (N. N. W.) from Beverley; containing 79 inhabitants. The living is a discharged rectory, valued in the king's books at £7; present net income, £252; patron, Earl of Egremont.

SCOREBY, a joint township with West Stamford-Bridge, in the parish of CATTON, union of YORK, wapentake of OUZE and DERWENT, East Riding of the county of YORK, 6½ miles (E. by N.) from York: the population is returned with West Stamford-Bridge.

SCORTON, a township, in the parish of CATTERICK, union of RICHMOND, wapentake of GILLING-EAST, North Riding of the county of YORK, 2½ miles (N. N. E.) from Catterick; containing 492 inhabitants. A free grammar school here is endowed with £200 a year, the bequest of Leonard Robinson, Esq. The school-house, erected in 1760, stands on the north side of a spacious green, around which the village, which is well built, is situated. The buildings on the east side are occupied by a religious community of thirty nuns, of the order of St. Clair, who emigrated to this country from Normandy, in 1795. There are also about 20 boarders, and they have a neat chapel belonging to the establishment. Within the parish is a spring, called St. Cuthbert's well, the water of which is efficacious in cutaneous and rheumatic disorders.

SCOSTHORPE, a township, in the parish of KIRKBY-IN-MALHAM-DALE, union of SETTLE, Western Division of the wapentake of STAINCLIFFE and EWCROSS, West Riding of the county of YORK, 6½ miles (S. E. by E.) from Settle; containing 95 inhabitants.

SCOTBY, a township, in the parish of WETHERAL, CUMBERLAND ward, Eastern Division of the county of CUMBERLAND, 3½ miles (E. by S.) from Carlisle; containing 397 inhabitants. The rail-road from Carlisle to Newcastle passes through the township. There is a meeting-house, with a burial ground, for the Society of Friends. A school is endowed with land, now producing £16 a year.

SCOTFORTH, a township, in the parish of LANCASTER, hundred of LONSDALE, south of the sands, Northern Division of the county palatine of LANCASTER, 1½ mile (S.) from Lancaster; containing 557 inhabitants. A school is partly supported by subscription.

SCOTHERN (*ST. GERMAN*), a parish, in the wapentake of LAWRESS, parts of LINDSEY, union and county of LINCOLN, 5¼ miles (N. E. by N.) from Lincoln; containing 497 inhabitants. The living is a discharged vicarage, valued in the king's books at £4. 5. 2½.; present net income, £102; patron, Earl of Scarborough, who, with the rector of Sudbrook, is the impropriator. There is a school in which six children are paid for by the parish, and ten are instructed at the expense of a lady.

SCOTTER (*ST. PETER*), a parish, in the union of GAINSBOROUGH, wapentake of CORRINGHAM, parts of LINDSEY, county of LINCOLN, 9¼ miles (N. E. by N.) from Gainsborough; containing 1043 inhabitants. The living is a rectory, valued in the king's books at £22. 4. 2.; present net income, £814; patron, Bishop of Peterborough. There is a place of worship for Wesleyan Methodists; also a National school. The river Eau runs through the parish, and falls into the Trent, which forms its north-west boundary. A charter for a market on Thursday, and a fair on July 10th, was granted by Richard I., but the former has been discontinued. There is a fair for horses and cattle on July 6th, called Scotter shew.

SCOTTLESTHORP, a hamlet, in the parish of EDENHAM, wapentake of BELTISLOE, parts of KESTEVEN, county of LINCOLN: the population is returned with the parish.

SCOTTON (*ST. GENEWYS*), a parish, in the union of GAINSBOROUGH, wapentake of CORRINGHAM, parts of LINDSEY, county of LINCOLN, 8¾ miles (N. E.) from Gainsborough; containing 494 inhabitants. The living is a rectory, valued in the king's books at £23; present net income, £636; patron, Sir Richard Frederick, Bart. There is a chapel of ease at East Ferry. Here is a place of worship for Wesleyan Methodists.

SCOTTON, a township, in the parish of CATTERICK, union of RICHMOND, wapentake of HANG-EAST, North Riding of the county of YORK, 3½ miles (S. S. E.) from Richmond; containing 138 inhabitants.

SCOTTON, a township, in the parish of FARNHAM, Lower Division of the wapentake of CLARO, West Riding of the county of YORK, 2¼ miles (N. W.) from Knaresborough; containing 312 inhabitants.

SCOTTOW (*ALL SAINTS*), a parish, in the union of AYLSHAM, hundred of SOUTH ERPINGHAM, Eastern Division of the county of NORFOLK, 10 miles (N. by E.) from Norwich; containing 460 inhabitants. This parish is situated on the turnpike road from Norwich to North Walsham. Near the church, on an elevated site, is the seat of Sir Henry Durrant, Bart. A fair is held annually on Easter-Tuesday. The living is a vicarage, annexed to the rectory of Belaugh, and valued in the king's books at £8. 13. 6½.; appropriator, Bishop of Norwich. The church is a neat building, with a lofty square embattled tower, and contains monumental inscriptions to the families of Hubbe, Hynde, Blake, Brougham, Robinson, and Durrant.

SCOTT-WILLOUGHBY, county of LINCOLN.—See WILLOUGHBY, SCOTT.

SCOULTON (*ALL SAINTS*), a parish, in the union and hundred of WAYLAND, Western Division of the county of NORFOLK, 4¼ miles (E.) from Watton; containing 328 inhabitants. The living is a discharged rectory, valued in the king's books at £10. 4. 2.; patron, John Weyland, Esq. The tithes have been commuted for a rent-charge of £450, subject to the payment of rates, which on the average have amounted to £43; the glebe comprises 53 acres, valued at £65 per annum. The church has a low tower, of which the basement is square, and the upper story octangular. At the end of each aisle there was formerly a chapel. Two schools are chiefly supported by the lord of the manor and principal proprietor.

SCRAFTON, WEST, a township, in the parish of COVERHAM, union of LEYBURN, wapentake of HANG-WEST, North Riding of the county of YORK, 4½ miles (S. W.) from Middleham; containing 145 inhabitants.

SCRAPTOFT (*ALL SAINTS*), a parish, in the union of BILLESDON, hundred of GARTREE, Southern Division of the county of LEICESTER, 4 miles (E. by N.) from Leicester; containing 126 inhabitants. The living is a vicarage, valued in the king's books at £8. 10.; present net income, £161; patron and impropriator, E. B. Hartopp, Esq. A National school has been established. Here is a curious and ancient stone cross.

SCRATBY (*ALL SAINTS*), a parish, in the hundred of EAST FLEGG, county of NORFOLK, 2½ miles (N. by W.) from Caistor: the population is returned with the parish of Ormsby-St. Margaret. The living is a discharged vicarage, united in 1548 to that of Ormsby-St. Margaret.

SCRAYFIELD (*ST. MICHAEL*), a parish, in the union of HORNCASTLE, hundred of HILL, parts of LINDSEY, county of LINCOLN, 3 miles (E. by S.) from Horncastle; containing 36 inhabitants. The living is a discharged rectory, united to that of Hameringham, and valued in the king's books at £4. 10. 4½.

SCRAYINGHAM (*St. Peter*), a parish, partly in the union of Pocklington, and partly in that of Malton, wapentake of Buckrose, East Riding of the county of York; containing 522 inhabitants, of which number, 164 are in the township of Scrayingham, 9½ miles (N. W. by N.) from Pocklington. The living is a rectory, valued in the king's books at £21. 11. 10½.; present net income, £661: it is in the patronage of the Crown. There is a chapel of ease at Leppingham. A school is chiefly supported by an allowance of £9 per annum from a lady and gentleman.

SCREDINGTON (*St. Andrew*), a parish, in the union of Sleaford, wapentake of Aswardhurn, parts of Kesteven, county of Lincoln, 4¼ miles (S. E. by S.) from Sleaford; containing 306 inhabitants. The living is a discharged vicarage, in the patronage of the Dean and Chapter of Lincoln (the appropriators), valued in the king's books at £6. 15. 4.; present net income, £80.

SCREMBY (*St. Peter and St. Paul*), a parish, in the union of Spilsby, Wold Division of the wapentake of Candleshoe, parts of Lindsey, county of Lincoln, 3½ miles (E. N. E.) from Spilsby; containing 204 inhabitants. The living is a discharged rectory, valued in the king's books at £16. 10. 2½.; present net income, £250; patron and incumbent, Rev. H. Brackenbury.

SCRENWOOD, a township, in the parish of Alnham, union of Rothbury, Northern Division of Coquetdale ward and of the county of Northumberland, 8 miles (N. W.) from Rothbury; containing 37 inhabitants.

SCREVETON (*St. Winifred*), a parish, in the union, and Northern Division of the wapentake, of Bingham, Southern Division of the county of Nottingham, 8½ miles (S. W. by S.) from Newark; containing 312 inhabitants. The living is a discharged rectory, valued in the king's books at £6. 19. 1.; present net income, £252; patrons, the Trustees of T. Hildyard, Esq. The church contains an altar-tomb and effigy to the memory of Gen. Whalley, the supposed executioner of Charles I., who commanded under Cromwell: figures of his three wives and twenty-two children are represented on the same monument. Dr. Thoroton, the antiquary and topographer, was born in an ancient mansion here belonging to his family.

SCRIVELSBY (*St. Benedict*), a parish, in the union of Horncastle, Southern Division of the wapentake of Gartree, parts of Lindsey, county of Lincoln, 2½ miles (S.) from Horncastle; containing 129 inhabitants. The living is a rectory, with that of Dalderby united in the year 1731, valued in the king's books at £12. 17. 6.; present net income, £562; patron, H. Dymoke, Esq. The Dymokes hold the manor of Scrivelsby by "the service of grand serjeantry, that, whenever any king of England is to be crowned, the lord of this manor for the time being, or, in case of sickness, some one for him, shall come well armed for battle, on a good horse, into the presence of our lord the king, at his coronation, and make proclamation that, if any one will say that our said lord the king has not a title to his kingdom and crown, he shall be ready and prepared to defend the right of the king and his kingdom, and the dignity of his crown, in his own person, against him and any other whatsoever."

SCRIVEN, a joint township with Tentergate, in the parish of Knaresborough, Lower Division of the wapentake of Claro, West Riding of the county of York, 1 mile (N. by W.) from Knaresborough; containing, with Tentergate, 1598 inhabitants.

SCROOBY (*St. Wilfred*), a parish, in the union of East Retford, and within the liberty of Southwell and Scrooby, though locally in the wapentake of Bassetlaw, Northern Division of the county of Nottingham, 1¾ mile (S.) from Bawtry; containing 281 inhabitants. The living is a discharged vicarage, united to that of Sutton. In this parish are some remains of an ancient palace of the Archbishops of York, converted into a farm-house; in the garden is a mulberry-tree, said to have been planted by Cardinal Wolsey.

SCROPTON (*St. Paul*), a parish, in the union of Burton-upon-Trent, hundred of Appletree, Southern Division of the county of Derby, 11½ miles (W. S. W.) from Derby; containing 500 inhabitants. The living is a perpetual curacy; net income, £19; patron and impropriator, J. Broadhurst, Esq. A day and Sunday school is partly supported by subscription. The river Dove runs through the parish.

SCRUTON (*St. Radegund*), a parish, in the union of Bedale, wapentake of Hang-East, North Riding of the county of York, 4¼ miles (N. E. by N.) from Bedale; containing 438 inhabitants. The living is a rectory, valued in the king's books at £14. 0. 5.; present net income, £515; patron, F. L. Coore, Esq.

SCULCOATES (*St. Mary*), a parish, and the head of a union, in the Hunsley-Beacon Division of the wapentake of Harthill, East Riding of the county of York, 1¼ mile (N.) from Kingston-upon-Hull; containing 13,468 inhabitants. This place is noticed in Domesday-book as one of the lordships granted to Ralph de Mortimer, a follower of the Conqueror. Its population, less than a century ago, did not exceed 100, but the southern part of the parish, since the construction of its dock on the western bank of the river Hull, in 1774, has been extensively built upon, and now forms a large and populous part of the environs of Hull. The petty sessions for the division are held here in a hall recently erected. The living is a discharged vicarage, valued in the king's books at £5. 6. 8.; present net income, £295: it is in the patronage of the Crown. The parish church was rebuilt in the year 1760. An act was obtained, in 1814, for the erection of an additional church, called Christchurch, which was consecrated in 1822: it is a handsome structure of white brick and Roche abbey stone, and cost upwards of £7000, part of which has been defrayed by subscription. By the same act the presentation was vested in subscribers of £100, and their survivors, till the number be reduced to eight, vacancies in which are to be filled up by the pew-holders, and these eight trustees, together with the vicar of Sculcoates, possess the patronage. The living is a perpetual curacy; net income, £169. Five schools are partly supported by subscription. The poor law union of Sculcoates comprises 18 parishes or places, under the care of 27 guardians, and, according to the census of 1831, contains a population of 29,238.

SCULTHORPE (*All Saints*), a parish, in the union of Walsingham, hundred of Gallow, Western Division of the county of Norfolk, 2 miles (N. W.) from Fakenham; containing 619 inhabitants. The living is a rectory, valued in the king's books at £16; patron,

Sir J. T. Jones, Bart. The tithes have been commuted for a rent-charge of £550, subject to the payment of rates, which on the average have amounted to £102. 11. 8.; the glebe comprises 71 acres, valued at £105. 5. per annum. The church was erected by Sir Robert Knollys, K. G., who, from a common soldier, rose to rank and eminence as a commander under Edward III., acquired an immense fortune, and, for his good services in subduing Wat Tyler's rebellion, received the freedom of the city of London. He died at the manor-house here in the 92nd year of his age, and was buried in the church of the Carmelites, Fleet-street, London.

SCUNTHORPE, a township, in the parish of FRODINGHAM, union of GLANDFORD-BRIDGE, Eastern Division of the wapentake of MANLEY, parts of LINDSEY, county of LINCOLN, 8½ miles (W. N. W.) from Glandford-Bridge; containing 240 inhabitants. There is a place of worship for Wesleyans.

SEA-BOROUGH, a parish, in the union of BEAMINSTER, hundred of CREWKERNE, Western Division of the county of SOMERSET, 2½ miles (S. by W.) from Crewkerne; containing 124 inhabitants. The living is a discharged rectory, valued in the king's books at £6. 15.; present net income, £150; patron, James Gear, Esq.

SEABRIDGE, a township, in the union of STOKE-UPON-TRENT, partly in the parish of SWINNERTON, but chiefly in that of STOKE-UPON-TRENT, Northern Division of the hundred of PIREHILL and of the county of STAFFORD, 1½ mile (S. by W.) from Newcastle-under-Lyme; containing 120 inhabitants.

SEABROOK, a joint hamlet with Horton, in the parish of IVINGHOE, hundred of COTTESLOE, county of BUCKINGHAM, 1½ mile (W. N. W.) from Ivinghoe: the population is returned with Horton.

SEACOMBE, a joint township with Poolton, in the parish of WALLASEY, union, and Lower Division of the hundred, of WIRRALL, Southern Division of the county of CHESTER, 11¾ miles (N. N. E.) from Great Neston; the population is returned with Poolton. It is bounded on the east by the river Mersey, and is situated opposite to the termination of the Leeds and Liverpool canal. There is a place of worship for Independents.

SEACOURT, a hamlet, in the parish of WYTHAM, union of ABINGDON, hundred of HORMER, county of BERKS; containing 25 inhabitants.

SEACROFT, a township, in the parish of WHITKIRK, Lower Division of the wapentake of SKYRACK, West Riding of the county of YORK, 4½ miles (E. N. E.) from Leeds; containing 918 inhabitants. There is a place of worship for Wesleyan Methodists.

SEAFORD (*St. Leonard*), a cinque-port and parish (formerly a representative borough and market-town), in the union of EAST-BOURNE, locally in the hundred of FLEXBOROUGH, rape of PEVENSEY, Eastern Division of the county of SUSSEX, 42 miles (E. by S.) from Chichester, and 59¼ (S. S. E.) from London; containing 1098 inhabitants. This was formerly a considerable town, and had four churches and chapels, until burnt by the enemy, and it is by some thought to have been the *Anderida Civitas* of the Romans; but it has greatly declined, being only resorted to for sea-bathing; it is defended by a small fort. Prawns of a large size and good flavour are caught here. The river Ouse, the estuary of which formerly constituted its harbour, now empties itself into the sea at Newhaven, about three miles westward. The market, which was on Saturday, is disused: fairs are held on March 15th and July 25th. Seaford was originally a member of the port of Hastings, but was made a port by charter of Henry VIII., who incorporated the inhabitants, under the style of " the bailiffs, jurats, and commonalty of the town, parish, and borough of Seaford:" the bailiff and other officers are chosen annually on September 29th; the jurats are twelve in number. The borough first sent barons to parliament in the 26th of Edward I., and continued to the 21st of Richard II., from which time a suspension took place until the reign of Edward IV., when the privilege was restored; but the borough was entirely disfranchised in the 2nd of William IV. The right of election was originally exercised by the freemen only, in number about five; but by a decision of the House of Commons, in 1792, it was subsequently vested in the inhabitant housekeepers paying scot and lot: the bailiff was the returning officer. The living is a discharged vicarage, annexed to that of Sutton, and valued in the king's books at £11. 15. The church is the nave of one of the old churches, with a tower, and a small chancel of later date; although the modern repairs and additions do not harmonize with the original style of architecture, the building still retains several vestiges of grandeur and beauty. A day and Sunday school is partly supported by subscription. Seaford gives the title of Baron to the family of Ellis, conferred on the present lord by patent, dated July 15th, 1826.

SEAFORTH, a chapelry, in the township of LITHERLAND, parish of SEFTON, hundred of WEST DERBY, Southern Division of the county palatine of LANCASTER, 5 miles (N.) from Liverpool: the population is returned with Litherland. The living is a perpetual curacy; net income, £87; patron, John Gladstone, Esq. The chapel, dedicated to St. Thomas, and containing about 800 sittings, is a neat edifice erected in 1815, at the sole expense of the present patron. Attached to the curacy is a good house, with about two acres of land.

SEAGRAVE (*All Saints*), a parish, in the union of BARROW-UPON-SOAR, hundred of EAST GOSCOTE, Northern Division of the county of LEICESTER, 3¼ miles (N. E. by E.) from Mountsorrel; containing 426 inhabitants. The living is a rectory, valued in the king's books at £19. 8. 11½.; present net income, £404; patrons, President and Fellows of Queen's College, Cambridge. Here is a National school, also a Sunday school, supported by the rector.

SEAGRY, LOWER and UPPER (*St. Mary*), a parish, in the union of CHIPPENHAM, hundred of MALMESBURY, Chippenham and Calne, and Northern, Divisions of the county of WILTS, 4¼ miles (S. S. E.) from Malmesbury; containing 234 inhabitants. The living is a discharged vicarage, valued in the king's books at £7. 13. 1½.; present net income, £173; patron and impropriator, Earl of Carnarvon. The church is a very ancient and irregular structure, combining portions in the Norman and early English styles of architecture; it contains a very ancient font; and under a niche in one of the walls is a recumbent effigy of a female.

SEAHAM (*St. Mary*), a sea-port town, and parish, in the union of EASINGTON, Northern Division of EASINGTON ward and of the county palatine of DURHAM, 4¾ miles (S. by E.) from Sunderland; containing 264 inhabitants, of which number, 130 are in the township of Seaham. Seaham harbour, on Seaham estate, then belonging to Sir Ralph Milbanke Noel, Bart., was first projected by his steward, W. Taylor, Esq., and planned by W. Chapman, Esq., engineer, of Newcastle, in 1820; but the sale of the estate delayed its commencement until 1823, by the present Marquess of Londonderry, who bought the estate in 1822, and carried into execution part of Mr. Taylor's projected harbour, through the medium of the above-named engineer. It appears to have been selected for the superior facility of its entrance, which is much safer than that of Sunderland; and from its situation on a prominent part of the shore, it has, with the exception only of Blyth, a better outlet to the south than any harbour on the coast, enabling vessels bound in that direction to clear the Yorkshire coast in north-east winds, when those from the river Tees are too deeply embayed to proceed in either direction. The harbour consists of three divisions, *viz.*, the Outer or North, the Inner, and the South harbour. The Outer harbour contains more than two acres and a half, and forms the general entrance to the other two; the outlet is 110 feet wide, and the depth at high water, with the exception of the north-west angle, in low neap tides, 12 feet, in common spring tides from 15 to 16 feet, and in extraordinary tides from 17 to 18 feet: in the north-west angle is a sloping beach for vessels to run upon, in gales that may require it; and on the southern part of it is an internal pier, where they may deliver their ballast. The inner harbour contains two acres and a half; the entrance, on the south side of which is a wooden jetty, is 30 feet wide at the base, and in 12 feet depth of water 33 feet; and the general depth the same as that of the outer harbour. Between the jetties will be a falling gate of seven feet in height, with a range of sluices, by means of which two acres and a half of water, seven feet in depth, may be sluiced off in spring tide low water, to scour the outer harbour from any deposit that might form in it to obstruct its entrance. On the south quay of this harbour are arrangements for four coal-loading berths, and on the west may be two more if requisite; the north quay wall will be sufficiently low for vessels to lie in a close tier, with their bowsprits over the quay, as at Yarmouth; at the head of the eastern quay, vessels will receive their cargoes of lime, and discharge their ballast near the entrance. The South harbour contains nearly eleven acres and a half; the entrance, between the eastern pier and the end of the wooden jetty of the inner harbour, is 110 feet wide: the sea face and south end of this harbour will be enclosed by a breakwater, of sufficient height to prevent any injurious agitation in gales from the sea, to be formed upon a ridge of rocks, parallel with the shore. The interior is divided into two portions, the northern division containing nearly seven acres, and the southern four acres and a half; on the northern extremity of the former a quay is being formed, 440 feet long, on which will be four coal-loading berths, and between this and the boundary pier will be two loading jetties, 360 feet long, on each of which will be six coal-loading berths; and on the north face of the pier will be others, making the whole number in this division nineteen. In the southern division, which will be nearly 350 feet in length, an intermediate jetty of the same length as the others will be constructed, on which and the two boundaries may be formed twelve coal-loading berths; altogether forming a very extensive vend of coal from the Marquess', and also from South Hetton collieries. The outward direction of the deepest channel leading to the harbour is S. $67\frac{1}{2}°$ E.; and on the starboard side, entering inwards, at the distance of 140 yards, N. 77° E., from the end of the north pier, will be placed a white buoy in six feet of water, low spring ebbs, nearly coinciding with the direction of the intended lighthouses on the ends of the piers, and pointing out the northern side of the channel. On the larboard side, entering inwards, S. 32° E. from the pier, and 180 yards distant from it, will be placed a black buoy in the same depth of water; between these there will be a clear channel, 165 yards in width. On the south side of the main channel is a shoal, called the Tangle rock, of which a small portion, about 55 yards S. 45° W. from the black buoy, has only three feet of water in low spring ebbs: it lies S. $27\frac{1}{2}°$ E. from the extremity of the north-east pier, at a distance of 195 yards. Still farther south, at the distance of a quarter of a mile from the mouth of the harbour, is a rocky ridge, projecting from the shore, called Liddell's Scars, which, like the Tangle rocks, is highly beneficial in breaking off the sea in heavy gales from the south-east; on the northern edge of this ridge, S. 15° E. from the end of the outer pier, and distant 520 yards, is a rock, dry at low spring ebbs, at 40 yards to the north-east of which will be placed a buoy striped black and white. At 150 fathoms E. N. E. from the mouth of the harbour there is good anchorage at three fathoms in low water at spring tides, in which light vessels may lie, when too late to enter the harbour, till the rise of the tide.

The town is pleasantly situated, and some plans have been contemplated for its extension, among which are the erection of a crescent, and of a north and south terrace fronting the beach. Conveyances run between Seaham and Sunderland two or three times daily. The living is a vicarage, endowed with the rectorial tithes, and valued in the king's books at £5. 0. 5.; present net income, £666; patron, F. Cresswell, Esq. A church was erected in 1837, containing 400 sittings, half of which are free, the Incorporated Society having granted £30 in aid of the expense. Here is a National school.

SEAL (*St. Peter*), a parish, in the union of Seven-Oaks, hundred of Codsheath, lathe of Sutton-at-Hone, Western Division of the county of Kent, $2\frac{1}{4}$ miles (N. E.) from Seven-Oaks; containing 1454 inhabitants. This parish abounds with rag-stone, which is quarried, and with layers of sand of various quality, some of which, consisting almost entirely of crystals of pure silex, are well adapted for the manufacture of glass; these layers are intersected in several instances by veins of iron-stone. The living is annexed to the vicarage of Kemsing. A school for eight girls is endowed with certain property in Lombard-street, London, bequeathed by Frances Bickerstaffe, in 1731: and there is also a National school, supported by subscription. A handsome school-house has been erected at Golden Green, by the Countess Amherst, at her own cost.

SEAL, a parish, in the hundred of Farnham, Western Division of the county of Surrey, $3\frac{1}{2}$ miles (E. by N.) from Farnham; containing 366 inhabitants. There are some very extensive chalk pits in this parish, which is intersected by a high ridge called the Hog's back, commanding a fine view over the surrounding country. The living is a perpetual curacy; net income, £44; patron, and appropriator, Archdeacon of Surrey. The church, an ancient structure in the early English style, has been within the last few years enlarged by the addition of a north chancel, in which are several monuments of the Long and Woodroffe families. A day and Sunday school is supported by subscription. The poor are maintained under Gilbert's Act.

SEAL, NETHER and OVER (*St. Peter*), a parish, in the union of Ashby-de-la-Zouch, hundred of West Goscote, Northern Division of the county of Leicester, 5¾ miles (S. W. by W.) from Ashby-de-la-Zouch; containing 1222 inhabitants. The living is a rectory, valued in the king's books at £17. 8. 11½.; present net income, £950; patron and incumbent, Rev William N. Gresley. Two schools are partly supported by subscription.

SEAMER (*St. Martin*), a parish, in the union of Stokesley, Western Division of the liberty of Langbaurgh, North Riding of the county of York, 2¼ miles (N. W. by W.) from Stokesley; containing 224 inhabitants. The living is a perpetual curacy; net income, £56; patron and impropriator, Col. Wyndham. The impropriate tithes have been commuted for a rent-charge of £367. 13. 8., and the perpetual curate's for £16. 10. 8., subject to the payment of rates. In the neighbourhood is a remarkable tumulus, and on the acclivity of a hill adjoining are vestiges of an ancient intrenchment, in the valley beneath which many human bones and warlike weapons have been discovered, conjectured to be relics of the great battle of Baden-hill, in which Prince Arthur overthrew the Saxons, in 492.

SEAMER (*St. Martin*), a parish, in the union of Scarborough, Pickering lythe, North Riding of the county of York; containing 981 inhabitants, of which number, 514 are in the township of Seamer, 4½ miles (S. W. by S.) from Scarborough. The living is a vicarage, with Cayton annexed, valued in the king's books at £18. 16. 5½.; present net income, £243; patron and impropriator, W. J. Denison, Esq. The church is a handsome cruciform structure. At East Ayton is a chapel of ease; and there is a place of worship for Wesleyan Methodists. An insurrection, headed by the parish clerk and two others, broke out here in the reign of Edward VI., in 1549, which had for its objects the restoration of the Roman Catholic religion, the abolition of monarchy, and the equalization of all ranks. The rebels, to the number of 3000, after committing great excesses, laid down their arms upon being offered the king's pardon; but the ringleaders were taken and executed at York, in Sept. of the same year. A fair is held annually on July 15th, and a market on the first Monday in every month, for cattle and sheep. A school, with a dwelling-house for the master, was built and endowed by the lord of the manor, in 1814.

SEARBY (*St. Nicholas*), a joint parish with Owmby, in the union of Caistor, Southern Division of the wapentake of Yarborough, parts of Lindsey, county of Lincoln, 4¾ miles (N. W.) from Caistor; containing, with Owmby, 252 inhabitants. The living is a discharged vicarage, with the vicarage of Owmby annexed, in the patronage of the Dean and Chapter of Lincoln, (the appropriators), valued in the king's books at £8; present net income, £172.

SEASALTER, LIBERTY (*St. Alphage*), a parish, in the union of Blean, hundred of Whitstable, lathe of St. Augustine, Eastern Division of the county of Kent, 5¼ miles (N. W. by N.) from Canterbury; containing 945 inhabitants. The living is a discharged vicarage, valued in the king's books at £11; present net income, £130; patrons and appropriators, Dean and Chapter of Canterbury. There is a place of worship for Independents. On the sea shore is an extensive oyster bed, called the Pollard, belonging to the Dean and Chapter of Canterbury, who let it to the Whitstable company of free dredgers. Four annual fairs were held here, but they have been long discontinued. Mrs. Frances Fagg, in 1794, bequeathed £800 three per cents. for the support of a school, in which seventeen children are educated.

SEASONCOTE (*St. Bartholomew*), a parish, in the union of Stow-on-the-Wold, Upper Division of the hundred of Kiftsgate, Eastern Division of the county of Gloucester, 2 miles (W. by S.) from Moreton-in-the-Marsh; containing 51 inhabitants. The living is a discharged rectory, united to the vicarage of Longborough, and valued in the king's books at £9. 12. 11. The church was demolished about 80 years since.

SEATHWAITE, a chapelry, in the parish of Kirkby-Ireleth, union of Ulverstone, hundred of Lonsdale, north of the sands, Northern Division of the county palatine of Lancaster, 8 miles (W. by N.) from Hawkeshead; containing 190 inhabitants. The living is a perpetual curacy; net income, £60; patrons, Devisees of the late R. Towers, Esq.

SEATON, a township, in the parish of Cammerton, union of Cockermouth, Allerdale ward below Derwent, Western Division of the county of Cumberland, 1¾ mile (N. E.) from Workington; containing 745 inhabitants. Here are extensive collieries and iron-works, near which the Derwent is crossed by a stone bridge, opposite to Workington.

SEATON (*St. Gregory*), a parish, in the hundred of Colyton, Honiton and Southern Divisions of the county of Devon, 2½ miles (S.) from Colyton; containing 1803 inhabitants. This place is situated on the sea coast, and is supposed to have been the *Moridunum* of Antoninus, and a landing-place of the Danes. Leland speaks of it as having been "a notable haven," and of the unsuccessful attempts of the inhabitants "to make a waul within the haven." The town has been much improved of late years, and is now a bathing-place: a pleasure fair is held on Whit-Tuesday. The living is a vicarage, valued in the king's books at £17. 0. 7½.; present net income, £206; patron and impropriator, Lord Rolle. There is a chapel of ease at Beer; and there are places of worship for Independents and Primitive Methodists; also three schools, one endowed with £30, and another with £15 per annum; the third contains 12 girls and is supported by a lady.

SEATON, a joint township with Slingley, in the parish of Seaham, union of Easington, Northern Division of Easington ward and of the county palatine of Durham, 4½ miles (S. by W.) from Sunderland; containing, with Slingley, 134 inhabitants.

SEATON (*All Saints*), a parish, in the union of Uppingham, hundred of Wrandike, county of Rutland, 2½ miles (E. by S.) from Uppingham; containing 435 inhabitants. The living is a rectory, valued in the king's books at £20. 7. 6.; present net income, £649; patron, Earl of Harborough. Here is a Sunday National school.

SEATON, a township, in the parish of Siggles-

THORNE, union of SKIRLAUGH, Northern Division of the wapentake of HOLDERNESS, East Riding of the county of YORK, 10½ miles (N. E. by E.) from Beverley; containing 288 inhabitants. A place of worship for Wesleyan Methodists was erected by subscription, in 1810.

SEATON-CAREW, a township, in the parish of STRANTON, union of STOCKTON, North-Eastern Division of STOCKTON ward, Southern Division of the county palatine of DURHAM, 10½ miles (N. E. by N.) from Stockton-upon-Tees; containing 333 inhabitants. The village is considerably resorted to during the bathing season, and contains respectable public, as well as private, accommodation for the visiters. The beach is smooth, and the sands firm and level to an extent of several miles, so that the convenience for bathing machines is exceedingly good. Here was a chapel dedicated to St. Thomas à Becket, but it has been long since demolished, there being no traces of it. A new church has been erected, containing 260 sittings, 150 of which are free, the Incorporated Society having granted £100 in aid of the expense. The living is a perpetual curacy; net income, £52: it is in the patronage of Mrs. Lawson; the impropriation belongs to Miss Smith. There is a place of worship for Wesleyan Methodists; and a National school has been established.

SEATON-DELAVAL, a township, in the parish of EARSDON, union of TYNEMOUTH, Eastern Division of CASTLE ward, Southern Division of the county of NORTHUMBERLAND, 6¼ miles (N. by W.) from North Shields; containing 271 inhabitants. Here are the ruins of one of the most magnificent mansions in the north of England; it was erected from a design by Sir John Vanbrugh, in 1707, by Admiral Delaval, and was destroyed by fire on Jan. 3rd, 1822. Near it is the site of the ancient castle of Seaton-Delaval, of which little remains, except the chapel, which is a fine specimen of Norman architecture, containing monuments of a Knight Templar and his lady, and ornamented with numerous escutcheons, banners, and pieces of ancient armour: divine service is still performed in it every Sunday. At a short distance from the chapel is a handsome mausoleum, erected by the late Lord Delaval, in memory of his son, the interior of which is fitted up as a chapel, having arched catacombs underneath for the reception of the dead.

SEATON-HOUSE, a joint township with Boulmer, in the parish of LONG HOUGHTON, union of ALNWICK, Southern Division of BAMBROUGH ward, Northern Division of the county of NORTHUMBERLAND, 6 miles (E.) from Alnwick: the population is returned with Boulmer.

SEATON, NORTH, a township, in the parish of WOODHORN, union, and Eastern Division of the ward, of MORPETH, Northern Division of the county of NORTHUMBERLAND, 6¾ miles (E.) from Morpeth; containing 150 inhabitants.

SEATON-ROSS (*St. Edmund*), a parish, in the union of POCKLINGTON, Holme-Beacon Division of the wapentake of HARTHILL, East Riding of the county of YORK, 7¼ miles (W. by S.) from Market-Weighton; containing 436 inhabitants. The living is a perpetual curacy; net income, £93; patron and impropriator, W. C. Maxwell, Esq. There is a place of worship for Wesleyan Methodists.

SEATON-SLUICE, or HARTLEY-PANS, a small sea-port, in the township of HARTLEY, parish of EARSDON, Eastern Division of CASTLE ward, Southern Division of the county of NORTHUMBERLAND, 6 miles (N.) from North Shields: the population is returned with Hartley. It is situated at the mouth of a rivulet, called Seaton-burn, where the late Sir Ralph Delaval, with great difficulty and expense, formed a new haven, and, to prevent its being choked with sand, constructed an immense sluice upon the brook, with flood-gates to retain the water from the flow of the tide till the ebb, when a sufficient body is collected, every twelve hours, to cleanse the bed of the harbour, and remove from it every impediment to its navigation. Considerable improvements upon the original plan were subsequently made by Lord Delaval, who also formed a second entrance, which is crossed by a drawbridge 900 feet in length. From twelve to fifteen vessels, of 300 tons' burden each, can now ride in safety at this port, and sail in or out with any wind. Coal is shipped here for the London and other markets, in very large quantities, from the neighbouring collieries. There are extensive glass-bottle works, malt-kilns, and a brewery, and there were formerly considerable manufactories for salt and copperas. A blockhouse and battery were erected during the late war, for the defence of the port, which is subordinate to that of Newcastle. The Presbyterians have a place of worship here. A whale, upwards of 50 feet long, was taken on this coast, in 1766.

SEAVINGTON (*St. Mary*), a parish, in the union of CHARD, hundred of SOUTH PETHERTON, Western Division of the county of SOMERSET, 3 miles (E.) from Ilminster; containing 366 inhabitants. The living is a perpetual curacy; net income, £50; patron and impropriator, Earl Poulett.

SEAVINGTON (*St. Michael*), a parish, in the union of CHARD, hundred of SOUTH PETHERTON, Western Division of the county of SOMERSET, 3½ miles (E.) from Ilminster; containing 397 inhabitants. The living is a rectory, valued in the king's books at £6. 15.; present net income, £290; patron, Earl Poulett.

SEBERGHAM (*Virgin Mary*), a parish, in the union of WIGTON, ward, and Eastern Division of the county, of CUMBERLAND; containing 840 inhabitants, of which number, 346 are in the division of Sebergham High bound, and 494 in that of Sebergham Low bound, the former 8¾, and the latter 6¼ miles (S. E. by E.) from Wigton. This parish is situated on the river Caldew, of which the south branch becomes subterraneous at Haltcliffe bridge, where it disappears under the high land for nearly three miles, and re-issues at Hives Hill mill. Near the church it is crossed by a bridge erected in 1689, by Alexander Denton, one of the justices of the Court of Common Pleas; and about a mile below is another bridge of one arch, built in 1772, near the site of a former structure, which was destroyed by a great flood the year before. A considerable quantity of limestone is quarried and burnt into lime, and there are extensive mines of coal and a powerful mineral spring in the parish. The living is a perpetual curacy; net income, £139; patrons and appropriators, Dean and Chapter of Carlisle. The church, which is a very neat structure, occupies the site of an ancient hermitage; it was repaired in 1774, and again in 1785. A National school has been lately rebuilt and is supported by subscription.

SECKINGTON (*All Saints*), a parish, in the union of TAMWORTH, Tamworth Division of the hundred of HEMLINGFORD, Northern Division of the county of WARWICK, $3\frac{3}{4}$ miles (N. E. by E.) from Tamworth; containing 129 inhabitants. The living is a rectory, valued in the king's books at £5. 16. $0\frac{1}{2}$.; present net income, £274; patron, Sir Francis Burdett, Bart. Near the church are vestiges of a large ancient encampment. In the neighbourhood is the site of a small priory, founded by William Burdett, in the reign of Henry II.

SEDBERGH (*St. Andrew*), a market-town and parish, in the Western Division of the wapentake of STAINCLIFFE and EWCROSS, West Riding of the county of YORK; containing 4711 inhabitants, of which number, 2214 are in the town of Sedbergh, 77 miles (W.N.W.) from York, and 260 (N. W. by N.) from London. This town is situated in a secluded vale, in a mountainous district, and contains two cotton-mills, in which several persons are employed. Coal is obtained from a mine rather more than two miles distant, near which the river Rother passes. The market, now almost disused, is on Wednesday; and fairs are held on March 20th, the Wednesday in Whitsun-week, and October 29th, chiefly for live stock. A constable is annually elected by the lay payers, and a court for the recovery of small debts has been recently instituted. The living is a discharged vicarage, valued in the king's books at £12. 8.; present net income, £184; patrons and impropriators, Master and Fellows of Trinity College, Cambridge. There are places of worship for the Society of Friends, Independents, and Wesleyan Methodists. The free grammar school was founded and endowed by Roger Lupton, D.D., Provost of Eton College, in the 5th of Edward VI., for all boys duly qualified to enter upon a course of classical instruction, without restriction: by letters patent it was ordained that there should be one master and one usher, and twelve of the inhabitants of Sedbergh were incorporated governors: his Majesty endowed the school with the rectory and church of Weston, and various messuages and lands, now producing a rental of £500, which sum is paid to the master, who allows the usher £100 per annum: the appointment to the mastership is vested in the Master and Fellows of St. John's College, Cambridge, who are visiters. Exhibitions to two fellowships and eight scholarships in St. John's College, Cambridge, were founded in favour of this school, by Dr. Lupton; one fellowship and two scholarships, in the same college, were also founded for boys from this school, by Henry Hebblethwayte, citizen and draper of London; and a further exhibition for one of the scholars, being a native of Sedbergh, to either of the Universities, is given by the governors, in appropriation of three bequests at their disposal. At Howgill, in this parish, is a school, erected near the chapel, and endowed with land by John Robinson, the income of which is £30 per annum.

SEDBURY, a hamlet, in the parish of TIDENHAM, hundred of WESTBURY, Western Division of the county of GLOUCESTER: the population is returned with the parish. This place is bounded on the east by the Severn, and on the west by the river Wye, by which it is separated from Chepstow. The principal estate is the property of George Ormerod, Esq., but the entire hamlet is within the manor of Tidenham, belonging to the Duke of Beaufort.

SEDGEBERROW, a parish, in the union of EVESHAM, Middle Division of the hundred of OSWALDSLOW, Pershore and Eastern Divisions of the county of WORCESTER, 4 miles (S. S. W.) from Evesham; containing 224 inhabitants. The living is a rectory, valued in the king's books at £13. 15. $7\frac{1}{2}$.; present net income, £228; patrons, Dean and Chapter of Worcester. The church has a small octagonal tower, surmounted by a spire, and exhibits portions in the decorated and later English styles of architecture.

SEDGEBROOK (*St. Lawrence*), a parish, in the union of NEWARK, wapentake of WINNIBRIGGS and THREO, parts of KESTEVEN, county of LINCOLN, 4 miles (W. N. W.) from Grantham; containing 252 inhabitants. The living is a rectory, in medieties, with East Allington united, one valued in the king's books at £7. 18. 9., and the other at £7. 4. 7.; present net income, £638: it is in the patronage of the Crown. Lady Margaret Thorold, in 1718, gave land, directing that £15 of the income arising therefrom should be applied for teaching fifteen poor children, £5 for apprenticing one boy, and £5 to the poor.

SEDGEFIELD (*St. Edmund*), a market-town, and parish, and the head of a union, in the North-Eastern Division of STOCKTON ward, Southern Division of the county palatine of DURHAM; containing 2178 inhabitants, of which number, 1429 are in the town of Sedgefield, 11 miles (S. E. by S.) from Durham, and $255\frac{1}{2}$ (N. by W.) from London. The town, which rather presents the appearance of a large village, occupies a gentle eminence commanding an extensive prospect to the south and south-east, and is remarkable for the peculiar salubrity of its atmosphere, and the longevity of its inhabitants, attributable, in a great degree, to the openness of its situation, and the fine gravel soil on which it stands: the inhabitants are supplied with water from springs. The centre of the town forms a spacious square, where the market, granted in 1312 by the charter of Bishop Kellaw, is held on Fridays, and is well supplied with provisions. A large fair, provincially called the Month-day, is held on the first Friday in every month, for the sale of hogs; and there are fairs for cattle on the first Fridays in April and October. This parish is a member of the great episcopal manor at Middleham, and is divided into seven constableries; that of Sedgefield includes the town itself, Hardwick, and Layton. A halmote court is held here, once in eighteen months, in rotation with Cornforth and Bishop's-Middleham, for the recovery of debts under 40*s*., at which the bishop's steward presides. The town has been made a polling-place for the southern division of the county. The living is a rectory, valued in the king's books at £73. 18. $1\frac{1}{2}$.; patron, Bishop of Durham. The tithes have been commuted for a rent-charge of £1481. 7. 2., subject to the payment of rates; the glebe comprises 385 acres, valued at £385 per annum. The glebe of Bradbury comprises 63 acres, valued at £45 per annum; and that of Embleton, 2 acres, valued at £7. 10. per annum. The church is a handsome cruciform structure, combining the early and later styles of English architecture, with a square embattled tower, crowned with pinnacles, at the west end; the windows are principally in the decorated style: the nave is separated from the aisles by pillars supporting pointed arches, and from the chancel by an oak screen of rich tabernacle-work; in the latter are some canopied stalls: the font is octagonal, and of black marble, and there are several brasses and mural tablets. At Embleton is a chapel of ease; and there is a place of worship for Wesleyan Methodists. A free grammar school, of unknown foundation and endowment, has an income of about £50 per annum, for which eight poor children are instructed; six are also educated and clothed from a benefaction of £400 three per cent. Bank Annuities, by Richard Wright, Esq., in the year 1790: the master is appointed by the rector and the vestry. The grammar school and master's house have been lately rebuilt, partly from the accumulated funds of the school at Bishop-Auckland, and partly by subscription, towards which £600 was given by the trustees of Bishop Barrington, £100 by the Rev. Viscount Barrington, £150 by the trustees of Lord Crewe, £50 by William Russel, Esq., and other donations amounting to about £30. William Soulsby, Esq., lately bequeathed the interest of £300 as an augmentation of the master's salary. In 1782, John Lowther, Esq., bequeathed £600 three per cent. Bank Annuities, for the instruction and clothing of eight poor girls. Here is a National school, also a parochial lending library. An almshouse for five poor men and five poor women, inhabitants of this town, was founded and endowed with £44 per annum, arising from land, by Thomas Cooper: additional benefactions were made by William Wrightson, Esq., and Thomas Foster, the latter of whom bequeathed the interest of £3435. 7. 9. three per cent. consols, for the benefit of the inmates. Upwards of 91 acres of land, under the superintendence of eight trustees, belong to a charity, founded by Lady Frevill, in 1630, for apprenticing poor children of the townships of Sedgefield and Bishop's-Middleham, and for purposes of general benevolence; in addition to which there are several minor charitable benefactions. The poor law union of Sedgefield comprises 23 parishes or places, containing a population of 5286, according to the census of 1831; the number of guardians is 24. Mrs. Elizabeth Elstob, the celebrated Saxon scholar, who died in 1756, was a native of Elstob, in this parish, of which the pious and erudite Bishop Lowth was rector, prior to his elevation to the see of London.

SEDGEFORD (*St. Mary*), a parish, in the union of Docking, hundred of Smithdon, Western Division of the county of Norfolk, 8½ miles (N. N. E.) from Castle Rising; containing 595 inhabitants. The living is a vicarage, in the patronage of the Dean and Chapter of Norwich (the appropriators), valued in the king's books at £8; present net income, £232. Here is a Sunday National school.

SEDGHILL (*St. Catherine*), a parish, in the union of Mere, hundred of Dunworth, Hindon and Southern Divisions of the county of Wilts, 4½ miles (S. W.) from Hindon; containing 235 inhabitants. The living is annexed to the rectory of Berwick-St. Leonard. The tithes have been commuted for a rent-charge of £250, subject to the payment of rates, which on the average have amounted to £29. A day and Sunday school is supported by subscription.

SEDGLEY (*All Saints*), a parish, in the union of Dudley, Northern Division of the hundred of Seisdon, Southern Division of the county of Stafford, 3 miles (S.) from Wolverhampton; containing 20,577 inhabitants. This populous parish is situated in the midst of a country abounding with coal, iron-stone, and limestone, the working of which furnishes employment to the greater part of the inhabitants. The village is supposed to occupy one of the highest sites in the kingdom, and the waters divide on the eminence, one portion running into the Trent and the other into the Severn, and flow into the sea at two opposite extremities of the island. The iron is manufactured in a variety of ways, both into pig-iron in furnaces, and into wrought or malleable in mills, which latter is again converted into bars, rods, hoops, hurdles, nails, coffee-mills, locks, &c. The Staffordshire and Worcestershire canal intersects the parish in various directions, affording a ready transit for these articles. A court leet is annually held, at which two constables and four deputies are chosen; and this has lately been made a polling-place for the southern division of the county. The living is a vicarage, endowed with a portion of the rectorial tithes, and valued in the king's books at £5. 12. 8½.; present net income, £503; patron, Lord Ward, who, with others, is the impropriator of the remainder of the rectorial tithes. The church is a beautiful edifice, in the purest style of English architecture, with side aisles, vaulted nave, and clerestory windows: it was completed in 1829, at the sole expense of the late Earl of Dudley. In the hamlet of Lower Gornall is a chapel dedicated to St. James; and a church dedicated to Our Saviour, in the ancient English style, with a tower, was erected in 1830, at Coseley, under the act of the 58th of George III., at an expense of £10,536. 19. 7., to which an ecclesiastical district has been annexed. The living is a perpetual curacy; net income, £138; patron, the Vicar; impropriator, Lord Ward. A church has been erected at Catchem's Corner, in the hamlet of Ettingsall, in this parish, with a parsonage-house and school, by subscription, through the exertions of the Rev. C. Girdlestone, Vicar; and Her Majesty's commissioners have approved of a plan for building a church at Upper Gornall, in this parish. There are two places of worship belonging to Particular Baptists, three to Wesleyan Methodists, one to Primitive Methodists, and one each to Independents, Presbyterians, and Roman Catholics; there is also a Roman Catholic chapel at Sedgley Park. A National school was erected by the late Earl of Dudley: there is also one at Gornall, and both are supported by subscription. A school for children of Dissenters, which has a permanent annual income of £30. 15. 9., was erected by subscription in 1753. The encrinite, and the singular fossil called the trilobite, or "Dudley locust," are found chiefly in this parish, the latter only in an isolated limestone rock, termed the Wren's Nest Hill.

SEDGWICK, a township, in the parish of Heversham, union and ward of Kendal, county of Westmorland, 4½ miles (S.) from Kendal; containing 204 inhabitants. A mill for the manufacture of gunpowder was established here about 1770. The river Kent and the Lancaster canal pass through the township. There is a place of worship for Independents; also a school supported by a private individual.

SEDLESCOMB (*St. John the Baptist*), a parish, in the union of Battle, hundred of Staple, rape of Hastings, Eastern Division of the county of Sussex, 3 miles (N. E.) from Battle; containing 732 inhabitants. The living is a rectory, valued in the king's books at

£9. 4. 2; present net income, £267: it is in the patronage of the Crown. The church is principally in the early style of English architecture, and contains 50 free sittings, the Incorporated Society having granted £50 in aid of the expense. There is a place of worship for Wesleyan Methodists. The Rev. George Barnsley, in 1723, bequeathed £150 for the education of children, which sum, added to subsequent gifts, was laid out in the purchase of an estate now producing £25 per annum, for which 20 boys are instructed. Here was formerly a preceptory of Knights Templars.

SEEND, a chapelry, in the parish, union, and hundred of MELKSHAM, Melksham and Northern Divisions of the county of WILTS, $3\frac{1}{2}$ miles (S. E. by E.) from Melksham; containing 1144 inhabitants. The chapel is dedicated to the Holy Cross. There is a place of worship for Wesleyan Methodists; also a school partly supported by subscription. The Kennet and Avon canal passes through the chapelry.

SEER GREEN, a hamlet, in the parish of FARNHAM-ROYAL, union of AMERSHAM, hundred of BURNHAM, county of BUCKINGHAM, $2\frac{1}{4}$ miles (N. N. E.) from Beaconsfield; containing 245 inhabitants.

SEETHING (*ST. MARGARET*), a parish, in the union of LODDON and CLAVERING, hundred of LODDON, Eastern Division of the county of NORFOLK, 3 miles (W.) from Loddon; containing 438 inhabitants. The living is a perpetual curacy; net income, £142; patrons and impropriators, Trustees of the Great Hospital, Norwich. The impropriate tithes have been commuted for a rent-charge of £458, subject to the payment of rates, which on the average have amounted to £56. A National school is partly supported by subscription.

SEFTON (*ST. HELEN*), a parish, in the union and hundred of WEST DERBY, Southern Division of the county palatine of LANCASTER; containing 4485 inhabitants, of which number, 403 are in the township of Sefton, 7 miles (N.) from Liverpool. The living is a rectory, valued in the king's books at £30. 1. 8.; present net income, £1378; patron and incumbent, Rev. R. R. Rothwell. The church was originally built in 1111, and partly rebuilt in the reign of Henry VIII., by the Rev. Anthony Molyneux, a distinguished preacher, then rector: it is partly Norman, and partly in the later style of English architecture, with a lofty spire: the interior is remarkably elegant; the chancel is separated from the nave by a magnificent screen, and contains sixteen richly-sculptured stalls, with numerous monuments of the family of Molyneux, of whom Sir William fought and performed signal acts of valour under the banner of the Black Prince, at Navaret, as did Sir Richard in the battle of Agincourt, and another Sir William in that of Flodden Field. This place confers the title of Earl on their descendants. A day and Sunday school is supported by the Earl of Sefton.

SEIGHFORD (*ST. CHAD*), a parish, in the Southern Division of the hundred of PIREHILL, union, and Northern Division of the county, of STAFFORD, $2\frac{3}{4}$ miles (W. N. W.) from Stafford; containing 898 inhabitants. This parish is intersected by the Grand Junction railway. The living is a discharged vicarage, valued in the king's books at £6; present net income, £119: it is in the patronage of the Crown; impropriator, F. Eld, Esq. Six children are instructed for £3. 3. a year, arising from a bequest by Dame Dorothy Bridgman.

SEISDON, a township, and the head of a union, in the parish of TRYSULL, Southern Division of the hundred of SEISDON and of the county of STAFFORD, 6 miles (W. S. W.) from Wolverhampton: the population is returned with the parish. The poor law union comprises 12 parishes or places, under the superintendence of 19 guardians, and contains, according to the census of 1831, a population of 11,170. Near Seisdon common is a large triangular stone, called the War stone; and at a short distance is a small square camp.

SELATTYN (*ST. MARY*), a parish, in the hundred of Oswestry, Northern Division of the county of Salop, $3\frac{1}{4}$ miles (N. N. W.) from Oswestry; containing 1142 inhabitants. The living is a rectory, valued in the king's books at £12. 9. $9\frac{1}{2}$.; patron, William Lloyd, Esq. The tithes have been commuted for a rent-charge of £800, subject to the payment of rates, which on the average have amounted to £45; the glebe comprises 83 acres, valued at £96 per annum. The church has been enlarged, and 100 free sittings provided, the Incorporated Society having granted £60 in aid of the expense. There are two free schools; one forming a part of the premises devised by Bishop Hanmer, in 1628, for the use of the poor; and the other erected in 1812, in a more distant part of the parish, on land given by G. H. Carew, Esq., the rector having contributed £200 on condition that both schools should be open to all the resident poor children; the remainder of the fund for their support arises from a bequest of £500 by Charles Morris, in 1721, with which 53 acres of land were purchased, yielding a rent of £60, out of which the master of the new school is paid £20, £15 is expended in clothing the children, and £5 is given annually towards an infants' school, the residue being distributed according to the will of the donor. James Wylding, one of the assembly of divines during the Interregnum, and the celebrated Dr. Henry Sacheverell, were rectors here. Offa's Dyke forms part of the western boundary of the parish, wherein formerly stood the ancient "Castle Brogyntyn," of which there are now scarcely any remains.

SELBORNE (*ST. MARY*), a parish, in the union of ALTON, hundred of SELBORNE, Alton and Northern Divisions of the county of SOUTHAMPTON, $4\frac{1}{2}$ miles (S. E. by S.) from Alton; containing 924 inhabitants. The living is a rectory, valued in the king's books at £8. 2. 1.; patrons, President and Fellows of Magdalene College, Oxford. The church is principally in the early style of English architecture: the altar-piece is ornamented with a fine painting, by Albert Durer, representing the offerings of the Magi, presented by the late Rev. Gilbert White, author of "The Natural History of Selborne," and other works on Natural History, who was born here in 1720, where he chiefly resided. In the time of the Saxons, Selborne was held in royal demesne. A fair is held at the village on May 29th. Gilbert White, in 1719, bequeathed £100, now producing an annual income of £9. 10., for teaching poor children. A priory of Black canons, in honour of the Blessed Virgin Mary, was founded here in 1233, by Peter de Rupibus, Bishop of Winchester; it was subsequently suppressed, and became part of the endowment of Magdalene College, Oxford. At Temple, in this parish, resided Sir Adam Gurdon, the celebrated freebooter in the time of Henry III.

SELBY (*ST. MARY AND ST. GERMAN*), a market-town and parish, and the head of a union, partly within the liberty of St. Peter's, East Riding, and partly in the Lower Division of the wapentake of Barkstone-Ash, West Riding, of the county of York, 14½ miles (S. by E.) from York, and 177 (N. by W.) from London; containing 4600 inhabitants. The ancient name of this place was *Salebeia*, whence its present appellation is obviously derived. The first remarkable event in its history is the foundation of a Benedictine abbey by William the Conqueror, in 1069, which was dedicated to St. German, and, in process of time, acquired such extensive possessions and immunities, as to render it equal in rank to the church of St. Peter at York; the superior of this establishment, with that of St. Mary's in that city, being the only mitred abbots north of the Trent; at the dissolution the revenue was valued at £819. 2. 6.: of this magnificent fabric the church is the only part remaining. In the early period of the great civil war, the town appears to have been held for the parliament, and, although subsequently taken by the royalists, it was eventually recaptured by Sir Thomas Fairfax, when the majority of the king's party were made prisoners, with several horses, pieces of ordnance, and a large quantity of ammunition.

The town is situated on the river Ouse, on the great road from London to Edinburgh: the streets are well paved, and lighted with oil, and the houses in general well built. A new street has been formed, called the Crescent, which consists of excellent and commodious houses, and adds considerably to the improved appearance of the town. The fertility of the surrounding district has been greatly increased by a process of irrigation, whereby the water of the rivers Ouse and Aire is detained upon the land until a sediment has been deposited, which forms excellent manure. A large quantity of woad, for the use of dyers, is produced in the vicinity, and formerly flax was cultivated and prepared to a considerable extent; this branch of trade, however, has greatly declined, owing to the importation of that article from France and the Netherlands, but flax-spinning is still carried on to some extent; there are also some iron-foundries, and manufactories for sail-cloth and leather. The general trade of the town has been much improved by means of a canal connecting the navigable rivers Ouse and Aire, thus opening a more direct communication with Leeds and the West Riding of Yorkshire, so that the greater quantity of the goods sent to that district is disembarked here; and a railway from Selby to Leeds has still more recently been formed, under an act obtained in 1830. The Hull and Selby railway joins the line between Selby and Leeds. A bridge of timber across the Ouse was opened in 1795, and is remarkable for the facility with which it can be turned round, though weighing 70 tons, being opened and closed within the space of a minute. A branch custom-house has been established, whereby vessels are enabled to clear out without touching at the port of Hull. The chief article exported is stone, which is sent coastwise: ships of 150 to 200 tons' burden navigate to Selby; steam-boats pass daily to and from Hull, and there are daily communications with London, and every port on the coast. Here is a ship-yard, in which vessels of considerable burden are built. The market is on Monday; and fairs are held on Easter-Tuesday, Monday after June 22nd, and on Michaelmas-day, for cattle, horses, cloth, &c.: in the centre of the market-place is a handsome cross, in the ancient style of English architecture. A petty session for the wapentake of Barkstone-Ash is held every alternate Monday; and courts leet and baron twice a year, by the lord of the manor, the Hon. E. R. Petre, who gave the site for the erection of a town-hall, which was built in 1825, at an expense of £800, raised by subscription; it is a neat edifice of brick, enclosed with an iron railing.

The living is a perpetual curacy; net income, £97; patron, Hon. E. R. Petre. The church, formerly conventual, and belonging to the abbey, was made parochial by letters patent of James I., dated March 20th, 1618. The ancient monastery stood on the west side of the river Ouse, and the principal buildings were on the west and south side of the church; the barn and granary are yet remaining, but the gateway was taken down about thirty years ago: over it was the abbot's court-house, with two rooms on the sides for the jury and the witnesses; on one side was the porter's lodge, and on the other a room in which to serve the poor. The appearance of this venerable pile is strikingly impressive: the magnificence, yet comparative simplicity, of the west front renders it deserving of particular notice, as its proportions and decorations merit remark from their singularity and elegance. The entrance is by a large and richly-ornamented Norman doorway, supported by six columns, with simply ornamented capitals. The triple arches above the doorway are in the English style, and the decorations partake in character with many found on the north and west doorways, and internal parts of the church. The centre arch forms the west window, being considerably larger than those at the sides, and filled with tracery. The walls of the nave and north transept are Norman, though few exterior arches of that character now remain, being mostly replaced by windows, &c., in the English style, at different periods. The most striking feature on this side is the porch, having circular and pointed arches indiscriminately introduced, composed of similar mouldings. Under it is a Norman doorway less enriched, but more elegantly proportioned than that at the west end. The nave is of massive and simple design; and the choir, of which the east window is highly enriched with tracery, is a perfect and splendid example of the early style of English architecture: on the south side of the choir are several stone stalls, and there are also some of wood enriched with tabernacle work. The upper part of the centre tower fell down, destroying the south transept and the roof of the western part of the south aisle, on March 30th, 1690. The present tower was probably rebuilt about the year 1700, but in a style by no means corresponding with the original. The chapter-house is a beautiful building attached to the south side of the choir: the room used for that purpose, now the vestry-room, appears, from its style and simplicity, to be of an earlier date: over it is a room now used as a school. Among the many striking architectural peculiarities which this magnificent edifice exhibits are two clusters of columns, or piers, supporting arches in the gallery, on the north side of the nave. The font is simple, with a beautiful and lofty cover of carved wood suspended from the second arch, on the north side of the nave. In 1826, a fine-toned organ was erected by private subscription, and adds considerably to the beauty of the choir. The church contains 320 free sittings, the Incorporated Society having granted £300 in aid of the expense. The only monuments of consequence are those of a knight and a lady, and two slabs, one for Abbot Selby, dated 1504, and the other for Abbot Berwick, 1526. The churchyard has been enlarged, and, by the removal of some houses which obstructed the view of the west end, the whole edifice has been thrown open to the market-place: these improvements were effected by the Hon. E. R. Petre, at an expense of not less than £2000: the whole has been since enclosed with an iron railing, 350 yards in length, at the cost of £600, which was defrayed by the voluntary contribution of the parishioners. There are places of worship for the Society of Friends, Independents, Wesleyan Methodists, Unitarians, and Roman Catholics. A Blue-coat school for boys is principally supported by voluntary contributions, augmented by a legacy of £100 from John Herbert, in 1775, which, with other donations, was vested in land; and £13 per annum is paid to the master by the trustees under the will of Joseph Rayner, who, in 1710, bequeathed £100 to be vested in land for the instruction of poor children: thirty-one boys are clothed and educated on this foundation; and £22. 10. per annum are likewise paid for the instruction of 20 children on the foundation of Leonard Chamberlain, in 1716, who also endowed an almshouse for seven poor widows. Another school for girls is supported by an endowment of £10 per annum, aided by subscription, for instructing and clothing 14 children; and in another, 50 children are paid for by an allowance of £35 per annum. The poor law union of Selby comprises 27 parishes or places, under the care of 29 guardians, and contains a population of 14,782, according to the census of 1831. Henry I., the youngest son of William the Conqueror, was born here, during the visit of that monarch and his queen, the year after the foundation of the abbey. Thomas Johnson, a botanist, who published the first local catalogue of plants in the kingdom, besides an improved edition of Gerard's Herbal, and who fell in a skirmish with the parliamentarians, in 1644, was also a native of this place.

SELBY'S-FOREST, a township, in the parish of Kirk-Newton, union, and Western Division of the ward, of Glendale, Northern Division of the county of Northumberland; containing 66 inhabitants. This extensive district consists chiefly of moors and mountains; the principal of the latter is the Cheviot, from which the celebrated range of hills so called derives its name; on its summit is a large lake, occasionally frozen in the summer.

SELHAM (*St. James*), a parish, in the union of Midhurst, hundred of Easebourne, rape of Chichester, Western Division of the county of Sussex, 3½ miles (W. by S.) from Petworth; containing 89 inhabitants. The living is a discharged rectory, valued in the king's books at £4. 15. 11½.; present net income, £150; patrons, Principal and Fellows of Brasenose College, Oxford. The church is in the early style of English architecture; the chancel is separated from the nave by a circular arch supported by slight columns with ornamented capitals. The Rother, or Arundel, navigation passes through the parish. Part of this parish is within the new boundary of the borough of Midhurst.

SELLACK (*St. Tesiliah*), a parish, in the union of Ross, Upper Division of the hundred of Wormelow, county of Hereford, 4¼ miles (N. W.) from Ross; containing 327 inhabitants. The living is a vicarage, with that of King's-Caple annexed, valued in the king's books at £16. 6. 8.; present net income, £420; patrons and appropriators, Dean and Chapter of Hereford. A day and Sunday school is supported by subscription.

SELLING or SELLYNG (*St. Mary*), a parish, in the union of Faversham, hundred of Boughton-under-Blean, Upper Division of the lathe of Scray, Eastern Division of the county of Kent, 4 miles (S. S. E.) from Faversham; containing 539 inhabitants. The living is a discharged vicarage, valued in the king's books at £6. 13. 4.; present net income, £255; patron and impropriator, Lord Sondes. The church is in the early style of English architecture. A fair is held on Whit-Tuesday. Children of this parish are eligible to the school in the parish of Sheldwich, founded and endowed by Lord Sondes. On Shottendon Hill is an ancient fortification of an irregular form, thought by some to be Roman, and by others Danish; but it seems probable that it was a work of the former, from the extensive Roman intrenchment still visible in a wood two miles to the south-east of it: there is also a tumulus in the neighbourhood.

SELLINGE (*St. Mary*), a parish, in the union of Elham, hundred of Street, lathe of Shepway, Eastern Division of the county of Kent, 5¼ miles (N. W. by W.) from Hythe; containing 451 inhabitants. The living is a discharged vicarage, valued in the king's books at £7. 4. 5.; present net income, £176: it is in the patronage of the Crown; appropriator, Archbishop of Canterbury. The church has an admixture of the various styles of English architecture.

SELMESTON (*St. Mary*), a parish, in the union of West Firle, hundred of Danehill-Horsted, rape of Pevensey, Eastern Division of the county of Sussex, 6¼ miles (E. S. E.) from Lewes; containing 189 inhabitants. The living is a discharged vicarage, valued in the king's books at £7. 5. 8.; present net income, £130; patron and appropriator, Prebendary of Heathfield in the Cathedral Church of Chichester. The church is in the early style of English architecture.

SELSEY (*St. Peter*), a parish, in the union of West Hampnett, hundred of Manhood, rape of Chichester, Western Division of the county of Sussex, 8 miles (S.) from Chichester; containing 821 inhabitants. Its name, according to Bede, is derived from the Saxon *Seals-ey*, signifying the island of Seals, for the resort of which it was anciently noted. When the aboriginal expedition was designed from the Belgic coast to take possession of the south-western region of Britain, it is certain that the first landing was made upon this peninsula. The establishment of a colony soon followed: and in the earliest annals mention is made of "Selsey" as among the more ancient of the Saxon establishments. Selsey was formerly inaccessible at flood tides except by a ferry, but by means of a raised causeway the peninsula is now accessible at all times. The parish is bounded on the east and south by Pagham harbour and the English channel. The western division is flat and low, and is frequently overflowed by the sea, from which it suffered very severely in the great storm of November 23rd, 1824, when nearly half the parish was under water. The village, consisting principally of one street of neatly built houses, occupies a dry gravelly site. There is an extensive fishery for prawns, lobsters, and crabs; on the coast, oysters in great quantities are taken in the winter season, also cod. The sands are remarkably dry, affording an exceedingly pleasant drive for about ten miles along the coast. A court baron is annually held; and there is a fair for toys, &c., on July 14th. The living comprises a discharged vicarage and a sinecure rectory united, the former valued in the king's books at £8, and the latter at £11. 3. 4.; present net income, £759; patron, Bishop of Chichester. The church is a stately pile, principally in the early style of English architecture: in the middle aisle are several coffin-shaped gravestones, with crosses and various other devices; and against the north wall of the chancel is a mural monument of Caen stone, with carved effigies of John and Agatha Lews, of the time of Henry VIII. There is a place of worship for Bryanites; also a National school. A school-house was erected for children of the parish by the late Rev. William Walker. Near the church is an intrenched mound, supposed to be a Roman military station. A monastery, dedicated to the Blessed Virgin, was founded here, about 681, by St. Wilfred, who, having previously converted many of the South Saxons to Christianity, and obtained of King Ædilwach the lands of this peninsula for its endowment, placed therein some religious who had been his companions in exile, of whom Eadbercht, abbot in 711, was consecrated first bishop of the South Saxons, and fixed his episcopal seat at this place. It remained an episcopal see till 1075, when William the Conqueror removed it to Chichester, and Stigand, the last bishop of Selsey, was appointed the first Bishop of Chichester. Vestiges of this ancient little city are mentioned, in old records, as being plainly visible at ebb-tide. Bones of large animals, trunks of trees, and fossil shells are occasionally found by the fishermen when dredging for oysters; several of the bones are deposited in the Chichester museum. The preventive service have four stations along the coast in this parish. Selsey gives the title of Baron to the family of Peachy.

SELSIDE, a chapelry, in the parish, union, and ward of KENDAL, county of WESTMORLAND, 4 miles (N. N. E.) from Kendal; containing 263 inhabitants. The living is a perpetual curacy; net income, £94; patrons, the Landowners; impropriators, Master and Fellows of Trinity College, Cambridge. The chapel, dedicated to Christ, was built about 1720, by the inhabitants, on a site given by William Thornburgh, Esq., the descendant of a Roman Catholic family, proprietors of Selside Hall, in consideration of his being allowed to devote the original chapel, attached to the hall of his ancestors, to his own religion, and which is now the kitchen of a farmhouse. The chapel was rebuilt and enlarged in 1837, at an expense of about £600, of which £80 is a grant from the Incorporated Society, and the remainder was raised by subscription; 60 free sittings have been provided. The free school, which was rebuilt by subscription in 1793, is supported from several sources, the principal being an estate left by John Kitching, in 1730, now producing an annual income of £50, for the education of all the poor children of the chapelry.

SELSTON (*ST. HELEN*), a parish, in the union of BASFORD, Northern Division of the wapentake of BROXTOW and of the county of NOTTINGHAM, 9 miles (S. W.) from Mansfield; containing 1580 inhabitants. The Mansfield and Pinxton railway intersects the parish, in the neighbourhood of which are extensive collieries; and several of the inhabitants are employed in frame-work knitting and the manufacture of lace. The living is a discharged vicarage, valued in the king's books at £5; present net income, £109; patron and impropriator, Sir W. W. Dixie, Bart.

SELWORTHY (*ALL SAINTS*), a parish, in the union of WILLITON, hundred of CARHAMPTON, Western Division of the county of SOMERSET, 4 miles (W.) from Minehead; containing 558 inhabitants. The living is a rectory, valued in the king's books at £12. 15. 5.; present net income, £221; patron, Sir T. D. Acland, Bart. The church is a neat edifice, in the decorated style, with a plain embattled tower; the roofs of the nave, chancel, and south aisle, are each divided into small square compartments, enriched with figures curiously carved in wood, and supported by two rows of pillars and arches of a peculiarly light and graceful construction. There are remains of two chapels, one at Tivington, now used as a school-room, the other at West Lynch, converted into a barn. A National school has been established. On a hill to the north-west of the church are vestiges of an ancient encampment, called Bury Castle; it is of an elliptical form, with a rampart of earth and stones, enclosing an area of about an acre and a half.

SEMER (*ALL SAINTS*), a parish, in the union and hundred of COSFORD, Western Division of the county of SUFFOLK, 2 miles (S. by E.) from Bildeston; containing 275 inhabitants, exclusively of the inmates of the union workhouse, which is in this parish. The living is a rectory, valued in the king's books at £11. 7. 1.; present net income, £385; patron and incumbent, Rev. James Young Cooke. The house of industry, erected in 1799, has been converted into a workhouse for the union, which comprises 28 parishes or places, under 36 guardians, and, according to the census of 1831, contains a population of 17,900.

SEMINGTON, a chapelry, in the parish of STEEPLE-ASHTON, union of MELKSHAM, hundred of WHORWELSDOWN, Melksham and Northern Divisions of the county of WILTS, 3 miles (N. E. by E.) from Trowbridge; containing 319 inhabitants. The chapel is dedicated to St. George. There is a place of worship for Wesleyan Methodists.

SEMLEY (*ST. LEONARD*), a parish, in the union of TISBURY, forming a detached portion of the hundred of CHALK, Hindon and Southern Divisions of the county of WILTS, $4\frac{3}{4}$ miles (S. by W.) from Hindon; containing 700 inhabitants. The living is a rectory, valued in the king's books at £17. 2. $8\frac{1}{2}$.; patrons and appropriators, Dean and Canons of Christ-Church, Oxford. The tithes have been commuted for a rent-charge of £492. 10., subject to the payment of rates, which on the average have amounted to £94; the glebe comprises 101 acres, valued at £135. 16. per annum. There is a place of worship for Baptists. Dr. William Thorn, a celebrated divine and Hebrew scholar, was born here, towards the close of the sixteenth century.

SEMPERINGHAM (*ST. ANDREW*), a parish, in the union of BOURNE, wapentake of AVELAND, parts of KESTEVEN, county of LINCOLN, $3\frac{1}{4}$ miles (E. S. E.) from Falkingham; containing 490 inhabitants. The living is a discharged vicarage, valued in the king's books at

£2. 15. 8.; present net income, £131; patron and impropriator, Earl Fortescue. The church appears to have been originally a larger structure, and is principally in the Norman style, with a plain tower of later date crowned with eight crocketed pinnacles. Gilbert de Sempringham, rector of this parish, and founder of the Gilbertine, or Sempringham, order, erected here, about 1139, a priory in honour of the Blessed Virgin Mary, for nuns and canons, whose revenue, at the dissolution, was valued at £359. 19. 7.: it was the superior establishment of the Sempringham order, where their general chapters were held: the buildings stood a little to the northward of the church, where the site only is discernible, being surrounded by a moat.

SEND (*St. Mary*), a parish, in the union of GUILDFORD, Second Division of the hundred of WOKEING, Western Division of the county of SURREY; containing, with the chapelry of Ripley, 1483 inhabitants. The living is a vicarage, valued in the king's books at £8. 18. 1½.; present net income, £260; patron, Earl Onslow. The church is principally in the early style of English architecture. At Ripley is a chapel of ease; and there is a place of worship for Baptists. Mrs. Ann Haynes gave £300, of which she appropriated the interest to the apprenticing of two poor children of this parish, which also participates in Henry Smith's charity; and General Evelyn bequeathed £20 per annum to keep his monument in repair, with residue to the poor. The Wey canal passes through the parish. A church and priory of Black canons, in honour of the Blessed Virgin Mary and St. Thomas à Becket, was founded in the time of Richard I., at Newark, in this parish, by Ruald de Calva, and Beatrix his wife, which at the dissolution possessed a revenue of £294. 18. 4.; there are some remains.

SENNEN (*St. Senan*), a parish, in the union of PENZANCE, Western Division of the hundred of PENWITH and of the county of CORNWALL, 8¼ miles (W. S. W.) from Penzance; containing 689 inhabitants. This parish is situated at the extreme western point of England, and includes the Land's End, a promontory 150 feet above the level of the sea, Whitsand bay, and Sennen Cove. King Stephen landed at Whitsand bay on his first arrival in England, as did also King John, on his return from the conquest of Ireland, and Perkin Warbeck in the reign of Henry VIII. The scenery along the coast is strikingly bold and magnificent; the Scilly islands, about nine leagues west-by-south from the Land's End, are distinctly seen in clear weather; and off the coast are several rocks, called the Longships, on one of which a lighthouse was erected in 1797, under the direction of the Master and Wardens of the Trinity House. A pilchard fishery is carried on at Sennen Cove, and great quantities of ling are cured and dried for the London and other markets. In the village of Churchtown is an inn, the sign of which has on one side the inscription—"the first inn in England;" and on the other—"the last inn in England." A coast guard is stationed at this place. The living is a rectory, united, with that of St. Levan, to the rectory of St. Burian. The tithes have been commuted for a rent-charge of £230, subject to the payment of rates. There are places of worship for Baptists and Wesleyan Methodists, and a cemetery for the Society of Friends; also a National school for this parish and that of St. Levan. Near Sennen Cove are the remains of an ancient chapel, and the site of Castle Mayon. In the village of Mayon is a large flat stone, called Table Maen, on which, according to Hals, seven Saxon kings dined together, when they came to visit the Land's End, towards the close of the sixth century. In 1807, 400 Roman coins of copper and plated metal were found between two flat stones under a large projecting rock near the Land's End; they were of Galienus, Victorinus, Porthumus, and others, and are now in the possession of James Trembath, Esq.

SERLBY, a township, in the parish of HAUGHTON, Hatfield Division of the wapentake of BASSETLAW, Northern Division of the county of NOTTINGHAM, 2¾ miles (S. S. W.) from Bawtry: the population is returned with the parish. Here was formerly a chapel of ease to the vicarage of Harworth, which has long been in ruins.

SESSAY (*St. Cuthbert*), a parish, in the union of THIRSK, wapentake of ALLERTONSHIRE, North Riding of the county of YORK, 6¾ miles (N. W. by W.) from Easingwould; containing 464 inhabitants. The living is a rectory, valued in the king's books at £17. 0. 2½.; present net income, £574; patron and incumbent, Viscount Downe, who partly supports a school.

SETCHEY, a parish, in the union and hundred of FREEBRIDGE-LYNN, Western Division of the county of NORFOLK, 5 miles (S.) from Lynn-Regis; containing 95 inhabitants. The living is a rectory, annexed to that of North Runcton.

SETMURTHEY, a chapelry, in the parish of BRIGHAM, union of COCKERMOUTH, ALLERDALE ward above Derwent, Western Division of the county of CUMBERLAND, 4 miles (E. N. E.) from Cockermouth; containing 182 inhabitants. The living is a perpetual curacy; net income, £54; patrons, the Inhabitants. A school-house was built by subscription in 1795, and is endowed with £22. 15. per annum.

SETTLE, a market-town, and the head of a union, in the parish of GIGGLESWICK, Western Division of the wapentake of STAINCLIFFE and EWCROSS, West Riding of the county of YORK, 59 miles (W. by N.) from York, and 234 (N. W. by N.) from London; containing 1627 inhabitants. The name of this town is derived from the Saxon word *Setl*, a seat: its situation is singular and picturesque, at the base of an almost perpendicular limestone rock, which rises to the height of 200 feet; it is neat and well built, the houses being chiefly of stone: the streets are partially paved, and the inhabitants are well supplied with water from numerous springs and wells. There is a subscription library and news-room. The surrounding vale consists of rich pastures, and is enclosed on each side by a long range of craggy mountains, including the lofty elevations of Pendle hill on the south, of Pennigant on the north, and Ingleborough on the north-west. Considerable business is done in the cotton trade, and there are several roperies and a paper-manufactory. The market is on Tuesday: fairs are held on April 26th, Whit-Tuesday, Aug. 18th and the two following days, and Tuesday after Oct. 27th; in addition to which there are fairs every alternate Tuesday from Easter to Whitsuntide, for lean cattle, and every second Monday in the year for fat cattle. A constable is appointed annually at a court baron of the lord of the manor, which is always held once, and sometimes twice a year. The town has been made a polling-place for the West Riding. A church was erected here in 1839. There are places of worship for Independents and Wesleyan Methodists; and a National school is supported by subscription. The poor law union of Settle comprises 31 parishes or places, under 33 guardians, and contains, according to the census of 1831, a population of 14,322.

SETTRINGTON (*All Saints*), a parish, in the union of MALTON, wapentake of BUCKROSE, East Riding of the county of YORK; containing 779 inhabitants, of which number, 527 are in the township of Settrington, 4 miles (E. S. E.) from New Malton. The living is a rectory, valued in the king's books at £42. 12. 6.; present net income, £1045: it is in the patronage of the Countess of Bridgewater. Two schools are chiefly supported by the lord of the manor and the rector. This has been made a polling-place for the East Riding.

SEVENHAMPTON (*St. Andrew*), a parish, in the union of NORTHLEACH, hundred of BRADLEY, Eastern Division of the county of GLOUCESTER, 4¼ miles (S.) from Winchcombe; containing 465 inhabitants. The living is a perpetual curacy; net income, £49; patrons, F. Craven and William Morris, Esqrs.

SEVENHAMPTON, a chapelry, in the parish of HIGHWORTH, hundred of HIGHWORTH, CRICKLADE, and STAPLE, Swindon and Northern Divisions of the county of WILTS, 1½ mile (S. E. by S.) from Highworth; containing 239 inhabitants. The chapel is dedicated to St. Andrew.

SEVEN-OAKS, a township, in the parish of GREAT BUDWORTH, union of RUNCORN, hundred of BUCKLOW, Northern Division of the county of CHESTER, 4 miles (N. W. by N.) from Northwich; containing 149 inhabitants.

SEVEN-OAKS (*St. Nicholas*), a market-town and parish, and the head of a union, in the hundred of CODSHEATH, lathe of SUTTON-AT-HONE, Western Division of the county of KENT; containing 4709 inhabitants, of which number, 2114 are in the town, 17½ miles (W.) from Maidstone, and 24 (S. E. by S.) from London. This place, which in the Textus Roffensis is written *Seovan Acca*, is supposed to have derived its name from seven large oaks which stood upon the eminence on which the town is built: the period of its origin is uncertain, and the only historical event connected with it is the defeat and death of Sir Humphrey Stafford, by Jack Cade and his followers, when sent to oppose the rebels by Henry VI., in 1450. The manor, formerly an appendage to Otford, and as such belonging to the see of Canterbury, was conveyed, about the time of the dissolution of the monastic establishments, by Archbishop Cranmer to Henry VIII., and subsequently became the property of the Dukes of Dorset. The town is situated on the ridge of hills which crosses the county, separating the Upland from the Weald, or southern part, near the river Darent, in a fertile and beautiful part of the country: it consists of two principal streets, and is well built and very respectably inhabited, being generally esteemed a most desirable place of residence. There were formerly some silk-mills in the neighbourhood, but they have been recently taken down. The market is on Saturday, principally for corn; and there is also a market, on the third Tuesday in every month, for cattle, which is very numerously attended: the fairs are on July 10th and Oct. 12th, the latter being also a statute fair. A bailiff, high constable, and several inferior officers, are chosen annually at a court leet. Petty sessions for the lathe of Sutton-at-Hone are held here; also a court of requests, for the recovery of debts not exceeding £5, the jurisdiction of which extends throughout the hundreds of Codsheath, Somerden, Westerham, and Wrotham, and the ville and liberty of Brastead. The town has lately been made a polling-place for the western division of the county. In the reign of Queen Elizabeth the assizes were held in the ancient market-house, near the middle of the High-street, and also two or three times at subsequent periods.

The living comprises a sinecure rectory and a vicarage; the rectory is valued in the king's books at £13. 6. 8., and the vicarage at £15. 3. 1½.; patron, incumbent, and impropriator, Rev. Thomas Curteis. The impropriate tithes have been commuted for a rent-charge of £543. 10., and the vicarial for £744. 10., subject to the payment of rates; the glebe comprises 15 acres, valued at £30. 15. per annum. The church is a spacious and handsome edifice, at the southern end of the town, and on so elevated a situation as to be a conspicuous object many miles around. Two chapels, with houses for the ministers, have been erected in this parish, by Earl Amherst and the late Multon Lambard, Esq., with the consent of the vicar, as patron; one, situated in the Weald, and built in 1821, contains about 200 sittings, and is endowed with 10 acres of land, and a parliamentary grant of £1800; the other, at Riverhead, erected in 1831, contains about 600 sittings, and is endowed with £220 Queen Anne's Bounty, and a rent-charge of £45 on the vicarial tithes, granted by the vicar: the livings are perpetual curacies, in the patronage of the vicar, after the decease of the founders. The Baptists have two places of worship, and the Supralapsarians and Wesleyan Methodists one each. The free grammar school was founded and endowed in 1432, by Sir William Sevenoake, usually written Sennocke, who, having been deserted by his parents, was brought up by some charitable persons, and apprenticed to a grocer in London, from which humble station he rose to be lord mayor of that city, and its representative in parliament, and left a portion of his wealth to found this school and an hospital for decayed elderly tradespeople. Queen Elizabeth granted a charter to the school, placing the management in the hands of the wardens and assistants of the town, who are elected annually, and appoint the master; and it is in consequence called "The Free Grammar School of Queen Elizabeth." It has seven scholarships; four of £15 per annum each, in any college at either of the Universities, founded by direction of the court of Chancery, in 1735, from the surplus of money received under this endowment in the hands of the trustees; two, formerly of £12 a year each, in Jesus' College, Cambridge, founded by Lady Boswell, which have greatly increased in value; and one of £4 per annum at either University, founded by Robert Holmden, and paid by the Leather-sellers' Company: in default of scholars from this school, that at Tonbridge has the right to appoint to the three last-named scholarships. The income at present derived from Sir William Sennocke's endowment, including some additions to it, particularly that of Anthony Pope, in 1571, is between

£700 and £800 per annum, of which sum, exclusively of repairs, &c., of the house, £50 a year is appropriated as a salary to the master, who has also a house and excellent premises: about eleven boys are generally on the foundation. Lady Margaret Boswell founded a school in 1675, for educating poor children of the town, with funds for apprenticing them: the present income is nearly £700 per annum, from which a considerable deduction is made for repairing the sea wall at Burnham Level, which, on an average of six years, amounted to upwards of £200 per annum: a new school-house was erected in 1827, on the site of the former, at an expense of about £2000, defrayed by savings from the income, and about 200 children are instructed on the National system: a premium of £12 is given as an apprentice-fee with each boy on leaving the school, the number of whom averages about six yearly. A school-house for boys and girls was erected in the Weald liberty, by the late M. Lambard, Esq.; another was also erected at Riverhead, by Earl Amherst; and an infants' school has been erected by Col. Austen, of Kippington, in this parish. In the almshouse founded by Sir William Sennocke are 32 persons; and sixteen out-pensioners receive an allowance from the endowment. The late Mr. Lambard erected eight almshouses in the Weald liberty, which he conveyed to trustees, for the residence of the out-pensioners of the town and parish. The poor law union of Seven-Oaks, to which Penshurst union has been added, now comprises 16 parishes or places, under 31 guardians, and contains, according to the census of 1831, a population of 19,607.

SEVERN-STOKE, county of WORCESTER.—See STOKE, SEVERN.

SEVINGTON (*St. Mary*), a parish, in the union of EAST ASHFORD, hundred of CHART and LONGBRIDGE, Upper Division of the lathe of SCRAY, Eastern Division of the county of KENT, 2¼ miles (S. E. by E.) fromAshford; containing 111 inhabitants. The living is a discharged rectory, valued in the king's books at £8. 14. 0½.; patron and incumbent, Rev. G. Norwood. The tithes have been commuted for a rent-charge of £230, subject to the payment of rates, which on the average have amounted to £2. 19. 2.; the glebe comprises 12 acres, valued at £24 per annum.

SEWARDSTONE, a hamlet, in the parish of WALTHAM-ABBEY, or HOLY CROSS, hundred of WALTHAM, Southern Division of the county of ESSEX, 1¼ mile (S.) from Waltham-Abbey; containing 825 inhabitants. This place is situated within the limits of the forest, and near the river Lea; it is said to have been formerly a distinct parish, and there is still in the vicinity a heap of rubbish, called "the ruins of the old church." The Wesleyan Methodists have a place of worship here.

SEWERBY, a joint township with Marton, in the parish and union of BRIDLINGTON, wapentake of DICKERING, East Riding of the county of YORK, 1¾ mile (E. N. E.) from Bridlington; containing, with Marton, 352 inhabitants.

SEWSTERN, a chapelry, in the parish of BUCKMINSTER, union of MELTON-MOWBRAY, hundred of FRAMLAND, Northern Division of the county of LEICESTER, 9 miles (E. by N.) from Melton-Mowbray; containing 368 inhabitants. The chapel is dedicated to St. Michael. The Independents and Wesleyan Methodists have each a place of worship.

SEXHOW, a township, in the parish of RUDBY-IN-CLEVELAND, union of STOKESLEY, Western Division of the liberty of LANGBAURGH, North Riding of the county of YORK, 5 miles (S. W. by W.) from Stokesley; containing 35 inhabitants.

SHABBINGTON (*St. Mary Magdalene*), a parish, in the union of THAME, hundred of ASHENDON, county of BUCKINGHAM, 2½ miles (W. by N.) from Thame; containing 298 inhabitants. This parish is bounded on the east and south by the river Thame. The living is a vicarage, endowed with a portion of the rectorial tithes, and valued in the king's books at £10. 9. 7.: it is in the patronage of Mrs. M. Wroughton; impropriators of the remainder of the rectorial tithes, the Landowners. The tithes have been commuted for a rent-charge of £380, subject to the payment of rates; the glebe comprises 90 acres.

SHACKERSTONE (*St. Peter*), a parish, in the union of MARKET-BOSWORTH, hundred of SPARKENHOE, Southern Division of the county of LEICESTER, 5 miles (N. W.) from Market-Bosworth; containing 432 inhabitants. The living is a discharged vicarage, valued in the king's books at £5. 2. 3½.; present net income, £150; patron and impropriator, T. S. Hall, Esq. A day and Sunday school, and an infants' school, are supported by Earl Howe. The Ashby-de-la Zouch canal passes through the parish.

SHADFORTH, a township, in the parish of PITTINGTON, Southern Division of EASINGTON ward, union, and Northern Division of the county palatine, of DURHAM, 4½ miles (E. by S.) from Durham; containing 236 inhabitants. A church has been built containing 500 sittings, 350 of which are free, the Incorporated Society having granted £200 in aid of the expense. Here is a National school. One of the towers of an ancient castellated mansion still remains.

SHADINGFIELD (*St. John the Baptist*), a parish, in the union and hundred of WANGFORD, Eastern Division of the county of SUFFOLK, 4¾ miles (S.) from Beccles; containing 198 inhabitants. The living is a discharged rectory, valued in the king's books at £12; present net income, £264; patron, Lord Braybrooke.

SHADOXHURST (*St. Peter and St. Paul*), a parish, in the union of WEST ASHFORD, hundred of BLACKBURN, Lower Division of the lathe of SCRAY, Western Division of the county of KENT, 5½ miles (S. W.) from Ashford; containing 239 inhabitants. The living is a discharged rectory, valued in the king's books at £7. 13.; present net income, £109: it is in the patronage of the Crown.

SHADWELL (*St. Paul*), a parish, in the union of STEPNEY, Tower Division of the hundred of OSSULSTONE, county of MIDDLESEX, 1½ mile (E. by S.) from London; containing 9544 inhabitants. This parish is situated on the northern bank of the river Thames, and comprises several streets, which are lighted with gas, and supplied with water from the East London water-works. It is within the jurisdiction of the New Police, the Thames Police Office, and a court of requests held in Whitechapel, for the recovery of debts under £5. There are some roperies in the parish. The living is a discharged rectory; net income, £352; patron, Dean of St. Paul's. The church is a handsome modern edifice, with a tower of stone surmounted by a small but elegant spire, erected on the site of the old structure. There are places of worship for Independents, and Primitive and Wesleyan Methodists. A parochial school, originally founded in 1699, for the maintenance and education of poor children, was rebuilt on an enlarged scale in 1837, and is now conducted on the National plan; there are at present 200 children, of whom 45 boys and 35 girls are clothed. The Protestant dissenters' original charity school was founded in 1712, and is situated in Shakspeare's walk: at present there are about 70 boys, who are educated and clothed by subscription, and afterwards apprenticed. A chapel is attached to the institution. Thirty-one almshouses for the widows of poor seamen, founded and endowed in 1713, by Captain James Cooke and Alice his wife, are situated in this parish.

SHADWELL, a township, in the parish of THORNER, Lower Division of the wapentake of SKYRACK, West Riding of the county of York, 5¾ miles (N. N. E.) from Leeds; containing 248 inhabitants. There is a place of worship for Wesleyan Methodists.

Arms.

SHAFTESBURY, or SHASTON, a borough and market-town, having separate jurisdiction, and the head of a union, locally in the hundred of Monckton-up-Wimborne, county of DORSET, 28 miles (N. N. E.) from Dorchester, and 101 (W. S. W.) from London, on the great western road from London to Exeter; containing 3061 inhabitants. The origin and derivation of the name of this town has given rise to much conjecture, it being supposed by some to have had existence even prior to the birth of Christ, and to have been called *Caer Calladwr* by the Britons: other periods have been assigned for its foundation, but that which appears to be most probable is the reign of King Alfred; in confirmation of which, Camden states, that in the time of William of Malmesbury was to be seen an old stone, brought from the ruins of a wall into the nuns' chapter-house, with an inscription purporting that King Alfred *built* this city (if we may so render "*fecit*") in 880, and in the eighth year of his reign. Its Saxon derivation from *Sceaft*, signifying the point of a hill, is supposed to be in allusion to the situation of the town. A Benedictine nunnery, founded here about the same period, has also been ascribed to various persons. Camden, following William of Malmesbury, attributes it to Elgiva, wife of Edmund, great grandson to King Alfred; but Leland and many other writers assert this latter monarch to be its founder, and that his daughter was the first abbess. To this abbey the remains of Edward the Martyr were removed after his murder at Corfe-Castle, and it appears to have been much resorted to by pilgrims, amongst whom was King Canute, who died here; and the extent of its endowments may be estimated from their value at its dissolution being £1166 per annum: the remains are very inconsiderable. The importance of the abbey naturally increased that of the town, which is reported at an early period to have contained ten parish churches. In the time of Edward the Confessor three mints were established here; and, according to a survey made shortly before the Norman Conquest, Shaftesbury contained 104 houses, and three mint-masters. The town is situated on a high hill, with a gradual rise on the east and south-east, but more precipitous on the west and south-west, at the extremity of the county of Dorset, and bordering on that of Wilts: it commands extensive views over both counties and also Somersetshire. The streets have been greatly widened and improved of late years, by the removal of obstructions and the erection of modern dwelling-houses and good shops. The inhabitants are amply supplied with water from wells of great depth on the hill: formerly they were chiefly supplied from the adjoining parish of Motcombe, which gave rise to a curious customary acknowledgment, called the Byzant, now discontinued. The manufacture of shirt buttons, formerly carried on to a considerable extent, has very much declined, but there is an extensive country trade. The market is on Saturday, and well supplied with all kinds of commodities; and there are fairs on the Saturday before Palm-Sunday, June 24th, and November 23rd.

Corporation Seal.

Obverse. *Reverse.*

This is a very ancient borough, being described as such in Domesday-book, but it was not incorporated till the reign of Queen Elizabeth, who, according to Hutchins, granted its first charter, appointing a mayor, recorder, twelve aldermen, a bailiff, and common council; but no charter can be found prior to that granted by James I., in 1604, which was followed by one of Charles II., vesting the government in a mayor and twelve capital burgesses, assisted by a recorder, town-clerk, coroner, and two sergeants-at-mace. The corporation now consists of a mayor, four aldermen, and twelve councillors, under the act of the 5th and 6th of William IV., cap. 76: the mayor and late mayor are justices of the peace, the county magistrates having concurrent jurisdiction. The borough first sent members to parliament in the 23rd of Edward I., and has since continued to do so without interruption; but under the act of the 2nd of William IV., cap. 45, it now sends only one. The right of election was formerly vested in the inhabitants paying scot and lot, about 400 in number; but an enlarged district has been added to the borough, which now comprises an area of 5644 acres, the limits of which are minutely described in the Appendix, No. II.: the municipal borough contains only 156 acres; the mayor is the returning officer. The corporation have power to hold a court of record weekly for debts under £10 contracted within the borough, but it has not been held for a long time past. Petty sessions for the division are held on the first Tuesday in every month; and the town has been made a polling-place for the county. A town-hall, which is a handsome building, has been recently erected, at an expense of about £3000, defrayed by the Marquess of Westminster.

The town comprises the parishes of *St. Peter, the Holy Trinity*, and *St. James*. The living of St. Peter's is a discharged rectory, valued in the king's books at £11. 10. 2½., and united to that of the Holy Trinity, valued at £4. 1. 10½., which union comprehends also the ancient parishes of St. Lawrence and St. Martin; present net income, £168; patron, Earl of Shaftesbury. St. Peter's church is of considerable antiquity, but has undergone many modern alterations; it contains a curiously carved font, and a very ancient monument, supposed to have been removed from the abbey. That of the Holy Trinity is also ancient, and is said to have been enlarged by Sir Thomas Arundel, in the early part of the reign of Queen Elizabeth; it has a square embattled tower, crowned with pinnacles; the churchyard is spacious, and adjoining it may be seen the remains of the wall of the abbey. The living of St. James' (a portion of which parish, in the liberty of Alcester, is without the borough,) is a rectory, valued in the king's books at £1. 11. 0½.; present net income, £286; patron, Earl of Shaftesbury. The church is a small and ancient fabric. The Society of Friends, Independents, and Wesleyan Methodists, have each places of worship here. The free school for educating, clothing, and apprenticing 20 poor boys, was founded and endowed by Mr. William Lush, in 1719. Here is also a National school. Spiller's spital, for ten poor men, was founded and endowed by Sir Henry Spiller in 1642; and an almshouse for sixteen poor women was founded by Matthew Chubb, in 1611, and endowed by him and several other benefactors. The poor law union of Shaftesbury comprises 19 parishes or places, under the care of 23 guardians, and contains, according to the census of 1831, a population of 12,239. On Castle hill, an eminence near the town, is a small mount, surrounded by a shallow ditch, which some have conjectured to be the site of a castle, but of which no mention can be found: by others it is supposed to have been a Roman intrenchment. The old city, which tradition reports to have existed prior to the time of Alfred, is said to have been near this mount. Shaftesbury is the birthplace of the Rev. James Granger, author of the Biographical History of England: it gives the title of Earl to the family of Ashley Cooper.

SHAFTO, EAST, a township, in the parish of HARTBURN, union of CASTLE ward, North-Eastern Division of TINDALE ward, Southern Division of the county of NORTHUMBERLAND, 11¾ miles (W. S. W.) from Morpeth; containing 41 inhabitants. Here was formerly a chapel. Behind the ancient hall rises a lofty verdant hill, termed Shafto Crag, adjacent to which there is a spacious cave formed in the solid rock.

SHAFTO, WEST, a township, in the parish of HARTBURN, union of CASTLE ward, North-Eastern Division of TINDALE ward, Southern Division of the county of NORTHUMBERLAND, 12½ miles (W. by S.) from Morpeth; containing 68 inhabitants.

SHAFTON, a township, in the parish of FELKIRK, wapentake of STAINCROSS, West Riding of the county of YORK, 5 miles (N. E.) from Barnesley; containing 248 inhabitants. There is a place of worship for Wesleyan Methodists.

SHALBOURN (*ST. MICHAEL*), a parish, in the union of HUNGERFORD, partly in the hundred of KINTBURY-EAGLE, county of BERKS, and partly in the hundred of KINWARDSTONE, Marlborough and Ramsbury, and Southern, Divisions of the county of WILTS, 4 miles (S. S. W.) from Hungerford; containing 922 inhabitants. The living is a vicarage, valued in the king's books at £14. 17. 6.; present net income, £271; patrons, Dean and Canons of Windsor; impropriator, Marquess of Aylesbury. The church is principally in the Norman style. In that part of the parish which is in the county of Wilts is an ancient chapel, in a dilapidated state, with a house attached to it, called West-court, and supposed to have been a retreat for the monks of Sarum, or rather a place of occasional relaxation from the austerities of the monastery. On the edge of the down, a continuance of Salisbury plain, is a tumulus, commanding very extensive prospects over several counties. Fragments of human skeletons and of horses, supposed to be the remains of those slain in the wars during the Heptarchy, are often met with in the neighbourhood. Wansdyke, the boundary between the kingdoms of Mercia and the West Saxons, runs along one side of the parish, on the north side of which is a chalybeate spring, formerly in great repute.

SHALBOURN, WEST, a township, in the parish of SHALBOURN, hundred of KINWARDSTONE, Marlborough and Ramsbury, and Southern, Divisions of the county of WILTS, 4 miles (S. S. W.) from Hungerford; containing 410 inhabitants.

SHALDEN (*ST. PETER AND ST. PAUL*), a parish, in the union of ALTON, hundred of ODIHAM, Alton and Northern Divisions of the county of SOUTHAMPTON, 2¾ miles (N. W.) from Alton; containing 167 inhabitants. The living is a rectory, valued in the king's books at £9. 15. 10.; present net income, £331: it is in the patronage of the Crown. A day and Sunday school is supported by the rector.

SHALDON, or SHALDON-GREEN, a township, in the parishes of STOKEINTINHEAD and ST. NICHOLAS, in a detached portion of the hundred of WONFORD, Teignbridge and Southern Divisions of the county of DEVON, 5¼ miles (E.) from Newton-Abbott: the population is returned with the parish. This place, which is composed of a tract of land recovered from the sea by an embankment, contains many genteel villas, and is pleasantly situated on the south bank of the river Teign across which a bridge has been erected, communicating with a new line of road leading from the west end of the township to Torquay; and a ferry boat is constantly passing between this and Teignmouth, so that a more direct communication is thus established, both for vehicles and foot passengers, with Teignmouth and Torquay. A church was erected in this township about 150 years ago, by the Carews of Haccombe. There are places of worship for Baptists, Independents, and Wesleyan Methodists.

SHALFLEET, a parish, in the liberty of WEST MEDINA, Isle of Wight Division of the county of SOUTHAMPTON, 3¾ miles (E. by S.) from Yarmouth; containing 1049 inhabitants. The living is a discharged vicarage, valued in the king's books at £18. 12. 1.; present net income, £127: it is in the patronage of the Crown; impropriator, — Wilkinson, Esq. The church is partly Norman and partly of later date, with a low tower of considerable magnitude and disproportionate dimensions, and a remarkable Norman doorway, with a rudely sculptured impost, or lintel, filling up the head of the arch, and said to represent a bishop, whose arms are extended, and the hands resting on animals resembling griffins. Here is a National school. The navigable river Newton bounds the parish on the north-east.

SHALFORD (*St. Andrew*), a parish, in the union of BRAINTREE, hundred of HINCKFORD, Northern Division of the county of ESSEX, 5 miles (N. N. W.) from Braintree; containing 701 inhabitants. This parish is supposed to have derived its name from an ancient ford over the river Blackwater, by which it is bounded on the east. It is about three miles in length, and two in breadth; the soil in some parts is a loam intermixed with sand, and in others a heavy wet loam on a substratum of brown clay. At the Domesday Survey the lands were held in royal demesne, and were subsequently divided into five manors. The living is a discharged vicarage, valued in the king's books at £7; present net income, £155; patron and appropriator, Prebendary of Shalford in the Cathedral Church of Wells. The church is an ancient edifice with a square embattled tower. Here is a Sunday National school.

SHALFORD (*St. Mary*), a parish, in the union of HAMBLEDON, First Division of the hundred of BLACKHEATH, Western Division of the county of SURREY, 1 mile (S. S. E.) from Guildford; containing 910 inhabitants. This place is situated on the road from Guildford to Brighton; the village is very neat. The living is a discharged vicarage, with Bramley annexed, valued in the king's books at £8. 4. 7½.; present net income, £240: it is in the patronage of the Crown; impropriator, Sir Henry Edmund Austen. The church was rebuilt in 1790, chiefly at the expense of Robert Austen, Esq., to whose memory there is a mural tablet in the chancel, by Bacon; he also presented the painted window at the east end. Four schools for girls are supported by four ladies; and the parish receives benefit from Henry Smith's charity. The Wey and Arun Junction canal passes through the parish.

SHALSTONE (*St. Edward*), a parish, in the union, hundred, and county of BUCKINGHAM, 4 miles (N. W. by W.) from Buckingham; containing 198 inhabitants. The living is a rectory, valued in the king's books at £8. 0. 5.; present net income, £199; patron, G. Jervoise, Esq.

SHAMBLEHURST, a tything, in the parish of SOUTH STONEHAM, hundred of MANSBRIDGE, Southampton and Southern Divisions of the county of SOUTHAMPTON, 4¼ miles (W. by S.) from Bishop's-Waltham; containing 921 inhabitants.

SHANGTON (*St. Nicholas*), a parish, in the union of MARKET-HARBOROUGH, hundred of GARTREE, Southern Division of the county of LEICESTER, 6¼ miles (N. by W.) from Market-Harborough; containing 39 inhabitants. The living is a rectory, valued in the king's books at £10. 13. 4.; present net income, £347; patron, Sir J. Isham, Bart. About three quarters of a mile from the church is "Gartre Bush," the spot where the hundred court was formerly held.

SHANKLIN, a parish, in the liberty of EAST MEDINA, Isle of Wight Division of the county of SOUTHAMPTON, 9½ miles (S. E.) from Newport; containing 255 inhabitants. The village occupies a sequestered site, sheltered by the lofty downs which nearly enclose it on two sides, yet sufficiently elevated to command a fine view of Sandown bay and the ocean: it contains several lodging-houses, and a small but picturesque hotel, for the accommodation of the visiters attracted hither by the delightful scenery of this secluded spot. A a short distance is a prodigious chasm, called Shanklin Chine, which, extending a considerable way inland from the coast, and being overgrown with trees, shrubs, and brushwood, contrasted at intervals with bold masses of rock or brown earth, forms a beautifully picturesque and extremely romantic scene. The living is annexed to the rectory of Bonchurch. In the church, which is a small edifice remarkable for the simplicity of its style, is an oak chest, curiously carved, with a Latin inscription, dated 1512, also the arms of the see, the gift of Thomas Silksted, Prior of Winchester.

SHAP (*St. Michael*), a parish, in WEST ward and union, county of WESTMORLAND, 6 miles (N. W.) from Orton; containing 1084 inhabitants. The town consists of one long straggling street on the high road between Penrith and Kendal. The river Lowther runs through the parish, which is bounded on the west by the lake Hawswater, and contains quarries of limestone and blue slate, and a remarkably fine range of red granite. The surrounding country is highly interesting to the geologist and antiquary, and this place is much resorted to for the efficacy of its mineral spring, the water of which closely resembles that of Harrogate. Shap well or spa is about three miles from the town, on the bank of a rivulet which separates the parish from that of Crosby-Ravensworth; and for the accommodation of the increased number of visiters, the Earl of Lonsdale has erected a spacious and handsome hotel, the grounds of which are enriched with thriving plantations, and tastefully laid out in walks and shrubberies. In 1687, a charter was obtained for a market on Wednesday, and three fairs on April 23rd, Aug. 1st, and Sept. 17th, each for two days, but they have been long in disuse. At present a small market is held on Monday, and a fair for cattle and pedlery on May 4th. This has been made a polling-place for the county. The living is a discharged vicarage, valued in the king's books at £8. 15. 7½.; present net income, £73; patron and impropriator, Earl of Lonsdale. A gallery has been erected in the church, and 110 free sittings provided, the Incorporated Society having granted £50 in aid of the expense. Thomas Jackson, in 1703, gave a messuage and land for the erection and support of a school; the annual income is £25, for which 20 children are instructed. The Earl of Lonsdale has recently erected a handsome school-house, which he has endowed. About one mile west from the town are the venerable ruins and tolerably perfect tower of Shap abbey, founded about 1150, by Thomas Fitz-Gospatrick, in honour of God and St. Mary Magdalene, for Premonstratensian canons, whom he caused to be removed hither from an abbey which he had previously established at Preston in Kendal: at the dissolution it contained 20 monks, whose revenue was estimated at at £166. 10. 6. In this parish is the site of a remarkable Druidical monument, or temple, upwards of half a mile in length, and from 20 to 30 yards in breadth: it is encompassed by huge masses of granite, many of them three or four yards in diameter, placed at irregular distances, having at the upper end a circus, or hippodrome, supposed to have been the place of sacrifice. At Hardendale, in this parish, Dr. John Mill, the learned editor of the Greek Testament, was born, in 1645; he died in 1701.

SHAPWICK (*St. Bartholomew*), a parish, in the union of Wimborne and Cranborne, hundred of Badbury, Wimborne Division of the county of Dorset, 4½ miles (S. E.) from Blandford-Forum; containing 462 inhabitants. The living is a vicarage, valued in the king's books at £7. 9. 4½.; present net income, £379; patron and impropriator, H. Bankes, Esq. The river Stour bounds the parish on the south. James Alexander, Esq., in 1818, gave a moiety of the dividends arising from £300 three per cents. for the education of poor children. Here was a small Carthusian priory, a cell to that of Sheen, in Surrey.

SHAPWICK (*St. Mary*), a parish, in the union of Bridg-water, hundred of Whitley, Western Division of the county of Somerset, 6¾ miles (W. by S.) from Glastonbury; containing 452 inhabitants. The living is a discharged vicarage, with the perpetual curacy of Ashcott annexed, valued in the king's books at £9. 13. 4.; present net income, £215; patron, incumbent, and impropriator, Rev. G. H. Templer.

SHARDLOW, a township, and the head of a union, in the parish of Aston-upon-Trent, hundred of Morleston and Litchurch, Southern Division of the county of Derby, 7 miles (S. E. by E.) from Derby: the population is returned with Wiln. The Trent and Mersey canal passes through the township. The poor law union comprises 46 parishes or places, some of which are in the counties of Leicester and Nottingham, and is under the superintendence of 57 guardians: the population, according to the census of 1831, amounts to 29,812.

SHARESHILL (*Assumption of the Virgin Mary*), a parish, in the union of Penkridge, Eastern Division of the hundred of Cuttlestone, Southern Division of the county of Stafford, 5¾ miles (N. N. E.) from Wolverhampton; containing 520 inhabitants. The living is a perpetual curacy; net income, £114; patron and impropriator, Lord Hatherton. The church, with the exception of the tower, is of modern erection, and contains several curious antique monuments, preserved at the demolition of the former edifice. The Staffordshire and Worcestershire canal crosses the north-western angle of the parish. On the north and south sides of the village are vestiges of two encampments, probably Roman.

SHARLSTON, a township, in the parish of Warmfield, union of Wakefield, Lower Division of the wapentake of Agbrigg, West Riding of the county of York, 4¼ miles (E. by S.) from Wakefield; containing 243 inhabitants.

SHARNBROOK (*St. Peter*), a parish, in the hundred of Willey, union and county of Bedford, 4 miles (N. E.) from Harrold; containing 754 inhabitants. The living is a discharged vicarage, valued in the king's books at £8; present net income, £144: it is in the patronage of the Crown; impropriator, J. Gibbard, Esq. The church is of early English architecture, with a spire. There is a place of worship for Baptists. This has been made a polling-place for the county.

SHARNFORD (*St. Helen*), a parish, in the union of Hinckley, hundred of Sparkenhoe, Southern Division of the county of Leicester, 4¼ miles (E. S. E.) from Hinckley; containing 545 inhabitants. The living is a rectory, valued in the king's books at £9. 18. 9.; present net income, £329: it is in the patronage of the Crown. The river Soar runs through the parish.

SHARPENHOE, a hamlet, in the parish of Streatley, union of Luton, hundred of Flitt, county of Bedford, 4 miles (S. S. W.) from Silsoe; containing 149 inhabitants. This hamlet contains a charity school for the instruction of eight children, founded by Richard Norton, in 1686, and endowed with a rent-charge of £10. Thomas Norton, a dramatic writer, was born here in the early part of the sixteenth century: he died about 1600.

SHARPERTON, a township, in the parish of Allenton, union of Rothbury, Western Division of Coquetdale ward, Northern Division of the county of Northumberland, 6½ miles (W. by N.) from Rothbury; containing 105 inhabitants.

SHARPLES, a township, in the parish and union of Bolton, hundred of Salford, Southern Division of the county palatine of Lancaster, 2¾ miles (N.) from Great Bolton; containing 2589 inhabitants. In this township are a large power-loom factory, and a bleaching establishment, affording employment to about 1300 persons. Coal abounds in the neighbourhood; and there are reservoirs for supplying the town of Bolton with water.

SHARRINGTON (*All Saints*), a parish, in the union of Walsingham, hundred of Holt, Western Division of the county of Norfolk, 4¼ miles (W. S. W.) from Holt; containing 252 inhabitants. The living is a discharged rectory, annexed to that of Saxlingham, and valued in the king's books at £10.

SHARROW, a township, in the parish and liberty of Ripon, West Riding of the county of York, 1½ mile (E. by N.) from Ripon; containing 103 inhabitants. A chapel was erected in 1824, containing 550 sittings, of which 280 are free, the Incorporated Society having granted £400 in aid of the expense. The living is a perpetual curacy; net income, £51: it is in the patronage of Mrs. Lawrence; appropriators, Dean and Chapter of Ripon. The Rev. Thomas Savage, in 1782, bequeathed £5 per annum for teaching eight poor children. The sum of £4. 10. a year is also paid out of the chapel rates, on account of sundry small gifts for the education of six others.

SHATTON, a joint hamlet with Brough, in the parish of Hope, union of Chapel-en-le-Frith, hundred of High Peak, Northern Division of the county of Derby, 6 miles (N. E. by N.) from Tideswell: the population is returned with Brough.

SHAUGH, a parish, in the union of Plympton-St. Mary, hundred of Plympton, Ermington and Plympton, and Southern, Divisions of the county of Devon, 6 miles (N.) from Earl's-Plympton; containing 570 inhabitants. This parish is situated in a district abounding with strikingly picturesque and romantic scenery. The living is a perpetual curacy, annexed to that of Samford-Spiney; net income, £107.

SHAVINGTON, a joint township with Gresty, in the parish of Wybunbury, union and hundred of Nantwich, Southern Division of the county of Chester, 4¼ miles (E.) from Nantwich; containing 320 inhabitants. The old manorial seat of the Wodenothes, of whom was descended John, the celebrated antiquary, born in 1624, was highly curious from its age, and the abundance of stained glass and other ancient relics it contained.

After remaining in the possession of that family for more than 500 years, the estate was sold in 1661; the old house was taken down, and a modern mansion, now belonging to — Turner, Esq., built upon the site, in which some of the ancient stained glass is preserved.

SHAW, a chapelry, in the parish of OLDHAM *cum* PRESTWICH, hundred of SALFORD, Southern Division of the county palatine of LANCASTER, 5 miles (S. E.) from Rochdale: the population is returned with the parish. The living is a perpetual curacy; net income, £220; patron, Rector of Prestwich. There is a place of worship for Wesleyan Methodists; also a National school.

SHAW *cum* DONNINGTON (*St. Mary*), a parish, in the union of NEWBURY, hundred of FAIRCROSS, county of BERKS, 1¼ mile (N. E.) from Newbury; containing 620 inhabitants. The living is a rectory, valued in the king's books at £12. 11. 8.; patron, Rev. Thomas Penrose, D.D. The impropriate tithes have been commuted for a rent-charge of £13. 7. 3., and the rectorial for £623. 7., subject to the payment of rates, which on the average, on the latter, have amounted to £108. 16.; the glebe comprises 28 acres, valued at £48 per annum. A school is supported by subscription. An attempt was made by a soldier of Cromwell's army, in 1644, to assassinate Charles I. at the manor-house, his usual place of rest, when journeying to the West of England: a brass plate fixed on the spot where the ball entered still records the event. In the second battle of Newbury this mansion was garrisoned for the king, and attacked by a large body of the enemy, which was repulsed with great loss. Several cannon balls, since found at intervals about the grounds, are preserved; as is also a bed on which Queen Anne reposed. Here are almshouses for twelve poor persons, founded about 1618, by Sir Richard Abberbury, Knt.

SHAWBURY (*St. Mary*), a parish, in the union of WEM, partly in the hundred of PIMHILL, partly in the liberties of SHREWSBURY, and partly in the Whitchurch Division of the hundred of NORTH BRADFORD, Northern Division of the county of SALOP; containing 1088 inhabitants, of which number, 336 are in the township of Shawbury, 7¼ miles (N. E.) from Shrewsbury. The living is a discharged vicarage, valued in the king's books at £7. 1. 5½.; patron, Sir Andrew Vincent Corbet, Bart., who, with Sir R. Hill, Bart., and W. Charlton, Esq., is the impropriator. The impropriate tithes have been commuted for a rent-charge of £436. 8.; and the vicarial for £394. 12., subject to the payment of rates; the glebe comprises 37 acres, valued at £76. 18. per annum. A school, supported partly by subscription and partly by a small endowment of Ralph Collins, is held in a building belonging to Sir A. Corbet: there is also a fund of £46 per annum, the rent of 25 acres of land devised by Dame Elizabeth Corbet, in 1702, and Robert Payne, in 1738, which is distributed in clothing the poor children of the school and apprenticing others with premiums from £6 to £12 each, and in bread for the poor.

SHAWDON, a township, in the parish of WHITTINGHAM, union of ALNWICK, Northern Division of COQUETDALE ward and of the county of NORTHUMBERLAND, 7½ miles (W.) from Alnwick; containing 80 inhabitants. Two ancient urns of common earthenware were found in the neighbourhood some years ago.

SHAWELL (*All Saints*), a parish, in the union of LUTTERWORTH, hundred of GUTHLAXTON, Southern Division of the county of LEICESTER, 3 miles (S.) from Lutterworth; containing 216 inhabitants. The living is a rectory, valued in the king's books at £9; present net income, £400: it is in the patronage of the Crown. A free grammar school for children of this parish and that of Newton, founded here by John Eckington, has an endowment of £20 per annum, with a house and garden for the master: the founder also erected an almshouse for six poor men. Another school is endowed with £2. 10. per annum, for which six children are taught.

SHEARSBY, a chapelry, in the parish of KNAPTOFT, union of LUTTERWORTH, hundred of GUTHLAXTON, Southern Division of the county of LEICESTER, 7 miles (N. E.) from Lutterworth; containing 354 inhabitants. The chapel is dedicated to St. Mary Magdalene. Here is a saline spring, which has been found serviceable in scorbutic affections.

SHEBBEAR (*St. Lawrence*), a parish, in the union of TORRINGTON, hundred of SHEBBEAR, Black Torrington and Shebbear, and Northern, Divisions of the county of DEVON, 7¾ miles (W. N. W.) from Hatherleigh; containing 1179 inhabitants. This parish is bounded on the west by the river Torridge. The living is a discharged vicarage, with the perpetual curacy of Sheepwash annexed, valued in the king's books at £11. 8. 4.; present net income, £243: it is in the patronage of the Crown; impropriator of Sheepwash, — Bendon, Esq.; impropriators of Shebbear, — Brent, — Brand, and — Snell, Esqrs. Here is a National school.

SHEDFIELD, a chapelry, in the parish of DROXFORD, hundred of BISHOP'S-WALTHAM, Droxford and Northern Divisions of the county of SOUTHAMPTON, 3 miles (S.) from Bishop's-Waltham: the population is returned with the parish. The living is a perpetual curacy, in the patronage of the Rector of Droxford; net income, £36.

SHEEN, a parish, in the union of LEEK, Southern Division of the hundred of TOTMONSLOW, Northern Division of the county of STAFFORD, 10 miles (E. by N.) from Winster; containing 366 inhabitants. The living is a perpetual curacy; net income, £63; patrons, alternately, J. Gould, Esq., and Captain Bateman. The church was nearly rebuilt in 1829, at the cost of £1100, raised by a parochial rate. Fifteen poor children are educated for £12 a year, the produce of sundry bequests.

SHEEN, EAST, a hamlet, in the parish of MORTLAKE, Western Division of the hundred of BRIXTON, Eastern Division of the county of SURREY, 6½ miles (S. W. by W.) from London: the population is returned with the parish.

SHEEPHALL, county of HERTFORD—See SHEPHALL.

SHEEPSHEAD (*St. Botolph*), a parish, in the union of LOUGHBOROUGH, hundred of WEST GOSCOTE, Northern Division of the county of LEICESTER, 4 miles (W.) from Loughborough; containing 3714 inhabitants, many of whom are employed in weaving stockings. The living is a discharged vicarage, endowed with a portion of the rectorial tithes, and valued in the king's books at £8. 10. 10.; present net income, £350; patron, C. M. Phillips, Esq.; the remainder of the rectorial tithes have been commuted for land. There are places of worship for Baptists, the Society of Friends, and Wesleyan Methodists. A National school is supported by subscription.

SHEEPSTOR, a parish, in the union of TAVISTOCK, hundred of ROBOROUGH, Midland-Roborough and Southern Divisions of the county of DEVON, 7 miles (S. E. by E.) from Tavistock; containing 154 inhabitants. The living is annexed to the vicarage of Bickleigh. Sheepstor rock is one of the most remarkable granite heaps upon Dartmoor, and is a conspicuous object from Roborough down: at the foot of it is situated the village, on the little river Mew. At Ailsborough, in this parish, a lofty eminence on Dartmoor, are very extensive tin mines.

SHEEPWASH (*ST. LAWRENCE*), a parish, in the union of TORRINGTON, hundred of SHEBBEAR, Black Torrington and Shebbear, and Northern, Divisions of the county of DEVON, 4 miles (W. N. W.) from Hatherleigh; containing 436 inhabitants. The living is a perpetual curacy, annexed to the vicarage of Shebbear. There is a place of worship for Baptists. The river Torridge runs through the parish. A market and three annual fairs were formerly held here. In 1743, the village was almost destroyed by fire. In this parish are the remains of a very large mansion, called Upcott Avenel, to which a chapel was formerly annexed; the few rooms still in being have been converted into a farm-house, and the fine old gateway was removed some years ago, for the purpose of building a handsome residence.

SHEEPWASH (*HOLY SEPULCHRE*), formerly a parish of itself, but now, by act of parliament passed in the latter part of the seventeenth century, forming with Ashington a township in the parish of BOTHAL, union, and Eastern Division of the ward, of MORPETH, Northern Division of the county of NORTHUMBERLAND, 4¾ miles (E.) from Morpeth; containing, with Ashington, 57 inhabitants. The living, now consolidated with Bothal, was a rectory, valued in the king's books at £3. 17. 1. The church has been demolished. The river Wansbeck, which is navigable for keels and small boats, is here crossed by a bridge.

SHEEPY, MAGNA (*ALL SAINTS*), a parish, in the union of ATHERSTONE, hundred of SPARKENHOE, Southern Division of the county of LEICESTER, 3 miles (N. E. by N.) from Atherstone; containing 627 inhabitants. The living is a rectory, consisting of the North and South medieties, with the rectory of Sheepy Parva annexed, valued in the king's books at £13. 4. 9½.; present net income, £835; patron and incumbent, Rev. T. C. Fell. There is a chapel of ease at Ratcliffe Culey; and at the Mythe, which formerly belonged to the monks of Merevale, are some slight remains of an ancient chapel.

SHEEPY, PARVA (*ALL SAINTS*), a parish, in the union of ATHERSTONE, hundred of SPARKENHOE, Southern Division of the county of LEICESTER, 3¼ miles (N. E.) from Atherstone; containing 87 inhabitants. The living is a rectory, annexed to that of Sheepy Magna, and valued in the king's books at £13. 4. 9½.

SHEERING (*ST. MARY*), a parish, in the union of EPPING, hundred of HARLOW, Southern Division of the county of ESSEX, 3 miles (N. E.) from Harlow; containing 547 inhabitants. This parish is bounded on the west by the river Stort; the soil is richly fertile, and the arable lands are under excellent cultivation. The living is a rectory, valued in the king's books at £13. 13. 4.; present net income, £433; patrons, Dean and Canons of Christ-Church, Oxford. The church is a small ancient edifice. A free chapel was formerly endowed here by Christiana de Valvines, the site of which is still called Chapel Field. Here is a National school.

SHEERNESS, a sea-port, market-town, and chapelry, in the parish of MINSTER, union of SHEPPY, having separate jurisdiction, locally in the liberty of the Isle of Sheppy, Upper Division of the lathe of SCRAY, Eastern Division of the county of KENT, 21 miles (N. E.) from Maidstone, and 50 (E. by S.) from London: the population is included in the return for the parish. This place, which is situated at the north-western point of the Isle of Sheppy, on the river Medway, at its junction with the Thames, was a mere swamp until the reign of Charles II., when the importance of its situation being appreciated, that monarch, early in 1667, directed the construction of a strong fort, and twice personally ascertained its progress. In the same year, before the new fortifications were in a very advanced state, the Dutch fleet entered the Thames, and made their memorable attack on the shipping in the Medway, having in their passage destroyed that portion of the works which was completed, and landed some men, who took possession of the fort. In consequence of this, a regular fortification, with a line of heavy artillery, and of smaller forts higher up, on each side of the Medway, was formed, to which other works have since been added. A garrison is kept up, under the command of a governor, lieutenant-governor, fort-major, and inferior officers; and the construction of a royal dock-yard, for repairing ships and building frigates and smaller vessels, has caused Sheerness to become a naval station of the first importance. In 1798, the mutiny of the fleet stationed at the Nore threatened this town with the most alarming consequences, and induced many of the inhabitants to make a precipitate retreat to Chatham and other places; but the fortunate suppression of this formidable insurrection saved the town from the apprehended danger. In 1827 it suffered from a dreadful fire, which destroyed 50 houses, and property of the value of £60,000; but the houses, which were before principally of wood, have been replaced by others built of brick.

Since the formation of the naval establishment here, Sheerness has grown up into a considerable town, consisting of two divisions, Blue Town and Mile Town: it has been recently much enlarged by the erection of some new streets. A pier and causeway extend from the town to low water mark, a distance of about a quarter of a mile, which are kept in good repair; and the town is paved, lighted, and cleansed under the authority of two acts of parliament passed in the 41st and 49th of George III. There was formerly a great scarcity of water, but it is now supplied, and of a very excellent quality, at one halfpenny per pail from four subscription wells, which have been sunk to a depth of 360 feet: the principal one, called King's well, is within the fort, and supplies the vessels at the Nore. Sheerness has latterly become the resort of much company, attracted by the facility of sea-bathing: the beach, on which the machines are stationed, is very clean, and forms a delightful promenade; and there are warm baths, both economical and well managed. On the cliffs leading from the beach towards Minster there is, perhaps, one of the most splendid and interesting views in the kingdom: the North Sea on the east; the rivers Thames and Medway, bearing innumerable vessels of all sizes, with the town and harbour of Sheerness, to the north and west; and the fertile valleys of Kent, with the Medway winding through them, and the towns and villages interspersed, towards the south, combine in presenting a diversity and sublimity of landscape rarely excelled. The harbour has of late years been much enlarged and improved, and is now safe and commodious, often presenting a splendid appearance from the number of vessels in it; passage boats ply with every tide, and a steam-boat twice a day, to and from Chatham; and there is daily communication by steam-boats with London.

The dock-yard has been greatly extended and improved within the last fifteen years, at a cost of about £3,000,000, and is now one of the finest in Europe: it covers an area of 60 acres, and is surrounded by an extremely well-built brick wall, which cost £40,000. The docks are sufficiently capacious to receive men of war of the first class, with all their guns, stores, equipments, &c., on board; and two steam-engines, each of 50-horse power, have been erected for the purpose of pumping them dry. There is a basin, with a depth of water of 26 feet, which will hold six ships of the first class; and two of a smaller size, for store ships and boats. The storehouse, which is the largest building in the country, is six stories high, with iron joists, beams, window frames, and doors, and will contain at least 30,000 tons of naval stores: there are also a handsome victualling storehouse, smithy, navy pay office, masthouses, &c. The superintendent and principal officers of the establishment have handsome houses in the yard; and a noble residence has been erected in the garrison for the port-admiral, in which are state rooms for the reception of the royal family, the lords of the admiralty, &c. The principal establishment of the ordnance department has been removed hence to Chatham, where the stores for the fleet at the Nore, &c., are kept, and the ground formerly occupied by it has been added to the dock-yard. An office connected with this department is still, however, retained here; and it is said that government contemplates the re-establishment of the department at this place on a very enlarged scale, and the extension of the present line of fortifications from the garrison point to the sea wall, called Queenborough wall; for which they have purchased 200 acres of land surrounding the town.

Considerable quantities of corn and seed, the produce of the isle, as well as oysters (of which the beds extend all along the coast, as far as Milton), are shipped for the London market. There are copperas works of considerable extent, within a few miles of the town: the pyrites, or copperas stones, are collected in heaps upon the beach, from the falling cliffs, and carried away in vessels. The market is on Saturday, but there is no regular market-place. The parish church is at Minster, about three miles distant; but a very neat chapel of ease, dedicated to the Holy Trinity, was erected at Miletown in 1835, containing 1070 sittings, of which 600 are free, the Incorporated Society having granted £700 in aid of the expense. At the east end of the dock-yard, outside the wall, there is also a spacious chapel, attached to the dock-yard: the chaplain is appointed by the Board of Admiralty. There are places of worship for Baptists, Independents, Primitive and Wesleyan Methodists, and Roman Catholics, also a synagogue. Connected with the church at Mile Town are three handsome schoolrooms, in which day and Sunday schools are held for 500 boys, girls, and infants; they were erected at an expense of £650; of which £250 was received from the Lords of the Treasury, and £150 from the National Society. A school in connection with the British and Foreign Society has been built at Sheerness for 500 children, at an expense of nearly £600, of which £275 was defrayed by the Lords of the Treasury. The Baptists, Independents, and Wesleyan Methodists have also their Sunday schools, in which more than 1000 children are instructed. There are several reading societies; also a mechanics' institute, with an extensive library, where lectures are delivered during six months of the year. In sinking the wells here, the workmen, at the depth of 200 feet, discovered a complete prostrate forest, through which they were obliged to burn their way. Stones well adapted for the composition of Roman cement, from being impregnated with copperas, are dredged up from the sea near the cliffs.

SHEET, a tything, in the parish and union of PETERSFIELD, hundred of FINCH-DEAN, Petersfield and Northern Divisions of the county of SOUTHAMPTON, ¾ of a mile (N. E. by E.) from Petersfield; containing 380 inhabitants.

Seal of the Town's Trust, or Sheffield Free Tenants.

SHEFFIELD (*St. Peter*), a newly enfranchised borough, market-town, and parish, comprising several chapelries and townships, and the head of a union, chiefly in the Southern, but partly in the Northern, Division of the wapentake of STRAFFORTH and TICKHILL, West Riding of the county of YORK; and containing 91,692 inhabitants, of which number, 59,011 are in the town of Sheffield, 55 miles (S. W. by S.) from York, and 163 (N. N. W.) from London. This place, which is of great antiquity, and was formerly called *Sheaffield*, derived its name from its situation on the river Sheaf, near its confluence with the Don, and forms the chief town of the extensive Saxon manor of Hallam, now called Hallamshire. The parish formed a portion of this large manor, which, although dismembered, and its jurisdiction dissolved, has imparted name to the more extensive tract called Hallamshire, the limits of which are undefined, but of which Sheffield may be considered the chief town. At the time of the Norman survey, the manor of Sheffield was held under Judith, niece of William the Conqueror, by Roger de Busli, and the widowed countess of the Saxon Earl Waltheof, who had been decapitated for entering into a conspiracy against William; and subsequently, with other manors, by the family of de Lovetot, of whom the first that owned the manor is supposed to have erected the ancient castle. On the death of William de Lovetot, without male issue, it was conveyed, with the other large possessions, to the Furnivals, in the early part of the reign of John, by the marriage of Maud, sole daughter and heir of William de Lovetot, with Gerard de Furnival, the younger, and continued in that family for many successions. On failure of issue male in this family, Joan, daughter and heir of William de Furnival, transferred it by marriage to Thomas de Neville, brother of Ralph, Earl of Westmorland, who thereupon was summoned to parliament as Lord Furnival. He died seised of this lordship, but leaving only two daughters and heirs, it came, by marriage of the eldest, to the celebrated John Talbot, who, for his distinguished civil and military services, was created Earl of Shrewsbury, and was succeeded in this lordship by several of his descendants. Gilbert, the last Earl of Shrewsbury who was lord of Hallamshire, left only three daughters, the youngest of whom married Thomas Howard, Earl of Arundel and Surrey; and her two sisters, the Countesses of Pembroke and Kent, having died without issue, the undivided possession of this lordship came, in consequence of the decease of her eldest son, to her grandson, and has since been enjoyed by the Dukes of Norfolk. Edward I. granted various privileges to the lords of the manor, who in turn releasing the inhabitants from the feudal tenure by which they held their estates, in consideration of a fixed annual payment, occasioned Sheffield to become a free town. Cardinal Wolsey, after his arrest in 1530, was detained in the manor-house here for eighteen days, in custody of the Earl of Shrewsbury; and Mary Queen of Scots was, with the exception of a few short intervals, held in captivity in the same place, or in the castle, for nearly fourteen years. During the civil war in the reign of Charles I., the inhabitants were in the interest of the parliamentarians, for whom they made only a feeble effort to maintain the town and castle; but the Earl of Newcastle, with a party of royalists, quickly gained possession for the king, and placed a garrison in the castle, under the command of Sir William Saville, who was appointed governor. The Earl of Manchester sent a force, under Major-General Crawford, to attempt its reduction, and, after a protracted siege, it was surrendered on honourable terms, and soon afterwards demolished by order of the parliament.

The town is pleasantly situated on a gentle eminence rising out of a spacious valley, which, with the exception of an opening to the north-east, is sheltered by a chain of lofty hills richly clothed with wood, and nearly surrounded by the rivers Don, Sheaf, and Porter. Over the river Don a stone bridge of three arches was erected in 1485, and called Lady bridge, from a religious house near it, dedicated to the Blessed Virgin, which was taken down, on widening the bridge, in 1768. An iron bridge of three arches has been also constructed over the same river; and, in 1828, an additional stone bridge of three arches was erected, for the purpose of affording an easier communication between the Rotherham and Barnsley roads, and the new corn and cattle markets: a bridge over the Sheaf, consisting of one arch, was rebuilt in 1769, by Edward Duke of Norfolk. The town extends for nearly a mile from north to south, and for three quarters of a mile from east to west, and consists of numerous streets, which, with the exception of one or two of the principal, are narrow and inconvenient: the houses, chiefly of brick, have obtained from the smoke of the numerous surrounding works, a sombre appearance, and are intermixed with many of very ancient character; the chief portion is within the angle formed by the rivers, but there are considerable ranges of building on the opposite banks. Considerable improvements have taken place, under the provisions of an act obtained in 1818, by which the town is partially paved, and lighted with gas by a company whose extensive works, situated at Shude Hill, near the bridge over the river Sheaf, were erected at an expense of £40,000, raised by a company of shareholders; a new company was formed in 1836, which, with a capital of £80,000, has erected extensive works on Blonk Island for affording a supply on more reasonable terms. The inhabitants were formerly supplied with water conveyed from springs in the neighbouring hills, by means of works situated on Crook's moor, erected by a few private individuals, in 1782; but the supply becoming inadequate to the increasing demands of the town, a company was formed in 1829, with a capital of £100,000, and incorporated by act of parliament. The environs abound with beautiful scenery and agreeable walks, and are interspersed with pleasant villages and ornamented with elegant villas. The barracks, forming an extensive range of building, and, with the parade ground, occupying a large tract of land to the north-east of the town, were erected in 1794, and are pleasantly situated on the bank of the river Don; they contain accommodation for two troops of cavalry. The public subscription library, originally in Surrey-street, was, in 1825, removed into a commodious room fitted up for it in the music hall; connected with the establishment is a reading-room, which is well supplied with periodical works. This library was commenced in 1771, and now comprises upwards of 7,000 volumes, in the various departments of literature and science. A copy of the public records, in 50 folio volumes, published by order of government, was deposited here in 1824 for the use of the inhabitants. A Literary and Philosophical Society was instituted in 1822, the members of which hold their meetings in an elegant apartment in the music hall, containing their apparatus, a collection of fossils, botanical specimens, and curiosities from the South Sea islands, and ornamented with a full-length portrait of Mr. Montgomery, the poet: the society deliver their public lectures in the saloon of the music hall. The Mechanics' library, established in 1824, has a collection of nearly 5,000 volumes, and the room is ornamented with a marble bust of James Watt, executed by Chantrey. The members of the Mechanics' institution, established in 1832, meet in George-street; but a hall is about to be erected by subscription. The public news-room is a neat stone building, forming part of the East Parade, recently erected: a news-room has been also established in the music hall; and another in the Commercial Buildings, a commodious structure, with a handsome Doric front, in High-street, built in 1834, by a company of proprietors, at a cost of £5000, raised in shares of £25 each. The ground story in the front forms a convenient post-office, and the upper story a spacious reading-room: the back part of the building is divided into offices let for various purposes. The Botanical and Horticultural Gardens, on the south-west side of the town, about $1\frac{1}{2}$ mile from the market-place, occupy an angular enclosure of about 18 acres of land, admirably adapted for the purpose, being a fertile and gently broken acclivity, with a southern aspect, in the picturesque vale of the river Porter, opposite the verdant and boldly rising banks of Sharrow, which are occupied by the general cemetery and many handsome villas. The grand entrance into these attractive gardens is at the north-east angle of the boundary wall in Clark-House-lane, through an elegant lodge gateway of the Ionic order, built of rubbed stone, and adapted to the model of the temple on the banks of the Ilyssus at Athens, but somewhat similar in the arrangement of its columns to the entrance at Hyde Park corner. The curator's house is a neat stone mansion near the south-west angle, and between it and the grand entrance is a magnificent range of conservatories, 100 yards in length, divided into five compartments, and having its long roof relieved in the centre and at the two ends by three lofty quadrangular glazed domes, constructed of metallic ribs, connected and surmounted by ornamental castings. The centre and terminating compartments form commodious greenhouses, and their fronts have Corinthian pillars alternating with the vertical sashes. From the front of the central conservatory a grand promenade path, 108 yards long, and 26 feet broad, descends to the margin of a circular tank, in which is a fountain, supplied by a dam in a higher part of the grounds. Below the fountain are tastefully laid out in artificial rock work and water, traversed by winding paths and romantic dells, small ponds, a rustic bridge, subterraneous archways, grottoes, a hermitage, aviaries, and a variety of other interesting objects. From the fountain, fine gravel walks diverge to all parts of the gardens, and one of them extends westward to the bear-pit, beyond which is an ancient wood, occupying the highest part of the grounds, and divided by the boundary wall. The lower lodge is in the style of a Swiss cottage, and has an approach from the Ecclesall new road. These delightful gardens, for recreation and the illustration of botanical science, are yet in their infancy; the land being purchased for its present use in 1834, at the cost of £200 per acre, and laid out and planted in 1835 and 1836. Being among the largest, so they are intended to form one of the most attractive, establishments of the kind in the kingdom; the spirited proprietors having not only provided ample space and suitable receptacles for specimens of all the British plants, and a great number of sxotics, but also a variety of zoological specimens, and epacious walks and grass plots, for the assemblage of the numerous visiters who throng hither on public promenade days, when commodious tents are erected, and the enchanting scene is enlivened by music. The company of proprietors have already expended about £16,000, raised in shares of £20 each. The gardens were first opened June 29th, 1836; and on the first four public days they were visited by more than 12,000 persons. The visitants were equally as numerous on September 14th and 15th to witness a grand Floricultural and Horticultural exhibition, which was open to competitors from all parts of the kingdom, and terminated in a distribution of prizes. The receipts for admission amounted, before the close of the year, to £1600, showing clearly that the public duly appreciate the liberality of the proprietors, and are disposed to grant them a remunerating profit on the capital expended.

Concerts are occasionally held, during the season, in the music hall, under the direction of the Philharmonic Society, which has been lately formed, and possesses one of the finest bands in Yorkshire, consisting of upwards 40 performers, and is supported by subscribers of £1. 1. annually. The Music hall is a spacious and elegant building, in the Grecian style of architecture, erected in Surrey-street, in 1824; it comprises on the ground-floor a room for the public library, a reading-room, and a saloon, and a spacious room for the Literary and Philosophical Society: the buildings also contain an elegant music-room, 99 feet in length and 38 feet wide, with a well-arranged orchestra; adjoining are a handsome saloon, with four recesses, two large refreshment-rooms, and the housekeeper's apartments. The theatre and assembly-rooms were erected in 1762, and form an extensive building of brick, handsomely ornamented with stone, and having a central portico supporting a pediment; the theatre is generally open from October to January, and the assembly-rooms, which are elegantly fitted up, are well attended during the season. The new circus and theatre is a spacious and handsome stone fabric, opposite the cattle-market, and adjoining Blonk-street bridge, close to the conflux of the Don and Sheaf, erected in 1836 and 1837 by a company of 50 proprietors, with a capital of £6000, raised in 240 shares of £25 each. The front is an elegant specimen of the Grecian Ionic order, and reflects much credit on its talented architect, Mr. James Harrison. It has a channelled rusticated basement, with three well-proportioned doors, and a neat pediment, above which is a receding balcony, with two pilasters and two fluted columns, having the enriched neckings of flowing foliage, and the involuted or spiral curves of the most elaborate examples of Grecian architecture. The cornice and frieze are continued along and returned on the wings, and finished with a bold blocking, except over the balcony, where the blocking forms a plinth for an attic, above which another cornice and blocking terminate this chaste and imposing elevation. The interior is handsomely fitted up, and adapted for both equestrian performances and dramatic representations, having a ring 42 feet in diameter, a stage 60 feet by 40, and stables for 14 horses. The boxes and galleries are approached by stone staircases, and from the ceiling is suspended a splendid chandelier. The public baths, on the road to Glossop, were erected in 1836, at an expense of £8000, by a company; they are neatly built and contain a public news-room and library, and connected with the baths are several handsome houses for families during the bathing season. There is also at Portobello an establishment of hot and medicated vapour baths, which are well attended. The Town's Trust has arisen from a grant made by a member of the ancient family of Furnival, about the year 1300, and consists of property in lands and tenements, shares in the river Don navigation, &c., producing about £1400 per annum, which is under the management of twelve resident trustees, elected by the freeholders, who have been incorporated under the title of the "Town's Trust," or "Sheffield Free Tenants:" the income is applied to the maintenance of Lady's bridge, the keeping in order of Barker-pool, the repair of the church and the highways, the payment of the stipendiary clergy, and other charitable and public uses.

This town appears to have been distinguished at a very early period for the manufacture of articles of cutlery, for which the numerous mines of coal and iron in the neighbourhood rendered its situation peculiarly favourable. Chaucer, in his Canterbury Tales, mentions the "Sheffield Thwytel, or Whittel," a kind of knife worn by such as had not the privilege of wearing a sword, for the making of which, as well as the iron heads for arrows, before the general use of fire-arms, Sheffield had, even at so early a period, become celebrated. From that time the principal articles manufactured were implements of husbandry, including scythes, sickles, shears, and other sharp instruments of steel, till the middle of the last century, when considerable improvements were introduced, and great ingenuity displayed in the finer articles of cutlery. The superintendence of the trade was entrusted to twelve master cutlers, appointed at the court leet of the lord of the manor, with power to enforce the necessary regulations for its protection and improvement. In 1624, the cutlers were incorporated by an act of parliament entitled "an act for the good order and government of the makers of knives, scissors, shears, sickles, and other cutlery wares, in Hallamshire, in the county of York, and parts near adjoining;" and the government was invested in a master, two wardens, six searchers, and 24 assistants, consisting of freemen only, in number about 600. The master, who, with the other officers of the company, is chosen annually by the whole corporation, on retiring from office, nominates the senior warden as his successor; but if the latter be rejected by the company, he nominates another member, till one is approved of by the body: the wardens are chosen by the officers of the company from among the searchers for the time being. The master, wardens, and assistants, have power to make by-laws for the regulation of the trade, and to inflict penalties for the neglect of them; and the jurisdiction of the company, which is restricted exclusively to affairs relating to the trade, extends throughout the whole district of Hallamshire, and all places within six miles of it. By an act obtained in 1814, permission is given to all persons, whether sons of freemen or not, and without their having served an apprenticeship, or obtained from the company a mark for their goods, to carry on business any where within the limits of Hallamshire. The privilege thus bestowed has been a great means of advancing the trade to its present state of perfection, by affording encouragement to men of genius from every part of the country to settle in this town; and the competition thus produced has furnished exquisite specimens of workmanship, in the finer branches of the trade, which abound in the show-rooms of the principal manufacturers, particularly in those of Messrs. Rodgers and Sons, and excite the admiration of the spectator. The cutlery trade employs from 8000 to 10,000 persons. The principal articles are table knives and forks; pen and pocket knives of every description; scissors; razors; surgical, mathematical, and optical instruments; engineers' and joiners' tools; scythes, sickles, and files, of which great quantities are manufactured and exported; and an endless variety of steel wares, which may be considered the staple trade of the town, though various other branches of manufacture have been subsequently introduced and carried to a high degree of perfection.

Seal of the Cutlers' Company.

Connected in some degree with the cutlery, but embracing a great variety of other objects, are ivory articles of every description; but the principal branches of manufacture which have more recently been established, and for which the town has obtained an unrivalled superiority, are spoons, tea and coffee pots, candlesticks, and a great variety of articles of Britannia metal, which are made in great quantities, and of every pattern; likewise silver-plated goods of every kind, among which are dessert knives and forks plated upon steel, tureens, épergues, and services for the table, candelabras, ice pails, urns, and a variety of similar articles, of the most elegant patterns, and of the richest workmanship, being generally known by the name of "Sheffield plate with silver edges." The manufacture of silver plate in all its branches, from the most minute to the most massive articles, is also carried on to a considerable extent, and has obtained deserved celebrity. The Assay office, erected in 1795, is a plain substantial edifice. The most ingenious and highly-finished specimens of cutlery displayed in the principal shops in the metropolis, and in those of the principal towns in England, notwithstanding their being stamped with the venders' names, are manufactured here; and so highly are the manufactures of this town esteemed, that they are found in every market in Europe, and exported in great quantities to every part of the globe. The making of buttons and button-moulds, wire-drawing, and the refining of silver, are also carried on; and along the banks of the rivers are numerous iron and steel works, in which the heavier castings are produced, and extensive works for slitting and preparing the iron and steel for the use of the manufacturers: among the manufactured iron goods are boilers for steam-engines, stove-grates (of most elegant design and exquisite workmanship), fenders, fire-irons, and various smaller articles. There are also extensive factories for the weaving of carpets, and of horse-hair seating for chairs. In 1806, a type-foundry was established with considerable success; and another was commenced in 1818, the proprietors having purchased the business of a house in London. The parish abounds with coal, of which the principal mines at present wrought are in the Park, and in the township of Attercliffe: the upper strata being nearly exhausted, new pits have been opened for procuring the lower beds, which are of good quality, and in the working of which very powerful engines are employed; but the whole produce is very insufficient for the supply of the numerous and extensive manufactories, and great quantities of coal are obtained from the surrounding districts. Iron-stone is also abundant, but the quality is not well adapted to the general purposes of the manufacturers, and consequently great quantities of iron, especially for the use of the cutlers, are imported from Sweden, Germany, and Russia. Sand-stone and firm grit-stone are quarried in several parts of the parish, and in others a grey slate of good quality is found. The trade of the town is greatly facilitated by its advantageous line of inland navigation. The river Don was, in 1751, made navigable to Tinsley, about three miles from the town; and in 1815, an act was obtained for enabling the proprietors of the Sheffield canal to connect the Don, at Tinsley, with the town, by means of a navigable cut, which was accomplished in 1819, forming a direct communication with the North Sea. Adjoining the basin of this canal, at the eastern extremity of the town, is a commodious wharf, where vessels can load and unload under cover; and there are spacious warehouses and offices for the transaction of business. The basin is capable of containing more than 40 vessels of about 50 tons' burden, which arrive here from Hull, York, Gainsborough, Manchester, Leeds, Liverpool, and Thorn, at which last place vessels from London generally unload goods intended for Sheffield. The Sheffield and Rotherham railway was opened on the 31st of October, 1838; its chief object is to make available to the manufactories of Sheffield the produce of the coal fields of Kimberworth, Greasborough, and Rawmarsh. The market was granted in 1296, to Thomas, Lord Furnival; the market-days are Tuesday and Saturday: the former, chiefly for corn, is held in the Corn Exchange, a large and elegant building, erected under an act of parliament obtained in 1827, by the Duke of Norfolk, on the site of the Shrewsbury hospital, (which has been removed) and situated in the Park, between the Sheaf and canal bridges, having in its principal front a handsome semicircular portico, supported by 16 massive pillars. Behind it is the cheese, poultry, and fish market; and in the area in front, enclosed with a neat iron railing, is the market for the sale of corn in sacks. Opposite this enclosure is the large open space where the fairs, and the hay and wholesale vegetable and fruit markets are held; and a little to the north of this is the spacious new cattle market. The market for butchers' meat is held in a convenient situation near Market-street, and the shambles are extensive and commodious, having covered walks in front of the rows of butchers' stalls; at the lower end is a convenient market for butter, eggs, and poultry; and round its exterior are shops for the sale of fruit, vegetables, &c. It is approached by several gateways, one of which opens into the market for shoes, tin ware, &c., and another into the vegetable and fruit market, on the opposite side of King-street. A market for earthenware is held every Tuesday in Paradise-square. The fairs are on the Tuesday in Trinity week and November 28th, for cattle and toys; and a cheese fair, held at the same time with the latter, has been established within the last few years, in which are sold many hundred tons of cheese from the counties of Derby, Stafford, Chester, and Lancaster.

By the act of the 2nd of William IV., cap. 45, Sheffield has been constituted a borough, with the privilege of sending two members to parliament, the right of election being vested in the £10 householders; the borough comprises an area of 19,307 acres, of which the limits are minutely described in the Appendix, No. II.: the Master Cutler is the returning officer. The town is within the jurisdiction of the magistrates for the district, who meet in the town-hall every Tuesday and Friday, for the determination of petty causes: the October sessions for the West Riding are also held here by adjournment. In the 29th of George II. (1756), an act of parliament was passed, by which the proceedings in personal actions in the courts baron of Sheffield and Ecclesall were regulated. In 1791 the Ecclesall court was revived, after having been discontinued many years, by Earl Fitzwilliam, who gave a piece of land on which a gaol was built by private subscription. The act of George II. having proved inefficient, was repealed by another in 1808, by which commissioners are appointed for determining cases in which the property in question does not exceed in value £5. The court for the manor of Sheffield is held every Thursday at the town-hall; that for the manor of Ecclesall is held on the Monday in every third week at the court-house in Ecclesall: the jurisdiction of each court extends for several miles round the parish. This town has been made a polling-place for the West Riding. The town-hall, a spacious and commodious building of plain architecture, with a cupola and clock, was erected in 1808, at the foot of the hay-market, on the site of an old building which was taken down: it was considerably enlarged at the expense of the Town Trust in 1833. The ground-floor consists of a large entrance-hall and offices, below which is a watch-house used by the police commissioners, who have their office here and hold their meetings in one of the court-rooms above, where are also held the quarter and petty sessions and public meetings. The sessions-hall is one of the largest in the county, and is handsomely fitted up and lighted by an elegant dome. Behind the building is the gaoler's house, and under it are cells for the temporary confinement of prisoners, who, after conviction, are generally sent for various periods to the Wakefield house of correction, except for capital offences, when they are conveyed to York. The present Cutlers' Hall, in Church-street, was completed in 1833, at a cost of £6500: it has a handsome stone front of the Corinthian order, ornamented with two fluted columns and four pilasters supporting a pediment, in the tympanum of which are the Cutlers' Arms sculptured in bold relief. The vestibule leads to a double staircase which ascends to the saloon; the banquet-room is a spacious apartment with a handsome dome light, and in front is the assembly-room: the rooms are adorned with five portraits and three busts, including a bust of the late Dr. Browne, by Chantrey.

Seal of the Twelve Capital Burgesses.

The living is a vicarage, valued in the king's books at £12. 15. $2\frac{1}{2}$.; present net income, £1285; patron, Duke of Norfolk; impropriators, P. Gell, Esq., and another. The vicarage-house is at the corner of St. James'-street, and the glebe land in its vicinity is covered with buildings. Three stipendiary clergymen, who, since the year 1819, have had an income of £250 each, are appointed to assist the vicar by twelve burgesses of the town and parish, incorporated by charter of Queen Mary, who hold certain lands and estates in trust, for the payment of the stipendiary assistants, and for the repairs of the church, bridges, &c., and relief of the needy poor: they are called the "Twelve Capital Burgesses," and hold their meetings in a room over the vestry-room of the church, vacancies in their number being filled up by vote among themselves. The church, erected in the reign of Henry I., is a spacious cruciform structure, with a central tower and spire, most probably in the Norman style of architecture; but it has been so altered by repairs, that, with the exception of part of the tower and spire, and a few small portions of the interior, very little of its original character can be distinguished. The chancel contains the first production from the chisel of Chantrey, a mural tablet, with the bust of the Rev. James Wilkinson, late vicar, canopied with drapery, in Carora marble, erected at the public expense, as a tribute of respect to his memory. Many illustrious persons have been interred in this church, among whom were Mary, Countess of Northumberland; Elizabeth, Countess of Lennox, mother of the unfortunate Lady Arabella Stuart; Lady Elizabeth Butler; four of the Earls of Shrewsbury; and Peter Roflet, French secretary of Mary, Queen of Scots. St. Paul's chapel was erected in 1720, by subscription, towards which Mr. R. Downes, silversmith, contributed £1000: it is a handsome edifice, in the Grecian style of architecture, with a tower surmounted by a well-proportioned dome, and a cupola of cast-iron; the interior is light, and elegantly ornamented, and contains a bust by Chantrey of the Rev. Alex. Mackenzie, with emblematical sculpture finely executed. The living is a perpetual curacy; net income, £136; patron, the Vicar. St. James' chapel, a neat structure, in the Grecian style of architecture, with a campanile turret, was erected by subscription in 1788: the interior is well arranged, and the east window is embellished with a beautiful painting of the Crucifixion, by Peckett. The living is a perpetual curacy; net income, £160; patron, the Vicar; impropriator, Duke of Norfolk. St. George's church, on an eminence at the western extremity of the town, erected in 1824, by grant from the parliamentary commissioners, at an expense of £14,819, is a very handsome structure, in the later style of English architecture, with a lofty square embattled tower at the west end, rising to the height of 139 feet, and crowned with pinnacles: the interior is handsomely finished, and contains about 2000 sittings, 1000 of which are free: the large altar-piece is an admirable representation of Christ blessing little children, painted and presented by Mr. Paris, in 1831. The living is a perpetual curacy; net income, £365; patron, the Vicar. St. Philip's church, near the infirmary, was erected in 1827, by grant from the parliamentary commissioners, at an expense of £13,970. 16.: it is a neat edifice, in the later English style of architecture, with a square embattled tower crowned with pinnacles. The living is a perpetual curacy; net income, £135; patron, the Vicar; impropriator, Duke of Norfolk. St. Mary's church, in Brammall-lane, of which the first stone was laid by the Countess of Surrey, in 1826, is a very handsome structure in the later style of English architecture, enriched externally with a profusion of grotesque heads and other ornaments, and presenting within, walls of polished stone, and a lofty groined roof, supported by two rows of light and elegant clustered columns, with a tower and a porch of beautiful design: it was erected by grant from the parliamentary commissioners, at an expense of £12,650: the site and the cemetery were given by His Grace the Duke of Norfolk. The living is a perpetual curacy; net income, £190; patron, the Vicar. St. John's church, on Park-hill, was erected by subscription in 1837, on a site of three acres presented by the Duke of Norfolk, at an expense of nearly £4000; it is a neat edifice, with a tower surmounted by a slender spire, and contains 1200 free sittings. There are also other churches in the townships belonging to the parish, which are noticed under their proper heads. The dissenting places of worship consist of eleven belonging to the various denominations of Methodists, six to Independents, and one each to Baptists, the Society of Friends, and Presbyterians, also a Roman Catholic chapel. A general cemetery has been formed on an acclivity of Sharrow Vale, in the south-western suburbs of the town, by a company of shareholders, which occupies more than 5½ acres of land, including the entrance road, and was opened on the 30th of July, 1836. The principal approach is by a level road to the entrance lodge, a handsome stone edifice of the Doric order, situated at the foot of the acclivity, from which a broad gravel walk leads to the lower range of catacombs, parallel with which, and forming a stage above it, is a longer series of catacombs, with a terrace-walk in front, and surmounted by a parapet with Egyptian balustrades. On a higher elevation stands the chapel, a neat stone edifice of the Doric order, with a portico of fluted columns approached by an Egyptian flight of steps; the door and windows are also of an elegant Egyptian character. Behind it is the quarry from which nearly the whole of the stone was dug, and of which one side is occupied by a range of catacombs. Above this, on the higher part of the grounds, stands the minister's house, a handsome building of the Doric order, commanding a fine prospect; and behind it is the boundary wall, in which are two arched Egyptian gateways. The nature of the situation has furnished some of the chief attractions of this beautiful cemetery, and in laying out the grounds nearly all the old trees have been preserved, and broad gravel walks formed leading to the buildings, terraces, &c.

The free grammar school was founded in 1603, by Thomas Smith, of Crowland, in the county of Lincoln, but a native of Sheffield, who endowed it with £30 per annum for a master and usher. In the following year, letters patent were obtained from James I., incorporating the church burgesses and the vicar, under the style and title of the "Governors of the Goods, Possessions, and Revenues of the Free Grammar School of James, King of England, &c.," with a common seal. The old school occupied a low situation in Townhead-street: the present handsome edifice, in Charlotte-street, was built in 1825, at a cost of £1600, of which £1400 was raised by subscription. By augmentation of the original bequest, the endowment now consists of a farmstead and 61¼ acres of land, at Wadsley, yielding a rental of £140 per annum; and two houses with ten acres of land at Gilberthorpe, let for £20. 10. per annum, and bequeathed by James Hill, in 1709: the master's salary is £100, and the usher's, £50, the remainder being reserved for repairs. The school is free for classical instruction to 25 boys of the parish, but for other branches of education the pupils pay each £2. 2. per quarter. The boys' charity school, at the north-east corner of the parish churchyard, was established in 1706; and the present school-house, a neat and commodious edifice of stone, was built in 1825, on the site of the original building, at the cost of £3000, and in 1830 was raised, and a large play-ground or gymnasium formed under the roof and extended over the court behind, by means of an iron gallery resting in the walls of the projecting wings. The school has an income arising from a benefaction of £5000 by Mr. Parkins, in 1766, aided by other donations, including one from Mr. T. Hanby, which maintains six boys on the establishment, at an expense of upwards of £60 per annum, the past masters of the Cutlers' Company being his principal trustees: the whole revenue is upwards of £600 per annum, with which, and annual subscriptions, 90 boys (of whom the greater part are clothed in blue, and the remainder in green) are maintained, clothed, educated, and apprenticed. At the opposite corner of the churchyard is a similar school, in which 70 girls are maintained, clothed, and educated, and afterwards placed out in service: a convenient school-house was erected, in 1786, at an expense of £1500, by voluntary subscription. A school for reading, writing, and arithmetic, has also been established here, in pursuance of the will of Mr. William Birley, who, in 1715, bequeathed £900 in trust for the purchase of an estate, of the rental of which, one-third was to be appropriated to the establishment of the school, one-third towards the maintenance of indigent tradesmen, or their widows, and the remainder towards the support of a minister to officiate in the chapel of the hospital. This school is situated in School-croft, a little below the site of the old grammar school, and was rebuilt in 1827: the entire income of the charity is about £200 per annum, arising from a large farm in the county of Derby, and other sources. A Lancasterian school for boys, established in 1809, and a similar institution for girls, established in 1815, are supported by subscription. There are a central and six other National schools, with numerous Sunday and infants' schools.

The Earl of Shrewsbury's hospital was projected by Gilbert Earl of Shrewsbury, in 1616, and completed in pursuance of his will, by the Earl of Norfolk, Earl Marshal of England, and the buildings erected in 1673: it is amply endowed for eighteen men and eighteen women: the original buildings were recently taken down to make room for the market-place, and the erection of the corn exchange; and a neat range of buildings, in the later style of English architecture, has been erected on the southern side of the town, in the centre of which is a chapel. Hollis' hospital was founded in 1703, by Mr. Thomas Hollis, a native of the town, who, with others of his descendants, endowed it for sixteen aged women, widows of cutlers, or of persons connected with the trade; a part of the funds is applied to the support of a school. The general infirmary was first opened for the reception of patients in 1797, and, in a manufacturing town, where so many artisans are continually exposed to accidents, and their health materially injured by the processes of many of the trades in which they are employed, has been deservedly regarded as an object entitled to the most liberal patronage and support. The premises, occupying an extensive site about a mile to the north-west of the town, and guarded against the too near approach of other buildings by the purchase of 31 acres of surrounding land, were erected by public subscription, at an expense of nearly £20,000, including the purchase of the land: they are handsomely built of stone, and form a conspicuous ornament in the principal approaches to the town. In front of the building is a neat portico, ornamented with statues of Hope and Charity, finely sculptured; and the grounds are enclosed by an iron palisade, with a central gateway, and a porter's lodge on each side. The internal arrangements are extensive and complete, and the institution is supported by an income arising from numerous donations and bequests, and by annual subscription.

Several extensive charitable benefactions have been made for the benefit of the inhabitants. Mr. Thomas Hanby left £8000, of which the interest of £3000 was for the benefit of the boys' charity school, and that of the remaining £5000 for distribution among the poor and creditable housekeepers, members of the church of England, and not under fifty years of age, two-thirds of the number to be men, and one-third women: the nomination is in the Master and Wardens of the Cutlers' Company, the past masters, the vicar and churchwardens, and the Town's Trust. Mrs. Eliza Parkins bequeathed £10,000, one-half of which is appropriated to the support of the boys' charity school, and the interest of the remainder divided annually among such poor persons as the vicar, the three assistant ministers, and the churchwardens, shall select. Mrs. Mary Parsons bequeathed £1500 to be invested in the funds, and the proceeds annually divided among 48 aged and infirm silver-platers; Mr. John Kirby left £400, the interest of which is annually divided between two poor widows; and Mr. Joseph Hudson, of London, gave £200 in trust to the Cutlers' Company, to divide the proceeds annually among sixteen of the most needy file-makers. There are also various other charitable bequests for distribution among the indigent; a humane society for recovering persons apparently drowned; a society for ameliorating the condition of the poor, and various others. The poor law union of Sheffield comprises three townships of the parish, together with the parish of Handsworth, containing in the whole, according to the census of 1831, a population of 74,058; and is under the care of thirteen guardians. On Spital Hill, near the town, was an hospital, founded in the reign of Henry II. by William de Lovetot, and dedicated to St. Leonard, of which there is no vestige; and of the ancient manor-house, in which Cardinal Wolsey and Mary Queen of Scots were confined, the ruins can but faintly be traced. In 1761, two thin plates of copper were ploughed up on a piece of land, called the Lawns, each containing an inscription commemorating the manumission of some Roman legionaries, and their enrolment as citizens of Rome. From the prevalence of iron-ore, the waters of Sheffield have a slight chalybeate property. The Rev. Dr. Robert Saunderson, Regius Professor of Divinity in the University of Oxford, and Bishop of Lincoln; and the Rev. Mr. Balguay, Prebendary of North Grantham in the Cathedral Church of Salisbury, and an eminent disputant in the Bangorian controversy, were natives of this place: and Chantrey, the celebrated sculptor, was born at Norton, a village about three miles from the town. Sheffield gives the title of Baron and Earl to the family of Holroyd.

SHEFFORD, a chapelry (formerly a market-town), in the parish of Campton, union of Biggleswade, hundred of Clifton, county of Bedford, 9½ miles (S. E. by S.) from Bedford; containing 763 inhabitants. The market, which was on Friday, has been discontinued. The river Ivel has been rendered navigable to Biggleswade, by the assistance of a canal recently cut. The chapel is dedicated to St. Michael, and has been enlarged by the addition of 200 free sittings, the Incorporated Society having granted £250 in aid of the expense. There is also a Roman Catholic chapel. Robert Bloomfield, the poet, died here in 1823.

SHEFFORD, LITTLE, or EAST, a parish, in the union of HUNGERFORD, hundred of KINTBURY-EAGLE, county of BERKS, 5¾ miles (N. E.) from Hungerford; containing 67 inhabitants. The living is a rectory, valued in the king's books at £9. 11. 3.; present net income, £400; patron, R. Harbert, Esq. The church contains some interesting monuments.

SHEFFORD - HARDWICKS, an extra - parochial liberty, in the hundred of WIXAMTREE, county of BEDFORD; containing 16 inhabitants.

SHEFFORD, GREAT, or WEST (*St. Mary*), a parish, in the union of HUNGERFORD, hundred of KINTBURY-EAGLE, county of BERKS, 5 miles (N. E. by N.) from Hungerford; containing 559 inhabitants. The living is a rectory, valued in the king's books at £14. 3. 4.; patrons, Principal and Fellows of Brasenose College, Oxford. The tithes have been commuted for a rent-charge of £833. 18. 6., subject to the payment of rates, which on the average have amounted to £111; the glebe comprises 110 acres, valued at £168 per annum. The church is principally in the Norman style, with a circular tower at the west end. Near the north door is a niche for the Virgin, enriched with pinnacles, &c., and the large font is richly and curiously carved with foliage: in the churchyard is the shaft of an ancient cross. There is a place of worship for Wesleyan Methodists; also a school partly supported by the Marquess of Downshire and the rector. Charles I. took up his quarters here on Nov. 19th, 1644.

SHEINTON (*St. Peter and St. Paul*), a parish, in the union of ATCHAM, hundred of STOTTESDEN, though locally in that of Condover, Southern Division of the county of SALOP, 3½ miles (N. by W.) from Much Wenlock; containing 133 inhabitants. The living is a discharged rectory, valued in the king's books at £6. 9. 2.; present net income, £288; patron, John Hodgson, Esq. This parish is bounded on the north by the river Severn.

SHELDING, a township, in the parish of RIPON, Lower Division of the wapentake of CLARO, West Riding of the county of YORK, 6½ miles (W. by S.) from Ripon; containing 49 inhabitants.

SHELDON, a chapelry, in the parish and union of BAKEWELL, hundred of HIGH PEAK, Northern Division of the county of DERBY, 3 miles (W.) from Bakewell; containing 148 inhabitants. The living is a perpetual curacy; net income, £99; patron, Vicar of Bakewell; impropriators, Dean and Chapter of Lichfield and Duke of Devonshire. The chapel is dedicated to All Saints. There are lead mines in the neighbourhood. Mary Frost, in 1756, gave £200 for apprenticing poor boys.

SHELDON (*St. James*), a parish, in the union of HONITON, hundred of HAYRIDGE, Cullompton and Northern Divisions of the county of DEVON, 7 miles (E. by N.) from Cullompton; containing 185 inhabitants. The living is a perpetual curacy; net income, £250; patron, Edward Simeon Drewe, Esq. The church contains 70 free sittings, the Incorporated Society having granted £20 in aid of the expense.

SHELDON (*St. Giles*), a parish, in the union of MERIDEN, Birmingham Division of the hundred of HEMLINGFORD, Northern Division of the county of WARWICK, 4¾ miles (S. W.) from Coleshill; containing 422 inhabitants. The living is a rectory, valued in the king's books at £8. 10. 10.; patron, Earl Digby. The tithes have been commuted for a rent-charge of £450, subject to the payment of rates, which on the average have amounted to £55; the glebe comprises 38 acres, valued at £63. 17. per annum. The church is principally in the decorated style, with a fine tower of later architecture.

SHELDWICH (*St. James*), a parish, in the union and hundred of FAVERSHAM, Upper Division of the lathe of SCRAY, Eastern Division of the county of KENT, 2¾ miles (S. by W.) from Faversham; containing 497 inhabitants. The living is a discharged vicarage, valued in the king's books at £6. 16. 8.; present net income, £109; patrons and appropriators, Dean and Chapter of Canterbury. The church is principally in the decorated style of English architecture. Here is a National school.

SHELF, a township, in the parish and union of HALIFAX, wapentake of MORLEY, West Riding of the county of YORK, 3½ miles (N. E. by N.) from Halifax; containing 2614 inhabitants, who are principally employed at some extensive iron-works here. There is a place of worship for Primitive Methodists. A Lancasterian school has been established for children of all persuasions.

SHELFANGER (*All Saints*), a parish, in the union of GUILTCROSS, hundred of DISS, Eastern Division of the county of NORFOLK, 2½ miles (N. by W.) from Diss; containing 435 inhabitants. The living is a rectory, valued in the king's books at £17; present net income, £440; patron, Duke of Norfolk. There is a place of worship for Baptists.

SHELFORD (*St. Peter and St. Paul*), a parish, in the union of BINGHAM, Southern Division of the wapentake of BINGHAM and of the county of NOTTINGHAM, 8 miles (E. N. E.) from Nottingham; containing 704 inhabitants. This parish forms part of the vale of Trent, which river bounds it on the west and north, and the Fosse-road touches its south-eastern boundary. The manor-house was garrisoned by Col. Stanhope, son of the first Earl of Chesterfield, for Charles I., and taken by storm by Col. Hutchinson, for the parliament, after a gallant resistance, during which Col. Stanhope and most of his men were slain. The living is a perpetual curacy; net income, £60; patron, Earl of Chesterfield. The church is the burial-place of the noble family of Stanhope, and contains the remains of Philip, the accomplished Earl of Chesterfield, who died in 1773. A school is supported by the Earl of Chesterfield. A priory, in honour of the Blessed Virgin Mary, was founded here in the time of Henry II., by Ralph Hanselyn, which, at the dissolution, had a revenue of £151. 14. 1. An hospital, called the Bede Houses, was founded and endowed in 1694, by Sir William Stanhope, for the reception and support of six of his decayed tenants, but the number has been reduced to four.

SHELFORD, GREAT (*St. Mary*), a parish, in the union of CHESTERTON, hundred of THRIPLOW, county of CAMBRIDGE, 4½ miles (S. by E.) from Cambridge; containing 812 inhabitants. This parish is situated on the river Granta, and on the road from London to Cambridge. There are extensive flour and oil-cake mills, driven by the stream of the Granta, and employing about 20 persons. The living is a discharged vicarage, valued in the king's books at £13. 6. 8.; present net income, £102; patron, Bishop of Ely; impropriators, Master and Fellows of Jesus' College, Cambridge. The church is said to have been built by Bishop Fordham, who died in 1425; the steeple was blown down by the great storm in 1703, and again, in 1798; it has since been rebuilt by subscription. In the chancel is a monument to Dr. Redman, Bishop of Norwich. There is a place of worship for Baptists, built by Mr. James Nutter, in 1823. A school is supported by subscription. On a farm, called Grannams, the property of St. John's College, are some remains of a Roman intrenchment. The Rev. Robert Hall, the eminent dissenting divine, was for three years a resident of this parish.

SHELFORD, LITTLE (*All Saints*), a parish, in the union of CHESTERTON, hundred of THRIPLOW, county of CAMBRIDGE, 5½ miles (S. by E.) from Cambridge; containing 483 inhabitants. This parish is situated on the river Granta, which separates it from the parish of Great Shelford, and on the high road from London to Cambridge, which forms a junction with the old road at Chesterford. The living is a rectory, valued in the king's books at £15. 9. 7.; present net income, £370; patron and incumbent, Rev. H. Finch. In the chancel of the church is a monument to Sir John de Treville, a Knight Templar and lord of the manor, with his figure in a recumbent position, with a lion at the feet. Near the altar a skeleton encased in lead was dug up in 1824, the hair of which was in a perfect state. There is a place of worship for Independents. A small school for girls is supported by subscription. Near the bridge over the Granta was anciently a hermitage.

SHELL, a hamlet, in the parish of HIMBLETON, Middle Division of the hundred of OSWALDSLOW, Droitwich and Eastern Divisions of the county of WORCESTER; containing 43 inhabitants.

SHELLAND, a parish, in the union and hundred of STOW, Western Division of the county of SUFFOLK, 3½ miles (W. N. W.) from Stow-Market; containing 126 inhabitants. The living is a donative; net income, £40; patron and impropriator, C. Tyrrell, Esq. The church was appropriated to that of Haughley in the 3rd of Edward III.: the present building bears date 1767.

SHELLEY (*St. Peter*), a parish, in the union and hundred of ONGAR, Southern Division of the county of ESSEX, 1½ mile (N.) from Chipping-Ongar; containing 163 inhabitants. This parish is intersected by a small stream, on the banks of which are luxuriant meadows, and the general beauty of the surrounding scenery renders it a desirable place of retirement. The living is a rectory, valued in the king's books at £9. 15.; patron, J. Tomlinson, Esq. The tithes have been commuted for a rent-charge of £180, subject to the payment of rates, which on the average have amounted to £30; the glebe comprises 35 acres, valued at £65 per annum. The church is a neat edifice of brick, erected in 1811 on the foundation of the ancient structure. The parsonage, a handsome ancient mansion of timber frame-work and plaster, was for some time the retreat of Dr. Thomas Newton, Bishop of Bristol, and author of a dissertation on the Prophecies. Here is a National school. The Rev. H. Soames, historian of the "Reformation," and author of other theological works, is the present rector.

SHELLEY (*All Saints*), a parish, in the hundred of SAMFORD, Eastern Division of the county of SUFFOLK, 2½ miles (S.) from Hadleigh; containing 142 inhabitants. The living is a perpetual curacy; net income, £72; patrons and impropriators, Heirs of Sir W. B. Rush, Knt. The steeple of the church is placed in a very unusual situation; it is on the north side of the nave, serving for a porch.

SHELLEY, a township, in the parish of KIRK-BURTON, union of HUDDERSFIELD, Upper Division of the wapentake of AGBRIGG, West Riding of the county of YORK, 6 miles (S. E.) from Huddersfield; containing 1319 inhabitants. There are places of worship for Independents and Wesleyan Methodists; also a school endowed with £12. 10. per annum.

SHELLINGFORD, county of BERKS.—See SHILLINGFORD.

SHELLOW-BOWELS (*St. Peter and St. Paul*), a parish, in the union of ONGAR, hundred of DUNMOW, Northern Division of the county of ESSEX, 6¼ miles (N. E.) from Chipping-Ongar; containing 143 inhabitants. This small parish, which consists only of one manor, is supposed to have been formerly much more extensive. The ancient manor-house is near the church. The living is a discharged rectory, consolidated with that of Willingale-Doe, and valued in the king's books at £7. 13. 4. The tithes have been commuted for a rent-charge of £120, subject to the payment of rates, which on the average have amounted to £18; the glebe comprises 13 acres, valued at £15 per annum. The church is a handsome edifice of brick, erected on the site of the former structure in 1752.

SHELSLEY-BEAUCHAMP (*All Saints*), a parish, in the union of MARTLEY, partly in the Lower and partly in the Upper Division of the hundred of DODDINGTREE, Hundred-House and Western Divisions of the county of WORCESTER, 8¼ miles (S. W.) from Stourport; containing 553 inhabitants. The living is a rectory, valued in the king's books at £9. 4. 4½.; present net income, £376; patron, Lord Foley. The free school in the churchyard was founded and endowed with the gift of £100 by the Rev. Owen Plwy, in 1681, and it subsequently received benefactions in land from Caleb Avenant and others, yielding in the whole £60 per annum, out of which the master is paid £52 annually for instructing 25 poor children, and supplying them with books and stationery; the remaining £8 being applied to apprenticing boys, according to the will of one of the donors The Rev. Thomas Webb, in 1703, bequeathed an estate called Hay-Oak Farm, the rent to be applied in apprenticing children of poor relatives of the donor, or, in default of these, poor children belonging to the parish.

SHELSLEY, KING'S, a hamlet, in the parish of SHELSLEY-BEAUCHAMP, union of MARTLEY, Upper Division of the hundred of DODDINGTREE, Hundred-House and Western Divisions of the county of WORCESTER, 9½ miles (S. W. by W.) from Stourport; containing 282 inhabitants.

SHELSLEY-WALSH (*St. Andrew*), a parish, in the union of MARTLEY, Upper Division of the hundred of DODDINGTREE, Hundred-House and Western Divisions of the county of WORCESTER, 9 miles (S. W.) from Stourport; containing 44 inhabitants. The living is a discharged rectory, valued in the king's books at £3. 8. 9.; present net income, £90; patron, Lord Foley.

SHELSWELL (*St. Ebbe*), a parish, in the union of BICESTER, hundred of PLOUGHLEY, county of OXFORD, 6 miles (N. N. E.) from Bicester; containing 49 inhabitants. The living is a rectory, annexed to that of Newton-Purcell, and valued in the king's books at £4. The church is in ruins.

SHELTON (*St. Mary*), a parish, in the union of ST. NEOTS, hundred of STODDEN, county of BEDFORD, 4 miles (W. by N.) from Kimbolton; containing 132 inhabitants. The living is a rectory, valued in the king's books at £13; present net income, £190; patron, Lord St. John. This parish has the privilege of sending five boys to the free school in the adjoining parish of Nether Dean.

SHELTON (*St. Mary*), a parish, in the union and hundred of DEPWADE, Eastern Division of the county of NORFOLK, 2½ miles (S. E. by S.) from Stratton St. Mary; containing 253 inhabitants. The living is a rectory, with that of Hardwick annexed, valued in the king's books at £8; patron and incumbent, Rev. E. Frank. The tithes have been commuted for a rent-charge of £628, subject to the payment of rates, which on the average have amounted to £149; the glebe comprises 42 acres, valued at £52. 10. per annum. There is also a rent-charge of £10. 8. payable to the rector out of the tithes of the parish of Ringstead, Northamptonshire. The church was built by Sir Ralph Shelton, who also erected the manor-house, now in ruins: it was a spacious castellated structure, with a chapel attached.

SHELTON (*St. Mary*), a parish, in the union of BINGHAM, Southern Division of the wapentake of NEWARK and of the county of NOTTINGHAM, 6½ miles (S. by W.) from Newark; containing 113 inhabitants. The living is a rectory, valued in the king's books at £6. 14. 4½.; present net income, £322; patron, Rev. R. Ffarmerie.

SHELTON, a township and chapelry, in the parish, union, and newly erected borough, of STOKE-UPON-TRENT, Northern Division of the hundred of PIREHILL and of the county of STAFFORD, 2 miles (E. N. E.) from Newcastle-under-Lyme; containing, with the hamlet of Etruria, and part of Cobridge, 9267 inhabitants. This place, which is adjoining to the township of Hanley, like many other places in this part of the county, has arisen from the very extensive potteries carried on in the vicinity. It is amply supplied with water; the foot-paths are paved with brick; and under the superintendence of commissioners appointed by an act of parliament obtained in 1815, and amended in 1828, for its better regulation and improvement, jointly with Hanley, it is lighted with gas. An act for the establishment and regulation of the market, and for the improvement of the market-place, was obtained in 1813, under the provisions of which the rents, tolls and duties are vested in trustees for the purposes of the act; and the surplus is directed to be appropriated from time to time to the promotion and aid of any public works, institutions, or establishments within the townships of Hanley and Shelton. A mechanics' institute was established here in 1826, for the potteries at large, under the patronage of the Marquess of Stafford, Josiah Wedgwood, Esq., and others. Concerts, generally for the benefit of some charitable institution, take place occasionally; and public meetings of the inhabitants are also held in the same room. Races, which are in general well attended, have been lately established in the neighbourhood. The principal articles of manufacture are porcelain and earthenware, affording employment to more than 3000 men, women, and children. Several of these manufactories are situated on the banks of the Trent and Mersey and Caldon canals, which pass through the township; and on the banks of which also are some extensive gas works. In the hamlet of Etruria are the extensive potteries and handsome mansion of Josiah Wedgwood, Esq., the latter remarkable for the beauty of its situation and style of architecture, and for the many splendid Etruscan vases with which it is ornamented: these elegant specimens of art, produced here under his own superintendence, are imitations of original vases found in Italy, to the discovery of which that gentleman was chiefly indebted for the elegance of form and purity of taste which he introduced into the manufacture of porcelain, china, and stone ware, for which this place is so deservedly celebrated, and which, by the use of flint in the composition of these articles, also introduced by Mr. Wedgwood, has, under his auspices, been progressively brought to its present state of perfection. The coal and iron-stone mines, in this and part of the adjoining township of Hanley, belong to the crown, and are extensively worked by Earl Granville, the lessee. The market for both townships is held at Hanley weekly on Wednesday and Saturday, and a cattle market is held there on the second Tuesday in every month. In 1813, an act was obtained for establishing and regulating the market, and for enlarging and improving the market-place, principally through the exertions of John Tomlinson, Esq., of Cliffville. Part of the site is held by lease for 21 years under the lords of the manor, at the yearly rent of £10, and perpetually renewable before the end of every 18th year, on payment of a fine of £50. Considerable additions have been made by purchase, and a handsome and capacious range of stone buildings has been recently erected by the trustees, at an expense of more than £5000, solely for a butchers' market, which is now one of the largest and best in the county. The buildings, rents, and tolls are vested in the trustees for the purposes of the act, the surplus to be applied to purposes of public utility, in Hanley and Shelton, which are in this respect united. A large public fund has thus been created, progressively increasing with the extension of these two towns, and already producing more than £1200 per annum, whilst, in the memory of persons now living, there was only one butcher in both places. The police establishment is under the superintendence of the commissioners; and a chief bailiff is annually elected, by whom all public meetings in both townships are convened, and at which he presides. The township is in the honour of Tutbury, duchy of Lancaster, and within the jurisdiction of a court of pleas held at Tutbury every third Tuesday, for debts under 40*s*.

Under the provisions of an act of parliament passed in 1827, relating to the rectory of Stoke-upon-Trent, this chapelry is to be separated from that parish, and to be made a distinct district rectory, to be endowed with not less than £10,000, nor more than £15,000, at the option of the patron, from the proceeds of tithes authorised to be sold and invested in land. A handsome and spacious church, in the early English style of architecture, with a square embattled tower crowned with pinnacles, has been erected by her Majesty's commissioners, at an expense of £9311, towards defraying which George IV. gave £250 from the revenues of the duchy of Lancaster, and further donations were made by Earl Granville, John Tomlinson, Esq., and others, to the amount of upwards of £500: it was consecrated on the 19th of June, 1834, and dedicated to St. Mark. In the chancel is a beautiful painted window, representing the Nativity and Ascension, and in the central apartment a full-length figure of St. Mark writing his gospel, presented as the joint gift of the patron, John Tomlinson, Esq., and the rector, the Rev. J. W. Tomlinson. A powerful organ has also been erected, at the sole expense of the patron. The church, till its endowment and conversion into a district church, will be a chapel of ease to the mother church; and the curate, whose present income is derived from the pew rents, will continue to be appointed by the Rector of Stoke-upon-Trent. The late rector, Dr. Woodhouse, gave £1000, which, with its accumulations, he appropriated to the erection of a parsonage-house, besides allotting a portion of the income arising from his munificent donation of £3000 for the support of a National school, which has also a permanent endowment from land given by Mrs. Hannah Bagnall, of Newcastle-under Lyme. There are places of worship for Baptists, Independents, Wesleyan Methodists of the Old and New Connection, and Unitarians. A British and Foreign school was established in 1821, and is supported by subscription; and there are various Sunday schools, in which several thousand children receive instruction. In this township is the North Staffordshire Infirmary, a noble institution, erected in 1816, and since very much enlarged; including the fever wards, which occupy one of the wings, it is capable of accommodating more than 100 in-patients: this institution has received donations and bequests to a very liberal extent, is further supported by annual subscriptions, and has also an accumulating fund originated by John Tomlinson, Esq., of Cliffville, which commenced with donations and subscriptions to the amount of £1142 exclusively for this purpose, and in fourteen years up to Michaelmas 1833, had accumulated more than threefold; the process of accumulation is to continue till the fund amounts to £20,000, when the entire annual produce of that sum is to be applied for ever as a permanent income for the support of the institution. Elijah Fenton, the poet, was born here in 1683.

SHELVE (*All Saints*), a parish, in the union of Clun, hundred of Chirbury, Southern Division of the county of Salop, 7¾ miles (N. by E.) from Bishop's-Castle; containing 71 inhabitants. The living is a discharged rectory, valued in the king's books at £2. 13. 4.; present net income, £70; patron, Robert Bridgeman More, Esq. The chapel has been repewed, and 100 free sittings provided, the Incorporated Society having granted £50 in aid of the expense. Here are numerous veins of lead-ore, which is considered to vie in richness with any in England: one of the mines was worked by the Romans in the time of Adrian, as is evident from an inscription on a pig of lead found in the vicinity. A market on Friday, and a fair on the festival of the Invention of the Cross, were granted by Henry III.

SHELWICK, a township, in the parish of Holmer, hundred of Grimsworth, union and county of Hereford, 2½ miles (N. E. by N.) from Hereford: the population is returned with the parish.

SHENFIELD (*St. Mary*), a parish, in the union of Billericay, hundred of Barstable, Southern Division of the county of Essex, 1 mile (N. E. by N.) from Brentwood; containing 665 inhabitants. There are several good houses in the village, which is pleasantly situated on the road to Colchester. The Eastern Counties railway intersects the parish: there is a small toy fair on Whit-Monday. The living is a rectory, valued in the king's books at £14. 18. 4.; patron, Lord de Grey. The tithes have been commuted for a rent-charge of £575, subject to the payment of rates, which on the average have amounted to £80; the glebe comprises 77 acres, valued at £92 per annum. The church is an ancient edifice with a shingled spire, and contains a much admired monument to Mrs. Robinson. Here is a National school, erected and supported by the present incumbent.

SHENINGTON (*Holy Trinity*), a parish, in the union of Banbury, Upper Division of the hundred of Tewkesbury, county of Gloucester, 6 miles (W. N. W.) from Banbury; containing 433 inhabitants. This parish is bounded by the counties of Oxford and Warwick, and is several miles distant from any other part of Gloucestershire: by the Reform Act it has been annexed to Oxfordshire as regards the election of parliamentary representatives. The living is a rectory, valued in the king's books at £15. 3. 4.; present net income, £321; patron, Earl of Jersey. The church contains 90 free sittings, the Incorporated Society having granted £60 in aid of the expense.

SHENLEY (*St. Mary*), a parish, in the union of Newport Pagnell, partly in the hundred of Cottesloe, but chiefly in that of Newport, county of Buckingham, 3½ miles (N. W. by W.) from Fenny-Stratford; containing 484 inhabitants. The living is a rectory, valued in the king's books at £22. 9. 7.; present net income, £424; patron and incumbent, Rev. P. Knapp. The chancel of the church is a fine specimen of the transitional style from the early to the later style of Norman architecture: in the south transept is a handsome monument to Sir Thomas Stafford, who lived in the reign of James I., in an old mansion now taken down, at Tattenhoe, in the parish of Waddon, and who founded, in 1626, an almshouse here, with an endowment of £35 per annum, for four widowers and two widows.

SHENLEY (*St. Botolph*), a parish, in the union of Barnet, hundred of Dacorum, county of Hertford, 6 miles (N. W.) from Chipping-Barnet; containing 1167 inhabitants. The living is a rectory, valued in the king's books at £16. 8. 1½.; present net income, £1244; patron and incumbent, Rev. T. Newcome. The church is built with flints, and has a wooden tower on the south side. Three schools are supported by the ladies of the parish. Sir Richard Cox, in 1633, bequeathed property now producing £24. 10. per annum, which is distributed in small sums to the poor; and the parish is entitled to send one poor widow into the Pemberton almshouses at St. Albans. A chapel is supposed to have formerly stood on a moated site in the park belonging to the house called Colney Chapel.

SHENSTONE (*St. John the Baptist*), a parish, in the union of LICHFIELD, Southern Division of the hundred of OFFLOW and of the county of STAFFORD, 3½ miles (S. by W.) from Lichfield; containing 1827 inhabitants. This parish is very extensive, comprising several hamlets and gentlemen's seats, and about 9000 acres of fertile land, watered by several rivulets abounding with excellent trout. The living is a discharged vicarage, valued in the king's books at £6. 5. 8.; present net income, £488; patron, Rev. John Peel; impropriators, various proprietors of land. The church exhibits specimens of the various styles of English architecture: a gallery has been erected, containing 70 free sittings, the Incorporated Society having granted £30 in aid of the expense. There is a chapel of ease at Upper Stonnal, containing 180 free sittings, the Incorporated Society having granted £200 in aid of the expense. A fair for cattle is held on the last Monday in February. Here is a National school, chiefly supported by William Leigh, Esq. There was once a castle or fortification at Upper Stonnal, of which only the remembrance is preserved in the name of Castle field.

SHENTON, a chapelry, in the parish and union of MARKET-BOSWORTH, hundred of SPARKENHOE, Southern Division of the county of LEICESTER, 2½ miles (S. W. by S.) from Market-Bosworth; containing 200 inhabitants. The Ashby-de-la-Zouch canal crosses the north-eastern angle of the chapelry.

SHEPHALL (*St. Mary*), a parish, in the union of HITCHIN, hundred of CASHIO, or liberty of ST. ALBANS, county of HERTFORD, 2¼ miles (S. E. by S.) from Stevenage; containing 217 inhabitants. The living is a vicarage, valued in the king's books at £9. 5. 10.; present net income, £193; the patronage and impropriation belong to the Crown. There is a fund of £24. 16. 6. per annum, arising from the bequests of Elizabeth Nodes, in 1730, and George Nodes, in 1737, which is annually distributed among the poor families in the parish.

SHEPLEY, a township, in the parish of KIRK-BURTON, union of HUDDERSFIELD, Upper Division of the wapentake of AGBRIGG, West Riding of the county of YORK, 7 miles (S. E. by S.) from Huddersfield; containing 893 inhabitants.

SHEPPERTON (*St. Nicholas*), a parish, in the union of STAINES, hundred of SPELTHORNE, county of MIDDLESEX, 2¼ miles (E. by S.) from Chertsey; containing 847 inhabitants. The living is a rectory, valued in the king's books at £26; present net income, £499; patron, S. H. Russell, Esq. Here is a National school.

SHEPRETH (*All Saints*), a parish, in the union of ROYSTON, hundred of WETHERLEY, county of CAMBRIDGE, 5¾ miles (N. by W.) from Royston; containing 345 inhabitants. The living is a discharged vicarage, valued in the king's books at £6. 11. 1.; present net income, £97; patron and impropriator, James Wortham, Esq. A school is supported by subscription. The river Cam runs through the parish.

SHEPSCOMB, a chapelry, in the parish of PAINSWICK, hundred of BISLEY, Eastern Division of the county of GLOUCESTER, 4 miles (N.) from Stroud; containing 803 inhabitants. This place is pleasantly situated in a retired vale remarkable for the variety and picturesque beauty of its scenery. To the east of the village, on the road from Stroud, is Shepscomb House; and on the acclivity of a wood-crowned hill, at the distance of a mile to the west, is Ebworth Park, from which is a beautiful view of a chain of hills stretching in the form of an amphitheatre, and richly clothed with beech trees of luxuriant growth. Extending along the eastern side of the vale is Loncheridge wood, comprising about 400 acres of beech and other trees. The manufacture of woollen cloth is carried on here to a considerable extent; there are two establishments, affording employment to a considerable number of persons in the making of Saxony broad cloths, for which this place has been celebrated for some years. The living is a perpetual curacy; net income, £45; patron, Vicar of Painswick. The chapel was built in 1819 by subscription: it contains 60 free sittings, the Incorporated Society having granted £60 in aid of the expense. A school is supported by subscription.

SHEPTON-BEAUCHAMP (*St. Michael*), a parish, in the union of CHARD, hundred of SOUTH PETHERTON, Western Division of the county of SOMERSET, 3¾ miles (N. E. by E.) from Ilminster; containing 648 inhabitants. The living is a rectory, valued in the king's books at £14. 8. 11½.; present net income, £373; patrons, — Nash and — Clark, Esqrs. The poor children of this parish are instructed for £10 a year, arising from the rent of certain land bequeathed by Thomas Rich, in 1723, and the interest of £100, the gift of Elizabeth Morgan, in 1763, which is applied in aid of a National school. Mrs. Morgan also bequeathed £200, the interest to be applied in apprenticing children.

SHEPTON-MALLET (*St. Peter and St. Paul*), a market-town and parish, and the head of a union, in the hundred of WHITESTONE, Eastern Division of the county of SOMERSET, 14 miles (N. E.) from Somerton, and 125 (W. by S.) from London; containing 5330 inhabitants. The origin of this town is not comparatively of very remote antiquity, the charter for its market having been granted by Edward II., in the 11th year of his reign. The manor, at the time of the Norman survey, was subordinate to that of Pilton, which had been granted by King Ina to the abbot of Glaston; and its pastures, from the sheep fed on which it is supposed to derive its name, are noticed in that work; the additional and distinguishing appellation having been received from the Barons Mallet, lords of Shepton in the reigns of Henry I. and II. The consequences of the Duke of Monmouth's rebellion were severely felt in this part; thirteen persons of the town, having been convicted at the "bloody western assizes," suffered here for their participation in that political enterprise. The town is situated chiefly on the southern bank of a deep valley, and consists of a number of streets and lanes, the principal of which, crossing the valley from north to south, is spacious, and well built, but the others are mostly narrow and irregular: the recent erection of a bridge, and the opening of a new road, have materially improved the town. It is well supplied with water, and a stream runs through the bottom of the valley, turning several mills in its course. The manufacture of woollen goods, silk, lace, stockings, sail-cloth, and hair-seating, is carried on to a considerable extent. The parish comprises a portion of the Mendip range of hills, prior to the enclosure of which lead-ore was obtained in it. The market days are Tuesday and Friday, the latter a very large one for all kinds of agricultural produce. The market cross, erected by Walter and Agnes Buckland, in 1500, is a fine old structure: it originally consisted of five arches, but it has lately undergone a thorough renovation (funds having been left to keep it in repair by the founders), and a sixth arch has been added: elevated above two rows of steps is an hexagonal pillar, supporting a flat roof, surmounted by a pyramidal spire, and ornamented with niches. The fairs are on Easter-Monday, 18th of June, and the 8th of August. The management of the local affairs is vested in a high constable and subordinate officers, who are chosen at a court leet held annually in October, by the householders generally; and a court for the recovery of debts under £2 has been held here from time immemorial. This town has been made a polling-place for the eastern division of the shire.

The living is a rectory, valued in the king's books at £33. 12. 1.; present net income, £533: it is in the alternate patronage of the Queen, in right of the duchy of Cornwall, and the Rev. Provis Wickham. The church is a venerable cruciform pile of building, to which are attached two small chapels; the roof of the nave is curiously wrought, and the pulpit and font, which are of stone, are much admired: it contains some ancient monuments, and has been enlarged and 700 free sittings provided, the Incorporated Society having granted £500 in aid of the expense. There are places of worship for Baptists, Independents, Wesleyan Methodists, and Roman Catholics. The nuns of the order of the Visitation have a convent here, containing about 30 inmates. The free school, founded by Sir George Strode and others, in 1639, is endowed with property at present producing about £75 per annum, principally arising from the rectorial tithes of the parish of Meare, subject to the master's keeping the school-house, and the chancel of the parish church of Meare, in repair, and paying a crown rent of £10. 18. per annum to the Earl of Radnor, for the rectory of Meare: there is no scholar on the foundation. Four poor boys are educated, and an apprentice-fee of £7 given with each, by means of a charity founded by Mr. John Curtis, in 1730, now producing about £20 per annum; and sixteen poor girls are clothed and educated from the produce of a bequest made by Mrs. Mary Gapper, in 1783. There is also a National school, and a school belonging to dissenters is supported by subscription. Almshouses for four poor men were founded and endowed in 1699, by Mr. Edward Strode, with property now yielding about £360 per annum, of which amount about £80 per annum is appropriated to the repairs of the almshouses, and allowances to the inmates, and about £200 to the purchase of bread for weekly distribution among the poor of the parish, for whose benefit there are several other small benefactions. The county bridewell, or house of correction, is in this town: it is capable of receiving from 200 to 300 prisoners, and comprises eighteen wards, nineteen day-rooms, nine work-sheds, two separate apartments for the sick, and tread-mills: the building was erected by prison labour, the stone being quarried within the walls. The poor law union of Shepton-Mallet comprises 25 parishes or places, containing a population of 18,040, according to the census of 1831, and is under the of 31 guardians. The Roman Fosse-way to Ilchester passes through the parish to the eastward of the town. Shepton-Mallet is the birthplace of Hugh Inge, Chancellor of Ireland, who died in 1528; of Walter Charlton, an eminent physician, author of a work on Stonehenge, and other productions; and of Simon Browne, a celebrated dissenting minister.

SHEPTON-MONTAGUE (*St. Peter*), a parish, in the union of Wincanton, hundred of Norton-Ferris, Eastern Division of the county of Somerset, 2½ miles (S.) from Burton; containing 452 inhabitants. The living is a perpetual curacy, valued in the king's books at £8. 15.; present net income, £62; patron and impropriator, Earl of Ilchester.

SHERATON, a township, in the parish of Monk-Hesleton, union of Easington, Southern Division of Easington ward, Northern Division of the county palatine of Durham, 11½ miles (N.) from Stockton-upon-Tees; containing 110 inhabitants.

SHERBORNE (*St. Mary*), a market-town and parish, and the head of a union, in the hundred of Sherborne, Sherborne Division of the county of Dorset, 18 miles (N. by W.) from Dorchester, and 117 (W. S. W.) from London; containing 4075 inhabitants. This place, though of remote antiquity, does not appear to have emerged from comparative insignificance until the Saxon era; the name, anciently written *Schiraburn*, *Schireburn*, and *Scyreburn*, and of which its present appellation is a corruption, is derived from the Saxon words *Sirce* clear, and *Burn*, a spring, or fountain, and was usually written in old Latin records *Fons clarus*. In 670, a house was founded here for Secular canons, by Cenwalh, King of the West Saxons, and others; and in the year 704, Sherborne was made the head of an episcopal see, which included the counties of Dorset, Somerset, Wilts, Devon, and Cornwall, by Ina, of which his kinsman, Aldhelm, was the first bishop. About 998, the Secular canons were displaced, and a society of Benedictines established under licence from Ethelred, by Wlfin, bishop of the see, who also rebuilt the monastery, and dedicated it to St. Mary: this institution became richly endowed, and at the dissolution its revenue was valued at £682. 14. 7.: the only remains of the convent are the cloister, the arched gateway of the abbey barn (the barn itself having been converted into a dwelling-house of modern erection, but of ancient style of architecture), and the ancient refectory, now used as a silk-manufactory. About 1103, Sherborne is stated to have been burnt by a detachment of the Danish invaders, and the entire destruction of the town and its ecclesiastical buildings, although doubtful, is a matter of great probability. The see continued for more than three centuries, when it was removed first to Wilton, afterwards to Old Sarum, and finally to New Sarum, or Salisbury; this event contributed much to depress the prosperity of Sherborne, and for a long period afterwards it was in comparative obscurity. It is evident that a castle was built here at a very early period, but the founder and the time of its erection and demolition are unknown. Previously to the time of Henry I., however, another had been built by Roger, the third Bishop of Salisbury, and became an episcopal palace: it was an octagonal structure, situated on a hill eastward of the town, and fortified by a moat and several drawbridges: having been seized by Stephen, it remained in the possession of the crown for some time, but about 1350 was recovered by Bishop Wyvil, and reverted to the bishoprick. During the civil war in the reign of Charles I., it was garrisoned in the royal interest, and, although gallantly defended and one of the last that yielded, it was eventually taken by the parliamentary forces under the command of Fairfax, and was demolished in 1645. Considerable portions of the ruins are remaining: the present mansion, called Sherborne Castle, the seat of Earl Digby, and standing in a very fine park, was built by Sir Walter Raleigh. The town is situated principally on a gradual slope near the border of the White Hart Forest, and the vale of Blackmore, and is divided by a small stream into two parts, of which one is called Castle Town: it is well paved, lighted, and amply supplied with water. The woollen trade, which formerly flourished, was succeeded by the making of buttons, haberdashery, and lace. In 1740 the first silk-mill was erected, and the various branches of this manufacture, especially the making of silk twist and buttons, afford employment to a great number of the working class. Markets are held on Tuesday, Thursday, and Saturday, the principal day being Thursday; and fairs on May 22nd, July 18th and 26th, and the first Monday after October 10th. The town is within the jurisdiction of the county magistrates; and formerly a court for the recovery of debts under 40*s.* was held by the steward of the hundred, but has been for some years discontinued. It has been made a polling-place for the county.

The living is a vicarage, valued in the king's books at £20. 4. 7.; present net income, £258: it is in the patronage of the Crown; impropriator, Earl Digby. The church, anciently the cathedral of the bishops of Sherborne, and originally erected by Bishop Aldhelm, stands south-west of the town: it was partially destroyed by fire in the reign of Henry VI., and having been rebuilt by the abbots during the three succeeding reigns, it consequently exhibits specimens of different styles of architecture: the semicircular arches and zigzag mouldings in the porch, transept, west end, and north side of the building, are Norman; the upper part of the nave, the tower, east end, aisles, and some of the chapels, are in the later style of English architecture: the roof is supported by groins springing from the aisles, and at the intersection of the tracery is a variety of arms and emblematical devices. It is a magnificent cruciform structure of freestone, with a central tower 154 feet high, containing eight bells, the largest of which, weighing more than three tons, was the gift of Cardinal Wolsey, and is considered the largest bell ever rung in a peal: the tower underwent considerable repair in 1830. The Saxon kings, Ethelbald and Ethelbert, and many Saxon nobles, bishops, and abbots, have been interred here: the church contains 100 free sittings, the Incorporated Society having granted £50 in aid of the expense; also some very ancient monuments, including a handsome one of the Digby family. There are places of worship for the Society of Friends, Independents, and Wesleyan Methodists. The free grammar school was founded by Edward VI., who endowed it with property belonging to the several chantries in the churches of Gillingham, Lytchett-Matravers, and Marnhull, in the county of Dorset, and of Martock in the county of Somerset, producing an income of £1200 per annum, and placed it under the government of 20 of the inhabitants, whom he incorporated, and appointed the bishop of Bristol special visiter: the masters are required to be clergymen, and graduates of one of the Universities. There are four exhibitions of £60 per annum each to either of the Universities, tenable for four years by boys on the foundation only, of whom the present number is 40. One of the principal benefactors to this town was Mr. Benjamin Vowell, who by will gave the dividends of £1000 three per cent. consols. to be annually distributed in clothing, besides two sums of £300, and one of £400, to various benefit societies. The Blue-coat school was founded in 1640, by Richard Foster, who gave land for the instruction and clothing of ten boys and ten girls, directing £5 per annum from the surplus rents to be applied towards maintaining one of the boys at the University, if required. A charity school for boys was founded in 1717, by John Woodman, who gave £250 to the vicar and churchwardens, for the instruction of poor boys, which sum was vested in land for that purpose. In 1743, William Lord Digby gave land for teaching and clothing thirteen poor girls. The churchwardens and overseers have the right of sending three boys to Christchurch Hospital, in London, for whose support Giles Russel gave lands, in the year 1670. A Lancasterian and a National school are supported by subscription. The almshouse, originally an hospital of the order of St. Augustine, was, by licence from Henry VI., refounded and dedicated to St. John the Baptist and St. John the Evangelist, for 20 brethren, twelve poor men, four poor women, and a chaplain; and was governed by a master and trustees: it now contains sixteen men and eight women, under the superintendence of a master and nineteen brethren, elected from among the inhabitants of the town; a chaplain officiates daily. There is a very considerable fund for the relief of the poor, arising from land and houses given for that purpose, in 1448, by Robert Neville, Bishop of Sarum, and others; also a small sum for apprenticing poor children, given by Agnes Broughton, in 1633. The poor law union of Sherborne comprises 30 parishes or places, containing a population of 11,243, according to the census of 1831, and is under the care of 34 guardians.

SHERBORNE (*St. Mary Magdalene*), a parish, in the union of Northleach, Lower Division of the hundred of Slaughter, Eastern Division of the county of Gloucester, 6 miles (N. W. by W.) from Burford; containing 767 inhabitants. The living is a discharged vicarage, with that of Windrush united, valued in the king's books at £15. 6. 8.; present net income, £194; patron and impropriator, Lord Sherborne. Here is a National school. James Bradley, D.D., Regius Professor of Astronomy, and Astronomer Royal, was born here in 1692: he died in 1762, and was buried at Minchinhampton. Sherborne gives the title of Baron to the family of Dutton.

SHERBORNE, ST. JOHN (or EAST) (*St. Andrew*), a parish, in the union and hundred of Basingstoke, Basingstoke and Northern Divisions of the county of Southampton, 2¾ miles (N. N. W.) from Basingstoke; containing 702 inhabitants. The living comprises a sinecure rectory and a vicarage, the former valued in the king's books at £9. 8. 1½., and of the present annual value of £352; patron, W. L. W. Chute,

Esq.: and the latter valued at £7; patron, the Rector: present net income, £139. There is a private chapel at the Vine, the seat of Mrs. Chute, containing a tomb in memory of Chaloner Chute, Esq., Speaker of the House of Commons in Richard Cromwell's parliament, and the purchaser of this noble mansion, which was erected in the reign of Henry VIII., by the first Lord Sandys.

SHERBORNE, MONK (or WEST) (*All Saints*), a parish, in the union, and partly in the hundred, of Basingstoke, but chiefly in the hundred of Chutely, Basingstoke and Northern Divisions, of the county of Southampton, $3\frac{1}{2}$ miles (N. W. by N.) from Basingstoke; containing 522 inhabitants. The living is a discharged vicarage, valued in the king's books at £8. 0. $7\frac{1}{2}$.; present net income, £50; patrons and impropriators, Provost and Fellows of Queen's College, Oxford. The chapel of a Benedictine priory still remains, and service is performed in it every Sunday; it contains an altar-tomb with the recumbent figure of a Knight Templar, carved in solid oak, supposed to be the effigy of Sir John de Port, the founder. A National school has been established. The priory was dedicated to St. Mary and St. John, and was a cell to the abbey of Cerasy in Normandy; at its suppression it was given by Henry VI. to Eton College, but was finally granted by Edward IV. to the hospital of St. Julian in Southampton, and is enjoyed by the Provost and Fellows of Queen's College, as masters of that hospital.

SHERBOURNE (*All Saints*), a parish, in the union of Warwick, Snitterfield Division of the hundred of Barlichway, Southern Division of the county of Warwick, $2\frac{3}{4}$ miles (S. W. by S.) from Warwick; containing 241 inhabitants. The living is a perpetual curacy, with the rectory of Fulbrook united; net income, £110; patron, Samuel Ryland, Esq.

SHERBURN, a township, in the parish of Pittington, Southern Division of Easington ward, union, and Northern Division of the county palatine, of Durham, $2\frac{3}{4}$ miles (E.) from Durham; containing 337 inhabitants. A branch of the Clarence railway extends to the lime and coal works here. A National school has been established.

SHERBURN (*St. Hilda*), a parish, in the union of Scarborough, wapentake of Buckrose, East Riding of the county of York, $11\frac{1}{4}$ miles (E. N. E.) from New Malton; containing 536 inhabitants. The living is a discharged vicarage, valued in the king's books at £6. 0. $2\frac{1}{2}$.; present net income, £120; patron, Sir W. Strickland, Bart.; impropriator, Hon. M. Langley. There is a place of worship for Wesleyan Methodists; also a day and Sunday school, partly supported by subscription.

SHERBURN (*All Saints*) a market-town and parish, partly in the liberty of St. Peter's, East Riding, and partly in the Upper Division of the wapentake of Barkstone Ash, West Riding, of the county of York; containing 3068 inhabitants, of which number, 1155 are in the town of Sherburn, 15 miles (S. W. by S.) from York, and 184 (N. by W.) from London. The name of this place appears to be of Saxon derivation, being compounded of *Scire*, clear or pure, and *Burn*, a spring or fountain, with reference to a small stream by which it is watered. It was anciently of more importance than it is at present: King Athelstan had a palace here, which was subsequently given to the Archbishop of York, but only the site remains, part of the materials having been used in the erection of the present church. In the great civil war, Sherburn was the scene of a remarkable conflict between Lord Digby, Lieutenant-General of the forces north of the Trent, and Colonel Copley, an officer in the interest of the parliament, when, owing to a misunderstanding as to the relative position of the parties, the army of the former was broken up and discomfited, and his baggage and cabinet papers fell into the enemy's hands. The town is situated on the direct road from Tadcaster to Ferrybridge, and the vicinity abounds with fine orchards; a species of plum, called the Winesoms, is abundantly prolific and fine in this soil. Flax is cultivated to some extent, and is sent to the Leeds market; teasel also forms a prominent article of trade. About a mile and a half from the town is a quarry of fine stone, and upon a stream called Bishop Dyke are several corn-mills. The Leeds and Selby railroad passes close to this place. The market, now almost disused, is on Friday; and there is a fair on Sept. 25th. The living is a perpetual curacy, valued in the king's books at £10. 17. 1.; present net income, £191; patron, Archbishop of York; appropriator, Prebendary of Fenton in the Cathedral Church of York. The church is a spacious and handsome structure, the nave presenting a rare and beautiful specimen of ancient architecture. There are chapels of ease at Lotherton and Micklefield; and a place of worship for Wesleyan Methodists. Sherburn hospital and school-house were erected in 1619, in pursuance of the will of Robert Hungate, Esq., who gave a rent-charge of £225. 6. 8. for the maintenance, clothing, and education of 24 male orphans, but the funds having become inadequate, the present number is only six. A charity school for the maintenance and education of six female orphans was founded and endowed in 1731, with certain land, and the sum of £1450, by the Rev. Samuel Duffield. A rich and elegant cross was found, some years ago, in the churchyard, in digging amongst the foundations of an old chapel: traces of a Roman road from this place to Aberford are yet visible.

SHERBURN-HOUSE, an extra-parochial liberty, in the Southern Division of Easington ward, union, and Northern Division of the county palatine, of Durham, $2\frac{1}{2}$ miles (E. by S.) from Durham; containing 59 inhabitants. An hospital for lepers was founded here previously to 1181, by Hugh Pudsey, Bishop of Durham: it was dedicated to St. Mary Magdalene, and its revenue, in the reign of Henry VIII., was certified as of the value of £142. 0. 4., the society consisting of a master, several priests, and 65 lepers. It is yet in being, having been incorporated in 1585, by Queen Elizabeth, for a master and 30 brethren, and is still subject to the regulations then adopted: the Bishop of Durham appoints the master, who must be in holy orders, and of the degree of M.A., at least; and the master nominates the brethren, who each receive a handsome yearly stipend, besides being comfortably lodged, fed, and clothed. At present this is one of the most richly endowed charitable foundations in the North of England, its income amounting to several thousand pounds per annum. The hospital was enlarged in 1819, by fifteen additional lodging-houses, for the accommodation of as many out-brethren, before which period there were only fifteen inmates.

The building, to which is attached a chapel and apartments for the master, is of a quadrangular form, situated in an agreeable vale on the eastern side of Sherburn water.

SHERE, or SHIRE (*St. James*), a parish, in the union of GUILDFORD, Second Division of the hundred of BLACKHEATH, Western Division of the county of SURREY, 6 miles (E. by S.) from Guildford; containing 1190 inhabitants. This parish is pleasantly situated on the road from Guildford to Dorking. Netley Place is a handsome residence commanding extensive views. Courts leet and baron are occasionally held here, by Edward Bray, Esq., lord of the manor. The living is a rectory, valued in the king's books at £26. 1. 5½.; patron, Rev. Mr. Delafosse. The church is principally in the early style of English architecture, with a tower and spire rising from the centre; in the windows are some fine remains of ancient stained glass, and there are several ancient brasses. There are two places of worship for Independents. Thomas Gatton, Esq., in 1758, bequeathed £400 for teaching poor children, who are sent to the school at Albury. Edward Woods, Esq., in 1837, left £500 to poor widows; and there are several small bequests to the poor of this parish, which derives also £30 per annum from Henry Smith's charity. William Bray, Esq., the antiquary and county historian, was born and is buried here.

SHEREFORD (*St. Nicholas*), a parish, in the union of WALSINGHAM, hundred of GALLOW, Western Division of the county of NORFOLK, 2¼ miles (W.) from Fakenham; containing 110 inhabitants. The living is a discharged rectory, valued in the king's books at £9; present net income, £227; patrons, Representatives of Marquess Townshend.

SHERFIELD-ENGLISH (*St. Leonard*), a parish, in the union of ROMSEY, hundred of THORNGATE, Romsey and Southern Divisions of the county of SOUTHAMPTON, 4¾ miles (W. N. W.) from Romsey; containing 338 inhabitants. The living is a rectory, valued in the king's books at £6. 10. 2½.; present net income, £148; patron, R. Bristow, Esq.

SHERFIELD-UPON-LODDON (*St. Leonard*), a parish, in the union of BASINGSTOKE, hundred of ODIHAM, Basingstoke and Northern Divisions of the county of SOUTHAMPTON, 4 miles (N. E. by N.) from Basingstoke; containing 599 inhabitants. The living is a rectory, valued in the king's books at £11. 3. 6½.; present net income, £544; patron and incumbent, Rev. W. Eyre. James Christian, in 1735, gave £100 to build a school-house, and £25 a year for the education of the children of this parish, including four from Stratfield-Saye.

SHERFORD (*St. Martin*), a parish, in the union of KINGSBRIDGE, hundred of COLERIDGE, Stanborough and Coleridge, and Southern, Divisions of the county of DEVON, 3¼ miles (E.) from Kingsbridge; containing 511 inhabitants. The living is annexed to the vicarage of Stokenham. The church contains some good screen-work. Attached to an old farm-house at Kennedon are some remains of the manorial seat of Justice Hals, who lived in the reign of Henry V.

SHERIFF-HALES (*St. Mary*), a parish, in the union of SHIFFNALL, partly in the Newport Division of the hundred of SOUTH BRADFORD, Northern Division of the county of SALOP, but chiefly in the Western Division of the hundred of CUTTLESTONE, Southern Division of the county of STAFFORD, 3 miles (N. by E.) from Shiffnall; containing 1109 inhabitants. The living is a vicarage, valued in the king's books at £11. 1. 8.; present net income, £614; patron and impropriator, Duke of Sutherland. The church is a neat stone edifice, seated on an eminence above a small stream that parts it from Shropshire. There is a chapel of ease at Woodcote; and a place of worship for Wesleyan Methodists, also a school supported by subscription. A milky vitriolic water is found among the iron mines in the neighbourhood.

SHERIFF-HUTTON, county of YORK. — See HUTTON, SHERIFF.

SHERMANBURY (*St. Giles*), a parish, in the union of STEYNING, hundred of WINDHAM and EWHURST, rape of BRAMBER, Western Division of the county of SUSSEX, 7½ miles (N. E. by N.) from Steyning; containing 345 inhabitants. The living is a rectory, valued in the king's books at £4. 19. 4½.; patron and incumbent, Rev. J. G. Challen, D.D. The tithes have been commuted for a rent-charge of £381. 15., subject to the payment of rates, which on the average have amounted to £102; the glebe comprises 14 acres, valued at £18 per annum. The river Adur, which divides into two branches in the parish, is navigable hither from Shoreham. The springs here are strongly impregnated with iron, and there are some of a saline quality. Here are the groined gateway and some other remains of a castellated mansion, surrounded by a moat, called Ewhurst, and anciently a seat of the Lords De la Warr.

SHERMANS-GROUNDS, an extra-parochial district, in the hundred of WEST GOSCOTE, Northern Division of the county of LEICESTER; containing 23 inhabitants.

SHERNBORNE (*St. Peter and St. Paul*), a parish, in the union of DOCKING, hundred of SMITHDON, Western Division of the county of NORFOLK, 6¼ miles (N. E. by N.) from Castle-Rising; containing 140 inhabitants. The living is a discharged vicarage, valued in the king's books at £8; present net income, £69; patron and appropriator, Bishop of Ely. The church was built by one Thorpe, lord of Shernborne, when Felix, Bishop of the East Angles, came to convert the inhabitants to Christianity, and is said to be the second founded in that kingdom; it has no tower, and the chancel has been long in ruins.

SHERRINGHAM (*All Saints*), a parish, in the union of ERPINGHAM, hundred of NORTH ERPINGHAM, Eastern Division of the county of NORFOLK 5¼ miles (W.) from Cromer; containing 899 inhabitants. The living is a vicarage; net income, £87; patron and appropriator, Bishop of Ely. A National school has been established. Here was formerly a monastery of Black canons, a cell to Nutley abbey, Buckinghamshire.

SHERRINGTON (*St. Laud*), a parish, in the union of NEWPORT-PAGNELL, hundred of NEWPORT, county of BUCKINGHAM, 1¾ mile (N. N. E.) from Newport-Pagnell; containing 804 inhabitants. The living is a rectory, valued in the king's books at £20. 0. 2½.; present net income, £631; patron, Bishop of Lincoln.

SHERRINGTON (*St. Michael*), a parish, in the union of WARMINSTER, forming a detached portion of the hundred of BRANCH and DOLE, Warminster and Southern Divisions of the county of WILTS, 5¼ miles

(N. E. by N.) from Hindon; containing 179 inhabitants. The living is a rectory, valued in the king's books at £11; present net income, £238; patron, A. B. Lambert, Esq. Here is a Sunday National school.

SHERSTON, MAGNA (*Holy Cross*), a parish, in the union of Malmesbury, and in a detached portion of the hundred of Chippenham, Malmesbury and Kingswood, and Northern, Divisions of the county of Wilts, 5¾ miles (W. by S.) from Malmesbury; containing 1361 inhabitants. This place, by the Saxons, was called Scarston, or Scaurston, signifying "the town on a rock:" from its situation near the Consular way, and from the coins of Antoninus, Faustinus, Gordianus, Flavius Julianus, and others, found here, it was evidently occupied by the Romans, but under what name cannot be ascertained; a very perfect formation of an ancient encampment is still to be seen on the cliff, at the rear of the village, with a remarkably deep well. In the neighbourhood are the foundations and fragments of three stone crosses; and here was fought, in 1016, an obstinate battle between Edmund Ironside and Canute the Great. Two small streams, forming the river Avon, unite in this parish. The living is a discharged vicarage, with the rectory of Sherston-Parva and the perpetual curacy of Alderton united, valued in the king's books at £10. 2.; present net income, £126; patrons, Dean and Chapter of Gloucester; impropriators, Rev. H. Cresswell, J. Neeld, Esq., and the Churchwardens of Cirencester, as lessees under the Dean and Chapter. The church exhibits portions of Norman and of the several styles of English architecture: it is a large structure, with a lofty tower rising from the centre. A day school is endowed with £10, and a Sunday school with £5, per annum.

SHERSTON, PARVA, or SHERSTON-PINKNEY, a parish, in the union of Malmesbury, and in a detached portion of the hundred of Chippenham, Malmesbury and Kingswood, and Northern, Divisions of the county of Wilts, 4¾ miles (W.) from Malmesbury; containing 122 inhabitants. The living is a discharged rectory, united, with the perpetual curacy of Alderton, to the vicarage of Sherston-Magna, and valued in the king's books at £3. 14. 4½. The church has long been demolished, no institution having taken place since 1640, when the patronage was in the Crown.

SHERWILL (*St. Peter*), a parish, in the union of Barnstaple, hundred of Sherwill, Braunton and Northern Divisions of the county of Devon, 4 miles (N. E.) from Barnstaple; containing 688 inhabitants. The living is a rectory, valued in the king's books at £30. 3. 11½.; patron, Sir Arthur Chichester, Bart. The tithes have been commuted for a rent-charge of £545, subject to the payment of rates; the glebe comprises 91 acres, valued at £90 per annum. A charity school was built and is supported by Lady Chichester; there are also six almshouses for aged persons.

SHEVINGTON, a township, in the parish of Standish, union of Wigan, hundred of Leyland, Northern Division of the county palatine of Lancaster, 3¼ miles (N. W. by W.) from Wigan; containing 899 inhabitants.

SHEVIOCK (*St. Mary*), a parish, in the union of St. Germans, Southern Division of the hundred of East, Eastern Division of the county of Cornwall, 3 miles (S. by E.) from St. Germans; containing 453 inhabitants. The living is a rectory, valued in the king's books at £26. 14. 7.; present net income, £412; patron, Rt. Hon. R. P. Carew. The church is an ancient structure, containing a sumptuous monument to the memory of Sir Edward and Lady Courtenay, and several curious tombs of the family of Dawnay. The parish is bounded on the north by the Lynher river, and on the south by the English Channel. At Wrinkle Cove is an ancient pier, and off the coast a considerable pilchard fishery.

SHIELDS, NORTH, a sea-port and market-town, partly in the township of North Shields, partly in that of Tynemouth, partly in that of Preston, and partly in that of Chirton, in the parish, union, and borough of Tynemouth, locally in the Eastern Division of Castle ward, and in the Southern Division of the county of Northumberland, 8 miles (E. N. E.) from Newcastle, and 276 (N. by W.) from London. The township of North Shields contains 6744 inhabitants, and the present population of the town is supposed to be about 30,000. The earliest notice of this place occurs in the reign of Edward I.; from which it appears that in the preceding reign the only buildings on the site of the present town consisted of fishermen's huts, or *Shielings*, of which the name is supposed to be a corruption. About the same period, the prior and monks of Tynemouth erected houses here, formed a harbour, established a market, and encouraged the settlement of traders. This exciting the jealousy of the burgesses of Newcastle, who possessed the exclusive trade of the Tyne, they commenced proceedings against the prior, who, by a Judgment of the Court of King's Bench, was driven from the trade he was encouraging and carrying on within the port of Tyne, on which he withdrew to the precincts of his priory, and formed an excavation as a harbour for his trading vessels, which is now called the Prior's Haven; consequently the town returned to its former obscurity, in which state it remained until the middle of the seventeenth century, when Cromwell endeavoured to remove the restrictions which prevented it from assuming that station as a commercial town, to which, from its situation, it was entitled, by causing an act to be passed for the erection of quays, &c., and the establishment of a market twice a week; but his death interfered with these plans, and it was not until the close of the seventeenth century, that any of the restrictions to its trade were removed. Since that period, the facilities afforded to the freedom of trade having caused the abolition of the remaining restrictions, North Shields has advanced rapidly to its present importance, and the immense augmentation of its commerce may be estimated from the increase of its population since the commencement of the present century, by more than four-fifths of the whole number of inhabitants of that time.

The town is situated on the northern shore of the river Tyne, at its confluence with the North Sea, and opposite to South Shields, which is on the southern bank of the river. The part which immediately adjoins the river, containing the township of North Shields, consists chiefly of narrow lanes and alleys, but the more recent additions comprise several spacious streets, three handsome squares, and many houses of a very superior description, both in the town and its immediate vicinity. Nearly opposite Trinity church a new entrance from

Newcastle has been lately made. The town is partly lighted with gas from gas-works at the Low Lights, and partly from gas-works in Hudson-street, established in 1836. Water, which formerly was difficult to be procured in dry seasons, is now abundantly supplied from the reservoirs at Percy Main, Whitley, and Waterville, and conveyed by pipes into the town by a company incorporated in 1786. A subscription library was established in 1802, and in 1807 the valuable collection of books was deposited in a handsome building erected for the purpose. The members of this library established the Tynemouth Literary and Philosophical Institution in 1835, when £60 was subscribed for providing apparatus for lectures on philosophical and literary subjects, and for the increase of the library of circulation and reference previously formed, which now comprises about 3600 volumes. A Natural History Society was established in 1836, by incorporating the members of the Philosophical and Mechanics' Institutions,—the latter originally formed in 1825, and re-established in 1835; the museum, consisting chiefly of mineralogical and geological specimens, was removed to Church-street. The Tyne news-room, in Dockwray-square, was instituted in 1836; in Tyne-street is the Commercial news-room; and a third has been lately established at the New Quay. The theatre, a brick building, fitted up in a neat and appropriate manner, is open at different times in the year.

The harbour is capable of containing 2000 sail of vessels, and in spring tides ships of 500 tons' burden can pass the bar at its mouth in safety. Coal, the principal article of trade of the Tyne, is exported from staiths up the river, in large quantities, to London, and places on the eastern coasts of England and Scotland, and employs the greater part of the vessels belonging to the port: on account of the general adoption of gas, and the great extension of steam-navigation, the coal trade to France and the Mediterranean has lately much increased. Many vessels are also engaged in the American and Baltic trades, and in the Greenland and Davis' Straits fishery. Arrangements are in progress for the erection of a quay from the part called Shepherd's Quay, on the west, to the Union road on the east, adjoining the Low Lights shore, being the line recommended several years since by the late Mr. Rennie: it will be formed by a solid stone wall extending 2365 feet, and will be chiefly filled up with ballast from vessels coming hither for coal; a frontage of 20 feet depth will be left free for public use, and the remainder of the enclosure will be attached to the respective dwelling houses adjacent; the estimated expense is about £9000. The houses adjoining the place called Custom-house Quay are about to be taken down, for the purpose of constructing a double and a single dock for repairing vessels. The principal manufactures are such as are connected with the shipping: there are two establishments for building ships, and several for boats and other small vessels; several chain-cable and anchor manufactories, two rope-walks, and a sail-cloth manufactory. There are also two salt-works, two manufactories for tobacco, and coarse earthenware, starch, and glove and hat manufactories. In the immediate vicinity are the Burdon Main coal staith, near Milburn place, and the Whitley coal and lime staith, near the Low lighthouse. Near the latter are the Tyne chain-works of Messrs. Tyzack, Dobinson and Co., where chain cables of every description are made; anchors and patent windlasses are also manufactured here. Near the former is the large iron-foundry of Messrs. Harrison and Co. There is also a large forge belonging to Messrs. Flynn and Co., with machinery for the manufacture of scrap iron; they have also an establishment for making chains and anchors.

Much inconvenience is sustained from ships engaged in the foreign trade being obliged to clear out at the custom-house at Newcastle: here is, however, an office belonging to that establishment, which facilitates the clearing out of vessels sailing coastwise; and at Clifford's Fort is a watch-house, where officers of the customs are stationed, with a lighthouse called the Low Light adjoining to it; there being another lighthouse, called the High Light, on the Bank Top, immediately to the west of it. The present lighthouses, which are under the control of the Trinity House at Newcastle, were erected about 30 years since, on account of the shifting of the bar at the entrance of the harbour: the old lighthouses, which give name to that part of the town called the Low Lights, have been appropriated as almshouses for aged seamen. A life-boat is kept here, which was presented to the town by the late Duke of Northumberland, in 1798, with a donation of £20 towards keeping it in repair. Steam-boats ply regularly every hour between this town and Newcastle; and a more intimate connection between North and South Shields has within the last few years been established, by means of a steam ferry across the river (which is here of considerable breadth), by an incorporated company. Carriages, horses, and passengers may cross in safety, every quarter of an hour, by the vessels now used, and, by means of drawbridges on each side of the vessel, the greatest facility is afforded for the embarkation and landing of carriages and other vehicles. A more direct communication with Newcastle has been effected by the Newcastle and North Shields railway, which has a branch to Tynemouth. The market, on Friday, is held in a commodious market-place, constructed in 1804, at the expense of the late Duke of Northumberland; immediately in front of it is the Northumberland or New Quay. Statute fairs are held on the first Fridays in March and November. A meeting of magistrates takes place every Tuesday; and a court leet and baron, for making presentments and for the recovery of debts under 40*s*., is held at Easter and Michaelmas, by the steward of the manor of Tynemouth, which belongs to the Duke of Northumberland. The town has also lately been made a polling-place for the southern division of the county. A handsome building in the Elizabethan style, with ornamental stone gables, has been lately built in Howard-street, intended for a register-office, also for the meetings of the guardians of the Tynemouth union under the new poor law act, and for the magistrates' meetings.

The parochial church, called Christ Church, is situated on the north-west side of the town. Another, called Trinity church, has been erected at the west end, and was consecrated October 27th, 1836: it is a handsome edifice, in the early English style, with a tower, the upper part of which is octangular and crowned with pinnacles: the cost of its erection was £3760, which was partly defrayed by the church commissioners and partly by subscription, towards which the present Duke of Northumberland contributed £350, and also presented the site. In 1834 a cemetery was formed on the west side of the town, with a handsome entrance gateway of four finely sculptured pillars, on the Newcastle road. There are two places of worship for Methodists in the New Connection, of which that called Salem chapel was erected in 1836; two for Independents, and one each for Particular Baptists, the Society of Friends, Primitive and Wesleyan Methodists, (the latter a spacious edifice, erected in 1807, at an expense of £2500), Presbyterians (called the Scotch church,—a handsome modern building), United Secession church and Roman Catholics; the last is a very handsome, Gothic edifice.

A free school, with houses for the master and mistress, was erected by subscription in 1810, in commemoration of the royal jubilee; it is supported by subscriptions, and nearly 300 children of both sexes are instructed on the Lancasterian plan. Another school was founded and liberally endowed by Mr. Thomas Kettlewell, in 1825, for which a handsome and commodious house has been erected, at an expense of £1200. There are also a National school, a Roman Catholic school for girls, and an infants' school. The dispensary was established in 1802, and there is a society for the relief of lying-in women. An asylum for decayed master mariners has been erected on the road to Tynemouth: the buildings, which are of a quadrangular form, and occupy about an acre of ground, are in the Elizabethan style, and consist of nine houses, each having accommodation for four pensioners; the site was presented to the society by the Duke of Northumberland. In addition to the institutions already noticed, there are several benefit societies. Margaret Richardson, in 1788, bequeathed £466. 13. 4. East India Annuities, directing the interest to be applied to purposes of general relief to the poor.

SHIELDS, SOUTH, a sea-port, market-town, and chapelry, a newly enfranchised borough, and the head of a union, in the parish of JARROW, Eastern Division of CHESTER ward, Northern Division of the county palatine of DURHAM, 20 miles (N. N. E.) from Durham, and 278 (N. N. W.) from London. The township of South Shields contains 9074 inhabitants, and the present population of the town is supposed to exceed 17,000. Although this town has only attained its present size and importance in very recent times, its origin appears to lay claim to considerable antiquity; many remains of the Romans, consisting of altars, an hypocaust, coins, and other memorials discovered at the Lawe, between the town and the mouth of the river, indicating that it was one of the stations occupied by that people, though the name by which it was then known has not been ascertained. The military road, called the "Wreken Dyke," terminated here, and an elevated pavement at the western end of the town appears to be a Roman work. The town is situated on the southern bank of the Tyne, at its junction with the North Sea, nearly opposite to North Shields: the old part consists principally of long, narrow, and inconvenient streets, extending a considerable distance along the shore of the river; but the more modern part comprises many spacious and well-built houses; buildings have much increased on the east side of the town. It is well supplied with water by pipes from springs, by a company of proprietors who obtained an act in 1788; and is lighted with gas supplied by works completed in 1824, at an expense of £4000. A more intimate connection, by means of the steam ferry, has been established with North Shields (*which see*). In the market-place, a large square near the centre of the town, is the town-hall, erected about 1768, by the Dean and Chapter of Durham, in which the petty sessions for this part of the Eastern Division of Chester ward are held, on the second and fourth Thursdays in each month; and a court leet and court baron by the Dean and Chapter, as lords of the manor, for making presentments, and for the recovery of small debts: it is also used as an exchange by the merchants, and as a public news-room; the under part consists of a colonnade, in which the market for butter, eggs, and poultry, is held. A subscription library was established in 1803, and a Literary, Scientific, and Mechanics' Institute, in 1825; the latter, in addition to its library, has apparatus for mechanical and scientific experiments. The theatre, erected in 1791, is on the Bank Top.

This town owed much of its importance to the salt trade, which was established about 1499, and, during the reigns of Elizabeth, James, and Charles I., attracted many strangers, who settled here. When in its most flourishing state, in 1696, it employed nearly 200 salt-pans, and many hundred individuals; but it has so much declined, that the number of pans does not now exceed two, producing about two tons of salt weekly. This trade has, however, been succeeded by other branches of business, from which the town derives much greater benefit, the principal being the coal trade, which is rapidly increasing. In addition to the coal brought down the river in keels and shipped here, two pits are worked in the immediate neighbourhood; connected with them are staiths on the banks of the river, for facilitating the loading of vessels, which also are the staiths belonging to the Stanhope and Tyne Railway Company. This company was established in 1833, and in two years completed a railway from the town of Stanhope, in the western part of the county, to South Shields, a distance of 34 miles, at an expense of about £250,000. From South Shields to Fatfield the line is traversed by locomotive engines, the remainder being worked by fixed engines and self-acting inclined planes. The staiths for shipping the coal at Shields are constructed on the most effective and scientific principles, and are capable of loading a vessel of 700 tons from each of the drops in about six hours. In 1836, 100,959 tons of coal were shipped at these staiths from the company's coal mines; and they convey annually about 166,500 tons from other collieries. A considerable quantity of lime of very superior quality is brought by this railway, and chiefly distributed through an extensive agricultural district; a portion is also shipped for Scotland from the staiths. The Stanhope and Tyne railway crosses King-street by a viaduct, and terminates at the company's staiths. The Durham Junction railway begins near Moorsley, where it joins the Hartlepool railway, and ends in the township of Usworth, where it is connected with the Stanhope and Tyne railway. The Brandling Junction railway, connects South Shields with Monk-Wearmouth on the south, and Gateshead on the west, a locomotive conveyance which it is expected will much increase the trade of this place. It is in contemplation to construct two large wet docks, of 20 acres each, in Jarrow Slake, (a large tract at present covered at high water,) on the west side of the town, and at the terminus of this railway, where staiths will be erected, and ships will at once take in their cargoes of coal.

The coal trade employs the majority of the vessels belonging to the port, about 350 in number, of an aggregate registered burden of about 77,000 tons; a few are engaged in the American, Baltic, and Indian trades: the number of seamen is nearly 3000. Ships are principally insured by mutual assurance associations, and are employed in the proportion of about two-thirds in the coasting and one-third in the foreign trade. The great inconvenience arising from the want of a custom-house, as described in the article on North Shields, applies equally to this place. Here are ten dry docks, which will contain fourteen large vessels, and extensive yards for ship-building are attached to each. During the late war, when the trade of the Tyne was most flourishing, 30 vessels have been launched in a year: at present, ship-building is not so extensively carried on, the business being principally confined to the repairing of vessels, for which there are two patent slips. The principal manufactures, which are chiefly exported, are plate, flint, and crown glass, and bottles; alkali, soda, salts, and soap; chain cables, anchors, and steam-boilers. There are nine glass-houses in the neighbourhood, and some mills for grinding it, altogether employing about 800 persons. The plate-glass works were established in 1827: the article is polished at Newcastle, and chiefly exported to London. Prior to the late reduction of the duty, the amount paid on glass manufactured here exceeded £120,000 per annum. The Jarrow alkali works, lately established, are said to be the most extensive in the kingdom; they employ more than 500 men, and produce every preparation of soda, salts, alum, sulphate of copper, sulphuric acid, and other minor articles. Chain cables and anchors are manufactured; and there are also a large manufactory for steam-engines, and one for steam-boilers, with some smaller ones; a paint manufactory, worked by steam; six breweries, and five rope-walks, in some of which patent cordage is made. The market is on Wednesday; and fairs, granted by charter of Bishop Trevor in 1770, are held on June 24th and September 1st, but are only indifferently attended.

The borough has the privilege of sending one member to parliament, chosen by the £10 householders of the townships of South Shields and Westoe, together constituting the borough, and comprising an area of 1845 acres. The number of voters registered in October 1836 was 644: the returning officer is annually appointed by the sheriff. It has also been made a polling-place for the northern division of the county. Nearly the whole of the land and property within the borough belongs to the Dean and Chapter of Durham, by whom the ground is let on building leases of 21 years, renewable every seven years on payment of a fine. In 1832 the Dean and Chapter obtained an act enabling them to appropriate certain properties named in the schedule, producing about £3000 per annum, towards the establishment of a university at Durham; a portion of this property has since been enfranchised.

The living is a perpetual curacy; net income, £330; patrons and appropriators, Dean and Chapter of Durham. The chapel, dedicated to St. Hilda, is of great antiquity, although very little remains of the original edifice, it having been, with the exception of the old tower, modernized and nearly rebuilt in 1810, at an expense of nearly £5000: it contains some fine old monuments. The incumbent receives a rent-charge of £249 out of the tithes of Harton, in the parish of Jarrow. A church, or chapel of ease, was erected in 1818, in that part of the town which is in the township of Westoe (anciently Wivestow), at an expense of about £2400, raised by subscription, towards which £1000 was given by the Dean and Chapter of Durham, and £500 by the trustees of the Right Rev. Lord Crewe. The ground floor of the building is used for a National school. Trinity church was erected in 1834, in the Western Commercial-road, at a cost of £3350, chiefly defrayed by the Dean and Chapter, and endowed by them with a net income of £350: it is a neat edifice, containing 1200 sittings, 800 of which are free, the Incorporated Society having granted £500 in aid of the expense. Another church has been recently built, containing 840 sittings, 500 of which are free, the Incorporated Society having granted £300 in aid of the expense. A National school is attached, and supported by subscription. There are two places of worship each belonging to the Baptists, Wesleyan Methodists, and Presbyterians, and one each to Independents, Methodists of the New Connection, Primitive Methodists, and the United Secession Church, nearly all of which have Sunday schools attached.

Christopher Maughan, in 1749, and Anne Aubone, in 1760, bequeathed property for the foundation of a charity school, which was established in 1769: the endowment, augmented by Ralph Redhead and others, produces at present £82. 2. 10. per annum. The school is held in a building of which the lower story is appropriated to the girls, the second to the boys, and the third, with the exception of two small rooms for the mistress, is let as a public library: about 85 boys and 40 girls are taught reading, writing, and arithmetic, and the girls needlework, by a master and mistress. Another school, called the Union school, was established in 1835, and is chiefly supported by subscription, by which means the school-house, a neat building, was erected. The dispensary, established in 1821, has been productive of great benefit, having afforded relief to nearly 3000 poor persons: there are several other benevolent and benefit societies in the town. The poor law union of South Shields comprises six parishes or places, containing a population of 24,427, according to the census of 1831, and is under the care of 25 guardians. The invention of the life-boat originated with a few benevolent individuals of this town, who assisted Mr. Greathead to construct one, which was found effectual for its purpose, and to whom parliament voted a reward of £1200. To Mr. Marshall, another native, the seamen of the eastern coast are indebted for the construction of the floating light off Newarp Sand, on the coast of Norfolk.

SHIFFNALL (*St. Andrew*), a market-town and parish, and the head of a union, in the Shiffnall Division of the hundred of Brimstree, Southern Division of the county of Salop; containing 4779 inhabitants, of which number, 1699 are in the town of Shiffnall, $17\frac{1}{2}$ miles (E. by S.) from Shrewsbury, and 143 (N. W.) from London.

This place, formerly called Idsall, appears to have been of greater note than it is at present, although the origin of its name and history is involved in obscurity. It belonged to Earl Morcar prior to the Conquest, and, at a period considerably later, it was the property of the family of Dunstanville, one of whom, Walter de Dunstanville, by the special command of Henry III., resided in the marches of Wales, to protect them against the ravaging incursions of the Welsh. It afterwards came into the possession of the Badlesmeres, who obtained from Edward I. a market for two days in the week, and two yearly fairs. Bartholomew de Badlesmere having been executed for his participation in the battle of Boroughbridge, it subsequently became the property of various persons of distinction, among whom were the families of Bohun, Tiptoft, Ab Rees, Mortimer, Talbot, &c. The town is supposed to have been destroyed by fire, and subsequently built on its present site, to the eastward of the church, having been, prior to its destruction, situated to the westward; and a book printed towards the end of the 15th century, entitled "The Burnynge of the Town of Idsall, alias Shiffnall," is said to be in existence, though very scarce. It is situated on the high road from London to Holyhead, in a country abounding with coal and iron ore, and is indifferently paved and not lighted, but the inhabitants are supplied with good water from wells: there are two paper-manufactories, and the coal and iron mines in the neighbourhood are worked on an extensive scale by a company at Priors-Lee. A subscription library has been established. The market is on Tuesday, and there are fairs on the first Monday in April, August 5th, and November 23rd, for hops, horses, and cattle of different kinds. A petty session for the division is held monthly by the magistrates, and a court leet annually; and the town has lately been made a polling-place for the southern division of the county.

The living is a vicarage, valued in the king's books at £15. 6. 8.; present net income, £450; patron, George Broke, Esq. The church is a large ancient cruciform structure, with a tower in the centre; the prevailing style of architecture is the later Norman, with many modern alterations, and the four pointed arches supporting the tower are good specimens of that style: the chancel, in which are two round-headed windows, (now blocked up,) with slender-shafted columns and decorated capitals, is evidently of earlier date, and is separated from the tower by a large semicircular arch, which is a fine specimen of the early Norman style: the roof of the nave, of oak richly carved, is concealed by a plaster ceiling, added in 1810, when the church underwent a thorough repair: there are some ancient monuments, and a tablet to the memory of one William Wakeley, stating that he died in 1714, having lived in the reigns of eight kings and queens. The Baptists and Independents have each places of worship here.

A free school, founded in 1595, by John Aron, had, from other endowments, an income of £13. 7. 4., which was paid to the master until 1816, when an addition was made from a fund raised by subscription, making the income £30 per annum, and the National system was adopted; about 130 children of both sexes are instructed. Three exhibitions to Christ-church College, Oxford, founded in 1689, by Edward Careswell, are attached to this school; but the course of education pursued not qualifying the scholars for the University, the benefit of them is enjoyed by a private classical school, the master of which is nominally classical master of the free school. Several small sums, called Dole charities, have been left by different persons for the benefit of the poor, and are periodically distributed agreeably to the intentions of the testators. The poor law union of Shiffnall comprises 15 parishes or places, containing a population of 10,577, according to the census of 1831, and is under the care of 20 guardians. In a field near the town are the remains of a military station, consisting of a circular mound with a ditch. Shiffnall is the birthplace of Dr. Beddoes, a physician, eminent as well for his literary attainments as for professional skill.

SHIFFORD, a chapelry, in the parish and hundred of BAMPTON, union of WITNEY, county of OXFORD, 6 miles (S. E.) from Witney; containing 47 inhabitants. The chapel is an ancient structure.

SHILBOTTLE (*St. James*), a parish, in the union of ALNWICK, Eastern Division of COQUETDALE ward, Northern Division of the county of NORTHUMBERLAND; containing 1195 inhabitants, of which number, 557 are in the township of Shilbottle, 4½ miles (S. by E.) from Alnwick. The living is a discharged vicarage, endowed with a portion of the rectorial tithes, and valued in the king's books at £4. 14. 8.; present net income, £222: it is in the patronage of the Crown; the impropriation of the remainder of the rectorial tithes belongs to various persons. The church was thoroughly repaired about 1793. Coal of excellent quality is obtained here. The parochial school, in which are about 40 boys and 20 girls, is partly aided by the vicar, who allows the master the Easter offerings (for the instruction of nine poor children); and also the surplice dues, and the interest of £100 left by Francis Strother, to the resident minister; partly by Messrs. T. and H. Taylor, lessees of Shilbottle colliery, who pay the master £4. 4. per annum for teaching five children; and partly by Miss Gallon, who contributes £2. 10. per annum. An evening and a Sunday school is attended by boys who work in the colliery, the lessees of which pay the master £6 per annum; and another Sunday school is supported by subscription.

SHILDON, a township, in the parish of ST. ANDREW-AUCKLAND, union of AUCKLAND, North-western Division of DARLINGTON ward, Southern Division of the county palatine of DURHAM, 3½ miles (S. E. by S.) from Bishop-Auckland; containing 867 inhabitants. Here is an extensive depôt for goods, on the line of the railway from Witton Park to Darlington and Stockton. A chapel has been recently erected: it occupies a conspicuous and commanding situation on the summit of the rising ground, east of the old village, and contains 600 sittings, 450 of which are free, the Incorporated Society having granted £300 in aid of the expense. The Earl of Eldon has subscribed 100 guineas towards building a parsonage-house. A school, erected by Mrs. Edward Whalton, is endowed with £25 a year, and a house and land, occupied by the master; it is on the National system.

SHILLINGFORD (*St. Faith*), a parish, in the union of FARRINGDON, hundred of GANFIELD, county of BERKS, 2¾ miles (S. E. by E.) from Great Farringdon; containing 246 inhabitants. The living is a rectory, valued in the king's books at £17. 8. 11½.; present net income, £497; patron, T. M. Goodlake, Esq. The church is partly Norman, and partly in the early style of English architecture: it contains some ancient and curious monuments, particularly an altar-tomb to the memory of John de Blewberry, a priest, who died in 1372, and one to the late Lord Ashbrook and his father, who resided and were interred here.

SHILLINGFORD (*St. George*), a parish, in the union of St. Thomas, hundred of Exminster, Wonford and Southern Divisions of the county of Devon, 3½ miles (S. S. W.) from Exeter; containing 89 inhabitants. The living is a discharged rectory, recently consolidated with that of Dunchideock, and valued in the king's books at £9; present net income of Shillingford, £191. The church contains an old monument to one of the Courteney family. Here is a National school.

SHILLINGSTONE, or SHILLING-OKEFORD (*HolyRood*), a parish, in the union of Sturminster, hundred of Cranborne, Sturminster Division of the county of Dorset, 5¾ miles (N. W.) from Blandford-Forum; containing 473 inhabitants. The living is a rectory in medieties, the first mediety valued in the king's books at £7. 9. 9½., and the second at £6. 16. 5½.; patron, J. Thompson, Esq. The tithes have been commuted for a rent-charge of £370, subject to the payment of rates; the glebe comprises 71 acres, valued at £144 per annum. The church has an embattled tower, crowned with pinnacles: it contains a small altar-tomb, erected, it is said, to the memory of the founder. The parish is bounded on the north by the river Stour.

SHILTON, a parish, in the union of Witney, partly in the hundred of Farringdon, county of Berks, and partly in the hundred of Bampton, county of Oxford, 2½ miles (S. S. E.) from Burford; containing 290 inhabitants. The living is a discharged vicarage, valued in the king's books at £5. 5. 5.; present net income, £112; patron, Rev. Thomas Neate; impropriator, J. Gwynne, Esq. The church is in Oxfordshire, the greater part of the parish in Berkshire, and the vicarage-house upon the boundary of the two counties. A day and Sunday school is supported by charity funds producing £8. per annum.

SHILTON (*St. Andrew*), a parish, in the union of Foleshill, Kirby Division of the hundred of Knightlow, Northern Division of the county of Warwick, 5¾ miles (N. E.) from Coventry; containing 460 inhabitants. The living is a perpetual curacy; net income, £76: it is in the patronage of the Crown; the impropriation belongs to Mrs. Gunman.

SHILTON, EARL, county of Leicester.—See EARL-SHILTON.

SHILVINGTON, a township, in the parish of Morpeth, union and Western Division of Castle ward, Southern Division of the county of Northumberland, 5 miles (S. W. by S.) from Morpeth; containing 101 inhabitants.

SHIMPLING (*St. George*), a parish, in the union of Depwade, hundred of Diss, Eastern Division of the county of Norfolk, 3¾ miles (N. E.) from Diss; containing 227 inhabitants. The living is a discharged rectory, valued in the king's books at £10. 13. 4.; present net income, £222; patron and incumbent, Rev. H. Harrison. The church was erected early in the thirteenth century, but the steeple appears to be of more ancient date; a representation of St. George and the Dragon, and the arms of the Shimplings, are carved on the front of it.

SHIMPLING (*St. George*), a parish, in the union of Sudbury, hundred of Babergh, Western Division of the county of Suffolk, 4 miles (W. N. W.) from Lavenham; containing 496 inhabitants. This parish comprises several estates with manorial rights; Chadacre Hall, the principal, was the seat of the family of Plampin, of which the late admiral was the last descendant, and is now the property of — Halifax, Esq., banker in London, who has built an elegant mansion on the site. Shimpling Thorn is a good mansion-house, late the property of the ancient family of Fiske; and Shimpling Hall, with 400 acres of land, and Gifford's Hall, are the property of Melford Hospital, to which they were bequeathed by ancient proprietors. The living is a rectory, valued in the king's books at £16. 17. 1.; patrons, Heirs of the late Rev. Thomas Fiske. The church, a commodious edifice, contains monuments to the Plampins.

SHINCLIFFE, a chapelry, in the parish of St. Oswald, Durham, union of Durham, Southern Division of Easington ward, Northern Division of the county palatine of Durham, 1¾ mile (S. E.) from Durham; containing 302 inhabitants. The living is a vicarage; net income, £98; patrons, Dean and Chapter of Durham; appropriator, Prebendary of Shincliffe in the Cathedral Church of Durham. The chapel was built and endowed in 1826, by the Dean and Chapter; and a National school has been established.

SHINETON, county of Salop.—See SHEINTON.

SHINFIELD (*St. Mary*), a parish, in the union of Wokingham, partly in the hundred of Charlton, and partly in that of Theale, county of Berks; and partly in the hundred of Amesbury, county of Wilts, 3 miles (S. by E.) from Reading; containing 1100 inhabitants. This parish is situated on the road from Reading to Basingstoke. The living is a vicarage, with that of Swallowfield annexed, valued in the king's books at £20. 3. 1½.; patrons and appropriators, Dean and Chapter of Hereford. The appropriate tithes have been commuted for a rent-charge of £930, and the vicarial for £200, subject to the payment of rates, which on the average have respectively amounted to £111 and £33; the appropriate glebe comprises 15 acres, and the vicarial 29 acres, respectively valued at £23. 12. 6. and £43. 10. per annum. There is a place of worship for Independents. A free school was founded here, in 1707, by Richard Piggot, who endowed it with land and houses, producing at present £57. 16. 2. per annum, for teaching and clothing 20 boys. There is another school, endowed by Mary Spicer in 1697, in which 20 children are taught for £11 a year, arising from the rent of a house and land.

SHINGAY (*St. Mary*), a parish, in the union of Royston, hundred of Armingford, county of Cambridge, 6½ miles (N. W. by N.) from Royston; containing 112 inhabitants. The living is annexed to the vicarage of Wendy. A preceptory of the Knights Hospitallers of St. John of Jerusalem was founded here in 1140, the revenue of which, at the suppression, was estimated at £175. 4. 6.

SHINGHAM (*St. Botolph*), a parish, in the union of Swaffham, partly in the hundred of South Greenhoe, but chiefly in that of Clackclose, Western Division of the county of NORFOLK, $4\frac{3}{4}$ miles (S. W. by W.) from Swaffham; containing 61 inhabitants. The living is a discharged rectory, annexed to that of Beechamwell All Saints, and valued in the king's books at £4. 6. 8.

SHIPBORNE (*St. Giles*), a parish, in the union of MALLING, hundred of WROTHAM, lathe of AYLESFORD, Western Division of the county of KENT, $3\frac{3}{4}$ miles (N.) from Tonbridge; containing 470 inhabitants. The living is a donative, in the patronage of John Simpson, Esq. A day and Sunday school is supported by subscription. A fair is held on Sept. 1st, being the festival of St. Giles the abbot, to whom the church is dedicated. Christopher Smart, a poet and miscellaneous writer of some eminence, was born here in 1722, and died in a state of derangement in 1771.

SHIPBROOK, a township, in the parish of DAVENHAM, union and hundred of NORTHWICH, Southern Division of the county of CHESTER, 3 miles (S. E.) from Northwich; containing 83 inhabitants. The Grand Trunk canal passes through the township.

SHIPDEN, formerly a parish, in the Northern Division of the hundred of ERPINGHAM, Eastern Division of the county of NORFOLK, adjacent to Cromer. The living was a rectory, but the church, which was dedicated to St. Peter, having been destroyed by an inundation of the sea, the parochial rights of Shipden have, for a very long period, been lost.

SHIPDHAM (*All Saints*), a parish, in the union of MITFORD and LAUNDITCH, hundred of MITFORD, Western Division of the county of NORFOLK, $1\frac{3}{4}$ miles (S. W. by W.) from East Dereham; containing 1889 inhabitants. The living is a rectory, valued in the king's books at £27. 7. 6.; present net income, £1120; patron and incumbent, Rev. B. Barker. The church is a stately pile, with a strong western tower, embattled and crowned with a handsome turret. Thomas Bullock, in 1735, bequeathed land producing £60 per annum, of which about £33 is paid to a master for teaching the poor children of the parish; and a National school has been established. There was anciently a hermitage within the parish, with a chapel dedicated to St. Thomas à Becket, for the repair of which the Bishop of Ely, in 1487, granted 40 days indulgence to all who might contribute.

SHIPHAM (*St. Leonard*), a parish, in the union of AXBRIDGE, hundred of WINTERSTOKE, Eastern Division of the county of SOMERSET, 3 miles (N. E. by N.) from Axbridge; containing 691 inhabitants. The living is a discharged rectory, valued in the king's books at £10. 13. 11.; present net income, £157; patrons, Dean and Chapter of Wells. There are lead and calamine works in operation in the parish. Here is a National school.

SHIPLAKE (*St. Peter and St. Paul*), a parish, in the union of HENLEY, hundred of BINFIELD, county of OXFORD, $2\frac{3}{4}$ miles (S.) from Henley-upon-Thames; containing 515 inhabitants. The living is a discharged vicarage, valued in the king's books at £7. 1.; present net income, £147; patrons and appropriators, Dean and Canons of Windsor. The church is in the early English style, with a tower at the west end of the north aisle, covered with ivy; it contains monuments to the families of Blundell and Plowden, the latter proprietors of the ancient manor-house built on the site of an old edifice near the church; on taking it down, a spacious crypt with a groined roof was discovered. In the south aisle of the church is a memorial of the Rev. James Grainger, author of the Biographical History of England, and vicar of the parish.

SHIPLEY, a township, in the parish of HEANOR, union of BASFORD, hundred of MORLESTON and LITCHURCH, Southern Division of the county of DERBY, $9\frac{1}{2}$ miles (N. E. by E.) from Derby; containing 632 inhabitants. The Nutbrook canal and several railways communicate with the coal mines here.

SHIPLEY, a township, in the parish of EGLINGHAM, union of ALNWICK, Southern Division of BAMBROUGH ward, Northern Division of the county of NORTHUMBERLAND, $4\frac{1}{2}$ miles (N. W. by N.) from Alnwick; containing 95 inhabitants.

SHIPLEY (*St. Mary*), a parish, in the union of HORSHAM, hundred of WEST GRINSTEAD, rape of BRAMBER, Western Division of the county of SUSSEX, 6 miles (S. S. W.) from Horsham; containing 1180 inhabitants. The soil is generally clayey, producing remarkably fine oak trees and excellent wheat; the system of agriculture has been greatly improved by the introduction of the draining plough by Sir Charles M. Burrell, Bart., through whose exertions also the parish is furnished with excellent roads. A branch of the river Adur intersects the parish, and there is a navigable canal from that river to the ancient castle of Kirap. This ancient castle, which appears to have been founded in an early period of the Norman era, was visited by King John in 1206 and 1215: it was garrisoned during the parliamentary war. Part of the keep, with a fine Norman arch, is still remaining in the grounds of Sir Charles M. Burrell, who has erected a magnificent castellated mansion within half a mile of the old ruin: the mansion contains a fine collection of paintings and many stately apartments; the grounds are enriched with much beautiful scenery and enlivened with a lake of 100 acres. The living is a perpetual curacy; net income, £98; patron and impropriator, Rev. L. Vernon Harcourt. The church is principally in the Norman style, with highly enriched circular arches at the west entrance of the nave and the chancel; it was repaired and enlarged in 1831, by the addition of 354 sittings, of which 276 are free, in consideration of a grant from the Incorporated Society. In the chancel is a fine monument of variegated marble to Sir Thomas Caryll, his lady and family, with effigies of himself as an armed knight, his lady in a recumbent posture, and three children kneeling. A National school for boys has been established, which has an endowment of £40 per annum by Mrs. Sarah Andrews, in 1825. Two other schools are partly supported by subscription. The union workhouse for children is in this parish.

SHIPLEY, a parochial district, in the parish and union of BRADFORD, wapentake of MORLEY, West Riding of the county of YORK, $3\frac{1}{4}$ miles (N. N. W.) from Bradford; containing 1926 inhabitants. The living is a perpetual curacy; net income, £50: it is in the patronage of the Society for Purchasing Livings. The church was founded in 1825, under the authority of His Majesty's Commissioners, and in the year 1828 it was constituted a district church; the estimate, including incidental expenses, was £7622. 7. There are places of worship for Baptists and Wesleyan Methodists. A branch of the Leeds and Liverpool canal passes through the district, in which the manufacture of worsted, woollen cloth, paper, &c., is carried on to some extent.

SHIPMEADOW (*St. Bartholomew*), a parish, in the union and hundred of Wangford, Eastern Division of the county of Suffolk, 3 miles (W. by S.) from Beccles; containing 133 inhabitants, exclusively of the inmates of the house of industry for the hundred, which is in this parish. The living is a discharged rectory, valued in the king's books at £10; present net income, £214; patrons, certain Trustees. The navigable river Waveney bounds the parish on the north.

SHIPPON, a chapelry, in the parish of St. Helen, Abingdon, hundred of Hormer, county of Berks, 1 mile (W. N. W.) from Abingdon; containing 151 inhabitants.

SHIPSTON-UPON-STOUR (*St. Edmund*), a market-town and parish, and the head of a union, forming, with the parishes of Tidmington and Tredington, a detached portion of the Upper Division of the hundred of Oswaldslow, Blockley and Eastern Divisions of the county of Worcester, being locally in the Kington Division of the hundred of Kington, county of Warwick, 16 miles (S. by W.) from Warwick, and 83 (N. W. by W.) from London; containing 1632 inhabitants. This place was formerly a township in the parish of Tredington, from which it was separated by an act of the 6th of George I., and made a distinct parish. The town is said to derive its name from having been formerly one of the largest markets for sheep in the kingdom; it is situated on the river Stour, in a fertile and rather hilly country, about two miles distant from the Stratford-upon-Avon and Moreton railway, to form a communication with which, by means of a branch railway, an act was obtained in 1833. There was formerly a large manufacture of shag, which has quite declined, and it has now little trade of any description. The Dean and Chapter of Worcester, who possess the manorial rights, hold a court annually, at which a constable is appointed. This town has been made a polling-place for the eastern division of the shire. The market is on Saturday, and there are fairs on the third Tuesday in April, June 22nd, the last Tuesday in August, and the Tuesday after October 10th. The living is a rectory, with that of Tidmington annexed, valued in the king's books at £33. 5. 10.; present net income, £700; patrons, Dean and Chapter of Worcester, and Principal and Fellows of Jesus' College, Oxford, the former presenting to every third vacancy. The Baptists, Society of Friends, and Wesleyan Methodists, have each a place of worship; and there is a Roman Catholic chapel at Foxcote. John Pittway, in 1706, bequeathed property, now producing £91. 17. 6. per annum, for teaching, apprenticing, and partly clothing six poor boys; and George Marshall, in 1747, bequeathed South Sea Stock, producing £39 per annum, for teaching boys and girls. These benefactions were united till 1822, when they were consolidated with a National school which was then formed in the parish. There are various small bequests distributed in bread and clothing among the poor of the town. The poor law union of Shipston includes 37 parishes or places, extending into the counties of Worcester, Warwick, and Gloucester, and contains a population of 19,030, according to the census of 1831; it is under the care of 44 guardians.

SHIPTON (*St. James*), a parish, in the union of Church-Stretton, and within the liberties of the borough of Wenlock, Southern Division of the county of Salop, 7 miles (S. W. by S.) from Much-Wenlock; containing 154 inhabitants. This was formerly a chapelry in the parish of Wenlock. The living is a donative curacy; net income, £3; patron, Thomas Mytton, Esq., who, with others, is the impropriator. A school is partly supported by the clergyman.

SHIPTON, a chapelry, in the parish of Market-Weighton, union of Pocklington, partly within the liberty of St. Peter's, and partly in the Holme-Beacon Division of the wapentake of Harthill, East Riding of the county of York, 1¼ mile (N. W. by W.) from Market-Weighton; containing 348 inhabitants. There is a school for the education of ten boys, endowed with a rent-charge of £5. 14., by John Hutchinson, in 1714, and another of £2 by John and Elizabeth Barker, in 1742.

SHIPTON, a township, in the parish of Overton, wapentake of Bulmer, North Riding, and extending into the liberty of St. Peter's, East Riding, of the county of York, 5½ miles (N. W. by N.) from York; containing 364 inhabitants. There are places of worship for Calvinistic and Wesleyan Methodists. A free grammar school was founded in 1655, by Ann Middleton, who endowed it with £40 per annum; a National school is also partly supported by endowment, and another school is supported by C. Sedgwick, Esq.

SHIPTON-BELLINGER (*St. Mary*), a parish, in the union of Andover, hundred of Thorngate, Andover and Northern Divisions of the county of Southampton, 4¼ miles (S. W. by S.) from Ludgershall; containing 287 inhabitants. The living is a discharged vicarage, valued in the king's books at £8; present net income, £116; patrons and impropriators, J. and G. Pothecary and J. Gilbert, Esqrs.

SHIPTON-GEORGE, a parochial chapelry, in the union of Bridport, hundred of Godderthorne, Bridport Division of the county of Dorset, 3 miles (E. by S.) from Bridport; containing 316 inhabitants. The living is a curacy, attached to the rectory of Burton-Bradstock. The chapel, dedicated to St. Martin, and situated on high ground, is a small edifice with a low embattled tower. Near the chapel, on the south-west, are slight remains of the ancient manor-house, still called the court-house.

SHIPTON-LEE, a hamlet, in the parish of Quainton, union of Aylesbury, hundred of Ashendon, county of Buckingham, 5½ miles (S. W. by S.) from Winslow; containing 104 inhabitants. Here was formerly a chapel, now demolished.

SHIPTON-MOYNE (*St. John the Baptist*), a parish, in the union of Tetbury, hundred of Longtree, Eastern Division of the county of Gloucester, 2¼ miles (S. by E.) from Tetbury; containing 389 inhabitants. The living is a rectory, valued in the king's books at £18. 1. 10½.; patron, T. G. B. Estcourt, Esq. The tithes have been commuted for a rent-charge of £345, subject to the payment of rates, which on the average have amounted to £51; the glebe comprises 173 acres, valued at £90 per annum. Here is a National school.

SHIPTON-OLLIFFE (*St. Oswald*), a parish, in the union of Northleach, hundred of Bradley, Eastern Division of the county of Gloucester, 6¼ miles

(N. W. by W.) from Northleach; containing 229 inhabitants. The living is a discharged rectory with Shipton-Sollers, united in 1776, valued in the king's books at £7. 5. 9.; present net income, £412; patrons, alternately, W. G. Peachy, Esq., lord of the manor of Shipton-Sollers, and W. P. Chapeau, Esq., lord of the manor of Shipton-Olliffe. The church has undergone a thorough repair, since which the service, which was previously once on every Sunday alternately at each of the churches of these united parishes, has been solely performed here. These parishes adjoin each other, but none of the inhabitants can precisely ascertain their respective boundaries; they are both situated on the north of the turnpike-road to London, and this parish abounds with limestone. The common field lands were enclosed in 1793, when 453 acres of arable and pasture land were allotted to the rector, in lieu of tithes.

SHIPTON-SOLLERS (*St. Mary*), a parish, in the union of Northleach, hundred of Bradley, Eastern Division of the county of Gloucester, 6¼ miles (W. N. W.) from Northleach; containing 98 inhabitants. The living is a discharged rectory, united to Shipton-Olliffe in 1776, and valued in the king's books at £7. 3. 4.

SHIPTON-UNDER-WYCHWOOD (*St. Mary*), a parish, in the union of Chipping-Norton, hundred of Chadlington, county of Oxford, 4 miles (N. N. E.) from Burford; containing 2454 inhabitants. The living is a vicarage, valued in the king's books at £16; present net income, £335; patron and impropriator, Professor of Civil Law in the University of Oxford. The impropriate tithes of Ramsden, a chapelry in this parish, have been commuted for a rent-charge of £100; and the vicarial for £27, subject to the payment of rates, which on the average have respectively amounted to £37 and £9. The church is an ancient structure in the early English style, with some portions of the Norman, and a lofty tower surmounted by a spire; the south porch is enriched with niches, containing mutilated statues, and there is a fine Norman doorway with zigzag mouldings; the pulpit is of stone exquisitely sculptured, and the font, which is octagonal, is ornamented with the arms of the Warwick family and with tracery; at the west end of the nave is a painting of the Resurrection, and in the north aisle an altar-tomb, with the recumbent effigy of a female, rudely sculptured, and supposed to be the founder. A National school is supported by subscription. At Langley was anciently a royal residence, and the remains of what is said to have been King John's palace may still be traced in the walls of one of the houses. From its proximity to Wychwood forest, it continued till the time of Charles I. to be the residence of the royal family, while taking the diversion of the chace. There are remains of three religious houses, which have not been noticed by any writer. Another ancient building has long been converted into the Crown Inn; and three singular stone vessels were found in digging the quarries at Milton.

SHIPTON-UPON-CHERWELL (*St. Mary*), a parish, in the union of Woodstock, hundred of Wootton, county of Oxford, 2¼ (E.) from Woodstock; containing 148 inhabitants. This parish is bounded on the east by the river Cherwell, and intersected by the Oxford canal. The living is a rectory, valued in the king's books at £11. 9. 4½.; present net income, £310; patron, William Turner, Esq. The church was rebuilt in 1832, at the sole cost of W. Turner, Esq., of Shipton House; it is a neat edifice in the later English style, with a square embattled tower. A Sunday school is supported by the patron and the rector.

SHIRBURN (*All Saints*), a parish, in the union of Thame, hundred of Pirton, county of Oxford, 4 miles (S. by E.) from Tetsworth; containing 325 inhabitants. This place was originally the property of Richard, Earl of Cornwall, but passed to Alice, wife of Warine de L'Isle, whose descendant of the same name obtained from Edward III. licence to embattle his house at Shirburn. Shirburn Castle, the seat of the Earl of Macclesfield, is surrounded by a moat, over which is a drawbridge; the interior contains a noble hall, an armoury, and a suite of splendid apartments; there is also a fine collection of paintings, amongst which is a portrait of Catherine Parr, wife of Henry VIII. The living is a discharged vicarage, valued in the king's books at £10. 16. 0½.; present net income, £112; patron and impropriator, Earl of Macclesfield. A school is partly supported by the minister, and a Sunday school by the Earl of Macclesfield.

SHIREBROOK, a chapelry, in the parish of Langwith, hundred of Scarsdale, Northern Division of the county of Derby, 5½ miles (N. by W.) from Mansfield: the population is returned with the parish. The living is annexed to the rectory of Pleasley.

SHIREHAMPTON, a chapelry, in the parish of Westbury-upon-Trym, Lower Division of the hundred of Henbury, Western Division of the county of Gloucester, 5 miles (N. W. by W.) from Bristol; containing 420 inhabitants. King-road and Hung-road, two noted anchorages for ships, are within the precincts of this chapelry. The chapel is dedicated to St. Michael, and contains 90 free sittings, the Incorporated Society having granted £200 in aid of the expense. There is a place of worship for Wesleyan Methodists.

SHIREHEAD, a chapelry, in the parish of Cockerham, hundred of Amounderness, Northern Division of the county palatine of Lancaster, 4 miles (N. W. by N.) from Garstang: the population is returned with the parish. The living is a perpetual curacy; net income, £93; patron, Vicar of Cockerham; impropriators, several Proprietors of land. The chapel contains 80 free sittings, the Incorporated Society having granted £70 in aid of the expense.

SHIRE-NEWTON (*St. Thomas à Becket*), a parish, in the union of Chepstow, Upper Division of the hundred of Caldicott, county of Monmouth; containing 791 inhabitants, of which number, 333 are in the village of Shire-Newton, 4½ miles (W.) from Chepstow. The living is a discharged rectory, valued in the king's books at £9. 8. 1½.; present net income, £304: it is in the patronage of the Crown. Here is a National school.

SHIREOAKS, a chapelry, in the parish of Worksop, wapentake of Bassetlaw, Northern Division of the county of Nottingham, 3½ miles (W. N. W.) from Worksop. The living is a perpetual curacy; net income, £50; patron, Duke of Norfolk.

SHIRLAND (*St. Leonard*), a parish, in the union of Chesterfield, hundred of Scarsdale, Northern Division of the county of Derby, 2 miles (N. by W.) from Alfreton; containing 1212 inhabitants. The living is a rectory, valued in the king's books at £7. 15. 5.; present net income, £215; patron, Earl of Thanet.

The church contains several ancient monuments of the De Greys. This was formerly a market-town, but the market was discontinued about 1785; a fair, chiefly for cattle, is still held on the Wednesday after New Year's Day. At Hatfield-gate is a charity school, endowed by Edward Revell, Mr. Stocks, and others, with about £25 per annum, and a house for the master, who educates 27 poor children.

SHIRLEY (*St. Michael*), a parish, in the hundred of Appletree, Southern Division of the county of Derby, $4\frac{1}{4}$ miles (S. E. by S.) from Ashbourn; containing 602 inhabitants. The living is a discharged vicarage, endowed with the rectorial tithes, with the chapelry of Yeaveley, and valued in the king's books at £6. 13. 4.; patron, Earl Ferrers. The tithes have been commuted for a rent-charge of £153. 17. 6., subject to the payment of rates; the glebe comprises 9 acres, valued at £25 per annum. The church is a small building with a wooden tower. A school is supported by subscription. Part of the old manor-house of the Shirleys, who settled here in the reign of Henry III., still remains, attached to a farm-house.

SHIRLEY-STREET, a hamlet, in the parish of Solihull, Solihull Division of the hundred of Hemlingford, Northern Division of the county of Warwick, 6 miles (S.) from Birmingham: the population is returned with the parish. A chapel of ease has been erected here, and endowed with a rent-charge of £50 upon the rectory of Solihull; the living is a perpetual curacy, in the patronage of E. B. Clive, Esq. Here is a National school.

SHITLINGTON (*All Saints*), a parish, in the union of Ampthill, partly in the hundred of Clifton, and partly in that of Flitt, county of Bedford, 4 miles (E. S. E.) from Silsoe; containing 1307 inhabitants. The living is a discharged vicarage, with the rectory of Lower Gravenhurst united, valued in the king's books at £18; present net income, £128; patrons and impropriators, President and Fellows of Trinity College, Cambridge. The church is a large and handsome edifice; the tower was rebuilt by the parishioners in 1750.

SHITLINGTON, a township, in the parish of Thornhill, union of Wakefield, Lower Division of the wapentake of Agbrigg, West Riding of the county of York, $5\frac{3}{4}$ miles (S. W. by W.) from Wakefield; containing 1893 inhabitants.

SHITLINGTON, HIGH, a township, in the parish of Wark, union of Bellingham, North-Western Division of Tindale ward, Southern Division of the county of Northumberland, 3 miles (W.) from Wark; containing 108 inhabitants.

SHITLINGTON, LOW, a township, in the parish of Wark, union of Bellingham, North-Western Division of Tindale ward, Southern Division of the county of Northumberland, $2\frac{1}{2}$ miles (W. by N.) from Wark; containing 58 inhabitants.

SHITTERTON, a tything, in the parish and hundred of Beer-Regis, Wareham Division of the county of Dorset; containing, with Beer Heath, 170 inhabitants.

SHOBDON (*St. John the Evangelist*), a parish, in the union of Leominster, hundred of Stretford, county of Hereford, $5\frac{1}{2}$ miles (E. S. E.) from Presteign; containing 536 inhabitants. The living is a rectory, valued in the king's books at £5. 7. 11.; present net income, £764; patron, William Hanbury, Esq. The church is the burial-place of the Bateman family; it was partially rebuilt in 1757, by John, Viscount Bateman. A school is supported by Mr. Hanbury. The rent of several acres of land, and the proceeds of some minor benefactions, are distributed among the poor by the rector and churchwardens. Near the church is a mount, called Castle Hill, encompassed with a moat, supposed to be the remains of a Roman, or Danish, fortification.

SHOBROOKE, a parish, in the union of Crediton, hundred of West Budleigh, Crediton and Northern Divisions of the county of Devon, 2 miles (E. N. E.) from Crediton; containing 644 inhabitants. The living is a rectory, annexed to the bishopric of Exeter, and valued in the king's books at £36. There is a place of worship for Independents. Various bequests, producing about £4 a year, are applied to the education of poor children.

SHOBY, a chapelry, in the parish of Saxelby, hundred of East Goscote, Northern Division of the county of Leicester, 5 miles (W. by N.) from Melton-Mowbray; containing 15 inhabitants.

SHOCKLACH (*St. Edith*), a parish, in the union of Wrexham, Higher Division of the hundred of Broxton, Southern Division of the county of Chester; containing 431 inhabitants, of which number, 140 are in the township of Church-Shocklach, $4\frac{1}{2}$ miles (N. W. by W.), and 216 in that of Oviatt-Shocklach, $3\frac{1}{2}$ miles (W. N. W.) from Malpas. This parish is bounded on the west by the river Dee, which is crossed by a bridge at Castletown, where is the moated site of Shocklach castle. The living is a perpetual curacy; net income, £107; patron, Sir R. Puleston, Bart. The church is a small ancient building, with an enriched Norman door.

SHOEBURY, NORTH (*St. Mary*), a parish, in the union and hundred of Rochford, Southern Division of the county of Essex, $3\frac{1}{4}$ miles (E. N. E.) from Southend; containing 226 inhabitants. This parish is bounded on the south by the parish of South Shoebury, and the villages of both are nearly contiguous. The living is a discharged vicarage, valued in the king's books at £9: it is in the patronage of the Crown; the impropriation belongs to Mrs. Jones. The impropriate tithes have been commuted for a rent-charge of £347, and the vicarial for £163. 5., subject to the payment of rates, which on the latter have on the average amounted to £13; the glebe comprises 10 acres, valued at £20 per annum. The church, a small ancient edifice with a tower and spire, contains a handsome monument to John Ibbotson, Secretary to the Admiralty.

SHOEBURY, SOUTH (*St. Andrew*), a parish, in the union and hundred of Rochford, Southern Division of the county of Essex, 4 miles (E.) from Southend; containing 202 inhabitants. This parish is situated on the river Thames, near its influx into the sea, and nearly opposite to the Nore; at its southern extremity is a small promontory called Shoebury Ness, on which a signal station has been established. The living is a rectory, valued in the king's books at £14. 13. 4.; patron, R. Bristow, Esq. The tithes have been commuted for a rent-charge of £410, subject to the payment of rates, which on the average have amounted to £71; the glebe conprises 7 acres, valued at £10 per annum. The church is a small edifice, with a tower of flint surmounted with a spire; between the nave and chancel is a richly ornamented arch.

SHOPLAND (*St. Mary Magdalene*), a parish, in

the union and hundred of ROCHFORD, Southern Division of the county of ESSEX, 2 miles (N. E. by E.) from Prittlewell; containing 48 inhabitants. The living is a discharged vicarage, valued in the king's books at £9; present net income, £89; patron and incumbent, Rev. T. Quarington. The church is a small edifice, consisting only of a nave and chancel.

SHOREDITCH (*St. Leonard*), a parish, in the Tower Division of the hundred of OSSULSTONE, county of MIDDLESEX, adjoining the north-eastern portion of the metropolis, and, with Haggerstone and Hoxton, which, by a recent act of parliament, have been constituted distinct parishes, containing 68,564 inhabitants. This parish, in ancient records called *Sordig, Soresdich,* and *Shordych,* appears to have derived its name from the great common sewer, or ditch, which passed through it, and to have given name to the family of Sir John de Sordig, lord of the manor, and one of the ambassadors of Edward III. to Philip of France, more than a century prior to the time of Jane Shore, from whom, according to a legendary tradition, it is supposed to have been originally derived. The Roman military way leading from London-wall to the ford at Hackney passed through part of the churchyard; and there are still some vestiges of the old artillery ground, anciently a Roman *Campus Martis,* which was subsequently celebrated for archery and other military exercises practised there by the citizens of London, but now covered with houses. The parish, which is very extensive, consists of numerous streets connecting it with the metropolis, and of several ranges of building on the roads to Kingsland, Hackney, and Bethnal-Green; it is well paved, lighted with gas, and amply supplied with water. There are some remains of ancient houses, but by far the greater number are modern. The only branches of manufacture carried on in the parish are such as are connected with the silk factories of the adjoining parish of Spitalfields, and there are several breweries, and some extensive foundries for church bells. The parish is within the jurisdiction of a court of requests for the Tower Hamlets, established under an act passed in the 23rd of George II., and 19th of George III., for the recovery of debts under 40*s.*, and within the limits of the new police act.

The living is a vicarage, valued in the king's books at £17; present net income, £656; patron and appropriator, Archdeacon of London. The church, rebuilt in 1740, is a handsome edifice, in the Grecian style of architecture, with a tower, from which rises an open turret surrounded with Corinthian pillars, supporting an elliptical dome surmounted by a small but well-proportioned spire; the western entrance is through a stately portico of four columns of the Doric order, supporting an enriched entablature and cornice, and surmounted by a triangular pediment. The interior is well arranged; the east window is embellished with stained glass, and there are numerous ancient monuments, among which may be noticed an altar-tomb, with recumbent effigies, of Sir John Elrington and his lady; a monument of Sir Thomas Leigh, in a kneeling posture; one for four ladies of the Rutland family, whose figures are represented kneeling at an altar, two on each side, in a recess; some erect statues, and various other memorials. A new church in the Curtain-road, capable of accommodating 1200 persons, was consecrated July 4th, 1839. There are places of worship for Baptists, Independents, and Wesleyan and other Methodists. The charity school for the maintenance, clothing, and education of boys was established in 1705, and a school-house erected by subscription in 1722: a similar institution for girls was established in 1709, and the school-house was erected in 1723; the former has an annual income of £100, and the latter one of £160, arising from rents and personal estates; they are further supported by subscription. There are National and Sunday schools in connection with the established church and the various dissenting congregations. On the north side of Old-street road are the Weavers almshouses, consisting of three houses, each containing four rooms, for twelve widows of freemen belonging to that company. Adjoining these are Walters' almshouses for eight widows of freemen of the Drapers' Company, who place in them two widows of freemen; the remaining six are appointed by the parish. Adjoining these are eight rooms built by Mr. Porter, and given to the parish for aged widows. On the south side of Old-street road are almshouses founded by Judge Fuller, in 1591, and endowed by him with £50 per annum for twelve widows. In Kingsland road are the Drapers' almshouses, containing twelve rooms, of which six are occupied by freemen of that company or their widows, and the other six by aged widows appointed by the parish. Beyond these are the Ironmongers' almshouses, founded in 1703 by Sir Robert Geffery for freemen of that company or their widows; the buildings form three sides of a quadrangle, of which the area is laid down in turf, and comprise fourteen houses of four rooms each, with a neat chapel in the centre of the principal range. The chaplain resides in one of the houses, and another is occupied by the matron; the remaining twelve are given to the almspeople, who have one room each, an allowance of £1. 10. per quarter from Geffery's endowment, and £1. 5. per quarter from the company, with a chaldron and a half of coal annually. Mr. Hawood left in trust to the Ironmongers' Company certain property, one-half of which he appropriated to the redemption of Algerine captives, and one-half for the benefit of charity schools in the vicinity of London, and the inmates of these almshouses; that portion of the property relating to the redemption of slaves has accumulated and now exceeds £100,000, and the whole matter is in Chancery. Beyond these, on the same road, are the almshouses of the Company of Frame work Knitters, consisting of twelve tenements for freemen of that company or their widows, who receive each £9 per annum and three sacks of coal. In Gloucester-street are almshouses founded by Mrs. Fuller, for sixteen aged widows, who receive £12 per annum each, and a chaldron of coal. There are also some almshouses founded by Egbert Guede, of Overyssel, for four poor men belonging to the Dutch church in Austin Friars. The Refuge for the Destitute, a spacious establishment in this parish, consists of two separate buildings, one for males, situated in Hoxton Old Town, and the other for females, situated in the Hackney-road. The management of the affairs of the poor of the parish is regulated under a local act.

SHOREHAM (*St. Peter and St. Paul*), a parish, in the union of SEVEN-OAKS, hundred of CODSHEATH, lathe of SUTTON-AT-HONE, Western Division of the county of KENT, 4½ miles (N.) from Seven-Oaks; containing 1015 inhabitants. A very elegant Palladian villa was commenced about the beginning of the last century, but being left unfinished, became infected with the dry rot, which induced the present Lord Ashburton to take it down and replace it by a mansion in the Elizabethan style. The living, a discharged vicarage, is one of the three which constitute the deanery of the Arches, and is valued in the king's books at £14. 6. 8.; present net income, £371; patrons, Dean and Chapter of Westminster. The rectory of Shoreham, with the curacy of Otford, is valued in the king's books at £34. 9. 9½., and is an appropriation belonging to the Dean and Chapter of Westminster, a certain stipend being allowed to the curate. The church is an ancient structure, containing several elegant monuments. Castle farm was built with the remains, and upon the site, of Shoreham castle. There are three almshouses for aged widows.

Corporation Seal of New Choreham.

Obverse. *Reverse.*

SHOREHAM, NEW (*St. Mary*), a borough, market-town, sea-port, and parish, in the union of STEYNING, hundred of FISHERGATE, rape of BRAMBER, Western Division of the county of SUSSEX, 23 miles (E.) from Chichester, and 56 (S. by W.) from London; containing 1503 inhabitants. This town is indebted for its origin to the decay of Old Shoreham, situated not far distant, which, though anciently a place of importance, is now an inconsiderable village. In ancient history it is chiefly remarkable for having been built on the spot where Ælla, the Saxon, landed with supplies from Germany, in aid of his countrymen, Hengist and Horsa. The town is situated about one mile from the English Channel, on the river Adur, across which is a long wooden bridge, on the main road between Brighton and Portsmouth, which was erected by annuity subscription, but with a reversion to the Duke of Norfolk, in 1781. A suspension bridge, on a design similar to that at Hammersmith, has been constructed over the river, at the western entrance into the town, at the expense of His Grace the Duke of Norfolk, by which the distance between Shoreham and Worthing has been reduced about two miles. Shoreham is noted for its ship-building, and vessels of 700 tons have been launched here. From its proximity to Brighton snd Worthing, the trade and importance of the port has, of late years, rapidly increased: its revenue, within the last 20 years, has been augmented five-fold, having been, in 1810, about £7000, and now amounting to about £35,000. The harbour, which is a tide harbour, is very commodious: in spring tides it has about nineteen feet of water, in common ones about fourteen feet, and not more than three feet at ebb: it was constructed in 1816, by subscription on shares, and has proved a very profitable undertaking. The river runs by the side of the town, parallel with the sea, with which it communicates about half a mile eastward, and is frequented by ships of considerable burden. The imports consist principally of timber, deals, merchandise from France, wine, spirits, coal, cheese and butter from Holland, &c.; and considerable quantities of oak timber are exported. Shoreham has lately been approved as a warehousing port for West India, Mediterranean, and other produce, for the reception of which large and commodious warehouses have been built. The amount of duties paid at the custom-house, for the year ended Jan. 5th, 1837, was about £23,000. Cement manufactories have been established here. The custom-house, erected in 1830, under the direction of Mr. Sydney Smirke, is an elegant building, in the Grecian style, situated in the centre of the town. A large market for corn is held every fortnight, and a fair on July 25. Shoreham is a borough by prescription, and is governed by a high constable appointed by the lord of the manor. It has sent two members to parliament since the first of Edward I.; the right of election being in the inhabitants paying scot and lot, and the freeholders of the rape of Bramber, subject now, however, to the preliminary condition of registering, as enjoined by the act of the 2nd of William IV., cap. 45.: the high constable is the returning officer. At the election in 1791, a majority of the electors having formed themselves into a society called the Christian Club, the real object of which was to sell their votes to the best bidder, an act of parliament was passed, disfranchishing every member of the society, and extending the votes for New Shoreham to the whole rape of Bramber: the polling-places are New Shoreham and Cowfold.

The living is a discharged vicarage, annexed to that of Old Shoreham, and valued in the king's books at £6. 1. 8.; present net income, £127; patrons, President and Fellows of Magdalene College, Oxford. The church is an extremely interesting specimen of Norman architecture, and contains 460 free sittings, the Incorporated Society having granted £250 in aid of the expense. It was originally cruciform, and one of the largest in the neighbourhood, as well as perhaps the most elegant; the architectural details within being still remarkable for their richness and diversity. The Independents and Wesleyan Methodists have each a place of worship. There is a National school adjoining the churchyard. Here was anciently a priory for Carmelites, or White friars, founded by Sir John Mowbray, Knt.; also an hospital, dedicated to St. James, but no remains of either are now discernible.

SHOREHAM, OLD (*St. Nicholas*), a parish, in the union of STEYNING, hundred of FISHERGATE, rape of BRAMBER, Western Division of the county of SUSSEX, ½ mile (N. W. by N.) from New Shoreham; containing 231 inhabitants. The living is a discharged vicarage, with that of New Shoreham annexed, valued in the king's books at £7. 18. 6.; present net income, £58; patrons and impropriators, President and Fellows of Magdalene College, Oxford. The church is a very ancient building, chiefly in the Norman style of architecture: it consists of a nave, chancel, two transepts, and a central tower; the northern transept is in ruins. Here was an ancient hospital, dedicated to St. James, which was valued, in the reign of Elizabeth, at £1. 6. 8. per annum.

SHORESWOOD, a township, in the parish of NOR-

HAM, otherwise Norhamshire, union of BERWICK-UPON-TWEED, county palatine of DURHAM, though locally to the northward, and for electoral purposes annexed to the Northern Division, of the county of Northumberland, 6½ miles (S. W. by S.) from Berwick-upon-Tweed; containing 279 inhabitants, who are chiefly employed in the adjacent coal mines. This township belongs to the Dean and Chapter of Durham.

SHORNCUTT (*All Saints*), a parish, in the union of CIRENCESTER, hundred of HIGHWORTH, CRICKLADE, and STAPLE, Cricklade and Northern Divisions of the county of WILTS, 5¾ miles (W. N. W.) from Cricklade; containing 29 inhabitants. The living is a discharged rectory, valued in the king's books at £4. 7. 6.; present net income, £130: it is in the patronage of the Crown.

SHORNE (*St. Peter and St. Paul*), a joint parish with Merston, in the union of NORTH-AYLESFORD, hundred of SHAMWELL, lathe of AYLESFORD, Western Division of the county of KENT, 3¼ miles (S. E.) from Gravesend; containing 730 inhabitants. The living is a vicarage, valued in the king's books at £13. 1. 8.; present net income, £358; patrons and appropriators, Dean and Chapter of Rochester. The church contains a fine monument to the memory of Sir Henry de Cobham. A National school is supported partly by subscription, but principally by the dividends arising from £1000 three per cent. consols., the bequest of the Rev. R. G. Ayerst in 1812. A battery for the protection of this part of the Medway was constructed in 1796.

SHORTFLATT, a township, in the parish of BOLAM, union of CASTLE ward, North-Eastern Division of TINDALE ward, Southern Division of the county of NORTHUMBERLAND 10¾ miles (W. S. W.) from Morpeth; containing 22 inhabitants.

HORTHAMPTON, a chapelry, in the parish of CHARLBURY, hundred of CHADLINGTON, county of OXFORD, 5 miles (S. by E.) from Chipping-Norton: the population is returned with the tything of Chilson. The living is annexed to the vicarage of Charlbury. The chapel is dedicated to All Saints.

SHORWELL (*St. Peter*), a parish, in the liberty of WEST MEDINA, Isle of Wight Division of the county of SOUTHAMPTON, 5 miles (S. W. by S.) from Newport; containing 699 inhabitants. The living comprises a sinecure rectory, valued in the king's books at £20. 0. 2½.; and a discharged vicarage, united to the rectory of Mottiston, and valued at £17. 16. 0½.; impropriator, S. Dowell, Esq. Here is a National school.

SHOSTON, a township, in the parish of BAMBROUGH, union of BELFORD, Northern Division of BAMBROUGH ward and of the county of NORTHUMBERLAND, 8 miles (E. by S.) from Belford; containing 89 inhabitants.

SHOTFORD, a hamlet (formerly a chapelry), in the parish of MENDHAM, hundred of EARSHAM, Eastern Division of the county of NORFOLK, 1 mile (S. E.) from Harleston: the population is returned with the parish. The chapel is desecrated, having been converted into a malt-house.

SHOT-HAUGH, a joint township with East and West Thriston, in the parish of FELTON, Eastern Division of MORPETH ward, Northern Division of the county of NORTHUMBERLAND: the population is returned with East and West Thriston.

SHOTESHAM (*All Saints*), a parish, in the union and hundred of HENSTEAD, Eastern Division of the county of NORFOLK, 4¾ miles (N. E. by N.) from Stratton-St.-Mary; containing 558 inhabitants. The living is a discharged vicarage, with the vicarage of St. Mary-Shotesham, and with the vicarage of St. Botolph and the rectory of St. Martin-Shotesham consolidated, (both of which were formerly distinct parishes, the latter being valued in the king's books at £4), and is valued in the king's books at £6. 13. 4.; patron, Robert Fellowes, Esq. A school is endowed with £3 per annum, and is further supported by subscription. This parish has the right of appointing eight poor men, aged 56, to the Lesser Hospital at East Greenwich, founded by Henry Howard, Earl of Northampton.

SHOTESHAM (*St. Mary*), a parish, in the union and hundred of HENSTEAD, Eastern Division of the county of NORFOLK, 4¾ miles (N. N. E.) from Stratton-St. Mary; containing, with St. Martin-Shotesham, 367 inhabitants. The living is a discharged vicarage, consolidated with the vicarages of All Saints and St. Botolph, and the rectory of St. Martin-Shotesham, and valued in the king's books at £5. The church of St. Botolph has been demolished; and the living, a discharged vicarage, was annexed in 1311 to this parish.

SHOTLEY (*St. Andrew*), a parish, in the union of HEXHAM, Eastern Division of TINDALE ward, Southern Division of the county of NORTHUMBERLAND; containing 1104 inhabitants, of which number, 590 are in the township of Shotley, 10½ miles (S. E.) from Hexham. The living is a perpetual curacy; net income, £139; patrons, Bishop Crewe's trustees. The church is situated upon an eminence about a mile and a half from the village of Shotley: in the cemetery is an elegant mausoleum in memory of the Hopper family. A chapel was erected in 1835, containing 210 free sittings, the Incorporated Society having granted £100 in aid of the expense. Here are several extensive and productive lead and coal mines; also a manufactory for the finest kinds of writing paper, in which twelve engines and two paper machines are impelled by water. A National school is supported by subscription; and there is an income of £14. 4., arising from bequests by Bishop Crewe, and by Ann Young in 1796, for educating 17 children.

SHOTLEY (*St. Mary*) a parish, in the hundred of SAMFORD, Eastern Division of the county of SUFFOLK, 8¼ miles (S. E. by S.) from Ipswich; containing 410 inhabitants. This parish is situated at the confluence of the navigable rivers Orwell and Stour, opposite to the town of Harwich, with which there is a communication by a regular ferry. The village is of remote antiquity, and was once the seat of the Filney family, of whom Sir Frederick Filney was knighted by Richard Cœur de Lion at the siege of Acre. Here was anciently a hamlet called Kirketon, and the parish is now called, in the king's books, Shotley *alias* Kirketon. The living is a rectory, valued in the king's books at £20; present net income, £604; patron, Marquess of Bristol. The church is remarkable for its elegance, which it owes to a former incumbent, the Hon. Hervey Aston, D.D., who completely pewed, paved, and beautified it, in 1745.

SHOTOVER, an extra-parochial liberty, in the union of HEADINGTON, hundred of BULLINGTON, county of OXFORD, 4¾ miles (E. by N.) from Oxford; containing 149 inhabitants.

SHOTTESBROOK (*St. John the Baptist*), a parish,

in the union of COOKHAM, hundred of BEYNHURST, county of BERKS, 5 miles (S. W.) from Maidenhead; containing 138 inhabitants. The line of the Great Western railway intersects the parish. The living is a vicarage not in charge, endowed with a portion of the rectorial tithes, with the vicarage of White Waltham united in 1744; net income, £513; patron, and impropriator of the remaining portion of the rectorial tithes of Shottesbrook, A. Vansittart, Esq.; impropriator of White Waltham, J. Sawyer, Esq. The church, though small, is an elegant cruciform structure, principally in the decorated style, with a tower and spire rising from the intersection: it was erected in 1337, and contains three stone stalls under trefoil arches, with a piscina adjoining, an octangular font, enriched with crocketed pinnacles, and several interesting monuments and inscriptions. In the chancel lie the remains of the learned Henry Dodwell, first Camden Professor of History at Oxford, and an able chronologist and historian, who wrote his celebrated work "De cyclis Veterum" at Smewins, in the parish of White Waltham, once a hunting seat belonging to Prince Arthur, eldest son of Henry VII., and now a farm-house. A chantry or college for a warden, five priests, and two clerks, was founded here in 1337, by Sir William Trussell, Knight, the revenue of which at the dissolution was estimated at £42. 2. 8.

SHOTTISHAM (*ST. MARGARET*), a parish, in the union of WOODBRIDGE, hundred of WILFORD, Eastern Division of the county of SUFFOLK, $5\frac{1}{4}$ miles (S. E.) from Woodbridge; containing 280 inhabitants. The living is a discharged rectory, valued in the king's books at £4. 16. $0\frac{1}{2}$.; present net income, £248: it is in the patronage of Mrs. Elizabeth Darby and Miss Mary Kett. The navigable river Deben runs on the western side of the parish, where there is an inlet called Shottisham creek. Crag or shell pits, supposed to be diluvial remains, abound here.

SHOTTLE, a joint township with Postern, in the parish of DUFFIELD, union of BELPER, hundred of APPLETREE, Southern Division of the county of DERBY, $2\frac{1}{4}$ miles (W. by S.) from Belper; containing 556 inhabitants. The sum of £7 per annum, arising partly from a bequest by Ralph Dowley in 1740, is paid to a schoolmaster for teaching the poor children of the township, in a school-room erected by subscription among the inhabitants.

SHOTTON, a joint township with Langley-Dale, in the parish of STAINDROP, South-Western Division of DARLINGTON ward, Southern Division of the county palatine of DURHAM, $5\frac{1}{2}$ miles (N. E. by E.) from Barnard-Castle: the population is returned with Langley-Dale.

SHOTTON, a township, in the parish and union of EASINGTON, Southern Division of EASINGTON ward, Northern Division of the county palatine of DURHAM, 9 miles (E. by S.) from Durham; containing 272 inhabitants. A free school, with a house for the master, was founded here in 1768, in pursuance of the will of Edward Walton, one of the Society of Friends; the annual income is £30, of which sum £22 is paid for the instruction of 22 children, £5 is appropriated for books and repairs, and the remainder for apprenticing the children. There is another school, in which six girls are instructed at the expense of a private individual.

SHOTTON, a joint township with Foxton, in the parish and union of SEDGEFIELD, North-Eastern Division of STOCKTON ward, Southern Division of the county palatine of DURHAM, $8\frac{1}{2}$ miles (N. W.) from Stockton-upon-Tees; containing 73 inhabitants.

SHOTTON, a joint township with Plessey, in the parish of STANNINGTON, Western Division of CASTLE ward, Southern Division of the county of NORTHUMBERLAND, $6\frac{1}{4}$ miles (S. by E.) from Morpeth: the population is returned with the parish.

SHOTTSWELL (*ST. LAWRENCE*), a parish, in the union of BANBURY, Burton-Dassett Division of the hundred of KINGTON, Southern Division of the county of WARWICK, $4\frac{1}{2}$ miles (N. N. W.) from Banbury; containing 302 inhabitants. The living is a discharged vicarage, valued in the king's books at £5. 13. 4.; present net income, £157: it is in the patronage of Lady G. North. There is a place of worship for Wesleyan Methodists; also a school chiefly supported by Lady G. North.

SHOTWICK (*ST. MICHAEL*), a parish, in the union of GREAT BOUGHTON, Higher Division of the hundred of WIRRALL, Southern Division of the county of CHESTER; containing 713 inhabitants, of which number, 96 are in the township of Shotwick, 6 miles (N. W.) from Chester. The living is a perpetual curacy; net income, £88; patrons and appropriators, Dean and Chapter of Chester. The church has a curious Norman door, and some portions in the later style of English architecture.

SHOTWICK-PARK, an extra-parochial liberty, in the union of GREAT BOUGHTON, Higher Division of the hundred of WIRRALL, Southern Division of the county of CHESTER, $4\frac{1}{4}$ miles (N. W.) from Chester; containing 18 inhabitants. This was the site of a castle formerly belonging to the crown. Henry II. is said to have lodged here on his journey to and from Ireland; and Edward I. occupied it in 1278. The castle was standing in Leland's time, and there were some remains in 1622, since which period every vestige has been gradually removed.

SHOULDEN (*ST. NICHOLAS*), a parish, in the union of EASTRY, hundred of CORNILO, lathe of ST. AUGUSTINE, Eastern Division of the county of KENT, $1\frac{1}{2}$ mile (W.) from Deal; containing 356 inhabitants. This parish is situated on the road from Sandwich to Deal, and bounded by the sea. Sandon castle, built by Henry VIII. for the defence of the coast, is within its limits. The inhabitants are within the jurisdiction of a court of requests held at Deal. The living is a vicarage, annexed to Northbourne. A gallery has been erected in the church, and 90 free sittings provided, the Incorporated Society having granted £15 in aid of the expense. Fragments of Roman urns, with several coins, chiefly of the Emperor Gallienus, were found in 1832, on removing some land near Sandon castle.

SHOULDHAM (*ALL SAINTS*), a parish, in the union of DOWNHAM, hundred of CLACKCLOSE, Western Division of the county of NORFOLK, $6\frac{1}{2}$ miles (N. E.) from Downham-Market; containing 725 inhabitants. The living is a perpetual curacy, with those of St. Margaret and Shouldham-Thorpe united; net income, £121; patron and impropriator, Sir Thomas Hare, Bart. The church of St. Margaret was standing in 1512, but after the dissolution was suffered to go to decay. There is a place of worship for Wesleyan Methodists. A Gilbertine priory, in honour of the Holy Cross and the

Blessed Virgin, was founded in the time of Richard I., by Jeffrey Fitz-Piers, Earl of Essex, for canons and nuns under the government of a prior which at the dissolution possessed a revenue of £171. 6. 8.: the site was granted, in the 7th of Edward IV., to Thomas Mildmay.

SHOULDHAM-THORPE (*St. Mary*), a parish, in the union of Downham, hundred of Clackclose, Western Division of the county of Norfolk, 5 miles (N. E.) from Downham-Market; containing 300 inhabitants. The living is a perpetual curacy, united to that of Shouldham.

SHOWELL, a chapelry, in the parish of Swerford, hundred of Chadlington, county of Oxford, 2½ miles (E. N. E.) from Chipping-Norton: the population is returned with the parish.

SHRAWARDINE, a township, in that part of the parish of Abberbury, which is in the hundred of Ford, Southern Division of the county of Salop, 6½ miles (N. W.) from Shrewsbury: the population is returned with the parish.

SHRAWARDINE (*St. Mary*), a parish, in the union of Atcham, hundred of Pimhill, Northern Division of the county of Salop, 6¾ miles (W. N. W.) from Shrewsbury; containing 189 inhabitants. The living is a rectory, valued in the king's books at £9. 12. 6.; present net income, £380; patron, Earl of Powis. Shrawardine castle was built by Alan, a follower of the Conqueror, and ancestor of the celebrated Fitz-Alans, who held it under the crown for many ages, to check the invasions of the Welsh. After having been the scene of many remarkable events, it was, in the reign of Elizabeth, purchased by Lord Chancellor Bromley: the site and remains, together with other estates in the parish, are now the property of the Earl of Powis. Here is a Sunday school which is supported by the minister.

SHRAWLEY (*St. Mary*), a parish, in the union of Martley, Lower Division of the hundred of Doddingtree, Hundred-House and Western Divisions of the county of Worcester, 4¼ miles (S. by W.) from Stourport; containing 497 inhabitants. The living is a rectory, valued in the king's books at £9. 17. 1.; patron, Thomas Taylor Vernon, Esq. The tithes have been commuted for a rent-charge of £340, subject to the payment of rates, which on the average have amounted to £35; the glebe comprises 63 acres, valued at £110. 5. per annum. The river Severn passes through the parish, in which there are vestiges of some ancient intrenchments. A school is partly supported by the rector; and Edward Burlton, in 1699, gave the interest of £40 for teaching poor children. Thomas Vernon, in 1711, bequeathed £1000 for providing clothing and fuel for the poor of Hanbury and Shrawley: the portion belonging to this parish was invested, in 1768 or 1770, in the purchase of an estate in the parish of Claines, near Worcester, consisting of about eighty acres; and now let for £100 per annum, which amount is annually distributed in clothing and fuel among the poor, with the concurrence of the Trustees.

SHREWLEY, a chapelry, in the parish of Hatton, union of Warwick, Snitterfield Division of the hundred of Barlichway, Southern Division of the county of Warwick, 4 miles (N. W. by W.) from Warwick; containing 264 inhabitants.

Arms.

SHREWSBURY, a borough and market-town, having separate jurisdiction, locally in the liberties of Shrewsbury, Northern Division of the county of Salop, of which it is the chief town, 154 miles (N. W.) from London; containing, exclusively of the parish of Meole-Brace, which is within the liberties, 21,297 inhabitants. This ancient borough is said to have arisen from the ruins of *Uriconium*, now Wroxeter, a celebrated Roman station on the line of the Watling-street, which, passing through the present town in a direction from east to west, divides it into two nearly equal parts. From its situation on two hills, richly covered with shrubs and trees, it obtained from the Britons the appellations of *Pengwerne* and *Smwithic*, or *Y Mwythig*, and was by the Saxons called *Scrobbes-byrig*, from which, written in Domesday-book *Sciropesberie*, its present name is derived: from what source it obtained the appellation of *Salopesberie*, by which it is mentioned in some ancient records, and from which it has, with the county, been denominated "Salop," has not been satisfactorily ascertained. During the Heptarchy, it was the capital of the district called Powysland, which comprised a portion of the Saxon and British frontier territories; and the residence of the princes of Powys, whom, in 778, Offa, King of Mercia, expelled from their possessions, which he added to his own kingdom, and, to secure his conquest, raised that stupendous barrier still called Offa's Dyke. In the reign of Alfred the Great, this town was numbered among the principal cities of Britain: it had a mint, which it retained till the reign of Henry III., and there are still extant several of the coins struck in the reigns of Athelstan, Edgar, Athelred, Canute, Edward the Confessor, and Harold II., besides several between the years 1066 and 1272. When Canute was pursuing his conquests through the northern parts of the kingdom, the inhabitants revolted in his favour and surrendered the town, which, in 1016, Edmund Ironside, a short time previously to the partition of the kingdom, recovered from the Danes, inflicting signal vengeance on the townsmen for their treachery. At the time of the Conquest, Shrewsbury, with nearly the whole of the county, was bestowed by William on his kinsman, Roger de Montgomery, whom he created Earl of Shrewsbury, Chichester, and Arundel, and who built here a formidable castle for his baronial residence. In 1069, the town was besieged by Edric Sylvaticus, and Owain Gwynedd, Prince of Wales, but was relieved by King William, who advanced from York, and defeated the assailants with great slaughter. In 1102, Robert de Belesme, son of Earl Roger, having espoused the cause of Robert Duke of Normandy, and commenced measures for raising him to the throne of England, in opposition to his brother Henry I., that monarch marched against the town with an army of 60,000 men; and the earl, although he had previously fortified it with a wall on each side of the castle, across the isthmus formed by the river Severn, submitted on the approach of the king, acknowledged his treasonable conduct, and was banished to Normandy; his estates were thus forfeited, and the castle became a royal fortress.

The importance of Shrewsbury as a frontier town has rendered it the scene of many and various transactions of historical interest. In the year 1116, the nobles of the realm are said to have assembled here to do homage, and take the oaths of allegiance, to William, son of the empress Matilda; but some historians state that this assembly took place at Salisbury. Stephen, in 1138, laid siege to the castle, while Fitz-Alan, the governor, was absent in forwarding the claims of the empress; and having taken it by storm, hanged several of the garrison. The frequent inroads of the Welsh induced John to assemble a council here, in order to concert measures for suppressing them; and, in 1215, Llewelyn, who had married Joan, natural daughter of that monarch, appeared before Shrewsbury with a numerous army, and obtained possession of the town and the castle. Henry III. soon dispossessed him of his capture, and drove him back to his own territory; but in the war with the barons, Richard Earl of Pembroke retired into Wales, and, being assisted by that prince, laid waste the intermediate district, and plundered and burnt the town after having put many of the inhabitants to the sword. Simon de Montfort, whilst prosecuting the war against Henry III., obtained possession of the town, which he held only for a short time. In 1241, and 1267, the same monarch assembled an army here for the invasion of Wales, but was diverted from his purpose by the submission of Llewelyn, with whom he subsequently concluded a treaty of peace. About this time the king recommended the inhabitants to complete the fortifications of the town, of which only one side was defended, but, notwithstanding the aid of royal bounty, the work was not accomplished in less than 30 years. The continued incursions of the Welsh upon the English frontier induced Edward I., in 1277, to fix his residence in Shrewsbury, to which he removed the courts of King's Bench and Exchequer, and in 1283, assembled the parliament here; the king and his court were accommodated at Acton-Burnell, the seat of Bishop Burnell, the Lord High Chancellor; the lords held their sittings in the castle, and the commons, who for the first time had any voice in the national councils, assembled in a building near the castle. This monarch having sent a force against the Welsh without success, took the field in person, at the head of a numerous army, and an engagement took place at the foot of Snowdon, in which they were completely routed, Llewelyn slain, and his brother Davydd, who had instigated him to this insurrection, taken prisoner, and, after a short confinement in Rhuddlan castle in Flintshire, brought to Shrewsbury, where, having been tried by the parliament, he was condemned and executed as a traitor, with a degree of degradation and severity previously unknown in this country, and which, till a very late period, furnished a precedent for the punishment of treason. Edward II. was received in this town with the greatest pomp in 1322, where, in the same year, he celebrated a grand tournament, which was attended by a numerous assemblage of knights and noblemen. In 1397, Richard II. adjourned the parliament from Westmorland to Shrewsbury, gave a splendid entertainment to the lords and commons, and created several peers, who at this time first assumed their seats in parliament: this, from the number of noblemen and others who attended, and from the importance of the state affairs transacted at it, was called the Great Parliament; but the measures enacted, though ratified by the pope's bull, were repealed during the following reign, and the king's conduct while in this town was made the subject of one of those charges which subsequently led to his deposition.

In 1403, a sanguinary battle was fought in the immediate vicinity, between the forces of Henry IV. and those of the Earl of Northumberland, who had rebelled against the king, assisted by a considerable body of Scottish troops under the command of Earl Douglas, amounting to 14,000 men. After a severe and protracted conflict, the victory was decided in favour of Henry: 2300 knights and gentlemen, among whom was Hotspur, son of Earl Percy, after performing prodigious exploits of valour, and 6000 common soldiers, were slain on both sides; the dead were interred on the spot, which has since been called Battlefield, where a church was afterwards erected by the king, in memory of his victory. Owain Glyndwr, who had raised an army to co-operate with the insurgents, marched with his advanced guard to Shelton, two miles from Shrewsbury, and on perceiving the battle decided, retreated into Wales. During the contest between the houses of York and Lancaster, the inhabitants embraced the cause of the former; and on the defeat of Richard Plantagenet, Duke of York, at the battle of Wakefield, in which he was slain, his son Edward Earl of March, afterwards Edward IV., levied in this town and neighbourhood a powerful army, with which he avenged the death of his father at the battle of Mortimer's Cross, where he gained a signal victory. Edward, on his elevation to the throne, selected Shrewsbury as an asylum for his queen during the agitation of the times; and in the convent of the Dominican friars, in which her Majesty resided, the princes Richard and George were born, the latter of whom died in childhood, and the former, with his elder brother, Prince Edward, was inhumanly murdered in the Tower of London, by their uncle, the Protector, afterwards Richard III. The Earl of Richmond, on landing at Milford Haven, proceeded to this town, where he was proclaimed king, and having strengthened his army with considerable reinforcements raised in the neighbourhood, advanced into Leicestershire, where he gained the battle of Bosworth Field, which terminated in the death of Richard III., and his own elevation to the throne, under the title of Henry VII. This monarch on his accession visited the town, with his queen and Prince Arthur; and after celebrating the festival of St. George in the church of St. Chad, granted the inhabitants several privileges, in acknowledgment of the alacrity with which they had supported his claims to the crown.

On the breaking out of the parliamentary war, Charles I. came to Shrewsbury, where he was received with every demonstration of loyalty by the inhabitants, and was soon afterwards joined by Prince Rupert, Prince Charles, the Duke of York, and several noblemen and gentlemen: the king kept his court in an ancient building, called the Council-house, and having established a mint for the supply of his exigencies, the inhabitants liberally presented their plate to be melted and coined into money for his use, of which considerable sums were expended in extending and strengthening the fortifications of the town. In 1664, Col. Mytton made two attempts to obtain possession of the town and castle for the parliament, and was repulsed in both with considerable loss; but having obtained a reinforcement, he made a third effort, in which he carried the place by storm, and was appointed governor. In 1651, Charles II. summoned it to surrender, but, on the refusal of the governor, marched on to Worcester; and after the disastrous battle there he took refuge in the Royal Oak at Boscobel, on the confines of this county. During that monarch's retreat on the Continent, a plan was formed by a party of royalists to besiege the castle; but their scheme was frustrated, and several of them were punished. James II. visited the town in 1687, and, attended by the nobility and gentry of the county, kept his court for several days at the council-house. During this reign the castle was dismantled, and all its ammunition and military stores removed. This castle, originally of such extent and formidable strength that, to make room for its erection, Earl Roger pulled down nearly one-fifth of the town, was a fortress of very great importance till the final subjugation of Wales, after which period it was entrusted to a constable, generally the sheriff, who made it the county prison: its importance as a frontier garrison having ceased, it fell into decay, and was repaired during the parliamentary war as a garrison for the king: after it came into the possession of the parliament, Cromwell erected an additional fort, called Roushill, which is among the most entire of the remaining portions. The remains are situated at the northern entrance into the town, on the summit of a bold eminence overlooking the Severn, by which it is nearly surrounded, and are composed principally of the keep, a spacious modernized structure of red stone, consisting of two round embattled towers connected by a quadrangular building, 100 feet in length; the walls of the inner court; and the great arch of the interior gateway: these include a grassy area, in which, though now private property, the knights of the shire, according to immemorial usage, are girt with their swords, on their election to serve in parliament. On the south side of the court is a lofty mount rising abruptly from the river: the summit is surrounded with a wall, and in one angle of the enclosure was a barbican, which has been converted into a summer-house, called Laura tower, after the name of Miss Pulteney, for whose use it was so perfected; it commands an extensive, varied, and picturesque view of the surrounding country. The ramparts formerly environing the town, together with the towers by which they were defended, have, with the exception of one of the towers on the south side of the town, been demolished. Adjoining the castle precinct, and formerly within its walls, are the remains of the ancient council-house, where the courts for the Marches of Wales were occasionally held, and which afforded a temporary residence to several of the English monarchs.

The town is pleasantly situated on two eminences rising gently from the river Severn, which, by its windings, forms a peninsula: it consists of several streets irregularly formed, and, with some exceptions, inconveniently narrow. Various improvements have been made under the provisions of an act obtained in 1821, and others are in progress, for removing numerous obstructions arising from the style of building, and widening the approaches and streets. It is well paved, lighted with gas by a company established in 1820, and supplied with water from a remarkably fine spring, called Bradwell, about two miles distant, and also from the river Severn, by a company established in 1827. Over this river are two bridges of stone, one called the English bridge, a handsome structure of Grinshill freestone of seven circular arches, crowned with a balustrade, erected in 1774, at an expense of £16,000, defrayed by public subscription, and connecting the suburb of Abbey Foregate with the town; the other, called the Welsh bridge, a neat plain structure of five spacious arches, erected in 1795, at a cost exceeding £8000, affording a passage into Wales. Near the Abbey Foregate is the military depôt, a handsome brick building, erected in 1806, from a design by Wyatt, at an expense of £10,000: it was discontinued as a depôt several years since, and the armoury removed to Chester castle; the building now belongs to Lord Berwick. At the entrance into the town from the London road is a lofty column of the Grecian Doric style, rising from a base, ornamented at the angles with lions couchant, to the height of 132 feet, and supporting on its summit a well-executed statue of Lieut.-Gen. Rowland Lord Hill, in honour of whose achievements in the late Continental war it was erected, by general subscription, in 1814. The public subscription library, near St. John's Hill, contains more than 5000 volumes in various departments of literature: attached to it is a news-room well supplied with periodical publications. A mechanics' institute was formed in 1825; and a museum was established about five years since. Extensive public buildings are now in course of erection. The ancient theatre was formerly part of the palace of the Princes of Powysland, of which it retained some vestiges, though materially altered by its appropriation to dramatic uses: a new theatre has been lately built, presenting a neat and extensive front occupied by shops. Assemblies are held monthly, during the season, in a suite of rooms well fitted up; and races annually in September, for three days, on a course adjoining Abbey Foregate. The river Severn, in addition to the salmon for which it is celebrated, and with which it formerly abounded to a much greater extent, produces trout, pike, perch, carp, eels, shad, flounders, lampreys, &c. On the south-western side of the town is a beautiful walk, called the Quarry, comprising about twenty acres, and extending along the winding margin of the Severn for 500 yards in length, forming a noble avenue of full-grown lime-trees, from which diverge three other walks leading to the town. In the vicinity also are numerous pleasant walks and rides, through a country abounding with beautiful and picturesque scenery.

The trade, which was formerly of considerable extent and importance, has been materially diminished by the growth of other places: but the town has, notwithstanding, always maintained a considerable share of internal commerce. Its ancient traffic in Welsh cloths and flannel was formerly the principal source of its opulence, and at present, though not restricted to the Drapers' Company as before, produces no inconsiderable profit: the greater portion of those made in the counties of Montgomery and Merioneth, and part of Denbighshire, is sent to Shrewsbury. An extensive manufactory for thread, linen yarn, and canvas, situated near the castle, adjoining the suburb of Castle Foregate, affords employment to a considerable number of persons; and on the banks of the river, in Coleham, are extensive iron-foundries, in which the immense chains that support the stupendous bridge over the Menai straits, and the iron-work in many similar erections, were cast. This town is also noted for its brawn, and for a particular kind of sweet cakes, called Shrewsbury cakes. The river affords a convenient transit for goods of every description to Worcester, Gloucester, Bristol, and other towns; and considerable quantities of grain, in which the trade is extensive, and of lighter manufactured articles, are forwarded by a junction canal, opened four years ago, the traffic on which is greatly increasing. The Shrewsbury canal, which is the great medium of supplying the town with coal, terminates near the Castle Foregate, where convenient wharfs have been constructed by the canal company, for the use of persons connected with the coal-works on the line of the canal, which, with the Birmingham and Liverpool junction canal, has opened a new and extensive species of traffic for the town: it was constructed under an act obtained in 1793. The market days are Wednesday and Saturday, the latter for grain: the general market is held in a stone building, erected in 1819; and that for corn, in the area under a spacious building of stone, erected in 1595, in the later style of English architecture: over the ground-floor is a room, appropriated formerly to the Drapers' Company, for the sale of flannel, and now used as a warehouse.

This town has received a succession of charters of incorporation, from the time of William the Conqueror to the reign of James II.: the earliest preserved in the archives of the corporation is dated November 11th, 1st of Richard I., under which, as extended and confirmed by succeeding sovereigns, and remodelled by Charles I., the government was vested in a mayor, twenty-four aldermen, and forty-eight assistants, constituting the common council; besides which there were a recorder, two chamberlains, town-clerk, and other officers. The mayor, ex-mayor, the bishop and the chancellor of the diocese, the recorder, steward, and three senior aldermen, were justices of the peace. The corporation now consists of a mayor, ten aldermen, and thirty councillors, under the act of the 5th and 6th of William IV., cap. 76, for an abstract of which see the Appendix, No. I. The borough is divided into five wards, and the municipal boundaries are co-extensive with those for parliamentary purposes. The freedom is acquired by birth, or obtained by apprenticeship to a member of one of the Incorporated Companies, of which there were sixteen, the Drapers' being the principal, but they are now much reduced in number. It was formerly the custom for the several companies to celebrate the festival of Corpus Christi in St. Chad's church, after which three days were devoted to festivity and recreation. All persons above the age of twenty-one, born in the town, and all persons having served an apprenticeship in it, may demand their freedom on payment of £6. 19. 1.; and all sons of burgesses, on paying £1. 3. 6. The borough has exercised the elective franchise from the 23rd of Edward I., and has regularly returned two members to parliament. The right of election was formerly vested in the resident burgesses paying scot and lot, and not receiving alms, in number about 1200; but, by the act of the 2nd of William IV., cap. 45, it has also been extended to the £10 householders of an enlarged district, which, by the act of the 2nd and 3rd of William IV., cap. 64, has been made to constitute the new borough, comprising an area of 3080 acres, the limits of which are minutely described in the Appendix, No. II.: the old borough contained only 1254 acres: the mayor is the returning officer. The recorder holds quarterly courts of session, on the Monday previous to the county quarter sessions, for all offences not capital; and the mayor, assisted by some of the other magistrates, holds a session every week, for the determination of petty causes: the recorder also holds a court of record, the jurisdiction of which extends over the liberties, every Tuesday, for the recovery of debts to any amount. A court of requests is held, every third Wednesday, by commissioners appointed under an act passed in the 23rd of George III., for the recovery of debts exceeding 2*s.* and under 40*s.*; and a court leet is held annually in May and October, at the latter of which constables and other officers for the town are appointed. The assizes and general quarter sessions for the county are held here; and the town has been made a polling-place and the place of election for the northern division of the county. The old town and shire hall, a spacious, handsome, and commodious building of stone, erected in 1785, has recently been taken down, and a new edifice built, from a design by Mr. Smirke. The town and county gaol, and house of correction for the county and the several boroughs therein, an extensive building of brick, pleasantly situated on the bank of the Severn, was erected in 1793, at an expense of £30,000: the entrance is through a freestone gateway, over which is a bust of the celebrated Howard, and on each side is a lodge for the inspection of prisoners previously to their admission: it contains a house for the governor, a chapel, infirmary, and twenty-six wards, nine work-rooms, and twenty-six day-rooms and airing-yards, and is admirably adapted to the classification, employment, and reformation of prisoners, of whom it is capable of receiving five hundred.

Corporation Seal.

Shrewsbury comprises the parishes of *St. Alkmond, St. Chad, Holy Cross, St. Julian,* and *St. Mary.* The living of St. Alkmond's is a vicarage, valued in the king's books at £6; present net income, £219: it is in the patronage of the Crown. The church was made collegiate by King Edgar, who endowed it for the support of ten canons, one of whom acted as dean; but the society was dissolved on the establishment of Lilleshull abbey, to which its revenue was appropriated. The old edifice, a cruciform structure of great antiquity, was, with the exception of the tower and spire, which are 184 feet in height, taken down, from an apprehension of insecurity, and rebuilt in 1795: the east window is embellished with a painting by Eginton, in stained glass, emblematical of Faith. The living of St. Chad's is a vicarage; present net income, £350: it is in the patronage of the Crown. The church, erected in 1792, at an expense of nearly £20,000, in lieu of an older edifice, which, while undergoing repair, fell down in 1788, is a handsome circular building, in the Grecian style of architecture, with a square rustic tower supporting an octagonal belfry, surmounted by a dome resting on eight Corinthian pillars: the body of the church forms a rotunda 100 feet in diameter, surrounded by a range of duplicated Ionic pillars between the lofty arched windows, rising from the basement, and supporting a handsome cornice surmounted by a balustrade: the entrance is through a stately portico of four Doric columns, supporting a triangular pediment: the interior has a rich and pleasing effect; the galleries are supported by a duplicated range of Ionic pillars, from which rise Corinthian pillars supporting the roof; the chancel is adorned with a painting of the Resurrection, in stained glass, by Eginton, from a design by West, removed from Lichfield cathedral. The remains of the ancient church, formerly collegiate, and once a royal free chapel, consist only of the south aisle of the chancel, containing portions in the Norman, early English, and decorated styles of architecture: it was fitted up for the performance of the funeral service, and is at present appropriated to the use of the charity school. A church in the ancient English style, with a tower, was built in this parish in 1831, at an expense of £3663. 19. 6., under the act of the 58th of George III. The living of the parish of the Holy Cross is a vicarage, with the chapel of St. Giles, valued in the king's books at £8; present net income, £323; patron and impropriator, Lord Berwick. The church, occupying a low site in the eastern suburb, to which it gives name, and surrounded on the south and west by the river Rea, commonly called Meole brook, is part of the conventual church of a splendid abbey, founded for Benedictine monks, by Roger de Montgomery, in 1083 (on the site of a religious institution established prior to the Conquest, with the revenue of which it was partly endowed), and dedicated to St. Peter and St. Paul: it was a mitred abbey, and the abbots exercised episcopal authority in their house, being in some respects exempt from the jurisdiction of the diocesan: at the dissolution, in 1513, its revenue was estimated at £615. 4. 3. The king intended to make Shrewsbury the seat of a diocese, and to raise the abbey church into a cathedral, Dr. Bourchier, the last abbot of Leicester, having been actually nominated bishop; but pecuniary exigencies compelled him to abandon the design. The abbey was further distinguished by the resort of many pilgrims to the shrine of St. Winifred, whose remains had been removed hither from Gwytherin, in Denbighshire. The walls of this establishment included ten acres, and the buildings, principally in the Norman style of architecture, were extensive and magnificent: the principal remains are the western tower, the north porch, the nave and aisles of the abbey, now the parish church, besides some small portions of the conventual buildings: the former retains several features of its ancient grandeur, though many alterations have been made, particularly the introduction of a large window of seven lights, of elegant tracery, and emblazoned with armorial bearings in stained glass, in the later English style, over the west doorway, which was originally a handsome circular arch, within which, at a much later period, a painted one has been placed, on each side of which are niches, one of them containing a statue of St. Peter, and the other a statue of St. Paul. The interior has a solemn grandeur of effect; the roof is finely vaulted, and supported on circular arches and massive piers, and in other parts the slender clustered column and the pointed arch prevail: the east window is enriched with armorial bearings, including those of Lord Berwick, by whom it was presented; and, in the central compartment, with paintings of St. Peter and St. Paul, in stained glass, by Mr. D. Evans, of Shrewsbury: there are various altar-tombs and ancient monuments, and within an arch, which formerly led to the south aisle of the transept, is an ancient figure in armour, supposed to be that of its founder, Earl Roger, who, having assumed the cowl toward the close of his life, died and was buried here. Among the ruins of the conventual buildings is a fragment supposed to be part of the refectory, on which is an exquisitely beautiful octagonal structure of stone, resting partly on a corbel, projecting from the wall, and supposed to have been an oratory, or pulpit, from which one of the monks, according to their custom, read to his brethren while at dinner: it is an unrivalled specimen of the decorated style of English architecture, ornamented with lofty and finely pointed windows, divided only by enriched mullions rising from the corbel, and crowned with trefoiled arches deeply moulded; the spaces between the three northern arches are filled up to the height of four feet with stone panels, in which are enshrined figures, and the exterior is crowned with an obtuse dome almost concealed by the ivy which has overspread the building: the interior is six feet in diameter, and the roof is elaborately groined, and ornamented in the centre, where the ribs unite, with an alto relievo of the Crucifixion. The chapel of St. Giles, which was originally attached to the hospital belonging to the abbey church, stands at the eastern extremity of the Abbey Foregate, and divine service is performed in it: it is a small ancient building, with a diminutive turret and an elegant eastern window of stained glass, and has been repewed and fitted up, at the expense of the Rev. Richard Scott. The living of St. Julian's is a perpetual curacy; net income, £159; patron and impropriator, Earl of Tankerville. The church, with the exception of the tower, which is in the Norman style of architecture, was rebuilt of brick in 1750; the interior, which is neatly arranged, is decorated with some relics of the ancient structure; in the east wall of the chancel is a small female figure, enshrined in rich tabernacle work, probably representing St. Juliana, the patroness, and in the ceiling is preserved a considerable portion of the ancient fret-work; the east window is embellished with a painting of St. James, in stained glass, brought from Rouen during the French revolution of 1792, above which are some armorial bearings; among the monuments is a slab of coarse alabaster, inscribed with Longobardic characters. The living of St. Mary's is a perpetual curacy; net income, £312: it is in the patronage of the Corporation, conjointly with the Bishop of Lichfield and others; the impropriation belongs to the free grammar school. The church is an ancient cruciform structure, partly in the Norman, and partly in the early English, style of architecture, with a western tower surmounted by a lofty spire of beautiful proportion; the lower part of the tower and the south porch are of the Norman style: the interior is well arranged, and, from its frequent enlargement and alteration, comprises specimens of various styles; the nave, of which the oak roof is finely panelled and carved, and supported on circular arches and massive piers, is lighted by a double range of clerestory windows, and separated from the chancel by a highly enriched pointed arch; the windows at the ends of the transepts, and on the north side of the chancel, are early English, and the east window of the latter is embellished with stained glass, formerly in the old church of St. Chad, representing the genealogy of Christ from the root of Jesse, and containing in each of the numerous oval compartments a king, or patriarch, of the ancestry of Joseph, the husband of the Virgin Mary. Among the monuments is one to the memory of Robert Cadman, who, attempting a second time to perform his descent from the summit of the spire, by means of a rope, the other end of which was secured in a field on the opposite side of the river, was killed by the accidental breaking of the rope. A chapel of ease to St. Mary's, dedicated to St. Michael, has been built near the Castle Foregate, by public subscription, aided by a grant of £500 from the Incorporated Society, for which 630 free sittings have been provided: the living is a perpetual curacy; net income, £59; patron, Incumbent of St. Mary's. A chapel of ease to the parish of St. Chad, now called St. George's, to which a district has been assigned, has also been erected in Frankwell, the living of which is a perpetual curacy; net income, £118; patron, Vicar of St. Chad's. A new church in the suburb of Coleham, called Trinity Church, was consecrated August 25th, 1837; it was built by subscription, aided by grants from the Diocesan and Incorporated Societies, and contains 500 free sittings. The patronage is in the Incumbent of St. Julian's. There are places of worship for Baptists, the Society of Friends, Independents, Wesleyan and Welsh Methodists, Sandemanians, Unitarians, and Roman Catholics.

The royal free grammar school was founded by Edward VI., in 1553: its endowment, augmented by Queen Elizabeth, produces an annual income of £2740: it is under the superintendence of the Bishop of Lichfield, as visiter, and thirteen trustees, the mayor of Shrewsbury, who presides at the several meetings, being one. The school is open for gratuitous classical instruction to all sons of burgesses so soon as they are qualified to commence the Latin accidence, and provided they have attained their sixth year; and it has maintained for many years a distinguished rank among the public schools of the country. There are usually 260 scholars, of whom about 45 are on the foundation, the remainder being private boarders either with the first or second master. It is conducted by a head master, appointed by the Master and Fellows of St. John's College, Cambridge, a second master, an usher, and a writing master, besides assistant masters who are paid by the head master. Belonging to the institution are four exhibitions of £70 per annum each, and four of £15 per annum each, to St. John's College, Cambridge; four of £60 per annum each to Christ-Church College, Oxford; and two of £25 per annum each, and one of £23 per annum, to either of the Universities; four scholarships of £63 per annum each, and two of £40 each, in Magdalene College, Cambridge; a by-fellowship in the same college, of £126 per annum, and three contingent exhibitions. The premises, in the later style of English architecture, occupy two sides of a quadrangle, with a square turret crowned with pinnacles in the angles, and comprise spacious school-rooms, with residences for the masters contiguous, and a chapel, over which is a fine library, rebuilt in 1815, at an expense of £1860, and containing an extensive and valuable collection of books and manuscripts, to which is annexed a museum of antiquities from Wroxeter, and fossils peculiar to this part of the country. Among the eminent persons who have received the rudiments of their education in this school are, Sir Philip Sydney; Sir Fulke Greville (Lord Brooke); Dr. John Thomas, Bishop of Salisbury; the Rev. Dr. John Taylor, a learned critic and philologist; Dr. Waring, Lucasian Professor of Mathematics in the University of Cambridge; W. Wycherly and Ambrose Phillips, poets; William Clark, a learned divine and antiquary; and various others. Mr. John Allat, formerly chamberlain of the borough, in 1792, bequeathed property amounting to upwards of £13,000, a portion of which was to be applied to clothing, instructing, and apprenticing poor children of parents not receiving parochial relief; and the remainder to an annual distribution of coats and gowns among a considerable number of aged men and women. Of the dividends, amounting to £393 per annum, about one-third is appropriated to the supply of clothing to the adults, and two-thirds to the purposes of the school, in which there are 25 boys and 25 girls, who are completely clothed, supplied with books, and, at the age of 14, are apprenticed. A handsome freestone building was erected for this charity in 1800, at an expense of £2000, with a convenient house for the master and mistress. A school for instructing, clothing, and apprenticing 18 boys and 12 girls of the parish of St. Julian was founded, and a neat brick building erected for it, in 1724, by Mr. Thomas Bowdler, alderman and draper. The public subscription charity school, near the abbey church, was established in 1778, at which time houses for the master and mistress were built. The royal Lancasterian school was established, and a commodious building, comprising two school-rooms, with apartments for the master and the mistress, was erected, by subscription, in 1812. Here are three National schools, partly supported by subscription. Sunday schools have also been instituted in connection with the established church, and with the several dissenting places of worship. St. Chad's almshouses were founded in 1409, by Mr. Bennet Tupton, with a small endowment; there were originally thirteen, but for want of funds two have fallen into decay. St. Mary's almshouses, sixteen in number, were founded in 1460 by Mr. Degory Watur, draper: the old almshouses were taken down in 1823, and a new building, consisting of 16 tenements, each containing two rooms, has been erected opposite St. Mary's church. St. Giles's almshouses, four in number, are inhabited by aged persons nominated by the Earl of Tankerville.

The house of industry, situated on an eminence adjoining Kingsland, on the south bank of the Severn, was erected in 1765, at an expense of £12,000, by the governors of the Foundling Hospital in London, as a branch establishment; on the relinquishment of that design, it was closed, and afterwards opened as a woollen manufactory for the employment of the children of the poor: it was subsequently rented by Government for the confinement of prisoners during the American war, and on the incorporation of the parishes for the maintenance of their poor, in 1784, it was purchased by the guardians and appropriated to its present use: the buildings, in addition to the inhabited apartments, comprise a chapel, school-rooms, work-rooms, an infirmary, &c., with about twenty acres of land adjoining. The general infirmary, established in 1745, was the second formed in the kingdom, that of Winchester being the first; it is supported by subscription: the premises, originally of brick, being found too small for the increased population of the town and neighbourhood, were taken down in 1827, and have been handsomely rebuilt of stone, upon a much more extensive scale, at an expense of £18,735. 17. 10., of which £13,044. 1. 3. was raised by subscription. In 1734, James Millington bequeathed property, now let for £1227 per annum, for the erection and endowment of an hospital in the suburb of Frankwell: this institution comprises schools for twenty-five boys and twenty-five girls, natives of Frankwell; and provision for twelve resident, and ten out, hospitallers, to be chosen from decayed housekeepers of Frankwell, or that part of the parish of St. Chad which is contiguous, the latter of whom, on vacancies occurring, have the preference of appointment to a residence. The children are clothed, educated, and apprenticed: there is a chaplain on the establishment, with a stipend of £50 per annum. Two exhibitions of £40 per annum each, to Magdalene College, Cambridge, were given by the same founder, to which boys educated in the hospital have the first claim, and which, in default of such, lapse to boys born in Frankwell, and educated in the free grammar school. Shrewsbury is one of the towns entitled to a share of the charities of Sir Thomas White and Henry Smith: and there is also a considerable sum, the produce of various other bequests, annually distributed in coal and clothing, and other relief. Among the monastic institutions anciently existing here were, a convent of Grey friars, founded in the reign of Henry III., by Hawise, wife of John de Charleton, Lord of Powis, of which there are some remains; a convent of Dominican friars, founded by Lady Genevile, of which there is not a vestige, the foundations having been lately dug up; and a convent of Augustine friars, founded by one of the family of Stafford, of which some small portions are remaining. Of the numerous chapels, the only one of which there are any remains is that of St. Nicholas, near the old Council-house, now converted into a stable. Among the eminent natives of this town were, Richard and George Plantagenet, sons of Edward IV.; Ralph of Shrewsbury, Bishop of Bath and Wells; Robert, Bishop of Bangor; Thomas Bower, and John Thomas, Bishops of Salisbury; Edward Wooley, Bishop of Clonfert; Sneyd Davies; Lord Chief Justice Jones; Richard Onslow, Speaker of the House of Commons; the Rev. Job Orton; George Costard, a distinguished mathematician; Thomas Churchyard, the poet; Vice-Admiral Benbow; Dr. John Taylor, a learned critic, and editor of Demosthenes; Hugh Farmer, an eminent divine; and Dr. Charles Burney, a celebrated musician. Ordericus Vitalis, one of the best of early English historians, born at Atcham, in 1074, was educated in the abbey. Shrewsbury gives the title of Earl to the family of Talbot.

SHREWTON (*St. Mary*), a parish, in the union of Amesbury, hundred of Branch and Dole, Salisbury and Amesbury, and Southern, Divisions of the county of Wilts, 5¾ miles (W. N. W.) from Amesbury; containing 491 inhabitants. The living is a discharged vicarage, valued in the king's books at £8; present net income, £197; patron, Bishop of Salisbury. There are two places of worship for Baptists. Ann Estcourt, of Newnton, in 1704, bequeathed a rent-charge upon her property now amounting to £34. 9. 8., for apprenticing yearly six poor boys belonging to the parishes of Newnton, Rolston, and Shrewton.

SHRIGLEY, POTT, county of Chester.—See POTT-SHRIGLEY.

SHRIPPLE, a tything, in the parish of Idmiston, hundred of Alderbury, Salisbury and Amesbury, and Southern, Divisions of the county of Wilts, 7 miles (E. by N.) from Salisbury; containing 56 inhabitants.

SHRIVENHAM (*St. Andrew*), a parish (formerly a market-town), in the union of Farringdon, hundred of Shrivenham, county of Berks, 5 miles (S. W. by S.) from Great Farringdon; containing 2113 inhabitants. The living is a vicarage, valued in the king's books at £20; present net income, £676: it is in the patronage of the Crown; impropriator, Viscount Barrington. The church is a very large structure, principally in the Norman style, with a tower rising from the centre, and contains a monument to Admiral Barrington, by Flaxman. There is a chapel of ease at Longcot, in this parish. A chantry was founded here in 1336, by John de Burghton and Agnes, his wife. William de Valence obtained a charter, in 1257, for a weekly market on Thursday, and an annual fair on the festival of St. Mary Magdalene, which were confirmed by another charter in 1383, but both have been long disused. The Wilts and Berks canal, and the line of the Great Western railway, pass through the parish. Thomas Stratton, in 1703, gave a rent-charge of £4, for which sum ten boys are educated in a school-house purchased by the parish, with aid from private subscriptions. In 1788, the materials of an old chapel at Watchfield were sold, and the produce vested in the purchase of £264. 4. 1. three per cents., the dividends arising from which are applied for teaching six poor children, and six others are instructed by a schoolmistress, for the annual proceeds of £100, left by Richard Smith, in 1818. A school has been erected at the expense of Archdeacon Berens, and is supported by him. Here are eight almshouses, founded in 1642, by Sir Henry Marten, with an endowment, including an augmentation by Mrs. Elizabeth Sadler, amounting to about £80 per annum, which is divided among the inmates and two out-pensioners.

SHROPHAM (*St. Peter*), a parish, in the union of Wayland, hundred of Shropham, Western Division of the county of Norfolk, 4 miles (N. by W.) from East Harling; containing 507 inhabitants. This place, which gave name to the hundred, was anciently a town of some importance, though it is now only a small village. The living is a discharged vicarage, valued in the king's books at £8. 13. 9.; patrons and impropriators, Mayor and Corporation of Norwich, but the advowson must be sold under the Municipal Reform Act. The tithes have been commuted for a rent-charge of £270, subject to the payment of rates, which on the average have amounted to £27; the glebe comprises 45 acres, valued at £39. 5. per annum.

SHROPSHIRE, an inland county, bounded on the north by Cheshire, and a detached portion of the Welsh county of Flint; on the east by Staffordshire; on the south-east by Worcestershire; on the south by Herefordshire; and on the south-west, west, and north-west, respectively, by the counties of Radnor, Montgomery, and Denbigh, in Wales. It extends from 52° 20′ to 53° 4′ (N. Lat.) and from 2° 17′ to 3° 14′ (W. Lon.), and comprises an area of upwards of 1341 square miles, or about 858,240 statute acres. The population amounts to 222,938. The name has been corrupted from the Saxon *Scrob-scire*, a contraction of *Scrobbes-byrig-scyre*, meaning the shire of *Scrobbes-byrig*, the Saxon name for Shrewsbury. The aboriginal inhabitants of this district were of the tribes called the Cornavii and the Ordovices, the former occupying the country on the north-eastern side of the Severn, the latter the opposite shores of that river, and the south-western tracts. Little is known of the Cornavii, but the Ordovices, joined with the Silures, under Caractacus, in defending their territories against the Roman invaders: and it is thought by some that the battle in which the Britons under that leader were finally defeated, by Ostorius Scapula, was fought within the limits of this county. Gough supposes it to have been at the hill called Caer Caradoc, or the Gaer, near the junction of the small rivers Clun and Teme, on the point of which are the remains of a very large and strongly-fortified camp. The annals of Tacitus place the camp of Caractacus at the Breyddin Chain, where in all probability that celebrated leader had his last fatal conflict with the Romans. Under the Roman dominion, Shropshire was included in the division called *Flavia Cæsariensis.* After that people had abandoned Britain, this county was the theatre of numerous sanguinary contests between the Britons and the Saxons, by the former of whom it was held as part of the kingdom of Powysland, of which Shrewsbury, called by them *Sengwerne*, was the capital. Though the British princes long disputed the possession of this territory, they were ultimately obliged to retreat; and in 777, their seat of royalty was transferred to Mathrafael, among the mountains of Powys, and Shropshire became part of the kingdom of Mercia. They still, however, made frequent inroads; and the warlike Saxon monarch, Offa, partly to avert the evils attendant upon these hostilities, caused a deep dyke and rampart to be made, which extended 100 miles along the mountainous border of Wales, from the Clwyddian hills to the mouth of the Wye, crossing the westernmost parts of Shropshire; but the Welsh continued their incursions far within this boundary, and in their hasty retreats often carried off immense booty. In the ninth century, when the Danes invaded the island, this part of Mercia, although it suffered less than some others, experienced much calamity, and its chief city, *Uriconium*, was destroyed. Shrewsbury then sprang up, and flourished in consequence; and Alfred, having subdued these ravagers, ranked it among his principal cities, and gave its name to the shire, of which it was the capital. In 1016, Shrewsbury was taken by Edmund Ironside, who severely punished the inhabitants for having taken part with Canute, in opposition to his father, Ethelred. The Welsh continued their incursions both before and after this event with great fierceness, particularly in the time of Edward the Confessor, under their reigning prince, Grufydd. Harold, afterwards king of England, undertook an expedition against this prince, both by land and sea, and harassed the Welsh so much, that they sent him the head of their chief, in token of subjection: he afterwards endeavoured to secure the advantages thus gained by a decree, forbidding any Welshman to appear on the eastern side of Offa's dyke, on pain of losing his right hand.

At the period of the Norman Conquest, nearly the whole of Shropshire, together with extensive possessions in other parts of England, was bestowed on Roger de Montgomery, a relation of William's, and one of his chief captains, in reward for his services. But the hostilities of the Welsh disturbed this warrior in the enjoyment of his good fortune; and, in 1067, Owain Gwynedd, their prince, in alliance with Edric Sylvaticus, or Edric the Forester, the Saxon Earl of Shrewsbury, laid siege to that town, with a force so formidable as to require the presence of the Conqueror, who repulsed the assailants with great slaughter, and bestowed the title of Earl of Shrewsbury upon Roger de Montgomery. This county, in like manner, was frequently the scene of contest, or of preparation for military enterprise, so long as the ancient British inhabitants of Wales maintained their independence. William the Conqueror, and his more immediate successors, for the purpose of subduing the resolute Britons, issued grants to certain noblemen of all the lands they should be able to wrest from them; and hence originated the seignories and jurisdictions of the lords marchers. The precise extent of the territory designated as the Marches it is difficult to determine, the word meaning, in a general sense, the borders between the Welsh and the English: but the western border of Shropshire certainly formed a principal portion. The tenure by which these lords marchers held under the king was, in case of war, to serve with a certain number of vassals, to furnish their castles with strong garrisons and with sufficient military implements and stores for defence, and to keep the king's enemies in subjection: to enable them to perform this, they were allowed to exercise, in their respective territories, absolute power. For their better security they fortified old castles and built new ones, garrisoning them with their own retainers; and thus it was that the greater part of the numerous castles on the Welsh border were erected. They had particular laws in their baronies, termed *Sngletheria* and *Waltheria*, where all suits between them and their tenants were commenced and determined; but if a question arose concerning the barony and its title, it was referred to the king's courts. There was also, so early as the reign of John, a lord warden of the marches, whose jurisdiction resembled that of a lord-lieutenant.

Shropshire is included in the several dioceses of Hereford, Lichfield, and St. Asaph: a detached portion, containing the parishes of Claverley, Hales-Owen, and Worfield, is in that of Worcester: the whole is contained in the province of Canterbury. But by the act of the 6th and 7th of William IV., c. 77, the entire county is to be divided between the dioceses of Chester and Hereford, and that part of it which is to be transferred to the former, will then be in the province of York. That part of the diocese of Hereford which is in Shropshire, forming about one-half of the county, is almost wholly included in the archdeaconry of Salop, which also comprises most of that part of the diocese of Lichfield contained in this county. Shropshire contains the deaneries of Burford, Clun, Ludlow, Marchia, Newport, Pontesbury, Salop, Stottesden, and Wenlock: the number of parishes is 214. For purposes of civil government it is divided into fifteen hundreds, or districts answering thereto, *viz.*, the hundreds of North Bradford, comprising the Drayton and Whitchurch divisions; South Bradford, comprising the Newport and Wellington divisions; and Brimstree, comprising the Hales-Owen and Shiffnall divisions; and the hundreds of Chirbury, Condover, Ford, Munslow, Oswestry, Overs, Pimhill, Purslow, (with which that of Clun has been incorporated), and Stottesden; the liberty of Shrewsbury; and the franchise of Wenlock. It contains the borough and market-towns of Shrewsbury, Bridgenorth, Ludlow, and Wenlock; and the market-towns of Bishop's-Castle, Broseley, Cleobury-Mortimer, Clun, Drayton-in-Hales, Ellesmere, Hales-Owen, Newport, Oswestry, Shiffnall, Church-Stretton, Wellington, Wem, and Whitchurch. By the act of the 2nd of William IV., cap. 45, the county has been divided into two portions for elective purposes, called the Northern and Southern divisions, each sending two members to parliament: the place of election for the Northern Division is Shrewsbury, and the polling-places are Shrewsbury, Oswestry, Whitchurch, Wellington, Ellesmere, Newport, Market-Drayton, and Wem; that for the Southern Division is Church-Stretton, and the polling-places Church-Stretton, Bridgenorth, Ludlow, Bishop's-Castle, Wenlock, Cleobury-Mortimer, Pontesbury, Shiffnall, and Clun. Two representatives are returned for each of the boroughs. Shropshire is included in the Oxford circuit: the assizes and general quarter sessions are held at Shrewsbury, where is the county gaol. That town is the seat of judicature for all suits of the inhabitants of North Wales, commenced by *Quo Minus* in the Exchequer at Westminster, which are tried at the assizes there, as they are for those of South Wales at Hereford.

The form of the county is an irregular parallelogram. It possesses almost every variety of fine scenery; bold and lofty mountains; woody and secluded valleys; fertile and widely-cultivated plains; a majestic river, which divides it into two nearly equal portions; and sequestered lakes. Though no part of the surface is absolutely flat, yet the north-eastern districts are comparatively so, as contrasted with the hills on the southern and western borders, approaching the Welsh mountains, and form an important part of the immense plain, or vale, which also includes the whole of the county of Cheshire, and the southern part of Lancashire, and is bounded on the east by the hills of Staffordshire, Derbyshire, and the western borders of Yorkshire; and on the west by the mountains of North Wales, and by the sea. The plain of Salop is about 30 miles long from north to south, or from Whitchurch to Church-Stretton, and 28 miles broad, from Oswestry to Colebrook-dale: it is divided into two unequal portions by the Severn. The famed Wrekin mountain, celebrated for the magnificent and extensive prospects which it commands, rises singly out of the plain, to the height of nearly 1200 feet above the level of the Severn, near which it is situated: north of it are excrescences of rock and partial swells. To the south-west the hills are more frequent; and on the western and south-western borders of the county is a striking succession of mountainous elevations, divided by beautiful valleys: some of the highest ground in the county is considered to be the summits of the hills in the vicinity of Oswestry. Lying to the east of the Wrekin, and on the eastern border of the county, the coal district of Colebrook-dale, which extends from north-east to south-west, about eight miles in length, and two in breadth, is considerably above the level of the plain of Shropshire, more especially its southern parts. South-west of the Severn, the limestone ridge of hills, which commences at Lincoln hill in Colebrook-dale, proceeds in a south-westerly direction towards Church-Stretton, near which place it turns southward from the hills, around Hope-Bowdler, and descends nearly in a direct line to Ludlow, on the southern border of the county. Westward is a vale about two miles broad, and nearly fifteen in length, from Colebrook-dale to the Stretton valley. Its western side is bounded by the line of low hills ranging, without any intermediate valley, along the base of a much more elevated ridge, of which the Wrekin forms the northern extremity: this chain is continued on the south-western side of the Severn, in a line with the Wrekin, and constitutes the Acton-Burnell hills, the Frodgesley hills, the Lawley, Caer Caradoc, and the Hope-Bowdler hills; all which have craggy summits, ascend abruptly from the plain, at an angle of about 60°, and command remarkably fine prospects. The vale in which Church-Stretton is situated separates from these the singular mass of hills called the Longmynd, which ascends gradually from the plain to a height much superior to that of the Wrekin, and then stretches, with a level and unvaried summit for several miles, towards Bishop's-Castle. Following the mountainous line that forms the boundary of the plain of Salop, a high and rocky district occurs between the high road from Shrewsbury to Bishop's-Castle and the vale of Montgomery. The most elevated peak of this assemblage of lofty hills is called the Stiperstones, its summit being extremely craggy, and overspread with enormous loose blocks of quartz, which at a distance, look like the ruins of some great fortress. This hill is somewhat higher than the Wrekin, and forms the abrupt termination of a line of mountains that hence extends south-westward into Radnorshire. From the Stiperstones a range of low hills stretches, in a north-easterly direction, as far as Shrewsbury, under the names of Lyth hill, Baystone hill, and the Sharpstones. In the southern parts of the county, the Clee hills, like the Wrekin, have their bases projecting towards the low lands which accompany the course of the Severn: the Brown Clee hill, and the Titterston Clee hill, are amongst the highest in Shropshire, and have flat tops, but very irregular sides, and, like many others similarly situated, have vestiges of ancient fortifications upon their summits. Of the Berwyn mountains only a small portion, the slate mountain of Selattyn, is within the boundary of Salop. The views obtained from many of the heights are remarkably grand and beautiful. The lakes, though neither numerous nor of great extent, form a variety in the landscape rarely met with in the midland counties: that adjoining Ellesmere covers 116 acres, and there are several others in the neighbourhood, but of smaller extent.

The variations of soil are as great as those of surface; and the different kinds are so intermingled as to render it difficult to define the limits of each. There is nearly an equal quantity of wheat and turnip land, though the proportions of the former somewhat preponderate. That part of the county which lies north-eastward of the Severn consists chiefly of a turnip soil under tillage, intermingled with large portions of meadow and pasture. The flat lands on the banks of that river, which are frequently overflowed, form rich meadows. On its south-western side, from Alderbury, in a district about eight miles wide, down to Cressage, the soils are chiefly of a good quality, but very variable; comprising pasture, wheat, and turnip land, each in small quantities. From Cressage, in a tract about six miles in width, to Bridgenorth, and thence to Cleobury and Ludlow, is chiefly a mixed soil upon clay, in some places very thin. The remaining south-western portion of the county has soils of almost every species, except chalk and flint: they are generally thin, some resting upon clay, others upon hard rock of different kinds, and compose extensive tracts of hilly wastes. In the hundred of Oswestry, on the northern side of the Severn, and occupying the north-western extremity of the county, is a considerable quantity of deep loam and gravelly soil, with some marl, in the parish of that name; while in that of West Felton, in the vicinity of the same town, is a large tract of black peaty bog. Towards the south-east the soil becomes sandy; and Pimhill hundred, which adjoins it on the east, contains a mixture of boggy land and of sand lying over a red sandstone, with a still greater proportion of sound wheat land. North Bradford, forming the north-eastern part of the county, has some low land of a peaty nature, good meadow land, a considerable quantity of sand lying upon a red sand-stone, and some gravelly soils. South of this, and on the eastern border, the hundreds of South Bradford and Brimstree consist for the most part of sandy loam. The franchise of Wenlock, immediately to the west of the latter, is chiefly occupied by pale-coloured clays, locally called dye-earth, which at a considerable depth are blue, but near the surface become a pale yellow: there is also some light sand, and a considerable quantity of soil formed chiefly of the decomposed matter upon which it rests. In the hundreds of Stottesden, Overs, and Munslow, which adjoin the south-eastern and southern borders of the county, there is also much clay, and a varying stone soil upon a substratum of limestone: they have also a shallow rocky loam lying upon freestone, and sometimes slate marl; sands covering red sand stone, particularly in the vicinity of Bridgenorth, and some clays of a reddish colour, especially near Ludlow. In Condover hundred, nearly in the centre of the county, and on the southern side of the Severn, is a good deal of gravelly loam, sand, and clay, frequently blending in very small beds. In the liberties of Shrewsbury, around that town, and in the hundred of Ford, which lies on the south side of the Severn, between it and the western border of the county, there is also much pebbly loam: north of Shrewsbury is some reddish clay upon red rock, and on the northern border of Ford hundred are some lighter-coloured clays upon limestone: the southern part of this division consists for the most part of a deep clayey soil, and, proceeding still southward and westward, becomes gravelly, rocky, and uneven. The small hundred of Chirbury, southward of this, is still more rugged, but has plains of a deep light-coloured loam, or clay. In the hundred of Purslow, forming the south-western extremity of the county, although the surface is very uneven, yet several of the hills are smooth, and there are some pale-coloured clays, and a considerable quantity of lighter soils.

The red soils are in general very productive. On the arable lands the courses of husbandry vary; but there is not much difference between the culture of the sandy and gravelly soils of this county, and that practised on similar soils in Norfolk, or any other light soils where turnips are grown. The crops most common are, wheat, barley, oats, peas, and turnips. In the southern part of the county, bordering on Worcestershire, are about 250 acres of hop plantations. The principal artificial grasses are the broad-leafed clover, Dutch clover (both red and white), trefoil, and rye-grass. Grass land for hay is seldom manured: on the banks of the Severn, and in other flat tracts, intersected by smaller streams that occasionally overflow their banks, are natural meadows regularly mown. In the vales of the south-western parts of the county, the grass lands are very good: the pasture lands are not, however, on the whole, of the richest kind. The meadows, in different parts, are irrigated by means of levels preserved from the natural streams, and continued along their banks. Lime and marl are very extensively employed as manures: with the former, which is most generally used, every part of the county is tolerably well supplied.

The neat cattle are considered to be for the most part of the same kind as those which prevail in Warwickshire and Staffordshire: the old Shropshire ox was remarkable for a large dewlap. Many are also reared of the improved breeds of Lancashire, Cheshire, Leicestershire, and Staffordshire, and upon the southern confines the Herefordshire breed prevails. In the vicinity of Bishop's-Castle the cattle are good and uniform in shape and colour, the latter being a dark red. The neat cattle on the north-eastern side of the Severn are an inferior sort of the Lancashire long-horned breed, in general for the dairy. The old Shropshire sheep are horned, and have black or mottled faces and legs: their size is about that of the South-Down sheep, but their necks are rather longer, and the carcass not quite so compact. On the Longmynd is a horned breed of sheep with black faces, apparently indigenous to the tract they inhabit. Upon the hills nearer Wales the flocks are without horns and with white faces; they have rather shorter legs and heavier and coarser fleeces than the Longmynd sheep, and are of about the same weight. Perhaps in no county of equal size are reared or fattened so many hogs as in this: the original hog of Shropshire was a high-backed, long-eared animal, but this is now rarely to be seen unmixed with other breeds.

This county has been cleared at different times of much of its timber, great supplies having been sent to Bristol, for ship-building; but it still retains more fine woods of oak than most other counties, there being sufficient for the home consumption, and a considerable surplus for exportation. The coppice-woods are extensive, and consist chiefly of oak. On the side of Shropshire towards Bewdley in Worcestershire, is a large tract of coppices. There are many modern plantations, generally of various kinds of fir and pine, intermingled with different deciduous trees: indeed, there are few trees which do not flourish in the soil. Exclusively of the heathy mountainous tracts before described, which are chiefly sheepwalks, there are some flat open heaths in the north-eastern part of the county, and in the parishes of Worfield and North Cleobury, in the vicinity of Bridgenorth. Clun Forest, an extensive sheepwalk, contains above 12,000 acres, and is a fine extent of smooth turf, with every variation of swelling banks and retired dingles: a part of the Longmynd has been enclosed. There are several large mosses and a great number of smaller ones: the largest district of swampy moorland surrounds the village of Kinnersley.

The mineral productions are various and considerable; the principal are coal, iron, lead, and stone of different kinds. The coal district of Colebrook-dale is about six miles long from north-east to south-west, and two miles broad: it commences on the south-western side of the Severn, in the parishes of Barrow and Much-Wenlock, and runs across that river through those of Broseley, Madeley, Little Wenlock, Wellington, Dawley, Malins-Lea, Shiffnall, Lilleshall, and some others. Red sand-stone is found immediately to the eastward of this tract, and the coal strata dip rapidly towards it. The whole of this "independent coal formation" is composed of the usual members, *viz.*, quartzose sandstone, indurated clay, clay porphyry, slate clay, and coal, alternating with each other without much regularity, except that each bed of coal is always immediately covered by indurated, or slaty, clay. Trop, or greenstone, appears in some places between the coal formation and the limestone upon which it rests, and which rises to the west of it. The strata are found most complete in Madeley colliery, where a pit has been sunk to the depth of 729 feet, through all the beds, 86 in number, that constitute the formation. The sand-stones, which make part of the first 30 strata, are fine-grained, and often contain thin plates, or minute fragments, of coal: and the 31st and 33rd strata are coarse grained sand-stone, entirely penetrated by petroleum. Below these, at different depths, are three thick beds of sand-stone, varying in quality: vegetable impressions are found in most of them, excepting those nearest the surface. The clay porphyry occurs but once, at the depth of 73 feet from the surface. The indurated clay is, in some beds, compact, dull, and smooth, when it is termed *clod;* in others it is glossy, unctuous, and tending to a slaty texture, and is then called *clunch;* it encloses beds of clay iron-stone, in the form of compressed balls, or broad flat masses, and some vegetable impressions and a few shells: the beds of iron-ore are five or six in number, and in the iron-stone nodules, vulgarly called ball-stone, impressions of various ferns are common. The slaty clay, called by the miners *basses*, is of a blueish black colour, usually containing pyrites, and is always either mixed with coal, or combined with petroleum. The first bed of coal occurs at the depth of 102 feet from the surface; it is not more than four inches thick, and is very sulphureous: nine other beds of a similar nature, but somewhat thicker, lie between this and the depth of nearly 400 feet; they are used only in burning lime. The first bed capable of being worked is 5 feet thick, and occurs at the depth of nearly 500 feet: below it are thirteen other beds of different quality and thickness: some possess the quality of caking: they are usually a mixture of slate-coal and pitch-coal, but rarely of cannel coal. Of these numerous beds, some are wanting in the neighbouring collieries, and the thickness of others varies considerably. The whole formation rests either upon die-earth, which is of a greyish colour, and contains petrifactions, chiefly of the Dudley fossil kind, and is so named because the beds of coal *die,* or cease, beneath it; or upon limestone, except in one place, where the greenstone trop interposes between the coal and the limestone. The principal faults, or breaks, in the strata run nearly north-east and south-west: two of these have thrown the strata on both sides of the district, from 100 to 200 feet lower than they are in the middle of it, but have not at all affected the surface; and it is in this middle tract, on account of the greater facility of working the mines, that by far the greatest quantity of coal and iron-ore is raised. The limestone formation consists of beds of limestone and sandstone, forming two mountain ranges westward of this, in the direction of north-east and south-west: at the northern extremity of one range, where it comes in contact with the coal formation, the limestone contains cavities, some of which are lined with, and some full of, petroleum. The combination of coal, iron-ore, &c., together with the advantages of water-carriage which it possesses, renders Colebrook-dale the centre of some of the most extensive iron-works in the kingdom, which consume by far the greater part of the coal raised there. In the Clee hills, from 20 to 30 miles southward of Colebrook-dale, are other coal-works, where the strata consist also of both coal and iron-stone, and dip towards the centre of the hills: this coal and iron-stone is in some places covered by a thick bed of basalt, which forms irregular ridges, higher than any other parts of the hills: the strata of coal in Brown Clee hill are much thinner than in the Tillerston, where the principal stratum is six feet thick: cannel coal is also found in this hill, and the strata are, as elsewhere, disarranged by faults. There are coal fields at Billingsley, two or three miles north-eastward of these, where a stratum of spatous iron-ore has been found; and valuable coal-works lie southward of the Clee hills, some of which produce cannel coal: coal is also found in most other parts of the hundred of Stottesden. Some miles westward of the first-mentioned coal district, pits have been sunk with success; and, indeed, out of the fifteen civil divisions of this county, ten are known to produce this valuable mineral: it is chiefly the south-western districts that are deficient of it. Nearly parallel with the Welsh border is a bank of coal strata, extending from the Dee to the Severn, and a portion of these is worked in the western and north-western parts of Shropshire, the coal having the caking quality of the Newcastle coal, and yielding a powerful heat; the principal works are near Chirk bridge: a stratum of coal seven feet thick has here been met with: spatous iron-ore, and common argillaceous iron-stone, are also found. The strata dip towards the east, and rest upon an irregular band of limestone, which in some places rises to a great height above the plain, but in others scarcely appears above the surface. In many parts, especially near Oswestry, this limestone changes into a kind of marble, and small quantities of both lead and copper are found through its whole extent.

There are mines of lead-ore of a good quality adjoining the Stiperstones, and in their vicinity; in the western parts of the county, the veins are in argillaceous schistus, and produce sulphuret of lead, both galena and steel-ore (which latter contains silver), carbonate of lead chrystallized, red lead-ore, and blende, or black-jack: the Bog mine has been worked to the depth of

150 yards, and a ton of the ore raised here yields 15 cwt. of pure lead: the ore of the White grit mine does not yield so much. At Snailbach, a vein, which is in some parts four yards in width, has been worked to the depth of 180 yards: calamine is also here met with. Ancient tools, judged to be Roman, have been found in these mines. The lead-ore is reduced at Minsterley and other places near the mines, whence it is sent by land-carriage to Shrewsbury, and there shipped, together with the raw calamine, in barges, for Bristol. There are appearances both of lead and copper in different other parts of the county. The various beds of stone are too numerous and diversified for minute description: besides those of limestone, a long range extends from Colebrook-dale, by Wenlock, to Ludlow, on the southern border of the county: the lime-works at Lilleshall are very considerable: limestone is also found in the Clee hills, and to a limited extent in the south-western districts, and in many places south of Shrewsbury, as well as in the parishes of Cardeston and Alberbury to the west of that town. At the eastern extremity of the Wrekin, and at some other lime-works, is produced a red lime, which sets very hard in water; in some parts of Shropshire limestone is found under very thick argillaceous strata. At Grinshill, seven miles north of Shrewsbury, is a noted quarry of white sand-stone, the bed of which is 60 feet thick: great use has been made of this stone in the more modern edifices of Shrewsbury and the vicinity. There are large extents of red sand-stone in the same neighbourhood, and westward and south-westward, as well as in the eastern parts of the county; and, near Bridgenorth, beds of red sand-stone are found under white sand-stone, and *vice versa.* Further south, sand-stone prevails; and at Orton bank is a stratum similar to the Bath and Portland stone, enclosed in strata of limestone. In the western district is a siliceous grit, difficult to work; but the more common stone is argillaceous. A kind of stony slate is found in the parish of Bettws, on the south-western verge of the county, and is used for covering roofs; and good flag-stones are obtained in Corndon-hill, west of Bishop's-Castle. In the Swinney mountain, near Oswestry, is a superior white sand-stone; and the same is found at Bowden quarry, in the hundred of Munslow. At Soudley, in the parish of Eaton and franchise of Wenlock, is a very good flag-stone for floors. At Pitchford, about seven miles south-east of Shrewsbury, is a red sandstone, approaching the surface in many places, from which exudes a mineral pitch: from this rock is extracted an oil, called "Betton's British oil," used medicinally. Clay slate occurs as the supporting rock of the trop formation of this county, which extends from the Wrekin to Church-Stretton. Lawley hill is in part formed of a kind of granite; but a still greater part consists of toadstone: Caer Caradoc hill is composed of a shivery kind of schistus: the Wrekin chiefly comprises reddish chert-stone, with granulated quartz imbedded in it: the hills near Oswestry are of coarse-grained sand-stone: this county affords throughout a rich field of enquiry for the mineralogist.

The rich stores of iron and lead ores, coal, and stone; the increasing manufactures, and the agricultural improvements of the district, have raised Shropshire to a high position in the scale of national importance; while its inland navigation has rendered it the emporium of the trade between England and Wales, and a grand centre of communication with the inland counties. The chief manufacture is that of iron: the number of blast furnaces for this metal between Ketley and Willey, in the great eastern coal district, in a space of about seven miles, exceeds that in any other tract of equal extent in the kingdom. The quantity of coal annually raised is nearly 300,000 tons: in Colebrook-dale, coked coal was first employed, on an extensive scale, as a substitute for charcoal, in the manufacture of iron. Various branches of the flannel manufacture are carried on near Shrewsbury; and there are mills at different places for dyeing woollen cloth. A considerable manufacture of gloves is carried on at Ludlow, chiefly for the London market, at which place paper is also made. Near Coalport, on the Severn, coloured china of all sorts, and of exquisite taste and beauty, is made: and at the same newly-formed town is a manufacture of earthenware, in imitation of that made at Etruria, commonly called Wedgwood ware. Glass is made at Donnington: earthenware, pipes, bricks and tiles, and nails, at Broseley; and at Coalport, in the neighbourhood of the last-mentioned place, is a china manufacture of great excellence: at Coalport are also manufactures of ropes and chains for the mines. There is a manufacture of carpets at Bridgenorth; paper and horse-hair seating are made at Drayton; and at nearly all the towns in the county the malting business is carried on to a very considerable extent. The staple trade of Shrewsbury is in fine flannels and Welsh webs, but it has very much declined.

The Severn, which, among British rivers, is next in magnitude and importance to the Thames, runs nearly through the centre of the county, in an irregular bending course of between 60 and 70 miles, and in a general direction of from north-west to south-east. During the whole of its course through Shropshire, this river is navigable for barges of from 20 to 80 tons' burden, which are towed up it; and for vessels called trows, which are larger, and navigate the ports lower down the river: by far the greater part of the barges are employed in exporting downwards the produce of the mines near Colebrook-dale; wines, groceries, &c., are brought up the Severn, for the consumption of this county, that of Montgomery, and others; and besides the exports of coal and iron by means of it, are those of lime, lead, flannel, grain, and cheese, with some others of minor importance. The inconveniences and interruptions attendant on the navigation of this river are very great, and are chiefly occasioned by the frequent fords and shoals that occur in its channel, which has a rapid fall; by a deficiency of water in times of drought; and by the superabundance of it in rainy seasons, when it rushes down with irresistible force. The fish found in it, within the limits of Shropshire, are salmon, flounders, a few pike, trout, graylings, perch, eels, shad, bleak, gudgeons, chub, roach, and dace (in great abundance), carp, a few lampreys, and ruff. The fishermen very commonly use a kind of canoe, being a very short wide boat, made of osiers covered with hides, and worked with a paddle, answering exactly to the description of the boats of the Britons in the time of Cæsar, and called a coracle: this bark is so light that the fisherman, on quitting the river, carries it upon his back, one end being pulled over his head, in the manner of a large basket. By the statute of the 30th of Charles II., cap. 9, the conservancy of the Severn, within the county, is vested in the county magistrates, with power to appoint one or more under-conservators. The smaller streams and brooks are extremely numerous, and the waters of almost all of them finally reach the Severn; its most important tributaries are, the Camlet, the Vyrnyw, the Tern, the Clun, the Ony, and the Teme.

The want of a navigable canal for conveying the produce of the more remote coal and iron mines of the eastern districts to the river Severn was long experienced, owing to the peculiar unevenness of the surface of the country, over which it must pass, and the impossibility of obtaining a sufficient quantity of water for lockage; until, at last, the remedy for these obstacles was supplied by a canal from the neighbourhood of the Oaken gates to the iron works at Ketley, a distance of about a mile and a half, with a fall of 73 feet, in which, instead of lockage, an inclined plane was formed. An act of parliament was then obtained for the Shropshire canal, which was finished in 1792. Immediately after the completion of this, the Shrewsbury canal was projected, for supplying that town with coal, which was conveyed thither, before its completion, only by an expensive land-carriage of about fourteen miles. The Ellesmere canal, or rather system of canals, which unites the Severn, the Dee, and the Mersey, crosses the river Ceiriog into the north-western parts of Shropshire, by an aqueduct 200 yards in length and 65 in height. At Frankton common, a branch strikes off eastward, which, after having passed close by the town of Ellesmere, proceeds by Welsh-Hampton to Fensmoss, were it divides, one branch proceeding to the town of Whitchurch, the other terminating at Prees heath, near the village of Prees. At Hordley also is a branch from the Ellesmere canal, in a south-westerly direction, which joins the Montgomeryshire canal. A canal, formed by the late Duke of Sutherland, commences at Donnington-Wood, and proceeds on a level to Pavé-lane, near Newport, a distance of seven miles: there is a branch from this to his grace's lime-works at Lilleshall. Iron railways, to convey heavy articles, have been adopted to a considerable extent in this county: the whole of the extensive iron and coal tract in the vicinity of Colebrook-dale is intersected by numerous tram-roads leading from the coal-works to the different foundries, and the wharfs on the banks of the canal and the river Severn.

The relics of antiquity are numerous and diversified. Remains of encampments, supposed by antiquaries to have been of early British formation, are to be seen in Brocard's Castle, near Church-Stretton; Bury ditches, on Tongley hill, near the village of Basford; on the Clee hills; on the hills called Caer Caradoc, two miles and a half from Church-Stretton, and the Caer Caradoc, or Gear, near Clun; at Old Port, near Oswestry, and on the Wrekin. The principal Roman stations were *Uriconium*, or *Viroconium*, now Wroxeter, which was a chief city of the Cornavii, fortified by the Romans; and *Rutunium*, at Rowton; but of the exact site of the last there is a difference of opinion: there were also *Bravinium* at Rushbury; *Sariconium*, at Bury hill; and *Usacona*, at Sheriff-Hales. The Roman station *Mediolanum* is by some fixed near Drayton, but with more probability at Meivod. Vestiges of Roman encampments and fortifications are found in the Bury walls, near Hawkstone; the Walls, near Chesterton; and the remains of the ancient city of *Uriconium*, near Wroxeter. A great Roman road enters Shropshire on the east between Crackley bank and Weston, and passes through it in a bending line, in the vicinity of Church-Stretton, which town derives its name from it, to Leintwardine in Herefordshire, on the southern border of this county: there are besides numerous minor vestiges of that people. Part of Offa's Dyke may be traced in the south-western part of Shropshire, which it enters from Knighton, in Radnorshire, and quits for Montgomeryshire, between Bishop's Castle and Newton; it is again visible in this county near Llanymynech, on the western border, whence it proceeds across the race-course, near Oswestry, and then descends to the river Ceiriog, the north-western boundary of the county, near Chirk, where it again enters Wales. There are the remains of a Danish camp near Cleobury-Mortimer. A very singular cave, in which were human bones, was discovered in 1809, in digging at the bottom of a rock, at Burncote, near Worfield: Kynaston's Cave, in the almost perpendicular side of Nesscliffe rock, and the traditions connected with it, are worthy of notice.

The number of religious houses, including collegiate establishments and hospitals, was about 47. The remains of some of them are interesting either for beauty or antiquity: the principal are those of the abbeys of Buildwas, Hales-Owen, Haughmond, Lilleshall, Much Wenlock, Shrewsbury, and White abbey, near Alberbury; and of the priories of Bromfield, Chirbury, and White Ladies. Of the ancient castles contained within the limits of the county, the great number of which has before been accounted for, some of the most remarkable that still remain, wholly, or in part, are those of Acton-Burnell; Alberbury; Bridgenorth, which was founded so far back as the year 912, by Ethelfleda, daughter of Alfred the Great; Caus; Clun; Hopton; Ludlow, so long the seat of the Lords President of the Marshes; Middle; Moreton-Corbet; Oswestry; Red Castle, Shrewsbury; Sibdon; Stoke; Wattlesborough, and Whittington. Among the most remarkable ancient mansions are those of Boscobel, where Charles II. was concealed after the battle of Worcester; White Hall, and Bellstone House: Shrewsbury Council-house is also remarkable for its antiquity. Of the more modern residences of nobility and gentry this county includes considerably more than a hundred. Shropshire contains numerous medicinal springs of various properties. At Kingley Wick, about two miles to the west of Lilleshall hill, is a strong spring of impure brine, from which salt was formerly made. There are medicinal springs of different qualities at Smeithmore and Moreton-Say, in the hundred of North Bradford; at Broseley, and at Admaston near Wellington, besides others near Ludlow; between Welbatch and Pulley common, in the vicinity of Wenlock, and on Prolley moor: that best known is Sutton Spa, about two miles to the south of Shrewsbury, and close to the village of Sutton, the waters of which are saline and chalybeate, and somewhat resemble those of Cheltenham: near Colebrook-dale is a bituminous spring of fossil tar. Numerous fossils are found among the strata of this county, particularly in the Colebrook-dale coal district. The *reseda luteola*, or dyers' weed, which affords a beautiful yellow dye, grows wild in many parts of the county; and the *berberis vulgaris*, or common barberry, is also occasionally found in a similar uncultivated state.

SHROTON, county of DORSET.—See IWERNE-COURTNAY.

SHUCKBURGH, LOWER (*ST. JOHN THE BAPTIST*), a parish, in the union of SOUTHAM, Burton-Dassett Division of the hundred of KINGTON, Southern Division of the county of WARWICK, 5 miles (E.) from Southam; containing 165 inhabitants. The living is a perpetual curacy, annexed, with that of Priors'-Marston, to the vicarage of Priors'-Hardwick. The Oxford canal passes through the parish.

SHUCKBURGH, UPPER (*ST. JOHN THE BAPTIST*), a parish, in the union of SOUTHAM, Southam Division of the hundred of KNIGHTLOW, Southern Division of the county of WARWICK, 5½ miles (E.) from Southam; containing 40 inhabitants. The living is a perpetual curacy; net income, £30; patron, Sir F. Shuckburgh, Bart. The church contains some finely-executed monuments to the Shuckburgh family; and the chancel window is embellished with a figure of St. John, painted by Mr. Eginton, of Birmingham. On the western boundary of the parish is Beacon Hill, a lofty elevation commanding, in a clear atmosphere, fine views of the Malvern hills and the Wrekin.

SHUDY-CAMPS (*ST. MARY*), a parish, in the union of LINTON, hundred of CHILFORD, county of CAMBRIDGE, 4¼ miles (E. S. E.) from Linton; containing 418 inhabitants. The living is a discharged vicarage, valued in the king's books at £9; present net income, £146; patrons and impropriators, Master and Fellows of Trinity College, Cambridge. Here is a Sunday National school.

SHURDINGTON, GREAT (*ST. PAUL*), a parish, in the union of CHELTENHAM, Upper Division of the hundred of DUDSTONE and KING'S-BARTON, Eastern Division of the county of GLOUCESTER, 3 miles (S. W.) from Cheltenham; containing 99 inhabitants. The living is annexed to the vicarage of Badgeworth. The impropriate tithes have been commuted for a rent-charge of £509. 10. 5., and the vicarial for £339. 13. 7., subject to the payment of rates, which on the average have respectively amounted to £59. 10. 4. and £39. 13. 7.; the glebe comprises 30 acres, valued at £37. 5. per annum. There is also a rent-charge of £6. 9. payable to the rector of Oddington. The church has a very handsome steeple, and a north aisle, called the Hatherly aisle. The village is pleasantly situated on the new road from Cheltenham, through Painswick and the vale of Rodborough, to Bath. The soil is a deep clay, and the land is chiefly in pasture. On opening a large tumulus here a stone coffin was found, at the depth of sixteen feet, which contained the body of a man, with a helmet almost consumed by rust.

SHURDINGTON, LITTLE, a hamlet, in the parish of BADGEWORTH, Upper Division of the hundred of DUDSTONE and KING'S-BARTON, Eastern Division of the county of GLOUCESTER: the population is returned with the parish.

SHURLACH, a township, in the parish of DAVENHAM, union and hundred of NORTHWICH, Southern Division of the county of CHESTER, 1¾ mile (E. S. E.) from Northwich; containing 98 inhabitants. The Grand Trunk canal passes in the vicinity.

SHUSTOCK (*ST. CUTHBERT*), a parish, in the union of MERIDEN, Atherstone Division of the hundred of HEMLINGFORD, Northern Division of the county of WARWICK, 2¾ miles (E. N. E.) from Coleshill; containing 634 inhabitants. The living is a discharged vicarage, valued in the king's books at £5. 7.; present net income, £258: it is in the patronage of the Crown; impropriator, Lord Leigh. Thomas and Charles Huntback, in 1714, gave certain houses and land for the endowment of a school for poor children, and an almshouse for six widows. There is another school, in which twelve children are instructed at the expense of a lady. Blythe Hall was the residence of the celebrated antiquary, Sir William Dugdale, who purchased that manor of Sir Walter Aston, in the 1st of Charles I., and here compiled "The Antiquities of Warwickshire:" he died on the 10th of February, 1685.

SHUTE (*ST. MICHAEL*), a parish, in the union of AXMINSTER, hundred of COLYTON, Honiton and Southern Divisions of the county of DEVON, 2 miles (N.) from Colyton; containing 617 inhabitants. The living is annexed, with that of Monkton, to the vicarage of Colyton. The church contains a memorial of Charles Bickford Templer, Esq., who was lost in the wreck of the Halsewell East Indiaman, in 1786.

SHUTFORD, EAST, a chapelry, in the parish of SWALCLIFFE, union and hundred of BANBURY, county of OXFORD, 5 miles (W. by N.) from Banbury; containing 31 inhabitants. The chapel is dedicated to St. Martin.

SHUTFORD, WEST, a township, in the parish of SWALCLIFFE, union and hundred of BANBURY, county of OXFORD, 5½ miles (W.) from Banbury; containing 431 inhabitants.

SHUTTINGTON, a parish, in the union of TAMWORTH, Tamworth Division of the hundred of HEMLINGFORD, Northern Division of the county of WARWICK, 3¾ miles (E. by N.) from Tamworth; containing 147 inhabitants. The living is a perpetual curacy; net income, £318; patron, Earl of Essex; impropriators, certain Trustees. The Coventry canal crosses the south-western angle of the parish.

SHUTTLEHANGER, a chapelry, in the parish of STOKE-BRUERNE, union of TOWCESTER, hundred of CLELEY, Southern Division of the county of NORTHAMPTON, 2¾ miles (E. N. E.) from Towcester; containing 325 inhabitants.

SIBBERTOFT (*ST. HELEN*), a parish, in the union of MARKET-HARBOROUGH, hundred of ROTHWELL, Northern Division of the county of NORTHAMPTON, 5 miles (S. W.) from Market-Harborough; containing 402 inhabitants. The living is a vicarage, annexed to that of Welford, and valued in the king's books at £6. 4. 9½. There is a place of worship for Wesleyan Methodists. A spot here, called Castle-yard, is supposed to be the site of an ancient castle.

SIBBERTSWOLD (*ST. ANDREW*), a parish, in the union of DOVOR, hundred of BEWSBOROUGH, lathe of ST. AUGUSTINE, Eastern Division of the county of KENT, 6¼ miles (N. W.) from Dovor; containing 363 inhabitants. The living is a vicarage, with that of Coldred annexed, valued in the king's books at £6; present net income, £255; patron, Archbishop of Canterbury. The church is principally in the early style of English architecture. A day and Sunday school is supported by subscription. Three Barrow Down, in this parish, is so named from three large tumuli or barrows, connected with each other by deep trenches, and occupying the hill between Denhill-terrace and the edge of Barham Downs. To the east of Long Lane farm are also other lines of intrenchment, with similar barrows or tumuli, supposed to be of Roman origin.

SIBDON-CARWOOD, a parish, in the union of CHURCH-STRETTON, hundred of PURSLOW, Southern Division of the county of SALOP, 7 miles (S. E. by E.) from Bishop's-Castle; containing 63 inhabitants. The living is a perpetual curacy; net income, £30; patron and impropriator, James Baxter, Esq.

SIBFORD-FERRIS, a hamlet, in the parish of SWALCLIFFE, union of BANBURY, hundred of BLOXHAM, county of OXFORD, 7¼ miles (W. S. W.) from Banbury; containing 248 inhabitants.

SIBFORD-GOWER, a hamlet, in the parish of SWALCLIFFE, union of BANBURY, hundred of BLOXHAM, county of OXFORD, 7¾ miles (W. by S.) from Banbury; containing 507 inhabitants. A chapel has been built here containing 500 sittings, 340 of which are free, the Incorporated Society having granted £300 in aid of the expense. One-third of the rents of the poor's estate is paid to a schoolmaster for teaching 40 children.

SIBSEY (*St. Margaret*), a parish, in the union of BOSTON, Western Division of the soke of BOLINGBROKE, parts of LINDSEY, county of LINCOLN, 5¼ miles (N. N. E.) from Boston; containing 1364 inhabitants. The living is a discharged vicarage, valued in the king's books at £11. 11. 3.; present net income, £315: it is in the patronage of the Crown. There is a place of worship for Wesleyan Methodists. A school is endowed by the Monson family; and four others are supported from the rents of lands in the parish.

SIBSON, a hamlet, in the parish of STIBBINGTON, hundred of NORMAN-CROSS, county of HUNTINGDON, 1½ mile (S. E.) from Wansford: the population is returned with the parish.

SIBSON, or SIBSTONE (*St. Botolph*), a parish, in the hundred of SPARKENHOE, Southern Division of the county of LEICESTER, 4 miles (W. S. W.) from Market-Bosworth; containing 427 inhabitants. The living is a rectory, valued in the king's books at £15. 18. 11½.; present net income, £962; patrons, Master and Fellows of Pembroke College, Oxford.

SIBTHORPE (*St. Peter*), a parish, in the union of BINGHAM, Southern Division of the wapentake of NEWARK and of the county of NOTTINGHAM, 6¾ miles (S. S. W.) from Newark; containing 144 inhabitants. The living is a donative, in the patronage of the Duke of Portland; net income, £20. The church was originally much larger than it is at present: the north and south aisles have been taken down, so that the pillars and lofty arches of the nave are now worked into the outer wall of the building. In the reign of Edward II., Thomas de Sibthorpe founded a chantry in the church, and subsequently erected it into a college for a warden, nine chaplains, three clerks, and four choristers; he also added four chapels, in honour of St. Anne, St. Katharine, St. Margaret, and St. Mary; the revenue, at the dissolution, was estimated at £31. 1. 2. Thomas Secker, D.D., Archbishop of Canterbury, was born here in 1693: he died in 1678.

SIBTON (*St. Peter*), a parish, in the union and hundred of BLYTHING, Eastern Division of the county of SUFFOLK, 2 miles (N. W. by W.) from Yoxford; containing 498 inhabitants. The living is a discharged vicarage, valued in the king's books at £8. 8. 4.; present net income, £200; patron, R. Sayer, Esq.; impropriators, the Landowners. Here are extensive remains of a Cistercian abbey, founded in 1149, by William de Cayneto; it was dedicated to the Blessed Virgin Mary, and at the dissolution possessed a revenue of £250. 15. 7. There was an hospital at the gate of the abbey.

SICKLINGHALL, a township, in the parish of KIRKBY-OVERBLOWS, Upper Division of the wapentake of CLARO, West Riding of the county of YORK, 3 miles (W.) from Wetherby; containing 212 inhabitants. There is a place of worship for Wesleyan Methodists.

SIDBURY (*St. Giles*), a parish, in the union of HONITON, hundred of EAST BUDLEIGH, Woodbury and Southern Divisions of the county of DEVON, 2½ miles (N. N. E.) from Sidmouth; containing 1725 inhabitants. The living is a vicarage, in the patronage of the Dean and Chapter of Exeter (the appropriators), valued in the king's books at £28; present net income, £476. There is a place of worship for Independents. This is a decayed market-town: there are still two annual fairs for cattle, on the Tuesday before Ascension-day and at Michaelmas. A National school has been established. On the Barton, or manor, of Sand, is an old mansion, with the inscription *Hortus Johannis Capelli* over the garden door.

SIDBURY (*Holy Trinity*), a parish, in the union of BRIDGENORTH, hundred of STOTTESDEN, Southern Division of the county of SALOP, 5¼ miles (S. S. W.) from Bridgenorth; containing 103 inhabitants. The living is a discharged rectory, valued in the king's books at £4. 17. 8½.; present net income, £227; patron, Earl of Shrewsbury.

SIDCUP, a hamlet, in the parish of FOOT'S-CRAY, union of BROMLEY, hundred of RUXLEY, lathe of SUTTON-AT-HONE, Western Division of the county of KENT, 3½ miles (S. E.) from Eltham: the population is returned with the parish. Here is a paper-mill; and there are several gentlemen's seats in the neighbourhood.

SIDDINGTON, a chapelry, in the parish of PRESTBURY, union and hundred of MACCLESFIELD, Northern Division of the county of CHESTER, 5 miles (N. by W.) from Congleton; containing 479 inhabitants. The living is a perpetual curacy; net income, £106; patron and impropriator, E. D. Davenport, Esq. The chapel is partly built of wood and plaster, and partly of brick. John Fowden, in 1712, founded a school, and endowed it with £8 per annum; and there is a school for girls, in which 17 are instructed by an endowment.

SIDDINGTON (*St. Mary and St. Peter*), a parish, in the union of CIRENCESTER, hundred of CROWTHORNE and MINETY, Eastern Division of the county of GLOUCESTER, 1¾ mile (S. S. E.) from Cirencester; containing 409 inhabitants. This parish is intersected by the river Churn and the Thames and Severn canal, and was formerly celebrated for the manufacture of pottery, which was carried on extensively, but within the last 30 years has been discontinued: the site of the ancient works is still called the "Pottery Court." The living of St. Mary's is a rectory, with which the discharged vicarage of St. Peter is united, the former valued in the king's books at £8. 12. 1., and the latter at £5. 12. 3½.; present net income of the united livings, £429: they are in the patronage of the Crown. The church, dedicated to St. Peter, combines several portions of the various English styles of architecture with some Norman details, of which latter the south door and the arch leading into the chancel are fine specimens. A day and Sunday school is supported by subscription. The celebrated Dr. George Bull, Bishop of St. David's, was for nearly thirty years incumbent of St. Peter's, during which period he composed the principal part of his writings: he was born at Wells, in 1634, and died at St. David's in 1709.

SIDE (*St. Mary*), a parish, in the union of Cirencester, hundred of Rapsgate, Eastern Division of the county of Gloucester, 7 miles (E.) from Painswick; containing 50 inhabitants. The living is a rectory, valued in the king's books at £3. 18. 4.; patron, W. Lawrence, Esq. The tithes have been commuted for a rent-charge of £83. 10., subject to the payment of rates, which on the average have amounted to £6. 5.; the glebe comprises 30 acres, valued at £22. 10. per annum.

SIDESTRANDS (*St. Michael*), a parish, in the union of Erpingham, hundred of North Erpingham, Eastern Division of the county of Norfolk, 3 miles (S. E. by E.) from Cromer; containing 160 inhabitants. The living is a discharged rectory, valued in the king's books at £5. 10.: it is in the patronage of the Crown, in right of the Duchy of Lancaster, and the Heirs of Gen. Carpenter. The tithes have been commuted for a rent-charge of £106. 4., subject to the payment of rates, which on the average have amounted to £25; the glebe comprises 1 acre, valued at £1. 13. per annum. The rector receives a rent-charge of £7. 10. out of the tithes of Trimingham.

SIDLESHAM (*St. Mary*), a parish, in the union of West Hampnett, hundred of Manhood, rape of Chichester, Western Division of the county of Sussex, 4 miles (S.) from Chichester; containing 1002 inhabitants. The parish is situated on the road to Selsey from Chichester, and is bounded on the south by Pagham harbour. It has a convenient quay, near which is a superior tide-mill, which for justness of principle is equal to any in the kingdom: it was erected at a considerable expense, by the late Mr. Woodroffe Drinkwater, under the direction of Benjamin Basle, the inventor and constructor of the machinery, which will grind a load of corn in an hour. The living is a discharged vicarage, endowed with a portion of the great tithes, and valued in the king's books at £7. 10. 10.; present net income, £186; patron, Prebendary of Sidlesham in the Cathedral Church of Chichester; appropriators of the remainder of the great tithes, Prebendaries of Sidlesham and Heighley. The church is in the early style of English architecture, and consists of a nave, transepts, and side aisles, with an embattled tower at the west end. There is a place of worship for Wesleyan Methodists.

SIDMONTON, county of Southampton. — See Sydmonton.

SIDMOUTH (*St. Nicholas*), a sea-port, market-town, and parish, in the union of Honiton, hundred of East Budleigh, Woodbury and Southern Divisions of the county of Devon, 13½ miles (E. S. E.) from Exeter, and 158 (W. S. W.) from London; containing 3126 inhabitants. The earliest account of this place is in the time of William the Conqueror, who bestowed the manor on the monastery of St. Michael in Normandy, from which, during the subsequent wars with France, it was alienated to the abbey of Sion, and has since belonged to various proprietors. In the reign of Edward III., the town appears to have been governed by a portreeve, and to have furnished that monarch, in his attack on Calais, with two vessels and 25 seamen. It is said to have been formerly famous for its fishery, and to have traded with Newfoundland; but the harbour, which then existed, is supposed, from the discovery of an old anchor and fragments of vessels, to have been in the Ham meadow, near the town; it has been choked up with sand and pebbles, and only boats and fishing-smacks can now approach the shore. In 1836 an act was passed for making and maintaining a harbour and other works here. To its great attractions as a watering-place is its present prosperity owing, the extent of which may be estimated by the circumstance of the population having increased more than 1000 since 1821. The town is situated at the entrance of a narrow valley, on a small stream called the Sid, from which it derives its name: the surrounding country is remarkably picturesque and beautiful, and the hills bounding it on the east and west sides are of very great altitude, and extremely precipitous, terminating abruptly on the shore, and, in addition to their scenic beauty, affording great shelter to the town, which, though irregularly built, is very neat, and skirted with numerous detached residences, altogether occupying a site of considerable extent. The inns and boarding-houses are of the best description, and every accommodation is provided for persons requiring sea-bathing. On the beach is a public walk more than half a mile in length, fronting which are the warm baths, public rooms, library, &c. Assemblies and concerts take place during the season. The markets, on Tuesday and Saturday, are well supplied; and there are fairs annually on Easter Monday and Tuesday, and the third Monday in September. Petty sessions are held on the first Monday in every month, by two resident magistrates, who include Salcombe, Branscombe, Sidbury, and Sidford, within their jurisdiction; and, at a court leet and baron held annually by the lord of the manor, two constables and tythingmen are appointed.

The living is a vicarage, valued in the king's books at £18. 15. 5.; present net income, £481; patron, incumbent, and impropriator, Rev. William Jenkins. The church is an ancient structure, with a well-built tower, and contains 160 free sittings, the Incorporated Society having granted £200 in aid of the expense. Among the monuments is one to the memory of Dr Currie, the distinguished biographer of Robert Burns. A new chapel has been built here, the patronage of which is vested in Sir John Kennaway, Bart., and others. There are places of worship for Baptists, Independents, and Unitarians. A National school is supported partly by a small endowment, and another partly by subscription. The poor of the parish are relieved by bequests from Anthony Isaack, in 1639, and John Minshull, in 1663, and others of small amount; and some charitable institutions are supported by donations and subscriptions from the inhabitants and visiters. A fraternity of Augustine monks is said to have once existed near Sidmouth, and there are still the remains of a building, which tradition affirms to have been a chapel of ease at a period when Sidmouth belonged to the parish of Otterton, on the road to which place there is an ancient stone cross. A fort, mounting four pieces of ordnance, formerly stood near the town. At Woolbrook cottage, in the vicinity, His Royal Highness the late Duke of Kent died. Sidmouth gives the title of Viscount to the family of Addington.

SIGGLESTHORNE (*St. Lawrence*), a parish, in the union of Skirlaugh, Northern Division of the wapentake of Holderness, East Riding of the county of York; containing 578 inhabitants, of which number, 204 are in the township of Sigglesthorne, 10 miles (E. N. E.) from Beverley. The living is a rectory, valued in the king's books at £31. 1. 3.; present net income, £685: it is in the patronage of the Crown. Marmaduke Constable, in 1810, gave £200 to be applied for the education of poor children: the annual income is £14., and fifteen children are instructed; the school is further aided by subscription.

SIGHILL, a township, in the parish of Earsdon, union of Tynemouth, Eastern Division of Castle ward, Southern Division of the county of Northumberland, $6\frac{3}{4}$ miles (N. W.) from North Shields; containing 985 inhabitants.

SIGNET, a joint hamlet with Upton, in the parish of Burford, union of Witney, hundred of Bampton, county of Oxford, $1\frac{1}{4}$ mile (S. by W.) from Burford: the population is returned with Upton.

SIGSTON, KIRBY (*St. Lawrence*), a parish, in the union of North-Allerton, wapentake of Allertonshire, North Riding of the county of York; containing 343 inhabitants, of which number, 131 are in the township of Kirby-Sigston, $3\frac{1}{2}$ miles (E. by N.) from North-Allerton. The living is a rectory, valued in the king's books at £12. 13. 4.

SILCHESTER (*St. Mary*), a parish, in the union of Basingstoke, hundred of Holdshott, Basingstoke and Northern Divisions of the county of Southampton, $7\frac{1}{2}$ miles (N.) from Basingstoke; containing 414 inhabitants. This place, situated near the border of Berkshire, was the *Caer Seiont*, or *Segont*, of the Britons, and the *Vinconum* of the Romans, having been one of the principal stations of the latter in the south of England. The usurper Constantine was invested with the purple in that city, in the year 407. About 493, it was destroyed by the Saxon chief, Ælla, on his march to Bath, from the coast of Sussex, where he had made his landing. The enclosed area is in the form of an irregular octagon, nearly a mile and a half in circumference. The walls are most perfect on the south side, being in some places nearly 20 feet high. About 150 yards from the north-east angle of the walls is a Roman amphitheatre, now covered with trees; and about a mile and a half to the north-west, near a village called the Soak, are some remains of a camp. The living is a rectory, valued in the king's books at £9. 6. $0\frac{1}{2}$.; present net income, £308; patron, Duke of Wellington. The church is partly in the early, and partly in the later, style of English architecture. Here is a National school. Silchester confers the title of Baron upon the family of Pakenham, Earls of Longford.

SILEBY (*St. Mary*), a parish, in the union of Barrow-upon-Soar, hundred of East Goscote, Northern Division of the county of Leicester, $1\frac{1}{2}$ mile (E.) from Mountsorrel; containing 1491 inhabitants, many of whom are employed in the manufacture of hosiery. The living is a discharged vicarage, valued in the king's books at £8. 15. 5.; present net income, £158; patron and impropriator, W. Pochin, Esq. The church has a highly enriched tower. There are places of worship for Baptists and Wesleyan Methodists. The river Soar bounds this parish on the west. The Rev. William Staveley, in 1702, founded a small free school, with an endowment of £5 per annum, to which George Pochin, in 1706, bequeathed £50.

SILFIELD, a township, in the parish of Wymondham, hundred of Forehoe, Eastern Division of the county of Norfolk, $1\frac{1}{2}$ mile (S. E.) from Wymondham; containing 593 inhabitants.

SILKSTONE (*All Saints*), a parish, in the wapentake of Staincross, West Riding of the county of York; containing 16,519 inhabitants, of which number, 1010 are in the township of Silkstone, $4\frac{1}{4}$ miles (W. by S.) from Barnsley. The living is a vicarage, valued in the king's books at £17. 13. 4.; present net income, £272; patron, Archbishop of York. The church is partly Norman, and partly of later date, with some elegant screen-work. There is a place of worship for Wesleyan Methodists. In this parish the various kinds of linen are manufactured; wire-drawing is also carried on, and there are extensive iron-foundries and collieries in the neighbourhood. The Rev. John Clarkson, in 1734, assigned his interest in the profits arising from certain messuages and lands, under lease for 500 years, for the education of poor children of the township of Silkstone; the income is £28. 15. per annum.

SILKSWORTH, a township, in the parish of Bishop-Wearmouth, union of Houghton-le-Spring, Northern Division of Easington ward, and of the county palatine of Durham, 3 miles (S. W. by S.) from Sunderland; containing 252 inhabitants.

SILK-WILLOUGHBY, county of Lincoln.—See WILLOUGHBY, SILK.

SILPHO, a joint township with Harwood-Dale, in the parish of Hackness, union of Scarborough, liberty of Whitby-Strand, North Riding of the county of York, $6\frac{1}{4}$ miles (N. W. by W.) from Scarborough: the population is included in the return for Harwood-Dale.

SILSDEN, a chapelry, in the parish of Kildwick, Eastern Division of the wapentake of Staincliffe and Ewcross, West Riding of the county of York, 4 miles (N. by W.) from Keighley; containing 2137 inhabitants. The living is a perpetual curacy; net income, £121; patron and impropriator, Earl of Thanet. The chapel is dedicated to St. James. There is a place of worship for Wesleyan Methodists; a school is partly supported by the Earl of Thanet, a small endowment, and £8 from the poor's rate; there is also a Sunday National school. The manufacture of woollen cloth, calico, and nails, is carried on.

SILSOE, a chapelry (formerly a market-town), in the parish of Flitton, union of Ampthill, hundred of Flitt, county of Bedford, 10 miles (S. by E.) from Bedford; containing 726 inhabitants. The chapel, dedicated to St. James, has a tower in the decorated style, with a wooden spire, and over the altar a representation of the Adoration of the Shepherds, painted and presented by Mrs. Mary Lloyd. Two day schools and an infants' school are supported by Earl de Grey. A market, now disused, was granted in 1319; also a fair, which is held on the festival of St. Peter and St. James, besides which there is another on Sept. 21st.

SILTON (*St. Nicholas*), a parish, in the union of Mere, hundred of Redlane, Shaston Division of the county of Dorset, 3 miles (S. S. W.) from Mere; containing 396 inhabitants. The living is a rectory, valued in the king's books at £7. 9. 7.; patron, H. C. Sturt, Esq. The tithes have been commuted for a rent-charge of £330, subject to the payment of rates, which on the average have amounted to £45; the glebe comprises 61 acres, valued at £122 per annum. A school is supported by the rector.

SILTON, NETHER, a chapelry, in the parish of Leak, union of North-Allerton, wapentake of Birdforth, North Riding of the county of York, 8 miles (N by E.) from Thirsk; containing 179 inhabitants. The living is a perpetual curacy, annexed to the vicarage of Leak. There is a place of worship for Wesleyan Methodists; also a school partly supported by F. Hicks, Esq.

SILTON, OVER (*All Saints*), a parish, in the union of North-Allerton, wapentake of Birdforth, North Riding of the county of York; containing 263 inhabitants, of which number, 111 are in the township of Over Silton, 8½ miles (N. by E.) from Thirsk. The living is a perpetual curacy; net income, £69; patrons and impropriators, Master and Fellows of Trinity College, Cambridge.

SILVERDALE, a chapelry, in the parish of Warton, hundred of Lonsdale, south of the sands, Northern Division of the county palatine of Lancaster, 4¾ miles (W. by S.) from Burton-in-Kendal; containing 240 inhabitants. This place is situated on the bay of Morecambe. The living is a perpetual curacy; net income, £47; patron, Vicar of Warton. A chapel was erected in 1830, containing 190 sittings, 100 of which are free, the Incorporated Society having granted £75 in aid of the expense.

SILVERLEY (*All Saints*), a parish, in the union of Newmarket, hundred of Cheveley, county of Cambridge, 3¾ miles (E.) from Newmarket: the population is returned with the parish of Ashley. The living is a vicarage, united to the rectory of Ashley, and valued in the king's books at £7. 17. 3½.; impropriator, Marquess of Bute.

SILVERSTONE (*St. Michael*), a parish, in the union of Towcester, hundred of Greens-Norton, Southern Division of the county of Northampton, 3 miles (S. by W.) from Towcester; containing 947 inhabitants. The living is annexed, with that of Whittlebury, to the rectory of Greens-Norton. There is a place of worship for Wesleyan Methodists.

SILVERTON (*St. Mary*), a parish, in the union of Tiverton, hundred of Hayridge, Cullompton and Northern Divisions of the county of Devon, 5¼ miles (S. W. by W.) from Cullompton; containing 1389 inhabitants. The living is a rectory, valued in the king's books at £51. 8. 4.; present net income, £589; patrons, Earl of Ilchester and the Hon. P. C. Wyndham. The church is a handsome specimen of the later style of English architecture: adjoining it are some slight remains of an ancient chapel, dedicated to the Virgin Mary. There is a place of worship for Wesleyan Methodists. A market was formerly held here weekly: there are cattle fairs on the second Thursday in February, the first Thursdays in March and July, and a pleasure fair on September 4th. The free school was founded in 1724, by John Richards, who gave £1200 for its erection and support; the present annual income is £90, and the number of boys taught about 70. Forty girls are also instructed in a school supported by voluntary contributions and an annuity of £2. 10., the gift of the Rev. Richard Troyte in 1730.

SILVINGTON (*St. Michael*), a parish, in the union of Cleobury-Mortimer, hundred of Overs, Southern Division of the county of Salop, 6 miles (N. W.) from Cleobury-Mortimer; containing 30 inhabitants. The living is a discharged rectory, valued in the king's books at £3. 6. 8.; present net income, £100: it is in the joint patronage of Lucy Fowler and Theophilus Salwey, Esq.

SIMONBURN (*St. Simon*), a parish, in the union of Hexham, North-Western Division of Tindale ward, Southern Division of the county of Northumberland; containing 1135 inhabitants, of which number, 600 are in the township of Simonburn, 9 miles (N. W. by N.) from Hexham. This parish was formerly the largest in the county, but in 1814 was divided, pursuant to an act obtained in 1811, into six parishes and rectories, the livings of all which are in the gift of the Governors of Greenwich Hospital, to which institution the manor of the ancient parish belongs, and from its funds the churches were erected. None but chaplains in the navy, who have served ten years, or lost a limb in the service, can be inducted to these benefices; they are not allowed to hold any other preferment, but by an act passed in 1820, may receive their half-pay. The living is a rectory, valued in the king's books at £34. 6. 3., for the ancient parish; present net income, £426; patrons, Governors of Greenwich Hospital. The church was repaired and beautified in 1821: it contains monuments to the family of Allgood. The ancient parish was about 33 miles in length and 14 in breadth, diversified with mountains and valleys of picturesque character: the Roman wall passed on the northern side of it, and within its limits coal is abundant, and iron-ore was formerly obtained. A very small portion of its surface is in tillage, the land being chiefly applied to depasturing sheep, and Scotch and Irish cattle. Giles Heron left an estate, now let for £180 per annum, for teaching and apprenticing poor children, and affording relief to adults belonging to the parishes of Simonburn and Wark: 80 boys are instructed gratuitously by a master who has a salary of £45, and the use of a residence. The castle was entirely destroyed in expectation of finding some hidden treasure, but part of the west end was rebuilt in 1766. In 1735, a stone inscribed VLPI. SABI., to Ulpius and Sabinus, Roman lieutenants in Britain, was found in taking down part of the rectory-house.

SIMONDSLEY, a township, in the parish and union of Glossop, hundred of High Peak, Northern Division of the county of Derby, 8 miles (N. by W.) from Chapel-en-le-Frith; containing 454 inhabitants.

SIMONSTONE, a township, in the parish of Whalley, union of Burnley, Higher Division of the hundred of Blackburn, Northern Division of the county palatine of LANCASTER, 4¾ miles (W. by N.) from Burnley; containing 440 inhabitants.

SIMONSWOOD, a township, in the parish of WALTON-ON-THE-HILL, union of ORMSKIRK, hundred of WEST DERBY, Southern Division of the county palatine of LANCASTER, 5 miles (S. by E.) from Ormskirk; containing 411 inhabitants.

SIMONWARD, county of CORNWALL.— See BREWARD, ST.

SIMPSON (*ST. NICHOLAS*), a parish, in the union of NEWPORT-PAGNELL, hundred of NEWPORT, county of BUCKINGHAM, 1½ mile (N. by E.) from Fenny-Stratford; containing 470 inhabitants. This parish is bounded on the south-east by a branch of the river Ouse, and is intersected by the Grand Junction canal. The living is a rectory, valued in the king's books at £17. 6. 8.; patron, Sir John Hanmer, Bart. The church is a handsome structure, in the Norman style of architecture; in the chancel is a fine monument to Sir T. Salden Hanmer, Bart., and his lady, with several others to different members of that family, who are buried in the church. Thomas Pigot, in 1573, bequeathed property now producing a rental of about £50, which is appropriated to the relief of the poor.

SINDERBY, a township, in the parish of PICKHILL, union of THIRSK, wapentake of HALLIKELD, North Riding of the county of YORK, 6¼ miles (W. by S.) from Thirsk; containing 93 inhabitants.

SINGLEBOROUGH, a hamlet, in the parish of GREAT HORWOOD, union of WINSLOW, hundred of COTTESLOE, county of BUCKINGHAM, 3 miles (N.) from Winslow; containing 110 inhabitants.

SINGLETON, a chapelry, in the parish of KIRKHAM, union of the FYLDE, hundred of AMOUNDERNESS, Northern Division of the county palatine of LANCASTER, 3 miles (E. by S.) from Poulton; containing 499 inhabitants. The living is a perpetual curacy; net income, £110; patron, H. Hornby, Esq. The chapel, consecrated in 1754, is dedicated to St. Anne. There is a place of worship for Roman Catholics; also a National school. A great fair for sheep and cattle is held on September 21st.

SINGLETON, a parish, in the union of WEST HAMPNETT, hundred of WESTBURN and SINGLETON, rape of CHICHESTER, Western Division of the county of SUSSEX, 6 miles (N.) from Chichester; containing 563 inhabitants. The village is situated on the road from London to Chichester, by way of Midhurst. The living is a discharged rectory, annexed to the vicarage of West Dean and chapelry of Binderton, and valued in the king's books at £6. 13. 4. The church is in the later style of English architecture, and consists of a nave, chancel, and side aisles, with a square tower at the west end. There are schools for boys and girls, the latter solely supported by Mrs. Colonel Wyndham. Henry Smith, about 1640, left land to the poor now producing about £60 per annum. St. Roche's, or Rook's, Hill, an elevation 702 feet above the level of the sea, skirts the southern boundary of the parish; and near its summit is an ancient encampment, known by the name of the Trundle, a corruption of Roundal, indicating its circular form. It includes an area of about five acres, has a deep fosse, and an outer and inner vallum: the inner vallum is raised to the height of about four feet all round the edge of the enclosure: in the centre are remains of a cell, now level with the ground, the walls of which are composed of flints cemented with mortar so hard as to render them almost immoveable: its size is 14 feet by 11. This was one of the stations fixed upon by the Board of Ordnance, in 1791, for taking the trigonometrical survey of the coast. The view is extensive, embracing the English Channel, Isle of Wight, Portsmouth, Hayling, and Thorney Islands, the beautiful country in the foreground, with the city of Chichester and lofty spire of the cathedral forming a conspicuous object.

SINNINGTON, a parish, in the union of PICKERING, partly in PICKERING lythe, and partly in the wapentake of RYEDALE, North Riding of the county of YORK; containing 584 inhabitants, of which number, 340 are in the township of Sinnington, 4¾ miles (W. N. W.) from Pickering. The living is a perpetual curacy; net income, £84; patron and impropriator, the Master of Hemsworth school. There is a place of worship for Wesleyan Methodists. An annuity of £5 is paid from the estate of Lady Lumley, for the free instruction of the poor.

SINWELL, a tything, in the parish of WOTTON-UNDER-EDGE, Upper Division of the hundred of BERKELEY, Western Division of the county of GLOUCESTER: the population is returned with the parish.

SISLAND (*ST. MARY*), a parish, in the union of LODDON and CLAVERING, hundred of LODDON, Eastern Division of the county of NORFOLK, 6¼ miles (N. by E.) from Bungay; containing 85 inhabitants. The living is a discharged rectory, valued in the king's books at £4. 3. 9.; patron and incumbent, Rev. William Hobson. The tithes have been commuted for a rent-charge of £132. 15., subject to the payment of rates, which on the average have amounted to £21; the glebe comprises 17 acres, valued at £17. 4. 4. per annum.

SISTON (*ST. ANNE*), a parish, in the union of KEYNSHAM, hundred of PUCKLECHURCH, Western Division of the county of GLOUCESTER, 6½ miles (E. by N.) from Bristol; containing 973 inhabitants, many of whom are employed in making pins. The living is a rectory, valued in the king's books at £5. 14. 4½.; present net income, £323; patron, F. Trotman, Esq. Here is a National school.

SITHNEY (*ST. SITHNEY*), a parish, in the union of HELSTON, hundred of KERRIER, Western Division of the county of CORNWALL, 2 miles (N. W.) from Helston; containing 2772 inhabitants. This parish, which is situated on the shore of the English Channel, is bounded on the south-west by Mount's bay, and includes the chief part of Porthleven fishing cove, and nearly the whole of Loe pool. The living is a discharged vicarage, valued in the king's books at £19. 11. 5½.; present net income, £368; patron, Bishop of Exeter; impropriator, S. T. Spry, Esq. The church is an ancient structure, and contains some fine remains of stained glass. There are places of worship for Baptists and Wesleyan Methodists; also two schools supported by subscription, and another partly supported by the curate. At Truthal are the remains of an ancient chapel; and there was formerly an hospital, dedicated to St. John, of which there are no remains. On Longston Downs is a rude pile of stones, one of which was formerly a logan rock, called Mên-Amber; it is 11 feet long, 6 feet wide, and 4 feet thick. Several stone battle-axes were found at Venton Vedna, in the year 1799.

SITTINGBOURNE (*ST. MICHAEL*), a parish (formerly a corporate and market-town), in the union and hundred of MILTON, Upper Division of the lathe of SCRAY, Eastern Division of the county of KENT, 10 miles (E. N. E.) from Maidstone; containing 2182 inhabitants. The only incident worthy of notice in the ancient history of this town is the fact that Henry V. was entertained at the Red Lion here, by one John Northwood, a gentleman then resident in the vicinity, at the small expense of nine shillings and ninepence; and several other English monarchs have occasionally conferred a like distinction on the place. The town is situated on the main road from Canterbury to London, and consists of one long wide street. By means of Milton creek, which bounds the parish on the north, and is navigable at Crown quay, hoys sail hence to London. In the neighbourhood are a manufactory for oil cake and a cement mill. A weekly market and two annual fairs were granted by charter of Elizabeth; the latter are held on Whit-Monday and October 10th, for linen and woollen goods, hardware, &c.; and a great market is held every three months. By the same charter the inhabitants were incorporated, under the style of "Guardian and Free Tenants," which was subsequently changed by another charter into that of "Mayor and Jurats;" they likewise had the privilege of sending two members to parliament, but all these were merely nominal, as it seems they were never exercised. The town has been made a polling-place for the eastern division of the county. The living is a vicarage, valued in the king's books at £10; late net income, £212 per annum, to which £40 has been recently added by the patron and appropriator, the Archbishop of Canterbury. The church was, with the exception of the walls, destroyed by fire in 1762; the present edifice exhibits specimens in the decorated and later styles of English architecture, and contains a fine enriched octagonal font, and some curious ancient monuments; there are 175 free sittings, the Incorporated Society having granted £45 in aid of the expense. There is a place of worship for Wesleyan Methodists; and a National school is supported by voluntary contributions. In the vicinity are some remains of ancient fortifications. Within the last ten years, several bodies were discovered in digging for clay in a field in this parish: they appear to have been buried here after some sanguinary conflict, as swords, javelins, and other weapons were found near them, with urns containing beads and ashes; several fibulæ were also found here, adorned with precious stones. About a quarter of a mile from the field are the remains of the ancient castle of Bayford, erected for the protection of the adjoining country.

SIX-HILLS (*ALL SAINTS*), a parish, in the union of CAISTOR, Eastern Division of the wapentake of WRAGGOE, parts of LINDSEY, county of LINCOLN, 5 miles (E. S. E.) from Market-Rasen; containing 169 inhabitants. The living is a discharged vicarage, valued in the king's books at £6; present net income, £67; patron and impropriator, George F. Heneage, Esq. There is a place of worship for Roman Catholics. A Gilbertine priory of nuns and canons, in honour of the Blessed Virgin, was founded here by one Grella, or Greslei, which at the dissolution had a revenue of £170. 8. 9.: the site was granted to Sir Thomas Heneage.

SIZEWELL, a hamlet, in the parish of LEISTON, hundred of BLYTHING, Eastern Division of the county of SUFFOLK, 4¼ miles (N. by E.) from Aldborough: the population is returned with the parish. Here was formerly a chapel, dedicated to St. Nicholas, but it has been demolished. Sizewell gap, a small bay on the coast, was formerly a notorious place for smuggling.

SKECKLING (*ALL SAINTS*), a parish, in the Southern Division of the wapentake of HOLDERNESS, East Riding of the county of YORK, 10½ miles (E.) from Kingston-upon-Hull: the population is included in the return for Burstwick. The living is a perpetual curacy, with Burstwick annexed, valued in the king's books at £7; present net income, £219; patron and impropriator, Sir T. A. Clifford Constable, Bart. In the chancel of the church is a fine painting of the Lord's Supper. There is no village named Skeckling, but there is a township, which, with those of Ryhill and Camerton, was returned in the last census with Burstwick parish.

SKEEBY, a township, in the parish of EASBY, union of RICHMOND, wapentake of GILLING-WEST, North Riding of the county of YORK, 2¼ miles (E. N. E.) from Richmond; containing 183 inhabitants.

SKEFFINGTON (*ST. THOMAS à BECKET*), a parish, in the union of BILLESDON, hundred of EAST GOSCOTE, Northern Division of the county of LEICESTER, 9½ miles (W. by N.) from Uppingham; containing 180 inhabitants. The living is a rectory, valued in the king's books at £12. 13. 9.; present net income, £446; patron, T. R. Davenport, Esq.

SKEFFLING (*ST. HELEN*), a parish, in the union of PATRINGTON, Southern Division of the wapentake of HOLDERNESS, East Riding of the county of YORK, 4½ miles (S. E. by E.) from Patrington; containing 204 inhabitants. The living is a perpetual curacy, valued in the king's books at £5; present net income, £53; patron and impropriator, Rev. H. T. Holme. The church is principally in the later style of English architecture. There is a place of worship for Wesleyan Methodists.

SKEGBY, a parish, in the union of MANSFIELD, Northern Division of the wapentake of BROXTOW and of the county of NOTTINGHAM, 3 miles (W.) from Mansfield; containing 656 inhabitants. The living is a perpetual curacy; net income, £78; patron and impropriator, Duke of Portland. The church is a small stone edifice, in a very dilapidated state, situated on an eminence some distance from the village. The manufacture of earthenware is carried on here, and on Skegby moor are extensive collieries.

SKEGNESS (*ST. CLEMENT*), a parish, in the union of SPILSBY, Marsh Division of the wapentake of CANDLESHOE, parts of LINDSEY, county of LINCOLN, 3¾ miles (E. by S.) from Burgh; containing 185 inhabitants. The living is a discharged rectory, valued in the king's books at £15. 6. 8.; present net income, £103; patron, Earl of Scarborough. According to Leland, here was once a considerable town, having a haven and a castle, and surrounded by walls, which was swallowed up by the sea. From its situation on an advantageous part of the coast, this has become a place for sea-bathing, much resorted to by visiters.

SKELBROOKE, a chapelry, in the parish of SOUTH KIRBY, Upper Division of the wapentake of OSGOLDCROSS, West Riding of the county of YORK, 7¼ miles (N. W. by N.) from Doncaster; containing 113 inhabitants. The living is a perpetual curacy; net income,

£60; patron and impropriator, G. Neville, Esq. The chapel is dedicated to St. Michael. The ancient Forest of Barnsdale, part of which is comprised in this chapelry, is said to have been one of the haunts of the famous Robin Hood, whose name is given to a well not far distant. It is further remarkable for the meeting which took place, in 1541, between Henry VIII. and the clergy of York, headed by the archbishop, who on their knees presented the king with £600.

SKELDIN, a township, in the parish and liberty of RIPON, West Riding of the county of YORK, $6\frac{1}{2}$ miles (W. by S.) from Ripon: the population is returned with Grantley.

SKELLINGTHORPE (*St. Lawrence*), a parish, in the Lower Division of the wapentake of BOOTHBY-GRAFFO, parts of KESTEVEN, union and county of LINCOLN, $5\frac{1}{2}$ miles (W.) from Lincoln; containing 417 inhabitants. The living is a discharged vicarage, valued in the king's books at £6. 18. 9.; present net income, £31; patron and impropriator, Master of Spital Hospital. There is a place of worship for Wesleyan Methodists; also a school partly supported by the Governors of Christ's Hospital and partly by the parish.

SKELLOW, a township, in the parish of OWSTON, union of DONCASTER, Upper Division of the wapentake of OSGOLDCROSS, West Riding of the county of YORK, $5\frac{1}{2}$ miles (N. N. W.) from Doncaster; containing 181 inhabitants. A school is partly supported by P. D. Cooke, Esq.

SKELMANTHORPE, a township, partly in the parish of EMLEY, Lower Division, partly in the chapelry of CUMBERWORTH, parish of KIRK-BURTON, Upper Division, of the wapentake of AGBRIGG, and partly in the parish of HIGH HOYLAND, wapentake of STAINCROSS, West Riding of the county of YORK, $7\frac{3}{4}$ miles (S. E. by E.) from Huddersfield: the population is returned with Cumberworth. There is a place of worship for Wesleyan Methodists.

SKELMERSDALE, a chapelry, in the parish and union of ORMSKIRK, hundred of WEST DERBY, Southern Division of the county palatine of LANCASTER, 4 miles (E. S. E.) from Ormskirk; containing 676 inhabitants. The living is a perpetual curacy; net income, £142; patron, Vicar of Ormskirk. There is a small endowment, bequeathed by Evan Swift, in 1720, for a school. Skelmersdale gives the title of Baron to the family of Bootle-Wilbraham.

SKELSMERGH, a township, in the parish, union, and ward of KENDAL, county of WESTMORLAND, $2\frac{1}{2}$ miles (N. by E.) from Kendal; containing 263 inhabitants. It is bounded on every side, except on the east, by the small rivers Kent, Mint, and Sprint, upon which there are corn, worsted, bobbin, and dye-wood mills. Here are remains of a chapel, dedicated to St. John the Baptist; and at Doddington Green there is one for Roman Catholics.

SKELTON (*St. Mary*), a parish, in the union of PENRITH, LEATH ward, Eastern Division of the county of CUMBERLAND; containing 854 inhabitants, of which number, 348 are in the township of Skelton, $6\frac{3}{4}$ miles (N. W. by W.) from Penrith. The living is a rectory, valued in the king's books at £43. 3. $6\frac{1}{2}$.; present net income, £294; patrons, President and Fellows of Corpus Christi College, Oxford. The church is an ancient structure: it was thoroughly repaired in 1794, and formerly contained a richly-endowed chantry. Freestone and limestone are obtained in the parish. A free school, erected in 1750, by Isaac Milner, was endowed in 1817, by the Rev. Joseph Nelson, with £1000, now producing upwards of £32 a year, for which 28 children are instructed. Another school for girls is partly supported by Joseph Cooper, Esq.

SKELTON, a township, in the parish and union of HOWDEN, wapentake of HOWDENSHIRE, East Riding of the county of YORK, 2 miles (S. E. by S.) from Howden; containing 228 inhabitants.

SKELTON (*All Saints*), a parish, partly in the liberty of ST. PETER'S, East Riding, but chiefly in the wapentake of BULMER, North Riding, of the county of YORK, $3\frac{3}{4}$ miles (N. W. by N.) from York; containing 291 inhabitants. The living is a rectory not in charge; net income, £80; patron, J. Hepworth, Esq. Here is a National school.

SKELTON (*All Saints*), a parish, in the union of GUILSBROUGH, Eastern Division of the liberty of LANGBAURGH, North Riding of the county of YORK; containing 1241 inhabitants, of which number, 781 are in the township of Skelton, 4 miles (N. E. by N.) from Guilsbrough. The living is a perpetual curacy, with that of Brotton annexed; net income, £137; patron and appropriator, Archbishop of York. There is a place of worship for Wesleyan Methodists; also a school endowed for the instruction of 10 boys. A castle was built here by Robert de Brus, a Norman baron, who came over with the Conqueror, and from whom descended some of the kings of Scotland and the present family of Bruce, Marquesses of Ailesbury. There are few remains of the ancient building, the whole having been entirely modernized, or renovated, in 1794. A market, formerly held on Sunday, but afterwards changed to Saturday, and a fair at Whitsuntide, have been discontinued.

SKELTON, a chapelry, in the parish and liberty of RIPON, West Riding of the county of YORK, 4 miles (S. E. by E.) from Ripon; containing 383 inhabitants. The living is a perpetual curacy; net income, £77; patrons, Dean and Chapter of Ripon. The chapel is small, but handsomely built in the early style of English architecture. A day and Sunday school is supported by the Earl and Countess de Grey.

SKELWITH, a joint township with Monk-Coniston, in the parish of HAWKSHEAD, union of ULVERSTON, hundred of LONSDALE, north of the sands, Northern Division of the county palatine of LANCASTER, $2\frac{1}{2}$ miles (S. W. by W.) from Ambleside: the population is returned with Monk-Coniston. A chapel was erected and endowed by Mr. Redmayne, of Brathey, on the Brathey Hall estate, in 1835, and consecrated the year following by the Bishop of Chester; it is a neat edifice, situated on the road from Hawkshead to Ambleside, near Brathey bridge, near which are two pleasing cascades and some fine views.

SKENDLEBY (*St. Peter*), a parish, in the union of SPILSBY, Wold Division of the wapentake of CANDLESHOE, parts of LINDSEY, county of LINCOLN, 4 miles (N. E. by N.) from Spilsby; containing 253 inhabitants. The living is a discharged vicarage, valued in the king's books at £4. 0. 5.; present net income, £155; patron and impropriator, Lord Willoughby de Eresby. There is a place of worship for Wesleyan Methodists.

SKENFRETH (*St. Bridget*), a parish, in the Upper Division of the hundred of Skenfreth, union and county of Monmouth, 6¾ miles (N. N. W.) from Monmouth; containing 609 inhabitants. The parish is intersected by the river Monnow, over which a bridge was erected in 1825, at an expense to the county of £1000, the road from London to Milford Haven being thus shortened by seven miles. The living is a discharged vicarage, valued in the king's books at £5. 16. 10½.; present net income, £80: it is in the patronage of Mrs. S. Pugh; impropriators, Vicar of Skenfreth and others. The church is a large handsome structure, the windows of which are enriched with fine stained glass. Of the ancient castle, which stood on an eminence rising gently from the bank of the Monnow, nothing remains but the outer wall: it was defended by six towers and a moat supplied from the river.

SKERNE, a parish, in the union of Driffield, Bainton-Beacon Division of the wapentake of Harthill, East Riding of the county of York, 2¾ miles (S. E. by S.) from Great Driffield; containing 201 inhabitants. The living is a discharged perpetual curacy; net income, £71; patron and impropriator, R. Arkwright, Esq.

SKERTON, a township, in the parish of Lancaster, hundred of Lonsdale, south of the sands, Northern Division of the county palatine of Lancaster, ¾ of a mile (N.) from Lancaster; containing 1351 inhabitants. This is a large village, separated from the town of Lancaster by the river Lune, in which there is a considerable salmon fishery. A new church has been erected here under the act of the 58th of George III.; it was consecrated on the 14th of November, 1833. A free school was erected by Jane Jephson, and endowed with £10 a year by Henry Williamson, in 1767, for 18 children; there is also a Sunday National school.

SKETCHLEY, a hamlet, in the parish of Aston-Flamville, hundred of Sparkenhoe, Southern Division of the county of Leicester, 1¼ mile (S. by W.) from Hinckley: the population is returned with the chapelry of Burbage. Here was formerly a chapel, now in ruins.

SKEWSBY, a township, in the parish of Dalby, union of Easingwould, wapentake of Bulmer, North Riding of the county of York, 8¾ miles (E. by N.) from Easingwould: the population is returned with the parish.

SKEYTON (*All Saints*), a parish, in the union of Aylsham, hundred of South Erpingham, Eastern Division of the county of Norfolk, 3¾ miles (E. by S.) from Aylsham; containing 317 inhabitants. The living is a discharged rectory, with that of Oxnead and the vicarage of Buxton annexed, valued in the king's books at £9. 10.; present net income, £646; patron, S. Bignold, Esq.

SKIDBROOK (*St. Botolph*), a parish, in the union of Louth, Marsh Division of the hundred of Louth-Eske, parts of Lindsey, county of Lincoln, 10½ miles, (N. E. by E.) from Louth; containing 362 inhabitants. The living is a discharged vicarage, valued in the king's books at £11. 3. 6.; patron and incumbent, Rev. J. M. Phillips; impropriator, — Ward, Esq. The impropriate tithes have been commuted for a rent-charge of £115. 16. 10., and the vicarial for £377. 3. 6., subject to the payment of rates, which on the average on the latter have amounted to £57; the glebe comprises 4 acres, valued at £6 per annum. A school is partly supported by an allowance of £8 per annum from the parish.

SKIDBY, a chapelry, in the union of Beverley, Hunsley-Beacon Division of the wapentake of Harthill, East Riding of the county of York, 4 miles (S. by W.) from Beverley; containing 315 inhabitants. The living is a perpetual curacy, annexed to the vicarage of Cottingham. The chapel is dedicated to St. Michael.

SKILGATE (*St. John the Baptist*), a parish, in the union of Dulverton, hundred of Williton and Freemanners, Western Division of the county of Somerset, 6½ miles (W. by S.) from Wiveliscombe; containing 227 inhabitants. The living is a rectory, valued in the king's books at £9. 9. 4½.; present net income, £192; patron, Rev. Richard Bere.

SKILLINGTON (*St. James*), a parish, in the union of Grantham, wapentake of Beltisloe, parts of Kesteven, county of Lincoln, 3 miles (N. W. by W.) from Colsterworth; containing 389 inhabitants. The living is a discharged vicarage, valued in the king's books at £4. 19. 4½.; present net income, £126; patrons and appropriators, Dean and Chapter of Lincoln. There is a place of worship for Wesleyan Methodists.

SKINBURNESS, a village, in the parish of Holme-Cultram, Allerdale ward below Derwent, Western Division of the county of Cumberland, 11½ miles (N. W. by W.) from Wigton: the population is returned with Low Holme. This place had anciently a market and a fair, granted to the abbot of Holme-Cultram; it was of considerable importance as a depôt, from which the army employed against the Scots was supplied with stores. About 1303, the town was washed away by an irruption of the sea, and the abbot, in consequence, having obtained licence to erect a church, or chapel, at Arlosh, a new town was also built, and called Newton-Arlosh. Skinburness is now a pleasant village and respectable sea-bathing place, affording public and private accommodation to its numerous visiters, and commanding a most extensive view over the Solway Firth and of the Scottish mountains beyond. A very productive herring fishery is carried on here.

SKINNAND, a parish, in the Higher Division of the wapentake of Boothby-Graffo, parts of Kesteven, union and county of Lincoln, 11½ miles (N. W.) from Sleaford; containing 24 inhabitants. The living is a discharged rectory, valued in the king's books at £5. 13. 11½.; present net income, £85; patron, S. Nicholls, Esq. The church is in ruins.

SKINNINGROVE, a township, in the parish of Brotton, union of Guilsbrough, Eastern Division of the liberty of Langbaurgh, North Riding of the county of York, 8 miles (N. E.) from Guilsbrough; containing 63 inhabitants. It has a small fishing village, situated on a creek, almost secluded from view by the lofty heights which closely environ it on every side.

SKIPLAM, a township, in the parish of Kirkdale, union of Helmsley-Blackmoor, wapentake of Ryedale, North Riding of the county of York, 5½ miles (N. E. by E.) from Helmsley; containing, with Bransdale West Side, 124 inhabitants.

SKIPSEA (*All Saints*), a parish, chiefly in the union of Bridlington, but partly in that of Skirlaugh, Northern Division of the wapentake of Holderness, East Riding of the county of York; containing 726 inhabitants, of which number, 386 are in the township of Skipsea, 11¼ miles (E. by S.) from Great Driffield. This parish is bounded on the east by the North Sea. The living is a discharged vicarage, valued in the king's books at £9. 16.; present net income, £96; patron and appropriator, Archbishop of York. The church is principally in the later style of English architecture. There are places of worship for Independents and Wesleyan Methodists. A lofty mount, termed Skipsea Brough, was anciently the site of a baronial castle of the lords of Holderness.

SKIPTON, a township, in the parish of TOPCLIFFE, union of THIRSK, wapentake of BIRDFORTH, North Riding of the county of YORK, 5 miles (S. W. by W.) from Thirsk; containing 114 inhabitants. There is a place of worship for Wesleyan Methodists. The Rev. Mr. Day, in 1764, bequeathed a small sum towards the support of a school.

SKIPTON (*HOLY TRINITY*), a market-town and parish, and the head of a union, chiefly in the Eastern Division of the wapentake of STAINCLIFFE and EWCROSS, but partly in the Upper Division of the wapentake of CLARO, West Riding of the county of YORK; containing 6193 inhabitants, of which number, 4181 are in the town of Skipton, 44 miles (W.) from York, and 211 (N. N. W.) from London. The name, which is variously spelt in Domesday-book as *Sciptone, Sceptone*, or *Sceptetone*, was probably acquired from the vast number of sheep anciently fed in the vicinity. On the left of the road leading to Knaresborough stands the ancient and magnificent castle, the residence of the Tuftons, Earls of Thanet, whose ancestors, the Veteriponts and Cliffords, inhabited it from the time of the Conquest. It was built by Robert de Romelie in 1066, and has been several times demolished in different wars, and as often rebuilt by its noble possessors. This old baronial residence now bears upon it the marks of time, and in some parts is falling to decay, although it contains several commodious and magnificent apartments, in which are many family relics and portraits. The town is situated in a valley of great fertility and beauty, near the river Aire, and is skirted on the south-west by the Leeds and Liverpool canal, which partly passes through it, affording great facility for the conveyance of goods. It consists chiefly of two long and wide streets, one crossing the other at its termination, nearly at right angles, which are partly paved: the town is well supplied with water, conveyed through pipes from a spring that rises upon Rumbles moor; and the houses are neatly built of stone, obtained in the immediate vicinity. The adjacent vale is celebrated for its productiveness, and is a fine grazing district: from the surrounding hills are some beautiful and picturesque views. There is a small subscription library. The situation of the town is highly advantageous to the purposes of trade, of which the spinning and weaving of cotton form the principal branch: there are several cotton-mills upon the neighbouring streams, and an extensive porter and ale brewery. The market is on Saturday, when a small quantity of grain is brought for sale; and there is a large market every second Monday for fat cattle and sheep: fairs are held on March 25th, Saturdays before Palm and Easter Sundays, the first and third Tuesdays after Easter, on Whitsun-eve, August 5th, and Nov. 23rd, chiefly for sheep, horned cattle, pedlery, &c.; and Sept. 23rd for horses. The local affairs are under the superintendence of a constable, who is appointed annually at the manorial court leet: the general quarter sessions for the West Riding are held here at Midsummer, in the town-hall, which is a neat stone building, situated on the east side of Sheep-street Hill. This town has been made a polling-place for the West Riding.

The living comprises a rectory and a discharged vicarage, the former valued in the king's books at £4. 0. 10.; and the latter, at £10. 12. 6.; present net income, £185; patrons and appropriators, Dean and Canons of Christ-Church, Oxford. The church immediately adjoins the castle, and is the work of different periods, but principally in the later style of English architecture; four stone seats, with pointed arches and cylindrical columns, in the south wall of the nave, are supposed to be the only remaining parts of the original edifice; the tower, which stands at the west end, was repaired by Lady Clifford, Countess Dowager of Pembroke, in the year 1655: there are several monuments to different members of the family of Clifford, Earls of Cumberland; and beneath the altar is a family vault, which was their place of interment from the dissolution of Bolton priory to the death of the last Earl of Cumberland. Another church was built in 1837, containing 630 sittings, 360 of which are free, the Incorporated Society having granted £350 in aid of the expense. There are places of worship for the Society of Friends, Independents, and Wesleyan Methodists. The free grammar school was founded in 1548, and endowed with lands in Addingham, Eastby, and Skipton, by the Rev. William Ermystead, the annual proceeds of which amount to upwards of £500: the master is appointed by the vicar and churchwardens, and has an agreeable residence, with pleasure grounds and gardens attached: boys are admissible from all parts, and there are about 60 in the school. They are eligible to the exhibitions of Lady Elizabeth Hastings, in Queen's College, Oxford; and the school has two exhibitions in Christ's College, Cambridge, founded by William Petyt, Esq., who gave £200 to the college for that purpose. The Clerk's school was originally endowed by the Rev. William Ermystead, with lands of great value at Wike, near Harewood, in Yorkshire, which, by mismanagement of the trustees, have been lost, and a prescriptive payment, amounting to £12 per annum, is made to the master of the boys' National school, which, with another for girls, is further supported by voluntary contributions. Sylvester Petyt, Esq., Principal of Bernard's Inn, and a native of this parish, bequeathed a library for the use of the parishioners, which is preserved in the church; and the sum of £24,048 South Sea Annuities, for various charitable purposes, especially for the payment of £20 per annum to Christ's College, Cambridge, for an augmentation of exhibitions for the free grammar school, for apprenticing twelve poor children of parishioners annually, the interest of £100 for the librarian, and of £50 to buy books for poor boys at the grammar school. The poor law union of Skipton comprises 42 parishes or places, under the care of 44 guardians, and contains, according to the census of 1831, a population of 25,562. The principal remains of the old castle are the western doorway, and several round towers. George Holmes, an eminent antiquary, who republished the first seventeen volumes of Rymer's Fœdera, was a native of this town.

SKIPWITH (*St. Helen*), a parish, in the union of Selby, wapentake of Ouse and Derwent, East Riding of the county of York; containing 648 inhabitants, of which number, 304 are in the township of Skipwith, 5½ miles (N. N. E.) from Selby. The living is a discharged vicarage, valued in the king's books at £10. 11. 3.; present net income, £300: it is in the patronage of the Crown. The Rev. Joseph Nelson, in 1813, bequeathed £400 for the education of the poorest children of the parish; the annual income is upwards of £13. 10., for which about fourteen children are instructed in a school founded by Dorothy Wilson, in 1717.

SKIRBECK (*St. Nicholas*), a parish, in the union of Boston, partly in the wapentake of Kirton, but chiefly in that of Skirbeck, parts of Holland, county of Lincoln, 1 mile (S. E. by S.) from Boston; containing 1578 inhabitants. The living is a rectory, valued in the king's books at £34. 17. 8½.; present net income, £737; patron, Rev. William Volans. An ancient hospital for ten poor persons, founded in honour of St. Leonard, was given, in 1230, by Sir Thomas Multon, Knt., to the knights hospitallers of St. John of Jerusalem, who settled here, and dedicated it anew to St. John the Baptist. In the time of Edward II., its revenue was sufficient for the maintenance of four priests, also 20 poor people in the infirmitory, and for the daily relief of 40 more at the gate.

SKIRBECK-QUARTER, a hamlet, in the parish of Skirbeck, union of Boston, wapentake of Kirton, parts of Holland, county of Lincoln; containing 323 inhabitants.

SKIRCOAT, a chapelry, in the parish and union of Halifax, wapentake of Morley, West Riding of the county of York, 1¾ mile (S. S. W.) from Halifax; containing 4060 inhabitants. The living is a donative; patron and incumbent, Rev. J. Akroyd: the permanent annual payments out of the proceeds of the living are equal to the income, which is £15. The manufacture of cotton and woollen goods is carried on here to a considerable extent. Here is a free grammar school, founded and endowed by Queen Elizabeth.

SKIRLAUGH, NORTH, a chapelry, in the parish of Swine, union of Skirlaugh, Northern Division of the wapentake of Holderness, East Riding of the county of York, 9 miles (N. N. E.) from Kingston-upon-Hull; containing, with the township of Rowton, and part of that of Arnold, 210 inhabitants. The chapel, dedicated to St. Augustine, was built by Walter Skirlaw, Bishop of Durham; it is an elegant specimen of the later English style. There is a place of worship for Wesleyan Methodists.

SKIRLAUGH, SOUTH, a chapelry, in the parish of Swine, union of Skirlaugh, Middle Division of the wapentake of Holderness, East Riding of the county of York, 8¾ miles (N. N. E.) from Kingston-upon-Hull; containing 228 inhabitants. Marmaduke Langdale, about 1609, bequeathed £200, directing the interest to be applied for teaching and apprenticing children, and other charitable purposes.

SKIRLINGTON, a township, in the parish of Atwick, union of Skirlaugh, Northern Division of the wapentake of Holderness, East Riding of the county of York, 13½ miles (E. S. E.) from Great Driffield: the population is returned with the parish.

SKIRPENBECK, a parish, in the union of Pocklington, partly in the wapentake of Buckrose, and partly in the liberty of St. Peter's, East Riding of the county of York, 7¾ miles (N. W. by N.) from Pocklington; containing 214 inhabitants. The living is a rectory, valued in the king's books at £14. 7. 8½.; present net income, £232: it is in the patronage of the Crown.

SKIRWITH, a township, in the parish of Kirkland, union of Penrith, Leath ward, Eastern Division of the county of Cumberland, 7½ miles (E.N.E.) from Penrith; containing 296 inhabitants. There is a place of worship for Wesleyan Methodists. A National school was erected by subscription in 1828, towards which Lady le Fleming and W. Parker, Esq., each contributed £20. The mansion of the latter is supposed to occupy the site of a preceptory of Knights Templars.

SKUTTERSKELFE, a township, in the parish of Rudby-in-Cleveland, union of Stokesley, Western Division of the liberty of Langbaurgh, North Riding of the county of York, 2 miles (W. by S.) from Stokesley; containing 38 inhabitants. Near the village is an excellent landmark, called Folly Hill, which is sometimes discernible 20 leagues at sea.

SLAD, or SLADE, a hamlet, partly in the parish of Painswick, and partly in that of Stroud, hundred of Bisley, Eastern Division of the county of Gloucester, 2 miles (N. E. by E.) from Stroud: the population is returned with the respective parishes. This hamlet, which is divided by a stream of water that separates the parishes of Painswick and Stroud into two parts, called respectively Painswick Slad and Stroud Slad, is pleasantly situated in a beautiful valley, through which passes the road to Cheltenham. The surrounding scenery is agreeably diversified; and near the village is a splendid mansion, lately the residence of D. H. Rucker, Esq. In the vale are several clothing manufactories, and in Stroud Slad is one of the largest establishments in the west of England. An elegant episcopal chapel was erected in 1831, through the exertions of Mrs. Rucker, on a site given by the lord of the manor: the want of funds for its endowment has hitherto retarded its consecration; but there is every probability of its being very soon opened for the performance of divine service. Here is a National school; and an infants' school has been established for the instruction of such children as are too young to work in the factories.

SLAIDBURN (*St. Andrew*), a parish, in the union of Clitheroe, Western Division of the wapentake of Staincliffe and Ewcross, West Riding of the county of York; containing 2409 inhabitants, of which number, 920 are in the township of Slaidburn, 8 miles (N. by W.) from Clitheroe. The living is a rectory, valued in the king's books at £32; present net income, £336; patron and incumbent, Rev. H. Wigglesworth. There is a chapel of ease, dedicated to St. Peter; also a place of worship for Wesleyan Methodists. A free grammar school was founded here in 1717, by John Brannord, who bequeathed an estate in trust to apply £200 for the erection of a school-house, with £50 a year to a master, and £30 to an usher, for teaching the classics, or English grammar, to all the children of the parish. A chantry was founded in 1332, by Stephen de Hamerton, in the chapel of St. Mary, then existing on his manor of Hamerton, for a Secular chaplain, to celebrate mass for the repose of the souls of himself, his father, and his mother.

SLAITHWAITE, a chapelry, in the parish and union of HUDDERSFIELD, Upper Division of the wapentake of AGBRIGG, West Riding of the county of YORK, 5 miles (W. S. W.) from Huddersfield; containing 2892 inhabitants. The living is a perpetual curacy; net income, £192; patron, Vicar of Huddersfield. The chapel, rebuilt in 1784, will accommodate upwards of 1500 persons. The canal, and the new line of road from Huddersfield to Manchester, pass through this place. The manufacture of woollen and cotton goods is carried on to a great extent, and the prosperity of the village is likely to be promoted by the recent discovery of an excellent spa, thought to be equal in its chalybeate properties to the spring at Harrogate. A free school was founded and endowed here, in 1721, by the Rev. Robert Meek; the income, with subsequent benefactions, amounts to £42 per annum.

SLALEY, a parish, in the union of HEXHAM, Eastern Division of TINDALE ward, Southern Division of the county of NORTHUMBERLAND, 5½ miles (S. E. by S.) from Hexham; containing 616 inhabitants. The living is a perpetual curacy; net income, £103; patron, T. W. Beaumont, Esq.; impropriators, R. Trevelyan and — Witham, Esqrs. The church has been enlarged, and 150 free sittings provided, the Incorporated Society having granted £100 in aid of the expense. At Dukesfield, in this parish, is a large mill for smelting and refining lead-ore, which is brought from Wardle, in the county of Durham. Ochre is obtained and manufactured on Slaley fell. A school in connection with the National society is endowed with £3. 10. per annum.

SLAPTON (*HOLY CROSS*), a parish, in the union of LEIGHTON-BUZZARD, hundred of COTTESLOE, county of BUCKINGHAM, 3¼ miles (N. by W.) from Ivinghoe; containing 360 inhabitants. The living is a rectory, valued in the king's books at £14. 9. 7.; present net income, £172; patrons, certain Trustees. There is a place of worship for Wesleyan Methodists. The Grand Junction canal passes through the parish, and the London and Birmingham railway within a mile of the church. A charity school is supported by voluntary subscription.

SLAPTON (*ST. MARY*), a parish, in the union of KINGSBRIDGE, hundred of COLERIDGE, Stanborough and Coleridge, and Southern, Divisions of the county of DEVON, 6 miles (S. W.) from Dartmouth; containing 665 inhabitants. The living is a discharged perpetual curacy; net income, £96; patron, William Paige, Esq.; impropriators, the Landowners. The church contains 86 free sittings, the Incorporated Society having granted £15 in aid of the expense. This place formerly belonged to Sir Guy de Brien, Knt., standard-bearer to Edward III., whom he attended at the battle of Calais, in 1349, in which, having greatly distinguished himself by his intrepidity, he was rewarded with a grant of 200 marks per annum, payable out of the exchequer during his life. Sir Guy founded in the church a chantry for a rector and four priests, which he endowed with £10 per annum in land, and with the advowson of the living. After the dissolution it was granted to Thomas Arundel, afterwards to Sir William Petre, and was subsequently sold to Captain Sir Richard Hawkins, R.N. John and Charles Kelland, about 1690, bequeathed £150, which, with accumulations, now produces about £35 per annum, applied to the instruction and apprenticing of poor children. Thomas Knyghton, in 1629, bequeathed an estate, now producing about £30 per annum, for the repair of the town-houses, and distribution among the resident poor, and paying the expenses of the funerals of labourers, their wives, or children.

SLAPTON (*ST. BOTOLPH*), a parish, in the union of TOWCESTER, hundred of GREENS-NORTON, Southern Division of the county of NORTHAMPTON, 3¾ miles (W. S. W.) from Towcester; containing 197 inhabitants. The living is a rectory, valued in the king's books at £9. 9. 9½.; patron, Rev. Thomas C. Welch. There is a place of worship for Wesleyan Methodists. Three poor children are educated by means of a trifling bequest from Thomas Knight, in 1723.

SLAUGHAM (*ST. MARY*), a parish, in the union of CUCKFIELD, hundred of BUTTINGHILL, rape of LEWES, Eastern Division of the county of SUSSEX, 4½ miles (S. W.) from Crawley; containing 740 inhabitants. This parish is situated on the road from London, through Crawley, to Brighton; and the river Ouse has its source in the grounds of Ashford, forming in its course a lake of about 30 acres at Slaugham mills. The surface is diversified with hill and dale; but the soil is sandy and generally poor. There are considerable remains of the old manor-house of Slaugham Place, the ancient seat of the Covert family; the grand staircase was given by the late Colonel Sergison to the proprietor of the Star Inn, at Lewes, and has been erected in that house. The living is a rectory, valued in the king's books at £10. 19. 2.; present net income, £282: it is in the patronage of Mrs. A. Sergison. The church is in the decorated style of English architecture, and was enlarged in 1837 by the addition of 200 free sittings, the Incorporated Society having granted £100 in aid of the expense: it contains a splendid monument to the Covert family. There is a place of worship for Baptists.

SLAUGHTER, LOWER, a parish, in the union of STOW-ON-THE-WOLD, Lower Division of the hundred of SLAUGHTER, Eastern Division of the county of GLOUCESTER, 3 miles (S. W.) from Stow-on-the-Wold; containing 258 inhabitants. The living is annexed to the rectory of Bourton-on-the-Water.

SLAUGHTER, UPPER (*ST. PETER*), a parish, in the union of STOW-ON-THE-WOLD, Lower Division of the hundred of SLAUGHTER, Eastern Division of the county of GLOUCESTER, 3¼ miles (S. W. by W.) from Stow-on-the-Wold; containing 260 inhabitants. The living is a rectory, valued in the king's books at £14. 14. 2.; present net income, £131; patron and incumbent, Rev. W. E. Witts. Two brooks run through the parish, and afterwards fall into another, all which uniting form the river Windrush. A school is supported by an allowance of £6 per annum from charitable funds; and there is a small endowment by the late rector, the Rev. F. T. Travell, for the support of a Sunday school.

SLAUGHTERFORD (*ST. NICHOLAS*), a parish, in the union and hundred of CHIPPENHAM, Chippenham and Calne, and Northern, Divisions of the county of WILTS, 5½ miles (W. by N.) from Chippenham; containing 115 inhabitants. The living is a perpetual curacy, annexed to the rectory of Biddestone. The church contains 100 free sittings, the Incorporated Society having granted £100 in aid of the expense.

SLAWSTON (*ALL SAINTS*), a parish, in the union

of UPPINGHAM, hundred of GARTREE, Southern Division of the county of LEICESTER, $5\frac{3}{4}$ miles (N. E. by N.) from Market-Harborough; containing 243 inhabitants. The living is a discharged vicarage, valued in the king's books at £6. 5. $7\frac{1}{2}$.; present net income, £174; patron, Earl Cadogan; impropriators, — Tailby, Esq., and others.

SLEAFORD, NEW (*St. Denis*), a market-town and parish, and the head of a union, in the wapentake of FLAXWELL, parts of KESTEVEN, county of LINCOLN, 18 miles (S. S. E.) from Lincoln, and 116 (N. by W.) from London; containing 2857 inhabitants. The name in ancient records is written *La Ford* and *Eslaforde*, which has been corrupted into Sleaford, and the epithet *New* given to distinguish it from Old Sleaford, an adjoining parish. A castle appears to have been erected here at an early period, but of its history there are few records, and of the building only some trifling remains. The town is situated on the main road from London to Lincoln; it is of respectable appearance, and is gradually improving in buildings and importance; it is well paved and lighted, and the inhabitants are supplied with water from an adjacent spring, called Bully, or Boiling, wells. A small theatre was erected in 1824. A canal connects this town with Boston, Lincoln, and the Trent navigation, and thus promotes the prosperity of its general trade. The market is on Monday; and fairs are on Plough-Monday, Easter-Monday, Whit-Monday, August 11, and October 20th, for horses, cattle, sheep, and provisions. The quarter sessions for the parts of Kesteven are held here by adjournment from Bourne; and the town has been made a polling-place for the parts of Kesteven and Holland. The old town-hall, being greatly dilapidated, has been pulled down, and a handsome edifice, in the later style of English architecture, erected.

The living is a discharged vicarage, valued in the king's books at £8; present net income, £170; patron and appropriator, Prebendary of Lafford, or New Sleaford, in the Cathedral Church of Lincoln. The church exhibits some fine specimens of almost every style of English architecture, and consists of a nave, aisles, a large chapel on the north side, and a chancel: at the west end is a tower, erected about 1150, which is by far the most ancient part of the building; it is in the early style of English architecture, and is surmounted by a spire of later date, 144 feet high; an enriched spiral staircase leads to it on the south side. In the chancel are three fine stalls, in the later style; and at the entrance are the screen and canopy of the ancient rood-loft: there are several ancient monuments, chiefly belonging to the family of Carr, formerly lords of the manor. There are places of worship for those in the Connection of the late Countess of Huntingdon, Independents, and Wesleyan Methodists. The free grammar school was founded in 1604, by Robert Carr, Esq., who endowed it with £20 per annum: the master, who must be a graduate of one of the Universities, is appointed by the Marquess of Bristol, as owner of the "late fair castle of Sleaford," for which he pays to the crown £40 per annum. A school was endowed with lands by William Alvey, in 1729, for the instruction of poor children: 20 boys and 20 girls are educated. An hospital, for a chaplain and twelve poor men, was founded and endowed by Sir Robert Carr, Bart., in 1636. The poor law union of Sleaford comprises 56 parishes or places, under the superintendence of 59 guardians, and contains, according to the census of 1831, a population of 19,832. The Bishop of Lincoln had anciently a magnificent palace here, in which King John sojourned a few days, but no part remains except the foundations: it is supposed to have been destroyed by Cromwell. A branch of the Ermin-street passes through this parish and that of Old Sleaford.

SLEAFORD, OLD (*St. Giles*), a parish, in the union of SLEAFORD, wapentake of ASWARDHURN, parts of KESTEVEN, county of LINCOLN, 1 mile (S. E.) from New Sleaford; containing 272 inhabitants. The living was a vicarage, valued in the king's books at £4. 10.; patron, Marquess of Bristol. The church has been demolished upwards of 200 years, for which period there has been no presentation, the vicarage being now supposed to have merged into the impropriation: the inhabitants attend divine service at Quarrington, or New Sleaford. The Roman Ermin-street passes through the parish.

SLEAGILL, a township, in the parish of MORLAND, WEST ward and union, county of WESTMORLAND, $2\frac{1}{2}$ miles (S. W.) from Morland; containing 184 inhabitants. Coal is obtained here. The sum of £6 per annum, the produce of land, is paid to a schoolmaster for teaching children.

SLEAP, a township, in the parish of WEM, hundred of PIMHILL, Northern Division of the county of SALOP, $1\frac{3}{4}$ mile (W. S. W.) from Wem; containing 27 inhabitants.

SLECKBURN, EAST, a township, in the parish of BEDLINGTON, a detached portion of the Eastern Division of CHESTER ward, county palatine of DURHAM, locally on the east side, and for electoral purposes annexed to the Northern Division, of the county of Northumberland, $6\frac{3}{4}$ miles (E. S. E.) from Morpeth: the population is returned with the parish.

SLECKBURN, WEST, a township, in the parish of BEDLINGTON, a detached portion of the Eastern Division of CHESTER ward, county palatine of DURHAM, locally on the east side, and for electoral purposes annexed to the Northern Division, of the county of Northumberland, $5\frac{3}{4}$ miles (E. by S.) from Morpeth: the population is returned with the parish.

SLEDDALE, LONG, a chapelry, in the parish, union, and ward of KENDAL, county of WESTMORLAND, 8 miles (N. by W.) from Kendal; containing 199 inhabitants. The living is a perpetual curacy; net income, £69; patrons, the Landowners. The chapel was rebuilt in 1712. Here are quarries of a fine blue slate, situated amid mountain scenery of the most romantic and picturesque character.

SLEDMERE (*St. Mary*), a parish, in the union of DRIFFIELD, wapentake of BUCKROSE, East Riding of the county of YORK, $7\frac{1}{2}$ miles (N. W. by W.) from Great Driffield; containing 480 inhabitants. The living is a perpetual curacy; patron, Sir Tatton Sykes, Bart. A day and Sunday school is partly supported by subscription.

SLEEP, a hamlet, in the parish of ST. PETER, ST. ALBANS, hundred of CASHIO, or liberty of ST. ALBANS, county of HERTFORD; containing 722 inhabitants.

SLENINGFORD, a joint township with North

Stanley, in the parish and liberty of RIPON, West Riding of the county of YORK, 5 miles (N. W. by N.) from Ripon: the population is returned with North Stanley. It takes its name from an ancient ford over the river Ure.

SLIMBRIDGE (*ST. JOHN THE EVANGELIST*), a parish, in the union of DURSLEY, Upper Division of the hundred of BERKELEY, Western Division of the county of GLOUCESTER, 4 miles (N. by W.) from Dursley; containing 923 inhabitants. This parish, of which the greater part belongs to the Berkeley family, is bounded on the north by the Severn, and intersected by the Gloucester and Berkeley canal. The living is a rectory, valued in the king's books at £28. 2. 11.; present net income, £601; patrons, President and Fellows of Magdalene College, Oxford. The church is a handsome structure with a fine spire. There is a place of worship for Independents.

SLINDON, a township, in the parish of ECCLESHALL, Northern Division of the hundred of PIREHILL and of the county of STAFFORD; containing 135 inhabitants.

SLINDON (*ST. MARY*), a parish, in the hundred of ALDWICK, rape of CHICHESTER, Western Division of the county of SUSSEX, 4¼ miles (W. by N.) from Arundel; containing 539 inhabitants. The living is a rectory, valued in the king's books at £14. 13. 1½.; present net income, £219; patron, Earl of Newburgh. The church is in the later style of English architecture. There is a Roman Catholic chapel, and a school is endowed for instructing children of Roman Catholics. A National school has been established.

SLINFOLD, or SLYNFOLD (*ST. PETER*), a parish, in the union of HORSHAM, partly in the hundred of EAST EASWRITH, rape of BRAMBER, but chiefly in the hundred of WEST EASWRITH, rape of ARUNDEL, Western Division of the county of SUSSEX, 4 miles (W. by N.) from Horsham; containing 682 inhabitants. This parish is intersected by the road from Horsham to Guildford: the soil, which is a deep clay resting upon slate, is very favourable for the growth of oak timber, of which many stately trees adorn Strood Park, the seat of J. Commerell, Esq. There are some good stone quarries, producing blocks of enormous size. Two branches of the river Arun unite in this parish, and various rare botanical plants are found here. The living comprises a sinecure rectory and a vicarage united, valued jointly in the king's books at £12. 14. 2.; present net income, £472; patron, Bishop of Chichester. The church is an ancient edifice, with a low tower; in the chancel is the figure of a female rudely sculptured. There are several small bequests for the poor. The Roman road from Regnum to London passed for about two miles through the parish, and Roman swords and ornaments of brass have been found. The Rev. James Dallaway, author of the Topography of the Rape of Arundel, was rector of this parish.

SLINGLEY, a joint township with Seaton, in the parish of SEAHAM, union of EASINGTON, Northern Division of EASINGTON ward and of the county palatine of DURHAM, 5¾ miles (S. by W.) from Sunderland: the population is returned with Seaton.

SLINGSBY (*ALL SAINTS*), a parish, in the union of MALTON, wapentake of RYEDALE, North Riding of the county of YORK, 6 miles (W. N. W.) from New Malton; containing 562 inhabitants. The living is a rectory, valued in the king's books at £12. 1. 10½.; present net income, £557; patron, Earl of Carlisle. The church is partly in the early and partly in the later style of English architecture. The Rev. R. Wood, in 1712, in exercise of a power reserved to him, gave a rent-charge of £5 for teaching ten poor children; and a National school for girls has been erected. Here is a fine bed of limestone, abounding with organic fossil remains. A Roman road from Malton to the westward passes through the parish; and here was a castle belonging to the family of Lacy before the Conquest, and afterwards to that of Mowbray, which Richard III. subsequently held till his death: it was partly rebuilt, in the later style of English architecture, by Sir C. Cavendish, in 1603, but the walls only are remaining.

SLIPTON (*ST. JOHN THE BAPTIST*), a parish, in the union of THRAPSTONE, hundred of HUXLOE, Northern Division of the county of NORTHAMPTON, 3¼ miles (W. by N.) from Thrapstone; containing 155 inhabitants. The living is a discharged vicarage, valued in the king's books at £5. 12. 3½.; present net income, £104; patron and impropriator, Duke of Dorset. A day and Sunday school is supported by Mr. and Mrs. Germain.

SLOLEY (*ST. BARTHOLOMEW*), a parish, in the hundred of TUNSTEAD, Eastern Division of the county of NORFOLK, 3½ miles (N. N. E.) from Coltishall; containing 267 inhabitants. The living is a discharged rectory, valued in the king's books at £5. 6. 8.; patron, B. Cubitt, Esq. The tithes have been commuted for a rent-charge of £240. 10., subject to the payment of rates, which on the average have amounted to £32; the glebe comprises 24 acres, valued at £42 per annum. Here is a National school.

SLOUGH, a village, partly in the parish of STOKE-POGES, and partly in that of UPTON, hundred of STOKE, county of BUCKINGHAM, 41 miles (S. E. by S.) from Buckingham, and 21 (W.) from London: the population is returned with the parishes. A cattle market is held on Tuesday. One of the stations of the Great Western Railway is situated at this village, and is much frequented by visiters going to Windsor Palace and Eton College. Sir William Herschell, the celebrated astronomer, resided at this place, and here he constructed his powerful telescope: he was born in Hanover, in 1738, and died in 1822.

SLYNE, a joint township with Hest, in the parish of BOLTON-LE-SANDS, hundred of LONSDALE, south of the sands, Northern Division of the county palatine of LANCASTER, 2½ miles (N.) from Lancaster; containing 286 inhabitants. At Hest bank, in this township, a breakwater was constructed in 1820, along the side of which vessels from Liverpool and Glasgow load and unload their cargoes; and, by means of a canal extending to within a short distance of the shore, a considerable trade is carried on with Kendal and other inland places. The road across the sands to Ulverstone commences at Hest; and the great road to Kendal, Carlisle, and Glasgow, passes through the village of Slyne. Courts leet and baron are held here. There are traces of salt-works which formerly existed in the neighbourhood.

SMALESMOUTH, a township, in the parish of GREYSTEAD, union of BELLINGHAM, North-Western Division of TINDALE ward, Southern Division of the county of NORTHUMBERLAND, 8 miles (W. by N.) from Bellingham; containing 173 inhabitants.

SMALLBURGH (*St. Peter*), a parish, in the hundred of TUNSTEAD, Eastern Division of the county of NORFOLK, 5¼ miles (N. E. by E.) from Coltishall; containing 699 inhabitants. The living is a discharged rectory, valued in the king's books at £10. 4.; patron, Bishop of Norwich. The tithes have been commuted for a rent-charge of £420, subject to the payment of rates, which on the average have amounted to £50; the glebe comprises 28 acres, valued at £56. 15. per annum. The steeple of the church fell down in 1677, and occasioned great injury to the edifice. Here is a National school; and in the house of industry is another.

SMALLEY, a chapelry, in the parish of MORLEY, union of BELPER, hundred of MORLESTON and LITCHURCH, Southern Division of the county of DERBY, 7 miles (N. E.) from Derby; containing 792 inhabitants. The chapel is dedicated to St. John the Baptist. There are extensive collieries in the neighbourhood. John and Samuel Richardson, in 1712, conveyed property to trustees for the erection and support of a school for twelve poor boys, each to receive 8*d.* per week; the present annual income is £88, and 28 boys are educated.

SMALLFORD, a ward, partly in the parish of ST. STEPHEN, and partly in that part of the parish of ST. PETER, ST. ALBANS, which is in the hundred of CASHIO, or liberty of ST. ALBANS, county of HERTFORD: the population is returned with the parish. Here are almshouses for three widows.

SMALL-HYTHE, a chapelry, in the parish and hundred of TENTERDEN, Lower Division of the lathe of SCRAY, Western Division of the county of KENT, 2 miles (S. by E.) from Tenterden. The living is a donative, in the patronage of the Householders of Dumborne, in Tenterden; net income, £107. The chapel is dedicated to St. John the Baptist.

SMALLWOOD, a township, in the parish of ASTBURY, union of CONGLETON, hundred of NORTHWICH, Southern Division of the county of CHESTER, 3 miles (E. by S.) from Sandbach; containing 554 inhabitants. There is a place of worship for Wesleyan Methodists.

SMARDALE, a township, in the parish of KIRKBY-STEPHEN, EAST ward and union, county of WESTMORLAND, 2¾ miles (W. S. W.) from Kirkby-Stephen; containing 52 inhabitants. Smardale Hall, an ancient manor-house, formerly belonging to the Warrop and Dalston families, proprietors of the township, is now a farm-house. A chapel was anciently situated at a place now called Chapel Well.

SMARDEN (*St. Michael*), a parish (formerly a market-town), in the union of WEST ASHFORD, hundred of CALEHILL, Upper Division of the lathe of SCRAY, Eastern Division of the county of KENT, 8 miles (N. E. by E.) from Cranbrooke, and 46 (S. E. by E.) from London; containing 1177 inhabitants. The living is a rectory, valued in the king's books at £24. 2. 6.; present net income, £501; patron, Archbishop of Canterbury. There are places of worship for Baptists and Wesleyan Methodists: the former has a school attached. The market has fallen into disuse, but the market-house is still standing: a fair is held on Oct. 10th, for toys and pedlery. A free school was founded in 1716, by Stephen Dadson, who endowed it with property now producing upwards of £65 a year. A National school for boys has also been established.

SMEATON, GREAT, a parish, in the union of NORTH-ALLERTON, partly in the wapentake of ALLERTONSHIRE, and partly in GILLING-EAST, North Riding of the county of YORK, 6½ miles (N. by W.) from North Allerton; containing 510 inhabitants. The living is a rectory, with the perpetual curacy of Appleton-upon-Wisk annexed, valued in the king's books at £13. 13. 4.; present net income, £472; patron and incumbent, Rev. Henry Hewgill. The river Wisk runs through the parish, and the Tees flows along the northern side of it. A National school has been established.

SMEATON, KIRK (*St. Mary*), a parish, in the Upper Division of the wapentake of OSGOLDCROSS, West Riding of the county of YORK, 6½ miles (S. E.) from Pontefract; containing 318 inhabitants. The living is a rectory, valued in the king's books at £10. 1. 0½.; present net income, £370; patron, Earl Fitzwilliam. Here is a National school. There is a tram-road from Wentbridge, in this parish, to Heckbridge, for conveying the stone procured here for shipment to London.

SMEATON, LITTLE, a township, in the parish of BIRKBY, union of NORTH-ALLERTON, wapentake of ALLERTONSHIRE, North Riding of the county of YORK, 5½ miles (N. by W.) from North-Allerton; containing 67 inhabitants.

SMEATON, LITTLE, a township, in the parish of WOMERSLEY, Lower Division of the wapentake of OSGOLDCROSS, West Riding of the county of YORK, 6½ miles (S. E. by E.) from Pontefract; containing 222 inhabitants. Ann Jackson, in 1675, bequeathed £30 in support of a school.

SMEETH (*St. Mary*), a parish, in the union of EAST ASHFORD, and in the franchise and barony of BIRCHOLT, lathe of SHEPWAY, Eastern Division of the county of KENT, 4 miles (E. S. E.) from Ashford; containing 497 inhabitants. The living is annexed to the rectory of Aldington, in the patronage of the Archbishop of Canterbury. The church is principally in the Norman style of architecture. This was formerly a market-town; fairs are still held on May 12th and Michaelmas-day, for toys and pedlery. Timothy Bedingfield, in 1691, bequeathed an estate producing £111. 10. per annum, for teaching poor children of the parishes of Smeeth, Lyminge, and Dymchurch, also for apprenticing them or assisting them at the university, at the discretion of the trustees: from five to ten have been annually clothed and educated, and one is apprenticed every year; a fund is also reserved for such as may be sent to the university, or otherwise require aid.

SMEETON-WESTERBY, a township, in the parish of KIBWORTH-BEAUCHAMP, union of MARKET-HARBOROUGH, hundred of GARTREE, Southern Division of the county of LEICESTER, 5¼ miles (N. W.) from Market-Harborough; containing 475 inhabitants.

SMERRILL, a joint chapelry with Middleton, in the parish of YOULGRAVE, hundred of WIRKSWORTH, Southern Division of the county of DERBY, 5 miles (S. S. W.) from Bakewell: the population is returned with Middleton.

SMETHCOTT (*St. Michael*), a parish, in the union of CHURCH-STRETTON, hundred of CONDOVER, Southern Division of the county of SALOP, 9½ miles (S by W.)

from Shrewsbury; containing 366 inhabitants. The living is a discharged rectory, valued in the king's books at £4. 9.; present net income, £276; patrons, Trustees of Hulme's charity, Manchester. The Rev. Henry Fletcher, in 1810, gave £10 per annum, of which £5 is distributed among poor widows, and the remainder applied to the education of poor children.

SMETHWICK, a township, in the parish of BRERETON, union of CONGLETON, hundred of NORTHWICH, Southern Division of the county of CHESTER, 4 miles (N. E. by E.) from Sandbach: the population is returned with the parish.

SMETHWICK, a hamlet, in the parish of HARBORNE, union of KING'S-NORTON, Southern Division of the hundred of OFFLOW and of the county of STAFFORD, 3 miles (W. by N.) from Birmingham; containing 2676 inhabitants. This place is situated on the turnpike road from Birmingham through Oldbury to Dudley, and is intersected by the Old Birmingham canal, with its improved line of navigation, over which have been constructed six bridges and an aqueduct of cast-iron: one of them, called the Summit Bridge, has a span of 150 feet, and the crown of the arch is 60 feet above the level of the water. Several important works are carried on here, affording employment to the labouring classes; of these, the French Walls steel and iron works, belonging to the Bordesley Steel Company, employ 150 persons; the Smethwick soap and red-lead manufactory, 60 men; the foundry belonging to the Soho establishment, for the manufacture of steam-engines and boilers, 300; the Smethwick iron-works, 90; and the British Crown Glass Company's works afford employment to 250 men. A chapel, with a house for the minister, was founded by Mrs. Dorothy Parkes; it is endowed, and is independent of the mother church. The living is a perpetual curacy, in the patronage of Trustees. A chapel was erected in 1836, containing 770 sittings, 440 of which are free, the Incorporated Society having granted £400 in aid of the expense. There are places of worship for Independents and Wesleyan Methodists. Mrs. Dorothy Parkes also endowed a charity school with £9. 9. per annum and a house for a mistress.

SMISBY, a parish, in the union of ASHBY-DE-LA-ZOUCH, hundred of REPTON and GRESLEY, Southern Division of the county of DERBY, 2½ miles (N. N. W.) from Ashby-de-la-Zouch; containing 324 inhabitants. The living is a perpetual curacy; net income, £58; patron and impropriator, Marquess of Hastings. A day and Sunday school is supported by subscription.

SNAILWELL (*St. Peter*), a parish, in the union of NEWMARKET, hundred of STAPLOE, county of CAMBRIDGE, 2¾ miles (N.) from Newmarket; containing 236 inhabitants. The living is a rectory, valued in the king's books at £27. 11. 0½.; present net income, £535; patron, J. Thorp, Esq.

SNAINTON, a chapelry, partly in the parish of EBBERSTON, but chiefly in that of BROMPTON, union of SCARBOROUGH, PICKERING lathe, North Riding of the county of YORK, 9½ miles (S. W. by W.) from Scarborough; containing 636 inhabitants. The chapel is subordinate to the vicarage of Brompton, and contains 200 free sittings, the Incorporated Society having granted £200 in aid of the expense. There is a place of worship for Wesleyan Methodists.

SNAITH (*St. Mary*), a market-town and parish, in the union of GOOLE, chiefly in the Lower Division of the wapentake of OSGOLDCROSS, but partly in the Lower Division of the wapentake of BARKSTONE-ASH, West Riding of the county of YORK; containing 8530 inhabitants, of which number, 885 are in the town of Snaith, 23 miles (S. by E.) from York, and 175 (N. by W.) from London. The town is situated on a gentle declivity on the south bank of the river Aire: it is small and irregularly built; the streets are lighted with oil; the houses are chiefly of brick, and rather of mean appearance, but a few handsome and substantial dwellings have been lately erected; the inhabitants are well supplied with water from wells. Flax is cultivated in the neighbourhood to a considerable extent, and conveyed to the market at Leeds by the river Aire: the canal from Knottingley to Goole passes southward of the town. There is a tram-road from Heckbridge, in this parish, to Wentbridge. The market is on Thursday; and fairs are held on the last Thursday in April, and August 10th, for cattle, &c. Courts are occasionally held for the manor. This town has been made a polling-place for the West Riding. The living is a vicarage; net income, £479; patron, Nicholas Edmund Yarburgh, Esq. The church is an ancient and spacious structure, in the later style of English architecture, with a low square tower surmounted with pinnacles, and a belfry of wood. There is a place of worship for Wesleyan Methodists. The free grammar school is of unknown foundation: in 1741, Nicholas Waller gave a rent-charge of £30 for its endowment, but classical instruction has been long discontinued, and the endowment is now applied in aid of a National school. There are almshouses for six poor persons, founded by the Yarburgh family; and others for six poor widows, which were rebuilt in 1802, by Viscount Downe.

SNAPE (*St. John the Baptist*), a parish, in the union and hundred of PLOMESGATE, Eastern Division of the county of SUFFOLK, 2¾ miles (S. by E.) from Saxmundham; containing 514 inhabitants. The living is a discharged vicarage, consolidated with that of Friston, and valued in the king's books at £5. 5. 7½. The church contains an hexagonal font, much enriched in the later English style. Here is a National school. The parish is bounded on the south by the river Ore, which is crossed by a bridge, where there is a quay for shipping corn, &c. A society of Benedictine monks from the abbey of St. John at Colchester settled here in 1155, and in 1400 was exempted from all subjection to that house, and raised into a distinct priory; it was dedicated to the Blessed Virgin Mary, and at its suppression, in 1524, was granted to Cardinal Wolsey towards the endowment of his intended colleges, when its revenue was valued at £99. 1. 11.

SNAPE, a township, in the parish of WELL, union of BEDALE, wapentake of HANG-EAST, North Riding of the county of YORK, 3¼ miles (S.) from Bedale; containing 656 inhabitants, who are principally employed in wool-combing for the worsted-spinners in this part of the county. There is a place of worship for Wesleyan Methodists. An almshouse for eight aged persons, and free schools for children of each sex, have been founded and endowed by Lady Neville.

SNAREHILL-HOUSE, an extra-parochial district, in the hundred of GUILT-CROSS, Western Division of the county of NORFOLK, 1¾ mile (S.) from Thetford:

the population is returned with Rushford. This place and Thetford Lodge are all that remain of two villages, called Great and Little Snareshill.

SNARESTON (*St. Bartholomew*), a parish, in the union of Ashby-de-la-Zouch, hundred of Sparkenhoe, Southern Division of the county of Leicester, 7 miles (N. W.) from Market-Bosworth; containing 353 inhabitants. The living is annexed to the rectory of Swepstone. A school for boys is supported by endowment. The Ashby-de-la-Zouch canal passes through the parish.

SNARFORD (*St. Lawrence*), a parish, in the wapentake of Lawress, parts of Lindsey, union and county of Lincoln, 7 miles (S. W.) from Market-Rasen; containing 61 inhabitants. The living is a discharged rectory, valued in the king's books at £4; present net income, £181; patron, Subdean of Lincoln.

SNARGATE (*St. Dunstan*), a parish, in the union of Romney Marsh, and liberties of Romney Marsh and New Romney, though locally in the hundred of Aloesbridge, lathe of Shepway, Eastern Division of the county of Kent, 5½ miles (N. W. by W.) from New Romney; containing 85 inhabitants. The living is a rectory, about to be united to the rectory of Snave, valued in the king's books at £17. 6. 8.; present net income, £84; patron, Archbishop of Canterbury.

SNAVE (*St. Augustine*), a parish, in the union and liberty of Romney Marsh, though locally in the hundreds of Aloesbridge, Ham, and Newchurch, lathe of Shepway, Eastern Division of the county of Kent, 4½ miles (N. W. by N.) from New Romney; containing 91 inhabitants. The living is a rectory, valued in the king's books at £19. 7. 11.; present net income, £160; patron, Archbishop of Canterbury. The church is a spacious edifice of stone, with a handsome western tower.

SNEAD, a hamlet, in the parish of Rock, Lower Division of the hundred of Doddingtree, Hundred-House and Western Divisions of the county of Worcester, 5½ miles (S. W.) from Bewdley: the population is returned with the parish.

SNEATON, a parish, in the union of Whitby, liberty of Whitby-Strand, North Riding of the county of York, 2¼ miles (S. by W.) from Whitby; containing 230 inhabitants. The living is a rectory, valued in the king's books at £13. 12. 6.; present net income, £170: it was formerly in the patronage of the Crown, but is now in that of James Wilson, Esq., to whom, as lord of the manor, the advowson was granted in consideration of his having built, at his own expense, a church to replace the old parish church, which had been for some time in ruins. The present handsome structure is in the decorated style of English architecture, with a low tower surmounted by a small spire; the eastern end and the south porch are ornamented with buttresses terminating in richly crocketed finials. This parish and lordship is bounded by the river Esk, and is the property of James Wilson, Esq., by whom courts leet and baron are regularly held. The free school was established by the late Jas. Wilson, Esq., and is endowed with £10 per ann. out of the Sneaton estate, and £5 per ann. paid by the overseer.

SNELLAND (*All Saints*), a parish, in the Western Division of the wapentake of Wraggoe, parts of Lindsey, union and county of Lincoln, 5 miles (W. N. W.) from Wragby; containing 105 inhabitants. The living is a rectory, valued in the king's books at £3. 17. 6.; present net income, £227; patron, Earl Brownlow.

SNELSMORE, a tything, in the parish of Chieveley, hundred of Faircross, county of Berks, 3¾ miles (N.) from Newbury: the population is returned with the parish.

SNELSON, a township, in the parish of Rosthern, union and hundred of Macclesfield, Northern Division of the county of Chester, 5¼ miles (S. E.) from Nether Knutsford; containing 136 inhabitants.

SNELSTON (*St. Peter*), a parish, in the hundred of Appletree, Southern Division of the county of Derby, 2½ miles (S. W.) from Ashbourn; containing 484 inhabitants. The living is annexed to the rectory of Norbury. There is a place of worship for Wesleyan Methodists.

SNENTON (*St. Stephen*), a parish, in the union of Radford, Southern Division of the wapentake of Thurgarton and of the county of Nottingham, ¾ of a mile (E.) from Nottingham; containing 3605 inhabitants. The village, which a few years ago consisted of a few scattered houses, and contained not more than 250 inhabitants, now presents a most respectable appearance, some new streets and many elegant houses having been erected within the last few years: several of the inhabitants are employed in the manufacture of stockings and lace. The living is a perpetual curacy; net income, £227; patron and impropriator, Earl Manvers. A new church has been built, and contains 800 free sittings, the Incorporated Society having granted £700 in aid of the expense; it was consecrated on the 26th of September, 1839. Here is a Sunday National school. The county asylum for lunatics (described in the article on Nottingham) is in this parish. In the neighbourhood are some curious excavations in the stone rock, used as dwellings.

SNETTERTON (*St. Andrew*), a parish, in the union of Wayland, hundred of Shropham, Western Division of the county of Norfolk, 3 miles (N.) from East Harling; containing 247 inhabitants. The living comprises the consolidated rectories of All Saints and St. Andrew the Apostle, united to the rectory of Quiddenham, and valued in the king's books at £12. 17. 1. A school is chiefly supported by the rector.

SNETTISHAM (*St. Mary*), a parish, in the union of Docking, hundred of Smithdon, Western Division of the county of Norfolk, 6¾ miles (N. by E.) from Castle-Rising; containing 926 inhabitants. This place was anciently called Snetham; it had a market on Friday, which has been discontinued. The living is a discharged vicarage, valued in the king's books at £5. 6. 8.; present net income, £110; patron and impropriator, Henry L. Styleman, Esq. There is a place of worship for Wesleyan Methodists. A school is endowed for the instruction of 20 boys, and a National school has been established. Ancient brass instruments, in the shape of axe heads, with handles to them, usually termed celts, have been discovered in the neighbourhood.

SNEYD, a township, in the parish of Burslem, Northern Division of the hundred of Pirehill and of the county of Stafford: the population is returned with the parish. This township adjoins the town of Burslem on the west, of which it forms a portion. Extensive coal-works and mines of iron-stone are wrought here. There is a place of worship for Wesleyan Methodists.

SNIBSTON, a chapelry, in the parish of PACKINGTON, hundred of WEST GOSCOTE, Northern Division of the county of LEICESTER, 4½ miles (S. E.) from Ashby-de-la-Zouch: the population is returned with the parish, of which it forms a detached part, on the road from Leicester to Ashby-de-la-Zouch. The chapel is dedicated to St. Mary.

SNILESBY, a township, in the parish of HAWNBY, wapentake of BIRDFORTH, North Riding of the county of YORK, 10 miles (N. W.) from Helmsley; containing 116 inhabitants.

SNITTER, a township, in the parish and union of ROTHBURY, Western Division of COQUETDALE ward, Northern Division of the county of NORTHUMBERLAND, 2¾ miles (N. W. by W.) from Rothbury; containing 165 inhabitants.

SNITTERBY (*St. Nicholas*), a parish, in the union of CAISTOR, Eastern Division of the wapentake of ASLACOE, parts of LINDSEY, county of LINCOLN, 11¼ miles (N. W. by W.) from Market-Rasen; containing 182 inhabitants. The living is annexed to the rectory of Wadingham.

SNITTERFIELD (*St. James*), a parish, in the union of STRATFORD-UPON-AVON, Snitterfield Division of the hundred of BARLICHWAY, Southern Division of the county of WARWICK, 4 miles (N. by E.) from Stratford-upon-Avon; containing 770 inhabitants. The living is a vicarage, valued in the king's books at £8; present net income, £271; patron, Bishop of Worcester; impropriator, Robert Philips, Esq. The church exhibits portions in the early, decorated, and later styles of English architecture. Here is a National school.

SNITTERTON, a joint hamlet with Winsley, in the parish of DARLEY, union of BAKEWELL, hundred of WIRKSWORTH, Southern Division of the county of DERBY, 1½ mile (W. by N.) from Matlock: the population is returned with Winsley.

SNITTLEGARTH, a joint township with Bewaldeth, in the parish of TORPENHOW, ALLERDALE ward below Derwent, Western Division of the county of CUMBERLAND, 1½ mile (W. S. W.) from Ireby: the population is returned with Bewaldeth.

SNODLAND (*All Saints*), a joint parish with Paddlesworth, in the union of MALLING, hundred of LARKFIELD, lathe of AYLESFORD, Western Division of the county of KENT, 3¾ miles (N. E. by N.) from West Malling; containing, with Paddlesworth, 518 inhabitants. The living is a rectory, valued in the king's books at £20; present net income, £297; patron, Bishop of Rochester. There is a paper-mill on a stream tributary to the Medway. John May, Esq., in 1800, founded a school for 40 children, and vested the government of it in the magistrates for the lathe.

SNOREHAM (*St. Peter*), a parish, in the hundred of DENGIE, Southern Division of the county of ESSEX, 5 miles (S. S. E.) from Maldon; containing 222 inhabitants. The living is a rectory, valued in the king's books at £3; present net income, £103; patron, J. H. Strutt, Esq. There is not even a vestige of the church; the inhabitants attend that of Latchingdon, with which parish it is rated for the poor.

SNORING, GREAT (*St. Mary*), a parish, in the union of WALSINGHAM, hundred of NORTH GREENHOE, Western Division of the county of NORFOLK, 1¾ mile (S. S. E.) from Little Walsingham; containing 437 inhabitants. The living is a rectory, with that of Thursford annexed, valued in the king's books at £24; present net income, £584; patrons, Master and Fellows of St. John's College, Cambridge. A school is supported by subscription.

SNORING, LITTLE (*St. Andrew*), a parish, in the union of WALSINGHAM, hundred of GALLOW, Western Division of the county of NORFOLK, 3¼ miles (N. E. by E.) from Fakenham; containing 287 inhabitants. The living is a rectory, annexed to the vicarage of East Barsham, and valued in the king's books at £12: the amount of commutation for tithes is stated under the head of East Barsham.

SNOWSHILL, a parish, in the union of WINCHCOMB, Lower Division of the hundred of KIFTSGATE, Eastern Division of the county of GLOUCESTER, 6 miles (N. E.) from Winchcomb; containing 292 inhabitants. The living is annexed to the rectory of Stanton.

SNYDALE, a township, in the parish of NORMANTON, Lower Division of the wapentake of AGBRIGG, West Riding of the county of YORK, 4 miles (W. by S.) from Pontefract; containing 114 inhabitants.

SOBERTON, a parish, in the union of DROXFORD, hundred of MEON-STOKE, Droxford and Northern Divisions of the county of SOUTHAMPTON, 3¾ miles (E. by S.) from Bishop's-Waltham; containing 931 inhabitants. The living is annexed to the rectory of Meon-Stoke. The church is principally in the early style of English architecture. There is a National school in the parish.

SOCKBRIDGE, a township, in the parish of BARTON, WEST ward and union, county of WESTMORLAND, 3 miles (S. S. W.) from Penrith; containing 263 inhabitants. It is situated on the south bank of the river Eamont, and abounds with limestone. The ancient hall, a quadrangular building, with a tower, has been converted into a farm-house. A school is supported by subscription.

SOCKBURN (*All Saints*), a parish, in the union of DARLINGTON, partly in the South-Western Division of STOCKTON ward, county palatine of DURHAM, but chiefly in the wapentake of ALLERTONSHIRE, North Riding of the county of YORK, 7 miles (S. E.) from Darlington; containing 191 inhabitants. The living is a discharged vicarage, valued in the king's books at £3. 18. 1½.; present net income, £79; patrons and impropriators, Master and Brethren of Sherburn Hospital. The church, principally in the early English style, has been partly taken down, and a new church erected on the opposite side of the river; the old church contains some ancient monuments, one of which is that of the valorous Sir John Conyers, representing him with his feet resting upon a lion, that appears to be contending with a winged dragon. In an adjoining field is the Grey Stone, where, according to legendary story, the dauntless knight slew the "monstrous venomous and poysonous wyveron, ask, or worm, which overthrew and devoured many people in fight." In reference to this is the curious tenure of the manor of Sockburn, for an account of which, see NEASHAM. Here is a Sunday National school. The river Tees runs through the parish.

SOCK-DENNIS, county of SOMERSET. — See STOCK-DENNIS.

SODBURY, CHIPPING (*St. John the Baptist*),

a market-town and parish, and the head of a union, in the Lower Division of the hundred of GRUMBALD'S-ASH, Western Division of the county of GLOUCESTER, 28 miles (S. S. W.) from Gloucester, and 113 (W. by S.) from London; containing 1306 inhabitants. This town, which existed in the twelfth century, and was endowed by King Stephen with the same privileges as Bristol, is situated on the road from Bristol to Cirencester, at the foot of a hill near the source of the Little Avon. The market is on Thursday: fairs are held on May 23rd and June 24th, for cattle, cheese, and pedlery; and on the Friday before Lady-day and Michaelmas-day, both statute fairs. The town was governed by a bailiff until 1681, when the inhabitants were incorporated by charter of Charles II., which ordained that the municipal body should consist of a mayor, six aldermen, and twelve burgesses; but this grant was annulled by proclamation of James II., in 1688, at the request of the inhabitants: constables are now elected annually at the court leet of the lord of the manor. It has been made a polling-place for the western division of the shire. The living is a perpetual curacy; net income, £126; patron, Vicar of Old Sodbury. There are places of worship for Baptists and the Society of Friends. A free grammar school is endowed with £20 per annum from the funds of the town and church lands. The poor law union of Chipping-Sodbury comprises 23 parishes or places, under the care of 29 guardians, and contains, according to the census of 1831, a population of 17,931.

SODBURY, LITTLE (*St. Adeline*), a parish, in the union of CHIPPING-SODBURY, Lower Division of the hundred of GRUMBALD'S-ASH, Western Division of the county of GLOUCESTER, $2\frac{3}{4}$ miles (E. N. E.) from Chipping-Sodbury; containing 126 inhabitants. The living is a rectory, valued in the king's books at £6. 10. 10.; present net income, £235; patron, W. H. H. Hartley, Esq. On the brow of a hill, in this parish, are traces of an ancient camp, probably of Roman origin. A slight skirmish took place in its vicinity previously to the fatal battle of Tewkesbury, between the forces of Queen Margaret and the advanced guard of Edward IV., when several of the latter were taken prisoners. During the parliamentary war, Oliver Cromwell lodged for one night at the old manor-house.

SODBURY, OLD (*St. John the Baptist*), a parish, in the union of CHIPPING-SODBURY, Lower Division of the hundred of GRUMBALD'S-ASH, Western Divison of the county of GLOUCESTER, $1\frac{3}{4}$ mile (E.) from Chipping-Sodbury; containing 729 inhabitants. The living is a vicarage, valued in the king's books at £14. 8. $1\frac{1}{2}$.; present net income, £490; patrons and appropriators, Dean and Chapter of Worcester. Here is a Sunday National school.

SOFTLEY, a joint township with Lynesack, in the parish of ST. ANDREW-AUCKLAND, union of AUCKLAND, North-Western Division of DARLINGTON ward, Southern Division of the county palatine of DURHAM, 8 miles (S. S. E.) from Walsingham: the population is returned with Lynesack. This township, which is commonly called South Side, is bounded on the south by the river Gaunless, and contains several coal-works.

SOHAM (*St. Andrew*), a market-town and parish, in the union of NEWMARKET, hundred of STAPLOE, county of CAMBRIDGE, $5\frac{3}{4}$ miles (S. E.) from Ely, and 69 (N. N. E.) from London; containing 3667 inhabitants. This was a place of some note at a very early period. About 630, St. Felix, first Bishop of the East Angles, is said to have founded a monastery here, which he made the seat of his diocese prior to the removal of the see to Dunwich, and where his remains were interred: they were afterwards taken up and conveyed to Romney abbey, when the cathedral church was erected by Luttingus, a Saxon nobleman. This building, as well as the bishop's palace, was destroyed by fire, and the monks, who at that time were a flourishing society, were killed by the Danish army under the command of Inguar and Ubba, in 870. Before the draining of the fens, here was a large lake, or mere, over which was anciently a dangerous passage by water to Ely, but it was subsequently rendered more safe by the construction of a causeway through the marshes, at the expense of Hervey, Bishop of Ely. The town is situated on the east bank of the river Cam, on the verge of the county; the streets are irregularly built, and the houses of mean appearance. Horticulture is carried on to a considerable extent, especially in the article of asparagus; the dairies are abundant, and cheese of a most excellent quality, and very similar to that of Stilton, is made here. A market, formerly held on Thursday, has been disused for more than a century: the present market is on Saturday. Fairs are held on May 9th, for horses, cattle, and pedlery; and on the Monday before Midsummer, which is a pleasure fair; another, formerly held three days before Michaelmas, has been discontinued.

The living is a vicarage, valued in the king's books at £32. 16. $5\frac{1}{2}$.; patrons and impropriators, Master and Fellows of Pembroke Hall, Cambridge. The impropriate tithes have been commuted for a rent-charge of £672. 14., and the vicarial for £1653, subject to the payment of rates, which on the latter have, on the average, amounted to £203; the impropriate glebe comprises 78 acres, and the vicarial 4 acres, respectively valued at £160 and £7. 10. per annum. The church is a venerable cruciform structure, with a lofty square embattled tower, visible at a great distance; in the interior are several monuments. At Barraway, in this parish, is a chapel of ease; and there are places of worship for Baptists, Independents, Wesleyan Methodists, and Unitarians. The free school for boys is endowed with the profits of an estate of moor land, allotted for that purpose, on the division of the commons, in 1658, now producing £60 per annum; and children are apprenticed with a premium of £20 from funds given by Bishop Laney. A Sunday school is endowed with £10 per annum. Three almshouses were founded for poor widows, in 1502, by Richard Bond; and nine others, in 1581, by Thomas Peachey, but neither has any endowment, excepting an allowance for fuel. Some few vestiges of the ancient palace and cathedral church are yet visible, and several human bones were dug up at the east end of the street, near the church, a few years ago.

SOHAM, EARL (*St. Mary*), a parish, in the union of PLOMESGATE, hundred of LOES, Eastern Division of the county of SUFFOLK, $3\frac{1}{2}$ miles (W.) from Framlingham; containing 762 inhabitants. The living is a rectory, valued in the king's books at £10; present net income, £515; patron and incumbent, Rev. J. H. Groome. There is a place of worship for Baptists; also a Sunday National school. A fair is held on Aug. 4th, for lambs. Soham Lodge was the residence of the

Countess of Surrey, the wife of the poet; she died here, and was buried at Framlingham.

SOHAM, MONK (*St. Peter*), a parish, in the union and hundred of Hoxne, Eastern Division of the county of Suffolk, 6 miles (W. by N.) from Framlingham; containing 433 inhabitants. The living is a rectory, valued in the king's books at £19. 5. 2½.; present net income, £530; patron and incumbent, Rev. J. H. Groome.

SOHO, a hamlet, in the parish of Handsworth, Southern Division of the hundred of Offlow and of the county of Stafford, 2¼ miles (N. W.) from Birmingham, which see.

SOKEHOLME, a chapelry, in the parish of Warsop, union of Mansfield, Hatfield Division of the wapentake of Bassetlaw, Northern Division of the county of Nottingham, 4 miles (N. by E.) from Mansfield; containing 68 inhabitants.

SOLIHULL (*St. Alphege*), a market-town and parish, and the head of a union, in the Solihull Division of the hundred of Hemlingford, Northern Division of the county of Warwick, 13 miles (N. W.) from Warwick, and 105 (N. W.) from London; containing 2878 inhabitants. This town is situated on the road from Warwick to Birmingham, and consists principally of one street, with another branching off from the high road to the market-place: the houses in general are modern and well built, and many of them large and handsome; the inhabitants are well supplied with water from the river Blythe, which flows through the eastern extremity of the town, and from springs: the air is salubrious, and the surrounding scenery of a pleasing character. The Warwick and Birmingham, and the Stratford-on-Avon, canals pass through the parish. The market is on Wednesday; and fairs are held on April 29th, for cattle and horses; Sept. 11th, for horses and the hiring of servants; and Oct. 12th, for cattle. A court leet, at which a constable is appointed, is held occasionally in the town-hall, a neat modern brick building, beneath which is the market-place, and in the upper part assemblies sometimes take place. Petty sessions are also held every alternate Wednesday; and the town has lately been made a polling-place for the northern division of the county. The living is a rectory, valued in the king's books at £24. 18. 4.; patron, E. B. Clive, Esq. The tithes have been commuted for a rent-charge of £1500, subject to the payment of rates, which on the average have amounted to £175; the glebe comprises 91 acres, valued at £240. 10. per annum. The church is a large cruciform structure, partly in the later and partly in the decorated style of English architecture, with an embattled tower and octagonal spire rising from the intersection: the tracery, mouldings, and corbels, in the interior, are extremely elegant, and there are some fine specimens of tabernacle and screen work; near the west entrance is an ancient stone font of octagonal form, with round Norman pillars at the angles: in the chancel and transepts are several piscinæ in trefoil niches, with triangular canopies: the vestry-room was formerly a chapel, dedicated to St. Thomas à Becket. In 1757, the spire fell on the north transept, and broke the roof and some monuments; it was afterwards rebuilt. There is also a chapel at Shirley, dedicated to St. James, and containing 500 sittings, 300 of which are free, the Incorporated Society having granted £175 in aid of the expense: the living is a perpetual curacy; net income, £49; patron, the Rector. There are places of worship for Independents and Roman Catholics. Sundry donations prior to 1697 were directed by the court of Chancery to be applied, among other purposes, to the instruction of poor children of this parish; the annual income is upwards of £317, of which sum, the head-master, who must be a graduate of one of the Universities, receives £100 for teaching the classics, and an under-master £65, for conducting the English department. Shenstone, the poet, was educated here. Fifteen poor girls are instructed by a schoolmistress for £8 per annum, arising from the united bequests of Mrs. Martha Palmer and Mrs. Fisher, in 1746. Here is a National school. The poor law union of Solihull comprises 11 parishes or places, and contains, according to the census of 1831, a population of 11,433; it is under the care of 20 guardians. A Benedictine nunnery, in honour of St. Margaret, was founded at Hean wood, in this parish, in the time of Henry II., by Ketelburn de Langdon, the revenue of which was valued, at the dissolution, at £21. 2.

SOLPORT, a township, in the parish of Stapleton, union of Longtown, Eskdale ward, Eastern Division of the county of Cumberland; containing 354 inhabitants.

SOMBOURN, KING'S (*St. Peter and St. Paul*), a parish, in the union of Stockbridge, hundred of King's-Sombourn, Romsey and Southern Divisions of the county of Southampton, 3 miles (S.) from Stockbridge; containing 1046 inhabitants. This place, which, prior to the Conquest, was held in royal demesne, and now forms part of the duchy of Lancaster, was the residence of the celebrated John of Gaunt, of whose palace there are still some remains. What are thought to have been the ancient stables have been converted into a farm-house; and the gardens and pleasure grounds, with the park, fish-ponds, and an extensive bowling-green, encompassed by an earthwork about three feet high, may still be traced. The spinning of silk for the Winchester manufacturers formerly afforded employment to 50 women and children of the village; but it has been discontinued. Considerable quantities of chalk are sent from this neighbourhood to Redbridge, for the improvement of the strong clay soil in the New Forest, by the Andover canal, which passes through the parish, and is crossed by a bridge at a place called Horsebridge, on the line of the old Roman road from Winchester to Old Sarum. The living is a vicarage, with those of Little Sombourn and Stockbridge annexed, valued in the king's books at £21. 1. 10½.; present net income, £696; patron and impropriator, Rev. John Barker. The church is an ancient structure, containing some interesting details, among which is the figure of an ecclesiastic in robes within a trefoiled niche. There was formerly a chapel of ease at Compton, but it has long since gone to decay. On an eminence three miles to the north of the church are the remains of an encampment called the Ring, with a deep intrenchment enclosing an area of about 21 acres; and on the adjoining down, but within the parish of Stockbridge, are some of smaller dimensions, probably the outworks of the former.

SOMBOURN, LITTLE (*All Saints*), a parish, in the union of Stockbridge, hundred of King's-Sombourn, Winchester and Northern Divisions of the county of Southampton, 2 miles (S. E.) from Stockbridge; containing 84 inhabitants. The living is annexed, with that of Stockbridge, to the vicarage of King's-Sombourn.

SOMERBY (*All Saints*), a parish, in the union of Melton-Mowbray, forming, with the parishes of Cold Overton and Withcote, a detached portion of the hundred of Framland, Northern Division of the county of Leicester, 6 miles (S. by E.) from Melton-Mowbray; containing 377 inhabitants. The living is a discharged vicarage, valued in the king's books at £6. 16. 8.; present net income, £224: it is in the patronage of Mrs. M. Burnaby; impropriators, S. Smith, Esq., and others. The church is an ancient structure, with a tower and spire rising from the centre. There is a place of worship for Wesleyan Methodists; also a school endowed with £15 per annum.

SOMERBY, a chapelry, in the parish and wapentake of Corringham, parts of Lindsey, county of Lincoln, $2\frac{3}{4}$ miles (E.) from Gainsborough: the population is returned with the parish.

SOMERBY (*St. Mary Magdalene*), a parish, in the union of Grantham, wapentake of Winnibriggs and Threo, parts of Kesteven, county of Lincoln, 4 miles (S. E. by E.) from Grantham; containing 282 inhabitants. The living is a rectory, valued in the king's books at £11. 12. $3\frac{1}{2}$.; present net income, £645; patron, Lord Willoughby de Eresby. The church contains tablets to the Rev. John Myers, for 42 years rector of this parish; and to the widow of the late Robert Cheney, Esq., erected by Mrs. Myers and Miss Cheney in 1833. There is a chapel of ease at Great Humby.

SOMERBY (*St. Margaret*), a parish, in the union of Caistor, Southern Division of the wapentake of Yarborough, parts of Lindsey, county of Lincoln, $4\frac{1}{2}$ miles (E.) from Glandford-Bridge; containing 61 inhabitants. The living is a discharged rectory, valued in the king's books at £7. 7. 6.: it is in the patronage of the Crown. The tithes have been commuted for a rent-charge of £198. 2. 11., subject to the payment of rates, which on the average have amounted to £35; the glebe comprises 13 acres, valued at £18. 15. per annum.

SOMERCOATES, NORTH (*St. Peter*), a parish, in the union of Louth, Marsh Division of the hundred of Louth-Eske, parts of Lindsey, county of Lincoln, $8\frac{1}{4}$ miles (N. E.) from Louth; containing 753 inhabitants. The living is a vicarage, valued in the king's books at £9. 18. 4.; present net income, £394; patron, Chancellor of the Duchy of Lancaster; impropriator, G. Lister, Esq. There is a place of worship for Wesleyan Methodists; also a school for boys, endowed with a house and land producing £31 per annum.

SOMERCOATES, SOUTH (*St. Mary*), a parish, in the union of Louth, Marsh Division of the hundred of Louth-Eske, parts of Lindsey, county of Lincoln, $7\frac{1}{4}$ miles (N. E. by E.) from Louth; containing 320 inhabitants. The living is a rectory, valued in the king's books at £22. 6. 3.; present net income, £405: it is in the patronage of the Crown, in right of the Duchy of Lancaster.

SOMERFORD-BOOTHS, a township, in the parish of Astbury, union of Congleton, hundred of Macclesfield, Northern Division of the county of Chester, $2\frac{3}{4}$ miles (N. W. by N.) from Congleton; containing 297 inhabitants.

SOMERFORD, GREAT (*St. Peter and St. Paul*), a parish, in the union and hundred of Malmesbury, Malmesbury and Kingswood, and Northern, Divisions of the county of Wilts, 4 miles (S. E. by S.) from Malmesbury; containing 500 inhabitants. The living is a rectory, valued in the king's books at £12. 14. 7.; present net income, £347; patrons, Rector and Fellows of Exeter College, Oxford.

SOMERFORD-KEYNES (*All Saints*), a parish, in the union of Cirencester, hundred of Highworth, Cricklade, and Staple, Cricklade and Northern Divisions of the county of Wilts, $6\frac{1}{2}$ miles (W.) from Cricklade; containing 327 inhabitants. The living is a vicarage, valued in the king's books at £8; present net income, £261; patron and impropriator, G. S. Foyle, Esq. A day and Sunday school is supported by the lord of the manor and the vicar.

SOMERFORD, LITTLE (*St. John the Baptist*), a parish, in the union and hundred of Malmesbury, Malmesbury and Kingswood, and Northern, Divisions of the county of Wilts, $3\frac{1}{2}$ miles (S. E.) from Malmesbury; containing 376 inhabitants. The living is a rectory, valued in the king's books at £8. 7. 1.; present net income, £241; patron, Earl of Ilchester. A day and Sunday school is supported by subscription.

SOMERLEYTON (*St. Mary*), a parish, in the hundred of Mutford and Lothingland, Eastern Division of the county of Suffolk, $4\frac{1}{2}$ miles (N. W. by N.) from Lowestoft; containing 419 inhabitants. The navigable river Waveney bounds the parish on the west. The living is a rectory, valued in the king's books at £12; patron, Rev. Geo. Anguish. The tithes have been commuted for a rent-charge of £350, subject to the payment of rates, which on the average have amounted to £22; the glebe comprises 45 acres, valued at £68. 12. 4. per annum.

SOMERSALL-HERBERT (*St. Peter*), a parish, in the union of Uttoxeter, hundred of Appletree, Southern Division of the county of Derby, $3\frac{1}{2}$ miles (E. by N.) from Uttoxeter; containing 117 inhabitants. The living is a discharged rectory, valued in the king's books at £4. 18. 10.; present net income, £225; patron, Earl of Chesterfield.

SOMERSBY (*St. Margaret*), a parish, in the union of Horncastle, hundred of Hill, parts of Lindsey, county of Lincoln, 7 miles (N. W.) from Spilsby; containing 69 inhabitants. The living is a discharged rectory, valued in the king's books at £4. 16. $5\frac{1}{2}$.; present net income, £92; patron, Robert Burton, Esq.

SOMERSETSHIRE, a maritime county, bounded on the north-west by the Bristol Channel, on the south-west by Devonshire, on the south-east by Dorsetshire, on the east by Wiltshire, and on the north-east by Gloucestershire. It extends from 50° 48′ to 51° 27′ (N. Lat.), and from 2° 35′ to 4° 5′ (W. Lon.); and comprises an area of 1642 square miles, or 1,050,880 statute acres. The population amounts to 404,200. At the period of the Roman Conquest, the district now forming the county of Somerset was part of the territory of the Belgæ, a people of Celtic origin, who had migrated hither out of Gaul, about three centuries before the commencement of the Christian era. Between the native Britons and this tribe continued hostilities existed, in consequence of the efforts of the former to regain possession of the territory which had been taken from them. About 250 years after the first settlement of the Belgæ, Divitiacus, King of the Suessones, brought over to them from the continent a considerable army of their fellow-countrymen, and a treaty was concluded between the contending nations, in which a line of demarcation between the territories of each was agreed upon. This line consisted of a large and deep fosse defended by a rampart, called Wansdike, parts of which may still be traced: commencing at Andover in Hampshire, it traverses the county of Wilts, and, on approaching Somersetshire, crosses the Avon near Binacre, and again at Bathampton, whence it continues over Claverton down to Prior Park, Inglish-Combe, Stanton-Prior, Publow, Norton, and Long Ashton, and terminates on the shores of the Bristol Channel at Portishead, being 80 miles in length. Thus nearly the whole of Somersetshire was included in the territory of the Belgæ; and of the three chief cities of that people, two, Bath and Ilchester, were situated within its limits. In the Roman division of the kingdom it was included in Britannia Prima.

This county, except the parish of Bedminster, which has been transferred to the diocese of Gloucester and Bristol, is co-extensive with the diocese of Bath and Wells, in the province of Canterbury, and is divided into the archdeaconries of Bath, Wells, and Taunton, the first having no archidiaconal court, and in the two latter the bishop exercising jurisdiction concurrently with the archdeacons: the first contains the deaneries of Bath and Redcliffe; the second, those of Axbridge, Cary, Frome, Ilchester, Marston, and Pawlett, and the jurisdiction of Glastonbury; and the last those of Bridg-water, Crewkerne, Dunster, and Taunton: the total number of parishes is 469. For purposes of civil government it is divided into the hundreds of Abdick and Bulstone, Andersfield, Bath-Forum, Bempstone, Brent with Wrington, Bruton, Cannington, Carhampton, Catsash, Chew, Chewton, Crewkerne, North Curry, Frome, Glaston-Twelve-Hides, Hampton and Claverton, Hartcliffe with Bedminster, Horethorne, Houndsborough, Berwick and Coker, Huntspill and Puriton, Keynsham, Kilmersdon, Kingsbury (East and West), Martock, Mells and Leigh, Milverton, Norton-Ferris, North Petherton, South Petherton, Pitney, Portbury, Somerton, Stone, Taunton and Taunton-Dean, Tintinhull, Wellow, Wells-Forum, Whitestone, Whitley, Williton and Freemanners, and Winterstoke. It contains the cities of Bath and Wells, the borough, market, and sea-port town of Bridg-water; the borough and market-towns of Frome-Selwood and Taunton; the market and sea-port town of Watchet; the small sea-port town of Porlock; and the market-towns of Axbridge, Bruton, Chard, Crewkerne, Dulverton, Dunster, Glastonbury, Ilminster, Langport-Eastover, Milverton, Minehead, Shepton-Mallet, Somerton, Wellington, Wincanton, Wiveliscombe, and Yeovil. Under the act of the 2nd of William IV., cap. 45, the county has been divided into two portions, called the Eastern and the Western divisions, each sending two members to parliament: the place of election for the former is Wells, and the polling-places are Wells, Bath, Shepton-Mallet, Bedminster, Axbridge, Wincanton, Frome, Clutton, Congresbury, and Wedmore; that for the latter is Taunton, and the polling-places are Taunton, Bridg-water, Ilchester, Williton, Ilminster, Langport, Wiveliscombe, and Dunster. Two representatives are returned for each of the cities, and one for the newly-enfranchised borough of Frome: the old boroughs of Ilchester, Milborne-Port, and Minehead, have been disfranchised. Somersetshire is included in the western circuit: the Lent assizes are held at Taunton; the summer assizes at Bridg-water and Wells, alternately. The quarter sessions are held on Jan. 11th and April 19th, at Wells; on July 12th at Bridg-water; and on Oct. 18th at Taunton.

To describe the variety of surface with some degree of perspicuity, it is necessary to consider it as divided into three districts: the first comprehends the north-eastern portion of the county, included between the harbours of Uphill and King-road, on the west, and the towns of Bath and Frome on the east: the next and central division, which is much the largest, comprising the entire middle part of the county, from the borders of Wiltshire and Dorsetshire to the Bristol Channel, is bounded on the north-east by the Mendip hills, and on the south-west by the Quantock hills and the forest of Neroche: the third forms the remaining western part of the county. The general surface of the north-eastern district is finely varied by lofty hills, which command magnificent views over the fertile plains that lie beneath them: the western part of it, however, including the hundreds of Winterstoke and Portbury, consists of low moorlands, as they are called, which are subject to frequent inundation. In the parishes of Congresbury, Yatton, Banwell, Winscombe, Churchill, and Puxton, there are not less than 3000 acres, which, for the most part, discharge their waters into the small river Yeo, and are under the inspection of commissioners of sewers: at spring-tides the waters of the river rise 5 feet above the level of the adjacent lands. To the northward of these parishes lie nearly 4000 acres, in the parishes of Kenn, Kingston-Seymour, Clevedon, Nailsea, and Chelvey, alike subject to inundation, being secured from the sea by a wall of stone, elevated about 10 feet above the level of the lands within; this wall is sometimes overflowed by high tides, and, when strong westerly winds prevail at the equinoxes, it is frequently broken down by the impetuosity of the waves, so that many hundred acres are laid under water. This tract discharges its waters by two rivers, called the Little Yeos, at the mouths of which are sluices: it is also subject to frequent land-floods. South-westward of these parishes lie six others, liable to the same circumstance, and discharging their waters by a sluice at Uphill. Northward of these is Leigh down, a tract of elevated land of nearly 3000 acres, extending from Clevedon to the Hot-Wells, near Bristol, south-eastward of which is a vale of rich grass land, extending from Bedminster, on the north-east, in a south-westerly direction, to the low districts just mentioned. The extensive mountainous range, called the Mendip hills, stretches from Cottle's Oak, near the town of Frome, on the eastern side of the county, in a direction nearly west-north-west, immediately northward of Wells and Axbridge, to a place called Black Rock, on the Bristol Channel, near Uphill, a distance of more than 30 miles. In the middle division, on the borders of Wilts and Dorset, the lands are high and chiefly occupied either as sheepwalks, or in the production of corn. The country around Shepton, Bruton, Castle-Cary, Ilchester, Somerton, Langport, Petherton, and Ilminster, is exceedingly productive, both in corn and pasture, and abounds with good orchards and fine luxuriant meadows; westward of this extensive tract rise the Polden and Ham hills, with a bold aspect. A distinguishing feature in this division is its marshes, or fen lands, which are divided into two districts, called Brent marsh, and the Bridg-water or South marsh. By far the greater part of Brent marsh, about 20,000 acres, has been drained and converted into fine grazing and dairy lands: there yet remain considerable tracts of turf bog, upon which much improvement remains to be effected. The river Brue drains the greater portion of this marsh, and has a barrier against the tide, with sluices at Highbridge. The two principal bogs of this district, comprising several thousand acres, situated one on each side of this river, a little to the westward of Glastonbury, are five or six feet higher than the adjacent lands, and consist of a mass of porous earth, saturated with, and floating in, water: some parts of the drained lands are occasionally subject to land-floods. The Ax has no barrier against the tide; and the waters, both of this river and of the Brue, are much obstructed in their progress towards the sea by accumulations of mud. The divisions of property are here marked by ditches, which discharge their waters into the rivers. That part of the South marsh lying nearest the sea has a surface more elevated than the interior, owing to the great deposit of mud by the tides, in the course of successive ages: the same observation is also applicable to Brent marsh. The river Parret is the principal drain of this marsh; but it has no barrier against the tides, the consequence being that in rainy seasons many thousand acres are laid under water for a considerable time, rendering the herbage unwholesome, and the air unhealthy. These tracts having, in former times, been constantly subject to occasional inundations from the sea, it was found necessary to establish a Commission of Sewers, the members of which should examine and inspect the sea-banks, ditches, gutters, and sewers, and order the requisite cleansings and repairs. The first commission of this kind upon record was in 1304; and the like offices are continued to the present day. Part of this marsh, which has been more recently drained (about the end of the last century), is called King's Sedgemoor, and contains nearly 14,000 acres. There are other tracts similar to this, on the borders of the rivers Tone and Parret, nearly all of which were, in like manner, and about the same period, drained and improved; *viz.*, Normoor, near North Petherton; Stanmoor, Currymoor, West Sedgemoor, &c., near North Curry; West-moor, near Kingsbury; and Wet-moor, near Muchelney; amounting in the whole to about 10,000 acres, independently of many other low enclosed tracts, which are liable to occasional inundation.

The south-western division of the county has nearly an equal proportion of lofty hills and fertile slopes and vales. In the vale of Taunton-Dean, which comprises 30 parishes, and the market-towns of Taunton, Wellington, and Milverton, the prospect is agreeably relieved by a mixture of arable and pasture ground; but to the north-west are wild and mountainous tracts. The Quantock hills, extending nearly the whole of the distance between the town of Taunton and the sea; the Brandon hills, to the westward of these; and others in this part of the county, are noted for their wild and picturesque scenery. The highest point of the Quantock hills is 1270 feet above the level of the sea; the elevation called Dundry beacon, situated near the sea, is the highest point of land, being, according to the Ordnance survey, 1668 feet above the level of the sea. The mountainous parts of the county have a smooth, undulating, and rounded outline, seldom presenting cliffs or precipitous faces, except on the sea-shore. The extensive line of sea-coast is very irregular, in some places projecting in lofty and rocky promontories, and in others receding into fine bays, with low and level shores. From Stert-point, at the mouth of the Parret, northward, the shore is for a considerable distance entirely flat, and composed of vast sand banks, repelling the waters of the ocean, which anciently spread over these shoals, and covered the extensive district now called Brent Marsh. The general direction of the Somersetshire coast, from the western extremity until near the mouth of the Parret, is from west to east; here, however, commence the shores of the marshes of the middle district, which extending in a direction nearly from north to south, form, with the last-mentioned, the bay of Bridg-water, so called from the sea-port of that name, situated some miles up the river Parret. This bay is terminated on the north by the promontory formed by Breane down; beyond this are two smaller bays and promontories, between which and the mouth of the Avon the coast runs nearly in a north-easterly direction.

For its general fertility, Somersetshire is particularly eminent; and the variety of soil is so great, that almost every species may be found within its limits. That of the moorlands of the north-eastern division is, for the most part, a deep and rich mixture of clay and sand, a marine deposit. Leigh down has a thin gravelly soil, lying immediately upon limestone, and frequently not more than three inches from the surface, being therefore unfavourable for tillage: that of Broadfield down, a few miles further south, is of the same nature. The soil of the Mendip hills is, for the most part, deep and loamy; but tracts of an inferior quality, light, spongy, and black, occasionally intervene; and the loam is sometimes intermingled with pieces of stone, and with gravel, clay, or other substances, that alter its quality in different degrees. The soil of the Polden and Ham hills is of an inferior quality and a very thin staple. That of the marshes, or moors, of the middle district, is generally very fertile, and consists of four kinds, *viz.*, strong, dry, and fertile clay, of considerable depth, which is esteemed the most valuable; red earth, varying in depth from one foot to six feet, and covering the black moory earth; black moory earth, having a substratum of clay at various depths; and what is called turf-bog, which is of a light spongy texture, and so full of the fibrous roots of plants that it is with difficulty cut with a spade: under it is found a stratum of black earth, from one to two feet thick; and next occurs the peat, which is from three to fifteen feet in depth, full of flaggy leaves and the hollow stalks of rushes, together with bituminous matter: it is employed as the common fuel of the district. Southward of this extensive level is an elevated tract of great fertility, composed chiefly of sea sand and shells, well adapted for tillage. The soil of the fruitful vale of Taunton-Dean is a rich loam, interspersed in some places with clay, and in others with sand, and that of some other parts of the western division is little inferior; but the hills and forests are, for the most part, left in a state of nature; the Quantock hills have various thin soils, covering a thin shaly rock, and sometimes limestone. The soil of White-down is very various; that of Black-down consists of a thin surface of black earth on a bed of sand and gravel. On the sea-coast, for some distance westward of the mouth of the river Parret, the remains of a forest are discoverable at low water, which is described in the Geological Transactions, Vol. III., p. 380.

In the north-eastern district the proportion of arable land is very small; in the middle division it is greater, but almost wholly on the south-eastern side; in the vale of Taunton, in the western part of the county, there is much arable land, and on the northern side, approaching Watchet: the whole amounts to nearly 300,000 acres. The rotations of crops are various: those commonly cultivated are wheat, barley, oats, beans, and peas, the produce of which varies greatly. The wheat produced on the rising lands to the south of Bridgwater marsh is of very superior quality; and the best barley in the county is supposed to grow in the parishes of Chedzoy, Weston-Zoyland, Middlezoy, and Othery, in the same district: on the Mendip hills the favourite crop is oats, which are there produced in abundance and of good quality. The most common artificial grass is the broad clover: sainfoin is much cultivated in the north-eastern districts, as also are rye-grass, marl-grass, and White Dutch clover: the marl-grass grows spontaneously on the marl ground, and bears a striking resemblance to red, or broad clover. Potatoes are very extensively cultivated in different districts, more especially on the fertile soils in the vicinity of Castle-Cary, where 160 sacks per acre are a common produce. In the parishes of Wrington, Blagdon, Ubley, Compton-Martin, and Harptree, in the north-eastern district, teasel is extensively grown, chiefly on a strong rich clay; the produce is very uncertain. The head of this plant, which is composed of well-turned vegetable hooks, is used in dressing cloth; and the manufactures of Somersetshire and Wiltshire are for the most part supplied from these parishes: large quantities are also exported from Bristol to Yorkshire. Woad is also cultivated in this district, chiefly in the vicinity of Keynsham, the quality of which is much esteemed: three or four crops are commonly gathered in the season; the average produce per acre is about a ton and a half. In the rich tract extending from Wincanton, by Yeovil, to Crewkerne, a great deal of flax and hemp is grown. The arable lands are not near sufficient to supply the consumption of grain, many thousand quarters being annually brought from the counties of Wilts and Dorset. The grass lands are of very great extent; and the plains are remarkable for their luxuriant herbage, furnishing a supply of produce much more than sufficient for consumption; London, Bristol, Salisbury, and other markets, receiving great quantities of fat oxen, sheep, and hogs, besides cider, cheese, butter, and different other articles from this county. In the northern district, on the rich marsh land near the Bristol Channel, the grazing system prevails: in the vicinity of Bristol and Bath the meadows are almost universally mown, while, in the parts more remote from those towns, dairying is almost the only object. Nearly the whole of the rich marshes and low lands of the middle division are under grass, and applied partly to grazing and partly to dairying. In the western part of the county are some irrigated meadows of excellent quality, the greater part lying on steep declivities. Lime is the principal manure: marl is applied in those parts where it is found of good quality, and some of the marsh farmers on the river Brue cut openings in the banks of that river, in the winter, and thus irrigate their land with the muddy water descending from the hills.

The cattle of Somersetshire form an object of great importance in its agricultural economy. In the north-western district the cows, which are all for the dairy, are almost entirely of the short-horned breed. From Crewkerne, extending southward into Dorsetshire, is one of those deep large vales for which this county is remarkable, and in which commences a district about twenty miles square (one half in Dorsetshire, and the other in Somersetshire), noted for supplying the summer markets at Exeter with calves, which are there bought by the Devonshire farmers, and, after being pastured three or four years in that county, are sold to the Somersetshire graziers, who fatten them for the London market: these are of the kind called Devonshire cattle. The neat cattle of Taunton-Dean are of the North Devon breed, and are held in high esteem by the graziers; and, indeed, the oxen of the whole western district are remarkably well shaped: they perform the greater part of the agricultural labour. Lean cattle of the red Devonshire and Somersetshire breed are bought at the fairs of this county by the graziers of Leicestershire, Oxfordshire, Warwickshire, &c.; and besides the numerous fairs for the sale of cattle, both in the counties of Somerset and Devon, a large market is held every three weeks, during the summer months, at Somerton, on the southern side of the great central marshes, to which many lean cattle are brought, together with an immense number of lean sheep, the latter chiefly of the Dorsetshire breed: this is a market for fat cattle also. In the vicinity of Bath is a valuable breed of large sheep, some of which, when fattened, weigh from 30 to 40 lb. per quarter. The Mendip hills have a native and very hardy breed, with fine wool, which thrive on very scanty pasturage: their flesh is also much esteemed. In the south-eastern part of the middle division the sheep are an improved sort of the Dorsetshire breed; the number kept, chiefly in breeding flocks, is exceedingly great. In Taunton Dean they are of the Dorsetshire breed; and in this western portion of the county are two other kinds of sheep, one a native breed, without horns, somewhat resembling the Leicester sheep; the other, a small horned sort, which are bought when young at South Molton in Devonshire, and are kept on the Forest of Exmoor, or the adjoining hills, for two or three years, merely for the sake of their wool. A few Leicester sheep have been introduced into the county. The extraordinary number of hogs fattened in the north-eastern district are, for the most part, procured from the Bristol market, to which they are brought from Wales. The few bred are of different sorts, *viz.*, the native white breed, which has large ears and a long body; the Berkshire black and white kind; the Chinese breed; and a mixed sort. But few horses are bred in the county; the northern district is supplied by dealers, who bring them from the great fairs in the North of England. The northern district contains numerous orchards; those which have a northern aspect, and are sheltered from the westerly winds, are considered the most regular in bearing: the fruit produced at the northern base of the Mendip hills affords a strong and palatable cider: the favourite apple is here the Court of Wick pippin, which takes its name from the place where it was first cultivated. In the middle division are also many orchards, from which a considerable quantity of cider is made. In the vale of Taunton-Dean cider of the very finest quality is made with particular care.

The woods and plantations occupy about 20,000 acres; the north-eastern district is but partially covered, and according to the demand at the collieries, the wood it contains is cut at very irregular intervals. Kingswood, the timber of which is chiefly oak, covers about 230 acres. On the northern declivity of the Mendip hills are some good coppices, the principal of which are those of Blagdon, Hasel, and Ubley. On the opposite declivity are other coppices, of which Stokewood is the principal; these, from their exposed situation, are less productive. In the eastern part of this district are other woods, large and productive, such as those of Mells, Leigh, Edford, Harwich, Compton, Cameley, &c., being, from their vicinity to the coal-works, very valuable: in the same part are also many beautiful plantations. The valleys of the north-eastern district are richly adorned with elms. In the eastern part of the middle division is an extensive chain of woodland, several miles in length, from the parish of East Cranmore, through Downhead, Cloford, Whatley, Elm, &c., besides other woods of considerable size. On the borders of Wiltshire was the large forest of Selwood, extending from Penscellwood to within three miles of Frome, which was disafforested in the reign of Charles I.: it appears to have extended over a vale of about 20,000 acres, 18,000 of which have been cleared and converted into arable and pasture land, with a small portion of meadow: the remainder continues in coppice woods, the chief sorts of timber being oak and ash, while the underwood is principally hazel, ash, alder, willow, and birch: the chief natural defects of these woods are the coldness of the soil upon which they are situated, and their exposure, for the most part, to violent south-west winds: the coal-pits near Mendip are a constant market for the poles cut from the underwood. Numerous modern plantations have been made in this tract, in which the Scotch fir thrives best. The marsh lands have few trees of any kind. On the declivities of the Quantock and other hills, in the western part of the county, are many coppice-woods, chiefly of oak. This district does not otherwise abound with oak, but elms grow in the hedge-rows, sometimes to a considerable size.

This county has different uncultivated wastes: in the north-western district are several unenclosed commons, the principal of which are Broadfield down and Lansdown, the former containing about 2500 acres, the latter nearly 1000: the surface of Lansdown is perfectly smooth, and it is remarkable for its excellence in feeding sheep. The large open tract called Leigh down, to the west of Bristol, is also subject to a right of commonage, and is chiefly depastured with sheep. More than one half of the ancient royal forest of Mendip, on the hills of that name, is now enclosed; the remainder is covered to the extent of several miles with heath and fern, and furnishes pasturage for large flocks of sheep. In the middle division, the largest unenclosed upland common is the forest of Neroche, near Ilminster, containing 800 or 900 acres, and upon which different parishes have a right of commonage without stint: the next in size is White-down, near Chard: the low marshy wastes comprise several thousand acres. At the western extremity of the county, and partly in Devonshire (which see), is the great forest of Exmoor, extending from east to west for a distance of ten or twelve miles, and from north to south about eight, and containing nearly 20,000 acres. There are several hundred acres of uncultivated land on the Quantock and Brandon hills, and in some other parts. The wastes of that part of Black-down which lies within this county are supposed to exceed 1000 acres: the occupiers of estates contiguous to these hills stock them with young cattle in the summer months.

The chief mineral productions are coal, lead, calamine, limestone, freestone, and various other kinds of stone: fullers' earth, marl, and ochre, are also occasionally found. The coal beds are the nearest to London of any yet discovered, and constitute the most southern deposit of that mineral in England. The deposit is comparatively small, and lies northward of the eastern parts of the Mendip hills: it may be divided into the northern and the southern; the former including the parishes of High Littleton, Timsbury, Paulton, Radstock, and the northern part of Midsummer-Norton; the latter, the southern part of Midsummer-Norton, Stratton-on-the-Foss, Kilmersdon, Babington, and Mells. This southern division comprises what have long been known as the Mendip collieries, and it is probable that they were once within the verge of the extensive forest of Mendip, but they are now in the midst of old enclosures, and their ancient name has become obsolete. In the northern collieries the strata of coal dip about nine inches in the yard, their thickness varying from ten inches to upwards of three feet; they are seldom worked if less than fifteen inches. The southern division is worked on a more limited scale: the strata here dip from 18 to 30 inches in the yard, while in some places they descend perpendicularly: their number is 25, varying in thickness from 6 inches to 7 feet, but they are seldom worked when less than 18 inches. At Clapton, a village lying to the north-west of Leigh down, and west of Bristol, is a coal-work, possessing the advantage of a land level of 44 fathoms. Under the rich and extensive vale lying to the south-east of Leigh down are beds of coal thought to be inexhaustible: some thousands of bushels are now daily raised at the pits in this district. The principal stratum is 5 feet thick, sometimes rather more: this and the other strata generally dip towards the south, about nine inches in the yard, seldom more than a foot in the same extent: the rocks above the coal being full of fissures, considerable inconvenience is experienced from the influx of water. The Mendip hills, which consist chiefly of limestone of that kind called, in mineralogical language, mountain limestone, are famous for their mines, chiefly of lead and lapis calaminaris. Those of the former metal are nearly exhausted, or at least, the deep working is so incumbered with water, that little can be done in them. In former times, however, many thousand pounds have been annually paid to the see of Wells for the lord's share (one-tenth) of the lead dug in the forest, in the parish of Wells only: on Broadfield down also there are veins of lead. In the parishes of Rowberrow, Shipham, and Winscombe, are valuable mines of lapis calaminaris, which is sometimes found within a yard of the surface, and is seldom worked deeper than 30 fathoms: in the parishes of Compton-Martin and East-Harptree are also mines of the same kind. From these parishes, eastward, through the whole tract of Mendip, to Mells, at its eastern extremity, are likewise found marks and indications of calamine; and at Merchant's hill, in the parish of Binegar, several tons of it have been raised. The Mendip mines are governed by a set of laws and orders, commonly called Lord Choke's Laws, which were enacted in the time of Edward IV., when, on some disputes arising, that monarch sent Lord Choke, the Lord Chief Justice of England, down to his royal forest of Mendip, and the said laws and orders were agreed upon by the lords royal of Mendip, *viz.*, the Bishop of Bath, Lord Glaston, Lord Benfield, the Earl of Chewton, and my lord of Richmond, at a great meeting then held at a place called the Forge. According to these, the miners are allowed to turn upon the forest as many cattle in summer as they are able to keep in winter: but before becoming such they must crave licence of the lord of the soil, where they purpose to work, or in his absence, of his officers, after which they proceed to break the ground: a tenth part of the ore must be paid to the lord, and a tenth part of the lead also, if it be smelted on his territory: every lord of the soil ought to hold a minor court twice a year, and to swear twelve men of the same occupation, for the redress of misdemeanors: the lord or lords may issue arrests for strife between man and man, on account of their works and for obtaining the payment of their own duties: and if any miner should by misfortune meet his death by the earth falling in upon him, or by any other accident, the other miners are bound to fetch him out of the earth, and bring him to Christian burial, at their own costs and charges; nor shall any coroner or officer at large have to do with him in any respect. The mountain limestone formation near Bristol, forming a feature in English geology, constitutes the hills rising from beneath the red marl to the west of Bristol, and forms a range of considerable elevation, through which the Avon passes, in its course to the Severn. These hills consist of a prodigious number of strata, of very different natures, but chiefly of limestone of several varieties, the dip of which is about 45 degrees. Some of the limestone strata contain different organic remains; and an assemblage of numerous strata, called the Black Rock, from the colour of the limestone, which is here quarried for paving-stones, contains numerous fossils and rounded concretions, penetrated by petroleum, which sometimes exudes from the rock. Very few of these numerous beds of limestone are quarried, and many of them will not burn into good lime. Calamine, accompanied by heavy spar and galena, is found in veins of calcareous spar, crossing the limestone. Manganese is found in a vein of iron-stone, crossing the limestone. The strata alternating with the limestone are beds of clay of various kinds, which sometimes contain nodules of coral and geodes of iron-ore: thin beds of iron-stone and quartzose sand are also found; besides a bed of coal about two inches thick. The mountain limestone ranges round Bristol, in almost every direction, forming a kind of irregular basin, and reposes on the red sand-stone, which visibly passes beneath it. On the top of the limestone strata forming the cliffs, on the side of the Avon, lies a yellowish sandstone, which has sometimes the appearance of breccia, occurring also in some other parts of the same district. The red clay in the neighbourhood of Bristol contains gypsum, and abounds with sulphate of strontian, in veins and large beds. Red sand-stone is found under the limestone of the Mendip hills. The mountainous part of the western district of this county is formed of a series of rocks, differing much in mineralogical character, but a great proportion of them having the structure of sand-stones: some of the finest of these sand-stones graduate into a fine-grained slate, divisible into laminæ as thin as paper, and having a smooth, silky, and shining surface: their prevailing colours are reddish brown and greenish grey, with many intermediate mixtures; but some of the slaty varieties are of a purplish hue, occasionally spotted with green. In many places large beds of limestone, full of madrepores, are contained in the slate, which, towards the external parts of the beds, is interstratified with limestone. Copper, in the state of sulphuret and of malachite, and veins of hematite, are frequently met with; and nests of copper-ore, of considerable magnitude, have been found in the subordinate beds of limestone. The Quantock hills, Grabbist hill, Croydon hill, Brendon hill, and some others to the west of them, consist chiefly of the kind of stone called grey-wacké, in some places interstratified with limestone; the quarries of limestone in the eastern side of the Quantock hills are very numerous. North hill, extending along the sea shore from Minehead to Porlock, and forming a very bold and precipitous coast, is of grey-wacké; and the whole of the precipitous coast of the county presents a great variety of mineralogical strata. Granite, of small grain, occurs near the foot of a hill a few miles to the north-east of Taunton, where it is quarried to a small extent; the inhabitants of the neighbourhood call it pottle stone. The kind of limestone called by mineralogists *lias*, and which extends in a direction nearly north-east and south-west almost to the banks of the Humber, commencing in Dorsetshire a little to the west of Ilchester, passes by Bath, and occupies a large tract of this county; it is also found on the coast of Somerset, whence it extends for some distance inland; proceeding eastward, it first occurs a little distance to the west of Watchet, and forms some very high cliffs to the eastward of that place; the Polden hills consist chiefly of lias: this stone burns into a very strong quick lime, valuable, when made into mortar, for its increasing hardness under water. The greatest quantity of freestone is raised at Coombe down, where the ground has been undermined for several miles.

The principal manufactures are those of woollen and worsted goods at Frome, Taunton, Wellington, and Wiveliscombe; of gloves, at Yeovil, Stoke, and Martick; of lace, at Chard and Taunton; of silk, at Taunton, Bruton, and Shepton-Mallet; of crape, at Taunton; and of knit worsted stockings, at Shepton-Mallet. Upon the Avon are several mills for preparing iron and copper, and others for the spinning of worsted, and the spinning and weaving of cotton. Many of the lower classes derive cheap and wholesome food from the salmon and herring fisheries of Porlock, Minehead, and Watchet, which are carried on to a considerable extent; the other fish found off this coast, and which are occasionally taken at different places upon it, are tublin flounders, sand-dabs, hakes, pipers, soles, plaice, skate, conger-eels, shrimps, prawns, crabs, muscles, and star-fish.

The chief rivers are the Lower Avon, the Parret, the Tone, the Brue, and the Ax, of which the first, besides constituting the harbour of Bristol, is navigable for small craft as high as Bath, a distance of sixteen miles above that port. The Parret forms the harbour of Bridg-water, and falls into Bridg-water bay at Stert point: the navigable part of its course commences at Langport, whence to Stert point is a distance of about twenty miles. The Tone becomes navigable at Taunton to the Parret at Boroughbridge, about eight miles from Taunton, and near the centre of the county. The Brue is navigable up to Highbridge, a distance of two miles from its mouth. The smaller streams are very numerous; they all flow through fertile tracts, and the banks of many of them are adorned with extensive ornamented grounds belonging to the various seats of the nobility and gentry, with which this county abounds: some of the principal are the Yeo, the Cale, the Chew, the Frome, the Ivel, and the Barl. The Kennet and Avon canal enters the county from Bradford in Wiltshire, and joins the Lower Avon at Bath. The Somersetshire coal canal commences in the Kennet and Avon canal at Limpley-Stoke, near Bradford, and proceeds to Poulton; a railway branching from it in the parish of South Stoke extends to the collieries at Wilton and Clandown. The Grand Western canal enters this county near the parish of Thorn-St. Margaret's, and proceeds to Taunton. A canal has been formed from Taunton to Bridg-water, and it is in contemplation to construct a branch to Chard, the expense of which is estimated at £40,000. The Great Western railway enters the county at Bathford, and passes by Bath and Keynsham to Bristol. The line of the Bristol and Exeter railway also passes through the county.

The remains of antiquity that have been found are very various. The parish of Stanton-Drew, in the north-eastern district, is remarkable as containing the remains of four clusters of huge massive stones, forming two circles, an oblong and an ellipsis, which are supposed to have anciently constituted a Druidical temple. The ancient boundary called Wansdyke may be traced in several places; in the vicinity of its course, near Great Bedwin, celts and ancient instruments of war have been discovered. Besides the Roman cities of Bath and Ilchester, there are numerous places which, although their names have been changed or altered, since that remote period, still bear evident marks of Roman origin in the foundations of some of their walls, and in various remains that have from time to time been dug from them: such are Camalet, Hamden, Wellow, Coker, Chilcompton, Conquest, Wiveliscombe, Bath-Ford, Warley, Street, Long Ashton, Postlebury, South Petherton, Watergore, Wigborough, Yeovil, Putsham, Kilton, Stogumber, Edington, Inglish-Combe, &c. Among the many miscellaneous remains of this people which have been discovered, more especially at Bath, are included temples, sudatories, tessellated pavements, altars, hypocausts, and coins of different ages. Traces of ancient encampments are still visible at Blacker's hills, Bowditch, Brompton, Bury Castle, and Burwalls; in the parish of South Cadbury, near Chesterton, in the parish of Chew Magna, Cowes Castle, Doleberry, Douxborough Castle, Godshill, Hawkridge Castle, on Hampton down, Masbury, Mearknoll, Modbury, on the forest of Neroche, Newborough, Norton-Hautville, Stantonbury, Stokeleigh, Tedbury, Trendle Castle, Turk's Castle, Wiveliscombe Castle, and Worleberry. The principal Roman road was the Fosse-way, which extends across this county from Bath, in a south-westerly direction to Perry Street, on the confines of Devonshire. In a direction nearly parallel with this ran another road from the forest of Exmoor, through Taunton, Bridg-water, and Axbridge, to Portishead, whence there was a *trajectus* or ferry, across the Bristol Channel to the city of *Isca Silurum*, new Caerleon. On Salisbury hill are traces of the earthworks thrown up at the time of the siege of Bath by the Saxons. An encampment called Jack's Castle, near Wilmington, is supposed to have been of Danish formation. The intrenchments formed by the forces of Harold, near Porlock, in 1052, are still to be seen.

According to Tanner, the number of religious houses in the county, of all denominations, before the Reformation, including two Alien priories, was about 44. There are remains of the abbey in the Isle of Athelney, founded by King Alfred; of that of Banwell, founded in the same reign; of those of Bath, Bruton, Cliff, Glastonbury, Hinton, Keynsham, Muchelney, and Wells; of the priories of Barlinch, Barrow, Bath, Berkeley, Buckland, Sordrum, Cannington, Chewton, Dunster, Frome, Hinton-Charterhouse, Ilchester, Kewstoke, Montacute, Portbury, Stavordale, Stogursey, Taunton, Woodspring, and Yeanston; and of the nunneries of Nunney, Walton, and Whiteball. Remains still exist of the ancient castles of Bridg-water, Dunster, Montacute, Stoke-under-Hamdon, Stowey, Taunton, and Walton. Combe-Sydenham, near Stogumber, is a very ancient mansion, the seat of the family of the Sydenhams. The more modern seats of nobility and gentry are particularly numerous. Besides the celebrated waters of Bath, there are mineral springs of different properties at Alford, Ashill, Castle-Cary, East Chinnock, Glastonbury, Queen-Camel, Wellington, and Wells: at Nether Stowey is a petrifying spring. In the Mendip hills, and surrounded by wild and magnificent scenery, is Wokey Hole (so called from the neighbouring village of Wokey), an extensive natural cavern, the most celebrated in the West of England, in which the waters of the Ax take their rise, issuing from it in a clear and rapid stream: in the parish of Cheddar, in the same district, is an immense chasm in the hills, called Cheddar Cliffs, the scenery of which is particularly rugged and striking. Somersetshire abounds with rare and curious plants: on the hilly wastes occur the dwarf juniper, the cranberry, and the wortleberry, the last being here provincially called *hurts*. The rocks on the coast have great quantities of the *lichen marinus*, or sea-bread: in the low moors grows the *gale*, or candleberry myrtle. The county of Somerset gives the title of Duke to the family of Seymour.

SOMERSHAM (*St. John the Baptist*), a parish (formerly a market-town), in the union of St. Ives, hundred of Hurstingstone, county of Huntingdon, $8\frac{3}{4}$ miles (E. N. E.) from Huntingdon, and $64\frac{1}{4}$ (N.) from London; containing 1402 inhabitants. This town, formerly called *Summersum*, is supposed to have derived its name from an adjacent hill, which was the site of a summer camp of the Romans: it is situated in a fertile country, abounding with springs of remarkable purity, some of which were considered to possess medicinal qualities, but are now disused. Several of the inhabitants are employed in preparing wicks for rushlights, which are sent to various parts of the kingdom. The market, long since discontinued, was on Friday: fairs are held on June 23rd and November 12th, but they are very inconsiderable. The living is a rectory, with Colne and Pidley annexed, valued in the king's books at £40. 4. 7.; present net income, £1770: it is annexed to the Regius Professorship of Divinity in the University of Cambridge. The tithes of Somersham have been commuted for a rent-charge of £531. 10., subject to the payment of rates, which on the average have amounted to £46; the glebe comprises 20 acres, valued at £34 per annum. There is a place of worship for Baptists. A free school is endowed with the proceeds of £200, the bequest of Thomas Hammond, in 1746, and with some land assigned by the commissioners in 1765: the whole income is £26 per annum. The Feoffees' estate, yielding £55 annually, is applied to repairing and maintaining a bridge over a stream called Cranbrook, on the road from Somersham to Colne, also a causeway leading from the bridge to the church. The Bishops of Ely had formerly a palace here.

SOMERSHAM (*St. Mary*), a parish, in the union and hundred of BOSMERE and CLAYDON, Eastern Division of the county of SUFFOLK, 6 miles (N. W. by W.) from Ipswich; containing 446 inhabitants. The living is a rectory, valued in the king's books at £8: it is in the patronage of Mrs. Stubbin. The tithes have been commuted for a rent-charge of £270, subject to the payment of rates. There are places of worship for Independents and Baptists; also a Sunday National school.

SOMERS-TOWN, a chapelry, in the parish of ST. PANCRAS, Holborn Division of the hundred of OSSULSTONE, county of MIDDLESEX, 2 miles (N. W.) from St. Paul's Cathedral, London. The living is a perpetual curacy; net income, £400; patron, Vicar of St. Pancras. This place has, within the last 30 years, become a very populous neighbourhood; a particular description of it is given in the article on ST. PANCRAS.

SOMERTON (*St. James*), a parish, in the union of BICESTER, hundred of PLOUGHLEY, county of OXFORD, 3½ miles (S. E.) from Deddington; containing 392 inhabitants. The living is a rectory, valued in the king's books at £15. 1. 10½.; present net income, £225; patron, Rev. H. Wintle. On the north side of the tower of the church is represented, in stone, our Saviour between the two thieves, and over the communion table is a painting (after the manner of L. Da Vinci) of Christ and the Eleven Apostles at the Last Supper. In the churchyard is a handsome stone cross, having on its south side a fine crucifix in basso relievo. The river Cherwell and the Oxford and Birmingham canal pass through the parish. Here was a castle, probably built in the reign of Stephen, as appears from the will of Thomas Fermor, Esq., dated 1580, by which he bequeathed "the castle-yard and chapel therein" to his executors, who erected on its site a free school, with a house for the master, and endowed it with £10 per annum, for which about fifteen boys are instructed. There is also a girls' school, endowed by the Countess of Jersey with £20 a year. Some remains of the old mansion of the Fermors still exist, particularly the large western window, which gave light to the grand hall; and very recently an apartment could be traced, termed the Prince's chamber, from its having been once occupied by James II., who granted a charter for a fair at Somerton, which was held in a place now called Broad Pound. There is a powerful petrifying spring in the parish, forming a small cascade.

SOMERTON (*St. Michael*), a market-town and parish, in the union of LANGPORT, hundred of SOMERTON, Western Division of the county of SOMERSET, 5 miles (N. N. W.) from Ilchester, and 123 (W. S. W.) from London; containing 1786 inhabitants. This was anciently the chief town in the county. During the Saxon era, a castle was erected here, which became a royal residence; it was subsequently converted into a state prison, and was the place of custody of many distinguished persons, among whom was John, King of France, who was removed hither from Hertford castle by Edward III.: its site is now occupied by the gaol, in the erection of which the materials of the ancient edifice were used. The town is situated near the river Cary, over which is a stone bridge, and consists of several narrow streets. The market is on Tuesday: the fairs are on the Tuesday in Passion week, and the third, sixth, ninth, and twelfth Tuesdays following, for cattle; and September 30th and November 8th, for cattle, sheep, hogs, and pedlery. The town is governed by a bailiff and constables, annually chosen by the inhabitants; and the county magistrates hold petty sessions in the town-hall, which stands in the centre of the market-place. The living is a vicarage, valued in the king's books at £16. 0. 7½.; present net income, £259; patron and impropriator, Earl of Ilchester. The church is an ancient structure, with an octagonal embattled tower on the south side. There is a place of worship for Independents. The free school was founded in the 27th of Charles II., by Thomas Glover, who endowed it with an estate producing £10. 10. per annum, in addition to which is a rent-charge of £5, given by Alice Yates, for the education of twelve poor boys of Hurcot and Easter, in this parish. On the eastern side of the hill, above the village of Hurcot, are considerable quantities of fine white alabaster.

SOMERTON (*St. Margaret*), a parish, in the union of SUDBURY, hundred of BABERGE, Western Division of the county of SUFFOLK, 7 miles (N. E. by N) from Clare; containing 141 inhabitants. The living is a rectory, valued in the king's books at £6. 16. 8.; present net income, £207; patron, Marquess of Downshire. The church stands on high ground, and commands an extensive prospect; it has remains of Norman architecture, and contains the family vault of Viscount Blundell. A Sunday National school is supported by the rector.

SOMERTON, EAST (*St. Mary*), a parish, in the hundred of WEST FLEGG, Eastern Division of the county of NORFOLK, 6¼ miles (N. N. W.) from Caistor; containing 54 inhabitants. The living is annexed to the rectory of Winterton. The church has been long since demolished.

SOMERTON, WEST (*St. Mary*), a parish, in the

hundred of West Flegg, Eastern Division of the county of Norfolk, 6¼ miles (N. W. by N.) from Caistor; containing 243 inhabitants. The living is a perpetual curacy; net income, £56; patron, Thomas Grove, Esq. The church is thatched, and has a tower round at the base, and octangular above: it contains 60 free sittings, the Incorporated Society having granted £20 in aid of the expense.

SOMPTING, a parish, in the union of Steyning, hundred of Brightford, rape of Bramber, Western Division of the county of Sussex, 2 miles (E. N. E.) from Worthing; containing 519 inhabitants. The living is a discharged vicarage, valued in the king's books at £8. 7.; present net income, £150; patron and impropriator, E. Barker, Esq. The church is a cruciform structure, principally in the early English style, with a tower at the west end, apparently of more ancient date, having a pilaster, or moulding, running up the centre of each face; it contains 80 free sittings, the Incorporated Society having granted £70 in aid of the expense.

SONNING (*St. Andrew*), a parish, in the union of Wokingham, partly in the hundred of Binfield, county of Oxford; and partly in the hundred of Charlton, and partly in that of Sonning, county of Berks, 3½ miles (E. N. E.) from Reading; containing 2588 inhabitants. This parish, which is of considerable antiquity, is said, during the separation of Berkshire and Wiltshire from the ancient see of Sherborne, to have been the seat of a diocese; but the fact has not been satisfactorily established. The Bishops of Salisbury had a palace here, in which Isabel, queen of the deposed monarch Richard II., resided from the time of his imprisonment in Pontefract castle till his lamentable death in 1399. The village is situated on the banks of the river Thames, over which is a wooden bridge, and the surrounding scenery is agreeably diversified. The line of the Great Western railway intersects the parish. The living is a vicarage, valued in the king's books at £20. 7. 1.; present net income, £451; patrons, Dean and Chapter of Salisbury; impropriator, R. Palmer, Esq. There is a place of worship for Independents. Sir Thomas Rich, in 1766, founded a free school, which he endowed with an estate now producing £52. 10. per annum, for educating and clothing 29 poor boys, and for apprenticing three annually in London. Mr. Payne, in 1709, bequeathed a rent-charge of £5, for placing out an apprentice; and Dame Harriet Read, in 1804, bequeathed the dividends on £500 South Sea annuities for the education, clothing, and apprenticing of poor boys, at the discretion of the vicar, which, with additional benefactions, produces an income of £26. 13. 8., of which £10 per annum is paid in aid of the Sunday school and school of industry.

SOOTHILL, a township, in the parish and union of Dewsbury, Lower Division of the wapentake of Agbrigg, West Riding of the county of York, 6 miles (N. W. by W.) from Wakefield; containing 3849 inhabitants. An ancient building, now used as a malt-house, is supposed to have been originally a church, or chapel.

SOPLEY (*St. Michael*), a parish, in the union and hundred of Christchurch, Ringwood and Southern Divisions of the county of Southampton, 2¾ miles (N.) from Christchurch; containing 1012 inhabitants. The living is a vicarage, valued in the king's books at £12. 16. 10½.; present net income, £230; patron, H. Compton, Esq.; impropriator, W. Wyndham, Esq. There is a place of worship for Baptists; also a National school and a Lancasterian school, partly supported by subscription; and a day and Sunday school, supported by Lady Fane. The village is pleasantly situated on the left bank of the river Avon.

SOPWORTH (*St. Mary*), a parish, in the union of Malmesbury, and in a detached portion of the hundred of Chippenham, Malmesbury and Kingswood, and Northern, Divisions of the county of Wilts, 7½ miles (W. by S.) from Malmesbury; containing 222 inhabitants. The living is a rectory, valued in the king's books at £8. 10. 5.; patron, Duke of Beaufort. The tithes have been commuted for a rent-charge of £174, subject to the payment of rates, which on the average have amounted to £24, the glebe comprises 101 acres, valued at £84. 12. per annum. An infants' school is chiefly supported by the Duchess of Beaufort.

SOTBY (*St. Peter*), a parish, in the union of Horncastle, Eastern Division of the wapentake of Wraggoe, parts of Lindsey, county of Lincoln, 5 miles (E. by N.) from Wragby; containing 157 inhabitants. The living is a discharged rectory, valued in the king's books at £9. 0. 10.; present net income, £193: it is in the patronage of the Crown. A Sunday school is supported by the minister.

SOTHERTON (*St. Andrew*), a parish, in the union and hundred of Blything, Eastern Division of the county of Suffolk, 4¼ miles (E. N. E.) from Halesworth; containing 196 inhabitants. The living is a discharged rectory, annexed to that of Uggeshall, and valued in the king's books at £5. 6. 8. The tithes have been commuted for a rent-charge of £275, subject to the payment of rates, which on the average have amounted to £40; the glebe comprises 12 acres, valued at £17. 3. per annum.

SOTTERLEY (*St. Margaret*), a parish, in the union and hundred of Wangford, Eastern Division of the county of Suffolk, 4½ miles (S. E. by S.) from Beccles; containing 243 inhabitants. The living is a discharged rectory, valued in the king's books at £10; present net income, £257; patron, Frederick Barne, Esq., whose ancestor, Sir George Barne, was Lord Mayor of London in the time of Edward VI. The church contains 100 free sittings, the Incorporated Society having granted £20 in aid of the expense. A day and Sunday school is supported by subscription.

SOTWELL (*St. James*), a parish, in the union of Wallingford, hundred of Moreton, county of Berks, 1¾ mile (N. W. by W.) from Wallingford; containing 157 inhabitants. The living is annexed to the rectory of St. Leonard, Wallingford. The rectorial tithes have been commuted for a rent-charge of £119. 11. 4., and the Provost and Scholars of Queen's College, Oxford, and the Warden of God's House, Southampton, have a portion of the impropriate tithes, which have been commuted for a rent-charge of £102. 10. 10., subject to the payment of rates, which, on the average, including the rectorial, have amounted to £21; the remainder of the impropriate tithes have been commuted for a rent-charge of £50. 18. 9., subject to the payment of rates, which on the average have amounted to £6.

SOUGHTON, or SYCHDIN, a township, in the parish of Llansillin, hundred of Oswestry, Northern

Division of the county of SALOP, $3\frac{3}{4}$ miles (S. W. by W.) from Oswestry; containing 247 inhabitants.

SOULBURY (*All Saints*), a parish, in the union of LEIGHTON-BUZZARD, hundred of COTTESLOE, county of BUCKINGHAM, 5 miles (W. N. W.) from Leighton-Buzzard; containing 578 inhabitants. The living is a perpetual curacy; net income, £116: the patronage and impropriation belong to Lady Lovett. There is a place of worship for Wesleyan Methodists. Robert Lovett, in 1710, and the Rev. John Sambee, in 1728, founded and endowed a school with land, tithes and other property, now producing an annual income of £77. 7. 10. The master receives a salary of £40, and has the free use of the school premises, comprising a residence and garden, for which he instructs gratuitously 24 boys and girls, some of whom are annually apprenticed out of the remaining funds. There are also some other trifling bequests, which are appropriated to charitable purposes. The London and Birmingham railway intersects the parish, about a mile to the north-east of the church.

SOULBY, a township, in the parish of DACRE, LEATH, ward, Eastern Division of the county of CUMBERLAND, 5 miles (S. W.) from Penrith: the population is returned with the parish. The village is situated on the margin of the beautiful lake Ullswater.

SOULBY, a chapelry, in the parish of KIRKBY-STEPHEN, EAST ward and union, county of WESTMORLAND, 2 miles (N. W.) from Kirkby-Stephen; containing 256 inhabitants. The living is a perpetual curacy; net income, £78; patron, Sir George Musgrave, Bart.; impropriator of the rectorial tithes, J. Wakefield, Esq. The chapel was erected in 1663, at the expense of Sir Philip Musgrave, Bart. This is a considerable village, situated on the rivulet Scandale, which is here crossed by a bridge of three arches, erected in 1819. Two fairs for cattle and sheep are held on the Tuesday before Easter, and August 30th: they are of recent establishment, and are well attended. Six poor boys are educated for a trifling annuity, the gift of several individuals since 1768.

SOULDERN (*St. Mary*), a parish, in the union of BICESTER, hundred of PLOUGHLEY, county of OXFORD, 4 miles (E. by S.) from Deddington; containing 599 inhabitants. The Oxford and Birmingham canal passes through the parish, and the river Cherwell forms the western boundary. The living is a rectory, valued in the king's books at £8. 14. 2.; present net income, £453; patrons, Master and Fellows of St. John's College, Cambridge. The church is a curious ancient edifice, with a tower in the Norman style; the Norman arch which separated the nave from the chancel was destroyed in rebuilding the latter; and other alterations and repairs have defaced the original character of the building. There is a place of worship for Wesleyan Methodists. A National school for boys is supported partly by contributions, and partly by a sum of money, amounting to £200, left by Miss Westcarr for that purpose, and producing £6. 7. per annum.

SOULDROP (*All Saints*), a parish, in the hundred of WILLEY, union and county of BEDFORD, $5\frac{1}{2}$ miles (N. E. by N.) from Harrold; containing 242 inhabitants. The living is a discharged rectory, united in 1735 to that of Knotting, and valued in the king's books at £10. The church has been rebuilt, but the ancient steeple remains. A Sunday school is supported by the rector and the parish.

SOULTON, a township, in the parish of WEM, Whitchurch Division of the hundred of NORTH BRADFORD, Northern Division of the county of SALOP; containing 31 inhabitants.

SOUND, a township, in the parish of WYBUNBURY, union and hundred of NANTWICH, Southern Division of the county of CHESTER, 3 miles (S. W. by S.) from Nantwich; containing 255 inhabitants.

SOURTON (*St. Thomas à Becket*), a parish, in the union of OAKHAMPTON, hundred of LIFTON, Lifton and Southern Divisions of the county of DEVON, $4\frac{1}{2}$ miles (S. W.) from Oakhampton; containing 625 inhabitants. The living is annexed to the rectory of Bridestowe. The village is situated on the high road from Oakhampton to Tavistock. A school is partly supported by subscription.

SOUTH-ACRE (*St. George*), a parish, in the union of SWAFFHAM, hundred of SOUTH GREENHOE, Western Division of the county of NORFOLK, $3\frac{1}{2}$ miles (N. by W.) from Swaffham; containing 96 inhabitants. The living is a rectory, valued in the king's books at £10. 18. $1\frac{1}{2}$.; patron, A. Fountaine, Esq. The tithes have been commuted for a rent-charge of £510, subject to the payment of rates; the glebe comprises $44\frac{1}{2}$ acres, valued at £40 per annum. At Racheness, in this parish, there was, in the time of Henry II., an hospital for lepers, subordinate to the priory of Castle-Acre.

SOUTHALL, a chapelry, in the parish of HAYES, union of UXBRIDGE, hundred of ELTHORNE, county of MIDDLESEX, $9\frac{1}{2}$ miles (W.) from London. A considerable weekly market for fat cattle is held here every Wednesday. A chapel has been built and endowed at Southall Green, by Henry Dobbs, Esq., in whom and his heirs the patronage is vested. The Great Western railway has a station at this place.

SOUTHAM, a hamlet, in the parish of BISHOP'S-CLEEVE, union of WINCHCOMB, hundred of CLEEVE, or BISHOP'S-CLEEVE, Eastern Division of the county of GLOUCESTER, $2\frac{1}{2}$ miles (N. E.) from Cheltenham: the population is returned with Brockhampton.

SOUTHAM (*St. James*), a market-town and parish, and the head of a union, in the Southam Division of the hundred of KNIGHTLOW, Southern Division of the county of WARWICK, 10 miles (E. S. E.) from Warwick, and 84 (N. W.) from London; containing 1256 inhabitants. This town, anciently called *Suthau*, is a place of great antiquity, and had formerly a mint. In an old mansion near the centre of the town, which appears to have been built prior to the reign of Elizabeth, Charles I. and his two sons are said to have slept, on the night before the battle of Edge Hill, in which a son of the Earl of Pembroke, who was buried in the church, was slain. The parochial register for the year 1641 contains an entry of money paid to the king's footman for opening the church doors, which had been locked up and sealed by the king's order, as a punishment to the inhabitants for not ringing the bells on his entering the town. The monks of Coventry had a religious establishment here, and in Bury orchard, near the churchyard, foundations have been discovered, and many skeletons dug up. The town is pleasantly situated on an eminence rising from the eastern bank of the river Stowe, and consists of two streets:

the houses in general are modern and well built, the inhabitants are well supplied with water from springs, and the surrounding scenery is pleasingly diversified. The river Stowe is crossed by a neat stone bridge of two arches, at the lower extremity of the town; and on the rising ground on the opposite side an antique mansion forms a striking contrast with the other buildings. The market, formerly on Wednesday, is now held on Monday, and is well supplied with corn: fairs are held on Easter-Monday, the Monday after Holy Thursday, and July 10th, for cattle and horses; the last of these is a show fair, at which, in imitation of that at Coventry, the procession of Lady Godiva is celebrated. A constable and headborough are appointed annually at the court leet of the lord of the manor. The town is a polling-place for the southern division of the shire.

The living is a rectory, valued in the king's books at £22. 17. 6.; present net income, £534: it is in the patronage of the Crown. The church is a stately structure, principally in the decorated style of English architecture, with a fine tower surmounted by a lofty spire; in the roof of the nave, which is lighted by eight clerestory windows, enriched with tracery, is some tabernacle-work well carved in oak. There is a place of worship for Baptists. A free school was founded 1762, and endowed with lands previously given for the relief of the poor, the proceeds of which amount to about £30 per annum, and the funds have been further augmented with £30 per annum, from the rents of the town lands; and a National school has been established. There is also an endowment of £200 per annum for the relief of the poor and the repairs of the bridge. A self-supporting dispensary, for the relief of the sick poor, was established here, on a peculiar plan, by Mr. Smith, a resident surgeon, who has been the means of founding similar institutions in different parts of the country. An infirmary for curing diseases of the eye and ear, established by the same indefatigable gentleman in 1818, under the patronage of the nobility and gentry of the neighbourhood, is supported by annual subscriptions and donations. Here is a mineral spring of similar properties to the waters at Leamington, also another called Holywell. The Rev. Mr. Holyoak, author of the first collection of English words ever published in the form of a dictionary, was at one time rector of this parish.

SOUTH AMBERSHAM, county of SOUTHAMPTON.—See AMBERSHAM, SOUTH.—*And all places having a similar distinguishing prefix will be found under the proper name.*

Arms.

SOUTHAMPTON, a sea-port, borough, and market-town, and a county of itself, under the designation of "The Town and County of the Town of Southampton," locally in the Southern Division of the county of Hants, 75 miles (S. W. by W.) from London; containing 18,670 inhabitants. This place probably derives its name from the ancient British *Ant*, the original name of one of the rivers which empty themselves into its fine estuary. To the north-east of the present town, on the opposite bank of the Itchen, the Romans had a military station, called *Clausentum,* which was succeeded by the Saxon town of *Hantune,* on the site of the present Southampton. In 838, the Danes, with a fleet of 33 ships, effected a landing on the coast, but were repulsed with considerable loss by Wulphere, governor of the southern part of the county, under Ethelwolf; and in 860 they again penetrated into the county and burned the city of Winchester. In the reign of Athelstan, two mints were established here. In 981, a party of Danish pirates, having made a descent from seven large vessels, plundered the town, and laid waste the neighbouring coast. In the reign of Ethelred II., Sweyn, King of Denmark, and Olave, King of Norway, landed here with a considerable force, plundered and burned the town, massacred the inhabitants, and committed the most dreadful depredations in the surrounding country, till Ethelred purchased peace by the payment of £16,000, on the receipt of which the invaders retired to Hantune, where they embarked for their own kingdom. Canute, after his establishment on the throne, made this town his occasional residence; and it was whilst seated on the beach here, at the influx of the tide, that he took occasion to make that memorable reproof of his courtiers, for their gross flattery, which has been recorded by historians. Adjoining the walk round the marsh, on the present dock land, is to be seen the reputed remains of a double circular protection by posts and rails of the spot on which the occurrence took place; the tops of the posts a few inches above the mud only remain, and the progress of the docks will probably obliterate these, unless some record be placed to continue so interesting a remembrance. At the time of the Conquest, the town was so much reduced by the repeated incursions of the Danes, that at the Norman survey, the king had only 79 demesne tenants. Henry II. and his queen landed at this port, on their return from France, in 1174. In the reign of John, Adam de Port was governor of the castle; and in that of Edward III. the town was completely destroyed by the French and their allies, the Spaniards and Genoese, but they were repulsed, with the loss of the Prince of Sicily and other commanders. Richard II. enlarged the castle and strengthened the fortifications that had been erected for the defence of the town and harbour. Henry V., previously to the battle of Agincourt, marshalled his army here for his expedition against France, and during his stay in the town, detected a conspiracy formed against him by the Lords Cambridge and Scroop, and Sir Thomas Grey, who were here executed for treason, and buried in the chapel of an ancient hospital, still remaining, called God's House. In the reign of Edward IV., Southampton was the scene of a sanguinary contest between the partisans of the houses of York and Lancaster, in which the former having gained the victory, many of the Lancasterian chiefs were, by the king's order, executed with extreme barbarity. The town had increased materially in extent and importance, and its trade had become so flourishing, that, in the reign of Edward V., the Lord Mayor of London was appointed collector of the king's duties at this port. In 1512, Grey, Marquess of Dorset, embarked here with a force for the assistance of Ferdinand, King of Spain; and, ten years after, the Emperor, Charles V., sailed from it, on his return to his own dominions, after having visited Henry VIII. Edward VI., in his tour through the western and southern parts of the kingdom, for the benefit of his health, visited the town, and was sumptuously entertained by the mayor and corporation. Philip, King of Spain, on his arrival in England to espouse Queen Mary, landed at this port, and was entertained at the sheriff's house by the mayor and his brethren, who sent him a present of wine, which he received on board his ship, the *Grace de Dieu*, then lying in the harbour.

The town is beautifully situated on a peninsular tract of ground, rising with a gradual ascent from the north-eastern shore of Southampton water, and bounded on the east by the river Itchen, over which is a floating or steam bridge leading to Gosport, constructed under an act obtained in 1834; and on the south and west by the fine open bay formed by the confluence of the Itchen with the river Test. The shores of the bay, or estuary, are richly clothed with wood, and afford a succession of beautifully diversified scenery, the vicinity being studded with villages, mansions, and villas. Southampton water, about two miles broad at its entrance near Calshot castle, stretches north-westward nearly seven miles: on the eastern shore are the ruins of Netley abbey, forming an object romantically picturesque. The town, rising gradually from the margin of the water, is distinguished for the beauty of its situation; and the approach from the London road, through an avenue of stately elms and a well-built suburb, is striking. The principal entrance is through Bar gate, one of the ancient gates, on the north front of which are two gigantic figures representing Sir Bevois of Southampton and the giant Ascupart, whom, according to legendary tale, Bevois is said to have slain in combat. From this gate, which is embattled and machicolated, a spacious street, more than half a mile in length, leads directly to the quay, for the improvement of which the old Water gate was taken down about 28 years since. The ancient part of the town was formerly enclosed with walls nearly a mile and a quarter in circuit, of which, with their ruined circular towers, considerable portions are still entire, the principal being that reaching from the south-east of West gate along the shore northward. Of the ancient gates, the principal now remaining are West gate and South gate, in addition to Bar gate, in relation to which last, the more modern part of the town is distinguished by the appellation of Above Bar, from the other part, which is called Below Bar. In that part Above Bar are many fine ranges of building. A new street of handsome houses has been recently erected, leading from the street Above Bar to the western shore, with a terrace, commanding a fine view of the surrounding scenery. The town is well paved, lighted with gas, and supplied with excellent water, chiefly from springs collected on an adjoining common. An increased supply being required, an Artesian well is now being sunk on the common, to a depth of 560 feet, of which 76 feet are finished, cased with iron cylinders, of 13 feet diameter, to be continued to 150 of that size, and then gradually reduced. A Literary and Philosophical Society, established some years ago, by a proprietary of 30 members, is further supported by an unlimited number of annual subscribers of £1. 1.: lectures are given periodically. In a central part of the High-street, during six months of the year, there is an exhibition of paintings for sale, well known as the Hants picture gallery. A Medical Society was established in 1834. The principal library and reading-rooms are in the High-street, at which a book of arrivals is regularly kept: there are also two circulating libraries and several reading-rooms in other parts of the town, together with billiard-rooms, elegantly fitted up. A mechanics' institute, in Hanover-buildings, was established about ten years ago, and is very flourishing; there are nearly 400 members; it contains a museum, library, and lecture-room. Near the Platform is a subscription bowling-green. There are two assembly-rooms, one called the Long Rooms, erected on the west side of the town, in 1761; and the other, recently erected, called the Archery Rooms. The theatre, in French-street, is well arranged and tastefully decorated: the season commences in August. Races are held annually in the autumn, and continue two days: the course, which is well adapted to the purpose, is pleasantly situated on Southampton common, and was given to the town by the corporation. The Archery grounds, on the west bank of the Southampton water, form an agreeable promenade; and Mr. Page's botanic gardens adjoining contain a very extensive collection of indigenous and exotic plants, constantly keeping pace with the improved state of botanical science and discovery. There are two lodges of freemasons, which meet at the Masonic hall, Bugle street; and in 1837 a Provident Masonic Association was formed, being the first established in the provinces. An annual regatta takes place during the summer, in which prizes, given by subscription, are contested for by yachts and small vessels belonging to the fishermen of Itchen, on the Southampton river, than which none can be more favourably adapted to aquatic excursions, from the bay being so beautiful and finely sheltered. A Yacht club has been established, further to promote these sports.

The salubrity of the air, and the beauty of its situation, have made Southampton a resort for sea-bathing; and hot, cold, medicated, and vapour baths have been constructed. In addition to those previously established, a handsome and commodious building was erected, in the Grecian style of architecture, at an expense of £7000, near the Platform on the beach; this has been taken by the Dock Company, and is now the Southampton dock-house; it stands in the centre of their land, but the other baths have much improved. Numerous respectable lodging-houses are let for the accommodation of visiters. On the beach is a causeway planted with trees, extending above half a mile. On the Platform, which has been much enlarged, is an ancient piece of ordnance, presented by Henry VIII., and recently mounted on a handsome cast-iron carriage, the gift of John Fleming, Esq. The government have presented to the town six pieces of ordnance, to be used in saluting any of the royal family, and on public occasions of rejoicing; they are also placed on the Platform. The barracks, erected here during the late war, and occupying about two acres of land, were, in 1816, considerably enlarged, and converted into a military asylum, as a branch of the institution at Chelsea, under the patronage of the late Duke of York, for the orphan children of soldiers, and of those whose mothers are dead, and their fathers absent on service: the buildings are of brick, handsome and commodious, and are appropriated to the reception of female children only. At

Itchen Ferry, and on the western side of the town, are bathing-machines, with experienced guides. The environs are equally remarkable for the varied beauty of their scenery, and for the number of elegant mansions and villas. In addition to the numerous attractions which the town itself possesses, and the facilities afforded for aquatic excursions, there are, in various directions, extensive rides through a country abounding with objects of extreme interest, and enriched with a great variety of scenery.

The port, of which the jurisdiction extends from Langstone harbour, on the east, to Hurst castle on the west, and midway from Calshot castle to the Isle of Wight, carries on a considerable foreign trade: the imports are wine and fruit from Portugal; hemp, iron, and tallow, from Russia; pitch and tar from Sweden; and timber from other ports on the Baltic: it has also a considerable trade with Jersey and Guernsey. By act of parliament of Edward III., making Southampton one of the staple ports for the exportation of wool, all cargoes of that material, not originally shipped to those islands from this port, must either be re-landed here, or pay a duty at the custom-house. A coasting trade is also carried on with Wales, from which it imports iron and slates; with Newcastle, from which it imports coal, lead, and glass; and with various other places. The quay, on which stands a convenient custom-house, is accessible to vessels of 250 tons' burden, and is commodiously adapted to the despatch of business. A spacious stone-faced quay has been added on the eastern side for smaller craft. A new landing pier, for the convenience of passengers to and from the Isle of Wight, Guernsey, Jersey, and France, was constructed by act of parliament in 1832: it is 900 feet in length, curving at the eastern extremity for the accommodation of steam-packets; the carriage road is 20 feet wide, and on each side of it is a foot-path protected by railing; the pier, which is of timber, is lighted with gas, and forms an interesting and agreeable promenade. In 1837, it was discovered to have been nearly destroyed by submarine insects between high and low water mark; and, in consequence, all the piles have been replaced by others thickly studded with nails, to prevent the ingress of the insects. The first stone of the Southampton Docks was laid by Sir L. Curtis, Bart., October 12th, 1838. Belonging to the port are 179 vessels, averaging a burden of 45 tons. The amount of duties paid at the custom-house, for the year ended Jan. 5th, 1837, was £49,140. The harbour is spacious, and affords good anchorage for ships, which may ride at all times in security, being sheltered from all winds. Steam-vessels proceed regularly, all the summer and autumn, from this port to Havre, and to Jersey and Guernsey; and there are sailing packets on the same destination at all other seasons daily: steam-packets also afford a constant communication with the Isle of Wight and Portsmouth. The trade of the town principally arises from the wants of the inhabitants and visiters, and is facilitated by the Itchen canal navigation to Winchester, the river itself being navigable as far as Northam; and a 74-gun ship and several frigates were built in the docks here during the late war. A railway from London to Southampton, a distance of 77 miles, is nearly completed, under an act obtained in the year 1834. The market days are Tuesday, Thursday, Friday, and Saturday; the market on Friday is for corn: the markets are well supplied with fish, eggs, poultry, and provisions of every kind. The fairs are on May 6th and 7th, for cows and pigs, and on Trinity Monday and Tuesday: the latter, a very ancient fair, is proclaimed by the mayor with particular ceremony on the preceding Saturday, and continues till the Wednesday noon following, during which time the junior bailiff presides and entertains the corporation in a booth erected on the occasion: this fair, which is principally for horses, cattle, and pigs, is held on the eastern side of the town, near the site of an ancient hermitage, formerly occupied by William Geoffrey, to whom its revenue, arising from standings, &c., was originally granted; a court of pie-powder is attached to it, and during its continuance all persons are free from arrest for debt within the precincts of the borough.

Corporation Seal.

Obverse. *Reverse.*

The inhabitants were first incorporated in the reign of Henry I., whose charter was confirmed by Richard I., and by John, who assigned the customs of the port, together with those of Portsmouth, to the burgesses, for an annual payment of £200: their privileges were extended and confirmed by numerous subsequent sovereigns, and were modified by Charles I., by whose charter the government was vested in a mayor, sheriff, two bailiffs and an indefinite number of aldermen, assisted by a recorder, town-clerk, two coroners, four sergeants-at-mace, and subordinate officers. The corporation now consists of a mayor, 10 aldermen, and 30 councillors, under the act of the 5th and 6th of William IV., cap. 76.; for an abstract of which see the Appendix, No. I. The borough is divided into five wards; the council appoint a sheriff. The mayor and late mayor are justices of the peace, and ten others are appointed under a commission. The police force consists of a superintendent, 3 sergeants, and 23 constables. The borough exercised the elective franchise in the 23rd of Edward I., since which time it has regularly returned two members to parliament: the right of election was formerly vested in the burgesses, resident and non-resident, and in the inhabitants generally paying scot and lot, in number more than 1700; but the non-resident burgesses, except within seven miles, have been disfranchised. The borough comprises about 1970 acres; it was erected, with a surrounding district, into a county of itself by Henry VI. The number of voters now registered is 1454, of whom 32 are freemen; the mayor and bailiffs are the returning officers. The recorder holds quarterly courts of session for all offences not capital; and the corporation have the privilege of holding assizes, when the judges are travelling the western circuit, to try for capital crimes committed within the limits of the town and county of the town. A court of record is held every alternate Tuesday by the recorder, for the recovery of debts to any amount. Petty sessions are held daily, and a court leet annually. The inhabitants paying scot and lot have right of common on the Town Lands, adjoining the town, the most extensive of which is Southampton common, containing about 350 acres. The audit-house is a handsome building, erected about 50 years since, comprising in the upper story a spacious hall, in which the business of the corporation is transacted, and the records and regalia are deposited; among the latter, which are splendid, is a silver oar, borne before the mayor on public occasions, as the ensign of his admiralty jurisdiction. The guildhall is a spacious room above the arches of the ancient Bar gate, which is a beautiful and venerable structure in the Norman style of architecture: the principal archway is deeply moulded and enriched, and is flanked by circular embattled turrets, and the approach is ornamented with two lions sejant, cast in lead, presented to the corporation in 1744, in lieu of two which were decayed, by William Lee, Esq., on his being elected a burgess; the south side of the gateway is neatly faced with stone, with a niche in the centre, in which is a statue of George III., presented to the corporation by the late Marquess of Lansdowne, to replace a decayed figure of Queen Anne. The common gaol for the borough comprises four rooms for 50 prisoners, but does not admit of classification. The bridewell contains three rooms, capable of receiving ten prisoners, and a small chapel, in which divine service is performed once in the week. The sheriff's prison for debtors contains two wards, and is adapted to the reception of ten prisoners. This town has been made a polling-place and the place of election for the southern division of the county.

Southampton comprises the parishes of *All Saints*, *Holy Rood*, *St. John* and *St. Lawrence* united, *St. Mary*, and *St. Michael*. The living of All Saints is a discharged rectory, valued in the king's books at £8. 1. 10½.; present net income, £400: it is in the patronage of the Crown. The church, rebuilt on the enlarged site of an ancient structure, is in the Grecian style of architecture, with a turret at the east end rising from a square pedestal, and surrounded by six Corinthian columns, supporting a circular entablature surmounted by a dome. The area underneath the church is divided into arched catacombs, in one of which are deposited the remains of Captain Carteret, the celebrated circumnavigator, and of Bryan Edwards, author of the "History of the West Indies;" and on the north side of the altar is a mural tablet to the memory of the Rev. Dr. Mant, many years rector of the parish, and another to the memory of the Rev. Mr. Mears, late rector. The living of Holy Rood parish is a discharged vicarage, valued in the king's books at £12. 1. 10½.; present net income, £379; patrons, Provost and Fellows of Queen's College, Oxford. The church is an ancient structure in the High-street, with a tower and spire at the south-west angle, and has a portico in front: among the monuments is one by Rysbrach to Miss E. Stanley, sister of the Rt. Hon. Hans Stanley, with an epitaph written by the poet Thomson, who has immortalised her memory in his poem of the Seasons. The living of St. John's is a discharged rectory, united to that of St. Lawrence, and valued in the king's books at £6. 13. 4. The church has been demolished. The living of the parish of St. Lawrence is a discharged rectory, with that of St. John's annexed, valued in the king's books at £7. 10.; present net income, £148: it is in the patronage of the Crown. The church was a small ancient building, which has been taken down and a much larger and very excellent edifice erected, forming an ornamental feature in the High-street: it contains 160 free sittings, the Incorporated Society having granted £200 in aid of the expense. The living of St. Mary's is a rectory, in the precinct of the town, valued in the king's books at £37. 5. 5.; patron, Bishop of Winchester. The church is modern, and contains 880 sittings, half of which are free, the Incorporated Society having granted £350 in aid of the expense; it has a very extensive churchyard, which is the principal cemetery of the town. The living of St. Michael's is a discharged vicarage, valued in the king's books at £12. 11. 10½.; present net income, £145: it is in the patronage of the Crown. The church is an ancient and spacious structure, principally in the Norman style of architecture, with a tower between the nave and the chancel, surmounted by a lofty and well-proportioned octagonal spire: the massive circular columns that supported the roof have been replaced with lighter octangular pillars, and sharply pointed arches; the windows are of a later style, and the tracery of the large west window has been carefully restored, and the upper compartments embellished with stained glass; a new window of elegant design has been placed by the corporation in the chapel of this church, in which, from time immemorial, the mayors have been sworn into office: the ancient font is of Norman character and highly enriched: there are some ancient monuments, and in the chapel is an old cenotaph of Lord Chancellor Wriothesley, who in the reign of Henry VIII. passed sentence of death on Queen Anne Boleyn. St. Paul's, a proprietary chapel, in the parish of All Saints, a handsome edifice in the later style of English architecture, has been recently erected, under the superintendence of a committee, and has an east window embellished with stained glass. There are also a chapel dedicated to the Holy Trinity, and a free chapel dedicated to our Saviour, the livings of both of which are perpetual curacies: net income of the former, £110; patrons, certain Trustees;—and of the latter, £70; patron, Rev. W. Davies. The chapel dedicated to our Saviour has been enlarged, and 135 free sittings provided, the Incorporated Society having granted £100 in aid of the expense. There are a handsome and commodious place of worship for Independents, a neat one for Baptists, others for the Society of Friends and Wesleyan Methodists, and a neat Roman Catholic chapel of modern erection.

The free grammar school was founded in the reign of Edward VI.: the corporation have erected a convenient school-house, capable of accommodating 40 boarders, on the site of an ancient edifice, called West-hall: the endowment produces not more than about £30 per annum: among other eminent men who have been educated in this school was the celebrated Dr. Watts, a native of Southampton, whose father kept a boarding school in the town. A charity school was founded in 1760, for qualifying twenty boys for the sea service, by Alderman Taunton, of this town, who left considerable funds for charitable uses: the original number of scholars has, by a decree of the court of Chancery, been reduced to ten, who are permitted to choose any mechanical trade, if they prefer it: pensions of £10 per annum were also paid from these funds to six decayed persons of the town, of whom the number has been increased to sixteen by a bequest of the late Charles D'Aussy, Esq.; and £40 per annum is appropriated from the same funds, as a reward for female servants, and a portion on their marriage. A National and a Lancasterian school, are partly supported by subscription. There are also several infants' schools, and various Sunday schools in connection with the established church and the dissenting congregations. The ancient hospital of Domus Dei, or God's House, was originally founded in the reign of Henry III. partly as a convent for nuns, and as a chapel to a neighbouring friary, which was burned by the French in the reign of Edward III., by whom it was given to Queen's College, Oxford; after various changes it was established as an hospital for a warden, four brothers, and four sisters: the buildings are ancient, and retain much of their original character; the ancient chapel was long used as a place of worship by the French Protestants. The hospital of St. John, on the site of which the present theatre has been built, consisted of a master and six boys: it was sold in 1774, under an act of the 13th of George III., for £425, which was appropriated towards the erection of the present workhouse, in which is a school for 50 children, who are clothed, maintained, and instructed, in consideration of an annuity of £40 charged upon the property of the corporation. Thorner's almshouses, a neat and commodious range of building, receiving their name from the funds for erecting them having arisen from a bequest by Robert Thorner, Esq., in 1690, for gradual accumulation, were originally built in 1789, and have lately been enlarged: they now accommodate 26 widows. The same benefactor also bequeathed considerable funds for apprenticing poor children. Almshouses in St. Mary's parish were founded in 1565, by Richard Butler, mayor, and built on ground given by Thomas Lister, mayor, in 1545. Six small unendowed tenements, near the work-house, of which the origin is unknown, are appropriated by the corporation as residences for the poor. The penitentiary, or refuge for destitute females, supported by donations and subscription, is a spacious and commodious building, with a handsome chapel attached to it, recently erected in front of Kingsland-place. The public dispensary was established in 1823, and the ladies' lying-in charity in 1812. Miss Elizabeth Bird bequeathed £1400 three per cents. to the mayor and corporation, in trust for the annual payment of £5 each to six unmarried women, members of the church of England, and upwards of 60 years of age. There are a Royal Humane Society, and several benefit and friendly societies. The late Mr. Newman bequeathed nearly £3000 in the funds for the erection and maintenance of an infirmary, which was established in 1839 in a building rented for the purpose. Southampton is one of the 24 corporations entitled to a share of Sir Thomas White's lending charity; and there are various other charitable bequests both as lending funds and for distribution among the poor. At Bittern, about a mile and a quarter from the town, supposed to have been the old station *Clausentum,* numerous Roman antiquities have been discovered, among which were considerable vestiges of a fortification, and a portion of a Roman wall, coins from the reign of Claudius to those of Valentinian and Valens, tessellated pavements, bricks, fragments of pottery, urns, vases, and sculptured stones, on several of which were Roman inscriptions. Southampton gives the title of Baron to the family of Fitzroy.

SOUTHAMPTON, COUNTY of, a maritime county, on the southern coast, bounded on the east by the counties of Surrey and Sussex, on the north by that of Berks, on the west by Wiltshire and Dorsetshire, and on the south by the English Channel. Including the Isle of Wight, it extends from 50° 36′ to 51° 23′ (N. Lat.), and from 45′ to 1° 53′ (W. Lon.); and comprises an area of upwards of 1628 square miles, or 1,041,920 statute acres. The population amounts to 314,280.

At the period of the invasion of Britain by Cæsar, the southern parts of this district formed a portion of the territory of the Regni, and the more northern tracts part of that of the Belgæ, who had come over from Gaul, and violently dispossessed the former inhabitants. Under the Romans it was included in the division called *Britannia Prima.* The Isle of Wight, called by the Romans *Vectis,* is mentioned by Suetonius as having been conquered by Vespasian, about the year 43: no other traces of Roman occupation have, however, been at any time discovered in it than a few coins. On the establishment of the kingdom of Wessex, by Cerdic, a great part of the county was included within the limits of that kingdom, at the same time that a part of its southern shores, together with the Isle of Wight, was comprised in the Saxon kingdom of Kent. The ancient British name of this district was *Gwent,* or *Y Went,* a term descriptive of its open downs; and hence the appellation *Caer Gwent,* or the city of the Gwentians, now Winchester. When the Saxon dominions in Britain were divided into shires, this district received the name of *Hamtunscyre,* from the ancient name of the present town of Southampton: this was afterwards corrupted into *Hamptescyre,* and hence its modern appellations of Hampshire and Hants. The name of the Isle of Wight is considered by Mr. Whitaker and other antiquaries to have been derived from the British word *Guith,* or *Guict,* signifying the divorced or separated, and apparently indicating a supposition of its having once been connected with the main land: hence also arose its Roman name of *Vectis,* or the separated region: by the Saxons it was called Weet.

William the Conqueror, on his accession to the throne of England, granted the lordship of the Isle of Wight, with a palatine jurisdiction, to his kinsman, William Fitz-Osbert. It afterwards several times escheated to, and otherwise became vested in, the crown, and was as often granted to different noble families Sir Edward Widville, in the first of Henry VII., was made captain of the Isle of Wight; but whether he received a grant of the lordship of it, is uncertain; but since the period of his death it has remained in the possession of the Crown, although some lands annexed to the castle at Carisbrooke continue to be holden by the governor *jure officii.* From the time that Edward I. purchased this lordship of Isabella de Fortibus, the defence of the island was generally entrusted to some person nominated by the Crown, who was at first distinguished by the appellation of warden, afterwards by that of captain, and, in later times, by that of governor.

This county is included in the diocese of Winchester, and province of Canterbury: the archdeaconry of Winchester is co-extensive with the county, and comprises the deaneries of Alresford, Alton, Andover, Basingstoke, Droxford, Fordingbridge, Sombourn, Southampton, Isle of Wight, and Winchester, and contains 305 parishes. Under the act of the 2nd of William IV., cap. 45, the county has been divided into two portions, called the Northern and the Southern divisions, each sending two members to parliament; and the Isle of Wight has, for elective purposes, been constituted a county of itself, empowered to return one representative. The place of election for the Northern Division is Winchester, and the polling-places are Winchester, Alton, Andover, Basingstoke, Kingsclere, Odiham, Petersfield, and Bishop's-Waltham; that for the Southern Division is Southampton, and the polling-places are Southampton, Fareham, Lymington, Portsmouth, Ringwood, Romsey, Havant, and Gosport; and that for the Isle of Wight is Newport, and the polling-places, Newport, Ryde, and West Cowes. The Northern Division comprises the minor sessional divisions of Alton, Andover, Basingstoke, Droxford, Kingsclere, Odiham, Petersfield, and Winchester; and the Southern Division, those of Fareham, Lymington, Ringwood, Romsey, Southampton, and the town and county of the town of Southampton. The ancient hundreds are the following:—Alton, Andover, Barton-Stacey, Basingstoke, Bermondspit, Bishop's-Sutton, Bishop's-Waltham, Bosmere, Bountisborough, Buddlesgate, Christchurch, Chutely, Crondall, East Meon, Evingar, Fareham, Fawley, Finchdean, Fordingbridge, Hambledon, Holdshott, Kingsclere, King's-Sombourn, Mainsborough, Mansbridge, Meonstoke, Mitcheldever, New Forest (East), New Forest (North), Odiham, Overton, Pastrow, Portsdown, Redbridge, Ringwood, Selborne, Thorngate, Titchfield, and Wherwell. Besides these hundreds, there are the ancient liberties of Alresford, Alverstoke and Gosport, Beaulieu, Bentley, Breamore, Dibden, Havant, Lymington, Soke (Winchester), and Westover; and the liberties of East and West Medina, in the Isle of Wight. It contains the city of Winchester; the borough, market, and sea-port towns of Christchurch, Lymington, Newport, Portsmouth, and Southampton; the borough and market-towns of Andover and Petersfield; the sea-port and market-town of Yarmouth; the sea-port of Newtown, and the small sea-port towns of Emsworth (a dependency on the harbour of Portsmouth) and Brading; and the market-towns of New Alresford, Alton, Basingstoke, Bishop's-Waltham, Fareham, Fordingbridge, Gosport, Havant, Kingsclere, Odiham, Ringwood, Romsey, Stockbridge, and Whitchurch. Two representatives are returned for the city of Winchester, and two for each of the boroughs, except Christchurch and Petersfield, which, under the act above mentioned, now send only one each. Hampshire is included in the western circuit: the assizes and quarter sessions are held at Winchester.

The form of the county, exclusively of the Isle of Wight, approaches nearly to a square, having a triangular projection at its south-western corner. The Isle of Wight is separated from the main land by a strait of unequal breadth, formerly called the Solent Sea, now the Sound, or, more usually, the West Channel, the breadth of which, at its western extremity, is about a mile, and, towards its eastern end, as much as 7 miles. The form of the island is somewhat rhomboidal, the greatest diagonal being 23 miles from east to west, and the transverse diameter, from north to south, about 13 miles. The surface of the whole county is beautifully varied by gently rising hills and fruitful valleys, and, in some parts, with extensive tracts of woodland. In the southern districts, approaching the coast, the population is much more dense than elsewhere; the mildness of the seasons, the beauty of the landscapes, and the proximity to the ports, operating as strong inducements to the continued residence of many families, besides those engaged in commercial pursuits. The agricultural report drawn up by Charles Vancouver, Esq., for the consideration of the Hon. the Board of Agriculture, divides the main land into five districts. The first, called the woodland district, occupies the northern portion of the county, comprising an area of 103,944 acres, and includes the woodlands and wastes of Bagshot, &c. Its soil and substrata are various: in the eastern part of it are some darkish-coloured sands and a gravelly mould of good depth resting upon a dry subsoil, but intermingled with a stronger and wetter brown loam. The borders of the streams have narrow tracts of meadow and pasture land, the soil of which is, for the most part, a dark-coloured sandy or gravelly loam, lying on a variety of substrata of clay, loam, peat, and gravel, and abounding with springs, which render it of a wet and spongy nature, and cause it to produce herbage of inferior quality. The great mass of this district, however, has a strong brown and grey loam, resting upon a tough blue and yellow clay, having generally an excess of moisture, with numerous unsound and boggy places. Ascending from the woodland valley northward, the soil becomes of a rather lighter quality; but, still proceeding northward, this improvement is lost in a thin sandy and gravelly mould, lying upon deep beds of white, red, and yellow sand and gravel, and a wet hungry loam, upon a moist, loose, white and yellow clay, which constitute the soil of part of Bagshot heath and Frimley common, in the county of Surrey. Along the southern side of this district is a tract composed of a temperate mixed soil, situated between the heavy clays, just mentioned, and the chalk district, which will hereafter be described; upon this is found a large proportion of valuable grass land. Peat is got on the wastes, and in some of the enclosed grounds: a large quantity is annually dug on the commons of Cove, Farnborough, and Aldershot. The second district comprises the whole body of the county, from the borders of Wiltshire to those of Sussex and Surrey, and is computed to contain 454,295 acres: the higher parts of this large central district have much the appearance of an elevated plain, divided into many unequal portions, and intersected by deep hollows, through which the brooks and rivulets rising in these elevated tracts descend, for the most part, in a southerly course towards the sea. In these valleys are considerable tracts of meadow and pasture land, and by far the greater number of the habitations in this district are situated in them. The higher tracts are almost wholly in open and extensive sheep downs: the substratum is throughout a firm unbroken bed of chalk. The soil covering some of them is provincially called *hazel-mould*, being light, dry, friable, and arenaceous, resting upon a chalk rubble containing flint, and naturally producing a short, but excellent, pasturage for sheep: another kind of the down land consists of a black vegetable mould, resting on a nearly similar substratum: a third is a thin grey loam, lying immediately on a firm bed of chalk, and constituting a very great proportion of the downs: a fourth sort consists of a deep, strong, red, flinty loam, lying, at the various depths of from one to eight or ten feet, upon the firm chalk rock; this is usually found on the flat summits of the lesser eminences, of which the brows and acclivities are occupied by a fifth description, which is of a thin staple, and chiefly consists of decomposed chalk, being favourable to the production of turnips and sainfoin. Below these hills a deep, strong, grey loam frequently occurs, the tillage of which, together with that of the red loam, is very difficult: the crops of wheat produced upon it are great. In other parts of this district, soils of a darker colour, equally strong, are met with; and in the numerous hollows intersecting the whole, exclusively of the valleys traversed by running streams, the surface is formed of an assemblage of small flat flints, combined together by a proportionably small quantity of extremely tough loam. This is provincially termed *shrave*; and there is another sort of it, consisting of a coarse red pebbly gravel, mixed with a small proportion of tough red loam, or, more commonly, with a dry sand, or small gravel. The soil of the deeper valleys is a black vegetable mould, resting on a strong calcareous loam, in which occur large chasms occupied by masses of peat, which is occasionally dug for fuel, or to be burned for manuring the ground with its ashes; numerous trunks of trees are found in this earth. The third district is small, containing only 49,525 acres, and including the forests of Woolmer and Alice-Holt, the hills of Binfield, Great and Little Worldham, Selborne, and Empshot, together with all the lower sides of the chalk hills surrounding and forming the vale of Petersfield, the soil of which is, for the most part, a grey sandy loam of good staple, lying on a kind of soft sand rock, being provincially termed *malmy* land. In the vale of Petersfield, formed by the chalk downs and the heaths of Woolmer Forest and its appendages, and traversed by a small branch of the river Loddon, the soil consists of a tough, brown, flinty clay, found at a certain distance from the chalk, interspersed with tracts of light sandy loam. Ascending from this valley, in a north-easterly direction from the town, from which it takes its name, is an extensive tract of sandy and gravelly heath, part of which has been applied to the culture of Scotch fir, and which extends along the borders of Sussex, within and upon the confines of Woolmer Forest. Still further, in the same direction, is found a considerable tract of convertible sandy loam, which forms good turnip land: small quantities of stronger soil are also found, while the valleys have a thin moory soil upon a clay of different colours: peat is obtained here also, chiefly upon the wastes, and in Woolmer and Alice-Holt Forests. The fourth district includes the whole southern part of the county situated on the main land (excepting a tract of 26,895 acres, at its south-eastern extremity), and comprises an area of 333,489 acres. This large division, besides many extensive wastes and commons, comprehends the Forest of Bere, the New Forest, and Waltham Chace; its soils are various, but consist chiefly of light sandy and gravelly loams, intermixed with clay and brick earth, and resting on substrata of argillaceous and calcareous marl. The heaths and commons chiefly comprise the higher lands between Gosport and Titchfield, between Titchfield, Bursledon, and Botley, and between the two latter places and the river Itchen: the cultivated district lying north of Southampton, Millbrook, and Redbridge, is also much contracted by the extensive commons of Shirley and Southampton. Descending southward from the heaths, the surrounding country preserves a smooth and uniform appearance, until broken on the south-west by Hill common and Tatchbury Mount, beyond which occurs a considerable extent of flat low ground, including Netley marsh, and thence extending towards Eling, and, for some distance westward, into the New Forest. Along the confines of the New Forest, and bordering upon the western side of Southampton water, the surface is much broken by hills. On the western side of the Avon the country rises suddenly, and spreads into extensive heaths and commons, upon which many plantations of forest trees have been made: a portion of Poole heath, so called from the vicinity of the town of Poole in Dorsetshire, is included within the limits of Hampshire. Much peat and turf moor is found on the heaths, low grounds, and wastes. Hayling Island, forming the south-western extremity of the county, and Portsea Island, containing the town of Portsmouth, together with the tracts on the main land immediately opposite to them, constitute the fifth district, comprising an area of 26,895 acres. In the islands and low grounds of the main land, a strong flinty and a tender hazel-coloured loam prevail. The soil and substrata of Portsdown hill, in the different degrees of its elevation, are similar to those of the chalk district. A large extent of Portsea Island is occupied as garden ground, and is very productive: on the coasts of this island, particularly on its eastern side, as well as on those of Hayling Island, is much loamy marsh land, subject to occasional inundation by spring tides, and chiefly appropriated as salterns. Its southern side is generally sandy, and the bed of shingles on its south-eastern coast affords large supplies of ballast to the coasting vessels, as well as an inexhaustible store of materials, of a good quality, for making and repairing the roads between Fareham and Chichester.

Through the centre of the Isle of Wight, from east to west, extends a range of lofty hills, affording only pasturage for sheep, and commanding views over every part of the island, with the ocean on the south, and the beautiful shores of Hampshire on the north. Its surface is otherwise much diversified: on the coast the land is, in some parts, very high, particularly on the south, where the cliffs are very steep, and vast fragments of rock, which the waves have at some time undermined, lie scattered below; on the northern side the ground slopes to the water in easy declivities, excepting towards the Needles, or western extremity, where the rocks are bare, broken, and precipitous. The height of the cliffs, of which the Needles form the extreme point, is, in some places, 600 feet above the level of the sea; in some parts they are perpendicular, and in others overhanging; they contain many deep caverns. The Needles derive their name from a lofty pointed rock rising to the height of about 120 feet above low water mark, and severed, with others, from the main land by the force of the waves: part of this rocky projection, about 60 years ago, having been undermined by the sea, fell and totally disappeared. St. Catherine's hill, the highest point in the island, rises 750 feet above the level of high water mark, and commands magnificent prospects; as also do the Culver cliffs, at the eastern end of the island; Carisbrooke castle, and Bembridge down. The soil and substrata of the Isle of Wight are extremely various. It may be said, generally, however, that the soil of the enclosed and depastured marshes and low grounds bordering on the Yarmouth river, on the inlets of Shalfleet and Newtown, on the Medina river between Newport and Cowes, on the Wooton and Ryde rivers, and of the embanked marshes above Brading haven, is composed of a tender hazel-coloured loam, lying, in some places, upon a blue, or rather black, sea-clay, but from which it is most frequently separated by a bed of coarse sand, or fine gravel. The surface mould of the low grounds and meadows, bordering the higher parts of the courses of these streams, is variable, in proportion to the quantity and quality of the adventitious matter washed down from the surrounding hills. On the whole northern side of the island, and for a considerable distance along its southern shores, the prevailing character of the soil is a rough strong clay, of different colours, in which are a blueish argillaceous marl, and a pure white shell marl. In the other parts are chiefly found tender red sand, and a gravelly mould, with argillaceous and calcareous marl, chalk, and its usual accompaniments, red loam and flints; and though the superior fruitfulness of most of these soils generally corresponds with the high estimation in which they are held by the inhabitants, yet some heaths and commons occur, and there are several small tracts of morass, the chief of which is on the western branch of the Medina river. The chalk downs of Brading and Arreton form an unbroken range from Culver cliff, on the eastern coast, to the valley that separates them from Staple's heath: those of Gatcombe and Shorwell are isolated from the western range by a highly-cultivated valley, extending from Shorwell to Newport, and terminating northward in the waste called Parkhurst Forest. From the vale of Shorwell to the western extremity of the island the high chalk downs are broken only by three gaps, or carriage roads, one of which is the passage between the head of the Yarmouth river and the innermost cove of Freshwater bay. The tract of downs situated towards the southern extremity of the island terminates abruptly towards the sea, in a precipice of limestone rock, having the appearance, particularly when seen from a distance, of an immense stone wall, and overhanging the romantic tract called the Undercliff, which extends along the sea-shore for a distance of nearly six miles.

On the arable lands of the county, the rotations of crops are various: the grain and pulse generally cultivated consist of wheat, barley, oats, rye, peas, and beans. The most common artificial grasses are the common broad clover, rye-grass, trefoil, sainfoin, and lucern: burnet is a plant that forms a large portion of the herbage of the downs; a much larger and stronger species is found on many of the low grounds, and upon the cold clay loams, where, as upon the downs, it has every appearance of being indigenous. In the parish of Alton and its vicinity, upon the borders of Surrey, hops are grown to a great extent: the produce varies greatly, but may be estimated, on an average, at about five cwt. per acre: their culture has been much encouraged by the reputation of the Farnham hops, that town, in Surrey, being situated only at the distance of a few miles. The entire extent of hop plantations throughout the county at present occupies 1609 acres. Several of the manures employed are remarkable: the principal are, marl, found in various parts; *malme*, a kind of marl, or chalky clay, found at Timsbury, above Romsey, and many other places along the valley of the Test, as high as the parish of Romsey, both of a black and of a white kind, as also in one or two other places; and chalk, which is extensively used, more especially on the strong lands, within a convenient distance of the Southampton water, to which it is brought from Fareham, from Portsdown hill, and even from the coasts of Kent and Essex. Much chalk is also employed in the islands of Portsea and Hayling, and in the northern or woodland district; as are also turf, peat, and coal-ashes, in situations where they can be procured: on the sea-coast, *rack*, or sea weed, is sometimes used. The most extensive and valuable tracts of meadow land are upon the borders of the various rivers: this county is particularly noted for its irrigated meadows. The green sward of the southern side of the Isle of Wight is, in numerous places, of a very rich quality: pastures of the same description are situated on the Medina river, above Newport, and upon some of the principal branches of the Brading river. The embanked marshes of Brading and Yaverland, in this island, form a valuable tract of rich feeding land: almost all the other embanked lands on the coasts of the islands, as well as of the main land of the county, have been appropriated as salterns for the manufacture of sea and medicinal salts: many of these salt works are now abandoned, but the brine and bitumen with which the ground is saturated, prevent their being brought into tillage.

Hampshire possesses no particular breed of cattle: those of Sussex, Suffolk, Hereford, Glamorgan, and North and South Devon, are indiscriminately met with, and are generally preferred for draught; many cattle of the Leicester breed are also seen. Many cows are kept in different parts of the county, for the purpose of suckling calves to supply the markets of London, Portsmouth, Chichester, Winchester, Newbury, Reading, Salisbury, &c., with veal. The number of sheep kept is remarkably great. In the woodland district the heath sheep, or the Old Hampshire, or Wiltshire breed, were formerly the most common, but have now, in many places, given way to breeds crossed between the New Leicester and others. The South Down breed occupies a great extent of the county, particularly of the open downs of the chalk district: upon the chalk downs of the Isle of Wight it is equally prevalent, although in that island are found many of the Dorsetshire breed, sometimes intermingled with the New Leicester. Numerous hogs are fed for a few weeks, at the close of the autumn, upon the mast produced in the forest and other woodlands; and a superior mode of curing being practised, the Hampshire bacon has become famous for its excellence. The native hog is a coarse, raw-boned, flat-sided animal, now seldom met with, the common stock being either of the Berkshire breed, or of a mixed kind between that and different other breeds. Upon the heaths and forests vast numbers of light small horses are bred, generally about twelve hands high, and provincially termed *heath-croppers,* which propagate indiscriminately upon these wastes, where they succeed in maintaining an existence throughout the year. Gardening is carried on to a great extent in the vicinity of all the large towns, and Portsea island is considered to produce the finest brocoli in the kingdom. This county has long been celebrated for its honey, called heath honey and down honey, from the different districts in which the bees collect it; the latter being the more valuable. The woods are numerous and extensive: the coppices in different parts are cut at various ages: their produce is formed into hop poles, wattled hurdles, hoops, rafter poles, and bavins and faggots. From those of the southern district, a number of straight hoops are also exported to the West Indies: woods of this kind, in the northern woodland district, consist chiefly of birch, willow, alder, hazel, wild cherry, ash, and sometimes oak; in the chalk district, of hazel, willow, oak, ash, maple, white thorn, and a little beech and wild cherry; and, in the southern part of the county, of hazel, willow, alder, birch, holly, and some ash, beech, and wild cherry. The coppice wood of the north-eastern part of the Isle of Wight is overshadowed by a heavy growth of timber. In the northern district a very fine growth of oak is generally observed; but the annual produce of this timber, in the southern parts, has been greatly diminished. In almost every part of the chalk district beech woods and groves flourish with peculiar vigour, and the forests and other woodlands are found to contain large proportions of this timber. Elm is generally of scarce growth, although it is occasionally seen, more especially in the southern districts; the largest elms are much in demand at the Royal Dock-yard for keel-pieces. The abele and aspen poplar, the lime or linden tree, and the Turin, or Lombardy poplar, are frequently met with. In the Isle of Wight the woods of Swainston are of considerable extent, and those of Wooton and Quarr occupy an area of a thousand acres.

The New Forest comprises a very extensive tract in the south-western part of the county. Its ancient boundaries, according to the oldest perambulation extant, which is dated 8th of Edward I., were, the Southampton river, on the east; the Sound and the British Channel, on the south; and the river Avon, on the west: northward, it extended as far as North Charford, on the west; and to Wade and Ower bridge, on the east. Different other perambulations are on record; but, according to that made in the 22nd of Charles II., the forest extends from Godshill, on the north-west, south-eastward to the sea, a distance of about 23 miles; and from Hardley, on the east, to Ringwood on the west, about fifteen miles; and contains 92,365 statute acres. The extent of the woods and waste lands of this tract were, however, at that time, reduced to 63,845 acres, by several manors and other freeholds within the perambulation, amounting to 24,797 acres; by 625 acres of copyhold, or customary lands, belonging to the king's manor of Lyndhurst; by 1004 acres of leasehold, granted for certain terms of years; by 901 acres of purprestures, or encroachments on the forest; and by 1193 acres of enclosed lands held by the master-keepers and groom-keepers, with their respective lodges. The remainder belonged to the crown, as it does at present, and in each kind of the property above mentioned, as being included within the limits of the forest, the king has also various rights and interests: in the freeholds he has certain rights relative to the deer and game; the copyholds are subject to quit-rents and fines, and the timber trees upon such property belong to the crown. The encroachments consist chiefly of cottages built by poor persons, with small parcel of enclosed land adjoining: such, however, as had been made by the proprietors of neighbouring estates, and had been held without any acknowledgment, the crown is authorised, by a modern act of parliament, to grant on lease, for valuable considerations, and provision is made against future encroachment. The forest lands, containing, as before stated, 63,845 acres, are subject to certain rights of commonage, pasturage, pannage, and fuel, possessed by the proprietors of estates within, or adjacent to, the forest; which rights, and those of the crown, are defined by an act of the 9th and 10th of William III., for the increase and preservation of timber in the forest. By this act the crown was empowered to enclose 6000 acres, as a nursery for timber, until the trees should be past danger of being injured by the deer or cattle, when the same shall be thrown open, and an equal quantity might, in like manner, be enclosed, afterwards to be thrown open, in any other part of the forest. The crown has also the right of keeping deer in the unenclosed part of the forest, at all times and without limitation. In consequence of this act, the woodlands, which according to surveys made at different periods, had been long remaining in a neglected state, received for a while some portion of attention; but that, ere long, was withdrawn from them, when the superintendence of the surveyor-general of the crown lands ceased, and the whole fell by degrees under the sole direction of the surveyor-general of the woods. For local purposes the New Forest is divided into nine bailiwicks, which are subdivided into fifteen walks. It is under the government of a lord-warden, who is appointed by letters patent under the great seal, during His Majesty's pleasure; and by his patent are granted to him the manor of Lyndhurst, and hundred of Redbridge, with various other privileges and emoluments; he appoints a steward for the king's house at Lyndhurst. A riding forester is appointed in the same manner as the lord-warden, whose office it is to ride before the king, when he enters the forest. A bow-bearer and two rangers are appointed by the lord-warden, during pleasure. A woodward is appointed by letters patent from His Majesty, during pleasure, and acts by deputy. Four verderers, the judges of the swanimote and attachment courts, are chosen by the freeholders of the county, in pursuance of the king's writ. A high-steward, and an under-steward, are appointed by the lord-warden during pleasure: the duty of the latter is to attend at and enrol the proceedings of the courts of attachment and swanimote, and hold the court leet for the hundred of Redbridge, and the courts baron for the manor of Lyndhurst. Twelve regarders are chosen by the freeholders of the county. There are, besides, nine foresters or master-keepers, and an indefinite number of under-foresters, or groom-keepers, though commonly one to each walk. A purveyor of the navy acts also for the forest, whose office is a naval appointment, and whose duty is to assign timber for the navy, and to prevent any fit for that use from being cut for other purposes. The limits of the forest were finally ascertained, and disputed boundaries settled, by an act of parliament passed in the year 1800. Of the perquisites of the under-keepers, which, at the period of the investigation made in 1789, arose chiefly from deer, browse-wood, rabbits, and swine, only the first and last are now allowed them, and the rabbits are nearly destroyed. The forest courts are regularly held by the verderers, who preside in them at Lyndhurst. The scenery is remarkable for its sylvan beauty, presenting magnificent woods, extended lawns, and vast sweeps of wild heath, unlimited by artificial boundaries, together with numerous river views and the prospect of distant shores. In some parts also are extensive bogs, the most considerable of which is at a place called Longsdale, on the road between Brockenhurst and Ringwood, which extends for about three miles. The oaks seldom rise into lofty stems, and their branches, which are more adapted to what the ship-builders call knees and elbows, are commonly twisted into the most picturesque forms: this is supposed to be owing to the nature of the soil through which their roots have to penetrate, it being generally an argillaceous loam, tempered in different degrees with sand or gravel. The advantage of water-carriage to the various royal or private dock-yards, in which its produce is employed, is superior to that of any other forest in the kingdom.

The forest of Bere, situated in the south-eastern part of the county, and extending northward from the Portsdown hills, which, according to the perambulation made in 1688, is now considered the boundary, comprises about 16,000 acres, upwards of one-third being enclosed. It is divided into two walks, named East and West, to each of which are annexed several smaller divisions, called purlieus, all of them being subject to the forest laws. Its officers are a warden, four verderers, two master-keepers, two under-keepers, a ranger, a steward of the swanimote court, twelve regarders, and two agistors. North-westward of it is the Chace of Bishop's-Waltham, containing about 2000 acres, which belongs to the see of Winchester. The forest of Alice-Holt and Woolmer, on the eastern border of the county, approaching the confines of Surrey and Sussex, and to the north-east of Petersfield, is divided into two parts by intervening private property: its limits comprehend 15,493 acres, of which 8694 belong to the crown: the division called Alice-Holt contains about 2740 acres of crown land. Parkhurst, or Carisbrooke forest, lying at a short distance to the north-west of Newport, in the Isle of Wight, occurs in Domesday-book under the appellation of the King's park, and was afterwards called the King's forest: it includes about 3000 acres, nearly destitute of valuable trees. The total quantity of waste land in Hampshire, exclusively of the forests, falls little short of 100,000 acres.

Upon the inlets and southern coasts of the main land, as also of the Isle of Wight, are large tracts of sea mud, generally less elevated than the highest level of the spring tides, which consequently overflow them. A long range of this kind of marsh extends on the western side of Southampton river, through the parish of Fawley, and a like tract occurs between Calshot castle and the Salterns of Fawley: the extent of these, together with others situated higher up the Southampton river, is estimated at about 2000 acres. The rivers of Beaulieu and Lymington, and the harbour of Christchurch, present similar tracts; and about 4000 acres of this kind of mud are found along the shore, between Hurst castle and the mouth of the Beaulieu river. The inlet or harbour of Portsmouth comprises about 3000 acres, and about 5500 are included in the harbour of Langport, and the portion of that of Emsworth which is within the limits of this county. In the Isle of Wight, similar tracts of marsh are found bordering on the Yarmouth river, on the inlets of Shalfleet and Newtown, and in Brading haven, the area of which is stated at 750 acres. In this latter marsh, different small tracts have been at various periods embanked against the flow of the tide; and an attempt was made by Sir Bevis Thelwall and Sir Hugh Myddelton, the constructor of the New River, to exclude the sea entirely by an embankment thrown across its narrow outlet, which was completed; but in the wet season, when the inner part of the haven was full of water, and there was a high spring tide, the waters met under the bank and made a breach, which was never repaired, and these marshes still remain in their ancient state, notwithstanding that the outlet is, at low water, only about 20 yards in width, and its surface appears for the most part to be more elevated than the embanked marshes lying westward of it, in the parishes of Brading and Yaverland.

The mineral productions are not numerous. On the southern shores of the county, particularly near the mouth of the Beaulieu river, iron-stone is washed up by the sea, and was formerly gathered, and conveyed to the iron-works at Sowley, but these have been discontinued. It is also occasionally found in small quantities in different other parts of the county, particularly in the cliffs near Hordwell, which are upwards of 100 feet high, and abound with nodules of iron-ore, together with pebbles, or flints, many of them containing fossil shells, or their impressions, of various and scarce species, found in a blueish kind of clay or marl. The range of chalk hills crossing the county from east to west, and occupying the central part of it, forms a portion of the vast formation that constitutes so considerable a feature in the geology of England. The strata constituting the southern part of the main land, and the northern part of the Isle of Wight, lie upon a depressed portion of the chalk beds, which, in geological language, is termed the Chalk Basin of the Isle of Wight. The chalk raised in this county is of two kinds, white and grey, both of them burned into lime of a good quality, that from the latter being also particularly serviceable as a cement under water, for which it is extensively employed. Between Milton and Christchurch is found a hard reddish stone, of which several ancient structures in that part of the county are built. The numerous strata of various kinds and formations, and exhibiting great diversity of position, of which the Isle of Wight consists, form a remarkably rich field of study for the geologist. At Alum bay, at the north-western extremity of the island, is found a vein of white sand, in great demand for the glass-works of Bristol and Liverpool, as also for others situated on the western coasts of England and Scotland, and in Ireland. Eastward of this, along the northern foot of the downs, grist or quarry stone, of a yellowish grey colour, and very porous texture, is found in detached masses, and used for building. A strong liver-coloured building stone rising in cubical masses, encrusted with a brownish kind of ochre, and enclosing specimens of rich ironstone, occurs on the southern side of the island: a rough calcareous freestone is frequently found in the marl pits, in loose detached pieces. Eastward of Staple's heath, and northward of Arreton downs, a close grey limestone is raised, the beds of which are separated from each other by small layers of marine shells, cemented together by alum, that substance being well known to pervade the western parts of the island. Freestone is sometimes found under marl in the northern districts of it: a plum-pudding stone exists in large quantities near Sandown fort, and is much used for paving and flooring. Potters' clay occurs in great variety in different parts of the county; and ochres of divers colours in the Isle of Wight.

The manufactures are various, but not extensive; ship-building, however, in addition to the works of the royal dock-yard at Portsmouth, is extensively carried on in most of the numerous creeks and harbours. The other productions are chiefly woollen goods, bed-ticking, light silk articles, sacking, leather, and a coarse kind of earthenware. At Overton are very extensive silk-mills, and the young female peasantry in the vicinity are much employed in the platting of straw for bonnets, the straw-hat manufacture being carried on at many towns in the county. There are paper-mills in different parts of the county, those in the vicinity of Overton being of considerable importance. At Lymington is a manufacture of salt. The advantages for maritime commerce may be estimated from the following list of harbours and roadsteads; *viz.*, Hayling and Portsmouth harbours, with their dependencies; the inlet of Southampton, with the mouths of the Itchen and Test rivers, their ship-yards, and the smaller havens of Redbridge, Eling, Hythe, Cadland, and Fawley, and of Botley, Bursledon, and Hamble, on the river Hamble; the Beaulieu river, with its dependencies, slips, and private dock-yards; Lymington, or Boldre water, with a number of small creeks through the salterns, including Keyhaven; and the harbour of Christchurch, with its branches, forming the mouth of the Avon and Stour rivers. In the Isle of Wight are the harbours of Hythe, Cowes, South Yarmouth, and Brading. The roadsteads separating that island from the rest of the county are those of St. Helen's, the Motherbank, Spithead, Cowes, Southampton bay, and Yarmouth roads, the latter of which is terminated westward by the Needles, the passage or channel of which is contracted to the space of only about a mile, by a broad bank of shingles thrown up by the sea, which beats upon it with great violence: this bank, from the main land in the parish of Milton, projects south-eastward, and upon its furthest extremity stands Hurst castle, built in the reign of Henry VIII., and commanding the passage. The shores of this county, particularly of the Isle of Wight, are much resorted to during the summer for the purpose of sea-bathing, &c.: the most frequented places are, on the main land, Christchurch, Muddiford, Lymington, and Southampton; and in the Isle of Wight, Yarmouth, Cowes, Hythe, Brading, and Shanklin. In all the rivers and creeks that discharge their waters directly into the sea, salmon are caught: the fisheries of the Southampton water are particularly extensive, and the boats engaged in them often make long coasting voyages to procure other fish, which are taken thence to the markets of London, Oxford, Bath, &c. Several persons are employed on the flat and rocky shores of the Isle of Wight, in catching shrimps and prawns, and on its bolder shores, in taking crabs and lobsters.

The principal rivers are the Test, the Anton, the Itchen, the Avon, the Boldre water, and the Exe. The Test, below Redbridge, expands and forms the head of the Southampton water, an arm of the sea extending from the "Above Town" of Southampton to the Sound at Calshot castle, and rendered exceedingly picturesque by its woody and irregular shores: the general direction of the Southampton water is from north-west to south-east. The Itchen, also called the Arbre, was brought into a regular channel, and made navigable up to Winchester, by Godfrey de Lacy, Bishop of Winchester, in 1215: towards its mouth it expands considerably. The Avon, by an act passed in 1665, was made navigable up to Salisbury; but the works having been swept away by a flood, the navigation was destroyed. The Boldre water is formed by several small streams rising in the New Forest, most of which unite above Brockenhurst, thence proceeding southward, by Boldre and Lymington, to the sea. The Exe, frequently called the Beaulieu river, from similar sources in the same district, flows south-eastwardly, and, beginning to expand near Beaulieu, opens into a broad estuary to the sea, below Exbury. The principal river of the Isle of Wight is the Medina, anciently called the Mede, which rises near the bottom of St. Catherine's down, in the southern part of the island, and, flowing directly northward, divides it into two equal parts, each constituting a liberty, which derives its name from its position on the eastern or western side of this stream: passing on the eastern side of the town of Newport, the Medina mingles its waters with those of the sea in Cowes harbour. The other principal streams of this island are the Yar, the Wooton, and the Ear; its shores are also indented by various creeks and bays. A navigable canal has been made, along the valleys of the Test and Anton, from Andover to the head of the Southampton water: from Barlowes-Mill, near Andover, its course is by Stockbridge and Romsey to its termination at Redbridge, in the parish of Millbrook: from Redbridge a branch proceeds directly to Southampton, and a collateral branch extends from it in a westerly direction, up the valley between East Dean, Lockerley, and East Tytherley, to Alderbury common, within two miles of Salisbury, but neither of them is navigable. From Basingstoke a canal has been made, under the authority of an act of parliament obtained in 1778, to the river Wey in Surrey, by which the navigation is maintained to the Thames: the total length of this canal is 37 miles and a quarter: the cost of cutting it amounted to about £100,000, a large portion of which was expended in forming a tunnel through Grewell hill, near Odiham, which is arched with brick, and is nearly three quarters of a mile in length: the articles of traffic conveyed upon it are chiefly corn, flour, coal, and timber. The Winchester and Southampton canal is one of the oldest in the kingdom: the act for its construction was obtained in the reign of Charles I., but, from the want of a suitable trade upon it, does not appear to have realised the expectations of the projectors. The London and South-Western railway, now nearly completed, enters the county at Farnborough, and passes by Basingstoke and Winchester to Southampton.

Within the limits of the county were the stations of *Venta Belgarum,* supposed to have been at Winchester; *Vindonum,* at Silchester; *Clausentum,* at Bittern; *Brigæ,* at Broughton; and *Andaoreon,* at Andover. The principal remains of Roman occupation discoverable are at Silchester, approaching the confines of Berkshire, where gold coins and rings, Roman bricks, and pottery, &c., have been dug up. About three-quarters of a mile north of Lymington is Buckland Rings, the remains of a Roman camp. Traces of other encampments of different periods are visible in various parts; some of the most extensive and remarkable are those of the camp on Dancbury hill, to the west and north-west of which are several barrows, many more of these monuments being found in different parts of the county. Three Roman roads branch from Silchester, one of them proceeding to the northern gate of Winchester; another by Andover, to Old Sarum; and the third, northward, across Mortimer heath: from Winchester also was a road leading to Old Sarum. The number of ancient religious establishments was about fifty-three. There are still interesting remains of the abbeys of Hide, Netley, Beaulieu, and Quarr; as also of the hospital of St. Cross near Winchester. The castles of Hurst, Porchester, and Carisbrooke, in the Isle of Wight, are still standing; and there are also remains of the castles of Christchurch, Odiham, and Warblington. The modern seats of the nobility and gentry are extremely numerous, more especially the villas. Several chalybeate springs are found in different parts of the Isle of Wight: at Pitland is a spring impregnated with sulphur; and at Shanklin another, the water of which is slightly tinctured with alum. The waters of the streams in the northern woodland part of the county are of a strong chalybeate quality: that which issues from the bogs and swampy ground is charged with a solution of iron. In the strong loam, woodland clay, and chalk districts, the want of a regular supply of water during seasons of drought is severely felt. Fossil remains of different kinds are found in some of the strata of this county; among the natural curiosities of which may also be mentioned the immense chasms near the sea-shore in the Isle of Wight, called Blackgang Chine, Luccombe Chine, and Shanklin Chine; and there is a large natural cavern at Freshwater Gate, a small creek in the centre of Freshwater bay. Samphire grows plentifully on some of the high cliffs of the Isle of Wight, and is gathered by the inhabitants.

SOUTHBOROUGH, a chapelry, in the parish and lowey of TONBRIDGE, lathe of AYLESFORD, Western Division of the county of KENT, $2\frac{3}{4}$ miles (S. by W.) from Tonbridge, with which the population is returned. It is situated about midway between Tonbridge and the Wells, and consists of a number of scattered houses. A new district church, in the early style of English architecture, has been erected, and endowed at an expense of £8436, defrayed by subscriptions: the property is vested in five trustees, who have also the appointment of the minister. The living is a perpetual curacy; net income, £153. In 1785, premises for a school were erected by the executors of the Rev. Edward Holmes, and endowed with £1050 four per cents., for the education of 70 children: it is managed by trustees.

SOUTH-BURN, a township, in the parish of KIRKBURN, union of DRIFFIELD, Bainton-Beacon Division of the wapentake of HARTHILL, East Riding of the county of YORK, $3\frac{3}{4}$ miles (S. W.) from Great Driffield; containing 107 inhabitants.

SOUTHCHURCH, a parish, in the union and hundred of ROCHFORD, Southern Division of the county of ESSEX, 1 mile (N. N. E.) from South-End; containing 401 inhabitants. This parish, which is situated near the sea, is bounded on the south by the river Thames; a considerable portion of the shore is flat and overflowed by the tide. Large oyster beds have for many years been preserved here. The living is a rectory, in the patronage of the Archbishop of Canterbury, valued in the king's books at £27. 0. 10. The tithes have been commuted for a rent-charge of £800, subject to the payment of rates, which on the average have amounted to £80; the glebe comprises 60 acres, valued at £98 per annum. The church is a small edifice with a tower and spire, and is in good repair.

SOUTHCOATES, a township, in the parish of DRYPOOL, union of SCULCOATES, Middle Division of the wapentake of HOLDERNESS, East Riding of the county of YORK, $1\frac{1}{2}$ mile (N. E. by E.) from Kingston-upon-Hull; containing 1114 inhabitants.

SOUTHCOT, a tything, in the parish of ST. MARY, READING, union and hundred of READING, county of BERKS, $1\frac{1}{2}$ mile (W. S. W.) from Reading; containing 84 inhabitants.

SOUTH-COVE, county of SUFFOLK.—See COVE, SOUTH.

SOUTHEASE, a parish, in the union of NEWHAVEN, hundred of HOLMSTROW, rape of LEWES, Eastern Division of the county of SUSSEX, $3\frac{3}{4}$ miles (S. by E.) from Lewes; containing 142 inhabitants. The living is a rectory, valued in the king's books at £16. 0. 10.; present net income, £191; patron, Rev. — Todd. The church is principally in the early style of English architecture, with a circular tower. A school is partly supported by subscription.

SOUTHEND, a hamlet, in the parish of PRITTLEWELL, hundred of ROCHFORD, Southern Division of the county of ESSEX, $1\frac{3}{4}$ mile (S. S. E.) from Prittlewell, and 42 (E.) from London: the population is returned with the parish. This place is situated at the mouth of the Thames, directly opposite to the river Medway, and at the southern extremity of the county, from which circumstance it probably derives its name. Though formerly an inconsiderable hamlet, consisting only of a few fishermen's huts, it has within the last few years grown into repute for sea-bathing, and, being the nearest watering-place to London, is rapidly rising into importance. The village comprises the lower, or old, town, and the upper, or new, town; the former situated on the beach, and the latter on an eminence fronting the sea. The old town consists principally of an irregular line of houses facing the sea, to which have been recently added some handsome dwellings, and a parade has been formed and partly enclosed: nearly in the centre is a large and commodious inn, and there are also two others on a smaller scale; at the eastern extremity is a small but neatly arranged theatre, which is opened every season, and on the beach are several good bathing machines. The new town, which is considered as the more fashionable residence, is superior both in its situation and in the character and style of its buildings, and consists principally of the terrace, a handsome range of buildings occupying an elevated site facing the sea, and having in the front a fine promenade; adjoining which is an hotel, containing several good suites of apartments, a handsome assembly-room with card and refreshment rooms, and every requisite accommodation. The library, a neat building in the later English style, contains a good reading-room, and is well supported; and adjoining it is a billiard-room. In front of the terrace, and extending the whole length, is a shrubbery tastefully laid out in walks, commanding a fine view of the sea, and forming an interesting feature in front of the houses; within this enclosure are the baths, a neat building in the cottage style, containing warm baths, which are supplied by machinery and fitted up with every requisite accommodation. Near the baths are the bathing machines for the use of the visiters of the upper town. From the terrace, and from the promenade in front of it, are fine views extending over the channel, from the Isle of Thanet along the richly wooded hills in the opposite county of Kent, and up the Thames as far as Gravesend, including the Medway and the government works at Sheerness. On the east the view reaches from the Nore beyond Whitstable and St. Nicholas' bays to the Reculvers, and terminates in the distant line of coast near Margate. A pier of framework has been constructed by an incorporated company of proprietors, and in the vicinity are many pleasant walks and rides through a district abounding with richly diversified and picturesque scenery. A chapel of ease has been lately built by subscription, to which the Society for building new churches granted £350. There is a place of worship for Independents.

SOUTHERNBY-BOUND, a township, in the parish of CASTLE-SOWERBY, LEATH ward, Eastern Division of the county of CUMBERLAND, 11½ miles (N. W. by W.) from Penrith; containing 162 inhabitants.

SOUTHERY (*St. Mary*), a parish, in the union of DOWNHAM, hundred of CLACKCLOSE, Western Division of the county of NORFOLK, 6½ miles (S.) from Downham-Market; containing 739 inhabitants. The living is a rectory, valued in the king's books at £7. 10.; patron, Robert Martin, Esq. The tithes have been commuted for a rent-charge of £629. 10., subject to the payment of rates, which on the average have amounted to £99; the glebe comprises 104 acres, valued at £106. 16. 2. per annum. The church is a very ancient structure, with a wooden screen separating the nave and chancel, and contains 140 free sittings, the Incorporated Society having granted £150 in aid of the expense. A school is supported by subscription, and there is a Sunday National school.

SOUTH-FIELDS, a liberty, in the hundred of GUTHLAXTON, Southern Division of the county of LEICESTER; containing 1608 inhabitants, many of whom are employed in frame-work knitting.

SOUTHFLEET (*St. Nicholas*), a parish, in the union of DARTFORD, hundred of AXTON, DARTFORD, and WILMINGTON, lathe of SUTTON-AT-HONE, Western Division of the county of KENT, 3½ miles (S. W.) from Gravesend; containing 624 inhabitants. This was a place of importance during the Heptarchy, when it was called *Sudfleta:* from its proximity to the old Watling-street, its distance from the station *Durobrivis* (Rochester), and the numerous Roman relics found on the spot, it is supposed to occupy the site of the *Vagniacæ* of Antoninus. Among other antiquities, a Roman milliary and a stone tomb, containing two leaden coffins, have been discovered in the parish. The living is a rectory, valued in the king's books at £31. 15.; present net income, £523; patron, Bishop of Rochester. The church is principally in the decorated style of English architecture, and exhibits many marks of antiquity, including six stone stalls under pointed arches, a piscina, a window of stained glass, and a font much admired for its curious workmanship. A school-house has been erected, and is endowed with a rent-charge of £20, bequeathed by the family of Sedley.

SOUTHGATE, a chapelry, in the parish and hundred of EDMONTON, county of MIDDLESEX, 8 miles (N. by W.) from London: the population is returned with the parish. The name of this place is derived from its situation at the south gate, or entrance, of Enfield Chace, and it is still called South-street division; the Chace has been entirely enclosed, and is now in a good state of cultivation. The village contains many handsome houses: the New River runs at its extremity. The neighbourhood is well wooded: the Duke of Buckingham and Chandos has a residence here, in the grounds of which is a very fine old oak tree, that covers with its shade nearly an acre of ground. The chapelry is within the jurisdiction of a court of requests held at Enfield, for the recovery of debts under 40*s*. The living is a perpetual curacy, in the patronage of the Vicar of Edmonton; net income, £180. The chapel, built in 1615, at the sole expense of Sir John Weld, has been pulled down and rebuilt: it is endowed with an estate in Essex, called Orsett. There is a place of worship for Independents; and a National school has been erected near the Green in a very neat style; the school-house built by the late John Walker, being at an inconvenient distance from the village, is not now used. Mrs. Cowley left a sum of money to clothe nine boys and nine girls in this school. In an adjacent field, called 'Camp Field,' have been found several pieces of cannon, and a gorget belonging to Oliver Cromwell, having his initials handsomely inlaid with jewels, now in the British Museum; and in 1829, several ancient coins were dug up in the neighbourhood.

SOUTH-HAMLET, an extra-parochial liberty, in the Middle Division of the hundred of DUDSTONE and KING'S-BARTON, union, and Eastern Division of the county, of GLOUCESTER; containing 834 inhabitants. Here is a mineral spring.

SOUTH-HILL (*St. Sampson*), a parish, in the union of LISKEARD, Middle Division of the hundred of EAST, Eastern Division of the county of CORNWALL, 3¼ miles (N. W.) from Callington; containing 530 inhabitants. The living is a rectory, with the perpetual curacy of Callington annexed, valued in the king's books at £38; present net income, £748; patron, Lord Ashburton. John Knill, in the year 1747, gave £5 per annum for teaching children.

SOUTHILL (*All Saints*), a parish, in the union of BIGGLESWADE, hundred of WIXAMTREE, county of BEDFORD; containing 1267 inhabitants, of which number, 675 are in the township of Southill, 4 miles (S. W. by W.) from Biggleswade. The living is a discharged vicarage, with that of Old Warden annexed, valued in the king's books at £11. 15.; present net income, £384; patron and impropriator, W. H. Whitbread, Esq. The church contains monuments of several of the Byng family, among which are those of the celebrated naval officer, Sir George Byng, the first Viscount Torrington, and of his son, Vice-Admiral the Hon. John Byng, who was executed for alleged professional misconduct, though the inscription upon his coffin, engraved in brass, represents him to have fallen a martyr to political persecution. The Baptists have a place of worship here; and two schools are supported by subscription.

SOUTH-MEAD, an extra-parochial liberty, in the Middle Division of the hundred of DUDSTONE and KING'S-BARTON, Eastern Division of the county of GLOUCESTER, adjacent to the south side of the city of Gloucester.

SOUTHMERE (*ALL SAINTS*), formerly a parish, in the hundred of SMITHDON, Western Division of the county of NORFOLK, 5 miles (W. by S.) from Burnham-Westgate: the population is returned with Docking. This place, though anciently of some note, has now dwindled into a mere hamlet, called Summerfield, containing a few inhabitants, who are considered as connected with the parish of Docking. The living is a discharged rectory, valued in the king's books at £2. 11. 2.; patrons, Provost and Fellows of Eton College. There is no church.

SOUTHMINSTER, county of ESSEX.—See MINSTER, SOUTH.

SOUTHOE (*ST. LEONARD*), a parish, in the union of ST. NEOTS, hundred of TOSELAND, county of HUNTINGDON, 3¼ miles (N. by W.) from St. Neots; containing 283 inhabitants. The living is a vicarage, endowed with the rectorial tithes, with that of Hail-Weston annexed, and valued in the king's books at £14. 2. 3½.; present net income, £288; patron and incumbent, Rev. R. Pointer, who erected a school-room in 1818, in which five children are instructed at his expense. There are some mineral springs in this parish. Bishop Chadderton was interred here.

SOUTHOLT (*ST. MARGARET*), a parish, in the union, and hundred, of HOXNE, Eastern Division of the county of SUFFOLK, 5 miles (S. E. by S.) from Eye; containing 193 inhabitants. The living is annexed to the rectory of Worlingworth. The tithes have been commuted for a rent-charge of £237. 10., subject to the payment of rates, which on the average have amounted to £47. A Sunday school is supported out of the rents of the town lands.

SOUTHORP, a hamlet, in the parish of BARNACK, union of STAMFORD, soke of PETERBOROUGH, Northern Division of the county of NORTHAMPTON, 4 miles (S. E.) from Stamford; containing 137 inhabitants.

SOUTHORPE, an extra-parochial liberty, in the union of GAINSBOROUGH, wapentake of CORRINGHAM, parts of LINDSEY, county of LINCOLN, 7 miles (N. E.) from Gainsborough; containing 36 inhabitants.

SOUTHOVER, county of SUSSEX.—See LEWES.

SOUTHPORT, a chapelry, in the parish of NORTH MEOLS, union of ORMSKIRK, hundred of WEST DERBY, Southern Division of the county palatine of LANCASTER, 9 miles (N. W.) from Ormskirk, and 22 (N.) from Liverpool: the population is returned with the parish. This place, situated at the mouth of the Ribble, on the shore of the Irish Sea, has of late years been much resorted to for sea-bathing, and possesses most of the usual accommodations for visiters: it consists of one principal street of handsome houses of brick, with gardens in front, and is surrounded by meols, or sand hills, resembling small tumuli. The Victoria baths, built by subscription at an expense of £6000, form a handsome range of building with a colonnade in front; the principal bath is 55 feet long and 25 feet wide, and there are cold and tepid baths supplied with water by a steam-engine: the building comprises rooms for refreshment, and attached to it are gardens and a conservatory, with a fine terrace walk of great extent. A theatre, news-rooms, and libraries, with other places of public resort, supply the means of amusement and relaxation. An episcopal chapel, called Christ church, was built in 1820, the living of which is a perpetual curacy; net income, £107; patron, Sir P. H. Fleetwood, Bart. It was enlarged in 1830, and 90 free sittings provided, the Incorporated Society having granted £90 in aid of the expense. Trinity church, a neat edifice, in the early English style, was consecrated Nov. 8th, 1837; it was built by subscription, and will accommodate 500 persons. The living is a perpetual curacy, in the patronage of the Trustees. There are places of worship for Independents, Wesleyan Methodists, and Roman Catholics. A National school has been established. The Strangers' charity furnishes to the sick poor the means of obtaining the benefits of sea air and bathing; and a dispensary, erected in 1823, supplies the requisite medical aid.

SOUTHPORT, county of SOUTHAMPTON. — See PORTSEA.

SOUTHROP (*ST. PETER*), a parish, in the union of NORTHLEACH, hundred of BRIGHTWELL'S-BARROW, Eastern Division of the county of GLOUCESTER, 3 miles (N.) from Lechdale; containing 350 inhabitants. The living is a discharged vicarage, valued in the king's books at £5. 16. 8.; present net income, £192; patrons, Warden and Fellows of Wadham College, Oxford; impropriator, J. Tuckwell, Esq.

SOUTHROP, a tything, in the parish of HERRIARD, hundred of BERMONDSPIT, though locally in that of Odiham, Basingstoke and Northern Divisions of the county of SOUTHAMPTON, 5 miles (N. W.) from Alton: the population is returned with the parish.

SOUTHROPE, a township, in the parish of HOOK-NORTON, hundred of CHADLINGTON, county of OXFORD: the population is returned with the parish.

SOUTHSEA, county of SOUTHAMPTON. — See PORTSEA.

SOUTHTOWN, formerly a parish, but now commonly esteemed a hamlet, in the parish of GORLESTON, locally in the hundred of Mutford and Lothingland, Eastern Division of the county of SUFFOLK; containing 1304 inhabitants. This place, formerly called Little Yarmouth, forms a suburb to Great Yarmouth, with which it is connected by a bridge over the Yare; and, as regards franchise, and matters of trade and jurisdiction, was united to that borough by an act passed in the 16th and 17th of Charles II. It consists of two separate parts, about a mile and a half distant from each other, of which the south-eastern, overlooking the sea, and adjoining Gorleston High-street, is called, by way of distinction, Southtown-on-the-Hill: the other extends from Yarmouth bridge about half a mile southward, along the western bank of the Yare, one side of the road being occupied by handsome private houses, and the other by timber wharfs, docks, and yards for ship-building, which afford employment to a great number of shipwrights and others. Here is an armoury erected during the late war, but for the present disused. The living was a discharged rectory, consolidated in 1520 with the vicarage of Gorleston. The church, which was dedicated to St. Nicholas, has fallen into decay. A chapel, dedicated to St. Mary, was erected in 1831 by subscription, at an expense of £2300: the Earl of Lichfield, then Viscount Anson, gave the land on which it is erected, and £500 towards building. The living is a perpetual curacy; net income, £200; patrons, Trustees.

SOUTHWARK, county of SURREY.—See LONDON.

SOUTHWELL (*St. Mary*), a market-town and parish, and the head of a union, in the liberty of SOUTHWELL and SCROOBY, Southern Division of the county of NOTTINGHAM, 14 miles (N. E.) from Nottingham, and 132 (N. N. W.) from London; containing 3384 inhabitants. This place, which is of great antiquity, derived its name from one of many large springs, or wells, that formerly existed in the neighbourhood, few of which are now remaining. It was distinguished, at a very early period, by the foundation of one of the first Christian churches in this part of the country, by Paulinus, who, at the request of Ethelburga, wife of Edwin, King of Northumberland, had been sent over to England by Pope Gregory VII., to preach the doctrines of Christianity, which she had herself embraced, and who, having converted Edwin to the Christian faith, was made Archbishop of York, in 627. The history of the town relates chiefly to the progress of its religious establishment, which flourished, under a succession of prelates, till the Conquest, at which time the church had become collegiate, had ample revenues and contained ten prebends, the number of which was subsequently increased to sixteen. From the time of the Conquest till the period of the Reformation, the revenue of the church continued to increase, and the establishment to prosper, during the reigns of Henry I., Henry II., Henry III., Edward I., and other sovereigns, who contributed largely to its endowment. Popes Alexander III. and Urban III. were also munificent patrons; every succeeding archbishop was anxious to promote its independence, aud the zeal and liberality of its own members were constantly devoted to its improvement. Soon after the dissolution of the monasteries by Henry VIII., the archbishop, and the prebendaries of Southwell, surrendered the church into the possession of that monarch, by whom, at the request of Cranmer, Archbishop of Canterbury, the chapter was refounded, in 1541, and Southwell subsequently erected into a see, of which, in 1543, Dr. Cox, afterwards translated to Ely, was appointed bishop. Edward VI., soon after his accession to the throne, dissolved the chapter, and granted the prebendal estates to John Earl of Warwick, upon whose attainder, in 1553, they reverted to the crown, and were restored by Queen Mary, who re-established the chapter upon its ancient foundation, and the prebendal establishment was finally confirmed by Queen Elizabeth, in 1585, and a new code of laws instituted. During the parliamentary war, Charles I. was frequently at this town, and held his court generally at the archiepiscopal palace, and occasionally at the King's Arms Inn, now the Saracen's Head, at which latter place, on the 6th of May, 1646, he privately surrendered himself to the Scottish commissioners. The parliamentary troops, during their stay in the town, converted the church into a stable, broke the monuments, defaced the ornaments, demolished the episcopal palace, in which Cardinal Wolsey resided during the summer previous to his death, and destroyed all the ancient records, except the *Registrum Album*, or white book, which is still in the possession of the chapter, and contains most of the grants to the church, from the year 1109 to 1525: the lands belonging to the see were sold for £4061.

The town is pleasantly situated on a gentle eminence richly clothed with wood, and surrounded by an amphitheatre of hills of various elevation, near the small river Greet, which is noted for the red trout abounding in it, and comprises the districts of Burgage, the High Town, Easthorpe, and Westhorpe, together forming a considerable, though scattered, town: it is well paved and abundantly supplied with water: the houses are in general well built, and the town has a neat and prepossessing appearance. Assemblies are held in a commodious suite of rooms erected in 1806; a harmonic society, established in 1786, is well supported, and a small neat theatre was opened in 1816. A pleasant promenade has been formed on the north side of the churchyard, and planted with trees, called the Parade: the roads in the vicinity have been recently improved. The air is salubrious, and the environs afford some agreeable walks. There is not much trade carried on, and the only branch of manufacture is that of silk, for which a mill has been erected on the river Greet by a firm at Nottingham. The market is on Saturday: the fairs are on Whit-Monday, which is a pleasure fair, and Oct. 21st, a statute fair. The town is under two separate jurisdictions, called the Burgage and the Prebendage: the former, denominated the Soke of Southwell *cum* Scrooby, includes twenty townships, for which quarterly courts of session are held by a *Custos Rotulorum*, and justices of the peace, nominated by the Archbishop of York and the Chapter of Southwell, and appointed by a commission under the great seal for the trial of all but capital offenders. The prebendage includes 28 parishes, over which the chapter, by their vicar-general, exercise ecclesiastical jurisdiction, and all episcopal functions, except confirmation and ordination. The subdivisions of the town are under the superintendence of the constable, or bailiff, assisted by two thirdboroughs for each district. This town has been made a polling-place for the southern division of the county. The house of correction for the county, a spacious and commodious building, which, after having been several times enlarged, was completed in 1829, comprises 11 wards, 34 work and day-rooms, 11 airing-yards, and a tread-mill with four wheels, being well adapted to the classification of prisoners, who are employed in frame-work knitting, spinning yarn, shoe-making, &c., and receive a portion of their earnings on their discharge: the buildings consist of the governor's house, chapel, infirmary, baths, and the requisite offices, with a cook-house and observatory in the centre of the prison.

The living is a discharged vicarage, in the patronage of the Prebendary of Normanton in the Collegiate Church (the appropriator), valued in the king's books at £7. 13. 4.; present net income, £144. The church, which is both parochial and collegiate, is a spacious and magnificent cruciform structure, chiefly of Norman architecture, with some portions in the early, decorated, and later styles of English architecture, having a low central tower, and two others of the same height at the west end, richly ornamented, between which is the principal entrance, through a fine circular arch, with a large window above it, of the later style, highly enriched with elegant tracery. The nave and western transept are of Norman character; the former has a flat roof of panelled oak finely carved, and supported upon a range of low massive circular columns and arches, and is lighted by a range of clerestory windows of small dimensions, above a triforium of large and undivided arches. The roof of the aisles is finely groined in stone; the arches and piers supporting the central tower are strikingly beautiful, from the simplicity of their style, and the stateliness of their elevation: the choir and the eastern transepts are beautiful specimens of the early English style, perhaps unrivalled for their purity of design and fidelity of minute detail; the former is lighted by two tiers of lancet-shaped windows, and is fitted up as a parochial church, with galleries; the stalls and the screen are in the later period of the decorated style. There are few monuments deserving notice, except that of Archbishop Sandys, in the south transept, the principal having been destroyed during the occupation of the church by the parliamentarian troops. On the eastern side of the north transept was formerly a chantry, or singing school, which has been converted into a library for the college, containing a valuable and extensive collection of works, chiefly on divinity. On the north side of the church is the chapter-house, in the decorated style of English architecture: the entrance doorway, which is double, is elegantly enriched with foliage of a character not very prevalent in England; the tracery in the windows, and in the stalls under them, is very beautiful. The prebendal houses, and especially that for the residentiary prebendary, are handsome buildings. In the churchyard are some remains of the ancient collegiate buildings, the establishment of which is still retained, consisting of sixteen prebendaries, six vicars choral, an organist, six singing men, six choristers, and six boys as probationers, with a registrar, treasurer, auditor, and other officers: two annual synods are held, at which all the Nottinghamshire clergy attend, over whom a certain number of the prebendaries are appointed by the archbishop to preside. There are places of worship for Baptists and Wesleyan Methodists.

The free grammar school, which occupies the site of the college of the chantry priests, is under the superintendence of the chapter of the collegiate church, by which body the master is appointed, subject to the approval of the Archbishop of York; it has two scholarships founded in St. John's College, Cambridge, by the Rev. Dr. John Keyton, Canon of Salisbury, in the reign of Henry VIII., for boys who have been choristers in the collegiate church: the school, which has a small endowment, is open to all boys of the town for gratuitous instruction in the classics. There are also several other schools with small endowments, besides Sunday schools in connection with the established church and the dissenting congregations. The poor law union of Southwell comprises 60 parishes or places, under the care of 62 guardians, and contains, according to the census of 1831, a population of 23,235. Of the ancient episcopal palace there are considerable remains, which are overspread with ivy, and form an interesting and romantic ornament to the town: they consist chiefly of the chapel and great hall, which are almost entire, and have been fitted up as a modern residence: in this portion of the building is a room lighted by the great west window of the hall, which is appropriated to the holding of the sessions: the quadrangle, once surrounded with offices, has been converted into a garden. Vestiges of a Roman fosse are perceptible on the Burgage hill. Of the springs which formerly distinguished this vicinity, and from one of which the town derived its name, St. Catherine's well, at Westhorpe, celebrated for the cure of rheumatism, and South well, about half a mile to the south-east of the town, are still open. Near the ruins of the palace were found, in 1780, a large ring of pure gold, on the inside of which was inscribed *Mieu mouri que change ma foi*; and, in 1828, a large brass seal, with the device of a female kneeling, and holding in her right hand a tilting spear bearing a breast-plate, out of which rises a unicorn's head, and in her left hand a shield, with the arms of Cavendish and Kemp quartered, and on a scroll encircling the device the legend *Gorge Rygmayden*, in Saxon characters: the seal is preserved in the museum at York.

SOUTHWELL-PARK, in the parish of HARGRAVE, hundred of THINGOE, Western Division of the county of SUFFOLK, 7 miles (W. S. W.) from Bury-St. Edmund's: the population is returned with the parish.

SOUTHWICK, a township, in the parish of MONK-WEARMOUTH, union of SUNDERLAND, Eastern Division of CHESTER ward, Northern Division of the county palatine of DURHAM, $1\frac{3}{4}$ mile (N. W. by W.) from Sunderland; containing 1301 inhabitants. There is a place of worship for Wesleyan Methodists; also a National school. On the banks of the Wear, in this township, are several lime-kilns, ship-yards, and earthenware and glass manufactories. Human bones, and sometimes entire skeletons, have been found on removing the soil above the limestone quarries on Southwick hills.

SOUTHWICK (*ST. MARY*), a parish, in the union of OUNDLE, hundred of WILLYBROOK, Northern Division of the county of NORTHAMPTON, $2\frac{3}{4}$ miles (N. N. W.) from Oundle; containing 154 inhabitants. The living is a discharged vicarage, valued in the king's books at £8. 7. 6.; present net income, £90; patron and impropriator, W. Lynn, Esq.

SOUTHWICK (*ST. JAMES*), a parish, in the union of FAREHAM, hundred of PORTSDOWN, Fareham and Southern Divisions of the county of SOUTHAMPTON, $3\frac{1}{4}$ miles (N. E. by E.) from Fareham; containing 723 inhabitants. The living is a donative, in the patronage of T. Thistlethwayte, Esq. A priory of Black canons, founded by Henry I., and originally established at Porchester, in 1133, was soon after removed hither, and flourished till the dissolution, when its revenue was valued at £314. 17. 10. per annum: it acquired some historical celebrity from its having been the scene of the marriage of Henry VI. with Margaret of Anjou; and there are still some small remains of the monastic buildings in Southwick Park. The manor-house is a large building of some antiquity, having two wings terminating in gables, and embattled. Charles I. was on a visit to the owner of this mansion at the time when the Duke of Buckingham, whom he had accompanied thus far from London, was assassinated by Felton, at Portsmouth: George I. was also entertained here. The publicans at Southwick enjoy the privilege, under a charter of Queen Elizabeth, of having no soldiers billeted upon them, nor quartered in their houses. A fair for horses is held on April 5th; and here was formerly a market, granted to the priory in 1235, but it has long been disused. A Sunday National school has been established.

SOUTHWICK (*St. Michael*), a parish, in the union of STEYNING, hundred of FISHERGATE, rape of BRAMBER, Western Division of the county of SUSSEX, $1\frac{1}{2}$ mile (E.) from New Shoreham; containing 502 inhabitants. The living is a discharged rectory, valued in the king's books at £9. 13. $9\frac{1}{2}$.; present net income, £207: it is in the patronage of the Crown. The church is principally in the Norman style, though the upper part of the tower and some smaller portions are of later date: it has been enlarged, and 200 free sittings provided, the Incorporated Society having granted £130 in aid of the expense.

SOUTHWICK, a chapelry, in the parish of NORTH BRADLEY, union of WESTBURY and WHORWELSDOWN, hundred of WHORWELSDOWN, Whorwelsdown and Northern Divisions of the county of WILTS, $2\frac{1}{2}$ miles (S. W. by S.) from Trowbridge; containing 1452 inhabitants. The living is a perpetual curacy; patron, Vicar of North Bradley. There are places of worship for Baptists and Wesleyan Methodists; also a National school for boys, and an infants' school partly supported by private charity. The manufacture of broad cloth and kerseymeres is carried on here.

Arms.

SOUTHWOLD (*St. Edmund*), a sea-port, incorporated market-town, and parish, having separate jurisdiction, in the union and hundred of Blything, Eastern Division of the county of SUFFOLK, 36 miles (N. E.) from Ipswich, and 104 (N. E.) from London; containing 1875 inhabitants. The ancient names of this place were *Suwald, Suwalda, Sudholda*, and *Southwood*, probably derived from an adjacent wood, the western confines still retaining the appellation of Wood's-end Marshes and Wood's-end Creek. In the year 1659, a dreadful conflagration took place, which in a few hours consumed the town-hall, market-house, prison, several other buildings, and 238 dwelling-houses, at an estimated loss of more than £40,000; the court baron rolls were all destroyed, and in consequence the copyholders under the corporation became freeholders. Another remarkable event was the memorable sea-fight between the English, under the command of His Royal Highness the Duke of York, and the Dutch under Admiral de Ruyter, which took place in Sole bay, to the east of the town, on the 26th of May, 1672, in which, though the former proved victorious, many brave and distinguished officers were slain, among whom was the Earl of Sandwich, second in command. In 1747, the haven, which is formed by the mouth of the river Blyth, and which had become choked up with sand, was cleared out by act of parliament. In 1749, a pier was erected on the north side; and in 1751, another on the south side, by the same authority. The town is pleasantly situated on a hill overlooking the North Sea, and is rendered peninsular by the sea and a creek, called the Buss creek, which runs into the river Blyth, over which there is a bridge, anciently called Myght's bridge, and formerly a drawbridge, leading into the town: it consists principally of one paved street; the houses are mostly well built and of modern appearance, and the inhabitants are well supplied with water from numerous excellent springs. On St. Edmund's, commonly called Gun Hill are six eighteen-pounders, presented by His Royal Highness the Duke of Cumberland, who landed here from the Netherlands, October 17th, 1745. There was formerly an ancient fort, probably in the possession of the Danes, on their invasion of the country in 1010, of which the foss is still discernible. From the nature of its situation and the convenience of the beach, Southwold is admirably adapted for sea-bathing, and has for several years been much resorted to for that purpose. There is a good promenade; also a reading-room, called the *Casino*, on the Gun Hill, with an assembly-room: races are held annually. To counteract the encroachments of the sea, a breakwater has been made under Gun Hill cliff, extending upwards of 300 yards. The trade of the town consists in the home fishery, which employs several small boats; and in the preparation and exportation of salt, for which there is a manufactory; and of red herrings, red sprats, and malt. The imports are coal and cinders: the coasting trade is chiefly in timber, lime, wool, and corn. The entrance into the haven is on the south side of the town: the superintendence of it is vested in commissioners, its revenue being about £1400 per annum, which is expended in repairing the piers and haven. The amount of duties paid at the custom-house, in 1838, was £245. 18. 1. The river Blyth is navigable to Halesworth, and, besides the bridge crossing it at Blythburgh, there is a ferry to Walberswick. The market is on Thursday; and a fair is held on Trinity-Monday.

Corporation Seal.

The first charter of incorporation was granted by Henry VII., and confirmed, with extended privileges, by Henry VIII. and subsequent sovereigns: that granted by William and Mary vested the government in two bailiffs, a high steward, town-clerk, coroner, and other officers: the bailiffs and high steward were justices of the peace, and held regular sessions for the trial of felons and other offenders. The corporation now consists of a mayor, 4 aldermen, and 12 councillors, under the act of the 5th and 6th of William IV., cap. 76, for an abstract of which see the Appendix, No. I. The mayor and late mayor are justices of the peace. A court of record for the recovery of debts to any amount was formerly held every week, but has been disused for nearly a century; and a court of admiralty, having jurisdiction over the town and fort, was held until the passing of the Municipal Reform Act, by which it was abolished. The guildhall was erected by the corporation, at an expense of £800; and the old gaol having been taken down, a new one was built, in the year 1819, which is now converted into a dwelling-house. The living is a perpetual curacy, annexed to Reydon; net income, £60; patron, Earl of Stradbroke. The church, which was completed about 1460, is a very elegant structure, in the later style of English architecture, with a large and lofty tower, surmounted with a spire, and constructed of freestone intermixed with flint of various colours. At each angle of the east end of the chancel is a low hexagonal embattled tower, decorated with crosses: the south porch is very elegant, and above the clerestory roof is a light open lantern: the ceiling was, in former times, handsomely painted, and the interior very richly ornamented, as appears by the carved work of the rood-loft, screen, and the seats of the magistrates. It was enlarged in 1836, and 100 free sittings provided, the Incorporated Society having granted £35 in aid of the expense. On the south side of the churchyard are three gravestones, in memory of Thomas Gardner, the historian of Dunwich and Southwold, and his two wives and daughter, on which are some singular inscriptions. There are places of worship for Baptists, Independents, and Wesleyan Methodists. The corporation have under their control for the building and maintenance of bridges, sea-walls, &c., and of the townhall, for the payment of the salaries of the corporation officers, and for general purposes of improvement, the following estates; *viz.*, the manorial rights of Southwold, granted by Henry VII., in the fifth year of his reign; an estate in Southwold, consisting of about 300 acres, the principal part of which was devised by William Godell, in 1509; a windmill, a house, and several shops in the town, now severally let to tenants at rack-rents; and an undivided moiety of 20 acres of land in Reydon, taken in exchange from Sir Thomas Gooch, Bart., in lieu of a parcel of land in Benacre, purchased by the corporation about the year 1642. John Sayer, in 1816, bequeathed £200 four per cent. consols. towards the support of the Burgh school; but that institution having been relinquished, the dividends are now applied, according to the will of the donor, to the relief of poor widows of Trinity pilots and masters of vessels belonging to the port. On a hill, called Eye cliff, at a small distance from the town, are vestiges of ancient encampments, and, in many parts, of circular tents, now called Fairy hills, most probably of Danish origin. Fossil remains of the elephant and mammoth are frequently found in the cliffs.

SOUTHWOOD (*St. Edmund*), a parish, in the union and hundred of Blofield, Eastern Division of the county of Norfolk, 3¾ miles (S. by W.) from Acle; containing 54 inhabitants. The living is a discharged rectory, with the vicarage of Limpenhoe annexed; net income, £163; patron, J. F. Leathes, Esq.

SOUTHWOOD, or SOUTHWOOD PARK, an extra-parochial district adjoining to Hargrave, in the hundred of Thingoe, Western Division of the county of Suffolk, consisting of about 300 acres.

SOUTHWORTH, a joint township with Croft, in the parish of Winwick, union of Warrington, hundred of West Derby, Southern Division of the county palatine of Lancaster, 3¼ miles (S. E. by E.) from Newton-in-Mackerfield; containing 1329 inhabitants. The hall once belonged to the Roman Catholic college of Stonyhurst, and part of it is still used as a chapel by persons of that persuasion.

SOW, or SOWE (*St. Mary*), a parish, in the union of Foleshill, partly in the Kirby Division of the hundred of Knightlow, but chiefly in the county of the city of Coventry, Northern Division of the county of Warwick, 3 miles (E. N. E.) from Coventry; containing 1414 inhabitants, of which number, 426 are in the Kirby Division of the hundred of Knightlow, and 988 in the county of the city of Coventry. Considerable coal-works are in operation in this parish, and many of the inhabitants are engaged in the riband manufacture, in connection with the trade of Coventry. The living is a vicarage, annexed to that of Stoke; impropriator, Earl of Craven. The church has been enlarged, and 160 free sittings provided, the Incorporated Society having granted £100 in aid of the expense. Here is a National school.

SOWERBY, a joint township with Inskip, in the parish of St. Michael, union of Garstang, hundred of Amounderness, Northern Division of the county palatine of Lancaster, 4¾ miles (S. S. W.) from Garstang: the population is returned with Inskip.

SOWERBY, a chapelry, in the parish and union of Thirsk, wapentake of Birdforth, North Riding of the county of York, 1 mile (S.) from Thirsk; containing 756 inhabitants. The living is a perpetual curacy; net income, £34; patron and appropriator, Archbishop of York.

SOWERBY, a chapelry, in the parish and union of Halifax, wapentake of Morley, West Riding of the county of York, 4 miles (W. S. W.) from Halifax; containing 6457 inhabitants. The living is a perpetual curacy; net income, £199; patron, Vicar of Halifax. The chapel, dedicated to St. Peter, contains a fine statue, erected about half a century ago, to the memory of John Tillotson, D.D., Archbishop of Canterbury, who was born at Haugh-End, (his father having been a manufacturer,) in 1630; he died in 1694. There are places of worship for Independents and Wesleyan Methodists. The manufacture of woollen and cotton goods is extensively carried on here. Paul Bairslow, in 1711, bequeathed £16 per annum for the support of a school; and another is partly supported by subscription. In 1678, a considerable number of Roman coins was ploughed up in the neighbourhood.

SOWERBY-BRIDGE, a chapelry, in the parish of Halifax, wapentake of Morley, West Riding of the county of York, 2¾ miles (S. W. by W.) from Halifax, with which the population is returned. Woollen and cotton goods are manufactured here to a considerable extent. There are also three large iron-foundries, and a great number of corn-mills on the Calder, which river and the Rochdale canal pass through this district. Great quantities of stone are obtained in the vicinity. The living is a perpetual curacy; net income, £166; patron, Vicar of Halifax. The old chapel being insufficient to accommodate the increased population of the place, a more commodious structure has been built, containing nearly 1000 sittings, of which 307 are free, the Incorporated Society having granted £800 in aid of the expense. There is a place of worship for Wesleyan Methodists; also a National school.

SOWERBY, CASTLE (*St. Kentigern*), a parish, in the union of Penrith, Leath ward, Eastern Division of the county of CUMBERLAND, $3\frac{1}{4}$ miles (S. E. by E.) from Hesket-Newmarket; containing 961 inhabitants. The living is a discharged vicarage, valued in the king's books at £17. 10. 5; present net income, £98; patrons and appropriators, Dean and Chapter of Carlisle. A school is endowed with £10 per annum for the instruction of 10 poor children. At Birksceugh are the remains of a chapel, formerly called Lady chapel; and in the township of How-bound is Castle Hill, the site of an ancient castle, from which the parish derives its distinguishing prefix.

SOWERBY, TEMPLE, a chapelry, in the parish of KIRKBY-THORE, EAST ward, and union, county of WESTMORLAND, 7 miles (N. W.) from Appleby; containing 438 inhabitants. The living is a perpetual curacy; net income, £96; patron, Earl of Thanet. The chapel, dedicated to St. James, is a handsome structure of red freestone, with a square tower and portico, rebuilt and enlarged in 1770, at the expense of the late Sir William Dalston. There is a place of worship for Independents. The village is situated on the river Eden, which is crossed by a bridge erected in 1823, at an expense of £3700, to replace a former bridge, which was destroyed by a flood in 1822; it is considered to be the finest bridge in the county. The village consists of two spacious streets of well-built houses, with several commodious inns, and near it are many handsome villas inhabited by genteel families. Fairs for sheep and cattle are held here on the last Thursdays in Jan., Feb., March, June, July, Aug., and Oct. and on the second Thursday in May. The Knights Templars had a preceptory here, which, when suppressed in 1312, was given to the Knights Hospitallers.

SOWERBY-UNDER-COTLIFFE, a township, in the parish of KIRBY-SIGSTON, union of NORTH-ALLERTON, wapentake of ALLERTONSHIRE, North Riding of the county of YORK, 3 miles (E. by S.) from North Allerton; containing 67 inhabitants.

SOWTON (*St. Michael*), a parish, in the union of ST. THOMAS, partly in the hundred of EAST BUDLEIGH, but chiefly in the hundred of WONFORD, Wonford and Southern Divisions of the county of DEVON, 4 miles (E.) from Exeter; containing 391 inhabitants. The living is a rectory, valued in the king's books at £11. 16. 3.; patron, Bishop of Exeter. The tithes have been commuted for a rent-charge of £264, subject to the payment of rates; the glebe comprises 19 acres, valued at £40 per annum. Five poor children are instructed for a small annuity left by Thomas Weare, in 1691.

SOYLAND, a township, in the parish and union of HALIFAX, wapentake of MORLEY, West Riding of the county of YORK, $5\frac{1}{2}$ miles (S. W. by W.) from Halifax; containing 3589 inhabitants, many of whom are employed in the manufacture of cotton and woollen goods.

SPALDING (*St. Mary and St. Nicholas*), a market-town and parish, and the head of a union, in the wapentake of ELLOE, parts of HOLLAND, county of LINCOLN, 44 miles (S. E. by S.) from Lincoln, and 100 (N.) from London; containing 6497 inhabitants. This place, which is said to have derived its name from a *Spa*, or spring of chalybeate water, in the market-place, is of great antiquity, as appears from the remains of Roman embankments in the neighbourhood. It is mentioned at an early period in the Saxon annals as one of the points on the boundary line of the estate belonging to Crowland abbey; and as the residence of Thorold de Buckenhale, the last Saxon governor of the province of Mercia, who in 1051 founded here a cell for a prior and five monks subordinate to that monastery. At the Norman Conquest the manor was presented to Ivo Talbois, Earl of Angiers and nephew of the Conqueror, who built a castle here, by which the religious society were so harassed that they abandoned their convent, which, falling into the hands of the earl, was, together with the church of St. Mary and the manor, given in 1074 to the abbey of St. Nicholas at Angiers; it became an Alien priory to that monastery, and was inhabited by monks of the Benedictine order. At the time of the suppression of Alien priories, this establishment was exempted: it was subsequently raised to the dignity of an abbey, and flourished till the dissolution, when its revenue was estimated at £878. 18. 3. The town is situated on the river Welland, in a fenny district, but now remarkably well drained; the streets are well paved, and lighted with gas, and there are many wells of excellent water for the supply of the inhabitants; the houses are in general well built and of neat appearance, and several of them are very handsome. An Antiquarian Society was established many years since by Mr. Maurice Johnson, a native of the town, of which Sir Isaac Newton, Sir Hans Sloane, Dr. Stukeley, and several other distinguished persons, were members; many of the valuable books, some manuscripts, relics of antiquity, and natural curiosities, are still preserved. The theatre, a small neat building, is opened annually for three weeks in the month of September. The land in the vicinity is extensively appropriated to grazing, and wool forms a material article in the trade of the town; a very considerable trade is also carried on in corn, coal, and timber; the river Welland is navigable to Stamford, and sloops of from 50 to 70 tons' burden can come up to the centre of the town, which maintains a regular coasting trade with London, Hull, Lynn, and other places. The port is a member of that of Boston; there is a quay for landing goods, with spacious storehouses for their reception, and this place may be considered one of the most thriving towns on the eastern coast. The market, which is one of the largest in the kingdom for fat cattle, is on Tuesday; fairs are held on April 27th and June 30th, by letters patent of George I.; and on August 28th, September 25th, and December 6th, by prescription, chiefly for live stock. This town has, for many centuries, been the principal seat of jurisdiction for the parts of Holland; in the Saxon times, the courts of law were held here by the earls, and subsequently to the Conquest the priors were invested with judicial authority, and possessed the power of life and death. At present the quarter sessions for the parts of Holland are held here and at Boston, and petty sessions for the wapentake are held here every week. Courts of sewers, and of requests for the recovery of debts not exceeding £5, for the hundred of Elloe, and the parishes of Gosberton and Surfleet, and also courts leet and baron, at which the steward presides, are held in the town. It has been made a polling-place for the parts of Kesteven and Holland. The town-hall, situated at the north-west end of the market-place, was erected at the expense of Mr. John Hobson, about the year 1620: the lower part is let for shops, and the rents are given to the poor, according to the will of the donor. A new house of correction for the parts of Holland was built in 1824: it is an airy and commodious edifice, and contains seven wards, day-rooms, and airing-yards, a tread-wheel, and infirmaries for the male and female sick.

The living is a perpetual curacy; net income, £950: it is in the patronage of certain Trustees, who are seised of the rectory in trust for the incumbent. The church, which was erected in 1284, when the old conventual church was taken down, is principally in the later style of English architecture, with a fine tower surmounted by a crocketed spire: considerable additions were made to it in 1466, among which is the beautiful north porch. There are places of worship for Baptists, the Society of Friends, Independents, and Wesleyan Methodists. The free grammar school was well endowed by John Blanche and Matthew Gamlyn, and latterly by Mr. Atkinson; and, by letters patent of the 30th of Elizabeth, the four trustees are incorporated, and have a common seal: the Master of St. John's College, Cambridge, is visiter, and the learned Dr. Bentley was once head master. The Petit school was founded in 1682, by Thomas Willesley, and well endowed with land. The Blue-coat school, founded by one of the Gamlyn family, is supported by a small endowment in land, aided by voluntary contributions. There is also a National school for boys. An almshouse for 22 poor persons was founded and endowed, in 1590, by Sir Matthew Gamlyn; and another was founded in 1709, for eight poor widows, by Mrs. Elizabeth Sparke. There are also considerable estates, called "Town Husbands," vested in trustees for the benefit of the poor. The poor law union of Spalding comprises 9 parishes, under the care of 27 guardians, and contains a population of 17,338, according to the census of 1831. A portion of the abbey buildings is yet remaining, which has been partly converted into tenements, and is partly in ruins. Many relics of antiquity have been found in the neighbourhood at different times, and several have been taken out of the river Welland.

SPALDINGTON, a township, in the parish of Bubwith, union of Howden, Holme-Beacon Division of the wapentake of Harthill, East Riding of the county of York, 4½ miles (N. by E.) from Howden; containing 352 inhabitants. Spaldington Hall, formerly the seat of the ancient family of Vavasour, is a fine specimen of the Elizabethan style of architecture. There is a place of worship for Wesleyan Methodists.

SPALDWICK (*St. James*), a parish, in the hundred of Leightonstone, union and county of Huntingdon, 3¾ miles (N. E. by N.) from Howden; containing 388 inhabitants. The living is a discharged vicarage, in the patronage of the Prebendary of Longstowe in the Cathedral Church of Lincoln, valued in the king's books at £12. 0. 10.; present net income, £84; appropriators, Dean and Chapter of Lincoln. There are places of worship for Baptists and Independents. Fairs for cattle are held on Whit-Monday and Nov. 28th.

SPALFORD, a hamlet, in the parish of North Clifton, union of Newark, Northern Division of the wapentake of Newark, Southern Division of the county of Nottingham, 7 miles (E. by S.) from Tuxford; containing 80 inhabitants.

SPANBY (*St. Nicholas*), a parish, in the union of Sleaford, wapentake of Aveland, parts of Kesteven, county of Lincoln, 4¼ miles (N. N. E.) from Falkingham; containing 84 inhabitants. The living is a rectory, annexed to the vicarage of Swaton.

SPARHAM (*St. Mary*), a parish, in the union of Mitford and Launditch, hundred of Eynsford, Eastern Division of the county of Norfolk, 3 miles (S. W.) from Reepham; containing 355 inhabitants. The living is a rectory, valued in the king's books at £9. 17. 11.; present net income, £548; patron, Edward Lombe, Esq.

SPARKFORD (*St. Mary Magdalene*), a parish, in the union of Wincanton, hundred of Catsash, Eastern Division of the county of Somerset, 4¼ miles (S. W. by S.) from Castle-Cary; containing 257 inhabitants. The living is a rectory, valued in the king's books at £12. 16. 3.; patron, Rev. Henry Bennett. The tithes have been commuted for a rent-charge of £245. 6. 8., subject to the payment of rates, which on the average have amounted to £23; the glebe comprises 40 acres, valued at £75 per annum. A day and Sunday school is chiefly supported by the Rev. H. Bennett.

SPARSHOLT (*Holy Cross*), a parish, in the union of Wantage, partly in the hundred of Shrivenham, but chiefly in that of Wantage, county of Berks, 3¼ miles (W.) from Wantage; containing 874 inhabitants. The living is a discharged vicarage, valued in the king's books at £20. 2. 3½.; present net income, £363; patrons and impropriators, Provost and Fellows of Queen's College, Oxford. The church is principally in the Norman style: it contains three stone stalls, and a piscina, highly enriched with trefoil ornaments and crocketed pinnacles. There is a place of worship for Presbyterians at Kingston-Lisle. The Wilts and Berks canal, and the line of the Great Western railway, pass through the parish, and the Ikeneld road through the vale of White Horse, to the south of the village. Abraham Atkins, in the year 1788, gave a school-house at Kingston-Lisle, and endowed it with a moiety of the rents arising from a certain estate: the annual income is about £63, for which the master instructs all the children of the parish who apply. Eight others are taught for £3. 10. a year, the gift of Richard Edmondson, in 1713. A National Sunday school was built here in 1838.

SPARSHOLT (*St. Stephen*), a parish, in the union of New Winchester, hundred of Buddlesgate, Winchester and Northern Divisions of the county of Southampton, 3¾ miles (W. N. W.) from Winchester; containing 357 inhabitants. The living is a vicarage, valued in the king's books at £16. 10. 2½.; present net income, £230: it is in the patronage of the Crown; impropriator, Sir W. Heathcote, Bart.

SPAUNTON, a township, in the parish of Lastingham, union of Pickering, wapentake of Ryedale, North Riding of the county of York, 7½ miles (N. W.) from Pickering; containing 138 inhabitants.

SPAXTON (*St. Margaret*), a parish, in the union of Bridg-water, hundred of Cannington, Western Division of the county of Somerset, 5 miles (W.) from Bridg-water; containing 963 inhabitants. The living is a rectory, valued in the king's books at £24. 8. 9.; patron and incumbent, Rev. William Gordon. The tithes have been commuted for a rent-charge of £650, subject to the payment of rates, which on the average have amounted to £64; the glebe comprises 65 acres, valued at £130 per annum. The Rev. Joseph Cook, in

1708, bequeathed lands, producing a liberal income, for the maintenance of six poor persons in an hospital, also £6 per annum, for teaching poor children.

SPECTON, a chapelry, in the parish and union of BRIDLINGTON, wapentake of DICKERING, East Riding of the county of YORK, 5½ miles (N. N. W.) from Bridlington; containing 111 inhabitants. The living is a perpetual curacy; net income, £50; patron, W. T. Denison, Esq.

SPEEN (*St. Mary*), a parish, in the union of NEWBURY, partly in the hundred of KINTBURY-EAGLE, and partly in that of FAIRCROSS, county of BERKS, 1 mile (W.) from Newbury; containing 3044 inhabitants. This was the *Spinæ* of the Romans, a station on the road from Gloucester to Silchester. To the north of the church, traces of an agger, or fortification, are distinctly visible; and on Speen moor a large urn was found under a tumulus of earth eight feet high; a Roman altar, consecrated to Jupiter, was also discovered, in 1730, at Fulsham, in this neighbourhood. This was the principal scene of the second battle of Newbury, fought Oct. 27th, 1644, between what is now the castle and the village. A market was formerly held on Monday, but it has been long disused. The parish is bounded on the south by the river Kennet and the Kennet and Avon canal, and on the north by the river Lambourn. The living is a vicarage, with the chapelry of Speenhamland, valued in the king's books at £14. 0. 10.; present net income, £424; patron and appropriator, Bishop of Salisbury. The church contains some curious monumental figures. A new church has been erected and endowed by the vicar, the Rev. H. W. Majendie, in the hamlet of Stockcross: it was consecrated in Nov. 1839. The communion service and organ were also presented by the vicar. A parochial school is supported by subscription.

SPEEN, CHURCH, a tything, in the parish of SPEEN, hundred of KINTBURY-EAGLE, county of BERKS, 1 mile (W.) from Newbury: the population is returned with the parish.

SPEEN, WOOD, a joint township with Bagnor, in the parish of SPEEN, hundred of FAIRCROSS, county of BERKS, 2¼ miles (N. W.) from Newbury: the population is returned with the parish.

SPEENHAMLAND, a chapelry, in the parish of SPEEN, hundred of FAIRCROSS, county of BERKS, adjoining the town of Newbury: the population is included in that of the parish. A chapel of ease was erected in 1831, chiefly by subscription, and contains 1000 sittings. A school-house for 100 boys and girls was erected and endowed by Mr. and Mrs. Page, of Goldwell House. An almshouse was founded here in 1664, by Mrs. Anne Watts, for two poor widows, with an allowance of two shillings each per week.

SPEKE, a township, in the parish of CHILDWALL, union of PRESCOT, hundred of WEST DERBY, Southern Division of the county palatine of LANCASTER, 7 miles (S. S. W.) from Prescot; containing 514 inhabitants.

SPELDHURST (*St. Mary*), a parish, in the union of TONBRIDGE, partly in the hundred of SOMERDEN, lathe of SUTTON-AT-HONE, but chiefly in the hundred of WASHLINGSTONE, late of AYLESFORD, Western Division of the county of KENT, 3 miles (N. W.) from Tonbridge-Wells; containing 2640 inhabitants. This parish comprises 3919 statute acres, of which 2843 are arable and in good cultivation. Iron ore abounds, and the springs are consequently more or less chalybeate. The well from which the water of Tonbridge well is drunk is situated in this parish, which is intersected by a branch of the river Medway, on which are several mills; and there are quarries of good building stone. Fairs for cattle are held at Groombridge on May 17th and Sept. 25th. The living is a rectory, valued in the king's books at £15. 5.; present net income, £380: it is in the patronage of Mrs. Harbroe. The church was struck by lightning and burned down in 1791, and was rebuilt in the following year: it contains two curious epitaphs on Sir Walter and Lady Anne Waller, and some monuments to the Bacon family. There is a private chapel at Groombridge. A certain number of children of this parish are admitted into the free school at Tonbridge, and a National school has been established here, and another is partly supported by subscription. The Duke of Orleans was detained prisoner at Groombridge in the reign of Henry V., and built the porch of the old church.

SPELSBURY (*All Saints*), a parish, in the union of CHIPPING-NORTON, hundred of CHADLINGTON, county of OXFORD, 5 miles (S. E. by S.) from Chipping-Norton; containing 609 inhabitants. The living is a vicarage, valued in the king's books at £9. 8. 9.; present net income, £211; patrons and appropriators, Dean and Canons of Christ-Church, Oxford. The church was founded by the Beauchamp family, and had formerly a lofty and elegant spire, which in 1772 was taken down from the insufficiency of the tower to sustain its weight, and other alterations have destroyed the original character of the edifice. In the north aisle is the sepulchral chapel of the family of Lee, which contains the remains of Henry Lord Wilmot, and of his son John, the celebrated Earl of Rochester. A school is chiefly supported by the vicar. On an eminence near the village is an extensive triangular intrenchment, called Castle Ditches, enclosing a space of about 24 acres.

SPENITHORN (*St. Michael*), a parish, in the union of LEYBURN, wapentake of HANG-WEST, North Riding of the county of YORK; containing 848 inhabitants, of which number, 198 are in the township of Spenithorn, 1 mile (N. E. by N.) from Middleham. The living is a rectory, valued in the king's books at £20. 10. 5.; present net income, £425; patron, Marmaduke Wyvill, Esq. The impropriate tithes of Harmby have been commuted for a rent-charge of £10. 16. 6., subject to the payment of rates. John Hutchinson, a philosophical writer, was born here, in 1667.

SPERNALL (*St. Leonard*), a parish, in the union of ALCESTER, Alcester Division of the hundred of BARLICHWAY, Southern Division of the county of WARWICK, 4 miles (N.) from Alcester; containing 95 inhabitants. The living is a discharged rectory, valued in the king's books at £3. 18. 1½.; present net income, £154; patron and incumbent, Rev. John Chambers.

SPETCHLEY (*All Saints*), a parish, in the union of PERSHORE, Lower Division of the hundred of OSWALDSLOW, Worcester and Western Divisions of the county of WORCESTER, 3¾ miles (E. by S.) from Worcester; containing 117 inhabitants. The living is a discharged rectory, valued in the king's books at £6. 11. 3.; present net income, £109; patron, Robert Berkeley, Esq. The church contains several monuments worthy of notice, erected to the memory of the Berkeley family. Here is a Roman Catholic chapel.

SPETISBURY (*St. John the Baptist*), a parish, in the union of Blandford, hundred of Loosebarrow, Blandford Division of the county of Dorset, 3 miles (S. E. by S.) from Blandford-Forum; containing 667 inhabitants. The living is a rectory, with Charlton-Marshall annexed, valued in the king's books at £28. 18. 1½.; patron, I. S. W. S. E. Drax, Esq. The tithes have been commuted for a rent-charge of £440, subject to the payment of rates; the glebe comprises 51 acres, valued at £63 per annum. Here is a Roman Catholic chapel. The river Stour runs close past the village. In 1728, a school was founded and endowed with land, given by Dr. Sloper and Bishop Hall, producing £110 per annum, of which £50 is paid to the master. Three other schools are partly supported by the rector. Here was a priory, a cell to the abbey of Preaux in Normandy, founded in the time of Henry I., but afterwards considered part of the cell of Monks' Toft in Norfolk, belonging to the same house. In the neighbourhood are the remains of an ancient British encampment, called Spetisbury Rings, in which coins and other relics of the Saxons have been found.

SPEXHALL (*St. Peter*), a parish, in the union and hundred of Blything, Eastern Division of the county of Suffolk, 2 miles (N. by W.) from Halesworth; containing 197 inhabitants. The living is a discharged rectory, valued in the king's books at £14: it is in the patronage of the Crown. The tithes have been commuted for a rent-charge of £286. 4., subject to the payment of rates, which on the average have amounted to £51; the glebe comprises 45 acres, valued at £60 per annum.

SPILSBY (*St. James*), a market-town and parish, and the head of a union, in the Eastern Division of the soke of Bolingbroke, parts of Lindsey, county of Lincoln, 31 miles (E.) from Lincoln, and 133 (N.) from London; containing 1384 inhabitants. The town is situated upon an elevated spot of ground which commands an extensive south-easterly view of a tract of marsh and fen land, bounded by Boston deeps and the North Sea, and within twelve miles of Skegness, on the best part of the Lincolnshire coast, which is much frequented by visiters. It consists of four principal streets diverging from a spacious square, forming the market-place, which is ornamented on its east side by a cross, consisting of a plain octagonal shaft rising from a quadrangular base, and resting on five steps. A subscription library and news-room is connected with the principal inn. The market is on Monday; and fairs are held on the Monday before, and the two next after, Whit-Monday (when Whitsuntide falls in May, otherwise there is no fair on the latter day), and on the third Monday in July, for cattle and wearing apparel: a market for fat stock, recently established, is held every fortnight. The general quarter sessions for the South Division of the parts of Lindsey are held here twice a year, in January and July; and the town is a polling-place for the parts of Lindsey. A court house and house of correction, begun in June, 1824, were completed within two years, at an expense of £25,000: the latter contains nineteen day-rooms, eight airing-yards, and a tread-wheel, with an infirmary and yards for the prisoners, so arranged that the governor's house commands a complete view of the whole: the site occupies about two acres of ground, and is surrounded by a brick wall, in which, in front of the building, is a handsome Doric portico.

The living is a perpetual curacy; net income, £109; patron and impropriator, Lord Willoughby de Eresby. The church is an ancient irregular stone edifice, with a handsome embattled tower at the west end, supposed to have been erected about the time of Henry VII., at a much later date than the body of the building. Amongst several ancient monuments is one in memory of the celebrated Lord Willoughby de Eresby, who, in the reign of Elizabeth, commanded 4000 English troops despatched to France, in aid of Henry IV., King of Navarre: he died in 1601, and was interred here. There are places of worship for Independents and Wesleyan Methodists, with Sunday schools attached. The grammar school, rebuilt in 1826, is endowed for the gratuitous education of 30 scholars. In 1735, the Duke of Ancaster and others endowed a school for the education and clothing of 20 poor boys. The poor law union of Spilsby comprises 66 parishes or places, under 69 guardians, and contains a population of 23,316, according to the census of 1831. At Eresby, near this town, are extensive remains of the foundations of a chapel, made collegiate in 1349, for a master and twelve priests, by Sir John Willoughby, and dedicated to the Holy Trinity. At the same place was formerly an elegant mansion belonging to the late Duke of Ancaster, which, in 1769, was destroyed by fire, while repairing the lead on the roof; one gateway pillar of exquisite brick work alone is remaining.

SPINDLESTONE, a township, in the parish of Bambrough, union of Belford, Northern Division of Bambrough ward and of the county of Northumberland, 3¼ miles (E. by S.) from Belford; containing 101 inhabitants. Here was anciently a military station of considerable extent, vestiges of mounds and intrenchments being conspicuous.

SPITALFIELDS (*Christchurch*), a parish, in the union of Whitechapel, Tower Division of the hundred of Ossulstone, county of Middlesex; containing 17,949 inhabitants. This parish, which is situated in the north-eastern part of the metropolis, was anciently called "Lolsenorth Field," and appears to have been selected as a place of sepulture by the Romans, during their occupation of London. On breaking up the ground in 1576, for clay to make bricks, numerous urns containing ashes and burnt bones were discovered, in each of which was found a brass coin of the emperor reigning at the time of the interment. Among these were some of the reigns of Claudius, Vespasian, Nero, Antoninus Pius, Trajan, and others; and vials, glasses, and pottery of red earth, were also found, with various other relics of Roman antiquity. The present name is derived from a priory of canons of the Augustine order, and an hospital for poor brethren, entitled "the new hospital of our Lady without Bishopsgate," founded in the year 1197, by Walter Brune, citizen, and afterwards sheriff, of London, and Roesia his wife, and dedicated to the Blessed Virgin. This establishment continued to flourish till the dissolution, when its revenue, according to Dugdale, was estimated at £478. 6. 6. From the time of the Reformation it was the custom for a bishop, a dean, and a doctor of divinity, to preach a sermon each on the Resurrection, on the Monday, Tuesday, and Wednesday in Easter week, in a pulpit cross in the churchyard of the priory; but, during the protectorate of Cromwell, the practice was discontinued and the cross destroyed. These sermons, from this circumstance called the "Spital Sermons," were revived after the restoration of Charles II., and preached in the parish church of St. Bridget, or St. Bride, Fleet-street, and are now delivered before the lord mayor and aldermen of the city of London, in the parish church of Christchurch, Newgate-street. Though undistinguished by any other features for many years, this place became the seat of the silk manufacture, originally established at Canterbury and other places by the refugees, who, after the revocation of the edict of Nantz, in the reign of Louis XIV., had found an asylum in this country, from which time it began to increase in importance, and is now one of the most populous parishes in the vicinity of the metropolis. It was originally a hamlet, in the parish of St. Dunstan, Stepney, from which it was separated by act of parliament in 1729, and erected into a distinct parish. In Church-street and several others are some spacious and well-built houses; the other streets are inhabited chiefly by weavers and other persons connected either immediately or remotely with the silk trade, who work in their own dwellings. This district has long been celebrated as the principal seat of the silk manufacture, which is carried on to a very considerable extent; and within the parish and the immediate vicinity are numerous houses in the trade, employing from 200 to 1500 persons each; and including the adjacent parishes of Bethnal-Green and Shoreditch, and the hamlet of Mile-End New Town, and its neighbourhood, not less than 15,000 looms are in operation, affording employment to more than 50,000 persons, exclusively of others engaged in other departments of the trade, which in all its branches is computed to employ from 130,000 to 150,000 in the surrounding district. The principal articles manufactured are broad silks and plain and figured velvets of the best quality. Connected with the silk manufacture are numerous dyeing establishments, some of which are on a large scale. In Brick-lane is the very extensive porter brewery of Messrs. Truman, Hanbury, and Buxton; the premises are capacious and substantially built, and a very considerable addition has been made within the last few years for the purpose of brewing ale. In Wheeler-street is a large soap-manufactory, in which about 40 persons are employed; and there are several manufactories of harp and violin strings, violins and double basses, and materials for colouring spirits and vinegar. In Montague-street is a very extensive timber-yard, in which is a large assortment of fancy mahogany and rosewood veneers; and in Bell-lane is a large timber and building-yard. The market, principally for fruit and vegetables, is very extensive, and has been for many years in high reputation for the abundance of its supply. The parish is within the jurisdiction of the Tower Hamlets court of requests, for the recovery of debts under £5.

The living is a rectory not in charge; net income, £445; patrons, Principal and Fellows of Brasenose College, Oxford, who pay a stipend of £120 to the curate. The church, built in 1729, under the provisions of an act of parliament in the reign of Anne, is a stately and massive structure, in the Roman style of architecture, with a tower surmounted by a pyramid of rather cumbrous appearance: the nave is separated from the aisles by round-shafted Corinthian columns, of which, near the east and west ends on each side, one is duplicated and connected with two square columns of the same order, forming a strong but elegant pier, supporting a highly enriched cornice, and surmounted by corresponding pilasters rising to the roof, which is panelled in compartments and lighted by a range of small clerestory windows; the roofs of the galleries are arched between the columns, ornamented with hexagonal panels and lighted by two tiers of small windows, of which the upper are circular: the chancel is separated from the nave by an elegant screen in the Grecian style, of four Corinthian pillars, extending also across the aisles and supporting an enriched frieze and cornice, above the centre of which are the royal arms finely sculptured, resting upon a tablet and displayed in beautiful relief by a fan-light above the east window, which is of Venetian character: a similar screen extends across the western end of the church, but it is broken in the centre of the cornice, for the reception of the organ, of which the gallery is supported on fluted Corinthian pillars of mahogany. On the north side of the chancel is a monument by Flaxman to Sir Robert Ladbroke, Knt., lord mayor of London, whose statue in his civic robes, with the sword and mace lying at his feet, is finely executed in marble; and on the south side is a monument to Edward Peek, Esq., one of the commissioners for building 50 new churches in the reign of Anne, and who laid the first stone of this church; his bust, resting on a sarcophagus, at each end of which is a weeping cherub, under a canopy of drapery, is well executed by Dunn: on the north side of the altar is a tablet of white marble, erected in 1828, by his parishioners, as a tribute of respect to the memory of the Rev. West Wheldale, who was for 24 years rector of the parish. Sir George Wheeler's chapel, in Chapel-street, was erected by that gentleman for the accommodation of his tenants, previously to the erection of the parish church, and for many years after continued in that family, but was subsequently purchased by the Tillard family, whose lands were contiguous to those of the founder: it is a proprietary episcopal chapel, licensed but not consecrated, and is in the patronage of the Rev. Richard Tillard, with the assent of the rector of Christchurch. There are places of worship for Independents and Calvinistic and Wesleyan Methodists.

The parochial school was founded in 1708, for clothing, instructing, and apprenticing children, and is endowed with the produce of successive benefactions, amounting to £241 per annum: the school-house is a neat building adjoining the churchyard, having over the entrance statues of a boy and a girl in the costume of the school. The National school, in Quaker-street, was built in 1819, at a total expense of £3300, raised by subscription, and is adapted to the reception of 1000 children; in the boys' school-room divine service is voluntarily performed every Sunday evening by the present rector of the parish. On the opposite side of Quaker-street is an infants' school, founded in 1820, and principally supported by Joseph Wilson, Esq.; the building, with an enclosed play-ground in front, and in a retired situation, is well adapted to the purpose: there are at present about 200 children in this school, in which have been trained many teachers of similar establishments. The Protestant Dissenters' charity school, in Wood-street, was established in 1717 by subscription, for clothing and instructing 50 boys and 50 girls of all religious denominations: the house is substantially built, with a good garden behind, and in one of the lower rooms is a library, with a philosophical apparatus for the members of the Eastern Mechanics' Institute, who hold their meetings here. The Jews' free school, in Bell-lane, originally founded in 1818 for 270 boys, and rebuilt on a larger scale in 1820, is supported by subscription and donations; it is a spacious building, with a neat stuccoed front, and, exclusively of apartments for the master and mistress and play-grounds for the children, contains two school-rooms. During the festival of the new year, and the period of the white fast, the larger school-room is fitted up as a synagogue for the accommodation of the poor who have no seats in the regular synagogues. Without the limits of the parish, in Whitechapel road, are almshouses founded by John Baker, Esq., exclusively for aged persons of this parish, who are nominated by the Master and Wardens of the Brewers' Company, as trustees.

SPITTAL-ON-THE-STREET, a chapelry, in the parish of GLENTWORTH, Western Division of the wapentake of ASLACOE, parts of LINDSEY, county of LINCOLN, 10 miles (E.) from Gainsborough. The chapel is dedicated to St. Nicholas. An hospital for poor women, with a chapel dedicated to St. Edmund, existed here in the reign of Edward II., was augmented in that of Richard II., and is now under the superintendence of the Dean and Chapter of Lincoln.

SPITTLE, a joint township with Poulton, in the parish of BEBINGTON, union and Lower Division of the hundred of WIRRALL, Southern Division of the county of CHESTER, 5½ miles (N. E. by N.) from Great Neston: the population is returned with Poulton.

SPITTLE, or SPITTAL, a considerable fishing and sea-bathing village, in the parish of TWEEDMOUTH, in ISLANDSHIRE, forming a part of the detached portion of the county palatine of DURHAM, locally northward, and for electoral purposes annexed to the Northern Division, of the county of Northumberland, 1 mile (S. E.) from Berwick-upon-Tweed: the population is returned with the parish. This place, which is situated on the coast of the North Sea, at the mouth of the river Tweed, consists of two principal streets. It was formerly inhabited by smugglers and others of disreputable character, but, since the enclosure of the adjacent common, these have gradually given place to honest and industrious fishermen. Here are six herring houses for curing red and white herrings, and good accommodation for persons who resort hither for sea-bathing and for drinking the water of a powerful chalybeate spring in the neighbourhood. On Sunnyside hill, half a mile from the village, is an extensive colliery, the property of the corporation of Berwick. There is a place of worship for Presbyterians.

SPITTLE, a township, in the parish of OVINGHAM, union of HEXHAM, Eastern Division of TINDALE ward, Southern Division of the county of NORTHUMBERLAND, 11 miles (W. by N.) from Newcastle-upon-Tyne; containing 7 inhabitants.

SPITTLE, a township, in the parish of FANGFOSS, union of POCKLINGTON, Wilton-Beacon Division of the wapentake of HARTHILL, East Riding of the county of YORK, 3½ miles (N. W.) from Pocklington: the population is returned with the parish.

SPITTLE-HILL, a township, in the parish of MITFORD, union, and Western Division of the ward, of MORPETH, Northern Division of the county of NORTHUMBERLAND, 1¾ mile (W. by N.) from Morpeth; containing 11 inhabitants. An hospital, dedicated to St. Leonard, was founded here, and endowed with lands by Sir Wm. Bertram; the site is occupied by a modern mansion.

SPITTLEGATE, a township, in the parish and union of GRANTHAM, wapentake of WINNIBRIGGS and THREO, parts of KESTEVEN, county of LINCOLN, 1 mile (S. by E.) from Grantham; containing 1063 inhabitants. This place, with Grantham and Little Gonerby, was lighted with gas in 1833, by a company established with a capital of £6000.

SPIXWORTH (*St. Peter*), a parish, in the union of ST. FAITH, hundred of TAVERHAM, Eastern Division of the county of NORFOLK, 4¾ miles (N. by E.) from Norwich; containing 54 inhabitants. The living is a discharged rectory, valued in the king's books at £6; patron, J. J. Longe, Esq. The tithes have been commuted for a rent-charge of £360, subject to the payment of rates, which on the average have amounted to £60; the glebe comprises 8 acres, valued at £12. 5. per annum. The church contains many monuments and inscriptions.

SPOFFORTH (*All Saints*), a parish, in the Upper Division of the wapentake of CLARO, West Riding of the county of YORK; containing 3233 inhabitants, of which number, 914 are in the township of Spofforth, 3¼ miles (N. W. by W.) from Wetherby. The living is a rectory, valued in the king's books at £73. 6. 8.; present net income, £1538; patron, Col. Wyndham. There is a place of worship for Wesleyan Methodists. This was for several ages, prior to Alnwick, or Warkworth, the seat of the Percy family, who had a princely castle here, which was demolished by the Yorkists after the battle of Towton, in which the Earl of Northumberland and Sir Charles Percy, his brother, were slain. The grand hall of this once magnificent mansion, though in ruins, still remains; it is nearly 76 feet in length, and about 37 in breadth, and is lighted by one of those large cathedral windows introduced subsequently to the reign of Edward I. A school for fifteen poor children is supported by an annuity of £10, the gift of the Earl of Egremont, in 1786, and another of £2. 10., by the Rev. Dr. Trip. There is also a school for girls, supported by the Hon. Mrs. W. Herbert.

SPONDON (*St. Mary*), a parish, in the union of SHARDLOW, hundred of APPLETREE, Southern Division of the county of DERBY, 3¼ miles (E. by S.) from Derby; containing 1867 inhabitants. This parish was formerly more extensive, the chapelries of Chaddesden, Lockhay, and Stanley, having been recently separated from it, and erected into distinct parishes. The village of Spondon, situated on a commanding eminence, overlooking the beautiful vale of Derwent, is of a considerable size, and the residence of several highly respectable families. The inhabitants are principally employed at the extensive cotton-mills established here, and in the manufacture of stockings, lace, and net, for the Nottingham market. The Derby canal passes through the parish. The living is a discharged vicarage, valued in the king's books at £6. 14. 7.; present net income, £162; patrons,

R. Holden and J. Harrison, Esqrs., and the Trustees under the will of the late W. D. Lowe, Esq.; the latter, with Sir Robert Wilmot, Bart., are the impropriators. The church is in the decorated style of English architecture, and has in the chancel three stone stalls; it has, within the last few years, undergone a thorough repair, at an expense of £1200, and contains 100 free sittings, the Incorporated Society having granted £65 in aid of the expense. In the churchyard is an antique stone, apparently Saxon. There is a place of worship for Wesleyan Methodists; also a school endowed with land producing £14 per annum.

SPOONBED, a tything, in the parish of PAINSWICK, hundred of BISLEY, Eastern Division of the county of GLOUCESTER; containing 899 inhabitants.

SPORLE (*St. Mary*), a parish, in the union of SWAFFHAM, hundred of SOUTH GREENHOE, Western Division of the county of NORFOLK, $2\frac{1}{2}$ miles (N. E. by E.) from Swaffham; containing, with Palgrave, 746 inhabitants. The living is a vicarage, with the rectory of Great and Little Palgrave, valued in the king's books at £10. 3. $6\frac{1}{2}$.; patrons, Provost and Fellows of Eton College; impropriator, W. Lucas, Esq. The tithes (including rent-charge on glebe), have been commuted for £929. 4. 6., subject to the payment of rates, which on the average have amounted to £219; the glebe comprises 106 acres, valued at £138 per annum. The church is an ancient and spacious building of flint, with a western tower, quoined and embattled with freestone, and a large embattled porch, with a dilapidated room over it, supposed to have been once inhabited by a recluse. In the chancel are many old gravestones, stripped of their brasses, and on the north side a small chapel, separated from the aisle by screen-work. A school is endowed with land, producing £9. 12. per annum. A priory of Black monks, a cell to the abbey of Saumers in Anjou, was founded here, as it is thought, by Henry II., and at the suppression was granted by Henry VI. towards the endowment of Eton College.

SPOTLAND, a chapelry, in the parish and union of ROCHDALE, hundred of SALFORD, Southern Division of the county palatine of LANCASTER, $1\frac{1}{4}$ mile (N. W.) from Rochdale; containing 15,325 inhabitants. This chapelry forms an extensive suburb to the town of Rochdale, and largely participates in the cotton and every other branch of trade and manufacture carried on there. The living is a perpetual curacy, in the patronage of the Vicar of Rochdale. A church has been erected in the later style of English architecture, with a campanile turret, at an expense of £4430. 9. 8., under the act of the 58th of George III. Samuel Taylor and Robert Jacques, in 1740, conveyed to trustees a schoolhouse and sundry other property, for the free education of children; the annual income is £20; the premises were rebuilt in 1819, at an expense of £400. Another school is endowed with £14. 10. per annum, and here is also a Sunday National school.

SPRATTON (*St. Luke*), a parish, in the union of BRIXWORTH, hundred of SPELHOE, Southern Division of the county of NORTHAMPTON, $6\frac{3}{4}$ miles (N. N. W.) from Northampton; containing 1012 inhabitants. The living is a vicarage, valued in the king's books at £15; present net income, £371; patron, R. J. Bartlett, Esq. The church has a Norman tower, and the body presents a mixture of that and the early and decorated styles of English architecture. Here is a manufactory for carpets on a small scale. Two schools are supported by Robert Ramsden, Esq.

SPREYTON (*St. Michael*), a parish, in the union of OAKHAMPTON, hundred of WONFORD, Crockernwell and Southern Divisions of the county of DEVON, 8 miles (E. by N.) from Oakhampton; containing 423 inhabitants. The living is a discharged vicarage, valued in the king's books at £10. 5. 8.; present net income, £135; patron, Rev. Richard Holland; impropriators, the Landowners. Two schools are partly supported by subscription.

SPRIDLINGTON (*St. Hilary*), a parish, in the Eastern Division of the wapentake of ASLACOE, parts of LINDSEY, union and county of LINCOLN, 9 miles (W. S. W.) from Market-Rasen; containing 250 inhabitants. The living is a rectory, valued in the king's books at £11. 10.; present net income, £454; patron and incumbent, Rev. Frederick Gildart, who, with the Rev. H. Neville, partly supports a school.

SPRINGFIELD (*All Saints*), a parish, in the union and hundred of CHELMSFORD, Southern Division of the county of ESSEX, 1 mile (N. E.) from Chelmsford; containing 1851 inhabitants. This parish, which is separated from that of Chelmsford by the river Chelmer, is supposed to have derived its name from the extraordinary number of springs within its limits. The surface rises gently from the banks of the river, and the soil is generally a fertile loam resting upon gravel. The village is pleasantly situated on elevated ground, commanding fine views, and has been much increased since the formation of the Chelmsford and Maldon navigation. The county gaol is situated here, and contains fourteen divisions, day-rooms, and airing-yards; and a building has also been lately erected for the reception of vagrants, consisting of four rooms and an airing-yard, capable of holding conveniently 48 prisoners. The living is a rectory in two portions, called Bosworth's and Richard's, but consolidated by Bishop Sherlock; the former portion is valued in the king's books at £11. 6. 8., and the latter at £11. 4. $9\frac{1}{2}$.; patron, Rev. Arthur Pearson. The tithes have been commuted for a rent-charge of £848, subject to the payment of rates; there are 50 acres of glebe. The church, an ancient edifice with an embattled tower, was fully repaired in 1837, when the lower part of a handsome window, which had been bricked up, was opened, and a carved oaken screen restored to its pristine beauty, by John Adey Repton, Esq.: it has an elegant font in the Norman style. Arrangements are now in progress for building an episcopal chapel. There is a place of worship for Wesleyan Methodists. A National and an infants' school have been established, to which an extensive lending library is attached; and a school is endowed for Dissenters by the trustees of the late Dr. Williams. Dr. Goldsmith is said to have composed his "Deserted Village" whilst residing at a farm-house nearly opposite the church. Joseph Strutt, the engraver and antiquary, was born here in 1749; he died in 1802.

SPRINGTHORPE (*St. George and St. Lawrence*), a parish, in the union of GAINSBOROUGH, wapentake of CORRINGHAM, parts of LINDSEY, county of LINCOLN, $4\frac{1}{2}$ miles (E. by S.) from Gainsborough; containing 194 inhabitants. The living is a rectory, valued in the king's books at £14. 3. 4.: it is in the patronage of the Crown. The tithes have been commuted for a rent-charge of £177, subject to the payment of rates, which on the average have amounted to £20; the glebe comprises 14¼ acres, valued at £18 per annum.

SPROATLEY (*All Saints*), a parish, in the union of Skirlaugh, Middle Division of the wapentake of Holderness, East Riding of the county of York, 7 miles (N. E. by N.) from Kingston-upon-Hull; containing 366 inhabitants. The living is a rectory, valued in the king's books at £7. 0. 10.; present net income, £230; patron, Sir T. Constable, Bart. The church was built in 1819, upon the site of the old edifice, which was dedicated to St. Swithin: it is of white brick, principally in the later style of English architecture, and contains 100 free sittings, the Incorporated Society having granted £150 in aid of the expense. In laying the foundation some antique tombstones were found, one of them bearing a Saxon inscription. There is a place of worship for Wesleyan Methodists. Bridget Biggs, in 1733, gave an estate for the erection and support of a school for twenty children, and for apprenticing them; the income is about £90 per annum, for which eighteen boys and eighteen girls are educated and apprenticed.

SPROSTON, a township, in the parish of Middlewich, union and hundred of Northwich, Southern Division of the county of Chester, 2¼ miles (E.) from Middlewich; containing 128 inhabitants.

SPROTBROUGH (*St. Mary*), a parish, in the union of Doncaster, Northern Division of the wapentake of Strafforth and Tickhill, West Riding of the county of York, containing 500 inhabitants, of which number, 322 are in the township of Sprotbrough, 2¾ miles (W. S. W.) from Doncaster. The living is a rectory, valued in the king's books at £44. 18. 9.; present net income, £685; patron, Sir J. Copley, Bart. The church contains monuments of the Fitzwilliams and Copleys: one of the former family founded an hospital here, before 1363, in honour of St. Edmund, the revenue of which, at the dissolution, was valued at £9. 13. 11. A school is partly supported by subscription.

SPROUGHTON (*All Saints*), a parish, in the hundred of Samford, Eastern Division of the county of Suffolk, 3 miles (W. by N.) from Ipswich; containing 524 inhabitants. The living is a rectory, valued in the king's books at £20. 18. 9.; patron, Marquess of Bristol. The tithes have been commuted for a rent-charge of £566, subject to the payment of rates, which on the average have amounted to £105; the glebe comprises 13 acres, valued at £20 per annum. There are some interesting monuments in the church, of which one is to the Rev. J. Waite, rector in 1670. A National school is supported by subscription. The Stow-Market and Ipswich navigation passes through the parish, a part of which is within the liberties of Ipswich.

SPROWSTON (*St. Mary and St. Margaret*), a parish, in the union of St. Faith, hundred of Taverham, Eastern Division of the county of Norfolk, 3 miles (N. E. by N.) from Norwich; containing 1179 inhabitants. The living is a perpetual curacy, in the patronage of the Dean and Chapter of Norwich, the appropriators; net income, £94. The church contains a mural monument of marble to the memory of Sir Miles Corbet, and Catharine his lady, a descendant of whom, Thomas Corbet, Esq., was one of the judges that signed the death warrant of Charles I., and, after the Restoration, was arrested and executed as a traitor, in 1661.

SPROXTON (*St. Bartholomew*), a parish, in the union of Melton-Mowbray, hundred of Framland, Northern Division of the county of Leicester, 8 miles (N. E. by E.) from Melton-Mowbray; containing 378 inhabitants. The living is a discharged vicarage, with that of Saltby consolidated, valued in the king's books at £7. 4. 4.; present net income, £282; patron and impropriator, Duke of Rutland. There is a place of worship for Wesleyan Methodists; also a school to which the parish allows £10 annually.

SPROXTON, a township, in the parish of Helmsley, union of Helmsley-Blackmoor, wapentake of Ryedale, North Riding of the county of York, 1¼ mile (S.) from Helmsley; containing 195 inhabitants.

SPURM-HEAD, or SPURN-HEAD, in the parish of Kilnsea, Southern Division of the wapentake of Holderness, East Riding of the county of York, 12 miles (S. S. E.) from Patrington: the population is returned with the parish. This place, which is identified with the "*Ocellum Promontorium*" of Ptolemy, is situated at the eastern extremity of the county, projecting into the mouth of the river Humber, near its influx into the North Sea, and forming a peninsular promontory, connected with the main land by a narrow isthmus, which is overflowed at high water. It is inhabited only by a few veteran seamen, who are pensioned by the Trinity House of Hull, and have the management of the life boats stationed here for the assistance of distressed mariners, who are frequently exposed to great hazards in navigating this part of the coast, to lessen which, two lighthouses have been erected, which are also under their care.

SPURSTOW, a township, in the parish of Bunbury, union of Nantwich, First Division of the hundred of Eddisbury, Southern Division of the county of Chester, 4¼ miles (S. E. by E.) from Tarporley; containing 588 inhabitants. A mineral spring here, called Spurstow Spa, was formerly much frequented, and baths were erected by Sir Thomas Mostyn, for the accommodation of visiters; but the waters are not at present in much repute.

STADHAMPTON (*St. John the Baptist*), a parish, in the union of Abingdon, hundred of Dorchester, county of Oxford, 5 miles (N.) from Bensington; containing 313 inhabitants. The living is a perpetual curacy; net income, £52; patron, Charles Peers, Esq. Sundry benefactions, producing £10 a year, are applied for the education of nine poor children. In the churchyard is a remarkably fine yew tree. There is a place of worship for Particular Baptists; and a school for 60 children is supported by Charles Peers, Esq. The Rev. John Owen, D.D., the celebrated and learned non-conformist, Dean of Christ-Church, and Vice-Chancellor of the University of Oxford, in the time of the Commonwealth, was born here; and Wilmot, Earl of Rochester, is supposed to have been a native of this parish.

STADMERSLOW, a township, in the parish of Wolstanton, Northern Division of the hundred of Pirehill and of the county of Stafford, 4 miles (N.) from Burslem; containing 290 inhabitants, who are located principally at the village of Harrisey Head, where there are a place of worship for Wesleyan Methodists and a school.

STAFFIELD, or STAFFOLD, a township, in the parish of KIRK-OSWALD, union of PENRITH, LEATH ward, Eastern Division of the county of CUMBERLAND, 1½ mile (N. N. W.) from Kirk-Oswald; containing 265 inhabitants. The village is situated in a deep vale on the north side of the river Croglin, and near it are the remains of an old border fortification, called Scarrow-manwick. Staffold Hall is distinguished for its walks and beautiful scenery.

STAFFORD, a hamlet, in the parish of BARWICK, hundred of HOUNDSBOROUGH, BERWICK, and COKER, Western Division of the county of SOMERSET, 2 miles (S. by E.) from Yeovil: the population is returned with the parish.

Arms.

STAFFORD, a borough and market-town, and the head of a union, consisting of the united parishes of *St. Mary* and *St. Chadd*, locally in the Southern Division of the hundred of Pirehill, Northern Division of the county of STAFFORD; containing 8512 inhabitants, of which number, 6956 are in the borough of Stafford, 136 miles (N. W.) from London, on the road to Chester. This place, which is of great antiquity, was originally called *Stadeford*, or *Stadford*, from the Saxon *Stade*, signifying a place on a river, and the *trajectus*, or ford, across the river Sow, on which it is situated. It is said to have been, in 705, the devotional retirement of St. Bertelin, the son of a Mercian king, upon whose expulsion from his hermitage, at a spot called Berteliney, and Betheney, meaning the island of Bertelin, several houses were built, which formed the origin of the present town. In 913, Ethelfleda, Countess of Mercia, erected a castle on the north side of the river, and surrounded the town with walls and a fosse, of which the only remains are one side of a groove for a portcullis, at the entrance to Eastgate-street. Edward the Elder, brother of Ethelfleda, about a year after the erection of the castle, built a tower, the site of which Mr. Pennant supposes to have been the mount called, by Speed, Castle hill. From this period to the Conquest, the town appears to have increased considerably in extent and importance, and though it had not received any charter of incorporation, it is, in Domesday-book, called a city, in which the king had eighteen burgesses in demesne, and the Earls of Mercia twenty mansions. William, out of all the manors in the county, reserved this only for himself, and built a castle to keep the barons in subjection, appointing as governor Robert de Toeni, the progenitor of the house of Stafford, on whom he bestowed all the other manors, with the title of Baron de Stafford. The castle, after having been rebuilt by Ralph de Stafford, a celebrated warrior, in the reign of Edward III., remained till the parliamentary war, when it was garrisoned by the royal forces under the Earl of Northampton, but was at length taken by the parliamentary troops under the command of Sir William Brereton, and subsequently demolished by order of the parliament. The remains consisted chiefly of the keep, and were situated on the summit of a lofty eminence, about a mile and a half to the south-west of the town; the walls were eight feet thick, and at each angle was an octagonal turret, with a tower similarly shaped on the south-west side. About 50 years since, the only visible remains were part of a wall, which the late Sir William Jerningham underbuilt, to prevent it from falling; in doing which it was discovered that the basement story lay buried under the ruins of the upper parts. Sir George Jerningham (now Lord Stafford) afterwards began to rebuild the castle on the old foundation, but has completed only the south front, flanked with two round towers, in which are deposited some ancient armour and other curiosities. The town is pleasantly situated on the north side of the river Sow, about six miles distant from its confluence with the Trent: the entrance from the London road is by a neat bridge over the river, near which was one of the ancient gates: the houses are in general well built of brick, and roofed with slate, and many of them are of modern erection; the streets are well paved, and lighted with gas, and the inhabitants are amply supplied with water. The environs are pleasant, abounding with noble mansions and elegant villas. There is a theatre, and assemblies are held in a suite of rooms in the town-hall: and races take place annually in September on Marston-field. The principal branch of manufacture is that of shoes and boots for supplying the London market, and for exportation; and the tanning of leather is carried on to a considerable extent. Stafford, in common with the neighbourhood, is also noted for the quality of its ale. The river Penk joins the Sow near Rutford bridge, an elegant structure of three arches, nearly a mile distant; the Staffordshire and Worcestershire canal passes near the town, and there is a station for the Grand Junction railway, which passes to the south-west of the town. The market is on Saturday: the fairs are on April 5th, May 14th, June 25th, October 3rd, and December 5th.

Corporation Seal.

The inhabitants first received a regular charter of incorporation in the fourth year of the reign of John, confirming all privileges previously enjoyed. After various confirmations and additions, in the subsequent reigns, it became forfeited in 1826, by the common council neglecting to fill up vacancies in the body corporate; and, on petition, a new charter was granted by George IV., in 1827, restoring and confirming all previous rights and privileges, with the exception only of exemption from serving on juries for the county. By this charter the government was vested in a mayor, ten aldermen, and ten principal burgesses, assisted by a recorder, deputy-recorder, town-clerk, sergeants-at-mace, and subordinate officers. The mayor, late mayor, two aldermen, and the recorder and his deputy, were justices of the peace. The corporation now consists of a mayor, six aldermen, and eighteen councillors, under the act of the 5th and 6th of William IV., cap. 76, for an abstract of which see the Appendix, No. I. The borough is divided into two wards, and the municipal and parliamentary boundaries are co-extensive. The freedom is inherited by birth, or obtained by servitude to a resident freeman. The borough first exercised the elective franchise in the 23rd of Edward I., since which time it has regularly returned two members to parliament. The right of election was formerly vested in all resident burgesses, upwards of 800 in number. The new borough comprises an area of 600 acres, the limits of which are minutely described in the Appendix, No. II.: the old borough contained only 370 acres: the mayor is the returning officer. The corporation have power to hold quarterly courts of session within the borough, for all offences not capital: but they transfer to the judges travelling the circuit all causes requiring the decision of a jury: they have power also to hold a court of record, for the recovery of debts to any amount, but no process has issued from it for the last fifty years. Petty sessions are held weekly. The custom of Borough English prevails within the town and liberties. The assizes and sessions for the county, which had previously been held here, were restored by Queen Elizabeth, the inhabitants having represented to her, on visiting the town in 1575, that to their removal its decay, at that time, was, among other causes, to be attributed. This town has been made the place of election and a polling-place for the northern division of the shire. The county hall is a spacious and handsome modern building of stone, in the centre of the High-street, and occupying nearly the whole of one side of a spacious square, appropriated as a market-place, over part of which is a room for 1000 stand of arms, for the Staffordshire militia; towards its erection the corporation contributed £1050: it is 120 feet in length, ornamented in the front with finely-sculptured figures of Justice and Peace, and contains several handsome apartments, with an assembly-room in the centre, elegantly fitted up, and occupying nearly the whole length of the front. The county gaol and house of correction is a spacious and substantial edifice, comprising nineteen wards, nineteen day-rooms, sixteen work-rooms and shops, twenty-one airing-yards, eight tread mills for grinding corn and supplying the prison with water, and a fulling-mill.

The living of St. Mary's is a rectory not in charge; net income, £221: it is in the patronage of the Crown. The church, formerly collegiate for a dean and thirteen prebendaries, is an ancient and spacious cruciform structure, in the early style of English architecture, with a lofty octagonal tower rising from the intersection, the upper part of which is of later date: the north entrance is richly ornamented with delicate shafts and bold hollows embellished with flowers and foliage: the interior is beautifully arranged; the piers and arches are of the early English, passing into the decorated, style, and to the east of the transepts diminish gradually in height; the windows are generally in the decorated style, though intermixed with others of the later English, of which the east window is an elegant specimen; the chancel is spacious, and the roof is supported on finely pointed arches, and piers of clustered columns; in the north transept is an ancient font of great beauty, and highly ornamented with sculptured figures and animals; there are many ancient and modern monuments, among which the most conspicuous are those of the family of Aston of Tixall. The living of St. Chad's is a perpetual curacy; net income, £85; patron and appropriator, Prebendary of Prees in the Cathedral Church of Lichfield. The church is a small edifice, originally in the Norman style of architecture, with a tower of the later English style between the chancel and the nave; the former is still in good preservation, and, with the exception of a modern east window, retains its original character; but the nave is of more recent date. The foundation stone of a new church was laid by the Earl of Harrowby in November 1837; it contains 600 sittings, half of which are free, the Incorporated Society having granted £300 in aid of the expense. In the parish of St. Mary is an endowed chapel, the living of which is a perpetual curacy; net income, £120; patron, the Rector. There are places of worship for Presbyterians, the Society of Friends, Independents, and Wesleyan Methodists of the Old and New Connection, also a Roman Catholic chapel; which last, in that part of the environs called Forebridge, is a small but handsome edifice, erected by the late Edward Jerningham, Esq., and contains several of the ancient oak stalls removed from Lichfield cathedral.

The free grammar school, which, according to Leland, was originally established by "Sir Thomas, Countre Parson of Ingestre by Heywodde, and Syr Randol, a chauntre preste of Stafford," and further endowed with subsequent benefactions, was, on petition of the inhabitants, refounded by Edward VI., who augmented the endowment, in 1550: the income is about £350 per annum, of which two-thirds are paid to the head-master, and the remainder to the usher: the school is open to all boys of the town, of whom there are sixteen on the foundation. School-rooms were built for a National school in 1825; and there are Sunday schools in connection with the established church and the dissenting congregations. The institution for the relief of the widows and orphans of poor clergymen of the county is supported by donations and annual subscription, and has also an income arising from property vested in old South Sea stock. The county infirmary was established in 1766, and the present building erected in 1772: the premises, situated in the Foregate, were considerably enlarged a few years since, and are well adapted to the reception of 78 patients: attached to it is a fever ward. This institution has lately received the name of hospital, and confers on pupils who have attended it the same advantages and privileges as the hospitals of London. The county general lunatic asylum was established in the year 1818, for patients from all parts of the kingdom, upon moderate terms, regulated according to their circumstances: the buildings, erected at an expense of £30,524. 16. 1., are capable of accommodating 212 inmates; the gardens and pleasure grounds comprise 30 acres. Almshouses for twelve aged and infirm persons were erected in 1640, by Sir Martin Noel, at an expense of £1000; twenty poor families at present reside in them. The poor law union of Stafford comprises 19 parishes or places, under the care of 27 guardians, and, according to the census of 1831, contains a population of 17,382. A priory of Black canons was founded by Richard Peeche, Bishop of Lichfield and Coventry, in 1181, and dedicated to St. Thomas à Becket, the revenue of which at the dissolution was £198. 0. 9.: a small portion of the buildings, now converted into a farmhouse, remains, about two miles east of the town. A house of Friars Eremites, of the order of St. Augustine, was founded in the suburb of Forebridge, by Ralph Lord Stafford, to which, on the suppression of the priory of Stone, the monuments of the Stafford family were removed; it continued till the dissolution, at which time these splendid monuments were destroyed. A priory of Franciscan friars was founded at the north end of the town walls by Sir James Stafford of Sandon, in the reign of Edward I., the revenue of which at the dissolution was £35. 13. 10. In addition to these were a free chapel, in the castle, dedicated to St. Nicholas; a free chapel, or hospital, of St. John, near the river, in Forebridge, for a master and poor brethren, the revenue of which at the dissolution was £10; and a free chapel, or hospital, dedicated to St. Leonard, of which the revenue was £4. 12. 4. Several silver coins, of a later date than the reign of Edward VI., a silver cross, the lower portion of an ancient font or piscina, a cannon ball, and two small millstones, were found, on repairing the walls of the castle, some few years since. Among eminent natives were, John de Stafford, a Franciscan monk; Edmund Stafford, Bishop of Exeter, and Chancellor of England, in the reigns of Richard II. and Henry IV.; Thomas Ashebourn, a strenuous opponent of Wickliffe; Thomas Fitz-Herbert, a learned Roman Catholic divine of the sixteenth and seventeenth centuries, and Principal of the English College at Rome; and Izaak Walton, the well-known author of the treatise on the art of angling. Stafford gives the title of Baron to the family of Jerningham.

STAFFORD, WEST, a parish, in the union of DORCHESTER, hundred of CULLIFORD-TREE, Dorchester Division of the county of DORSET, 2½ miles (S. E. by E.) from Dorchester; containing 184 inhabitants. The living is a rectory, with that of Frome-Billet united, valued in the king's books at £10. 8. 1½.; it is in the patronage of Mrs. Elizabeth Floyer. The tithes have been commuted for a rent-charge of £265, subject to the payment of rates, which on the average have amounted to £9; the glebe comprises 39 acres, valued at £75 per annum. The church, by the date 1640 over the porch, seems to have been rebuilt in that year. Here is a National school. At Bingham Court are the remains of a free chapel, which has been long desecrated. Frome-Billet was formerly a parish, but the church having been destroyed, and the parish becoming almost depopulated, the living was united, about the middle of the fifteenth century, to that of West Stafford: it contains an ancient mansion, which formerly belonged to the family of Gould, now the property of John Floyer, Esq.

STAFFORDSHIRE, an inland county, bounded on the north and north-west by Cheshire, on the west by Shropshire, on the south by Worcestershire and a detached portion of Shropshire, on the south-east by Warwickshire, and on the east and north-east by a small projecting portion of the county of Leicester, and by Derbyshire. It extends from 52° 23′ to 53° 14′ (N. Lat.), and from 1° 33′ to 2° 22′ (W. Lon.); and includes an area of 1148 square miles, or 734,720 statute acres. The population amounts to 410,512. Its ancient British inhabitants were the Cornavii, whose territory, on its subjection by the Romans, was included in the division called *Flavia Cæsariensis.* On the completion of the Anglo-Saxon Heptarchy, it was included in the powerful kingdom of Mercia, several of the principal towns of which were situated within its limits.

This county is in the diocese of Lichfield, and in the province of Canterbury: it forms an archdeaconry, containing the deaneries of Tamworth, Tutbury, Lapley, Treizull, Alveton, Leek, Newcastle-under-Lyme, and Stone, and comprises 146 parishes. For purposes of civil government it is divided into the hundreds of Cuttlestone (East and West), Offlow (North and South), Pirehill (North and South), Seisdon (North and South), and Totmonslow (North and South.) It contains the city of Lichfield; the borough and market-towns of Newcastle-under-Lyme, Stafford, Tamworth, Stoke-upon-Trent, Walsall, and Wolverhampton, the three last recently enfranchised; and the market-towns of Burslem, Burton-upon-Trent, Cheadle, Eccleshall, Hanley, Lane-end, Leek, Longnor, Penkridge, Rugeley, Stone, Uttoxeter, and Wednesbury. Under the act of the 2nd of William IV., cap. 45, the county has been divided into two portions, called the Northern and the Southern divisions, each sending two members to parliament: the place of election for the former is Stafford, and the polling-places are Stafford, Leek, Newcastle-under-Lyme, Cheadle, Abbot's-Bromley, Burton-upon-Trent, Uttoxeter, Eccleshall, and Stone; that for the latter is Lichfield, and the polling-places, Walsall, Lichfield, Wolverhampton, Penkridge, King's-Swinford, West Bromwich, Wednesbury, Rugeley, Tamworth, Brewood, Handsworth, Bilston, Sedgley, and Tipton. Two representatives are returned for the city of Lichfield, and two for each of the boroughs, except Walsall, which sends only one. This county is included in the Oxford circuit: the assizes and quarter sessions are held at Stafford.

Its surface is various. The northern part rises into hills, called the Moorlands, constituting the southern extremity of the long mountainous range which extends hence through the north of Derbyshire, and along the western confines of Yorkshire, towards the mountainous borders of Scotland. These moorlands are situated to the north of a supposed line drawn from Uttoxeter to Newcastle-under-Lyme, and comprise large tracts of waste and uncultivated land, appropriated almost entirely to the pasturage of sheep. A large portion of them has been enclosed with stone walls, almost the only fence to be seen in this part of the county; but these enclosures have not been subdivided, and large breadths have never undergone the least improvement. The pleasant vale in which is situated the town of Cheadle, in this part of the county, is bounded, in the vicinity of that town, by high barren hills, composed of huge heaps of gravel: the wastes upon these hills, and others equally dreary and barren, extending both northward and westward of Cheadle, are extensive; almost their only produce being heath, broom, whortleberries, and mountain cinquefoil. Eastward of this town also, approaching the borders of Derbyshire, are similar desolate wastes, one of which, near the banks of the Dove, is called Oak-moor, from its being nearly covered with dwarf oaks. A little to the north of this, commences an extensive tract of limestone country, included between the rivers Dove and Churnet, extending westward as far as Ipstones, and northward as far as Longnor, and comprising an area of 50 or 60 square miles: this is the most valuable part of the moorlands, the soil naturally producing a fine herbage: many of the hills, which are composed of immense masses of limestone, rise to a very considerable height, and in various places present huge perpendicular cliffs: the Weaver hills, in the southern part of it, of very considerable extent, rise, in common with some other of the highest peaks of the moorlands, to the height of 1000 feet and upwards above the level of the tide in the Thames at Brentford, and command remarkably extensive views, in which are included the Peak hills of Derbyshire; these are almost covered with irregular excrescences, clothed with moss or lichens. Large tracts of the other parts of the moorlands, notwithstanding their great superiority of elevation, are entirely wet peat moors, or mosses; such are Morrage, Axedge, the Cloud heath, High Forest, Leek Frith, and Mole Cop.

The middle and southern parts of the county are level, or diversified only by gently-rising eminences. The following tracts, however, are exceptions to this observation, *viz.*, the limestone hills of Dudley and Sedgley; the parish of Rowley-Regis, principally composed of an isolated mountain, terminating in various peaks, the loftiest of them, called Turner's Hill, being the highest spot of ground in the south of Staffordshire, rising to the height of 900 feet above the level of high water in the Thames at Brentford; the hills of Clent, in the detached portion of the county lying to the south of Stourbridge, in Worcestershire, and nearly equal in height to those of Rowley; Barbeacon, rising to the height of 653 feet, and many others of less elevation. The soil of the Rowley hills is a strong marly loam, through which the rocky substratum frequently rises in innumerable fragments; that of the Clent hills is of the kind commonly called stone-brash, the lower parts being of stronger staple than the summits: much of these latter hills consists of sheepwalks of good herbage. The lowest points of land are supposed to be the banks of the Severn at Upper Arely, which are only 60 feet above the level of high water at Brentford; and the banks of the Trent, at its junction with the Dove below Burton, about 100 feet above the same level: those of the Tame, at Tamworth, are 50 feet higher than the last-mentioned.

The arable soils have been distinguished by Mr. Pitt, in his general view of the agriculture of the county, into four species; *viz.*, first, strong or clayey soils, which are of two kinds, the harsh, stiff, and untractable, and the more mild and friable, both of them commonly resting on a substratum of marl, with a hard stone rock underneath. Lands of this nature occupy a very extensive tract, stretching across the middle of the county, from the border of Derbyshire to that of Shropshire, but excluding the extensive waste of Cannock heath, and the country to the east of Stafford, between that town and the Trent. That part of the county lying eastward of the river Tame has the same kind of soil, as also have several parishes in the south-western extremity approaching the banks of the Severn. Secondly, loose, light, sandy land, adapted to the culture of turnips, and occupying a tract bounded on the north by the Trent, on the east by the Tame, on the south by the confines of the county, and on the west by an imaginary line drawn from the village of Armitage, near the Trent, southward by Longdon, Hammerwich, Aldridge, and a short distance eastward of Walsall, to the verge of the county, near Birmingham: of the same kind also are a considerable tract lying to the south-west of Dudley and Wolverhampton, extending from the border of Worcestershire to that of Shropshire; and another, of much smaller size, extending westward from Brewood, and including the village of Sheriff-Hales. Thirdly, the calcareous soils, or those resting on a substratum of limestone; such are the soils of the great northern limestone district before described, as also of one of very small extent to the north and north-west of Dudley, and of another to the north-east of Walsall. Fourthly, and lastly, mixed soils and loams, formed of the above in different proportions, frequently with the addition of gravel and various adventitious matter, and occupying the remaining extensive portions of the county: the substrata are various, including sand, gravel, clay, marl, and stone of different kinds. In some of the uncultivated tracts, and in one or two other places, are found small pieces of ground of a thin light black earth, of a peaty nature, generally lying upon gravel. The meadow soils are in many places similar to those of the adjoining arable lands, with the addition, when within reach of natural inundation, of the accumulating sediment so deposited; and in others composed of peat-earth, varying in thickness, sometimes to the depth of several feet, and containing trunks of trees.

The quantity of land devoted to agricultural purposes is estimated at 600,000 acres, of which 500,000 are arable, the rest meadow, or pasture. Of the arable lands, 200,000 acres are of the clayey, or of the more friable of the mixed loams; an equal quantity is of gravelly or sandy loam, or of the calcareous soils, while the remaining 100,000 acres are, for the most part, of light sandy or gravelly loams, suitable for turnips. The courses of crops are various: it may be observed, however, that the Norfolk system, including the rotation of turnips, barley, clover, and wheat, is in common practice on the light soils. The crops of grain and pulse commonly cultivated are wheat, barley, oats, beans, and peas. On the moorlands oats are almost the only grain ever cultivated, and are commonly sown for three succeeding years, after which the ground is laid down for grass: a considerable quantity of oaten bread is eaten in the moorlands. Buck-wheat, here called French wheat, is sometimes cultivated, both as a crop and for ploughing under as manure. Hemp and flax are also grown, though upon a small scale; and many leases are subject to restrictions, to prevent the cultivation of these plants to any great extent. The common artificial grasses are red clover, white clover, trefoil, and rye-grass: burnet and rib-grass are also sown in considerable quantities. The low lands adjoining the rivers and brooks consist of meadows and pastures, as also do considerable tracts of flat land, which, by the backing on of water in former times, have acquired a stratum of peaty earth upon their surface. The meadows on the banks of the Dove, in the higher part of its course, before it is joined by the Churnet, are rendered proverbially fertile, by the calcareous particles deposited by the overflowing of that river, the plain within reach of which is in some places nearly a mile broad. There is a considerable extent of grass land in the vicinity of all the principal towns: the vale of the Trent is regarded as the richest tract in the county for its extent. Lime is extensively applied as manure upon all kinds of land: an extraordinary quantity of marl also is used for the same purpose, being found under the loamy soils, and in gravelly land.

The cattle are for the most part of the long-horned breed: the dairies, to which they are almost all appropriated, vary in size, containing from ten to forty, and, in some instances, to seventy cows each. Staffordshire contains the following distinct native breeds of sheep. First, the grey-faced, without horns, which is the native breed of Cannock heath, and the neighbouring commons. Secondly, the black-faced horned sheep, with fine wool, which are peculiar to the commons on the western side of the county, towards Drayton, in Shropshire. Thirdly, the white-faced polled breed, with long combing wool, peculiar to the eastern parts of the moorlands. Fourthly, the mixed waste-land breed, which is found upon the wastes and uncultivated enclosures of the western parts of the moorlands, and is much inferior to the last mentioned. The sheep on the commons of the southernmost part of the county are also of mixed breeds; their wool is tolerably fine, and of a clothing quality. Lastly, the pasture flocks, of various sorts and crosses, but chiefly of the Old and New Leicester breeds: those of the Cotswold, Wiltshire, and Dorsetshire, are also occasionally met with. The most esteemed breed of hogs is a cross between the large slouch-eared kind and a dwarf breed, which is finer boned, broad, and plump: part of the consumption of bacon, which is very great in this county, is supplied from Shropshire and North Wales. In the parish of Tettenhall, near Wolverhampton, great quantities of a peculiar kind of pear, called from the name of the place where it is produced, are grown: a great quantity of fruit is brought to the markets of this county from Worcestershire. The woods, wastes, and impracticable lands, are supposed to occupy an extent of upwards of 100,000 acres. The county is well stocked with almost every species of thriving English timber, growing on the numerous estates of the nobility and gentry. Plantations to a great extent have been made on various parts of the steep moorland hills, particularly those of Dilhorne, Kingsley, and Oakmoor: from the underwood of these, many rods and staves, to make crates for the use of the potteries, are cut. Needwood Forest, in the eastern part of the county, situated between the rivers Trent and Dove, before the passing of an act of enclosure about the commencement of the present century, was an entirely wild tract of nearly 10,000 acres, presenting much romantic and beautiful scenery, and affording pasturage to numerous herds of deer: it was also subject to a common right for cattle and horses. Of the wastes at present remaining, Cannock heath is by far the most extensive: it contains upwards of 25,000 acres, and is situated near the centre of the county, lying chiefly to the north and east of the small town of Cannock, whence it extends to the southern banks of the Trent. The northern and western parts of this common have for the most part a light soil, but the eastern and southern are of a cold, wet, gravelly nature; the best sheep land is a tract on the western side, towards Tiddesley, and another in the northern part of it, near Rugeley and Beaudesert Park, to which may also be added the vicinity of Hedgford: the southern and eastern parts are in a great measure barren, producing little besides heath, whortleberries, lichens, and mosses. Although now a bleak and dreary tract, entirely devoid of trees, this waste in former times is asserted to have been covered with a profusion of majestic oaks, and to have been a favourite chace of the Saxon monarchs of Mercia. In addition to Walsall wood, Whittington heath, and the Weeford Flats, the other principal waste lands are those of Swindon and Wombourne, and that near Stewponey, on the south; those of Morrage, Wetley moor, Stanton moor, Hollington heath, and Caverswall common, in the north; and, in the middle districts, those of Essington wood, Sneyd common, Wyrley and Pelsal commons, Tirley, Ashley, Maer, Swinnerton, Tittensor, and Shelton heaths, Houlton, Milwich, Hardwick, and Fradswell commons, and others of smaller extent, chiefly used as sheepwalks. Most of them bear evident marks of ancient cultivation.

The mineral productions are numerous and valuable, consisting chiefly of coal, iron, lead, copper, marble, alabaster, and stone of various kinds. The coal strata have been found to exist to a superficial extent of 50,000 acres. The coal district of the southern part of the county extends in length from Cannock heath, a part of which it includes, to near Stourbridge, in Worcestershire, and in breadth from Wolverhampton to Walsall. In the north of the county this mineral is raised in abundance, in the neighbourhood of Newcastle and the Potteries, near Lane-End and Hollybush, and again in the vicinity of Cheadle and Dilhorne. In the colliery of Birch hill, near Walsall, occurs a bed of trap, or greenstone, as it is here called, lying upon indurated sandstone, and covered by a bed of slaty clay: the effect it has upon the coal is that of depriving it of its bitumen, and thus reducing it to what is called blind coal: the whole of the strata of this mine are interrupted by faults, as the miners call them, of this greenstone. A singular species of coal, called peacock-coal, from the prismatic colours it exhibits, is raised at Hanley. In the district called the Potteries, the strata intervening between the beds of coal consist chiefly of clays of different kinds, some of which make excellent fire-bricks for building the kilns, and making the saggars, or cases, in which the ware is burned. Iron-ore abounds in all the coal-mines: the strata of it that are found in the neighbourhood of Wednesbury, Tipton, Bilston, in part of the parish of Sedgley, and to the west of Newcastle, are very extensive, lying generally under a vein of coal: the iron-ore found on Cannock heath is of a peculiar kind, called Cannock stone, of very little value. In the numerous mines of coal and iron, and in the foundries, blast-furnaces, slitting-mills, and other iron manufactories, an immense number of workmen are employed: the works on the banks of the Birmingham canal are particularly numerous and extensive. The other metallic ores obtained are those of copper and lead, of both which considerable quantities are raised at Ecton, near Warslow, approaching the north-eastern border of the county: a copper mine is also worked at Mixon, within a few miles of Leek; and a lead mine near Stanton moor, in the same part of the county. Limestone forms the substratum of a very great extent of the country already described: an immense quantity of it is raised for burning into lime; the limeworks on Caldon Low, and in the neighbourhood of the Weaver hills, are particularly extensive. Under several of the limestone hills, in the southern part of the county, that are perforated by tunnels, large caverns have been hollowed out, without disturbing the surface soil, some of them penetrating to a distance of 300 yards from the canal. The hills in the parishes of

Dudley and Sedgley are chiefly of limestone; that in Dudley is peculiarly compact and hard, and abounds with numerous shells and marine substances; immense excavations have been made to the depth of 200 feet under the Castle and the Wren's Nest hills, and vast quantities of limestone have been burnt into lime. The upper stratum of the limestone in the parish of Sedgley forms what is called the Black lime, which is much more porous and friable, and is found on the very surface; and in the parish of Rowley-Regis these hills are composed of an amazingly hard, rugged and angular stone, called the Rowley Rag, which in some places rises nearly to the surface. The limestone of this county, in different places, has some of the qualities of marble, and is susceptible of a high polish; in others it is composed, in a great measure, of petrified marine substances. The kind of marble called rance-marble, which is white, with red veins formed of shining gritty particles, and takes so good a polish as to be frequently used for chimney-pieces and monuments, is found in great abundance in Yelpersley Tor and the adjoining hills: there is a considerable quantity of grey marble at Stansop, and at Powke hill is obtained a very hard black marble. In the great limestone district, particularly on the banks of the river Dove, are some veins of alabaster, which is also dug between Needwood Forest and Tutbury: many of the moulds used in the potteries are composed of this material, after it has been ground. Extensive quarries of excellent freestone are numerous: at Bilston is obtained a particularly fine kind: Gornall, near Sedgley, has different quarries of a coarser sort; and among the numerous other places where freestone is obtained may be specified, more particularly Tixall, Wrottesley, Brewood park, and Pendeford. Clays of almost every description are found in this county: potters' clay, of several sorts, abounds chiefly in the vicinity of Newcastle-under-Lyme, where the pottery wares were formerly manufactured from it. At Amblecot, in the southern part of the county, is a clay of a dark blueish colour, of which glass-house pots of a remarkably superior quality are made, great quantities of them being sent to different parts of the kingdom, and many consumed in the neighbouring glass-works. Yellow and red ochre are also found in Staffordshire: and a blue clay, obtained at Darlaston near Wednesbury, is used by glovers. A kind of black chalk exists in beds of grey marble, in Langley-close; and a fine reddish earth, little inferior to the red chalk of France, is obtained near Himley Hall.

The manufactures are various and extensive: that of hardware, in the southern district, is very important, and affords employment to many thousand persons. At Wolverhampton, and in its vicinity, are made locks of every kind, edge-tools, files, augers, japanned goods, and a great variety of other articles of the same material. The town and neighbourhood of Walsall are famous for the manufacture of sadlers' ironmongery, such as bridle-bits, stirrup-irons, spurs, &c., sent thence to every part of the kingdom. The manufacture of nails employs many thousand persons in some of the most populous parishes in this part of the county, particularly in those of Sedgley, Rowley, West Bromwich, Smethwick, Tipton, Wombourne, and Pelsall, and in the Foreign of Walsall: many women and children are employed in making the lighter and finer sorts. The other kinds of hardware produced are chiefly plated, lackered, japanned, and some enamelled goods, toys, tobacco and snuff boxes, of iron and steel; and machinery for steam-engines. Some places also partake of the manufacture of guns; and there are several works for making brass, and for preparing tin plates, chiefly in the northern part of the county. In those parts of Staffordshire situated in the vicinity of Stourbridge and Dudley, in Worcestershire, are a number of large glass-houses, where the manufacture is carried on to a great extent. The manufacture of china and earthenware, in the north-western part of the county, is the most extensive and important in the kingdom: the district called the Potteries consists of numerous scattered villages occupying an extent of about ten square miles, and containing about 20,000 inhabitants; it is crossed by the Trent and Mersey, or Grand Trunk canal. This manufacture, though of very ancient establishment in this part of the country, was of inferior importance until the latter part of the eighteenth century, when, chiefly by the exertions of the late Josiah Wedgwood, Esq., it was raised to such a pitch of excellence, as confers honour upon that gentleman's ingenuity and taste; and, in consequence, several of the villages of this district, particularly Burslem and Hanley, have grown rapidly into populous market-towns. The several species of ware invented by Mr. Wedgwood, varied by the industry and ingenuity of the manufacturers into an infinity of forms, and differently painted and embellished, constitute nearly the whole of the fine earthenwares at present manufactured in England, which are the object of a very extensive traffic. They are chiefly the following; *viz.*, the Queen's ware, which is composed of the whitest clays from Derbyshire and Dorsetshire, and different other parts of England, mixed with a due proportion of ground flint; terra cotta, resembling porphyry, granite, and other stones of the silicious or chrystalline order; basaltes, or black ware; porcelain biscuit, of nearly the same properties as the natural stone of that name; a white porcelain biscuit, of a smooth wax-like surface, of the same properties as the preceding; jasper, a white porcelain biscuit of exquisite beauty and delicacy, possessing the general properties of the basaltes, together with the singular one of receiving, through its whole substance, from the mixture of metallic calces with other materials of the same colours which those calces communicate to glass, or enamels, in fusion, a property that no other porcelain or earthenware, of either ancient or modern composition, has been found to possess; bamboo, or cane-coloured porcelain biscuit, possessing the same qualities as the white porcelain biscuit; and a porcelain biscuit, remarkable for great hardness, little inferior to that of agate: the glazes are of vitreous composition. A very great number of persons are constantly employed in raising and preparing the raw materials for this manufacture in different parts of the kingdom, more especially upon the southern coasts, from Norfolk round to North Devon, and on the shores of Wales and Ireland. Vessels, which in the proper seasons have been employed in the Newfoundland fishery, convey these materials coastwise to the most convenient ports, whence they are forwarded by the inland small craft to the Potteries. Notwithstanding that almost every part of the kingdom receives supplies of pottery from this manufacture, yet by far the greater portion of its produce is exported to foreign countries.

The exports of earthenware and china to the United States alone amount to 60,000 packages annually. The quantity of wool manufactured is small, nearly the whole of the produce of the county being sold to the clothing and hosiery districts. The cotton manufacture is considerable; the works at Rochester and other places near the Dove are on a large scale, as are also those at Fazeley and Tutbury. The town of Leek and its neighbourhood has a considerable manufacture of silk and mohair, the articles being chiefly sewing-silk, twist, buttons, ribands, ferrets, shawls, and handkerchiefs. Tape is manufactured at Cheadle and Tean, affording employment to many of their inhabitants. Stafford has manufactures of shoes and boots, for exportation and home consumption; and tanning and hat-making are carried on to a great extent in several of the towns. This county is also celebrated for its ale, particularly that made at Burton-upon-Trent.

The principal rivers are the Trent, the Dove, the Tame, the Blythe, the Penk, and the Sow: the Severn also, though not considered a Staffordshire river, takes its navigable course by the parish of Upper Arely, at the south-western extremity of the county. The Trent, which ranks as the third largest river in England, becomes navigable at Burton, a little below which, being joined by the Dove, it enters Derbyshire, after a course, through this county and bordering upon it, of upwards of 50 miles. The Dove, which, throughout its course, forms the boundary between this county and that of Derby, not far from its source enters the beautiful and sequestered Dove Dale, flowing through it in a southerly direction, to the vicinity of Ashbourn in Derbyshire, whence it proceeds south-westward towards Uttoxeter, near which town it assumes a south-easterly direction, by Tutbury, to its junction with the Trent to the north-east of Burton: from the great inclination of the bed of this river its water flows with great rapidity, in some places dashing over rugged masses of rock, in others forming gentle cascades. Near the village of Ilam, in this county, the Dove is greatly augmented by the waters of the rivers Manifold and Hamps: the former, rising near the source of the Dove, takes a very circuitous route through a romantic vale situated in the north-eastern part of the county, and sinking into the earth to the south of Ecton hill, between the villages of Butterton and Wetton, emerges again at Ilam, shortly before its junction with the Dove, and at the distance of about four miles from the spot where it sinks into the ground: this stream is joined during its subterraneous transit by the Hamps, which in like manner passes under ground for a considerable distance. The extent of artificial navigation for the ready transport of the produce of the mines, manufactures, &c., is remarkably great. The Grand Trunk canal, which was planned, and in a great measure executed, by the celebrated engineer Brindley, enters this county from Cheshire, near Lawton, and almost immediately passes through the Harecastle tunnel, which is 2880 yards long: the highest level of this canal is at Harecastle, from which, on the south-eastern side, there is a fall of 316 feet. The Staffordshire and Worcestershire canal branches from this at Haywood, near the confluence of the rivers Sow and Trent, and quits the county, in its course to the Severn, a short distance to the south of Kinver: this canal, with the Grand Trunk, completes the communication between the ports of Bristol, Liverpool, and Hull. The Coventry and Oxford canal branches from the Grand Trunk at Fradley heath, and near Fazeley enters Warwickshire: from Fazeley a branch proceeds to Birmingham, and is called the Birmingham and Fazeley canal. The Wyrley and Essington canal, commencing at a place called Wyrley Bank, forms a junction with the Birmingham canal a little beyond the village of Wednesfield, near Wolverhampton: its branches are, one from the vicinity of Wolverhampton to Stow heath, another from Pool-Hayes to Ashmore Park, and a third from Lapley-Hayes to Ashmore Park. At Huddlesford commences a branch from the Coventry canal, called the Wyrley and Essington Extension, which forms a junction with the Wyrley and Essington canal near Bloxwich; on the western side of Cannock heath a branch is carried southward, by Walsall wood, to the limeworks at Hay-head: its whole extent, including the branches, is $34\frac{1}{2}$ miles; and from Cannock heath to the Coventry canal it has a fall of 264 feet. The Birmingham canal, from that town in Warwickshire, joins the Staffordshire and Worcestershire canal a little to the north of Wolverhampton, after a course of 22 miles. Of the very numerous branches of this canal, one proceeds northward, over Ryder's Green, to the collieries of Wednesbury, and the vicinity of Walsall; while another, commencing about a mile from Dudley, passes south-westward by Brierly Hill, and to the left of Brockmore Green joins a canal which, commencing in a large reservoir at Pensett's Chace, and passing nearly in a straight line by Wordsley, crosses the river Stour, and joins the Staffordshire and Worcestershire canal, a few miles to the west of Stourbridge in Worcestershire, to which town there is a small branch. The cut which connects the Dudley canal with that of Birmingham, called the Dudley Extension canal, has part of its course in this county. Sir Nigel Gresley's canal extends from the Grand Trunk, near Newcastle-under-Lyme, past that town, to the coal mines in Apedale. The Birmingham and Liverpool railway enters the county a little to the north-west of Birmingham, and passing by Wednesbury, Penkridge, and Stafford, quits it to the north-west of Madeley. The Birmingham and Derby Junction railway also crosses an angle of the county, on its eastern side.

Several large single stones at Cannock are supposed to be Druidical, as also are the eight upright stones, called "the Bridestones," near Biddulph, on the north-western boundary of the county; and on Drood, or Druid, heath, where there are several singular earthworks, Mr. Shaw, the historian of this county, considers the chief seat of the Arch-Druid of Britain to have been situated. Thyrsis, or Thor's house, a cavern, situated in the side of a lofty precipice in the vale of the Manifold, near the village of Wetton, is also supposed to have been the scene of Druidical rites. Some very ancient artificial caves have been discovered at Biddulph. The encampment of Billington, about three miles to the west of Stafford, and that on Castle hill, near Beaudesert park, in the vicinity of Rugeley, are of ancient British formation. Under the Roman dominion, the tract now constituting Staffordshire contained the stations of *Etocetum*, at Wall, near Lichfield; and *Pennocrucium*, now Penkridge. Sheriff-Hales, near the confines of Shropshire, is supposed by some antiquaries to have been the site of the station *Uxacona*, or *Usacona*. Two of the great prætorian ways also crossed Staffordshire: the Watling-street, entering it from Warwickshire, near Tamworth, proceeded westward across the southern part, and quitted it for Shropshire, to the west of the town of Brewood: the Ikeneld-street, which entered from Warwickshire, at the village of Handsworth, near Birmingham, proceeded thence, in a north-north-easterly direction, to a little beyond Shenstone, where it crossed the Watling-street, and afterwards pursued a north-easterly course, entering Derbyshire at Monks' Bridge, on the Dove. Roman domestic remains, and traces of their roads, are discoverable in different places; and Roman earthworks are visible at Arely wood, Ashton heath, Ashwood heath, near Kinver, at Oldbury, near Shareshill, and in Tiddesley park. Near Maer are intrenchments supposed to have been thrown up by Cenred, in the progress of his hostilities against Osred, King of Northumbria; and on Sutton-Coldfield there is a camp, considered to be of Danish formation. The number of religious houses in this county, including free chapels, hospitals, and colleges, was about 40. Remains of the abbeys of Burton and Croxden, and of the priories of Rowton, Stafford, and Stone, are still visible. The remains of ancient castles are chiefly those of Alveton, Caverswall, Chartley, Dudley, Healy or Heyley castle, Tamworth, and Tutbury. Among the most remarkable ancient mansions are Bentley Hall and Moseley Hall, in both which Charles II. remained concealed for some time after the battle of Worcester. Staffordshire contains numerous modern seats of the nobility and gentry, many of which are elegant, and several magnificent: among the most distinguished are Trentham, the property and residence of the Duke of Sutherland; and Beaudesert, that of the Marquess of Anglesey. Most modern houses of the ordinary class are built of brick, and roofed with tile or slate, the latter brought along the canals, chiefly from Wales and Westmorland: near Newcastle, and in one or two other places, large quantities of a peculiar kind of blue tiles are made, which, owing to their superior durability, are in great demand. Salt springs exist in different places; the principal are in the parish of Weston; and there are other mineral springs of various qualities, the most remarkable being that near Codsall, formerly famous for the cure of leprosies; St. Erasmus' well, between Ingestre and Stafford; and that at Willoughby. Numerous fossil remains occur in different parts of the strata of this county, more particularly in some of the limestone beds. At Bradley, to the east of Wolverhampton, a stratum of coal, about four feet thick, and eight or ten yards below the surface, having been set on fire, burned for about 50 years, and has reduced a considerable extent of land to a complete calx, used for the mending of roads: sulphur and alum are found in its vicinity. The original calendar of the Norwegians and Danes was, till lately, used by many inhabitants of this county, where it has the appellation of the Staffordshire Clog: this is a quadrangular piece of wood, on each of the four sides of which are contained three months of the year, the days being expressed by notches, to which are added the symbols of the several saints, to denote the day of their festival, &c. This county gives the inferior title of Marquess of Stafford to the Duke of Sutherland.

STAGBATCH, a hamlet, in the parish of LEOMINSTER, hundred of WOLPHY, county of HEREFORD, 2 miles (W. S. W.) from Leominster: the population is returned with the chapelry of Ivington.

STAGSDEN (*St. Leonard*), a parish, in the hundred of WILLEY, union and county of BEDFORD, 5½ miles (W. by S.) from Bedford; containing 597 inhabitants. The living is a discharged vicarage, valued in the king's books at £8; present net income, £185; patron and impropriator, Hon. G. R. Trevor. A school is endowed with £6 per annum, and is further aided by the Hon. G. R. Trevor.

STAINBROUGH, a chapelry, in the parish of SILKSTONE, wapentake of STAINCROSS, West Riding of the county of YORK, 3¼ miles (S. W.) from Barnsley; containing 304 inhabitants. Here is a school, founded by the family of Cutler, with a house for the master, and an endowment of £8 per annum, which, with an annuity of £2 from T. V. Wentworth, Esq., is applied for teaching twelve children.

STAINBURN, a township, in the parish of WORKINGTON, union of COCKERMOUTH, ALLERDALE ward above Derwent, Western Division of the county of CUMBERLAND, 1 mile (E.) from Workington; containing 174 inhabitants. Here was formerly an oratory, subordinate to the priory of St. Bees.

STAINBURN, a chapelry, in the parish of KIRKBY-OVERBLOWS, Upper Division of the wapentake of CLARO, West Riding of the county of YORK, 4¼ miles (N. E. by E.) from Otley; containing 290 inhabitants. The living is a perpetual curacy; net income, £66; patron and impropriator, Rev. Dr. Marsham.

STAINBY (*St. Peter*), a parish, in the union of GRANTHAM, wapentake of BELTISLOE, parts of KESTEVEN, county of LINCOLN, 2 miles (W.) from Colsterworth; containing 186 inhabitants. The village is situated on the south-western border of the county. One portion of the parish lies upon a red stone rock, and the other, which is thinner in soil, abounds generally with coarse limestone; the soil is well adapted for the growth of corn. There is an extensive quarry of freestone, now little used, which is supposed to have supplied the materials for building many of the beautiful churches in the fenny district of the county. The living is a rectory, with which that of Gunby was consolidated in 1773, valued in the king's books at £6. 6. 8.; present net income, £400; patron, Earl of Harborough. The church was neatly rebuilt in 1805, at the expense of the late earl. On the enclosure of the parish, in 1773, land was allotted to the rector in lieu of tithes and all payments, except mortuaries and surplice fees. The proceeds of 13 acres of land left by an unknown benefactor have been for many years applied to the repairs of the church and the liquidation of church rates. Near the river Witham, which forms the eastern boundary of the parish, are the remains of a Roman villa, where a sudatory, some tessellated pavements of ordinary character, and pieces of leaden pipes and tiles, have been found; and walls extending to a considerable distance have been traced. Near the village is an ancient fortification with some out-works, now called Tower Hill, but no record of it is extant, nor are any traces of masonry discernible. In the western portion of the parish are two tumuli, not far distant from each other, and supposed to be the graves of some slaughtered Danes.

STAINDROP (*St. Mary*), a market-town and parish, in the union of Teesdale, South-Western Division of Darlington ward, Southern Division of the county palatine of Durham; containing, exclusively of Cleatlam, 2527 inhabitants, of which number, 1478 are in the town of Staindrop, $5\frac{1}{2}$ miles (N. E. by E.) from Barnard-Castle, and 244 (N. N. W.) from London. This place, formerly called also Stainthorp, or the stony town, is of great antiquity, having been granted by King Canute, who had a mansion at Raby, in this parish, to the monastery at Durham. The town is pleasantly situated in a valley, and consists chiefly of one long well-built street. Here is a subscription library and news-room. In Langley-dale are very extensive works for smelting lead-ore. A weekly market on Saturday, and fairs annually on the Vigil of St. Thomas the Martyr and the two following days, were granted in 1378, by Bishop Hatfield, which, after a time, fell into disuse, but the market has been revived and is well supplied with provisions. The magistrates hold petty sessions every alternate Saturday; and a court leet and court baron for the lordship of Raby are held at Michaelmas by the lord of the manor, at which constables are sworn in at the former, and debts under 40*s*. are recoverable at the latter: the jurisdiction extends over the other townships in the parish. The living is a vicarage, annexed to the rectory of Cockfield; impropriator, Duke of Cleveland. The impropriate tithes of Hilton have been commuted for a rent-charge of £136. 1. 10., subject to the payment of rates, which on the average have amounted to £21. 3. 7. The church was formerly collegiate: it is a handsome structure, exhibiting portions in the early, decorated, and later styles of English architecture, with a square embattled tower rising from the centre, and contains some ancient and handsome monuments to the Neville family. The Society of Friends, Independents, Wesleyan Methodists, and Presbyterians, have each a place of worship. Here is a Sunday National school. A collegiate establishment was founded here in the reign of Henry IV., by Ralph Nevill, Earl of Westmorland, in honour of the Virgin Mary, for a master, six priests, six clerks, six decayed gentlemen, six poor officers, and other poor men: its revenue at the dissolution was £170. 4. 6.

STAINES (*St. Mary*), a market-town and parish, and the head of a union, in the hundred of Spelthorne, county of Middlesex, 10 miles (W. S. W.) from Brentford, and 17 (W. S. W.) from London; containing 2486 inhabitants. This place has by some been conjectured to derive its name from a Roman milliarium, said to have been placed here; and the traces of a Roman road pointing towards Staines' bridge, mentioned by Dr. Stukeley, who also describes the town as having been surrounded by a ditch, may in some degree strengthen this conjecture. But the more general opinion is, that its name is derived from a stone which, standing on the banks of the river near it, marks the extent of the jurisdiction of the Lord Mayor of London, as conservator of the Thames; the inscription on it bears date 1285. To perpetuate this relic of antiquity, the stone was raised upon a pedestal, erected upon the spot where it originally stood, in 1781. An army of Danes, on their way from Oxford (which they had burned) to their ships, crossed the river here, in 1009, on hearing that an army was marching from London to oppose them. The town, which has been much improved of late years, consists principally of one wide street, containing several good houses, terminating at the river Thames, across which was an iron bridge of one arch; but this being considered unsafe, a handsome stone bridge has been erected in lieu of it, which was opened in 1832, and a new street formed in a line with it, which avoids the sharp turn over the former, where many accidents have occurred. The town is lighted with gas, the works for which are situated on the opposite bank of the river, on the road to Egham. A handsome building was, in 1835, erected near the bridge for a Literary and Scientific Institution. The market is on Friday: the market-house is a small building surmounted by a spire. There are fairs on May 11th and September 19th. The living is a vicarage, with Laleham and Ashford annexed, valued in the king's books at £12. 3. 4.; present net income, £425: it is in the patronage of the Crown; impropriator, — Coussmaker, Esq. Attached to the vicarage are 59 acres of glebe in this parish, 16 acres in the parish of Laleham, and 26 in that of Ashford. The church is a neat structure, consisting of a nave, chancel, and north and south aisles, rebuilt in 1828, and a square embattled tower of brick, erected by Inigo Jones in 1631, and in 1829 raised 12 feet and surmounted with a battlement of stone, crowned with pinnacles; the chancel belongs to the lay impropriator: 340 free sittings have been provided, the Incorporated Society having granted £250 in aid of the expense. There are places of worship for Baptists, the Society of Friends, and Independents. Here is a school for boys, supported by subscription; also a Sunday National school. The poor law union of Staines comprises 13 parishes or places, under the care of 21 guardians, and contains a population of 12,644, according to the census of 1831. Duncroft House, in which King John is said to have slept the night after he signed Magna Charta at the neighbouring plain of Runymede, and now the seat of Col. Carmichael, is in this parish. A forest anciently extended from Staines to Hounslow, but part of it, consisting of about 300 acres, has been enclosed.

STAINFIELD, a chapelry, in the parish of Hacconby, wapentake of Aveland, parts of Kesteven, county of Lincoln, $3\frac{1}{2}$ miles (N. N. W.) from Bourne: the population is returned with the parish.

STAINFIELD, a parish, in the Western Division of the wapentake of Wraggoe, parts of Lindsey, union and county of Lincoln, 4 miles (S. W. by S.) from Wragby; containing 136 inhabitants. The living is a perpetual curacy; net income, £71; patron and impropriator, T. T. Drake, Esq. A school is supported by Mrs. Drake. A priory of Benedictine nuns was founded here, in the reign of Henry II., by Henry Percy, which at the dissolution possessed a revenue of £112. 5.

STAINFORTH, a township, in the parish of Giggleswick, union of Settle, Western Division of the wapentake of Staincliffe and Ewcross, West Riding of the county of York, 2 miles (N.) from Settle; containing 263 inhabitants. An application has been made to Her Majesty's Commissioners from Pudsey Dawson, Esq., and others, for the perpetual patronage of a new chapel which they propose to build and endow here.

STAINFORTH, a township, in the parish of Hatfield, union of Thorne, Southern Division of the wapentake of Strafforth and Tickhill, West Riding of the county of York, $3\frac{1}{4}$ miles (W. S. W.) from

Thorne; containing, with South Bramwith, 852 inhabitants. There are places of worship for Wesleyan Methodists and Unitarians. Henry Travers, in 1706, bequeathed land, now producing £15 per annum, for teaching poor children.

STAININGHALL, a parish, in the union of St. Faith, hundred of Taverham, Eastern Division of the county of Norfolk, 2 miles (S. W. by S.) from Coltishall: the population is returned with the parish of Horstead. The living is a discharged rectory, annexed to that of Frettenham, and valued in the king's books at £1. 13. 6½. The church is desecrated.

STAININGTON, a chapelry, in the parish of Eccleshall, Northern Division of the wapentake of Strafforth and Tickhill, West Riding of the county of York, 4 miles (W. by N.) from Sheffield. The chapel, erected under the act for promoting the building of additional churches, &c., was finished in Nov. 1829, at an expense of £2607. 19. 3., defrayed by the parliamentary commissioners: it is in the later style of English architecture, with a cupola.

STAINLAND, a township, in the parish and union of Halifax, wapentake of Morley, West Riding of the county of York, 4½ miles (S. by W.) from Halifax; containing 3037 inhabitants, who are extensively employed in the manufacture of worsted, woollen cloth, cotton, and paper. There is a place of worship for Independents. Various Roman coins have been found here.

STAINMORE, a chapelry, in the parish of Brough, East ward and union, county of Westmorland, 4 miles (E. S. E.) from Brough; containing 707 inhabitants. The living is a perpetual curacy; net income, £119; patron, Earl of Thanet. The chapel was erected as a school-house in 1594, consecrated for divine service in 1680, and repaired in 1699, by Thomas Earl of Thanet, who built the school-house adjoining, in which about 30 children are instructed for an endowment of £30 per annum. At a place called Maiden Castle is a Roman fort, aud there is another at Rere Cross, which, according to tradition, was erected in the first or second century, by Marius, a petty king of the Britons, in memory of a victory which he obtained there over the Picts.

STAINSBY, a township, in the parish of Ault-Hucknall, hundred of Scarsdale, Northern Division of the county of Derby, 5¾ miles (S. E.) from Chesterfield: the population is returned with the parish.

STAINSBY, a hamlet, in the parish of Ashby-Puerorum, hundred of Hill, parts of Lindsey, county of Lincoln: the population is returned with the parish.

STAINSIKER, or STAINSACRE, a joint township with Hawsker, in the parish and union of Whitby, liberty of Whitby-Strand, North Riding of the county of York, 1¾ mile (S. E.) from Whitby: the population is returned with Hawsker.

STAINTON, a township, in the parish of Stanwix, Cumberland ward, and Eastern Division of the county of Cumberland, 1½ mile (N. W.) from Carlisle; containing 67 inhabitants.

STAINTON, a township, in the parish of Dacre, Leath ward, Eastern Division of the county of Cumberland, 2¾ miles (S. W. by W.) from Penrith: the population is returned with the parish. Eight children of this township are instructed for the dividends arising from £100 three per cents., the gift of Mark Scott, in 1758.

STAINTON, a joint township with Streatlam, in the parish of Gainford, union of Teesdale, South-Western Division of Darlington ward, county palatine of Durham, 1¾ mile (N. E. by N.) from Barnard Castle; containing, with Streatlam, 324 inhabitants.

STAINTON, a township, in the parish of Urswick, hundred of Lonsdale, north of the sands, Northern Division of the county palatine of Lancaster, 1½ mile (S. E.) from Dalton: the population is returned with the parish.

STAINTON, a chapelry, in the parish of Heversham, union and ward of Kendal, county of Westmorland, 4 miles (S. by E.) from Kendal; containing 388 inhabitants. The living is a perpetual curacy; net income, £89; patron, Vicar of Heversham. The chapel, called Cross-Crake chapel, was founded in the reign of Richard II., by Anselm de Furness, son of the first Michael le Fleming; it was rebuilt in 1773, and had a burial-ground attached to it in 1823. There is a place of worship for Independents. The Lancaster canal passes through the chapelry; and on a rivulet, tributary to the Belo, are two flax-mills and a woollen-mill: the manufacture of bobbin is also carried on. At Helme are the remains of an encampment.

STAINTON, a township, in the parish of Downholme, union of Richmond, wapentake of Hang-West, North Riding of the county of York, 5½ miles (S. W. by W.) from Richmond; containing 44 inhabitants.

STAINTON (*St. Peter*), a parish, in the union of Stockton, Western Division of the liberty of Langbaurgh, North Riding of the county of York; containing 1000 inhabitants, of which number, 271 are in the township of Stainton, 5 miles (N. W. by N.) from Stokesley. The living is a vicarage, with the perpetual curacy of Thornaby annexed, valued in the king's books at £5. 14. 2.; present net income, £323; patron and appropriator, Archbishop of York. The church is an ancient structure, partially repaired and modernized about 1820.

STAINTON (*St. Winifred*), a parish, in the union of Doncaster, Southern Division of the wapentake of Strafforth and Tickhill, West Riding of the county of York, 7 miles (S. by W.) from Doncaster; containing 251 inhabitants. The living is a discharged vicarage, valued in the king's books at £5. 15.; present net income, £168; patron and impropriator, Earl of Scarborough.

STAINTON-BY-LANGWORTH (*St. John the Baptist*), a parish, in the Western Division of the wapentake of Wraggoe, parts of Lindsey, union and county of Lincoln, 5 miles (W.) from Wragby; containing 238 inhabitants. The living is a discharged vicarage, valued in the king's books at £4. 18. 4.; present net income, £183; patron, Earl of Scarborough, who, with Earl Manvers, is the impropriator.

STAINTON-DALE, a township, in the parish of Scalby, union of Scarborough, Pickering lythe, North Riding of the county of York, 8½ miles (N. W. by N.) from Scarborough; containing 252 inhabitants.

STAINTON, GREAT (*All Saints*), a parish, in the union of Sedgefield, North-Eastern Division of

STOCKTON ward, Southern Division of the county palatine of DURHAM; containing 248 inhabitants, of which number, 154 are in the township of Great Stainton, 7 miles (N. E. by N.) from Darlington. The living is a rectory, valued in the king's books at £12. 13. 4.; present net income, £279: it is in the patronage of the Crown. A school was founded and endowed in 1749, by the Rev. Thomas Nicholson, and has since received considerable donations, chiefly from Lord Crewe's charity. Mr. Hubbock, of Stainton-Grange, gave £60 for the instruction of three or four children from Little Stainton.

STAINTON-LE-VALE (*St. Andrew*), a parish, in the union of CAISTOR, Southern Division of the wapentake of WALSHCROFT, parts of LINDSEY, county of LINCOLN, 6 miles (N. E.) from Market-Rasen; containing 118 inhabitants. The living is a discharged rectory, valued in the king's books at £4. 17. 6.; present net income, £99; patron, J. Angerstein, Esq.

STAINTON, LITTLE, a township, in the parish of BISHOPTON, union of SEDGEFIELD, South-Western Division of STOCKTON ward, Southern Division of the county palatine of DURHAM, 5½ miles (N. E.) from Darlington; containing 54 inhabitants.

STAINTON, MARKET (*St. Michael*), a parish, in the union of HORNCASTLE, Northern Division of the wapentake of GARTREE, parts of LINDSEY, county of LINCOLN, 7 miles (E. by N.) from Wragby; containing 132 inhabitants. The living is a donative curacy; net income, £66; patrons, Representatives of the late J. Loft, Esq. Here was formerly a market on Monday, also an annual fair on Oct. 29th; the latter was removed to Horncastle in 1768, the sum of £200 having been paid in compensation to the lord of this manor.

ST. ALBANS, county of HERTFORD.—See ALBANS, ST.—*And all places having a similar distinguishing prefix will be found under the proper name.*

STALBRIDGE (*St. Mary*), a market-town and parish, in the union of STURMINSTER, hundred of BROWNSHALL, Sturminster Division of the county of DORSET, 7½ miles (E. by N.) from Sherborne, and 111 (W. S. W.) from London; containing 1773 inhabitants. The name of this place, in Domesday-book, is written *Staplebridge*, and at the time of the Conquest it belonged to the abbey of Sherborne. The town and the greater part of the parish are situated upon a rock, which supplies building materials for the neighbourhood; the streets are not regularly paved, but are partially lighted by subscription, and the inhabitants are well supplied with water. From the south end of the main street another diverges, and at the intersection is an ancient stone cross, 30 feet in height, including the pedestal (ornamented on the sides with sculptured emblematical figures), from which rises the frustum of a pyramid, twelve feet high, with fluted angles, and decorated on one of the faces with a figure of our Saviour with a lamb at his feet, and at the bottom with shields of arms, and surmounted with canopied shrines, in one of which is a representation of the Crucifixion; above these are enriched canopies, terminating in a crocketed pinnacle, formerly surmounted by a cross, the whole supported on three octagonal flights of steps, which diminish in the ascent. In the park formerly belonging to the manor-house the Anglesea cricket club is held: a building has been erected for the accommodation of the members, who meet weekly during the season, but the rest is converted to agricultural purposes, and is surrounded by a wall five miles in circumference. Stalbridge was formerly noted for the manufacture of stockings: several of the inhabitants are now employed in winding silk. A branch of the river Stour, and the Dorsetshire and Somersetshire canal, pass through the parish. In the reign of Edward I., a grant of a market and fair was made to the abbot of Sherborne; the present market is on Tuesday, and on every alternate Tuesday is a great market for cattle: fairs are held on May 6th and Sept. 4th. The living is a rectory, valued in the king's books at £27. 4. 7.; present net income, £888; patron, Senior Bachelor of Christ's College, Cambridge. The church is a spacious and ancient structure, with a lofty embattled tower, and contains some ancient monuments. There is a place of worship for Independents; also a National school supported by subscription.

STALHAM (*St. Mary*), a parish, in the hundred of HAPPING, Eastern Division of the county of NORFOLK, 7¾ miles (S. E. by E.) from North Walsham; containing 613 inhabitants. The living is a discharged vicarage, valued in the king's books at £5; present net income, £180; patron, impropriator, and incumbent, Rev. B. Cubitt. There is a place of worship for Wesleyan Methodists; also a National school.

STALISFIELD (*St. Mary*), a parish, in the union and hundred of FAVERSHAM, Upper Division of the lathe of SCRAY, Eastern Division of the county of KENT, 2¼ miles (N. N. E.) from Charing; containing 342 inhabitants. The living is a vicarage, valued in the king's books at £5. 6. 8.; present net income, £134; patron and appropriator, Archbishop of Canterbury. The church is a handsome and ancient cruciform structure.

STALLINGBOROUGH (*St. Peter and St. Paul*), a parish, in the union of CAISTOR, Eastern Division of the wapentake of YARBOROUGH, parts of LINDSEY, county of LINCOLN, 7¼ miles (W. N. W.) from Great Grimsby; containing 366 inhabitants. The living is a discharged vicarage, valued in the king's books at £11. 10. 10.; present net income, £127; patron and appropriator, Bishop of Lincoln. The church, with its tower, fell down in 1746; the chancel, and the burial-place of the Ayscough family, were afterwards rebuilt. There is a place of worship for Wesleyan Methodists.

STALLING-BUSK, a chapelry, in the parish of AYSGARTH, wapentake of HANG-WEST, North Riding of the county of YORK, 17 miles (W. by S.) from Middleham: the population is returned with the parish. The living is a perpetual curacy; net income, £91; patron, Vicar of Aysgarth. The river Ure here forms a fine cataract.

STALMINE, a chapelry, in the parish of LANCASTER, union of GARSTANG, hundred of AMOUNDERNESS, Northern Division of the county palatine of LANCASTER, 5 miles (N. N. E.) from Poulton; containing 504 inhabitants. The living is a perpetual curacy; net income, £267; patron, Vicar of Lancaster. The chapel is dedicated to St. James.

STALYBRIDGE, a market-town and chapelry, partly in the township of HARTSHEAD, parish and union of ASHTON-UNDER-LINE, hundred of SALFORD, Southern Division of the county palatine of LANCASTER, and partly in that of DUCKINFIELD, parish of STOCKPORT, and partly in that of STALYBRIDGE, parish of MOTTRAM-IN-LONGDEN-DALE, hundred of MACCLESFIELD,

Northern Division of the county of CHESTER, 8 miles (N. E. by N.) from Stockport; containing upwards of 12,000 inhabitants, included in the returns for the several townships in which it is situated. It is on the banks of the Tame, and derives its name, which was originally Staveleigh, from the family of the Staveleighs, who formerly had their residence here, which is still in existence, and the addition from the bridge, which has been recently rebuilt, at an expense of £4000, and, crossing the river here, connects the two palatine counties of Lancaster and Chester: it is partially paved and lighted with gas, and considerable improvements have been made, under an act of parliament for that purpose obtained in 1828; it is well supplied with water. The neighbourhood, formerly much covered with wood, presents some bold scenery; and from the "Wild Bank," which rises 1300 feet above the level of the sea, the prospect is extensive. The principal market day is Saturday, and there is a fair for pedlery on March 5th. The court of requests, held at Ashton-under-Line, for the recovery of debts under £5, comprises this place within its jurisdiction. The advance of the trade and population of the town has been singularly rapid; the first cotton-mill was erected in 1776, into which a steam-engine was introduced in 1795, since which period numerous factories, worked by steam-engines, have been established, giving employment to about 7000 persons. Upwards of 800 houses have been erected within the last few years. Large quantities of excellent fire-bricks are manufactured. The new road from Manchester to Sheffield runs on the north side of the town; and the Huddersfield canal passes in the vicinity. A temporary place of confinement, termed a "Lock-up," has been recently built.

The living is a perpetual curacy, subject to the rector of Ashton-under-Line; net income, £143; patron and impropriator, Earl of Stamford and Warrington. The chapel, dedicated to St. George, has been rebuilt on a new site, and was consecrated in Oct. 1839; it contains 500 free sittings, the Incorporated Society having granted £500 in aid of the expense. There are places of worship for General and Particular Baptists, Wesleyan Methodists, and Methodists of the New Connection, to which Sunday schools are attached, affording instruction to upwards of 2000 children. A Society for Mutual Instruction has been established here, with a library and apparatus for lecturing; and there are also a news-room and several benefit societies.

STAMBORNE (*St. Peter*), a parish, in the union of HALSTEAD, hundred of HINCKFORD, Northern Division of the county of ESSEX, 10 miles (N. W.) from Halstead; containing 475 inhabitants. This parish is about four miles in length and three in breadth. The living is a rectory, valued in the king's books at £15; present net income, £353: it is in the patronage of the Crown, in right of the Duchy of Lancaster. The church is an ancient edifice with a tower, and consists of a nave and chancel, with a north aisle extending the whole length of the building; the east window is embellished with some ancient stained glass. There is a place of worship for Independents. A National school is supported partly by an endowment from Mrs. Cole, who gave the fourth part of the rent of a farm, and partly by subscription. Sir John Fairwell, who served under William III., and was governor of the Tower, was interred here.

STAMBRIDGE, GREAT (*St. Mary and All Saints*), a parish, in the union and hundred of ROCHFORD, Southern Division of the county of ESSEX, 1¾ mile (E.) from Rochford; containing 405 inhabitants. The living is a rectory, valued in the king's books at £20; present net income, £600; patrons, Governors of the Charter-house, London. The church is situated on an eminence, and consists of a nave, south aisle and chancel, with a tower and shingled spire. The National school at Rochford is open to children of this parish.

STAMBRIDGE, LITTLE, a parish, in the union and hundred of ROCHFORD, Southern Division of the county of ESSEX, 1½ mile (N. E. by N.) from Rochford; containing 105 inhabitants. The extensive brewery of Mr. John English is in this parish. The living is a discharged rectory, valued in the king's books at £12: it is in the patronage of the Crown. The tithes have been commuted for a rent-charge of £167. 2. 8., subject to the payment of rates, which on the average have amounted to £24; the glebe comprises 30 acres, valued at £30 per annum. The church is an ancient edifice, consisting of a nave and chancel.

STAMFORD, a borough and market-town, having separate jurisdiction, and the head of a union, locally in the wapentake of Ness, parts of Kesteven, county of LINCOLN, 46 miles (S. by E.) from Lincoln, and 89 (N. by W.) from London; containing 5837 inhabitants. Its original name, *Seanforde*, signifying a stone ford, was derived from the passage across the river Welland being paved with stones; it was afterwards called Stanford, which was subsequently changed to its present name. The town is of very remote antiquity, its origin being ascribed by tradition to a period long before the Christian era; but the earliest authentic account respecting it is by Henry of Huntingdon, who records it as the place where the Picts and Scots, after having ravaged the country to Stamford, were defeated by the Britons, aided by the Saxons under the command of Hengist, who had been called to the assistance of the Britons by their king Vortigern. It was one of the five cities into which the Danes were distributed by Alfred the Great, when, after defeating them, he allowed that people, with Guthrum their prince, to settle in the kingdom, and who were thence called *Fif-burgenses*, or Five-burghers (the other places being Derby, Nottingham, Leicester, and Lincoln), and subsequently *Sefen-burgenses*, on the addition of two more cities, namely, Chester and York. A castle was erected by Edward the Elder, early in the tenth century, on the bank of the river, opposite the town, to check the incursions of the Danes, and of the Five-Burghers and other internal enemies, but every vestige of it has long since disappeared. Another castle, on the north-west of the town, the foundations of which are still visible, was fortified by Stephen, during the war with the Empress Matilda, but was captured by Henry of Anjou, her son, afterwards Henry II.; and the town appears to have been at this period surrounded by a wall, of

Seal and Arms.

which a few traces are discernible on its northern side, and of its gateways on the east and west sides: though there are no traces of a gate towards the north, the street is called Scot-gate, from the gate which formerly stood there. The barons met at Stamford in the 17th of John, to concert those measures which led to the signing of Magna Charta by that monarch. In the reign of Henry III., the Carmelites and members of other religious establishments here commenced giving lectures on divinity and the liberal arts, which being attended by a great number of youths of good family, led to the erection of colleges, and Stamford became celebrated as a place for education: insomuch that, from dissensions occurring, in the reign of Edward III., in the University of Oxford, amongst the students from the southern and those from the northern parts of England, a considerable number of the latter, with several professors, removed hither; but they soon returned to Oxford, in consequence of a royal proclamation, and statutes were passed by both Universities, by which any person taking a degree at either of them bound himself by oath not to attend any lectures at Stamford: a part of the gate of Brasenose College, standing in St. Paul's-street, is all that now remains of its university. This place suffered much during the war between the houses of York and Lancaster, a great portion of it having been burnt and otherwise destroyed about 1461, and it never afterwards regained its former importance.

The town is pleasantly situated on the side of a hill, rising gradually from the northern bank of the Welland, across which is a stone bridge of five arches connecting it with Stamford-Baron, or St. Martin's, in Northamptonshire: the houses are chiefly built of freestone, obtained from the neighbouring quarries of Ketton, Wettering, and Barnoak, and covered with slate: the streets are partially paved, and lighted with gas, the works for which were erected in 1824, at an expense of upwards of £9000: it is well supplied with water, which is raised by the Marquess of Exeter, who obtained an act in 1837 for better supplying the town with water, and great improvements have since been made at a very considerable expense. The surrounding country is finely varied, and the approach to the town from the south is pleasing and picturesque. The theatre, erected in 1768, is a neat and commodious edifice, lighted with gas; and there are assembly-rooms in St. George's square. Races are held in July, on a good course, a mile in circumference, on Wittering heath, near the town. On the banks of the river are excellent cold and hot water baths. The trade is principally in coal, rafts, malt, and beer, and is much promoted by the Welland being navigable hither for boats and small barges. There was formerly a school for spinning and winding raw silk, which has been discontinued. The market days are Monday and Friday, the latter being noted for corn: the butchers' and fish markets were erected in 1807, by the corporation. The fairs are on the Tuesday before Feb. 13th, Monday before Mid-Lent, Mid-Lent Monday, Monday before May 12th, Monday after the festival of Corpus Christi, and Nov. 8th and 9th.

At the time of the Conquest, Stamford was governed by *lagemen*, or aldermen, but was not incorporated by charter until the 1st of Edward IV. In the year 1663, a charter was granted by Charles II., in which the chief magistrate is first styled mayor; it was confirmed in the year 1685, by James II. The corporation consisted of a mayor, twelve aldermen, and twenty-four capital burgesses, who appointed a recorder, deputy recorder, coroner, town-clerk, and subordinate officers: the mayor, aldermen, recorder, and deputy recorder were justices of the peace. The government is now vested in a mayor, 6 aldermen, and 18 councillors, under the act of the 5th and 6th of William IV., cap. 76, for an abstract of which see the Appendix, No. I. The borough is divided into two wards, and the municipal boundaries are co-extensive with those for parliamentary purposes. The mayor and late mayor are justices of the peace, with four others appointed by commission. The police force consists of a chief constable and nine men. The freedom is acquired by birth and apprenticeship. The custom of Borough English is said to have prevailed here, by which, when the father dies intestate, the youngest son inherits the lands and tenements, to the exclusion of the elder branches of the family. This borough first sent members to parliament in the reign of Edward I., and continued to do so, with occasional intermissions, until that of Henry VIII. (1542), since which period it has exercised this privilege without interruption. The right of voting was formerly vested in the inhabitants of the borough paying scot and lot, and not receiving alms, in number about 700. The new borough comprises an area of 2399 acres, the limits of which are minutely described in the Appendix, No. II.: the old borough contained only 1623 acres. The number of voters now registered is 671, of whom 206 are freemen: the mayor is the returning officer. The recorder holds quarter sessions, and a court of record for the recovery of debts to the amount of £40, contracted within the limits of the borough, every Thursday. Petty sessions are held every Monday. The town-hall was rebuilt in 1776; it is a large and handsome detached building, standing in the main street, near the bridge, and containing a sessions-room, house of correction, gaol, guard-room, and other apartments.

Stamford formerly contained fourteen parish churches, but several of those in the liberties were destroyed by the northern soldiers, in 1461; and the number was again reduced, in 1538, at the dissolution of the monastic institutions: under an act of parliament passed in 1547, the parishes were consolidated, and five churches were allowed to remain. The living of *All Saints'* is a rectory, with that of *St. Peter's* consolidated, valued in the king's books at £12. 7. 8½.; present net income, £431: it is in the patronage of the Crown for one turn, and the Marquess of Exeter for two. The church is a large and handsome structure, combining some fine specimens of the early and later styles of English architecture, with a lofty embattled tower surmounted by an elegant octangular crocketed spire, in the later English style; it was built about 1465, at the expense of Mr. John Brown, a merchant at Calais, who was buried in it. The living of *St. George's* is a discharged rectory, with that of *St. Paul's* consolidated, valued in the king's books at £5. 3. 11½.; present net income, £124; patron, Marquess of Exeter. The church, a spacious plain edifice with a square embattled tower, was rebuilt in 1450, by William Bruges, the first Garter King at Arms. The living of *St. John's the Baptist* is a rectory, with that of *St. Clement's* consolidated, valued in the king's books at £8. 8. 6½.; present net income, £167: it is in the patronage of

R. Newcomb, Esq., for one turn, and the Marquess of Exeter for two. The church, rebuilt about the year 1452, is principally in the later English style; it has a neat embattled tower, adorned with pinnacles, and a handsome south porch; the screen separating the chancel from the nave and aisles, and the roof, are both very handsome. The living of *St. Mary's* is a discharged vicarage, valued in the king's books at £4. 18. 9.; present net income, £87; patron, Marquess of Exeter. This handsome church, supposed to have been built about the end of the thirteenth century, on the site of one so early as the Conquest, is considered the mother church; it is principally in the later English style, with some portions, particularly a very fine tower and spire, of early English architecture, highly deserving the attention of the antiquary. The living of *St. Michael's* is a discharged rectory, with the vicarage of *St. Andrew's* and the rectory of *St. Stephen's* consolidated, valued in the king's books at £18. 14. 2.; present net income, £136; patron, Marquess of Exeter. The church, situated near the centre of the town, was probably the oldest, having been built early in the thirteenth century; it had been much altered, and an embattled tower at the west end had been erected, in 1761; but in 1832, whilst the workmen were employed in improving the interior, by widening the span of the arches and diminishing the number of pillars, the walls gave way, and nearly the whole of the roof and the body of the church fell into a mass of ruins: it was rebuilt and consecrated Oct. 26th, 1836, and contains 400 free sittings, the Incorporated Society having granted £400 in aid of the expense. There are places of worship for Independents, Wesleyan Methodists, and Roman Catholics.

The Ratcliffe free school was founded by Alderman William Ratcliffe, about the year 1530, and endowed by him with estates now producing £547. 16. per annum, but of the yearly value of upwards of £800 on the expiration of the present leases. The mayor is, under an act of the 1st of Edward VI., sole trustee of the revenues, and is authorised, with the consent of the Master of St. John's College, Cambridge, to nominate and appoint the schoolmaster, with power to remove him for reasonable cause; and the form of instruction is to be approved and allowed by the master of the above college: the remains of the ancient church of St. Paul were assigned for the school-house. In 1608, a dwelling-house, garden, and orchard, nearly adjoining, were vested in feoffees for the residence of the schoolmaster; and in 1726 the dwelling-house was rebuilt by subscription. On the death of the late schoolmaster in 1832, measures were adopted by the mayor to restore the school to a degree of usefulness corresponding with its ample endowment. A claim to the right of appointing the master having, however, been advanced by the Master of St. John's College, the mayor consulted two eminent barristers, who gave their opinions in his favour; the corporation voted resolutions in his support; £564. 4. were subscribed towards his expenses; and a petition to parliament was prepared. But, that the town might not be deprived of the immediate benefit of the free school, the mayor put an end to the dispute, as to that vacancy, and bestowed the preferment upon a gentleman to whose appointment the Master of St. John's College gave consent. The schoolmaster receives the whole income of the estates, and employs such under-masters as he thinks fit. The free school is open to all boys of the town and of the adjoining parishes of St. Martin, Easton, Tinwell, Great Casterton, Little Casterton, Ryhall, Belmisthorpe, and Uffington, for instruction in the classics, and in writing, arithmetic, and the mathematics; the scholars are eligible from 8 to 12 years of age, and may remain till 15 years of age; and six of the head free scholars may continue till the age of 18, if intended for the university: the schoolmaster receives an unlimited number of boarders into his house. The school is entitled to one of the 24 scholarships at St. John's College, Cambridge, augmented by the first Lord Burghley; and Thomas Lord Exeter, in 1613, also founded three fellowships and eight scholarships at Clare Hall, Cambridge, with preference to candidates educated at Stamford school, provided they are equally qualified with their competitors. Mr. Thomas Truesdale, by will in 1700, left £50 for scholars going to the university; and £12 per annum was bequeathed by Mr. Marshall, for an exhibition for a scholar from the grammar school of Southwark, or Stamford. The Blue-coat school was established in 1704, by subscription among the inhabitants, with which an estate at Hogsthorpe was purchased, the proceeds, aided by other contributions, affording instruction and clothing to 40 boys. Wells', or the petty school, was endowed in 1604. A National school for girls was established in 1815, in conjunction with a lying-in charity, assisted by voluntary contributions; there is also a Sunday National school. An infants' school was established in 1833; and there is a Dorcas charity, founded in 1816.

The charitable institutions are numerous and liberally endowed. The principal is the hospital, or bead-house, founded and largely endowed by William Browne, in 1493, for a warden, confrater, twelve poor aged men, and two nurses, who are incorporated, and have a common seal. The edifice is a very neat structure, containing a house for the warden, apartments for the confrater, and rooms for the aged men and nurses, with a chapel in which prayers are read daily by the warden, or the confrater: the hospital is under the control of the dean of Stamford and the vicar of All Saints, who appoint the inmates. Truesdale's hospital, in Scot-gate, was founded in 1700, and rebuilt in 1833; eight poor men, with their wives, are lodged in it, and as, on the decease of any inmate, his widow must quit the hospital, the sum of five shillings a week was bequeathed by H. Fryer, Esq., to each widow so leaving it, for the remainder of her life. Snowden's hospital, endowed in 1604, and rebuilt in 1823, affords an asylum to eight poor women. Williamson's callis, or almshouse, has apartments for ten poor women. All Saints' callis, for men and women, is supported by incidental legacies, and by subscriptions from the corporation; and Peter's Hill callis, for an unlimited number of poor women, is endowed by the corporation with the interest of £200, arising from the Black Sluice drainage. The principal bequests for charitable purposes are one of £1800, by John Warrington, Esq., for the benefit, in equal proportions, of the poor widows of All Saints' callis and Snowden's hospital; £3000, left by Mr. Fryer, for the poor of Snowden's hospital and Peter's Hill callis; the rent of four houses left by Mrs. Williamson, to be paid, in sums of three shillings and sixpence a week, to six poor women; and an estate producing £50 per annum, left by Mr. W. Wells for the education of children under ten years of age belonging to the parish of All Saints. A handsome infirmary, for Stamford and the county of Rutland, capable-of receiving 32 patients, has been erected near the town, by subscription: towards its support upwards of £7000 stock was bequeathed by Mr. Henry Fryer, and £2000 collected by the ladies at a bazaar; it is further supported by voluntary contributions. The poor law union of Stamford extends into four counties, and comprises 37 parishes or places, containing a population of 15,411, according to the census of 1831; the number of guardians is 43. A Benedictine priory, dedicated to St. Leonard, and valued at the dissolution at £36. 17. per annum, is supposed to have been founded here in the seventh century, and refounded in the time of William the Conqueror, when it was made a cell to the monastery of Durham; the site is a small distance from the town, though formerly included within it; a portion of the conventual church still remains. Of the Carmelite friary, founded in 1291, the west gate exists, and is a handsome specimen of the architecture of that period; the infirmary occupies a portion of the site. Part of an outer wall, and a postern, are the only remains of the convent of Grey friars, founded by Henry III. A Dominican priory was founded prior to the year 1240; a Gilbertine priory in 1291; an Augustine priory before 1346; and an hospital, or house of lepers, in 1493. A custom, called bull-running, is practised here on St. Brice's day (Nov. 13th), and is said to have originated in William Earl of Warren having, in the reign of King John, granted a meadow for the common use of the butchers of the town, on condition that they should find a bull to be hunted and baited annually on that day. Stamford gives the title of Earl to the family of Grey of Groby.

STAMFORD, a township, in the parish of EMBLETON, union of ALNWICK, Southern Division of BAMBROUGH ward, Northern Division of the county of NORTHUMBERLAND, 5 miles (N. E. by N.) from Alnwick; containing 94 inhabitants.

STAMFORD-BARON, county of NORTHAMPTON.—See MARTIN'S (ST.) STAMFORD-BARON.

STAMFORD-BRIDGE, EAST, a township, in the union of POCKLINGTON, partly in the liberty of ST. PETER'S, but chiefly in the parish of CATTON, Wilton-Beacon Division of the wapentake of HARTHILL, East Riding of the county of YORK, 8 miles (E. N. E.) from York; containing 385 inhabitants.

STAMFORD-BRIDGE, WEST, a joint township with Scoreby, in the parish of CATTON, union of YORK, wapentake of OUZE and DERWENT, East Riding of the county of YORK, 7½ miles (E. N. E.) from York; containing, with Scoreby, 151 inhabitants. The river Derwent separates this township from that of East Stamford. There is a place of worship for Wesleyan Methodists. Christopher Wharton, in 1787, gave £600 in support of a school, for twelve boys and six girls; the annual income is about £22, and the expense of the school-house, erected in 1795, was defrayed out of previous accumulations. The celebrated battle between Harold and Tosti, in 1066, was fought near this place.

STAMFORD-HILL, county of MIDDLESEX.—See HACKNEY.

STAMFORDHAM (*ST. MARY*), a parish (formerly a market-town), in the North-Eastern Division of TINDALE ward, Southern Division of the county of NORTHUMBERLAND, 12½ miles (W. N. W.) from Newcastle-upon-Tyne; containing 1736 inhabitants. The living is a vicarage, valued in the king's books at £14. 18. 1½.; present net income, £574: it is in the patronage of the Crown; appropriator, Bishop of Durham. At Ryall, in this parish, is a chapel of ease; and there is a place of worship for Presbyterians. Lime and coal abound within the parish. The market has fallen into disuse, but the market cross, erected by Sir John Swinburne, Bart., in 1735, is still standing. Fairs for the sale of cattle, pigs, &c., are held on the second Thursday in April, and on Aug. 14th, if on Thursday; if not, on the Thursday following: there are also statute fairs for hiring servants, on the Thursdays before Old May-day and Nov. 14th, and on the last Thursday in February. This has been made a polling-place for the southern division of the shire. A free school was founded in 1663, by Sir Thomas Widdrington, Knt., who endowed it with 76 acres of land, now producing about £200 per annum, which is paid to the master, who also has the free use of a cottage and garden, with a detached close of about half an acre. A rent-charge of £12, left by Henry Paston, in 1698, is distributed among the poor.

STANALL, a township, in the parish of LANCASTER, union of GARSTANG, hundred of AMOUNDERNESS, Northern Division of the county palatine of LANCASTER, 4¼ miles (N. by E.) from Poulton: the population is returned with the chapelry of Stalmine.

STANBRIDGE, county of DORSET.—See HINTON, PARVA.

STANCILL, a joint township with Wellingley and Wilsick, in the parish of TICKHILL, union of DONCASTER, Southern Division of the wapentake of STRAFFORTH and TICKHILL, West Riding of the county of YORK, 3 miles (N. N. E.) from Tickhill; containing, with Wellingley and Wilsick, 66 inhabitants.

STAND, a hamlet, in the parish of PRESTWICH, hundred of SALFORD, Southern Division of the county palatine of LANCASTER: the population is returned with the parish. The living is a perpetual curacy; net income, £170; patron, Rector of Prestwich. Here is a National school.

STANDBRIDGE, a chapelry, in the parish and union of LEIGHTON-BUZZARD, hundred of MANSHEAD, county of BEDFORD, 3¼ miles (E. by S.) from Leighton-Buzzard; containing 416 inhabitants. The living is a perpetual curacy, annexed to the vicarage of Leighton-Buzzard. The chapel is dedicated to St. John the Baptist.

STANDERWICK, a parish, in the union and hundred of FROME, Eastern Division of the county of SOMERSET, 4 miles (E. N. E.) from Frome; containing 97 inhabitants. The living is a rectory, annexed to that of Beckington, and valued in the king's books at £2. 9. 7.

STANDFORD (*ALL SAINTS*), a parish, in the union of ELHAM, hundred of STOUTING, lathe of SHEPWAY, Eastern Division of the county of KENT, 3½ miles (N. W.) from Hythe; containing 243 inhabitants. The living is annexed, with that of Paddlesworth, to the rectory of Lyminge. The tithes have been commuted for a rent-charge of £170. 11., subject to the payment of rates, which on the average have amounted to £15; the glebe comprises nine acres, valued at £10 per annum. A school is supported by subscription. The ancient Stane-street passes through the village; and there are remaining the entrance gateway, the tower and gallery, with the garden walls and moat, of the ancient mansion of Westenhanger, in which Fair Rosamond, previously to her removal to Woodstock, and Queen Elizabeth and other sovereigns of England, are said to have resided.

STANDGROUND (*St. John the Baptist*), a parish, in the union of PETERBOROUGH, hundred of NORMAN-CROSS, county of HUNTINGDON, 1 mile (S. E. by S.) from Peterborough; containing 1242 inhabitants. The living is a vicarage, valued in the king's books at £6. 6. 10½.; present net income, £1299; patrons, Master and Fellows of Emanuel College, Cambridge; impropriators, the Landowners. There is a chapel of ease at Farcett. A small fund, arising from various bequests, is appropriated to apprenticing children and relieving adults.

STANDHILL, a hamlet, in the parish and hundred of PIRTON, county of OXFORD, 3 miles (W. S. W.) from Tetsworth: the population is returned with the parish.

STANDISH (*St. Nicholas*), a parish, in the union of WHEATENHURST, Upper Division of the hundred of WHITSTONE, Eastern Division of the county of GLOUCESTER, 4 miles (N. W.) from Stroud; containing 536 inhabitants. The living is a vicarage, with that of Hardwick consolidated, valued in the king's books at £44. 2. 8½.; present net income, £527; patron and appropriator, Bishop of Gloucester and Bristol. The church is principally in the decorated style of English architecture. Two schools are supported by subscription.

STANDISH (*St. Wilfrid*), a parish, in the unions of WIGAN and CHORLEY, hundred of LEYLAND, Northern Division of the county palatine of LANCASTER; containing 7719 inhabitants, of which number, 2407 are in the township of Standish with Langtree, 3¼ miles (N. W. by N.) from Wigan. The living is a rectory, valued in the king's books at £45. 16. 8.; present net income, £1874; patron, C. Standish, Esq. The church was built in 1584, by the Rev. Richard Moodie, the first Protestant rector; it is an elegant structure of the Tuscan order. A new church has been built at Adlington, in this parish. There is a Roman Catholic chapel at Standish Hall. Fairs for horses, cattle, toys, &c., are held on June 29th and November 22nd. The free grammar school, founded in 1603, by Mary Langton, is endowed with lands, &c., producing an annual income exceeding £100, for the support of a master and an usher. Mary Smalley, in 1794, bequeathed £1000 for the endowment of a school, in which 20 poor girls are taught and clothed: the income amounts to £50 per annum. Two of the twelve ancient castles of Lancashire, *viz.*, Standish and Penwortham, stood here, but their sites only can now be distinguished.

STANDLAKE (*St. Giles*), a parish, in the union of WITNEY, hundred of BAMPTON, county of OXFORD, 5½ miles (S. S. E.) from Witney; containing 669 inhabitants. The living is a rectory, valued in the king's books at £16. 10. 10.; present net income, £373; patrons, President and Fellows of Magdalene College, Oxford. The church is a handsome cruciform structure, in the early and decorated styles of English architecture, with a lofty octangular tower crowned with a pierced parapet, from within which rises a low octagonal spire; the arched timber roof is supported by springers resting on corbels ornamented with heraldic devices. Thirty poor children are instructed for about £25 per annum, arising from the united gifts of William Plaisterer, in 1711, and John Chambers, in 1732. Gaunt House, now occupied by a farmer, and said to have been originally built by John of Gaunt and Johan his wife, to whose memory there is a brass in the church, was garrisoned for Charles I. in 1643 and 1644, by Dr. Fell, Dean of Christ-Church, to whom it then belonged.

STANDLINCH, a parish, in the union of ALDERBURY, hundred of DOWNTON, Salisbury and Amesbury, and Southern, Divisions of the county of WILTS, 4¾ miles (S. E. by S.) from Salisbury; containing 31 inhabitants. It is bounded on the west and south-west by the river Avon. A chantry was founded here by Queen Elfrida, in expiation of the murder of Edward the Martyr, on the site of which a small chapel was erected in 1147, and rebuilt in 1677; but, though still in existence, no living is attached to it, nor is it used for divine service, except occasionally by the family of the lord of the manor, Earl Nelson, of Trafalgar House.

STANDON (*St. Mary*), a parish, (formerly a market-town), in the union of WARE, hundred of BRAUGHIN, county of HERTFORD, 8 miles (N. E.) from Hertford; containing 2272 inhabitants. The living is a discharged vicarage, valued in the king's books at £14. 13. 4.; patron and impropriator, R. P. Ward, Esq. The impropriate tithes have been commuted for a rent-charge of £1310, and the vicarial for £520, subject to the payment of rates, which on the average have respectively amounted to £218 and £108; the glebe comprises six acres, valued at £16 per annum. The church is a large ancient building, with a tower on the north side. There are places of worship for Baptists and Wesleyan Methodists. The market, which was on Friday, was granted by Charles II., together with two fairs; a pleasure fair is held on April 25th. About five miles from Ware, on the Cambridge road, in this parish, is St. Edmund's College, established on the expulsion of the English Roman Catholics from their college at Douay, at the commencement of the French revolution in 1789, for the education of the sons of the English nobility, clergy, and gentry, of the Roman Catholic religion. The building was erected in 1795, and consists of a range of buildings four stories high, and, with its two wings, 300 feet long: more than 80 students can be conveniently accommodated. A free school for poor children was endowed with £35 per annum, by Thomas Fisher, in 1612. There is also a school for Roman Catholics, supported by the Principal of the college. The ancient Ermin-street runs through this parish.

STANDON (*All Saints*), a parish, in the union of STONE, Northern Division of the hundred of PIREHILL and of the county of STAFFORD, 4 miles (N. N. W.) from Eccleshall; containing 420 inhabitants. The living is a rectory, valued in the king's books at £6. 18. 4.; patron and incumbent, Rev. Thomas Walker. The tithes have been commuted for a rent-charge of £470, subject to the payment of rates, which on the average have amounted to £2; the glebe comprises 89½ acres, valued at £179. 4. 3. per annum.

STANDON, or STONDON-MASSEY (*St. Peter and St. Paul*), a parish, in the union and hundred of ONGAR, Southern Division of the county of ESSEX, 2 miles (E. S. E.) from Chipping-Ongar; containing 299 inhabitants. This parish takes the adjunct to its name from the ancient proprietor of the manor. The living is a rectory, valued in the king's books at £13. 6. 8.; present net income, £461; patron, John Hubbard, Esq., of West Ham. The church is a small edifice with a belfry turret, surmounted by a spire of wood. Here is a parochial school.

STANE (*All Saints*), a parish, in the Marsh Division of the hundred of Calceworth, parts of Lindsey, county of Lincoln, 6½ miles (N.) from Alford: the population is returned with Witherne. The living is a rectory, united to that of Mablethorpe-St. Mary, and valued in the king's books at £5. 6. 8.

STANFIELD (*St. Margaret*), a parish, in the union of Mitford and Launditch, hundred of Launditch, Western Division of the county of Norfolk, 6 miles (N. W. by N.) from East Dereham; containing 234 inhabitants. The living is a discharged rectory, valued in the king's books at £6. 14. 2.; patrons, certain Trustees. The tithes have been commuted for a rent-charge of £258, subject to the payment of rates, which on the average have amounted to £52; the glebe comprises 20 acres, valued at £29. 1. per annum.

STANFORD, a hamlet, in the parish of Southill, hundred of Wixamtree, county of Bedford, 3¼ miles (S. W. by S.) from Biggleswade; containing 335 inhabitants.

STANFORD (*All Saints*), a parish, in the union of Swaffham, hundred of Grimshoe, Western Division of the county of Norfolk, 6 miles (S. W.) from Watton; containing 153 inhabitants. The living is a discharged vicarage, valued in the king's books at £5. 13. 1½.; present net income, £60; patron and appropriator, Bishop of Ely. The church is built of brick, now much decayed; it has a tower of flint at the west end, circular at the base, and octangular above.

STANFORD (*St. Nicholas*), a parish, in the union of Rugby, hundred of Guilsborough, Southern Division of the county of Northampton, 5 miles (S. E.) from Lutterworth; containing 24 inhabitants. The living is a discharged vicarage, valued in the king's books at £9. 10. 5.; present net income, £85; the patronage and impropriation belong to Mrs. O. Cave. The river Avon and the Grand Union canal pass through the parish.

STANFORD (*St. Mary*), a parish, in the union of Martley, Upper Division of the hundred of Doddingtree, Hundred-House and Western Divisions of the county of Worcester, 8 miles (N. E. by N.) from Bromyard; containing 198 inhabitants. The living is a rectory, valued in the king's books at £7. 4. 2.; present net income, £260; patron, Sir T. E. Winnington, Bart. The church was erected in 1768, in the later style of English architecture. There is a considerable quantity of limestone in the parish.

STANFORD, BISHOP'S (*St. James*), a parish, in the union of Bromyard, hundred of Broxash, county of Hereford, 3½ miles (S. E. by S.) from Bromyard; containing 362 inhabitants. The living is a perpetual curacy; net income, £58; patron, Vicar of Bromyard; appropriators, the three Portionists of Bromyard.

STANFORD-DINGLEY (*St. Denis*), a parish, in the union of Bradfield, hundred of Faircross, county of Berks, 8½ miles (E. N. E.) from Newbury; containing 139 inhabitants. The living is a rectory, valued in the king's books at £8. 1. 8.; it is in the patronage of Mrs. Valpy. The tithes have been commuted for a rent-charge of £270. 6. 4., subject to the payment of rates, which on the average have amounted to £32; the glebe comprises 18 acres, valued at £24 per annum. The church is principally in the Norman style of architecture. A school is supported by subscription.

STANFORD-IN-THE-VALE (*St. Denis*), a parish, in the union of Farringdon, partly in the hundred of Ock, but chiefly in that of Ganfield, county of Berks, 4 miles (E. S. E.) from Great Farringdon; containing 1016 inhabitants. The living is a discharged vicarage, valued in the king's books at £21. 1. 10½.; present net income, £337; patrons and appropriators, Dean and Chapter of Westminster. There is a chapel of ease at Goosey, in this parish. Here was formerly a market on Thursday, with a fair on the festival of St. Dionysius, or St. Denis, granted in 1230 to Ferrars, Earl of Derby, by Henry III. Here is a National school, in which 25 children are instructed for £6. 5. per annum, arising from the united bequests of John Hulton, in 1750, and William Shilton, in 1753. A quarry of stone in this parish is remarkable for its variety of fossil remains.

STANFORD, KING'S, a township, in the parish of Bishop's-Stanford, hundred of Broxash, county of Hereford, 3½ miles (S. S. E.) from Bromyard; containing 93 inhabitants. Here is a petrifying spring.

STANFORD-LE-HOPE (*St. Margaret*), a parish, in the union of Orsett, hundred of Barstable, Southern Division of the county of Essex, 1½ mile (S. E. by S.) from Horndon-on-the-Hill; containing 330 inhabitants. This parish takes its name from a stone ford across the stream, which separates it from the parishes of Mucking and Horndon-on-the-Hill; and its adjunct from a bay formed by the river Thames, called the Hope: the ancient ford has been superseded by a bridge, which is kept in repair at the joint expense of the three parishes. The living is a rectory, valued in the king's books at £12. 19. 9½.; present net income, £591; patrons, Trustees. The church, situated on the village green, is an ancient edifice, with a tower on the south side. Mrs. Elizabeth Davison bequeathed £950 three per cents., and £300 South Sea Annuities, for the instruction of poor children: the school supported by this charity was suspended in 1836, and its expected restoration has not yet been completed.

STANFORD-RIVERS (*St. Mary*), a parish, in the union and hundred of Ongar, Southern Division of the county of Essex, 2 miles (S. W. by W.) from Chipping-Ongar; containing 905 inhabitants. This parish derives its name from an ancient stone ford across the river Rodon, and its adjunct from the family of Rivers, to whom the manor at one time belonged. The living is a rectory, valued in the king's books at £26. 13. 4.: it is in the patronage of the Crown, in right of the duchy of Lancaster. The tithes have been commuted for a rent-charge of £1020, subject to the payment of rates, which on the average have amounted to £70; the glebe comprises 54 acres, valued at £81 per annum. The church is an ancient edifice, with a tower surmounted by a shingled spire. There is a place of worship for Independents; and a school is partly supported by a private individual. The work-house of the Ongar union is situated here. Dr. John Crayford, Master of University College, Oxford; Dr. Thomas Cole, Dean of Salisbury; Dr. Richard Montague, Bishop of Norwich; Dr.

Roger Manwaring, Bishop of St. David's; and the learned Drs. Richard Mulcaster and Nathaniel Lancaster, were rectors of this parish.

STANFORD-UPON-SOAR (*St. John the Baptist*), a parish, in the union of Loughborough, Southern Division of the wapentake of Rushcliffe and of the county of Nottingham, 2½ miles (N. by E.) from Loughborough; containing 129 inhabitants. The living is a rectory, valued in the king's books at £9. 7. 6.; present net income, £435; patron and incumbent, Rev. S. V. Dashwood. A day and Sunday school is supported by a private family. The river Trent enters the county here, and runs through the parish. Roman coins have been discovered.

STANGHOE, a township, in the parish of Skelton, Eastern Division of the liberty of Langbaurgh, North Riding of the county of York, 4½ miles (E.) from Guilsbrough; containing 122 inhabitants.

STANHOE (*All Saints*), a parish, in the union of Docking, hundred of Smithdon, Western Division of the county of Norfolk, 4 miles (S. W. by S.) from Burnham-Westgate; containing 436 inhabitants. The living is a rectory, valued in the king's books at £16; patron, D. Hoste, Esq. The tithes have been commuted for a rent-charge of £480, subject to the payment of rates, which on the average have amounted to £130; the glebe comprises 18 acres, valued at £22. 3. per annum. A school is endowed with £7. 10. per annum, and another is partly supported by subscription.

STANHOPE (*St. Thomas the Apostle*), a parish, in the union of Weardale, North-Western Division of Darlington ward, Southern Division of the county palatine of Durham; containing 9541 inhabitants, of which number, 2080 are in the township of Stanhope-Quarter, 5¾ miles (W. N. W.) from Wolsingham. On the incursion of the Scots in the reign of Edward III., that monarch encamped his forces in this neighbourhood; and on an eminence to the west of the village may still be seen the remains of an ancient fortress, demolished by the Scots upon that occasion. The village is situated on the northern bank of the river Wear, and near the western extremity is Stanhope Hall, an ancient mansion regularly defended by a curtain, and formerly the residence of the family of Featherstonhaugh, to whom the manor belonged. With the exception of that part of the parish which is situated on the banks of the river, the country is rugged and mountainous, and abounds with mineral wealth. Lead-ore is wrought here upon a very extensive scale in works belonging to the London Company, erected on the banks of the Wear: of these mines, one-tenth part of the produce belongs to the Bishop of Durham, and one-ninth to the rector of the parish. The neighbourhood abounds also with limestone of excellent quality, of which large quantities are procured. The market, granted by charter of Bishop Langley, in 1421, has been discontinued: fairs are held on the Wednesday before Easter, the second Friday in September, and December 21st, the last for cattle. A savings' bank has been established in the town: petty sessions for the division are held here on alternate Fridays, and a court baron is held occasionally at Frosterley, in this parish. Stanhope has been made a polling-place for the southern division of the county.

The living is a rectory, valued in the king's books at £67. 6. 8.; present net income, £4848; patron, Bishop of Durham. The church is situated on an eminence to the north of the town. There is a chapel of ease at Rookhope; and divine service is performed every Sunday afternoon in the National school in the hamlet of Eastgate. There is a place of worship for Wesleyan Methodists. The free school, in Westgate, was originally founded and endowed in 1681, by Richard Bambridge, and subsequently further endowed in 1724, by the Rev. Dr. Hartwell, and in 1829 by Mrs. Barbara Chapman: these bequests, altogether producing about £25 per annum, are now under the control of the Trustees of the Weardale school, who pay the master a salary, which far exceeds the annual income of the old foundation. The Rev. Dr. Hartwell likewise bequeathed a rent-charge of £80 for two exhibitioners from the schools of Durham and Newcastle, and for other purposes, with a further rent-charge of £7 for apprenticing two boys or girls yearly; and Mrs. Chapman also left an additional sum of £12 for placing four boys in the school at Frosterley, and distributing the residue among the poor at Christmas. The National school, to which a library is attached, was erected by Bishop Barrington, and endowed with a portion of £2000 given by that prelate for the support of six schools in this parish; it is also further supported by subscription, and a small quarterly payment by the children. Near the site of the ancient fortress several altars and other Roman antiquities have been found, and to the north are some natural excavations called Hetherburn Caves.

STANION (*St. Peter*), a parish, in the union of Kettering, hundred of Corby, Northern Division of the county of Northampton, 4¾ miles (S. E.) from Rockingham; containing 313 inhabitants. The living is annexed to the vicarage of Brigstock. There is a place of worship for Wesleyan Methodists.

STANLEY, a chapelry, in the parish of Spondon, union of Shardlow, hundred of Appletree, though locally in that of Morleston and Litchurch, Southern Division of the county of Derby, 6 miles (N. E. by E.) from Derby; containing 391 inhabitants. The living is a perpetual curacy; net income, £64; patron, Sir R. Wilmot, Bart.; impropriator, R. Bateman, Esq. The chapel is dedicated to St. Andrew.

STANLEY, a township, in the parish and union of Leek, Northern Division of the hundred of Totmonslow and of the county of Stafford, 5 miles (S. W. by W.) from Leek; containing 118 inhabitants. The Rev. Richard Shaw bequeathed £10 a year for teaching poor children.

STANLEY, a chapelry, in the parish and union of Wakefield, Lower Division of the wapentake of Agbrigg, West Riding of the county of York, 1¾ mile (N. N. E.) from Wakefield; containing, with the township of Wrenthorp, 5047 inhabitants. A district has been assigned to the chapel of this place, under the 16th section of the act of the 59th of George III. The living is a perpetual curacy; net income, £85; patron, Vicar of Wakefield. Here is a Sunday National school. This was the site of a Roman station, where a vast quantity of Roman crucibles, moulds, and silver and copper coins, has been found; of the latter, 40lb. weight was dug up in 1812. Many of these relics are deposited in the British Museum and other similar establishments. The scene of the battle fought by Robin Hood, Scarlet, and Little John, against the Pindar of Wakefield, is laid here, according to the ancient ballad. A lunatic asylum for paupers has been established in this township.

STANLEY, KING'S (*St. George*), a parish, in the union of STROUD, Lower Division of the hundred of WHITSTONE, Eastern Division of the county of GLOUCESTER, 3¼ miles (W. by S.) from Stroud; containing 2438 inhabitants. This place is supposed to have derived the adjunct to its name from its having been anciently the residence of some of the Mercian kings. The manufacture of woollen cloth appears to have been introduced into the parish at a very early period: in the reign of Elizabeth it was conducted by Richard Clotterbooke, who was interred in the church, and whose descendants for many generations carried on that business in various parts of the neighbourhood. The Stanley mills employ from 800 to 1000 persons in the manufacture of fine woollen cloths. The living is a rectory, valued in the king's books at £18. 15. 2½.; present net income, £312; patrons, Master and Fellows of Jesus' College, Cambridge. The church has been enlarged and 140 free sittings provided, the Incorporated Society having granted £150 in aid of the expense. There is a place of worship for Baptists; also a National school, partly supported by an endowment of £18 per annum. There are vestiges of an ancient encampment, near which eight Roman altars, a large brass of Alexander Severus, and other relics of antiquity, were found some years since.

STANLEY, ST. LEONARD (*St. Swithin*), a parish (formerly a market-town), in the union of STROUD, Lower Division of the hundred of WHITSTONE, Eastern Division of the county of GLOUCESTER, 4¼ miles (W. S. W.) from Stroud; containing 942 inhabitants. This place, before 1686, when a great fire destroyed most of its buildings, was a considerable town, with two fairs, on St. Swithin's day and November 6th, which are still held; but the market, which was on Saturday, under a grant of Edward II., renewed in 1620, has been discontinued. There is an extensive manufacture of woollen cloth in the village, the houses in which are now scattered and irregular. The living is a perpetual curacy; net income, £200; the patronage and impropriation belong to Mrs. Cumberland. The church is an ancient cruciform structure, partly in the early, and partly in the later, styles of English architecture, with a low tower in the centre, singularly constructed with double walls, and a passage and recesses between them: it formerly belonged to a priory of Benedictine monks, founded here in 1146, and dedicated to St. Leonard, as a cell to the abbey of St. Peter, Gloucester, which at the dissolution possessed a revenue of £126. 0. 8. There are considerable remains of the conventual buildings, of which the kitchen has been converted into a dairy. A school containing 14 children is supported by endowment: and there is also a Sunday school, erected and supported by voluntary contributions. Mrs. Rishton gave £200, which has been invested in land; and Daniel Badger gave £2 per annum, to the poor.

STANLEY, NORTH, a joint township with Sleningford, in the parish and liberty of RIPON, West Riding of the county of YORK, 4½ miles (N. W. by N.) from Ripon; containing, with Sleningford, 407 inhabitants.

STANLEY-PONTLARGE, a parish, in the union of WINCHCOMB, Lower Division of the hundred of KIFTSGATE, Eastern Division of the county of GLOUCESTER, 2¾ miles (N. W.) from Winchcomb; containing 62 inhabitants. The living is annexed to the vicarage of Toddington.

STANLEY, SOUTH, a parish, in the Lower Division of the wapentake of CLARO, West Riding of the county of YORK, 2¾ miles (N. E. by N.) from Ripley; containing 243 inhabitants. The living is a vicarage; net income, £75; patrons and impropriators, Horner Reynard and J. S. Browne, Esqrs.

STANLOW-HOUSE, an extra-parochial liberty, in the union of GREAT BOUGHTON, Higher Division of the hundred of WIRRALL, Southern Division of the county of CHESTER, 8 miles (N. by E.) from Chester; containing 13 inhabitants. This place is situated on the river Mersey, which is navigable here and forms its southern boundary. An abbey of Cistercian monks was founded here in 1178, by John Lacy, constable of Chester, which, on account of the inundations of the Mersey in 1296, was removed to Whalley in Lancashire, a cell being left at this place: some small remains of the conventual buildings are visible in a farm-house.

STANMER, a parish, in the union of NEWHAVEN, hundred of RINGMER, rape of PEVENSEY, though locally in that of Lewes, Eastern Division of the county of SUSSEX, 4 miles (N.N.E.) from Brighton; containing 123 inhabitants. The living is a rectory, in the patronage of the Archbishop of Canterbury, valued in the king's books at £16; present net income, £141. The church is an ancient structure, in the early English style. A school of industry, containing 21 boys, is supported by the Earl of Chichester.

STANMORE, GREAT (*St. John the Evangelist*), a parish, in the union of HENDON, hundred of GORE, county of MIDDLESEX, 10 miles (N. W.) from London; containing 1144 inhabitants. The living is a rectory, valued in the king's books at £10; patron, George Drummond, jun., Esq. The tithes have been commuted for a rent-charge of £430, subject to the payment of rates, which on the average have amounted to £63; the glebe comprises 41 acres, valued at £82 per annum. There is a place of worship for Independents; also a National school and an infants' school. The first meeting, after the conclusion of the late war, of the Prince Regent and his illustrious guests, the Emperor of Russia and the King of Prussia, with Louis XVIII., took place here. The celebrated Dr. Parr kept a school on the site of a house now belonging to Mr. Barren. Here is a monument in memory of Cassibelaunus, also a mound called Belmont, thrown up at the expense of the late Duke of Chandos.

STANMORE, LITTLE (*St. Lawrence*), a parish, in the union of HENDON, hundred of GORE, county of MIDDLESEX, ½ a mile (N. W.) from Edgware; containing 876 inhabitants. The living is a perpetual curacy; patron, incumbent, and impropriator, Rev. G. Mutter. The impropriate tithes have been commuted for a rent-charge of £36. 10., and the incumbent's for £415. 5. 10., subject to the payment of rates, which on the latter have on the average amounted to £70. The church stands about half a mile from the village, and was rebuilt, with the exception of the tower, about 1715, by the Duke of Chandos, whose splendid mansion of Canons was in this parish, but the internal decorations were not completed until 1720: the ceiling and walls were painted by Laguerre; on each side of the altar is a painting of the Nativity, and a dead Christ, by Belluchi; and behind it is a recess for the organ, supported by columns of the Corinthian order; in the back ground are paintings of Moses receiving the Law, and Christ preaching. Handel, who resided at Canons as chapel-master, is said to have composed his sacred drama of Esther for its consecration; the anthems used in it were composed by him, and the morning and evening services by Pepusch. On September 25th, 1790, a grand miscellaneous concert of sacred music, selected from Handel's works, was performed to his honour in this church. A vault was constructed on the north side of the chancel by the Duke of Chandos, for the interment of his family, and in a large chamber over it is a monument of his ancestor, James, first Duke of Chandos. The free grammar school was founded and endowed by Sir Lancelot Lake, in 1656; the income is derived from a field producing £50 per annum, of which £30 is paid to the master, and the remainder applied to charitable purposes. Almshouses were founded in 1640, by Dame Mary Lake, for seven poor persons, having an endowment of about £45 per annum; and this parish is entitled to send three poor persons to the almshouses in Edgware, founded in 1828, by Charles Day, Esq.

STANNEY, GREAT, an extra-parochial liberty, in the union of GREAT BOUGHTON, Higher Division of the hundred of WIRRALL, Southern Division of the county of CHESTER, 6¾ miles (N.) from Chester; containing 32 inhabitants. This liberty formerly belonged to the adjacent abbey of Stanlaw: it comprises some excellent arable and rich meadow land, in which is found marl of very good quality, composed of alluvial matter, and in great abundance; and in the meadows large trees have been dug up. The ancient mansion of the family of Bunbury, called Rake Hall, has been repaired by its present owner, Sir H. E. Bunbury, Bart.; several farm-buildings have been erected; and the roads have been very much improved. The Chester canal passes through this liberty.

STANNEY, LITTLE, a township, in the parish of STOAK, union of GREAT BOUGHTON, Higher Division of the hundred of WIRRALL, Southern Division of the county of CHESTER, 5½ miles (N.) from Chester; containing 201 inhabitants. A free school was founded by Sir Thomas Bunbury, Bart., who endowed it with £5 per annum.

STANNINGFIELD (*ST. NICHOLAS*), a parish, in the union of THINGOE, hundred of THEDWASTRY, Western Division of the county of SUFFOLK, 5¼ miles (S. by E.) from Bury-St.-Edmunds; containing 306 inhabitants. The living is a rectory, valued in the king's books at £8. 0. 2½.; patron, John Gage Rokewoode, Esq. The tithes have been commuted for a rent-charge of £345, subject to the payment of rates, which on the average have amounted to £63; the glebe comprises 44 acres, valued at £50. 12. per annum. The body of the church is of Norman architecture. Mrs. Inchbald, an ingenious novelist and dramatic writer, was a native of this place: she died at Kensington, in 1821.

STANNINGTON (*ST. MARY*), a parish, in the union and Western Division of CASTLE ward, Southern Division of the county of NORTHUMBERLAND, 5 miles (S. by E.) from Morpeth; containing 1021 inhabitants. The living is a vicarage, valued in the king's books at £5. 13. 4.; present net income, £355; patron, Bishop of Durham; impropriator, Sir M. W. Ridley, Bart. The church is an ancient structure: it had formerly a chantry, and the windows exhibit some fine old specimens of stained glass. There are places of worship for Wesleyan Methodists and Unitarians. The river Blyth runs through the parish, which abounds with coal. Here is an extensive manufacture of oil-cloth. Mrs. Grey, in 1720, gave a rent-charge of £2, and John Moor, in 1813, bequeathed the interest of £200, which, together amounting to £11 per annum, are applied for teaching eleven children.

STANNINGTON, a chapelry, in the parish of ECCLESFIELD, Northern Division of the wapentake of STRAFFORTH and TICKHILL, West Riding of the county of YORK, 4 miles (W.) from Sheffield: the population is returned with the parish. The chapel, which is in the later English style, with a tower and steeple, was erected in 1829, by the parliamentary commissioners, at an expense of £2607. 19. 3. The living is a perpetual curacy; patron, Vicar of Ecclesfield; impropriators, Rev. T. R. Ryder and others. Here is a National school. The manufacture of cutlery is extensively carried on here.

STANSFIELD (*ALL SAINTS*), a parish, in the union and hundred of RISBRIDGE, Western Division of the county of SUFFOLK, 5¼ miles (N. by E.) from Clare; containing 470 inhabitants. The living is a rectory, valued in the king's books at £11. 9. 4½.: it is in the patronage of the Crown. The tithes have been commuted for a rent-charge of £481. 6., subject to the payment of rates, which on the average have amounted to £122; the glebe comprises 68 acres, valued at £68 per annum. There is a small place of worship for Dissenters, of ancient date; and there are some small charitable bequests. Dr. Samuel Ogden, a learned divine, who died in 1778, was rector of this parish.

STANSFIELD, a township, in the parish of HALIFAX, union of TODMORDEN, wapentake of MORLEY, West Riding of the county of YORK, 4½ miles (S. W.) from Halifax; containing 8262 inhabitants. It is bounded on the south by the river Calder. At Crosstone, in this township, a new church, dedicated to St. Paul, has been lately erected. Here is the site of a manor-house, formerly belonging to the Warrens; and the rocky height above was once crowned with a castle. A school is supported by subscription.

STANSTEAD (*ST. MARY*), a parish, in the union of MALLING, hundred of WROTHAM, lathe of AYLESFORD, Western Division of the county of KENT, 2 miles (N.) from Wrotham; containing 262 inhabitants. The living is a vicarage, annexed to the sinecure rectory of Wrotham; net income, £500. A school is chiefly supported by Mr. William Hickson.

STANSTEAD (*ST. JAMES*), a parish, in the union of SUDBURY, hundred of BABERGH, Western Division of the county of SUFFOLK, 5¼ miles (N. E. by E.) from Clare; containing 353 inhabitants. The living is a rectory, valued in the king's books at £10; present net income, £247; patron and incumbent, Rev. S. Sheen. Here is a National school.

STANSTEAD, ABBOTS' (*ST. JAMES*), a parish, in the union of WARE, hundred of BRAUGHIN, county of HERTFORD, 2¾ miles (N. E. by E.) from Hoddesdon; containing 966 inhabitants. The parish is bounded on the west by the navigable river Lea, on the north by the Ashe, and on the east and south-east by the navigable river Stort, being nearly insulated. The Rye House, noted for the plot laid there, in 1683, against the lives of Charles II. and James, Duke of York, was built in the reign of Henry VI., by Adrew Ogard. It was formerly surrounded by a moat, but the only remains of the structure now existing is an embattled gate-house of brick, with a handsome stone doorway, long since converted into a workhouse for the parish. The living is a discharged vicarage, valued in the king's books at £10; present net income, £134; patron and incumbent, Rev. T. Feilde; impropriator, D. Hankin, Esq. The church, situated on an eminence one mile south-east from the village, was built in 1578, by Ralph Baesh, Esq. Here is a National school. Almshouses for six poor widows were founded in 1636, by Sir Edward Baesh, who endowed them with certain lands and a rent-charge of £25, for the payment of two shillings per week to each; the residue to be applied for apprenticing poor children. He also founded a free grammar school with an endowment of £20 per annum, and gave a small cottage for the use of the parish clerk.

STANSTEAD, ST. MARGARET'S, a parish, in the union of WARE, hundred and county of HERTFORD, 1½ mile (N. by E.) from Hoddesdon; containing 107 inhabitants. The living is a donative curacy; net income, £6; patron and impropriator, Rev. T. S. Pratt. The impropriate tithes have been commuted for a rent-charge of £87. 3. 8., subject to the payment of rates, which on the average have amounted to £16. A school for girls is supported by the lady of the manor. This parish is situated between the New River and the Lea. A college, or chantry, for a master and four Secular priests, was founded here in 1315, by Sir William de Goldington, Knt., in consequence of the impoverishment of the tithes, oblations, and other ecclesiastical rights of the church; but it was dissolved in 1431, in consequence of neglect and misapplication of its revenue.

STANSTED-MOUNTFITCHET (*St. Mary*), a parish, in the union of BISHOP'S-STORTFORD, partly in the hundred of CLAVERING, but chiefly in that of UTTLESFORD, Northern Division of the county of ESSEX, 3 miles (N. N. E.) from Bishop's-Stortford; containing 1560 inhabitants. This parish takes its name from a branch of the Roman road from Bishop's-Stortford to Colchester, which passes through it, and its adjunct from its possessor at the time of the Conquest, Robert Gernon, surnamed Montfitchet, who erected a castle here, of which there are still some remains. It is about forty miles in circumference, comprehending a great variety of surface, and is generally well cultivated. Stansted Hall, the old manor-house, has been taken down and converted into farm-buildings. The village, which is extensive and contains many well-built houses, is situated chiefly on the road to Newmarket, and partly on that to Takely. A fair is annually held here on the 12th of May. The living is a discharged vicarage, valued in the king's books at £13. 6. 8.; present net income, £251; patron and impropriator, E. F. Maitland, Esq. The church, a small ancient edifice with a tower of brick, formerly belonged to the priory of Thremhall; it contains 200 free sittings, the Incorporated Society having granted £270 in aid of the expense, and has a very ancient font rudely sculptured, some interesting architectural details, and several ancient monuments and brasses. There is a place of worship for Independents. A handsome school-house, containing spacious rooms for 100 girls and 100 infants, has been erected at a cost of £400 by the Rev. Josias Torriano, the present incumbent. Another school for girls, and a National Sunday school, are supported by subscription. About two miles from the church was the priory of Thremhall, founded by Richard de Montfichet, and dedicated to St. James.

STANTHORNE, a township, in the parish of DAVENHAM, union and hundred of NORTHWICH, Southern Division of the county of CHESTER, 1¼ mile (W. N. W.) from Middlewich; containing 149 inhabitants.

STANTON, a chapelry, in the parish of YOULGRAVE, union of BAKEWELL, hundred of HIGH PEAK, Northern Division of the county of DERBY, 3¾ miles (N.) from Winster; containing 744 inhabitants. A school is partly supported by W. P. Thornhill, Esq.

STANTON, a joint township with Newhall, in the parish of STAPENHILL, union of BURTON-UPON-TRENT, hundred of REPTON and GRESLEY, Southern Division of the county of DERBY, 3 miles (S. S. E.) from Burton-upon-Trent; containing, with Newhall, 1182 inhabitants.

STANTON (*St. Bartholomew*), a parish, in the union of WINCHCOMB, Lower Division of the hundred of KIFTSGATE, Eastern Division of the county of GLOUCESTER, 4¾ miles (N. E. by N.) from Winchcomb; containing 293 inhabitants. The living is a discharged rectory, with that of Snowshill annexed, valued in the king's books at £17. 11. 5½.; present net income, £377; patron and incumbent, Rev. R. Wynniatt.

STANTON, a township, in the parish of LONG HORSLEY, union, and Western Division of the ward, of MORPETH, Northern Division of the county of NORTHUMBERLAND, 6 miles (N. W. by W.) from Morpeth; containing 135 inhabitants. Here are a colliery, several quarries of limestone, and kilns for burning lime. The ancient manor-house has been converted into a poor-house: a chapel which stood a little to the northward of it has quite disappeared. From the foundations in the neighbourhood, Stanton seems to have been once a considerable place.

STANTON, a township, in the parish of ELLASTONE, Southern Division of the hundred of TOTMONSLOW, Northern Division of the county of STAFFORD, 3½ miles (W.) from Ashbourn; containing 371 inhabitants. Gilbert Sheldon, Archbishop of Canterbury, was born here in 1598, and died in 1677.

STANTON (*All Saints*), a parish, in the union of THINGOE, hundred of BLACKBOURN, Western Division of the county of SUFFOLK, 3¼ miles (N. E.) from Ixworth; containing, with Stanton-St.-John, 1035 inhabitants. The living is a discharged rectory, with Stanton-St.-John united, valued in the king's books at £9. 6.; present net income, £843; patron, R. E. Lofft, Esq. A school for girls is supported by the rector. A fund of £20 per annum, the rent of two parcels of land, left at a very remote period by some unknown benefactors, is appropriated to the repairs of the church; and some minor bequests are distributed annually among the poor.

STANTON (*St. John*), a parish, in the union of THINGOE, hundred of BLACKBOURN, Western Division of the county of SUFFOLK: the population is returned with the parish of Stanton-All-Saints. The living is a discharged rectory, united to Stanton-All-Saints, and valued in the king's books at £9. 4. 10.

STANTON, ST. BERNARD (*All Saints*), a parish, in the union of Devizes, hundred of Swanborough, Devizes and Northern Divisions of the county of Wilts, $5\frac{3}{4}$ miles (E. by N.) from Devizes; containing 319 inhabitants. The living is a discharged vicarage, valued in the king's books at £7; present net income, £222; patron and impropriator, Earl of Pembroke. The church is a neat edifice, lately rebuilt at an expense of £500, defrayed by subscription. A school which was enlarged and improved in 1832 is partly supported by the vicar.

STANTON-BURY (*St. Peter*), a parish, in the union of Newport-Pagnell, hundred of Newport, county of Buckingham, 3 miles (W. by S.) from Newport-Pagnell; containing 51 inhabitants. The living is a discharged vicarage, valued in the king's books at £7. 6. 8., present net income, £54; patron and impropriator, Earl Spencer. The church exhibits many Norman remains, particularly a fine and richly decorated arch between the nave and the chancel.

STANTON-BY-BRIDGE (*St. Michael*), a parish, in the union of Shardlow, hundred of Repton and Gresley, Southern Division of the county of Derby, $6\frac{3}{4}$ miles (S. by E.) from Derby; containing 215 inhabitants. The living is a rectory, valued in the king's books at £6. 12. $8\frac{1}{2}$.; present net income, £345; patron, Sir George Crewe, Bart. The church is partly in the Norman and partly in the decorated style of architecture. The distinguishing appellation of this place arises from the ancient bridge over the Trent, termed Swarkstone bridge, which connects these two parishes.

STANTON-BY-DALE (*St. Michael*), a parish, in the union of Shardlow, hundred of Morleston and Litchurch, Southern Division of the county of Derby, 9 miles (E. by N.) from Derby; containing 740 inhabitants. The living is a vicarage, endowed with the rectorial tithes, with the chapel of Dale-Abbey; net income, £195; patrons, Trustees appointed by the proprietors of the lordship of Stanton. A school is supported by subscription. The Erewash and the Nutbrook canals pass through the parish. Almshouses for six poor persons were founded in 1711, by a bequest from Joseph Middlemore, with an endowment of more than £100 per annum.

STANTON-DREW (*St. Mary*), a parish, in the union of Clutton, hundred of Keynsham, Eastern Division of the county of Somerset, $1\frac{1}{2}$ mile (W. by S.) from Pensford; containing 731 inhabitants. The living is a discharged vicarage, with that of Pensford-St.-Thomas annexed, valued in the king's books at £7. 2. $8\frac{1}{2}$.; patron and appropriator, Archdeacon of Bath. The appropriate tithes have been commuted for a rent-charge of £234. 18., and the vicarial for £237. 10., subject to the payment of rates; the appropriate glebe comprises 48 acres, valued at £108, and the vicarial one acre, valued at £4. 10., per annum. Richard Jones, in 1688, bequeathed two-fifths of the proceeds of an estate for educating and apprenticing poor children: the income is £60 a year. Six poor girls are instructed for £5 per annum, the gift of Elizabeth Lyde, in 1772. The neighbourhood abounds with various objects of interest to the antiquary, the most prominent of which are, *Maes Knoll Tump*, a stupendous barrow, and an extensive Druidical temple of three circles of stones, whose diameters are respectively 120, 43, and 32 yards, spreading itself over ten acres of ground: the stones, which are of amazing dimensions, were apparently brought from the neighbouring quarries; many of them, however, now lie prostrate on the ground. The name of Belton, or Belluton, a hamlet at nearly an equal distance from each of these British monuments, is thought to be a corruption of Belgeton, a town of the Belgæ, being situated on the line of Wansdyke, the ancient boundary of their territory.

STANTON, FEN, county of Huntingdon.—See FEN-STANTON.

STANTON-FITZWARREN (*St. Leonard*), a parish, in the union of Highworth and Swindon, hundred of Highworth, Cricklade, and Staple, Cricklade and Northern Divisions of the county of Wilts, $2\frac{1}{4}$ miles (S. W. by W.) from Highworth; containing 188 inhabitants. The living is a rectory, valued in the king's books at £10. 2. 6.; present net income, £174; patron and incumbent, Rev. Dr. Trenchard, who supports a school.

STANTON, ST. GABRIEL, a chapelry, in the parish and hundred of Whitchurch-Canonicorum, union of Bridport, Bridport Division of the county of Dorset, 4 miles (W. by S.) from Bridport; containing 101 inhabitants. This chapelry is bounded on the south by the English Channel, and on the shore, on one of the highest hills in the county, a signal staff has been erected. The living is annexed, with those of Chideock and Marsh-wood, to the vicarage of Whitchurch-Canonicorum.

STANTON HARCOURT (*St. Michael*), a parish, in the union of Witney, hundred of Wootton, county of Oxford, $4\frac{1}{2}$ miles (W.) from Oxford; containing 657 inhabitants. This place was granted by Adeliza, second queen of Henry I., to her kinswoman Milicent, wife of Richard de Camville, whose daughter Isabel married Robert de Harcourt, from whom it derived the adjunct to its name, and in whose descendants the manor has remained for more than 600 years. It is pleasantly situated near the confluence of the small river Windrush with the Thames, and abounds with interesting and picturesque scenery. The living is a discharged vicarage, valued in the king's books at £16. 13. 4.; present net income, £136; patron and appropriator, Bishop of Oxford. The church is a handsome and venerable cruciform structure, chiefly in the decorated style of English architecture, with a lofty square embattled tower, and some Norman portions: the principal entrance is through a large Norman door-case, contiguous to which is a smaller for women only; the windows in the lower part of the tower are Norman, and those in the upper stages of more recent date. The nave is Norman, and is lighted by a range of clerestory windows in that style; on the north side of the chancel are a rich altar-tomb, and a recessed monument to Maud, daughter of John Lord Grey, of Rotherfield, with her recumbent effigy in the costume of the time of Richard II. On the south side of the chancel is the sepulchral chapel of the Harcourt family, containing monuments to Sir Philip Harcourt and his wife Anne, daughter of Sir William Waller; Sir Robert Harcourt and his wife Margaret, daughter of Sir John Byron; and to several others of that family. The font is octangular, and embellished with a cross, and with shields of armorial bearings in compartments. There is a chapel of ease at

South Leigh, in this parish. A school is supported by the proceeds of various benefactions, amounting to £14 per annum. Only a small portion of the ancient mansion of the Harcourts is now remaining, and is in the occupation of a farmer: the chapel, with a chamber over it, and the adjoining tower, are in a very good state of preservation; the tower contains three apartments above each other, of which the uppermost is called Pope's study, from the poet having passed much of his time there while employed in his translation of Homer, during which he spent two summers at Stanton-Harcourt, where he was occasionally visited by Gay, who was then at a neighbouring seat of Lord Harcourt's, at Cokethorpe. On one of the panes of glass in the window of Pope's study, which is now preserved at Nuneham-Courtenay, is the following inscription, in his own hand-writing;—"In the year 1718, Alexander Pope finished here the fifth volume of Homer." The kitchen, which bears marks of remote antiquity, was repaired about the reign of Henry IV., and has a great resemblance to the abbot's kitchen at Glastonbury. Some ancient remains, called the Devil's Quoits, were probably placed here to commemorate a victory obtained by the Saxon king Cynegils, and his son Cwichelm, over the Britons, of whom more than 2000 were killed in battle.

STANTON, ST. JOHN (*St. John the Baptist*), a parish, in the union of Headington, hundred of Bullingdon, county of Oxford, 4½ miles (N. E. by E.) from Oxford; containing 470 inhabitants. This place takes the adjunct to its name from the family of St. John, who held the manor in the reign of Edward III. A hill in the neighbourhood, called Irondon hill, is supposed to have obtained its name from Ireton, who lived there after his marriage with the daughter of Cromwell. The living is a rectory, valued in the king's books at £16. 9. 4½.; present net income, £287; patrons, Warden and Scholars of New College, Oxford. The church is in the early English style, with a handsome square embattled tower; part of the north aisle is enclosed by a richly decorated screen, and is used as a vestry; and on the north side of the chancel is an enriched recess, apparently intended for a stone coffin. Lady Elizabeth Holford, in 1717, gave £500 in support of a charity school for the children of this parish and that of Forest Hill. About a mile to the north-east of the church is the hamlet of Woodpury, in which are some interesting remains of the ancient village and church of that name. Some Roman tiles and pottery, with two coins, have been dug up here.

STANTON, LACY (*St. Peter*), a parish, in the union of Ludlow, hundred of Munslow, Southern Division of the county of Salop, 3 miles (N. N. W.) from Ludlow; containing 1467 inhabitants. The living is a vicarage, valued in the king's books at £16; present net income, £485; patron and impropriator, Earl of Craven. The church has been repewed, and 56 free sittings provided, the Incorporated Society having granted £30 in aid of the expense.

STANTON, LONG (*All Saints*), a parish, in the union of Chesterton, hundred of Northstow, county of Cambridge, 6¼ miles (N. W. by N.) from Cambridge; containing 428 inhabitants. The living is a discharged vicarage, valued in the king's books at £13. 13. 4½.; present net income, £155; patron, Bishop of Ely; the impropriation belongs to the Hutton family. There is a place of worship for Wesleyan Methodists. The bishops of Ely anciently had a palace here, at which Queen Elizabeth was entertained on the day after her visit to the University of Cambridge, in Aug. 1564.

STANTON, LONG (*St. Michael*), a parish, in the union of Chesterton, hundred of Northstow, county of Cambridge, 5½ miles (N. W. by N.) from Cambridge; containing 127 inhabitants. The living is a rectory, valued in the king's books at £6. 12. 8½.; present net income, £237; patrons, Master and Fellows of Magdalene College, Cambridge. The church is a small thatched building.

STANTON, LONG (*St. Michael*), a parish, in the union of Bridgenorth, hundred of Munslow, Southern Division of the county of Salop, 7¾ miles (S. W. by S.) from Much Wenlock; containing 278 inhabitants. The living is a discharged vicarage, valued in the king's books at £7; present net income, £134; patrons and appropriators, Dean and Chapter of Hereford.

STANTON-ON-THE-WOLDS (*All Saints*), a parish, in the union of Bingham, Northern Division of the wapentake of Rushcliffe, Southern Division of the county of Nottingham, 7½ miles (S. E. by S.) from Nottingham; containing 125 inhabitants. The living is a discharged rectory, valued in the king's books at £2. 13. 4.; present net income, £109; patron and incumbent, Rev. G. Randolph. The parish is bounded on the east by the old Fosse-road.

STANTON-PRIOR (*St. Lawrence*), a parish, in the union and hundred of Keynsham, Eastern Division of the county of Somerset, 5 miles (W. S. W.) from Bath; containing 159 inhabitants. The living is a discharged rectory, valued in the king's books at £10. 1. 10½.; present net income, £182; patron, W. G. Langton, Esq. On a long isolated eminence, called Stanton Bury, are the remains of an ancient intrenchment, enclosing more than 30 acres: it has been thought a work of the Romans, some of their coins having been found near it; but being situated on the Wansdyke, it had probably a more remote origin, and might have been subsequently occupied by them. Gilbert Sheldon, Archbishop of Canterbury, was born here, in 1598; he died in 1677.

STANTON, ST. QUINTIN (*St. Giles*), a parish, in the union of Chippenham, hundred of Malmesbury, Chippenham and Calne, and Northern, Divisions of the county of Wilts, 4¼ miles (N. by W.) from Chippenham; containing 317 inhabitants. The living is a rectory, valued in the king's books at £10. 5. 7½.; present net income, £312; patron, Earl of Radnor. The church has been enlarged, and contains 80 free sittings, the Incorporated Society having granted £40 in aid of the expense. Good slate, limestone, and a very hard blue stone for building, abound within the parish. Here were formerly the ruins of an ancient castle, but they have entirely been removed within the last few years.

STANTON-STONEY (*St. Michael*), a parish, in the union of Hinckley, hundred of Sparkenhoe, Southern Division of the county of Leicester, 4¼ miles (E. by N.) from Hinckley; containing 549 inhabitants. The living is a rectory, valued in the king's books at £14. 13. 1½.; present net income, £348; patron, Marquess of Hastings.

STANTON-UNDER-BARDON, a chapelry, in the parish of THORNTON, union of MARKET-BOSWORTH, hundred of SPARKENHOE, Southern Division of the county of LEICESTER, 9 miles (W. N. W.) from Leicester; containing 295 inhabitants. There is a place of worship for Wesleyan Methodists.

STANTON-UPON-ARROW (*St. Peter*), a parish, in the union of KINGTON, partly in the hundred of STRETFORD, but chiefly in that of WIGMORE, county of HEREFORD, 5½ miles (E. N. E.) from Kington; containing 393 inhabitants. The living is a discharged vicarage, valued in the king's books at £5. 7. 10.; present net income, £220: it is in the patronage of the Crown; appropriator, Bishop of Hereford.

STANTON-UPON-HINE-HEATH (*St. Andrew*), a parish, in the union of WEM, Whitchurch Division of the hundred of NORTH BRADFORD, Northern Division of the county of SALOP; containing 722 inhabitants, of which number, 262 are in the township of Stanton, 5½ miles (S. E. by E.) from Wem. The living is a discharged vicarage, valued in the king's books at £5. 10. 10.; present net income, £200; patron and impropriator, Sir R. Hill, Bart. A school, in which fifteen poor children are instructed, was founded and endowed by Mrs. S. Baddeley, in 1721, with £5 per annum, which is paid to the parish clerk for that purpose, who has also the use of the house and garden, and permission to receive other scholars. Two other schools are partly supported by Lady Hill.

STANTON-WYVILLE, county of LEICESTER.—See STONTON WYVILLE.

STANWAY (*All Saints*), a parish, in the union of LEXDEN and WINSTREE, Colchester Division of the hundred of LEXDEN, Northern Division of the county of ESSEX, 4 miles (W. by S.) from Colchester; containing 665 inhabitants. This parish is about nine miles in circumference, forming part of a fertile and highly cultivated district, and, from the remains of a second church, appears to have been formerly divided into the two parishes of Stanway Magna and Stanway Parva. A fair is held in the village on the 23d of April. The living is a rectory, with the chapel of Albright, valued in the king's books at £10. 17. 6.; present net income, £723; patrons, President and Fellows of Magdalene College, Oxford. The church, situated near the London road, is a small ancient edifice, with a wooden turret, and contains 50 free sittings, the Incorporated Society having granted £50 in aid of the expense. The ruins of the other church, which belonged to Stanway Magna, bespeak it to have been a stately structure. Four schools are partly supported by subscription. The workhouse for the union of Lexden and Winstree is in this parish, and within 2½ miles from Colchester; it was built in 1837, at a cost of £6500 (including the purchase of the site), and will accommodate 400 paupers. The union comprises 35 parishes or places, under the care of 38 guardians, and, according to the census of 1831, contains a population of 19,811. A number of large bones and other remains, probably of elephants brought over by Claudius in 43, was found here in 1764, lying in a stratum of sea sand and shells.

STANWAY (*St. Peter*), a parish, in the union of WINCHCOMB, Upper Division of the hundred of TEWKESBURY, Eastern Division of the county of GLOUCESTER, 3½ miles (N. E.) from Winchcomb; containing 401 inhabitants. The living is a discharged vicarage, valued in the king's books at £9; present net income, £220; the patronage and impropriation belong to Lady Elcho. Limestone is obtained in the parish. The Earl of Wemyss supports two schools for the education of poor children; and a Sunday school is partly supported by endowment of the late Viscountess Hereford.

STANWELL (*St. Mary*), a parish, in the union of STAINES, hundred of SPELTHORNE, county of MIDDLESEX, 2¾ miles (N. E. by N.) from Staines; containing 1386 inhabitants. The living is a discharged vicarage, valued in the king's books at £9; present net income, £301: it is in the patronage of the Crown; impropriator, Sir John Gibbon, Bart., and others. The church is principally in the later style of English architecture. There is a place of worship for Independents. A school for boys was endowed in 1624, by Thomas, Lord Knevitt; there is also a National school for girls.

STANWICK (*St. Lawrence*), a parish, in the union of THRAPSTONE, hundred of HIGHAM-FERRERS, Northern Division of the county of NORTHAMPTON, 2¼ miles (N. N. E.) from Higham-Ferrers; containing 503 inhabitants. The living is a rectory, valued in the king's books at £12. 9. 4½.; present net income, £373: it is in the patronage of the Crown. The church exhibits portions in the early, decorated, and later styles of English architecture; it has an octangular tower of the former character, surmounted by an enriched spire, in the decorated style. There is a place of worship for Wesleyan Methodists; also a Sunday National school. Richard Cumberland, the dramatist, was born here in 1732.

STANWICK-ST. JOHN (*St. John the Baptist*), a parish, in the union of RICHMOND, wapentake of GILLING-WEST, North Riding of the county of YORK; containing, exclusively of a portion of the township of Stapleton, which is in this parish, 955 inhabitants, of which number, 73 are in the township of Stanwick-St. John, 7½ miles (N. by E.) from Richmond. The living is a vicarage, valued in the king's books at £6. 13. 4.; present net income, £61; patron, John Wharton, Esq. The church is very ancient, and contains two fine marble statues to the memory of Sir Hugh and Lady Smithson. In this parish is a most extensive intrenchment, enclosing an area of nearly 1000 acres: it has been ascribed to the Romans and also to the Scots; but Whittaker, the learned antiquary, considers it the site of an ancient camp, or city of the Britons.

STANWIX (*St. Michael*), a parish, in the union of CARLISLE, partly in CUMBERLAND ward, but chiefly in ESKDALE ward, Eastern Division of the county of CUMBERLAND; containing 1788 inhabitants, of which number, 545 are in the township of Stanwix, ½ a mile (N.) from Carlisle. The parish is bounded on the south by the river Eden, which is crossed by a handsome stone bridge, connecting it with the city of Carlisle: the village is beautifully situated on the northern bank of that river. A soft freestone abounds in the neighbourhood. The living is a vicarage, valued in the king's books at £9; present net income, £264; patron, Bishop of Carlisle, who, with the Dean and Chapter, is the appropriator. The old church was built upon the site and out of the ruins of the *Congavata* of the Romans, of which station Severus' Wall formed the northern rampart, and near which many altars and inscriptions have been found. A new church has been recently erected. Here is a National school.

STAPELEY, a township, in the parish of WYBUNBURY, union and hundred of NANTWICH, Southern Division of the county of CHESTER, 1¾ mile (S. E.) from Nantwich; containing 356 inhabitants.

STAPELEY, a joint tything with Rye, in the parish and hundred of ODIHAM, Odiham and Northern Divisions of the county of SOUTHAMPTON, 2½ miles (S. S. W.) from Hartford-bridge: the population is returned with the parish.

STAPENHILL (*St. Peter*), a parish, in the union of BURTON-UPON-TRENT, hundred of REPTON and GRESLEY, Southern Division of the county of DERBY, 1 mile (S. E.) from Burton-upon-Trent; containing 1926 inhabitants. The living is a discharged vicarage, valued in the king's books at £5. 6. 0½.; present net income, £373; patron, Marquess of Anglesey. A chapel has been erected at Newhall, which was consecrated on July 9th, 1832. As it was the first founded and endowed under the 1st and 2nd of William IV., it is entitled to peculiar advantages. The erection cost £2000, the gift of the Rev. Joseph Clay, curate of Stapenhill, John Clay and Miss Sarah Clay, who have thus become patrons, and who endowed it with £1000 for the minister, besides building a parsonage-house and a school. They also intend to grant a further sum of £2000 towards the endowment, in order to obtain an equal amount from the governors of Queen Anne's bounty. At Cauldwell, in this parish, is also a chapel of ease, and there is a place of worship for Baptists. John Hieron, an eminent non-conformist divine and critic, was born here in 1608.

STAPLE, a tything, in the parish of TISBURY, hundred of DUNWORTH, Hindon and Southern Divisions of the county of WILTS: the population is returned with the parish.

STAPLE-FITZPAINE (*St. Peter*), a parish, in the union of TAUNTON, hundred of ABDICK and BULSTONE, Western Division of the county of SOMERSET, 5 miles (S. E. by S.) from Taunton; containing 415 inhabitants. The living is a rectory, with Bickenhall annexed, valued in the king's books at £17. 14. 2.; patron, Lord Portman. The tithes have been commuted for a rent-charge of £183. 10., subject to the payment of rates, which on the average have amounted to £25; the glebe comprises 50 acres, valued at £64 per annum. The tithes of Bickenhall have been commuted for a rent-charge of £200, subject to the payment of rates, which on the average have amounted to £7; the glebe comprises 8 acres, valued at £12 per annum. A school is supported by Lord Portman.

STAPLE-NEXT-WINGHAM (*St. James*), a parish, in the union of EASTRY, hundred of DOWNHAMFORD, lathe of ST. AUGUSTINE, Eastern Division of the county of KENT, 1¾ mile (E. by S.) from Wingham; containing 502 inhabitants. This parish is chiefly inhabited by market-gardeners, who supply the watering-places in the Isle of Thanet with vegetables. The living is annexed to the rectory of Adisham. The church contains a very ancient font, and many handsome monuments to the Lynch family, formerly residing at Grove, an ancient mansion in this parish. Two schools are partly supported by subscription.

STAPLEFORD (*St. Andrew*) a parish, in the union of CHESTERTON, hundred of THRIPLOW, county of CAMBRIDGE, 5¼ miles (S. S. E.) from Cambridge; containing 464 inhabitants, of whom a considerable number are employed in the straw-plat manufacture. The living is a vicarage, valued in the king's books at ££7. 18. 9.; present net income, £181; patrons and appropriators, Dean and Chapter of Ely. Two schools are supported by Lady Godolphin.

STAPLEFORD (*St. Mary*), a parish, in the union, hundred, and county of HERTFORD, 3¼ miles (N. by W.) from Hertford; containing 237 inhabitants. The living is a rectory, valued in the king's books at £8. 8. 6½.; patron, Samuel Smith, Esq. The tithes have been commuted for a rent-charge of £250, subject to the payment of rates, which on the average have amounted to £38; the glebe comprises 20 acres, valued at £30 per annum. A school is supported by Abel Smith, Esq. The river Bean runs through the parish; and there is a large watercourse called the New Cut, made at the expense of S. Smith, Esq., to diminish the violence of the floods.

STAPLEFORD (*St. Mary Magdalene*), a parish, in the union of MELTON-MOWBRAY, hundred of FRAMLAND, Northern Division of the county of LEICESTER, 4 miles (E. by S.) from Melton-Mowbray; containing 185 inhabitants. The living is a discharged vicarage, consolidated with the rectory of Saxby, and valued in the king's books at £13. The church was erected in 1783, and contains some fine monuments to the Sherard family, among which is one by Rysbrach, in memory of the first Earl of Harborough. The river Wreake and the Melton-Mowbray and Oakham canal run through the parish. Here is an endowed hospital.

STAPLEFORD (*All Saints*), a parish, in the union of NEWARK, Lower Division of the wapentake of BOOTHBY-GRAFFO, parts of KESTEVEN, county of LINCOLN, 6½ miles (N. E. by E.) from Newark; containing 185 inhabitants. The living is a discharged vicarage, valued in the king's books at £5. 3. 4.; present net income, £68; patron and impropriator, Lord Middleton.

STAPLEFORD (*St. Helen*), a parish, in the union of SHARDLOW, Southern Division of the wapentake of BROXTOW, Northern Division of the county of NOTTINGHAM, 5¾ miles (W. S. W.) from Nottingham; containing 1533 inhabitants. The river Erewash bounds the parish on the west and north-west: the stocking manufacture is here carried on to a considerable extent. The living is a perpetual curacy; net income, £80: it is in the patronage of the Crown. The church underwent a thorough repair in 1785. There is a place of worship for Wesleyan Methodists; also a National school. An obelisk, apparently of Saxon construction, and a Druidical monument, called the Hemlock Stone, are the only remains of antiquity here. Stapleford Hall was the residence of that distinguished admiral, Sir John Borlace Warren.

STAPLEFORD (*St. Mary*), a parish, in the union of WILTON, hundred of BRANCH and DOLE, Salisbury and Amesbury, and Southern, Divisions of the county of WILTS, 4½ miles (N. N. W.) from Wilton; containing 337 inhabitants. The living is a discharged vicarage, valued in the king's books at £10; present net income, £105; patrons and appropriators, Dean and Canons of Windsor.

STAPLEFORD, ABBOT'S (*St. Mary*), a parish, in the union and hundred of Ongar, Southern Division of the county of Essex, $5\frac{3}{4}$ miles (S. E. by S.) from Epping; containing 507 inhabitants. This parish is separated from that of Stapleford Tawney by the river Rodon, over which was a ford that may have given rise to its name; it takes its distinguishing affix from its having belonged to the monastery of St. Edmondsbury, in the county of Suffolk. The fine old mansion of Albyns, the residence of the Abdy family, is situated here; it was erected by Inigo Jones, and contains some fine ceilings, mantel-pieces, and other enriched details, and a valuable collection of books, among which is a copy of Coverdale's Bible in good preservation. The living is a rectory, valued in the king's books at £16. 15.; present net income, £483: it is in the patronage of the Crown. In the east window of the church is a very ancient figure of Edward the Confessor in stained glass, removed by Dr. Pearce, Bishop of Rochester, from the palace of Havering-atte-Bower. There are several monuments to the family of Abdys; John Lord Fortescue, one of the Justices of the Court of Common Pleas; and his son, Dormer, the last Lord: and in the churchyard a monument to Sir H. Gould, also one of the Justices of the Court of Common Pleas. There is a school-house, endowed with three old cottages and £25 a year, by John Lord Fortescue, in 1734, in which 40 boys are instructed; and a National school has been established. Dr. Pearce, Bishop of Rochester, and Dr. Godfrey Goodman, Bishop of Gloucester, were rectors of this parish; and John Day, author of "Sandford and Merton," resided here.

STAPLEFORD, BRUEN, a township, in the parish of Tarvin, union of Great Boughton, Second Division of the hundred of Eddisbury, Southern Division of the county of Chester, $4\frac{1}{2}$ miles (W. N. W.) from Tarporley; containing 159 inhabitants.

STAPLEFORD, FOULK, a township, in the parish of Tarvin, union of Great Boughton, Lower Division of the hundred of Broxton, Southern Division of the county of Chester, $5\frac{1}{2}$ miles (W.) from Tarporley; containing 244 inhabitants. A school is endowed by Sir T. Moulson, Bart.

STAPLEFORD, TAWNEY, (*St. Mary*), a parish, in the union and hundred of Ongar, Southern Division of the county of Essex, 6 miles (S. E. by E.) from Epping; containing 297 inhabitants. This parish is of very small extent: within its limits is Suttons, the residence of the family of Smijth. The living is a rectory, with that of Mount-Thoydon united, valued in the king's books at £15. 8. 9.; patron, Sir John Smijth, Bart. The tithes of Stapleford-Tawney have been commuted for a rent-charge of £350, subject to the payment of rates, which on the average have amounted to £66; the glebe comprises $127\frac{1}{2}$ acres, valued at £150 per annum. The church is a small edifice, with a belfry turret and a spire of wood. Here is a National school.

STAPLEGATE, an extra-parochial district, forming the northern suburb of the city of Canterbury, in the union of Blean, hundred of Westgate, lathe of St. Augustine, Eastern Division of the county of Kent; containing 247 inhabitants.

STAPLEGROVE, a parish, in the union of Taunton, hundred of Taunton and Taunton-Dean, Western Division of the county of Somerset, $1\frac{3}{4}$ mile (N. W.) from Taunton; containing 457 inhabitants. This parish was separated from Taunton, and made a distinct parish in 1554. The living is a rectory; patron, Vincent Stuckey, Esq. The tithes have been commuted for a rent-charge of £205, subject to the payment of rates. A gallery has been erected in the church, and 60 free sittings provided, the Incorporated Society having granted £30 in aid of the expense. A school is supported by subscription.

STAPLEHURST (*All Saints*), a parish, in the union of Maidstone, partly in the hundred of Cranbrooke, and partly in that of Marden, Lower Division of the lathe of Scray, Western Division of the county of Kent, 4 miles (N. by E.) from Cranbrooke; containing 1484 inhabitants. The living is a rectory, valued in the king's books at £26. 5. 10.; patrons, Master and Fellows of St. John's College, Cambridge. The tithes have been commuted for a rent-charge of £1199. 15., subject to the payment of rates; the glebe comprises one acre, valued at £1. 10 per annum. There is a place of worship for Independents. A fair for cattle, corn, and hops, is held on Oct. 11th. Two schools are supported here for about £60 per annum, arising from the united bequests of Lancelot Bathurst in 1539, and John Gibbon, Esq., in 1707.

STAPLETON (*St. Mary*), a parish, in the union of Longtown, Eskdale ward, Eastern Division of the county of Cumberland; containing 1097 inhabitants, of which number, 447 are in the township of Stapleton, 10 miles (E. N. E.) from Longtown. The living is a discharged rectory, valued in the king's books at £1. 8. $11\frac{1}{2}$.; present net income, £90; patron, Earl of Carlisle. The church was rebuilt in 1829, and 440 free sittings provided, the Incorporated Society having granted £300 in aid of the expense. The river Line, which is here crossed by a bridge, runs through the parish, and on its northern bank are the ruins of Shank castle. Limestone abounds here, and a colliery has been opened in the vicinity. A school is endowed with £1. 10. per annum.

STAPLETON (*Holy Trinity*), a parish, in the union of Clifton, hundred of Barton-Regis, Western Division of the county of Gloucester, $2\frac{1}{2}$ miles (N. E. by N.) from Bristol; containing 2715 inhabitants. This parish is situated in the north-western angle of the South Gloucester and Somerset coal field, and is bounded on the north side by a range of hills from 150 to 200 feet in height, to which elevation the strata of the coal measures on the south side are lifted up by a mass of mill-stone grit. The turnpike road from Bristol to Wootton-under-Edge and Gloucester passes through the village of Stapleton; and another from the same city to Sodbury, Cirencester, and Oxford, passes through the village of Fishponds. From east to west the parish is traversed by the river Frome, flowing through a richly wooded glen, occasionally interspersed with precipitous rocks; and on the banks of the stream are several mills. Under the northern hills, immediately above the glen, which there become less wild, is the village of Stapleton, with its white-washed church, backed with the woods and pleasure grounds of Heath House, the property of Sir John Smyth, Bart.; and Stoke House, belonging to the Duke of Beaufort, which is finely situated. The parish abounds with valuable Pennant stone, of which considerable quarries are wrought; and there are several coal mines in operation, the produce of which contributes to the supply of Bristol. The manufacture of hats, formerly more extensive, is now on a small scale, and a very considerable number of the inhabitants are employed in the quarries and mines: the soil is a stone-brash. About a mile from Stapleton, on elevated ground rising from the southern bank of the Frome, is the village of Fishponds, so called from two extensive ponds which formerly existed there, and which is now a populous place. In this village is an establishment for the insane, called Fishponds Asylum, one of the first private establishments erected for that purpose. It was built by Dr. Mason in 1770, and has since that time been under the superintendence of Dr. Cox and Dr. Bompas, his lineal descendants: it is a spacious building covering a large plot of ground, in consequence of the system of classification being there carried to the fullest extent. A mild and humane plan of treatment was very early introduced into this establishment, and has been since continued under a conviction of its efficacy. The buildings are surrounded by extensive grounds for exercise, affording various amusements, and to such as are fond of those pursuits, opportunities of gardening and agriculture in the gardens and on the farm adjoining it. Attached to the building is a chapel for religious worship, in which service is regularly performed for the benefit of such invalids as may be in a fit state to attend. There are several respectable houses in the village, which is chiefly inhabited by individuals engaged in the quarries and in agriculture. The living is a perpetual curacy; net income, £112; patron and impropriator, Sir John Smyth, Bart. The church is a small neat edifice, with a tower crowned with pinnacles. A chapel of ease was built in the village of Fishponds by voluntary subscription, in 1817, containing 570 free sittings, the Incorporated Society having granted £500 in aid of the expense: the living is a perpetual curacy; net income, £80; patron, Rev. H. Shute. There are two places of worship for Baptists, and one for Wesleyan Methodists. Mary Webb, in 1729, bequeathed £450, which, with subsequent benefactions, produces an annual income of £42. 10., which is appropriated to the instruction of 30 children, and to the support of three poor women in an almshouse adjoining the school, both of which were founded by the testatrix. Two other schools are supported by subscription. The coal measures abound with specimens of vegetable fossils of the Cactus pine form, and other species. About half-way between Stapleton and Fishponds is an extensive building, formerly occupied as a depôt for French prisoners of war. The late Mrs. Hannah More was born in the house now occupied as the free school; and the Rev. John Foster, author of some Essays, resided in the village of Stapleton.

STAPLETON, a joint township with Frog-street, in the parish of PRESTEIGN, union of KNIGHTON, hundred of WIGMORE, county of HEREFORD; containing, with Frog-street, 156 inhabitants.

STAPLETON, a chapelry, in the parish of BARWELL, hundred of SPARKENHOE, Southern Division of the county of LEICESTER, 3 miles (N. by E.) from Hinckley; containing 249 inhabitants. The chapel is dedicated to St. Martin.

STAPLETON (*ST. JOHN*), a parish, in the union of ATCHAM, hundred of CONDOVER, Southern Division of the county of SALOP, 6 miles (S. by W.) from Shrewsbury; containing 235 inhabitants. The living is a rectory, valued in the king's books at £6. 7. 6.; patron, Hon. W. H. Powys. The tithes have been commuted for a rent-charge of £464, subject to the payment of rates; the glebe comprises 10½ acres, valued at £13. 8. per annum. A day and Sunday school is partly supported by subscription.

STAPLETON, a township, partly in the parish of ST. JOHN-STANWICK, wapentake of GILLING-WEST, but chiefly in the parish of CROFT, wapentake of GILLING-EAST, union of DARLINGTON, North Riding of the county of YORK, 2¼ miles (S. W.) from Darlington; containing 121 inhabitants.

STAPLETON, a township, in the parish of DARLINGTON, Upper Division of the wapentake of OSGOLDCROSS, West Riding of the county of YORK, 4¼ miles (S. E. by E.) from Pontefract; containing 107 inhabitants.

STARBOTTON, a township, in the parish of KETTLEWELL, union of SKIPTON, Eastern Division of the wapentake of STAINCLIFFE and EWCROSS, West Riding of the county of YORK, 17 miles (N. E. by N.) from Settle: the population is returned with the parish.

STARCROSS, a small port and chapelry, in the parish of KENTON, hundred of EXMINSTER, Wonford and Southern Divisions of the county of DEVON, 1½ mile (W. by N.) from Exmouth: the population is returned with the parish. This place is pleasantly situated on the western bank of the river Exe, near its influx into the English Channel, and a little to the south of Powderham Castle, a fortress erected in the reign of Edward I., by Isabella, the last of the powerful family of Rivers. The surrounding scenery is richly diversified, and embellished with the park, pleasure grounds, and plantations, of the castle, which are nearly ten miles in circumference. On an eminence in the neighbourhood is a conspicuous landmark, called Belvidere, erected in 1773 by the Earl of Devon, and consisting of a lofty triangular tower, with an hexagonal turret rising from each of the angles, commanding an extensive view over the channel and the country adjacent. From the advantages of its situation and the beauty of the surrounding scenery, the village is much frequented as a bathing-place. The trade of the port consists principally in the importation of coal and timber, for the landing of which a convenient quay has been constructed. A fair is held annually on the Wednesday in Whitsuntide. The living is a perpetual curacy; net income, £84; patrons and appropriators, Dean and Chapter of Exeter, and Dean and Chapter of Salisbury. The chapel is a modern structure. There is a place of worship for Wesleyan Methodists; and a National school has been established.

STARSTON (*ST. MARGARET*), a parish, in the union of DEPWADE, hundred of EARSHAM, Eastern Division of the county of NORFOLK, 1¼ mile (N. N. W.) from Harleston; containing 449 inhabitants. The living is a rectory, valued in the king's books at £15; patrons, certain Trustees. The tithes have been commuted for a rent-charge of £651, subject to the payment of rates, which on the average have amounted to £132; the glebe comprises 47½ acres, valued at £60 per annum. A school is partly supported by the rector.

STARTFORTH (*HOLY TRINITY*), a parish, in the union of TEESDALE, wapentake of GILLING-WEST,

North Riding of the county of York, $\frac{3}{4}$ of a mile (W. S. W.) from Barnard-Castle; containing 632 inhabitants. The living is a discharged vicarage, valued in the king's books at £4. 0. 10.; present net income, £150; patron, Earl of Lonsdale; impropriator, J. B. S. Morritt, Esq. The church is of great antiquity.

STATFOLD, a parish, in the union of Tamworth, Southern Division of the hundred of Offlow and of the county of Stafford, $3\frac{1}{4}$ miles (N. E.) from Tamworth; containing 41 inhabitants. The living is annexed to the vicarage of St. Mary, Lichfield. The church, now used only as a chapel for interment, contains some ancient monuments. This place, under its ancient name, Stotfold, is one of the prebends in the Cathedral Church of Lichfield, the revenue of which has been formed by act of parliament into a "Fabric Fund," for repairs, &c., no prebendary being appointed.

STATH DIVISION, a tything, in the parish of Stoke-St.-Gregory, hundred of North Curry, Western Division of the county of Somerset: the population s returned with the parish.

STATHERN (*St. Guthlake*), a parish, in the union of Melton-Mowbray, hundred of Framland, Northern Division of the county of Leicester, $8\frac{3}{4}$ miles (N. by E.) from Melton-Mowbray; containing 481 inhabitants. The living is a rectory, valued in the king's books at £16. 3. $1\frac{1}{2}$.; present net income, £566; patrons, Master and Fellows of St. Peter's College, Cambridge. The Grantham canal passes through the parish. There is a trifling endowment for a school, the gift of Joseph Westley, in 1786.

STAUGHTON, GREAT (*St. Andrew*), a parish, in the union of St. Neot's, hundred of Toseland, county of Huntingdon, $3\frac{1}{4}$ miles (S. E. by S.) from Kimbolton; containing 1191 inhabitants. The living is a vicarage, valued in the king's books at £20; present net income, £545; patrons, President and Fellows of St. John's College, Oxford; impropriator, D. Onslow, Esq. There are two small bequests in support of a school, by Lady Elizabeth Conyers, in 1707, and John Poachby, in 1727: the income arising from these bequests and an allotment of twelve acres, under an enclosure act in 1808, in lieu of other land granted for purposes of education, amounts to £18. 10., which is paid to a master for teaching ten poor boys. Under the same act an allotment of 24 acres of the town field was awarded for other parcels of land bequeathed for charitable uses, which produces £50 annually.

STAUGHTON, LITTLE (*All Saints*), a parish, in the union of St. Neot's, hundred of Stodden, county of Bedford, 4 miles (S. by E.) from Kimbolton; containing 455 inhabitants. The living is a rectory, valued in the king's books at £13. 8. 4.; present net income, £200; patrons, President and Fellows of Corpus Christi College, Oxford. There is a place of worship for Baptists.

STAUNTON (*All Saints*), a parish, in the union of Monmouth, hundred of St. Briavells, Western Division of the county of Gloucester, $3\frac{1}{2}$ miles (E. by N.) from Monmouth; containing 204 inhabitants. The living is a discharged rectory, valued in the king's books at £7; present net income, £241; patron, E. Machen, Esq. A National school has been established.

STAUNTON (*St. Mary*), a parish, in the union of Newark, Southern Division of the wapentake of Newark and of the county of Nottingham, $6\frac{3}{4}$ miles (S.) from Newark; containing 173 inhabitants. The living is a rectory, with that of Kilvington consolidated in 1826, valued in the king's books at £16. 13. $11\frac{1}{2}$.; present net income, £322; patron and incumbent, Rev. Dr. Staunton. There is a chapel of ease at Flawborough.

STAUNTON (*St. James*), a parish, in the union of Newent, and forming, with the parishes of Chaseley and Eldersfield, a distinct portion of the Lower Division of the hundred of Pershore, Upton and Western Divisions of the county of Worcester, 6 miles (N. E. by E.) from Newent; containing 348 inhabitants. The living is a rectory, valued in the king's books at £11. 5.; present net income, £404; patron and incumbent, Rev. Thomas Hill. The church is partly in the decorated, and partly in the later, style of English architecture, with a tower and spire. A rental of £30, the produce of several small benefactions, is annually appropriated to the supply of coal, flour, &c., to the poor.

STAUNTON-HARROLD, a chapelry, in the parish of Breedon, union of Ashby-de-la-Zouch, hundred of West Goscote, Northern Division of the county of Leicester, $3\frac{1}{2}$ miles (N. N. E.) from Ashby-de-la-Zouch; containing 342 inhabitants. The chapel, dedicated to the Holy Trinity, is a domestic chapel belonging to Earl Ferrers.

STAUNTON-UPON-WYE (*St. Mary*), a parish, in the union of Weobley, hundred of Grimsworth, county of Hereford, $8\frac{3}{4}$ miles (W. N. W.) from Hereford; containing 544 inhabitants. The living is a rectory, valued in the king's books at £13. 13. 4.; present net income, £340; patron, certain Trustees. Here is a National school supported by endowment.

STAVELEY (*St. John the Baptist*), a parish, in the union of Chesterfield, hundred of Scarsdale, Northern Division of the county of Derby, $4\frac{3}{4}$ miles (N. E. by E.) from Chesterfield; containing 2984 inhabitants. This was for many generations the chief seat of the Frechevilles. In the reign of Charles I., Sir John Frecheville, an active royalist, strongly fortified his house here with twelve pieces of cannon, but capitulated in August 1644. The river Rother runs through the parish; and the Chesterfield canal, and several railroads, pass between it and the neighbouring collieries. A considerable quantity of iron-stone, obtained here, is smelted near the village, where are two blast furnaces. The living is a rectory, valued in the king's books at £12. 7. 6.; present net income, £706; patron, Duke of Devonshire. The church contains 100 free sittings, the Incorporated Society having granted £120 in aid of the expense; there are several monuments of the Frecheville family, and the east window exhibits some stained glass, presented by Lord Frecheville, in 1676. A free grammar school was founded at Netherthorp, in 1537, by Judge Rodes, in support of which, and of two scholarships in St. John's College, Cambridge, he bequeathed £20 per annum. A new house was erected by subscription, in 1804, for the master, whose annual income, including the bequests of Margaret Frecheville, in 1599, Lord James Cavendish, in 1742, and the Rev Francis Gisborne, in 1796, is about £30 a year. There is also a school, in which 22 children are paid for by the parish.

An hospital for four aged persons of each sex was erected at Woodthorpe, in 1632, by Sir Peter Frecheville, who endowed it with £4 per annum to each of the inmates; but Richard Robinson, in 1777, having augmented the original endowment with a donation of £18 per annum, the allowances have since been doubled.

STAVELEY, a chapelry, in the parish of CARTMEL, union of ULVERSTONE, hundred of LONSDALE, north of the sands, Northern Division of the county palatine of LANCASTER, 9 miles (N. E.) from Ulverstone; containing 326 inhabitants. The living is a perpetual curacy; net income, £108; patron, Earl of Burlington.

STAVELEY (*All Saints*), a parish, in the Lower Division of the wapentake of CLARO, West Riding of the county of YORK, 3 miles (S. W. by S.) from Boroughbridge; containing 330 inhabitants. The living is a rectory, valued in the king's books at £8. 17. 11.; present net income, £354; patron and incumbent, Rev. Richard Hartley. The church has been rebuilt. Here is a National school.

STAVELEY, NETHER, a township, in the parish, union, and ward of KENDAL, county of WESTMORLAND, 4¾ miles (N. W. by N.) from Kendal; containing 190 inhabitants.

STAVELEY, OVER, a chapelry, in the parish, union, and ward of KENDAL, county of WESTMORLAND, 5 miles (N. W. by N.) from Kendal; containing 412 inhabitants. The living is a perpetual curacy; net income, £80; patron, Vicar of Kendal. The manufacture of woollens and bobbin is carried on to some extent. Edward III. granted a charter for a market on Friday, and an annual fair on the festival of St. Luke, but both have been long disused. George Jopson, in 1696, gave two tenements, now let for £60 per annum, to the minister, provided he should instruct all the children of the chapelry.

STAVERTON (*St. George*), a parish, in the union of TOTNES, hundred of HAYTOR, Teignbridge and Southern Divisions of the county of DEVON, 3¼ miles (N. by W.) from Totnes; containing 1055 inhabitants. This parish is bounded on the south and south-west by the river Dart, and is famous for its cider. In the neighbourhood are quarries of blue and grey marble, slate, and excellent limestone. The living is a vicarage, in the patronage of the Dean and Chapter of Exeter (the appropriators), valued in the king's books at £32. 14. 9½.; present net income, £366. The church contains some good screen-work. The Rev. Thomas Baker, in 1802, gave £200 for charitable uses, of the proceeds of which £8 is applied to the instruction of young children. A National school has been established.

STAVERTON (*St. John the Baptist*), a parish, in the union of CHELTENHAM, Lower Division of the hundred of DEERHURST, Eastern Division of the county of GLOUCESTER, 4¾ miles (W. by N.) from Cheltenham; containing 245 inhabitants. The living is a vicarage, with Boddington annexed, valued in the king's books at £12; present net income, £436; patron, J. Blagdon, Esq. A school is partly supported by subscription. The Gloucester and Cheltenham railroad passes through the parish.

STAVERTON (*St. Mary*), a parish, in the union of DAVENTRY, hundred of FAWSLEY, Southern Division of the county of NORTHAMPTON, 2 miles (W. S. W.) from Daventry; containing 475 inhabitants. The living is a discharged vicarage; net income, £549; patrons and appropriators, Dean and Canons of Christ-Church, Oxford. The Rev. Francis Baker, in 1767, gave certain land, now producing an annual income of £52, which is applied to the instruction of 25 children, in a house occupied by the master, and purchased with a bequest of £100 by Catherine Burbidge, in 1767. There is also a Sunday school, supported by the annual proceeds of a legacy of £200, bequeathed by the late Rev. Sir John Knightley.

STAVERTON, a chapelry, in the parish of TROWBRIDGE, hundred of MELKSHAM, Westbury and Northern Divisions, and Trowbridge and Bradford Subdivisions, of the county of WILTS, 2½ miles (N.) from Trowbridge, with which the population is returned. The chapel has been enlarged, and 100 free sittings provided, the Incorporated Society having granted £125 in aid of the expense. Here is a large cloth-manufactory.

STAWELL, a chapelry, in the parish of MOORLINCH, union of BRIDG-WATER, hundred of WHITLEY, Western Division of the county of SOMERSET, 5 miles (E. by N.) from Bridg-water; containing 214 inhabitants.

STAWLEY (*St. Michael*), a parish, in the union of WELLINGTON, hundred of MILVERTON, Western Division of the county of SOMERSET, 3 miles (S. W. by S.) from Wiveliscombe; containing 180 inhabitants. The living is a discharged rectory, valued in the king's books at £8. 8. 6½.; present net income, £150; patron, Richard Harrison, Esq.

STAXTON, a township, in the parish of WILLERBY, wapentake of DICKERING, East Riding of the county of YORK, 6½ miles (S. by W.) from Scarborough; containing 260 inhabitants. There is a place of worship for Wesleyan Methodists.

STAYLEY, a township, in the parish of MOTTRAM-IN-LONGDEN-DALE, union of ASHTON-UNDER-LINE, hundred of MACCLESFIELD, Northern Division of the county of CHESTER, 1 mile (N. E.) from Ashton-under-Line; containing, with a part of the town of Stalybridge (which is described under its own head), 2440 inhabitants. The Huddersfield canal passes through the township. Here is a Sunday National school.

STAYTHORPE, a township, in the parish of AVERHAM, union of SOUTHWELL, Northern Division of the wapentake of THURGARTON, Southern Division of the county of NOTTINGHAM, 4 miles (W.) from Newark; containing 61 inhabitants.

STEAN (*St. Peter*), a parish, in the union of BRACKLEY, hundred of KING'S-SUTTON, Southern Division of the county of NORTHAMPTON, 2¾ miles (N. W.) from Brackley; containing 24 inhabitants. The living is a discharged rectory, united to that of Hinton-in-the-Hedges, and valued in the king's books at £5. 9. 7.

STEARSBY, a hamlet, in the parish of BRANSBY, union of EASINGWOULD, wapentake of BULMER, North Riding of the county of YORK, 7½ miles (E. N. E.) from Easingwould: the population is returned with the parish.

STEBBING (*St. Mary*), a parish, in the union of DUNMOW, hundred of HINCKFORD, Northern Division of the county of ESSEX, 3¼ miles (N. E. by E.) from Great Dunmow; containing 1434 inhabitants. This parish is about twenty-five miles in circumference; the surface is elevated, and the soil for the greater portion dry and fertile; it is intersected by a stream on which are several mills, and there are two artificial mounts, on one of which there is said to have been formerly a castle. The village is pleasantly situated, and contains several well-built houses. A fair for cattle and fat calves is held on the 10th of July. The living is a vicarage, valued in the king's books at £12; present net income, £203: the patronage and impropriation belong to Mrs. Batt. The church is a spacious and lofty structure, situated on an eminence, and contains 140 free sittings, the Incorporated Society having granted £30 in aid of the expense. There is a place of worship for Independents. Some small bequests have been made for teaching poor children; and a National school was established in 1832, and is chiefly supported by subscription.

STEDHAM (*St. James*), a parish, in the union of MIDHURST, hundred of EASEBOURNE, rape of CHICHESTER, Western Division of the county of SUSSEX, 2 miles (W. N. W.) from Midhurst; containing 494 inhabitants. This parish is intersected by the river Rother, and under the Reform Act is partly within the new boundary of the borough of Midhurst. The living is a rectory, with that of Heyshot united, valued in the king's books at £17. 18. $6\frac{1}{2}$.; present net income, £386; patron, Rev. L. V. Harcourt. The church consists of a nave and chancel, with a tower rising from the centre; in the churchyard is a fine old yew tree, 28 feet in girth about 3 feet from the ground.

STEEL, a joint township with Prees, in the parish of PREES, Whitchurch Division of the hundred of NORTH BRADFORD, Northern Division of the county of SALOP, 3 miles (S.) from Whitchurch: the population is returned with Prees.

STEEP, a parish, in the union of PETERSFIELD, hundred of EAST MEON, Petersfield and Northern Divisions of the county of SOUTHAMPTON, $1\frac{3}{4}$ mile (N.) from Petersfield; containing 835 inhabitants. The living is annexed, with that of Froxfield, to the vicarage of East Meon. The church contains 117 free sittings, the Incorporated Society having granted £70 in aid of the expense.

STEEP-HOLMES ISLAND, in the parish of UPHILL, hundred of WINTERSTOKE, Eastern Division of the county of SOMERSET, 2 leagues (W. by N.) from Uphill. This island is a vast rock, about a mile and a half in circumference, rising perpendicularly out of the Bristol Channel to the height of 400 feet above the level of the sea, and inaccessible at all points except two. A few rabbits burrow here, and great numbers of sea-fowl build their nests within the recesses of its overhanging cliffs. A house was erected in 1776, for the accommodation of fishermen, who occasionally make this island their resort. A priory is supposed to have been founded here, about the reign of Edward II., by Maurice Lord Berkeley.

STEEPING, GREAT (*All Saints*), a parish, in the union of SPILSBY, Wold Division of the wapentake of CANDLESHOE, parts of LINDSEY, county of LINCOLN, 3 miles (E. S. E.) from Spilsby; containing 281 inhabitants. The living is a discharged vicarage, united to the rectory of Firsby, and valued in the king's books at £7. 18. 4. There is a place of worship for Wesleyan Methodists. This parish is bounded on the south by the river Steeping. The remains of an old mansion, surrounded by a moat, are now occupied as a farm-house, and a moated enclosure in the neighbourhood is said to have been the site of a monastery.

STEEPING, LITTLE (*St. Andrew*), a parish, in the union of SPILSBY, Eastern Division of the soke of BOLINGBROKE, parts of LINDSEY, county of LINCOLN, $3\frac{3}{4}$ miles (S. E. by E.) from Spilsby; containing 263 inhabitants. The living is a discharged rectory, valued in the king's books at £9. 19. 4.; present net income, £181; patron, Lord Willoughby de Eresby. A school is supported by an allowance of £10 per annum out of some charity lands in the parish, and a contribution of £15 from the rector.

STEEPLE (*St. Michael*), a parish, in the union of WAREHAM and PURBECK, hundred of HASILOR, Wareham Division of the county of DORSET, $4\frac{1}{4}$ miles (W. by S.) from Corfe-Castle; containing 237 inhabitants. The living is a rectory, with that of Tyneham, united by act of parliament in the 8th of George I., and valued in the king's books at £9. 15. 5.; present net income, £382; patron, John Bond, Esq. The church has a plain but lofty tower. Two schools are partly supported by the minister. The hamlet of West Creech, in this parish, formerly belonged to the abbey of Bindon, and had the privilege of a market and fair, granted by Henry III.

STEEPLE (*St. Lawrence and All Saints*), a parish, in the union of MALDON, hundred of DENGIE, Southern Division of the county of ESSEX, 3 miles (N. W. by N.) from Southminster; containing 497 inhabitants. This parish, which includes the island of Ramsey, is bounded on the north by the river Blackwater, on which a quay, the property of the governors of St. Bartholomew's Hospital, has been constructed, at which barges of 70 tons deliver cargoes of chalk, and return laden with corn: the hamlet of Stanesgate, in this parish, is nearly surrounded by water. Fairs are held in the village on the Wednesday in Whitsun-week, and the Wednesday after Michaelmas-day. The living is a discharged vicarage, with Stanesgate, valued in the king's books at £15. 18.; present net income, £195; patrons, alternately, Sir Brook W. Bridges, Bart., and another; the impropriation belongs to J. K. and T. Hunt, Esqrs., and Miss Hunt. The church is an ancient edifice. Here is a National school. At Stanesgate a priory of Cluniac monks, subordinate to that of Lewes, existed before 1176: it was dedicated to St. Mary Magdalene, and at the dissolution had a revenue of £38. 18. 3.: the only remains now form part of the walls of a barn.

STEEPLE-ASHTON, county of WILTS.—See ASHTON, STEEPLE.—*And all places having a similar distinguishing prefix will be found under the proper name.*

STEEPLETON-IWERNE, or PRESTON (*St. Mary*), a parish, in the union of BLANDFORD, hundred of PIMPERNE, Blandford Division of the county of DORSET, $4\frac{1}{4}$ miles (N. N. W.) from Blandford-Forum; containing 36 inhabitants. The living is a discharged rectory, valued in the king's books at £6. 18. 4.; present net income, £30; patron, Lord Rivers.

STEETON, a township, in the parish of BOLTON-PERCY, Ainsty of the city, and East Riding of the county, of YORK, $3\frac{1}{2}$ miles (E. by N.) from Tadcaster; containing 85 inhabitants.

STEETON, a joint township with Eastburn, in the parish of KILDWICK, union of KEIGHLEY, Eastern

Division of the wapentake of STAINCLIFFE and EWCROSS, West Riding of the county of YORK, 2¾ miles (N. W.) from Keighley; containing, with Eastburn, 859 inhabitants. There is a place of worship for Wesleyan Methodists.

STELLA, a township, in the parish of RYTON, union of GATESHEAD, Western Division of CHESTER ward, Northern Division of the county palatine of DURHAM, 7½ miles (W. by N.) from Gateshead; containing 482 inhabitants. A considerable English army was defeated here, Aug. 28th, 1640, by the Scots, who passed the Tyne under cover of several pieces of cannon, which they had planted in Newburn church. The navigable river Tyne runs past the village, where is a wharf, belonging to the London Lead Company, and a coal-staith. There is a Roman Catholic chapel at Stella Hall; and a school for Roman Catholic children is supported by subscription.

STELLING (*ST. MARY*), a parish, in the union of ELHAM, partly in the hundred of LONINGBOROUGH, but chiefly in that of STOUTING, lathe of SHEPWAY, Eastern Division of the county of KENT, 6 miles (S. by W.) from Canterbury; containing 313 inhabitants. The living is annexed to the rectory of Upper Hardres. The tithes have been commuted for a rent-charge of £257. 10., subject to the payment of rates. The parish is bounded on the west by the ancient Stane-street.

STELLING, a township, in the parish of BYWELL-ST.-PETER, union of HEXHAM, Eastern Division of TINDALE ward, Southern Division of the county of NORTHUMBERLAND, 8½ miles (E. by N.) from Hexham; containing 17 inhabitants.

STENIGOT (*ST. NICHOLAS*), a parish, in the union of LOUTH, Northern Division of the wapentake of GARTREE, parts of LINDSEY, county of LINCOLN, 5¾ miles (S. W. by W.) from Louth; containing 89 inhabitants. The living is a discharged rectory, valued in the king's books at £7. 12. 3½.; present net income, £222; patron and incumbent, Rev. M. Alington.

STENSON, a township, in the parish of BARROW, union of BURTON-UPON-TRENT, hundred of APPLETREE, Southern Division of the county of DERBY, 4¼ miles (S. S. W.) from Derby: the population is returned with the chapelry of Twyford. The inhabitants are entitled to the benefit of Alsop's school at Findern.

STEPHENS, ST., a parish, including the disfranchised borough of Newport, in the Northern Division of the hundred of EAST, Eastern Division of the county of CORNWALL, ¾ of a mile (N. N. W.) from Launceston; containing 1084 inhabitants. The living is a perpetual curacy; net income, £80: it is in the patronage of Feoffees and the Inhabitants. The church is an ancient structure, and contains some interesting details. This place derived its name from a collegiate church, dedicated to St. Stephen, which was founded prior to the Norman Conquest, for canons of the Augustine order, and which was subsequently occupied by Secular canons till 1126, when they were removed by Warlewast, Bishop of Exeter, to a priory which he had founded in the parish of St. Thomas. The village is pleasantly situated on the brow of a lofty hill immediately above the town of Newport, and commands some extensive views and interesting scenery. Fairs, chiefly for cattle, are held on May 12th, July 31st, and September 25th. John Horwell, in 1717, bequeathed property for boarding, clothing, instructing, and apprenticing six poor boys, which, in 1821, after paying for the auction of the premises, produced £6444, which was vested in the 3 per cent. consols., yielding an income of £192. 13. 4.

STEPHENS, ST., a parish, in the hundred of CASHIO, or liberty of ST. ALBANS, county of HERTFORD, 1 mile (S. W.) from St. Albans; containing 1746 inhabitants. The living is a discharged vicarage, valued in the king's books at £15; patron and incumbent, Rev. M. R. Southwell; impropriator, Rev. C. Lomax. The impropriate tithes have been commuted for a rent-charge of £1420. 9., and the vicarial for £500, subject to the payment of rates, which on the average on the latter have amounted to £24. The church, situated on the Roman Watling-street, occupies the site of that built in the reign of King Eldred, by Ulsinus, sixth abbot of St. Albans. A fine brass eagle with expanded wings, on an ornamented pedestal of the same metal, was dug up some years ago in the churchyard, and is now used as a stand in the chancel for Fox's Martyrology. A school is partly supported by subscription. The rivers Ver and Colne run through the parish, in which many Roman coins have been found.

STEPHENS, ST., county of KENT.—See HACKINGTON.

STEPHENS (ST.) BY SALTASH, a parish, in the union of ST. GERMANS, Southern Division of the hundred of EAST, Eastern Division of the county of CORNWALL, 1 mile (W. by S.) from Saltash; containing 3092 inhabitants. The living is a vicarage, valued in the king's books at £26; present net income, £139; patrons and appropriators, Dean and Canons of Windsor. Here is a National school. In this parish are considerable remains of the ancient castle of Trematon, erected before the Conquest, in a beautiful situation on the banks of the Lyner: the area covered more than an acre of ground, and was enclosed by embattled walls; the keep is situated on the summit of a conical elevation, and is approached by a circular arched doorway; the principal gateway consists of three arches, supporting a square embattled tower, containing a museum for natural curiosities.

STEPHENS (ST.) IN BRANNEL, a parish, in the union of ST. AUSTELL, Eastern Division of the hundred of POWDER and of the county of CORNWALL, 4½ miles (W. by N.) from St. Austell; containing 2477 inhabitants. This parish, which takes its name from the dedication of its church, is situated in the heart of a district abounding with mineral treasure. Several tin and copper mines have been opened in the vicinity, but they have not been found sufficiently productive to remunerate the proprietors, and have been discontinued, with the exception of one called the Strawberry mine, which is still worked, but not on a large scale. Moorstone of excellent quality, which is extensively used in building, is found in the parish, which also abounds with a fine white clay, procured in great quantities for the potteries. In the clay works are found a species of black spar, and some beautiful transparent chrystals of regular polygonal form, known by the appellation of Cornish diamonds. The living is a rectory, annexed, with the rectory of St. Dennis, to that of St.-Michael-Caerhays. The tithes have been commuted for a rent-charge of £780, subject to the payment of rates. The church is an ancient structure, principally in the Norman style of architecture, with some later details, and a square detached tower. There is a place of worship for Independents. A school for the gratuitous instruction of poor children is endowed with a small annnal income: there is also a National school. In 1711, Ellen Mabbott bequeathed a rent-charge of £35. 10. for poor widows not receiving parochial relief; and in 1726, James Buller endowed four almshouses for poor people. There are vestiges of a circular intrenchment comprising an area of about one acre, surrounded with a fosse.

STEPNEY (*St. Dunstan and All Saints*), a parish, and the head of a union, in the Tower Division of the hundred of Ossulstone, county of Middlesex, $2\frac{1}{2}$ miles (E.) from London; comprising the hamlets of Mile-End New Town, Mile-End Old Town, and Ratcliffe, and containing 51,023 inhabitants. This parish, called in various ancient records *Stebunhithe* and *Stebenhythe*, occurs in Domesday-book under the name *Stibenhede*, from which its present appellation is obviously deduced. It anciently included a widely-extended district, comprising, in addition to its present parochial limits, the hamlets of Stratford-le-Bow, Limehouse, Poplar and Blackwall, Shadwell, St. George's-in-the-East, Wapping, Spitalfields, Whitechapel, and Bethnal-Green, which, from their increased extent and importance, have been successively separated from it, and erected into distinct parishes, at present constituting some of the most populous districts in the vicinity of the metropolis. According to Stowe, Edward I. held a parliament at Stepney, in the mansion of Henry Walleis, mayor of London, in which he conferred several valuable privileges on the citizens. The manor was, in 1380, annexed to the see of London, and the bishops had a palace, called Bishop Hall, now included in the parish of Bethnal-Green, in which they continued to reside till 1550, when it was alienated from the see by Bishop Ridley, who gave it to Edward VI. In the rebellion under Jack Cade, in the reign of Henry VI., the insurgents who attacked the metropolis encamped for some time at the hamlet of Mile-End; and, in 1642, at the commencement of the parliamentary war, fortifications were constructed in this parish for the defence of the city.

From the then pleasantness of its situation, and the beauty of its scenery, which are noticed in a letter from Sir Thomas More to Dean Colet, Stepney was formerly the favourite residence of many persons of distinction. Isabel, Countess of Rutland, had a seat here in the latter part of the sixteenth century, at which time Sir Thomas Lake, secretary of state in the reign of James I., was also a resident; but there are no vestiges of the houses which they occupied. Henry, the first Marquess of Worcester, had a mansion near the parsonage-house, of which the gateway, handsomely built of brick, with a turret at one of the angles, is still remaining, and forms part of a house wherein Dr. Richard Mead was born, and where he resided for many years: the site of the ancient mansion is now occupied by an academy for the education of young men intended for ministers of the Baptist denomination. Sir Henry Colet, father of Dean Colet, the founder of St. Paul's school, lived in a spacious residence to the west of the church, called the Great Place, the site being now partly occupied by a place of public entertainment, called Spring Gardens: on two sides of the pleasure grounds, traces of the moat that surrounded the ancient mansion are still discernible. During part of the seventeenth century, Stepney suffered severely from the ravages of the plague, of which 2978 persons died in the year 1625; and in the year 1665, not less than 6583. In the course of the latter year, 116 sextons and grave-diggers belonging to this parish, died of the plague, and so greatly was the parish, then principally inhabited by sea-faring men, depopulated, that it is recorded, in the life of Lord Clarendon, that "there seemed an impossibility to procure seamen to fit out the fleet." In 1794, a most calamitous and destructive fire, occasioned by the boiling over of a pitch kettle in a barge-builder's yard, broke out, and consumed more than half of the hamlet of Ratcliffe, communicated to the shipping in the river, and destroyed several ranges of warehouses, among which was one belonging to the East India Company, containing more than 200 tons of saltpetre. Of 1200 houses then in that hamlet, only 570 escaped the conflagration; and 36 warehouses, chiefly stored with articles of combustion, were totally consumed. By this dreadful calamity several hundred families were reduced to the utmost distress, deprived of shelter, and made dependent for subsistence on the public benevolence. One hundred and fifty tents, furnished by government from the Tower, were pitched for their reception in an enclosed piece of ground near the churchyard, and provisions were daily supplied to them from the vestry-room of the church. A public subscription was opened at Lloyd's coffee-house, by which, together with the contributions of thousands who came to visit the extensive ruins caused by this desolating conflagration, more than £16,000 was collected for the relief of the sufferers.

The parish is situated on the northern bank of the Thames, and is chiefly inhabited by persons connected with the shipping: it extends for a considerable distance from the river to the principal road leading into Essex, and comprises many handsome ranges of building. The Commercial road, leading from Whitechapel to the East and West India docks, passes through the parish. On the south side of this road, a tram-road has been laid down; and the London and Blackwall railway is now in course of formation in connection with the docks. The basin, or dock, at the junction of the Regent's canal with the Thames, capable of containing 100 ships, occupies a portion of the east side of the hamlet of Ratcliffe. The parish is paved, lighted with gas, and supplied with water by the East London Company, from their works at Old Ford, the reservoir of which, excavated in 1827, and covering ten acres of ground, is situated to the north of the high road. On the banks of the Regent's canal, which crosses the Mile-End road under a stone bridge, are several coal and timber wharfs; in the hamlets of Mile-End Old Town and Mile-End New Town are some extensive breweries, a large distillery, an extensive floor-cloth manufactory, a manufactory for tobacco-pipes, and a very spacious nursery-ground; in the hamlet of Ratcliffe there are extensive manufactories for sail-cloth, sails, chain-cables, and mooring-chains, steam-engines, and machinery connected with the docks and shipping, and large establishments belonging to coopers for the West India trade, timber and hoop merchants, ship chandlers, sugar-bakers, rope-makers, and various other trades, for which its situation renders it peculiarly favourable. The market, granted by

Charles II., in 1664, is now held at Whitechapel; and the fair, granted at the same time, and originally held on Mile-End green, was afterwards removed to Stratford-le-Bow, and subsequently suppressed.

Stepney is within the jurisdiction of the county magistrates, who sit at the police-office in Lambeth-street, Whitechapel, for the despatch of business relating to the hamlets of Mile-End Old and New Towns; and at the Thames police-office, Wapping, for the hamlet of Ratcliffe. Its local affairs are under the superintendence of twelve trustees, who, pursuant to the provisions of an act passed in 1810, are annually elected by the inhabitants. It is within the limits of the new police establishment; and under the jurisdiction of the court of requests for the "Tower Hamlets," held in Osborne-street, Whitechapel, for the recovery of debts under £5. A notion has for many years been very generally entertained, that all persons born at sea are, from that circumstance alone, parishioners of Stepney: to counteract the influence of this error, which has subjected the parish to serious expense, the overseers, in 1813, applied for a criminal information against the magistrate of the county of Chester, for having removed a vagrant, who stated that he was born at sea, from the parish of Stockport, to Stepney. On this occasion Lord Ellenborough observed, that this was a delusion, and in the hope that the promulgation of his lordship's decision, that "certainly it must be understood, that all these sea-born persons are not to be marched off, at the pleasure of the magistrate, to the parish of Stepney," would produce the desired effect, the overseers forebore to press further proceedings.

The church of Stepney, together with the manor, was appropriated to the see of London, in 1380, and the bishops of that diocese appointed to the rectory, which was a sinecure, the rectors being patrons of the vicarage: in 1544, the great tithes were impropriated, and the impropriator presented both to the rectory and to the vicarage; in 1708, they were purchased by the Principal and Fellows of Brasenose College, Oxford, which purchase was afterwards confirmed by act of parliament, and they were annexed to the vicarage, subject to an annual payment of £40 to the college, and divided into moieties, of which the incumbents were styled portionists of Church-Stepney and Spitalfields-Stepney. After the separation of the several parishes, and the consequent diminution of the value of the benefice, the arrangement was altered, and the living became vested in one person. It is a rectory, valued in the king's books at £73. 6. 8.; present net income, £1190; patrons, Principal and Fellows of Brasenose College, Oxford. The church is a spacious structure of flint and stone, principally in the later style of English architecture, with a low broad tower, strengthened with buttresses, and surmounted by a turret crowned with a small dome. Near the western entrance is a bas-relief, indifferently executed and much decayed, representing the Virgin and Child, with a female figure in the attitude of supplication; and over the south door is a rudely-sculptured representation of the Crucifixion, in tolerable preservation. The nave is separated from the aisles by clustered columns and pointed arches; and on the south side of the chancel are two arched recesses. There are many ancient monuments in the church: on the north side of the chancel is the altar-tomb of Sir Henry Colet, Knt., under an arched canopy, finely groined, and near it a monument to Benjamin Kenton, Esq., on which is a finely sculptured representation of the Good Samaritan, by Westmacott: this benevolent individual, who died in 1800, at the advanced age of 83, bequeathed to different charitable institutions the sum of £63,550. On the east wall is a monument to Lady Dethic, and on the south a tablet to Sir Thomas Spert, Knt., founder and first master of the corporation of the Trinity: the church was repaired and beautified in 1828. The churchyard is spacious, and there are numerous monuments to distinguished persons who have been buried here, among whom were, the Rev. Matthew Mead, who was ejected from the living of Shadwell for non-conformity; Admiral Sir John Leake, Knt., a distinguished officer in the reign of Queen Anne; and various others. St. Philip's church, behind the London Hospital, a handsome edifice in the later English style, with minarets at the angles crowned with finials, was erected in 1822, towards the expense of which the parliamentary commissioners granted the sum of £3500: the living is a perpetual curacy; net income, £218; patrons, Principal and Fellows of Brasenose College, Oxford. A district has been assigned to this church under the 59th of George III., cap. 134. St. Thomas's church, in Arbour-square, a neat edifice of Suffolk brick, in the early English style, with two octangular turrets, was erected in 1837 by grant from the metropolis church-building fund: it contains 1100 sittings, of which 500 are free. St. Peter's church, in Mile-End Old Town, also a neat edifice of Suffolk brick, in the early Norman style, was was erected in 1839 by a grant from the same fund; it contains 1400 sittings, of which 750 are free. St. James's church, at Ratcliffe, a neat edifice of brick, in the early English style, with a tower and spire, was erected in 1838, by grant from the commissioners for building churches; it contains 1200 sittings, of which 350 are free. The livings of these three new churches are perpetual curacies, in the patronage of the Rector. There are places of worship for Baptists, the Society of Friends, those in the connection of the late Countess of Huntingdon, Calvinistic Methodists, and three for Independents, of one of which, near the church, founded by the lecturer, the Rev. William Greenhill, and built in 1674, the Rev. Matthew Mead became the first minister. Stepney College, in Mile-End Old Town, was founded in 1810 for the education of ministers of the Baptist denomination; the premises, which have been greatly enlarged, include part of an ancient building called King John's Tower; they contain private studies and sleeping-rooms for twenty-four students, with apartments for the head and other masters, and a chapel in which the students officiate.

The charity schools at Ratcliffe were established in 1710, and the school-house was erected in 1719; the endowment of this institution, originally for the clothing and instruction of 30 boys and 20 girls, consists of an estate at Edmonton, given to the school by Mr. Wakeling; a legacy of £500, by Edward Turner, Esq., and other benefactions amounting in the whole to £2000: the school was enlarged in 1814, and is conducted on the National system; 40 boys and 25 girls are annually clothed in green, and apprenticed. The charity school at Mile-End Old Town was established by subscription in 1714, and has been subsequently endowed with various benefactions, producing £150 per annum: it is on the National plan, and 60 boys and 40 girls are clothed; a school-room for the girls, and other apartments, were erected at Stepney Green, in 1786, behind which is a school for the boys. The Stepney Meeting charity school, in which 130 boys and 70 girls are instructed, partly on the National and partly on the Lancasterian plan, was founded in 1783, and the present building erected in 1828: it has an endowment of £188 per annum, arising from various benefactions, and is further supported by subscription. The charity school for Mile-End New Town was established by subscription, in 1785, for the instruction of 30 boys and 30 girls; the permanent income arises from £715, vested in the four per cents., the deficiency being supplied by subscription. For want of room in the parish church, the scholars were allowed to attend public worship in the Methodist meeting-house; and the school, together with its funds, is supposed to have merged into the school belonging to that place of worship. There are also Sunday schools in connection with the established church and the dissenting congregations.

In School-house-lane, Ratcliffe, are the almshouses of the Coopers' Company, founded in 1538 by Toby Wood, Esq., and Mr. Cloker, members of that company, for fourteen aged persons of both sexes. Adjoining them is a free grammar school, largely endowed by Nicholas Gibson, Esq., master of the company, and sheriff of London, in the reign of Henry VIII., for the instruction of 35 boys; in this school Bishop Andrews, and several other distinguished persons, received the rudiments of their education. These premises were destroyed by the fire of 1794, but were rebuilt in 1796, and the almshouses more liberally endowed by the company; they now afford an asylum to six men and eighteen women. The buildings occupy three sides of a quadrangle, with a chapel in the central range. The almshouses belonging to the Vintners' Company were taken down and rebuilt in 1802, upon a larger scale, in appropriation of a bequest of £2250 by Mr. Benjamin Kenton of Stepney: they consist of twelve separate tenements and a chapel, and are endowed for twelve widows of freemen of the Vintners' Company, who receive £36 per annum each; a chaplain performs divine service weekly, and has a stipend of £52. 10. per annum. The almshouses erected by the Brethren of the Trinity House comprise twelve sets of apartments, with a handsome chapel in the centre, in the front windows of which are some armorial bearings in stained glass. Francis Bancroft, in 1727, bequeathed in trust to the Drapers' Company property then worth £28,000 for the erection and endowment of 24 almshouses for aged men, members of that company, and a school for 100 boys, who are also boarded and clothed; the present income, arising from £40,800 three per cent. consols., £33,400 three per cent. reduced annuities, and from landed property, exceeds £4000 per annum; the almsmen receive each £20 per annum, and a chaldron and a half of coal, with a gown every third year. The buildings, in the Mile-End road, consist of two ranges of houses occupying two sides of a quadrangular area, of which the school-room, chapel, and other apartments form the third side. A chaplain, who is appointed by the Master and Wardens of the Drapers' Company, has an annual stipend of £31. 10. Mr. John Fuller, in 1592, founded twelve almshouses, which he endowed with £50 per annum, for twelve aged and unmarried men. Near the churchyard are the Mercers' almshouses, founded in 1691, by Dame Jane Mico, relict of Sir Samuel Mico, which she endowed for ten aged widows, who receive each £30 per annum. Mrs. Bowry, in 1715, bequeathed a leasehold estate and a sum of money in the South Sea annuities, amounting to £2636. 13., for the erection and endowment of eight almshouses between Mile-End and Stratford-le-Bow, for decayed seamen and their widows, of this parish. Capt. James Cook, and his widow, Dame Alice Row, founded four almshouses in the Grove-road, Mile-End, for widows of seamen of Stepney, to which the hamlet of Mile-End exclusively presents, on condition of keeping them in repair. Eight almshouses were erected in Mile-End Old Town, in 1698, by Mr. John Pennell, who endowed them for aged widows. The East London Institution for lying-in women, established by subscription, is well supported and judiciously regulated. At Mile-End Old Town is the Jews' hospital for aged poor, and for the education and employment of children, founded in 1806, and enlarged in 1818: the building, which is on the south side of the road, is spacious, and ornamented in front with a pediment supported by Ionic pilasters. Nearly opposite is the hospital for Spanish and Portuguese Jews, established in 1747, for the reception of sick poor and lying-in women, and intended also as an asylum for the aged and infirm. On the north side of the high road are two spacious cemeteries belonging to the Portuguese Jews, and a third for German or Dutch Jews, in which several of the Rabbins, and other eminent individuals of that class, are interred. The poor law union of Stepney comprises Limehouse, Shadwell, Mile-End Old Town, Ratcliffe, and Wapping, under the care of 23 guardians, and, according to the census of 1831, contains a population of 72,446.

STEPNEY MARSH, county of MIDDLESEX.—See DOGS, ISLE OF.

STEPPINGLEY (*St. Lawrence*), a parish, in the union of AMPTHILL, hundred of REDBORNESTOKE, county of BEDFORD, 2½ miles (S. W. by S.) from Ampthill; containing 348 inhabitants. The living is a discharged rectory, valued in the king's books at £6. 16. 3.; present net income, £289; patron, Duke of Bedford.

STERNDALE, EARL, a chapelry, in the parish of HARTINGTON, hundred of WIRKSWORTH, Northern Division of the county of DERBY, 5½ miles (S. E. by S.) from Buxton; containing 354 inhabitants. The living is a perpetual curacy; net income, £96; patron, Vicar of Hartington; impropriator, Duke of Devonshire. The chapel was erected in 1829, and contains 70 free sittings, the Incorporated Society having granted £70 in aid of the expense. The Peak Forest and Cromford rail-road passes through the chapelry. Four poor children are instructed for £2 a year, the gift of James Hill, in 1712.

STERNFIELD (*St. Mary Magdalene*), a parish, in the union and hundred of PLOMESGATE, Eastern Division of the county of SUFFOLK, 1½ mile (S. S. E.) from Saxmundham; containing 203 inhabitants. The living is a rectory, valued in the king's books at £8. 14. 4½.; present net income, £297; patron, William Long, Esq. A school is partly supported from the poor's rate.

STERSCOTE, county of STAFFORD.—See SYERSCOTE.

STERT (*St. James*), a parish, in the union of DEVIZES, hundred of SWANBOROUGH, Devizes and Northern Divisions of the county of WILTS, 2 miles (S. E.) from Devizes; containing 185 inhabitants. The living is annexed to the vicarage of Urchfont.

STETCHWORTH (*St. Peter*), a parish, in the union of NEWMARKET, hundred of RADFIELD, county of CAMBRIDGE, 2¾ miles (S. by W.) from Newmarket; containing 545 inhabitants. The living is a discharged vicarage, valued in the king's books at £10. 12. 1.; present net income, £174; patron and impropriator, Richard Eaton, Esq. The church contains a handsome monument in white marble, to the memory of the Hon. Henry Gorges. An almshouse for two poor persons of each sex was founded here in 1700, by Lord and Lady Gorges, who endowed it with £30 per annum.

STEVENAGE (*St. Nicholas*), a market-town and parish, in the union of HITCHIN, hundred of BROADWATER, county of HERTFORD, 12 miles (N. W. by N.) from Hertford, and 31 (N. N. W.) from London; containing 1859 inhabitants. The ancient name of this town was *Stigenhaght*, signifying the hills by the highway, evidently derived from six barrows, or hills, near the road side, half a mile south of the town: about the time of the Heptarchy it was called *Stigenhace*, and in Domesday-book *Stevenach*, or *Stevenadge*. It formed a part of the demesne of the Saxon kings, and was given, by Edward the Confessor, to the abbey of Westminster, on the suppression of which it was granted, by Edward VI., to the see of London, to which the manor still belongs. The town is pleasantly situated on the great North road from London to Edinburgh, and consists principally of one long and spacious street, with two or three smaller ones, comprising some well-built brick residences, and is amply supplied with water. The trade is principally that of carcass butchers, who dispose of the slaughtered cattle principally at Hertford, and in the London market. The platting of straw furnishes employment to many of the females in the town and its vicinity. In the reign of James I., Monteine, Bishop of London, procured the grant of a weekly market, and three fairs annually, which was confirmed, with liberty to alter the market day, by a charter of William and Mary; but, from the contiguity of other towns, in which large markets are held, that of Stevenage has fallen into disuse; and the fairs, except one on September 22nd, have also been nearly discontinued. Petty sessions for the division are held here, and a manorial court annually by the Bishop of London. The town has been made a polling-place for the county.

The living is a rectory, valued in the king's books at £33. 6. 8.; patron, William Baker, Esq. The tithes have been commuted for a rent-charge of £1023. 7. 3., subject to the payment of rates; the glebe comprises 26½ acres, valued at £43 per annum. The church is situated on a chalky eminence about half a mile from the town, approached by a fine avenue of trees; it is a neat well-built edifice, with a square tower at the west end, surmounted by a spire covered with lead; attached to the chancel are two small chapels. There are places of worship for Independents and Wesleyan Methodists. The Rev. Thomas Alleyn, in 1558, devised all his estates to the Master and Fellows of Trinity College, Cambridge, in trust for certain charitable uses, among which was the founding of free grammar schools at Stevenage, Uttoxeter, and Stowe, each with an annual income of £13. 6. 8. Shortly after the testator's demise, a free English school was founded by the inhabitants of Stevenage, and subsequently endowed with some land by Robert Gynne, in 1614, and a rent-charge of £12 by Edmond Woodward, in 1659; and this school was eventually placed under the management of the master of the grammar school, who is appointed by the Master and Fellows of Trinity College: the present income, arising from the above endowments, is £37. 6. 8. A National school also has been established. An almshouse for three poor persons, called "All Christians' Souls' House," was founded by Stephen Hellard, in 1501; the present rental of the charity is £31. 9., which is applied to the relief of poor persons, and the almshouse, having been let, is now occupied as a dwelling-house. There are various other bequests, amounting to about £20 per annum, distributed among the poor. Henry Trigg, an eccentric inhabitant of this town, by his will, dated in 1724, directed his body after death to be deposited on a floor, to be erected in one of the outbuildings of his house, leaving his property to his brother on condition that he complied with this direction, which was accordingly done, the corpse still remaining where it was deposited. The six barrows supposed to give name to the town have been conjectured to be sepulchral monuments, although in those that have been opened no human remains were discovered. It is generally supposed that they were erected by the Danes, several battles having been fought between them and the Saxons in this county; some fields, at the distance of about three-quarters of a mile, still retaining the name of Danes' Blood. In a wood about half a mile eastward of the barrows, called Humbley wood, are the apparent remains of an intrenched camp, or fortification, of unknown construction, consisting of a large and perfectly square area, surrounded with a deep moat containing water, with only one entrance on the north side. Richard de Stevenage, abbot of St. Albans at the dissolution, was a native of this place.

STEVENTON (*St. Michael*), a parish, in the union of ABINGDON, hundred of OCK, county of BERKS, 5 miles (S. W. by S.) from Abingdon; containing 691 inhabitants. The living is a vicarage, valued in the king's books at £9. 5. 2½.; present net income, £192; patrons and appropriators, Dean and Chapter of Westminster. There is a place of worship for Baptists; also a day and Sunday school, partly supported by endowments amounting to about £12 per annum. The Berks and Wilts canal, and the line of the Great Western railway, pass through the parish. In the village is an ancient cross, consisting of a tall shaft rising from a base of several steps. A castle was formerly erected here by Baldwin Wake, in 1281, of which there are no vestiges. A priory of Black monks, a cell to the abbey of Bec in Normandy, was founded in the time of Henry I., which, at the suppression of Alien houses, was bestowed upon the abbot and convent of Westminster.

STEVENTON (*St. Nicholas*), a parish, in the union and hundred of BASINGSTOKE, Kingsclere and Northern Divisions of the county of SOUTHAMPTON, 6¼ miles (E.) from Whitchurch; containing 197 inhabitants. The living is a rectory, valued in the king's books at £11. 4. 7.;

patron, Edward Knight, Esq. The tithes have been commuted for a rent-charge of £500, subject to the payment of rates. A school is supported by Mr. Knight. The manor house has a very antiquated appearance, and bears evident marks of former grandeur. The London and South-Western railway passes through the parish.

STEVINGTON (*St. Mary*), a parish, in the hundred of Willey, union and county of Bedford, $5\frac{1}{2}$ miles (N. W. by W.) from Bedford; containing 500 inhabitants. The living is a discharged vicarage, valued in the king's books at £12. 13. 4.; present net income, £108; patron, Duke of Bedford; impropriator, Earl Spencer.

STEWKLEY (*St. Mary*), a parish, in the union of Winslow, hundred of Cottesloe, county of Buckingham, 5 miles (W. N. W.) from Leighton-Buzzard; containing 1053 inhabitants. The living is a vicarage, valued in the king's books at £9. 9. 7.; present net income, £194; patron and appropriator, Bishop of Oxford. The church is one of the most enriched and complete specimens of the Norman style of architecture now remaining: it contains 130 free sittings, the Incorporated Society having granted £120 in aid of the expense. There is a place of worship for Wesleyan Methodists. The lace manufacture, formerly carried on here, is nearly extinct, and the principal part of the female population are employed in the manufacture of straw plat. There is a lime quarry, in which are occasionally found curious fossil antediluvian remains, especially some specimens of very large spiral fossil shells. The Grand Junction canal is distant only three miles. A fund of about £25 per annum, arising from various bequests, is distributed among the poor families of the parish on Lady-day.

STEWTON (*St. Andrew*), a parish, in the union of Louth, Wold Division of the hundred of Louth-Eske, parts of Lindsey, county of Lincoln, $2\frac{3}{4}$ miles (E.) from Louth; containing 69 inhabitants. The living is a discharged rectory, valued in the king's books at £7; patron, T. Heneage, Esq. The tithes have been commuted for a rent-charge of £200, subject to the payment of rates, which on the average have amounted to £11. 19.; the glebe comprises 11 acres, valued at £13 per annum.

STEYNING (*St. Andrew*), a market-town and parish (formerly a representative borough), and the head of a union, in the hundred of Steyning, rape of Bramber, Western Division of the county of Sussex, 24 miles (E. by N.) from Chichester, and $49\frac{1}{2}$ (S. by W.) from London; containing 1436 inhabitants. The name is supposed to be derived from the Steyne-street, an ancient road which passed through this part of the county from Arundel to Dorking. Camden considers it to have been mentioned in Alfred's will by the name of *Steyningham*. It appears in the Saxon age to have been a place of considerable note; a church, or monastery, having been here built, wherein St. Cadman was buried; and in the Catalogue of Religious Houses, ascribed to Gervase of Canterbury, in the time of Richard I., mention is made of a Dean and Secular canons. It is more certain that King Edward the Confessor gave lands to the monastery of Feschamp in Normandy, which included this place; these being taken away by Earl Godwin, and restored by William the Conqueror, some Benedictine monks were sent from that house, who erected an Alien priory here, which was given to the monastery of Sion by Edward IV., and continued part of its possessions till the dissolution. Speed says, the conventual church was dedicated to St. Mary Magdalene, and contained the sacred relics of St. Cuthman (Cadman), and Ethelwulph, father of Alfred the Great: here anciently was also a parochial church of St. Cuthman. Camden speaks of its market as well frequented in his time: the town afterwards became reduced, and is, in the Magna Britannia, a century later, mentioned as "a mean, contemptible place, with hardly a building fit to put a horse in," being said then to have contained not more than 150 families; but since that period it has been considerably enlarged. It stands at the foot of a lofty hill near the river Adur, over which is a bridge, and consists of four streets, crossing each other. It is supplied with water by a celebrated spring, issuing from a hill half a mile distant, its stream turning two mills belonging to the town. Great improvement in the buildings and general appearance of the town has been lately made, through the liberality of the Duke of Norfolk. Here were extensive barracks for infantry, which have been pulled down; and within one mile of the town, on the downs, is a race-course, but races have not been held for the last 50 years. The land in the vicinity is fertile, and the adjoining downs afford good pasturage for sheep. The chief traffic is in cattle, for which there is a monthly market: great numbers are also sold at the fairs. The market is on Wednesday: the fairs are on June 9th, September 20th, and October 10th; at the Michaelmas fair more than 3000 head of Welsh oxen alone have been disposed of, exclusively of other kinds, together with sheep, horses, hogs, wheat, seeds, &c. Steyning is a borough by prescription, under the authority of a constable, appointed at the court leet of the lord of the manor. It formerly sent two members to parliament, but was disfranchised by the act of the 2nd of William IV., cap. 45: the right of election, which has been frequently contested, was in the inhabitant householders paying scot and lot, about 80 in number; the constable was the returning officer. The members were formerly elected in conjunction with Bramber, but subsequently each town returned two representatives, although one part of Bramber is in the centre of Steyning. This town has been made a polling-place for the western division of the county.

The living is a vicarage, valued in the king's books at £15; present net income, £308; patron and impropriator, Duke of Norfolk. The church has been repewed, and contains 220 free sittings, the Incorporated Society having granted £150 in aid of the expense; it consists of a portion only of a larger cruciform structure, presenting beautiful specimens of the Norman style of architecture. The interior is magnificently enriched: four elaborately ornamented circular arches, each surmounted with a small roundheaded window on the south side of the nave, have, for their beauty and variety, been copied in the repairs of Arundel Castle; the side aisles are much and disproportionately lower. At the east end, where the transept is intersected, are clusters of columns and arches for supporting a central tower: a lofty Norman arch leads into the chancel. The present tower on the west, of more modern date, is of chequered flint and rubble stone, with angular buttresses. The free school was founded and endowed in 1614, by William Holland, a native and alderman of Chichester, who bequeathed for that purpose to trustees a garden and messuage, called "Brotherhood Hall," then used as a school-house, together with his manor of Festoes, &c., to pay from the proceeds of the latter £20 yearly to a schoolmaster, for the instruction of children of persons dwelling within Steyning and its liberty: the income is £81. 10. per annum, and ten boys receive a classical education. Brotherhood Hall is still standing, and most likely received its name from having been the hall of some guild, or fraternity, prior to the dissolution: it consists of a centre, with a large arched entrance, and two wings, the roofs split into five divisions and pointed, with large square windows of the time of Henry VIII. A National school is supported by subscription. The poor law union of Steyning comprises 23 parishes or places, under the care of 31 guardians, and contains a population of 11,071, according to the census of 1831. John Pell, the mathematician, was educated here.

STIBBARD (*All Saints*), a parish, in the union of Walsingham, hundred of Gallow, Western Division of the county of Norfolk, 4¼ miles (E. by S.) from Fakenham; containing 505 inhabitants. The living is a discharged rectory, annexed to that of Colkerk, and valued in the king's books at £11. 13. 4. A school is partly supported by the rector.

STIBBINGTON (*St. John the Baptist*), a parish, in the union of Stamford, hundred of Norman-Cross, county of Huntingdon, 1 mile (E. by S.) from Wansford; containing 456 inhabitants. The living is a rectory, valued in the king's books at £7. 13. 6½.; present net income, £431; patron, Duke of Bedford.

STICKFORD (*St. Helen*), a parish, in the union of Spilsby, Western Division of the soke of Bolingbroke, parts of Lindsey, county of Lincoln, 5½ miles (S. W.) from Spilsby; containing 425 inhabitants. The living is a discharged vicarage, valued in the king's books at £6. 3. 6.; present net income, £133; patron and appropriator, Bishop of Lincoln. There is a place of worship for Wesleyan Methodists; also a day and Sunday school, partly supported by subscription.

STICKNEY (*St. Luke*), a parish, in the union of Spilsby, Western Division of the soke of Bolingbroke, parts of Lindsey, county of Lincoln, 2½ miles (E. by N.) from Bolingbroke; containing 809 inhabitants. The living is a rectory, valued in the king's books at £13. 11. 3.; present net income, £356; patron, Thomas Coltman, Esq. There is a place of worship for Wesleyan Methodists; also a school supported by endowment.

STIDD, or STEDE, an extra-parochial liberty, in the Lower Division of the hundred of Blackburn, Northern Division of the county palatine of Lancaster, 7 miles (N. N. W.) from Blackburn: the population is returned with Ribchester. Here are the ruins of an ancient chapel, in the early style of English architecture, endowed with £25 a year, to preserve which stipend, service has occasionally been performed within its walls since the Reformation, by the vicar of Ribchester, to which place the living of Stidd, which is a perpetual curacy, is united. There is a Roman Catholic chapel at Stidd Lodge.

STIFFKEY (*St. John*), a parish, in the union of Walsingham, hundred of North Greenhoe, Western Division of the county of Norfolk, 3¼ miles (E.) from Wells; containing 460 inhabitants. The living comprises the united rectories of St. John and St. Mary, with the rectory of Morston annexed, valued in the king's books at £25; patron, Marquess Townshend. A school is partly supported by subscription. The old hall, having been in a state of dilapidation for some years, is now used as a farm-house. It was built in 1604, by Sir Nicholas Bacon, Knt., Lord Keeper of the Great Seal, and the west front, flanked by two embrasured towers, still remains. To the westward of the village is Warborough hill, on the summit of which is a circular camp.

STIFFORD (*St. Mary*), a parish, in the union of Orsett, hundred of Chafford, Southern Division of the county of Essex, 10 miles (S. E.) from Romford; containing 274 inhabitants. This parish is situated on the Horndon road, and takes its name from a ford over a stream that flows into the river Thames. Extensive chalk pits have been worked here for the last two centuries, the produce of which is sent by the Thames, together with considerable quantities of agricultural produce. The village is pleasantly situated, and contains many well-built houses. Stifford Lodge and Ford Place are handsome residences. The living is a rectory, valued in the king's books at £15; patrons, R. Wingfield, Esq., and the Embroiderers' Company in fee. The tithes have been commuted for a rent-charge of £450, subject to the payment of rates; the glebe comprises 26 acres, valued at £45 per annum. The church is a neat structure of the time of Edward I., with a spire: in the chancel is an altar-tomb, with a Norman inscription, to David Percy, in Saxon characters; there is also the effigy of a monk in brass, with an inscription in Latin to Randulph Peachey, formerly rector of this parish. Here is a Sunday National school.

STILLINGFLEET (*St. Helen*), a parish, in the union of York, partly in the Ainsty of the city of York, partly in the liberty of St. Peter's, but chiefly in the wapentake of Ouze and Derwent, East Riding of the county of York; containing 909 inhabitants, of which number, 406 are in the township of Stillingfleet with Moreby, 7½ miles (S. by W.) from York. The living is a discharged vicarage, valued in the king's books at £9. 7. 6.; present net income, £412; patrons, Dean and Chapter of York, who, as trustees of St. Mary's school, are the impropriators. The church is an ancient structure, exhibiting some portions in the Norman style; attached to it is a chapel, in which is a cross-legged figure in armour, of one of the ancient family of Moreby. There is a place of worship for Wesleyan Methodists; also a school partly supported by subscription.

STILLINGTON, a chapelry, in the parish of Redmarshall, union of Sedgefield, South-Western Division of Stockton ward, Southern Division of the county palatine of Durham, 8 miles (N. W. by W.) from Stockton-upon-Tees; containing 96 inhabitants.

STILLINGTON (*St. Nicholas*), a parish, in the union of Easingwould, liberty of St. Peter's, East Riding, though locally in the wapentake of Bulmer, North Riding, of the county of York, 4½ miles (E. S. E.) from Easingwould; containing 717 inhabitants. The living is a discharged vicarage, in the patronage of the Dean and Chapter of York, valued in the king's books at £4. 15. 5.; present net income, £120; impropriator, Col. Croft. There is a place of worship for Wesleyan Methodists; also a National school. The celebrated Laurence Sterne held this living, and resided at Sutton, in the neighbourhood.

STILTON (*St. Mary*), a parish (formerly a market-town), in the union of Peterborough, hundred of Norman-Cross, county of Huntingdon, 12½ miles (N. N. W.) from Huntingdon; containing 793 inhabitants. This place takes its name from *Stivecle*, signifying stiff clay, according to Stukeley, and is situated upon the Roman road, Ermin-street; though formerly a market-town, it has dwindled into comparative insignificance. It gives name to the famous cheese so called, great quantities of which are sold here, though it is made in Leicestershire, twenty miles off. This has been made a polling-place for the county. The living is a rectory, valued in the king's books at £11. 5. 10.; present net income, £355; patron, Bishop of Lincoln. There is a place of worship for Wesleyan Methodists. John Apreece, Esq., in 1821, bequeathed Navy stock, the dividends of which amount to £6. 3. 4., for the relief of poor widows. There is a fine spring about a quarter of a mile from the town, at one period celebrated for the cure of ulcerated legs, which properties are said to have ceased; and to the south-east are the remains of an ancient circular encampment.

STINCHCOMBE (*St. Cyril*), a parish, in the union of Dursley, Upper Division of the hundred of Berkeley, Western Division of the county of Gloucester, 2 miles (W. by N.) from Dursley; containing 352 inhabitants. The living is a perpetual curacy; net income, £65; patron, Bishop of Gloucester and Bristol; impropriators, P. B. Purnell, Esq., and others. A school is endowed, and further supported by a salary allowed by the parish.

STINSFORD (*St. Michael*), a parish, in the union of Dorchester, hundred of George, Dorchester Division of the county of Dorset, 1¼ mile (E. N. E.) from Dorchester; containing 382 inhabitants. The living is a vicarage, valued in the king's books at £12. 17. 1.; patron and impropriator, Earl of Ilchester. The impropriate tithes have been commuted for a rent-charge of £115, and the vicarial for £190, subject to the payment of rates, which on the average on the latter have amounted to £1. 12.; the glebe comprises one acre, valued at £3 per annum. In the chancel of the church is a stone recording the death of Wadham Strangeways, who was slain at Bridport, whilst opposing the rebellion of the Duke of Monmouth. A day school and a National Sunday school are supported by the clergyman. There are some remarkable circular sand-pits in this parish.

STIRCHLEY (*St. James*), a parish, in the union of Madeley, Wellington Division of the hundred of South Bradford, Northern Division of the county of Salop, 3 miles (W. by S.) from Shiffnall; containing 271 inhabitants. The living is a discharged rectory, valued in the king's books at £6. 5. 10.; present net income, £262; patrons and impropriators, Revel Phillips, Esq., and others. A school is supported by private contributions. The Shropshire canal passes through the parish.

STIRTON, a joint township with Thorlby, in the parish of Kildwick, union of Skipton, Eastern Division of the wapentake of Staincliffe and Ewcross, West Riding of the county of York, 1¼ mile (N. W.) from Skipton; containing, with Thorlby, 170 inhabitants.

STISTED (*All Saints*), a parish, in the union of Braintree, hundred of Hinckford, Northern Division of the county of Essex, 3½ miles (E. N. E.) from Braintree; containing 895 inhabitants. This parish, which is bounded by the river Blackwater, is about thirty miles in circumference, richly wooded, and abounds with variety of surface and scenery. A handsome modern mansion with an Ionic portico has been erected on the site of the old manor-house, Stisted Hall; and there are several other stately mansions in the parish. The village is extensive and beautifully situated on ground sloping gently from the river, and commanding richly varied prospects. The living is a rectory, in the patronage of the Archbishop of Canterbury, valued in the king's books at £22; present net income, £466. The church is an ancient structure in the Norman style, with a tower on the south side surmounted by a shingled spire, and contains some interesting monuments. Here is a Sunday National school.

STITHIANS (*St. Stedian*), a parish, in the union of Redruth, Eastern Division of the hundred of Kerrier, Western Division of the county of Cornwall, 4 miles (N. W. by W.) from Penryn; containing 1874 inhabitants. This parish, which includes a part of the village of Ponsnooth, occupies an agreeable district; and the views from the high grounds in Kennall Vale extend over Falmouth harbour, Tregothnan, St. Day, Roseland, Brown Whilly, Carn Marth, and other places. In Kennall Vale and in Cofawes Wood are extensive mills for making gunpowder, which is chiefly used in the mines: the works produce annually about 10,000 barrels. Above the Kennall powder-mills is a paper-mill. The living is a vicarage, with that of Perran-Arworthal annexed, valued in the king's books at £14. 0. 10.; present net income, £377; patron and impropriator, Earl of Falmouth. The church has a handsome embattled tower crowned with pinnacles. There are places of worship for Wesleyan Methodists; also a school partly supported by subscription. There are various ancient crosses enriched with sculpture in the parish; and in Cofawes Wood was an ancient chapel, dedicated to St. Mary Magdalene.

STITTENHAM, a township, in the parish of Sheriff-Hutton, union of Malton, wapentake of Bulmer, North Riding of the county of York, 8¼ miles (W. S. W.) from New Malton; containing 86 inhabitants.

STIVICHALL (*St. James*), a parish, in the union of Warwick, county of the city of Coventry, Northern Division of the county of Warwick, 1¾ mile (S. by W.) from Coventry; containing 103 inhabitants. The living is a perpetual curacy; net income, £90; patron and impropriator, Francis Gregory, Esq.

STIXWOULD (*St. Peter*), a parish, in the union of Horncastle, Southern Division of the wapentake of Gartree, parts of Lindsey, county of Lincoln, 6¾ miles (W. S. W.) from Horncastle; containing 221 inhabitants. The living is a discharged vicarage, valued in the king's books at £7. 10.; present net income, £70; patron and impropriator, C. Turner, Esq. A convent of Cistercian nuns, in honour of the Blessed Virgin, was founded here in the reign of Stephen, by the Countess Lucy, relict of Ranulph, first Earl of Chester, which at the dissolution possessed a revenue of £163. 1. 2.

STOAK, or STOKE, a parish, in the union of GREAT BOUGHTON, Higher Division of the hundred of WIRRALL, Southern Division of the county of CHESTER; containing 302 inhabitants, of which number, 101 are in the township of Stoak, 5½ miles (N. by E.) from Chester. The living is a perpetual curacy; net income, £130; patron and impropriator, Sir H. E. Bunbury, Bart. The church, which is the burial-place of the Bunbury family, has a Norman doorway, some ancient wooden screen-work, and a small chapel attached to the south side of the chancel: it was partially rebuilt in 1827, and 55 free sittings provided, the Incorporated Society having granted £50 in aid of the expense. This parish is intersected by the Ellesmere canal. There are some bequests to the poor, amounting to about £25 per annum, chiefly by the Bunbury family.

STOBOROUGH, a liberty, in the parish of the HOLY TRINITY, borough of WAREHAM, Wareham Division of the county of DORSET, ¾ of a mile (S.) from Wareham; containing 347 inhabitants. It was formerly governed by a mayor, chosen annually at Michaelmas; but the inhabitants declining to qualify themselves, when the Schism act came into operation, in 1714, the office no longer exists, but a bailiff is appointed in the same manner as the mayor was, *viz.*, by a jury at the court held by the lord of the manor.

STOCK (*ALL SAINTS*), a parish, in the union and hundred of CHELMSFORD, Southern Division of the county of ESSEX, 3 miles (E. by S.) from Ingatestone; containing 619 inhabitants. This parish is almost surrounded by that of Buttsbury, in which it is supposed to have been formerly a hamlet: the soil is in some parts fertile, and there is abundance of fine clay, of which bricks of superior quality are made. The living is a rectory, with that of Ramsden-Bellhouse annexed, valued in the king's books at £10; present net income, £658; patron, Rev. Edward Edison. The church is a large edifice of brick, with a turret of wood: in the south wall is a table monument with the recumbent figure of an armed warrior. Here is a place of worship for Independents; also a National school, erected in 1839. Richard Tweedy, of Boreham, in 1574, erected and endowed four almshouses for two poor men of this parish, and two of the parish of Boreham.

STOCK, a hamlet, in the parish of FLADBURY, union of DROITWICH, Middle Division of the hundred of OSWALDSLOW, Droitwich and Eastern Divisions of the county of WORCESTER, 6¾ miles (E. S. E.) from Droitwich: the population is returned with the chapelry of Bradley.

STOCK-CROSS, a hamlet, in the parish of SPEEN, hundred of KINTBURY, county of BERKS, 2½ miles (N. W.) from Newbury: the population is returned with the parish. A district church was erected in this hamlet in 1839, and endowed solely at the expense of the Rev. Henry William Majendie; it is a handsome cruciform edifice, in the early English style, containing 400 sittings, of which 300 are free, and, being situated on elevated ground, forms an interesting and picturesque feature in the landscape. A school is supported by subscription.

STOCK-DENNIS, a tything, in the parish and hundred of TINTINHULL, Western Division of the county of SOMERSET; containing 13 inhabitants. This was formerly a parish, but the church having been destroyed, and the place almost depopulated, it has long since lost its parochial rights. The living was a rectory, valued in 1294 at £20: it now belongs to the Rev. J. H. Wyndham, and yields a net income of £188.

STOCK-GAYLAND, a parish, in the union of STURMINSTER, hundred of BROWNSHALL, Sturminster Division of the county of DORSET, 7 miles (E. S. E.) from Sherborne; containing 63 inhabitants. The living is a discharged rectory, valued in the king's books at £5. 7. 1.; present net income, £194; patron, Rev. H. F. Yeatman. Here is a National school.

Arms.

STOCKBRIDGE (*ST. PETER*), a market-town and parish (formerly a representative borough), having separate jurisdiction, and the head of a union, locally in the hundred of King's-Sombourn, Andover and Northern Divisions of the county of SOUTHAMPTON, 18 miles (N. by W.) from Southampton, and 66 (W. S. W.) from London; containing 851 inhabitants. This small town is situated on the great western road from London to Exeter, and consists of one long street, which is intersected at the west end by the river Test, and at the east by the Andover and Redbridge canal, over each of which is a bridge; that over the former was rebuilt in 1799, and is a handsome structure: five smaller streams cross the street in the intermediate space, over which were formerly bridges of one arch, but these have been recently taken down, except one, and the whole has been arched over: the inhabitants are supplied with excellent water. On Houghton down, about two miles west of the town, was formerly a race-course; but a new one has been formed, immediately adjoining it, in the parishes of Wallop and Longstock, under Danebury hill, from the area and intrenchments of which the whole of it may be seen: a stand has been erected, which is also used by the members of the Bibury Racing club, lately removed hither from Gloucestershire: races are held in June, and, for some years past, a plate has been given by the Marquess of Westminster. The streams are particularly favourable for trout-fishing; the principal nobility and gentry of this and the adjoining counties meet here three or four times a year, and spend several weeks in this favourite recreation, during the season. The preparation of parchment and glue affords employment to a few persons. Some thousand bushels of peat ashes are annually disposed of to the neighbouring farmers for manure, that article being much used for fuel by the inhabitants, though they are also well supplied with coal by means of the canal. The market, on Thursday, is well attended; and a large and handsome market-room, adjoining the Grosvenor Arms, has been built, at the expense of the Marquess of Westminster. There were formerly three fairs, of which one only is now held, on July 10th, which is one of the largest in the county for lambs, several thousands being annually sold; and a fortnight cattle market has been lately established.

Stockbridge is a borough by prescription, under a bailiff and a constable, who are elected annually by the jury, at the court leet of the manor, held by the steward on Easter-Wednesday, the constable for the year preceding being generally made bailiff for the following year: the jury are summoned by a sergeant-at-mace. It first sent representatives to parliament in the 1st of Elizabeth, but was disfranchised by the act of the 2nd of William IV., cap. 45: the right of election was vested in the inhabitant householders paying scot and lot, in number about 100; the bailiff was the returning officer. Petty sessions are held monthly. The town-hall, a neat edifice, is situated near the centre of the town, and was rebuilt in 1810, on the site of the previous structure, at an expense of £1500, defrayed by the inhabitants. The living is annexed, with that of Little Sombourn, to the vicarage of King's Sombourn. There is a place of worship for Independents. The poor law union of Stockbridge comprises 15 parishes or places, under the care of 19 guardians, and, according to the census of 1831, contains a population of 6552. About two miles and a half from the town is Danebury hill, a circular intrenchment, in good preservation, enclosing an extensive area, with very high ramparts. On the north and west are several barrows, one of which is named Canute's barrow. On the east, at the distance of about one mile and a half, is another circular intrenchment, with a high rampart, enclosing an area of about twenty acres, called Woolberry, on the east side of which is the representation of a white horse, cut many years since, at the expense of W. P. Powlett, Esq., of Sombourn House. Robert, Earl of Gloucester, natural brother of the Empress Matilda, was taken prisoner in this town on his flight from Winchester: according to tradition, he took refuge in the church, after having effected the escape of the empress, who, feigning death, was conveyed thence in funeral procession through the besieging army, and, having arrived at a certain distance, mounted a horse and reached Gloucester in safety.

STOCKBURY (*St. Mary Magdalene*), a parish, in the union of Hollingbourn, hundred of Eyhorne, lathe of Aylesford, Western Division of the county of Kent, 3 miles (W. by S.) from Sittingbourne; containing 618 inhabitants. This parish contains several streets of straggling houses, called respectively Stockbury, Guilstead, and South streets, with North Dean and Hill Green, the last occupying elevated ground commanding a distant view of the sea. The lands are generally well cultivated. A fair, chiefly for toys, is held annually on the 2nd of August. The living is a vicarage, valued in the king's books at £9. 11. 0½.; present net income, £243; patrons and appropriators, Dean and Chapter of Rochester. The church is a spacious cruciform structure, in the early English style, the columns and arches of which, on the north side, are of Petworth marble, and peculiarly elegant. A great part of the chancel and north transept was destroyed by fire in 1836, and the whole was rebuilt, partly at the expense of the dean and chapter, and partly by subscription: the new pillars and arches are of Bethersdon marble, and the ancient carved work, of which some capitals on the south side were elaborately enriched, has been carefully restored. A vicarage-house was erected in 1834. Mrs. Jane Bentley, in 1752, bequeathed an annuity of £2. 10. a year, for teaching six children, and £2 every fourth year to purchase bibles and prayer-books for their use. A dreadful tempest, attended with the most destructive effects, happened here in 1746.

STOCKELD, a township, in the parish of Spofforth, Upper Division of the wapentake of Claro, West Riding of the county of York, 2 miles (W.) from Wetherby; containing 62 inhabitants. This township forms a manor, comprising 970 acres of land, the property of Peter Middleton, Esq., of Stockeld Hall, in the grounds of which is a lake, from the margin of which rises a rock of a peculiar form, 65 feet in girth, and in height 30, which probably gave name to the place, Stockheldt being the Dutch term for a misshapen figure of stone. There is a Roman Catholic chapel at Stockeld Hall.

STOCKERSTON (*St. Peter*), a parish, in the union of Uppingham, hundred of Gartree, Southern Division of the county of Leicester, 2¾ miles (W. S. W.) from Uppingham; containing 60 inhabitants. The living is a rectory, valued in the king's books at £13; present net income, £174; patrons, T. Walker, Esq., and two others. Near it, John Boyvile, Esq., in 1465, obtained leave of Edward IV. to erect, in honour of the Blessed Virgin, an almshouse for a chaplain and three poor persons, and to settle lands upon them in mortmain, of the annual value of £10.

STOCKHAM, a township, in the parish and union of Runcorn, hundred of Bucklow, Northern Division of the county of Chester, 3½ miles (N. E.) from Frodsham; containing 52 inhabitants.

STOCKHILL, a joint township with Sandholme, in the parish of St. John, union and liberties of the borough of Beverley, East Riding of the county of York, 1¾ mile (N. E.) from Beverley; containing, with Sandholme, 34 inhabitants.

STOCKHILL, a joint township with Middleton, in the parish of Ilkley, Upper Division of the wapentake of Claro, West Riding of the county of York, 6½ miles (N. W.) from Otley: the population is returned with Middleton. These two hamlets comprise about 2280 acres of land, the property of William Middleton, Esq., of Middleton Lodge, an ancient mansion in the Gothic style, near which is a neat Roman Catholic chapel, erected about 14 years since.

STOCKINGFORD, a chapelry, in the parish of Nuneaton, Atherstone Division of the hundred of Hemlingford, Northern Division of the county of Warwick, 1½ mile (W.) from Nuneaton: the population is returned with the parish. Here is a chapel, erected in 1824, and dedicated to St. Paul: it is a handsome building, consisting of a nave, aisles, and chancel, with a square embattled tower.

STOCKLAND (*St. Michael*), a parish, in the union of Axminster, and forming a detached portion of the hundred of Whitchurch-Canonicorum, Bridport Division of the county of Dorset, though locally in the county of Devon (to the Southern Division of which for electoral purposes it is annexed), 6¼ miles (N. E. by E.) from Honiton; containing 1206 inhabitants. The living is a vicarage, valued in the king's books at £15. 13. 11½.; present net income, £457; patrons, the Freeholders and Inhabitants; impropriators, the Landowners. The church is a large ancient structure, and contains 80 free sittings, the Incorporated Society having granted £50 in aid of the expense. There is a chapel of ease at Dalwood. A National school is supported by subscription, and an allowance of £10 a year from lands left for charitable purposes.

STOCKLAND-BRISTOL, a parish, in the union of

Bridg-water, hundred of Cannington, Western Division of the county of Somerset, 7 miles (N. W. by N.) from Bridg-water; containing 202 inhabitants. The living is a discharged vicarage, valued in the king's books at £6. 9. 4.; present net income, £161: it was in the patronage of the Mayor and Corporation of Bristol, the impropriators. The impropriate tithes have been commuted for a rent-charge of £60, and the vicarial for £151. 8., subject to the payment of rates, which on the latter have on the average amounted to £13; the glebe comprises 16 acres, valued at £20 per annum.

STOCKLEWATH-BOUND, a township, in the parish of Castle-Sowerby, Leath ward, Eastern Division of the county of Cumberland, 8 miles (S. by W.) from Carlisle; containing 260 inhabitants. Castle-Steads, a Roman camp, 188 yards long and 160 broad, is within this township: it has an inner and an outer rampart, and is placed in a triangular position with, and at an equal distance from, two other fortifications, called Whitestones and Stoneraise, the latter of which, it is supposed, was originally a burial-ground of the Druids, afterwards occupied by the Romans. About a mile from these are vestiges of a Druidical temple, where three stone coffins, containing human bones and other relics, have been found; and a little to the southward are fragments of a large rocking-stone, to which an avenue of stones seems to have once led.

STOCKLEY, a township, in the parish of Brancepeth, North-Western Division of Darlington ward, union, and Southern Division of the county palatine, of Durham, 4¾ miles (S. W. by W.) from Durham; containing 57 inhabitants.

STOCKLEY-ENGLISH (*St. Mary*), a parish, in the union of Crediton, hundred of West Budleigh, Crediton and Northern Divisions of the county of Devon, 5¼ miles (N. by E.) from Crediton; containing 144 inhabitants. The living is a rectory, valued in the king's books at £7; present net income, £116: it is in the patronage of the Crown.

STOCKLEY-POMEROY (*St. Mary*), a parish, in the union of Crediton, hundred of West Budleigh, Crediton and Northern Divisions of the county of Devon, 3½ miles (N. E. by E.) from Crediton; containing 238 inhabitants. The living is a rectory, valued in the king's books at £15. 6. 8.; present net income, £222; patron, Bishop of Exeter. The church is partly of Norman architecture, having an enriched doorway in that style. A day and Sunday school is partly supported by the lord of the manor.

STOCKLINCH, MAGDALENE (*St. Mary Magdalene*), a parish, in the union of Chard, hundred of Abdick and Bulstone, Western Division of the county of Somerset, 2¾ miles (N. E.) from Ilminster; containing 95 inhabitants. The living is a discharged rectory, valued in the king's books at £4. 4. 7.; present net income, £140; patron and incumbent, Rev. James Upton.

STOCKLINCH, OTTERSAY (*St. Mary*), a parish, in the union of Chard, hundred of Abdick and Bulstone, Western Division of the county of Somerset, 2½ miles (N. E.) from Ilminster; containing 120 inhabitants. The living is a discharged rectory, valued in the king's books at £6. 9. 2.; present net income, £148; patron, Jeffreys Allen, Esq.

STOCKPORT (*St. Mary*), a newly-enfranchised borough, market-town, and parish, and the head of a union, in the hundred of Macclesfield, Northern Division of the county of Chester; containing 66,610 inhabitants, of which number, 25,469 are in the town of Stockport, 39 miles (N. E. by E.) from Chester, and 179 (N. W. by N.) from London.

Arms.

This place, from its situation near a common centre, from which several Roman roads diverged, is supposed to have been a Roman military station, and the fort to have occupied the summit of Castle hill, on the site of which the Saxons subsequently erected a baronial castle; from which, expressive of its situation in the woods, the town derived its name Stokeport, or Stockport. Though not mentioned in Domesday-book, it is of considerable antiquity, and, till the Conquest, was a military station of some importance, most probably one of those laid waste by the Normans on their conquest of the island. In confirmation of this opinion may be adduced the name of an adjacent vill, called Portwood, also omitted in the survey, the first notice of it occurring in the records of the lands of the Baron of Dunham, under the name of Brinnington, or the burnt town. In 1173, the castle of Stokeport was held by Geoffrey de Costentyn, against Henry II., but whether in his own right or not, is uncertain. The first baron appears, from the best authority, to have been Ranulph le Dapifer, the progenitor of the family of the De Spencers, from whom it passed to Robert de Stokeport, who, in the reign of Henry III., made the town a free borough. In 1260, it obtained the grant of an annual fair for seven days, commencing on the festival of St. Wilfrid, and a weekly market on Friday. During the parliamentary war, it was garrisoned for the parliament; but Prince Rupert, advancing against it with a party of the royal troops, expelled the garrison, and took possession of it for the king; it was subsequently retaken by the parliamentarians, who retained it till the termination of the war. In 1745, Stockport was twice visited by the troops under the Pretender, on their approach to Derby, and in their retreat; on the latter occasion, the bridge over the Mersey was destroyed, and the rebels, with Prince Charles, were compelled to wade through the river, in order to effect their escape. Of the ancient castle not a vestige can be traced; a circular brick building was erected on the site by the late Sir George Warren, as a hall for the sale of muslin, for which article of manufacture it was his wish to make this town a mart; but since the failure of that project, the building has been converted into an inn.

Stockport is romantically situated on elevated ground, of irregular and precipitous ascent: the south-eastern portion is intersected by the rivers Goit and Tame, which at their confluence in the centre of the borough form the river Mersey. From the valleys through which these rivers run, the houses rise in successive tiers round the sides of the hill, from the base to the summit; and the numerous extensive factories elevated above each other, and spreading over the extent of the town, present, when lighted during the winter months, an appearance strikingly impressive. The most ancient part surrounds the church and market-place, on the high ground overlooking the Mersey, from the bank of which several steep streets, ascending the acclivity, lead into the market-place, whence various other streets diverge in different directions: many of the houses at the base of the hill have apartments excavated in the rock, which is of soft red sand-stone. The principal street, here called the Underbank, follows the direction of the old Roman road, leading southward to Buxton, and contains an ancient timber and brick mansion, formerly occupied by the family of Arderne of Harden and Alvanley, now a banking-house. On the summit of the hill is a range of houses surrounding the market-place; and to the north of the church is the site of the ancient castle, and of the Roman military works. The town extends, on the south, a considerable distance along the road to London; and on the north-east, by a bridge over the Mersey, to Portwood; on the west towards Cheadle, and towards Manchester by another bridge across the Mersey on the north, on which side of the river is the township of Heaton-Norris, forming part of this town, though in the county of Lancaster. To prevent the inconvenience and delay of travelling through the narrow and hilly part of the town, the trustees of the Manchester and Buxton turnpike-roads applied, in 1824, for an act of parliament to empower them to construct a new line of road from Heaton-Norris chapel, on the north side of the Mersey, to Bramall-lane, at the southern extremity of the town, through an open and airy situation, affording eligible sites for the erection of houses, and an admirable opportunity for improving and extending the town. This important work was completed under the superintendence of Mr. Thomas Broadhurst; its especial object was to cross the river without the necessity of descending from the high grounds on each side to the level of the vale of the Mersey, which has been accomplished by the construction of a noble bridge, of eleven arches, across the valley and the river, of which nine are on the Cheshire, and two on the Lancashire side of the Mersey. The arch over the river has a span of more than 90 feet, and is built of hard white stone from the Saddleworth and Runcorn quarries; the arches on the Cheshire side are carried over several of the streets, the thoroughfare being continued underneath, and others are closed up, forming commodious warehouses. From the last of them the road is carried for a considerable distance over an artificial embankment, formed of earth cut from the hill through which it passes, to its junction with the Warrington road, near which it again joins the old road at Heaviley, at the distance of three miles from its commencement. The whole expense of this work, which was completed in less than two years, was £40,000: it is called the Wellington road. Between the Wellington bridge, and the Lancashire bridge, a foot bridge, termed Vernon bridge, over the Mersey, forming an intermediate and more direct communication between the town and the township of Heaton-Norris, has been built by subscription, the first stone having been laid in 1828.

The town is well paved, and lighted with gas, and the inhabitants are amply supplied with water. An act of parliament for incorporating a gas company, and another for the construction of water-works, were obtained in 1825. In 1837, the corporation obtained a general improvement act, the council being the sole commissioners. Under this act they have purchased the property of the gas company, the profits from which are applied to the improvement of the borough. There are several private news-rooms and libraries; and there is a library in the Mechanics' Institute, to which purpose the theatre has been converted. The surrounding scenery is richly diversified with hill and dale, wood and water. The winding and throwing of silk, for which mills were first established here upon the Italian plan, have been nearly superseded by the introduction of the cotton manufacture, which has for some years been the staple trade of the town: of the former there are still some respectable factories; but the latter, since its introduction, has been rapidly increasing, and has attained, both for its extent, and the perfection to which it has been brought, a very high degree of celebrity. There are within the town, including Heaton-Norris and Portwood, not less than 50 cotton-factories, worked by steam-engines and water-wheels: the printing of calico is carried on to a very great extent; and there are many large establishments and dye-houses in the vicinity. The weaving of calico has spread over all the neighbouring villages, which in some instances have become virtually a part of the town. The manufacture of hats has been long established, and is carried on to a very considerable extent for the supply of the London, and many of the principal country markets: there are also several extensive thread manufactories. Connected with the various branches of manufacture, the construction of machinery affords employment to a great number of persons, and of several additional steam-engines, others again being used in grinding corn and other purposes. The importance of Stockport, as a manufacturing town, has been materially promoted by the facility and the abundance of its supply of coal from Poynton, Worth, and Norbury, and the neighbouring districts on the line of the Manchester and Ashton canal, which joins the Peak Forest canal, a branch of the latter extending to this town, and affords also a direct communication with the principal towns in the kingdom. The Manchester and Birmingham railway passes through the borough in a direction parallel with the Wellington-road; it crosses the valley of the Mersey by an immense viaduct, the centre arch over the river being 120 feet from the surface of the water, with a span of about 90 feet; this viaduct is two-thirds of a mile in length, and is one of the most magnificent works connected with modern railways. The market, on Friday, is more abundantly supplied with corn, meal, and cheese, than any other market in the county: in the higher part of the town (the Hill-gate), extensive and convenient shambles, covering an area of 2000 square yards, were built in 1827; but the inhabitants of the vicinity do not avail themselves of this market, which has fallen into disuse, preference being given to the general market in the centre of the town. The fairs are on March 4th and 25th, May 1st, and October 23rd, for cattle. Stockport was anciently incorporated: it retained the office of mayor, which, however, was little more than nominal, and it was within the jurisdiction of the county magistrates, until the passing of the Municipal Reform Act. The government is now vested in a mayor, 14 aldermen, and 42 councillors, under the act of the 5th and 6th of

William IV., cap. 76, for an abstract of which see the Appendix, No. I.: the municipal and parliamentary boundaries are co-extensive, and the borough is divided into six wards. The mayor and late mayor are justices of the peace, with eight others appointed by commission. The police force consists of a superintendent, sergeant, and 12 men. By the act of the 2nd of William IV., cap. 45, it has been constituted a borough, with the privilege of sending two members to parliament, the right of election being vested in the £10 householders: the borough comprises an area of 2505 acres, of which the limits are minutely described in the Appendix, No. II. The number of voters now registered is about 1230; the mayor is the returning officer. Petty sessions are held every Monday, Wednesday, and Saturday. Courts leet and baron are held twice in the year, at which the lord of the manor appoints two constables and other officers, to the number of 50, who are sworn into office at an adjourned court. The churchwardens are appointed by the four lords of the manors of Bramall, Bredbury, Brinnington, and Norbury, who from time immemorial have represented the parish in ecclesiastical matters. The ancient court baron, for the recovery of debts under 40*s*., has fallen into disuse; and a court for the recovery of debts not exceeding £5 has been established, by an act passed in the 46th of George III., the jurisdiction of which extends over the townships of Stockport and Brinnington, and the hamlets of Edgeley and Brinksway. This town has been made a polling-place for the northern division of the shire.

The living is a rectory, valued in the king's books at £70. 6. 8.; present net income, £1882: it is in the patronage of Lady Vernon. The ancient church, supposed to have been erected in the fourteenth century, of the soft red sandstone in the neighbourhood, having fallen to decay, was, with the exception of the chancel, rebuilt, at an expense of £30,000, by act of parliament passed in the 50th of George III., and an extensive cemetery added to it. The present structure, situated on the eastern side of the market-place, is a handsome building in the later style of English architecture, with a lofty square tower, crowned with a pierced parapet and pinnacles; the pillars of the nave are carried up to the roof, producing an unusual, but impressive effect, from the loftiness of their elevation (an arrangement affording ample accommodation for galleries, which the increasing population of the parish rendered highly necessary). The chancel, which was in the decorated style of English architecture, has undergone considerable alteration, but still retains some of the ancient stone stalls, which are of elegant design, and the original window has been removed only within the last few years. Several of the ancient monuments have been preserved, and are distributed in various parts of the church. The parish of St. Mary has been divided into two separate and distinct parishes, under the 16th section of the 58th of George III., cap. 45; the church of the new parish having been endowed out of the revenues of the mother church. St. Peter's chapel, a neat edifice of brick, was erected in 1768, at the sole expense of William Wright, Esq., of Mottram St.-Andrew, to whom a handsome mural monument has been erected in the centre of the north aisle. The living is a perpetual curacy; net income, £220; patron, Lawrence Wright, Esq. A district has been assigned to this chapel, under the 16th section of the 59th of George III., cap. 134. The church dedicated to St. Thomas was erected in 1825, by grant from the parliamentary commissioners, at an expense of £14,555. 13.: it is a handsome structure, in the Grecian style of architecture, with a tower surmounted by a cupola; the principal entrance is at the east end, through a noble portico of six lofty Ionic pillars: the interior is handsomely decorated; from the panelled pedestals that support the galleries rises a beautiful range of fluted Corinthian columns, sustaining the roof; corresponding with which is a series of pilasters, of the same order, supporting a handsome entablature and cornice round the walls; the ceilings are panelled in large compartments; and above the altar, which occupies the whole central breadth, is a pediment resting on Ionic pillars, and surmounted on the apex by a gilt cross. The living is a perpetual curacy; net income, £110; patron, Rector of Stockport. A church has been built at Dukinfield, by aid of a grant from Her Majesty's Commissioners; it contains 430 free sittings. There are three places of worship for Independents, three for Wesleyan Methodists, two for the New Connection of Methodists, and one each for the Society of Friends, Primitive Methodists, Warrenite Methodists, Unitarians, and Roman Catholics. A public cemetery has been established by a company; the grounds are several acres in extent, and a neat chapel has been erected.

The free grammar school was founded in 1482, by Sir Edmund Shaw, citizen and goldsmith of London, who endowed it with £10 per annum, to which several subsequent benefactions have been added. The Goldsmiths' Company, who are the patrons, have erected, on the Wellington-road, a handsome and extensive school-room, with a house for the master, in the later style of English architecture, at an expense of £4500, on a site of land presented by Lady Vernon: the master, who must be a graduate of Oxford or Cambridge, has a salary of £210, with £15 for fuel; and the usher a salary of £105: there are 150 boys in the school. The National school was established in 1826, and is partly supported by subscription: the school-rooms are a handsome and spacious edifice of brick, fronted with stone, and well adapted to the purpose; there are 2000 children of both sexes instructed in this establishment. The Stockport Sunday school, upon a very extensive and comprehensive plan, admitting children of all denominations, was established in 1805, and a very extensive building of brick, four stories high, was erected for its use, at an expense of £10,000, raised by subscription: there are 4000 children belonging to this institution, who are instructed by 300 gratuitous teachers: attached to it are four branch schools, in the vicinity of the town, erected at an expense of £6000, in which 1500 children are taught, who cannot conveniently attend the parent establishment. These and some others, all supported by subscription, afford instruction to more than 10,000 children. Sunday schools are also supported in connection with the established church and the several dissenting congregations. On the eastern side of the old churchyard are six almshouses, founded by an ancestor of the late Sir George Warren, in 1685, for six aged men. The allowance was augmented by Humphrey Warren, Esq., who died in the middle of the last century; and the late Lady Bulkeley bequeathed £1200, vested in trustees for the same purpose, and £1000 for the poor of Stockport. A dispensary was established, and a commodious building erected, by subscription, in 1797, to which nine fever wards were added in 1799: it is liberally supported, and affords relief to more than 2000 patients annually; but being inadequate to the wants of the population, an infirmary was erected in 1833, on the Wellington road, at an expense of £6300, raised by subscription; it is an elegant structure of stone, and forms a prominent feature in the approach to the town. There is a bequest by Mr. Wright, for apprenticing four children, besides some other bequests for distribution among the poor. The poor law union of Stockport comprises 17 parishes or places, under the care of 22 guardians, and, according to the census of 1831, contains a population of 70,886.

STOCKSFIELD-HALL, a township, in the parish of Bywell-St. Andrew, union of Hexham, Eastern Division of Tindale ward, Southern Division of the county of Northumberland, 9 miles (E. by S.) from Hexham; containing 35 inhabitants. It is bounded on the north by the Tyne.

STOCKTON, a township, in the parish of Malpas, union of Wrexham, Higher Division of the hundred of Broxton, Southern Division of the county of Chester, 1¾ mile (S. S. W.) from Malpas; containing 30 inhabitants.

STOCKTON (*St. Michael*), a parish, in the union of Loddon and Clavering, hundred of Clavering, Eastern Division of the county of Norfolk, 3 miles (N. W. by N.) from Beccles; containing 110 inhabitants. The living is a discharged rectory, valued in the king's books at £8; present net income, £275; patron, Duke of Norfolk.

STOCKTON (*St. Chad*), a parish, in the union of Shiffnall, Shiffnall Division of the hundred of Brimstree, Southern Division of the county of Salop, 4¾ miles (N. by E.) from Bridgenorth; containing 459 inhabitants. The living is a rectory, with Boningale annexed, valued in the king's books at £13. 11. 3.; patron, T. Whitmore, Esq. The tithes have been commuted for a rent-charge of £589. 15., subject to the payment of rates; the glebe comprises 184 acres, valued at £369. 6. 6. per annum. A day and Sunday school is supported by private charity.

STOCKTON (*St. Michael*), a parish, in the union of Southam, Southam Division of the hundred of Knightlow, Southern Division of the county of Warwick, 2¼ miles (N. E. by E.) from Southam; containing 380 inhabitants. The living is a rectory, valued in the king's books at £10. 7. 1.; present net income, £338; patrons, Warden and Fellows of New College, Oxford. The Warwick and Napton canal passes in the vicinity. The Rev. Charles Crane, in 1807, gave a house towards the support of a school, in which 50 children are instructed for £12 a year, arising from the church lands; and a National school has been established.

STOCKTON (*St. John the Baptist*), a parish, in the union of Warminster, forming a detached portion of the hundred of Elstub and Everley, Warminster and Southern Divisions of the county of Wilts, 6¼ miles (N. E.) from Hindon; containing 274 inhabitants. The living is a rectory, valued in the king's books at £18. 2. 1.; present net income, £493; patron, Bishop of Winchester. Two schools are partly supported by three ladies. An almshouse was founded in 1657, under the will of John Topp, Sen., who bequeathed £1000 for charitable uses, at the discretion of his trustees. It is occupied by eight poor unmarried people, chosen from the parishes of Stockton and Codford-St. Mary; the present annual revenue of the charity is £158. 10.

STOCKTON (*St. Andrew*), a parish, in the union of Martley, Lower Division of the hundred of Doddingtree, Hundred-House and Western Divisions of the county of Worcester, 7¼ miles (S. W.) from Bewdley; containing 133 inhabitants. The living is a discharged rectory, valued in the king's books at £5. 13. 11½.; present net income, £254; patron, John Bury, Esq.

STOCKTON-ON-THE-FOREST, a parish, in the union of York, partly within the liberty of St. Peter's, East Riding, and partly in the wapentake of Bulmer, North Riding, of the county of York, 5¼ miles (N. E.) from York; containing 319 inhabitants. The living is a perpetual curacy, in the patronage of the Prebendary of Bugthorpe in the Cathedral Church of York; net income, £140. There is a place of worship for Wesleyan Methodists.

Corporation Seal.

STOCKTON-UPON-TEES (*St. Thomas*), an incorporated market-town, inland port, and parish, and the head of a union, in the South-Western Division of Stockton ward, Southern Division of the county palatine of Durham; containing 7991 inhabitants, of which number, 7763 are in the town of Stockton, 20 miles (S. S. E.) from Durham, and 244 (N. by W.) from London. This place is of considerable antiquity, and the discovery of a Roman coin near the site of the castle, has led to the conjecture that it was a Roman station, but nothing farther to confirm this opinion is recorded: it formed a part of the possessions of the see of Durham at an early period, and the castle was occupied by Hugh de Pudsey, bishop of that diocese in the reign of Richard I. His successor, Philip de Poictou, entertained King John here in 1214, and the charter granted by that monarch to the burgesses of Newcastle bears date at Stockton. It continued to be the occasional residence of the Bishops of Durham, and seems to have escaped in a great measure the commotions and border feuds which then agitated this part of England, with the exception of an inroad of the Scots in 1322, who plundered and burnt the town. At the period of the parliamentary war the castle was taken possession of by the royalists, some importance being attached to its commanding the old passage of the Tees; it was afterwards surrendered to the parliamentary forces, and in 1645 was garrisoned by the Scots, but delivered by them to the English; in 1647 it was ordered by the parliament to be dismantled, and about five years afterwards its complete destruction was accomplished, no part of the structure now remaining, although the fosse may still be traced. The town suffered severely from the overflowing of the river in 1771, 1783, and 1822. It is situated on an eminence on the northern bank of the Tees, and has advanced rapidly in prosperity since the middle of the seventeenth century, at which period it consisted principally of mean hovels, the better houses being constructed with "post and pile," and not one built of brick. It is now one of the cleanest and handsomest towns in the northern part of the kingdom: the main street, which is about half a mile in length, is broad, and contains some good houses, chiefly of brick, the few of stone having been erected with the materials of the dilapidated castle: from this street smaller ones branch off towards the river, and on the western side of the town a great number of new houses has been recently built: the streets are well paved, and lighted with gas under the authority of an act of parliament passed in 1822. A handsome stone bridge over the river was commenced in 1764, and completed in 1769, at an expense of £8000, raised by subscription on shares; it has five elliptical arches, the span of the central arch being 72 feet, and its height from low-water mark 23. An annuity of £90, with £3 for every acre of land occupied by the road leading to the bridge, was directed, by the act of parliament under which it was constructed, to be paid to the Bishop of Durham, as a compensation for the tolls of the ferry which existed previously: land was purchased for the bishop in lieu of this annuity, and the debt having been paid off, the tolls ceased in 1820. The theatre, in Green-Dragon-yard, Finkle-street, is a neat building. A mechanics' institute and library was established in 1824; it contains upwards of 300 volumes, with apparatus for lecturing. There are also a subscription library and two news-rooms. Races are held annually in August, a week after those of York, at Tibbersley, about three miles from Stockton; and assemblies occasionally.

The situation of Stockton, on a river navigable eight miles above it, and about the same distance from the sea, affords it many commercial advantages; and the increased shipping, and amount of duties, evince the progressive extension of its mercantile interests. The port is a member of that of Newcastle-upon-Tyne; the dues, from the payment of which vessels belonging to the cinque-ports are exempt, are the property of the bishop, and are held on lease by the corporation. Ships of large size were formerly obliged to receive and unload their cargoes at Portrack, a mile down the river, or at Cargo-Fleet, or Cleveland Port, a mile lower on the Yorkshire side; but, in 1808, a company was incorporated by act of parliament, called the "Tees Navigation Company," and a canal was cut from Portrack to the town, capable of admitting vessels of 300 tons' burden, which has greatly benefited the trade of the town, and amply repaid the shareholders: another act was recently obtained for the extension of this canal to Newport, a distance of nearly two miles. In 1815, this port was made a bonding-port for certain goods: its principal trade coastwise is with London, Hull, Leith, Sunderland, &c., and comprises the exportation of most articles of agricultural produce, coal, linen and worsted yarn, and more particularly lead, of which many hundred tons, brought chiefly from Yorkshire and the borders of Durham and Northumberland, are annually shipped; it also forms the chief article of exportation in its foreign trade, which is with the Baltic, Holland, Hamburgh, and the British and American colonies, whence it receives in return materials for ship-building, timber for other purposes, tallow, &c. Two shipping companies in the London, and two in the foreign, trade have been established. The amount of duties paid at the custom-house, for the year ended Jan. 5th, 1838, was £64,515; and in that year 1,145,837 tons of coal, &c., were shipped coastwise. The principal branches of manufacture are those connected with the shipping; there are two ship-builders' yards, five manufactories for sail-cloth, two rope-walks, two iron-foundries, and a block and pump manufactory; there are also three breweries, some corn-mills, a mill for spinning yarn, and one for worsted. The fishery of the Tees was formerly a great source of prosperity to the town, but it has considerably declined; eastward of the bridge it belongs to the bishop, but is open to poor fishermen under certain regulations. A rail-road from Witton Park and other collieries, by Darlington, to this place, was constructed in 1825, and is productive of great advantage: by a recent act of parliament the line has been extended to Middlesbro', about four miles lower down the river, where commodious staiths have been erected. A branch of the Clarence railway, which terminates at Haverton Hill, on the river Tees, extends to this town. On the decline of Hartlepool, in 1680, Stockton was selected as the port for the establishment of the principal officers of the customs, and three legal or free quays were appointed. A new custom-house, a plain commodious building, has recently been erected; the out-stations are Hartlepool and Seaton. The market, granted by Bishop Anthony Beck, in 1310, is on Wednesday and Saturday, and is well attended; and the shambles, erected in 1825, in front of the town-hall, form a neat range of enclosed brick buildings: a handsome stone column, of the Doric order, 33 feet high, stands in the centre of the market-place. Fairs are held on the last Wednesday before May 13th, and Nov. 23rd, which are general and statute fairs, and there are cattle fairs on the last Wednesday in every month.

The period at which Stockton was incorporated is uncertain, but is supposed to be about the commencement of the thirteenth century; the last charter was granted by Bishop Cosin, in 1666. The corporation consisted of a mayor, eight aldermen, a recorder, town-clerk, sergeant-at-mace, and others: the mayor was a justice of the peace. The government is now vested in a mayor, six aldermen, and 18 councillors, under the act of the 5th and 6th of William IV., cap. 76, for an abstract of which see the Appendix, No. I. The borough is divided into two wards. The mayor and late mayor are justices of the peace, and three others are appointed by commission. The mayor's jurisdiction on the river, under the bishop, empowers him to levy a duty on all vessels entering the port; it extends from the bar to the Wathstead, betwixt Aislaby and Middleton-St. George, on the Yorkshire as well as the Durham side of the Tees. The town comprises two constablewicks, one called the Borough, including that part which is wholly freehold, and the other called the Town, consisting of that portion which is partly held by copy of court roll under the Bishop of Durham, and partly by long leases for years under the vicar and vestrymen of Stockton: they form, however, but one township, uniting in the maintenance of the poor. The bishop is lord of the borough, and holds courts leet and baron by his steward; suits of trespass and debts under 40*s.* are cognizable by these courts, the jurisdiction of which is limited to the borough. A halmote court is also held twice a year, in which similar causes are tried as in the court baron. Petty sessions for Stockton ward are holden here; and the town has been made a polling-place for the southern division of the county. The municipal business is transacted in the town-hall, built in 1735, and enlarged in 1744: it stands nearly in the centre of the main street, and is a handsome quadrangular brick building, surmounted by a light clock-tower and a spire, with a piazza stretching along the lower story on its north side; the upper part of the building contains a court-room, an assembly-room, and other apartments, and the lower part is disposed in shops, &c.

Stockton was formerly a chapelry in the parish of Norton, from which it was separated by an act of parliament obtained in 1713, and constituted a distinct parish. The living is a vicarage not in charge; net income, £247; patron, Bishop of Durham; impropriator, R. W. Myddleton, Esq. The ancient chapel, supposed to have been built about the year 1237, was taken down, and the building of the present church was completed in 1712, at an expense of about 1600: it is a neat and commodious edifice of brick, with a tower 80 feet high, at the western end; in the vestry-room is a small library, chiefly of theological works. A new church has recently been erected at the south end of the town, which has in consequence been divided into two ecclesiastical districts; it contains 400 free sittings, the Incorporated Society having granted £600 in aid of the expense. There are places of worship for Particular Baptists, the Society of Friends, Independents, Primitive and Wesleyan Methodists, Unitarians, and Roman Catholics, several of which have Sunday schools attached. A charity school was founded by subscription, in 1721, since which period it has been endowed with various bequests and donations, among which are £1100 by George Brown in 1811, £950 by George Sutton, in 1815, £700 by the Bishop of Durham in 1824, £200 by John Swainson, and £100 each by the Trustees of Lord Crewe's charity and Nicholas Swainson, altogether producing an annual income of £250, which is usually augmented to £300 by subscriptions. The present building, comprising separate school-rooms for boys and girls, and a dwelling-house for the master and the mistress, whose joint salaries are £100, was erected in 1819. The school is conducted on the National system, and affords instruction to about 200 boys and 60 girls, 40 of each being clothed. A school of industry for girls, instituted in 1803, is supported partly by endowment, and partly by subscription. In 1785, a room for a grammar school was erected by subscription: it possesses no endowment, but the corporation give a small annual stipend towards its support. Stockton, in conjunction with Norton, is entitled to a scholarship at Brasenose College, Oxford, with an endowment of £8 per annum, founded by Dr. Claymond, formerly vicar of Norton. The almshouses were originally erected about the year 1682, and were rebuilt in 1816, from a bequest of £3000 by George Brown, Esq., the benefactor to the charity school; they contain a committee-room and dispensary, and thirty-six apartments for the same number of inmates. Elizabeth Bunting, in 1765, bequeathed £300 for the benefit of poor persons not receiving parochial aid: this sum was invested in the purchase of £378. 13. 6. three per cent. consols.; and the dividends, amounting to £11. 7., are distributed according to the will of the donor. The dispensary, founded in 1792, and revived in 1815, occupies a part of the workhouse. The savings' bank, established in 1815, is held in a room in the almshouses. There are several benefit societies in the town. The poor law union of Stockton comprises 42 parishes or places, under the care of 55 guardians, and, according to the census of 1831, contains a population of 23,468. Stockton is the birthplace of Joseph Ritson, a refined critic, and author of "Ancient Songs and Metrical Romances;" of Brass Crosby, Lord Mayor of London at the period of the commotions occasioned by the prosecution of Wilkes, who was committed to the Tower for refusing to allow a warrant, issued by the Speaker of the House of Commons, to be executed in the city; and of Joseph Reed, a dramatic poet.

STOCKWELL, a chapelry, in the parish of LAMBETH, Eastern Division of the hundred of BRIXTON and of the county of SURREY, 3 miles (S. S. W.) from London: the population is returned with the parish. This place, which consists of ranges of handsome houses extending on both sides of the road from Kennington to Clapham, has within the last few years been greatly increased by the erection of numerous pleasing villas and elegant cottages; and, to meet the wants of the increasing population, it is in contemplation to erect a church, for which funds will be raised by subscription, aided by a grant from the Diocesan Society. The streets are partially lighted with gas, and the inhabitants are supplied with water from the South London water-works. Here is an extensive ale brewery; also a spacious and handsome assembly-room, in which concerts and public meetings are held. The chapelry is within the jurisdiction of the court of requests held in the borough of Southwark, and within the limits of the new police act. The chapel, erected by Archbishop Secker, is dependent on the mother church at Lambeth; it has been repaired within the last few years, and a district has been recently assigned to it. There is a place of worship for Independents. The Stockwell proprietary grammar school, in Park-road, Clapham-road, erected at an expense of £1658, from a design by Mr. J. Davies, architect of Highbury College, is a handsome building fronted with Bath stone, in the Elizabethan style of architecture, consisting of a centre and two wings extending 94 feet in length, and having in front an open corridor, communicating with a vestibule to the school-room in the centre, which rises above the wings and is flanked with square embattled turrets, between which is a large handsome window giving light to the school-room, which is additionally lighted by windows on each side; above this window is a canopied niche with a pedestal for a statue, surmounted by a pinnacled turret rising from the apex of the gable. A National school is supported by subscription; the school-house, of which the first stone was laid by Archbishop Sutton, was erected in 1818. A building formerly used as a place of worship for Baptists has been lately appropriated as an infants' school. A Sunday Reading Society, which is held in this school, is conducted by the minister of Stockwell for the instruction of elderly persons during the intervals of divine service, and is attended by about 70 members. By a will of the late Mr. Angel, the validity of which is disputed, it appears to have been his intention to found a college, with a chapel attached to it, for aged persons of respectable station in reduced circumstances.

STOCKWITH, EAST, a hamlet, in the parish and union of GAINSBOROUGH, wapentake of CORRINGHAM, parts of LINDSEY, county of LINCOLN, 3¾ miles (N. N. W.) from Gainsborough; containing 269 inhabitants. There is a place of worship for Wesleyan Methodists.

STOCKWITH, WEST, a chapelry, in the parish of MISTERTON, union of GAINSBOROUGH, North-Clay Division of the wapentake of BASSETLAW, Northern Division of the county of NOTTINGHAM, 4 miles (N. N. W.) from Gainsborough; containing 635 inhabitants. There is a place of worship for Wesleyan Methodists. Ten poor children are instructed for £10 a year, paid by the trustee of the late William Huntington.

STOCKWOOD (*ST. EDWOLD*), a parish, in the union of SHERBORNE, liberty of SUTTON-POINTZ, Sherborne Division of the county of DORSET, 8 miles (S. S. W.) from Sherborne; containing 33 inhabitants. The living is a discharged rectory, valued in the king's books at £5. 13. 4.; present net income, £160: it is in the patronage of Miss E. Bellamy.

STODDAY, a joint township with Ashton, in the parish of LANCASTER, hundred of LONSDALE, south of the sands, Northern Division of the county palatine of LANCASTER, 2 miles (S. S. W.) from Lancaster: the population is returned with Ashton. A National school has been established.

STODMARSH (*ST. MARY*), a parish, in the union of BRIDGE, hundred of DOWNHAMFORD, lathe of ST. AUGUSTINE, Eastern Division of the county of KENT, 4½ miles (E. N. E.) from Canterbury; containing 119 inhabitants. The living is a donative, endowed with the rectorial tithes; net income, £128; patron, Archdeacon of Canterbury. The church is in the early style of English architecture.

STODY (*ST. MARY*), a parish, in the union of ERPINGHAM, hundred of HOLT, Western Division of the county of NORFOLK, 3 miles (S. W. by S.) from Holt; containing 161 inhabitants. The living is a discharged rectory, with that of Hunworth united, valued in the king's books at £6. 3. 4.; present net income, £342: it is in the patronage of Lady Suffield. Three pounds per annum, the gift of a Mr. Symonds, is applied for teaching poor children.

STOGUMBER (*ST. MARY*), a parish, in the union of WILLITON, hundred of WILLITON and FREEMANNERS, Western Division of the county of SOMERSET, 7 miles (N. by E.) from Wiveliscombe; containing 1294 inhabitants. The living is a discharged vicarage, valued in the king's books at £11. 18. 7½.; present net income, £239; patrons and appropriators, Dean and Chapter of Wells. There is a place of worship for Baptists; and two schools and a Sunday National school are partly supported by subscription. A market was formerly held on Saturday, and fairs are still held on May 6th and August 1st. Roman coins have been discovered here.

STOGURSEY, or STOKE-COURCY (*ST. ANDREW*), a parish, in the union of WILLITON, hundred of CANNINGTON, Western Division of the county of SOMERSET, 8½ miles (N. W. by W.) from Bridg-water; containing 1496 inhabitants. The parish is bounded on the north by the Bristol Channel. The living is a vicarage, with that of Lilstock annexed, valued in the king's books at £16. 7. 6.; present net income, £389; patrons and impropriators, Provost and Fellows of Eton College. A school is supported by private charity. A Benedictine priory, a cell to the abbey of L'Onley in Normandy, was founded here in the reign of Henry II., which at the suppression was valued at £58 per annum, and granted by Henry VI. to Eton College.

STOKE, a township, in the parish of ACTON, union and hundred of NANTWICH, Southern Division of the county of CHESTER, 3½ miles (N. W.) from Nantwich; containing 124 inhabitants. The Chester canal passes through the township.

STOKE (*ST. MICHAEL*), a parish, in the union of FOLESHILL, county of the city of COVENTRY, Northern Division of the county of WARWICK, 1½ mile (E.) from Coventry; containing 848 inhabitants. The living is a vicarage, not in charge, with that of Sow or Walsgrave annexed; net income, £333: it is in the patronage of the Crown; impropriator of Stoke, W. Pridmore, Esq., and of Sow, Earl of Craven. The Coventry canal passes through the parish. A National school has been established.

STOKE, a township, in the parish of HOPE, union of BAKEWELL, hundred of HIGH PEAK, Northern Division of the county of DERBY, 1½ mile (N. E.) from Stoney-Middleton; containing 60 inhabitants.

STOKE (*ST. PETER*), a parish, in the union and hundred of Hoo, lathe of AYLESFORD, Western Division of the county of KENT, 8 miles (N. E.) from Rochester; containing 432 inhabitants. The living is a discharged vicarage, valued in the king's books at £8. 11. 8.; present net income, £180; patron and impropriator, Baldwin Duppa Duppa, Esq. Here is a Sunday National school.

STOKE (*HOLY CROSS*), a parish, in the union and hundred of HENSTEAD, Eastern Division of the county of NORFOLK, 5½ miles (N. E. by E.) from St. Mary-Stratton; containing 350 inhabitants. The living is a vicarage; net income, £162; patrons and appropriators, Dean and Chapter of Norwich. Here is a National school.

STOKE (*ST. MILBURGH*), a parish, in the union of LUDLOW, partly in the hundred of MUNSLOW, but chiefly in the liberty of the borough of WENLOCK, Southern Division of the county of SALOP, 7 miles (N. E. by N.) from Ludlow; containing 597 inhabitants. The living is a vicarage, endowed with a portion of the rectorial tithes, and valued in the king's books at £6. 13. 4.; present net income, £474; patron and incumbent, Rev. George Morgan; impropriators of the remainder of the rectorial tithes, the Landowners.

STOKE (*ST. GREGORY*), a parish, in the union of TAUNTON, hundred of NORTH CURRY, Western Division of the county of SOMERSET, 5 miles (W. by N.) from Langport; containing 1507 inhabitants. The parish is bounded on the north-east by the navigable river Parret, and on the north-west by the Tone, which latter is crossed by three bridges, one of them forming a connection with the Isle of Athelney, famous as the retreat of the renowned Alfred. The living is a perpetual curacy; net income, £61; patron, E. W. King Coker, Esq.; appropriators, Dean and Chapter of Wells. The appropriate tithes have been commuted for a rent-charge of £388, subject to the payment of rates, and the vicar of North Curry receives a rent-charge of £147 out of the tithes of this parish.

STOKE (*St. Mary*), a parish, in the union of Taunton, hundred of Taunton and Taunton-Dean, Western Division of the county of Somerset, 3¼ miles (S. E. by E.) from Taunton; containing 275 inhabitants. The living is a rectory and donative; patron, Lord Portman. The tithes have been commuted for a rent-charge of £110, subject to the payment of rates. There is a place of worship for Independents.

STOKE-ABBAS (*St. Mary*), a parish, in the union and hundred of Beaminster, Bridport Division of the county of Dorset, 2 miles (W. by S.) from Beaminster; containing 587 inhabitants. The living is a rectory, valued in the king's books at £9. 15.; present net income, £420; patrons, Warden and Fellows of New College, Oxford.

STOKE-ALBANY (*St. Botolph*), a parish, in the union of Market-Harborough, hundred of Corby, Northern Division of the county of Northampton, 5 miles (S. W. by W.) from Rockingham; containing 339 inhabitants. The living is a rectory, valued in the king's books at £13. 6. 8.; present net income, £216; patron, Lord Sondes.

STOKE-ASH (*All Saints*), a parish, in the union and hundred of Hartismere, Western Division of the county of Suffolk, 3½ miles (S. W.) from Eye; containing 392 inhabitants. The living is a rectory, valued in the king's books at £11. 1. 3.; present net income, £275; patron and incumbent, Rev. John Ward. There is a place of worship for Baptists.

STOKE-BARDOLPH, a township, in the parish of Gedling, union of Basford, Southern Division of the wapentake of Thurgarton and of the county of Nottingham, 5 miles (E. N. E.) from Nottingham; containing 181 inhabitants.

STOKE-BISHOP, a tything, in the parish of Westbury-upon-Trym, Lower Division of the hundred of Henbury, Western Division of the county of Gloucester, 2¼ miles (N. W. by N.) from Bristol; containing 2328 inhabitants.

STOKE, BISHOP'S (*St. Mary*), a parish, in the union of New Winchester, hundred of Fawley, Winchester and Northern Divisions of the county of Southampton, 6½ miles (W. by N.) from Bishop's-Waltham; containing 1026 inhabitants. The London and South-Western railway passes near the parish. The living is a rectory, valued in the king's books at £14. 17. 6.; present net income, £437, patron, Bishop of Winchester. Here is a National school for boys.

STOKE-BLISS, a parish, in the union of Tenbury, partly in the Upper Division of the hundred of Doddingtree, Worcester and Western Divisions of the county of Worcester, but chiefly in the hundred of Broxash, county of Hereford, 6¼ miles (N.) from Bromyard; containing 344 inhabitants. The living is a rectory, valued in the king's books at £6. 16. 8.; present net income, £374: it is in the patronage of the Crown.

STOKE-BRUERNE (*St. Mary*), a parish, in the union of Towcester, hundred of Cleley, Southern Division of the county of Northampton, 3½ miles (E. N. E.) from Towcester; containing 762 inhabitants. The living is a rectory, valued in the king's books at £30; present net income, £422; patrons, Principal and Fellows of Brasenose College, Oxford. The Grand Junction canal passes through a tunnel two miles long, partly in this parish, and partly in that of Blisworth.

STOKE-BY-CLARE (*St. Augustine*), a parish, in the union and hundred of Risbridge, Western Division of the county of Suffolk, 2¼ miles (S. W. by W.) from Clare; containing 792 inhabitants. The living is a perpetual curacy; net income, £130; it is in the patronage of Lady Rush. The navigable river Stour passes on the south of the parish. Sir Gervaise Elwes, Bart., in 1678, bequeathed a rent-charge of £10 for teaching poor children, which is applied in aid of a Sunday school; and there is a fund for apprenticing children, amounting to £33 per annum, the rent of twelve acres of land purchased with a bequest of £450 by Mary Barnes, in 1681, for that purpose. An almshouse, consisting of three cottages, occupied by six poor widows, was founded by Richard Brown, in 1526, and has an endowment of £8 per annum, arising from different bequests. Richard de Clare, Earl of Hereford, in 1124, removed the monks of Bec, whom his father had placed in the castle of Clare, to this village, first into the parish church of St. Augustine, and afterwards to a church built for them, and dedicated to St. John the Baptist. In 1415, Edmund Mortimer, Earl of March, its then patron, procured it to be changed into a college of Secular priests, for a dean, six prebendaries, eight vicars, and other officers. This was valued in the 26th of Henry VIII. at £324. 4. 1. per annum. Matthew Parker, Archbishop of Canterbury, was the last dean. A modern house now stands upon the site, which is the seat of J. P. Elwes, Esq., and was the estate and residence of the well-known miser, John Elwes, Esq.

STOKE-CANNON (*St. Mary Magdalene*), a parish, in the union of St. Thomas, hundred of Wonford, Wonford and Southern Divisions of the county of Devon, 3½ miles (N. N. E.) from Exeter; containing 446 inhabitants. The living is a perpetual curacy, in the patronage of the Dean and Chapter of Exeter, the appropriators; net income, £180. The church contains 255 free sittings, the Incorporated Society having granted £120 in aid of the expense; it has a curious ancient font. This parish is intersected by the rivers Exe and Culm: the church and manor were given by King Athelstan to the Cathedral Church of Exeter.

STOKE-CHARITY (*St. Michael*), a parish, in the union of New Winchester, hundred of Buddlesgate, Winchester and Northern Divisions of the county of Southampton, 6½ miles (S. by E.) from Whitchurch; containing 135 inhabitants. The living is a rectory, valued in the king's books at £15. 13. 6½.; patrons, President and Fellows of Corpus Christi College, Oxford. The tithes have been commuted for a rent-charge of £420, subject to the payment of rates, which on the average have amounted to £38; the glebe comprises 20 acres, valued at £26 per annum.

STOKE-CLIMSLAND, a parish, in the union of Launceston, Northern Division of the hundred of East, Eastern Division of the county of Cornwall, 3½ miles (N.) from Callington; containing 1608 inhabitants. This parish is bounded on the north by the river Inney, which runs into the Tamar on the east, and near the romantic and picturesque rocks of Cartharmartha. A fair for cattle is held on May 29th. The living is a rectory, valued in the king's books at £40; present net income, £621: it is in the patronage of the Crown, in right of the Duchy of Cornwall. The church is a very spacious structure, with a fine tower. There is a place of worship for Wesleyan Methodists. Two small sums, bequeathed by Ralph Tope, in 1718, and

Joan Clarke, in 1783, are applied in support of a school.

STOKE-D'ALBORNE (*St Mary*), a parish, in the union of EPSOM, Second Division of the hundred of ELMBRIDGE, Western Division of the county of SURREY, 1½ mile (S. E. by E.) from Cobham; containing 289 inhabitants. The living is a rectory, valued in the king's books at £13. 11. 3.; present net income, £418; patron, Rev. Hugh Smith. The church contains monuments to the Vincent family; the pulpit is richly embellished.

STOKE-DAMERALL, a parish, in the hundred of ROBOROUGH, Roborough and Southern Divisions of the county of DEVON, adjoining the borough of Plymouth, and containing 34,883 inhabitants. This parish, which includes Devonport and Morice Town, is one of the most extensive in the county; the village occupies an elevated site, and comprises several rows of excellent houses, a crescent, and some private mansions of more than ordinary beauty. Among the important public structures are, the immense reservoir of the Devonport Water Company, which supplies the government establishments and the neighbourhood in general; the military hospital, a spacious edifice of grey marble, erected in 1797, on the west side of Stonehouse creek, comprising four large square buildings, of similar size and form, connected by a piazza of 41 arches; and the Blockhouse, occupying an eminence north of the village, surrounded by a fosse and drawbridge, commanding a most magnificent prospect: in addition to the military use of this fortress, it forms an admirable landmark for ships entering the Sound. On the eastern bank of the Hamoaze is Morice Town, consisting of four principal streets, and so named from the former lord of the manor, now commonly called the New Passage, where a ferry was established in 1800, to communicate with Cornwall, at Torpoint, on the opposite shore; a floating bridge, worked by steam, and held in its course by chains across the bed of the river, was completed a few years since. The sides of the harbour are lined with various wharfs; and in the town is a large establishment, called the Tamar Brewery. At a short distance is the powder magazine, which, although it covers an area of five acres, was insufficient in time of war, when line-of-battle ships were fitted up as floating magazines. A fair is held here on Whit-Monday. In the vicinity, at Cross hill, is a very extensive quarry of slate of a durable quality. Stoke-Damerall, with the township of East Stonehouse, is included in the newly-enfranchised borough of Devonport. The living is a rectory, valued in the king's books at £18. 18. 9.; patron, Sir John St. Aubyn, Bart. The church is a mean but spacious edifice, with a low square tower. A new church has been erected. There are places of worship for Independents, and Calvinistic and Wesleyan Methodists. The Devonport public school for poor boys, and the Naval and Military school for the sons of seamen, soldiers and marines, fishermen, and watermen, are in this parish. Another school is supported by a parochial rate, and two others by subscription; there are also numerous Sunday schools.

STOKE-DOYLE (*St. Rumbald*), a parish, in the union of OUNDLE, hundred of NAVISFORD, Northern Division of the county of NORTHAMPTON, 1½ mile (S. W. by S.) from Oundle; containing 165 inhabitants. The living is a rectory, valued in the king's books at £20. 2. 11.; present net income, £142; patron, G. Capron, Esq.

STOKE, DRY (*St. Andrew*), a parish, in the union of UPPINGHAM, partly in the hundred of GARTREE, Southern Division of the county of LEICESTER, but chiefly in the hundred of WRANDIKE, county of RUTLAND, 3½ miles (S. S. W.) from Uppingham; containing 53 inhabitants. The living is a rectory, valued in the king's books at £11. 2. 1.; present net income, £385; patron, Marquess of Exeter.

STOKE, EARL, county of WILTS.—See EARL-STOKE.

STOKE, EAST (*St. Mary*), a parish, in the union of WAREHAM and PURBECK, hundred of WINFRITH, Wareham Division of the county of DORSET, 4 miles (W. by S.) from Wareham; containing 561 inhabitants. This parish occupies a pleasant situation, commanding a distant view of the Purbeck hills, and comprising a fertile valley watered by the river Frome. The surrounding scenery is finely varied, and greatly enriched by the plantations at Heffleton, the seat of James C. Fyler, Esq., for which improvement two gold medals were awarded by the Society of Arts to the late A. Bain, Esq., M.D., formerly proprietor of that estate; and among the interesting features within the limits of the parish are the beautiful and picturesque remains of Bindon Abbey. The living is a rectory, valued in the king's books at £14. 12. 11.; present net income, £135; patron, Sir W. Oglander, Bart. The church is a neat and simple edifice, rebuilt by subscription in 1827, at an expense of £1700; it contains 180 free sittings, the Incorporated Society having granted £100 in aid of the expense. A school-house has been built near the church, in which poor children are gratuitously instructed: there is also a National school. Bindon Abbey was founded in 1172, by Robert de Newburgh and Maud his wife, who endowed it for monks of the Cistercian order; it was dedicated to St. Mary, and at the dissolution its revenue was valued at £229. 2. 1.: the site was granted, in the 37th of Henry VIII., to Sir Thomas Poynings. The remains consist principally of an angle of the tower of the church, and part of the walls, with the foundations.

STOKE (EAST) NEAR NEWARK (*St. Oswald*), a parish, in the union of SOUTHWELL, Northern Division of the wapentake of THURGARTON, Southern Division of the county of NOTTINGHAM, 3¾ miles (S. W.) from Newark; containing 320 inhabitants. The living is a discharged vicarage, with those of Coddington and Syerston annexed, valued in the king's books at £8. 13.; present net income, £372; patron and appropriator, Chancellor of the Cathedral Church of Lincoln. There is a chapel of ease at Elston, in this parish. On Stoke field was fought, in 1487, the decisive battle between the armies of Henry VII. and John de la Pole, Earl of Lincoln, who had espoused the cause of the impostor Lambert Simnel, in which the earl and 4000 of his followers were slain: this is said to be the first action wherein cannon was used with success: human bones, fragments of armour, coins, &c., have been frequently ploughed up on the spot. The river Trent and the old Fosse-road pass through the parish. A school is partly supported by an individual. An hospital, dedicated to St. Leonard, was founded here before the time of Henry I., for a master and brethren, chaplains, and several sick persons, whose revenue at the dissolution was valued at £9.

STOKE-EDITH (*St. Mary*), a parish, in the hundred of RADLOW, union and county of HEREFORD, 7¼ miles (E.) from Hereford; containing 505 inhabitants. The living is a rectory, with the perpetual curacy of West Hide annexed, valued in the king's books at £15; present net income, £470; patron, Edward T. Foley, Esq. A school is supported by the lord of the manor. An ancient sword, some curious beads, several human skeletons with their faces downwards, and other relics, were found at Radlow Bush some years ago.

STOKE-FERRY (*All Saints*), a market-town and parish, in the union of Downham, hundred of Clackclose, Western Division of the county of Norfolk, 38 miles (W. by S.) from Norwich, and 88½ (N. N. E.) from London; containing 706 inhabitants. This town is situated on the banks of the river Wissey, on the turnpike-road from London to Newmarket. In the reign of Henry III. it obtained a grant for holding a weekly market and an annual fair, which was confirmed by Henry VI.: the market was for a long period disused, but has been recently revived, and is now held on Friday, principally for corn; and the fair on December 6th, for cattle. Many of the inhabitants are employed in the very extensive malting establishment of Messrs. Whitbread and Co., whose superintendent has successfully adopted a superior plan for drying the malt, and has obtained a patent for his improvements. The living is a perpetual curacy; net income, £71: it is in the patronage of the Crown; impropriator, G. Eyres, Esq. The church had formerly a square tower. Twenty-five boys of Stoke-Ferry and Wretton are gratuitously educated in a school founded by the late James Bradfield, Esq.; and a Sunday National school has been established.

STOKE-FLEMING (*St. Peter*), a parish, in the union of Kingsbridge, hundred of Coleridge, Stanborough and Coleridge, and Southern, Divisions of the county of Devon, 2 miles (S. S. W.) from Dartmouth; containing 725 inhabitants. The living is a rectory, valued in the king's books at £31. 6. 0½.; present net income, £649; patron, Charles Farwell, Esq. The church contains some interesting monuments. Three schools are partly supported by subscription.

STOKE-GABRIEL (*St. Gabriel*), a parish, in the union of Totnes, hundred of Haytor, Paignton and Southern Divisions of the county of Devon, 4 miles (S. E. by E.) from Totnes; containing 718 inhabitants. The living is a discharged vicarage, valued in the king's books at £16. 11. 10½.; present net income, £163; patrons, Sir Stafford H. Northcote, Bart., the Executors of the late Rev. John Templar, and the Rev. F. Belfield, alternately; appropriator, Chancellor of Exeter Cathedral. The church contains an ancient wooden screen; and at Watton there was formerly a chantry chapel. There is a place of worship for Baptists; also a National school. The navigable river Dart bounds the parish on the south. Capt. John Davies, the discoverer of Davies' Straits, was born here.

STOKE-GIFFORD (*St. Michael*), a parish, in the union of Clifton, Upper Division of the hundred of Henbury, Western Division of the county of Gloucester, 5 miles (N. N. E.) from Bristol; containing 441 inhabitants. The living is a discharged vicarage, valued in the king's books at £6; present net income, £60; patron and impropriator, Duke of Beaufort. John Silcocks, in 1741, bequeathed £200, directing the interest to be applied for teaching six poor children.

STOKE-GOLDING, a chapelry, in the parish of Hinckley, hundred of Sparkenhoe, Southern Division of the county of Leicester, 2¾ miles (N. W.) from Hinckley; containing 543 inhabitants. The Ashby-de-la-Zouch canal passes through the chapelry. The chapel is dedicated to St. Margaret. Here is a Sunday National school.

STOKE-GOLDINGTON (*St. Peter*), a parish, in the union of Newport-Pagnell, hundred of Newport, county of Buckingham, 4 miles (W. S. W.) from Olney; containing 912 inhabitants. The living is a rectory, united in 1736 to that of Gayhurst, and valued in the king's books at £14. 6. 3. The river Ouse runs through the parish, in which lime and other stone abound. There was formerly a chapel at Eakley, which is said to have been once a distinct parish. A National school has been established.

STOKE-HAMOND (*St. Mary*), a parish, in the union of Leighton-Buzzard, hundred of Newport, county of Buckingham, 2¾ miles (S.) from Fenny-Stratford; containing 323 inhabitants. The living is a rectory, valued in the king's books at £19. 9. 4½.; present net income, £249; patron, Bishop of Lincoln. There is a place of worship for Wesleyan Methodists. The London and Birmingham railway passes near the church.

STOKE-LACY (*St. Pete and St. Paul*), a parish, in the union of Bromyard, hundred of Broxash, county of Hereford, 4¼ miles (S. W. by S.) from Bromyard; containing 381 inhabitants. The living is a rectory, valued in the king's books at £8; present net income, £250; patron, T. Apperley, Esq. Limestone abounds in the neighbourhood. The late Archdeacon of Hereford bequeathed property for the erection and endowment of a free school for children of both sexes, which gift has been augmented by Mr. Brown, of Evington, with £20 per annum, and £50 towards building the school-house; it is on the National system.

STOKE-LANE, or STOKE (*St. Michael*), a parish, in the union of Shepton-Mallet, hundred of Whitestone, Eastern Division of the county of Somerset, 4 miles (N. E.) from Shepton-Mallet; containing 980 inhabitants. The living is a perpetual curacy; net income, £80; patron, Vicar of Doulting; impropriator, Richard Strachey, Esq. The church contains 210 free sittings, the Incorporated Society having granted £200 in aid of the expense. There is a place of worship for Wesleyan Methodists.

STOKE, LIMPLEY, a joint chapelry with Winsley, in the parish of Great Bradford, union and hundred of Bradford, Westbury and Northern Divisions, and Trowbridge and Bradford Subdivisions, of the county of Wilts, 3 miles (W. by S.) from Bradford: the population is returned with Winsley. The chapel is dedicated to St. Mary.

STOKE-LYNE (*St. Peter*), a parish, in the union of Bicester, hundred of Ploughley, county of Oxford, 4¼ miles (N. by W.) from Bicester; containing 593 inhabitants. The living is a discharged vicarage; net income, £173; patrons, Trustees of J. Bullock, Esq.; impropriator, — Coles, Esq. A school is partly supported by the minister.

STOKE-MANDEVILLE (*St. Mary*), a parish, in the union of Wycombe, hundred of Aylesbury, county of Buckingham, 2¾ miles (N. W. by W.) from Wendover; containing 461 inhabitants. The living is annexed, with those of Buckland and Quarrendon, to the vicarage of Bierton. Here is a National school.

STOKE-NEAR-NAYLAND (*St. Mary*), a parish, in the union of SUDBURY, hundred of BABERGH, Western Division of the county of SUFFOLK, 2 miles (N. E. by N.) from Nayland; containing 1447 inhabitants. In this parish is Gifford Hall, an ancient structure with a fine entrance gateway, built in the early part of the reign of Henry VIII.: the hall has a lofty roof of richly carved oak. Tendring Hall, the seat of Sir J. R. Rowley, Bart., was formerly the residence of the Dukes of Norfolk, and the Earl of Surrey wrote his poems here; the grounds contain some fine specimens of stately oak. The living is a vicarage, valued in the king's books at £19. 0. 10.; patron, Sir Joshua R. Rowley, Bart.; impropriator, P. Mannock, Esq. The impropriate tithes have been commuted for a rent-charge of £1254. 7. 6., and the vicarial for £305, subject to the payment of rates, which on the average on the latter have amounted to £42. The church is a spacious structure, in the later style of English architecture, with a finely-proportioned tower, and contains numerous monuments and some ancient brasses, noticed in Weever's Funeral Monuments. A chapel for the inhabitants of Leavenheath has lately been erected by subscription, and is endowed with three acres of land, and £1100 in the public funds: it contains 190 free sittings, the Incorporated Society having granted £100 in aid of the expense. There is a Roman Catholic chapel; and a National school has been established. Four cottages were endowed with 10*s.* per quarter from land in the parish of Higham, by Lady Anne Windsor; and five cottages were left in 1675, by Thos. Purslowe, for the benefit of the poor. A monastery existed here in the middle of the tenth century, to which Earl Alfgar, and his daughters, Æthelfled and Ægelfled, made considerable donations, it being the burial-place of that noble family. Sir John Capel, lord mayor of London in 1503, was a native of this place; and the Rev. William Jones, the well-known author, was vicar of the parish.

STOKE-NEXT-GUILDFORD (*St. John the Evangelist*), a parish, in the union of GUILDFORD, First Division of the hundred of WOKEING, Western Division of the county of SURREY, ¾ of a mile (N.) from Guildford; containing 1327 inhabitants. This parish, which is intersected by the Wey canal, is situated on the road by Kingston to Portsmouth, and part of it is now included within the limits of the borough of Guildford. There are a paper and a flour mill. The living is a rectory, valued in the king's books at £18. 0. 5.; patrons, certain Trustees. The tithes have been commuted for a rent-charge of £679, subject to the payment of rates. The church is in the later style of English architecture, and contains several neat monuments. A girls' school is supported by the Rev. Samuel Paynter, the rector. James Price bequeathed Bank annuities, producing a dividend of £96, to the poor; and three almshouses for six poor women above 60 years of age were founded and endowed by Henry and William Parsons, Esqrs. Mrs. Charlotte Smith, the celebrated novelist, was buried here.

STOKE, NORTH, a township, in the parish of SOUTH STOKE, union of GRANTHAM, wapentake of WINNIBRIGGS and THREO, parts of KESTEVEN, county of LINCOLN, 2¾ miles (N. by W.) from Colsterworth; containing 124 inhabitants.

STOKE, NORTH (*St. Mary*), a parish, in the union of WALLINGFORD, hundred of LANGTREE, county of OXFORD, 2½ miles (S.) from Wallingford; containing 199 inhabitants. The living is a vicarage, with that of Newnham Murren annexed, valued in the king's books at £14. 10.; present net income, £568; patrons and impropriators, Master and Fellows of St. John's College, Cambridge. There is a chapel of ease at Ipsden, in this parish.

STOKE, NORTH (*St. Martin*), a parish, in the union of KEYNSHAM, hundred of BATH-FORUM, Eastern Division of the county of SOMERSET, 4¼ miles (N. W.) from Bath; containing 128 inhabitants. The living is a discharged rectory, valued in the king's books at £5. 7. 6.: it is in the patronage of the Crown. The tithes have been commuted for a rent-charge of £100, subject to the payment of rates, which on the average have amounted to £6; the glebe comprises 26 acres, valued at £52 per annum. The parish is bounded on the west by the river Avon, the ground gradually rising from that river to the heights of Lansdown, which gives the title of Marquess to the family of Petty.

STOKE, NORTH, a parish, in the hundred of POLING, rape of ARUNDEL, Western Division of the county of SUSSEX, 3 miles (N. by E.) from Arundel; containing 86 inhabitants. The living is a perpetual curacy, valued in the king's books at £5. 14. 4½.; present net income, £57; patron and impropriator, Earl of Egremont. The river Avon separates this parish from that of South Stoke. Under an old drain lying in the course of an arm or tributary of this river was found, in March 1834, a canoe, or ancient vessel, formed out of the trunk of an oak tree, which was presented by the Earl of Egremont to the trustees of the British Museum: it is 35 feet 4 inches in length, nearly 2 feet in depth, and between 4 and 5 feet in breadth.

STOKE-ORCHARD, a chapelry, in the parish of BISHOP'S-CLEEVE, union of TEWKESBURY, hundred of CLEEVE, or BISHOP'S-CLEEVE, Eastern Division of the county of GLOUCESTER, 4¼ miles (S. E.) from Tewkesbury; containing 229 inhabitants.

STOKE-PERO, a parish, in the union of WILLITON, hundred of CARHAMPTON, Western Division of the county of SOMERSET, 6¾ miles (W. S. W.) from Minehead; containing 61 inhabitants. The living is a discharged rectory, valued in the king's books at £4. 10. 10.; present net income, £92; patron, John Quick, Esq. The rusty appearance of the water among the hills indicates the probability of iron-ore lying beneath. Dunkry Beacon, a large and lofty mountain, is partly in the parishes of Cutcombe, Luccombe, Wotton-Courtney, Stoke-Pero, and Exford: its base is about twelve miles in circuit, and its height above the sea at high water is 1770 feet, being the highest eminence in the western part of England: it serves as a distant landmark, but the summit is often obscured by clouds. On the top are many loose stones of large dimensions, and among them the ruins of three large fire-hearths, the remains of Beacons. Collinson says, when the air is serene and clear, the line which bounds the horizon cannot be less than 500 miles in circumference.

STOKE-POGES (*St. Giles*), a parish, in the union of ETON, hundred of STOKE, county of BUCKINGHAM,

2 miles (N.) from Slough; containing, with a part of the town of Slough, 1252 inhabitants. The living is a discharged vicarage, valued in the king's books at £7. 17.; present net income, £319; patron and impropriator, Lord Godolphin. There is a place of worship for Wesleyan Methodists. A fair is held here on Whit-Tuesday. A school, now conducted on the National system, was erected at the expense of the parishioners, in 1798, and is supported by sundry bequests, including one by Lady Elizabeth Hatton, in 1645, the whole producing an annual income of £30. Two other schools are supported by two individuals, who have built school houses. "The hospital of Stoke-Poges," for four poor men and two women, was founded in 1557, by Lord Hastings of Sloughborough, who endowed it with a rent-charge of about £53, for the support of a chantry priest and four bedesmen. It was originally built in Stoke park, and its noble founder, becoming one of its inmates, ended his days within its walls, and was buried in the chapel attached. The ancient building was pulled down, in 1765, in pursuance of an act obtained by Mr. Penn, who, with Lady Elizabeth Hatton, in 1645; and Richard Redding, in 1717; and the Rev. Arthur Bold, in 1830, also augmented the endowment; and the hospital was refounded on its present site: the revenue is £142. The east end of the present buildings contains the residence of the almspeople, comprising a kitchen, pantry, cellar, larder, wash-house, and brew-house on the ground floor, with six bed-rooms above: the west end contains the master's apartments and the chapel. The inmates are three brethren and two sisters, who have each an allowance in money, and are provided with board, lodging, clothing, and fuel; they all dine at a common table, with the master, who must be in priest's orders, and whose salary is about £23. The vicar is eligible to the mastership, and the visiters are the Dean of Windsor and the Provost of Eton. There are bequests by Lady Elizabeth Hatton, and others, for the poor. The churchyard is the scene of "Gray's Elegy," and contains the remains of the poet; and in the field adjoining a large sarcophagus was erected, in 1799, by the late Mr. Penn, of Stoke Park, to the memory of Mr. Gray, who died at Cambridge in 1771; his body was removed hither, and deposited, with those of his mother and aunt, in a vault constructed at his own expense.

STOKE-PRIOR, a parish, in the union of LEOMINSTER, hundred of WOLPHY, county of HEREFORD; containing 478 inhabitants, of which number, 335 are in the township of Stoke-Prior, 3 miles (S. E.) from Leominster. The living is a perpetual curacy, with that of Docklow annexed; net income, £132; patron, Vicar of Leominster.

STOKE-PRIOR (*St. Michael*), a parish, in the union of BROMSGROVE, Middle Division of the hundred of OSWALDSLOW, Droitwich and Eastern Divisions of the county of WORCESTER, 1¾ mile (S.) from Bromsgrove; containing 1100 inhabitants. The living is a discharged vicarage, with St. Godwald's chapel, valued in the king's books at £12; present net income, £270; patrons and appropriators, Dean and Chapter of Worcester. In the great civil war, the court-house was almost destroyed by the royalists. The Birmingham and Worcester canal passes through the parish. Here is a National school; and Mr. John Saunders gave £10 a year for apprenticing one poor boy annually.

STOKE-RIVERS (*St. Bartholomew*), a parish, in the union of BARNSTAPLE, hundred of SHERWILL, Braunton and Northern Divisions of the county of DEVON, 5 miles (E. by N.) from Barnstaple; containing 270 inhabitants. The living is a rectory, valued in the king's books at £14. 14. 7.; present net income, £231; patron and incumbent, Rev. Charles Hiern.

STOKE, RODNEY (*St. Leonard*), a parish, in the union of WELLS, hundred of WINTERSTOKE, Eastern Division of the county of SOMERSET, 5 miles (N. W. by W.) from Wells; containing 333 inhabitants. This was long the seat of the knightly family of Rodney, whose descendant, the distinguished admiral, was elevated to the peerage as Baron Rodney of Rodney-Stoke, in 1782, for the memorable victory he had achieved over the French fleet commanded by the Compte de Grasse. The living is a discharged rectory, valued in the king's books at £8. 12. 8½.; present net income, £339; patron, Bishop of Bath and Wells. The church is principally in the Norman style of architecture, and has a curious ancient font. Here is a small endowed almshouse.

STOKE, SEVERN (*St. Denis*), a parish, in the union of UPTON-UPON-SEVERN, Lower Division of the hundred of PERSHORE, Upton and Western Divisions of the county of WORCESTER, 3 miles (N. by E.) from Upton-upon-Severn; containing 745 inhabitants. The living is a rectory, valued in the king's books at £21. 17. 4.; present net income, £746; patron, Earl of Coventry. A market and a fair were granted to be held here by Edward II., but both of them have been long since disused.

STOKE, SOUTH, otherwise STOKE-ROCHFORD (*St. Andrew and St. Mary*), a parish, in the union and soke of GRANTHAM, parts of KESTEVEN, county of LINCOLN, 2 miles (N. by W.) from Colsterworth; containing 438 inhabitants. This parish derives the latter adjunct to its name from the family of Rochford, who were anciently proprietors of the lordship: it is pleasantly situated on the river Witham, and the surrounding scenery is enlivened by the seat of Christopher Turnor, Esq., in an extensive park of richly varied and highly picturesque beauty. The living is a rectory, formerly in medieties, which were united in 1776, valued jointly in the king's books at £18. 15.; present net income, £685; patron, Prebendary of South Grantham in the Cathedral Church of Salisbury. The church contains some monuments to the Rochford family, by whom it was partly built; to the Cholmeley family of Easton; and the Turnors of Stoke. An almshouse was founded in 1677, by Sir Edmund Turnor, who endowed it for six poor persons. In this parish are the remains of a Roman villa and baths.

STOKE, SOUTH (*St. Andrew*), a parish, in the union of WALLINGFORD, hundred of DORCHESTER, county of OXFORD, 4¼ miles (S. by W.) from Wallingford; containing 751 inhabitants. The living is a discharged vicarage, valued in the king's books at £12. 16. 0½.; present net income, £136; patrons, Dean and Canons of Christ-Church, Oxford. At Woodcote, in this parish, is a chapel of ease, dedicated to St. Leonard; and there is a place of worship for Independents. Ten poor children are taught for £18 a year, arising from a bequest by the Rev. Griffith Higgs, D.D., in 1659, who also bequeathed £100, which has been invested in land, and the rents divided among the poor according to his will. There is also a National school. This parish is entitled to send one poor man to the almshouse, and two children to the school, at Goring Heath, under the will of Mr. H. Allnutt in 1724. The Great Western railway passes through the parish.

STOKE, SOUTH (*St. James*), a parish, i the union of BATH, hundred of BATH-FORUM, Eastern Division of the county of SOMERSET, 2½ miles (S. by W.) from Bath; containing 266 inhabitants. The living is a discharged rectory, with Moncton-Combe and Combe-Down annexed, valued in the king's books at £7. 18. 9.; present net income, £163; patrons, Trustees of Charles Johnson, Esq. The river Avon and the Radford canal run through the parish; the former is crossed by a bridge near its junction with the Kennet.

STOKE, SOUTH, a parish, in the hundred of AVISFORD, rape of ARUNDEL, Western Division of the county of SUSSEX, 2½ miles (N. N. E.) from Arundel; containing 101 inhabitants. The living is a rectory, valued in the king's books at £11. 15. 10.; present net income, £162; patron, Earl of Albemarle.

STOKE-TALMAGE (*St. Mary Magdalene*), a parish, in the union of THAME, hundred of PIRTON, county of OXFORD, 2 miles (S. S. W.) from Tetsworth; containing 107 inhabitants. The living is a rectory, valued in the king's books at £12. 17. 1.; present net income, £248; patron, Earl of Macclesfield.

STOKE-TRISTER, a parish, in the union of WINCANTON, hundred of NORTON-FERRIS, Eastern Division of the county of SOMERSET, 2 miles (E.) from Wincanton; containing 428 inhabitants, a few of whom are employed in the manufacture of dowlas and ticking. The living is a discharged rectory, united to Cucklington, and valued in the king's books at £7. 15. 2½.

STOKE-UNDER-HAMDON (*St. Denis*), a parish, in the union of YEOVIL, hundred of TINTINHULL, Western Division of the county of SOMERSET, 5¾ miles (W. by N. from Yeovil; containing 1365 inhabitants. The living is a perpetual curacy, valued in the king's books at £5. 10. 2½.; present net income, £89; patron and impropriator, — Hawkesworth, Esq. Here is a considerable manufacture of gloves. A free chapel, or chantry, for a provost and four priests, in honour of St. Nicholas, was founded in 1304, by Sir John Beauchamp, Knt., in the ancient castle, of which, in the time of Leland, there were extensive remains near the village, as also in the chapel, many old monuments, statues, &c., without inscriptions, and a flat marble stone, with the effigy of Maheu de Gurney, dated 1406.

STOKE-UPON-TERN (*St. Peter*), a parish, in the union of DRAYTON, Drayton Division of the hundred of NORTH BRADFORD, Northern Division of the county of SALOP; containing 1030 inhabitants, of which number, 526 are in the township of Stoke, 6 miles (S. W. by S.) from Drayton-in-Hales. The living is a rectory, valued in the king's books at £20; patron, R. Corbet, Esq. The tithes have been commuted for a rent-charge of £939, subject to the payment of rates; the glebe comprises 50 acres, valued at £100 per annum. The church contains a handsome monument of alabaster to the memory of Sir Reginald Corbet, a judge of the Common Pleas in the reign of Elizabeth. A school is partly supported by the rector. The ancient mansion-house of the Corbets has been demolished, and a farm-house erected upon its site.

STOKE-UPON-TRENT (*St. Peter ad Vincula*), a newly-enfranchised borough, market-town, and parish, in the Northern Division of the hundred of PIREHILL and of the county of STAFFORD, 1¾ mile (E.) from Newcastle-under-Lyme, and 150 (N. W. by N.) from London; containing, in the parish, 37,240 inhabitants; in the borough, 52,946; and in the district forming the rectory, 36,079. This parish comprises about two-thirds of the populous district called the Staffordshire Potteries, and the town, in common with various others in the parish and in this part of the county, is indebted for its increase and importance to the numerous potteries established in the neighbourhood. It is pleasantly situated on the river Trent, is amply supplied with water, and, with the adjoining townships of Fenton and Longton, is lighted with gas from extensive works recently erected by subscription on the banks of the Trent and Mersey canal. Very considerable improvements have taken place within the last few years; many handsome houses have been built, and new streets formed opening into the glebe and other lands adjoining. A spacious and elegant pile of stone buildings has been erected for an additional market-house and town-hall, of which the first stone was laid in September 1834, by the late John Tomlinson, Esq., of Cliff ville, chairman of the subscribers to the undertaking; and beautiful tablets of porcelain commemorating the event, and containing the names of the subscribers, respectively manufactured and presented by Messrs. Copeland and Garrett, and the late Mr. Arthur Minton, were on that occasion deposited underneath it. The surplus of the tolls and rents of the market and buildings, after defraying the expenses, are to be for ever applied to public and charitable objects within the town and immediate district. The principal manufactures are china and earthenware in all their various branches, for which there are several very extensive establishments; the largest are those of Messrs. Copeland and Garrett, Messrs. Wm. Adams and Sons, and Messrs. Minton and Boyle. The Trent and Mersey canal, and a branch from it to Newcastle-under-Lyme, pass through the town, affording great facility of communication and other advantages; and on their banks are numerous wharfs, warehouses, mills and other buildings. In connection with the above is a railway to Lane-End, for the conveyance of goods. The market is on Saturday. The greater part of the property in Penkhull and the neighbourhood is of copyhold tenure within the manor of Newcastle-under-Lyme, which belongs to the Crown, and is held on lease by the Duke of Sutherland; and a manorial court is held monthly at Newcastle for passing surrenders and other business. In 1839, an act was obtained for establishing an effective police in this town, Fenton, Longton, and Trentham and for improving and cleansing the streets; the police is upon a similar footing to the metropolitan police. Commissioners with certain qualifications are appointed for carrying the act into operation, and out of their body a chief bailiff is appointed. Lewis Adams, Esq., was elected the first to fill that office. By the act of the 2nd of William IV., cap. 45, this place, with others, has been constituted a borough, with the privilege of sending two members to parliament: the right of election is vested in the £10 householders of a district comprising 7084 acres, of which the limits are minutely described in the Appendix, No. II. The number of voters registered in

1838 was 1661: the returning officer is annually appointed by the sheriff. The town-hall is a small neat structure.

The rectory was originally much more extensive, but has at different times been subdivided, and parts of it formed into distinct parishes and rectories. In the year 1807, an act was obtained for separating from it the chapelries of Newcastle-under-Lyme, Burslem, Whitmore, Bucknall with Bagnall, and Norton-on-the-Moors, which are all now distinct rectories, though Bucknall and Bagnall still form part of this parish for all except ecclesiastical purposes. In 1827 the late John Tomlinson, Esq., patron, obtained an act of parliament authorising the sale to the respective landowners of all tithes and rectorial dues belonging to the rectory, and for the endowment of two new churches, which have been built at Shelton and Longton: the latter has been purchased by John Carey, Esq., and is now a distinct rectory. The living of Stoke-upon-Trent is a rectory, valued in the king's books at £41. 0. 10.; present net income, £2717; the Rev. John Weekes Tomlinson, eldest son of the late patron, is now rector. The old church, a very ancient structure, is supposed to have been built before the Conquest, and is mentioned in the Taxation of Pope Nicholas in 1291, with its chapels annexed, and valued at 60 marks. Being not only too small for the increased population, but also in a state of decay, it was taken down, and in 1826 a new church was erected near its site, at an expense of more than £14,000, of which the greater part was raised by subscription, and the remainder by the sale of pews, and by parochial rates. Among the subscriptions were £3300 from Dr. Woodhouse, Dean of Lichfield, then rector, who also gave a painted window for the chancel; £500 from the late Josiah Spode, Esq.; £300 from John Tomlinson, Esq., who also presented a marble font; £500, the amount of contributions arising from extra labour, presented by the workmen within Stoke Proper; and £400 from the Incorporated Society, for which 500 free sittings were provided. It is a handsome structure, in the later style of English architecture; the east window is a fine specimen of stained glass, after the antique, containing fifteen well-executed figures of the Apostles and Evangelists; and in the four side windows are the arms of the Bishop, Archdeacon, Rector, and Patron, and also of some of the principal contributors. There are several handsome monuments in the chancel by eminent sculptors, and those of the late Josiah Wedgwood, Esq., of Etruria, and Mrs. Wedgwood, were removed hither from the old church. The churchyard contains nearly five acres, and is fenced with a stone wall and iron-railing. The parsonage-house, at a small distance from the church, has been enlarged and modernized from the funds of the rectory. There are places of worship for Baptists, the Society of Friends, Wesleyan and Primitive Methodists, and Methodists of the New Connection. The National school is supported by subscription and by an allotment of one-third of the proceeds arising from Dr. Woodhouse's permanent endowment. Under the Poor Law Amendment Act this parish is not united to any other, but is under the care of 24 guardians. Dr. John Lightfoot, an eminent Hebrew scholar, and one of the principal persons employed in finally arranging the Liturgy of the Church of England, was born here in 1602; and Fenton, the poet, was interred at this place.

STOKE-WAKE (*All Saints*), a parish, in the union of Sturminster, hundred of Whiteway, Sturminster Division of the county of Dorset, 10 miles (W.) from Blandford-Forum; containing 147 inhabitants. The living is a rectory, valued in the king's books at £8. 8. 9.; present net income, £156; patron, H. K. Seymour, Esq.

STOKE, WEST, a parish, in the union of West Hampnett, hundred of Bosham, rape of Chichester, Western Division of the county of Sussex, $3\frac{1}{2}$ miles (N. W.) from Chichester; containing 101 inhabitants. In this parish is the picturesque valley of Kingley Bottom, in which is a grove of yew trees of great size and luxuriance. The living is a discharged rectory, valued in the king's books at £9. 11.; present net income, £170: it is in the patronage of the Crown. The church, which is beautifully situated in Stoke park, is in the early English style; it contains a handsome monument to the Stoughton family. It is conjectured that this is the site of the dreadful slaughter of the Danes by the men of Chichester, about the year 900.

STOKEHAM, a parish, in the union of East Retford, South-Clay Division of the wapentake of Bassetlaw, Northern Division of the county of Nottingham, 5 miles (N. E. by N.) from Tuxford; containing 48 inhabitants. The living is annexed, with that of Askham, to the vicarage of East Drayton.

STOKEINTINHEAD (*St. Andrew*), a parish, in the union of Newton-Abbott, and forming, with Combintinhead, Haccombe, and Shaldon, a detached portion of the hundred of Wonford, Teignbridge and Southern Divisions of the county of Devon, 4 miles (E. by S.) from Newton-Bushell; containing 621 inhabitants. The living is a rectory, valued in the king's books at £36. 15. 10.; present net income, £467; patron, Bishop of Exeter. The church contains some ancient screen-work: it was formerly collegiate, for a warden and several chaplains, established in honour of the Virgin Mary and St. Andrew, by John de Stanford, in the reign of Edward III. Here is a National school.

STOKENCHURCH, (*St. Peter and St. Paul*), a parish, in the union of Wycombe, hundred of Lewknor, county of Oxford, 7 miles (W. N. W.) from High Wycombe; containing 1290 inhabitants. This parish is situated on the upper road from London to Oxford. The village is on one of the highest points of the Chiltern hills: the manufacture of common chairs is carried on to a considerable extent in the parish, principally for the London market. The living is annexed to the vicarage of Aston-Rowant. The church contains monuments to two members of the Morley family, who distinguished themselves in the wars of Edward III. and Richard II.; also a monument to Bartholomew Tipping. There is a place of worship for Independents. Twelve poor children are educated, clothed, and apprenticed for a rent-charge of £41, the bequest of B. Tipping, in 1675; there are also a National and a Sunday school.

STOKENHAM, otherwise STOKINGHAM (*St. Barnabas*), a parish, in the union of Kingsbridge, hundred of Coleridge, Stanborough and Coleridge, and Southern, Divisions of the county of Devon, $5\frac{1}{4}$ miles (E. by S.) from Kingsbridge; containing 1609 inhabitants. The living is a vicarage, with Chivelstone and Sherford annexed, valued in the king's books at £48. 7. $8\frac{1}{2}$.; present net income, £625: it is in the patronage of the Crown; impropriator, A. H. Holdsworth, Esq. The church has an ancient wooden screen. Here is a Sunday National school.

STOKESAY (*St. John the Baptist*), a parish, in the union of Ludlow, hundred of Munslow, Southern Division of the county of Salop, 7 miles (N. W.) from Ludlow; containing 529 inhabitants. The living is a discharged vicarage, valued in the king's books at £4. 13. 4.; present net income, £326; patron, R. Marston, Esq.; impropriator, H. O. Davies, Esq. A school is partly supported by an endowment of £7. 8. per annum, and partly by the Earl of Craven.

STOKESBY (*St. Andrew*), a parish, in the hundred of East Flegg, Eastern Division of the county of Norfolk, 2¼ miles (E.) from Acle; containing, with the parish of Herringby, 324 inhabitants. The living is a rectory, with that of Herringby united, valued in the king's books at £13. 6. 8.; present net income, £526; patron and incumbent, Rev. G. Lucas. There is a place of worship for Wesleyan Methodists; also a school partly supported by subscription.

STOKESLEY (*St. Peter*), a market-town and parish, and the head of a union, in the Western Division of the liberty of Langbaurgh, North Riding of the county of York; containing 2376 inhabitants, of which number, 1967 are in the town of Stokesley, 41 miles (N. by W.) from York, and 242 (N. by W.) from London. This place is situated on the northern bank of the river Leven, in the centre of the fruitful tract called Cleveland, which, at the distance of about five miles from the town, is bounded by the Cleveland hills, forming a vast and majestic amphitheatre. The houses are neatly built in a modern style, and are ranged chiefly in one spacious street, from east to west. The trade is principally in linen; one of the mills for its manufacture is worked by a powerful steam-engine. The market, which is held on Saturday, is abundantly supplied with provisions; and there are fairs for horses, cattle, &c., on the Saturdays before Trinity and Palm Sundays, and before Old Lammas-day. A court leet is held annually, and the magistrates hold petty sessions every week. The town has been made a polling-place for the North Riding. The living is a rectory, with the perpetual curacy of Westerdale annexed, valued in the king's books at £30. 6. 10½.; patron, Archbishop of York. The tithes of Stokesley and the several townships have been commuted separately, for an aggregate rent-charge of £956. 6. 8., subject to the payment of rates, which on the average have amounted to £81. 17. 10.; the glebe comprises 76 acres, valued at £250 per annum. The tithes of Westerdale have been commuted for a rent-charge of £250, subject to the payment of rates, which on the average have amounted to £26; the glebe comprises 11 acres, valued at £14 per annum. The impropriate tithes of the township of Newby have been commuted for a rent-charge of £3. 18., and there is also a rent-charge of £2. 18. payable to the perpetual curate of Seamer. There are places of worship for Independents and Primitive and Wesleyan Methodists. A National school has been established, and is partly supported by subscription, and partly by the dividends arising from £2780. 1. 5. five per cents., bequeathed in 1805 by John Preston. The Society for the Promotion of Christian Knowledge, in 1818, established a depository of books in the vestry-room of the church, from which many thousand volumes have been since distributed. The poor law union of Stokesley comprises 28 parishes or places, under the care of 29 guardians, and, according to the census of 1831, contains a population of 9618. The then Lord Stokesley was the patron of Bishop Hall, and effected his mission to the synod of Dort.

STONALL, OVER, a chapelry, in the parish of Shenstone, Southern Division of the hundred of Offlow, and of the county of Stafford, 6 miles (S. W.) from Lichfield: the population is returned with the parish. The living is a perpetual curacy; net income, £92; patron, Vicar of Shenstone.

STONAR (*St. Augustine*), a parish, in the union of the Isle of Thanet, hundred of Ringslow, or Isle of Thanet, lathe of St. Augustine, Eastern Division of the county of Kent, ¾ of a mile (N. by E.) from Sandwich; containing 52 inhabitants. It is supposed that the site of this place, in the time of the Romans, was entirely covered with water. On the sea retiring from Ebbs-fleet, it became a common landing-place, at an early period, and, in consequence, a town and port of considerable importance, having, in 1090, so increased, that the seignory of Stonar was claimed by the citizens of London as subject to that port. After sustaining repeated injuries from the Danes and other marauders, as well as from inundations of the sea, it began about the reign of Richard II. to decay. Leland, who wrote in the time of Henry VIII., describes it as "sometime a pretty town," but then "having only the ruin of the church, which some people call Old Sandwich." Abundant remains of former dwellings may be traced, though at present there are not more than ten inhabited houses in the entire parish. Salt-works are carried on near the site of the church, the produce of which serves all the purposes of bay salt. The living is a rectory, valued in the king's books at £3. 6. 8., and in the patronage of the Crown, by lapse. The church has been destroyed; and no presentation has lately been made to the living.

STONDON, LOWER, a hamlet, in the parish of Shitlington, union of Ampthill, hundred of Clifton, county of Bedford, 3 miles (S. by E.) from Shefford: the population is returned with the parish.

STONDON-MASSEY, county of Essex. — See STANDON-MASSEY.

STONDON, UPPER (*All Saints*), a parish, in the union of Biggleswade, hundred of Clifton, county of Bedford, 2¾ miles (S.) from Shefford; containing 37 inhabitants. The living is a rectory, valued in the king's books at £6. 6. 10½.; present net income, £125; patrons, J. and T. Smith, Esqrs.

STONE (*St. John the Baptist*), a parish, in the union and hundred of Aylesbury, county of Buckingham, 3 miles (W. S. W.) from Aylesbury; containing 773 inhabitants. This parish is separated from Waddesdon by the river Thame. The manufacture of lace, which was formerly more considerable, is still carried on. The living is a discharged vicarage, valued in the king's books at £9; present net income, £149; patron, Dr. Lee. The church is partly in the Norman, and partly in the early English, style of architecture. There are two places of worship for Methodists; and a National school is supported by subscription.

STONE, a chapelry, in the parish and Upper Division of the hundred of Berkeley, Western Division of the county of Gloucester, 2¾ miles (S. by W.) from Berkeley: the population is returned with the tything of Ham. The living is a perpetual curacy; net income, £54; patron, Vicar of Berkeley; appropriators, Dean and Chapter of Bristol. The chapel, dedicated to All Saints, is partly in the early, and partly in the later, style of English architecture. The turnpike road from Gloucester to Bristol passes through the village.

STONE (*St. Mary*), a parish, in the union of Tenterden, hundred of Oxney, lathe of Shepway, Eastern Division of the county of Kent, 6½ miles (S. E.) from Tenterden; containing 410 inhabitants. The living is a vicarage, valued in the king's books at £8. 14. 4½.; present net income, £345; patrons and appropriators, Dean and Chapter of Canterbury. The rectorial tithes have been commuted for a rent-charge of £929, subject to the payment of rates, which on the average have amounted to £85; the glebe comprises 7½ acres, valued at £16 per annum. The church is a spacious and handsome structure. A school is partly supported by the parish and partly by subscription. The Grand Military canal passes through the parish. A fair for pedlery is held here on Holy Thursday.

STONE (*St. Michael*), a market-town and parish, and the head of a union, in the Southern Division of the hundred of Pirehill, Northern Division of the county of Stafford, 7 miles (N. by W.) from Stafford, and 141 (N. W. by N.) from London; containing 7808 inhabitants. The name is traditionally reported to be derived from a monumental heap of stones, which, according to the custom of the Saxons, had been placed over the bodies of the princes Wulford and Rufinus, who were here slain by their father Wulfhere, King of Mercia, on account of their conversion to Christianity. The king himself becoming subsequently a convert, founded, in 670, a college of Secular canons, dedicating it to his children, in expiation of his crime, and to this establishment the town is supposed to owe its origin. The canons having been expelled, during the war with the Danes, the college fell into the possession of some nuns, who established themselves here. No mention is made of it in Domesday-book, but it appears to have been granted by Henry I. to Robert de Stafford, who displaced the nuns, and made it a cell to the monastery of Kenilworth, which it continued to be until 1260, when it became independent, with the exception of paying a small sum annually to that monastery, and an acknowledgment of its patronage; its revenue was valued, at the dissolution, at about £119. The town is situated on the high road from London to Liverpool, on the eastern bank of the river Trent, over which there is a bridge to Walton, and is paved, and well supplied with water: it consists chiefly of one long street, with several others branching off. Races are occasionally held in the neighbourhood, and assemblies sometimes in the town. The prevailing branch of manufacture is that of shoes, and there are two considerable breweries: on a stream which falls into the Trent are four corn mills. The Trent and Mersey canal (commonly called the Grank Trunk) passes through the town, running parallel for several miles with the river; and the principal office of the Company of Proprietors of this prosperous and important navigation is here. The Norton Bridge Station of the Grand Junction railway is within three miles of this place. The market, which is on Tuesday, was, about 50 years since, a great mart for corn, but it has very much declined, owing, probably, to the rapidly increasing population and additional markets in the neighbouring potteries. The fairs are on the Tuesday after Mid-Lent, Shrove-Tuesday, Whit-Tuesday, Aug. 5th. Petty sessions are held by the county magistrates every fortnight, and two constables are annually chosen at the court leet of the lord of the manor. The town has lately been made a polling-place for the northern division of the county.

The living is a perpetual curacy; net income, £214: it is in the patronage of the Crown. The church is a modern structure, in the later English style, with a square tower: the altar-piece is a fine painting, by Sir William Beechey, of St. Michael binding Satan: there is a marble monument surmounted by a bust, to the memory of the late Earl St. Vincent. The old church fell down about the middle of the last century, occasioned, it is said, by the undermining of one of the pillars in digging a vault; in consequence of which no interment is allowed to take place within the walls of the present edifice. There are places of worship for Independents and Wesleyan Methodists of the Old and New Connections; and a Roman Catholic chapel at Aston Hall, in this parish. The free school was founded and endowed with a small annual income by the Rev. Thomas Alleyn, in 1558. The Master and Fellows of Trinity College, Cambridge, are the trustees, and appoint the master: the school-house adjoins the churchyard, but there are no boys on the foundation. Here is a National school. A bequest of £100 per annum to ten poor widows, charged on the Stone Park estate, is paid by Earl Granville, though void by the mortmain act: there are other small charitable endowments. The poor law union of Stone comprises 10 parishes or places, under the care of 20 guardians, and, according to the census of 1831, contains a population of 17,871. The workhouse is a large and handsome brick building near the town. The remains of the abbey adjoin the churchyard, and consist of one perfect arch and rather extensive cloisters. On clearing away some of the rubbish, in 1828, a stone coffin was discovered under the foundation of one of the walls, with the remains enclosed in a tolerably perfect state. In a field, now allotted to the poor, at a short distance from the town, the army under the Duke of Cumberland was encamped, in 1745, expecting the Pretender to pass that way, but he avoided them by taking the route by Leek. The late celebrated naval commander, Earl St. Vincent, was born at Meaford, in this parish, and was buried in the churchyard here. At the small hamlet of Burston, one of the sons of Wulfere, King of Mercia, is said to have suffered martyrdom.

STONE (*St. Mary*), a parish, in the union of Kidderminster, Lower Division of the hundred of Halfshire, Kidderminster and Western Divisions of the county of Worcester, 2¼ miles (S. E. by E.) from Kidderminster; containing 551 inhabitants. This place was formerly a chapelry in the parish of Chaddesley-Corbet. The spinning of yarn, connected with the manufactures at Kidderminster, is carried on, for which there are two mills. The living is a vicarage, endowed with the rectorial tithes, and valued in the king's books at £15; present net income, £827: it is in the patronage of the Crown. The church contains, over the northern door, some small remains of Norman architecture.

The free school, founded pursuant to the will of the Rev. Mr. Hill, B. D., is endowed with 24 acres of land, let for £32 per annum, and a house for the master. This parish possesses a valuable charity of unknown origin, consisting of some land near Stourbridge, containing clay for making fire-bricks, and producing, on an average, nearly £700 per annum, which, with the dividends of about £5000 three per cent. stock, is applied to repairing the church, and for charitable purposes.

STONE-DELPH, a joint township with Almington, in the parish and union of TAMWORTH, Tamworth Division of the hundred of HEMLINGFORD, Northern Division of the county of WARWICK, 3 miles (S. E.) from Tamworth: the population is returned with Almington.

STONE-EASTON, a parish, in the union of CLUTTON, hundred of CHEWTON, Eastern Division of the county of SOMERSET, 6½ miles (N.) from Shepton-Mallet; containing 386 inhabitants. The living is annexed, with those of Emborrow and Paulton, to the vicarage of Chewton-Mendip. Here is a National school.

STONE-NEAR-DARTFORD (*St. Mary*), a parish, in the union of DARTFORD, hundred of AXTON, DARTFORD, and WILMINGTON, lathe of SUTTON-AT-HONE, Western Division of the county of KENT, 2 miles (E. by N.) from Dartford; containing 719 inhabitants. The living is a rectory, valued in the king's books at £26. 10.; present net income, £765; patron, Bishop of Rochester. The church is much admired as being a peculiarly fine specimen of the later style of English architecture: it contains several ancient stalls, remarkable for the elegance of their workmanship and the delicacy of their pillars, which are of crown marble. A day and Sunday school is partly supported by the rector. From the rents of the lands attached to the castle twenty-six sermons are annually preached, one on each Wednesday during summer, alternately at Gravesend and Dartford, agreeably to the will of Dr. Plume, founder of the Plumian Professorship at Cambridge. The river Thames bounds the parish on the north. Stone Castle stands to the south of the high Dovor road, and is said to be one of the 115 castles which were not dismantled in accordance with an express stipulation to that effect between Stephen and Henry II.

STONE - NEXT - FAVERSHAM, a parish, in the union and hundred of FAVERSHAM, Upper Division of the lathe of SCRAY, Eastern Division of the county of KENT, 2½ miles (W. by N.) from Faversham; containing 80 inhabitants. The living is a perpetual curacy.

STONEBECK, DOWN, a township, in the parish of KIRKBY-MALZEARD, union of PATELEY-BRIDGE, Lower Division of the wapentake of CLARO, West Riding of the county of YORK, 14 miles (W. by S.) from Ripon containing 494 inhabitants.

STONEBECK, UPPER, a township, in the parish of KIRKBY-MALZEARD, union of PATELEY-BRIDGE, Lower Division of the wapentake of CLARO, West Riding of the county of YORK, 16 miles (W. by N.) from Ripon; containing 332 inhabitants.

STONEFERRY, a township, in the parish of SUTTON, union of SCULCOATES, Middle Division of the wapentake of HOLDERNESS, East Riding of the county of YORK, 1½ mile (N. by E.) from Kingston-upon-Hull; the population is returned with the parish. There is a place of worship for Wesleyan Methodists. Ann Waters, in 1720, bequeathed property for the erection and endowment of almshouses for seven widows or poor old maids, who each receive £13 per annum: she also left an annuity of £5 to be paid to one of the almswomen for teaching ten poor girls.

STONEGRAVE, a parish, in the union of HELMSLEY BLACKMOOR, wapentake of RYEDALE, North Riding of the county of YORK; containing 327 inhabitants, of which number, 189 are in the township of Stonegrave, 4¾ miles (S. E. by S.) from Helmsley. The living is a rectory, valued in the king's books at £33. 6. 8.; present net income, £495: it is in the patronage of the Crown. The church is partly in the decorated, and partly in the later, styles of English architecture. A school is partly supported by an allowance of £5 per annum from the proprietor of the estate.

STONEHAM, NORTH (*St. Nicholas*), a parish, in the union of SOUTH STONEHAM, hundred of MANSBRIDGE, Southampton and Southern Divisions of the county of SOUTHAMPTON, 4¾ miles (N. N. E.) from Southampton; containing 766 inhabitants. The living is a rectory, valued in the king's books at £21. 9. 7.; present net income, £536; patron, John Fleming, Esq. The church contains the remains of the celebrated Admiral Lord Hawke, to whose memory there is a superb monument, composed of white and variegated marble, bearing the family arms and other appropriate emblems, with a sculptured representation of his victory over the French admiral, Conflans, in Quiberon bay. Two miles south of the village is an old mansion, formerly the residence of his lordship. The Itchen navigation, and the London and South-Western railway, pass through the parish. Here is a National school. Edmund Dummer, in 1720, gave £300 for erecting a school-house, and an annuity of £5, for which five boys are instructed.

STONEHAM, SOUTH (*St. Mary*), a parish, and the head of a union, partly in the county of the town of SOUTHAMPTON, but chiefly in the hundred of MANSBRIDGE, Southampton and Southern Divisions of the county of SOUTHAMPTON, 3 miles (N. N. E.) from Southampton; containing 2737 inhabitants. The river Itchen, which is navigable from Winchester to its influx into the Southampton water, passes through the parish, which is well situated for trade, and is intersected by the London and South-Western railway. At Wood Mills blocks and pumps were formerly manufactured for the supply of nearly the whole of the Royal Navy, but the factory was destroyed by fire some years since, and there is now a flour-mill upon its site. The living is a vicarage, valued in the king's books at £12; present net income, £250; patron and appropriator, Rector of St. Mary's, Southampton. A district chapel has been recently erected, containing 610 sittings, 380 of which are free, the Incorporated Society having granted £350 in aid of the expense. Here is a National school; and another is partly supported by the Rev. John Hurncell, and T. B. Hay, Esq. The poor law union of South Stoneham comprises 9 parishes or places, under the care of 16 guardians, and, according to the census of 1831, contains a population of 9447. At Swathling is a mineral spring.

STONEHOUSE (*St. Cyril*), a parish, in the union of STROUD, Lower Division of the hundred of WHITSTONE, Eastern Division of the county of GLOUCESTER,

3 miles (W.) from Stroud; containing 2469 inhabitants. This parish is situated in a pleasing and fertile vale on the turnpike-road from Gloucester to Bath, and is intersected by the river Frome and the Stroudwater canal, which flow through the village. The surrounding scenery is agreeably diversified, and the soil, of which the substratum is limestone, is favourable to the growth of apples for cider. The cloth manufacture appears to have been introduced here at an early period, as, in the reign of Henry VIII., a fulling-mill, established in this parish, formed part of the possessions of the abbey of Gloucester. During the 17th century, and the greater part of the 18th, this place was celebrated for its scarlet cloth, which was considered the finest in the kingdom, and its clothing establishments still rank among the most extensive and flourishing in the clothing districts. Fairs are held annually in the village on May 1st and Oct. 11th; and it has lately been made a polling-place for the eastern division of the shire. The living is a vicarage, endowed with the rectorial tithes, and valued in the king's books at £22; present net income, £510: it is in the patronage of the Crown. The church, though much modernized, retains some portions of its original Norman style of architecture, of which the north door is a good specimen. There are places of worship for Independents and Wesleyan Methodists. John Elliott and others, in 1774, subscribed £612. 10. for establishing a free school in the village of Stonehouse, and another in the hamlet of Ebley, of which the former has been converted into a National school; two handsome school-rooms were built in 1831, with apartments for the master and mistress; the income arising from the endowment is £47 per annum.

STONEHOUSE, EAST (*St. George*), a parish, in the suburbs of the borough of PLYMOUTH, Roborough and Southern Divisions of the county of DEVON; containing 9571 inhabitants. This place, originally called Hipperston, was in the reign of Henry III., the property of Joel de Stonehouse, from whom it derives its present name; it was then situated more southerly, but, after subsequent improvements and extension to the northward, the ancient buildings were allowed to fall into decay. It includes several good streets, which are mostly paved, and lighted with gas; the houses are of neat and respectable character; and the inhabitants are well supplied with water by means of pipes leading from the reservoir of the Devonport Water Company, situated in the parish of Stoke-Damerall, and from a fine stream brought into the town under an act passed in the 35th of Elizabeth. A very handsome quadrangle of Greek architecture, enclosing the new chapel of St. Paul, has been lately built in the south-western part of the town. A communication was made with Devonport by means of a stone bridge across Stonehouse creek, erected at the joint expense of the Earl of Mount-Edgecumbe and Sir John St. Aubyn: the tolls are let annually, at a public survey, and the income derived from them is very considerable. Higher up the creek, to the north, a bridge has been recently erected, affording a passage to Stoke. On the Devil's Point (which commands, perhaps, the finest prospect of Mount-Edgecumbe) is the picturesque ruin of a blockhouse, erected in the time of Elizabeth; and over this old edifice is a modern battery, occupied by the Royal Marine Artillery. At a short distance is Eastern King's battery, commanding the mouth of the Hamoaze: there is also a fort for the protection of the creek. The three towns of Stonehouse, Plymouth, and Devonport, are brilliantly lighted from the gas-works in this parish; the gasometer presents a conspicuous object from the road from Plymouth to Devonport. The road to the ferry at New Passage passes through this place. In Stonehouse pool are convenient quays for merchant vessels; and in addition to the general business arising from the maritime relations of this town, and its naval and military establishments, there are some large manufactories for varnish used in the dock-yards, soap, and tallow. A customary market is held on Wednesday, in a neat and convenient building in Edgecumbe-street; and there are fairs on the first Wednesday in May, and the second Wednesday in September. By the act to amend the representation, recently passed, the township of Stonehouse has been included within the limits of the newly-enfranchised borough of Devonport. The town is within the jurisdiction of the county magistrates, who hold their sessions in the town-hall at Devonport. A manorial court leet and baron is held annually.

Among the most important public establishments is the Royal Naval Hospital, for the reception of wounded seamen and marines, opened in 1762; it is situated on an eminence near the creek, and comprises ten buildings, each containing six wards, each ward affording accommodation for about 20 patients, with a chapel, store-room, operating-room, small-pox ward, and dispensary; they form an extensive quadrangle, ornamented on three sides with a piazza, and the entire edifice, with its spacious lawn, is said to occupy an area of 24 acres. In 1795, the government of this institution was vested in a post-captain; the other officers are, the first and second lieutenants, physician, surgeon, dispenser, chaplain, agent, and steward; the chapel is open to the public. The Royal Marine barracks, on the west shore of Mill bay, comprise a handsome range of buildings forming an oblong square, and are adapted for the accommodation of about 1000 men: the Long Room barracks, built chiefly of wood, will contain 900. A new victualling establishment has been erected at Devil's Point, upon a scale of great magnitude: it is approached through a granite gateway and double colonnade of singular beauty, and the various ranges of building are surprisingly magnificent. Among the more remarkable features of the work are, the removal of 300,000 cubic yards of limestone rock, and the erection of a granite sea-wall, 1500 feet in length, the foundation of which was laid by means of a diving bell. The water for the brewery is supplied, at the rate of 350 tons per day, from the Plymouth Leat: it first runs into a reservoir capable of receiving 2000 tons, and is thence conveyed through iron pipes into a second basin, of 6000 tons. The Royal Military Hospital is situated on the opposite side of the creek, in the parish of Stoke-Damerall; its government is similar to that of the Naval Hospital. Stonehouse was formerly a chapelry in the parish of St. Andrew, Plymouth. The living is a perpetual curacy; net income, £197; patron, Vicar of St. Andrew's, Plymouth; impropriators, Corporation of Plymouth. The church was built in 1787, when the old chapel was taken down. An additional church, in the later English style, with a tower, was erected in

1831, at an expense of £2899. 11. 6., under the act of the 58th of George III., and was consecrated and dedicated to St. Paul in 1833. There are places of worship for Baptists, Independents, Wesleyan Methodists, and Roman Catholics. A National school is supported by subscription. In the arrangements under the Poor Law Amendment Act, this parish is not united to any other, but is under the care of 10 guardians of its own.

STONELEIGH (*St. Mary*), a parish, in the union of Warwick, Kenilworth Division of the hundred of Knightlow, Southern Division of the county of Warwick, 3¾ miles (E. by N.) from Kenilworth; containing 1298 inhabitants. This place, anciently called Stanlei, from the rocky nature of the soil, is pleasantly situated in an elevated district, which is, notwithstanding, tolerably fertile. It was chiefly distinguished for its venerable abbey, founded in 1154 by Henry II., for monks of the Cistercian order, who were removed to this place from Radmore, in the county of Stafford. In 1245, the abbey suffered greatly from an accidental fire, and was repaired by Robert de Hockele, the 16th abbot, who in 1300 built the gateway tower and entrance, now almost the only portion remaining entire: the revenue of this establishment, at the dissolution, was valued at £178. 2. 5. The village, consisting chiefly of a few old cottages, occasionally interspersed with some modern houses, in detached situations, is intersected by the river Sowe, which, after passing under an ancient stone bridge of eight arches, unites with the Avon about half a mile beyond, and flows through the grounds of Stoneleigh Abbey, the elegant seat of Lord Leigh. This spacious mansion, erected on the site of the ancient monastery, is a stately structure of stone, in the Grecian style of architecture, situated about a mile from the village, in a widely extended and richly diversified domain, enclosed on one side by gently sloping hills, and sheltered on another by luxuriant woods; the lawn in front of the house slopes gently to the margin of the Avon, which within the grounds is crossed by an elegant bridge of stone: the entrance to the house is through the ancient arched gateway of the abbey, a fine specimen in the decorated style of English architecture. Of the ancient monastic buildings the chief remains are found in the cellars and domestic offices of the modern mansion, and consist chiefly of groined arches resting upon massive pillars, with richly ornamented capitals, and of numerous details in the latest and most finished period of the Norman style, which prevails generally throughout the lower parts of that splendid structure. The grounds are extensive, and are enriched with finely varied scenery. Opposite to the entrance lodge is another leading into the woods, which extend for some miles, and are in various parts ornamented with tasteful lodges, the residences of the keepers, the last of which forms an entrance from the suburbs of Kenilworth. The extensive range of stabling attached to the mansion, and erected in the ancient style of English architecture, forms a prominent and interesting feature in the grounds near the house; and the park, which is well stocked with deer, combines a variety of interesting objects, and is enriched with a profusion of stately and venerable oaks of luxuriant growth. A market and a fair, granted to the abbots by Henry II., were formerly held in the village; and on the south side of the river Sowe, opposite to the church, is Motstow Hill, where the tenantry used to assemble to render suit and service to the lord of the manor.

The living is a vicarage, valued in the king's books at £16. 15. 5.; present net income, £510: it is in the patronage of the Crown; impropriator, Lord Leigh. The church is an ancient and venerable structure, partly in the Norman style of architecture, with a low massive tower, strengthened with angular buttresses, and surmounted by another of smaller dimensions, crowned with pinnacles at the angles; on the north side is a handsome arched doorway in the Norman style, enriched with round and flat mouldings; the south porch was closed up when the church was re-pewed in 1821, and the present entrance made through the tower; the south aisle is separated from the nave by a range of three massive octagonal pillars supporting pointed arches, and the chancel by a large and richly ornamented Norman arch resting upon columns, of which the shafts and capitals are elaborately embellished; around the east end of the chancel is a series of small Norman arches, which for many years had been concealed by plaster, but have been lately restored; and near the altar is a splendid monument to Lady Alice Lee, Duchess of Dudley, and a recumbent figure of stone, which was found in an upright position when digging the foundation for the mausoleum of the Leigh family, and supposed to be that of Geoffrey de Muschamp, Bishop of Coventry and Lichfield in the reign of John. On the north side of the chancel is the mausoleum above mentioned, recently erected, of which the roof is elaborately groined: the ancient font is of cylindrical form, and ornamented with figures of the twelve apostles under Norman arches. A chapel of ease is about to be erected, towards the expense of which Lord Leigh has given £1000. The free school was founded in 1708, by Thomas, Lord Leigh, who endowed it with lands in the parish of Wintoft, for the instruction of 70 boys and 50 girls in reading, writing, and arithmetic; and is now in connection with the National Society. Almshouses for five aged men and five aged women were founded in 1576, by Dame Alice Leigh, whose endowment is augmented with £7 per annum from a charity at Bidford. This place gave the title of Baron to the family of Leigh, of whom there were five lords prior to 1784, when the title became extinct, but was revived in the person of the present Chandos Leigh, created, May 2nd, 1839, Baron Leigh, of Stoneleigh, in the county of Warwick.

STONERAISE, a joint township with Brocklebank, in the parish of Westward, Allerdale ward below Derwent, Western Division of the county of Cumberland, 2¼ miles (S. S. E.) from Wigton: the population is returned with Brocklebank. In this township are the ruins of Old Carlisle, where was a considerable Roman city, supposed by Horsley to have been the *Olenacum* of the Notitia.

STONESBY (*St. Peter*), a parish, in the union of Melton-Mowbray, hundred of Framland, Northern Division of the county of Leicester, 6 miles (N. E.) from Melton-Mowbray; containing 287 inhabitants. The living is a discharged vicarage, valued in the king's books at £5. 0. 7½.; present net income, £90; patron and impropriator, R. Norman, Esq. Here is a National school.

STONESFIELD (*St. James*), a parish, in the union

of WOODSTOCK, hundred of WOOTTON, county of OXFORD, 4¼ miles (W.) from Woodstock; containing 535 inhabitants. The living is a discharged rectory, valued in the king's books at £4. 19. 9½.; present net income, £139; patron, Duke of Marlborough. There is a place of worship for Wesleyan Methodists; also a day and Sunday National school. In the slate pits, which employ two-thirds of the male population, fossils are occasionally found.

STONEY-MIDDLETON, county of DERBY.—See MIDDLETON, STONEY. *And all places having a similar distinguishing prefix will be found under the proper name.*

STONHAM, ASPAL, or ANTEGAN (*St. Lambert*), a parish, in the union and hundred of BOSMERE and CLAYDON, Eastern Division of the county of SUFFOLK, 5 miles (N. E.) from Needham-Market; containing 612 inhabitants. The living is a rectory, valued in the king's books at £19. 10. 2½.; patron, Sir W. F. F. Middleton, Bart. The tithes have been commuted for a rent-charge of £666. 10., subject to the payment of rates, which on the average have amounted to £106; the glebe comprises 32 acres, valued at £40 per annum. In the churchyard is a fine, though mutilated, monument of alabaster, to the memory of Anthony Wingfield, Esq., who died in 1714. Here is a free school, endowed in 1612, by the Rev. John Metcalf, rector of this parish, with land producing £80 per annum.

STONHAM, EARL (*St. Mary*), a parish, in the union and hundred of BOSMERE and CLAYDON, Eastern Division of the county of SUFFOLK, 3 miles (N. N. E.) from Needham-Market; containing 757 inhabitants. The living is a rectory, valued in the king's books at £17. 2. 6.; patrons, Master and Fellows of Pembroke College, Cambridge. The tithes have been commuted for a rent-charge of £650, subject to the payment of rates, which on the average have amounted to £174; the glebe comprises 31 acres, valued at £39 per annum. There is a place of worship for Baptists. John Punchard, about 1475, gave a house for the use of a school, which George Reeve, in 1599, endowed with land, now producing more than £20 a year, for the education and clothing of eight poor children.

STONHAM, PARVA, or JERNEGAN (*St. Mary*), a parish, in the union and hundred of BOSMERE and CLAYDON, Eastern Division of the county of SUFFOLK, 4 miles (N. N. E.) from Needham-Market, containing 329 inhabitants. The living is a discharged rectory, valued in the king's books at £9. 18. 11½.; present net income, £360: it is in the patronage of Mrs. C. Bevan.

STONTON-WYVILLE (*St. Denis*), a parish, in the union of MARKET-HARBOROUGH, hundred of GARTREE, Southern Division of the county of LEICESTER, 5½ miles (N. by E.) from Market-Harborough; containing 106 inhabitants. The living is a rectory, valued in the king's books at £9. 18. 11½.; present net income, £190; patron, Earl of Cardigan.

STONYHURST, county palatine of LANCASTER.—See MITTON.

STOODLEY (*St. Margaret*), a parish, in the union of TIVERTON, hundred of WITHERIDGE, Cullumpton and Northern Divisions of the county of DEVON, 4 miles (S. W.) from Bampton; containing 524 inhabitants. The living is a rectory, valued in the king's books at £20. 0. 2½.; present net income, £341; patron, John Nicholas Fazakerley, Esq., who supports a school. On Warbrightsleigh hill, in this parish, are the remains of an ancient beacon, said to have been erected by Edward II.

STOPHAM (*St. Mary*), a parish, in the union of THAKEHAM, hundred of ROTHERBRIDGE, rape of ARUNDEL, Western Division of the county of SUSSEX, 4¼ miles (S. E. by E.) from Petworth; containing 129 inhabitants. The living is a discharged rectory, valued in the king's books at £5. 12. 8½.; patron, Walter Smith, Esq. The tithes have been commuted for a rent-charge of £146, subject to the payment of rates, which on the average have amounted to £21; the glebe comprises 28 acres, valued at £20 per annum. The church is partly in the early, and partly in the decorated, style of English architecture. The river Rother flows through the parish, which is bounded on the east by the Avon.

STOPSLEY, a hamlet, in the parish of LUTON, hundred of FLITT, county of BEDFORD, 2 miles (N. N. E.) from Luton; containing 510 inhabitants.

STORETON, a township, in the parish of BEBINGTON, union, and Lower Division of the hundred, of WIRRALL, Southern Division of the county of CHESTER, 4¾ miles (N. by E.) from Great Neston; containing 192 inhabitants.

STORITHS, a joint township with Hazlewood, in the parish and union of SKIPTON, Upper Division of the wapentake of CLARO, West Riding of the county of YORK, 7½ miles (E.) from Skipton: the population is returned with Hazlewood.

STORMORE, an extra-parochial liberty, in the union of RUGBY, hundred of GUTHLAXTON, Southern Division of the county of LEICESTER: the population is returned with Westrill.

STORRINGTON (*St. Mary*), a parish, in the union of THAKEHAM, hundred of WEST EASWRITH, rape of ARUNDEL, Western Division of the county of SUSSEX, 8½ miles (N. E.) from Arundel; containing 916 inhabitants. The living is a rectory, valued in the king's books at £18; present net income, £387; patron, Duke of Norfolk. A market was formerly held here on Wednesday, and there are still fairs on May 12th and November 11th. A National school is partly supported by an endowment of £25 per annum.

STORTFORD, BISHOP (*St. Michael*), a market-town and parish (formerly a borough), and the head of a union, in the hundred of BRAUGHIN, county of HERTFORD, 14 miles (E. N. E.) from Hertford, and 30 (N. N. E.) from London; containing 3958 inhabitants. This place derives its name from its situation on each side of a ford on the river Stort, now crossed by two bridges, and its prefix from having been bestowed by William, soon after the Conquest, upon Maurice, Bishop of London, and his successors. In the reign of Stephen, the Empress Matilda negotiated to obtain, by exchange, from the Bishop of London, the castle erected here by William the Conqueror, and not succeeding, threatened its demolition; it, however, remained till the eighth year of King John's reign, who, exasperated at the bishop's promulgation of the pope's menace of laying the kingdom under an edict, razed it to the ground, seized the town into his own hands, incorporated the inhabitants, and granted them the elective franchise, which they continued to exercise only from the fourth of Edward II. till the reign of Edward III., when their privileges ceased. In the reign of Mary, this town became the scene of religious persecution, and Bishop Bonner made use of a prison, formerly attached to the castle, for the confinement of convicted Protestants, of whom one was burnt on Goose Green adjoining. The town is situated on two gentle acclivities, called respectively Windhill and Hockerhill, in a fertile valley on the banks of the river Stort, and consists principally of four streets, in the form of a cross, of which Windhill is the western and Hockerhill the eastern extremity: the inhabitants are well supplied with water from springs. There is a public library, instituted in 1827. The trade consists chiefly in malt and other grain, of which considerable quantities are sent to London by the river, which is navigable, and a canal, on the banks of which are commodious wharfs and quays. The market, for which a very handsome and spacious market-house, or corn-exchange, was erected in 1828, by a company of proprietors, is on Thursday: this building is of the Ionic order, and the area of a semicircular form, with a colonnade of cast-iron pillars; it contains an assembly and coffee-rooms, and magistrates' chamber, on the first floor, and underneath a spacious hall, where the corn-exchange is held; at the southern extremity are the registrar's office for births and marriages under the late act, and the police station-house. Fairs are held annually on Holy Thursday, the Thursday after Trinity-Sunday, and on October 11th, for horses and cattle. The town is within the jurisdiction of the county magistrates, who hold a petty session every fortnight; and it has been made a polling-place for the shire.

The living is a vicarage, valued in the king's books at £12; present net income, £419; patron and appropriator, Precentor of St. Paul's Cathedral, London. The church is an elegant and spacious structure, standing at the south-west angle of the town, with a fine tower surmounted by a lofty spire: it was built in the reign of Henry VI., and partly rebuilt in 1820, and contains many ancient and curious monuments, among which are those of Charles Denny, grandson to Sir Anthony Denny, Knt., Privy Councillor to Henry VIII.; and of Sir George Duckett, who was the last surviving proprietor of the Stort navigation. There are places of worship for Baptists, the Society of Friends, Independents, and Methodists, with which are connected Sunday schools, affording instruction to several hundred children. Here was formerly a free grammar school, in High-street, near the church, to which an excellent library was presented by Thomas Leigh, Esq., and increased by the Rev. Thomas Leigh, vicar, and other benefactors, of which many valuable portions still remain preserved in the tower of the church: this school, in which Sir Henry Chauncey, a native of this town, and author of the History and Antiquities of Hertfordshire, was educated, has declined, nothing remaining of its former celebrity but the libraries of Leigh, the founder, and Tooke, the reviver of it on its former failure, together with the books presented by the boys on their leaving school. The National school, established in 1818, is supported by funds in the hands of trustees, and by subscription. There are five newly-erected almshouses for poor people, which have been established with the proceeds of the sale of two almshouses in Potter-street, endowed by Mr. R. Pilston, in 1572; and several estates, producing about £120 per annum, are appropriated to the apprenticing of poor children, the relief of the poor, and the repair of the church; to which latter purpose about £75 per annum, arising from the revenue of a dissolved chantry, and some ancient guilds, formerly established here, is also applied. The poor law union of Bishop-Stortford comprises 20 parishes or places, half of which are in the county of Essex, and contains a population of 18,012, according to the census of 1831; it is under the care of 27 guardians. There are some small remains of the castle, in the garden of which Roman coins have been found, and among them one of Marcus Aurelius Antoninus; and near the castle an ancient well, dedicated to St. Osyth, the water of which is esteemed beneficial in diseases of the eyes. Hoole, the translator of Tasso, was born here.

STORWOOD, a township, in the parish of THORNTON, union of POCKLINGTON, Holme-Beacon Division of the wapentake of HARTHILL, East Riding of the county of YORK, 8¼ miles (S. W. by W.) from Pocklington; containing 119 inhabitants.

STOTFOLD (*St. Mary*), a parish, in the union of BIGGLESWADE, hundred of CLIFTON, county of BEDFORD, 2½ miles (N. W.) from Baldock; containing 833 inhabitants. The living is a discharged vicarage, valued in the king's books at £5. 17. 1.; present net income, £185; patrons and impropriators, Master and Fellows of Trinity College, Cambridge. There is a place of worship for Wesleyan Methodists; and a school for 15 boys is endowed with £20 per annum.

STOTFORD, a township, in the parish of HOOTON-PAGNELL, union of DONCASTER, Southern Division of the wapentake of STRAFFORTH and TICKHILL, West Riding of the county of YORK, 7 miles (N. W.) from Doncaster; containing 9 inhabitants.

STOTTESDEN (*St. Mary*), a parish, in the union of CLEOBURY-MORTIMER, partly in the hundred of WOLPHY, county of HEREFORD, but chiefly in the hundred of STOTTESDEN, Southern Division of the county of SALOP, 5¼ miles (N.) from Cleobury-Mortimer; containing 1579 inhabitants. The living is a vicarage, valued in the king's books at £15. 10. 10.; present net income, £670; patron and impropriator, Duke of Cleveland. The church was rebuilt by Robert de Belesme, Earl of Shrewsbury, who gave it to the abbey of that place: it was repewed in 1838, and 215 free sittings provided, the Incorporated Society having granted £100 in aid of the expense. There is a chapel of ease at Farlow, in this parish; and here is a Sunday National school.

STOUGHTON, a chapelry, in the parish of THURNBY, union of BILLESDON, hundred of GARTREE, Southern Division of the county of LEICESTER, 4 miles (E. S. E.) from Leicester; containing 139 inhabitants. The chapel is dedicated to St. Mary.

STOUGHTON (*St. Mary*), a parish, in the union of WEST-BOURNE, hundred of WEST-BOURNE and SINGLETON, rape of CHICHESTER, Western Division of the county of SUSSEX, 8½ miles (N. W.) from Chichester; containing 570 inhabitants. This parish is bounded on the west by the county of Southampton; the range of lofty downs, called Bowhill, extends along the southern boundary, and the village is situated in a valley to the north of it. Henry IV. granted a charter for holding a weekly market and three fairs, but both have fallen into disuse. The living is a discharged vicarage, valued in the king's books at £8. 10.; present net income, £160: it is in the patronage of the Crown; appropriators, Dean and Chapter of Chichester. The church is a cruciform structure, in the early and later styles of English architecture; the chancel is separated from the nave by a fine circular arch. Three schools are partly supported by gentlemen of the parish; and Henry Smith, in 1628, bequeathed land, now producing £30 per annum, to the poor, and for apprenticing children. Standsted, with its extensive forest, is chiefly in this parish, and has been honoured by several royal visits. Queen Elizabeth was entertained there in one of her progresses. George, Prince of Wales (afterwards George II.), on the 20th of September, 1716; his father, George I., on the 31st of August, 1722; and George III. and Queen Charlotte also lunched here on their way from Portsmouth. The mansion, situated in a well-wooded park, of about 900 acres, pleasingly diversified with hill and dale, and commanding most extensive land and sea views, was erected about the close of the 17th century, by Richard, Earl of Scarborough. Besides carvings by Grindley Gibbons, there is a suite of Arras tapestry, representing the Battle of Namur, the largest of six sets wrought at Arras for the Duke of Marlborough and five of his generals; it was brought from Flanders by the first Lord Scarborough. The Rev. Lewis Way, a former proprietor, in 1819, converted a portion of the north-west side of the old mansion into a beautiful chapel in the Gothic style, and embellished the windows with the arms of the Fitz-Alans and subsequent proprietors, and endowed the same. Charles Dixon, Esq., the present proprietor of Stansted, and who has very much improved it, presents to this chapel, which is open to the public.

STOULTON (*St. Edmund*), a parish, in the union of Pershore, Lower Division of the hundred of Oswaldslow, Worcester and Western Divisions of the county of Worcester, 3¾ miles (N. W.) from Pershore; containing 312 inhabitants. The living is a perpetual curacy; net income, £100; patron, Earl Somers; appropriators, Dean and Chapter of Worcester. The appropriate tithes have been commuted for a rent-charge of £218, subject to the payment of rates, which on the average have amounted to £34; the glebe comprises 11 acres, valued at £16. 10. per annum. Here is a Sunday National school, commenced in 1827 and supported by subscription.

STOURBRIDGE, a chapelry, in the parish of St. Andrew the Less, or Barnwell, hundred of Flendish, county of Cambridge, 1½ mile (N. E. by N.) from Cambridge. This place is remarkable for its celebrated fair, one of the largest in the kingdom: it is held in a field to the eastward of Barnwell, and commences on September 18th, on which day it is proclaimed by the vice-chancellor, doctors, and proctors of the University of Cambridge, and by the mayor and aldermen of that borough, and continues more than three weeks: the staple commodities exposed for sale are, leather, timber, cheese, hops, wool, cattle, and on the 25th, horses. The hospital of St. Mary Magdalene, for lepers, was anciently at the disposal of the burgesses of Cambridge; but, about 1245, Hugh, Bishop of Ely, possessed the patronage of it, which was also enjoyed by his successors, till the suppression, in 1497: its chapel, called St. Mary's chapel, has been converted into a barn. Here is a National school.

STOURBRIDGE, a market-town, and the head of a union, in the parish of Old Swinford, Lower Division of the hundred of Halfshire, Stourbridge and Eastern Divisions of the county of Worcester, 21 miles (N. by E.) from Worcester, and 124 (N. W.) from London; containing 6148 inhabitants. This place, originally called *Bedcote*, which name the manor still retains, derives its present appellation from the erection of a bridge, about the time of Henry VI., across the small river Stour, which here separates the counties of Worcester and Stafford. The surrounding country abounds with coal and iron-stone, the mines of which appear, by a manuscript in the possession of the Lyttleton family, to have been worked so early as the reign of Edward III.; and the manufacture of glass was established here in 1557, about the period it was introduced into this country from Lorraine. The town consists chiefly of one long street, called the High-street, which is well flagged and Macadamized, and the lower part spacious, and containing some good houses; the upper is somewhat narrow. A subscription library was established in 1790, which contains upwards of 3000 volumes, and of which Parkes, the self-taught and celebrated chymist, was the first president. There are races on two days in the last week in August, during which the theatre, a small and mean-looking building, is open; assemblies are held monthly during the season. The principal branches of trade and manufacture are those of glass, iron, and fire-bricks: the first is now carried on to a very great extent, there being twelve houses in the immediate neighbourhood, in which the different varieties of flint, crown, bottle, and window glass are manufactured, besides several cutting-mills. The flourishing state of this branch of manufacture is chiefly owing to the plentiful supply of fuel, and to the existence, near the town, of that superior species of clay used in making glass-house pots, crucibles, and fire-bricks, which is found here in large quantities, and furnishes a considerable article of export, by the name of "Stour-bridge fire-clay:" the best lies at about 150 feet below the surface of the earth, in strata of three or four feet thick, in the compass of about 200 acres, near the town: large quantities of these fire-bricks are made, and sent to London and other places. The manufacture of iron forms also a most important branch of the trade of this town and neighbourhood, and the manufactories are generally on a most extensive scale, particularly that of Bradley and Co., which covers nearly four acres, and gives employment usually to about 1000 men: nearly every article in wrought or cast iron is here manufactured, comprehending, in the foundry, steam-engines, boilers, gasometers, and every description of heavy machinery; the bearers, roofs, and fire-proof guards belonging to the custom-house, the new post-office, and the recently erected portion of the British Museum, were cast here: in the wrought-iron manufactory are made merchant, wire, and sheet, iron; hoops; nail-rods; small rounds and squares, &c. In the other manufactories are made the various articles of hammered iron, besides scythes, spades, anvils, and vices, plantation tools, chains, called gearing, &c.; but that branch of the trade which is carried on to the largest extent is the making of nails, which, in the town and neighbourhood, affords employment to some thousand men, women, and children. The trading interests are greatly benefited by a canal, which, running from the town to the Staffordshire and Worcestershire canal, connects it with that extensive line of inland navigation which spreads in various branches over the mining and manufacturing districts of the country, and also with the Severn, affording an opening for the transit of goods to all parts of the kingdom. The market, granted in 1486, by Henry VII., is on Friday, and is well attended: the market-house, erected at an expense of about £15,000, is a handsome modern brick building; the principal front, towards the High-street, is stuccoed, and of the Doric order of architecture: it consists of a spacious triangular area, having on each side an arcade; the centre is open, and that portion of the front not occupied by the entrance is disposed in shops. The fairs are on March 29th and September 8th; the former, which continues seven days, is a celebrated horse fair; the latter is for horses, horned cattle, sheep, and pedlery. A court of requests for the recovery of debts under 40*s*., the jurisdiction of which extends over the parish of Old Swinford, is held here; and by the act of the 2nd and 3rd of William IV., cap. 64, the town is a polling-place for the eastern division of the county.

An episcopal chapel, dedicated to St. Thomas, was erected by subscription among the inhabitants, under an act of parliament obtained about a century since: it is a neat brick edifice, with a square tower, and is not within the jurisdiction of the bishop; net income, £134; patrons, the Inhabitant Householders. There are places of worship for Baptists, the Society of Friends, Independents, Wesleyan Methodists, Presbyterians, and Roman Catholics. The free grammar school was founded and endowed by letters patent, granted in 1553, by Edward VI.: it has a very considerable endowment, from which £150 is paid to the head master, and £90 to the second master; the remainder, according to a special clause in the deed of endowment, after deducting for repairs of the school, &c., is divided between them; each master has also a good rent-free residence. The government is vested in eight of the principal inhabitants, who are a body corporate, and elect the masters. For several years there were no pupils on the foundation, but at present there are a few: attached to the school is a library of ancient books. Dr. Johnson received the rudiments of his education in this school until he was fifteen years of age; but the report of his having been an unsuccessful candidate for the head-mastership is void of truth. A National school was erected in 1815, and is supported by subscription: it is also used as a Sunday school. This town has the privilege of sending four boys to a noble institution near it, called the Blue-coat hospital, founded by Thomas Foley, Esq.; and it also participates in the advantages of the endowments of John Wheeler and Henry Glover; but as these institutions are for the benefit of the parish generally, an account of them will be found in the article on Old Swinford. The poor law union of Stourbridge embraces portions of the counties of Worcester, Salop, and Stafford, and comprises 14 parishes or places, containing a population of 35,911, according to the census of 1831; it is under the care of 24 guardians. In a sandy tract of ground to the westward of the town, numerous detached portions of jasper, porphyry, rock-salt, granite, chalcedony, agate, cornelian, and several varieties of marble, supposed to be diluvial remains, have been discovered.

STOURMOUTH (*All Saints*), a parish, in the union of Eastry, hundred of Bleangate, lathe of St. Augustine, Eastern Division of the county of Kent, 8 miles (E. N. E.) from Canterbury; containing 257 inhabitants. The living is a rectory, valued in the king's books at £19; present net income, £399; patron, Bishop of Rochester. The tithes have been commuted for a rent-charge of £416, subject to the payment of rates; the glebe comprises 12 acres, valued at £31. 19. 1. per annum. The navigable river Stour passes through the parish.

STOURPAIN (*Holy Trinity*), a parish, in the union of Blandford, hundred of Pimperne, Blandford Division of the county of Dorset, 3 miles (N. W. by N.) from Blandford-Forum; containing 594 inhabitants. The living is a discharged vicarage, in the patronage of the Dean and Chapter of Salisbury (the appropriators), valued in the king's books at £7. 18. 6½.; present net income, £130. The river Stour runs on the west and south of this parish. Here is a National school. Lacerton, formerly a distinct parish, was united, in 1431, to Stourpain, to which it is now only a hamlet. In a field, called Chapel Close, adjoining a farm-house, the foundations of its ancient church, which was dedicated to St. Andrew in 1331, may be still traced. On an eminence called Hod hill are the remains of a Danish camp, in the form of the letter D, defended by a double rampart and fosse, which, on the north and south sides, are almost inaccessible: there are five entrances, and within the area, which comprises several acres, are many circular trenches, four and five yards in diameter, and some round pits, contiguous to each other, supposed to have been so deep and numerous, at one period, as to be capable of concealing a large army.

STOURPORT, a market-town, in the chapelry of Mitton, parish and union of Kidderminster, Lower Division of the hundred of Halfshire, Kidderminster and Western Divisions of the county of Worcester, 4 miles (S. S. W.) from Kidderminster, and 130 (W. N. W.) from London: the population is returned with Mitton. This place, which is of modern origin, derives its name from its situation on the river Stour, near its confluence with the Severn, and from being a principal depôt for the manufactures and agricultural and mineral produce of the adjoining counties, which are hence transmitted to the various commercial towns in the kingdom. Prior to the formation of the Staffordshire and Worcestershire canal, in 1770, it consisted only of a few scattered cottages forming the lower part of the hamlet of Mitton; but since that period, from the advantages of its situation affording a communication with most parts of the kingdom, by means of the Grand Trunk canal, which connects the Severn with the Trent, it has risen into commercial importance and become an inland port of considerable trade. The town is neat and well built; the chief streets are paved, and lighted partly with gas. A handsome iron bridge of one arch, 150 feet in span and 50 in height, with several land arches affording a free course for the water in case of floods, has been constructed over the Severn, connecting the town with Arely-King's, and replacing a bridge of three arches, which was swept away by a flood after a sudden thaw. A subscription library was established in 1821; and there are three reading societies, which are well supported. The trade consists principally in the conveyance by canal navigation of the produce of the adjoining counties, for the reception of which extensive warehouses have been erected; and basins on a large scale have been formed, with wharfs for the loading and unloading of the craft employed in the trade; the building of boats and barges, for which several small docks have been constructed, is also carried on extensively. A canal from this town to Kington, in Herefordshire, was projected some time since, but it has been completed only as far as Mamble. There is a considerable iron-foundry belonging to Messrs. Baldwin. The market, which is amply supplied, is on Wednesday, and in 1833 was made a corn market: there is also a market on Saturday; both are well supplied with meat, poultry, vegetables, and fruit. The market-house, a commodious building, was erected upon a site purchased by a proprietary, who receive the tolls, and a room over it has been erected for the transaction of the public business of the town. A great quantity of hops was formerly sold here, but the trade has very much declined. Fairs are held annually on the first Tuesday in April and the second Tuesday in October, and are abundantly supplied with sheep and cattle. This town has been included in the borough of Bewdley, and has been also made a polling-place for the western division of the county. The inhabitants attend divine service in the chapel of Mitton, which has been enlarged, and 330 free sittings provided, the Incorporated Society having granted £250 in aid of the expense. There is a place of worship for Wesleyan Methodists. Spacious and airy Sunday school-rooms have been erected, in which 180 children of both sexes are instructed, and there is also a Sunday school attached to the Wesleyan meeting-house.

STOURTON (*St. Peter*), a parish, in the union of Mere, partly in the hundred of Norton-Ferris, Eastern Division of the county of Somerset, but chiefly in the hundred of Mere, Hindon and Southern Divisions of the county of Wilts, 2½ miles (W. N. W.) from Mere; containing 653 inhabitants. This place, which is of considerable antiquity, was the scene of some memorable events during the earlier periods of English history. In 656, Cenwalh, King of the West Saxons, here encountered an army of the Britons, which he defeated with great slaughter, and compelled them to retreat to Petherton, on the river Parret: and in 879, Alfred the Great, issuing from his retreat in the Isle of Athelney, erected his standard on an eminence in this parish, since called Kingsettle hill, while on his route to Edington, where he obtained a signal victory over the Danes. In 1001, an obstinate and sanguinary battle was fought near Kingsettle hill, between the Danes and the Saxons under the command of Cola and Edsigus, in which the latter were defeated: and, in 1016, another engagement took place between the Danes, under Canute, and Edmund Ironside, in which the latter was victorious. A castle was built here by John de Stourton, to whom the manor belonged, on the site of which a spacious and elegant mansion has been erected, in the Italian style of architecture, by the Hoare family, the present owner being Sir H. H. Hoare, Bart. Within the domain, and on the spot where Alfred erected his standard, a lofty tower was raised to the honour of that monarch by Henry Hoare, Esq.; on the pedestal of a statue of that hero is the following inscription,—"Alfred the Great, A. D. 879, on this summit erected his standard against the Danish invaders;" to which is added a tribute to the numerous virtues of that illustrious monarch. The living is a rectory, valued in the king's books at £17; present net income, £520; patron, Sir H. H. Hoare, Bart. The church is partly in the decorated and partly in the later styles of English architecture, and contains some monuments to the ancient family of Stourton. Henry Hoare, Esq., in 1724, gave £2000 to be appropriated to the erection of charity schools and workhouses generally; £2000, the produce to be applied yearly to supplying the poor of this parish with Bibles and other books; and £1300 for other charitable purposes. There is a fund of about £35 per annum, arising from bequests by Jane Hoare, Nicholas King and others, which is annually distributed in blankets and other relief among the poor. At the south-western extremity of the parish, in the county of Somerset, there is a wide boggy tract, containing many curious excavations, called Pen Pits, of which there are not less than several thousands scattered over a surface of nearly 700 acres. This place gives the title of Baron to the family of Stourton, created in 1448.

STOUTING (*St. Mary*), a parish, in the union of Elham, hundred of Stouting, lathe of Shepway, Eastern Division of the county of Kent, 8 miles (E. by S.) from Ashford; containing 254 inhabitants. The living is a rectory, valued in the king's books at £7. 17. 11.; present net income, £252; patron, Rev. Jacob Geo. Wrench, D.C.L. The church is principally in the early style of English architecture: it has been recently repaired and enlarged by the addition of a gallery. There is a place of worship for Wesleyan Methodists. A school-house, with a residence for a mistress, was erected in 1838, at the sole expense of the rector: the school is supported by subscription. This parish is bounded on the east by the Roman Stane-street, and a branch of the river Stour rises here. In the neighbourhood is a mound overgrown with wood, around which was a double moat. Some urns and Roman coins have been discovered in the parish.

STOVEN (*St. Margaret*), a parish, in the union and hundred of Blything, Eastern Division of the county of Suffolk, 5¼ miles (N. E. by E.) from Halesworth; containing 112 inhabitants. The living is a perpetual curacy; net income, £69; patron and impropriator, Rev. G. O. Leman. The tithes have been commuted for a rent-charge of £200, subject to the payment of rates, which on the average have amounted to £24. The church contains two Norman arches of great beauty.

STOW, a hamlet, in the parish of Threckingham, union of Sleaford, wapentake of Aveland, parts of Kesteven, county of Lincoln, 2¼ miles (N. E. by E.) from Falkingham: the population is returned with the parish.

STOW (*St. Mary*), a parish, in the union of Gainsborough, wapentake of Well, parts of Lindsey, county of Lincoln, 7¾ miles (S. E.) from Gainsborough; containing 808 inhabitants. The living is a perpetual curacy; net income, £102; patrons, alternately, Prebendaries of Corringham and Stow in the Cathedral Church of Lincoln. The church is a large structure, principally in the Norman style, with the upper part of the tower, the west window, and a few other portions, of later date.

It was founded for Secular priests by Eadnorth, Bishop of Dorchester, its revenue having been greatly augmented by Earl Leofric and his Lady Godiva. After the Conquest these religious became Benedictine monks, under the government of an abbot, and Bishop Remigius obtained for them, from William Rufus, the then desolate abbey of Eynsham in Oxfordshire, where they soon afterwards settled. There is a place of worship for Wesleyan Methodists. Courts leet and baron are annually held here, and a fair for horses and cattle on Oct. 10th. A school for the education of poor children is endowed with £12 per annum. The ancient Watling-street passes near this place, which is supposed to be the *Lidnacester* of the Romans.

STOW (*St. Michael*), a parish, in the union of Knighton, hundred of Purslow, Southern Division of the county of Salop, 1½ mile (N. E.) from Knighton; containing 147 inhabitants. The living is a discharged vicarage, valued in the king's books at £4. 7. 4.; present net income, £188: it is in the patronage of the Crown.

STOW-BARDOLPH (*Holy Trinity*), a parish, in the union of Downham, hundred of Clackclose, Western Division of the county of Norfolk, 2 miles (N. N. E.) from Downham-Market; containing 760 inhabitants. The living is a discharged vicarage, with the rectory of Wimbotsham annexed, valued in the king's books at £6. 6. 8.; present net income, £400; patron, Sir T. Hare, Bart. The church has a large square tower with brick buttresses; and on the north side of the chancel is a chapel, the burial-place of the Hare family. Near a bridge which crosses the Ouse, about two miles from the village, a fair is held for horses and cows, on the eve of the festival of the Holy Trinity. South of the church are the remains of an ancient hermitage, of flint and brick, now converted into a farm-house.

STOW-BEDON (*St. Botolph*), a parish, in the union and hundred of Wayland, Western Division of the county of Norfolk, 4¾ miles (S. E. by S.) from Watton; containing 303 inhabitants. The living is a discharged vicarage, endowed with the rectorial tithes, and valued in the king's books at £4. 19. 4½.; present net income, £295; patron, C. Eade, Esq. The church was anciently appropriated to Marham abbey, and had a guild founded in honour of the Virgin Mary.

STOW *cum* QUY (*St. Mary*), a parish, in the union of Chesterton, hundred of Staine, county of Cambridge, 5 miles (N. E.) from Cambridge; containing 400 inhabitants. The living is a perpetual curacy; net income, £52; patron and appropriator, Bishop of Ely. The tithes have been commuted for a rent-charge of £530, subject to the payment of rates; the glebe comprises 64½ acres. Jeremy Collier, the celebrated non-juring divine, was born here in the year 1650; he died in 1730.

STOW-LANGTOFT (*St. George*), a parish, in the union of Stow, hundred of Blackbourn, Western Division of the county of Suffolk, 2½ miles (S. E.) from Ixworth; containing 204 inhabitants. The living is a rectory, valued in the king's books at £8. 7. 8½.; present net income, £307; patron, J. Wilson, Esq. A school is supported by the chief landed proprietor, who allows £20 per annum to the master. There is an almshouse, consisting of four tenements, occupied by poor widows, with an acre of land adjoining, left by an unknown benefactor. Here was the seat of Sir Symonds D'Ewes, Bart., an eminent antiquary; and Tillemans, the Dutch painter, was buried in the church.

STOW, LONG, a parish, in the union of Caxton and Arrington, hundred of Longstow, county of Cambridge, 2¾ miles (S. S. E.) from Caxton; containing 231 inhabitants. The living is a rectory, valued in the king's books at £4. 8. 4.; patron and incumbent, Rev. H. Holloway. A day and Sunday school is supported by the rector. An hospital for poor sisters was founded here, and dedicated to the Blessed Virgin, in the reign of Henry III., by Walter, the then vicar.

STOW, LONG (*St. Botolph*), a parish, in the union of St. Neot's, hundred of Leightonstone, county of Huntingdon, 2½ miles (N. by E.) from Kimbolton; containing 180 inhabitants. The living is a perpetual curacy; net income, £70; patron, Prebendary of Long Stow in the Cathedral Church of Lincoln.

STOW-MARIES (*St. Mary and St. Margaret*), a parish, in the union of Maldon, hundred of Dengie, Southern Division of the county of Essex, 7 miles (S. by W.) from Maldon; containing 242 inhabitants. This parish, which is situated on the river Crouch, takes the adjunct to its name from the family of Marey, to whom the lands at one time belonged. A fair is held annually on the 24th of June; it is chiefly a pleasure fair. The living is a rectory, valued in the king's books at £18. 6. 8.; present net income, £439; patron, Rev. T. H. Storie. The church is an ancient edifice. A school is supported by the rector.

STOW-MARKET (*St. Peter and St. Mary*), a market-town and parish, and the head of the union of Stow, in the hundred of Stow, Western Division of the county of Suffolk, 12 miles (N. N. W.) from Ipswich, and, by way of that town, through which the mail travels, 81 (N. E.) from London, but only 75 through Sudbury; containing 2672 inhabitants. This town, which received the adjunct to its name in order to distinguish it from other parishes of the same name, is the most central in the county. It is situated at the confluence of three rivulets, which form the river Gippen, on the high road from Ipswich to Bury and Cambridge, and consists of several streets, which are, for the most part, regularly built and paved; many of the houses are handsome, especially those near the market-place; and the inhabitants are well supplied with water from land-springs and wells. The commercial interests of the town are essentially promoted by its locality, and have been much improved by making the Gippen navigable to Ipswich, which was effected under an act obtained in 1790. From the basin of the navigable river Orwell extends a pleasant walk, about a mile in length, chiefly through the hop plantations in the neighbourhood. The trade consists chiefly in the making of malt, for which there are more than 20 houses, and which is rapidly increasing; and the exportation of corn, to a considerable extent, to London, Hull, Liverpool, and other places: there are also small manufactories for rope, twine, and sacking, and an iron-foundry. By means of the navigation to Ipswich, grain and malt are conveyed thither, the returns consisting of timber, deals, coal, and slate, for the supply of the central parts of the county. The market is on Thursday, for corn, cattle, and provisions. Fairs are held on August 12th, for cattle; July 10th, a pleasure fair; and in September for hops, cheese, butter, and cattle. The meetings for the nomination of the county members, and other general county meetings, were, from its central situation, generally held in this town, which is now a polling-place for the western division only. The county magistrates hold a petty session every alternate Monday, and a manorial court baron is held annually.

The living is a discharged vicarage, with that of Stow-Upland annexed, valued in the king's books at £16. 15.; present net income, £281; patron, — Wilcox, Esq.; impropriator, C. R. Freeman, Esq. The church has been repewed, and contains 480 free sittings, the Incorporated Society having granted £150 in aid of the expense: it is a spacious and handsome edifice, in the centre of the town, partly in the decorated, and partly in the later, style of English architecture, with a square tower surmounted by a slender wooden spire of tasteful appearance, 120 feet in height; the latter was erected from the proceeds of a legacy left for that purpose in the reign of Anne. There are places of worship for Baptists and Independents; and a National school is supported by subscription. There are also several benevolent institutions for the relief of the poor, the funds arising from the same source. The poor law union of Stow comprises 34 parishes or places, under the superintendence of 35 guardians, and contains, according to the census of 1831, a population of 18,308. Abbot's Hall, the seat of J. Rust, Esq., was so called from having formerly belonged to the abbey of St. Osyth, in the county of Essex. In a stone pit near the entrance to the town, the tusks and bones of a species of elephant have been found. Here is a spring slightly impregnated with iron. Dr. Young, tutor to the poet Milton, and Master of Jesus' College, Cambridge, was vicar of this parish from 1630 to 1655, and was interred here.

STOW-MARKET, a hamlet, in the chapelry of GIPPING, union and hundred of STOW, Western Division of the county of SUFFOLK: the population is returned with Gipping.

STOW-ON-THE-WOLD (*St. Edward*), a market-town and parish, and the head of a union, in the Upper Division of the hundred of SLAUGHTER, Eastern Division of the county of GLOUCESTER, 25 miles (E. by N.) from Gloucester, and 82 (W. N. W.) from London; containing 1810 inhabitants. This place was the scene of a battle between the royalists and the parliamentary forces in the great civil war, on which occasion the former were put to flight. In old records the town is called Stow-St. Edward: it is situated on the summit of a steep elevation; the houses in general are of stone, but low, irregularly built, and of ancient appearance; and it is so indifferently supplied with fuel and water, and having no common field attached, that it is vulgarly remarked it has only one of the four elements, namely air. A charter for a market was procured in the reign of Edward III., by the abbot of Evesham, then lord of the manor: it is held on Thursday; and fairs are held on May 12th and October 24th, for the sale of hops, cheese, and sheep, of which last 20,000 are said to have been sold at one fair. The inhabitants were incorporated by Henry VI., but at present the town is governed by two bailiffs, who are appointed annually at the manorial court leet. It has lately been made a polling-place for the eastern division of the shire. The living is a rectory, valued in the king's books at £18; present net income, £525; patron, Rev. H. Hippisley. The church is a spacious edifice, in the ancient English style, erected at different periods in the fourteenth and fifteenth centuries: the tower is conspicuous at a great distance. There is a place of worship for Baptists. A school is endowed with £13. 9. per annum for teaching Latin, but no children attend. A National school is supported by subscription. An almshouse for nine poor persons, on the south side of the churchyard, was founded in the sixteenth of Edward IV., under the will of William Chestre, and subsequent endowments have been given for the maintenance of its inmates. The poor law union comprises 28 parishes or places, containing a population of 9105, according to the census of 1831; it is under the care of 30 guardians. A park, house, and garden, named St. Margaret's chapel, at a place called Merke, in this parish, constituted part of the estates of Charles I. and his queen. The Fosse-way intersects the northern part of the parish.

STOW-UPLAND (*St. Mary*), a parish, in the union and hundred of STOW, Western Division of the county of SUFFOLK, adjoining Stow-Market, and containing 826 inhabitants. The living is a discharged vicarage, annexed to that of Stow-Market. There is neither church nor chapel; the parishioners consequently attend divine service in the church of Stow-Market.

STOW, WEST (*St. Mary*), a parish, in the union of THINGOE, hundred of BLACKBOURN, Western Division of the county of SUFFOLK, $5\frac{1}{4}$ miles (N. N. W.) from Bury-St. Edmund's; containing 266 inhabitants. The living is a discharged rectory, with that of Wordwell united, valued in the king's books at £9. 17. $3\frac{1}{2}$.; patron, R. B. de Beauvoir, Esq. The tithes have been commuted for a rent-charge of £191. 5., subject to the payment of rates; the glebe comprises 29 acres. The church contains numerous memorials of the ancient family of Croft, of Weston Hall. The remains of Weston Hall convey some idea of its former magnificence; the gateway entrance is a fine specimen of brick-work of the time of Henry VIII. The Rev. John Boys, one of the learned divines employed in the translation of the Bible, was rector of this parish.

STOW-WOOD, a parish, in the union of HEADINGTON, hundred of BULLINGDON, county of OXFORD, 4 miles (N. E.) from Oxford; containing 26 inhabitants. There being no church, the inhabitants resort to the adjoining parish church of Beckley.

STOWE (*St. Mary*), a parish, in the union, hundred, and county of BUCKINGHAM, $2\frac{1}{2}$ miles (N. N. W.) from Buckingham; containing 490 inhabitants. The living is a vicarage, valued in the king's books at £11. 14. 7.; present net income, £95; patron and impropriator, Duke of Buckingham. There is a place of worship for Wesleyan Methodists; also a day and Sunday school, in which 50 children of both sexes are educated and clothed, and provided with breakfast and dinner on Sundays, at the expense of the Duchess of Buckingham. This place is celebrated for the princely mansion of the Duke of Buckingham. Hammond, the elegiac poet, died whilst on a visit here, in 1742.

STOWE (*St. John the Baptist*), a parish, in the union of STAMFORD, wapentake of NESS, parts of KESTEVEN, county of LINCOLN, $2\frac{3}{4}$ miles (W. N. W.) from Market-Deeping; containing 25 inhabitants. The living is a discharged vicarage, united in 1772 to that of Barholme, and valued in the king's books at £4. 3. 9.

STOWE (*St. John the Baptist*), a parish, in the Southern Division of the hundred of Pirehill, union and Northern Division of the county of Stafford, 7 miles (N. E. by E.) from Stafford; containing 1283 inhabitants. The living is a perpetual curacy; net income, £61; patron, John C. Browne, Esq.; impropriator, John Fitzgerald, Esq. The church is an ancient building, containing a monument to the Devereux family. Fourteen children are educated for about £15 a year, arising from land given by an individual whose name is unknown.

STOWE-NINE-CHURCHES (*St. Michael*), a parish, in the union of Daventry, hundred of Fawsley, Southern Division of the county of Northampton, 6 miles (S. E. by E.) from Daventry; containing 404 inhabitants. This parish, which is pleasantly situated at a short distance to the west of the turnpike road from London to Holyhead, obtained the adjunct to its name from the circumstance of the manor having nine advowsons appended to it in the reign of Henry VII. It was for some time in the possession of Sir John Danvers, one of the principal leaders of the parliamentary faction, and also one of those who signed the warrant for the execution of Charles I. The Grand Junction canal, and the London and Birmingham railway, pass through the parish. The living is a rectory, valued in the king's books at £18; present net income, £705; patron and incumbent, Rev. J. L. Crawley. The church, which is situated on the brow of a steep declivity, contains a sumptuous monument to the memory of Elizabeth, fourth daughter of John Lord Latimer. Here is a National school. The Roman Watling-street forms the ancient boundary of the parish.

STOWELL (*St. Leonard*), a parish, in the union of Northleach, hundred of Bradley, Eastern Division of the county of Gloucester, 2 miles (W. S. W.) from Northleach; containing 43 inhabitants. The living is a discharged rectory, annexed, in 1660, to that of Hampnett, and valued in the king's books at £5. 17. 1. Sir William Scott, late Judge of the court of Admiralty, was created Baron Stowell, of Stowell Park, in 1821: the title is now extinct.

STOWELL (*St. Mary Magdalene*), a parish, in the union of Wincanton, hundred of Horethorne, Eastern Division of the county of Somerset, 5 miles (S. S. W.) from Wincanton; containing 123 inhabitants. The living is a discharged rectory, valued in the king's books at £6. 15.; present net income, £193; patron, H. M. Dodington, Esq. The tithes have been commuted for a rent-charge of £169, subject to the payment of rates, which on the average have amounted to £11; the glebe comprises 27 acres, valued at £40 per annum.

STOWELL, a tything, in the parish of Overton, hundred of Elstub and Everley, Everley and Pewsey and Southern Divisions of the county of Wilts, 6½ miles (S. W. by S.) from Marlborough: the population is returned with the chapelry of Alton-Priors.

STOWER, EAST (*St. Mary*), a parish, in the union of Shaftesbury, hundred of Redlane, Shaston Division of the county of Dorset, 4¼ miles (W.) from Shaftesbury; containing 531 inhabitants. The living is annexed, with those of Motcomb and West Stower, to the vicarage of Gillingham. Here is a National School. Henry Fielding, Esq., the celebrated novelist, resided for some time on his estate in this parish.

STOWER-PROVOST (*St. Michael*), a parish and liberty, in the union of Shaftesbury, Shaston Division of the county of Dorset, 4½ miles (W. by S.) from Shaftesbury; containing 870 inhabitants. The living is a rectory, to which that of Todbere was annexed in 1746, valued in the king's books at £16. 4. 9½.; present net income, £655; patrons, Provost and Fellows of King's College, Cambridge. Here is a National school, with an endowment of £6 per annum. In the reign of William the Conqueror, a cell to the nunnery of St. Leger de Pratellis, or Preaux, in Normandy, was founded here, which at the suppression was granted to Eton College, and then to King's College, Cambridge.

STOWER, WEST (*St. Mary*), a parish, in the union of Shaftesbury, hundred of Redlane, Shaston Division of the county of Dorset, 5¼ miles (W.) from Shaftesbury; containing 219 inhabitants. The living is annexed, with those of Motcomb and East Stower, to the vicarage of Gillingham. William Watson, M. D., author of some theological productions, was a native of this place, where, though he regularly graduated as a physician, and was distinguished for knowledge of his profession, he practised as a quack.

STOWERTON, a hamlet, in the parish of Whichford, union of Shipston-upon-Stour, Brails Division of the hundred of Kington, Southern Division of the county of Warwick, 4 miles (S. E.) from Shipston-upon-Stour; containing 197 inhabitants.

STOWEY (*St. Mary*), a parish, in the union of Clutton, hundred of Chew, Eastern Division of the county of Somerset, 3½ miles (S. S. W.) from Pensford; containing 234 inhabitants. The living is a discharged vicarage, endowed with the rectorial tithes, and valued in the king's books at £6. 12.; present net income, £183; patron, Bishop of Bath and Wells. A school is chiefly supported by subscription.

STOWEY, NETHER (*St. Mary*), a market-town and parish, in the union of Bridg-water, hundred of Williton and Freemanners, Western Division of the county of Somerset, 8 miles (W. N. W.) from Bridgwater, and 147 (W. by S.) from London; containing 778 inhabitants. This place, which is situated on a stream tributary to the river Parret, consists of three streets diverging obliquely from the market-place. The town is neat and well built, but is neither paved nor lighted: at the western extremity is a hill said to have been the site of an ancient castle, of which nothing but a circular earthwork remains; this eminence commands a fine view of the channel, with the Mendip hills and the surrounding country, which is agreeably diversified. The manufacture of silk is carried on to a limited extent. The market is on Saturday; but, from its proximity to Bridg-water, very little business is transacted: the market-house is a rude building, situated in the centre of the town. A fair for cattle is held annually on Sept. 18th; and a court leet and baron is held at Michaelmas, when constables and other officers are appointed. The living is a vicarage, valued in the king's books at £5. 2. 8½.; present net income, £334; patrons and appropriators, Dean and Canons of Windsor. The church is situated at the entrance into the town from Bridg-water. There is a place of worship for Independents. A National school has been established; and a Sunday school is partly supported by a small endowment.

STOWEY, OVER (*St. Mary Magdalene*), a parish, in the union of Bridg-water, hundred of Cannington, Western Division of the county of Somerset, 8 miles (W. by N.) from Bridg-water; containing 592 inhabitants. The living is a discharged vicarage, valued in the king's books at £7. 1. 5½.; present net income, £153; patron, Bishop of Bath and Wells; impropriators, the Corporation of Bristol. The impropriate tithes have been commuted for a rent-charge of £130, and the vicarial for £165, subject to the payment of rates; the glebe comprises 1½ acre, valued at £3 per annum. The manufacture of silk is here carried on to a small extent. A National Sunday school is endowed with the interest of £100.

STOWFORD, a parish, in the union of Tavistock, hundred of Lifton, Lifton and Southern Divisions of the county of Devon, 8 miles (E. by N.) from Launceston; containing 463 inhabitants. The living is a rectory, valued in the king's books at £11. 12. 6.; patron and incumbent, Rev. John Wollocombe. The tithes have been commuted for a rent-charge of £240, subject to the payment of rates; the glebe comprises 50 acres, valued at £80 per annum. In the church is a monument, on which are marble statues of Christopher Harris, Esq., in the ancient Roman costume, and his wife Mary. Margaret Doyle, in 1777, bequeathed the interest of £200 for teaching 20 poor children. On the north side of the high road to Exeter are the remains of an ancient circular encampment. Dr. John Prideaux, a learned divine, was born here in 1578; he died in 1650.

STOWICK, a tything, in the parish of Henbury, Lower Division of the hundred of Henbury, Western Division of the county of Gloucester; containing 568 inhabitants.

STRADBROOK (*All Saints*), a parish, in the union and hundred of Hoxne, Eastern Division of the county of Suffolk, 5¾ miles (E.) from Eye; containing 1527 inhabitants. The living is a discharged vicarage, valued in the king's books at £9. 18. 6½.; present net income, £712, including the great tithes, which the vicar holds of the bishop upon a lease, renewable every 21 years, at the small rent of £8 per annum, originally granted in 1661, as an augmentation of the living; patron, Bishop of Ely. The church is a large building; the steeple, which is handsome, appears to have been built by the De la Poles, Earls of Suffolk. There is a place of worship for Baptists. William Grendling, in 1599, bequeathed land to be applied, among other purposes, in support of a school, for 5 poor children; and Michael Wentworth, in 1587, gave the town-house for the use of the poor, with a chamber for a school. Mary Warner also, in 1746, left an annuity of £15 for teaching 12 children. Robert Grostete, Bishop of Lincoln, who died in 1253, was a native of this parish.

STRADISHALL (*St. Margaret*), a parish, in the union and hundred of Risbridge, Western Division of the county of Suffolk, 5 miles (N. by W.) from Clare; containing 393 inhabitants. The living is a rectory, valued in the king's books at £9. 11. 0½.; present net income, £325; patron, Sir Robert Harland, Bart. Dr. Valpy, the master of Reading school, was long rector of this parish. Stradishall Place, the seat of William Rainor, Esq., lord of the manor, is a handsome residence, situated in a small park.

STRADSETT (*St. Mary*), a parish, in the union of Downham, hundred of Clackclose, Western Division of the county of Norfolk, 3¾ miles (E. N. E.) from Downham-Market; containing 183 inhabitants. The living is a discharged vicarage, valued in the king's books at £3. 6. 8.; present net income, £108; patron and impropriator, W. Bagge, Esq. The east window of the church exhibits the arms of the see of Ely, those of the East Angles, and of Bury and Dereham abbeys, in stained glass; the north window also is decorated with various emblems. Two day schools and a Sunday school are supported by a private family.

STRAGGLESTHORPE (*St. Michael*), a parish, in the union of Newark, wapentake of Loveden, parts of Kesteven, county of Lincoln, 8 miles (E. by S.) from Newark; containing 82 inhabitants. The living is annexed to the rectory of Beckingham. A day and Sunday school is partly supported by Sir W. E. Welby, Bart.

STRAMSHALL, a township, in the parish of Uttoxeter, Southern Division of the hundred of Totmonslow, Northern Division of the county of Stafford, 1¾ mile (N. N. W.) from Uttoxeter: the population is returned with the parish. St. Modwenna, on her arrival from Ireland, early in the ninth century, founded a nunnery here, and presided as abbess in it.

STRANTON (*All Saints*), a parish, in the union of Stockton, North-Eastern Division of Stockton ward, Southern Division of the county palatine of Durham; containing 736 inhabitants, of which number, 381 are in the township of Stranton, 5 miles (S. W. by W.) from Hartlepool. The village is pleasantly situated on the south side of Hartlepool harbour. Limestone abounds within the parish, and is quarried and burnt into lime. The living is a discharged vicarage, valued in the king's books at £17. 16. 0½.; present net income, £280; patron, Sir M. W. Ridley, Bart; impropriators, John Stephenson, Esq., and others. The church exhibits specimens of various styles of architecture. There is a place of worship for Wesleyan Methodists; also a National school. An immense quantity of human bones was discovered in draining a morass, supposed to have been those of the Scots who fell at the siege of Hartlepool, in 1644.

STRATFIELD-MORTIMER (*St. Mary*), a parish, in the union of Bradfield, partly in the hundred of Holdshott, Basingstoke and Northern Divisions of the county of Southampton, but chiefly in the hundred of Theale, county of Berks, 7½ miles (S. W. by S.) from Reading; containing 1208 inhabitants. The living is a discharged vicarage, valued in the king's books at £8. 19. 4½.; present net income, £176; patrons and impropriators, Provost and Fellows of Eton College. The impropriate tithes have been commuted for a rent-charge of £941, and the vicarial for £244, subject to the payment of rates; the impropriate glebe comprises 83 acres, and the vicarial 32 acres, respectively valued at £50 and £35 per annum. There is a place of worship for Independents. Parochial schools are supported by subscription for about 150 children. A fair for cattle is annually held on the 7th of November. There are some remains of a Roman amphitheatre, attached to the ancient Roman station of Silchester.

STRATFIELD-SAYE (*St. Mary*), a parish, in the union of Basingstoke, partly in the hundred of Reading, county of Berks, but chiefly in the hundred of

HOLDSHOTT, Basingstoke and Northern Divisions of the county of SOUTHAMPTON, 7¾ miles (N. E. by N.) from Basingstoke; containing 808 inhabitants. The living is a rectory, valued in the king's books at £24. 13.; present net income, £669; patron, Duke of Wellington. In this parish is the noble mansion belonging to the Duke of Wellington: the estate was formerly the property of Lord Rivers, from whom it was purchased by Government, and presented to His Grace. Lora Pitt and others, in 1739, erected a school-house, and endowed it with £400, now producing an annual income of about £18. 18., for which 36 children are instructed in connection with the National Society. There is also an annuity of £5, the bequest of James Christmas, for the education and relief of the poor. A Benedictine priory, in honour of St. Leonard, was founded here in 1170, by Nicholas de Stoteville, as a cell to the abbey of Vallemont in Normandy, and at the suppression was granted to Eton College.

STRATFIELD-TURGIS (*All Saints*), a parish, in the union of BASINGSTOKE, hundred of HOLDSHOTT, Basingstoke and Northern Divisions of the county of SOUTHAMPTON, 5 miles (N. W. by W.) from Hartford-Bridge; containing 232 inhabitants. The living is a rectory, valued in the king's books at £6. 10. 2½.; present net income, £204; patron, Duke of Wellington.

STRATFORD (*St. Andrew*), a parish, in the union and hundred of PLOMESGATE, Eastern Division of the county of SUFFOLK, 3 miles (S. W.) from Saxmundham; containing 200 inhabitants. The living is a discharged rectory, valued in the king's books at £5; present net income, £137: it is in the patronage of the Crown, in right of the Duchy of Lancaster. Ranulph de Glanville, Justiciary of England in the reign of Henry II., was born here.

STRATFORD (*St. Mary*), a parish, in the hundred of SAMFORD, Eastern Division of the county of SUFFOLK, 5¼ miles (W. N. W.) from Manningtree; containing 630 inhabitants. The living is a discharged rectory, valued in the king's books at £13; present net income, £296: it is in the patronage of the Crown, in right of the Duchy of Lancaster. The river Stour is navigable on the west of this parish, and also on the south, where it is crossed by a bridge. The parish is entitled to send one child to each of the schools of Dedham and East Bergholt; and £7 per annum, from land in the latter parish, is applied to teach 8 poor children. Stratford Hall, in 1672, was, by purchase, the seat of Major-Gen. Phillip Skippon. Dr. William Nicholson, Bishop of Gloucester, who died in 1672, was a native of this parish.

STRATFORD, ST. ANTHONY (*St. Mary*), a parish, in the union of ALDERBURY, hundred of CAWDEN and CADWORTH, Salisbury and Amesbury, and Southern, Divisions of the county of WILTS, 4 miles (S. W. by W.) from Salisbury; containing 125 inhabitants. The living is a rectory, valued in the king's books at £12; patrons, President and Fellows of Corpus Christi College, Oxford. The tithes have been commuted for a rent-charge of £242, subject to the payment of rates, which on the average have amounted to £23; the glebe comprises 48 acres, valued at £153 per annum. There is a school in which 9 children are instructed at the expense of a private individual.

STRATFORD, FENNY, a market-town, and chapelry, in the union of NEWPORT-PAGNELL, partly in the parish of BLETCHLEY, and partly in that of SIMPSON, in the three hundreds of NEWPORT, county of BUCKINGHAM, 13½ miles (E.) from Buckingham, and 45 (N. W.) from London; containing 635 inhabitants. The distinguishing prefix is derived from the nature of the surrounding land; the town itself stands on an eminence. In 1665, it was much depopulated by the plague, on account of the ravages of which the inns were shut up, and the road turned in another direction: it contains two streets. The Grand Junction canal crosses the high road at the bottom of the town, and about three quarters of a mile distant, is a station of the London and Birmingham railway. Lace-making employs a considerable number of poor females. The market, which has never flourished since the time of the plague, is on Monday; and fairs for cattle are held on April 19th, July 18th, Oct. 10th, and Nov. 28th. The living is a perpetual curacy; patron, John Willis, Esq. The chapel, dedicated to St. Martin, and situated in Bletchley, having been dilapidated since the reign of Elizabeth, was rebuilt by subscription, through the exertions of Mr. Browne Willis, the antiquary, who resided here: the first stone was laid by him on St. Martin's day, in 1724. The remains of Mr. Willis are interred within the rails of the communion-table: he bequeathed a benefaction for a sermon to be preached on St. Martin's day, and requested that the rector of Bletchley may never have the cure of Fenny-Stratford, but directed that, if he would contribute £6 per annum towards the salary of the curate, he should have the appointment; this has never been done. There are places of worship for Baptists and Wesleyan Methodists. A Sunday National school was erected in 1817.

STRATFORD - LANGTHORNE, a ward, in the parish of WEST HAM, hundred of BEACONTREE, Southern Division of the county of ESSEX, 4 miles (N. E. by E.) from London: the population is returned with the parish. The village is situated on the high road to Harwich, and on the banks of the river Lea, over which is a bridge connecting it with the village of Stratford-le-Bow; it is well lighted with gas by the trustees of the road, and supplied with water from the East London water-works. The printing and dyeing of calico and silk are extensively carried on. There are also two chymical establishments, and a porter brewery, on the river Lea, which is navigable to the Thames. This has lately been made a polling-place for the southern division of the county. A church, in the early style of English architecture, with a tower and spire, was erected in 1833, at an expense of £7100. 10., under the act of the 58th of George III. There are places of worship for Independents, Wesleyan Methodists, and Roman Catholics. A charity school was founded in 1802, for clothing and educating 30 girls; and there is also a National school. John Hiett, in 1719, bequeathed £5 a year for apprenticing one poor boy. About 1135, a Cistercian abbey was founded here by William of Montfichet, which, from its low situation in the marshes, being damaged by the floods, the society removed to a cell at Burghstead, near Billericay; but on its repair, they returned, and continued till the dissolution, at which period the revenue was valued at £573. 15. 6.

STRATFORD-LE-BOW, county of MIDDLESEX.—See BOW.

STRATFORD, OLD, a hamlet, partly in the parishes of COSGROVE, FURTHO, PASSENHAM, and POTTERS-PURY, hundred of CLELEY, Southern Division of the county of NORTHAMPTON, ¼ of a mile (N. W.) from Stony-Stratford: the population is returned with the parishes. At Chapel Close there formerly stood a hermitage and free chapel.

STRATFORD, OLD (*HOLY TRINITY*), a parish, in the union of STRATFORD-UPON-AVON, Stratford Division of the hundred of BARLICHWAY, Southern Division of the county of WARWICK; comprising the town of Stratford-upon-Avon, and containing 5171 inhabitants. The living is a discharged vicarage, valued in the king's books at £20; present net income, £239: it is in the patronage of the Countess of Plymouth; impropriators, Mayor and Corporation of Stratford-upon-Avon, which see. Three schools are partly supported by subscription.

STRATFORD, STONY, a market-town, comprising the united parishes of *St. Giles* and *St. Mary Magdalene* (commonly called West Side and East Side), in the union of POTTERS-PURY, three hundreds of NEWPORT, county of BUCKINGHAM, 8 miles (N. E.) from Buckingham, and 51 (N. W.) from London, and containing about 1700 inhabitants. At or near this spot appears to have been the boundary of King Alfred's kingdom, running from Bedford along the river Ouse and ending at the Watling-street; and also the *Lactodorum* of the Itinerary. Camden is of opinion that the latter was at this town, because the derivation of *Lactodorum*, in the ancient British language, agrees with the present name, both signifying "a river forded by means of stones." Dr. Stukely supposes it was at Old Stratford, on the Northamptonshire side of the river; and Dr. Salmon, at Calverton, an eminence close by, which led to the ford at Passenham, and an adjoining parish, where the army of Edward the Elder was stationed, whilst he fortified Towcester. Through Stratford passes the Roman road, Watling-street, in a direct line through the county from Dunstable. One of the crosses of Eleanor, queen of Edward I., was erected here, in memory of the body resting at this place in its way from Lincolnshire to Westminster, but it was demolished in the great civil war. At an inn in this town, Richard III., when Duke of Gloucester, accompanied by the Duke of Buckingham, seized the unfortunate young prince, Edward V., and in his presence arrested Lord Richard Grey, Sir Thomas Vaughan, and Sir Richard Hawt. In 1736, an accidental fire destroyed 53 houses; and, in 1742, a similar catastrophe consumed 113, and the church of St. Mary Magdalene, which has never been rebuilt: the tower, however, escaped the flames, and is yet standing. The damage was estimated at £10,000, of which £7000 was raised for the sufferers by a brief and subscriptions. The town, which is situated on the parliamentary road to Birmingham, Holyhead, Chester, and Liverpool, consists of one long street, which is Macadamized, with a good market-square and two back streets: the houses are principally built of brick. It consisted originally of only a few inns, and was a noted place of rendezvous for pack-horses, prior to the introduction of waggons, for the conveyance of goods to London. The bridge over the Ouse is supposed to have been built by the Romans: it consists of five arches, and, having been partially destroyed during the civil war of the 17th century, is now in a very dilapidated state. In 1834, an act was obtained enabling the justices of the two counties of Buckingham and Northampton to take down and rebuild the bridge on an enlarged plan. The manufacture of bone lace was formerly carried on here to a considerable extent, but has greatly declined, very little being now made. The Grand Junction canal passes about two miles north-east of the town, where it is carried over the river Ouse, across Wolverton valley, by a large embankment and aqueduct of cast iron: there were formerly three arches, which fell in 1806. About half a mile from the aqueduct terminates a branch canal from Buckingham, constructed under the authority of an act obtained in 1794, which extends along the north side of the valley of the Ouse, passing by Old Stratford, to its junction with the main line at Cosgrove. The London and Birmingham railway also passes within two miles, on which part of its course is the largest viaduct throughout the entire extent of the line, also the Wolverton station. Henry III., in 1257, granted to Hugh de Veer, Earl of Oxford, a fair to be held here on the eve, day, and morrow of St. Giles; and Edward I., in 1290, granted another fair, to be held on the eve and festival of St. Mary Magdalene. Charles II., in 1648, granted to Simon Bennett, Esq., four annual fairs, to be held in the west part of the town, *viz.*, on the Friday next before the feast of St. Michael the Archangel, on the Feast of All Saints, April 9th, and the Wednesday next before Whitsuntide. By the last charter was also granted a market, which is held on Friday, with a court of pie poudre. The magistrates for the two counties of Buckingham and Northampton hold a petty session here on alternate Fridays.

The livings of the two parishes, having been united, form a perpetual curacy; net income, £130; patron and appropriator, Bishop of Lincoln. The church, dedicated to *St. Giles*, formerly a chantry, was erected in 1451, and endowed in 1482, but rebuilt, except the tower, in 1776, by Mr. Irons of Warwick: it was formerly considered a chapel to the mother church of Calverton, on the west side of the street, whilst that of *St. Mary Magdalene* belonged to Wolverton, on the east side. There are places of worship for Baptists, Independents, and Wesleyan Methodists. A free school was founded and endowed with property now producing £28 per annum, by Michael Hipwell, in 1610; in 1819 it was incorporated with a National school then established, and the master receives the amount of the permanent endowment, subject to a deduction of £5 for the poor of the two parishes. John Whally, in 1670, devised an estate, now consisting of a farm containing nearly 169 acres in the parish of Hartwell, for apprenticing poor boys born in this town, and whose parents have lived in it five years as housekeepers, each with a premium not exceeding £10; the same sum also to be given to each boy who has honestly and truly served his apprenticeship, to set him up in business: by a decree of the court of Chancery the trustees are now empowered to pay any sum not exceeding £25 as an apprentice fee. Edmund Arnold, Esq., devised the manor of Furtho, and all his lands there, in trust, among other things, to pay £20 per annum for apprenticing poor children of this town, and towards setting them up in business; £5 per annum towards the relief of the poor; and £20 per annum towards the support of an orthodox minister to perform certain religious duties in one of the churches or chapels of Stony-Stratford, to be elected by twelve of the most substantial Protestant housekeepers within the town, with the assistance and advice of the rectors or ministers of Furtho, Cosgrove, Passenham, and Calverton: the bequest for placing out apprentices has been augmented by the court of Chancery, and about £100 a year is now appropriated out of this estate for that purpose. Mr. Serjeant Piggott in 1519, John White in 1674, and John Mashe, gave several estates in trust for the purpose of maintaining and keeping in repair the bridge and causeways, and the highways on both sides of the town: these estates were sold under an act passed in the 41st of George III., and the proceeds were invested in the purchase of a farm now let for £240 per annum. By the act for rebuilding the bridge, that portion of the funds which was applicable to its repair is to be given up to the justices, in consideration of the inhabitants being in future released from all charge for its maintenance. There is also a fund of about £60, arising from estates devised by Sir Simon Bennett, James Barnes, and others, which is applicable to the relief of the poor and setting them to work, and for providing clothing, bread, &c.

STRATFORD-UNDER-THE-CASTLE (*St. Lawrence*), a parish, in the union of Alderbury, hundred of Underditch, Salisbury and Amesbury, and Southern, Divisions of the county of Wilts, 1¾ mile (N. W. by W.) from Salisbury; containing 374 inhabitants. The living is a perpetual curacy; net income, £80; patrons and appropriators, Dean and Chapter of Salisbury.

Seal aud Arms.

STRATFORD-UPON-AVON, an incorporated market-town, and the head of a union, in the parish of Old Stratford, having separate jurisdiction, though locally in the Stratford Division of the hundred of Barlichway, Southern Division of the county of Warwick, 8 miles (S. W.) from Warwick, and 94 (N. W.) from London, on the road through Oxford to Shrewsbury; containing 3488 inhabitants. This place, originally called *Streat-ford* and *Stretford*, derived its name from its situation on the great north road, and from a Saxon ford on the river Avon, at the entrance to the town. It was a place of considerable importance prior to the Conquest, and was distinguished for its monastery, founded in the reign of Ethelred, on or near the site of the present church. In 1197, Richard I. granted the inhabitants a weekly market; and, during the succeeding reigns, various other privileges were conferred upon the town. In the 36th and 37th of Elizabeth it suffered materially from accidental fires, which destroyed the greater part of it; and again, in 1614, it experienced a similar calamity. In 1588, both ends of the bridge over the Avon were carried away by a flood that inundated the lower part of the town. During the parliamentary war, a party of royalists stationed here was driven out by a superior force of parliamentarians, under the command of Lord Brooke, in 1642; but the inhabitants still maintained their adherence to the royal cause, and, in the following year, Henrietta Maria, queen of Charles I., at the head of 3000 infantry, 1500 cavalry, and with a train of artillery and 150 waggons, advanced to the town, where she was met by Prince Rupert; and, after remaining for three days at New Place, then the residence of Shakspere's daughter, where she was hospitably entertained by the family, proceeded to Kington, to meet the king, whom she accompanied to Oxford. The parliamentarians, having subsequently obtained possession of the town, demolished one of the arches of the bridge, over the deepest part of the river, to prevent the approach of the royalists.

The town is beautifully situated on the south-west border of the county, on an eminence rising gently from the west bank of the Avon. The entrance from the London road is over a handsome stone bridge of fourteen pointed arches, built by Sir Hugh Clopton, in the reign of Henry VII., and widened by act of parliament in 1814. Nearly parallel with it is another of nine cycloidal arches, built of brick, and exclusively used as a rail-road to the wharfs at this extremity of the town. The town consists of several spacious streets, intersecting each other, some at right angles, and others crossing obliquely: the houses in that part which is called the Old Town, though rather of ancient appearance, are commodious and well built, occasionally interspersed with modern buildings of large dimensions and handsome appearance, and in some of the streets are smaller houses of frame-work timber and plaster; among these, part of the ancient house in which Shakspere was born is still preserved in its antique state, and is an object of much interest. The house in which he lived in retirement, for a few years previously to his decease, was originally the mansion of the Clopton family, and was purchased by the bard, who, after repairing and improving it, called it "New Place:" it has been taken down by a late proprietor, who also cut down the mulberry tree planted by Shakspere in the gardens. The town is partially paved, and lighted with gas, and the inhabitants are amply supplied with water from pumps attached to their houses. The public library and reading-rooms are supported by subscription; the Shaksperian library, also supported by private subscription, was established in the year 1810, and is a permanent and useful institution. The theatre is a neat building of brick, within the precincts of Shakspere's garden: assemblies are held occasionally, during the winter, at the town-hall. To the south of the town is a race-course, where races took place as early as 1691, and were in general well attended; but since 1786 they have been discontinued. A jubilee, in honour of Shakspere, was instituted by Garrick, in 1769, when the town-hall, which had been recently rebuilt, was dedicated to the poet; and his statue, finely sculptured, and presented to the town by Garrick, at the close of the ceremony, was placed in a niche in the north end of the building; this festival has been recently revived, and is celebrated every third year. The environs, abounding with diversified scenery and objects of considerable interest, afford many beautiful walks; and the salubrity of the air, and its central situation in a neighbourhood enlivened with the elegant villas of respectable families, and the noble mansions of the wealthy, make this town eligible as a place of residence. There is not much trade carried on, the inhabitants being principally employed in agriculture. The Stratford canal, passing close to the north of the town, and joining the Birmingham, Warwick, and Oxford canals, connects them with the Avon, which is navigable for barges of 40 tons to Tewkesbury, where it joins the Severn, thus affording a line of inland navigation to the principal towns in the kingdom: near the bridge are some extensive wharfs for lime, timber, coal, and other articles of merchandise. A railway has also been constructed from this town to Moreton-in-the-Marsh, in the county of Gloucester, from which a branch has been formed to Shipston-upon-Stour. The market, which was formerly on Thursday, is now, by charter granted in the 59th of George III., held on Friday, and is very considerable for corn and other grain, and for cattle. The fairs, to which are attached courts of pie-poudre, are on May 14th and the three following days, for cattle, horses, and toys; and September 25th, for cattle and cheese; besides these there are great cattle markets on the third Monday in February, the Friday after the 25th of March, the last Monday in July, the second Friday after the 25th of September, and on the second Monday in December: there is also a statute fair on the morrow after Old Michaelmas. The corn market is held in the area near the town-hall; the poultry market in a neat stuccoed building erected at the east end of Wood-street, near the site of the ancient cross, and surmounted by a cupola and vane, representing a falcon grasping a tilting spear, Shakspere's family crest; and the cattle market in Rother-street.

The town received its first regular charter of incorporation from Edward VI., in 1553, which, reciting and confirming former grants of privileges to the "Bailiff and Burgesses of Stratford-on-Avon," was extended by James I., in 1611, and subsequently by Charles II., in the 16th and 26th years of his reign. Under this last charter the government was vested in a mayor, high steward, recorder, two chamberlains, twelve aldermen, and twelve capital burgesses, assisted by a steward of the borough court, clerk of the peace, two sergeants-at-mace, and subordinate officers. The mayor, the late mayor, the high steward, the recorder, and the two senior aldermen, were justices of the peace, and, by the second charter of Charles II., the mayor, recorder, and the senior aldermen, were also justices of the peace for part of the parish of Old Stratford not otherwise within the jurisdiction of the borough, which includes Old Town, and the church and churchyard. The corporation now consists of a mayor, four aldermen, and twelve councillors, under the act of the 5th and 6th of William IV., cap. 76, for an abstract of which see the Appendix, No. I. The mayor and late mayor are justices of the peace by virtue of their office; a commission of the peace has also been granted to the borough, by which two additional justices are appointed. The police consists of a superintendent, one day constable, and three night constables. The corporation are, by their charter, empowered to hold a court of record for the recovery of debts not exceeding £40; but it has fallen into disuse. This town has been made a polling-place for the southern division of the shire. The guildhall, in which the courts were held, and the business of the corporation is now transacted, is an ancient building, possessing few claims to architectural notice: it occupies the west side of a small quadrangular area, of which the chapel of the ancient guild of the Holy Cross forms the north side, the vicar's and schoolmaster's houses the east, and the entrance to the school the south side; above the hall are rooms appropriated to the use of the free grammar school. The town-hall was rebuilt in 1768, by the corporation, assisted by the nobility and gentry of the neighbourhood, on the site of the former, of which the upper room, having been used during the civil war as a magazine, by an accidental explosion was destroyed, and the building greatly damaged. The present building is a plain and substantial structure, of the Tuscan order, on piazzas: at the north angle are two small cells for the temporary confinement of prisoners, and the rest of the area is appropriated to the use of the market: on the west front are the arms of the corporation, and in a niche at the north end of the building is the statue of Shakspere, presented by Garrick. The upper story comprises a handsome banquetting-room, decorated with paintings, among which are a full-length portrait of Shakspere, sitting in an antique chair, by Wilson, and, at the opposite end, one of Garrick, reclining against a bust of the poet, by Gainsborough; besides several smaller apartments, which are also ornamented with paintings: the larger meetings of the corporation, and the town meetings, are held here; and the celebration of the jubilee, concerts, and assemblies, take place in this suite of rooms.

The parochial church, which was formerly collegiate, is situated within the town; it is a spacious and venerable cruciform structure, in the early style of English architecture, with a square embattled tower rising from the centre, and surmounted by a lofty octagonal spire; the west entrance is through a richly-moulded and deeply-recessed archway, above which is a large window in the later style, having the lower central compartment filled up with three richly-canopied shrines: an avenue of lime trees, with their branches entwined, forms a pleasing approach to the north porch, over which is an apartment originally lighted by a window, now covered by a tablet. The nave, of which the fine oak roof is richly carved, and supported on clustered pillars and pointed arches, is very lofty, and is lighted by a range of twelve large clerestory windows enriched with tracery of the later style. In the south aisle, which is in the decorated style, is a chapel dedicated to St. Thomas à Becket; and in the north aisle is a sepulchral chapel, separated by a richly-carved stone screen, containing several altar-tombs, with recumbent figures of the Clopton family, finely sculptured in marble, and painted to represent the natural complexion of the persons. In the transepts are several ancient and some handsome modern monuments, and at the extremity of each is a large enriched window. Massive piers of clustered columns and lofty arches support the tower and separate the chancel from the nave. The chancel, parted off by an oak screen, which has been glazed, has a fine roof of richly-carved oak, and is lighted by a handsome range of five windows on each side, in the later style, and a large east window of rich tracery, in which are placed several portions of stained glass that have been preserved: on the south side, near the altar, is a piscina, and near it are stone stalls of elegant design. On a slab at the entrance to the altar, covering the ashes of the bard, is an inscription written by Shakspere; and on the north wall is his monument, in which is his bust, representing him in the act of composing, with a pen in the right hand, and the left arm resting upon a scroll on a cushion: this bust, which is a well-attested likeness, and was originally painted with strict resemblance to the complexion, and colour of the eyes and hair, of the poet, has, by the direction of his commentator Malone, been painted to resemble stone, and forms a lamentable contrast to the complexioned monuments of the Clopton family, and others in the church. The ancient stone font in which Shakspere is supposed to have been baptized having been removed, to make room for a modern one of marble, was preserved by the late Captain Saunders, of Stratford, who placed it on the pedestal of the ancient market-cross, and, upon the erection of the new market-house, removed it into his garden. The church is undergoing a thorough repair, in which a due regard to its original character will be preserved, and a complete restoration of its ancient elegance, under the superintendence of a committee appointed by the Shaksperian Club, at an estimated expense of £3000, raised by subscription, aided by grants from the National and Incorporated Societies: it will contain 320 free sittings. The soil of the churchyard, which had accumulated to the height of three feet above the natural level of the ground, has been removed; the organ gallery, which closed up the east end of the nave, and excluded the chancel and transepts, has been also removed, and the entire view from the western entrance to the east window of the chancel thrown open; the flat ceiling of the chancel has been taken away, and the original roof restored, into the spandrels of which have been introduced the armorial shields of many of the nobility and gentry of the county; and the thick coats of whitewash, which had obscured the more delicate and minute details of the stone stalls and other elegant sculptures, have been also removed, and the whole has been restored to its original purity and elegance: the renovation of the chancel and its monuments, which was peculiarly the work of the committee, was effected at an expense of £1100, raised in subscriptions not exceeding £1 each. The chapel of ease, dedicated to the Holy Trinity, is a handsome edifice, in the later style of English architecture: it formerly belonged to the ancient guild of the Holy Cross, and was rebuilt by Sir Hugh Clopton, in the reign of Henry VII. It has a square embattled tower, and a beautiful north porch, of which the entrance is a deeply-recessed and highly-enriched arch, surmounted by a canopy embellished with scrolls and flowers; the nave is lofty, and is lighted on each side by a range of four windows; the chancel appears from frequent alteration and repair, which have been made at different times without due regard to the prevailing style of the structure, to have lost its original character: in repairing the chapel, the walls were found to have been originally decorated with various legendary paintings of great antiquity. The master of the grammar school is usually appointed chaplain, to whom the corporation pay a stipend of £50 per annum. There are places of worship for Baptists, Independents, and Wesleyan Methodists.

The free grammar school was founded in 1482, by Thomas Jolyffe, a native of the town, and one of the brethren of the ancient guild of the Holy Cross: at the dissolution the estate was seized by Henry VIII., but was afterwards restored to the corporation by charter of Edward VI. By that pious monarch the school, which is open to all inhabitants of the town for gratuitous instruction, was refounded: the number of scholars on the foundation is twelve. The income arising from the endowment is about £115 per annum, which is paid by the corporation to the master, who is appointed by the lord of the manor; in addition to which he has £30 per annum for a house, and the chaplainship of the Holy Trinity, as assistant to the vicar. In this school Shakspere received his education, but he was removed at an early age. The National school is supported by subscription; a Lancasterian school is supported by the Independents; and an infants' school was founded by Miss Mason, at whose expense a neat building, capable of receiving 200 children, was erected. The almshouses nearly adjoining the guildhall, and in a similar style of building, were refounded and endowed under the charter of Edward VI., for twelve men and twelve women: there are numerous other charitable bequests for distribution among the poor. The poor law union of Stratford-upon-Avon comprises 36 parishes or places, containing a population of 18,745, according to the census of 1831; it is under the care of 44 guardians.

About a mile to the west of the town, in the hamlet of Bishopton, is a mineral spring, which, having been analysed by Dr. Perry, in 1744, was found to be of a saline quality, strongly impregnated with sulphur, in its properties like the water of Leamington. The water has been recently analysed by Professor Daniel, of King's College, London, and since by Professor Phillips, of St. Thomas's Hospital; it contains sulphuric, muriatic and carbonic acids; soda, magnesia, lime, carbonate and muriate of lime; sulphates of soda and magnesia, and chloride of sodium; with a portion of sulphuretted hydrogen gas, which last ingredient was found in the analysis made by Professor Phillips, and in smaller proportion than in the Harrogate waters. An elegant pump-room has been erected at the spring; and for the accommodation of invalids, to whom the distance from the town may be inconvenient, a handsome and commodious hotel has been built, affording every requisite comfort for visiters of rank. The spa, which is designated the Victoria Spa, is a tasteful erection in the embellished rustic style; and the grounds and walks are laid out with great beauty and variety. It is sheltered from the north, north-east, and north-west, by richly wooded hills, and the immediate neighbourhood abounds with pleasingly diversified scenery. At Welcombe, about one mile to the north of the town, are the remains of a military intrenchment: in the neighbourhood are several tumuli, in which human bones, spear-heads, and other military weapons, have been found: in opening one of these, in 1795, the proprietor discovered a human skull, transfixed with a spear, which appeared to be the gilded head of a standard pike. On the surface of Borden hill, about a mile to the west, astroites, or star stones, are found in profusion, in small columns apparently formed of successive layers, which are easily separated: the soil is calcareous, and in the region of limestone; to the north-west, large specimens of testaceous fossils are found: the star stone is found also at Shuckburgh and Southam, and the arms of the family of Shuckburgh are three of these stones of five points, parted by a chevron. Of the ancient monastery, or of the college that succeeded it, not the slightest vestige is discernible. So intimately is the name of Shakspere associated with every recollection and description of this place, that every circumstance connected with his memory is deemed worthy of being recorded. It is singular that, though a letter addressed to him has been discovered, which is now in the possession of Mr. Wheler, no traces of his hand-writing, nor any thing that was ever known to have belonged to him, have ever been fully authenticated, until recently, when an autograph was discovered which sold for £100. In 1810, a large gold seal ring, which had evidently lain there many years, was found near the churchyard, bearing the initials W. S., tied together with a string and tassels, according to the fashion of his time: this, which with great probability is supposed to have been his signet, is in the possession of Mr. Wheler, author of the history of Stratford. Stratford is eminently distinguished as the place where Shakspere was born, on the 23rd of April, 1564, and in which, after having lived a few years in retirement, he ended his days, on the anniversary of his birth, in 1616, in the 52nd year of his age. Among other eminent natives were John de Stratford, Lord Treasurer in the reign of Edward II., and Lord Chancellor in that of Edward III., who promoted him to the see of Canterbury; Robert de Stratford, his brother, Archdeacon of Canterbury, and afterwards Lord Chancellor, on the translation of his brother to the primacy, and who, together with him and the Bishop of Lichfield and Coventry, was committed to the Tower, on a charge of having detained the supplies for the war with France, but was subsequently liberated and promoted to the see of Chichester; Ralph de Stratford, Bishop of London; John Huckell, educated in the free grammar school, author of a poem on the Avon, and who assisted Garrick in the composition of the Ode and other poetical addresses, delivered at the celebration of the jubilee, in 1769; and Francis Ainge, a memorable instance of longevity, who was baptized on the 23rd of Aug., 1629, left England in his youth, and died in North America, on the 13th of April, 1767, having attained the extraordinary age of 137 years. The small village of Shottery, a mile west of Stratford, is supposed to have been the residence of Ann Hathaway, the wife of Shakspere, prior to her marriage; and the cottage is still shown in which she dwelt.

STRATFORD, WATER (*St. Giles*), a parish, in the union, hundred, and county of Buckingham, 3 miles (W. by N.) from Buckingham; containing 186 inhabitants. The living is a rectory, valued in the king's books at £7. 0. 5.; present net income, £306; patron, Duke of Buckingham. The church is partly Norman.

STRATTON, a joint hamlet with Holme, in the parish and hundred of Biggleswade, county of Bedford, $\frac{3}{4}$ of a mile (E. by S.) from Biggleswade, with which the population is returned.

STRATTON (*St. Andrew*), a market-town and parish, and the head of a union, including the small sea-port of Bude, in the hundred of Stratton, Eastern Division of the county of Cornwall, $17\frac{1}{2}$ miles (N. N. W.) from Launceston, and 223 (W. by S.) from London; and containing 1613 inhabitants. This place was the scene of a great victory obtained, in the early part of the civil war, by the royalist forces over the parliamentarians: and in consideration of the eminent services rendered by Sir Ralph Hopeton, on this occasion, he was created Lord Hopeton of Stratton, in 1643; and after his death in 1654, Sir John Berkley, to whose prowess and courage the victory at Stratton was mainly owing, was, in 1658, created Baron Berkley, of Stratton, by Charles II., who was then in exile. The town is situated in a flat country, and the streets are but indifferently paved. The Bude canal passes within a mile of it, and extends to Draxton bridge, about three miles north of Launceston: upon it are six inclined planes, worked by very powerful machinery, particularly that near Bude. The market is on Tuesday, and fairs are on May 19th, Nov. 8th, and Dec. 11th. A court leet is held annually by the lord of the manor, and a court baron by the lord of the manor of Efford: petty sessions for the hundred are also held on the first Tuesday in every month; and the town has been made a polling-place for the eastern division of the county. The living is a discharged vicarage, valued in the king's books at £10. 11. 8.; present net income, £129: it is in the patronage of the Crown, in right of the Duchy of Cornwall; impropriators, — James and — Shepherd, Esqrs. The church is in the later style of English architecture, with a lofty square embattled tower crowned with pinnacles. In the north aisle is the effigy of a Knight Templar, supposed to be that of Ranulph de Blanchminster, constable of Ennour castle in Scilly; and at the east end of the south aisle is a tomb of black marble, on the lid of which are the effigies in brasses of Sir John Arundle, Knight, his two wives, and their thirteen children. A new chapel has been built and endowed at Bude by Sir Thomas Dyke Acland, Bart., in whom the patronage is vested. There is a place of worship for Wesleyan Methodists. Here is a small charitable donation for the education of fifteen boys and ten girls; and another school is partly supported by subscription. Some lands, now let for about £115 per annum, are vested in feoffees for the benefit of the poor of this parish. About half a mile to the west of the town are the remains of Binhammy castle, the occasional residence of Ranulph de Blanchminster, occupying an elevated site surrounded by a deep foss. There are vestiges of Roman roads in this parish, and several coins and tessellated pavements have been discovered.

STRATTON (*St. Mary*), a parish, in the union of Dorchester, hundred of George, Dorchester Division of the county of Dorset, $3\frac{1}{2}$ miles (N. W.) from Dorchester; containing 310 inhabitants. The living is a perpetual curacy, annexed to that of Charminster. The church has a lofty tower, but no chancel, the latter having been pulled down in 1547. A Roman road from Dorchester to Ilchester passes through the parish.

STRATTON (*St. Peter*), a parish, in the union of Cirencester, hundred of Crowthorne and Minety, Eastern Division of the county of Gloucester, $1\frac{3}{4}$ mile (N. W.) from Cirencester; containing 468 inhabitants. The living is a rectory, valued in the king's books at £12. 7. 6.; present net income, £300: it is in the patronage of Jane Masters. The church is a small ancient structure, with a low slated tower rising from between the nave and the chancel. A day and Sunday school is supported by subscription. The ancient Erming-street passes through the parish.

STRATTON (*St. Michael*), a parish, in the union and hundred of Depwade, Eastern Division of the county of NORFOLK, 1 mile (N. by E.) from Stratton-St. Mary; containing 203 inhabitants. The living is a rectory, with that of St. Peter consolidated, valued in the king's books at £6. 12. 8½.; present net income, £388; patrons, Warden and Fellows of New College, Oxford. The church of St. Peter has been long since demolished.

STRATTON, an extra-parochial liberty, in the hundred of COLNEIS, Eastern Division of the county of SUFFOLK, adjoining the parish of Levington, and containing but one house, the ancient hall. In Chapel-field, between Levington and Trimley, are the ruins of a church or chapel.

STRATTON (*St. Margaret*), a parish, in the union of HIGHWORTH and SWINDON, hundred of HIGHWORTH, CRICKLADE, and STAPLE, Swindon and Northern Divisions of the county of WILTS, 2¾ miles (N. E. by N.) from Swindon; containing 924 inhabitants. The living is a discharged vicarage, valued in the king's books at £8. 12. 3½.; present net income, £216; patrons, Warden and Fellows of Merton College, Oxford (the impropriators), on the nomination of the Bishop of Salisbury. John Herring, in 1720, bequeathed lands producing a small income for teaching poor children; and here is a National school. An Alien priory was founded here soon after the Conquest, which, at the dissolution, was given by Henry VI. to King's College, Cambridge. The line of the Great Western railway passes through the parish.

STRATTON-AUDLEY (*St. Mary*), a parish, in the union of BICESTER, partly in the hundred and county of BUCKINGHAM, but chiefly in the hundred of PLOUGHLEY, county of OXFORD, 3 miles (N. E. by N.) from Bicester; containing 360 inhabitants. The living is a perpetual curacy; net income, £89; patrons and appropriators, Dean and Canons of Christ-Church, Oxford. The church contains monuments to Sir — Borlase, Bart., who died in 1688; and to Admiral Sir John Borlase Warren, Bart. A day and Sunday school is partly supported by subscription. This place is supposed by Bishop Kennet to have derived its name from a Roman road or street, an opinion strengthened by the discovery of Roman coins and arms in the vicinity.

STRATTON, EAST (*All Saints*), a parish, in the union of NEW WINCHESTER, hundred of MITCHELDEVER, Winchester and Northern Divisions of the county of SOUTHAMPTON, 6 miles (N. N. W.) from New Alresford; containing 386 inhabitants. The living is annexed, with those of Northington and Popham, to the vicarage of Mitcheldever. A day and Sunday school is supported by Sir Thomas Baring, Bart.

STRATTON, LONG (*St. Mary*), a parish, in the union and hundred of DEPWADE, Eastern Division of the county of NORFOLK, 10½ miles (S. by W.) from Norwich; containing 721 inhabitants. This parish, which is situated on the turnpike road from Norwich to London, by way of Ipswich, was, during Wat Tyler's rebellion, chosen as a place of meeting for the magistrates and gentry of the county. It has for many centuries been the place for holding the petty sessions for the hundreds of Depwade and Henstead, and has been made a polling-place for the Eastern Division of the county. A fair was granted by King John to Roger de Stratton, in 1207, which is now disused. Two other fairs are annually held here. The living is a rectory, valued in the king's books at £10; present net income, £345; patrons, Master and Fellows of Gonville and Caius College, Cambridge. The church, in old records called Stratton *cum turri*, was built about the year 1330; it contains monuments to Sir Roger de Bourne, to Judge Reve and his lady, to the Rev. Randall Burroughes and his lady (the heiress of William Ellis, Esq.), to the Rev. Ellis Burroughes, and others of that family. There is a place of worship for Wesleyan Methodists. Part of the ancient manor-house, encircled by a deep moat, is still remaining. The Roman road leading to the station *Ad Tuam*, or Tasburgh, passes through the parish. There was anciently a hermitage, with an oratory attached to it. Several Roman urns, of which one was curiously ornamented, were found in 1773, on opening a gravel pit, near which a sepulchral hearth has been since discovered.

STRATTON-ON-THE-FOSS (*St. Vigor*), a parish, in the union of SHEPTON-MALLET, hundred of KILMERSDON, Eastern Division of the county of SOMERSET, 6 miles (N. N. E.) from Shepton-Mallet; containing 407 inhabitants. The living is a discharged rectory, valued in the king's books at £9. 11. 5½.; present net income, £100: it is in the patronage of the Crown, in right of the Duchy of Cornwall. The village is situated on the ancient Fosse-way, now part of the turnpike-road from Bath to Shepton-Mallet. On the eastern side of the road is the Roman Catholic college of Downside. Coal, iron-stone, and marl, are found in abundance. The Bath market is principally supplied with butter from the dairy farms in this neighbourhood.

STRATTON-STRAWLESS (*St.. Margaret*), a parish, in the union of AYLSHAM, hundred of SOUTH ERPINGHAM, Eastern Division of the county of NORFOLK, 4¾ miles (S. S. E.) from Aylsham; containing 218 inhabitants. The living is a discharged rectory, valued in the king's books at £8. 8.; patron, R. Marsham, Esq. The tithes have been commuted for a rent-charge of £280, subject to the payment of rates, which on the average have amounted to £40; the glebe comprises 32½ acres, valued at £37. 7. 6. per annum. The church contains numerous monuments and inscriptions, and the windows some curious specimens of ancient stained glass. Thomas Bulwer, in 1693, left property, now producing, with a rent-charge by Henry Marshall, in 1683, £16. 9. 10. per annum, for the poor.

STRATTON, UPPER, a tything, in the parish of STRATTON-ST.-MARGARET, hundred of HIGHWORTH, CRICKLADE, and STAPLE, Swindon and Northern Divisions of the county of WILTS, 4¼ miles (S. W.) from Highworth: the population is returned with the parish.

STRATTON, WEST, a tything, in the parish and hundred of MITCHELDEVER, Winchester and Northern Divisions of the county of SOUTHAMPTON, 6¾ miles (N. W. by N.) from New Alresford: the population is returned with the parish.

STREATHAM (*St. Leonard*), a parish, in the union of WANDSWORTH, Eastern Division of the hundred of BRIXTON and of the county of SURREY, 6 miles (S. by W.) from London; containing 5068 inhabitants. This parish, which derives its name from its situation near the great Roman road from Arundel to London, extends along the principal road to Brighton for nearly three miles. The houses, which are mostly modern, are well built, and interspersed with several detached villas and stately mansions, particularly in the neighbourhood of the common, between which and the lower part of the village was an ancient stately mansion of red brick, for some time the residence of Lord William Russell, but now taken down, Streatham park, where Dr. Johnson spent much of his time, was formerly the residence of Mrs. Thrale, afterwards Madame Piozzi. The neighbourhood is richly wooded, and is diversified with hills and valleys, and the surrounding scenery is finely varied: the air, which is considered particularly salubrious and invigorating, combining with other local advantages, has rendered this village a favourite residence of many opulent families. Among the attractions is a mineral spring, which was discovered in 1660, and is still held in esteem, being highly efficacious in scorbutic eruptions, and in many other cases. The only branch of manufacture is that of silk, recently introduced. The London and Croydon railway passes through the parish. Streatham is within the jurisdiction of the court of requests for the Eastern Division of the hundred of Brixton, held in the borough of Southwark, for the recovery of debts under £5, and within the limits of the new police establishment. The living is a rectory, valued in the king's books at £18. 13. 9.; patron, Duke of Bedford. The ancient church, with the exception of the tower, which is of flint, and surmounted by a spire of shingles, forming a picturesque object in the distant landscape, was taken down in 1830, and handsomely rebuilt upon an enlarged scale in the later English style. A chapel has been erected at Upper Tooting, within the last few years, of which the living is a perpetual curacy, in the patronage of the rector. It is in contemplation to erect an additional church near Brixton Hill, in this parish, for which purpose nearly £4000 has been subscribed. There are places of worship for Independents and Wesleyan Methodists. National schools are supported by subscription, and a handsome and commodious school-house has been erected, with a residence attached for the master and mistress. Four handsome almshouses for four aged women have been lately erected in the Elizabethan style, by a bequest of the late Mrs. Henry Thrale, of Streatham Park. Mrs. Elizabeth Howland, in 1716, bequeathed £20 per annum for clothing and educating ten girls; and Mrs. Dorothy Appleby, in 1681, left £5 per annum for apprenticing a poor child. The celebrated Dr. B. Hoadley, Bishop of Bangor, was for several years rector of this parish.

STREATLAM, a joint township with Stainton, in the parish of GAINFORD, union of TEESDALE, South-Western Division of DARLINGTON ward, Southern Division of the county palatine of DURHAM, $2\frac{3}{4}$ miles (N. E. by E.) from Barnard-Castle: the population is returned with Stainton. In the neighbourhood are extensive quarries, from which stone has been raised for the erection of the principal buildings in this part of the county; among which is Streatlam castle, a stately structure, built in the seventeenth century, on the site of the old castle, by Sir William Bowes. Here was anciently a chapel.

STREATLEY (*St. Margaret*), a parish, in the union of LUTON, hundred of FLITT, county of BEDFORD, 5 miles (N. by W.) from Luton; containing 339 inhabitants. The living is a discharged vicarage, valued in the king's books at £6. 15. 2.; present net income, £79; patron, Sir G. P. Turner, Bart.; impropriators, Messrs. Smyth and others. Richard Norton, in 1686, gave a rent-charge of £10 in support of a school.

STREATLEY (*St. Mary*), a parish, in the union of BRADFIELD, hundred of MORETON, county of BERKS, $5\frac{1}{2}$ miles (S. by W.) from Wallingford; containing 582 inhabitants. This place is supposed to have taken its name from its situation on the ancient Ikeneld-street, or Ickleton Long, as it is here called, which crosses the Thames from this place to Goring in Oxfordshire. The living is a vicarage, valued in the king's books at £10. 7. 6.; present net income, £276; patron, Bishop of Salisbury; impropriator, P. Pusey, Esq. In 1837, an act was passed for building a bridge here over the Thames. Four children are taught for an annuity of £2. 5., the bequest of Mr. R. Tull. Here was formerly a Dominican convent.

STREET (*Holy Trinity*), a parish, in the union of WELLS, hundred of WHITLEY, Western Division of the county of SOMERSET, $1\frac{3}{4}$ mile (S. S. W.) from Glastonbury; containing 899 inhabitants. The living is a rectory, with that of Walton annexed, valued in the king's books at £24. 12. $3\frac{1}{2}$.; present net income, £675; patron, Marquess of Bath. There are places of worship for Baptists and the Society of Friends; also a National school and an infants' school. A large fair for all kinds of cattle is held at Christmas. Blue lias and limestone, abounding with marine impressions, are found here.

STREET, a tything, in the parish and hundred of CHRISTCHURCH, Ringwood and Southern Divisions of the county of SOUTHAMPTON: the population is returned with the parish.

STREET, a parish, in the union of CHAILEY, hundred of STREET, rape of LEWES, Eastern Division of the county of SUSSEX, $6\frac{1}{4}$ miles (N. W.) from Lewes; containing 168 inhabitants. The living is a discharged rectory, valued in the king's books at £6. 19. 7.: it is in the patronage of Mrs. Lane. The tithes have been commuted for a rent-charge of £198, subject to the payment of rates, which on the average have amounted to £32; the glebe comprises 31 acres, valued at £23 per annum. The church is an ancient structure of flint, containing 238 free sittings, the Incorporated Society having granted £150 in aid of the expense; there are several monuments to the Dobell family and others.

STREETHALL, a parish, in the union of SAFFRON-WALDEN, hundred of UTTLESFORD, Northern Division of the county of ESSEX, 4 miles (W. N. W.) from Saffron-Walden; containing 41 inhabitants. This small parish, which is not more than two miles in length and one mile in breadth, is situated on elevated ground commanding a richly diversified prospect over the surrounding country. The living is a discharged rectory, valued in the king's books at £13; present net income, £155; patron and incumbent, Rev. William Forbes Raymond. The church is a substantial edifice of stone, and contains several ancient monuments.

STREETHAY, a hamlet, in the parish of ST. MICHAEL, LICHFIELD, union of LICHFIELD, Northern Division of the hundred of OFFLOW and of the county of STAFFORD, $2\frac{1}{4}$ miles (N. E. by E.) from Lichfield; containing 112 inhabitants.

STRELLY (*All Saints*), a parish, in the union of BASFORD, Southern Division of the wapentake of BROXTOW, Northern Division of the county of NOTTINGHAM,

$4\frac{1}{2}$ miles (W. N. W.) from Nottingham; containing 426 inhabitants. The living is a discharged rectory, valued in the king's books at £6. 4. 8.; present net income, £90; patron, T. Webb Edge, Esq. The church is a large handsome cruciform structure, with a lofty tower; the nave is separated from the chancel by a richly carved oaken screen: it contains several tombs of the Strelley family, and the windows exhibit some ancient stained glass in good preservation. A school is endowed by the Earl of Stamford with £5 per annum. The ancient hall, though it has been much modernized, still retains slight traces of the style of Edward III. In the park is an extensive area, surrounded by a moat; and in the neighbourhood are considerable coal mines.

STRENSALL (*St. Mary*), a parish, in the union of York, and within the liberty of St. Peter's, East Riding, though locally in the wapentake of Bulmer, North Riding, of the county of York, 6 miles (N. N. E.) from York; containing 398 inhabitants. The living is a discharged vicarage, with that of Haxby annexed, in the patronage of the Prebendary of Strensall in the Cathedral Church of York (the appropriator), valued in the king's books at £4. 13. 4.; present net income, £195. Robert Wilkinson, in 1718, gave by deed £14. 6. per annum in support of a school; the present annual income, arising from this and other sources, is £28.

STRENSHAM (*St. John the Baptist*), a parish, in the union, and Upper Division of the hundred, of Pershore, Pershore and Eastern Divisions of the county of Worcester, $4\frac{1}{2}$ miles (S. W. by S.) from Pershore; containing 328 inhabitants. This place, which is pleasantly situated on the river Avon, between the hills of Malvern and Bredon, is renowned in history for the siege it sustained against the parliamentary forces, and for the signal bravery displayed here by the then lord of the manor, Sir William Russel, as well as in the memorable battle of Worcester. Blue stone abounds in every part of the parish, and, in some places, fossils and minerals are met with. The living is a rectory, valued in the king's books at £12; present net income, £200; patron, John Taylor, Esq. The church is a noble structure: it contains many memorials of the Russel family, among which are some fine specimens of Italian sculpture, in Parian and other marbles. Lady Ann Russel bequeathed a rent-charge of £10 for teaching poor children in a school founded by her daughter, Lady Ann Guise, in 1709, but it has been allowed to fall into a ruinous state. There are nine almshouses, endowed by the same lady and her father, Sir Francis Russel: they consist each of a sitting-room and bedroom, with gardens attached, occupied by six women and three men; the total income amounts to about £43 per annum. Samuel Butler, author of Hudibras, was born here in 1612.

STRETFORD (*St. Peter*), a parish, in the union of Weobley, hundred of Stretford, county of Hereford, $4\frac{3}{4}$ miles (S. W. by W.) from Leominster; containing 44 inhabitants. The living is a discharged rectory, valued in the king's books at £6. 19. 8.; patron, John Wall, Esq. The tithes have been commuted for a rent-charge of £92, subject to the payment of rates, which on the average have amounted to £2. 15.: the glebe comprises 13 acres, valued at £35 per annum.

STRETFORD, a hamlet, in the parish of Leominster, hundred of Wolphy, county of Hereford, $2\frac{1}{2}$ miles (E. by S.) from Leominster: the population is returned with the township of Broadward.

STRETFORD, a chapelry, in the parish of Manchester, union of Chorlton, hundred of Salford, Southern Division of the county palatine of Lancaster, 4 miles (S. W.) from Manchester; containing 2463 inhabitants. The living is a perpetual curacy; net income, £124; patrons, Warden and Fellows of the Collegiate Church of Manchester. The chapel has been enlarged, and 50 free sittings provided, the Incorporated Society having granted £20 in aid of the expense. A school is endowed by Mrs. Hind with £35 per annum; the children are also clothed. This is a celebrated mart for pigs, of which from 600 to 700 are sent weekly to the Manchester market.

STRETHAM (*St. James*), a parish, in the hundred of South Witchford, union and Isle of Ely, county of Cambridge, $4\frac{1}{4}$ miles (S. W. by S.) from Ely; containing 1173 inhabitants. The living is a rectory, in the patronage of the Bishop of Ely, valued in the king's books at £22; present net income, £756. At Thetford, in this parish, is a chapel of ease; and there are places of worship for Baptists and Wesleyan Methodists; also two free schools.

STRETTON, a township, in the parish of Tilston, union of Great Boughton, Higher Division of the hundred of Broxton, Southern Division of the county of Chester, $4\frac{1}{2}$ miles (N. W. by N.) from Malpas; containing 105 inhabitants.

STRETTON, a chapelry, in the parish of Great Budworth, union of Runcorn, hundred of Bucklow, Northern Division of the county of Chester, $4\frac{1}{2}$ miles (S. by E.) from Warrington; containing 324 inhabitants. In 1827 a chapel, with a tower, in the early English style, was erected, towards defraying the expense of which the Incorporated Society granted £1800. The living is a perpetual curacy; net income, £63; patron, Rector of Great Budworth.

STRETTON, a township, in the parish of North Wingfield, union of Chesterfield, hundred of Scarsdale, Northern Division of the county of Derby, $4\frac{1}{2}$ miles (N. by W.) from Alfreton; containing 439 inhabitants. There is a small bequest for teaching poor children to read.

STRETTON (*St. Nicholas*), a parish, in the union of Oakham, hundred of Alstoe, county of Rutland, $8\frac{1}{4}$ miles (N. E. by E.) from Oakham; containing 208 inhabitants. The living is a rectory, valued in the king's books at £7. 17. 1.; patron, Sir G. Heathcote, Bart. The tithes have been commuted for a rent-charge of £299. 10., subject to the payment of rates, which on the average have amounted to £15; the glebe comprises 3 acres, valued at £6 per annum. A school is supported by the Hon. Mrs. Heathcote, and another by the clergyman and the parish.

STRETTON, a chapelry, in the parish and union of Penkridge, Western Division of the hundred of Cuttlestone, Southern Division of the county of Stafford, 3 miles (S. W. by W.) from Penkridge; containing 268 inhabitants. This place, which is now an obscure hamlet, is supposed to occupy the site of the *Pennicrocium* of the Romans, with the situation of which, as laid down by Antoninus in his Itinerary, it perfectly agrees; the supposition is further strengthened by the discovery of several coins, and other relics of Roman antiquity. The Grand Trunk canal passes in the vicinity. The living is a perpetual curacy; net income, £96; patron and impropriator, Lord Hatherton. The chapel is dedicated to St. John. A school is supported by George Monckton, Esq.

STRETTON, a township, in the parish and union of BURTON-UPON-TRENT, Northern Division of the hundred of OFFLOW and of the county of STAFFORD, 2½ miles (N.) from Burton-upon-Trent; containing 373 inhabitants. This township is bounded on the north by the river Dove, and on the east by the Trent. A chapel has been built, and 160 free sittings provided, the Incorporated Society having granted £100 in aid of the expense.

STRETTON-BASKERVILLE (*ALL SAINTS*), a parish, in the union of HINCKLEY, Kirby Division of the hundred of KNIGHTLOW, Northern Division of the county of WARWICK, 3 miles (E. by S.) from Nuneaton; containing 59 inhabitants. The living is a sinecure rectory, valued in the king's books at £6, and in the patronage of Miss Pinchin and Mrs. Wilcox. The church is in ruins. Here is a National school.

STRETTON, CHURCH (*ST. LAWRENCE*), a market-town and parish, and the head of a union, in the hundred of MUNSLOW, Southern Division of the county of SALOP, 13 miles (S. by W.) from Shrewsbury, and 153 (N. W.) from London; containing 1302 inhabitants. This place, which by its adjunct is distinguished from its townships as the seat of the parish church, derived its name Stretton, or Street-town, from its situation within a quarter of a mile of the ancient Watling-street, which passes in a direction parallel with the road from Shrewsbury to Ludlow. The town is romantically situated in a rich and fertile vale, enclosed on one side by a bold range of mountains, among which are the Caer Caradoc, the lofty and precipitous retreat of Caractacus; the Lawley; and the Raglish; and on the other by the extensive chain of hills called the Longmynd, flat on the summit, but deeply indented, on the south-eastern acclivity, with numerous valleys, from which many mountain streams descend with impetuosity. It consists of one street only, in the wider part of which is the market-house, erected in 1617, an antique building of timber and plaster, consisting of two upper rooms now used for storing wool, and supported on pillars of wood resting on stone plinths, affording a sheltered area for the use of the market. The houses are in general built of brick, and of neat and modern appearance, occasionally interspersed with handsome dwellings and many small cottages: the inhabitants are amply supplied with water by pumps attached to the more respectable houses, and from a stream which, descending from the Longmynd, flows through one extremity of the town. The secluded and romantic situation of the place, its proximity to scenes of deep interest, its fine mountain scenery, and various other attractions, render it a place of resort for parties from the neighbouring towns. But little trade is carried on: a manufactory for flannel was established in 1816, which is now flourishing; large flocks of sheep are depastured on the neighbouring hills, and a fair for wool was established in 1819. The market is on Thursday, chiefly for provisions: the fairs are March 10th, for cattle, horses, and sheep; May 14th, a statute fair; July 3rd, a great wool fair; September 25th, a very large sheep fair; and the last Thursday in November for cattle, sheep, and horses. The county magistrates hold a court of petty sessions at the Talbot hotel, on the third Thursday in every month; and two constables for each township are annually appointed at the court leet held in the old manor-house, now an inn; at which also, under the steward, who must be a lawyer, a court of requests for the recovery of debts under 40*s.* was formerly held by letters patent of Charles II., granted to the Marquess of Bath, then lord of the manor. This town has been made a polling-place and the place of election for the southern division of the county.

The living is a rectory, valued in the king's books at £15. 10.; patron and incumbent, Rev. Robert Norgrave Pemberton. The tithes have been commuted for a rent-charge of £500, subject to the payment of rates; the glebe comprises 68 acres, valued at £136 per annum. The church is an ancient and venerable cruciform structure, principally in the early style of English architecture, with a square embattled tower rising from the centre, strengthened by buttresses and crowned with pinnacles: in the buttress at the south angle is a figure of St. Lawrence, and in other parts of the tower are groups of figures well sculptured. The south porch and the entrance on the north are of Norman character, and the interior contains several portions in the Norman style, with insertions in the decorated style of English architecture. The nave, chancel, and transepts are separated by four lofty clustered columns and pointed arches, which support the tower; the chancel is beautifully ornamented with richly carved oak in antique devices, collected from ancient manorial and ecclesiastical edifices, and put up by the present rector, who has bestowed much care and expense on the embellishment of the church; and in the central compartment of the altar is an elegant and well-carved representation of a dead Christ in the lap of the Virgin. The windows, principally in the decorated style, with rich and flowing tracery, are embellished with stained glass; and in the south transept, the ancient oak roof, finely carved, is carefully preserved. A large stone coffin with a lid, now broken, and an alabaster slab, with an illegible inscription, were taken from under a low arch in the south transept. The triennial visitation is held in this church by the bishop of the diocese, in August, and in the intermediate years by the archdeacon in May. The rectory-house, a handsome mansion, is beautifully situated at the foot of the Longmynd. The grounds have been laid out by the Rev. Mr. Pemberton with a due regard to the characteristic features of the surrounding scenery; Alpine bridges have been thrown over the various mountain streams, and walks have been formed leading to those parts of the mountain from which the most interesting views are obtained.

The free school was endowed by successive benefactors, in addition to whose legacies, of which a portion is appropriated to apprenticing the children, it has an endowment of 27 acres of land, under a late enclosure act. The building, which was erected in 1779, upon the site of an old school, is neat, and comprises two schoolrooms, with apartments for the master and the mistress; the school is free for all children of the parish, and combines the objects of a National school with the advantages of a Sunday school: the annual income is about £43. Almshouses are appropriated as residences for four poor people, who pay a rent of one shilling per quarter: the rental of four acres of land in the parish, and several charitable bequests, are distributed among the poor on Easter-day by the rector and churchwardens. The poor law union of Church-Stretton comprises 14 parishes or places, under the care of 15 guardians, and contains, according to the census of 1831, a population of 5703. On Caer Caradoc are the remains of a large encampment, defended on the steepest acclivities with one, and on the more accessible ascents with two, and in some places with three, intrenchments, hewn out of the solid rock; this was probably an exploratory station of Caractacus, from whom the hill received its name. On the summit of Longmynd a pole has been erected, denoting the highest point in that extensive range of hills, commanding a panoramic view of a wide extent of country in every direction: the prospect includes on the west, the Stiperstones, the Welsh mountains, the Sugar Loaf in Abergavenny, the Table mountain, the Cader Idris, and the intervening range from that mountain to Snowdon; on the south-east, the Edgewood, between Wenlock and Ludlow, the Wrekin, and the Clee and Malvern hills; and the Radnorshire hills on the south-west. On the Longmynd are many low tumuli and cairns of stones; and on one of the eminences, called Bodbury, is a large intrenchment of earth: this mountain was the scene of many battles between the Romans and the Britons, and afterwards between the Welsh and the English. In 1825, one of the tumuli was opened, under the superintendence of the Rev. Mr. Pemberton: on the level of the base was found a circular enclosure of loose stones, appearing to have endured the action of fire, and several pieces of bone were discovered in a calcined state, supposed to have been parts of the bodies burnt there according to the rites of Roman sepulture: this tumulus was surrounded by the trench, from the excavation of which it had been raised. On an eminence at Minton is a very lofty tumulus, supposed to be one of those mounts upon which, in the earlier times of the Britons, justice was administered to the people. One mile to the south-west of Church-Stretton was Brockard's Castle, of which the site, the intrenchments, the moat, and foundations, with the approaches from the Watling-street, may be traced. Among the eminent natives of this town were William Thynne, Receiver of the Marches, in 1546; Sir John Thynne, Knight, who founded Longleat House in the county of Wilts; and Dr. Roger Mainwaring, vicar of St. Giles's in the Fields, London, and chaplain to Charles I., who, for preaching two sermons called "Religion" and "Allegiance," was censured by parliament, and imprisoned and suspended for three years; being afterwards by the king made Bishop of St. David's, he retained that dignity till the abolition of the episcopacy, when he again underwent various persecutions till his death, in 1653.

STRETTON-EN-LE-FIELDS (*St. Michael*), a parish, in the hundred of Repton and Gresley, Southern Division of the county of Derby, though locally in the Western Division of the hundred of Goscote, county of Leicester, 5 miles (S. W.) from Ashby-de-la-Zouch; containing 109 inhabitants. The living is a rectory, valued in the king's books at £9. 10. 5.; present net income, £196; patron, Sir William Browne Cave, Bart. The church contains some ancient tombs of ecclesiastics.

STRETTON-GRANDSOME (*St. Lawrence*), a parish, in the union of Ledbury, hundred of Radlow, county of Hereford, 7¾ miles (N. W.) from Ledbury; containing 168 inhabitants. The living is a vicarage, endowed with the rectorial tithes, with that of Ashperton annexed, and valued in the king's books at £9. 4. 2.; present net income, £479; patron and incumbent, Rev. J. Hopton. Here is a National school with a small endowment.

STRETTON, MAGNA (*St. John the Baptist*), a parish, in the union of Billesdon, hundred of Gartree, Southern Division of the county of Leicester, 5½ miles (S. E. by E.) from Leicester; containing 27 inhabitants. The living is annexed to the vicarage of Glen Magna. The church contains monuments to the Hewitt family. Stretton Hall, the property of Sir Geo. Robinson, Bart., is a fine mansion, embellished with plantations of oak. The Roman *Via Devana* passes through the parish.

STRETTON-ON-THE-FOSS (*St. Peter*), a parish, in the union of Shipston-upon-Stour, forming a detached portion of the Brails Division of the hundred of Kington, Southern Division of the county of Warwick, 3 miles (W. S. W.) from Shipston-upon-Stour; containing 455 inhabitants. The living is a rectory, with that of Ditchford annexed in 1642, valued in the king's books at £11: it is in the patronage of Mrs. Jervoise. The tithes have been commuted for a rent-charge of £184, subject to the payment of rates, which on the average have amounted to £44. The old Roman Fosse-way passes through the parish, also an unfinished rail-road. There is a spring in the neighbourhood, the water of which is slightly impregnated with salt. Ditchford friary is divided into three farms, but there are no remains of its ancient chapel.

STRETTON, PARVA, a chapelry, in the parish of King's-Norton, union of Billesdon, hundred of Gartree, Southern Division of the county of Leicester, 6 miles (E. S. E.) from Leicester; containing 96 inhabitants.

STRETTON-SUGWAS (*St. Mary Magdalene*), a parish, in the hundred of Grimsworth, union and county of Hereford, 3¾ miles (N. W. by W.) from Hereford; containing 155 inhabitants. The living is a discharged rectory, valued in the king's books at £9. 7. 1.; present net income, £184; patrons, Governors of Guy's Hospital, London.

STRETTON-UNDER-FOSS, a hamlet, in the parish of Monks-Kirby, union of Lutterworth, Kirby Division of the hundred of Knightlow, Northern Division of the county of Warwick, 6¼ miles (N. W. by N.) from Rugby; containing, with Newbold-Revel, 304 inhabitants. The Oxford canal is crossed by the old Fosse-road to the westward of this place. There is a place of worship for Independents.

STRETTON-UPON-DUNSMORE (*All Saints*), a parish, in the union of Rugby, Rugby Division of the hundred of Knightlow, Northern Division of the county of Warwick, 6 miles (S. E. by E.) from Coventry; containing 817 inhabitants. This place derives its name from its situation on the Roman Fosse-way, and nearly in the centre of what was formerly Dunsmore heath. The parish extends for about two miles and a half on the London and Holyhead road, and the village is situated about half a mile to the south-west of it. There is no manufacture appropriated to this place, but many of the women and children are employed in preparing the silk for the riband-weavers of Coventry. Plaster of Paris is also made here from the gypsum of which a considerable stratum is found in the parish; and large quantities of lime are burnt from limestone which abounds here. The nearest canal is at Brinklow, about five miles distant. The living is a vicarage; net income, £438; patrons, Rev. H. T. Powell, the present Vicar, for one turn, and — Fauquier, Esq., for two; impropriators, several Proprietors of land. The late Rev. William Daniel, formerly vicar of the parish, bequeathed £4000, subject to the life of his wife, to the Bishop of Lichfield and Coventry and to the Archdeacon of Coventry, in trust for building a new church, which has been erected. A National school is supported from the proceeds of land bequeathed by Mr. William Herbert, in 1694, for charitable uses, of which he appropriated £10. 10. to a master, and the remainder for apprenticing one poor boy every two years; and a Sunday school for children of Stretton and Princethorpe is supported by subscription. Several sums have been left by unknown individuals for repairing the church, for poor widows, and for distribution among the poor generally. There is a spring strongly impregnated with lime, which will incrust rough substances with limestone formation if left in the water for some length of time. In this parish was Brandon Castle, supposed to have been founded either by Geoffrey de Clinton, or by Norman de Verdune, who had married the daughter of that nobleman. During the rebellion of Simon de Montfort, Earl of Leicester, who then held Kenilworth Castle against Henry III., his followers, understanding that John de Verdune had a commission from the king to raise forces in Worcestershire to oppose them, assaulted and destroyed his castle. It was, however, rebuilt by Theobald de Verdune, his successor, who had the privilege of court leet and gallows, with assize of bread and beer, to all of which, in the reign of Edward I., he claimed prescriptive right, which was allowed: the only remains of the castle are the moat and some heaps of rubbish. On the summit of a hill in the hamlet of Princethorpe, by the side of the ancient Fosse-way, several Roman coins have been found, but none of them of any particular note. At Knightlow Hill, on the boundary of the parish, is an ancient stone called Knightlow Cross, one of the oldest memorials of feudal tenure existing. On this stone certain fines are annually paid by the surrounding parishes, among which is a rent called Wroth money or Swart penny, due to the lord of the hundred, which must be paid every Martinmas-day before sun-rise; the party paying it must go three times round the cross, saying "the Wroth money," and then deposite it in the hole of the cross before good witnesses, under a penalty of forfeiting 30 shillings and a white bull.

STRICKLAND, GREAT, a township, in the parish of MORLAND, WEST ward and union, county of WESTMORLAND, 3 miles (W. N. W.) from Morland; containing 245 inhabitants. This place takes its name from the ancient family of the Stricklands, who were lords of the manor, and formerly resided here. There is a meeting-house belonging to the Society of Friends, with a burial-ground attached.

STRICKLAND-KETTLE, a township, in the parish, union, and ward of KENDAL, county of WESTMORLAND, 2 miles (N. by W.) from Kendal; containing 386 inhabitants. It is bounded on the east by the Kent river. The chapel and part of the village of Burneside is within this township. A school is endowed with £24 per annum.

STRICKLAND, LITTLE, a township, in the chapelry of Thrimby, parish of MORLAND, WEST ward and union, county of WESTMORLAND, 3 miles (N. E.) from Shap; containing 121 inhabitants. The chapel and school-house are both situated in this township.

STRICKLAND-ROGER, a township, in the parish, union, and ward of KENDAL, county of WESTMORLAND, 4 miles (N.) from Kendal; containing 326 inhabitants. It is bounded on the west by the river Kent, and on the east by the Sprint, and contains part of the village of Burneside. Near Garnet-bridge is a mill for the manufacture of bobbin, and at Cowen Head there is a paper-mill. At a place called Hundhow was anciently a chapel, called Chapel-en-le-Wood.

STRINGSTON, a parish, in the union of WILLITON, hundred of CANNINGTON, Western Division of the county of SOMERSET, 10¼ miles (W. N. W.) from Bridg-water; containing 128 inhabitants. The living is a vicarage, united to the rectory of Kilve. There is a place of worship for Independents. In the neighbourhood is an ancient fortification, called Danes-burrow, or Douse-borough, Castle, with a double embankment and wide ditch; it is about three quarters of a mile in circumference, and is wholly covered with oak coppice wood, among which a prætorium may be distinctly traced.

STRIXTON (*St John the Baptist*), a parish, in the union of WELLINGBOROUGH, hundred of HIGHAM-FERRERS, Northern Division of the county of NORTHAMPTON, 4¼ miles (S. by E.) from Wellingborough; containing 69 inhabitants. The living is a discharged rectory, consolidated with the vicarage of Bozeat, and valued in the king's books at £7. The church is small, but affords a good specimen of the early style of English architecture.

STROOD (*St. Nicholas*), a parish, in the union of NORTH AYLESFORD, partly within the jurisdiction of the city of ROCHESTER, and partly in the hundred of SHAMWELL, lathe of AYLESFORD, Western Division of the county of KENT, ½ a mile (N. W.) from Rochester; containing 2722 inhabitants. The village consists of one principal street, on the high road from London to Rochester, to which latter place it is joined by a bridge over the Medway, at its eastern extremity: the houses are irregularly built, and equally destitute of uniformity and respectability of appearance; but since the last act of parliament for paving, watching, and lighting the village, it has been considerably improved: the adjoining heights command interesting and extensive prospects. The inhabitants are principally engaged in maritime pursuits, in the fisheries on the Medway, and in dredging for oysters, of which large quantities, as well as shrimps, are sent to the London and other markets: the trade principally arises from the resort of sea-faring men, from its situation as a thoroughfare, and, more particularly, from its proximity to Rochester. The fair is on August 26th and two following days, by grant of King John, and has become very considerable. That part of the parish, called Strood Extra, which is not within the city of Rochester, is under the jurisdiction of the county magistrates, and within that of the court of requests held at Rochester, for the recovery of debts not exceeding £5. The living is a perpetual curacy; gross income, about £240; patrons, Dean and Chapter of Rochester; appropriator, Bishop of Rochester. The old church, situated at the western extremity of the village, was taken down and rebuilt in 1812 at the expense of the parishioners. There is a place of worship for Independents. Francis Barrel, Esq., residuary legatee of Sir John Hayward's estate, in 1718, bequeathed £1100 for the endowment of three charity schools, to be called Sir John Hayward's schools, two of which were to be in the parish of St. Nicholas, Rochester, and one in Strood, for instructing 30 children, of which number, 20 were to be of this parish; and, in 1721, Mr. William Turner gave a rent-charge of £2 to the school here. Mr. Phillips, in 1740, gave a sum of money producing an interest of £5. 13. 4., and Mr. Hulkes gave a sum producing £2. 10. per annum, to be annually distributed in bread to the poor. Of Strood Temple, originally a preceptory founded for Knights Templars, and valued at the dissolution at £52. 6. 10., there are some interesting remains on the Temple farm; and of Strood hospital, founded by Bishop Gilbert de Glanville, in the reign of Richard I., for infirm and indigent travellers, the almonry, which has been converted into a stable, and some other portions, are remaining. About two miles from Strood, on the London road, is Gadshill, celebrated by Shakspere as the scene of Falstaff's valorous exploits.

STROUD, a tything, in the parish of CUMNER, hundred of HORMER, county of BERKS; containing 72 inhabitants.

STROUD, or STROUDWATER (*ST. LAWRENCE*), a newly enfranchised borough, market-town, and parish, and the head of a union, in the hundred of BISLEY, Eastern Division of the county of GLOUCESTER, 10 miles (S. by E.) from Gloucester, and 102 (W. by N.) from London; containing 8607 inhabitants. The first notice of this place in any records extant occurs in an agreement in 1304, between the Rector of Bisley and the inhabitants of La Stroud, which, at the time of the Norman survey, formed part of that parish. It derives its name from its situation on the Slade, or Stroud water, near its confluence with the Frome. It stands on a considerable declivity, in the midst of a most beautiful country, and consists principally of a long street extending up the side of the hill, with another diverging from it at the base. The streets are paved, and lighted with gas; the town contains many handsome houses; and the inhabitants are well supplied with water conveyed by pipes from two springs in the neighbourhood. Stroud has long been famous as the centre of the woollen manufacture in Gloucestershire, and is supposed to owe much of its prosperity to the peculiar properties of the stream called the Stroud water, which is admirably adapted for dyeing scarlet, and which, consequently, was the means of attracting, at an early period, many clothiers and dyers to its banks. It possesses great advantages in water-carriage, the Thames and Severn canal passing close to the south of the town. The inhabitants of the neighbourhood and surrounding villages, of which Bowbridge, Brimscomb-Port, and Thrupp, are noticed under their respective heads, are employed in different processes of this manufacture, several thousand pieces of broad and narrow cloth being annually made, and conveyed by the canal to different parts of the empire. At the distance of a mile from the town, on the Bath and Birmingham road, are Light Pool Mills, an extensive establishment for the manufacture of solid-headed pins. The building consists of five stories, each 100 feet long, and filled with machinery put in motion by a large water wheel of 40-horse power, and ingeniously adapted to the entire making of these pins without any manual assistance. The various processes are simultaneously performed by an intricate mechanical combination of apparatus, each appropriated to its particular use, from the drawing of the wire to the finishing of the pins, which, when completed, fall from the machine into a receiver placed beneath it. The peculiarity of these pins consists in the formation of a portion of the shank into a solid head, by which the possibility of its slipping is entirely prevented; there are about 100 of these machines at work, each producing from 40 to 50 pins every minute, and in the aggregate about two tons' weight weekly. The town has been greatly improved in consequence of an act of parliament obtained, within a few years, for paving, lighting, and widening the streets; and many new roads have been formed, extending in various directions, to connect it more closely with the contiguous towns. The market, which is on Friday, is well supplied; and there are fairs on May 10th and August 21st, for cattle, sheep, and pigs. Stroud has been constituted a borough, with the privilege of sending two members to parliament, the right of election being vested in the £10 householders of a large manufacturing district, which has been formed into a borough for elective purposes, comprising an area of 42,356 acres, the limits of which are minutely described in the Appendix, No. II.: the returning officer is annually appointed by the sheriff. The petty sessions for the hundred are held here, on the first and third Fridays in every month: and the town is within the jurisdiction of the court of requests, for the recovery of debts under 40*s.*, held at Cirencester, on Thursday every three weeks; and in that of a court baron held annually by the lord of the manor of Bisley. It has been made a polling-place for the eastern division of the county.

The parish was separated from that of Bisley, and made a distinct parish, in the reign of Edward II. The living is a perpetual curacy; net income, £132; patron, Bishop of Gloucester and Bristol; impropriator, —Goodlake, Esq. There is an endowed lectureship, in the gift of the parishioners. The church is a large building, erected and enlarged at several different periods, with a tower at its western end, surmounted by a lofty octangular steeple. A new church has been built and consecrated at Stroudshill; it contains 700 free sittings, the Incorporated Society having granted £500 in aid of the expense. Her Majesty's commissioners have also approved of a plan for building a church at White hill. There are places of worship for Particular Baptists, Independents, and Wesleyan Methodists. Thomas Webb, in 1642, gave an endowment, now amounting to about £54 per annum, by means of which four poor boys are boarded, clothed, and educated; and, in 1734, Henry Windowe bequeathed £21, for the maintenance and clothing of two more: there are other small endowments for educating and apprenticing poor boys, and several hundred children are instructed in the National and Sunday schools. The parochial school, instituted in 1700, by the Rev. William Johns, is supported by subscription. Many endowments also provide relief for the poor. The poor law union of Stroud comprises 15 parishes or places, under the care of 31 guardians, and contains, according to the census of 1831, a population of 40,767. Stroud is the birthplace of John Canton, F. R. S., a celebrated natural philosopher, who died in 1772; and Joseph White, D.D., Professor of Arabic at Oxford, who died in 1814; both these distinguished men were the sons of weavers.

STROUD-END, a tything, in the parish of PAINSWICK, hundred of BISLEY, Eastern Division of the county of GLOUCESTER; containing 838 inhabitants.

STROXTON (*ALL SAINTS*), a parish, in the union of GRANTHAM, wapentake of WINNIBRIGGS and THREO, parts of KESTEVEN, county of LINCOLN, $3\frac{3}{4}$ miles (S. S. W.) from Grantham; containing 124 inhabitants. The living is a discharged rectory, valued in the king's books at £3. 8. $6\frac{1}{2}$.; present net income, £250; patron, Sir W. E. Welby, Bart.

STRUBBY (*ST. OSWALD*), a parish, in the union of LOUTH, Wold Division of the hundred of CALCEWORTH, parts of LINDSEY, county of LINCOLN, 4 miles (N.) from Alford; containing 201 inhabitants. The living is a discharged vicarage, in the patronage of the Dean and Chapter of Lincoln (the appropriators), valued in the king's books at £4. 13. 4.; present net income, £150.

STRUMPSHAW (*ST. PETER*), a parish, in the union and hundred of BLOFIELD, Eastern Division of the county of NORFOLK, $3\frac{1}{2}$ miles (S. W. by W.) from Acle; containing 374 inhabitants. The living is a discharged rectory, with that of Bradseton united, valued in the king's books at £8; present net income, £474; patron and incumbent, Rev. Thomas Woodward. Here is a windmill, standing on the highest ground in the county, and forming a conspicuous landmark.

STUBBS, a joint township with Hamphall, in the parish of ADWICK-LE-STREET, Northern Division of the wapentake of STRAFFORTH and TICKHILL, West Riding of the county of YORK, 7 miles (N. W.) from Doncaster: the population is returned with Hamphall.

STUBBY-LANE, a hamlet, in the parish of HANBURY, Northern Division of the hundred of OFFLOW and of the county of STAFFORD, $4\frac{1}{2}$ miles (S. E.) from Uttoxeter; containing 173 inhabitants.

STUBLACH, a township, in the parish of MIDDLEWICH, union and hundred of NORTHWICH, Southern Division of the county of CHESTER, 3 miles (N. by E.) from Middlewich; containing 66 inhabitants.

STUBTON (*ST. MARTIN*), a parish, in the union of NEWARK, wapentake of LOVEDEN, parts of KESTEVEN, county of LINCOLN, $6\frac{3}{4}$ miles (S. E. by E.) from Newark; containing 182 inhabitants. The living is a rectory, valued in the king's books at £12. 3. 9.; patron, Sir Robert Heron, Bart. The tithes have been commuted for a rent-charge of £270, subject to the payment of rates, which on the average have amounted to £39; the glebe comprises 44 acres, valued at £80 per annum.

STUDHAM (*ST. MARY*), a parish, in the union of LUTON, partly in the hundred of DACORUM, county of HERTFORD, but chiefly in the hundred of MANSHEAD, county of BEDFORD; containing 821 inhabitants, of which number, 173 are in the hamlet of Studham, $3\frac{3}{4}$ miles (W. by S.) from Market-Street. The living is a discharged vicarage, valued in the king's books at £9; present net income, £129: it is in the patronage of the Crown; impropriators, Rev. J. Wheeldon and others.

STUDLAND (*ST. NICHOLAS*), a parish, in the union of WAREHAM and PURBECK, hundred of ROWBARROW, Wareham Division of the county of DORSET, $5\frac{1}{2}$ miles (E. by N.) from Corfe-Castle; containing 435 inhabitants. This parish, which includes Brownsea and several smaller islands, is bounded on the north by Poole harbour, on the east by Studland bay, and by Swanwich bay on the south-east, where there is a signal station, on a hill called Ballard down. The bay, though an open roadstead, affords excellent anchorage for ships drawing fourteen or fifteen feet of water. Brownsea island is of an oval form, about three miles in circumference, and contained anciently a hermitage and chapel, dedicated to St. Andrew, of which there are now no remains. The castle, at its eastern extremity, was built in the reign of Elizabeth, by the inhabitants of Poole, for the defence of that port: adjoining it is a platform, upon which, in time of war, a few pieces of ordnance are mounted. There is also a quay, where vessels of considerable burden can lie conveniently for taking in, or discharging, their cargoes. The living is a discharged rectory, valued in the king's books at £7. 10. 5.; present net income, £126; patron, Edmund Morton Pleydell, Esq. The church is supposed to have been built about the time of the Conquest. Two schools are partly supported by Mrs. Banks. On Studland common there are many ancient barrows, which must be either British or Danish: the principal of them is 90 feet in perpendicular height, and is called Agglestone, or Stone Barrow, from its being surmounted by an enormous circular red sandstone, eighteen feet high, and computed to weigh 400 tons.

STUDLEY, a hamlet, in the parish of BECKLEY, union of HEADINGTON, chiefly in the hundred of ASHENDON, county of BUCKINGHAM, and partly in that of BULLINGDON, county of OXFORD, 5 miles (N.) from Wheatley; containing, with the hamlet of Horton, 405 inhabitants. A priory of Benedictine nuns, in honour of the Blessed Virgin Mary, was founded here in the reign of Henry II., by Bernard de S. Walerico, which at the dissolution had a revenue of £102. 6. 7. The remains of a Roman villa were discovered here in a wood belonging to Sir Alexander Croke, and various pieces of masonry, as window slabs, apparently parts of some ancient edifice, have been found.

STUDLEY (*ST. MARY*), a parish, in the union of ALCESTER, Alcester Division of the hundred of BARLICHWAY, Southern Division of the county of WARWICK, $4\frac{3}{4}$ miles (N. by W.) from Alcester; containing 1903 inhabitants. This place is situated on the river Arrow, and has a large manufacture of needles and fish-hooks. The living is a vicarage, valued in the king's books at £8; present net income, £103; patron, R. Knight, Esq. There is a place of worship for Wesleyan Methodists. Eight children are taught free, and two are annually clothed, from a small income arising from bequests by William Mortiboys, in 1733, and William Ayres, in 1739: the school-house was built by subscription in 1810. There are considerable remains of a priory founded in honour of St. Mary, early in the reign of Henry II., by Peter de Studley, who translated hither a society of Augustine canons, which he had previously established at Wicton in Worcestershire: this house, at the dissolution, had a revenue of £181. 3. 6., and at its gate William de Cantilupe erected an hospital for the relief of poor impotent people.

STUDLEY-ROGER, a township, in the parish of RIPON, Lower Division of the wapentake of CLARO, West Riding of the county of YORK, 1¾ mile (W. S. W.) from Ripon; containing 157 inhabitants. Here is a National school.

STUDLEY-ROYAL, a hamlet, in the parish of RIPON, Lower Division of the wapentake of CLARO, West Riding of the county of YORK, 2½ miles (W. S. W.) from Ripon; containing 60 inhabitants.

STUKELEY, GREAT (*St. Bartholomew*), a parish, in the hundred of HURSTINGSTONE, union and county of HUNTINGDON, 2½ miles (N. W.) from Huntingdon; containing 397 inhabitants. The living is a discharged vicarage, valued in the king's books at £6. 14. 2.; present net income, £124; patrons, Master and Fellows of Trinity Hall, Cambridge; impropriator, J. Heathcote, Esq. The church is principally in the Norman style of architecture. Two allotments, comprising together eight acres, and yielding £14 per annum, have been assigned in lieu of other land appropriated to the repairs of the church, and the income is applied to that purpose and occasionally to the relief of the poor. There are two almshouses on the road from Stukeley to Huntingdon, occupied by poor families appointed by the family of Torkington, of this parish.

STUKELEY, LITTLE (*St. Martin*), a parish, in the hundred of HURSTINGSTONE, union and county of HUNTINGDON, 3¼ miles (N. W. by N.) from Huntingdon; containing 413 inhabitants. The living is a rectory, valued in the king's books at £13. 13. 1½.; present net income, £252: it is in the patronage of Lady Olivia Sparrow. A day school and a Sunday school are chiefly supported by the rector. There are a few small allotments of land assigned to poor persons at moderate rents, in order to cultivate them by spade husbandry.

STUNTNEY, a chapelry, in the parish of the HOLY TRINITY, ELY, hundred and Isle of ELY, county of CAMBRIDGE, 1½ mile (S. E.) from Ely: the population is returned with the parish. The living is a perpetual curacy; net income, £77; patrons and appropriators, Dean and Chapter of Ely. The tithes have been commuted for a rent-charge of £580, subject to the payment of rates, which on the average have amounted to £93; the glebe comprises 32 acres. The chapel is in the Norman style of architecture.

STURBRIDGE, county of CAMBRIDGE.—See STOURBRIDGE.

STURMERE, a parish, in the union of RISBRIDGE, hundred of HINCKFORD, Northern Division of the county of ESSEX, 1 mile (S. E.) from Haverhill; containing 320 inhabitants. This parish takes its name from a lake, or mere, comprising about 20 acres, which extended from the river Stour, by which the parish is bounded on the north. It is about twenty miles in circumference, and well watered by a rivulet which flows through it; and though now an obscure place, was formerly of considerable importance. It extended into the counties of Suffolk and Cambridge, and included the parishes of Haverhill and Kedington, each now exceeding it in population. Sturmere Hall, situated in a fine park, is the residence of John Purkiss, Esq., who has greatly improved the lands by extensive plantations. The living is a rectory, valued in the king's books at £8. 10.; present net income, £189; patron, Duke of Rutland. The church, an ancient structure of flint and rubble stone, is partly in the Norman and early English styles, with a rich Norman arch on the south side. Here is a National school. Numerous Roman coins of Antoninus Pius and of the Lower Empire have been found here; and in widening the road, in 1820, several skeletons, of gigantic size, were discovered.

STURMINSTER-MARSHALL (*St. Mary*), a parish, in the union of WIMBORNE and CRANBORNE, hundred of COGDEAN, Wimborne Division of the county of DORSET, 5 miles (W.) from Wimborne-Minster; containing 803 inhabitants. This place derives its name from the situation of its church on the river Stour, and its adjunct from the Earl of Pembroke, then Earl Marshall, to whom it anciently belonged, and who, in the reign of Henry I., obtained for it the grant of an annual fair. The parish is bounded on the north-east by the river Stour, over which is a bridge of eight arches, called Whitmill bridge. In the centre of the village is an open space, still called the market-place, though no market has been held there within the memory of man. The living is a vicarage, endowed with the rectorial tithes of Lytchett-Minster, Corfe-Mullen, and Hamworthy, with the perpetual curacy of Lytchett-Minster annexed, and valued in the king's books at £31. 5.; present net income, £920; patrons and impropriators, Provost and Fellows of Eton College. The church has an embattled tower and a remarkably large chancel: at the west end of the north aisle a space is partitioned off, in which the royal peculiar court of Sturminster-Marshall is held. In 1799, William Mackrell endowed two schools with the interest of £1200 three per cent. consols., for the education of 52 children. A national school is supported by subscription, aided by an annual grant of £10 by the society. On Cogdean Elms, an eminence in this parish, on which the courts of the hundred to which it gives name were formerly held, are some stately elm trees, and near it are several barrows.

STURMINSTER-NEWTON-CASTLE (*St. Mary*), a market-town and parish, and the head of a union, in the hundred of STURMINSTER-NEWTON-CASTLE, Sturminster Division of the county of DORSET, 9 miles (N. W.) from Blandford, and 108 (W. S. W.) from London; containing 1831 inhabitants. This place, which derives its name from the river on the northern bank of which it is situated, and from the minster, or church, is supposed to be the *Anicetis* of Ravennas, and was known to the Saxons at an early period. Alfred the Great gave some lands here to his son Ethelwald; and, in 968, Edgar gave the manor of Sturre, or Stour, to the abbey of Glastonbury, which grant was confirmed by Edmund Ironside. In the Norman survey this place was included in Newenton, or Newton, from which it appears to have derived the adjunct to its name. At the dissolution it was given by Henry VIII. to Catherine Parr, and, after her death, by Edward VI. to his sister Elizabeth, who devised it to Sir Christopher Hatton, from whom it passed to the family of Lord Rivers. In 1645, some hundred clubmen of Dorsetshire and Wiltshire forced the quarters of the parliamentary troops here, and, after some slaughter on both sides, were victorious, taking sixteen dragoons, with several horses and arms. In

1681, and 1729, the town suffered by conflagrations, having sustained damage at the latter period to the amount of £13,000. Sturminster Newton comprises the two townships of Sturminster and Newton, occupying different sides of the river Stour, and connected by a causeway and bridge of six arches; the latter has been widened and improved, and the former raised, to prevent the inundation to which it was previously subject. The streets are in general narrow, and the houses low and indifferently built, except in the market-place, where there is a large oblong market-house, with ware-rooms above and shambles below. A turnpike-road, lately completed, runs through this town to Sherborne. Some trade is carried on with Newfoundland, and the little manufacture in the town consists of baizes, though woollen goods were formerly made. The market is held on Thursday, and on every alternate Thursday is a large market for cattle: fairs are on May 12th, and October 24th. A court leet is held annually, at which the constable for the hundred, and tythingmen, are appointed. The living is a vicarage, valued in the king's books at £16. 16. 8.; present net income, £104; patron, Lord Rivers; impropriator, Rev. T. H. Lane Fox. The church, a handsome edifice, situated on the south side of the town, was originally built by John Selwood, abbot of Glastonbury, and has been lately repaired and enlarged at the sole expense of the Rev. Thomas H. Lane Fox; it consists of a chancel, nave, and two aisles, with an embattled tower, and contains a painted window, which cost 400 guineas. A chapel of ease, which stood at Bagbere in this parish, has fallen into decay. There is a place of worship for Wesleyan Methodists. A National school has been erected by the Rev. T. H. L. Fox, by whom it is principally supported, and by whom an infants' school and a school of industry for girls have been established. The poor law union of Sturminster comprises 19 parishes or places, under the superintendence of 22 guardians, and contains, according to the census of 1831, a population of 9553. The principal object of interest is an ancient fortification, or camp, called the Castle, situated on an eminence in Newton, near the south bank of the river, supposed to have been constructed by the Romans, or not later than the Saxon era: it consists of a vallum and deep foss, in the shape of the Roman letter D, and on the top, near the centre, is a small artificial mount, or keep, near which are the ruins of an ancient house, where the courts were formerly held. On enclosing a common, in 1828, a considerable quantity of silver coins of Elizabeth, and of gold coins of Edward III., was discovered.

STURRY (*St. Nicholas*), a parish, in the union of Blean, hundred of Bleangate, lathe of St. Augustine, Eastern Division of the county of Kent, 2½ miles (N. E.) from Canterbury; containing 925 inhabitants. The living is a vicarage, valued in the king's books at £13. 1. 8.; present net income, £219; patron, Archbishop of Canterbury; impropriator, — Greville, Esq. The church is in the early style of English architecture, with a tower surmounted by a spire. Here is a Sunday National school. The river Stour runs through the parish, and is crossed by a bridge at the village, which is large and well built, on the road between Canterbury and the Isle of Thanet. A fair is held here on Whit-Monday.

STURSTON, a hamlet, in the parish of Ashbourn, hundred of Appletree, Southern Division of the county of Derby, 1 mile (E.) from Ashbourn; containing 578 inhabitants.

STURSTON (*Holy Cross*), a parish, in the union of Thetford, hundred of Grimshoe, Western Division of the county of Norfolk, 5¼ miles (S. W. by S.) from Watton; containing 49 inhabitants. The living is a perpetual curacy; net income, £28; patron, Lord Walsingham. The church is in ruins.

STURTON, a township, in the parish of Scawby, union of Glandford-Bridge, Eastern Division of the wapentake of Manley, parts of Lindsey, county of Lincoln, 2¾ miles (S. W.) from Glandford-Bridge: the population is returned with the parish.

STURTON, a joint township with Bransby, in the parish of Stow, union of Gainsborough, wapentake of Well, parts of Lindsey, county of Lincoln, 8¼ miles (S. E.) from Gainsborough; containing 318 inhabitants. There is a place of worship for Wesleyan Methodists. The Countess Dowager of Warwick, in 1626, gave an annuity of £5 for teaching poor children; and Edward Burgh subsequently bequeathed property producing about £7 per annum, for a like purpose.

STURTON (*St. Peter*), a parish, in the union of East Retford, North-Clay Division of the wapentake of Bassetlaw, Northern Division of the county of Nottingham, 6 miles (E. N. E.) from East Retford; containing 638 inhabitants. The living is a vicarage, valued in the king's books at £5. 7. 3½.; present net income, £282; patrons and appropriators, Dean and Chapter of York. The Roman road from Lincoln to Doncaster passes through the parish, and the river Trent forms its southern boundary. A National school, erected in 1830, is partly supported by a rental of £6. 14., the gift of George Green, in 1710, for teaching eight poor children. Several coins were lately discovered here, and among them one of Louis XIII., and a leaden seal of Pope Innocent III.

STURTON-GRANGE, a township, in the parish of Warkworth, union of Alnwick, Eastern Division of Coquetdale ward, Northern Division of the county of Northumberland, 2½ miles (W. N. W.) from Warkworth; containing 88 inhabitants.

STURTON-GRANGE, a township, in the parish of Aberford, Lower Division of the wapentake of Skyrack, West Riding of the county of York, 7 miles (E.) from Leeds; containing 74 inhabitants.

STURTON, GREAT (*All Saints*), a parish, in the union of Horncastle, Northern Division of the wapentake of Gartree, parts of Lindsey, county of Lincoln, 5½ miles (N. W. by N.) from Horncastle; containing 138 inhabitants. The living is a discharged vicarage, valued in the king's books at £8; present net income, £116: it is in the patronage of the Crown; impropriator, R. J. Loft, Esq.

STUSTON, or STURSTON (*All Saints*), a parish, in the union and hundred of Hartismere, Western Division of the county of Suffolk, 3 miles (N. by W.) from Eye; containing 212 inhabitants. The living is a discharged rectory, valued in the king's books at £6. 16. 8.; present net income, £174; patron, Sir Edward Kerrison, Bart. Elizabeth Bosworth, in 1710, bequeathed four acres of land, in trust for the rector to preach and read prayers in the parish church on Good Friday. A school is partly supported by subscription.

STUTCHBURY (*St. John the Baptist*), a parish, in the union of Brackley, hundred of King's Sutton, Southern Division of the county of Northampton, 5 miles (N. by W.) from Brackley; containing 29 inhabitants. The living is a rectory, valued in the king's books at £3. 6. 8.; present net income, £5: it is in the patronage of the Crown. The church is in ruins.

STUTTON (*St. Peter*), a parish, in the hundred and union of Samford, Eastern Division of the county of Suffolk, $6\frac{1}{4}$ miles (S. by W.) from Ipswich; containing 475 inhabitants. The river Stour, which is navigable for vessels of considerable burden, separates this parish from Essex, and, at high tides, is here from two to three miles broad. The gateway and other remains of Stutton Hall are good specimens of the domestic style of architecture prevalent in the reign of Elizabeth: it was formerly the residence of the ancient family of Jermy, and is now occupied by a tenant. The living is a rectory, valued in the king's books at £12. 17. 6.; present net income, £550; patron and incumbent, Rev. Thomas Mills. The parsonage-house is delightfully situated, and the grounds tastefully embellished. A National school is supported by the incumbent. Several fossil remains have been dug up here.

STUTTON, a joint township with Hazlewood, in the parish of Tadcaster, Upper Division of the wapentake of Barkstone-Ash, West Riding of the county of York, $1\frac{1}{2}$ mile (S. by W.) from Tadcaster; containing 330 inhabitants. The hamlet of Stutton is situated in the vale of the Cock rivulet, and comprises about 600 acres. Hazlewood, which is extra-parochial, contains about 2000 acres, the property of the Hon. Sir Edward Marmaduke Vavasour, Bart., of Hazlewood Hall, an ancient mansion, pleasantly situated on a lofty eminence commanding extensive views. Near the hall is a handsome and venerable Roman Catholic chapel. In the neighbourhood are quarries of excellent limestone, and several limekilns.

STYDD, a township, in the parish of Shirley, hundred of Appletree, Southern Division of the county of Derby, $4\frac{3}{4}$ miles (S. by W.) from Ashbourn; containing 29 inhabitants.

STYFORD, a township, in the parish of Bywell-St.-Andrew, union of Hexham, Eastern Division of Tindale ward, Southern Division of the county of Northumberland, 7 miles (E. by S.) from Hexham; containing 65 inhabitants. It is bounded on the south by the river Tyne.

STYRRUP, a township, in the parishes of Blyth, Harworth, and Houghton, union of Worksop, Hatfield Division of the wapentake of Bassetlaw, Northern Division of the county of Nottingham, $3\frac{1}{2}$ miles (W. S. W.) from Bawtry; containing 510 inhabitants.

SUCKLEY (*St. John the Baptist*), a parish, in the union of Martley, Upper Division of the hundred of Doddingtree, Worcester and Western Divisions of the county of Worcester, $5\frac{1}{2}$ miles (E. S. E.) from Bromyard; containing 1196 inhabitants. The living is a rectory, valued in the king's books at £26. 19. $4\frac{1}{2}$.; present net income, £634: it is in the patronage of the Crown. At Alfrick and Lulsley, in this parish, are chapels of ease; and there is a place of worship for Wesleyan Methodists. Courts leet and baron are annually held here. There is a free school, which is endowed with about £10. 10. per annum, arising from bequests by J. Palmer, in 1683, and an unknown benefactor; eight boys are taught free. Thomas Freeman, in 1794, left £1000, which was vested in land, now producing £46 per annum, which sum is distributed among the poor not receiving parochial relief, in quarterly payments. There is a further sum, arising from sundry minor bequests, appropriated to the general relief of the poor.

SUDBORNE, or SUDBOURN (*All Saints*), a parish, in the union and hundred of Plomesgate, Eastern Division of the county of Suffolk, $1\frac{1}{2}$ mile (N. by E.) from Orford; containing 631 inhabitants. This parish is bounded on the east by the river Ore and the North Sea. Sudbourn Hall, formerly the seat of the Viscounts Hereford, is now the occasional residence of the Marquess of Hertford. The living is a rectory, with that of Orford annexed, valued in the king's books at £33. 6. 8.; present net income, £577: it is in the patronage of the Crown. A school is partly supported by the rector. Dr. Pretyman Tomline, Bishop of Winchester, was rector of this parish.

SUDBOROUGH (*All Saints*), a parish, in the union of Thrapstone, hundred of Huxloe, Northern Division of the county of Northampton, $4\frac{1}{4}$ miles (N. W.) from Thrapstone; containing 346 inhabitants. The living is a rectory, valued in the king's books at £10. 5. 10.; patron, Bishop of London. The tithes have been commuted for a rent-charge of £359, subject to the payment of rates, which on the average have amounted to £62; the glebe comprises 14 acres, valued at £65 per annum. The church is in the early, decorated, and later English styles of architecture, and contains 50 free sittings, the Incorporated Society having granted £20 in aid of the expense. A Sunday school was founded by the Marchioness of Bath, in 1788, and endowed with £366. 13. 4. three per cent. reduced annuities.

SUDBROOK (*Holy Trinity*), a parish, in the Upper Division of the hundred of Caldicott, county of Monmouth, $5\frac{1}{4}$ miles (S. W. by S.) from Chepstow. The living is a discharged rectory, annexed, with that of St. Pierre, to the rectory of Portscuett, and valued in the king's books at £4. 14. 7. The church is in ruins; and the parish has greatly declined in population and importance.

SUDBROOKE (*St. Edward*), a parish, in the wapentake of Lawress, parts of Lindsey, union and county of Lincoln, $4\frac{3}{4}$ miles (N. E.) from Lincoln; containing 84 inhabitants. The living is a discharged rectory, valued in the king's books at £7. 10.; patron, Bishop of Lincoln. The tithes have been commuted for a rent-charge of £56, subject to the payment of rates, which on the average have amounted to £5.

SUDBURY (*All Saints*), a parish, in the union of Uttoxeter, hundred of Appletree, Southern Division of the county of Derby, 5 miles (E. by S.) from Uttoxeter; containing 642 inhabitants. The living is a rectory, valued in the king's books at £14. 13. $1\frac{1}{2}$.; present net income, £747; patron, Lord Vernon. The church contains some very ancient monuments. An infants' school, and two day and Sunday schools, are supported by Lord and Lady Vernon. Here are almshouses for seven poor persons.

Arms.

SUDBURY, a borough and market-town, and the head of a union, locally in the hundred of Babergh, Western Division of the county of SUFFOLK, 22 miles (W. by S.) from Ipswich, and 56 (N. E. by N.) from London; containing 4677 inhabitants. This place, which was originally called *South Burgh*, is of great antiquity, and at the period of the Norman survey was of considerable importance, having a market and a mint. A colony of the Flemings, who were introduced into this country by Edward III., for the purpose of establishing the manufacture of woollen cloth, settled here, and that branch of trade continued to flourish for some time, but at length fell to decay. The town is situated on the river Stour, which is crossed by a bridge leading into Essex. For some years after the loss of the woollen trade, it possessed few attractions, the houses belonging principally to decayed manufacturers, and the streets being very dirty: it has, however, within the last few years, been greatly improved, having been paved and lighted in 1825, under an act obtained for the purpose, and some good houses built. The town-hall, erected by the corporation, in the Grecian style of architecture, is a great ornament to the town, in which is also a neat theatre. The trade principally consists in the manufacture of silk, crape, and buntings used for ships' flags: that of silk was introduced by the manufacturers from Spitalfields, in consequence of disputes with their workmen, and now affords employment to a great number of persons, about 1500 being engaged in the silk, and 400 in the crape and bunting, business. The river Stour, navigable hence to Manningtree, affords a facility for the transmission of coal, chalk, lime, and agricultural produce. The statute market is on Saturday, and the corn market on Thursday: fairs are held on March 12th and July 10th, principally for earthenware, glass, and toys.

The first charter of incorporation was granted by Queen Mary, in 1554, and confirmed by Elizabeth, in 1559: another was given by Oliver Cromwell, but that under which the corporation derived its power was bestowed by Charles II. The government was vested in a mayor, six aldermen, and 24 capital burgesses, with a recorder, town-clerk, bailiff, chief constable, and subordinate officers; the mayor and late mayor were justices of the peace. The government is now vested in a mayor, four aldermen, and twelve councillors, under the act of the 5th and 6th of William IV., cap. 76, (for an abstract of which see the Appendix, No. I.:) the municipal and parliamentary boundaries are co-extensive. The freedom is obtained by birth or apprenticeship. The borough first sent members to parliament in the commencement of the reign of Elizabeth, when the elective franchise was vested in the body corporate; but it was subsequently decided to be in the freemen, about 800 in number. The privilege is now extended to the £10 householders of the old borough, and of the hamlet of Ballingdon-*cum*-Brunden, in the county of Essex, which have been constituted the new borough, and comprise 1685 acres: the old borough contained only 925 acres; the mayor is the returning officer. The recorder holds courts of quarter session; and a court of record is held every Monday, for the recovery of debts to the amount of £20.

Corporation Seal.

Sudbury comprises the parishes of *All Saints*, *St. Gregory*, and *St. Peter*. The living of All Saints is a discharged vicarage, valued in the king's books at £4. 11. 5½.; present net income, £119; patron and impropriator, J. Sperling, Esq. The living of St. Gregory's is a perpetual curacy, with that of St. Peter's annexed; net income, £160; patron and impropriator, Sir Lachlan Maclean. The churches are all of considerable antiquity, and are spacious and handsome structures, mostly in the later English style of architecture, of which they present some fine specimens, though generally much defaced. In the church of All Saints is a curious monument to the Eden family, whose pedigree is painted on the walls. The church of St. Peter contains 130 free sittings, the Incorporated Society having granted £30 in aid of the expense. St. Gregory's, which is the most ancient, was formerly collegiate, until Henry VIII. granted its site and other possessions, for the sum of £1280, to Sir T. Paston, Knt.: it contains 80 free sittings, the Incorporated Society having granted £20 in aid of the expense: the font is very magnificent, and in a niche in the wall of the vestry-room, enclosed with an iron grating, is a human head, supposed to be that of Symon de Theobald, alias de Sudbury, Archbishop of Canterbury in the time of Richard II., a native of this town, who was beheaded by the mob in Wat Tyler's rebellion. The free grammar school was founded in 1491, under the will of William Wood, warden of Sudbury College, who endowed it with a farm, called the School farm, in the parish of Little Maplestead, in the county of Essex, worth about £100 per annum; a good house is provided for the master, with a large school-room, and about one acre of land, for which he pays a moderate rent: there are twelve boys on the foundation, whose parents must be inhabitants of Sudbury. There is also a National school, with a small endowment. The hospital of St. Leonard, for lepers, was founded by John Colneys, and endowed by Simon Theobald de Sudbury, with about five acres of land, a chapel, and a dwelling-house: it is now in the possession of the corporation of the poor, and is applied towards their maintenance. From a bequest by Thomas Carter, in 1706, 50 poor men receive coats and 50 poor women gowns annually on St. Thomas's day, and there are several other smaller charities for the benefit of the poor. The poor law union of Sudbury extends into Essex, and comprises 42 parishes or places, under the care of 46 guardians; the population, according to the census of 1831, amounts to 27,896. The college of St. Gregory, for Secular priests, founded by Simon de Theobald, was richly endowed, and valued, at the period of the dissolution, at £122. 18. 3. per annum: its only remains are the gateway, and portions of a wall now forming a part of the workhouse. A gateway, part of a monastery of Augustine friars, standing in Friars'-street, also exists. An hospital was founded here, in the reign of King John, by Amicia, Countess of Clare, which was afterwards given to the monks of Stoke; and there was also a Benedictine cell to the abbey of Westminster, founded in the reign of Henry II. About half a mile from the town is a spring of pure water, which, from its supposed efficacy in curing many diseases, is called by the inhabitants "Holy water." Sudbury is the birthplace of Gainsborough, the celebrated painter. It gives the inferior title of Baron to the Duke of Grafton.

SUDELEY-MANOR (*St. Mary*), a parish, in the union of Winchcomb, Lower Division of the hundred of Kiftsgate, Eastern Division of the county of Gloucester, 1 mile (S. S. E.) from Winchcomb; containing 84 inhabitants. The living is a rectory, valued in the king's books at £6. 11. 5½.; present net income, £45; patron, Lord Rivers. The church, which has remained in a dilapidated state ever since the injury it sustained in the great civil war, was the burial-place of Queen Catharine Parr, and of several of the family of Bridges. The ancient castle is said to have been built, *ex spoliis Gallorum*, by Boteler, Lord Sudeley, a celebrated warrior in the reigns of Henry V. and VI., who sold it to Edward VI., for fear of confiscation. It was granted by Edward VI. to his uncle, Lord Seymour, who espoused Queen Catherine Parr. Mary bestowed it upon Sir John Bridges, created by her Baron Chandos of Sudeley, whose grandson, the third Lord Chandos, here entertained Queen Elizabeth, in 1592. George, the sixth lord, having embraced the cause of Charles I., the castle was twice besieged by the parliamentary forces, who reduced it to its present state of ruin: the remains are considerable and interesting. Charles Hanbury Tracy, Esq., was created Baron Sudeley in 1838.

SUDELEY-TENEMENTS, a hamlet, in the parish of Winchcomb, Lower Division of the hundred of Kiftsgate, Eastern Division of the county of Gloucester: the population is returned with the parish.

SUFFIELD (*St. Margaret*), a parish, in the union of Erpingham, hundred of North Erpingham, Eastern Division of the county of Norfolk, 3¼ miles (W. by N.) from North Walsham; containing 272 inhabitants. The living is a discharged rectory, valued in the king's books at £14; patron, Lord Suffield, who derives his title of Baron from this place. The tithes have been commuted for a rent-charge of £350, subject to the payment of rates, which on the average have amounted to £75; the glebe comprises 10½ acres, valued at £15. 13. 6. per annum. A school is partially endowed by the late Dowager Lady Suffield, but principally supported by the present Lord Suffield, who built the school-house. Thomas Bulwer, in 1693, bequeathed property, now let for £12. 10. per annum, for distribution among the poor of the parish; and the Rev. Thomas Symonds, in 1682, left land, producing £15 per annum, to be divided among six poor widows.

SUFFIELD, a joint township with Everley, in the parish of Hackness, union of Scarborough, liberty of Whitby-Strand, North Riding of the county of York, 5 miles (W. N. W.) from Scarborough; containing, with Everley, 124 inhabitants.

SUFFOLK, a maritime county, bounded on the east by the North Sea, or German Ocean, on the north by the county of Norfolk, on the west by that of Cambridge, and on the south by that of Essex. It extends from 51° 56′ to 52° 36′ (N. Lat.), and from 23′ to 1°44′ (E. Lon.), and comprises an area of about 1512 square miles, or 967,680 statute acres. The population amounts to 296,317.

At the period of the Roman invasion, this county formed part of the territory inhabited by the Iceni, or Cenomanni, who, according to Whitaker, were descended from the Cenomanni of Gaul. Under the Roman dominion it was included in the division called *Flavia Cæsariensis.* After the withdrawal of the Roman legions, Cerdic, one of the earliest Saxon invaders, and founder of the kingdom of Wessex, landed, in 495, at a place afterwards called Cerdic Sand, in the hundred of Mutford and Lothingland, forming the north-eastern extremity of the county, and, after gaining some advantages over the opposing Britons, set sail for the western parts of the island. During the succeeding invasions of the Saxons, the territory now comprised in the counties of Suffolk, Cambridge, and Norfolk, was erected by Uffa, about the year 575, into the kingdom of East Anglia, in which the relative position of this district obtained for its inhabitants the name of *Suth-folc*, or southern people (in contradistinction to those of Norfolk, who were called the *North-folc*, or northern people), whence, by contraction, its modern name. The Christian religion was permanently established in this kingdom by King Sigebert, who brought over with him a Burgundian ecclesiastic, named Felix, whom he made bishop of East Anglia, and who fixed his seat at Dunwich, in this county, where he died in 647. Bisus, or Bosa, on succeeding to this see, in 669, divided it into two bishopricks, the seat of one of which was fixed at North Elmham, in Norfolk, the other remaining at Dunwich; but these were re-united about the year 870, when North Elmham became the sole seat of the diocese. In 655, during the struggles for independence maintained by East Anglia against the powerful kingdom of Mercia, then under the sway of Penda, a battle was fought at Bulcamp, near Dunwich, in which Anna, monarch of the East Angles, and his son Ferminus, were slain. East Anglia was again the scene of desolation, at the period of its subjugation by Offa, King of Mercia, who had basely assassinated its king, Ethelbert. It remained tributary to Mercia until, in the reign of Egbert, the kingdom of Wessex obtained a preponderating influence in the Heptarchy: under that monarch it continued to have its own sovereigns, until the reign of the East Anglian king, Edmund, who, after being barbarously murdered by the Danes under Inguar and Ubba, was surnamed the Martyr. These marauders directing their early attacks chiefly upon this part of the island, possessed themselves of the whole of East Anglia before the death of Egbert, making lamentable ravages in their progress. After the total defeat of the Danes by Alfred, in the West of England, East Anglia was one of the principal portions of territory allotted to them, for their limited residence, by that monarch, and was put under the government of Guthmund, who had his residence at Hadleigh, and, on his conversion to Christianity, took the name of Athelstan.

Under the act of the 6th and 7th of William IV., cap. 77, Suffolk is partly in the diocese of Norwich, and partly in that of Ely, in the province of Canterbury:

it is divided into the archdeaconries of Suffolk and Sudbury. By the new ecclesiastical division of the county, in 1837, the deaneries of Hartismere and Stow, with the parish of Rickinghall Inferior, have been separated from the archdeaconry of Sudbury, and annexed to that of Suffolk, and the archdeaconry of Sudbury, with the above exception, has been annexed to and forms part of the diocese of Ely. The total number of parishes is 504. For purposes of civil government it is divided into the hundreds of Babergh, Blackbourn, Blything, Bosmere and Claydon, Carlford, Colneis, Cosford, Hartismere, Hoxne, Lackford, Loes, Mutford and Lothingland, Plomesgate, Risbridge, Samford, Stow, Thedwastry, Thingoe, Thredling, Wangford, and Wilford. It contains the borough, market-town, and sea-port of Ipswich; the borough and market-towns of Bury-St. Edmund's, Eye, and Sudbury; the market-towns and sea-ports of Lowestoft, Southwold, and Woodbridge; the sea-ports of Aldborough and Dunwich; and the market towns of Beccles, Bungay, Clare, Debenham, Framlingham, Hadleigh, Saxmundham, and Stow-Market. By the act of the 2nd of William IV., cap. 45, the county has been divided into two portions, called the Eastern and Western divisions, each sending two members to parliament: the place of election for the Eastern Division is Ipswich, and the polling-places, are Ipswich, Needham, Woodbridge, Framlingham, Saxmundham, Halesworth, and Beccles; and that for the Western Division is Bury-St. Edmund's, and the polling-places, Bury-St. Edmund's, Wickham-Brook, Lavenham, Stow-Market, Botesdale, Mildenhall, and Hadleigh. Two representatives are returned for each of the boroughs, except Eye, which was deprived of one by the act of the 2nd of William IV., cap. 45, by which also the boroughs of Aldborough, Dunwich, and Orford were totally disfranchised. Suffolk is included in the Norfolk circuit: the assizes are held alternately at Bury and Ipswich; and the general quarter sessions at Beccles, Woodbridge, Ipswich, and Bury, each for its respective district. The county gaols and houses of correction are at Bury and Ipswich; and there are houses of correction at Beccles and Woodbridge.

The two grand civil divisions are, the franchise, or liberty, of Bury-St. Edmund's, and the remaining part, or body of the county, as it is termed; each, at the county assizes, furnishing a distinct grand jury. In its civil government Suffolk is also divided into the Geldable portion, in which the issues and forfeitures are paid to the king; and the Franchises, in which they are paid to the lords of the liberties. The former comprises the hundreds of Blything, Bosmere and Claydon, Hartismere, Hoxne, Mutford and Lothingland, Samford, Stow, and Wangford: the sessions for Blything, Mutford and Lothingland, and Wangford, are held at Beccles; and for the rest of the Geldable hundreds at Ipswich. The franchises are three in number; first, the franchise, or liberty, of St. Ethelred, which formerly belonged to the prior and convent, now to the Dean and Chapter, of Ely, containing the hundreds of Carlford, Colneis, Loes, Plomesgate, Thredling, and Wilford, the sessions for which are held at Woodbridge. Secondly, the franchise, or liberty, of St. Edmund, given to the abbot of Bury, by Edward the Confessor, and comprising the hundreds of Babergh, Blackbourn, Cosford, Lackford, Risbridge, Thedwastry, and Thingoe, the sessions for which are held at Bury. And thirdly, the liberty of the duchy of Norfolk, granted by letters patent from Edward IV., dated Dec. 7th, 1468, in which the Duke has the returning of all writs, and the right of appointing a special coroner, and of receiving all fines and amerciaments: it comprises, within the limits of this county, his manors of Bungay, Kelsale, Carlton, Peasenhall, the three Stonhams, Dennington, Brundish, the four Ilketshalls, and Cratfield. The counties of Suffolk and Norfolk formed only one shrievalty, until the year 1576, when a sheriff for each was first appointed. With regard to the government and management of the poor, the most remarkable circumstance is the incorporation of several hundreds for the erection and support of houses of industry, on a very large scale, and in situations chosen for their pleasantness and salubrity. Thus, Colneis and Carlford hundreds were incorporated in the 29th of George II., and have their house of industry in the parish of Nacton; the hundred of Blything was incorporated in 1764, and has its house of industry in the hamlet of Bulcamp, near Blythburgh; that of Mutford and Lothingland, in the same year, having its house of industry in the parish of Oulton, near Lowtestoft; and that of Wangford, also in 1764, which has its house of industry at Shipmeadow, between Bungay and Beccles. The following hundreds were incorporated in 1765: *viz.*, those of Loes and Wilford (since dis-incorporated), which had their house of industry in the parish of Melton, and which has been since converted into a lunatic asylum; that of Samford, in the parish of Tattingstone; and that of Bosmere and Claydon, in the parish of Barham. The hundred of Cosford, and the parish of Polstead, which have theirs in the parish of Semer, were incorporated in the 19th of George III. The hundred of Stow was incorporated in the 20th of George III., and has its house of industry in the parish of Onehouse.

The soils are various, but the limits of each may be clearly traced. Strong clayey loams, on a substratum of clay marl, occupy the largest tract, which is generally called High Suffolk, and extends from the confines of Cambridgeshire and Essex, on the south-west, across the central parts of the county, to those of Norfolk, on the north-east: on the north-west this district is bounded by an irregular imaginary line, passing from the western border of the county, near Dalham, by Barrow, Little Saxham, the vicinity of Bury-St. Edmund's, Rougham, Pakenham, Ixworth, and Honington, to the northern boundary at Knettishall; and on the south-east by another, drawn from the banks of the Waveney, near North Cove, a few miles to the east of Beccles, southward by Wrentham and Wangford, and then south-westward by Blythford, Holton, Bramfield, Yoxford, Saxmundham, Campsey-Ash, Woodbridge, Culpho, Bramford, Hadleigh, and along the high lands bordering on the western side of the river Bret, to the confluence of that stream with the Stour. The bottoms of the vales traversed by running streams, which are numerous, and the slopes descending to them, are of a soil superior in quality to the rest of this district, generally consisting of a rich friable loam. Rich loams, of various qualities, occupy that portion of the county included between the south-eastern part of the strong loams and the estuaries of the rivers Stour and Orwell, lying to the south of a line drawn from Ipswich to Hadleigh: some of these are of a sandy quality, others much stronger: from Stratford and Higham, on the borders of the Stour, eastward across the Orwell, to the banks of the river Deben, near its mouth, extends a tract of friable and putrid vegetable mould of extraordinary fertility, more especially at Walton, Trimley, and Felixstow. In the projecting north-eastern district, lying between the river Waveney and the ocean, is much land of the same rich quality; but as it is interspersed with many sandy tracts, and on the sea-coast is of a sandy character throughout, it may be considered to form part of the great sandy maritime district extending from the river Orwell, between the clayey loams and the sea, to the north-eastern extremity of the county: the lands in this district, which is called the Woodlands, are generally of excellent staple, and are among the best-cultivated in England; although, in the country lying between the towns of Woodbridge, Orford, and Saxmundham, and north-eastward, as far as Leiston, there is a large extent of poor, and in some places even *blowing*, sands, which have caused this south-eastern part of the county to receive the name of "Sandlings," or "Sandlands:" the substratum of the eastern district, though sometimes marl, is generally sand, chalk, or crag; which last is a singular mass, consisting of cockle and other shells, found in numerous places, from Dunwich, southward, to the Orwell, and even beyond that river. Another district of sand occupies the whole extent between the clayey soils and the fenny tract, which latter forms the north-western angle of the county, and may be separated from the sand by an irregular line drawn from near where the river Lark begins to form the western boundary of Suffolk, to the Little Ouse, a short distance below Brandon: these western sands, unlike much of the last-mentioned, are seldom of a rich loamy quality, but comprise numerous warrens and poor sheepwalks: much of that now under tillage is apt to *blow*, that is, to be driven by the wind, and consequently ranks among the worst soils: the chief exceptions to the general inferiority of this district lie to the south-east of a line drawn from Barrow to Honington, and at Mildenhall: the substratum is throughout a perfect chalk, at various depths. Of the Fens, it is only necessary to observe, that the surface, to the depth of from one foot to six, consists of the ordinary peat of bogs, some of which is very solid and black; but in other places it is more loose, porous, and of a reddish colour: the substratum is generally a white clay, or marl.

By far the greater part of the county is under tillage: the modes of culture on the clayey loams are various; but, on the rich loam and lighter soils, the Norfolk system has been generally introduced. On the sand, turnips are everywhere employed as a preparative crop before corn or grass. Paring and burning, hitherto practised only in the Fens, have lately been adopted on the heavy plough lands with great advantage: in the former, the course of crops, after this operation, is generally cole-seed; then oats twice in succession, with the last crop of which are sown rye-grass and clover, under which the land remains for six or seven years, and afterwards is again pared and burned. The crops commonly cultivated are wheat, barley, oats, beans, peas, buck-wheat, turnips, cabbages, carrots, potatoes, beet, tares, cole-seed, red and white clover, trefoil, sainfoin, hemp, and hops. The culture of carrots in the Sandlings is of very ancient practice, great quantities having been formerly sent from that district by sea to the London market; but the chief object for which they are now grown is as food for horses. In the fen district, cole-seed constitutes one of the principal crops; and the cultivation of sainfoin is particularly extensive in the chalky subsoils. The most remarkable agricultural implement is the well-known light Suffolk swing plough; drilling is much practised.

The grass lands were remarkable for their richness, but the best have been ploughed up, and the extent occupied by dairy farms is not so great as formerly, though large quantities of butter are still sent to the London market. Large tracts of grass land are also mown for the supply of the towns with hay: the herbage, which springs up after the gathering of the hay crop, is here called *rowings*. Clay and marl are extensively employed as manures; as also is chalk, which the inhabitants of the hundreds of Colneis and Samford obtain from Kent and Essex, by means of the corn-hoys; besides these substances, the shell marl, or crag, as it is provincially termed, that is found in the Sandlings, is much used for this purpose, and considerable quantities of manure are brought from London. The Suffolk cows have long been famous for the great quantity of their milk, which, however, is not remarkable for richness: they are universally polled, and not of a large size. The number of sheep kept is very great, the South Down breed being most prevalent. The only remarkable breed of hogs is a particularly good one, which is found in the dairy district. Poultry is exceedingly plentiful, especially turkeys, for which this county is nearly as famous as Norfolk. The Suffolk breed of horses is as celebrated as that of cows: it is found in the greatest perfection in the tract included between the coast and the towns of Woodbridge, Debenham, and Eye, extending as far as Lowestoft.

The woods are of very small extent, and are not generally of luxuriant growth; the strong loams formerly bore considerable quantities of fine oak, a great proportion of which has been cleared off, and various plantations made, but only with a view to ornament. The most important tracts of waste land are those occupying nearly all the country from Newmarket, on the borders of Cambridgeshire, to the confines of Norfolk, near the towns of Thetford and Brandon; and those lying between Woodbridge, Orford, and Saxmundham, in the eastern part of the county; besides which, numerous heaths of smaller extent are scattered in every quarter of it; the chief use of these wastes is as sheepwalks. The manufactures and commerce are very inconsiderable, in comparison with those of many other counties. The chief manufacture is the combing and spinning of wool, in a great measure for the Norwich manufacturers, which is carried on, though not to any great extent, in most parts of the county, excepting the hemp district before mentioned, where the latter material was extensively spun and woven into linen, but this manufacture is almost extinct. At Sudbury are manufactories for silk and woollen goods: there is also a silk-manufactory at Mildenhall, a branch of an extensive concern at Norwich, and also at Glimsford. The imports are merely the ordinary supplies of foreign articles for the inhabitants: the chief exports are corn and malt. The principal fishery on the coast is that of herrings, which is the chief support of the town of

Lowestoft, where about 40 boats, of 40 tons' burden each, are engaged in it: the season commences about the middle of Sept., and lasts until towards the end of Nov. The town also partakes in the mackarel fishery, in which the same boats are employed, the season commencing about the end of May, and continuing until the end of June. In the Orford river there is a considerable oyster fishery.

This is a well-watered county: the principal rivers are the Stour, the Gippen or Orwell, the Deben, the Ore, the Waveney, the Little Ouse or Brandon river, and the Lark; besides which the smaller streams are exceedingly numerous. The Stour first meets the tide at Manningtree in Essex, and begins to expand into a broad estuary, which at high water has a beautiful appearance; but at low water the river shrinks into a narrow channel, bordered by extensive mud banks. Proceeding eastward, it is joined near Harwich by the Orwell, and their united waters having formed the port of Harwich, discharge themselves into the North Sea, between that town, in Essex, and Landguard fort at the south-eastern extremity of this county: this river is navigable up to Sudbury. The Gippen is formed by the confluence of three rivulets at Stow-Market, from which place it was made navigable in 1793: below Ipswich it assumes the name of Orwell, expands into an estuary, and continues its course to its junction with the Stour opposite Harwich: the Orwell is navigable for ships of considerable burden up to Ipswich, and the scenery on its banks is beautiful. The Deben rises near Debenham, and at Woodbridge expands into an estuary, and proceeds thence in a southerly direction to the North Sea: towards its mouth it takes the name of Woodbridge haven, which joins the sea about ten miles below that town, to which it is navigable for vessels of considerable burden. The Ore expands into an estuary as it approaches Aldborough, where, having arrived within a very short distance of the sea, it suddenly takes a southerly direction, and discharges its waters into the North Sea below Orford: it is navigable to a short distance above Aldborough. The Waveney joins the river Yare at the head of Bredon-water, an expansion formed by these united rivers, which, contracting again near Yarmouth, pursues a nearly southerly course to the sea, below that town: this river, the meadows on the banks of which are among the richest in England, is navigable for barges as high as Bungay bridge. The little Ouse, or Brandon river, is navigable up to Thetford; the Lark is navigable to within a mile of Bury-St. Edmund's; and the Blythe, to Halesworth. The only artificial navigation is that in the channel of the Gippen, from Stow-Market to Ipswich: it is 16 miles and 40 rods long, and has 15 locks, each 60 feet in length and 14 in width: this canal was opened in the year 1793: the total expense of its formation was about £26,380. The line of the Eastern Counties railway passes through the county.

Within the limits of the county were comprised the Roman stations *Ad Ansam*, at Stratford, on the border of Essex; *Cambretonium*, at Brettenham, or Icklingham; *Garianonum*, at Burgh Castle (though some fix it at Caistor, near Yarmouth); and *Sitomagus*, probably at Dunwich. Remains of Roman military works exist at Burgh Castle, Brettenham, Icklingham, Stow-Langtoft, and Stratford, on the banks of the Stour. Numerous domestic and sepulchral relics of that people have also been dug up in different places, such as pavements, coins, medals, urns, rings, &c. That stupendous work of human labour, called the Devil's Ditch, on Newmarket heath, is supposed to have served as the line of demarcation between the kingdoms of Mercia and East Anglia. Near Barnham, on the borders of the Little Ouse, is a range of eleven tumuli, on a spot supposed to have been the scene of one of the conflicts between the Danes, under Inguar, and the forces of Edmund, King of East Anglia: others occur in different places, the most remarkable group being that called the Seven Hills, at Fornham-St. Geneveve, near Bury. The number of religious houses, of all denominations, was about 59, including four Alien priories. There are remains of the abbeys of Bury-St. Edmund's, Leiston, and Sibtow; of the priories of Blythburgh, Butley, Clare, Herringfleet, Campsey-Ash, Dodnash, Gorleston, Kersey, Ixworth, Orford, Wangford, Ipswich, Mendham, and Sudbury; and of the nunneries of Bungay and Redlingfield. The remains of ancient fortresses are chiefly those of the castles of Bungay, Clare, Framlingham, Haughley, Lidgate, Mettingham, Orford, and Wingfield. Ancient mansions are seen in different parts of the county; the most remarkable is Hengrave Hall. There are many elegant seats, among the most distinguished of which is Euston Park, the property and residence of the Duke of Grafton; Heveningham Hall, the seat of Lord Huntingfield; Flixton Hall, and Kentwell Hall. Suffolk gives the title of Earl to the family of Howard.

SUGLEY, a township, in the parish of NEWBURN, union and Western Division of CASTLE ward, Southern Division of the county of NORTHUMBERLAND, $3\frac{3}{4}$ miles (W.) from Newcastle; containing 255 inhabitants. The extensive manufactory termed the Tyne Iron Works is within this township.

SUGNALL, MAGNA, a township, in the parish of ECCLESHALL, Northern Division of the hundred of PIREHILL and of the county of STAFFORD, $2\frac{1}{2}$ miles (N. W. by W.) from Eccleshall; containing 130 inhabitants.

SUGNALL, PARVA, a township, in the parish of ECCLESHALL, Northern Division of the hundred of PIREHILL and of the county of STAFFORD, 3 miles (N. W.) from Eccleshall; containing 61 inhabitants.

SULBY, an extra-parochial district, in the union of MARKET-HARBOROUGH, hundred of ROTHWELL, Northern Division of the county of NORTHAMPTON, $6\frac{1}{4}$ miles (S. W.) from Market-Harborough; containing 78 inhabitants. An abbey of the Premonstratensian order, in honour of the Blessed Virgin Mary, was founded here about 1155, by Robert de Querceto, Bishop of Lincoln, and its possessions were so much increased by Sir Robert de Paveley, Knt., that, at the dissolution, its revenue was estimated at £305. 8. 5.

SULGRAVE (*St. James*), a parish, in the union of BRACKLEY, hundred of CHIPPING-WARDEN, Southern Division of the county of NORTHAMPTON, $8\frac{1}{2}$ miles (N. by W.) from Brackley; containing 576 inhabitants. The living is a discharged vicarage, valued in the king's books at £9. 17.; present net income, £231; patron and incumbent, Rev. W. Harding; impropriator, C. F. Annesley, Esq. Near the church, to the westward, is Castle hill; and about a mile to the northward is an artificial mount, called Burrough hill, crowned with an ancient fortification, 40 feet square, and commanding a most extensive prospect, nine counties being visible from its summit. A school, open to all the poor children of this parish, was founded by John Hodges, in 1722, who endowed it with £4 per annum, which, with £5 annually left by Robert Gardiner, in the year 1763, for teaching six poor boys, and a subscription raised for the purpose, is paid to the master: the last-named individual also bequeathed £500, in 1776, for apprenticing one of his charity boys, clothing others, and for distribution among the poor.

SULHAM (*St. Nicholas*), a parish, in the union of Bradfield, hundred of Theale, county of Berks, 1 mile (S. S. E.) from Pangbourn; containing 72 inhabitants. This parish consists of a narrow slip of land extending from the river Thames to the river Kennet, on the roads from London to Bristol and from Reading to Oxford, the former towards the Kennet, and the latter towards the Thames. The Thames navigation for coal, and the Great Western railway, pass through the parish. The living is a rectory, valued in the king's books at £6. 4. 2.; present net income, £147; patron, Frederick Wilder, Esq. The church has been recently rebuilt. A school was commenced in 1830, and is supported from private funds.

SULHAMPSTEAD-ABBOTTS (*St. Bartholomew*), a parish, in the union of Bradfield, hundred of Reading, county of Berks, 7 miles (S. W. by W.) from Reading; containing, with the tything of Grageby, 423 inhabitants. The Avon and Kennet navigation passes through the parish, and the Great Western railway within five miles of it. The living is a rectory, to which, in 1782, that of Sulhampstead-Bannister was annexed, valued in the king's books at £10. 6. 0½.; present net income, £600; patrons, Provost and Fellows of Queen's College, Oxford. There is a day and Sunday school, supported by endowment, with a house for the master and mistress.

SULHAMPSTEAD-BANNISTER (*St. Michael*), a parish, in the union of Bradfield, hundred of Theale, county of Berks, 6¾ miles (S. W. by W.) from Reading; containing 289 inhabitants. The living is a rectory, annexed to that of Sulhampstead-Abbotts, and valued in the king's books at £6. 5. A Sunday school is supported by endowment. The river Kennet runs through the parish.

SULLINGTON (*St. Mary*), a parish, in the union of Thakeham, hundred of East Easwrith, rape of Bramber, Western Division of the county of Sussex, 5½ miles (W. by N.) from Steyning; containing 320 inhabitants. The living is a rectory, valued in the king's books at £12. 17. 6.; present net income, £296; patron and incumbent, Rev. G. Palmer. The church is principally in the early style of English architecture.

SUMMERFORD, a township, in the parish of Astbury, union of Congleton, hundred of Northwich, Southern Division of the county of Chester, 1¾ mile (N. W.) from Congleton; containing 112 inhabitants. Here is a school, in which are about 75 children, 38 of whom are paid for by a lady, and the remainder by their parents; also a Sunday school, which is wholly supported by the same lady, who provides books for both schools.

SUMMERHOUSE, a township, in the parish of Gainford, union of Darlington, South-Western Division of Darlington ward, Southern Division of the county palatine of Durham, 6¾ miles (N. W. by W.) from Darlington; containing 192 inhabitants. This place derived its name from having been anciently the summer residence of the Lords of Raby, of whose mansion, surrounded by a moat, there are still some vestiges in the southern part of the village. There is a place of worship for Wesleyan Methodists; also a National school.

SUMMER-TOWN, formerly a hamlet, in the parish of St. Giles, Oxford, (with which the population is returned), now a district parish, for which a church was erected in 1838, and a district annexed to it, under the 21st section of the 58th of George III.

SUNBURY (*St. Mary*), a parish, in the union of Staines, hundred of Spelthorne, county of Middlesex, 15 miles (S. W. by W.) from London; containing 1863 inhabitants. The living is a vicarage, valued in the king's books at £13. 6. 8.; present net income, £336; patrons, Dean and Chapter of St. Paul's, London; the impropriation belongs to Mrs. Fish, and — Edwards and — Taylor, Esqrs. Here is a National school, partly supported by the interest of bequests, amounting to £4. 1. 2. per annum, and partly by subscription. There is also a Sunday school, in which 20 children of each sex are taught gratuitously by congregational dissenters.

SUNDERLAND, a township, in the parish of Isall, union of Cockermouth, Allerdale ward below Derwent, Western Division of the county of Cumberland, 6 miles (N. E.) from Cockermouth; containing 77 inhabitants.

SUNDERLAND (*Holy Trinity*), a sea-port, newly enfranchised borough, market-town, and parish, and the head of a union, in the Northern Division of Easington ward and of the county palatine of Durham, on the southern bank of the river Wear, 13 miles (N. E.) from Durham, and 269 (N. by W.) from London; containing (exclusively of Bishop-Wearmouth and Monk-Wearmouth, the population of which is stated under their respective heads) 19,000 inhabitants, including seamen. The early history of this place is interwoven with that of Bishop-Wearmouth, until the end of the twelfth century, when, in a charter of privileges and free customs, similar to those then enjoyed at Newcastle, granted to it under the name of Wearmouth, by Bishop Pudsey, we meet with the first authentic record of its distinct maritime and commercial character as a port. The etymology and date of its present name are somewhat obscure; the more probable conjecture is, that it was intended to designate its original peninsular situation, occasioned by the influx of the river Wear into the sea, and a deep ravine at Hendon Dene, which separated it from the main land. Its prosperity having been essentially promoted by the provisions of the above-mentioned charter, it gradually increased in population and importance, and, in the reign of Henry VIII., was becoming a place of considerable note. At the commencement of the 17th century, some Scottish families and foreign merchants came to reside here, and the charter of Bishop Morton raised it to the dignity of a corporate town. Soon after the Conquest, Malcolm, King of Scotland, when traversing the eastern coast, in one of his destructive incursions, met with Edgar Atheling, heir to the English crown, his sister Margaret, the future queen of Scotland, and a train of distressed Saxons, whilst waiting in the harbour of Sunderland for a wind and tide favourable to their escape from the victorious Normans into Scotland. During the great civil war the inhabitants were entirely devoted to the interests of the parliament, and, in 1642, the town was garrisoned in its behalf, in consequence of the seizure of Newcastle by the royalists, and the exportation of coal from that port to London being prohibited; a parliamentary commissioner also was sent to reside here until the surrender of that town. In 1644 and 1645, repeated skirmishes occurred in Sunderland and the neighbourhood, between the contending parties, during which period the resident Scots suffered greatly from want of provisions, owing to the wreck of some vessels laden with supplies from Scotland, and the capture of others by the royalists, in the river Tyne, whither adverse winds had most inopportunely driven them.

The town consists of many streets, of which High-street, the principal, is broad and handsome, extending upwards of a mile, and is well paved and the foot-paths flagged. In general the houses are well built, except in the lower part of the town: by virtue of an act of parliament, passed in 1809, considerable improvements were made, by the removal of nuisances; and, in 1823, a company was formed for lighting the streets with gas, supplied from two gasometers, which are calculated to contain 25,000 cubic feet, the works having been built at an expense of £8000. The inhabitants are supplied with water from a large well, at the head of Bishop-Wearmouth, by means of a steam-engine, which raises it into two reservoirs, at the rate of 150 gallons per minute, whence it is conveyed to the houses through pipes: the expense of these works, estimated at £5000, was defrayed by shares of £25 each. The facilities and accommodations afforded to bathers, during the season, have rendered the town a place of fashionable resort: in 1800, hot and cold baths were established at Hendon; and, in 1821, a suit of hot, cold, vapour, and medicated baths, was erected near the town moor: on the sands, which are favourable for bathing, are several machines for the purpose. A new entrance to the town from the Stockton road has been lately made, at a considerable expense, by cutting through an eminence called Building Hill: it enters at the new street called Fawcett-street, being a continuation of Bridge-street, which is also of recent date; the former contains a new Methodist chapel, and the latter a Roman Catholic chapel, two of the handsomest buildings in the town. A Mechanics' Institute, established in 1825, was for some years in a flourishing condition, but having of late been broken up, it has been re-established by aid of a donation of £50 from the Earl of Durham: it is now situated in Bridge-street, and comprises a library and a reading and news-room. The other news-rooms are the exchange, the borough, and one recently formed in Bedford-street. A subscription library, originally instituted in 1795, is kept in the upper part of a handsome edifice completed in 1802, and consists of more than 9000 volumes. The Methodists have a select library, kept in a house in which a law library has also been established. A Literary and Philosophical Institution and Natural History Society have been lately formed. There are a commodious theatre in Drury-lane, and an assembly-room. Barracks were erected on the town moor in 1794, and in 1828 a portion of the buildings was removed and the remainder cased with brick: they now afford accommodation for 800 men and stabling for 10 horses, and there is an hospital adapted to the reception of 20 patients. The Monk-Wearmouth and the opposite batteries, constructed for the protection of the port, are now in decay. Of all the improvements which the town has received during the last 50 years, the cast-iron bridge across the Wear, which connected it with Monk-Wearmouth, may be considered the chief. Previously to 1792, the river had been crossed by means of two ferries: the first project entertained was that of erecting a stone bridge, which was abandoned for the present structure, of which the first stone was laid September 24th, 1793, and the work was completed in 1796, at a total expense of £33,400, of which sum, £30,000 was advanced by Rowland Burdon, Esq., M.P. for the county, to whom the invention of the bridge is ascribed; the tolls are let annually for about £3000. The bridge was built under the direction of Mr. Thomas Wilson, of Bishop-Wearmouth, and consists of one magnificent arch, 236 feet in the span, and 100 feet in height from low water mark, admitting vessels of considerable burden to pass underneath without striking their topgallant-masts. The whole weight of iron is 260 tons, 46 of which are malleable and the remainder cast.

Sunderland is mainly indebted for its present importance to its advantageous situation on the coast, and near the influx of a navigable river, which flows through a district abounding with coal and limestone. The export of coal, which is its staple article of commerce, appears to have commenced so early as the reign of Elizabeth, or the beginning of that of James. The river dues are collected by the commissioners or conservators of the Wear, under the Sunderland Improvement Act, and applied to the cleansing and improving of the harbour: the amount expended for several years has been, on the average, £12,000. The metropolis and the western coast of England are its principal marts; large quantities of coal are also shipped to Holland, France, and other parts of the continent. The coal shipped here from the staiths of the different collieries, during the year 1837, amounted to 932,135 tons. The principal coal staiths are those of the Earl of Durham, the Hetton Coal Company, and the Durham and Sunderland Railway Company. The last-named company conveys the coal of various collieries in the vicinity of the railway, which is also a public conveyance for goods and passengers: at Quarrington it joins the Clarence railway to Stockton, and at Haswell, the Hartlepool railway. The Hetton Company conveys its own coal exclusively; but the greater part of the Marquess of Londonderry's coal is conveyed to the Lambton staiths for shipment, the remainder being brought by keels from Pensher, a few miles above the town.

Next in importance to the coal trade is the trade in lime, of which upwards of 30,000 chaldrons are annually shipped for the ports of Yorkshire and the eastern coast of Scotland, employing from 25 to 30 vessels of from 40 to 100 tons burden. The remainder of the export trade is supplied by the numerous manufactories in the town and its vicinity, which consist of several extensive establishments for manufacturing flint and crown glass and bottles, alkali and copperas works, several extensive potteries; many roperies, four of which are patent and worked by steam power; chain cable and anchor manufactories; four large foundries, one of which employs upwards of 300 men and boys daily; brass foundries; sail-cloth manufactories; a block manufactory worked by steam, lately established; and two paper mills. There are also several flour and saw mills, impelled by steam; and grindstone quarries. The patent ropery at Deptford was worked by a steam-engine of 16-horse power, and was capable of furnishing 500 tons of cordage in the year, within the usual hours of labour; it was lately destroyed by fire, but has been rebuilt on a more extensive scale. The principal imports are flour, wine, spirituous liquors, timber, tallow, iron, flax, and all articles of Baltic produce. Ship-building is here carried on more extensively than at any other port in the empire: on the shores of the river from 80 to 100 vessels are sometimes on the stocks at once. There are 30 yards for building ships, 5 for building boats, four dry docks, and 11 floating docks. The largest vessel ever constructed here was the Lord Duncan, launched from Southwick quay in 1798: its extreme length was 144 feet, the breadth 39 feet, and the burden 925 tons. The salmon fishery was formerly extensive, 72 having been taken at one draught so late as the year 1788; it has been wholly abandoned, the supplies now consisting of cod, ling, turbot, haddock, skate, herrings, crabs, and lobsters, all of which are caught in abundance. The general trade of the port has increased, within the last fifty years, to a height and with a rapidity almost incredible; the population and shipping having within that period become doubled in number. The exports and imports are annually greater, and it may be fairly inferred that the port will rise still higher and more rapidly in commercial and maritime importance. A capacious wet dock, containing an area of nearly eight acres, has been lately completed and opened at Monk-Wearmouth Shore, on the northern side of the river, and near the entrance to the harbour: this dock is connected with the Newcastle and Carlisle railway by means of the Brandling Junction railway, thus establishing a communication between the east and west seas. During the year ending January 5th, 1837, 187 vessels entered the port from foreign parts, of which 108 were British and 79 foreign; and 902 cleared outwards, of which 542 were British and 360 foreign. The number of coasting vessels that entered, during the same period, was 874, (exclusively of 57 entering "for warehouse,") and that cleared outwards, 5095. There are 685 vessels belonging to the port, of the aggregate burden of 131,471 tons, and navigated by 5626 seamen. The value of the shipping insured in the mutual insurance "policies" of the town amounts to £850,000, besides which there is nearly one-fourth of that amount either not insured or insured at other places. The amount of custom-house duties received in the year ended January 5th, 1838, was £91,406.

The entrance of the harbour is formed by two piers, of which the northern is now being rebuilt a little to the north of its present site, thus widening and improving the entrance. The depth of water at the highest spring tide is from 15 to 20 feet, and at the lowest neap tides from 10 to 11; vessels of 400 tons' burden can enter it. In 1669, letters patent were obtained from Charles II., by Edward Andrew, Esq., empowering certain commissioners to levy contributions for the purpose of cleansing the harbour, and erecting a pier and light-house. Under this authority the southern pier had been partially constructed, at an expense of £50,000, and, in 1765, its completion was expected to require as much more: its present length is about 626 yards, and a tide light is placed near its extremity. The northern pier was begun in 1787, and is now 1850 feet in length. Near its extremity is an elegant octagonal lighthouse of freestone, 68 feet high from the pier to the under side of the cap, lighted with gas by means of nine argand lamps and reflectors, each of the latter being 18 inches in diameter. A gasometer, and two neat houses for the residence of the men who attend to the lights, have been erected on the pier: the disasters incident to this rocky and dangerous coast are additionally provided against by having in constant readiness three life-boats of peculiar construction. All the improvements connected with the harbour were placed under the superintendence of commissioners appointed, at first, for 21 years, by an act passed in the 3rd of George I., of which the provisions have been subsequently enlarged and confirmed, the last renewal having taken place in 1830; their jurisdiction extends up the river as far as to Biddick Ford, and from Souter Point to Ryhope Dene on the coast.

The exchange, a handsome structure in the High-street, erected in 1814 at an expense of £8000, advanced on shares of £50 each, comprises a large room for public meetings, commercial and news rooms, merchants' walk, and justice-room: in the news-room, which is a handsome and spacious apartment, is a full-length portrait of Sir Henry Vane Tempest, Bart., presented by the Marquess of Londonderry. The custom-house for some time past having occupied a very inconvenient site near the town moor, a more commodious structure has been lately erected in an excellent situation near to and fronting the river, at an expense of £5000, raised by a company of subscribers: the building has been taken by government on a renewable lease of 21 years, and was opened on Oct. 10th, 1837. The excise-office is in East-cross-street. The market, formerly held on Friday, is now on Saturday; there is a cattle market once a fortnight, and fairs for toys and pedlery are held on May 13th and 14th, and Oct. 12th and 13th; there is also a statute fair twice a year for the hiring of servants. A site for a new market was purchased in 1830, for £4200, and soon after extensive ranges of shambles and stalls for butchers' meat, poultry, butter, &c., were erected: the entrance from the High-street is through a handsome arcade, over which is a spacious room for auctions, exhibitions, &c. The Durham Junction railway, about five miles in length, extends from the Stanhope and Tyne railway at North Biddick to the foot of Rainton Hill, where it joins the Seabam railway; it has a branch of 1½ mile to Houghton-le-Spring. Near Coxgreen, about five miles above Sunderland, it crosses the valley of the Wear by a magnificent bridge of hewn stone, from designs by Mr. T. E. Harrison, and consisting of four nearly semicircular arches, of which the principal or waterway arch is 160 feet in the span; the adjoining one on the north, 144 feet; and the two other dry arches, 100 feet each. The roadway of the bridge, which is the highest stone bridge in the kingdom, is 130 feet above the level of high water: the total length is 810¾ feet, and the width 23 feet. From the spring of the arches rise circular pillar towers, the main pier and tower being 160 feet from the foundation.

The first charter of privileges granted to the inhabitants in the 12th century, by Bishop Pudsey, conferred upon them similar immunities to those enjoyed by the burgesses of Newcastle, likewise the right to determine all pleas arising within the borough, excepting those of the crown; other privileges and appointments were made by subsequent Bishops of Durham. Previously to 1634, the town was governed by a bailiff appointed by letters patent from the bishop; but in that year the inhabitants were incorporated by charter of Bishop Morton, under the style of mayor, twelve aldermen, and commonalty of the borough of Sunderland, which also stated that it had been a borough from time immemorial, known as the new borough of Wearmouth. The practical operation of this charter having ceased some time after it was granted, a body of the influential tradesmen and resident gentlemen have, for some years past, formed a private corporation, under the title of "Freemen and Stallingers," and in that capacity have claimed, and are in possession of, the town moor as their corporate property; they have not interfered in the regulation or government of the town, and their right to hold the moor is disputed by the new corporation, by which, under the act of the 5th and 6th of William IV., c. 76, (for an abstract of which see the Appendix, No. I.), the town is now governed. The corporation is styled the "Mayor, Aldermen, and Burgesses," and consists of a mayor, 14 aldermen, and 42 councillors, forming the council of the borough, which is divided into seven wards. The mayor is a justice of the peace during his year of office and one year after; there are also four other justices appointed by the Crown. By the act of the 2d of William IV., cap. 45, the town was constituted a borough, with the privilege of returning two members to parliament, the right of election being vested in the £10 householders of a very populous district, comprising an area of 4761 acres, of which the extent is described in the Appendix, No. II.: the number of registered voters is about 1580; the mayor is the returning officer. The municipal borough comprises the parish of Sunderland, and the townships of Monk-Wearmouth, Monk-Wearmouth-Shore, Bishop-Wearmouth-Pans, and so much of that of Bishop-Wearmouth as is included within a circle of one mile from the centre of the bridge: the parliamentary borough contains, in addition, the township of Southwick and the remainder of Bishop-Wearmouth township. The various interests in the parish are let on two leases by the Bishop of Durham; one includes the borough courts, fairs, market tolls, anchorage, and beaconage, and the office of water-bailiff; the other comprises the ferry-boats, and metage and tolls of fruits, herbs, and roots. The bishop's lessee is empowered to hold courts leet and baron annually, or oftener if need be, but no court has been held for many years. The borough magistrates hold petty sessions every Monday and Thursday at the justice-room in the exchange; a petty session of the county magistrates is also held at the same place every Saturday; and the town is a polling-place for the northern division of the county.

Prior to 1719, Sunderland formed a portion of the parish of Bishop-Wearmouth, but, in that year, an act was obtained to make it a distinct parish. The living is a rectory; net income, £241; patron, Bishop of Durham. The church was erected in 1719, and repaired in 1803: it stands in the upper part of the town, and is a commodious and handsome brick edifice, with a square tower: the altar is placed in a recess covered with a dome opening into the nave, under fluted pilasters, with Corinthian capitals. The parochial business is transacted by 24 vestrymen, who are elected by the parishioners, and continue in office three years: they choose churchwardens, overseers, and constables, and fix the church-rate, &c., but have no connection with the guardians under the new poor law act. St. John's chapel, a spacious building with a square tower, was erected in 1769, chiefly at the expense of John Thornhill, of Thornhill, Esq., and contains 580 free sittings, the Incorporated Society having granted £200 in aid of the expense: the site, at the head of Barrack-street, was given by Marshall Robinson, of Herrington, Esq. The living is a perpetual curacy; net income, £288. 6. 4.: it is in the patronage of the Bishop, being generally held by the rector of Sunderland, who, after paying the curates' stipends, derives about £280 per annum from his incumbency. Another church was erected in John-street, in Bishop-Wearmouth, in 1829, at an expense of £4879. 7. 9., by grant of the parliamentary commissioners. At Deptford, in Bishop-Wearmouth, is a new place of worship in connection with the Established Church, and another at Ryhope, both in the patronage of the rector; a third, at Hylton Ferry, is in the patronage of the Rev. R. Gray: the two latter have cemeteries. Attached to these chapels are national schools; that of Hylton Ferry has a handsome school-house, erected in 1836. There are two places of worship for Wesleyan, and two for Primitive Methodists, of which a handsome meeting-house for the former was erected in the later English style, in Fawcett-street, in 1836, at an expense of about £3000: in the front, which is supported by buttresses crowned with crocketed pinnacles, is a large window of stained glass, but the interior is entirely surrounded with a gallery, which does not harmonize with the style of the building; it has a good organ. There are also a place of worship each for Baptists; Scotch Burghers; the Society of Friends; Independents, called Bethel chapel, in Villiers-street, a handsome stone-fronted edifice; Methodists of the New Connection; Presbyterians, called St. George's chapel or Scotch church, also in Villiers-street, a handsome Doric structure, with a pediment in front supported by four pilasters; Unitarians, a neat stone-fronted building in Bridge-street; Roman Catholics, and Jews. The Roman Catholic chapel, in Bridge-street, erected in 1836 at a cost of about £4000, is a chaste and beautiful edifice in the early English style, from a design by Mr. Buonomi, of Durham: the interior is of corresponding character, and is embellished with a beautiful altar and altar-piece, a copy of that in Durham cathedral; it is furnished with an excellent organ. A new cemetery is about to be formed in the "Rector's Gill."

A school for instructing and clothing poor girls of the parish was endowed in 1764, under the will of Elizabeth Donnison, widow, with £1500; the management is vested in eight trustees, of whom the rector of Sunderland is one: the property has been vested in the three per cents. and allowed to accumulate, and the interest now amounts to £120. 7. 10. per annum; about 20 poor girls are completely clothed and instructed, and in part boarded; the mistress has a salary of £30 per annum, with a residence built by Mrs. Eliz. Woodcock, in 1827. A school for children of members of the Society of Friends was established, pursuant to the will of Edward Walton, in 1768; out of the income, which amounts to about £36 per annum, £20 is paid to a schoolmaster in Bishop-Wearmouth, who instructs fifteen children. A school of industry in Sans-street, for females, is supported by subscription. A National school, established in Vine-street, in 1822, is supported by the interest of £1000 3½ per cent. stock, given by Mrs. Elizabeth Woodcock in 1823, £20 per annum from Lord Crewe's trustees, and by annual contributions from the Marchioness of Londonderry and the rector: the building consists of two stories, and cost the sum of £1750, defrayed by the National School Society, Mrs. Woodcock, the Diocesan Society, Bishop Barrington, Lord Crewe's trustees, and the rector: residences for the master and mistress were built by Mrs. Woodcock in 1829. In Nicholson-street is a school under the British and Foreign School Society, chiefly supported by subscription; and an infants' school was established in 1833, near St. John's chapel, under the patronage of the rector, for which a handsome building has been erected by subscription. In addition to these are numerous Sunday schools.

The almshouses in Assembly Garth, for the residence of 38 superannuated seamen and their widows, belonging to the "Muster Roll," were purchased in 1750, by the trustees of the "Seamen's Fund," appointed by virtue of an act passed in the 20th of George II., which obliges the masters of all vessels to levy sixpence per month from each sailor, towards this provident institution, to be under the management of fifteen trustees, who are elected annually: upwards of 700 individuals derive benefit from it. The Sunderland and Bishop-Wearmouth Maritime Institution, for ten widows, or unmarried daughters, of master-mariners, was founded and endowed in 1820, by Mrs. Elizabeth Woodcock; each candidate must have passed her 56th year: each inmate receives an annuity of £10 for life. Almshouses in Church-lane, Bishop-Wearmouth, for the maintenance of twelve poor women, were built and endowed by a legacy of £1400, the bequest of Mrs. Jane Gibson, in 1725; the appointment of the inmates is vested in the family of Mowbray. On Wearmouth Green are almshouses for twelve poor persons, erected about 1712, under the will of John Bowes, rector of Bishop-Wearmouth, and endowed in 1725, by Thomas Ogle, with a bequest of £100; the Dean and Chapter of Durham are trustees. In Church-street are almshouses for eight poor widows, who are appointed by the old freemen. The infirmary, comprising also a dispensary, each instituted at different periods, but now combined in one establishment in Durham-lane, was erected in 1822, at an expense of £3000. A Humane Society has been established here for many years: there are five stations on the south and three on the north side of the river, where the drags and other necessary apparatus are kept. A penitentiary has been lately opened. Amongst the several friendly societies are three lodges of freemasons, of which the Phœnix Lodge has a masonic hall in Queen-street, erected in 1785. Here are also numerous benevolent societies for the benefit of the sick and poor of the town and its vicinity. The poor law union of Sunderland comprises 11 chapelries and townships, in the two parishes of Bishop-Wearmouth and Monk-Wearmouth, containing a population of 42,664, according to the census of 1831, and is under the care of 34 guardians. Sunderland confers the inferior title of Earl upon the family of Churchill, Dukes of Marlborough.

SUNDERLAND-BRIDGE, a township, in the chapelry of CROXDALE, parish of ST. ANDREW-AUCKLAND, union of DURHAM, South-Eastern Division of DARLINGTON ward, Southern Division of the county palatine of DURHAM, 3½ miles (S. S. W.) from Durham; containing 283 inhabitants. It is situated between the river Wear and Croxdale water, the former being crossed by a bridge on the great north road. Here is a day and Sunday National school. On the manor of Butterby there are saline and sulphureous springs.

SUNDERLAND, NORTH, a township, in the parish, and Northern Division of the ward of BAMBROUGH, union of BELFORD, Northern Division of the county of NORTHUMBERLAND, 8½ miles (E. by S.) from Belford; containing 860 inhabitants. This township has the North Sea on the east, and possesses a small port, subject to Berwick, from which corn, fish, and lime are exported; of the latter article considerable quantities are burned at the kilns in the neighbourhood. A new church, in the purest Norman style, with a parsonage-house corresponding, has been erected here, which was opened in June 1833: the expense was defrayed by the trustees of Lord Crewe's charities, who also endowed it from the same fund. There is a place of worship for Presbyterians. Seven children are educated for £5 a year, with a house and garden for the master, allowed by the trustees of Lord Crewe's charities; and there is a National School.

SUNDERLAND-WICK, a township, in the parish of HUTTON-CRANSWICK, union of DRIFFIELD, Bainton-Beacon Division of the wapentake of HARTHILL, East Riding of the county of YORK, 2½ miles (S. S. W.) from Great Driffield; containing 35 inhabitants.

SUNDON (*St. Mary*), a parish, in the union of LUTON, hundred of FLITT, county of BEDFORD, 4¾ miles (N. W. by N.) from Luton; containing 408 inhabitants. The living is a discharged vicarage, valued in the king's books at £8. 6. 8.; present net income, £83; patron and impropriator, Sir G. P. Turner, Bart. The church is partly in the decorated style of architecture. A market and fair, formerly held by royal grant in 1316, have been long disused. Here is a Sunday school, supported by subscription.

SUNDRIDGE, a parish, in the union of SEVEN-OAKS, hundred of CODSHEATH, lathe of SUTTON-AT-HONE, Western Division of the county of KENT, 4 miles (W. by N.) from Seven-Oaks; containing 1268 inhabitants. The river Darent flows through the parish, part of which lies below the great ridge of sand hills in the Weald. The manufacture of paper is carried on here. The living is a rectory, in the patronage of the Archbishop of Canterbury, valued in the king's at £22. 13. 4.; present net income, £615. The church has some fine windows in the later English style. The glebe-house has 10 acres of land attached to it. Here are two day and Sunday schools, one of which is endowed with £20 per annum, arising from the interest of £663. 13. 4. invested in government securities, and further supported by voluntary contributions; this school has been long established, but was permanently endowed and placed on the present plan in 1824. The other was built by subscription in 1825, and is supported by the same means. There is a National school for 100 children. H. Hyde, Esq., left £6 per annum for a parochial school. Bishop Porteus resided in this parish, to which he bequeathed £1600 for charitable uses; he was interred in the churchyard. Sundridge gives the English title of Baron to the Duke of Argyll.

SUNK-ISLAND, an extra-parochial district, in the union of PATRINGTON, Southern Division of the wapentake of HOLDERNESS, East Riding of the county of YORK, 20 miles (S. E. by E.) from Kingston-upon-Hull; containing 242 inhabitants. This island has been gradually recovered from the Humber; a century ago it comprised only 800 acres, but it now contains 5000, in a high state of cultivation, and more is expected to be embanked within a very short period. It was originally two miles from the opposite shore, and vessels formerly passed through the channel, which is now so narrow as to be crossed by a bridge to the main land. There is a small chapel: the living is a perpetual curacy, in the patronage of the Crown; net income, £250. There is also a place of worship for Wesleyan Methodists.

SUNNINGHILL, (*ST. MICHAEL*), a parish, in the union of WINDSOR, hundred of COOKHAM, county of BERKS, 6 miles (S. S. W.) from New Windsor; containing 1520 inhabitants. The living is a vicarage; net income, £328; patrons and impropriators, Master and Fellows of St. John's College, Cambridge. The church has been lately rebuilt, at the expense of £3000: in the churchyard is a yew tree, supposed to have been planted before the Conquest. A school-room has been appropriated as a place of worship for Wesleyan Methodists. A National school has been established, and is supported by an endowment amounting to £43 per annum, of which £40 per annum was given by Augustus Schutz, Esq., and £3 by Miss Dawson, and by subscription. Two chalybeate springs, in the garden of an inn called Sunning Wells, were formerly in great repute, and adjoining them is a room where public breakfasts have been given: the water has properties similar to those of Tunbridge Wells. At a place called Bromehall there was formerly a small convent of Benedictine nuns, founded before the reign of John, which was deserted by the sisters in 1522. The noted race-course of Ascot Heath is situated in this vicinity.

SUNNINGWELL (*ST. LEONARD*), a parish, in the union of ABINGDON, hundred of HORMER, county of BERKS, 2½ miles (N.) from Abingdon; containing 339 inhabitants. The living is a rectory, valued in the king's books at £12. 14. 7.; present net income, £318; patron, Sir. G. Bowyer, Bart. The church is an ancient structure, of a singular form. There is a chapel of ease at Kennington. Here are two Sunday schools, one of which is supported by the minister, the other by subscription.

SURFLEET (*ST. LAWRENCE*), a parish, in the union of SPALDING, wapentake of KIRTON, parts of HOLLAND, county of LINCOLN, 4 miles (N.) from Spalding; containing 871 inhabitants. The living is a perpetual curacy, valued in the king's books at £11; present net income, £65; patrons and impropriators, J. and T. Pickworth, Esqrs. The church is partly in the later style of English architecture, and partly of earlier date, with a tower and spire. Here are two endowed schools. There is a canal, called the Glen, by which the waters of Pinchbeck are conveyed to the Welland river; and another, termed the Grand Sluice, that conveys the waters of the fen to Boston. This parish contains one of the largest heronries in England.

SURLINGHAM (*ST. MARY*), a parish, in the union and hundred of HENSTEAD, Eastern Division of the county of NORFOLK, 5½ miles (E. S. E.) from Norwich; containing 399 inhabitants. The living is a vicarage, with the perpetual curacy of St. Saviour's annexed, valued in the king's books at £6. 13. 4.; present net income, £40; patrons, Bishop of Norwich and Rev. W. Collet, the latter of whom is the impropriator and incumbent. A National school is partly supported by subscription. Here is a ferry across the Yare, which is much frequented.

SURRENDRAL, a tything, in the parish of HULLAVINGTON, hundred of CHIPPENHAM, Chippenham and Cale, and Northern, Divisions of the county of WILTS, 5¾ miles (S. W.) from Malmesbury; containing 38 inhabitants.

SURREY, an inland county, bounded on the north by the river Thames, which separates it from Middlesex and the south-eastern extremity of Buckinghamshire; on the north-west, by Berkshire; on the west, by Hampshire; on the south, by Sussex; and on the east by Kent. It extends from 51° 5′ to 51° 31′ (N. Lat.), and from 3′ (E. Lon.) to 51′ (W. Lon.); and comprises an area of 758 square miles, or about 485,120 statute acres. The population amounts to 486,334. The most ancient British inhabitants of this district, of whom we have authentic information, were the *Segontiaci*, or, as they are called by Ptolemy, the *Regni*, a people who had been expelled from Hampshire by the invading Belgæ. Cæsar, in his exploratory invasion of Britain, crossed the north-eastern part of Surrey, from the county of Kent to the Thames, which he is supposed to have passed at a place now called Cowey Stakes, at Walton-on-Thames, into the territory of Cassivellaunus, though the Britons endeavoured to prevent his passage by driving stakes into the bed of the river. Under the Roman dominion Surrey was included in the division called *Britannia Prima*. On the complete establishment of the Saxon kingdom of Wessex, it appears to have included the greater part of this county. In the year 568, Ethelbert, in defence of his own kingdom, having invaded the territories of Ceawlin, king of Wessex, a great battle was fought between them at Wimbledon, in which the former was defeated with considerable loss: this was the first battle fought between the Saxon kings. The county suffered severely from the ravages of the Danes, who entered it in 852, after sacking London, but were defeated with great slaughter at Ockley, near its southern border, by Ethelwulph and his son Ethelbald.

Surrey is included in the diocese of Winchester, and province of Canterbury, excepting the exempt deanery of Croydon, which contains nine parishes, and is in the peculiar jurisdiction of the Archbishop of Canterbury. Under the new ecclesiastical arrangements provided by the act of the 6th and 7th of William IV., cap. 77, the parish of Croydon will remain in the diocese of Canterbury, to which also the parish of Addington, and the district of Lambeth Palace, will be annexed; and the borough of Southwark, and the parishes of Battersea, Bermondsey, Camberwell, Christchurch, Clapham, Lambeth, Rotherhithe, Streatham, Tooting-Graveney, Wandsworth, Merton, Kew, Richmond, St. Mary Newington, Barnes, Putney, Mortlake, and Wimbledon, will be annexed to the diocese of London. It forms an archdeaconry, in which are the deaneries of Ewell, Southwark, and Stoke: the total number of parishes is 141. For purposes of civil government it is divided into the hundreds of Blackheath, Brixton, Copthorne and Effingham, Elmbridge, Farnham, Godalming, Godley, Kingston, Reigate, Tandridge, Wallington, Wokeing, and Wotton, all of them having first and second divisions, except Brixton, which is divided into east and west, and Farnham, which has no division. It contains the borough and market-towns of Southwark, Guildford, and Reigate; the newly enfranchised borough of Lambeth, with the populous suburban parishes of Rotherhithe, Bermondsey, Newington, Camberwell, Clapham, and Battersea, of which the first two are included within the electoral limits of the borough of Southwark, and Newington and the greater part of Camberwell in that of Lambeth; the market-towns of Chertsey, Croydon, Dorking, Farnham, Godalming, Haslemere, and Kingston; and the large and elegant village of Richmond. By the act of the 2nd of William IV., cap. 45, this county has been divided into two portions, called the Eastern and Western divisions, each sending two members to parliament: the place of election for the former is Croydon, and the polling-places are Croydon, Reigate, Camberwell, Kennington Common, Bermondsey, Wandsworth, and Kingston; that for the latter is Guildford, and the polling-places, Guildford, Dorking, Chertsey, Farnham, Godalming, Epsom, and Chobham. Two representatives are returned to parliament for each of the boroughs, except Reigate, which, under the above-named act, sends only one. The old boroughs of Bletchingley, Gatton, and Haslemere have been disfranchised. Surrey is included in the Home circuit: the lent assizes are held at Kingston, and the summer assizes at Guildford and Croydon alternately: the winter assizes have been discontinued since the establishment of the Central Criminal Court held at the Old Bailey. The county gaol is in Horsemonger-lane, in the parish of Newington. The winter quarter sessions are held at the New Sessions House, Newington; the spring, at Reigate; the summer, at Guildford; the autumn, at Kingston. According to the earliest authentic accounts, Surrey had always a sheriff of its own until the beginning of the reign of John, when it was placed under the same shrievalty with Sussex; and though, during the reigns of succeeding monarchs, it was occasionally under a distinct sheriff, yet the regular appointment of a separate sheriff for it did not commence until 1615.

The form of the county is nearly oblong, except that the northern border is rendered extremely irregular by the devious course of the Thames. The scenery, celebrated for its beauty, possesses also great variety, presenting in some parts wild and naked heaths, which form a powerful contrast with the adjoining highly cultivated and ornamented districts. The surface, for the most part, is gently undulating, excepting the Weald, a district of about 30 miles in length, and varying from three to five in breadth, which extends along the whole southern border, and forms, with the Wealds of Kent and Sussex, an immense plain, the flat surface of which is of very inferior elevation: some of the hills, however, rise to a very considerable height, and command rich and extensive views. The middle of the county is crossed from east to west by the Downs, which rise with a gentle acclivity from the north, but on the south are broken into precipitous cliffs of great height and romantic irregularity. Southward of the Downs rise the hills that overhang the Weald, in the vicinities of Oxted, Godstone, Reigate, and Dorking. Approaching the western side of the county, this range becomes of greater extent, and near Wonersh, Godalming, and Pepper-Harrow, is covered with rich woods, and intersected by pleasing valleys, watered by streams tributary to the Wey; the whole forming one of the most picturesque portions of the county. The largest tracts of the very extensive heaths lie in the western part: from Egham, on the banks of the Thames, south-south-westward as far as the village of Ash, the district consists, with little exception, of heath and moor; as also does that extending in a transverse direction from Bagshot, on the north-western confines by Chobham and Byfleet, to Cobham, Ripley, and Oatlands: the whole south-western angle is of the same barren character, from Haslemere to Farnham, in one direction, and from Elstead to Frensham in the other.

The soils, which are extremely various, are by no means so clearly discriminated as in many other counties, the different species lying in small patches much intermixed: they may, however, be reduced under the four general heads of clay, loam, chalk, and heath. The most extensive tract bearing a uniform character is that denominated the Weald of Surrey, where the surface consists of a pale, cold, retentive clay, upon a subsoil of the same nature, in which iron is found: still deeper occurs a white clay of a slaty texture, the laminæ of which are extremely thin. In some places, particularly on the northern border of the Weald, where the soil is more of a loamy character, rag-stone is the prevailing substratum: the difficulties experienced in the cultivation of this soil are extremely great. Northward of it lies a district of sandy loam, stretching across the whole county, and forming the range of hills overlooking the Weald: this soil is every where of great depth, though variable in colour and fertility, and rests on a sandstone, veined with oxyde of iron; the richest portion of it lies around Godalming. Although this district, in the eastern part of the county, is in few places more than half a mile broad, yet, proceeding westward, in the neighbourhood of Godalming, it expands to a breadth of five or six miles. The chalky soil of the Downs adjoins this on the north, the most fertile portions of it being a pure hazel loam of various depths, while the substratum of the whole is chalk rock. A less friable soil, mixed with flints, intervenes between Croydon and Godstone, and is found on all the declivities of the chalk hills. Along the elevated summit of the Downs, particularly about Walton and Headley, is a large extent of heathy land, consisting of ferruginous and barren sand. A peculiar kind of clay, of a blueish-black colour, here called "black land," extends in a long narrow tract along the southern side of the chalk hills, from the vicinity of Reigate, into the county of Kent. Proceeding northward from the eastern extremity of the Downs, a variety of soils occurs, chiefly strong clays, interspersed with tracts of sandy loam, and these again with patches of gravel, which continue nearly to Dulwich, whence, to the north-eastern extremity of the county, a strong pure clay occupies the whole tract. Northward from the chalk, at Banstead Downs, is also a long tract of clay, continuing by Sutton, Morden, and the eastern side of Merton, to the sandy loams of Wimbledon, Putney heath, and Mortlake. A similar soil is found on descending northwards from any part of the Downs, between Banstead and Clandon, near Guildford; but the breadth of clay separating the sandy loam from the chalk continues decreasing as it proceeds westward. The rich sandy soil lying between the clay and the Thames is intermingled, especially on the banks of the Mole and the Wey, with loams of various qualities, and even with clay: the subsoil of nearly the whole of these northern sands is a yellow silicious gravel. North-westward from Guildford, sandy loams first occur in the vicinity of Stoke, and afterwards strong retentive clays, which extend beyond Warplesdon, where they unite with the heaths. The soil on the banks of the Thames consists partly of a sandy loam and partly of a rich strong loam.

The proportion of arable land greatly exceeds that of meadow and pasture: the drill husbandry is general in the western part of the county, about Bagshot, Esher, Send, Cobham, Ripley, &c. The corn and pulse crops are wheat, barley, oats, beans, and peas: wheat is cultivated to a great extent; the barley, being of a superior quality, is generally appropriated to the making of malt. The field varieties of peas and beans are extensively cultivated, as an agricultural crop, in most parts of the county, especially on the chalk hills, while the finer sorts are grown in the vicinity of the metropolis, and on the sandy loams in the vale of the Thames, chiefly for the supply of the London market. The cultivation of turnips and cabbages is now carried on to a great extent, partly for the supply of the metropolitan markets, and partly for the consumption of cattle. Great quantities of carrots are grown in the northern part of the county, to the west of the river Mole, and parsnips on the rich deep lands in the district lying between Wandsworth and Kingston, for the London markets. Red clover has long been in general cultivation; trefoil, white clover, and rye-grass are occasionally sown; and large tracts of chalky soil are occupied by sainfoin, most of which is made into hay. The Farnham hops are cultivated to a considerable extent on the border of Hampshire, and have long been celebrated for their excellent quality, always bringing a higher price than any other hops in the kingdom: the entire extent of hop plantations in the county occupies at present 1170 acres. Woad flourishes on the chalk hills about Banstead Downs, where it is generally sown with barley. By far the most extensive and valuable tracts of meadow are situated along the banks of the Thames, in the north-western part of the county, and on the banks of the Wey, near Godalming: there is also a small extent of meadow in its north-eastern angle, near the metropolis. Of dairy pastures there are scarcely any: the greatest extent lying together is on the estate of the Duke of Norfolk, in the parishes of Newdigate and Charlwood, on the southern border. Large quantities of manure of different kinds are carried from the metropolis into the northern parts of Surrey; in the more remote districts lime and chalk are procured from the quarries and works on both sides of the range of chalk downs.

A great variety of cattle is kept in this county. The large Wiltshire, the South Down, and the Dorsetshire, are the principal breeds of sheep, the last-named being kept chiefly for the rearing of early house lambs, for which Surrey was formerly much celebrated, though the number sent to the London market has greatly decreased: many grass lambs are prepared for the butcher in April and May. Great numbers of hogs are fed at the distilleries and starch-manufactories, whither they are brought chiefly from Berkshire, Shropshire, and the East Riding of Yorkshire; those from the first-named district are preferred. Budgwick, on the border of Sussex, is remarkable for a breed of swine that fatten to an enormous size. Numerous flocks of geese are kept on the commons, especially in the Weald: the Dorking breed of fowls is much celebrated. The quantity of garden ground employed in raising vegetables for the London market is very considerable: the parishes of Mortlake and Battersea are particularly distinguished for the cultivation of asparagus of excellent quality; in the latter much ground is also occupied in the cultivation of vegetables for seed. It is considered that a greater extent of land is employed in the cultivation of medicinal plants in this county than in any other in England: those grown in the largest quantity are peppermint, lavender, wormwood, chamomile, aniseed, liquorice, and poppies, which, with a few others for the druggists and perfumers, occupy a large portion of land in the parishes of Mitcham and Tooting.

The part most remarkable for its woods is the Weald, on the southern side of the county, which, there is every reason to believe, was formerly wholly covered with wood, much of which has been cleared off at no very remote period. The coppices consist chiefly of oak, birch, ash, chesnut, sallow, hazel, and alder; and their produce is formed into hoops, poles for the hop plantations, hurdles, and faggots; great quantities are also made into charcoal, for gunpowder and other purposes. The woodlands in the other parts of the county, particularly on the chalk hills, have a greater proportion of coppice, and fewer timber trees, than those in the Weald. The box in this county, and chiefly on Box hill, near Dorking, attains a considerable size; its wood is bought principally by the mathematical instrument makers, and by the turners in London and Tonbridge. Surrey is noted for the great number of yew-trees that are scattered in a wild state over its chalk hills, and for the size which some of those that have been artificially planted have attained. Besides forming a considerable portion of the underwoods, the birch flourishes on the heaths: great quantities of brooms are made of its small branches, and sold chiefly at Southwark. Extensive plantations of fir and larch have been made on the heathy lands in the western part of the county. In the western and northern parts the osier and willow are much cultivated, particularly about Byfleet, Chertsey, &c.; the common furze is also cultivated in different places for fuel.

Under the early Norman sovereigns a large portion of Surrey was reserved as part of the royal demesne. By Henry II. the limits of Windsor Forest were gradually extended until he had afforested nearly the whole of this county; but Richard, his son and successor, in the first year of his reign, consented to disafforest all the county lying eastward of the river Wey, and southward of Guildford Down, his charter to that effect having been confirmed by King John. That part still enclosed in the forest, according to the provisions of this charter, was called the Bailiwick of Surrey, being exempt from the jurisdiction of the sheriff, and subject only to that of its own bailiff: it contained the parishes and townships of Ash, Bisley, Byfleet, Chobham, Horsell, Purbright, Pyrford, Stoke, Tongham, Wanborough, Worplesdon, Windlesham, and Woking: within the same district also lay Chertsey, Egham, and Thorpe, but these being estates of the abbey of Chertsey, were not subject to the bailiff's jurisdiction. Notwithstanding the charter of forests granted by King John, this part of the county remained in forest until after the granting of another charter by Henry III., in the ninth year of his reign, which disafforested the whole, excepting only the park of Guildford. Edward I. and Edward II. attempted to set aside this grant, but without effect; and Edward III., in the first year of his reign, fully confirmed it. The royal pretensions were revived in the 7th of Charles I., but without success; and from that period, the district known, since the time of Richard I., as the Bailiwick of Surrey, has been regarded as only a purlieu of the forest of Windsor; and the king has still a right and property over the deer escaping into it, which may be destroyed by none, except the owners of woods, or lands, in which they may be found, which are especially exempted from the operation of the forest laws: it is, however, so far free and open to all owners of land within its limits, that, under certain restrictions, they may chase and kill any of the deer actually found therein. For the better preservation of the deer escaping out of the forest into this purlieu, the king has a ranger, appointed by letters patent, whose office it is to chase them back again, and to whom, in his official capacity, belongs Fangrove Lodge, near Chertsey.

It appears surprising, that a county so near the metropolis should contain so large a quantity of waste land. About the commencement of the present century it was computed, that one-sixth lay in a wild and uncultivated state; and though this extent has been greatly lessened by numerous enclosures, there yet remain in heaths about 48,000 acres, and in commons about 17,000. The principal heaths are, that of Bagshot, on the western border of the county, which, with Romping Downs, and the wastes of Purbright, Windlesham, &c., contains upwards of 30,000 acres, almost entirely covered with short heath; Frensham, Thursley, and Witley heaths, comprising together 5800 acres; Hindhead heath, which, with the last-mentioned tracts, is situated on the south-western confines of Surrey, and occupies an extent of upwards of 3000 acres; the heaths of Farnham and Crooksbury, containing about 3700 acres; Blackheath, about four miles south of Guildford, comprising about 1000 acres; and Headley heath, containing about 900 acres. The most extensive commons still remaining unenclosed are, those of Leith hill and Hurtwood, containing upwards of 3000 acres; those of Walton, Kingswood, and Banstead, occupying about 1500 acres; and those of Epsom, Leatherhead, and Ashtead, extending over 1200 acres. Those in the more immediate vicinity of the metropolis are Wimbledon and Putney, containing about 1000 acres; Barnes common about 200; Wandsworth common, about 350; Battersea and Clapham commons, together about 50; Streatham common, about 250; and Kennington common, about 20.

A sand-stone, commonly called rag-stone, containing oxyde of iron, abounds along the line of junction of the Weald with the sand hills, which skirt that tract on the north: the oxyde of iron sometimes prevails so greatly, as to have been formerly worked and smelted as iron-ore; but in consequence of the high price of fuel, these works have long been totally abandoned. Stone of a similar kind to the above is found in smaller quantities about Send and Chobham; and iron ore appears in the sand in the vicinities of Puttenham and Godstone. At Purbright, and in many parts of the surrounding country, are found loose blocks of stone bearing a strong resemblance, both in quality and appearance, to those termed the Grey Wethers, on the downs of Berkshire and Wiltshire. In the neighbourhoods of Godstone, Gatton, Merstham, Reigate, and Bletchingley, are extensive quarries of a peculiar kind of stone, which, when first dry, is soft and unable to bear the action of a damp atmosphere; but after being kept under cover for a few months, its texture becomes so firm and compact, that it can resist the heat of an ordinary fire, and is, in consequence, in great demand for fire-places. On the white hills near Bletchingley this stone is softer than elsewhere, and is now chiefly dug for the glass-manufacturers, who, by means of it, have been enabled to produce plate-glass of much larger dimensions than they formerly could: great quantities are taken by water to Liverpool and the north of England. Limestone of a blueish-grey colour, containing a very small portion of flint, is extensively quarried near Dorking, and affords lime of great purity and strength, particularly serviceable in works under water, having been employed in the construction of the West India and Wapping docks: limestone is also dug and burned at Guildford, Sutton and Carshalton. Some of the most extensive chalk-pits are those at Croydon, Sutton, Epsom, Leatherhead, Bookham, Effingham, Horsley, Clandon, Stoke, Guildford, and Puttenham, on the northern side of the Downs; and those of Godstone, Catterham, Reigate, Merstham, Buckland, and Betchworth, on the southern side. The sand about Tandridge, Reigate, and Dorking, is in great request for hour-glasses, writing, and a variety of other purposes; that about Reigate, more especially, is considered unequalled in the kingdom for purity and colour; great quantities are sent to London. Fullers' earth is found in very extensive beds about Nutfield, Reigate, and Bletchingley, to the south of the Downs; and some, though of an inferior quality, near Sutton and Croydon, to the north of them: it is of two kinds, blue and yellow, the latter, which is the most valuable, being chiefly employed in fulling the finer cloths of Wiltshire and Gloucestershire, while the former is sent into Yorkshire, for the coarser manufactures: this earth is sent in waggons to the railway commencing at Merstham, along which it is conveyed to the Thames, where it is shipped. Brick-earth is found in most parts of the county, though not of the finest quality: at Nonsuch, in the parish of Cheam, however, there is a particularly valuable bed, from which a kind of brick, capable of resisting intense heat, is made.

Though Surrey cannot be regarded as a manufacturing county, yet its vicinity to the metropolis, and the convenience of its streams for the working of mills, have caused several manufactures of importance to be established in it. On the Wandle is situated a great number of flour, paper, snuff, and oil mills, besides mills for preparing leather and parchment, and for grinding logwood: upon its banks also, chiefly in the parishes of Croydon and Mitcham, are large calico, bleaching, and printing works. This river, which is usually not more than three feet deep and eight broad, is remarkable for turning 90 mills in a course of only ten miles. On the Mole are several flour-mills, some iron-mills at Cobham, and several flatting-mills at Ember. There are extensive powder-mills near Malden, to the north of Ewell; and several paper-mills on the different tributary branches of the Wey. At Godalming are considerable factories for the weaving of all kinds of stockings and the making of patent fleecy hosiery; at the same place are also establishments for the combing of wool, and the manufacture of worsteds, blankets, tilts, and collar-cloths. At Stoke, near Guildford, is a sawing-mill for staves, ship-pins, &c.; and at Mortlake a manufacture of delft and stone ware. In the neighbourhood of London, particularly at Battersea and Lambeth, are several distilleries on a very extensive scale; and at the latter place, manufactories for patent shot and artificial stone: the manufactures carried on in Southwark, and its immediate vicinity, are of different kinds, being chiefly such as are connected with the varied trade of the port of London. This north-eastern extremity of the county has a very large share in the vast commerce of the port of London; and, besides its numerous wharfs and quays on the banks of the Thames, it possesses various large commercial docks, among which may be noticed, more particularly, the Grand Surrey docks (Outer and Inner), connecting the Grand Surrey canal with the river.

The principal rivers are the Thames, the Wey, and the Mole. The Thames, forming the entire northern boundary of the county, first touches it at its north-western extremity, above Egham, whence it takes a south-easterly course, by the town of Chertsey, to the confluence of the Wey, where it assumes an irregular north-easterly direction to the village of Kew, passing by the town of Kingston (to which and the village of Thames-Ditton it makes an extensive sweep on the south-east), the village of Petersham, and the bold heights of Richmond: at Kew it takes a winding easterly course, which it pursues by Mortlake, Barnes, Putney, Wandsworth, and Battersea: then forming Chelsea Reach, it pours its majestic stream through the spacious arches of the six magnificent bridges which connect the cities of London and Westminster with the borough of Southwark and the southern suburbs of the metropolis, and immediately eastward of the last of them forms the pool or harbour of London; between Rotherhithe and Deptford it quits Surrey, as it approaches the superb pile of Greenwich hospital: above London bridge it is navigable for barges of large burden during the whole of its course past this county; the tide flows to Richmond bridge. The Wey enters Surrey on its south-western border, near Frensham, and becomes navigable at Godalming; it falls into the Thames at Harn Haw, near Weybridge. The Mole is famed for, and is supposed to derive its name from, the circumstance of a part of its waters pursuing a subterraneous passage; this is occasioned by the porous and cavernous nature of the soil over which the river runs during several miles of its course below Dorking: when its waters are at their ordinary height, no particular irregularity in its stream is here observable; but in seasons of drought its current is wholly carried through the swallows, as the subterraneous passages are called, and its ordinary channel, similar to that of any other river of the same size, is left dry, except here and there a stagnant pool: by the bridge at Thorncroft it rises again, and thenceforward the current is uninterrupted.

Under the head of canals it may be proper to observe, that the navigation of the Wey is artificial, and has locks upon it, which are supposed to have been the first constructed in the kingdom; the navigable channel also is, in some places, wholly separate from the natural course of the river: the bill for the formation of this navigation up to Guildford was passed in 1651, but the work was not carried into execution until towards the close of the century: it was extended to Godalming in 1760, and by its means timber, flour, and paper, are now exported to London. The Basingstoke canal enters Surrey from Hampshire near Dradbrook, crossing the river Loddon, from which it derives its chief supply of water; from Dradbrook to its junction with the navigable channel of the Wey, a distance of about fifteen miles, it has a fall of 195 feet; from Hook common there is a branch to Turgis Green, six miles long, and on the same level: the act for the formation of this navigation was passed in 1778, and it was completed in 1796: the chief articles of traffic upon it are timber, corn, and coal. The Grand Surrey canal, the act for constructing which was obtained in 1801, commencing a little to the west of the road from London to Camberwell, is carried eastward across the Kent road, and then northward to the Grand Surrey docks, through which it communicates with the Thames. The Surrey and Sussex canal forms a junction between the navigable channel of the Arun in Sussex, and that of the Wey, a little above Guildford. The Croydon canal, the act for making which was obtained in 1801, was sold to the London and Croydon Railway Company, who have formed their line of railway along the greater portion of its bed. This line is continued by the Brighton railway, now in progress, which proceeds to Reigate, where the South-Eastern railway branches from it to Dovor: the Brighton railway quits the county at Black Corner. The London and South-Western railway crosses the northern part of the county, which it quits a little beyond Purbright.

Surrey contained the Roman station of *Noviomagus*, situated at Woodcote, near Croydon; besides two others, supposed to have been respectively at Kingston-on-Thames and Walton-on-the-Hill. It was traversed by the roads leading from the southern and eastern coasts to the capital, which met in St. George's Fields, near Southwark, the principal of them being the Ermin-street, which ran nearly parallel to, and at a very short distance to the eastward of, the present turnpike road through Clapham, Tooting, Merton, Ewell, and Epsom to Ashtead, thence proceeding, nearly in a southerly direction, to Dorking, where it took a westerly course, about a mile southward of Guildford, to Farnham, beyond which town it soon entered Hampshire; the Stane-street, which, branching from the Ermin-street at Dorking, proceeded southward, through the parish of Ockley, into Sussex; and another Stane-street, which from the metropolis passed through Streatham, Croydon, Coulsdon, Catterham, and Godstone, also into Sussex; and the Watling-street, from Dover, which crossed its north-eastern extremity to London. Remains of ancient encampments, supposed to be Roman, may be seen at Bottle hill, in the parish of Warlingham; on Castle hill, in that of Hascomb; near Chelsham; on Holmbury hill, in the parish of Ockley; at Ladlands and Oatlands; and on St. George's hill, near Walton-on-Thames. Foundations of Roman edifices have been discovered at Walton-on-the-Hill, and on Blackheath, in the parish of Albury, both surrounded by intrenchments. Other remains of buildings, supposed to be of the like origin, have been traced in the vicinities of Wollington, Carshalton, and Beddington; and, near Kingston-on-Thames, Roman sepulchral urns, coins, earthenware, and foundations of buildings, have been found. Many Roman coins and pavements have also been found in St. George's Fields, Southwark. Different ancient encampments, the date of the formation of which is uncertain, exist in various places, besides those above mentioned; that at the south-western angle of Wimbledon common is supposed by Camden to mark the site of the battle fought in 568; while those of Hanstie Bury, on a projection of Leith hill, and War Coppice hill, in the parish of Catterham, are attributed to the Danes.

The number of religious houses of all denominations, prior to the general dissolution, was about 28. Remains yet exist of the abbeys of Chertsey and Waverley, and of the priories of Merton, Newark or Newstead, and Southwark. There are extensive remains of the castles of Farnham and Guildford. The most remarkable ancient residence is Lambeth palace: there are also remains of the ancient palace of the Archbishops of Canterbury at Croydon. Few counties in England can vie with Surrey in the number and elegance of its seats of nobility and gentry, and certainly none not exceeding it in size: this circumstance is owing chiefly to its vicinity to the metropolis, and the superior pleasantness of its scenery. Of its parks, the royal one of Richmond, to the south-east of that elegant place of resort, is the most extensive, being nearly eight miles in circumference, and containing upwards of 2250 acres. The mineral springs are numerous, and were formerly in high repute and much frequented, more particularly those of Epsom. On the northern side of the chalk hills, and in the valleys by which they are traversed, in the eastern part of the county, copious streams of water, in the shape of remarkably powerful springs, provincially called bourns, are periodically discharged. In sinking wells on the chalk it is often found necessary to bore to the depth of 300 feet; and the thick beds of clay lying in the north-eastern part of the county must, for the same purpose, be bored entirely through, frequently to an equal depth. There are several remarkable excavations, of unknown date, in the chalk hill upon which Guildford castle stands. Surrey gives the inferior title of Earl to the family of Howard, Dukes of Norfolk.

SUSSEX, a maritime county, bounded on the west by Hampshire, on the north by Surrey, on the north-east and east by Kent, and on the south by the English Channel. It extends from 50° 44′ to 51° 9′ (N. Lat.), and from 50′ (E. Lon.) to 57′ (W. Lon.), and comprises an area of upwards of 1463 square miles, or about 936,320 statute acres. The population amounts to 272,340. At the period of the invasion of Britain by the Romans, Sussex formed part of the territory of the Regni. The reduction of this part of the island was effected by Flavius Vespasian, who was commissioned by the Emperor Claudius, about the year 47, to establish the Roman dominion in the maritime provinces, which he accomplished without much difficulty, and fixed his head-quarters near the site of the present city of Chichester. This territory was included in the division called *Britannia Prima*. No particular mention of it occurs in history until after the departure of the Romans from Britain, when, in 477, a Saxon chieftain, named Ælla, landed, with his three sons and a considerable number of followers, at West Wittering, a village about eight miles south-west of Chichester: they soon made themselves masters of the adjacent coasts, but were too weak to penetrate into the country, which was vigorously defended by its inhabitants. Hostilities appear to have been carried on for several years between Ælla and the Britons, the former occasionally receiving reinforcements; and, in 485, a sanguinary but indecisive battle was fought near *Mecreadesbourne*, in the vicinity of Pevensey. Ælla's forces having, however, been recruited by fresh arrivals of his countrymen, he undertook, in 490, the siege of *Anderida*, the capital of the Regni (the precise situation of which has not been ascertained), and at last succeeded in taking it by assault; as a punishment for the obstinacy of its defenders, he ordered them all to be put to the sword. From this period may be dated the foundation of the kingdom of the South Saxons, called in Saxon *Suth Seaxe*, of which Sussex is a contraction. Ælla, on the death of Hengist, founder of the Saxon kingdom of Kent, became the most influential of the Saxon chieftains in Britain, which he continued to be until his death, in 504 or 505. Cissa, the only surviving son of Ælla, succeeded his father in the government of the South Saxons, and employed much of his time and treasure in rebuilding and improving the capital of his kingdom, to which he gave the name of *Cissa-ceaster*, now Chichester. About the year 950, Adelwalch succeeded to the throne of Sussex, and was attacked, vanquished, and made prisoner by Wulfhere, monarch of the more powerful kingdom of Mercia. Having at the court of the latter embraced the Christian religion, Adelwalch was reinstated in his dominions, and made every exertion to propagate the same faith among his subjects, receiving into his dominions Bishop Wilfrid, who had been expelled from Northumbria, at the same time assigning the peninsula of Selsea as his abode, and granting to him and his companions that and other lands for their maintenance. Ceadwalla, a prince of the blood royal of Wessex, having failed in an attempt to usurp the supreme authority in that kingdom, fled to the great forest of *Anderida*, which occupied the Weald of Sussex, Surrey, and Kent, where he succeeded in maintaining himself for some time at the head of a band of freebooters: Adelwalch attacked and expelled him from his territories; but Ceadwalla, having undertaken an expedition against Kent, which proved unsuccessful, in his retreat from that kingdom, again encountered Adelwalch, whom he defeated and slew. On the death of the king, Berthun and Anthun, two South Saxon nobles, rallied their countrymen around them, and compelled the invader, Ceadwalla, to retire. The latter, however, on the death of the reigning monarch of Wessex, soon after succeeded peaceably to the throne of that kingdom, and renewing his contests with the South Saxons, entered their country with a powerful army, defeated them, and made great devastations throughout their whole territory: the final subjugation of Sussex was, however, left for Ceadwalla's successor, who, in 728, united it to his other dominions. Bishop Wilfrid returned to Northumbria about the year 658, and after his departure it appears that the ecclesiastical affairs of Sussex were under the government of the bishops of Winchester, until the year 711, when Eadbert, abbot of Selsea, was appointed bishop of the South Saxons, and for more than three succeeding centuries Selsea was the episcopal see, until it was removed to Chichester, in the reign of the Conqueror, about the year 1082.

Sussex is co-extensive with the diocese of Chichester, in the province of Canterbury, and is divided into the two archdeaconries of Chichester and Lewes, the former containing the deaneries of Arundel, Boxgrove, Chichester, Midhurst, and Storrington, and locally that of Pagham; the latter those of Dallington, Hastings, Lewes, and Pevensey, and locally that of South Malling: all the parishes comprised in the exempt deaneries of Pagham and South Malling, with those of All Saints at Chichester, and St. Thomas in the Cliffe at Lewes, are in the peculiar jurisdiction of the Archbishop of Canterbury; the total number is 300. The great civil divisions are called *rapes*, a term peculiar to this county: they are six in number, *viz.*, Arundel, comprising the hundreds of Arundel, Avisford, Bury, Poling, Rotherbridge, and West Easwrith; Bramber, comprising those of Brightford, Burbeach, East Easwrith, Fishergate, Patching, Singlecross, Steyning, Tarring, Tipnoak, West Grinstead, and Windham and Ewhurst; Chichester, those of Aldwick, Bosham, Box and Stockbridge, Dumpford, Easebourne, Manhood, and Westbourn and Singleton; Hastings, those of Baldslow, Battle, Bexhill, Foxearle, Goldspur, Gostrow, Guestling, Hawkesborough, Henhurst, Netherfield, Ninfield, Shoyswell, and Staple; Lewes, those of Barcomb, Buttinghill, Dean, Fishergate, Holmstrow, Lewes, Poynings, Preston, Street, Swanborough, Whalesbone, and Younsmere; and Pevensey, those of Alciston, Bishopstone, Burley-Arches or Burarches, Danehill-Horsted, Dill, Eastbourne, East Grinstead, Flexborough, Hartfield, Longbridge, Loxfield-Dorset, Loxfield-Pelham, Ringmer, Rotherfield, Rushmonden, Shiplake, Totnore, and Willingdon, and the lowey of Pevensey. It contains the city and port of Chichester; the following members of the cinque-ports, *viz.*, Hastings, Rye, Seaford, and Winchelsea, all which have markets except Seaford; the borough, market, and sea-port towns of Brighton and Horsham, the former of which has been lately enfranchised; the borough and market-towns of Arundel, Lewes, Midhurst, and Shoreham; the market-town and sea-port of Hastings; and the market-towns of Cuckfield, East Grinstead, Hailsham, Little Hampton, Petworth, Steyning, and Worthing. Under the act of the 2nd of William IV., cap. 45, the county has been divided into two portions, called the Eastern and Western divisions, each sending two members to parliament: the place of election for the former is Lewes, and the polling-places are Lewes, East Grinstead, Battle, Mayfield, Brighton, Hastings, Rye, and Cuckfield; that for the latter is Chichester, and the polling-places, Chichester, Steyning, Petworth, Horsham, Arundel, Midhurst, and Worthing. Two citizens are returned to parliament for the city of Chichester; two barons for Hastings, and one for Rye, the latter having been deprived of one by the act above named, by which also Seaford and Winchelsea were entirely disfranchised; and two burgesses for each of the boroughs, except Midhurst, Horsham, and Arundel, which now return only one each. Bramber, East Grinstead, and Steyning have also been disfranchised. This is one of the counties forming the Home Circuit: the Lent assizes are held at Horsham, and the summer and winter assizes at Lewes: the county gaols are at Lewes and Horsham, a portion of the former being used as the county house of correction. The quarter sessions are held at Petworth, Horsham, and Chichester, for the western division, and at Lewes for the eastern.

The most remarkable feature in the surface and scenery is the bold and open range of chalk hills, called the South Downs, extending into this county from Hampshire, and stretching in nearly an easterly direction for the greater part of its length, gradually approaching the sea: their northern declivity is precipitous, but on the south their descent is gradual, except in the vicinity of Brighton, where they form a shore broken into stupendous cliffs, terminated on the east by the bold promontory of Beachy Head, which rises perpendicularly above the strand to the height of 564 feet, and is the most elevated point on the southern coast of England. The rest of the coast is flat, excepting the vicinity of Selsea Bill, where a few rocks present themselves, and the rocks of Hastings. The district generally understood to constitute the South Downs consists only of the chalk hills lying to the east of Shoreham: many parts of the Downs westward of the river Arun are overgrown with much beech wood, chiefly of a dwarf size, furze, &c., so that the herbage is much inferior to that covering them further eastward. Southward of the chalk hills, extending from their base to the sea, lies a fertile and richly-cultivated vale, which towards its eastern extremity, between Brighton and Shoreham, is, for the most part, less than a mile in breadth: proceeding westward, between the rivers Adur and Arun, this is increased to three miles; and from the Arun to the borders of Hampshire its breadth varies from three to seven miles: its length is about 36. Extensive tracts of marsh land lie adjacent to the coast, between the eastern extremity of the South Downs at Beachy Head and the confines of Kent, in the vicinity of Rye; others also are situated on the lower part of the course of the rivers Ouse, Adur, and Arun. "The Weald" comprises a tract of country, the exact limits of which are not defined: in a legal sense it means the large woodland district contained in the three counties of Sussex, Surrey, and Kent, in which the woods pay no tithe. The large portion of it within the limits of this county, called the

Weald of Sussex, comprises nearly the whole of the level tract lying to the north of the Downs, together with the range of hills running the whole length of the county, at a short distance from its northern and north-eastern boundaries, a great part of which is completely barren. Such is the quantity of timber and other trees in the low plains of the Weald, that, when viewed from the chalk hills, they present to the eye the appearance of one mass of wood; this is, in part, owing to the common practice, at the period that this tract was first reclaimed from its condition of a wild forest, of leaving a shaw of wood, several yards in width, around each enclosure, as a nursery for timber.

The different soils of chalk, clay, sand, loam, and gravel, are found in this county. The first is the soil of the South Downs, consisting, in its natural state, of a rich light hazel mould, lying upon a substratum of loose chalk, or chalk rubble, which covers the more compact rock of the same nature: when brought under cultivation it becomes intermixed with a great portion of chalk by the action of the plough. In some places along the summit of the Downs, nearly the whole surface consists of flints covered by a natural turf: descending from these hills the soil becomes of a deeper staple, and near their base is every where of a good and sufficient depth for the plough. Westward of the river Arun this soil is very gravelly, and contains large flints, and between the Adur and the Ouse has a thin substratum of reddish sand. Along the northern base of the chalk hills, throughout nearly their whole length, and lying between them and the clay of the Weald, is a narrow tract of rich and strong arable land, consisting of an excessively stiff, deep, calcareous loam, on a substratum of pure clay, which, from the admixture of chalk, has generally a whitish appearance. The soil of the fertile vale lying to the south of the Downs consists of a rich loam, in some places rather stiff, more particularly in the projecting south-western angle of the county, and in the peninsula of Selsea Bill, though more commonly light and sandy, and resting upon a substratum of brick-earth, or gravel; the latter, in the western part of it, being by far the most prevalent. Between this and the chalk hills extends a narrow tract, inferior to the former in richness, but consisting of excellent land for turnips; its breadth is greatest on the confines of Hampshire, the flints lying upon its surface in such abundance as entirely to hide the mould: vegetation, however, flourishes through these beds of stone with singular luxuriance. The prevailing soil of the Weald, namely, that of its low plains, is a very stiff clayey loam, on a substratum of brick clay, and that again upon one of sand-stone. Upon the hills before mentioned, as traversing it in a direction nearly from east to west, the soil is in some places a sandy loam, resting upon a sandy gritstone; and in others, a poor, black, sandy, vegetable mould on a soft clay. A long line of these poor sandy tracts, in an unimproved state, crosses the northern part of this county from Kent, and extends into Hampshire: these wastes are, however, wholly separated from each other by extensive districts of clay and others under cultivation. The soil of the marsh lands is composed of decayed vegetable substances, intermixed with sand and other matter, deposited by inundations: in the rape of Lewes this vegetable mould is about twelve inches thick; while in that of Pevensey it is many feet deep, and rests upon a heavy black silt, or sea-sand, containing various kinds of shells: stumps of trees, and timber of large size, have been dug from both these tracts.

The rich arable lands lying to the south of the Downs, and those at the foot of their northern declivity, amount to about 100,000 acres: of Down land there is about 68,000 acres, of which a great proportion is under its native green sward: the arable and grass lands of the Weald, which are of nearly equal extent, amount together to about 425,000 acres. The rotations of crops are regulated chiefly by the nature and properties of the soil, and are therefore various: the common system in the Weald is a fallow, two crops of corn, and one of clover. The corn and pulse commonly grown are wheat, barley, oats, and peas. A valuable species of wheat, called the *Chidham white*, or hedge wheat, takes the former name from the village of Chidham, in this county, where it was first cultivated, the plant which produced the original seed having been found growing wild in a hedge in that neighbourhood. The soil of the extensive tract of the Weald being generally too heavy for the culture of barley, the quantity produced is consequently inferior. Oats are grown on the largest scale in the Weald. Peas are very extensively cultivated, especially on the South Downs, and in the maritime districts: beans are very little grown. Coleseed, barley, and rye are sown, and are in great esteem among the flock-masters of the Downs, as green food for their sheep. Potatoes are extensively and very successfully grown, particularly in the vicinities of Battle, East Bourne, and Chichester: they are chiefly applied to the fattening of cattle.

The principal artificial grasses are, red and white clover, trefoil, and rye-grass. In the eastern and north-eastern parts, hops are very extensively cultivated. The meadow lands are mown every year, and afterwards grazed. It is only in the western part of the county that there are any extensive tracts of irrigated meadows; these are chiefly on the course of the small river Lavant. The marshes may be classed among the finest and most profitable of their kind, having undergone great improvement: they occupy about 30,000 acres, and are wholly employed in the feeding of cattle and sheep; the level of Pevensey is preferred for the cattle, while the marshes about Winchelsea and Rye, adjoining the western side of Romney-Marsh, in Kent, not possessing fresh water, are better calculated for sheep. An act of parliament was obtained, about the commencement of the present century, for widening the channel of the Ouse, near Lewes, and making a shorter cut to the sea, the execution of which design has been of essential benefit to the Lewes and Laughton levels. The great extent of Down land having its native green sward is applied to the feeding of numerous flocks of sheep: the herbage is short, sweet, and aromatic, and of an excellent kind, peculiar to these hills, which is supposed to give to the flesh of the sheep fed upon them that firmness and exquisite flavour for which it is so remarkable. Several of the manures are peculiar: chalk is used in immense quantities, as also is lime burned from it; and as the chalk hills extend no further eastward than East Bourne, it is shipped in sloops from Holywell pits at Beachy Head, and conveyed to the kilns at Bexhill, Hastings, and Rye: lime is also burned from a kind of limestone dug in the Weald. The lands in the maritime district are extensively manured with marl, which is found almost every where on the south side of the Downs, at the depth of only a few feet from the surface: great quantities are also dug from pits on the sea-shore, which are generally covered at high water. Sea mud, provincially called *sleech*, is frequently used as manure near the coast; wood ashes are employed in the Weald.

The chief object of the cattle system is the breeding and rearing of stock, for the purposes of working, and fattening for the butcher, the dairy being only a secondary object. The native cattle are ranked among the best in the kingdom: their colour is universally red, and they have a great disposition to fatten; they yield but very little milk, but the quality of it is peculiarly rich: many oxen are employed in the labours of the farm. Besides its valuable native breed of cattle, Sussex has a breed of sheep, the South Down, one of the most celebrated and numerous in the kingdom: this breed has of late years been introduced in great numbers into various parts of England, more particularly the southern and midland counties; and its excellence is every where acknowledged. The wool is of a very fine carding kind, and its lightness is in proportion to its fineness. The hogs bear much affinity to those of Berkshire; some of them are of a mixed breed, between this and a smaller species. Most of what are called Dorking fowls are bred in the Weald of Sussex, the chief market for them being at Horsham.

In the western parts of the county are some considerable orchards, from which cider is made: it is only to the South of the Downs that orchards for cider are much attended to, though the vicinity of Petworth is considered to produce the best in the county. Sussex has, from the remotest period of antiquity, been celebrated for its fine growth of timber, chiefly oak: the extent of its woodlands cannot be estimated at less than 1700 acres, nearly all included within the Weald, the timber produced in which is preferred by the navy contractors to that of any other district. In the Saxon times here appears to have been one continued forest, stretching from Hampshire into Kent, which, at the time of the Norman Conquest, was valued, not according to the quantity of timber it produced, but the number of hogs that could be fed on the acorns. The waste lands are chiefly situated on the northern side of the county, occupying an extent of about 100,000 acres: their soil has a discouraging aspect, generally consisting of a poor blackish sand, frequently very wet; and their chief value is as rabbit warrens.

The chief mineral productions are the various descriptions of limestone obtained in the Weald: one of these is the Sussex marble, which, when cut and polished, is equal in beauty to most marbles; it is found in the highest degree of perfection in the neighbourhood of Petworth. The limestone and the iron-stone in contact with it often rise to within a very few feet of the surface. Alternate strata of sand-stone and iron-stone occur every where in the Weald; and under these, at a considerable depth, are numerous strata of limestone, which, when burned, makes the finest cement in the kingdom. Anciently the iron-stone of this district was very extensively worked as ore, until the successful establishment of the great iron and coal works in the midland and northern districts of the kingdom occasioned the works in the Weald, the fuel of which was supplied by the extensive surrounding woodlands, to be wholly abandoned. Fuller's earth is found at Tillington, and used in the neighbouring fulling-mills; red ochre is obtained at Graffham, Chidham, and several other places on the coast, some of it being sent to London. The manufacture of charcoal, chiefly for gunpowder, has been of considerable importance in this county, from which large quantities have been annually sent to London over land. At Chichester a small woollen manufacture is carried on: sacks, blankets, linen and worsted yarn, cotton and stuff goods, and other articles, are made in the workhouses. There are paper-mills at Iping and a few other places. Potash is made at Bricksill hill, near Petworth, for the soap-makers of that town. Brick-making is common in many parts of the county: near Petworth are kilns for the burning of bricks and tiles to be exported to the West Indies. Ship and boat building is carried on in some of the small harbours of Sussex; yet, notwithstanding the great extent of sea-coast, its maritime commerce is of nearly as little importance as its manufactures. Corn is exported in different directions, much of it being sent to Portsmouth. A considerable quantity of timber is exported; as are charcoal, cord-wood, and oak-bark; horned cattle and sheep, hides, and wool, are among its agricultural exports. There are several fisheries upon the coast, chiefly for herrings, mackarel, and flat-fish, much of the produce being sent to London. In the Weald are very numerous ponds for feeding fresh-water fish for the London markets: these are chiefly carp; though tench, perch, eels, and pike, are also kept: many of the ponds were originally formed for the purpose of working the machinery of the iron-manufactories, long since abandoned. The most fashionable places of resort for sea-bathing are Brighton, Worthing, and Hastings.

All the principal rivers rise in the Weald, within the limits of the county, and take a tolerably direct course to the English Channel, so that their length is not great. They are, the Arun, with its tributary, the Rother; the Ouse; and the Adur. The Arun, with the aid of several artificial cuts, has been made navigable up to Newbridge, near Guillenhurst; and the Rother, with the like assistance, as high as the town of Midhurst. A small canal also branches from the Rother to the village of Haslingbourne, within half a mile of Petworth. The largest barges navigating these rivers are of 30 tons' burden: the tide flows up the Arun, a distance of seventeen miles, to the vicinity of the village of Amberley. The Ouse is formed by the junction of two streams, one of which rises in the forest of Worth, the other in that of St. Leonard, uniting near Cuckfield: it has been made navigable beyond Lewes to within five miles of Cuckfield. The Adur, sometimes called the Beeding, is navigable for ships of considerable burden to Shoreham, and for barges to the vicinity of the village of Ashurst. The Lavant, a much smaller stream than any of the above, becomes navigable for ships some distance below Chichester, and expands into an estuary, which opens into the sea between the village of Wittering and the south-eastern point of Hayling island in Hampshire: remarkably fine lobsters are bred in this river, near its mouth. The shores of the south-western part of the county are rendered very irregular by several other arms of the sea, one of which separates Thorney island from the rest of the county. The Portsmouth and Arundel canal, the act for the formation of which was obtained in 1815, commencing from the river Arun, a little below the latter town, proceeds westward, in nearly a direct line, to the broad estuary of the Lavant, below Chichester, to which city there is a short branch northward: from the Lavant the navigation is continued through the channels which separate Thorney and Hayling islands from the main land to the eastern side of Portsea island, where the artificial navigation recommences, and proceeds westward to Portsmouth. The London and Brighton railway enters the county at Black Corner, and proceeds past Balcomb, and to the east of Cuckfield, to Brighton.

This county is supposed to have contained the Roman stations of *Anderida Civitas*, at Seaford or East Bourne; *Anderida Portus*, at Pevensey; *Cilindunum*, at Slindon; *Mida*, at Midhurst; *Mantuantonis*, or *Mutuantonis*, at Lewes; *Portus Adurni*, at Aldrington; and *Regnum*, at Chichester. The present roads from Portsmouth, from Midhurst, and from Arundel, to Chichester, are considered to have been originally of Roman formation; and from this city the Roman road, commonly called the Stane-street, proceeded in a north-easterly direction towards Dorking in Surrey, where it fell into the Ermin-street, being traceable in many parts of its course. Various Roman domestic remains have been dug up in different places, particularly at Chichester, Bognor, and East Bourne, including tesselated pavements and baths: coins of the Lower Empire have been found in various other places. The number of ancient encampments upon the Downs and in other parts of the county, near the sea, evince that they have been frequently the scene of military operations: some of these ancient fortifications are supposed to have been made by the Romans, others by the Saxon and the Danish invaders, while one on Mount Caburn, about a mile and a half from Lewes, on the northern edge of the Downs, is thought to be British. Near the western confines of the county, to the west of Chichester, are, the encampment of the Broile, of an oblong form, about half a mile in length and a quarter in breadth; and another, called Gonshill, of the same form. On the northern brow of the Downs, overlooking the Weald, and proceeding from west to east, occur the following camps; Chenkbury, about two miles west of Steyning, which is circular; eight miles further, above Poynings, a large camp of an oval form; Wolstenbury, about three miles further, on a hill projecting from the rest like a bastion, nearly circular; Ditchling, three miles from the last, nearly square; and lastly, that on Mount Caburn, already mentioned, which is round; and a quarter of a mile westward of it is another fortification, much larger, but not so perfect. Those on the southern side of the Downs, proceeding in the same direction, are, that on St. Roche's, or St. Rook's hill, which is circular; that on High Down, four miles eastward of Arundel, a small square; and Cissbury, four miles south-west of Steyning. The only one on the central heights of the chalk hills is Hollingbury, two miles north of Brighton, which is square, and the area about five acres. About a mile eastward of the same town, on the top of a hill near the sea, is a camp with a triple ditch and bank, also square, except that the angles are rounded off. In the parish of Telscombe, about five miles from the last, are two other squares; and at Newhaven, on the point of a hill overlooking the mouth of the Ouse, is a strong fortification of an oval form. About a mile eastward of Seaford is another work of the same kind, also situated close to the sea, but of a semicircular form; and lastly, about three miles east of Cuckmere haven, near Burling Gap, are other earthworks, enclosing a hill, of a semi-elliptical form: traces of several more earthworks of less extent and importance are discoverable in other places.

The number of religious houses in this county, before the general dissolution, including hospitals and colleges, was about 58. There are yet extensive remains of the magnificent abbey of Battle, and of that of Bayham, on the confines of Kent; and considerable relics of the priories of Boxgrove, Hardham, Lewes, Michelham, and Shelbred, about four miles north of Midhurst. The most considerable remains of ancient castles are those of Amberley, Arundel, Bodiham, Bramber, Eridge, in the parish of Frant, Hastings, Hurstmonceaux, Ipres at Rye, Lewes, Pevensey, Scotney, and Winchelsea. The most remarkable ancient mansion is that of Cowdry House, now in ruins. Several of the modern seats of the nobility and gentry are magnificent; and the Pavilion at Brighton is distinguished as one of the residences of the sovereign. Some of the seats most worthy of notice are, Petworth Park, the residence of Colonel Wyndham; Arundel Castle, that of the Duke of Norfolk; the episcopal palace of Chichester; Eridge Castle, the residence of the Earl of Abergavenny; Goodwood, that of the Duke of Richmond, the lord-lieutenant of the county; Parham Park, that of Lord de la Zouche; Penshurst Place, that of Sir John Shelley Sidney, Bart.; Sheffield Park, that of the Earl of Sheffield; Slindon House, that of the Earl of Newburgh; and Stanmer Park, that of the Earl of Chichester. There is a chalybeate spring at Brighton, and another at East Bourne: near Hastings is a singular dropping well, and, in the same vicinity, a fine waterfall of 40 feet perpendicular. The title of the Duke of Sussex is borne by His Royal Highness Prince Augustus Frederick, sixth son of George III., upon whom it was conferred in the year 1801.

SUSTEAD (*St. Peter and St. Paul*), a parish, in the union of Erpingham, hundred of North Erpingham, Eastern Division of the county of Norfolk, $4\frac{1}{4}$ miles (S. W.) from Cromer; containing 162 inhabitants. The living is a perpetual curacy; net income, £34; patron, Admiral Windham. Here is a National school.

SUTCOMBE (*St. Andrew*), a parish, in the union of Holsworthy, hundred of Black Torrington, Holsworthy and Northern Divisions of the county of Devon, 5 miles (N. by E.) from Holsworthy; containing 491 inhabitants. The living is a rectory, valued in the king's books at £17. 10. $7\frac{1}{2}$.; present net income, £170; patron, Rev. W. Cohern. The church has a Norman doorway, but is mostly of later date: it contains some neat monuments to the family of Davie: the Prideaux family presented the communion plate. There is an almshouse for six poor persons, founded and endowed by Sir William Morris, Secretary of State to Charles II.

SUTTERBY (*St. John the Baptist*), a parish, in the union of Spilsby, Wold Division of the wapentake of Candleshoe, parts of Lindsey, county of Lincoln, $4\frac{3}{4}$ miles (W. S. W.) from Alford; containing 34 inhabitants. The living is a rectory, valued in the king's books at £5. 10. 2½.; present net income, £125: it is in the patronage of the Crown.

SUTTERTON (*St. Mary*), a parish, in the union of Boston, wapentake of Kirton, parts of Holland, county of Lincoln, 8¼ miles (N. W. by N.) from Holbeach; containing 1093 inhabitants. The living is a vicarage, valued in the king's books at £23. 3. 4.; present net income, £885: it is in the patronage of the Crown. The church is principally in the later style of English architecture, with a tower, surmounted by an elegant crocketed spire. Here is a National school.

SUTTON (*All Saints*), a parish, in the union and hundred of Biggleswade, county of Bedford, 1¾ mile (S.) from Potton; containing 386 inhabitants. The living is a rectory, valued in the king's books at £20; present net income, £362; patrons, President and Fellows of St. John's College, Oxford. There is a National school, endowed with £11 per annum. Here were the seat and royalty of the celebrated John of Gaunt, Duke of Lancaster, who conferred Sutton and Potton upon Sir Roger Burgoyne and his heirs, by a curious laconic deed in doggerell verse, which is preserved among the ancient records in the Arches, Doctors' Commons. The manor-house was burnt down in 1826. There is a fine chalybeate spring near the parsonage-house. The learned Bishop Stillingfleet was rector of Sutton, about the middle of the seventeenth century, where he wrote his *Origines Sacræ*.

SUTTON (*St. Andrew*), a parish, in the hundred of South Witchford, union and Isle of Ely, county of Cambridge, 6¼ miles (W. by S.) from Ely; containing 1362 inhabitants. The living is a vicarage, united to the rectory of Mepal, and valued in the king's books at £10; appropriators, Dean and Chapter of Ely. The appropriate tithes have been commuted for a rent-charge of £450, and the vicarial for £1175, subject to the payment of rates, which on the average have respectively amounted to £66 and £92; the appropriate glebe comprises 72 acres, and the vicarial 45 acres, valued at £130. 6. and £80. 7. per annum. The church was built by Barnet, Bishop of Ely, who died in 1373: it is a beautiful specimen of the decorated style of English architecture. There are places of worship for Baptists and Wesleyan Methodists. A school is endowed with £15 per annum, the rent of poor's land; another is supported by subscription, and there is a Sunday National school. This place had anciently a market and a fair, granted to the first abbot of Ely. In 1634, some labourers discovered near this place, several ancient coins and gold rings, and three silver plates, one of which had a curious inscription engraven upon it.

SUTTON, a township, in the parish and union of Runcorn, hundred of Bucklow, Northern Division of the county of Chester, 2 miles (N. E. by E.) from Frodsham; containing 237 inhabitants.

SUTTON, a township, in the union and borough of Macclesfield, parish of Prestbury, Northern Division of the county of Chester; containing 5856 inhabitants. There are several extensive cotton and silk mills in this township. A district chapel, dedicated to St. George, was erected and consecrated in 1834, the living of which is a perpetual curacy; net income, £225; patrons, Trustees. Two small bequests, by Catherine Nixon, in 1689, and John Upton, are applied to the instruction of poor children. Here is a Sunday National school. This was the seat of the family of Holinshed, the historian, and is supposed to have been his birth-place.

SUTTON, a township, in the parish of Middlewich, union and hundred of Northwich, Southern Division of the county of Chester, 1¼ mile (S.) from Middlewich; containing 18 inhabitants.

SUTTON (*St. Mary*), a parish, in the union of Chesterfield, hundred of Scarsdale, Northern Division of the county of Derby, 4 miles (E. S. E.) from Chesterfield; containing 700 inhabitants. The living is a discharged rectory, with the vicarage of Duckmanton annexed, valued in the king's books at £12. 16. 0½.; present net income, £309; patron, R. Arkwright, Esq., the impropriator of Duckmanton. The tithes have been commuted for a rent-charge of £300, subject to the payment of rates, which on the average have amounted to £30; the glebe comprises 58 acres, valued at £54. 7. per annum. The windows of the church exhibit some remains of ancient stained glass. Nicholas Deincourt, Earl of Scarsdale, in 1643, fortified the hall, which he had previously erected here, for Charles I.; it was taken by assault, and the works demolished by Sir John Gell, and, some time afterwards, was plundered by the parliamentarian garrison of Bolsover; this mansion is situated in an extensive and beautiful park. Eighteen poor children are instructed for about £18 per annum, arising from the rent of sixteen acres of land allotted by the lord and freeholders of the manor.

SUTTON, a parish, in the union and hundred of Rochford, Southern Division of the county of Essex, 1½ mile (S. E. by S.) from Rochford; containing 96 inhabitants. This parish derives its name, originally South town, from its relative position with respect to Rochford. The living is a rectory, valued in the king's books at £11; patron, J. Aitkin, Esq. The tithes have been commuted for a rent-charge of £270, subject to the payment of rates, which on the average have amounted to £40; the glebe comprises 10 acres, valued at £30 per annum. The church is a small ancient edifice with a stone tower, and consists only of a nave and chancel.

SUTTON (*St. Michael*), a parish, in the hundred of Broxash, union and county of Hereford, 4¼ miles (N. N. E.) from Hereford; containing 98 inhabitants. The living is a perpetual curacy; net income, £64; patrons and impropriators, — Allen and — Unett, Esqrs.

SUTTON (*St. Nicholas*), a parish, in the hundred of Broxash, union and county of Hereford, 4¼ miles (N. E. by N.) from Hereford; containing 234 inhabitants. The living is a rectory, valued in the king's books at £8. 1. 8.; present net income, £195; patron, J. Johnston, Esq. A school is endowed with £6. 13. 4., and a house and garden.

SUTTON, a township, in the parish and union of Prescot, hundred of West Derby, Southern Division of the county palatine of Lancaster, 1½ mile (S. E. by E.) from St. Helen's; containing 3173 inhabitants, who are extensively employed in the manufacture of crown and flint glass, earthenware, and watch movements. The Liverpool and Manchester railway passes through the township. A school is endowed with £20 per annum; and another is endowed by J. Nattall, Esq., for 40 girls.

SUTTON, a township, in the parish of Wymondham, hundred of Forehoe, Eastern Division of the county of

NORFOLK, 1½ mile (S. W. by S.) from Wymondham; containing 739 inhabitants.

SUTTON (*St. Michael*), a parish, in the hundred of HAPPING, Eastern Division of the county of NORFOLK, 8¼ miles (E. N. E.) from Coltishall; containing 313 inhabitants. The living is a discharged rectory, valued in the king's books at £6. 16. 8.; present net income, £292; patron, Earl of Abergavenny.

SUTTON, a chapelry, in the parish of CASTOR, union and soke of PETERBOROUGH, Northern Division of the county of NORTHAMPTON, 1¼ mile (E. by S.) from Wansford; containing 118 inhabitants. The river Nene runs through the chapelry, in which there is a fine stone quarry, producing stone resembling that at Ketton. The chapel is dedicated to St. Michael.

SUTTON, a hamlet, in the parish of GRANBY, Northern Division of the wapentake of BINGHAM, Southern Division of the county of NOTTINGHAM, 14 miles (E. by S.) from Nottingham: the population is returned with the parish.

SUTTON (*St. Bartholomew*), a parish, in the union of EAST RETFORD, liberty of SOUTHWELL and SCROOBY, though locally in the wapentake of BASSETLAW, Northern Division of the county of NOTTINGHAM, 3¼ miles (N. N. W.) from East Retford; containing 801 inhabitants. The living is a discharged vicarage, with that of Scrooby annexed, valued in the king's books at £10; present net income, £185; patron and impropriator, Duke of Portland. The river Idle runs through the parish, in which there is a very ancient mansion of singular appearance, said to have been formerly much larger than at present, and the country residence of some of the ancestors of Earl Fitzwilliam. A school, erected by subscription in 1783, is endowed with the proceeds of £70, the gift of Richard Taylor, in 1737, and with two allotments of the waste lands, enclosed in 1773: the annual income is about £28; it is in connection with the National Society.

SUTTON, a township, in the parish of DIDDLEBURY, hundred of MUNSLOW, Southern Division of the county of SALOP, 6 miles (N.) from Ludlow: the population is returned with the parish.

SUTTON (*St. John*), a parish, in the union of ATCHAM, liberties of the borough of SHREWSBURY, Northern Division of the county of SALOP, 2¼ miles (S. S. E.) from Shrewsbury; containing 81 inhabitants. The living is a discharged rectory, valued in the king's books at £3; present net income, £17; patron and impropriator, Lord Berwick. Sutton Spa, a fine mineral spring issuing from a stratum of ash-coloured clay, close to the village, is nearly similar in its properties to sea water.

SUTTON (*All Saints*), a parish, in the union of WOODBRIDGE, hundred of WILFORD, Eastern Division of the county of SUFFOLK, 3½ miles (S. E. by E.) from Woodbridge; containing 680 inhabitants. The living is a discharged vicarage, valued in the king's books at £8. 2. 1.; present net income, £299; patron and incumbent, Rev. Robert Field. There is a place of worship for Baptists. The navigable river Deben bounds the parish on the west, where there is a ferry to Woodford.

SUTTON (*St. Nicholas*), a parish, in the union of EPSOM, Second Division of the hundred of WALLINGTON, Eastern Division of the county of SURREY, 5 miles (E. N. E.) from Epsom; containing 1121 inhabitants. The living is a rectory, valued in the king's books at £16. 18. 4.; present net income, £660; patron, Rev. Thomas Hatch. The church is partly in the decorated style of English architecture; it had formerly a wooden tower, which has been replaced by one of brick, and contains, among other handsome monuments, chiefly of the Talbots, one to the memory of Lady Dorothy Brownlow, and also one to Mrs. Sarah Glover. In Domesday-book two churches are mentioned to have existed in this place. There is a place of worship for Independents. Nine poor children are educated for £6 per annum, the gift of William Beck, in 1789; and the interest of a bequest of £100 by Susannah Bentley, in 1823, is paid in aid of the National school, which is further supported by voluntary contributions. There are also two Sunday schools, endowed by Mrs. Lucy Manners with £700 three per cent. consols. Mary Gibson left £500 three per cent. annuities, directing the interest to be distributed annually in the following manner; £5 to the rector, £5 to the poor, £2 each to the two churchwardens, and £1 to the parish clerk, provided that, on the 12th of August, the mausoleum of her family in the churchyard be opened and inspected by them, and that they then repair to the church, to hear a sermon preached by the rector. There is a large chalk pit in the parish, in which a variety of curious fossils has been found.

SUTTON (*St. John*), a parish, in the hundred of ROTHERBRIDGE, rape of ARUNDEL, Western Division of the county of SUSSEX, 5 miles (S.) from Petworth; containing 379 inhabitants. The living is a rectory, valued in the king's books at £15. 0. 10.; patron, Col. Wyndham. The tithes have been commuted for a rent-charge of £340, subject to the payment of rates, which on the average have amounted to £68: the glebe comprises 25 acres, valued at £44 per annum. The church is partly in the early, and partly in the decorated, style of English architecture.

SUTTON, a hamlet, in the parish of TENBURY, Upper Division of the hundred of DODDINGTREE, Hundred-House and Western Divisions of the county of WORCESTER, 2½ miles (S. by E.) from Tenbury; containing 185 inhabitants.

SUTTON, a township, in the parish of NORTON, wapentake of BUCKROSE, East Riding of the county of YORK, 1 mile (S. by E.) from New Malton; the population is returned with the parish.

SUTTON (*St. James*), a parish, in the union of SCULCOATES, Middle Division of the wapentake of HOLDERNESS, East Riding of the county of YORK, 3½ miles (N. N. E.) from Kingston-upon-Hull; containing 4383 inhabitants. The living is a perpetual curacy; net income, £98; patron and impropriator, H. Broadley, Esq. The church had formerly a chantry of six priests, endowed by John of Sutton, and valued at the dissolution at £13. 18. 8. per annum. Many of the most opulent merchants of Hull have residences in this neighbourhood. John Marshall, in 1803, bequeathed £150 for the instruction of ten poor children. In the village are two hospitals; one of them founded by Leonard Chamberlain, and rebuilt in 1800, for the maintenance of two poor aged widowers and eight widows; the other erected in 1819, by the trustees of the late Mrs. Watson, for the reception of the widows and daughters of poor clergymen deceased. A house of White friars existed here in the time of Edward I.

SUTTON, a township, in the parish of KIRKLINGTON, wapentake of HALLIKELD, North Riding of the county of YORK, $5\frac{3}{4}$ miles (N.) from Ripon; containing 121 inhabitants. There is a place of worship for Wesleyan Methodists.

SUTTON, a joint township with Healey, in the parish of MASHAM, union of LEYBURN, wapentake of HANG-EAST, North Riding of the county of YORK, $6\frac{1}{4}$ miles (S. W.) from Bedale: the population is returned with Healy.

SUTTON, a township, in the parish of BROTHERTON, partly within the liberty of ST. PETER'S, East Riding, and partly in the Lower Division of the wapentake of BARKSTONE-ASH, West Riding, of the county of YORK, 1 mile (N. E. by E.) from Ferry-Bridge; containing 57 inhabitants.

SUTTON, a township, in the parish of CAMPSALL, union of DONCASTER, Upper Division of the wapentake of OSGOLDCROSS, West Riding of the county of YORK, $6\frac{1}{4}$ miles (N. by W.) from Doncaster; containing 134 inhabitants.

SUTTON, a township, in the parish of KILDWICK, union of KEIGHLEY, Eastern Division of the wapentake of STAINCLIFFE and EWCROSS, West Riding of the county of YORK, 5 miles (W. N. W.) from Keighley; containing 1153 inhabitants, several of whom are employed in the manufacture of cotton goods and worsted. There is a place of worship for Baptists.

SUTTON-AT-HONE (*St. John the Baptist*), a parish, in the union of DARTFORD, hundred of AXTON, DARTFORD, and WILMINGTON, lathe of SUTTON-AT-HONE, Western Division of the county of KENT, $2\frac{1}{2}$ miles (S. by E.) from Dartford; containing 1012 inhabitants. This extensive parish, from which the lathe derives its name, is pleasantly situated on the river Darent, by which it is bounded on the east, and is intersected by the road from Dartford to Seven-Oaks. The village, which lies on the bank of that river, has an interesting and picturesque appearance, and the surrounding scenery is agreeably diversified and enlivened with some good houses, among which are Sutton Place and St. John's, the latter occupying the site of an ancient commandery; and in the hamlet of Hawley, near the northern extremity of the parish, is Hawley House, a mansion of considerable antiquity. The living is a vicarage, valued in the king's books at £10; present net income, £519; patrons and appropriators, Dean and Chapter of Rochester. The church, an ancient structure, with some portions in the decorated style of English architecture, was partly burnt down in 1615. An almshouse for four poor persons was founded by Katherine Wrott in 1596, with a small endowment, and another tenement has been added to it. Jeffrey Fitz-Piers, Earl of Essex, in the reign of Richard I., or of John, gave his estates in this parish to the Archdeacon of Taunton, for the foundation and endowment of an hospital for three chaplains and thirteen brethren. About the same time Robert Basinge gave the manor to the Knights Hospitallers of St. John of Jerusalem, who had a commandery here, of which the ancient refectory forms the scullery of the present modern house, and in the gardens are still remaining the fish stews belonging to the establishment.

SUTTON-BASSETT (*St. Mary*), a parish, in the union of MARKET-HARBOROUGH, hundred of CORBY, Northern Division of the county of NORTHAMPTON, $3\frac{1}{4}$ miles (N. E.) from Market-Harborough; containing 139 inhabitants. The living is a vicarage, united to that of Weston-by-Welland.

SUTTON-BENGER (*All Saints*), a parish, in the union of CHIPPENHAM, hundred of MALMESBURY, Chippenham and Calne, and Northern, Divisions of the county of WILTS, 5 miles (N. N. E.) from Chippenham; containing 443 inhabitants. The living is a discharged vicarage, valued in the king's books at £6. 3. 4.; present net income, £285; patrons and appropriators, Dean and Chapter of Salisbury. There is a place of worship for Independents. The line of the Great Western railway passes through the parish.

SUTTON-BINGHAM, a parish, in the union of YEOVIL, hundred of HOUNDSBOROUGH, BERWICK, and COKER, Western Division of the county of SOMERSET, $3\frac{1}{2}$ miles (S. by W.) from Yeovil; containing 78 inhabitants. The living is a rectory, valued in the king's books at £4. 15. 10.; present net income, £214; patron, W. Helyar, Esq. The church is principally in the early style of English architecture.

SUTTON, BISHOP'S (*St. Nicholas*), a parish, in the union of ALRESFORD, hundred of BISHOP'S-SUTTON, Alton and Northern Divisions of the county of SOUTHAMPTON, $1\frac{3}{4}$ mile (E. S. E.) from New Alresford; containing 527 inhabitants. The living is a vicarage, with that of Ropley annexed, valued in the king's books at £19. 10. $2\frac{1}{2}$.; present net income, £350; joint patrons, Sir Thomas Baring, Bart., and John Deacon, Esq.; impropriators, several proprietors of land. There is a place of worship for Independents. The Bishops of Winchester had anciently a palace here, the remains of which have been converted into a malt-house. Fairs are held on the Thursday after the festival of the Holy Trinity, and on Nov. 6th.

SUTTON-BONNINGTON, a parish, in the union of LOUGHBOROUGH, Southern Division of the wapentake of RUSHCLIFFE and of the county of NOTTINGHAM, 2 miles (S. E. by E.) from Kegworth; containing 1136 inhabitants. The living consists of the united rectories of *St. Anne* and *St. Michael*, the former valued in the king's books at £4. 17. 6.; present net income, £462: it is in the patronage of the Crown:—the latter is valued at £15. 2. 1.; present net income, £509; patrons, Dean and Chapter of Bristol. There is a place of worship for Wesleyan Methodists; also a school endowed with land producing £40 per annum.

SUTTON-BOURNE, a hamlet, in the parish of LONG SUTTON, union of HOLBEACH, wapentake of ELLOE, parts of HOLLAND, county of LINCOLN, 5 miles (E. by N.) from Holbeach; containing 706 inhabitants.

SUTTON-BY-DOVOR (*St. Peter and St. Paul*), a parish, in the union of EASTRY, hundred of CORNILO, lathe of ST. AUGUSTINE, Eastern Division of the county of KENT, 4 miles (S. W. by W.) from Deal; containing 164 inhabitants. The living is a perpetual curacy, consolidated in 1835 with the rectory of Little Mongeham, according to act of parliament; net income, £47; appropriator, Archbishop of Canterbury. The church is a small ancient structure, in the early English style, with a circular east end, but no tower.

SUTTON-CHART, county of KENT.—See CHART, SUTTON.

SUTTON-CHENEY, a chapelry, in the parish of

Market-Bosworth, hundred of Sparkenhoe, Southern Division of the county of Leicester, 2 miles (S. S. E.) from Market-Bosworth; containing 335 inhabitants. The chapel is dedicated to St. James, and contains 120 free sittings, the Incorporated Society having granted £60 in aid of the expense. There is a place of worship for Wesleyan Methodists. The Ashby-de-la-Zouch canal passes in the vicinity.

Seal and Arms.

SUTTON-COLDFIELD (*Holy Trinity*), a market-town and parish, having separate jurisdiction, in the union of Aston, locally in the Birmingham Division of the hundred of Hemlingford, Northern Division of the county of Warwick, 26 miles (N. W. by N.) from Warwick, and 110 (N. W. by N.) from London; containing 3684 inhabitants. This town, formerly called *Sutton-Colville* and *King's-Sutton*, is of considerable antiquity, having been of some note in the Saxon times. During the reign of Edward the Confessor, the manor was in the possession of Edwin Earl of Mercia, but subsequently William the Conqueror held it in his own hands, and Henry I. exchanged it with the Earl of Warwick for other manors. In later times, the town, having nearly fallen into decay, was indebted to the attachment of Vesey, Bishop of Exeter, and chaplain to Henry VIII., who was a native of this place, for that munificence which led to its revival, and laid the foundation of its future prosperity. It occupies a bleak and exposed situation on rising ground of steep acclivity, and consists principally of one long street; the houses are mostly modern, well built, and of handsome appearance, and the inhabitants are amply supplied with water from springs. Adjacent to it is a very extensive park finely wooded and adorned with several large sheets of water, in which the inhabitants have the privilege of pasturage, for a small payment to the corporation: it is crossed by the Ikenild-street, which is distinctly traceable for two miles, entering the park near a small artificial mount, called King's Standing on the Coldfield, from the circumstance of Charles having harangued his troops from Shropshire on this spot, and taking thence a direction into the Lichfield road. Here is a medicinal spring, called Rounton Well; another, possessing sulphureous qualities, is now disused. The principal occupation is the manufacture of spades, saws, axes, and other implements: mills for grinding gun-barrels are worked by streams of water issuing from the pools in the park, of which one covers from 30 to 35 acres. The Birmingham and Fazeley canal passes through the parish. The market is held on Monday; and fairs are on Trinity-Monday and November 8th, for cattle, sheep, and pedlery. The town is governed by a corporation, which obtained its charter from Henry VIII., at the instance of Bishop Vesey, consisting of a warden, two capital burgesses, and 22 other corporate members. The warden, who is chosen annually, and the capital burgesses for life, by the corporation, from their own body, are justices of the peace by virtue of their office: the warden acts as coroner for the town, manor, and lordship of Sutton. The corporation are lords of the manor, and elect a lord high steward and park-keepers: the former appoints his deputy, who must be a lawyer, and presides at the courts leet and baron. The other members are also chosen by the corporation; the inhabitants are free and eligible by residence. Under the charter they were empowered to hold courts of Oyer and Terminer, and of gaol delivery, which, from disuse, have been transferred to the county town, the corporation paying a quota towards the county rate: they hold a petty session at the town-hall, on the Friday in the week of general quarter sessions; and a court of record. The town has lately been made a polling-place for the northern division of the county. In the town-hall, a neat brick building, are the arms of Bishop Vesey, emblazoned on a shield surmounted with a mitre.

The living is a rectory, valued in the king's books at £33. 9. 2.; patron, Rev. William Riland Bedford, by whose ancestors the advowson has been held since the reign of Elizabeth, when it was sold by the crown. The church is a fine ancient structure, built probably in the thirteenth century, though combining different styles of English architecture: the aisles were added by Bishop Vesey, whose effigy, in a recumbent posture, with a mitre on his head and a crosier in his right hand, is in the chancel: part of the nave fell down about 70 years ago, and was rebuilt by the corporation, at the expense of £1500. A chapel of ease erected at Mere Green, in this parish, was consecrated in 1835; it contains 560 sittings, 440 of which are free, the Incorporated Society having granted £200 in aid of the expense. A Roman Catholic chapel, capable of containing nearly 500 persons, has been erected; also a Roman Catholic college, at an expense of nearly £60,000. The free grammar school was founded in the reign of Henry VIII., and endowed with land in the parish, by Bishop Vesey: the salary of the master is from £300 to £400 per annum, and a handsome house was erected for him, chiefly at the expense of the corporation, on the condition of his teaching 24 poor boys additionally in reading, writing, and arithmetic, but which obligation, in consequence of his having given up certain property to the corporation, has ceased. Six schools, in which about 350 children of both sexes are educated, and of whom 40 are clothed, are supported from funds belonging to the corporation. Almshouses for five aged men and five aged women, with gardens attached, were built and are supported by the corporation. Among various charitable benefactions, four marriage portions, of £24 each, are allowed annually to four poor maidens, natives or long resident. Near Driffold House, so called from the custom of driving and folding the cattle of the parishioners, a farm-house occupies the site of the old manor-house, formerly a royal palace of great strength, of which a few remains are still visible.

SUTTON-COURTNEY (*All Saints*), a parish, in the union of Abingdon, hundred of Ock, county of Berks, 2 miles (S. by E.) from Abingdon; containing 1284 inhabitants. The living is a vicarage, valued in the king's books at £18. 13. 4.; present income, £148; patrons and appropriators, Dean and Canons of Windsor. The church is very ancient: it has a wooden rood-loft, also a Norman font surrounded with pillars and enriched with sculptured foliage, &c. There is a chapel of ease at Appleford: and at Sutton-Courtney is a place of worship for Independents. The Wilts and

Berks canal passes through the parish. A paper-mill employs about 25 persons. Edmund Bradstock, in 1607, bequeathed a house and lands, of the present annual value of £80, for the education of children: the premises are now in the occupation of a schoolmaster, who teaches twenty boys. An almshouse was erected in 1820, pursuant to the will of Francis Elderfield, Esq., who endowed it for six widows. Abbey House was formerly a residence for the monks of Abingdon.

SUTTON, EAST (*St. Peter and St. Paul*), a parish, in the union of HOLLINGBOURN, hundred of EYHORNE, lathe of AYLESFORD, Western Division of the county of KENT, 6 miles (S. E.) from Maidstone; containing 379 inhabitants. This parish is crossed by the ridge of hills bounding the Weald, of which its southern side forms a part. The living is annexed to the vicarage of Sutton-Valence. The church is a handsome edifice, and contains some interesting monuments. A school is partly supported by subscription.

SUTTON, ST. EDMUND'S, a chapelry, in the parish of LONG SUTTON, union of HOLBEACH, wapentake of ELLOE, parts of HOLLAND, county of LINCOLN, 10 miles (E. by N.) from Crowland; containing 626 inhabitants. The living is a perpetual curacy; net income, £195; patron and appropriator, Vicar of Long Sutton. A school for boys is supported by endowment.

SUTTON, FULL, a parish, in the union of POCKLINGTON, Wilton-Beacon Division of the wapentake of HARTHILL, East Riding of the county of YORK, 5¼ miles (N. W. by N.) from Pocklington; containing 140 inhabitants. The living is a discharged rectory, valued in the king's books at £10. 12. 8½.; present net income, £150; patron, Lord Feversham.

SUTTON-GRANGE, a township, in the parish and liberty of RIPON, West Riding of the county of YORK, 3 miles (N. W. by N.) from Ripon; containing 83 inhabitants.

SUTTON, GREAT, a township, in the parish of EASTHAM, union, and Higher Division of the hundred, of WIRRALL, Southern Division of the county of CHESTER, 7 miles (N. N. W.) from Chester; containing 162 inhabitants.

SUTTON, GUILDEN, a parish, in the union of GREAT BOUGHTON, Lower Division of the hundred of BROXTON, Southern Division of the county of CHESTER, 3¼ miles (E. N. E.) from Chester; containing 132 inhabitants. The living is a perpetual curacy; net income, £50; patron and impropriator, Lord Stanley.

SUTTON-IN-ASHFIELD (*St. Mary*), a parish, in the union of MANSFIELD, Northern Division of the wapentake of BROXTOW and of the county of NOTTINGHAM, 3½ miles (W. S. W.) from Mansfield; containing 5746 inhabitants. The village is situated on an eminence, and comprises several streets, covering a considerable extent of ground. The inhabitants are chiefly engaged in the manufacture of cotton hose and lace, which are carried on to a great extent; and a large factory for spinning cotton, and making checks and nankeens, has long been conducted here. A few of the inhabitants also find employment in making a coarse kind of red pottery ware. The Mansfield and Pinxton railway passes through the parish. Limestone of excellent quality abounds in the vicinity. A book society has been established for several years. A small customary market, for provisions, is held on Saturday. The living is a perpetual curacy; net income, £118; patron, Duke of Devonshire; impropriator, Duke of Portland. There are places of worship for General and Particular Baptists, Independents, and Primitive and Wesleyan Methodists. A National school for boys is principally supported by subscription, excepting about £10 per annum arising from two small benefactions in land given by James Mason and Elizabeth Root. Joseph Whitehead, a framework-knitter, eminent for his attainments in astronomy and mechanics, who constructed an orrery upon Ferguson's principle, and other complicated pieces of machinery, and was also an excellent musician, was a native of this place; he died in 1811, at the early age of 27 years.

SUTTON-IN-THE-ELMS, a township, in the parish of BROUGHTON-ASTLEY, hundred of GUTHLAXTON, Southern Division of the county of LEICESTER, 6½ miles (E.) from Hinckley: the population is returned with the parish. There is a place of worship for Baptists.

SUTTON-IN-THE-MARSH (*St. Clement*), a parish, in the union of SPILSBY, Marsh Division of the hundred of CALCEWORTH, parts of LINDSEY, county of LINCOLN, 6½ miles (N. E. by E.) from Alford; containing 183 inhabitants. The living is a discharged vicarage, valued in the king's books at £6. 13. 4.; present net income, £68; patron and appropriator, Prebendary of Sutton-in-Marisco in the Cathedral Church of Lincoln.

SUTTON, ST. JAMES, a chapelry, in the parish of LONG SUTTON, union of HOLBEACH, wapentake of ELLOE, parts of HOLLAND, county of LINCOLN, 5½ miles (S. E. by S.) from Holbeach; containing 391 inhabitants. The living is a perpetual curacy; net income, £66; patron and appropriator, Vicar of Long Sutton. The chapel was built of a large sort of brick; but the chancel and the steeple, composed of brick and stone, are the only remains. Near it is a remarkable stone, called Ivy Cross.

SUTTON, KING'S (*St. Peter*), a parish, in the union of BRACKLEY, hundred of KING'S-SUTTON, Southern Division of the county of NORTHAMPTON, 6 miles (W. by S.) from Brackley; containing 1270 inhabitants. The living is a discharged vicarage, valued in the king's books at £5. 6. 8.; present net income, £83; patron and impropriator, William Willes, Esq. The church is a beautiful specimen of the later English style of architecture, and the tower is surmounted by a lofty crocketed spire. There are places of worship for Independents and Wesleyan Methodists; also a National school. At Astrop there is a mineral spring, called St. Rumbald's well, which formerly attracted many visiters.

SUTTON, LITTLE, a township, in the parish of EASTHAM, union, and Higher Division of the hundred, of WIRRALL, Southern Division of the county of CHESTER, 7¾ miles (W. N. W.) from Chester; containing 387 inhabitants.

SUTTON, LONG, or ST. MARY'S, a parish, in the union of HOLBEACH, wapentake of ELLOE, parts of HOLLAND, county of LINCOLN, 4¾ miles (E. by S.) from Holbeach; containing 5233 inhabitants. The living is a vicarage, valued in the king's books at £40; present net income, £600; patron, impropriator, and incumbent, Rev. T. L. Bennett. The church is a fine structure, with an ancient stone steeple, and a lofty spire covered with lead. There is a place of worship for Independents; also a National school.

SUTTON, LONG (*Holy Trinity*) a parish, in the union of Langport, hundred of Somerton, Western Division of the county of Somerset, 2¾ miles (S. S. W.) from Somerton; containing 957 inhabitants. The living is a discharged vicarage, valued in the king's books at £8. 18.; present net income, £229; impropriator, Earl of Burlington. An infants' school is partly supported by subscription. The river Yeo, or Ivel, is navigable along the southern boundary of the parish. Roman coins, pateræ, and other antiquities, have been found in the neighbourhood.

SUTTON, LONG (*All Saints*), a parish, in the hundred of Crondall, Odiham and Northern Divisions of the county of Southampton, 2½ miles (S.) from Odiham; containing 326 inhabitants. The living is a perpetual curacy; net income, £40; patrons and impropriators, Master and Brethren of the Hospital of St. Cross.

SUTTON-MADDOCK (*St. Mary*), a parish, in the union of Shiffnall, Shiffnall Division of the hundred of Brimstree, Southern Division of the county of Salop, 6 miles (N.) from Bridgenorth; containing 384 inhabitants. The living is a discharged vicarage, annexed to the rectory of Kemberton, and valued in the king's books at £5. The Severn bounds the parish on the west, and the Shropshire canal forms a junction with that river near the china-manufactory established on its banks.

SUTTON-MALLET, a chapelry, in the parish of Moorlinch, union of Bridg-water, hundred of Whitley, Western Division of the county of Somerset, 5¼ miles (E.) from Bridg-water; containing 153 inhabitants. The chapel has been enlarged, and 80 free sittings provided, the Incorporated Society having granted £50 in aid of the expense.

SUTTON-MANDEVILLE (*All Saints*), a parish, in the union of Tisbury, hundred of Cawden and Cadworth, Hindon and Southern Divisions of the county of Wilts, 7 miles (W. S. W.) from Wilton; containing 256 inhabitants. The living is a rectory, valued in the king's books at £13. 6. 8.; patron, W. Wyndham, Esq. The tithes have been commuted for a rent-charge of £240, subject to the payment of rates, which on the average have amounted to £26; the glebe comprises 50 acres, valued at £70 per annum.

SUTTON-MONTIS, or MONTAGUE (*Holy Trinity*), a parish, in the union of Wincanton, hundred of Catsash, Eastern Division of the county of Somerset, 5¼ miles (S.) from Castle-Cary; containing 178 inhabitants. The living is a rectory, valued in the king's books at £6. 12. 1.; patron, R. Leach, Esq. The tithes have been commuted for a rent-charge of £134. 7., subject to the payment of rates; the glebe comprises 37 acres.

SUTTON-NEAR-SEAFORD, a parish, in the hundred of Flexborough, rape of Pevensey, Eastern Division of the county of Sussex, ¾ of a mile (N. E. by E.) from Seaford, with which the population is returned. The living is a discharged vicarage, with that of Seaford annexed; net income, £167; patrons, Prebendaries of Sutton and Seaford in Chichester Cathedral; impropriators, Earl of Chichester and Prebendaries of Seaford and Bargham in Chichester Cathedral. The church is desecrated.

SUTTON-ON-THE-FOREST (*All Saints*), a parish, in the union of Easingwould, wapentake of Bulmer, North Riding of the county of York; containing 1019 inhabitants, of which number, 493 are in the township of Sutton-on-the-Forest, 8½ miles (N. by W.) from York. The living is a vicarage, valued in the king's books at £17. 3. 4.; present net income, £390; patron and appropriator, Archbishop of York. The church is a very handsome structure. The celebrated Lawrence Sterne was vicar of this parish. There is a place of worship for Independents.

SUTTON-ON-THE-HILL (*St. Michael*), a parish, in the union of Burton-upon-Trent, hundred of Appletree, Southern Division of the county of Derby, 8 miles (W. by S.) from Derby; containing 574 inhabitants. The living is a vicarage, valued in the king's books at £4. 16. 8.; present net income, £225; patron and incumbent, Rev. R. R. Ward; impropriator, Thomas Cox, Esq. Fourteen poor children are educated and apprenticed for an annuity of £20, bequeathed in 1722, by Anne Jackson. The school-house was erected by subscription in 1736. All the lands in the township, except one farm, were given by Humphrey Chetham, Esq., to the Blue-coat hospital at Manchester.

SUTTON-POINTZ, a tything, in the parish of Preston, liberty of Sutton-Pointz, Dorchester Division of the county of Dorset, 4 miles (N. N. E.) from Melcombe-Regis; containing 340 inhabitants. Here was formerly a chapel, dedicated to St. Giles, some remains of which are still visible.

SUTTON-SCOTNEY, a chapelry, in the parish of Wonston, hundred of Buddlesgate, Winchester and Northern Divisions of the county of Southampton, 5¾ miles (S.) from Whitchurch: the population is returned with the parish.

SUTTON-UNDER-BRAILS (*St. Thomas à Becket*), a parish, in the union of Shipston-upon-Stour, Upper Division of the hundred of Westminster, county of Gloucester, though locally in the hundred of Kington, county of Warwick, 4¾ miles (S. E.) from Shipston-upon-Stour; containing 239 inhabitants. The living is a rectory, valued in the king's books at £13. 13. 4.; present net income, £313; patron, Bishop of London. This parish, for electoral purposes, is wholly annexed to the southern division of the county of Warwick. A Sunday school is supported by the rector.

SUTTON-UNDER-WHITESTONE-CLIFFE, a township, in the parish of Felix-Kirk, union of Thirsk, wapentake of Birdforth, North Riding of the county of York, 3½ miles (E. by N.) from Thirsk; containing 328 inhabitants. There is a place of worship for Calvinistic Methodists.

SUTTON-UPON-DERWENT (*St. Michael*), a parish, in the union of Pocklington, Wilton-Beacon Division of the wapentake of Harthill, East Riding of the county of York, 6½ miles (W. by S.) from Pocklington; containing 417 inhabitants. The living is a rectory, valued in the king's books at £14. 14. 7.; present net income, £509; patron, Sir T. Clarges, Bart. There is a place of worship for Wesleyan Methodists. The village is pleasantly situated on the banks of the Derwent, which is here crossed by a substantial stone bridge, and near it is a spring, strongly impregnated with iron. A day and Sunday school is chiefly supported by Sir T. Clarges and the rector.

SUTTON-UPON-TRENT (*All Saints*), a parish, in

the union of SOUTHWELL, Northern Division of the wapentake of THURGARTON, Southern Division of the county of NOTTINGHAM, 5½ miles (S. E.) from Tuxford; containing 1002 inhabitants. The living is a discharged vicarage, valued in the king's books at £5. 6. 8.; present net income, £200; patron, Sir Charles Hulse, Bart.; impropriator, J. E. Denison, Esq. The church exhibits a mixture of various styles of architecture. There are places of worship for Baptists and Wesleyan Methodists. Six poor children are instructed in the parochial school-house for £6 per annum, the bequest of Mary Sprigg, in 1816. This has lately been made a polling-place for the southern division of the county.

SUTTON-VALENCE (*St. Mary*), a parish, in the union of HOLLINGBOURN, hundred of EYHORNE, lathe of AYLESFORD, Western Division of the county of KENT, 4½ miles (S. E. by S.) from Maidstone; containing 1144 inhabitants. The living is a vicarage, with that of East Sutton annexed, valued in the king's books at £7. 9. 7.; present net income, £318; patrons and appropriators, Dean and Chapter of Rochester. The church has been lately rebuilt, in a plain substantial style, with a square tower, and contains 280 free sittings, the Incorporated Society having granted £100 in aid of the expense. There is a place of worship for Independents. The village, called Town Sutton, is situated below the ridge of hills bounding the Weald, and was anciently distinguished for a strong castle, of which part of the keep still remains: it is a highly picturesque ruin, being overgrown with ivy, and having branches of trees sprouting from its walls. Fruit is grown in large quantities, for the supply of the London market. A free grammar school, founded here pursuant to letters patent of the 18th of Elizabeth, by which the master and four wardens of the Clothworkers' Company were constituted governors, is endowed with a rent-charge of £30, by William Lambe and John Franklin, in support of a master and an usher; another of £5, bequeathed in 1713, by George Maplisden, for the usher; and with £200, the gift of Francis Robins, in 1721 (who also bequeathed £3 per annum for the poor), to found two exhibitions, of £10 a year each, in St. John's College, Cambridge. Mr. Lambe also left £4 per annum for a visitation, and the master occupies a house rent free. Thirty boys receive an English education, and instruction in the classics when required. There is a National school for children of this and the adjoining parishes, supported by subscription. Mr. Lambe founded and endowed almshouses for six widows, appointed by the Clothworkers' Company.

SUTTON-VENEY (*St. Leonard*), a parish, in the union and hundred of WARMINSTER, Warminster and Southern Divisions of the county of WILTS, 2 miles (W. S. W.) from Heytesbury; containing 848 inhabitants. The living is a rectory, valued in the king's books at £21; present net income, £800; patron, G. W. Heneage, Esq. The church has been enlarged, and 75 free sittings provided, the Incorporated Society having granted £75 in aid of the expense.

SUTTON-WALDRON (*St. Bartholomew*), a parish, in the union of SHAFTESBURY, hundred of REDLANE, Shaston Division of the county of DORSET, 5½ miles (S.) from Shaftesbury; containing 236 inhabitants. The living is a rectory, valued in the king's books at £9. 9. 4½.; present net income, £169; patron, H. C. Sturt, Esq.

SUTTON-WICK, a township, in the parish of SUTTON-COURTNEY, union of ABINGDON, hundred of OCK, county of BERKS, 1¾ mile (S. by W.) from Abingdon; containing 271 inhabitants.

SWABY (*St. Nicholas*), a parish, in the union of LOUTH, Marsh Division of the hundred of CALCEWORTH, parts of LINDSEY, county of LINCOLN, 5½ miles (W. by N.) from Alford; containing 396 inhabitants. The living is a discharged rectory, valued in the king's books at £12. 1. 10.; present net income, £330; patrons, President and Fellows of Magdalene College, Oxford. There is a place of worship for Wesleyan Methodists; and a National school has been established.

SWADLINCOTE, a chapelry, in the parish of CHURCH-GRESLEY, union of BURTON-UPON-TRENT, hundred of REPTON and GRESLEY, Southern Division of the county of DERBY, 4¾ miles (S. E. by E.) from Burton-upon-Trent; containing 645 inhabitants. It has lately been made a polling-place for the southern division of the county. Here are extensive potteries for the manufacture of yellow earthenware: coal is obtained in the neighbourhood. There is a place of worship for Wesleyan Methodists.

SWAFFHAM (*St. Peter and St. Paul*), a market-town, parish, and the head of a union, in the hundred of SOUTH GREENHOE, Western Division of the county of NORFOLK, 28 miles (W. by N.) from Norwich, and 95 (N. N. E.) from London; containing 3285 inhabitants. This ancient town is situated on an eminence, commanding an extensive view of the surrounding country, and is remarkable for the salubrity of its air, and the longevity of its inhabitants: it consists of four principal and several inferior streets; the houses in general are well built, and the inhabitants are well supplied with water from springs. A book-club is supported by subscriptions among the clergy and gentry in the town and neighbourhood: a neat theatre has been erected; and an elegant assembly-room, on the market-hill, repaired and modernized, at a considerable expense; subscription assemblies are held monthly. On the north-west side of the town is a fine heath, of some thousand acres, admirably adapted for the diversions of racing and coursing: greyhounds are annually entered here for the latter amusement, in the month of November, being subject to the same restrictions as race horses. A charter for a market and two annual fairs was granted by King John to one of the Earls of Richmond, who were anciently lords of the manor. The market is on Saturday; and fairs are held on May 12th for sheep, July 21st, and November 3rd, for sheep and cattle. The market-place, a fine area, surrounded by handsome buildings, contains a beautiful cross, erected in 1783 by Lord Orford, consisting of a circular dome, covered with lead, supported on eight pillars, and crowned with a figure of Ceres, the whole being enclosed with palisades. The county magistrates hold a weekly petty session: the general quarter sessions are held here at Midsummer only, by adjournment from Norwich; and manorial courts leet and baron are held annually in April or May. This town has been made the place of election and a polling-place for the western division of the county. The town having formerly been held in royal demesne, the inhabitants are exempt from toll, and from being empanelled on juries, or any recognizances, except in the court of the manor. Anciently the Earl of Richmond had a prison in the town; and a house of correction, or bridewell, was erected in the 41st of Elizabeth, for the convenience of several adjoining hundreds. The New Bridewell was built about 1787, and will hold upwards of 100 prisoners: attached to it is a chapel, the chaplain, who has a stipend of £200 per annum, being elected by the justices: a treadmill was erected in 1822, and in 1825 a handsome residence for the governor.

The living is a vicarage, with the rectory of Threxton annexed, valued in the king's books at £14. 5. 10.; present net income, £738; patron, Bishop of Norwich; appropriators, Dean and Chapter of Westminster. The church contains 100 free sittings, the Incorporated Society having granted £50 in aid of the expense; it is a splendid and spacious cruciform structure, in the later style of English architecture, with a stately embattled tower crowned with turrets and surmounted by a well-proportioned spire: the nave is separated from the aisles by lofty ranges of slender clustered columns supporting the roof, which is richly ornamented with figures of angels carved in Irish oak; in the transepts are three chapels. The interior is lighted by a fine range of clerestory windows; and there are several handsome monuments and some brasses, bearing a variety of inscriptions: in a library attached to the church is a curious missal. The north aisle is commonly reported to have been built by John Chapman, a tinker of this town; concerning which circumstance there is a curious monkish legend, and there are various devices of a pedler, and others representing a person keeping a shop, in different parts of the church, which are, in all likelihood, only rebuses on the name of Chapman, a conceit very prevalent in former times; the founder having probably been a person of that name, who was churchwarden in 1462. Here was anciently a free chapel, dedicated to St. Mary; and about half a mile distant, in a hamlet formerly called Guthlac's Stow, now Goodluck's Close, stood another, dedicated to St. Guthlac. There are places of worship for Baptists and Wesleyan Methodists. The grammar school was founded by Nicholas Hammond, Esq., who by will bequeathed £500 for erecting a school-house, and £500 for the instruction of 20 boys. A National school is supported by subscription. Several houses have been given, at different periods, as rent free residences for the poor, for whom there is also a workhouse, which was the residence of the rector before the impropriation took place. The poor law union of Swaffham comprises 29 parishes or places, under the care of 42 guardians, and contains, according to the census of 1831, a population of 12,474. John de Swaffham, a man of great learning, and a strenuous opponent of Wickliffe, who was raised to the see of Bangor by Pope Gregory II., was a native of this town.

SWAFFHAM-BULBECK (*St. Mary*), a parish, in the union of Newmarket, hundred of Staine, county of Cambridge, 6 miles (W. by S.) from Newmarket; containing 727 inhabitants. This parish is partly bounded by the Cam, from which river there is a cut called Swaffham Lode, navigable to the village. Here is a quarry of chalk marl, which is extensively worked for building purposes. The living is a discharged vicarage, valued in the king's books at £16. 10.; present net income, £219; patron and appropriator, Bishop of Ely. A charity school was founded here in 1721, by Mrs. Frances Towers, who gave £50 towards it, and the Rev. Mr. Hill added as much as purchased 12 acres of land, producing £20 per annum; the school is conducted on the National plan. Here are the remains of a Benedictine nunnery, founded before the reign of John, by one of the Bolebecs, and dedicated to St. Mary. At the dissolution its revenue was estimated at £46. 18. 10.; and the house is now occupied by paupers.

SWAFFHAM-PRIOR (*St. Cyriat*), a parish, in the union of Newmarket, hundred of Staine, county of Cambridge, 5½ miles (W. by N.) from Newmarket; containing 1102 inhabitants. This parish, which includes part of Newmarket heath, is bounded on the north by the Cam; and several navigable drains, or lodes, communicating with that river, pass through it. A market and fair, anciently granted to the prior of Ely, have been long disused. The living consists of the consolidated vicarages of *St. Cyriac* and *St. Mary*, the former valued in the king's books at £16. 18. 11½., and the latter at £14. 12. 11.; present net income, £301; patrons, alternately, the Bishop, and the Dean and Chapter, of Ely, the latter the appropriators. Here were formerly two churches in the same cemetery; that of St. Mary has fallen to ruin, except the tower, which, from the peculiarity of its situation, still forms an interesting object. The church dedicated to St. Cyriac has been lately rebuilt, and contains 50 free sittings, the Incorporated Society having granted £15 in aid of the expense. There are an endowed school, and a National school.

SWAFIELD (*St. Nicholas*), a parish, in the hundred of Tunstead, Eastern Division of the county of Norfolk, 1½ mile (N. by E.) from North Walsham; containing 155 inhabitants. The living is a discharged rectory, valued in the king's books at £6; present net income, £200: it is in the patronage of the Crown, in right of the Duchy of Lancaster.

SWAINBY, a joint township with Allarthorp, in the parish of Pickhill, union of Bedale, wapentake of Hallikeld, North Riding of the county of York, 6 miles (E. S. E.) from Bedale; containing 27 inhabitants. Butchers' knives, and some other articles of cutlery, are manufactured. There are places of worship for Primitive and Wesleyan Methodists. A Premonstratensian abbey, founded here by Hellewise, daughter of Ranulph de Glanville, in the time of Henry II., was afterwards removed to Coverham.

SWAINSCOE, a joint township with Blore, in the parish of Blore, Northern Division of the hundred of Totmonslow and of the county of Stafford, 4 miles (W. N. W.) from Ashbourn: the population is returned with Blore.

SWAINSTHORPE, a parish, in the union of Henstead, hundred of Humbleyard, Eastern Division of the county of Norfolk, 4¼ miles (N. N. E.) from Stratton-St. Mary; containing 180 inhabitants. The living consists of the united rectories of *St. Mary* and *St. Peter*, with the rectory of Newton-Flotman, valued in the king's books at £12. 13. 4.; present net income, £422; patron and incumbent, Rev. R. C. Long. The tithes have been commuted for a rent-charge of £245, subject to the payment of rates, which on the average have amounted to £49; the glebe comprises 45 acres, valued at £63 per annum. The church is a small ancient structure, with a steeple, circular at the basement and hexangular above; that of St. Mary was pulled down at the Reformation.

SWAINSWICK (*St. Mary*), a parish, in the union of Bath, hundred of Bath-Forum, Eastern Division of the county of Somerset, 3 miles (N. by E.) from Bath; containing 427 inhabitants. The living is a discharged rectory, valued in the king's books at £9. 17. 8.; patrons, Provost and Fellows of Oriel College, Oxford. The impropriate tithes have been commuted for a rent-charge of £55. 19., and the rectorial for £190, subject to the payment of rates, which on the average on the latter have amounted to £2. 12. 6.; the glebe comprises 15 acres, valued at £45 per annum. The church contains the remains of the celebrated William Prynne, barrister-at-law, an active statesman and public writer during the disturbed reign of Charles I.: he was born at this place in 1600, and died in 1669. Here is a school on the National plan.

SWALCLIFFE (*St. Peter and St. Paul*), a parish, in the union of Banbury, partly in the hundred of Banbury, partly in that of Bloxham, and partly in that of Dorchester, county of Oxford, 6 miles (W. S. W.) from Banbury; containing 1962 inhabitants. The living is a vicarage, valued in the king's books at £7. 9. 4½.; present net income, £209; patrons and impropriators, Warden and Fellows of New College, Oxford. There are chapels of ease at Epwell and East Shutford. One-third of the sum of £72. 15. per annum, arising from the rent of certain land bequeathed by an unknown individual, is paid in support of a school, and the residue for other charitable purposes.

SWALECLIFFE (*St. John the Baptist*), a parish, in the union of Blean, hundred of Bleangate, lathe of St. Augustine, Eastern Division of the county of Kent, 6½ miles (N.) from Canterbury; containing 133 inhabitants. The living is a rectory, valued in the king's books at £11. 9. 4½.; patron, Earl Cowper. The tithes have been commuted for a rent-charge of £315, subject to the payment of rates, which on the average have amounted to £48. 10.; the glebe comprises 9½ acres. A new parsonage-house was erected by the late incumbent. The church contains some costly monuments to the families of Wykeham, Loggin, and Duncombe. A Sunday school is endowed with the interest of £120 three per cent. consols. The village is delightfully situated on rising ground, and was the occasional residence of William of Wykeham; and in the old parsonage-house is a mantel-piece said to have been designed by him.

SWALLOW (*Holy Trinity*), a parish, in the union of Caistor, wapentake of Bradley-Haverstoe, parts of Lindsey, county of Lincoln, 4 miles (E. N. E.) from Caistor; containing 168 inhabitants. The living is a rectory, valued in the king's books at £7. 10. 10.; present net income, £408; patron, Lord Yarborough.

SWALLOWCLIFFE, a parish, in the union of Tisbury, hundred of Dunworth, Hindon and Southern Divisions of the county of Wilts, 6½ miles (S. E.) from Hindon; containing 278 inhabitants. The living constitutes the endowment of a prebend in the church of Heytesbury, in the patronage of the Dean of Salisbury, as Dean of Heytesbury, valued in the king's books at £8. 13. 4.; present net income, £38. There is an allotment of common land of about 20 acres belonging to the poor, and now let for £20 per annum.

SWALLOWFIELD (*All Saints*), a parish, in the union of Wokingham, partly in the hundred of Charlton, county of Berks, but chiefly in the hundred of Amesbury, Wokingham and Southern Divisions of the county of Wilts, 6 miles (S. by E.) from Reading; containing 1106 inhabitants. The living is annexed to the vicarage of Shinfield. There is a place of worship for Wesleyan Methodists; and three schools are supported by subscription. A fair is held here on June 9th. The celebrated Lord Chancellor Clarendon, after his retirement from public life, resided at the manor-house, then the property of his son, where he wrote "The History of the Rebellion."

SWALWELL, a township, in the parish of Whickham, Western Division of Chester ward, Northern Division of the county palatine of Durham, 4¾ miles (W. by S.) from Gateshead; containing 1372 inhabitants, the greater number of whom are employed at the extensive iron-works of Messrs. Crawley, Millington, and Co., where anchors of the largest size, chain cables, pumps, and cylinders for steam-engines, with every other description of cast and wrought iron articles, are produced. This factory was founded about 1690, by Sir Ambrose Crawley, who was originally a blacksmith: he also benevolently established schools for instructing the children of his workmen, provided asylums for their widows and orphans, and adopted the most laudable regulations for orderly conduct among them. An annual festival is held, on May 22nd, in the village, which is of considerable size: a freemasons' lodge is occasionally held at the Queen's Head Inn. There are places of worship for Presbyterians and Wesleyan Methodists. William Shield, the celebrated musical composer, was a native of this place.

SWANAGE, county of Dorset.—See SWANWICH.

SWANBOURNE (*St. Swithin*), a parish, in the union of Winslow, hundred of Cottesloe, county of Buckingham, 2¼ miles (E.) from Winslow; containing 668 inhabitants. The living is a vicarage, valued in the king's books at £9. 9. 7.; present net income, £158: it is in the patronage of the Crown; impropriators,—Graves and—Lamb, Esqrs. There is a place of worship for Baptists. Nicholas Godwin, in 1712, bequeathed a rent charge of £15 for teaching poor children: this income is now carried to the account of a National school.

SWANLAND, a township, in the parish of North Ferriby, union of Sculcoates, county of the town of Kingston-upon-Hull, 6¾ miles (W. by S.) from Kingston-upon-Hull; containing 478 inhabitants. There is a place of worship for Independents.

SWANNINGTON, a chapelry, in the parish of Whitwick, union of Ashby-de-la-Zouch, hundred of West Goscote, Northern Division of the county of Leicester, 4½ miles (E. by S.) from Ashby-de-la-Zouch; containing 549 inhabitants. There is a place of worship for Wesleyan Methodists. A railway has been constructed to Leicester, for conveying coal and limestone.

SWANNINGTON (*St. Margaret*), a parish, in the union of St. Faith, hundred of Eynsford, Eastern Division of the county of Norfolk, 3½ miles (S. E.)

from Reepham; containing 370 inhabitants. The living is a discharged rectory, with the vicarage of Wood-Dalling annexed, valued in the king's books at £6. 11. 5.; present net income, £376; patrons, Master and Fellows of Trinity Hall, Cambridge.

SWAN, OLD, a chapelry, in the parish of WALTON-ON-THE-HILL, union and hundred of WEST DERBY, county palatine of LANCASTER, 3 miles (E.) from Liverpool: the population is returned with the parish. This place is situated on the turnpike road to Manchester and Birmingham. Some extensive glass-works are carried on, and here is one of the largest roperies in the kingdom. The Liverpool cattle market is held here. The living is a perpetual curacy; the income, valued at £150 per annum, arises chiefly from the rents of the pews; patron and incumbent, Rev. Thomas Gardner. The chapel, dedicated to St. Ann, is a plain neat edifice, erected in 1831 by the late Thomas Gardner, Esq., at an expense of £2000, aided by a grant of £400 from the Incorporated Society. Here is a Roman Catholic chapel. Two schools in connection with the Established Church, and in which are about 120 children, are supported by subscription.

SWANSCOMBE (*ST. PETER AND ST. PAUL*), a parish, in the union of DARTFORD, hundred of AXTON, DARTFORD, and WILMINGTON, lathe of SUTTON-AT-HONE, Western Division of the county of KENT, 4 miles (E.) from Dartford; containing 1166 inhabitants. The name of this place, anciently "Swenes-Camp," is supposed to have been derived from the encampment of Sweyn, King of Denmark, who, on his arrival in England, landed his forces here. It is also celebrated as the spot where the Kentish men, carrying boughs of trees in their hands and prepared for battle, surrounded William the Conqueror, from whom they obtained a confirmation of their ancient privileges, particularly of the law of gavelkind, the existence of which, confined at that time almost exclusively to this county, appears, in the opinion of Camden, to confirm the authenticity of that fact, which otherwise rests only on the authority of a monkish historian. The parish is bounded on the north by the river Thames, from which the village, surrounded with woods, has an interesting and picturesque appearance. From Greenhithe large quantities of chalk and lime are sent to the neighbouring ports. The living is a rectory, valued in the king's books at £25. 13. 4.; present net income, £612; patrons, Master and Fellows of Sydney Sussex College, Cambridge. The church is principally in the early style of English architecture.

SWANSCOMBE-CROSS, a hamlet, in the parish of SWANSCOMBE, hundred of AXTON, DARTFORD, and WILMINGTON, lathe of SUTTON-AT-HONE, Western Division of the county of KENT: the population is returned with the parish. In this hamlet are a large manufactory of Roman cement and some extensive chalk works, affording employment to more than 100 persons.

SWANTHORPE, a joint tything with Crondall, in the parish and hundred of CRONDALL, Odiham and Northern Divisions of the county of SOUTHAMPTON, 3 miles (S. E.) from Odiham: the population is returned with Crondall.

SWANTON-ABBOTT (*ST. MICHAEL*), a parish, in the union of AYLSHAM, hundred of SOUTH ERPINGHAM, Eastern Division of the county of NORFOLK, 1 mile (N.) from Scottow; containing 448 inhabitants. This parish is situated on the turnpike road from Cromer to Norwich; and some of the poorer inhabitants are employed in the weaving of cotton, which is carried on here to a small extent. The living is a discharged rectory, valued in the king's books at £6. 10.; patron, Rev. W. Jex Blake. The tithes have been commuted for a rent-charge of £274. 11., subject to the payment of rates, which on the average have amounted to £80; the glebe comprises $1\frac{1}{2}$ acre, valued at £2 per annum. The church has a square embattled tower, and contains an ancient font; there are also a brass with the effigy of the Rev. Stephen Multon, rector of the parish in the fifteenth century, who was interred in the chancel; and monumental inscriptions to Gallant, a former rector, and to the Blakes. There is a place of worship for Wesleyan Methodists; and a National school is supported by subscription. A fund arising from various bequests is distributed in clothing among the poor.

SWANTON-MORLEY (*ALL SAINTS*), a parish, in the union of MITFORD and LAUNDITCH, hundred of LAUNDITCH, Western Division of the county of NORFOLK, $3\frac{3}{4}$ miles (N. E.) from East Dereham; containing 837 inhabitants. The living is a rectory, with that of Worthing annexed, valued in the king's books at £15. 10. $2\frac{1}{2}$.; present net income, £920; patron, Edward Lombe, Esq. The church was erected in 1379, on an eminence in the centre of the village. Near it stood the ancient manor-house, surrounded by a moat. A school is supported by the rector. Here is an extensive and long-established paper-manufactory. William Small, in 1651, bequeathed a rent-charge of £11 for teaching and apprenticing children; and the town lands yield a rental of £84, of which a moiety is appropriated to the repair of the body of the church, and the remainder to the poor.

SWANTON-NOVERS (*ST. EDMUND*), a parish, in the union of WALSINGHAM, hundred of HOLT, Western Division of the county of NORFOLK, $6\frac{1}{4}$ miles (S. W.) from Holt; containing 377 inhabitants. The living is a discharged rectory, annexed to that of Wood-Norton, and valued in the king's books at £4. 15. $2\frac{1}{2}$. The tithes have been commuted for a rent-charge of £131, subject to the payment of rates, which on the average have amounted to £30; the glebe comprises 19 acres, valued at £24 per annum.

SWANWICH (*ST. MARY THE VIRGIN*), a market-town and parish, in the union of WAREHAM and PURBECK, hundred of ROWBARROW, Wareham Division of the county of DORSET, 7 miles (E. S. E.) from Corfe-Castle, and 122 (S. W. by W.) from London; containing 1734 inhabitants. In the Saxon Chronicle this place is called *Swanawic;* Asser Menevensis names it *Swanavine* and *Gnavewic*, and in Domesday-book it is written *Snanwic* and *Sonwic*. The earliest and principal historical circumstance which we find on record connected with it is the destruction, by a violent storm in 877, of a Danish fleet, on its way from Wareham to the relief of Exeter, in the bay on which the town stands; and a similar disaster is said to have befallen another of their fleets, after its defeat by Alfred, in the same place and year. The town, which is situated on the small bay of the same name, consists principally of one street about a mile long, containing many neat houses, built and roofed with stone; and the bay having of late years become a place of resort for sea-bathing, has led to the erection of some new houses in the town, among which are a library containing more than 2000 volumes, and the Royal Victoria Hotel, a spacious building upon arches (precluding the possibility of damp), sheltered in its different aspects from all winds, and furnished with a warm air apparatus similar to that in the British Museum, by which any required temperature may be maintained throughout the winter. In the house are hot and cold sea-water baths, supplied by pipes; and there are suites of apartments for the reception of from six to eight families. The sea sands afford pleasant walks, rides, and drives; considerable improvements have taken place in the neighbourhood; and the mildness and salubrity of the air, possessing all the advantages of a southern climate, render this place peculiarly desirable as a winter residence. The manufacture of straw-plat employs many of the females, but the chief occupation of the inhabitants is derived from working the numerous quarries in the parish, which produce great quantities of the freestone called Purbeck stone, which is conveyed in carts to boats, and by them to the larger vessels in the bay, for transmission to various parts of the kingdom, a small quantity being also sent abroad. The bay affords a tolerable harbour for vessels of 300 tons' burden. In addition to other public works, Ramsgate pier was constructed of this stone, 50,000 tons of which were conveyed thither for the purpose. The quarry-men are governed by local laws, or regulations, by which none but their sons, who must serve an apprenticeship of seven years, are allowed to work. The market is on Tuesday and Friday.

The living is a rectory, valued in the king's books at £27. 9. 9½.; present net income, £550; patron, John Hales Calcraft, Esq. The church is an extremely ancient structure, with a very large chancel and a lofty tower: it was formerly a chapel to the vicarage of Worth-Matravers, but was made parochial in 1500. There are places of worship for Independents and Wesleyan Methodists; and a school is partly supported by subscription. Fossils of different fish, particularly bream, are frequently found in the quarries, and there are also some mineral springs in the parish. Among various bones discovered in the isle of Purbeck are large vertebræ and two bones of the iguanodon, a fragment of a femur, bones of large and small crocodiles, of the plesiosaurus, and of various reptiles.

SWANWICK, a hamlet, in the parish of ALFRETON, hundred of SCARSDALE, Northern Division of the county of DERBY, 1½ mile (S. by W.) from Alfreton, with which the population is returned. In the neighbourhood are extensive collieries. There are places of worship for Baptists and Wesleyan Methodists. A free school, erected in 1740, was endowed by Mrs. Elizabeth Turner with £500, which was laid out in the purchase of a house and lands of the annual value of £60, now in the occupation of the schoolmaster, who, for this endowment, instructs 40 poor children.

SWARBY (*St. Mary and All Saints*), a parish, in the union of SLEAFORD, wapentake of ASHWARDHURN, parts of KESTEVEN, county of LINCOLN, 5½ miles (N. W. by N.) from Falkingham; containing 142 inhabitants. The living is a discharged vicarage, valued in the king's books at £6; present net income, £60; patron and impropriator, Sir T. Whichcote, Bart. The church is a handsome structure, principally in the later English style.

SWARDESTON (*St. Andrew*), a parish, in the union of HENSTEAD, hundred of HUMBLEYARD, Eastern Division of the county of NORFOLK, 4½ miles (S. S. W.) from Norwich; containing 371 inhabitants. The living is a discharged vicarage, valued in the king's books at £6; patron, Rev. J. H. Steward.

SWARKESTONE (*St. James*), a parish, in the union of SHARDLOW, hundred of REPTON and GRESLEY, Southern Division of the county of DERBY, 5¾ miles (S. by E.) from Derby; containing 308 inhabitants. This place was distinguished during the civil war of the seventeenth century, by the efforts of Col. Hastings, in 1643, to secure the passage of the Trent for the royalists, for which purpose he threw up some works at the bridge, and also placed a garrison in the house of Sir John Harpur, which he fortified for that purpose; but Sir John Gell, marching hither with Sir George Gresley's troops, after an obstinate defence, succeeded in driving the garrison from their post, and securing the pass of the river for the parliamentarians. The village is pleasantly situated on the river Trent, over which is a bridge, 1304 yards in length, comprising additional arches beyond the span of the river, which is 138 yards in breadth, to secure a passage over the low grounds, which are usually flooded in winter; this bridge, which originally was not more than twelve feet in breadth, was widened a few years since, to allow carriages to pass each other. The Trent and Mersey canal, which is here joined by the Derby canal, passes through the parish. The living is a rectory, valued in the king's books at £5; present net income, £182; patron, Sir George Crewe, Bart. The church is principally in the Norman style of architecture, but much disfigured by the insertion of modern windows: it contains 70 free sittings, the Incorporated Society having granted £75 in aid of the expense.

SWARLAND, a township, in the parish of FELTON, union of ALNWICK, Eastern Division of COQUETDALE ward, Northern Division of the county of NORTHUMBERLAND, 8¼ miles (S. by W.) from Alnwick; containing 210 inhabitants. An obelisk of white freestone, erected by the late Alexander Davison, Esq., to the memory of Admiral Lord Nelson, stands near Swarland Hall, and close to the great road between Morpeth and Alnwick.

SWARRATON, a parish, in the union of ALRESFORD, hundred of BOUNTISBOROUGH, Winchester and Northern Divisions of the county of SOUTHAMPTON, 4 miles (N. N. W.) from New Alresford; containing 120 inhabitants. A stream from one of the sources of the river Itchen runs through the valley. Here is Grange House, built after the model of the Temple of Theseus at Athens, and the splendid seat of Lord Ashburton. The living is a discharged rectory, valued in the king's books at £4. 5. 2½.; patron, Lord Ashburton. The tithes have been commuted for a rent-charge of £110, subject to the payment of rates. A school is supported by Lady Ashburton.

SWATON (*St. Michael*), a parish, in the union of SLEAFORD, wapentake of AVELAND, parts of KESTEVEN, county of LINCOLN, 5¾ miles (N. E.) from Falkingham; containing 311 inhabitants. The living is a vicarage, with the rectory of Spanby annexed, valued in the king's books at £12. 7. 1.; present net income, £514: the patronage and impropriation belong to Mrs. Knapp.

SWAVESEY (*St. Andrew*), a parish, in the union of St. Ives, hundred of Papworth, county of Cambridge, 5¼ miles (E. S. E.) from St. Ives; containing 1115 inhabitants. The living is a vicarage, valued in the king's books at £7. 6. 8.; present net income, £428; patrons, Master and Fellows of Jesus' College, Cambridge; appropriator, Bishop of Ely. The appropriate tithes have been commuted for a rent-charge of £750, and the vicarial for £265, subject to the payment of rates, which on the average on the latter have amounted to £64; the appropriate glebe comprises 72 acres, valued at £52. 10. per annum. The church anciently belonged to an alien priory of Black monks, founded here, soon after the Conquest, as a cell to the abbey of St. Sergius and St. Bachus, and that of St. Briocus, at Angiers; at the suppression it was given by Richard II. to the priory of St. Anne, Coventry, and some slight remains of the monastic buildings are still visible. There is a place of worship for Baptists. A market and fair were granted in 1243, to the family of Zouch, the site of whose ancient castle is about half a mile south-west from the church.

SWAY, a hamlet, in the parish of Boldre, hundred of Christchurch, Lymington and Southern Divisions of the county of Southampton, 3½ miles (N. W.) from Lymington: the population is returned with the parish. It is in contemplation to build a chapel here. There is a place of worship for Baptists.

SWAYFIELD (*St. Nicholas*), a parish, in the union of Bourne, wapentake of Beltisloe, parts of Kesteven, county of Lincoln, 2 miles (S. by W.) from Corby; containing 260 inhabitants. The living is a rectory, valued in the king's books at £11. 2. 11.; present net income, £391: it is in the patronage of the Crown.

SWEETHOPE, a township, in the parish of Thockrington, union of Bellingham, North-Eastern Division of Tindale ward, Southern Division of the county of Northumberland, 9¾ miles (E. by S.) from Bellingham; containing 18 inhabitants. It is occupied as an extensive sheep farm.

SWEFFLING (*St. Mary*), a parish, in the union and hundred of Plomesgate, Eastern Division of the county of Suffolk, 2¾ miles (W. N. W.) from Saxmundham; containing 336 inhabitants. The living is a rectory, valued in the king's books at £9. 2. 8½.; present net income, £262; patron, Thomas Williams, Esq.

SWELL (*St. Catherine*), a parish, in the union of Langport, hundred of Abdic and Bulstone, Western Division of the county of Somerset, 4 miles (W. S. W.) from Langport; containing 87 inhabitants. The living is a discharged vicarage, endowed with the rectorial tithes, annexed to that of Fivehead, and valued in the king's books at £5. 10. 5. The tithes have been commuted for a rent-charge of £168. 9. 6., subject to the payment of rates, which on the average have amounted to £8; the glebe comprises 28 acres, valued at £35 per annum.

SWELL, LOWER (*St. Mary*), a parish, in the union of Stow-on-the-Wold, Upper Division of the hundred of Slaughter, Eastern Division of the county of Gloucester, 1 mile (W.) from Stow-on-the-Wold; containing 298 inhabitants. The living is a discharged vicarage, valued in the king's books at £6. 12. 3½.; present net income, £100; patrons and appropriators, Dean and Canons of Christ-Church, Oxford.

SWELL, UPPER, a parish, in the union of Stow-on-the-Wold, Upper Division of the hundred of Kiftsgate, Eastern Division of the county of Gloucester, 1¼ mile (N. W.) from Stow on-the-Wold; containing 95 inhabitants. The living is a discharged rectory, valued in the king's books at £6. 14. 6.; present net income, £85; patron, Charles Pole, Esq. A school is supported by the proprietor and the clergyman of the parish. The Roman Fosse-way bounds the parish on the east.

SWEPSTONE (*St. Peter*), a parish, in the union of Ashby-de-la-Zouch, hundred of West Goscote, Northern Division of the county of Leicester, 4¾ miles (S. by E.) from Ashby-de-la-Zouch; containing, with Newton, 627 inhabitants. The living is a rectory, with that of Snareston annexed, valued in the king's books at £21. 18. 4.; present net income, £894; patron, — Charnell, Esq. The river Mease, and the Ashby-de-la-Zouch canal, run through the parish.

SWERFORD (*St. Mary*), a parish, in the union of Chipping-Norton, hundred of Chadlington, county of Oxford, 4¼ miles (N. E. by E.) from Chipping-Norton; containing 441 inhabitants. The living is a rectory, valued in the king's books at £15. 7. 1.; present net income, £496; patrons, President and Fellows of Magdalene College, Oxford. On a hill to the north of the church, called Castle Hill, are some remains indicating the existence of ancient military works; according to tradition there was a camp, but it is not noticed by any writer.

SWETTENHAM, a parish, in the union of Congleton, hundred of Northwich, Southern Division of the county of Chester; containing 421 inhabitants, of which number, 247 are in the township of Swettenham, 5 miles (N. W.) from Congleton. The living is a rectory, valued in the king's books at £5. 1. 3.; present net income, £255; patron, Rev. J. D'Arcey. The church is of brick, with a tower, forming a conspicuous object in the romantic scenery on the banks of the Dane.

SWILLAND (*St. Mary*), a parish, in the union and hundred of Bosmere and Claydon, Eastern Division of the county of Suffolk, 6 miles (N. by E.) from Ipswich; containing 272 inhabitants. The living is a discharged vicarage, endowed with the rectorial tithes and valued in the king's books at £7. 8. 4½.; present net income, £227: it is in the patronage of the Crown. The church is an ancient edifice, with a richly ornamented Norman arch leading into it from the porch. Here is a National school.

SWILLINGTON (*St. Mary*), a parish, in the Lower Division of the wapentake of Skyrack, West Riding of the county of York, 6½ miles (E. S. E.) from Leeds; containing 523 inhabitants. The living is a rectory, valued in the king's books at £16. 1. 8.; present net income, £510; patron, Sir John Lowther, Bart. The church is partly in the decorated, and partly in the later, style of English architecture. Sir William Lowther, Bart., and others, gave £92 for the erection of a school, which the former endowed with land producing a small annuity for teaching six poor children.

SWIMBRIDGE (*St. James*), a parish, in the union of Barnstaple, hundred of South Molton, South

Molton and Northern Divisions of the county of DEVON, $4\frac{1}{4}$ miles (S. E. by E.) from Barnstaple; containing 1511 inhabitants. The living is a perpetual curacy, annexed to that of Landkey. The church is a fine specimen of the later style of English architecture, with a spire: it contains a stone pulpit, enriched with figures of saints, and several monuments to the Chichester family: the nave and chancel are separated by a handsome wooden screen. There is a place of worship for Wesleyan Methodists. The village is situated in a hollow surrounded by verdant hills of singular formation. Limestone is found here, enclosed in a strata of hard blueish building stone. Sixteen children are instructed for £6 per annum, arising from the parish estate.

SWINBROOK (*St. Mary*), a parish, in the union of WITNEY, hundred of CHADLINGTON, county of OXFORD, $2\frac{1}{2}$ miles (E.) from Burford; containing 222 inhabitants. The living is a perpetual curacy; net income, £57; patron and appropriator, Chancellor of the Cathedral Church of Salisbury. The church is partly Norman, and partly of later date, with a remarkable tower open by an arch to the west; it has some remains of a rood-loft and wooden screen-work. In the chancel, which is separated from the nave by a finely pointed arch resting upon columns with beautifully ornamented capitals, are numerous memorials, and some costly monuments to the family of Fettiplace, who resided in a mansion here for more than four centuries. The river Windrush runs through the parish, which possesses a very considerable right in the forest of Wychwood. Mrs. Anne Pytts, in 1715, endowed a school with £40 per annum, for teaching boys of Swinbrook and the parish of Widford.

SWINBURN, a joint township with Colwell, in the parish of CHOLLERTON, North-Eastern Division of TINDALE ward, Southern Division of the county of NORTHUMBERLAND, 7 miles (N.) from Hexham: the population is returned with Colwell. It is bounded on the west by a rivulet of the same name, tributary to the North Tyne. There is a domestic Roman Catholic chapel at Swinburn Castle, a handsome stone structure belonging to Ralph Riddell, Esq.

SWINBURN, LITTLE, a joint township with Whiteside-Law, in the parish of CHOLLERTON, North-Eastern Division of TINDALE ward, Southern Division of the county of NORTHUMBERLAND, $9\frac{1}{2}$ miles (N. by E.) from Hexham: the population is returned with the township of Chollerton.

SWINCOMB (*St. Botolph*), a parish, in the union of HENLEY, hundred of EWELME, county of OXFORD, $5\frac{1}{4}$ miles (E. by N.) from Wallingford; containing 367 inhabitants. The living is a rectory, valued in the king's books at £7. 9. $4\frac{1}{2}$.; present net income, £325; patron, Rev. Chas. Edmund Keene. A school is partly supported by endowment, aided by subscription and the children's work.

SWINDALE, a chapelry, in the parish of SHAP, West ward, county of WESTMORLAND, 12 miles (W. N. W.) from Orton: the population is returned with the parish. The living is a perpetual curacy; net income, £56; patron, Vicar of Shap. The chapel was built at the expense of the inhabitants, in 1749. Near it is a school, founded in 1703, by Mr. Baxter, and endowed with a rent-charge of £25.

SWINDEN, a township, in the parish of GISBURN, union of SETTLE, Western Division of the wapentake of STAINCLIFFE and EWCROSS, West Riding of the county of YORK, $7\frac{1}{2}$ miles (S. E.) from Settle; containing 36 inhabitants.

SWINDERBY (*All Saints*), a parish, in the union of NEWARK, Lower Division of the wapentake of BOOTHBY-GRAFFO, parts of KESTEVEN, county of LINCOLN, $8\frac{1}{2}$ miles (S. W. by W.) from Lincoln; containing 449 inhabitants. The living is a vicarage, valued in the king's books at £3. 19. $9\frac{1}{2}$.; present net income, £148; patron and incumbent, Rev. W. C. Kendall; impropriator, — Burgham, Esq. There is a place of worship for Wesleyan Methodists; also a school, endowed with £5 per annum, for which twelve children are instructed. Half-Way Houses, in this parish, has been made a polling-place for the parts of Kesteven and parts of Holland.

SWINDON (*St. Lawrence*), a parish, in the union and hundred of CHELTENHAM, Eastern Division of the county of GLOUCESTER, $2\frac{1}{4}$ miles (N. N. W.) from Cheltenham; containing 225 inhabitants. The living is a discharged rectory, valued in the king's books at £13. 1. $0\frac{1}{2}$.; present net income, £339; patron, Rev. W. Raymond. The church contains 50 free sittings, the Incorporated Society having granted £30 in aid of the expense. A day and Sunday school is supported by subscription.

SWINDON (*Holy Rood*), a market-town and parish, in the union of HIGHWORTH and SWINDON, hundred of KINGSBRIDGE, Swindon and Northern Divisions of the county of WILTS, 41 miles (N.) from Salisbury, and 81 (W.) from London; containing 1742 inhabitants. This place is mentioned in Domesday-book, but nothing further connected with its ancient history is on record. The town is pleasantly situated on the summit of a considerable eminence, commanding extensive and beautiful views of parts of Berkshire and Gloucestershire: the principal street is wide, and contains some good houses, and the general aspect of the town is prepossessing: there is a good supply of water, which is of excellent quality. No branch of manufacture is carried on. Some very extensive quarries are worked in the immediate vicinity, the stones raised from which are usually very large, and of an excellent quality. The Wilts and Berks canal passes about half a mile from the town; and a reservoir, covering about 70 acres, for its supply in dry seasons, has been constructed about a mile and a half from it, and is partly in this parish, adding much to the beauty of the scenery. The line of the Great Western railway passes through the parish. The market is on Monday, for corn, &c., and on every second Monday for cattle; the latter is termed the great market. Fairs are held on the Monday before April 5th, the second Monday after May 12th, the second Monday after Sept. 11th, and the second Monday in Dec., for cattle of all kinds, pedlery, &c. The petty sessions for the Swindon Division of the hundred are held here; and the town has been made a polling-place for the northern division of the county, and for the borough of Cricklade, of which Swindon forms a portion. The living is a vicarage, valued in the king's books at £17; present net income, £302: it is in the patronage of the Crown; impropriator, Col. Vilett. The church, situated at the south-eastern extremity of the town, is a small unadorned edifice, with a low tower; the interior is neatly fitted up. There are places of worship for Independents and Wesleyan Methodists. The free school, which was established in 1764, was founded by the gentry of the town and neighbourhood, and is supported partly by an endowment of about £40 per annum; there is a house for the master. In 1837, a National school was erected, and the two establishments were united.

SWINDON, a township, partly in the parish of PANNALL, Lower Division, and partly in that of KIRKBY-OVERBLOWS, Upper Division, of the wapentake of CLARO, West Riding of the county of YORK, 6 miles (W. by S.) from Wetherby; containing 46 inhabitants.

SWINE (*St. Mary*), a parish, in the union of SKIRLAUGH, chiefly in the Middle, but partly in the Northern, Division of the wapentake of HOLDERNESS, East Riding of the county of YORK; containing 1603 inhabitants, of which number, 231 are in the township of Swine, 6½ miles (N. N. E.) from Kingston-upon-Hull. The living is a discharged vicarage, valued in the king's books at £8; present net income, £102; patron, W. Wilberforce, Esq.; impropriator, Earl of Shaftesbury. The church is partly in the early, and partly in the later, style of English architecture. There is a chapel of ease at Skirlaugh. Six poor children are educated for £6 a year, the bequest of Mrs. Lamb; and five more are instructed at the expense of the Earl of Shaftesbury. A nunnery of the Cistercian order, dedicated to the Blessed Virgin Mary, was founded here in the reign of Stephen, by Robert de Verli, which at the dissolution possessed a revenue of £134. 6. 9.

SWINEFLEET, a chapelry, in the parish of WHITGIFT, union of GOOLE, Lower Division of the wapentake of OSGOLDCROSS, West Riding of the county of YORK, 4¾ miles (S. S. E.) from Howden; containing 1055 inhabitants. The living is a perpetual curacy; net income, £127; patron, Vicar of Whitgift; impropriator, A. R. Worsop, Esq. There is a place of worship for Wesleyan Methodists; also a school partly supported by an allowance of £15 per annum from the town's land.

SWINESHEAD (*St. Nicholas*), a parish, in the union of ST. NEOTS, hundred of LEIGHTONSTONE, county of HUNTINGDON, though locally in the hundred of Stodden, county of Bedford, 3½ miles (S. W. by W.) from Kimbolton; containing 262 inhabitants. The living is a rectory, valued in the king's books at £12. 13. 6½.; patron, Duke of Manchester.

SWINESHEAD (*St. Mary*), a decayed market-town and parish, in the union of BOSTON, wapentake of KIRTON, parts of HOLLAND, county of LINCOLN, 7 miles (W. by S.) from Boston; containing 1994 inhabitants. An abbey for Cistercian monks was founded here by Robert de Greslie, in 1134, the revenue of which at the dissolution was valued at £175. 19. 10.; many valuable coins and several skeletons have, at various periods, been dug up near the spot, and, in 1825, on sinking a well, one of the latter was discovered, which measured six feet four inches. King John, in passing the Cross Keys wash, near this place, lost his carriages and baggage, and escaped to the monastery only with his life, where he died. The ruins of the monastery have entirely disappeared, but its site is still visible, and a mansion, recently modernized, was erected with a portion of its materials, about 220 years since. The sea formerly flowed up to the town, and near the market-place was a harbour. About 30 years ago, a bridge was taken down, which crossed a river then navigable for small craft, but now choked up. The market, nearly disused, is on Thursday; and a fair is held on Oct. 2nd. The living is a discharged vicarage, valued in the king's books at £14. 9.; present net income, £240; patrons and impropriators, Master and Fellows of Trinity College, Cambridge. The church is a handsome edifice, with a lofty spire. A free school was founded in 1720, by Thomas Cowley, Esq., who endowed it with certain lands producing £35 per annum, with a small surplus for clothing the poor, who also receive from the interest of various charitable bequests the benefit of a distribution amounting to £200 per annum. About a quarter of a mile north-westward of the town is a circular Danish encampment, called the Man-war-rings, about 60 yards in diameter, and surrounded by a double fosse.

SWINETHORP, an extra-parochial liberty, in the Higher Division of the wapentake of BOOTHBY-GRAFFO, parts of KESTEVEN, union and county of LINCOLN, 7 miles (W. by S.) from Lincoln; containing 54 inhabitants. Coal may be obtained here, though the mines have not yet been worked.

SWINFEN, a hamlet, in the parish of WEEFORD, union of LICHFIELD, Southern Division of the hundred of OFFLOW and of the county of STAFFORD, 2¼ miles (S. E. by S.) from Lichfield; containing 120 inhabitants. Swinfen Hall is a magnificent structure, built by the late Mr. Wyatt, father of the present celebrated architect; it stands in a large and well-wooded park, with a fine lawn and lake, and commands an extensive view of the country around Lichfield.

SWINFORD, a tything, in the parish of CUMNER, hundred of HORMER, county of BERKS; containing 38 inhabitants.

SWINFORD (*All Saints*), a parish, in the union of LUTTERWORTH, hundred of GUTHLAXTON, Southern Division of the county of LEICESTER, 3½ miles (S. E. by S.) from Lutterworth; containing 438 inhabitants. The living is a discharged vicarage, valued in the king's books at £5. 7. 11.; present net income, £216; the patronage and impropriation belong to Mrs. O. Cave. The church contains 100 free sittings, the Incorporated Society having granted £50 in aid of the expense.

SWINFORD, KING'S (*St. Mary*), a parish, in the union of STOURBRIDGE, Northern Division of the hundred of SEISDON, Southern Division of the county of STAFFORD, 3 miles (N. by W.) from Stourbridge; containing 15,156 inhabitants. The situation of this parish, in a country abounding with coal and iron mines, has given rise to the establishment of its manufactures, the principal of which are iron and glass, both carried on to a considerable extent, and for the conveyance of which great facilities are afforded by the Stourbridge and the Staffordshire and Worcestershire canals passing through the parish; a rail-road from some of the principal mines to the latter canal was constructed at a great expense, by the late Earl of Dudley. There are also, in various parts of the parish, numerous potteries of stone and coarse black ware, large brick and tile yards, a wire-mill, and several chain and nail manufacturers. A court leet and a court baron are held annually, and the inhabitants claim an exemption from tolls, under a charter granted by Queen Elizabeth, and confirmed by Charles I. A copyhold court is held occasionally; and the village has been made a polling-place for the southern division of the shire. The living is a rectory, valued in the king's books at £17. 13. 4.; present net income, £961; patron, Lord Ward. The church is an ancient edifice with a massive tower, and contains several monumental inscriptions to the families of Corbyn, Scott, Hodgett, and Bendy. An additional church, in the later English style, with a tower, was erected in 1831, at an expense of £7414. 5., under the act of the 58th of George III., at Wordsley, in this parish, which, by a recent act of parliament, will become the mother church. The living is a perpetual curacy; net income, £400; patron, Lord Ward; appropriator, Rector of King's-Swinford. There is a chapel at Brierly hill, containing 700 sittings, half of which are free, the Incorporated Society having granted £700 in aid of the expense. There are places of worship for Wesleyan and Primitive Methodists, and Independents. Holbeach House, in this parish, then the residence of Sir Stephen Littleton, who had fled for concealment to a private house at Rowley-Regis, was occupied by Catesby and the other conspirators in the gunpowder plot, and defended as their last retreat against the sheriff of Worcester. By the blowing up of their powder several were dreadfully burnt; Catesby, Piercy, and two more were killed on the spot; four were taken and conveyed to London, and executed; and Sir Stephen Littleton, after being concealed for some time, was apprehended and hanged, with some others in the country. A National school is supported by subscription. There are the remains of a Roman encampment on Ashwood heath; and the spa of Lady well is partly in this parish, and partly in that of Dudley.

SWINFORD, OLD (*St. Mary*), a parish, in the union of Stourbridge, partly in the Southern Division of the hundred of Seisdon, in the Southern Division of the county of Stafford, but chiefly in the Lower Division of the hundred of Halfshire, Stourbridge and Eastern Divisions of the county of Worcester, 1 mile (S. S. E.) from Stourbridge; containing 13,874 inhabitants, many of whom are engaged in the making of nails, exclusively of the trade connected with the town of Stourbridge, which see. The living is a rectory, valued in the king's books at £26. 6. 8.; present net income, £781; patron, Lord Foley. The church is partly in the decorated, and partly in the later, style of English architecture, with some Norman remains of good design, and a lofty and well-proportioned spire; it is now about to be entirely rebuilt. A chapel was built at the Lye, in this parish, by the late Thomas Hill, Esq., and endowed by his representatives. The patronage has been vested in certain Trustees, and a district is proposed to be assigned to it. The Blue-coat hospital, in this parish, was founded by Thomas Foley, Esq., ancestor of the noble family of that name, and endowed by him with estates now producing nearly £2300 per annum: it is a commodious brick edifice, somewhat in the style of a college, pleasantly situated on the road to Bromsgrove, and contains on the ground floor an excellent school and dining hall, for the boys, and a parlour for the use of the trustees, above which is a dormitory extending along the whole front, containing beds for 60 boys; over these are apartments for the assistants, the younger boys and the sick; and at the back are rooms for the accommodation of the master and his family, with kitchen, &c. The original number of boys in this noble institution was 60, but the increase of income has enabled the feoffees to add ten more: they are chosen from the following parishes and towns, *viz.*, Old Swinford, three; town of Stourbridge, four; Kidderminster, six (three from the borough, and three from the foreign); Bewdley, four; Dudley, four; Great Whitley, King's-Swinford, Kinver, Harbourn, Hales-Owen, West Bromwich, Bromsgrove, Rowley-Regis, Wednesbury, and Sedgeley, two each; Hagley, Little Whitley, Alvechurch, Pedmore, and Wombourn, one each; and the remaining 24 are appointed by the representative of the founder. Each boy must, on admission, be between seven and eleven years of age, free from disease, and the son of parents who have never received parochial relief; he is annually provided with a suit of clothes (similar to that worn at Christ's Hospital), a cap, and four pair of shoes and stockings; is boarded and instructed in reading, writing, arithmetic, mensuration, and land-surveying; is supplied with books and stationery, and, when 14 years of age, is apprenticed with a premium of £4, and, on producing a certificate of good conduct during his apprenticeship, receives at its expiration a gratuity of £15. The master is appointed by the trustees, with a salary of £75, and is allowed to appoint an assistant with a salary not exceeding £25, and board and lodging; there are a steward with a salary of £31. 10., and a housekeeper (who is at present the master's wife) with a salary of £25; all these, with the master's family, three servants, and a farming man, are also maintained from the funds of the charity. The establishment is under the direction of nineteen feoffees, who are noblemen, dignitaries of the church, or gentlemen of large landed property in this or the adjoining counties, and who elect the master, stewards, &c. The school at Red-hill is supported with endowments by John Wheeler, Esq., and Henry Glover, Esq., the former of whom granted property for the instruction of 20 poor boys, and for furnishing them with books and stationery; and the latter bequeathed £400, since laid out in lands, for the instruction of six poor boys, in reading, writing, and arithmetic, for six years each, providing them with books and stationery, and apprenticing one of them annually, with a premium of £5: two boys have since been added by the trustees, who are the governors of the Stourbridge free grammar school, and the boys on the foundation of both these charities are instructed by the same master. The remainder of Henry Glover's endowment, after all necessary charges for the school are deducted, is distributed amongst the poor of that part of the parish which is in the county of Worcester, who do not receive parochial relief, in sums of not more than one shilling per week to each person. There are some other small endowments for teaching poor children; and a school has also been established in connection with the National Society. There are also several benefactions for distributing bread, clothing, &c.

SWINGFIELD (*St. Peter*), a parish, in the union of Elham, hundred of Folkestone, lathe of Shepway, Eastern Division of the county of Kent, 5 miles (N.) form Folkestone; containing 282 inhabitants. The living is a perpetual curacy; net income, £52. 10.; patron and impropriator, Sir John W. E. Brydges, Bart.

A preceptory of Knights Templars was founded here before 1190, to which Sir Waresius de Valoniis and others were considerable benefactors: it subsequently became part of the possessions of the Knights of St. John of Jerusalem, and, at the dissolution, had a revenue of £87. 3. 3. On Swingfield common, during the agitations of 1745, the neighbouring nobility, gentry, and yeomen, to the number of several thousands, accoutred with arms and ammunition, assembled, to oppose an expected invasion on the coast of Kent.

SWINHOE, a township, in the parish and Northern Division of BAMBROUGH ward, union of BELFORD, Northern Division of the county of NORTHUMBERLAND, 9½ miles (S. E. by E.) from Belford; containing 110 inhabitants.

SWINHOPE (*St. Helen*), a parish, in the union of CAISTOR, wapentake of BRADLEY-HAVERSTOE, parts of LINDSEY, county of LINCOLN, 7¾ miles (S. E. by E.) from Caistor; containing 126 inhabitants. The living is a discharged rectory, valued in the king's books at £4. 17. 8½.; present net income, £117; patron and incumbent, Rev. M. Alington.

SWINNERTON (*St. Mary*), a parish, in the union of STONE, Northern Division of the hundred of PIREHILL and of the county of STAFFORD, 3½ miles (W. N. W.) from Stone; containing 791 inhabitants. The living is a rectory, valued in the king's books at £10. 2. 6.; patron, Rev. Christopher Dodsley. The church is an ancient edifice; in the south aisle, which is used as the parish school, is a colossal figure of our Saviour, sitting and pointing to the wound in his side, which was discovered buried at a short distance from its present situation. There is a Roman Catholic chapel at Swinnerton Park.

SWINSTEAD (*St. Mary*), a parish, in the union of BOURNE, wapentake of BELTISLOE, parts of KESTEVEN, county of LINCOLN, 2 miles (S. E.) from Corby; containing 402 inhabitants. The living is a discharged vicarage, valued in the king's books at £6. 19. 7.; present net income, £80; patron and impropriator, Lord Willoughby de Eresby, who chiefly supports three schools. There is a place of worship for Wesleyan Methodists.

SWINTON, a chapelry, in the township of WORSLEY, parish of ECCLES, hundred of SALFORD, Southern Division of the county palatine of LANCASTER, 4½ miles (W.) from Manchester: the population is returned with Worsley. The living is a perpetual curacy; net income, £126; patron, Vicar of Eccles. The chapel, which is dedicated to St. Peter, contains 350 free sittings, the Incorporated Society having granted £300 in aid of the expense: it has a Sunday school attached, attended by about 300 children.

SWINTON, a joint township with Warthermask, in the parish of MASHAM, union of BEDALE, partly in the liberty of ST. PETER'S, East Riding, and partly in the wapentake of HANG-EAST, North Riding, of the county of YORK, 1 mile (S. W.) from Masham; containing, with Warthermask, 207 inhabitants. Many relics of antiquity have been discovered in this neighbourhood, among which were the handle of a shield of gold, and a Roman battle-axe of brass.

SWINTON, a chapelry, in the parish of APPLETON-LE-STREET, union of MALTON, wapentake of RYEDALE, North Riding of the county of YORK, 2¼ miles (N. W. by W.) from New Malton; containing 333 inhabitants. There is a place of worship for Wesleyan Methodists.

SWINTON, a chapelry, in the parish of WATH-UPON-DEARN, union of ROTHERHAM, Northern Division of the wapentake of STRAFFORTH and TICKHILL, West Riding of the county of YORK, 4¾ miles (N. N. E.) from Rotherham; containing 1252 inhabitants. The living is a perpetual curacy; net income, £160; patrons, Earl Fitzwilliam and the Vicar of Wath-upon-Dearn; appropriators, Dean and Canons of Christ-Church, Oxford. The chapel, dedicated to St. Mary, has a fine Norman door. There is a place of worship for Wesleyan Methodists; and a Sunday National school has been established. A considerable manufacture of earthenware is carried on in this chapelry.

SWITHLAND (*St. Leonard*), a parish, in the union of BARROW-UPON-SOAR, hundred of WEST GOSCOTE, Northern Division of the county of LEICESTER, 2¼ miles (S. W. by W.) from Mountsorrel; containing 352 inhabitants. The living is a rectory, valued in the king's books at £10. 4. 7.; present net income, £300: it is in the patronage of the Crown. A day and Sunday school is partly supported by subscription. There are quarries of slate in the parish.

SWYRE (*Holy Trinity*), a parish, in the union of BRIDPORT, hundred of UGGSCOMBE, Dorchester Division of the county of DORSET, 5½ miles (S. E.) from Bridport; containing 226 inhabitants. This parish is bounded on the south by the English Channel, and the village is situated about one mile from the coast. A fair was granted in the 56th of Henry VIII.; and a wake is annually kept on Trinity Monday. The living is a discharged rectory, valued in the king's books at £7. 0. 5.; patron, Duke of Bedford. The tithes have been commuted for a rent-charge of £160, subject to the payment of rates, which on the average have amounted to £35; the glebe comprises 25 acres, valued at £30 per annum. The church was consecrated in 1503; it has a lofty tower, and north and south porches. A day and Sunday school is supported by the Duke of Bedford and the parish. *Cornua Ammonis* and the *Lapis Judaicus*, the latter exactly resembling the half of a peascod, and of a faint green colour, are found in this parish; and in different parts of it is dug a grey coarse marble full of shells, which is of a black colour when polished.

SYDE, county of GLOUCESTER.—See SIDE.

SYDENHAM, a chapelry, in the parish and union of LEWISHAM, hundred of BLACKHEATH, lathe of SUTTON-AT-HONE, Western Division of the county of KENT, 8½ miles (S. S. E.) from London: the population is returned with the parish. This place, which formerly consisted only of a few scattered dwellings, was first brought into notice by the discovery, in 1640, of a saline chalybeate spring, the waters of which, similar in their properties to those of Epsom, made it the occasional resort of invalids; and though the wells have fallen almost into disuse, the salubrity of the air, the pleasantness of its situation, and its proximity to the metropolis, have made it the permanent residence of numerous families of respectability, who have erected in the vicinity many handsome seats and elegant villas. The village is well built, and contains many genteel houses, with detached cottages of pleasing appearance. The upper part of the common commands extensive and richly-varied prospects, and the surrounding scenery possesses much rural beauty: the neighbourhood affords many pleasant walks, and the adjoining woods are much frequented by parties from the metropolis on excursions of pleasure. The London and Croydon railway intersects the chapelry, near the church, where a station has been established. This place is within the jurisdiction of the court of requests held at Bromley and at Greenwich, for the recovery of debts not exceeding £5. An annual fair, chiefly for pleasure, is held on Trinity Monday, and is in general well attended. The proprietary episcopal chapel, of which the Rev. P. A. French appoints the minister, was originally a meeting-house, of which Dr. John Williams, author of a Greek concordance, was the minister for many years: it is a convenient edifice, but not in any respect entitled to architectural notice. A church, dedicated to St. Bartholomew, to which a district has been assigned, was erected in 1831, at an expense of £9485. 18., under the act of the 58th of George III.: it is a handsome structure of white Suffolk brick, ornamented with stone, in the later style of English architecture, with a square embattled tower crowned with pinnacles, and contains 1000 sittings, of which 500 are free: the nave is lighted by a handsome range of clerestory windows, and separated from the aisles by lofty piers and arches of graceful elevation. The living is a perpetual curacy; net income, £240; patron, Vicar of Lewisham. There are places of worship for Independents and Wesleyan Methodists; and a National school is supported by subscription.

SYDENHAM (*St. Mary*), a parish, in the union of THAME, hundred of LEWKNOR, county of OXFORD, 2¾ miles (E.) from Tetsworth; containing 423 inhabitants. The living is annexed, with those of Towersey and Tetsworth, to the vicarage of Thame. On a headland in Sydenham field was found the umbo of an Anglo-Saxon shield.

SYDENHAM-DAMAREL (*St. Mary*), a parish, in the union of TAVISTOCK, and forming, with the parish of Lamerton, a distinct portion of the hundred of LIFTON, Lifton and Southern Divisions of the county of DEVON, 5¼ miles (W. by N.) from Tavistock; containing 296 inhabitants. The living is a rectory, valued in the king's books at £10. 6. 8.; present net income, £207; patron, John Carpenter, Esq. A school is supported by endowment. The river Tamar separates the parish from Cornwall. A copper mine was formerly worked, but has been discontinued.

SYDERSTONE (*St. Mary*), a parish, in the union of DOCKING, hundred of GALLOW, Western Division of the county of NORFOLK, 6¼ miles (S.) from Burnham-Westgate; containing 421 inhabitants. The living is a rectory, valued in the king's books at £13. 13. 4.; present net income, £534; patron, Marquess of Cholmondeley. Here is a Sunday National school.

SYDLING (*St. Nicholas*), a parish and liberty, in the union of CERNE, Cerne Division of the county of DORSET, 8 miles (N. W. by N.) from Dorchester; containing 617 inhabitants. The living is a discharged vicarage, valued in the king's books at £13. 1. 0½.; present net income, £169; patrons and impropriators, Warden and Fellows of Winchester College. The church is a neat structure, with a high embattled tower; the chancel has been elegantly rebuilt by the late Sir William Smith, who also constructed a large family vault within it, where his remains were deposited in the year 1752. At Hilfield, in this parish, is a chapel of ease. There is a place of worship for Independents; and a National school is supported by endowment, and another school by subscription. Near the ancient mansion-house of the family of Hardy, at Up-Sydling, there was before the Reformation a chapel of ease, but there are no vestiges of it.

SYDMONTON, a chapelry, in the parish, union, and hundred of KINGSCLERE, Kingsclere and Northern Divisions of the county of SOUTHAMPTON, 7 miles (N. by E.) from Whitchurch; containing 170 inhabitants. The ancient manor-house, a spacious mansion, built prior to the time of Elizabeth, was entirely remodelled in 1837. The chapel, dedicated to St. Mary, is situated in the park; the chancel is separated from the nave by a fine Norman arch.

SYERSCOTE, otherwise STERSCOTE, a liberty, in the parish and union of TAMWORTH, Northern Division of the hundred of OFFLOW and of the county of STAFFORD, 3 miles (N. N. E.) from Tamworth; containing 34 inhabitants.

SYERSTON (*All Saints*), a parish, in the union of SOUTHWELL, Southern Division of the wapentake of NEWARK and of the county of NOTTINGHAM, 5¾ miles (S. W.) from Newark; containing 138 inhabitants. The living is annexed, with that of Coddington, to the vicarage of East Stoke. The old Fosse-road passes through the parish, which is partly bounded by the river Trent.

SYKEHOUSE, a chapelry, in the parish of FISHLAKE, union of THORNE, Southern Division of the wapentake of STRAFFORTH and TICKHILL, West Riding of the county of YORK, 5½ miles (N. W. by W.) from Thorne; containing 617 inhabitants. The living is a perpetual curacy; net income, £50; patron, certain Trustees. The chapel is dedicated to St. Peter. A school is partly supported by endowment.

SYLEHAM (*St. Mary*), a parish, in the union and hundred of HOXNE, Eastern Division of the county of SUFFOLK, 3½ miles (S. W.) from Harleston; containing 391 inhabitants. The living is a perpetual curacy; net income, £88; patron and impropriator, L. Press, Esq. Mr. Anthony Barry, by his will in 1678, settled the impropriate tithes on the minister, and that settlement was afterwards confirmed by his son, Mr. Christopher Barry, but not the fee simple thereof. The church is an ancient edifice with a round tower and other ancient details. There was formerly a chapel at Esham, a hamlet of this parish.

SYMONDSBURY (*St. John the Baptist*), a parish, in the union of BRIDPORT, hundred of WHITCHURCH-CANONICORUM, Bridport Division of the county of DORSET, 1½ mile (W. N. W.) from Bridport; containing 1147 inhabitants. The living is a rectory, valued in the king's books at £36. 3. 4.; present net income, £969; patron, Rev. G. Raymond. The church is a large cruciform structure, partly in the early, and partly in the later style of English architecture, with a tower rising from the intersection, and contains some monuments to the family of Syndercombe. Three schools are supported by the rector. There are several springs in the neighbourhood slightly impregnated with iron.

SYMONDS-HALL, a tything, in the parish of WOTTON-UNDER-EDGE, Upper Division of the hundred of BERKELEY, Western Division of the county of GLOU-

CESTER, 3 miles (N. E. by E.) from Wotton-under-Edge, with which the population is returned.

SYNFIN, a joint liberty with Arleston, in the parish of BARROW, hundred of APPLETREE, Southern Division of the county of DERBY, $2\frac{1}{2}$ miles (S. by W.) from Derby; containing 71 inhabitants.

SYRESHAM (*St. James*), a parish, in the union of BRACKLEY, hundred of KING'S-SUTTON, Southern Division of the county of NORTHAMPTON, $4\frac{3}{4}$ miles (N. E.) from Brackley; containing 895 inhabitants. The living is a rectory, valued in the king's books at £13; present net income, £152; patron, Sir C. C. Dormer, Knight. There is a place of worship for Wesleyan Methodists. The Rev. George Hammond, in 1755, bequeathed the interest of £300 for teaching ten poor children; and in augmentation of the master's salary, Conquest Jones, in 1773, left the interest of £100.

SYSONBY, a parish, in the union of MELTON-MOWBRAY, hundred of FRAMLAND, Northern Division of the county of LEICESTER, 1 mile (W.) from Melton-Mowbray; containing 81 inhabitants. The living is annexed to the vicarage of Melton-Mowbray.

SYSTON (*St. Peter*), a parish, in the union of BARROW-UPON-SOAR, Eastern Division of the hundred of GOSCOTE, Northern Division of the county of LEICESTER, $5\frac{1}{4}$ miles (N. N. E.) from Leicester; containing 1349 inhabitants, several of whom are engaged in framework knitting. The living is a discharged vicarage, valued in the king's books at £7. 2. 7.; present net income, £115; patrons and impropriators, Vice-Chancellor, Masters, and Scholars of the University of Oxford. The church is extremely beautiful, and contains monuments to the Thorold family; over the altar is a fine painting of the Nativity, presented by the late Sir John Thorold, and in the churchyard is a curious arch. A National and an infants' school are partly supported by subscription. About a mile south-west of the village is a tumulus, on the eastern side of a Roman road passing through the vicinity.

SYSTON (*St. Mary*), a parish, in the union of NEWARK, wapentake of WINNIBRIGGS and THREO, parts of KESTEVEN, county of LINCOLN, 4 miles (N. N. E.) from Grantham; containing 203 inhabitants. The living is a vicarage; net income, £83; patron and impropriator, Sir J. C. Thorold, Bart. A school is endowed with £5 per annum, and there is also a Sunday National school.

SYWELL (*St. Peter and St. Paul*), a parish, in the union of WELLINGBOROUGH, hundred of HAMFORDSHOE, Northern Division of the county of NORTHAMPTON, 5 miles (W.) from Wellingborough; containing 216 inhabitants. The living is a rectory, valued in the king's books at £11. 1. $5\frac{1}{2}$.; present net income, £492; patron, Earl Brownlow.

T

TABLEY, INFERIOR, a township, in the parish of GREAT BUDWORTH, union of ALTRINCHAM, hundred of BUCKLOW, Northern Division of the county of CHESTER, 3 miles (S. W. by W.) from Nether Knutsford; containing 134 inhabitants. Here was formerly a chapel, the site of which is still called Chapel field. An infants' school is supported by Lord de Tabley. Tabley confers the title of Baron on the family of Leicester, created July 16th, 1826, at which period Sir John Fleming Leicester, Bart., a gentleman distinguished for his munificent patronage of the fine arts, and encouragement of native artists, was raised to the peerage by that title; he died in June 1827.

TABLEY, SUPERIOR, a township, in the parish of ROSTHERN, union of ALTRINCHAM, hundred of BUCKLOW, Northern Division of the county of CHESTER, 2 miles (W. N. W.) from Nether Knutsford; containing 442 inhabitants. Here are the ruins of an ancient chapel, formerly called, from its situation, "The chapel in the street."

TACHBROOK, BISHOP'S (*St. Chad*), a parish, in the union of WARWICK, partly in the Kenilworth Division of the hundred of KNIGHTLOW, but chiefly in the Warwick Division of the hundred of KINGTON, Southern Division of the county of WARWICK, $3\frac{3}{4}$ miles (S. E.) from Warwick; containing 674 inhabitants. The living is a vicarage, valued in the king's books at £5. 13. 4.; present net income, £293; patron, Bishop of Lichfield; appropriator, Prebendary of Tachbrook in the Cathedral Church of Lichfield. A school was erected and endowed in 1771, by subscription of Sir William Bagot, the Earl of Warwick, and others, in which about 100 children are instructed on the National system: the income is £39 per annum, with a good house and garden for the master.

TACHBROOK-MALLORY, a hamlet, in the parish of BISHOP'S-TACHBROOK, union of WARWICK, Kenilworth Division of the hundred of KNIGHTLOW, Southern Division of the county of WARWICK, $3\frac{1}{2}$ miles (S. E. by E.) from Warwick: the population is returned with the parish.

TACKLEY (*St. Nicholas*), a parish, in the union of WOODSTOCK, hundred of WOOTTON, county of OXFORD, $3\frac{1}{4}$ miles (N. E.) from Woodstock; containing 564 inhabitants. The living is a rectory, valued in the king's books at £19. 9. $4\frac{1}{2}$.; present net income, £742; patrons, President and Fellows of St. John's College, Oxford. The church is an ancient cruciform structure, chiefly in the early English style; the north aisle appears to have been destroyed by fire, and in the chancel is a triple lancet-shaped window. There is a place of worship for Wesleyan Methodists; also a National school. Earth of a peculiar quality, used for flooring barns, cottages, &c., abounds here. The Roman Akeman-street passes through the parish, and separates the two manors of the Duke of Marlborough and Sir Henry Dashwood, Bart.: on that of the latter the two gateways of the ancient mansion built by the Harborne family still remain.

TACOLNESTON (*All Saints*), a parish, in the union and hundred of DEPWADE, Eastern Division of the county of NORFOLK, $4\frac{1}{2}$ miles (N. W. by W.) from Stratton-St. Mary; containing 486 inhabitants. The living is a rectory, valued in the king's books at £12; present net income, £498: it is in the patronage of Mrs. Warren. Here is a Sunday National school. John Tasephans, who was prior of the Carmelite friary at Norwich, was born here in 1404: he was a learned and pious divine, and a powerful orator, but his intolerant zeal brought much persecution upon the Lollards. The hall, a fine brick mansion, is a good specimen of the domestic style of architecture prevalent in the seventeenth century; it is said to have been built in 1670, by the Browne family, who then held the estate.

TADCASTER (*St. Mary*), a market-town and parish, chiefly in the Upper Division of the wapentake of Barkstone-Ash, West Riding, but partly in the Ainsty of the city of York, East Riding, of the county of York; containing 2855 inhabitants, of which number, 2403 are in the town of Tadcaster, 10 miles (S. W.) from York, and 189½ (N. N. W.) from London. This place was the Roman station *Calcaria*, so named from the nature of the soil, which abounds with calx, or limestone, and one of the out-ports, or gates, on the Consular way, to their chief military station, *Eboracum* (York). Under the name *Calca-cester* Bede relates that Heina, the first female who assumed the habit of a nun in this country, retired hither, where she built a residence. In all the great civil wars of England, it was regarded as a post of considerable importance, and the possession of it was repeatedly contested. On the appointment of the Earl of Newcastle to the command of the royal army, in 1642, he advanced from York towards this town, with 4000 men and seven pieces of cannon, and commenced his attack on the enemy's works, which lasted without intermission from eleven in the morning to five in the afternoon, when, his ammunition being exhausted, he desisted from the assault, in expectation of a fresh supply from York, before the following morning; but during the night, Sir Thomas Fairfax, who was posted here with 700 men, drew them off to Cawood and Selby, and left the royalists in possession. The town, which is a great thoroughfare, is situated on the river Wharf, over which is a very handsome stone bridge, considered the finest in the county, erected in the beginning of the last century; its centre marks the divisions of the jurisdictions of the West Riding and the Ainsty of the city of York. On the banks of the river, which is navigable up to the town, are several flour-mills. The streets are arranged on each side, and the houses are neat and modern. The walks on the banks of the river are highly interesting, and have of late been greatly improved. In the immediate neighbourhood are stone quarries, one of which, called Jack-daw-Cragg (singularly interesting and romantic, and in the possession of the ancient family of Vavasour), supplied stone for the erection of the magnificent cathedral at York, and recently furnished materials for its repair, after the partial conflagration in 1829. The market is on Wednesday; and fairs are held on the last Wednesdays in the months of May and Oct., for cattle and sheep; and in Nov., for hiring servants.

The living is a discharged vicarage, valued in the king's books at £8. 4. 9½.; present net income, £240; patron, Col. Wyndham; impropriator,—Sharan, Esq. The church is a handsome structure, in the later style of English architecture, with a fine tower. There are places of worship for Independents, Inghamites, and Primitive and Wesleyan Methodists; the last has a school-room attached, capable of containing 300 children. The grammar school, and an hospital for four poor men, were founded and endowed with lands and the sum of £600, by Dr. Oglethorpe, Bishop of Carlisle, and confirmed by licence in the 5th of Philip and Mary; and the annual income is £145. Forty girls are also instructed by four poor women, almshouse pensioners, on the foundation of Mrs. Henrietta Dawson, who bequeathed £15 per annum to ten widows, and £10 per annum to ten spinsters, with an additional £5 per annum to each of the four women for teaching the children. A Sunday school, in connection with the established church, was built by subscription in 1788, on a plot of ground given by the late William Hill, Esq., whose daughter, the present Miss Hill, has ever since contributed materially towards its support, and has assigned to it a permanent endowment of £20 per annum. Several Roman coins have been found here at different times; and there are some vestiges of a trench, surrounding part of the town, which is supposed to have been thrown up at the time of the civil war in the reign of Charles I.

TADDINGTON, a chapelry, in the parish and union of Bakewell, hundred of High Peak, Northern Division of the county of Derby, 3½ miles (S. S. W.) from Tideswell; containing, with the township of Priestcliffe, 391 inhabitants. The living is a perpetual curacy; net income, £87; patron, Vicar of Bakewell; appropriators, Dean and Chapter of Lichfield. The chapel, dedicated to St. Michael, is fast going to decay: near it is the mutilated shaft of an ancient cross. There is a place of worship for Baptists. Twelve poor children are taught in a school, erected by subscription in 1805, and supported with a rent-charge of £15, the bequest of Michael White, in 1798.

TADLEY (*St. Peter*), a parish, in the union of Kingsclere, hundred of Overton, Kingsclere and Northern Divisions of the county of Southampton, 6 miles (N. N. W.) from Basingstoke; containing 683 inhabitants. The living is annexed to the vicarage of Overton. There is a place of worship for Independents.

TADLOW (*St. Giles*), a parish, in the union of Caxton and Arrington, hundred of Armingford, county of Cambridge, 4½ miles (E. S. E.) from Potton; containing 176 inhabitants. The living is a discharged vicarage, valued in the king's books at £6. 17.; present net income, £120; patrons and impropriators, Master and Fellows of Downing College, Cambridge.

TADMARTON (*St. Nicholas*), a parish, in the union of Banbury, hundred of Bloxham, county of Oxford, 4¾ miles (W. S. W.) from Banbury; containing 355 inhabitants. The living is a rectory, valued in the king's books at £13. 11. 0½.; present net income, £307; patrons, Provost and Fellows of Worcester College, Oxford. The Danes, who in 914 plundered this part of the county, advanced with great havoc to Hook-Norton, where they killed many of the Saxons, and are supposed to have, on that occasion, raised the works called Tadmarton Castle and Hook-Norton Barrow, of which there are some vestiges.

TAKELEY (*Holy Trinity*), a parish, in the union of Dunmow, hundred of Uttlesford, Northern Division of the county of Essex, 2¾ miles (S. E. by E.) from Stansted-Mountfitchet; 4 miles (W.) from Dunmow; containing 1099 inhabitants. This parish is about three miles in length and two in breadth; it had formerly a very extensive forest, and is still richly wooded. The living is a vicarage, valued in the king's books at £11; present net income, £207; patron and appropriator, Bishop of London. The church contains 160 free sittings, and is an ancient edifice of stone, consisting of a nave and chancel, with a south aisle, in which is a sepulchral chapel belonging to the Bassingbourne estate. A small priory was founded here in the reign of Henry I., as a cell to the abbey of St. Valery in Picardy. Dr. Robert Fowler, Archbishop of Dublin, was buried here October 19th, 1801. Here is a Sunday school on the National plan.

TALATON (*St. James*), a parish, in the union of HONITON, hundred of HAYRIDGE, Cullompton and Northern Divisions of the county of Devon, $4\frac{1}{2}$ miles (N. W. by N.) from St. Mary-Ottery; containing 479 inhabitants. The living is a rectory, valued in the king's books at £32. 3. $1\frac{1}{2}$.; present net income, £518; patron and incumbent, Rev. R. P. Welland. The church has an elegant wooden screen in good preservation. At Escot House, which was destroyed by fire in 1808, George III. and three of the princesses were entertained by Sir G. Young, Bart., Aug. 14th, 1780. A National school has been established. Dr. Thomas Sprat, Bishop of Rochester, an historian and poet, was born here in 1636; he died in 1713. Southcote, in this parish, was the occasional residence of Sir William Pole, the antiquary.

TALK-O'-TH'-HILL, a chapelry, in the parish of AUDLEY, Northern Division of the hundred of PIREHILL, and of the county of STAFFORD, 5 miles (N. N. W.) from Newcastle-under-Lyme; containing 1196 inhabitants. The living is a perpetual curacy; net income, £118; patron, Vicar of Audley; impropriator, G. Tollet, Esq. The chapel is a small brick building, surmounted by a cupola, and contains 100 free sittings, the Incorporated Society having granted £150 in aid of the expense. There is a place of worship for Wesleyan Methodists. The great north road formerly passed through the village, which is situated upon an eminence commanding a view into nine counties, with the mountains of North Wales in the distance. In the centre is a stone cross, where a market was formerly held. A free school was erected by subscription in 1760, in which fourteen children are instructed for £15 a year. Adjacent to the village is a spring, the water of which is of a blue milky colour, strongly impregnated with sulphur, and much in request for cutaneous diseases.

TALKIN, a township, in the parish of HAYTON, union of BRAMPTON, ESKDALE ward, Eastern Division of the county of CUMBERLAND, 3 miles (S. E. by S.) from Brampton; containing 376 inhabitants. It is bounded on the west by the river Gelt, and contains quarries of freestone and limestone, and collieries. A school is supported by small annual subscriptions. Three valuable gold clasps were discovered in 1790, on Netherton farm, where a battle was formerly fought.

TALLAND (*St. Tallan*), a parish, in the union of LISKEARD, hundred of WEST, Eastern Division of the county of CORNWALL, 2 miles (S. W. by W.) from West Looe; containing, with the borough of West Looe, 1434 inhabitants. The parish is bounded on the south by the English Channel, and includes the decayed market-town of West Looe, and part of the small fishing town of Polperro. The living is a discharged vicarage, valued in the king's books at £10; present net income, £110; patron and incumbent, Rev. N. Kendall; impropriator, J. Graves, Esq. Mary Kendall, in 1710, left £4 per annum for teaching poor girls; and Charles Kendall, in 1746, gave £6 a year for the instruction of boys.

TALLENTIRE, a township, in the parish of BRIDEKIRK, union of COCKERMOUTH, ALDERDALE ward below Derwent, Western Division of the county of CUMBERLAND, $3\frac{3}{4}$ miles (N. by W.) from Cockermouth; containing 237 inhabitants. Limestone is quarried and burned in the vicinity. Here is a free school, with a trifling endowment.

TALLINGTON (*St. Lawrence*), a parish, in the union of STAMFORD, wapentake of NESS, parts of KESTEVEN, county of LINCOLN, $3\frac{1}{2}$ miles (W. by S.) from Market-Deeping; containing 220 inhabitants. The living is a discharged vicarage, valued in the king's books at £8. 9. 8.; present net income, £200; patron and impropriator, Earl of Lindsey. Sixteen children are instructed from an annual payment of £5 out of a charity belonging to the poor of the parish.

TALWORTH, a hamlet, in the parish of LONG DITTON, union, and Second Division of the hundred, of KINGSTON, Eastern Division of the county of SURREY, $2\frac{1}{4}$ miles (S. S. E.) from Kingston-upon-Thames; containing 264 inhabitants.

TAMERTON-FOLLIOTT (*St. Mary*), a parish, in the union of PLYMPTON-ST. MARY, hundred of ROBOROUGH, Midland-Roborough and Southern Divisions of the county of DEVON, 5 miles (N. by W.) from Plymouth; containing 1061 inhabitants. This place, supposed by Camden to be the ancient *Tamara*, is delightfully situated on a creek of the river Tamar, and is inhabited by several respectable families. Maristow, the property of Sir Ralph Lopes, Bart., whose uncle, in 1789, had the honour of entertaining here George III. and three of the princesses, is a noble mansion with a chapel attached, in which divine service is regularly performed: the domain is extensive, and enriched with pleasingly diversified scenery. Warlegh, the ancient seat of the Bampfylde family, is also in this parish. The living is a vicarage, valued in the king's books at £12. 7. $8\frac{1}{2}$.; present net income, £315: it is in the patronage of the Crown; impropriators, G. Leach, Esq., and others. The church has a remarkably fine tower, and contains, among several handsome memorials of the Coplestone, Bampfylde, and Radcliffe families, an ancient altar-tomb, with the figures of an armed knight and his lady. A free school for twenty boys was founded, and liberally endowed with land and money, by Mary Deane, in 1734; the income is about £120 a year, for which they are clothed and educated: a school-room was built in 1831, at an expense of £130, in which children of both sexes are gratuitously instructed. An almshouse for four widows was erected in 1669, by Sir C. Bampfylde, who endowed it with property producing about £12 per annum.

TAMERTON, NORTH (*St. Denis*), a parish, in the union of HOLSWORTHY, hundred of STRATTON, Eastern Division of the county of CORNWALL, 5 miles (S. S. W.) from Holsworthy; containing 517 inhabitants. The living is a donative curacy; net income, £230; patrons, R. P. Coffin, Esq., and the heirs of the late Colonel I'Ans. A school is partly supported by subscription. The river Tamar and the Bude canal run through the parish, in a parallel direction, from north to south. There is a dilapidated chapel at Hornacot.

TAMHORN, a hamlet, in the parish of WHITTING-

TON, Northern Division of the hundred of OFFLOW and of the county of STAFFORD, 3 miles (N. W. by N.) from Tamworth; containing 7 inhabitants. The Birmingham and Fazeley canal passes in the vicinity.

Corporation Seal.

TAMWORTH (*St. Edith*), a borough, market-town, and parish, and the head of a union, partly in the Northern, and partly in the Southern, Division of the hundred of OFFLOW, and of the county of STAFFORD, and partly in the Tamworth Division of the hundred of HEMLINGFORD, Northern Division of the county of WARWICK; containing 7182 inhabitants, of which number, 1711 are in the borough of Tamworth, 22 miles (S. E. by E.) from Stafford, 27 (N. by W.) from Warwick, and 112 (N. W. by N.) from London. This town, which is considered the most ancient in the county, derives its name from the river Tame, on which it is situated, and *Waert*, or *Worthidge*, a water farm. At a very early period it was the site of a Mercian fortification and royal residence, and was the seat of government under Offa, Cenwulf, Beornwulf, and others, at which period it had also a mint. Having been nearly destroyed by the Danes, it was rebuilt early in the tenth century, by Ethelfleda, daughter of Alfred the Great, who also erected a castle for its defence, which, having undergone recent repairs, is now a private residence; and the ancient fosse which surrounded the town, called the King's Dyke, is still visible. The town, which is about equally divided between the counties of Stafford and Warwick, though commonly considered a Staffordshire town, is situated near the confluence of the rivers Tame and Anker, both crossed by bridges about a mile distant from the Coventry canal, and consists of some good streets. The manufacture of lace, cotton, tapes, and patten-ties, affords employment to several persons. Many veins of coal have been found, and are worked in the vicinity; and bricks and tiles of great durability are made from a clay which abounds here. The Birmingham and Derby Junction railway passes near the town. There is a permanent library, under the direction of a respectable committee. The market is on Saturday: fairs are held by charter on May 4th, July 26th, and Oct. 24th, for cattle and merchandise, and there are five new fairs for the sale of cattle only. The town was, till the passing of the Municipal Reform Act, governed under a charter granted by Charles II., upon the surrender of a former one, which had been conferred by Elizabeth, authorising the appointment of a high steward, two bailiffs, recorder, twenty-four capital burgesses, a town-clerk, and other officers, of whom the high steward, bailiffs, recorder, and town-clerk, were justices of the peace. The government is now vested in a mayor, four aldermen, and twelve councillors, under the act of the 5th and 6th of William IV., cap. 76, for an abstract of which see the Appendix, No. I. The mayor and late mayor are justices of the peace, and four other magistrates are appointed; the county magistrates also have concurrent jurisdiction. The borough returns two members to parliament: the elective franchise was formerly in the inhabitants paying scot and lot, and not receiving alms, in number about 600; but it has been extended also to the £10 householders of the entire parish, which has been made to constitute the new elective borough, comprising an area of 11,000 acres: the old borough contained only 83 acres. The number of voters now registered is 501; the mayor is the returning officer. The corporation hold courts leet and baron, and are empowered also to hold a court of record for the recovery of debts, but this right has not been exercised for many years. Under the provisions of the Municipal Reform Act, the quarter sessions have been discontinued, and trials take place at the assizes and sessions for the county. Petty sessions for the borough are held every alternate Wednesday; and the town has lately been made a polling-place for the southern division of the county. The town-hall is a handsome building in the market-place.

The living is a vicarage; net income, £170; patron, Captain A'Court, R. N. The church, situated in that part of the parish which is in the county of Stafford, is a spacious and handsome edifice, with a fine tower, in which are two remarkable spiral staircases, communicating with separate floors, their respective entrances being within and without the church: beneath the edifice is a small crypt: the building combines the decorated and later styles of English architecture. It was formerly collegiate, and occupies the site of an ancient monastery: the foundation of the college, which consisted of a dean and six prebendaries, is uncertain, but is attributed, with the greatest probability, to the Marmions, who were successively owners of the castle. Some fine tesselated pavement, now placed in front of the communion-table, was discovered a few years since, when the church was undergoing repair. There are places of worship for Baptists, the Society of Friends, Independents, and Wesleyan Methodists; also a Roman Catholic chapel. The free grammar school was refounded in the reign of Edward VI., and the stipend of £10. 13. 2¼. was confirmed to the master, and made payable from the revenues of the Crown. In the reign of Elizabeth the bailiffs were incorporated governors, and in 1677 the school-room was rebuilt: the revenue has been subsequently increased by various benefactors, and now amounts to £33. 11. 3. Boys from this school are eligible to a scholarship at Catherine Hall, Cambridge, founded by Mr. Frankland; and a native of this town to a fellowship in St. John's College, Cambridge, on the foundation of Mr. Bailey. A free school for twelve boys and ten girls has an income of £20 per annum, partially arising from a charitable bequest: a National school has been established; and the Sunday schools are endowed with the interest of £50, the gift of Mrs. Mary Done. In 1686, the Rev. John Rawlett bequeathed lands and houses for teaching and apprenticing poor children. An almshouse for fourteen poor men and women was endowed in 1678, by Thomas Guy, Esq., of London, founder of Guy's hospital in the borough of Southwark, who represented this borough in seven parliaments, and, in 1701, rebuilt the town-hall. The poor law union of Tamworth comprises 22 parishes or places, and contains a population of 12,175, according to the census of 1831; the number of guardians is 27. A new bridge has been erected at Fazeley, over the Tame, along which passes the ancient Watling-street. Edward Lord Thurlow was a representative of this borough until his elevation to the peerage, and recorder until his death. Tamworth confers the inferior title of Viscount on Earl Ferrers.

TAMWORTH-CASTLE, a liberty, adjoining the borough, and in the parish and union of TAMWORTH, Tamworth Division of the hundred of HEMLINGFORD, Northern Division of the county of WARWICK; containing 19 inhabitants.

TANDRIDGE (*St. Peter*), a parish, in the union of GODSTONE, First Division of the hundred of TANDRIDGE, Eastern Division of the county of SURREY, 2 miles (E. by S.) from Godstone; containing 478 inhabitants. This parish is situated on the road from Guilford, by Godstone, to Maidstone; and the South-Eastern railway will pass through it to the south of Tilbusta Hill. The living is a perpetual curacy; net income, £80; patron and impropriator, Sir W. Clayton, Bart. The church is an ancient edifice with a tower surmounted by a spire of wood: in the churchyard is an ancient yew tree, 30 feet in girth at the height of five feet from the ground. A National school for the children of Tandridge, Godstone, and Oxted, is supported by subscription, and a school-room in the rustic cottage style has been built by the same means. A priory of Augustine canons, in honour of St. James, to which Odo de Damartin was a great benefactor, was founded in the time of Richard I., and at the dissolution had possessions valued at £86. 7. 6. per annum. In the grounds of the Priory are the lids of two stone coffins, which were dug up here; and in 1828 several silver and copper coins of Julius Cæsar and other Roman Emperors were found.

TANFIELD, a chapelry, in the parish of CHESTER-LE-STREET, union of LANCHESTER, Middle Division of CHESTER ward, Northern Division of the county palatine of DURHAM, $6\frac{3}{4}$ miles (S. W.) from Gateshead: the population is returned with the townships of Beamish and Lintz-Green. The living is a perpetual curacy; net income, £133; patron and impropriator, Lord Ravensworth. The chapel, dedicated to St. Margaret, was rebuilt by subscription in 1749, with the exception of a portion of the chancel, in the southern wall of which there is a piscina. There are two paper-mills and extensive collieries in the neighbourhood. Tanfield Arch, a magnificent stone structure, 130 feet in the span, springing from abutments 9 feet high, to the height of 60 feet, was erected upon the site of a wooden arch that had recently fallen, by certain coal-owners, at the expense of £12,000, to expedite the passage of their waggons. Elizabeth Davidson, in 1762, bequeathed £500, and Robert Robinson, in 1730, a rent-charge of £6, towards the support of a free school: the latter, and the interest of the former, are paid to a master, who instructs 18 boys. The same individuals bequeathed also property now producing, with other benefactions, £12 per annum, which is distributed among the poor. Here is a Sunday National school.

See a singular Law case in Serjt. Barnardiston's Repts. v. 1. p. 318 remarkable as to the early introduction of wooden railways Pitt v Lady Clavers.

TANFIELD, EAST, a township, in the parish of KIRKLINGTON, wapentake of HALLIKELD, North Riding of the county of YORK, $6\frac{1}{2}$ miles (N. N. W.) from Ripon; containing 35 inhabitants.

TANFIELD, WEST (*St. Nicholas*), a parish, in the wapentake of HALLIKELD, North Riding of the county of YORK, $6\frac{1}{2}$ miles (N. W. by N.) from Ripon; containing 693 inhabitants. The living is a rectory, valued in the king's books at £13. 0. 5.; patron, Marquess of Ailesbury. The tithes have been commuted for a rent-charge of £415. 10., subject to the payment of rates, which on the average have amounted to £35; the glebe comprises 63 acres, valued at £63 per annum. The church is an ancient structure, containing many curious old monuments: attached to it is the chantry of Maud Marmion, founded in the time of Henry III., for a master, warden, and two brothers, to pray for the souls of Lord and Lady Marmion. There is a place of worship for Wesleyan Methodists. Eleven poor children are educated for £8 per annum, left by Diana, Countess of Oxford, and a smaller annuity bequeathed by Catherine Allen, in 1769. On the banks of the Ure, which is here of considerable width and crossed by a bridge, are the remains of Tanfield castle, the origin and history of which are involved in obscurity.

TANGHAM, a hamlet, in the parish of BUTLEY, hundred of PLOMESGATE, but locally in that of Wilford, Eastern Division of the county of SUFFOLK: the population is returned with the parish.

TANGLEY (*St. John the Baptist*), a parish, in the union of ANDOVER, hundred of PASTROW, Andover and Northern Divisions of the county of SOUTHAMPTON, $5\frac{1}{2}$ miles (N. N. W.) from Andover; containing 283 inhabitants. The living is annexed to the rectory of Faccombe. The tithes have been commuted for a rent-charge of £329. 12., subject to the payment of rates, which on the average have amounted to £79; the glebe comprises $1\frac{1}{2}$ acre, valued at £2 per annum. There is a place of worship for Wesleyan Methodists. A fair for sheep is held on April 15th.

TANGMERE (*St. Andrew*), a parish, in the union of WEST HAMPNETT, hundred of ALDWICK, rape of CHICHESTER, Western Division of the county of SUSSEX, 3 miles (E. by N.) from Chichester; containing 197 inhabitants. The living is a rectory, valued in the king's books at £13. 5.; present net income, £282; patron, Duke of Richmond. The church is in the early style of English architecture, with a spire; it contains a Norman font; and in the churchyard is a venerable yew tree, 20 feet in girth at a height of 3 feet from the ground. Here is a Sunday National school.

TANKERSLEY (*St. Peter*), a parish, in the union of WORTLEY, wapentake of STAINCROSS, West Riding of the county of YORK; containing 1596 inhabitants, of which number, 678 are in the township of Tankersley, $5\frac{1}{4}$ miles (S.) from Barnesley. The living is a rectory, valued in the king's books at £26. 0. $9\frac{1}{2}$.; present net income, £474; patron, Earl Fitzwilliam. The church is principally in the later style of English architecture. There is an estate, comprising a farm-house and about 17 acres of land, understood to have been left by a Countess of Devonshire, the rental of which, amounting to £29 per annum, is applied to the relief of decayed housekeepers and other necessitous persons.

TANNINGTON (*St. Ethelbert*), a parish, in the union and hundred of HOXNE, Eastern Division of the county of SUFFOLK, $4\frac{3}{4}$ miles (N. W.) from Framlingham; containing 264 inhabitants. The living is a discharged vicarage, with that of Brundish annexed, valued in the king's books at £12. 10. $2\frac{1}{2}$.; present net income, £196; patron and appropriator, Bishop of Rochester. The rents of the town lands, producing £60 per annum, are applied to the repair of the church, and to the clothing of the poor, and supplying them with coal; and £5 are assigned towards the support of a Sunday school.

TANSHELF, a township, in the parish of PONTEFRACT, Upper Division of the wapentake of OSGOLDCROSS, West Riding of the county of YORK, ¼ of a mile (W. by S.) from Pontefract; containing 423 inhabitants.

TANSLEY, a hamlet in the parish of CRICH, hundred of WIRKSWORTH, Southern Division of the county of DERBY, 1½ mile (E.) from Matlock; containing 507 inhabitants. Her Majesty's commissioners have agreed to afford facilities for obtaining a site for a chapel here, and the Incorporated Society have granted £120 in aid of the expense, for which 120 free sittings will be provided. There is a place of worship for Wesleyan Methodists.

TANSOR (*ST. MARY*), a parish, in the union of OUNDLE, hundred of WILLYBROOK, Northern Division of the county of NORTHAMPTON, 2¼ miles (N. N. E.) from Oundle; containing 225 inhabitants. The living is a rectory, valued in the king's books at £13. 12. 11.; present net income, £283; patrons, Dean and Chapter of Lincoln. The church is a small structure, partly in the Norman, and partly in the early English style, with some screen-work and ancient monuments. This parish is partly bounded by the river Nen: at Cotterstock, on its opposite bank, a fine Roman tesselated pavement has been discovered. Two schools are partly supported by the clergyman. On the enclosure in 1777, fifteen acres, now let for £24 per annum, were allotted, in lieu of other lands, for the support of a Sunday school and for the purchase of coal and clothing for the poor; and £5 per annum from Bellamy's charity is given for apprenticing a child.

TANWORTH (*ST. MARY MAGDALENE*), a parish, in the union of SOLIHULL, Warwick Division of the hundred of KINGTON, Southern Division of the county of WARWICK, 4¼ miles (N. W. by N.) from Henley-in-Arden; containing 2201 inhabitants. The living is a discharged vicarage, valued in the king's books at £6. 13. 4.; present net income, £380; the patronage and impropriation belong to the Countess of Plymouth. A chapel has been built here, containing 230 free sittings, the Incorporated Society having granted £150 in aid of the expense. The Stratford-on-Avon canal passes through the parish. Two schools are supported out of the proceeds of several charitable bequests, amounting to £80 per annum.

TAPLOW (*ST. NICHOLAS*), a parish, in the union of ETON, hundred of BURNHAM, county of BUCKINGHAM, 1 mile (E. N. E.) from Maidenhead; containing 647 inhabitants. This parish is separated from the county of Berks by the river Thames, on which is an extensive paper-mill; and the Great Western railway passes within half a mile of the church. Cleifden was the residence of the Prince and Princess of Wales, during the infancy of their son, George III. It formerly belonged to a member of the Hamilton family, who fought under the celebrated Duke of Marlborough, and who, on his return from the continent, indulged the curious fancy of figuring the battle of Blenheim, by plantations of trees, now in full vigour. The living is a rectory, valued in the king's books at £11. 18. 9.; present net income, £329: it is in the patronage of the Crown. The church is a neat structure of brick, lately erected at some distance from the site of the former edifice, which was taken down, with the exception of part of the chancel, including the east window, and part of the west end of the nave, now forming a picturesque ruin: in the chancel of the former were interred the remains of Sarah Milton, the mother of the immortal poet, who resided here for some years. Francis Sharp, in 1797, gave £100 stock, the dividends on which, amounting to £3. 13. 6., are distributed in bread among the poor; and there is an allotment of waste land, awarded under an enclosure act in 1779, and now let for £5, which sum is applied to the supply of coal among the poor inhabitants.

TAPTON, a township, in the parish and union of CHESTERFIELD, hundred of SCARSDALE, Northern Division of the county of DERBY, 1½ mile (N. E. by E.) from Chesterfield; containing 171 inhabitants.

TARBOCK, a township, in the parish of HUYTON, union of PRESCOT, hundred of WEST DERBY, Southern Division of the county palatine of LANCASTER, 3½ miles (S. by W.) from Prescot; containing 755 inhabitants, of whom a considerable number are employed at the collieries here.

TARDEBIGG (*ST. BARTHOLOMEW*), a parish, in the union of BROMSGROVE, partly in the Alcester Division of the hundred of BARLICHWAY, Southern Division of the county of WARWICK, but chiefly in the Upper Division of the hundred of HALFSHIRE, Droitwich and Eastern Divisions of the county of WORCESTER, 3 miles (E. S. E.) from Bromsgrove; containing 4145 inhabitants. This parish is situated on the turnpike road from Birmingham to Pershore, which passes directly through the town of Redditch, celebrated for its manufacture of needles and fish-hooks, and of which a description is given under its own head. Hewell Park, the seat of the late Earl of Plymouth, is in this parish: the mansion is a beautiful structure, and the demesne is highly embellished. The remains of the late earl were deposited in the church, where is the family vault. There are quarries of excellent building stone, from which was raised the stone for Tardebigg church and the chapel of Redditch. The Birmingham and Worcester canal passes through the parish, within two miles of Redditch. Fairs are annually held here on the first Monday in August and the third Monday in September, for cattle. The living is a vicarage, valued in the king's books at £8; present net income, £031; the patronage and impropriation belong to the Countess of Plymouth. The church is an elegant structure, in the Grecian style of architecture, with a very beautiful spire; it was rebuilt by act of parliament in 1776. There are places of worship at Redditch for Arminians, Independents, and Wesleyan Methodists, and a Roman Catholic chapel. A National day and Sunday school is supported by the Countess of Plymouth; and another school is also partly supported by private charity. The site of Bordesley Abbey, supposed to have been originally founded by Maud, daughter of Henry I., is in this parish; its revenue at the dissolution was estimated at £392. 8. 6.; the site and remains were granted by Henry VIII. in exchange to Lord Windsor, one of the ancestors of the late Earl of Plymouth.

TARLETON (*ST. MARY*), a parish, in the union of ORMSKIRK, hundred of LEYLAND, Northern Division of the county palatine of LANCASTER, 8½ miles (N. by E.)

from Ormskirk; containing 1886 inhabitants. This was formerly a chapelry in the parish of Croston. The living is a perpetual curacy; patron, Rector of Croston. The church was consecrated in 1719. A free school, the building for which was erected in 1650, is endowed with £30 per annum, for which 20 children are gratuitously instructed. Here is also a National school.

TARNICAR, a joint township with Upper Rawcliffe, in the parish of St. Michael, union of Garstang, hundred of Amounderness, Northern Division of the county palatine of Lancaster, $4\frac{1}{4}$ miles (S. W.) from Garstang: the population is returned with Upper Rawcliffe.

TARPORLEY (*St. Helen*), a market-town and parish, in the union of Nantwich, First Division of the hundred of Eddisbury, Southern Division of the county of Chester; containing 2391 inhabitants, of which number, 995 are in the town of Tarporley, $10\frac{1}{2}$ miles (E. S. E.) from Chester, and 172 (N. W.) from London. This town, which is situated on the great road from Chester to London, has a neat appearance, and consists of one long street, which is well paved, and terminated at the southern extremity by the ancient manor-house. At the close of the thirteenth century, a grant of a market and fair was obtained by Hugh de Thorpley, then proprietor of the manor. The market is on Thursday; and fairs are held on May 1st, the first Monday after August 24th, and December 11th. The town was governed by a mayor from 1297 to 1348, but now only two constables are appointed. The living is a rectory, valued in the king's books at £20. 3. 4.: it is in the joint patronage of the Dean and Chapter of Chester, Lord Alvanley, and the Rev. Sir P. G. Egerton, Bart. The tithes of Tarporley have been commuted for a rent-charge of £190. 15., subject to the payment of rates; the glebe comprises 12 acres, valued at £27. 11. 6. per annum. The tithes of the other townships have been commuted for £509. 5., subject to the payment of rates. The church contains 100 free sittings, the Incorporated Society having granted £80 in aid of the expense; it is an ancient structure of red stone, and there are some good monuments. There is a place of worship for Wesleyan Methodists. A school, situated in the churchyard, was endowed with £20 per annum by Lady Jane Done, who left also a small bequest for apprenticing poor children. By a subsequent augmentation of the endowment, the master's salary is £55: the school-house has been recently enlarged, and a girls' school added, which is supported by subscription and conducted on the National plan. Almshouses were founded by Sir John Crewe, in 1704, for four poor widows, each of whom receives thirty shillings per annum.

TARRABY, a township, in the parish of Stanwix, Eskdale ward, Eastern Division of the county of Cumberland, $1\frac{3}{4}$ mile (N. N. E.) from Carlisle; containing 138 inhabitants.

TARRANT, CRAWFORD, county of Dorset.—See CRAWFORD-TARRANT.

TARRANT-GUNVILLE (*St. Mary*), a parish, in the union of Blandford, hundred of Cranborne, Blandford Division of the county of Dorset, 6 miles (N. N. E.) from Blandford-Forum; containing 502 inhabitants. The living is a rectory, valued in the king's books at £19. 7. 11.; present net income, £448; patrons, Master and Fellows of University College, Oxford. The church, which was consecrated in 1503, consists of a nave, aisles, and chancel, with an embattled tower crowned with pinnacles; the nave is raised above the aisles.

TARRANT-HINTON, county of Dorset.—See HINTON, TARRANT.

TARRANT-KEYNSTON (*All Saints*), a parish, in the union of Blandford, hundred of Pimperne, Blandford Division of the county of Dorset, $3\frac{1}{2}$ miles (S. E. by E.) from Blandford-Forum; containing 277 inhabitants. The living is a rectory, valued in the king's books at £7. 17. $8\frac{1}{2}$.; patron, Rev. John Austen. The tithes have been commuted for a rent-charge of £383, subject to the payment of rates; the glebe comprises 37 acres, valued at £60 per annum. The church stands on the western bank of the small river Tarrant, which falls into the Stour on the southern side of the parish. A day and Sunday school is supported by subscription.

TARRANT-LAUNCESTON, a parish, in the union of Blandford, hundred of Pimperne, Blandford Division of the county of Dorset, $5\frac{1}{4}$ miles (N. E. by E.) from Blandford-Forum; containing 72 inhabitants. The living is annexed to the vicarage of Tarrant-Monckton.

TARRANT-MONCKTON, county of Dorset.—See MONCKTON, TARRANT.

TARRANT-RAWSTON (*St. Mary*), a parish, in the union of Blandford, hundred of Pimperne, Blandford Division of the county of Dorset, $4\frac{1}{2}$ miles (E. by N.) from Blandford-Forum; containing 48 inhabitants. The living is a discharged rectory, valued in the king's books at £8. 9. 2.; patron, Sir J. W. Smith, Bart. The tithes have been commuted for a rent-charge of £90, subject to the payment of rates, which on the average have amounted to £6; the glebe comprises 33 acres, valued at £33 per annum.

TARRANT-RUSHTON (*St. Mary*), a parish, in the union of Blandford, hundred of Cranborne, Wimborne Division of the county of Dorset, $3\frac{3}{4}$ miles (E.) from Blandford-Forum; containing 226 inhabitants. The living is a rectory, valued in the king's books at £4. 19. 2.; present net income, £219; patron, Rev. George E. Saunders. A day and Sunday school is supported by subscription. Here was formerly an hospital or chantry, dedicated to St. Leonard, which was granted to the prior of Christchurch-Twynham in the 7th of Edward III.

TARRETBURN, a township, in the parish and union of Bellingham, North-Western Division of Tindale ward, Southern Division of the county of Northumberland, 3 miles (N. W.) from Bellingham; containing 265 inhabitants.

TARRING-NEVILLE (*St. Mary*), a parish, in the union of Newhaven, hundred of Danehill-Horsted, rape of Pevensey, Eastern Division of the county of Sussex, $2\frac{1}{2}$ miles (N.) from Newhaven; containing 80 inhabitants. The living is a rectory, united to that of Heighton in 1660, and valued in the king's books at £7. The church is a neat structure, in the early English style, with a remarkably large chancel.

TARRING, WEST (*St. Andrew*), a parish, in the hundred of Tarring, rape of Bramber, Western Division of the county of Sussex, $1\frac{1}{2}$ mile (N. W.) from Worthing; containing 626 inhabitants. This was anciently a place of much importance, and appears to have had a church or monastery, built to the honour of St.

Andrew in the time of Offa, King of Mercia, some remains of which might be traced in a free chapel, or peculiar jurisdiction, which continued here to the time of Edward III. Henry VI. granted the inhabitants a market, long since discontinued. The living consists of a sinecure rectory, valued in the king's books at £22. 13. 4., and a vicarage consolidated with the rectory of Patching, valued at £8. 13. 4., and in the patronage of the Archbishop of Canterbury, to whom belongs the sinecure rectory, the present net income of which is £576. The church is principally in the early English style of architecture: it consists of a nave, side aisles, and chancel, with a lofty tower surmounted by a handsome octagonal spire. The ancient parsonage-house was formerly of much greater extent, and is thought to have been a manor-house or palace, occasionally inhabited by Thomas à Becket. A school is partly supported by the rector. John Selden, the celebrated antiquary and historian, was born at the hamlet of Salvington, in this parish, in 1584; he died in 1654, and was interred with great pomp in the Temple church, London. The house still remains, very little larger than a cottage, with some curious old wooden frame-work at the end.

TARRINGTON (*St. James*), a parish, in the union of LEDBURY, hundred of RADLOW, county of HEREFORD, 7 miles (W. N. W.) from Ledbury; containing 540 inhabitants. The living is a vicarage, valued in the king's books at £5. 0. 2½.; present net income, £300; patron and impropriator, E. T. Foley, Esq. The church has been enlarged, and 90 free sittings provided, the Incorporated Society having granted £100 in aid of the expense. A day and Sunday school is supported by the chief landed proprietor.

TARSET, WEST, a township, in the parish of THORNEYBURN, union of BELLINGHAM, North-Western Division of TINDALE ward, Southern Division of the county of NORTHUMBERLAND, 4 miles (W. N. W.) from Bellingham; containing 149 inhabitants.

TARVIN (*St. Andrew*), a parish, in the union of GREAT BOUGHTON, partly in the Lower Division of the hundred of BROXTON, but chiefly in the Second Division of the hundred of EDDISBURY, Southern Division of the county of CHESTER; containing 3415 inhabitants, of which number, 1020 are in the township of Tarvin, 6 miles (E. by N.) from Chester. About the middle of the sixteenth century, Sir John Savage, lord of the manor, procured a charter for a market and a fair to be held here, which have been long disused. During the great civil war, Tarvin was a considerable military post, often taken and retaken by each party, till September 1644, when it fell into the power of the parliament, and so remained to the end of the war. The living is a vicarage, valued in the king's books at £19. 11. 0½.; present net income, £563; patron, Bishop of Lichfield; appropriators, Dean and Chapter of Lichfield. The church is in the later style of English architecture, with a fine tower considerably enriched with sculpture, though now much mutilated. A chapel was erected in 1833, which contains 100 free sittings, the Incorporated Society having granted £150 in aid of the expense. A grammar school was founded in 1600, with a house for the master, by John Pickering, who endowed it with £200, which was laid out in lands now producing an annual income of £16, for which 20 children are instructed, in connection with a National school: a diocesan school has also been established, and is partly supported by subscription. The celebrated calligrapher, John Thomason, was master of the grammar school; he died in 1740.

TASBURGH (*St. Mary*), a parish, in the union and hundred of DEPWADE, Eastern Division of the county of NORFOLK, 2 miles (N.) from Stratton-St. Mary; containing 479 inhabitants. The living is a rectory, with Rainsthorpe, valued in the king's books at £8; present net income, £275; patron and incumbent, Rev. G. Preston. The church, which was very recently damaged by lightning, stands on a lofty eminence, within the area of a square intrenchment containing 24 acres, which Gale considers the Roman station *Ad Tuam*; it was an advantageous position for the defence of the river Tesse, running hence to Caistor, and here forming a convenient harbour. Coins, fibulæ, and other relics of antiquity, have been found.

TASLEY, a parish, in the union of BRIDGENORTH, hundred of STOTTESDEN, Southern Division of the county of SALOP, 1¾ mile (N. W. by W.) from Bridgenorth; containing 102 inhabitants. The living is a discharged rectory, valued in the king's books at £5. 6. 8.; patron, E. F. Acton, Esq. The tithes have been commuted for a rent-charge of £210, subject to the payment of rates; the glebe comprises 8½ acres, valued at £12. 15. per annum.

TATENHILL (*St. Michael*), a parish, in the union of BURTON-UPON-TRENT, Northern Division of the hundred of OFFLOW and of the county of STAFFORD, 3¾ miles (W. S. W.) from Burton-upon-Trent; containing 2180 inhabitants. The living is a rectory, annexed, with the prebend of Abdaston, to the deanery of Lichfield, and valued in the king's books at £26. 1. 8. The tithes have been commuted for a rent-charge of £1337. 6., subject to the payment of rates; the glebe comprises 123½ acres, valued at £200 per annum. The Grand Trunk canal passes through the parish. Here is a National school.

TATHAM (*St. James*), a parish, in the hundred of LONSDALE, south of the sands, Northern Division of the county palatine of LANCASTER; containing 853 inhabitants, of which number, 744 are in the township of Tatham, 11 miles (N. E. by E.) from Lancaster. The living is a rectory, valued in the king's books at £12. 5.; present net income, £195; patron, Devisee of the late John Marsden, Esq. The church is a handsome modern structure. An old Roman road passes through the parish, in which there is an extensive colliery. A fair for cattle is held annually on March 12th, in the village of Lowgill. A school is endowed with an estate producing £26 a year, which is applied for the instruction of all the poor children of the lower division of the parish.

TATHAM-FELL, a chapelry, in the parish of TATHAM, hundred of LONSDALE, south of the sands, Northern Division of the county palatine of LANCASTER, 12½ miles (E. N. E.) from Lancaster: the population is returned with the parish. The living is a perpetual curacy; net income, £125; patron, Rector of Tatham.

TATHWELL (*St. Vedast*), a parish, in the union of LOUTH, Wold Division of the hundred of LOUTH-ESKE, parts of LINDSEY, county of LINCOLN, 3¼ miles (S. by W.) from Louth; containing 338 inhabitants.

The living is a discharged vicarage, valued in the king's books at £10; present net income, £227; patron, C. Chaplin, Esq. The church contains monuments to the Hanby and Chaplin families. Tathwell Hall, the seat of George Chaplin, Esq., is beautifully situated. On a high hill in this parish are six barrows, in a line running from east to west; and on Dagarth Hill, about half a mile from the barrows, are the remains of two Roman encampments.

TATSFIELD, or TATTESFIELD, a parish, in the union of GODSTONE, Second Division of the hundred of TANDRIDGE, Eastern Division of the county of SURREY, 7 miles (N. E. by E.) from Godstone; containing 166 inhabitants. The living is a discharged rectory, valued in the king's books at £5. 0. 5.; present net income, £150; patron, William L. Gower, Esq. The church, principally in the early style of English architecture, was almost entirely rebuilt in 1838, by private subscription; the south porch and the tower, which are elegant specimens of that style, were built at the expense of the Rev. S. Streatfield, the rector, by whom the parochial charity school is principally supported.

TATTENHALL (*St. Alban*), a parish, in the union of GREAT BOUGHTON, Lower Division of the hundred of BROXTON, Southern Division of the county of CHESTER; containing 1141 inhabitants, of which number, 978 are in the township of Tattenhall, 5¾ miles (S. W. by W.) from Tarporley. The living is a rectory, valued in the king's books at 13. 17. 6.; present net income, £277; patron, Bishop of Chester. The church contains 100 free sittings, the Incorporated Society having granted £30 in aid of the expense. There are places of worship for Independents and Wesleyan Methodists. Dr. Paploe, rector of this parish, who died in 1781, gave money, vested in the purchase of £334 three per cents., for the education of twelve children; and a National school has been established, which is partly supported from the same funds, aided by subscription.

TATTENHOE, a parish, in the union of WINSLOW, hundred of COTTESLOE, county of BUCKINGHAM, 3¾ miles (W.) from Fenny-Stratford; containing 13 inhabitants. The living is a donative curacy, holden by institution as a rectory; net income, £50; patron and impropriator, W. S. Lowndes, Esq. The church was rebuilt in 1540, but the parish containing only a few inhabitants, it fell into disuse and desecration, until the rector of Shealey claimed the tithes of the parish, in 1636, when it was consecrated anew, and the living was presented to as a rectory.

TATTERFORD (*St. Margaret*), a parish, in the union of WALSINGHAM, hundred of GALLOW, Western Division of the county of NORFOLK, 4 miles (W. by S.) from Fakenham; containing 75 inhabitants. The living is a discharged rectory, consolidated with that of Tatterset, and valued in the king's books at £6. 6. 8.

TATTERSET (*St. Andrew*), a parish, in the union of WALSINGHAM, hundred of GALLOW, Western Division of the county of NORFOLK, 5 miles (W.) from Fakenham; containing 118 inhabitants. The living is a discharged rectory, with that of Tatterford consolidated, valued in the king's books at £11. 1. 8.; present net income, 685; patron, Sir Charles Chadd, Bart.

TATTERSHALL (*Holy Trinity*), a market-town and parish, in the union of HORNCASTLE, Southern Division of the wapentake of GARTREE, parts of LINDSEY, county of LINCOLN, 30 miles (S. E. by E.) from Lincoln, and 125 (N.) from London; containing 599 inhabitants. This place, anciently a Roman military station, as two encampments at Tattershall park in its immediate neighbourhood indicate, was granted at the Conquest to Eudo, one of William's followers, whose descendants erected a castle here about 1440, of which some remains are yet visible south-westward from the town: it stood on a moor, and was surrounded by two fosses, which received the waters of the Bane, but the principal part was demolished during the parliamentary war: the north-west tower, a rectangular brick structure 100 feet high, flanked by four embattled octangular turrets, was built by Sir Ralph Cromwell, treasurer of the Exchequer in the reign of Henry VI., and still remains; he likewise erected a lofty tower, with a spiral staircase leading to its summit, about four miles northward, as an appendage to the larger structure, but this is now in a very dilapidated state. The town, situated on the river Bane, near its junction with the Witham, is much decayed, and the trade inconsiderable: a canal from the Witham to Horncastle passes through it. The market, originally granted by King John to Robert Fitz-Eudo, was formerly on Friday, but is now on Thursday: fairs are on May 15th and Sept. 25th. The parish is within the jurisdiction of the Bolingbroke and Horncastle court of requests for debts not exceeding £5. The living is a donative; net income, £110; patron and impropriator, Earl Fortescue. The church is situated on the eastern side and in the outer moat of the castle, and was made collegiate, in the time of Henry VI., for seven chaplains (one of whom was master), six clerks, and six choristers: at the dissolution its revenue was estimated at £348. 5. 11. The collegiate buildings have been taken down, and the church alone remains, which is a beautiful and venerable cruciform structure, consisting of a nave, transept, and choir, of which the last was once much admired for its magnificent painted windows, but since their removal to the chapel of Burleigh, the seat of the Marquess of Exeter, this part of the edifice has been allowed to fall into decay. There is a place of worship for Wesleyan Methodists. A National school is held in the south side of the transept of the church; and an almshouse for thirteen poor persons, originally established by the licence which raised the church into a college, still remains, with a small endowment for its support.

TATTINGSTONE (*St. Mary*), a parish, in the hundred of SAMFORD, Eastern Division of the county of SUFFOLK, 5½ miles (S. W. by S.) from Ipswich; containing 356 inhabitants, exclusively of 310 in the house of industry for the hundred, which is in this parish. The living is a rectory, valued in the king's books at £6. 13. 4.; patron, Rev. John Garwood Bull. The tithes have been commuted for a rent-charge of £402, subject to the payment of rates, which on the average have amounted to £53; the glebe comprises 38 acres, valued at £55 per annum. There is a place of worship for Wesleyan Methodists; and a National school has been established. A spot, on the estate of T. Burch Western, Esq., whose mansion, called the Place, is in this parish, is remarkable for a very thick deposit of marine shells. This is the head of a union containing 28 parishes, governed by a peculiar act of parliament, and not subject to the poor law commissioners.

TATTON, a township, in the parish of ROSTHERN, union of ALTRINCHAM, hundred of BUCKLOW, Northern Division of the county of CHESTER, 2 miles (N.) from Nether Knutsford; containing 69 inhabitants.

TATWORTH, a joint tything with Forton, in the parish of CHARD, Eastern Division of the hundred of KINGSBURY, Western Division of the county of SOMERSET, 1¾ mile (S.) from Chard: the population is returned with the parish.

TAUNTON, a borough and market-town, and the head of a union, in the hundred of TAUNTON and TAUNTON-DEAN, Western Division of the county of SOMERSET, 11 miles (S. by W.) from Bridg-water, and 144 (W. by S.) from London; containing 11,139 inhabitants. This place, which was called by the Saxons *Tantun*, and subsequently *Tawnton* and *Thoneton*, from its situation on the river Thone, or Tone, is of great antiquity; and the discovery of several urns, containing Roman coins, in the neighbourhood, has led to the conjecture that it existed in the time of that people. But the earliest authentic accounts refer to the period of the Heptarchy, when a castle was built here for a royal residence, by Ina, King of the West Saxons, about the year 700, in which he held his first great council. This castle was afterwards demolished by his queen Ethelburga, after expelling Eadbricht, King of the South Saxons, who had seized it. The town and manor are supposed to have been granted to the church of Winchester in the following reign; and another castle is said to have been built, on the site of the first, by the Bishops of Winchester, in the reign of William the Conqueror, in which they principally resided for some years. At this period Taunton had a mint, some of the coins bearing the Conqueror's effigy being still in existence. In the reign of Henry VII., in 1497, Perkin Warbeck seized the town and castle, which he quickly abandoned, on the approach of the king's troops. In 1645, it again participated in civil war, being celebrated for the long siege it sustained, and the defence it made under Colonel (afterwards the renowned Admiral) Blake, who held it for the parliament against 10,000 royalist troops under Lord Goring, until relieved by Fairfax. On this memorable occasion a public thanksgiving was appointed by the Commons, who voted £500 to the Colonel, and £1000 to the men under his command; but the inhabitants incurred the displeasure of the king, who, on his restoration, suspended their charter, and ordered the walls to be razed to the ground, which was so effectually done, that even their site is not now known. James Duke of Monmouth was proclaimed king on the Cornhill of Taunton, June 21st, 1685; and many of his followers, amongst whom were some of the inhabitants of Taunton, were, after his defeat at Sedgmoor, inhumanly put to death, on the same spot, by the brutal Kirke, without form of trial, besides those who were condemned by the merciless Judge Jeffreys, at the "bloody assize" which he held here in the following September.

The town is situated in a central part of the singularly beautiful and luxuriant vale of Taunton-Dean, and is upwards of a mile in length: the principal streets, which terminate in the market-place, are spacious, well paved, and lighted with gas by a company established in 1821: the houses, mostly built of brick, are generally commodious and handsome, and well supplied with excellent water. The respectability of the town, combined with the beauty of the surrounding country, has rendered it very attractive as a place of residence, and many recent improvements have been effected, amongst which are the erection of a neat crescent and terrace, and the removal of some old houses at East Gate, which has rendered the entrance from London more spacious. In 1833, an act was obtained for better regulating the market, cleansing the streets, and preventing nuisances. The Parade, in the centre of the town, is a fine open triangular space, enclosed with iron posts and chains; and on the east side of it is a wide street, erected by the late Sir Benjamin Hammet, which forms a handsome approach to the beautiful church of St. Mary. A substantial stone bridge of two arches crosses the Tone, and connects the town with the village of North-town, or Nurton. Several detached villas commanding beautiful views, have also been erected in the suburbs of Wilton, Staplegrove, West Monckton, and adjoining parishes. The Taunton and Somerset Institution, established in 1823, has a small but valuable library, particularly works of reference, and a museum containing various specimens in mineralogy, ornithology, zoology, &c.: it has likewise a noble and spacious public reading and news-room. The theatre, in Silver-street, is usually open two months in the year; and balls and concerts occasionally take place.

Taunton was formerly noted for its woollen manufacture, being one of the first places into which that branch of trade was introduced, but it has long since given place to the silk trade, which was begun here in 1778, and is now carried on to a considerable extent: the chief articles manufactured are crapes, persians, sarsnets, and mixed goods; and, as nearly every cottage has a silk-loom, the trade furnishes employment to a great number of persons, principally females. Two patent lace-manufactories have been established. The river Tone is navigable, but its course to Bridg-water being circuitous, and the navigation frequently interrupted, a canal, called the Taunton and Bridg-water canal, has been constructed, which has given increased activity to trade, considerable quantities of Welsh coal being brought to the town, and, in return, the produce of the vale of Taunton being exported to Bristol and other parts of England. The Grand Western canal, forming a communication with the river Exe, terminates here; and the line of the Bristol and Exeter railway passes by the town. The markets are on Wednesday and Saturday, the latter being the principal; they are well supplied with fish from both channels, with every other kind of provisions, and with fruit in abundance. The old market-house at the south end of the Parade, is a lofty brick building, supported on each side by an arcade, one of which is used as a corn market, and the other by various tradesmen: it contains the guildhall, and a handsome assembly-room, in which is a full-length portrait of George III. in his robes, presented by the late Sir Benjamin Hammet. On the west side of the Parade is a handsome building of freestone, erected in 1821, in the lower part and rear of which, and on the northern side, are the markets for meat, fish, pork, poultry, and dairy produce; the upper being used as the library and reading-room of the institution before mentioned: it is of the Ionic order, the entablature supported by four handsome columns, and forms a great ornament to this part of the town. The last Saturday in every month is called the great market, including the sale of live stock. This is one of the towns which, by act of parliament, are obliged to make weekly returns of corn sold; the tolls of the markets are let for about £1400 per annum. There is an annual fair in the town on June 17th, and another in the suburb called North-town on July 7th, for horses and cattle.

The town was for several centuries under the jurisdiction of portreeves and bailiffs, chosen at the courts of the Bishops of Winchester, as lords of the manor, which was formerly very extensive and valuable; the rental, at the time of the Conquest, appearing, from a document found amongst the court rolls, to amount to nearly £700 per annum: it was, however, divided by the Conqueror, and portions of it distributed among his favourites. By the custom of the manor the wife is considered her husband's heir, and succeeds before the children, the youngest son before the eldest, and also in other relations, the younger succeeding first. It continued in the possession of the Bishops of Winchester until the year 1822, when it was sold by Bishop Tomline to Thomas Southwood, Esq., and is now the property of Robert Mattock, Esq., at whose annual courts, held in the castle, two portreeves, who collect the lord's rents, two bailiffs, two constables, and six tythingmen, are chosen. A charter was granted in 1627, by Charles I., vesting the civil power in a mayor, justice, aldermen, and burgesses; this was suspended by Charles II., but subsequently confirmed by him, and existed until the year 1792, when, in consequence of the corporation having suffered a majority of the members to die without filling up vacancies, and a majority being required to swear in the officers, none could be legally elected; whereby the charter became forfeited, and a fruitless application having been made to the Privy Council for its renewal, the town is now under the jurisdiction of the county magistrates, who hold a petty session on Wednesdays and Saturdays at the guildhall. The bailiffs usually convene and preside at public meetings; and the constables have the distribution of most of the public charities. This is a borough by prescription, and first sent members to parliament in the reign of Edward I., in 1295. The right of election was formerly in the potwallers, or persons who boiled their own pot, and had resided six months within the limits of the borough, and not receiving alms, in number about 600; but the borough has been enlarged, and comprises an area of 742 acres, the limits of which are minutely described in the Appendix, No. II.: the old borough contained only 158 acres; the present number of voters is nearly 800; the bailiffs are the returning officers. The Lent assizes for the county are held in the castle, as are also the Michaelmas general quarter sessions; and the town has been made the place of election, and a polling-place, for the western division of the county. A court for the recovery of debts under 40*s.* is holden, in the castle, for the manor and hundred, and a similar court for the borough at the guildhall, alternately, every week. The castle, supposed to be part of a stately edifice erected by William Giffard, Bishop of Winchester, in the reign of Henry I., was thoroughly repaired by Bishop Langton, towards the end of the 15th century; and in addition to other improvements, the present assize hall was built, by Bishop Horne, in 1577, since which period, various sums have been expended for keeping it in repair. The building consists of a south front, with a gateway in the centre, over which are two escutcheons, one bearing the arms of Henry VII., with the motto *Vive le roi Henri;* the other the inscription *Laus tibi Xte.*, and *T. Langto Winto*, 1495, both in Saxon characters: at the east end is a circular tower. The inner court-yard is an irregular quadrangle, the east side being the shorter, and on the north side are the county courts, grand jury room, &c.: the access to it is through an open court, called Castle Green, formerly enclosed with two gates, but one only remains, over which is what was the porter's lodge, now occupied as a dwelling-house; besides these is another apartment, called the Exchequer chamber, where the rolls of the manor are preserved; the walls are of great strength, and the grooves for the portcullis quite perfect. The moat was filled up, and the drawbridge removed, in 1785. There is a small place of temporary confinement, called the Nook, used for malefactors until they can be taken before a magistrate; and closely adjoining the town is the house of correction, at Wilton, which was erected in 1754, and enlarged in 1815, being now capable of receiving 80 prisoners. Debtors, as well as criminals, are sent to the county gaol at Ilchester, and the latter also to Wilton and Shepton-Mallet prisons.

Taunton comprises the parishes of *St. James* and *St. Mary Magdalene*, but many of the houses extend into the adjoining parishes of Wilton and Bishop's Hull. The living of St. James's is a perpetual curacy; net income, £254; patron and impropriator, Sir T. B. Lethbridge, Bart. The church, which was formerly the conventual church of the priory, has lately been considerably enlarged, through the exertions of the Rev. James Cottle, the present incumbent, and is now an elegant and commodious structure, capable of containing 1400 sittings, about 500 of which are free, the Incorporated Society having granted £450 in aid of the expense. The living of St. Mary Magdalene's is a vicarage; patron, Lord Ashburton. The church, standing near the centre of the town, was originally a chapel to the conventual church of St. James, but was made parochial in 1308, under Walter Huselshaw, then Bishop of Bath and Wells: it is a spacious and magnificent edifice, in the decorated and later styles of English architecture, consisting of a chancel, nave, and four aisles, two of the latter having been probably added at a later date, separated by four rows of clustered columns supporting pointed arches: the quadrangular tower at the west end is an elegant structure in four compartments, containing thirteen windows, which, by the variety of their ornaments, add much to its lightness and beauty; it is 121 feet in height, exclusively of its pinnacles of 32 feet, which are richly adorned with carved work, and the top crowned with most exquisitely delicate battlements. Another church is in course of erection, the Incorporated Society having granted £500 in aid of the expense. There is a place of worship each for Baptists, the Society of Friends, Independents, and Unitarians, and two for Wesleyan Methodists, one of them erected under the immediate direction of their founder, the Rev. John Wesley. The Roman Catholics have a handsome chapel, with a portico supported by two Ionic pillars, and a façade ornamented with windows and pilasters of the same order; they have also a convent of Franciscan nuns, who emigrated from Brussels during the French Revolution in the last century, and first settled at Winchester, until they became possessed of their present residence, which is a noble building at the east end of the town, near the entrance from London, originally intended for a public hospital.

The free grammar school was founded in 1522, by Richard Fox, Bishop of Winchester, and endowed in 1554, by William Walbee, clerk, with property now producing about £36 per annum: the school-house is on the south side of the Castle green, and adjoining it is a house for the master, whose appointment is vested in the Warden of New College, Oxford: there are no boys on the foundation. The school of industry. in which 80 boys and 50 girls are clothed and instructed, is supported by subscription; and there are also a National school, a Lancasterian school, and two infants' schools; a school-room for one of the latter was erected near St. James's church in 1828. The almshouses at East Gate, for ten poor women and seven men, were founded in 1635, by Robert Gray, Esq., a native of this town, and endowed by him with £2000, since augmented by additional benefactions. Huish's almshouses, on the north side of Hammet-street, were founded and endowed by Richard Huish, Esq., for thirteen poor men, one of whom is appointed president, and reads prayers daily in the chapel attached to the building. Henley's and Pope's almshouses, and seventeen other tenements, are also appropriated to the rent-free residence of the poor, but are not endowed. Of the other charities, the principal is that arising from the Town Lands, consisting of property to which no claimant appeared after a plague had raged in Taunton, which, together with some land and houses purchased under bequests of John Meredith and Margery Acland, is vested in feoffees; the annual produce is about £360, and, after some deductions for expenses, the income from the Town Lands is distributed in money among the poor of the parish of St. Mary Magdalene; that from John Meredith's bequest, in clothing to poor men and women; and that from Margery Acland's, in small sums to poor widows. The Taunton and Somerset hospital was founded in 1809, in commemoration of George III. entering upon the fiftieth year of his reign: it was opened on the 25th of March, 1812, and contains four wards, two for men and two for women, being capable of accommodating about 34 patients, and is about to be enlarged to double its present size. An eye infirmary, established in 1816, is supported by voluntary contributions; and there is a society for the relief of lying-in women. The poor law union of Taunton comprises 38 parishes or places, containing a population of 31,378, according to the census of 1831, and is under the care of 51 guardians. Taunton is the birthplace of Samuel Daniel, the poet, born in 1562; and of the Rev. Henry Grove, in 1683, an eminent dissenting minister, who, in addition to other works, contributed some excellent papers to the Spectator. Amongst the bishops of Winchester who made it their occasional residence are the names of Cardinals Beaufort and Wolsey.

TAVERHAM (*St. Edmund*), a parish, in the union of St. Faith, hundred of Taverham, Eastern Division of the county of Norfolk, $5\frac{3}{4}$ miles (N. W.) from Norwich; containing 191 inhabitants. The living is a rectory, formerly in medieties, now united, each valued in the king's books at £4. 2. $8\frac{1}{2}$.; present net income, £300; patron, Bishop of Norwich.

TAVISTOCK (*St. Eustachius*), a borough, market-town, and parish, and the head of a union, in the hundred of Tavistock, Tavistock and Southern Divisions of the county of Devon, 33 miles (W. by S.) from Exeter, and 204 (W. S. W.) from London; containing 5602 inhabitants. This place, which takes its name from its situation on the river Tavy, was anciently the residence of Orgar, Earl of Devonshire, whose daughter Elfrida, surreptitiously obtained in marriage by Athelwold, favourite of King Edgar (for whom he had been sent to negotiate), on the subsequent discovery of his treachery to his sovereign, became the wife of that monarch. The town appears to have derived its origin from the erection of an abbey of Black monks, begun in 961, by Orgar, who, according to tradition, had been admonished in a dream to found a monastery here, which was completed in 981 by his son Ordulf, by whom it was endowed with ample possessions, and dedicated to St. Mary the Virgin and St. Rumon. After having been destroyed by the Danes, the monastery was restored by the contributions of the neighbouring families, of whom the De Eggecombes were munificent benefactors, and was rebuilt with greater magnificence. Henry I. granted to the abbots the entire jurisdiction of the hundred of Tavistock, and a weekly market and annual fairs, and invested them with other privileges; and in 1513 Henry VIII. conferred the right of a seat among the peers upon Abbot Banham, who also procured from Pope Leo X. an exemption from all episcopal and metropolitical jurisdiction. Soon after the introduction of printing into England, a printing press was established in the monastery, from which issued a code of the Stannary laws, and a translation of Boëthius by Walton, printed by Dan Thomas Rychard, one of the monks, of which perfect copies are still preserved in the library of Exeter College, Oxford. The monastery continued to flourish till the year 1539, when it was surrendered to the king by the last abbot, John Peryn, on whom was settled a pension of £100 per annum for life. The revenue at the dissolution was £902. 5. 7.; and the site, together with the borough and town, was assigned to Lord John Russell, ancestor of the Duke of Bedford. A school for the study of Saxon literature was established here at a very early period, under the patronage of the abbots, and continued to flourish till the time of the Reformation. While the plague raged at Exeter, in 1591, the summer assizes were held in this town, and thirteen criminals were executed on the Abbey green; and during its ravages in London, a market and a fair were held above Merivale bridge, about three miles distant from the town, the memorial of which is still preserved by three long rows of stones, which indicate the spot. After the defeat of the parliamentarians on Bradock Down, in 1643, the royalists were quartered here; and Charles I. visited the town on

Arms.

his route to Cornwall, after his unsuccessful attempt on Plymouth.

The town is pleasantly situated in a valley, through which the river Tavy rushes with tumultuous impetuosity over an uneven and rocky bed, and which combines some of the most beautiful and picturesque scenery in this justly admired county; it is irregularly built, partly in the vale and partly on the acclivities by which it is enclosed; and was first lighted with gas in the year 1832. The approaches are easy and commodious, that from Plymouth being remarkably fine, and those from the east of Cornwall, and from the roads over Dartmoor, having undergone considerable improvement, under the auspices of the late Duke of Bedford, in 1839. On the right of the entrance into the town from the Plymouth road, and opposite to the church, are various embattled and turreted buildings originally belonging to the abbey, of which a part has been converted into the Bedford Hotel, which has an extensive façade in the ancient English style. The public library is held in a building over the grand archway of the ancient abbey; and adjoining it is a building in which the members of a literary and scientific institution deliver lectures once a fortnight during the winter months: the library was fitted up, and the building for the literary institution was erected, by the late Duke of Bedford, in lieu of a building in the Grecian style of architecture, for both those purposes, which was built by subscription, and which, as not harmonising with the venerable remains of the abbey, his Grace was anxious to remove. Over the Tavy are two ancient bridges within the town, and a third of modern construction about a quarter of a mile on the Plymouth road, near which is a bridge over the Tavistock canal, which extends to the town in a direction parallel with the river. Races are held on Whitchurch Down. The manufacture of serge and coarse woollen cloths, which formed the principal employment of the inhabitants, has been long on the decline; and the mining trade, which was formerly carried on to a considerable extent, has also materially diminished. An extensive iron-foundry is conducted in the town; and at a place called Crowndale, at the distance of a mile from it, is a tin-smelting establishment. The neighbourhood abounds with mineral productions, and in the section of a mining field between the rivers Tavy and Tamar, considerable quantities of porphyritic rock, in alternate layers, called Elvan, are found. From the mines near the town grey and ruby copper are produced; and in the mine called Wheal Friendship, native rich yellow, red, and chrystallised pyrites are found in profusion. Lead abounds in the district, and silver, tin, manganese, iron, and the loadstone are also found. The Tavistock canal, constructed to form a junction with the Tamar at Morwell Ham quay, with which it is connected by an inclined plane 240 feet high, flows under a tunnel at Morwell down, one mile and three quarters in length; it was opened in June 1817, having been completed at an expense of £68,000. The boats employed on this canal are of iron, and the principal articles conveyed by them are ore, coal, and lime. The market, which is noted for its ample supply of corn, is on Friday; the fairs were on Jan. 17th, May 6th, Sept. 9th, Oct. 10th, and Dec. 14th, but are now held on the second Wednesday in each of those months respectively; there are also great cattle markets on the second Wednesday in March, July, Aug., and Nov.

Seal of the Court of the Lordship.

The inhabitants never received a regular charter of incorporation. The town, which is one of the four chief stannary towns, is governed by a portreeve elected at the annual court leet of the lord of the manor. The borough, which exists by prescription, first sent members to parliament in the reign of Edward I. The elective franchise was formerly vested in the resident freeholders, in number about 30; but it has been extended to the £10 householders of the parish (except the detached manor of Cudliptown), which has been constituted the new elective borough, comprising an area of 11,112 acres: the old borough contained only 336 acres; the portreeve is the returning officer. Among the most distinguished of its representatives were John Pym, the great opposer of Charles I., and William Lord Russell, who was beheaded in the reign of Charles II. A court of pie-poudre, and a court for the recovery of debts under £20, formerly held here, have fallen into disuse. The town has been made a polling-place for the southern division of the county.

The living is a discharged vicarage, valued in the king's books at £10. 17. 6.; present net income, £298; patron and impropriator, Duke of Bedford. The church is a neat, spacious, and ancient structure, with a lofty tower supported on arches, affording a thoroughfare underneath it for carriages: it contains some good monuments, especially those to Sir John Fitz and Sir John Glanville, the latter of whom was judge of the Common Pleas, and died in 1600; there is also a neat and expensive monument to the family of Carpenter, of this parish, put up many years since by the father of J. Carpenter, Esq., of Mount Tavey. There are places of worship for the Society of Friends, Independents, Wesleyan Methodists, and Unitarians. The grammar school is of very ancient though uncertain foundation, and during the existence of the monastery, was, under the auspices of the abbots, for many years pre-eminently distinguished; it contained accommodation within the building for 100 classical scholars, and many eminent men, whose families were unconnected with the town, were educated here. In 1552, John Earl of Bedford granted, for 200 years, the amount of dues claimed by him within the borough for the support of this school; and in 1649 Sir John Glanville, Knight, Speaker of the House of Commons, gave an estate at South Brentor, producing £25 per annum, for the better maintenance of a poor scholar at either of the universities. Since the expiration of the Earl of Bedford's grant, his successors have allowed the master a residence, school-house, and garden rent-free, and an annual voluntary stipend of £20: a handsome school-house was erected by the late Duke, in 1838; and the school, which previously had fallen almost into disuse, has again begun to flourish: for several years no claimant has been found for Sir John Glanville's grant, which has accumulated to about £450, since which time an allowance of £4. 4. from the

Ford-street charity has been discontinued. A free school, conducted on the Lancasterian plan, was erected at the expense of the Duke of Bedford, and is chiefly supported by subscription. In 1674, Nicholas Watts bequeathed lands and houses, the rents of which together amount to £65. 18. 8., for the benefit of poor persons not receiving parochial relief, and a part to be appropriated to the assistance of a poor scholar of Tavistock, in maintaining himself at the university. An almshouse was founded by Lord Courtenay for four aged widows, among whom an annuity of £8. 12. is divided. The amounts of several benefactions called the Ford-street charity, producing £120 per annum, were settled by act of parliament and vested in the Duke of Bedford, for various purposes, in fulfilment of which an almshouse has been erected for fifteen poor persons, who receive each £3 per annum in quarterly payments; and from the same fund the sum of £5 is paid quarterly as a portion to a poor maiden on the day of her marriage; two poor boys receive annually £7. 10. each as an apprentice fee; £30 is distributed among the poor; £4. 4. are payable to the master of the free grammar school; and the residue is expended in repairs. The general dispensary was established in 1832. The poor law union of Tavistock comprises 24 parishes or places, containing a population of 20,630, according to the census of 1831, and is under the care of 35 guardians. The principal remains of the ancient monastery are the gateway, the refectory (now used as a place of worship for Unitarians), traces of the boundary walls, and an entire gateway near the canal bridge, probably forming a private entrance to the gardens and orchard of the abbey; they are principally in the later style of English architecture, and being in many parts mantled with ivy, have an interesting and picturesque appearance. Within the parish are the remains of Old Morwell House, the ancient hunting-seat of the abbots; in the woods attached to the mansion is a precipitous cliff, from the summit of which is a fine view of the river Tamar winding through a valley of great beauty. Within a mile of the town, in the parish of Whitchurch, is Holwell House, the ancient seat of the Glanville family, of which the last male representative of the elder branch, by whose father the property was alienated, died in 1830: the family arms are emblazoned over the mantel-piece of one of the state rooms on the ground floor of a house in the centre of the town: the general appearance of the mansion, which is in good preservation, bears testimony to its original magnificence. Among the eminent natives of this place were Sir Francis Drake; Judge Glanville; his son, Sir John Glanville; and William Browne, author of "Britannia's Pastorals," the "Shepherd's Pipe," and other works. This place gives the inferior title of Marquess to the Duke of Bedford.

TAVY (*St. Mary*), a parish, in the union of Tavistock, hundred of Lifton, Tavistock and Southern Divisions of the county of Devon, 4 miles (N. E. by N.) from Tavistock; containing 1123 inhabitants. The living is a rectory, valued in the king's books at £14. 5. 7½.; present net income, £224; patron, John Buller, Esq. In this parish is an extensive copper mine, called Wheal Friendship. A school is partly supported by subscription.

TAVY (*St. Peter*), a parish, in the union of Tavistock, partly in the hundred of Lifton, but chiefly in the hundred of Roborough, Tavistock and Southern Divisions of the county of Devon, 3½ miles (N. E.) from Tavistock; containing 444 inhabitants. The living is a rectory, valued in the king's books at £17. 1. 8.; patron, Bishop of Exeter. The church contains a monument to the Rev. Mr. Pocock, a former rector, and, with the churchyard, forms a strikingly picturesque feature in the scenery of the place. Here is a National school for girls. At Wilsworthy there was formerly a chantry chapel, which has been converted into a cow-house.

TAWNEY-STAPLEFORD, county of Essex.—See STAPLEFORD, TAWNEY.

TAWSTOCK (*St. Peter*), a parish, in the union of Barnstaple, hundred of Fremington, Braunton and Northern Divisions of the county of Devon, 2 miles (S.) from Barnstaple; containing 1348 inhabitants. The living is a rectory, valued in the king's books at £69. 12. 1.; present net income, £783; patron, Sir B. Wrey, Bart. There are places of worship for Independents and Roman Catholics. Here is a National school, with an endowment of £2 per annum; it was built at the expense of the Rev. B. W. Wrey. The manor-house, which was garrisoned by Sir T. Fairfax, in February 1646, was almost consumed by fire in 1787: it was rebuilt by the late Sir B. Wrey, Bart., except the ancient gateway, which still remains, bearing date 1574.

TAWSTOCK, county of Suffolk.—See TOSTOCK.

TAWTON, BISHOP'S (*St. John the Baptist*), a parish, in the union of Barnstaple, hundred of South Molton, Braunton and Northern Divisions of the county of Devon, 2 miles (S. by E.) from Barnstaple; containing 1641 inhabitants. The living is a vicarage, valued in the king's books at £21; present net income, £440; patron and appropriator, Dean of Exeter. The church is a neat ancient structure, with a handsome stone spire, and contains some ancient monuments to the Chichester family. On the division of the West Saxon see of Sherborne, this was made the seat of the Devonshire diocese by Werstan, its first Bishop, soon after his consecration in 905. He was succeeded by Putta, and afterwards by Eadulphus, who was installed at Crediton, to which place he removed the see, and who died in 931. Some remains of the episcopal palace are still discernible, and in the churchyard are the ruins of the deanery. A National school has been established. This parish contains the populous village of Newport, which see.

TAWTON, NORTH (*St. Peter*), a parish, in the union of Oakhampton, hundred of North Tawton, South Molton and Northern Divisions of the county of Devon, 12 miles (W. by N.) from Crediton; containing 1788 inhabitants. This place was anciently called Cheping Tawton, *i. e.* "a market-town on the Taw," which river runs through, and forms a boundary to, the parish. Its market charter was confirmed in the year 1270, at which period it was a borough town, being still governed by a portreeve, elected annually at the manorial court. The market was discontinued about 1720; but cattle fairs are held on the third Tuesday in April, Oct. 3rd, and December 18th. Here was once an extensive woollen manufacture, and there is still a spinning-mill, employing 200 persons in spinning yarn, which is woven at Crediton and other places. The parish contains a quarry of good freestone: the scenery is pleasantly varied. Ashridge, one of the most ancient demesnes in the county, comprises nearly 100 acres of woodland, containing a vast quantity of fine oak trees. The living is a rectory, valued in the king's books at £32. 4. 7.; present net income, £751; patron and incumbent, Rev. George Hole. There is a place of worship for Independents. Fifteen children are educated for about £14. 14. a year, part of the produce of a messuage and lands, the gift of the Rev. Richard Hole, in 1783. There were formerly chapels at Crook-Burnell, Nichols-Nymet, and Bath-Barton, in this parish. The last is the birthplace of Henry de Bathe, who had been, by Henry III., in 1238, made one of the justices of the common pleas, and in 1240 constituted one of the justices itinerant for many of the counties: in 1251, he fell under the king's displeasure, and was arraigned for high treason, but so great was the influence and the power of his family that he was, after a time, restored to all his former dignities; he died in 1262. Henry Tozer, expelled from Exeter College for his loyalty, in 1648, was a native of this place; he was the author of "Directions for a Devotional Life," which passed through ten editions. In the neighbourhood, a small brook sometimes issues out of a large pit ten feet deep, called Bathe Pool, and continues running for several days together, like that called Woodbourne, in Hertfordshire.

TAWTON, SOUTH (*St. Andrew*), a parish, in the union of Oakhampton, hundred of Wonford, Crockernwell and Southern Divisions of the county of Devon, $3\frac{1}{4}$ miles (E. by S.) from Oakhampton; containing 1937 inhabitants. The living is a vicarage, valued in the king's books at £10; present net income, £150; patrons and appropriators, Dean and Canons of Windsor. Here is a National school.

TAXALL (*St. James*), a parish, in the union and hundred of Macclesfield, Northern Division of the county of Chester; containing 587 inhabitants, of which number, 184 are in the township of Taxall, $8\frac{1}{4}$ miles (N. E. by E.) from Macclesfield. The living is a discharged rectory, valued in the king's books at £9. 2. 6.; present net income, £250; patron and incumbent, Rev. J. Swain. The church was, with the exception of the tower, taken down and rebuilt on a larger scale, in 1825; it contains 200 free sittings, the Incorporated Society having granted £200 in aid of the expense. Against the north wall is a monument to Michael Heathcote, Esq., gentleman of the pantry to George II.; and in the chancel are several memorials to the Shallcross family, who were patrons of the living in the early part of the last century, and resided at Shallcross Hall, in Derbyshire, which is on the east bank of the river, immediately opposite the church. The village occupies a pleasing situation on the banks of the river Goyt, which separates it from Derbyshire.

TAYNTON (*St. Lawrence*), a parish, in the union of Newent, hundred of Botloe, Western Division of the county of Gloucester, $3\frac{1}{2}$ miles (S. S. E.) from Newent; containing 555 inhabitants. The living is a rectory, valued in the king's books at £9. 6. 8.; present net income, £321; patrons, Dean and Chapter of Gloucester. The church was rebuilt during the protectorate of Cromwell, by an ordinance passed in January 1647. Robert Aldridge, in 1737, bequeathed a rent-charge of £2. 10. for teaching four children; and William Guilding bequeathed the interest of £100 for apprenticing them. A Sunday National school has been established.

TAYNTON (*St. John*), a parish, in the union of Witney, hundred of Chadlington, county of Oxford, $1\frac{3}{4}$ mile (N. W.) from Burford; containing 371 inhabitants. The living is a discharged vicarage, valued in the king's books at £7. 9. $4\frac{1}{2}$.; present net income, £56; patron and impropriator, Lord Dynevor. The church is an elegant edifice, in the later English style; the ancient font is highly enriched, and, from initials interwoven with other ornaments, appears to have been dedicated to St. John. The river Windrush runs through the parish, in which are considerable quarries of excellent freestone. A school is partly supported by the interest of £100, bequeathed by John Collier, in 1725.

TEALBY (*All Saints*), a parish, in the union of Caistor, Southern Division of the wapentake of Walshcroft, parts of Lindsey, county of Lincoln, $4\frac{1}{2}$ miles (E. N. E.) from Market-Rasen; containing 824 inhabitants. The living is a discharged vicarage, valued in the king's books at £6. 16. 8.; present net income, £120; patron and impropriator, G. Tennyson, Esq. There is a place of worship for Wesleyan Methodists.

TEAN, a hamlet, in the parish of Checkley, union of Cheadle, Southern Division of the hundred of Totmonslow, Northern Division of the county of Stafford, $7\frac{1}{4}$ miles (N. W. by W.) from Uttoxeter: the population is returned with the parish. The manufacture of tape, supposed to be the most extensive in Europe, was established at Upper Tean, in 1748, at which, and in the adjoining bleaching grounds, several hundred persons find employment. The proprietors support schools for the children of their workpeople. In the neighbourhood are several fine mansions and elegant villas. There are places of worship for Independents and Wesleyan Methodists. Two fairs are held in the village on Easter-Tuesday and November 10th.

TEATH, ST., a parish, in the union of Camelford, hundred of Trigg, Eastern Division of the county of Cornwall, 3 miles (S. W. by W.) from Camelford; containing 1260 inhabitants. This parish is bounded on the west by the Bristol Channel, and comprises the extensive and valuable slate quarry of Delabole; also a lead mine, employing 80 persons, in the ore of which a great proportion of silver is found. The living is a vicarage, valued in the king's books at £12; present net income, £226; patron, Bishop of Exeter; impropriator, E. P. Lyon, Esq. The church was formerly collegiate for two prebendaries, or portionists. There are places of worship for Bryanites and Wesleyan Methodists.

TEDBURN (*St. Mary*), a parish, in the union of St. Thomas, hundred of Wonford, Crockernwell and Southern Divisions of the county of Devon, $4\frac{1}{2}$ miles (S. W. by S.) from Crediton; containing 821 inhabitants. The living is a rectory, valued in the king's books at £8. 6. 3.; patron and impropriator, Rev. Charles Burne. The tithes have been commuted for a rent-charge of £400, subject to the payment of rates, which on the average have amounted to £33; the glebe comprises $38\frac{1}{2}$ acres. A school is partly supported by subscription. A cattle fair is held here on the Monday before Michaelmas-day. At Hackworthy, in this parish, there was formerly a chapel of ease.

TEDDINGTON (*St. Mary*), a parish, in the union

of KINGSTON, hundred of SPELTHORNE, county of MIDDLESEX, 11 miles (S. W. by W.) from London; containing 895 inhabitants. The village stands on the western bank of the Thames, on the road from London, through Isleworth, to Hampton Court. Bushy Park, the usual country residence of William IV. and his queen Adelaide, before their accession to the throne, is partly in this parish. Here are the wax bleaching grounds and candle manufactory of Messrs. Barclay, the largest and most complete establishment of the kind in the kingdom. During the summer months, nearly four acres of ground are covered with wax, of which about 200,000lbs. are annually bleached, and in winter formed into candles by hand. Connected with this manufactory is a very extensive one of spermaceti, chiefly carried on in Leicester-square by the same firm. The living is a donative curacy; net income, £91; patron and impropriator, Earl of Bradford. The church has been repewed, and 80 free sittings provided, the Incorporated Society having granted £100 in aid of the expense: it is principally in the later style of English architecture, and contains the remains of Sir Orlando Bridgeman, who died in 1674, and of Dr. Stephen Hall, Clerk of the Closet to the Princess of Wales, mother of George III., and 51 years minister of this parish, to which he was a most liberal benefactor: he died in 1761. Her Majesty the Queen Dowager has presented £100 to the parish, towards the erection of a parsonage-house. Twelve girls are instructed for £20 a year, the rent of certain cottages and lands purchased with £40 left by Dame Dorothy Bridgeman, and a smaller sum from the parish funds. Here is also a National school.

TEDDINGTON, a chapelry, in the parish of OVERBURY, union of TEWKESBURY, Middle Division of the hundred of OSWALDSLOW, Pershore and Eastern Divisions of the county of WORCESTER, 5 miles (E. by N.) from Tewkesbury; containing 129 inhabitants. The chapel is dedicated to St. Nicholas.

TEDSTONE-DELAMERE (*St. James*), a parish, in the union of BROMYARD, hundred of BROXASH, county of HEREFORD, $4\frac{1}{2}$ miles (N. E. by E.) from Bromyard; containing 230 inhabitants. The living is a discharged rectory, valued in the king's books at £6. 13. 4.; present net income, £238; patrons, Principal and Fellows of Brasenose College, Oxford. There is a petrifying spring in the parish.

TEDSTONE-WAFER, a parish, in the union of BROMYARD, hundred of BROXASH, county of HEREFORD, $3\frac{3}{4}$ miles (N. E. by N.) from Bromyard; containing 91 inhabitants. The living is a rectory, united to that of Edvin-Loach, and valued in the king's books at £1. 10. The tithes have been commuted for a rent-charge of £80, subject to the payment of rates, which on the average have amounted to £6; the glebe comprises 2 acres, valued at £2. 5. per annum. Limestone abounds in the neighbourhood.

TEETON, a hamlet, in the parish of RAVENSTHORPE, union of BRIXWORTH, hundred of NEWBOTTLE-GROVE, Southern Division of the county of NORTHAMPTON, $7\frac{3}{4}$ miles (N. W. by N.) from Northampton; containing 73 inhabitants.

TEFFONT-EVIAS, a parish, in the union of TISBURY, hundred of DUNWORTH, Hindon and Southern Divisions of the county of WILTS, $6\frac{1}{2}$ miles (W.) from Wilton; containing 176 inhabitants. The living is a rectory, valued in the king's books at £8; present net income, £148; patron, John Thomas Mayne, Esq. The church has been rebuilt, and contains 100 free sittings, the Incorporated Society having granted £100 in aid of the expense. A National school has been established. The river Nadder, and a beautifully clear stream, which rises near the adjoining village of Teffont-Magna, run through the parish; the latter forms a lake, covering two acres, well stocked with trout. There is a fine freestone quarry, besides very extensive excavations, from which the stone used in building Salisbury cathedral was taken. The manor-house, a handsome structure in the later style of English architecture, was the birthplace of Henry, Earl of Marlborough, Lord High Treasurer and Chancellor of England in the time of James II.

TEFFONT-MAGNA, a parish, in the union of TISBURY, forming a distinct portion of the hundred of WARMINSTER, being locally in that of Dunworth, Hindon and Southern Divisions of the county of WILTS, $5\frac{1}{4}$ miles (E.) from Hindon; containing 213 inhabitants. The living is annexed to the vicarage of Dinton.

TEIGH (*Holy Trinity*), a parish, in the union of OAKHAM, hundred of ALSTOE, county of RUTLAND, 5 miles (N.) from Oakham; containing 176 inhabitants. The living is a rectory, valued in the king's books at £14. 2. 11.; present net income, £349; patron, Earl of Harborough.

TEIGNMOUTH, a sea-port and market-town, comprising two parishes, called *East* and *West Teignmouth*, in the union of NEWTON-ABBOT, hundred of EXMINSTER, Teignbridge and Southern Divisions of the county of DEVON, 15 miles (S. by E.) from Exeter, and $187\frac{3}{4}$ (W. S. W.) from London; containing 4688 inhabitants, of which number, 2878 are in West Teignmouth. This place was originally an insignificant village, and is stated to have been the first landing-place of the Danes, in 787, on being sent to reconnoitre the British coast; who, having slain the governor, were encouraged by this omen of success to pursue their warlike purpose throughout the island. The town has been twice destroyed by fire, first by a French pirate, in 1340, and subsequently, on July 26th, 1690, when the French, having effected a landing, proceeded to ransack the churches, and burnt 116 houses, with a number of ships and small craft lying within the harbour. In commemoration of this calamitous event, one of the streets still retains the appellation of French-street; and the original brief granted for the relief of the sufferers is now in the possession of the Jordan family. Alarmed at the threat of a similar attack, in 1744, the inhabitants obtained permission to erect a small fort on the beach of East Teignmouth, and petitioned the Admiralty for the requisite supply of ordnance. In Camden's time the eastern town was called Teignmouth-Regis, and the other Teignmouth-Episcopi, the manor of the latter having belonged to the see of Exeter, until alienated by Bishop Vesey: it is now the property of Lord Clifford. The town, as its name implies, is situated on the navigable river Teign, at its influx into the sea: it occupies a gentle declivity at the foot of a chain of hills, by which it is sheltered on the north and west, the two parts being separated by a small rivulet, called the Tame. East Teignmouth, which is the more modern, is almost entirely appropriated as a watering-place, in which respect it is considered equal, if not superior, in magnitude and fashionable repute to any on the Devonshire coast: its situation is beautiful, and in the vicinity are prospects, particularly from Little Haldon, of great and deserved celebrity: the cliffs are of a reddish colour, and of considerable height, and at the southern side of the river's mouth is a singular elevation, called the Ness. On the strand fronting the sea are spacious carriage drives, promenades, and an extensive lawn. The public rooms, recently erected by subscription, form the centre of a crescent, and comprise spacious assembly-rooms, with apartments for refreshments, cards, and billiards; the façade of the building is decorated with an Ionic portico over a Doric colonnade. There are also a public library, bathing establishments, and a small theatre. A regatta takes place annually, about the month of August. West Teignmouth is the port and principal seat of business: in this respect it had risen to some importance at an early period, having sent members to the great council for maritime affairs, and contributed seven ships, with 120 men, towards the expedition against Calais, in 1347. The town is roughly paved, and irregularly built, and, with its quay and dock-yard, is situated on the curve formed by the sudden expansion of the river. A post-road, passing through it from Exeter to Torquay, crosses a modern bridge over the Teign, said to be the longest in England: it is constructed of wood and iron, with a drawbridge at one end, for the passage of vessels, and was erected by subscription. The quay was constructed in 1820, by G. Templer, Esq.; and, in the small dock-yard, sloops of war and vessels of upwards of 200 tons' burden have been built. The harbour is safe and commodious, though somewhat difficult to enter, on account of a moveable bar, or sand bank, which shifts with the wind. In the middle of the last century, a large number of vessels, of from 50 to 200 tons' burden each, were employed in the trade with Newfoundland, and some business of this description is still carried on, but it is on the decline: coal and culm are imported in large quantities, and the home fishery at present occupies a considerable number of the inhabitants. By means of a rail-road and canal, which latter joins the Teign at Newton-Abbot, and is navigable thence to the sea at Teignmouth, a communication has been effected between the granite quarries at Haytor and the clay-pits of Bovey, which greatly facilitates the exports of granite and pipe and potters' clay. A grant of a market and a fair was obtained in the reign of Henry III., by the Dean and Chapter of Exeter, for East Teignmouth, where there is a commodious market-house, which belongs to the Earl of Devon, lord of the manor. The market is on Saturday, principally for provisions: fairs are held on the third Tuesday in January, the last Tuesday in February, and the last Tuesday in September. The government of West Teignmouth is vested in a portreeve, who is annually elected by a jury of twelve, sworn in at the court leet and baron held by Lord Clifford, the lord of the manor, at which court also the town-clerk, four constables, two bailiffs, and other officers, are appointed: in East Teignmouth, a reeve and two constables are elected by the court, and two constables by the parish.

The living of East Teignmouth is a perpetual curacy; net income, £127; patron, Vicar of Dawlish; appropriators, Dean and Chapter of Exeter. The church, which is dedicated to *St. Michael*, was almost rebuilt in 1821, and 200 free sittings provided, the Incorporated Society having granted £500 in aid of the expense. West Teignmouth is annexed ecclesiastically to Bishop's-Teignton: the church, which is dedicated to *St. James*, is a spacious modern octagonal structure, with a tower at the west side, and surmounted in the centre by a lantern. There are places of worship for Baptists, Independents, and Calvinistic Methodists, at West Teignmouth. In East Teignmouth, thirteen poor children are instructed from the proceeds of a joint benefaction, made in 1731, by Captains John and Thomas Coleman; and there is another small school founded by means of a benefaction from Sir John Elwill, for the education of twelve poor children. A National school and an infants' school are partly supported by subscription. Teignmouth confers the title of Baron on the family of Shore.

TEIGNTON, BISHOP'S (*St. John the Baptist*), a parish, in the union of Newton-Abbot, hundred of Exminster, Teignbridge and Southern Divisions of the county of Devon, $1\frac{3}{4}$ mile (W. by N.) from West Teignmouth; containing 1085 inhabitants. The river Teign forms a boundary of the parish, in which are extensive quarries of limestone, affording also compact blocks of various-coloured marble. The living is a discharged vicarage, endowed with the rectorial tithes, with West Teignmouth annexed, and valued in the king's books at £25. 8. 10.; present net income, £334; patron and incumbent, Rev. John Comyns. The church, which has been lately entirely renovated and re-pewed, is principally in the Norman style of architecture, with an enriched western doorway in excellent preservation, and a low massive tower between the nave and the chancel. Near it are the remains of an ancient chapel. There is a charity school, with a house for the master, founded in 1729, and endowed by Christopher Coleman and others, the money having been vested in the purchase of an estate in Bovey-Tracey, yielding £37 per annum, for teaching and clothing poor children. In the upper part of the parish are the remains of a palace and chapel of Bishop Grandison, and an hospital founded by that prelate for poor decayed clerks. There was formerly a chapel at Venn, in this parish.

TEIGNTON, DREWS (*Holy Trinity* , a parish, in the union of Oakhampton, hundred of Wonford, Crockernwell and Southern Divisions of the county of Devon, $4\frac{1}{2}$ miles (N. N. W.) from Moreton-Hampstead; containing 1313 inhabitants. The name of this place is supposed to signify "the Druids' Town on the Teign," which river here pursues its rapid course through scenery of the wildest description, and is crossed by Fingle bridge, in a romantic valley. The living is a rectory, valued in the king's books at £40. 13. 4.; present net income, £776; patrons, Messrs. Ponsford. The church is an interesting ancient structure, with a beautiful window of stained glass at the east end, and contains 70 free sittings, the Incorporated Society having granted £40 in aid of the expense. An almshouse was founded by Richard Eggecomb, in 1542. On the Shilston estate is a cromlech, consisting of three supporting stones, each about six feet and a half high, with a covering stone, twelve feet long and nine feet across the widest part. On the bank of the Teign is one of the celebrated logan, or rocking, stones, which are thought by some to be a work of design, by others of nature; and at Preston Bury are the remains of an encampment.

TEIGNTON, KING'S (*St. Michael*), a parish, in the union of Newton-Abbot, hundred of Teignbridge, Teignbridge and Southern Divisions of the county of Devon, 2 miles (N. E. by N.) from Newton-Bushell; containing 1288 inhabitants. This parish is situated on the navigable river Teign, and in the neighbourhood are large beds of potters' clay, of very superior quality, and a chalybeate spring. The living is a vicarage, with the perpetual curacy of Highweek annexed, valued in the king's books at £28. 13. 9.; present net income, £396; patron and appropriator, Prebendary of King's-Teignton in the Cathedral Church of Salisbury. The church was enlarged in 1824, and 80 free sittings provided, the Incorporated Society having granted £30 in aid of the expense: it contains a monument bearing a singular epitaph to the memory of the Rev. Richard Adlam, vicar of the parish in 1669. There is a place of worship for Independents; and two schools are partly supported by the vicar. Theophilus Gale, a learned non-conformist divine, was born here, in 1628; he died in 1687.

TEINGRACE (*St. Mary*), a parish, in the union of Newton-Abbot, hundred of Teignbridge, Teignbridge and Southern Divisions of the county of Devon, $2\frac{1}{4}$ miles (N. by W.) from Newton-Bushell; containing 160 inhabitants. The living is a rectory, valued in the king's books at £5. 9. $4\frac{1}{2}$.; patron, Duke of Somerset. The tithes have been commuted for a rent-charge of £160, subject to the payment of rates, which on the average have amounted to £11; the glebe comprises 65 acres, valued at £70 per annum. The church is a handsome edifice, rebuilt in 1787 by James Templer, Esq., George Templer, Esq., and the Rev. John Templer, brothers. Among other monuments of that family, it contains one to the memory of Charles Templer, who perished in the wreck of the Halsewell East Indiaman, on the Dorsetshire coast, in 1786. The Stover canal and railway, constructed by the Templer family, facilitate the exportation of potters' clay found in the neighbourhood, and of granite from the extensive quarries near High Tor. There is a small school, supported partly by the minister, and the feoffees of another charity.

TELLISFORD (*All Saints*), a parish, in the union of Frome, hundred of Wellow, Eastern Division of the county of Somerset, 6 miles (N. N. E.) from Frome; containing 162 inhabitants. The living is a discharged rectory, valued in the king's books at £9. 1. $0\frac{1}{2}$.; patron and incumbent, Rev. C. W. Baker. The tithes have been commuted for a rent-charge of £150. 6. 8., subject to the payment of rates; the glebe comprises 59 acres, valued at £76 per annum. This parish is separated from Wiltshire by the river Frome, which is here crossed by a bridge. There is a fulling-mill. About a third part of the village was destroyed by fire in 1785. A school is endowed with the interest of £143 three per cents., given by Edward Crabb, a native of this place.

TELSCOMBE (*St. Lawrence*), a parish, in the union of Newhaven, hundred of Holmstrow, rape of Lewes, Eastern Division of the county of Sussex, $3\frac{1}{2}$ miles (N. W. by W.) from Newhaven; containing 121 inhabitants. The living is a discharged vicarage, valued in the king's books at £13. 13. 4.; present net income, £231; patron and incumbent, Rev. J. Hutchins. The church is a small structure, principally in the Norman style. The Rev. Josiah Povey, in 1727, bequeathed land, &c., yielding about six guineas a year, for teaching poor children. Henry Smith, Esq., bequeathed an estate in this parish, called Court Farm, directing the rents to be applied for the relief of the poor of seven parishes in the north of England.

TEMPLE, a parish, in the union of Bodmin, hundred of Trigg, Eastern Division of the county of Cornwall, $6\frac{1}{4}$ miles (N. E. by E.) from Bodmin; containing 29 inhabitants. The living is a donative, in the patronage of Sir B. Wrey, Bart.; net income, £21. The church is quite dilapidated. The extensive moors, which lie between Bodmin and Launceston, take their name from this parish, in which they are partly situated.

TEMPLE-BREWER, an extra-parochial liberty, in the wapentake of Flaxwell, parts of Kesteven, county of Lincoln, $6\frac{3}{4}$ miles (N. W. by N.) from Sleaford; containing 73 inhabitants. A preceptory of the Knights Templars was founded here before 1185, which afterwards belonged to the Hospitallers, and had possessions, at the dissolution, valued at £184. 6. 8. per annum.

TEMPLE-GRAFTON, county of Warwick.—See GRAFTON, TEMPLE.—*And other places having a similar distinguishing prefix will be found under the proper name.*

TEMPLE-HALL, an extra-parochial liberty, in the hundred of Sparkenhoe, Southern Division of the county of Leicester, $2\frac{1}{2}$ miles (W. by S.) from Market-Bosworth; containing 28 inhabitants.

TEMPLE-NEWSOM, a township, in the parish of Whitkirk, Lower Division of the wapentake of Skyrack, West Riding of the county of York, 4 miles (E. by S.) from Leeds; containing 1458 inhabitants. The Knights Templars had a preceptory here, which, at the suppression of their order, was granted to Sir John D'Arcy. A school is endowed with £10 per annum, and two others have an allowance of £5 annually from the Marquess of Hertford.

TEMPLETON (*St. Margaret*), a parish, in the union of Tiverton, hundred of Witheridge, Cullompton and Northern Divisions of the county of Devon, 5 miles (W. by N.) from Tiverton; containing 222 inhabitants. The living is a rectory, valued in the king's books at £8. 15.; present net income, £162; patron, Sir W. T. Pole, Bart. A school is partly supported by the rector and curate.

TEMPSFORD (*St. Peter*), a parish, in the union and hundred of Biggleswade, county of Bedford, $6\frac{1}{2}$ miles (N. N. W.) from Biggleswade; containing 535 inhabitants. This is a place of great antiquity; it was occupied by the Danes before 921, when they were expelled by the Saxons, but they returned in 1010, and reduced it to ashes. The village is situated on the river Ivel, which is navigable through the parish, and falls into the Ouse as it passes along the western boundary. The living is a rectory, valued in the king's books at £24; present net income, £227: it is in the patronage of the Crown. There is a place of worship for Wesleyan Methodists; and a day and Sunday school is partly supported by subscription.

TENBURY (*St. Mary the Virgin*), a market-town and parish, and the head of a union, in the Upper Division of the hundred of Doddingtree, Hundred-House and Western Divisions of the county of Worcester; containing 1768 inhabitants, of which number, 1093 are in the town of Tenbury, 22 miles (N. W. by W.) from Worcester, and 134 (N. W. by W.) from London. The town is situated on the southern bank of the river Teme, from which it derives its name, having been originally called *Temebury*: it consists of two streets, crossing nearly at right angles, and partially paved; the houses in general are but indifferently built. Races are held annually in June, on a good course about a mile south of the town. The surrounding country is rich, and very productive of hops and apples; great quantities of cider and perry are made, forming a principal source of trade: there are also a considerable malting trade, and a tannery. A canal, commenced in 1794, originally intended to extend from Leominster to Stourport, but not completed the whole distance, passes within half a mile of the town. The river Teme, which here separates Worcestershire from Shropshire, is crossed, at the northern entrance into the town, by a bridge of six arches. The market, granted by Henry III. in 1249, is on Tuesday; the building for the corn market is an ancient structure, but the butter-cross is more recent. Fairs are on April 22nd, May 1st, July 18th, and Sept. 26th. Petty sessions are held once in two months, and a court leet and court baron by the lord of the manor. The town has been made a polling-place for the western division of the county. The living is a vicarage, endowed with a portion of the rectorial tithes, with that of Rochford annexed, and valued in the king's books at £21; present net income, £607; patron and incumbent, Rev. George Hall; impropriator of the remainder of the rectorial tithes, R. Bagnall, Esq., and others. The church was rebuilt in 1777, the old building having been swept away by a flood, in Nov. 1770: it is a spacious and neat edifice, and had formerly a chantry attached to it, which was valued at the period of the suppression at £5. 0. 6. per annum. There is a place of worship for Particular Baptists. The National school was founded in 1816, and is supported by subscription: there is also a day and Sunday school, supported by the trustees of the late Mr. E. Goff. The principal charity arises from some houses and lands given by Mr. Philip Baylis, which having been let by the feoffees in 1767, on lease for 99 years, at a rental of £7. 10. per annum, little benefit is derived from it. The poor law union of Tenbury extends also into the counties of Salop and Hereford, comprising 19 parishes or places, and containing a population of 7109, according to the census of 1831; it is under the superintendence of 22 guardians. A saline spring has recently been discovered here, the water of which is similar in its properties to that of Cheltenham, but more strongly impregnated with saline matter; a pint of it in evaporation producing one ounce of crystalline salt.

TENBURY-FOREIGN, a hamlet, in the parish of Tenbury, Upper Division of the hundred of Doddingtree, Hundred-House and Western Divisions of the county of Worcester; containing 325 inhabitants.

TENDRING (*St. Edmund*), a parish, and the head of a union, in the hundred of Tendring, Northern Division of the county of Essex, $6\frac{1}{2}$ miles (S. S. E.) from Manningtree; containing 758 inhabitants. This parish, which is situated in the centre of the hundred, is about 10 miles in circumference; the surface is elevated, the soil generally light but fertile, and in some parts strong and heavy. A fair is held in the village on the 14th of September. The living is a rectory, valued in the king's books at £16; present net income, £734; patrons, Master and Fellows of Balliol College, Oxford. The church is an ancient edifice with a belfry turret of wood, and contains some ancient and interesting monuments. Two schools are partly supported by subscription. The poor law union of Tendring comprises 30 parishes or places, containing a population of 21,002, according to the census of 1831, and is under the care of 35 guardians. The workhouse, situated on the heath in this parish, was erected in 1838, for 400 paupers, at an expense of £6500, including the purchase of the site.

Corporation Seal of Tenterden.

Obverse. *Reverse.*

TENTERDEN (*St. Mildred*), a market-town and parish, within the cinque-port liberties, having separate jurisdiction, and the head of a union, locally in the hundred of Tenterden, Lower Division of the lathe of Scray, Western Division of the county of Kent, 18 miles (S. E. by S.) from Maidstone, and 53 (S. E. by E.) from London; containing 3177 inhabitants. This place, of which the present name appears to be a corruption of *Theinwarden*, or the ward of Thanes, that is, the guard in the valley, was one of the first places in which the woollen manufacture was established, in the reign of Edward III. It became a scene of opposition to the church of Rome, at an early period prior to the Reformation, when, in the time of Archbishop Warham, 48 inhabitants of the town and its vicinity were publicly accused of heresy, and five of them condemned to be burned. The town stands upon a pleasant eminence, surrounded by some fine plantations of hops: the houses are well built and of respectable appearance. The streets are paved and lighted with gas, under the provisions of a general act, by which it has been recently much improved; and the trade, which consists chiefly in supplying the adjoining grazing districts, of which it is the chief centre, has greatly increased. The town-hall was built in 1792, the former having been destroyed by fire, and contains a commodious room occasionally used for public assemblies. The Royal Military canal passes within six miles of the town. The market, principally for corn, is held on Friday; and there is a fair for horses, cattle, and pedlery, on the first Monday in May. A stock market was established on the 28th of June, 1839, and is held on the Fridays before the first and third Tuesdays in each month. The inhabitants were incorporated, by the style of "The Bayliffe and

Commonaltie of the Town and Hundred of Tenterden," and the town annexed as a member to the town and port of Rye, by Hen. VI.; a new charter was granted in the 42nd of Elizabeth, changing the style to "Mayor, Jurats, and Commons," according to the provisions of which the town was governed by a mayor, twelve jurats, and an indefinite number of admitted freemen, who were the commons, assisted by a town-clerk, chamberlain, and two sergeants-at-mace. The corporation now consists of a mayor, four aldermen, and twelve councillors, under the act of the 5th and 6th of William IV., cap. 76, for an abstract of which see the Appendix, No. I. The mayor and late mayor are justices of the peace. The recorder holds a court of quarter sessions, with power to try for all offences not capital; and has also power to hold a court of record every fifteen days, for the recovery of debts to any amount, and for trying all pleas, its jurisdiction extending over the hundred, which includes only this parish and part of Ebony, but this court is at present in disuse. The town has lately been made a polling-place for the western division of the county.

The living is a vicarage, valued in the king's books at £33. 12. 11.; present net income, £177; patrons and appropriators, Dean and Chapter of Canterbury. The church is spacious and handsome, with a lofty tower at the west end, to which a beacon was formerly attached; it contains 150 free sittings, the Incorporated Society having granted £100 in aid of the expense. There are places of worship for Baptists, Wesleyan Methodists, and Unitarians. At Smallhythe, in this parish, is a chapel, erected about 1509, dedicated to St. John the Baptist, and licensed by faculty from Archbishop Warham: it is repaired and maintained out of lands in this parish and that of Wittersham, vested in feoffees, and the chaplain is appointed by the inhabitants. It appears that, at the time of the erection of this chapel, the sea came up to Smallhythe, as power was then given to inter in the chapelyard the bodies of shipwrecked persons cast on shore. The free grammar school, founded at an early period by an ancestor of the late Sir Peter Hayman, was endowed in 1521, by William Marshall, with a rent-charge of £10, and, in 1702, by John Mantel, with the sum of £200, which was laid out in land: the present income is £52 per annum, which sum is appropriated to the support of a National school. In 1660, Dame Jane Maynard bequeathed land for apprenticing children and the maintenance of poor widows, for which purpose Mrs. Anne Shilton also gave land, in 1674; and in 1834, John Butler Pomfret gave £10 per annum, charged on lands, for the benefit of poor widows, under the same trustees. Dame Frances Norton, in 1719, gave an estate, the proceeds to be divided between this parish and that of Hollingbourne. Dr. Edward Curteis, in 1797, bequeathed property, now producing £101. 6. 10. per annum, for the clothing and instruction of 10 poor girls, for the distribution of bread to the poor, and other charitable uses, reserving a portion of the gift for an accumulating fund; the original endowment has been considerably increased, and its benefit already extended to several other parishes within the Weald, and will ultimately include the whole. The management is vested in the mayor of Tenterden, the rectors of Halden and Biddenden, and the Vicar of Bennenden. The poor law union of Tenterden comprises 11 parishes or places, under the care of 11 guardians, and contains a population of 10,478, according to the census of 1831. Hoole, the translator of Tasso, resided at this place. Tenterden confers the title of Baron on the family of Abbot; Sir Charles Abbot, late Lord Chief Justice of the Court of King's Bench, having been raised to the peerage by that title, on the 30th of April, 1827.

TENTERGATE, a joint township with Scriven, in the parish of KNARESBROUGH, Lower Division of the wapentake of CLARO, West Riding of the county of YORK: the population is returned with Scriven.

TERLING (*ALL SAINTS*), a parish, in the union and hundred of WITHAM, Northern Division of the county of ESSEX, $3\frac{3}{4}$ miles (W.) from Witham; containing 892 inhabitants. This parish is bounded on the west by the hundred of Chelmsford, and is situated at no great distance from the ancient Roman stations of Colchester, Maldon, and Pleshey. On making a new road here, in 1824, about 300 gold and silver coins were dug up, and on further search a jar was discovered, containing two large rings and 30 small pieces of gold, with some silver coins of the twelve Roman emperors, in regular succession, from Constantius to Honorius, and as bright as if just taken from the mint. In 1269, the Bishop of Norwich had a palace and park here, which subsequently became the residence of Henry VIII.: the chapel attached to it possessed the privilege of sanctuary, and as such afforded shelter to the celebrated Hubert de Burgh, when under the indignation of Henry III. A fair, chiefly for pleasure, is held on Whit-Monday. The living is a discharged vicarage, valued in the king's books at £10; present net income, £226; patron and impropriator, J. H. Strutt, Esq., of Terling Hall. The church is a spacious edifice, with a tower of brick, replacing one of stone, which had fallen down; it consists of a nave, south aisle, and chancel, and has been elegantly restored and fitted up by Mr. Strutt. There is a place of worship for Wesleyan Methodists. Benjamin Jocelyne, Esq., in 1775, bequeathed an annuity of £10 for teaching 10 poor boys; 10 others are paid for by the vicar, and 10 by Mr. Strutt. Another school for girls is partly supported by subscription.

TERRINGTON (*ST. CLEMENT'S*), a parish, in the union of WISBEACH, hundred of FREEBRIDGE-MARSHLAND, Western Division of the county of NORFOLK, $5\frac{1}{2}$ miles (W. by N.) from Lynn-Regis; containing 1466 inhabitants. Though not noticed in Domesday-book, Terrington was an extensive place, and had considerable salt-works in the time of the Saxons, as appears from a grant of Godric, brother to Ednoth, abbot of Ramsey, about 970. The living is a vicarage, with that of Terrington-St. John's united, valued jointly in the king's books at £23. 6. 8.; present net income, £502: it is in the patronage of the University of Cambridge; appropriator, Bishop of Peterborough: the rectory is annexed to the Margaret Professorship at Cambridge. The church is a handsome cruciform structure, in the later style of English architecture, having on the battlements of the south aisle many shields with different armorial bearings. Here was also a chapel, dedicated to St. James. There is a place of worship for Wesleyan Methodists; also a National school, partly supported by endowment.

TERRINGTON (*ST. JOHN'S*), a parish, in the union of WISBEACH, hundred of FREEBRIDGE-MARSHLAND,

Western Division of the county of NORFOLK, 6 miles (W. S. W.) from Lynn-Regis; containing 595 inhabitants. The living is a vicarage, united to that of Terrington-St. Clement's; the rectory is, with that of St. Clement's, annexed to the Margaret Professorship in the University of Cambridge. The church is a regular pile of building, having, at its south-west angle, a square tower crowned with pinnacles. Here is a National school.

TERRINGTON (*All Saints*), a parish, in the union of MALTON, partly within the liberty of ST. PETER'S, East Riding, but chiefly in the wapentake of BULMER, North Riding, of the county of YORK; containing 759 inhabitants, of which number, 649 are in the township of Terrington with Wigginthorpe, 8 miles (W. by S.) from New Malton. The living is a rectory, valued in the king's books at £23. 18. 6½.; present net income, £571; patron and incumbent, Rev. C. Hall. There is a place of worship for Wesleyan Methodists.

TERWICK, a parish, in the union of MIDHURST, hundred of DUMPFORD, rape of CHICHESTER, Western Division of the county of SUSSEX, 4½ miles (W. N. W.) from Midhurst; containing 97 inhabitants. This parish is situated on the road from Midhurst to Petersfield, and is bounded on the south by the river Rother. Sandstone is everywhere abundant. The living is a discharged rectory, valued in the king's books at £5. 0. 5.; patrons, Trustees of the late Sir Charles Paget, Knt. The tithes have been commuted for a rent-charge of £173, subject to the payment of rates; the glebe comprises 10 acres.

TESTERTON (*St. Remigius*), a parish, in the union of WALSINGHAM, hundred of GALLOW, Western Division of the county of NORFOLK, 2¾ miles (S. E. by S.) from Fakenham; containing 18 inhabitants. The living is a discharged rectory, valued in the king's books at £5; present net income, £13; patron, P. M. Case, Esq. The church has been destroyed.

TESTON (*St. Peter and St. Paul*), a parish, in the union of MAIDSTONE, hundred of TWYFORD, lathe of AYLESFORD, Western Division of the county of KENT, 4 miles (W. by S.) from Maidstone; containing 255 inhabitants. The living is a discharged vicarage, valued in the king's books at £6. 10.; present net income, £233; patron and impropriator, Lord Barham. The church, which was a remarkably small structure, has been repaired, considerably enlarged, and beautified, by subscription, towards which the vicar largely contributed: it stands on the banks of the Medway, over which there is a fine bridge of seven arches. There is a National school, for which a very good house has been built.

TETBURY (*St. Mary*), a market-town and parish, and the head of a union, in the hundred of LONGTREE, Eastern Division of the county of GLOUCESTER, 20 miles (S. by E.) from Gloucester, and 99 (W. by N.) from London; containing 2939 inhabitants. The town is pleasantly situated on an eminence at the southern verge of the county, bordering on Wiltshire, and near the source of the river Avon, over which is a long bridge or causeway, leading into the main road to Malmesbury: it consists principally of a long street, crossed at right angles by two shorter ones, with a spacious market-house near one of the intersections. An act was obtained, in 1817, for paving and lighting the town, the expense of which was defrayed out of the funds in the hands of trustees appointed in 1814, under the act "for enclosing certain common fields and waste grounds within the parish:" £1000 was also appropriated from the same fund for the repair of the market-house. The poor are chiefly employed by woolstaplers, and the market was formerly noted for the sale of woollen yarn, but the introduction of machinery has put an end to the trade. The market is on Wednesday; and fairs are held on Ash-Wednesday, the Wednesday before and after April 5th, and July 22nd for corn, cheese, horses, and cattle. A bailiff and a constable are elected annually at the court leet of the feoffees of the manor. Petty sessions for the town and hundred are held here and at Horseley and Rodborough alternately.

The living is a vicarage, valued in the king's books at £36. 13. 4.; present net income, £771; patrons, Trustees of Tetbury charity; appropriators, Dean and Canons of Christ-Church, Oxford. The appropriate tithes have been commuted for a rent-charge of £240, and the vicarial for £800, subject to the payment of rates, which on the average have respectively amounted to £34 and £107; the glebe comprises 67 acres. The church having been undermined by a flood in 1770, was, with the exception of the tower, which is surmounted by a fine modern spire, rebuilt in 1781, in the early style of English architecture, at an expense of £6000. There are places of worship for Baptists and Independents. A grammar school was endowed by Sir William Romney, a native of this town, and an alderman and sheriff of London in the reign of James I., who bequeathed to certain trustees a lease for years of the weights of wool and yarn, tolls, and other profits within the town, with the proceeds of which lands have since been purchased, and out of the rents £40 per annum was paid to the master; but since the year 1800 no payments have been made, and the affairs of the institution are under investigation. Here is a school partly supported by an endowment of £30 per annum, bequeathed by Elizabeth Hodges, in 1723, and partly by subscription. There is also a National school. The Sunday school, opened to all the poor children of the parish, is supported by bequests of £100 each from Ann Wright, in 1788, Sarah Paul, in 1795, and Ann Gastrell, in 1797. An almshouse for eight poor persons was founded and endowed by the above-mentioned Sir William Romney. The poor law union of Tetbury comprises 13 parishes or places, under the care of 15 guardians: the population, according to the census of 1831, amounts to 5797. In Maudlin meadow, which belongs to Magdalene College, Oxford, and is situated north of the town, is a petrifying spring, impregnated with calcareous earth. A castle is said to have been built here, long before the invasion of Britain by the Romans; and ancient British coins and fragments of weapons have been found within the area of a camp in the vicinity, of which all traces are now obliterated. Roman coins of the Lower Empire have also been discovered.

TETCHWORTH, a hamlet, in the parish of LUDGERSHALL, hundred of ASHENDON, county of BUCKINGHAM, 10 miles (W. N. W.) from Aylesbury: the population is returned with the parish.

TETCOTT (*Holy Cross*), a parish, in the union of HOLSWORTHY, hundred of BLACK TORRINGTON, Holsworthy and Northern Divisions of the county of DEVON, 5 miles (S. by W.) from Holsworthy; containing 293 inhabitants. The living is a discharged rectory, valued in the king's books at £13. 6. 8.; patron, Sir W. Molesworth, Bart. The tithes have been commuted for a rent-charge of £140, subject to the payment of rates; the glebe comprises 40 acres, valued at £40 per annum. The church contains some interesting monuments to the Arscott family. A school is partly supported by subscription.

TETFORD (*St. Mary*), a parish, in the union of HORNCASTLE, hundred of HILL, parts of LINDSEY, county of LINCOLN, 6 miles (N. E. by E.) from Horncastle; containing 690 inhabitants. The living is a discharged rectory, valued in the king's books at £5. 0. 10.; present net income, £373: it is in the patronage of Miss Harrison. The church contains 60 free sittings, the Incorporated Society having granted £40 in aid of the expense. There is a place of worship for Wesleyan Methodists; also a National school.

TETNEY (*St. Peter and St. Paul*), a parish, in the union of LOUTH, wapentake of BRADLEY-HAVERSTOE, parts of LINDSEY, county of LINCOLN, 10½ miles (N.) from Louth; containing 647 inhabitants. The living is a discharged vicarage, valued in the king's books at £7. 18. 4.; present net income, £250; patron and appropriator, Bishop of Lincoln. There is a place of worship for Wesleyan Methodists.

TETSWORTH (*St. Giles*), a parish, in the union and hundred of THAME, county of OXFORD, 11½ miles (E. S. E.) from Oxford; containing 530 inhabitants. The living is annexed, with those of Sydenham and Towersey, to the vicarage of Thame. The church is an ancient edifice, consisting only of a nave and chancel, separated by a Norman arch: above the south entrance is a circular moulding, under which are a mitred figure, having a crosier in the left hand, and the right hand raised as in the act of benediction; and the figure of a priest, with a book in the left hand, and the right hand pointing above to the paschal lamb, with a banner. There is a place of worship for Wesleyan Methodists.

TETTENHALL-REGIS (*St. Michael*), a parish, in the union of SEISDON, and partly in the Northern, and partly in the Southern, Division of the hundred of SEISDON, Southern Division of the county of STAFFORD, 1¾ mile (N. W.) from Wolverhampton; containing 2618 inhabitants. It is chiefly inhabited by persons engaged in the manufactures of the surrounding district, and more especially in that of locks. The living is a perpetual curacy; net income, £196; patron and impropriator, Lord Wrottesley. The church, which is in the early, decorated, and later styles of English architecture, was made collegiate before the Conquest for a dean and four prebendaries: it was enlarged in 1825, and 200 free sittings provided, the Incorporated Society having granted £300 in aid of the expense. The eastern window contains an ancient painting on glass, representing the archangel trampling on a dragon: the font is curiously ornamented with sculpture. There is a place of worship for Wesleyan Methodists; also a National school. The Worcestershire and Staffordshire canal passes near the village.

TETTON, a township, in the parish of WARMINGTON, union of CONGLETON, hundred of NORTHWICH, Southern Division of the county of CHESTER, 3 miles (W. N. W.) from Sandbach; containing 181 inhabitants.

TETWORTH, a joint parish with Everton, in the union of ST. NEOTS, hundred of TOSELAND, county of HUNTINGDON, 3 miles (N. by W.) from Potton; containing, exclusively of Everton, 183 inhabitants. See EVERTON.

TEVERSALL (*St. Catherine*), a parish, in the union of MANSFIELD, Northern Division of the hundred of BROXTOW and of the county of NOTTINGHAM, 5¼ miles (W. by N.) from Mansfield; containing 400 inhabitants. The living is a rectory, valued in the king's books at £9. 19. 2.; present net income, £510; patron, Lord Porchester. The church is principally in the Norman style, and contains several old monuments of the Greenhalghe, Molyneux, and Babington families. Coal and limestone abound in the parish, but neither are now worked. South of the church are the extensive ruins of the ancient mansion-house, built by Gilbert Greenhalghe in the reign of Henry VII., and the remains of a hanging garden, on a very magnificent scale.

TEVERSHAM (*All Saints*), a parish, in the union of CHESTERTON, hundred of FLENDISH, county of CAMBRIDGE, 3½ miles (E.) from Cambridge; containing 197 inhabitants. The living is a rectory, in the patronage of the Bishop of Ely, valued in the king's books at £19. 16. 0½.; present net income, £352. A school is supported by endowment.

TEW, GREAT (*St. Michael*), a parish, in the union of CHIPPING-NORTON, hundred of WOOTTON, county of OXFORD, 3¾ miles (N. N. E.) from Neat-Enstone; containing 440 inhabitants. The living is a vicarage, valued in the king's books at £6. 13. 4.; present net income, £134; patron and impropriator, M. R. Boulton, Esq. The church has been enlarged, and 260 free sittings provided, the Incorporated Society having granted £200 in aid of the expense: it is a handsome Norman structure, beautifully situated, and contains some ancient brasses and sculptured effigies. T. E. Freeman, in 1781, gave an estate producing £31 per annum for the education of ten children; the school is on the National system.

TEW, LITTLE, a parish, in the union of CHIPPING-NORTON, hundred of WOOTTON, county of OXFORD, 2½ miles (N. by E.) from Neat-Enstone; containing 216 inhabitants. The parishioners pay church-rates to the parish of Great Tew, to which the living is annexed, and attend divine service in the church of that parish. A school-house has been recently built in connection with the National Society.

TEWIN (*St. Peter*), a parish, in the union, hundred, and county of HERTFORD, 3 miles (E. S. E.) from Welwyn; containing 474 inhabitants. The living is a rectory, valued in the king's books at £14; patrons, Master and Fellows of Jesus' College, Cambridge. The tithes have been commuted for a rent-charge of £460, subject to the payment of rates, which on the average have amounted to £87; the glebe comprises 42 acres, valued at £45 per annum. The church has a square embattled tower, with a low spire. The Rev. Henry Yarborough, in 1788, bequeathed the rents of four tenements, for teaching ten children, and increasing the parish-clerk's income; and Lady Cathcart, in the same year, gave an annuity of £5 in further support of the school. There are sundry other bequests for the relief of the poor, and apprenticing children.

Seal and Arms.

TEWKESBURY (*St. Mary*), a borough, market-town, and parish, having separate jurisdiction, and the head of a union, locally in the Lower Division of the hundred of Tewkesbury, Eastern Division of the county of GLOUCESTER, 10 miles (N. N. E.) from Gloucester, and 103 (W. N. W.) from London; containing 5780 inhabitants. This place, which is of great antiquity, is supposed to have derived its name from *Theot*, a Saxon recluse, who, during the latter period of the Heptarchy, founded a hermitage here, where he lived in solitude and devotion, and from whom it was called *Theotisberg*, from which its present appellation is deduced. In 715, a monastery was founded here by the two brothers Odo and Dodo, Dukes of Mercia, and dedicated to the Blessed Virgin Mary, which, after having experienced great injury during the Danish wars, became a cell to the abbey of Cranborne in Dorsetshire. After the Conquest, Robert Fitz-Hamon, who attended William in his expedition to Britain, enlarged the buildings of this monastery, and so amply augmented its possessions, that the monks of Cranborne removed, in 1101, to Tewkesbury, which they made the principal seat of their establishment: it was subsequently raised into an abbey of Benedictine monks, and continued to flourish till the dissolution, at which time its revenue was estimated at £1598. 1. 3. The last decisive battle between the Yorkists and the Lancastrians took place within half a mile of this town, in 1471; on this memorable occasion, many of the principal nobility were slain on both sides, and not less than 3000 of the Lancastrian troops. Queen Margaret, who headed her own forces, was intrenched on the summit of an eminence, called the Home Ground, at the distance of a mile from the town, on the east side of the road to Gloucester; and the troops of Edward IV., who advanced against his opponents by way of Tredington, occupied the sloping ground to the south, called the Red Piece: the victory was decisive in favour of the Yorkists, the defeat of the Lancastrians having been ascribed to the treacherous inactivity of Lord Wenlock, one of their generals, whom the chief commander, the Duke of Somerset, struck dead on the field with his battle-axe. After their defeat, the Duke of Somerset, with about 20 other distinguished persons, took shelter in the church, from which they were dragged with violence, and immediately beheaded. At the commencement of the great civil war in the reign of Charles I., Tewkesbury was occupied by the parliamentarians, who were afterwards driven out, and the town was taken by the royalists, by whom it was again lost and retaken, till, in 1644, it was surprised and captured by Col. Massie, governor of Gloucester, for the parliamentarians, in whose possession it remained till the conclusion of the war.

The town is pleasantly situated in the northern part of the luxuriant vale of Gloucester, and on the eastern bank of the river Avon, near its confluence with the Severn: it is nearly surrounded by the small rivers Carron and Swilgate, both which fall into the Avon; over the latter are two bridges of stone, and over the former is a handsome stone bridge recently erected. It is handsome and well built, consisting of three principal streets, lighted with gas and well paved, and is amply supplied with water: the houses are in general of brick, but occasionally interspersed with ancient timber and brick buildings. Considerable improvements have taken place, under the provisions of an act obtained in 1786, among which may be noticed the ranges of building, erected in 1810, to the east of the High-street, on a tract of land called Oldbury, and the formation of a new street. An elegant cast-iron bridge has been constructed over the river Severn, near the hamlet of Mythe, within half a mile of the town, at an expense of £36,000, subscribed in shares of £100 each, opening a direct communication between London and Hereford: it consists of one noble arch, 172 feet in span, with a light iron balustrade, and was opened to the public in 1826. Near the division of the Worcester and Pershore roads is an ancient bridge of several arches over the Avon, from which a level causeway has been raised, extending to the iron bridge. A subscription library, with a news-room, was established in 1828, and is well supported: it contains at present more than 1000 volumes. A small theatre is occasionally opened by the Cheltenham company. The races, established in 1825, take place on the Ham, a large meadow near the town; and assemblies are held at the town-hall.

About the beginning of the fifteenth century, this place seems to have carried on a considerable trade upon the Severn; and a petition was forwarded to the House of Peers, in the 8th of Henry VI., stating that the inhabitants had been accustomed "to ship all manner of merchandise down the Severn to Bristol," and complaining of the disorderly conduct of the people of the forest of Dean, who are reported to have stopped and plundered their ships as they passed by the coasts near the forest. For the redress of these grievances an act was passed in the same year; and, in 1580, Queen Elizabeth made Tewkesbury an independent port, which grant was afterwards revoked, on a petition from the inhabitants of Bristol. Tewkesbury formerly enjoyed a considerable trade in woollen cloth, and was celebrated for the manufacture of mustard of superior quality; the principal branch of trade at present is the stocking frame-work knitting. The manufacture of cotton-thread lace was established at Oldbury in 1825; a considerable trade is carried on in malt, and some in leather, and there is a large manufactory for nails. An extensive distillery and a rectifying establishment were opened in 1770; the former has been abandoned, but the latter is still conducted on an advantageous scale. A very considerable carrying trade centres here, in connection with the Avon and the Severn, goods being conveyed by land and water to all parts of the kingdom: on the bank of the Avon are extensive corn-mills, formerly belonging to the abbey. The market days are Wednesday and Saturday; the former for corn, sheep, and pigs; the latter for poultry and provisions. The fairs are on the second Monday in March, the second Wednesday in April, May 14th, the first Wednesday after Sept. 4th, and Oct. 10th, for cattle, leather, and pedlery; fairs were also held in June and Dec., but they have been recently discontinued: statute fairs are held on the Wednesday before, and the Wednesday after,

Old Michaelmas-day, and there are great cattle markets on the second Wednesday in June, Aug., and Dec. The market-house, erected by a company, to whom the corporation have mortgaged the tolls for 99 years, is a handsome building, with Doric columns and pilasters, supporting a pediment in front.

Tewkesbury, though a borough by prescription, was first incorporated in 1574, by Elizabeth, whose charter was confirmed by James I., in the third year of his reign; and when that monarch sold the manor to the corporation, in 1609, he granted a new charter, with extended privileges, which being lost or destroyed, during the parliamentary war, an exemplification of it was obtained under the great seal in the reign of Charles II.: this was surrendered in 1685, to James II., who, in the following year, incorporated the inhabitants, under the title of "Mayor, Aldermen, and Common-Councilmen;" but the functions of the municipal body having ceased in 1692, the town remained without a corporation till 1698, when William III. granted a charter, extending the jurisdiction of the magistrates over the whole of the parish. The government, under this last charter, was vested in two bailiffs, a high steward, recorder, 24 principal burgesses, or common-councilmen, and 24 assistants, with a town-clerk, coroner, chamberlain, sergeants-at-mace, and subordinate officers. The bailiffs, with four principal burgesses, appointed annually on the second Thursday in Oct., were justices of the peace within the borough. By the act of the 5th and 6th of William IV., cap 76 (for an abstract of which see the Appendix, No. I.), the corporation now consists of a mayor, four aldermen, and twelve councillors. There were formerly several trading companies incorporated under the charter, but the only one now in existence is that of Cordwainers. The freedom of the borough is acquired by birth, or servitude to a resident freeman. The eldest son of a freeman, born after his father's admission, is entitled to the freedom on his father's decease, but if the son die first, the right does not devolve upon any other. The borough first received the elective franchise in the 7th of James I., since which time it has continued to return two members to parliament. The right of election was formerly vested in the freemen generally, and in all proprietors of freehold houses within the ancient limits of the borough, in number about 600; but the privilege has been extended to the £10 householders of the entire parish, comprising an area of 2388 acres: the old borough contained only 53 acres; the mayor is the returning officer. The recorder holds quarterly courts of session, for all offences within the borough not capital; a court of record is held every Friday, for the recovery of debts not exceeding £50; and the corporation, as lords of the manor, hold a court leet, the jurisdiction of which extends over all the parishes in the hundred of Tewkesbury. The town-hall, in which the courts are held, and the public business of the corporation is transacted, is a handsome building, erected in 1788, by Sir William Codrington, Bart., at an expense of £1200; the lower part is appropriated to the use of the courts, and the upper part contains a hall for the meetings of the corporation, and an assembly-room. The common gaol, house of correction, and penitentiary for the borough, was built in 1816, at the northern extremity of the High-street, at an expense of £3420, raised by a rate on the inhabitants, and was subsequently enlarged and improved: it contains four wards for the classification of prisoners. The county magistrates hold here a petty session for the division every alternate Wednesday; and the town has been made a polling-place for the eastern division of the county.

The living is a vicarage; net income, £313: it is in the patronage of the Crown; impropriators, several proprietors of land. The church, situated at the south-western part of the town, and formerly the collegiate church of the ancient monastery, is a spacious and venerable cruciform structure, principally in the Norman style of architecture, with a noble and richly ornamented tower rising from the centre. The nave and choir, which latter was repaired in 1796, at an expense of £2000, are separated from the aisles by a noble range of cylindrical columns and circular arches, highly enriched with mouldings and other ornaments peculiar to the Norman style; the former is lighted by a range of clerestory windows in the later style, inserted in the Norman arches of the triforium, and the latter by an elegant range of windows in the decorated style, with rich tracery, and embellished with considerable portions of ancient stained glass: the windows of the aisles and transepts are in the decorated and later styles; and the large west window, in the later style, is inserted in a very lofty Norman arch of great depth, with shafts and mouldings richly ornamented: the roof is finely groined, and embellished, at the intersections of the ribs, with figures of angels playing on musical instruments: the east end of the choir is hexagonal, and contains several beautiful chantry chapels, in the decorated style: the Lady chapel and the cloisters have been destroyed, but the arches which led to them may be traced on the outside of the building; and on the north side are the remains of the chapter-house, now used for a school. The church has been repewed, and a gallery built, by which means 500 free sittings have been provided, the Incorporated Society having granted £200 in aid of the expense. It contains a fine series of monuments, from the earliest period of the decorated, to the latest period of the later, style of English architecture, among which are several to the early patrons of the abbey, and to those who fell in the battle of Tewkesbury. In a light and elegant chapel on the north side of the choir, erected by Abbot Parker, in 1397, is the tomb of Robert Fitz-Hamon, the founder, who was killed at Falaise in Normandy, in 1107, and whose remains, after having been interred in the chapter-house, were removed into the church, in 1241: an altar-tomb, enclosed with arches surmounted by an embattled cornice, on which are the figures of a knight and his lady, is supposed to have been erected for Hugh le Despenser and his wife Elizabeth, daughter of William Montacute, Earl of Salisbury. Near this is a beautiful sepulchral chapel, built by Isabel, Countess of Warwick, for her first husband, Richard Beauchamp, Earl of Worcester, who was killed at the siege of Meaux, in 1421; it is profusely ornamented, and the roof, which is richly embellished with tracery, was supported on six pillars of blue marble, of which only two are remaining. Among the modern monuments is one, by Flaxman, to Anne, wife of Sir S. Clarke, Bart., of finely executed sculpture. Another church was erected in 1837; it is constructed of red brick, with stone dressings. There are places of worship for Baptists, the Society of Friends, Independents, and Wesleyan Methodists, all of which have Sunday schools annexed; also a Roman Catholic chapel.

The free grammar school was founded in 1576, and endowed with £20 per annum, by Mr. Ferrers, payable out of the manor of Skillingthorpe, in the county of Lincoln, also with lands purchased with money left by Sir Dudley Digges, and with some chief rents: the room appropriated to it is supposed to have been the chapter-house of the abbey. The Blue-coat school is endowed with one-twelfth part of the rents of a farm in Kent, devised for charitable uses by Lady Capel, in 1721, and with £2. 10. per annum given by Mr. Thomas Merret, in 1724, being further supported by subscription: 40 boys are clothed and instructed in it. The National school, under the superintendence of the same master, was established in 1813; and a building for its use, and also for that of the Blue-coat school (the two establishments having been incorporated), was erected adjoining the churchyard, in 1817, at an expense of £1345. 8. 3¼. A Lancasterian school was established in 1813, for which a building had been previously erected, at the cost of more than £600, raised by contributions; the ground was given by N. Hartland, a member of the Society of Friends: these schools are supported by subscription. In the churchyard are some unendowed almshouses for ten poor widows. A dispensary, established in 1815; a lying-in charity, in 1805; and a society for the distribution of blankets among the poor, in 1817, are supported by subscription; and there are various charitable bequests for the poor. Near the entrance into the town from Gloucester is the house of industry, a large brick building. The late Samuel Barnes, Esq., of this place, erected an extensive structure in the Oldbury, as an almshouse for 24 poor parishioners, which he endowed with land for their support; but the bequest being void by the statute of mortmain, the property devolved to his nieces, the Misses Mines, who, by deed conformably with the provisions of that statute, assigned it in trust to the corporation, to be applied to the purpose for which it was originally intended. The poor law union of Tewkesbury comprises 23 parishes or places, containing a population of 14,193, according to the census of 1831; it is under the care of 30 guardians.

Of the monastic buildings, with the exception of the church, there are few remains: the principal is the gateway of the ancient monastery, which appears to have been erected in the fifteenth century; it is surmounted with an embattled parapet rising above the cornice, from which are projecting figures, and below it is a canopied niche between two square-headed windows. Roman coins have been frequently dug up in the vicinity, and, in 1828, several were found near the abbey church. At Walton, near the town, is a mineral spring, the water of which resembles that at Cheltenham. The hamlet of Mythe, remarkable for the beauty of its situation, is on the north-west side of the town, near the confluence of the Severn and the Avon. On the south-west side is a tumulus, from which the descent to the Severn is precipitous and abrupt. George III. visited this spot in 1788, since which time it has obtained the name of Royal Hill: in the immediate vicinity are some handsome seats. Southwick, another hamlet, situated, as its name implies, to the south of the town, is mentioned in the Norman survey, under the name Sudwick, as containing three hides of land. Alan of Tewkesbury, a monk of the abbey, the friend and biographer of Thomas à Becket; and Estcourt, the celebrated dramatist, who was contemporary with Steele and Addison, were natives of this town. Tewkesbury gave the title of Baron to George I., previously to his accession to the throne, which at present is held by the Earl of Munster.

TEY, GREAT (*St. Barnabas*), a parish, in the union of Lexden and Winstree, Colchester Division of the hundred of Lexden, Northern Division of the county of Essex, 4 miles (N. E. by E.) from Great Coggeshall; containing 682 inhabitants. This parish, which is about 17 miles in circumference, has a great variety of soil, and the lands, which are principally arable, are in a high state of cultivation. The living is a discharged vicarage, valued in the king's books at £7; patron, Rev. R. S. Dixon, who is also patron and incumbent of the sinecure rectory, which is valued at £18. The rectorial tithes have been commuted for a rent-charge of £547. 14., and the vicarial for £232. 6., subject to the payment of rates, which on the average have respectively amounted to £112 and £18; the vicarial glebe comprises 18 acres, and the rectorial, 7 acres, valued at £20 and £8 per annum. The church is a very ancient edifice, originally cruciform, with a central tower supported on four arches, and side aisles separated from the nave by massive columns: in 1829, the tower was found to have pressed the pillars of the nave so much out of the perpendicular, that it became necessary to take down all the building to the west of it; since then divine service has been performed in the old chancel and transept, and a small erection has been raised on the site of the nave, forming the vestry-room and organ gallery: it contains 120 free sittings, the Incorporated Society having granted £200 in aid of the expense. A National school was built by the late rector, the Hon. R. L. Melville. Great Tey is the mother parish of Pontisbright, commonly called Chapel: the former contains 2480, and the latter about 1500 acres.

TEY, LITTLE (*St. James*), a parish, in the union of Lexden and Winstree, Witham Division of the hundred of Lexden, Northern Division of the county of Essex, 2¾ miles (E. by N.) from Great Coggeshall; containing 58 inhabitants. This parish is one of the smallest in the county, comprising only 448 acres; the soil, though heavy, is fertile, and the lands are in good cultivation. The living is a discharged rectory, valued in the king's books at £4; present net income, £120; patron, Bishop of London. The church is a small ancient edifice, with a belfry turret of wood.

TEY, MARKS (*All Saints*), a parish, in the union of Lexden and Winstree, Witham Division of the hundred of Lexden, Northern Division of the county of Essex, 4 miles (E. by N.) from Great Coggeshall; containing 363 inhabitants. This parish takes the adjunct to its name from the family of Marks, or Merks, to whom it anciently belonged; it is also in some documents called Tey *ad ulmos*, from the number of elm trees with which it formerly abounded, and for the growth of which the soil is peculiarly favourable. The living is a rectory not in charge; net income, £234; patrons, Master and Fellows of Balliol College, Oxford. In the chancel of the church is a window, containing the arms of Dr. Compton, Bishop of London, in painted glass.

TEYNHAM (*St. Mary*), a parish, in the union of Faversham, hundred of Teynham, Upper Division of the lathe of SCRAY, Eastern Division of the county of KENT, 4¼ miles (E.) from Sittingbourne; containing 753 inhabitants. The living is a vicarage, valued in the king's books at £10; present net income, £179; patron and appropriator, Archdeacon of Canterbury. The church is a handsome cruciform structure, principally in the early style of English architecture; it contains many brasses and other ancient memorials, and the windows exhibit fragments of old stained glass. An accession has lately been made to this parish, by the embankment of the island of Fowley. Conyer creek, an inlet of the sea, is terminated by a quay, to which vessels of 250 tons' burden come up and discharge their cargoes of coal, for the supply of the inhabitants, taking in the produce of the neighbourhood for the London and other markets. The parish abounds with cherry orchards, and there are a few plantations of hops. Here are vestiges of a Roman encampment, and the ruins of a palace formerly belonging to the Archbishops of Canterbury. Teynham confers the title of Baron on the family of Curzon.

THAKEHAM (*St. Mary*), a parish, and the head of a union, in the hundred of EAST EASWRITH, rape of BRAMBER, Western Division of the county of SUSSEX, 6½ miles (N. W.) from Steyning; containing, with the house of industry for other parishes, 597 inhabitants. The living is a rectory, valued in the king's books at £14. 9. 9½.; present net income, £585; patron, Duke of Norfolk. The church is partly in the early, and partly in the later, style of English architecture. A National school has been established. The poor law union comprises 14 parishes or places, under the superintendence of 16 guardians, and contains, according to the census of 1831, a population of 7311.

THAME (*St. Mary*), a market town and parish, and the head of a union, in the hundred of THAME, county of OXFORD, 13 miles (E.) from Oxford, and 44½ (N. W. by W.) from London; containing 2885 inhabitants. This town is evidently of Roman origin, and is first mentioned, as a place of some importance, at the commencement of the tenth century, when Wulfhere, King of Mercia, granted a charter, dated "in the vill called Thames:" in the year 970, Osketyl, Archbishop of York, died here. It suffered much from the Danish invasions, particularly in 1010, during which period a fortification was erected. At the Conquest it belonged to the Bishop of Lincoln, and formed part of the extensive possessions of the succeeding prelates, in this county, till the reign of Edward VI.: among the many benefits conferred by them on the town was the diverting through it the road which previously passed on its side. In 1138, a monastery for Cistercian monks was founded at Thame Park, in honour of the Virgin Mary, the revenue of which at the dissolution was valued at £256. 13. 7.: the site is occupied by the modern mansion of Lady Wenman. About the reign of Edward IV., an hospital for destitute persons was founded and endowed with lands by Richard Quatremain, a member of an ancient family of high repute. In the civil war of the sixteenth century, Thame was the centre of military operations, and experienced much consequent distress: during the late contest with France, it became one of the depôts for prisoners of war. The town derives its name from its situation on a gentle declivity on the bank of the river Thame, which here separates the counties of Oxford and Bucks, and over which is a bridge of considerable length. It consists principally of one long and spacious street, with a commodious market-place in the centre, over which is the town-hall, a handsome and commodious building. The manufacture of lace is carried on, but the poor are chiefly employed in husbandry. The market, which is of great antiquity, is on Tuesday, and is well supplied with corn and cattle. Fairs are held on Easter-Tuesday, the Tuesday before Whit-Sunday, the first Tuesday in Aug., and a statute fair on Oct. 11th.

The living, anciently a prebend in the Cathedral Church of Lincoln, valued in the king's books at £82. 12. 3½., but impropriated and dissolved in 1547, is now a discharged vicarage, with those of Sydenham, Tetsworth, and Towersey annexed, valued in the king's books at £18; present net income, £300; patron, Rev. John Peers: the impropriation belongs to Lady Wenman. The church, built in 1138, is a large and handsome cruciform structure, in the decorated style of English architecture, with an embattled tower rising from the intersection, supported on four massive pillars, and surmounted by an octagonal turret of nearly equal height. The interior, which in 1839 was thoroughly restored and beautified, at an expense of £500, is divided by columns and pointed arches, and is entered by a stone porch with an elegant canopied niche, in which was formerly a statue of the tutelar saint. In the chancel is a tomb of white marble, to the memory of Lord Williams, with the recumbent effigies of himself and his lady, in the costume of the time of Elizabeth; against the south wall is a curious brass, with a kneeling effigy of Sir John Clerke, of Weston, who, according to the legend, took prisoner Louis of Orleans, Duke of Longueville, in the reign of Henry VIII. The north transept is the burying-place of the Dormer family, and the south transept is the sepulchral chapel of the Quatremains: both contain handsome monuments. Lord Williams, in 1558, bequeathed estates for the foundation of a free grammar school, erected by his executors in 1574, near the church, and the maintenance of a master and an usher: the Warden and Fellows of New College, Oxford, are trustees of the school, and are empowered to nominate the master, who must be a clergyman and a graduate of one of the Universities, subject to the approbation of the Earl of Abingdon, as heir of the founder: the school is open to all boys of the parish. Hampden, the patriot; Dr. Fell; Justice Sir George Croke; Pocock, the learned orientalist; King, Bishop of Chichester; Anthony à Wood, the antiquary; and the celebrated John Wilkes, were educated in this school. A free school was established by bequests from the second Earl of Abingdon and others, the income amounting to £26 annually, in which 24 boys are instructed. Here is also a National school. There are several small annuities for apprenticing poor boys, and other benefactions, amounting to £150 per annum, for the poor. An almshouse for five poor men and one woman was also founded and endowed by Lord Williams, upon the dissolved foundation of Richard Quatremain. The poor law union of Thame includes a part of Buckinghamshire, and comprises 34 parishes or places, containing, according to the census of 1831, a population of 14,540: it is under the care of 38 guardians. A little to the north of the church are the remains of the prebendal house, originally attached to the monastery at Thame Park, and which, till 1837, consisted of nearly three sides of a quadrangle, but in that year, Mr. Chas. Stone converted the remains into a mansion-house, retaining the original character and grandeur of the ancient edifice, and in 1840, the same gentleman restored the chapel, at the east end of which is a triple lancet window circumscribed by a circular arch. George Hetheridge, an eminent Hebraist and Grecian in the reign of Elizabeth, and Regius Professor of Greek at Corpus Christi College, Oxford; and Lord Chief Justice Holt, were natives of this town.

THAMES - DITTON, county of SURREY. — See DITTON, THAMES.

THANINGTON (*St. Nicholas*), a parish, in the union of BRIDGE, hundred of WESTGATE, lathe of ST. AUGUSTINE, Eastern Division of the county of KENT, $1\frac{3}{4}$ mile (S. W. by W.) from Canterbury; containing 316 inhabitants. The living is a perpetual curacy; net income, £98; patron, Archbishop of Canterbury; impropriator, G. Gipps, Esq. The impropriate tithes have been commuted for a rent-charge of £610, subject to the payment of rates, which on the average have amounted to £71; the glebe comprises 3 acres, valued at £3 per annum. The ancient road, called Stane-street, passes through the parish. In Wincheap-street, a suburb of Canterbury, extending into this parish, are some small remains of the hospital of St. James, founded in the reign of John, by Archbishop Walter, for female lepers: its revenue, at the dissolution, was £46. 6. 3. Here is a school, in which 18 children are clothed and educated at the expense of Mrs. Bell.

THARSTON (*St. Mary*), a parish, in the union and hundred of DEPWADE, Eastern Division of the county of NORFOLK, $1\frac{1}{2}$ mile (N. W.) from Stratton-St.-Mary; containing 369 inhabitants. The living is a discharged vicarage, valued in the king's books at £5. 1. 8.; present net income, £118; patron and appropriator, Bishop of Ely. A school is partly supported by Lady Harvey.

THATCHAM (*St. Luke*), a parish (formerly a market-town), in the union of NEWBURY, partly in the hundred of FAIRCROSS, but chiefly in that of READING, county of BERKS, 3 miles (E.) from Newbury; containing 3912 inhabitants. This place, according to the Norman survey, appears to have been once a town of some importance; tradition has assigned to it the rank of a borough, but there is no proof that it ever sent representatives to parliament. A market on Sunday was confirmed by charter of Henry II. to the abbot or monks of Reading, then possessors of Thatcham, which was changed to Thursday in 1218, by Henry III.; but it has long been discontinued: the remains of the butter-cross still exist. The town is pleasantly situated near the navigable river Kennet, on the Bath road: the inhabitants are well supplied with water. The Kennet and Avon canal passes a little to the southward. There is a paper-mill at Colthrop, which affords employment to 80 persons. A statute fair is held on the first Tuesday after October 12th.

The living is a vicarage, valued in the king's books at £20; patron, J. Hanbury, Esq.; impropriators, various proprietors of land. The tithes have been commuted for a rent-charge of £700, subject to the payment of rates. The church has some portions in the early, and some in the later, style of English architecture: at the south entrance is a fine Norman arch, and in the interior are, an altar-tomb to the memory of William Danvers, Chief Justice of the Court of Common Pleas; and a mural monument to Nicholas Fuller, Esq., barrister of Gray's Inn. At Greenham and Midgham, in this parish, are chapels of ease; and at Crookham or Crokeham, was formerly another, of which there are no remains. There is a place of worship for Independents. A free school was founded in 1707, by Lady Frances Winchcomb, who gave a rent-charge of £53 for the education of 30 poor boys of the parishes of Thatcham, Bucklebury, and Little Shefford, and apprenticing some of them. It was opened about the year 1713, but continued only for a few years, in consequence of the attainder of Lord Bolingbroke, who was the owner of the estate charged, and also the only surviving trustee of the school, &c.; the affairs of the charity subsequently came under the direction of the court of Chancery. In 1741, arrears were recovered sufficient to purchase £1406. 9. 7. old South Sea annuities, since which period the funds have continued to increase, the amount of stock being now upwards of £5000, exclusively of the rent-charge of £53, which is regularly received. The school was re-opened in June 1794, under the regulations of the decree of the court of Chancery, and 40 boys are educated and clothed, five or six of them being annually apprenticed, with premiums varying from £12 to £20 each. This school is now united with a National school, in which about 80 boys and 40 girls are instructed, and which is supported partly by Lady Winchcomb's endowment, and partly by subscription. There is an almshouse for nine widows, founded by means of bequests from the Rev. Mr. Herdsman and John Hunt, besides various minor charitable benefactions.

THAXTED (*St. Mary*), a parish, in the union and hundred of DUNMOW, Northern Division of the county of ESSEX, 19 miles (N. N. W.) from Chelmsford; containing 2293 inhabitants. This parish, which is of considerable extent, is situated on the river Chelmer, near its source, and on the high road from Chelmsford to Cambridge. The village is large and pleasantly situated on the banks of the Chelmer, and contains several well-built houses: it was formerly a town of considerable importance, and had a charter of incorporation from Philip and Mary, vesting the government in a mayor, recorder, two bailiffs, and a council of 20 principal burgesses; but on a writ of *Quo Warranto* issued by James II., the burgesses resigned their functions, and the market, which was on Thursday, was discontinued. The market was subsequently revived, but it never recovered its former importance. Fairs are held on the 27th of May and the 10th of August, the latter for cattle. The living is a vicarage, valued in the king's books at £24; present net income, £450; patron and impropriator, Viscount Maynard, whose ancestor gave £2000 in augmentation of the vicarage. The church is a spacious embattled structure, strengthened by buttresses terminating in canopied niches, in the later style of English architecture, and having a tower and crocketed spire 183 feet high, the exact length of the church; the south porch is much enriched. There are places of worship for Baptists, the Society of Friends, and Independents. A free grammar school for the instruction of 30 boys, founded by Thomas Yardley, is now merged into a National school. Near the church are almshouses for 16 poor persons, who are partly supported by donations from various charity funds. Some Roman coins, and a beautiful amphora, were discovered in this parish some years ago.

THEAKSTONE, a township, in the parish of Burneston, union of Bedale, wapentake of Hallikeld, North Riding of the county of York, $3\frac{1}{4}$ miles (S. E. by E.) from Bedale; containing 82 inhabitants.

THEALE, a parish, in the hundred of Reading, county of Berks, $4\frac{1}{4}$ miles (W. by S.) from Reading: the population is included in the return for the parish of Tilehurst, within which this place was formerly a chapelry, but was separated from it by act of parliament, and made a distinct parish. An elegant church, in the later style of English architecture, has been erected, at the sole expense of Mrs. Sophia Sheppard; and, under the provisions of the act, the living of Tilehurst will be divided, and a portion appropriated to this church, to be attached to the headship of Magdalene College, Oxford. The Rev. Thomas Sheppard, D.D., bequeathed £20 per annum for the establishment of a school, which is conducted on the National system.

THEALE, a chapelry, in the parish of Wedmore, hundred of Bempstone, Eastern Division of the county of Somerset, 7 miles (S. S. E.) from Axbridge: the population is returned with the parish. The living is a perpetual curacy; net income, £100; patron and appropriator, Dean of Wells.

THEARNE, a township, in the parish of St. John, Beverley, union and liberties of the borough of Beverley, East Riding of the county of York, $3\frac{3}{4}$ miles (S. E. by E.) from Beverley; containing 67 inhabitants.

THEBERTON (*St. Peter*), a parish, in the union and hundred of Blything, Eastern Division of the county of Suffolk, 4 miles (N. E. by E.) from Saxmundham; containing 537 inhabitants. The living is a discharged rectory, valued in the king's books at £26. 13. 4.; present net income, £354: it is in the patronage of the Crown. The church has a round tower and other ancient details of Norman character. Here is a National school.

THEDDINGWORTH (*All Saints*), a parish, in the union of Market-Harborough, partly in the hundred of Rothwell, Northern Division of the county of Northampton, but chiefly in the hundred of Gartree, Southern Division of the county of Leicester, $4\frac{1}{2}$ miles (W. by S.) from Market-Harborough; containing 283 inhabitants. The living is a discharged vicarage, valued in the king's books at £8. 15. 7.; present net income, £137; patron, J. Cook, Esq.; impropriator, Earl Spencer. The Grand Union canal passes through the parish.

THEDDLETHORPE (*All Saints*), a parish, in the union of Louth, Marsh Division of the hundred of Calceworth, parts of Lindsey, county of Lincoln, $10\frac{1}{2}$ miles (N. N. E.) from Alford; containing 266 inhabitants. The living is a vicarage, valued in the king's books at £7. 5. $2\frac{1}{2}$.; present net income, £98; patron and impropriator, J. Alcock, Esq. The impropriate tithes have been commuted for a rent-charge of £400, subject to the payment of rates, which on the average have amounted to £47.

THEDDLETHORPE (*St. Helen's*), a parish, in the union of Louth, Marsh Division of the hundred of Calceworth, parts of Lindsey, county of Lincoln, $9\frac{3}{4}$ miles (N. by E.) from Alford; containing 275 inhabitants. The living is a rectory, with that of Mablethorpe-St.-Peter united in 1745, valued in the king's books at £18. 10. $2\frac{1}{2}$.; present net income, £498; patron, Lord Willoughby de Eresby. There is a place of worship for Wesleyan Methodists.

THELBRIDGE (*St. David*), a parish, in the union of Crediton, hundred of Witheridge, South Molton and Northern Divisions of the county of Devon, $7\frac{1}{4}$ miles (E. by S.) from Chulmleigh; containing 219 inhabitants. The living is a rectory, valued in the king's books at £10. 6. $5\frac{1}{2}$.; present net income, £198; patron, G. Tanner, Esq.

THELNETHAM (*St. Nicholas*), a parish, in the union of Thetford, hundred of Blackbourn, Western Division of the county of Suffolk, $3\frac{1}{2}$ miles (N. W.) from Botesdale; containing 553 inhabitants. The living is a discharged rectory, valued in the king's books at £16. 18. 4.; present net income, £508; patron and incumbent, Rev. S. Colby. There is a manor belonging to the rectory. An allotment of 28 acres of land, now let for £32. 12. per annum, was appropriated for parochial purposes under an enclosure act in 1821, together with another allotment of about 40 acres in lieu of the right of cutting turf.

THELVETON (*St. Andrew*), a parish, in the union of Depwade, hundred of Diss, Eastern Division of the county of Norfolk, $1\frac{3}{4}$ mile (N.) from Scole; containing 175 inhabitants. The living is a discharged rectory, valued in the king's books at £9; present net income, £249: it is in the patronage of the Crown.

THELWALL, a chapelry, in the parish and union of Runcorn, hundred of Bucklow, Northern Division of the county of Chester, $3\frac{1}{2}$ miles (E. S. E.) from Warrington; containing 332 inhabitants. This was formerly a considerable town, but is now only an obscure village. The Duke of Bridgewater's canal passes in the vicinity, and the river Mersey forms its northern boundary, on the south bank of which river are some gunpowder-mills. The living is a perpetual curacy; net income, £83; patron, T. A. Pickering, Esq. A day and Sunday school is supported by three families, who subscribe about £14 per annum.

THEMELTHORPE (*St. Andrew*), a parish, in the union of Aylsham, hundred of Eynsford, Eastern Division of the county of Norfolk, $1\frac{3}{4}$ mile (E. by S.) from Foulsham; containing 89 inhabitants. The living is a discharged rectory, annexed to that of Bintree, and valued in the king's books at £4. 2. $8\frac{1}{2}$. The tithes have been commuted for a rent-charge of £131, subject to the payment of rates, which on the average have amounted to £30; the glebe comprises 19 acres, valued at £24 per annum.

THENFORD (*St. Mary*), a parish, in the union of Brackley, hundred of King's-Sutton, Southern Division of the county of Northampton, $5\frac{1}{2}$ miles (N. W. by W.) from Brackley; containing 231 inhabitants. The living is a discharged rectory, valued in the king's books at £10; present net income, £120: it is in the patronage of the Crown. A school is supported by Mr. Severne. There is a mineral spring in the parish.

THERFIELD (*St. Mary*), a parish, in the union of Royston, hundred of Odsey, county of Hertford, $2\frac{1}{2}$ miles (S. W. by S.) from Royston; containing 974 inhabitants. The living is a rectory, valued in the king's books at £50; present net income, £937; patrons,

Dean and Chapter of St. Paul's, London. Here is a National school, chiefly supported by the rector.

THETFORD, a chapelry, in the parish of STRETHAM, hundred of SOUTH WITCHFORD, union and Isle of ELY, county of CAMBRIDGE, 2¼ miles (S. by W.) from Ely; containing 257 inhabitants. The chapel is dedicated to St. George.

Arms.

THETFORD, a borough and market-town, having exclusive jurisdiction, and the head of a union, locally in the hundred of Shropham, Western Division of the county of NORFOLK, but partly in the hundred of LACKFORD, Western Division of the county of SUFFOLK, 30 miles (S. W.) from Norwich, and 80 (N. N. E.) from London; containing 3462 inhabitants. This ancient place, called *Theodford* by the Saxons, evidently derives its name from the river Thet, which unites its stream with the Lesser Ouse at this spot; the latter then passes through the town, separates the two counties, and is navigable hence to Lynn. The majority of antiquaries consider it to be the site of the celebrated *Sitomagus* of the Romans, who possessed it in 435, and it is known to have been the metropolis of East Anglia; on which account, and from its proximity to the North Sea, it was frequently, during the Heptarchy, desolated by the Danes, who, having retained possession of the town for fifty years, totally destroyed it by fire in the ninth century. In 1004, it sustained a similar calamity from their king, Sweyn, who had invaded East Anglia; and in 1010 it became, for the third time, the scene of plunder and conflagration by these marauders, into whose hands it again fell, after a signal victory which they had obtained over the Saxons. In the reign of Canute, Thetford began to recover from the effects of these repeated calamities, and in that of Edward the Confessor, had nearly regained its former prosperity, containing not less than 947 burgesses, who enjoyed divers privileges. In the time of the Conqueror, the episcopal see of North Elmham was transferred hither, and hence to Norwich, by Herbert de Losinga, in the following reign; but Henry VIII. made it the seat of a bishop suffragan to Norwich, which it continued during his reign. From the time of Athelstan to that of John here was a mint, in which coins of Edmund and Canute were struck. The ancient extent and importance of this town may be gathered from the fact that, in the reign of Edward III., it comprised twenty-four principal streets, five market-places, twenty churches, six hospitals, eight monasteries, and other religious and charitable foundations, of all which there are comparatively but few remains. Thetford has been honoured with the presence and temporary residence of several British sovereigns, particularly Henry I., Henry II., and Elizabeth, who rebuilt the ancient mansion of the Earls of Warren, on its lapse to the crown, and occasionally resided in it, as did also James I. for the purpose of hunting; and it is still called the King's House.

The town, which has of late been much improved, comprises five principal streets, which are partly paved, and is connected with the few remaining houses on the Suffolk side by a handsome iron bridge over the Ouse, erected in 1829: the modern buildings are plain and neat, and the inhabitants are supplied with water from wells and springs. Assemblies are occasionally held: a small theatre was formerly open during the Lent assizes, but has been converted into a dwelling-house; and there is a subscription library. In addition to a very extensive paper-mill, there are a large iron-foundry, an agricultural machine manufactory, three good breweries, and several malting establishments. The navigation of the river, in its course to Lynn, having been improved between this place and Brandon, a brisk business is carried on in corn, wool, coal, and other articles. The market is on Saturday; the market-house is a neat and commodious building, covered with cast iron, with a portico and palisades in front. Fairs are held on May 14th, Aug. 2nd, and Aug. 16th, for sheep; Sep. 25th, for cattle; and there is a wool fair in July.

Corporatian Seal.

The charter of incorporation, granted by Elizabeth in 1573, was surrendered to the crown in the 34th of Charles II., and a very imperfect one obtained in its stead, which, in 1692, was annulled, and the original charter restored, by a decree in Chancery. By this charter the government was vested in a mayor, recorder, ten aldermen, and twenty common-councilmen, assisted by a town-clerk, sword-bearer, sergeant-at-mace, and inferior officers. The mayor, (who was clerk of the market, and acted as coroner after his year of office), the recorder, and the coroner, were justices of the peace, and held quarter sessions for the borough, over which they possessed exclusive jurisdiction; also a court of record for the recovery of debts to the amount of £50. The corporation now consists of a mayor, four aldermen, and twelve councillors, under the act of the 5th and 6th of William IV., cap. 76, for an abstract of which see the Appendix, No. I.: the mayor, late mayor, and recorder, are justices of the peace. This borough sends two members to parliament: the right of election was formerly vested exclusively in the corporation, but the privilege has been extended to the £10 householders of a new district: the present number of electors is about 240: the mayor is the returning officer. There has been a re-grant of a court of quarter session for the borough. The county assizes, which had been held here, in Lent, ever since the year 1234, were removed a few years since. The town is a polling-place for the western division of the county. The guildhall is a fine old building, erected at the expense of Sir Joseph Williamson, Knt., Secretary of State to Charles II., in which the assizes are held. The gaol is a plain edifice of flint and white brick, commodiously arranged, and capable of holding 100 prisoners: on these buildings many thousand pounds have been expended by the inhabitants.

Thetford comprises the parishes of *St. Cuthbert, St. Peter*, and *St. Mary the Less*, in the patronage of the Duke of Norfolk. The living of St. Cuthbert's is a discharged perpetual curacy, with the rectory of the Holy Trinity united; net income, £50: the church has a square embattled tower. The living of St. Peter's is a discharged rectory, with that of St. Nicholas' united, valued in the king's books at £5. 1. 5½.; present net income, £55. The church is commonly called "the black church," being constructed chiefly of flint: the tower and part of the body of the church were rebuilt in 1789. The living of the parish of St. Mary the Less is a perpetual curacy, valued in the king's books at £1. 13. 6½.; present net income, £83; impropriator, Duke of Norfolk. The church, which stands in the county of Suffolk, has a square tower. There were formerly many more parishes in this town, the churches of which have all been demolished. There are places of worship for the Society of Friends, Independents, Wesleyan Methodists, and Roman Catholics. A free grammar school, and an hospital for two poor men and two poor women, were founded in the reign of James I., under the will of Sir Richard Fulmerston, who died in 1566, having bequeathed property now producing £200 per annum, for the erection of a free school and other buildings, with adequate salaries for the master and usher, and remuneration to a clergyman for the performance of certain prescribed duties; it was therefore decreed, by act of parliament, that this should be a free grammar school and hospital for ever, and that the master, usher, and the four poor people, should be incorporated under the title of "The Master and Fellows of the School and Hospital at Thetford." A National school is supported by subscription. Almshouses for six poor men were erected in 1680, at the expense of William Harbord, Esq., and endowed with £30 per annum for a limited term, which expired about 25 years since, after having been renewed: the inmates participate in the proceeds of a bequest of £30 per annum, left in 1679, by his father, Sir Charles Harbord, Knt., Surveyor-General to Charles I. A certain number of boys and girls, children of the inhabitants, are apprenticed from a fund of £2000, bequeathed by Sir J. Williamson. In 1818, Mr. P. Sterne, of this place, bequeathed £1000 for the benefit of the poor: there are several minor charitable benefactions. The poor law union of Thetford comprises 34 parishes or places, under the care of 43 guardians, and contains, according to the census of 1831, a population of 16,198.

The relics of antiquity consist chiefly of fragments of the nunnery, founded in the reign of Canute, by Urius, the first abbot of Bury-St. Edmund's; some of the walls, buttresses, and windows, with a fine arch and cell, are still visible, the conventual church having been converted into a barn, and a farm-house built with the other ruinous portions. Of the priory, or abbey, founded on the brink of the river, in 1104, by Roger Bigod, for Cluniac monks, which at the dissolution was valued at £418. 16. 3., the gateway, constructed with freestone and black flint, and parts of the church, alone remain. Of the monastery of St. Sepulchre, founded in 1109, by the Earl of Warren, and additionally endowed by Henry II., the church has been converted into a barn; and the site of St. Augustine's friary, founded in 1387, by John of Gaunt, for mendicants of that order, still bears the name of Friars' Close: of the rest, no certain traces can be distinguished. At the eastern extremity of the town are remains of an ancient Danish fortification, which consisted of a large keep and double rampart, erected on an artificial mount, called Castle Hill, of which the height is 100 feet, and the circumference of the summit 81 feet, and of the base 984; the remains of the ramparts are 20 feet high, and the surrounding fosse 70 feet wide: it is somewhat singular, that no trace of any steps, or path, by which military stores could be conveyed up the very steep ascent to the fortress, is visible. Among the various fossils found in the vicinity is a perfect nautilus, which has been deposited in the British Museum. A mineral spring, the properties of which are similar to those of Tonbridge, was discovered here, about 80 years ago, by Matthew Manning, Esq., M.D., and about that time was much resorted to: it was afterwards shut up for many years, but in 1819 was re-opened, and the waters having been again analysed, were found to be very effectual in strengthening the stomach. A handsome pump-room was erected, to which hot and cold baths are attached: it is situated near the river side, and is approached by pleasant sheltered walks. Thomas Martin, F.A.S., and author of the "History of Thetford," was born here, in 1696, and educated at the free school, of which his father was master, and also rector of the parish of St. Mary. The notorious Thomas Paine, author of the "Age of Reason," "Rights of Man," &c., was also born at a house in White Hart-street, and educated at the school.

THICKLEY, EAST, a township, in the parish of St. Andrew-Auckland, union of Auckland, North-Western Division of Darlington ward, Southern Division of the county palatine of Durham, 4¼ miles (S. E.) from Bishop-Auckland; containing 35 inhabitants. The Stockton and Darlington railway passes in the vicinity.

THIMBLEBY (*St. Margaret*), a parish, in the union and soke of Horncastle, parts of Lindsey, county of Lincoln, 1¼ mile (W. by N.) from Horncastle; containing 364 inhabitants. The living is a discharged rectory, valued in the king's books at £13. 10. 10.; present net income, £441; patron, T. Hotchkin, Esq. A school is chiefly supported by the rector.

THIMBLEBY, a township, in the parish of Osmotherley, union of North-Allerton, wapentake of Allertonshire, North Riding of the county of York, 5½ miles (E. N. E.) from North-Allerton; containing 185 inhabitants, many of whom are employed in the manufacture of worsted. A school is partly supported by the rector.

THINGDON, county of Northampton.—See FINEDON.

THINGWELL, a township, in the parish of Woodchurch, union, and Lower Division of the hundred, of Wirrall, Southern Division of the county of Chester, 5½ miles (N. by W.) from Great Neston; containing 77 inhabitants.

THIRKLEBY, a township, in the parish of Kirby-Grindalyth, union of Malton, wapentake of Buckrose, East Riding of the county of York, 10 miles (E. by S.) from New Malton; containing 44 inhabitants.

THIRKLEBY (*All Saints*), a parish, in the union of Thirsk, wapentake of Birdforth, North Riding of the county of York, 4 miles (S. E. by E.) from Thirsk; containing 317 inhabitants. The living is a discharged vicarage, valued in the king's books at £6; present net income, £210; patron and appropriator, Archbishop of York. The church was rebuilt in the year 1722, by Sir Thomas Frankland, Bart.

THIRLBY, a township, in the parish of Felix-Kirk, union of Thirsk, wapentake of Birdforth, North

Riding of the county of York, 5 miles (E. N. E.) from Thirsk; containing 131 inhabitants.

THIRLWALL, a township, in the parish and union of Haltwhistle, Western Division of Tindale ward, Southern Division of the county of Northumberland, 4 miles (W. N. W.) from Haltwhistle; containing 328 inhabitants. The ancient mansion of Wardrew, situated on the eastern side of the river Irthing, has been handsomely fitted up, to accommodate visiters who resort to the adjacent spa of Gilsland. On the western bank of the Tippal bourn, which is here crossed by the great Roman wall, are the ruins of the once strong castle of Thirlwall, occupying the summit of a rocky precipice. The walls of this fortress were nine feet thick, vaulted within, and defended by an outer wall of great strength.

THIRN, a township, in the parish of Thornton-Watlass, union of Leyburn, wapentake of Hang-East, North Riding of the county of York, 4 miles (S. W. by W.) from Bedale; containing 142 inhabitants.

THIRNE, a parish, in the hundred of West Flegg, Eastern Division of the county of Norfolk, 4½ miles (N. by E.) from Acle; containing 138 inhabitants. The living is a rectory, annexed to that of Ashby, and valued in the king's books at £5. Here is a school, in which six children are paid for by charity.

THIRNTOFT, a township, in the parish of Ainderby-Steeple, union of North-Allerton, wapentake of Gilling-East, North Riding of the county of York, 3½ miles (W. by S.) from North-Allerton; containing 170 inhabitants.

THIRSK (*St. Mary*), a borough, market-town, and parish, and the head of a union, partly within the liberty of St. Peter's, East Riding, but chiefly in the wapentake of Birdforth, North Riding, of the county of York; containing 3829 inhabitants, of which number, 2829 are in the borough of Thirsk, 23 miles (N. W. by N.) from York, and 223 (N. N. W.) from London. The name of this place is supposed to be derived from *Tre Isk*, two ancient British words, signifying a town and river or brook. A strong and extensive castle was erected here, about 979, by the ancient family of Mowbray, on which Roger de Mowbray, in the time of Henry II., having become a confederate of the King of Scotland, erected his standard against his lawful sovereign: on the suppression of that revolt, this fortress, with many others, was entirely demolished by order of the king. In the reign of Henry VII., during a popular commotion, Henry Percy, Earl of Northumberland, and lieutenant of this county, is said to have been put to death here, beneath a very ancient elm-tree, which formerly grew on Elm Green. The town is situated on the road from York to Edinburgh, nearly in the centre of the vale of Mowbray, a tract of country remarkable for the fertility of its soil, and the picturesque beauty and richness of its scenery: it consists of the Old and the New towns, which are separated by a small stream, called Cod-beck, over which are two substantial stone bridges. A neat gravel walk across the fields leads to the adjacent village of Sowerby; it commands a fine prospect of the surrounding country, terminated by the Hambleton hills, and is the favourite promenade of the inhabitants. At the south-western extremity of the town, the moat and rampart, together with some subterranean vaults, and the site of the court-yard of the castle, still exist. Within the precincts of this ancient fortress New Thirsk is situated, with its spacious and commodious market-place in the centre. The Old Town, which alone is included within the limits of the borough, is on the north-east side of the stream, and consists of a long range of cottages on each side of the turnpike road leading from York to Yarm and Stockton, and two squares surrounded by similar buildings, one called St. James's Green, where the cattle fairs are held, the other formerly comprising an ancient church, dedicated to St. James, of which there are no vestiges. This place is called Elm Green, from the ancient elm that formerly grew there, beneath which the members of parliament for the borough were usually elected. A small quantity of coarse linen and sacking is manufactured. Coal, which is partly brought from the county of Durham, in small carts containing from 18 to 22 bushels each, is sold at a very high price, although the supply has been somewhat increased by the extension of the Darlington railway to Croft Bridge, 21 miles distant. The market is on Monday, and is a large one for provisions, of which great quantities purchased here are carried for sale to Leeds and other places. Fairs are held on Shrove-Monday and April 4th and 5th, for cattle, sheep, leather, &c.; Easter-Monday and Whit-Monday, for woollen cloth, toys, &c.; and August 4th and 5th, October 28th and 29th, and the first Tuesday after December 11th, for cattle, sheep and leather. The municipal regulations of the town are vested in a bailiff, chosen by the burgage-holders, and sworn in before the stewart of the lord of the manor, who holds a court leet annually at Michaelmas, for that and other purposes. Old Thirsk is a borough by prescription, and first sent members to parliament in the 23rd of Edward I., but made no other return till the last parliament of Edward VI.: under the act of the 2nd of William IV., cap. 45, it now sends only one. The elective franchise was formerly vested in the owners of burgage tenements, 50 in number, of whom only six are resident within the borough; but the privilege has been extended to the £10 householders of an enlarged district, which has been made to constitute the new elective borough, comprising an area of 8570 acres, of which the limits are minutely described in the Appendix, No. II.: the bailiff is the returning officer. This town has been made a polling-place for the North Riding.

The living is a perpetual curacy; net income, £143; patron and appropriator, Archbishop of York. The church, situated at the northern extremity of the New Town, is a spacious and handsome structure, in the later style of English architecture, with a lofty embattled tower at the west end, and is supposed to have been constructed from the ruins of the castle. There are places of worship for the Society of Friends, Independents, and Wesleyan Methodists. In 1769, Jane Day bequeathed £100 for the instruction of poor children. There is a school-house under the chancel of the church; also a school of industry, for clothing and educating poor girls; and a dispensary. The poor law union of Thirsk comprises 40 parishes or places, under the care of 41 guardians, and contains, according to the census of 1831, a population of 12,013.

THIRTLEBY, a township, in the parish of Swine, union of Skirlaugh, Middle Division of the wapentake of Holderness, East Riding of the county of York, 6 miles (N. E.) from Kingston-upon-Hull; containing 59 inhabitants.

THISTLETON, a joint township with Greenhalgh, in the parish of KIRKHAM, union of the FYLDE, hundred of AMOUNDERNESS, Northern Division of the county palatine of LANCASTER, $4\frac{1}{4}$ miles (N. N. W.) from Kirkham: the population is returned with Greenhalgh.

THISTLETON (*St. Nicholas*), a parish, in the union of OAKHAM, hundred of ALSTOE, county of RUTLAND, 8 miles (N. N. E.) from Oakham; containing 151 inhabitants. The living is a rectory, valued in the king's books at £3. 11. $0\frac{1}{2}$.; present net income, £118; patron, G. Fludyer, Esq. The Rev. Henry Foster, in 1692, bequeathed land producing upwards of £10 per annum, in support of a school, now conducted on the National system.

THIXENDALE, a township, in the parish of WHARRAM-PERCY, union of POCKLINGTON, wapentake of BUCKROSE, East Riding of the county of YORK, $8\frac{3}{4}$ miles (S. S. E.) from New Malton; containing 207 inhabitants.

THOCKRINGTON, a parish, in the union of BELLINGHAM, North-Eastern Division of TINDALE ward, Southern Division of the county of NORTHUMBERLAND; containing 203 inhabitants, of which number, 71 are in the township of Thockrington, $11\frac{1}{4}$ miles (N. by E.) from Hexham. The living is a perpetual curacy, in the patronage of the Rev. R. Afferk, the impropriator; net income, £48.

THOLTHORP, a township, in the parish of ALNE, union of EASINGWOULD, partly within the liberty of ST. PETER'S, East Riding, and partly in the wapentake of BULMER, North Riding, of the county of YORK, $4\frac{1}{2}$ miles (S. W.) from Easingwould; containing 265 inhabitants.

THOMAS-CLOSE, a township, in the parish of HUTTON-IN-THE-FOREST, union of PENRITH, LEATH ward, Eastern Division of the county of CUMBERLAND, $8\frac{3}{4}$ miles (N. W. by N.) from Penrith; containing 106 inhabitants.

THOMAS (ST.) STREET, an extra-parochial hamlet, locally in the parish of St. Thomas the Apostle, union of LAUNCESTON, Northern Division of the hundred of EAST, Eastern Division of the county of CORNWALL; containing 378 inhabitants.

THOMAS (ST.) THE APOSTLE, a parish, in the union, and adjoining the borough of LAUNCESTON, Northern Division of the hundred of EAST, Eastern Division of the county of CORNWALL; containing 626 inhabitants. The living is a perpetual curacy; net income, £83; patrons, the Inhabitants. The great tithes of the parish of St. Clether were purchased for this curacy with Queen Anne's bounty. A priory was founded here by Bishop Warlewast, who in 1126 removed to it the establishment of Secular canons which had previously existed at St. Stephen's: at the dissolution its revenue was estimated at £354. 0. 11. At Kestelwood are vestiges of ancient earthworks.

THOMAS (ST.) THE APOSTLE, a parish, and the head of a union, in the hundred of WONFORD, Wonford and Southern Divisions of the county of DEVON, $\frac{1}{2}$ a mile (S. by W.) from Exeter; containing 4203 inhabitants. The parish is bounded on the east by the river Exe, from which the Exeter canal passes to the southward. The living is a vicarage, valued in the king's books at £11. 2. $8\frac{1}{2}$.; present net income, £237; patron and impropriator, James W. Buller, Esq. Twenty-four children are educated for an annuity of £10, bequeathed by William Gould, and four for £1. 10. a year, the gift of Robert Pate. The poor law union comprises 49 parishes or places, under the superintendence of 61 guardians, and contains, according to the census of 1831, a population of 42,155. A small priory of Black canons, a cell to that of Plympton, founded in the time of Henry III., in honour of the Blessed Virgin, stood partly in this parish, and partly in that of Alphington.

THOMPSON (*St. Martin*), a parish, in the union and hundred of WAYLAND, Western Division of the county of NORFOLK, 3 miles (S. S. E.) from Watton; containing 478 inhabitants. The living is a perpetual curacy; net income, £49; patron and impropriator, S. E. Emsworth, Esq. Sir Thomas de Shardelow, Knt., and his brother, about 1349, founded, in honour of the Blessed Virgin and All Saints, a chantry, or college, for a master and five chaplains, whose revenue, at the dissolution, was valued at £52. 15. 7.

THOMPSON'S-WALL, a joint township with Couldsnouth, in the parish of KIRKNEWTON, Western Division of GLENDALE ward, Northern Division of the county of NORTHUMBERLAND, $8\frac{1}{2}$ miles (W. by N.) from Wooler; containing 41 inhabitants.

THOMPSON (*St. Andrew*), a parish, in the union of BLANDFORD, hundred of COOMBS-DITCH, Blandford Division of the county of DORSET, 7 miles (S. by E.) from Blandford-Forum; containing 41 inhabitants. The living is a discharged rectory, valued in the king's books at £4. 8. 9.; patron, H. Bankes, Esq. The tithes have been commuted for a rent-charge of £83, subject to the payment of rates, which on the average have amounted to £6 per annum; the glebe comprises about one acre, valued at £1. 5. per annum. The church is a small brick building without a tower, circular at the east end: it was wholly rebuilt and pewed by Archbishop Wake.

THONG, NETHER, a township, in the parish of ALMONDBURY, union of HUDDERSFIELD, Upper Division of the wapentake of AGBRIGG, West Riding of the county of YORK, $5\frac{1}{4}$ miles (S. by W.) from Huddersfield; containing 1004 inhabitants, of whom many are employed in the manufacture of woollen cloth. A chapel, in the later English style, with a cupola, has been erected by the parliamentary commissioners, at an expense of £2869. 12. 1.: the living is a donative; net income, £57; patron, Vicar of Almondbury; impropriators, Governors of Clitheroe school. A day and Sunday school is supported by subscription.

THONG, UPPER, a township, in the parish of ALMONDBURY, union of HUDDERSFIELD, Upper Division of the wapentake of AGBRIGG, West Riding of the county of YORK, $6\frac{1}{4}$ miles (S. S. W.) from Huddersfield; containing 1648 inhabitants. The manufacture of woollen goods is here carried on to a considerable extent. There is a place of worship for Wesleyan Methodists.

THORALBY, a township, in the parish of AYSGARTH, wapentake of HANG-WEST, North Riding of the county of YORK, $8\frac{1}{2}$ miles (W. by S.) from Middleham; containing 272 inhabitants.

THORESBY, NORTH (*St. Helen*), a parish, in the union of LOUTH, wapentake of BRADLEY-HAVERSTOE, parts of LINDSEY, county of LINCOLN, $8\frac{1}{2}$ miles

(N. by W.) from Louth; containing 544 inhabitants. The living is a rectory, valued in the king's books at £24. 10. 10.; present net income, £273; patron and incumbent, Rev. H. Bassett. There is a place of worship for Wesleyan Methodists. Dr. Robert Mapletoft, in 1676, bequeathed land producing £30 per annum, for teaching poor children.

THORESBY, SOUTH (*St. Andrew*), a parish, in the union of Louth, Marsh Division of the hundred of Calceworth, parts of Lindsey, county of Lincoln, 4 miles (W. by N.) from Alford; containing 142 inhabitants. The living is a discharged rectory, valued in the king's books at £6. 3. 6½.; present net income, £214; patron, Chancellor of the Duchy of Lancaster.

THORESTHORPE, a hamlet, in the parish of Saleby, union of Louth, Wold Division of the hundred of Calceworth, parts of Lindsey, county of Lincoln, ¾ of a mile (N. N. E.) from Alford: the population is returned with the parish.

THORESWAY (*St. Mary*), a parish, in the union of Caistor, Southern Division of the wapentake of Walshcroft, parts of Lindsey, county of Lincoln, 5 miles (S. E.) from Caistor; containing 158 inhabitants. The living is a discharged rectory, valued in the king's books at £8. 10. 10.; present net income, £493: it is in the patronage of the Crown.

THORGANBY (*All Saints*), a parish, in the union of Caistor, Southern Division of the wapentake of Walshcroft, parts of Lindsey, county of Lincoln, 6¼ miles (E. S. E.) from Caistor; containing 108 inhabitants. The living is a discharged rectory, held by sequestration, and valued in the king's books at £6. 0. 10.; present net income, £47; patron, Lord Yarborough.

THORGANBY (*St. Elen*), a parish, in the union of York, wapentake of Ouse and Derwent, East Riding of the county of York, 8½ miles (N. E. by N.) from Selby; containing 342 inhabitants. The living is a perpetual curacy; net income, £53; patron, J. D. Jefferson, Esq. There is a place of worship for Wesleyan Methodists. Thomas Dunnington, in 1733, gave a school-house, with a residence for the master, and an annuity of £2 towards his support; which, with the annual sums of £2 by Robert Blythe, £2 by Thomas Bradford, and £10. 10. by Robert Jefferson, is paid for teaching 20 children.

THORINGTON (*St. Peter*), a parish, in the union and hundred of Blything, Eastern Division of the county of Suffolk, 4 miles (S. E.) from Halesworth; containing 159 inhabitants. The living is a discharged rectory, valued in the king's books at £7; present net income, £239; patron, Lieutenant-Colonel Bence, who has his seat here. The church has a round tower, and other Norman details.

THORLBY, a joint township with Stirton, in the parish of Kildwick, union of Skipton, Eastern Division of the wapentake of Staincliffe and Ewcross, West Riding of the county of York, 1¾ mile (N. W. by W.) from Skipton: the population is returned with the parish.

THORLEY (*St. James*), a parish, in the union of Bishop's-Stortford, hundred of Braughin, county of Hertford, 2½ miles (S. S. W.) from Bishop's-Stortford; containing 414 inhabitants. The living is a rectory, valued in the king's books at £16. 13. 4.; present net income, £455; patron, Bishop of London. The church has an embattled tower, surmounted by a lofty spire at the west end, and a Norman doorway on the south.

THORLEY (*St. Mary*), a parish, in the liberty of West Medina, Isle of Wight Division of the county of Southampton, 1 mile (E. S. E.) from Yarmouth; containing 146 inhabitants. The living is a discharged vicarage, valued in the king's books at £6. 18. 9.; present net income, £100; patrons, Rev. Dr. Walker and Edward Roberts, Esq. The church has a belfry over the porch, but no tower.

THORMANBY (*St. Mary*), a parish, in the union of Easingwould, wapentake of Bulmer, North Riding of the county of York, 4¼ miles (N. W. by N.) from Easingwould; containing 133 inhabitants. The living is a discharged rectory, valued in the king's books at £8. 2. 11.; present net income, £216; patron and incumbent, Viscount Downe.

THORN (*St. Margaret*), a parish, in the union of Wellington, hundred of Milverton, Western Division of the county of Somerset, 3 miles (W.) from Wellington; containing 165 inhabitants. The living is a perpetual curacy; net income, £113; patron and appropriator, Archdeacon of Taunton,

THORN-COFFIN (*St. Andrew*), a parish, in the union of Yeovil, hundred of Tintinhull, Western Division of the county of Somerset, 2½ miles (N. W. by W.) from Yeovil; containing 101 inhabitants. The living is a discharged rectory, valued in the king's books at £5. 5. 2½.; present net income, £200; patron and incumbent, Rev. Alfred Tooke.

THORN-GUMBALD, a chapelry, in the parish of Paul, union of Patrington, Southern Division of the wapentake of Holderness, East Riding of the county of York, 2 miles (S. E.) from Hedon; containing 266 inhabitants. There is a place of worship for Independents.

THORNABY, a chapelry, in the parish of Stainton, union of Stockton, Western Division of the liberty of Langbaurgh, North Riding of the county of York, 1¾ mile (S. S. E.) from Stockton-upon Tees; containing 301 inhabitants. The living is a perpetual curacy, united to the vicarage of Stainton. The weaving of linen yarn is carried on here to a limited extent.

THORNAGE, a parish, in the union of Erpingham, hundred of Holt, Western Division of the county of Norfolk, 2¾ miles (S. W. by W.) from Holt; containing 332 inhabitants. The living is a rectory, with that of Brinton annexed, valued in the king's books at £6. 18. 4.; patron, Sir J. D. Astley, Bart. The tithes have been commuted for a rent-charge of £321, subject to the payment of rates, which on the average have amounted to £63; the glebe comprises 33 acres, valued at £35. 5. 3. per annum.

THORNBOROUGH (*St. Mary*), a parish, in the union, hundred, and county of Buckingham, 3½ miles (E.) from Buckingham; containing 673 inhabitants. The living is a discharged vicarage, valued in the king's books at £8. 17.; present net income, £187; patron, Sir H. Verney, Bart.; impropriator of a portion of the great tithes, Duke of Buckingham, the remainder having been given to the different proprietors on the inclosure of waste lands in 1804. The chancel of the church belongs to W. F. Lowndes Stone and John Clark, Esqrs., who keep it in repair. On opening an ancient barrow, about

25 feet high, at Thornborough field, in November 1839, there were discovered near the base, on a layer of rough limestone, various bronze ornaments, in a state of excellent preservation, amongst which were a curious lamp beautifully shaped, two large and elegant jugs, a large dish, a bowl, and the hilt of a sword, also a small ornament of pure gold, with a figure of Cupid elegantly chased upon it, and a large glass, which contained the bones and ashes of the individual there interred, supposed to have been a person of distinction: the whole were deposited at Stow House.

THORNBOROUGH, a township, in the parish of CORBRIDGE, union of HEXHAM, Eastern Division of TINDALE ward, Southern Division of the county of NORTHUMBERLAND, 5¾ miles (E.) from Hexham; containing 81 inhabitants. Considerable quantities of limestone are quarried and burned in this township. A lead mine anciently wrought here was re-opened in 1801, but the speculation proving unsuccessful, it was soon after abandoned.

THORNBROUGH, a township, in the parish of SOUTH KILVINGTON, union of THIRSK, wapentake of BIRDFORTH, North Riding of the county of YORK, 2¾ miles (E. N. E.) from Thirsk; containing 21 inhabitants.

THORNBURY (*St. Peter*), a parish, in the union of HOLSWORTHY, hundred of BLACK TORRINGTON, Holsworthy and Northern Divisions of the county of DEVON, 4½ miles (N. E. by E.) from Holsworthy; containing 546 inhabitants. The living is a rectory, valued in the king's books at £11. 3. 11½.; present net income, £198: it is in the patronage of Mrs. Spencer. The church has a Norman door, and contains a monument of an armed knight and his lady, with several monumental effigies of the Edgecumbe family.

THORNBURY (*St. Mary*), a market-town and parish, and the head of a union, in the Lower Division of the hundred of THORNBURY, Western Division of the county of GLOUCESTER, 24 miles (S. W.) from Gloucester, and 124 (W. by N.) from London; containing 4375 inhabitants. This town, which is of considerable antiquity, is situated on the banks of a small rivulet, 2 miles east of the Severn, in the vale of Berkeley, and consists of three principal streets. The chief object worthy of notice is the remains of an old castle at the end of the town, begun by Edward, Duke of Buckingham, in 1511, but left in an unfinished state; the outer wall is still in good preservation, and over the arched gateway, which formed the principal entrance, and is greatly admired, is an inscription in raised letters, recording the date of its erection: these ruins command a fine view of the river Severn, which flows on the western side of the parish, and the remote landscape of South Wales. Henry VIII. and Anne Boleyn were sumptuously entertained here for ten days, in 1539. The clothing business formerly flourished, but has been long discontinued, and there is now no particular branch of trade. The market is on Saturday. Fairs are held on Easter-Monday, August 15th, and the Monday before December 21st, for cattle and pigs. The corporation, now merely nominal, consists of a mayor and twelve aldermen, with a sergeant-at-mace and two constables. A manorial court leet is held annually, and occasionally a court baron; also a manor court for the surrender of admission to copyholds. A court for the recovery of debts under 40*s*., for the hundred, is held once in three weeks, on Thursday; a court of record, for pleas to any amount, for the honour of Gloucester, is also held every three weeks, on Tuesday; and the town has been made a polling-place for the western division of the shire.

The living is a vicarage, valued in the king's books at £25. 15. 10.; present net income, £500; patrons and appropriators, Dean and Canons of Christ Church, Oxford. The church is a spacious and handsome cruciform structure, in the later style of English architecture, with a lofty tower, with open worked battlements and eight pinnacles: the north and south doors are of much earlier date. At Falfield and Oldbury upon Severn, in this parish, are chapels of ease; and there are places of worship for Baptists, the Society of Friends, Independents, and Wesleyan Methodists. A free grammar school was founded and endowed in 1648, by William Edwards; its funds having been augmented by subsequent benefactors, the present income is £57. 3. 6., and twelve boys are instructed on the foundation. Another free school was founded in 1729, by means of a bequest of £500 from John Atwells, and endowed with lands in 1789; the income is £70 per annum, and twenty-four boys and twelve girls are educated: there is also a National school. Here are six almshouses for fifteen poor people, founded by Sir John Stafford, Thomas Slimbridge, and Katherine Rippe: there are also other benefactions for the poor. The poor law union of Thornbury comprises 21 parishes or places, under the care of 26 guardians, and contains, according to the census of 1831, a population of 15,422.

THORNBURY, a parish, in the union of BROMYARD, hundred of BROXASH, county of HEREFORD, 4¼ miles (N. N. W.) from Bromyard; containing 212 inhabitants. The living is a rectory, valued in the king's books at £5. 6. 8.; present net income, £183: it is in the patronage of Mrs. Pytts. Wall Hill camp, in this parish, has a triple intrenchment, almost perfect, and is supposed to be a work of the ancient Britons. At Netherwood, Robert Devereux, Earl of Essex, who was beheaded in 1601, was born in 1567.

THORNBY (*St. Helen*), a parish, in the union of BRIXWORTH, hundred of GUILSBOROUGH, Southern Division of the county of NORTHAMPTON, 10½ miles (N. N. W.) from Northampton; containing 198 inhabitants. The living is a rectory, valued in the king's books at £13; present net income, £364; patron and incumbent, Rev. N. Cotton.

THORNCOMBE (*St. Mary*), a parish (formerly a market-town), in the union, and forming a detached portion of the hundred, of AXMINSTER, Honiton and Southern Divisions of the county of DEVON, 6½ miles (S. W. by W.) from Axminster; containing 1368 inhabitants. The living is a vicarage, valued in the king's books at £15. 18. 9.; present net income, £457; patron and impropriator, John Bragge, Esq. A fair is held on Easter Tuesday. Here is a free school, founded by the Rev. Thomas Cooke, in 1734, with a small endowment for the instruction of twelve poor children. At Ford, in this parish, an abbey for Cistercian monks was founded by Adelign, daughter of Baldwin de Brioniis, in 1140, the revenue of which was valued at £381. 10. 8½.; the remains are considerable, consisting partly of the entrance tower, the old abbey walls, and various other parts now used as a private mansion. The chapel has a groined roof in the early English style, and some arches of late Norman character: the hall and cloisters are in the later English. The possessor of this abbey holds a court at Holditch.

THORNCOTE, a hamlet, in the parish of NORTHILL, hundred of WIXAMTREE, county of BEDFORD, $3\frac{1}{4}$ miles (N. W.) from Biggleswade; containing, with Brookend, Budnor, Hatch, and a part of Beeston, 268 inhabitants.

THORNDON (*All Saints*), a parish, in the union and hundred of HARTISMERE, Western Division of the county of SUFFOLK, $3\frac{1}{4}$ miles (S. by W.) from Eye; containing 696 inhabitants. The living is a rectory, valued in the king's books at £24. 11. $10\frac{1}{2}$.; present net income, £600; patron and incumbent, Rev. Thomas Howes. The church is said to have been built by Robert de Ufford, Earl of Suffolk, in 1358. A school is partly supported by the rector. Dr. John Bale, Bishop of Ossory, was rector of this parish.

THORNE (*St. Nicholas*), a market-town and parish, and the head of a union, in the Southern Division of the wapentake of STRAFFORTH and TICKHILL, West Riding of the county of YORK, 29 miles (S. by E.) from York, and 165 (N. by W.) from London; containing 3779 inhabitants. This place, in Leland's time only a small village, with a castle near it, the foundations of which are still visible, has become a neat and flourishing town; it is situated on the verge of the moors, near Marshland, a fenny district, supposed to have been once a forest, from the numerous fossil trees, &c., which have been discovered here; the streets are paved, and many of the houses well built. On the moor, large quantities of pcat are obtained, and conveyed by means of a canal to the town and other places, to be used as a substitute for coal. The inhabitants carry on a considerable trade in grain and other commodities with London; rope is made to some extent. At Hangman's hill, about a mile distant is the quay, where all merchandise is shipped and landed. Vessels for the coasting trade are built here, and, being launched at spring tides, are conveyed down the river Don to Hull, to be rigged and otherwise completed. A canal from this river to the Trent passes westward of the town, by which its trade is greatly promoted. The market, originally granted by Richard Cromwell, and renewed by Charles II., is on Wednesday; and fairs, chiefly for horses, cattle, and pedlery, are held on the Monday and Tuesday next after June 11th and October 11th.

The living is a perpetual curacy; net income, £117; patron and impropriator, Heir of Sir H. Hetherington. The church is principally in the later style of English architecture, with a square tower surmounted by pinnacles. There are places of worship for the Society of Friends, Independents, Primitive and Wesleyan Methodists, Unitarians, and the followers of Joanna Southcote. The free school was endowed with land by William Brook, in 1705, for the perpetual maintenance of a schoolmaster, and the instruction of ten of the poorest boys within the town: the annual income is £148. 19. 9. Another free school was founded and endowed with land in 1706, by Henry Travis, the income of which is about £35. The poor law union of Thorne comprises 13 parishes or places, the greater number included in Lincolnshire, and contains a population of 14,918, according to the census of 1831; it is under the superintendence of 19 guardians. The Rev. Abraham de la Pryme, F.R.S., a celebrated antiquary and historian, was some time minister of Thorne, and died in 1704, at the early age of 34. At Crowtrees, near this town, resided Sir Cornelius Vermuyden, who, having expended £400,000 in draining Hatfield Chace, and an additional sum in litigation, died in indigent circumstances.

THORNE-FALCON (*Holy Cross*), a parish, in the union of TAUNTON, hundred of NORTH CURRY, Western Division of the county of SOMERSET, $3\frac{1}{2}$ miles (E. by S.) from Taunton; containing 273 inhabitants. The living is a rectory, valued in the king's books at £14. 10.; patrons, E. and J. Batten, Esqrs. The tithes have been commuted for a rent-charge of £140, subject to the payment of rates, which on the average have amounted to £26; the glebe comprises 73 acres, valued at £126 per annum.

THORNER (*St. Peter*), a parish, in the Lower Division of the wapentake of SKYRACK, West Riding of the county of YORK; containing 1220 inhabitants, of which number, 804 are in the township of Thorner, $6\frac{1}{2}$ miles (S. by W.) from Wetherby. The living is a discharged vicarage, valued in the king's books at £8. 3. 4.; present net income, £143: it is in the patronage of the Crown; impropriator, Earl of Mexborough. There is a place of worship for Wesleyan Methodists. A school was erected by subscription in 1787, and endowed with 14 acres of land from the waste, the annual income arising from which, being £15. 10., is applied to the instruction of six children. In the neighbourhood is a fine spring of water, called St. Sykes's Well.

THORNES, a chapelry, in the parish and union of WAKEFIELD, Lower Division of the wapentake of AGBRIGG, West Riding of the county of YORK, 1 mile (S. by W.) from Wakefield: the population is returned with the chapelry of Alverthorpe. A church of the Roman Doric order, with a cupola, was erected here in 1830, at an expense of £2038. 17. 6., under the act of the 58th of George III. The living is a perpetual curacy; net income, £43; patron, Vicar of Wakefield; impropriator, Heir of the late Sir J. Ramsden, Bart.

THORNEY (*St. Helen*), a parish, in the union, and Northern Division of the wapentake, of NEWARK, Southern Division of the county of NOTTINGHAM, $8\frac{3}{4}$ miles (E. by N.) from Tuxford; containing 308 inhabitants. The living is a discharged vicarage, valued in the king's books at £4. 7. 6.; present net income, £160; patron, C. Neville, Esq. The Fosse-dyke canal borders on the parish.

THORNEY ABBEY (*St. Botolph*) a market-town and parish, in the hundred of WISBEACH, Isle of ELY, county of CAMBRIDGE, 35 miles (N. W.) from Cambridge, and 86 (N.) from London; containing 2055 inhabitants. This place derived its original name of *Ankeridge* from a monastery for hermits, or anchorites, founded here in 662, by Saxulphus, abbot of Peterborough, who became its first prior; the edifice having been destroyed by the Danes, the site lay waste until 972, when Ethelwold, Bishop of Winchester, founded upon it a Benedictine abbey, in honour of the Virgin, which became so opulent that, at the dissolution, its revenue was valued at £508. 12. 5.: of this abbey, which was a mitred one, the only remains are portions of the parish church, a gateway, and some fragments of the old walls. A Literary Society was established in 1823, which possesses a good library. The market, granted in 1638, is on Thursday; and fairs are held on July 1st and Sept. 21st, for horses and cattle, and on Whit-Monday is a pleasure fair. Upwards of 3000 sheep are sent annually from this district to the London market. The petty sessions are held here. The living is a donative, in the patronage of the Duke of Bedford, the impropriator; net income, £220. The church, originally the nave of the conventual church, and built about 1128, is partly in the Norman style of architecture, with portions in the later English: in the churchyard are several tombs of the French refugees, of whom a colony settled here about the middle of the sixth century, having been employed, by the Earl of Bedford, in draining the fens. A school-house was erected by a member of the illustrious house of Russell, and the present Duke of Bedford allows the master a salary of £20 per annum for the instruction of poor children: ten or twelve poor families also are supported in some almshouses by the munificence of his Grace.

THORNEY, WEST, a parish and island, in the union of WEST BOURNE, hundred of BOSHAM, rape of CHICHESTER, Western Division of the county of SUSSEX, 7½ miles (W. by S.) from Chichester; containing 104 inhabitants. This place, called also Thorney Island, is situated nearly in the centre of the great estuary called Chichester harbour, communicating with the small port of Emsworth by a causeway passable at low water for horses and carriages; it is about 5 miles in circumference, and comprises about 1500 acres of arable, meadow, and pasture land; the soil is rich and highly favourable for the production of wheat. About a furlong to the south is Pilsey Island, also within the parish, comprising about 18 acres of good land, and a house has recently been built here; this island has for many years been the resort of almost every species of marine wild fowl which frequents the English Channel. By an act of inclosure, in 1812, about 960 acres of open land were inclosed, of which one-fifth part of arable and one-eighth part of pasture were allotted to the rector in lieu of tithes, besides the glebe, which amounts to about 48 acres. The living is a rectory, valued in the king's books at £10. 8. 4.; present net income, £330; patron, P. Lyne, Esq. The church is an ancient edifice, chiefly in the early English style, with a fine tower of the Norman at the west end; the chancel is separated from the nave by an ancient screen and rood-loft; and among numerous interesting details, contains a Norman font of cylindrical form; the interior was wholly restored in 1839, chiefly at the expense of the Rev. P. Lyne, the rector. On the exterior of the north wall are three large circular arches, now stopped up, evidently an aisle or chantry chapel, supposed to have belonged to an ancient religious house near the church, the remains of which are now incorporated in a farm-house, erected on the spot. Nothing of the history of this establishment is known. Cædmon, a celebrated Saxon poet, was born here in 660. A school is chiefly supported by the rector.

THORNEYBURN, a parish, in the union of BELLINGHAM, North-Western Division of TINDALE ward, Southern Division of the county of NORTHUMBERLAND; containing 334 inhabitants, of which number, 185 are in the township of Thorneyburn, 5 miles (N. W. by W.) from Bellingham. This is one of the five new parishes anciently forming part of the extensive parish of Simonburn: it is a wild and mountainous district, extending from the North Tyne river to Reedsdale, and is bounded on the east by the Tarset bourn. Coal is obtained within the parish. The living is a rectory, valued in the king's books at £4. 5.; present net income, £113; patrons, Governors of Greenwich Hospital, who, in 1818, at the expense of £4000, erected the church, which is a neat structure, situated in a field formerly called Draper Croft.

THORNFORD (*St. Mary Magdalene*), a parish, in the union and hundred of SHERBORNE, Sherborne Division of the county of DORSET, 3½ miles (S. W.) from Sherborne; containing 383 inhabitants. The living is a discharged rectory, valued in the king's books at £6. 17. 3.; present net income, £200; patron, Rev. G. H. Templer. The church was anciently a chapel dependent on Sherborne abbey. Here is a Sunday National school.

THORNGRAFTON, a township, in the parish and union of HALTWHISTLE, Western Division of TINDALE ward, Southern Division of the county of NORTHUMBERLAND, 5½ miles (E. by N.) from Haltwhistle; containing 263 inhabitants. House Steads, in this township, was the site of the Roman station *Borcovicus*, adjacent to which passed the Roman wall. It stands on the brow of a rocky eminence, on the western declivity of which are several terraces, one above another. The area of this fort, on the north side, is level, but on the south it exhibits vast and confused heaps of ruins. In the neighbourhood are foundations of houses and traces of streets, squares, baths, &c., extending over several acres, and to the distance of two miles and a half; and on Chapel hill, a little to the south, are the remains of a temple of the Doric order, among which have been found altars, sepulchral inscriptions, and curiously carved figures in relief.

THORNHAM (*St. Mary*), a parish, in the union of HOLLINGBOURN, hundred of EYHORNE, lathe of AYLESFORD, Western Division of the county of KENT, 4 miles (N. E. by E.) from Maidstone; containing 571 inhabitants. The living is a vicarage, valued in the king's books at £8. 0. 10.; present net income, £392; patron, and incumbent, Rev. J. McMahon Wilder; impropriator, Sir E. Dering, Bart. The church is principally in the decorated style of English architecture. A vicarage-house has recently been built. There is a place of worship for Wesleyan Methodists. A vein of white sand, known by the name of Maidstone sand, though discovered in this parish, is said to have caused the first improvement in the manufacture of glass in this country: it was first worked by experienced Italians, and soon became of infinite importance in the trade: the pits are remarkable for their vast subterranean caverns, which are curiously arched. The ruins of Thornham, or Godard's, castle still exist on the brow of a hill, forming part of the great range of chalk hills; the walls, which are more than thirteen feet high, and three feet thick, enclose an area of a quarter of an acre, including the keep mount. Urns and other vestiges of a Roman station have been found here.

THORNHAM, a township, in the parish of MIDDLETON, union of OLDHAM, hundred of SALFORD, Southern Division of the county palatine of LANCASTER, 3½ miles (S.) from Rochdale; containing 1455 inhabitants.

THORNHAM (*All Saints*), a parish, in the union of DOCKING, hundred of SMITHDON, Western Division of the county of NORFOLK, $6\frac{1}{4}$ miles (W. by N.) from Burnham-Westgate; containing 668 inhabitants. The living is a discharged vicarage, with that of Holme-near-the-Sea annexed, valued in the king's books at £10; present net income, £428; patrons and impropriators, Bishop of Norwich and H. Styleman, Esq. Here is a National school.

THORNHAM-MAGNA (*St. Mary*), a parish, in the union and hundred of HARTISMERE, Western Division of the county of SUFFOLK, $3\frac{1}{4}$ miles (W. S. W.) from Eye; containing 380 inhabitants. The living is a discharged rectory, with which that of Thornham-Parva was consolidated in 1744, valued in the king's books at £7. 11. 3.; present net income, £497; patron, Lord Henniker, whose seat, Major House, is in this parish. A National school has been established. Here was formerly an ancient chapel, dedicated to St. Eadburga, and called St. Arborough's chapel, in which an anchorite resided; it appears to have been standing in the reign of Elizabeth.

THORNHAM-PARVA, a parish, in the union and hundred of HARTISMERE, Western Division of the county of SUFFOLK, $2\frac{3}{4}$ miles (W. by S.) from Eye; containing 206 inhabitants. The living is a discharged rectory, consolidated with that of Thornham-Magna, and valued in the king's books at £4. 14. $4\frac{1}{2}$. The church is a small Norman edifice covered with thatch.

THORNHAUGH (*St. Andrew*), a parish, in the union of STAMFORD, soke of PETERBOROUGH, Northern Division of the county of NORTHAMPTON, $1\frac{1}{4}$ mile (N.) from Wansford; containing 271 inhabitants. The living is a rectory, with that of Wansford annexed, valued in the king's books at £17. 1. 3.; present net income, £483; patron, Duke of Bedford. The church exhibits portions in the various styles of English architecture.

THORNHILL, a hamlet, in the parish of HOPE, union of CHAPEL-EN-LE-FRITH, hundred of HIGH PEAK, Northern Division of the county of DERBY, $6\frac{3}{4}$ miles (N. E. by N.) from Tideswell; containing 135 inhabitants.

THORNHILL, a tything, in the parish of STALBRIDGE, hundred of BROWNSHALL, Sturminster Division of the county of DORSET, 2 miles (S.) from Stalbridge; containing 257 inhabitants.

THORNHILL (*St. Michael*), a parish, in the union of DEWSBURY, Lower Division of the wapentake of AGBRIGG, West Riding of the county of YORK; containing 6271 inhabitants, of which number, 2371 are in the township of Thornhill, 6 miles (W. by S.) from Wakefield. The river Calder runs through the parish, which abounds with the most beautiful and picturesque scenery. In the extensive park sloping to the banks of the river, and ornamented with aged woods, stood the castellated mansion of the Thornhills, which was surrounded by a moat, and garrisoned by Sir George Saville, a descendant of that family, for Charles I., but was taken and destroyed by the parliamentarians. Thornhill, though now only a manufacturing village, was formerly a place of considerable importance, indications of which are still discernible. It had a market and a fair, granted by charter of Edward II., in 1320, now discontinued. The manufacture of woollen cloth, plaids, shawls, and thread, is carried on; there are also brass and iron works, at the latter of which bar and sheet iron, anchor-palms, piston-rods, boiler-plates, and various other articles, are made. The living is a rectory, valued in the king's books at £40. 0. $7\frac{1}{2}$.; present net income, £988; patron, Earl of Scarborough. The church is principally in the early style of English architecture, and contains 190 free sittings, the Incorporated Society having granted £100 in aid of the expense. There are places of worship for Baptists and Wesleyan Methodists. The Rev. Charles Greenwood, in 1642, bequeathed £500 for erecting and endowing a free school; the income is £20 a year, for which all children of the parish who apply receive an English education. There is also a free school, founded and endowed about 1712, by Richard Walker, with a residence for the master, and an annual income of £40, which is applied to the instruction of eighty children of both sexes; and the Sunday school is aided by an annuity of £4. 10., the bequest of the same benefactor.

THORNHOLM, a township, in the parish of BURTON-AGNES, union of BRIDLINGTON, wapentake of DICKERING, East Riding of the county of YORK, $4\frac{3}{4}$ miles (S. W. by W.) from Bridlington; containing 93 inhabitants.

THORNLEY, a township, in the parish of KELLOE, union of EASINGTON, Southern Division of EASINGTON ward, Northern Division of the county palatine of DURHAM, $6\frac{1}{4}$ miles (S. E. by E.) from Durham; containing 50 inhabitants. An extensive colliery is carried on here, of which the produce is shipped at Hartlepool; and limestone is quarried in the neighbourhood. Thornley Hall, now a farm-house, is supposed to occupy the site of a castle, which was strongly fortified in 1140, when Bishop Barbara and his adherents fled hither from William Comyn, who had usurped the see of Durham. A National school has been established.

THORNLEY, a joint township with Wheatley, in the parish of CHIPPING, union of CLITHEROE, Lower Division of the hundred of BLACKBURN, Northern Division of the county palatine of LANCASTER, $7\frac{3}{4}$ miles (W.) from Clitheroe; containing, with Wheatley, 516 inhabitants.

THORNSETT, a hamlet, in the parish of GLOSSOP, union of HAYFIELD, hundred of HIGH PEAK, Northern Division of the county of DERBY, $5\frac{3}{4}$ miles (N. W. by N.) from Chapel-en-le-Frith; containing 685 inhabitants.

THORNTHWAITE, a chapelry, in the parish of CROSTHWAITE, ALLERDALE ward above Derwent, Western Division of the county of CUMBERLAND, 4 miles (W. N. W.) from Keswick; containing 174 inhabitants. The living is a perpetual curacy; net income, £59; patron, Vicar of Crosthwaite. The chapel has been enlarged, and 60 free sittings provided, the Incorporated Society having granted £50 in aid of the expense. The village, in which the manufacture of woollen cloth is carried on, commands most romantic views of Bassenthwaite lake and Skiddaw. There is a smelting mill, though it has not been in operation since the neighbouring lead mine was discontinued.

THORNTHWAITE, a chapelry, in the parish of HAMPSTHWAITE, union of PATELEY-BRIDGE, Lower Division of the wapentake of CLARO, West Riding of the county of YORK, $7\frac{1}{2}$ miles (W. by S.) from Ripley; containing, with the township of Padside, 304 inhabitants. The living is a perpetual curacy; net income, £109; patron, Vicar of Hampsthwaite. Flax-spinning and the manufacture of linen are carried on here. Francis Day, in 1748 and 1757, gave land producing an annual income of £20, for teaching poor children of Thornthwaite, Padside, Menwith Hill, and Darley.

THORNTON (*St. Michael*), a parish, in the union, hundred, and county of BUCKINGHAM, 4½ miles (E. N. E.) from Buckingham; containing 94 inhabitants. The living is a rectory, valued in the king's books at £11. 16. 3.; present net income, £197; patron, Sir T. C. Sheppard, Bart. William Bredon, who died rector of this parish in 1638, was celebrated for his skill in calculating nativities, and had a share in composing Sir Christopher Haydon's Judicial Astrology.

THORNTON (*St. Mary*), a parish, in the union of GREAT BOUGHTON, Second Division of the hundred of EDDISBURY, Southern Division of the county of CHESTER; containing 914 inhabitants, of which number, 181 are in the township of Thornton-in-the-Moors, 6½ miles (N. N. E.) from Chester. The living is a rectory, valued in the king's books at £24. 7. 8½.; present net income, £508; patron, Lord Berwick. The church is partly in the later style of English architecture, with a handsome tower. A National school is endowed with £8. 15 per annum; the school-house was rebuilt about 1790.

THORNTON, a joint hamlet with Aston, in the parish of HOPE, hundred of HIGH PEAK, Northern Division of the county of DERBY, 6½ miles (N. N. E.) from Tideswell; containing 104 inhabitants.

THORNTON, a tything, in the parish of MARNHULL, hundred of REDLANE, Sturminster Division of the county of DORSET, 3½ miles (N. N. E.) from Sturminster-Newton-Castle. It is now almost depopulated, though formerly a distinct parish, and was united to Marnhull at the Reformation, when the church, dedicated in 1464 to *St. Martin*, was desecrated, and is now used as a stable.

THORNTON, a township, in the parish of NORHAM, otherwise NORHAMSHIRE, union of BERWICK-UPON-TWEED, county palatine of DURHAM, though locally to the northward, and for parliamentary purposes included in the Northern Division, of the county of Northumberland, 5¼ miles (S. W.) from Berwick-upon-Tweed; containing 192 inhabitants.

THORNTON, a township, in the parish of POULTON, union of the FYLDE, hundred of AMOUNDERNESS, Northern Division of the county palatine of LANCASTER, 1¾ mile (N. by E.) from Poulton; containing 842 inhabitants. A chapel was built here in 1835, and 190 free sittings provided, the Incorporated Society having granted £250 in aid of the expense. There is a place of worship for Wesleyan Methodists. James Baines, in 1717, bequeathed land, now producing an annual income of £40, for teaching poor children. The schoolroom has been rebuilt by subscription, and there is a house for the master.

THORNTON, a township, in the parish of SEFTON, union and hundred of WEST DERBY, Southern Division of the county palatine of LANCASTER, 6½ miles (N.) from Liverpool; containing 342 inhabitants.

THORNTON (*St. Peter*), a parish, in the union of MARKET-BOSWORTH, hundred of SPARKENHOE, Southern Division of the county of LEICESTER, 6 miles (N. E.) from Market-Bosworth; containing 1078 inhabitants. The living is a discharged vicarage, valued in the king's books at £6. 10. 2.; present net income, £202; patron, Viscount Maynard, who, with the Duke of Rutland, is the impropriator. There are chapels of ease at Bagworth and Stanton-under-Bardon, in this parish. Here is a National school.

THORNTON (*St. Wilfrid*), a parish, in the union of HORNCASTLE, Southern Division of the wapentake of GARTREE, parts of LINDSEY, county of LINCOLN, 1¾ mile (S. W.) from Horncastle; containing 216 inhabitants. The living is a discharged vicarage, valued in the king's books at £5. 12. 1.; present net income, £160; patrons and appropriators, Dean and Chapter of Lichfield.

THORNTON, a parish, in the union of POCKLINGTON, partly in the Wilton-Beacon, but chiefly in the Holme-Beacon, Division of the wapentake of HARTHILL, East Riding of the county of YORK, 4¼ miles (S. W.) from Pocklington; containing 791 inhabitants. The living is a discharged vicarage, with Allerthorpe annexed, valued in the king's books at £7. 5. 10.; present net income, £210; patron and appropriator, Dean of York.

THORNTON, a joint township with Baxly, in the parish of COXWOLD, union of EASINGWOULD, wapentake of BIRDFORTH, North Riding of the county of YORK, 3 miles (N. by E.) from Easingwould; containing, with Baxly, 67 inhabitants.

THORNTON, a chapelry, in the parish and union of BRADFORD, wapentake of MORLEY, West Riding of the county of YORK, 4½ miles (W.) from Bradford; containing 5968 inhabitants. The living is a perpetual curacy; net income, £155; patron, Vicar of Bradford. The chapel, dedicated to St. James, is principally in the later style of English architecture. There are quarries of freestone in the neighbourhood, and the manufacture of worsted is carried on to a considerable extent. A school, erected by subscription, is endowed with £45 per annum, arising from the produce of divers benefactions, the principal of which are those of George Ellis and Samuel Sunderland: there is also a National school.

THORNTON (*St. Mary*), a parish, in the union of SKIPTON, Eastern Division of the wapentake of STAINCLIFFE and EWCROSS, West Riding of the county of YORK, 5¾ miles (W. S. W.) from Skipton; containing 2246 inhabitants. The living is a rectory, valued in the king's books at £19. 5. 2½.; present net income, £248; patron, Sir Lister Kaye, Bart. A chapel has been built here, containing 260 sittings, 200 of which are free, the Incorporated Society having granted £150 in aid of the expense. Here was formerly a market on Thursday, granted to Walter de Muncey, in the 28th of Edward I., with a fair, for five days, commencing on the eve of the festival of St. Thomas the Martyr. At a short distance from the village is a huge rocky cliff, called Thornton Scar, partly clothed with wood, and rising to the height of 300 feet. Thornton Force is a beautiful cataract, rushing from an aperture in a precipitous rock, and having a fall of 90 feet in one sheet of water 16 feet wide. A free school was erected in 1633, at an expense of £100, the bequest of Robert Windle, who also endowed it with a rent-charge of £20 for a master, who has also a dwelling and garden attached to the schoolhouse.

THORNTON, BISHOP, a chapelry, in the parish and liberty of RIPON, West Riding of the county of

YORK, 3½ miles (N. N. W.) from Ripley; containing 614 inhabitants. The living is a perpetual curacy; net income, £95; patrons and appropriators, Dean and Chapter of Ripon. The chapel has been rebuilt, and 200 free sittings provided, the Incorporated Society having granted £200 in aid of the expense. Here is a Roman Catholic chapel; also a National school.

THORNTON-BRIDGE, a township, in the parish of BRAFFERTON, wapentake of HALLIKELD, North Riding of the county of YORK, 4¼ miles (N. E. by N.) from Boroughbridge; containing 47 inhabitants.

THORNTON-CHILDER, a township, in the parish of EASTHAM, union, and Higher Division of the hundred, of WIRRALL, Southern Division of the county of CHESTER, 5 miles (E.) from Great Neston; containing 296 inhabitants.

THORNTON-CURTIS (*St. Lawrence*), a parish, in the union of GLANFORD-BRIDGE, Northern Division of the wapentake of YARBOROUGH, parts of LINDSEY, county of LINCOLN, 5 miles (S. E. by E.) from Barton-upon-Humber; containing 362 inhabitants. The living is a discharged vicarage, valued in the king's books at £5. 18. 4.; present net income, £123; patron and impropriator, C. Winn, Esq. There is a school in which ten children are paid for by the parish. A priory of Black canons, in honour of the Blessed Virgin, was founded here in 1139, by William le Gros, Earl of Albemarle and Lord of Holderness, which at the dissolution had a revenue of £730. 17. 2. Henry VIII. applied the greater part of its possessions to the erection of a sumptuous college, in honour of the Holy and Undivided Trinity, for a dean and nineteen prebendaries, which was dissolved in the 1st of Edward VI., when its site was granted to the Bishop of Lincoln. It occupied an extensive square area, encompassed by a deep fosse and strong ramparts, within which an avenue of large ash trees led to the church, the ruins of which, particularly the chapter-house, are very fine. There are some remains of the gate-house, approached by a bridge flanked with embattled walls, and arches with loop-holes, supporting two round towers. Various other portions of these once magnificent buildings are still standing, and exhibit good specimens of the decorated and later styles of English architecture. Opposite the entrance are four small mounds, called Butts, supposed to be tumuli.

THORNTON-DALE (*All Saints*), a parish, in PICKERING lythe and union, North Riding of the county of YORK; containing 1368 inhabitants, of which number, 937 are in the township of Thornton-Dale, 2½ miles (E. by S.) from Pickering. The living is a rectory, valued in the king's books at £20; present net income, £396; patron, R. Hill, Esq. There is a place of worship for Wesleyan Methodists. A school was endowed in 1657, by Viscountess Lumley, who also erected twelve alms-houses for as many poor people: two other schools are partly supported by private charity.

THORNTON, EAST, a township, in the parish of HARTBURN, union, and Western Division of the ward, of MORPETH, Northern Division of the county of NORTHUMBERLAND, 6 miles (W.) from Morpeth; containing 59 inhabitants.

THORNTON-IN-LONSDALE (*St. Oswald*), a parish, in the union of SETTLE, Western Division of the wapentake of STAINCLIFFE and EWCROSS, West Riding of the county of YORK; containing 1152 inhabitants, of which number, 441 are in the township of Thornton, 11½ miles (N. W.) from Settle. The living is a vicarage, valued in the king's books at £28. 13. 1½.; present net income, £99; patrons and appropriators, Dean and Chapter of Worcester. The manufacture of cotton is carried on here to a limited extent. One of the sources of the river Greta, which bounds the parish to the south and south-east, is in the valley of Kingsdale, the stream flowing through most romantic scenery, and forming several beautiful cascades in its descent. The celebrated Yorda's cave is at the northern extremity of the vale. Ralph Redmayne, Esq., in 1702, founded a free school, and endowed it with £200, which, having been vested in land, produces annually £55, for teaching all the poor children of the parish, in a school-room built by subscription.

THORNTON-LE-BEANS, a township, in the parish of NORTH OTTERINGTON, union of NORTH-ALLERTON, wapentake of ALLERTONSHIRE, North Riding of the county of YORK, 3½ miles (S. E.) from North-Allerton; containing 219 inhabitants. There is a place of worship for Wesleyan Methodists; also a school endowed with £4 per annum, for which six children are instructed.

THORNTON-LE-FEN, a township, in the union of BOSTON, soke of HORNCASTLE, parts of LINDSEY, county of LINCOLN, 8½ miles (N. W.) from Boston; containing 156 inhabitants. Here is a chapel, the living of which is a perpetual curacy; net income, £84; patrons, certain Trustees.

THORNTON-LE-MOOR (*All Saints*), a parish, in the union of CAISTOR, Northern Division of the wapentake of WALSHCROFT, parts of LINDSEY, county of LINCOLN, 5½ miles (S. W. by W.) from Caistor; containing 99 inhabitants. The living is a discharged rectory, valued in the king's books at £9. 10. 10.; present net income, £319; patron, Bishop of Ely.

THORNTON-LE-MOOR, a township, in the parish of NORTH OTTERINGTON, union of THIRSK, wapentake of BIRDFORTH, North Riding of the county of YORK, 5 miles (N. W. by N.) from Thirsk; containing 337 inhabitants. Here was formerly a chapel of ease, the remains of which have been converted into a school-room: the school is endowed with £14 per annum, for which 16 boys are taught, and some additional poor children are taught from the proceeds of a donation of £25 by John Haroksly Ackerley, of Bath, in 1830.

THORNTON-LE-STREET (*St. Leonard*), a parish, in the union of THIRSK, wapentake of ALLERTONSHIRE, North Riding of the county of YORK; containing 226 inhabitants, of which number, 162 are in the township of Thornton-le-Street, 3 miles (N. N. W.) from Thirsk. The living is a discharged vicarage, valued in the king's books at £4; present net income, £60; patrons and appropriators, Dean and Canons of Christ-Church, Oxford. A school is partly supported by subscription.

THORNTON-MAYOW, a township, in the parish of NESTON, union, and Higher Division of the hundred, of WIRRALL, Southern Division of the county of CHESTER, 2½ miles (N. N. E.) from Great Neston; containing 144 inhabitants.

THORNTON-RUST, a township, in the parish of AYSGARTH, wapentake of HANG-WEST, North Riding of the county of YORK, 10 miles (W.) from Middleham;

containing 158 inhabitants. A school was established and endowed in conjunction with a Calvinistic chapel, in 1827.

THORNTON-STEWARD (*St. Oswald*), a parish, in the union of Leybourn, wapentake of Hang-West, North Riding of the county of York, $3\frac{1}{2}$ miles (E. by S.) from Middleham; containing 310 inhabitants. The living is a discharged vicarage, valued in the king's books at £6. 13. $11\frac{1}{2}$.; present net income, £234; patron and appropriator, Bishop of Chester. The church is principally in the Norman style. A neat school-house was erected in 1815, at the expense of George Horn, Esq.: it has an endowment of £10 per annum for the education of the poor children of the parish.

THORNTON-UPON-CLAY, a township, in the parish of Foston, union of Malton, wapentake of Bulmer, North Riding of the county of York, 11 miles (N. N. E.) from York; containing 205 inhabitants.

THORNTON-WATLASS (*St. Mary*), a parish, in the union of Bedale, wapentake of Hang-East, North Riding of the county of York; containing 448 inhabitants, of which number, 185 are in the township of Thornton-Watlass, $2\frac{3}{4}$ miles (S. W.) from Bedale. The living is a rectory, valued in the king's books at £6. 10. 10.; present net income, £475; patron, M. Millbank, Esq. Eight poor children are instructed for £4 per annum, the gift of an ancestor of the Rev. Frederic Dodsworth, D.D., in 1756.

THORNTON, WEST, a township, in the parish of Hartburn, union, and Western Division of the ward, of Morpeth, Northern Division of the county of Northumberland, $7\frac{1}{4}$ miles (W. by N.) from Morpeth; containing 53 inhabitants. This is the supposed site of a Roman camp, and it is recorded that, at the commencement of the eighteenth century, vestiges of a town, intersected by a military way, were discernible.

THORNVILLE, a township, in the parish of Whixley, Lower Division of the wapentake of Claro, West Riding of the county of York, $5\frac{1}{4}$ miles (S. by E.) from Boroughbridge; containing 17 inhabitants.

THORNWOOD, a hamlet, in the parish of North Weald, hundred of Harlow, Southern Division of the county of Essex, $2\frac{1}{2}$ miles (N. N. E.) from Epping: the population is returned with the parish.

THOROTON (*St. Elena*), a parish, in the union, and Northern Division of the wapentake, of Bingham, Southern Division of the county of Nottingham, 8 miles (S. S. W.) from Newark; containing 143 inhabitants. The living is annexed, with that of Scarrington, to the vicarage of Orston. The church is a small edifice.

THORP, a township, in the parish of Rothwell, union of Wakefield, Lower Division of the wapentake of Agbrigg, West Riding of the county of York, $4\frac{1}{2}$ miles (N. by W.) from Wakefield; containing 62 inhabitants.

THORP-ACRE, a parish, in the union of Loughborough, hundred of West Goscote, Northern Division of the county of Leicester, $1\frac{1}{4}$ mile (W. N. W.) from Loughborough; containing 366 inhabitants. The living is a donative curacy, united to that of Dishley. A school of industry for girls is partly supported by an allowance from the trustees of charity land.

THORP-ARCH (*All Saints*), a parish, in the Ainsty of the city, and East Riding of the county, of York, $2\frac{3}{4}$ miles (S. E. by E.) from Wetherby; containing 316 inhabitants. The living is a discharged vicarage, endowed with the rectorial tithes, and valued in the king's books at £3. 15. 5.; present net income, £356: it is in the patronage of Mrs. Wheeler. The church is a handsome structure. There is a place of worship for Wesleyan Methodists. A mineral spring, sulphureous and chalybeate, and in high repute, was discovered here, in 1744, the water of which, in some disorders, has a decided superiority over that of Harrogate. The rapid river Wharf separates the village from that of Boston, and forms a picturesque cascade, as viewed through the arches of the bridge by which it is crossed. Here are three good inns, and a number of private lodging-houses, erected for the accommodation of visiters, of whom many have derived considerable benefit from the spa. A school was founded in 1738, by Lady Elizabeth Hastings, who endowed it with £15 a year; but from other sources the annual income has been increased to £40.

THORP-AUDLING, a township, in the parish of Badsworth, Upper Division of the hundred of Osgoldcross, West Riding of the county of York, $4\frac{1}{2}$ miles (S. S. E.) from Pontefract; containing 355 inhabitants.

THORP-BASSETT (*All Saints*), a parish, in the union of Malton, wapentake of Buckrose, East Riding of the county of York, 5 miles (E. by N.) from New Malton; containing 206 inhabitants. The living is a rectory, valued in the king's books at £12; present net income, £328; patron, Earl Fitzwilliam. Ten boys are instructed for the dividends arising from £200, the gift of the Rev. James Graves, in 1804.

THORP-STAPLETON, a township, in the parish of Whitkirk, Lower Division of the wapentake of Skyrack, West Riding of the county of York, $3\frac{3}{4}$ miles (S. E.) from Leeds; containing 19 inhabitants.

THORP-SUB-MONTEM, a joint township with Burnsall, in the parish of Burnsall, union of Skipton, Eastern Division of the wapentake of Staincliffe and Ewcross, West Riding of the county of York, $8\frac{1}{2}$ miles (N. by E.) from Skipton; containing 242 inhabitants.

THORP-UNDERWOODS, a township, in the parish of Little Ouseburn, Lower Division of the wapentake of Claro, West Riding of the county of York, $6\frac{1}{2}$ miles (S. E.) from Aldborough; containing 144 inhabitants.

THORPE (*St. Leonard*), a parish, in the hundred of Wirksworth, Southern Division of the county of Derby, $3\frac{1}{4}$ miles (N. W. by N.) from Ashbourn; containing 189 inhabitants. The living is a discharged rectory, valued in the king's books at £6. 1. 6.; present net income, £129; patron, Dean of Lincoln. The church is partly in the Norman style of architecture. A day and Sunday school is partly supported by the clergyman. To the northward of the village is a remarkable conical hill of limestone, called Thorpe Cloud; its altitude is 300 feet above the bed of the river Dove, which flows at its base.

THORPE (*St. Peter*), a parish, in the union of Spilsby, Eastern Division of the soke of Bolingbroke, parts of Lindsey, county of Lincoln, $1\frac{1}{2}$ mile (N. W.) from Wainfleet; containing 498 inhabitants. The living is a discharged vicarage, valued in the king's books at

£20. 19. 4.; present net income, £313; patron and impropriator, W. Hopkinson, Esq.

THORPE, a hamlet, in the parish of West Ashby, union of Horncastle, Southern Division of the wapentake of Gartree, parts of Lindsey, county of Lincoln; containing 284 inhabitants.

THORPE (*St. Lawrence*), a parish, in the union of Southwell, Southern Division of the wapentake of Newark and of the county of Nottingham, 3¼ miles (S. W.) from Newark; containing 105 inhabitants. The living is a rectory, valued in the king's books at £8; present net income, £280: it is in the patronage of the Crown. The church exhibits portions in the several styles of English architecture. The Trent bounds the parish on the north-west, and the old Fosse road passes between that river and the village. A fine tesselated pavement, coins, and many other Roman relics, have been discovered.

THORPE, a parish, in the union of Loughborough, Southern Division of the wapentake of Rushcliffe and of the county of Nottingham, 6¾ miles (N. E.) from Loughborough; containing 39 inhabitants. The living is a discharged rectory, valued in the king's books at £12. 9. 4½.; patron, Lord Rancliffe. The church has been destroyed.

THORPE, a hamlet, in the parish of Aldringham, union and hundred of Blything, Eastern Division of the county of Suffolk, 3¼ miles (N. by E.) from Aldborough: the population is returned with the parish. Here was formerly a chapel, dedicated to St. Mary, the ruins of which have been recently removed.

THORPE, a hamlet, in the parish of Ashfield, union of Bosmere and Claydon, hundred of Thredling, Eastern Division of the county of Suffolk, 2 miles (E. S. E.) from Debenham: the population is returned with the parish. The chapel, dedicated to St. Peter, was rebuilt by George Pitt, Esq., in 1739, but the tower, which is circular, must have belonged to a more ancient edifice; the chapelyard is extensive, but no interments have taken place here, the churchyard of Ashfield having been solely used for that purpose.

THORPE (*St. Mary*), a parish, in the union of Windsor, Second Division of the hundred of Godley, Western Division of the county of Surrey, 2 miles (S. W. by S.) from Staines; containing 471 inhabitants. This parish is situated on the river Thames, and between the lines of the Great Western and London and South-Western railways, from each of which it is about five miles distant. The living is a vicarage, valued in the king's books at £5. 13. 4.; present net income, £141: it is in the patronage of the Crown; impropriator, Rev. Henry Leigh Bennett. The church is a very old edifice, with a tower of brick covered with ivy, and contains some ancient monuments, one with an inscription in brass. There is a licensed room for Baptists. Parochial and Sunday schools are supported by subscription.

THORPE, a township, in the parish and union of Howden, wapentake of Howdenshire, East Riding of the county of York, 1¼ mile (N. by E.) from Howden; containing 44 inhabitants.

THORPE, a township, in the parish of Wycliffe, union of Teesdale, wapentake of Gilling-West, North Riding of the county of York, 1½ mile (N. E. by E.) from Greta-Bridge: the population is returned with the parish. This township is situated on the banks of the Tees.

THORPE, a joint township with Whitcliff, in the parish and liberty of Ripon, West Riding of the county of York, 1½ mile (S. by E.) from Ripon: the population is returned with Whitcliff.

THORPE-ABBOTS (*All Saints*), a parish, in the union of Depwade, hundred of Earsham, Eastern Division of the county of Norfolk, 2¼ miles (E.) from Scole; containing 272 inhabitants. The living is a discharged rectory, valued in the king's books at £6; present net income, £305; patron, G. Read, Esq.

THORPE-ACHURCH (*St. John the Baptist*), a parish, in the union of Oundle, hundred of Navisford, Northern Division of the county of Northampton, 4¼ miles (N. N. E.) from Thrapstone; containing 240 inhabitants. The living is a rectory, with the vicarage of Lilford annexed, valued in the king's books at £14 16. 3.; present net income, £420; patron, Lord Lilford, the impropriator of Lilford. An allotment of common land, let for £13. 10., was awarded for parochial purposes under an enclosure act.

THORPE-ARNOLD (*St. Mary*), a parish, in the union of Melton-Mowbray, hundred of Framland, Northern Division of the county of Leicester, 1½ mile (E. N. E.) from Melton-Mowbray; containing 117 inhabitants. This parish contains 1740 acres, of which 23 are glebe land; and is intersected by the Melton-Mowbray and Oakham canal. The living is a vicarage, with the chapelry of Brentingby annexed, valued in the king's books at £6. 17. 8½., patron, Duke of Rutland.

THORPE, BISHOP'S (*St. Andrew*), a parish, in the union of Blofield, partly in the city of Norwich, but chiefly in the hundred of Blofield, Eastern Division of the county of Norfolk, 2¾ miles (E. by S.) from Norwich; containing 2151 inhabitants. The living is a rectory, valued in the king's books at £8; present net income, £420; patron, W. L. W. Chute, Esq. The village is delightfully situated on the southern declivity of a hill, at the base of which flows the navigable river Wensum: the vicinity is ornamented with rich plantations, finely interspersed with the handsome villas of many of the opulent citizens of Norwich. The county lunatic asylum (described in the article on Norwich) is in this parish.

THORPE-BRANTINGHAM, a township, in the parish of Brantingham, Hunsley-Beacon Division of the wapentake of Harthill, East Riding of the county of York, 3 miles (S. S. W.) from North Cave; containing 66 inhabitants.

THORPE-BULMER, a township, in the parish of Monk-Hesleton, union of Stockton, Southern Division of Easington ward, Northern Division of the county palatine of Durham, 12½ miles (N. by E.) from Stockton-upon-Tees; containing 28 inhabitants.

THORPE-BY-IXWORTH (*All Saints*), a parish, in the union of Thingoe, hundred of Blackbourn, Western Division of the county of Suffolk, ½ mile (N. W. by N.) from Ixworth; containing 128 inhabitants. The manor was part of the endowment of Ixworth priory, and at the dissolution was granted, with the Priory, to Richard and Elizabeth Codyngton. The living is a donative; net income, £21; patron and impropriator, Sir C. M. Lamb, Bart.

THORPE-BY-WATER, a hamlet, in the parish of SEATON, union of UPPINGHAM, hundred of WRANDIKE, county of RUTLAND, 5 miles (S. E. by S.) from Uppingham; containing 89 inhabitants.

THORPE-CONSTANTINE (*St. Constantine*), a parish, in the union of TAMWORTH, Northern Division of the hundred of OFFLOW and of the county of STAFFORD, 5 miles (N. E.) from Tamworth; containing 49 inhabitants. The living is a rectory, valued in the king's books at £5. 5. 5.; present net income, £389; patron, William Phillips Inge, Esq.

THORPE, EAST, a hamlet, in the parish of LONDESBOROUGH, Holme-Beacon Division of the wapentake of HARTHILL, East Riding of the county of YORK, 2¾ miles (N.) from Market-Weighton: the population is returned with the parish.

THORPE-IN-BALNE, a township, in the parish of BARNBY-UPON-DON, union of DONCASTER, Southern Division of the wapentake of STRAFFORTH and TICKHILL, West Riding of the county of YORK, 6½ miles (N. N. E.) from Doncaster; containing 121 inhabitants. A rent-charge of £5, the gift of James Fretwell, in 1751, and sundry smaller bequests, are applied to the instruction of poor children.

THORPE-IN-THE-STREET, a township, in the parish of NUN-BURNHOLME, union of POCKLINGTON, Holme-Beacon Division of the wapentake of HARTHILL, East Riding of the county of YORK, 2½ miles (N. W. by W.) from Market-Weighton; containing 31 inhabitants.

THORPE-LE-SOKEN (*St. Mary*), a parish, in the union and hundred of TENDRING, Northern Division of the county of ESSEX, 12 miles (E. S. E.) from Colchester; containing 1173 inhabitants. The three parishes of Thorpe-le-Soken, Kirby-le-Soken, and Walton-le-Soken, form a manor, termed "the liberty of the Soken," having within its limits two or three reputed manors of smaller extent. It was given to the church of St. Paul, London, by King Athelstan, before 941, and belonged to the canons of St. Paul's at the time of the Norman survey. The Dean and Chapter held the manor, with the three advowsons, as their peculiars, until deprived of them by Henry VIII.; and Mary, by letters patent dated March 2nd, 1554, placed them under the visitation of the Bishop of London. Edward VI. granted the manors and advowsons, with all their peculiar privileges, to Sir Thomas D'Arcy, vice-chamberlain of his household, on his advancement to the barony, from whose descendants, the Lords D'Arcy, they passed to the Barons Rivers, thence to the Earls of Rochford, who sold them to Richard Rigby, Esq., and they have since had various owners, the advowsons being now separate from the manors. The lord of the manor, who styles himself "lord of the liberty, franchise, dominion, and peculiar jurisdiction of the Soken, in the county of Essex," appoints a commissary, by the title of "Official Principal, and Vicar-general in spiritual causes," who holds a court in Thorpe church annually, or at other times as often as occasion requires, and proves wills, and grants marriage licences, &c.: the wills and records are deposited at the residence of the registrar of the court, who is also steward of the manor, at Harwich. The lord holds his court annually on St. Anne's day, at Kirby: he also appoints a coroner, and other officers in the liberty, which has the privilege, though not exercised, that no bailiff, except his own, can arrest within its limits. This parish is 17 miles in circumference; the soil is various, but in general fertile, producing fair average crops, and the lands are in good cultivation. A creek, or arm of the sea, runs up to Landermere, a small hamlet in the parish, where there is a convenient wharf, at which vessels are laden with corn for the London market, and discharge their cargoes of coal, manure, &c. A small customary market is held on Wednesday evenings, and there are fairs on the Monday before Whitsuntide and September 29th. The petty sessions for the division are held here on Mondays, alternately with Mistley. This has been made a polling-place for the northern division of the county. The living is a discharged vicarage, consolidated with that of Walton-le-Soken and the vicarage of Kirby-le-Soken, (which are held as one benefice,) and valued in the king's books at £16; impropriator, John Martin Leake, Esq., of Thorpe Hall. The church, an ancient structure, was entirely repewed in 1827, when a gallery was erected, and 240 additional sittings provided, 150 of which are free, the Incorporated Society having granted £100 in aid of the expense. An action was afterwards commenced against the commissary for removing a pew, for which a faculty was claimed as granted by his predecessor, and the matter, on trial, being left to arbitration, it was awarded that the pew should remain in its altered state. The chancel contains several monuments to the family of Leake, and one to a member of the Wharton family; and there was formerly in the south aisle a sculptured figure of a warrior, with a lion at his feet, now preserved in the vestry-room. There is a place of worship for Baptists; and adjoining the churchyard is a good school-room, at present used for a Sunday school. Here is a house called the Abbey, with some land attached, but no evidence of its having ever been a religious house has been discovered. A number of French refugees formerly settled and had a chapel here, of which there are now no remains.

THORPE, LITTLE (*St. Mary*), a parish, in the union of DEPWADE, hundred of DISS, Eastern Division of the county of NORFOLK, ½ a mile (E.) from Scole: the population is returned with the parish of Scole. The living is a rectory, annexed to that of Billingford, and valued in the king's books at £4. The church is now in ruins.

THORPE-LUBENHAM, a township, partly in the parish of LUBENHAM, hundred of GARTREE, Southern Division of the county of LEICESTER, but chiefly in the parish of MARSTON-TRUSSEL, hundred of ROTHWELL, Northern Division of the county of NORTHAMPTON, 2 miles (W. by S.) from Market-Harborough; containing 3 inhabitants.

THORPE-MALSOR (*All Saints*), a parish, in the union of KETTERING, hundred of ROTHWELL, Northern Division of the county of NORTHAMPTON, 2 miles (W. N. W.) from Kettering; containing 297 inhabitants. The living is a rectory, valued in the king's books at £11. 14. 2.; present net income, £255; patron, P. Marmsell, Esq. A day and Sunday school is partly supported by subscription. Robert Talbot, an early English antiquary, was born here, about the close of the fifteenth century.

THORPE-MANDEVILLE (*St. John the Baptist*), a parish, in the union of BRACKLEY, hundred of KING'S-SUTTON, Southern Division of the county of NORTHAMP-

THO

TON, 6 miles (N. E. by E.) from Banbury; containing 175 inhabitants. The living is a rectory, valued in the king's books at £10. 2. 11.; present net income, £281; patron, R. P. Humfrey, Esq.

THORPE, MARKET (*St. Margaret*), a parish, in the union of Erpingham, hundred of North Erpingham, Eastern Division of the county of Norfolk, $4\frac{3}{4}$ miles (N. W. by N.) from North Walsham; containing 254 inhabitants. The living is a discharged vicarage, with the donative mediety of Bradfield annexed, valued in the king's books at £5. 11. 3.; patron and impropriator, Lord Suffield. The impropriate tithes have been commuted for a rent-charge of £155. 10., and the vicarial for £75, subject to the payment of rates; the glebe comprises 33 acres, valued at £33 per annum. The church, rebuilt a few years ago, at the expense of Lord Suffield, is an elegant structure of flint and freestone, having at each angle a turret, and each side being terminated by a gable, with a stone cross: the windows are adorned with modern stained glass; the chancel is separated from the nave by a light oaken screen, and at the west end is another of similar workmanship.

THORPE-MORIEUX (*St. Mary*), a parish, in the union and hundred of Cosford, Western Division of the county of Suffolk, $4\frac{1}{4}$ miles (N. W.) from Bildeston; containing 412 inhabitants. The living is a rectory, valued in the king's books at £18. 14. $4\frac{1}{2}$.; present net income, £500; patron, Rev. Thomas Harrison.

THORPE-NEXT-HADDISCOE (*St. Matthias*), a parish, in the union of Loddon and Clavering, hundred of Clavering, Eastern Division of the county of Norfolk, $5\frac{3}{4}$ miles (N. by E.) from Beccles; containing 79 inhabitants. The living is a discharged rectory, valued in the king's books at £3. 6. 8.; present net income, £175; it is in the patronage of the Crown and Lord Calthorpe.

THORPE-ON-THE-HILL (*All Saints*), a parish, in the Lower Division of the wapentake of Boothby-Graffo, parts of Kesteven, union and county of Lincoln, 6 miles (S. W.) from Lincoln; containing 273 inhabitants. The living is a rectory, valued in the king's books at £9. 10.; present net income, £247; patrons, Dean and Chapter of Lincoln. There is a place of worship for Wesleyan Methodists.

THORPE-SALVIN (*St. Peter*), a parish, in the union of Worksop, Southern Division of the wapentake of Strafforth and Tickhill, West Riding of the county of York, $5\frac{3}{4}$ miles (W. by N.) from Worksop; containing 233 inhabitants. The living is a perpetual curacy; net income, £63; patron and appropriator, the Prebendary in York Cathedral. The church has a remarkably fine Norman arch, and a font of the same character. There is a place of worship for Wesleyan Methodists; also a school in which 10 children are instructed at the expense of the Duke of Leeds, who allows £5. 5. per annum.

THORPE-SATCHVILLE, a chapelry, in the parish of Twyford, union of Melton-Mowbray, hundred of East Goscote, Northern Division of the county of Leicester, $5\frac{1}{2}$ miles (S. by W.) from Melton-Mowbray; containing 163 inhabitants. The chapel is dedicated to St. Michael.

THORPE-TINLEY, a township, in the parish of Timberland, union of Sleaford, First Division of the wapentake of Langoe, parts of Kesteven, county of Lincoln, $10\frac{1}{2}$ miles (N. N. E.) from Sleaford; containing 127 inhabitants.

THORPE-UNDERWOOD, a hamlet, in the parish and hundred of Rothwell, Northern Division of the county of Northampton, $1\frac{1}{4}$ mile (W.) from Rothwell; containing 18 inhabitants.

THORPE, WEST, a parish, in the wapentake of Lawress, parts of Lindsey, union and county of Lincoln, $7\frac{1}{2}$ miles (N. W. by N.) from Lincoln; containing 69 inhabitants. The living is a discharged vicarage, annexed to the rectory of Aisthorpe, and valued in the king's books at £5. 7. 6.

THORPE-WILLOUGHBY, a township, in the parish of Brayton, union of Selby, Lower Division of the wapentake of Barkstone-Ash, West Riding of the county of York, $2\frac{1}{2}$ miles (W. S. W.) from Selby; containing 148 inhabitants.

THORPLAND, a hamlet, in the parish of Wallington, union of Downham, hundred of Clackclose, Western Division of the county of Norfolk, $3\frac{3}{4}$ miles (N.) from Downham-Market: the population is returned with the parish. Here was formerly a chapel, dedicated to St. Thomas.

THORPLAND, a hamlet, in the parish of Fakenham-Lancaster, hundred of Gallow, Western Division of the county of Norfolk, 2 miles (N.) from Fakenham: the population is returned with the parish. Here was formerly a chapel, long since destroyed.

THORRINGTON (*St. Mary Magdalene*), a parish, in the union and hundred of Tendring, Northern Division of the county of Essex, $8\frac{1}{4}$ miles (S. E. by E.) from Colchester; containing 431 inhabitants. This parish is about seven miles in circumference: the situation is low, and the soil light and much intermixed with sand. The living is a rectory, united to that of Frating, and valued in the king's books at £16. The church is an ancient edifice with a tower of flint and stone, and consists of a nave and chancel, with a north aisle extending the whole length of the building. There is a place of worship for Wesleyan Methodists; also a National school.

THORROCK, a hamlet, in the parish of Gainsborough, parts of Lindsey, county of Lincoln: the population is returned with the parish. Thorrock Hall, the seat of Henry Bacon Hickman, Esq., lord of the manor of Gainsborough, is pleasantly situated in a fine park in this hamlet. Here are the remains of an ancient fortification, called the Danish camp; it is supposed to have been occupied by the army of Sweyn, King of Denmark, who made from this place his incursions into Oxfordshire, on his return from which he was killed by an unknown assassin in the midst of his festivities.

THORVERTON (*St. Thomas à Becket*), a parish, in the union of Tiverton, hundred of Hayridge, Cullompton and Northern Divisions of the county of Devon, 7 miles (W. S. W.) from Cullompton; containing 1455 inhabitants. The living is a vicarage, valued in the king's books at £18. 12. $8\frac{1}{2}$.; present net income, £507; patrons and appropriators, Dean and Chapter of Exeter. There is a place of worship for Presbyterians. A school is endowed with £18 per annum, and a Lancasterian school is supported by subscription. The parish is bounded on the west by the river Exe. At East Raddon there was formerly a chapel, dedicated to St. John the Baptist, the remains of which have been converted into a dwelling-house, called "No Man's chapel."

THOYDON-BOIS (*St. Mary*), a parish, in the union of Epping, hundred of Ongar, Southern Division of the county of Essex, 3½ miles (S.) from Epping; containing 676 inhabitants. This parish, which is the least extensive of the three parishes of that name, and is partly included within the limits of Epping Forest, takes its distinguishing epithet from the abundance of woodlands within its boundaries. The living is a perpetual curacy; net income, £68; patron, R. W. H. Dare, Esq. The church is a small edifice with a belfrey tower of wood, surmounted with a shingled spire.

THOYDON-GARNON, or COOPER-SAIL (*All Saints*), a parish, in the union of Epping, hundred of Ongar, Southern Division of the county of Essex, 2½ miles (S. S. E.) from Epping; containing 841 inhabitants. This parish takes the adjunct to its name from the family of Gernon, who were anciently its proprietors. The living is a rectory, valued in the king's books at £17; patron, J. R. H. Abdy, Esq. The tithes have been commuted for a rent-charge of £634. 6., subject to the payment of rates, which on the average have amounted to £108. 18.; the glebe comprises 64 acres, valued at £39 per annum. The church is an ancient edifice with a massive square tower, and contains some interesting monuments. On the steeple is an inscription commemorating the bounty of Sir John Crosbie, the founder of Crosbie Hall, London, who contributed largely towards its erection. Here is a National school. Lady Ann Sydney Fitzwilliam, in 1602, bequeathed a small rent-charge towards the foundation of an almshouse for four poor widows. Baron Dimsdale, the celebrated inoculator for the small-pox, was born here.

THOYDON, MOUNT (*St. Michael*), a parish, in the union and hundred of Ongar, Southern Division of the county of Essex, 4 miles (S. E.) from Epping; containing 249 inhabitants. This parish derives the adjunct to its name from its situation in the most elevated portion of the ancient district of Thoydon. The living is a rectory, annexed to that of Tawney-Stapleford, and valued in the king's books at £13. 6. 8. The tithes have been commuted for a rent-charge of £301, subject to the payment of rates, which on the average have amounted to £43; the glebe comprises 19½ acres, valued at £33. 10. 6. per annum. The church, which is a handsome edifice, contains many fine monuments to the family of Smyth, among which is one to the memory of Sir Thomas Smyth, Chancellor of the Garter and Principal Secretary of State, in the reigns of Edward VI. and Elizabeth; opposite this is one to Sir William Smyth and his lady, with their three sons and four daughters. A National school is supported by the Smyth family.

THRANDESTON (*St. Margaret*), a parish, in the union and hundred of Hartismere, Western Division of the county of Suffolk, 3 miles (N. W.) from Eye; containing 358 inhabitants. The living is a rectory, valued in the king's books at £13. 6. 8.; present net income, £391; patron, Sir E. Kerrison, Bart. A considerable fair is held here on the 31st of July. There are some cottages and land, derived from an unknown source, the rental of which, amounting to about £26, is applied to parochial purposes.

THRAPSTON, or THRAPSTONE (*St. James*), a market-town and parish, and the head of a union, in the hundred of Navisford, Northern Division of the county of Northampton, 21 miles (N. E. by E.) from Northampton, and 74 (N. N. W.) from London; containing 1014 inhabitants. This small town is delightfully situated in a rich and luxuriant valley, on the eastern bank of the river Nen, over which is a handsome wooden bridge of several arches, built in 1795, in lieu of an old stone bridge, which was swept away by the inundation of that year: the houses are neat and regularly built, and the inhabitants are well supplied with excellent water: the vicinity is adorned with numerous residences of the nobility and gentry. The town appears to have been more extensive than it is at present, as several traces of buildings destroyed by fire are visible. The principal articles of manufacture are whips and bobbin-lace: on the river are corn-mills and a paper-mill; and in the vicinity are some stone and other quarries, which yield a beautiful white sand, much used for domestic purposes. Some trade is carried on in conveying grain, by means of the Nen, which was made navigable in 1737, to Northampton, Peterborough, Lynn, and other places; and bringing back timber, coal, and other commodities. The market is on Tuesday, for corn and seed, and is the largest hog market in the county. Fairs are held on the first Tuesday in May, for cattle and sheep; Aug. 5th, for the hiring of servants, and for shoes and pedlery; and the first Tuesday after Michaelmas, a very large fair for cattle. There is a resident magistrate in the town; and subordinate officers are appointed at the manorial court, the court of the honour of Gloucester, and that for Navisford hundred, all of which are held here.

The living is a rectory, valued in the king's books at £14. 5. 5.; present net income, £348: it is in the patronage of the Crown. The church is a cruciform structure, with a western tower and spire, combining the early, decorated, and later styles of English architecture; it was extensively repaired in 1810: in the chancel are three stone stalls, with rich mouldings and crocketed canopies, and in the churchyard is an ancient and very curious monument. There is a place of worship for Baptists. Here is a National school, in which are 80 children, aided by a bequest of Mrs. Mary Ekins, in 1794. An "Institution for bettering the condition of the poor," established in 1826, is supported by voluntary contributions. The poor law union of Thrapston comprises 26 parishes or places, under the care of 30 guardians, and contains, according to the census of 1831, a population of 11,099. There are 24 villages within five miles of this town, and, from an adjacent eminence, 32 churches may be seen.

THREAPLAND, a township, in the parish of Torpenhow, Allerdale ward below Derwent, Western Division of the county of Cumberland, 6¾ miles (N. N. E.) from Cockermouth: the population is returned with the township of Bothel.

THRECKINGHAM (*St. Peter*), a parish, in the union of Sleaford, wapentake of Aveland, parts of Kesteven, county of Lincoln, 2 miles (N. E. by N.) from Falkingham; containing 191 inhabitants. The living is a discharged vicarage, valued in the king's books at £6. 8. 9.; present net income, £144; patron and impropriator, Sir G. Heathcote, Bart. The church has a lofty tower and spire, and exhibits a curious admixture of the Norman, early English, and decorated styles of architecture. In the chancel is an elegant stall; the font is circular, with early English panelling, and there are some old monuments and good screen-works. An infants' school and a Sunday school are supported by subscription.

THREE-FARMS, a township, in the parish of ECCLESHALL, Northern Division of the hundred of PIREHILL, and of the county of STAFFORD; containing 67 inhabitants.

THRELKELD, a chapelry, in the parish of GREYSTOCK, union of PENRITH, LEATH ward, Eastern Division of the county of CUMBERLAND, 4 miles (E. N. E.) from Keswick; containing 320 inhabitants. The living is a perpetual curacy; net income, £53; patron, Earl of Lonsdale. The chapel, dedicated to St. Mary, was rebuilt by subscription upwards of 50 years since. A school is partly supported by an endowment of £4 per annum.

THRESHFIELD, a township, in the parish of LINTON, union of SKIPTON, Eastern Division of the wapentake of STAINCLIFFE and EWCROSS, West Riding of the county of YORK, 9 miles (N.) from Skipton; containing 212 inhabitants. Here is a school endowed with £30 per annum.

THREXTON, a parish, in the union of SWAFFHAM, hundred of WAYLAND, Western Division of the county of NORFOLK, $2\frac{1}{4}$ miles (W. by S.) from Watton; containing 29 inhabitants. The living is a discharged rectory, annexed to the vicarage of Swaffham, and valued in the king's books at £7. 9. $4\frac{1}{2}$. The tithes have been commuted for a rent-charge of £171. 2. 8., subject to the payment of rates, which on the average have amounted to £18. The church has a low round steeple, and in the south window are various armorial bearings in stained glass.

THRIBERGH (*St. Leonard*), a parish, in the union of ROTHERHAM, Southern Division of the wapentake of STRAFFORTH and TICKHILL, West Riding of the county of YORK, 3 miles (N. E.) from Rotherham; containing 332 inhabitants. The living is a rectory, valued in the king's books at £12. 11. $5\frac{1}{2}$.; present net income, £329; patron, John Fullerton, Esq. The church is principally in the later style of English architecture. Elizabeth Finch, in 1760, bequeathed money producing ten guineas a year, and the Rev. W. Hodges, £9 annually, for teaching poor children. A new school-room has been erected by Mr. Fullerton, and the former converted into a residence for the master. Another school is partly supported by ladies.

THRIGBY (*St. Mary*), a parish, in the hundred of EAST FLEGG, Eastern Division of the county of NORFOLK, $4\frac{1}{2}$ miles (W.) from Caistor; containing 43 inhabitants. The living is a discharged rectory, valued in the king's books at £6; patron, Thomas Browne, Esq. The tithes have been commuted for a rent-charge of £210. 17., subject to the payment of rates, which on the average have amounted to £20; the glebe comprises $4\frac{1}{2}$ acres, valued at £9 per annum.

THRIMBY with LITTLE STRICKLAND, a chapelry, in the parish of MORLAND, WEST ward, and union, county of WESTMORLAND, 3 miles (N. by W.) from Shap; containing 81 inhabitants. The living is a perpetual curacy; net income, £53; patron, Vicar of Morland; appropriators, Dean and Chapter of Carlisle. The chapel, dedicated to St. Mary, was consecrated in 1814, having been rebuilt at Little Strickland, together with a school-house, by the Earl of Lonsdale. The school was founded in 1684, by Thomas Fletcher, Esq., who endowed it with a rent-charge of £10, for the teacher and preacher.

THRINGSTONE, a chapelry, in the parish of WHITWICK, hundred of WEST GOSCOTE, Northern Division of the county of LEICESTER, $4\frac{3}{4}$ miles (E.) from Ashby-de-la-Zouch; containing 1267 inhabitants. The living is a perpetual curacy; net income, £44; patron, Vicar of Whitwick; impropriator, Sir G. Beaumont, Bart. The chapel is dedicated to St. George, and contains 380 free sittings, the Incorporated Society having granted £450 in aid of the expense. There is a place of worship for Wesleyan Methodists; and a National school has been established.

THRIPLOW or TRIPLOW (*All Saints*), a parish, in the union of ROYSTON, hundred of THRIPLOW, county of CAMBRIDGE, 6 miles (N. N. E.) from Royston; containing 417 inhabitants. The living is a discharged vicarage, in the patronage of the Bishop of Ely, valued in the king's books at £4. 9. 2.; present net income, £129; impropriators, Master and Fellows of Peter-House, Cambridge. The church is an ancient cruciform structure, and contains several monuments to the family of Lucas. There is a place of worship for Independents. A grand rendezvous of the parliamentarian army, commanded by Fairfax and Cromwell, took place on Thriplow heath, in July 1647; instruments of warfare are frequently found here.

THRISLINGTON, a township, in the parish of BISHOP'S MIDDLEHAM, union of SEDGEFIELD, North-Eastern Division of STOCKTON ward, Southern Division of the county palatine of DURHAM, $7\frac{1}{2}$ miles (S. S. E.) from Durham; containing 15 inhabitants.

THRISTON, EAST and WEST, a joint township with Shot-haugh, in the parish of FELTON, Eastern Division of MORPETH ward, Northern Division of the county of NORTHUMBERLAND, $9\frac{3}{4}$ miles (S. by E.) from Alnwick; containing, with Shot-haugh, 307 inhabitants.

THROAPHAM, a township, in the parish of LAUGHTON-EN-LE-MORTHEN, union of WORKSOP, Southern Division of the wapentake of STRAFFORTH and TICKHILL, West Riding of the county of YORK, $6\frac{1}{2}$ miles (S. S. W.) from Tickhill; containing 70 inhabitants and 1080 acres.

THROCKING (*Holy Trinity*), a parish, in the union of BUNTINGFORD, hundred of EDWINSTREE, county of HERTFORD, 2 miles (W. N. W.) from Buntingford; containing, exclusively of the extra-parochial liberty of Wakely, 76 inhabitants. The living is a rectory, valued in the king's books at £8; present net income, £252; patron, J. Ray, Esq.

THROCKLEY, a township, in the parish of NEWBURN, union and Western Division of CASTLE ward, Southern Division of the county of NORTHUMBERLAND, $6\frac{1}{4}$ miles (W. by N.) from Newcastle-upon-Tyne; containing 208 inhabitants. A school is partly supported by an allowance of £5 per annum from the Governors of Greenwich Hospital.

THROCKMORTON, a chapelry, in the parish of FLADBURY, union of PERSHORE, Middle Division of the hundred of OSWALDSLOW, Pershore and Eastern Divisions of the county of WORCESTER, 3¼ miles (N. E.) from Pershore; containing 159 inhabitants.

THROPPLE, a township, in the parish of MITFORD, union, and Western Division of the ward, of MORPETH, Northern Division of the county of NORTHUMBERLAND, 4 miles (W.) from Morpeth; containing 78 inhabitants. At Whittle Hill are slight remains of a Roman camp; the intrenchments have been almost obliterated by the plough, and the stones removed for repairing the roads. Coins have been found near it, in an ancient barrow, called Money Hill.

THROPTON, a township, in the parish and union of ROTHBURY, Western Division of COQUETDALE ward, Northern Division of the county of NORTHUMBERLAND, 2 miles (W. by N.) from Rothbury; containing 218 inhabitants. The village is pleasantly situated near the confluence of the river Coquet and the Snitter-burn, over the latter of which a substantial bridge was erected by subscription in 1810, the old bridge having fallen down several years before. There is a cross road at both ends of the village, and a stone cross is situated at each intersection. There are places of worship for Presbyterians and Roman Catholics.

THROSTON, a township, in the parish of HART, union of STOCKTON, North-Eastern Division of STOCKTON ward, Southern Division of the county palatine of DURHAM, 3¾ miles (W.) from Hartlepool; containing 70 inhabitants.

THROWLEY, a parish, in the union of OAKHAMPTON, hundred of WONFORD, Wonford and Southern Divisions of the county of DEVON, 6¾ miles (E. S. E.) from Oakhampton; containing 460 inhabitants. The living is a rectory, valued in the king's books at £19. 6. 10½.; present net income, £200: it is in the patronage of the Crown. The church is a small plain building. The river Teign forms one of the boundaries of the parish. There are some remains of a chapel at Walland Hill.

THROWLEY (*St. Mary*), a parish, in the union and hundred of FAVERSHAM, Upper Division of the lathe of SCRAY, Eastern Division of the county of KENT, 4½ miles (S. S. W.) from Faversham; containing 675 inhabitants. The living is a vicarage, valued in the king's books at £7. 11. 8.; present net income, £200; patron and appropriator, Prebendary of Rugmere in the Cathedral Church of St. Paul's, London. Sir Thomas Sondes, in 1592, endowed a free school for educating fourteen boys: this school no longer exists, but Lord Sondes, the present proprietor of the estate at Throwley, contributes, in lieu of the above endowment, to the support of a National school, established in 1814, which is open to the children of the parishes of Throwley, Sheldwich, Selling, and Baddlesmere. There are three almshouses, founded by the same family. The vicarage-house occupies the site of a priory, founded as a cell to the abbey of St. Bertin, at St. Omers in Artois, and granted, in the 22nd of Henry VI., to the abbey of Sion.

THROWLEY, a hamlet, in the parish of ILAM, Northern Division of the hundred of TOTMONSLOW and of the county of STAFFORD, 7¾ miles (N. W. by W.) from Ashbourn: the population is returned with the parish. Throwley Hall, now a farm-house, is an ancient mansion, once occupied by the family of Oliver Cromwell.

THROXENBY, a township, in the parish of SCALBY, union of SCARBOROUGH, PICKERING lythe, North Riding of the county of YORK, 2½ miles (W.) from Scarborough; containing 54 inhabitants.

THRUMPTON (*All Saints*), a parish, in the union of BASFORD, Northern Division of the wapentake of RUSHCLIFFE, Southern Division of the county of NOTTINGHAM, 4 miles (N. N. E.) from Kegworth; containing 132 inhabitants. This place is situated on the river Trent, on the banks of which is a fine old mansion in the style of architecture prevalent in the reign of Elizabeth. The living is a perpetual curacy; net income, £77; patron and impropriator, J. E. Wescomb, Esq. A day and Sunday school is supported by endowment.

THRUP, a hamlet, in the parish of KIDLINGTON, union of WOODSTOCK, hundred of WOOTTON, county of OXFORD, 1½ mile (E. S. E.) from Woodstock; containing 84 inhabitants.

THRUPP, a joint tything with Wadley, otherwise Littleworth, in the parish of GREAT FARRINGDON, hundred of SHRIVENHAM, county of BERKS, 2 miles (N.) from Great Farringdon: the population is returned with the parish.

THRUPP, THE, a hamlet, in the parish of STROUD, hundred of BISLEY, Eastern Division of the county of GLOUCESTER, 1 mile from Stroud: the population is returned with the parish. The village is pleasantly situated in a vale through which passes the Thames and Severn canal, affording a facility of conveyance for the produce of the several works carried on within its limits. Of these the principal is the manufacture of superfine woollen cloths, chiefly black, for which there are two very extensive and several smaller establishments, together employing about 2000 persons. There are also in the hamlet a large wool-stapling business, an extensive iron and brass foundry, and a general engineering establishment, in which from 40 to 50 persons are regularly employed. Here is an endowed school in connection with the established church, in which an unlimited number of children are gratuitously instructed.

THRUPPWICK, a liberty, in the parish of RADLEY, hundred of HORMER, county of BERKS, 1½ mile (E.) from Abingdon; containing 31 inhabitants.

THRUSHELTON (*St. George*), a parish, in the union of TAVISTOCK, hundred of LIFTON, Lifton and Southern Divisions of the county of DEVON, 10 miles (S. W. by W.) from Oakhampton; containing 353 inhabitants. The living is annexed to the vicarage of Mary-Stow. By an ancient deed, dated 1504, the parishioners on certain conditions were allowed to have a cemetery near their chapel, to avoid the difficulty they experienced from inundation, in conveying their dead to the churchyard of the mother church at Mary-Stow. A day and Sunday school is supported by subscription.

THRUSSINGTON (*Holy Trinity*), a parish, in the union of BARROW-UPON-SOAR, hundred of EAST GOSCOTE, Northern Division of the county of LEICESTER, 8¼ miles (N. N. E.) from Leicester; containing 454 inhabitants. The living is a discharged vicarage, valued in the king's books at £6; present net income, £240; patron and incumbent, Rev. C. B. Woolley. Thomas Hayne, in 1640, bequeathed an annuity of about £7. for teaching poor children; and a National school has been established. In this parish are the kennels of the Melton Hunt, erected at an expense of £12,000 by the late Sir Harry Goodricke, Bart.; they contain stabling for 60 horses and kennels for 300 hounds, with offices for 20 servants.

THRUXTON (*St. Bartholomew*), a parish, in the union of Dore, hundred of Webtree, county of Hereford, 6¼ miles (S. W. by W.) from Hereford; containing 59 inhabitants. The living is a discharged rectory, with the vicarage of Kingstone united, valued in the king's books at £4. 8. 4.; present net income, £252; patron, Dean of Hereford; appropriators, Dean and Chapter of Hereford.

THRUXTON (*Holy Rood*), a parish, in the union and hundred of Andover, Andover and Northern Divisions of the county of Southampton, 5 miles (W.) from Andover; containing 269 inhabitants. The living is a rectory, valued in the king's books at £15. 12. 11.; patron and incumbent, Rev. Donald Baynes. The tithes have been commuted for a rent-charge of £387, subject to the payment of rates; and there are 50 acres of glebe, valued at £70 per annum. There is a place of worship for Wesleyan Methodists. Here is a beautiful Roman pavement, nearly perfect.

THUNDERLEY, a hamlet (formerly a distinct parish), in the parish of Wimbish, hundred of Uttlesford, Northern Division of the county of Essex, 2 miles (S. E. by S.) from Saffron-Walden: the population is returned with the parish. The vicarage has been consolidated with that of Wimbish. The church is in ruins.

THUNDERSLEY (*St. Peter*), a parish, in the union of Billericay, partly in the hundred of Rochford, but chiefly in the hundred of Barstable, Southern Division of the county of Essex, 2¼ miles (S. W. by W.) from Rayleigh; containing 526 inhabitants. This parish is about two miles in length, and a mile and a half in breadth. The village is small, but pleasantly situated on elevated ground, and the surrounding scenery is pleasingly diversified. The living is a discharged rectory, valued in the king's books at £14. 13. 4.; patron and incumbent, Rev. G. Hemming. The tithes have been commuted for a rent-charge of £570, subject to the payment of rates, which on the average have amounted to £76; the glebe comprises 40 acres, valued at £40 per annum. The church is a venerable structure, in the later Norman and early English styles of architecture, with a tower and spire; the nave is separated from the aisles by low round and octagonal pillars with enriched capitals, supporting a series of pointed arches. Here is a small National school.

THUNDRIDGE (*St. Mary and All Saints*), a parish, in the union of Ware, hundred of Braughin, county of Hertford, 2½ miles (N. N. E.) from Ware; containing 588 inhabitants. The living is a vicarage, annexed to that of Ware, and valued in the king's books at £6. The church has an embattled tower with a lofty spire: it had formerly a Norman arch between the nave and the chancel, which having been enlarged, in recently repairing the edifice, its original character has been destroyed. Two day and Sunday schools are partly supported by subscription.

THURCASTON (*All Saints*), a parish, in the union of Barrow-upon-Soar, hundred of West Goscote, Northern Division of the county of Leicester, 4 miles (N. by W.) from Leicester; containing 1241 inhabitants, a few of whom are employed in frame-work knitting. The living is a rectory, with Anstey consolidated, valued in the king's books at £23. 7. 8½.; present net income, £676; patrons, Master and Fellows of Emanuel College, Cambridge. The Rev. Richard Hill, in 1730, bequeathed land, now producing £21 per annum, for teaching 22 children; and a National school has been established. The zealous and venerable reformer, Dr. Hugh Latimer, Bishop of Worcester, was born here, about 1480. Doctor Hurd, also Bishop of Worcester, was for some time rector of this parish.

THURCROSS, a chapelry, in the parish of Fewston, union of Pateley-bridge, Lower Division of the wapentake of Claro, West Riding of the county of York, 9 miles (W. by S.) from Ripley; containing 601 inhabitants.

THURGARTON (*All Saints*), a parish, in the union of Erpingham, hundred of North Erpingham, Eastern Division of the county of Norfolk, 4¾ miles (S. W. by S.) from Cromer; containing 247 inhabitants. The living is a discharged rectory, valued in the king's books at £9. 6. 8.; present net income, £206; patron, Bishop of Norwich. The church is a small thatched building without a tower. A day and Sunday school is partly supported by the curate. Here are slight remains of an ancient hall.

THURGARTON (*St. Peter*), a parish, in the union of Southwell, Southern Division of the wapentake of Thurgarton and of the county of Nottingham, 3¼ miles (S. by W.) from Southwell; containing 329 inhabitants. The living is a perpetual curacy; net income, £56; patrons and impropriators, Master and Fellows of Trinity College, Cambridge. There is a school in which 20 boys are educated from funds accruing from Trinity College lands; and 20 girls are instructed at the expense of an individual. An Augustine priory, in honour of St. Peter, was founded here in the year 1130, by Ralph de Ayncourt, which, at the dissolution, had a revenue of £359. 15. 10.

THURGOLAND, a township, in the parish of Silkstone, union of Wortley, wapentake of Staincross, West Riding of the county of York, 3¾ miles (E. S. E.) from Penistone; containing 1147 inhabitants, many of whom are employed in the manufacture of woollen cloth and wire. There is a place of worship for Wesleyan Methodists.

THURLASTON (*All Saints*), a parish, in the union of Blaby, hundred of Sparkenhoe, Southern Division of the county of Leicester, 6 miles (N. E. by E.) from Hinckley; containing 636 inhabitants. The living is a rectory, valued in the king's books at £13. 19. 7.; present net income, £400; patron, R. Arkwright, Esq. The chancel of the church is the property of — Gundry, Esq., lord of the manor. Here is a National school.

THURLASTON, a hamlet, in the parish of Dunchurch, union of Rugby, Rugby Division of the hundred of Knightlow, Northern Division of the county of Warwick, 1 mile (W.) from Dunchurch; containing 281 inhabitants.

THURLBEAR (*St. Thomas*), a parish, in the union of Taunton, hundred of North Curry, Western Division of the county of Somerset, 3¼ miles (S. E.) from Taunton; containing 202 inhabitants. The living is a rectory and donative; patron, Lord Portman. The tithes have been commuted for a rent-charge of £150, subject to the payment of rates; the glebe comprises 36 acres, valued at £41. 15. per annum. The church was anciently a chapel to the vicarage of St. Mary Magdalene, in Taunton, but the tithes were restored by Sir Thomas Petman, Bart.

THURLBY (*St. German*), a parish, in the union of Newark, Lower Division of the wapentake of Boothby-Graffo, parts of Kesteven, county of Lincoln, 9½ miles (S. W.) from Lincoln; containing 145 inhabitants. This parish lies between the rivers Trent and Witham, the latter of which is celebrated for its eels and pike. The living is a perpetual curacy; net income, £55; patron and appropriator, Prebendary of Carlton *cum* Thurlby in the Cathedral Church of Lincoln. The appropriate tithes have been commuted for a rent-charge of £204. 1. 10., subject to the payment of rates, which on the average have amounted to £22; the glebe comprises one acre, valued at £2 per annum. The church is principally in the later style of English architecture.

THURLBY, a hamlet, in the parish of Bilsby, union of Spilsby, Wold Division of the hundred of Calceworth, parts of Lindsey, county of Lincoln, 2¼ miles (E.) from Alford: the population is returned with the parish.

THURLBY (*St. Firmin*), a parish, in the union of Bourne, wapentake of Ness, parts of Kesteven, county of Lincoln, 5¼ miles (N. N. W.) from Market-Deeping; containing 632 inhabitants. The living is a discharged vicarage, valued in the king's books at £10. 9. 4½.; present net income, £252; patrons and impropriators, Provost and Fellows of Eton College. The church is an ancient but handsome structure: the ancient Roman canal, Carr Dyke, passes close by it, where it is plainly discernible.

THURLEIGH (*St. Peter*), a parish, in the hundred of Willey, union and county of Bedford, 6½ miles (N.) from Bedford; containing 538 inhabitants. The living is a discharged vicarage, valued in the king's books at £9; present net income, £142; patron and impropriator, S. Crawley, Esq. Ten boys are instructed for £10 per annum, the bequest of George Franklyn, in 1618; and a National school has been established. Here is the moated site of the ancient mansion of Blackbull Hall; and on Bury hill are vestiges of a circular camp.

THURLESTONE, a parish, in the union of Kingsbridge, hundred of Stanborough, Stanborough and Coleridge, and Southern, Divisions of the county of Devon, 4½ miles (W. by S.) from Kingsbridge; containing 466 inhabitants. This parish is bounded on the north-west by the river Avon, and on the south-west by the English Channel. The living is a rectory, valued in the king's books at £25. 10.; present net income, £321; patron, Sir J. B. Y. Buller, Bart.

THURLESTONE, a township, in the parish of Penistone, union of Wortley, wapentake of Staincross, West Riding of the county of York, 8½ miles (W. by S.) from Barnesley; containing 1599 inhabitants. The manufacture of woollen cloth and hair cloth is extensively carried on, and there are fulling and scribbling mills in the neighbourhood. There is a place of worship for Wesleyan Methodists.

THURLOW, GREAT (*All Saints*), a parish, in the union and hundred of Risbridge, Western Division of the county of Suffolk, 5 miles (N. by E.) from Haverhill; containing 425 inhabitants. The living is a vicarage, valued in the king's books at £10. 11. 5½.; present net income, £300: the Queen presents to the vicarage, on the nomination of the proprietor of the manor; the impropriation belongs to Lady Harland. An hospital, or free chapel, dedicated to St. James, and subordinate to that of Hautpays, or *De Alto Passu*, was founded here in the time of Richard II., which, at the suppression of Alien houses, was valued at £3 per annum, and granted by Edward IV. to the Maison de Dieu in Cambridge, now part of King's College. Great Thurlow Hall was the seat of the Vernon family.

THURLOW, LITTLE (*St. Peter*), a parish, in the union and hundred of Risbridge, Western Division of the county of Suffolk, 5½ miles (N. by E.) from Haverhill; containing 464 inhabitants. Little Thurlow Hall has been the seat of the family of Soame since the reign of Elizabeth; it was built by Sir Stephen Soame, Knt., who was lord mayor of London, died in 1619, and was buried in the church of this parish under a handsome monument. The living is a rectory, valued in the king's books at £7. 10. 5.; present net income, £401; patrons, Representatives of the late R. C. Barnard, Esq. Sir Stephen Soame, in 1618, founded and endowed an almshouse for eight poor unmarried men or women, and a school for twenty boys: the former contains eight apartments, with a room in the centre intended for a reader, which are occupied by poor persons who have resided 24 years in this parish, or, in default thereof, in those of Great Thurlow or Wratting: the school premises consist of a school-room, with apartments over it for the use of the master.

THURLOXTON (*St. Giles*), a parish, in the union of Bridg-water, hundred of North Petherton, Western Division of the county of Somerset, 5 miles (S. S. W.) from Bridg-water; containing 229 inhabitants. The living is a rectory, valued in the king's books at £6. 15. 10.; patron, Lord Portman. The tithes have been commuted for a rent-charge of £93. 14. 2., subject to the payment of rates, which on the average have amounted to £10.

THURLTON (*All Saints*), a parish, in the union of Loddon and Clavering, hundred of Clavering, Eastern Division of the county of Norfolk, 5½ miles (N.) from Beccles; containing 416 inhabitants. The living is a discharged rectory, valued in the king's books at £6. 13. 4.: it was in the patronage of the Mayor and Corporation of Norwich, but the advowson must be sold under the Municipal Reform Act.

THURLTON (*St. Botolph*), a parish, within the liberty of the borough of Ipswich, Eastern Division of the county of Suffolk, 2½ miles (N. N. W.) from Ipswich: the population is returned with the parish of Whitton. The living is a rectory, annexed to that of Whitton.

THURMASTON, NORTH, a chapelry, partly in the parish of Barkby, and partly in that of Belgrave, union of Barrow-upon-Soar, hundred of East Goscote, Northern Division of the county of Leicester, 3¼ miles (N. N. E.) from Leicester; containing 184 inhabitants. The petty sessions for the hundred of East Goscote are held here. The walls of the ancient chapel, dedicated to St. John the Evangelist, are still remaining. There is a charity of £1 per annum to the poor, payable out of the lands of Mr. Simpkin. The manor is held by the trustees of — Pochin, Esq.

THURMASTON, SOUTH, a chapelry, in the parish of Belgrave, union of Barrow-upon-Soar, hundred of East Goscote, Northern Division of the county of LEICESTER, 3 miles (N. N. E.) from Leicester; containing 947 inhabitants. In consequence of a benefaction of £200 by Dr. Percy, Bishop of Dromore, and the surrender of the yearly sum of £22 by the vicar of Belgrave, as part of his stipend paid by the bishop, this chapelry was severed from the mother church of Belgrave by the Governors of Queen Anne's bounty in 1798, and the patronage of the perpetual curacy was vested in Bishop Percy and his heirs, who sold it to William Pochin, Esq., of Barkby, from whom it descended to the present family of that name: the net income is £100. The tithe farm, containing upwards of 140 acres, and valued at £207. 7. 4. per annum, belongs to the see of Lichfield. The chapel is dedicated to St. Michael, and contains many ancient monuments to the family of Simons. There is a place of worship for Wesleyan Methodists; also a National school. The most ancient Roman *milliarium* known in Britain was found here; it is 3½ feet high, and 7½ inches in circumference, and now stands on a pillar in a conspicuous situation in the town of Leicester. The Leicester canal passes through this lordship, and joins the Melton-Mowbray canal near the village, which is intersected by the Roman Fosse-way. The chapelry comprises 1100 acres of land, the soil of which is in general light, the substratum consisting of loam, gravel, and clay.

THURNBY (*St. Luke*), a parish, in the hundred of GARTREE, Southern Division of the county of LEICESTER, 4 miles (E. by S.) from Leicester; containing 383 inhabitants. The living is a vicarage, valued in the king's books at £11; present net income, £73; patron and impropriator, G. A. Leigh-Keck, Esq. There is a chapel of ease at Stoughton; and a National school has been established.

THURNHAM, a township, in the parish of LANCASTER, hundred of LONSDALE, south of the sands, Northern Division of the county palatine of LANCASTER, 4¾ miles (S. S. W.) from Lancaster; containing 526 inhabitants. This place, which is situated on the river Lune, contains the modern harbour of Lancaster, called Glasson Dock, which was constructed in the year 1787, and is capable of receiving about 25 merchant vessels, of which the cargoes are discharged at this place and forwarded by small craft to Lancaster. A considerable traffic is carried on here, affording employment to most of the inhabitants; and a canal has been opened connecting Glasson Dock with the Preston and Lancaster canal, and greatly facilitating the trade of the place. There is a Roman Catholic chapel. A school is partly supported by the Ditton family, and a day and Sunday school partly by subscription. Within this township are the venerable remains of Cockersand Abbey, founded originally as an hospital for a prior and infirm brethren, in the reign of Henry II., by William of Lancaster, and afterwards endowed for Premonstratensian canons: at the dissolution its revenue was £157. 14. The remains consist chiefly of the octagonal chapter-house, with the roof supported on a single central column, now the sepulchral chapel of the Dalton family; the building is romantically situated on a neck of land projecting into the sea, near the Cocker sands, from which the abbey derived its name.

THURNING (*St. James*), a parish, in the union of OUNDLE, partly in the hundred of POLEBROOKE, Northern Division of the county of NORTHAMPTON, but chiefly in that of LEIGHTONSTONE, county of HUNTINGDON, 4½ miles (S. E.) from Oundle; containing 130 inhabitants. The living is a rectory, valued in the king's books at £11. 4. 2.; patrons, Master and Fellows of Emanuel College, Cambridge. The tithes have been commuted for a rent-charge of £180, subject to the payment of rates, which on the average have amounted to £24; the glebe comprises 60 acres, valued at £45 per annum.

THURNING (*St. Andrew*), a parish, in the union of AYLSHAM, hundred of EYNSFORD, Eastern Division of the county of NORFOLK, 4¾ miles (N. by W.) from Reepham; containing 140 inhabitants. The living is a discharged rectory, valued in the king's books at £7; patrons, Master and Fellows of Corpus Christi College, Cambridge. The tithes have been commuted for a rent-charge of £367, subject to the payment of rates, which on the average have amounted to £73; the glebe comprises 18 acres, valued at £18 per annum. Here is a National school for girls.

THURNSCOE (*St. Helen*), a parish, in the union of DONCASTER, Northern Division of the wapentake of STRAFFORTH and TICKHILL, West Riding of the county of YORK, 8 miles (E.) from Barnesley; containing 223 inhabitants. The living is a rectory, valued in the king's books at £11. 7. 8½.; present net income, £341; patron, Earl Fitzwilliam.

THURROCK, GRAYS (*St. Peter and St. Paul*), a market-town and parish, in the union of ORSETT, hundred of CHAFFORD, Southern Division of the county of ESSEX, 12 miles (S. E.) from Romford, 22 miles (S. S. W.) from Chelmsford, and 20½ (E. by S.) from London; containing 1248 inhabitants. The town, consisting of a single street irregularly built, is situated on the verge of the river Thames, from which branches a creek navigable for small craft: on the north bank of this river is a wharf communicating with some lime-works by means of a railway. Bricks are here manufactured to a considerable extent, affording employment to about 300 persons, for the London builders, and are conveyed hence in barges. Great facility of communication with the metropolis is afforded by the London and Gravesend steamers, which perform the distance in two hours. The market is on Thursday; and fairs for cattle and hardware are held on May 23rd and Oct. 20th. The parish, which is about two miles in length, and nearly of equal breadth, takes its distinguishing epithet from the noble family of Grey, who were proprietors of the manor for more than three centuries. The living is a discharged vicarage, valued in the king's books at £5. 0. 10.; present net income, £160; patron, Rev. R. S. Hele. The church is a handsome cruciform structure, with a tower rising from the north transept. The old market-house is used as a place of worship for Dissenters. A free school, situated in the churchyard, and now united with a National school, was founded and endowed by William Palmer, in 1706, in which 20 children are instructed, and four are clothed on the old foundation. Rare and valuable fossils are found frequently in the chalk pits.

THURROCK, LITTLE (*St. Mary*), a parish, in the union of ORSETT, hundred of BARSTABLE, Southern Division of the county of ESSEX, 1 mile (E.) from Grays-Thurrock; containing 302 inhabitants. The living is a rectory, valued in the king's books at £13. 15.; present net income, £505; patron, Rev. E. Bowlby.

In the south wall of the church are some arched recesses. This parish is bounded on the south-west by the Thames. In a wood near the highway leading to tifford are some ancient excavations, termed "Danes' Holes."

THURROCK, WEST (*St. Clement*), a parish, in the union of ORSETT, hundred of CHAFFORD, Southern Division of the county of ESSEX, 1 mile (W.) from Grays-Thurrock; containing 804 inhabitants. This parish, which includes the populous village of Purfleet, is the smallest of the three parishes of that name; it is bounded on the south by the river Thames, where is a landing-place opposite to Greenhithe, and is about three miles in length and two miles and a half in breadth. The living is a discharged vicarage, valued in the king's books at £15. 13. 4.; present net income, £306; patron, W. H. Whitbread, Esq., who, with — Montgomery, Esq., is the impropriator. The church is an ancient structure of stone, with a massive square tower, and consists of a nave and chancel, with north and south aisles to each. There is a chapel of ease at Purfleet.

THURSBY (*St. Andrew*), a parish, in the union of WIGTON, ward, and Eastern Division of the county, of CUMBERLAND; containing 564 inhabitants, of which number, 373 are in the township of High Thursby, 6½ miles (S. W.) from Carlisle. The living is a discharged vicarage, valued in the king's books at £11. 10. 5.; present net income, £160; patrons and appropriators, Dean and Chapter of Carlisle. This place is supposed to have derived its name from Thor, the Saxon deity, to whose honour a temple is said to have been erected at Woodrigs, in this neighbourhood. A school, founded in 1740 by subscription, was endowed in 1798, by Thomas Tomlinson, Esq., with the interest of £354. A pillar of coarse stone, inscribed to Philip the Emperor and his son, A. D. 248, dug up near the military way at Wigton, is carefully preserved here.

THURSFIELD, a chapelry, in the parish of WOLSTANTON, Northern Division of the hundred of PIREHILL and of the county of STAFFORD, 6½ miles (N. by E.) from Newcastle-under-Lyme; containing 389 inhabitants. Dr. Robert Hulme, in 1708, bequeathed certain lands, now producing an annual income of £73, for the instruction of children. James Brindley, of Turnhurst, the celebrated engineer, was interred here in 1772; a plain altar-tomb has been erected to his memory.

THURSFORD (*St. Andrew*), a parish, in the union of WALSINGHAM, hundred of NORTH GREENHOE, Western Division of the county of NORFOLK, 3½ miles (S. E. by E.) from Little Walsingham; containing 392 inhabitants. The living is a discharged rectory, annexed to that of Great Snoring, and valued in the king's books at £8.

THURSLEY (*St. Michael*), a parish, in the union of HAMBLEDON, Second Division of the hundred of GODALMING, Western Division of the county of SURREY, 5 miles (S. W. by W.) from Godalming; containing 708 inhabitants. This parish abounds with ironstone, and there were anciently several large iron-foundries, of which the only memorials are four large ponds called "Hammer Ponds." The silk manufacture was subsequently introduced here, but has been for some years discontinued. The living is annexed to the vicarage of Witley. A parochial school is supported by subscription; and the parish partakes of Henry Smith's charity.

THURSTASTON (*St. Bartholomew*), a parish, in the union, and Lower Division of the hundred, of WIRRALL, Southern Division of the county of CHESTER, 5 miles (N. W. by N.) from Great Neston; containing 92 inhabitants. The living is a discharged rectory, valued in the king's books at £6. 13. 6.; present net income, £242; patrons, Dean and Chapter of Chester. The church has been lately rebuilt, and a rectory-house has been erected partly at the expense of the present incumbent. The river Dee passes by this place to Chester.

THURSTON (*St. Peter*), a parish, in the union of STOW, hundred of THEDWASTRY, Western Division of the county of SUFFOLK, 5½ miles (E. by N.) from Bury-St. Edmund's; containing 462 inhabitants. The situation of this parish is particularly healthy; the soil is generally light, with gravel and sandpits: a large portion of the great tithes has been redeemed by the principal landowners. The living is a discharged vicarage, valued in the king's books at £6. 13. 4.; present net income, £250; patron, Charles Tyrell, Esq.; impropriator, Gill Stedman, Esq. The church is a remarkably fine structure, and the pillars of the nave are peculiarly light and airy. There is some land, arising from various bequests and an allotment under an enclosure act, the rental of which, amounting to about £50, is appropriated to parochial and charitable purposes.

THURSTONLAND, a township, in the parish of KIRK-BURTON, union of HUDDERSFIELD, Upper Division of the wapentake of AGBRIGG, West Riding of the county of YORK, 5½ miles (S. by E.) from Huddersfield; containing 1098 inhabitants. There is a place of worship for Wesleyan Methodists.

THURTON (*St. Ethelbert*), a parish, in the union of LODDON and CLAVERING, hundred of LODDON, Eastern Division of the county of NORFOLK, 8 miles (S. E.) from Norwich; containing 193 inhabitants. The living is a perpetual curacy; net income, £70; patron and impropriator, Sir W. B. Proctor, Bart. The impropriate tithes have been commuted for a rent-charge of £220, subject to the payment of rates, which on the average have amounted to £52.

THURVASTON, a joint township with Osleston, in the parish of SUTTON-ON-THE-HILL, hundred of APPLETREE, Southern Division of the county of DERBY, 7¾ miles (W. by N.) from Derby: the population is returned with Osleston. There is a place of worship for Wesleyan Methodists.

THUXTON (*St. Paul*), a parish, in the union of MITFORD and LAUNDITCH, hundred of MITFORD, Western Division of the county of NORFOLK, 4 miles (N. by E.) from Hingham; containing 83 inhabitants. The living is a discharged rectory, valued in the king's books at £4. 6. 3.; present net income, £211; patron and incumbent, Rev. W. Castell.

THWAITE (*All Saints*), a parish, in the union of AYLSHAM, hundred of South Erpingham, Eastern Division of the county of NORFOLK, 4¾ miles (N.) from Aylsham; containing 142 inhabitants. The living is a discharged rectory, valued in the king's books at £7; present net income, £110; patron, Bishop of Norwich. The church has a fine Norman entrance on the south.

THWAITE (*St. Mary*), a parish, in the union of LODDON and CLAVERING, hundred of LODDON, Eastern Division of the county of NORFOLK, 3½ miles (N.) from Bungay; containing 107 inhabitants. The living is a discharged rectory, valued in the king's books at £4; present net income, £143; patron, Duke of Norfolk.

THWAITE (*St. George*), a parish, in the union and hundred of HARTISMERE, Western Division of the county of SUFFOLK, 4¾ miles (S. W. by S.) from Eye; containing 175 inhabitants. The living is a discharged rectory, valued in the king's books at £6. 3. 5½.; present net income, £193; patrons, Executors of the late John Wilson Sheppard, Esq. In 1832, a large quantity of Saxon coins of silver were found here, in removing an old tree.

THWAITS, a chapelry, in the parish of MILLOM, ALLERDALE ward above Derwent, Western Division of the county of CUMBERLAND, 10 miles (S. E.) from Ravenglass; containing 324 inhabitants. The living is a perpetual curacy; net income, £99; patrons, four Landowners in the chapelry. The chapel was rebuilt in 1715, and dedicated to St. Anne in 1724.

THWING (*All Saints*), a parish, in the union of BRIDLINGTON, wapentake of DICKERING, East Riding of the county of YORK, 8¾ miles (W. N. W.) from Bridlington; containing 350 inhabitants. The living is a rectory in medieties, each valued in the king's books at £8. 12. 1.; present net income, £520: it is in the patronage of the Crown. The church was repaired and beautified, and a splendid east window of painted glass added, at the expense of Robert Prickett, Esq., lord of the manor, who partly supports a school.

TIBBENHAM (*All Saints*), a parish, in the union and hundred of DEPWADE, Eastern Division of the county of NORFOLK, 5 miles (W. S. W.) from Stratton-St. Mary; containing 650 inhabitants. The living is a discharged vicarage, valued in the king's books at £7. 6. 8.; present net income, £335; patron and appropriator, Bishop of Ely. Here is a National school. About a mile south-east of the church is Chanons Hall, occupying the site of the ancient manor-house of Chanons, which was a very extensive building surrounded by a moat.

TIBBERTO (*Holy Trinity*), a parish, in the union of NEWENT, duchy of LANCASTER, Western Division of the county of GLOUCESTER, 4¼ miles (S. E.) from Newent; containing 307 inhabitants. This parish formed part of the hundred of Botloe until the 50th of Edward III., when Lancashire being made a county palatine, all the estates of the Duke of Lancaster in this county, of which Tibberton was one, were erected into a new hundred, called the hundred of the duchy of Lancaster. The Herefordshire and Gloucestershire canal passes on the eastern side of the parish. The living is a discharged rectory, valued in the king's books at £7. 16. 0½.; patron, James Scott, Esq. The tithes have been commuted for a rent-charge of £327, subject to the payment of rates, which on the average have amounted to £32; the glebe comprises 5 acres.

TIBBERTON, a chapelry, in the parish of EDGMOND, union of DROITWICH, Newport Division of the hundred of SOUTH BRADFORD, Northern Division of the county of SALOP, 4¼ miles (W. by N.) from Newport; containing 351 inhabitants. The chapel is dedicated to All Saints.

TIBBERTON (*St. Nicholas*), a parish, in the union of NEWPORT, Middle Division of the hundred of OSWALDSLOW, Worcester and Western Divisions of the county of WORCESTER, 4 miles (E. N. E.) from Worcester; containing 337 inhabitants. The living is a discharged vicarage, in the patronage of the Dean and Chapter of Worcester (the appropriators), valued in the king's books at £3. 15. 10.; present net income, £123. The Birmingham and Worcester canal passes through the parish.

TIBERTON (*St. Mary*), a parish, in the union of DORE, hundred of WEBTREE, county of HEREFORD, 9 miles (W.) from Hereford; containing 118 inhabitants. The living is annexed to the vicarage of Madley.

TIBSHELF (*St. John the Baptist,*) a parish, in the union of MANSFIELD, hundred of SCARSDALE, Northern Division of the county of DERBY, 4 miles (N. E. by N.) from Alfreton; containing 759 inhabitants. The living is a discharged vicarage, valued in the king's books at £4. 5. 3.; present net income, £172; patron, — Lord, Esq. The church was rebuilt in 1729. The manufacture of stockings is carried on, and there are extensive collieries in the parish.

TIBTHORP, a township, in the parish of KIRKBURN, union of DRIFFIELD, Bainton-Beacon Division of the wapentake of HARTHILL, East Riding of the county of York, 5¼ miles (W. S. W.) from Great Driffield; containing 227 inhabitants. There is a place of worship for Wesleyan Methodists.

TICEHURST (*St. Mary*), a parish, and the head of a union, in the hundred of SHOYSWELL, rape of HASTINGS, Eastern Division of the county of SUSSEX, 4 miles (E. S. E.) from Wadhurst; containing 2314 inhabitants. The living is a vicarage, valued in the king's books at £18. 7. 6.; present net income, £350; patrons, Dean and Chapter of Canterbury; impropriator, Rev. J. Constable. The church is principally in the later style of English architecture. A chapel has been built at Flimwell, containing 325 sittings, 225 of which are free, the Incorporated Society having granted £200 in aid of the expense. There is a very extensive lunatic asylum, beautifully situated at the extremity of the village, the grounds of which are tastefully laid out. A National school has been established. The poor law union comprises 8 parishes or places, under the superintendence of 18 guardians, and contains, according to the census of 1831, a population of 13,347.

TICHFIELD, county of SOUTHAMPTON. — See TITCHFIELD.

TICKENCOTE (*St. Peter*), a parish, in the union of OAKHAM, hundred of EAST, county of RUTLAND, 3 miles (N. W. by W.) from Stamford; containing 128 inhabitants. The living is a discharged rectory, valued in the king's books at £6. 5. 8.; patron, John Wingfield, Esq. The tithes have been commuted for a rent-charge of £162. 9., subject to the payment of rates; the glebe comprises 3½ acres, valued at £6 per annum. The church was in the earliest style of Norman architecture, but it has been partially rebuilt. Stukely says, "it is the most venerable church extant, and was the entire oratory of Prince Peada, founder of Peterborough abbey." Here is a National school.

TICKENHAM (*St. Quiricus and Julietta*), a parish, in the union of BEDMINSTER, hundred of PORTBURY, Eastern Division of the county of SOMERSET, $9\frac{1}{4}$ miles (W. by S.) from Bristol; containing 427 inhabitants. The living is a discharged vicarage, united to that of Portbury, and valued in the king's books at £8. 15. 5. Here is a Sunday National school. About a mile north of the church are the remains of a double intrenched Roman camp; the ramparts, constructed of limestone, enclose an area of about an acre.

TICKHILL (*St. Mary*), a parish, in the union of DONCASTER, Southern Division of the wapentake of STRAFFORTH and TICKHILL, West Riding of the county of York; containing 2804 inhabitants, of which number, 2108 are in the town of Tickhill, 45 miles (S.) from York, and 157 (N. by W.) from London. This manor was given by William the Conqueror to Roger de Busli, who erected or rebuilt a castle, which, with the honour of Tickhill, being subsequently forfeited, was given by King Stephen to the Count of Eu, in Normandy; it afterwards reverted to the crown, and was granted by Richard I. to his brother, Prince John. In the reign of Henry III., it was restored to the Count of Eu, but, after several changes, became again vested in the crown, in the reign of Henry IV. At the commencement of the great civil war, the castle, then considered a very strong fortress, was garrisoned for the king, and, after a siege of two days, was surrendered to the assailants, and soon after dismantled by order of parliament. The town is situated in a fertile valley, on the border of the county of Nottingham: the streets, which intersect at right angles, are neat and spacious; the houses are in general of respectable appearance, but built in a straggling manner, and the inhabitants are well supplied with water. The trade in malt was formerly extensive, but at present there are not more than three kilns: a small paper-manufactory affords employment to a few persons. The market, formerly held on Friday, is entirely disused. A fair is held on Aug. 21st, for cattle, and various articles of merchandise. The market-cross is a circular building of stone, erected in 1776, and situated in the centre of the town. Manorial courts leet and baron are held annually. The living is a discharged vicarage, valued in the king's books at £7. 2. 6.; present net income, £261; patron and impropriator, G. S. Foljambe, Esq. The church is a handsome structure, in the later style of English architecture, having a fine tower with pinnacles; it was greatly injured by lightning in 1825, but has undergone an entire repair, at an expense of £1950. In the chancel is an altar tomb, ornamented at the sides with large quatrefoils, to the memory of William Estfield, seneschal of the lordship of Holderness, and of the honour of Tickhill, who died in 1386: at the east end of the south aisle is an alabaster monument, with the effigies of a knight and his lady. There are places of worship for Independents and Wesleyan Methodists. A National school is supported by subscription, aided by an annuity of £4. 8. 8. from the duchy of Lancaster. Near the church is a *Maison de Dieu*, comprising fourteen almshouses for poor widows, of uncertain foundation. The remains of the castle, on the south-east side of the town, consist of the mound, on which the foundations of the keep are visible; the ditch, with part of the external walls, and a dilapidated Norman gateway. The northern part has been converted into a modern residence, and the ground within the walls is formed into gardens and shrubberies. The ruins of an ancient Augustine priory, founded in the reign of Henry III., and situated in an adjacent vale, have been converted into a farm-house. John of Gaunt, Duke of Lancaster, resided at Tickhill castle.

TICKNALL (*St. Thomas à Becket*), a parish, in the union of ASHBY-DE-LA-ZOUCH, hundred of REPTON and GRESLEY, Southern Division of the county of DERBY, $5\frac{3}{4}$ miles (N. by W.) from Ashby-de-la-Zouch; containing 1274 inhabitants, many of whom are employed in the extensive limeworks in the parish. The living is a perpetual curacy; net income, £97; patron and impropriator, Sir George Crewe, Bart. The church is partly in the early and partly in the later style of English architecture, and has lately been repaired and embellished. There is a place of worship for Wesleyan Methodists. A school-house was erected by Dame Catherine Harpur, who, in 1744, conveyed for its support land now producing an annual income of £25: the premises were rebuilt in 1825, at the expense of Sir George Crewe. An hospital for seven decayed housekeepers of Ticknall and Calke was founded in 1771, by Charles Harpur, Esq., who gave £500 for building it, and endowed it with £2000: seven aged women at present enjoy the benefits of the charity.

TICKTON, a joint township with Hull-Bridge, in the parish of ST. JOHN, union, and liberties of the borough, of BEVERLEY, East Riding of the county of YORK, $2\frac{1}{4}$ miles (N. E.) from Beverley; containing, with Hull-Bridge, 110 inhabitants.

TIDCOMBE (*St. Michael*), a parish, in the union of HUNGERFORD, hundred of KINWARDSTONE, Everley and Pewsey, and Southern, Divisions of the county of WILTS, $6\frac{1}{4}$ miles (N. N. E.) from Ludgershall; containing 243 inhabitants. The living is a perpetual curacy, valued in the king's books at £6. 13. 4.; present net income, £77; patrons and appropriators, Dean and Canons of Windsor. A day and Sunday school is partly supported by the lord of the manor.

TIDDESLEY-HAY, an extra-parochial liberty, comprising about 3500 acres, in the Eastern Division of the hundred of CUTTLESTONE, Southern Division of the county of STAFFORD, $2\frac{1}{4}$ miles (N. E.) from Penkridge; and containing 50 inhabitants. This was a royal chace adjoining that of Cannock, till the reign of Elizabeth, who granted it jointly to the Earls of Warwick and Leicester, by whom it was sold to Sir Edward Littleton, of Pillaton Hall: there were at that time no other enclosures upon it than two ancient parks, and in that state it continued till recently, when it was wholly enclosed by Lord Hatherton, its present proprietor.

TIDDINGTON, a hamlet, in the parish of ALBURY, union of THAME, hundred of BULLINGDON, county of OXFORD, $3\frac{1}{4}$ miles (N. N. W.) from Tetsworth; containing 198 inhabitants.

TIDENHAM (*St. Mary*), a parish, in the union of CHEPSTOW, and forming, with the parish of Wollaston, a detached portion of the hundred of WESTBURY, Western Division of the county of GLOUCESTER, 2 miles (N. E. by N.) from Chepstow; containing, with Lancaut, 1180 inhabitants. This place was granted, in the year 956, by King Edwy, to the abbey of Bath, and after the Conquest was vested in the Earl of Hereford and the Count D'Eu, by which families it was shortly afterwards forfeited for treason, and was granted to the family of Clare. From the Clares the manor descended to the Duke of Beaufort, who also inherited from his paternal ancestors the rights of Tintern Abbey with respect to this parish, being thus invested with nearly all the rights of the Norman Lords of Strigul or Chepstow, with the exception of the church, which had been granted by them to the convent of De Leira in Normandy, which subsequently merged into that of Shene. After the dissolution, the rectory and advowson remained in the crown till the time of James I.; the latter now belongs to Mrs. Burr, the former is in severalties. This parish, with that of Wollaston, was originally in Norman Gloucestershire; in the hundred rolls it is noticed as an exclusive liberty of the Earls Marshal: and subsequently, as part of the paramount lordship marcher of Strigul, was in the jurisdiction of the marches, from which it was separated, and, by statute of the 27th of Henry VIII., re-annexed to the county of Gloucester. It is situated at the extremity of the Forest peninsula, commanding from its hills extensive and richly diversified views. The Severn is here crossed by the Old Passage ferry, which has been lately much improved, and forms the principal communication with South Wales. Sedbury Park, the property and residence of George Ormerod, Esq., is within the parish: the mansion has been nearly rebuilt from designs by Sir Robert Smirke; and within the grounds is the southern termination of Offa's Dyke, which passes through the estate, and over Buttindon Hill, to a lofty cliff overhanging the river Severn, near its confluence with the Wye. The living is a discharged vicarage, valued in the king's books at £7. 14.; present net income, £441: it is in the patronage of Mrs. Mary Burr; impropriators, J. Buckle, Esq., and others. A new chapel has been built at Beachley. A school is partly supported by James Jenkins, Esq. Among the relics of antiquity in the parish are the ruins of St. Tecla's chapel, on a small rocky island near the confluence of the rivers Wye and Severn; the Akeman-street, crossing Sedbury in its line from Oldbury to Caerwent; and some Roman and Danish camps on the line of Offa's Dyke, in the hamlets of Churchend and Wibdon, of which some were occupied as stations during the parliamentary war in the reign of Charles I.

TIDESWELL (*St. John the Baptist*), a market-town and parish, in the union of Bakewell, hundred of High Peak, Northern Division of the county of Derby; containing 2807 inhabitants, of which number, 1553 are in the town of Tideswell, 33 miles (N. N. W.) from Derby, and 160 (N. W. by N.) from London. The first account of this place is in Domesday-book, in which, under the name *Tiddeswall*, it is described as a royal demesne, having a chapel, which in 1215 was given by King John to the canons of Lichfield. The town is situated in a valley, surrounded by some of the most barren lands in the county, on the road from Chesterfield to Manchester. The houses in general are of very mean appearance, but the inhabitants are supplied with good water, by means of a small stream which flows through the town. The chief branches of trade are calico-weaving and mining. A market and two fairs were granted by Henry III., and confirmed by subsequent sovereigns: the market is on Wednesday; and fairs are held on March 24th, May 15th, the last Wednesday in July, the second Wednesday in September, and October 29th, for cattle and sheep. A court leet and court baron are held twice a year.

The living is a discharged vicarage, in the patronage of the Dean and Chapter of Lichfield (the appropriators), valued in the king's books at £7. 0. 7½.; present net income, £109. The church contains 360 free sittings, the Incorporated Society having granted £180 in aid of the expense: it is a remarkably fine cruciform structure, principally in the decorated style of English architecture, having an embattled tower at the west end, with crocketed pinnacles: the chancel is separated from the nave by a light screen of carved oak, and from the vestry-room by an embattled stone screen enriched with tracery. In the south transept is a tombstone to the memory of John Foljambe, who contributed largely to the erection of the church, in 1358: in the chancel is an altar-tomb, ornamented with brasses, to the memory of Sampson Meverell, who served under the Duke of Bedford in France, and was knighted upon the field at St. Luce: another altar-tomb records the death of Robert Pursglove, a native of this town, Prior of Gisburn abbey, Prebendary of Rotherham, and Bishop of Hull, who died May 2nd, 1579. There are places of worship for Wesleyan Methodists and Roman Catholics. The free grammar school was founded in 1560, under letters patent from Queen Elizabeth, by the above-mentioned Robert Pursglove, and endowed with lands at Priestcliff and Taddington, in this county, and at Colmworth in Bedfordshire; also with £2 per annum, chargeable on the estate of Earl Manvers; to be called "The school of Jesus," and to be open to all boys of this parish. The master is appointed by the Dean and Chapter of Lichfield; he and the vicar and churchwardens constitute a body corporate. The income arising from the lands is £227 per annum, one-fourth of which has generally been distributed among the poor of Tideswell on New Year's eve. Here is a Sunday National school.

TIDMARSH (*St. Lawrence*), a parish, in the union of Bradfield, hundred of Theale, county of Berks, 6 miles (W. N. W.) from Reading; containing 143 inhabitants. The living is a rectory, valued in the king's books at £5. 2. 6.; present net income, £223; patron, Robert Hopkins, Esq. The church is partly Norman, and partly in the early style of English architecture, the doorway being a particularly fine specimen of the former; the ceiling of the chancel is of panelled oak, and there are two slabs of blue marble, with ancient brasses.

TIDMINGTON, a parish, in the union of Shipston-upon-Stour, and forming, with Shipston-upon-Stour and Tredington, a detached portion of the Upper Division of the hundred of Oswaldslow, Blockley and Eastern Divisions of the county of Worcester, being locally in the Kington Division of the hundred of Kington, county of Warwick, 1½ mile (S. by E.) from Shipston-upon-Stour; containing 76 inhabitants. Tidmington and Shipston-upon-Stour were formerly townships in the parish of Tredington, from which they were separated by act of parliament, in the 6th of George I., and made distinct parishes; on this occasion the rectory of Tredington was divided into three parts. The living is a rectory, annexed to that of Shipston-upon-Stour. The church is partly in the early style of English architecture, and partly of later date.

TIDWORTH, NORTH (*Holy Trinity*), a parish,

in the union of ANDOVER, hundred of AMESBURY, Everley and Pewsey, and Southern, Divisions of the county of WILTS, 2½ miles (S. W. by W.) from Ludgershall; containing 392 inhabitants. The living is a rectory, valued in the king's books at £11. 17. 1.; present net income, £266: it is in the patronage of the Crown. An almshouse was founded and endowed with a rent-charge of £21 by Dr. Thomas Price, in 1689; and is occupied by four poor unmarried people. North-west of the village, on the summit of an isolated hill, is the large earthwork called Chidbury Camp, in form resembling a heart, and enclosing an area of seventeen acres. Robert Maton, a celebrated divine, was born here about 1607.

TIDWORTH, SOUTH (*St. Mary*), a parish, in the union and hundred of ANDOVER, Andover, and Northern Divisions of the county of SOUTHAMPTON, 2¾ miles (S. W. by S.) from Ludgershall; containing 217 inhabitants. The living is a rectory, valued in the king's books at £11. 15. 2½.; present net income, £379; patron, T. A. Smith, Esq. A school is supported by Mrs. Smith.

TIFFIELD (*St. John*), a parish, in the union and hundred of TOWCESTER, Southern Division of the county of NORTHAMPTON, 2¾ miles (N. by E.) from Towcester; containing 131 inhabitants. The living is a rectory, valued in the king's books at £9. 9. 7.; present net income, £175; patron, J. Flesher, Esq. The old Roman Watling-street passes through the parish.

TILBROOK (*All Saints*), a parish, in the union of ST. NEOTS, hundred of STODDEN, county of BEDFORD, 1½ mile (N. W. by W.) from Kimbolton; containing 295 inhabitants. The living is a rectory, valued in the king's books at £13. 10.; present net income, £388; patron, Lord St. John. Charles Higgins bequeathed £300 in support of a Sunday school, which sum, together with subscriptions, produces £20 a year.

TILBURY, EAST (*St. Margaret*), a parish, in the union of ORSETT, hundred of BARSTABLE, Southern Division of the county of ESSEX, 5¼ miles (E. by S.) from Grays-Thurrock; containing 245 inhabitants. This parish is bounded on the south-east by part of the Thames, called the Hope, where was an ancient ferry, said to be the place where Claudius passed the Thames in pursuit of the Britons; on Hope point is a battery for the defence of the river below Tilbury Fort. The living is a discharged vicarage, valued in the king's books at £13. 6. 8.; present net income, £169: it is in the patronage of the Crown; impropriator, Rev. E. Lloyd. The lofty tower of the manor-house of Gossalyne was battered down by the Dutch fleet, which ascended the Thames in the reign of Charles II.

TILBURY-JUXTA-CLARE, a parish, in the union of HALSTEAD, hundred of HINCKFORD, Northern Division of the county of ESSEX, 4 miles (N. N. W.) from Castle-Hedingham; containing 236 inhabitants. This parish, which derives the affix to its name from its proximity to the parish of Clare, in the county of Suffolk, is about five miles in circumference, and is intersected by a rivulet which has its source in the adjoining parish of Ridgwell. The soil is moderately fertile, and the lands in a good state of cultivation. The living is a rectory, consolidated with that of Ovington, and valued in the king's books at £8. The church, with the exception of the tower, which is of brick, is an ancient edifice of stone.

TILBURY, WEST (*St. James*), a parish, in the union of ORSETT, hundred of BARSTABLE, Southern Division of the county of ESSEX, 3¾ miles (E.) from Grays-Thurrock; containing 276 inhabitants. According to Bede, Tilbury, or Tillaburgh, was the seat of Bishop Cedda, when, about 630, he was engaged in baptizing the East Saxons. The parish is bounded on the south by the Thames, is about three miles in length and one mile in breadth, and lies directly opposite to Gravesend, with which town and the interior of Kent there is a constant traffic, by means of the ferry-boats stationed here for the conveyance of foot passengers, cattle, carriages, and merchandise. Tilbury Fort, partly in this parish, and partly in that of Chadwell adjoining, was originally a block-house, built in the reign of Henry VIII.; but after the memorable attack of the Dutch fleet, in 1667, upon the English shipping in the Medway, it was converted into a regular fortification, to which considerable additions have since been made. It is encompassed by a deep wide fosse, and its ramparts present several formidable batteries of heavy ordnance, particularly towards the river. It contains comfortable barracks, and other accommodations for the garrison, which at present consists of a fort-major and a detachment of invalids. The living is a rectory, valued in the king's books at £20; present net income, £558: it is in the patronage of the Crown. The church is an ancient edifice of stone, and had originally a lofty square embattled tower, which fell down some years since, and has been replaced with a belfry turret and spire of wood. Here is a National school. In a chalk hill near the village are several caverns, termed Danes' Holes, curiously constructed of stone, being narrow at the entrance, and very spacious at the depth of 30 feet; and some traces of the camp formed in the neighbourhood, to oppose the invasion of the Spanish Armada, are still visible. A medicinal spring was discovered in 1737.

TILDESLEY, county of LANCASTER.—See TYLDERSLEY.

TILEHURST (*St. Michael*), a parish, in the union of BRADFIELD, hundred of READING, county of BERKS, 2¾ miles (W.) from Reading; containing 1878 inhabitants. This extensive parish, which has the river Thames on the north, and the Kennet on the south, and is intersected by the line of the Great Western railway, was by act of parliament divided into two parts, of which one constitutes the present parish of Theale. The living is composed of a rectory and vicarage, united in 1586, valued in the king's books at £21. 15. 2½., and in the patronage of Mrs. Sophia Sheppard. The church is a plain brick structure, containing some ancient brasses, and a sumptuous monument to the memory of Sir Peter Vanlore, Knt., who died in 1627. There is a place of worship for Wesleyan Methodists; also a school endowed by the Rev. Dr. Sheppard and Mrs. Sheppard: Richard Lloyd, the learned Bishop of Worcester, was born here, in 1627; he died in 1717.

TILFORD, a joint tything with Culverlands, in the parish and hundred of FARNHAM, Western Division of the county of SURREY, 3 miles (S. E.) from Farnham: the population is returned with Culverlands.

TILLEY, a township, in the parish of WEM, Whitchurch Division of the hundred of NORTH BRADFORD, Northern Division of the county of SALOP, 1 mile (S.) from Wem; containing 296 inhabitants.

TILLINGHAM (*St. Nicholas*), a parish, in the union of Maldon, hundred of Dengie, Southern Division of the county of Essex, $2\frac{1}{4}$ miles (S. by W.) from Bradwell-near-the-Sea; containing 970 inhabitants. This parish, situated on the shore of the North Sea, which washes its eastern boundary, is about two miles in length and one mile in breadth; the surface rises gradually from the marshes till it attains a considerable elevation. The lands are watered by numerous fine springs, and the soil is generally fertile. The living is a vicarage, in the patronage of the Dean and Chapter of St. Paul's, London (the appropriators), valued in the king's books at £25. 3. 9. The appropriate tithes have been commuted for a rent-charge of £665, and the vicarial for £335, subject to the payment of rates, which on the average have respectively amounted to £135 and £57; the glebe comprises 12 acres, valued at £18 per annum. The church was rebuilt, at the expense of the inhabitants, in 1708. Here is a National school.

TILLINGTON, a township, in the parish of Burghill, hundred of Grimsworth, union and county of Hereford, 5 miles (N. W. by N.) from Hereford; containing 419 inhabitants.

TILLINGTON, a township, in the parish of St. Mary and St. Chad, Stafford, Southern Division of the hundred of Pirehill, Northern Division of the county of Stafford, $1\frac{1}{2}$ mile (N. N. W.) from Stafford; containing 42 inhabitants.

TILLINGTON, a parish, in the union of Midhurst, hundred of Rotherbridge, rape of Arundel, Western Division of the county of Sussex, 1 mile (W.) from Petworth; containing 806 inhabitants. This parish is bounded on the south by the Rother navigation, and contains some very extensive quarries of stone of good quality for building and other purposes. The living is a rectory, valued in the king's books at £13. 10.; patron, Col. Wyndham. The tithes have been commuted for a rent-charge of £740, subject to the payment of rates, which on the average have amounted to £180; the glebe comprises 22 acres. A school is supported by subscription. The church, principally in the decorated style of English architecture, with trifling insertions of a later date, was almost entirely rebuilt in 1837, at the sole expense of the late Earl of Egremont, with the exception of the tower, which was also built by the late Earl in 1808, after the style of the tower of St. Dunstan's in the East, London; the interior is elegantly arranged, and contains monuments to the Rev. Dr. James Stanier Clark, to the Mitford family, Lieut. John Ayling, and others. In the hamlet of River was formerly a chapel, and some years since a stone coffin was dug up near the site, which is now used as a trough for water. Two handsome school-rooms were erected in 1838 by the Misses Mitford, formerly of Pitts Hill, in this parish, and now of Bath, in connection with a National school for 60 children of each sex, supported by subscription. An almshouse for six poor persons was built chiefly from a bequest of the Styles family, now extinct; and in 1839, Col. Wyndham built two almshouses for two poor persons, and endowed them with £20 per annum. The late Rev. Dr. James Stanier Clark, rector of this parish, and chaplain and librarian at Carlton-House, was editor of the Stuart Papers, and author of the Life of Nelson, with whom he was present at the battle of Trafalgar, as chaplain in the navy.

TILMANSTONE (*St. Andrew*), a parish, in the union and hundred of Eastry, lathe of St. Augustine, Eastern Division of the county of Kent, 6 miles (W. by S.) from Deal; containing 407 inhabitants. The living is a discharged vicarage, valued in the king's books at £7. 12. 6.; present net income, £217; patron and appropriator, Archbishop of Canterbury.

TILNEY, a joint parish with Islington, in the union of Wisbeach, hundred of Freebridge-Marshland, Western Division of the county of Norfolk, $4\frac{1}{4}$ miles (S. W.) from Lynn-Regis: the population is returned with Islington, which see.

TILNEY (*All Saints*), a parish, in the union of Wisbeach, hundred of Freebridge-Marshland, Western Division of the county of Norfolk, $4\frac{1}{2}$ miles (W. S. W.) from Lynn-Regis; containing 420 inhabitants. The living is a vicarage, with that of Tilney-St.-Lawrence annexed, valued in the king's books at £30; present net income, £280; patrons and impropriators, Master and Fellows of Pembroke Hall, Cambridge. The church is principally in the later style of English architecture. Tilney Smeath, a common in this parish, though no more than three miles long and one broad, is said to have been so remarkably fertile, as to constantly feed 30,000 sheep, and all the horned cattle belonging to seven villages.

TILNEY (*St. Lawrence*), a parish, in the union of Wisbeach, hundred of Freebridge-Marshland, Western Division of the county of Norfolk, 6 miles (S. W. by W.) from Lynn-Regis; containing 672 inhabitants. The living is a vicarage, annexed to that of Tilney-All Saints.

TILSHEAD (*St. Thomas à Becket*), a parish, in the union of Amesbury, hundred of Branch and Dole, Devizes and Southern Divisions of the county of Wilts, 4 miles (S. S. E.) from East Lavington; containing 465 inhabitants. The living is a discharged vicarage, valued in the king's books at £7. 16.; present net income, £216; it is in the patronage of the Crown; impropriator, G. W. Taylor, Esq.

TILSOP, a joint chapelry with Nash and Weston, in the parish of Burford, union of Tenbury, hundred of Overs, Southern Division of the county of Salop, $4\frac{1}{2}$ miles (W. S. W.) from Cleobury-Mortimer: the population is returned with Nash.

TILSTOCK, a chapelry, in the parish of Whitchurch, Whitchurch Division of the hundred of North Bradford, Northern Division of the county of Salop, $2\frac{1}{2}$ miles (S.) from Whitchurch: the population is returned with the parish. The living is a perpetual curacy; net income, £87; patron and appropriator, Rector of Whitchurch. The chapel is dedicated to St. Giles, and has been rebuilt by a bequest of Francis, late Earl of Bridgewater, who was rector of Whitchurch. Here is a National school.

TILSTON (*St. Mary*), a parish, in the union of Great Boughton, Higher Division of the hundred of Broxton, Southern Division of the county of Chester; containing 873 inhabitants, of which number 395 are in the township of Tilston, 3 miles (N. W. by N.) from Malpas. The living is a rectory, valued in the king's books at £12. 2. 11.; present net income, £333; patrons, Marquess of Cholmondeley and T. T. Drake, Esq. A National school is supported partly by an endowment of £16 per annum, and partly by subscription. The

Chester canal passes through the parish and close to the village.

TILSTON-FERNALL, a township, in the parish of BUNBURY, union of NANTWICH, First Division of the hundred of EDDISBURY, Southern Division of the county of CHESTER, $2\frac{3}{4}$ miles (S. S. E.) from Tarporley; containing 170 inhabitants. A school is supported by subscription.

TILSWORTH (*All Saints*), a parish, in the union of WOBURN, hundred of MANSHEAD, county of BEDFORD, $3\frac{1}{2}$ miles (N. W. by W.) from Dunstable; containing 275 inhabitants. The living is a discharged vicarage, valued in the king's books at £8; present net income, £60; patron and impropriator, Sir G. O. Page Turner, Bart. The church contains several old monuments of the Fowler family, one to the memory of Sir Henry Chester, K.B., and an ancient altar-tomb, with an inscription in French, and the effigy of Adam de Tullesworth, in sacerdotal robes.

TILTON (*St. Peter*), a parish, in the union of BILLESDON, partly in the hundred of GARTREE, but chiefly in that of EAST GOSCOTE, Northern Division of the county of LEICESTER, $8\frac{1}{4}$ miles (W. S. W.) from Oakham; containing, with part of the liberty of Whatborough, 361 inhabitants. The living is a vicarage, valued in the king's books at £12. 16. 8.; patron, Rev. George Greaves. The church is partly in the later style of English architecture. Here was an ancient hospital, which Sir William Burdet annexed to Burton-Lazars, in the time of Henry II.

TILTS, a joint township with Langthwaite, in the parish and union of DONCASTER, Northern Division of the wapentake of STRAFFORTH and TICKHILL, West Riding of the county of YORK, 4 miles (N.) from Doncaster: the population is returned with Langthwaite.

TILTY (*St. Mary*), a parish, in the union and hundred of DUNMOW, Northern Division of the county of ESSEX, 3 miles (S. by W.) from Thaxted; containing 82 inhabitants. The living is a donative; net income, £30; patron and impropriator, Viscount Maynard. The church constitutes the remains of an abbey church, a fine specimen of the decorated style of English architecture; the east and north windows are ornamented with remarkably elegant tracery, and there are some rich stalls in the chancel, and several ancient and interesting monuments. The abbey was founded about 1152, by Robert Ferrers, Earl of Derby, and Maurice Fitz-Jeffery, for White monks, whose revenue, at the dissolution, was valued at £177. 9. 4.

TIMBERLAND (*St. Andrew*), a parish, in the union of SLEAFORD, First Division of the wapentake of LANGOE, parts of KESTEVEN, county of LINCOLN, 10 miles (N. N. E.) from Sleaford; containing 1278 inhabitants. The living is a discharged vicarage, valued in the king's books at £12. 2. 11.; present net income, £216; patron and impropriator, Sir T. Whichcote, Bart. There is a place of worship for Wesleyan Methodists.

TIMBERSCOMBE (*St. Michael*), a parish, in the union of WILLITON, hundred of CARHAMPTON, Western Division of the county of SOMERSET, $2\frac{1}{2}$ miles (W. S.W.) from Dunster; containing 453 inhabitants. The living is a discharged vicarage, in the patronage of the Prebendary of Timberscombe in the Cathedral Church of Wells (the appropriator), valued in the king's books at £6. 10.; present net income, £170. The church has an embattled tower, surmounted by a low spire. Richard Ellsworth, in 1714, bequeathed £200 towards building a school-house, and an annuity of £20 for clothing and educating poor children: it was not erected till 1824, and the original endowment having accumulated to £50 per annum, about 60 children are instructed and clothed. He also left £40 per annum to Balliol College, Oxford, for the endowment of two scholarships, to be enjoyed for seven years, by boys of Timberscombe, Cutcombe, Selworthy, Wooton-Courtney, Winchead, or Dunster, in default of which, by two from any other part of the county, to be chosen by the Master and Fellows of that College.

TIMBLE, GREAT, a township, in the parish of FEWSTON, Lower Division of the wapentake of CLARO, West Riding of the county of YORK, $5\frac{1}{4}$ miles (N. by W.) from Otley; containing 218 inhabitants.

TIMBLE, LITTLE, a township, in the parish of OTLEY, Upper Division of the wapentake of CLARO, West Riding of the county of YORK, $5\frac{1}{2}$ miles (N.) from Otley; containing 56 inhabitants.

TIMPERLEY, a township, in the parish of BOWDON, union of ALTRINCHAM, hundred of BUCKLOW, Northern Division of the county of CHESTER, $1\frac{3}{4}$ mile (N. E. by E.) from Altrincham; containing 752 inhabitants. A chapel has been built here, containing 500 sittings, 340 of which are free, the Incorporated Society having granted £300 in aid of the expense. A school is supported by the interest of £300 given by Mrs. Jane Houghton, who also gave £100 for purchasing bibles and prayer-books for the scholars.

TIMSBURY (*St. Mary*), a parish, in the union of CLUTTON, hundred of CHEW, Eastern Division of the county of SOMERSET, 5 miles (S. E.) from Pensford; containing 1367 inhabitants. The living is a rectory, valued in the king's books at £11. 19. $9\frac{1}{2}$.; patrons, Master and Fellows of Balliol College, Oxford. The tithes have been commuted for a rent-charge of £283, subject to the payment of rates; the glebe comprises 64 acres, valued at £113 per annum. The church has been enlarged, and 280 free sittings provided, the Incorporated Society having granted £250 in aid of the expense. There is a place of worship for Wesleyan Methodists. The Radford canal passes through the parish, in which there are coal mines that supply the city of Bath. A National school has been established.

TIMSBURY (*St. Andrew*), a parish, in the union of ROMSEY, hundred of KING'S-SOMBOURN, Andover and Northern Divisions of the county of SOUTHAMPTON, $2\frac{1}{4}$ miles (N. by W.) from Romsey; containing 165 inhabitants. The living is a vicarage; net income, £64; patrons, J. Fleming and William Chamberlayne, Esqrs. The church contains 280 free sittings, the Incorporated Society having granted £250 in aid of the expense. There is a place of worship for Wesleyan Methodists. The Andover canal passes through the parish.

TIMWORTH (*St. Andrew*), a parish, in the union of THINGOE, hundred of THEDWASTRY, Western Division of the county of SUFFOLK, $4\frac{1}{4}$ miles (N. by E.) from Bury-St.-Edmund's; containing 216 inhabitants. The living is a rectory, consolidated with the rectory of Ingham, and valued in the king's books at £9. 17. 11. A school is partly supported by the minister.

TINCLETON, a parish, in the union of DORCHES-

TER, hundred of PIDDLETOWN, Dorchester Division of the county of DORSET, 5¼ miles (E.) from Dorchester; containing 171 inhabitants. The living is a perpetual curacy, valued in the king's books at £5. 11. 8.; present net income, £92; patron, C. H. Sturt, Esq. The church is a small structure, the burial-place of the Baynards of Cliff, of which family it contains several sepulchral memorials. This parish is bounded on the south by the river Frome.

TINGEWICK (*St. Mary*), a parish, in the union, hundred, and county of BUCKINGHAM, 2¾ miles (W. by S.) from Buckingham; containing 866 inhabitants. The living is a rectory, valued in the king's books at £12. 16. 3.; present net income, £359; patrons, Warden and Fellows of New College, Oxford. The church exhibits some remains of Norman architecture, and contains a curious brass to the memory of Erasmus Williams, who died rector of this parish in 1608. Here was formerly a market on Tuesday, granted in 1246 to the abbey *De Monte Rothomago* in Normandy, to which the manor had previously been given by the family of Finmore. The Rev. Francis Edmonds, in 1751, endowed a charity school with £15 per annum, for teaching and clothing twelve boys, appointing the warden and Fellows of New College, Oxford, the trustees: it is now conducted on the National system. Charles Longland, in 1688, bequeathed property now producing £11 per annum, which is annually distributed among poor persons of the place.

TINGRITH (*St. Nicholas*), a parish, in the union of WOBURN, hundred of MANSHEAD, county of BEDFORD, 4¼ miles (E. by S.) from Woburn; containing 162 inhabitants. The living is a discharged rectory, valued in the king's books at £9; patron, R. Trevor, Esq. The tithes have been commuted for a rent-charge of £240, subject to the payment of rates; the glebe comprises 5½ acres, valued at £6. 2. 6. per annum.

TINHEAD, a tything, in the parish of EDINGTON, hundred of WHORWELSDOWN, Whorwelsdown and Northern Divisions of the county of WILTS, 1 mile (N. by E.) from Edington; containing 472 inhabitants.

TINSLEY, a chapelry, in the parish and union of ROTHERHAM, Southern Division of the wapentake of STRAFFORTH and TICKHILL, West Riding of the county of YORK, 2¾ miles (S. W. by W.) from Rotherham; containing 368 inhabitants. The living is vicarial; net income, £124; patron and impropriator, Earl Fitzwilliam. The chapel contains 100 free sittings, the Incorporated Society having granted £100 in aid of the expense. A school is endowed by Earl Fitzwilliam with £10 per annum, for which 10 children are instructed.

TINTAGEL (*St. Symphorina*), a parish, in the union of CAMELFORD, hundred of LESNEWTH, Eastern Division of the county of CORNWALL; comprising the disfranchised borough of Bossiney, and with it containing 1006 inhabitants. This parish is situated on the shore of the Bristol Channel, by which it is bounded on the north, and was distinguished at an early period for a castle, of which the foundation is attributed to King Arthur. This fortress, which was built partly on a stupendous craggy rock, surrounded by the sea, and partly on the precipitous cliff which skirts the main land, consisted of two divisions separated by a frightful chasm, 300 feet deep, over which was a drawbridge affording means of communication. It was occupied occasionally by several of the English princes, and in 1245, Richard, Earl of Cornwall, entertained in this castle his nephew, Davydd, Prince of Wales, during his rebellion against Henry III. In subsequent reigns, till within a few years of that of Elizabeth, it continued to be a royal castle, under a governor appointed by the crown, and was used as a state prison for the duchy of Cornwall. The surrounding scenery is strikingly picturesque: on the Trevillet estate is a deep romantic vale of considerable length, in some parts richly wooded, in others alternated with spiral rocks and overhanging precipices, and terminating on the south-east with a lofty and picturesque cascade. On the cliffs, which are romantically bold, are several slate quarries, from which 200 cargoes are annually procured, and shipped at a wharf erected for the purpose, near the remains of Arthur's Castle. In these quarries are found those beautifully transparent and regular polygonal chrystals called "Cornish diamonds." A fair for cattle is held at Trevenna, in this parish, on the Monday after October 19th. The living is a vicarage, valued in the king's books at £8. 11. 3.; present net income, £220; patrons, Dean and Canons of Windsor; impropriator, Lord Wharncliffe. The church is an ancient structure, with some interesting details, and contains a curious Norman font: there were formerly two chapels in the parish, of which one was dedicated to St. Piran, and the other to St. Denis. There is a place of worship for Wesleyan Methodists. On the Trevillet estate are some remains of ancient earthworks, called Condolden Burrows; in the churchyard are three barrows, and also in the town of Bossiney is a barrow, on which the writ for the election of members for that borough was formerly read. Near the town is an ancient cross. The remains of King Arthur's Castle consist chiefly of large scattered masses of the broken towers, and walls pierced for the discharge of arrows: in Leland's time the keep was remaining; it stood on the peninsula, and, according to that writer, contained "a praty chapel, with a tumbe on the left syde."

TINTERN, LITTLE (*St. Michael*), a parish, in the union of CHEPSTOW, Upper Division of the hundred of RAGLAND, county of MONMOUTH, 4¾ miles (N.) from Chepstow; containing 313 inhabitants. The living is a discharged rectory, valued in the king's books at £2. 1. 5½.; present net income, £162; patron, W. Gale, Esq. Here is a National school. Philip Hacket, in 1634, bequeathed property now producing £36 per annum, which is distributed in moieties among the poor of Chapel-Hill and Little Tintern.

TINTINHULL (*St. Margaret*), a parish, in the union of YEOVIL, hundred of TINTINHULL, Western Division of the county of SOMERSET, 2¼ miles (S. W.) from Ilchester; containing 473 inhabitants. The living is a perpetual curacy; net income, £90; patron and impropriator, Hon. Hugh Arbuthnot. The old Roman Fosse-way passes through the parish, which is bounded on the north by the navigable river Ivel. Stock-Dennis, now a tything in this parish, was anciently a very populous place, but at present it contains only two families.

TINTWISTLE, a township, in the parish of MOTTRAM-IN-LONGDENDALE, union of ASHTON-UNDER-LYNE, hundred of MACCLESFIELD, Northern Division of the county of CHESTER, $9\frac{1}{2}$ miles (N. E. by E.) from Stockport; containing 1820 inhabitants, who are mostly employed in the manufacture of cotton and woollen goods, and at the quarries of stone in the neighbourhood. The village is situated on an acclivity rising from the western bank of the river Etherow: fairs for cattle are held on May 2nd and November 1st. This was anciently a borough, and had a court leet of its own; it is now only a member of the lordship of Mottram-in-Longdendale, and is within the jurisdiction of a court of requests held there, for the recovery of debts under 40*s*. A chapel has been erected in this township, and was consecrated October 18th, 1837. There is a place of worship for Calvinistic Methodists. A school is endowed with £5. 5. per annum.

TINWELL (*All Saints*), a parish, in the union of STAMFORD, hundred of EAST, county of RUTLAND, $1\frac{1}{2}$ mile (S. W. by W.) from Stamford; containing 262 inhabitants. The living is a rectory, valued in the king's books at £12. 10. 5.; present net income, £303; patron, Marquess of Exeter. The church contains a monument to Elizabeth Cecil, sister of Lord Treasurer Burghley. There is an endowment of £14 per annum, given by the Marquess of Exeter, for the instruction of children; and a National school has been established.

TIPTON (*St. Martin*), a parish, in the union of DUDLEY, Southern Division of the hundred of OFFLOW, and of the county of STAFFORD, $1\frac{1}{2}$ mile (N. E.) from Dudley; containing 14,951 inhabitants. This place, which is situated nearly in the centre of an extensive and rich mining district, has progressively risen from an inconsiderable village to its present extent and importance, from the abundant and apparently exhaustless mines of coal and iron-stone which are found under almost every acre of its surface. The former is of superior quality, and is found in strata of 30 feet in thickness, and the latter is wrought to a very considerable extent, affording together employment to an immense population, of whose dwellings, with the exception of some few of a superior class, the village principally consists. There are not less than nine forges, with blast furnaces and other apparatus, for the manufacture of pig-iron, of which, on an average, more than 70 tons per week are made, and in which the weekly consumption of coal is not less than 600 tons. Nails and hinges of every kind, fenders, fire-irons, and boilers for steam-engines, are manufactured to a great extent; and there are also manufactories for tinned plates, soap, muriatic potash, and red lead. The trade is much facilitated by the Birmingham canal, and several of its collateral branches, which intersect the parish, whereby a communication has been established with almost every line of inland navigation, and the produce of its mines and manufactures is conveyed to many of the principal towns in the kingdom. This place is supplied with gas from the extensive works at West Bromwich, $2\frac{1}{2}$ miles distant. The river Trent has its source within a few hundred yards of the western boundary of the parish. Tipton is within the jurisdiction of the county magistrates, and also within that of a court of requests for the parishes of Hales-Owen, Rowley-Regis, West Bromwich, Harborne, and the manor of Bradley, in the parish of Wolverhampton, in the counties of Worcester, Salop, and Stafford, respectively, established by an act passed in the 47th of George III., for the recovery of debts not exceeding £5. Officers for the internal regulation of the parish are annually appointed at the court leet of the lord of the manor; and it has lately been made a polling-place for the southern division of the county. The living is a perpetual curacy; net income, £419; patron, J. S. Hellier, Esq.; appropriator, Prebendary of Prees, or Pipa Minor, in the Cathedral Church of Lichfield. The ancient church, having become greatly dilapidated many years since, a neat and commodious new church of brick was erected; and an additional church, erected at an expense of £3700, of which £2000 was a grant from the Commissioners, and the remainder raised by subscription, was opened for divine service on May 14th, 1839: it contains 1300 sittings, of which 770 are free, the Incorporated Society having granted £300 in aid of the expense. There are six places of worship for Wesleyan Methodists, and one each for Primitive Methodists, Baptists, Independents, and Kilhamites. In 1796, Mr. Solomon Woodall gave £650 for the foundation and endowment of a school, and several subsequent donations and benefactions have been made, which, aided by subscriptions, are appropriated to the support of several National schools, in which more than 700 children receive instruction: there are also Sunday schools in connection with the Established Church and the dissenting congregations, supported by subscription.

TIRLEY (*St. Matthew*), a parish, in the union of TEWKESBURY, partly in the Lower Division of the hundred of WESTMINSTER, and partly in that of the hundred of DEERHURST, Eastern Division of the county of GLOUCESTER, $5\frac{1}{4}$ miles (S. W. by W.) from Tewkesbury; containing 498 inhabitants. The navigable river Severn flows through the parish, and is crossed at Haw by a handsome stone bridge, completed in 1824, on the new line of road leading from Cheltenham into Herefordshire, Monmouthshire, and South Wales. The living is a discharged vicarage, valued in the king's books at £9. 6. 8.; present net income, £375: it is in the patronage of the Crown; impropriator, Earl of Coventry. The church is partly in the decorated and partly in the later style of English architecture. There is a place of worship for Wesleyan Methodists. A National school is supported by a small endowment, aided by annual subscriptions.

TISBURY (*St. John the Baptist*), a parish, and the head of a union, in the hundred of DUNWORTH, Hindon and Southern Divisions of the county of WILTS, $3\frac{1}{2}$ miles (S. E.) from Hindon; containing 2259 inhabitants. A castle, of which the origin is remote and the history obscure, appears to have been erected here prior to the reign of Edward III., since which time it has been successively the seat of the families of St. Martin, Touchet, Audley, and Willoughby de Broke, and subsequently of Sir John Arundel, whose son Thomas was, by James I., created Lord Arundel of Wardour, by which name the castle was distinguished. In the civil war of the seventeenth century, the castle was besieged by a detachment of the parliamentarian army, consisting of 1300 men, under the command of Sir Edward Hungerford, and defended in the absence of Lord Arundel by his wife, the Lady Blanch, with a garrison of only 25 men, for nearly a week, when it surrendered on May 8th, 1643, upon honourable terms, which, however, were not fulfilled by the captors. It was, in the course of the same summer, retaken by the royalists under Lord Arundel and Sir Francis Doddington, after a siege of several weeks, from the celebrated Edmund Ludlow, who had been made governor by the parliament, and who, in his memoirs, accuses the royalists of the same disregard of the terms of capitulation which had been shown by the parliamentarians. From the great injury which the castle received, especially on the latter occasion, it became totally unfit either for the purposes of a fortress or a residence; and, since the year 1776, the family of Arundel have erected a magnificent mansion, called Wardour Castle, consisting of a centre and two wings projecting in a curvilinear form, from a design by Mr. Paine; it is a handsome structure of freestone, beautifully situated within a mile of the original castle. The living is a vicarage, valued in the king's books at £18. 10. 10.; present net income, £306; patron, Lord Arundel; appropriators, Dean and Chapter of Bristol. The appropriate tithes have been commuted for a rent-charge of £880, and the vicarial for £440, subject to the payment of rates, which on the average have amounted to £247 and £87; the appropriate glebe comprises 12 acres, and the vicarial, 3 acres, respectively valued at £24 and £5 per annum. There is also a rent-charge of £67. 12. payable to the rector of Compton-Chamberlayne, and a rent-charge of £50 to other impropriators, subject to the payment of rates, which on the average have respectively amounted to £12 and £10. The church is a spacious structure, in the Norman style of architecture, and contains numerous monuments to the noble family of Arundel. There is a place of worship for Independents. Alice Combes, in 1731, bequeathed £400, directing the interest to be appropriated to the instruction of poor children. In consequence of an increase in the funds of the charity, its original object has been extended, and there are now two masters and two mistresses, appointed in different parts of the parish. Another school is supported by subscription. An estate at Budbush, now let for £70 per annum, was purchased with bequests by Sir Matthew Arundel, in 1598; Albinus Davies, in 1703; and Sir John Davies; and the rental is appropriated partly to apprenticing poor children, and partly to the relief of the poor. The poor law union of Tisbury comprises 20 parishes or places, under the care of 21 guardians, and, according to the census of 1831, contains a population of 9763. The remains of the ancient castle are situated under a range of hills in the form of an amphitheatre, richly crowned with wood, and consist principally of the hexagonal court which formed the centre of the buildings, and in which a deep well was sunk by Ludlow for the supply of the garrison during the siege. Almost contiguous to the castle are the remains of the mansion occupied by the Arundel family, after the destruction of the castle till the completion of their present seat. Sir Nicholas Hyde, Chief Justice of the King's Bench and Lord Treasurer in the reign of James I., was born in Wardour Castle, about the year 1750; and about the same time Sir John Davies, eminent as a lawyer, a poet, and a political writer, was born in the hamlet of Chisgrove, in this parish.

TISSINGTON (*St. Mary*), a parish, in the hundred of Wirksworth, Southern Division of the county of Derby, 3¾ miles (N.) from Ashbourn; containing 459 inhabitants. Dovedale, in this parish, abounds with strikingly romantic scenery. Thorpe Cloud on the right, and a towering pile of massive rocks on the left of the entrance, form natural ramparts of majestic elevation, between which the river Dove winds through the vale with varied course, sometimes rushing with tumultuous effort along the bases of stupendous cliffs, and at others expanding into a smooth and placid surface, reflecting the luxuriant verdure of its wood-crowned banks. At various intervals, rude rocky masses of grotesque form, which have been fancifully denominated my Lady's Chair, Dovedale Castle, the Church, the Twelve Apostles, the Lion's Head, the Sugar Loaves, and the Lover's Leap, rise in succession throughout this enchanting dale, in which the simpler and the sublimer beauties of nature, in all their variety, are richly and strikingly combined. The living is a perpetual curacy; net income, £97; patron, Sir H. Fitzherbert, Bart., who, with J. G. Johnson, Esq., is the impropriator. The church is partly Norman and partly of later date: it stands on an eminence overlooking the village, where are five springs of the purest water, which at a remote period are said to have furnished the only supply of the neighbourhood for several miles round. In grateful commemoration of this, an ancient custom, termed the "Floralia," prevails among the villagers, annually on Holy Thursday; *viz.*, that of decorating these fountains with the choicest flowers, as offerings to the Naïads: they afterwards repair to the church, where the ritual of the day is performed, and a sermon preached: the springs are then visited by the minister, choristers, and people; the Psalms, Epistle, and Gospel are read, and a hymn sung, and the remainder of the day is spent in rural festivity. Catherine Port, in 1722, bequeathed £5 a year, and Francis Fitzherbert, in 1735, gave an annuity of £4, for teaching poor children.

TISTED, EAST (*St James*), a parish, in the union of Alton, hundred of Selborne, Alton and Northern Divisions of the county of Southampton, 5 miles (S. by W.) from Alton; containing 278 inhabitants. The living is a rectory, valued in the king's books at £16; present net income, £333; patron, James Scott, Esq. The Rev. Philip Valois, in 1760, bequeathed £300 in support of a school; and the Rev. John Williams, in 1822, gave £400 three per cents. for a like purpose: the united income is about £20, since increased by £20 more: a school-house and dwelling were built by Mr. Scott, in 1837, at a cost of more than £200; the school is attended by 40 children, and is supported by the endowment, by Mr. Scott, and the rector.

TISTED, WEST, a parish, in the union of Alresford, hundred of Bishop's-Sutton, Alton and Northern Divisions of the county of Southampton, 4¾ miles (S. E. by E.) from New Alresford; containing 264 inhabitants. The living is a perpetual curacy; net income, £58; patrons and impropriators, President and Fellows of Magdalene College, Oxford. The church contains 44 free sittings, the Incorporated Society having granted £20 in aid of the expense.

TITCHBOURN (*St. Andrew*), a parish, in the union of Alresford, hundred of Fawley, Winchester and Northern Divisions of the county of Southampton, 2½ miles (S. W. by S.) from New Alresford; containing 363 inhabitants. The living is annexed, with that of Kilmeston, to the rectory of Cheriton. A school is partly supported by Lady Tichborne, the rector, and curate.

TITCHFIELD (*St. Peter*), a parish, in the union of Fareham, hundred of Titchfield, Fareham and Southern Divisions of the county of Southampton, $2\frac{1}{4}$ miles (W.) from Fareham; containing 3712 inhabitants. This is a small well-built town, pleasantly situated at the mouth of Southampton water, about two miles to the west of the Titchfield river. A customary corn market is held on Saturday; and fairs are on the Saturday fortnight before Lady-day, May 14th, September 25th (for hiring servants), and on the Saturday fortnight before December 21st. A court baron is held twice a year, and a court leet annually, the latter having jurisdiction in all pleas of debt under 40*s*. The living is a vicarage, valued in the king's books at £6. 17. $3\frac{1}{2}$.; present net income, £266; patron, H. P. Delmé, Esq.; appropriators, Dean and Chapter of Winchester. The appropriate tithes have been commuted for a rent-charge of £2886, and the vicarial for £35, subject to the payment of rates; the glebe comprises 7 acres, valued at £25 per annum. The north aisle of the church was built by William of Wykeham: the chancel, which is kept in repair by the Duke of Portland, contains a handsome monument to Henry, the first Earl of Southampton. A district chapel has been built at Sarisbury, containing 440 sittings, 320 of which are free; the Incorporated Society having granted £300 in aid of the expense. At Crofton, in this parish, is a chapel of ease; and there is a place of worship for Independents. Twelve girls are educated from funds arising out of the rental of land and premises demised, in 1620, by Henry Earl of Southampton, for charitable uses. In 1703, Richard Godwin bequeathed a rent-charge of £4, for teaching twelve poor children; and a National school has been established. At a short distance north of the town are the remains of Titchfield Place, or House, erected by the first Earl of Southampton, on the site and with the materials of an abbey for Premonstratensian canons, founded by Peter de Rupibus, in 1231, the revenue of which, at the suppression, was valued at £280. 19. 10. In this mansion Charles I. was concealed after his escape from Hampton Court, in 1647, and again previously to resigning himself to Colonel Hammond, who conducted him to Carisbrooke castle, in the Isle of Wight. The building is now in a state of ruin, the entrance gateway being the only part standing; the old stables were taken down a few years since. It is asserted that the nuptials of Henry VI. with Margaret of Anjou were celebrated at this place. Rachel, wife of William Lord Russell, who was beheaded in the reign of Charles II., was born here. Titchfield confers the title of Marquess on the family of Bentinck.

TITCHMARSH (*St. Mary*), a parish, in the union of Thrapstone, hundred of Navisford, Northern Division of the county of Northampton, 2 miles (E. N. E.) from Thrapstone; containing 843 inhabitants. The living is a rectory, valued in the king's books at £45; present net income, £782; patron, Lord Lilford. Here are a National and an infants' school. An allotment of about 28 acres was awarded under an enclosure act in 1778, in lieu of an estate purchased with a bequest by Edward Pickering, in 1697: the rental, amounting to £36. 10., is distributed in money among poor persons not receiving parochial aid. Dorothy Elizabeth Pickering and Frances Byrd, in 1756, founded and endowed an almshouse for eight poor unmarried women not possessing property above £8 per annum: the income of the charity is £165.

TITCHWELL (*St. Mary*), a parish, in the union of Docking, hundred of Smithdon, Western Division of the county of Norfolk, 5 miles (W. by N.) from Burnham-Westgate; containing 159 inhabitants. The living is a rectory, valued in the king's books at £12; present net income, £354; patrons, Provost and Fellows of Eton College. There is a place of worship for Wesleyan Methodists.

TITLEY (*St. Peter*), a parish, in the union of Kington, hundred of Wigmore, county of Hereford, 3 miles (N. E. by E.) from Kington; containing 328 inhabitants. The living is a perpetual curacy; net income, £231; patrons and impropriators, Warden and Fellows of Winchester College. The church was erected about 60 years ago, on the site of that which belonged to a priory of Benedictine monks, founded as a cell to the abbey of Tyrone in France, of which there are no vestiges, except the moat that encompassed it, and a remarkably fine spring of water, still called the Priory well. A National school was established here before 1826, when Lady Coffin Greenby erected a new schoolhouse, with the aid of £80 granted by the Christian Knowledge Society; the school is endowed with £5 per annum.

TITLINGTON, a township, in the parish of Eglingham, union of Alnwick, Northern Division of Coquetdale ward and of the county of Northumberland, $7\frac{1}{4}$ miles (W. by N.) from Alnwick; containing 78 inhabitants.

TITSEY, a parish, in the union of Godstone, Second Division of the hundred of Tandridge, Eastern Division of the county of Surrey, 5 miles (N. E. by E.) from Godstone; containing 202 inhabitants. This parish is situated on the road from Croydon to Maidstone, and within its limits is one of the sources of the river Medway. Lime of superior quality is made here from the chalk-pits of Botley Hill. The living is a rectory, valued in the king's books at £7. 17. $3\frac{1}{2}$.; present income, £290; patron, W. L. Gower, Esq. The ancient church, which stood near the mansion-house of Titsey Place, was taken down and a new church erected, in 1776, by Sir John Graham, who removed all the ancient tombs and monuments, with the exception of the tomb of the Staples family, now in the grounds of Titsey Place: in the north wall of the chancel is a stone with brass effigies of William Graham and family. A parochial school is supported by Mrs. Gower, and there is also a National school. The parish derives some benefit from Henry Smith's charity.

TITTENHANGER, a hamlet, in the parish of St. Peter, borough of St. Albans, hundred of Cashio, or liberty of St. Albans, county of Hertford, $2\frac{1}{2}$ miles (S. E. by E.) from St. Albans; containing 1038 inhabitants.

TITTENLEY, a township, in the parish of Audlem, union of Drayton, hundred of Nantwich, Southern Division of the county of Chester; containing 30 inhabitants.

TITTISWORTH, a township, in the parish and union of Leek, Northern Division of the hundred of Totmonslow and of the county of Stafford, 2 miles (N. E. by N.) from Leek; containing 447 inhabitants.

TITTLESHALL (*St. Mary*), a parish, in the union

of MITFORD and LAUNDITCH, hundred of LAUNDITCH, Western Division of the county of NORFOLK, 6¼ miles (S. S. W) from Fakenham; containing, with the parish of Godwick, 570 inhabitants. The living is a rectory, with those of Godwick and Wellingham annexed, valued in the king's books at £9. 12. 8½; patron, Earl of Leicester. The tithes have been commuted for a rent-charge of £665, subject to the payment of rates, which on the average have amounted to £105; the glebe comprises 58 acres, valued at £58 per annum. The church contains, among several other monuments, an altar-tomb and effigy in white marble, of the learned and celebrated Sir Edward Coke, in his judicial costume, with an elaborate inscription.

TIVERTON, a township, in the parish of BUNBURY, union of NANTWICH, First Division of the hundred of EDDISBURY, Southern Division of the county of CHESTER, 1¾ mile (S.) from Tarporley; containing 618 inhabitants. At Four-lane Ends, in this township, is an old established corn market, held every Monday. The Chester canal passes in the vicinity. There is a place of worship for Wesleyan Methodists.

Corporation Seal.

TIVERTON (*ST. PETER*), a borough, market-town, and parish, possessing exclusive jurisdiction, and the head of a union, locally in the hundred of Tiverton, Cullompton and Northern Divisions of the county of DEVON, 14 miles (N. by E.) from Exeter, and 175 (W. by S.) from London; containing 9766 inhabitants. This place, formerly called *Twy-ford, Twy-ford-ton*, or *Two-ford-ton*, derives its name from its situation between two rivers, anciently called Fords, the Exe and the Lowman, and was known as the village of Twyford so early as 872. A castle was erected here in 1106, by Rivers, Earl of Devon, which continued for many ages the head of the barony, and which, with the lordship of the hundred and the manor, is now the property of Sir W. P. Carew, Bart. In 1200, the town had a market and three annual fairs; and, in 1250, it was supplied with water by means of a stream, called the Leat, at the expense of Isabel, Countess of Westmorland. In the year 1353, the wool trade was introduced, and in the year 1500 the inhabitants were extensively engaged in the manufacture of baizes, plain cloths, and kerseys, for which, in the time of Elizabeth, the town enjoyed considerable repute; but, in the year 1591, the plague greatly checked its prosperity, destroying nearly 600 of the inhabitants. Notwithstanding this event, however, and a destructive fire in 1598, Tiverton was regarded, in 1612, as the chief manufacturing town in the West of England; but about this time a second fire destroyed 600 houses, and occasioned very great distress among the inhabitants. During the contest between Charles and the parliament, the townsmen were much divided; in 1643, they were for a time subject to the king, but in 1645 the parliamentary forces effected the entire subjugation of the place, and the castle, church, and outworks were taken, together with the governor and 200 men. In 1731, a third fire destroyed 300 houses; and, ten years after, one-twelfth of the population was cut off by a severe epidemic fever. In 1745, the introduction of Norwich stuffs, and the subsequent establishment of a manufactory at Wellington, occasioned the decay of the woollen trade, which, in 1815, was entirely superseded by the patent net manufacture, now the staple trade of the place.

The parish is divided into four portions, or quarters: *viz.*, Clare, Pitt, Pryors, and Tidcombe. The town is pleasantly situated on elevated ground between the rivers Exe and Lowman, which unite their streams a little to the south of it, and consists of several streets of respectable appearance, which are paved throughout, under an act obtained in 1794, for its improvement, and it is now lighted with gas, by private subscription. Some of the private mansions are spacious and handsome, and the inhabitants are well supplied with water, as mentioned above. At its eastern extremity is a wharf, whence a canal extends to Burlescombe, passing in its course near the rocks of Canonsleigh, which yield excellent limestone: the lofty manufactories on the west side of the river have an imposing effect. The Exe is crossed by a handsome stone bridge originally erected in 1590, by the munificence of Mr. Walter Tyrrel, a linen-draper of this town, and lately rebuilt, from which is a fine view of the castle and church. A subscription reading-room, theatre, and assembly-room, are the chief sources of amusement. About 1500 persons are employed in the lace manufacture. The markets are on Tuesday and Saturday, the former being the principal; and there are also four great markets for cattle during the year. Fairs are held on the second Tuesday after Whit-Sunday and on Michaelmas-day. The first charter of incorporation was granted by James I., in 1615; but, in 1723, the mayor absconding on the day of election, it became forfeited, and a second was granted by George I., in 1737, under which the municipal body consisted of a mayor, twelve capital burgesses, and twelve assistants, with a recorder and town-clerk, and subordinate officers: the mayor, late mayor, and recorder, were justices of the peace, with exclusive jurisdiction within the borough. The corporation now consists of a mayor, six aldermen, and 18 councillors, under the act of the 5th and 6th of William IV., cap. 76, for an abstract of which see the Appendix, No. I. The borough is divided into three wards, and the municipal boundaries are co-extensive with those for parliamentary purposes. The mayor, the late mayor, and recorder, are justices of the peace, and three others are appointed by commission. This borough returns two representatives to parliament: the elective franchise was formerly vested exclusively in the twenty-five members of the corporation, but it has been extended to the £10 householders of the borough and parish, which are co-extensive, and contain an area of 17,500 acres. The number of voters now registered is 496: the mayor is the returning officer. The recorder holds a court of session quarterly, and a court of record occasionally for all pleas not exceeding £100. Petty sessions are held every alternate week. The police establishment consists of 16 constables. The town has lately been made a polling-place for the northern division of the county. The bridewell, a commodious edifice, was rebuilt about 40 years since: the other principal public buildings are the guildhall, and a spacious new market-place, erected in 1830.

At the close of the thirteenth century, the living was

divided, by Hugh Courtenay, Baron of Oakhampton and Earl of Devon, into the portions of Clare, Pitt, Tidcombe, and Pryors: the last was given to the monastery of St. James, at Exeter, and, having been subsequently assigned, with the monastery, to the Provost and Fellows of King's College, Cambridge, that society, as owners of the impropriate rectory, appoint the curate: the Clare portion is valued in the king's books at £27; the Pitt portion, with Cove chapelry annexed, at £36; and the Tidcombe portion, at £27: these three, which are rectorial, are in the patronage of the Earl of Harrowby, Sir W. P. Carew, Bart., Sir R. Vyvyan, Bart., and the Rev. John Spurway: present net income of the Clare portion, £452; of the Pitt portion, £675; and of the Tidcombe portion, £735. The church has been rebuilt on an enlarged plan, and contains 237 free sittings, the Incorporated Society having granted £200 in aid of the expense: the altar-piece, of which the subject is "The Deliverance of St. Peter from Prison," was painted and presented by the celebrated Mr. Cosway, a native of this town: the churchyard occupies a commanding elevation, and forms an agreeable promenade. A handsome edifice, in the modern style of architecture, has been erected as a chapel of ease: it is dedicated to St. George, and each of the four portionists officiates in turn. There are places of worship for Baptists, Independents, and Wesleyan Methodists.

The free grammar school was founded and amply endowed in 1604, pursuant to the will of Peter Blundell, a clothier of Tiverton, who gave £2400 for the purchase of ground and the erection of the building; and for its maintenance devised all his lands in Devonshire to 27 trustees, directing his executors to apply £2000 of the proceeds in the establishment and perpetual maintenance of six students at either of the Universities: certain exhibitions were added by John Ham, in 1678; by R. Downe, in 1806; and one to Balliol College, Oxford, by John Newte, in 1715; there are likewise two other exhibitions of £30 per annum each, endowed with the dividends on stock bequeathed by Benjamin Gilberd, in 1783. The upper and under master have each a house rent-free: the number of boys educated, the majority of whom are boarders, has varied considerably at different times: the whole income is upwards of £1100 per annum. The building is a venerable edifice, having its north front cased with freestone; the façade exhibits two porches, and is of considerable extent, with a spacious quadrangular court opposite. The free English school, in Peter-street, was founded in 1611, by Robert Comyn, alias Chilcot, who gave £400 for its erection, and an annuity of £20 for the master's salary. The charity school, which is situated in the churchyard, was originally founded by subscription, in 1713, and has been since extensively endowed with various benefactions; it is conducted on the National plan. A school of industry for girls is supported by subscription; another school is endowed with £10 per annum from Newte's charity; and a Sunday school belonging to Independents is endowed with £3 per annum. Almshouses for nine poor men, situated in Gold-street, were founded by John Greenway, in 1529; a chapel is attached, which contains some good carved work. The Western almshouse, which has also a small chapel, for eight poor men, was founded in 1579, by John Waldron; and another, in Peter-street, for six aged women, in 1613, by George Slee. A charitable fund was established pursuant to the will of Mary Rice, in 1697, under the management of trustees, from which 67 poor persons receive life annuities. The other charitable benefactions belonging to Tiverton, too numerous to particularise, are expended in apprenticing poor children, and contributing, in various ways, to the comfort of the poor, and the general improvement and welfare of the town. The poor law union comprises 27 parishes or places, under the care of 38 guardians, and, according to the census of 1831, contains a population of 31,229. Some few remains of the boundary wall of the old castle, with its flanking and angular towers, are still perceptible; particularly part of the grand east entrance, and fragments on the south-west: the site, which occupies about an acre of ground, is on a level with the churchyard, and overhangs the river. Mrs. Cowley, the dramatic writer, was a native of this town.

TIVETSHALL (*St. Margaret*), a parish, in the union of Depwade, hundred of Diss, Eastern Division of the county of Norfolk, 5¾ miles (N. E. by N.) from Diss; containing 376 inhabitants. The living is a rectory, annexed to that of Tivetshall-St. Mary.

TIVETSHALL (*St. Mary*), a parish, in the union of Depwade, hundred of Diss, Eastern Division of the county of Norfolk, 5¼ miles (N. E. by N.) from Diss; containing 313 inhabitants. The living is a rectory, with that of Tivetshall-St. Margaret annexed, valued in the king's books at £20; present net income, £760; patron, Earl of Orford. The church is a venerable structure, erected before the Conquest. Here is a Sunday National school.

TIXALL (*St. John the Baptist*), a parish, in the Southern Division of the hundred of Pirehill, union and Northern Division of the county of Stafford, 3¾ miles (E. by S.) from Stafford; containing 176 inhabitants. The living is a discharged rectory, valued in the king's books at £8. 0. 8.; present net income, £200; patron, Sir Clifford Constable, Bart. There is a Roman Catholic chapel in the parish; and a school is endowed with £35 per annum. Immense quantities of freestone are quarried in the neighbourhood of Tixhall Hall, which fine old mansion was built of that found upon the spot. Much of it has been used also in the construction of the bridges and locks of the Staffordshire and Worcestershire canal, which passes through the parish, and of the Trent and Mersey canal in the vicinity, the stone being peculiarly adapted for resisting the action of water.

TIXOVER (*St. Mary Magdalene*), a parish, in the union of Stamford, hundred of Wrandike, county of Rutland, 7½ miles (E. by N.) from Uppingham; containing 108 inhabitants. The living is annexed to the vicarage of Ketton.

TOCKENHAM (*St. John*), a parish, in the union of Cricklade and Wootton-Bassett, hundred of Kingsbridge, Swindon and Northern Divisions of the county of Wilts, 2¾ miles (S. W.) from Wootton-Bassett; containing 164 inhabitants. The living is a rectory, valued in the king's books at £6. 13. 4.; it is in the patronage of the Crown. The tithes have been commuted for a rent-charge of £245, subject to the payment of rates, which on the average have amounted to £35; the glebe comprises 36 acres,

TOCKETTS, a township, in the parish and union of

GUILSBOROUGH, Eastern Division of the liberty of LANGBAURGH, North Riding of the county of YORK, 1¾ mile (N. by E.) from Guilsbrough; containing 35 inhabitants.

TOCKHOLES, a chapelry, in the parish, union, and Lower Division of the hundred of BLACKBURN, Northern Division of the county palatine of LANCASTER, 3 miles (S. S. W.) from Blackburn; containing 1124 inhabitants. The living is a perpetual curacy; net income, £95; patron, Vicar of Blackburn. The chapel, dedicated to St. Michael, was rebuilt in the later style of English architecture, in 1833, at an expense of £2567. 0. 6., under an act of the 58th of George III. There is a place of worship for Independents; and a National school has been established.

TOCKINGTON, LOWER, a tything, in the parish of ALMONDBURY, Lower Division of the hundred of LANGLEY and SWINEHEAD, Western Division of the county of GLOUCESTER, 3¾ miles (S. by E.) from Thornbury; containing 327 inhabitants.

TOCKINGTON, UPPER, a tything, in the parish of OLVESTON, Lower Division of the hundred of LANGLEY and SWINEHEAD, Western Division of the county of GLOUCESTER, 3 miles (S. by W.) from Thornbury; containing 729 inhabitants. Three schools are partly supported by subscription.

TOCKWITH, a township, in the parish of BILTON, Ainsty of the city, and East Riding of the county, of YORK, 5¾ miles (N. E.) from Wetherby; containing 547 inhabitants. There is a place of worship for Wesleyan Methodists.

TODBERE, a parish, in the union of SHAFTESBURY, hundred of REDLANE, Sturminster Division of the county of DORSET, 4½ miles (S. W. by W.) from Shaftesbury: containing 119 inhabitants. The living is a discharged rectory, united in 1746 to that of Stower-Provost, and valued in the king's books at £5. 19. 4. The church was considered a chapel to Gillingham till 1434, when it was made parochial: the inhabitants, by ancient custom, bury at Stower.

TODBURN, a township, in the parish of LONGHORSLEY, union of ROTHBURY, Western Division of MORPETH ward, Northern Division of the county of NORTHUMBERLAND, 8 miles (N. W. by N.) from Morpeth; containing 32 inhabitants.

TODDENHAM (*St. Thomas à Becket*), a parish, in the union of SHIPSTON-UPON-STOUR, Upper Division of the hundred of WESTMINSTER, Eastern Division of the county of GLOUCESTER, 3¾ miles (N. E.) from Moreton-in-the-Marsh; containing 481 inhabitants. The living is a rectory, valued in the king's books at £18. 19. 9½.; present net income, £254; patron, Bishop of London; impropriator, A. Pole, Esq. The church is a handsome structure, with a tower and spire, and in the chancel are some canopied stone stalls. Here is a National school. The old Fosse-way bounds the parish on the west.

TODDINGTON (*St. George*), a market-town and parish, in the union of WOBURN, hundred of MANSHEAD, county of BEDFORD, 15 miles (S.) from Bedford; containing 1926 inhabitants. This place, which is of remote antiquity, was distinguished as the scene of a battle between the Romans under Aulus Plautius, who encempted his forces on Conger hill, near the church, and the Britons, commanded by their prince, Togodumnus, in which the latter were defeated, with the loss of their leader. In the reign of Henry III., the manor, which was a free warren, was granted by that monarch to Sir Paulinus Peyvre, who obtained for the inhabitants the grant of a market, and other privileges. The grand manor-house, built by Hugh Wadlowe, and rebuilt by Sir Paulinus Peyvre, steward to Henry III., was situated at the distance of one mile from Toddington, on the road to Westoning and Ampthill, and was the residence of his descendants (amongst whom was Sir John Broughton, Lord Cheney, Chamberlain to Edward VI., and Queen Elizabeth), who are all buried in Toddington church. Queen Elizabeth visited this place in 1563, and passed some time in the manor-house, which was also honoured by a visit from James I., in 1608. During the civil war of the 17th century, the parliamentary general called Hudibras, with his army, was encamped at this place; and the king, who had posted himself on Sundon hills, occupied a house at Woodend, in this parish, the site of the encampment and the moat surrounding it being still visible. The town is pleasantly situated on an eminence; the houses are chiefly of ancient appearance and irregularly built, and the poorer inhabitants are principally employed in the manufacture of straw plat, which is carried on here to a considerable extent. The market, granted by charter of Henry III.. is on Saturday, but has greatly declined; it is still considerable for straw plat, drapers' goods, and vegetables. The fairs are on St. George's day, the first Monday in June, September 4th, November 2nd, and December 16th. The ancient market-house, which was very spacious, was demolished in 1799. The parish contains about 6500 acres of good arable and pasture land.

The living is a rectory, valued in the king's books at £29. 2. 11.; present net income, £829; patrons, Heirs of the late Lady Louisa Conolly. The church is partly in the later style of English architecture, with earlier portions; the exterior is ornamented with grotesque sculptures of various animals; and the interior contains several ancient and interesting monuments, among which are some to the descendants of Sir Paulinus Peyvre; a very costly monument to the memory of Henrietta, Baroness Wentworth, who is said to have died of grief, a few months after the execution of James Duke of Monmouth, to whom she had been betrothed; and another to Lady Maria Wentworth. There is a place of worship for Wesleyan Methodists. Toddington manor-house, about one mile from the town, on the road to Milton-Bryant and Woburn, was the residence of the Duke of Cleveland, and of Thomas Wentworth, Earl of Strafford; in this house James Duke of Monmouth was concealed for some time after the battle of Sedgmoor. In digging gravel in a field on the estate of Mr. William Harbett, in 1829 and 1830, great quantities of human bones and skulls, several urns containing small bones, the head of a spear, a sword blade, some beads, and other relics of antiquity, were discovered.

TODDINGTON (*St. Leonard*), a parish, in the union of WINCHCOMB, Lower Division of the hundred of KIFTSGATE, Eastern Division of the county of GLOUCESTER, 3 miles (N. by E.) from Winchcomb; containing 290 inhabitants. The living is a discharged vicarage, with that of Stanley-Pontlarge annexed, valued in the king's books at £7. 15. 4.; patron, Lord Sudeley. A school is supported by Lady Sudeley.

TODMORDEN, a chapelry, and the head of a union, in the parish of ROCHDALE, hundred of SALFORD, Southern Division of the county palatine of LANCASTER, 20 miles (N. E.) from Manchester, and 207 (N. W. by N.) from London; containing, with part of the market-town of Todmorden, and the township of Walsden, 6054 inhabitants: the remaining part of the town, which contains a much greater population than is here specified, is included within the extensive parish of Halifax, county of York. This town, anciently called *Todmaredene*, is situated in a picturesque and fertile district, denominated the vale of Todmorden, or "the valley of the fox mere, or lake," through which flows the river Calder, here separating the counties of Lancaster and York. The manufactories for calico, fustian, dimities, satteen, and velveteen, are numerous and extensive, worked by water-mills on the river, and by steam-engines. Great facility of conveyance is afforded by the Rochdale canal, which skirts the town on the south, and other navigations which connect this place with the eastern and western oceans. The railway from Manchester to Leeds passes by Todmorden. The great prosperity of the local manufactures has essentially contributed to the extension and improvement of the town. The market is on Thursday, the first Thursday in every month being noted for the sale of cattle. Fairs are held on the Thursday before Easter and Sept. 27th, the latter continuing three days. The living is a perpetual curacy; net income, £134; patron, Vicar of Rochdale. The chapel, rebuilt about 1770, on the site of a more ancient one, is dedicated to St. Mary, and is situated on an eminence near the centre of the town. A church, in the later English style, with a tower, was erected in 1831, at an expense of £3494. 8., under the act of the 58th of George III., to supersede the old chapel. There are places of worship for Baptists, the Society of Friends, Independents, Wesleyan Methodists and those of the New Connection, and Unitarians. A school, adjoining the chapelyard, was endowed in 1713 with the sum of £100, contributed by the Rev. Richard Clegg, and £50 voluntary subscriptions. The poor law union of Todmorden comprises six chapelries or townships, under the care of 18 guardians, and, according to the census of 1831, contains a population of 23,397.

TODRIDGE, a township, in the parish of HARTBURN, union, and Western Division of the ward, of MORPETH, Northern Division of the county of NORTHUMBERLAND; containing 4 inhabitants.

TODWICK (*St. Peter and St. Paul*), a parish, in the union of WORKSOP, Southern Division of the wapentake of STRAFFORTH and TICKHILL, West Riding of the county of YORK, 8½ miles (S. E. by S.) from Rotherham; containing 210 inhabitants. The living is a rectory, valued in the king's books at £6. 14. 7.; present net income, £121; patron, Duke of Leeds.

TOFT (*St. Andrew*), a parish, in the union of CAXTON and ARRINGTON, hundred of LONGSTOW, county of CAMBRIDGE, 5 miles (E. by S.) from Caxton; containing 279 inhabitants. The living is a rectory, with the vicarage of Caldecote annexed, valued in the king's books at £6. 16. 10½.; present net income, £287; patrons, Master and Fellows of Christ's College, Cambridge. A National school is endowed by the Rev. John Preston with the interest of £500.

TOFT, a township, in the parish of KNUTSFORD, union of ALTRINCHAM, hundred of BUCKLOW, Northern Division of the county of CHESTER, 1¾ mile (S.) from Nether Knutsford; containing 200 inhabitants.

TOFT, a joint hamlet with Lound, in the parish of WITHAM-ON-THE-HILL, union of BOURNE, wapentake of BELTISLOE, parts of KESTEVEN, county of LINCOLN, 3 miles (S. W.) from Bourne; containing, with Lound, 194 inhabitants.

TOFT, a hamlet, in the parish of DUNCHURCH, Rugby Division of the hundred of KNIGHTLOW, Northern Division of the county WARWICK, ½ a mile (S. W.) from Dunchurch: the population is returned with the parish.

TOFT, MONKS' (*St. Margaret*), a parish, in the union of LODDON and CLAVERING, hundred of CLAVERING, Eastern Division of the county of NORFOLK, 3½ miles (N.) from Beccles; containing 333 inhabitants. The living is a discharged rectory, annexed to that of Haddiscoe, and valued in the king's books at £8. An Alien priory, a cell to the abbey of St. Peter and St. Paul, at Preaux in Normandy, was founded here in the time of Henry I., the revenue of which, at the suppression, was annexed by Henry V. to the Carthusian monastery at Witham, by Henry VI. to Eton College, and by Edward IV. to King's College, Cambridge.

TOFT-NEXT-NEWTON (*St. Peter and St. Paul*), a parish, in the union of CAISTOR, Northern Division of the wapentake of WALSHCROFT, parts of LINDSEY, county of LINCOLN, 4 miles (W.) from Market-Rasen; containing 74 inhabitants. The living is a rectory, valued in the king's books at £9. 10. 10.; present net income, £111: it is in the patronage of the Crown.

TOFT, TREES (*All Saints*), a parish, in the union of WALSINGHAM, hundred of GALLOW, Western Division of the county of NORFOLK, 2½ miles (S. W.) from Fakenham; containing 78 inhabitants. The living is a discharged vicarage, valued in the king's books at £7. 18. 6.; present net income, £168; patron and impropriator, Lord C. Townshend.

TOFT, WEST (*St. Mary*,) a parish, in the union of THETFORD, hundred of GRIMSHOE, Western Division of the county of NORFOLK, 5¼ miles (N. E.) from Brandon-Ferry; containing 182 inhabitants. The living is a discharged rectory, valued in the king's books at £8. 6.; present net income, £110; patron, Sir R. Sutton, Bart. The church is an ancient building of flint and stone, with a large square tower, erected early in the reign of Edward IV., and coped and embattled with freestone; the nave and chancel are separated by a screen, and in the latter is a piscina. In 1720, an oaken coffin was discovered here, containing, among other relics, human bones, the representation of a face cut in jet, a blue cypher, and several beads.

TOGSTON, a township, in the parish, of WARKWORTH, union of ALNWICK, Eastern Division of MORPETH ward, Northern Division of the county of NORTHUMBERLAND, 10 miles (S. E. by S.) from Alnwick; containing 149 inhabitants. Coal is obtained here.

TOLLAND (*St. John the Baptist*), a parish, in the union of TAUNTON, hundred of TAUNTON and TAUNTON-DEAN, Western Division of the county of SOMERSET, 3¼ miles (N. N. E.) from Wiveliscombe; containing 121 inhabitants. The living is a rectory, valued in the king's books at £7; it is in the patronage of the Crown. The tithes have been commuted for a rent-charge of £140, subject to the payment of rates; the glebe comprises 40 acres, valued at £75 per annum. A day and Sunday school is supported by the rector.

TOLLARD-ROYAL (*St. Peter*), a parish, in the union of TISBURY, partly in the hundred of CRANBORNE, Shaston Division of the county of DORSET, but chiefly in the hundred of CHALK, Hindon and Southern Divisions of the county of WILTS, 6¾ miles (S. E. by E.) from Shaftesbury; containing 286 inhabitants. The living is a rectory, valued in the king's books at £16; patron, J. Austin, Esq. The impropriate tithes have been commuted for a rent-charge of £10, and the rectorial for £560, subject to the payment of rates, which on the latter have on the average amounted to £70; the glebe comprises 55 acres, valued at £55 per annum. In this parish is an old farm-house, called King John's hunting seat, thought to be the remains of an ancient royal residence for hunting in Cranborne Chace.

TOLLER-FRATRUM (*St. Basil*), a parish, in the union of DORCHESTER, hundred of TOLLERFORD, Dorchester Division of the county of DORSET, 9 miles (N. W. by W.) from Dorchester; containing 56 inhabitants. The living is a discharged vicarage, with Winford-Eagle annexed, valued in the king's books at £10. 6.; present net income, £161; patron, F. J. Browne, Esq., who, with Lord Wynford, is the impropriator. This parish formerly belonged to the brethren of the order of St. John of Jerusalem, whence it derived its distinguishing appellation. Near the road leading to Maiden-Newton are slight traces of an ancient intrenchment, upon an eminence called White Sheet; and on Farn down a barrow was opened, many years since, which contained seventeen urns, full of firm bones and black ashes. George Brown, in 1772, left a rent-charge of £21, to be applied in support of a school for the children of this parish and that of Toller-Porcorum.

TOLLER-PORCORUM (*St. Peter*), a parish, in the union of DORCHESTER, partly in the hundred of BEAMINSTER-FORUM and REDHONE, Bridport Division, but chiefly in the hundred of TOLLERFORD, Dorchester Division, of the county of DORSET, 10 miles (W. N. W.) from Dorchester; containing 540 inhabitants. This place is said to have derived its distinguishing name from the great number of swine formerly bred here. The living is a vicarage, valued in the king's books at £5; present net income, £180; patron and impropriator, F. J. Browne, Esq. A National school has been established. The parish partakes, with Toller-Fratrum, in the benefit of a school founded in 1772, by George Browne.

TOLLERTON (*St. Peter*), a parish, in the union of BINGHAM, Southern Division of the wapentake of BINGHAM and of the county of NOTTINGHAM, 4¾ miles (S. E. by S.) from Nottingham; containing 149 inhabitants. The living is a rectory, valued in the king's books at £15. 9. 4½.; present net income, £435; patron, Pendock Barry, Esq. The church is a small structure, with a tower surmounted by eight pinnacles with vanes: the interior is peculiarly neat, though not pewed.

TOLLERTON, a township, in the parish of ALNE, partly within the liberty of ST. PETER'S, East Riding, but chiefly in the wapentake of BULMER, North Riding, of the county of YORK, 4½ miles (S. by W.) from Easingwould; containing 529 inhabitants.

TOLLESBURY (*St. Mary,*) a parish, in the union of MALDON, hundred of THURSTABLE, Northern Division of the county of ESSEX, 8 miles (E. N. E.) from Maldon; containing 1066 inhabitants. This parish is bounded on the south by the bay and river Blackwater, and the Creek of Southfleet is navigable to the village for vessels drawing six feet of water. It is supposed to have derived its name from its having been the place where customs or tolls were formerly paid by ships entering the bay. The living is a vicarage, valued in the king's books at £16. 6. 3.; present net income, £484; patron, — Lawson, Esq.; impropriator, Richard Benyon de Beauvoir, Esq. The church is an ancient edifice with a stone tower. There is a place of worship for Independents; and a National school has been established.

TOLLESHUNT, D'ARCY (*St. Nicholas*), a parish, in the union of MALDON, hundred of THURSTABLE, Northern Division of the county of ESSEX, 7 miles (N. E. by E.) from Maldon; containing 690 inhabitants. This parish, which is bounded on the south-east by the river Blackwater and Northfleet Creek, derives the adjunct to its name from the family of D'Arcy, who were anciently lords of the manor. Corn is sent to Maldon to be shipped, and great quantities of fish manure are landed in the parish. The living is a discharged vicarage, valued in the king's books at £18. 10.; present net income, £150; patron and impropriator, Lieut.-General Rebow. The church has a square embattled tower of stone, and contained a chapel of the D'Arcy family (now a pew), and some monumental brasses to that family. Here is a National school. New House, or White House farm, in this parish, was purchased in 1635, by the trustees of the charity of Henry Smith, Esq., who, besides his great munificence to almost every town and village in Surrey, left money to buy lands, directing the rents to be distributed among the poor of fourteen parishes, of which this is one.

TOLLESHUNT, KNIGHTS' (*All Saints*), a parish, in the union of MALDON, hundred of THURSTABLE, Northern Division of the county of ESSEX, 7½ miles (N. E.) from Maldon; containing 374 inhabitants. This parish is pleasantly situated, and contains some ancient mansions. The village is neatly built, and an annual fair is held there on the 29th of June. The living is a rectory, valued in the king's books at £16. 13. 4.; present net income, £488: it is in the patronage of the Crown. The church is a very ancient edifice, with a belfry turret of wood, and contains a monument of a Knight Templar. Near the manor-house of Barnewalden, some Roman pavements were discovered a few years ago.

TOLLESHUNT, MAJOR, or BECKINGHAM (*St. Nicholas*), a parish, in the union of MALDON, hundred of THURSTABLE, Northern Division of the county of ESSEX, 5½ miles (N. E. by E.) from Maldon; containing 428 inhabitants. The living is a discharged vicarage, valued in the king's books at £8; present net income, £150; patron, Rev. Charles William Carwardine; impropriators, The New England Company. The church is an ancient edifice, consisting of a nave and chancel, on the north side of which was a chapel, now destroyed, and the arched entrance walled up.

TOLPUDDLE, a parish, in the union of DORCHESTER, hundred of PIDDLETOWN, Dorchester Division of the county of DORSET, 7 miles (E. N. E.) from Dorchester; containing 349 inhabitants. The living is a vicarage, valued in the king's books at £15. 7. 3½.; present net income, £240; patrons and appropriators, Dean and Canons of Christ Church, Oxford. The church is a small ancient fabric, built of rubble. There is a place of worship for Wesleyan Methodists; also a Sunday National school.

TONBRIDGE, or TUNBRIDGE (*St. Peter and St. Paul*), a market-town and parish, and the head of a union, in the lowey of TONBRIDGE, late of AYLESFORD, Western Division of the county of KENT, 14 miles (W. S. W.) from Maidstone, and 30 (S. E.) from London; containing, with part of the chapelry of Tonbridge-Wells, 10,380 inhabitants. This place is supposed to have been originally called "Town of Bridges," from the stone bridges crossing the five streams into which the river Medway here branches, of which the present name is a contraction. The town probably owes its origin to a castle of formidable strength, supposed by some to have existed before the Conquest, but more generally thought to have been erected, very soon after that period, by Richard Earl of Clare, a relation of the Conqueror's. This castle, which was frequently an object of contention, was besieged by William Rufus, soon after his accession to the throne, the proprietor having declared in favour of Robert Duke of Normandy: it was afterwards taken by King John, in his war with the barons; and subsequently was besieged by Prince Edward, son of Henry III., on which occasion the town was burned by the garrison, to prevent its giving shelter to the assailants. Edward having ascended the throne, was sumptuously entertained here by Gilbert Earl of Clare; and during his absence in Flanders, his son, afterwards Edward II., when administering the government of the kingdom, resided in this castle, and, having been crowned king, took possession of it, in consequence of the rebellion of its owner, after which it became, with three others, the depository of the records of the kingdom. The lordship, some time after, was the property of the family of Stafford; and on the attainder of the Duke of Buckingham (the last powerful member of that family), in the reign of Henry VIII., it was seized by the crown, with his other possessions, and the castle was suffered to fall into decay.

The town consists principally of one long and spacious street, containing some good houses, and its situation, on the declivity of a hill, contributes greatly to its cleanliness: it is paved and lighted with gas. The only public buildings are the town-hall and market-house, a plain brick edifice. The principal bridge was erected in 1775, by Mr. Milne, at an expense of £1100. The chief articles manufactured are Tonbridge ware and gunpowder, but both to a less extent than formerly. The river Medway, on which convenient wharfs have been erected for the accommodation of the trade, which is very considerable, was made navigable to this town about the middle of the last century, and a considerable quantity of coal and timber is brought by it from Maidstone. The line of the South-Eastern railway passes near the town, where one of the principal stations is established. The weekly market, on Friday, is now discontinued. There is a cattle market on the first Tuesday in every month, which is very numerously attended; and a fair is held on October 12th. The county magistrates meet on the second and fourth Wednesdays in each month; and a high constable and borsholder are appointed annually at the court leet of the lord of the manor. A court of requests, for the recovery of debts under £5, is held on the third Monday in the month, comprehending within its jurisdiction the hundreds of Brenchley and Horsemonden, Codsheath, Somerden, Washlingstone, Westerham, and Wrotham, the lowey of Tonbridge, and the ville and liberty of Brasted. Two representatives were sent to parliament from this town in the 23rd of Edward I., but it has not since exercised the elective franchise: it is a polling-place for the western division of the county.

The living is a vicarage, valued in the king's books at £20. 3. 4.; present net income, £763; patron, John Deacon, Esq.; impropriators, various proprietors of land. The church is a spacious and handsome structure, with a square embattled tower; it contains some good monuments, and was some years since repaired and enlarged. There are places of worship for Calvinistic and Wesleyan Methodists; and a National school has been established. The free grammar school was founded by Sir Andrew Judd, alderman of London, in the 7th of Edward VI.; and, by letters patent of that monarch, it was ordained that, after the death of the founder, the management of the school should be vested in the "Master, Wardens, and Commonalty of the Mystery of Skinners, of London," who should appoint the master, and, with the advice of the Warden and Fellows of All Souls' College, Oxford, should make statutes and ordinances for the due government of the school. The rental of the estates, with some small additional bequests, amounting in 1819 to upwards of £4500 per annum, a suit was instituted in Chancery, respecting its application, and in 1825 the Lord Chancellor approved of a scheme, and decreed that the Skinners' Company might, with the advice of the Society of All Souls' College, Oxford, make such alterations as they should think fit, provided they did not interfere with the plan of such scheme, under the authority of which a salary of £500 per annum is paid to the head master, and £200 to the under-master, both having also rent-free residences. Sixteen exhibitions of £100 a year each, to continue for four years, were also founded from the income, for boys going to either of the Universities. The school is open to boys residing in the town, or within ten miles of it, free of charge, and to boys from any part of the United Kingdom, on payment of £7. 10. per annum to the master, and £3 per annum to the under-master; and the exhibitions are open to all the boys in the school, with preference to those on the foundation. The masters are allowed to take boarders, and any housekeeper of the town, having a licence from the Skinners' Company, which is granted on testimonials as to their character from the master, may receive scholars as boarders, the number not to exceed 30. In addition to the exhibitions founded from the endowments, the pupils are also eligible to a fellowship at St. John's College, Oxford, founded by Sir Thos. Whyte; to six exhibitions of £10 per annum each, tenable at any college in either University, founded by Sir Thomas Smith; to a scholarship of £17. 9. 6. per annum, at Brasenose College, Oxford, founded by Mr. Henry Fisher; to an exhibition of £2. 13. 4. per annum, at either of the Universities, founded by Mr. Thomas Lampard; to two exhibitions, of £6 per annum each, at St. John's College, Cambridge, founded by Mr. Worrall; to an exhibition originally £4, now £8, per annum, at either University (in default of scholars from Seven-Oaks school), founded by Mr. Robert Holmedon; and to two exhibitions, of £75 per annum each, at Jesus'

College, Cambridge (also in default of scholars from Seven-Oaks school), founded by Lady Mary Boswell. The school premises have been repaired and enlarged, and form an elegant range of building, with a frontage of about 130 feet: attached to them is a play-ground of about 12 acres in extent, and adjoining it a commodious dwelling-house for the master. The National school is supported by voluntary contributions; and the boys who were educated from the funds of the town charity school, which amounted to about £36, are now instructed in this school. The poor law union of Tonbridge comprises 10 parishes or places, containing a population of 21,159, according to the census of 1831, and is under the care of 11 guardians. The remains of the once celebrated castle consist only of the entrance gateway, which is flanked by two round towers and an artificial mount, on which the keep stood. At some distance, on the opposite side of the river, are the ruins of a priory of Black canons, founded by Richard de Clare, about the end of the reign of Henry I.: at its dissolution, in 1525, the revenue, amounting to £169. 10. 3., was intended to form a part of the endowment of Wolsey's colleges at Ipswich and Oxford; but the cardinal's disgrace occurring before the grant was confirmed, it became vested in the crown: the foundation is still visible, but little remains besides the refectory, or hall, which has been converted into a barn. About a mile from the town is a well of mineral water of the same quality as that of Tonbridge Wells.

TONBRIDGE, or TUNBRIDGE, WELLS, a market-town and chapelry, partly in the parish and lowey of TONBRIDGE, union of TONBRIDGE, and partly in the parish of SPELDHURST, hundred of WASHLINGSTONE, lathe of AYLESFORD, Western Division of the county of KENT; and partly in the parish of FRANT, hundred of ROTHERFIELD, rape of PEVENSEY, Eastern Division of the county of SUSSEX, 20 miles (S. W.) from Maidstone, and 36 (S. E. by S.) from London: the population within the district of a local act in 1838 was 9154. This attractive and fashionable watering-place owes its importance to its medicinal springs, which were first discovered in 1606, by Dudley, Lord North, then staying at Eridge House, for the benefit of his health; and in consequence of the benefit he derived from the use of them, Lord Abergavenny, who resided at Eridge, was induced to fit up the wells, and make such improvements as might lead to their becoming a place of public resort. The springs soon acquired such celebrity, that Henrietta Maria, queen of Charles I., retired hither to enjoy the benefit of the waters, after the birth of her eldest son, Prince Charles; and there being no suitable residence, she and her suit were lodged in tents upon Bishop's Down. Their increasing reputation continuing to attract many visiters, various retail dealers constructed standings, on which they exhibited their wares, under a row of trees in the road by which the company usually passed to the Wells, and finally lodging-houses were erected. Soon after the Restoration, in 1664, the place was visited by Catherine, queen of Charles II., who, residing here for some time, with the gay court of that monarch, gave it additional attraction. It was also a very favourite residence of Queen Anne, prior to her occession to the throne, and has continued ever since to attract a great concourse of company during the season, which is from May to November. The waters, which are chalybeate, are of nearly equal strength with those of the German Spa, and are considered very efficacious in cases of weak digestion, or where tonics are necessary. The town is irregularly but beautifully built, consisting of clusters of houses in different situations, and is lighted with gas and watched under the provisions of a local act obtained in 1815. In 1835, also, an act was obtained for lighting, watching, cleansing, regulating, and otherwise improving the town, and for regulating the supply of water, and establishing a market within it. The Well is situated in a sort of dingle, on a sandy bottom, surrounded by hills; and the water, which rises into a stone basin, is served by women, called "Slippers," who receive a certain sum for the season from each person drinking it. About 45 years since, a marble basin was substituted for that of stone, and, in consequence of the dirt which had accumulated in the latter, a fixed cover was added, and the water drawn off by a spout; but some of the visiters, not experiencing the usual benefit in the succeeding season, fancifully imagined that the marble cover had neutralised the effects of the water, and it was consequently removed, and a stone basin replaced. Near the Well, which is 300 feet above the level of the sea, are the principal shops and places of amusement. A spacious handsome building, called the Bath House, has been erected, and contains both hot and cold mineral baths. The Parade, which is broad and handsome, is bounded on one side by the assembly-rooms, libraries, and by shops in which Tonbridge ware and fancy articles of every kind are sold, and in front of which is a piazza extending nearly the whole length; and on the opposite side is a row of trees, with an orchestra in the midst, where a band usually plays during a portion of each day in the season: with the Parade is connected what are called the Upper and Lower walk, divided by palisades of iron. The other parts of the town are situated on detached eminences, at short distances from the Wells, called Mount Ephraim, Mount Sion, Mount Pleasant, and Bishop's Down; which, being interspersed with shrubberies and pleasure grounds, and connected with the Wells by beautiful walks regularly disposed, present a combination of interesting scenery. The inns and boarding and lodging houses are generally of a superior description. Some rocks of considerable height, surrounded with wood, about a mile and a half south-west from the town, are much visited and admired. A Literary and Scientific Institution, and a Horticultural Society, have been established; there is a small neat theatre near the Wells; and races are held, annually in August, on the common. The manufacture of wooden toys and articles for domestic use, commonly denominated "Tonbridge ware," is carried on to a considerable extent. A handsome market-place has been erected by John Ward, Esq., near that part of the town called Calverley Park; it is an elegant range of building, with an area in front, in the centre of which is a fountain, and contains an elegant and spacious room for assemblies and public meetings. The government is vested in commissioners appointed under the local act of 1835, which embraces a district of one mile beyond the town; and constables are appointed at the court leet for the "hundred of Southborough and manor of Rusthall."

The living is a perpetual curacy, in the patronage of certain Trustees. The chapel, dedicated to King Charles the Martyr, was built about 150 years since, by subscription amongst the visiters, on ground given by the lady of the manor, and is supported by an annual collection after a sermon and by subscriptions: it is a plain Grecian building, fitted up and wainscoted with fine old oak, which, with its ornamented ceiling, is much admired. Trinity church, to which a district has been attached under the 21st section of the act of the 58th of George III., was erected in 1829, at an expense of £12,000, by subscription, aided by a grant of £6000 from His Majesty's Commissioners for building new churches, in that part of the town which is in the parish of Tonbridge: it is a handsome structure, in the later style of English architecture. The Incumbent is appointed by the Vicar of Tonbridge during his life, after which the living will be a perpetual curacy, in the gift of the Patron of Tonbridge. An additional church, to be called Christ-church, was erected in 1838, in the Norman style of architecture, but is not yet consecrated or licensed. There are places of worship for those in the connection of the late Countess of Huntingdon, Independents, and Wesleyan Methodists, and a Roman Catholic chapel was erected in 1838. A charity school, which has a small endowment, adjoins the chapel, in which about 80 boys are instructed; and one has been established for girls: both are on the National system, and supported by voluntary contributions. On the 29th of Sept., 1834, the first stone of a building, to be called the Victoria National school, was laid in Calverley-lane, by their Royal Highnesses the Duchess of Kent and the Princess Victoria. There is a female British school in connection with the Independents; and an infants' school, a lying-in charity, a dispensary, a savings' bank, and a mendicity society, have also been established. The late Richard Cumberland, Esq., the celebrated dramatist, was for many years a resident on Mount Sion, frequently attracting hither some of the most eminent literary characters of the day.

TONG (*St. Giles*), a parish, in the union and hundred of Milton, Upper Division of the lathe of Scray, Eastern Division of the county of Kent, 2 miles (E. by N.) from Sittingbourne; containing 226 inhabitants. The living is a vicarage, valued in the king's books at £8. 6. 8.; patron, W. Baldwin, Esq.; appropriator, Archbishop of Canterbury. The appropriate tithes have been commuted for a rent-charge of £522. 10., and the vicarial for £205, subject to the payment of rates, which on the former have on the average amounted to £107; the glebes comprise respectively seven and two acres. The church has a tower steeple on the south side. In Tong castle Hengist surprised King Vortigern and his nobles, the latter of whom he massacred, and the former he kept a prisoner till he surrendered his kingdom: of this ancient fortress the ditch and keep-mount still remain, at a short distance to the south of the church. William Housson, in 1779, bequeathed £200, directing the interest to be applied in teaching poor boys of Tong, Bapchild, and Murston. At Pukeshall, in this parish, there was anciently an hospital, dedicated to St. James.

TONG (*St. Bartholomew*), a parish, in the union of Shiffnall, Shiffnall Division of the hundred of Brimstree, Southern Division of the county of Salop, 3¼ miles (E. by S.) from Shiffnall; containing 510 inhabitants. The living is a perpetual curacy; net income, £83; patron and impropriator, George Durant, Esq. The impropriate tithes have been commuted for a rent-charge of £391. 1. 10., subject to the payment of rates; the glebe comprises 2 acres, valued at £1. 10. per annum. The church is in the decorated style, with a handsome spire rising from the centre: it originally belonged to the abbey of Shrewsbury, and was purchased, in 1410, by Isabel, relict of Sir Fulk Pembridge, Knt., and others, who rebuilt and made it collegiate for a warden, four chaplains, priests, and two clerks, with an hospital for thirteen poor persons, whose revenue at the dissolution was valued at £45. 9. 10. Tong castle, a magnificent and extensive mansion, the seat of George Durant, Esq., was erected in the last century, upon the site of the ancient structure, which was then demolished: it is crowned with numerous turrets, pinnacles, and two lofty domes, producing a grand and striking effect. Near the church, and within the demesne of the castle, are considerable remains of the old hospital; new almshouses have been founded in its stead at the village. There are benefactions by Lady Harris, Lord and Lady Pierrepont, and the Rev. Lewis Petier, producing £45 a year, for educating and clothing poor boys and girls.

TONG, a chapelry, in the parish of Birstall, union of Bradford, wapentake of Morley, West Riding of the county of York, 4¾ miles (E. S. E.) from Bradford; containing 2067 inhabitants. The living is a perpetual curacy; net income, £166; patron and impropriator, J. P. Tempest, Esq. The manufacture of woollen cloth, worsted, rope, and twine, is here carried on. Six poor children are instructed for £6 a year, bequeathed in 1739 by Sir George Tempest, who erected the schoolhouse.

TONGE, a joint township with Haulgh, in the parish of Bolton, hundred of Salford, Southern Division of the county palatine of Lancaster, 1¼ mile (E. N. E.) from Great Bolton; containing 2201 inhabitants, who are chiefly employed in the manufacture of cotton and counterpanes, and in the extensive bleaching grounds, spinning-mills, and paper-mills, established here. The ingenious Crompton resided at Hull-i'-th'-Wood, the ancient seat of the Starkie family, where he completed his invention of the spinning-mill, which he sold for not more than £100. He, however, received a grant from parliament of £5000, and a subscription was opened, by the cotton spinners and others of Bolton and Manchester, for the purchase of an annuity, which he enjoyed during the remainder of his life. The commissioners of enclosure have awarded land, now let for £5 a year, towards the foundation of a school. A limited number of boys of this place are annually appointed to the High Style school, founded by Henry Mather, at Kearsley.

TONGE, a township, in the parish of Oldham *cum* Prestwich, union of Oldham, hundred of Salford, Southern Division of the county palatine of Lancaster; containing 1800 inhabitants. This township adjoins Middleton, and forms a populous part of the environs of that town. A church has been built by aid from Her Majesty's commissioners, at an expense of £1773; it contains 350 free sittings.

TONGHAM, a hamlet, in the parish of Seal, hundred of Farnham, Western Division of the county of Surrey, 3½ miles (N. E. by E.) from Farnham: the population is returned with the parish.

TOOLEY, a hamlet, in the parish of PECKLETON, hundred of SPARKENHOE, Southern Division of the county of LEICESTER, 5 miles (N. E.) from Hinckley: the population is returned with the parish.

TOOTHOG, a township, in the parish of CWMYOY, hundred of EWYASLACY, county of HEREFORD, 10 miles (N. N. W.) from Abergavenny; containing 127 inhabitants. Here is a Sunday school, supported by subscription.

TOOTING, LOWER, or TOOTING-GRAVENEY (*St. Nicholas*), a parish, in the union of WANDSWORTH, Western Division of the hundred of BRIXTON, Eastern Division of the county of SURREY, 7 miles (S. S. W.) from London; containing 2063 inhabitants. This village, consisting of two streets, is situated on the road from London to Brighton, through Reigate; it is lighted with oil, and supplied with water from wells produced by boring. The atmosphere is considered very salubrious, and the environs are studded with elegant cottages and villas. Assemblies are occasionally held during the winter months. The parish is within the jurisdiction of a court of requests held at Wandsworth, for the recovery of debts under £5, and is also within the superintendence of the new police. The living is a rectory, valued in the king's books at £8. 8. 6½.; present net income, £374; patron, Rev. Richard Greaves. The church was rebuilt in 1833, by subscription, by a sale of part of Tooting common, and by a grant of £350 from the Incorporated Society, for which 440 free sittings have been provided: it contains monuments to Sir John Hebden, ambassador to Russia in the reign of Charles I.; and to Sir James Bateman, and others. There are places of worship for Independents and Wesleyan Methodists. The National school is endowed with the interest of a bequest of £400 in the five per cent. consols., by Mr. John Avarn, in 1809; and by another bequest, from William Powell, Esq., in 1823, of £210, and a share of the residue of his estate, amounting to £1113, which were together laid out in the purchase of £1441. 18. 11. three per cent. consols.: it is further supported by voluntary contributions; 30 of the girls are clothed. The school-rooms, and separate apartments for the master and mistress, were erected on the site of some former ones, in 1828, at an expense of £1800. In 1718, Sir James Bateman, Knt., bequeathed £100 for apprenticing children.

TOOTING, UPPER, a hamlet, in the parish of STREATHAM, Eastern Division of the hundred of BRIXTON and of the county of SURREY, 6¾ miles (S. S. W.) from London: the population is returned with the parish. This village, which is also designated Tooting-Beck, is well sheltered from the north winds; and the salubrity of the air, the purity of the water, and its dry gravelly soil, have made it the residence of several respectable families. In that part of the hamlet adjoining Balham Hill, also a hamlet in the parish of Streatham, is a proprietary episcopal chapel, erected by the inhabitants of these two places, at the expense of nearly £7000, about the year 1806, and since greatly enlarged by the addition of two wings: it has not been consecrated, but is merely licensed by the bishop, and is under the management of a committee of twelve persons, chosen from among the proprietary, who, on a vacancy occurring in the incumbency, present three candidates to the rector of Streatham, one of whom he appoints to be the officiating minister: it will accommodate about 1000 persons: over the altar is a painted window. Two school-houses, with houses for the master and mistress, have been erected by subscription at Balham Hill; the schools are supported by voluntary contributions.

TOPCLIFFE (*St. Columb*), a parish, in the union of THIRSK, partly in the wapentake of BIRDFORTH, and partly in the wapentake of HALLIKELD, North Riding of the county of YORK; containing 2592 inhabitants, of which number, 590 are in the township of Topcliffe, which extends within the liberty of St. Peter's, East Riding, 5½ miles (S. S. W.) from Thirsk. The living is a vicarage, valued in the king's books at £19. 19. 2.; present net income, £600; patrons and appropriators, Dean and Chapter of York. The church is a building of great antiquity. There is a place of worship for Wesleyan Methodists. Here are slight vestiges of the ancient baronial mansion of the Percy family, called Maiden Bower, in which Henry, the fourth Earl of Northumberland, was murdered by the populace, in 1520, for enforcing a tax imposed in the reign of Henry VII. Charles I. was confined in it, and the sum of £200,000, for giving him up to the parliament, was here paid to the Scottish commissioners. John Hartforth, in 1588, gave land and money in support of a free grammar school, which, with the subsequent smaller bequests of William Robinson and Henry Roper, produce £70 a year.

His Monumt. in Beverley Minster

TOPCROFT (*St. Margaret*), a parish, in the union of LODDON and CLAVERING, hundred of LODDON, Eastern Division of the county of NORFOLK, 4¾ miles (E. by S.) from Stratton-St. Mary; containing 463 inhabitants. The living is a rectory, valued in the king's books at £10. 13. 4.; present net income, £366; patron, Bishop of Norwich. Near Topcroft Hall there was formerly a free chapel, dedicated to St. Giles. Here is a Sunday National school.

TOPPESFIELD (*St. Margaret*), a parish, in the union of HALSTEAD, hundred of HINCKFORD, Northern Division of the county of ESSEX, 8 miles (N. W.) from Halstead; containing 1088 inhabitants. The living is a rectory, valued in the king's books at £26; present net income, £579: it is in the patronage of the Crown. The church is an ancient structure, with a handsome modern tower of brick, supplying the place of the original tower of stone, which was burnt down in 1700, and rebuilt by the Rev. Mr. Wilde, then rector. A gallery has been erected, and 80 free sittings provided, the Incorporated Society having granted £50 in aid of the expense: under an arch in the south wall of the chancel is a very ancient tomb, and there are several interesting monuments. Here is a place of worship for Independents. Robert Edwards, in 1730, left £10 per annum for teaching children: a National school has been established. A silver coin of Edward VI. was found here in 1837.

TOPSHAM (*St. Margaret*), a market-town and parish, in the union of ST. THOMAS, hundred of WONFORD, Wonford and Southern Divisions of the county of DEVON, 3½ miles (S. E.) from Exeter, and 170 (W. S. W.) from London; containing 3184 inhabitants. In the civil war of the seventeenth century, the Earl of Warwick brought some ships up the river Exe, and captured a small fort here; but the vessels being left upon the sands, on the ebbing of the tide, two were captured and one burnt by the army under Fairfax, who remained here some time. This neat little town is situated near the influx of the river Exe into the sea, and is within the limits of the port of Exeter; it is lighted with gas. The river expands here to a considerable width, forming, at high tides, a noble sheet of water. About a mile to the south, on the opposite side of it, are the sea locks, opening into the canal leading to Exeter: the prosperity of the commercial interests of Topsham is dependent on the foreign and coasting trade and the manufactures of that city. Ship-building is carried on extensively; and chain cables, anchors, ropes, twine, and sacking are manufactured: there is a large paper-manufactory, in which 200 persons are employed; and a considerable trade is also carried on in coal and timber. A quay, built about 1313, by Hugh Courtenay, was purchased by the Chamber of Exeter, in 1778, and is capable of receiving vessels of 200 tons' burden: it is large and commodious. On the strand are some neat residences, fronted with gardens extending to the water's edge, the view being justly admired for its variety and extent. In 1257, an annual fair for three days was granted to the inhabitants, and, together with a market on Saturdays, confirmed to them by Edward I. The market is still held on Saturday; and there is a small fair on the first Wednesday in August. The living is a perpetual curacy, in the patronage of the Dean and Chapter of Exeter (the appropriators); net income, £227. The church contains 250 free sittings, the Incorporated Society having granted £150 in aid of the expense: there are some good monuments by Chantry, among which is one to Sir John Duckworth, Bart.: the view from the churchyard is considered very fine. There are places of worship for Independents and Wesleyan Methodists. Sundry benefactions for the instruction of poor children produce an income of about £15 per annum, £5 of which is paid to the National school, and £10 to another school. Capt. Burgess, R.N., who was killed at the battle of Camperdown, and to whose memory a public monument has been erected in St. Paul's Cathedral, was a native of this place; and Capt. Watson, who lost his life in the West Indies, under Admiral Rowley, resided here for some time.

TORBRIAN (*Holy Trinity*), a parish, in the union of Newton-Abbot, hundred of Haytor, Teignbridge and Southern Divisions of the county of Devon, 4 miles (S. W. by S.) from Newton-Bushell; containing 257 inhabitants. The living is a rectory, valued in the king's books at £20. 14. 7.; present net income, £286; patron, John Wolston, Esq. The church contains three sepulchral chapels, and has an elegant wooden screen, an enriched pulpit of wood, an ancient font, and a piscina: there is a monument to the memory of John Petre, who bequeathed £1 each per annum to this and nineteen other parishes. The porch is ornamented with sculptured angels; and in the churchyard is an ancient cross in good preservation. This parish abounds with limestone of excellent quality.

TORKINGTON, a township, in the parish and union of Stockport, hundred of Macclesfield, Northern Division of the county of Chester, 4½ miles (S. E.) from Stockport; containing 284 inhabitants.

TORKSEY (*St. Peter*), a parish, in the union of Gainsborough, wapentake of Lawress, though locally in the wapentake of Well, parts of Lindsey, county of Lincoln, 8 miles (S. by E.) from Gainsborough; containing 484 inhabitants. This place is situated at the junction of the Fosse-Dyke with the river Trent, and formerly enjoyed many privileges, on condition that the king's ambassadors, when travelling this way, should be conveyed by the inhabitants, in their own barges, down the Trent to York. A priory of Black canons, in honour of St. Leonard, was founded here by King John, which at the dissolution was valued at £27. 2. 8. per annum. The living is a perpetual curacy; net income, £42; patron and impropriator, Sir A. Hume, Bart.

TORLETON, a hamlet (formerly a chapelry), partly in the parish of Coates, hundred of Crowthorne and Minety, and partly in that of Rodmarton, hundred of Longtree, Eastern Division of the county of Gloucester, 5 miles (W. by S.) from Cirencester: the population is returned with the respective parishes. The chapel is desecrated.

TORMARTON (*St. Mary*), a parish, in the union of Chipping-Sodbury, Lower Division of the hundred of Grumbald's-Ash, Western Division of the county of Gloucester, 4 miles (E. S. E.) from Chipping-Sodbury; containing 402 inhabitants. The living is a rectory, with the vicarage of Acton-Turville united, valued in the king's books at £27; present net income, £800; patron, N. Castleton, Esq.; impropriator, Duke of Beaufort. There is a chapel of ease at West Littleton.

TOR-MOHUN, or TOR-MOHAM, a parish, in the union of Newton-Abbot, hundred of Haytor, Paignton and Southern Divisions of the county of Devon, ¾ of a mile (N. W.) from Torquay; containing 3582 inhabitants. The living is a perpetual curacy, with that of Cockington annexed; net income, £270; patron, Rev. R. Mallock, the impropriator of Cockington; impropriators of Tor-Mohun, Sir L. V. Palk, Bart., and H. G. Cary, Esq. The church has an elegant wooden screen, formerly painted and gilt, also an ancient stone font. In 1835 an act was passed for lighting, watching, and improving this place. Here is a National school. Of 32 Premonstratensian monasteries in England, that of Torre, founded and endowed by William de Brewer, in 1196, was by far the richest: it was dedicated to our Holy Saviour, the Virgin Mary, and the Holy Trinity, and, at the dissolution, had a revenue of £396. 0. 11. The situation of the abbey is most beautiful; and the remains of the church (which is said to have been richly furnished with cloth of gold), the chapter-house, &c., evince the former magnificence of the buildings: the old refectory was, many years since, converted into a Roman Catholic chapel, still existing; and of the three gateways mentioned by Leland, the only one now remaining is much admired for the beauty of its architectural proportions. The modern mansion of Torre Abbey is the seat of H. G. Cary, Esq., in whose family it has continued since 1662. On a hill, about half a mile from the church, are the remains of a chapel, which was dedicated to St. Michael.

TORPENHOW (*St. Michael*), a parish, in the union of Wigton, Allerdale ward below Derwent, Western Division of the county of Cumberland; containing 1032 inhabitants, of which number, 317 are in the township of Torpenhow with Whitrigg, 2½ miles (W. by N.) from Ireby. This parish, which is bounded on the north by the river Ellen, abounds with freestone and limestone. The living is a vicarage, valued in the king's books at £33. 6. 8.; patron, Bishop of Carlisle. The church is principally in the Norman style; the roof of carved oak is painted and curiously embellished. The ancient free school at Bothel was founded and endowed by subscription, and was rebuilt about 35 years ago; the annual income is £42.

TORPOINT, a chapelry, in the parish of ST. ANTHONY, Southern Division of the hundred of EAST, Eastern Division of the county of CORNWALL, 3 miles (W.) from Devonport: the population is returned with the parish. The village occupies a peninsula, formed by the river Tamar, the Lynher, and St. John's lake, from which the inhabitants derive an abundance of fish. Though small, it is highly respectable; and in the vicinity are many genteel seats, of which Trematon Castle is the most distinguished. The living is a perpetual curacy; net income, £124; patron, Hon. R. P. Carew. There are places of worship for Independents and Wesleyan Methodists. Sir Coventry Carew founded a free school, for teaching and clothing ten children: there is also a National school, supported by subscription.

TORQUAY, a chapelry, in the parish of TOR-MOHUN, hundred of HAYTOR, Paignton and Southern Divisions of the county of DEVON, 7 miles (S. E. by S.) from Newton-Bushell, and 23 (S.) from Exeter: the population is returned with the parish. This place, about 45 years since, was an insignificant fishing hamlet, but is now a fashionable and attractive watering-place: it is situated in the most northerly cove of Torbay, and occupies a somewhat irregular, but singularly beautiful, site. The first great improvement was the erection of a pier and quay, for which an act of parliament was obtained by Sir Lawrence Palk, to whom the town is greatly indebted: it was commenced in 1803, and completed in 1807; and another pier has since been constructed, forming a secure basin, 500 feet long and 300 broad. A considerable portion of the town, consisting of neat and comfortable residences (principally lodging-houses), and shops of the best description, is built at the sides of the basin and on the strand; there are also two excellent hotels, warm and cold baths, and a library, with billiard and news-rooms. On the north, east, and west sides the town is completely sheltered by hills of very considerable elevation, on the declivities of which are detached houses and terraces, some of them very handsome buildings; and the heights on which they are situated being richly clothed with wood, their appearance from the pier head is strikingly beautiful. An annual regatta takes place about August, which is well attended; the principal prize is a splendid gold challenge cup, of the value of £100, with an accumulated fund added. The assembly-room, erected in 1826, is much frequented during the season, which is from September to May. The salubrity and mildness of the air, arising from its contiguity to the sea and its sheltered situation, renders this a most desirable winter residence for persons of a consumptive habit, or those for whom a mild climate is necessary; and it is usually, at this period of the year, very full of company: it is well supplied with water. Torquay has a trifling share in the Newfoundland trade; and, in addition to several coasting vessels employed in the importation of coal and other commodities, it has a weekly communication by water with London, and the advantage of steam-boats passing four times in the week. There is a small but very convenient market-place, well supplied with provisions at the customary markets, which are on Tuesday and Friday: a fair is held annually at Easter. The living is a perpetual curacy; net income, £104; patron, Perpetual Curate of Tor-Mohun. The chapel being found insufficient to accommodate the increasing population of the town, and, from its confined situation, incapable of enlargement, another has recently been erected; they are both handsome modern structures. There are also places of worship for Calvinistic and Wesleyan Methodists; and a National school. In the cliffs in this neighbourhood are several remarkable fissures, or openings, particularly that called Kent's Hole, which is of extraordinary magnitude, comprising numerous caves of various elevations, to which are several openings, one of them 93 feet deep, 100 wide, and 30 in height, containing many interesting specimens, both stalactital and organic, and fossil remains of the elephant and several other animals. Druidical knives have also been discovered.

TORRINGTON, BLACK (*ST. MARY*), a parish, in the union of HOLSWORTHY, hundred of BLACK TORRINGTON, Holsworthy and Northern Divisions of the county of DEVON, 5¼ miles (W. by N.) from Hatherleigh; containing 1083 inhabitants. This parish, which is intersected by the river Torridge, comprises the ancient manor of the Cohans, who were proprietors of it in 1066. The living is a rectory, valued in the king's books at £22. 8. 9.; present net income, £303; patron, Lord Poltimore. There is a place of worship for Baptists. A National school has been established.

TORRINGTON, EAST (*ST. MICHAEL*), a parish, in the union of CAISTOR, Western Division of the wapentake of WRAGGOE, parts of LINDSEY, county of LINCOLN, 4 miles (N. N. E.) from Wragby; containing 87 inhabitants. The living is a discharged rectory, with the vicarage of Wragby united in 1735, valued in the king's books at £7. 10. 10.; present net income, £327; patron, C. Turnor, Esq.

Arms.

TORRINGTON, GREAT (*ST. MICHAEL*), an incorporated market-town and parish, having separate jurisdiction, and the head of a union, locally in the hundred of FREMINGTON, Great Torrington and Northern Divisions of the county of DEVON, 34 miles (N. W.) from Exeter, and 202 (W. by S.) from London; containing 3093 inhabitants. The name of this place is derived from its situation on the river Torridge; and its antiquity as a market-town is evident from various old records, in which it occurs under the appellation of "Cheping Toriton." At a very early period it gave the title of Baron to its lords, who had the power of life and death throughout the lordship; and, in 1340, Richard de Merton, in whose possession it then was, erected a castle here, of which the chapel was remaining about the close of the last century. In 1484, Bishop Courtenay was tried at the sessions held here, on a charge of treason against Richard III.; and, in 1590, the county sessions were held at this place, on the appearance of the plague at Exeter, which malady afterwards extended to Great Torrington. During the civil war of the 17th century, Colonel Digby, who had fortified himself in this place, was attacked, in 1643, by a party of the parliamentary forces, strengthened by the garrisons of Barnstaple and Bideford, whom he defeated and put to flight. In 1646, the royalists, under Lords Hopton and Capel, and Sir John Digby, having taken possession of and fortified the town, were attacked by the parliamentary forces under Sir Thomas Fairfax, who, after a severe contest, drove them from their post, and obtained a victory which put an end to the power of the royalists in this part of the country. This achievement was celebrated by a thanksgiving sermon preached in the market-place by the noted Hugh Peters, whose eloquence was considered to be very effectual in promoting the parliamentarian interest. Sir Thomas Fairfax was frustrated in his intention of prolonging his stay here, by the accidental explosion of 80 barrels of gunpowder, deposited in the church, by which the south-west angle of that building was destroyed, and 200 prisoners who were confined in it, together with the soldiers on guard, perished. In 1724, this place suffered severely from an accidental fire, by which about 80 houses were destroyed, and the records of the corporation were burnt. The town occupies a singularly bold and picturesque situation on the summit and declivity of a lofty cliff, washed at its base by the river Torridge, over which is a bridge connecting this parish with that of Little Torrington: it is lighted with gas, and consists of several good houses surrounding the market-place, and of two streets respectively on the ridge and on the declivity of the cliff, with gardens sloping down towards the river, the banks of which are crowned with finely-varied scenery; in its winding course, a little above the town, the stream passes beneath some of the richest hanging woods in the kingdom. The bowling-green, which occupies the site of the ancient castle, on the highest portion of the cliff, commands an extensive and beautiful prospect. The woollen trade, which was formerly considerable, is now confined to the manufacture of a few serges, blankets, and some coarse woollen cloths; and the principal trade at present is the making of kid, chamois, beaver, and other gloves, for the London and foreign markets; the beaver gloves made here are the same as those called Woodstock, and the preparation of the leather affords employment to a considerable number of men; great quantities of gloves are also sewn here by commission, and the trade together affords employment to 3000 girls in the town and neighbourhood. There are two tan-yards in the town, and on the river is a large corn-mill. Some veins of lead-ore have been found in the neighbourhood, which is supplied with coal, lime, and timber, by a canal extending from the town to the sea lock near Bideford, and running in a direction nearly parallel with the river, which at that place becomes navigable for sloops. This line of canal owes its construction wholly to Lord Rolle, the owner of the manor, through whose lands and a portion of the common it passes, and at whose expense it was executed at a cost of more than £40,000. About two miles to the north-west of the town, it is carried over the river by a noble aqueduct, of which the first stone was laid on August 11th, 1824, and nearer the sea lock it has a descent by an inclined plane. New turnpike-roads have lately been made to Exeter and Bideford, from which many rich and romantic views are obtained. The market, held by prescription, is on Saturday, and that on the third Saturday in March is also one of the largest cattle markets in the West of England; there is also a smaller cattle market in November: the fairs are on May 4th, July 5th, and October 10th.

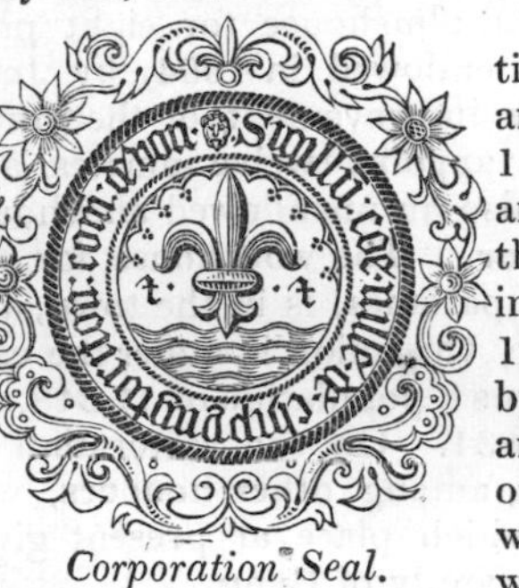

Corporation Seal.

By charters of incorporation granted to the inhabitants by Philip and Mary, in 1554, by James I. in 1617, and by James II. in 1686, the government was vested in a mayor, 8 aldermen, and 12 capital burgesses, assisted by a recorder, town-clerk, and other officers: the mayor, ex-mayor, and recorder, were justices of the peace, with exclusive jurisdiction within the borough. The corporation at present consists of a mayor, 4 aldermen, and 12 councillors, under the act of the 5th and 6th of William IV., cap. 76, for an abstract of which see the Appendix, No. I. The mayor and late mayor are justices of the peace, and hold a court of petty sessions for the borough every three weeks. A court of record, which was directed to be held every three weeks for the recovery of debts to the amount of £50, has fallen into disuse. The county magistrates hold petty sessions for the division every Saturday. This place sent representatives to thirteen parliaments, in the reigns of Edward I. and succeeding sovereigns, but the inhabitants were released on their own petition. They enjoy the right of pasturage on an extensive common near the town, consisting of about 500 acres, which was granted to the occupiers of ancient messuages by one William Fitz-Robert, lord of the manor of Great Torrington: of this tract, 50 acres were enclosed a few years since, under a general act of parliament, to be allotted for cultivation by the poor. The town-hall is a neat modern edifice of brick, ornamented with stone, supported on arches, affording a covered area underneath; and there is also a small prison. The town has been made a polling-place for the northern division of the shire.

The living is a perpetual curacy, with that of St. Giles'-in-the-Wood annexed, valued in the king's books at £20; present net income, £162; patrons and appropriators, Dean and Canons of Christ Church, Oxford. The church, after its partial destruction by gunpowder in 1646, was rebuilt in 1651; and the present structure, which is in the Tuscan order in the interior, includes such portions of the old church as escaped destruction: it previously contained a north transept, and, in 1831, a south transept was erected, at the expense of £130, on the site of the old steeple; and a new western tower, surmounted by a spire, was built, at an expense of £1600, of which £700 was defrayed by the feoffees of the town lands, and the remainder by a rate. Over the communion table is a large painting of the Ascension, presented by Lady Rolle, in 1812. There are places of worship for Baptists, Independents, and Wesleyan Methodists. A grammar school was formerly held in the town-hall, but it has been discontinued for many years. The Blue school, in Well-street, was founded in 1709 by Denys Rolle, Esq., who endowed it with a messuage and £200 in money, which sum was increased by the Rolle family to £950, from the produce of which 22 boys are instructed and clothed, of whom two are apprenticed annually; and a National school is partly supported by subscription: a new school-house has been built for the latter by Lord Rolle. A district Diocesan school is about to be established. An almshouse for eight poor persons was founded and endowed in 1604, by John Huddle; but, owing to the improvement of the funds, the feoffees are enabled to appoint twelve inmates, and sometimes more: there is also an unendowed almshouse for the residence of the poor. The workhouse for the union, which comprises 23 parishes, is in the town, and is adapted for 200 paupers; the union is under the care of 28 guardians, and contains a population of 17,348, according to the census of 1831. On the restoration of Charles II., General Monk, among other honours, was made Earl of Torrington, which place at present gives the title of Viscount to the family of Byng.

TORRINGTON, LITTLE, a parish, in the union of TORRINGTON, hundred of SHEBBEAR, Black Torrington and Shebbear, and Northern, Divisions of the county of DEVON, 2 miles (S.) from Great Torrington; containing 572 inhabitants. The living is a rectory, valued in the king's books at £14. 18. 11½.: it is in the patronage of Lord Rolle, — Buckingham, Esq., and Mrs. Stephens. The tithes have been commuted for a rent-charge of £460, subject to the payment of rates, which on the average have amounted to £63; the glebe comprises 44 acres, valued at £50 per annum. At Taddiport, in this parish, is an hospital, with a chapel attached, appropriated to the poor of both parishes.

TORRINGTON, WEST (*St. Mary*), a parish, in the union of HORNCASTLF, Western Division of the wapentake of WRAGGOE, parts of LINDSEY, county of LINCOLN, 2¾ miles (N. by E.) from Wragby; containing 126 inhabitants. The living is a discharged vicarage, valued in the king's books at £4; present net income, £80; patron and impropriator, Sir R. S. Ainslie, Bart., by whom a school is supported.

TORRISHOLME, a township, in the parish of LANCASTER, hundred of LONSDALE, south of the sands, Northern Division of the county palatine of LANCASTER, 2 miles (N. W.) from Lancaster; containing 188 inhabitants.

TORTINGTON, a parish, in the hundred of AVISFORD, rape of ARUNDEL, Western Division of the county of SUSSEX, 2½ miles (S. W.) from Arundel; containing 76 inhabitants. The living is a vicarage not in charge, endowed with the rectorial tithes; net income, £158; patron, Duke of Norfolk. The church is principally in the early style of English architecture. A priory of Augustine canons, in honour of St. Mary Magdalene, was founded by the Lady Hadwisa Corbet, before the reign of John, which at the dissolution possessed a revenue of £101. 4. 1.

TORTWORTH (*St. Leonard*), a parish, in the union of THORNBURY, Upper Division of the hundred of GRUMBALD'S-ASH, Western Division of the county of GLOUCESTER, 4 miles (W) from Wotton-under-Edge; containing 266 inhabitants. The living is a rectory, valued in the king's books at £16. 3. 9.; present net income, £428; patrons, Provost and Fellows of Oriel College, Oxford.

TORVER, a chapelry, in the parish and union of ULVERSTONE, hundred of LONSDALE, north of the sands, Northern Division of the county palatine of LANCASTER, 6 miles (W. S. W.) from Hawkeshead; containing 224 inhabitants. The living is a perpetual curacy; net income, £59; patron, T. R. G. Braddyll, Esq. Three children are educated for the interest of £200, the gift of John Fleming, in 1777, in support of a free grammar school, for which a school-house has been built by subscription. John Middleton, in 1695, bequeathed £50; and Thomas Woodall, in 1729, gave £15, for teaching poor children.

TORWORTH, a township, in the parish of BLYTH, union of EAST RETFORD, Hatfield Division of the wapentake of BASSETLAW, Northern Division of the county of NOTTINGHAM, 4¾ miles (N. W. by N.) from East Retford; containing 205 inhabitants.

TOSELAND (*St. Mary*), a parish, in the union of ST. NEOT'S, hundred of TOSELAND, county of HUNTINGDON, 4¾ miles (E. N. E.) from St. Neot's; containing 161 inhabitants. The living is annexed, with that of Little Paxton, to the vicarage of Great Paxton.

TOSSEN, GREAT, a township, in the parish and union of ROTHBURY, Western Division of COQUETDALE ward, Northern Division of the county of NORTHUMBERLAND, 2 miles (W S. W.) from Rothbury; containing 195 inhabitants, who are chiefly employed in the manufacture of woollen cloth, and at the limestone quarries in the neighbourhood. The village was formerly a considerable place: there are still the remains of an ancient tower.

TOSSEN, LITTLE, a township, in the parish and union of ROTHBURY, Western Division of COQUETDALE ward, Northern Division of the county of NORTHUMBERLAND, 2½ miles (W. S. W.) from Rothbury; containing 29 inhabitants.

TOSSIDE, or TOSSET, a chapelry, in the parish of GISBURN, union of SETTLE, Western Division of the wapentake of STAINCLIFFE and EWCROSS, West Riding of the county of YORK, 7½ miles (S. W. by S.) from Settle: the population is returned with the parish. The living is a perpetual curacy; net income, £50; patrons, Executors of the late Lord Ribblesdale. The chapel is dedicated to St. Bartholomew.

TOSSIDE ROW, an extra-parochial liberty, in the wapentake and liberty of STAINCLIFFE, West Riding of the county of YORK, 8 miles (S.) from Settle: the population is returned with the parish of Gisburn.

TOSTOCK (*St. Andrew*), a parish, in the union of STOW, hundred of THEDWASTRY, Western Division of the county of SUFFOLK, 7 miles (E.) from Bury-St.-Edmund's; containing 283 inhabitants. This parish, which is situated on the high road from Bury to Ipswich, comprises about 1000 acres of profitable land; the soil is of a mixed quality, but generally fertile. Gravel abounds in the parish, and the roads are generally good. The living is a discharged rectory, valued in the king's books at £6. 8. 6½.; present income, £250; patron, Rev. William Gilbert Tuck. The church is an ancient edifice, and contains some old richly-carved benches for free seats, which have been much defaced, probably by Cromwell's agents during the interregnum. There was formerly an ancient mansion, the residence of Lords North and Grey, but no part of it remains. The late George Brown, Esq., built a very handsome residence here.

TOTHAM, GREAT (*St. Peter*), a parish, in the

union of MALDON, hundred of THURSTABLE, Northern Division of the county of ESSEX, 3 miles (N. N. E.) from Maldon; containing 696 inhabitants. The surface is generally much elevated, and some parts of the parish are supposed to be the highest land in the county. The living is a discharged vicarage, valued in the king's books at £10; present net income, £95; patrons and impropriators, Trustees of the late W. P. Honeywood, Esq. The church contains several ancient monuments. A National school has been established.

TOTHAM, LITTLE (*All Saints*), a parish, in the union of MALDON, hundred of THURSTABLE, Northern Division of the county of ESSEX, 3½ miles (N. E.) from Maldon; containing 306 inhabitants. This parish, which is situated on the shore of Blackwater bay, is about five miles in circumference; the situation is low and uninviting, the soil light and gravelly. Some salt-works are carried on here in a creek of the river Blackwater. The living is a rectory; net income, £331: it is in the patronage of the Executrix of the late Rev. W. Westcomb. The church is a small ancient edifice, with a tower of flint and stone, surmounted with a spire. Here is a National school.

TOTHILL (*St. Mary*), a parish, in the union of LOUTH, Marsh Division of the hundred of CALCEWORTH, parts of LINDSEY, county of LINCOLN, 5¼ miles (N. W. by N.) from Alford; containing 67 inhabitants. The living is a discharged rectory, valued in the king's books at £6. 17.; present net income, £173; patron, Lord Willoughby de Broke.

TOTLEY, a hamlet, in the parish of DRONFIELD, union of ECCLESALL-BIERLOW, hundred of SCARSDALE, Northern Division of the county of DERBY, 3½ miles (W. N. W.) from Dronfield; containing 351 inhabitants. Ten children are instructed for an annuity of £6, the gift of the Rev. Robert Turie, in a school-room built by subscription, in 1821.

TOTMONSLOW, a village, in the parish of DRAYCOTT-IN-THE-MOORS, hundred of TOTMONSLOW, Northern Division of the county of STAFFORD: the population is returned with the parish.

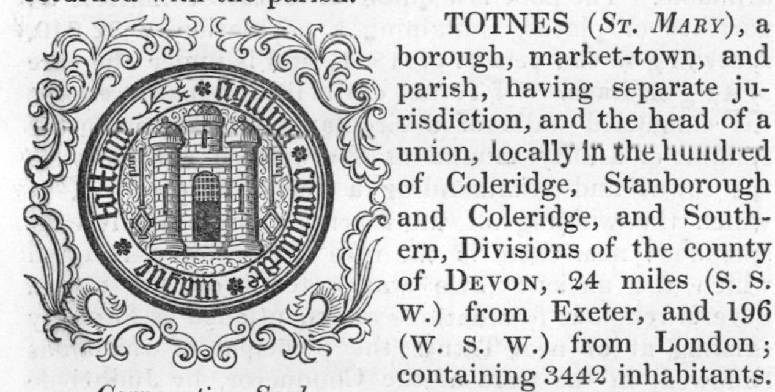

Corporation Seal.

TOTNES (*St. Mary*), a borough, market-town, and parish, having separate jurisdiction, and the head of a union, locally in the hundred of Coleridge, Stanborough and Coleridge, and Southern, Divisions of the county of DEVON, 24 miles (S. S. W.) from Exeter, and 196 (W. S. W.) from London; containing 3442 inhabitants. It is variously denominated in ancient records: in Domesday-book it is called *Totneis;* Camden speaks of it as having once been named *Totonese;* and Risdon alludes to it under the name of *Toutaness,* by contraction *Totnes,* or *Totness.* The latter author accedes to the opinion of Leland, who imagines the name to be a modernization of *Dodonesse,* signifying a rocky town, its situation rendering this supposition probable. The antiquity of the place is attested by Venerable Bede, who describes it as the station where the British troops assembled under Ambrosius and Pendragon, prior to their successful attack upon the tyrant Vortigern. The manor of Great Totnes, having been a royal demesne in the time of the Confessor, was bestowed by William I. upon Judhel, one of his nobles, who took the title of "de Totneis," and erected the castle at the north-western extremity of the town. It is probable that Totnes was fortified at a very early period, having (according to Risdon) undergone alteration under the Romans, Saxons, Danes, and Normans. Of the present town, which is divided into the Higher, Middle, and Lower quarters, the Middle quarter was included within the ancient boundary wall, in which were three gateways, *viz.*, the East, West, and North. At the time of the Norman survey, Totnes was rated when Exeter was rated, and, if there was any expedition by land or water, Totnes, Barnstaple, and Lidford, paid as much as Exeter: in that record it is described as containing ninety-five burgesses within the borough, and fifteen without. During the civil war of the seventeenth century, this place became the temporary station of General Goring: Fairfax subsequently halted here, on his way to and from the town of Dartmouth.

The town is neatly built and of highly respectable appearance, containing many good shops and substantial private residences: it occupies a situation of much beauty and salubrity, on the western bank of the river Dart, over which is a handsome bridge of three arches, completed in 1828, at an expense of about £12,000. It consists chiefly of one long street, descending from Bridgetown, on the east, to the bridge, from the foot of which it rises gradually in a western direction till it reaches a considerable elevation near the site of the castle: this street is crossed midway by the East gateway belonging to the old fortifications; and many of the fronts of the houses beyond are supported by pillars, affording a spacious covered way for foot passengers: the inhabitants are well supplied with water. The general aspect of the town, from the bridge, is picturesque, the church tower appearing on the right of the ascent, and the ivied ruins of the castle crowning the summit of the hill. The surrounding country, particularly as seen from the castle and the hills, is extremely fine; and the course of the Dart between Totnes and its influx into the channel is through diversified and interesting scenery. The town is fast increasing; several villas and handsome houses have been finished in Bridgetown, and many new houses are now in progress of erection on the Plymouth and other roads, which have been lately much improved. There are two libraries, a small theatre, and an assembly-room; and races are held annually, in July or August, on a good course. This town has been noted for its serge manufacture, and there is still some weaving carried on, but the trade is on the decline. The Dart is navigable to the bridge, above which, at a short distance, is a weir, from which a Leat was cut in the reign of Elizabeth, terminating a little above the bridge, for supplying the fulling and corn-mills with water: salmon are caught in great quantities in the Dart, and the town is also plentifully supplied with other kinds of fish. During spring tides only, vessels of 100 tons' burden could come up to the quay, but the river has lately been deepened at an expense of £8000, by which means vessels can approach the quay at all times of the tide, a convenience which much facilitates the commercial intercourse with London and Plymouth, and will greatly improve the trade. Cider is the chief article of exportation: coal, grain, and culm (the last chiefly used for the burning of lime, which abounds in the neighbourhood) are the principal imports. A steam-packet and several other boats proceed daily to Dartmouth. A customary market is held on Saturday; and there is a great cattle market, which is one of the best in the West of England, on the first Tuesday in every month. Fairs for cattle are held on May 12th and October 28th.

The burgesses obtained a charter of privileges from King John, which was confirmed by Edward I., in whose reign, it is understood, Totnes first sent members to parliament: the burgesses obtained the right of electing their mayor in the reign of John. Queen Elizabeth, in the 30th of her reign, granted a charter, whereby the government of the town was entrusted to three magistrates, *viz.*, the mayor, recorder, and a justice: there were fourteen masters and councillors, or aldermen (of which body the mayor was one), and about fifty resident burgesses; also a town-clerk, portreeve, two sergeants-at-mace, and twelve constables, including the two sergeants; the other ten constables were chosen, not under the charter, but at the annual court leet, held by the corporation for the manor of Great Totnes. The corporation at present consists of a mayor, four aldermen, and twelve councillors, under the act of the 5th and 6th of William IV., cap. 76 (for an abstract of which see the Appendix, No. I.); and the municipal boundaries have been made co-extensive with those for parliamentary purposes. The mayor and late mayor are justices of the peace within the borough, and the county magistrates have concurrent jurisdiction. The police establishment consists of twelve constables, appointed by the watch committee under the Municipal Act, and twelve at the court leet. The borough sends two representatives to parliament, who were formerly elected by the masters and burgesses at large; but the privilege has been extended to the £10 householders of the entire parish of Totnes and the adjoining manor of Bridgetown, in the parish of Berry-Pomeroy, which, by the act of the 2nd and 3rd of William IV., cap. 64, have been constituted the new borough, comprising 1162 acres: the old borough contained only 118 acres. The number of voters now registered is 352, of whom 30 are freemen: the mayor is the returning officer. The magistrates of the corporation, previously to the passing of the Municipal Reform Act, held quarter sessions for all but capital offences arising within the borough; and a court of requests was formerly held, but it is now in disuse. There are a guildhall and chamber, and a town prison. The corporation claim many privileges, such as freedom from quayage and wharfage throughout the kingdom, except the port of London, and exemption from serving on juries, except in the borough, for all inhabitants of the borough and parish, whether members of the corporation or not.

The living is a discharged vicarage, valued in the king's books at £12. 8. 9.; present net income, £150: it is in the patronage of the Crown; impropriator, Duke of Somerset. The church has been enlarged, and 350 free sittings provided, the Incorporated Society having granted £250 in aid of the expense: it is in the later style of English architecture, with a handsome embattled tower surmounted by octagonal pinnacles of lighter-coloured material than the remainder of the building, which is composed of a red stone strongly resembling brick; and contains an elegant stone screen, with the remains of the ancient rood-loft and steps; a curious stone pulpit, enriched with tracery; a handsome altar-piece; and a library, in which are many old and valuable books. An episcopal chapel has been erected at Bridgetown, at an expense of £7000, by the Duke of Somerset. There are places of worship for Independents and Wesleyan Methodists, and a Roman Catholic chapel at Follaton. The grammar school was founded in 1554, and endowed in 1658 with lands now worth £70 a year, by Sir John Maynard, trustee of Elizeus Hele, Esq., who left considerable property for charitable purposes. The schoolmaster is appointed by the corporation, who have the right of presenting two boys for gratuitous education: and an unlimited number of day scholars are admitted, at the annual charge of £8. 8.: the old school-room is used for a charity school on Dr. Bell's plan. The charity school is endowed with lands, of which the proceeds amount to about £40 per annum, but is chiefly supported by voluntary contributions; 70 children are educated, of whom about 30 are annually clothed. Among the numerous charitable donations were some for re-establishing decayed tradesmen in business, and others for apprenticing children, but these funds have been long since exhausted. There is an almshouse, recently rebuilt on a different site, occupied by about 20 people, supposed to be an enlargement of a foundation by Mr. Norris, who bequeathed £250 for its erection, in the year 1635. Here was formerly a lazar-house, the lands appertaining to which now yield an income of about £4. 10., which is applied to the repairs of the church, &c.: the remains of the building were incorporated in an edifice fitted up, in 1832, for the reception of cholera patients, and now inhabited, with permission of the parishioners, by some poor people. Trustees for the several charities have been appointed by the Lord Chancellor for carrying into effect the various purposes, for which ample funds had been bequeathed, and property to a considerable amount will be rendered available. The poor law union of Totnes comprises 28 parishes or places, containing a population of 32,240, according to the census of 1831, and is under the care of 44 guardians. Of Totnes castle little remains, except the embattled walls of a circular keep, occupying the summit of a lofty mound at the western extremity of the town, and commanding a delightful prospect, in which the windings of the Dart are prominently conspicuous: near them is the ruin of a gateway, through which the ancient town was entered on the north. Several religious foundations are mentioned as formerly existing at or near Totnes, the principal of which was endowed, in the time of the Conqueror, by Judhel de Totneis: it was of the Benedictine order, dedicated to St. Mary, and formed an appendage to an abbey at Angiers: the site is occupied by a dwelling-house, called "the Priory." Here are some remains of an ancient chapel. Leland mentions a Roman Fosse-way, commencing in this vicinity. Chrystallised rhomboidal carbonate of lime has been found on grey limestone in the Peto quarry, about three miles to the west of the town, on the north of the Plymouth road. Dr. Philip Furneaux, a learned non-conformist divine; Benjamin Kennicott, a learned biblical critic, and who, it may be noticed, was in early life master of the charity school above mentioned; and Edward Lye, a celebrated lexicographer, were natives of Totnes.

TOTON, a hamlet, in the parish of ATTENBOROUGH, union of SHARDLOW, Southern Division of the wapentake of BROXTOW, Northern Division of the county of NOTTINGHAM, 5¾ miles (S. W. by W.) from Nottingham; containing 202 inhabitants.

TOTTENHAM (*All Saints*), a parish, in the union and hundred of EDMONTON, county of MIDDLESEX, 4 miles (N. by E.) from London; containing 6937 inhabitants. This place, written in Domesday-book *Toteham*, and now frequently called Tottenham High Cross, is a genteel village, consisting chiefly of one long street, formed by houses irregularly arranged, on the line of road from London to Cambridge: it is lighted with gas, and well supplied with water from several fountains produced by boring; and the immediate vicinity is adorned with numerous handsome villas. Near Tottenham Green a cross has stood for many years: the present structure, superseding the original one of wood, is an octangular brick column, erected in 1600 by Dean Wood: it was repaired, covered with cement, and decorated with various architectural embellishments, in 1809, by subscription. At the entrance of Page Green, on the east side of the high road, is a remarkable circular clump of elm trees, called "the Seven Sisters:" in the centre was formerly a walnut tree, which, according to tradition, never increased in size, though it continued annually to bear leaves: these trees appear to have been at their full growth in 1631, but no authentic account of their being planted is extant. Within a short distance from the high road is Bruce Castle, a mansion rebuilt, in the seventeenth century, on the site of an ancient castellated edifice, erected in the reign of Henry VIII., and honoured, in the year 1516, with the presence of that monarch, who came hither to meet his sister, Margaret, Queen of Scots; in 1578, Elizabeth also honoured it with her presence. The original castle was the residence and property of Robert de Bruce, father of Robert, King of Scotland; the present building has been converted into a school for young gentlemen: a detached brick tower, which covers a deep well, is the only vestige of the ancient edifice. In the parish is a well, of which the water is similar in its properties to that at Cheltenham; also a spring, called Lady's Well, of reputed efficacy for disorders in the eyes; it is said that this water never freezes. Here are extensive flour and oil mills, the former having been established time immemorially; also a pottery for coarse brown ware, and a brewery: a large silk-manufactory is unoccupied. Near the entrance of the village, and on an ancient stream now called the Moselle, are the works of the London Caoutchouc company, for the manufacture of India rubber solution for rendering waterproof cloth and garments of every kind, also tie-bands, ropes, cables, webs, and various other articles to which the use of India rubber has been appropriated; the machinery is driven by two steam-engines, of the aggregate power of 55 horses; and is applied to the cutting of India rubber into threads and sheets; the works afford employment to 130 persons. The navigable river Lea passes through the parish. Tottenham is within the jurisdiction of a court of requests held at Enfield, for the recovery of debts under 40*s*.

The living is a vicarage, valued in the king's books at £14; present net income, £978; patrons and appropriators, Dean and Chapter of St. Paul's, London. The church stands about a quarter of a mile west of the high road, and is in the later style of English architecture, with a square embattled and ivy-mantled tower: on the summit was formerly a lofty wooden cross (whence, according to some, the adjunct to the name of the village), which was destroyed during the civil war: on the south side of the church is a large brick porch, erected prior to 1500; and over it a room, called "a church house," formerly used for the transaction of parochial business, and now occupied as a Sunday school. At the east end of the north aisle is a vestry of circular form, surmounted by a dome, erected in 1696 by Lord Henry Coleraine, and repaired in 1790; underneath this is the family vault. The eastern window, divided into eight compartments, and containing representations of the Evangelists and some of the Prophets, in fine old painted glass, was given to the parish, in 1807, by the late John Eardley Wilmot, Esq.: the font is curious and of great antiquity: many ancient monuments adorn the interior, of which one in white marble, to the memory of the family of Sir Robert Barkham, is worthy of especial notice. This church was repaired in 1816, at an expense of £3000. A new church, dedicated to the Holy Trinity, in the later style of English architecture, with turrets at each angle, and pinnacles over the aisles, was erected in 1829, on Tottenham Green, by aid of the parliamentary commissioners, who granted £4893. 11. 6., the remainder of the expense having been defrayed by subscription. The living is a perpetual curacy; net income, £309; patron, Vicar of Tottenham. There are places of worship for Baptists, the Society of Friends, Independents, Wesleyan Methodists, and Roman Catholics. The grammar school, founded by means of a bequest from Nicholas Reynardson, alderman of London, in 1685, was endowed in the following year by Sarah Duchess Dowager of Somerset, with £250 for enlarging the buildings, and £1100 for extending the benefits of the institution to all children of such inhabitants of the parish as were not possessed of real property to the amount of £20 per annum; several small bequests have been made since that period. The Blue school, instituted in 1735, in which are 50 poor girls; and the Green school, established in 1792, in which are 40, are supported by voluntary contributions; all the children are clothed, and instructed on the Madras system. Two Lancasterian schools, and a school for Roman Catholic children, are similarly supported. Almshouses for four poor men and four poor women were founded and endowed, about 1600, with a small rent-charge by Balthasar Sanches, first pastry-cook to Philip of Spain, with whom he came over to this country, and was the first who exercised that trade in London. An almshouse for six poor men and six poor women, with a small chapel in the centre, was founded and endowed with £2000 by Nicholas Reynardson, Esq., in 1685; some valuable augmentations have been made to the funds, and, in 1828, the buildings were repaired at an expense of £450, defrayed by voluntary contributions. Some almshouses on the high road are occupied by four poor women, placed there by the parishioners. The savings bank in this parish was one of the first established in England.

TOTTENHILL (*St. Botolph*), a parish, in the union

of DOWNHAM, hundred of CLACKCLOSE, Western Division of the county of NORFOLK, 6½ miles (N. N. E.) from Downham-Market; containing 358 inhabitants. The living is a perpetual curacy; net income, £62; patron and appropriator, Bishop of Ely.

TOTTERIDGE (*St. Andrew*), a parish, in the union of BARNET, and forming a detached portion of the hundred of BROADWATER, though locally in that of Cashio, or liberty of St. Albans, county of HERTFORD, 2 miles (S. by W.) from Chipping-Barnet; containing 595 inhabitants. The living is annexed to the rectory of Bishop's-Hatfield. The church, which was rebuilt in 1798, has a latticed square tower, with a spire. There is a place of worship for Independents. A school for 10 girls is supported by Mrs. Arrowsmith, and another is partly supported by Mrs. Pugett. A Sunday school is partly supported by an endowment of £400 stock, by Elizabeth Williams, in 1789.

TOTTERNHOE (*St. Giles*), a parish, in the union of LUTON, hundred of MANSHEAD, county of BEDFORD, 2 miles (W. S. W.) from Dunstable; containing 515 inhabitants. The living is a discharged vicarage, valued in the king's books at £10; present net income, £120: it is in the patronage of the Countess of Bridgewater, to whom, with the Master and Fellows of Trinity College, Cambridge, the impropriation belongs. On the north side of the church passes the Roman Ikeneld-street, skirting the downs, upon which are the remains of Totternhoe castle, overhanging the village of Stanbridge: the keep mount is lofty, and encompassed by a circular fosse within another that is square, the latter enclosing the entire breadth of the ridge. Near this fortification is an ancient quadrangular camp; and to the eastward are extensive quarries of freestone and limestone, below which, at a great depth, is a bed of clay.

TOTTINGTON (*St. Andrew*), a parish, in the union and hundred of WAYLAND, Western Division of the county of NORFOLK, 3¾ miles (S. S. W.) from Watton; containing 313 inhabitants. The living is a discharged vicarage, valued in the king's books at £6. 14. 9½.; present net income, £89; patrons and impropriators, Trustees of Chigwell free schools. The church is a large ancient structure: the churchyard wall is coped with coffin stones having crosses on them, supposed to have enclosed the remains of the ancient vicars and other religious buried here.

TOTTINGTON, HIGHER, a township, in the parish of BURY, union of HASLINGDEN, hundred of SALFORD, Southern Division of the county palatine of LANCASTER, 3 miles (N. W.) from Bury; containing 2572 inhabitants. Here is a National school.

TOTTINGTON, LOWER, a chapelry, in the parish and union of BURY, hundred of SALFORD, Southern Division of the county palatine of LANCASTER, 3 miles (N. W. by W.) from Bury; containing 9280 inhabitants. Here are very extensive establishments for the printing and bleaching of cotton, in which more than 2500 persons are employed. A fair is held on October 12th. The living is a perpetual curacy; net income, £145; patrons, Trustees. The chapel is dedicated to St. Anne. A school was erected in 1715, by Thomas Nuttall, who endowed it with a rent-charge of £3; in 1773, the building was enlarged by subscription among the inhabitants, and with the subsequent bequests of Peter and Ann Baron, the income has been augmented to £24 a year, for which 20 children are gratuitously instructed. Dr. Wood, the celebrated mathematician, was born at this place.

TOUCHEN, a division, in the parish and hundred of BRAY, county of BERKS: the population is returned with the parish.

TOULSTON, a township, in the parish of NEWTON-KYME, Upper Division of the wapentake of BARKSTONE-ASH, West Riding of the county of YORK, 2¾ miles (W. by N.) from Tadcaster: the population is returned with the parish.

TOWCESTER (*St. Lawrence*), a market-town and parish, and the head of a union, in the hundred of TOWCESTER, Southern Division of the county of NORTHAMPTON, 8½ miles (S. W. by S.) from Northampton, and 60 (N. W.) from London; containing 2671 inhabitants. This place, the name of which in Domesday-book is written *Tovecestre*, "a city, or fortified place, on the river Tove," is considered to have been a Roman station, from the discovery of numerous coins, especially on an artificial mount north-eastward of the town, called Berry-mont hill, and in many of the gardens and homesteads of the town: of these, Mr. Deacon, of this place, has a considerable collection, the greater part of which are thought to have been found in the fields to the north-west of the town. On the north-west side are vestiges of a fosse, and the ruins of a tower, supposed to be Saxon. Some antiquaries have thought that the station of *Lactodorum* should be placed here, in preference to Stony-Stratford. During the Saxon era, the town appears to have been so well defended as to have offered a protracted and effectual resistance to the attacks of the Danes: about the year 921, a mandate was issued by Edward, for rebuilding and fortifying it, and it was surrounded by a stone wall, of which some vestiges are yet discernible. In the reign of Henry VI., a college and chantry were founded here by William Sponne, Archdeacon of Norfolk, the revenue of which, at the dissolution, was valued at £19. 6. 8. per annum. The town, which is situated on the river Tove, consists principally of one long street, composed of well-built houses, and paved under the direction of the trustees of the charities of Archdeacon Sponne, who devised the Tabart Inn, and certain lands producing about £150 per annum, for that purpose; the inhabitants are well supplied with water. The manufactures consist of bobbin lace, boots, and shoes; and great advantages are derived from the situation of the town on the great road from London to Holyhead. The market is on Tuesday; and fairs are held on Shrove-Tuesday, May 12th, and October 29th, for cattle: on October 10th is a statute fair for hiring servants. A manorial court is held at Michaelmas, at which the constables for the parish are chosen; and the town has been made a polling-place for the southern division of the county.

The living is a discharged vicarage; net income, £217; patron and appropriator, Bishop of Lichfield. The church contains 410 free sittings, the Incorporated Society having granted £300 in aid of the expense: it is a neat building of the eleventh century, in the early style of English architecture, and contains the monument of Archdeacon Sponne, who held the living in the time of Henry VI. Among the various incumbents was Pope Boniface VIII., at the time of his promotion to the pontificate, in 1294. Abthorpe, which was formerly a chapelry in this parish, was separated from it by act of parliament, about 1756. There are places of worship for Baptists, Independents, and Wesleyan Methodists. The grammar school was founded in 1552, by the trustees of Sponne's charity, who, on the dissolution of the college and chantry, purchased and converted them to this use, with a house and garden for the master: the income, arising from bequests and donations, is £57. Three almshouses were founded and endowed in 1695, by Thomas Bickerstaff, of this place; and there are a few other bequests for the poor. The poor law union of Towcester comprises 23 parishes or places, containing a population of 12,142, according to the census of 1831, and is under the superintendence of 31 guardians. In the vicinity is a petrifying spring. The Roman Watling-street passed along the site of the town. Sir Richard Empson, once proprietor of the manor, and a celebrated lawyer, who was promoted to the chancellorship of the duchy of Lancaster, in the time of Henry VII., and beheaded on Tower-hill, in the succeeding reign, in the year 1509, was the son of a sieve-maker in this town. About a mile and a half from Towcester, at Easton-Neston, is the seat of Earl Pomfret, formerly celebrated for its splendid collection of paintings and statues, presented in 1756 to the University of Oxford, by the then Countess of Pomfret.

TOWEDNACK (*St. Twinnock*), a parish, in the union of Penzance, Western Division of the hundred of Penwith, and of the county of Cornwall, 3 miles (S. W. by W.) from St. Ives; containing 737 inhabitants. The living is a vicarage, annexed to that of Uny-Lelant. There is a place of worship for Wesleyan Methodists; and two schools are supported by subscription. This parish contains the mine called Wheal Durla, and in various parts are vast rocks of fine granite.

TOWER HAMLETS, one of the newly enfranchised metropolitan boroughs, in the union of Whitechapel, comprising the Liberty of the Tower, and part of the Tower Division of the hundred of Ossulstone, forming the north-eastern part of the suburbs of the metropolis, and containing by estimation 302,519 inhabitants. This borough, of which the limits are minutely described in the Appendix, No. II., comprises an area of 3954 acres, and returns two members to parliament: the number of voters is about 10,000; the chief bailiffs of the liberty are the returning officers. Separate courts held for this liberty are noticed in the article on London.

TOWERSEY (*St. Catharine*), a parish, in the union of Wycombe, hundred of Ashendon, county of Buckingham, $2\frac{1}{4}$ miles (E. S. E.) from Thame; containing 403 inhabitants. The living is annexed, with those of Sydenham and Tetsworth, to the vicarage of Thame. This parish has an interest in the charity of Mrs. Katherine Pye, described in the account of Risborough (Prince's).

TOWNGREEN, a township, in the parish of Wymondham, hundred of Forehoe, Eastern Division of the county of Norfolk; containing 1052 inhabitants.

TOWNSTALL, county of Devon.—See DARTMOUTH.

TOWTHORPE, a township, in the parish of Wharram-Percy, union of Driffield, wapentake of Buckrose, East Riding of the county of York, 9 miles (W. N. W.) from Great Driffield; containing 48 inhabitants.

TOWTHORPE, a township, in the union of York, partly in the parish of Strensall, within the liberty of St. Peter's, East Riding, and partly in that of Huntingdon, wapentake of Bulmer, North Riding, of the county of York, 5 miles (N. by E.) from York; containing, in the latter portion, 70 inhabitants, the former part being returned with Strensall.

TOWTON, a township, in the parish of Saxton, Upper Division of the wapentake of Barkstone-Ash, West Riding of the county of York, 3 miles (S.) from Tadcaster; containing 115 inhabitants. Between this place and Saxton is the celebrated Towton-field, where was fought, on Palm-Sunday, 1461, the most important battle between the houses of York and Lancaster, which lasted from nine in the morning till seven in the evening, and ended in the defeat of the latter: in this bloody conflict, it is recorded, 110,000 Englishmen were engaged, of whom 36,776 were slain.

TOXTETH-PARK, an extra-parochial district, in the union and hundred of West Derby, Southern Division of the county palatine of Lancaster, adjoining the town of Liverpool, and containing 24,007 inhabitants. This place, which is delightfully situated on the banks of the Mersey, was anciently a park belonging to the Dukes of Lancaster, and afterwards passed to the Molyneaux family. In consequence of its proximity to Liverpool, it has become the residence of numerous merchants, manufacturers, retired tradesmen, &c., and several new streets have been formed within the last few years. It is supposed to have formerly been included in the parish of Walton-on-the-Hill, to which it still pays tithes, though commonly deemed extra-parochial. It possesses two chapels, dedicated respectively to St. James and St. Michael, the latter consecrated in 1818: the livings are perpetual curacies, in the patronage of the Rector of Walton; net income of St. James's, £188; net income of St. Michael's, £210. A new church, dedicated to St. John, in the later English style, with a tower, was erected in 1831, at an expense of £6648. 1. 6., under the act of the 58th of George III., to which a district has been attached. There is a place of worship for Unitarians; and National schools have been established.

TOYNTON (*All Saints*), a parish, in the union of Spilsby, Eastern Division of the soke of Bolingbroke, parts of Lindsey, county of Lincoln, $1\frac{1}{2}$ mile (S.) from Spilsby; containing 475 inhabitants. The living is a discharged perpetual curacy, valued in the king's books at £5. 11. 3.; present net income, £243; patron and impropriator, Lord Willoughby de Eresby. Two schools are partly supported by subscription.

TOYNTON (*St. Peter's*), a parish, in the union of Spilsby, Eastern Division of the soke of Bolingbroke, parts of Lindsey, county of Lincoln, $2\frac{1}{4}$ miles (S. by E.) from Spilsby; containing 372 inhabitants. The living is a discharged rectory, valued in the king's books at £12. 0. 2.; present net income, £199; patron. Lord Willoughby de Eresby.

TOYNTON, HIGH (*St. John the Baptist*), a parish, in the union and soke of Horncastle, parts of Lindsey, county of Lincoln, $1\frac{1}{2}$ mile (E.) from Horncastle; containing 164 inhabitants. The living is a perpetual curacy; net income, £53; patron and appropriator, Bishop of Carlisle. There is a place of worship for Wesleyan Methodists.

TOYNTON, LOW (*St. Peter*), a parish, in the union and soke of Horncastle, parts of Lindsey, county of Lincoln, 1¼ mile (N. E.) from Horncastle; containing 108 inhabitants. The living is a discharged rectory, valued in the king's books at £11. 1. 8.; present net income, £332; patron, Lord Willoughby de Eresby. There is a place of worship for Wesleyan Methodists.

TRAFFORD, BRIDGE, a township, in the parish of Plemonstall, union of Great Boughton, Second Division of the hundred of Eddisbury, Southern Division of the county of Chester, 4½ miles (N. E. by N.) from Chester; containing 38 inhabitants.

TRAFFORD, MICKLE, a township, in the parish of Plemonstall, union of Great Boughton, Lower Division of the hundred of Broxton, Southern Division of the county of Chester, 3½ miles (N. E.) from Chester; containing 333 inhabitants. Here is a school, endowed with £4. 5. per annum.

TRAFFORD, WIMBOLDS, a township, in the parish of Thornton, union of Great Boughton, Second Division of the hundred of Eddisbury, Southern Division of the county of Chester, 6 miles (N. E. by N.) from Chester; containing 118 inhabitants.

TRANMERE, a township, in the parish of Bebington, union, and Lower Division of the hundred, of Wirrall, Southern Division of the county of Chester, 7¼ miles (N. N. E.) from Great Neston; containing 1168 inhabitants. The river Mersey is here crossed by a ferry, and the Birkenhead and Chester railway passes through the township. A neat and commodious church, or chapel, was erected in 1831, which contains a fine painting of the Resurrection by Le Brun, presented by T. Warrington, Esq.: the living is a perpetual curacy; net income, £105; patron, Rector of Bebington. A National school is supported by subscription.

TRANWELL, a joint township with High Church, in the parish and union of Morpeth, Western Division of Castle ward, Southern Division of the county of Northumberland, 2 miles (S. S. W.) from Morpeth; containing, with High Church, 64 inhabitants.

TRAWDEN-FOREST, a township, in the parish of Whalley, union of Burnley, Higher Division of the hundred of Blackburn, Northern Division of the county palatine of Lancaster, 1¾ mile (S. E.) from Colne; containing 2853 inhabitants. This district, like other forests in the kingdom, was rejected as of little value, at the time of the original distribution of land: it comprises ten square miles, and manufactures, similar to those carried on in the neighbouring towns and villages, have been introduced, and are progressively increasing. There are places of worship for the Society of Friends and Wesleyan Methodists; also two schools supported by subscription.

TRAYFORD, county of Sussex.—See TREYFORD.

TREALES, a joint township with Roseacre and Wharles, in the parish of Kirkham, union of the Fylde, hundred of Amounderness, Northern Division of the county palatine of Lancaster, 1¼ mile (N. E.) from Kirkham; containing 756 inhabitants. A school, now conducted on the National plan, was established in 1814, from the surplus funds of an estate of William Grumbaldson, bequeathed in 1725, for charitable purposes.

TREBOROUGH (*St. Peter*), a parish, in the union of Williton, hundred of Carhampton, Western Division of the county of Somerset, 6¼ miles (S. by E.) from Dunster; containing 105 inhabitants. The living is a discharged rectory, valued in the king's books at £7. 10. 5.; present net income, £137; patron, Sir J. Trevelyan, Bart. A National school has been established.

TREDEGAR, a market-town, in the township of Ushlawrcoed, parish of Bedwelty, Lower Division of the hundred of Wentllogge, county of Monmouth, 12 miles (W. by S.) from Abergavenny; containing upwards of 6000 inhabitants. This place, which, previously to the year 1800, contained only three houses, has since that period rapidly increased in extent and importance, and is become a populous and flourishing market-town. For its present prosperity it is indebted to the persevering efforts of Messrs. Samuel Homfray, the late Richard Fothergill, William Thompson, William Foreman, and the Rev. Matthew Monkhouse, who, appreciating its advantageous situation in a district abounding with coal and iron-stone, established some extensive iron-works, which have been progressively enlarged and improved. Coal of excellent quality is obtained on the spot; the mines of iron-stone are of vast extent, and appear, from scoria frequently found on the sides of the hills, to have been worked at an early period; a little below the town are slight remains of an ancient furnace, with heaps of half-smelted scoria scattered round it. Limestone also is found in the adjoining parish of Llangynidr, in the county of Brecon. A tram-road was constructed in 1802 from the above works, leading down the valley of Sirhowy to Newport, a distance of 24 miles. At Risca it is joined by several other tram-roads, and is carried across the valley and the river by a viaduct of 32 arches, nearly a mile and a quarter in length, from which the road is continued in a direction parallel with the canal for some distance. On this line of road are locomotive steam-engines belonging to the Tredegar iron-works, and the Ebbw Vale Company, who, in 1832, carried a tunnel through the centre of the Manmoel mountain, nearly two miles in length, connecting their works with those of Tredegar and Sirhowy, all in the parish of Bedwelty; the produce of all of which is now conveyed down the Sirhowy rail-road to Newport, and thence shipped to all parts of the world. An excellent turnpike-road, adjoining the tram-road, has been constructed from Tredegar to Newport; and the mail-coach road from Merthyr to Abergavenny, constructed about 25 years since, and leading through a singularly wild and romantic tract of country, passes close to the town.

The town is situated on the west bank of the Sirhowy river, and consists of a square, from which diverge four principal streets, a long and wide street leading to the Sirhowy works, several smaller streets, and some detached houses. The whole is built on land belonging to the Tredegar Iron Company, by whom it was arranged and the buildings were erected; and who, under the direction of Samuel Homfray, Esq., of Bedwelty House, the only resident proprietor, built a very handsome town-hall, market-house, shambles, and slaughter-houses, which were completed in 1833. The market is on Saturday; it is well supplied with provisions and necessaries of all kinds, and is numerously attended by the inhabitants of the surrounding populous mining district. Fairs are annually held at present on the 19th of April, Sept. 23rd, and Nov. 18th, but it is in contemplation to alter these days. The parish church is eight miles distant; but a church has been built in the town, towards the erection of which the parliamentary commissioners granted £1000, the Incorporated Society £450, and upwards of £600 was subscribed by private individuals; it contains 1020 sittings, half of which are free. There are places of worship for English and Welsh Baptists, Independents, and English and Welsh Wesleyan and Calvinistic Methodists. Near the town are some very powerful balance machines, by which coal is raised from the mines without the aid of horses, water-wheels, or steam; they were introduced in the year 1820, by Samuel Homfray, Esq.

TREDINGTON, a parish, in the union, and Lower Division of the hundred of TEWKESBURY, Eastern Division of the county of GLOUCESTER, 3 miles (S. E. by S.) from Tewkesbury; containing 132 inhabitants. The living is a perpetual curacy; net income, £51; patron, Bishop of Gloucester and Bristol. The church is very small, but the churchyard is remarkable for containing a lofty pillar of one single stone, resting upon a basis of four steps.

TREDINGTON (*St. Gregory*), a parish, in the union of SHIPSTON-UPON-STOUR, and forming, with the parishes of Shipston-upon Stour and Tidmington, a distinct portion of the Upper Division of the hundred of OSWALDSLOW, Blockley and Eastern Divisions of the county of WORCESTER, locally in the Kington Division of the hundred of Kington, county of Warwick, 2¼ miles (N.) from Shipston-upon-Stour; containing 606 inhabitants. This parish, which is intersected by the river Stour, was divided under an act passed in the 6th of George I., when the townships of Shipston and Tidmington were separated from it, and constituted distinct parishes; and in 1833 an act was obtained for separating from it, and forming into a distinct parish, the hamlets of Newbold and Armscott. The living is a rectory, in two portions, jointly valued in the king's books (for the whole of the ancient parish) at £99. 17. 6.; present net income of the first portion, £415, and of the second, £539; patrons, Principal and Fellows of Jesus' College, Oxford. John Jordan, in 1830, bequeathed annuities of £50, £10, and £5, for founding and endowing free schools respectively at Armscott, Cropredy, in the county of Oxford, and Little Compton, in the county of Gloucester. Another school is endowed by Thomas Eden, with £12 per annum, accruing from houses in the city of Bristol. Here was formerly a religious house, the remaining part of which is now the rectory-house.

TREDUNNOCK (*St. Andrew*), a parish, in the union of NEWPORT, Upper Division of the hundred of USK, county of MONMOUTH, 4½ miles (S.) from Usk; containing 158 inhabitants. The living is a rectory, valued in the king's books at £10. 0. 5.; patron, Capel Hanbury Leigh, Esq. The tithes have been commuted for a rent-charge of £180, subject to the payment of rates; the glebe comprises 46 acres, valued at £55 per annum. The church contains the monument of a Roman soldier of the second legion; this stone, a kind of blue slate, was discovered fastened by four pins to the foundation of the church, and is now fixed in a similar manner to the wall, near the font.

TREETON (*St. Helen*), a parish, in the union of ROTHERHAM, Southern Division of the wapentake of STRAFFORTH and TICKHILL, West Riding of the county of YORK; containing 680 inhabitants, of which number, 345 are in the township of Treeton, 4 miles (S. by E.) from Rotherham. The living is a rectory, valued in the king's books at £12; present net income, £674; patron, Duke of Norfolk. A cottage near the churchyard has long been occupied rent-free by a schoolmistress, for teaching six poor children.

TREGARE (*St. Mary*), a parish, in the Lower Division of the hundred of RAGLAND, union and county of MONMOUTH, 5½ miles (W. S. W.) from Monmouth; containing 326 inhabitants. The living is annexed to the vicarage of Dingestow. Hugh Watkins bequeathed money and land now producing £18 per annum, which sum, together with some minor bequests, is annually distributed among the poor of this place.

TREGAVETHAN, a manor, in the parish of KEA, union of TRURO, Western Division of the hundred of POWDER and of the county of CORNWALL; containing 59 inhabitants.

TREGONEY-CUM-ST. JAMES, a parish, comprising the disfranchised borough and market-town of Tregoney, in the union of TRURO, Western Division of the hundred of POWDER and of the county of CORNWALL, 41½ miles (S. W.) from Launceston, and 248 (S. W. by W.) from London, and containing 1127 inhabitants. The original town, situated at the base of the hill on which the present town is built, was of very great antiquity, and formerly a place of considerable importance: it was distinguished for its castle, supposed to have been erected in the reign of Richard I., the site of which, and the moat by which it was surrounded, are plainly discernible. A priory is supposed by some writers to have existed here, as a cell to the convent of Merton, in Surrey; but this opinion rests solely, according to Bishop Tanner, on the erroneous mention of the advowson of the *priory* of Tregoney, having been appropriated to the prior and convent of Merton, instead of the advowson of the *rectory* of St. James in Tregoney, which was actually so appropriated by the abbot of de Valle in Normandy, to whom it previously belonged. The present town, which has materially decreased in importance since the increase of Truro, is pleasantly situated on the turnpike road from St. Austell to St. Mawes, and consists of one principal street; it is watered by the river Fal, which was formerly navigable for a mile above the town, for small barges, and over which a neat bridge has been erected. The market, which is well supplied with meat and provisions, is on Saturday; and fairs are held on Shrove Tuesday, May 3rd, July 25th, Sept. 1st, and Nov. 6th. The inhabitants received a charter of incorporation from James I., in 1620, by which the government of the borough is vested in a mayor, recorder, and eight capital burgesses or aldermen. The mayor is elected annually on the first Tuesday after the festival of St. Michael the Archangel, and he and the senior aldermen are justices of the peace within the borough. The borough first returned members to parliament in the reign of Edward I., and, after having discontinued for many years, regained the elective franchise in 1559, which it continued to exercise till, by the late bill for amending the representation, it was totally disfranchised: the right of election was vested in the potwallopers, in number about 300. The magistrates have power to hold courts of session, but they have not exercised it for many years. The living is a rectory, with the vicarage of Cuby annexed, valued in the king's books at £10. 4. 2.; present net income, £311; patron, J. A. Gordon, Esq., the impropriator of Cuby. The church, though situated close to the town, is within the parish of Cuby. The church dedicated to St. James, which stood in a meadow below the town, and near the site of the original town, has been taken down more than 50 years since; human bones of large dimensions have been frequently found near the spot, on the site of the ancient churchyard. There are places of worship for Independents and Wesleyan Methodists. A school for the gratuitous instruction of children, conducted on the British system, has been established, for which a handsome building, capable of receiving 300 children, was erected in 1829, at an expense of £500, towards defraying which J. A. Gordon, Esq., contributed £300, and the Rev. John Lloyd Lugger, the rector of the parish, by whom the school is principally supported, the remainder. An hospital for decayed housekeepers was founded here in 1696, by Hugh Boscawen, Esq., who endowed it with lands producing at present £90 per annum, for decayed housekeepers of this parish.

TRELLECK (*St. Nicholas*), a parish, chiefly in the union of Monmouth, but partly in that of Chepstow, Upper Division of the hundred of Ragland, county of Monmouth; containing 1110 inhabitants, of which number, 140 are in the township of Trelleck, 5 miles (S.) from Monmouth. This place derives its name, anciently written "Tre lech," from three massive upright stones, which, according to an inscription on the pedestal of a sun-dial near the churchyard gate, were raised in commemoration of a victory obtained here by Harold over the Britons. Though called by the inhabitants Harold's stones, they are more generally supposed to be the remains of some Druidical monuments, of which there are other vestiges in the vicinity. The village, which is situated on the high road from Monmouth to Chepstow, contains a few handsome houses, and the surrounding scenery is strikingly diversified with features of wildly romantic and pleasingly picturesque beauty. The living is a vicarage, endowed with the rectorial tithes, with that of Penalth annexed, and valued in the king's books at £8; present net income, £430: it is in the patronage of the Crown, in right of the duchy of Cornwall. The church is an ancient structure, and contains several monuments of interest. Zacharias Babington, in 1689 and 1691, bequeathed property for keeping the church clock in repair, maintaining a school, and establishing a lectureship. In the school about 40 boys and girls are gratuitously taught upon the National system, by a master who occupies the school premises, and receives about £15 per annum; the lecturer, who is the present curate, receives about £30 per annum. There are also some small bequests for the poor. Near the village are several chalybeate springs strongly impregnated with iron, which, from the cinders of ancient blomeries, appears to have been wrought here at an early period. In the garden of a house in the village is a large tumulus surrounded by a deep fosse, about 450 feet in circumference, supposed by some to have been the site of a castle formerly belonging to the Earls of Clare, and by others to be a barrow raised over the bodies of the Britons slain in some ancient battle, which took place near the spot, and to have been subsequently occupied by the English, previously to the battle of Craig y Dorth, in which Owain Glyndwr defeated the royal forces and pursued them to the gate of Monmouth.

TRELLECK-GRANGE, a chapelry, in the parish of Trelleck, union of Chepstow, Upper Division of the hundred of Ragland, county of Monmouth, 6 miles (N. W. by N.) from Chepstow; containing 170 inhabitants. The living is a perpetual curacy; present net income, £74; patron and impropriator, Duke of Beaufort.

TREMAYNE, a parish, in the union of Launceston, Northern Division of the hundred of East, Eastern Division of the county of Cornwall, 6¾ miles (W. N. W.) from Launceston; containing 118 inhabitants. The living is a perpetual curacy, annexed to that of Egloskerry. The tithes have been commuted for a rent-charge of £83. 14. 3., subject to the payment of rates.

TRENDLE, a tything, in the parish of Pitminster, hundred of Taunton and Taunton-Dean, Western Division of the county of Somerset: the population is returned with the parish.

TRENEGLOS (*St. Werburgh*), a parish, in the union of Launceston, hundred of Lesnewth, Eastern Division of the county of Cornwall, 7½ miles (N. E. by E.) from Camelford; containing 183 inhabitants. The living is a vicarage, with that of Warbstow annexed, valued in the king's books at £9. 9. 7.; present net income, £187: it is in the patronage of the Crown, in right of the duchy of Cornwall; impropriator, Earl of St. Germans. On the moors in this parish are several ancient barrows.

TRENT (*St. Andrew*), a parish, in the union of Sherborne, hundred of Horethorne, Eastern Division of the county of Somerset, 3 miles (N. E. by E.) from Yeovil; containing 449 inhabitants. The living is a rectory, valued in the king's books at £23. 5. 5.; present net income, £433; patrons, President and Fellows of Corpus Christi College, Oxford. The church has a tower at the south-east corner, surmounted by an hexagonal spire. John Young, in 1678, bequeathed £1000 for the erection and endowment of a free school; the annual income is about £95, for which fifteen poor boys are instructed

TRENTHAM (*St. Mary*), a parish, in the union of Stone, Northern Division of the hundred of Pirehill and of the county of Stafford; containing 2344 inhabitants, of which number, 631 are in the township of Trentham, 4 miles (S. S. E.) from Newcastle-under-Lyme. The ancient parish, which is situated on the London and Liverpool road, has been divided into three distinct and separate parishes, under the 16th section of the act of the 58th of George III. The Trent and Mersey canal passes through the parish, in the neighbourhood of which there is a considerable manufacture of remarkably good bricks and tiles, of a dark blue colour, much in request at Northampton and the intervening places, in which about 200 persons are employed, and for the conveyance of which, and other produce, there is a commodious wharf. The living is a perpetual curacy; net income, £113; patron and impropriator, Duke of Sutherland. The church is an ancient structure without a tower, which was taken down about a century ago; it is the only remaining portion of an Augustine priory. A chapel, erected at Hanford, was opened in July 1828; it contains 350 free sittings, the Incorporated Society having granted £200 towards defraying the expense: there is also a chapel at Blurton; both are endowed. A chapel at More-Heath has also been recently licensed by the Bishop. Lady Catherine Leveson, in 1670, bequeathed an annuity of £20 for teaching poor children, and a like sum for apprenticing them. Two schools, one at Trentham, and one at More-Heath, are partly supported by subscription. Here was anciently a nunnery, of which St. Werburgh, in the seventh century, was appointed abbess, by her brother, King Ethelred. In the reign of Henry I., Randal, Earl of Chester, converted it into a priory of Augustine canons, in honour of the Blessed Virgin Mary and all Saints, which, at the dissolution, had a revenue of £121. 3. 2., and was granted to Charles Duke of Suffolk. Trentham gives the inferior title of Viscount to the Duke of Sutherland, whose noble mansion is in this parish.

TRENTISHOE (*St. Peter*), a parish, in the union of Barnstaple, hundred of Braunton, Braunton and Northern Divisions of the county of Devon, $10\frac{1}{2}$ miles (E. by N.) from Ilfracombe; containing 128 inhabitants. The living is a discharged rectory, valued in the king's books at £8. 8. 4.; present net income, £118: it is in the patronage of Mrs. A. W. Griffiths.

TREPRENAL, a township, in the parish of Llanymynech, hundred of Oswestry, county of Salop, 5 miles (S.) from Oswestry: the population is returned with the parish.

TRESCOTT, a joint hamlet with Pirton, in the parish of Tettenhall, Northern Division of the hundred of Seisdon, Southern Division of the county of Stafford, 4 miles (W. by S.) from Wolverhampton: the population is returned with Pirton.

TRESHAM, a chapelry, in the parish of Hawkesbury, Upper Division of the hundred of Grumbald's-Ash, Western Division of the county of Gloucester, $3\frac{1}{2}$ miles (S. E. by E.) from Wotton-under-Edge: the population is included in the return for Killcott.

TRESMEER (*St. Nicholas*), a parish, in the union of Launceston, Northern Division of the hundred of East, Eastern Division of the county of Cornwall, $6\frac{1}{2}$ miles (W. by N.) from Launceston; containing 171 inhabitants. The living is a perpetual curacy; net income, £85: it is in the patronage of the Crown.

TRESWELL (*St. John the Baptist*), a parish, in the union of East Retford, South-Clay Division of the wapentake of Bassetlaw, Northern Division of the county of Nottingham, $5\frac{1}{4}$ miles (E. by S.) from East Retford; containing 224 inhabitants. The living is a rectory, formerly in two portions, which were united in 1764; the eastern portion is valued in the king's books at £8. 1. 4.; and the western at £9. 15. 8.; present net income, £254; patrons, Dean and Chapter of York. There is a place of worship for Wesleyan Methodists; also a school partly supported by the rector.

TRETIRE (*St. Mary*), a parish, in the union of Ross, Lower Division of the hundred of Wormelow, county of Hereford, $5\frac{3}{4}$ miles (W.) from Ross; containing 120 inhabitants. The living is a rectory, with the rectory of Michael-Church united, valued in the king's books at £6. 1. 8.; present net income, £182; patrons and impropriators, Governors of Guy's Hospital, London.

TREVALGA, a parish, in the union of Camelford, hundred of Lesnewth, Eastern Division of the county of Cornwall, $1\frac{1}{2}$ mile (N. E.) from Bossiney; containing 192 inhabitants. The living is a discharged rectory, valued in the king's books at £7. 6. $0\frac{1}{2}$.; present net income, £146; patrons, Dean and Chapter of Exeter. There are quarries of slate in the parish.

TREVENA, county of Cornwall.—See BOSSINEY with TREVENA.

TREVETHAN (*St. Cadocus*), a parish, in the union of Pont-y-pool, Upper Division of the hundred of Abergavenny, county of Monmouth, 2 miles (N.) from Pont-y-pool; containing, with the market-town of Pont-y-pool, 10,280 inhabitants. The Monmouthshire and Brecon canals, and numerous railroads, pass through the parish. The inhabitants are employed in extensive mines of iron and coal, with which the neighbourhood abounds, in the burning of lime, and in the large iron works established at Pont-y-pool and in its vicinity. The British Mining Company have established furnaces at the Varteage, three miles from Pont-y-pool; and buildings for the residence of the overseers and workmen have been erected in almost every direction. The living is united, with that of Mamilad, to the vicarage of Llanover. The church, a very ancient building, has been enlarged, and contains 180 free sittings, the Incorporated Society having granted £100 towards defraying the expense. A church in the later English style, with a campanile turret, was erected in 1832, at an expense of £1764, under the act of the 58th of George III. Two schools, called the "Varteage Hill Iron Company's schools," are partly supported by the proprietors; a school of industry for girls is partly supported by subscription; and an infants' school is supported by Capel Hanbury Leigh, Esq. Charles Price, in 1826, bequeathed £200, the interest to be appropriated in supplying bread to the poor.

TREVILLE, an extra-parochial liberty, in the union of Dore, Upper Division of the hundred of Wormelow, county of Hereford, $6\frac{1}{2}$ miles (N. W. by N.) from Ross; containing 66 inhabitants.

TREWEN (*St. Michael*), a parish, in the union of Launceston, Northern Division of the hundred of East, Eastern Division of the county of Cornwall, $5\frac{1}{4}$ miles (W. by S.) from Launceston; containing 213 inhabitants. The living is a joint vicarage with South Petherwin. Fairs for colts, sheep, and lambs, are held on May 1st and Oct 10th.

TREWHITT, HIGH and LOW, a township, in the parish and union of Rothbury, Western Division of Coquetdale ward, Northern Division of the county of Northumberland; containing 137 inhabitants. High Trewhitt is $4\frac{1}{4}$ miles (N. W.), and Low Trewhitt $4\frac{1}{2}$ miles (N. W. by N.), from Rothbury.

TREWICK, a township, in the parish of Bolam, union and Western Division of Castle ward, Southern Division of the county of Northumberland, $7\frac{1}{2}$ miles (S. W.) from Morpeth; containing 30 inhabitants. It is bounded on the south by the river Blyth.

TREYFORD, a parish, in the union of Midhurst,

hundred of DUMPFORD, rape of CHICHESTER, Western Division of the county of SUSSEX, 4 miles (S. W. by W.) from Midhurst; containing 130 inhabitants. The living is a rectory, with the vicarage of Didling annexed, valued in the king's books at £7. 12. 1.; present net income, £144; patron, Rev. L. V. Harcourt. Contiguous to the Downs are several circular and conical barrows.

TRIMDON (*St. Mary Magdalene*), a parish, in the union of SEDGEFIELD, Southern Division of EASINGTON ward, Northern Division of the county palatine of DURHAM, 9 miles (S. E.) from Durham; containing 276 inhabitants. The living is a perpetual curacy; net income, £96; patron and impropriator, William Beckwith, Esq. A lectureship was endowed here, before 1730, with £21. 5. a year, by John Smith, Esq. Some large specimens of lead-ore have been dug up in the neighbourhood, but no mine has yet been opened. Henry Airey, in 1680, gave a rent-charge of £5 for the maintenance of a free school, which is conducted on the National plan, and is further endowed with £7 per annum; the school-room was built by subscription, in 1823. The produce of an estate, purchased with the amount of various bequests, and now let for £32 per annum, is applied partly in apprenticing children, and partly to the relief of the poor.

TRIMINGHAM (*St. John the Baptist*), a parish, in the union of ERPINGHAM, hundred of NORTH ERPINGHAM, Eastern Division of the county of NORFOLK, 5 miles (S. E. by E.) from Cromer; containing 168 inhabitants. The living is a discharged rectory, valued in the king's books at £6; it is in the patronage of the Crown, in right of the Duchy of Lancaster. The tithes have been commuted for a rent-charge of £134. 2. 9., subject to the payment of rates, which on the average have amounted to £20; the glebe comprises 3 acres, valued at £4. 4. per annum. There is also a rent-charge of £7. 10. payable to the rector of Sidestrand. A day and Sunday school is supported by private charity. In papal times it was pretended that the head of St. John the Baptist, the patron saint of the church, was deposited here, to which numerous pilgrimages and rich offerings were made.

TRIMLEY (*St. Martin*), a parish, in the union of WOODBRIDGE, hundred of COLNEIS, Eastern Division of the county of SUFFOLK, 8½ miles (S. E. by E.) from Ipswich; containing 514 inhabitants. The living is a discharged rectory, with Alleston, or Alkeston, consolidated (of the church of which there are no remains), valued in the king's books at £12. 0. 5.; present net income, £423; patron, Rev. C. Waller. The church is situated in the same churchyard with that of Trimley-St. Mary, and contains a mausoleum of the family of Sir John Barker, Bart.; the walls of the churchyard were repaired with the stones of Felixton castle and priory, of which the former was anciently a Roman station; the sites have been lost by encroachment of the sea. Here is a National school. The navigable river Orwell bounds the parish on the south. Grimston Hall, the site of which is now occupied by a farm-house, was the seat of Thomas Cavendish, celebrated as the first English circumnavigator, who was born here.

TRIMLEY (*St. Mary*), a parish, in the union of WOODBRIDGE, hundred of COLNEIS, Eastern Division of the county of SUFFOLK, 8¾ miles (S. E. by E.) from Ipswich; containing 401 inhabitants. The living is a discharged rectory, valued in the king's books at £16. 13. 4.; it is in the patronage of the Crown. The tithes have been commuted for a rent-charge of £470, subject to the payment of rates. The church, of which the steeple and part of the nave are in ruins, is in the same churchyard with that of Trimley-St. Martin, and service is performed once in each every Sunday. The navigable river Orwell bounds the parish on the south; and it is stated that there was anciently a considerable town here, which gave name to the haven and the river, and which is recorded in history to have been plundered by the Danes. A school is supported by subscription, aided by an annual grant from the Suffolk society. The rent of an allotment of land awarded under an enclosure act in 1808, amounting to £10, is annually expended in coal for distribution among the poor.

TRING (*St. Peter and St. Paul*), a market-town and parish, in the union of BERKHAMPSTEAD, hundred of DACORUM, county of HERTFORD, 30 miles (W. by N.) from Hertford, and 31 (N. W. by W.) from London; containing 3488 inhabitants. The origin of this town is of remote antiquity: at the time of the division of the county by Alfred, it was considered of sufficient importance to give name to the hundred in which it was situated, being then called *Treung*. Antiquaries have attributed the derivation of its name to the form of the town, which they suppose to have been originally triangular. The opinion that it is of Roman origin receives confirmation from the fact that the Ikeneld-way from Dorchester to Colchester passed in its vicinity; part of it is included within the Chiltern hundreds. The town consists of two principal streets, the larger being crossed at the top by the other, both containing good houses, generally of modern style. Contiguous to it is the elegant mansion of Tring Park (built by Charles II., for his favourite mistress, Eleanor Gwynn, and since modernized), with the hills rising in the back ground, clothed with fine beech trees. The general appearance is exceedingly neat, the atmosphere very salubrious, and the inhabitants are amply supplied with water. A silk-mill, worked partly by water and partly by steam, gives employment to upwards of 300 persons, and is in progress of great improvement: the manufacture of canvas and straw-plat is also carried on. The Grand Junction canal passes within about a mile of the town; and in the parish are four large reservoirs, to supply any loss of water to that navigation. At Wilstone, also within the parish, is one of the sources of the river Thames. About two miles distant is the Tring station of the London and Birmingham railway. The market, granted by charter of Charles II. to Henry Guy, Esq., in 1681 (to whom that monarch had, the year before, granted the manor), is on Friday, for straw-plat, corn, meat, and pedlery; and cattle fairs are held on Easter-Monday and Old Michaelmas-day. The market-house, the property of the lord of the manor, is situated on the north side of the principal street. The living is a perpetual curacy; net income, £157; patrons and appropriators, Dean and Canons of Christ-Church, Oxford. The church, situated near the market-house, is a handsome embattled structure, in the ancient English style, with a large tower at the west end, surmounted with a low spire: the font is in the later style of English architecture, highly enriched. At Long Marston, in this parish, is a chapel of ease. There are places of worship for

Baptists and Independents. Two allotments of land, containing together about 110 acres, and now let for £71 per annum, were awarded to the parish in 1793, under an enclosure act: the rent, together with the produce of some other small bequests, is appropriated to supplying the poor with coal, &c. A Roman helmet was found in digging the Grand Junction canal, near Northcote hill, between this town and Berkhampstead, of which a drawing was published by the Society of Antiquaries of London, in 1819. Robert Hill, a remarkable self-taught linguist, was born here in 1699; he died in 1777.

TRIPPLETON, a joint township with Whitton, in the parish of LEINTWARDINE, hundred of WIGMORE, county of HEREFORD; containing 63 inhabitants.

TRITLINGTON, a township, in the parish of HEBBURN, union, and Western Division of the ward, of MORPETH, Northern Division of the county of NORTHUMBERLAND, 4¾ miles (N.) from Morpeth; containing 82 inhabitants. A school is partly supported by the Duke of Portland and the rector.

TROSTON (*St. Mary*), a parish, in the union of THINGOE, hundred of BLACKBOURN, Western Division of the county of SUFFOLK, 7 miles (N. N. E.) from Bury-St.-Edmund's; containing 399 inhabitants. The living is a discharged rectory, valued in the king's books at £10. 4. 7.; present net income, £332: it is in the patronage of the Crown. There is a place of worship for Wesleyan Methodists. An allotment of land, containing about 15 acres, was awarded to the parish under an enclosure act in 1800: it is let for £[illegible] per annum, and the rental is expended in coal for distribution among the poor. Capel Lofft had a seat here.

TROSTREY (*St. David*), a parish, in the union of PONT-Y-POOL, Upper Division of the hundred of USK, county of MONMOUTH, 2½ miles (N. by W.) from Usk; containing 202 inhabitants. The living is a discharged perpetual curacy, valued in the king's books at £3. 8. 11½.; present net income, £72; patron and impropriator, Sir S. Fludyer, Bart.

TROTTERSCLIFFE, or TROSLEY (*St. Peter and St. Paul*), a parish, in the union of MALLING, hundred of LARKFIELD, lathe of AYLESFORD, Western Division of the county of KENT, 2 miles (N. E. by E.) from Wrotham; containing 310 inhabitants. The living is a rectory, valued in the king's books at £10. 2. 11.; present net income, £287: it is in the alternate patronage of the Crown and the Bishop of Rochester. The Rev. Paul Baristow, in 1711, bequeathed land producing £8 a year, for teaching poor children.

TROTTON (*St. George*), a parish, in the union of MIDHURST, partly in the hundred of EASEBOURNE, but chiefly in that of DUMPFORD, rape of CHICHESTER, Western Division of the county of SUSSEX, 3½ miles (W. by N.) from Midhurst; containing 416 inhabitants. The living is a rectory, valued in the king's books at £9; present net income, £296; patron, S. Twyford, Esq. The church is principally in the decorated style of architecture, and contains a beautiful monument of brass inlaid with Sussex marble, to Lord and Lady Camois, and in the nave is another brass to a lady of that family; there are also several neat memorials to the families of Twyford, Alcock, and Aylevin. There is a chapel of ease at Milland, near the border of Hampshire. Under the Municipal Reform Act this parish is within the new boundary of the borough of Midhurst. A day and Sunday school is partly supported by private charity. Thomas Otway, the poet, was born here in 1651; he died in 1685.

TROUGH, a township, in the parish of STAPLETON, union of LONGTOWN, ESKDALE ward, Eastern Division of the county of CUMBERLAND, 9½ miles (N. E. by E.) from Longtown; containing 169 inhabitants.

TROUGHEND-WARD, a township, in the parish of ELSDON, union of BELLINGHAM, Southern Division of COQUETDALE ward, Northern Division of the county of NORTHUMBERLAND, 7¼ miles (N. N. E.) from Bellingham; containing 327 inhabitants.

TROUTBECK, a chapelry, in the parish of WINDERMERE, union and ward of KENDAL, county of WESTMORLAND, 5½ miles (S. E. by E.) from Ambleside; containing 349 inhabitants. This chapelry is intersected by a rivulet, from which it derives its name. In the neighbourhood are extensive quarries of fine blue slate. The living is a perpetual curacy, net income, £43; patron, Rector of Windermere. The chapel, called Jesus' chapel, was consecrated in 1562: adjoining is a school-house, built in 1639, with an endowment of £8 per annum. There were formerly two cairns, supposed to be British, on the removal of one of which a rude stone chest was discovered, enclosing a quantity of human bones.

TROUTSDALE, a township, in the parish of BROMPTON, union of SCARBOROUGH, PICKERING lythe, North Riding of the county of YORK, 8 miles (W.) from Scarborough; containing 59 inhabitants.

TROWAY, a township, in the parish of ECKINGTON, hundred of SCARSDALE, Northern Division of the county of DERBY, 6½ miles (N. by E.) from Chesterfield: the population is returned with the parish.

TROWBRIDGE (*St. James*), a market-town and parish, in the union and hundred of MELKSHAM, Westbury and Northern Divisions, and Trowbridge and Bradford Subdivisions, of the county of WILTS, 30 miles (N. W.) from Salisbury, and 99 (W. by S.) from London; containing 10,863 inhabitants. The origin of this place, and the etymology of its name, are involved in much obscurity: Camden says it was called by the Saxons *Truthabrig*, a strong and faithful town. It is not mentioned in Domesday-book; but a place called Little Trowle, now a hamlet in the parish, is therein recorded, and hence the present name is by many supposed to be a corruption of *Trowlebridge*, by which it is described by Geoffrey of Monmouth; Leland writes it *Throughbridge*, or *Thorough-bridge*. Trowbridge was formerly a royal manor, forming part of the duchy of Lancaster, having been granted by the crown to John of Gaunt: it afterwards reverted to the crown, and was given by Henry VIII., in the 28th year of his reign, to Sir Edward Seymour, Knight, Viscount Beauchamp. Having again lapsed to the crown, Queen Elizabeth, in the 24th of her reign, assigned it, with the profits of the fairs and markets, &c., to Edward Earl of Hertford: it afterwards became the property of the Duke of Rutland, who alienated it to Thomas Timbrell, Esq., father of the present proprietor. The earliest historical circumstance relating to the town is its defence against King Stephen, by Humphrey de Bohun, who held it for the Empress Matilda, at which period its castle is supposed to have existed, though some writers ascribe its erection to John of Gaunt, Duke of Lancaster: it was demolished previously to the time of Henry VIII., as, when Leland wrote, it was in ruins, only two of its seven towers remaining: not a vestige of it now exists, its site being occupied by other buildings.

The town is situated upon a rocky hill, near the river Biss, across which is a stone bridge, and is very irregularly built, the houses being mostly of stone: the principal street is spacious, and contains some excellent houses, but the others are generally narrow, and the houses old, and of rather a mean appearance: it is paved, lighted with gas, and tolerably well supplied with water. The manufacture of woollen cloth was introduced here at an early period, and must have very soon become a thriving branch of trade, as Camden mentions that Trowbridge was then famous for the clothing trade: the articles made are chiefly kerseymeres, with some superfine broad cloth. The Kennet and Avon canal passes about a mile north of the town, by which a communication is opened with London and Bristol; and a branch of the Great Western railway will be made to the town. The markets are on Tuesday, Thursday, and Saturday, the last being the principal market day, and are well supplied with provisions: there is a fair on the 5th of August, which lasts two days, for cattle, cheese, woollen goods, &c. The town is under the government of the county magistrates, and a petty session is held, by those resident in it, on the first Tuesday in the month, for the transaction of business connected with the parish. A court of requests, for the recovery of debts not exceeding £5, is holden here every Tuesday three weeks, comprehending within its jurisdiction the hundreds of Melksham, Bradford, and Whorwelsdown. A court leet and a court baron are also held by the lord of the manor at Easter, at the former of which constables, tythingmen, a crier, and cornets of the market, are appointed and sworn.

The living is a rectory, valued in the king's books at £20. 12. 8½.; patron, Duke of Rutland. The tithes have been commuted for a rent-charge of £609. 6., subject to the payment of rates, which on the average have amounted to £219; the glebe comprises 54 acres. There is also a rent-charge of £16. 16. payable to the lay impropriator of Haverton-Fishing, subject to the payment of rates, which on the average have amounted to £5. The church is called the New church, in consequence of a more ancient one having existed about 70 yards to the south-east of the former; it is a large building, with a tower at the west end, surmounted by a lofty spire: the walls of the nave and aisles are crowned with battlements and crocketed pinnacles; the nave has a flat ceiled roof, ornamented with flowers, &c., and is separated from the aisles by five pointed arches on each side, springing from clustered columns, with rich capitals; there are fragments of painted glass in some of the windows; the font is lofty, and is covered with a profusion of tracery and panelling. Attached to the eastern extremities of the two aisles are two chapels, that on the south belonging to the lord of the manor, and that on the north to the family of John Clark, Esq. A church has also been erected, of which the first stone was laid May 1st, 1837: it contains 1000 sittings, half of which are free, the Incorporated Society having granted £430 towards defraying the expense; and a district has been formed under the act of the 58th of George III. At Staverton, in this parish, there is a district chapel. There are four places of worship for Particular Baptists, one for General Baptists, one for Independents, two for Wesleyan Methodists, and one for Presbyterians. A National school, in which 50 boys are instructed, is supported by property left for charitable purposes: it stands in the churchyard, and the master is provided with a house. A Lancasterian school was established in 1832, and a substantial building was erected at an expense of £1800, by subscription: and there are also ten Sunday schools, in which about 2200 children are taught to read; and an infants' school, partly supported by subscription. In an almshouse in Hilperton-lane, founded by a person named Yerbury, six poor widows are lodged, and receive a weekly allowance. The poor enjoy the benefit of the moiety of a small estate at Hoddesdon, in Hertfordshire, the rental of which, with the produce of other funds, is distributed in bread, annually at Christmas, among such as do not receive parochial relief. George Keate, a poetical and miscellaneous writer of some celebrity, was born here, in the year 1730: and the Rev. George Crabbe, the late well-known poet, was instituted to the rectory in 1814, which he held till his death in 1832. Trowbridge formerly gave the title of Baron to the family of Seymour, Dukes of Somerset, one of whom is buried here, which title is extinct.

TROWELL (*St. Helen*), a parish, in the union of Basford, Southern Division of the wapentake of Broxtow, Northern Division of the county of Nottingham, 5¾ miles (W.) from Nottingham; containing 402 inhabitants. The living is a rectory in two portions, each valued in the king's books at £4. 14. 4½.; present net income, £440; patron, Lord Middleton, who partly supports a school. The Nottingham canal passes through the parish, which is separated from Derbyshire by the river Erewash. There are coal mines in the neighbourhood.

TROWLE, a tything, in the parish of Great Bradford, union and hundred of Bradford, Westbury and Northern Divisions, and Trowbridge and Bradford Subdivisions, of the county of Wilts; containing 290 inhabitants.

TROWSE (*St. Andrew*), a parish, in the union and hundred of Henstead, Eastern Division of the county of Norfolk, 1¾ mile (S. E. by S.) from Norwich; containing 583 inhabitants. The living is a vicarage, with that of Lakenham annexed, in the patronage of the Dean and Chapter of Norwich (the appropriators), valued in the king's books at £5; present net income, £314. Trowse-Newton Hall, an ancient building with a chapel, erected by the priors of Norwich, has been converted into a farm-house.

TRUDOX-HILL, a hamlet, in the parish of Nunney, hundred of Frome, Eastern Division of the county of Somerset, 4 miles (S. W.) from Frome: the population is returned with the parish. Here was formerly a chapel, long since desecrated.

TRULL (*All Saints*), a parish, in the union of Taunton, hundred of Taunton and Taunton-Dean, Western Division of the county of Somerset, 2 miles (S. S. W.) from Taunton; containing 506 inhabitants. The living is a perpetual curacy; net income, £98; patron and impropriator, Sir F. G. Cooper, Bart. John Wyatt, in 1756, gave the proceeds of £210 in support of a school; the annual income is £24, which is applied for teaching eighteen poor children.

TRUMPINGTON (*St. Mary and St. Michael*), a parish, in the union of Chesterton, hundred of Thriplow, county of Cambridge, 2½ miles (S.) from Cambridge; containing 722 inhabitants. The living is a vicarage, valued in the king's books at £5. 6. 8.; present net income, £241; patrons and impropriators, Master and Fellows of Trinity College, Cambridge. William Austin, in 1679, gave fourteen acres of land, now producing £18 per annum, for teaching eight poor children. There are still some remains of the mill, celebrated by Chaucer, in his Reeve's Tale. At Dam Hill, near the river Cam, which runs through the parish, several beautiful vases and pateræ, urns containing calcined human bones, and other relics of antiquity, have been discovered. Christopher Anstey, author of the poetical "Bath Guide," was born here in 1724.

TRUNCH (*St. Botolph*), a parish, in the union of Erpingham, hundred of North Erpingham, Eastern Division of the county of Norfolk, 3 miles (N. by E.) from North Walsham; containing 441 inhabitants. The living is a rectory, valued in the king's books at £10. 13. 4.; present net income, £350; patrons, Master and Fellows of Catherine Hall, Cambridge. There is a place of worship for Wesleyan Methodists.

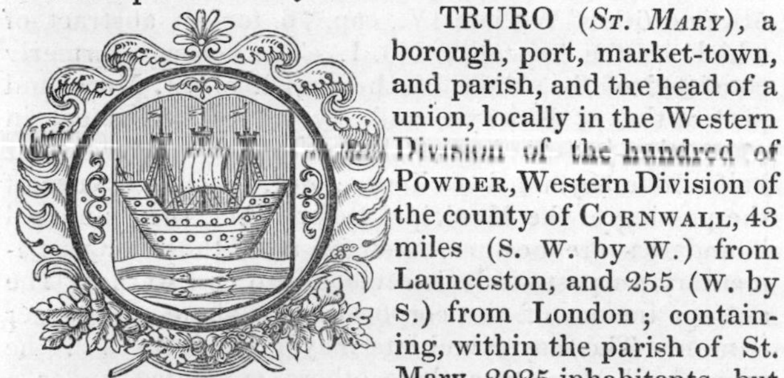

Seal and Arms.

TRURO (*St. Mary*), a borough, port, market-town, and parish, and the head of a union, locally in the Western Division of the hundred of Powder, Western Division of the county of Cornwall, 43 miles (S. W. by W.) from Launceston, and 255 (W. by S.) from London; containing, within the parish of St. Mary, 2925 inhabitants, but within the whole town (which extends into the parishes of *St. Clement* and *Kenwyn*), about 8468. This place, called in ancient records *Triueru*, *Treuru*, *Truru*, and *Truruburgh*, and in a receipt given for the payment of a fine to the king in the 15th of Henry VII., the *ville de Truro*, all of similar import, is supposed to have derived its name from the three streets of which the town originally consisted. The manor, in 1161, belonged to Richard de Luci, chief justice of England and lord of Truro, who probably built the castle, of which the memorial is preserved in the name of its site, still called Castle Hill, and invested the inhabitants with numerous privileges, which were subsequently confirmed by Reginald Fitz-Henry, Earl of Cornwall, natural son of Henry I., one of the witnesses to whose charter was Robert de Dunestanville, ancestor of the late Lord de Dunestanville, through intermarriage with whose family Sir Francis Bassett, on his elevation to the peerage, assumed that title. In 1410, a petition was presented to the parliament by the inhabitants of the town, praying that the rent payable to the crown, which had been reduced by Richard II. from £12. 1. 10. to £2. 10., for a term of years, in consequence of their sufferings from war and pestilence, might be continued in perpetuity; and stating that, instead of rebuilding their houses, the inhabitants were about to leave the town, which might be considered as the defence of that part of the country from the attempts of the enemy. Here, after the defeat at Naseby, the remains of the royalist army surrendered to Sir Thomas Fairfax; and whilst the negociations were pending, Lord Hopton, their general, and the Prince of Wales, afterwards Charles II., Sir Edward Hyde, afterwards Lord Clarendon, Lord Capel, and other royalists of distinction, made their escape and embarked at Falmouth for Scilly and thence to Jersey.

The town is pleasantly situated at the confluence of the rivers Kenwyn and St. Allen, which here fall into a creek from the river Fal, forming together an estuary sufficient to enable vessels of 100 tons' burden to approach the town at spring tides; and in the centre of a rich and extensive mining district, to which it is principally indebted for its commercial importance. The surrounding scenery is pleasingly diversified, and at high tides a beautiful lake, nearly two miles in length, is formed above Mopus, from which the tide never recedes. A considerable increase has recently taken place in the number of its houses, and great improvements have been made in the principal streets and approaches; it has consequently become a handsome, well-built town, paved and lighted with gas, and amply supplied with water by streams flowing through the principal streets. These improvements were effected by commissioners authorized by an act passed in 1790; but the town having outgrown their jurisdiction, a new act was obtained in 1835; and to the exertions of the commissioners the town is indebted for its general neatness and cleanliness, and comparative exemption from disease. The Royal Institution of Cornwall, to the support of which Her Majesty subscribes £50 per annum, is established here: it possesses a museum, handsomely fitted up and enriched with specimens of objects of natural history not only belonging to the county, but from all parts of the globe: it also contains a very respectable number of geological and mineralogical specimens, antiquities, coins, and various productions of the artists of China, America, Africa, the South Sea Islands, &c. Monthly meetings are held for reading papers and discussing subjects tending to elucidate questions submitted to the society relating to the history, antiquities, and manufactures of the county. In the same building is the county library, also very liberally supported, and containing at present about 4000 volumes. The Truro Institution holds its meetings in the lecture-room of the Royal Institution, the object of which is to promote the diffusion of knowledge by means of lectures on literary and scientific subjects: these lectures have been very attractive, and the society is in a very flourishing state. The Royal Horticultural Society of Cornwall has its museum and library in a part of the same building. It has three exhibitions in the year, one at Falmouth, one at Redruth, and one at Truro. The Cornwall Agricultural Society also holds its annual exhibition at Truro, and an attempt has been recently made to combine to a certain extent these societies, by extending horticultural experiments to the practical test of agriculture. There is a very handsome assembly-room, which is occasionally converted into a theatre, and to which an elegant subscription billiard-room is now added. At the top of Lemon-street a handsome Doric column of granite has been erected to commemorate the discovery of the termination of the river Niger, or Quorra, in the sea at the Bight of Benin, unsuccessfully attempted by Park,

Clapperton, and others. John and Richard Lander were natives of Truro; both are dead.

The port exercises jurisdiction over the several creeks of Newham, Tresillian, Restronguet, Devoran, Tregoney, Pill, and Mylor. The principal exports are tin and copper-ore; the former, which is made into blocks weighing $3\frac{1}{2}$ cwt., ingots from 60 to 70 lb., and bars from 4 to 6 oz. each, is shipped at this port chiefly to France, the Mediterranean, and the Baltic; the copper-ore, principally from the neighbourhood of Redruth, is shipped at Devoran, where there is a ferry for horse and foot passengers, making the distance from Truro to Falmouth only $7\frac{1}{2}$ miles. The imports are principally iron, coal, timber, and other commodities. The amount of duties paid at the custom-house, for the year ended January 5th, 1837, was £48,552. 13. 7. There are 25 vessels belonging to the port, averaging a burden of 80 tons each, chiefly employed in the coasting trade. An extensive carpet and woollen manufactory has been established here for more than 40 years. There are also an iron-foundry, several tanneries, and two small potteries for the coarser kinds of earthenware. The smelting of tin is carried on extensively; there are now four smelting-houses in the town and its immediate vicinity, viz., at Calenick, on the Falmouth-road, where the best crucibles in Europe for assaying are made; at Carvedras, containing four reverberating furnaces, with a chimney 110 feet high, with which the flues from the furnaces communicate; a third near Garras Wharf, on the south side of the town; and a fourth recently established at the eastern entrance to the town, where an elegant chimney, 120 feet high, has been erected, forming one of the chief ornaments of the neighbourhood. The coinage hall, in which the tin formerly received the duchy stamp, is an ancient edifice, at the east end of Boscawen-street; and the quantity of tin stamped there, in 1831, was 7274 blocks, each block weighing about 4 cwt.; the duty was 4*s.* per cwt. This town and Helston and Penzance were the principal stannary towns in the county: the custom of coining the tin, as it was called, has been recently abolished. The jurisdiction of the ancient stannary courts having been confined to cases wherein tin or tinners were concerned, and this having been found a serious inconvenience to persons engaged in raising other minerals, by an act passed in the 6th and 7th of William IV., cap. 106, it is declared expedient to unite the court of equity of the Vice-Warden with the courts of common law of the Stewards of the Stannaries, and to extend the jurisdiction of the court to and over all metals and metallic minerals in the said stannaries, and to and over all transactions connected therewith in the county of Cornwall. This court is held quarterly at Truro, and has proved to be of the greatest utility to the mining interests of the county. It has also been found of essential service to the trading community generally, from the facilities afforded for the recovery of small debts whenever the plaintiff or defendant can be shown to be connected with transactions relating to the raising or purifying of metals or metallic minerals. The markets are on Wednesday and Saturday, the former for corn, and both are abundantly supplied with provisions of all kinds; a cattle market has been recently established, on the first Wednesday in every month. Fairs for cattle are also held on the Wednesday after Mid-Lent Sunday, Wednesday after Whit-Sunday, November 19th, and December 8th: that in November belongs to the high lords of the manor, and is held in a place near the church, called the Cross; the others are held on the Castle Hill.

The original charter granted to the inhabitants by Earl Reginald has no date, but it must have been granted between 1140 (5th of Stephen) and 1176 (22nd of Henry II.): other charters were granted by Edward I., in 1284; Edward III., in 1369; Henry IV., in 1402; Edward Prince of Wales, as Duke of Cornwall (afterwards Edward V.), in 1477; Henry VII., in 1488; and by Elizabeth, in 1589; all which charters are among the muniments of the corporation. James II., in the first year of his reign, also granted a charter to the inhabitants, which was abandoned on the abdication of that monarch, from its unconstitutional provisions, by which the king reserved the right of dismissing the capital burgesses and of appointing others in their stead, so as to enable him to influence the election of members of parliament. The government of the borough was vested in a mayor, four aldermen, and twenty capital burgesses (including the mayor), assisted by a recorder, town-clerk, coroner, and town-steward, elected by the corporation: the mayor and recorder were justices of the peace. The corporation now consists of a mayor, six aldermen, and 18 councillors, under the act of the 5th and 6th of William IV., cap. 76, for an abstract of which see the Appendix, No. I. The borough formerly consisted of the whole of the parish of St. Mary and part of that of Kenwyn, and, by act of parliament in 1790, the jurisdiction of the justices was extended for half a mile beyond these limits; but, in consequence of the passing of the Municipal Reform Act, the municipal boundaries are now co-extensive with those for parliamentary purposes. It is divided into two wards. The average income of the corporation is about £1200 per annum. The mayor and late mayor are justices of the peace, and there are three others appointed under a commission of the crown. The police force consists of an inspector and five constables. The borough first sent members to parliament in the reign of Edward I. The right of election was formerly vested in the mayor, aldermen, and capital burgesses; but it has been extended to the £10 householders of an enlarged district, comprising 1235 acres, the limits of which are minutely described in the Appendix, No. II. The number of voters now registered is 644: the mayor is the returning officer. It was formerly usual, at the election of a mayor, to deliver the town mace to the lord of the manor, who retained it till sixpence had been paid for certain houses within the borough; this custom is now discontinued, but sixpence is still paid by such tenants as occupy these ancient houses, under the appellation of "Smoke Money." The charter of Elizabeth describes the mayor of Truro to be also mayor of Falmouth, and as such he exercised jurisdiction over Falmouth harbour, the customs and dues of which he received; but this claim was in part successfully resisted by the inhabitants of that town, and the mayor has now jurisdiction only over a small part of the harbour, which is preserved by the practice of arresting there, once in seven years, when the water boundaries of the port of Truro are renewed, in the presence of the members of the corporation, an inhabitant for a nominal debt of £999, who is immediately liberated on bail: this was last done in June

1839. The Easter quarter sessions for the county were held at this place, but at the Midsummer sessions held at Bodmin in 1839, the magistrates voted that for the future they should be held there. The petty sessions for the Western Division of the hundred are held on the first Thursday in every month; and, by the act of the 2nd and 3rd of William IV., cap. 64, the town has been made the place of election and a polling-place for the western division of the county.

The living is a discharged rectory, valued in the king's books at £16; present net income, £135; patron, Earl of Mount-Edgecumbe, one of the high lords of the manor. The church, a handsome structure, partly of granite and partly of freestone, in the later style of English architecture, was built in 1518; but the tower, which is surmounted by a spire, was not erected until 1769. It contains some remains of ancient stained glass, and some elegant monuments to the families of Robarts, Vivian, Pendarves, and others. A church, dedicated to St. John, in the Grecian style of architecture, with a campanile turret, was erected in Lemon-street by subscription of the inhabitants, aided by a grant of £700 from the parliamentary commissioners: the living is a perpetual curacy, in the patronage of the Vicar of Kenwyn. There are places of worship for Baptists, Bryanites, the Society of Friends, Independents, and Methodists of the Old and New Connection. Near the site of the castle is a cemetery with a chapel, for the performance of the funeral service. The free grammar school, which is of uncertain foundation, is under the management of the corporation, who allow the master £25 per annum: it has two exhibitions of £30 per annum each to Exeter College, Oxford, founded by the Rev. St. John Elliot, formerly rector of St. Mary's, who died in 1760. Sir Humphrey Davy, the celebrated experimental chymist and natural philosopher; Lord Exmouth, Sir Hussey Vivian, Polwhele, Henry Martyn, and many other distinguished characters, received the rudiments of their education in this school. A National and a British school have been established. John White, Esq., bequeathed £10 per annum for apprenticing two poor children. An hospital for ten poor people was founded in 1631, by Mr. Henry Williams, who endowed it with lands now producing about £200 per annum; it is under the management of the charity trustees, who appoint females only The county infirmary, situated on an elevated and healthy spot near the town, was opened in 1799, under the patronage of George IV., then Duke of Cornwall, and is liberally supported by subscription. The poor law union of Truro comprises 24 parishes or places, containing a population of 39,114, according to the census of 1831, and is under the care of 45 guardians. A convent of Grey friars was established here in the latter part of the reign of Henry III., by an ancestor of Rauf Reskmyer, who was a great benefactor to the establishment in the reign of Edward IV.; it flourished till the dissolution, and in the reign of Edward VI. the site was granted to Edward Aglianby, and is now occupied by a tanyard, in Kenwyn-street, in sinking the pits of which, about 30 years since, many stone coffins, in which were bones and urns containing various coins, were discovered. Samuel Foot, of dramatic celebrity, was born, in 1721, in the house now called the Red Lion hotel; and the Rev. Richard Polwhele, author of the histories of Cornwall and Devon, and other works, was born here in 1760, and died here in March 1838.

TRUSHAM, a parish, in the union of NEWTON-ABBOT, hundred of EXMINSTER, Teignbridge and Southern Divisions of the county of DEVON, 2¼ miles (N. N. W.) from Chudleigh; containing 207 inhabitants. The living is a rectory, valued in the king's books at £9. 4. 9½.; patron, Sir W. T. Pole, Bart. The tithes have been commuted for a rent-charge of £120, subject to the payment of rates, which on the average have amounted to £7; the glebe comprises 56 acres, valued at £70 per annum. In the church is a very rich wooden screen. Ten poor children are instructed for a rent-charge of £5, purchased with the sum of £100, raised by subscription.

TRUSLEY (*All Saints*), a parish, in the union of BURTON-UPON-TRENT, hundred of APPLETREE, Southern Division of the county of DERBY, 7 miles (W.) from Derby; containing 101 inhabitants. The living is a discharged rectory, valued in the king's books at £5. 6. 8.; patron, John Coke, Esq. The tithes have been commuted for a rent-charge of £94, subject to the payment of rates, which on the average have amounted to £6. 14. 11.; the glebe comprises 44 acres, valued at £68 per annum.

TRUSTHORPE (*St. Peter*), a parish, in the union of LOUTH, Marsh Division of the hundred of CALCEWORTH, parts of LINDSEY, county of LINCOLN, 7 miles (N. E.) from Alford; containing 286 inhabitants. The living is a discharged rectory, valued in the king's books at £19. 10. 8½.; present net income, £212; patron and incumbent, Rev. H. Rycroft. There is a place of worship for Wesleyan Methodists.

TRYSULL (*All Saints*), a parish, in the union, and Southern Division of the hundred, of SEISDON, Southern Division of the county of STAFFORD, 5 miles (S. W.) from Wolverhampton; containing 562 inhabitants. This place takes its name from John de Tressel, or Trysull, to whom the manor, together with that of Seisdon, which gives name to the hundred, belonged in the reign of Edward II. The living is a vicarage not in charge, annexed to that of Wombourne. The church is a small ancient edifice, having the figure of a bishop carved upon its tower. Thomas Rudge bequeathed £200 to purchase land, now producing, with other bequests, £16. 10. per annum, for which eleven children are gratuitously instructed.

TUBNEY, a parish, in the union of ABINGDON, hundred of OCK, county of BERKS, 4¼ miles (W. by N.) from Abingdon; containing 167 inhabitants. The living is a sinecure rectory, valued in the king's books at £3. 1. 10½.; present net income, £120; patrons, President and Fellows of Magdalene College, Oxford. The church has been entirely demolished: on the induction of a rector, the ceremony takes place in the open air. The parishioners repair to Fyfield church for the performance of ecclesiastical rites.

TUDDENHAM (*St. Martin*), a parish, in the union of WOODBRIDGE, hundred of CARLFORD, Eastern Division of the county of SUFFOLK, 3¼ miles (N. E. by N.) from Ipswich; containing 369 inhabitants. The living is a discharged vicarage, valued in the king's books at £10. 13. 4.; present net income, £50; patron and impropriator, Rev. C. W. Fonnereau. The impropriate tithes have been commuted for a rent-charge of £220, and the vicarial for £110, subject to the payment of rates, which on the average on both have amounted to £73: there are three acres of glebe, valued at £4. 10. per annum. The north door of the church is a richly moulded Norman arch, and the font has the date 1363 inscribed on it.

TUDDENHAM (*St. Mary*), a parish, in the union of MILDENHALL, hundred of LACKFORD, Western Division of the county of SUFFOLK, 3 miles (S. E. by S.) from Mildenhall; containing 388 inhabitants. The living is a rectory, valued in the king's books at £10. 17. 6.; patron, Marquess of Bristol. The tithes have been commuted for a rent-charge of £360, subject to the payment of rates, which on the average have amounted to £70; the glebe comprises 17 acres, valued at £11. 18. per annum. John Cockerton, in 1723, founded and endowed with an estate, now producing a rental of £70, a free school for instructing all the poor children of the parish.

TUDDENHAM, EAST (*All Saints*), a parish, in the union of MITFORD and LAUNDITCH, hundred of MITFORD, Western Division of the county of NORFOLK, 6¾ miles (E. S. E.) from East Dereham; containing 587 inhabitants. The living is a discharged vicarage, annexed to that of Honingham, and valued in the king's books at £7. 6. 0½. A day and Sunday school is partly supported by the minister and the parish.

TUDDENHAM, NORTH (*St. Mary*), a parish, in the union of MITFORD and LAUNDITCH, hundred of MITFORD, Western Division of the county of NORFOLK, 4¼ miles (E. by S.) from East Dereham; containing 399 inhabitants. The living is a rectory, valued in the king's books at £10. 5. 5.; present net income, £716; patron and incumbent, Rev. J. Day.

TUDELY, or TUDELY *cum* CAPEL (*All Saints*), a parish, in the union of TONBRIDGE, partly in the hundred of TWYFORD, but chiefly in that of WASHLINGSTONE, lathe of AYLESFORD, Western Division of the county of KENT, 2½ miles (E. S. E.) from Tonbridge; containing 575 inhabitants. The living is a vicarage, held jointly with Capel, and valued in the king's books at £4. 16. 0½.; present net income, £238: it is in the patronage of the Baroness le Despencer. The church is a small building of stone with a square tower of brick, and a spire, built about 70 years since. The line of the South Eastern railway passes through the parish. There are mineral springs, having the same properties as those of Tonbridge Wells. Here was formerly a seat of the Earls of Westmorland, now a farm-house.

TUDERLEY, EAST and WEST, county of SOUTHAMPTON.—See TYTHERLEY.

TUDHOE, a township, in the parochial chapelry of WHITWORTH, South-Eastern Division of DARLINGTON ward, union, and Southern Division of the county palatine, of DURHAM, 5 miles (S. by W.) from Durham; containing 237 inhabitants. A small school on the National plan has been established.

TUDY, ST., a parish, in the union of BODMIN, hundred of TRIGG, Eastern Division of the county of CORNWALL, 6¼ miles (N.) from Bodmin; containing 658 inhabitants. The living is a rectory, valued in the king's books at £31; present net income, £700; patrons, Dean and Canons of Christ-Church, Oxford. The church contains several ancient monuments to the family of Nichols, of Penrose, in this parish. There is a place of worship for Wesleyan Methodists. Fairs for sheep and cattle are held on May 20th and Sept. 14th. A school on the National plan is partly supported by subscription. At Tintern and Kelly Green are the remains of two ancient chapels. Dr. Richard Lower, an eminent physician in the time of Charles II., who first brought into notice the mineral water at Astrop, in Northamptonshire, and who is mentioned in Dr. Good's "Study of Medicine," as having either discovered or brought to perfection the practice of transfusing blood, was born at Tremere, in this parish, about 1631; he died in London in 1690, and was interred at his native place.

TUFFLEY, a hamlet, in the parish of ST.-MARY-DE-LODE, city of GLOUCESTER, Middle Division of the hundred of DUDSTONE and KING'S-BARTON, union, and Eastern Division of the county, of GLOUCESTER, 2¼ miles (S. by W.) from Gloucester; containing 109 inhabitants.

TUFTON, or TUCKINGTON (*St. Mary*), a parish, in the union of WHITCHURCH, hundred of WHERWELL, Kingsclere and Northern Divisions of the county of SOUTHAMPTON, ½ a mile (S. W.) from Whitchurch; containing 197 inhabitants. The living is annexed, with that of Bullington, to the vicarage of Wherwell.

TUGBY (*St. Thomas à Becket*), a parish, in the union of BILLESDON, partly in the hundred of GARTREE, but chiefly in the hundred of EAST GOSCOTE, Northern Division of the county of LEICESTER, 7½ miles (W. by N.) from Uppingham; containing 226 inhabitants. The living is a vicarage, with the perpetual curacy of East Norton annexed, valued in the king's books at £11. 8. 4.; present net income, £284: it is in the patronage of the Crown. The church has been repewed, and 130 free sittings provided, the Incorporated Society having granted £30 in aid of the expense. Robert Wilson, in 1726, bequeathed land, directing the produce to be applied for teaching children and the relief of the poor: a National school has been established.

TUGFORD (*St. Catherine*), a parish, in the union of LUDLOW, hundred of MUNSLOW, Southern Division of the county of SALOP, 10 miles (N. N. E.) from Ludlow; containing 188 inhabitants. The living is a discharged rectory, united to that of Holdgate, and valued in the king's books at £4. 13. 4.

TUGGAL, a township, in the parish, and Northern Division of the ward, of BAMBROUGH, union of BELFORD, Northern Division of the county of NORTHUMBERLAND, 9½ miles (N. by E.) from Alnwick; containing 102 inhabitants.

TUMBY, a hamlet, in the parish of REVESBY, Western Division of the soke of BOLINGBROKE, parts of LINDSEY, county of LINCOLN, 2½ miles (N. by W.) from Bolingbroke; containing 6 inhabitants.

TUMBY, a township, in the parish of KIRKBY-UPON-BAIN, union of HORNCASTLE, Southern Division of the wapentake of GARTREE, parts of LINDSEY, county of LINCOLN, 2½ miles (N. E. by N.) from Tattershall; containing 322 inhabitants. There is a place of worship for Wesleyan Methodists.

TUNBRIDGE, county of KENT. — See TONBRIDGE.

TUNSTALL, a township, in the parish of BISHOP-WEARMOUTH, union of SUNDERLAND, Northern Division of Easington ward and of the county palatine of Durham, $2\frac{3}{4}$ miles (S. by W.) from Sunderland; containing 75 inhabitants. Limestone abounds here, imbedded in which have been found fossils, and a considerable quantity of iron-ore. On the Tunstall hills are some vestiges of a Druidical circle; and a rude sepulchre, constructed with fragments of limestone, was discovered in 1814.

TUNSTALL (*St. John the Baptist*), a parish, in the union and hundred of Milton, Upper Division of the lathe of Scray, Eastern Division of the county of Kent, $1\frac{1}{2}$ mile (S. W. by W.) from Sittingbourne; containing 171 inhabitants. The living is a rectory, valued in the king's books at £14. 8. 4.; present net income, £479; patron, Archbishop of Canterbury. The church is principally in the later style of English architecture, and contains several handsome monuments. Edward Rowe Mores, Esq., a distinguished antiquary, was born here, in 1730; he published the history and antiquities of his native parish, and various other works, and was the original projector of the Equitable Society for insurance on lives and survivorships by annuities.

TUNSTALL (*St. John the Baptist*), a parish, in the hundred of Lonsdale, south of the sands, Northern Division of the county palatine of Lancaster; containing 862 inhabitants, of which number, 142 are in the township of Tunstall, $3\frac{3}{4}$ miles (S.) from Kirkby-Lonsdale. The living is a discharged vicarage, valued in the king's books at £6. 3. $11\frac{1}{2}$.; present net income, £332; patron and impropriator, R. T. North, Esq. The church is very ancient, and occupies a retired situation. Twenty four children receive a gratuitous education for about £26 a year, arising from sundry bequests.

TUNSTALL (*St. Peter and St. Paul*), a parish, in the union of Blofield, hundred of Walsham, Eastern Division of the county of Norfolk, $2\frac{3}{4}$ miles (S. S. E.) from Acle; containing 101 inhabitants. The living is a perpetual curacy; net income, £48; patron and appropriator, Bishop of Norwich.

TUNSTALL, a chapelry, in the parish of Abdaston, Northern Division of the hundred of Pirehill and of the county of Stafford; containing 98 inhabitants.

TUNSTALL (*St. Michael*), a parish, in the union and hundred of Plomesgate, Eastern Division of the county of Suffolk, $4\frac{1}{2}$ miles (E. by S.) from Wickham-Market; containing 733 inhabitants. The living is a discharged rectory, with that of Dunningworth annexed, valued in the king's books at £21. 0. 5.; present net income, £352; patron and incumbent, Rev. T. G. Ferrand. There is a place of worship for Baptists; and two day and Sunday schools are partly supported by the rector.

TUNSTALL (*All Saints*), a parish, in the union of Patrington, partly within the liberty of St. Peter's, but chiefly in the Middle Division of the wapentake of Holderness, East Riding of the county of York, $14\frac{1}{2}$ miles (E. by N.) from Kingston-upon-Hull; containing 172 inhabitants. The living is a perpetual curacy, in the patronage of the Dean and Chapter of York, the appropriators; net income, £52. Here is a National school.

TUNSTALL, a township, in the parish of Catterick, union of Richmond, wapentake of Hang-East, North Riding of the county of York, 2 miles (S. W.) from Catterick; containing 312 inhabitants.

TUNSTALL-COURT, a considerable modern town in the Staffordshire potteries, in the parish of Wolstanton, union of Wolstanton and Burslem, Northern Division of the hundred of Pirehill, and of the county of Stafford, 4 miles (N. by E.) from Newcastle-under-Lyme; containing 3673 inhabitants, according to the census of 1831, since which time the population has increased to more than 6000, who are chiefly employed in the china and earthenware manufactures (of which here are nearly 20 establishments), and in the collieries and brick and tile works, the latter being extensively carried on and producing articles of superior hardness and quality, which are in great demand in Lancashire and the northern parts, whither these and the other manufactured goods are forwarded by the Grand Trunk canal, which passes on its summit level near the west side of the town, and is conducted into Cheshire in two collateral tunnels under Harecastle-hill, within half a mile (N. W.) of the town: these tunnels are 2888 yards in length. The town has very much increased and improved within the present century; the population, in 1811, having been 1677 only. It is a member of the parliamentary borough of Stoke-upon-Trent: the number of registered voters, in 1838, was 125. In 1816, a market-place was set out and a town-hall erected in shares by the principal inhabitants, in conjunction with the lord of the manor, who gave the land. The town-hall, or court-house, is a neat building of brick in the centre of the market-place. A market has since been established, which, though not at present sanctioned by act of parliament, is held on Saturdays and Mondays. A church was erected here in 1831 by the parliamentary commissioners, at an expense of £4000, including £1000 raised by subscription, and will accommodate about 1000 persons; the site was given by Ralph Sneyd, Esq. It is a plain building of stone, with a spire, and has been lately made a district church for the townships of Tunstall, Oldcott, and Ranscliff. A vicarage-house is in course of erection by aid from Queen Anne's bounty and private subscriptions. There are places of worship for Wesleyan, Primitive, and New Conection of Methodists; those belonging to the two former are large and handsome structures, with spacious school-rooms attached. The society of Primitive Methodists had their origin here in 1808. Tunstall has at present no regular police establishment, but is governed by a chief constable appointed at the manorial court, with subordinate constables: it is, however, entitled to the benefit of an act lately passed for appointing a stipendiary magistrate for the borough of Stoke-upon-Trent. The township of Tunstall contains 795 acres, but the manor, of which Ralph Sneyd, Esq., of Keele Hall, is lord, comprehends 12 other contiguous townships, including Burslem.

TUNSTEAD (*St. Mary*), a parish, in the hundred of Tunstead, Eastern Division of the county of Norfolk, $3\frac{1}{4}$ miles (N. E. by E.) from Coltishall; containing 498 inhabitants. The living is a discharged vicarage, with that of South Ruston annexed, valued in the king's books at £18. 9. 7.; present net income, £286; patron, — Mack, Esq., who, with — Johnson, Esq., is the impropriator. Here is a Sunday National school.

TUNWORTH (*All Saints*), a parish, in the union and hundred of Basingstoke, Basingstoke and Northern Divisions of the county of Southampton, $3\frac{3}{4}$ miles (S. E.) from Basingstoke; containing 122 inhabitants. The living is a rectory, valued in the king's books at £8. 18. 9.; patron, G. P. Jervoise, Esq. The tithes have been commuted for a rent-charge of £175, subject to the payment of rates, which on the average have amounted to £35; the glebe comprises 38 acres, valued at £30 per annum. A day and Sunday school is supported by subscription.

TUPHOLME, a parish, in the union of HORNCASTLE, Western Division of the wapentake of WRAGGOE, parts of LINDSEY, county of LINCOLN, 9 miles (W. by S.) from Horncastle; containing 68 inhabitants. The living is a discharged vicarage, valued in the king's books at £2. 10. 10.; present net income, £89; patron and appropriator, Bishop of Lincoln. An abbey of Premonstratensian canons, in honour of the Blessed Virgin Mary, was founded here, in the time of Henry II., by Allan and Gilbert de Nevill, which, at the dissolution, possessed a revenue of £119. 2. 8.

TUPSLEY, a township, in the parish of BISHOP-HAMPTON, hundred of GRIMSWORTH, union and county of HEREFORD, 2 miles (E. S. E.) from Hereford; containing 512 inhabitants.

TUPTON, a township, in the parish of NORTH WINGFIELD, union of CHESTERFIELD, hundred of SCARSDALE, Northern Division of the county of DERBY, 4 miles (S.) from Chesterfield; containing 201 inhabitants.

TURKDEAN (*All Saints*), a parish, in the union of NORTHLEACH, hundred of BRADLEY, Eastern Division of the county of GLOUCESTER, $2\frac{1}{4}$ miles (N. by W.) from Northleach; containing 237 inhabitants. The living is a discharged vicarage, with the perpetual curacy of Aldsworth annexed, valued in the king's books at £10; present net income, £208; patrons and appropriators, Dean and Canons of Christ-Church, Oxford.

TURNASTONE (*St. Mary*), a parish, in the union of DORE, hundred of WEBTREE, county of HEREFORD, 10 miles (E. S. E.) from Hay; containing 54 inhabitants. The living is a discharged rectory, valued in the king's books at £2. 14. 2.; present net income, £73: it is in the patronage of Lady Boughton.

TURNDITCH (*All Saints*), a chapelry, in the parish of DUFFIELD, union of BELPER, hundred of APPLETREE, Southern Division of the county of DERBY, $3\frac{3}{4}$ miles (W. by S.) from Belper; containing 370 inhabitants. The living is a perpetual curacy; net income, £63; patron, Vicar of Duffield; impropriator, Lord Beauchamp. There are places of worship for Baptists and Primitive Methodists. A school is supported by subscription.

TURNERS-PUDDLE (*Holy Trinity*), a parish, in the union of WAREHAM and PURBECK, hundred of HUNDREDSBARROW, Wareham Division of the county of DORSET, $7\frac{1}{2}$ miles (N. W.) from Wareham; containing 82 inhabitants. The living is a discharged rectory, valued in the king's books at £7. 13. 4.; patron, J. Frampton, Esq. The tithes have been commuted for a rent-charge of £168, subject to the payment of rates; the glebe comprises $3\frac{1}{2}$ acres, valued at £3 per annum. The church was partly blown down in 1758, and rebuilt in 1759.

TURNHAM-GREEN, a hamlet, in the parish of CHISWICK, Kensington Division of the hundred of OSSULSTONE, county of MIDDLESEX, 5 miles (W. by S.) from London: the population is returned with the parish. The great western road passes through this village, which contains many handsome houses occupied by genteel families; it is lighted with gas, and supplied with water from the West London water-works. On the south side is the Horticultural Society's garden (see Chiswick) of which the principal entrance is from the green here. This hamlet is within the jurisdiction of a court of requests held in Kingsgate-street, Holborn, for the recovery of debts under 40*s*., and is under the superintendence of the new police. There is a place of worship for Wesleyan Methodists. A National school for the parish of Chiswick is situated here: about 100 boys are instructed, and 20 are clothed.

TURNWORTH (*St. Mary*), a parish, in the union of BLANDFORD, hundred of CRANBORNE, Blandford Division of the county of DORSET, 5 miles (W. by N.) from Blandford-Forum; containing 78 inhabitants. The living is a rectory, valued in the king's books at £10. 12. 3.; present net income, £135; patron, Bishop of Salisbury.

TURTON, a chapelry, in the parish and union of BOLTON, hundred of SALFORD, Southern Division of the county palatine of LANCASTER, $4\frac{1}{4}$ miles (N.) from Great Bolton; containing 2563 inhabitants. This chapelry is bounded by two rivulets tributary to the Irwell, which supply the power for various cotton-spinning, bleaching, dyeing, and printing works, of which the most extensive are the Egerton spinning and dye-mills, worked by a powerful water-wheel: at these establishments about 1000 persons are employed. The weaving of cotton, by hand-looms, is extensively carried on by the cottagers. A manorial court is held twice a year; and there are fairs for cattle, horses, &c., on Sept. 4th and 5th, at the village of Chapel Town. The living is a perpetual curacy; net income, £155; patron, G. N. Hoare, Esq. The chapel, dedicated to St. Bartholomew, has been rebuilt, and 560 free sittings provided, the Incorporated Society having granted £400 towards defraying the expense. Her Majesty's commissioners have also agreed to afford facilities for obtaining a site for a new church. There is a place of worship for Unitarians. Humphrey Cheetham, Esq., in 1746, endowed a small school for clothing and teaching ten boys: a school-room, with a house for the master, had previously been erected, the former of which was rebuilt and enlarged by subscription, about 30 years ago. He also made provision for ten poor boys of this township at Manchester College. Another school was founded by Abigail Cheetham, who endowed it with property now let for £28 a year, for which six poor boys are clothed and educated in a school-room built by subscription, the master of which has a dwelling-house rent-free. A small sum is also appropriated, from Mrs. Smalley's charity, to the education of poor children at Eagley Bridge school, which was established by subscription, in 1794. A Roman road passed through this chapelry, in which the remains of a Druidical temple, and the copper head of an old British standard, have been discovered. Turton Tower, an embattled structure, four stories high, has been the residence of the Orrells, the Cheethams, and the Greames, but is now occupied as a farm-house.

TURVEY (*All Saints*), a parish, in the hundred of WILLEY, union and county of BEDFORD, 8 miles (W. N. W.) from Bedford; containing 988 inhabitants. This parish is situated on the border of Buckinghamshire, and is intersected by the turnpike road from Olney to Bedford. The living is a rectory, valued in the king's books at £16; patron, T. C. Higgins, Esq.; appropriators, Bishop of Ely and the Rector. The appropriate tithes have been commuted for a rent-charge of £253. 4., and the rectorial for £458. 9., subject to the payment of rates, which on the average have respectively amounted to £50 and £105; the rector of Carleton also receives a rent-charge of £10. 10. out of the tithes, subject to the payment of rates, which on the average have amounted to £2. The church contains several fine monuments of the ancient and noble family of Mordaunt, and the remains of the celebrated Earl of Peterborough are deposited in the family vault. There are places of worship for Independents and Wesleyan Methodists. A National school has been established; there is also a Sunday school, supported by endowment.

TURVILLE (*St. Mary*), a parish, in the union of WYCOMBE, hundred of DESBOROUGH, county of BUCKINGHAM, 6 miles (N. W. by W.) from Great Marlow; containing 442 inhabitants, several of whom are employed in the manufacture of lace. The living is a discharged vicarage, valued in the king's books at £9. 9. 9½.; present income, under commutation, £90, and there are 40 acres of glebe; patron, Joseph Bailey, Esq.; impropriators, the landed proprietors. Parochial schools, in which about 400 children of both sexes are taught, are supported by subscription. The celebrated French general, Dumourier, resided at this place during the last two or three years of his life, where he died, and was buried at Henley.

TURWESTON (*St. Mary*), a parish, in the union of BRACKLEY, hundred and county of BUCKINGHAM, ¾ of a mile (E.) from Brackley; containing 371 inhabitants. The living is a rectory, valued in the king's books at £12. 16. 3.; present net income, £300; patrons, Dean and Chapter of Westminster. A school is endowed with £5 per annum, the interest of £100 bequeathed by the Rev. William Fairfax, in 1762.

TURWICK, county of SUSSEX.—See TERWICK.

TUSHINGHAM, a joint township with Grindley, in the parish of MALPAS, union of NANTWICH, Higher Division of the hundred of BROXTON, Southern Division of the county of CHESTER, 3½ miles (E.S.E.) from Malpas; containing 328 inhabitants.

TUSMORE, a parish, in the union of BICESTER, hundred of PLOUGHLEY, county of OXFORD, 6 miles (N. by W.) from Bicester: the population is returned with the parish of Hardwicke. The living is a discharged rectory, valued in the king's books at £3. 5.; present net income, £15; patron, Sir H. Dashwood, Bart. The church has been destroyed.

TUTBURY (*St. Mary*), a parish (formerly a market-town), in the union of BURTON-UPON-TRENT, Northern Division of the hundred of OFFLOW and of the county of STAFFORD, 5¼ miles (N. W. by N.) from Burton-upon-Trent; containing 1553 inhabitants. On the division of lands after the Conquest, Tutbury was included in the domain allotted to Henry de Ferrars, a Norman nobleman, who either built the castle, or received it in gift from the Conqueror. His descendant, Robert, joining Leicester in the rebellion against Henry III., was fined £50,000, and, being unable to pay so large a sum, forfeited his castle to the king, who granted it to his son, Edmund Earl of Lancaster. After the attainder of Thomas Earl of Lancaster, who, with the Earl of Hereford, had attempted the dethronement of Edward II., the fortress was suffered to fall to ruin, and so remained till the year 1350, when John of Gaunt becoming its possessor, he rebuilt the greater part of it, with the gatehouse, and surrounded it on three sides by a wall, the precipitous declivity on the fourth rendering further security unnecessary. Mary Queen of Scots was for some time imprisoned in this castle; and, at the commencement of the parliamentary war, it was garrisoned for the king; but, by order of parliament, it was nearly demolished in 1646: the ruins, however, are still sufficient to indicate its former extent and magnificence, and exhibit good specimens of the early and later styles of English architecture. The town, situated on the west bank of the river Dove, which is crossed by a stone bridge of five arches, of modern erection, was, at a very early period, erected into a free borough, and possessed many valuable privileges, though it never had the right of sending members to parliament. Fairs for horses and cattle are held on February 14th, August 15th, and December 1st. On the river are extensive corn and cotton-spinning mills, and there is also a considerable cut-glass manufactory. The country between Tutbury and Needwood Forest abounds with alabaster. The manor, or honour, of Tutbury belongs to the Crown, in right of the Duchy of Lancaster; its jurisdiction extends over a great portion of Staffordshire, and into several of the neighbouring counties, *viz.*, those of Derby, Leicester, Nottingham, and Warwick; and, in Her Majesty's name, courts leet and baron are here held once a year; also a court of pleas, every third Tuesday, for all debts under 40*s.* contracted within the honour.

The living is a discharged vicarage, valued in the king's books at £7; present net income, £131; patron, Vicar of Bakewell; impropriators, Executors of R. Stone, Esq. The church is the nave of a more extensive structure, and a fine specimen of the Norman style of architecture: it has been enlarged, newly pewed, and greatly improved, at an expense of nearly £2000, of which £250 was contributed by the Incorporated Society, for which 320 free sittings have been provided. There are places of worship for Independents, and Primitive and Wesleyan Methodists. A free school was founded by Richard Wakefield, who, about 1730, endowed it with lands producing about £40 per annum, for the education of thirty poor children: the school-house was rebuilt in 1789. Richard Wakefield also, by his will in 1773, devised land and tithes, now producing about £450, to trustees, for charitable uses. On the declivity of the commanding eminence upon which the castle stood, a Benedictine priory, in honour of the Blessed Virgin, was founded in 1080, by Henry de Ferrars, which, though a cell to the abbey of St. Peter super Divam, in Normandy, survived till the general dissolution, when its revenue was valued at £244. 16. 8. Among other curious customs that formerly prevailed here was a minstrel fête, given by the Duke of Lancaster on Assumption-day, to which all the itinerant musicians of the neighbourhood were invited. There was also a sport termed "Bull running," which consisted in chasing a bull with a soaped tail, and, if caught in the county, he was conducted to the market-place and there baited, otherwise he remained the property of the Duke of Devonshire, who held the priory on condition of furnishing a bull annually for the purpose. His Grace has, however, compounded with the minstrel king and his subjects, and the bailiff, having purchased his right to the animal, sends him to the manor at Hardwick, for a Christmas feast to the poor. Ann Moore, who professed the ability to live without food, resided here during the period of her imposture.

On the 1st of June, 1831, some workmen who had been employed by Mr. Webb, the proprietor of the cotton-mills, to obtain a greater fall for the water by which the works are put in motion, while digging a quantity of gravel out of the bed of the river for that purpose, discovered a few pieces of silver coin, about 60 yards below the bridge. As they advanced up the river, they continued to find more; on the 7th of June they found several thousand coins, and on the following day discovered the deposit from which they had been washed down the river, about 30 yards below the bridge and from four to five feet below the surface of the gravel. The coins were so abundant that 5000 of them were collected by two of the men employed on that day, and sold to the by-standers at 8 shillings per 100. The officers of the crown at length asserted the king's right, and a commission was issued from the chancellor of the duchy of Lancaster, to institute a farther search on behalf of the crown, when more than 1500 more were found and forwarded to the chancellor. The total number of coins found was on a moderate computation at least 100,000. They were chiefly sterlings of the empire of Brabant, Lorraine, and Hainault; several Scottish coins of Alexander III., John Baliol, and Robert Bruce; and a complete English series of coins of Edward I., from the mints of London, York, Canterbury, Chester, Durham, Lincoln, Bristol, Exeter, Berwick, Bury-St.-Edmund's, Kingston, and Newcastle; and also of Dublin, Waterford, and Cork. There were also specimens of all the prelatical coins of the reigns of Edward I. and II.; of Beck, Keller, and Beaumont, bishops of Durham; some others supposed to have been struck by the abbot of Bury-St. Edmund's, bearing the inscription "Rob. de Hadley," and a few from the archiepiscopal see of York. There were also many of the reign of Henry III., both of his first and second coinage, and a few of the early part of that of Edward II. The English coins, with the exception only of one of the reign of Edward I., about the size of a half-crown piece and of very beautiful design, were all of the same size and value; and among them was found a ring rudely chased, with the motto "*spreta vivant*" within the circle. These coins were the contents of the military chest of the Earl of Lancaster, left at Tutbury castle on his retreat from that place, then threatened by the army of Edward II., to his castle of Pontefract, in the county of York; and which, with his baggage entrusted to his treasurer, was lost in the river Dove, on his attempting to cross it at a high flood, in the darkness of the night and with a panic-struck guard.

TUTNAL, a joint hamlet with Cobley in the parish of TARDEBIGG, union of BROMSGROVE, Alcester Division of the hundred of BARLICHWAY, Southern Division of the county of WARWICK, 2 miles (E. S. E.) from Bromsgrove; containing, with Cobley, 518 inhabitants.

TUTTINGTON (*ST. PETER AND ST. PAUL*), a parish, in the union of AYLSHAM, hundred of SOUTH ERPINGHAM, Eastern Division of the county of NORFOLK, 2¾ miles (E.) from Aylsham; containing 228 inhabitants. The living is a discharged vicarage, valued in the king's books at £5. 0. 7½.; present net income, £124; patron and appropriator, Bishop of Ely.

TUXFORD (*ST. NICHOLAS*), a market-town and parish, in the union of EAST-RETFORD, South-Clay Division of the wapentake of BASSETLAW, Northern Division of the county of NOTTINGHAM, 30 miles (N. E. by N.) from Nottingham, and 139 (N. by W.) from London, on the great north road; containing 1113 inhabitants. This place, often denominated Tuxford-in-the-Clay, to designate its situation, is a small town of modern appearance, having been rebuilt since 1702, when the old town was destroyed by fire. The only branch of trade, which is somewhat extensive, is in hops, of which large quantities are grown in the neighbourhood. The market is on Monday; fairs are on May 12th, for cattle, sheep, swine, and poultry; and on Sept. 28th, for hops. The living is a discharged vicarage, valued in the king's books at £4. 14. 7.; present net income, £260; patrons and impropriators, Master and Fellows of Trinity College, Cambridge. The church contains portions in various styles of architecture. There is a place of worship for Wesleyan Methodists. The free grammar school was founded in 1670, by Charles Read, Esq., who bequeathed £200 for the erection of the buildings, endowed it with lands, and directed a salary of £20 per annum to be paid to the master, and £5 per annum towards the maintenance of four boys, "being the sons of poor widows of ministers, and of decayed gentlemen and their widows," from the age of seven to sixteen years. A National school has been established; and there is also an infants' school, partly supported by the vicar.

TWAMBROOK, a township, in the parish of GREAT BUDWORTH, union and hundred of NORTHWICH, county of CHESTER, ¼ of a mile (E.) from Northwich: the population is returned with the chapelry of Witton.

TWEEDMOUTH (*ST. BARTHOLOMEW*), a parish, in the union of BERWICK-UPON-TWEED, ISLANDSHIRE, county palatine of DURHAM, though locally northward, and for electoral purposes connected with the Northern Division, of the county of Northumberland, adjoining Berwick-upon-Tweed, and containing 4971 inhabitants. In 1203, King John made an attempt to fortify Tweedmouth to repel the Scots, who twice interrupted the design, and the works were entirely demolished by William, surnamed the Lion, who then occupied Berwick. This place, situated on the southern bank of the Tweed, forms a handsome suburb to the town of Berwick-upon-Tweed, to which it is joined by an elegant stone bridge. The inhabitants are principally employed in manufactures, ship-building, and in a very productive salmon-fishery. Petty sessions are held here every Saturday. The living is a perpetual curacy; net income, £144; patrons and appropriators, Dean and Chapter of Durham. The church, formerly a chapel of ease to Holy Island, was rebuilt in 1780. There is a place of worship for Scotch Presbyterians, also a National school. An ancient hospital formerly existed here. Near the village of East Ord, on the bank of the river, are vestiges of an ancient British intrenchment, where many fragments of military weapons have been found. In this parish is Spittle, a sea-bathing place, which see.

TWEMLOW, a township, in the parish of SANDBACH, union of CONGLETON, hundred of NORTHWICH, Southern Division of the county of CHESTER, 5¼ miles (E. N. E.) from Middlewich; containing 152 inhabitants.

TWICKENHAM (*St. Mary*), a parish, in the union of BRENTFORD, hundred of ISLEWORTH, county of MIDDLESEX, 9 miles (W. S. W.) from London; containing 4571 inhabitants. The name of this place, formerly written *Twicknam*, is said to refer to its situation between two streams, or brooks, that flow into the Thames at each end of the village, which occupies a most delightful position on its western bank, on the road from London, through Isleworth, to Hampton Court. Twickenham is deservedly admired for the beauty of its scenery, enlivened by the windings of the Thames; it has been the favourite retreat of the statesman and the poet, and is embellished with handsome seats and tasteful villas. At the southern extremity of the village, fronted by a lawn sloping to the verge of the river, is Pope's villa; the house has been much enlarged, and the celebrated grotto erected under the immediate superintendence of the poet has lost nearly all its original character. Strawberry Hill, formerly the residence of Sir Robt. Walpole, forms an interesting object as seen from the river, in the centre of which, nearly opposite to the church, is an island, called Twickenham Ait, comprising about eight acres, the chief part of which is laid out in pleasure ground. The Eel Pie house has been noted for the last two centuries, as a favourite resort for refreshment and recreation to water parties, and persons repairing hither for the amusement of fishing; the old house was taken down in 1830, and a very handsome and commodious edifice, comprising a good assembly-room, measuring 50 feet by 15, has been erected on the site by the present proprietor. There are powder and oil mills in the parish. Fairs are held on Holy Thursday and Aug. 9th and 10th. The inhabitants are under the jurisdiction of a court of requests, for the recovery of debts under 40s., held at Brentford during the summer half year, and at Uxbridge during the winter.

The living is a vicarage, valued in the king's books at £11; present net income, £717; patrons, Dean and Canons of Windsor; impropriator, — Pownall, Esq. The church, which stands near the river, is a neat structure of brick, ornamented with stone, of the Doric order, with an ancient embattled tower of the eleventh century; it was rebuilt in 1714: in the interior is a monument to the memory of Pope, erected by Bishop Warburton; and another to Mrs. Clive, the actress. Midway between Twickenham and Richmond is a chapel of ease, erected about 1721. There are places of worship for Independents and Wesleyan Methodists. A National school was formed in 1809, by the union of three schools, and the appropriation of some small endowments belonging to them, amounting to £133 per annum: 100 boys and 70 girls are educated, of whom about 30 boys and 24 girls are clothed annually. Another school is partly supported by the vicar, and in two others 22 children are paid for by charitable individuals. Six boys and one girl of this parish are eligible for instruction and apprenticeship, or to be put to service, on the foundation of John and Frances West, who conveyed estates in trust to the Governors of Christ's Hospital for that purpose; £20 being paid with each boy, and £5 with each girl. One man or woman of the parish receives £5 per annum also from a benefaction by Frances West of £2600, to be laid out in land for the use of the Clothworkers' Company, to pay to each of ten blind men and ten blind women that sum annually.

TWIGMOOR, a hamlet, in the parish of MANTON, union of GLANDFORD-BRIDGE, Eastern Division of the wapentake of MANLEY, parts of LINDSEY, county of LINCOLN, 5½ miles (W.) from Glandford-Bridge; containing 25 inhabitants.

TWIGWORTH, a hamlet, in the parish of ST. CATHERINE, GLOUCESTER, Upper Division of the hundred of DUDSTONE and KING'S-BARTON, union, and Eastern Division of the county, of GLOUCESTER, 2 miles (N. N. E.) from Gloucester; containing 87 inhabitants.

TWINEHAM (*St. Peter*), a parish, in the union of CUCKFIELD, hundred of BUTTINGHILL, rape of LEWES, Eastern Division of the county of SUSSEX, 5 miles (S. W.) from Cuckfield; containing 337 inhabitants. The living is a rectory, valued in the king's books at £10. 15. 5.; patron, Sir C. F. Goring, Bart. The tithes have been commuted for a rent-charge of £400, subject to the payment of rates, which on the average have amounted to £135; the glebe comprises 3 acres, valued at £4. 10. per annum.

TWINING (*St. Mary Magdalene*), a parish, in the union of TEWKESBURY, Lower Division of the hundred of KIFTSGATE, though locally in the Lower Division of that of Tewkesbury, Eastern Division of the county of GLOUCESTER, 2¾ miles (N. by E.) from Tewkesbury; containing 942 inhabitants. The living is a discharged vicarage, valued in the king's books at £7. 9. 7.; present net income, £127; patrons and appropriators, Dean and Canons of Christ-Church, Oxford. The church contains 100 free sittings, the Incorporated Society having granted £40 towards defraying the expense; it exhibits portions in the Norman style of architecture. This parish is separated from Worcestershire by the navigable river Avon, across which there is a ferry.

TWINSTEAD, a parish, in the union of SUDBURY, hundred of HINCKFORD, Northern Division of the county of ESSEX, 5½ miles (N. E. by N.) from Halstead; containing 205 inhabitants. It is of small extent; the situation is pleasant, the surface pleasingly diversified, and the soil fertile. Twinstead Hall, the ancient manor-house, has been partly modernized, but still retains much of its original character, and part of the moat by which it was surrounded, and the ancient bridge across it, are still remaining. The living is a discharged rectory, valued in the king's books at £6; present net income, £250: it is in the patronage of the Crown. The church, originally a small ancient edifice, has been rebuilt and enlarged. Here is a National school.

TWISTON, a township, in the parish of WHALLEY, union of CLITHEROE, Higher Division of the hundred of BLACKBURN, Northern Division of the county palatine of LANCASTER, 5 miles (E. N. E.) from Clitheroe; containing 222 inhabitants.

TWITCHEN (*St. Peter*), a parish, in the union and hundred of SOUTH MOLTON, South Molton and Northern Divisions of the county of DEVON, 6½ miles (N. E. by E.) from South Molton; containing 170 inhabitants. It is a perpetual curacy, annexed to the vicarage of North Molton. A school is aided by Lady Morley.

TWIVERTON (*St. Michael*), a parish, in the union of BATH, hundred of WELLOW, Eastern Division of the county of SOMERSET, 1¾ mile (W. S. W.) from Bath; containing 2478 inhabitants. The living is a discharged vicarage, endowed with a portion of the rectorial tithes, and valued in the king's books at £5. 18. 1½.; present net income, £395; patrons, Provost and Fellows of Oriel College, Oxford; impropriator of the remainder of the rectorial tithes, G. Langton, Esq. The church has been enlarged, and 100 free sittings provided, the Incorporated Society having granted £170 towards defraying the expense. There is a place of worship for Wesleyan Methodists, also a National school. The river Avon runs through the parish, from east to west, and turns numerous mills; and the line of the Great Western railway also passes through it.

TWIZELL, a township, in the parish of NORHAM, otherwise Norhamshire, union of BERWICK-UPON-TWEED, county palatine of DURHAM, though locally to the northward, and for electoral purposes annexed to the Northern Division, of the county of Northumberland, 10 miles (S. W.) from Berwick; containing 292 inhabitants. It is situated on the river Till, which is here crossed by a stone bridge of one arch, ninety feet and three-quarters in the span. Twizell castle, a fine, though unfinished, castellated mansion of the Blakes, is seated on a rocky precipice, surrounded by scenery extremely picturesque. Near it is Tillmouth House, the present residence of the family, of which was the celebrated Admiral Blake, who died in 1657: it has a small chapel attached. Here is a National school; and in the neighbourhood are the remains of an ancient chapel, dedicated to St. Cuthbert.

TWIZELL, a township, in the parish of MORPETH, union and Western Division of CASTLE ward, Southern Division of the county of NORTHUMBERLAND, $6\frac{1}{4}$ miles (S. W. by S.) from Morpeth; containing 50 inhabitants. Here is a National school.

TWYCROSS (*ST. JAMES*), a parish, in the union of MARKET-BOSWORTH, hundred of SPARKENHOE, Southern Division of the county of LEICESTER, $4\frac{3}{4}$ miles (W. N. W.) from Market-Bosworth; containing 319 inhabitants. The living is annexed to the vicarage of Orton-on-the-Hill. Charles Jennings, in 1773, bequeathed £333. 6. 8., directing the interest to be applied in support of a school for poor children, which is further supported by the Earl and Countess Howe, at whose expense a Sunday school is wholly maintained.

TWYFORD, a chapelry, in the parish of HURST, partly in the hundreds of CHARLTON and SONNING, county of BERKS, and partly in that of AMESBURY, Southern Division of the county of WILTS, 8 miles (S. W.) from Maidenhead: the population is returned with the parish. A battle was fought near this place in 1688, between the partisans of James II. and those of the Prince of Orange, afterwards William III. This village, which is neatly built and populously inhabited, is situated on the Bath and Bristol road, by which it is divided into two parts, one in the county of Berks and the other in that of Wilts. The river Thames flows at a short distance, and the Great Western railway passes close by it. Silk-throwing is extensively carried on; and a fair for horses and other cattle is held on the 15th of July, but is very indifferently attended. The chapel, dedicated to St. Swithin, was erected and endowed with £30 per annum by Mr. Edward Polehampton, who died in 1721. The living is a donative, in the patronage of three Trustees, appointed by the will of the founder. There is a place of worship for Independents. Edward Polehampton, in 1721, also bequeathed a rent-charge of £10, with a dwelling-house for the master, to teach ten boys. An hospital was founded here in 1640, by Lady Frances Winchcombe, for the maintenance of eight poor single women. Edward Polehampton was born in this village, and interred in the church at Hurst; he is said to have made these endowments in acknowledgment of the hospitable assistance his mother received here while travelling through the village.

TWYFORD (*ST. MARY*), a parish, in the union, hundred, and county of BUCKINGHAM, $5\frac{1}{4}$ miles (S. W. by S.) from Buckingham; containing 660 inhabitants. The living is a rectory not in charge, annexed to the Rectorship of Lincoln College, Oxford; net income, £725. Two schools are supported by the Rev. Dr. Tatham.

TWYFORD, a chapelry, in the parish of BARROW, union of BURTON-UPON-TRENT, hundred of APPLETREE, Southern Division of the county of DERBY, $5\frac{1}{2}$ miles (S. S. W.) from Derby; containing, with the township of Stenson, 219 inhabitants. The Trent and Mersey canal passes in the vicinity. The chapel is dedicated to St. Andrew. John Harpur and others, in 1696, bequeathed a rent-charge of £15, to be applied for teaching and apprenticing poor children, or towards their support at the University.

TWYFORD (*ST. ANDREW*), a parish, in the union of MELTON-MOWBRAY, hundred of EAST GOTCOTE, Northern Division of the county of LEICESTER, $6\frac{1}{4}$ miles (S. by W.) from Melton-Mowbray; containing 512 inhabitants. The living is a discharged vicarage, united to that of Hungerton, in 1732, and valued in the king's books at £8. 8. 6.

TWYFORD, a township, in the parish of COLSTERWORTH, hundred of BELTISLOE, parts of KESTEVEN, county of LINCOLN, $\frac{3}{4}$ of a mile (S.) from Colsterworth: the population is returned with the parish.

TWYFORD, an extra-parochial liberty, in the Kensington Division of the hundred of OSSULSTONE, county of MIDDLESEX, 6 miles (W. N. W.) from London; containing 43 inhabitants. There is a private chapel at Twyford Abbey.

TWYFORD (*ST. NICHOLAS*), a parish, in the union of MITFORD and LAUNDITCH, hundred of EYNSFORD, Eastern Division of the county of NORFOLK, $\frac{3}{4}$ of a mile (W.) from Foulsham; containing 82 inhabitants. The living is a discharged rectory, valued in the king's books at £4. 19. $9\frac{1}{2}$.; patron and incumbent, Rev. John Spurgeon. The tithes have been commuted for a rent-charge of £151, subject to the payment of rates, which on the average have amounted to £26; the glebe comprises 11 acres, valued at £19. 5. per annum.

TWYFORD (*ST. MARY*), a parish, in the union of NEW WINCHESTER, hundred of FAWLEY, Winchester and Southern Divisions of the county of SOUTHAMPTON, 3 miles (S.) from Winchester; containing 1177 inhabitants. The living is a vicarage, valued in the king's books at £12. 12. $8\frac{1}{2}$.; present net income, £213: it is in the patronage of Lady Mildmay, on the nomination of the Master and Fellows of Emanuel College, Cambridge; impropriators, Governors of the Hospital of St. Cross. The Itchin navigation passes through the parish. A National school has been established here; and 24 poor children are educated for the proceeds of £500, bequeathed by Richard Wooll, in 1780. Here was formerly a Roman Catholic seminary, in which the celebrated poet, Alexander Pope, received part of his education.

TWYWELL (*St. Nicholas*), a parish, in the union of Thrapstone, hundred of Huxloe, Northern Division of the county of Northampton, 3½ miles (W. by S.) from Thrapstone; containing 199 inhabitants. The living is a rectory, valued in the king's books at £9; present net income, £280; patron and incumbent, Rev. W. Alington.

TYDD (*St. Giles*), a parish, in the union and hundred of Wisbeach, Isle of Ely, county of Cambridge, 5½ miles (N. W. by N.) from Wisbeach; containing 967 inhabitants. The Bedford Level canal, which is 100 feet wide and 30 feet deep, passes through the centre of the parish, by the construction of which 30,000 acres of fenny land, belonging to the Duke of Bedford, have been rendered arable, and seven alluvial deposits have been discovered. Woad for dyeing cloth is prepared here. The living is a rectory, in the patronage of the Bishop of Ely, valued in the king's books at £21. 13. 1½.; present net income, £653. The church and steeple are widely detached, probably in consequence of the percolating soil. There is a place of worship for Independents. A school, in which are 60 children, is supported by the Rev. J. H. Watson, the rector; and there is also a school for children of Dissenters.

TYDD (*St. Mary's*), a parish, in the union of Holbeach, wapentake of Elloe, parts of Holland, county of Lincoln, 6 miles (N. by W.) from Wisbeach; containing 960 inhabitants. The living is a rectory, valued in the king's books at £17. 6. 5½.: it is in the patronage of the Crown. The tithes have been commuted for a rent-charge of £1295, subject to the payment of rates, which on the average have amounted to £144; the glebe comprises 35 acres, valued at £72 per annum. This parish is bounded on the east by the river Nene. Martha Trafford, in 1740, gave certain land, the produce of which is applied in teaching poor children, in a school-house built by subscription; and there is another school, in which 4 girls are taught and clothed from a bequest of £10 per annum by the late Dr. Wills, a former rector. Nicholas Breakspear, who was raised to the papal dignity, as Adrian IV., was rector of this parish.

TYLDERSLEY, or TILDESLEY *cum* SHAKERLEY (*St. George*), a parochial district, in the union of Leigh, hundred of West Derby, Southern Division of the county palatine of Lancaster, 2½ miles (E. N. E.) from Leigh; containing 5038 inhabitants. The living is a perpetual curacy; net income, £135; patron, Lord Lilford. The church was erected by the commissioners for promoting the building of additional churches, at an expense of more than £12,000, and will accommodate 2000 persons: it is a chaste and handsome structure, designed by Smirke, in the later style of English architecture, with a spire rising to the height of 150 feet, and was consecrated in September 1825. The site was presented by the late Thomas Johnson, Esq.; and the munificence of George Ormerod, Esq., has supplied the enclosure of the cemetery, a peal of six fine-toned bells, three beautifully painted windows, an organ, an elegant communion cloth, &c.; the communion plate was the gift of Mrs. Ormerod. There are places of worship for those in the connection of the late Countess of Huntingdon and Wesleyan Methodists. In 1827, the township of Tyldesley was erected into a distinct parish, as regards ecclesiastical affairs. Cotton-spinning is extensively carried on, and affords employment to about 1000 persons; the remainder of the labouring class are employed in weaving, in agriculture, and in the neighbouring collieries, which are very considerable: there are several cotton-mills, and one for the making of machinery. A National school, erected in 1827, at an expense of £650, on a site given by George Ormerod, Esq., adjacent to the church, is a neat and substantial stone building of two stories. A subscription library was established in 1828. Of the several antique mansions in the neighbourhood, there are considerable remains of Dam House, a very old brick building, with bay windows and gables; and, near it, the ruins of another, still more ancient.

TYNEHAM (*St. Mary*), a parish, in the union of Wareham and Purbeck, hundred of Hasilor, Wareham Division of the county of Dorset, 6½ miles (W. by S.) from Corfe-Castle; containing 247 inhabitants. This parish is bounded on the south by the English Channel, on the coast of which is a circular battery, for the defence of Worbarrow bay. The living is a rectory, united, by an act passed in the 8th of George I., to that of Steeple, and valued in the king's books at £11. 0. 10. The church was repaired in 1744; it has no tower. Two schools are supported by the rector. There was formerly a chapel at Povington, and another, dedicated to St. Margaret, at North Egleston. Here was an Alien priory, subordinate to the abbey of Bec, in Normandy, which, at the suppression, was given by Henry VI. to St. Anthony's hospital, London; by Edward IV. to Eton College, and afterwards to the Dean and Prebendaries of Westminster.

TYNEMOUTH (*St. Oswin*), a parish and newly-enfranchised borough, and the head of a union, in the Eastern Division of Castle ward, Southern Division of the county of Northumberland; comprising the whole of the town and suburbs of North Shields, and containing 24,778 inhabitants, of which number, 10,182 are in the township of Tynemouth, 8½ miles (E. N. E.) from Newcastle-upon-Tyne. This parish is of great extent, occupying the south-eastern section of the county, and locally called "Tynemouthshire;" the river Tyne bounds it on the south, in its course to the ocean, which forms the eastern limit. It abounds with coal; there is also some iron-stone, and the only magnesian limestone strata in the county slightly occur here. In the time of the ancient Britons the village was denominated *Penbal Crag*, or "the head of the rampart on the rock:" from remains discovered in the year 1783, it is conjectured that the Romans had a strong fortress here. A chapel of wood was built by Edwin, King of Northumberland, in 625; it was afterwards rebuilt with stone by Oswald, in the eighth century, and dedicated to St. Mary, and, having been repeatedly plundered by the Danes, was refounded by Tostig, Earl of Northumberland: in 1074, it was annexed to the monastery of Jarrow, and both institutions were made cells to the abbey of Durham. In 1090, it was elevated into a priory for Black canons, by Earl Mowbray, who converted it into a fortress during his conspiracy against William Rufus, when it was again nearly demolished, but rebuilt in 1110. After other ravages, it became the occasional residence of the queens of Edward I. and II., was afterwards plundered by the Scots, and eventually surrendered in 1539, when its revenue was valued at £511. 4. 1. After the dissolution, the church continued parochial until the year 1657, when having become ruinous, a new parish church was built at North Shields. Of the once magnificent priory there are still some interesting and venerable remains, consisting chiefly of an arch, the eastern part of the church, and other parts now converted into a magazine for military stores: they stand within the walls of the garrison, near the east end of the village, on a peninsula of stupendous rocks at the mouth of the Tyne. The cemetery attached to these beautiful ruins is still used by the parishioners. The approach from the west is by a gateway and tower of square form, with an exploratory turret at the north-east corner: this tower has been modernized and converted into a barrack, capable of containing about 250 soldiers. At the eastern extremity of the garrison-yard is a lighthouse with a revolving light: the whole is defended by a double wall extending to the shore. This fortress was defended by the Earl of Newcastle, in 1642, but eventually captured in 1645: a governor and deputy-governor are still appointed to it, but the office will be abolished on the next avoidance. Immediately south of the abbey-yard is an artificial haven, now used as a bathing-place, and called the Prior's Haven, having been excavated by one of the priors when his attempt to open a trade in the port of Tyne was frustrated by the burgesses of Newcastle (see Shields, North); and between it and the mouth of the Tyne is a small battery, called the Spanish Battery, which in time of war is mounted with a few guns. In 1672, Clifford's fort was constructed at the mouth of the river, where a guard from the garrison is constantly mounted; and, in 1758, barracks were erected near the village for the accommodation of 1000 men; but, at the general peace, they were sold, and have since been converted into dwelling-houses, which now constitute Percy-square.

The village of Tynemouth, which is about a mile from North Shields, consists principally of one main street and a minor one, extending nearly parallel with it; it is much resorted to in the bathing season, having some neat and commodious baths, erected in the Prior's Haven in 1807. The Newcastle and North Shields railway extends nearly to the baths. The haven is sheltered by an amphitheatre of rocks, and has an excellent strand, near which the baths are situated. Cullercoates, anciently Caller Cots, is a small fishing village and much frequented bathing-place, about a mile to the north of Tynemouth, to which it is attached by a fine sandy beach, called the Long Sand; and at the Sands is a mineral spring, which is in considerable repute. By the act of the 2nd of William IV., cap. 45, Tynemouth has been constituted a borough, with the privilege of returning one member to parliament, the right of election being in the £10 householders of the townships of Tynemouth, North Shields, Chirton, Preston, and Cullercoates, comprising 4754 acres: 552 voters polled in July 1837; the returning officer is annually appointed by the sheriff. The house of correction contains a day-room, two airing-yards, and several separate cells; it has been lately enlarged.

The living is a discharged vicarage, valued in the king's books at £24. 19. 4.; present net income, £298; patron, Duke of Northumberland, who, with the Guardians of the poor of the parish, is the impropriator. A chapel has been lately built here, containing 500 sittings, half of which are free, the Incorporated Society having granted £300 towards defraying the expense. There is a place of worship for Wesleyan Methodists; and at Cullercoates is a chapel used by three different sects, also an ancient burial-ground, called the Quaker's burial-ground. A free school was opened about 1827, under the will of Thomas Kettlewell, who erected the school premises, and endowed them with property since invested in the purchase of £2000 New Four per cents. and £2000 three per cent. consols.: about 200 boys, appointed by the trustees from the children of residents in, or parishioners of, Tynemouth, preference being given to orphans, are instructed by a master who receives a salary of £60, with £10 for house-rent. Here is also a school for girls, children of Roman Catholics, supported by subscription. George Crawford, of King's-Langley, in 1811, bequeathed £700 three per cent. consols, since vested in the purchase of £420 five per cent. annuities, and producing a dividend of £17. 12. 9., for annual distribution among the poor of the village of Tynemouth only. Sir Mark Milbanke, Bart., of Halnaby, in the county of York, gave a moiety of the corn tithes, now yielding from £70 to £80, for distribution among the poor of the eight townships in Tynemouthshire. The poor law union of Tynemouth comprises 24 parishes or places, containing a population of 47,715, according to the census of 1831, and is under the care of 47 guardians.

TYRLEY, county of STAFFORD.—See BLOORE-IN-TYRLEY.

TYRRINGHAM (*St. Peter*), a joint parish with Filgrove, in the union of NEWPORT-PAGNELL, hundred of NEWPORT, county of BUCKINGHAM, $2\frac{1}{4}$ miles (N. N. W.) from Newport-Pagnell; containing, with Filgrove, 227 inhabitants. The living is a rectory, with that of Filgrove united, valued in the king's books at £13. 6. $10\frac{1}{2}$.; patron, William Praed, Esq. The tithes have been commuted for a rent-charge of £450, subject to the payment of rates, which on the average have amounted to £50; the glebe comprises 15 acres, valued at £24 per annum. A school, containing 10 boys and 10 girls, is supported by Mrs. Praed.

TYSOE (*St. Mary*), a parish, in the union of SHIPSTON-UPON-STOUR, Kington Division of the hundred of KINGTON, Southern Division of the county of WARWICK, 5 miles (S. by E.) from Kington; containing 1007 inhabitants. The living is a discharged vicarage, with the rectory of Compton-Wyniates united, valued in the king's books at £10; present net income, £266; patron, Marquess of Northampton, who receives the tithes of Compton-Wyniates in consideration of £50 per annum paid to the incumbent; impropriators of Tysoe, several proprietors of land. There is a place of worship for Wesleyan Methodists. Thirty-six poor boys are educated for £26 per annum, arising from certain property bequeathed to the parish, in 1541, by John Middleton and Edward Richards: the school-room and master's residence were built a few years ago. Opposite to the church, on the side of a hill, is cut the figure of a horse, which, from the colour of the soil, is called the Red Horse, and gives to the adjacent low lands the name of the Vale of Red Horse. This figure is supposed to have been designed to commemorate the well-known act of Richard Neville, Earl of Warwick, in killing his horse on the day of the battle of Towton, fought on Palm-Sunday, 1461; on which day annually it has been customary for the country people to assemble, for the purpose of clearing the figure of the horse from whatever has grown upon it in the course of the year, which is locally termed "scouring the horse."

TYTHBY (*Holy Trinity*), a parish, in the union, and Southern Division of the wapentake, of Bingham, Southern Division of the county of Nottingham, 9 miles (E. S. E.) from Nottingham; containing 695 inhabitants. The living is a perpetual curacy; net income, £102; patron and impropriator, J. Musters, Esq. This parish is intersected by the Grantham canal, in the south-west portion of it, where it is crossed by the Fosse-road.

TYTHERINGTON, a township, in the parish of Prestbury, union and hundred of Macclesfield, Northern Division of the county of Chester, 1 mile (N.) from Macclesfield; containing 427 inhabitants.

TYTHERINGTON (*St. James*), a parish, in the union of Thornbury, partly in the Upper Division of the hundred of Henbury, but chiefly in the Lower Division of that of Thornbury, Western Division of the county of Gloucester, 3 miles (S. E.) from Thornbury; containing 476 inhabitants. The living is a discharged vicarage, valued in the king's books at £9. 11. 7.; present net income, £300; patron and impropriator, T. Hardwick, Esq.

TYTHERINGTON (*St. James*), a parish, in the union of Warminster, hundred of Heytesbury, Warminster and Southern Divisions of the county of Wilts, 4¼ miles (S. E. by S.) from Warminster; containing 132 inhabitants. The living is a perpetual curacy, in the patronage of the Dean of Salisbury, as Dean of Heytesbury.

TYTHERLEY, otherwise TUDERLEY, EAST (*St. Peter*), a parish, in the union of Stockbridge, hundred of Thorngate, Romsey and Southern Divisions of the county of Southampton, 7 miles (S. W.) from Stockbridge; containing 294 inhabitants. The living is a donative; net income, £40; patron and impropriator, J. L. Goldsmid, Esq. Sarah Rolle, in 1736, conveyed lands, &c., in support of a schoolmaster and schoolmistress; the income is about £208 a year, for which all the poor children of this parish and Lockerley, who apply, are gratuitously instructed, and of the present number about twenty or thirty girls are partly clothed and boarded: the school premises consist of a house containing two school-rooms and a residence for the master and mistress, of whom the former has £50 and the latter £25 per annum, with a garden and orchard.

TYTHERLEY, otherwise TUDERLEY, WEST, a parish, in the union of Stockbridge, hundred of Thorngate, Romsey and Southern Divisions of the county of Southampton, 7¾ miles (S. W. by W.) from Stockbridge; containing 489 inhabitants. The living is a rectory, valued in the king's books at £8. 5. 10.; patron, C. B. Wall, Esq. The tithes have been commuted for a rent-charge of £355, subject to the payment of rates, which on the average have amounted to £65; the glebe comprises 30 acres, valued at £32 per annum. The church has been enlarged, and 120 free sittings provided, the Incorporated Society having granted £120 towards defraying the expense. Here is a National school, supported by the rector and C. B. Wall, Esq.

TYTHERTON-KELLAWAYS, a tything, in the parish of Bremhill, union and hundred of Chippenham, Chippenham and Calne, and Northern, Divisions of the county of Wilts, 3¼ miles (E. N. E.) from Chippenham: the population is returned with the parish. This place merits notice from the peculiar circumstances attending its origin and progressive improvement. An individual name Connicker, a native of Reading, having embraced the original doctrines of Whitfield and Wesley, at the period of their first promulgation, became so zealous a devotee, that he expended his patrimonial estates in building meeting-houses in different parts of the country, one of which he erected at Tytherton, and attached to it a burying ground, &c. Here he fixed his own residence, and propagated his opinions with great success during several years; but, on the schism between Wesley and Whitfield, he joined the Moravians, and induced most of his followers at Tytherton to do the same. Accordingly, two cottages, adjoining each other, were purchased, and converted into a house for the reception of the young unmarried women of the sect. A house for young men was also attempted to be established, but without success. In this situation, Tytherton settlement continued till about 30 years ago, when the society, having grown more numerous and more wealthy, built a new chapel and sister-house, with a neat residence for their pastor. Since that period, they have erected a large school-house, into which female children of every persuasion are received as boarders, for the purpose of instruction in morality and the elements of knowledge.

TYTHERTON-LUCAS, a chapelry, in the parish and hundred of Chippenham, Chippenham and Calne, and Northern, Divisions of the county of Wilts, 3¼ miles (N. E. by E.) from Chippenham, with which the population is returned.

TYTHERTON-STANLEY, a joint tything with Nethermore, in the parish and hundred of Chippenham, Chippenham and Calne, and Northern, Divisions of the county of Wilts, 2 miles (E. by S.) from Chippenham, with which the population is returned.

TYWARDRETH (*St. Andrew*), a parish, in the union of St. Austell, Eastern Division of the hundred of Powder and of the county of Cornwall, 3¾ miles (W. N. W.) from Fowey; containing 1238 inhabitants. This parish is bounded on the south by the English Channel, where, on Greber Head, is a signal station. The petty sessions for the district are held here, on the third Monday in every month. The living is a vicarage, valued in the king's books at £9. 6. 8.; present net income, £135; patron, W. Rashleigh, Esq. The church has been repewed, and 150 free sittings provided, the Incorporated Society having granted £60 towards defraying the expense. A chapel has been erected by W. Rashleigh, Esq., about half a mile from his seat, Menabilly House. There is a place of worship for Wesleyan Methodists; also a school, in which 20 children are instructed at the expense of Mr. Rashleigh; and there are two Sunday schools, one in connection with the Established Church, and the other appertaining to Wesleyan Methodists. Here was a Benedictine priory, a cell to the monastery of St. Sergius and St. Bacchus in Normandy, supposed to have been founded before 1169, by Ricardus Dapifer, steward of the household of the Earl of Cornwall. This house, which was dedicated to St. Andrew, continued till the general dissolution, when its revenue was estimated at £151. 16. 1.: the site is now occupied by a farm-house. There is an almshouse for four poor widows, founded by one of the Rashleigh family.

U

UBBESTON (*St. Peter*), a parish, in the union and hundred of Blything, Eastern Division of the county of Suffolk, 5¾ miles (S. W. by W.) from Halesworth; containing 199 inhabitants. This parish is watered by the river Blyth, which has its source in the adjoining parish of Laxfield. Ubbeston Hall, for generations the ancient seat of the family of Kemp, has been taken down, and a farm-house built on the site. The living is a discharged vicarage, endowed with the rectorial tithes, and valued in the king's books at £6. 13. 4.; patron, Lord Huntingfield. The tithes have been commuted for a rent-charge of £315, subject to the payment of rates, which on the average have amounted to £62; the glebe comprises 6 acres, valued at £10. 2. 6. per annum. The church has a fine Norman arched doorway on the north side.

UBLEY, or OBLEIGH (*St. Bartholomew*), a parish, in the union of Clutton, hundred of Chewton, Eastern Division of the county of Somerset, 9 miles (N. by W.) from Wells; containing 340 inhabitants. The living is a discharged rectory, valued in the king's books at £11. 11. 5½.: it is in the patronage of the Crown. The tithes have been commuted for a rent-charge of £194, subject to the payment of rates; the glebe comprises 70 acres, valued at £40 per annum. There is a place of worship for Wesleyan Methodists.

UCKERBY, a township, in the parish of Catterick, union of Richmond, wapentake of Gilling-East, North Riding of the county of York, 3½ miles (N. by E.) from Catterick; containing 50 inhabitants.

UCKFIELD (*Holy Cross*), a parish, and the head of a union, in the hundred of Loxfield-Dorset, rape of Pevensey, Eastern Division of the county of Sussex, 8 miles (N. E. by N.) from Lewes; containing 1261 inhabitants. The parish is bounded on the west by the river Ouze: in the neighbourhood are two powerful chalybeate springs. About a mile from the village, which is a well-built, respectable place, is a manufactory for raw and refuse silk. Fairs are held on May 14th and Aug. 29th. The living is annexed to the rectory of Buxted. The church is principally in the later style of English architecture: the nave has been rebuilt, and 480 free sittings provided, the Incorporated Society having granted £350 towards defraying the expense. There is a place of worship for Baptists. Dorothy Ellis, in 1706, bequeathed £4. a year, for the instruction of ten children; and in the same year, Dr. Anthony Saunders left a messuage, school-house, and land, in trust for the establishment of a free grammar school for six boys of this parish, and six of Buxted; he also gave his library for the use of the school: the master receives £10. per annum, and the residue, amounting to £20 a year, is applied in apprenticing poor boys of Buxted. This school is now incorporated with a National school, supported by subscription. The poor law union of Uckfield comprises 11 parishes or places, containing a population of 16,109, according to the census of 1831, and is under the superintendence of 18 guardians. In a house once occupied by Bishop Christopherson, confessor to Queen Mary, are massive rings and other vestiges of popery. Dr. Edward Clarke, the celebrated traveller, and librarian to the University of Cambridge, passed much of the early part of his life at Uckfield.

UCKINGTON, a chapelry, in the parish of Elmstone-Hardwicke, union of Cheltenham, Lower Division of the hundred of Deerhurst, Eastern Division of the county of Gloucester, 2¾ miles (N. W.) from Cheltenham; containing 175 inhabitants.

UDIMORE (*St. Mary*), a parish, in the union of Rye, hundred of Gostrow, rape of Hastings, Eastern Division of the county of Sussex, 3½ miles (W. N.W.) from Winchelsea; containing 454 inhabitants. This parish is bounded on the south by Brede Channel. The living is a perpetual curacy, valued in the king's books at £8. 5. 2.; present net income, £100; patron, Earl of Burlington. The impropriate tithes have been commuted for a rent-charge of £400, subject to the payment of rates, which on the average have amounted to £99; the glebe comprises 5 acres, valued at £6 per annum. The church is principally in the early style of English architecture. A school is supported by subscription.

UFFCULME (*St. Mary*), a parish, in the union of Tiverton, hundred of Bampton, Cullompton and Northern Divisions of the county of Devon, 4¾ miles (N. E.) from Cullompton; containing 2082 inhabitants. This is a decayed market-town; fairs are still held on the Wednesday in Passion week, June 29th, and the middle Wednesday in Sept. During the last century, a great quantity of serges was made here, and there are still some flannels manufactured. The living is a vicarage, in the patronage of the Prebendary of Uffculme in the Cathedral Church of Salisbury (the appropriator), valued in the king's books at £18. 0. 2½.; present net income, £350. The church has a rich wooden screen. There are places of worship for Baptists and Independents. The free grammar school was founded in 1701, by Nicholas Ayshford, who gave £1200 for its erection and endowment, of which sum, £400 was expended in building the school-house and master's residence. Two boys of Uffculme, and two of Burlescombe, or Holcombe-Rogus, are entitled to gratuitous education for the dividends arising from the residue, which amount to about £46 per annum. Bradfield Hall, in this parish, is a perfect ancient mansion, containing several curious apartments, and to which a chapel was formerly attached. On a common in the neighbourhood is a place called Pixy Garden, an ancient earthwork.

UFFINGTON (*St. Mary*), a parish, in the union of Farringdon, hundred of Shrivenham, county of Berks, 4¼ miles (S. S. E.) from Great Farringdon; containing 1019 inhabitants. The living is a discharged vicarage, valued in the king's books at £21; present net income, £369; patron and impropriator, C. Eyre, Esq. The church is a handsome cruciform structure, in the early style of English architecture: the spire was destroyed by lightning, about 1750. There are chapels of ease at Baulking and Wolstone. The Wilts and Berks canal, and the line of the Great Western railway, pass through the parish. Thomas Saunders, in 1636, founded and endowed a free school for twelve boys of this parish, and six from Wolstone: the rents applied for its support amount to £40, and thirty boys are educated. Another school is partly supported by subscription. On White Horse hill, just above the village, is Uffington

Castle, a large encampment, surrounded by a double vallum, the inner one very high: it is 700 feet from east to west, 500 from north to south, and is supposed to be a work of the Britons, afterwards occupied by the Romans. This hill takes its name from the rude figure of a horse, 374 feet in length, cut in the turf, near the summit, said to be commemorative of the victory which Alfred obtained over the Danes in this neighbourhood, though some consider it a British work. Lands were formerly held here by cleaning, or rather cutting, away the turf, to render the figure more visible: for which purpose a custom still prevails among the inhabitants of assembling to scour, as it is termed, the horse, on which occasion they are entertained by the lord of the manor, and spend the day in festivity. To the westward of Uffington Castle is a large tumulus or cromlech, called Wayland Smith's cave; and there are various other tumuli scattered on these downs, particularly between Uffington and Lambourn, the most considerable of which are those called the Seven Barrows. Uffington gives the inferior title of Viscount to the Earl of Craven.

UFFINGTON (*St. Michael*), a parish, in the union of Stamford, wapentake of Ness, parts of Kesteven, county of Lincoln, $2\frac{1}{4}$ miles (E. by N.) from Stamford; containing 481 inhabitants. The living is a rectory, valued in the king's books at £21. 5. $2\frac{1}{2}$.; present net income, £837; patron, Earl of Lindsey, whose splendid mansion is in this parish. The church is a handsome structure, partly in the early and partly in the later style of English architecture, with a fine tower and spire, and some fragments of ancient stained glass. A National school is endowed by Albermarle, the last Earl of Lindsey, with £20 per annum. An hospital, or priory, of Augustine canons, in honour of the Virgin Mary, was founded in the reign of Henry III., or his predecessor, by William de Albini, which at the dissolution had a revenue of £42. 1. 3.

UFFINGTON (*Holy Trinity*), a parish, in the union of Atcham, Wellington Division of the hundred of South Bradford, Northern Division of the county of Salop, $3\frac{1}{4}$ miles (E. N. E.) from Shrewsbury; containing 343 inhabitants. The living is a perpetual curacy; net income, £59; the patronage and impropriation belong to Mrs. Corbett, who chiefly supports a day and Sunday school. The Shrewsbury canal passes through the parish.

UFFORD, a tything, in the parish and hundred of Crediton, Crediton and Northern Divisions of the county of Devon, $3\frac{1}{2}$ miles (W. by S.) from Crediton; containing 288 inhabitants.

UFFORD (*St. Andrew*), a parish, in the union of Stamford, soke of Peterborough, Northern Division of the county of Northampton; containing 480 inhabitants, of which number, 183 are in the hamlet of Ufford, 5 miles (S. E.) from Stamford. The living is a rectory, valued in the king's books at £26. 13. 4.; present net income, £688; patrons, Master and Fellows of St. John's College, Cambridge. There is a chapel of ease at Bainton. The river Welland runs through the parish, and the Roman road from the station at Castor, leading towards Lincoln, bounds it on the east.

UFFORD (*St. Mary*), a parish, in the union of Woodbridge, hundred of Wilford, Eastern Division of the county of Suffolk, $2\frac{1}{2}$ miles (N. E.) from Woodbridge; containing 661 inhabitants. The living is a discharged rectory, valued in the king's books at £8. 5.; present net income, £299; patron, F. C. Brooke, Esq., whose seat, Ufford Place, is in the parish. The church contains a font, with a curious cover, which has been engraved at the expense of the Society of Antiquaries. Here was anciently a chapel, called Sogenho, of which there are no remains. An hospital for four poor men was founded by Thomas Wood, D.D., Bishop of Lichfield and Coventry, who endowed it with £15 per annum. William Otley, Lord Mayor of London in 1434, was born here. The Earls of Suffolk took their name from this place.

UFTON, or UPTON-NERVET (*St. Peter*), a parish, in the union of Bradfield, hundred of Theale, county of Berks, $7\frac{1}{4}$ miles (S. W. by W.) from Reading; containing 357 inhabitants. The living is a rectory, valued in the king's books at £11. 3. $1\frac{1}{2}$.; present net income, £426; patrons, Provost and Fellows of Oriel College, Oxford. There are slight remains of a church, which formerly belonged to Upton-Greys, once a distinct parish, but consolidated with this in 1442. A school is chiefly supported by the lord of the manor and the rector.

UFTON (*St. Michael*), a parish, in the union of Southam, Kenilworth Division of the hundred of Knightlow, Southern Division of the county of Warwick, $2\frac{1}{2}$ miles (W. by N.) from Southam; containing 166 inhabitants. The living is a perpetual curacy, in the patronage of the Prebendary of Ufton in the Cathedral Church of Lichfield, the appropriator; net income, £46.

UGBOROUGH, a parish, in the union of Totnes, hundred of Ermington, Ermington and Plympton, and Southern, Divisions of the county of Devon, 3 miles (N. N. E.) from Modbury; containing 1467 inhabitants. The living is a discharged vicarage, endowed with a portion of the rectorial tithes, and valued in the king's books at £20; present net income, £260; patrons, Master and Wardens of the Grocers' Company; impropriator of the remainder of the rectorial tithes, Lady Carew. The church contains a Norman font and some remains of ancient screen-work. There was formerly a chapel at Earlscomb. A fair for cattle is held on the third Tuesday in every month. Here is a school, in which 75 children are paid for from the parish funds. Sir John Kempthorn, a distinguished admiral, was born at Widescomb, in this parish, in 1620.

UGGESHALL (*St. Mary*), a parish, in the hundred of Blything, Eastern Division of the county of Suffolk, $4\frac{3}{4}$ miles (N. W.) from Southwold; containing 303 inhabitants. The living is a rectory, with that of Sotherton annexed, valued in the king's books at £13. 6. 8.; patron, Earl of Stradbroke. The tithes have been commuted for a rent-charge of £380, subject to the payment of rates, which on the average have amounted to £74; the glebe comprises 43 acres, valued at £64. 10. per annum.

UGGLEBARNBY, a chapelry, in the parish and union of Whitby, liberty of Whitby-Strand, North Riding of the county of York, 4 miles (S. S. W.) from Whitby; containing 426 inhabitants. The living is a perpetual curacy, annexed to that of Eskdale-side. The chapel was built in 1137, by Nicholas, abbot of Whitby.

UGLEY (*St. Peter*), a parish, in the union of Bishop-Stortford, hundred of Clavering, Northern Division of the county of Essex, 3½ miles (N.) from Stansted-Mountfitchet; containing 318 inhabitants. This parish is pleasantly situated in the north-eastern portion of the hundred, and is intersected by the high road from London to Cambridge. Ugley Hall is an ancient mansion near the church; and Orford House, a handsome residence of brick, takes its name from the Earl of Orford, by whom it was built. The living is a discharged vicarage, with the perpetual curacy of Berdon annexed, valued in the king's books at £14. 13. 4.; patrons and impropriators, Governors of Christ's Hospital, London. The impropriate tithes have been commuted for a rent-charge of £360. 12. 9., and the vicarial for £99. 8., subject to the payment of rates, which on the average have respectively amounted to £76 and £19; the glebes comprise 48 acres and 3 acres, valued at £48 and £3. 15. per annum. The church is a small edifice, with a belfry turret surmounted by a cupola; on the south side of the chancel is a chapel belonging to Bollington Hall. Here is a National school.

UGTHORPE, a township, in the parish of Lythe, union of Whitby, Eastern Division of the liberty of Langbaurgh, North Riding of the county of York, 7½ miles (W.) from Whitby; containing 260 inhabitants. A Roman Catholic chapel was erected here about 1812; and a day and Sunday school is partly supported by an allowance of £10 per annum from a Catholic fund.

ULCEBY (*All Saints*), a parish, in the union of Spilsby, Wold Division of the hundred of Calceworth, parts of Lindsey, county of Lincoln, 3½ miles (S. W.) from Alford; containing 218 inhabitants. The living is a discharged rectory, valued in the king's books at £19. 16. 8.; present net income, £519; patron and incumbent, Rev. J. Robinson. There is a place of worship for Wesleyan Methodists. The Bull's Head, a lofty hill in this parish, is a noted landmark.

ULCEBY (*St. Nicholas*), a parish, in the union of Glandford-Bridge, Northern Division of the wapentake of Yarborough, parts of Lindsey, county of Lincoln, 7¼ miles (S. E.) from Barton-upon-Humber; containing 694 inhabitants. The living is a discharged vicarage, valued in the king's books at £11. 18. 4.; present net income, £146: it is in the patronage of the Crown; impropriators, W. D. Field, Esq., and others. A school is partly supported by endowment, for which ten children are instructed.

ULCOMBE (*All Saints*), a parish, in the union of Hollingbourn, hundred of Eyhorne, lathe of Aylesford, Western Division of the county of Kent, 7¼ miles (S. E. by E.) from Maidstone; containing 761 inhabitants. This parish lies partly in the Weald, and is intersected by several small streams, which empty themselves into the Medway. The living is a rectory, valued in the king's books at £16. 5. 10.; patron, Hon. C. B. C. Wandesford. The tithes have been commuted for a rent-charge of £950, subject to the payment of rates; the glebe comprises 79 acres, valued at £100 per annum. The church, principally in the later style of English architecture, was formerly collegiate, for an archpresbyter, two canons, a deacon, and one clerk: it contains some very ancient monuments to several of the St. Leger family, many to the family of Stringer, and, of more recent date, one to the memory of the Marquess and Marchioness of Ormonde, and another to Lady Sarah Wandesford. A National school has been established.

ULDALE, a parish, in the union of Wigton, Allerdale ward below Derwent, Western Division of the county of Cumberland, 2½ miles (S. S. E.) from Ireby; containing 344 inhabitants. The living is a rectory, valued in the king's books at £17. 18. 1½.; present net income, £151; patron and incumbent, Rev. Joseph Cape. The church was rebuilt by the parishioners in 1730. The river Ellen has its source here, in two small lakes, well stocked with various kinds of fish; about a mile and a half south-east from which, a brook, tumbling from a lofty mountain down several rocky precipices, forms a beautiful cascade, termed White Water Dash. Coal, freestone, limestone, and peat, abound here. A large fair for sheep, established in 1791, is held annually on Aug. 29th. The free school was founded in 1726, by Matthew Caldbeck, who endowed it with £100, and the like sum having been raised by subscription, both were expended in the purchase of freehold property, which, with £350 bequeathed by Thomas Tomlinson, produces an annual income of about £47; it is further supported by subscription.

ULEY (*St. Giles*), a parish, in the union of Dursley, Upper Division of the hundred of Berkeley, Western Division of the county of Gloucester, 2½ miles (E. by N.) from Dursley; containing 2641 inhabitants. This parish is situated in a district abounding with picturesque scenery, and with objects of interest. The manufacture of woollen cloth was formerly carried on here extensively, but has of late considerably declined. The living is a rectory, valued in the king's books at £13. 3. 4.; it is in the patronage of the Crown. The tithes have been commuted for a rent-charge of £236, subject to the payment of rates; the glebe comprises 16 acres, valued at £30 per annum. There are places of worship for Baptists, Independents, and Wesleyan Methodists. Two schools, one for infants, are partly supported by subscription. On an eminence, north-west of the village, is an ancient encampment, called Uley Bury, where various Roman coins have been found.

ULGHAM, a parochial chapelry, in the union, and Eastern Division of the ward, of Morpeth, Northern Division of the county of Northumberland, 5 miles (N. E. by N.) from Morpeth; containing 359 inhabitants. The living is a perpetual curacy, annexed to the rectory of Morpeth. The chapel, which is a plain sto[illegible] edifice, is dedicated to St. John. Coal is found her[illegible] the banks of the river Line. A market was fo[illegible] held, and the ancient market-cross still remair[illegible] centre of the village.

ULLENHALL, a chapelry, in the par[illegible] ton-Wawen, Henley Division of the h[illegible] lichway, Southern Division of the co[illegible] 2½ miles (N. W. by W.) from Henle[illegible] pulation is returned with Apsle[illegible] chapel is dedicated to St. Ma[illegible] in 1733, bequeathed a rent-c[illegible] six poor children.

ULLESKELF, a tow[illegible] Wharfe, liberty of S[illegible] locally in the Up[illegible] Barkstone-Ash, W[illegible]

$4\frac{1}{2}$ miles (S. E.) from Tadcaster; containing 339 inhabitants.

ULLESTHORPE, a hamlet, in the parish of CLAYBROOKE, union of LUTTERWORTH, hundred of GUTHLAXTON, Southern Division of the county of LEICESTER, $3\frac{1}{4}$ miles (N. W.) from Lutterworth; containing 599 inhabitants. There are places of worship for Baptists and Independents.

ULLEY, a township, partly in the parish of ASTON, but chiefly in that of TREETON, union of ROTHERHAM, Southern Division of the wapentake of STRAFFORTH and TICKHILL, West Riding of the county of YORK, $4\frac{1}{2}$ miles (S. E. by S.) from Rotherham; containing 193 inhabitants.

ULLINGSWICK, a parish, in the union of BROMYARD, hundred of BROXASH, county of HEREFORD, 5 miles (S. W.) from Bromyard; containing 293 inhabitants. The living is a rectory, with that of Little Cowarne annexed, valued in the king's books at £9; patron, Bishop of Hereford. The tithes have been commuted for a rent-charge of £195, subject to the payment of rates, which on the average have amounted to £18; the glebe comprises 20 acres, valued at £24 per annum.

ULLOCK, a joint township with Pardsey and Dean-Scales, in the parish of DEAN, ALLERDALE ward above Derwent, Western Division of the county of CUMBERLAND, $5\frac{1}{2}$ miles (S. W. by S.) from Cockermouth; containing, with Pardsey and Dean-Scales, 356 inhabitants.

ULNES-WALTON, a township, in the parish of CROSTON, union of CHORLEY, hundred of LEYLAND, Northern Division of the county palatine of LANCASTER, $5\frac{1}{4}$ miles (W. by N.) from Chorley; containing 537 inhabitants.

ULPHA, a chapelry, in the parish of MILLOM, union of BOOTLE, ALLERDALE ward above Derwent, Western Division of the county of CUMBERLAND, 9 miles (E. by S.) from Ravenglass; containing 405 inhabitants. This chapelry extends along the western bank of the river Duddon to the mountains Hard Knot and Wrynose, where is a stone called three-shire stone, marking the boundaries of Cumberland, Lancaster, and Westmorland. A Roman road crosses both these mountains; and about half way up the former are the remains of Hard Knot Castle, a fortress anciently of great importance, though the period of its erection is involved in obscurity. There are quarries of excellent blue slate, of which about 1400 tons are annually raised. Copper mines were formerly worked, and zinc is known to exist here. The coppices, with which this district abounds, produce a large supply of wood for making hoops and bobbins, the former being disposed of at Liverpool, and the latter to the manufacturers of cotton, woollen, linen, and silk in other towns. A fair for sheep is held on the first Monday in September, and there were formerly fairs for cloth and yarn, on the Monday before Easter and July 9th, but these are now only resorted to for pleasure. Ulpha Hall, which bears marks of high antiquity, has been converted into a farm-house: adjoining it is a well, termed "Lady's Dub," where it is said a lady was surprised and killed by one of the wolves that anciently infested the neighbourhood. The living is a perpetual curacy; net income, £49; patron, Vicar of Millom. The chapel is dedicated to St. John. There is a place of worship for Baptists, also a National school.

ULPHA, a joint township with Methop, in the parish of BEETHAM, union and ward of KENDAL, county of WESTMORLAND, 11 miles (S. S. W.) from Kendal: the population is returned with Methop. It is bounded on the south by the estuary of the Kent.

ULROME, a chapelry, in the union of BRIDLINGTON, partly in the parish of BARMSTON, but chiefly in that of SKIPSEA, Northern Division of the wapentake of HOLDERNESS, East Riding of the county of YORK, $6\frac{1}{4}$ miles (N. N. W.) from Hornsea; containing 166 inhabitants. The living is a perpetual curacy, valued in the king's books at £3. 19. 2.; present net income, £71; patrons, Executors of J. Lockwood, Esq.; impropriator, Rev. J. Brown. The chapel is very ancient.

ULTING (*ALL SAINTS*), a parish, in the union and hundred of WITHAM, Northern Division of the county of ESSEX, $4\frac{1}{2}$ miles (S. S. W.) from Witham; containing 158 inhabitants. This parish, which is bounded on the south by the river Chelmer, is about six miles in circumference; the soil is generally fertile and the lands well cultivated. The living is a discharged vicarage, valued in the king's books at £7. 4. 2.; patron and impropriator, R. Nicholson, Esq. The impropriate tithes have been commuted for a rent-charge of £187. 17. 6., and the vicarial for £164. 12. 6., subject to the payment of rates. A day and Sunday school is supported by subscription. The church is a small ancient edifice of stone, with a turret of wood surmounted by a shingled spire; and is beautifully situated on the bank of the river.

ULVERCROFT, an extra-parochial liberty, in the union of BARROW-UPON-SOAR, hundred of WEST GOSCOTE, Northern Division of the county of LEICESTER, 6 miles (W. by S.) from Mountsorrel; containing 100 inhabitants. At this place, in a deep sequestered valley of Charnwood forest, are the ruins of a church, anciently belonging to an Augustine priory, dedicated to the Blessed Virgin, which was founded by Robert Blanchmains, Earl of Leicester, in the reign of Henry II., and had, at the dissolution, a revenue of £101. 3. 10.

ULVERSTONE (*ST. MARY*), a market-town and parish, and the head of a union, in the hundred of LONSDALE, north of the sands, Northern Division of the county palatine of LANCASTER; containing 7741 inhabitants, of which number, 4876 are in the town of Ulverstone, 22 miles (N. W.) from Lancaster, and 271 (N. W. by N.) from London. This place derives its name (written in old records *Olvestonam*) from *Ulpha*, a Saxon lord; and was conferred, in 1127, on the abbey of Furness, by Stephen, afterwards King of England. It was subsequently granted to Gilbert, who had succeeded to the barony of Kendal, and who released the inhabitants from their state of feudalism, granting them a charter, which was augmented and confirmed by his successors. It afterwards reverted to the crown; and being, in 1609, divided into moieties, was eventually purchased, in 1736, by the Duke of Montague, for £490, and is at present vested in the Duke of Buccleuch. A charter was granted by Edward I., for a market and an annual fair; but it continued to be merely nominal until the dissolution of the abbey of Furness, near Dalton, the capital of that district, from which event the prosperity of Ulverstone may be dated. The town is situated near the estuary of the rivers Crake and Leven, and consists principally of four spacious streets; the houses are chiefly of stone. There are a news-room, and two subscription libraries, one of which is general, founded in 1797, under the auspices of Thomas Sutherland, Esq., and contains 3000 volumes; the other clerical, instituted by the associates of Dr. Bray, and greatly augmented by the donations and exertions of the Rev. Dr. Stonard, the learned commentator on the Prophecies, and by the contributions of other members. The theatre and public rooms, erected by subscription in 1796, were considerably improved in 1828, and during the hunt week in November are genteelly attended. The peninsular situation of the town led to the appointment of mounted guides to direct travellers across the sands, who are paid by government, and directed to be in attendance from sunrise to sunset, when the channel is fordable; but this arrangement has been partially superseded by the construction of a new road from Carnforth to Ulverstone, under an act of parliament.

The prevailing branches of manufacture are those of cotton, linen, check, canvas for sails, sacking, candlewicks, hats, axes, adzes, spades, hoes, and sickles. The chief articles of export, in addition to some of the above, are iron and copper ores, pig and bar iron of the finest quality, the best blue and green slates, and limestone, wool, grain, malt, butter, gunpowder, pyroligneous acid, leather, hoops, basket-rods, brush-sticks, baskets called swills, brooms, crate and wheel-spoke wood, laths, and oak and larch poles: these are principally sent coastwise, the intercourse with foreign countries being limited. There is a yard for ship-building, and the aggregate registry of ships belonging to the place is nearly 3000 tons: two or three vessels are employed in the American timber trade; and from other ports a few belonging to this town are engaged in the West India trade. Ulverstone is a creek within the limits of the port of Lancaster, and from this circumstance it has, with the liberty of Furness, been released from the heavy duty on coal carried coastwise. In 1793, an act of parliament was obtained for making a canal, by means of which ships of 400 tons' burden are safely moored in a capacious basin with extensive wharfs, and discharge their cargoes close to the town. The market, granted to Roger de Lancaster, in the 8th of Edward I., is on Thursday: fairs are held on the Tuesday before Easter-Sunday, April 29th, Holy Thursday, October 7th, and the first Thursday after October 23rd. Manorial courts leet and baron are held on the Monday next after October 24th. The court baron for the liberty of Furness, for the recovery of debts under 40*s*., is held here every Saturday three weeks; and the baronial court for the manor of Bolton with Adgarley, annually. The petty sessions for the hundred of Lonsdale, north of the sands, are also held here; and the town has been made a polling-place for the northern division of the shire.

Ulverstone anciently formed part of the parish of Dalton. The living is a perpetual curacy; net income, £149; patron and impropriator, Rev. T. R. G. Braddyll. The church, situated on the south side of a hill, about a quarter of a mile from the town, is of very ancient and obscure foundation; it was rebuilt in the time of Henry VIII., and again, with the exception of the tower and a Norman doorway, at the commencement of the present century. In the east window is some fine stained glass, the designs from Rubens; the altar-piece, representing the Descent from the Cross, with the three Cardinal graces, was designed by T. R. G. Braddyll, Esq., after the manner of Sir Joshua Reynolds: in the interior are several elegant and sumptuous monuments. An additional church, in the later English style, with a tower and spire, and dedicated to the Holy Trinity, was erected in 1831, at an expense of £5301. 3., under the act of the 58th of George III., to which a district has been assigned. The living is a perpetual curacy; net income, £143; patron, Rev. T. R. G. Braddyll. There are places of worship for Independents and Wesleyan Methodists, and Roman Catholics; and about a mile to the south-west is another for the Society of Friends (the first possessed by that community in England), built under the superintendence of George Fox, founder of the sect, who resided at Swartmoor Hall, in this neighbourhood. Townbank school was erected by subscription, and having been endowed with various bequests, has an income of £24 per annum. The school-rooms, of which the upper is appropriated to the classics, and the lower to instruction in English, writing, and arithmetic, were rebuilt about 1781. A National school is supported by subscription, and there is a school for children of Roman Catholics. The poor law union of Ulverstone comprises 27 parishes or places, containing a population of 22,563, according to the census of 1831, and is under the superintendence of 36 guardians. Conishead priory, in this parish, was founded by Gamel de Pennington, for Black canons, and at the dissolution its revenue was valued at £124. 2. 1.: the building was then dismantled, and the materials were sold for £333. 6. 3½., but some remains of the cemetery, pillars of the transept, the foundation walls of the church, with several skeletons, were discovered in the year 1823: the site is occupied by a modern edifice, in the English style of architecture. Richard de Ulverstone, a monk of considerable eminence, and author of a work entitled "Articles of Faith," was born here in 1434.

UNDERBARROW, a chapelry, in the parish, union, and ward of KENDAL, county of WESTMORLAND, 2¾ miles (W.) from Kendal; containing 526 inhabitants. The living is a perpetual curacy; net income, £92; patron, Vicar of Kendal.

UNDERMILBECK, a township, in the parish of WINDERMERE, union and ward of KENDAL, county of WESTMORLAND, 8 miles (W. by N.) from Kendal; containing, with the chapelry of Winster, 854 inhabitants. A school is supported by endowment; in another, nine children are paid for by T. Bolton, Esq.; and a day and Sunday school is partly supported by subscription.

UNDER-SKIDDAW, a township, in the parish of CROSTHWAITE, union of COCKERMOUTH, ALLERDALE ward below Derwent, Western Division of the county of CUMBERLAND, 6 miles (N. N. W.) from Keswick; containing 477 inhabitants. There is a school at this place, free to the parishioners of Crosthwaite, supported by sundry donations amounting to nearly £100 a year. A handsome school-house has been erected at High Hill, in this township, at the sole expense of James Stanger, Esq., in which 80 girls are taught on the National system.

UNDERWOOD, a joint liberty with Offcoat, in the parish of ASHBOURN, hundred of WIRKSWORTH, Southern Division of the county of DERBY: the population is returned with Offcoat.

UNDY, a parish, in the union of CHEPSTOW, Lower

Division of the hundred of CALDICOTT, county of MONMOUTH, 7 miles (E. by S.) from Newport; containing 291 inhabitants. The living is a discharged vicarage, valued in the king's books at £4. 10. 7½.; present net income, £183: the patronage and impropriation belong to the Chapter of Llandaff.

UNERIGG, a joint township with Ellenborough, in the parish of DEARHAM, union of COCKERMOUTH, ALLERDALE ward below Derwent, Western Division of the county of CUMBERLAND, 1½ mile (S. E.) from Maryport: the population is returned with Ellenborough. A school was founded and endowed in 1718, by Ewan Christian and others, with about £30 per annum.

UNSTONE, a township, in the parish of DRONFIELD, union of CHESTERFIELD, hundred of SCARSDALE, Northern Division of the county of DERBY, 4 miles (N. by W.) from Chesterfield; containing 586 inhabitants. A day and Sunday school is partly supported by an endowment of £16 per annum.

UNSWORTH, a chapelry, in the parish of OLDHAM *cum* PRESTWICH, hundred of SALFORD, Southern Division of the county palatine of LANCASTER, 3 miles (S. S. E.) from Bury: the population is returned with the parish. The living is a perpetual curacy; net income, £63; patron, Rector of Prestwich. The chapel, dedicated to St. George, was consecrated in 1730. There is a place of worship for Wesleyan Methodists. James Lancaster, in 1737, bequeathed certain property, now producing £12. 12. per annum, for teaching ten poor children.

UNTHANK, a township, in the parish of SKELTON, LEATH ward, Eastern Division of the county of CUMBERLAND, 5½ miles (N. W.) from Penrith; containing 235 inhabitants.

UNTHANK, a township, in the parish of ALNHAM, union of ROTHBURY, Northern Division of COQUETDALE ward and of the county of NORTHUMBERLAND, 8¼ miles (N. N. E.) from Rothbury; containing 34 inhabitants.

UPCHURCH (*ST. MARY*), a parish, in the union and hundred of MILTON, Upper Division of the lathe of SCRAY, Eastern Division of the county of KENT, 5½ miles (E. by S.) from Chatham; containing 456 inhabitants. This parish is bounded on the north by the Medway, where is Otterham creek and quay, at which corn produced in the neighbourhood is shipped for exportation. By the survey made in the reign of Elizabeth, it appears that twelve vessels belonged to this place. The living is a vicarage, valued in the king's books at £11; present net income, £155; patrons and impropriators, Warden and Fellows of All Souls' College, Oxford. The church is a handsome structure, partly in the decorated, and partly in the later, style of English architecture, with a lofty spire, noted as a landmark, and some remains of stained glass.

UP-EXE, a tything, in the parish of REWE, hundred of HAYRIDGE, Wonford and Northern Divisions of the county of DEVON, 6 miles (S. W. by W.) from Cullompton; containing 100 inhabitants.

UPHAM, a parish, in the union of DROXFORD, hundred of BISHOP'S-WALTHAM, Winchester and Northern Divisions of the county of SOUTHAMPTON, 2¾ miles (N. W. by N.) from Bishop's-Waltham; containing 511 inhabitants. The living is a rectory, valued in the king's books at £11. 2. 1.; present net income, £625; patron, Bishop of Winchester. The church contains 30 free sittings, the Incorporated Society having granted £20 towards defraying the expense. There is a chapel of ease at Durley; and here is a National school. Dr. Edward Young, author of the "Night Thoughts," whose father was rector of this parish, was born here in 1681; he died in 1707.

UPHAVEN (*ST. MARY*), a parish, in the union of PEWSEY, hundred of SWANBOROUGH, Everley and Pewsey, and Northern, Divisions of the county of WILTS, 4 miles (S. W. by S.) from Pewsey; containing 498 inhabitants. The living is a discharged vicarage, valued in the king's books at £7. 16. 8.; present net income, £112; it is in the patronage of the Crown; impropriators, Provost and Fellows of King's College, Cambridge. There is a place of worship for Particular Baptists. A market was granted by Henry III. to Peter de Mauley; and in the reign of Edward I., Hugh de Spencer procured a charter of free warren and two annual fairs, one of which, as well as the market, is discontinued. A Benedictine priory, a cell to the abbey of Fontanelle in Normandy, was founded here about the commencement of the reign of Henry I., and, at its suppression, was granted by Henry VI. to the monastery of Ivy-Church, in exchange for lands, &c., in Clarendon park.

UPHILL (*ST. NICHOLAS*), a parish, in the union of AXBRIDGE, hundred of WINTERSTOKE, Eastern Division of the county of SOMERSET, 8 miles (N. W. by W.) from Axbridge; containing 306 inhabitants. The river Axe bounds the parish on the south, and falls into the Bristol Channel at the village, the proximity of which to Weston-Super-Mare (of late become a fashionable bathing-place) has induced capitalists to purchase a considerable portion of land in the neighbourhood, with a view to erect houses upon it. Fuel is very abundant and cheap here. The living is a discharged rectory, valued in the king's books at £11. 7.; present net income, £184; patron, John Fisher, Esq. The church has a central tower, and occupies the summit of a lofty eminence south of the village. There is a place of worship for Baptists; also a National school. A cave was discovered a few years ago, similar to those in the same ridge of hills at Burrington and Banwell.

UP-HOLLAND, county palatine of LANCASTER.—See HOLLAND, UP.

UPLEADON (*ST. MARY*), a parish, in the union of NEWENT, hundred of BOTLOE, Western Division of the county of GLOUCESTER, 3¾ miles (E. by N.) from Newent; containing 241 inhabitants. This parish takes its name from the river Leadon, by which it is intersected. The living is a perpetual curacy; net income, £82; patron, Bishop of Gloucester and Bristol; appropriators, Dean and Chapter of Gloucester. The church has a Norman entrance on the north side.

UPLEATHAM, a parish, in the union of GUILSBROUGH, Eastern Division of the liberty of LANGBAURGH, North Riding of the county of YORK, 3¼ miles (N. E. by N.) from Guilsbrough; containing 265 inhabitants. The village is pleasantly situated on a declivity, and commands a fine prospect of Skelton castle, and the rich vale below. The living is a perpetual curacy; net income, £57; patron and appropriator, Archbishop of York. The church has been rebuilt, and contains 50 free sittings, the Incorporated Society having granted £75 towards defraying the expense. There is a place of worship for Wesleyan Methodists. A school for boys and another for girls are respectively supported by the Earl and Countess of Zetland.

UPLOWMAN (*St. Peter*), a parish, in the union of Tiverton, partly in the hundred of Halberton, but chiefly in that of Tiverton, Cullompton and Northern Divisions of the county of Devon, 4¾ miles (E. N. E.) from Tiverton; containing 335 inhabitants. The living is a rectory, valued in the king's books at £21. 0. 10.; patron and incumbent, Rev. Sydenham Pidsley. The tithes have been commuted for a rent-charge of £450, subject to the payment of rates; the glebe comprises 36 acres, valued at £78 per annum. Two schools are partly supported by subscription.

UPLYME (*St. Peter and St. Paul*), a parish, in the union and hundred of Axminster, Honiton and Southern Divisions of the county of Devon, 1¼ mile (N. W.) from Lyme-Regis; containing 975 inhabitants. The living is a rectory, valued in the king's books at £20. 8. 11½.; present net income, £386; patron and incumbent, Rev. C. W. Ethelston. The church has been enlarged, and 50 free sittings provided, the Incorporated Society having granted £45 towards defraying the expense. Two schools are partly supported by the minister; and there is a Sunday National school. In this parish are extensive beds of blue and white lias, replete with organic marine remains, and applicable to building, paving, or burning into lime: there is also a manufactory for woollen cloth.

UPMINSTER (*St. Lawrence*), a parish, in the union of Romford, hundred of Chafford, Southern Division of the county of Essex, 4 miles (E. S. E.) from Romford; containing 1033 inhabitants. This parish is about seven miles in length and about one mile in average breadth; the surface towards the north is considerably elevated; the soil in the uplands is clayey and in the low lands light and sandy. The scenery is finely varied, and enlivened with numerous handsome residences and flourishing plantations. The living is a rectory, valued in the king's books at £26. 13. 4.; present net income, £960; patrons, Trustees of the late J. R. Holden, Esq. The church is a handsome structure, with a tower and spire; on the north side of the chancel is a chapel belonging to Gaines Hall. There is a place of worship for Independents; also a National school. Here is a mineral spring. Dr. Derham, author of "Physico-Theology," &c., was rector of the parish from 1689 to 1735.

UP-OTTERY (*St Mary*), a parish, in the union of Honiton, hundred of Axminster, Honiton and Southern Divisions of the county of Devon, 5¼ miles (N. E. by N.) from Honiton; containing 940 inhabitants. The living is a vicarage, valued in the king's books at £15. 5. 7½.; present net income, £392; patrons and appropriators, Dean and Chapter of Exeter. The church has been enlarged, and 100 free sittings provided, the Incorporated Society having granted £100 towards defraying the expense. There are places of worship for Baptists and Calvinistic Methodists; also a National school supported by Lord Sidmouth and the vicar. Fairs for cattle are held here on March 17th and Oct. 24th. At Roridge, in this parish, was anciently a chapel.

UPPER ALLITHWAITE, county of Lancaster.—See ALLITHWAITE, UPPER.—*And all places having a similar distinguishing prefix will be found under the proper name.*

UPPERBY, a township, in the parish of St. Cuthbert, Carlisle, Cumberland ward, Eastern Division of the county of Cumberland, 1¾ mile (S. E. by S.) from Carlisle; containing 393 inhabitants, who are chiefly employed in the manufacture of linen.

UPPINGHAM (*St. Peter and St. Paul*), a market-town and parish, and the head of a union, in the hundred of Martinsley, county of Rutland, 6 miles (S.) from Oakham, and 89 (N.N.W.) from London; containing 1757 inhabitants. The name of this place is descriptive of its elevated situation: the town consists principally of one good street, with a square area in the centre, and is tolerably well paved: the houses are commodious and well-built, and the inhabitants are supplied with water from a spring in the upper part of the town. The air, though keen, is pure and salubrious, and the surrounding country is pleasingly diversified. The market, granted by Edward I., in 1280, to Peter de Montfort, is held on Wednesday, and is well supplied with corn and cattle. Fairs are held on March 7th and July 7th, chiefly for horses, horned cattle, and sheep, and also for coarse linen cloth. By statute of the 11th of Henry VII., the standard of weights and measures for the county is kept at this place. The living is a rectory, valued in the king's books at £20. 0. 10.; present net income, £661; patron, Bishop of London. The church, situated on the south side of the square, is a spacious structure, in the ancient style of English architecture, with a tower surmounted by a lofty spire. There are places of worship for Independents and Calvinistic and Wesleyan Methodists. The free grammar school adjoining the churchyard, and an hospital for poor men, were founded in 1584, by the Rev. Robert Johnson, Archdeacon of Leicester, and Rector of North Luffenham in this county, who also founded a similar school and hospital at Oakham, all of which he endowed with alienated ecclesiastical property, now producing an income of £3000 per annum, which is appropriated to the maintenance of these establishments at both places, the details of which are given in the article on Oakham (which see). Many eminent persons have been educated in these schools, among whom, in the school of this place, were Dr. Charles Manners Sutton, late Archbishop of Canterbury; Lord Manners, late Chancellor of Ireland; Dr. Henry Ferne, Bishop of Chester; and Dr. Bramston, Roman Catholic Bishop of the London District. A National school is supported by subscription; and commodious rooms, with a house, for the master and mistress, have been erected on a site given by Sir Gerard Noel, by subscription, aided by a grant of £70 from the National, and £113 from the District, Societies. The poor law union of Uppingham extends into the counties of Leicester and Northampton, and comprises 35 parishes or places, containing a population of 11,027, according to the census of 1831; it is under the superintendence of 36 guardians. The celebrated Jeremy Taylor, Bishop of Down and Connor, was rector of this parish.

UPPINGTON (*Holy Trinity*), a parish, in the union of Atcham, Wellington Division of the hundred of South Bradford, Northern Division of the county of Salop, 3½ miles (W. S. W.) from Wellington; containing 117 inhabitants. The living is a donative; net income, £70; patron and impropriator, Duke of Cleveland. This parish has an interest in the grammar school at Doddington, in the parish of Wroxeter.

UPSALL, a township, in the parish of SOUTH KILVINGTON, union of THIRSK, wapentake of BIRDFORTH, North Riding of the county of YORK, $3\frac{3}{4}$ miles (N. N. E.) from Thirsk; containing 114 inhabitants. There are some remains of a castle of the Mowbrays, which subsequently became the residence of the Scroops.

UPSALL, a township, in the parish of ORMSBY, union of GUILSBROUGH, Eastern Division of the liberty of LANGBAURGH, North Riding of the county of YORK, 3 miles (W.) from Guilsbrough: the population is returned with Ormsby.

UPSHIRE, a hamlet, in the parish of WALTHAM-ABBEY, or HOLY-CROSS, hundred of WALTHAM, Southern Division of the county of ESSEX; containing 745 inhabitants. There is a place of worship for Wesleyan Methodists.

UPSLAND, a joint township with Kirklington, in the parish of KIRKLINGTON, union of BEDALE, wapentake of HALLIKELD, North Riding of the county of YORK: the population is returned with Kirklington.

UPTON, a chapelry, in the parish of BLEWBERRY, union of WANTAGE, hundred of MORETON, county of BERKS, $4\frac{3}{4}$ miles (N. N. E.) from East Ilsley; containing, with the liberty of Nottingham-Fee, 254 inhabitants. The chapel is a very ancient edifice.

UPTON (*St. Lawrence*), a parish, in the union of ETON, hundred of STOKE, county of BUCKINGHAM, 3 miles (N. W. by W.) from Colnbrook; containing 1502 inhabitants. This parish is situated on the great Bath road, and about a mile from the river Thames; the Great Western railway passes through it, and has a station at Slough. The living is a discharged vicarage, valued in the king's books at £6. 17.; present net income, £220: it is in the patronage of the Crown; impropriator, W. Bousey, Esq. The church, which is a very ancient structure, said to have been built more than 1200 years since, has a fine Norman doorway, and is principally in that style of architecture. A new church has been erected at Chalvey, in this parish, towards which the late King gave £100, the Queen Dowager £50, the Impropriator £200, the Incumbent £200, and the Curate £10: it is a handsome edifice in the Norman style, and is adapted for a congregation of 800. There is a place of worship for Independents. A parochial school is supported by subscription; and a British school has been recently erected. Benjamin Lane, in 1720, bequeathed a rent-charge of £20 for clothing six poor men and six poor women, and distributing six bibles annually. The late Sir William Herschel, the celebrated astronomer, resided and was buried here; and his son, the present Sir John Herschel, Bart., was born and resides in the parish.

UPTON, a township, in the parish of ST. MARY, CHESTER, union of GREAT BOUGHTON, Lower Division of the hundred of BROXTON, Southern Division of the county of CHESTER, $2\frac{1}{4}$ miles (N.) from Chester; containing 289 inhabitants.

UPTON, a township, in the parish of PRESTBURY, union and hundred of MACCLESFIELD, Northern Division of the county of CHESTER, $1\frac{1}{2}$ mile (N. W.) from Macclesfield; containing 64 inhabitants.

UPTON, or OVER-CHURCH, a parish, in the union, and Lower Division of the hundred, of WIRRALL, Southern Division of the county of CHESTER, $7\frac{3}{4}$ miles (N. by W.) from Great Neston; containing 191 inhabitants. The living is a perpetual curacy; net income, £52; patron, William Webster, Esq.; impropriator, Sir T. S. M. Stanley, Bart. The impropriate tithes have been commuted for a rent-charge of £145, subject to the payment of rates; the glebe, which belongs to the incumbent, comprises $19\frac{1}{2}$ acres, valued at £35 per annum. There is a rent-charge of £4. 6. payable to the rector of Woodchurch. At Upton (from which the parish church, called Over-Church, is distant half a mile) a market was held so late as 1662: there are still two fairs for cattle. A court leet and baron is held annually. Mr. Webster, who is lord of the manor, has erected a school, and pays a yearly salary to the master.

UPTON, a tything, in the parish of HAWKESBURY, Upper Division of the hundred of GRUMBALD'S-ASH, Western Division of the county of GLOUCESTER; containing 696 inhabitants. There is a place of worship for Wesleyan Methodists.

UPTON (*St. Leonard's*), a parish, in the Middle Division of the hundred of DUDSTONE and KING'S-BARTON, union, and Eastern Division of the county, of GLOUCESTER, $3\frac{1}{4}$ miles (S. E. by S.) from Gloucester; containing 898 inhabitants. The living is a perpetual curacy; net income, £86; patron and appropriator, Bishop of Gloucester and Bristol. The church is principally Norman, but the tower and some of the details are in the later style of English architecture; it has been enlarged, and 168 free sittings provided, the Incorporated Society having granted £160 towards defraying the expense. Here is a National school.

UPTON, (*St. Margaret*), a parish, in the hundred of LEIGHTONSTONE, union and county of HUNTINGDON, $6\frac{1}{2}$ miles (N. W.) from Huntingdon; containing 150 inhabitants. The living is a rectory, with that of Coppingford consolidated; net income, £160; patron, Lord Montagu. The church is partly in the early style of English architecture, with a curious ancient font. A day and Sunday school is supported by Mr. Heathcote.

UPTON, a township, in the parish of SIBSON, union of MARKET-BOSWORTH, hundred of SPARKENHOE, Southern Division of the county of LEICESTER, $3\frac{3}{4}$ miles (S. W.) from Market-Bosworth; containing 148 inhabitants. Here was formerly a chapel, now in ruins.

UPTON (*All Saints*), a parish, in the union of GAINSBOROUGH, hundred of WELL, parts of LINDSEY, county of LINCOLN, 5 miles (S. E. by E.) from Gainsborough; containing 460 inhabitants. The living is a discharged vicarage, valued in the king's books at £7. 4. 2.; present net income, £131; patron, Sir W. A. Ingilby, Bart. There is a place of worship for Wesleyan Methodists.

UPTON (*St. Margaret*), a parish, in the union of BLOFIELD, hundred of WALSHAM, Eastern Division of the county of NORFOLK, $1\frac{3}{4}$ mile (N.) from Acle; containing, with the parish of Fishley, 510 inhabitants. The living is a discharged vicarage, united to that of Ranworth, and valued in the king's books at £5.

UPTON (*St. Michael*), a parish, in the hundred of NEWBOTTLE-GROVE, union, and Southern Division of the county, of NORTHAMPTON, 2 miles (W.) from Northampton; containing 48 inhabitants. The living is annexed, with that of Kingsthorpe, to the rectory of St. Peter's, Northampton. Here are still the remains of a castle founded by Simon de St. Liz. James Harrington, an eminent political writer in the time of the Commonwealth, was born at Upton Hall, in 1611.

UPTON, a chapelry, in the parish of Castor, union and soke of Peterborough, Northern Division of the county of Northampton, $2\frac{1}{4}$ miles (E. N. E.) from Wansford; containing 122 inhabitants. The chapel is dedicated to St. John the Baptist. In the neighbourhood is a stone quarry, resembling that of Ketton.

UPTON (*St. Peter*), a parish, in the union of Southwell, and in that part of the liberty of Southwell and Scrooby which separates the Northern from the Southern Division of the wapentake of Thurgarton, Southern Division of the county of Nottingham, $2\frac{3}{4}$ miles (E.) from Southwell; containing 533 inhabitants. The living is a discharged vicarage, in the patronage of the Chapter of the Collegiate Church of Southwell (the appropriators), valued in the king's books at £4. 11. $5\frac{1}{2}$.; present net income, £91. The church is endowed with lands of the annual value of £20, for keeping it in repair, the surplus to be given to poor soldiers travelling through the place. Here is a considerable manufactory for starch.

UPTON, a hamlet, in the parish of Headon, union of East Retford, South-Clay Division of the wapentake of Bassetlaw, Northern Division of the county of Nottingham, $3\frac{1}{4}$ miles (N. by E.) from Tuxford: the population is returned with the parish.

UPTON, a joint hamlet with Signet, in the parish of Burford, union of Witney, hundred of Bampton, county of Oxford, $1\frac{1}{4}$ mile (W.) from Burford; containing, with Signet, 246 inhabitants.

UPTON (*St. James*), a parish, in the union of Dulverton, hundred of Williton and Freemanners, Western Division of the county of Somerset, $4\frac{1}{4}$ miles (E. by N.) from Dulverton; containing 344 inhabitants. The living is a perpetual curacy; net income, £50; patron and impropriator, Executor of the late R. Bere, Esq. The impropriate tithes have been commuted for a rent-charge of £241. 15., subject to the payment of rates. A school is partly supported by charitable donations.

UPTON, a township, in the parish of Ratley, Burton-Dassett Division of the hundred of Kington, Southern Division of the county of Warwick, 4 miles (S. E. by S.) from Kington: the population is returned with the parish.

UPTON, a township, in the parish of Badsworth, Upper Division of the wapentake of Osgoldcross, West Riding of the county of York, $6\frac{1}{4}$ miles (S. by E.) from Pontefract; containing 229 inhabitants. There are limeworks in the neighbourhood. Here are two schools, in which 14 children are paid for by a private family.

UPTON BISHOP'S (*St. John the Baptist*), a parish, in the union of Ross, hundred of Greytree, county of Hereford, 4 miles (N. E. by E.) from Ross; containing 626 inhabitants. The living is a vicarage, valued in the king's books at £8. 17. 6.; present net income, £708; patrons and appropriators, Dean and Chapter of Hereford. Here is a school endowed with £4 per annum, and a cottage for the residence of the teacher; it is further supported by subscription.

UPTON-CRESSETT (*St. Michael*), a parish, in the union of Bridgenorth, hundred of Stottesden, Southern Division of the county of Salop, 5 miles (W. by S.) from Bridgenorth; containing 43 inhabitants. The living is a discharged rectory, valued in the king's books at £4. 15. $2\frac{1}{2}$.; present net income, £125; late patron, J. C. Pelham, Esq.

UPTON-GRAY, a parish, in the union of Basingstoke, hundred of Bermondspit, Basingstoke and Northern Divisions of the county of Southampton, 4 miles (W. S. W.) from Odiham; containing 388 inhabitants. The living is a perpetual curacy; patrons and impropriators, Provost and Fellows of Queen's College, Oxford. The tithes have been commuted for a rent-charge of £490, subject to the payment of rates. A National school is supported principally by Mrs. Beaufoy and partly by subscriptions.

UPTON-HELLIONS (*St. Mary*), a parish, in the union of Crediton, hundred of West Budleigh, Crediton and Northern Divisions of the county of Devon, $2\frac{1}{2}$ miles (N. N. E.) from Crediton; containing 152 inhabitants. The living is a rectory, valued in the king's books at £10. 6. 8.; present net income, £223; patron and incumbent, Rev. W. Wellington.

UPTON-LOVELL, a parish, in the union of Warminster, hundred of Heytesbury, Warminster and Southern Divisions of the county of Wilts, 2 miles (S. E. by E.) from Heytesbury; containing 249 inhabitants. The living is a rectory, valued in the king's books at £17. 18. $11\frac{1}{2}$., and in the patronage of the Crown. The tithes have been commuted for a rent-charge of £320, subject to the payment of rates, which on the average have amounted to £39; the glebe comprises 29 acres, valued at £55 per annum. The Rev. John Crouch, in 1794, bequeathed £500 three per cent. consols., the interest to be applied to the instruction of six poor children during the week, and all such children as desire it on the Sunday: a salary of £15 is paid to the master. On Upton-Lovell down, about two miles north of Heytesbury, is a single intrenchment, called Knook Castle, including about two acres. On the summit of a hill, to the north-west of Elder Valley, is a large tumulus, called Bowls Barrow, which has been found to contain fourteen human skeletons. There is also, in the neighbourhood of Knook Castle, and near the north bank of the Wily, another large barrow, which, from the number of gold ornaments discovered in it, has been called Golden Barrow.

UPTON-MAGNA (*St. Lucia*), a parish, in the union of Atcham, Wellington Division of the hundred of South Bradford, Northern Division of the county of Salop, $5\frac{1}{4}$ miles (E.) from Shrewsbury; containing, with the extra-parochial liberty of Haughmond Demesne, 512 inhabitants. The living is a rectory, valued in the king's books at £12; present net income, £546; patrons, Devisees of J. Corbet, Esq. The Shrewsbury canal passes through the parish. Here is a National school, the master of which receives £7. 10. per annum, a portion of a bequest by Mr. Thomas Blakeway, in 1767.

UPTON-NOBLE, a parish, in the union of Shepton-Mallet, hundred of Bruton, Eastern Division of the county of Somerset, 4 miles (N. N. E.) from Bruton; containing 282 inhabitants. The living is a perpetual curacy, annexed to the rectory of Batcombe. There is a place of worship for Wesleyan Methodists.

UPTON-PYNE, a parish, in the union of St. Thomas, hundred of Wonford, Wonford and Southern Divisions of the county of Devon, $3\frac{1}{2}$ miles (N. by W.) from Exeter; containing 514 inhabitants. The living is a rectory, valued in the king's books at £23. 6. 8.; patron, Sir S. H. Northcote, Bart. The tithes have been commuted for a rent-charge of £400, subject to the payment of rates, which on the average have amounted to £19; the glebe comprises 35 acres, valued at £126 per annum. The church contains a good painting of the Last Supper, the monument of a crusader, and some remains of ancient stained glass. A day and Sunday school is supported by Lady Northcote. This parish is bounded on the south by the river Exe, and contains some mines of manganese.

UPTON-SCUDAMORE (*St. Mary*), a parish, in the union and hundred of WARMINSTER, Warminster and Southern Divisions of the county of WILTS, 2 miles (N.) from Warminster; containing 393 inhabitants. The living is a rectory, valued in the king's books at £16. 7. 1.; patrons, Provost and Fellows of Queen's College, Oxford. The rector's tithes have been commuted for a rent-charge of £480, subject to the payment of rates, which on the average have amounted to £38; his glebe comprises nearly 23 acres, valued at £29. 10. per annum. The appropriate tithes have been commuted for a rent-charge of £20 and £35 (respectively payable to the Dean and Chapter of Salisbury and the Prebendary of Luxford), subject to the payment of rates, which on the average have amounted to £5. There is also a rent-charge of £50 payable to an impropriator, subject to the payment of rates, which on the average have amounted to £5; the impropriate glebe comprises 23½ acres, valued at £26 per annum. Here is a National school.

UPTON-SNODSBURY (*St. Kenelme*), a parish, in the union, and Upper Division of the hundred, of PERSHORE, Worcester and Western Divisions of the county of WORCESTER, 6½ miles (E. by S.) from Worcester; containing 316 inhabitants. The living is a discharged vicarage, valued in the king's books at £8; present net income, £95; patron and incumbent, Rev. Henry Green.

UPTON-UPON-SEVERN (*St. Peter and St. Paul*), a market-town and parish, and the head of a union, in the Lower Division of the hundred of PERSHORE, Upton and Western Divisions of the county of WORCESTER, 10 miles (S.) from Worcester, and 109 (N. W. by W.) from London; containing 2343 inhabitants. According to Dr. Stukeley, this was the *Upoessa* of Ravennas; and the probability of its having been a Roman station is strengthened by the discovery of some ancient armour in the neighbourhood. During the parliamentary war, a bridge of six arches, erected pursuant to act of parliament in the reign of James I., was broken down, and a battery placed in the churchyard, to prevent the approach of Cromwell and his forces; but the plan was ineffectual, and the parliamentary forces entered the town. Upton is situated on the right bank of the river Severn, which is here navigable for vessels of 100 tons' burden, and is crossed by a bridge erected in 1606: it is neatly built, and the streets are well paved. There is a subscription library. The surrounding country is in a state of high cultivation, and the scenery is varied and picturesque. A considerable quantity of cider, brought from Herefordshire and other places, is shipped here for conveyance to different parts of England: there is a harbour for barges, also a wharf on the river for the convenience of loading and discharging. The market is on Thursday: a handsome market-house, including an assembly-room and apartments for the meetings of the magistrates, has been erected by subscription. Fairs are held on Mid-Lent and Whitsun Thursdays, July 10th, and the Thursday before Oct. 2nd. A manorial court is held occasionally, and petty sessions once a fortnight; and, by the act of the 2nd and 3rd of William IV., cap. 64, the town has been made a polling-place for the western division of the county.

The living is a rectory, valued in the king's books at £27; present net income, £917; patron, Bishop of Worcester. The church is a handsome structure, completed in 1758: the ancient tower was once surmounted by a spire, which, from an apprehension of insecurity, was taken down, and a wooden cupola substituted. There are places of worship for Baptists and Wesleyan Methodists. A charity school for 20 poor girls was endowed in 1718, by Richard and Anne Smith, with property producing at present £28 per annum, which was augmented with a bequest of £5 per annum, in 1824, by Mrs. Sarah Husband: a boys' school was added to it in 1797, by a bequest from George King, of property vested in the purchase of £100 three per cents., and £100 four per cent. consols.: and these are now incorporated into a National school, which is further supported by voluntary contributions. Edward Hall, in 1578, bequeathed an estate, now producing a rental of about £80 a year, for maintaining a bridge over the Severn at this place. Thomas Morris, alias Woodward, in 1675, bequeathed a legacy of £185, which was invested in land, &c., now producing a rental of £35. 10. 6., which sum is appropriated to parochial purposes. The poor law union comprises 22 parishes or places, under the care of 28 guardians, and contains a population of 15,496, according to the census of 1831. Dr. John Dee, a celebrated astrologer in the reign of Elizabeth, was a native of this town: the Rev. J. Davison, B. D., author of some highly-esteemed theological works, was succeeded in the incumbency of this parish by the Rev. H. J. Tayler, B. D., the present rector.

UPTON-WARREN (*St. Michael*), a parish, in the union of DROITWICH, Upper Division of the hundred of HALFSHIRE, Droitwich and Eastern Divisions of the county of WORCESTER, 3½ miles (N. E. by N.) from Droitwich; containing 474 inhabitants. The living is a rectory, valued in the king's books at £11. 2. 3½.; patron, Earl of Shrewsbury. The tithes have been commuted for a rent-charge of £650, subject to the payment of rates, which on the average have amounted to £51; the glebe comprises 72 acres, valued at £150 per annum. The church has been partly rebuilt, and the interior is neatly fitted up. A National school is endowed with £10. 10. per annum: a house for the master and a school-room are given by the Earl of Shrewsbury, at a nominal rent. There is also an annuity of £10, bequeathed by a person named Saunders, to be paid by the Grocers' Company, for apprenticing, in London, a boy born of poor parents of this parish, or, in default, to one from the parishes of Stoke-Prior or Chaddesley.

UPTON-WATERS (*St. Michael*), a parish, in the union of WELLINGTON, Wellington Division of the hundred of SOUTH BRADFORD, Northern Division of the county of SALOP, 5½ miles (N. by W.) from Wellington; containing 193 inhabitants. The living is a discharged rectory, valued in the king's books at £3. 17. 3½.; it is in the patronage of the Crown. The tithes have been commuted for a rent-charge of £135, subject to the payment of rates; the glebe comprises $32\frac{1}{2}$ acres, valued at £60 per annum.

UP-WALTHAM, county of SUSSEX.—See WALTHAM, UP.

UPWAY (*ST. LAWRENCE*), a parish, in the union of WEYMOUTH, comprising the liberty of Weybey-house, the tything of Stottingway in the hundred of CULLIFORD-TREE, and that of Elwell in the liberty of WYKE-REGIS and ELWELL, Dorchester Division of the county of DORSET, $4\frac{1}{2}$ miles (S. W. by S.) from Dorchester; and containing 618 inhabitants. The living is a rectory, valued in the king's books at £18. 3. $1\frac{1}{2}$.; patron, Bishop of Salisbury. The tithes have been commuted for a rent-charge of £380, subject to the payment of rates; the glebe comprises $46\frac{1}{2}$ acres, valued at £68. 12. per annum. The church has an embattled tower, crowned with pinnacles, and has been enlarged, and 200 free sittings provided, the Incorporated Society having granted £50 towards defraying the expense. The liberty of Weybey-house and the manor of Upway belong to the Rev. George Gould, whose ancestors have been seated here since the reign of James I.: part of the ancient manor-house is still remaining, but the family have for some years chiefly resided at Fleet, in this county. On this estate are some excellent quarries, from which was taken the stone for building the new church at Fleet. The manor of Stottingway belongs to the vicars-choral in the cathedral church of Salisbury. Near the church, from the foot of a steep hill, rises the small river Way, which runs through the parish, and falls into the sea at Weymouth. On Ridgway down are numerous barrows, extending from that part of the ridge opposite Sutton-Pointz to beyond Longbridy, a distance of nearly six miles, in a direction parallel to the ancient Roman road called *Via Iceniana*.

UPWELL (*ST. PETER*), a parish, in the union of WISBEACH, partly in the hundred of WISBEACH, Isle of ELY, county of CAMBRIDGE, and partly in the hundred of CLACKCLOSE, Western Division of the county of NORFOLK, $5\frac{3}{4}$ miles (S. E. by S.) from Wisbeach; containing 4176 inhabitants. The village is intersected by the river Nene, by which the productions of the large garden-grounds here are conveyed to the markets of the various towns situated upon its banks. The living is a rectory, valued in the king's books at £16; present net income, £3855; patron, R. G. Townley, Esq. The church, with the greater part of the parish, is in Norfolk. At Welney, in this parish, is a chapel of ease. There is a place of worship for Wesleyan Methodists; also a school supported by Marshall's charity, and another by the rector. In that part of the parish which is in Cambridgeshire are the sites of two ancient religious houses, one of which, at Mirmound, was a small priory of Gilbertines, dedicated to the Blessed Virgin Mary, a cell to the priory of Sempringham, in Lincolnshire, and valued at the dissolution at £13. 6. 1. per annum.

UPWOOD (*ST. PETER*), a parish, in the hundred of HURSTINGSTONE, union and county of HUNTINGDON, $2\frac{3}{4}$ miles (S. W. by W.) from Ramsey; containing 326 inhabitants. The living is a perpetual curacy, with that of Great Raveley annexed; net income, £78: it is in the patronage of Miss Bickerton. The impropriate tithes have been commuted for a rent-charge of £340, subject to the payment of rates. Two schools for girls are supported by subscription. Robert Gorden and Anthony Ashton, in 1660, gave several small parcels of land, containing together between 10 and 11 acres, and now let for £10 per annum, which is applied to parochial purposes.

URCHFONT (*ST. MICHAEL*), a parish, in the union of DEVIZES, hundred of SWANBOROUGH, Devizes and Northern Divisions of the county of WILTS, $2\frac{3}{4}$ miles (N. E.) from East Lavington; containing 1389 inhabitants. The living is a discharged vicarage, with that of Stert annexed, valued in the king's books at £15. 15. 10.; present net income, £237; patrons and appropriators, Dean and Canons of Windsor. Here is a National school.

URMSTON, a township, in the parish of FLIXTON, hundred of SALFORD, Southern Division of the county palatine of LANCASTER, $5\frac{1}{2}$ miles (S. W. by W.) from Manchester; containing 706 inhabitants. John Collier, commonly called Tim Bobbin, the author of the "Lancashire Dialect," was born here.

URPETH, a township, in the parish and union of CHESTER-LE-STREET, Middle Division of CHESTER ward, Northern Division of the county palatine of DURHAM, 9 miles (N. by W.) from Durham; containing 716 inhabitants, who are mostly employed in the extensive coal mines adjacent.

URSWICK (*ST. MARY*), a parish, in the union of ULVERSTONE, hundred of LONSDALE, north of the sands, Northern Division of the county palatine of LANCASTER, 3 miles (S. W. by S.) from Ulverstone; containing, with the hamlet of Little Urswick, 752 inhabitants. The living is a discharged vicarage, valued in the king's books at £7. 17. 6.; present net income, £86; patrons, the Landowners. The church, which was re-pewed in 1826, is situated between the villages of Great and Little Urswick. At Bolton are the remains of an ancient chapel, near which Roman coins and a brass vessel upon three feet have been found.

URSWICK, LITTLE, a hamlet, in the parish of GREAT URSWICK, hundred of LONSDALE, north of the sands, Northern Division of the county palatine of LANCASTER: the population is returned with the parish. The village, which adjoins the village of Great Urswick, is pleasantly situated, and is distinguished for a fine circular lake, about half a mile in diameter, and abounding with tench, roach and other fish; the sole right of fishing is vested in the Earl of Burlington. A school was founded in the village in 1580, by Wm. Marshall, who endowed it with a rent-charge of £15, for which 40 boys are instructed.

USHAW, a hamlet, in the chapelry of ESH, parish of LANCHESTER, ward of CHESTER, county palatine of DURHAM, 4 miles (W.) from Durham: the population is returned with Esh. This place derives its name from the abundance of yew trees which formerly grew in the neighbourhood. The lands belong to the Roman Catholic College, which was established here in 1808, and which owed its origin to the dissolution of the English Roman Catholic College of Douay, in French Flanders, by the arms of the French republic in 1794. The majority of the professors and students having made their escape to their native land, established themselves at Crook Hall, in this county. That building, however, was soon found too small, and by the liberal support of the Catholic clergy and laity, they were enabled to raise their present extensive edifice. The building, which comprises a spacious quadrangle, is adapted to the reception of 150 students, with apartments for a president, vice-president, and professors: it has a valuable library of more than 12,000 volumes of biblical and general literature, with numerous splendidly illuminated MSS. Dr. Lingard and Dr. Wiseman were educated in this college.

USHLAWRCOED, a hamlet, in the parish of Bedwelty, union of Abergavenny, Lower Division of the hundred of Wentllooge, county of Monmouth; containing 5359 inhabitants.

USK (*St. Mary*), a market-town and parish, in the union of Pont-y-pool, partly in the Lower, but chiefly in the Upper, Division of the hundred of Usk, county of Monmouth, 13 miles (S. W.) from Monmouth, and 144 (W. by N.) from London; containing 1775 inhabitants. This place, which derives its name from the Gaelic *Ysc*, signifying water, is of remote antiquity, and is generally admitted by antiquaries to have been the *Burrium* of the Romans. The ancient castle, erected on an eminence overlooking the town, experienced repeated assaults during the wars between the Welsh chieftains and the Anglo-Norman lords, especially in the time of the celebrated Owain Glyndwr; and, in the civil commotions in the reign of Charles I., it was, with the town, partly demolished by the parliamentary forces. The town is agreeably situated on the river Usk, which is crossed by a stone bridge, and consists of several streets, composed of detached houses, with intervening gardens and orchards. Some of the inhabitants are engaged in husbandry, and others in the salmon fishery, and there is a small manufactory for japanned tin, or Pont-y-Pool ware. The market is on Friday, and a cattle market is held on the first Monday in each month. Fairs are held on April 20th, a large one for wool; June 20th; Oct. 29th; and on the Monday before Christmas-day. The town is governed by a corporation, consisting of a portreeve, recorder, and burgesses, assisted by four constables: the portreeve is elected from among the burgesses, who are chosen by the recorder; and the constables are appointed at a borough-leet, held annually: on retiring, the portreeve becomes an alderman. This borough, of which the old limits include 327 acres, conjointly with those of Monmouth and Newport, returns one member to parliament: the right of election was formerly vested in the burgesses at large, but the privilege has been extended to the £10 householders of an enlarged district, which has been made to constitute the new elective borough, comprising an area of 522 acres, and the limits of which are minutely described in the Appendix, No. II.: the number of qualifying tenements under the new law is 90, but they are not all at present rated in proportion to their value. The portreeve possesses magisterial authority, the county magistrates having concurrent jurisdiction. The quarter sessions for the county, and the petty sessions for the division, are held here; also a court leet once a fortnight, at which the portreeve and recorder preside; and the town has been made a polling-place for the county. The town-hall is a handsome edifice over the market-place, built at the expense of the Duke of Beaufort. The prison has been enlarged, and a tread-mill erected, by the county, at an expense of about £600.

The living is a discharged vicarage, valued in the king's books at £10. 10.; present net income, £250; patron, William Addams Williams, Esq.; impropriator, Duke of Beaufort. The church, which was formerly conventual, appears to be of Anglo-Norman origin, and was originally cruciform, but has undergone numerous alterations: it contains several ancient monuments, and a modern one, erected in 1822, to commemorate the extended benevolence of Mr. Roger Edwards, the founder of the grammar school. There are places of worship for Independents, Wesleyan Methodists, and Roman Catholics. Roger Edwards, in 1621, bequeathed property now producing a yearly rental of £412, to found and endow a free grammar school; to support an almshouse previously built by him at Llangeview, for 12 poor persons, chosen from the parishes of Llangeview, Gwernesney, and Llangwym: and for other charitable purposes. There are now two separate schools held in premises adjoining the church: that which is at present called the grammar school is held in the lower room; the master, who is a graduate of Oxford, has a salary of £60, with the use of a good dwelling-house, two gardens, and an orchard. In the other, called the writing school, held in the upper rooms, about 40 younger children are instructed in reading, writing, and accounts, by a master in holy orders, who receives a salary of £70. The almshouses at Llangeview, originally erected in 1612, were rebuilt in 1826: they consist of a range of 12 apartments, each with a closet for a bed and a pantry, forming two sides of a quadrangle, with a room fitted up as a chapel in the angle. The same benefactor also founded and endowed a scholarship with £5 per annum in the University of Oxford, for a boy educated at the above school. A National school is supported by subscription. Almshouses for 24 poor persons were erected in 1826, near the Priory, upon the site of some old ones, the founder of which is unknown. The remains of the castle, standing on an abrupt eminence to the east of the river, consist of the exterior walls and a tower gateway, with several apartments, amongst which is the baronial hall: the area is of considerable extent, and is flanked by square and round towers. To the south-east of the church are a few remains of a priory founded here by one of the Earls of Clare.

USSELBY (*St. Margaret*), a parish, in the union of Caistor, Northern Division of the wapentake of Walshcroft, parts of Lindsey, county of Lincoln, $3\frac{1}{4}$ miles (N. by W.) from Market-Rasen; containing 84 inhabitants. The living is a perpetual curacy; net income, £44; patron and impropriator, G. Tennyson, Esq.

USWORTH, GREAT and LITTLE, a chapelry, in the parish of Washington, union of Chester-le-Street, Eastern Division of Chester ward, Northern Division of the county palatine of Durham, $4\frac{1}{4}$ miles (S. E.) from Gateshead; containing, with North Biddick, 1477 inhabitants. In 1834, an act was obtained for constructing a railway from the Hartlepool railway, near Moorsley, to the Stanhope and Tyne rail-road, in this chapelry. A chapel was built here in 1831, and 300 free sittings provided, the Incorporated Society having granted £100 towards defraying the expense. There is a place of worship for Wesleyan Methodists. The late Mrs. Penrith, in 1814, built a commodious school-house, and endowed it with £30 per annum; and a National school has been established.

UTKINTON, a township, in the parish of Tarporley, union of Nantwich, First Division of the hundred of EDDISBURY, Southern Division of the county of CHESTER, $1\frac{1}{2}$ mile (N. by W.) from Tarporley; containing 564 inhabitants.

UTON, a tything, in the parish and hundred of CREDITON, Crediton and Northern Divisions of the county of DEVON, 2 miles (W. S. W.) from Crediton; containing 259 inhabitants.

UTTERBY (*St. Andrew*), a parish, in the union of LOUTH, wapentake of LUDBOROUGH, parts of LINDSEY, county of LINCOLN, $4\frac{3}{4}$ miles (N. by W.) from Louth; containing 198 inhabitants. The living is a discharged vicarage, valued in the king's books at £5. 6. 8.; patrons and impropriators, J. Staunton, Esq., and others, as trustees of the Rev. L. E. Towne. The impropriate tithes have been commuted for a rent-charge of £200, and the vicarial for £125, subject to the payment of rates; the glebe is valued at £1. 8. per annum. The church contains monuments to the Harrold family, of which several members were buried here. Utterby House, the seat of the Rev. Henry Bristowe Benson, is beautifully situated; and the grounds comprehend some pleasingly picturesque scenery; over the entrance are the armorial bearings of the Sapsford family. The Roman road, Barton-Street, bounds the parish on the west, and, according to tradition, here was a Roman encampment.

UTTOXETER (*St. Mary*), a market-town and parish, and the head of a union, in the Southern Division of the hundred of TOTMONSLOW, Northern Division of the county of STAFFORD, 13 miles (N. E. by E.) from Stafford, and 135 (N. W. by N.) from London; containing 4864 inhabitants. Uttoxeter, anciently called *Uttokeshather*, is a place of great antiquity, and is supposed to have derived its name from the Saxon words *Uttoc*, a mattock, and *Hather*, heath; it was afterwards called *Utoc Cestre* and *Utcester*. One of its late commons, called the High Wood (a moiety of which was seized by the crown within the last two centuries), anciently constituted, with other lands, one of the wards of the late Forest of Needwood. The manor heretofore formed part of the possessions of the duchy of Lancaster, and formerly belonged to the Peverills of the Peak, Lords of Nottingham. Having come, by marriage, into the possession of William de Ferrars, Earl of Derby, it was forfeited to the crown, together with the other large possessions of that family, by Earl Robert, in the reign of Henry III., and given to Edmund Earl of Lancaster, the king's second son. In 1308, Thomas Earl of Lancaster, son of Edmund, obtained for it the grant of a market, and a fair on the eve, day, and morrow of St. Mary Magdalene. The manor reverted to the crown, as parcel of the duchy of Lancaster, in the person of Henry IV., son of John of Gaunt, Duke of Lancaster, who obtained it by marriage with Blanche, daughter and co-heiress of Henry Earl of Lancaster, nephew of Earl Thomas. Charles I., in the first year of his reign, granted it and the demesne to Robert Dixon and William Walley, as trustees for Henry Viscount Manderville, afterwards Earl of Manchester; and it is now vested, in twelve shares, in Earl Talbot and other proprietors: the market and fairs were sold at the same time, and are now the property of Earl Talbot. During the civil war of the seventeenth century, from its proximity to Tutbury castle, it was alternately the head-quarters of the royalist and the parliamentary forces.

The town stands upon an eminence, rising from the western bank of the river Dove, across which is an ancient stone bridge of six arches, connecting the counties of Stafford and Derby: it consists of several spacious streets, and a good central market-place: the houses in general are well built, and several of them are handsome. It has long been noted for the manufacture of clock cases and movements; there are also several maltsters, tanners, fell-mongers, nail-makers, bendware-manufacturers, wool-staplers, rope and twine spinners, timber-merchants, &c., and a large brewery. The local trade in cheese, corn, and other articles, is benefited by the communication with the Potteries, by means of the Caldon branch of the Trent and Mersey canal, which comes up to a wharf at the northern end of the High-street. The land near the town, and in the vicinity of the Dove, is very fertile in pasturage; and the neighbouring rivers and brooks afford trout, grayling, and other kinds of fish. Near the town is found a pure red brick clay, from one to five yards below the surface, in irregular masses. The market, which is well attended, is held on Wednesday, every alternate Wednesday being a large market for cattle, merchandise, &c. Fairs for cattle are held on the Tuesday before Old Candlemas, May 6th, July 31st, Sept. 1st and 19th, and Nov. 11th and 27th; those on May 6th and Sept. 19th are the principal. The first charter was granted in the 36th of Henry III., by William de Ferrars, Earl of Derby, which conferred on the burgesses all the privileges of a free borough. Uttoxeter, though a manor, with power to hold a court baron, was subject to the jurisdiction of the officers of the courts held for the honour of Tutbury; but, in 1636, an order of the court of the duchy chamber was made, discharging the inhabitants from further attendance at the courts for the honour. Petty sessions for the southern division of the hundred of Totmonslow are held here, once a fortnight, by the county magistrates, who appoint surveyors of the highways, and also constables, head-boroughs, &c., in cases where the lords of the different courts leet in the neighbourhood neglect to hold their courts, and make the appointments. The town has lately been made a polling-place for the northern division of the shire.

The living is a discharged vicarage, valued in the king's books at £27. 1. 8.; present net income, £136; patrons, Dean and Canons of Windsor, who hold courts for the rectorial manor. The church has been rebuilt, with the exception of the ancient tower and beautiful and lofty spire, and has received an additional number of sittings, of which 420 are free: the spire was damaged by lightning in 1814, and has been partly rebuilt. The chantries of St. Mary and the Holy Trinity, founded in this church, were endowed with houses and lands in the neighbourhood. There are places of worship for Independents, the Society of Friends, Wesleyan Methodists, and Roman Catholics. A free grammar school, situated in Bridge-street, was founded by the Rev. Thomas Allen, a celebrated mathematician in the sixteenth century: the management is vested in the Master, Fellows, and Scholars of Trinity College, Cambridge; fifteen scholars are instructed, and the master's salary is £13. 6. 8. per annum. A National school is supported by subscription. There are almshouses for twelve poor persons, with small endowments; and a fund, amounting to about £60 per annum, for apprenticing poor children. The poor law union of Uttoxeter comprises 16 parishes or places, containing a population of 12,837, according to the census of 1831; it is under the care of 22 guardians. Thomas Allen, the mathematician; Sir Simon Degge, the antiquary; and the distinguished Admiral Gardner, were natives of this place.

UXBRIDGE, a market-town and chapelry, and the head of a union, in the parish of HILLINGDON, hundred of ELTHORNE, county of MIDDLESEX, 15 miles (W. by N.) from London; containing 3043 inhabitants. The most ancient name of this place was *Oxebreuge*, or *Woxbrigge*, probably of Saxon origin, which has passed through the several variations of *Waxbridge, Woxbridge*, and *Oxbridge*, whence its present name. The town, which was probably founded about the time of Alfred, was surrounded by a ditch, and the whole site comprised about 85 acres: in feudal times it was an important station as a frontier town, and appears to have been fortified at an early period. It afterwards had a regular garrison; and, during the civil war of the seventeenth century, it was the scene of the memorable, but unsuccessful, negociation between the king and his parliament: sixteen commissioners on each side held a conference here, which commenced on the 30th of Jan., 1645, and continued about three weeks, in an ancient brick mansion, situated at the west end of the town, still designated as the Treaty House, which has undergone various alterations, and is now the Crown Inn: two of the principal rooms used on this occasion still present specimens of the ancient and curious wainscot, in a fine state of preservation. This edifice was occupied by the Earl of Northumberland, and a mansion in its vicinity was the temporary residence of the Earl of Pembroke: the royal commissioners selected the Crown Inn, which formerly stood opposite the present White Horse; and the parliamentary commissioners, the George, which, although materially diminished in size, yet remains. In 1647, the head-quarters of the parliamentary army were fixed here; and there was a garrison so late as 1689.

The town is situated on the high road from London to Oxford, called the Uxbridge road, occupying a gentle declivity on the banks of the river Colne: it is paved, lighted, and supplied with water from numerous wells, and consists of one principal street, about a mile in length, called London or High-street, which runs south-east and north-west, with another diverging from it, in the direction towards Windsor. The common, which is surrounded by rich and beautiful scenery, has been reduced by enclosures to a space of fifteen acres, called the Recreation Ground. Vine-street, branching to the south-east, defines the limits of what was formerly denominated the borough, in that direction; and although the town extends considerably beyond it, eastward, this part, which is called Hillingdon End, is within the parish of Hillingdon, and is neither paved nor lighted. The Grand Junction canal passes through the town. A library and reading-room, called the Uxbridge Book Society, and containing about 1300 volumes, is supported by subscription. An assembly-room is attached to one of the inns. The facilities afforded by the river Colne for the erection of water-mills, and of water-carriage by the canal to Paddington, and the Thames, have rendered Uxbridge remarkable for an extensive flour trade. At the western extremity there are three large flour-mills, and within three or four miles up and down the river, ten more, which are supposed, in the aggregate, to supply upwards of 3000 sacks of flour per week, a great part of it being sent to the metropolis: there are also two small breweries. South-east of the town is a fine soil of brick-earth, which extends several miles, and has been sold at £500 or £600 per acre; the burning of bricks on these fields employs several hundred persons. The general trade of the town is very considerable; and manufactories for the implements of husbandry and Windsor and garden chairs, are carried on to a considerable extent. The Colne is crossed by two bridges; over its principal branch is High-bridge, which is of brick, and was built about 50 years since, at the joint expense of the counties of Buckingham and Middlesex, replacing an ancient one which had existed from the time of Henry VIII.: over the smaller branch is a short bridge at Mercer's mill. There is likewise a bridge across the Grand Junction canal, on the banks of which are warehouses and wharfs for the convenience of trade. The market, granted in the reign of Henry II., is on Thursday, and is one of the largest markets in the kingdom for corn, which is pitched in considerable quantities: there is another market on Saturday, for meat, poultry, eggs, butter, &c. Fairs are held annually on March 25th, July 31st, Sept. 29th, and Oct. 11th; the two latter are now used as statute fairs. The old market-house, erected in 1561, was removed, by act of parliament, in 1785, and the present erected, at an expense of nearly £3000: it is a commodious building of brick, supported on 51 wooden columns, with spacious apartments used for various purposes.

Uxbridge was part of the manor of Colham till 1669, when it became a separate property, and in 1729 was vested, by purchase and survivorship, in Edmund Baker and Edmund Blount, who conveyed it to trustees and their successors, who must be inhabitants and housekeepers in the town, for charitable uses, certain rights being reserved to the lord of the manor of Colham. The town was anciently a borough, and until the close of the seventeenth century was governed by bailiffs: it is now under the superintendence of two constables, four headboroughs, and two ale-conners, who are elected annually, with a beadle and town-crier. In the 13th of Edward I. it was ordained that the high constable for the Uxbridge division, who generally resides in the town, should be chosen by the justices in quarter session. A petty session for the town and eleven adjoining parishes is held by the magistrates, on the first and third Mondays in every month; and there is a county court of requests, for debts under 40*s*., on the first Tuesday in every month. The town has been made a polling-place for the county.

The living is a perpetual curacy; net income, £111; patrons, Trustees of the late G. Townsend, Esq., who are to present a Fellow of Pembroke College, Oxford. The chapel, dedicated to St. Margaret, and built about 1447, on the site and partly from the materials of an old chapel, which stood here in the thirteenth century, is in the later English style, and contains 300 free sittings, the Incorporated Society having granted £200 towards defraying the expense: it is constructed of brick and flint, and consists of a chancel, nave, and two aisles, separated by octagonal pillars and pointed arches, with a low square tower at the north-west angle: in the interior are an ancient octagonal stone font, decorated with quatrefoils and roses, and several fine monuments. A church has been built at Uxbridge-Moor, containing 450 sittings, 250 of which are free, the Incorporated Society having granted £250 for this purpose. There are places of worship for Baptists, the Society of Friends, and Independents. The free school, for the education of boys, was founded in 1809, principally through the benevolent exertions of Thomas Truesdale Clarke, Esq.: it is held in a spacious apartment over the market-place, and is supported by donations and annual subscriptions, together with the interest of £600 given by Mr. J. Hall, to be divided between this school and the school of industry: the lords of the manor and borough lately subscribed £50 per annum. The school of industry for Uxbridge and its vicinity, established in 1809, by uniting two small schools on the improved system of education, is for girls only, and is held in a building in George-yard, erected in 1816, by subscription among the inhabitants. The profits arising from the manor and borough are, by the trustees, according to the deed, appropriated to the payment of £20 per annum amongst six poor men or women, inhabitants of the town, to be nominated by the overseers; and £10 a year by weekly payments, the donation of John Clarke; the remainder to be appropriated, at the discretion of a majority of the trustees, "for the benefit and advantage of the town of Uxbridge only." In consequence of an accident, which occasioned the death of a boy, as Lord Osselton's carriage was passing through the town, his lordship gave £100 for the purchase of land, directing the rental to be applied in apprenticing poor boys. In addition to these are several benefactions to the poor by different persons. The poor law union of Uxbridge comprises 10 parishes or places, under the care of 20 guardians, and contains, according to the census of 1831, a population of 12,663. About four miles from the town, at Breakspear, the seat of J. A. Partridge, Esq., some remains of Roman sepulchres have been discovered. Uxbridge gives the inferior title of Earl to the Marquess of Anglesey.

V

VANGE (*All Saints*), a parish, in the union of Billericay, hundred of Barstable, Southern Division of the county of Essex, 5 miles (N. E. by E.) from Horndon-on-the Hill; containing 165 inhabitants. This parish, which is about two miles and a half in length, and about two miles in breadth, is bounded on the south by a creek from the river Thames, and comprehends a low tract of marshy land, called Bower's Marsh, and a portion of Canvey island. The living is a discharged rectory, valued in the king's books at £14; present net income, £302; patron, Sir C. Smith, Bart. The church, a small ancient edifice, consisting of a nave and chancel, has been recently repaired and enlarged, by the addition of a gallery, at the expense of the parishioners, and 70 free sittings have been provided, the Incorporated Society having granted £35 towards defraying the expense. Here is a Sunday National school.

VAULTERSHOME, a tything, in the parish of Maker, hundred of Roborough, Roborough and Southern Divisions of the county of Devon, $1\frac{1}{2}$ mile (S. by W.) from Devonport; containing 1092 inhabitants.

VAUXHALL, county of Surrey—See LAMBETH.

VEEP, ST. (*St. Cyricius*), a parish, in the union of Liskeard, hundred of West, Eastern Division of the county of Cornwall, 3 miles (N. N. E.) from Fowey; containing 697 inhabitants. This parish is bounded on the west by the navigable river Fowey, on the north by the Leryn, and on the south by Penpol creek. In the civil war of the seventeenth century, the royalist cavalry were quartered here previously to the capitulation of the Earl of Essex, in the year 1644. A fair for cattle and sheep is annually held on the first Wednesday after June 16th. The living is a vicarage, valued in the king's books at £5. 0. $7\frac{1}{2}$.; present net income, £215; patron and impropriator, David Howell, Esq. There are some remains of the chapel of the small priory founded by one of the Earls of Cornwall as a cell to that of Montacute, in the county of Somerset, and dedicated to St. Cyric and St. Juliett. Walter de Exon, author of a history of Guy Earl of Warwick, in the latter part of the thirteenth century, was an inmate of this priory, in which he was interred.

VENN-OTTERY (*St. Gregory*), a parish, in the union of Honiton, hundred of East Budleigh, Woodbury and Southern Divisions of the county of Devon, $3\frac{1}{4}$ miles (S. W. by S.) from Ottery-St. Mary; containing 133 inhabitants. The living is a vicarage, endowed with the rectorial tithes, and annexed to that of Harpford. This parish is bounded on the east by the river Otter.

VERNHAM-DEAN, a parish, in the union of Andover, hundred of Pastrow, Andover and Northern Divisions of the county of Southampton, 8 miles (N. by W.) from Andover; containing 694 inhabitants. The living is annexed to the vicarage of Hurstbourn-Tarrant. The church occupies a lonely and romantic situation, in a delightfully sequestered spot. There is a place of worship for Wesleyan Methodists.

VERYAN (*St. Symphoriana*), a parish, in the union of Truro, Western Division of the hundred of Powder and of the county of Cornwall, 4 miles (S. by W.) from Tregoney; containing 1525 inhabitants. This parish is situated on a bay to which it gives name, in the English Channel, by which it is bounded on the south, and includes the fishing cove of Portloe, in the trade of which the inhabitants are principally employed. The living is a vicarage, valued in the king's books at £19; present net income, £339; patrons and appropriators, Dean and Chapter of Exeter. The church contains several neat monuments and an ancient font enriched with sculpture. A National school is supported by subscription, and the dividends arising from £562 three per cent. consols., purchased with the bequests of Richard Thomas and John Kempe, Esqrs., and of Admiral Kempe, Mrs. Sarah James, and others, and augmented with a bequest of £166. 13. 4., in the same stock, by Mrs. Collins. Two school-rooms, to one of which is attached a dwelling-house for the master and mistress, were built at the expense of the Rev. Jeremiah Trist, the late vicar. Within a mile of the church is a very large barrow, called "the Beacon," appearing to have been surrounded by a moat, from which a fine view is obtained over the surrounding country, particularly towards the west; and on the road to Gwenda is a singular mound on the side of a hill, surrounded by a fosse.

VIRGINSTOW (*St. Bridget*), a parish, in the union of the Holsworthy, hundred of Lifton, Lifton and Southern Divisions of the county of Devon; $6\frac{1}{4}$ miles (N. E. by N.) from Launceston; containing 136 inhabitants. The living is a discharged rectory, valued in the king's books at £5. 6. 8.; present net income, £103: it is in the patronage of the Crown.

VIRLEY (*St. Mary*), a parish, in the union of Lexden and Winstree, hundred of Winstree, Northern Division of the county of Essex, $8\frac{1}{4}$ miles (S. S. W.) from Colchester; containing 65 inhabitants. The living is a discharged rectory, valued in the king's books at £7. 13. 4.; present net income, £140: it is in the alternate patronage of the families of Abdy and Glover. Virley creek is navigable on the east to the North Sea.

VOWCHURCH (*St. Bartholomew*), a parish, in the union of Dore, hundred of Webtree, county of Hereford, $11\frac{1}{2}$ miles (W. by S.) from Hereford; containing 371 inhabitants. The living is a discharged vicarage, valued in the king's books at £5. 9.; present net income, £200; patron and appropriator, Prebendary of Putson Major in the Cathedral Church of Hereford. In the neighbourhood is an ancient square encampment.

VOWMINE, a township, partly in the parish of Clifford, hundred of Huntington, and partly in the parish of Dorstone, hundred of Webtree, county of Hereford, 4 miles (E. by S.) from Hay: the population is returned with the parishes.

W

WABERTHWAITE (*St. John*), a parish, in the union of Bootle, Allerdale ward above Derwent, Western Division of the county of Cumberland, $1\frac{1}{2}$ mile (E. S. E.) from Ravenglass; containing 139 inhabitants. This parish forms an inclined plane from the mountains to the river Esk, which bounds it on the north-west. The living is a discharged rectory, valued in the king's books at £3. 11. 8.; present net income, £131; patron, Lord Muncaster.

WACKERFIELD, a township, in the parish of Staindrop, union of Teesdale, South-Western Division of Darlington ward, Southern Division of the county palatine of Durham, 7 miles (S. W. by S.) from Bishop-Auckland; containing 112 inhabitants.

WACTON, a parish, in the union of Bromyard, hundred of Broxash, county of Hereford, $4\frac{1}{2}$ miles (N. W. by W.) from Bromyard; containing 112 inhabitants. The living is a perpetual curacy; net income, £61; patron, Vicar of Bromyard; appropriators, the Portionists of Bromyard.

WACTON-MAGNA (*All Saints*), a parish, in the union and hundred of Depwade, Eastern Division of the county of Norfolk, $1\frac{1}{4}$ mile (W. by S.) from Stratton-St. Mary; containing 242 inhabitants. The living is a discharged rectory, with the sinecure rectory of Wacton-Parva annexed, valued in the king's books at £5; present net income, £302; patron, Rev. Ellis Burroughes. The church contains a monument to one of the Knyvet family. A Sunday school is supported by the rector and the patron.

WACTON-PARVA (*St. Mary*), a parish, in the hundred of Depwade, Eastern Division of the county of Norfolk, $1\frac{1}{2}$ mile (S. W.) from Stratton-St. Mary. The living is a discharged sinecure rectory, annexed to the rectory of Wacton-Magna, and valued in the king's books at £2. 13. 4. The church is in ruins.

WADBOROUGH, a hamlet, in the parish of the Holy Cross, Pershore, union, and Upper Division of the hundred, of Pershore, Pershore and Eastern Divisions of the county of Worcester, $3\frac{1}{2}$ miles (W. N. W.) from Pershore; containing 198 inhabitants.

WADDESDON (*St. Michael*), a parish, in the union of Aylesbury, hundred of Ashendon, county of Buckingham, $5\frac{1}{2}$ miles (W. N. W.) from Aylesbury; containing 1734 inhabitants. The living is a rectory, in three portions, each valued in the king's books at £15, and all in the patronage of the Duke of Marlborough: the three portionists officiate alternately; net income of the first portion, £178; of the second, £202; and of the third, £152. There are places of worship for Baptists and Wesleyan Methodists. Eighteen boys, sons of poor parishioners, are gratuitously educated at a school in Waddesdon; and one boy, also the son of a parishioner, is apprenticed out of funds amounting to about £15 per annum, bequeathed by Lewis Fetto and John Beck. Almshouses for six aged widows were founded and endowed with a rent-charge of £30, by Arthur Goodwin, in 1645. William Turner, in 1784, bequeathed £3265. 11. 6., three per cent. consols., the dividends of which, £97. 19. 4., are distributed among the poor of the whole parish in money, clothing, food, &c. The milk of one cow, kept for that purpose by a tenant of the Duke of Marlborough, is given daily in rotation to twenty-two parishioners: it is called the Alms' Cow, but no document exists relating to the origin of the custom.

WADDINGTON (*St. Michael*), a parish, in the union, and within the liberty of the city of Lincoln, parts of Lindsey, county of Lincoln, $4\frac{1}{2}$ miles (S.) from Lincoln; containing, with Meer Hospital, 768 inhabitants. The living is a rectory, with Meer annexed, valued in the king's books at £20. 16. 8.; present net income, £566; patrons, Rector and Fellows of Lincoln College, Oxford. The church is principally in the Norman style. Connected with this parish was the ancient chapelry of Meer, of which the chapel, dedicated to St. James, is in ruins. There is a place of worship for Wesleyan Methodists. This parish participates, with others within the limits of the city and county of the city of Lincoln, in the benefits of the Blue-coat school and hospital founded, in 1602, by Richard Smith, M.D.; and the Jersey school, established in 1693, pursuant to the will of Henry Stone, Esq. A house of Knights Templars, with an hospital, near Danston, was founded in 1246 by Simon de Poppele, which latter was suffered to continue after the dissolution of similar establishments.

WADDINGTON, a chapelry, in the parish of Mitton, union of Clitheroe, Western Division of the wapentake of Staincliffe and Ewcross, West Riding of the county of York, $1\frac{3}{4}$ mile (N. W. by N.) from Clitheroe; containing 624 inhabitants, many of whom are employed in the spinning and manufacture of cotton.

The living is a perpetual curacy; net income, £122; patron and impropriator, T. Parker, Esq. The chapel, dedicated to St. Helen, is principally in the later style of English architecture. There is a place of worship for Wesleyan Methodists; also an hospital for twenty-four poor widows, founded by Robert Parker, with a chapel attached.

WADDINGWORTH (*St. Margaret*), a parish, in the union of Horncastle, Southern Division of the wapentake of Gartree, parts of Lindsey, county of Lincoln, 6 miles (W. by N.) from Horncastle; containing 63 inhabitants. The living is a discharged rectory, valued in the king's books at £7. 0. 10.: it is in the patronage of the Crown. The tithes have been commuted for a rent-charge of £132. 10., subject to the payment of rates, which on the average have amounted to £7; the glebe comprises 25 acres, valued at £17. 10. per annum.

WADEBRIDGE, a market-town, partly in the parish of St. Breock, hundred of Pyder, and partly in that of Egloshayle, hundred of Trigg, Eastern Division of the county of Cornwall, 5 miles (E. S. E.) from Padstow: the population is returned with the respective parishes. This place is chiefly remarkable for its noble bridge of 17 arches, nearly 320 feet long, over the navigable river Camel: it was erected about 1485, at a cost of £22,500, of which £16,500 was raised in shares, and £6000 by mortgage at £5 per cent.; and, in the reign of James I., it was made a county bridge: there are certain estates vested in trustees, with the rents of which, aided by tolls (from which the inhabitants of the two parishes are exempt), it is kept in repair. A railway has been completed from this place to Bodmin, with a branch to Ruthern Bridge: it is intended principally for the conveyance of a marine sand obtained at Padstow, which is in high repute as a manure. A discharging dock has also been constructed, which, with the quay, is capable of containing five vessels; and another dock, for the reception of sand barges, has been formed, at the sole expense of Sir W. Molesworth, Bart. The trade of the town principally consists in the exportation of corn, in vessels not exceeding 150 tons' burden. The market, which is of ancient establishment, though inconsiderable, is still held on Friday; and there are fairs on May 12th, June 22nd, and October 10th. There are places of worship for Independents and Wesleyan Methodists.

WADENHOE (*St. Michael*), a parish, in the union of Oundle, hundred of Navisford, Northern Division of the county of Northampton, 4¼ miles (S. W.) from Oundle; containing 252 inhabitants. The living is a rectory, valued in the king's books at £11; present net income, £186; patron and incumbent, Rev. Dr. Roberts. A day and Sunday school is supported by subscription.

WADHURST (*St. Peter and St. Paul*), a parish, in the union of Ticehurst, hundred of Loxfield-Pelham, rape of Pevensey, Eastern Division of the county of Sussex, 5 miles (S. E.) from Tonbridge Wells; containing 2256 inhabitants. The living is a vicarage, valued in the king's books at £5. 1. 0½.; present net income, £659; patrons, Warden and Fellows of Wadham College, Oxford; impropriator, S. Playsted, Esq. The church is partly in the early, and partly in the later, style of English architecture. There is a place of worship for Baptists; also a day and Sunday National school, commenced in 1824.

WADINGHAM, a parish, in the union of Caistor, Eastern Division of the wapentake of Manley, parts of Lindsey, county of Lincoln, 8½ miles (S. by W.) from Glandford-Bridge; containing 523 inhabitants. The living consists of the united rectories of St. Mary and St. Peter, with that of Snitterby annexed, valued in the king's books at £29. 6. 8.; present net income, £955: it is in the patronage of the Crown. There is a place of worship for Wesleyan Methodists; also a school, endowed with £25 per annum, for the instruction of 20 children. A new cut, or canal, called the river Ancholme, passes through the parish.

WADLEY, or LITTLEWORTH, a joint tything with Thrup, in the parish of Great Farringdon, hundred of Farringdon, county of Berks, 1¾ mile (N. E. by E.) from Great Farringdon: the population is returned with the parish.

WADSLEY, a hamlet, in the township and parish of Ecclesfield, wapentake of Strafforth and Tickhill, West Riding of the county of York, 3 miles (N. W.) from Sheffield: the population is returned with the parish. A church has been erected here, at the sole expense of the Misses Harrison, of Weston House; it is a neat building of stone, calculated to accommodate 1000 persons. There is a National school.

WADSWORTH, a township, in the parish of Halifax, union of Todmorden, wapentake of Morley, West Riding of the county of York, 7½ miles (W. N. W.) from Halifax; containing 5198 inhabitants, who are extensively employed in the manufacture of cotton and worsted goods.

WADWORTH (*St. Mary*), a parish, in the union of Doncaster, Southern Division of the wapentake of Strafforth and Tickhill, West Riding of the county of York, 5¼ miles (S.) from Doncaster; containing 690 inhabitants. The living is a discharged vicarage, valued in the king's books at £4. 2. 6.; present net income, £110; patron, W. Walker, Esq.; impropriators, several Proprietors of land. The church contains 84 free sittings, the Incorporated Society having granted £70 towards defraying the expense: in the interior are two altar-tombs of the Fitzwilliam family, with effigies of a knight and his lady. There is a place of worship for Wesleyan Methodists. A school is endowed by E. S. Foljambe, Esq., with six guineas annually.

WAGHEN, or WAWN (*St. Peter*), a parish, in the union of Beverley, partly within the liberty of St. Peter's, but chiefly in the Middle Division of the wapentake of Holderness, East Riding of the county of York; containing 338 inhabitants, of which number, 255 are in the township of Waghen, 5 miles (S. E. by E.) from Beverley. The living is a discharged vicarage, in the patronage of the Chancellor of the Cathedral Church of York, valued in the king's books at £7. 0. 10.; present net income, £49; appropriators, Dean and Chapter of York. The church is partly in the decorated style, with a tower of later date: there are three stalls in the chancel. Here is a National school.

WAINFLEET, a market-town, in the union of Spilsby, Marsh Division of the wapentake of Candleshoe, parts of Lindsey, county of Lincoln, 39½ miles (E. S. E.) from Lincoln, and 128 (N. by E.) from

London; containing 1795 inhabitants. In the time of the Romans, the whole province is said to have been supplied from this place with salt made from the sea-water; and a road across the fens, still called the Salters' road, is supposed to have been the Roman road between *Bannovallium* and *Lindum.* Wainfleet returned one burgess to the grand council summoned in the 11th of Edward III., and, in 1359, it supplied two ships of war for the armament prepared for the invasion of Brittany. The town is situated on a small creek in a marshy district; and, in consequence of the enclosure of the east fen, the waters of the haven have been carried off by a wide drain to Boston Scalf, which has so reduced them as to preclude the entrance of any but small craft. Previously to this event, it is believed that the town was situated higher up the creek, chiefly because the old church of All Saints stood at High Wainfleet, about a mile and a half from the town: it was taken down in 1820. The market is on Saturday; and fairs are held on the third Saturday in May for cattle, and October 24th for sheep, the latter being also a pleasure fair. This place is within the jurisdiction of the Bolingbroke and Horncastle court of requests, for the recovery of debts not exceeding £5.

The town comprises the parishes of *All Saints*, *St. Mary*, and *St. Thomas.* The living of All Saints is a rectory, valued in the king's books at £16. 3. 6½., and in the patronage of the Crown. The tithes have been commuted for a rent-charge of £406. 9. 7., subject to the payment of rates, which on the average have amounted to £68; the glebe comprises 23 acres, valued at £40. 10. per annum. The church has been rebuilt at an expense of £3000. The living of St. Mary's is a perpetual curacy, valued in the king's books at £8. 13. 4.; present net income, £201; patrons and impropriators, Governors of Bethlehem Hospital, London. The living of St. Thomas's is also a perpetual curacy, in the patronage of — Barnes, Esq. There are places of worship for the Society of Friends and Wesleyan Methodists. The free grammar school was founded in 1424, by William Patten, generally known as William of Waynflete, Bishop of Winchester, Lord High Chancellor of England in the reign of Henry VI., and founder of Magdalene College, Oxford, from the President and Fellows of which the master receives £11. 6. 8. per annum, and has, in addition, seventeen acres of land and a rent-free residence. Another school, called the "Bethlem Free school," is supported by the Governors of Bethlehem Hospital.

WAITBY, a township, in the parish of KIRKBY-STEPHEN, East ward and union, county of WESTMORLAND, 1¾ mile (W.) from Kirkby-Stephen; containing 41 inhabitants. A free school was erected in 1680, by James Highmore, citizen of London, who endowed it with £400, now producing £40 per annum, for children of this township and of Smardale; he also bequeathed £5. 5. annually to the poor of Waitby.

WAITH (*ST. MARTIN*), a parish, in the union of LOUTH, wapentake of BRADLEY-HAVERSTOE, parts of LINDSEY, county of LINCOLN, 6¾ miles (S. by E.) from Great Grimsby; containing 31 inhabitants. The living is a discharged vicarage, valued in the king's books at £2. 14. 2.; present net income, £86: the patronage and impropriation belong to William Haigh, Esq., and Mrs. Haigh.

WAKEFIELD (*ALL SAINTS*), a newly-enfranchised borough, market-town, and parish, and the head of a union, in the Lower Division of the wapentake of AGBRIGG, West Riding of the county of YORK; containing 24,538 inhabitants, of which number, 12,232 are in the town of Wakefield, 32 miles (S. W. by W.) from York, and 178 (N. N. W.) from London. The discovery of many coins and other relics of the Romans, in the neighbourhood, indicates its existence in the time of that people. In the reign of Edward the Confessor, it formed part of the royal demesne, and was subsequently transferred to the Conqueror: in Domesday book it is denominated *Wachefeld.* This was the scene of the celebrated battle, at the close of the year 1459, between Richard Duke of York and Margaret of Anjou, queen of Henry VI., in which the former lost his life; and the spot of ground where he was buried, a short distance from the town, is still parted off from the adjoining field; his gold ring was found here some years ago. Wakefield again suffered from the calamities of the civil war between Charles I. and the parliament, falling, at different periods, into the hands of each party, according to the various chances of war. The manor, or lordship, which is one of the largest in the kingdom, extending from east to west a distance of more than 30 miles, and now comprising a population of upwards of 120,000, was granted by Henry I., about the year 1107, to William Earl of Warren, in whose family it continued until the reign of Edward III., when, by default of heirs, it lapsed to the Crown, and so remained until granted by Charles I. to Henry Earl of Holland; and, after passing through several other families, it was, in 1700, purchased by the Duke of Leeds, to whose descendant, the present duke, it still belongs.

Arms.

The town is principally situated on the side of an eminence sloping to the river Calder, in the midst of a fertile and picturesque country, and consists of spacious and regular streets, with many well-built and handsome brick houses: it is paved and flagged, lighted with gas, and well supplied with water. Very great improvements have been made, of late years more especially, on its northern side, where some handsome rows of houses have been erected, which, with a few detached mansions standing amidst shrubberies and pleasure-grounds, form a great ornament to this part of the town, which is called St. John's. A handsome building, of the Ionic order, situated in Wood-street, near the court-house, has been erected by subscription, and contains a library and news-room, the upper part being used for concerts, assemblies, and other public amusements. A Literary and Philosophical Society, established in 1827, meets every fortnight at the court-house; and there are also an Horticultural Society, a museum, and public baths, over the latter of which are spacious assembly-rooms; and a Temperance Hall has been lately built. The theatre, in Westgate, was erected by the late celebrated Tate Wilkinson, Esq., and is usually opened in Aug. by the York company. The river, which was made navigable in 1698, is crossed by a handsome stone bridge of nine arches, built in the reign of Edward III., on the eastern side of which is a chapel, supposed to have been founded by that monarch, dedicated to the Virgin Mary, and endowed with £10 per annum, for two chaplains to perform divine service. The present structure was rebuilt by Edward IV., in memory of his father, the Duke of York, and his followers, who fell with him at the battle of Wakefield; it is about ten yards in length and six in width, and is in the later style of English architecture: the eastern window, overhanging the river, is ornamented with rich and delicate tracery, and the western front has much architectural decoration; it is divided by buttresses into compartments, with lofty pedestals, and pointed arches in relief, the spandrels being covered with sculptured ornaments; over these is an entablature, with five shorter compartments, in relief, representing scriptural subjects, the whole surmounted by battlements: the revenue ceased at the dissolution of monastic establishments, and the chapel has been long desecrated, being now used as a counting-house by a corn-factor.

The manufacture of woollen cloth and worsted yarn was formerly carried on to a great extent, insomuch that Leland, in speaking of the town, says, "it standeth now al by clothying;" but these branches of trade have greatly declined, and corn and wool are now the staple commodities. The Tammy Hall, a building of two stories, about 70 yards long and 10 broad, erected by subscription many years since, for the sale of the lighter woollen stuffs, has been converted into a manufactory for stuff pieces by the use of power-looms, the shares of the original proprietors having been bought up. There are also, in the town and its vicinity, large dye-houses, several roperies, iron-foundries, breweries, copperas-works, a bleach-yard, starch-works, several malting establishments, and boat and sloop building yards. Coal is procured in great abundance in the surrounding country, and is brought to the town by railways from the collieries, and conveyed in barges down the Calder and up the Ouse to York, or by the Humber to Hull. The trade in corn has greatly increased of late years, and warehouses for storing it have been built on a most extensive scale. At the foot of the bridge is the soke mill, where persons living in the soke of Wakefield, which comprehends Stanley, Sandal, Alverthorpe, Ossett, Horbury, and Crigglestone, are obliged to have their corn ground. Much barley is grown in the neighbourhood, and a considerable quantity is here converted into malt: great quantities of wool are sold, being brought from distant parts of the country, to be disposed of to the neighbouring manufacturers. The Aire and Calder Navigation Company have their principal establishment and wharf near the bridge, whence fly-boats leave for Huddersfield every day; and, by means of their navigation, and the Barnsley and other canals connected with it, a direct line of communication is opened with Hull, Lincolnshire, and Lancashire, which affords great facilities to the trade of the town. The North Midland or Leeds and Derby railway, and the Manchester and Leeds railway, pass near the town. The market, on Friday, is well supplied with provisions: the market cross, built by subscription more than 120 years since, is a handsome structure of the Doric order, consisting of an open colonnade supporting a dome; a spiral staircase leads to a spacious room, lighted by a lantern at the top, in which business relating to the town is generally transacted: in consequence of the confined space of the market-place, the corn market has been removed to the broad street called Westgate, where a new and elegant corn exchange has been recently built on an extensive scale. There is also a very large market for fat cattle and sheep every alternate Wednesday, commenced in 1765, which is attended by graziers and butchers from a very considerable distance; it is held in a spacious area on the west side of the town, suitably fitted up with pens, and comprising about 3½ acres. Fairs, chiefly for horses, horned cattle, and pedlery, are held on July 4th and 5th, and Nov. 11th and 12th. The town is under the superintendence of a chief constable, who is appointed by the inhabitants, and sworn by the steward of the lord of the manor, at a court leet held at the moot-hall, in Kirk-gate. By the act of the 2nd of William IV., cap. 45, Wakefield has been constituted a borough, with the privilege of returning one member to parliament, the right of election being in the £10 householders: the limits of the borough comprehend by estimation 1036 acres, and are minutely described in the Appendix, No. II.: the returning officer is appointed annually by the sheriff. The Christmas quarter sessions for the West Riding are holden, by adjournment from Wetherby, at the court-house, a handsome and appropriate edifice, erected in 1806; and a petty session for the district is held weekly, by the county magistrates, on Monday. The manor court, for petty causes and the recovery of debts under £5, is held by the steward, once in three weeks, at the moot-hall. The register-office, established in 1704, a handsome edifice of stone, which was enlarged by the erection of a fire-proof wing in 1829, and the office of clerk of the peace for the West Riding, are both in this town; and it has been made a polling-place for that division of the county. The house of correction for the West Riding is an extensive pile of building, situated near the bottom of Westgate, and constructed on the improved plan of county prisons, comprising 36 work-rooms, 19 day-rooms, 19 airing-yards, separate apartments for the sick, a tread-mill for grinding corn, a chapel, a school for juvenile offenders, and 307 separate sleeping cells: the prisoners are employed in weaving coarse cloth, calico, and linsey.

The living is a vicarage, valued in the king's books at £29. 19. 2.; present net income, £537: it is in the patronage of the Crown; impropriators, Heir of the late Sir J. Ramsden, Bart., and others. The church is a spacious and handsome edifice of early English architecture, erected in the reign of Henry III., and occupying an elevated site in the centre of the town, but, from the repairs and improvements it has undergone, little of the original character remains. It consists of a nave, separated from the aisles by two rows of clustered columns supporting pointed arches, and from the chancel by a lofty screen: the pulpit and reading-desk are of carved oak; and the font, which bears the date 1611, and the initials of Charles II., was replaced in its present situation at the west end of the nave (from which it had been removed), in 1821, with the addition of a beautifully wrought canopy. The square tower is adorned with battlements and pinnacles, and surmounted by an octagonal spire, the height of both being about 237 feet, exceeding that of any other in the county: in 1715, the vane and about one-third of the spire were blown down and only partially rebuilt. In 1802, its dilapidated condition exciting alarm as to its safety, it was secured by iron bands; and, in 1823, a portion of it was taken down and rebuilt, so that it is now perfectly secure. There are two lectureships; that in the afternoon, founded in 1652, under the will of Lady Camden, with an income of £100 per annum, is in the gift of the Master and Wardens of the Mercers' Company; and that in the evening, established in 1801, is supported by voluntary subscriptions, and in the patronage of seven trustees, of whom the vicar is one. The church dedicated to *St. John*, standing in that part of the town called St. John's, was commenced in 1792, and completed in 1795, at an expense of about £10,000: the site, and £1000 for the support of the officiating minister, were bequeathed by Mrs. Newstead. It was made parochial, jointly with the church of All Saints, by an act obtained in 1815; net income, £118; patron, Vicar of All Saints, who has also the patronage of the chapelries of Horbury, Alverthorpe, and Stanley, to the two latter of which districts have been assigned under the 16th section of the act of George III. There are two places of worship for Independents, three for Wesleyan Methodists, and one each for Baptists, the Society of Friends, Primitive Methodists, Unitarians, and Roman Catholics.

The free grammar school was founded in 1592, by Queen Elizabeth, and is endowed with property given by numerous benefactors, and producing rather more than £320 per annum; about 40 scholars receive a classical education. It is under the government of fourteen trustees, who are a corporate body, and appoint the two masters: the first with a salary of £160, and the other with one of £80, per annum. There is also a writing school, where boys are instructed in writing and arithmetic, on paying a certain sum quarterly to the master. The children of all residents, both in the town and parish, are admissible, and are eligible to the several exhibitions from this school to the Universities; two of which, founded by Thomas Cave, are to Clare Hall, Cambridge; one, founded by Lady Elizabeth Hastings, to Queen's College, Oxford: and three, founded by John Stone, to either of the Universities, each of the value of about £50 per annum. Of the candidates, preference is given, first, to the natives of the town; next, to those born in the parish; and, in failure of these, to residents; but each candidate must have been at the school for three years. The school-room is a commodious building, with a good library attached to it. Several eminent persons have been educated here, amongst whom are Richard Bentley, D.D., who was born in the neighbourhood; Dr. John Potter, Archbishop of Canterbury, a distinguished author, born here in 1674; and Dr. John Radcliffe, the munificent benefactor to the University of Oxford, and founder of the Radcliffe library, who was also a native of the town. The West Riding Proprietary School is a large and handsome edifice, built in 1833, by a company of proprietors, with a capital of £15,000, raised in 600 shares of £25 each; it is attended by about 200 students, and is under the management of a president, vice-presidents, and directors, and the superintendence of a principal, vice-principal, and several other masters. The Green-coat charity school was founded about 1707, by the trustees of the poor, from the charity estates; and the present income, amounting to upwards of £600 per annum, and subject to certain charges, arises from the ground on which the fortnight cattle market is held, together with various donations, including a grant by John Storie, in 1674, of certain lands, now by exchange, and the appropriation of the site for the cattle market, producing more than £500 per annum: in this school 70 boys and 50 girls are clothed and instructed. There are also National and Lancasterian schools, a school of industry, and an infants' school, partly supported by subscription.

The almshouses, in Almshouse-lane, founded by Cotton Horne, in 1646, and endowed with lands producing nearly £300 per annum, are appropriated to ten poor women. Adjoining are almshouses, endowed by William Horne, in which are ten poor men. There are also almshouses at Brooksbank, founded by Leonard Bate, for five poor widows of this parish. The management of all these establishments is vested in the governors of the grammar school, who have also the distribution of a bequest made, in 1722, by John Bromley, of property now producing upwards of £700 per annum, which is applied to the clothing and apprenticing of poor boys, and the relief of poor housekeepers; and of various other benefactions and bequests for the relief of the poor, producing upwards of £800 per annum. The West Riding pauper lunatic asylum, about a mile north-east of the town, was erected about 1817, and is calculated to contain nearly 400 patients: it is a handsome and commodious edifice, and is conducted in a manner admirably adapted both for bodily comfort and mental relief of its inmates. A dispensary and fever ward are supported by voluntary contributions. The poor law union of Wakefield comprises 17 parishes or places, under the care of 22 guardians, and contains, according to the census of 1831, a population of 37,032. A mineral spring at Stanley, and another at Horbury, each within about two miles of the town, possess medicinal qualities somewhat similar to those of the waters at Harrogate or Cheltenham. In addition to those natives of the town who are mentioned as having been educated at the grammar school, the following were also born here: Dr. Thomas Zouch, a learned divine; Joseph Bingham, M.A., author of "*Origines Ecclesiasticæ*;" and Dr. John Burton, author of the "*Monasticon Eboracense*."

WAKELEY, an extra-parochial liberty, formerly a distinct parish, in the hundred of EDWINSTREE, county of HEREFORD, 2 miles (S. W.) from Buntingford; containing 7 inhabitants.

WAKERING, GREAT (*St. Nicholas*), a parish, in the union and hundred of ROCHFORD, Southern Division of the county of ESSEX, 4½ miles (E. N. E.) from Southend; containing 834 inhabitants. This parish is situated near the mouth of the Thames, where is a small but convenient haven, and is traversed by the road to Foulness Island. The living is a vicarage, valued in the king's books at £20. 13. 4.; patron, Bishop of London; impropriator, T. Clough, Esq. The impropriate tithes have been commuted for a rent-charge of £640, and the vicarial for £290, subject to the payment of rates, which on the average have respectively amounted to £120 and £49; the impropriate glebe comprises 60 acres, and the vicarial, 2 acres. The church is a neat substantial structure, with a tower and spire. There is a place of worship for Independents; a National school has been established; and there are almshouses for six inmates, but without endowment.

WAKERING, LITTLE (*St. Mary*), a parish, in the union and hundred of ROCHFORD, Southern Division of the county of ESSEX, 4½ miles (N. E. by E.) from Southend; containing 297 inhabitants. This parish, which is bounded on the south by the parish of Great Wakering, includes the Island of Potten, which is formed by the river Bromhill and the haven of Wakering, and comprises a tract of very rich land. The living is a vicarage, valued in the king's books at £12; present net income, £190; patrons and impropriators, Governors of St. Bartholomew's hospital, London. The church is a small ancient edifice with a tower, on which are the armorial bearings of Bishop Wakering.

WAKERLEY (*St. Mary*), a parish, in the union of UPPINGHAM, hundred of CORBY, Northern Division of the county of NORTHAMPTON, 6¾ miles (E.) from Uppingham; containing 218 inhabitants. The living is a rectory, valued in the king's books at £11. 12. 6.; present net income, £100; patron, Marquess of Exeter. The Rev. Matthew Snow, in 1797, bequeathed £100 three per cent. consols., the interest to be expended in Bibles and Prayer-books for distribution among poor children of the parish.

WALBERSWICK (*St. Andrew*), a parish, in the union and hundred of BLYTHING, Eastern Division of the county of SUFFOLK, 1¾ mile (S. W. by S.) from Southwold; containing 279 inhabitants. The living is a perpetual curacy, held with that of Blythburgh; net income, £41; patron, Sir C. Blois, Bart. The church is in ruins, but a part of the south aisle and nave has been fitted up for the performance of Divine service; from the extent of the ruins, it is probable that this place was formerly of much greater importance than at present. The navigable river Blyth runs through the parish, and falls into the North Sea on its east side.

WALBERTON (*St. Mary*), a parish, in the union of WEST HAMPNETT, hundred of AVISFORD, rape of ARUNDEL, Western Division of the county of SUSSEX, 3½ miles (W. S. W.) from Arundel; containing 616 inhabitants. The living is a discharged vicarage, with that of Yapton united, valued in the king's books at £10. 19. 2.; present net income, £468; patron and appropriator, Bishop of Chichester. The church is principally in the early style of English architecture. John Nash, in 1732, bequeathed a house and land, with a rent-charge of £12, for teaching eighteen poor children; and a National school has been established in connection with this charity.

WALBURN, a township, in the parish of DOWNHOLME, union of RICHMOND, wapentake of HANG-WEST, North Riding of the county of YORK, 5 miles (S. W.) from Richmond; containing 26 inhabitants.

WALBY, a township, in the parish of CROSBY-UPON-EDEN, ESKDALE ward, Eastern Division of the county of CUMBERLAND, 4 miles (N. E. by N.) from Carlisle; containing 52 inhabitants. This village appears to have derived its name from its situation near the Roman wall.

WALCOT (*St. Nicholas*), a parish, in the union of SLEAFORD, wapentake of AVELAND, parts of KESTEVEN, county of LINCOLN, 1½ mile (N. W.) from Falkingham; containing 183 inhabitants. The living is a vicarage; net income, £159; patron and impropriator, Sir G. Heathcote, Bart. The church is principally in the decorated style of English architecture, with a tower surmounted by a fine crocketed spire: the clerestory is in the later style. On the edge of the fens is a powerful mineral spring. The monks of Sempringham had formerly a prison here.

WALCOTE, a hamlet (formerly a chapelry), in the parish of MISTERTON, hundred of GUTHLAXTON, Southern Division of the county of LEICESTER, 1¾ mile (E. by S.) from Lutterworth: the population is returned with the parish. The chapel, dedicated to St. Martin, has been destroyed.

WALCOTT, a chapelry, in the parish of BILLINGHAY, union of SLEAFORD, First Division of the wapentake of LANGOE, parts of KESTEVEN, county of LINCOLN, 8¾ miles (N. E. by N.) from Sleaford; containing 514 inhabitants. The living is a perpetual curacy, annexed to the vicarage of Billinghay. The chapel is dedicated to St. Oswald. There is a place of worship for Wesleyan Methodists.

WALCOTT (*All Saints*), a parish, in the hundred of HAPPING, Eastern Division of the county of NORFOLK, 5¼ miles (E. by N.) from North Walsham; containing 129 inhabitants. The living is a perpetual curacy, valued in the king's books at £30; present net income, £43; patron and appropriator, Bishop of Norwich.

WALCOTT (*St. Swithin*), a parish, in the union of BATH, partly within the city of BATH, and partly in the hundred of BATH-FORUM, Eastern Division of the county of SOMERSET; containing 26,023 inhabitants. This parish includes all those parts of the city of Bath lying on the north, north-east, and north-west sides of the parish of St. Michael; also some handsome ranges of buildings on the declivities of Lansdown and Beacon hills. A church in the later style of English architecture, with a tower, was erected in 1831, at an expense of £6383. 9. 6., by the act of the 58th of George III.; and another church has been subsequently erected. For a more detailed account see the article on BATH.

WALCOTT, a joint hamlet with Membris, in the parish of HOLY CROSS, PERSHORE, union, and Upper Division of the hundred, of PERSHORE, Pershore and Eastern Divisions of the county of WORCESTER, 2 miles (N. by W.) from Pershore; containing, with Membris, 375 inhabitants.

WALDEN, a joint township with Burton, in the parish of AYSGARTH, wapentake of HANG-WEST, North Riding of the county of YORK, 10 miles (W. S. W.) from Middleham: the population is returned with Burton.

WALDEN, KING'S (*St. Mary*), a parish, in the union of HITCHIN, hundred of HITCHIN and PIRTON, county of HEREFORD, 4¼ miles (S. S. W.) from Hitchin; containing 1004 inhabitants. The living is a donative curacy, in the patronage of W. Hale, Esq.; net income, £57. On the north side of the chancel is a chapel, the burial-place of the family of Hale, erected by William Hale, Esq., who died in 1648. About £12 per annum, arising from the bequests of Richard Hale, in 1616, and William Smith, in 1771, is distributed among the necessitous poor.

WALDEN, ST. PAUL'S (*All Saints*), a parish, in the union of HITCHIN, hundred of CASHIO, or liberty of ST. ALBANS, county of HERTFORD, 5¼ miles (N. N. W.) from Welwyn; containing 1058 inhabitants. The living is a vicarage, valued in the king's books at £10; present net income, £142; patrons and appropriators, Dean and Chapter of St. Paul's, London. There are places of worship for Baptists and Independents; and a National school has been established. This parish is entitled to a share of Henry Smith's bequest, and of the charity of Thomas Chapman, of Stevenage.

Corporation Seal.

WALDEN, SAFFRON, (*St. Mary*), an incorporated market-town and parish, possessing separate jurisdiction, and the head of a union, locally in the hundred of Uttlesford, Northern Division of the county of Essex, 27 miles (N. N. W.) from Chelmsford, and 40 (N. N. E.) from London; containing 4762 inhabitants. The name of Walden is said to be derived from the Saxon words *Weald* and *Den*, signifying a woody valley. At a later period the place was called *Waldenburgh;* and in the reign of Stephen, when Geoffrey de Mandeville, Earl of Essex, procured from the Empress Maud the grant of a market, previously held at Newport, the town took the appellation of *Cheping-Walden.* The present designation owes its origin to the culture of saffron in the neighbourhood, which is supposed to have been introduced into England in the time of Edward III., but has long since been discontinued: the device of the seal of the corporation is a rebus on the name, being *three saffron flowers walled in.* The Earl of Essex, above mentioned, was the grandson of Geoffrey de Mandeville, a Norman chief, and one of the most distinguished followers of William I.: he founded a Benedictine priory, near the south-western extremity of the parish, which was richly endowed, and, in 1190, converted into an abbey; its revenue, at the time of the suppression, amounted, according to Speed, to £406. 5. 11. In 1537, the abbey was surrendered, with all its possessions, to the king, who granted them to Sir Thomas Audley, K.G., afterwards Lord Chancellor, and created Baron Audley of Walden. Upon the site of the monastic buildings, and partly out of the ruins, Thomas, first Earl of Suffolk, in 1603, erected a stately fabric, which he called Audley End, in honour of his maternal grandfather, the chancellor; but of this magnificent house, which occupied thirteen years in completing, and was considered the largest mansion within the realm, one court only remains, and even this comparatively small portion of the original building forms a splendid residence. Upon the death of Henry, tenth Earl of Suffolk, in 1745, without issue, the Audley End estates were divided between George William, Earl of Bristol, (who had a half share) and Elizabeth, Countess of Portsmouth, and Ann Griffin, wife of William Whitwell, Esq., (who had a quarter share each) as representatives of the daughters and coheirs of James, third Earl of Suffolk. Lady Portsmouth gave her share of the property, together with the house, in 1762, to her nephew, Sir John Griffin Griffin, K.B., who, in 1784, established his claim, in the female line, to the ancient barony of Howard de Walden; and, dying in 1797, bequeathed his estates to Richard Aldworth Griffin, Lord Braybrooke, the father of the present possessor of Audley End, who has greatly improved the estate.

The town, which is beautifully situated in a district abounding with interesting scenery, contains several good streets, and a spacious market-place, in which there is a neat town-hall. The old houses are principally built of lath and plaster, and some of them are very ancient, but the more modern ones are of brick, and the recent improvements have materially altered the general appearance of the place. A bridge has been built over the Slade, and some pleasant promenades have been opened for the use of the inhabitants. A Scientific and Literary Institution has been established, and there are horticultural and other societies. The situation of the town is thus emphatically described by Dr. Stukeley: "A narrow tongue of land shoots itself out like a promontory, encompassed with a valley, in the form of a horse-shoe, enclosed by distant and delightful hills. On the bottom of the tongue, towards the east, stand the ruins of the castle, and on the top, or extremity, the church, the greater part of which is seen above the surrounding houses." The trade in malt and barley is very considerable. The market is on Saturday. Fairs are held annually on Mid-Lent Saturday and November 1st; and a fair for sheep and lambs is also held on the 3rd and 4th of August, which is much frequented. By the first charter of incorporation, in 1549, the government of the town was vested in twenty persons, a treasurer and two chamberlains being chosen annually from this number; but the corporation was re-modelled by the charter of William and Mary, and consisted of a mayor, recorder, twelve aldermen, deputy-recorder, coroner, town-clerk, and other officers. The mayor and his immediate predecessor, and the two senior aldermen, were magistrates *ex officio:* the county magistrates had concurrent jurisdiction. The borough court of sessions appears to have had the power of inflicting capital punishment, which was exercised more than once in the seventeenth century. Under the act of the 5th and 6th of William IV., cap. 76 (for an abstract of which see the Appendix, No. I.), the corporation at present consists of a mayor, four aldermen, and twelve councillors: the mayor, ex-mayor, and recorder, are justices of the peace, and two others have been appointed under a separate commission. The sessions are held quarterly, under a grant from William IV.; and a court of record is also held, every three weeks, for the recovery of debts and the determination of pleas to any amount, at which the recorder presides. The courts leet and baron for the manors of Brook and Chipping-Walden, belonging to the owner of Audley End, take place at stated times. Constables are appointed by the town council under the Municipal Reform Act. The magistrates for the division also hold their sessions in the town, once a fortnight; and it has been made a polling-place for the northern division of the county.

The living is a vicarage, valued in the king's books at £33. 6. 8.; present net income, £237; patron and impropriator, Lord Braybrooke. The church, which was erected in the reigns of Henry VI. and VII., is a spacious and elegant structure, in the later English style, with a lofty square embattled tower, strengthened with double buttresses of five stages, terminating in minarets rising above the battlements, and surmounted by a lofty crocketed spire of recent erection. The western front is of imposing grandeur, having over the central doorway a handsome window of three, and at the extremities of the side aisles windows of five, lights of rich and elegant design, and at the angles of the building enriched buttresses terminating in crocketed pinnacles; the exterior walls are embattled and pinnacled, and the interior of the church is beautifully arranged. The nave is lighted by a range of clerestory windows, and is separated from the aisles by clustered columns supporting the roof, which, like that of the chancel and aisles, is richly groined: the altar is embellished with a fine painting of the Holy Family, after Correggio. The churchyard has recently been enlarged. The middle and south chancels were erected by Chancellor Audley, and the north by the inhabitants, aided by John Leche, who was vicar from 1489 to 1521, and whose tomb may still be seen near the north chancel door: the chancellor's monument, of black marble, is placed under the east window of the south chancel. There are places of worship for General Baptists, the Society of Friends, Independents, Wesleyan Methodists, and Unitarians. The school, in which the classics were formerly taught, owed its foundation to John Leche, before mentioned, and his sister, Johane Bradbury. The learned Sir Thomas Smith, Secretary to Edward VI., a native of Walden, is said to have received his early education here, and through his interest the school was advanced to a royal foundation. The income of the original foundation is now reserved with the intention of appropriating it to its legitimate purpose. A charity school was established by subscription, and subsequently endowed with benefactions producing £100 per annum; the children are now taught in the National school. There is also a school for 100 girls, similarly conducted, near the eastern end of the church, adjoining the Bury, or Castle Hill; and another school for boys, upon the plan of the British and Foreign school Society, has been erected in East-street. A range of almshouses was built in 1829, at the south-west end of the town, to replace those founded by Edward VI., for the reception of sixteen decayed housekeepers of each sex, which had been long considered too much dilapidated to admit of reparation. The elevation of the new buildings, which cost nearly £5000, is handsome and appropriate, and adds much to the general appearance of the town, as well as to the comforts of the inmates: the income of this charity exceeds £900 per annum; the number of inmates is about 30. These and the remaining charities belonging to the parish are all properly administered. This was the first town in which the system of allotment for the poor and working classes was introduced; about 40 acres are thus appropriated, much to the benefit of nearly 800 of the population. It is also the head of a union, comprising 24 parishes, with a population of 17,987, according to the census of 1831, and under the care of 31 guardians; a spacious and commodious workhouse has been erected. The late Jabez Gibson, Esq., in 1836, sunk an Artesian well to the depth of 1004 feet: a fine spring was discovered at the depth of 400 feet, which rose to within 18 feet of the surface; the chief strata were chalk marl, with many shells. Between the town and Audley End Park are the remains of an old embankment, called "The Battle Ditches," respecting which there is no clear or satisfactory tradition. Dr. Stukeley found the south bank to be 730 feet long, 20 feet high, 50 broad at the base, and 8 at the top: the length of the western bank is 588 feet: both banks and ditches are well preserved and extremely bold. The ruins of the castle, erected soon after the Conquest, by Geoffrey de Mandeville, are only remarkable for the thickness of the walls and the rude character of the building. These ruins, and the hill on which they stand, are held by trustees, under lease from Lord Braybrooke, for the benefit of the town. A museum was built within the grounds, in 1835, which contains many rare specimens of Zoology and other departments of natural history; a spacious hall has been added to the building by Lord Braybrooke for the agricultural society of the town and vicinity. The hamlet of Little Walden, containing a few straggling houses, stands a mile and a half from the town, on the Linton road. The parish contains 7296 acres. Lord Thomas Howard, afterwards Earl of Suffolk, in 1597, took his title of baron from this town, which has descended to the present Lord Howard de Walden, in right of his mother, who was granddaughter to Frederic, fourth Earl of Bristol; the barony, being a female honour, has, at different periods, been disunited from each of the above earldoms.

WALDEN-STUBBS, a township, in the parish of Womersley, Lower Division of the wapentake of Osgoldcross, West Riding of the county of York, 7½ miles (S. E. by E.) from Pontefract; containing 139 inhabitants.

WALDERSHARE (*All Saints*), a parish, in the union and hundred of Eastry, lathe of St. Augustine, Eastern Division of the county of Kent, 4½ miles (N. by W.) from Dover; containing 67 inhabitants. The living is a discharged vicarage, valued in the king's books at £5. 8.; present net income, £133; patron and appropriator, Archbishop of Canterbury. The church contains some handsome monuments. A fair for toys and pedlery is held on Whit-Tuesday. Here is a National school for this and the adjoining parishes, supported by the Earl of Guilford, whose seat is in this parish, and who has erected a handsome school-house, with residences for the master and mistress.

WALDINGFIELD, GREAT (*St. Lawrence*), a parish, in the union of Sudbury, hundred of Babergh, Western Division of the county of Suffolk, 3¼ miles (N. E. by E.) from Sudbury; containing 679 inhabitants. The living is a rectory, valued in the king's books at £21. 6. 8.; patrons, Master and Fellows of Clare Hall, Cambridge. The tithes have been commuted for a rent-charge of £710, subject to the payment of rates, which on the average have amounted to £210; the glebe comprises 20 acres, valued at £35 per annum. Three schools are partly supported by private contributions. Roger Spencer, Lord Mayor of London in 1594, was a native of this parish.

WALDINGFIELD, LITTLE (*St. Lawrence*), a parish, in the union of Sudbury, hundred of Babergh, Western Division of the county of Suffolk, 4½ miles (N. E. by E.) from Sudbury; containing 403 inhabitants. The living is a discharged vicarage, valued in the king's books at £4. 18. 11½.; present net income, £112; patron, incumbent, and impropriator, Rev. B. B. Syer. There is a place of worship for Wesleyan Methodists. Holbrook Hall, in this parish, is the seat of a branch of the Hanmer family.

WALDLEY, a village, in the parish of Marston-Montgomery, hundred of Appletree, Southern Division of the county of DERBY: the population is returned with the parish.

WALDRIDGE, a township, in the parish and union of CHESTER-LE-STREET, Middle Division of CHESTER ward, county palatine of DURHAM, 5½ miles (N. by W.) from Durham; containing 104 inhabitants.

WALDRINGFIELD (*All Saints*), a parish, in the union of WOODBRIDGE, hundred of CARLFORD, Eastern Division of the county of SUFFOLK, 3½ miles (S. by E.) from Woodbridge; containing 166 inhabitants. The living is a discharged rectory, valued in the king's books at £4. 17. 11.; present net income, £187; patron, Rev. William Edge. There is a place of worship for Baptists.

WALDRON (*All Saints*), a parish, in the union of UCKFIELD, hundred of SHIPLAKE, rape of PEVENSEY, Eastern Division of the county of SUSSEX, 5¾ miles (E. S. E.) from Uckfield; containing 997 inhabitants. The living is a rectory, valued in the king's books at £13. 4. 7.; present net income, £455; patrons, Rector and Fellows of Exeter College, Oxford. The church is partly in the early, and partly in the later, style of English architecture: a gallery has been erected, and 50 free sittings provided, the Incorporated Society having granted £30 in aid of the expense. There is a place of worship for Wesleyan Methodists.

WALES (*St. John*), a parish, in the union of WORKSOP, partly within the liberty of ST. PETER'S, East Riding, but chiefly in the Southern Division of the wapentake of STRAFFORTH and TICKHILL, West Riding of the county of YORK, 8 miles (S. S. E.) from Rotherham; containing 226 inhabitants. The living is a perpetual curacy, in the patronage of the Prebendary in the Cathedral Church of York, the appropriator; net income, £100. Six poor children are taught to read for an annuity of £6, the gift of a Mr. Turie; and a National school has been established.

WALESBY (*All Saints*), a parish, in the union of CAISTOR, Southern Division of the wapentake of WALSHCROFT, parts of LINDSEY, county of LINCOLN, 3¼ miles (N. E.) from Market-Rasen; containing 247 inhabitants. The living is a rectory, valued in the king's books at £23. 18. 1½.; present net income, £441; patron, J. Angerstein, Esq.

WALESBY (*St. Edmund*), a parish, in the union of SOUTHWELL, Hatfield Division of the wapentake of BASSETLAW, Northern Division of the county of NOTTINGHAM, 3 miles (N. E.) from Ollerton; containing 340 inhabitants. The living is a discharged vicarage, valued in the king's books at £6. 1. 3.; present net income, £158; patron, Earl of Scarborough. The church is in the Norman style of architecture, with a low tower surmounted by a pyramidical roof. The Rev. Richard Jackson, in 1760, bequeathed a rent-charge of £2 for teaching poor children; nine are educated for £5 a year, being the rental of land received in lieu of the annuity.

WALFORD (*St. Leonard*), a parish, in the union of ROSS, hundred of GREYTREE, county of HEREFORD, 2¾ miles (S. S. W.) from Ross; containing 1155 inhabitants. The living is a discharged vicarage, with that of Ruardean annexed, valued in the king's books at £13. 2. 1.; present net income, £218; patron, the Precentor in the Cathedral Church of Hereford. A school is endowed with £5 per annum; a day and Sunday school, appertaining to Baptists, is supported by the trustees of the late Edward Goff, Esq.; and another school is partly supported by a donation of £20 per annum from a private individual.

WALFORD, a joint township with Letton and Newton, in the parish of LEINTWARDINE, union of KNIGHTON, hundred of WIGMORE, county of HEREFORD, 13 miles (N. W. by N.) from Leominster; containing, with Letton and Newton, 212 inhabitants.

WALGHERTON, a township, in the parish of WYBUNBURY, union and hundred of NANTWICH, Southern Division of the county of CHESTER, 3¾ miles (S. E. by E.) from Nantwich; containing 213 inhabitants.

WALGRAVE (*St. Peter*), a parish, in the union of BRIXWORTH, hundred of ORLINGBURY, Northern Division of the county of NORTHAMPTON, 7¼ miles (N. W. by W.) from Wellingborough; containing 575 inhabitants. The living is a rectory, with that of Hannington annexed, valued in the king's books at £22. 4. 7.; patron, Bishop of Lincoln. There is a place of worship for Baptists. Montague Lane, in 1670, bequeathed £200, directing the interest to be applied for teaching poor children; and £12 per annum, on account of this bequest, is now paid to the master of a National school. Here were formerly some almshouses, which have long since fallen into decay.

WALHAM-GREEN, a chapelry, in the parish of FULHAM, Kensington Division of the hundred of OSSULSTONE, county of MIDDLESEX, 3 miles (S. W. by W.) from London: the population is returned with the parish. The living is a perpetual curacy; net income, £230; patron, Vicar of Fulham. The chapel, dedicated to St. John, was erected in 1829, at an expense of £9683. 17. 9., raised by subscription, and a grant from the parliamentary commissioners.

WALKER, a township, in the parish of LONG BENTON, Eastern Division of CASTLE ward, Southern Division of the county of NORTHUMBERLAND, 3¼ miles (E.) from Newcastle-upon-Tyne: the population is returned with the parish. It is bounded on the south by the river Tyne, along the banks of which are extensive manufactories and coal-staiths.

WALKERINGHAM (*St. Mary Magdalene*), a parish, in the union of GAINSBOROUGH, North-Clay Division of the wapentake of BASSETLAW, Northern Division of the county of NOTTINGHAM, 4 miles (N. W. by W.) from Gainsborough; containing 529 inhabitants. The living is a discharged vicarage, valued in the king's books at £7. 11. 4.; present net income, £204; patrons and impropriators, Master and Fellows of Trinity College, Cambridge. There is a place of worship for Wesleyan Methodists. The Chesterfield canal passes through the parish, and the river Trent forms the eastern boundary, where there is a ferry. Robert Woodhouse, in the year 1719, bequeathed a rent-charge of £15 for teaching poor children, and another of £1 for providing books.

WALKERITH, a hamlet, in the parish and union of GAINSBOROUGH, wapentake of CORRINGHAM, parts of LINDSEY, county of LINCOLN, 2½ miles (N. W. by N.) from Gainsborough; containing 65 inhabitants.

ALKERN (*St. Mary*), a parish, in the hundred of BROADWATER, union and county of HERTFORD, 4¾ miles (E. by N.) from Stevenage; containing 771 inhabitants. The living is a rectory, valued in the king's books at £20. 1. 10½.; patrons, Provost and Fellows of King's College, Cambridge. The impropriate tithes have been commuted for a rent-charge of £75. 7., and the rectorial for £588. 13., subject to the payment of rates, which on the average have amounted to £11 and £88; the glebes comprise respectively 100 and 26 acres, valued at £100 and £23. 8. per annum. The church contains a curious monument of a Knight Templar. There is a place of worship for Independents; also a day and Sunday school supported by subscription. A fair for cattle is held on Nov. 5th.

WALKHAMPTON, a parish, in the union of TAVISTOCK, hundred of ROBOROUGH, Midland-Roborough and Southern Divisions of the county of DEVON, 4½ miles (S. E. by E.) from Tavistock; containing 691 inhabitants. The living is a vicarage, valued in the king's books at £9. 14. 7.; present net income, £125; patron and impropriator, Sir R. Lopes, Bart. The church is situated on the verge of Dartmoor Forest. The Plymouth railway passes through the parish. Lady Modyford, in 1719, gave a school-house, with the rents and profits of certain premises, now producing about £161 a year, for the education of poor children; and an infants' school is partly supported by the minister.

WALKINGHAM-HILL, an extra-parochial liberty, in the Upper Division of the wapentake of CLARO, West Riding of the county of YORK, 4 miles (N.) from Knaresborough; containing 25 inhabitants.

WALKINGTON (*All Hallows*), a parish, in the union of BEVERLEY, partly in the Hunsley-Beacon Division of the wapentake of HARTHILL, but chiefly in the wapentake of HOWDENSHIRE, East Riding of the county of YORK, 2¾ miles (S. W. by W.) from Beverley; containing 558 inhabitants. The living is a rectory, valued in the king's books at £24. 13. 4.; present net income, £676; patron and incumbent, Rev. D. Ferguson. There is a place of worship for Wesleyan Methodists.

WALKINSTEAD, county of SURREY.—See GODSTONE.

WALKMILL, a township, in the parish of WARKWORTH, union of ALNWICK, Eastern Division of COQUETDALE ward, Northern Division of the county of NORTHUMBERLAND; containing 7 inhabitants.

WALL, a chapelry, in the parish of ST.-JOHN-LEE, union of HEXHAM, Southern Division of TINDALE ward and of the county of NORTHUMBERLAND, 3¾ miles (N. by W.) from Hexham; containing 495 inhabitants. The living is a perpetual curacy; patron, T. R. Beaumont, Esq. The impropriate tithes have been commuted for a rent-charge of £274, subject to the payment of rates, which on the average have amounted to £8. The chapel, dedicated to St. Oswald, was erected by the monks of Hexham, upon the spot where King Oswald, who was afterwards canonized, raised the standard of the cross, and defeated the Britons under Cadwalla. A silver coin of the former was found when the chapel underwent repair, and a mutilated Roman altar lies in the cemetery; adjoining which is a field, where human skulls and fragments of military weapons have been often turned up by the plough.

WALL, a hamlet, in the parish of ST. MICHAEL, LICHFIELD, union of LICHFIELD, Southern Division of the hundred of OFFLOW and of the county of STAFFORD, 2½ miles (S. S. W.) from Lichfield; containing 93 inhabitants. The Rev. Burnes Floyer has given a piece of land for the site of a church, and John Smith, Esq., has given £500 towards building it. This hamlet is intersected by Watling-street, and is the ancient Roman station of *Etocetum,* of which many vestiges may still be traced in the walls.

WALL-TOWN, a township, in the parish and union of HALTWHISTLE, Western Division of TINDALE ward, Southern Division of the county of NORTHUMBERLAND, 3 miles (N. W. by W.) from Haltwhistle; containing 96 inhabitants. The Roman wall passed through the village; and in this township were the stations *Vindolana,* now called Little Chesters, and *Æsica,* called Great Chesters, the ramparts of which, particularly the latter, where there are also considerable traces of a town, are in a better state of preservation than those of any other on the whole line of the wall. Roman baths, altars, tombstones, inscriptions, curious pieces of sculpture, and numerous other relics of antiquity, have been found in both; and in a neighbouring hill, called Chapel-Steads, many urns have been discovered. Near the military road connecting the two stations are tumuli, termed the "Four Lawes;" and, on an adjoining hill, a rude monument of three large stones, vulgarly called the "Mare and Foals." Part of the ruins of a castellated mansion, formerly the residence of the Ridleys, has been removed for building a modern seat, and the remainder has been converted into a farm-house.

WALLASEA, ISLE OF, partly in the parishes of CANEWDON, EASTWOOD, PAGLESHAM, GREAT STAMBRIDGE, and LITTLE WAKERING, hundred of ROCHFORD, Southern Division of the county of ESSEX, 6 miles (E. N. E.) from Rochford. It is now a peninsula, formed by the rivers Crouch and Broomhill, and joined to the main land by a causeway, kept up at the expense of the several parishes to which it belongs.

WALLASEY (*St. Hilary*), a parish, in the union, and Lower Division of the hundred, of WIRRALL, Southern Division of the county of CHESTER; containing 2737 inhabitants, of which number, 558 are in the township of Wallasey, 11¾ miles (N. by E.) from Great Neston. This parish is situated in the north-west corner of the county: it is a peninsula of a triangular form, bounded on the west by the Irish Sea, on the north-east by the Mersey, and on the south-east by a branch of the Mersey, called Wallasey Pool: there are sand-hills bordering on the sea, which form a natural barrier against its encroachments. Many handsome houses and marine villas have been erected on the banks of the Mersey, this place being much frequented for sea-bathing. The principal house in the village is an ancient mansion by the sea side, denominated Mockbeggar Hall, or, more properly, Leasowe Castle, formerly a seat of the Egertons, which has been converted by its proprietor, Col. Edward Cust, into a commodious hotel for the accommodation of visiters. A handsome pillar near Poolton, with an inscription, has been erected to the memory of the colonel's mother-in-law, Mrs. Boodie, who was thrown out of her carriage and killed on the spot. On the Black rock, at the north-west point of the parish, is a very strong fort, mounting fifteen large guns; and, further in the sea, a small lighthouse, on the plan of the Eddystone lighthouse, has been lately erected. The masses of sand-stone near the Black rock, called the "Red Noses," well merit the attention of the naturalist, being worn, by the action of the sea, into a variety of caverns of the most romantic forms. Between the village and the sea-shore is an enclosure (formerly a common), called the Leasowe, where races were held, which were of very early origin: here the unfortunate Duke of Monmouth ran his horse, in the reign of Charles II., won the plate, and presented it to the daughter of the mayor of Chester: the races were discontinued in the year 1760, and the ground is now under cultivation. Steam-boats cross every half-hour from Egremont and Seacomb ferries to Liverpool, which is directly opposite. At Liscard, on the banks of the river, is a magazine, where all ships entering the port of Liverpool deposit their gunpowder, prior to admission into the docks. The living is a discharged rectory, valued in the king's books at £11. 0. 2½.; present net income, £393; patron, Bishop of Chester. The church was rebuilt about 70 years ago, excepting the tower, which bears the date 1560: it stands in the centre of the parish, on a hill composed of red sandstone, used for building. There was another church, prior to the dissolution, appropriated to Birkenhead abbey, but there are no traces of it: a path near its site is still called the "Kirkway." Major Henry Meols built a school-house and endowed it in 1656; it was further endowed by his brother William, with £125, which was laid out in the purchase of a house and land; a close of land was subsequently given to it by Mr. Henry Young, and a legacy of £100 by the Rev. George Briggs, a late rector. The present endowment consists of 38 acres of land, producing £89. 11. per annum, which, with the interest of the legacy, constitutes the income of the school; the salary of the master, who is appointed by the rector and churchwardens, is £80 per annum, with a house and the privilege of taking boarders. The old school house, which was inconveniently situated near the church, was pulled down and rebuilt on another site, in the year 1799. The interest of several small sums, and of £100 by the Rev. G. Briggs, is distributed among the poor.

WALLBOTTLE, a township, in the parish of NEWBURN, union and Western Division of CASTLE ward, Southern Division of the county of NORTHUMBERLAND, 4¾ miles (W. by N.) from Newcastle-upon-Tyne; containing 688 inhabitants, many of whom are employed in an extensive colliery at this place.

WALLCOTT, a hamlet, in the parish of CHARLBURY, hundred of BANBURY, county of OXFORD, 5½ miles (S. S. E.) from Chipping-Norton; containing 9 inhabitants.

WALLDITCH (*ST. MARY*), a parish, in the union of BRIDPORT, hundred of GODDERTHORNE, Bridport Division of the county of DORSET, 1½ mile (E. by S.) from Bridport; containing 164 inhabitants. The living is a perpetual curacy; net income, £54; patrons and impropriators, Lord Rolle and J. Bragge, Esq. The impropriate tithes have been commuted for a rent-charge of £60, and the incumbent's for £33, subject to the payment of rates, which on the average on the latter have amounted to £3. 5.; the impropriate glebe comprises 28 acres, valued at £45 per annum. The church was formerly a free chapel, or chantry. An infants' school and a Sunday school are supported by subscription.

WALLERSCOAT, a township, in the parish of WEAVERHAM, union of NORTHWICH, Second Division of the hundred of EDDISBURY, Southern Division of the county of CHESTER, 1½ mile (W.) from Northwich; containing 10 inhabitants.

WALLERTHWAITE, a joint township with Markington, in the parish and liberty of RIPON, West Riding of the county of YORK, 4 miles (N. N. E.) from Ripley: the population is returned with Markington.

WALLINGFORD, a borough and market-town, having exclusive jurisdiction, and the head of a union, locally in the hundred of Moreton, county of BERKS, 15 miles (N. N. W.) from Reading, and 46 (W. by N.) from London; containing, with the extra-parochial liberty of the Castle, but exclusively of the liberty of Clapcot, which is in the parish of Allhallows, 2467 inhabitants. The name is derived from the ancient British word *Guallen*, or the Roman *Vallum*, each signifying "an old fort," and from a ford over the Thames: subsequently to the Roman invasion, it was converted into a strong fortification by that people, and is supposed to have been the principal station of the Attrebatii. On the arrival of the Saxons, it became one of their principal forts, and continued to be a place of considerable repute, until it was burnt by the Danes, in 1006: from the effects of this calamity it speedily recovered, and, in the reign of Edward the Confessor, had risen to the dignity of a royal prescriptive borough. At the Conquest, William, having arrived with his army, received here the homage of Stigand, Archbishop of Canterbury, and many other prelates and barons. During the civil war between Stephen and the Empress Matilda, the castle was occupied and held for the latter; it was subsequently the place of meeting between John and the barons. The honour, having become vested in the crown, was given by Richard I. to his brother John; and Henry III., on being elected King of the Romans, entertained all the prelates and barons in the castle. Having been subsequently annexed, by act of parliament, to the duchy of Cornwall, on the reversion of these estates to the crown, the castle and manor were granted to Cardinal Wolsey, who conferred them on his then newly-erected college of Christ-Church, Oxford; and, in Camden's time, part of the castle was used, as an occasional retreat in time of sickness, by the students of that college: a portion of these buildings, called the "Priests' Chambers," has been converted into a malthouse. At the commencement of the parliamentary war, it was repaired and garrisoned for the king, and was not surrendered till nearly the close of the war; and, about four years afterwards, in 1653, it was completely demolished, insomuch that, at present, part of a wall towards the river is all that remains of this ancient and celebrated structure.

Seal.

The town is situated on the road between Reading and Oxford, and has a remarkably neat and clean appearance: it consists principally of a handsome market-place and two streets, well paved and lighted with gas, under an act obtained in 1795, and is abundantly supplied with water. Across the river Thames, which passes on the eastern side of the town, is a fine stone bridge of several arches, about 300 yards in length, constructed in 1809, in lieu of a dilapidated structure which was supposed to have been built five centuries ago: there is a rent-charge of £42 per annum on houses for its repair, under the management of trustees appointed by the 49th of George III. Some business is done in malting, but it is not so extensive as formerly. A line of communication has been opened with Birmingham, Bath, and Bristol, by means of a canal navigation, running into the Thames, by which river coal is brought hither, and corn and flour are conveyed to London and other places. The Great Western railway passes through the town. The market is on Friday; and a statute and pleasure fair is held on September 29th. Wallingford is a borough by prescription, and has received charters from various sovereigns: by that bestowed by Charles II., in the 15th year of his reign, the government was vested in a mayor, recorder, town-clerk, 6 aldermen, 2 bailiffs, a chamberlain, 18 assistants, and 2 sergeants-at-mace: besides these, but not named in the charter, were a high steward, two bridgemen, and other subordinate officers: the mayor, recorder, and aldermen, were justices of the peace, with exclusive jurisdiction. By the act of the 5th and 6th of William IV., cap. 76, (for an abstract of which see the Appendix, No. I.), the corporation now consists of a mayor, 4 aldermen, and 12 councillors. The borough formerly returned two members to parliament, but it now sends only one. The right of election was vested in the corporation, and in the inhabitants at large paying scot and lot, above 300 in number; but it has been extended to the £10 householders of an enlarged district, which has been made to constitute the new elective borough, and the limits of which are minutely described in the Appendix, No. II. The old borough comprised 435 acres; the new borough comprehends by estimation 16,352. The number of electors now registered is 366, including householders and scot and lot voters; the mayor is the returning officer. In former times, criminals convicted capitally in this borough, for the first time, had their lives spared on certain conditions; and, in the 45th of Henry III., a return made by the jurors declared, that no person belonging to the borough ought to be executed for one offence. The corporation are empowered, by charter, to hold a court for the recovery of small debts; but this right is seldom exercised, though debts to any amount may be recovered by a process from the town-clerk. Petty sessions for the division are held every Friday.

Wallingford comprises the parishes of *All Hallows*, *St. Leonard*, *St. Mary-le-More*, and *St. Peter*. The living of All Hallows is a sinecure rectory, in the patronage of the Master and Fellows of Pembroke College, Oxford: the church was demolished in 1648. The living of St. Leonard's is a discharged rectory, with the perpetual curacy of Sotwell annexed, valued in the king's books at £7. 12. 6.; present net income, £153: it is in the patronage of the Crown. The church is a very ancient structure, but destitute of claim to architectural description, with the exception of some few Norman remains. The living of St. Mary-le-More is a discharged rectory, valued in the king's books at £4; present net income, £137: it is in the patronage of the Crown. The church is a very handsome edifice, situated in the space near the market-house, with a square embattled tower, ornamented with pinnacles, and on which is the figure of an armed knight on horseback, supposed to represent King Stephen; the tower, which bears the date of 1658, was built by the corporation, with materials said to have been taken from the ruins of the castle. The living of St. Peter's is a discharged rectory, valued in the king's books at £6. 1. 3.; present net income, £100; patron, W. S. Blackstone, Esq. The church is a very handsome structure, bearing date 1769, with a square tower, surmounted by an elegant spire of Portland stone, supported on pillars and arches, and erected in 1777, by voluntary subscriptions, to which the learned Sir William Blackstone, who was an inhabitant of the town, and whose remains are deposited in the church, was a liberal contributor. There are places of worship for Baptists, the Society of Friends, Independents, Calvinists, and Wesleyan Methodists. The free school was founded by Walter Bigg, alderman of London, in 1659, by whom it was endowed with £10 per annum, for six boys' who are elected by the aldermen. A school, combining an infants' and day school, is partly supported by subscription; the master and mistress have a house and garden rent free; and another school, containing 24 boys, is supported by W. S. Blackstone, Esq. An almshouse for six poor widows was founded and endowed with £34 per annum, in 1681, by William Angier and Mary his sister; the endowment has been augmented by subsequent benefactions. The poor law union of Wallingford comprises 29 parishes or places, nearly half of which are in the county of Oxford, and contains a population of 13,085, according to the census of 1831; it is under the care of 29 guardians. On Wittenham hill (the ancient *Sinodun*), in this neighbourhood, are some remains of a Roman camp, where numerous coins have been found. Richard de Wallingford, abbot of St. Alban's, a celebrated mathematician and mechanic; and John de Wallingford, a monk of that abbey, are supposed to have been natives of this town: the former invented, and presented to the abbey church, an ingenious clock, that showed not only the course of the sun, moon, and principal stars, but also the ebbing and flowing of the sea. Joan, the fair maid of Kent, and widow of the Black Prince, died here in the year 1385. Wallingford formerly conferred the title of Viscount on the Earl of Banbury, but it has now merged in the Earldom of Abingdon.

WALLINGTON (*St. Mary*), a parish, in the union of Buntingford, hundred of Odsey, county of Hertford, 3 miles (E.) from Baldock; containing 213 inhabitants. The living is a rectory, valued in the king's books at £16. 15. 2½.; present net income, £398; patron, Master of Emanuel College, Cambridge. The church is an ancient structure, with an embattled tower, surmounted by a short spire: attached to the north side of the chancel are several mutilated altar-tombs, and other sepulchral remains. The Rev. John Browne, in 1736, bequeathed £100, the interest of which, amounting to £3. 7. 3., is paid towards instructing poor children: the same benefactor also gave the interest of £20 to be distributed on Easter-Monday to the needy poor.

WALLINGTON (*St. Margaret*), a parish, in the union of Downham, hundred of Clackclose, Western Division of the county of Norfolk, 3½ miles (N. by E.) from Downham-Market; containing 47 inhabitants. The living is a rectory, united, with that of Holme, to the rectory of South Runcton. The church is in ruins.

WALLINGTON, a hamlet, in the parish of BEDDINGTON, union of CROYDON, Second Division of the hundred of WALLINGTON, Eastern Division of the county of SURREY, $2\frac{3}{4}$ miles (W. by S.) from Croydon; containing 933 inhabitants. Here was formerly a chapel, now in ruins. This place gives name to the hundred, the whole of which is within the jurisdiction of the court of requests, held at Croydon, for the recovery of debts under £5. Eight schools are partly supported by public and private subscription.

WALLINGTON-DEMESNE, a township, in the parish of HARTBURN, union of MORPETH, Northern Division of TINDALE ward, Southern Division of the county of NORTHUMBERLAND, $12\frac{1}{2}$ miles (W. by S.) from Morpeth; containing 193 inhabitants. In pulling down the remains of Fenwick tower, in 1775, several hundred gold nobles, of the coinage of Edward III., were found in an open stone chest; it is supposed that they were concealed on the invasion of David, King of Scotland, in 1360, who made prisoners the two sons of Sir John Fenwick, then owner of the castle.

WALLINGSWELL, an extra-parochial liberty, in the Hatfield Division of the wapentake of BASSETLAW, county of NOTTINGHAM, $3\frac{3}{4}$ miles (N. by W.) from Worksop; containing 21 inhabitants. A Benedictine nunnery, in honour of the Virgin Mary, was founded here in the reign of Stephen, by Ralph de Cheroulcourt, which at the dissolution had a revenue of £88. 11. 6.: it is now the residence of Sir Thomas Woollaston White, Bart. In excavating near the house, in 1829, several stone coffins were found, and amongst them that of Dame Margery Dourant, second abbess of the convent, who died in the reign of Richard I.; on opening it, the body appeared nearly perfect, but, on exposure, soon suffered decomposition; her shoes were entire, as was also a silver chalice: these relics were again deposited, with the ashes, in the same receptacle, and re-interred.

WALLOP, NETHER (*St. Andrew*), a parish, in the union of STOCKBRIDGE, hundred of THORNGATE, Andover and Northern Divisions of the county of SOUTHAMPTON, 4 miles (W. by N.) from Stockbridge; containing 900 inhabitants. The living is a discharged vicarage, valued in the king's books at £13. 13. 4.; patrons and appropriators, Vicars Choral of York. The vicarial tithes have been commuted for a rent-charge of £346. 18. 4., subject to the payment of rates; the glebe comprises two acres. There is a place of worship for Wesleyan Methodists; also a National school, partly supported by endowment. On a point, or head, of an elevated ridge, called Danebury Hill, or Bill, are remains of a circular fortification, with lofty ramparts, enclosing an extensive area: a short distance to the westward is an outwork, for the defence of that side; but on the east and north sides, where the ground is more steep, it is protected by a single ditch only: the entrance is by a winding course, strengthened by embankments. There are several barrows near this camp, one of which, two miles distant, is called Canute's barrow.

WALLOP, OVER (*St. Peter*), a parish, in the union of STOCKBRIDGE, hundred of THORNGATE, Andover and Northern Divisions of the county of SOUTHAMPTON, 5 miles (W. N. W.) from Stockbridge; containing 478 inhabitants. The living is a rectory, valued in the king's books at £27. 5. $2\frac{1}{2}$.; patron, Earl of Portsmouth. The tithes have been commuted for a rent-charge of £820, subject to the payment of rates; and there are 9 acres of glebe. A school is supported by Mrs. Brownjohn, aided by £2 a year, the gift of Mr. Smith, in 1786; and 12 children are taught at the rector's expense.

WALLSEND (*Holy Cross*), a parish, in the union of TYNEMOUTH, Eastern Division of CASTLE ward, Southern Division of the county of NORTHUMBERLAND, $3\frac{1}{2}$ miles (E. N. E.) from Newcastle-upon-Tyne; containing 5510 inhabitants. The name of this parish is obviously derived from its situation at the extremity of the wall of Severus, on the east: it contained the Roman station *Legedunum*, so called from its situation, and from having been a magazine for corn, whence other stations in the interior were supplied. It was garrisoned by the first cohort of the *Lergi*, who were stationed here for the defence of their shipping, of which the Romans kept, in the rivers on the borders of their settlements, ships called *lusoriæ*, which were always on the alert, either to protect them from the invasion of the inhabitants of the neighbouring coasts, or to assist them in extending their empire, by the invasion of the territories of their neighbours: and many altars, coins, and urns, with other curious relics, have been discovered upon the spot. Beyond this point the wall does not appear to have been continued; the Tyne itself, near its influx into the ocean, forming, by its great breadth and depth, a sufficient barrier against the incursions of those enemies from whose frequent depredations it was originally erected to protect the inhabitants of this part of the island. The ruins of a quay still further evince that it was anciently a considerable trading colony of the Romans, who, more than 1000 years since, discharged their freights where now are numerous staiths, projecting from the northern bank of the Tyne, whence vessels employed in the coal trade are continually taking in immense quantities of the excellent coal termed "Wallsend," for the London and other markets. The village is large and well built, situated near the Shields road, and contains many good houses, with a spacious green in the centre, crossed by a raised causeway. Here are several yards for ship-building, extensive limekilns, and manufactories for copperas and earthenware. The Newcastle and Shields railway passes through the village. The living is a perpetual curacy; net income, £289; patrons, Dean and Chapter of Durham; impropriators, Bishop of Bristol and Rev. J. S. Ogle. The church, dedicated to *St. Peter*, is a neat stone building with a spire, situated at some distance from the village: it was erected at the expense of nearly £5000, of which about £3300 was raised by tontine; the first stone was laid in 1807, and it was consecrated in August, 1809. The old church, which was dedicated to the *Holy Cross*, has been pulled down. There are three places of worship for Methodists, and one for Anti-Burghers. At the eastern extremity of the village is a school-room, with a house and garden for the master, given in 1748 by Mrs. Stewart and Mrs. Muncaster; and a National school has been established. The Sunday schools are attended by about 500 children. Wallsend is the birth-place of the two brothers, John and William Martin, the first distinguished as historical painter to the late king, the other as an ingenious inventor of several useful machines.

WALMER (*St. Mary*), a parish, and a member of the cinque-port liberty of SANDWICH, in the union of EASTRY, locally in the hundred of Cornilo, lathe of ST. AUGUSTINE, Eastern Division of the county of KENT, 2 miles (S.) from Deal; containing 1779 inhabitants. Walmer-street, on the turnpike-road from Deal to Dovor, is neatly built, being interspersed with genteel houses, marine villas, &c., and, partly on account of its convenient situation as regards those two towns, is much frequented during the season for sea-bathing. It is noted for the salubrity of its air, and for the fine prospects, in its vicinity, over the Downs and the straits of Dovor to the French coast; but chiefly for the celebrated fortress, Walmer Castle, erected by Henry VIII., at the same period with those of Deal and Sandown, for the defence of the coast, and now appropriated to the Lord Warden of the cinque-ports, for whose residence the principal apartments were fitted up some years ago, and the fosse was converted into a garden. Since the appropriation of the castle as a residence for the Lord Warden, many handsome marine villas have been erected in the vicinity, and an esplanade has been formed, consisting of numerous pleasing residences, occupied by highly respectable families. Bathing machines are in constant attendance, and a complete establishment has been formed of hot, vapour, and shower baths, with reading-rooms and every accommodation for visiters. From the esplanade is a delightful promenade from Walmer to Deal Castle (the principal part of which is in this parish), commanding a splendid view of the sea, with the shipping in the Downs. In the village is a large brewery and malting establishment. By the act of the 2nd of William IV., cap. 45, this place was united with Deal and Sandwich in the exercise of the elective franchise. His late Majesty and the present Queen Dowager, when Duke and Duchess of Clarence, resided at Walmer Castle in the summer of 1822; and the Princess Amelia resided for many years in an old mansion in the village. The living is a perpetual curacy, endowed with the vicarial tithes; net income, £154; patron and appropriator, Archbishop of Canterbury. The church has been repaired, and the nave considerably enlarged; 700 additional seats have been obtained, of which 500 are free. The western entrance is under a highly-enriched Norman arch, and there is a similar arch between the nave and chancel. In the churchyard are two remarkably fine yew trees. A National school and an infants' school have been established. Near the church is a deep fosse, with other vestiges of ancient intrenchments; and in the churchyard several stone coffins were discovered about 50 years since, supposed to have belonged to the ancient family of Crowl, of whom Sir Nicholas, in the reign of Edward I., erected a mansion in the village, of which there are still some remains. The North and South Infantry barracks, and the Naval hospital, which are partly within this parish, are described under the head of DEAL.

WALMERSLEY, a township, in the parish and union of BURY, hundred of SALFORD, Southern Division of the county palatine of LANCASTER, $2\frac{1}{4}$ miles (N. by E.) from Bury; containing 3456 inhabitants, who are chiefly employed in the extensive spinning-mills on the river Irwell, which runs through it. A chapel was erected in 1837, containing 670 sittings, 340 of which are free, the Incorporated Society having granted £250 for that purpose. There is a place of worship for Independents; also a National school endowed with £10 per annum.

WALMSGATE, a parish, in the union of LOUTH, hundred of HILL, parts of LINDSEY, county of LINCOLN, $6\frac{1}{2}$ miles (S. S. E.) from Louth; containing 72 inhabitants. This parish is situated on the high road from Louth to London; the surrounding scenery is pleasing, and the seat of James Whiting Yorke, Esq., commands some finely varied prospects. The whole of the land is tithe free. The living is annexed to the vicarage of Burwell. The church has fallen to ruins, and the inhabitants attend the parish church of Burwell, about a mile distant, in which they have a right of pews. The poor children of the parish are instructed in the National school at Burwell, which is supported by subscription.

WALMSLEY, a chapelry, in the parish of BOLTON, hundred of SALFORD, Southern Division of the county palatine of LANCASTER, 4 miles (N.) from Great Bolton: the population is returned with the parish. The living is a perpetual curacy; net income, £69; patron, Vicar of Bolton-in-the-Moor. The chapel has been enlarged, and 260 free sittings provided, the Incorporated Society having granted £300 in aid of the expense. There is a place of worship for Unitarians. Ten poor children are instructed in a school-house erected by subscription, upon land given by Miles Lonsdale, in 1716: the school is partly supported with the interest of £50, the bequest of James Lancashire, in 1737. Here is also a Sunday National school.

WALNEY, ISLE OF, a chapelry, in the parish of DALTON-IN-FURNESS, hundred of LONSDALE, north of the sands, Northern Division of the county palatine of LANCASTER, 5 miles (S. W.) from Dalton: the population is returned with the parish. Walney, which is insular at high water, is ten miles in length, and about one in breadth, having a lighthouse on its southern extremity, a short distance north from which is a rocky islet, termed the Pile of Fouldrey, *i. e.*, the island of fowls, where are the venerable ruins of a strong castle, built by an abbot of Furness. There are several other small isles in the group, the principal of which is Old Barrow, lying between this and the main land, opposite the small village and port of Barrow. On Walney are some remarkable intermitting springs of fresh water. It is stated to have been once covered with wood, of which it is now extremely barren. West, in his "Antiquities of Furness," describes it as lying on a bed of moss, which is found by digging through a layer of sand and clay, and in which trees have been discovered. To prevent the encroachments of the sea, the abbots of Furness kept up a dyke, which, since the dissolution, has been neglected, and the sea has made considerable inroads. The living is a perpetual curacy; net income, £94; patron, Vicar of Dalton.

WALPOLE (*St. Andrew*), a parish, in the union of WISBEACH, hundred of FREEBRIDGE-MARSHLAND, Western Division of the county of NORFOLK, $8\frac{3}{4}$ miles (W. by S.) from Lynn-Regis; containing 514 inhabitants. The living is a discharged vicarage, endowed with the rectorial tithes, and valued in the king's books at £26. 13. 4.; present net income, £1259; patron, Rev. C. H. Townshend. The church is a regular, well-built structure, entirely covered with lead. This place derives its name from the great wall raised by the Romans to defend it from the encroachments of the sea, and from an extensive pool of water near it. In a garden at the foot of this embankment, many Roman bricks, and the remains of an aqueduct, formed of 26 earthen pipes, have been found. The estuary, called Cross Keys Wash, in the neighbourhood, may be passed on foot, at the reflux of the tides, to Long Sutton, in Lincolnshire.

WALPOLE (*St. Peter*), a parish, in the union of Wisbeach, hundred of Freebridge-Marshland, Western Division of the county of Norfolk, 9 miles (W. by S.) from Lynn-Regis; containing 1237 inhabitants. The living is a rectory, valued in the king's books at £21; present net income, £925: it is in the patronage of the Crown. The church, which was built in the reign of Henry VI., is esteemed one of the most beautiful parochial structures in England; it has thirteen clerestory windows on each side, exhibiting fine specimens of stained glass. There is a place of worship for Wesleyan Methodists; also a school endowed with 60 acres of land, which is free to the children of this parish and Walpole-St. Andrew.

WALPOLE (*St. Mary*), a parish, in the union and hundred of Blything, Eastern Division of the county of Suffolk, 2 miles (S. W.) from Halesworth; containing 658 inhabitants. The living is a perpetual curacy; net income, £82; patron and impropriator, Rev. B. Philpot. The church is an ancient edifice with a fine Norman door-way from the porch into the interior. There is a place of worship for Independents. Thomas Neale, in the year 1704, bequeathed an annuity of £2. 10. for teaching poor children.

WALRIDGE, a township, in the parish of Stamfordham, North-Eastern Division of Tindale ward, Southern Division of the county of Northumberland, 5½ miles (N. by W.) from Durham; containing 7 inhabitants.

Seal and Arms.

WALSALL (*St. Matthew*), a parish, and the head of a union, in the Southern Division of the hundred of Offlow and of the county of Stafford; comprising the market-town and newly enfranchised borough of Walsall, which is 18 miles (S. E. by S.) from Stafford, and 118 (N. W.) from London, and the township of Walsall-Foreign; and containing 15,066 inhabitants, of which number, 6401 are in the township of the borough, but there are 10,796 in the entire town, which in many parts has stretched in numerous streets into what is properly called the Foreign. This place is supposed to have derived its name (in various ancient records written *Waleshall* and *Walshale*) from its situation in or near an extensive forest, resorted to by the Druids for the celebration of their religious rites, and in which the Saxons subsequently erected a temple to their god Woden; from which also the town of Wednesbury, in the vicinity, is supposed to have derived its name. In the early part of the tenth century, it was fortified by Ethelfleda, daughter of Alfred, and Countess of Mercia, probably about the same time that she built a castle at Stafford, and surrounded the town with walls. At the time of the Conquest, it was retained by William, and continued to be a royal demesne for nearly 20 years, till it was given by the Conqueror to Robert, son of Asculfus, who accompanied him to Britain. Walsall is not connected with any events of historical interest: Queen Elizabeth, in one of her tours through the country, visited it, and affixed the royal seal and signature at *Walshale*, on the 13th of July, in the 28th year of her reign, to a deed preserved in the archives of the corporation, containing a grant of certain lands to the town. In 1643, Henrietta Maria, queen of Charles I., remained here for a short time previously to joining the king at Edgehill; and Charles II., on his road from Boscobel to the coast, found an asylum at Bently Hall, about a mile distant.

The town is pleasantly situated on the summit and acclivities of a rock of limestone, and is watered by a small brook called by Erdeswick "Walsal water," which falls into the river Tame, a little below the town: it consists of several regular and spacious streets, in some of which are handsome houses of modern erection, many of them being of a superior description. The environs are interesting, and contain some pleasant villas, and much beautiful and varied scenery. The town is well paved and lighted with gas, under the superintendence of commissioners, appointed by act of parliament passed in 1824, and amply supplied with water. A subscription library was established in 1800, the plan of which was, a few years since, enlarged; and a splendid edifice, containing reading and news rooms, ornamented by a Doric colonnade 30 feet high, has been erected. The principal hotel, a very spacious and handsome building, has been enlarged and beautified at a considerable expense; the handsome portico is formed of pillars formerly belonging to Fisherwick, the noble mansion of Lord Donegall. The principal articles of manufacture are bridle-bits, stirrups, spurs, saddle-trees and every kind of saddlers' ironmongery; buckles, snuffers, spoons, and various other kinds of hardware; coach harness and furniture, plated ware, locks, chain curbs, dog chains, and other articles, many of which are brought into the town and sold by factors. Many mercantile houses have been established in the town, carrying on an extensive trade with America and other countries, and a considerable home trade. A manufactory for Hebert's patent progressive corn-mills has lately been erected within four miles of the town, where one of these mills is in operation. There are several brass and iron foundries, of which the iron foundry at Goscote is the most extensive, as well as the oldest, in this district; steam-engines of every power, cylinders, and cannon, besides the various smaller articles of cast iron, are founded here upon the most improved principles. A considerable trade is also carried on in malt; and in the vicinity are large quarries of limestone. Several extensive mines of coal and iron-stone, with which the neighbourhood abounds, have lately been opened at the Birchills and near Bloxwich, in consequence of which the population of the town has been and is increasing more rapidly than any other in the county. The situation of the town, in the north-eastern part of an extensive mining and manufacturing district, abundantly supplied with coal, is peculiarly favourable to its manufactures; and a branch of the Old Birmingham canal, which comes up to the west end of the town, and the Wyrley and Essington canal, which passes within a mile north of it, afford every facility of inland navigation. The junction of these canals has been agreed upon and is now in progress. From the wharf at Park-street, fly-boats ply to Birmingham daily. About one mile from the town is the Bescot-Bridge station of the Grand Junction railway. The market is on Tuesday and Saturday; and the fairs are on Feb. 24th; Whit-Tuesday, a pleasure fair; and the Tuesday before Michaelmas-day, chiefly for horses, cattle, and cheese.

The inhabitants enjoy several privileges and immunities by prescription: Henry I. granted them exemption from toll throughout England, and from serving upon juries out of the limits of the "borough and foreign," into which Walsall is divided; and the guilds of St. John the Baptist, and of Our Lady, appear to have been ancient establishments, exercising various rights and privileges. The earliest existing charter of incorporation was granted in the 3rd of Charles I. and confirmed by Charles II. in the 13th of his reign; by it the government was vested in a mayor, recorder, and twenty-four capital burgesses (including the mayor), assisted by a town-clerk, two sergeants-at-mace, and other officers. The mayor, late mayor, recorder, and two senior burgesses, were justices of the peace. The government of the borough is now vested in a mayor, six aldermen, and 18 councillors, under the act of the 5th and 6th of William IV., cap. 76, for an abstract of which see the Appendix, No. I. The borough is divided into three wards. The mayor and late mayor are justices of the peace, with four others appointed by commission. By the act of the 2nd of William IV., cap. 45, Walsall has been constituted a parliamentary borough, with the privilege of returning one member, the right of election being in the £10 householders of the parish, with the exception of a small detached part, which continues for elective purposes annexed to the southern division of the county: the limits of the franchise comprise about 7080 acres; the mayor is the returning officer. The recorder holds quarterly courts of session for all offences not capital; and a court of record, under the charter of Charles II., as often as may be requisite, for the recovery of debts above 40*s.*, and not exceeding £20. A police force, organized under a superintendent sent from London, was established during the excitement produced by the election of 1832, and is supported by the council. The hundred court is held here, for the recovery of debts under 40*s.*, before a steward appointed by the high sheriff of the county; and the lord of the manor holds an annual court leet, at which constables and other officers are appointed. The town-hall, where the several courts are held, and the public business of the corporation is transacted, is a handsome, though rather ancient, building, well adapted to its purpose. The common gaol for the town is a small building, calculated to receive only ten prisoners. A new gaol is about to be erected.

The living is a vicarage, valued in the king's books at £10. 19. 7.; present net income, £368; patron, Earl of Bradford, who, with Col. Walhouse, is the impropriator. The church, an ancient and spacious cruciform structure, with several chapels in the aisles, was, with the exception of the tower and chancel, which latter has undergone several alterations, taken down and rebuilt in the later style of English architecture, in 1821, at an expense of £20,000: it occupies a commanding situation on the summit of the rock on which the town is built, and the tower, which is in fine proportion, and surmounted by a lofty spire, forms a conspicuous object in the distant view of the town. St. Paul's chapel, a handsome edifice, in the Grecian style of architecture, was erected by the governors of the free grammar school, who, having sold some mines under part of the land belonging to that establishment, in 1797, obtained an act of parliament for applying part of the purchase money to the erection of the chapel, which was completed in 1826: it contains 750 free sittings, the Incorporated Society having granted £2000 in aid of the expense. The living is a perpetual curacy, in the patronage of the governors, who appoint the head master of the school to the office of minister; net income, £50. A new church was erected at Walsall Wood, and consecrated on the 22nd of Aug., 1837: it contains 330 free sittings, the Incorporated Society having granted £200 in aid of the expense. A church for the accommodation of 1100 persons has lately been erected at the end of Stafford-street; there are also two other churches and schools now in contemplation. There are places of worship for Independents, Wesleyan Methodists, and Unitarians, and two Roman Catholic chapels, one of which is a handsome edifice, in the Grecian style of architecture.

The free grammar school, in Park-street, was founded in 1557, by Queen Mary, who endowed it with lands belonging to the guilds and chantries which existed here previously to the dissolution, and placed it under the control of certain governors, whom she incorporated: the income is about £780 per annum: the salary of the head master, including his stipend as minister of the chapel, is £220 per annum; and there are three other masters, whose salaries respectively are £100, £80, and £60: the head, second, and third masters have houses rent-free and the privilege of taking boarders, and the school is open to all boys of the parish. Bishop Hough received the rudiments of his education in this school: the premises, built a few years since, are handsome and commodious. An English school is supported from the same funds, in the old school buildings in the churchyard, in which about 90 boys are instructed gratuitously. The Blue-coat charity school, which was endowed with £14 per annum, for the instruction of 25 children of each sex, has been incorporated with a National school, which is principally supported by subscription. There is also a National school at Walsall Wood, partly supported by an annual grant of £35 from the governors of the free grammar school; and a school for Roman Catholics is partly supported by subscription. Some almshouses, founded by Mr. John Harper, in the reign of James I., and endowed with land producing £40 per annum, were rebuilt in 1790, by the Rev. Mr. Rutter, then vicar, for the reception of six aged widows, among whom £10 per quarter is divided, in equal shares. Almshouses were, in 1825, erected and endowed for eleven aged widows, five from the Borough and five from the Foreign, of Walsall, and one from the parish of Rushall; to which purpose a dole of one penny, paid by the corporation to every person in the parishes of Walsall and Rushall, on the eve of Epiphany, was appropriated. In the reign of Henry VI., Mr. Thomas Mollesley gave to the corporation a manor and estates in the county of Warwick, which now constitute part of their extensive possessions. There are also numerous charitable bequests for apprenticing children, and for distribution among the poor. The poor law union of Walsall comprises 8 parishes or places, under the care of 19 guardians, and contains, according to the census of 1831, a population of 24,931. A custom has prevailed, from time immemorial, of throwing apples and nuts from the town-hall into the street, to be scrambled for by the populace. On a farm belonging to Lord Bradford, near the town, is a powerful chalybeate spring, called Alum Well, on the site of the ancient manor-house, of which the moat is still remaining.

WALSALL-FOREIGN, a township, in the parish and union of WALSALL, Southern Division of the hundred of OFFLOW and of the county of STAFFORD; comprising the hamlets of Great and Little Bloxwich, Birchills, Coldmore, Horden, Walsall Wood, and the Windmill-streets, in the manor of Walsall; and Goscote, which is a manor of itself; and containing 8665 inhabitants, a great proportion being within the Town, which has extended in many directions into the Foreign.

WALSDEN, a township, in the parish of ROCHDALE, union of TODMORDEN, hundred of SALFORD, Southern Division of the county palatine of LANCASTER: the population is returned with the chapelry of Todmorden.

WALSHAM-LE-WILLOWS (*St. Mary*), a parish, in the union of STOW, hundred of BLACKBOURN, Western Division of the county of SUFFOLK, 4½ miles (E. by N.) from Ixworth; containing 1167 inhabitants. The living is a perpetual curacy; net income, £93; patron, S. Golding, Esq. There are places of worship for Baptists and Wesleyan Methodists; and a National school has been established.

WALSHAM, NORTH (*St. Mary*), a market-town and parish, in the union of ERPINGHAM, hundred of TUNSTEAD, Eastern Division of the county of NORFOLK, 15 miles (N. N. E.) from Norwich, and 124 (N. E. by N.) from London; containing 2615 inhabitants. In the year 1600, nearly the whole of this town was destroyed by a fire, which, although it continued but three hours, consumed property of the value of £20,000. It is situated on the high road to Norwich, and consists of three streets diverging from a central area, in which stands the church: the inhabitants are well supplied with water. A neat theatre has been erected, and is opened for performances once in two years. A navigable canal passes through the parish, a short distance north-east of the town, in its course from Antingham to Yarmouth; and a silk-manufactory has been recently established. The market is on Thursday: and a fair is held on the day before Holy Thursday, for cattle: statute fairs for hiring servants are held twice a year. The market cross, erected by Bishop Thirlby, in the reign of Edward III., was repaired after the great fire in 1600, by Bishop Redman. Two courts baron are held annually, one by the Bishop of Norwich, and the other by Lord Suffield. The magistrates for the hundred meet here every week; and the town has been made a polling-place for the eastern division of the county.

The living is a vicarage, with the rectory of Antingham-St. Margaret annexed, valued in the king's books at £8; present net income, £336: it is in the patronage of the Crown; appropriator, Bishop of Norwich. The church is a spacious structure; but the tower, having fallen in the year 1724, is in ruins: in the chancel is an elegant mural monument to the memory of Sir William Paston, Knt., a native of this town, and founder of the grammar school: it was erected during his life, and is surmounted by a recumbent statue in armour. There are places of worship for the Society of Friends, Independents, and Primitive and Wesleyan Methodists. The free grammar school was founded, in 1606, by the above-mentioned Sir William Paston, for the education of the sons of residents in either of the hundreds of North Erpingham, Happing, Tunstead, and Flegg, and endowed by him with the rents of certain estates at Horsey and Walcot, to the amount of about £300 per annum, together with a small endowment of £4. 15. by a person unknown: the master has a salary of £70, and an allowance of £30 for an usher, with the free use of a residence, and liberty to take pay scholars, for which consideration he is required to instruct the foundation pupils gratuitously in Latin, Greek, algebra, mathematics, English grammar and history, writing and arithmetic: the school contains a good library, bequeathed by the Rev. Richard Berney, in 1787; and a monthly lecturer receives £12. 12. per annum out of the funds of the charity. Admiral Lord Nelson received part of his education at this school. A National school is supported partly by a small rental bequeathed by Mary Scarburgh, but chiefly by voluntary subscription. About £30 per annum, the rent of an allotment of waste land, is expended in coal for distribution among the poor. About a mile south of the town is a stone cross, erected to commemorate a victory obtained, in 1382, by Spencer, Bishop of Norwich, over some rebels, headed by a dyer named Leytester.

WALSHAM, SOUTH, a village, in the union of BLOFIELD, hundred of WALSHAM, Eastern Division of the county of NORFOLK, 2¾ miles (N. W. by W.) from Acle; comprising the parishes of *St. Lawrence* and *St. Mary*, and containing 575 inhabitants. The living of St. Lawrence's is a rectory, valued in the king's books at £13. 6. 8.; present net income, £376; patrons, President and Fellows of Queen's College, Cambridge: that of St. Mary's is a discharged vicarage, valued in the king's books at £5; present net income, £195; it was in the patronage of the Corporation of Norwich, the impropriators, but the advowson must be sold under the Municipal Reform Act. The church of St. Mary only remains. Richard Harrold, in 1718, bequeathed property, now let for about £20 per annum, for apprenticing poor children; and £34 per annum, the rental of some waste land awarded under an enclosure act in the 41st of George III., is expended in coal for distribution among the poor.

WALSHFORD, a joint township with Great Ribston, in the parish of HUNSINGORE, Upper Division of the wapentake of CLARO, West Riding of the county of YORK, 3¼ miles (N. by E.) from Wetherby: the population is returned with Great Ribston.

WALSINGHAM, county palatine of DURHAM.—See WOLSINGHAM.

WALSINGHAM, GREAT, a parish, in the union of WALSINGHAM, hundred of NORTH GREENHOE, Western Division of the county of NORFOLK, 1 mile (N. by E.) from Little Walsingham; comprising the united parishes of *All Saints* and *St. Peter*, and containing 434 inhabitants. This place, also called Old Walsingham, was formerly of considerable importance, having contained three churches. The living is a donative, with that of Little Walsingham united; net income, £168;

patron and impropriator, H. Lee Warner, Esq. The church is remarkable for the fine proportions of its architecture, which is in the English style. A school for girls is supported by the minister.

WALSINGHAM, LITTLE (*St. Mary*), a parish (formerly a market-town), and the head of a union, in the hundred of North Greenhoe, Western Division of the county of Norfolk, 28 miles (N. W.) from Norwich, and 114 (N. N. E.) from London; containing 1004 inhabitants. This place, also denominated New Walsingham, was of great celebrity, for many centuries, as the site of a shrine of the Virgin, or Our Lady of Walsingham, constructed of wood, after the plan of the Sancta Casa at Nazareth, and founded, in 1061, by the widow of Ricoldie Faverches, whose son confirmed her endowment, and added a monastery for Augustine canons, with a conventual church: this institution became immensely rich, and at the dissolution its revenue was valued at £446. 14. 4., exclusively of the valuable offerings of the numerous devotees of all nations who had visited the shrine, which are said to have equalled those presented at the shrine of Our Lady of Loretto, and that of St. Thomas à Becket at Canterbury. Among the illustrious visitants were several of the kings and queens of England, especially Henry VIII., who, in the second year of his reign, walked hither barefoot from Barsham, to present a valuable necklace to the image of the Virgin. During the prevalence of superstition, the credulous were taught to believe that the galaxy, or milky way, in the heavens was the peculiar residence of the Virgin, whence it obtained the name of "Walsingham way." The venerable remains of this once noble and stupendous pile are situated, in the midst of a grove of stately trees, in the pleasure grounds of H. Lee Warner, Esq., and contiguous to a fine stream of water, over which that gentleman has built a handsome bridge: they chiefly consist of the great western portal, a lofty and magnificent arch, 60 feet high, which formed the east end of the conventual church; the spacious refectory, 78 feet by 27, with walls 26½ feet in height; a portion of the cloisters, and a stone bath with two wells, called St. Mary's, or the "Wishing Wells," near which is a Saxon arch with zigzag mouldings, removed hither from the mansion as an ornamental object. The devotees who had permission to drink of these wells were taught to believe that, under certain restrictions, they should obtain whatever they might desire. The cross, resting upon a platform on which the pilgrims knelt to receive the water, is still visible. The other relics are, the abbey, a stone pulpit belonging to the refectory, and the ruins of a fine window. Here was also a house of Grey friars, founded in 1346, by Elizabeth de Burgo, Countess of Clare. The town is situated in a vale, surrounded by bold heights presenting diversified scenery, near a small stream which, within a few miles, falls into the sea: the inhabitants are supplied with water from wells. A fair is held annually a fortnight after Whit-Monday. The general quarter sessions for the county are held here by adjournment. The bridewell, or house of correction, formerly an hospital for lepers, founded in 1486, has been considerably enlarged, and contains six wards and day-rooms, three airing-yards, two work-rooms, a room for the sick, a chapel, and a tread-mill. This place was formerly noted for the growth of saffron, which has been discontinued for some years. The living is a donative, united to that of Great Walsingham. The church is a spacious structure, and contains a very ancient and beautiful font, of octagonal form, resting on a plinth of four ornamented steps, and representing, in compartments, the seven Sacraments of the church of Rome and the Crucifixion: there is also an ancient monument, erected to the memory of a Roman Catholic bishop; and a very fine one to that of Sir Henry Sidney. Here is a place of worship for Wesleyan Methodists. A free grammar school was founded by Richard Bond, Esq., in 1639, and endowed with an estate at Great Snoring, in this county, producing £110 per annum, vested in feoffees, for the maintenance of a master and an usher, to teach thirty boys. A day and Sunday school is supported by subscription. There are eight almshouses, occupied by paupers having large families. The poor law union of Walsingham comprises 50 parishes or places, under the care of 53 guardians, and contains, according to the census of 1831, a population of 20,866. The remains of a Danish encampment are visible towards the sea. This place confers the title of Baron on the family of De Grey.

WALSOKEN (*All Saints*), a parish, in the union of Wisbeach, hundred of Freebridge-Marshland, Western Division of the county of Norfolk; containing 1856 inhabitants. The village is pleasantly situated adjoining the town of Wisbeach, and extending a mile from it; and the pleasant walks in its vicinity are much frequented by the inhabitants of that town. The living is a rectory, valued in the king's books at £30. 13. 4.; present net income, £706; patron, — Watson, Esq. The church is a handsome structure, with a lofty spire.

WALTERSTONE (*St. Mary*), a parish, in the union of Dore, hundred of Ewyaslacy, county of Hereford, 15 miles (S. W. by W.) from Hereford; containing 149 inhabitants. The living is a perpetual curacy; net income, £136; patron, Edmund Higginson, Esq.; impropriator, Incumbent of Rollstone.

WALTHAM (*St. Bartholomew*), a parish, in the union of Bridge, hundred of Bridge and Petham, lathe of St. Augustine, Eastern Division of the county of Kent, 7 miles (S. S. W.) from Canterbury; containing 572 inhabitants. The living is a vicarage, with that of Petham annexed, valued in the king's books at £7. 15. 5.; present net income, £535; patrons, alternately, Archbishop of Canterbury and Sir J. C. Honywood, Bart.; appropriator of Waltham, Archbishop of Canterbury, and of Petham, Rev. J. K. S. Brooke, and the Archbishop of Canterbury of a small portion. The church is in the early style of English architecture. There are some remains of a chapel and castle at Ashenfield, and of a chapel at Waddenhall, in this parish. The parish is bounded on the east by the Roman Stane-street.

WALTHAM (*All Saints*), a parish, in the union of Caistor, wapentake of Bradley-Haverstoe, parts of Lindsey, county of Lincoln, 3¾ miles (S. by W.) from Great Grimsby; containing 545 inhabitants. The living is a rectory, valued in the king's books at £15. 10. 10.; present net income, £331: it is in the patronage of the Chapter of the Collegiate Church of Southwell. There is a place of worship for Wesleyan Methodists.

WALTHAM-ABBAS, county of Berks.—See WALTHAM, WHITE.

WALTHAM - ABBEY, or **HOLY - CROSS** (*Holy Cross and St. Lawrence*), a parish, in the union of Edmonton, hundred of Waltham, Southern Division of the county of Essex; comprising the market-town of Waltham-Abbey, and the hamlets of Holyfield, Sewardstone, and Upshire; and containing 4104 inhabitants, of which number, 2202 are in the town of Waltham-Abbey, 23½ miles (W. by S.) from Chelmsford, and 12½ (N. by E.) from London. The name of this place is compounded of the Saxon words *Weald* and *Ham*, signifying a residence in or near a wood; the adjunct is a term of distinction derived from an ancient abbey which was founded here. The town derived its origin, in the time of Canute the Great, from the facility and inducement for hunting afforded by the neighbourhood, which led Ralph de Toni, standard-bearer to that monarch, to build a few houses. A church was soon afterwards erected, principally for the preservation of the holy cross, to which many legends of miraculous efficacy were attached; and, upon a lapse of the property to the crown, Harold, to whom it had been given by Edward the Confessor, founded, in 1062, a monastery for Secular canons; for which, in 1177, Henry II. substituted monks of the order of St. Augustine, and dedicated it to the Holy Cross. At the dissolution, the revenue was valued at £1079. 12. 1.: within the choir, or eastern chapel, the body of Harold, who was slain in the battle of Hastings, with those of his brothers, Gurth and Leofwin, was entombed. In a place called Romeland, adjoining the abbey, was a house at which Henry VIII. occasionally resided; and to a conversation held here, on the important subject of the king's divorce, Dr. Cranmer was eventually indebted for the royal favour, and his ultimate elevation to the see of Canterbury. The town, which is spacious and irregularly built, consisting chiefly of one long street, is situated on the banks of the river Lea, which here divides into many streams, and separates the two counties of Essex and Herts about half a mile to the west, and also the parishes of Cheshunt and Waltham-Abbey: the inhabitants are well supplied with water. The gunpowder-mills belonging to government are situated here, and at present afford employment to nearly 200 persons, but in time of war from 400 to 500 were engaged. About 100 persons are occupied in printing silk handkerchiefs, and some business is done in the manufacture of pins, though it is by no means so extensive as formerly: here are also a brewery, flour-mill, and two malt-kilns. In the hamlet of Sewardstone is an extensive factory for throwing and spinning silk, in which between 200 and 300 persons are employed: at the west end of the town is the new cut from the river Lea. The line of the Northern and Eastern railway passes within half a mile of the town. The market is on Tuesday: fairs are held on May 14th and September 25th, for horses and cattle; and on the 26th is a statute fair for hiring servants. Courts leet and baron are held on Whit-Monday. A town-hall is now being erected adjoining the Leverton school-house, by voluntary contribution.

The living is a donative curacy; net income, £100; patrons and impropriators, Trustees of the Earl of Norwich. The church, which comprises only the nave of the old abbey church, is yet a spacious structure, in the Norman style of architecture, with a tower of later date; on the south side is the Lady chapel, now used as a vestry and school-room. In the interior are three tiers of semicircular arches, enriched with zigzag ornaments, supported on circular massive piers, some of which are also decorated with waving lines; the windows are of various kinds: beneath the Lady chapel is a fine crypt, now arranged in vaults. Among the various monuments and sepulchral tablets the principal is one to the memory of Sir Edward Denny, who died in 1599; under an arch of veined marble are recumbent effigies of him and his lady, and, beneath, those of their children, in a kneeling posture, surmounted by an appropriate inscription: a slab near the communion table retains the impression of an abbot with his crosier, the brass having been taken away. A district church, dedicated to St. Paul, for which a site and a house for the minister was given by Captain Sotheby, lord of the manor, has been erected in the hamlet of Sewardstone, by subscription, aided by a grant from the church-building commissioners; it was consecrated December 20th, 1837. There are places of worship for Baptists and Wesleyan Methodists. A free school, in which 20 boys and 20 girls receive instruction, is supported by voluntary contributions; and four boys are also instructed from an endowment by Mr. John Edmondson, who, in 1708, bequeathed for this purpose land now producing £18 per annum. The Leverton school, in which 20 boys and 20 girls are clothed and instructed, was founded about 1824, by the late Mrs. Rebecca Leverton, and is supported by an endowment bequeathed by Thomas Leverton, Esq.; the endowment was augmented by a bequest of £1000 by George Fawbert, Esq., in 1819. Green's almshouses, for four poor widows, originally built, and endowed with an orchard and barn now let for £20 per annum, by — Green, Esq., in 1626, were rebuilt and enlarged, in 1818, with four additional rooms, by means of a bequest from Robert Mason, in 1807. Eight poor widows reside in them: those occupying the additional rooms receive a weekly allowance of two shillings and sixpence each, arising from the interest of £1350 given in 1826, by Mowbray Woolard. The only remains of the venerable abbey, exclusively of the church, are a fine gate with a postern, the bridge leading to it, and some dilapidated walls.

WALTHAM, BISHOP'S (*St. Peter*), a market-town and parish, in the union of Droxford, hundred of Bishop's-Waltham, Droxford and Northern Divisions of the county of Southampton, 10 miles (E. N. E.) from Southampton, and 65 (S. W. by W.) from London; containing 2181 inhabitants. The river Hamble has its source about half a mile from the village, and passes through the piece of water termed Waltham Pond, which formerly deserved the appellation, given it by historians, of "a large and beautiful lake," but is now contracted by the encroachments of alluvial soil and rushes. On its banks are the remains of the once magnificent palace of the Bishops of Winchester, built in 1135, by Bishop Henry de Blois, brother of King Stephen, and greatly embellished by Wykeham: it continued to be the principal episcopal residence till the parliamentary war, when it was destroyed by the army under Waller: the extensive park in which it stood was afterwards converted into farms by Bishop Morley. The market is on Friday; fairs are held on the second Friday in May, for horses and toys; July 30th, for cheese and pedlery; and the first Friday after Old Michaelmas-day, for horses, stockings, and toys. A bailiff is appointed at the court of the manor, held by the Bishop of Winchester. This town has been made a polling-place for the northern division of the shire. The living is a rectory, valued in the king's books at £26. 12. 8½.; present net income, £915; patron, Bishop of Winchester. In addition to the parochial church, a chapel, dedicated to St. Peter, has been erected on Curdridge common, to which a district has been assigned: it contains 340 sittings, 220 of which are free, the Incorporated Society having granted £230 for this purpose. A free school was founded by Bishop Morley, who endowed it with an annuity of £10, which sum has been augmented, by subsequent benefactions, to £38, for which 36 boys are instructed in a National school, which is further supported by subscription.

WALTHAM, BRIGHT, commonly called BRICKLETON, or BRIGHT-WOLTON (*All Saints*), a parish, in the union of WANTAGE, hundred of FAIRCROSS, county of BERKS, 5 miles (W. S. W.) from East Ilsley; containing 442 inhabitants. The living is a rectory, valued in the king's books at £11. 15.; present net income, £700; patron, Bartholomew Wroughton, Esq. Here is a school in which 12 boys and 18 girls are educated and clothed at the expense of the rector.

WALTHAM-COLD, a parish, in the union of THAKEHAM, hundred of BURY, rape of ARUNDEL, Western Division of the county of SUSSEX, 5½ miles (S. E.) from Petworth; containing 449 inhabitants. The living is a vicarage, in the patronage of the Bishop of Chichester, the appropriator; net income, £65. This parish is bounded on the north by the Rother, and on the east by the Arun.

WALTHAM-CROSS, a ward, in the parish of CHESHUNT-ST. MARY, hundred and county of HERTFORD, 9 miles (S. by E.) from Hertford: the population is returned with the parish. This place received the adjunct to its name from a noble cross erected, on the eastern side of the high road, by Edward I. to his beloved consort Eleanor, whose corpse rested at Waltham abbey, on its way from Lincolnshire to London: it is hexangular, and highly enriched with tabernacle work and foliage, having pendant shields bearing the devices of England, Castile, Leon, and Ponthieu, and crowned statues of the queen, the left hand holding a cordon, and the right a sceptre, or globe. This beautiful monument having suffered much from mutilation, was, in 1757, at the instance of the Society of Antiquaries, enclosed by a brick wall, at the expense of Lord Monson, then lord of the manor; and one of the statues has recently been replaced, and the cross perfectly restored, and surrounded by an iron pallisade, by voluntary subscription of the inhabitants. Courts leet are held annually at Whitsuntide, and a court baron in October. The river Lea separates this ward from the parish of Waltham-Holy-Cross, and the New River runs through the western portion of the ward. On the road leading from the village to the abbey the Northern and Eastern railway is carried over a dry bridge. At a short distance from the village is the mansion of Sir George B. Prescott, Bart., now in the occupation of Sir Henry Meux, Bart., it is built near the site of the palace of Theobalds, and is pleasantly situated in an extensive park. A chapel has been erected by voluntary contribution, aided by parliamentary grant, the living of which is a perpetual curacy, in the patronage of the Vicar of Cheshunt. There is a place of worship for Independents. Almshouses for four poor widows, founded and endowed by Beaumont Spital, were taken down in 1830, and have been lately rebuilt in the decorated style of English architecture.

WALTHAM, GREAT (*St. Mary and St. Lawrence*), a parish, in the union and hundred of CHELMSFORD, Southern Division of the county of ESSEX, 4½ miles (N. by W.) from Chelmsford; containing 2013 inhabitants. This parish, which is situated in a fertile district, and intersected by the river Chelmer, is about seven miles in length, and amply supplied with excellent water from numerous springs. The soil is rich and favourable to the growth of wheat, and the lands are in a high state of cultivation. The living is a vicarage, valued in the king's books at £18. 13. 4.; present net income, £294; patrons and appropriators, President and Fellows of Trinity College, Oxford. The church is a spacious edifice of brick, with an octangular tower surmounted by a spire, and contains 165 free sittings, the Incorporated Society having granted £70 in aid of the expense: there are several splendid monuments. Near the western gateway of the churchyard is an ancient building, called the Guildhall. Here is a Sunday National school.

WALTHAM, ST. LAWRENCE, a parish, in the union of COOKHAM, hundred of WARGRAVE, county of BERKS, 5¼ miles (S. W.) from Maidenhead; containing 739 inhabitants. The living is a vicarage, valued in the king's books at £7. 6. 8.; present net income, £211; patron, Lord Braybrooke. The church contains a fine monument in memory of Sir Henry Neville, one of the gentlemen of the privy chamber of Edward VI., who died in 1593. A fair is held on August 11th: a court baron takes place annually at Wargrave. Michael Wandesford, in 1712, bequeathed land now let for £6 a year, which, with a smaller bequest from Richard How, in 1652, is applied to the instruction of poor children. A school is supported by Lord Braybrooke, the lord of the manor, who has an ancient residence at Billingbear, in this parish. The line of the Great Western railway passes through the northern part of the parish. In a field between the church and the Bath road was a Roman station, where coins, urns, and tiles, have frequently been dug up.

WALTHAM, LITTLE (*St. Martin*), a parish, in the union and hundred of CHELMSFORD, Southern Division of the county of ESSEX, 4 miles (N. by E.) from Chelmsford; containing 674 inhabitants. This parish is situated on the river Chelmer, and on the turnpike road to Norwich through Sudbury and Bury; it comprises about 2000 acres of profitable land in good cultivation, and is rich in agricultural produce. The village is pleasantly situated and neatly built; there are two flour-mills driven by water. The living is a rectory, valued in the king's books at £11. 10.; patron, T. L. Hodges, Esq. The tithes have been commuted for a rent-charge of £670, subject to the payment of rates, which on the average have amounted to £111; the glebe comprises 12 acres, valued at £24. 12. 6. per annum. The church is a small edifice with an embattled tower, and contains several interesting monuments. There is a place of worship for Independents. Mr. Roger Poole, in the reign of Philip and Mary, gave property for the support of a parochial school; and John Aleyn, in 1660, gave £500 to be vested in land for apprenticing children with premiums of £15 each. A Sunday National school is supported chiefly from Poole's bequest, and partly by annual subscriptions.

WALTHAM, NORTH (*St. Michael*), a parish, in the union of Basingstoke, hundred of Overton, Kingsclere and Northern Divisions of the county of Southampton, 6 miles (S. W. by W.) from Basingstoke; containing 458 inhabitants. The living is a rectory, valued in the king's books at £15. 13. 4.; present net income, £379; patron, Bishop of Winchester. The church has been enlarged, and 120 free sittings provided, the Incorporated Society having granted £40 in aid of the expense. Here is a National school.

WALTHAM-ON-THE-WOLDS (*St. Mary Magdalene*), a parish, (formerly a market-town), in the union of Melton-Mowbray, hundred of Framland, Northern Division of the county of Leicester, 5¼ miles (N. E.) from Melton-Mowbray; containing 653 inhabitants. The living is a rectory, valued in the king's books at £19. 5.; present net income, £481; patron, Duke of Rutland. The church is principally in the decorated style, with portions of earlier date; it has three enriched stalls, and the font presents a curious admixture of the Norman and early English styles. Joseph and George Noble, in 1776, and Thomas Baker, left sums for the instruction of children, now producing £12. 13. 7. per annum, which is applied in aid of a National school.

WALTHAM, UP, a parish, in the union of West-Hampnett, hundred of Box and Stockbridge, rape of Chichester, Western Division of the county of Sussex, 6 miles (S. S. W.) from Petworth; containing 95 inhabitants. The village is situated on the road from London to Chichester, by Petworth. The living is a discharged rectory, valued in the king's books at £6. 2. 11.; patron, Colonel Wyndham. The tithes have been commuted for a rent-charge of £128, subject to the payment of rates, which on the average have amounted to £30; the glebe comprises 4¼ acres, valued at £10 per annum. The church is in the early style of English architecture, with a circular east end.

WALTHAM, WHITE, or WALTHAM-ABBAS (*St. Mary*), a parish, in the union of Cookham, hundred of Beynhurst, county of Berks, 4 miles (S. W.) from Maidenhead; containing 902 inhabitants. The living is a vicarage, united to the rectory of Shottesbrook, in 1744, and valued in the king's books at £10. 13. 4.; patron, Arthur Vansittart, Esq. There are places of worship for Independents and those of Lady Huntingdon's Connection. Here is a National school. The Great Western railway passes through the parish. Smewin's house, now occupied by a farmer, is surrounded by a moat, and is said to have been a hunting seat of Prince Arthur's, eldest son of Henry VII.: it was also the retreat of the learned Dodwell, first Camden Professor of Ancient History at Oxford, and a celebrated writer on ecclesiastical antiquity. The vicarage-house was partly paved with Roman bricks, and many Roman tiles, coins, and other relics, have been found near the church. There was formerly a chapel of ease at Feens, in this parish. Thomas Hearne, the antiquary, was born here, in 1678.

WALTHAMSTOW (*St. Mary*), a parish, in the union of West Ham, hundred of Becontree, Southern Division of the county of Essex, 6 miles (N. E. by N.) from London; containing 4258 inhabitants. The manor, according to the Norman survey, wherein it is called *Welannestun*, was in the possession of Judith, niece to the Conqueror; and having subsequently belonged to the Earls of Warwick, on the attainder and execution of Earl Thomas, in 1396, it lapsed to the crown. The name appears to be of Saxon origin, consisting of *weald*, a wood, and *ham*, a dwelling; the adjunct *stowe*, a place, being intended to distinguish this from other *Walthams* within the county; and the entire name being accurately descriptive of the village, which consists of numerous dwelling-houses and mansions, detached and encompassed with trees and woodland, and pleasantly situated on the borders of Epping Forest, through which a new road has been cut to Woodford, in order to form a nearer communication with the great road from London to Newmarket. The parish is separated from the county of Middlesex by the navigable river Lea, over which is a bridge, and on its banks are extensive copper and flour-mills, and an oil-mill. Courts leet and baron, for the manors within it, are held as occasion requires. The government of the parish was entrusted to a select vestry of seventeen persons, besides the minister and churchwardens, according to a grant of Bishop Montaigne, in 1624, which does not appear to have been otherwise acted upon. The living is a vicarage, valued in the king's books at £13. 6. 8.; patron and incumbent, Rev. W. Wilson, B.D.; impropriator, R. Orlebar, Esq. The church, situated on an eminence, is a neat structure, originally built of flint and stone, with a tower at the west end, which was partly rebuilt by Sir George Monox, who also built the chapel at the end of the north aisle, in 1535. The church was enlarged, repaired, and beautified in 1817, at an expense of about £2000: in the chancel is a circular window, divided into compartments, of stained glass, representing a "*Gloria*;" it was originally intended for Southampton castle, but was presented to this parish by Miss Russell. Among the various sepulchral memorials which adorn the interior are those of Sir George Monox, Lord Mayor of London in 1514, and his lady; a splendid monument of white marble, with figures as large as life, to Sigismond Trafford, his wife, and their infant daughter; and another in memory of Lady Lucy Stanley, erected by her husband, Sir Edward Stanley. In the churchyard is a white marble tomb, by Chantry, in memory of Jesse Russell, Esq., father of the above-named benefactress. At Chapel End, in this parish, a chapel of ease, dedicated to St. John, was erected, at an expense of £1800, raised by subscription. There are places of worship for Independents and Unitarians.

The almshouses and free school on the north side of the churchyard were founded in 1542, by Sir George Monox, and endowed with a rent-charge upon property in the city of London of £42. 17. 4., which has been augmented by subsequent bequests in rent-charges and dividends, from Edward Alford, Sir Henry Maynard, Richard Banks, John Harman, and William Bedford, to an annual income of about £155. The almshouses are occupied by eight men, married or single, and five single women. The schoolmaster's emoluments are about £85 per annum, for which he is required to give gratuitous instruction to boys of the parish not exceeding 38 in number; but the indefinite terms of the will have created a doubt as to whether a classical or English education was intended by the testator, and consequently each, in turn, has been adopted. A National school is supported by subscription and bequests, amounting to £17. 10. per annum: it was established in 1815, and the building was enlarged by subscription in 1825: a Sunday school is attached to it. An infants' school, in which are 170 children, was established in 1823, and is supported in a similar manner: the building adjoins the churchyard, and comprises a large school-room, with separate houses for the master and mistress. In a school belonging to the Independents 30 girls receive instruction, of whom 20 are clothed. Almshouses for six poor widows were built and endowed by Mrs. Mary Squires, in 1795, with stock producing an annual dividend of £87. Henry Maynard, one of the benefactors to the free school, in 1686, bequeathed for various charitable purposes property now producing an annual net income of about £200. Mary Newell, in 1810, bequeathed property in the funds, the dividends on which amount to £15, two-thirds being appropriated to apprenticing one boy annually, and one-third towards the support of the Sunday school. The churchwardens and other members of the vestry have under their discretionary control a fund of £273 per annum, arising from the bequests of Edward Corbett, in 1674, Thomas Colby, in 1633, and several others: it is chiefly distributed in coal: and another fund of £61. 10. per annum, bequeathed by James Holbrook, in 1805, and others, is appropriated towards supplying bread to the poor. George Gascoigne, a poet of considerable repute, and author of several dramatic pieces, was a native of this village; he died in 1578. The Rev. William Piers, D.D., Bishop of Bath and Wells, lies interred in the chancel of the church; he died at the advanced age of 94, and was at the time the oldest Bishop in Christendom, both with respect to years and date of consecration. Edward Rowe Mores, an eminent scholar and antiquary, and one of the principal agents in forming the Equitable Society for Assurance on Lives, was buried here in 1778. Thomas Cartwright, afterwards Bishop of Chester, and Edmund Chishall, a learned antiquary and divine, and author of Travels in Turkey, and antiquities of Asia before the Christian Era, were respectively vicars of the parish.

WALTON (*St. Michael*), a parish, in the union of Newport-Pagnell, hundred of Newport, county of Buckingham, 2 miles (N. by E.) from Fenny-Stratford; containing 114 inhabitants. The living is a rectory, valued in the king's books at £8. 9. 7.; present net income, £232; patron and incumbent, Rev. Valentine Ellis.

WALTON, a parish, in the union of Brampton, Eskdale ward, Eastern Division of the county of Cumberland; containing 481 inhabitants, of which number, 168 are in High Walton, 10½ miles (N. E. by E.), and 313 in Low Walton, 10 miles (N. E. by E.) from Carlisle. The living is a perpetual curacy; net income, £131; patrons, Heirs of the late C. Dacre, Esq.; impropriator, W. P. Johnson, Esq. The ancient Roman wall crossed the parish, which contained the station *Petriana*, the site of which is now called Castle Steads, and out of its ruins several houses have been built. From the blackness of the stones, it is thought that the ancient buildings had suffered greatly from fire: numerous inscriptions and other relics of antiquity have been discovered. A charity school, with a small endowment, was founded by J. Boustead, Esq.; and a Sunday school is supported by Mr. Johnson.

WALTON, a chapelry, in the parish and union of Chesterfield, hundred of Scarsdale, Northern Division of the county of Derby, 3 miles (S. W. by W.) from Chesterfield; containing 935 inhabitants.

WALTON, or DEERHURST-WALTON, a hamlet, in the parish of Deerhurst, Lower Division of the hundred of Westminster, Eastern Division of the county of Gloucester, 3¼ miles (S.) from Tewkesbury: the population is returned with the parish.

WALTON, a township, in the parish of Bishop's-Froome, hundred of Radlow, county of Hereford, 4½ miles (S.) from Bromyard; containing 101 inhabitants.

WALTON, a hamlet, in the parish of Knaptoft, union of Lutterworth, hundred of Guthlaxton, Southern Division of the county of Leicester, 4 miles (N. E. by E.) from Lutterworth; containing 234 inhabitants.

WALTON, a hamlet, in the parish of Paston, union and soke of Peterborough, Northern Division of the county of Northampton, 2¾ miles (N. N. W.) from Peterborough; containing 160 inhabitants.

WALTON (*Holy Trinity*), a parish, in the union of Wells, hundred of Whitley, Western Division of the county of Somerset, 3½ miles (S. W. by W.) from Glastonbury; containing 732 inhabitants. The living is annexed to the rectory of Street. The church has a tower rising from the centre. There is a place of worship for Wesleyan Methodists; and a National school has been established.

WALTON, a township, in the parish of Baswich, Eastern Division of the hundred of Cuttlestone, Southern Division of the county of Stafford, 2¾ miles (S. E. by E.) from Stafford: the population is returned with the parish.

WALTON, a township, in the parish of Eccleshall, Northern Division of the hundred of Pirehill and of the county of Stafford; containing 92 inhabitants.

WALTON (*St. Mary*), a parish, in the union of Woodbridge, hundred of Colneis, Eastern Division of the county of Suffolk, 10 miles (S. E. by E.) from Ipswich; containing 887 inhabitants. This parish is bounded on the north-east by the river Deben, on the south-west by the harbour of Harwich, and on the south by the North Sea. On the shore is a Martello tower for the defence of the coast; and there are still some small remains of Walton castle, a stronghold of the Bigods, in the parish of Felixtow, anciently a Roman station, of which nearly the whole has been washed away by the sea: it had the privilege of a mint, and large quantities of Roman coins have been found on the site. Languard Fort is within this parish. The living is a discharged vicarage, endowed with the rectorial tithes, with that of Felixtow annexed, and valued in the king's books at £4. 6. 8.; present net income, £290; patron, — Richards, Esq. There is a place of worship for Baptists; also a school in which 20 children are paid for by subscription. A cell of Benedictine monks, subordinate to the monastery of Rochester, was founded here in the reign of William Rufus, which continued till 1528, when it was given to Cardinal Wolsey towards the endowment of his intended colleges: there are very considerable remains of the ancient buildings.

WALTON (*St. Peter*), a parish, in the Ainsty of the city, and East Riding of the county, of York, $2\frac{1}{2}$ miles (E. by S.) from Wetherby; containing 237 inhabitants. The living is a perpetual curacy; net income, £75; patrons and impropriators, C. A. Fischer, Esq., and another. There is a place of worship for Wesleyan Methodists. The old Roman Watling-street crosses the river Wharf at a place called St. Helens, and passes through the parish to that part of the wall now called Redgate.

WALTON, a township, in the parish of Great Sandall, union of Wakefield, Lower Division of the wapentake of Agbrigg, West Riding of the county of York, 3 miles (S. E. by S.) from Wakefield; containing 376 inhabitants. A school is endowed by Mrs. Neville with £6. 6. per annum.

WALTON-CARDIFF (*St. James*), a parish, in the union, and Lower Division of the hundred, of Tewkesbury, Eastern Division of the county of Gloucester, $1\frac{1}{4}$ mile (E. S. E.) from Tewkesbury; containing 57 inhabitants. The living is a perpetual curacy; net income, £53; patrons, Warden and Fellows of All Souls' College, Oxford; impropriator, J. Merrell, Esq. The church was built in 1658.

WALTON-DEIVILE (*St. James*), a parish, in the Warwick Division of the hundred of Knightlow, Southern Division of the county of Warwick, $3\frac{3}{4}$ miles (W. N. W.) from Kington: the population is returned with the parish of Wellesbourn-Hastings. The living is a rectory, annexed to the vicarage of Wellesbourn-Hastings, and valued in the king's books at £4. 13. 4. The church was rebuilt by Sir C. Mordaunt, about 80 years ago.

WALTON, EAST (*St. Mary*), a parish, in the union and hundred of Freebridge-Lynn, Western Division of the county of Norfolk, 7 miles (N. W.) from Swaffham; containing 220 inhabitants. The living is a discharged vicarage, valued in the king's books at £6. 3. 4.; present net income, £157; patron, Andrew Hamond, Esq.; appropriator, Bishop of Ely.

WALTON, INFERIOR, a township, in the parish and union of Runcorn, hundred of Bucklow, Northern Division of the county of Chester, 2 miles (S.) from Warrington; containing 340 inhabitants. The Mersey and Irwell canal passes in the vicinity.

WALTON-IN-GORDANO (*St. Paul*), a parish, in the union of Bedminster, hundred of Portbury, Eastern Division of the county of Somerset, $11\frac{1}{2}$ miles (W.) from Bristol; containing 297 inhabitants. The living is a discharged rectory, valued in the king's books at £9. 15. 5.; patron, P. J. Miles, Esq. The tithes have been commuted for a rent-charge of £174, subject to the payment of rates, which on the average have amounted to £14; the glebe comprises 29 acres, valued at £54 per annum. The church is a plain modern building: there are some remains of a more ancient structure at the foot of the hill occupied by Walton castle, an octangular pile, embattled and crowned at each angle with a turret: the principal entrance is on the east, and the keep, which is octangular, is situated in the centre of the area.

WALTON-LE-DALE, a chapelry, in the parish and Lower Division of the hundred of Blackburn, union of Preston, Northern Division of the county palatine of Lancaster, 2 miles (S. E.) from Preston; containing 5767 inhabitants. It is distinguished as the scene of a great battle, fought Aug. 17th, 1648, between Cromwell and the Duke of Hamilton; also for a gallant achievement performed in 1715, by General, or Parson, Wood, and his congregation, in defending the passage of the Ribble against the Scottish rebels. In 1701, the Duke of Norfolk, the Earl of Derwentwater, and other leaders of the jacobites, incorporated themselves by the style of the "Mayor and Corporation of the ancient borough of Walton," and held their meetings in a small public-house here, concealing their real motives under the guise of ludicrous transactions; they kept a register, a mace, a sword of state, and other mock insignia of office: the society, notwithstanding the diminution of the number of its members by the unsuccessful rebellion of 1715, existed till about 40 years ago, when it was entirely dissolved. This place is situated on an eminence which commands fine views of the vale of Ribble on one side, and of the vale of Derwent on the other. Both these valleys are extremely picturesque, the banks of their respective rivers being steep and richly clothed with wood. The back ground of the Ribble is formed by the high and extensive ranges of Longridge and Pendle; and that of the Derwent by Billinge hill, and an abrupt elevation crowned with the ruins of Houghton Tower, the ancient baronial residence of the family of that name. Here are three large cotton-manufactories and several printing establishments, affording employment to the greater portion of the inhabitants. The living is a perpetual curacy; net income, £150; patron, Vicar of Blackburn. The chapel, dedicated to St. Leonard, is principally in the later style of English architecture. Her Majesty's commissioners have made a conditional grant for building a church. There is a National day and Sunday school, supported by subscription. Another school-house was built by subscription among the inhabitants, in 1672; it is endowed with about £16 per annum, for teaching all children who apply.

WALTON-LE-SOKEN, or WALTON-ON-THE-NAZE (*All Saints*), a parish, in the union and hundred of Tendring, Northern Division of the county of Essex, $13\frac{1}{2}$ miles (S. E. by E.) from Manningtree; containing 469 inhabitants, according to the census of 1831, since which period the population has been rapidly increasing. This parish, which is bounded on three sides by the sea, forms a noted promontory, called the Naze, from the Saxon term, signifying a nose of land. Imbedded in the clay, which composes the basis of the cliffs, are found, among various alluvial remains, some curious fossils, the tusks of elephants, with the horns, bones, and teeth of other huge animals, which have been usually discovered after the ebbing of very strong tides; and in the cliffs are found a variety of shells and fossil remains. The shore abounds with pyrites, chiefly of wood, of which immense quantities have been here manufactured into the crystal commonly called green copperas, or sulphate of iron; but the works having gone to decay and been entirely removed, the pyrites are at present collected and sent to other places, to undergo the like process. Nodules of argillaceous clay, which continually fall from the cliffs and harden into stone, are gathered and conveyed to London and Harwich, for making Roman cement. The beach is a delightful promenade, and affords superior facilities for bathing, the ebb tides leaving a firm smooth sand several miles in extent; which advantages have, of late years, occasioned a number of persons from this and the adjoining counties, as well as from the metropolis, to resort hither for the benefit of cold and warm sea-bathing, for whose accommodation convenient machines and baths are in constant readiness; and a highly respectable hotel, with a bazaar, reading and billiard-rooms attached, and many lodging-houses of various sizes and descriptions, have been erected. A handsome row of houses, called Walton Terrace, has been built overlooking the sea; and various detached lodges are in course of erection. Coaches run daily to Colchester, 17 miles distant, and the Ipswich and Harwich steamers call regularly during the season. Adjoining the hall is a square tower, built by the corporation of the Trinity House, as a mark to guide ships passing or entering the port of Harwich; and there were two Martello towers within the parish, one of which has very lately been taken down.

This district, comprising three parishes, takes the distinguishing appellation "Le Soken" from the Saxon word *Soc*, or *Soca*, signifying "power, authority, or liberty to minister justice and execute laws." The living is a discharged vicarage, consolidated, with that of Thorpe-le-Soken, with the vicarage of Kirby-le-Soken, in the jurisdiction of the peculiar court of the Sokens, and subject, by letters patent granted by Queen Mary, to the visitation of the Bishop of London: it is valued in the king's books at £9; impropriators, Hope Insurance Company, London. The wills and records are deposited at the residence of the registrar of the peculiar court, who is also steward of the manor, at Harwich. The church was consecrated and erected by Bishop Porteus, in 1804, the ancient structure having, a few years previously, been entirely swept away by the sea, as well as the churchyard and every house near it but one of the old village. It was built to contain only about 150 sittings, and consequently being found much too small for the increasing population, particularly during the visiting season, the south side was taken out, and the building enlarged twofold in 1832: this being also inadequate, a further enlargement took place in 1835, by which means accommodation has been provided for 600 persons. The expenses, amounting to about £1000, were almost entirely defrayed by voluntary contributions. Here was formerly an endowment of a prebend in the cathedral of St. Paul's, but it has long since been consumed by the encroachment of the sea: the dignity is now held as "*Prebenda Consumpta per Mare.*" School-rooms for infants' and Sunday schools are in course of erection; and there are about 35 acres of land, left chiefly by Mr. John Sadler, in 1563, for the benefit of the poor, the proceeds of which are managed and distributed by trustees.

WALTON-ON-THE-HILL (*St. Mary*), a parish, in the union and hundred of West Derby, Southern Division of the county palatine of Lancaster; containing 22,575 inhabitants, of which number, 1400 are in the township of Walton-on-the-Hill, 3 miles (N. by E.) from Liverpool. In consequence of its proximity to Liverpool, this place has greatly advanced in population, and has become the residence of numerous merchants, retired tradesmen, &c. There is a house of correction at Kirkdale, at which place are held the adjourned general quarter sessions for the hundred, and the petty sessions for the Kirkdale division thereof. The living is a vicarage, endowed with the rectorial tithes, and valued in the king's books at £6. 13. 4.; present net income, £294; patron, J. S. Leigh, Esq.: the rectory is valued in the king's books at £69. 16. 10½. The church has been partially rebuilt, and is partly in the decorated style of English architecture: it contains 60 free sittings, the Incorporated Society having granted £40 in aid of the expense. Up to 1698 it was the mother church of Liverpool, and the present church of St. Nicholas there was a chapel of ease under the Vicar of Walton. The church of St. James, in Toxteth Park, is dependent on Walton. A chapel has been erected at Poplar Grove, containing 1100 sittings, 640 of which are free, the Incorporated Society having granted £400 in aid of the expense; and I. Gladstone, Esq., has applied to Her Majesty's commissioners for the perpetual patronage of a chapel which he proposes to build and endow at Toxteth Park. A school containing 47 boys is supported by endowment.

WALTON-ON-THE-HILL (*St. Peter*), a parish, in the union of Reigate, First Division of the hundred of Copthorne, Western Division of the county of Surrey, 4¼ miles (S. by E.) from Epsom; containing 352 inhabitants. The living is a discharged rectory, valued in the king's books at £12. 6. 5½.; present net income, £346; patron, Sir B. H. Carew. The body of the church was rebuilt in 1826 by the parishioners, aided by a grant from the Incorporated Society, for which 60 additional sittings are free; and an elegant octagonal tower was erected at the expense of Mrs. Anne Paston Gee, the late patroness. The church contains a curious ancient leaden font of nine compartments, in each of which is a figure in a sitting posture; and the windows of the chancel contain some remains of ancient stained glass. Here is a National school. Walton Place, the old manor-house, now occupied as a farm-house, bears evident marks of having been once strongly fortified. Roman tiles and pottery have been dug up in an enclosure on Walton Heath, an ancient earthwork, where also a brass figure of Æsculapius was found. There are some springs in the parish, the water of which is of a mineral quality.

WALTON-ON-THE-WOLDS (*St. Mary*), a parish, in the union of Barrow-upon-Soar, hundred of East Goscote, Northern Division of the county of Leicester, 4 miles (E.) from Loughborough; containing 289 inhabitants. The living is a rectory, valued in the king's books at £15; present net income, £403; patron and incumbent, Rev. A. Hobart, who supports an infants' school.

WALTON-ON-TRENT (*St. John the Baptist*), a parish, in the union of Burton-upon-Trent, hundred of Repton and Gresley, Southern Division of the county of Derby, 4¼ miles (S. W.) from Burton-upon-Trent; containing 408 inhabitants. The living is a rectory, with that of Rosliston annexed, valued in the king's books at £17. 2. 8½.; present net income, £828; patron, Lord C. Townshend. The church was, a few years since, repaired at a considerable expense, defrayed by subscription: it contains several ancient monumental tombs. Edward II. forded the Trent at this place, in pursuit of Thomas Earl of Lancaster, and the disaffected barons. In 1833, an act was obtained for building a bridge over the Trent to Barton-under-Needwood, in Staffordshire. A school-room has been erected by voluntary contributions, in which all the poor children of the parish are instructed on the National system: the school is partly supported by subscription, and partly with £20 a year arising from land bequeathed, in 1760, in support of a school, by two ladies named Levett and Bailey, who also gave a dwelling-house for the master.

WALTON, SUPERIOR, a township, in the parish and union of Runcorn, hundred of Bucklow, Northern Division of the county of Chester, 2¾ miles (S. S. W.) from Warrington; containing 238 inhabitants. The Duke of Bridgewater's canal passes in the vicinity.

WALTON-UPON-THAMES (*St. Mary*), a parish, in the union of Chertsey, First Division of the hundred of Elmbridge, Western Division of the county of Surrey, 3 miles (N. W.) from Esher, and 18 (S. W. by W.) from London; containing 2035 inhabitants. This place probably derived its name from the formidable Roman works yet visible within its precincts, the principal of which, on St. George's hill, is called the camp of Cæsar, who here gave battle to Cassivelaunus, at the head of the Britons: that chieftain, having first taken the precaution of driving stakes into the bed of the Thames, successfully opposed the vigorous attempts of the Romans to force the passage of the river, at a place still called Coway Stakes. Strata of iron-ore appear in different parts of the parish. The village is pleasantly situated on the banks of the river Thames, and is much frequented by anglers; it derives some importance from the many noble mansions in its immediate neighbourhood, and the elegant villas by which it is surrounded. Ashley Park, one of the many mansions built by Cardinal Wolsey, is now the residence of Sir Henry Fletcher, Bart.; the entrance-hall is a magnificent apartment recently restored to its original character; the grounds are embellished with stately trees, among which are some fine specimens of firs and cedars. Oatlands, formerly the property of the late Duke of York, and now the seat of Lord Francis Egerton, is partly in this parish and partly in that of Weybridge, the boundary line passing through the house. Apps Court, the residence of Richard Sharp, Esq., was once the residence of Cardinal Wolsey; but the ancient building has given place to a more modern and elegant mansion, the proprietor of which is subject to an old custom of distributing annually a quarter of wheat made into bread, and a barrel of small beer, amongst such travellers as may happen to present themselves on Nov. 13th; he enjoys the privilege of nominating four poor widows of Walton to a charity derived from the parish of Effingham. A farm-house is also mentioned as having been the seat of Bradshaw, who presided at the trial of Charles I., and which was afterwards occupied by Judge Jeffreys: some of the old carving is still preserved in several of the apartments. A curious wooden bridge, of three arches, over the Thames, was built about 1750, by Samuel Dicker, Esq.; and, more recently, another of brick and stone, of fifteen arches, across the low meadows, was added to it; but the former, falling to decay, was replaced by the present structure, built uniform with that which remained, and both now appear as one bridge of considerable length and beauty. The line of the London and South-Western railway intersects the parish, and a station has been established here. A fair for cattle, granted by Henry VIII., is held on Wednesday and Thursday in Easter week. The living is a discharged vicarage, valued in the king's books at £12. 13. 4.; present net income, £209: it is in the patronage of the Crown; impropriator, J. W. Spicer, Esq. The church is a handsome structure of considerable antiquity, containing many fine monuments, of which the most conspicuous is one by Roubilliac, to the memory of Richard Boyle, Viscount Shannon, who distinguished himself at the memorable battle of the Boyne: it contains the remains of several members of the Rodney family, and of many other eminent persons. A chapel of ease was built at Hersham, on a site given by Mr. William Holmes, at an expense of £1800, of which £500 was a grant from the Diocesan, and £250 a grant from the Incorporated Society; it is a neat edifice, in the early Norman style, containing 472 sittings, half of which are free; it has been endowed with £1000 by Sir H. Fletcher, Bart., and was consecrated on the 8th of November, 1839. There is a place of worship for Independents; and a National school is supported by subscription. Thomas Fenner, in 1635, bequeathed a messuage in the parish of St. Helen's, Bishopsgate, producing £210 per annum, which is appropriated to the relief of 20 poor families, and the apprenticing of poor boys. There are also some almshouses; and considerable benefit is derived from Henry Smith's charity. William Lilly, the celebrated astrologer, was interred here.

WALTON, WEST (*St. Mary*), a parish, in the union of Wisbeach, hundred of Freebridge-Marshland, Western Division of the county of Norfolk, 2¾ miles (N. by E.) from Wisbeach; containing 905 inhabitants. The living is a rectory in medieties, called West Walton Lewis and West Walton Elien; the former is valued in the king's books at £16. 13. 4.; present net income, £747; patron, Rev. C. H. Townshend: the latter is valued in the king's books at £16; present net income, £404: it is in the patronage of the Crown. A school is partly supported by an endowment of £22. 6. 2. per annum.

WALTON, WOOD (*St. Andrew*), a parish, in the hundred of Norman-Cross, union and county of Huntingdon, 7 miles (N. by W.) from Huntingdon; containing 305 inhabitants. The living is a rectory, valued in the king's books at £11; present net income, £471; patron, Admiral Hussey.

WALWICK, an extra-parochial liberty, locally in the parish of Warden, North-Western Division of Tindale ward, Southern Division of the county of Northumberland, 5¼ miles (N. W. by N.) from Hexham: the population is returned with Warden. This place, situated on the western bank of the North Tyne, and on the line of Severus' Wall, was the *Cilurnum* of the Romans, and the station of the *Ala Secunda Asturum*; its extent, which may still be traced, was, from east to west, 570 feet, and from north to south 400. Among the numerous relics discovered are a spacious vault, a mutilated statue of Europa neatly sculptured in freestone, and a curious tablet commemorative of the rebuilding of some edifice by the second wing of the *Astures*. Walwick Grange, formerly the seat of the Errington family, built out of an old tower, has been converted into a farm-house; and in Homer's-lane are fragments of an ancient cross.

WALWORTH, a township, in the parish of HEIGHINGTON, union of DARLINGTON, South-Eastern Division of DARLINGTON ward, Southern Division of the county palatine of DURHAM, $4\frac{1}{2}$ miles (N. W.) from Darlington; containing 155 inhabitants.

WALWORTH, a hamlet, in the parish and union of ST. MARY NEWINGTON, Eastern Division of the hundred of BRIXTON and of the county of SURREY, 2 miles (S.) from London: the population is returned with the parish. This place consists of a continued line of modern houses on the road to Camberwell, and also extends on the west to Kennington, and on the east to the Kent-road, several streets in each direction having been built within a few years. The Royal Surrey Zoological gardens were opened in August 1831; they occupy an area of about 17 acres attached to the manor-house, which have been enclosed and tastefully laid out in parterres of flowers, lawns, and shrubberies, intersected by gravel walks leading to the various objects of attraction within the grounds. A church, dedicated to St. Peter, was erected in 1825, at an expense of £19,126. 13., of which sum, one moiety was granted by the parliamentary commissioners, and the other advanced on loan for eight years without interest, to be repaid by a rate on the inhabitants: it is a spacious and handsome edifice of brick ornamented with stone, having at the western entrance a receding portico of four Ionic columns supporting a cornice and balustrade, with a slender square tower, ornamented at the quoins with pillars of the Corinthian order, and surmounted by a circular campanile turret, surrounded with Corinthian pillars and crowned with a conical dome. The churchyard, which is spacious, is enclosed with handsome iron palisades, and planted with trees. The living, which will eventually become a district incumbency, is a perpetual curacy, in the patronage of the Rector of Newington. There are places of worship for Baptists in East-street, Horseley-street, and Lion-street; for Independents, in York-street, Beresford-street, and West-street; for Wesleyan Methodists, in Walworth-road, Portland-street, and Providence-street; and a meeting-house for the followers of Johanna Southcote, in Amelia-street. A charity school for girls, established in 1723, and a school of industry in 1796, were united in 1818, and a neat and commodious building was erected for their use in Mount-street; in this establishment, which is supported by subscription among the dissenting congregations, 100 girls are clothed and instructed. Mrs. Clayton's charity school, in York-street, in which 60 girls are instructed, of whom 20 are also clothed, was founded in 1810, and is supported by the congregation of Independents attending that place of worship. The charity school in South-street, in which 100 girls are taught on the plan of the British and Foreign Society, was established in 1810, and is supported by subscription; and a similar school for boys, in Flint-street, was established in 1816; a spacious school-room, capable of receiving 400 boys, and a house for the master, were built by subscription; there are at present 320 boys in the school.

WAMBROOK (*St. Mary*), a parish, in the union of CHARD, hundred of BEAMINSTER, Bridport Division of the county of DORSET, $1\frac{3}{4}$ mile (S. W.) from Chard; containing 217 inhabitants. The living is a rectory, valued in the king's books at £8. 7. 1.; present net income, £262: it is in the patronage of Mrs. Martha Edwards. The church was anciently a chapel to the vicarage of Chardstock.

WAMPOOL, a township, in the parish of AIKTON, CUMBERLAND ward, and Eastern Division of the county of CUMBERLAND, $4\frac{3}{4}$ miles (N. by W.) from Wigton; containing 127 inhabitants.

WANBOROUGH, a parish, in the union of GUILDFORD, First Division of the hundred of WOKEING, Western Division of the county of SURREY, $4\frac{1}{4}$ miles (W.) from Guildford; containing 111 inhabitants. This parish, which comprises about 1800 acres, is intersected by a high ridge of land called the Hog's Back, which commands an extensive and richly diversified view of the surrounding country. The soil is light, and there are several chalk-pits in the parish, which is tithe free and exempt from ecclesiastical jurisdiction. The church, which was presented to by the abbot of Waverley, has fallen into decay: part of it was converted into a mausoleum for the Mangles family, by the late James Mangles, Esq., who was interred here in 1838; and the parishioners attend divine service in the parish church of Puttenham. Here is a school, supported by Captain Charles Mangles.

WANBOROUGH (*St. Andrew*), a parish, in the union of HIGHWORTH and SWINDON, hundred of KINGSBRIDGE, Swindon and Northern Divisions of the county of WILTS, $3\frac{1}{2}$ miles (E. by S.) from Swindon; containing 1016 inhabitants. The living is a vicarage, valued in the king's books at £21. 10. $7\frac{1}{2}$.; present net income, £375; patrons and appropriators, Dean and Chapter of Winchester. There is a place of worship for Wesleyan Methodists.

WANDSWORTH (*All Saints*), a parish, and the head of a union, in the Western Division of the hundred of BRIXTON, Eastern Division of the county of SURREY, 6 miles (S. W.) from London; containing 6879 inhabitants. The name of this place is derived from its situation on the river Wandle, which falls into the Thames in this parish. It consists chiefly of one street, occupying the declivities of two hills, on each of which are several mansions of a superior description: the inhabitants are supplied with water from springs. The manufactures comprise scarlet-dyeing, established for more than a century; hat-making, introduced by some French emigrants who settled here in the time of Louis XIV.; the making of bolting cloths, the printing of kerseymeres and silk handkerchiefs, the whitening and pressing of stuffs, and calico-printing. There are also three corn-mills and mills for the preparation of iron, white-lead, and linseed oil, now on the decline; vinegar works, distilleries, and a large brewery; the whole furnishing employment to several hundred persons. A tram-road extends from the basin, near the junction of the Wandle with the Thames, through Mitcham to Croydon; and the line of the London and South-Western railway passes through the parish, on which is a station about a mile from the church. A fair is held on Whit-Monday, for cattle, horses, and pigs; and there is a pleasure fair on the two following days. The town is within the jurisdiction of the new police. Petty sessions for the Western Division of the hundred of Brixton are held every Saturday; and a court of requests, for the recovery of debts under £5, comprises within its jurisdiction the parishes of Barnes, Battersea, Lower Tooting, Mortlake, Merton, Putney, Wandsworth, and Wimbledon. It has also lately been made a polling-place for the eastern division of the county.

The living is a vicarage, valued in the king's books at £15. 5. 5.; present net income, £840; patron, W. Borradaile, Esq.; impropriators, Trustees of Marshall's charity. The church is a plain brick structure, rebuilt in 1780, with the exception of a square tower at the west end; it contains several monuments. A new church, dedicated to St. Anne, was erected, a few years since, at an expense of £14,600, which was defrayed by Her Majesty's commissioners for building new churches: the living is a perpetual curacy; net income, £162; patron, Vicar of Wandsworth. There are places of worship for Baptists, the Society of Friends, Independents, and Wesleyan Methodists. The free, or Green-coat, school was founded and endowed in 1710, under the will of William Wicks; of late years it has been incorporated with the National school, in which about 200 children are educated, 25 of the boys and 30 of the girls being also clothed: the produce of the old endowment is appropriated exclusively to the use of 35 boys on the original foundation. The school of industry, founded in 1805, in which 40 girls are instructed in knitting, spinning, &c., besides apportioning rewards for good behaviour at service, is partly supported by endowment; and another school is partly supported by subscription. The Society of Friends have two schools, at one of which the eminent citizen, Sir John Barnard, was educated. Fifteen watermen of the parish receive £4 per annum each, the produce of bequests by Nicholas Tonnett, Sir Alan Broderick, and Sir Francis Millington. Here are some small funds for apprenticing poor children, and relieving the poor, for whose benefit a parochial library was instituted in 1826. Amongst the miscellaneous charities, those of the famous Alderman Smith, commonly called Dog Smith, who was born and buried here, deserve particular notice, extending not only to Wandsworth, but to most of the parishes in the county. The poor law union of Wandsworth and Clapham comprises 8 parishes, under the care of 19 guardians, and contains, according to the census of 1831, a population of 33,090. The first Presbyterian congregation established in the kingdom was at this place, in the year 1572. In Garratt-lane, between Wandsworth and Tooting, a mock election used to be held after every parliamentary election, to which Foote's dramatic production of the Mayor of Garratt has given celebrity.

WANGFORD (*St. Peter*), a parish, in the union and hundred of Blything, Eastern Division of the county of Suffolk, $3\frac{1}{2}$ miles (N. W. by N.) from Southwold; containing 792 inhabitants. The living is a perpetual curacy; net income, £79; patron and impropriator, Earl of Stradbroke. The church has the steeple at the north-east angle of the building, and is the place of sepulture of his lordship's family. A school containing 20 children is supported by the Countess of Stradbroke. A Cluniac priory, a cell to Thetford, was founded here before 1160 by Doudo Asini, steward to the king's household, or, as some think, by Eudo Dapifer, and at the suppression had a revenue of £30. 9. 5.

WANGFORD (*St. Denis*), a parish, in the union of Mildenhall, hundred of Lackford, Western Division of the county of Suffolk, 3 miles (S. W. by W.) from Brandon Ferry; containing 53 inhabitants. The living is a discharged rectory, annexed to that of Brandon, and valued in the king's books at £9. 11. $8\frac{1}{2}$. Robert Wright, and Joanna his widow, about 1644, bequeathed the respective rent-charges of £30 and £10 towards the establishment of a school for poor children of this parish and those of Brandon, Downham, and Weeting: the former gave a school-room and house for the master, and the latter certain land for repairs.

WANLIP (*St. Nicholas*), a parish, in the union of Barrow-upon-Soar, hundred of West Goscote, Northern Division of the county of Leicester, $3\frac{1}{4}$ miles (S. E. by S.) from Mountsorrel; containing 91 inhabitants. The living is a rectory, valued in the king's books at £14. 4. $4\frac{1}{2}$.; patron, Sir G. J. Palmer, Bart. The tithes have been commuted for a rent-charge of £275, subject to the payment of rates, which on the average have amounted to £8; the glebe comprises 26 acres. The church was founded in the reign of Richard II., by Sir Thomas Walsh, over whose tomb there is a brass plate. A school is supported by Lady Palmer. The river Soar, or the Leicester and Melton-Mowbray navigation, runs through the parish, and is crossed by a bridge. The sum of £5 per annum is annually divided amongst the poor. Near the old Fosse-road, which passes in the vicinity, a Roman tesselated pavement, coins, broken urns, and other relics, have been found.

WANSFORD (*St. Mary*), a parish, in the union of Stamford, soke of Peterborough, Northern Division of the county of Northampton, 36 miles (N. E.) from Northampton; containing 179 inhabitants. The living is annexed to the rectory of Thornhaugh. The church exhibits specimens of various styles of architecture. A school is endowed with £17. 13. 8. per annum for the instruction of ten children.

WANSFORD, a township, in the parish of Nafferton, union of Driffield, wapentake of Dickering East Riding of the county of York, $3\frac{1}{4}$ miles (E. S. E.) from Great Driffield; containing 152 inhabitants, many of whom are employed in the manufacture of cotton goods and carpets; the latter establishment, which is situated on the navigable river Hull, is the only one of the kind in this part of the kingdom. There is a place of worship for Wesleyan Methodists.

WANSTEAD (*St. Mary*), a parish, in the union of West Ham, hundred of Becontree, Southern Division of the county of Essex, 7 miles (N. E.) from London; containing 1403 inhabitants. This parish, which is separated from that of Barking by the river Rodon, is about 25 miles in circumference, and exclusively of the land situated within the limits of the forest of Waltham, has about 600 acres in high cultivation. The village is situated on the borders of Waltham Forest, near the main road from London to Cambridge, and is principally worthy of note as the site of that once princely mansion, Wanstead House, built in 1715, by Sir Richard, son of Sir Josiah Child, created Viscount Castlemain in 1718, and Earl of Tylney in 1731, and considerably extended and embellished by his descendants. This splendid mansion was surrounded by a very extensive and beautiful park, laid out with great taste, and interspersed with gardens, pleasure grounds, and grottos: it was the temporary residence of the Prince of Condé, but having come, by marriage, into the possession of the Hon. W. P. T. L. Wellesley, it was sold and demolished in 1822, since which time the park has been let out in portions for the grazing of cattle; of the buildings, nothing remains but the stables and out-offices. Snaresbrook, a hamlet in this parish, and situated on the borders of the forest, contains several handsome houses, the residences of respectable families; the scenery is beautifully picturesque, and enlivened with a fine sheet of water; and the immediate neighbourhood abounds with handsome seats and pleasing villas. The living is a rectory, valued in the king's books at £6. 13. 9.; present net income, £616; patron, Hon. W. P. T. L. Wellesley. The church was rebuilt about the year 1790; it is a handsome edifice of brick and Portland stone, with a fine Doric portico, and, at the west end, a cupola supported on eight Ionic pillars: the interior is of light and elegant appearance, the aisles being separated from the nave by columns of the Corinthian order: in the chancel is a window of beautifully stained glass, by Eginton, representing Christ bearing the Cross, in imitation of the altar-piece in the chapel of Magdalene College, Oxford; also a superb monument to the memory of Sir Josiah Child, Bart., who died in 1699, embellished with a marble effigy of the deceased. A free school, in connection with the National Society, in which 100 children are educated, 40 of whom are also clothed, is partly supported by the proceeds of £200 three per cents., the bequest of George Bowles, Esq., in 1805. About the year 1735, a tesselated pavement of considerable dimensions, brass and silver coins, fragments of urns, and other relics of antiquity, were dug up on the south side of Wanstead Park.

WANSTROW (*St. Mary*), a parish, in the union and hundred of Frome, Eastern Division of the county of Somerset, 5 miles (N. N. E.) from Bruton; containing 410 inhabitants. The living is a rectory, valued in the king's books at £13. 9. 9½.; patron, John Shore, Esq. The tithes have been commuted for a rent-charge of £320, subject to the payment of rates; the glebe comprises 57 acres, valued at £83 per annum. A new road has been formed in this parish, which has added considerably to its local advantages.

WANTAGE (*St. Peter and St. Paul*), a market-town and parish, and the head of a union, in the hundred of Wantage, county of Berks; containing 3282 inhabitants, of which number, 2507 are in the town of Wantage, 9 miles (S. W. by W.) from Abingdon, 24 miles (W. N. W.) from Reading, and 60 (W.) from London. This town, celebrated as the birthplace of Alfred the Great, and as a royal residence in the time of the West Saxons, was made a borough after the Conquest, through the influence of Fulk Fitz-Warren, who had obtained a grant of the manor from Bigod, Earl Marshal of England. It is situated at the edge of the Vale of White Horse, on a branch of the river Ock: the streets are irregular, but contain many good houses. The town is lighted, paved, and watched, under a local act; and the inhabitants are supplied with water from wells, and from a brook which runs into the river. The principal branches of trade and manufacture are those of sacking, twine, malt, and flour; coal is brought hither, and corn, flour, and malt sent to different parts, by means of a branch of the Wilts and Berks canal, which comes up to the town, affording a communication with Bath, Bristol, and London; and the line of the Great Western railway passes within two miles to the north of the town. The market is held every alternate Saturday, chiefly for corn, also for pigs and cattle. Fairs are held on the first Saturdays in March and May, for cattle and cheese; July 18th, for cherries; and October 18th, a statute fair: a cheese fair is also held on the first Saturday in every month. The town-hall, which is in the centre of the market-place, was erected in 1835. The petty sessions for the division are held here every Saturday: a manorial court is held annually; and this has been made a polling-place for the county.

The living is a vicarage, in the patronage of the Dean and Canons of Windsor (the appropriators), valued in the king's books at £35. 2. 8½. The tithes have been commuted for a rent-charge of £750, subject to the payment of rates. The church is a spacious and handsome cruciform structure, with a square embattled tower rising from the intersection; it is said to have been built by some of the Fitzwarrens, to different members of which family there are several monuments. A district for a chapelry has been formed out of this parish. There are places of worship for Baptists, Independents, and Wesleyan Methodists. About £400 per annum, the proceeds of certain town lands, bequeathed in the reign of Henry VI. and VII., and in 1598, and vested in twelve governors for the support of a free grammar school, is now appropriated to the purchase of cloth to the amount of £100, the payment of £160 in pensions to aged widows, the maintenance of apartments for poor persons, and to the support of an English school, in which 24 boys and 18 girls are educated. Twelve almshouses were founded and endowed by Richard Styles, in 1680, with land in Hampshire, producing about £70 per annum. The poor law union of Wantage comprises 34 parishes or places, under the superintendence of 37 guardians, and contains, according to the census of 1831, a population of 16,120. King Alfred, whose memory is here retained by a petrifying spring, called "Alfred's Well," was born in 849, and died in 901. Dr. Joseph Butler, Bishop of Durham, and author of "The Analogy," was born here in 1692; as was also the Rev. Isaac Kimber, a learned theological writer: the former died in 1752, and the latter in 1755.

WANTISDEN (*St. John the Baptist*), a parish, in the union and hundred of Plomesgate, Eastern Division of the county of Suffolk, 4¾ miles (N. W. by W.) from Orford; containing 125 inhabitants. The living is a perpetual curacy; net income, £64; patron and impropriator, N. C. Barnardiston, Esq. The church has many Norman details.

WAPLEY (*St. Peter*), a parish, in the union of Chipping-Sodbury, Lower Division of the hundred of Grumbald's-Ash, Western Division of the county of Gloucester, 2½ miles (S. S. W.) from Chipping-Sodbury; containing 253 inhabitants. The living is a discharged vicarage, valued in the king's books at £7. 18.; present net income, £400; patrons and appropriators, Dean and Chapter of Bristol.

WAPLINGTON, a township, in the parish of Allerthorpe, union of Pocklington, Wilton-Beacon Division of the wapentake of Harthill, East Riding of the county of York, 2¾ miles (S. W.) from Pocklington; containing 18 inhabitants.

WAPPENBURY (*St. John the Baptist*), a parish, in the union of Warwick, Southam Division of the hundred of Knightlow, Southern Division of the county of Warwick, $6\frac{3}{4}$ miles (N. N. W.) from Southam; containing 252 inhabitants. The living is a vicarage, valued in the king's books at £8; present net income, £60; patron and impropriator, Lord Clifford.

WAPPENHAM (*St. Mary*), a parish, in the union of Towcester, hundred of King's-Sutton, Southern Division of the county of Northampton, 5 miles (W. S. W.) from Towcester; containing 568 inhabitants. The living is a rectory, valued in the king's books at £21. 9. $9\frac{1}{2}$.; present net income, £354; patron, Bishop of Lincoln. There is a place of worship for Wesleyan Methodists; also a school partly supported by the parish. At Astwell is an ancient mansion-house, formerly the seat of the Earls Ferrars.

WAPPING (*St. John the Evangelist*), a parish, adjoining the eastern portion of the city of London, in the union of Stepney, Tower Division of the hundred of Ossulstone, county of Middlesex; containing 3564 inhabitants. This place, originally overflowed by the Thames, was first recovered from inundation, and denominated *Wapping Wash*, in the reign of Elizabeth, under whose auspices it was enclosed and defended by walls. In the early part of the reign of Charles II., it comprised only one long street, which extended from within a quarter of a mile from the Tower, along the northern bank of the Thames, to the entrance of St. Katherine's Docks. In the reign of William and Mary it was made a parish, by act of parliament. About the end of the last century, upwards of 60 houses and other buildings were destroyed by fire, and several lives were lost, from the explosion of some barrels of gunpowder: the damage sustained on this occasion was estimated at more than £200,000. On the abdication of James II., the notorious judge Jeffreys, who had fled in order to escape the probable effects of popular rage, assumed the disguise of a sailor, and concealed himself for a short time in an obscure part of Wapping, but was at last discovered and committed to the Tower, where he died in a few days. The parish, part of which is in the precincts of Wellclose, in the liberty of the Tower, consists of several streets, which are well paved and lighted with gas, the main street having been widened in several places within the last few years; and the inhabitants are well supplied with water. It is within the jurisdiction of the court of requests for the Tower Hamlets, for the recovery of debts under £5. The business transacted is chiefly of a maritime and commercial character, to the growth of which the construction of the London docks has materially contributed: they occupy more than twenty acres of ground, extend nearly as far as Ratcliffe-Highway, are enclosed by a wall, and contain numerous and extensive warehouses and cellars; the larger dock, called St. George's, is capable of receiving 500 ships, and the smaller, denominated Shadwell dock, about 50: they are entered from the Thames by means of three basons, and from corresponding stairs, called Hermitage, Old Wapping, and Old Shadwell stairs. The largest tobacco warehouse is 762 feet by 160, and the smallest 250 feet by 200. The first stone of the entrance bason, and those of the respective warehouses, were laid June 26th, 1802; and the docks were opened, in the presence of the Chancellor of the Exchequer, and some of the principal officers of state, with appropriate ceremonies, at the commencement of the year 1805. The whole of this immense establishment is under the control of the officers of the customs, and the capital of the Dock Company is estimated at £1,200,000. The living is a rectory not in charge; net income, £258; patrons, Principal and Fellows of Brasenose College, Oxford. The church contains a very fine monument, by Roubilliac. There is a place of worship for Roman Catholics. The free school was established by subscription, in 1704; in 1822, its funds were augmented by a bequest of £5000 from Samuel Troutbeck, of Madras, Esq., and it is further supported by voluntary contributions; 80 boys and 50 girls are educated and clothed. A Roman Catholic school is supported by subscription, and another is partly supported by the British and Foreign Sailors' Society. Thomas Dilworth, author of the spelling-book, and system of arithmetic, was master of the parochial free school.

WARBLETON (*St. Mary*), a parish, in the union of Hailsham, hundred of Hawkesborough, rape of Hastings, Eastern Division of the county of Sussex, $6\frac{1}{2}$ miles (N. by E.) from Hailsham; containing 1225 inhabitants. The living is a rectory, valued in the king's books at £13. 6. 8.; patrons, Trustees of Smith's charities. The tithes have been commuted for a rent-charge of £930, subject to the payment of rates, which on the average have amounted to £300; the glebe comprises 38 acres, valued at £52 per annum. Here is a National school.

WARBLINGTON, a parish, in the union of Havant, hundred of Bosmere, Fareham and Southern Divisions of the county of Southampton, $\frac{3}{4}$ of a mile (S. E. by E.) from Havant; containing 2118 inhabitants. This parish is bounded on the south by Langstone harbour. The living is a rectory, valued in the king's books at £19. 9. $4\frac{1}{2}$.; patron and incumbent, Rev. W. Norris. The tithes have been commuted for a rent-charge of £740, subject to the payment of rates, which on the average have amounted to £138; the glebe comprises 35 acres, valued at £90 per annum. The church is partly in the Norman and partly in the early English style of architecture, with an oratory at the end of each aisle: several stone coffins are deposited in both aisles. Here is a National school. The ruins called Warblington castle are the remains of a quadrangular mansion of the Montacutes, of which there are only the gateway and tower, the whole being surrounded by a deep fosse.

WARBOROUGH *cum* SHILLINGFORD (*St. Lawrence*), a parish, in the union of Wallingford, hundred of Ewelme, county of Oxford, 2 miles (N. E.) from Bensington; containing 681 inhabitants. This parish is partially bounded by the rivers Thame and Thames. The living is a perpetual curacy; net income, £350; patrons and impropriators, President and Fellows of Corpus Christi College, Oxford. The church is an ancient edifice, with a tower, built in 1666: the font is of lead on an octagonal stone shaft, and there are some interesting monuments, among which is one of marble to Francis Randolph, Margaret Professor of Divinity, and for some time Principal of St. Alban's Hall, Oxford. There is a place of worship for the Society of Friends; and a National school, in which 120 children are instructed, was established in 1838.

WARBOYS (*St. Mary Magdalene*), a parish, in the union of St. Ives, hundred of Hurstingstone, county of HUNTINGDON, $4\frac{1}{2}$ miles (S. S. E.) from Ramsey; containing 1550 inhabitants. The living is a rectory, valued in the king's books at £27. 10.; present net income, £1250; patron, T. Daniel, Esq. The church has been enlarged, and 150 free sittings provided, the Incorporated Society having granted £350 in aid of the expense. There is a place of worship for Baptists; and a National school has been established. The Rev. Robert Fowler, in 1824, bequeathed £200, which was invested on the security of the rates and taxes of the Fourth Fen district, and the interest, together with a small rental arising from the poor's lands, is distributed among the poor.

WARBRICK, a joint township with Layton, in the parish of BISPHAM, hundred of AMOUNDERNESS, Northern Division of the county palatine of LANCASTER, $2\frac{1}{2}$ miles (W. S. W.) from Poulton: the population is returned with Layton.

WARBSTOW (*St. Werburgh*), a parish, in the union of LAUNCESTON, hundred of LESNEWTH, Eastern Division of the county of CORNWALL, $8\frac{1}{2}$ miles (N. E.) from Camelford; containing 481 inhabitants. The living is a vicarage, annexed to that of Treneglos. The church has a curious Norman font. There are places of worship for Bryanites and Wesleyan Methodists; also a National school. Here is a remarkable ancient fortification, called Warbstow Barrow.

WARBURTON (*St. Werburgh*), a parish, in the union of ALTRINCHAM, hundred of BUCKLOW, Northern Division of the county of CHESTER, $6\frac{1}{2}$ miles (E. by N.) from Warrington; containing 510 inhabitants. The living is a perpetual curacy, annexed to the second mediety of the rectory of Lymm. The tithes have been commuted for a rent-charge of £249. 18. 6., subject to the payment of rates; the glebe comprises 3 acres. Here is a National school. The rivers Mersey and Botling run through the parish. Here was anciently a monastery of Premonstratensian canons, dedicated to St. Werburgh.

WARCOP (*St. Columba*), a parish, in EAST ward and union, county of WESTMORLAND, 3 miles (W. by N.) from Brough; containing 680 inhabitants. The living is a discharged vicarage, valued in the king's books at £9. 5. $1\frac{1}{2}$.; present net income, £194; patron, incumbent, and impropriator, Rev. W. M. S. Preston. There is a place of worship for Wesleyan Methodists. The river Eden runs through the parish, in the mountainous part of which there is said to be lead, but no mines have been opened. In the village is an ancient cross, which was recently brought from the common, at the expense of the lord of the manor, who holds his court annually in June, or July. Castle hill, near its south-east end, is supposed to be the site of an ancient castle; and Kirksteads, near it, that of a chapel.

WARD-END, a hamlet, in the parish of ASTON, Birmingham Division of the hundred of HEMLINGFORD, Northern Division of the county of WARWICK, 3 miles (S. E.) from Birmingham: the population is returned with the parish. Here is an ancient chapel of ease to Aston, supposed to have been founded about the year 1516, and which, having fallen into decay since the Reformation, has been restored by subscription: it contains 180 free sittings, the Incorporated Society having granted £150 in aid of the expense. The living is a perpetual curacy, in the patronage of the Vicar of Aston.

WARDEN (*St. James*), a parish, in the union of SHEPPEY, and within the liberty of the Isle of SHEPPEY, Upper Division of the lathe of SCRAY, Eastern Division of the county of KENT, $6\frac{3}{4}$ miles (E.) from Queenborough; containing 27 inhabitants. The living is a discharged rectory, valued in the king's books at £4. 17. $8\frac{1}{2}$.; present net income, £90; patron, V. B. Simpson, Esq. The tower of the church was built at the expense of Delmark Banks, Esq., in 1834, with part of the materials of the old London bridge.

WARDEN (*St. Michael*), a parish, in the union of HEXHAM, North-Western Division of TINDALE ward, Southern Division of the county of NORTHUMBERLAND; containing, with the extra-parochial liberty of Walwick, 2286 inhabitants, of which number, 540 are in the township of Warden, $2\frac{1}{2}$ miles (N. W. by N.) from Hexham. The living is a vicarage, with Newbrough annexed, valued in the king's books at £8. 16. 3.; present net income, £504; patron, T. W. Beaumont, Esq.; impropriators, Governors of Greenwich Hospital, R. R. Allgood, Esq., and others. The church is a cruciform structure, in the early style of English architecture. There is a chapel of ease at Haydon Bridge. A National school has been established. This parish lies between the wall of Severus and the North and South Tyne rivers, and near their junction is a petrifying spring. On an eminence, called Castle Hill, are vestiges of a circular British fortification, which was defended by a rampart of rough stone; it was subsequently occupied by the Romans, by whom additional earthworks were raised, and surrounded by a fosse: within the area the foundations of buildings, and several hand-mills, have been discovered. Not far from the vicarage-house are traces of another fort, termed Castle Hill.

WARDEN, CHIPPING (*St. Peter and St. Paul*), a parish (formerly a market-town), in the union of BANBURY, hundred of CHIPPING-WARDEN, Southern Division of the county of NORTHAMPTON, $7\frac{3}{4}$ miles (N. N. E.) from Banbury; containing 500 inhabitants. The living is a rectory, valued in the king's books at £26. 10.; present net income, £277: it is in the patronage of the Rt. Hon. Lady Susan North. Here is a National school. The Rev. William Smart, in 1466, bequeathed an estate, the present rental of which, amounting to about £80 per annum, is distributed in coal, clothing, and money among the poor. In the neighbourhood are some Saxon, or Danish, intrenchments, called Arberry Banks.

WARDEN-LAW, a township, in the parish and union of HOUGHTON-LE-SPRING, Northern Division of EASINGTON ward and of the county palatine of DURHAM, $8\frac{1}{2}$ miles (N. E.) from Durham; containing 54 inhabitants. It comprises a lofty eminence, crossed by a railway having a steam-engine on its summit, for drawing up and letting down coal-waggons employed in conveying coal from the Heaton pits.

WARDEN, OLD (*St. Leonard*), a parish, in the union of BIGGLESWADE, hundred of WIXAMTREE, county of BEDFORD, $3\frac{3}{4}$ miles (W. by S.) from Biggleswade; containing 660 inhabitants. The living is a discharged vicarage, united to that of Southill. In the cemetery is the mausoleum of Lord Ongley. A day and Sunday school is supported by two subscribers. A market and fair, granted in 1218, were formerly held, but both have been long disused. An abbey for Cistercian monks from Rivaulx was founded here, in 1135, by Walter L'Espec: it was dedicated to the Blessed Virgin Mary, and at the dissolution had a revenue of £442. 11. 11.

WARDINGTON, a chapelry, in the parish of CROPREDY, union and hundred of BANBURY, county of OXFORD, 5 miles (N. E. by N.) from Banbury; containing 824 inhabitants, some of whom are employed in the manufacture of plush and girth-webbing. The chapel is dedicated to St. Mary Magdalene. A school is supported by subscription, and a day and Sunday school is partly supported by endowment.

WARDLE, a township, in the parish of BUNBURY, union of NANTWICH, First Division of the hundred of EDDISBURY, Southern Division of the county of CHESTER, 4¼ miles (N. W.) from Nantwich; containing 144 inhabitants.

WARDLE, a joint township with Wuerdale, in the parish and union of ROCHDALE, hundred of SALFORD, Southern Division of the county palatine of LANCASTER, 3 miles (N. N. E.) from Rochdale: the population is returned with Wuerdale. There is a place of worship for Wesleyan Methodists.

WARDLEWORTH, a township, in the parish and union of ROCHDALE, hundred of SALFORD, Southern Division of the county palatine of LANCASTER, 1½ mile (N. by E.) from Rochdale; containing 9360 inhabitants. The principal part of the town of Rochdale is within this township. A day and Sunday school is supported by subscription.

WARDLEY (*St. Mary*), a parish, in the union of UPPINGHAM, soke of OAKHAM, county of RUTLAND, 2¾ miles (W. by N.) from Uppingham; containing 50 inhabitants. This parish is bounded on the south by the river Eye, which separates it from Leicestershire. The living is a discharged rectory, with the vicarage of Belton annexed, valued in the king's books at £10. 16.; present net income, £287: it is in the patronage of the Crown; impropriators of Belton, Dr. Bishop and J. Eagleton, Esq.

WARDLOW, a township, in the union of BAKEWELL, partly in the parish of HOPE, but chiefly in that of BAKEWELL, hundred of HIGH PEAK, Northern Division of the county of DERBY, 2 miles (E. by S.) from Tideswell; containing 149 inhabitants. In making a turnpike road through the village, in 1759, a circular heap of stones was opened, and found to contain the remains of about seventeen human bodies, interred in rude cells, or coffins of stone, apparently brought from a quarry about a quarter of a mile distant: by some they are supposed to have been the bodies of persons slain during the war between the houses of York and Lancaster, but others think that the tomb was a family burial-place.

WARE (*St. Mary*), a market-town and parish, and the head of a union, in the hundred of BRAUGHIN, county of HERTFORD, 2¼ miles (E. N. E.) from Hertford, and 21 (N.) from London; containing 4214 inhabitants. This place, anciently called *Guare*, derived both its origin and name from a *weare*, or dam, constructed on the river Lea, and strongly fortified by the Danes in 894, in order to defend their vessels; but Alfred, by draining the bed of the river, is said to have stranded them and destroyed the fort. His son Edward built a town here, which was of no importance till the reign of John, when Sayer de Quincy forced the thoroughfare of the bridge over the river Lea, by breaking the chain placed there until toll was paid to the king's bailiff at Hertford. This led to the diversion of the northern road through this town, instead of Hertford, which essentially conduced to its prosperity. In the reign of Henry III., a tournament was held here by Gilbert Marshall, Earl of Pembroke, in which he was slain; and, in the same reign, a Benedictine priory was founded by Margaret, Countess of Leicester, as a cell to the monastery of Ebralf, at Uttica, in Normandy, which was eventually bestowed by Henry V. on the Carthusian monastery of Sheen in Surrey: here was also a house of Franciscan friars. The town is situated in a valley, on the east side of the navigable river Lea, and consists of several streets, the principal extending about a mile along the high road from London to Cambridge: it is lighted, well supplied with both river and spring water, and is in a state of general improvement. A public library was established in 1795. The place was formerly subject to floods, but, from diverting into the river the water that flowed through Baldock-street to near the centre of the town, the inconvenience has been removed. The trade is chiefly in malt, which is made to a very great extent, and most of the London breweries are supplied from this town: there are 70 malting establishments in the town, and others are in progress of erection. The river Lea is navigable hence to Hertford and London, furnishing ample facilities for the conveyance of malt and corn to the metropolis, and for bringing back coal and manure. The market is on Tuesday; and fairs are held on the last Tuesday in April, and on the Tuesday before September 21st, for cattle. A market-house, erected by subscription, supported on sixteen arches, and containing an elegant assembly-room, was completed in 1827; the site was given by the lord of the manor. The town is under the superintendence of four constables and three headboroughs: the county magistrates hold a petty session every alternate Tuesday, and a court baron is held annually.

The living is a vicarage, with that of Thundridge annexed, valued in the king's books at £20. 10.; present net income, £333; patrons and impropriators, Master and Fellows of Trinity College, Cambridge. The church has been repewed, and 330 free sittings provided, the Incorporated Society having granted £125 in aid of the expense: it is situated in the centre of the town, and is an ancient cruciform edifice, with two sepulchral chapels and a west tower, surmounted by a low spire; in the interior is an antique font, in the later style of English architecture. In the churchyard is an ancient tombstone, bearing the following inscription:—"To the memory of William Mead, M.D., who departed this life on the 28th day of October, 1652, aged 148 years, 9 months, 3 weeks, and 4 days." A chapel has been erected at English Hall, containing 500 sittings, 380 of which are free, the Incorporated Society having granted £400 in aid of the expense. There are two places of worship each for Independents and Wesleyan Methodists, and one for the Society of Friends; also a Roman Catholic chapel at Old Hall Green. Humphrey Spencer gave a school-house, and funds for keeping it in repair and for the salary of the master, to which additions have since been made: 10 boys are educated free. A National school for boys is supported by voluntary contributions; a British school is partly supported by subscription; and a school for girls is maintained by two ladies. Here is an old school-house belonging to the Governors of Christ's Hospital, also a range of buildings for the accommodation of the nurses and children. There are seventeen almshouses for widows and other poor persons, some of which have small endowments: bequests to the amount of about £300 per annum have been left for the poor, for whose further relief a Lying-in Society and a Friendly Institution have been established. The poor law union of Ware comprises 15 parishes or places, under the superintendence of 21 guardians, and contains, according to the census of 1831, a population of 14,654. Near the town are two springs of excellent water, of which one, called "the Chadwell Spring," is also denominated the "New River Head," and the other the "Amwell Spring;" these, under the superintendence of the New River Company, supply the metropolis. In the grounds of Amwell House is a beautiful grotto. The great bed of Ware, sufficiently capacious to accommodate six couple, is of uncertain and conjectural origin: at the head is carved the date 1453. Four stone coffins were found in a field called Bury Field, at the south-west corner of the town, in 1802, supposed to have been the burial-place of the priory.

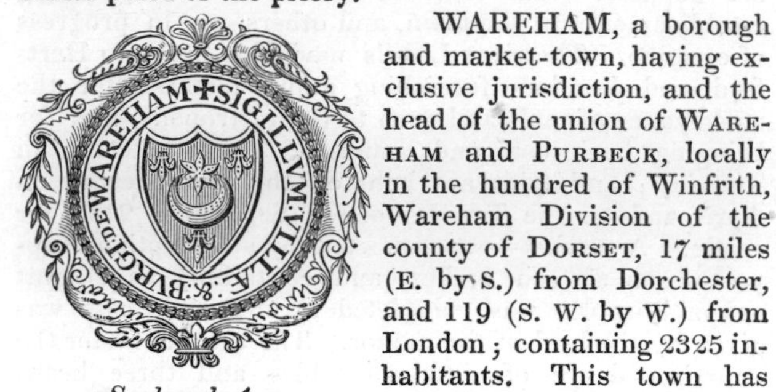

Seal and Arms.

WAREHAM, a borough and market-town, having exclusive jurisdiction, and the head of the union of WAREHAM and PURBECK, locally in the hundred of Winfrith, Wareham Division of the county of DORSET, 17 miles (E. by S.) from Dorchester, and 119 (S. W. by W.) from London; containing 2325 inhabitants. This town has been of great note in history; it existed in the time of the Britons, and was called *Durngueis*; by the Saxons it was named *Vœpham*, *Vepham*, *Veapham*, and *Thornsæta*, and in ancient records it is designated *Werham* and *Varama*, a compound of *Var-Ham*, "a habitation on a fishing shore." It has been supposed to occupy the site of the *Morionium*, or *Moriconium*, of Ravennas, but this is doubtful. That it was known to the Romans is demonstrated by the existence of a Roman way proceeding to Dorchester, and by the discovery of coins in the vicinity. In the Saxon times it was of some importance, and the burial-place of Brithric, the West Saxon king, about the year 800. The Danes soon afterwards massacred the inhabitants, and reduced the town to ruins; but it had so recovered in the time of Athelstan, that he established two mints in it. In 978, the body of Edward the Martyr, after his assassination at Corfe Castle, was temporarily interred here, and was removed by St. Dunstan, with much ceremony, to Shaftesbury. After the lapse of twenty years more, the town was again ravaged by the Danes, who, making the Isle of Wight their general place of rendezvous, proceeded thence to the mouth of the river Frome, and kept Wareham in a state of continual alarm. In 1138, the castle and town were seized for the Empress Maud, by Robert de Lincoln, but retaken and burnt by Stephen. On the intended expedition of John against France, in 1205, that monarch landed here, and three years afterwards garrisoned the town, which, in 1213, became the scene of the cruel execution of Peter of Pomfret, a religious enthusiast, and his son, because the former had foretold the deposition of that monarch. During the parliamentary war, Wareham was alternately possessed by the king and the parliament, but was finally given up to the former, on the surrender of Corfe Castle. In 1762, two-thirds of it were destroyed by fire; but, by a liberal subscription and an act procured for its restoration, it was, within two years, completely rebuilt.

The town, which has been greatly improved and is lighted with gas, is pleasantly situated on an eminence between the mouths of the Frome and the Piddle, which commands a prospect of Poole harbour, and in form resembles a parallelogram, occupying an area of about 100 acres, enclosed, except on the south, by a high wall, or rampart of earth; the intervening space between the wall and the town is laid out in large garden grounds, divided into regular squares by lanes, which still exhibit traces of some ancient buildings. The four principal streets, as well as the minor streets and lanes, diverge at right angles, and the former are open and spacious, corresponding with the cardinal points of the compass. The south and north entrances are formed by bridges over the Frome and the Piddle; the former is a handsome modern stone structure of five arches, erected in 1775, in lieu of an old bridge, which had stood from the time of William Rufus; the latter has three arches; from both bridges are raised stone causeways, that from the south leading to Stobborough, and 800 paces in length; the other to North Port, on the London road. Wareham was formerly a noted port, and in the time of Edward III., furnished three ships and 59 men for the siege of Calais; but the retreat of the sea from its harbour has long destroyed its importance, and withdrawn its commercial traffic: although, at very high tides, the water flows up nearly five miles to Holme bridge: the quay is on the south side of the town. The river Frome was anciently a celebrated salmon fishery, of which the profits formed part of the dowry granted by Henry VII. to his queen; but the fishery has long since declined, very few being now caught. The manufacture of shirt buttons and straw-plat, and the knitting of stockings, employ a great number of females; pipe-clay is obtained in large quantities from pits in the neighbourhood, and considerably more than 10,000 tons are annually shipped at Poole, for the Staffordshire potteries, also to London, Liverpool, Hull, and Glasgow, for the manufacture of tobacco pipes: coal, manufactured goods, and grocery, are imported. The gardens within the town produce a sufficient quantity of vegetables for the supply of its own market, and also for those of Poole and Portsmouth. The market is on Saturday; and fairs are held on April 17th and September 11th, for cattle, cheese, and hogs: the toll of the market and fairs belongs to the mayor: of late years six cattle markets have been held during the spring, and are well attended. This is a borough by prescription, and the inhabitants have had their privileges confirmed by several charters: the last, under which the town is now governed, was granted by Queen Anne, in 1703. The municipal body consists of a mayor, and six capital and twelve assistant burgesses, with a recorder, town-clerk, and inferior officers. The borough returned two members to parliament from the time of Edward I. to the 2nd of William

IV., when it was deprived of one by the act then passed to amend the representation. The right of election was formerly vested in the mayor, magistrates, and inhabitants paying scot and lot, and in the freeholders of lands and tenements who had been *bona fide* in the occupancy or receipt of the rents and profits thereof for the space of one whole year next before the election; except the same had come to such freeholders by descent, devise, marriage settlement, or appointment to some benefice in the church. By the 2nd of William IV., cap. 45, the privilege has been extended to the £10 householders of an enlarged district, comprehending 13,950 acres, which has been constituted the new elective borough, and the limits of which are minutely described in the Appendix, No. II.: the old borough comprised only 300 acres. The present number of registered voters is about 400: the mayor is the returning officer. The mayor, who is a justice of the peace, and coroner for the town and the isles of Purbeck and Brownsea, and the capital burgesses, hold quarter sessions of the peace, having exclusive jurisdiction: a court of record is held on the first Monday in every month, for the recovery of debts under £40; and a court baron is held annually by the lord of the manor. The town is a polling-place for the county.

Wareham comprises the parishes of the *Holy Trinity* (Within and Without), *St. Martin* (Within and Without), and *Lady St. Mary* (Within and Without). The living of Holy Trinity parish is a rectory, to which those of St. Martin's and St. Mary's were united in the year 1678, valued in the king's books at £7. 5. 5., and in the patronage of John Hales Calcraft, Esq.; the church has been appropriated for a school on the National plan. The living of St. Martin's is valued in the king's books at £8. 2. 6.: only the burial service is read in this church. The living of St. Mary's is a rectory not in charge: the church is a spacious and ancient structure, containing early and decorated portions: it is believed to have been conventual, and attached to a priory founded here before 876, when the monastery was destroyed by the Danes, and to have been rebuilt about the period of the Conquest: over a small north door is a rude piece of sculpture, representing the Crucifixion, surmounted by a Norman arch. In a small south chapel, of which the ceiling is richly groined, are the recumbent effigies of two warriors in complete mail: in this chapel the remains of Mr. Hutchins, rector of this place, and author of the "History and Antiquities of the County of Dorset," are deposited; and in the chancel are several mural monuments to the members of the Calcraft family. There were formerly two other parochial churches, St. Peter's and St. Michael's. There are places of worship for Independents, Wesleyan Methodists, and Unitarians: that for the Independents was erected in 1670, and its first minister was one of the confessors of Bartholomew-day.

The free school, situated in the parish of Lady St. Mary, was founded by George Pitt, of Strathfieldsaye, Esq., with a salary of £20 per annum for a schoolmaster; the endowment was further augmented by a bequest from Henry Harbin, in 1703, who left £200 for the purchase of land, now producing £10 per annum, which is paid to a mistress for teaching poor children to read. Another school, an infants' and day school combined, is partly supported by subscription. An almshouse, opposite St. Peter's church, was founded by John Streche, of Exeter, Esq., for six aged men and five women; it was rebuilt, in the year 1741, by Henry Drax and John Pitt, Esqrs., and by the corporation, and valued in the chantry roll at £11. 13. 10. The poor law union of Wareham and Purbeck comprises 27 parishes or places, under the care of 31 guardians, and contains a population of 14,579, according to the census of 1831. The antiquities of Wareham comprise the walls, supposed to have been built by the Britons. Bloody-bank was the place of execution, in 1684, of Mr. Baxter, Holman, and others, for their attachment to the Duke of Monmouth. Of the castle, situated in the south-west angle of the town, and originally supposed to have been built by the Romans, and rebuilt by the Conqueror, only the mound, or keep, called Castle hill, remains. The relics of the priory have been converted into a dwelling-house. At Stobborough, on opening a barrow, called King Barrow, in 1767, a large hollow trunk of an oak was discovered, in which were human bones wrapped up in a large covering composed of several deer skins, and a small vessel of oak, in the shape of an urn, conjectured by Mr. Hutchins to have been the drinking cup of the deceased, who, in the opinion of Mr. Gough, was some Saxon, or Danish, chieftain. Dr. John Chapman, tutor to the great Lord Camden; and Horace Walpole, Earl of Orford, were natives of this town.

WAREHORNE (St. Matthew), a parish, in the union of East Ashford, hundred of Ham, lathe of Shepway, Eastern Division of the county of Kent, 7¼ miles (S. by W.) from Ashford, containing 439 inhabitants. The living is a rectory, valued in the king's books at £19; present net income, £236: it is in the patronage of the Crown. Here is a National school. The Grand Military canal from Hythe to Rye passes through the parish; and the river Rother has its source here. Fairs are held on May 14th, for toys; and October 2nd and 3rd, for cattle; the former on Ham-Street green, the latter on Warehorne green.

WARESLEY, (St. Andrew), a parish, in the union of St. Neots, hundred of Toseland, county of Huntingdon, 4¼ miles (N. N. E.) from Potton; containing 241 inhabitants. The living is a discharged vicarage, valued in the king's books at £8. 16. 5½.; present net income, £303; patrons and impropriators, Master and Fellows of Pembroke Hall, Cambridge. A day and Sunday school is supported by subscription.

WARFIELD (St. Michael), a parish, in the union of East Hampstead, hundred of Wargrave, county of Berks, 6 miles (E. N. E.) from Wokingham; containing 1207 inhabitants. This parish once formed part of Windsor Forest: it is thickly studded with gentlemen's seats. At New Bracknell fairs are held on April 25th, August 22nd, and October 1st. The living is a vicarage, valued in the king's books at £13. 6. 8.; present net income, £150; patron, Maxwell Windle, Esq.; impropriators, the Landowners. The church contains some handsome monuments; and in a chapel (the burial-place of the Stavertons) attached to the north side of the chancel, is an ancient brass, with an effigy of one of that family. The sum of £200, bequeathed by the Hon. Gen. Wm. Hervey, has been expended in the erection of premises for a National school, on land given by the late Lord Braybrooke, the lord of the manor; the school is supported by voluntary contributions. Here are some remains of an intrenchment, called Cæsar's Camp, where many Roman coins have been found.

WARFORD, GREAT, a township, in the parish of ALDERLEY, union and hundred of MACCLESFIELD, Northern Division of the county of CHESTER, 5 miles (E. by S.) from Nether Knutsford; containing 349 inhabitants. There is a place of worship for Baptists; also a school partly supported by the rector.

WARFORD, LITTLE, a joint township with Martall, in the parish of ROSTHERN, union of ALTRINCHAM, hundred of BUCKLOW, Northern Division of the county of CHESTER, 4½ miles (E. S. E.) from Nether Knutsford: the population is returned with Martall.

WARGRAVE (*St. Mary*), a parish, in the union of WOKINGHAM, hundred of WARGRAVE, county of BERKS, 6½ miles (N. E. by E.) from Reading; containing 1423 inhabitants. This parish is bounded on the north by the river Thames. A market, granted in 1218 to Peter de Rupibus, Bishop of Winchester, which was held here on Monday, has long been disused. The living is a vicarage, valued in the king's books at £13. 13. 6½.; present net income, £226; patron, Lord Braybrooke, as lord of the manor, and impropriator of the great tithes; to whose ancestor, Sir Henry Nevill, the Billingbear estates, and the hundred of Wargrave, formerly attached to the see of Winchester, were granted by Edward VI. The church has a tower in the later style of English architecture. The first stone of a district church, to be dedicated to St. Peter, was laid at Knowl Hill on the 20th of July, 1839, by Lady Clayton East; it will contain 436 sittings, and near it will be a house for the minister, both erected by subscription on a site given by Thomas Wethered, Esq. It has been endowed with a rent-charge of £40 by Sir E. G. Clayton East, Bart., and £10 per annum by Lord Braybrooke. Richard Aldworth, in 1692, bequeathed an annuity of £5 for teaching four poor children; and Robert Pigot, Esq., in 1796, left by will £6700 three per cent. stock, directing the interest to be applied towards instructing and clothing 20 boys and 20 girls: these bequests have been applied to the support of a National school.

WARHAM, a village, in the union of WALSINGHAM, hundred of NORTH GREENHOE, Western Division of the county of NORFOLK, 2 miles (S. E. by E.) from Wells; containing 451 inhabitants. It comprises the parishes of *All Saints*, a discharged rectory, in the patronage of the Crown, valued in the king's books at £16; and *St. Mary*, and *St. Mary Magdalene*, united rectories, with that of Waterden annexed, valued jointly at £11. 6. 8.; patron, Earl of Leicester. The tithes of All Saints have been commuted for a rent-charge of £245, subject to the payment of rates, which on the average have amounted to £25; the glebe comprises 8 acres, valued at £8. 15. per annum. The tithes of St. Mary's have been commuted for a rent-charge of £373. 10., subject to the payment of rates, which on the average have amounted to £20. The amount of commutation for Waterden is stated under the head of that parish. There are small remains of an old baronial mansion, surrounded by a moat, and of some ancient fortifications.

WARK, a parish, in the union of BELLINGHAM, North-Western Division of TINDALE ward, Southern Division of the county of NORTHUMBERLAND; containing 861 inhabitants, of which number, 417 are in the township of Wark, 4½ miles (S. S. E.) from Bellingham. This parish is one of the six into which the late extensive parish of Simonbourn was divided, in 1814, under the authority of an act of parliament obtained in 1811: it is bounded on the east by the North Tyne, across which there is a ferry. The village is ancient, and was considerably improved, a few years since, by the erection of a handsome row of houses, with stone taken from some extensive ruins. The living is a rectory not in charge; net income, £240; patrons, Governors of Greenwich Hospital, who have erected a handsome church, the first stone of which was laid in October 1815, and it was opened for divine service on Aug. 10th, 1818; they also built the parsonage-house. There is a place of worship for Presbyterians; and a school is endowed with £45 per annum. About half a mile to the north of the village are the remains of an old church, a tumulus, and a cairn, in which urns and other relics have been found. Within the parish are vestiges of several ancient fortifications, said to have been thrown up by Edward III.; and on the bank of the river is Moat Hill, formerly occupied by a tower, and more recently used as an observatory to watch the movements of an enemy.

WARKLEY (*St. John*), a parish, in the union and hundred of SOUTH MOLTON, South Molton and Northern Divisions of the county of DEVON, 5½ miles (W. S. W.) from South Molton; containing 283 inhabitants. The living is a rectory, valued in the king's books at £14. 4. 7.; present net income, £215; patron, J. Gould, Esq. A school is supported by subscriptions.

WARKSBURN, a township, in the parish of WARK, union of BELLINGHAM, North-Western Division of TINDALE ward, Southern Division of the county of NORTHUMBERLAND, 5¼ miles (S.) from Bellingham; containing 278 inhabitants. Near the farm-house of Roses Bower is a medicinal spring, called Holy Well.

WARKTON (*St. Edmund*), a parish, in the union of KETTERING, hundred of HUXLOE, Northern Division of the county of NORTHAMPTON, 2¼ miles (E. N. E.) from Kettering; containing 301 inhabitants. The living is a rectory, valued in the king's books at £18. 16. 3.; present net income, £293; patron, Duke of Buccleuch. The church is chiefly remarkable for containing sumptuous monuments of the Montague family, two of which are by Roubilliac.

WARKWORTH (*St. Mary*), a parish, in the union of BANBURY, hundred of KING'S-SUTTON, Southern Division of the county of NORTHAMPTON, 2 miles (E.) from Banbury; containing, with the hamlets of Grimsbury and Nethercote (which form part of the adjoining parish of Banbury, in the county of Oxford, which extends into this county), 521 inhabitants. The living is annexed to the vicarage of Marston-St. Lawrence.

WARKWORTH (*St. Lawrence*), a parish, in the union of ALNWICK, partly in the Eastern Division of COQUETDALE ward, and partly in the Eastern Division of MORPETH ward, Northern Division of the county of NORTHUMBERLAND; containing 2478 inhabitants, of which number, 614 are in the township of Warkworth, 7 miles (S. E.) from Alnwick. This parish abounds with excellent coal, freestone, limestone, and whinstone. The village is situated a short distance westward from the sea, and is almost surrounded by the river Coquet, which is here crossed by an ancient bridge of two arches, having at the south end a tower gateway, formerly defended by an iron gate, through which the road passes: some valuable gems and pebbles are frequently found in its bed. This place consists of one main street leading from the castle to the bridge, and a short street opening to the church, of good modern houses, and contains two very commodious inns. A harbour is in course of formation. At the ancient cross, a market, granted by King John, was formerly held for provisions, but it has been long disused. Fairs for cattle are proclaimed on the first Thursday in May, and on the Thursdays before August 18th and November 23rd, one being held also on the last-mentioned day. Warkworth was anciently a borough, having been probably so constituted by King Ceolwulph, in the time of the Saxons. The Duke of Northumberland holds a court leet within 28 days after Michaelmas, at which a boroughreeve, two moorgrieves, three constables, and other officers, are annually chosen. The living is a vicarage, valued in the king's books at £18. 5. 7½.; present net income, £528; patron and appropriator, Bishop of Carlisle. The church is a handsome structure, said to have been originally founded by Ceolwulph, King of Northumbria, in 736, and rebuilt at a later period; it has a spire nearly 100 feet high: in its south-western extremity is a monument to the memory of Sir Hugh de Morwick, a Knight Templar, who gave the common to the inhabitants. There are places of worship for Scottish Seceders and Wesleyan Methodists. A building, now sometimes called the "Town Hall," was erected in 1736, by Mr. G. Lawson, for a school-house, and is used as such; but a more commodious one was erected by subscription in 1821, the children in which are educated on the National system: there are also a girls' National school, and an infants' school. Nearly adjoining the churchyard are some remains of a small Benedictine priory, founded by Nicholas de Farnham, Bishop of Durham, who died in 1257. The venerable and magnificent ruins of Warkworth castle occupy a fine elevation, rising from the margin of the river, south of the village, the moat by which it is surrounded enclosing more than five acres; the area is of an oblong form, on the north side of which stands the keep, on a lofty mound, encompassed by a wall 35 feet high, both in good preservation; but of the grand entrance to this once stately fortress only a few portions remain. It is not recorded by whom the castle was erected; the arms of the Percy family, however, appear to have been inserted in different parts of the building, at a much later period than that of its foundation. About a mile westward is an ancient hermitage, with a neat small chapel, hewn out of the solid rock, constructed without beams or timber, and curiously adorned, in the early style of English architecture, containing an altar and various other devotional emblems, with the representation of a recumbent female on a table monument placed in a niche near the altar, and that of a hermit standing over it, in a mournful attitude. This interesting retreat is celebrated in the beautiful poem entitled "The Hermit of Warkworth," published by Dr. Percy, Bishop of Dromore, in 1771. Warkworth gives the inferior title of Baron to the Duke of Northumberland.

WARLABY, a township, in the parish of AINDERBY-STEEPLE, union of NORTH ALLERTON, wapentake of GILLING-EAST, North Riding of the county of YORK, 2¾ miles (S. S. W.) from North Allerton; containing 76 inhabitants.

WARLEGGON (*St. Bartholomew*), a parish, in the union of BODMIN, hundred of WEST, Eastern Division of the county of CORNWALL, 5¾ miles (E. N. E.) from Bodmin; containing 274 inhabitants. The living is a discharged rectory, valued in the king's books at £5. 17. 6.; present net income, £125; patron, G. W. F. Gregory, Esq.

WARLEY, a township, in the parish and union of HALIFAX, wapentake of MORLEY, West Riding of the county of YORK, 2½ miles (W.) from Halifax; containing 5685 inhabitants, many of whom are employed in the manufacture of cotton, worsted, and stuffs. The chapelry of Sowerby-Bridge is in this township. There is a place of worship for Independents; also a school partly supported by subscription. In the neighbourhood is one of those remarkable rocking-stones, supposed to be of Druidical origin.

WARLEY, GREAT (*St. Mary*), a parish, in the union of ROMFORD, hundred of CHAFFORD, Southern Division of the county of ESSEX, 2 miles (S.) from Brentwood; containing 424 inhabitants. This parish, which is about seven miles in length, and one mile in average breadth, is separated from the parish of Little Warley by a rivulet which flows into the Thames. During the first revolutionary war with France, the common was a celebrated encampment. The surface is elevated, and the soil strong and heavy, resting on a substratum of clay, but under good cultivation producing average crops. The village is about two miles in length, and consists of well-built houses widely detached from each other. The line of the Eastern Counties railway passes through the north-eastern extremity of the parish. The living is a rectory, valued in the king's books at £14; present net income £404; patrons, Master and Fellows of St. John's College, Cambridge. The master of Ilford Hospital is impropriator of two-thirds of the great tithes of 903 acres, which have been commuted for a rent-charge of £90, subject to the payment of rates. The rectorial tithes have been commuted for a rent-charge of £520, subject to rates, which on the average have amounted to £61; the glebe comprises 10 acres. The church is an ancient edifice of brick, with a belfry turret of wood surmounted by a small spire. A National school is supported by the rector. Dr. Fulke, a puritan divine, and author of annotations on the Rhemish Testament, was rector of this parish; and Mr. Day, author of Sandford and Merton, was born here.

WARLEY, LITTLE (*St. Peter*), a parish, in the union of BILLERICAY, hundred of CHAFFORD, Southern Division of the county of ESSEX, 3½ miles (S. by E.) from Brentwood; containing 163 inhabitants. This parish is bounded on the west by a stream which flows into the Thames. A barrack for horse artillery, capable of receiving two troops, was erected here in 1804. The Eastern Counties railway passes within a mile of the parish. The living is a rectory, valued in the king's books at £11. 3. 9.; patron, Rev. John Pearson. The tithes have been commuted for a rent-charge of £280, subject to the payment of rates, which on the average have amounted to £42; the glebe comprises 32 acres, valued at £35 per annum. The church is a small ancient edifice with a tower, which has been rebuilt of brick; it contains two fine monuments to the Strutt family. A parochial school is supported by subscription. A small farm, with a house, producing £25 per annum, was bequeathed to the poor by Hugh Chappington, Esq., about a century since.

WARLEY-WIGORN, a hamlet, in the parish of HALES-OWEN, union of WEST BROMWICH, Lower Division of the hundred of HALFSHIRE, Dudley and Eastern Divisions of the county of WORCESTER, 3 miles (N. E. by E.) from Hales-Owen; containing 921 inhabitants. John Moore, in 1797, bequeathed £100, now producing £5 a year, for which fourteen poor children are educated.

WARLINGHAM *cum* CHELSHAM (*ALL SAINTS*), a parish, in the union of GODSTONE, Second Division of the hundred of TANDRIDGE, Eastern Division of the county of SURREY, 4¾ miles (S. S. E.) from Croydon; containing 454 inhabitants. The living is a vicarage, endowed with the rectorial tithes, united with Chelsham, and valued in the king's books at £11. 12. 11.; present net income, £471; patron, A. D. Wigsell, Esq. The church is principally in the early style of English architecture. A parochial school is supported by subscription. H. Atwood, Esq., in 1675, bequeathed two annuities of £10 each, chargeable on Warlingham and Chelsham court, to the curate of this parish; and erected an almshouse, which he endowed with £20 per annum, for four aged persons, two of whom were to be of this parish, and one each from the parishes of Chelsham and Sanderstead. The parish participates to a small extent in Henry Smith's charity.

WARMFIELD (*ST. PETER*), a parish, in the union of WAKEFIELD, Lower Division of the wapentake of AGBRIGG, West Riding of the county of YORK; containing 995 inhabitants, of which number, 752 are in the township of Warmfield with Heath, 3½ miles (E.) from Wakefield. The living is a vicarage, endowed with a moiety of the rectorial tithes, and valued in the king's books at £5. 4. 2.; present net income, £148; patrons, Master and Fellows of Clare Hall, Cambridge, the impropriators of the other moiety of the rectorial tithes. Lady Mary Bowles, in 1660, conveyed to trustees a building, to be used as a school-house for 10 poor boys, and a rent-charge of £20, of which £4 was to be appropriated in apprenticing one of the scholars annually. John Smyth, Esq., in 1729, left three houses and an annuity of £3, for educating six children; and John Smith, Jun., in 1731, bequeathed the interest of £50 for clothing six poor children of Heath, Kirkthorpe, and Warmfield annually. Othoneus Sagar, in 1558, founded and endowed, with a rent-charge of £12 upon his estate at Castleford, an almshouse for four poor women of this parish: it is situated at Kirkthorpe, and the inmates are benefited by two other small annuities. There is also an almshouse at Kirkthorpe, containing a common hall with seven apartments, and an adjoining cottage for a nurse, founded and endowed with a rent-charge of about £30 per annum, by John Freestone, in 1592, for seven poor, aged, unmarried men, of the parishes of Warmfield and Normanton.

WARMINGHAM (*ST. LEONARD*), a parish, partly in the union of NANTWICH, but chiefly in that of CONGLETON, hundred of NORTHWICH, Southern Division of the county of CHESTER; containing 1167 inhabitants, of which number, 372 are in the township of Warmingham, 3¾ miles (W.) from Sandbach. The living is a rectory, valued in the king's books at £12. 4. 7.; present net income, £631; patrons, Trustees of the late Lord Crewe. A free school, founded by Thomas Minshull, has an endowment in land of the annual value of £20.

WARMINGHURST, a parish, in the union of THAKEHAM, hundred of EAST EASWRITH, rape of BRAMBER, Western Division of the county of SUSSEX, 6 miles (N. W.) from Steyning; containing 113 inhabitants. The living is not endowed with any portion of the tithes, the minister receiving a voluntary stipend from the Duke of Norfolk.

WARMINGTON (*ST. MARY*), a parish, in the union of OUNDLE, chiefly in the hundred of POLEBROOK, but partly in that of WILLYBROOK, Northern Division of the county of NORTHAMPTON, 3¼ miles (N. E.) from Oundle; containing 617 inhabitants. The living is a discharged vicarage, valued in the king's books at £13. 6. 8.; present net income, £107; patron, Earl of Westmorland; impropriator, T. Gardner, Esq. The church is a beautiful structure, principally in the early English style, with an enriched tower and spire. There is a place of worship for Wesleyan Methodists; and two small bequests, by Mr. Blowfield and Mr. Elmes, are applied in support of a National school. The water of Chadwell Spring, in the neighbourhood, possesses some mineral properties.

WARMINGTON (*ST. MICHAEL*), a parish, in the union of BANBURY, Burton-Dassett Division of the hundred of KINGTON, Southern Division of the county of WARWICK, 5 miles (N. W. by N.) from Banbury; containing 470 inhabitants. The living is a rectory, valued in the king's books at £16. 3. 11¾.; present net income, £450; patrons, Trustees of Mr. Hulme's exhibitions to Brasenose College, Oxford. Captain Alexander Gordon, who was killed in the battle of Edge Hill, was buried in the churchyard. There is a place of worship for Wesleyan Methodists. A school-house was built by the Rev. H. B. Harrison, D.D., the late incumbent, and the school is partly supported by subscription. A Benedictine priory, subordinate to the abbey of St. Peter and St. Paul de Pratellis, or Preaux, in Normandy, was founded here in the time of Henry I., which, after the suppression of Alien houses, was granted by Henry VI. to the Carthusian priory at Witham, in Somersetshire. Nadbury camp, in this vicinity, where some fix the ancient *Tripontium*, is of a square form, rounded at the angles, and comprises about twelve acres.

WARMINSTER (*ST. DENIS*), a market-town and parish, and the head of a union, in the hundred of WARMINSTER, Warminster and Southern Divisions of the county of WILTS, 21 miles (W. N. W.) from Salisbury, and 97 (W. S. W.) from London; containing 6115 inhabitants. Antiquaries are at variance respecting the etymology of its name: according to Camden this was the *Verlucio* of the Romans, and he considers the first syllable to be a corruption of that of its ancient appellation: others deduce it from the *wears* near which the town stood, and from a minster, or monastery, stated to have been once situated in its vicinity; which conjecture receives confirmation from the supposed site being still called "The Nunnery," and a walk upon the neighbouring hill, "Nuns' Path." At the Conquest it was denominated *Guermistre*, and, according to the Norman survey, was then possessed of many privileges: at a later period it became celebrated for its corn market, which, in the time of Henry VIII., appears to have been of considerable note. The town is situated on the river Willey, near the south-western extremity of Salisbury Plain, and consists principally of one street, nearly a mile long, well paved by the commissioners of the roads, and of clean appearance: it is considered one of the most healthy towns in England, and has been remarkable for the longevity of its inhabitants. The malt trade was formerly carried on here to a greater extent than in any other town in the West of England, and it is still a considerable branch of trade; the manufacture of broad cloths and kerseymeres was also much more extensive than it is at present: the silk business has been recently introduced, and affords employment to many females and children. The market, which is on Saturday, is very considerable for the sale of corn: the whole is previously warehoused in the town, and a sack from every load is pitched in the market-place. Fairs are held on April 22nd, August 10th, and October 26th; the last is pre-eminently called "The Great Fair," for sheep, cattle, and cheese. A high constable, deputy constables, and tythingmen, are chosen annually at the manorial court of the Marquess of Bath. The county sessions of the peace for the summer quarter are holden here annually in July, and petty sessions are held monthly by the neighbouring magistrates. A court of requests, for the recovery of debts not exceeding £5, is held every fortnight, on Tuesday, at Warminster and Westbury alternately, the jurisdiction of which extends over the hundreds of Warminster, Westbury, and Heytesbury. The town-hall was pulled down a few years since, and the Marquess of Bath has erected, at his own expense, a noble building in the centre of the market-place, in the same elegant style of architecture as his lordship's own mansion at Longleat, comprising every accommodation for holding the sessions, and a handsome suite of apartments for assemblies, public meetings, &c. The town is a polling-place for the southern division of the shire.

The living is a vicarage, valued in the king's books at £18. 0. 2½.; present net income, £324; patron, Bishop of Salisbury; appropriators, Dean and Chapter of Salisbury. The parochial church is situated on the Bath road, near the north-western extremity of the town, and is a spacious and handsome structure, of various styles of architecture, with a tower rising from the centre, originally built about the time of Edward III.; but the body and aisles were rebuilt on the old foundation, in 1724. A proprietary chapel, founded in the reign of Edward I., and dedicated to St. Lawrence, stands near the market-place: it was endowed by two maiden sisters, named Hewett, and is vested in feoffees; the original tower remains, but the body of the chapel was rebuilt in 1725, and has lately been repaired and beautified. Christ Church, to which a district has been assigned, was built in 1831, at an expense of £3114. 11., defrayed by voluntary subscriptions, aided by a grant from the parliamentary commissioners; it occupies an elevated site, and forms an interesting object in the view of the town: the living is a perpetual curacy; net income, £100; patron, Vicar of Warminster. There are places of worship for Baptists, Independents, Wesleyan Methodists, and Unitarians. A free grammar school was built and endowed by the first Viscount Weymouth: the master is appointed by the Marquess of Bath, and receives a salary of £30 per annum, for which 20 boys are instructed. National schools for boys and girls, and a Lancasterian school for girls, are supported by voluntary contributions. The poor law union of Warminster comprises 22 parishes or places, under the care of 31 guardians, and contains, according to the census of 1831, a population of 17,150. Various Roman coins, both of silver and brass, have been found here. In the vicinity are many British tumuli, and several remains of Roman encampments, particularly Battlesbury, a strong earthwork with double sides, where spear-heads and other weapons have been occasionally ploughed up. Near this intrenchment, on the edge of the river Willey, a beautiful tesselated pavement, and the foundations of a Roman villa, with its hypocaust, sudatory, &c., were discovered, in 1786: among other portraits was a figure of Diana, with a hare; the former was too much injured to be removed; the latter is carefully preserved at Longleat House. On the west side of the town is Clay hill, a steep and conical eminence surmounted by a tumulus: it is nearly 900 feet above low-water mark at Bristol, and was formerly used as a beacon. The environs are rich in fossil remains, many of which have been deposited in the British Museum. In the year 1816, a toad and a newt, both living, were found imbedded in a thick stratum of rock, which had not the smallest crack, or orifice; an account of this discovery was published in the 38th volume of the Medical and Physical Journal, for 1817. Dr. Huntingford, Bishop of Hereford; and Dr. Samuel Squire, Bishop of St. David's, an able and learned writer, were natives of this town.

WARMSWORTH (*St. Peter*), a parish, in the union of Doncaster, Southern Division of the wapentake of Strafforth and Tickhill, West Riding of the county of York, 3¼ miles (S. W.) from Doncaster; containing 362 inhabitants. The living is a discharged rectory, valued in the king's books at £6. 10. 10.; present net income, £168; patron, W. Wrightson, Esq. At the distance of half a mile from the church is a steeple, containing a bell. An infants' school is partly supported by William Adam, Esq. There are some limekilns in the neighbourhood. George Fox, founder of the Society of Friends, held his first meetings at this place.

WARMWELL, a parish, in the union of Dorchester, hundred of Winfrith, Dorchester Division of the county of Dorset, 5½ miles (S. E.) from Dorchester; containing 87 inhabitants. The living is a rectory, with which that of Poxwell was united in 1749, valued in the king's books at £15; present net income, £350: it is in the patronage of the Trenchard family.

WARNBOROUGH, NORTH, a tything, in the parish and hundred of Odiham, Odiham and Northern Divisions of the county of Southampton, 1 mile (N. W.) from Odiham: the population is returned with the parish.

WARNBOROUGH, SOUTH (*St. Andrew*), a parish, in the union of Hartley-Wintney, hundred of Bermondspit, Odiham and Northern Divisions of the county of Southampton, 2¾ miles (S. W. by S.) from Odiham; containing 374 inhabitants. The turnpike road from Odiham to Alton runs through the parish. The living is a rectory, valued in the king's books at £14. 12. 3½.; present net income, £594; patrons, President and Fellows of St. John's College, Oxford. The church is very ancient, having a fine Norman arch at the entrance; in the chancel is a curious monument to Sir Thomas White and his family. Tradition says that

Queen Elizabeth, then occasionally residing at Odiham, rode over to the manor-house, and after partaking of breakfast with this gentleman, knighted him in his own saloon. There is a singular mound in the churchyard, as if the bones of combatants had been deposited under it. Peter Mews, afterwards Bishop of Winchester, and Peter Heylin, the cosmographer, were rectors of this parish. The Rev. John Duman, D.D., rector, in 1785, gave £200 in the 3 per cents., the interest to be applied in support of a school. A school-room has been erected for a Sunday school by the present rector, at a cost exceeding £160.

WARNDON (*St. Nicholas*), a parish, in the union of Droitwich, Lower Division of the hundred of Oswaldslow, Worcester and Western Divisions of the county of Worcester, 2½ miles (E. N. E.) from Worcester; containing 171 inhabitants. The living is a rectory, valued in the king's books at £10. 0. 2½.; present net income, £151; patron, R. Berkeley, Esq. The Birmingham and Worcester canal passes through the parish.

WARNFORD, or WARRINGTON, a township, in the parish of Bambrough, union of Belford, Northern Division of Bambrough ward and of the county of Northumberland, 3½ miles (S. E. by S.) from Belford; containing 35 inhabitants. This was formerly a considerable village, but there are now only a few houses. A Presbyterian chapel, built in 1750, was rebuilt by subscription in 1824, and has a handsome house attached, for the residence of the minister.

WARNFORD, a parish, in the union of Droxford, hundred of Meon-Stoke, Droxford and Northern Divisions of the county of Southampton, 13 miles (S. W. by S.) from Alton; containing 418 inhabitants. The living is a rectory, valued in the king's books at £21. 9. 4½.; present net income, £502; patron, W. Abbott, Esq. The church is in the early style of English architecture, with a Norman tower; it was founded by Wilfrid, and rebuilt by Adam de Portu, in the reign of the Conqueror. There is a place of worship for Independents; also a school partly supported by subscription. Near the church is a very ancient and curious ruin of a house, called King John's, a corruption of St. John's, which family formerly possessed it: the walls, composed of flints set in grout-work, were four feet thick, with semicircular arched windows and doors; and it was divided into two apartments, under a vaulted roof, which, though fallen in, appears to have been once supported by four slender, well proportioned columns, the bases and capitals of which are still entire, and by four half-columns worked into the east and west walls.

WARNHAM (*St. Margaret*), a parish, in the union of Horsham, hundred of Singlecross, rape of Bramber, Western Division of the county of Sussex, 2 miles (N. N. W.) from Horsham; containing 952 inhabitants. This parish is bounded on the north by the county of Surrey: the surface is diversified with hill and dale and enriched with wood; near Warnham Mill is a large sheet of water, comprising about 100 acres. Warnham Court is a spacious mansion in the Elizabethan style, on an elevated site commanding extensive views. On Oldhouse farm is a large quarry of stone, much used for paving. The living is a vicarage, valued in the king's books at £10. 1. 0½.; present net income, £191; patrons and appropriators, Dean and Chapter of Canterbury. The church is an ancient structure; in the north chancel are the effigies of Sir John Caryll and his lady, with their children, in a kneeling position. Two day and Sunday schools are partly supported by subscription. Percy Bysshe Shelley, the poet, was born at Field Place, the residence of Sir Timothy Shelley, Bart.

WARNINGCAMP, an ancient chapelry, annexed to the parish of Leominster, hundred of Poling, rape of Arundel, Western Division of the county of Sussex, 1¾ miles (E.) from Arundel; containing 104 inhabitants.

WARPSGROVE (*St. James*), a parish, in the union of Thame, hundred of Ewelme, county of Oxford; containing 36 inhabitants. The living is a rectory, valued in the king's books at £2. 11. 10½.: it is in the patronage of the Crown. There is no church; the parishioners attend the parish church of Chalgrove.

WARRENTON, a township, in the parish, and Northern Division of the ward, of Bambrough, union of Belford, Northern Division of the county of Northumberland, 3 miles (S.) from Belford; containing 158 inhabitants.

WARRINGTON, a hamlet, in the parish of Olney, union of Newport-Pagnell, hundred of Newport, county of Buckingham, 1¾ mile (N. by E.) from Olney: the population is returned with the parish.

WARRINGTON (*St. Helen*), a newly enfranchised borough, market-town, and parish, and the head of a union, in the hundred of West Derby, Southern Division of the county palatine of Lancaster: containing 19,155 inhabitants, of which number, 16,018 are in the town of Warrington, 52 miles (S. by E.) from Lancaster, and 188 (N. W. by N.) from London. This place, which is of very great antiquity, is by Mr. Whitaker, in his history of Manchester, supposed to have been originally a British town, and on the invasion of the Romans under Agricola, in the year 79, to have been converted into a Roman station. This opinion rests chiefly on the circumstance of three Roman roads leading respectively from the stations of *Condate*, *Coccium*, and *Mancunium*, to a ford here over the Mersey; the vestiges of a castrum and fosse, still discernible; and in the discovery of some Roman relics, consisting of coins found on both sides of the river, near the ancient ford, and other antiquities, which have been subsequently dug up. On its occupation by the Saxons, it obtained the name of *Weringtun*, from the Saxon *Wæring*, a fortification, and *tun*, a town, from which its present appellation is derived. It was at that time of sufficient importance to give name to a wapentake, which afterwards merged into the hundred of West Derby, and formed part of the demesne of Edward the Confessor; it had been previously the head of a deanery, of which the jurisdiction still remains. In Domesday-book it is noticed under the name of *Wallintun*; and in the reign of Edward I., it was in the possession of William le Boteler, who obtained for it the grant of a market, and other privileges. From the earliest period the river Mersey at this place was passed only by the ancient ford, till the close of the fifteenth century, when Thomas, first Earl of Derby, in compliment to Henry VII., on his visit to Latham and Knowsley, in 1496, erected the first bridge of stone, soon after which the passage of the river by the ford ceased. In the reign of Henry VIII., Leland, speaking of Warrington, says, "it is a pavid towne of a prety bignes, the paroche church is at the tayle of the towne; it is a better market than Manchestre." Nothing of importance is recorded of it from this period till the commencement of the parliamentary war, when the inhabitants openly declared in favour of the royal cause, and the town was garrisoned for the king. In 1643, a detachment of the parliamentary forces, stationed at Manchester, laid siege to it, on which occasion the royalists under Colonel Norris, the governor, took refuge in the church, and, fortifying that edifice against the assailants, obstinately resisted their attack for five days; but the enemy having erected a battery, which they brought to bear upon it, the royalists were compelled to surrender. Their number was 1600, of whom 300 were taken prisoners; and ten pieces of ordnance, with a large quantity of arms and ammunition, fell into the hands of the enemy. The royalists seem, however, to have soon regained possession of the town, for in less than three months it was again attacked by the parliamentarians, who carried it by storm, when the former lost 600 men and eight pieces of cannon. In 1648, a numerous body of Scottish troops, under the command of the Duke of Hamilton, on their retreat after the battle of Ribbledale, rallied at Warrington; and, after an obstinate but unsuccessful encounter with the parliamentarian forces under General Lambert, in which 1000 men were slain, the remainder, consisting of 2000, surrendered themselves prisoners of war. The same general, in 1651, encountered and repulsed the Scottish army, under the command of the young king, near this town.

Towards the close of the Interregnum, Sir George Booth, knight of the shire, who had been a strenuous supporter of the parliament, being dissatisfied with the conduct of public affairs, and anxious for the re-establishment of a free parliament under a legitimate head, raised a considerable force, in 1658; but, after a severe engagement with the forces under General Lambert, at Winnington bridge, near Delamere Forest, he was defeated, and part of his army retreating to Warrington, the men were arrested in their flight by the parliamentary garrison stationed in that town: the services of Sir George, on this occasion, procured for him, after the Restoration, the title of Baron Delamere of Dunham Massey. Since the erection of the bridge over the Mersey, Warrington, as a military station, was regarded as commanding the entrance into the county of Lancaster; and in 1745, on the approach of the army under Prince Charles Edward, the young pretender, who was advancing from Wigan, the central arches were demolished by the Liverpool Blues, who, having thus intercepted their progress, captured part of the rebel army, whom they sent prisoners to Chester castle: the bridge was repaired in 1747, but afterwards becoming much dilapidated, it was taken down, and a wooden bridge on stone piers was erected in 1812, at the joint expense of the counties of Chester and Lancaster; this has since been replaced by the present substantial stone bridge.

The town, which is pleasantly situated on the river Mersey, consists of four principal streets diverging from the centre, and intersected by several smaller; they are in general narrow and inconvenient, but have undergone considerable improvement, under the superintendence of commissioners appointed by an act of parliament obtained in 1813: the houses are, for the greater part, of indifferent appearance, but interspersed with numerous respectable modern edifices, which form a striking contrast. Prior to the construction of the railroad between Liverpool and Manchester, it was the great thoroughfare between these two places, and 70 stage coaches passed through it daily, but only one now continues to run. The town is well paved, under the provisions of the act; lighted with gas by a company incorporated in 1822, whose works, on a very extensive scale in Mersey-street, were erected at an expense of £15,000, advanced on shares of £20 each; and amply supplied with water by public works. The public subscription library was established in 1760, and is well supported. A Floral and Horticultural Society are now united, the former of which originated in 1817, and in 1824 was extended, to embrace the objects of the latter; a mechanics' institute was established in 1825; a neat and well-arranged theatre is opened occasionally; and a handsome suite of assembly-rooms has been erected.

Warrington has been long celebrated as a place of trade: until the early part of the eighteenth century, the principal branches of manufacture were coarse linen and checks, to which succeeded that of sail-cloth, which was carried on so extensively that one-half of the sail-cloth for the use of the British navy is computed to have been made here: the raw material, imported from Russia to Liverpool, was conveyed up the Mersey. On the decline of this branch of business after the peace, the spinning of cotton was introduced, together with the manufacture of muslin, calico, velveteen, and other cotton goods, which, with sail cloth on a less extensive scale, constitute a very great portion of the trade of the town, and for the sale of which two cloth-halls have been erected. There are several considerable pin-manufactories, which is the principal staple trade; and the making of files, for which the artificers have obtained a high degree of reputation, and other articles of hardware, employs a great number of men. The manufacture of glass and glass bottles is also extensively carried on, there being several large establishments, of which the Bank Quay Glass Company is the chief. The manufacture of flint-glass has been subject to considerable fluctuation: at one period the export trade in this article to America was exceedingly great, but it is now almost annihilated; the Americans, in order to encourage their own manufactures, having imposed a tariff on English glass, which operates in many instances as a prohibition. A considerable trade is also carried on in malt, and there are several tanneries, soap manufactories and breweries: the ale of this place is in high repute. The soil in the neighbourhood is extremely fertile, and productive of early vegetables for the supply of the neighbouring markets: formerly large quantities of potatoes were exported to the countries bordering on the Mediterranean, but since the increase of the population in the neighbouring districts, that species of traffic has been discontinued. The Mersey and Irwell navigation affords a direct communication with Manchester, and the other districts with which that town is connected by various canals. The Sankey canal, commencing at the river Mersey, about one mile westward of the town, and approaching very near its northern extremity, was the first formed in the county, the act for its construction having been obtained in 1755: it extends about twelve miles to the collieries near St.

Helen's. In 1830, a railway, with two collateral branches, was constructed from this town to join the one between Manchester and Liverpool, at Newton-in-Mackerfield, and subsequently an act was obtained to extend it across the latter, until it reached the branch from the borough of Wigan. Here is also a station of the Grand Junction railway, which passes through the parish.

On the river Mersey was formerly a valuable fishery, which, in 1763, was let for £400 per annum: it abounded with salmon and smelts of very superior quality, but has now greatly declined, not only in the quantity, but also in the size and flavour, of the fish. At spring tides the water in this river rises to a height varying from ten to twelve feet at Warrington bridge, at which time vessels of from 70 to 100 tons' burden can sail up to the quay, a little above the town, where warehouses and other accommodations have been erected. The town, from its situation as regards Liverpool and Manchester, enjoys considerable traffic, having a constant communication, both by land and water, with those places. The market days are Wednesday and Saturday; the former, which is the principal, is abundantly supplied with corn: there is also a large cattle market every alternate Wednesday. The fairs are on July 18th, and Nov. 30th, each continuing ten days, for the sale of woollen cloth and other goods, and for horses, cattle, sheep and pigs. The market hall, a neat and convenient building in the market-place, used on market days for the sale of corn, is, during the fairs, let to different tenants, for the sale of flannels and other goods; over it is a good suite of rooms, in which the assemblies were formerly held. Adjoining the market-house is the principal cloth-hall, occupying three sides of a quadrangle; the ground floor is divided into shops for the sale of linen-drapery, fustians, hardware, and toys, during the fairs, and the upper part is used for the sale of cloths. There is also a cloth-hall, on a smaller scale, in Buttermarket-street. The town is within the jurisdiction of the county magistrates, who hold a petty session for the division on the first and third Wednesdays in every month; and constables and other officers are appointed, annually in October, at the court leet of the lord of the manor. By the act of the 2nd of William IV., cap. 45, Warrington has been constituted a borough, with the privilege of returning one member to parliament, the right of election being in the £10 householders: the limits of the borough comprise by estimation 5657 acres, and are described in the Appendix, No. II. The number of voters now registered is 631; the returning officer is annually appointed by the sheriff. The town-hall, a neat building, in which the petty sessions are held, was erected in 1820, at an expense of £2800, raised by subscription.

The living is a rectory, valued in the king's books at £40; patron, Lord Lilford. The ancient church, dedicated to St. Elfin, which was of Saxon origin, existed at the time of the Conquest; of this there are few remains, the site being occupied by the present church, dedicated to St. Helen, a spacious and ancient cruciform structure, of various styles of architecture, with a central tower, which, with the piers and arches supporting it, and the chancel, are parts of the original building, and a fine specimen of the decorated style of English architecture: the windows of the chancel, particularly the east window, are enriched with elegant tracery of beautiful design; the north transept is in the later style of English architecture, of an inferior character, and the nave and south transept are modern additions: it contains 400 free sittings, the Incorporated Society having granted £240 towards defraying the expense. Two of the ancient sepulchral chapels are remaining, in one of which is the magnificent tomb of Sir Thomas Boteler and his lady, with their effigies, the former in armour, and both surrounded by various sculptured figures; in the other, which formerly belonged to the family of Massey, are several monuments to the Patten family, one of which is embellished with an elegant specimen of Italian sculpture to the memory of Thomas Wilson Patten, Esq., who died at Naples in 1819. Trinity chapel, in Sankey-street, is a neat and commodious edifice: the living is a perpetual curacy; net income, £130; patron, Thomas Legh, Esq. A church, dedicated to St. Paul, was erected in Bewsey-street, in 1830, at an expense of £5347. 2. 6., under the act of the 58th of George III.: the living is a perpetual curacy; net income, £110; patron, Rector of Warrington. There are places of worship for Baptists, the Society of Friends, those in the late Countess of Huntingdon's Connection, Independents, Independent and Wesleyan Methodists, Unitarians, and Roman Catholics; also one for the last-named at Garswood.

The free grammar school was founded and amply endowed in 1526, by a member of the Boteler family: by a decree of the court of Chancery, in 1820, the trustees were ordered to pay the master a salary of £300 per annum, with the gratuitous use of the school-house, garden, and land adjoining, and to increase it when the number of scholars should exceed 30; the same order directs the usher's salary to be from £60 to £100 per annum, and the writing-master's from £40 to £60: the school, which is under the patronage of Lord Lilford, who appoints the master, is open to all boys of the parish who have attained the age of seven: the late Rt. Hon. George Tierney was educated here. The Blue-coat school, in Winwick-street, was established in 1677, and is supported partly by subscription, and partly by an income arising from benefactions and legacies vested in the purchase of land, producing more than £200 per annum: this establishment has also the reversion of an estate at Sankey, worth £6000, granted by John Watkins, Esq., in 1797, and the reversionary interest of an estate in the county of Bedford, given, in 1685, by Arthur Borron, Esq.: 150 boys and 40 girls are instructed in it, of which number, 24 boys and 16 girls are also clothed. A charity school, established in 1814, in the same street, and in which 150 girls are taught to read and sew; and an infants' school, opened in 1826, are supported by subscription. A National school also has been erected. A society for the relief of widows and orphans of clergymen of the archdeaconry of Chester was established in the early part of the last century, under the patronage of the bishop of the diocese, and is liberally supported. The dispensary was established in 1810, and an appropriate building erected for its use in 1818, at an expense of £1030: a branch of the Royal Humane Society has been annexed to it. A ladies' society, for affording relief to lying-in women, was established in 1819: there are also various other institutions and provident societies for promoting the religious instruction and the comfort of the poor, and divers funds for charitable uses. The poor law union of Warrington comprises 15 chapelries and townships, in the parishes of Warrington, Winwick and Prescot, containing a population of 27,757, according to the census of 1831, and is under the care of 18 guardians. Orford Hall, about a mile from the town, was the residence of John Blackburne, Esq., a celebrated botanist: he was the second person that succeeded in raising the pine-apple in England; he was also successful in the cultivation of the cotton tree, which he brought to a considerable degree of perfection, and died in 1796, at the advanced age of 96. Litherland, the inventor of the patent lever watch, was a native of this town. Warrington formerly gave the title of Earl to the family of Booth, which, after becoming extinct, was revived in 1796, in the family of Grey, who are now Earls of Stamford and Warrington.

WARSILL, a township, in the parish and liberty of RIPON, West Riding of the county of YORK; containing 86 inhabitants. There is a place of worship for Independents.

WARSLOW, a chapelry, in the parish of ALLSTONEFIELD, union of LEEK, Northern Division of the hundred of TOTMONSLOW and of the county of STAFFORD, $7\frac{1}{4}$ miles (E. N. E.) from Leek; containing 595 inhabitants. The living is a perpetual curacy; net income, £105; patron, Vicar of Allstonefield; impropriator, Sir George Crewe, Bart. The chapel, dedicated to St. John the Baptist, is a neat modern structure. The school, erected by subscription in 1728, is endowed with about £17 per annum, for which twenty children are instructed.

WARSOP (*St. Peter and St. Paul*), a parish, in the union of MANSFIELD, Hatfield Division of the wapentake of BASSETLAW, Northern Division of the county of NOTTINGHAM, $5\frac{1}{4}$ miles (N. N. E.) from Mansfield; containing 1281 inhabitants. The village, which is a considerable place, has been formed into divisions, called Church-Warsop and Market-Warsop. Fairs for cattle and horses are held on May 21st and Nov. 17th. The living is a rectory, valued in the king's books at £22. 15. $2\frac{1}{2}$.; present net income, £1020; patron, H. G. Knight, Esq. There is a chapel of ease at Sokeholme. Thomas Whiteman, in 1811, bequeathed £400, now producing £15. 15. per annum, for which 25 children are instructed. Dr. Samuel Halifax, Bishop of St. Asaph, a prelate of deep erudition, died also rector of this parish, in 1790.

WARTER (*St. James*), a parish, in the union of POCKLINGTON, Bainton-Beacon Division of the wapentake of HARTHILL, East Riding of the county of YORK, $4\frac{1}{2}$ miles (E. by N.) from Pocklington; containing 470 inhabitants. The living is a discharged vicarage, valued in the king's books at £4; present net income, £100; patron and impropriator, Lord Muncaster. There is a place of worship for Wesleyan Methodists. A priory of Black canons, in honour of St. James, was founded here in 1132, by Geoffry Fitz-Pain, which, at the period of the dissolution of the monasteries, possessed a revenue of £221. 3. 10.

WARTHERMASK, a joint township with Swinton, in the parish of MASHAM, union of BEDALE, partly within the liberty of ST. PETER'S, East Riding, and partly in the wapentake of HANG-EAST, North Riding, of the county of YORK, $7\frac{3}{4}$ miles (S. W. by S.) from Bedale: the population is returned with Swinton.

WARTHILL (*St. Mary*), a parish, in the union of YORK, partly in the wapentake of BULMER, North Riding, and comprising the township of Warthill within the liberty of ST. PETER'S, East Riding, of the county of YORK; containing 162 inhabitants, of which number, 127 are in the township of Warthill, $5\frac{1}{2}$ miles (N. E. by E.) from York. The living is a discharged vicarage, in the patronage of the Prebendary of Warthill in the Cathedral Church of York (the appropriator), valued in the king's books at £3. 1. 8.; present net income, £100.

WARTLING (*St. Mary Magdalene*), a parish, in the union of HAILSHAM, hundred of FOXEARLE, rape of HASTINGS, Eastern Division of the county of SUSSEX, $4\frac{1}{2}$ miles (E. by S.) from Hailsham; containing 948 inhabitants. The living is a vicarage, valued in the king's books at £10. 0. $2\frac{1}{2}$.; present net income, £307; it is in the patronage of the Rev. — Hayley and Miss Rosarn.

WARTNABY, a chapelry, in the parish of ROTHLEY, union of MELTON-MOWBRAY, hundred of EAST GOSCOTE, Northern Division of the county of LEICESTER, $4\frac{1}{2}$ miles (N. W.) from Melton-Mowbray; containing 86 inhabitants. The chapel is dedicated to St. Michael.

WARTON, a chapelry, in the parish of KIRKHAM, union of the FYLDE, hundred of AMOUNDERNESS, Northern Division of the county palatine of LANCASTER, 3 miles (S. S. W.) from Kirkham; containing 531 inhabitants. The living is a perpetual curacy; net income, £86; patron, Vicar of Kirkham. The chapel, which is dedicated to St. Paul, was consecrated in 1725. Here is a National school, built by subscription, and supported by various endowments, producing an annual income of nearly £100.

WARTON (*Holy Trinity*), a parish, in the hundred of LONSDALE, south of the sands, Northern Division of the county palatine of LANCASTER; containing 2159 inhabitants, of which number, 558 are in the township of Warton with Lindeth, 7 miles (N. by E.) from Lancaster. The living is a vicarage, valued in the king's books at £74. 10. $2\frac{1}{2}$.; present net income, £187; patrons, Dean and Chapter of Worcester. The Lancaster canal passes through the parish. A free grammar school and an hospital were founded and endowed in 1594, by Matthew Hutton, Archbishop of York, and further endowed with bequests from Robert Lucas and others. A National Sunday school has been established. The remains of a Roman encampment may still be traced; and under Warton Cragg there is a copper mine, but it is not now worked.

WARTON, a township, in the parish and union of ROTHBURY, Western Division of COQUETDALE ward, Northern Division of the county of NORTHUMBERLAND, $3\frac{1}{4}$ miles (W. by N.) from Rothbury; containing 59 inhabitants.

WARWICK (*St. Leonard*), a parish, in the union of CARLISLE, partly in CUMBERLAND ward, and partly in ESKDALE ward, Eastern Division of the county of CUMBERLAND; containing 686 inhabitants, of which number, 266 are in the township of Warwick, $4\frac{3}{4}$ miles (E. by N.) from Carlisle. The parish is bounded on the north by the river Eden, and on the west by the Irthing;

and, from the large earthworks still remaining, is supposed to be the site of the ancient *Virosidum,* where the sixth cohort of the Nervii was stationed. The village is pleasantly situated on the western bank of the river Eden, which is crossed by a bridge of four arches, near the base of an eminence, on which are the remains of trenches, probably thrown up to guard the pass during the border feuds. The living is a perpetual curacy, annexed to that of Wetheral. The church is built of stone, in the Norman style of architecture, with a semicircular east end, and appears to have been formerly much larger. Here is a National school; and a building for a Sunday school was erected by the late Thomas Parker, Esq.

Corporation Seal.

WARWICK, a borough and market-town, having separate jurisdiction, and the head of a union, locally in the Warwick Division of the hundred of Kington, Southern Division of the county of WARWICK, of which it is the chief town, 90 miles (N. W.) from London; containing 9109 inhabitants. This place is said by Rous, the historian of the county, to have been a British town of considerable importance prior to the Roman invasion, and this statement is confirmed by Camden, Dugdale, and other writers. The same author relates that, after its devastation by the frequent incursions of the Picts, it was rebuilt by Caractacus, on whose defeat by Claudius, in the year 50, the Romans, in order to secure their conquests in Britain, erected several fortresses on the banks of the Severn and Avon, of which latter, Warwick castle was one; but this is very doubtful, the nearest Roman station having, probably, been that at Chesterton. Upon the establishment of the Saxons in the island, this town, included in the kingdom of Mercia, fell under the dominion of Warremund, who rebuilt it, and, after his own name, called it *Warre-wyke:* it appears, however, from a coin of Hardicanute, that its Anglo-Saxon name was *Werhica*, but from either of these sources its present name may be derived. Warwick was subsequently destroyed by the Danes, and, according to the most authentic records, Ethelfleda, daughter of Alfred, and Countess of Mercia, restored it, about the year 913, and built a fort, which evidently forms the most ancient part of the present castle. At the time of the Conquest, this fortress was considerably enlarged, and the town was surrounded with walls and a ditch, of which there are still some vestiges, and a memorial is preserved in the appellation of a certain part of the town, called "Wall-dyke." In the reign of Edward I., the fortifications were repaired by Guy Earl of Warwick, who, in 1312, with the Earl of Lancaster, having taken Piers Gavestone, the favourite of Edward II., on his route to Wallingford, brought him to this castle, where he was secured for the night under the barons' guard, and in the morning removed to Blacklow hill, about a mile from the town, where he was tried and beheaded. In 1571, Robert Dudley, Earl of Leicester, celebrated in St. Mary's church the ceremony of the order of St. Michael, which, by permission of Elizabeth, had been conferrred upon him by Charles XI. of France. William Parr, brother of Catherine, the last consort of Henry VIII., assisted at this ceremony, and, dying soon after, was buried in the chancel of the church. Queen Elizabeth visited Warwick, in 1572, on her route to Kenilworth castle, where she remained for two or three days; and where also, in 1575, she was sumptuously entertained by the Earl of Leicester for seventeen days, and amused with costly pageants and magnificent spectacles, at an expense of not less than £1000 per day; and in 1617, James I. visited the town, and was splendidly entertained in the great hall of the Earl of Leicester's hospital, in commemoration of which, a tablet, with an appropriate inscription, was put up on one of the walls of that building. During the great civil war, in the reign of Charles I., Robert Greville, Lord Brooke, embraced the cause of the parliament, and defended the castle against the king. Having occasion to repair to London, in order to procure a supply of arms and ammunition, he deputed Sir Edward Peto governor during his absence. The supply being obtained, he was met on his return by the Earl of Northampton, with a considerable force, near Edge Hill; but an accommodation taking place, Lord Brooke deposited his artillery and ammunition in Banbury castle, and returned to London. After his departure, the earl, having attacked Banbury castle, and taken the military stores, advanced to Warwick, and laid siege to the castle, which was defended by the governor for fourteen days, till Lord Brooke, on his return from London, after a successful skirmish with the earl near Southam, came to his assistance, and compelled the royalists to abandon the siege. William III., in 1695, visited the town, of which, in the course of the preceding year, more than one-half was destroyed by a dreadful conflagration, occasioned by a spark, from a lighted piece of wood in the hand of a boy, communicating with the thatched roof of a dwelling-house; a great quantity of goods, probably in a state of ignition, having been removed for safety into the collegiate church of St. Mary, set fire to that venerable pile, which, with the exception of the chancel, the Beauchamp chapel, and the chapter-house, was destroyed. In a few years after, the town was rebuilt, in consequence of a national contribution, amounting to £110,000, of which, £1000 was bestowed by Queen Anne.

The town is pleasantly situated on a rock of freestone, rising gently from the north side of the river Avon, which winds round its base: the approaches on every side are beautiful, and the surrounding scenery is richly diversified. The entrance from the Banbury road is strikingly picturesque; a handsome stone bridge, of one noble arch 100 feet in the span, leads into the town, which rises gradually from the bank of the river, and presents in succession the venerable castle on the left, the spire of St. Nicholas' church in the lower ground, and the lofty tower of St. Mary's in the distance. The entrance from the Birmingham road, after passing through the suburb called Saltisford, commands a view of the priory, the county hall, and the fine tower of St. Mary's church. The entrance from the Stratford road is through a long ancient arched gateway, with a lofty tower on the west; and that from the Emscote road through an archway, which supports the chapel of St. Peter. The streets are spacious and regularly formed, consisting chiefly of two principal ones running east and west, crossed by another inclining to the centre of the town: the houses are in general modern and well built, interspersed with elegant mansions, and others affording occasional specimens of the style which prevailed before the fire: the town is well paved, lighted with gas, and amply supplied with water from springs about half a mile distant. The castle, which is situated on the south side of the town, is one of the most splendid and entire specimens of feudal grandeur in the kingdom, and is not less remarkable for its stately magnificence than for the elegance of its architecture and the beauty of its situation: it encloses within the walls an area of nearly three acres, the plot surrounded by the moat being more than five acres and a half. A winding road cut through the solid rock, the sides of which are covered with ivy and skirted with shrubs, leads from the outer lodge to a massive gateway flanked with two towers connected by an embrasure above, and formerly defended by a portcullis, which leads into the inner court, in the north angle of which is a lofty octangular tower, with a projecting and embattled parapet resting upon corbels; in the south angle is Guy's tower, of a duodecagonal form, but more ancient, and having an exploratory turret rising from within the battlements; on the north-east side are two low embattled towers, in one of which bears were anciently kept, for the purpose of baiting. The range of state apartments on the east, as viewed from this side of the castle, is strikingly magnificent; the windows are in fine proportion, and every part is in the highest preservation. At the western extremity, and commanding, from its elevated situation, an extensive view of the surrounding country, is the keep, erected by Ethelfleda, as a place of security against any sudden irruption of the Danes, and also as an exploratory tower, from which their movements might be observed: the ascent is by a winding path, now richly planted with forest trees, among which are some beautiful cedars of Lebanon, and laid out in walks and shrubberies. The façade of the castle, rising from the river Avon, is a long line of flat masonry relieved only by the number and variety of its windows: the broken arches of an ancient bridge, which formerly led into the town, are still preserved, and add greatly to the picturesque beauty of the scene. The state-rooms, the armoury, and the other various apartments, are preserved in a style of appropriate grandeur: the lawns and gardens are tastefully laid out, and in the greenhouse, built expressly for its reception, is the beautiful Grecian vase of Lysippus, which was dug from the ruins of Adrian's palace, at Tivoli, near Rome, and brought to England by Sir William Hamilton, under the direction and at the expense of his nephew, the late Earl of Warwick. This celebrated specimen of ancient sculpture is of white marble, and is placed upon a pedestal of the same material: its form is nearly that of a hemisphere with reverted rim: two intertwining vines, of which the stems form the handles, wreathe their tendrils, with fruit and foliage, round the upper part of the exterior, and the central part is ornamented with antique heads, and enveloped with a panther's skin with the head and paws, the thyrsus of Bacchus, the lituus of the augurs, and other embellishments: it is of large dimensions, being capable of containing 163 gallons. Assemblies are held in the town-hall, and for larger meetings, and during the races, in the county-hall: the theatre is opened during the race week, and occasionally at other times, by the Cheltenham company. The races take place in the first week of Sept., and continue for three days. The course is a fine level, with a little rising ground in one part, and has undergone such improvement as to make it one of the best in the kingdom: the grand stand is handsome and commodious.

Very little trade is carried on beyond what is necessary for the supply of the inhabitants: the cotton manufacture, which was formerly introduced, has entirely declined; and a worsted manufactory, subsequently established, is decreasing. There are several large malting-houses, and lime, timber, and coal wharfs on the banks of the Warwick and Birmingham, and Warwick and Napton canals, which form a junction at Warwick, and come up to the northern part of the town, and, communicating with the Oxford and Birmingham canal, afford every facility of inland navigation. From the wharfs at Saltisford fly-boats start daily, conveying goods to all the intermediate counties, on their way to London; others proceed to Birmingham and Wolverhampton, Manchester, and Liverpool, and, on Tuesday and Friday, to Oxford and Banbury. The market, which is abundantly supplied with corn and provisions of every kind, is on Saturday: the fairs are on Jan. 21st, Feb. 11th and 23rd, April 1st, May 13th, June 3rd, July 5th, Aug. 12th, Sept. 4th, Oct. 12th (which is a pleasure and a statute fair, and during which an ox is generally roasted in the market-place), Nov. 8th, and Dec. 16th. The market-place is an extensive area surrounded by respectable houses, in the centre of which is the market-house, a neat substantial building of stone, supported on arches; the upper story, which is surmounted by a cupola and dome, is now occupied by the interesting museum of the Warwickshire Natural History and Archaiological Society.

Warwick was first incorporated in the 37th of Henry VIII., and made a "mayor town" by Queen Mary, in 1553: charters were afterwards granted in the 16th and 35th of Charles II.; and by charter of William III., in 1694, the government was vested in a mayor, recorder, deputy recorder, twelve aldermen, and twelve assistant burgesses, aided by a town-clerk, a sergeant-at-mace, a yeoman sergeant, and subordinate officers. The mayor, recorder, late mayor, and the three senior aldermen, were justices of the peace, with exclusive jurisdiction. The corporation now consists of a mayor, six aldermen, and 18 councillors, under the act of the 5th and 6th of William IV., cap. 76, for an abstract of which see the Appendix, No. I. The borough is divided into two wards, and the municipal and parliamentary boundaries are co-extensive. The mayor and late mayor are justices of the peace, with five others appointed by commission. The borough first exercised the elective franchise in the 23rd of Edward I., since which time it has regularly returned two members to parliament. The right of election is vested in the £10 householders, the scot and lot right being reserved: the limits of the ancient borough (which comprise 5273 acres) have not been extended. The number of voters now registered is 977: the mayor is the returning officer. The recorder holds quarterly courts of session, for all offences not capital; and a court of record every Wednesday, except in the Christmas, Easter, and Whitsun weeks, for the recovery of debts not exceeding £40, at which the town-clerk generally presides. A court leet is held annually before the town-clerk, as steward. Petty sessions are held every Monday. The court-house, in which the borough sessions and courts of record are held, is a handsome stone building in High-street, ornamented with fluted Corinthian pilasters, and having over the entrance a sculptured figure of Justice, surmounted by the arms of the borough. In the upper story is an elegant assembly-room: the walls are decorated with fluted Corinthian pilasters supporting an entablature and cornice, with an orchestra at one end: from the ceiling, which is lofty, three large and brilliant chandeliers of cut glass are suspended; adjoining it is a card-room, in which is a good portrait of Charles II. The assizes and general quarter sessions of the peace for the county are held at Warwick, as the county town; and it has been made the place of election and a polling-place for the southern division of the shire. The county-hall, in Northgate-street, is an elegant building of freestone, in the Grecian style of architecture; the façade is embellished with pilasters of the Corinthian order, and with a central portico of Corinthian columns supporting a triangular pediment. On the left of the county-hall is the judges' mansion, a neat stone building with a handsome portico, and having communication with the hall; and on the right hand is the county gaol, a large stone building of the Doric order, with massive columns in front, having a central entrance to the office of the clerk of the peace. The entrance to the prison is through a spacious gateway, over which is the platform for the execution of criminals: the interior is divided into ten wards (exclusively of the female debtors' yard) for the classification of prisoners, with day-rooms, work-rooms, and airing-yards, in one of which is a tread-mill with three compartments: the cells are ranged round the governor's house in the centre, which commands a distinct view of each of the wards: the chapel, capable of containing 400 persons, is divided by screens, with a view to preserve the same classification. Opposite to the side entrance of the gaol is the county bridewell, enclosed within a high stone wall: having been enlarged at different times, the arrangement is rather inconvenient; the same regard to classification, order, and cleanliness, prevails here as in the county gaol; a flour-mill, worked by hand, and employing 24 men, who relieve each other at intervals, grinds a sufficient quantity to supply the county gaol and bridewell, and bread sufficient for the supply of both establishments is made here, and baked in an oven large enough to hold 400 loaves at once: the boys and the women are employed in heading pins for the manufacturers at Birmingham, and the men in drawing and preparing the wire for that purpose, and in other occupations: the tread-wheel is applied to working a triple pump, which, from an excellent spring, raises seven gallons of water in a minute, and supplies the whole prison.

The town comprises the parishes of *St. Mary* and *St. Nicholas*. The living of St. Mary's is a vicarage, valued in the king's books at £20; present net income, £280, besides surplice fees, being the vicar's stipend, and an assistant minister is paid £120 per annum out of charity estates belonging to the borough: it is in the gift of the Crown; the impropriation belongs to the Corporation. The church, formerly collegiate, of which the tower and the greater part were destroyed in the conflagration, and rebuilt in 1704, though comprising an incongruous mixture of styles, blending the Roman and later English architecture, is, notwithstanding, a very stately and magnificent structure; the tower, which rises in successive stages, variously embellished, to the height of 130 feet, is supported on four pointed arches, affording a spacious passage underneath, and crowned with lofty pinnacles at the angles, and with others in the centre, of less elevation. The exterior, in many parts, is strikingly handsome, but the eastern part, in particular, is elaborately embellished with panelled and richly-canopied buttresses. The chancel, which is in its original state, is an elegant and highly enriched specimen of the later style of English architecture, and contains a fine altar-tomb to the memory of Thomas Beauchamp, Earl of Warwick, and his lady, Catherine, daughter of Roger Mortimer, first Earl of March. The nave, separated from the aisles by lofty clustered columns, is spacious, and well-lighted by a range of clerestory windows, and the windows in the aisles and transepts are of large dimensions, but totally destitute of beauty in the details. In the south transept is the entrance to the chapel of St. Mary, erected by Richard Beauchamp, Earl of Warwick, and thence called the Beauchamp chapel: it is an elegant and highly enriched edifice, in the later style of English architecture, and, both in its external and internal embellishment, is inferior only to the chapel of Henry VII. at Westminster: the roof is elaborately groined and enriched with fan tracery; the altar is adorned with a well-executed representation of the Salutation, in basso relievo, by Collins; behind it is an apartment within the buttresses, said, but on insufficient authority, to have been the library of John Rous, the historian; and on the north side is a chantry, from which an ascent of four stone steps, deeply worn, leads into an apartment supposed to have been formerly used as a confessional. In the centre of the chapel is the splendid monument of the founder, in gilt brass, in which his effigy, recumbent on an altar-tomb decorated with shields of armorial bearings and numerous figures, and surmounted by a canopy, is finely executed; on the north side is a large monument, in the Elizabethan style, to the memory of Robert Dudley, Earl of Leicester. On the north side of the church is the ancient chapter-house, the interior of which is entirely occupied by the stately monument of Sir Fulke Greville, the first Lord Brooke; the vault was used as a place of sepulture by the Greville family till the whole space was entirely filled, and the chapel above it is now appropriated to the use of the National school. The living of St. Nicholas' is a vicarage, valued in the king's books at £13. 6. 8.; present net income, £220, besides surplice fees: it is in the patronage of the Countess of Warwick, by purchase from the Corporation, who are the impropriators. The church was rebuilt in 1780, the tower and spire having been rebuilt about 40 years previously: it is a neat edifice, in the later style of English architecture; the roof is groined, and supported on clustered columns; the interior is lighted with three handsome windows on each side, and the altar is placed in a recess at the east end. There are places of worship for Baptists, the Society of Friends, Independents, Wesleyan Methodists, and Unitarians; also a Roman Catholic chapel at Hampton Cottage, Grove Park.

The free grammar school, situated on the Butts (a place formerly set apart for the young men of the town to exercise themselves in the use of the bow, prior to the invention of gunpowder), was founded by Henry VIII., and endowed with a portion of the lands of the dissolved monasteries: the master is appointed by the Crown, with a salary of £135 per annum. There are two exhibitions, of £70 per annum each, to any of the colleges at Oxford, founded by Mr. Fulk Weale, of Warwick; and the school is also entitled to two exhibitions to Trinity College, Cambridge, in failure of candidates from Combrook school, founded by Lady Verney. The premises occupy a quadrangle, with a cloister on two sides, built by Richard Beauchamp, Earl of Warwick, for the dean and canons of the church of St. Mary, in which he founded a collegiate establishment. The charity school now held in the ancient chapel of St. Peter was endowed by Lady Greville, Lord Brooke, and Mr. T. Oaken, for the instruction of 39 boys and 36 girls. Forty of this number are completely clothed, and the remainder receive each a coat and a pair of shoes annually: the master's salary is £70 per annum, out of which he pays a mistress for teaching the girls to sew. Another school for boys, and National schools, are supported by subscription. The school of industry, in Castle-street, was established by the Countess of Warwick, and is chiefly supported by the Earl of Warwick and the family of Mr. Wise: in this establishment, to which the house formerly occupied by Mr. Oaken, a great benefactor of the town, is appropriated, 40 girls are completely clothed, and provided, at a moderate charge, with dinner in the school-room daily, from Michaelmas to Lady-day: the school is conducted on the National system.

The hospital, founded by Robert Earl of Leicester, occupies the buildings formerly used by the ancient guild of St. George, which after being united in the reign of Richard II. with the guild of the Blessed Virgin and the Holy Trinity, became, at the dissolution, vested in the corporation, by whom the buildings were conveyed to the earl, who converted them into an hospital, which he endowed for a master and twelve aged brethren, especially such as had been maimed or wounded in the service of their country: the increase in the rental having produced £130 per annum to each of the brethren, and the master's salary, by the deed of endowment, being limited to £50 per annum, an act of parliament was obtained for augmenting it to £400 per annum, and for reducing the yearly stipend of the brethren to £80, till, by the application of the difference, the great hall should be converted into dwellings for ten additional inmates, after which time the revenue is to be equally divided among the brethren. The almsmen wear a blue gown bearing the crest of the founder, a bear and ragged staff, on the left sleeve, without which they are not permitted to appear in public: the appointment of the master, who must be a clergyman, and of the brethren, is vested in the heirs of the founder, and now belongs to Sir John Shelley Sidney, Bart.; and the presentation to the vicarage of Hampton-in-Arden is in the gift of the Master and Brethren. The premises, near the west end of High-street, occupy a quadrangle, on one side of which is the great hall, on another are the master's apartments, the two remaining sides being occupied by the brethren, who have separate dwellings, and a common kitchen. The chapel of St. James, over the west gate of the town, is annexed to, and forms part of, the hospital; it is neatly fitted up, and the altar is embellished with a painting of the Ascension, by Millar, a pupil of Sir Joshua Reynolds: behind the quadrangle is a spacious and well-planted garden, bounded on one side by part of the ancient walls of the town. Those parts of the building which were ornamented in the time of the guilds were, during the Commonwealth, concealed with a covering of lath and plaster, to preserve them from mutilation by the emissaries of the parliament; but in 1833, part of this covering, having fallen into decay, was blown down, and on the discovery of the ornamented parts, the original exterior of the building was restored by the master and brethren. Warwick is one of the towns included in Sir Thomas White's charity, by which young tradesmen are assisted with a free loan of £100 for nine years, to enable them to commence business. There are not less than 40 almshouses in various parts of the town, chiefly for aged women. Large funds for charitable uses and for distribution among the poor, are now vested in the charity Trustees. The poor law union of Warwick comprises 34 parishes or places, under the care of 45 guardians, and contains, according to the census of 1831, a population of 28,924.

About a mile from Warwick, on the road to Kenilworth, is Guy's Cliff, the solitary retreat, for some years prior to his death, of the celebrated Guy Earl of Warwick, of whom so many legendary tales are recorded: the cave in which he is said to have lived in retirement and devotion, and in which he was buried, is hewn in the rock near the bank of the Avon. Near it is a range of cells, having the appearance of a nunnery, and some cloisters hewn in the rock, and rudely arched, called Phillis' Cloisters, after the countess, who survived him only a few days, and was buried near him. Under a Roman arch, built by the late proprietor to sustain an ancient pointed one, which was falling to decay, are preserved two stone basins, called Guy's Well, covered with moss, into which a fine spring of clear water is constantly flowing. On this cliff, Richard de Beauchamp, Earl of Warwick, built a chapel, dedicated to St. Margaret, in which he erected a colossal statue of Guy in armour, in the attitude of drawing his sword; the right arm is wanting, and the left bears a shield: the chapel, now dismantled, is in the later style of English architecture, with a very beautiful porch, the roof of which, like that of the chapel, is richly groined. The mansion built on this cliff by the late Mr. Greatheed, and now the seat of the Honourable Charles Bertie Percy, is a handsome modern structure, with a stately avenue of noble fir-trees in front; the Avon winds beautifully round the base of the cliff, and through the grounds, in which is a water-mill for grinding corn, erected prior to the Conquest. Nearly opposite to Guy's Cliff, on the other side of the road, is Blacklow hill, a rocky eminence planted with forest trees: in the hollow part of the rock, which appears to have been quarried, Piers Gavestone was beheaded, in commemoration of which event a monument of four slender upright shafts, resting upon a pedestal, with a suitable inscription, and supporting a flat stone surmounted by a cross, has been erected on the summit.

Numerous monastic establishments anciently existed

in the town: the priory was founded by Henry de Newbury, Earl of Warwick, and completed by his son Roger, in the reign of Henry I., for canons Regular of the order of the Holy Sepulchre, and its revenue at the dissolution was £49. 13. 6.: the remains have been converted into a private mansion, but retain very considerable portions of the ancient architecture, and are situated at the entrance into the town from Birmingham. The hospital of St. John the Baptist was founded in the reign of Henry II., by William, Earl of Warwick, for the reception of strangers and pilgrims, and at the dissolution had a revenue valued at £19. 17. 3.: the building, which is a fine specimen of the architecture of the time, is now occupied as a private boarding-school, and is situated near the extremity of the town, on the road to Leamington. Within the precincts of the castle was the collegiate church of All Saints, of which John Rous relates, that St. Dubricius made it an episcopal seat, about the latter end of the sixth century, the Secular priests, or canons, of which establishment were, in 1125, united to the college of St. Mary. In the north-west part of the town was an abbey, which was destroyed by Canute in 1016, who also reduced to ashes a nunnery, occupying the site of St. Nicholas' churchyard. In the north suburb was the chapel of St. Michael, to which was annexed an hospital, founded about the close of the reign of Henry I., or the beginning of that of Stephen, by Roger, Earl of Warwick, for a master and leprous brethren, the revenue of which, at the dissolution, was £10. 19. 10.; the remains are appropriated as an almshouse for aged women. Of the hospital of St. Thomas, stated by Rous to have been founded by William, Earl of Warwick, not even the site is known. The convent of Dominican friars, which was situated in the western suburbs, was founded in the reign of Henry III., by the Botelers, Lords Studley, and the Mountforts; the revenue, at the time of the dissolution, was £4. 18. 6. Attached to the chapel of St. James, over the west gate of the town, now forming part of the Leicester hospital, was a college for four Secular priests, founded in the reign of Richard II., which continued till the dissolution. There were also numerous churches in the town, of which only St. Mary's and St. Nicholas' are remaining; these being found sufficient for the accommodation of the inhabitants, the others, which were greatly dilapidated, were suffered to fall into decay. Edward Plantagenet, son of George, Duke of Clarence, and the last male heir of that family, was born in Warwick castle: he was kept a close prisoner in the reigns of Edward IV., Richard III., and Henry VII., and, attempting to effect his escape from the Tower, during the reign of the last-named monarch, was beheaded in 1499. Warwick gives the title of Earl to the family of Greville.

WARWICKSHIRE, an inland county, bounded on the east by Leicestershire and Northamptonshire, on the south by Oxfordshire and Gloucestershire, on the west by Worcestershire, and on the north-west and north by Staffordshire. It extends from 51° 58′ to 52° 42′ (N. Lat.), and from 1° 10′ to 1° 57′ (W. Lon.); and comprises an area of 902 square miles, or 577,280 statute acres. The population, including that of the city and county of the city of Coventry, amounts to 336,610.

At the period of the invasion of Britain by Julius Cæsar, this county was included partly in the territory of the Cornavii, and partly in that of the Wigantes, or Wiccii; the former occupying the northern, and the latter the southern, part of it. It was first subjected to Roman sway by Ostorius Scapula, the second Roman governor of Britain, who entered it with his forces about the year 50, and constructed a line of intrenched camps along the banks of the Avon; and it was afterwards included in the province called *Flavia Cæsariensis.* On the complete establishment of the Saxon Heptarchy, it formed part of the powerful kingdom of Mercia, the sovereigns of which selected Warwick, Tamworth, and Kingsbury, as occasional places of residence.

Warwickshire was formerly partly in the diocese of Lichfield and Coventry, and partly in that of Worcester; but under the new ecclesiastical arrangements made pursuant to the act of the 6th and 7th of William IV., cap. 77, it is now entirely within the latter diocese, in the province of Canterbury: it contains the deaneries of Arden, Coventry, Marton, and Stonely, in the archdeaconry of Coventry; and the deaneries of Kington and Warwick, in the archdeaconry of Worcester. For purposes of civil government it is divided into four hundreds; *viz.*, Barlichway, having the divisions of Alcester, Henley, Snitterfield, and Stratford; Hemlingford, having those of Atherstone, Birmingham, Solihull, and Tamworth; Kington having those of Brails, Burton-Dassett, Kington, and Warwick; and Knightlow, having those of Kenilworth, Kirby, Rugby, and Southam. The county of the city of Coventry comprises an extent of 18,161 acres, and includes nine parishes. Warwickshire contains the city of Coventry (locally), the borough and market-town of Warwick, the large manufacturing market-town and lately enfranchised and incorporated borough of Birmingham, and the market-towns of Alcester, Atherstone, Coleshill, Henley-in-Arden, Kenilworth, Kington or Kineton, Leamington-Priors, Nuneaton, Rugby, Southam, Stratford-upon-Avon, and Sutton-Coldfield. Under the act of the 2nd of William IV., cap. 45, the county has been divided into two electoral portions, called the Northern and Southern Divisions, each empowered to send two members to parliament. The place of election for the Northern Division is Coleshill, and the polling-places are Coleshill, Nuneaton, Coventry, Birmingham, Dunchurch, Sutton-Coldfield, Solihull, Atherstone, and Polesworth. The place of election for the Southern Division is Warwick, and the polling-places, Warwick, Kington, Stratford-upon-Avon, Henley-in-Arden, Long Compton, and Southam. Two citizens are returned for the city of Coventry, and two burgesses for each of the boroughs of Birmingham and Warwick. Warwickshire is in the Midland circuit: the assizes and quarter sessions for the county are held at Warwick, where stand the common gaol and house of correction; and those for the city and county of the city of Coventry, at Coventry.

A considerable part of the county, on its south-western border, is separated from the rest by a detached portion of Worcestershire: and a smaller isolated district, lying at a short distance beyond its western confines, is surrounded by Worcestershire. The general surface is undulating, and, though seldom presenting romantic scenery, has, for the most part, a rich and pleasing appearance, greatly heightened by numerous small tracts of woodland. The banks of the Avon, though in some places flat and uninteresting, are in many, particularly near Warwick, highly beautiful and picturesque. The soils are generally fertile, but various, comprehending almost every kind, except such as contain chalk or flints. The southern and south-eastern part of the county has, nearly throughout, a strong clay loam, resting on limestone; and a similar soil occupies its north-eastern extremity, bordering on the course of the small river Anker. The soil of an extensive tract of Barlichway hundred, reaching from the vicinity of Warwick to the western border of the county, near Tamworth, and in the neighbourhood of Salford, and including the towns of Henley-in-Arden and Alcester, is also a strong clay loam resting on marl and limestone. A little westward from Warwick commences a considerable tract of strong clay on limestone, which extends south-westward to the confines of the county; and the large detached portion of the county to the west of its southern extremity has a soil of similar quality. A light sandy soil, in several places mixed with sharp gravel, and well adapted for turnip husbandry, occupies the tract lying between the town of Rugby and the village of Grandborough, bounded on the east by the confines of the county, and on the west by the road from Southam to Coventry. A soil of a similar kind, poor, but also well adapted for the same purpose, extends from the vicinity of the village of Meriden, northward, to the boundary of the county, chiefly bordering on the valleys of the Blythe and the Tame. Between this and the northern tract of clay and limestone land is a considerable extent of various poor and moory soils, which also occupy another large district on the west side of these light lands, in the vicinity of Sutton-Coldfield. The remaining extensive portions of the county consist chiefly of a red sandy loam, and a red clay loam, resting on freestone or limestone, and sometimes on sharp gravel: some of the sand is well suited for turnip husbandry. Coventry is also surrounded by a rich deep sandy loam, resting on marl and freestone, which, in different directions, soon becomes intermingled with some of the other soils above mentioned. On the south side of the river Avon, in the vicinity of Castle Bromwich, a good red clayey loam is found extending eastward to Coleshill, and westward towards Birmingham; near the latter town, and in the vicinities of Aston and Hockley brook, a light, dry, red sandy soil prevails. A considerable district around Sylihull, and various parishes to the north and east of that town, are composed of a strong marl clay soil on a wet clay substratum. In the vale of the Avon the soil is remarkable for its fertility, and is excellently adapted for the culture of turnips.

The courses of crops are various: those commonly cultivated are wheat, barley, oats, peas, beans, turnips, potatoes, and tares or vetches. Rye is seldom sown, except on the light, poor, sandy soils, and then chiefly for spring food for sheep. The principal artificial grasses are red and white clover, and rye-grass. The permanent meadow and pasture lands amount by computation to 235,000 acres, and the quantity of land under artificial grasses to 60,000, making a total of 295,000 acres. This is a noted grazing county: the most extensive tracts of permanent pasture are towards the eastern and north-eastern confines of the county, bordering on Leicestershire. Towards Oxfordshire, at Radway, Warmington, Avon-Dassett, Shotswell, &c., and along the great road thence towards Warwick, as far as Gaydon, is a large extent of very rich pastures; on each bank of the Avon, during the whole of its course through this county, there is also much rich meadow and grazing land; and numerous other parts abound with fine old pastures. The principal manures are lime, marl (which abounds in the western districts, but instead of which lime has, of late years, been much employed), soap ashes, horn scrapings, malt dust, and soot. Many different breeds of cattle are common in the county, although the Scotch and Herefordshire oxen, and the long-horned heifers and cows, are always preferred for grazing. The cows and oxen bred are chiefly of the long-horned Lancashire sort: the other breeds that have been introduced are principally the Devonshire, the Herefordshire, the Yorkshire, the Tees-water, the Scotch, and the Welsh: of these, the Yorkshire and long-horned and Durham breeds are most esteemed for the dairy. The sheep bred are chiefly of the Old Warwickshire and the New Leicester breeds: the former are a large polled kind, but they have been much intermingled with the New Leicester sheep, which are here bred in great perfection. The middle, western, and northern parts of the county are those most abounding with timber, of which a large proportion is oak of remarkably fine growth, those parts having been formerly occupied by the extensive forest of Arden: there are numerous thriving plantations of different kinds of forest trees in various parts. The extent of unenclosed land is inconsiderable: the commons of Sutton-Coldfield and Sutton-Park are the most extensive.

The chief mineral productions are coal, limestone, freestone, and a blue flag-stone. The coal of the best quality is found at Bedworth, between Coventry and Nuneaton, where the seam varies in thickness from three to four feet, and is worked to a considerable extent. Large quantities of coal are also raised at Griff-hollow, Chilvers-Coton, Nuneaton common, Hunts hall, and Oldbury, lying to the north of the first-mentioned place; and the same vein extends still further northward, by Merevale, to Polesworth and Wilnecote. Limestone is found to a great extent, and is quarried at numerous places, where it is also burned into lime: it is generally of a close texture and a dark blue colour, and produces strong lime. Abundance of freestone exists in the neighbourhoods of Warwick, Leamington, Kenilworth, Coventry, and many other places, chiefly where the soil is light and sandy: the blue flag-stone, used for paving and flooring, is found in many places, and is quarried in the neighbourhoods of Bidford and Wilnecote. Ironstone exists at Oldbury and Merevale, near the former of which it was anciently worked. The western part of the county abounds with marl of different colours and qualities, much of which is strong and excellent; and a peculiar kind of blue clay, having some of the properties of soap, exists in great quantities in the eastern part of it.

The hardware manufactures of Birmingham and its vicinity are the principal in the county. The next in importance is that of silk, ribands, &c., at Coventry and the surrounding villages: that city is also noted for the making of watches. There are considerable flax-mills at Berkeswell and Balsall, and in the vicinity of Tamworth, where much linen yarn is spun; at Kenilworth horn combs of all descriptions are manufactured; at Alcester are made fish-hooks and needles; and at

Atherstone are several manufactories for hats and ribands; the latter article is also manufactured at Nuneaton.

The principal rivers are the Avon and the Tame: the former, called the Upper Avon, to distinguish it from the river Avon, which forms the harbour of Bristol, was made navigable for vessels of 40 tons' burden up to Stratford, in 1637. This county has an extensive artificial navigation. Birmingham is the grand centre from which most of the important lines of communication radiate, enabling that town to send the produce of its manufactures, by a direct and easy water carriage, to the four great ports of the kingdom. The Birmingham old canal affords a medium for the conveyance of coal and iron to Birmingham and other places, from the numerous mines on its banks, and the manufactured goods of that town to Liverpool, Manchester, &c. The Birmingham and Worcester canal was formed principally for the conveyance of coal, and for opening a more direct communication between Birmingham and the Severn. The Dudley Extension canal branches from this a little before it enters the county near Birmingham; and the Stratford-on-Avon canal commences in it at King's-Norton in Worcestershire, and proceeds through this county to its termination in the navigable channel of the Avon at Stratford: this canal has a short branch to the village of Tamsworth, and a longer one to the Grafton limeworks; it also communicates by a short cut with the Warwick and Birmingham canal, near Lapworth-street. The Birmingham and Fazeley canal, commencing in the Coventry canal at Whittington brook, was formed chiefly for the conveyance of the produce of the Birmingham manufactures towards London and Hull, and for supplying Birmingham with grain and other commodities. The Coventry canal forms an important line in the communication between London, Birmingham, Manchester, Liverpool, &c., and, by means of it, great quantities of coal are conveyed from the pits in its vicinity, chiefly to the city of Coventry; it has a branch, about a mile in length, extending to the Griff collieries, and another, from which are several minor branches, to those near Lees-wood, Pool, and Bedworth. The Ashby-de-la-Zouch canal commences in the Coventry canal at Marston bridge, near Nuneaton, and, taking an irregular north-easterly course, soon quits the county near Hinckley. The Oxford canal commences in the Coventry canal at Longford, about four miles from Coventry, and finally quits for Oxfordshire a little to the south of Wormleighton; the Grand Junction canal commences in the last-mentioned at Braunston, on the eastern border of Warwickshire, but in the county of Northampton. The Warwick and Birmingham canal, commencing in the Digbeth cut of the Birmingham and Fazeley canal at Digbeth, near Birmingham, proceeds south-eastward near Solihull to Warwick, whence the navigation is continued by the Warwick and Napton canal, which terminates in the Oxford canal near Napton-on-the-hill: these two canals, with the few miles of the Oxford canal lying between Napton and Braunston, where the Grand Junction canal commences, form an important part of the most direct line of water communication between Birmingham and London, and the former supplies the town of Warwick with coal. The London and Birmingham railway enters this county to the south-east of Rugby, and passes by Coventry to Birmingham, where the Grand Junction railway commences. The Birmingham and Derby Junction railway has a branch from the former at Hampton-in-Arden, and another near Birmingham, which unite near Kingsbury, the railway then proceeding to Tamworth, where it quits the county for Staffordshire. A rail-road has been constructed from Stratford-upon-Avon to Moreton-in-the-Marsh, in the county of Gloucester, and an act was obtained in 1833 for making a branch to Shipston-upon-Stour.

Warwickshire contained the Roman station of *Manduessedum*, situated on the Watling-street, at Mancetter; and that of *Alauna*, at Alcester; while another was probably fixed at Chesterton. It was traversed by the Watling-street, the Fosse-way, the Ryknield-street, and the Ridge-way; and several vicinal ways diverged from each extremity of the great roads. The Roman camps are not very numerous; the principal are situated along the course of the Fosse-way, and on the banks of the river Avon: in the vicinity of these and of the roads are found many tumuli and coins, and various other vestiges of Roman occupation have been discovered in almost every part of the county. On Welcombe hills to the west of the village of Alveston, are extensive earthworks, called the Dingles, supposed to be of Saxon origin.

The number of religious houses, including hospitals and colleges, was about 57. There yet exist the remains of the abbey of Merevale, comprising some interesting specimens of Saxon or early Norman architecture; of the priories of Coventry, Kenilworth, and Maxstoke; and of the nunneries of Nuneaton, Pindley, and Polesworth. There are remains of the castles of Astley, Brandon, Kenilworth, Maxstoke, Tamworth, and Warwick, the last of which are particularly extensive, and form the chief part of the present magnificent residence of the Earl of Warwick. The most remarkable ancient mansions are Clopton House, Compton-Wyniates House, and Aston Hall near Birmingham; and some of the most distinguished of the more modern seats of the nobility and gentry, besides Warwick Castle, are Ragley Hall, Combe Abbey, Packington Hall, and Stoneleigh Abbey. There are various chalybeate springs, particularly at Birmingham, Ilmington, and Newnham-Regis: but the mineral waters of Leamington are by far the most celebrated, and their reputation has of late years converted this formerly obscure village into a place of fashionable resort.

WARWICK-BRIDGE, a township, in the parish of WITHERAL, ESKDALE ward, Eastern Division of the county of CUMBERLAND, $5\frac{1}{4}$ miles (E.) from Carlisle: the population is returned with the township of Great Corby. The river Eden is here crossed by a fine stone bridge of four arches to the opposite village of Warwick. A strong party of royalists, stationed to defend the passage of this bridge, in June 1648, was put to the rout by General Lambert. Extensive cotton-mills and bleaching-grounds, established by Messrs. Dixon and Sons, employ more than 500 persons: these gentlemen support a Sunday school. There is a Roman Catholic chapel at this place.

WASDALE, or NETHER-WASDALE, a chapelry, in the parish of ST. BEES, union of WHITEHAVEN, ALLERDALE ward above Derwent, Western Division of the county of CUMBERLAND, 7 miles (N. N. E.) from Ravenglass; containing 185 inhabitants. The living is a perpetual curacy; net income, £66; patron, Incumbent of St. Bees. The beautiful lake Wast-water is three miles long, half a mile broad, and forty-five fathoms deep, or about fifteen fathoms below the level of the sea, which disproportion as to its extent and depth accounts, perhaps, for its never having been known to freeze. A fair for sheep is held on the first Monday in September.

WASDALE-HEAD, a joint chapelry with ESKDALE, in the parish of ST. BEES, union of BOOTLE, ALLERDALE ward above Derwent, Western Division of the county of CUMBERLAND, 11 miles (S. W. by S.) from Keswick: the population is returned with Eskdale. This place is pleasantly situated at the head of Wast-water lake, in a narrow valley almost surrounded by lofty hills. The living is a perpetual curacy; net income, £49; patron, Incumbent of St. Bees. The chapel is very small, and, having no burial-ground attached, the inhabitants inter their dead at Nether-Wasdale.

WASHAWAY, a hamlet, in the parish of EGLOSHAYLE, hundred of TRIGG, Eastern Division of the county of CORNWALL, 3 miles (N. W.) from Bodmin: the population is returned with the parish. The petty sessions for the division are held here, on the last Monday in every month.

WASHBOURN, GREAT (*ST. MARY*), a parish, in the union of WINCHCOMB, Upper Division of the hundred of TEWKESBURY, Eastern Division of the county of GLOUCESTER, $5\frac{1}{2}$ miles (N. N. W.) from Winchcomb; containing 87 inhabitants. The living is a perpetual curacy; net income, £59; patron, J. B. Smith, Esq.

WASHBOURN, LITTLE, a chapelry, in the parish of OVERBURY, union of WINCHCOMB, Middle Division of the hundred of OSWALDSLOW, Pershore and Eastern Divisions of the county of WORCESTER, $6\frac{1}{2}$ miles (E. by N.) from Tewkesbury; containing 51 inhabitants.

WASHBROOK (*ST. MARY*), a parish, in the hundred of SAMFORD, Eastern Division of the county of SUFFOLK, 4 miles (W. by S.) from Ipswich; containing 418 inhabitants. The living is a discharged vicarage, annexed to the rectory of Copdock, and valued in the king's books at £8. 6. 8. The impropriate tithes have been commuted for a rent-charge of £191. 10., and the vicarial for £217, subject to rates, which on the average have amounted to £37 and £42; the glebe comprises 26 acres, valued at £39. 8. per annum. The church contains several ancient stalls, which have been recently renovated; and over the communion-table a window of stained glass has been placed, at the expense of Lord Walsingham. Felchurch was formerly a chapel to Washbrook, but has been for many years destroyed.

WASHFIELD (*ST. MARY*), a parish, in the union of TIVERTON, hundred of WEST BUDLEIGH, Cullompton and Northern Divisions of the county of DEVON, $2\frac{1}{2}$ miles (N. N. W.) from Tiverton; containing 453 inhabitants. The living is a rectory, valued in the king's books at £19. 7. 6.; patron, John Francis Worth, Esq. The tithes have been commuted for a rent-charge of £400, subject to the payment of rates, which on the average have amounted to £12; the glebe comprises 25 acres, valued at £46 per annum. The church contains an ancient oak screen, which has been painted white, and a curious monument with brasses to the family of Worth. Ancient swords and other military weapons have been found upon the site of what is supposed to have been a Roman encampment.

WASHFORD-PINE (*ST. PETER*), a parish, in the union of CREDITON, hundred of WITHERIDGE, South Molton and Northern Divisions of the county of DEVON, $8\frac{1}{2}$ miles (N. by W.) from Crediton; containing 174 inhabitants. The living is a discharged rectory, valued in the king's books at £6. 0. $2\frac{1}{2}$.; present net income, £144; patron, William Comyns, Esq. There was formerly a chapel at Whenham, in this parish.

WASHINGBOROUGH (*ST. JOHN THE EVANGELIST*), a parish, in the Second Division of the wapentake of LANGOE, parts of KESTEVEN, union and county of LINCOLN, 2 miles (N. E.) from Lincoln; containing 1124 inhabitants. This parish is bounded on the north by the navigable river Witham. The living is a rectory, valued in the king's books at £26. 13. 4.; present net income, £1554; patron, Sir W. A. Ingilby, Bart. The church is a large and handsome structure, with a lofty tower. There is a school for young children, with an endowment of £15. 10. per annum, arising from the bequests of Timothy Pike and others, in 1728. The free grammar school at Heighington was founded in 1619, by Thomas Garrett, who endowed it with lands and houses of the present annual value of £134; he was, in other respects, a great benefactor to the poor. In 1701, Sir Thomas Clack left land now producing £70 per annum, for apprenticing poor children.

WASHINGLEY, a parish, in the union of PETERBOROUGH, hundred of NORMAN-CROSS, county of HUNTINGDON, $1\frac{1}{2}$ mile (W.) from Stilton; containing 81 inhabitants. The living is a rectory, united to that of Lutton. There being no church, the inhabitants attend at Lutton.

WASHINGTON, a parish, in the union of CHESTER-LE-STREET, Eastern Division of CHESTER ward, Northern Division of the county palatine of DURHAM; containing 2673 inhabitants, of which number, 1123 are in the township of Washington, $5\frac{1}{2}$ miles (S. E.) from Gateshead. The living is a rectory, valued in the king's books at £18; present net income, £953; patron, Bishop of Durham. A gallery has been erected in the church, and 100 free sittings provided, the Incorporated Society having granted £40 in aid of the expense. At Usworth, in this parish, is a chapel of ease; and there is a place of worship for Wesleyan Methodists. The parish abounds with coal, and there are considerable cast-iron and rope works. A National school is partly supported by the rector.

WASHINGTON, a parish, in the union of THAKEHAM, hundred of STEYNING, rape of BRAMBER, Western Division of the county of SUSSEX, $4\frac{1}{2}$ miles (W. N. W.) from Steyning; containing 793 inhabitants. The living is a discharged vicarage, valued in the king's books at £9. 10.; present net income, £63; patrons and impropriators, President and Fellows of Magdalene College, Oxford. The church is in the early style of English architecture. Four schools for poor children are chiefly supported by private charity.

WASING (*ST. NICHOLAS*), a parish, in the union of NEWBURY, hundred of FAIRCROSS, county of BERKS, $7\frac{1}{2}$ miles (E. S. E.) from Newbury; containing 79 inhabitants. The living is a discharged rectory, valued in the king's books at £3. 13. 4.; present net income, £100; patron, W. Mount, Esq.

WASPERTON (*St. John the Baptist*), a parish, in the union of Warwick, Warwick Division of the hundred of Kington, Southern Division of the county of Warwick, 4 miles (S. S. W.) from Warwick; containing 292 inhabitants. The living is a discharged vicarage, valued in the king's books at £5; present net income, £230; patron and appropriator, Rector of Hampton-Lucy. A school is partly supported by the vicar.

WASTE-LANDS, an extra-parochial liberty, locally in the parish of Swineshead, wapentake of Kirton, parts of Holland, county of Lincoln, 6½ miles (W. by S.) from Boston: the population is returned with the parish.

WATCHETT, a sea-port and market-town, in the parish of St. Decuman, hundred of Williton and Freemanners, Western Division of the county of Somerset, 5 miles (E.) from Dunster, and 154 (W. by S.) from London: the population is returned with the parish. This place, anciently called *Weced-poort*, suffered severely from the Danes in 886: it is agreeably situated in a pleasant valley, on a creek of the Bristol Channel, and consists chiefly of four paved streets. It was once a place of extensive trade, and noted for its herring-fishery: vessels are now employed in the coasting trade, and in the importation of coal from Newport and Swansea; and two packets ply between this place and Bristol every fortnight. There is a small manufacture of woollen cloth and paper. A pier, originally erected by the Wyndham family, was repaired by Sir William Wyndham previously to 1740. The cliffs in the vicinity abound with alabaster and limestone. The market is on Saturday; and a fair is held on Nov. 17th. Manorial courts are held annually. There are places of worship for Baptists and Wesleyan Methodists; and a National school has been established.

WATCHFIELD, a township (formerly a chapelry), in the parish and hundred of Shrivenham, union of Farringdon, county of Berks, 4½ miles (S. W. by S.) from Great Farringdon; containing 341 inhabitants. The chapel was taken down about 1770.

WATERBEACH (*St. John*), a parish, in the union of Chesterton, hundred of Northstow, county of Cambridge, 5¼ miles (N. E. by N.) from Cambridge; containing 1146 inhabitants. The living is a discharged vicarage, valued in the king's books at £5. 15. 7½.; patron and appropriator, Bishop of Ely. The appropriate tithes have been commuted for a rent-charge of £9. 6. 6., and the vicarial for £73, subject to the payment of rates, which on the average have respectively amounted to 17*s.* and £6; the glebes comprise allotments of land consisting of 332 and 288 acres, respectively valued at £536 and £366 per annum. A charity school, now conducted on the National plan, was founded in the year 1687, and endowed with lands, by Grace Clarke and Dorothy Staines: the master's salary is upwards of £40 per annum, with a good house and garden, and the number of scholars is limited to eighteen. An almshouse for six poor widows was founded in 1628, by a bequest from Mr. John Yaxley, alderman of Cambridge, and endowed with £12 per annum; to which a rent-charge of £15 was added by Mrs. Jane Brigham, in 1705. About the year 1160, a cell to the monastery of Ely was established in a small island, called Elmeneye, but was shortly after removed to Denney, both in this parish: in the following century it was occupied by the Knights Templars, who then possessed the manor of Waterbeach. In 1293, an abbey for minoresses of the order of St. Clare was founded by Dionysia de Mountchensi, which, in 1338 (the order of the Templars being then abolished), was transferred to their house at Denney: at the dissolution there were 25 nuns, and the annual value of the lands was estimated at £172: the abbey house and the demesne have been many years rented as a farm, and the refectory has been converted into a barn.

WATERDEN (*All Saints*), a parish, in the union of Docking, hundred of Brothercross, Western Division of the county of Norfolk, 4 miles (W. by S.) from Little Walsingham; containing 24 inhabitants. The living is a discharged rectory, annexed to that of Warham-St. Mary, and valued in the king's books at £5. 6. 8. The tithes have been commuted for a rent-charge of £190, subject to the payment of rates, the glebe comprises 18 acres, valued at £16. 4. per annum.

WATER-EATON, a township, in the parish of Bletchley, union of Newport-Pagnell, hundred of Newport, county of Buckingham, ¾ of a mile (S.) from Fenny-Stratford; containing 243 inhabitants. There is an ancient manor-house, with a chapel, in which divine service is performed every Sunday.

WATER-EATON, a chapelry, in the parish of Kidlington, union of Woodstock, hundred of Wootton, county of Oxford, 3¾ miles (N.) from Oxford; containing 102 inhabitants.

WATER-EATON, a township, in the parish of Penkridge, Eastern Division of the hundred of Cuttlestone, Southern Division of the county of Stafford, 2¾ miles (S. W. by S.) from Penkridge, with which the population is returned.

WATER-EATON, a township, in the parish of Eisey, hundred of Highworth, Cricklade, and Staple, Cricklade and Northern Divisions of the county of Wilts, 2¼ miles (E. S. E.) from Cricklade: the population is returned with the parish.

WATERFALL (*St. James*), a parish, in the Southern Division of the hundred of Totmonslow, Northern Division of the county of Stafford, 7 miles (E. S. E.) from Leek; containing 531 inhabitants. The river Hamps surrounds the parish; it enters the ground at the Waterhouses, and pursues a subterraneous course of about three miles to Ilam, where it emerges and joins the river Manifold. Limestone (a considerable quantity of which is burned), grit-stone, and lead-ore, are found in the neighbourhood; and at the adjoining hamlet of Winkshill are two paper-mills, a flax-mill, and an iron forge and foundry. The living is a perpetual curacy; net income, £65: it is in the patronage of Mrs. Jane Wilmott; and the impropriation belongs to Mrs. Townsend. There is a place of worship for Wesleyan Methodists. Eight poor children are instructed in a schoolhouse built by subscription, for £6. 10. a year, arising from four acres and a half of land given by the freeholders about 60 years since.

WATERGALL, an extra-parochial liberty, in the Southam Division of the hundred of Knightlow, Southern Division of the county of Warwick, 4 miles (S.) from Southam; containing 13 inhabitants.

WATERHEAD, a township, in the parish of Lanercost-Abbey, union of Brampton, Eskdale ward, Eastern Division of the county of Cumberland, 24 miles (N. E.) from Carlisle; containing 177 inhabitants.

WATERINGBURY (*St. John the Baptist*), a parish (formerly a market-town), in the union of West Malling, hundred of Twyford, lathe of Aylesford, Western Division of the county of Kent, 5 miles (W. by S.) from Maidstone; containing 1109 inhabitants. The living is a discharged vicarage, valued in the king's books at £5; present net income, £727; patrons and appropriators, Dean and Chapter of Rochester. The church, an ancient edifice in the early English style, with a spire, formerly exhibited a profusion of stained glass, with portraits of Edward III. and his consort Philippa: it has been enlarged, and contains 320 free sittings, the Incorporated Society having granted £380 in aid of the expense. A day and Sunday school is supported by subscription. The scenery is agreeably enlivened by several gentlemen's seats in the neighbourhood of the village. Fruit is extensively raised here for the supply of the London market. On a monument in the cemetery, to the memory of Sir Oliver Style, Bart., it is recorded that, in the height of an entertainment given to a party of his friends at Smyrna, every individual, himself excepted, perished by an earthquake.

WATER-MILLOCK, a chapelry, in the parish of Greystock, union of Penrith, Leath ward, Eastern Division of the county of Cumberland, 7 miles (S. W.) from Penrith; containing 429 inhabitants. This place, sometimes called Newchurch (from the present chapel, which was built in 1558, on a more convenient site than the former), is situated on the north side of Ullswater lake, in a district abounding with diversified scenery, the natural beauties of which have been heightened and improved by the erection of several handsome private residences, with pleasure grounds tastefully laid out. In a deep glen in Gow-Barrow Park, rushing impetuously through the thick foliage of full-grown trees, is Airey Force, a beautiful cataract, which, dashing violently from rock to rock, emits a considerable spray. The discharge of a gun here produces, from the reverberation of the hills, an effect somewhat like thunder, and one or two French horns that of an harmonious concert of musical instruments. The living is a perpetual curacy; net income, £55; patron, Rector of Greystock. A school for boys is endowed with £525 in the 3½ per cent. consols.; and a school containing 16 girls is partly supported by Mrs. Marshall, whose late husband built the school-house in 1832.

WATER-OAKLEY, a division in the parish and hundred of Bray, county of Berks: the population is returned with the parish.

WATER-OVERTON, a chapelry, in the parish of Aston, Birmingham Division of the hundred of Hemlingford, Northern Division of the county of Warwick, 2½ miles (N. W.) from Coleshill: the population is returned with the parish. The living is a perpetual curacy; net income, £115; patrons and impropriators, Trustees. The chapel is dedicated to St. Peter and St. Paul.

WATERPERRY (*St. Mary*), a parish, in the union of Thame, hundred of Bullingdon, county of Oxford, 2 miles (S. W.) from Wheatley; containing 243 inhabitants. The living is a discharged vicarage, valued in the king's books at £8. 1. 5½.; present net income, £60; patron and impropriator, Joseph Henley, Esq. The church consists of a nave, south aisle, and chancel, with a wooden tower of singular construction, and contains some fine brasses to the Curzon family, whose ancient mansion near it is now the property of Joseph Henley, Esq.; in the south aisle is an altar-tomb, with the recumbent effigy of a crusader, supposed to be one of the family of Ledwell; and in the chancel is a splendid monument, by Chantrey, to Mrs. Greaves. A parochial school is supported by Joseph Henley, Esq.

WATERSTOCK (*St. Leonard*), a parish, in the union and hundred of Thame, county of Oxford, 3 miles (E.) from Wheatley; containing 142 inhabitants. The living is a rectory, valued in the king's books at £10. 16. 0½.; present net income, £58; patron, William Henry Ashurst, Esq. The chancel of the church is of modern erection, and contains a monument to the memory of Sir George Crook, one of the judges of the Court of King's Bench in the reign of Charles I. A school is chiefly supported by Mr. Ashurst.

WATFORD (*St. Mary*), a market-town and parish, and the head of a union, in the hundred of Cashio, or liberty of St. Albans, county of Hertford; containing 5293 inhabitants, of which number, 2960 are in the town of Watford, 20 miles (W. S. W.) from Hertford, and 15 (N. W.) from London. This town, situated on the river Colne, derives its name from the Watling-street, which passes in the vicinity, and from a ford over the river, to which latter its origin also is attributed: it consists of one principal street, about a mile in length and irregularly built, and is supplied with water by a forcing pump, erected by subscription. By means of the Grand Junction canal, which passes a mile to the westward, a communication is maintained with the metropolis and the northern part of the kingdom. The manufacture of straw-plat, and three silk-throwsting mills, employ a considerable number of persons; there are likewise eight malt-kilns, and two extensive breweries. The London and Birmingham railway intersects the southern part of the parish, and at a short distance from the town is a station: about 2¼ miles beyond the town is the Watford tunnel, one mile in length. The market, granted by Henry I., is held on Tuesday: the market-house is an indifferent building, supported on wooden pillars, with granaries over it, and its situation is very confined. Fairs are held on the Tuesday after Whit-Tuesday, and on August 29th and 30th, for cattle and pedlery; the latter, originally granted by Edward III., in 1335, had fallen into disuse, but was revived in 1827; there is also a statute fair for hiring servants in Sept. A meeting of magistrates is held every Tuesday, and there is a weekly court of requests, for the recovery of debts under 40*s*., the jurisdiction of which is co-extensive with the liberty of St. Albans; this court is constituted of commissioners, whose qualification to act is the possession of £800 real or personal property. The town has lately been made a polling-place for the election of the county representatives.

The living is a vicarage, valued in the king's books at £21. 12. 1.; present net income, £730; patron and impropriator, Earl of Essex. The church, situated in the centre of the street, on the south side of the town, has two chapels annexed, and a tower at the west end. There are places of worship for Baptists, those in the Connection of the late Countess of Huntingdon, and Wesleyan Methodists. The free school was founded in

1708, by Elizabeth Fuller, who endowed it with a rent-charge of £52, which has been subsequently augmented by the following bequests; namely, a rent-charge of £21 from Silvester Chilcot, in 1715; £1100 from Cornelius Denne, in 1805; £1000 from Elizabeth Whittingstall, in 1824; and about £1000 more from three other benefactors, altogether producing a revenue of £178. The school-house is a handsome structure, at the south-west corner of the churchyard, containing a school-room for the boys, and another for the girls, a room for the trustees, apartments for the master and the mistress, and domestic offices. Forty boys and twenty girls, children of poor inhabitants of Watford, are clothed and instructed. A parochial free school, in which eleven boys are taught, was founded in 1641, and endowed with a rent-charge of £10, by Francis Coombe, who also bequeathed to this parish an estate, the rent of which, together with the produce of various bequests from the Villiers family and others, amounting in the aggregate to about £100 per annum, is distributed in clothing among the poor on the first Monday in each year. A National Sunday school is chiefly supported by the Earl of Essex. Almshouses for eight poor widows of Watford, Cheneys, and Langley, were founded by Francis Earl of Bedford and his Countess, in 1580, and were endowed by Charles Morrison, in 1583, Lady Mary Morrison, in 1629, and Mary Newman, in 1789, with property now producing an annual income of £72. The almshouses in Lote's-lane were erected in 1824, in lieu of a building given by Lady Dorothy Morrison, in 1614, as a free residence for a lecturer and four poor widows: it was endowed, in 1627, by Sir Charles Morrison, Bart., the son of the founder, with a rent-charge of £50, and in 1791 with a bequest of £350 from Hannah Pocock: the present annual income is £55. 10., and the lecturer receives about £100 per annum, arising from the rent of a corn-mill given by Lady Elizabeth Russell, in 1610. Lady Mary Morrison, above mentioned, bequeathed, in 1629, a rent-charge of £50, for apprenticing children; and Sarah Ewer, in 1765, gave stock now producing a dividend of £16. 2. 9. for the same purpose. The annual rent of the church lands is £151. 6. The poor law union of Watford comprises 6 parishes or places, containing a population of 15,379, according to the census of 1831, and is under the superintendence of 16 guardians.

WATFORD (*St. Peter and St. Paul*), a parish, in the union of Daventry, hundred of Guilsborough, Southern Division of the county of Northampton, $4\frac{3}{4}$ miles (N. N. E.) from Daventry; containing 353 inhabitants. The living is a vicarage, valued in the king's books at £11. 7. $8\frac{1}{2}$.; present net income, £296: it is in the patronage of the Crown; impropriator, Lord Henley. The union canal passes through the parish, which is bounded on the west by the ancient Watling-street. Sarah Clarke in 1702, gave £400, now producing £33 a year, which is applied in aid of a National school. Here are some springs strongly impregnated with iron.

WATH (*St. Mary*), a parish, partly in the wapentake of Allertonshire, and partly in the wapentake of Hallikeld, North Riding of the county of York; containing 730 inhabitants, of which number, 196 are in the township of Wath, $4\frac{1}{4}$ miles (N. by E.) from Ripon. The living is a rectory, valued in the king's books at £17. 17. 1.; present net income, £981; patron, Marquess of Ailesbury. The Rev. Peter Samwaise, in 1690, founded a free school, and endowed it with lands and houses producing about £75 a year, for which 40 boys are educated.

WATH, a township, in the parish of Hovingham, union of Malton, wapentake of Ryedale, North Riding of the county of York, 8 miles (W. by N.) from New Malton; containing 21 inhabitants.

WATH-UPON-DEARNE (*All Saints*), a parish, in the union of Rotherham, Northern Division of the wapentake of Strafforth and Tickhill, West Riding of the county of York; containing 6927 inhabitants, of which number, 1149 are in the township of Wath-upon-Dearne, $5\frac{3}{4}$ miles (N.) from Rotherham. There are extensive potteries, furnaces, and collieries in the parish. The living is a discharged vicarage, with the perpetual curacy of Adwick-upon-Dearne annexed, valued in the king's books at £15. 10. $2\frac{1}{2}$.; present net income, £315; patrons and appropriators, Dean and Canons of Christ-Church, Oxford. There is a place of worship for Wesleyan Methodists. The Rev. Thomas Wombwell, in 1663, gave £30 towards the erection of a school-house, which is endowed with £17 per annum, for which 30 children are taught. The Roman Ikeneld-street passed through the parish.

WATLESBOROUGH, a township, in the parish of Cardeston, hundred of Ford, Southern Division of the county of Salop, 6 miles (W.) from Shrewsbury: the population is returned with the parish.

WATLINGTON (*St. Peter and St. Paul*), a parish, in the union of Downham, hundred of Clackclose, Western Division of the county of Norfolk, $5\frac{1}{2}$ miles (N.) from Downham-Market; containing 500 inhabitants. The living is a rectory, valued in the king's books at £14. 16. 8.; patron, C. B. Plastow, Esq. The church, which has been highly decorated, contains some fine old fragments of stained glass, a curious ancient font, and several monuments and inscriptions.

WATLINGTON (*St. Leonard*), a market-town and parish, in the union of Henley, hundred of Pirton, county of Oxford, 15 miles (E. S. E.) from Oxford, and 43 (W. by N.) from London; containing 1833 inhabitants. The name is supposed to have been derived from the Saxon *Watelar*, hurdles or wattles, alluding to the way in which the Britons are described to have built their towns, "as groves fenced in with hewn trees." About three miles from this place is Chalgrove Field, where a battle was fought in 1643, between the royalists and parliamentarians, in which the royalists were victorious, and Hampden was killed. It is traditionally said that a military chest of money was left at the house of Robert Parslow, of this town, and never afterwards reclaimed, in consequence of which he bequeathed a liberal donation to the poor of this parish. On the hill immediately above the town is an obelisk, from the summit of which is obtained a view over nine counties. The town is situated between the two high roads leading from London to Oxford, about half a mile from the line of the Iknield-street; it is irregularly built, and consists of some narrow streets, the houses, with a few exceptions, being but of mean appearance; water is supplied from an adjacent brook, which rises in one of the Chiltern hills, above the town, and on which are four corn-mills. A few females are employed in lace-making; a school, in which from 30 to 40 girls attend, having been established to teach them the art. The market, granted in the reign of Richard I., is on Saturday: a substantial market-house was built in 1666, by Thomas Stonor, Esq., and over it is a room where the public business of the town is transacted. A fair is held on April 5th, and on the Saturday before and after Old Michaelmas is a statute fair. Two courts leet are held annually, and the petty sessions for the hundred take place once a fortnight. The living is a discharged vicarage, valued in the king's books at £12; present net income, £175: the patronage and impropriation belong to Miss Tilson. The church stands on the north-western side of the town; in the chancel is a burial-place of the Horne family, also some interesting monuments. There are places of worship for Independents and Wesleyan Methodists, and one for the latter also on Christmas common. The free grammar school, once a noted classical school, but now confined to English instruction, was founded in 1664, and endowed with a rent-charge of £10, by Thomas Stonor, Esq., which sum has been augmented by subsequent benefactions: the master, who must be a graduate of one of the Universities, receives a salary of £20 per annum; it is held in a room over the market-place, and 20 boys are instructed. Twenty poor men are annually provided with coats from Parslow's bequests, and there are several other charities for the repair of the church and for distribution among the poor not receiving parochial relief. On Bretwell hill there are some remains of trenches, indicating the site of an ancient encampment. Of Watlington castle, which stood south-east of the church, there are only some traces of the moat by which it was surrounded, now filled with water.

WATTISFIELD (*St. Margaret*), a parish, in the union of Stow, hundred of Blackbourn, Western Division of the county of Suffolk, 2 miles (W. S. W.) from Botesdale; containing 592 inhabitants. The living is a discharged rectory, valued in the king's books at £8. 11. 8.; present net income, £336: it is in the patronage of Mrs. Morgan. There is a place of worship for Independents. A rental of £71. 17., derived from the town lands, is appropriated to general parochial purposes. Mr. Thomas Harmer, author of "Observations on Divers Passages of Scripture," resided in this parish more than 50 years.

WATTISHAM (*St. Nicholas*), a parish, in the union and hundred of Cosford, Western Division of the county of Suffolk, 2 miles (N. E.) from Bildeston; containing 202 inhabitants. The living is a perpetual curacy; net income, £100; patrons and impropriators, Provost and Fellows of King's College, Cambridge. In the church is a tablet recording a singular calamity which happened to a poor family, by which six persons lost their feet by mortification, and of which an account is published in the Philosophical Transactions for 1762. There is a place of worship for Baptists; also a National school.

WATTLEFIELD, a division, in the parish of Wymondham, hundred of Forehoe, Eastern Division of the county of Norfolk, $2\frac{3}{4}$ miles (S. by W.) from Wymondham; containing 451 inhabitants.

WATTON (*St. Mary and St. Andrew*), a parish, in the hundred of Broadwater, union and county of Hertford, $4\frac{3}{4}$ miles (N. N. W.) from Hertford; containing 830 inhabitants. The living is a rectory, valued in the king's books at £19. 8. $6\frac{1}{2}$.; patron, Abel Smith, Esq. The tithes have been commuted for a rent-charge of £686, subject to the payment of rates, which on the average have amounted to £98; the glebe comprises 68 acres, valued at £87 per annum. The church has a square western tower embattled, and a chapel attached to the north side of the chancel. The river Beane runs through the parish, which was also intersected by one of the Roman vicinal ways, on the supposed line of which there is still a large stone, apparently of high antiquity, and several coins have been found in the vicinity. Sir William and Maurice Thompson, in 1662, founded a free school for 20 children, and endowed it with property now producing a rental of £39. 12. 6.: of this sum, £20 is now paid towards a National school, which was established in the parish in 1818, a school-room having been erected at the sole expense of Samuel and Abel Smith, Esqrs.; and the remaining portion of the bequest is appropriated to apprenticing children.

WATTON (*St. Mary*), a market-town, and parish, in the union and hundred of Wayland, Western Division of the county of Norfolk, 24 miles (W. by S.) from Norwich, and 94 (N. N. E.) from London; containing 1027 inhabitants. This town, which suffered severely, in 1673, from an extensive fire that destroyed property to the amount of £10,000, is situated in the centre of the hundred, on the verge of that part of the county called *Filand*, or "the open country:" the streets are lighted with oil, and the inhabitants are supplied with water from springs: the neighbourhood is noted for supplying the metropolis with large quantities of butter, called Cambridge butter. The market is on Wednesday; and the ancient fairs are on July 10th, Oct. 11th, and Nov. 3rd; two others, of modern establishment, are held on the first Wednesday in July and the first Wednesday after Old Michaelmas-day. A manorial court is held annually, and petty sessions for the hundred every month. The living is a discharged vicarage, valued in the king's books at £7. 0. 4.; present net income, £187; patron and impropriator, W. H. Hicks, Esq. The church appears to have been originally built in the time of Henry I., and to have been re-dedicated in that of Henry VI.; the tower is circular at the base, and octangular above; the north porch is surmounted by a much admired crucifix, now partly dilapidated; there are a few brasses. In the centre of the town is a small building, with a clock and one bell, the latter rung on Sundays, preparatory to divine service: the lower part is used as a lock-up house. There is a place of worship for Independents, a short distance from the town. A National school was erected in 1819, by William Lane Robinson, Esq., and is supported by voluntary contributions. Almshouses for four poor widows were founded and endowed with a small rent-charge by Mr. Goff. In Wayland wood, near this town, supposed to be the scene of the tale of the Babes in the Wood, and which gives name to the hundred, the sheriff's torn, or court, was anciently held.

WATTON (*St. Mary*), a parish, in the union of Driffield, Bainton-Beacon Division of the wapentake of Harthill, East Riding of the county of York, $5\frac{1}{2}$ miles (S.) from Great Driffield; containing 345 inhabitants. The living is a perpetual curacy; net income, £60; patron, R. Bethell, Esq., who chiefly supports two schools. A nunnery of the Sempringham order, in honour of the Blessed Virgin Mary, was founded here in 1150, upon the site of a more ancient priory, which existed in the year 686: at the dissolution, its revenue was valued at £453. 7. 8., when its site and remains, which are still discernible, were granted to John Earl of Warwick.

WAULDBY, a township, in the parish of ROWLEY, union of SCULCOATES, Hunsley-Beacon Division of the wapentake of HARTHILL, East Riding of the county of YORK, $4\frac{1}{4}$ miles (E. by S.) from South Cave; containing 50 inhabitants.

WAVENDON (*ST. MARY*), a parish, in the union of NEWPORT-PAGNELL, hundred of NEWPORT, county of BUCKINGHAM, $3\frac{1}{2}$ miles (N. E.) from Fenny-Stratford; containing 802 inhabitants. The living is a rectory, valued in the king's books at £26. 6. $10\frac{1}{2}$.; present net income, £593; patron, Sir H. Hugh Hoare, Bart. There is a place of worship for Wesleyan Methodists. George Wells, in 1714, bequeathed £800, and his niece, Beatrice Miller, subsequently added £200, for founding and endowing a free school: these sums were invested in land now producing a clear rental of £84. 18. 8., out of which, £24 is paid to a master, who has also a residence and garden free, for instructing ten poor boys, who are clothed and occasionally apprenticed from the remaining funds of the charity. A National school for boys has also been established. The Duke of Bedford annually supplies coal to the amount of £150 for distribution among the poor, in lieu of an allotment of waste land awarded under an enclosure act in 1791. The parish is in possession of some property known as the Town lands, the rents of which, amounting to about £20, together with the produce of some minor charities, are annually distributed in bread among the poor.

WAVERLEY, an extra-parochial liberty, in the hundred of FARNHAM, Western Division of the county of SURREY, 2 miles (S. E. by E.) from Farnham; containing 74 inhabitants. An abbey of Cistercian monks, in honour of the Blessed Virgin Mary, was founded here in 1128, by William Giffard, Bishop of Winchester, which, at the dissolution, had a revenue of £196. 13. 11.: it is said to have been the first of that order established in England. The remains are still considerable, and have an interesting effect, much enhanced by the luxuriant ivy with which they are overgrown. This property has been purchased by G. T. Nicholson, Esq., who has greatly improved it, and it is now one of the most magnificent seats in the county.

WAVERTON (*ST. PETER*), a parish, in the union of GREAT BOUGHTON, Lower Division of the hundred of BROXTON, Southern Division of the county of CHESTER; containing 720 inhabitants, of which number, 324 are in the township of Waverton, $4\frac{1}{2}$ miles (E. S. E.) from Chester. The living is a perpetual curacy; patron, Bishop of Chester; the rectory is valued in the king's books at £23. 6. 8., and is annexed to the bishopric of Chester. The tithes have been commuted for a rent-charge of £150, subject to the payment of rates, which on the average have amounted to £15; the glebe comprises 20 acres, valued at £41. 10. per annum. The tithes of the townships of Hatton and Huxley have been respectively commuted for rent-charges of £150 and £135, subject to the payment of rates. The Chester canal passes through the parish.

WAVERTON, HIGH and LOW, a township, in the parish and union of WIGTON, ward and Eastern Division of the county of CUMBERLAND, $2\frac{1}{2}$ miles (W. S. W.) from Wigton; containing 487 inhabitants. The river Waver intersects the township, dividing it into what is termed High and Low Waverton.

WAVERTREE, a chapelry, in the parish of CHILDWALL, union and hundred of WEST DERBY, Southern Division of the county palatine of LANCASTER, $2\frac{1}{2}$ miles (E. by S.) from Liverpool; containing 1932 inhabitants. Its proximity to Liverpool has made it the residence of several respectable families. The living is a perpetual curacy; net income, £117; patrons, certain Trustees. A National school is supported by subscription. Here is a well, at which contributions were formerly received by some monks; it bears a curious Latin inscription, and the date 1414.

WAXHAM (*ST. JOHN AND ST. MARGARET*), a parish, in the hundred of HAPPING, Eastern Division of the county of NORFOLK, 12 miles (E. S. E.) from North Walsham; containing 59 inhabitants. The living is a discharged rectory, with the vicarage of Palling annexed, valued in the king's books at £6. 13. 4.; patron, — Conyers, Esq. The tithes have been commuted for a rent-charge of £343, subject to the payment of rates, which on the average have amounted to £27; the glebe comprises $1\frac{1}{2}$ acres, valued at £2. 5. per annum.

WAXHOLME, a township, in the parish of OWTHORNE, union of PATRINGTON, Middle Division of the wapentake of HOLDERNESS, East Riding of the county of YORK, $15\frac{1}{2}$ miles (E.) from Kingston-upon-Hull; containing 68 inhabitants.

WAYFORD, a parish, in the union of CHARD, hundred of CREWKERNE, Western Division of the county of SOMERSET, $2\frac{1}{2}$ miles (S. W.) from Crewkerne; containing 219 inhabitants. The living is a discharged rectory, valued in the king's books at £5. 1. $5\frac{1}{2}$.; present net income, £132; patron, John Pinney, Esq. Elizabeth Bragge, in 1719, gave a rent-charge of £2. 10. for teaching eight children.

WEALD, a chapelry, in the parish and union of SEVEN-OAKS, hundred of CODSHEATH, lathe of SUTTON-AT-HONE, Western Division of the county of KENT, 2 miles (S.) from Seven-Oaks: the population is returned with the parish. The living is a perpetual curacy, in the patronage of Earl Amherst, with reversion to the Vicar of Seven-Oaks. The church, an elegant structure, in the early English style, was erected in 1820, and the tower was added in 1839 by subscription. A National school is supported by subscription; and a neat range of almshouses for eight persons has been built here in connection with those at Seven-Oaks.

WEALD, a joint hamlet with Greenhill, in the parish of HARROW-ON-THE-HILL, hundred of GORE, county of MIDDLESEX, $2\frac{1}{2}$ miles (N.) from Harrow-on-the-Hill: the population is returned with the parish.

WEALD, NORTH, or NORTH-WEALD-BASSET, (*ST. ANDREW*), a parish, in the union of EPPING, partly in the hundred of HARLOW, but chiefly in that of ONGAR, Southern Division of the county of ESSEX, $3\frac{1}{4}$ miles (N. E. by E.) from Epping; containing 887 inhabitants. This parish is situated near the northern extremity of the hundred; the soil is heavy, but under good management has been rendered abundantly productive. The living is a vicarage, valued in the king's books at £13. 6. 8.; present net income, £353; patrons, Bishop of London and R. P. Ward, Esq.; impropriator, J. King, Esq. The impropriate tithes have been commuted for a a rent-charge of £426, and the vicarial for £446, subject to the payment of rates, which on the average on each have amounted to £65; the impropriate glebe comprises 24 acres, and the vicarial 11, respectively valued at £48 and £22 per annum. The church is a small ancient edifice, with a substantial tower of brick. A day and Sunday school is endowed with £10 per annum.

WEALD, SOUTH (*St. Peter*), a parish, in the union of Billericay, hundred of Chafford, Southern Division of the county of Essex; containing 2825 inhabitants. This parish, which from its name is supposed to have been that portion of the forest which was first inhabited, comprises about 6000 acres, including the hamlets of Brentwood and Brook-street. The living is a vicarage, valued in the king's books at £26. 13. 4.; patron, Bishop of London; impropriator, C. T. Tower, Esq. The impropriate tithes have been commuted for a rent-charge of £13, and the vicarial for £680, subject to the payment of rates, which on the latter have on the average amounted to £16. 10. 6.; the glebe comprises 18 acres, valued at £27. 18. per annum. The church contains 130 free sittings, the Incorporated Society having granted £100 towards defraying the expense; it is a handsome structure of stone, with a massive and lofty tower, forming a conspicuous object; and there is a chapel at Brentwood (which see). A National school has been established; and there are five almshouses for the poor. In front of the ancient hall is a mild chalybeate spring, much resorted to in summer, the water possessing properties somewhat similar to those of sea water.

WEARDALE, ST. JOHN, or ST. JOHN'S CHAPEL, a market-town and chapelry, in the parish of Stanhope, North-Western Division of Darlington ward, Southern Division of the county palatine of Durham, 6¼ miles (W. N. W.) from Stanhope: the population is returned with the parish. This is a small thriving town, situated in the Vale of Wear, through which runs the river of that name: its chief support is derived from the neighbouring lead mines. A customary market, on Saturday, has been established for more than a century: the market-cross was erected at the expense of the late Sir Ralph Milbank, Bart. The living is a perpetual curacy; net income, £186; patron, Rector of Stanhope. The chapel, which is a handsome structure, was rebuilt at the expense of the late Sir William Blackett, Bart., aided by a bequest of £50 from Dr. Hartwell. Dr. Shute Barrington, the late Bishop of Durham, also erected another chapel, which is presented to by the rector; and, about the same time, a National school. There are places of worship for Independents and Primitive and Wesleyan Methodists.

WEARDLEY, a township, in the parish of Harewood, Upper Division of the wapentake of Skyrack, West Riding of the county of York, 6½ miles (E.) from Otley; containing 169 inhabitants.

WEARE (*St. Gregory*), a parish, in the union of Axbridge, hundred of Bempstone, Eastern Division of the county of Somerset, 2½ miles (S. W.) from Axbridge; containing 764 inhabitants. The living is a vicarage, endowed with the rectorial tithes, and valued in the king's books at £12. 1. 5½.; present net income, £350; patrons, Dean and Chapter of Bristol. A Sunday school is supported by the vicar and the curate. The river Ax is crossed by an ancient bridge, and runs through that part of the parish termed Nether Weare; which place, among many other privileges granted by different monarchs, enjoyed that of sending members to parliament in the 34th and 35th of Edward I., and had a weekly market on Wednesday, with an annual fair.

WEAR-GIFFORD (*Holy Trinity*), a parish, in the union of Torrington, hundred of Shebbear, Great Torrington and Northern Divisions of the county of Devon, 2¾ miles (N. N. W.) from Great Torrington; containing 547 inhabitants. This parish takes its name from the ancient family of the Giffords, to whom the manor belonged till the reign of Henry III., when it passed by female heirs into other hands. It again reverted to the descendants of Sir Walter Gifford, from whom it passed to the Denzells, and from them by marriage to Martin, son of Sir John Fortescue, chancellor to Henry VI., and ancestor of the present proprietor, Earl Fortescue. The ancient manor-house, built by the Denzells in the fifteenth century, though now occupied as a farm-house, retains much of its original character; the hall is ceiled with oak richly carved, and the walls are curiously plastered and ornamented with escutcheons of armorial bearings; the original oak screen at the east end is still remaining, and in good preservation. The living is a rectory, valued in the king's books at £13. 5.; patron, Earl Fortescue. The tithes have been commuted for a rent-charge of £172, subject to the payment of rates; the glebe comprises 11½ acres, valued at £30 per annum. The church contains the figures of a cross-legged knight and his lady, carved in stone, and now placed in an erect position against the wall, though formerly recumbent on an altar-tomb, and most probably the effigies of some of the Gifford family. Sixteen poor children are taught to read for £15 per annum, the bequest of John Lovering, in 1671.

WEARMOUTH, BISHOP (*St. Michael*), a parish, adjoining the town of Sunderland, chiefly in the union of Sunderland, but partly in that of Houghton-le-Spring, Northern Division of Easington ward and of the county palatine of Durham; containing 16,590 inhabitants, of which number, 14,462 are in the township of Bishop-Wearmouth. This parish derived its distinguishing appellation from having belonged to the Bishop of Durham, in contradistinction to Monk-Wearmouth on the opposite side of the river, which belonged to the monks of his cathedral. It is situated on the south-western side of Sunderland, and, prior to 1719, comprised that town within its limits: the two places are now connected by a continuation of the principal street of Sunderland uniting with the High-street of this town, and may be said to form one large town and port. The rector is lord of the manor, for which he occasionally holds courts. On the side of an eminence, called Building Hill, is a quarry of stone, which, on the division of lands in 1649, was reserved for the free use of the copyholders within the manor: and the inhabitants, from time immemorial, have enjoyed the privilege of bleaching their linen, &c., on a small piece of ground, called Burnfields, a little westward. The living is a rectory, valued in the king's books at £89. 18. 1½.; present net income, £2899; patron, Bishop of Durham. The church was rebuilt in the year 1807, on the site of the ancient edifice, which existed from the time of Athelstan. Another church was erected in the early English style, in 1829, by the parliamentary commissioners, at an expense of £4879. 7. 9.; and there are three other churches or chapels in the parish. There are places of worship for Methodists of the Old and New Connection. Near the church is a National school, and schools are also attached to the chapels of ease. For a more detailed account, see SUNDERLAND.

WEARMOUTH, MONK (*St. Peter*), a parish, in the union of SUNDERLAND, Eastern Division of CHESTER ward, Northern Division of the county palatine of DURHAM; containing 9428 inhabitants, of which number, 1498 are in the township of Monk-Wearmouth, ½ a mile (N.) from Sunderland. This place is situated on the northern bank of the Wear, near its mouth, and takes its distinguishing prefix from having belonged to the monks of Durham. A monastery was founded here about 674, by Biscopius, a Saxon nobleman attached to the court of Oswy, King of Northumberland, and dedicated to St. Peter. In the reign of Ethelred, this establishment was completely destroyed by the Danes: accounts differ with respect to the period of its re-erection, but, in 1083, the majority of the monks were removed to Durham, and this institution became a cell subordinate to the monastery of St. Cuthbert, in that city: at the dissolution, the revenue was valued only at £26. 9. 9. The town consists chiefly of two long streets, stretching east and west, situated on the declivity of an eminence, at the base of which is the river Wear; and of some irregular buildings on the shore, once comprising merely a few fishermen's huts, but now containing the major part of the population. It is situated on the north bank of the river, opposite to Sunderland, with which it is joined by the magnificent iron bridge described in the article on that town, and the inhabitants are engaged in ship-building and the various branches of trade connected with the shipping interests of that port. Gas works were established here in 1836. The Brandling Junction railway connects Gateshead with South Shields and Monk-Wearmouth. About ¼ of a mile north-west of Sunderland-bridge is the Pemberton coal-pit, which occupied upwards of seven years in sinking, and with reference to the level of the sea is supposed to be the deepest shaft in the world, being 1680 feet in perpendicular depth from the surface, which is scarcely 100 feet above the level of the river Wear at high water. The living is a perpetual curacy; net income, £225; patron and impropriator, Sir H. Williamson, Bart. The church is of considerable antiquity, believed to have been built about 634; in the tower, which is supported on four circular arches, are some Norman windows: during some repairs a stone coffin was found, also many curious specimens of ancient carving in stone, now deposited in the library of the Dean and Chapter of Durham. A chapel of ease was established about five years since, the incumbency of which is in the patronage of the Dean and Chapter: the building formerly belonged to the Independents. There are places of worship for Baptists, Independents, Presbyterians, and Wesleyan Methodists: that for the Presbyterians, called the Scotch church, is a handsome building in the new street, named North Bridge-street. A free school for boys is supported by an annual subscription of £25 from Lady Williamson, and by voluntary contributions: the number of children is about 100. Here is a Sunday National school. A savings bank has been opened here: for an account of other charities, see SUNDERLAND. A skeleton of gigantic stature, and some relics of stags' horns, have been found in the parish. Some remains of the monastery are still visible near the church. Venerable Bede passed the early part of his monastic life in this establishment, whence he removed to Jarrow: some consider Wearmouth, and others Monkton, to have been his native place; he was born in 672, and died in 735.

WEARMOUTH-PANS, BISHOP, a township, in the parish of BISHOP-WEARMOUTH, union of SUNDERLAND, Northern Division of EASINGTON ward and of the county palatine of DURHAM, containing 363 inhabitants: it forms part of the borough of Sunderland, which see.

WEARMOUTH-SHORE, MONK, a township, on the northern bank of the river Wear, adjoining the town of Sunderland, in the parish of MONK-WEARMOUTH, union of SUNDERLAND, Eastern Division of CHESTER ward, Northern Division of the county palatine of DURHAM; containing 6051 inhabitants. This place owes its origin to the extensive yards for ship-building constructed here during the late continental war, and to the increased commerce of the port of Sunderland. A school of industry for girls is partly supported by subscription; the school and house for the mistress were built from the proceeds of a bazaar, in 1832, and other donations; a surplus of £300 remains, which has been vested on mortgage as an endowment for the school. Another school for boys is partly, and one for girls wholly, supported by subscription.

WEASENHAM (*All Saints*), a parish, in the union of MITFORD and LAUNDITCH, hundred of LAUNDITCH, Western Division of the county of NORFOLK, 7¾ miles (S. W.) from Fakenham; containing 313 inhabitants. The living is a discharged vicarage, with that of Weasenham-St. Peter's annexed, valued in the king's books at £15. 10.: it is in the patronage of the Crown.

WEASENHAM (*St. Peter's*), a parish, in the union of MITFORD and LAUNDITCH, hundred of LAUNDITCH, Western Division of the county of NORFOLK, 7¼ miles (S. W.) from Fakenham; containing 309 inhabitants. The living is a discharged vicarage, annexed to that of Weasenham-All Saints.

WEATHERSFIELD, county of ESSEX.—See WETHERSFIELD.

WEAVERHAM (*St. Mary*), a parish, in the union of NORTHWICH, Second Division of the hundred of EDDISBURY, Southern Division of the county of CHESTER; containing 2321 inhabitants, of which number, 818 are in the township of Weaverham *cum* Milton, 3¼ miles (W. by N.) from Northwich. The living is a vicarage, valued in the king's books at £12. 11. 10½.; present net income, £325; patron and appropriator, Bishop of Chester. The free school for 50 children of the lordship of Weaverham was endowed by Mr. William Barker; and two other schools for girls are supported by subscription. The interest of £100, left by Mary Barker, is applied for apprenticing poor children, as is also a rent-charge of £2, bequeathed by Mr. Mobberley. Here is a charity for six poor decayed housekeepers and their wives, or for deserving widowed or maiden women, selected by the vicar. The Grand Junction railway passes through the parish.

WEAVERTHORPE (*All Saints*), a parish, in the union of DRIFFIELD, partly within the liberty of ST. PETER'S, but chiefly in the wapentake of BUCKROSE,

East Riding of the county of YORK; containing 753 inhabitants, of which number, 403 are in the township of Weaverthorpe, $10\frac{1}{2}$ miles (N. N. W.) from Great Driffield. The living is a vicarage, in the patronage of the Dean and Chapter of York (the appropriators), valued in the king's books at £9. 6. $0\frac{1}{2}$.; present net income, £168. At West Lutton, in this parish, is a chapel of ease. There is a place of worship for Wesleyan Methodists; also a school partly supported by the neighbouring gentlemen.

WEDDINGTON (*St. James*), a parish, in the union of NUNEATON, Atherstone Division of the hundred of HEMLINGFORD, Northern Division of the county of WARWICK, $1\frac{1}{4}$ mile (N.) from Nuneaton; containing 69 inhabitants. The living is a rectory, valued in the king's books at £8. 10. $7\frac{1}{2}$.; present net income, £278; patron, S. B. Heming, Esq.

WEDGWOOD, a township, in the parish of WOLSTANTON, Northern Division of the hundred of PIREHILL and of the county of STAFFORD, 3 miles (N. E.) from Burslem; containing 125 inhabitants. This place, which comprises 431 acres of arable land, is supposed to have been originally the residence of the ancestors of the Wedgwood family, of whom several individuals have been eminent for their improvements in the earthenware and porcelain manufacture.

WEDHAMPTON, a tything, in the parish of URCHFONT, hundred of SWANBOROUGH, Devizes and Northern Divisions of the county of WILTS, $3\frac{3}{4}$ miles (N. E.) from East Lavington; containing 221 inhabitants.

WEDMORE (*St. Mary*), a parish, in the union of AXBRIDGE, hundred of BEMPSTONE, Eastern Division of the county of SOMERSET, $4\frac{3}{4}$ miles (S. by E.) from Axbridge; containing 3557 inhabitants. This place, originally called *Wet-moor*, which appellation it retained till a late period, was the residence of the West Saxon monarchs. Few places have undergone such rapid and extensive improvement; since, within memory, the immediate neighbourhood was usually under water nine months in the year. At present the situation of the village is extremely pleasant, being considerably elevated above the subjacent level, which, from the extensive drainage effected during the last half century, has been rendered valuable land. This ancient borough, by which distinction a part of it is still known, is under the superintendence of a portreeve, chosen annually at the manorial court, with water-bailiffs, constables, and other officers. It has lately been made a polling-place for the eastern division of the county. The living is a discharged vicarage, in the patronage of the Dean and Chapter of Wells (the appropriators), valued in the king's books at £20. 8. $6\frac{1}{2}$.; present net income, £304. The church is a handsome cruciform edifice, in the early style of English architecture, with a stately tower at the intersection, crowned with balustrades: on each side of the chancel there is a chapel, and annexed to the south aisle is another of smaller dimensions. Over the porch is a library, the gift of the Rev. — Andrews, a former vicar. A chapel has been built at Theale containing 340 sittings, 210 of which are free, the Incorporated Society having granted £250 in aid of the expense. There are places of worship for Baptists and Wesleyan Methodists; and a National school is supported by subscription.

WEDNESBURY (*St. Bartholomew*), a market-town and parish, in the union of WEST BROMWICH, Southern Division of the hundred of OFFLOW and of the county of STAFFORD, 19 miles (S. S. E.) from Stafford, and 117 (N. W.) from London; containing 8437 inhabitants. This place, denominated by the Saxons *Weadesbury*, and now commonly called *Wedgebury*, was fortified, in 916, against the Danes, by Ethelfleda, daughter of Alfred the Great: at the Conquest it was held in royal demesne. The town is lighted with gas from the extensive works at West Bromwich, three miles distant. The trade consists principally in the manufacture of articles of iron, both cast and wrought, such as screws, hinges, gun-locks, gun-barrels, coach ironmongery, agricultural implements, apparatus for gas-lights, &c., many of which are prepared for exportation. In the vicinity of the town are numerous collieries, yielding a superior species of coal, which, from its great heat, is admirably adapted for the forges; and a species of iron is here manufactured, termed Damascus iron, of which the best gun barrels are made; it passes through several processes, and when finished throws up a beautiful figure on the surface of the barrel by some chymical application. On a small rivulet near the town are an extensive manufactory for edge-tools, and some corn-mills. A branch of the Walsall and Birmingham canal extends to the western extremity of the town; and about a mile from it is the Bescot-bridge station of the Grand Junction railway, which passes through the parish. The market is on Friday; and fairs are held on May 6th and Aug. 3rd, for cattle. The town is governed by a constable chosen at the manorial court held annually in October; and it has lately been made a polling-place for the southern division of the county. A court of requests for the townships of Bilston and Willenhall, and the parishes of Wednesbury and Darlaston, in the county of Stafford, excepting the manor of Bradley, is held here occasionally, for the recovery of debts not exceeding £5. The living is a discharged vicarage, valued in the king's books at £4. 3. 4.; present net income, £301: it is in the patronage of the Crown; impropriators, Sir E. D. Scott, Bart., and E. T. Foley, Esq. The church, occupying an elevated site, supposed to be that of the ancient castle, and commanding a beautiful prospect, is a fine structure, principally in the later style of English architecture, with an octagonal east end, and contains some ancient wooden seats, and monuments to several families of eminence: it has undergone complete repair, at an expense of £5600, towards which the Incorporated Society gave £500, for which 460 free sittings have been provided. The organ cost £500, and was the gift of Benjamin Wright, Esq., of Birmingham. There are places of worship for Independents and Primitive and Wesleyan Methodists; also a National day and Sunday school, and a Lancasterian school for boys. An almshouse, erected and endowed by Thomas Parkes, in 1602, has received some subsequent benefactions: the same benefactor also bequeathed a house and land, called Clay Pit Leasow, and his son a small sum, for the education of poor children, but at present neither the almshouses, school, nor land can be recognised. William, the first Lord Paget, Secretary of State to Henry VIII., was a native of this town.

WEDNESFIELD, a chapelry, in the parish and union of WOLVERHAMPTON, Southern Division of the hundred of OFFLOW and of the county of STAFFORD, 2 miles (N. E. by E.) from Wolverhampton; containing 1879 inhabitants. The living is a perpetual curacy; net income, £113; patron, John Gough, Esq. The chapel, dedicated to St. Giles, was built in 1750. There is a place of worship for Wesleyan Methodists; also a National school. Edward the Elder, in 911, here defeated the Danes, when two of their kings, two earls, and nine other chiefs, were slain. There were formerly two barrows on the supposed site of the battle, one of which has been levelled. The Essington and Wyrley canal passes through the parish; and the Grand Junction railway passes about a mile to the south-west of the church. The manufacture of locks, keys, traps, &c., constitutes the principal trade of the place.

WEEDON, a hamlet, in the parish of HARDWICKE, union of AYLESBURY, hundred of COTTESLOE, county of BUCKINGHAM, 2¾ miles (N.) from Aylesbury; containing 405 inhabitants.

WEEDON, or WEEDON-BEC (*ST. PETER AND ST. PAUL*), a parish, in the union of DAVENTRY, hundred of FAWSLEY, Southern Division of the county of NORTHAMPTON, 4 miles (S. E. by E.) from Daventry; containing 1439 inhabitants. This place is by Camden and other antiquaries supposed to have been the *Beneventa* of the Romans, which station is now generally referred to Borough Hill, near Daventry. Wulfhere, the first Christian King of Mercia, had a palace here, which after his death was converted by his daughter Werburgh into a nunnery, of which she became abbess, and which was destroyed by the Danes in the ninth century. The village, which is pleasantly situated in a valley, is divided into Upper and Lower Weedon, of which the latter is partly on the Holyhead road, at its junction with the turnpike road from Northampton to Daventry. The Royal Military Depôt, one of the most magnificent establishments of the kind in Europe, is situated above the village, and consists of a handsome centre with two detached wings, forming the residence of the governor and of the principal officers of the establishment; and on the summit of the hill are barracks for 500 men, where troops are always stationed for the protection of the place. At the bottom of the lawn, to the south of the governor's house, are eight storehouses and four magazines, capable of containing 240,000 stand of small arms, exclusively of a proportionate quantity of artillery and ammunition. Between the two ranges of building is a cut communicating with the Grand Junction canal, and affording a facility of conveyance for stores to any part of the kingdom; attached to the buildings are shops for artificers of every kind connected with the establishment, and an hospital, with accommodation for 40 patients. Courts leet are held occasionally, and a court baron annually; and near Dodford Mill is a spot called "Gallows Furlong," where criminals were anciently executed. The London and Birmingham railway passes through the parish, and a station has been established here. The living is a discharged vicarage, valued in the king's books at £11; present net income, £180; patron and impropriator, T. R. Thornton, Esq. The church was, with the exception of the tower, taken down and rebuilt in 1825, under the superintendence of the Rev. Mr. Hunt, the present incumbent; it contains 260 free sittings, the Incorporated Society having granted £500 towards defraying the expense. The parsonage-house occupies the site of the ancient palace of Wulfhere. There are places of worship for Independents and Wesleyan Methodists. Nathaniel Billing, in 1712, bequeathed certain houses to be sold, and the money to be appropriated to the establishment of a school, for which purpose also the Rev. John Rogers, in 1736, bequeathed £76; the annual income is about £100, for which 20 boys are clothed and educated. The Roman Watling-street passed through the parish.

WEEDON-LOYS (*ST. PETER AND ST. MARY*), a parish, in the union of TOWCESTER, hundred of GREENS-NORTON, Southern Division of the county of NORTHAMPTON, 6¼ miles (W. by S.) from Towcester; containing 528 inhabitants. The living is a vicarage, valued in the king's books at £6. 17. 6.; present net income, £462; patrons and impropriators, Provost and Fellows of King's College, Cambridge. In the neighbourhood is a mineral spring, called St. Loys, or St. Lewis's well.

WEEFORD (*ST. MARY*), a parish, in the union of LICHFIELD, Southern Division of the hundred of OFFLOW and of the county of STAFFORD, 4 miles (S. S. E.) from Lichfield; containing 470 inhabitants. This place is supposed to have taken its name from a ford on the line of the Roman Watling-street, called Wayford. Within the parish is the lowe, termed Offlow, which gives name to the hundred; it is erroneously stated to have been the burial-place of Offa, who was interred at Bedford. The living is a perpetual curacy; net income, £63; patron, Bishop of Lichfield; appropriator, Prebendary of Alrewas and Weeford in the Cathedral Church of Lichfield. A day and Sunday school is supported by Lady Wenlock.

WEEK (*ST. MARY*), a parish, in the union and hundred of STRATTON, Eastern Division of the county of CORNWALL, 7 miles (S.) from Stratton; containing 769 inhabitants. This place appears to have been formerly of more importance than it is at present; in all ancient records it is denominated the borough of Week-St.-Mary, and the occupiers of certain fields in the parish are still called Burgage-holders. The surface is undulating, and some of the higher grounds command pleasing views of the surrounding scenery, which is finely varied, and of the adjacent country, which abounds with interesting features. Morris, formerly the seat of Lord Rolle, and now the property of Cecil Bray, Esq., is an ancient mansion containing several apartments, the ceilings of two of which are exquisitely ornamented in plaster. Fairs for bullocks and sheep are annually held in May, July, September, and December. The living is a rectory, valued in the king's books at £17; present net income, £388; patrons, Master and Fellows of Sydney Sussex College, Cambridge. The church, situated on elevated ground, is an ancient building with a stately tower, and consists of a nave with north and south aisles, and a chancel. A parochial school is supported by subscription. A chantry, with a free school, was founded and endowed here in the reign of Henry VIII., by Dame Thomasine Percival, but was suppressed in the reign of Edward VI.: some portions of the building may be traced, and the well is still remaining. Adjoining the churchyard is the site of an ancient fortress, called Castle Hill.

WEEK (*ST. MARY*), a parish, in the union of NEW WINCHESTER, hundred of BUDDLESGATE, Kingsclere and Northern Divisions of the county of SOUTHAMPTON,

1 mile (N. W. by W.) from Winchester; containing 182 inhabitants. The living is a rectory, valued in the king's books at £12. 19. 2.; patron, Bishop of Winchester. The tithes have been commuted for a rent-charge of £250, subject to the payment of rates, which on the average have amounted to £19; the glebe comprises 1 acre, valued at £2 per annum. Part of the parish is bounded by the old castle walls of the city of Winchester.

WEEKE-CHAMPFLOWER, county of SOMERSET.—See WYKE-CHAMPFLOWER.

WEEKLEY (*St. Mary*), a parish, in the union of KETTERING, hundred of CORBY, Northern Division of the county of NORTHAMPTON, $2\frac{1}{4}$ miles (N. E. by N.) from Kettering; containing 273 inhabitants. The living is a discharged vicarage, valued in the king's books at £9. 0. 5.; present net income, £94; patron, Duke of Buccleuch. The church contains a monument to the memory of Lord Chief Justice Montague. A school, in which 24 children of Weekley and Warkton are instructed, is endowed with land producing £17 per annum, bequeathed by the Rev. Nicholas Latham, in 1619, and the master is also allowed £3 per annum from the two parishes. Near the south side of the church is an hospital for seven poor men, founded and endowed with property of the value of £130 per annum, by Lord Chief Justice Montague, in 1614.

WEEL, a township, in the parish of St. John, Beverley, union, and liberties of the borough, of Beverley, East Riding of the county of York, $2\frac{1}{4}$ miles (E.) from Beverley; containing 136 inhabitants.

WEELEY (*St. Andrew*), a parish, in the union and hundred of TENDRING, Northern Division of the county of ESSEX, $9\frac{1}{2}$ miles (S. S. E.) from Manningtree; containing 573 inhabitants. This parish is about eight miles in circumference; the situation is pleasant and the soil fertile. Here were formerly extensive barracks, the removal of which has reduced the population to nearly one half. The living is a discharged rectory, valued in the king's books at £12; present net income, £375; patron, Bishop of London. The church has an embattled tower built of remarkably large bricks. There is a place of worship for Wesleyan Methodists. A National school is supported partly by an endowment of £9. 4. by Mrs. Jefferson, widow of Archdeacon Jefferson, formerly rector of the parish.

WEETHLEY (*St James*), a parish, in the union of ALCESTER, Alcester Division of the hundred of BARLICHWAY, Southern Division of the county of WARWICK, $3\frac{1}{4}$ miles (S. W. by W.) from Alcester; containing 62 inhabitants. The living is annexed to the rectory of Kinwarton.

WEETING, a parish, in the union of THETFORD, hundred of GRIMSHOE, Western Division of the county of NORFOLK, $1\frac{1}{2}$ mile (N.) from Brandon-Ferry; containing 357 inhabitants. The living comprises the united rectories of *All Saints* and *St. Mary*, valued jointly in the king's books at £18. 9. $9\frac{1}{2}$.; present net income, £470; patrons, Master and Fellows of Gonville and Caius College, Cambridge. The church of St. Mary is in ruins. A day and Sunday school is supported by J. Angerstein, Esq.

WEETON, a township, in the parish of KIRKHAM, union of the FYLDE, hundred of AMOUNDERNESS, Northern Division of the county palatine of LANCASTER, $3\frac{1}{4}$ miles (N. W. by W.) from Kirkham; containing 477 inhabitants. A fair for cattle and pedlery is held on Trinity Monday and the following day.

WEETON, a township, in the parish of HAREWOOD, Upper Division of the wapentake of CLARO, West Riding of the county of YORK, 6 miles (E. N. E.) from Otley; containing 322 inhabitants. There is a place of worship for Wesleyan Methodists.

WEETSTED, a township, in the parish of LONG BENTON, Eastern Division of CASTLE ward, Southern Division of the county of NORTHUMBERLAND: the population is returned with the parish.

WEEVER, a township, in the parish of MIDDLEWICH, union, and First Division of the hundred, of EDDISBURY, Southern Division of the county of CHESTER, $4\frac{1}{4}$ miles (W. S. W.) from Middlewich; containing 196 inhabitants.

WEIGHTON, MARKET (*All Saints*), a market-town and parish, in the union of POCKLINGTON, partly within the liberty of St. PETER's, but chiefly in the Holme-Beacon Division of the wapentake of HARTHILL, East Riding of the county of YORK; containing 2169 inhabitants, of which number, 1821 are in the town of Market-Weighton with Arras, 19 miles (E. S. E.) from York, and 190 (N. by W.) from London. This town, which is situated at the western foot of the Wolds, near a branch of the river Foulness, on the high road from York to Beverley, is progressively improving, its trade having been considerably increased by the construction of a canal to the Humber. The market is on Wednesday; and fairs are held on May 14th and September 25th, for horses, cattle, and sheep. The living is a discharged vicarage, in the patronage of the Prebendary of Weighton in the Cathedral Church of York (the appropriator), valued in the king's books at £4. 13. 9.; present net income, £176. The church is an ancient edifice. At Shepton, in this parish, is a chapel of ease; and there are places of worship for Independents and Wesleyan Methodists. Near the town are some tumuli, which have been found to contain human bones, and remains of ancient armour, supposed to be Danish.

WELBECK, an extra-parochial liberty, in the Hatfield Division of the wapentake of BASSETLAW, Northern Division of the county of NOTTINGHAM, $3\frac{1}{2}$ miles (S. W. by S.) from Worksop; containing 63 inhabitants. An abbey for Premonstratensian canons, in honour of St. James, was founded here in 1153, by Thomas le Flemangh, which at the dissolution had a revenue of £298. 4. 8.

WELBORNE (*All Saints*), a parish, in the hundred of FOREHOE, Eastern Division of the county of NORFOLK, $6\frac{1}{2}$ miles (N. N. W.) from Wymondham; containing 231 inhabitants. The living is a discharged rectory, annexed to that of Yaxham, and valued in the king's books at £5. 18. 4. The tithes have been commuted for a rent-charge of £225, subject to the payment of rates, which on the average have amounted to £31; the glebe comprises 44 acres, valued at £70 per annum.

WELBOURN (*St. Chad*), a parish, in the union of SLEAFORD, Higher Division of the wapentake of BOOTHBY-GRAFFO, parts of KESTEVEN, county of LINCOLN, $9\frac{1}{2}$ miles (N. W.) from Sleaford; containing 494 inhabitants. The living is a rectory, valued in the king's books at £19. 16. $0\frac{1}{2}$.; present net income, £493: it is in the patronage of the Countess of Buckinghamshire. The church exhibits fine specimens of the early, decorated, and later styles of English architecture: the tower, which is of very early date, is surmounted by a crocketed spire, supported by flying buttresses springing from the angles. A school is partly supported by the Countess of Buckinghamshire, and another partly by subscription.

WELBURN, a township, in the parish and wapentake of BULMER, union of MALTON, North Riding of the county of YORK, 5¾ miles (S. W. by W.) from New Malton; containing 391 inhabitants. A school is partly supported by the Earl of Carlisle.

WELBURN, a township, in the parish of KIRKDALE, union of HELMSLEY-BLACKMOOR, wapentake of RYEDALE, North Riding of the county of YORK, 4¾ miles (E. by S.) from Helmsley; containing 112 inhabitants.

WELBURY (*St. Leonard*), a parish, in the union of NORTH ALLERTON, wapentake of BIRDFORTH, North Riding of the county of YORK, 6¾ miles (N. N. E.) from North Allerton; containing 233 inhabitants. The living is a rectory, valued in the king's books at £7. 2. 11.; present net income, £360: it is in the patronage of the Crown, in right of the duchy of Lancaster. The church has been lately rebuilt. A school is partly supported by the rector.

WELBY, a chapelry, in the parish and union of MELTON-MOWBRAY, hundred of FRAMLAND, Northern Division of the county of LEICESTER, 2¼ miles (N. W. by W.) from Melton-Mowbray; containing 44 inhabitants.

WELBY (*St. Bartholomew*), a parish, in the union of GRANTHAM, wapentake of WINNIBRIGGS and THREO, parts of KESTEVEN, county of LINCOLN, 4½ miles (E. by N.) from Grantham; containing 399 inhabitants. The living is a rectory, valued in the king's books at £10. 6. 3.; present net income, £350; patron, Prebendary of South Grantham in the Cathedral Church of Salisbury. A school is endowed with £15 per annum by a late rector, the Rev. W. Dodwell, and is further supported by a yearly donation of £11 from Sir W. E. Welby, Bart.

WELCHES-DAM, an extra-parochial liberty, in the union of NORTH WITCHFORD, hundred of SOUTH WITCHFORD, Isle of ELY, county of CAMBRIDGE; containing 137 inhabitants.

WELDON, GREAT (*St. Mary*), a parish (formerly a market-town), in the union of OUNDLE, hundred of CORBY, Northern Division of the county of NORTHAMPTON, 4¼ miles (E. S. E.) from Rockingham; containing 778 inhabitants. The living is a rectory, valued in the king's books at £13. 6. 8.; present net income, £209; patron, Earl of Winchelsea. There is a place of worship for Independents; also a National school. The market was on Wednesday; and there were four fairs, three of which are still held, *viz.*, on the first Thursdays in Feb., May, and Nov., but the fourth is disused. The market-house, erected at the expense of Viscount Hatton, over which were the sessions-chambers, supported by pillars of the Tuscan order, was pulled down about ten years ago. The houses are built of rag-stone from extensive quarries in the neighbourhood. In an enclosure called Chapel field, the pavements of a Roman villa, forming a double square, measuring 100 feet by 50, with the foundations of a stone wall, and a great number of coins of the Lower Empire, besides some of Constantine, Constans, Magnentius, and Constantine, jun., were discovered in 1738: higher up the hill are the remains of an ancient town.

WELDON, LITTLE, a hamlet, in the parish of GREAT WELDON, union of OUNDLE, hundred of CORBY, Northern Division of the county of NORTHAMPTON, 4 miles (E. S. E.) from Rockingham; containing 440 inhabitants.

WELFORD (*St. Gregory*), a parish, in the union of NEWBURY, partly in the hundred of KINTBURY-EAGLE, but chiefly in that of FAIRCROSS, county of BERKS, 5 miles (N. W.) from Newbury; containing 1061 inhabitants. The living is a rectory, valued in the king's books at £35. 15. 5.; patron, Rev. William Nicholson. The tithes have been commuted for a rent-charge of £1353, subject to the payment of rates, which on the average have amounted to £213. 7. 2.; the glebe comprises 189½ acres, valued at £200 per annum. The church is principally in the decorated style of English architecture, with a tower of which the lower part is circular, and in the Norman, and the upper square, and in the early English style, surmounted by a spire. Three schools are supported by subscription. There is a chapel of ease at Wickham, dedicated to St. Swithin, the tower and chancel of which are in the Norman style. At the time of the Norman survey, there was a church in the hamlet of Weston.

WELFORD (*St. Mary*), a parish, in the union of LUTTERWORTH, hundred of GUILSBOROUGH, Southern Division of the county of NORTHAMPTON, 8¾ miles (S. W. by W.) from Market-Harborough; containing 1011 inhabitants. The living is a discharged vicarage, with that of Sibbertoft united, valued in the king's books at £8; present net income of Welford, £244, and of Sibbertoft, £462; patron and appropriator, Bishop of Oxford. There are places of worship for Independents and Wesleyan Methodists. The Grand Union canal passes through the parish. The premises of the free school were purchased out of funds arising from the church and poor's lands; the annual income is £24. 10., for which 20 children are instructed.

WELFORD (*Holy Trinity*), a parish, in the union of STRATFORD-UPON-AVON, partly in the Stratford Division of the hundred of BARLICHWAY, Southern Division of the county of WARWICK, but chiefly in the Upper Division of the hundred of DEERHURST, Eastern Division of the county of GLOUCESTER, 4¼ miles (W. S. W.) from Stratford-upon-Avon; containing 669 inhabitants. The living is a rectory, valued in the king's books at £29. 15. 10.; present net income, £442: it is in the patronage of Countess Amherst. The church is principally in the Norman style, with a lofty tower crowned with pinnacles. There is a place of worship for Wesleyan Methodists; also a school endowed with £3 per annum for the education of eight boys.

WELHAM (*St. Andrew*), a parish, in the union of MARKET-HARBOROUGH, hundred of GARTREE, Southern Division of the county of LEICESTER, 4¼ miles (N. E. by N.) from Market-Harborough; containing 73 inhabitants. The living is a vicarage, valued in the king's books at £6. 3. 4.; present net income, £98: it is in the patronage of the Crown; impropriators, Proprietors of land.

WELHAM, a township, in the parish of NORTON, wapentake of BUCKROSE, East Riding of the county of

YORK, $1\frac{3}{4}$ mile (S.) from New Malton: the population is returned with the parish.

WELL (*St. Margaret*), a parish, in the union of SPILSBY, Wold Division of the hundred of CALCEWORTH, parts of LINDSEY, county of LINCOLN, $2\frac{1}{4}$ miles (S. S. W.) from Alford; containing 76 inhabitants. The living is a discharged rectory, with the vicarage of Claxby united; net income, £372; patron, Bateman Dashwood, Esq. The church has been rebuilt in the form of an elegant Grecian temple. Near this place, in 1725, two urns, containing 600 Roman coins, were found; in the neighbourhood are three Celtic barrows, contiguous to each other.

WELL (*St. James*), a parish, in the union of BEDALE, wapentake of HANG-EAST, North Riding of the county of YORK; containing 1062 inhabitants, of which number, 406 are in the township of Well, $4\frac{1}{2}$ miles (S.) from Bedale. This place derives its name from a celebrated well, dedicated to St. Michael, which at all times of the year is supplied with water by a spring issuing from a rock. An hospital, in honour of St. Michael the Archangel, for a master, two priests, and 24 poor brethren and sisters, was founded here in 1342, by Sir Ralph de Neville, Lord of Middleham, which at the dissolution had a revenue of £42. 12. 3. The living is a discharged vicarage, valued in the king's books at £8. 13. 7.; present net income, £120; patron and impropriator, Charles Chaplin, Esq. The church contains several monuments of the lords of Snape. Thomas, Earl of Exeter, in 1605, established a charity, called Neville's workhouse, for the maintenance of a master and mistress and twelve poor girls, the latter of whom are also educated. A school for boys, and another for girls, were founded here, and two others at Snape, in 1788, and are supported from these funds, which amount to about £100 per annum.

WELL-HAUGH, a township, in the parish of FALSTONE, union of BELLINGHAM, North-Western Division of TINDALE ward, Southern Division of the county of NORTHUMBERLAND, $12\frac{1}{4}$ miles (W. N. W.) from Bellingham; containing 272 inhabitants.

WELLAND (*St. James*), a parish, in the union of UPTON-UPON-SEVERN, Lower Division of the hundred of OSWALDSLOW, Upton and Western Divisions of the county of WORCESTER, 3 miles (W. by S.) from Upton-upon-Severn; containing 490 inhabitants. The living is a discharged vicarage, valued in the king's books at £8. 2. 11.; present net income, £378: the patronage and impropriation belong to the Crown. Here is a Sunday National school.

WELLCOMBE (*St. Nictan*), a parish, in the union of BIDEFORD, hundred of HARTLAND, Great Torrington and Northern Divisions of the county of DEVON, $5\frac{3}{4}$ miles (S. W. by S.) from Hartland; containing 258 inhabitants. The living is a perpetual curacy; net income, £71; patron, Lord Clinton; impropriator, W. Heddon, Esq.

WELLESBOURN-HASTINGS (*St. Peter*), a parish, in the union of STRATFORD-UPON-AVON, Warwick Division of the hundred of KINGTON, Southern Division of the county of WARWICK, $4\frac{3}{4}$ miles (N. W.) from Kington; containing, with Walton-Deivile and Wellesbourn-Montford, 1357 inhabitants. The living is a discharged vicarage, with the rectory of Walton-Deivile annexed, valued in the king's books at £7. 11. 8.; present net income, £422; it is in the patronage of the Crown; impropriators, Sir J. Mordaunt, Bart., and others. The church is partly in the Norman, and partly in the early English, style of architecture, with a tower of later character: it contains a monument to the memory of Sir Thomas le Strange, lord-lieutenant of Ireland in the reign of Henry VI. About 50 boys and 60 girls receive gratuitous instruction in schools founded in 1723, by the Rev. Richard Boyse, who endowed them with land and houses now producing nearly £60 per annum, which sum has been augmented by subscriptions to about £82, for the maintenance of a master and a mistress: these schools are in connection with the National Society.

WELLESBOURN-MONTFORD, a hamlet, in the parish of WELLESBOURN-HASTINGS, union of STRATFORD-UPON-AVON, Warwick Division of the hundred of KINGTON, Southern Division of the county of WARWICK, 5 miles (N. W. by W.) from Kington; containing 660 inhabitants.

WELLHOUSE, a tything, in the parish of HAMPSTEAD-NORRIS, hundred of FAIRCROSS, county of BERKS, $3\frac{1}{2}$ miles (N.) from Newbury: the population is returned with the parish.

WELLING, a village, partly in the parish of BEXLEY, and partly in that of EAST WICKHAM, hundred of LESSNESS, lathe of SUTTON-AT-HONE, Western Division of the county of KENT, $2\frac{1}{2}$ miles (W. by N.) from Crayford. This place, which is of modern origin, is situated on the great road from London to Canterbury and Dovor, and contains some good posting-houses.

WELLINGBOROUGH (*All Saints*), a market-town and parish, and the head of a union, in the hundred of HAMFORDSHOE, Northern Division of the county of NORTHAMPTON, 10 miles (N. E. by E.) from Northampton, and 67 (N. N. W.) from London; containing 4688 inhabitants. The name is derived from the wells, or springs, that abound here, of which, that denominated Red Well was formerly in such repute for its medicinal properties, that in 1626, Charles I. and his queen resided in tents during a whole season, for the purpose of drinking its salubrious water at the source. In 1738, the town was nearly destroyed by fire, and rebuilt on the slope of a hill nearly a mile northward from the navigable river Nen: it consists of several streets lighted and pitched, the principal of them meeting in the market-place: the houses, built of red sand-stone, which abounds in the vicinity, are of modern style and handsome appearance. The chief articles of manufacture are boots and shoes, and bobbin-lace: the former was very extensive during the war, and is still considerable; and the latter, though on the decline, employs many females and children: a silk-mill has been recently established. The market was granted by King John, at the request of the monks of Croyland abbey, the proprietors of the manor, which was subsequently held by Queen Elizabeth, after the dissolution; it is on Wednesday, and is a very considerable corn market. Fairs are on the Wednesdays in Easter and Whitsun weeks, and Oct. 29th; the last is a large one for live stock. Manorial courts are held annually in Oct.; and petty sessions for the division, every week, by the county magistrates, who assemble at the town-hall, which has been recently erected by the feoffees, and is used for vestries and other public meetings. The town has been made a polling-place for the northern division of the shire.

The living is a vicarage, valued in the king's books at £24. 1. 8.; present net income, £400; patron and impropriator, Quintus Vivian, Esq. The church is a spacious and handsome structure, combining specimens of the different styles of English architecture, with an elegant tower and spire; on the south side is a Norman door, and in the interior are some ancient screen-work and stalls; the east window is richly ornamented with sculpture and tracery. There are three places of worship for Independents, and one each for Baptists, the Society of Friends, and Wesleyan Methodists. A free grammar school, adjoining the churchyard, was founded in the 2nd of Edward VI., and endowed with a revenue of a guild of the Blessed Virgin, formerly attached to the church, and subsequently with an estate at Burton-Latimer, and three-eighths of the rental of 55 acres of land under the will of Richard Fisher, in 1711: the master is elected by the inhabitants assessed to the land-tax, and receives an annual income of £130, out of which he provides an usher for the lower school. In the upper school, sons of the inhabitants of Wellingborough are instructed in Latin gratuitously, and in English, writing, and arithmetic, upon payment of four guineas per annum each; and in the lower school about 100 children are gratuitously taught reading, writing, and arithmetic. A copyhold house was bequeathed by John Freeman, in 1711, to be used as a charity school. Here is a National school, in which 60 children are educated: it is endowed with one-half of the rental of Fisher's estate above mentioned, the remaining eighth of which is appropriated to two aged inhabitants not receiving parochial relief, and with legacies of £100 each from Mary Roane, Samuel Knight, and John Robinson, altogether producing £77. 8. 7. per annum. The Town estate, which is under the management of sixteen feoffees, produces an annual income of £350. 18. 2.: out of this fund, £54 is paid to the master of the grammar school, as the rental of that portion of the school endowment which is included in this estate, and forming a part of his salary of £130; and the remainder is appropriated to purposes of general improvement in the town, and the relief of poor inhabitants. There is also a fund of £53. 4., arising from bequests by Mrs. Anne Glasbrook, in 1790, and others, which is annually distributed in bread and money. The poor law union of Wellingborough embraces a small part of Bedfordshire, and comprises 27 parishes or places, containing a population of 18,571, according to the census of 1831; it is under the care of 36 guardians.

WELLINGHAM (*St. Andrew*), a parish, in the union of Mitford and Launditch, hundred of Launditch, Western Division of the county of Norfolk, 6¼ miles (S. W. by S.) from Fakenham; containing 165 inhabitants. The living is a discharged rectory, united, with that of Godwick, to the rectory of Tittleshall, and valued in the king's books at £5. 8. 6½. The impropriate tithes have been commuted for a rent-charge of £1. 12., and the rectorial for £265, subject to the payment of rates, which on the average on the latter have amounted to £35. 8. 9.; the glebe comprises 3 acres, valued at £4. 3. 4. per annum.

WELLINGLEY, a joint township with Stancill and Wilsick, in the parish of Tickhill, union of Doncaster, Southern Division of the wapentake of Strafforth and Tickhill, West Riding of the county of York, 2 miles (N. by W.) from Tickhill: the population is returned with Stancill.

WELLINGORE (*All Saints*), a parish, in the union of Sleaford, Higher Division of the wapentake of Boothby-Graffo, parts of Kesteven, county of Lincoln, 9 miles (N. W.) from Sleaford; containing 752 inhabitants. The living is a discharged vicarage, in the patronage of the Dean and Chapter of Lincoln (the appropriators), valued in the king's books at £11. 10.; present net income, £206. There is a place of worship for Wesleyan Methodists.

WELLINGTON (*St. Margaret*), a parish, in the hundred of Grimsworth, union and county of Hereford, 5¼ miles (N.) from Hereford; containing 630 inhabitants. The living is a discharged vicarage, valued in the king's books at £6. 13. 4.; present net income, £280; patron and appropriator, Prebendary of Wellington in the Cathedral Church of Hereford. Here is a National school.

WELLINGTON (*All Saints*), a market-town and parish, and the head of a union, in the Wellington Division of the hundred of South Bradford, Northern Division of the county of Salop, 11 miles (E.) from Shrewsbury, and 151 (N. W.) from London; containing 9761 inhabitants. During the great civil war, this was the first place of rendezvous of Charles I., who, on September 19th, 1642, mustered his forces near the town, and having commanded his military orders to be read, delivered in person the remarkable address mentioned by Clarendon. The town occupies a low situation, near the ancient Roman Watling-street, about two miles southward from the Wrekin, which rises from the plain to a height of about 1100 feet above the bed of the Severn, embraces an horizon of from 350 to 400 miles in circumference, and is surmounted by an ancient fortification: a part of the parish is bounded by the river Tern. The streets are mostly narrow, but have been much improved, and are now Macadamized and lighted with gas, and many of the houses are of modern and respectable appearance. There are two valuable springs at Admaston, about a mile and a half from the town, called the Upper and Lower, the former chalybeate, and the latter sulphureous. A very comfortable inn and baths have been erected here. The waters have been found highly efficacious, particularly in rheumatic complaints, and it has become a favourite watering-place, being frequented by persons from various parts of the kingdom. The mineral productions of the parish, consisting of coal, iron-stone, and limestone, form the basis of its trade, which chiefly consists in the different branches of iron manufacture, especially that of nails: several companies of iron-masters have establishments in the neighbourhood, amongst which are, the Hadley, Ketley, Lawley, and Lilleshall companies. There are also a glass-manufactory, corn-mills, and malt-kilns, and some business is transacted in timber. The various articles of manufacture and commerce are conveyed by the Shrewsbury and Shropshire canals, which communicate with the navigable river Severn, and the midland counties. The market, granted to Hugh Burnel, in the 11th of Edward I., is on Thursday, and is on a very extensive scale: fairs, chiefly for live stock and butter and cheese, are held on March 29th, June 22nd, September 29th, and November 17th. The town is under a mayor and constables, and two clerks are appointed to regulate the market: a manorial court is held in November, at which these officers are appointed. Petty sessions for the hundred take place weekly; and a court of record, for the recovery of debts under £20, is held on certain specified days. It has been made a polling-place for the northern division of the county.

The living is a vicarage, with the rectory of Eyton-on-the-Wild-Moors annexed, valued in the king's books at £9. 5.; present net income, £842; patron, T. Eyton, Esq. The church is a light and elegant modern edifice of freestone. A church was erected here in 1838, containing 1140 sittings, 740 of which are free, the Incorporated Society having granted £400 in aid of the expense. At Ketley, a church, built and endowed at the sole expense of the Duke of Sutherland, was consecrated in August, 1839; it is an elegant structure, in the later English style, and is built on an eminence commanding extensive views. There are places of worship for Baptists, Independents, and Wesleyan Methodists. A free school for children of both sexes is chiefly supported by voluntary contributions; and there is an almshouse for poor women. A National school is held in a building in the churchyard. The poor law union of Wellington comprises 11 parishes or places, under 19 guardians, and contains, according to the census of 1831, a population of 17,945. Several curious petrifactions of plants and shells are found occasionally in some of the iron mines in the vicinity. Dr. Withering, author of a "Botanical Arrangement of British Plants," and some medical treatises, was born here, in 1741.

WELLINGTON (*St. John the Baptist*), a market-town and parish, and the head of a union, forming, with the parish of West Buckland, one of the two unconnected portions which comprise the Western Division of the hundred of Kingsbury, Western Division of the county of Somerset, 24 miles (W. S. W.) from Somerset, and 149 (W. S. W.) from London; and containing 4762 inhabitants. This town is situated on the main road from Bath to Exeter, and of late years has been much improved, many of the streets having been paved, and a few of the old houses removed. The manufacture of druggets and serges was formerly carried on to a considerable degree, and still prevails on a less extended scale. The Grand Western canal, from Bridg-water to Tiverton, passes near the town, and affords considerable facility for the increase of its trade; and the line of the Bristol and Exeter railway passes through the parish. During the possession of the manor by the Bishops of Wells, a charter was obtained for a market and two fairs; the former is held on Thursday, principally for corn; the latter on the Thursdays before Easter and Whitsuntide. The market-house being in a very dilapidated condition, and not affording suitable accommodation for the market people, is thought to have caused a partial decline in the market of late years; to obviate which, His Grace the Duke of Wellington, lord of the manor, granted a lease for 99 years, and the inhabitants have erected a new edifice by subscription on shares. The government of the town is in a bailiff and subordinate officers, chosen at the annual court leet held for the manor. The living is a vicarage, with West Buckland annexed, valued in the king's books at £15. 10. 2½.; present net income, £894; patron, Rev. W. P. Thomas; appropriator, Dean of Wells. The church is a handsome edifice, with an embattled tower crowned with pinnacles, and has two sepulchral chapels, in one of which is a splendid monument to the memory of Sir John Popham, Knt., Lord Chief Justice of England in the reigns of Elizabeth and James I., ornamented with a profusion of effigies and carved work. The Rev. Mr. Thomas has erected an elegant chapel, at his own expense, near the west end of the town, which is dedicated to the Holy Trinity. There are places of worship for Baptists, the Society of Friends, Independents, and Wesleyan Methodists; also two infants' schools, partly supported by subscription. Almshouses for six poor men and six poor women were founded in 1604, and endowed with land by Sir John Popham, the master and matron to instruct poor children. The poor law union of Wellington embraces a small part of Devonshire, and comprises 24 parishes or places, under the care of 34 guardians; it contains, according to the census of 1831, a population of 20,985. Wellington confers the titles of Viscount, Earl, Marquess, and Duke, on that distinguished military commander, Arthur Wellesley, Prince of Waterloo; the first created September 4th, 1809; the second, February 28th, 1812; the third, August 18th, of the same year; and the fourth, May 3rd, 1814. At a short distance from the town a magnificent pillar has been erected, by public subscription, in commemoration of the signal victory obtained by his Grace on the plain of Waterloo, in 1815.

WELLOW (*St. Swithin*), a parish, in the union of Southwell, South-Clay Division of the wapentake of Bassetlaw, Northern Division of the county of Nottingham, 1½ mile (S. E. by E.) from Ollerton; containing 473 inhabitants. The living is a perpetual curacy; net income, £66; patron, Earl of Scarborough; appropriator, Bishop of Lincoln. Here is a school, with a small endowment.

WELLOW (*St. Julian*), a parish, in the union of Bath, hundred of Wellow, Eastern Division of the county of Somerset, 4¼ miles (S.) from Bath; containing 960 inhabitants. The living is a discharged vicarage, valued in the king's books at £20. 6. 10½.; present net income, £380; patron, Charles Jones, Esq.; impropriator, W. G. Langton, Esq. The church was built by Sir Walter Hungerford, about 1732. There is a place of worship for Wesleyan Methodists. A railway from the Welton collieries, communicating with the Avon and Kennet, and the Radford coal canals, passes through the parish. Ten poor children are educated for £10 a year, the bequest of Rachael Coles, in 1756. Among numerous other Roman relics discovered in the neighbourhood, a tesselated pavement was found in 1644, another in 1670, and a third in 1685, with altars, pillars, and fragments of pateræ, and other vessels. At the extremity of the parish is an immense barrow, called Woodeborough, and another smaller one has been found to contain several stone coffins.

WELLOW, EAST (*St. Margaret*), a parish, in the union of Romsey, hundred of Thorngate, Romsey and Southern Divisions of the county of Southampton, 3¾ miles (W.) from Romsey; containing 318 inhabitants. The living is a vicarage, valued in the king's books at £5; patron, W. E. Nightingale, Esq., who, with — Harvey, Esq., is the impropriator. The vicarial tithes have been commuted for a rent-charge of £245. 5., subject to the payment of rates, which on the average have amounted to £24. 10.; the glebe comprises 23 acres, valued at £35 per annum. The impropriate tithes have been commuted for a rent-charge of £317. 10. 4., subject to the payment of rates. There is a place of worship for Wesleyan Methodists; also a school, partly supported by Mr. Nightingale, at whose expense a school-room was built in 1828.

WELLOW, WEST, a tything, in the parish of EAST WELLOW, union of ROMSEY, hundred of AMESBURY, Salisbury and Amesbury, and Southern, Divisions of the county of WILTS, $4\frac{3}{4}$ miles (W. by N.) from Romsey; containing 394 inhabitants.

WELLS (*ST. PETER*), a sea-port town and parish, in the union of WALSINGHAM, hundred of NORTH GREENHOE, Western Division of the county of NORFOLK, 33 miles (N. W. by N.) from Norwich, and 120 (N. N. E.) from London; containing 3624 inhabitants. This place, called in Domesday-book *Guella*, is situated on a creek about a mile from the North Sea, and consists principally of two streets, partially paved: it is well supplied with water. There is a theatre, a plain brick building, neatly fitted up within; and a subscription library has been established: races were formerly held here, but they have been discontinued. The magistrates of the hundred hold their sittings once a fortnight; and courts leet and baron are held annually by the lord of the manor, at which the steward presides. The trade consists in the exportation of grain and malt, and the importation of coal, deals, tiles, bark, linseed and rapeseed cakes, tar, and wine: oysters of an excellent quality are also found, and the fishery furnishes employment to many persons. The harbour, from the accumulation of sand, is rather difficult of access, but considerable improvements have been made in it by the Harbour Commissioners. The custom-house is a brick building, situated on the quay, and the establishment consists of a collector, comptroller, land and tide-waiter, &c. A fair is held annually on Shrove-Tuesday. The living is a rectory, valued in the king's books at £26. 13. 4.; present net income, £738; patron, J. R. Hopper, Esq. The church is a handsome spacious edifice of flint, with a lofty embattled tower. The Society of Friends, Independents, and Wesleyan Methodists, have each a place of worship. There are free schools for 30 boys and 30 girls, endowed with land producing £24 per annum; and two other schools are supported by theCountess of Leicester and William Nettleton, Esq.

Seal and Arms.

WELLS, a city, having separate jurisdiction, and the head of a union, locally in the hundred of Wells-Forum, Eastern Division of the county of SOMERSET, 19 miles (S. W.) from Bath, 19 (S.) from Bristol, and 120 (W. by S.) from London; containing, with that part of the parish of St. Cuthbert which is without the limits of the city, 6649 inhabitants. This place derives its name from the numerous springs with which it abounds, more particularly from St. Andrew's well, the water of which, rising near the episcopal palace, flows through the south-western part of the city: it owes its origin to Ina, King of the West Saxons, who, in 704, founded a collegiate church, which he dedicated to St. Andrew the Apostle. This establishment was subsequently endowed by Cynewulf, one of his successors, with considerable estates in the vicinity, in 766, and continued to flourish till 905, when, in pursuance of an edict of Edward the Elder, for the revival of religion which, from the frequent incursions of the Danes, had almost fallen into disuse, several new bishops were consecrated by Pligmund, Archbishop of Canterbury, of whom Aldhelm, formerly abbot of Glastonbury, was appointed to preside over Wells, which was then erected into a see, having jurisdiction over the entire county of Somerset. After a succession of twelve bishops, Giso, chaplain to Edward the Confessor, was appointed to the see, to which that monarch gave the extensive possessions of Harold, Earl of Wessex, whom, with his father, Godwin, Earl of Kent, he had banished from the kingdom. Harold, during his exile, made an incursion into this part of Somersetshire, raised contributions on his former tenantry, despoiled the church of its ornaments and treasure, expelled the canons, and converted their possessions to his own use. Giso, on his return from Rome, where he had been consecrated, obtained some compensation for these injuries from the queen, who was Harold's sister; but that prince, on his restoration to favour, procured the banishment of Giso, and, upon his subsequent accession to the throne, resumed all the estates granted by Edward to the church, and greatly impoverished the see. Bishop Giso remained in exile till the Conquest, when he was reinstated; and William, in the second year of his reign, restored to the bishopric all Harold's estates, with the exception of some small portions which had been granted to the monastery of Glastonbury, adding, in lieu of them, two other manors. Giso exerted himself in augmenting the revenue of his see: he increased the number of canons, over whom he appointed a provost, built a cloister, hall, and dormitory, and enlarged and embellished the choir of the cathedral: these buildings, however, were demolished by his successor, John de Villula, who erected a palace on their site. This prelate removed the seat of the diocese of Bath, and assumed the title of Bishop of Bath, in which he was followed by his two next successors. Great disputes arising between the inhabitants of both cities, each claiming to be regarded as the head of the diocese, the matter was at length referred to the arbitration of the bishops, who decided that the prelates should take the title of Bishops of Bath and Wells, that their election should be made by an equal number of delegates from both places, and that the ceremony of installation should take place in both churches. Reginald Fitz-Jocelyne, who was bishop in the reign of Richard I., granted the town a charter of incorporation, and made it a free borough; and during the captivity of that monarch in Austria, Savaricus, who succeeded Fitz-Jocelyne in the see, and was nearly allied to the emperor, obtained, through his influence, a promise from Richard, as a condition of his restoration, that the abbacy of Glastonbury, then vacant, should be annexed to the see of Bath and Wells: this prelate afterwards removed the seat of his diocese to Glastonbury, and assumed the title of Bishop of Glastonbury. After his death, in 1205, the monks, under his successor, Jocelyne de Walles, petitioned the court of Rome that they might be restored to their ancient government by an abbot, which indulgence they obtained, on condition of their relinquishing to the bishop a considerable portion of their revenue, and Jocelyne assumed the style of Bishop of Bath and Wells, which the prelates of the see have ever since retained. After the death of Jocelyne, disputes arose in the election of his successors, the monks of Bath frequently exercising that right without the concurrence of the canons of Wells; but an appeal having been made to the pope, the union of the churches appears to have subsequently remained without interruption. At the time of the Reformation, the monastery of Bath was suppressed; and, though the name of the see was retained, the ecclesiastical authority, and the right of electing the bishops, were vested in the Dean and Chapter of Wells, then constituted the sole chapter of the diocese. The revenue of the monastery of Wells, at the period of its dissolution, was valued at £1939. 12. 8.

The city, which appears to have grown up around the ancient ecclesiastical establishment, and to have flourished in proportion to its prosperity, is pleasantly situated on the south side of the Mendip hills, in a fertile plain lying at their base, and is sheltered from the north winds by that mountainous range of richly-wooded eminences, and open on the south side to an extensive tract of fine meadow land. The houses are well built, and of respectable appearance; several of them are ancient, having been erected for ecclesiastical residences, and many are of modern and elegant structure. The grandeur of its cathedral, the beauty of its church, and the character of the conventual buildings, give it an air of peculiar interest. It is divided into four verderies by four principal streets, from which they take their name, and is well paved, and amply supplied with water from a public conduit of great beauty, built by Bishop Beckington, and filled by pipes leading from an aqueduct near the source of St. Andrew's well. The environs, which abound with diversified and picturesque scenery, contain many handsome seats, and afford a variety of pleasing walks and rides. Races are held annually a short distance east of the city, beyond the limits of its liberties. The principal branch of manufacture is the knitting of stockings; and at Wookey, about two miles distant, are several paper-mills, where, from the excellent quality of the water, paper of the best kind is made. The market days are Wednesday and Saturday for provisions; and on every fourth Saturday a large market is held for corn, cattle, and cheese. The fairs are on Jan. 6th, May 14th, July 6th, Oct. 25th, and Nov. 30th, for cattle, horses, and pedlery. The market-place, on the east side of the city, is a fine spacious area, on the north side of which is a handsome range of twelve houses of stone, built by Bishop Beckington, for twelve priests, now inhabited by townsmen: at the eastern extremity is an ancient gateway, communicating with the Cathedral Close, and, fronting the street, another leading to the episcopal palace, both erected by the same bishop, who intended to rebuild the whole area. Near the site of the ancient cross, which was taken down in 1780, formerly stood the city conduit, an elegant hexagonal structure, in the later style of English architecture, erected by the same bishop, in 1450, richly embellished with canopied niches and delicate ornaments, and crowned with a conical dome; but this being considered an obstruction, it was taken down about 30 years since, and soon afterwards removed to Stowerhead, then the seat of Sir R. C. Hoare, Bart., and a very handsome one was erected on the site of the old cross: in the south-eastern angle is the town-hall and market-house, a plain but commodious building.

The charter granted by Reginald Fitz-Jocelyne was confirmed by King John, who entrusted the government to a master and commonalty of the borough of Wells. Queen Elizabeth gave the inhabitants a new charter, in the 31st of her reign, under which the government was vested in a mayor, recorder, seven masters, and sixteen capital burgesses, assisted by a town-clerk and other officers: the mayor, recorder, and one of the masters, were justices of the peace. The corporation now consists of a mayor, four aldermen, and twelve councillors, under the act of the 5th and 6th of William IV., c. 76, for an abstract of which see the Appendix, No. I.; and the municipal and parliamentary boundaries are co-extensive. The freedom is inherited by the eldest son of a freeman, and obtained by servitude. The inhabitants first exercised the elective franchise in the 23rd of Edward I., since which time they have regularly returned two members to parliament. The right of election was formerly vested in the mayor, masters, and burgesses generally; but the privilege has been extended to the £10 householders of an enlarged district, which has been constituted the new borough, and comprises 715 acres, of which the limits are minutely detailed in the Appendix, No. II.; the old borough contained 692 acres; the mayor is the returning officer. The corporation held quarterly courts of session for all offences not capital, arising within the city, but no prisoners have been tried for several years; and a court of record formerly held, for the recovery of debts to any amount, has been for many years discontinued. The assizes for the county are held here every alternate year, and the Epiphany and Easter quarter sessions annually; and by the act of the 2nd and 3rd of William IV., cap. 64, this city has been made the place of election and a polling-place for the eastern division of the shire.

Arms of the Bishopric.

The present ecclesiastical establishment, as refounded by Henry VIII., on the dissolution of the monastery, consists of a bishop, dean, precentor, chancellor, three archdeacons, treasurer, sub-dean, 49 prebendaries, four priest-vicars, eight lay-vicars, organist, six choristers, and other officers. The cathedral church, dedicated to *St. Andrew*, is a magnificent cruciform structure, principally in the early style of English architecture, with partial insertions of the decorated and later styles: the foundation was laid by Wiffeline, second bishop of the diocese, and the edifice was completed and improved by Bishop Jocelyne, in 1239. The west front is a striking and superb combination of stately grandeur and splendid embellishment; the whole of it, together with the buttresses, by which it is divided into compartments, is replete with elaborate sculpture, from the base to the summit, in successive tiers of richly-canopied shrines, containing the statues of kings, popes, bishops, cardinals, and abbots; the mullions of the west window and the lower stages of the western towers are similarly enriched; the canopies of the niches in which these figures are enshrined are supported by slender-shafted pillars of polished marble, and the intermediate spaces between the several series are filled with architectural ornaments of elegant design and appropriate character. In the upper range of the central compartment are the statues of the twelve Apostles, in a series of lofty niches separated by slender shafts, and in the range immediately beneath them are figures of the hierarchs, below which is a sculptured representation of the Resurrection, in alto relievo. The entrance, which is through a deeply-recessed arch, is flanked by the western towers, of which the lower stages are comprised in the general design of the front, and the upper, which are wreathed with pierced parapets, are relieved by fine windows, and with lofty canopies rising from the buttresses, and terminating in crocketed finials. The central tower is crowned with a pierced parapet of elegant design, and decorated with lofty angular pinnacles surmounted by vanes, and with smaller pinnacles in the intervals: though of large dimensions, it has an airy appearance, from the proportionate size and elegance of the windows. The interior displays some specimens of the early English style, which are of unfrequent occurrence, and equally remarkable for simplicity and elegance. Of this character are the nave and transepts: the former is separated from the aisles by a beautiful range of clustered columns and finely-pointed arches, above which are a triforium of lancet-shaped arches, and a fine range of clerestory windows, in which elegant tracery, in the later English style, has been inserted; the roof is finely groined, and the great west window is embellished with ancient stained glass of great brilliancy. The choir is in the decorated style, and of very elegant character, and beyond it is the Lady chapel, both forming parts of one general arrangement, which, for beauty of design, and richness of architectural embellishment, is perhaps unequalled. There are numerous chapels in various parts of the cathedral, some of which are enclosed with screens of beautiful design, and in one is an ancient clock, removed from Glastonbury, with an astronomical dial, and a train of figures of knights in armour, which, by the machinery, are moved round in circular procession: in the south transept is an ancient font of the same date as that part of the building. There are many interesting and ancient monuments of the bishops who were interred within its walls, among which are, the tomb of Bishop Beckington, in a chapel in the presbytery, with his effigy in alabaster; the gravestone of Bishop Jocelyne in the middle of the choir, marking the spot where an elegant marble monument, bearing his effigy in brass, formerly stood; that of King Ina, who was interred in the centre of the nave, and many others. The cloisters form three sides of a quadrangle south of the cathedral: the western range, in which are the school and the treasury, was built by Bishop Beckington, who also began the south side, which was finished by Thomas Henry, treasurer of Wells, and archdeacon of Cornwall; and the eastern range, containing a chapel and a library, was erected by Bishop Bubwith. The chapter-house is an elegant octagonal structure: the roof, which is finely groined, is supported on an elegant clustered column of Purbeck marble in the centre, and the interior is lighted by windows of handsome design: beneath it is a crypt of good character, with a roof displaying a fine specimen of plain groining, from which a staircase of singular construction leads into the chapter-room, and to several other parts of the adjacent buildings. To the south of the cathedral is the episcopal palace, an ancient castellated mansion, surrounded with walls, enclosing nearly seven acres of ground, and defended by a deep moat, which is supplied from the water of St. Andrew's well: on the north side is a venerable gateway tower leading over a bridge into the outer court, on the east side of which is the palace, containing several spacious and magnificent rooms, and a chapel; opposite the entrance are the remains of the great hall, now in ruins, having been demolished in the reign of Edward VI., for the sake of the materials. The vicar's close was originally built by Walter de Hull, canon of Wells, and archdeacon of Bath, and improved, in 1348, by Bishop Ralph de Salopia, who erected a new college for the residence of the vicars and choristers, which he endowed with lands of his own, in addition to what were given by Walter de Hull: it was subsequently enlarged and its endowment augmented, by Bishop Beckington, who erected the gateways, of which that on the east, adjoining the cathedral buildings, has a long gallery communicating with the church and the vicar's close, with a large flight of steps at each end: at the south end is a hall, with a buttery and other conveniences, under which is an arched gateway: at the north end are the chapel and library, and on the east and west sides are handsome ranges of dwelling-houses. This college, the revenue of which, in the 26th of Henry VIII., was £72. 10. 9½., escaped the general dissolution, and was afterwards refounded by Queen Elizabeth, who appointed the number of vicars to be not less than fourteen, nor more than twenty. The deanery is a spacious and handsome structure, erected by Dean Gunthorp, in allusion to whose name the walls are ornamented with several guns, carved in stone: in this mansion the founder entertained Henry VII., on his return from the West of England. Near the deanery is the west gate, a plain ancient edifice, forming the principal entrance into the city from Bath.

The city comprises only the in-parish of *St. Cuthbert*, which surrounds the cathedral precincts: the several hamlets, which are without the limits of the city, extending for seven miles in circuit, form the out-parish of St. Cuthbert. The living is a vicarage, valued in the king's books at £33. 13. 6.; present net income, £564; patrons and appropriators, the Dean and Chapter. The church of St. Cuthbert is a spacious and handsome structure, in the later style of English architecture, with a lofty square embattled tower, strengthened with angular buttresses, and crowned with pinnacles, forming one of the most beautiful specimens of a tower in that style of architecture. Though of large dimensions, it has a degree of lightness from the judicious distribution of its ornaments, and the relief afforded by niches of elegant design; the belfry windows are lofty, and, from the beauty of their composition, give to the tower above the roof the character of a magnificent lantern, and the west door, and the large window over it, are also richly embellished. The interior consists of a nave, aisles, and choir, and contains several sepulchral chapels, among which are traces of an earlier style of architecture than the church: the walls are adorned with several ancient monuments and mural tablets. A chapel has been erected at East Horrington, containing 260 sittings, 170 of which are free, the Incorporated Society having granted £150 in aid of the expense. Another chapel has been built at Coxley, in this parish, by aid of a grant from Her Majesty's Commissioners. There are places of worship for Baptists, Independents, and Wesleyan Methodists. The collegiate grammar school contains 26 boys, 8 of whom, choristers of the cathedral, are paid for by the Dean and Chapter. The United charity school, founded in 1654, by Mrs. Mary Barkham, Mr. Adrian Hickes, and Mr. Philip Hodges, the last of whom erected a school-house, is endowed with property producing above £500 per annum, which is appropriated to the instruction of 34 boys and 20 girls, 20 of each being completely clothed; the boys, on leaving the school, are apprenticed, with a premium of £10, and an additional sum of £10 on the completion of the fourth year; at the expiration of their term, upon producing a certificate of good conduct, each receives a present of £5; the girls are also taught needlework, and placed out in service. There is a central National school. On the north side of the churchyard is an hospital, founded and endowed by Bishop Bubwith, who died in 1424, for twelve aged men, twelve aged women, and a chaplain, to which six more aged men were added, in 1607, by Bishop Still, who augmented the endowment for that purpose; including the previous augmentation by Bishops Beckington and Bourne, the present income is about £400 per annum: the buildings are neat, and comprise separate apartments for each, with a common room, and a small chapel at the east end. The almshouses in Priest's Row were founded in 1614, by by Mr. Henry Llewellyn, who endowed them for six aged women: the income arising from the endowment is about £170 per annum, from which fund also a weekly allowance is paid to four aged widows not in the almshouses. An almshouse for decayed burgesses was founded in 1638, by Mr. Walter Brick, who placed it under the inspection of the Bishop. Almshouses were founded in 1711, by Mr. Archibald Harper, who endowed them with property now producing about £70 per annum, for five decayed wool-combers of the city: the buildings comprise five apartments for the men, with a committee-room for the meeting of the trustees, which, at other times, is appropriated as a common room for the inmates. There are numerous other charitable bequests and funds, at the disposal of the charity commissioners, for distribution among the poor. The poor law union of Wells comprises 18 parishes or places, under the care of 27 guardians, and contains, according to the census of 1831, a population of 19,197. In the verdery of Southover are the remains of the priory of St. John, founded in 1206, by Hugh, Archdeacon of Wells, afterwards Bishop of Lincoln, and subsequently augmented by Bishop Jocelyne; the revenue at the dissolution was £41. 3. 6.: the buildings have been converted into a wool-comber's shop. The neighbourhood, especially on the side of the Mendip hills, abounds with geological interest. Among the eminent prelates of the see were Cardinal Wolsey and Archbishop Laud; the celebrated historian, Polydore Virgil, was archdeacon in the sixteenth century; and the learned and pious Dr. George Bull, Bishop of St. David's, was born in the city, in the year 1634; he died in 1709.

WELNETHAM, GREAT, a parish, in the union of THINGOE, hundred of THEDWASTRY, Western Division of the county of SUFFOLK, $3\frac{1}{2}$ miles (S. E. by S.) from Bury-St.-Edmunds; containing 422 inhabitants. The living is a rectory, valued in the king's books at £9. 15. $7\frac{1}{2}$.; present net income, £314; patron, F. Wing, Esq. Mrs. Mary Green, in 1814, bequeathed a legacy of £200, which was invested in the funds, and the interest is annually expended in coal for the poor. Here was a priory of Crouched, or Crossed, friars, subordinate to the principal house of that order, near the Tower of London. Numerous remains of Rōman antiquities have been dug up. Sir Richard Gipp, Knt., a great collector of Suffolk antiquities, resided and was buried here.

WELNETHAM, LITTLE (*St. Mary*), a parish, in the union of THINGOE, hundred of THEDWASTRY, Western Division of the county of SUFFOLK, $3\frac{1}{2}$ miles (S. E.) from Bury-St.-Edmunds; containing 180 inhabitants. The living is a discharged rectory, valued in the king's books at £4. 13. 4.; patron, Marquess of Bristol. The tithes have been commuted for a rent-charge of £155, subject to the payment of rates, which on the average have amounted to £25; the glebe comprises 25 acres.

WELNEY, a chapelry, in the parish of UPWELL, union of DOWNHAM, partly in the hundred of WISBEACH, Isle of ELY, county of CAMBRIDGE, and partly in the hundred of CLACKCLOSE, Western Division of the county of NORFOLK, 8 miles (E. S. E.) from March; containing 805 inhabitants. A day and Sunday school is endowed with £28 per annum. Roman coins have been found here in urns, &c., turned up by the plough.

WELTON (*St. Mary*), a parish, in the wapentake of LAWRESS, parts of LINDSEY, union and county of LINCOLN, 6 miles (N. N. E.) from Lincoln; containing 516 inhabitants. This place constitutes the endowment of five prebends in the cathedral church of Lincoln, called respectively Welton-Rivall, Welton-Bekall, Welton-Brinkhall, Welton-Painshall, and Welton-Westhall. The living is a discharged vicarage, valued in the king's books at £7. 6. 8.; present net income, £150; patrons, the five prebendaries; appropriators, Dean and Chapter of Lincoln. There is a place of worship for Wesleyan Methodists; also a school partly supported by subscription.

WELTON (*St. Andrew*), a parish, in the union of DAVENTRY, hundred of FAWSLEY, Southern Division of the county of NORTHAMPTON, $2\frac{1}{2}$ miles (N. N. E.) from Daventry; containing 600 inhabitants. The living is a discharged vicarage, valued in the king's books at £7; present net income, £193: it is in the patronage of the Crown. The Grand Junction and Union canals meet at the south-eastern extremity of the parish, and the latter is previously crossed by Watling-street, which skirts the eastern boundary. The sum of £8 a year is paid from the receipts of the charity lands, for teaching poor children, in connection with the National Society; and a Sunday school is supported with £5. 4. per annum from the same fund.

WELTON, a township, in the parish of OVINGHAM, union of HEXHAM, Eastern Division of TINDALE ward,

Southern Division of the county of NORTHUMBERLAND, $9\frac{1}{4}$ miles (E. N. E.) from Hexham; containing 69 inhabitants. This was the seat of King Oswy, and the place where the kings Peada and Segbert received the rites of baptism from Finan, Bishop of Lindisfarne: the Roman wall passed in the vicinity.

WELTON (*St. Helen*), a parish, in the union of SCULCOATES, wapentake of HOWDENSHIRE, East Riding of the county of YORK; containing 805 inhabitants, of which number, 672 are in the township of Welton, $3\frac{1}{2}$ miles (S. E.) from South Cave. The living is a vicarage, endowed with the rectorial tithes, and valued in the king's books at £25; present net income, £383: it is in the patronage of the Crown. The church is supposed to have been built by William Rufus. There is a place of worship for Wesleyan Methodists; and a National school has been established, also an infants' school.

WELTON-IN-THE-MARSH (*St. Martin*), a parish, in the union of SPILSBY, Wold Division of the wapentake of CANDLESHOE, parts of LINDSEY, county of LINCOLN, 6 miles (E. N. E.) from Spilsby; containing 363 inhabitants. The living is a perpetual curacy, valued in the king's books at £14. 8. 9.; present net income, £122; patrons and impropriators, P. and M.A. Massingberd, Esqrs.

WELTON-LE-WOLD (*St. Martin*), a parish, in the union of LOUTH, Wold Division of the hundred of LOUTH-ESKE, parts of LINDSEY, county of LINCOLN, $3\frac{3}{4}$ miles (W.) from Louth; containing 241 inhabitants. The living is a rectory, valued in the king's books at £11. 12. 1.; present net income, £448: it is in the patronage of the Crown. The church contains 90 free sittings, the Incorporated Society having granted £15 in aid of the expense.

WELWICK (*St. Mary*), a parish, in the union of PATRINGTON, Southern Division of the wapentake of HOLDERNESS, East Riding of the county of YORK, 2 miles (S. E. by E.) from Patrington; containing 401 inhabitants. The living is a discharged vicarage, valued in the king's books at £6. 13. 4.; present net income, £104: it is in the patronage of the Crown; impropriator, W. Fewson, Esq. The church is principally in the decorated style of English architecture, and contains the remains of a once splendid monument, said to have been removed from Burstall abbey, and bearing marks of high antiquity. There are places of worship for the Society of Friends and Wesleyan Methodists.

WELWYN (*St. Mary*), a parish, and the head of a union, in the hundred of BROADWATER, county of HERTFORD, 8 miles (W. N.W.) from Hertford; containing 1369 inhabitants. The village is situated on the small river Mimram, and on the great north road from London to York; it consists of one principal street, with a smaller leading to Stevenage, and contains several genteel residences, besides which there are many others in the immediate vicinity. In Mill-lane is a fine chalybeate spring, formerly in considerable repute: here is also an assembly-room. The living is a rectory, valued in the king's books at £21; patrons, Warden and Fellows of All Souls' College, Oxford. The tithes have been commuted for a rent-charge of £620, subject to the payment of rates, which on the average have amounted to £64; the glebe comprises 85 acres, valued at £125 per annum. The church contains 150 free sittings, the Incorporated Society having granted £100 in aid of the expense. Over the altar is a piece of embroidery, with a suitable inscription, by Lady Betty Young, wife of Dr. Edward Young, author of the "Night Thoughts," who was many years rector of this parish: he died on Good Friday, 1765, and was buried by the side of his lady, under the communion table. There are places of worship for Huntingtonians and Wesleyan Methodists. Dr. Young, in 1760, founded a school for clothing and educating sixteen poor boys, and endowed it with £1500 Old South Sea Annuities, which was augmented in 1810 by a bequest of £200 from Daniel Spurgeon. In 1830, the school-room was rebuilt upon an enlarged plan, chiefly from the funds of the charity, but partly by subscription; and in addition to the sixteen boys upon the original foundation, upwards of 50 are educated on the National system. John Bexfield, in 1570, bequeathed some land, the rent of which, now amounting to £13. 10. per annum, is distributed, together with the produce of some other bequests, among the resident poor in bread, &c. The poor law union only embraces the four parishes of Welwyn, Digswell, Ayot-St. Lawrence, and Ayot-St. Peter, containing a population of 1970, according to the census of 1831, and is under the superintendence of five guardians.

WEM (*St. Peter and St. Paul*), a market-town and parish, and the head of a union, partly in the hundred of PIMHILL, but chiefly in the Whitchurch Division of the hundred of NORTH BRADFORD, Northern Division of the county of SALOP; containing 3973 inhabitants, of which number, 1932 are in the township of Wem, 11 miles (N. by E.) from Shrewsbury, and 172 (N. W.) from London. It has been conjectured by Horsley that this place occupies the site of the ancient *Rutunium*, but there is no authentic account of it prior to the Conquest, at which period William Pandulph, who held 28 manors of Earl Roger de Montgomery, made it the head of a barony, and fixed his residence here; and, on the forfeiture of the estates of Robert de Belesme, son of Earl Roger, for rebellion in the reign of Henry I., Pandulph held it immediately of the crown, and thence became a baron of the realm. After continuing for several generations in this family, and passing through the hands of other proprietors, the barony was, in 1665, purchased by Daniel Wycherley, father of the poet, and by him sold to the unprincipled Judge Jeffreys, who was created Baron of Wem in 1685, being the first who enjoyed that dignity by patent, but at the death of his son the title became extinct. Wem was the first town in the county which declared for the parliament, in 1643, in which year, a party of the king's troops, under Lord Capel, attempted to capture it by storm, but were repulsed by the small garrison, aided, it is said, by the active exertions of the women: in the following year it was reconnoitred by Prince Rupert, who deemed it unworthy of any effort to capture. Under the government of Major-General Mytton, the garrison plundered the possessions of the neighbouring royalists, and the booty brought by them into the town caused it to flourish at that time more than at any antecedent or subsequent period. In 1677, it suffered from a dreadful fire, which consumed the church, market-house, and whole ranges of building, destroying property of the value of upwards of £23,000.

The town, situated in a level district, on the northern bank of the river Roden, consists principally of one spacious street, called High-street, from which several smaller streets and lanes diverge, and is well supplied with good water. Tanning and malting are carried on to a very considerable extent. The Ellesmere and Chester canal skirts the north-western boundary of the parish. The market was granted by King John, in 1205, to be held on Sunday, at that time a usual circumstance; since the 24th of Edward III., it has been held on Thursday. The market-house, on the south side of High-street, is a small neat edifice of brick, with stone quoins, commenced in 1702, but not completed until 1728; in the room over it the courts leet are held. The fairs are on March 4th and May 6th, for linen cloth; May 20th and June 29th, for cattle; and Sept. 30th and Nov. 22nd, chiefly for swine. Wem appears to have been incorporated, though at what period the charter was granted is not known; but, from a copy of court roll, dated 9th of Edward VI., it must have been prior to that period. The principal officers are two bailiffs, appointed annually at the court leet held after Michaelmas, one by the lord's steward, and the other by the borough jury. Their authority is now very limited, their duties consisting chiefly in the returning of the jury to attend the steward at courts leet, in preventing fraud by the use of false weights and measures, and in being present at public proclamations. The burgesses are the holders of burgage tenements, about 80 in number. It has lately been made a polling-place for the Northern Division of the county.

The living is a rectory, valued in the king's books at £26. 4. 4½.; present net income, £1767; patron, Duke of Cleveland. The church is a spacious edifice with a lofty tower: it appears to have been built at an early period, but the subsequent alterations and repairs it has undergone have left little of the original style. An elevated spot at the north-west corner of the churchyard, now converted into gardens, is supposed to have been the site of the ancient castle. There is a chapel of ease at Edstaston, in this parish. Baptists and Presbyterians have each a place of worship. The free grammar school, in Noble-street, was founded and endowed for three masters, in 1650, by Thomas Adams, Esq., who was born here in 1586, and became a wealthy trader and active magistrate of the city of London, having been created a baronet in 1660: the present school premises were erected in 1670. Its management is vested in feoffees, who appoint the masters; and, to increase the original endowment, the statutes of the founder direct that the school shall not be open to the children of those parents who, having the ability, do not contribute towards its support: several subsequent bequests have been added, making the present gross annual income £331. 2. The school enjoys the benefit of two exhibitions, founded by Mr. Careswell, for an account of which see BRIDGENORTH. Here is also a National school. A sum of about £20, arising from various bequests, is annually distributed in clothing and bread. The poor law union of Wem comprises 12 parishes or places, under the care of 16 guardians, and contains a population of 11,353, according to the census of 1831. Mr. John Ireland, author of "Hogarth Illustrated," was born in this parish.

WEMBDON (*St. George*), a parish, in the union of BRIDG-WATER, hundred of NORTH PETHERTON, Western Division of the county of SOMERSET, 1½ mile (N. W.) from Bridg-water; containing 289 inhabitants. The living is a discharged vicarage, valued in the king's books at £9. 16. 10.; present net income, £612; patron, C. K. K. Tynte, Esq.; impropriator, J. Credland, Esq. A day and Sunday school is partly supported by the clergyman. The river Parret is navigable on the east of this parish.

WEMBURY (*St. Werburgh*), a parish, in the union of PLYMPTON-ST. MARY, hundred of PLYMPTON, Ermington and Plympton, and Southern, Divisions of the county of DEVON, 5¼ miles (S. by W.) from Plympton-Earls; containing 652 inhabitants. This parish is bounded on the south and west by the English Channel, and on the east by the river Yealm. The surface is boldly undulated, and the views over the channel and the adjacent country are interesting and extensive. Nearly opposite to the church, from which it is about two miles distant, bearing west-south-west, and at the entrance of Plymouth Sound, is the small island called by mariners the Mew Stone, the property of Charles Biggs Calmady, Esq., of Langdon Hall, in this parish. The living is a perpetual curacy; net income, £83; patrons and appropriators, Dean and Canons of Windsor. The appropriate tithes have been commuted for a rent-charge of £380, subject to the payment of rates. The church, situated on the brow of a bold eminence on the shore, is principally in the later style of English architecture, with the exception of the north aisle, which is of an earlier period, and is substantially built of granite: in the chancel is a curious monument to Sir John Hele, serjeant-at-law in the reigns of Elizabeth and James I.; and in the south aisle is a massive tomb inscribed to the memory of Lady Narborough, and dated 1678: there are also various memorials of the family of Calmady by whom the communion plate was presented, the Heles and other families. A school is partly supported by subscription. An almshouse for ten poor people was founded and endowed, in 1625, by Sir Warwick Hele.

WEMBWORTHY (*St. Michael*), a parish, in the union of CREDITON, hundred of NORTH TAWTON, South Molton and Northern Divisions of the county of DEVON, 3½ miles (S. S. W.) from Chulmleigh; containing 378 inhabitants. This parish is situated nearly in the centre of the county; the surrounding scenery is agreeably diversified, and enlivened with the handsome residence of Eggesford, the seat of the Hon. Newton Fellowes. Near Eggesford is a circular encampment surrounded by a fosse. The living is a rectory, valued in the king's books at £11. 13. 4.; patrons, Rev. P. Johnson and others. The tithes have been commuted for a rent-charge of £165. 12. 6., subject to the payment of rates, which on the average have amounted to £3. 10.; the glebe comprises 40 acres. The church has been almost entirely rebuilt, at the expense of the Hon. Newton Fellowes. A day and Sunday school is supported by the rector. Dr. Burton, author of the "Pentalogia," and other learned works, was a native of this place.

WENDLEBURY (*St. Giles*),) a parish, in the union of BICESTER, hundred of PLOUGHLEY, county of OXFORD, 2½ miles (S. W.) from Bicester; containing 196 inhabitants. The living is a rectory, valued in the king's books at £11. 9. 4½.; present income, £250; patrons, Dean and Canons of Christ-Church, Oxford. The church, with the exception of the tower, which had stood for 700 years, was rebuilt in 1761. A school containing 10 children is supported by private donation. The Rev. Robert Welborne, rector of the parish from 1730 to 1764, bequeathed 61 folio volumes as the foundation of a theological library.

WENDLING (*St. Peter and St. Paul*), a parish, in the union of Mitford and Launditch, hundred of Launditch, Western Division of the county of Norfolk, 4¼ miles (W.) from East Dereham; containing 347 inhabitants. The living is a perpetual curacy; net income, £52; patron, Earl of Leicester; impropriator, T. Smith, Esq. An abbey of Premonstratensian canons, in honour of the Blessed Virgin Mary, was founded here before 1267, by Willian de Wendling, which at the dissolution had a revenue of £55. 18. 4.: part of the church was lately standing, but it has been removed for the purpose of mending the roads.

WENDON-LOFTS (*St. Dunstan*), a parish, in the union of Saffron-Walden, hundred of Uttlesford, Northern Division of the county of Essex, 5¼ miles (W. by N.) from Saffron-Walden; containing 54 inhabitants. This parish, which is situated in a pleasant open country everywhere presenting interesting scenery, is supposed to have derived the adjunct to its name from a former proprietor of the manor. The living is a discharged rectory, with the vicarage of Elmdon annexed, valued in the king's books at £9. 10. 10.; present net income, £470; patron and impropriator, John Wilkes, Esq. The church is a small ancient edifice near the manor-house, and in good repair; in the chancel are some ancient brasses and monumental inscriptions.

WENDONS-AMBO (*St. Mary*), a parish, in the union of Saffron-Walden, hundred of Uttlesford, Northern Division of the county of Essex, 2½ miles (S. W. by W.) from Saffron-Walden; containing 333 inhabitants. This parish, which is about seven miles in circumference, appears to have derived its affix from the consolidation of both parishes consequent on the destruction of the parish church of Little Wendon. The living is a discharged vicarage, with the rectory of Little Wendon united, valued jointly in the king's books at £17; present net income, £165; patron, Marquess of Bristol. The church is an ancient structure in the early English style, with a low square tower; the chancel is separated from the nave by a richly carved screen of oak. Here is a Sunday National school. The river Cam has its source in the parish.

WENDOVER (*St. Mary*), a market-town and parish (formerly an unincorporated borough), in the union of Wycombe, hundred of Aylesbury, county of Buckingham, 23 miles (S. E. by S.) from Buckingham, and 35 (N. W. by W.) from London; containing 2008 inhabitants. The manor was given by Henry II. to Faramus de Boulogne, and it was subsequently in the possession of the Fiennes; Sir John Molins; Alice Perrers, a favourite of Edward III.; Thomas Holland, Earl of Kent; Edward, Duke of York, in 1338 (between which period and 1564 it was held either by the queen or some branch of the royal family); and Sir Francis Knollys and Catherine his wife: in 1660 it was purchased by the Hampden family, and continued in their possession until the decease of the late lord. It then became the property of the Earl of Buckinghamshire, who sold it to Samuel Smith, Esq., in 1828, and it is now the property of Abel Smith, Esq., who represented the borough till its disfranchisement by the act of the 2nd of William IV. The town, situated at the foot of the Chiltern hills, near the entrance to the Vale of Aylesbury, is indifferently built, containing but few good houses; the inhabitants are well supplied with water from wells. Many of the females are engaged in lace-making. A branch of the Grand Junction canal extends to the town, and affords a medium of conveyance for coal from Staffordshire; it passes through a reservoir in the neighbourhood, extending over 70 acres. A market was granted in 1403, and confirmed in 1464, with two fairs: the former is on Monday; and the latter are held on May 13th and October 2nd, chiefly for cattle. Wendover, which was a borough by prescription, returned members to parliament from the 28th of Edward I. to the 2nd of Edward II.; at which period the right ceased, and, after a discontinuance of more than 400 years, was restored through the exertions of Mr. Hakeville, a barrister, who, on examining the parliamentary writs in the Tower, in the 21st of James I., discovered that Amersham, Wendover, and Great Marlow, had all sent representatives. Petitions were accordingly presented from these places; and, notwithstanding the opposition of the monarch, who declared that "he was troubled with too many burgesses already," the commons decreed the renewal of the privilege; but the borough was finally disfranchised in the 2nd of William IV. Hampden, the celebrated patriot, represented this borough in five successive parliaments. Petty sessions are held once a fortnight, and courts leet and baron occasionally. The living is a discharged vicarage, valued in the king's books at £12. 6. 1.; present net income, £271: it is in the patronage of the Crown; impropriator, Abel Smith, Esq. The church stands about a quarter of a mile from the town. An ancient chapel, dedicated to St. John, was taken down to afford a site for an infants' school, which was erected a few years since. There are places of worship for Baptists and Independents. Joan Bradshaw, in 1578, bequeathed property now producing a rental of £31. 10., half of which is distributed, together with £32. 13. 4., arising from various other bequests, among the poor of this parish; the other moiety being appropriated to a like purpose in the parish of Halton; and William Hill, in 1723, bequeathed an estate, now let for £145 per annum, for the support of National schools in the parishes of Bierton and Wendover, and for the distribution of coal to poor men in the above and four other parishes. The National school is situated near the market-house. The infants' and another school are supported by Mr. and Mrs. Smith. Roger de Wendover, historiographer to Henry II.; and Richard, Bishop of Rochester, in the reign of Henry III., were natives of this place.

WENDRON (*St. Wendron*), a parish, in the union of Helston, comprising the borough and market-town of Helston (which has a separate jurisdiction), and partly in the Western Division of the hundred of Kerrier, Western Division of the county of Cornwall; and containing 8073 inhabitants. This parish, which is situated near the coast of the English Channel, is rich in mineral treasure, and tin and copper mines have been opened within its limits, which afford employment to many of the inhabitants. The living is a vicarage, endowed with the rectorial tithes, with the rectory of Helston annexed, and valued in the king's books at £26. 19. 4½.; present net income, £878; patrons, Provost and Fellows of Queen's

College, Oxford. There are places of worship for Baptists, Bryanites, and Wesleyan Methodists. A National school is partly supported by an endowment of £3 per annum, and partly by subscription. On the summit of a hill, called Caer Bonalas, is a circle of upright stones, enclosing an intrenchment thirty-five feet in diameter, in the centre of which are four thin flat stones placed on each other, the uppermost of which is nineteen feet in diameter; on the same hill are two barrows, one of which is enclosed by a wall about five feet high; and between the village and Redruth are nine upright stones, called the "nine maidens." Roman coins have been found at a place called Golvaduck Barrow; and at Trehill is an ancient well.

WENDY, a parish, in the union of ROYSTON, hundred of ARMINGFORD, county of CAMBRIDGE, 6¼ miles (N. N. W.) from Royston; containing 125 inhabitants. The living is a discharged vicarage, with Shingay annexed, valued in the king's books at £5. 10. 10.; present net income, £200; patron and impropriator, Representative of the late Hon. T. Windsor. A school is supported by a rent-charge of £30, given by the late Hon. T. Windsor, who erected the school-house.

WENHAM, GREAT, or COMBUST (*St. John*), a parish, in the hundred of SAMFORD, Eastern Division of the county of SUFFOLK, 8 miles (S. W. by W.) from Ipswich; containing 181 inhabitants. The living is a discharged rectory, valued in the king's books at £8.13.4.; present net income, £211; patron, Rev. D. C. Whalley.

WENHAM, LITTLE, a parish, in the hundred of SAMFORD, Eastern Division of the county of SUFFOLK, 5 miles (S. E. by E.) from Hadleigh; containing 88 inhabitants. The living is a discharged rectory, consolidated with that of Capel-St. Mary, and valued in the king's books at £5. 8. 11½.; present net income, £250. The church contains memorials of several individuals of the family of Brewes. Here are the remains of an old castellated mansion, the seat of that ancient family, by whom it appears to have been erected in 1569; it has been converted into a granary.

WENHASTON (*St. Peter*), a parish, in the union and hundred of BLYTHING, Eastern Division of the county of SUFFOLK, 2¼ miles (E. S. E.) from Halesworth; containing 1070 inhabitants. The navigable river Blythe bounds this parish on the north-east; and the family of Leman, of whom many are buried in the church, had a seat here. The living is a discharged vicarage, valued in the king's books at £6. 0. 10.; present net income, £110: it is in the patronage of the Crown; impropriator, Earl of Gosford. William Pepyn, in 1562, and Reginald Lessey, in 1563, bequeathed land for the support of a free school for poor children: 20 boys are instructed gratuitously from this endowment.

WENLOCK, LITTLE (*St. Lawrence*), a parish, within the liberties of the borough of WENLOCK, union of MADELEY, Southern Division of the county of SALOP, 3¼ miles (S.) from Wellington; containing 1057 inhabitants. The living is a rectory, valued in the king's books at £11. 13. 4.; present net income, £550; patron, Lord Forester. The church has been enlarged, and 500 free sittings provided, the Incorporated Society having granted £200 in aid of the expense. There are coal and iron mines, with extensive quarries of limestone, in the parish.

WENLOCK, MUCH (*Holy Trinity*), a borough, market-town, and parish, having separate jurisdiction, and the head of a liberty, in the union of MADELEY, Southern Division of the county of SALOP, 12 miles (S. E.) from Shrewsbury, and 148 (N. W.) from London; containing 2424 inhabitants. This town is of considerable antiquity; its British name was *Llan Meilien*, or "St. Milburgh's-Church;" and in the Monasticon it is denominated *Winnica*, or "the windy place." Its early importance was derived from the establishment of a convent, about 680, by Milburga, daughter of King Merwald and niece of Wulphere, King of Mercia, who presided as abbess, and at her death was interred here. Having been destroyed by the Danes, it was restored by Leofric, Earl of Mercia, in the time of Edward the Confessor, after which it fell into decay. It was rebuilt, or repaired, by Roger de Montgomery, soon after the Conquest, who largely endowed it, converted it into a priory for Cluniac monks, and dedicated it to St. Milburga: at the dissolution, the revenue was valued at £434. 1. 2. The ruins, which are situated on the south side of the town, are extensive, and present every variety of the most finished specimens of the latest Norman, and the early and decorated styles of English architecture. Of the church, the south transept is in the most perfect state: the end and side walls, including the triforium and clerestory windows, are standing, and exhibit the purest specimens of elegant design and elaborate execution; one wall of the north transept also is remaining, in which is a continuation of the same details: the bases of the four massive piers which supported the tower, and of those which separated the aisles from the nave and choir, are still uncovered by turf, and mark out the ground plan of a cathedral, which, for its magnificence and elegant decoration, scarcely had its equal in the kingdom. Three beautiful Norman arches, highly ornamented, form an entrance to the chapter-house, the walls of which are embellished with successive series of intersecting arches, with clustered columns of exquisite design. Two of the cloisters also remain in a very perfect state; one is of the lighter decorated style, with a lofty ceiling richly groined, and ornamented with slender shafts terminating in corbels on the walls; the other of the more massive, but finished Norman style, with low clustered pillars ranged upon circular plinths.

Corporation Seal.

The town is situated in a pleasant vale, and consists principally of one long street, from which another diverges at right angles; the houses are in general of brick and well built, several of them being modern and handsome, with many cottages of stone, with thatched roofs; the streets are Macadamized, but not lighted; and the inhabitants are supplied with water by pumps attached to the houses. In the time of Richard II., this place was noted for lime quarries and copper mines; the former are still extensive; the latter are not now worked. The market, originally granted to the prior and brethren, is on Monday. Fairs are held on the second Monday in March, and May 12th, for horned cattle, horses, and sheep, and for hiring servants; July 5th, for sheep; and October 17th and December 4th, for horned cattle, horses, sheep, and swine. Much-Wenlock enjoys many peculiar privileges, with a jurisdiction extending over the seventeen parishes of Beckbury, Badger, Broseley, Barrow, Benthall, Deuxhill, Ditton-Prior, Hughley, Linley, Madeley, Monk-Hopton, Skipton, Stoke-St. Milborough, Wenlock (Little), Wenlock (Much), Willey, and Eaton, and the extra-parochial district of Posenall, which constitute the liberty; but in consequence of the parishes of Beckbury, Badger, Deuxhill, part of Skipton, Stoke-St. Milborough, and Eaton, being disjoined, they no longer form a part of the municipal, though still included within the limits of the parliamentary, borough. By a charter of incorporation granted by Edward IV., in the seventh year of his reign, confirmed and extended by subsequent sovereigns, the corporation consisted of a bailiff, recorder, and an unlimited number of bailiff's peers; the bailiff, recorder, and two of the peers, were justices of the peace: there were also a coroner, treasurer, town-clerk, sergeant-at-mace, and subordinate officers. The corporation at present consists of a mayor, six aldermen, and 18 councillors, under the act of the 5th and 6th of William IV., cap. 76, for an abstract of which see the Appendix, No. I.; and the borough is divided into three wards. The mayor and the ex-mayor are justices of the peace, and seven others are appointed by the Crown, of whom only five act. The freedom is obtained by birth after the father has been sworn, and by servitude. This borough was the first that possessed the right of parliamentary representation by virtue of a charter from the crown: the elective franchise was granted in 1478, by Edward IV., when it returned one member: at present it sends two. The right of election was formerly vested in the burgesses, in number from 500 to 600; but the privilege has been extended to the £10 householders of the borough, which is considered to be co-extensive with the liberty or franchise, and comprises an area of 47,589 acres. The present number of registered voters is 850; the mayor is the returning officer. The corporation were empowered to hold a court of common pleas, every Tuesday fortnight; a court of assize for trying criminals, with the power of life and death, which now, by the refusal of a borough sessions under the Municipal Reform Act, is taken away; and a court of record, for the recovery of debts to any amount, but which, as there is no longer a recorder appointed, is consequently discontinued. A court of requests is held under the 22nd of George III., for the recovery of debts under 40*s.*, the jurisdiction of which extends over the parishes of Broseley, Benthall, Madeley-Barrow, Linley, Willey, Little Wenlock, and the extra-parochial place called Posenall. Manorial courts are held at Easter and Michaelmas; at the latter, constables are appointed. This town has been made a polling-place for the southern division of the county. The guildhall is an ancient building of timber frame-work, resting on piazzas, more remarkable for its antiquity than the beauty of its architecture.

Wenlock is the head of a deanery: the living is a discharged vicarage, valued in the king's books at £12. 9. 7.; present net income, £180; patron and impropriator, Sir W. W. Wynn, Bart. The church is a venerable structure, with a square tower surmounted by a spire: it partakes, in a very remote degree, of the style of the abbey, being partly of the Norman, and partly of the decorated English style: the interior consists of a chancel, nave, and aisles, separated by clustered piers and obtusely pointed arches. A small theological library, left by one of the vicars for the use of the clergy, was, about 40 years since, extended by subscription into a circulating library for the use of the inhabitants. There is a place of worship for Wesleyan Methodists. The free school, endowed with £14. 5. 10. per annum, by the Rev. Francis Southern and others, is further supported by subscription: fourteen boys are educated on the foundation. Another school is supported by Lady Wenlock. There are almshouses for four poor widows. Paul Beilby Thompson, Esq., was created Baron Wenlock on the 2nd of May, 1839.

WENN, ST., a parish, in the union of ST. COLUMB-MAJOR, Eastern Division of the hundred of PYDER and of the county of CORNWALL, 4 miles (N. E. by E.) from St. Columb-Major; containing 649 inhabitants. This parish, which derives its name from the dedication of its church, is intersected in the northern part by the river Camel, a few miles to the south of its influx into the Bristol Channel. Fairs for cattle are annually held at Tregonetha on April 25th, May 6th, and August 1st. The living is a vicarage, valued in the king's books at £16. 6. 8.; patron, W. Rashleigh, Esq. The church, with the exception of the tower, was rebuilt in 1825. There is a place of worship for Wesleyan Methodists; also a school endowed by the Rev. J. P. Gilbert, with £5 per annum.

WENNINGTON (*ST. PETER*), a parish, in the union of ROMFORD, hundred of CHAFFORD, Southern Division of the county of ESSEX, 2 miles (N. by W.) from Purfleet; containing 127 inhabitants. This parish, which is about three miles and a half in length, and little more than one mile in breadth, is bounded by the river Thames, from the banks of which extends a considerable tract of marsh land. The living is a rectory, valued in the king's books at £8; present net income, £373; patron, Bishop of London. The church is a handsome ancient structure, with a square embattled tower: near the altar is an ancient stone with some obsolete letters or characters. Here is a Sunday National school.

WENNINGTON, a township, in the parish of MELLING, hundred of LONSDALE, south of the sands, Northern Division of the county palatine of LANCASTER, 6½ miles (S. by E.) from Kirkby-Lonsdale; containing 155 inhabitants.

WENSLEY (*HOLY TRINITY*), a parish, in the union of LEYBURN, wapentake of HANG-WEST, North Riding of the county of YORK; containing 2266 inhabitants, of which number, 288 are in the township of Wensley, 3 miles (N. W. by W.) from Middleham. The living is a rectory, valued in the king's books at £49. 9. 9½.; present net income, £1337; patron, Lord Bolton. In the church is some fine screen work, which is said to have belonged to the abbey of St. Agatha at Richmond. Here is a National school. The river Ure runs through the parish, and is crossed by an ancient bridge of three or four arches, which was erected about the commencement of the fourteenth century, and has been lately widened and repaired, at the expense of the Riding. Above Wensley are the ruins of Bolton castle, built in the reign of Richard II., by Richard Lord Scroop, Lord High Chancellor of England. According to Leland it consisted of four principal towers, and was eighteen years in building, the expense having amounted to 1000 marks yearly, or £12,000 sterling in the whole. The timber was brought from Inglewood Forest, in Cumberland, the conveyance of which was the chief cause of the great expense incurred in the building.

WENSLEY-FOLD, a township, in the parish and Lower Division of the hundred of BLACKBURN, Northern Division of the county palatine of LANCASTER, $1\frac{3}{4}$ mile (W. by N.) from Blackburn; containing 1047 inhabitants, upwards of 300 of whom are employed in an extensive spinning establishment here. It is bounded on the south by the river Darwent.

WENTNOR (*ST. MICHAEL*), a parish, in the union of CLUN, hundred of PURSLOW, Southern Division of the county of SALOP, $5\frac{1}{2}$ miles (N. E. by E.) from Bishop's-Castle; containing 707 inhabitants. The living is a rectory, valued in the king's books at £7. 2. 11.; present net income, £189; patrons, Dean and Canons of Christ-Church, Oxford. A school, in which twelve poor children are instructed, is endowed with £100 New Four per cents., and a house and garden.

WENTWORTH, or WINGFORD, a parish, in the hundred of SOUTH WITCHFORD, union and Isle of ELY, county of CAMBRIDGE, $4\frac{1}{2}$ miles (W. S. W.) from Ely; containing 144 inhabitants. The living is a rectory, valued in the king's books at £10; present net income, £286; patrons, Dean and Chapter of Ely. A school is partly supported by subscription.

WENTWORTH, a chapelry, in the parish of WATH-UPON-DEARNE, union of ROTHERHAM, Northern Division of the wapentake of STRAFFORTH and TICKHILL, West Riding of the county of YORK, $5\frac{1}{2}$ miles (N. W. by N.) from Rotherham; containing 1394 inhabitants. The living is a perpetual curacy; net income, £125; patron, Earl Fitzwilliam; appropriators, Dean and Canons of Christ-Church, Oxford. The chapel, dedicated to the Holy Trinity, is principally in the later style of English architecture, and has been lately enlarged at the expense of Earl Fitzwilliam: it contains several monuments of the Wentworth family. A school was erected here in 1716, by Thomas Wentworth, who endowed it with land and a rent-charge of £65, towards teaching and clothing 50 poor children: the master occupies the school-house and land rent-free, and 4 children have been added in consequence of a bequest of £2 per annum from Ann Pickles. A National school for girls and an infants' school were built in 1837, by Earl Fitzwilliam. Almshouses for six men and six women were built in 1697, pursuant to the will of William Wentworth, Earl of Strafford, who endowed them with £36 a year: they were repaired and improved by the late Earl Fitzwilliam, at a cost of £500.

WEOBLEY (*ST. PETER AND ST. PAUL*), a market-town and parish (formerly an unincorporated borough), and the head of a union, in the hundred of STRETFORD, county of HEREFORD, 12 miles (N. W.) from Hereford, and 145 (W. N. W.) from London; containing 819 inhabitants. This ancient town consists of one principal street on the main road from Hereford to Knighton. The market is on Thursday; and fairs are held on Holy Thursday and three weeks after. The elective franchise was granted in the reign of Edward I., and renewed or confirmed by Charles I.; but was withdrawn in the 2nd of William IV. The right of election was vested in "the inhabitants of the ancient vote houses of 20*s*. per annum, resident during 40 days previous to the election, and paying scot and lot, also in such owners of ancient vote houses, paying scot and lot, as were resident in such houses at the time of election:" the number of voters was about 93; and the constables were the returning officers. A manorial court is held annually in October, the jurisdiction of which extends to the recovery of debts under 40*s*. The petty sessions for the hundred take place here. The living is a discharged vicarage, valued in the king's books at £9. 1.; patron and appropriator, Bishop of Hereford. The appropriate tithes have been commuted for a rent-charge of £358, and the vicarial for £250, subject to the payment of rates. The free grammar school was founded in 1655, by William Crother, citizen of London, for the education of children born in the parishes of Weobley, Wormesley, and in the village of Wooton, in the parish of King's-Pion, and endowed with £20 per annum; he likewise bequeathed £100 to build a school-house. A National school also is partly supported by endowment. The poor law union of Weobly comprises 26 parishes or places, under the superintendence of 29 guardians, and contains, according to the census of 1831, a population of 8340. On the south side of the town are the remains of an ancient castle, which was taken by Stephen in the war between him and the Empress Matilda, for whom it had been kept by William Talbot.

WEONARD'S, ST., a parish, in the union of Ross, Lower Division of the hundred of WORMELOW, county of HEREFORD, $7\frac{1}{4}$ miles (W. by N.) from Ross; containing 564 inhabitants. The living is annexed, with those of Little Dewchurch, Hentland, and Llangarran, to the vicarage of Lugwardine. There is a place of worship for Wesleyan Methodists; also a school supported by subscription.

WEREHAM (*ST. MARGARET*), a parish, in the union of DOWNHAM, hundred of CLACKCLOSE, Western Division of the county of NORFOLK, $1\frac{3}{4}$ mile (N. W.) from Stoke-Ferry; containing 575 inhabitants. The living is a perpetual curacy, with that of Wretton annexed; net income, £109; patron, Edward R. Pratt, Esq., who, with G. R. Eyres, Esq., is the impropriator. Here is a National school. A Benedictine priory, in honour of St. Winwaloe, or St. Guenolo, was founded here, about the beginning of the reign of John, by the Earl of Clare, as a cell to the abbey of Mounstroll in France; it was given, in 1321, to the abbey of West Dereham, and at the dissolution had a revenue of £7. 2. 8.

WERNETH, a township, in the parish and union of STOCKPORT, hundred of MACCLESFIELD, Northern Division of the county of CHESTER, 4 miles (N. E. by E.) from Stockport; containing 3462 inhabitants. This place is situated to the east of Stockport, on the right bank of the river Etherow, and upon the edge of Derbyshire. It is noticed in the Norman survey as "Warnet," by which name Stockport is also described, which, after an intermediate alienation to the Lords of Longdendale, by whom it was held in the 16th of Edward III., passed to the Stokeports, and from them to the Davenports of Henbury; the manor is now the property of Thomas William Tatton, Esq., of Withenshaw, to whom it passed from the Bretlands, who probably purchased it from the representatives of the Davenports. The township comprises the hamlets of Gee Cross and Compstal Bridge;

the former of which takes its name from the ancient family of Gee, who erected here a cross of stone, which has disappeared; it consists of one wide street, half a mile in length, through which passes the turnpike-road from Stockport to Mottram-in-Longdendale. The cotton manufacture, calico-printing, and the making of hats, are extensively carried on, and coal mines and stone quarries abound in this township. Fairs for cattle are annually held at Gee Cross, on April 28th and November 20th. The Peak Forest canal passes through the township; and there is a daily post from Manchester to this place. A court baron for the manor of Werneth is held annually at Gee Cross. At that place is a meeting-house for Unitarians, with an extensive burying-ground, also a Sunday school. In the immediate vicinity of Gee Cross rises the celebrated Werneth Loe, which, though of very considerable height, is enclosed and cultivated to its summit; this hill commands a most extensive and varied prospect.

WERRINGTON (*St. Martin and St. Giles*), a parish, in the union of LAUNCESTON, hundred of BLACK TORRINGTON, Lifton and Northern Divisions of the county of DEVON, 2½ miles (N. by W.) from Launceston; containing 661 inhabitants. The living is a donative curacy; net income, £229; patron, Earl of Buckinghamshire. A day and Sunday school is partly supported by subscription.

WERRINGTON, a chapelry, in the parish of PASTON, union and soke of PETERBOROUGH, Northern Division of the county of NORTHAMPTON, 3½ miles (N. N. W.) from Peterborough; containing 537 inhabitants. The interest of £100, bequeathed by John Goodwin in 1755, is distributed at Easter and Christmas among poor widows of this place.

WERVIN, a township, in the parish of ST. OSWALD, city of CHESTER, union of GREAT BOUGHTON, Lower Division of the hundred of BROXTON, Southern Division of the county of CHESTER, 4¼ miles (N. by E.) from Chester; containing 64 inhabitants. The Ellesmere, or Wirrall, canal bounds it on the west.

WESHAM, a joint township with Medlar, in the parish of KIRKHAM, union of the FYLDE, hundred of AMOUNDERNESS, Northern Division of the county palatine of LANCASTER, 1¼ mile (N. by W.) from Kirkham; the population is returned with Medlar.

WESSINGTON, a township, in the parish of CRICH, union of CHESTERFIELD, hundred of SCARSDALE, Northern Division of the county of DERBY, 3½ miles (N. W. by W.) from Alfreton; containing 465 inhabitants.

WEST-ACOMB, county of NORTHUMBERLAND.—See ACOMB, WEST.—*And other places having a similar distinguishing prefix will be found under the proper name.*

WEST-ACRE (*All Saints*), a parish, in the hundred of FREEBRIDGE-LYNN, Western Division of the county of NORFOLK, 5¼ miles (N. W. by N.) from Swaffham; containing 415 inhabitants. The living is a donative; net income, £31; patron and impropriator, A. Hammond, Esq. A priory of Black canons, in honour of St. Mary and All Saints, was founded here, in the time of William Rufus, by Ralph de Toney, which at the dissolution had a revenue of £308. 19. 11.: the remains of this once celebrated house exhibit specimens of the early and later styles of English architecture.

WESTANSWICK, a township, in the parish of STOKE-UPON-TERN, Drayton Division of the hundred of NORTH BRADFORD, Northern Division of the county of SALOP; containing 181 inhabitants.

WESTBEER (*All Saints*), a parish, in the union of BLEAN, hundred of BLEANGATE, lathe of ST. AUGUSTINE, Eastern Division of the county of KENT, 3½ miles (N. E. by E.) from Canterbury; containing 219 inhabitants. The living is a rectory, valued in the king's books at £7: it is in the patronage of the Crown. The tithes have been commuted for a rent-charge of £274, subject to the payment of rates, which on the average have amounted to £37; the glebe comprises two acres. The Archbishop of Canterbury is appropriator of the great tithes of a district called Rushbourne, which have been commuted for a rent-charge of £90, subject to the payment of rates. A piece of land comprising nearly two acres was left to this parish by an unknown benefactor for supplying every person resident in it with a roll of bread, a piece of cheese, and part of 28 gallons of beer annually, which the tenants paid in lieu of rent, and the land was thence called the Bread and Cheese Field. On the demise of the last tenant, the parish took possession of this land, and let it out in small allotments to the poor at a trifling rent, which has been productive of great benefit.

WESTBOROUGH (*All Saints*), a parish, in the union of NEWARK, wapentake of LOVEDEN, parts of KESTEVEN, county of LINCOLN, 7 miles (N. W. by N.) from Grantham; containing 215 inhabitants. The living is a rectory, in medieties, in the patronage and incumbency of the Rev. R. Hall; the first mediety, with the vicarage of Dry Doddington annexed, is valued in the king's books at £20; present net income, £76: the second mediety is valued at £6. 13. 4.; present net income, £477.

WESTBOURNE, county of SUSSEX.—See BOURNE, WEST.

WEST-BROMWICH, county of STAFFORD.—See BROMWICH, WEST.

WESTBROOK, a tything, in the parish of BOXFORD, hundred of KINTBURY-EAGLE, county of BERKS, 3¼ miles (N. W. by N.) from Speenhamland: the population is returned with the parish.

WESTBURTON, a tything, in the parish and hundred of BURY, rape of ARUNDEL, Western Division of the county of SUSSEX, 4 miles (N.) from Arundel: the population is returned with the parish.

WESTBURY (*St. Augustine*), a parish, in the union of BRACKLEY, hundred and county of BUCKINGHAM, 4¾ miles (W. N. W.) from Buckingham; containing 391 inhabitants. The living is a vicarage, valued in the king's books at £9. 17. 1.; present net income, £106; patron and impropriator, Benjamin Price, Esq. The church has been enlarged, and 330 free sittings provided, the Incorporated Society having granted £300 for this purpose. The rent of an allotment of land, awarded under an enclosure act in 1764, amounting to about £11 per annum, is distributed among the poor, who have also the right of cutting furze for fuel on this land.

WESTBURY (*St. Mary*), a parish, in the union of ATCHAM, hundred of FORD, Southern Division of the county of SALOP; containing 2228 inhabitants, of which number, 1495 are in the township of Westbury (including Westley and Yockleton townships), 8¾ miles

(W. by S.) from Shrewsbury. The living is a rectory in two portions; Westbury *in Dextera Parte*, valued in the king's books at £13. 9. 4½.; present net income, £643;—and Westbury *in Sinistra Parte*, valued at £11. 12. 8½.; present net income, £556; they are both in the patronage of E. W. S. Owen, Esq. An extensive colliery is worked in the neighbourhood. Petty sessions for the division are held here during the winter months. The Rev. John Earl, in 1716, gave land producing about £30 per annum, £10 of which is paid to a schoolmaster for teaching 24 poor children, and the remainder is appropriated to supplying the children with stationery and apprenticing them. Another school is partly supported by subscription.

WESTBURY (*St. Lawrence*), a parish, in the union of Wells, hundred of Wells-Forum, Eastern Division of the county of Somerset, 4 miles (N. W. by W.) from Wells; containing 681 inhabitants. The living is a discharged vicarage, with that of Priddy annexed, valued in the king's books at £11. 4. 9½.; patron and appropriator, Bishop of Bath and Wells. The appropriate tithes have been commuted for a rent-charge of £150, and the vicarial for £183, subject to the payment of rates, which on the latter on the average have amounted to £2. 10.; the glebe comprises 12 acres, valued at £20 per annum. There is also a rent-charge of £5 payable to the Subdean of Wells.

WESTBURY, a joint hamlet with Peak, in the parish of East Meon, hundred of Meon-Stoke, Petersfield and Northern Divisions of the county of Southampton, 6¼ miles (W.) from Petersfield, containing 226 inhabitants.

Seal and Arms.

WESTBURY (*All Saints*), a borough and parish, forming the hundred of Westbury, and the head of the union of Westbury and Whorwelsdown, Westbury and Southern Divisions, and Trowbridge and Bradford Subdivisions, of the county of Wilts; and containing 7324 inhabitants, of which number, 2495 are in the town of Westbury, 24 miles (N. W. by W.) from Salisbury, and 98 (W. by S.) from London. This place is of very great antiquity, and is generally supposed to have been a British settlement, and to occupy the site of the Roman station *Verlucio*. The name is of Saxon origin, being intended to designate the importance, or relative position, of the town: here, according to tradition, was a palace belonging to the West Saxon kings. The town is situated under Salisbury Plain, and consists of three principal streets, irregularly built, branching off towards Frome, Bradford, and East Lavington: the inhabitants are supplied with water from springs, and a small stream which falls into the Avon. The clothing trade formerly flourished here, one house alone employing 1000 persons: the principal manufuctures are broad cloth and kerseymere, there being in and near the town eight manufactories, and several others within the parish: a considerable quantity of malt is made. The market, now merely nominal, is on Tuesday, for pigs only; and fairs are held on the first Friday in Lent and Whit-Monday for pedlery; and on Easter-Monday and Sept. 24th, for cattle, horses, and cheese. The charter of incorporation was granted by Henry IV.: the municipal body consists of a mayor, recorder, twelve aldermen, and burgesses, with subordinate officers, none of them possessing magisterial authority. Courts leet are held by the mayor in Nov., and by the steward of the lord of the manor in May. There is also a court of requests, for the recovery of debts under £5, the jurisdiction of which is co-extensive with the hundreds of Westbury, Warminster, and Heytesbury; it is held here and at Warminster alternately, every fortnight, on Tuesday. Two high constables are appointed at the manorial court. This borough constantly returned two members to parliament from the 27th of Henry VI. to the 2nd of William IV., when it was deprived of one by the act then passed to amend the representation. The right of election was formerly in the resident occupiers of burgage tenements, in fee or for lives, or 99 years determinable on lives, or by copy of court roll, in number 61; but the privilege has been extended to the £10 householders of the entire parish, which has been constituted the new elective borough, and comprises 11,536 acres; the old borough contained only 7 acres: the mayor is the returning officer. A handsome town-hall, in the centre of the town, was erected in 1815, at the sole expense of the late Sir Manasseh Masseh Lopes, Bart.

The living is a discharged vicarage, in the patronage of the Precentor of the Cathedral Church of Salisbury (the appropriator), valued in the king's books at £44. 16. 0½.; present net income, £238. The church is a spacious and handsome structure, with a central tower, supposed to have been built about 900 years ago; in the interior are several handsome monuments. There are two places of worship for Independents, three for Baptists, and one for Wesleyan Methodists in the town, besides some others in the parish. John Matravers, an opulent clothier of this place, and a member of the Society of Friends, in 1814, gave £1000, to found a free school, and £1000 for clothing 20 poor women at Christmas; under this bequest upwards of 170 boys are taught according to the British system, upon payment of one penny per week each: the school is held in a room supplied by a member of the testator's family. Out of the clothing fund, 25 poor women are now annually provided with clothing. The sum of £17. 1. 10., being the dividends of a bequest of £500 by John Gibbs, in 1772, is appropriated to supplying clothing to six poor men. Westbury is entitled to a share of Henry Smith's charity, and every fourth year to a sum of about £30, the rent of an estate bequeathed in 1615, by Thomas Ray, to the several towns of Westbury, Trowbridge, Chippenham, and Marlborough, for the relief of poor clothiers. The poor law union of Westbury and Whorwelsdown comprises 10 parishes or places, under the care of 17 guardians, and contains, according to the census of 1831, a population of 13,164. Roman coins have been found here in great abundance. William de Westbury, one of the puisne judges of the Court of Common Pleas, and James Ley, Earl of Marlborough, are interred within the church. Bryan Edwards, historian of the British colonies in the West Indies; and Dr. Philip Withers, a writer of some eminence about the close of the last century, were natives of this town.

WESTBURY-UPON-SEVERN (*St. Peter and St. Paul*), a parish, and the head of a union, in the hundred of Westbury, Western Division of the county of Gloucester, $2\frac{1}{2}$ miles (N. E. by E.) from Newnham; containing 2032 inhabitants. This parish is bounded on the east and south by the river Severn, which is here crossed by a ferry to Framilode: it was the scene of divers military transactions during the civil war of the 17th century. The living is a vicarage, valued in the king's books at £20. 2. $8\frac{1}{2}$.; present net income, £261; it is in the patronage of the Custos of the College of Vicars-Choral in the Cathedral Church of Hereford, the appropriators. There is a place of worship for Wesleyan Methodists; also a day and Sunday school, endowed with £10 per annum. The poor law union of Westbury comprises 13 parishes or places, containing a population of 8760, according to the census of 1831, and is under the superintendence of 18 guardians.

WESTBURY-UPON-TRYM (*Holy Trinity*), a parish, in the union of Clifton, partly in the Lower Division of the hundred of Henbury, Western Division of the county of Gloucester, and partly in the county of the city of Bristol, 3 miles (N. N. W.) from Bristol; containing 4263 inhabitants. This parish is bounded on the south-west by the river Avon, and furnishes the celebrated stone called Cotham-stone; lead-ore has also been found. About a mile and a half north-east of the village is a prodigious cavern, called Pen Park Hole. The living is a perpetual curacy; net income, £630; patron and impropriator, Rev. Charles Vivian. There are chapels of ease at Redland and Shirehampton, in this parish. Here is a place of worship for Wesleyan Methodists; also a National school, and an infants' school, partly supported by subscription. A monastery existed here early in the ninth century, which was refounded near the close of the eleventh; it was dedicated to the Blessed Virgin, and made a cell to the priory of Worcester, but was dissolved in the reign of Henry I. About 1288, it became a college for a dean and canons, in honour of the Holy Trinity: in 1443, it was rebuilt, and its possessions augmented by Wm. Canning, a merchant, and Dr. Carpenter, Bishop of Worcester, who styled himself Bishop of Worcester and Westbury. Its revenue at the dissolution was estimated at £232. 14.; and the house, which remained till the reign of Charles I., was burned by Prince Rupert, to prevent its falling into the power of the parliament, but some traces of it are still visible in a mansion erected on its site.

WESTBY, a joint parish with Basingthorpe, wapentake of Beltisloe, parts of Kesteven, county of Lincoln, $3\frac{3}{4}$ miles (N. W.) from Corby: the population is returned with Basingthorpe. There is a place of worship for Roman Catholics.

WESTBY, a joint township with Plumptons, in the parish of Kirkham, union of the Fylde, hundred of Amounderness, Northern Division of the county palatine of Lancaster, $2\frac{1}{2}$ miles (W.) from Kirkham; containing, with Plumptons, 686 inhabitants.

WESTCOTE (*St. Mary*), a parish, in the union of Stow-on-the-Wold, Upper Division of the hundred of Slaughter, Eastern Division of the county of Gloucester, 4 miles (S. E. by S.) from Stow-on-the-Wold; containing 188 inhabitants. The living is a rectory, valued in the king's books at £9. 7. $3\frac{1}{2}$.; present net income, £209; patron and incumbent, Rev. T. P. Pantin. The church contains 70 free sittings, the Incorporated Society having granted £50 in aid of the expense.

WESTCOTE, a township, in the parish of Tysoe, Kington Division of the hundred of Kington, Southern Division of the county of Warwick, 5 miles (S. E. by E.) from Kington: the population is returned with the parish.

WESTCOTT, a hamlet, in the parish of Waddesdon, union of Aylesbury, hundred of Ashendon, county of Buckingham, 7 miles (W. N. W.) from Aylesbury; containing 242 inhabitants.

WESTEND, a township, in the parish of Burgh-upon-the-Sands, ward, and Eastern Division of the county, of Cumberland; containing 457 inhabitants.

WESTEND, a tything, in the parish of Worplesdon, First Division of the hundred of Wokeing, Western Division of the county of Surrey: the population is returned with the parish.

WESTENHANGER, or OSTENHANGER (*St. Thomas à Becket*), anciently a parish, now a manor, in the parish of Stanford, hundred of Stouting, lathe of Shepway, Eastern Division of the county of Kent, 3 miles (N. W.) from Hythe: the population is returned with Standford. The living was a rectory, valued in the king's books at £7. 12. 6., and in the patronage of the Crown. The church has been long demolished.

WESTERDALE, a parish, in the union of Guilsbrough, Eastern Division of the liberty of Langbaurgh, North Riding of the county of York, $7\frac{1}{2}$ miles (S. S. E.) from Guilsbrough; containing 281 inhabitants. The living is a perpetual curacy, annexed to the rectory of Stokesley. The tithes have been commuted for a rent-charge of £250, subject to the payment of rates, which on the average have amounted to £26; the glebe comprises 11 acres, valued at £14 per annum. Fifteen children are instructed for £15 per annum, arising from a rent-charge of £3, given by Jane Duck, in 1734, and certain land left by Mary Fish, in 1741.

WESTERFIELD (*St. Mary Magdalene*), a parish, in the union of Ipswich, partly within the borough of Ipswich, and partly in the hundred of Bosmere and Claydon, Eastern Division of the county of Suffolk, $2\frac{1}{2}$ miles (N. N. E.) from Ipswich; containing 327 inhabitants. The living is a discharged rectory, valued in the king's books at £11. 10. $7\frac{1}{2}$.; present net income, £292; patron, Bishop of Ely. Bridget Collett, in 1662, bequeathed land producing about £10 per annum, in support of a day and Sunday school. Francis Brooke, Esq., appropriated one moiety of the interest of £300 for clothing and books for the scholars, and the other moiety for coal for the poor.

WESTERGATE, a hamlet, in the parish of Aldingbourn, hundred of Box and Stockbridge, rape of Chichester, Western Division of the county of Sussex, $4\frac{1}{4}$ miles (E. by N.) from Chichester: the population is returned with the parish.

WESTERHAM (*St. Mary*), a market-town and parish, in the union of Seven-Oaks, hundred of Westerham, lathe of Sutton-at-Hone, Western Division of the county of Kent, 22 miles (W.) from Maidstone, and 21 (S. S. E.) from London; containing 1985 inhabitants. The name of this town implies its situation on the western border of the county. Two remarkable phenomena, called land slips, occurred here on the southern escarpment of the Sand hill in 1596 and 1756; in the former, nine acres of ground continued in motion for eleven days, and in the latter, two acres and a half, some parts sinking into pits, and others rising into hills. The town stands on the northern declivity of the same formation, and is of neat and clean appearance; about the centre is the market-house. The river Darent rises in the parish, and, after watering the ancient park of Squerries, takes a north-eastern direction. The market, which was granted in the 25th of Edward III. to the abbot of Westminster, who possessed the manor, is on Wednesday; and a cattle fair is held on May 3rd. This place is within the jurisdiction of a court of requests held for the hundred, for the recovery of debts under £5. The living is a vicarage, with that of Edenbridge annexed, valued in the king's books at £19. 19. $4\frac{1}{2}$.; present net income, £608; patron and incumbent, Rev. Richard Board; impropriators, H. V. Bodicoate, Esq., and others. The church is a large and venerable structure. There is a place of worship for Dissenters; and a National school is supported by voluntary contributions. Bishop Hoadley and the celebrated General Wolfe were both natives of this town; in the church is a simple tablet, with the well-known elegant tribute to the memory of the latter; and in the grounds of Squerries is a pillar, erected for the like purpose.

WESTERLEIGH (*St. James*), a parish, in the union of Chipping-Sodbury, hundred of Puckle-Church, Western Division of the county of Gloucester, 3 miles (S. W. by W.) from Chipping-Sodbury; containing 1709 inhabitants. This place anciently formed part of the parish of Puckle Church, and was not invested with parochial rights, nor had it a church, till the 14th century. Coal is procured in considerable quantities, and is conveyed by a railroad to Bristol. The living is united, with that of Abson, to the vicarage of Puckle-Church. The church is a handsome structure, in the later English style, with a lofty tower and stone pulpit. Sir John Smythe, Bart., in 1715, gave an annuity of £20 in support of two schools, in which ten boys and ten girls are educated; and there are two Sunday schools. Edward Fowler, Bishop of Gloucester, a theological writer of the seventeenth century, was born here.

WESTERTON, a township, in the parish of St. Andrew-Auckland, union of Auckland, South-Eastern Division of Darlington ward, Southern Division of the county palatine of Durham, $2\frac{1}{2}$ miles (E. by N.) from Bishop-Auckland; containing 85 inhabitants.

WESTFIELD (*St. Andrew*), a parish, in the union of Mitford and Launditch, hundred of Mitford, Western Division of the county of Norfolk, $2\frac{1}{2}$ miles (S.) from East Dereham; containing 127 inhabitants. The living is a discharged rectory, united to that of Whinbergh, and valued in the king's books at £4. 4. 2. The tithes have been commuted for a rent-charge of £145, subject to the payment of rates, which on the average have amounted to £24; the glebe comprises 20 acres, valued at £30 per annum.

WESTFIELD (*St. John the Baptist*), a parish, in the union of Battle, hundred of Baldslow, rape of Hastings, Eastern Division of the county of Sussex, $4\frac{1}{4}$ miles (E. by S.) from Battle; containing 938 inhabitants. The living is a vicarage, valued in the king's books at £11. 6. 8.; present net income, £372; patron and appropriator, Bishop of Chichester. The church is in the early style of English architecture. A day and Sunday school is partly supported by subscription.

WESTGATE, a township, in the parish of St. John, Newcastle, union of Newcastle-upon-Tyne, Western Division of Castle ward, Southern Division of the county of Northumberland; containing 2996 inhabitants. It forms the north-western suburb of the town of Newcastle, and several streets, containing many handsome residences, have lately been erected.

WESTHALL (*St. Andrew*), a parish, in the union and hundred of Blything, Eastern Division of the county of Suffolk, 4 miles (N. E.) from Halesworth; containing 442 inhabitants. The living is a discharged vicarage, valued in the king's books at £10. 2. $3\frac{1}{2}$.; present net income, £195; patrons and appropriators, Dean and Chapter of Norwich. The church has a Norman tower with elaborately enriched arches. Two small rent-charges, amounting together to £2. 18., are applied for teaching poor children. The family of the Bohuns had a seat here for many generations.

WESTHAM (*St. Mary*), a parish, in the union of Eastbourne, lowey and rape of Pevensey, Eastern Division of the county of Sussex, $5\frac{3}{4}$ miles (S. E.) from Hailsham; containing 752 inhabitants. The living is a vicarage, valued in the king's books at £21. 10. 10.; patron and impropriator, Earl of Burlington. The impropriate tithes have been commuted for a rent-charge of £302. 15. 9., and the vicarial for £570, subject to the payment of rates, which on the latter, on the average, have amounted to £133; the glebe comprises $1\frac{1}{2}$ acres, valued at £5 per annum. The church is partly in the later style of English architecture, and partly of earlier date. A National school has been established; and there is an almshouse, containing four tenements, called the hospital of St. John, endowed with 30 acres of land, granted, as it is supposed, by one of the religious societies of Layney and Priest Hawes, the remains of which have been converted into farm-buildings: £20 per annum has been granted from the revenue of the hospital for the support of the National school.

WESTHAMPNETT, county of Sussex.—See HAMPNETT, WEST.

WESTHORPE (*St. Margaret*), a parish, in the union and hundred of Hartismere, Western Division of the county of Suffolk, $7\frac{3}{4}$ miles (N.) from Stow-Market; containing 263 inhabitants. The living is a discharged rectory, valued in the king's books at £4. 18. $1\frac{1}{2}$.; present net income, £293; patron and incumbent, Rev. R. Hewitt, D.D. The Hall, formerly the residence of Charles Brandon, Duke of Suffolk, was a noble mansion; it was taken down about the middle of the last century. His royal consort, Mary, died here in 1533.

WESTINGTON, a joint hamlet with Combe, in the parish of Chipping-Campden, Upper Division of the hundred of Kiftsgate, Eastern Division of the county of Gloucester, $\frac{1}{2}$ a mile (S.) from Chipping-Campden; containing, with Combe, 128 inhabitants.

WESTLETON (*St. Peter*), a parish, in the union and hundred of Blything, Eastern Division of the county of Suffolk, $2\frac{3}{4}$ miles (E.) from Yoxford; containing 884 inhabitants. This parish is bounded on the south by a stream of which the mouth, in ancient records, is called the Port of Mismere, from a large sheet of water formerly near it; and it contained the hamlet of Dingle, which now consists only of a single farm. The living is a discharged vicarage, with the rectory of Middleton, annexed to the rectory of Fordley, and valued in the king's books at £8; patron and incumbent, Rev. Harrison Packard. In the chancel of the church are some elegant stone seats.

WESTLEY, a township, in the parish of WESTBURY, hundred of FORD, Southern Division of the county of SALOP, 10 miles (W. S. W.) from Shrewsbury: the population is returned with the township of Westbury.

WESTLEY (*St. Thomas à Becket*), a parish, in the union and hundred of THINGOE, Western Division of the county of SUFFOLK, 2 miles (W.) from Bury-St.-Edmund's; containing 132 inhabitants. The living is a rectory, annexed to that of Fornham-All-Saints, and valued in the king's books at £9. 15. 5. The tithes have been commuted for a rent-charge of £302. 10., subject to the payment of rates, which on the average have amounted to £39; the glebe comprises 31½ acres, valued at £38 per annum. There is a rent-charge of £17. 10. payable to the rector of Horringer, subject to the payment of rates, which on the average have amounted to £2. 9. 8. The old church, of which the tower fell down in 1774, having become completely dilapidated, a new and remarkably handsome church was erected in 1837, by the liberal contributions of the Marquess of Bristol, the patrons, and the parishioners; it contains 55 free sittings, the Incorporated Society having granted £100 in aid of the expense.

WESTLEY-WATERLESS (*St. Mary*), a parish, in the union of NEWMARKET, hundred of RADFIELD, county of CAMBRIDGE, 5 miles (S. S. W.) from Newmarket; containing 158 inhabitants. The living is a rectory, valued in the king's books at £10. 5.; present net income, £326; patron, R. Chapman, Esq. The church has a circular tower. A school is partly supported by the rector.

WESTMANCOATE, a hamlet, in the parish of BREDON, Middle Division of the hundred of OSWALDSLOW, Pershore and Eastern Divisions of the county of WORCESTER, 4¾ miles (N. E.) from Tewkesbury: the population is returned with the parish. There is a place of worship for Baptists, the minister of which receives £15 per annum, the produce of a bequest by John Haydon in 1781; and for which he gratuitously instructs 15 children of that sect.

WESTMESTON (*St. Martin*), a parish, in the union of CHAILEY, hundred of STREET, rape of LEWES, Eastern Division of the county of SUSSEX, 5¾ miles (N. W. by W.) from Lewes; containing 494 inhabitants. The living is a rectory, valued in the king's books at £22. 4. 2.; patron, G. Courthope, Jun., Esq. The tithes have been commuted for a rent-charge of £585, subject to the payment of rates, which on the average have amounted to £117; the glebe comprises 17 acres, valued at £20 per annum. The church is principally in the early style of English architecture, with a plain Norman arch between the nave and the chancel, decorated with the remains of a painting of the signs of the Zodiac: it contains several ancient monumental slabs, and a rudely-constructed circular stone font. At the east end is an ancient chapel, the burial-place of the Marten family. At East Chiltington, in this parish, is a chapel of ease. A school is partly supported by subscription, and an endowment of £3. 2. 8. per annum. Anthony Shirley, who acquired some celebrity as a traveller and writer, in the time of James I., was born here. A charter for an annual fair on Martinmas-day was granted by Edward II.

WESTMILL (*St. Mary*), a parish, in the union of BUNTINGFORD, hundred of BRAUGHIN, county of HERTFORD, 1½ mile (S. by E.) from Buntingford; containing 418 inhabitants. The living is a rectory, valued in the king's books at £20; present net income, £474; patron, Earl of Hardwicke. Here is a day and Sunday National school, commenced in 1827, and partly supported by an endowment of £20 per annum by Lord Hardwicke, and partly by small annual subscriptions.

WESTMINSTER, county of MIDDLESEX.—See LONDON.

WESTMORLAND, an inland county, bounded on the north and west by Cumberland, on the south-west and south by Lancashire, on the south-east and east by Yorkshire, and on the north-east by the county of Durham. It extends from 54° 11′ 30″ to 54° 42′ 30″ (N. Lat.), and from 2° 20′ to 3° 12′ (W. Lon.); and includes an area of 763 square miles, or 488,320 statute acres. The population amounts to 55,041. The ancient British inhabitants of the territory included within the limits of this county were of two tribes of the *Brigantes*, called the *Voluntii* and the *Sistuntii*, the former occupying the eastern parts of it, the latter the western. Under the Roman dominion it was included in the division called *Maxima Cæsariensis*; and, at the period of the Saxon Heptarchy, it formed part of the extensive and powerful kingdom of Northumbria: from its Saxon conquerors it received the name of *West-moringa-land*, or land of the western moors, since contracted into Westmorland.

This county is partly in the diocese of Chester, and partly in that of Carlisle, in the province of York, but under the act of the 6th and 7th of William IV., c. 77, it is to be wholly included in the latter diocese: the former comprises the barony of Kendal, which is divided between the two deaneries of Kendal and Kirkby-Lonsdale, both of which extend into the adjoining parts of Lancashire; the barony of Westmorland, forming the remaining portion of the county, is in the diocese of Carlisle, and constitutes the deanery of Westmorland: the total number of parishes is 32. Its great civil divisions are the two baronies of Kendal and Westmorland, the former containing the wards of Kendal and Lonsdale, and the latter, which has in later ages been occasionally styled the "barony of Appleby," and is often called the "Bottom of Westmorland," comprising the East and West wards. The county contains the newly enfranchised borough and market-town of Kendal, the small market-town and sea-port of Milnthorpe, and the market-towns of Ambleside, Appleby, Brough, Burton-in-Kendal, Kirkby-Lonsdale, Kirkby-Stephen, and Orton: two knights are returned to parliament for the shire, and one representative for the borough of Kendal: the borough of Appleby was disfranchised by the act of the 2nd of William IV., cap. 45, passed to amend the representation. The polling-places in county elections are Appleby, Kirkby-Stephen, Shap, Ambleside, Kendal, and Kirkby-Lonsdale; and the place of election is Appleby. It is included in the northern circuit: the assizes, and the Easter and Michaelmas quarter sessions, are held at Appleby, and the Epiphany and Midsummer sessions at Kendal.

In the later periods of the Saxon dominion, when the ancient kingdom of Northumbria was divided into six shires, one of these was called Appleby-scyre: this however does not seem to have included the barony of Kendal, which, according to various records, appears, for some ages after the Norman Conquest, to have continued to form part of the hundred of Lonsdale, county palatine of Lancaster. In Domesday-book many places in the barony of Kendal are noticed, while Westmorland, properly so called, is, with Cumberland, Durham, Northumberland, and part of Lancashire, wholly omitted, those counties having been excluded from the survey. Lands in this county were, for centuries after the Norman Conquest, held by services similar to those of the border counties of Cumberland and Northumberland. The barony of Westmorland was granted by the Conqueror to Ranulph de Maschines, from whom, in a few generations, the possessions attached to it descended through the families of Trevers, Engain, and Morville to that of the Veteriponts, from whom it passed to the Cliffords, and from the latter, in the seventeenth century, to the Tuftons, Earls of Thanet, in which family it still remains; the present Earl of Thanet being hereditary sheriff of the county, as owner of that barony.

The county is in general so mountainous, that the soil of a great portion of it must necessarily for ever remain undisturbed by the plough. The mountains are separated by pleasant and fertile valleys, requiring only a greater number of trees and hedge-rows to complete the beauty of their appearance. The most extensive vales are, that of the Eden, reaching from about ten miles south-east of Kirkby-Stephen, north-westward by Appleby, towards Penrith; and that of Kendal, more particularly southward and westward of that town. Loose masses of rock, of various sizes and descriptions, are scattered over all the lower hills and the champaign parts of the county; and on the southern side of Shap, along the road towards Kendal, different streams, and especially Wasdale-beck, force their passage amidst stupendous blocks of rounded granite. Cross-fell, at the north-eastern extremity of the county, which is the highest of the chain of mountains extending along the eastern borders of Westmorland and Cumberland, rises to the height of 2901 feet above the level of the sea. The other greatest elevations, included wholly or partly within its limits, are Helvellyn, 3055 feet high; Bow-fell, 2911 feet high; Rydal-head, about the same height as the last-mentioned; and the High Street, which is about 2730 feet high, and derives its name from an ancient road that runs along its summit, and on which the people of the neighbourhood have annual horse-races and other sports, on July 10th. All these mountains command magnificent and extensive prospects, and from Rydal-head are seen the lakes Windermere, Elter-water, Grassmere, and Rydal-water.

Many beautiful lakes adorn the numerous romantic and sequestered dales, and, together with those of Cumberland, have afforded an abundant theme for description, and have been the subjects of some of the finest efforts of landscape painting. The principal of those in Westmorland are, Ullswater, Windermere, Grassmere, Haws-water, Elter-water, Broad-water, and Rydal-water. *Ullswater* is on the north-western side of the county: the higher part of it is wholly within the limits of this county, while its lower part is divided between it and Cumberland: it is about nine miles long, its breadth varying from a quarter of a mile to two miles: the lower end is called Ousemere: its depth varies from six to thirty-five fathoms. The shores of this lake are extremely irregular, and from its making different bold sweeps, only parts of it are seen at once: the lower extremity is bordered by pleasant enclosures, interspersed with woods and cottages, scattered on the sides of gently rising hills; but, advancing upwards towards Patterdale, the enclosures are of smaller extent, and the hills more lofty and rugged, until their aspect becomes wholly wild and mountainous: in its highest expanse are a few small rocky islands. Place-fell, on the east, projects its barren and rugged base into the lake; and on the west rise several rocky hills, one of which, called Stybarrow Crag, is clothed with oaks and birches: these and the other surrounding hills are furrowed with glens and the channels of torrents, causing remarkable echoes. When the sky is uniformly overcast and the air perfectly calm, this lake, in common with some others, has its surface overspread by a smooth oily appearance, provincially called a *keld,* which term is also applied to the places that are longest in freezing: it contains abundance of fine trout, perch, skellies, and eels, some char, and a species of trout, called grey trout, almost peculiar to it, which frequently attains the weight of 30lb. *Windermere* is ten miles and a half long, and lies on the western border of the county, which it separates, for the greater part of its length, from Lancashire, in which county its lower extremity is wholly included: its breadth is from one to two miles, and its area is computed at 2574 acres, including thirteen islands, occupying a space of about 40 acres, the largest of which is now called Curwen's Island, and contains 27 acres. The Westmorland margin of this lake is bordered by enclosures rising gently from the water's edge, adorned with numerous woody and rocky knolls of various elevations and sizes; the Lancashire shore is higher and more abrupt, and is clothed with wood, though not to the summit; and a simple magnificence is the chief characteristic of the whole surrounding scenery. Its fisheries, which are rented of the crown, are chiefly for common and grey trout, pike, perch, skellies, and eels, and more particularly for char, its most remarkable produce, of which there are two sorts, called, from the difference of their colour, silver and golden char, the former of which is considered the most delicious, and is potted for the London market: great numbers of water fowl resort to this lake, and to a few of the smaller ones. *Grassmere* is a particularly beautiful small lake, situated at the lower end of a valley bearing its name: in the centre of it is one small island, and its head is adorned by the church and village of Grassmere. *Haws-water* is situated in a narrow vale, called Mardale, and is about three miles long, and from a quarter to half a mile broad: near the centre it is nearly divided into two by a low enclosed promontory; and the mountains which environ its head are steep, bold, and craggy, but are skirted at their feet by enclosures. On its northern side is Naddle Forest, a steep mountainous ridge, in the form of a bow, and in the centre of which rises Wallow Crag, a mass of upright rocks: the other portions of its scenery are equally picturesque. The char and trout of this lake are in great esteem; besides these, it produces perch, skellies, and eels. *Elter-water*, at the bottom of Great Langdale, which is rather larger than Grassmere, is inferior to none of the smaller lakes in the variety and beauty of its scenery. *Broad-water*, about a mile above the head of Ullswater, is environed by high and rugged mountains, and is viewed to great advantage from a spot called Hartsop-high-field. *Rydal-water*, on the course of the Rothay, is shallow, and has several picturesque woody islands: it is about a mile in length. The principal of the smaller lakes, most commonly called *tarns*, are Ais-water, about a mile south-west of Hartsop, and about a mile northward of which is Angle-tarn; Grisedale-tarn, at the head of Grisedale; Red-tarn, under the eastern side of Helvellyn, and westward of which lies Kepel-cove-tarn, Red-tarn and Small-water, at the head of Riggindale, the highest branch of Mardale; Skeggles-water, in the mountains between Long Sleddale and Kentmere; Kentmere, in the valley of the Kent; Sunbiggin-tarn, in the parish of Orton; and Whinfell-tarn, in the parish of Kendal. Along the chain of mountains extending from Cross-fell, in a southern direction, to Stainmore near Brough, a distance of about 20 miles, occurs a singular phenomenon, called the Helm Wind, which blows at various times of the year, but most commonly from October to April. A light-coloured cloud covers the summit of the mountain, and extends nearly half-way down, above which the blue sky generally appears, and above that another cloud somewhat darker than the former: the latter is called, by the country people, the Burr, or Bar, from a popular notion that it represses the fury of the storm. During the time the Helm is forming, a noise is heard something like the distant roaring of the sea, and when it assumes the appearance of one continued and unbroken cloud, with a tremulous motion, the phenomenon is said to be completely formed. The wind then rushes down the mountain with incredible fury, extending its influence in a westerly direction for about three miles, beyond which the air is often quite calm and occasionally a wind is even found blowing in an opposite direction. Over the top, or eastern side, of the mountain, the cloud extends no farther downwards than on the western side; and after it has passed through, the air there also is often quite calm. The Helm, therefore, is purely a local wind: it is more or less violent at different times of the year, and is most severely felt about the villages of Dufton, Murton, and Hilton, where it occasionally does considerable damage, tearing up trees by the roots, and unroofing houses, and when it unfortunately occurs in harvest time, destroying the crops within its influence.

The most prevailing soil is a dry gravelly mould. Sand and hazel-mould appear in various places, but chiefly towards the eastern and north-eastern confines. Clay is found on a few farms near the Eden, and bordering on the eastern mountains, and a heavy moist soil in others in the northern parts of the county. Peat moss occurs in small patches in many of the valleys, and abounds on the tops of several of the higher mountains, which, however, are in general covered with a dry soil upon a hard blue rock, provincially termed rag. The soil resting on a limestone bottom is every where esteemed the best. Notwithstanding the numerous enclosures and improvements that have taken place since the commencement of the present century, the cultivated lands hardly amount to one-half of the whole extent of the county. Upon these the oldest system of husbandry is, when the pastures have become very full of moss, to have, first, a crop of oats, then one of barley manured, and lastly, oats again, after which they become grass land as before: the farmer does not usually sow seeds with his last crop of oats, the land of itself producing a tolerable herbage, which is greatly encouraged by the humidity of the climate; but, in a few years, it again becomes of little value, on account of the increase of moss, when it is again brought under a similar course of tillage. This system, however, is nearly exploded, and the turnip and clover husbandry now chiefly prevails, particularly in the Bottom of Westmorland, and in the parishes of Heversham, Burton, and Kirkby-Lonsdale, where considerable quantities of wheat are grown. Hence it is obvious, that the greater part, amounting to about three-fourths, of the enclosed lands, are always under grass, particularly in high situations; and as the farmers, during the summer months, can keep almost any quantity of cattle on the commons, &c., at a very little expense, their chief object is to get as much hay as possible from their enclosed lands against the approach of winter. The artificial irrigation of meadows lying on the borders of streams is practised in many parts, but generally on a small scale. Lime is extensively employed as manure, limestone being found in almost inexhaustible quantities in most parts of the county: rock-marl obtained from Bolton common is also used, as well as peat ashes. Paring and burning is much practised on the moorlands and the rough pastures.

The cattle formerly bred in the county were chiefly of the long-horned breed, and many farmers, particularly about Kendal and the neighbourhood, are still partial to them; but, of late, the Durham, or short-horned breed, has almost superseded the former, being generally considered much superior. There are few counties where, in proportion to their size, more milch cows are kept than in this, and where the produce of the dairy is an object of greater importance: this is chiefly butter, of which great quantities are annually sent to the London market, in firkins containing 56lb. net. Not less than 10,000 Scotch cattle are annually brought to Brough Hill fair, whence great numbers are driven towards the rich pastures of the more southern parts of England, though many are retained and fattened within the limits of Westmorland. The breeds of sheep kept on the mountains and commons are either native or have been intermingled with Scotch sheep: they are horned, and have dark or grey faces, thick pelts, and coarse hairy wool. Silverdale, a small tract in the southern part of the county, in the neighbourhood of Milnthorpe, gives name to a peculiar breed found in the surrounding districts: they are said to be native, and are in every respect superior to the common sort. In Westmorland it is not unusual for the proprietor of the land to be the owner of the sheep upon it, in which case the farmer is little more than a shepherd; and any difference in the value of the flock between the time of his entering upon the farm and that of his quitting it, must be accounted for by either party in money. The hogs, though not large, are considered of a good kind:

farmers, butchers, and others, who kill swine, often dispose of the hams to persons who make a trade of curing them, in which state they are highly prized. They are packed in hogsheads, with straw, or the husks of oats, and sent to London, Lancaster, and Liverpool, in such quantities as to form one of the principal articles of export.

In some parts, considerable tracts are covered with coppices, consisting chiefly of oak, ash, alder, birch, and hazel: these underwoods, particularly in the barony of Kendal, are usually cut every sixteenth year, hardly any trees being left for timber, and their produce converted partly into hoops, which are made in the county, and sent coastwise to Liverpool; and partly into charcoal, which is in demand for the neighbouring ironworks. Timber is chiefly found in the plantations, which are numerous, and, at Whinfield Forest and around Lowther Hall, extensive: the larch is generally the most flourishing tree, though indeed most of the woods spring with a degree of vigour hardly to be expected from the bleak and exposed situations which many of them occupy. The extensive wastes are partly subject to common rights, constituting a great part of the value of many farms, to which they are attached, and partly in severalties and stinted pastures. A few of them consist of extensive commons in low situations, possessing a good soil; but by far the greater number is composed of large mountainous tracts, called by the inhabitants *fells* and moors, which produce little besides a very coarse grass, heath, and fern, provincially called *ling* and *brackens*: the soil of these is generally a poor hazel-mould and peat moss. The higher wastes are principally applied to the pasturage of large flocks of sheep, which, during the winter, are all brought down to the enclosures. By the end of April, they are sent back to the wastes. Numerous herds of black cattle are likewise seen on the lower commons; a few are of the breed of the county; the rest are Scotch.

The mineral productions are various, and some of them valuable: they consist chiefly of lead, coal, marble, slate (the finest in England), limestone, freestone, and gypsum; every part of the county presents an interesting field of study to the geologist. The principal lead mines are those at Dunfell, which are considered to be nearly exhausted; at Dufton, where they are unusually rich; at Eagle Crags, in Grisdale, a branch of the vale of Patterdale; and at Greenside, near Patterdale. A small quantity of this metal is also annually procured in the hills above Staveley, and large loose masses of ore have been found in different other situations: a very rich and productive vein at Hartley ceased to be worked about the commencement of the last century. Copper has been wrought to a limited extent at Limbrig, Asby, and Rayne, and is found in small quantities in many other parts. Coal is neither abundant nor of good quality: it is wrought only in the south-eastern extremity of the county, chiefly on Stainmore heath, and in the neighbourhood of Shap: in the vale of Mallerstang a kind of small coal, chiefly used for burning limestone, is procured. Bordering upon the river Kent, about three miles below Kendal, a bed of beautiful white marble, veined with red and other tints, was discovered in 1793, and quarries were immediately opened. Near Ambleside, and between that town and Penrith, is found a marble of a dusky green colour, veined with white; a black sort is also found near Kirkby-Lonsdale, and another species at Kendal Fell. The western mountains produce vast quantities of slate, all the various kinds of which are used in the surrounding districts for covering the roofs of buildings, while the best of them are conveyed by sea to Liverpool, London, Lynn, Hull, &c., and by land into Cumberland, Northumberland, Durham and Lancashire. The most general colour is blue, of many different shades, sometimes having a greenish cast; one kind is purple; and another, used to make writing slates, is nearly black: the best kinds are obtained at the greatest depth. The prevailing strata in the southern and eastern parts of the county are limestone and freestone, together with a soft laminous schistus, horizontally stratified. The western and north-western mountains, besides the slate before mentioned, consist of masses of the trap genera, chiefly basalt, commonly called whinstone. Around the head of Windermere, and for some distance eastward of it, lies a stratum of dark grey limestone, which is occasionally burned into lime, or polished for tomb-stones and chimney-pieces. Wasdale Crag is a mass of coarse flesh-coloured granite; higher up the dale, a greenish-coloured granite, of a finer and harder texture, is found: a very coarse species of granite also appears in many other parts of the county. A vein of red porphyry crosses the road between Kendal and Shap; and at Acorn-bank, near Kirkby, there is one of gypsum, which is used for laying floors. In many parts are also detached round pieces of blue ragstone, of granite, and of a very hard composite stone, called by the masons *callierde*. In Knipe Scar are found talky fibrous bodies, opaque and of an ash colour, which burn for a considerable time without any sensible diminution. Fossil remains exist only in the strata of the southern and eastern parts of the county: coralloid bodies are very common, some of them being beautifully variegated.

The manufactures are but of minor importance, and consist chiefly of coarse woollen cloths, called Kendal *cottons* (supposed to be corrupted from *coatings*), linseys, knit-stockings, waistcoat-pieces, flannels, and leather. Nor is the commerce extensive: the principal exports are, the coarse cloths manufactured at Kendal, stockings, slates, tanned-hides, gunpowder, hoops, charcoal, hams, bacon, wool, sheep, and cattle: the imports are, grain, and Scotch cattle and sheep. Much fish from the river and lakes is annually sent to Lancaster and Liverpool, and some even to London.

The principal rivers are the Eden, Eamont, Lowther, Lune, and Kent. This county derives considerable benefit from the Lancaster canal, which, commencing at Kendal, proceeds for some distance parallel with the course of the Kent, and afterwards across that of the Betha, to the vicinity of Burton, where it enters Lancashire, in the southern part of which county it communicates with the Leeds and Liverpool canal, &c.

A singular collection of huge stones, called Penhur-rock, now nearly destroyed, and a Druidical circle of stones near Oddendale, both in the parish of Crosby-Ravensworth, are supposed to have been British; as also are the rude circle of stones at the head of the stream called the Ellerbeck; that on the waste of Moorduvock, called the Druid's Cross; that of Mayborough, on a gentle eminence on the western side of Eamont bridge; and that about a mile north-eastward of Shap, called the Druids' Temple. Various other relics of this people have been discovered, including several cairns and encampments. Westmorland was traversed by a variety of Roman roads of minor importance, and contained the stations of *Verteræ*, which has been fixed at Brough; *Brovacum*, at Brougham castle; *Galacum*, at the head of Windermere; and another at Natland, the name of which is uncertain. A branch of the great Roman road, called the Watling-street, passed through it from Stainmore to Brougham castle, and several parts of it between Brough and Kirkby-Thore are still tolerably perfect. From this the Maiden-way branched off at Kirkby-Thore, and passed over the lower extremity of Cross-fell, by Whitley castle, into Northumberland: this road may still be clearly traced, being uniformly about seven yards broad, and formed of large loose stones. Other vestiges of Roman occupancy are also very numerous, including altars, urns, coins, bricks, tesselated pavements, foundations of buildings, &c., which have been found on the sites of the stations, and in a few other places. There are also, a Roman camp, about 100 yards southward of Borrowbridge, in Borrowdale, now called Castlehows; others, called Castlesteads and Coney-beds, near the station at Natland; and several between Crackenthorpe and Cross-fell; besides Maiden Castle, upon Stainmore, a very strong square fort, about five miles from Brough; and several other remarkable intrenchments. Near Shap is a stupendous monument of antiquity, called Carl-lofts, supposed to be Danish, consisting of two long lines of huge obelisks of unhewn granite, with different other masses of the same material, arranged in various forms. The religious houses were only the Premonstratensian abbey of Shap, and a monastery of White friars at Appleby, together with an hospital for lepers near Kirkby-in-Kendal: there are some remains of the abbey of Shap. Remains of more modern fortifications are numerous and extensive, comprising the ruins of the castles of Appleby, Beetham, Brough, Brougham, Bewley, Howgill, Kendal, and Pendragon; Arnside tower, Helsback tower, and several other ancient castellated buildings. Of ancient mansions, the most remarkable specimens are Sizergh Hall and Levens Hall, together with the ruins of Old Calgarth Hall and Preston Hall. Of the more modern seats of the nobility and gentry, those most worthy of notice are Lowther Castle, the residence of the Earl of Lonsdale, lord-lieutenant of the county; and Appleby Castle, that of the Earl of Thanet, hereditary high sheriff. The small freeholds are very numerous. The enclosed fields are generally very small, and are fenced partly by hedges, and partly by stone walls. The inhabitants, owing to their secluded situation, have, until recently, been distinguished for their adherence to several antiquated customs: *haver-bread*, from the oat sometimes called *haver*, made into unleavened cakes, is still in common use; shoes with wooden soles bound with small iron plates, called clogs, are worn by the common people of both sexes, especially in the winter season. There are mineral springs of various qualities in several places; the principal being that near the village of Clifton, at which a great number of people annually assemble, on the first day of May, to drink its waters; that called Gonsdike, a little to the south of Rounthwaite, which continually casts up small metallic spangles; Shap wells, much resorted to in the summer season by persons afflicted with scorbutic complaints, and by lead-miners from Alston and Arkingartdale; the numerous petrifying springs on the borders of the river Kent; and a petrifying well in the cave called Pate-hole. The most remarkable cascades on the numerous mountain streams are, Leven's Park waterfall, on the Kent; another on the Betha, below Betham—the *Caladupæ* of Camden; and Gillforth spout, in Long Sleddale, which has an unbroken fall of 100 feet. Pate-hole, before mentioned, is a very curious and extensive cavern in a limestone rock near Great Asby, from which, in rainy seasons, powerful streams of water issue. Westmorland gives the title of Earl to the family of Fane; and Baron Vipont of Westmorland is one of the titles borne by the noble family of Clifford.

WESTOE, a chapelry, in the parish of JARROW, union of SOUTH SHIELDS, Eastern Division of CHESTER ward, Northern Division of the county palatine of DURHAM; containing 9682 inhabitants. This is a populous suburb to South Shields, the market-place and many of the principal streets of which are comprised within the chapelry. The living is a perpetual curacy; net income, £220; patrons, Incumbent of St. Hilda and certain Trustees; appropriators, Dean and Chapter of Durham. There is a school attached to the chapel, which is now conducted on the National system, and supported by subscription, to which the Dean and Chapter liberally contribute.

WESTON, a township, in the parish and union of RUNCORN, hundred of BUCKLOW, Northern Division of the county of CHESTER, $3\frac{1}{4}$ miles (N. N. W.) from Frodsham; containing 532 inhabitants. The Weston canal passes in the vicinity, parallel with the river Mersey.

WESTON, a township, in the parish of WYBUNBURY, union and hundred of NANTWICH, Southern Division of the county of CHESTER, 6 miles (E.) from Nantwich; containing 401 inhabitants. A school is partly supported by an allowance of £10 per annum from Sir John Delves Broughton's charity, for which 20 girls are instructed.

WESTON, a tything, in the parish of STALBRIDGE, hundred of BROWNSHALL, Sturminster Division of the county of DORSET; containing 225 inhabitants.

WESTON (*HOLY TRINITY*), a parish, in the union of HITCHIN, hundred of BROADWATER, county of HERTFORD, $4\frac{1}{2}$ miles (N. E. by N.) from Stevenage; containing 1046 inhabitants. The living is a discharged vicarage, valued in the king's books at £10. 6. 8.; present net income, £197; patron and impropriator, William Hale, Esq. The church has been repewed and repaired, and 145 free sittings provided, the Incorporated Society having granted £80 towards defraying the expense: it is partly Norman, and partly of later date. There is a place of worship for Wesleyan Methodists; also a National school.

WESTON (*ST. MARY*), a parish, in the union of SPALDING, wapentake of ELLOE, parts of HOLLAND, county of LINCOLN, $3\frac{1}{4}$ miles (N. E. by E.) from Spalding; containing 567 inhabitants. The living is a vicarage not in charge, in the patronage of the Crown; impropriator, Sir J. Trollope, Bart. The impropriate tithes have been commuted for a rent-charge of £970, and the vicarial for £163. 10., subject to the payment of rates; the glebe comprises nearly two acres, valued at £4. 10. per annum.

WESTON, county of NORFOLK.—See WESTON-LONGVILLE.

WESTON, a hamlet, in the parish of LOYS-WEEDON, hundred of GREEN'S-NORTON, Southern Division of the county of NORTHAMPTON, 7 miles (W. by S.) from Towcester: the population is returned with the parish. There is a place of worship for Baptists. A chalybeate spring in the neighbourhood was formerly much esteemed, but has fallen into disrepute.

WESTON (*ALL SAINTS*), a parish, in the union of SOUTHWELL, Northern Division of the wapentake of THURGARTON, Southern Division of the county of NOTTINGHAM, 3 miles (S. E.) from Tuxford; containing 395 inhabitants. The living is a rectory, valued in the king's books at £19. 2. 11.; present net income, £168; patron, Earl Manvers. The church exhibits specimens of various styles of architecture. Richard Hawksworth, in 1736, bequeathed £50 for erecting a school-house, and £100 towards its endowment, for the education of ten poor children.

WESTON, a joint chapelry with Nash and Tilsop, in the parish of BURFORD, union of TENBURY, hundred of OVERS, Southern Division of the county of SALOP, 6 miles (E. S. E.) from Ludlow: the population is returned with Nash.

WESTON (*ALL SAINTS*), a parish, in the union of BATH, hundred of BATH-FORUM, Eastern Division of the county of SOMERSET, 1¾ mile (N. W. by W.) from Bath; containing 2560 inhabitants. The parish is bounded on the south by the river Avon, a stream tributary to which has its source in Lansdown Hill, flows through the village, and is crossed by a stone bridge of one arch on the high road from Bath to Bristol. The living is a vicarage, endowed with the rectorial tithes, and valued in the king's books at £10. 1. 8.; present net income, £468: it is in the patronage of the Crown. The church has been enlarged, and 220 free sittings provided, the Incorporated Society having granted £300 towards defraying the expense. Another church was erected in 1836, which contains 300 free sittings, in consequence of a grant of £200 from the Incorporated Society. There is a place of worship for Wesleyan Methodists. A National school has been established.

WESTON, a hamlet, in the parish of WANSTROW, hundred of FROME, Eastern Division of the county of SOMERSET, 5¼ miles (S. W.) from Frome; the population is returned with the parish.

WESTON, a tything, in the parish of BURITON, hundred of FINCH-DEAN, Petersfield and Southern Divisions of the county of SOUTHAMPTON, 1¼ mile (S. S. W.) from Petersfield: the population is returned with the parish. John Goodyer, in 1664, bequeathed premises now let for £79 a year, which is applied to the education of children and the relief of the poor.

WESTON (*ST. PETER*), a parish, in the union and hundred of WANGFORD, Eastern Division of the county of SUFFOLK, 2¾ miles (S.) from Beccles; containing 233 inhabitants. The living is a discharged rectory, valued in the king's books at £13. 6. 8.; present net income, £260: it is in the patronage of the Crown. Weston Hall, the ancient seat of the family of Rede, was a handsome mansion in the Elizabethan style; part of it has been within a few years taken down, and the remainder converted into a farm-house.

WESTON, a joint hamlet with Ember, in the parish of THAMES-DITTON, Second Division of the hundred of ELMBRIDGE, Western Division of the county of SURREY: the population is returned with Ember.

WESTON, a hamlet, in the parish of BULKINGTON, Kirby Division of the hundred of KNIGHTLOW, Northern Division of the county of WARWICK, 3½ miles (S. S. E.) from Nuneaton: the population is returned with the parish.

WESTON (*ALL SAINTS*), a parish, in the Upper Division of the wapentake of CLARO, West Riding of the county of YORK; containing 521 inhabitants, of which number, 121 are in the township of Weston, 2 miles (N. W. by W.) from Otley. The living is a discharged vicarage, valued in the king's books at £6. 11. 5½.; present net income, £51; patron and impropriator, Representative of the late W. Vavasour, Esq. Here is a Sunday National school.

WESTON, ALCONBURY, county of HUNTINGDON.—See ALCONBURY WESTON.

WESTON-BAMFYLD (*HOLY CROSS*), a parish, in the union of WINCANTON, hundred of CATSASH, Eastern Division of the county of SOMERSET, 5¾ miles (S. S. W.) from Castle-Cary; containing 123 inhabitants. The living is a rectory, valued in the king's books at £8. 15. 10.; patron, Rev. J. Goldesbrough. The tithes have been commuted for a rent-charge of £169. 14., subject to the payment of rates, which on the average have amounted to £12. 11.; the glebe comprises 22 acres, valued at £39 per annum.

WESTON-BEGGARD (*ALL SAINTS*), a parish, in the hundred of RADLOW, union and county of HEREFORD, 5 miles (E.) from Hereford; containing 281 inhabitants. The living is a discharged vicarage, valued in the king's books at £5. 15. 3.; present net income, £135; patrons, Dean and Chapter of Hereford; impropriator, Warden of St. Catherine's Hospital, Ledbury. Here is a National school.

WESTON-BIRT (*ST. CATHERINE*), a parish, in the union of TETBURY, hundred of LONGTREE, Eastern Division of the county of GLOUCESTER, 3¾ miles (S. W. by S.) from Tetbury; containing 138 inhabitants. The living is a discharged rectory, valued in the king's books at £6. 2.; present net income, £226; patron, R. Holford, Esq. A school is partly supported by charity.

WESTON-BY-WELLAND (*ST. MARY*), a parish, in the union of MARKET-HARBOROUGH, hundred of CORBY, Northern Division of the county of NORTHAMPTON, 4¼ miles (N. E.) from Market-Harborough; containing 208 inhabitants. The living is a vicarage, with that of Sutton-Bassett united, valued in the king's books at £11. 17. 1.; present net income, £260; patron, Lord Sondes. Here is a National school.

WESTON, COLD (*ST. MARY*), a parish, in the union of LUDLOW, hundred of MUNSLOW, Southern Division of the county of SALOP, 6¾ miles (N. E. by N.) from Ludlow; containing 25 inhabitants. The living is a discharged rectory, valued in the king's books at £2. 8. 4.; present net income, £100; patron, — Cornwall, Esq.

WESTON-COLLEY, a tything, in the parish and hundred of MITCHELDEVER, Winchester and Northern Divisions of the county of SOUTHAMPTON, 8 miles (N. by E.) from Winchester: the population is returned with the parish.

WESTON-COLVILLE (*ST. MARY*), a parish, in the

union of LINTON, hundred of RADFIELD, county of CAMBRIDGE, 6 miles (N. E. by N.) from Linton; containing 444 inhabitants. The living is a rectory, valued in the king's books at £21. 13. 6½.; present net income, £200; patron, John Hall, Esq. A day and Sunday school is supported by subscription.

WESTON-CONEY (*St. Mary*), a parish, in the union of THETFORD, hundred of BLACKBOURN, Western Division of the county of SUFFOLK, 6¼ miles (N. N. E.) from Ixworth; containing 255 inhabitants. The living is a discharged rectory, annexed to that of Barningham, and valued in the king's books at £13. 0. 5. A school is partly supported by the parish.

WESTON-CORBETT, an extra-parochial liberty, in the hundred of BERMONDSPIT, Basingstoke and Northern Divisions of the county of SOUTHAMPTON, 4 miles (S. E.) from Basingstoke; containing 17 inhabitants.

WESTON-COYNEY, a joint township with Hulme, in the parish of CAVERSWALL, Northern Division of the hundred of TOTMONSLOW and of the county of STAFFORD, 5 miles (W.) from Cheadle; containing 619 inhabitants.

WESTON, EDITH, county of RUTLAND. — See EDITH-WESTON.

WESTON-FAVELL (*St. Peter*), a parish, in the union and hundred of SPELHOE, Southern Division of the county of NORTHAMPTON, 2½ miles (E. N. E.) from Northampton; containing 443 inhabitants. This parish is situated on the turnpike road from Northampton to Peterborough. The living is a rectory, valued in the king's books at £16. 16. 3.; present net income, £236; patron and incumbent, Rev. R. H. Knight. There is a place of worship for Baptists. A free school for 15 boys and 9 girls was founded and endowed with £22. 8. 10. per annum, by Harvey Ekins, Esq., and Elizabeth his wife, who also, at the desire of their daughter, endowed a charity for apprenticing one poor boy yearly, which is called Gertrude Ekins' charity. Thomas Green, in 1739, gave certain lands for the further endowment of the school. The Rev. James Harvey, M. A., author of the "Meditations," was rector of this parish for many years; he rebuilt the rectory-house on an enlarged scale, but did not inhabit it, having died on the 25th of Dec., 1758, in the forty-fifth year of his age, before it was completed; he was buried in the church.

WESTON-IN-GORDANO (*St. Paul*), a parish, in the union of BEDMINSTER, hundred of PORTBURY, Eastern Division of the county of SOMERSET, 10 miles (W. by N.) from Bristol; containing 124 inhabitants. The living is a discharged rectory, valued in the king's books at £6. 3.; patron, P. John Mills, Esq. The tithes have been commuted for a rent-charge of £112, subject to the payment of rates, which on the average have amounted to £7. 15.; the glebe comprises 22 acres, valued at £33 per annum. A National school has been established.

WESTON-JONES, a township, in the parish of NORBURY, union of NEWPORT, Western Division of the hundred of CUTTLESTONE, Southern Division of the county of STAFFORD, 3¼ miles (N. N. E.) from Newport; containing 113 inhabitants.

WESTON, KING, county of SOMERSET. — See KINGWESTON.

WESTON, KING'S, a tything, in the parish and Lower Division of the hundred of HENBURY, Western Division of the county of GLOUCESTER, 4½ miles (N. W.) from Bristol; containing 107 inhabitants.

WESTON, LAWRENCE, a tything, in the parish of HENBURY, Lower Division of the hundred of BERKELEY, Western Division of the county of GLOUCESTER, 5¼ miles (N. W. by N.) from Bristol; containing 329 inhabitants.

WESTON-LONGVILLE (*All Saints*), a parish, in the union of ST. FAITH, hundred of EYNSFORD, Eastern Division of the county of NORFOLK, 5¼ miles (S.) from Reepham; containing 406 inhabitants. The living is a discharged rectory, valued in the king's books at £8. 18. 1½.; present net income, £583; patrons, Warden and Fellows of New College, Oxford. A school is partly supported by Mrs. H. T. Custance.

WESTON, MARKET (*St. Mary*), a parish, in the union of THETFORD, hundred of BLACKBOURN, Western Division of the county of SUFFOLK, 7 miles (S.) from East Harling; containing 312 inhabitants. The living is a discharged rectory, valued in the king's books at £8. 19. 7.; present net income, £242; patron, Rev. H. G. Wilkinson. The church was repaired and improved in 1838. The seat of the family of Thruston is in the parish.

WESTON, OLD (*St. Swithin*), a parish, in the union of THRAPSTONE, hundred of LEIGHTONSTONE, county of HUNTINGDON, 7¼ miles (N.) from Kimbolton; containing 356 inhabitants. The living is united, with that of Bythorn, to the rectory of Brington. There is a school in which six children are paid for by the rector.

WESTON-ON-THE-GREEN (*St. Mary*), a parish, in the union of BICESTER, hundred of PLOUGHLEY, county of OXFORD, 4½ miles (S. W. by W.) from Bicester; containing 494 inhabitants. The living is a discharged vicarage; net income, £148; patron and impropriator, Hon. — Peregrine Bertie. The church is a Grecian structure, erected in 1743, at the sole expense of Norreys Bertie, Esq., on the site of the ancient edifice, which had fallen into decay. Near it is the ancient manor-house, in which are several portraits of the families of Norreys and Bertie. Here are quarries of stone of good quality for building. Numerous Roman coins of the Lower Empire have been found.

WESTON-PATRICK (*St. Lawrence*), a parish, in the union of BASINGSTOKE, hundred of ODIHAM, Basingstoke and Northern Divisions of the county of SOUTHAMPTON, 4¼ miles (S. W. by W.) from Odiham; containing 210 inhabitants. The living is a perpetual curacy; net income, £48; patron and appropriator, Chancellor of Salisbury Cathedral.

WESTON-PEVEREL, or PENNY-CROSS, a chapelry, in the parish of ST. ANDREW, PLYMOUTH, union of PLYMPTON-ST.-MARY, hundred of ROBOROUGH, Roborough and Southern Divisions of the county of DEVON, 2¾ miles (N. by W.) from Plymouth; containing 274 inhabitants. The chapel is dedicated to St. Pancras.

WESTON-RHYN, a joint township with Bron-y-garth, in the parish of ST. MARTIN, hundred of OSWESTRY, Northern Division of the county of SALOP: the population is returned with Bron-y-garth.

WESTON, SOUTH (*St. Lawrence*), a parish, in the union of THAME, hundred of PIRTON, county of

Oxford, $2\frac{3}{4}$ miles (S. by E.) from Tetsworth; containing 118 inhabitants. The living is a rectory, valued in the king's books at £9. 2. 6.; present net income, £200; patrons, Provost and Fellows of Queen's College, Oxford. There is a place of worship for Wesleyan Methodists.

WESTON-SUB-EDGE (*St. Lawrence*), a parish, in the union of Evesham, Upper Division of the hundred of Kiftsgate, Eastern Division of the county of Gloucester, $1\frac{3}{4}$ mile (W. N. W.) from Chipping Campden; containing 367 inhabitants. The living is a rectory, valued in the king's books at £31; present net income, £811; patron and incumbent, Rev. H. Smith. A school is supported by endowment.

WESTON-SUPER-MARE (*St. John*), a parish, in the union of Axbridge, hundred of Winterstoke, Eastern Division of the county of Somerset, $9\frac{3}{4}$ miles (N. W. by W.) from Axbridge; containing 1310 inhabitants. This parish, which is situated on the margin of Uphill Bay, near the Bristol Channel, has within the last few years more than doubled its population, from the construction of a bathing establishment at Knightstone, since which time it has become a fashionable and well-frequented watering-place. The bathing-house contains commodious apartments for the residence of invalids; and contiguous to the baths are furnished lodging-houses for the reception of families, and also several good inns. The establishment comprises an open swimming bath, cold plunging, and warm, and hot and cold shower baths, in separate wings of the building, appropriated to each sex; and to obviate the inconvenience of unpleasant odours, detached rooms are fitted up for the application of the dry sulphureous vapour bath, the sudatorium, the Ioduretted warm bath, and for simple and medicated humid vapour baths. The requisite apparatus is also provided for the administration of the Douche, and the superintendent has been instructed in the process of shampooing. The establishment, which comprises also a public reading-room, may be heated to any required temperature by a steam apparatus detached from the building. A few of the inhabitants of the parish are engaged in the sprat and herring fishery, which is carried on off this coast; and a convenient market-house has been recently erected at the expense of Richard Parsley, Esq. The line of the Bristol and Exeter railway passes near the parish. The living is a discharged rectory, valued in the king's books at £14. 17. 11.; present net income, £264; patron, Bishop of Bath and Wells. The church is a neat modern edifice. There are two places of worship for dissenters; and a day and Sunday school is supported by subscription. At Worteberry, above the village, is a rampart of stones, 20 feet high, with ditches; and a well in the parish possesses the unusual property of being empty at high water, and full when the tide is at its ebb.

WESTON-TURVILLE (*St. Mary*), a parish, in the union and hundred of Aylesbury, county of Buckingham, $2\frac{1}{4}$ miles (N. by W.) from Wendover; containing 637 inhabitants. The living is a rectory, valued in the king's books at £22. 0. 10.; present net income, £484; patrons, Warden and Fellows of All Souls' College, Oxford. A Sunday National school is supported by subscription.

WESTON-UNDER-LIZARD (*St. Andrew*), a parish, in the union of Shiffnall, Western Division of the hundred of Cuttlestone, Southern Division of the county of Stafford, $5\frac{3}{4}$ miles (N. E. by E.) from Shiffnall; containing 257 inhabitants. This parish takes the adjunct to its name from Lizard, a hill in Shropshire, to distinguish it from that of Weston-upon-Trent. The living is a discharged rectory, valued in the king's books at £6. 7. $8\frac{1}{2}$.; present net income, £503; patron, Earl of Bradford.

WESTON-UNDER-PENYARD (*St. Lawrence*), a parish, in the union of Ross, hundred of Greytree, county of Hereford, $2\frac{1}{4}$ miles (E. S. E.) from Ross; containing 639 inhabitants. The living is a rectory, valued in the king's books at £18; patron, Bishop of Hereford. The tithes have been commuted for a rent-charge of £619. 10., subject to the payment of rates, which on the average have amounted to £57. 14. 5.; the glebe comprises 2 acres, valued at £4 per annum. Here is a National school.

WESTON-UNDER-RED-CASTLE, a chapelry, in the parish of Hodnet, union of Wem, Drayton Division of the hundred of North Bradford, Northern Division of the county of Salop, 4 miles (E.) from Wem; containing, with Wixhill, 328 inhabitants. A day and Sunday school is supported by Sir Rowland Hill, Bart.

WESTON-UNDER-WEATHERLY (*St. Michael*), a parish, in the union of Warwick, Southam Division of the hundred of Knightlow, Southern Division of the county of Warwick, 6 miles (N. E. by E.) from Warwick; containing 208 inhabitants. The living is a discharged vicarage, valued in the king's books at £5. 0. 2.; present net income, £90; patron and impropriator, Lord Clifford. A day and Sunday school is supported by subscription.

WESTON-UNDERWOOD (*St. Lawrence*), a parish, in the union of Newport-Pagnell, hundred of Newport, county of Buckingham, $1\frac{3}{4}$ mile (W. S. W.) from Olney; containing 441 inhabitants. The parish is bounded on the south by the river Ouse, and contains the ancient seat, now uninhabited, of the Throckmorton family, who have also a neat Roman Catholic chapel here, with a handsome residence for the priest. The living is a perpetual curacy; net income, £51; patron and impropriator, Robert Throckmorton, Esq., who endowed a school in 1826 for educating 24 children. Charles Higgins, in 1792, bequeathed £500, the dividends of which, now amounting to £20. 8. 4. per annum, are expended in the purchase of clothing for 10 poor old women; and an annual sum of about £55, arising from several small bequests, is appropriated to purposes of general relief among the poor. In this pleasant village Cowper resided for several years during the latter part of his life; and the neighbourhood is supposed to have furnished many of his descriptions of rural scenery.

WESTON-UNDERWOOD, a township, in the parish of Stanton-by-Dale, union of Belper, hundred of Morleston and Litchurch, Southern Division of the county of Derby, $5\frac{3}{4}$ miles (N. W. by N.) from Derby; containing 272 inhabitants.

WESTON-UPON-AVON (*All Saints*), a parish, in the union of Stratford-on-Avon, partly in the Alcester Division of the hundred of Barlichway, Southern Division of the county of Warwick, but chiefly in the Upper Division of the hundred of Kiftsgate, Eastern Division of the county of Gloucester, $4\frac{1}{2}$

miles (S. W. by W.) from Stratford-on-Avon; containing 108 inhabitants. The living is a discharged vicarage, valued in the king's books at £7. 14. 7.; present net income, £84; the patronage and impropriation belong to Countess Amherst. A school is endowed with £15 per annum.

WESTON-UPON-TRENT (*St. Mary*), a parish, in the union of SHARDLOW, hundred of MORLESTON and LITCHURCH, Southern Division of the county of DERBY, 7 miles (S. E. by S.) from Derby; containing 387 inhabitants. The living is a rectory, valued in the king's books at £11. 16. 3.; present net income, £594; patron, Sir Robert Wilmot, Bart. The Trent and Mersey canal passes through the parish.

WESTON-UPON-TRENT, a parish, in the Southern Division of the hundred of PIREHILL, union and Northern Division of the county of STAFFORD, 4½ miles (N. E.) from Stafford; containing 608 inhabitants. The Grand Trunk canal passes through the parish. Extensive salt-works have been established at the village: the brine is raised in the parish of Ingestre, by means of machinery worked by the waters of the Trent, and is conveyed across that river and under the canal, through pipes, to certain reservoirs, whence it runs into iron pans, is heated, and becomes chrystallized for use. The living is a vicarage; net income, £96; patrons, alternately, J. N. Lane and W. Inge, Esqrs.; impropriator, William Moore, Esq. The church is an ancient structure, with a large tower and spire: it was partly rebuilt in 1685, when the north aisle was taken down, and was not restored till 1825, when the chancel was also rebuilt. Two of the windows are decorated with stained glass, and in 1829 the spire was rebuilt at the expense of the parishioners.

WESTON-ZOYLAND (*St. Mary*), a parish, in the union of BRIDG-WATER, hundred of WHITLEY, Western Division of the county of SOMERSET, 4 miles (E. S. E.) from Bridg-water; containing 937 inhabitants. This parish is bounded on the south by the navigable river Parrett. The living is a vicarage, valued in the king's books at £14. 6. 8.; present net income, £284; patron and appropriator, Bishop of Bath and Wells. The church is a cruciform structure, with a stately western tower, highly enriched and crowned with pinnacles.

WESTONING (*St. Mary Magdalene*), a parish, in the union of AMPTHILL, hundred of MANSHEAD, county of BEDFORD, 4 miles (S. by W.) from Ampthill; containing 627 inhabitants. The living is a discharged vicarage, valued in the king's books at £9. 17.; present net income, £195; patrons, Executors of the late J. Everitt, Esq.; the impropriation belongs to Mrs. Penyston. There is a place of worship for Baptists.

WESTOVER, a tything, in the parish and hundred of WHERWELL, Andover and Northern Divisions of the county of SOUTHAMPTON, 2½ miles (S. by W.) from Andover: the population is returned with the parish.

WESTOW, county of SUFFOLK. — See STOW, WET.

WESTOW (*St. Mary*), a parish, in the union of MALTON, wapentake of BUCKROSE, East Riding of the county of YORK; containing 606 inhabitants, of which number, 389 are in the township of Westow, 5½ miles (S. S. W.) from New Malton. The living is a discharged vicarage, valued in the king's books at £3. 18. 4.; present net income, £173; patron and appropriator, Archbishop of York. There is a place of worship for Wesleyan Methodists. Three poor children are educated for an annuity of £2. 10., the gift of Mrs. Elizabeth Sugar.

WEST-FEN, county of LINCOLN.—See FRITHVILLE.

WEST-PARK, a joint tything with Cole, in the parish and hundred of MALMESBURY, Malmesbury and Kingswood, and Northern, Divisions of the county of WILTS: the population is returned with Cole.

WESTPORT (*St. Mary*), a parish, in the union and hundred of MALMESBURY, Northern Division of the county of WILTS, adjacent to the north-west side of Malmesbury; containing 1286 inhabitants. The living is a vicarage, with those of Brokenborough and Charlton annexed, valued in the king's books at £16. 17. 8½.; present net income, £310: it is in the patronage of the Crown; impropriator, Earl of Suffolk.

WESTRILL, an extra-parochial liberty, in the union of RUGBY, hundred of GUTHLAXTON, Southern Division of the county of LEICESTER; containing, with Stormore, 7 inhabitants.

WESTROP, a tything, in the parish of HIGHWORTH, hundred of HIGHWORTH, CRICKLADE, and STAPLE, Swindon and Northern Divisions of the county of WILTS; containing 644 inhabitants.

WEST-VILLE, a township, in the union of BOSTON, Western Division of the soke of BOLINGBROKE, parts of LINDSEY, county of LINCOLN; containing 118 inhabitants. This township, with six others, is not dependent on any parish, and was created such by act of parliament, in 1812, on the occasion of a very extensive drainage of Wildmore, and the East and West Fens: the inhabitants attend the chapel at Carrington, which was consecrated in 1818.

WESTWARD, a parish, in the union of WIGTON, ALLERDALE ward below Derwent, Western Division of the county of CUMBERLAND, 2¾ miles (S. E. by S.) from Wigton; containing 1253 inhabitants. This place derives its name from its situation in the great forest of Inglewood, of which it formed the western ward, under the charge of the forester. The parish is bounded on the east by the Wampool river, and on the south by the branches of the river Waver; it abounds with limestone, red freestone, and slate, all of excellent quality, of which there are extensive quarries, affording employment to many of the labouring class; and several seams of cannel and other coal have been found within its limits. The living is a perpetual curacy; net income, £99; patrons and appropriators, Dean and Chapter of Carlisle. The church is situated on an eminence called Church-hill, in the township of Stoneraise. A chapel has been built here, containing 280 free sittings, the Incorporated Society having granted £70 towards defraying the expense. Seven poor children are gratuitously instructed, for which purpose John Jefferson, in 1744, bequeathed £60, which has been invested in land now produring £6. 18. per annum. There are also some small charitable bequests for distribution among the poor. In the township of Stoneraise, about one mile and a half to the north of the church, and on the Roman road from *Lugovallum* (Carlisle) to *Volantium* or *Virosidum* (Ellenborough), are the remains of the station, Old Carlisle, a considerable Roman city, which Horsley supposes to have been the *Olenacum* of the Notitia, where the *Ala Herculea* and *Ala Augusta* were stationed. Antiquaries, however, differ with respect to the right name of this important station, which with its appendages occupied many acres of ground; its site is still overspread with the ruins and foundations of numerous buildings, fragments of altars, equestrian statues, images, inscriptions, and many other relics. The walls enclosed a quadrilateral area, 170 yards long and 120 yards broad, with obtuse angles, and an entrance on each side, and were surrounded by a double ditch. Near a place called the Heights, in another part of the parish, vestiges of several square and circular intrenchments may be traced, though many of them, since the enclosure of the lands, have been levelled with the surface of the ground. Ilekirk Hall, anciently called Hildkirk, from a hermitage, dedicated to St. Hilda, which was granted by John, in the 16th of his reign, to the abbey of Holme-Cultram, is in the township of Stoneraise, and now a farm-house; it was for some time the residence of the celebrated Richard Barwise, a man of extraordinary stature and prodigious strength.

WESTWATER, a tything, in the parish and hundred of AXMINSTER, Honiton and Southern Divisions of the county of DEVON, 2 miles (N. W.) from Axminster: the population is returned with the parish.

WESTWELL (*ST. MARY*), a parish, in the union of WEST ASHFORD, hundred of CALEHILL, Upper Division of the lathe of SCRAY, Eastern Division of the county of KENT, 2½ miles (E. S. E.) from Charing; containing 861 inhabitants. The living is a vicarage, in the patronage of the Archbishop of Canterbury (the appropriator), valued in the king's books at £13; present net income, £235. The Earl of Thanet has recently built two school-rooms, one for boys and the other for girls; the schools are on the National plan, and are principally supported by his lordship.

WESTWELL (*ST. MARY*), a parish, in the union of WITNEY, hundred of BAMPTON, county of OXFORD, 2 miles (W. S. W.) from Burford; containing 162 inhabitants. The living is a rectory, valued in the king's books at £5. 3. 9.; present net income, £159; patrons, Dean and Canons of Christ-Church, Oxford. The church is situated on an eminence, and is in the Norman style of architecture; on the north and south sides are circular-arched doorways with zigzag ornaments; in the nave is a mural monument to Charles Trindor, and on the south side is a recumbent effigy of an ecclesiastic of the time of Elizabeth.

WESTWICK, a hamlet, in that part of the parish of OAKINGTON which is in the hundred of CHESTERTON, county of CAMBRIDGE, 5¼ miles (N. N. W.) from Cambridge; containing 47 inhabitants.

WESTWICK, a township, in the parish of GAINFORD, union of TEESDALE, South-Western Division of DARLINGTON ward, Southern Division of the county palatine of DURHAM, 2 miles (S. E.) from Barnard-Castle; containing 98 inhabitants. It is bounded on the south by the river Tees, over which is a bridge connecting it with the parish of Rokeby.

WESTWICK (*ST. BOTOLPH*), a parish, in the hundred of TUNSTEAD, Eastern Division of the county of NORFOLK, 2¾ miles (S.) from North Walsham; containing 210 inhabitants. The living is a discharged rectory, valued in the king's books at £9. 13. 9.; patron, J. Petre, Esq. The tithes have been commuted for a rent-charge of £171. 11. 5., subject to the payment of rates, which on the average have amounted to £29; the glebe comprises 16 acres, valued at £15 per annum. At a short distance from Westwick House is an obelisk, 90 feet high, with a neat room at the top, commanding beautiful and extensive views of the sea-coast on one hand, and of a richly diversified country on the other.

WESTWICK, a township, in the parish and liberty of RIPON, West Riding of the county of YORK, 3½ miles (W. by S.) from Boroughbridge; containing 30 inhabitants.

WESTWOOD, a township, in the parish of THORNBURY, hundred of WOLPHY, county of HEREFORD, 4½ miles (N. W. by N.) from Bromyard: the population is returned with the parish.

WESTWOOD, a parish, in the union of BRADFORD, and forming a detached portion of the hundred of ELSTUB and EVERLEY, locally in that of Bradford, Westbury and Northern Divisions, and Trowbridge and Bradford Subdivisions, of the county of WILTS, 2 miles (S. W.) from Bradford; containing 390 inhabitants. The living is annexed to the vicarage of Great Bradford.

WESTWOOD PARK, an extra-parochial liberty, in the Upper Division of the hundred of HALFSHIRE, Droitwich and Eastern Divisions of the county of WORCESTER, 2¼ miles (W. N. W.) from Droitwich; containing 10 inhabitants. A priory, dedicated to the Blessed Virgin, for six nuns of the order of Fontevrault, was founded here in the reign of Henry II., and at the dissolution had a revenue of £75. 18. 11.

WETHERAL (*HOLY TRINITY*), a parish, in the union of CARLISLE, chiefly in CUMBERLAND ward, but partly in ESKDALE ward, Eastern Division of the county of CUMBERLAND; containing 2864 inhabitants, of which number, 607 are in the township of Wetheral, 5 miles (E. by S.) from Carlisle. The living is a perpetual curacy, with that of Warwick annexed; net income, £108; patrons and appropriators, Dean and Chapter of Carlisle. The church was built in the reign of Henry VIII.; and a handsome chapel was attached to it, as a burial-place, by Henry Howard, Esq., in 1791. Here are three schools; one is endowed with £20. 16., another with £16, and a third with land producing £4. 5. per annum. The river Eden runs through the parish, in which there are quarries of red freestone and alabaster. A priory of Benedictine monks, dedicated to the Holy Trinity, St. Mary, and St. Constantine, was founded here by Ranulph de Meschines, as a cell to the abbey of St. Mary at York; at the dissolution its revenue was estimated at £128. 5. 3. Of the conventual buildings, the gatehouse still remains, and near the site are three ancient cells, excavated in the rock, at the height of 40 feet above the river Eden, which flows at its base.

WETHERBY (*ST. JAMES*), a market-town and chapelry, in the parish of SPOFFORTH, Upper Division of the wapentake of CLARO, West Riding of the county of YORK, 12½ miles (W. by S.) from York, and 194 (N. N. W.) from London; containing 1321 inhabitants. The Saxon name of this town, whence the present is obviously deduced, was *Wederbi*, intended to designate its situation on a bend of the river Wharf. During the civil war of the seventeenth century, it was garrisoned for the parliament, and successively repulsed two attacks made upon it by Sir Thomas Glemham. About three miles and a half below it is St. Helen's ford, where the Roman military way crossed the Wharf. The town consists chiefly of one long street, behind which is the market-place. Over the river is a handsome stone bridge, and a little above this a weir, formed for the benefit of some mills for grinding corn, extracting oil from rape-seed, and pulverising logwood for the use of clothiers and dyers. Many old houses have been recently removed and new ones erected, under the direction of the lord of the manor. The market is on Thursday; and fairs are held on Holy Thursday, August 5th, October 10th, and the first Thursday after November 22nd; there are also fortnight fairs for the sale of cattle. The quarter sessions for the West Riding are held here at Christmas, in rotation with Knaresborough, Skipton, and Pontefract, and courts leet and baron on Lady-day and Michaelmas-day. The living is a perpetual curacy; net income, £101; patron and impropriator, Hon. and Rev. William Herbert. The foundation stone of a new church, to be called Christ-Church, was laid on Easter-Monday, 1839. There are places of worship for Independents and Wesleyan Methodists; also a Sunday National school.

WETHERDEN (*St. Mary*), a parish, in the union and hundred of Stow, Western Division of the county of Suffolk, 4¼ miles (N. W.) from Stow-Market; containing 487 inhabitants. The living is a discharged rectory, valued in the king's books at £6. 13. 4.; present net income, £371: it is in the patronage of the Crown. The aisle of the church is ornamented on the outside with numerous armorial bearings of the owners of the Hall; and many of the Sulyard family, its ancient proprietors, are buried in the church. The sum of £5. 5. out of the rents of the town lands is applied towards the support of a Sunday school.

WETHERINGSETT (*All Saints*), a parish, in the union and hundred of Hartismere, Western Division of the county of Suffolk, 2¼ miles (E. N. E.) from Mendlesham; containing 1001 inhabitants. The living is a rectory, valued in the king's books at £33. 9. 2.; present net income, £604; late patron, A. Steward, Esq. There is a chapel of ease at Brockford, in this parish. The church is a spacious and handsome structure. The Rev. John Sheppard left a trifling legacy for a dinner to poor people. The Rev. Richard Hackluyt, compiler of English voyages, was rector of this parish.

WETHERSFIELD (*St. Mary Magdalene*), a parish, in the union of Braintree, hundred of Hinckford, Northern Division of the county of Essex, 7 miles (N. N. W.) from Braintree; containing 1698 inhabitants. This parish is bounded by the river Blackwater, on which are several extensive flour-mills, and having been formerly a royal manor, the proprietor has the privilege of holding a manorial court annually, and a pleasure fair is held in July. The living is a vicarage, valued in the king's books at £12; present net income, £239; patrons, Master and Fellows of Trinity Hall, Cambridge; appropriator, Bishop of London. The church is an ancient edifice with a tower and spire, and contains some interesting monuments. There is a place of worship for Independents. Thomas Fitch, in 1702, bequeathed land producing £20 per annum for teaching and clothing 20 boys; he also founded a second school for 10 boys and 10 girls, for which he gave £5 per annum, with a house for the master and mistress; and Dorothy Mott gave land producing £76 per annum, of which £18 is paid for teaching and clothing 20 girls. There is also a National school.

WETTENHALL, a chapelry, in the parish of Over, union of Nantwich, First Division of the hundred of Eddisbury, Southern Division of the county of Chester, 5½ miles (E. by S.) from Tarporley; containing 272 inhabitants. The living is a perpetual curacy; net income, £75; patron, Vicar of Over.

WETTON (*St. Margaret*), a parish, in the Southern Division of the hundred of Totmonslow, Northern Division of the county of Stafford, 7½ miles (N. W. by N.) from Ashbourn; containing 497 inhabitants. The living is a perpetual curacy; net income, £90; patron, M. Burgoyne, Esq.; impropriator, Duke of Devonshire. The church was rebuilt in 1820, except the tower, which is very ancient, at the cost of £600. The river Manifold runs through the parish, as far as Wetton-mill, then suddenly disappears through the fissures of its limestone bed, and, continuing a subterraneous course for about five miles, emerges within a few yards of the place where the river Hamps re-appears in like manner from its channel underground. At Ecton hill there is a copper mine, which was first wrought in the seventeenth century, and for many years produced a yearly profit of £30,000 to the Duke of Devonshire; but the ore having become scarce, it was given up by His Grace some years ago, and let to a small company of working miners, who still find a tolerable remuneration for their labours. On the opposite side of the hill was a prolific lead mine, now exhausted. Twelve poor children are instructed for an annuity of £4. Within this parish is a remarkable cavern of large dimensions, termed Thor's House, in which the Druids, it is believed, sacrificed to their god Thor. Here are quarries of excellent marble.

WETWANG (*St. Michael*), a parish, in the union of Driffield, partly within the liberty of St. Peter's, and partly in the wapentake of Buckrose, East Riding of the county of York; containing 621 inhabitants, of which number, 482 are in the township of Wetwang, 5¾ miles (W. by N.) from Great Driffield. The living is a discharged vicarage, in the patronage of the Prebendary of Wetwang in the Cathedral Church of York (the appropriator), valued in the king's books at £9. 7. 8½.; present net income, £220. There is a place of worship for Wesleyan Methodists. Lady Sykes allows a small sum to a school for teaching poor children.

WEXHAM (*St. Mary*), a parish, in the union of Eton, hundred of Stoke, county of Buckingham, 3 miles (N. W.) from Colnbrook; containing 181 inhabitants. The living is a rectory, valued in the king's books at £5. 15.; present net income, £200: it is in the patronage of the Crown. In the church is a vault in which several of the Godolphin family are interred. Rag-stone abounds in the neighbourhood. The learned Fleetwood, before his elevation to episcopal dignity, was rector of this parish from 1705 to 1708, during which period he published his *Chronicon Pretiosum*.

WEYBOURNE (*All Saints*), a parish, in the union of Erpingham, hundred of Holt, Western Division of the county of Norfolk, 3¾ miles (N. E.) from Holt; containing 273 inhabitants. The living is a donative, in the patronage of the Earl of Orford.

WEYBREAD (*St. Andrew*), a parish, in the union and hundred of Hoxne, Eastern Division of the county of SUFFOLK, 1¾ mile (S. S. W.) from Harleston; containing 708 inhabitants. The living is a discharged vicarage, valued in the king's books at £4. 15.; present net income, £102; patron and incumbent, Rev. J. E. Daniel; impropriators, the Landowners. The church is an ancient structure with a round tower.

WEYBRIDGE (*St. Nicholas*), a parish, in the union of CHERTSEY, First Division of the hundred of ELMBRIDGE, Western Division of the county of SURREY, 3 miles (S. W.) from Chertsey, and 20 miles (S. W. by W.) from London; containing 930 inhabitants. The parish is bounded on the north by the Thames, where it receives the Wey, which is crossed by a bridge, and thus gave name to the place. The Wey canal commences a little to the westward of the village; and the London and South-Western railway, which passes through the parish, has a station on Weybridge Common. The neighbourhood is adorned with many elegant seats; the principal of these is Oatlands, which was the country residence of His Royal Highness the late Duke of York, occupying the brow of an eminence, near a fine sweep of the Thames. A pillar has been erected in the village to the Duchess of York, as a mark of respect to her memory, by the inhabitants; it is a plain Tuscan column, rising from a square pedestal, on which is inscribed a tribute to the benevolence of her character, and surmounted by a coronet. The living is a rectory, valued in the king's books at £7. 0. 5.; present net income, £292: it is in the patronage of the Crown. The church is a small neat edifice, and contains several ancient and modern monuments, among which is one to Her Royal Highness the late Duchess of York, who was interred here: it is a plain tablet, on which is a well-sculptured figure, by Chantry, of the duchess kneeling in the attitude of prayer, with her hands crossed upon her breast, and a suitable inscription. James Taylor, Esq., in 1836, built a Roman Catholic chapel, with a a house for the clergyman, near his own residence, at a cost of £2000. Charles Hopton, in 1739, bequeathed £100, the interest to be applied in support of a school erected by his sister Elizabeth, and which is now merged into a National school. The parish, in which is included part of St. George's Hill, partakes of Henry Smith's charity. Among the various relics of antiquity found here, several curious wedges, or celts, were discovered, in 1725, at Oatlands, about 20 feet below the surface of the earth; which circumstance seems to sanction the opinion that Julius Cæsar attacked the Britons at the place now called Coway Stakes, a short distance from his camp at Walton.

WEYHILL (*St. Michael*), a parish, in the hundred of ANDOVER, Andover and Northern Divisions of the county of SOUTHAMPTON, 2¾ miles (W. by N.) from Andover; containing 429 inhabitants. The village is celebrated for a great fair commencing Oct. 10th, for horses and sheep, of the latter of which it is estimated that more than 140,000 are sold on the first day; it continues the five following days, and is visited by persons from all parts of the kingdom; cheese, hops, and leather, are also sold in considerable quantities. The living is a rectory, valued in the king's books at £26; patrons, Provost and Fellows of Queen's College, Oxford. The tithes have been commuted for a rent-charge of £500, subject to the payment of rates, which on the average have amounted to £60; the glebe comprises 20 acres, valued at £40 per annum. The interior of the church was mutilated by Cromwell's soldiers. Chaucer, the poet, had the manor and advowson, which were afterwards given by Charles I. to Queen's College, Oxford, for "services rendered during the civil wars." Twelve children of this parish attend a school at Penton-Mewsey, which is endowed with £6 per annum. Richard Taunton, in 1759, left the interest of £200 for bread for labourers not receiving parochial relief.

WEYMOUTH and MELCOMBE-REGIS, a sea-port, borough, and market-town, having separate jurisdiction, and the head of a union, in the Dorchester Division of the county of DORSET, 8 miles (S. by W.) from Dorchester, and 129 (S. W. by W.) from London; containing 7655 inhabitants, of which number, 2529 are in Weymouth, and 5126 in Melcombe-Regis. This borough comprises the towns of Weymouth and Melcombe-Regis, forming opposite boundaries of the harbour, in the conveniences of which they had their origin; and, to terminate their mutual rivalry for the exclusive possession of which, they were united into one borough, in the 13th of Elizabeth. Weymouth, which derives its name from its situation at the mouth of the river Wey, is the more ancient, and was probably known to the Romans; as, in the immediate neighbourhood, there are evident traces of a vicinal way, leading from one of the principal landing stations connected with their camp at Maiden Castle to the *via Iceniana*, where the town of Melcombe-Regis now stands. The earliest authentic notice of it occurs in a grant by Athelstan, in 938, wherein he gives to the abbey of Milton "all that water within the shore of Waymouth, and half the stream of that Waymouth out at sea, a saltern, &c." It is also noticed in the Norman survey, with several other places, under the common name of Wai, or Waia, among which it is clearly identified by the mention of the salterns exclusively belonging to it.

Arms of Weymouth and Melcombe-Regis.

The ports of Weymouth and Melcombe, with their dependencies, were, by the charters of Henry I. and II., granted to the monks of St. Swithin, in Winchester, from whom, by exchange, Weymouth passed into the possession of Gilbert de Clare, Earl of Gloucester, who, in the reigns of Henry III. and Edward I., held it with view of frankpledge and other immunities. His successor, Lionel Duke of Clarence, obtained many privileges for the town, which he made a borough, and which, through his heir, Edward IV., subsequently reverted to the crown, and formed part of the dowry of several queens of England. In the reign of Edward II. it received the staple of wine, and collectors were appointed, in the 4th and 6th years of that reign, to receive the duties. Weymouth, in the 10th of Edward III., had become a place of some importance, and, with Melcombe and Lyme, contributed several ships towards the equipment of that monarch's expedition to Gascony. In the year 1347, it furnished 20 ships and 264 mariners towards the fleet destined for the siege of Calais: in this subsidy, Melcombe, though not mentioned, was probably included. In 1471, Margaret of Anjou, with her son, Prince Edward, landed at this port from France, to assist in restoring her husband, Henry VI., to the throne of England; and, in the 20th of Henry VII., Philip King of Castile, on his voyage from Zealand to Spain, with a fleet of 80 ships, on board of which was his queen, being driven by a storm on the English coast, put into it for safety, intending to re-embark, after having refreshed himself from his toils, before his arrival could be known to the English monarch; but Sir Thomas Trenchard and Sir John Carew, who, fearing some hostile attack, had marched with their forces to the town, detained him till he might have an interview with the king, and for that purpose conducted him to Woolveton, the seat of Sir Thomas Trenchard. This port, in 1588, contributed six ships to oppose the armada of Spain, and one of the enemy's vessels, having been taken in the English Channel, was brought into Weymouth harbour. Melcombe-Regis, on the north side of the harbour, derived its name from being situated in a valley, in which was an ancient mill; and its adjunct from its having formed part of the demesnes of the crown. It is not mentioned in Domesday-book, being included in the parish of Radipole, which at that time belonged to Cerne abbey; but it passed from the monks into the possession of the crown at an earlier period than Weymouth, and, in the reign of Edward I., became the dowry of Queen Eleanor, on which account it obtained many valuable and extensive privileges. In the reign of Edward III., it was made one of the staple towns for wool, and flourished considerably; but, in the following reign, having been burnt by the French, it became so greatly impoverished, that the inhabitants petitioned the king to be excused from the payment of their customs. Edward IV., in order to afford relief, granted them a new charter, conferring the same privileges as were enjoyed by the citizens of London.

In the reign of Elizabeth, the lords of the council, wearied by the continual disputes of these two towns, which were both boroughs, and endowed with extensive privileges, by the advice of Cecil, Lord Treasurer, united them into one borough by an act of parliament, which was afterwards confirmed by James I., under the designation of "The United Borough and Town of Weymouth and Melcombe Regis," from which time their history becomes identified. Weymouth afterwards gradually fell into decay, and suffered greatly during the parliamentary war, having been alternately garrisoned for both parties. In 1644, it was evacuated by the royalists, on which occasion several ships, and a great quantity of arms, fell into the hands of the parliament, who obtained possession of it. The royalists soon afterwards attempted to recover it, but the garrison sustained the attack for eighteen days, and finally obliged them to raise the siege. An additional fort was built, in 1645, on the Weymouth side of the harbour, to defend it from the incursions of the Portlanders; and, four years after, the corporation petitioned for an indemnification for the destruction of their bridge and chapel (the latter, from its commanding situation, having been converted into a fort), and for assistance in the maintenance of the garrison, which application appears to have been disregarded; but, in 1666, a brief was granted to repair the damage, and, in 1673, another was bestowed for the collection of £3000, to repair the injury which the town had received from an accidental fire, whereby a considerable portion of it was destroyed. The rise of the town of Poole, which was rapidly growing into importance, the decay of the haven, and the loss of its trade, with various other causes, contributed powerfully to the decline of the town, which, from an opulent and commercial port, had almost sunk into a mere fishing town, when Ralph Allen, Esq., of Bath, in 1763, first brought it into notice as a bathing-place; and the subsequent visits of the Duke of Gloucester, and afterwards of George III. and the royal family, with whom it was a favourite place of resort, laid the foundation of its present prosperity.

The town is beautifully situated on the western shore of a fine open bay in the English Channel, and separated into two parts by the river Wey, which, expanding to a considerable breadth, in its progress to the bay, forms a small, but secure and commodious, harbour, on the south side of which is Weymouth, at the foot of a high hill near the mouth of the river; and, on the north side, Melcombe-Regis, on a peninsula, connected with the main land by a narrow isthmus, which separates the waters of the bay from those formed by the estuary of the river, called the Backwater. A long and handsome stone bridge of two arches, with a swivel in the centre, to admit small vessels into the upper part of the harbour, has been erected, by act of parliament in the 1st of George IV., and connects the two parts of the town. In building it, the workmen, on clearing the foundation of some ancient premises, discovered an urn, covered with a thin piece of sheet-iron, containing a great number of half-crowns, shillings, and sixpences, of the reigns of Elizabeth, James I., and Charles I.: in taking down an old house, nearly opposite the bridge, a richly-gilt crucifix of brass, about four inches in length, was found, which, with the exception of a part of the gilding, was in a very perfect state; and on removing another, which tradition reports to have been the council-house of Melcombe, and in which the charter of Queen Elizabeth was granted to the borough, a complimentary inscription was discovered painted on the wall, in black letter, with the date 1577. Since the town has become a place of fashionable resort for sea-bathing, various handsome ranges of building, and a theatre, assembly-rooms, and other places of public entertainment, have been erected. Among the former, Belvidere, the Crescent, Gloucester-Row, Royal-Terrace, Chesterfield-Place, York-Buildings, Charlotte-Row, Augusta-Place, and Clarence, Pulteney, and Devonshire Buildings, are conspicuous; to which may be added, Brunswick-Buildings, a handsome range of houses at the entrance of the town, and numerous detached villas in the vicinity. From the windows of these buildings, which front the sea, a most extensive and delightful view is obtained, comprehending, on the left, a noble range of hills and cliffs extending, for many miles, in a direction from west to east, and of the sea in front, with the numerous vessels, yachts, and pleasure-boats, which are continually entering and leaving the harbour. The town, especially on the Melcombe side of the harbour, is regularly built, and consists partly of two principal streets, parallel with each other, intersected by others at right angles; it is well paved and lighted, under the provisions of an act passed in the year 1766, and is supplied, by a public company incorporated by another act, with excellent water, conveyed by pipes from the Boiling Rock, in the parish of Preston, a distance of two miles. The houses, excepting such as have been erected for the accommodation of visiters, are in general built of stone and roofed with tiles, and are low and of indifferent appearance.

About a mile to the south-west are the remains of Weymouth, or Sandsfoot, castle, erected by Henry VIII., in the year 1540, on the threatened invasion of the Pope, and described by Leland as "a right goodly and warlyke castle, having one open barbicane." It is quadrangular in form: the north front has been nearly destroyed, the masonry with which it was faced having been removed; the apartments on this side are all vaulted, and appear to have been the governor's residence; at the extremity was a tower, on the front of which were the arms of England, having a wyvern and unicorn for supporters. The greater part of the south front fell into the sea in 1837. On this side is a low building, broader than the castle, and flanking its east and west sides, in which are embrasures for great guns, and loop-holes for small arms: the walls, in some parts, are of amazing thickness, but in a very dilapidated state, and rapidly falling to decay. On the south of the town are the cavalry barracks, a neat and commodious range of building. The Esplanade, a beautiful terrace 30 feet broad, rising from the sands, and secured by a strong wall extending in a circular direction, parallel with the bay, a mile in length, and commanding an extensive and beautiful view of the sea and the mountainous range of cliffs by which the bay is enclosed, is the finest marine promenade in the kingdom. Among the buildings that adorn it is the Royal Lodge, where George III. and the royal family resided while visiting this place, comprising several houses of handsome, though not of uniform, appearance. There are several flights of steps, of Portland stone, leading to the sands, to which also is a gently sloping descent from one extremity of the Esplanade to the other: in the centre is the principal public library, elegantly furnished. The assembly and card rooms form part of the Royal hotel, a handsome range of building, with commodious stabling and other appendages, and occupying an area 600 feet in length and 250 in breadth, erected at an expense of £6000, advanced on shares of £100 each, and are in every respect well adapted to the meetings which are held there during the season, under the superintendence of a master of the ceremonies. The theatre, of which the box entrance is in Augusta Place, is a neat and well-arranged edifice, handsomely fitted up; it will accommodate 300 persons in the boxes, and is open four nights in the week during the season. Races were established in 1821, which take place in August, and are generally well attended; among the prizes contended for are the queen's plate of 100 guineas, the members' of 50 guineas, and the ladies and tradesmen's plates: the course, about a mile from the town, is conveniently adapted to the purpose.

About the time of the races, a splendid regatta is celebrated in the bay, which has a fine circular sweep of nearly two miles, and, being sheltered from the north and north-east winds by a continuous range of hills, the water is generally calm and transparent. The sands are smooth, firm, and level; and so gradual is the descent towards the sea, that, at the distance of 300 feet, the water is not more than five feet deep. Numerous bathing-machines are in constant attendance, and on the South Parade is an establishment of hot salt-water baths, furnished with dressing-rooms and every requisite accommodation. At the south entrance of the harbour are the piers: two new quays have been recently erected, and the harbour has been deepened. Part of the ground over which the sea formerly flowed has been embanked, and is now covered with buildings; and other parts are enclosed with iron railings, which form a prominent feature on the Esplanade. On the Weymouth side are the Look-out and the Nothe, affording extensive and interesting prospects: on the latter was a battery, formerly mounted with six pieces of ordnance, which, on the dismantling of the fort, were removed into Portland castle: within the walls a signal post has been established, which communicates with several other stations, and apartments have been built for the accommodation of a lieutenant and a party of men. The bay almost at all times affords ample facilities for aquatic excursions, its tranquil surface being never disturbed, except by violent storms from the south or south-east; yachts and pleasure-boats are always in readiness, the fares of which are under strict regulations. The air is so mild and pure, that the town is not only frequented during the summer, but has been selected, by many opulent families, as a permanent residence; and the advantages which it possesses in the excellence of its bay, the beauty of its scenery, and the healthfulness of its climate, have contributed to raise it from the low state into which it had fallen, from the depression of its commerce, to one of the most flourishing towns in the kingdom.

The port formerly carried on an extensive trade with France, Spain, Norway, and Newfoundland, in the fishery of which last place it employed 80 vessels; but the war with France, after the Revolution, put an end to its commerce with that country; the trade with Newfoundland was, in a great measure, transferred to Poole; and the accumulation of sand in the harbour, operating with other causes, considerably diminished its importance as a port. A few vessels are still employed in the American and Mediterranean trade, in addition to which there is a tolerable coasting trade. The principal imports are coal, timber, wine, brandy, geneva, tobacco, and rice, for which it was made a bonding port by an order of council, in 1817: the chief exports are Portland-stone, pipe-clay, Roman cement, bricks, tiles, slates, corn, and flour. The amount of duties paid at the custom-house, for the year ended Jan. 5th, 1837, was £13,120. Ship-building is carried on to some extent; and many persons are employed in the manufacture of ropes, twine, and cordage, and in the making of sails. The quay, on which is the custom-house, a neat and commodious building, is well adapted to the loading and unloading of goods, but, from the accumulation of sand in the harbour, it is not accessible to ships of large burden. Two Post-Office steam-packets sail regularly, on Wednesday and Saturday, for Guernsey, Jersey, and the neighbouring islands; and arrangements have been recently made for establishing a communication by steam with Cherbourg, on the coast of France, twice a week. The market days are Tuesday and Friday: the town is abundantly supplied with fish of every description, with the small mutton from the isle of Portland, and with provisions of all kinds.

Arms.

Weymouth and Melcombe-Regis, which had been distinct boroughs, and had returned members to parliament, the latter since the 8th, and the former since the 12th, of Edward II., were united into one borough, by charter of Elizabeth, confirmed by James I., in the 14th year of his reign, who granted a charter of incorporation, which was amended by the charter of the 21st of George II.; but it having been allowed to expire by neglect in filling up vacancies in the corporate body, George III., on petition, granted a new charter in the 44th year of his reign. Under this charter the government was vested in a mayor, recorder, two bailiffs, an indefinite number of aldermen (not less than eight), and 24 principal burgesses, assisted by a town-clerk, two sergeants-at-mace, and subordinate officers. The mayor, recorder, and bailiffs, were justices of the peace. The corporation now consists of a mayor, 6 aldermen, and 18 councillors, under the act of the 5th and 6th of William IV., cap. 76, for an abstract of which see the Appendix, No. I.: the borough is divided into two wards, and the municipal and parliamentary boundaries are now co-extensive. It formerly comprised 505 acres, and, since its union, continued to return four members to parliament until the 2nd of William IV., when it was deprived of two. The right of election was formerly vested in the corporation and freeholders generally not receiving alms, in number about 600, every elector being entitled to vote for four candidates: but the privilege has been extended to the £10 householders of an enlarged district, which has been constituted the new borough, comprising an area of 812 acres, the limits of which are minutely described in the Appendix, No. II.: the mayor is the returning officer. The corporation were empowered to hold a court of session quarterly, for offences not capital, which court was held only at Michaelmas. A court of record is held every Tuesday, for the recovery of debts to any amount, and separate courts leet were held for the two places. A handsome town-hall, situated in the market-place, has been recently erected, the old one having become dilapidated; under it are a small prison and watch-house.

Weymouth is a chapelry to Wyke-Regis: the chapel, dedicated to *St. Nicholas,* was situated on the top of the hill, but has long since disappeared; the site, called Chapel-Hay, is distinctly marked by large stones at the four corners. Under the hill, and nearly adjoining this site, a new church or chapel has been built, from a design by Mr. P. Wyatt, at the sole expense of the Rev. George Chamberlaine, rector of Wyke-Regis and Weymouth; underneath it are spacious catacombs, capable of containing upwards of 1000 bodies. Melcombe, previously to the reign of James I., was a chapel of ease to Radipole, from which it was separated in 1605, when a new church was built on the site of the former chapel, and made parochial in 1606: the living is a rectory, with Radipole annexed, valued in the king's books at £11. 5. 5.; present net income, £298; patron, W. Wyndham, Esq. The church, dedicated to *St. Mary,* having become greatly dilapidated, an act of parliament was obtained in the 55th of George III., for rebuilding it, which was completed in 1817; it is a spacious neat edifice, containing upwards of 2000 sittings, including 500 free sittings purchased by the Rev. George Chamberlaine, at an expense of £500, for the exclusive use of the poor on both sides of the water: it is appropriated to the use of the inhabitants of Weymouth and Melcombe-Regis; the interior is neatly fitted up, and the altar-piece is embellished with a painting of the Last Supper, by Sir James Thornhill. There are two places of worship for Independents, and one each for Baptists and Wesleyan Methodists; also a Roman Catholic chapel. A Lancasterian school was erected in 1813, at an expense of more than £1000, which sum, from the failure of subscriptions, the committee being unable to liquidate, the committee of a National school then forming took the debt upon themselves, and in 1819 obtained from the National Society a grant of £400, of which £200 was appropriated to the payment of the arrears on the original school, and £200 towards the erection of a girls' school adjoining, which was completed in 1820, at an expense of £600. The schools are built on ground granted on lives by Sir G. F. Johnson, at a nominal rent of 5*s.* per annum. There is also a British school, supported by subscription; and there are several small bequests for the education of children, especially one of £70 per annum, and another of £28, for teaching six boys, left by Mr. Taylor, in 1753. Sir James Thornhill built an almshouse, in St. Mary's-street, for decayed seamen, but, having no endowment, it fell to decay; a small portion only is remaining, the greater part having been taken down. The poor law union of Weymouth comprises 18 parishes or places, under the care of 22 guardians, and contains, according to the census of 1831, a population of 16,947. At Nottington, about two miles and a half distant, on the Dorchester road, is a mineral spring, the water of which is considered efficacious in scrofula; and about one mile from the town is Radipole Spa, discovered in 1830, by John Henning, Esq. In the centre of the town was a priory of Black canons, dedicated to St. Winifred, and founded by some member of the family of Rogers, of Bryanston: the buildings occupied a quadrangular area of nearly one acre, but they have been entirely removed, and several small houses erected on the site: in digging the foundations, a great quantity of human bones was found. The burning cliff at Holworth, five miles from Weymouth, was first introduced to public notice by Mr. George Frampton, in 1827, and has since attracted the notice of naturalists. Certain masses of septaria, which, when sawn asunder, exhibit beautiful specimens of spar, cornua ammonis, &c., were discovered a few years since in the rear of Melcombe. Sir James Thornhill, the celebrated painter, was a native of Melcombe-Regis, and represented that borough in parliament in the reign of George I. The late Mr. John Harvey, of Weymouth, projected the plan of a breakwater for Portland Roads, which has been matured and improved by his son, the present post-master of this place; for a detailed account of this proposed undertaking, see PORTLAND (ISLE OF). Melcombe conferred the title of Baron on Bubb Doddington, with whom it became extinct; Weymouth gives that of Baron to the family of Thynne.

WHADDON (*ST. MARY*), a parish, in the union of

Winslow, hundred of Cotteslobe, county of Buckingham, 4¼ miles (S. by E.) from Stony-Stratford; containing 889 inhabitants. The living is a vicarage, valued in the king's books at £10; present net income, £152; patrons and impropriators, Warden and Fellows of New College, Oxford. A small priory of Benedictine monks, in honour of St. Leonard, was founded at Snelleshall, prior to the time of Henry III., by Ralph Martel, which, at the dissolution, had a revenue of £24. The prior obtained, in 1227, a grant of a weekly market on Thursday, long since disused. Whaddon Hall was once the seat of Arthur Lord Grey, who was honoured by a visit from Queen Elizabeth, in 1568, then on her Buckinghamshire progress. Spencer, the poet, his lordship's secretary, was frequently here: it was afterwards purchased and occupied by Browne Willis, the antiquary. A charity school was founded here by a Mr. Coare, who endowed it with £10 per annum for teaching 20 children: he also erected an almshouse, but died before fulfilling his intention of endowing it. Dr. Richard Cox, Bishop of Ely, an eminent champion of the Reformation, and one of the principal composers of the Liturgy, was born in this parish, in 1499. This place gave the title of baron, the first conferred upon him, to Villiers Duke of Buckingham, the favourite of James I. and Charles I.

WHADDON (*St. Mary*), a parish, in the union of Royston, hundred of Armingford, county of Cambridge, 4¼ miles (N.) from Royston; containing 339 inhabitants. The living is a discharged vicarage, valued in the king's books at £7. 2. 3½.; present net income, £166; patrons and appropriators, Dean and Canons of Windsor. A school is partly supported by subscription.

WHADDON (*St. Margaret*), a parish, in the Middle Division of the hundred of Dudstone and King's-Barton, union and Eastern Division of the county of Gloucester, 3¼ miles (S. by W.) from Gloucester; containing 152 inhabitants. The living is a perpetual curacy; net income, £46; patron, J. Pitt, Esq.

WHADDON, a parish, in the union and hundred of Melksham, Melksham and Northern Divisions of the county of Wilts, 2¾ miles (N. E. by N.) from Trowbridge; containing 58 inhabitants. The living is a discharged rectory, valued in the king's books at £8. 4. 4½.; present net income, £159; joint patrons, R. and J. Long, Esqrs. The river Avon, and the Kennet and Avon canal, pass through the parish.

WHALEY, a joint township with Yeardsley, in the parish of Taxall, union and hundred of Macclesfield, Northern Division of the county of Chester, 9¼ miles (S. E.) from Stockport; containing, with Yeardsley, 403 inhabitants. There is a place of worship for Wesleyan Methodists. The Peak Forest canal passes through the parish.

WHALLEY (*All Saints*), a parish, in the union of Clitheroe, comprising the borough aud market-town of Clitheroe, the market-towns of Burnley, Colne, and Haslingden, with several chapelries and townships, chiefly in the Higher Division, but partly in the Lower Division, of the hundred of Blackburn, Northern Division of the county palatine of Lancaster; partly in the Second Division of the hundred of Eddisbury, Southern Division of the county of Chester; and partly in the Western Division of the wapentake of Staincliffe and Ewcross, West Riding of the county of York; and containing 97,868 inhabitants, of which number, 1151 are in the township of Whalley, 4 miles (S. by W.) from Clitheroe. This parish is about 30 miles in length by 15 in breadth, though but little more than half its ancient extent, which included also the present parishes of Blackburn, Chipping, Mitton, Ribchester, Rochdale, and Slaidburn, which have been separated from it at different times. The rivers Calder and Ribble form a junction at the western extremity of the parish. The village is chiefly celebrated for the venerable ruins of its abbey, which exhibit portions in the early, decorated, and later, styles of English architecture: they are still considerable, and possess much interest. This house was founded in 1296, by Henry Lacy, Earl of Lincoln, in honour of the Blessed Virgin Mary, for monks of the Cistercian order, whose revenue, at the dissolution, was estimated at £551. 4. 6. Here are manufactures of cotton, rope, and nails. The living is a vicarage, valued in the king's books at £6. 3. 9.; present net income, £137; patron, Archbishop of Canterbury; impropriators, Earl Howe and Lord Ribblesdale. The church is a large structure, principally in the early English style, of which the chancel is a very fine specimen: the interior contains eighteen ancient stalls, and some considerable remains of good screen-work, brought from the old abbey. A chapel at Goodshaw was rebuilt in 1828, and contains 300 free sittings, the Incorporated Society having granted £300 in aid of the expense. Her Majesty's Commissioners have agreed to afford facilities for obtaining a site for a church at Chatburn, in this parish. There is a place of worship for Wesleyan Methodists. The free grammar school, founded by Queen Elizabeth, was rebuilt by subscription in 1725, with a dwelling-house for the master, who receives an annuity of £12. 8. 2. from the crown rents, and another of £4. 14. arising from the bequests of John Chewe and Sir Edmund Assheton: it is open for the classical instruction of all boys of the township, and, with those of Burnley and Middleton, has an interest in thirteen scholarships, founded in Brasenose College, Oxford, by Dr. Nowell, in the year 1572. A National school was erected by subscription, in 1819, at Rossendale Forest, in this parish.

WHALTON, a parish, in the union and Western Division of Castle ward, Southern Division of the county of Northumberland; containing 548 inhabitants, of which number, 311 are in the township of Whalton, 6 miles (S. W. by W.) from Morpeth. The living is a rectory, valued in the king's books at £13. 8. 1½.; present net income, £753; patron, R. Bates, Esq. The church is an ancient structure; it was repaired in the year 1783, when parapets and pinnacles were added to the tower. Here is a National school, with an endowment of £2 per annum. A little to the eastward of the village are the remains of considerable earthworks, supposed to enclose the site of the castle of the ancient barons of Whalton; and about a mile and a half south of it are some slight remains of Ogle castle, surrounded by a double fosse, part of which is in fine preservation.

WHAPLODE (*St. Mary*), a parish, in the union of Holbeach, wapentake of Elloe, parts of Holland, county of Lincoln, 2 miles (W.) from Holbeach; containing 1998 inhabitants. The living is a vicarage, valued in the king's books at £16. 14. 9½.; present net income, £309: it is in the patronage of the Crown; impropriators, Trustees of Uppingham school. The church has been enlarged, and contains 130 free sittings, the Incorporated Society having granted £200 in aid of the expense. Elisha Wilson, and Frances his wife, in 1708, bequeathed land producing £10 per annum for the support of a school.

WHAPLODE-DROVE, a chapelry, in the parish of WHAPLODE, union of HOLBEACH, wapentake of ELLOE, parts of KESTEVEN, county of LINCOLN, 5¾ miles (E. N. E.) from Crowland; containing 580 inhabitants. The living is a perpetual curacy; net income, £380; patrons and impropriators, certain Trustees. The chapel is dedicated to St. John the Baptist. Six children are instructed for £5 per annum from the church rates.

WHARLES, a joint township with Treales and Roseacre, in the parish of KIRKHAM, union of the FYLDE, hundred of AMOUNDERNESS, Northern Division of the county palatine of LANCASTER, 2¾ miles (N. E. by N.) from Kirkham; containing 756 inhabitants.

WHARRAM-LE-STREET (*St. Mary*), a parish, in the union of MALTON, wapentake of BUCKROSE, East Riding of the county of YORK, 6¾ miles (S. E. by E.) from New Malton; containing 150 inhabitants. The living is a discharged vicarage, valued in the king's books at £6; patron and impropriator, Lord Middleton. The tithes have been commuted for a rent-charge of £157. 10., subject to the payment of rates; the glebe comprises 3½ acres, valued at £5 per annum.

WHARRAM-PERCY, a parish, partly in the unions of MALTON, DRIFFIELD, and POCKLINGTON, wapentake of BUCKROSE, East Riding of the county of YORK; containing 350 inhabitants, of which number, 30 are in the township of Wharram-Percy, 7¼ miles (S. E.) from New Malton. The living is a discharged vicarage, valued in the king's books at £11. 13. 4.; present net income, £60; patron and impropriator, Lord Middleton.

WHARTON, a township, in the parish of DAVENHAM, union and hundred of NORTHWICH, Southern Division of the county of CHESTER, 2½ miles (W. N. W.) from Middlewich; containing 1060 inhabitants. The Grand Junction railway passes through the township.

WHARTON, a township, in the parish of BLYTON, union of Gainsborough, wapentake of CORRINGHAM, parts of LINDSEY, county of LINCOLN, 3¾ miles (N. E.) from Gainsborough: the population is returned with the parish.

WHARTON, a township, in the parish of KIRKBY-STEPHEN, East ward and union, county of WESTMORLAND, 2¼ miles (S. by W.) from Kirkby-Stephen; containing 76 inhabitants. The hall, once a large quadrangular building, with a tower at each angle, was the princely residence of Philip, the celebrated Duke of Wharton, and his ancestors, but is now occupied as a farm-house. The ancient village was demolished many years ago for the enlargement of the park, when the inhabitants settled at Wharton Dikes, on the southern boundary of the park. The hall, with all the estates and manorial rights of the Wharton families, is now possessed by the Earl of Lonsdale.

WHASHTON, a township, in the parish of KIRKBY-RAVENSWORTH, union of RICHMOND, wapentake of GILLING-WEST, North Riding of the county of YORK, 4 miles (N. by W.) from Richmond; containing 159 inhabitants.

WHATBOROUGH, a liberty, in the parish of TILTON, union of BILLESDON, hundred of EAST GOSCOTE, Northern Division of the county of LEICESTER, 12 miles (E. by N.) from Leicester; containing 19 inhabitants.

WHATCOMBE, a tything, in the parish of FAWLEY, hundred of KINTBURY-EAGLE, county of BERKS, 6 miles (S.) from Wantage: the population is returned with the parish.

WHATCOTT (*St. Peter*), a parish, in the union of SHIPSTON-UPON-STOUR, Brails Division of the hundred of KINGTON, Southern Division of the county of WARWICK, 4¼ miles (N. E.) from Shipston-upon-Stour; containing 219 inhabitants. The living is a rectory, valued in the king's books at £12. 17. 3½.; present net income, £213; patron, Sir S. Graham, Bart.

WHATCROFT, a township, in the parish of DAVENHAM, union and hundred of NORTHWICH, Southern Division of the county of CHESTER, 3 miles (N. W. by N.) from Middlewich; containing 56 inhabitants.

WHATFIELD (*St. Margaret*), a parish, in the union and hundred of COSFORD, Western Division of the county of SUFFOLK, 3¼ miles (S. E.) from Bildestone; containing 377 inhabitants. The living is a rectory, valued in the king's books at £15. 0. 5.; present net income, £393; patrons, Master and Fellows of Jesus' College, Cambridge. The Rev. John Clubb, rector of this parish, published, in 1753, the "History and Antiquities of the Ancient Villa of Wheatfield," intended as a satire on antiquaries and conjectural etymologists.

WHATLEY (*St. George*), a parish, in the union and hundred of FROME, Eastern Division of the county of SOMERSET, 2¾ miles (W. by S.) from Frome; containing 386 inhabitants. The living is a rectory, valued in the king's books at £12. 17. 1.; patron, T. S. Horner, Esq. The tithes have been commuted for a rent-charge of £226, subject to the payment of rates; the glebe comprises 13 acres, valued at £18 per annum. The church is situated on an eminence, separated from the parish of Mells by a deep ravine, the sides of which are clothed with thick woods, and at the bottom runs a rivulet: it is an ancient structure, of which the nave was rebuilt about 9 years since, and contains the recumbent effigy of a crusader, whose armorial bearings are on the tower. There are two places of worship for Wesleyan Methodists, and one for Independents. A parochial school is supported by subscription. This parish abounds with freestone, under which is limestone rock to a considerable depth: in the former are imbedded fossil shells, which are also found in great abundance over the entire surface. On a bold height, at the western extremity of the parish, are vestiges of a Roman encampment; and in 1838 was discovered what, from the figures of dolphins, is supposed to have been a Roman bath, consisting of an apartment, 30 feet long and 15 feet wide, the floor of which is a finely tessellated pavement in excellent preservation; there is also a smaller apartment, in the centre of which is the head of a female, supposed to be Cybele; the *tesseræ* are very perfect, and are wrought into elegant devices.

WHATLINGTON, a parish, in the union and hundred of BATTLE, rape of HASTINGS, Eastern Division of the county of SUSSEX, 2 miles (N. by E.) from Battle; containing 286 inhabitants. The living is a discharged rectory, valued in the king's books at £7. 4. 6.; present net income, £160; patron, Duke of Dorset.

WHATTON (*St. John of Beverley*), a parish, in the union, and Northern Division of the wapentake, of BINGHAM, Southern Division of the county of NOTTINGHAM, 2¾ miles (E.) from Bingham; containing 677 inhabitants. The living is a discharged vicarage, valued in the king's books at £5. 6. 8.; present net income, £212; patron, G. S. Foljambe, Esq.; impropriators, T. Hall, Esq., and others. The church contains an effigy of a Knight Templar in armour, and a monumental tablet in memory of Thomas Cranmer, father of the celebrated Archbishop Cranmer, who was born at Aslacton, in the year 1489. There is a place of worship for Wesleyan Methodists.

WHATTON, LONG (*All Saints*), a parish, in the union of LOUGHBOROUGH, hundred of WEST GOSCOTE, Northern Division of the county of LEICESTER, 4¼ miles (N. W. by W.) from Loughborough; containing 855 inhabitants. The living is a rectory, valued in the king's books at £13. 6. 8.; present net income, £275: it is in the patronage of the Crown.

WHEATACRE (*All Saints*), a parish, in the union of LODDON and CLAVERING, Eastern Division of the county of NORFOLK, 4¾ miles (N. E. by E.) from Beccles; containing 186 inhabitants. The living is a discharged rectory, with that of Barnby and the vicarage of Mutford annexed, valued in the king's books at £6. 6. 5½.; present net income, £660; patrons, Master and Fellows of Gonville and Caius College, Cambridge.

WHEATACRE-BURGH, or BURGH (*St. Peter*), a parish, in the union of LODDON and CLAVERING, hundred of CLAVERING, Eastern Division of the county of NORFOLK, 6¾ miles (E. N. E.) from Beccles; containing 316 inhabitants. The living is a discharged rectory, valued in the king's books at £7. 6. 8.; patron and incumbent, Rev. W. Boycatt. The tithes have been commuted for a rent-charge of £370, subject to the payment of rates, which on the average have amounted to £21; the glebe comprises 15½ acres, valued at £23. 7. per annum. There is a rent-charge of £3 payable to the rector of Monks' Toft.

WHEATENHURST, or WHITMINSTER (*St. Andrew*), a parish, and the head of a union, in the Lower Division of the hundred of WHITSTONE, Eastern Division of the county of GLOUCESTER, 7 miles (N. W. by W.) from Stroud; containing 423 inhabitants. The living is a perpetual curacy, valued in the king's books at £7. 12. 3½.; present net income, £135; the patronage and impropriation belong to Mrs. Hawkins. The impropriate tithes have been commuted for a rent-charge of £266. 10., subject to the payment of rates; the glebe comprises 11½ acres, valued at £23 per annum. A day and Sunday school is supported by C. O. Cambridge, Esq., who built the school-house. The Gloucester and Berkeley, and the Thames and Severn, canals pass through the parish; and the river Severn is navigable on its north-western boundary. The great road from Gloucester to Bristol also passes through the parish. Wheatenhurst poor law union comprises 14 parishes or places, containing a population of 7770, according to the census of 1831, and is under the superintendence of 18 guardians.

WHEATFIELD, a parish, in the union of THAME, hundred of PIRTON, county of OXFORD, 2¼ miles (S.) from Tetsworth; containing 105 inhabitants. The living is a rectory, valued in the king's books at £9. 10. 10.; present net income, £237; patrons, Trustees of C. V. Spencer, Esq. The manor-house was destroyed by fire in 1814, and is now in ruins.

WHEATHAMPSTEAD (*St. Helen*), a parish, in the union of ST. ALBANS, hundred of DACORUM, county of HERTFORD, 4¼ miles (W. S. W.) from Welwyn; containing 1666 inhabitants. The living is a rectory, with that of Harpenden annexed, valued in the king's books at £42. 1. 10½.; present net income, £1356; patron, Bishop of Lincoln. The church is an ancient cruciform structure: the font is a curious specimen of the early decorated style. There is a place of worship for Independents. The rebellious barons here assembled their forces against Edward II., in 1311, on which occasion two nuncios, sent by the pope, endeavoured to restore peace between the contending parties, when the papal authority was rejected by the former. The St. Albans races are held on ground, called Noman's land, which extends into this parish. A National school, erected in 1814, is supported by subscription. James Marshall, of Harpenden, in 1719, bequeathed property in the parishes of Harpenden and Luton, the rental of which, now amounting to £184. 15. per annum, is equally divided between the parishes of Wheathampstead and Harpenden, and appropriated to apprenticing poor children. There are some other trifling bequests for the general relief of the poor. John Bostock, abbot of St. Albans, a learned divine and poet in the time of Henry VI., was born here, and was commonly called John of Wheathampstead.

WHEATHILL (*Holy Trinity*), a parish, in the union of CLEOBURY-MORTIMER, hundred of STOTTESDEN, Southern Division of the county of SALOP, 9½ miles (N. E. by E.) from Ludlow; containing 123 inhabitants. The living is a rectory, valued in the king's books at £7. 5. 7½.; patron and incumbent, Rev. John Churton. The tithes have been commuted for a rent-charge of £201. 4. 9., subject to the payment of rates, which on the average have amounted to £14; the glebe comprises 93 acres, valued at £105 per annum. A weekly market and an annual fair were granted to this place by Edward I., both which have been long disused.

WHEATHILL (*St. John the Baptist*), a parish, in the union of WINCANTON, hundred of WHITLEY, Western Division of the county of SOMERSET, 4 miles (W. by S.) from Castle-Cary; containing 56 inhabitants. The living is a discharged rectory, valued in the king's books at £4. 5. 2½.; present net income, £105: it is in the patronage of Mrs. Harbin.

WHEATLEY, a joint township with Thornley, in the parish of CHIPPING, union of CLITHEROE, Lower Division of the hundred of BLACKBURN, Northern Division of the county palatine of LANCASTER, 8½ miles (W. by S.) from Clitheroe: the population is returned with Thornley.

WHEATLEY, a chapelry, in the parish of CUDDESDEN, union of HEADINGTON, hundred of BULLINGDON, county of OXFORD, 5½ miles (E. by S.) from Oxford; containing 976 inhabitants. The living is a perpetual curacy; net income, £120; patron and appropriator, Bishop of Oxford. The chapel is dedicated to St. Mary. Dr. Moss, Bishop of Oxford, in 1811, bequeathed £3000 in trust for the foundation of a National school here for children of Cuddesden, Wheatley, Denton, and Chippenhurst, and for other charitable uses; in pursuance of which, two school-rooms, and a dwelling-house for the master and mistress, have been provided, and £1500 given by the trustees as a permanent endowment, producing, with subscriptions, £100 per annum. Sir Sebastian Smith gave a rent-charge of £5 issuing out of Simons Close; Lady Curzon, in 1692, gave lands producing £15 per annum for the apprenticing of poor children; Dr. Cyril Jackson, in 1816, gave £166. 13. 4. three per cent. consols. for clothing the poor; and the rental of the town meadow, amounting to £26. 10., is applied to the general relief of the poor. A post-office has been established in the village.

WHEATLEY, a joint township with Long Sandal, in the parish, union, and soke, of DONCASTER, West Riding of the county of YORK, 2 miles (N. E. by N.) from Doncaster: the population is returned with Long Sandal. A school for teaching poor children, and almshouses for twelve aged persons, were erected and are liberally supported by the family of Cooke.

WHEATLEY-CARR, a township, in the union of BURNLEY, parish of WHALLEY, Higher Division of the hundred of BLACKBURN, Northern Division of the county palatine of LANCASTER, $3\frac{3}{4}$ miles (W. S. W.) from Colne; containing 58 inhabitants.

WHEATLEY, NORTH (*St. Peter*), a parish, in the union of EAST RETFORD, North-Clay Division of the wapentake of BASSETLAW, Northern Division of the county of NOTTINGHAM, $5\frac{1}{4}$ miles (N. E.) from East Retford; containing 435 inhabitants. The living is a discharged vicarage, valued in the king's books at £3. 18. $11\frac{1}{2}$.; present net income, £246; patron and impropriator, Lord Middleton. There is a place of worship for Wesleyan Methodists; also a National school. The Roman road from Doncaster to Lincoln passes through the parish.

WHEATLEY, SOUTH (*St. Helen*), a parish, in the union of EAST RETFORD, North-Clay Division of the wapentake of BASSETLAW, Northern Division of the county of NOTTINGHAM, $5\frac{1}{2}$ miles (N. E. by E.) from East Retford; containing 35 inhabitants. The living is a discharged rectory, in the patronage of the Chapter of the Collegiate Church of Southwell, valued in the king's books at £6. 14. 2.; present net income, £140.

WHEATON-ASTON, a chapelry, in the parish of LAPLEY, union of PENKRIDGE, Western Division of the hundred of CUTTLESTONE, Southern Division of the county of STAFFORD, $5\frac{1}{4}$ miles (W. by S.) from Penkridge: the population is returned with the parish. The chapel has been enlarged, and 320 free sittings provided, the Incorporated Society having granted £250 in aid of the expense. Here are places of worship for Independents and Primitive Methodists. Fairs for cattle, &c., are held on April 20th and November 1st.

WHEDDICAR, a township, in the parish of ST. BEES, union of WHITEHAVEN, ALLERDALE ward above Derwent, Western Division of the county of CUMBERLAND, $2\frac{3}{4}$ miles (E. by S.) from Whitehaven; containing 55 inhabitants.

WHEELOCK, a township, in the parish of SANDBACH, union of CONGLETON, hundred of NORTHWICH, Southern Division of the county of CHESTER, $1\frac{1}{2}$ mile (S. S. W.) from Sandbach; containing 440 inhabitants. The Grand Trunk canal passes through the parish, and on its banks are commodious wharfs and warehouses. Cotton is manufactured here, and there is a brewery; but the chief trade of the place is in salt, of which large quantities are extracted from the brine found, at the depth of 60 yards, on both sides of the river Wheelock. There is a place of worship for Wesleyan Methodists.

WHEELTON, a township, in the parish and hundred of LEYLAND, union of CHORLEY, Northern Division of the county palatine of LANCASTER, 4 miles (N. E. by N.) from Chorley; containing 1519 inhabitants.

WHELDRAKE (*St. Helen*), a parish, in the union of YORK, partly within the liberty of ST. PETER'S, but chiefly in the wapentake of OUZE and DERWENT, East Riding of the county of YORK, $7\frac{1}{2}$ miles (S. E.) from York; containing 691 inhabitants. The living is a rectory, valued in the king's books at £25. 17. $3\frac{1}{2}$.; present net income, £474; patron, Archbishop of York. The church was rebuilt in 1789. There are places of worship for Wesleyan Methodists and Methodists of the New Connection; also a National school, partly supported by an endowment of £17. 8. per annum. A manorial court is occasionally held here, for the recovery of small debts, at which a bailiff is appointed.

WHELPINGTON, KIRK (*St. Bartholomew*), a parish, in the union of BELLINGHAM, North-Eastern Division of TINDALE ward, Southern Division of the county of NORTHUMBERLAND; containing 789 inhabitants, of which number, 260 are in the township of Kirk-Whelpington, 11 miles (E. by N.) from Bellingham. The river Wansbeck has its source in this parish, which is a hilly and extensive district, for the most part composed of sheep and dairy farms. Limestone and sandstone are plentiful, and the moors afford an almost inexhaustible supply of peat for fuel. The living is a vicarage, valued in the king's books at £7. 3. 4.; present net income, £288; patron, Bishop of Durham; impropriator, Sir J. E. Swinburne, Bart. The church is ancient, and constitutes the remains of a much larger structure. Here is a National school. There is a spring, the water of which is impregnated with sulphur, and has been found efficacious in chronic disorders. In various parts of the parish are traces of circular and rectilinear earthworks, probably thrown up in the border wars, for the protection of cattle from the moss-troopers. Whelpington tower, now the vicarage-house, was anciently fortified.

WHELPINGTON, WEST, a township, in the parish of KIRK-WHELPINGTON, union of BELLINGHAM, North-Eastern Division of TINDALE ward, Southern Division of the county of NORTHUMBERLAND, $15\frac{1}{2}$ miles (W.) from Morpeth; containing 72 inhabitants. Horn's castle, situated on a commanding eminence in this township, has been converted into a farm-house.

WHENBY (*St. Martin*), a parish, in the union of EASINGWOULD, wapentake of BULMER, North Riding of the county of YORK, $9\frac{1}{4}$ miles (E.) from Easingwould; containing 115 inhabitants. The living is a discharged vicarage, valued in the king's books at £4. 8. 4.; present net income, £121; patron and impropriator, W. Garforth, Esq.

WHEPSTEAD, a parish, in the union and hundred of THINGOE, Western Division of the county of SUFFOLK, $4\frac{1}{4}$ miles (S. S. W.) from Bury-St. Edmund's;

containing 618 inhabitants. The living is a rectory, valued in the king's books at £14. 4. 2.; present net income, £468; patron and incumbent, Rev. T. Image. Thomas Sparke, in 1721, bequeathed land, now producing an annual income of about £21, for which ten poor children are educated. The parish is entitled to a share of the bequests of Sir Robert Drury and Sir Robert Jarvis, which, together with £8, the interest of £200 given by J. W. Allen, Esq., are distributed among the poor at Christmas. Plumpton, an ancient house in this parish, is the seat of Lieut.-Gen. Sir Thomas Hammond.

WHERSTEAD (*St. Mary*), a parish, in the hundred of Samford, Eastern Division of the county of Suffolk, 2¾ miles (S. by W.) from Ipswich; containing 233 inhabitants. This parish is situated on the river Orwell, adjoining the town of Ipswich; and at a very early period there was a small religious foundation, united to the priory of St. Peter and St. Paul at Ipswich. Wherstead Lodge, a handsome mansion, is the seat of Sir Robert Harland, Bart. There are several crag pits in the parish. The living is a discharged vicarage, valued in the king's books at £5. 6. 8.; present net income, £194: it is in the patronage of the Crown; impropriator, Sir R. Harland, Bart. A school is supported by Lady Harland, and a Sunday school by the vicar.

WHERWELL (*Holy Cross*), a parish, in the union of Andover, hundred of Wherwell, Andover and Northern Divisions of the county of Southampton, 3¾ miles (S. S. E.) from Andover; containing 686 inhabitants. The living is a vicarage, with those of Bullington and Tufton annexed, valued in the king's books at £14; present net income, £301; patron, Colonel Iremonger, as rector of the sinecure rectory, which was a prebend in the nunnery of Wherwell, and is valued in the king's books at £44. 11. 0½. A day and Sunday school is partly supported by the curate. The rivers Test and Ande run through the parish; the latter falls into the Redbridge and Andover canal. A fair for sheep is held on Sept. 24th. A Benedictine nunnery was founded and amply endowed here, about 986, by Elfrida, Queen Dowager, in expiation of the murders of her first consort, Athelwold, and her step-son, Edward the Martyr; she spent the latter part of her life in it, and was buried within its walls. It was dedicated to the Holy Cross and St. Peter, and at the dissolution had a revenue of £403. 12. 10. There is a very extensive wood in this parish, in a recess of which is a stone cross, with the following inscription on its base: "About the year of our Lord DCCCCLXIII Upon this spot, beyond the time of memory Called Dead Man's Plack, Tradition reports that Edgar (Sirnamed the Peaceable) King of England, in the ardour of Youth, Love and Indignation, Slew with his own hand his treacherous and ungrateful Favourite, Earl Athelwold, owner of this Forest of Harewood, in resentment of the Earl's having basely betrayed his Royal confidence, and perfidiously married his intended Bride The beautious Elfrida, Daughter of Ordgar, Earl of Devonshire, after Wife to King Edgar and by him Mother of King Etheldred the 2nd, which Queen Elfrida, after Edgar's Death, murdered, his eldest Son, King Edward the Martyr, and founded the Nunnery of Whorwell."

WHESSOE, a chapelry, in the parish of Haughton-le-Skerne, union of Darlington, South-Eastern Division of Darlington ward, Southern Division of the county palatine of Durham, 2½ miles (N. by W.) from Darlington; containing 123 inhabitants. The Stockton and Darlington railway passes through the chapelry.

WHESTON, a hamlet, in the parish of Tideswell, hundred of High Peak, Northern Division of the county of Derby; containing 75 inhabitants.

WHETMORE, a joint chapelry with Buraston, in the parish of Burford, union of Tenbury, hundred of Overs, Southern Division of the county of Salop, 2½ miles (N. N. E.) from Tenbury: the population is returned with the parish.

WHETSTONE, a hamlet, in the parish of Tideswell, union of Bakewell, hundred of High Peak, Northern Division of the county of Derby, 1¼ mile (W.) from Tideswell; containing 75 inhabitants.

WHETSTONE (*St. Peter*), a parish, in the union of Blaby, hundred of Guthlaxton, Southern Division of the county of Leicester, 5¼ miles (S. S. W.) from Leicester; containing 903 inhabitants, a few of whom are employed in frame-work knitting. The living is annexed to the vicarage of Enderby.

WHETSTONE, a hamlet and chapelry, partly in the parish of East Barnet, hundred of Cashio, or liberty of St. Albans, county of Hertford, and partly in that of Fryern-Barnet, Finsbury Division of the hundred of Ossulstone, county of Middlesex, 8 miles (N. N. W.) from London: the population is returned with the respective parishes. The living is a perpetual curacy; net income, £50; patrons, Bishop of London and the Trustees. The chapel is dedicated to St. John. Here is a National school.

WHICHAM (*St. Mary*), a parish, in the union of Bootle, Allerdale ward above Derwent, Western Division of the county of Cumberland, 10 miles (S. S. E.) from Ravenglass; containing 285 inhabitants. The living is a rectory, valued in the king's books at £8. 15.; patron, Earl of Lonsdale. The tithes have been commuted for a rent-charge of £160, subject to the payment of rates; the glebe comprises 75 acres, valued at £90 per annum. At High and Low Team, in this parish, are extensive iron-works. An annuity of £16, supposed to have been granted by Queen Elizabeth from the Crown revenues of the county, and payable out of the Exchequer, is applied towards the support of a grammar school for the parishes of Whicham and Millom. The school-house was erected at Churchgate, pursuant to a decree of Chancery in 1688.

WHICHFORD (*St. Michael*), a parish, in the union of Shipston-upon-Stour, Brails Division of the hundred of Kington, Southern Division of the county of Warwick, 6 miles (S. E.) from Shipston-upon-Stour; containing 638 inhabitants. The living is a rectory, valued in the king's books at £19. 8. 6½.; present net income, £623; patron, Earl Beauchamp. Here is a National school.

WHICKHAM (*St. Mary*), a parish, in the union of Gateshead, Western Division of Chester ward, Northern Division of the county palatine of Durham; containing 3848 inhabitants, of which number, 873 are in the township of Whickham, 3½ miles (W. S. W.) from Gateshead. The village is pleasantly situated on an eminence, and contains several very neat and respectable houses. There are extensive coal mines in the parish; also a bed of calcined earth, produced by the English having set fire to their camp, when pressed by the Scottish army under Leslie, which communicated with a seam of coal that burnt with great fury for some years. It has been made a polling-place for the northern division of the county. The living is a rectory, valued in the king's books at £20. 8. 11½.; present net income, £663; patron, Bishop of Durham. A charity school was erected about 1711, by Robert Tomlinson, D.D., then rector of the parish; it is supported by various bequests subsequently made by him and others, and from the rental of certain galleries and pews constructed in the church at the founder's expense: the building was enlarged in 1825, by Archdeacon Bouyer, and the school is conducted on the National system. Another school, chiefly for girls, is partly supported by the Bishop of Hereford. John Hewett, in 1738, left a small fund for apprenticing children; and about £30 per annum, the produce of some bequests by Bishop Wood, Sir James Clavering, and others, is annually distributed among the poor in money and bread. See SWALWELL.

WHIDHILL, a township, in the parish of ST. SAMPSON, borough of CRICKLADE, hundred of HIGHWORTH, CRICKLADE, and STAPLE, Cricklade and Northern Divisions of the county of WILTS, 2¾ miles (S. E.) from Cricklade: the population is returned with the parish.

WHILE, a joint parish with Puddlestone, in the hundred of WOLPHY, county of HEREFORD, 5½ miles (E. by N.) from Leominster: the population is returned with Puddlestone. The living is a rectory, united to that of Puddlestone.

WHILLYMOOR, a township, in the parish of ARLECDON, ALLERDALE ward above Derwent, Western Division of the county of CUMBERLAND, 5½ miles (E. N. E.) from Whitehaven: the population is returned with the parish.

WHILTON (*ST. ANDREW*), a parish, in the union of DAVENTRY, hundred of NEWBOTTLE-GROVE, Southern Division of the county of NORTHAMPTON, 5 miles (E. N. E.) from Daventry; containing 397 inhabitants. The living is a rectory, valued in the king's books at £12. 16. 3.; present net income, £328; patron, William Rose, Esq. The old Watling-street, the Grand Junction canal, and the London and Birmingham railway, pass through the parish. Jonathan Emery, in 1768, bequeathed £500, and Judith Worsfold £1000 three per cent. consols., producing together £40 per annum, which is applied in aid of a National school.

WHIMPLE (*ST. MARY*), a parish, in the union of ST. THOMAS, hundred of CLISTON, Woodbury and Northern Divisions of the county of DEVON, 4½ miles (W. N. W.) from Ottery-St. Mary; containing 739 inhabitants. The living is a rectory, valued in the king's books at £30; present net income, £357; patron, Duke of Bedford. A day and Sunday school is supported by subscription. A fair for sheep is held on the Monday before Michaelmas-day.

WHINBERGH (*ST. MARY*), a parish, in the union of MITFORD and LAUNDITCH, hundred of MITFORD, Western Division of the county of NORFOLK, 3½ miles (S. S. E.) from East Dereham; containing 219 inhabitants. The living is a discharged rectory, with that of Westfield united, valued in the king's books at £6. 18. 6½.; present net income, £283; patron, Sir W. Clayton, Bart.

WHINFELL, a township, in the parish of BRIGHAM, union of COCKERMOUTH, ALLERDALE ward above Derwent, Western Division of the county of CUMBERLAND, 3¼ miles (S.) from Cockermouth; containing 122 inhabitants.

WHINFELL, a township, in the parish, union, and ward of KENDAL, county of WESTMORLAND, 6½ miles (N. E. by N.) from Kendal; containing 214 inhabitants. The school-room was rebuilt by the late Mr. Shepherd, but there is no endowment.

WHIPPINGHAM (*ST. MILDRED*), a parish, in the liberty of EAST MEDINA, Isle of Wight Division of the county of SOUTHAMPTON, 3½ miles (N. by E.) from Newport; containing 2229 inhabitants. The parish lies on the east side of the navigable river Medina, and is bounded on the north-east by the Motherbank: it contains the populous hamlet of East Cowes, which is separated from West Cowes by the river. On the brow of a neighbouring hill is East Cowes castle, erected by Mr. Nash for his own residence, and commanding some fine sea views; it has one square, and two circular, embattled towers. Old Castle point, on this coast, is the site of a fort constructed in the reign of Henry VIII. The living is a rectory, valued in the king's books at £19. 1. 5½.; present net income, £757: it is in the patronage of the Crown. The church is a large structure, principally in the later English style, with a tower and spire. Here is a Sunday National school. At Barton, in this parish, was an oratory of Augustine monks, founded in 1282 by John de Insula, the remains of which have been converted into a farm-house.

WHIPSNADE (*ST. MARY MAGDALENE*), a parish, in the union of LUTON, hundred of MANSHEAD, county of BEDFORD, 3 miles (S. W.) from Dunstable; containing 204 inhabitants. The living is a discharged rectory, valued in the king's books at £7. 13. 4.; present net income, £156: it is in the patronage of the Crown.

WHISBY, a chapelry, in the parish of DODDINGTON, Lower Division of the wapentake of BOOTHBY-GRAFFO, parts of KESTEVEN, union and county of LINCOLN, 6 miles (S. W. by W.) from Lincoln; containing 58 inhabitants.

WHISSENDINE (*ST. ANDREW*), a parish, in the union of OAKHAM, hundred of ALSTOE, county of RUTLAND, 4½ miles (N. W. by N.) from Oakham; containing 800 inhabitants. The living is a discharged vicarage, valued in the king's books at £7. 1.; present net income, £155; patron, Earl of Harborough; impropriator, W. Bissill, Esq. A school is chiefly supported by the Earl of Harborough.

WHISSONSETT (*ST. MARY*), a parish, in the union of MITFORD and LAUNDITCH, hundred of LAUNDITCH, Western Division of the county of NORFOLK, 4¾ miles (S.) from Fakenham; containing 628 inhabitants. The living is a discharged rectory, with that of Horningtoft united, valued in the king's books at £10. 3. 4.; present net income, £714; patron, F. R. Reynolds, Esq. In the chancel of the church are gravestones, of grey marble, with effigies of some members of the ancient family of Bozoun.

WHISTLEY-HURST, a liberty, in the parish of HURST, union of WOKINGHAM, hundred of CHARLTON, county of BERKS, 5¼ miles (E. by N.) from Reading; containing 867 inhabitants.

WHISTON, a township, in the parish and union of

PRESCOT, hundred of WEST DERBY, Southern Division of the county palatine of LANCASTER, 1¼ mile (S.) from Prescot; containing 1468 inhabitants, most of whom are employed in the extensive collieries here.

WHISTON (*St. Mary*), a parish, in the union of HARDINGSTONE, hundred of WYMMERSLEY, Southern Division of the county of NORTHAMPTON, 6¾ miles (E. by S.) from Northampton; containing 64 inhabitants. The living is a rectory, to which a portion of the rectory of Denton is annexed, valued in the king's books at £14. 11. 0½.; present net income, £286; patron, Lord Boston. The church was built about 1530, by Anthony Catesby, Esq.: it is remarkable for the beauty of its proportions, and is in the later style of English architecture, with a lofty and elegant tower crowned with rich pinnacles; the font is octagonal, with panelled sides handsomely executed. In the chancel is a monument to the founder, and there are several to the Boston family, of which, one to the memory of the first Lord Boston, and his secretary, is finely executed by Nollekens. The river Nene runs through the parish, in which are the remains of a moated building, said to have been the residence of King John. Limestone abounds here.

WHISTON, a township, in the parish of PENKRIDGE, Eastern Division of the hundred of CUTTLESTONE, Southern Division of the county of STAFFORD, 2 miles (W.) from Penkridge: the population is returned with the township of Penkridge. Here is a place of worship for Methodists.

WHISTON, a township, in the parish of KINGSLEY, union of CHEADLE, Southern Division of the hundred of TOTMONSLOW, Northern Division of the county of STAFFORD, 3¾ miles (N. E.) from Cheadle; containing 549 inhabitants.

WHISTON (*St. James*), a parish, in the union of ROTHERHAM, Southern Division of the wapentake of STRAFFORTH and TICKHILL, West Riding of the county of YORK, 2½ miles (S. E.) from Rotherham; containing 927 inhabitants. The living is a rectory, valued in the king's books at £10; present net income, £868; patron, Lord Howard of Effingham. A gallery has been erected in the church, and 40 free sittings provided, the Incorporated Society having granted £25 in aid of the expense. There is a place of worship for Wesleyan Methodists. Francis Mansel, in 1728, bequeathed a rent-charge of £6; and Joseph Hammond, in 1794, gave £300, since increased to £443, for teaching poor children. Here is also a Sunday National school.

WHISTONES, a tything, in the parish of CLAINES, Lower Division of the hundred of OSWALDSLOW, union, and Worcester and Western Divisions of the county, of WORCESTER, adjacent to the north side of the city of Worcester; containing 2518 inhabitants. A priory of White nuns, in honour of St. Mary Magdalene, was founded here before 1255, by a bishop of Worcester, which at the dissolution had a revenue of £56. 3. 7. An hospital, dedicated to St. Oswald, said to have been founded by Bishop Oswald, for a master and poor brethren, existed before 1268, and at the dissolution was valued at £15. 18. per annum, when it was granted to the Dean and Chapter of Worcester: it was demolished in the reign of Elizabeth, but after the Restoration was rebuilt by Bishop Fell, who recovered most of its ancient possessions, and it now affords an asylum for twelve poor men. A new gaol has been erected here.

WHITACRE, NETHER (*St. Giles*), a parish, in the union of MERIDEN, Atherstone Division of the hundred of HEMLINGFORD, Northern Division of the county of WARWICK, 3¼ miles (N. E.) from Coleshill; containing 413 inhabitants. The living is a perpetual curacy, endowed with the rectorial tithes; net income, £333; patron, Earl Howe. There is a place of worship for Wesleyan Methodists. Charles Jennins, Esq., in 1775, bequeathed one-third of the interest of £1000 in support of a school; and a Sunday school is supported by the rector.

WHITACRE, OVER (*St. Leonard*), a parish, in the union of MERIDEN, Atherstone Division of the hundred of HEMLINGFORD, Northern Division of the county of WARWICK, 3½ miles (E. N. E.) from Coleshill; containing 288 inhabitants. The living is a donative; net income, £142; patron, Earl Digby. Six poor children are educated for the interest of £40, bequeathed by the Rev. Thomas Morrall, in 1740.

WHITBECK (*St. Mary*), a parish, in the union of BOOTLE, ALLERDALE ward above Derwent, Western Division of the county of CUMBERLAND, 8¾ miles (S. S. E.) from Ravenglass; containing 234 inhabitants. This parish is situated between the mountain of Black-Comb and the sea. In the former is a cavity, similar to the crater of a volcano, several hundred yards in diameter and depth; the inside is lined with vitrified and chrystallised matter, having at the bottom a fine spring of water. On the west side of the mountain is a cascade, and on the shore a mineral spring, formerly in repute for the cure of gravel and scurvy. Trunks of oak and fir, of an immense size, have been found in the peat mosses, considerably below the surface. The living is a perpetual curacy; net income, £76; patron, Earl of Lonsdale. The impropriate tithes have been commuted for a rent-charge of £88. 17. 4., subject to the payment of rates, which on the average have amounted to £3. 15. The perpetual curate has 4 acres of glebe, which, with the house, is valued at £10 per annum. An hospital, built by the parishioners in 1632, is endowed with a rent-charge of £24, purchased with the sum of £400 bequeathed by Henry Parke, a native of this place. Here are the remains of three Druidical temples; one, termed Standing-stones, consists of eight massive stones disposed in a circle; Kirkstones, of thirty, in two circles, like Stonehenge; and the third of twelve stones: there is also a large cairn, encompassed at the base by a circle of huge stones.

WHITBOURNE (*St. John the Baptist*), a parish, in the union of BROMYARD, hundred of BROXASH, county of HEREFORD, 6 miles (E. by N.) from Bromyard; containing 899 inhabitants. The living is a rectory, valued in the king's books at £14. 14. 9½.; present net income, £533; patron, Bishop of Hereford, whose predecessors had anciently a palace here, the remains of which are occupied as a farm-house. Ten children are taught in a school for the use of a house and garden belonging to the parish; here is also a Sunday National school.

WHITBURN, a parish, in the union of SOUTH SHIELDS, Eastern Division of CHESTER ward, Northern Division of the county palatine of DURHAM, 3½ miles (N.) from Sunderland; containing 1001 inhabitants. This parish is bounded on the east by the North Sea; and on the shore several copper coins of Constantine,

Licinius, Maxentius, and Maximian, have been found. Limestone is quarried to a great extent, and conveyed up the river Tees into Yorkshire: coal also is obtained here, though it lies at a great depth. The living is a rectory, valued in the king's books at £39. 19. 4½.; present net income, £1113; patron, Bishop of Durham. There is a place of worship for Wesleyan Methodists. A National school is endowed with £10 per annum by Lord Crewe's Trustees. Dr. Triplett, in 1664, bequeathed a rent-charge of £18, which has been increased to £61. 4. per annum, which sum is appropriated to apprenticing boys and girls of the parishes of Whitburn, Washington, and Woodhorn. There are several chalybeate springs in the neighbourhood, the water of which is in great request among the inhabitants.

WHITBY, a township, partly in the parish of EASTHAM, and partly in that of STOAK, union, and Higher Division of the hundred, of WIRRALL, Southern Division of the county of CHESTER, 6¾ miles (N.) from Chester; containing 234 inhabitants. The Ellesmere, or Wirral, canal passes through the parish, and communicates with the river Mersey, where a port with extensive docks and wharfs is about to be formed.

WHITBY (*ST. MARY*), a sea-port, newly enfranchised borough, market-town, and parish, and the head of a union, in the liberty of WHITBY-STRAND, North Riding of the county of YORK; containing 11,725 inhabitants, of which number, 10,429 are in the town of Whitby, 48 miles (N. N. E.) from York, and 241 (N. by W.) from London. This place was originally called by the Saxons *Streoneshalh,* which Bede translates *Sinus fari,* the bay of the lighthouse, or tower, probably from a tower built there by the Romans. At the time of the Conquest, either from the colour of the houses, or from its conspicuous situation, it was called *Whitteby,* or the White Town, of which its present name is a contraction; and a certain portion of the town, or an appendage belonging to it, was then called *Prestby,* or Priests' Town. Its origin may be ascribed to the monastery founded by Oswy, King of Northumberland, in fulfilment of a vow which he made in 655, prior to his encountering Penda, King of Mercia, who had invaded his dominions, and whom he defeated and killed. In fulfilment of the same vow, his daughter Ælfleda became a nun in this monastery, which was placed under the superintendence of Hilda, grand niece of Edwin, King of Northumberland, who, in 658, came from Hartlepool to commence this establishment, of which she was made the first abbess. A national synod, at which Oswy presided, was held here in 664, for the regulation of some minor observances in ecclesiastical affairs, about which considerable differences at that time prevailed. Under Lady Hilda, and her successor Ælfleda, the monastery became celebrated as a seat of learning, where several bishops and many eminent men received their education; and several cells, or smaller convents, were erected as appendages to the abbey, of which the principal was that of Hackness, founded in 679, by Hilda, who died in the following year, and whose virtues subsequently obtained for her the honour of canonization. The Danes, in 867, completely destroyed the monastery, laid waste the town, and massacred the inhabitants with the most horrible barbarity: the abbot previously escaped, and is said to have carried with him the relics of St. Hilda to Glastonbury: but so complete was the devastation of these merciless invaders, that the town remained in ruins till its ancient name was lost. Towards the close of the Saxon period, Whitby began to revive, and at the time of the Conquest had become so considerable, that the manor, including its dependencies, was valued in the Norman survey at £112, being the richest in the north-east part of the county of York. The Conqueror granted the site of the town and monastery to his nephew, Hugh Earl of Chester, who assigned it to William de Percy, by whose assistance the monastery was rebuilt by Reinfrid, a Benedictine monk from the abbey of Evesham, soon after 1074. William de Percy afterwards endowed it with 240 acres of land, and its revenues were subsequently augmented by the Earl of Chester, who also granted it a charter conferring many privileges. After having been plundered and greatly injured by a band of pirates, it was again restored and dedicated to St. Peter and St. Hilda, by William de Percy, who appointed his brother Serlo prior. The monastery, notwithstanding the repeated attacks of pirates, to which it was continually exposed, increased in wealth and importance: it was constituted an abbey by Henry I., its privileges were greatly extended by royal charters and pontifical decrees, and it continued to flourish till the dissolution, at which period its revenue was £437. 2. 9. Tradition relates that Robin Hood and Little John paid a visit to Richard de Waterville, who was then abbot, and, to give a proof of their dexterity in archery, shot each of them an arrow from the tower of the abbey to the distance of more than a mile; and that pillars were erected by the abbot on the spots where the arrows fell, to commemorate the event. The enclosures are still called Robin Hood and Little John's fields; and about six miles from the town is Robin Hood's bay, where that celebrated outlaw is said to have kept a small fleet, to assist his escape when pursued by an enemy.

The town is situated on the shore of the North Sea, at the mouth of the river Esk, by which it is divided into two nearly equal parts, and consists of several streets, of which the greater number are narrow, and some steep; the houses are partly of stone and partly of brick, and several of the modern buildings, of both kinds, are spacious and elegant: it is paved under the provisions of an act of parliament obtained in 1837, and is lighted with gas from the works of a company established in 1825. The approaches have been greatly improved; the environs, in which are many gentlemen's seats and pleasant villas, abound with pleasing and picturesque scenery, and the view of the town derives considerable beauty from the ruins of the abbey on the east bank of the river. A handsome stone building of three stories was erected by subscription on the west pier, in 1826, of which the lower story is occupied by baths, with spacious ante-rooms, dressing-rooms, and every requisite accommodation; the middle, or principal, story is occupied by the public subscription library, established in 1776; and the upper story by the museum of the Literary and Philosophical Society, instituted in 1823. The museum contains an extensive collection of petrifactions and fossil organic remains, which abound in the rocks in the vicinity, especially in the alum rock. The fossils form a rich variety of plants, shells, and fishes, including some animals of the larger size, among which are the ichthyosaurus, or lizard fish, and the plesiosaurus. The finest specimen in the collection, and perhaps the noblest fossil in the world, is the great crocodile found in the alum rock near Whitby, in 1824; it is nearly entire, and, when alive, must have measured eighteen feet in length. There are also several varieties of ammonites, or snake stones, which are found here in great abundance, and are fabled to have been snakes deprived of their heads, and petrified by St. Hilda: three white snakes on a blue shield were the ancient arms of the abbots of Whitby. A handsome news-room was erected by subscription, in Haggersgate, in 1814; and a tradesmen's news-room was also established on the opposite side of the river. A commodious theatre was built by subscription, in 1784, but was destroyed by fire in 1823, and has not been since rebuilt. Assemblies are held in a handsome room at the Angel Inn.

The origin of the commercial prosperity and consequent importance of Whitby may be attributed to the discovery of the alum mines, in the latter part of the reign of Elizabeth. Early in the seventeenth century these mines were worked, and very soon became the source of an extensive trade, of which the necessary supply of coal requisite for conducting the works formed an additional branch; ship-building was introduced about 30 years after, and great improvements were made in the harbour, by removing rocks and other obstructions. In 1632 the west pier was rebuilt, principally through the exertions of Sir Hugh Cholmeley, under whose sanction a subscription of £500 was raised for promoting the work, and the town began to assume a considerable degree of maritime importance. An act of parliament was obtained in 1702, imposing certain dues for the improvement of the piers and harbour, which, being in force only for a limited time, was, on its expiration, renewed from time to time, with certain modifications, till 1827, when it was made perpetual. The duties collected under the authority of this act consist of a halfpenny per chaldron (Newcastle measure) on all coal shipped at that port and at Sunderland, with their dependencies, and of some trifling payments on coal, grain, salt, and other commodities delivered in the harbour. By these means the piers and the harbour have been very greatly improved; the former, with the exception of those at Ramsgate, are perhaps the finest in England; there are also two inner piers to break the force of the waves in stormy weather, the Burgess pier and the Fish pier. A very handsome bridge, with horizontal leaves, was erected across the Esk in 1835, at an expense of £10,000: it will allow vessels of 600 tons' burden to pass into the inner harbour, which is capable of receiving a large fleet, and affords secure shelter in stormy weather. Adjoining the inner harbour are spacious dock-yards, and commodious dry docks for the building and repairing of ships. Ship-building was carried on here to a very considerable extent in the time of the late war, during which the extensive demand for transports was a source of great emolument to the ship-builders and owners: the return of peace, and other causes, occasioned for a period the decline of this trade, and of all the branches connected with it; but within the last few years the trade has greatly revived, and is now carried on as briskly as ever. The shipping belonging to the port consists of 299 vessels, whose aggregate burden is 45,625 tons. The larger vessels are employed in the foreign trade, and the smaller coastwise, especially in the coal trade. The Greenland and Davis' Straits fisheries, which formerly were of great benefit to the town, and in 1819 employed twelve large ships, are now altogether abandoned; and the manufacture of alum, which once constituted a principal branch of the trade, has also very much declined; several of the works have been discontinued, and those which are still carried on are not in full operation. The works at Kettleness were totally destroyed, in Dec. 1829, by the falling of the rock under which they were situated, but have been lately rebuilt. The export trade in alum to France, Holland, and other parts of the continent, was formerly very considerable, but it has long been extinct, and the principal part of the alum now manufactured is sent coastwise to London, Hull, and other towns. The sail-cloth manufactured at Whitby is in great repute; and during the last two years the quantity produced has been greater than at any former period.

The fishery on the coast, which formed originally the principal employment of the inhabitants, was very much injured by the late war, which afforded more advantageous opportunities for the employment of capital and labour. The fish generally taken are cod, ling, halibut, soles, and haddocks; salmon and salmon-trout have been very scarce for several years, though at one time they were plentiful in the Esk, and were a considerable article of trade; salmon-trout has been lately taken on the coast by a particular method of fishing, and sometimes in considerable quantities. Since 1833 the herring fishery has been carried on here on an extensive scale, principally through the exertions of the Herring Company. Oats, butter, and bacon, were formerly sent to London in great cargoes, but this trade has been much reduced within the last 20 years. The foreign exports are inconsiderable: the principal imports are timber from British America, and timber, wooden wares, hemp, and flax, from the Baltic. The principal articles sent coastwise are freestone, whinstone, and ironstone, of which great quantities are procured from the quarries at Aislaby, Grosmont, &c., by the Whitby Stone Company. The establishment of this, and of other trading and manufacturing associations in the place, must be ascribed to the formation of the Whitby and Pickering railway, completed in 1836; an undertaking which, though hitherto unprofitable to the proprietors, has been eminently beneficial to the public. The chief articles brought coastwise are groceries, salt, and coal. The amount of duties paid at the custom-house, for the year ended January 5th, 1837, was £1106. In 1839, this port obtained the privilege of bonding generally, with the usual restriction on tobacco. The depth of water in the harbour in common neap tides is about 10 feet, and in spring tides from 15 to 16 feet, and sometimes more. The great swell occasioned by tempestuous weather renders the harbour difficult of access, and many disastrous accidents had occurred before it attained its present state of security. In November 1710, the shipping was greatly damaged by a violent storm, in which it suffered injury to the amount of £40,000; and in the night of the 24th of December, 1787, a newly erected quay, on which was a range of houses, 80 feet above the level of the sea, from the insecurity of the foundation, suddenly fell, involving the greater number of them in one common ruin, and leaving 196 families, who had been forewarned of their danger in time to escape, destitute of a home. The custom-house is a commodious building in Sandgate; a small office for the governor of the pilots has been erected on the pier; and there are four marine associations for mutual insurance. Whitby is a station for the preventive coast guard; and regular trading vessels to and from London, Hull, and Newcastle, sail from two wharfs in Church-street. The market, granted by Henry VI., is held on Saturday; and there are fairs on August 25th and Martinmas-day; the former was granted in 1168, by Henry II., to the abbot, but they are both inconsiderable.

By the act of the 2nd of William IV., cap. 45, Whitby has been constituted a borough, with the privilege of sending one member to parliament, the right of election being in the £10 householders of the townships of Whitby, Ruswarp, and Hawsker *cum* Stainsacre, together comprising an area of 5132 acres. The number of registered voters is 445: the returning officer is annually appointed by the sheriff. The town is under the superintendence of the magistrates of the North Riding, who meet every Tuesday and Saturday. A court for the recovery of small debts, every third Monday, and a court leet, annually at Michaelmas, are held in the town-hall, a substantial building of the Tuscan order, situated in the market-place, and erected in 1788, by Nathaniel Cholmeley, Esq. This town has been made a polling-place for the North Riding.

The living is a perpetual curacy; net income, £206; patron, Archbishop of York. The church has been enlarged, and 300 free sittings provided, the Incorporated Society having granted £300 in aid of the expense: it is supposed to have been built by William de Percy, but has undergone so much alteration that little of the original structure remains. It is situated on the summit, and near the verge, of a high cliff, and is approached by 191 stone steps. Her Majesty's commissioners have made a conditional grant of £700 in aid of building a new church, and a subscriptiou is now in progress for that purpose. In the western division of the town is a neat chapel of ease, erected in 1778. There are two places of worship for Wesleyan Methodists, and one each for the Society of Friends, Independents, Presbyterians, Primitive Methodists, Unitarians, and Roman Catholics. Lancasterian or public schools have been established for 30 years; and within the last two years, infants' schools have also been introduced. A National school has been erected at Sleights, in this parish; and here is a school for the children of Roman Catholics, endowed with £20 per annum. The dispensary was established in 1786. The Seamen's hospital, erected in 1676, affords a comfortable asylum to 42 seamen's widows and their children: there are numerous small benefactions and donations applied annually to the relief of the poor, and several societies for administering to their wants, principally formed by the ladies. The poor law union of Whitby comprises 22 parishes or places, under the care of 24 guardians, and contains, according to the census of 1831, a population of 19,882.

The remains of Whitby abbey, which stand a small distance from the church, on a cliff 200 feet above the level of the sea, present one of the most interesting ruins in the country, and consist of the choir, or eastern part of the church; the north transept, which is nearly entire; and considerable portions of the north wall of the nave, and of the western wall, or front, of the building: the beauty of these ruins has been much impaired by the fall of the central tower on June 25th, 1830, occasioned by one of the four massive clustered columns by which it was supported having given way. About half a mile west of the pier is a chalybeate spring, celebrated formerly, when baths existed near the spot, but the latter were demolished about a century ago, and the former has fallen into disuse. The neighbourhood abounds with petrifactions, which have been frequently investigated by learned naturalists and other scientific inquirers, and are particularly described in Young's "Geological Survey of the Yorkshire Coast." Some lands near Whitby are still held by the tenure formerly called Horngarth, thought to be that now called the Penny Hedge: this service, which is supposed to have been originally intended for the repairs of the quays and piers of the harbour, at that time constructed of wood with stones thrown in between, is still continued by a family of the name of Herbert; it consists in the erection of a small hedge of stake and yether in the harbour, on the eve of Ascension-day.

WHITCHBURY, county of WILTS.—See WHITSBURY.

WHITCHESTER, a township, in the parish of HEDDON-ON-THE-WALL, union of CASTLE ward, Eastern Division of TINDALE ward, Southern Division of the county of NORTHUMBERLAND, 9½ miles (W. N. W.) from Newcastle-upon-Tyne; containing 57 inhabitants. Here was the site of a Roman station, defended on almost every side by deep ravines.

WHITCHURCH (*ST. JOHN THE EVANGELIST*), a parish, in the union of AYLESBURY, hundred of COTTESLOE, county of BUCKINGHAM, 4¾ miles (N. by W.) from Aylesbury; containing 928 inhabitants. The living is a discharged vicarage, valued in the king's books at £8. 17.; present net income, £61; it is in the patronage of the Crown. There is a place of worship for Wesleyan Methodists. A market on Monday, and a fair on the festival of St. John the Evangelist, granted in 1245, were formerly held here. John Westcar, in 1833, bequeathed £500 to each of the parishes of Whitchurch and Cublington, in the county of Bucks, and Souldern, in the county of Oxford, desiring the interest to be appropriated to a supply of clothing for the poor.

WHITCHURCH (*ST. ANDREW*), a parish, in the union of TAVISTOCK, hundred of ROBOROUGH, Tavistock and Southern Divisions of the county of DEVON, 1¼ mile (S. E.) from Tavistock; containing 791 inhabitants. The living is a vicarage, valued in the king's books at £16. 5. 5.; present net income, £195; patron, incumbent, and impropriator, Rev. Peter Sleeman. Walreddon House, the property of William Courtenay, Esq., a descendant of the Courtenays, Earls of Devon, is an ancient mansion of the time of Edward VI., whose arms in the hall are still in good preservation. Holwell House, also in this parish, was, until within a recent period, the property and residence of the Glanville family; it is now the property of John Scobell, Esq. The Tavistock races are held annually on Whitchurch Down. A chantry chapel was founded in 1300, by the abbot of Tavistock. Francis Pengelly, Esq., in 1719, left £6 per annum for teaching six poor children; also £4 per annum, to be given in wool to poor women not receiving parochial relief.

WHITCHURCH (*St. Dubritius*), a parish, in the union of Monmouth, Lower Division of the hundred of Wormelow, county of Hereford, 6½ miles (S. W. by S.) from Ross; containing 885 inhabitants. The navigable river Wye runs through the parish, in which was formerly an extensive iron forge, for some years discontinued. The living is a rectory, with that of Ganerew annexed, valued in the king's books at £6. 0. 2½.; present net income, £300; patron, Josiah Pyrke, Esq. There is a place of worship for Wesleyan Methodists.

WHITCHURCH (*St. Mary*), a parish, in the union of Bradfield, hundred of Langtree, county of Oxford, 6½ miles (N. W.) from Reading; containing 745 inhabitants. The living is a rectory, valued in the king's books at £16. 2. 8½.; present net income, £456: it is in the patronage of the Crown. Two day and Sunday schools are supported by subscription, and another school for girls is supported by a gentleman of the parish.

WHITCHURCH (*St. Alkmund*), a market-town and parish, chiefly in the Whitchurch Division of the hundred of North-Bradford, Northern Division of the county of Salop, but partly in the hundred of Nantwich, Southern Division of the county of Chester, 20 miles (N. by E.) from Shrewsbury, and 160 (N. W. by N.) from London; containing 5819 inhabitants. This place was anciently called *Album Monasterium* and *Blancminster*, which have the same signification as its present name, and appear to imply the existence of a monastery, of which, however, there is no account; an hospital was standing here in the reign of Henry III., and was endowed by the lord of the manor with the whole town of Wylnecot, for the relief of the poor at its gate. In 1211, King John assembled his forces here, prior to attacking the Welsh, on which occasion he penetrated to the foot of Snowdon. At the commencement of the civil war of the 17th century, the inhabitants appear to have taken an active part in favour of the king, and to have raised a regiment in support of his cause. Of the foundation and history of the ancient castle, a portion of the ruined walls of which was standing in 1760, nothing is known. The town is situated on elevated ground, in a rich and picturesque country, and contains some neat streets and respectable houses: in its neighbourhood are three fine lakes, called Osmere, Blackmere, and Brown Moss-water, and several brooks, one of which, called Red Brook, is the boundary between England and Wales; another separates this county from that of Chester. The trade is principally in malt and hops: shoes are manufactured for the Manchester market; and near the town is an establishment for making oak acid, also several limekilns and brick ovens. A branch of the Ellesmere canal extends to the town, by means of which boats ply to London and many intervening towns on Saturday, and to Manchester and Shrewsbury, from the canal wharf. The market is on Friday; and there are fairs on the second Friday in April, Whit-Monday, Friday after August 2nd, and October 28th. A high steward, who superintends the affairs of the town, is appointed by the lord of the manor, and presides at courts baron and leet held in October, at the town-hall, which is the depository for the rolls and archives of the lordship; and the town has been made a polling-place for the northern division of the county.

The living is a rectory, with Marbury annexed, valued in the king's books at £44. 11. 8,; present net income, £1458: it is in the patronage of the Countess of Bridgewater. The church, erected in 1722, on the site of an ancient edifice, is a noble structure of the Tuscan order, built of freestone, with a square embattled western tower: it contains several handsome monuments of the Shrewsbury family, and amongst them an effigy in alabaster of the renowned John Talbot, Earl of Shrewsbury, who was killed in France, in 1453, and who, for his remarkable prowess, was called the English Achilles. A chapel of ease and parsonage-house were built at Ash in 1835; and the Bridgewater church, dedicated to St. Catherine, an elegant structure, built and endowed by the Countess of Bridgewater, was consecrated August 31st, 1837; it cost upwards of £8000. There are places of worship for Baptists, Independents, Wesleyan Methodists, and Unitarians. The free grammar school, situated at Bargates, was founded in 1550, by Sir John Talbot, who was incumbent of the parish, and endowed by him with £200, which was augmented by a bequest from William Thomas in 1662, and others, now, with the original endowment, producing an income of £454 per annum: the Earl of Shrewsbury is hereditary visiter. The school is open to all boys, whether resident in the parish or not, for gratuitous instruction in Greek and Latin and mathematics: two of the scholars are also clothed. A charity school for children of both sexes, and an almshouse for six decayed housekeepers, were endowed by Samuel Higginson, in 1697, and Jane Higginson, in 1707, with property now producing about £250 per annum. A National school was also established in 1827, at Highgate; and there is a school in connection with the Presbyterians, which was founded and endowed by Thomas Beuyon, in 1707. Another school is partly supported by a lady. The interest of £2200, arising from bequests by Elizabeth Turton in 1794, and others, is distributed at the commencement of every year among persons in reduced circumstances; and a considerable sum is likewise annually distributed in bread. In 1828, the late Earl of Bridgewater, who was rector of the parish, bequeathed £2000 to the overseer and churchwarden, to be applied to such charitable use in the parish as they should think proper. At the northern extremity of the town is an extensive house of industry, built and principally supported from the funds of several bequests left for general purposes of relief, and which were vested in certain trustees appointed under an act of the 32nd of George III.: there are some vestiges of a foss in its vicinity. Whitchurch is the birthplace of Dr. Bernard, chaplain and biographer of Archbishop Usher, and of Abraham Wheelock, a celebrated linguist, who died in 1654.

WHITCHURCH, otherwise FELTON (*St. Gregory*), a parish, in the union and hundred of Keynsham, Eastern Division of the county of Somerset, 3 miles (N.) from Pensford; containing 423 inhabitants. The name Filton, or Felton, is derived from a very ancient town, situated to the north-west of the present village, in a forest, or chace, once called Filwood; a church having been erected on the site of an ancient chapel, dedicated to St. Whyte, the inhabitants of Filton gradually removed into its vicinity, upon which the new village and the parish assumed their present designation. The living is a perpetual curacy; net income, £88; patrons and impropriators, Sir J. Smyth, Bart., and W. G. Langton, Esq.

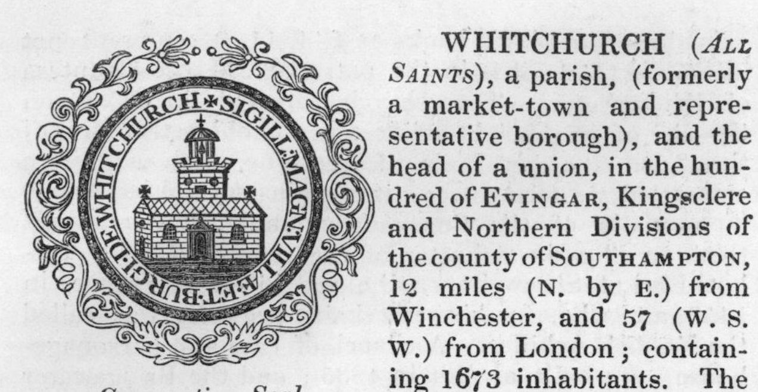

Corporation Seal.

WHITCHURCH (*All Saints*), a parish, (formerly a market-town and representative borough), and the head of a union, in the hundred of Evingar, Kingsclere and Northern Divisions of the county of Southampton, 12 miles (N. by E.) from Winchester, and 57 (W. S. W.) from London; containing 1673 inhabitants. The town, which is small and irregularly built, is situated on the river Test, on very low ground, under a range of chalk hills. Many of the inhabitants are employed in silk-weaving, and two silk-mills also furnish employment to about 100 persons: there are also several corn-mills on the river. A pleasure fair is held on the third Thursday in June, and there is one on October 19th and 20th, for cattle, pigs, &c. Whitchurch is a borough by prescription, and is governed by a corporation consisting of a titular mayor and bailiff, who do not now exercise any authority within the borough: they are chosen, with a constable, at the court leet of the lords of the manor, held annually in October at the town-hall, a neat building, erected about 50 years since; and another court is held at the manor farm, in May, under the Dean and Chapter of Winchester, as lords of the manor. It first sent members to parliament in the 27th of Queen Elizabeth, and was deprived of its franchise by the act passed to amend the representation in the 2nd of William IV. The right of election was vested in the freeholders (in their own right, or that of their wives) of land and tenements not divided since the act of the 7th and 8th of William III. The number of voters was about 84; and the mayor was the returning officer. This place is within the jurisdiction of the Cheyney Court held at Winchester every Thursday, for the recovery of debts to any amount. The living is a vicarage, valued in the king's books at £13. 12. 8½.; present net income, £120; patron, Bishop of Winchester; impropriator, J. Portal, Esq. The church, which is a low and plain structure, with a tower, contains a library, chiefly of theological works, bequeathed by the Rev. William Wood, to which access is obtained by permission of the vicar. There are places of worship for Baptists, Independents, and Wesleyan Methodists. A school for girls is supported by Mrs. Portar, and another is partly supported by subscription. Here is also a Sunday National school. A quantity of clothing and bedding, of the value of about £80, is annually distributed amongst the poor of this parish, arising from the produce of a bequest made by Richard Wollaston, Esq., in the year 1688. The poor law union of Whitchurch comprises 7 parishes or places, under the care of 10 guardians, and contains a population of 5175, according to the census of 1831.

WHITCHURCH (*St. Mary*), a parish, in the union of Stratford-upon-Avon, in a detached portion of the Kington Division of the hundred of Kington, Southern Division of the county of Warwick, 5½ miles (S. S. E.) from Stratford-upon-Avon; containing 261 inhabitants. The living is a rectory, valued in the king's books at £20. 17. 3½.; present net income, £186; patron, James West, Esq.

WHITCHURCH-CANONICORUM (*Holy Cross*), a parish, in the union of Bridport, hundred of Whitchurch-Canonicorum, Bridport Division of the county of Dorset, 5 miles (W. N. W.) from Bridport; containing 1399 inhabitants. This parish, which is one of the most ancient and most extensive in the county, derives its name from the original dedication of its church to St. Candida, or White, in honour of whom a monastery was founded here, which was called *Album Monasterium*, and at the time of the Norman survey belonged to the abbey of St. Wandragasil in Normandy. The grant of a market and a fair obtained in the reign of Henry III. was confirmed in the 4th of Edward II. Chidiock, a hamlet in this parish, was distinguished for an ancient castle, the residence of the Chidiocks and Arundels, which, during the civil war of the 17th century, became a powerful check upon the garrison of Lyme, and was alternately in the possession of the contending parties. Marshwood Vale is an extensive tract, in which were formerly two parks, and some stately mansions, of which only the ruins are remaining. The soil of the vale, which principally affords pasture for cattle, is a cold chalky clay, but in the parish generally the soil is deep and fertile. The living is a vicarage, endowed with a portion of the rectorial tithes, with those of Chidiock, Stanton-St. Gabriel, and Marshwood annexed, and valued in the king's books at £32. 6. 3.; present net income, £739; patron, Bishop of Bath and Wells; appropriators of the remaining portion of the rectorial tithes, the Dean and Chapter of Wells and of Sarum, in moieties. The church, originally dedicated to *St. Candida*, and afterwards to the *Holy Cross*, is an ancient structure with a tower, and contains some interesting monuments, among which are one to Sir John Jeffery, Knt., with his effigy in armour, and another to one of the Hemley family; the pulpit is curiously carved. A school is supported by subscription; and in another, 11 children are paid for by an allowance of £5 per annum from the poor's rate. Here is a Sunday National school.

WHITCLIFF, a joint township with Thorpe, in the parish and liberty of Ripon, West Riding of the county of York, 1½ mile (S.) from Ripon; containing, with Thorpe, 198 inhabitants.

WHITCOMBE, a parish, in the union of Dorchester, hundred of Culliford-Tree, Dorchester Division of the county of Dorset, 2¼ miles (S. E.) from Dorchester; containing 64 inhabitants. The living is a donative; net income, £13; patron, Hon. G. L. D. Damer.

WHITECHAPEL, a chapelry, in the parish of Kirkham, hundred of Amounderness, Northern Division of the county palatine of Lancaster, 5½ miles (S. E. by E.) from Garstang: the population is returned with the parish. The living is a perpetual curacy; net income, £104; patron, Vicar of Kirkham.

WHITECHAPEL (*St. Mary*), a parish, and the head of a union, in the Tower Division of the hundred of Ossulstone, county of Middlesex, adjoining the city of London, and containing 30,733 inhabitants. This populous parish extends in an eastern direction from Aldgate to Mile-End, a continuous line nearly a mile in length, and including Whitechapel High-street and Whitechapel-road, the former a noted market for butchers' meat, and the latter containing numerous manufacturing establishments. On the south side of

Whitechapel-road is the extensive and long-established bell-foundry of Mr. Mears. In Fieldgate-street, nearly adjoining, but within the hamlet of Mile-End Old Town, in the parish of Stepney, is the large iron-foundry belonging to Mr. Ford, to which is attached an extensive manufactory of gun carriages and wheelwrights' work, which latter department of that concern is in this parish. In Great Garden-street, on the north of the road, is a large brass-foundry belonging to Messrs. Glascott and Co.; and nearly opposite is the manufactory of Messrs. Schooling and Everet, for every kind of furnishing ironmongery, smoke and wind-up jacks, scales and scale beams, and other articles, upon a very extensive scale. Near this extremity of the parish, and partly bordering on that of Bethnal-Green, is the very extensive distillery of Mr. Smith, for British spirits and compounds, established in that family for nearly a century: the premises, which have been recently rebuilt on a commodious plan, occupy a large extent of ground; and the works, in which are two powerful steam-engines, afford employment to about 200 persons. In Thomas-street are the starch-works of Messrs. Joseph Lescher, Son and Co., which have been conducted by the Leschers for half a century: the works, in which a steam-engine of sixteen-horse power is applied to the grinding of wheat and other purposes connected with the manufactory, afford employment to 40 men, and from 800 to 900 hogs are usually fed on the premises. In Osborne-place is a large establishment belonging to Mr. Kirke, for the dyeing of woollen cloth: on the premises a spring of remarkably soft water has been opened by boring to the depth of 160 feet. In a southern direction the parish extends to Well-Close square, one half of which is within its limits; this portion of it comprises Goodman's Fields and several spacious and well-built streets, among which are Great Prescot-street, Leman-street, Great and Little Aylie-street, and others, in the neighbourhood of which, in almost every direction, are numerous and extensive establishments for the refining of sugar, which constitutes the principal trade of the place. In Church-lane is the proof-house of the City of London Company of Gunmakers, originally erected by the company in 1757, and handsomely rebuilt in 1818. There are several large manufactories of floor-cloth in Whitechapel-road, and some extensive establishments of coach and coach harness makers, with various other works in different parts of the parish. The Royal pavilion theatre, on the north side of the road, is a commodious and well-arranged building, with a handsome front and principal entrance between four Ionic pillars supporting a cornice. In Leman-street is the Royal Garrick theatre, which is neatly fitted up. The court of requests for the Tower Hamlets, situated in Osborne-street, established by act of the 23rd of George II., for the recovery of debts under £2, and of which the power has been extended to debts not exceeding £5, by an act of the 3rd of William IV., has jurisdiction over the parishes of St. Mary Whitechapel, Christchurch Spitalfields, St. Leonard Shoreditch, St. John Hackney, St. Matthew Bethnal-Green, St. Mary Stratford-Bow, Bromley St. Leonard, St. Ann and St. Paul Shadwell, St. George and St. John Wapping, St. Botolph without Aldgate (commonly called East Smithfield), Trinity Minories, Mile-End Old Town, Mile-End New Town; Poplar; Radcliffe and Blackwall, in the parish of Stepney; the precincts of the Tower Without, of St. Catherine's, of Well-Close, and of the Old Artillery Ground; and the liberty of Norton-Falgate.

The living is a rectory, valued in the king's books at £31. 17. 3½.; present net income, £868; patrons, Principal and Fellows of Brasenose College, Oxford. The church, previously to the year 1329, was a chapel of ease to St. Dunstan's Stepney, the rector of which parish, in that year, made this incumbency a rectory; the ancient building was taken down, and the present church erected of brick in the year 1673, by private subscription; it has a small tower at the west end, with an illuminated dial, surmounted by a cupola. The interior is handsomely arranged, and the roof, which is partly arched, is supported on Corinthian columns pleasingly disposed. Near the altar is a mural monument by Banks, erected by the parishioners to the Rev. R. Markham, D.D., formerly rector of the parish; the figure of a weeping female kneeling on the pedestal of a sarcophagus, of which one side has a sculptured representation of the Good Samaritan, is finely executed: near this is a monument to Luke Flood, Esq., justice of the peace for the county, on which is the figure of a weeping cherub with an inverted torch in the left hand, and in various parts of the church and in the churchyard there are several others. St. Mark's church, on the Tenter Ground, was erected from funds granted by the Metropolitan Church-building Society, and consecrated in May 1839; it is a neat edifice of brick in the early English style, with a square tower surmounted by an octagonal spire, and contains 1800 sittings, of which 500 are free. There are places of worship for Baptists, Independents, Wesleyan Methodists, and other dissenters. In Little Aylie-street is the German Lutheran church, dedicated to St. George, a neat building with a campanile turret, and adjoining is a German and English school, supported by subscription: in Hooper's square is also a German Calvinistic chapel. The Danish church in Well-Close square is at present in the occupation of Mr. Smith, the original projector of the Sailors' Home. In Church-lane is a private burial-ground belonging to Mr. Sheen, comprising about an acre and a half neatly laid out and planted. The parochial school, originally founded and endowed by the Rev. R. Davenant, rector of the parish in 1680, for the clothing, instruction, and apprenticing of 30 boys and 30 girls, and handsomely rebuilt in 1818, and enlarged for the reception of 100 children of each sex, has an annual income of £600, arising from subsequent benefactions and annual subscriptions; the children are instructed and clothed. The free school in Gowers walk, established in 1806, under the immediate superintendence of the late Dr. Bell, and the first formed on that system, was founded by Mr. William Davis, who gave the ground and erected the building at his own cost, and endowed the institution with £2400 three per cents, for the instruction of children in useful knowledge, and training them to habits of industry; the same benefactor also gave a complete printing apparatus for the use of the boys, who are instructed in that trade; the school, in which there are at present 110 boys and 80 girls, is partly maintained by the profits of the printing establishment, and the sums received for needlework; what the boys earn by printing and the girls by needlework is paid monthly to each, and by them deposited in the school fund, from which they receive the whole amount on leaving the school: the premises comprise, on one side, the printing-office and boys' school, and on the other the girls' school, with apartments in the centre for the superintendent of the office and the master and mistress of the school: the annual income of the institution, including the profits of the printing-office and the needlework, is about £1200. The Whitechapel Society's Institution in Whitechapel-road was established in 1814, in union with the National Society: it is a spacious and neat building of brick, with a cupola at the west end; the school-room is internally arranged as a chapel, and consecrated, in which two regular services are performed every Sunday by the chaplain and superintendent of the institution, under whose direction the school is conducted. There is also another National school for boys in St. Mary-street. The Infants' school, in Great Garden-street, was founded and is partly supported by John Coope, Esq., who gave the premises, comprising a large school-room with an enclosed play-ground in front, and apartments for the master and mistress. The almshouses in Whitechapel-road were founded and endowed in 1658, by William Meggs, Esq., for twelve aged widows; the endowment, including subsequent benefactions, amounts to £149 per annum. Eight almshouses, founded by Thomas Baker, Esq., who endowed them for eight widows, form a neat range of building in the Elizabethan style.

The London Hospital owes its origin to Mr. John Harrison, surgeon, who, having conducted a small establishment of the kind near Upper Moorfields, removed it to Prescott-street, Goodman's-Fields, in the year 1740, under the designation of the "London Infirmary;" an appropriate building upon a larger scale having been subsequently erected in the Whitechapel-road, the institution was removed to that place in the year 1758, and the conductors incorporated by the name of the Governors of the London Hospital. The buildings have been progressively enlarged, and are now adapted to the reception of 370 patients: the medical department is under the superintendence of 3 physicians and 3 assistant physicians, 3 surgeons and 3 assistant surgeons, and an apothecary and assistant apothecary; the average number of in-patients is about 320, and of out-patients 7000 annually; and the average annual income, including annual contributions from public bodies and private subscriptions, is about £9000. The Sailors' Home, or Brunswick Maritime Establishment, is intended for the reception of unemployed sailors belonging to the port of London, to provide them with board and lodging at a moderate charge, and with religious and moral instruction while on shore; to procure for them employment in the Navy or Merchants' service, and to furnish such as are needy with the necessary outfits for the voyage: the plan of the institution also comprises a general register-office for seamen. The building occupies the site of the late Brunswick theatre, in Well-street, London docks, and has accommodation for 500 men; the first stone was laid in June 1830. The poor law union of Whitechapel comprises also the parishes of Christchurch, Spitalfields; St. Botolph without Aldgate; and the Holy Trinity, Minories; with the districts of Norton-Falgate, and Old Artillery Ground, the precincts of St. Katherine, and the Tower, within and without, and the hamlet of Mile-end New Town; and contains altogether a population of 64,141, according to the census of 1831; it is under the care of 25 guardians. There are workhouses in Whitechapel and Christ-church parishes.

WHITEFIELD, a joint hamlet with Apperley, in the parish of DEERHURST, Lower Division of the hundred of WESTMINSTER, Eastern Division of the county of GLOUCESTER, 4¼ miles (S. S. W.) from Tewkesbury: the population is returned with the parish.

WHITEGATE, otherwise NEWCHURCH (*St. Mary*), a parish, in the First Division of the hundred of EDDISBURY, Southern Division of the county of CHESTER, 3¼ miles (S. W.) from Northwich; containing 909 inhabitants. This place, which once formed part of the parish of Over, was separated from it and made a distinct parish, in 1541: it is bounded and partly intersected by the river Weaver. During the confinement at Hereford of Prince Edward, afterwards Edward I., while a prisoner in the hands of the barons, the monks of the neighbouring monastery of Dore visited and consoled him: he afterwards removed them, about the year 1273, to Dernhall, in this parish. A few years subsequently, the king, having resolved to build them a new abbey on a neighbouring spot, gave it the name of Vale-Royal, and laid the first stone of the new monastery, in August 1277, wherein the monks took up their abode in 1330, at which period £32,000 had been issued from the royal treasury for defraying the expense: the solemnity of the removal was observed with much magnificence, being attended by a great concourse of prelates, nobility, and gentry: at the dissolution the revenue was estimated at £518. There are still some small remains of this house in the doorways of the modern mansion that now occupies its site, which, in the great civil war, was plundered and partly destroyed. The living is a vicarage; net income, £163; patron and impropriator, Lord Delamere. The school, which is conducted on the National plan, was built at the expense of Lord Delamere, who pays the master £20 per annum, and Lady Delamere also pays for the instruction of 20 girls.

WHITEHAVEN, a sea-port, market-town, and newly enfranchised borough, and the head of a union, in the parish of ST. BEES, ALLERDALE ward above Derwent, Western Division of the county of CUMBERLAND, 40 miles (S. W.) from Carlisle, and 320 (N. W.) from London; containing 11,393 inhabitants, exclusively of about 800 seamen belonging to registered vessels, or, including these and the suburb of Preston Quarter, nearly 17,000 inhabitants. The derivation of the name is very uncertain: in the record of a trial between the abbot and monks of St. Mary's at York and the crown, relative to a claim to the wreck of the sea in the manor of St. Bees, the place is called *Whitothaven*: by some it is supposed to have derived its name from the light-coloured rocks which surmount the bay, though others derive it from the circumstance of a fisherman, named White, having been the first who frequented the bay, and built a cottage here; but the last account is generally discredited. In the reign of Henry I., the manor formed part of the possessions of the monastery of St. Mary near the walls at York, to which the priory of St. Bees belonged: and, so late as that of Elizabeth, the town consisted of only a few small huts inhabited by fishermen. In 1599, the manor of St. Bees was purchased from Sir Thomas Chaloner, Knt., by Gerard

Lowther and Thomas Wybergh, Esqrs.; and the whole having come into the possession of Sir John Lowther, Bart., in the year 1644, Whitehaven, under his auspices, advanced rapidly in prosperity. Having obtained from Charles II. a grant of land which was estimated at 150 acres, lying between high and low water mark, to the extent of two miles northward, Sir John materially improved the harbour, extended the collieries, and erected a mansion near the town, which, aided and improved by the patronage of his family, subsequently created Earls of Lonsdale, continued to increase until it has become one of the most populous and flourishing places in Cumberland.

The town is situated on a creek of the Irish Sea, and consists of several spacious and well-built streets, intersecting each other at right angles, and paved with pebbles: it is lighted with gas; supplied with water partly from wells, and partly by means of carts, in which it is brought into the town; and is watched under the superintendence of police. The ground, on three approaches to it, rises abruptly and precipitously, and the entrance from the north is under a fine arch of red sandstone, with a rich entablature bearing the arms of the Lowther family. On the south-east is the castle of the Earl of Lonsdale, a quadrangular building, with square projections at the angles, and a circular bastion in the centre, having fine meadow land to the south, and commanding an extensive prospect of the harbour. The theatre, erected in 1769, in Roper-street, is a handsome and commodious structure. Races are occasionally held in the neighbourhood. The subscription library, established in 1797, occupies a neat building, erected by the Earl of Lonsdale, in Catherine-street, and contains about 3000 volumes: the subscription news-room, also fitted up by his lordship, is well supplied with newspapers. The mechanics' institute and library, in Lowther-street, was established in 1825. Cold, warm, and shower salt-water baths have been fitted up in a building erected near the old platform. The harbour has always been an object of importance with those interested in the trade of the town, and many important improvements have been effected in it. Several stone piers extend, some in a diverging and some in a parallel direction, into the harbour; and another bends in an angular manner towards the north-west, on which is a battery. A watchhouse and a lighthouse have been built on the pier called the Old Quay, which was constructed in the time of Charles II., or previously, affording protection to the shipping in the harbour, which is capable of sheltering several hundred sail of vessels. At high water, in spring tides, there were about twenty feet, and in neap tides about twelve feet, of water; but at low water the harbour was dry. To remedy this, a new west pier, 20 yards in thickness, has been lately constructed to the north-west; it was commenced in 1824, on a plan by Mr. John Rennie. The estimated expense was £80,000, but this sum having been found insufficient, the trustess were empowered to borrow £180,000, to complete the undertaking. The harbour was defended by four batteries, mounting together nearly 100 guns; but, since the termination of the late war, many of the guns have been removed. At the entrance of the harbour are two lighthouses, that already mentioned, and another on the New Quay, which has a revolving light; and a life-boat, stationed here since 1803, has frequently been instrumental in saving life.

Whitehaven is a place of very considerable trade, of which coal forms the chief article: in addition to this, it exports lime, freestone, alabaster, and grain: the imports consist chiefly of American, Baltic, and West Indian produce, linen and flax from Ireland, fruit from the Levant, and wine from Spain and Portugal. The principal manufactures are linen, linen yarn, sail-cloth, checks, ginghams, cordage, earthenware, copperas, colours, anchors, and nails; soap and candles are also made for the West India market and for home consumption. The coal mines, which are of a magnitude only inferior to those of Newcastle and Sunderland, furnish the principal employment of the inhabitants; some have been sunk to a depth of upwards of 150 fathoms, and extend to a considerable distance under the sea. They are worked by means of shafts formed at great expense, and to some are entrances, called Bear Mouths, which opening at the bottom of a hill, lead through passages, by a steep descent, to the bottom of the pit, by which horses are taken into the mines: the coal, after being raised, is carried to the harbour in waggons on railways, the progress of which is aided by the declivity of the ground, and shipped by means of an inclined plane and wooden spouts, called *hurries*, placed sloping over the quays. These mines have suffered occasionally from fire-damp, but the safety-lamp has removed much of this danger. The coal lies in seams varying in thickness from two to eleven feet, in proportion to the depth. A quantity of a very rich iron-ore is sent from the mines here to the iron-works in South Wales. The herring fishery was formerly carried on to a great extent, but now very few of the inhabitants are employed in it. There are several ship-builders' yards, the ships being distinguished for their durability, and for drawing little water. A patent slip was erected in 1821, by the Earl of Lonsdale, which will admit vessels of 700 tons' burden, and, with great convenience, four vessels of 150 tons' burden each, and by which a few men can draw a large vessel into the yard to be repaired. A communication with Liverpool, Dublin, Carlisle, the Isle of Man, Dumfries, Annan, and Garliestown, is maintained by steam-boats, which sail regularly for those places. The jurisdiction of the port extends from Maryport, northward, to mid-stream in the river Duddon, southward, and to the intermediate coast and ports of Workington, Harrington, Ravenglass, and Millom, and as far into the sea as ten fathoms of water. The amount of duties paid at the custom-house, for the year 1838, was £102,939; and in 1837, 409,493 tons of coal were shipped from this port. The custom-house, erected in 1811, is a commodious structure; the establishment consists of a collector and comptroller, with the usual subordinate officers. There are three weekly markets, on Tuesday, Thursday, and Saturday, that on Thursday being the principal, and they are all well supplied with provisions: the fair, held on the 12th of August, has nearly fallen into disuse. The market-place is a handsome area, containing a neat market-house, designed by Smirke, in 1813, for the sale of poultry, eggs, and dairy produce; and there is another, erected in 1809, at the expense of the Earl of Lonsdale, for fish, of which there is a good supply: there are also shambles, called the Low and George's markets, for butchers. The regulation of the affairs of the town and harbour was, by acts of parliament passed in the 7th and 11th of Queen Anne, confirmed by subsequent acts, vested in 21 trustees, of whom, seven are appointed by the lord of the manor (himself being one), and the remaining fourteen are elected triennially by ballot; such of the inhabitants as pay harbour dues, or possess one-sixteenth share of a vessel belonging to the port, and the masters of vessels, being the electors. The constables of the town are nominated by the trustees, and appointed by the justices of the peace, who meet at the public office, in Lowther-street, on Thursday and Saturday, for the despatch of business. By the act of the 2nd of William IV., cap. 45, this town has been constituted a borough, with the privilege of sending one member to parliament, the right of election being vested in the £10 householders: the borough comprises an area of 1778 acres, of which the limits are minutely described in the Appendix, No. II.: the returning officer is annually appointed by the sheriff. A court leet is held annually, and a court baron monthly; the latter is for the recovery of debts under 40*s*.

Whitehaven contains three chapels, to which districts have been assigned; the livings are perpetual curacies, in the patronage of the Earl of Lonsdale, the impropriator. St. James' stands on an eminence at the eastern extremity of the town, and was rebuilt in 1753; it is a neat structure with a square tower, surmounted with pinnacles: net income, £200. St. Nicholas' was erected in 1693, and is a plain building of good proportion, with a square tower; the interior is decorated with paintings of the Last Supper, and of Moses and Aaron, by Matthias Reed, an artist of some merit, who came from Holland in the fleet with the Prince of Orange, and settled in this town: net income, £188: the ecclesiastical courts are held in this chapel. That of the Holy Trinity stands near the southern extremity of the town, at the head of Roper-street; it is a plain building with a lofty tower: net income, £250. There are two places of worship for Presbyterians, and one each for Particular Baptists, the Society of Friends, Independents, Primitive and Wesleyan Methodists, and Roman Catholics. The Marine school, near St. James' church, or chapel, in which 60 boys are instructed, was endowed by Matthew Piper, Esq., with the interest of £2000; the site was given, and the building erected, by the Earl of Lonsdale. The National school, a large and commodious stone building, was erected in 1824. The interest of £1000 was bequeathed by Matthew Piper, Esq., also, for the purchase of soup, to be distributed, during winter, among the poor; for whose benefit there are also some minor bequests and donations, besides several charitable institutions for providing them with clothing, &c. A dispensary was established in 1783, and a house of recovery in 1819: about the commencement of the year 1830, a spacious mansion in Howgill-street was purchased, and fitted up for the purpose of an infirmary, which establishment includes the dispensary and house of recovery. A savings bank was instituted in 1818, and, from the accumulation of interest beyond what was paid to the depositors, a new and elegant edifice has been erected in Lowther-street. The poor law union of Whitehaven comprises 23 parishes or places, under the care of 32 guardians, and contains a population of 27,816, according to the census of 1831. Dean Swift, when a child, resided with his attendant in a house in Roper-street, during the disturbance in Ireland about the time of the Revolution; and Dr. Brownrigs, who by his publications first attracted the notice of strangers to the beauties of Keswick and the surrounding scenery, for many years practised as a physician in the town.

WHITEPARISH (*All Saints*), a parish, in the union of Alderbury, hundred of Frustfield, Salisbury and Amesbury, and Southern, Divisions of the county of Wilts, 8 miles (S. E. by E.) from Salisbury; containing, with the extra-parochial liberty of Earldoms, 1254 inhabitants. The living is a discharged vicarage, valued in the king's books at £13. 7. 2.; present net income, £126; patron and impropriator, Robert Bristow, Esq. A free school for boys was founded by James Lynch, in 1639, and endowed by him with lands now let for £40 per annum: it received a further endowment of £200 three per cent. consols from Henry Eyre, about 1720, making the total annual income of the school £46; about 35 boys are taught gratuitously. A free school for girls was founded in 1722 by several members of the Hitchcock family, who endowed it with property now producing an annual income of £17. 1. 6., which is paid to a schoolmistress, who teaches 30 girls. The same family bequeathed also £2 per annum, which, with £4 arising from a bequest by Sir John Evelyn, in 1684, is distributed in clothing among the poor.

WHITE-ROOTHING, county of Essex. — See ROOTHING, WHITE.

WHITESIDELAW, a township, in the parish of Chollerton, North-Eastern Division of Tindale ward, Southern Division of the county of Northumberland, 7½ miles (N. N. E.) from Hexham: the population is returned with the township of Chollerton.

WHITEWELL, a chapelry, in the parish of Whalley, Western Division of the wapentake of Staincliffe and Ewcross, West Riding of the county of York, 7 miles (N. W. by W.) from Clitheroe: the population is returned with the parish. The living is a perpetual curacy; net income, £88; patron, Vicar of Whalley. A gallery has been erected in the church, and 70 free sittings provided, the Incorporated Society for building churches and chapels having granted £25 in aid of the expense.

WHITFIELD, a township, in the parish and union of Glossop, hundred of High Peak, Northern Division of the county of Derby, 8½ miles (N. by W.) from Chapel-en-le-Frith; containing 1734 inhabitants. There is a place of worship for Wesleyan Methodists. A school-house was erected about 1786, by Joseph Hague, Esq., who endowed it with land and houses of the present annual value of £40; he also left the interest of £1000 to be expended in clothes for twelve poor men and as many women of this and other townships in the parish.

WHITFIELD (*St. Mary*), a parish, in the union of Dovor, hundred of Bewsborough, lathe of St. Augustine, Eastern Division of the county of Kent, 3¼ miles (N. N. W.) from Dovor; containing 199 inhabitants. The living is a perpetual curacy, in the patronage of the Archbishop of Canterbury (the appropriator), valued in the king's books at £5. 18. 8.; present net income, £109.

WHITFIELD (*St. John the Evangelist*), a parish, in the union of Brackley, hundred of King's-Sutton, Southern Division of the county of Northampton, 2 miles (N. E. by N.) from Brackley; containing 328 inhabitants. The living is a rectory, valued in the king's books at £8. 15.; present net income, £258; patrons, Provost and Fellows of Worcester College, Oxford. Here is a National school, endowed with £6 a year, arising from land purchased with sundry donations; and another school is supported by William Morriss, Esq.

WHITFIELD, a parish, in the union of Haltwhistle, Western Division of Tindale ward, Southern Division of the county of Northumberland, 11½ miles (W. S. W.) from Hexham; containing 388 inhabitants. The living is a discharged rectory, valued in the king's books at £8; patron, William Ord, Esq. The tithes have been commuted for a rent-charge of £260, subject to the payment of rates, which on the average have amounted to £17. 17.; the glebe comprises 43½ acres, valued at £82 per annum. The church was rebuilt about the year 1784. The East and West Allen rivers join their streams at Cupola, in this parish, at which place lead-ore, obtained from a mine at Limestone-Cross, was formerly smelted. A new line of road from Alston to Haydon-bridge has been formed through the parish. Mr. Ord allows a schoolmaster the use of a house and garden, with £25 a year, for teaching poor children; and a school is conducted on the National plan. At Redmires there is a chalybeate spring.

WHITGIFT (*St. Mary Magdalene*), a parish, in the union of Goole, Lower Division of the wapentake of Osgoldcross, West Riding of the county of York; containing 2252 inhabitants, of which number, 310 are in the township of Whitgift, 6½ miles (S. E.) from Howden. The living is a perpetual curacy; net income, £287; patron, N. E. Yarburgh, Esq.; impropriator, — Worsop, Esq. Here is a place of worship for Primitive Methodists.

WHITGREAVE, a township, in the parish of St. Mary and St. Chad, Stafford, Southern Division of the hundred of Pirehill, union, and Northern Division of the county, of Stafford, 3½ miles (N. N. W.) from Stafford; containing 195 inhabitants.

WHITKIRK (*St. Mary*), a parish, in the Lower Division of the wapentake of Skyrack, West Riding of the county of York, 4 miles (E.) from Leeds; containing 2564 inhabitants. The living is a discharged vicarage, valued in the king's books at £13. 5. 7½.; present net income, £202; patrons and impropriators, Master and Fellows of Trinity College, Cambridge. In a school, built by subscription, six children are instructed for an annuity of £10, bequeathed by Richard Brooke, in 1702. Here is a National school.

WHITLEY, a tything, in the parish of Cumner, hundred of Hormer, county of Berks, 5 miles (W. by S.) from Oxford; containing 38 inhabitants.

WHITLEY, a hamlet, in the parish of St. Giles, Reading, union and hundred of Reading, county of Berks, 2 miles (S.) from Reading; containing 363 inhabitants. This place is lighted with gas.

WHITLEY, a chapelry, in the parish and union of Tynemouth, Eastern Division of Castle ward, Southern Division of the county of Northumberland, 2½ miles (N. by E.) from North Shields; containing 632 inhabitants. The village is pleasantly situated near the sea, and contains many well-built houses. Iron-stone abounds in the neighbourhood; and there are extensive mines of coal, and quarries of limestone, the productions of which are conveyed, by means of a railroad, to Shields for exportation. The North Shields Water-works Company have a reservoir here. The living is a perpetual curacy; net income, £124; patron and incumbent, Rev. J. Hewetson.

WHITLEY, a township, in the parish of Kellington, Lower Division of the wapentake of Osgoldcross, West Riding of the county of York, 5½ miles (W. by S.) from Snaith; containing 310 inhabitants. Here is a place of worship for Wesleyan Methodists.

WHITLEY-BOOTHS, a joint township with Barley, in the parish of Whalley, union of Burnley, Higher Division of the hundred of Blackburn, Northern Division of the county palatine of Lancaster, 4 miles (W. by N.) from Colne; containing, with Barley, 707 inhabitants.

WHITLEY, LOWER, a chapelry, in the parish of Great Budworth, union of Runcorn, hundred of Bucklow, Northern Division of the county of Chester, 4¾ miles (N. W. by N.) from Northwich; containing 237 inhabitants. The living is a perpetual curacy; net income, £108; patron, Sir John Chetwode, Bart.; appropriators, Dean and Canons of Christ-Church, Oxford. There is a place of worship for Wesleyan Methodists.

WHITLEY, LOWER, a township, in the parish of Thornhill, union of Dewsbury, Lower Division of the wapentake of Agbrigg, West Riding of the county of York, 5¼ miles (S. W.) from Wakefield; containing 1012 inhabitants. There are two scribbling-mills and a tan-yard in the neighbourhood.

WHITLEY, OVER, a township, in the parish of Great Budworth, union of Runcorn, hundred of Bucklow, Northern Division of the county of Chester, 5½ miles (N. N. W.) from Northwich; containing 283 inhabitants.

WHITLEY, UPPER, a township, in the parish of Kirk-Heaton, union of Huddersfield, Upper Division of the wapentake of Agbrigg, West Riding of the county of York, 6 miles (E. by N.) from Huddersfield; containing 885 inhabitants.

WHITLINGHAM (*St. Andrew*), a parish, in the union and hundred of Henstead, Eastern Division of the county of Norfolk, 2½ miles (E. S. E.) from Norwich; containing 45 inhabitants. The living is a perpetual curacy, in the patronage of — Hare, Esq. The church has fallen into ruins.

WHITMORE, a parish, in the union of Newcastle-under-Lyme, Northern Division of the hundred of Pirehill and of the county of Stafford, 4 miles (S. W.) from Newcastle-under-Lyme; containing 281 inhabitants. The living is a rectory not in charge; net income, £470; patron, E. Mainwaring, Esq. Here is a station of the Grand Junction railway, which passes through the parish. Two day and Sunday schools are supported by Mrs. and Miss Mainwaring.

WHITNASH (*St. Margaret*), a parish, in the union of Warwick, Kenilworth Division of the hundred of Knightlow, Southern Division of the county of

Warwick, 3 miles (E. S. E.) from Warwick; containing 260 inhabitants. The living is a rectory, valued in the king's books at £5. 9. 9½.; patron, Lord Leigh. The tithes have been commuted for a rent-charge of £280, subject to the payment of rates, which on the average have amounted to £43; the glebe comprises 77 acres. A day and Sunday school is supported by subscription.

WHITNEY (*St. Peter and St. Paul*), a parish, in the union of Hay, hundred of Huntington, county of Hereford, 5 miles (N. E.) from Hay; containing 254 inhabitants. The living is a discharged rectory, valued in the king's books at £5. 8.; present net income, £200; patron, Tomkin Dew, Esq.

WHITREY, a tything, in the parish of Willand, hundred of Halberton, Cullompton and Northern Divisions of the county of Devon, 2¼ miles (N. N. E.) from Cullompton; containing 122 inhabitants.

WHITRIDGE, a township, in the parish of Hartburn, union, and Western Division of the ward, of Morpeth, Northern Division of the county of Northumberland, 9½ miles (W. by N.) from Morpeth; containing 11 inhabitants.

WHITRIGG, a joint township with Torpenhow, in the parish of Torpenhow, union of Wigton, Allerdale ward below Derwent, Western Division of the county of Cumberland, 1 mile (S.) from Torpenhow: the population is returned with the parish. Here is a National school. On a hill, called Caer Mot, are the remains of a square double intrenchment, intersected by the old road from Keswick to Old Carlisle; near it is a smaller encampment, defended by a rampart and fosse: there are also the remains of a beacon.

WHITSBURY, or WHITCHBURY (*St. Leonard*), a parish, in the union of Fordingbridge, hundred of Cawden and Cadworth, Salisbury and Amesbury, and Southern, Divisions of the county of Wilts, 3½ miles (N. by W.) from Fordingbridge; containing 183 inhabitants. The living is a discharged vicarage, endowed with the rectorial tithes, and valued in the king's books at £5. 13. 4.; present net income, £284; patrons, Trustees of the late Admiral Purvis. The church is situated at the extremity of the parish, and the chancel is within the county of Southampton. This parish anciently formed part of the possessions of the priory of Breamore, founded by Baldwin de Redveriis in the reign of Henry I.; and appears, from discoveries which have been recently made, to have been formerly connected with events of historical importance, the particulars of which are not recorded. A school is partly supported by the vicar. Charles Delafaye, in 1762, bequeathed £200 South Sea annuities, the interest of which is applied to furnishing medical aid to persons who may be disabled by accident or sickness, and to apprenticing children. Within the parish is a Roman encampment, occupying an elevated and extensive area surrounded by a trench, and commanding a fine view of Salisbury cathedral and the castle of Old Sarum. In 1823, a barrow was opened on the estate of Sir Lucius Curtis, Bart., and in it were found a human skeleton, some spear heads, the top of a helmet, and a jaw bone of gigantic size; and, in opening the ground for a vault on the north side of the chancel, 21 human skulls were found, supposed to be those of persons killed in some battle that had taken place near the spot. Traces of a Roman road are discernible towards Clarebury Riggs.

WHITSON, county of Monmouth.—See WITSTON.

WHITSTABLE (*All Saints*), a parish, in the union of Blean, hundred of Whitstable, lathe of St. Augustine, Eastern Division of the county of Kent, 5¾ miles (N. N. W) from Canterbury; containing 1926 inhabitants. This parish lies near the entrance to the East Swale, opposite the Isle of Sheppey. On the shore by Tankerton are several establishments, where considerable quantities of copperas, or green vitriol, are manufactured. Whitstable bay is frequented by a number of colliers, from which Canterbury and the surrounding places are supplied with coal, by means of the Canterbury and Whitstable railway. It is also the station of hoys, which sail to and from London alternately, every week, with goods and passengers. Many boats are employed in the fisheries, Whitstable being a royalty of fishery, or oyster dredging, appendant to the manor; and for the due regulation of the trade a court is held annually on the second Thursday in July. There are fairs on Thursday before Whitsuntide, near the water side; on Midsummer-day, at Church-street; and on St. James' day, on Greensted-green, in Whitstable-street, which is a thriving and populous village, containing shops well stored with every necessary article of consumption for those engaged in the extensive traffic here carried on. The living is a perpetual curacy; net income, £100; patron, Archbishop of Canterbury; impropriator, T. Foord, Esq. There are places of worship for Independents and Wesleyan Methodists. A house and two school-rooms, built on land given by Mr. Nott, are occupied rent-free by a master, who has also £10 per annum for teaching 17 boys named by the feoffees of the poor's estate, which they have greatly improved. Great quantities of Roman pottery have been found in dredging for oysters round a rock, now called the Pudding-pan, which is supposed by some to have been the island *Caunos*, mentioned by Ptolemy, though now covered by the sea.

WHIT-STAUNTON (*St. Andrew*), a parish, in the union of Chard, and forming a distinct portion of the hundred of South Petherton, being locally in that of Kingsbury, Western Division of the county of Somerset, 3¼ miles (W. N. W.) from Chard; containing 318 inhabitants. The living is a rectory, valued in the king's books at £14. 2. 11.; patron, Isaac Elton, Esq. The tithes have been commuted for a rent-charge of £215, subject to the payment of rates; the glebe comprises 54 acres, valued at £65 per annum. Robert Somerhays bequeathed £10. 10. per annum, for the instruction of 11 boys and 10 girls.

WHITSTONE, a parish, in the union and hundred of Stratton, Eastern Division of the county of Cornwall, 7 miles (S. S. E.) from Stratton; containing 481 inhabitants. This parish is situated in the north-eastern part of the county, bordering upon Devonshire, and is intersected by the Bude canal, in its course to Launceston. The living is a discharged rectory, valued in the king's books at £14. 11. 0½.; present net income, £231; patron and incumbent, Rev. J. Kingdon. A school is partly supported by Lord de Dunstanville, and partly by subscription.

WHITSTONE (*St. Catherine*), a parish, in the

union of St. Thomas, hundred of Wonford, Wonford and Southern Divisions of the county of Devon, 3¾ miles (W. N. W.) from Exeter; containing 643 inhabitants. The living is a rectory, valued in the king's books at £19. 3. 4.; present net income, £641; patron and incumbent, Rev. Charles Brown. The surrounding scenery is pleasingly diversified and enriched with wood. John Spratt, in 1753, bequeathed £20 per annum for teaching the poor children of the parish. He also founded almshouses for five poor people. Another school, containing 25 girls, is supported by the wife of the rector.

WHITTERING (*All Saints*), a parish, in the union of Stamford, soke of Peterborough, Northern Division of the county of Northampton, 2½ miles (N. N. W.) from Wansford; containing 216 inhabitants. The living is a discharged rectory, valued in the king's books at £8. 0. 10.; present net income, £101; patron, Marquess of Exeter.

WHITTINGHAM, a township, in the parish of Kirkham, hundred of Amounderness, Northern Division of the county palatine of Lancaster, 5¾ miles (N. N. E.) from Preston; containing 710 inhabitants.

WHITTINGHAM (*St. Bartholomew*), a parish, in the union of Rothbury, Northern Division of Coquetdale ward and of the county of Northumberland; containing 1790 inhabitants, of which number, 611 are in the township of Whittingham, 8½ miles (W. by S.) from Alnwick. The living is a vicarage, valued in the king's books at £19. 11. 3.; present net income, £540; patrons and appropriators, Dean and Chapter of Carlisle. The church has been enlarged, and 140 free sittings provided, the Incorporated Society having granted £100 in aid of the expense. A school is partly supported by Lord Ravensworth, who allows the master a dwelling-house and garden, with £10 per annum, for the instruction of 10 poor children; and Lady Ravensworth allows a mistress £12 per annum, with a room and garden, for instructing 12. In this parish is a vaulted tower, that has often afforded refuge and defence to the inhabitants during the border warfare.

WHITTINGTON (*St. Bartholomew*), a parish, in the union of Chesterfield, hundred of Scarsdale, Northern Division of the county of Derby, 2¼ miles (N.) from Chesterfield; containing 740 inhabitants. A public-house here is distinguished by the name of the Revolution House, from the adjournment to it of a select meeting of friends to liberty and the Protestant religion, held on the moor early in 1688, at which the Earl (afterwards Duke) of Devonshire, the Earl of Derby (afterwards Duke of Leeds), Lord Delamere, and Mr. John Darey, eldest son of the Earl of Holderness, attended. When the centenary anniversary of that glorious event was commemorated in Derbyshire, in 1788, the committee dined on the preceding day at this house; and on the anniversary, a sermon was preached in the parish church by Dr. Pegge, the celebrated antiquary, then rector, before the descendants of these illustrious revolutionists, and a large assemblage of the most distinguished families of the county, who afterwards went in procession to take refreshment at the Revolution House, and then proceeded to Chesterfield to dinner. A subscription was then opened for erecting a column on Whittington moor, in memory of the Revolution, but the design was abandoned, in consequence, as it is supposed, of the revolution which so speedily followed in France. The Chesterfield races are held on this moor. The manufacture of earthenware is here carried on. The living is a rectory, valued in the king's books at £7. 10. 10.; present net income, £302; patron, Dean of Lincoln. The chancel of the church was built in 1827. A free school was founded in 1674, by Peter Webster, who, in 1678, gave £200 to purchase land; and Joshua Webster, in 1681, gave land in Whittington to be applied for teaching ten children: the annual income is now £35. 15., for which 17 children are taught. A chalybeate spring here was formerly much resorted to, and, for the convenience of visiters, a cold bath was erected in 1769.

WHITTINGTON, a parish, in the union of Northleach, hundred of Bradley, Eastern Division of the county of Gloucester, 4½ miles (E. S. E.) from Cheltenham; containing 247 inhabitants. The living is a rectory, valued in the king's books at £13. 6. 8.; patron, Walter Lawrence Lawrence, Esq. The tithes have been commuted for a rent-charge of £295. 14. 8., subject to the payment of rates, which on the average have amounted to £33; the glebe comprises 29 acres, valued at £50 per annum. Here is a National school, endowed with £33 per annum.

WHITTINGTON, a parish, in the hundred of Lonsdale, south of the sands, Northern Division of the county palatine of Lancaster, 2 miles (S. S. W.) from Kirkby-Lonsdale; containing 542 inhabitants. The living is a rectory, valued in the king's books at £13. 9. 9½.; present net income, £415; patron, E. Hornby, Esq. William Margison, in 1762, left £1000 for building and endowing a school, in which six children are now educated.

WHITTINGTON (*St. John the Baptist*), a parish, in the hundred of Oswestry, Northern Division of the county of Salop, 3 miles (E. N. E.) from Oswestry; containing 1788 inhabitants. Lloyd, in his "Archæologia," imagines this place to have been celebrated, under the name *Drêv Wen*, or the White Town, by Llywarch Hen, a noble British bard, who flourished about the close of the sixth century; and describes it as the place where *Condolanus*, a British chieftain, was slain, in an attempt to expel some Irish invaders. According to the bards, it was subsequently the property and chief residence of Tudor Trevor. After the Conquest it was given to Roger Earl of Shrewsbury, and, on the defection of his son, Earl Robert, and the confiscation of that nobleman's immense estates, in the reign of Henry I., the castle and barony were granted to the Peverells, from whom, by the marriage of Mellet, second daughter of William Peverell, to Guarine de Mets, who received her hand as the reward of his distinguished prowess in a tournament held at the castle in the Peak, in Derbyshire, they passed to the illustrious race of Fitz-warine, whose feats of chivalry and valorous exploits have furnished a subject for romance, and, in modern times, have been beautifully illustrated in a poem by J. F. M. Dovaston, Esq., of West Felton, in the vicinity. The Fitz-warines were lords of the place for nearly 400 years, and every heir, for nine descents, preserved the Christian name of Fulk. The castle then became a border fortress, and the neighbourhood the frequent scene of battle between the lords retainers and the Welsh: in these conflicts the building, probably, sustained considerable injury, since licence was granted by Henry III. to the renowned Fulk Fitz-warine, for repairing and fortifying it. The remains consist of one large tower, with traces of four others, and the exterior gateway, which is inhabited by a farmer. On the green, annually at Midsummer, a gay assemblage of the young people of the vicinity, called the Whittington club, similar to those at Ellesmere and Oswestry, takes place. A court leet and baron is annually held in a modern portion of the castle, rebuilt a few years ago by William Lloyd, Esq., the present lord of the manor. Coal is thought to lie beneath the surface of some parts of the parish, but no mines have yet been opened. The river Perry runs through it; also the Ellesmere canal, which here divides into four branches, called the Chester, Llangollen, Montgomeryshire, and Weston canals. The living is a rectory, valued in the king's books at £25. 4. 2.; patron, W. Lloyd, Esq. The impropriate tithes have been commuted for a rent-charge of £285, and the vicarial for £1041. 8., subject to the payment of rates, which on the average have respectively amounted to £8 and £47: the glebe comprises 58 acres, valued at £101. 10. per annum. The church is supposed to have been built in the reign of Henry II., by Fulk Fitz-warine, lord of the manor, who procured a market and fair to be held here, both which have been long disused: it was rebuilt in 1806. Robert Jones, in 1679, bequeathed two cottages and five acres of land, directing the rents to be applied in support of a school; and, in 1706, Griffith Hughes left seventeen acres of land, one-half to be given to the school here, and the other to that of Ruabon. A day and Sunday school is partly supported by a small endowment, and partly by the rector.

WHITTINGTON (*St. Giles*), a parish, in the union of LICHFIELD, Northern Division of the hundred of OFFLOW and of the county of STAFFORD, 2½ miles (E. by S.) from Lichfield; containing 773 inhabitants. The living is a perpetual curacy; net income, £251; patron and impropriator, T. Levett, Esq. The church has a lofty spire, and was rebuilt in 1762. The Coventry canal passes near the village. Sarah Neal, in 1741, gave a house and land towards the education of poor children; and, in 1800, the Rev. Richard Levett bequeathed £200 for the like purpose: the income is £8. 6. 7. per annum.

WHITTINGTON, a hamlet, in the parish of GRENDON, Tamworth Division of the hundred of HEMLINGFORD, Northern Division of the county of WARWICK, 2 miles (N. W.) from Atherstone: the population is returned with the parish.

WHITTINGTON, a chapelry, in the parish of ST. PETER, WORCESTER, union of PERSHORE, Lower Division of the hundred of OSWALDSLOW, Worcester and Western Divisions of the county of WORCESTER, 2½ miles (S. E. by E.) from Worcester; containing 279 inhabitants. The chapel, dedicated to St. Philip and St. James, is an ancient structure of wood, with some curious tracery in the windows.

WHITTINGTON, GREAT, a township, in the parish of CORBRIDGE, union of HEXHAM, Eastern Division of TINDALE ward, Southern Division of the county of NORTHUMBERLAND, 7 miles (N. E.) from Hexham; containing 209 inhabitants. James Kirsopp, Esq., left £5 a year for the education of ten poor children, in a school built by subscription in 1825.

WHITTINGTON, LITTLE, a township, in the parish of CORBRIDGE, union of HEXHAM, Eastern Division of TINDALE ward, Southern Division of the county of NORTHUMBERLAND, 6½ miles (N. E.) from Hexham; containing 11 inhabitants.

WHITTLE, a hamlet, in the parish of GLOSSOP, union of HAYFIELD, hundred of HIGH PEAK, Northern Division of the county of DERBY, 6¼ miles (N. W.) from Chapel-en-le-Frith; containing 2266 inhabitants.

WHITTLE, a township, in the parish of SHILBOTTLE, union of ALNWICK, Eastern Division of COQUETDALE ward, Northern Division of the county of NORTHUMBERLAND; 5 miles (S.) from Alnwick; containing 53 inhabitants.

WHITTLE, a township, in the parish of OVINGHAM, union of HEXHAM, Eastern Division of TINDALE ward, Southern Division of the county of NORTHUMBERLAND, 11 miles (W.) from Newcastle-upon-Tyne; containing 20 inhabitants.

WHITTLE-LE-WOODS, a township, in the parish and hundred of LEYLAND, union of CHORLEY, Northern Division of the county palatine of LANCASTER, 2 miles (N.) from Chorley; containing 2015 inhabitants. Here are several valuable millstone quarries; and a lead mine was formerly worked with great success. A church in the later English style of architecture was built here in 1830, by grant from the parliamentary commissioners, at an expense of £2756. 5. The living is a perpetual curacy; net income, £40; patron, Vicar of Leyland; impropriators, Sir H. Hoghton, Bart., and — Silvester, Esq. A school, erected by subscription in 1769, was endowed by Samuel Crooke, in 1770, with the interest of £220; another, for the children of Roman Catholics, is supported by subscription. Here is also a school on the National plan.

WHITTLE, WELSH, a township, in the parish of STANDISH, union of CHORLEY, hundred of LEYLAND, Northern Division of the county palatine of LANCASTER, 3 miles (S. W.) from Chorley; containing 147 inhabitants.

WHITTLEBURY (*St. Mary*), a parish, in the union of TOWCESTER, hundred of GREENS-NORTON, Southern Division of the county of NORTHAMPTON, 3¾ miles (S. by W.) from Towcester; containing 670 inhabitants. The living is annexed, with that of Silverstone, to the rectory of Greens-Norton. There is a place of worship for Wesleyan Methodists. The manufacture of bone-lace affords employment to the female population. A National school has been established, and is endowed with the rent of land in the parish of Slapton, producing £14 per annum, and partly supported by subscription. Several Roman tiles and ornaments were found under the roots of trees which had been felled in a field belonging to Mr. Cooke, adjoining the churchyard, in 1822; and several Roman coins have also been found here in high preservation, some of which are in the possession of Mr. Baker, the county historian.

WHITTLESEY, a village (formerly a market-town), containing the parishes of *St. Andrew* and *St. Mary*, forming a union of itself, in the hundred of NORTH WITCHFORD, Isle of ELY, county of CAMBRIDGE, 6 miles (E. by S.) from Peterborough; containing 6019 inhabitants. This place, called *Witesie* in Domesday-book, is supposed to have been a Roman station, from the traces of a military way, and the numerous relics of antiquity discovered in the neighbourhood. The village, which is bounded on the north and south by branches of the river Nene, is still a large and respectable place, though its market, formerly held on Friday, has been for some years disused: the market-house still remains, and there is a fair for horses on June 13th. At the Falcon, the principal inn, courts leet and baron are held twice a year; also a court of requests, for the recovery of debts under 40*s*., on the third Friday in every month. A public library and news-room have been established by subscription. Adjoining this place, but in the county of Huntingdon, is an expanse of water, termed Whittlesey Mere; it bears also the appellation of the White Sea, and abounds with a variety of fish, a considerable quantity of which is sent to the metropolis. Whittlesey has been made a polling-place for the Isle of Ely. The living of St. Andrew's is a discharged vicarage, valued in the king's books at £4. 13. 4.; present net income, £62: it is in the patronage of the Crown; impropriators, Earl Waldegrave and others. The church is a handsome structure, with a stately tower crowned with turrets. The living of St. Mary's is a discharged vicarage, valued in the king's books at £19. 13. 9.; present net income, £222; patron and impropriator, Earl Waldegrave. The church is a handsome edifice, with a lofty tower of peculiar elegance, surmounted by a slender enriched spire of good proportions. Her Majesty's Commissioners have made a conditional grant in aid of building another church in this parish. Within the limits of the two parishes are places of worship for Baptists, Independents, and Calvinistic and Wesleyan Methodists. There are two endowed schools, one of them founded in 1735, by Adam Kelfull, and endowed with £15 per annum, also £12 from the rental of the town lands; and the other in 1815, by John Sudbury, with £20 per annum; and a Sunday National school has been established. William de Whittlesey, Archbishop of Canterbury, was born here in 1367.

WHITTLESFORD (*St. Mary and St. Andrew*), a parish, in the union of Linton, hundred of Whittlesford, county of Cambridge, 6¼ miles (W. by N.) from Linton; containing 524 inhabitants. The living is a discharged vicarage, valued in the king's books at £10; present net income, £169; patrons, Master and Fellows of Jesus' College, Cambridge; impropriator, H. J. Thurnall, Esq. A market and a fair were formerly held here, but both have been long disused. William Westley, in 1723, bequeathed lands now let for £50 a year, for teaching poor children. At Whittlesford bridge are the remains of an ancient hospital, said to have been founded before the time of Edward I., by William Colvill, and dedicated to St. John the Baptist.

WHITTON, a township, in the parish of Grindon, union of Stockton, North-Eastern Division of Stockton ward, Southern Division of the county palatine of Durham, 5½ miles (N. W. by W.) from Stockton-upon-Tees; containing 75 inhabitants.

WHITTON, a joint township with Trippleton, in the parish of Leintwardine, hundred of Wigmore, county of Hereford; containing, with Trippleton, 63 inhabitants.

WHITTON (*St. John the Baptist*), a parish, in the union of Glandford-Bridge, Northern Division of the wapentake of Manley, parts of Lindsey, county of Lincoln, 11 miles (W. N. W.) from Barton-upon-Humber; containing 245 inhabitants. The living is a discharged rectory, united to the vicarage of Aukborough, and valued in the king's books at £6. 10. A day and Sunday school is partly supported by Marmaduke Constable, Esq.

WHITTON, a township, in the parish and union of Rothbury, Western Division of Coquetdale ward, Northern Division of the county of Northumberland, ½ a mile (S.) from Rothbury; containing 104 inhabitants. Whitton Tower, formerly a very strong fortress, is still a commodious edifice, occupied by the rector of the parish: near it is a circular observatory, built by the late Dr. Sharp.

WHITTON, a chapelry, in the parish of Burford, union of Tenbury, hundred of Overs, Southern Division of the county of Salop, 3¾ miles (N. W. by N.) from Tenbury; containing 76 inhabitants.

WHITTON (*St. Mary*), a parish, in the union and within the liberty of the borough of Ipswich, Eastern Division of the county of Suffolk, 2½ miles (N. N. W.) from Ipswich; containing 346 inhabitants. The living is a rectory, with that of Thurlton annexed, valued in the king's books at £6. 11. 5½.; patron, Bishop of Ely. The manor and impropriation of Thurlton were granted to Cardinal Wolsey by Henry VIII., and now belong to the Rev. Edward Woolnough. The tithes have been commuted for a rent-charge of £448, subject to the payment of rates. Three schools are partly supported by subscription. The Stow-Market and Ipswich navigation passes through the parish.

WHITTONSTALL, a chapelry, in the parish of Bywell-St. Peter, union of Hexham, Eastern Division of Tindale ward, Southern Division of the county of Northumberland, 10 miles (S. E. by E.) from Hexham; containing 175 inhabitants. The living is a perpetual curacy; net income, £45; patrons, Dean and Chapter of Durham. The chapel, dedicated to St. Philip and St. James, has been rebuilt; it contains 120 free sittings, the Incorporated Society having granted £50 in aid of the expense. The Governors of Greenwich Hospital, of which this place forms part of the endowment, allow £15 a year to a schoolmaster for teaching poor children. The Roman Watling-street passes through the chapelry.

WHITWELL (*St. Lawrence*), a parish, in the union of Worksop, hundred of Scarsdale, Northern Division of the county of Derby, 10¾ miles (E. N. E.) from Chesterfield; containing 1007 inhabitants. This place, together with some of the neighbouring villages, has been on the decline since the opening of the Chesterfield canal; frame-work knitting is still carried on to a small extent. A statute fair for hiring servants, formerly held on Nov. 1st, is now in disuse. The ancient hall has been converted into a farm-house. The living is a rectory, valued in the king's books at £20. 3. 4.; patron, Duke of Portland. The tithes have been commuted for a rent-charge of £642, subject to the payment of rates; the glebe comprises 143 acres, valued at £137 9. 5. per annum. The church has a Norman tower. A school for boys is principally supported by the Duke of Portland, and another for girls by the Duchess. At Steetly, said to have been at one period a distinct parish, is a desecrated church, exhibiting a curious and good specimen of the later and more enriched style of Norman architecture; it is an interesting ruin, and is preserved with great care.

WHITWELL, a parish, in the union of AYLSHAM, hundred of EYNSFORD, Eastern Division of the county of NORFOLK, 1¼ mile (S. W.) from Reepham; containing 483 inhabitants. The living is a discharged vicarage, united to the rectory of Hackford.

WHITWELL (*St. Michael*), a parish, in the union of OAKHAM, hundred of ALSTOE, county of RUTLAND, 7 miles (W. by N.) from Stamford; containing 124 inhabitants. The living is a rectory, valued in the king's books at £5; patron, Representative of the late Sir G. Noel Noel, Bart. The tithes have been commuted for a rent-charge of £140, subject to the payment of rates, which on the average have amounted to £10; the glebe comprises 54 acres, valued at £78 per annum. The water of this place is slightly impregnated with iron.

WHITWELL (*St. Radegund*), a parish, in the liberty of EAST MEDINA, Isle of Wight Division of the county of SOUTHAMPTON, 8 miles (S. by E.) from Newport; containing 556 inhabitants. The living is a perpetual curacy, united, with the vicarage of Godshill, to the rectory of Niton.

WHITWELL, a township, in the parish, union, and ward, of KENDAL, county of WESTMORLAND, 4½ miles (N. by E.) from Kendal: the population is returned with the chapelry of Selside. This was an extensive common previously to 1825, when it was enclosed by act of parliament.

WHITWELL, a chapelry, in the parish of CATTERICK, union of NORTH ALLERTON, wapentake of GILLING-EAST, North Riding of the county of YORK, 3 miles (E.) from Catterick; containing 86 inhabitants.

WHITWELL-HOUSE, an extra-parochial liberty, in the Southern Division of EASINGTON ward, Northern Division of the county palatine of DURHAM, 2¾ miles (E. S. E.) from Durham; containing 32 inhabitants.

WHITWELL-ON-THE-HILL, a township, in the parish of CRAMBE, union of MALTON, wapentake of BULMER, North Riding of the county of YORK, 5¾ miles (S. W.) from New Malton; containing 227 inhabitants.

WHITWICK (*St. John the Baptist*), a parish, in the union of ASHBY-DE-LA-ZOUCH, hundred of WEST GOSCOTE, Northern Division of the county of LEICESTER, 5½ miles (E. by S.) from Ashby-de-la-Zouch; containing 3368 inhabitants, many of whom are employed in the manufacture of hosiery. The living is a discharged vicarage, valued in the king's books at £9. 14. 7.; present net income, £179: it is in the patronage of the Crown, in right of the Duchy of Lancaster; impropriator, Sir G. Beaumont, Bart. There is a place of worship for Wesleyan Methodists; and a Roman Catholic chapel was erected in 1837. Here is a National school, and another at Coalville.

WHITWOOD, a township, in the parish of FEATHERSTONE, Lower Division of the wapentake of AGBRIGG, West Riding of the county of YORK, 4¾ miles (N. W. by W.) from Pontefract; containing 306 inhabitants. There is an extensive manufacture of earthenware at Mere pottery, also large glass-bottle works, in this township.

WHITWORTH, a parochial chapelry, partly in the union of AUCKLAND, and partly in that of DURHAM, South-Eastern Division of DARLINGTON ward, Southern Division of the county palatine of DURHAM; containing 341 inhabitants, of which number, 104 are in the township of Whitworth, 5¼ miles (N. E. by N.) from Bishop-Auckland. The living is a perpetual curacy, in the patronage of the Dean and Chapter of Durham, the appropriators; net income, £243. The chapel was originally a chapel of ease to the vicarage of Merrington. In the cemetery, among other ancient sepulchral memorials, are a monument of a knight in armour, and the effigies of two ladies.

WHITWORTH, a chapelry, in the parish of ROCHDALE, hundred of SALFORD, Southern Division of the county palatine of LANCASTER, 2¾ miles (N. by W.) from Rochdale: the population is returned with the parish. The living is a perpetual curacy; net income, £256; it is in the patronage of Mrs. Langton, Mrs. Hornby, and J. Starky, Esq. The chapel is dedicated to St. Bartholomew. Twelve poor children are taught to read for £14. 10. a year, arising from the rents of certain cottages bequeathed by James Starky, in 1724. Here is a Sunday National school.

WHIXHALL, a chapelry, in the parish of PREES, Whitchurch Division of the hundred of NORTH BRADFORD, Northern Division of the county of SALOP, 3¾ miles (N. by E.) from Wem; containing 957 inhabitants. The living is a perpetual curacy; net income, £152; patron, Vicar of Prees. William Higgins, in 1737, bequeathed a rent-charge of £2 for teaching poor children.

WHIXLEY, a parish, partly in the Lower and partly in the Upper Division of the wapentake of CLARO, West Riding of the county of YORK; containing 968 inhabitants, of which number, 621 are in the township of Whixley, 6½ miles (S. S. E.) from Aldborough. The living is a perpetual curacy, valued in the king's books at £7. 17. 1.; present net income, £82; patrons and impropriators, Governors of the Tancred charities. There is a place of worship for Wesleyan Methodists. C. Tancred, Esq., whose family were long seated at the hall, at his death in 1754, left his house to be converted into an hospital for 12 decayed gentlemen, and endowed it with estates which, in 1815, were let for £2,480 per annum. He also endowed it with £1282. 15. 9. three per cent. consols., but directed part of the income to be appropriated for 12 exhibitions, 4 at Christ's College for divinity, 4 at Caius College for physic, and 4 at Lincoln's-Inn for law. The hospital is a spacious brick building, consisting of a centre and wings.

WHIXOE, county of SUFFOLK.—See WIXOE.

WHORLTON, a chapelry, in the parish of GAINFORD, union of TEESDALE, South-Western Division of DARLINGTON ward, Southern Division of the county palatine of DURHAM, 4 miles (E. S. E.) from Barnard-Castle; containing 311 inhabitants. The living is a perpetual curacy; net income, £107; patron, Vicar of Gainford. The chapel stands near the edge of a steep cliff above the river Tees. Here is a National school. Limestone abounds in the chapelry, in which also there are some petrifying springs.

WHORLTON, a township, comprising East and West Whorlton, in the parish of NEWBURN, union and Western Division of CASTLE ward, Southern Division of the county of NORTHUMBERLAND, 4¾ miles (N. W. by W.) from Newcastle-upon-Tyne; containing 59 inhabitants.

WHORLTON (*Holy Cross*), a parish, in the union of Stokesley, Western Division of the liberty of Langbaurgh, North Riding of the county of York; containing 915 inhabitants, of which number, 585 are in the township of Whorlton, 5½ miles (S. W. by S.) from Stokesley. The living is a perpetual curacy; net income, £84; patron and impropriator, Marquess of Ailesbury. The church is remarkable for a beautiful ivy tree, which flourishes in the interior. At Scarth, in this parish, in the time of Henry I., a cell of Augustine canons, subordinate to the monastery of Gisburn, was founded by Stephen Meinil. The lofty gateway tower of a castle, supposed to have been built in the reign of Richard II., still remains, and bears the arms of D'Arcy, Meynell, and Gray, its ancient possessors.

WIBSEY, a chapelry, in the parish of Bradford, wapentake of Morley, West Riding of the county of York, 2¼ miles (S. W. by S.) from Bradford, with which the population is returned. The living is a perpetual curacy; net income, £157; patron, Vicar of Bradford. The chapel, dedicated to the Holy Trinity, has been enlarged; it contains 300 free sittings, the Incorporated Society having granted £250 in aid of the expense. The manufacture of worsted is extensively carried on here.

WIBTOFT, a chapelry, in the parish of Claybrooke, union of Lutterworth, Kirby Division of the hundred of Knightlow, Northern Division of the county of Warwick, 5½ miles (W. N. W.) from Lutterworth; containing 104 inhabitants. The old Watling-street and Fosse-way meet at a Roman fort on the Leicestershire boundary, north of this place.

WICHAUGH, a township, in the parish of Malpas, union of Wrexham, Higher Division of the hundred of Broxton, Southern Division of the county of Chester, 5½ miles (N. W.) from Whitchurch; containing 35 inhabitants.

WICHENFORD (*St. Lawrence*), a parish, in the union of Martley, Lower Division of the hundred of Oswaldslow, Worcester and Western Divisions of the county of Worcester, 6¼ miles (N. W. by N.) from Worcester; containing 355 inhabitants. The living is a vicarage, endowed with the rectorial tithes, and valued in the king's books at £9. 10.; patrons, Dean and Chapter of Worcester. The tithes have been commuted for a rent-charge of £420. 6. 8., subject to the payment of rates, which on the average have amounted to £46. 5.; the glebe comprises 8 acres, valued at £8 per annum.

WICHLING, county of Kent.—See WITCHLING.

WICHNOR, a chapelry, in the parish of Tatenhill, union of Burton-upon-Trent, Northern Division of the hundred of Offlow and of the county of Stafford, 6½ miles (N. E.) from Lichfield; containing 157 inhabitants. The living is a perpetual curacy; net income, £71; patron, T. Levett, Esq. The chapel is dedicated to St. Leonard. A Sunday school is supported by Mrs. Levett. The Grand Trunk canal passes through the chapelry, and communicates with the neighbouring ironworks. James I. visited this place August 21st, 1621, and held a court at the hall; he dined there again, August 19th, 1624. Many Roman coins have been found in the neighbourhood, and in the park are vestiges of an encampment.

WICK and ABSON (*St. James*), a parish, in the union of Chipping, hundred of Puckle-Church, Western Division of the county of Gloucester, 7¼ miles (E. by N.) from Bristol; containing 824 inhabitants. This place is distinguished as the scene of a memorable victory obtained near Toghill, in this parish, in 1642, by the royalists, commanded by the Marquess of Hertford, over the parliamentarians under Sir William Waller. In this battle was slain the brave Sir Beville Granville, whose son was created Viscount Lansdown and Earl of Bath, and whose descendant erected on Lansdown, near Bath, a monument in commemoration of the event. The village is situated at the foot of a rocky hill rising to the height of more than 200 feet, and consisting of alternate beds of limestone and petrosilex. The parish abounds with coal, which is raised in great quantities; and lead-ore is also found here. It has lately been made a polling-place for the western division of the shire. The living is united, with that of Westerleigh, to the vicarage of Puckle-Church. Henry Burrow, in 1718, bequeathed £500 for charitable uses, of which £9 per annum is applied in aid of a school. Here is an ancient camp, supposed to be of British origin; and several Roman coins, urns, and bricks, have been dug up in the parish, which abounds with antique remains and natural curiosities.

WICK (*St. Lawrence*), a parish, in the union of Axbridge, hundred of Winterstoke, Eastern Division of the county of Somerset, 8½ miles (N. N. W.) from Axbridge; containing 281 inhabitants. This parish is situated on the shore of the Bristol Channel. The living is annexed to the vicarage of Congresbury, to which the church was formerly a chapel of ease. Here is a National school.

WICK, a tything, in the parish of Kemble, hundred of Malmesbury, Northern Division of the county of Wilts, 7½ miles (N. E. by N.) from Malmesbury: the population is returned with the parish.

WICK-EPISCOPI, a township, in the parish of St. John Bedwardine, Lower Division of the hundred of Oswaldslow, Worcester and Western Divisions of the county of Worcester: the population is returned with the parish.

WICK-NEAR-PERSHORE, a chapelry, in the parish of St. Andrew, Pershore, union, and Upper Division of the hundred, of Pershore, Pershore and Eastern Divisions of the county of Worcester, 1½ mile (E. S. E.) from Pershore; containing 280 inhabitants. The living is a perpetual curacy; net income, £105; patrons and appropriators, Dean and Chapter of Westminster. The chapel is dedicated to St. Lawrence. An Augustine priory was founded here, early in the reign of Stephen, by Peter de Corbezon, who, a few years afterwards, removed it to Studley in Warwickshire.

WICKEN (*St. Lawrence*), a parish, in the union of Newmarket, hundred of Staploe, county of Cambridge, 7 miles (S. S. E.) from Ely; containing 892 inhabitants. The living is a perpetual curacy; net income, £56: it is in the patronage of Mrs. Rayner. The Buckingham canal passes through the parish. Here is a National school with an endowment, also an almshouse for widows. At Spinney there was a priory founded by Sir Hugh de Malebisse, in the reign of Henry III., for three Augustine canons.

WICKEN (*St. John the Evangelist*), a parish, in the union of Potters-pury, hundred of Cleley, Southern Division of the county of Northampton, 3½ miles (W. S. W.) from Stony-Stratford; containing 536 inhabitants. The living is a rectory, with that of Wyke-Hamon consolidated, valued in the king's books at £15. 1. 10½.; present net income, £405; patron, Sir J. Mordaunt, Bart. The church of Wyke-Hamon has been long demolished. A National school is supported by Sir J. Mordaunt, Bart.

WICKEN-BONANT (*St. Margaret*), a parish, in the union of Saffron-Walden, hundred of Uttlesford, Northern Division of the county of Essex, 4¾ miles (S. W. by S.) from Saffron-Walden; containing 134 inhabitants. This parish is about a mile in length and of nearly equal breadth; the situation is low, but the lands are in good cultivation. The living is a rectory, valued in the king's books at £11; present net income, £213; patron, A. George, Esq. The church is a small ancient edifice of stone, with a tower of wood. A school is chiefly supported by the incumbent.

WICKENBY (*St. Peter and St. Lawrence*), a parish, in the Western Division of the wapentake of Wraggoe, parts of Lindsey, union and county of Lincoln, 5½ miles (N. W.) from Wragby; containing 137 inhabitants. The living is a rectory, valued in the king's books at £6. 17. 6.; present net income, £312; patron, C. Nevile, Esq. A school is supported by the rector.

WICKERSLEY (*St. Alban*), a parish, in the union of Rotherham, Southern Division of the wapentake of Strafforth and Tickhill, West Riding of the county of York, 3½ miles (E. by S.) from Rotherham; containing 527 inhabitants. The living is a rectory, valued in the king's books at £8. 0. 2½.; present net income, £345; patron, Henry Kater, Esq. The church has a tower, with north and south aisles, and was erected soon after the Norman Conquest: in the east window are some remains of painted glass. A school is endowed with £3. 10. per annum. Here are quarries of excellent stone for grindstones, from which the Sheffield manufacturers are supplied.

WICKFORD, a parish, in the union of Billericay, hundred of Barstable, Southern Division of the county of Essex, 6 miles (E. by S.) from Billericay; containing 402 inhabitants. This parish lies low, and the soil is generally wet and heavy: in the Vale of Wickford there is some rich land, producing excellent crops of wheat. The living is a rectory, valued in the king's books at £14; present net income, £381; patron, R. B. de Beauvoir, Esq. The church is situated on an eminence, and is a small edifice consisting of a nave and chancel. There is a place of worship for Independents.

WICKHAM, a chapelry, in the parish of Welford, hundred of Kintbury-Eagle, county of Berks, 5½ miles (N. W. by N.) from Speenhamland: the population is returned with the parish. The chapel is dedicated to St. Swithin.

WICKHAM, a chapelry, in the parish of Spalding, wapentake of Elloe, parts of Holland, county of Lincoln, 3¼ miles (N. E. by N.) from Spalding: the population is returned with the parish. The living is a donative curacy; net income, £33; patrons, Governors of Spalding Free Grammar School. The chapel, dedicated to St. Nicholas, is in ruins.

WICKHAM, a hamlet, in the parish and hundred of Banbury, county of Oxford, 1½ mile (W.) from Banbury, with which the population is returned.—See Banbury.

WICKHAM (*St. Nicholas*), a parish, in the union of Fareham, hundred of Titchfield, Fareham and Southern Divisions of the county of Southampton, 4 miles (S. by E.) from Bishop's-Waltham; containing 1106 inhabitants. The living is a rectory, valued in the king's books at £8. 2. 8½.; present net income, £578; patron, W. Rashleigh, Esq. Courts leet and baron are held annually: and there is a fair for cattle on May 20th. Here is a National school. The village, which is situated on the high road from London to Gosport, is remarkable as the birthplace, in 1324, of the distinguished and munificent prelate, William of Wykeham; and as the residence of Dr. Joseph Warton, the poet, who died here in 1800.

WICKHAM, BISHOP'S, a parish, in the union of Witham, hundred of Thurstable, Northern Division of the county of Essex, 2½ miles (S. by E.) from Witham; containing 549 inhabitants. This parish derives the affix to its name from the appropriation of the manor to the Bishops of London, who had formerly a palace here. The living is a rectory, valued in the king's books at £12. 3. 4.; present net income, £393; patron, Bishop of London. The church is a small ancient edifice, with a belfry turret of wood. Here is a National school.

WICKHAM, CHILDS (*St. Mary*), a parish, in the union of Evesham, Lower Division of the hundred of Kiftsgate, Eastern Division of the county of Gloucester, 5¼ miles (W. by S.) from Chipping-Campden; containing 415 inhabitants. The living is a discharged vicarage, valued in the king's books at £7. 16. 10.; present net income, £105; patron and incumbent, Rev. H. Pruen; impropriator, Sir T. Phillips, Bart. A day and Sunday school is supported by subscription.

WICKHAM, EAST (*St. Michael*), a parish, in the union of Dartford, hundred of Lessness, lathe of Sutton-at-Hone, Western Division of the county of Kent, 3½ miles (W. N. W.) from Crayford; containing 399 inhabitants. The living is annexed to the vicarage of Plumstead. Part of the lands and tithes of this parish were given by the famous Admiral, Sir John Hawkins, in the reign of Elizabeth, to the hospital for distressed mariners founded by him at Chatham, to which they still belong. William Forster, in 1727, gave lands in trust, among other purposes, to erect and endow a school; the income is £68 a year, for which about 40 children are educated in a National school.

WICKHAM-MARKET (*All Saints*), a parish, (formerly a market-town), in the union of Plomesgate, hundred of Wilford, Eastern Division of the county of Suffolk, 12½ miles (N. E.) from Ipswich; containing 1202 inhabitants. The village occupies an elevated and pleasant site, rising from the river Deben, and as its name implies, was formerly a market-town; attempts have been lately made to revive the market. It had also a shire-hall, where the general quarter sessions were usually held, but the building was, a few years since, taken down by the lord of the manor. The living is a discharged vicarage, valued in the king's books at £6. 16. 8.; present net income, £208: it is in the patronage of the Crown; the impropriation belongs to Pemberton's charity at Ipswich. The church is situated on an eminence commanding a most extensive prospect, including no less than 50 churches, and the spire is a conspicuous landmark; the archdeacon holds his spiritual courts here. There is a place of worship for Independents, who have also a school. Ann Barker, in 1730, bequeathed one-third of the produce of £300, which, with the rents arising from certain lands given by an unknown individual, is applied for teaching poor children: the annual income is £26. Here is a National school. Four poor men of this parish partake of the benefit of Bishop Wood's hospital at Ufford, but have no house to reside in; the endowment is £21 a year, which is divided equally among them. A union workhouse for Plomesgate hundred has lately been erected here: the union comprises 40 parishes or places, under the care of 41 guardians, and, according to the census of 1831, contains a population of 20,703. Mr. John Kirby, compiler of the Suffolk Traveller, was for some time resident in this parish, but removed to Ipswich, where he died.

WICKHAM, ST. PAUL (*All Saints*), a parish, in the union of SUDBURY, hundred of HINCKFORD, Northern Division of the county of ESSEX, 4½ miles (N. by E.) from Halsted; containing 388 inhabitants. This parish is about six miles in circumference; the soil is light, but very fertile in some parts. The village is pleasantly situated, and consists of a number of neat, well-built houses, ranged around an extensive green. The living is a rectory, in the patronage of the Dean and Chapter of St. Paul's, London, valued in the king's books at £9. The tithes have been commuted for a rent-charge of £400, subject to the payment of rates, which on the average have amounted to £77; the glebe comprises 20 acres, valued at £30 per annum. The church is a neat substantial edifice, with a square embattled tower. A National Sunday school is supported by the rector. On the village green is an almshouse for one inmate, but without any endowment. The rent of a field of 4 acres, amounting to £11. 10., is given to the poor at Christmas.

WICKHAM-SKEITH (*St. Andrew*), a parish, in the union and hundred of HARTISMERE, Western Division of the county of SUFFOLK, 2¾ miles (N.) from Mendlesham; containing 556 inhabitants. The living is a discharged vicarage, valued in the king's books at £5. 8. 1½.; present net income, £125; patrons, Executors of the late T. Bond, Esq.; impropriators, the Landowners.

WICKHAM, WEST (*St. Mary*), a parish, in the union of LINTON, hundred of CHILFORD, county of CAMBRIDGE, 4¾ miles (N. E. by E.) from Linton; containing 529 inhabitants. The living is a perpetual curacy; net income, £88; patron and impropriator, Earl of Hardwicke. Here is a Sunday National school.

WICKHAM, WEST (*St. John the Baptist*), a parish, in the union of BROMLEY, hundred of RUXLEY, lathe of SUTTON-AT-HONE, Western Division of the county of KENT, 2¾ miles (S. S. W.) from Bromley; containing 614 inhabitants. This was formerly a market-town; a fair for cattle is held on Easter-Monday. The manor-house is a curious square structure, with angular towers, of the time of Henry VII. The living is a rectory, valued in the king's books at £11. 10. 10.: it is in the patronage of the Dowager Lady Farnaby. The tithes have been commuted for a rent-charge of £495, subject to the payment of rates; the glebe comprises 35 acres. The chancel windows of the church are beautifully ornamented with stained glass. There is a place of worship for Wesleyan Methodists. A National school is supported by subscription. The learned Gilbert West, the friend of Gray, the poet, long resided in the village, where he was visited by Lyttelton and Pitt; he was the author of "Poems," and "Observations on the Resurrection," and translated some of the Odes of Pindar; he was buried here, in 1756.

WICKHAMBREUX (*St. Andrew*), a parish, in the union of BRIDGE, hundred of DOWNHAMFORD, lathe of ST. AUGUSTINE, Eastern Division of the county of KENT, 5 miles (E. by N.) from Canterbury; containing 486 inhabitants. The living is a rectory, valued in the king's books at £29. 12. 6.; present net income, £790; patron, J. P. Plumtree, Esq. The tithes have been commuted for a rent-charge of £749. 11. 6., subject to the payment of rates; the glebe comprises 20½ acres, valued at £50 per annum. The Rev. John Smith, B.D., in 1756, gave a house and school-room in trust to the minister and parish officers, for the education of children.

WICKHAMBROOK (*All Saints*), a parish, in the union and hundred of RISBRIDGE, Western Division of the county of SUFFOLK, 6¼ miles (N. by W.) from Clare; containing 1400 inhabitants. The living is a vicarage, valued in the king's books at £8. 6. 10½.; present net income, £210: it is in the patronage of the Crown; impropriator, N. W. Bromley, Esq. The church is a large and handsome building, and contains some good monuments. There is a place of worship for Independents. Two day schools are partly supported by subscription; and here is a Sunday National school. This has been made a polling-place for the western division of the county. Near the church is an almshouse consisting of six tenements under the same roof, with a garden attached, which was founded in 1615, by Anthony Sparrow, who endowed it with £3 per annum, paid out of a mill and land in Stansfield: the inmates are six aged maids or widows. Mrs. Anne Warner, in 1785, bequeathed £400 three per cent. reduced annuities, the dividends of which, together with the rental of the poor estate and some other bequests, altogether amounting to £30. 5. per annum, are distributed among the poor; and Elizabeth Chinery, by will in 1818, gave £250 three per cents., directing the dividends to be applied to purchasing linen for the poor. Some Roman remains have been found in the parish.

WICKHAMFORD (*St. John the Baptist*), a parish, in the union of EVESHAM, Upper Division of the hundred of BLACKENHURST, Pershore and Eastern Divisions of the county of WORCESTER, 2¼ miles (E. S. E.) from Evesham; containing 136 inhabitants. The living is a discharged perpetual curacy, valued in the king's books at £2. 4. 4½.; present net income, £41; patrons and appropriators, Dean and Canons of Christ-Church, Oxford. Limestone in abundance is obtained in the neighbourhood.

WICKHAMPTON (*St. Andrew*), a parish, in the union of BLOFIELD, hundred of WALSHAM, Eastern Division of the county of NORFOLK, 4½ miles (S. S. E.) from Acle; containing 122 inhabitants. The living is a discharged rectory, valued in the king's books at £4; present net income, £150; patron, J. F. Leathes, Esq. The arms of the ancient family of Gerbrigges are in various parts of the church, though much defaced by time. There is a place of worship for Wesleyan Methodists.

WICKLEWOOD, a parish, in the hundred of FOREHOE, Eastern Division of the county of NORFOLK, 3 miles (W. N. W.) from Wymondham; containing 787 inhabitants. The living comprises the rectory of *St. Andrew*, consolidated with the discharged vicarage of *All Saints*, and valued in the king's books at £6. 3. 11½.; present net income, £127; the patronage and impropriation belong to Elizabeth Darby, Mary Kett, and R. Heber, Esq. The church of St. Andrew has been long demolished: it stood in the same churchyard as that of All Saints, which is a spacious structure, with a large handsome tower, in the later style of English architecture. A market and two fairs, granted in 1440 by Henry VI., were formerly held here.

WICKMERE, a parish, in the union of AYLESHAM, hundred of SOUTH ERPINGHAM, Eastern Division of the county of NORFOLK, 5 miles (N. N. W.) from Aylsham; containing 319 inhabitants. The living is a discharged rectory, annexed to that of Wolterton, and valued in the king's books at £9. A school for girls is supported by the Countess of Orford.

WICKTON, a joint township with Risbury, in the parish of STOKE-PRIOR, hundred of WOLPHY, county of HEREFORD, 4½ miles (S. E.) from Leominster; containing, with Risbury, 143 inhabitants.

WICKWAR (*HOLY TRINITY*), a market-town and parish, in the union of CHIPPING-SODBURY, Upper Division of the hundred of GRUMBALD'S-ASH, Western Division of the county of GLOUCESTER, 24 miles (S. S. W.) from Gloucester, and 111 (W.) from London; containing 1000 inhabitants. This town, which consists principally of one spacious and well-built street, is pleasantly situated on the nearest and best road from Bath to Gloucester and the north of England, and is watered by two small streams. The surrounding scenery is highly picturesque, and the air proverbially pure and salubrious; great improvements have taken place, and a new road has been formed to Wotton-under-Edge, by which the distance to that place has been shortened two miles. The clothing trade was formerly carried on here to a considerable extent, but it has for some time been discontinued; coal abounds in the adjacent waste lands. The market is on Monday, and fairs are held on April 6th and July 2nd, for horses and horned cattle. The town, under a charter granted by Charles I., is governed by a mayor and an indefinite number of aldermen, consisting of all who have served the office of mayor. A manorial court leet is held triennially in October. The living is a rectory, valued in the king's books at £18; patron, Lord Ducie. The tithes have been commuted for a rent-charge of £430, subject to the payment of rates, which on the average have amounted to £70; the glebe comprises 14½ acres. The church, a spacious structure, with a lofty and handsome tower, has been enlarged, and 70 free sittings provided, the Incorporated Society having granted £60 in aid of the expense: there are places of worship for Independents and Wesleyan Methodists. The free school was founded in 1683, by Alexander Hosea, a native of the town, who endowed it with property now producing £126 per annum, for a classical and a writing master, who are appointed by the trustees, and have dwelling-houses rent-free; for some time there have been no boys in the classical school; the average number attending the writing school is about 25. Here is a National school; and there are several small endowments for the poor.

WIDCOMBE (*ST. THOMAS à BECKET*), a joint parish with Lyncombe, in the union of BATH, hundred of BATH-FORUM, Eastern Division of the county of SOMERSET, 1½ mile (S. E. by E.) from Bath; containing 8704 inhabitants. The living is a vicarage, annexed to the rectory of St. Peter and St. Paul, Bath. The manufacture of fine woollen cloth is carried on here to a considerable extent. A fair is held annually on May 14th. At Holloway, in this parish, John Cantlow, Prior of Bath, towards the close of the fifteenth century, built a small chapel, in honour of St. Mary Magdalene; and adjoining it an hospital for lunatics, which latter was rebuilt in 1761.

WIDCOMBE, a tything, in the parish of CHEWTON-MENDIP, union of CLUTTON, hundred of CHEWTON, Eastern Division of the county of SOMERSET, 5 miles (S. W. by S.) from Pensford; containing 160 inhabitants.

WIDDINGTON (*ST. MARY*), a parish, in the union of SAFFRON-WALDEN, hundred of UTTLESFORD, Northern Division of the county of ESSEX, 5 miles (W. by N.) from Thaxted; containing 386 inhabitants. The living is a rectory, valued in the king's books at £25; patron, W. J. Campbell, Esq. The tithes have been commuted for a rent-charge of £570, subject to the payment of rates, which on the average have amounted to £103; the glebe comprises 42 acres, valued at £63. 10. per annum. The church is a small ancient edifice of stone, partly rebuilt with brick; it retains several details of the Norman style. Here is a National school.

WIDDINGTON, a township, in the parish of LITTLE OUSEBURN, Upper Division of the wapentake of CLARO, West Riding of the county of YORK, 8¼ miles (S. E. by E.) from Aldborough; containing 30 inhabitants.

WIDDRINGTON, a parochial chapelry, in the union, and Eastern Division of the ward, of MORPETH, Northern Division of the county of NORTHUMBERLAND, 8 miles (N. E. by N.) from Morpeth; containing 395 inhabitants. This was long the seat of the family of Widdrington, of whom many have at various periods distinguished themselves against the Scots. Sir William, in 1642, was expelled from the House of Commons, for raising forces in defence of Charles I., who, in the following year elevated him to the dignity of Baron Widdrington of Blankney. After the battle of Marston Moor he left the kingdom, when his estates were confiscated by the parliament; but returning in the service of Charles II., he was slain at the battle of Wigan. His son and successor, William, Lord Widdrington, was attainted in 1715, and his property, to the amount of £100,000, was sold for the public use. The ancient castle, which stood in a noble park of 600 acres, was burned down more than fifty years ago, and the present edifice, which occupies the site of the former, is much out of repair, and now uninhabited. Widdrington was separated from the parish of Woodhorn, and invested with distinct parochial rights, in 1768. A small colliery is worked near the village. The living is a perpetual curacy; net income, £67: it is in the patronage of Lord and Lady Vernon; impropriators, Mercers' Company, and Incumbent of Hampstead, Middlesex. The church is ancient, and appears to have been once much larger. A Scotch church was erected here in 1765. There is a school-room, with a house and garden occupied by the master, whose salary of £25 a year is paid by the lady of the manor, for teaching the poor children of the parish, who are also clothed.

WIDECOMBE-IN-THE-MOOR (*St. Pancras*), a parish, in the union of Newton-Abbott, and forming, with the parish of Buckland-in-the-Moor, a detached portion of the hundred of Haytor, being locally in that of Lifton, Teignbridge and Southern Divisions of the county of Devon, 5¾ miles (N. W. by N.) from Ashburton; containing 959 inhabitants. The parish is intersected by three rivulets, tributary to the river Dart, which bounds it on one side. It comprises several long valleys, enclosed by rugged hills, bordering upon Dartmoor, and exhibits, in many places, marks of streamworks: tin has long been found in the neighbourhood. The living is a vicarage, valued in the king's books at £25. 13. 9.; present net income, £268; patrons and appropriators, Dean and Chapter of Exeter. The church was greatly injured by lightning during the performance of divine service, on October 21st, 1638, when portions of the stone and wood work fell in. There are four small schools in different parts of the parish, supported by subscription and the trifling endowments of William Culling and Miss Smith.

WIDFORD (*St. Mary*), a parish, in the union and hundred of Chelmsford, Southern Division of the county of Essex, 1½ mile (S. W.) from Chelmsford; containing 157 inhabitants. This parish is supposed to have derived its name from a ford over the river Chelmer. It comprises 800 acres; the soil is rich, and around the village the lands are in a high state of cultivation. The living is a discharged rectory, valued in the king's books at £8; patron and incumbent, Rev. W. Warner. The tithes have been commuted for a rent-charge of £250, subject to the payment of rates, which on the average have amounted to £49; the glebe comprises 20 acres, valued at £36 per annum. The church is a small ancient edifice, partly in the early, and partly in the decorated, style of English architecture; it is situated on the west side of the high road from London to Chelmsford.

WIDFORD (*St. Oswald*), a parish, in the union of Witney, Lower Division of the hundred of Slaughter, Eastern Division of the county of Gloucester, 1½ mile (E. S. E.) from Burford; containing 51 inhabitants. This parish is entirely surrounded by the county of Oxford, to which it has been united as regards parliamentary representation. The living is a discharged rectory, valued in the king's books at £3. 14. 2.; present net income, £90; patron, Lord Redesdale.

WIDFORD (*St. John the Baptist*), a parish, in the union of Ware, hundred of Braughin, county of Hertford, 3¾ miles (E. by N.) from Ware; containing 506 inhabitants. The living is a rectory, valued in the king's books at £12. 13. 4.; present net income, £242; patron, W. P. Hamond, Esq. The church has a square embattled tower, with a tall slender spire, and occupies a considerable eminence. A rent-charge of £5, from an unknown benefactor, is applied for teaching three boys; and a National school has been established.

WIDLEY (*St. Mary*), a parish, in the union of Fareham, hundred of Portsdown, Fareham and Southern Divisions of the county of Southampton, 5½ miles (E. by N.) from Fareham; containing 512 inhabitants. The living is a rectory, with the vicarage of Wymering annexed, valued in the king's books at £14. 11. 10½.; present net income, £678; patron, T. Thistlethwaite, Esq. The church has been enlarged, and 100 free sittings provided, the Incorporated Society having granted £100 in aid of the expense. Two schools are supported by subscription.

WIDMER-POOL (*St. Peter*), a parish, in the union of Bingham, Southern Division of the wapentake of Rushcliffe and of the county of Nottingham, 9 miles (S. S. E.) from Nottingham; containing 180 inhabitants. The living is a rectory, valued in the king's books at £14. 16. 0½.; present net income, £232; patron, F. Robinson, Esq. The church was damaged by lightning, but was restored in 1836, the Incorporated Society having granted £300 in aid of the expense. A day and Sunday school is supported by subscription. The parish is bounded on the west by the old Fosse-road.

WIDNESS, a joint township with Appleton, in the parish and union of Prescot, hundred of West Derby, Southern Division of the county palatine of Lancaster, 6¼ miles (W. by S.) from Warrington; containing 1986 inhabitants.

WIDWORTHY (*St. Cuthbert*), a parish, in the union of Honiton, hundred of Colyton, Honiton and Southern Divisions of the county of Devon, 3½ miles (E. by S.) from Honiton; containing 278 inhabitants. The living is a rectory, valued in the king's books at £11. 16. 0½.; patron, E. Marwood Elton, Esq. The tithes have been commuted for a rent-charge of £200, subject to the payment of rates; the glebe comprises 31 acres, valued at £50 per annum. The church contains the effigy of a knight in armour, and a fine monument, by Bacon, to the memory of James Marwood, Esq., a liberal benefactor to the parish. Benedictus Marwood, Esq., in 1742, gave £100, and the Rev. Joseph Somaster, in 1770, left £50 to be applied for the education of poor children: the latter also left £50, directing the proceeds to be distributed in bread among the poor. In 1831, the Rev. W. J. Tucker, then rector, gave £200 to his successor for similar purposes. At Wilmington a fair is held on the morrow of St. Matthew's day. Near the church is an old earthwork, and in the north-east part of the parish are vestiges of an ancient intrenchment.

WIELD (*St. James*), a parish, in the union of Alton, hundred of Fawley, Alton and Northern Divisions of the county of Southampton, 6 miles (W.) from Alton; containing 248 inhabitants. The living is a perpetual curacy; net income, £64; patron and impropriator, Earl of Portsmouth. The church is very ancient, and contains a marble monument to Sir Richard Wallop, an ancestor of the Earls of Portsmouth. A school is supported by subscription.

WIGAN (*All Saints*), a parish, comprising the borough and market-town of Wigan (which has separate jurisdiction), and several chapelries and townships, and the head of a union, chiefly in the hundred of West Derby, but partly in the hundred of Salford, Southern Division of the county palatine of Lancaster; and containing 44,486 inhabitants, of which number, 20,774 are in the town of

Corporation Seal.

Wigan, 18 miles (W. N. W.) from Manchester, and 199 (N. W. by N.) from London. This place is stated by Camden to have been originally called *Wibiggin:* the vicinity is said to have been the scene of some sanguinary battles between the Britons, under their renowned King Arthur, and the Saxons; and the discovery, about the middle of the 18th century, of a large quantity of human bones, and the bones and shoes of horses, over an extensive tract of ground near the town, tends to confirm this opinion. During the great civil war, several battles were fought here by the contending parties, it being the principal station of the king's troops commanded by the Earl of Derby, who was defeated and driven from the town by the parliamentary forces under Sir John Smeaton, early in 1643; and shortly afterwards, in the same year, he was again defeated by Colonel Ashton, who, in consequence of the devotion of the inhabitants to the royal cause, ordered the fortifications of the town to be demolished. From this time Wigan remained tranquil (with the exception of Oliver Cromwell pursuing through it, in 1648, the Scotch army under the Duke of Hamilton, whom he had driven from Preston) until 1651, when the Earl of Derby, having been summoned from the Isle of Man by King Charles II., was again defeated here by a very superior force under Colonel Lilburne. To record the courage and loyalty of Sir Thomas Tildesley, who was slain in this action, a monumental pillar was erected in 1679, by Alexander Rigby, Esq., then high sheriff of the county, on the spot where he fell, at the northern end of the town. In the year 1745, Prince Charles Edward marched through Wigan on his route from Preston to Manchester.

The town is situated on the eastern bank, and near the source, of the river Douglas, and is described by Leland as "a paved town, as big as Warrington, but better builded;" a patent for paving it and building a bridge over the Douglas was granted so early as the 7th of Edward III. The old and greater part of the town consists of irregular streets and mean houses, but great improvements have been made, and two new streets formed, which contain some well-built houses. It is lighted with gas by a company established in 1823, and supplied with good water by works erected by a company formed under the authority of an act in 1761, but now affording a very inadequate supply. The manufacture of calicoes, fustians, and other cotton goods, linens, and checks, and the spinning of cotton yarn, are extensively carried on; and there are brass and iron foundries, pewter-works, several manufactories for spades and edge-tools, and some corn-mills, on the river. The Douglas, under the authority of an act obtained in 1720, was made navigable to its junction with the Ribble. The Leeds and Liverpool canal passes the town, and by its branches and various communications with Manchester and Kendal on one side, and the North Sea on the other, affords every facility for the conveyance of the manufactures and of the coal abounding in this neighbourhood, among which is cannel coal, to all parts of the kingdom: fly-boats ply daily upon it to Manchester and Liverpool. A branch from the Liverpool and Manchester railway, seven miles long, has been formed to this town; and a line, 17 miles in length, to Preston, has been recently completed. The market is on Monday and Friday, that on the latter day being the principal; and there are fairs on Holy Thursday, June 27th, and Oct. 28th, on which days the commercial hall, a commodious brick building in the market-place, erected in 1816, is open for the use of the clothiers.

The first charter of incorporation was granted by Henry III., and the privileges which it bestowed have been confirmed and augmented by succeeding monarchs; but that under which the corporation acted, previously to the passing of the Municipal Reform Act, was granted by Charles II., and vested the government of the borough in a mayor, 12 aldermen, two bailiffs, a recorder, town-clerk, and subordinate officers. The mayor, the ex-mayor, and six senior aldermen, were justices of the peace. The corporation now consists of a mayor, 10 aldermen, and 30 councillors, under the act of the 5th and 6th of William IV., cap. 76, for an abstract of which see the Appendix, No. I. The borough is divided into five wards, and the municipal and parliamentary boundaries are made co-extensive. The mayor and late mayor are justices of the peace, with ten others appointed by commission. The police force consists of seven constables. The borough first sent members to parliament in the 23rd of Edward I., and again in the 35th of the same reign, after which period this privilege was not exercised until the 1st of Edward VI. The right of election was formerly vested in the free burgesses; but the privilege has been extended to the £10 householders within the borough, which is co-extensive with the township of Wigan, and comprises an area of 1640 acres. The present number of voters is 564, of whom 32 are burgesses: the mayor is the returning officer. The corporation are authorized by their charter to try for all civil actions, and hold a court of sessions quarterly for felonies not capital, committed within the borough. Petty sessions for the Warrington Division of the hundred are held here; and by the act of the 2nd and 3rd of William IV., cap. 64, the town has been made a polling-place for the southern division of the shire. The town-hall was rebuilt in 1720, by the Earl of Barrymore and Sir Roger Bradshaigh, then members for the borough. The gaol is used only for temporary confinement, prisoners being committed to the county gaol.

The living is a rectory, valued in the king's books at £80. 13. 4.; present net income, £3000; patron, Earl of Bradford. The parochial church, dedicated to All Saints, is an ancient and handsome edifice. St. George's was erected, as a chapel of ease, in 1781: the living is a perpetual curacy; net income, £118: it is in the alternate patronage of the owners of pews and the rector, the latter of whom is patron of the perpetual curacies of Billinge, Hindley, and Upholland, in this parish. Her Majesty's Commissioners have also approved of a plan for building a church at Scholes. There are two places of worship each for Baptists, Independents, and Roman Catholics, and one each for Presbyterians and Wesleyan Methodists; there are also Roman Catholic chapels at Ashton-in-the-Willows, Brinn, and Writington Hall, in the vicinity. The free grammar school, at Millgate, appears to have been founded in the 16th year of the reign of James I., but by whom is uncertain; the earliest recorded benefaction to it is one of £6. 13. 4. per annum, in 1619, by Mr. James Leigh. A considerable increase in the income having arisen from various subsequent donations and benefactions, an act of parliament was passed in 1812, incorporating fifteen members of the corporation governors of the institution, who appoint a master and an usher; the number of scholars is limited to eighty, and the income exceeds £200 per annum. A Blue-coat school, wherein 40 boys were clothed and instructed, was established in 1773, but a building for a National school having been erected in 1825, the former has been united to it. Two schools for children of Roman Catholics are supported by subscription. The dispensary was established in 1798, and the building erected in 1801. There are many minor bequests and benefactions for the poor, amounting in the aggregate to a considerable sum annually. A clothing society, for furnishing them with warm clothing and bedding during the winter months, was formed in 1817; a savings bank was established in 1821; a Bible society in 1825; and a mechanics' institute and library in 1825. The poor law union of Wigan comprises 20 townships or chapelries, under the care of 28 guardians, and contains a population of 58,402, according to the census of 1831. A spring was discovered near Scholes bridge, some years since, the water of which possessed nearly the same medicinal properties as those at Harrogate; and a handsome building was erected for the convenience of persons who wished to drink, or use it for a bath; but it has ceased to be resorted to, the water having lost much of its medicinal virtue, owing, it is supposed, to its being mixed with the water of the neighbouring coal-pits.

WIGBOROUGH, GREAT (*St. Stephen*), a parish, in the union of Lexden and Winstree, hundred of Winstree, Northern Division of the county of Essex, 7 miles (S. S. W.) from Colchester; containing 434 inhabitants. This parish, which is bounded on the south by a creek of the river Blackwater, called the Verley, is of considerable antiquity, and was the scene of a great battle, probably with the northern pirates, to whose incursions it was, from its situation, peculiarly exposed; and there is near the church a large tumulus, supposed to have been raised over the bodies of those who were slain on that occasion. The village is situated on the road from Maldon to Colchester, and though now of little note, was formerly of much greater importance, as is evident from several green lanes still retaining the appellation of streets. There were formerly extensive salt-works in the immediate neighbourhood, from which circumstance the hamlet where they were carried on is called Salcot-Wigborough, where a fair is held annually on the 24th of August. The living is a rectory, valued in the king's books at £18. 17. 6.; present net income, £591; patrons, H. Bewes, Esq., and the Rev. William Fookes. The church is situated on a considerable eminence commanding extensive views of the sea and adjacent country. There is a place of worship for Independents. A quantity of ancient coins, inclosed in an earthen jar, was found in the marshes about 45 years since.

WIGBOROUGH, LITTLE (*St. Nicholas*), a parish, in the union of Lexden and Winstree, hundred of Winstree, Northern Division of the county of Essex, 7¼ miles (S. by W.) from Colchester; containing 123 inhabitants. This parish is bounded on the north by a creek of the river Blackwater, called Mersey channel, and on the south by another called Verley channel; it comprises about 700 acres of land in good cultivation. The living is a discharged rectory, valued in the king's books at £10; patrons, Governors of the Charter-house, London. The tithes have been commuted for a rent-charge of £220, subject to the payment of rates, which on the average have amounted to £9; the glebe comprises 20 acres, valued at £30 per annum. The church is a small ancient edifice, with a square tower, romantically situated on the sea-shore.

WIGGENHALL (*St. Germans*), a parish, in the union of Downham, hundred of Freebridge-Marshland, Western Division of the county of Norfolk, 4¼ miles (S. S. W.) from Lynn-Regis; containing 552 inhabitants. The living is a discharged vicarage, valued in the king's books at £6; present net income, £107; patrons and appropriators, Dean and Chapter of Norwich. The appropriate tithes have been commuted for a rent-charge of £272. 5., and the vicarial for £133. 15., subject to the payment of rates, which on the average on both have amounted to £90; the appropriate glebe comprises 25 acres, and the vicarial 2 acres, respectively valued at £50 and £5 per annum. The church is situated on the east bank of the Ouse. A school is partly supported by a donation of £12 per annum from the trustees of St. German's bridge. There is also a Sunday National school.

WIGGENHALL (*St. Mary*), a parish, in the union of Downham, hundred of Freebridge-Marshland, Western Division of the county of Norfolk 5¾ miles (S. W. by S.) from Lynn-Regis; containing 206 inhabitants. The living is a discharged vicarage, valued in the king's books at £12. 10.; present net income, £94: it is in the patronage of the Crown; impropriator, W. Bagge, Esq. The impropriate tithes have been commuted for a rent-charge of £380, and the vicarial for £85, subject to the payment of rates, which on the average have amounted respectively to £55 and £13; the glebe comprises 2½ acres, valued at £5. 3. per annum. The church is a stately structure; at the east end is an oaken screen, separating what was once a chapel from the rest of the interior. The windows over the arches of the nave exhibit some remains of ancient stained glass, representing the twelve Apostles, and various portions of sacred history. It contains a fine altar-tomb, bearing the arms of Kervile and Plowden, with the effigies of a knight in armour, his lady, and two children. This parish is bounded on the west by the river Ouse.

WIGGENHALL (*St. Mary Magdalene*), a parish, in the union of Downham, hundred of Freebridge-Marshland, Western Division of the county of Norfolk, 6 miles (S. S. W.) from Lynn-Regis; containing 576 inhabitants. The living is a discharged vicarage, valued in the king's books at £5. 15. 10.; present net income, £336; patron and impropriator, W. Franks, Esq. The church is a regular structure, with a square stone tower at the west end, and chapels at the termination of the north and south aisles: various parts of the building are ornamented with the arms of Kervile, Scales, Berney, and several other ancient families. In 1833, an act was obtained for draining and preserving certain fen lands, and for other purposes, in this parish. On the west bank of the Ouse was a hermitage, dedicated to St. John the Evangelist, which in 1181 was appropriated, by the prior and convent of Reynham, to nuns of the order of St. Augustine, subject to Castle-Acre, whose revenue at the dissolution was valued at £31. 16. 7.

WIGGENHALL (*St. Peter*), a parish, in the union of Downham, hundred of Freebridge-Marshland, Western Division of the county of Norfolk, 5 miles

(S. by W.) from Lynn-Regis; containing 114 inhabitants. The living is a discharged vicarage, valued in the king's books at £6; present net income, £85: it is in the patronage of the Crown; impropriator, J. Hatt, Esq. The church stands on the east side of the Ouse: on the font are two cross keys, an emblem of the patron saint.

WIGGESLEY, a hamlet, in the parish of THORNEY, union, and Northern Division of the wapentake, of NEWARK, Southern Division of the county of NOTTINGHAM, 8½ miles (E. by S.) from Tuxford; containing 86 inhabitants.

WIGGINTHORPE, a joint township with Terrington, in the parish of TERRINGTON, union of MALTON, partly within the liberty of ST. PETER'S, East Riding, and partly in the wapentake of BULMER, North Riding, of the county of YORK, 9 miles (W.) from New Malton: the population is returned with Terrington.

WIGGINTON (*St. Bartholomew*), a parish, in the union of BERKHAMPSTEAD, hundred of DACORUM, county of HERTFORD, 1¼ mile (S. E.) from Tring; containing 536 inhabitants. The living is a perpetual curacy; net income, £69; patrons and appropriators, Dean and Canons of Christ-Church, Oxford. The London and Birmingham railway passes nearly within a mile to the north-east of the church. In this parish is an extensive common, which, according to tradition, has been the scene of frequent military achievements, not only during the parliamentary war, but even in the time of the Romans; in support of the latter opinion an almost perfect specimen of a Roman camp may still be distinctly traced.

WIGGINTON (*St. Giles*), a parish, in the union of BANBURY, hundred of BLOXHAM, county of OXFORD, 5¼ miles (W. N. W.) from Deddington; containing 327 inhabitants. The living is a rectory, valued in the king's books at £17. 2. 8½.; present net income, £290; patrons, Principal and Fellows of Jesus' College, Oxford. A National school has been established. To the south-east of the church are some vestiges of a Roman villa, extending over a considerable space: on digging here, an octagonal apartment, with a tessellated pavement, was discovered, and to the south-east of it, another of square form, with tesseræ of coarse and imperfect character; on the floor was a human skeleton: these apartments were heated by flues under the floor, and some Roman coins were also found.

WIGGINTON, a chapelry, in the parish and union of TAMWORTH, Southern Division of the hundred of OFFLOW and of the county of STAFFORD, 1¾ mile (N.) from Tamworth; containing 737 inhabitants. The living is a perpetual curacy; net income, £92; patron, Vicar of Tamworth. The chapel, dedicated to St. Leonard, was enlarged in 1830, and contains 100 free sittings, the Incorporated Society having granted £100 in aid of the expense. Thomas Barnes, in 1717, gave property, of the value of about £5 per annum, in support of a school.

WIGGINTON, a parish, in the wapentake of BULMER, union and North Riding of the county of YORK, 5 miles (N.) from York; containing 359 inhabitants. The living is a rectory, valued in the king's books at £14. 13. 4.; present net income, £297: it is in the patronage of the Crown. Here is a National school.

WIGGLESWORTH, a township, in the parish of LONG PRESTON, union of SETTLE, Western Division of the wapentake of STAINCLIFFE and EWCROSS, West Riding of the county of YORK, 6½ miles (S. S. W.) from Settle; containing 443 inhabitants. The free school was founded by Lawrence Clark, in 1789, and is endowed with £1135. 19. 8. three per cent. consols.

WIGGONBY, a township, in the parish of AIKTON, ward, and Eastern Division of the county, of CUMBERLAND, 4¾ miles (N. E.) from Wigton; containing 175 inhabitants. Margaret Hodgson, in 1792, left land now let for about £175 a year in support of a school. Near Down Hall, within the township, is an encampment 60 yards square, now planted with fir, and surrounded by a deep ditch.

WIGGONHOLT, a parish, in the hundred of WEST EASWRITH, rape of ARUNDEL, Western Division of the county of SUSSEX, 8 miles (N.N.E.) from Arundel; containing 37 inhabitants. This parish is bounded on the west by the river Arun, and the road from Petworth to Brighton passes through it. The living is a rectory, with that of Greatham consolidated, valued in the king's books at £7. 4. 4½.; patron, Hon. Robert Curzon. The tithes have been commuted for a rent-charge of £102. 14. 8., subject to the payment of rates, which on the average have amounted to £2; the glebe comprises 9 acres, valued at £18 per annum. The church is small, and was repewed and repaired in 1839, at the expense of the Hon. Robert Curzon. The rectory-house was enlarged and altered in 1838 by the incumbent, and is now a spacious residence in the Elizabethan style. The parish is in the Preston incorporation under Gilbert's Act for the maintenance of the poor. A great number of Roman urns was found here in 1827; they were of red pottery, beautifully figured, but, from the unprotected situation in which they were deposited, few of them are in a perfect state. Several Roman coins of the emperors Nero, Vespasian, Claudius, Adrian, and Marcus Antoninus, were also discovered.

WIGHILL (*All Saints*), a parish, in the Ainsty of the city, and East Riding of the county, of YORK, 2½ miles (N. by W.) from Tadcaster; containing 276 inhabitants. The living is a discharged vicarage, endowed with a portion of the rectorial tithes, and valued in the king's books at £5. 3. 6½.; present net income, £114; patron, and impropriator of the remainder of the rectorial tithes, R. F. Wilson, Esq. The church is situated on an eminence rising from the margin of the river Wharfe. A school is partly supported by Richard Yorke, Esq.

WIGHT, ISLE OF.—See article on the county of SOUTHAMPTON.

WIGHTERING, EAST and WEST, county of SUSSEX.—See WITTERING, EAST and WEST.

WIGHTON (*All Saints*), a parish, in the union of WALSINGHAM, hundred of NORTH GREENHOE, Western Division of the county of NORFOLK, 2¼ miles (N. by E.) from Little Walsingham; containing 542 inhabitants. The living is a discharged vicarage, valued in the king's books at £11. 11. 8.; patrons, Dean and Chapter of Norwich. A school is endowed with £6 per annum.

WIGLAND, a township, in the parish of MALPAS, union of WREXHAM, Higher Division of the hundred of BROXTON, Southern Division of the county of CHESTER, 1¾ mile (S. S. E.) from Malpas; containing 265 inhabitants. At Dirtwich, in this township, are brine springs, from which salt is made. In 1643, the works were destroyed by a detachment of the parliamentary army, but they were soon restored.

WIGMORE (*St. James*), a parish, in the union of

LUDLOW, hundred of WIGMORE, county of HEREFORD, 10 miles (N. W. by N.) from Leominster; containing 476 inhabitants. The living is a discharged vicarage, valued in the king's books at £8; present net income, £120; patron and appropriator, Bishop of Hereford. Limestone abounds here, and it is supposed that coal may be obtained in the neighbourhood. A court leet is occasionally held; and there are fairs for cattle, sheep, &c., on May 6th and August 5th. On a commanding elevation, a little to the westward of the village, are the ivy-mantled ruins of Wigmore castle, the outer works of which are the most perfect: the massive fragments of the keep occupy the summit of a lofty artificial mound, and present an appearance highly grand and picturesque: the founder of this once stately edifice is now unknown, but it is recorded that Edward the Elder caused it to be repaired. It was taken from Edric, Earl of Shrewsbury, by Ranulph de Mortimer, who came over with the Conqueror, and made it his principal seat. The same nobleman, in 1100, established in the parish church a small college of three prebendaries, which continued till 1179, when his son Hugh founded, in honour of St. James, a noble abbey for monks of the order of St. Augustine, about one mile distant from the castle, and endowed it so amply that, at the dissolution, its revenue was estimated at £302. 12. 3. An Alien priory, a cell to that of Aveney in Normandy, is said to have existed, at an early period, at Limebrook, in this parish; but it is more certain that a priory of nuns of the order of St. Augustine was founded there by the Mortimers, some time in the reign of Richard I., which at the suppression was valued at £23. 17. 8. In the neighbourhood are traces of a Danish camp.

WIGSTHORPE, a hamlet, in the parish of LILFORD, union of OUNDLE, hundred of HUXLOE, Northern Division of the county of NORTHAMPTON, 4 miles (S. by E.) from Oundle: the population is returned with the parish. Here was anciently a chapel.

WIGSTON, MAGNA (*ALL SAINTS*), a parish, in the union of BLABY, hundred of GUTHLAXTON, Southern Division of the county of LEICESTER, $3\frac{1}{2}$ miles (S. S. E.) from Leicester; containing 2174 inhabitants. The living is a discharged vicarage, valued in the king's books at £9. 8. 9.; present net income, £107; patrons, alternately, Master and Wardens of the Haberdashers' Company, and the Governors of Christ's Hospital, London, who are the impropriators. There is a place of worship for Independents. A school for girls is endowed with £8 per annum; and an infants' school is partly supported by subscription. This place was formerly called Wigston Two Steeples, from its having two churches; one of these, now in a very dilapidated state, is used as a school-room, where poor children are taught to read and write at the expense of the parishioners. At a place called the Gaol Close, during the civil war in the reign of Charles I., a temporary prison was erected, to which the prisoners were removed from the county gaol at Leicester: the royal army lay near this place some few days. The Leicester canal runs through the parish. The village is pleasantly situated on the high road between Welford and Leicester, and is chiefly inhabited by persons employed in the manufacture of stockings. Here are a lunatic asylum, and an hospital, or almshouse, for six poor widows and as many widowers; the latter was endowed by a Miss Clarke.

WIGSTON, PARVA, a chapelry, in the parish of CLAYBROOKE, union of LUTTERWORTH, hundred of GUTHLAXTON, Southern Division of the county of LEICESTER, $6\frac{1}{2}$ miles (N. W. by W.) from Lutterworth; containing 69 inhabitants. The chapel is dedicated to St. Mary. Here is a Sunday National school.

WIGTOFT (*ST. PETER AND ST. PAUL*), a parish, in the union of BOSTON, wapentake of KIRTON, parts of HOLLAND, county of LINCOLN, 3 miles (S. E. by S.) from Swineshead; containing 697 inhabitants. The living is a discharged vicarage, with that of Quadring united, valued in the king's books at £11. 5.; present net income, £412; patron and appropriator, Bishop of Lincoln. A free school for 20 children is endowed with land producing £26. 15. per annum.

WIGTON (*ST. MARY*), a market-town and parish, and the head of a union, in the ward, and Eastern Division of the county, of CUMBERLAND; containing 6501 inhabitants, of which number, 4885 are in the town of Wigton, 11 miles (S. W. by W.) from Carlisle, and 305 (N. N. W.) from London. Of the early history of this place little is recorded: the barony was given by William de Meschines to Waldeof, Lord of Allerdale, and by him to Odoard, who lived about the period of the Norman Conquest, and assumed the name of De Wigton. The town was burnt by the Scots when they plundered the abbey of Holme-Cultram, in 1322; and during the civil war, in 1648, the van of the Duke of Hamilton's army was quartered here. It consists principally of one spacious street, with a narrower extending transversely at one end of it, containing some handsome, well-built houses; it is pitched with pebbles, and supplied with water from wells, the property of individuals, and a public well and pump, erected near the centre of the town. There are a public subscription and a circulating library: races formerly took place annually in the month of August, but they have been discontinued. The principal articles of manufacture are checks, muslins, and ginghams, which are made to a considerable extent; and a large establishment for calico printing and dyeing also affords employment to many of the inhabitants. Coal is obtained within three miles, and copper-ore is found within five miles, of the town. The market days are Tuesday and Friday, the former only for corn, of which a great quantity is pitched in the market-place. Fairs are held on February 20th, a very large horse fair; April 5th, for horned cattle; and on December 21st, called Wallet fair, for cattle, butchers' meat, apples, and honey; there are also statute fairs at Whitsuntide and Martinmas. The county magistrates hold here a petty session every month; and constables are appointed at the court leet and baron of the lord of the manor, which is held annually in September. This town has been made a polling-place for the eastern division of the county.

The living is a discharged vicarage, valued in the king's books at £17. 19. $0\frac{1}{2}$.; present net income, £120; patron, Bishop of Carlisle; impropriators, the Landowners. The church is said to have been originally erected by Odoard, with materials brought from the neighbouring Roman station, called Old Carlisle, and it subsequently belonged to the abbey of Holme-Cultram; it was taken down in 1788, and the present edifice, a light and handsome building, erected on its site: attached to it is a library for the use of the clergy, presented by Dr.

Bray. There are places of worship for the Society of Friends, Independents, Wesleyan Methodists, and Roman Catholics: a large and handsome meeting-house for the Independents, with school-rooms, &c., attached, has been lately opened. The free grammar school, at Market Hill, near the entrance of the town, was founded in 1730, by certain of the inhabitants, who agreed to subscribe towards its erection £1 for every penny for which their houses were assessed to the purvey, on condition of having the privilege of sending their children; this privilege is retained by their successors in the several houses which have thus contributed. In 1787, the sum of £1000 three per cent. stock was bequeathed by John Allison to this school, for the education of four boys belonging to the parish of Wigton, and not living in free tenements; and, in 1798, £355 was bequeathed to it by Thomas Tomlinson, Esq., who also left £100 for the establishment of a public library: the present income is about £68 per annum. The master and usher are elected by those inhabitants whose tenements are entitled to the freedom of the school, and the former has a good dwelling-house. The Rev. John Brown, D.D., author of the tragedy of *Barbarossa,* received his early education in this school. A National Sunday school has been erected. At Brookfield, near the town, is a school for 60 boys, founded by the Society of Friends, in 1825. The Sunday school, erected in 1820, is a neat and spacious building of freestone, capable of receiving 500 children; the general number attending it is about 450: there are also Sunday schools supported by the dissenting congregations. An hospital for six widows of beneficed clergymen, or curates of two years' standing, of the county of Cumberland, or that part of Westmorland which is in the diocese of Carlisle, or of Rothbury in the county of Northumberland, was founded in 1725, by the Rev. John Tomlinson, who endowed it with a rent-charge of £45. 12., to which other benefactions have subsequently been added. About a mile south of Wigton, on an eminence, are the remains of a Roman station, called Old Carlisle, where a great variety of antiquities have been dug up, consisting of coins, altars, statues, and inscriptions, which prove that the *Ala Augusta* was stationed here, in the reign of the Emperor Gordian. Ewan Clarke, the well-known Cumberland poet; Joseph Rooke, a self-taught genius, who has become a distinguished mathematician and philosopher, excelling in his knowledge of music, mechanics, optics, and botany; R. Smirke, R. A., the celebrated historical painter; and Mr. George Barnes, professor of mathematics, were natives of this town.

WIGTON, a township, in the parish of HAREWOOD, Upper Division of the wapentake of SKYRACK, West Riding of the county of YORK, 5½ miles (N. by E.) from Leeds; containing 168 inhabitants. A school is partly supported by subscription.

WIKE, a township, in the parish of BIRSTALL, union of BRADFORD, wapentake of MORLEY, West Riding of the county of YORK, 3½ miles (S. by W.) from Bradford; containing 1918 inhabitants. There is a place of worship for Independents.

WIKE, a township, in the parish of HAREWOOD, Upper Division of the wapentake of SKYRACK, West Riding of the county of YORK, 6¼ miles (N. N. E.) from Leeds; containing 142 inhabitants. A school is endowed for the instruction of ten children.

WILBARSTON (*ALL SAINTS*), a parish, in the union of MARKET-HARBOROUGH, hundred of CORBY, Northern Division of the county of NORTHAMPTON, 4½ miles (S. W. by W.) from Rockingham; containing 681 inhabitants. The living is a vicarage, valued in the king's books at £7. 17. 1.; present net income, £187; patron and impropriator, Lord Sondes. There is a place of worship for Independents; and a National school has been established. Mrs. Catherine Palmer bequeathed £200 for educating 22 girls, 12 of the parish of Wilbarston, and 10 of the parish of Cottingham; £100 for establishing a school at Burton-Overy; £100 for employing the aged poor of Wilbarston in spinning flax, &c.; and £100 to the poor of each of the parishes of Cottingham, Foxton, and Illston. Some other trifling bequests are annually distributed in bread and money among the poor.

WILBERFOSS (*ST. JOHN THE BAPTIST*), a parish, in the union of POCKLINGTON, Wilton-Beacon Division of the wapentake of HARTHILL, East Riding of the county of YORK; containing 580 inhabitants, of which number, 352 are in the township of Wilberfoss, 5½ miles (W. N. W.) from Pocklington. The living is a perpetual curacy; net income, £67; patron, Colonel Wyndham, who, with others, is the impropriator. A Benedictine nunnery, in honour of the Blessed Virgin Mary, was founded here before 1153, by Alan de Catton, which at the dissolution had a revenue of £28. 8. 8.

WILBRAHAM, GREAT (*ST. NICHOLAS*), a parish, in the union of CHESTERTON, hundred of STAINE, county of CAMBRIDGE, 7¼ miles (E. by S.) from Cambridge; containing 510 inhabitants. The living is a discharged vicarage, valued in the king's books at £11. 18. 4.; present net income, £203; patron and impropriator, Edward Hicks, Esq. The church is a cruciform structure, with a tower at the west end; it had originally a tower rising from the centre. The manor-house, an ancient building formerly belonging to the Knights Templars, is still called the Temple.

WILBRAHAM, LITTLE (*ST. JOHN THE EVANGELIST*), a parish, in the union of CHESTERTON, hundred of STAINE, county of CAMBRIDGE, 7¼ miles (E.) from Cambridge; containing 315 inhabitants. The living is a rectory, valued in the king's books at £19. 16. 8.; present net income, £326; patrons, President and Fellows of Corpus Christi College, Cambridge. An infants' school is supported by Johnson's charity.

WILBURTON (*ST. PETER*), a parish, in the hundred of SOUTH WITCHFORD, union and Isle of ELY, county of CAMBRIDGE, 6½ miles (S. W.) from Ely; containing 471 inhabitants. The living is a perpetual curacy; net income, £68; patron, Archdeacon of Ely; impropriator, Heir of the Hon. Sir Albert Pell. The tithes have been commuted for a rent-charge of £519, subject to the payment of rates; the glebe comprises 170 acres, valued at £220 per annum. The church is a handsome structure. There is a place of worship for Baptists; also a Sunday National school. The parsonage-house was anciently the seat of the Archdeacons of Ely, at which Henry VII. and his son, Prince Henry, were entertained for several days, when that sovereign came to visit the shrine of St. Ethelreda.

WILBY (*ALL SAINTS*), a parish, in the union of GUILTCROSS, hundred of SHROPHAM, Western Division of the county of NORFOLK, 3½ miles (N. E. by

E.) from East Harling; containing 123 inhabitants. The living is a discharged rectory, with that of Hargham annexed, valued in the king's books at £7. 4. 7½.; present net income, £380: it is in the patronage of Ann Hare. A school is partly supported by subscription.

WILBY (*St. Mary*), a parish, in the union of Wellingborough, hundred of Hamfordshoe, Northern Division of the county of Northampton, 2¼ miles (S. W. by W.) from Wellingborough; containing 386 inhabitants. The living is a rectory, valued in the king's books at £13. 19. 4½.; present net income, £273; patron, W. Stockdale, Esq. The church is partly in the early and partly in the decorated style of English architecture, with a beautiful spire steeple of later date.

WILBY (*St. Mary*), a parish, in the union and hundred of Hoxne, Eastern Division of the county of Suffolk, 6 miles (E. S. E.) from Eye; containing 649 inhabitants. The living is a rectory, valued in the king's books at £26. 6. 10½.; patron and incumbent, Rev. George Mingay. The tithes have been commuted for a rent-charge of £630, subject to the payment of rates, which on the average have amounted to £202; the glebe comprises 51 acres, valued at £87. 10. per annum.

WILCOT (*Holy Cross*), a parish, in the union of Pewsey, hundred of Swanborough, Everley and Pewsey, and Northern, Divisions of the county of Wilts, 1¾ mile (W. N. W.) from Pewsey; containing 677 inhabitants. The living is a discharged vicarage, valued in the king's books at £6. 17.; present net income, £143; patron and impropriator, Lieut.-Col. George Wroughton Wroughton. The Kennet and Avon canal passes through the parish. The manor-house is said to have been anciently a monastery, of which there are no further particulars.

WILCOTE (*St. Peter*), a parish, in the union of Witney, hundred of Wootton, county of Oxford, 4 miles (N. by E.) from Witney; containing 10 inhabitants. The living is a discharged rectory, valued in the king's books at £2. 13. 4.: it is in the patronage of Mrs. Pickering.

WILDBOAR-CLOUGH, a township, in the parish of Prestbury, union and hundred of Macclesfield, Northern Division of the county of Chester, 6½ miles (S. E. by E.) from Macclesfield; containing 476 inhabitants. There is a place of worship for Wesleyan Methodists.

WILDEN (*St. Nicholas*) a parish, in the hundred of Barford, union and county of Bedford, 5¼ miles (N. E. by N.) from Bedford; containing 411 inhabitants. The living is a rectory, valued in the king's books at £18. 7. 1.; present net income, £250; patron, Duke of Bedford. The church has been repewed, and 80 free sittings provided, the Incorporated Society having granted £40 in aid of the expense. John and Thomas Rolle, in 1624, bequeathed land now producing an annual income of £22. 7. 6., for teaching poor children.

WILDON-GRANGE, a township, in the parish of Coxwold, union of Easingwould, wapentake of Birdforth, North Riding of the county of York, 6½ miles (N. by W.) from Easingwould; containing 27 inhabitants.

WILDSWORTH, a hamlet, in the parish of Laughton, union of Gainsborough, wapentake of Corringham, parts of Lindsey, county of Lincoln, 7½ miles (N.) from Gainsborough; containing 132 inhabitants. A chapel was erected here in 1836, containing 150 sittings, 50 of which are free, the Incorporated Society having granted £50 towards defraying the expense.

WILERICK, or WILLCRICK, a parish, in the union of Newport, Lower Division of the hundred of Caldicott, county of Monmouth, 4½ miles (E. S. E.) from Caerleon; containing 33 inhabitants. The living is a discharged rectory, annexed to that of Llanmartin, and valued in the king's books at £2. 10. 2½.

WILFORD (*St. Wilfrid*), a parish, in the union of Basford, Northern Division of the wapentake of Rushcliffe, Southern Division of the county of Nottingham, 1¾ mile (S. by W.) from Nottingham; containing 602 inhabitants. The living is a rectory, valued in the king's books at £18. 17. 6.; present net income, £574; patron, Sir R. Clifton, Bart. The river Trent is here crossed by a ford and a ferry.

WILKESLEY, a joint township with Dodcot, partly in the parish of Wrenbury, but chiefly in that of Audlem, union and hundred of Nantwich, Southern Division of the county of Chester, 3¾ miles (S. W. by W.) from Audlem: the population is returned with Dodcot.

WILKSBY (*All Saints*), a parish, in the union and soke of Horncastle, parts of Lindsey, county of Lincoln, 5¾ miles (S. S. E.) from Horncastle; containing 67 inhabitants. The living is a discharged rectory, valued in the king's books at £4. 4. 2.; present net income, £120; patron, Henry Dymoke, Esq.

WILLAND (*St. Mary*), a parish, in the union of Tiverton, hundred of Halberton, Cullompton and Northern Divisions of the county of Devon, 2¼ miles (N. N. E.) from Cullompton; containing 321 inhabitants. The living is a discharged rectory, valued in the king's books at £7. 10. 5.; present net income, £110; patron, — Salter, Esq. There is a place of worship for Wesleyan Methodists; also a National school.

WILLASTON, a township, in the parish and hundred of Nantwich, Southern Division of the county of Chester, 20 miles (S. E. by E.) from Chester; containing 99 inhabitants.

WILLASTON, a township, in the parish of Wybunbury, union and hundred of Nantwich, Southern Division of the county of Chester, 1½ mile (E. by N.) from Nantwich; containing 122 inhabitants.

WILLASTON, a township, in the parish of Neston, union, and Higher Division of the hundred, of Wirrall, Southern Division of the county of Chester, 2¾ miles (E.) from Great Neston; containing 276 inhabitants.

WILLEN (*St. Mary Magdalene*), a parish, in the union of Newport-Pagnell, hundred of Newport, county of Buckingham, 1½ mile (S.) from Newport-Pagnell; containing 83 inhabitants. The living is a vicarage, valued in the king's books at £7. 10.; present net income, £115; patrons, Trustees of the late Dr. Busby, who nominate a Westminster student of Christ-Church College, Oxford. The church was erected in 1680, at the expense of Dr. Busby, head master of Westminster school, who endowed it with the rectorial tithes, and gave a library for the use of the vicar: he also appointed 22 lectures on the catechism to be preached annually: the vicar now receives a stipend in lieu of the rectorial tithes.

WILLENHALL, a chapelry, in the parish and union of WOLVERHAMPTON, Southern Division of the hundred of OFFLOW and of the county of STAFFORD, 3¼ miles (E.) from Wolverhampton; containing 5834 inhabitants. This place, at the period of the Norman survey, was called *Winehala*, the Saxon term for victory, probably from the great battle fought near it in 911. The village began to thrive in the reign of Elizabeth, when, from the extensive mines of iron-stone and coal in the neighbourhood, the iron manufacture was first established here: at present it is noted for its collieries and its flourishing trade in locks, the latter of which it possesses to a greater extent than any other place of its size in Europe. As an instance of the great ingenuity of the workmen, it is recorded that, in 1776, one Lees, aged 63, produced a padlock and key, wrought by himself, of less weight than a silver twopenny coin. Many other articles of hardware are made, particularly currycombs, gridirons, screws, &c. The Wyrley and Essington canal, and the Grand Junction railway (which has a station here) pass through the parish. Courts leet and baron are annually held; and there is a court of requests on three Mondays in every alternate month, for the recovery of debts under £5. In the neighbourhood are the remains of an old hall, formerly the seat of the maternal ancestors of the Duke of Cleveland. The living is a perpetual curacy; net income, £300; patrons, the Inhabitants. The chapel, dedicated to St. Giles, was built about 1748. There are places of worship for Baptists and Wesleyan Methodists; also five Sunday schools, supported by subscription.

WILLENHALL, a hamlet, in the parish of HOLY TRINITY, city of COVENTRY, union of FOLESHILL, Kirby Division of the hundred of KNIGHTLOW, Northern Division of the county of WARWICK, 2¾ miles (S. E.) from Coventry; containing 120 inhabitants.

WILLERBY (*ST. PETER*), a parish, in the union of SCARBOROUGH, wapentake of DICKERING, East Riding of the county of YORK; containing 356 inhabitants, of which number, 38 are in the township of Willerby, 7½ miles (S. S. W.) from Scarborough. The living is a discharged vicarage, valued in the king's books at £9. 0. 7½.; present net income, £116: it is in the patronage of the Crown; impropriator, W. J. Dennison, Esq.

WILLERBY, a township, in the union of SCULCOATES, partly in the parish of COTTINGHAM, Hunsley-Beacon Division of the wapentake of HARTHILL, and partly in that of KIRK-ELLA, county of the town of KINGSTON-UPON-HULL, East Riding of the county of YORK, 5½ miles (W. N. W.) from Kingston-upon-Hull: the latter part contains 189 inhabitants, and the former is returned with the parish of Cottingham.

WILLERSEY (*ST. PETER*), a parish, in the union of EVESHAM, Upper Division of the hundred of KIFTSGATE, Eastern Division of the county of GLOUCESTER, 3 miles (W.) from Chipping-Campden; containing 327 inhabitants. The living is a rectory, valued in the king's books at £13. 2. 6.; present net income, £162; patron, W. Preedy, Esq. The church is a cruciform structure of various dates, with a tower at the intersection, crowned with pinnacles. Here is a Sunday National school. On the top of the hill above the village is a large camp, enclosing about 60 acres, supposed to have been formed during the incursions of the Danes: from this camp is a fine view of the vale below.

WILLERSLEY (*ST. MARY MAGDALENE*), a parish, in the union of KINGTON, hundred of HUNTINGTON, county of HEREFORD, 7½ miles (E. N. E.) from Hay; containing 13 inhabitants. This parish is bounded on the south by the river Wye, the banks of which are adorned with much picturesque scenery. The living is a discharged rectory, valued in the king's books at £3. 6. 8.; present net income, £79: it is in the patronage of Mrs. Jane Lilly.

WILLESBOROUGH (*ST. MARY*), a parish, in the union of EAST ASHFORD, hundred of CHART and LONGBRIDGE, Upper Division of the lathe of SCRAY, Eastern Division of the county of KENT, 2 miles (S. E. by E.) from Ashford; containing 472 inhabitants. The living is a vicarage, valued in the king's books at £8. 16. 8.; present net income, £167; patrons and appropriators, Dean and Chapter of Canterbury. The church is principally in the decorated style of English architecture.

WILLESDEN, or WILSDON (*ST. MARY*), a parish, in the union of HENDON, Kensington Division of the hundred of OSSULSTONE, county of MIDDLESEX, 5 miles (W. N. W.) from London; containing 1876 inhabitants. The living is a vicarage, in the patronage of the Dean and Chapter of St. Paul's, London (the appropriators), valued in the king's books at £14; present net income, £130. The church is principally in the later style of English architecture. Here is a National school. The London and Birmingham railway passes through the southern part of the parish.

WILLESLEY (*ST. THOMAS*), a parish, in the union of ASHBY-DE-LA-ZOUCH, and forming, with the parishes of Measham and Stratton-en-le-Fields, a detached portion of the hundred of REPTON and GRESLEY, Southern Division of the county of DERBY, locally in the Western Division of the hundred of Goscote, county of Leicester, 2½ miles (S. W. by S.) from Ashby-de-la-Zouch; containing 63 inhabitants. The living is a perpetual curacy; net income, £62; the patronage and impropriation belong to Lady Hastings. The Ashby-de-la-Zouch canal skirts the south-western boundary of the parish, whence a railroad passes to that town.

WILLEY, a township, in the parish of PRESTEIGN, union of KNIGHTON, hundred of WIGMORE, county of HEREFORD, 2¾ miles (N.) from Presteign; containing 147 inhabitants.

WILLEY (*ST. JOHN THE BAPTIST*), a parish, within the liberties of the borough of WENLOCK, union of MADELEY, Southern Division of the county of SALOP, 4¾ miles (N. W. by N.) from Bridgenorth; containing 155 inhabitants. The living is a discharged rectory, with the perpetual curacy of Barrow annexed, valued in the king's books at £5. 6. 3.; patron, Lord Forester. The tithes have been commuted for a rent-charge of £245, subject to the payment of rates; the glebe comprises 27 acres, valued at £45 per annum.

WILLEY (*ST. LEONARD*), a parish, in the union of LUTTERWORTH, Kirby Division of the hundred of KNIGHTLOW, Northern Division of the county of WARWICK, 3½ miles (W.) from Lutterworth; containing 96 inhabitants. The living is a rectory, valued in the king's books at £8. 6. 0½.; present net income, £191: it is in the patronage of the Crown.

WILLIAMSCOTT, or WILLSCOTT, a hamlet, in

the parish of CROPREDY, hundred of BANBURY, county of OXFORD, 3¾ miles (N. N. E.) from Banbury: the population is returned with the chapelry of Wardington. Walter Calcott, in 1575, gave a rent-charge of £13, and John Ditchfield, in 1708, an annuity of £2, to be applied for teaching poor children; and a National school has been established. There are some remains of an ancient house, in which Charles I. slept a night or two prior to the battle of Cropredy bridge.

WILLIAN (*All Saints*), a parish, in the union of HITCHIN, hundred of BROADWATER, county of HERTFORD, 2¾ miles (E. by N.) from Hitchin; containing 313 inhabitants. The living is a vicarage, endowed with the rectorial tithes, and valued in the king's books at £5; patron, Francis Pym, Esq. The tithes have been commuted for a rent-charge of £593. 12. 6., subject to the payment of rates, which on the average have amounted to £50; the glebe comprises 20 acres, valued at £20 per annum. The church has been recently beautified, at the expense of £250.

WILLINGALE-DOE (*St. Christopher*), a parish, in the union of ONGAR, hundred of DUNMOW, Northern Division of the county of ESSEX, 4½ miles (N. E.) from Chipping-Ongar; containing 466 inhabitants. This is the larger of the two parishes of the name. Three separate constables are appointed, one for the township of Willingale, one for the hamlet of Torrels Hall, and one for that of Birds Green. The living is a rectory, with that of Shellow-Bowels consolidated, valued in the king's books at £16; patron, T. W. Bramston, Esq. The tithes have been commuted for a rent-charge of £489. 5., subject to the payment of rates, which on the average have amounted to £87; the glebe comprises 31 acres, valued at £40 per annum. The church, consisting of a nave and chancel, with a square embattled tower, stands in the same churchyard with that of Willingale-Spain: the parishes are much intermixed, though quite distinct as to ecclesiastical and civil concerns. Robert Cole, in 1733, gave an annuity of £4 for teaching six poor children: there is also a small sum, the bequest of the Rev. Mr. Walker, for purchasing books; and a National school has been established.

WILLINGALE-SPAIN (*All Saints*), a parish, in the union of ONGAR, hundred of DUNMOW, Northern Division of the county of ESSEX, 4¼ miles (N. E.) from Chipping-Ongar; containing 239 inhabitants. This parish derives the adjunct to its name from the family of Hervey de Spain, to whom it belonged at the time of the Norman Survey. Spain Hall, the ancient manor-house, is near the church. The living is a rectory, valued in the king's books at £7. 13. 4.: it is in the patronage of the Crown, on the nomination of the Bishop of London. The tithes have been commuted for a rent-charge of £322. 12. 6., subject to the payment of rates, which on the average have amounted to £50; the glebe comprises 29½ acres, valued at £50 per annum. The church has a handsome altar-piece, the gift of William Brocket, Esq.—See WILLINGALE-DOE.

WILLINGDON, a parish, in the union of EAST-BOURNE, hundred of WILLINGDON, rape of PEVENSEY, Eastern Division of the county of SUSSEX, 2¼ miles (N. by W.) from East Bourne; containing 603 inhabitants. The living is a discharged vicarage, valued in the king's books at £12; present net income, £67; patrons, Dean and Chapter of Chichester; impropriators, J. Thomas and R. Newman, Esqrs. The church is principally in the early style of English architecture. A school for girls is supported by subscription.

WILLINGHAM (*St. Mary and All Saints*), a parish, in the union of CHESTERTON, hundred of PAPWORTH, county of CAMBRIDGE, 6¼ miles (E. by S.) from St. Ives; containing 1403 inhabitants. The living is a rectory, valued in the king's books at £18. 8. 1½.; present net income, £667; patron, Bishop of Ely. On the north side of the chancel of the church is a chapel, in the decorated style of English architecture, with a stone roof of singular construction. There is a place of worship for Baptists. Much of the cheese which takes its name from the neighbouring village of Cottenham is made in this parish, where about 1200 milch cows are usually kept. A charity school was founded by subscription, in 1593, and an estate purchased for its endowment, which now produces £20 per annum; it is also further endowed with a rent-charge of £10, bequeathed in 1700 by Dr. Saywell, Master of Jesus' College, Cambridge; the number of children is limited to 30; in another school the curate instructs about 50 children gratuitously. An almshouse for four widows was founded in 1616, by William Smith, Provost of King's College, Cambridge, and endowed with £18 per annum.

WILLINGHAM, a chapelry, in the parish of CARLTON, hundred of RADFIELD, county of CAMBRIDGE, 5½ miles (S. by E.) from Newmarket: the population is returned with the parish. The chapel is dedicated to St. Matthew.

WILLINGHAM (*St. Helen*), a parish, in the union of GAINSBOROUGH, wapentake of WELL, parts of LINDSEY, county of LINCOLN, 6 miles (S. E.) from Gainsborough; containing 392 inhabitants. The living is a rectory, valued in the king's books at £18. 6. 8.; present net income, £352; patron, Rev. J. Peel. There is a place of worship for Wesleyan Methodists.

WILLINGHAM (*St. Mary*), a parish, in the union and hundred of WANGFORD, Eastern Division of the county of SUFFOLK, 4 miles (S.) from Beccles; containing 158 inhabitants. The living is a rectory, annexed to that of North Cove, and valued in the king's books at £6. 13. 4. The church was standing in 1529, but is now in ruins.

WILLINGHAM, CHERRY (*St. Peter*), a parish, in the wapentake of LAWRESS, parts of LINDSEY, union and county of LINCOLN, 4 miles (E. by N.) from Lincoln; containing 102 inhabitants. The living is a discharged vicarage, valued in the king's books at £6. 13. 4.; present net income, £95; patrons and impropriators, — Cock, — Gordon, and — Ellis, Esqrs.

WILLINGHAM, NORTH, a parish, in the union of CAISTOR, Southern Division of the wapentake of WALSHCROFT, parts of LINDSEY, county of LINCOLN, 4 miles (E. by S.) from Market-Rasen; containing 223 inhabitants. The living is a discharged vicarage, valued in the king's books at £5. 4. 4½.; present net income, £69; patron and impropriator, Ayscoghe Boucherett, Esq. A day and Sunday school is partly supported by subscription.

WILLINGHAM, SOUTH (*St. Martin*), a parish, in the union of LOUTH, Eastern Division of the wapentake of WRAGGOE, parts of LINDSEY, county of LINCOLN, 5¼ miles (E. N. E.) from Wragby; containing 212 inhabitants. The living is a discharged vicarage, valued in the king's books at £13. 10. 10.; present net income, £389; patron, G. F. Heneage, Esq. A school is partly supported by the incumbent.

WILLINGTON (*St. Lawrence*), a parish, in the hundred of Wixamtree, union and county of Bedford, 4 miles (E) from Bedford; containing 332 inhabitants. The living is a discharged vicarage, valued in the king's books at £7. 17.; present net income, £194; patron and impropriator, Duke of Bedford. The church is principally in the later style of English architecture, and contains some old monuments to the Gostwicke family. The navigable river Ouse bounds the parish on the north.

WILLINGTON, a township, in the parish of Whalley, union of Great Boughton, Second Division of the hundred of Eddisbury, Southern Division of the county of Chester, 3 miles (N. N. W.) from Tarporley; containing 115 inhabitants. This township is deemed to be part of the parish of Whalley, it having formerly belonged to the abbey there, though for the performance of ecclesiastical rites the inhabitants resort to the church of St. Oswald, Chester, and pay a portion of the great tithes to the rectors of Wem and Tarvin. There is a place of worship for Unitarians; also a school supported by Major and the Misses Tomkinson.

WILLINGTON (*St. Michael*), a parish, in the union of Burton-upon-Trent, hundred of Morleston and Litchurch, Southern Division of the county of Derby, 5 miles (N. E.) from Burton-upon-Trent; containing 402 inhabitants. The living is a discharged vicarage, valued in the king's books at £4. 17. 3.; present net income, £82; patrons and impropriators, Governors of Etwall hospital and Repton Grammar School. A school is supported by Mrs. Spilsbury. The river Trent (over which there is a bridge of five arches to Repton), the Grand Trunk canal, and the Birmingham and Derby Junction railway, pass through the parish.

WILLINGTON, a township, in the parish of Brancepeth, North-Western Division of Darlington ward, union, and Southern Division of the county palatine, of Durham, 4 miles (N.) from Bishop-Auckland; containing 216 inhabitants. It is situated on the north side of the river Wear, and is intersected by the great Roman road. There is a place of worship for Wesleyan Methodists; also a school partly supported by subscription.

WILLINGTON, a township, in the parish of Wallsend, Eastern Division of Castle ward, Southern Division of the county of Northumberland, 3 miles (W. by S.) from North Shields: the population is returned with the parish. It is situated on the river Tyne, where are a fine quarry, several coal-staiths, a patent ropery, and a corn-mill, worked by steam. There are extensive collieries within the township.

WILLINGTON, a hamlet, in the parish of Barcheston, Brails Division of the hundred of Kington, Southern Division of the county of Warwick, 1¼ mile (S. S. E.) from Shipston-upon-Stour: the population is returned with the parish.

WILLISHAM (*St. Mary*), a parish, in the union and hundred of Bosmere and Claydon, Eastern Division of the county of Suffolk, 3 miles (S. S. W.) from Needham-Market; containing 224 inhabitants. The living is a perpetual curacy; net income, £56; patrons and impropriators, Trustees of the late T. Myers, Esq. The sum of £4. 10., arising from land, is distributed among the poor of the parish.

WILLITOFT, a joint township with Gribthorpe, in the parish of Bubwith, union of Howden, Holme-Beacon Division of the wapentake of Harthill, East Riding of the county of York, 5¼ miles (N.) from Howden: the population is returned with Gribthorpe.

WILLITON, a chapelry, in the parish of St. Decuman, hundred of Williton and Freemanners, Western Division of the county of Somerset, 6¼ miles (E. S. E.) from Dunster: the population is returned with the parish. The living is a perpetual curacy, in the patronage of the Prebendary of St. Decuman's in Wells Cathedral; net income, £53. The chapel is dedicated to St. Peter. There are places of worship for Baptists and Wesleyan Methodists. Williton has been constituted a polling-place for the western division of the county.

WILLOUGHBY with SLOOTHBY (*St. Helen*), a parish, in the union of Spilsby, Wold Division of the hundred of Calceworth, parts of Lindsey, county of Lincoln, 3½ miles (S. S. E.) from Alford; containing 557 inhabitants. The living is a rectory, valued in the king's books at £39. 10. 2½.; patron, Lord Willoughby de Eresby. The tithes have been commuted for a rent-charge of £1020, subject to the payment of rates, which on the average have amounted to £185; the glebe comprises 50 acres, valued at £81. 5. per annum. The church contains an altar-tomb, with the recumbent effigy of a Knight Templar. There is a place of worship for Wesleyan Methodists. Anthony Barnes, in 1728, bequeathed land now producing more than £25 per annum, for teaching and apprenticing poor children.

WILLOUGHBY (*St. Nicholas*), a parish, in the union of Rugby, Rugby Division of the hundred of Knightlow, Northern Division of the county of Warwick, 3 miles (S. by E.) from Dunchurch; containing 376 inhabitants. This place, in the neighbourhood of which many Roman antiquities have been discovered, is in Domesday-book called *Wilbere* and *Wilebei*, from which its present name is derived. It was a royal demesne in the reign of Henry I., who gave it to Wigan, one of his servants, by one of whose descendants it was granted, in the reign of Henry III., to the hospital of St. John, founded by that monarch without the east gate at Oxford, in the 17th year of his reign. On the dissolution of that establishment in the reign of Henry VI., William de Wainfleet, Bishop of Winchester, having obtained a grant of the site for the foundation of Magdalene College, procured from the master and brethren the manor of Willoughby, which at present forms part of the endowment of that institution. The parliamentarian army, in their retreat from the battle of Edge Hill, passed through the village, and fastened a rope round the ancient cross, with the intention of pulling it down, from which they were dissuaded by the vicar. Willoughby was formerly a place of much more importance than it is at present, having enjoyed a market and fairs, to which, from the name of a small hamlet in the parish, called "Pie Court," probably a court of pie-poudre was attached. Part of the foundation of a gaol was discovered by some labourers, in digging for gravel near the church. The village is situated on the high road from London to Holyhead, from which it extends, in a westerly direction, for nearly three quarters of a mile: the houses, which are chiefly of stone with thatched roofs, occasionally interspersed with a few of more modern construction, have a very rural appearance, and the whole village has a pleasing air of tranquillity and retirement. The lands in the neighbourhood are fertile and in a high state of cultivation, and the environs abound with pleasing scenery and with various objects of interest. Within the last few years it has been growing into notice from the discovery of some powerful sulphureous and saline springs, the properties of which have been found similar to those at Harrogate. Two small bathing establishments have been erected, one called Willoughby Lodge Spa, situated in a field at the western extremity of the village, and the New Sulphureous and Saline baths, on the high road, opposite to the inn called the Four Crosses. Some neat cottages have been erected as lodging-houses; hot, cold, and shower baths have been constructed, and there is a pump-room for drinking the water, which is efficacious in all cases of scrofula, and in scorbutic and cutaneous diseases: separate baths are provided for the poor.

The living is a discharged vicarage, valued in the king's books at £9. 4. 4.; present net income, £217; patrons and impropriators, President and Fellows of Magdalene College, Oxford. The church is a spacious and neat structure, in the later style of English architecture, with a low square embattled tower, strengthened with angular buttresses; the exterior, which is plain, is relieved by a north and south porch of good design, and above the western entrance is a large window enriched with tracery. The interior, which is of appropriate character, consists of a nave, aisles, and chancel, the last of which was rebuilt in 1779, and is separated from the nave by an obtusely pointed arch; the nave is separated from the aisles by three clustered columns and arches of a similar character, and from the tower by a lofty arch of good proportion, plainly moulded; the font, which is placed in the south aisle, is a large cylindrical vase of stone, supported on a square pedestal, and is slightly ornamented: the church contains several ancient monuments and brasses, among which is an altar-tomb of the family of Clerke. There is a place of worship for Primitive Methodists. A school was founded in 1816, and a school-house, with accommodation for a master and mistress, was erected at an expense of £430, paid by the trustees of property amounting to £458 per annum, bequeathed by various benefactors for pious and charitable uses: the school is on the National system.

WILLOUGHBY-IN-THE-WOLDS (*St. Mary and All Saints*), a parish, in the union of LOUGHBOROUGH, Southern Division of the wapentake of RUSHCLIFFE and of the county of NOTTINGHAM, 7½ miles (N. E. by E.) from Loughborough; containing 465 inhabitants. The living is a discharged vicarage, valued in the king's books at £6. 18. 6½.; present net income, £87; patrons, W. Melville and — Garton, Esqrs. There is a place of worship for Wesleyan Methodists; also a school endowed with £2. 10. per annum. According to Horsley, this was the Roman station *Vernometum*, but Gale and Stukeley place *Margidunum* here. In a field called Herrings, or Black field, are traces of an old town, where many coins, pavements, and other relics of antiquity have been found. In the centre of the village stands a cross, the shaft consisting of one entire stone, fifteen feet high, resting on four steps. Near this place, in the great civil war, a battle was fought, which is commonly termed the battle of Willoughby Field. The old Fosse-road bounds the parish on the west, near which is a tumulus, called Cross Hill, where an annual revel is held.

WILLOUGHBY, SCOTT (*St. Andrew*), a parish, in the union of SLEAFORD, wapentake of AVELAND, parts of KESTEVEN, county of LINCOLN, 4 miles (N. W. by N.) from Falkingham; containing 24 inhabitants. The living is a discharged rectory, valued in the king's books at £7. 1. 3.; patron, Earl Brownlow. The tithes have been commuted for a rent-charge of £156, subject to the payment of rates, which on the average have amounted to £5. 18. 8.; the glebe comprises 3 acres, valued at £4. 11. per annum.

WILLOUGHBY, SILK (*St. Denis*), a parish, in the union of SLEAFORD, wapentake of ASWARDHURN, parts of KESTEVEN, county of LINCOLN, 2¼ miles (S. S. W.) from Sleaford; containing 193 inhabitants. The living is a rectory, valued in the king's books at £14. 8. 1½.; present net income, £591; patron, Lord Huntingtower. The church is a handsome structure, with a well-proportioned tower and spire: the body is principally in the decorated style of English architecture, and the chancel of later date; the latter contains three stalls, some fine screen-work of wood, and fragments of ancient stained glass in the east window: the font is circular, with Norman shafts. A day and Sunday school is partly supported by the rector.

WILLOUGHBY-WATERLESS (*St. Mary*), a parish, in the union of LUTTERWORTH, hundred of GUTHLAXTON, Southern Division of the county of LEICESTER, 5¾ miles (N. N. E.) from Lutterworth; containing 327 inhabitants. The living is a rectory, with the vicarage of Peatling-Magna united in 1729, valued in the king's books at £11. 11. 3.; present net income, £347; patron and incumbent, Rev. John Miles; impropriator of Peatling-Magna, J. R. Swindall, Esq.

WILLOUGHTON (*St. Andrew*), a parish, in the union of GAINSBOROUGH, Western Division of the wapentake of ASLACOE, parts of LINDSEY, county of LINCOLN, 8½ miles (E. by N.) from Gainsborough; containing 477 inhabitants. The living is a discharged vicarage, valued in the king's books at £7. 4. 2.; present net income, £192; patrons, alternately, Provost and Fellows of King's College, Cambridge, and the Earl of Scarborough, the latter of whom is the impropriator. There is a place of worship for Wesleyan Methodists. An Alien priory, a cell to the abbey of St. Nicholas at Angiers, is said to have existed here. Roger de Buslei and Simon de Canci, in the time of Stephen, gave a moiety of the church, and the greatest part of the town, to the Knights Templars, who had a preceptory here, which from that order came to the hospitallers, and at the dissolution its revenue was estimated at £219. 19. 8.

WILLSBOROUGH, an extra-parochial liberty, in the hundred of SPARKENHOE, Southern Division of the county of LEICESTER, 2½ miles (W. by S.) from Market-Bosworth: the population is returned with Temple-Hall.

WILLS-PASTURES, an extra-parochial liberty, in the Southam Division of the hundred of KNIGHTLOW, Southern Division of the county of WARWICK; containing 7 inhabitants.

WILLSWORTHY, a hamlet, in the parish of St.-Peter Tavy, hundred of LIFTON, Tavistock and Southern Divisions of the county of DEVON, 6 miles (N. E. by N.) from Tavistock: the population is returned with the parish.

WILMINGTON (*St. Michael*), a parish, in the union of DARTFORD, hundred of AXTON, DARTFORD, and WILMINGTON, lathe of SUTTON-AT-HONE, Western Division of the county of KENT, 1 mile (S.) from Dartford; containing 724 inhabitants. The living is a vicarage, valued in the king's books at £6. 17. 6.; present net income, £340; patrons and appropriators, Dean and Chapter of Rochester. The church occupies the summit of a hill near the high road, and has a handsome spire steeple: it contains 115 free sittings, the Incorporated Society having granted £100 in aid of the expense. The celebrated Earl of Warwick, in the reign of Edward IV., resided at the manor-house in the village, which is remarkable for the beauty of its situation.

WILMINGTON (*St. Mary*), a parish, in the union of EAST-BOURNE, hundred of LONGBRIDGE, rape of PEVENSEY, Eastern Division of the county of SUSSEX, 4½ miles (S. W.) from Hailsham; containing 328 inhabitants. The living is a discharged vicarage, valued in the king's books at £8; patron and impropriator, Earl of Burlington. The impropriate tithes have been commuted for a rent-charge of £65, and the vicarial for £51. 16., subject to the payment of rates, which on the average have respectively amounted to £7 and £6; the glebe comprises 4 acres, valued at £7 per annum. The church is principally in the Norman style of architecture. A Sunday school is supported by the vicar, who built the school-house, and endowed it with £50, the interest to be applied in repairs. A Benedictine priory, a cell to the abbey of Grestein, in Normandy, was founded here in the time of William Rufus, which, at its suppression, was valued at 240 marks per annum, and sold by licence of Henry IV. to the Dean and Chapter of Chichester, to whom it was confirmed by Henry V., towards founding a chantry of two priests in the cathedral church. Wilmington gives the inferior title of Baron to the Marquess of Northampton.

WILMSLOW (*St. Bartholomew*), a parish, in the union of ALTRINCHAM, hundred of MACCLESFIELD, Northern Division of the county of CHESTER, 7 miles (N. W. by N.) from Macclesfield; containing 4296 inhabitants. The whole of this parish was held under the Earl of Chester, by the Fittons of Pownall, from whom it passed by marriage to the Venables, of Kinderton, and from them, in default of heirs male, to the families of Dunham and Trafford. The parish is intersected by the small river Bollin, on the banks of which, about a quarter of a mile to the east of the church, is Bolling Hall, the seat of the Fitton family. On the same river are two cotton-mills and one silk-mill; the former, which are at Styal, afford, on the average, employment to 400 persons. A court baron is held by the Trafford family, in Pownall-Fee, the jurisdiction of which extends also over the townships of Chorley, Hough, and Morley; and another, by the Earl of Stamford and Warrington. The living is a rectory, valued in the king's books at £32. 15.; present net income, £955; patron, Thomas Joseph Trafford, Esq. The church is a handsome structure, in the decorated and later English styles, with a square tower; it comprises a nave, chancel, and two aisles, of which the east end of one and the west end of the other are enclosed as sepulchral chapels, for the families of Dunham and Trafford. Near the altar are brasses with inscriptions to Sir Robert Booth, of Dunham, and Douce Venables, his wife, and the figure of a divine, with an inscription to Henry Treffort, Rector, 1537: in the north chapel are two altar-tombs sunk in the wall, on which are two figures representing the Newtons, of Newton and Pownall; there is also a chapel of more recent date, in which are several tombs of the family of Leigh, of Hawthorn Hall, near Wilmslow, by one of whom it was built. There are places of worship for Calvinistic and Wesleyan Methodists. A school was founded by the Rev. Henry Hough, who endowed it with £5 per annum for teaching ten children; it is now conducted on the National plan. A workhouse was established about 1780, on Lindon common; and land, now producing more than £200 per annum, was assigned for its support. There are some remains of an ancient chapel, now forming part of a farm-house; and on a common, near the village, have been found many trees, of which some were very large, at a depth of from 2 to 5 yards below the surface of the ground.

WILNE, GREAT (*St. Chad*), a parish, in the union of SHARDLOW, hundred of MORLESTON and LITCHURCH, Southern Division of the county of DERBY, 7¾ miles (S. E.) from Derby; containing, with the inmates of a large poor-house belonging to several townships, 1091 inhabitants. The living is a perpetual curacy, annexed to the vicarage of Sawley. A school is partly supported by subscription.

WILNE, LITTLE, a chapelry, in the parish of SAWLEY, hundred of MORLESTON and LITCHURCH, Southern Division of the county of DERBY, 7¾ miles (E. S. E.) from Derby: the population is returned with the liberty of Draycott.

WILNECOTE, a chapelry, in the parish and union of TAMWORTH, Tamworth Division of the hundred of HEMLINGFORD, Northern Division of the county of WARWICK, 2¾ miles (E. E. by S.) from Tamworth; containing, with the hamlet of Dosthill, 688 inhabitants. The living is a perpetual curacy; net income, £90; patron, Vicar of Tamworth. The chapel is dedicated to the Holy Trinity, and was rebuilt in 1821. Collieries and brick and lime-kilns have been established in this chapelry of late years.

WILPSHIRE, a township, in the parish, union, and Lower Division of the hundred, of BLACKBURN, Northern Division of the county palatine of LANCASTER, 3¼ miles (N. by E.) from Blackburn; containing 337 inhabitants.

WILSDEN, a township, in the parish and union of BRADFORD, wapentake of MORLEY, West Riding of the county of YORK, 5 miles (S. E. by S.) from Keighley; containing 2252 inhabitants. The spinning of worsted and cotton is extensively carried on, and there are also manufactories for cotton and linen goods. A neat edifice for a mechanics' institute was erected in 1827. The first stone of a new chapel was laid in 1823, which was finished in 1825, at an expense of £7710. 13. 6., defrayed by the parliamentary commissioners: it was made a district church in 1828. The living is a perpetual curacy; net income, £46; patron, Vicar of Bradford. There are places of worship for Independents and Wesleyan Methodists. A day and Sunday school is partly supported by subscription, and there is also a Sunday National school.

WILSFORD (*St. Mary*), a parish, in the union of Sleaford, wapentake of Winnibriggs and Threo, parts of Kesteven, county of Lincoln, 4½ miles (W. S. W.) from Sleaford; containing 393 inhabitants. The living is a rectory, valued in the king's books at £10; present net income, £500; patron and incumbent, Rev. C. Brackenbury. The church has a tower and spire, and exhibits an admixture of the early and decorated styles of English architecture: the font, which is octagonal, with concave sides, is of later date. Here is a Sunday National school. A Benedictine priory, a cell to the abbey of Bec, in Normandy, was founded in the reign of Stephen; at the suppression of Alien houses, it was settled upon the abbey of Bourn, in this county, and at the dissolution was granted to Charles, Duke of Suffolk.

WILSFORD (*St. Michael*), a parish, in the union of Amesbury, hundred of Underditch, Salisbury and Amesbury, and Southern, Divisions of the county of Wilts, 1¾ mile (S. W. by W.) from Amesbury; containing 119 inhabitants. The living is a vicarage, with that of Woodford consolidated, in the patronage of the Prebendary of Wilsford and Woodford, in the Cathedral Church of Salisbury, the appropriator; net income, £168. A school is partly supported by a lady. The parish is bounded on the east by the river Avon. The manor-house of Lake is a remarkably fine specimen of the Elizabethan style of architecture.

WILSFORD-DAUNTSEY (*St. Nicholas*), a parish, in the union of Pewsey, hundred of Swanborough, Devizes and Northern Divisions of the county of Wilts, 4½ miles (W. S. W.) from Pewsey; containing 512 inhabitants. The living is a vicarage, valued in the king's books at £8. 17. 11.; present net income, £242; patron and impropriator, Master of the hospital of St. Nicholas, Salisbury.

WILSHAMPSTEAD (*All Saints*), a parish, in the hundred of Redbornestoke, union and county of Bedford, 4 miles (S. by E.) from Bedford; containing 753 inhabitants. The living is a vicarage, endowed with one-third of the rectorial tithes, and valued in the king's books at £9. 9. 7.; present net income, £280; patron, Lord Carteret; impropriators of the remainder of the rectorial tithes, J. C. Crook, Esq., and another. There is a place of worship for Wesleyan Methodists. A small charity school here is endowed with land producing £8 per annum.

WILSICK, a joint township with Stancill and Wellingley, in the parish of Tickhill, union of Doncaster, Southern Division of the wapentake of Strafforth and Tickhill, West Riding of the county of York, 6 miles (S. by W.) from Doncaster: the population is returned with Stancill.

WILSTHORPE, a chapelry, in the parish of Greatford, union of Stamford, wapentake of Ness, parts of Kesteven, county of Lincoln, 5 miles (N. W.) from Market-Deeping; containing 69 inhabitants.

WILSTROP, a township, in the parish of Kirk-Hammerton, Ainsty of the city, and West Riding of the county, of York, 7½ miles (W. by N.) from York; containing 112 inhabitants.

WILTON (*St. James*), a parish, in the union of Thetford, hundred of Grimshoe, Western Division of the county of Norfolk, 4¼ miles (W.) from Brandon-Ferry: the population is returned with the parish of Hockwold. The living is a discharged vicarage, united to the rectory of Hockwold, and valued in the king's books at £6. 7. 6. The church is built of flint and boulder stones, with a massive embattled tower, surmounted by an octangular spire of freestone: on the north side of the building is a curious arch, probably constructed for the burial-place of the founder. There is a place of worship for Wesleyan Methodists.

WILTON, a tything, in the parish of Midsummer-Norton, hundred of Chewton, Eastern Division of the county of Somerset, 8½ miles (S. W.) from Bath: the population is returned with the parish.

WILTON (*St. George*), a parish, in the union of Taunton, hundred of Taunton and Taunton-Dean, Western Division of the county of Somerset; containing 795 inhabitants. The living is a perpetual curacy; net income, £104; patron, Rev. E. T. Halliday; impropriator, — Willment, Esq. The church has been enlarged, and 100 free sittings provided, the Incorporated Society having granted £120 in aid of the expense: it was formerly a chapel to the vicarage of St. Mary Magdalene, in Taunton, to which town Wilton forms an extensive suburb. The house of correction contains fourteen wards, day-rooms, and airing-yards, two work-shops and yard, two tread-wheels, a chapel, and infirmaries for both sexes; it is capable of containing 175 prisoners. Here was formerly an hospital, built by one of the Bishops of Winchester.

WILTON (*St. Mary*), a borough (formerly a market-town) and parish, having exclusive jurisdiction, and the head of a union, locally in the hundred of Branch and Dole, Salisbury and Amesbury, and Southern, Divisions of the county of Wilts, 3 miles (W. by N.) from Salisbury, and 85 (W. S. W.) from London; containing 1997 inhabitants. This town, which derives its name from the river Wily, on which it is situated, is of great antiquity, and is supposed by Baxter to have been the *Caer-Guilo*, or capital of the British Prince, Caroilius, and subsequently a seat of the West Saxon kings. It was a place of great importance for several centuries preceding the Norman Conquest, possessing an eminent religious establishment, and giving name to the county in which it is situated: it had also a mint. Wilton is stated by Camden and other writers to have been originally called *Ellandune*, and the scene of a sanguinary battle, fought between Egbert, King of the West Saxons, and Beorwulf, the Mercian king, in which the latter was defeated; but later writers have controverted this opinion, and the battle is now supposed to have been fought at a place called *Ellendune*, situated in another part of the county. An engagement took place here in 871, between King Alfred and the Danes, in which the latter, though ultimately successful, were obliged to sue for peace. The celebrated monastery was commenced in the year 800, by Wulstan, Earl of Wiltshire, who, having defeated Ethelmund, the Mercian king, founded a chantry, or oratory, and repaired the old church of St. Mary at Wilton, which had been destroyed by the Danes, and placed in it a college of Secular priests. About 30 years after Earl Wulstan's death, his widow, Alburga, sister to King Egbert,

Corporation Seal.

induced that monarch to convert the oratory into a priory of thirteen sisters, of which she was the first prioress, and hence Egbert has been commonly reputed its founder. Immediately on granting peace to the Danes, King Alfred, at the solicitation of his queen, Ealswitha, built a nunnery on the site of the palace, and transferred to it the thirteen sisters of the priory, adding to them an abbess and twelve nuns; his successors were great benefactors to this establishment, particularly Edgar, who enlarged its buildings and augmented its revenue; his natural daughter, Editha, having been abbess, and, after her death, being canonized, became its patron saint. Editha, daughter of Earl Godwin, and queen of Edward the Confessor, who was educated in this nunnery, rebuilt it in a magnificent manner with stone, it having been originally constructed of wood; and Matilda, queen of Henry I., was also educated in it, under her aunt, the abbess Christina. Early in the tenth century Wilton became the seat of the diocese of Wiltshire, and continued so during the lives of eleven successive bishops, the last of whom, Hermannus, having been also appointed to the see of Sherborne, united the two bishoprics, and removed to Old Sarum, where he founded a cathedral, which continued the seat of the see until its transfer, in 1217, to Salisbury. After the Conquest the town continued to flourish, until the year 1143, when King Stephen took possession of it, intending to convert the nunnery into a place of defence; but being surprised by Robert, Earl of Gloucester, with the troops of the Empress Matilda, who set fire to the town on all sides, the king was obliged to flee, leaving behind his troops and baggage. Wilton recovered from this disaster; but in the succeeding reign it began to decline, in consequence of the foundation of New Sarum or Salisbury, and the change in the direction of the great western road, which quickly followed. Its monastic institution, however, continued of importance until the dissolution, when it was granted to Sir William Herbert, afterwards Earl of Pembroke, its revenue being at that time estimated at upwards of £600. A house of Black friars, and two hospitals, dedicated to St. Mary Magdalene and St. John, also existed here at the period of the dissolution. Wilton was visited by Queen Elizabeth, in September, 1579, and it became the residence of the court for a short time, in October, 1603.

The town, consisting principally of two streets which cross at nearly right angles, is situated in a broad and fertile valley, near the confluence of the rivers Nadder and Wily, and is partially paved, and well supplied with water. The manufacture of carpets, for which Wilton has been so much celebrated, was introduced by a former Earl of Pembroke, who brought over workmen from France for that purpose, this being the first place in England where this manufacture commenced; and the manufacture of carpets at Axminster having been recently discontinued, those splendid articles called Axminster carpets are now manufactured at Wilton Fancy cloth waistcoatings also formed, at one time, a considerable branch of manufacture, but this is nearly extinct. The market days were on Wednesday and Saturday, but, since the decline of the manufactures, no regular market has been held. The fairs are on May 4th and Sept. 12th, the former for cattle and sheep, and the latter one of the largest sheep fairs in the West of England, the number sold often exceeding 100,000. Wilton is a borough by prescription, its ancient rights and franchises having been confirmed by charters of various monarchs, from the time of Henry I. to that of Henry VI.: it is governed by a mayor, recorder, high steward, five aldermen, and an unlimited number of burgesses, appointed occasionally by the corporation, with a town-clerk, two sergeants-at-mace, and four constables. The mayor is chosen by the corporation at large, on the first Thursday after Michaelmas-day, from three persons previously nominated by such members of the corporation as have served the office of mayor, and is sworn into office, with the other officers who are then chosen, on Oct. 13th, at the court leet of the lord of the manor, held at the town-hall, an ancient plain brick building, which was repaired and improved, about ten years since, by the corporation. The mayor and recorder are justices of the peace, with exclusive jurisdiction. The borough first sent members to parliament in the 23rd of Edward I., and has since done so without interruption; but at present, under the act of the 2nd of William IV., cap. 45, it sends only one. The right of voting was formerly vested in the members of the corporation, resident or non-resident, in number between 40 and 50; but the privilege has been extended to the £10 householders of an enlarged district, which has been made to constitute the new elective borough, comprising an area of 32,150 acres, of which the limits are minutely described in the Appendix, No. II.: the mayor is the returning officer.

The living is a rectory, with that of Bulbridge and the vicarage of Ditchampton united, valued in the king's books at £12. 16. 3.; present net income, £450; patron, Earl of Pembroke. The church is an ancient edifice. At Netherhampton, in this parish, is a chapel of ease. The Independents and Methodists have each a place of worship. The free school, situated in North-street, was founded in 1714, under the will of Walter Dyer, who, in 1706, bequeathed £600 for that purpose: part of this bequest was expended in the erection of the school premises, to which additions have been made, at different times, by the trustees; and the residue, with a legacy of £1000 Bank stock, producing by accumulation £2090, from Richard Uphill, in 1716, was laid out in an estate at East Knoyle, the rental of which, amounting to £120 per annum, is applied in support of the school, which is also entitled to the interest of £1000, part of a sum of £4200 three per cent. consols., bequeathed in 1775, by Mr. Robert Sumption, for various purposes. The master has a salary of £40 per annum, and a dwelling-house, and 20 boys of the parish are clothed and educated, and an apprentice-fee of £8. 10. given with each: the school is on the National system. Of the remainder of Mr. Sumption's bequest, the interest of £1000 is given annually, as a marriage portion, to poor young women, natives of and resident in the town; and the interest of £2000 is appropriated to the benefit of five poor men and as many women, who receive £6 per annum each, to which an addition is made from a bequest of £500 New Four Per Cents. by the late James Swayne, Esq.: there are also several minor charitable bequests for annual distribution. In 1816, Thomas Mease gave to the lord high steward and corporation of Wilton, on the death of his wife, £4000 Navy five per cents., directing that the dividends should be allowed to accumulate for 20 years after her decease; the accumulations to be applied in improving the parish church, and the annual interest of the principal to various charitable purposes. The poor law union of Wilton comprises 22 parishes or places, under the care of 24 guardians, and, according to the census of 1831, contains a population of 10,270. The hospital of St. John, supposed to have been founded by Hubert, who was bishop of Salisbury in 1189 and Archbishop of Canterbury in 1193, is endowed for a master, or prior, who is a clergyman, nominated by the Dean of Salisbury, and two poor men and two women, chosen by the prior: the tenements are going to decay, and the pensioners are lodged in an adjoining cottage; the chapel, which had fallen into decay, has been repaired and enlarged, at the expense of the prior; and divine service is now performed every Sunday evening and every alternate Friday evening, by a chaplain appointed by the prior, who receives a stipend of £44. On the site of the celebrated nunnery, Sir William Herbert, to whom it was granted, commenced the erection of that princely pile, now the residence of his descendants, the Earls of Pembroke: it was designed by Holbein and Inigo Jones, and contains a collection of paintings, statues, and various antiquities, not excelled by any in the kingdom: in this house the distinguished Sir Philip Sidney, whose sister Mary, the celebrated Countess of Pembroke, was the wife of the earl, composed his heroic romance of "Arcadia."

WILTON, a chapelry, in the parish of KIRK-LEATHAM, union of GUILSBROUGH, Eastern Division of the liberty of LANGBAURGH, North Riding of the county of YORK, 3½ miles (N. N. W.) from Guilsbrough; containing 411 inhabitants. The living is a perpetual curacy; net income, £136; patron, Sir J. Lowther, Bart.; the impropriation belongs to Lady Turner. The chapel is dedicated to St. Cuthbert. At the west end of the village is Wilton castle, recently erected upon the site of the ancient baronial castle of the Bulmers, which family possessed it for many generations, till Sir John Bulmer, Knt., was attainted of high treason, when his estates were confiscated.

WILTON, a chapelry, in the parish of ELLERBURN, PICKERING lythe, and union, North Riding of the county of YORK, 3¾ miles (E. by S.) from Pickering; containing 192 inhabitants.

WILTON, BISHOP (*ST. EDITH*), a parish, in the union of POCKLINGTON, partly within the liberty of ST. PETER'S, and partly in the Wilton-Beacon Division of the wapentake of HARTHILL, East Riding of the county of YORK; containing 831 inhabitants, of which number, 622 are in the township of Bishop-Wilton, with Belthorpe, 4½ miles (N.) from Pocklington. The living is a discharged vicarage, valued in the king's books at £7. 3. 6½.; present net income, £148; patron and impropriator, Sir Tatton Sykes, Bart. There is a place of worship for Wesleyan Methodists. Here was formerly a palace, built in the reign of Edward IV., by Bishop Neville, and encompassed with a moat, which still remains.

WILTSHIRE, an inland county, bounded on the north and north-west by Gloucestershire, on the west by Somersetshire, on the south-west and south by Dorsetshire, on the south-east and east by Hampshire, and on the north-east by Berkshire. It extends from 50° 55′ to 50° 42′ (N. Lat.), and from 1° 30′ to 2° 22′ (W. Lon.); and comprises an area of 882,560 statute acres, or about 1379 square miles. The population amounts to 240,156. A large portion of this county was occupied, in the time of Cæsar, by the Belgæ: the Hedui inhabited the north-western parts of it, and the Carvilii another district; the Cangi are also supposed, either at this period or soon after, to have possessed some territory within its northern limits. On the second invasion of the Romans, during the reign of the Emperor Claudius, in the year 44, the Belgæ were found to have subdued nearly the whole, which they occupied as far north as the rude barrier of the Wansdyke, beyond which the Cangi are, by some writers, supposed to have preserved their dominion. Under the Roman government, Wiltshire was comprised in the division called *Britannia Prima.* After the withdrawal of the Roman forces, Cerdic, the founder of the kingdom of the West Saxons, who had been engaged in an arduous warfare, for upwards of 20 years, with the Romanized Britons near the place of his landing, on the coast of Hampshire, at last penetrated into this territory, in the year 520, but was defeated in a great battle by the British hero Arthur, and the Saxons did not return hither for upwards of 30 years. In 554, Cynric, son of Cerdic, and his successor in the sovereignty of Wessex, advanced with his army towards *Sorbiodunum,* or Old Sarum, and defeated a British army opposed to his progress near that place, of which he immediately after took possession. Four years afterwards another decisive battle was fought, at "Beranbyrig," or Barbury Castle, near Marlborough, in which the Britons were again routed, and Wiltshire shortly became incorporated in the kingdom of Wessex.

It derives its name from Wilton, which for a long period anterior to the Norman Conquest, and for a considerable time subsequent to that event, was its principal town, and still continues the county town. In the early period of Christianity among the West Saxons, it was included in the diocese of Winchester, and was afterwards, in the reign of Ina, annexed to that of Sherborne, and so remained for a long succession of years. Soon after the year 905, however, a bishopric was erected co-extensive with the county, the seat of which was placed successively at Ramsbury and at Wilton. Hermannus, who held it at the Conquest, and who had obtained the bishopric of Sherborne to be united to it, soon after fixed the seat of his diocese at Old Sarum; whence, in the early part of the thirteenth century, it was removed to its present situation at New Sarum, or Salisbury. Wiltshire is now in the diocese of Salisbury, and province of Canterbury; but by the act of the 6th and 7th of William IV., cap. 77, the deaneries of Cricklade and Malmesbury will be annexed to and form part of the diocese of Gloucester and Bristol. It forms the two archdeaconries of Sarum and Wilts, the former comprising the deaneries of Amesbury, Chalk, Potterne, Salisbury, Wilton, and Wily; the latter, those of Avebury, Cricklade, Malmesbury and Marlborough: the total number of parishes is 295. For purposes of civil government it is divided into the following hundreds, *viz.*, Alderbury, Amesbury, Bradford, Branch and Dole, Calne, Cawden and Cadworth, Chalk, Chippenham, Damerham (North and South), Downton, Dunworth, Elstub and Everley, Frustfield, Heytesbury, Kingsbridge, Kinwardstone, Malmesbury, Melksham, Mere, Potterne and Cannings, Ramsbury, Selkley, Swanborough, Underditch, Warminster, Westbury, Whorwelsdown, and Highworth, Cricklade, and Staple. It contains the city of Salisbury, or New Sarum; the borough and market-towns of Calne, Chippenham, Cricklade, Devizes, Malmesbury, Marlborough, and Westbury; the borough of Wilton; and the market-towns of Amesbury, Great Bradford, Hindon, Market-Lavington, Melksham, Mere, Swindon, Trowbridge, Warminster and Wootton-Bassett. By the act of the 2nd of William IV., cap. 45, the county has been divided into two portions, called the Northern and Southern divisions, each empowered to send two members to parliament: the place of election for the former is Devizes, and the polling-places are, Devizes, Melksham, Malmesbury, and Swindon; that for the latter is Salisbury, and the polling-places, Salisbury, Warminster, East Everley, and Hindon. Two citizens are returned for Salisbury; two representatives for each of the boroughs of Chippenham, Cricklade, Devizes, and Marlborough; and one each for Calne, Malmesbury, Westbury, and Wilton, these four last having been each deprived of one by the above-named act, by which also the old boroughs of Old Sarum, Great Bedwin, Downton, Heytesbury, Hindon, Ludgershall, and Wootton-Bassett, were entirely disfranchised. This county is included in the Western circuit: the Lent assizes are held at Salisbury, and the summer assizes at Devizes; the quarter sessions are held at Devizes in the winter, at Salisbury in the spring, at Warminster in the summer, and at Marlborough in the autumn. The county gaol is at Fisherton-Anger, the county house of correction at Devizes, and the bridewells at Devizes and Marlborough.

The form of the county is nearly an ellipse, the transverse diameter of which bears north and south. It is common to consider it as divided into North Wiltshire and South Wiltshire, by a line passing through it from east to west, at or near Devizes; but the natural division is into South-east Wiltshire and North-west Wiltshire, by an irregular line passing from the confines of Berkshire, near Bishopston, south-westward to those of Somersetshire, near Maiden-Bradley. South-east Wiltshire, containing nearly 500,000 acres, thus comprehends, and is almost entirely occupied by, the whole of the Wiltshire Downs, with their intersecting valleys, forming the western division of the ranges of chalk hills which occupy so great a portion of Hampshire, and a smaller extent of Berkshire. At a distance, this portion of the county presents the appearance of a large elevated plain; but on a nearer approach its surface is found to be broken by numerous and frequently extensive valleys, and to possess an almost constant series of gentle eminences, but no where a mountainous elevation: the declivities on one side of some of the ridges are very abrupt, while on the other they sink gently, in irregular gradation, sometimes into a perfect flat. The two grand divisions of the chalk hills are into Marlborough Downs, being those to the north of the Kennet and Avon canal, and Salisbury Downs, or Plain, occupying nearly all the county southward of that line: these great districts are separated by the vale of Pewsey, and the only difference in their general appearance is, that the eminences of the former are more abrupt and elevated than those of the latter. The most extensive level prevails around Stonehenge, where the scenery is peculiarly tame. On Marlborough Downs are scattered many of the singular masses of stone, called "grey wethers," and when broken, "Sarsden-stones," or, by contraction, "Sarsons." The principal valleys, which display scenes of rich meadow and arable lands, adorned with seats, villages, and occasionally woods, are traversed by streams of excellent water: those descending from Salisbury Plain take a direction towards Wilton and Salisbury. The north-western division of the county presents a remarkably different appearance, being a rich tract of vale land, extending from the base of the Downs to the northern and western confines of the county, and generally so flat that few deviations from the ordinary level are perceptible: approaching the Cotswold hills of Gloucestershire, however, the surface becomes gradually more elevated. This low plain is so well wooded, that, when viewed from any of the surrounding hills, it appears like a vast plantation. The most remarkable eminences in the county, and some of those which command the finest prospects, are, Beacon hill, near Amesbury, which rises to the height of 600 feet above the level of the sea; Bidcombe hill, near Maiden-Bradley; Codford Hill; the high grounds near Standlinch House; Old Sarum hill, 339 feet high; and Westbury Down, 775 feet high.

The soil of the downs, though varying considerably in quality, is uniformly calcareous: the hills consist of chalk, with its usual accompaniment of flint; and the soil on their sides, where hardly any flints are found, consists of a chalky loam, or rather decomposed chalk, called "white land." The flatter parts have a flinty loam; and in the centre of the principal valleys is a bed of broken flints, covered with black vegetable earth, washed from the hills above: in some of the more extensive valleys are beds of black peat. These hills are skirted by narrow tracts of very fertile soil, called by geologists green sand, from its containing little round masses of a green substance, frequently prevailing to such an extent as to give a green hue to the aggregate of which they form a part. The substratum of the north-westernmost part of the county, separated from the rest of it by a line passing from the border of Somersetshire, near Bradford, through the neighbourhood of Malmesbury, to the confines of Gloucestershire, near Cirencester, consists of a loose irregular mass of the broken strata of a light-coloured, calcareous, and sandy stone, provincially called *corn-grate*, the soil resting upon which is chiefly a reddish calcareous loam, mixed with flat stones, and is called *stonebrash*. The quality of this soil varies much, according to its depth above the stone rock, and the absence or presence of an occasional intervening stratum of blue clay, which has the appearance of marl, but is devoid of its good qualities: its presence is denoted by a spontaneous growth of oak, and that of the drier substratum by a natural and luxuriant growth of elm. About Chippenham, and thence southward by Melksham and Trowbridge, is a much more fertile district, having a greater depth of soil without any of the clay. From Melksham, by Chippenham, to Cricklade, runs a stratum of gravel, in general covered to a good depth with a rich loam: the greatest extent of it is, however, from Tytherton, through Christian-Malford and Dauntsey, to Somerford: this is considered to be the richest soil in the county. In the north-western district are also two tracts of sand of a sharp, loose, gravelly nature, one of which runs from Rodborne, by Seagry, Draycot, and Sutton-Benger, to Langley-Burrel, near Chippenham; the other from

Charlcot, through Bremhill, to Bromham: of the latter, there are also detached parts at Rowde and Seend, and the stratum of sand appearing in different places to the north of it is supposed to be the same. The remaining parts of North-west Wiltshire, particularly those extending from Highworth, by Wootton-Bassett, to Clack, have a cold retentive soil, on a hard, close, rough kind of limestone. Braden Forest, between Cricklade and Malmesbury, has a soil peculiar to itself, being a cold iron clay, proverbially sterile. Strong clays and clayey loams, of a tolerably rich quality, are found to a small extent in various places on the skirts of the south-eastern district. South-eastern Wiltshire is chiefly appropriated to tillage and to sheepwalks; the north-western district comprises a very rich and extensive tract of grazing and dairy land, on the borders of the Thames and the Lower Avon, while the rest of it is chiefly arable.

The courses of crops are very various: in the South-eastern district the red wheat is most commonly sown. Barley is a favourite crop in the chalk district, but hardly in any other part of the county: oats are no where sown to a great extent: a few peas and beans are grown in the north-western parts of the county: rye is often sown as spring food for sheep, but is seldom suffered to stand for a corn crop. Turnips are extensively cultivated on the chalky and stonebrash soils: rape, or coleseed, is grown to a great extent on the Downs; as also are vetches in this and in the North-western district. Potatoes are much cultivated, particularly on the rich sands adjoining the chalk. The principal artificial grasses are, rye-grass, broad clover, marl-grass, or Dutch clover, and trefoil. The grass lands of the North-eastern district are of the richest quality, and are partly occupied by dairies, and partly in the fattening of cattle. The cheese, which is the only produce of the numerous dairies, excepting the poor kind of butter made from the whey, was, for many years, sold in the London market as Gloucestershire cheese; but it is now well known and much esteemed there under the name of "North Wiltshire." Bordering on the streams of the downs are continued narrow tracts of meadow land, under an excellent system of irrigation, which became general about the commencement of the last century: the quantity is estimated at about 20,000 acres. With the grass lands of Wiltshire may be classed its spacious downs, which are unenclosed, and subject to common rights; and though a portion of them is always under tillage, yet by far the most extensive tracts are covered with a fine native sward, affording food to no less than 500,000 sheep and lambs during the summer and autumn. Several parts of the downs, usually the most level and valuable, are pastured with common herds of cattle, which are brought upon them early in May, and remain until the end of harvest, when they are taken to the stubble-fields, and the down then becomes common to the sheep. These lands derive almost their only manure from having the sheep folded upon them, which practice also bestows upon the arable lands in every other part of the county the chief part of that which they receive: the other manures most frequently employed, besides the ordinary ones, are chalk, lime burned from chalk, soot, coal-ashes, peat-ashes, woollen rags, and soap-ashes. The dairy cows of the northern parts of the county are nearly all of the long-horned breed, and are chiefly obtained from the more northern counties of England; some Devonshire cattle have also been introduced: great numbers of calves are annually sent for the supply of the Bath and London markets with veal. The principal breeds of sheep are the native Wiltshire and the South-Down; the former, however, on the downs, has been much intermixed with a larger breed, and is now, in its pure state, only occasionally seen in the northern districts. The native hog is large, white, and long-eared; but the prevailing kind is a mixed breed between this and a small black species: the county is famous for its excellent bacon, which is prepared in very large quantities in the dairy districts.

Wiltshire was anciently well-wooded, but its present woodlands are of comparatively small extent. Different parts near its border are occupied by valuable woods, generally in a thriving condition, though much injured by cattle, to which they are common. The only forest still remaining in a well-wooded state is Savernake forest, the property of the Earl of Ailesbury, which is about sixteen miles in circumference, and is situated to the south-east of Marlborough; it contains many majestic oaks, and exhibits some fine and interesting scenery; it is also well stocked with deer. Cranborne Chace occupied a long narrow tract on the extreme southern verge of the county, and contained six lodges, with walks appropriated to each, the whole under the care of a ranger deputed by Lord Rivers, as lord of the chace; but it has lately been disfranchised, his lordship receiving an annual payment from the owners of the woods in it, and the lands adjoining. Vernditch Chace, belonging to the Earl of Pembroke, adjoins the latter on the east, and is now nearly all under cultivation. Grovely Forest, now generally called Grovely woods, also belongs to Lord Pembroke: it occupies a long narrow tract of the high ground between the valleys of the Nadder and the Wily. The ancient forest of Penchett, or Ponsett, near Salisbury, is now better known as Clarendon park and woods, the property of Sir F. H. Bathurst. The wastes are comparatively trifling, and consist chiefly of small marshy commons, most of them in the north-western part of the county, where there are also a few small heaths.

The mineral productions are of little importance. The chalk, forming the substratum of nearly all the extensive south-eastern district, is, in some places, extremely hard, though more frequently of a soft marly texture; the finest kind is found at Sidbury hill, which furnishes a supply to several of the western counties. Sandstone is obtained in the low grounds both of North and South Wiltshire: the "corn-grate" is frequently found in masses so thin as to be employed in the roofing of houses; it is also used for building and paving: a more regular stratified sandstone is found under the sandy surface at Swindon, and is in much request for paving, for cisterns, and for tombstones. The inferior kind of limestone, found in the country between Highworth and Clack, is used only for the making and repairing of roads. On the western side of the county, bordering on Somersetshire, are numerous and extensive quarries of a fine kind of freestone: those at Box, near Bath, are among the most celebrated in the vicinity of that city, and produce a great variety of fossil shells and other marine exuviæ. Near Wootton-Bassett, in the blue clay, and near Grittleton, in the freestone strata, other singular fossil remains are found. The freestone quarries at Chilmark, Tisbury, and that neighbourhood, are extensive, and the stone is of a very superior quality.

The manufactures are of considerable extent and importance, particularly that of woollen goods. At Salisbury great quantities of flannel were made till within the last 20 years, and also fancy woollens; but the manufacture has gradually declined, and a very small quantity of flannel and linsey is now made: this city has also a manufacture of cutlery and steel goods of great excellence. Wilton has a manufacture of carpets and of kerseymere and linsey. Bradford, Trowbridge, Westbury, and all the adjacent towns and villages, from Chippenham to Heytesbury inclusive, carry on extensive woollen manufactures, chiefly of superfine broad cloth, kerseymere, and fancy cloths. At Mere and in its vicinity is a manufacture of linen, chiefly dowlas and bed-ticking; and at Aldbourn is one of cotton goods, chiefly fustians and thicksets. The manufacture of silk has been introduced at Devizes; and the parishes of Stourton and Maiden-Bradley, and others in their vicinity, participate to a small extent in the neighbouring linen-manufacture of Dorsetshire, and the silk-manufacture of Bruton in Somersetshire. Ale of a superior quality is brewed in some parts of the county, and a considerable quantity of it is sold in London, under the names of "Wiltshire" and "Kennet" ale. The commerce consists chiefly in the exportation of the agricultural and manufacturing produce: of the former there is a considerable surplus, principally wheat, barley, fat cattle, calves, sheep, hogs, and cheese, part of which is taken to the London market, and the rest to Bath, Bristol, and the eastern parts of Somersetshire. Wiltshire, besides supplying its own woollen manufactures with the raw material, also sends a considerable quantity to other counties. The principal imports are the ordinary articles of merchandise, coal, and cows for the supply of the dairy districts.

The rivers and streams are very numerous, and all of them rise either within the county, or near its borders: the principal are the Isis, or Thames, the Lower Avon, the Kennet, and the Salisbury, or Wiltshire and Hampshire Avon, not one of which is navigable within its limits. The Kennet and Avon canal, which crosses the centre of the county from west to east, and connects the navigation of the Lower Avon with that of the Kennet and the Thames, passes through a tunnel, near Burbage, to the valley in which Great Bedwin is situated, and down which it proceeds to the banks of the Kennet, near Hungerford, with which river it quits the county. The first act of parliament for its formation was obtained in 1794, and several other acts for the alteration of the course originally designed for it, and for the raising of additional funds, were afterwards passed: the whole line, however, was not completed and opened until the end of the year 1809. The Wilts and Berks canal, branching from the Kennet and Avon at Semington, about two miles to the west of Devizes, passes northward, by Melksham, to the vicinities of Chippenham and Calne, to each of which towns it has a short branch: the first act for its formation was obtained in 1795, but the work experienced many delays. The North Wilts canal, executed under an act obtained in 1813, commences in the Wilts and Berks canal near Swindon, and terminates in the Thames and Severn canal at Weymoor bridge, in the parish of Latton, being eight miles and three furlongs in length. The Salisbury and Southampton canal was designed to have commenced at Salisbury, and to have proceeded, first south-eastward and then eastward, into Hampshire; but, after a considerable portion of the work had been completed, it was at last abandoned, on encountering an extensive quicksand. The Thames and Severn canal crosses only the northern extremity of the county, passing the northern bank of the Isis, and near the town of Cricklade. The line of the Great Western railway enters the county a little to the east of Stratton-St. Margaret's, and passing to the north of Swindon, and by Wootton-Bassett, Chippenham, and Corsham, quits it a little beyond the last-named town. A new line of turnpike-road, about 17 miles in length, has been projected from Amesbury to Kennet, following the course of the river Avon, and passing through the centre of the county, to open a direct intercourse between the Northern and Southern Divisions of Wilts.

This county contained the Roman stations of *Sorbiodunum*, at Old Sarum; *Verlucio*, in the vicinity of Heddington; and *Cunetio*, a little to the east of Marlborough: this people had also several other permanent settlements in Wiltshire, particularly at Easton-Grey, Wanborough, near Heytesbury, and Littlecot. The principal of the Roman roads which traversed it was a continuation of the *Julia Strata*, which, entering from Bath, proceeded north-eastward, by Medley and Spye Park, to the station of *Verlucio*, and thence by Colston and across the river Kennet to that of *Cunetio*, beyond which it stretched across the eastern confines of the county. The Fosse-way branched from the *Julia Strata* at Bath-Ford, and passed by Banner Down, Easton-Grey, and across the turnpike road between Tetbury and Malmesbury, to Cirencester in Gloucestershire: another great road entered from Cirencester, and passed south-eastward by Cricklade to Wanborough, at which latter place it separated into two branches, one proceeding by Baydon towards Speen in Berkshire, and the other by Ogbourne, Mildenhall, Manton, and Chute Park, towards Winchester. *Sorbiodunum* was connected with other stations by three roads, one of which passed by Bemerton, Stratford-St. Anthony, and Woodyates-Inn, towards Dorchester; another by Ford, Winterslow, Buckholt Farm, and Bossington, towards Winchester; and the third by Porton and Idmiston, towards Silchester, in the north of Hampshire. The Ridge-way, extending north-eastward from Avebury into the adjoining county of Berks, is also mentioned by Whitaker as a Roman road. Wiltshire is distinguished for remarkably numerous traces (chiefly in its south-eastern districts) of the nations which successively occupied it during the earlier periods of our history. Of these, the stupendous monument of Stonehenge, two miles westward of Amesbury, and that of Avebury, about five miles to the west of Marlborough, are entitled to primary notice. The vast earthwork of the Wansdyke is conjectured by some to have been the northern boundary of the Belgæ, and supposed to have intersected the whole county, from the north of Somersetshire to the north of Hampshire: though in the greater part of its course it can be distinctly traced only in detached spots, yet, throughout the range of hills to the south and west of Marlborough, it is still tolerably entire, and in one place is conspicuous, in a bold and connected line, for the distance of ten or twelve miles. The sepulchral mounds called barrows, or tumuli, are abundant, more particularly around Stonehenge and Avebury: the most remarkable is Silbury hill, near Avebury. There is a cromlech at Clatford Bottom, near the village of Clatford, and another at Littleton-Drew. The Roman roads may yet be distinctly traced in several places; and the Ridgeway is clearly visible on the high chalk ridge extending north-eastward from Avebury into Berkshire.

The encampments, which are very numerous, vary in the period of their formation, in their size, shape, and mode of construction, and in the peculiarities of their situations. Some of these are undoubtedly the work of British tribes and of the Belgæ; others of successive invaders, the Romans, the Saxons, and the Danes. The largest and most noted are, the vast fortifications of Old Sarum, enclosing an area of nearly 30 acres, the foundations of the walls of which are still visible; Chidbury Camp, to the north-west of Tidworth, comprising an area of seventeen acres; and Vespasian's Camp, as it is commonly called, to the westward of Amesbury, enclosing an area of 39 acres. There are many others nearly equal in extent, and scarcely less interesting to the antiquary: the principal are situated at Whitesheet hill, to the north-west of Mere; Clay hill, near Warminster; Warminster Down; Whiten hill, near Longbridge-Deverill; Cottley hill, to the north-west of Heytesbury; Knighton Down; Pewsey heath; Oldbury hill, near Calne; Roundway hill, near Devizes; Martinsall hill, near Marlborough; Chidbury hill; Blunsden hill, near Highworth; Beacon hill, near Amesbury; Southley wood, to the south of Warminster; Barberry Castle, near Marlborough; Liddington Castle; Hays, on the western border of the county, near the Lower Avon; Bratton, two miles from Eddington; Battlesbury, near Warminster; Scratchbury, near Cottley; Yarnborough, at an angle formed by the old trackway from Salisbury to Bath and the present turnpike road from Amesbury to Mere; Banbury, near Wily; Groveley Castle, Rolston; Casterley, near Shrewton; the vicinity of Berwick-St. John; Haydon hill, near Chute; Ogbury, near Great Durnford; Newton-Toney; Alderbury; Whitchbury; Clearbury, near Downton; Broad-Chalk; Chiselbury, near Fovant; Old Camp, on Boreham Down, near Warminster; Dinton, and Little Path hill. These intrenchments were evidently formed for purposes of military defence, but there is a variety of other earthworks spread over Salisbury Plain and Marlborough Downs, the uses of which are unknown; some of them are considered the sites of British villages, others as denoting places consecrated to religion. Many less vestiges of antiquity, such as tessellated pavements, coins, urns, &c., of the Romans, and fragments of sculpture, daggers, shields, gold and silver ornaments, and a great variety of other articles of British, Saxon, Danish, or Norman manufacture, have been discovered at different periods. The number of religious houses, including colleges and hospitals, was about 57. There are remains of the abbeys of Kingswood, Laycock, and Malmesbury; of the priory of Bradenstoke; and of the nunnery of Kington-St. Michael. There yet exist extensive remains of the ancient castles of Castle-Combe, Devizes, Farley, Ludgershall, Malmesbury, Marlborough, and Wardour. Amongst the principal of the numerous seats of the nobility and gentry, the most splendid are Bowood, the residence of the Marquess of Lansdowne, lord-lieutenant of the county; Charlton House, that of the Earl of Suffolk; Stowerhead, of Sir Henry Hugh Hoare, Bart.; Longleat, of the Marquess of Bath; Tottenham Park, of the Earl of Ailesbury; Wardour Castle, of Lord Arundel; Wilton House, of the Earl of Pembroke; Longford Castle, of the Earl of Radnor; and Corsham House, of Lord Methuen. There is a chalybeate spring at Chippenham, also a chalybeate and saline aperient spring near Melksham, and mineral springs of different other qualities at Heywood, Holt, and Middle Hill Spa, near Box. Wiltshire gives the inferior title of Earl to the Marquess of Winchester.

WILY, or WYLY (*St. Mary*), a parish, in the union of Wilton, hundred of Branch and Dole, Hindon and Southern Divisions of the county of Wilts, 7 miles (E. N. E.) from Hindon; containing 476 inhabitants. The living is a rectory, valued in the king's books at £21. 14. 2.; present net income, £492; patron, Earl of Pembroke. This parish is situated on the high western road, and contains an inn, called the Deptford Inn, which affords accommodation to the most distinguished families. A large sheep fair is held here on the 4th of October. Lady Elizabeth Mervyn, in 1581, bequeathed $7\frac{1}{2}$ bushels of wheat and 100 ells of linen and cloth, to be provided by the possessor of her estates, on Good Friday in each year, and distributed among the poor of Wily, Steeple-Langford, Upton, Padworth, and Tisbury. Christopher Willoughby, in 1678, gave £200 to be distributed among the poor of Wily and Marlborough. Here is a National school. About one mile from the village is a large British encampment, called Badbury-Rings, or Wily Camp, which occupies a point of down projecting from the principal ridge, and encloses an area of more than seventeen acres. Two miles from Deptford Inn is Yarnbrough Castle, a large and ancient Roman encampment.

WIMBISH (*All Saints*), a parish, in the union of Saffron-Walden, hundred of Uttlesford, Northern Division of the county of Essex, $4\frac{1}{4}$ miles (E. S. E.) from Saffron-Walden; containing, with Thunderley, 921 inhabitants. This parish, including the hamlet of Thunderley, is about sixteen miles in circumference, and the scenery is much enriched with timber of stately growth. The living is a vicarage, with that of Thunderley united in 1425, valued in the king's books at £8; present net income, £190; patron and incumbent, Rev. J. Raymond. The rectory is a sinecure, valued at £12; present net income, £435; patron and incumbent, Rev. J. Dolignon. The church is an ancient structure of stone, with a tower of brick replacing the original tower, which had fallen down. Sarah Barnard, in 1774, bequeathed an annuity of £4 in support of a school. Here is a National school for girls.

WIMBLEDON (*St. Mary*), a parish, in the union of Kingston, Western Division of the hundred of Brixton, Eastern Division of the county of Surrey, 7 miles (S. W.) from London; containing 2195 inhabitants. The name of this place, anciently written *Wymbandune*, *Wymbaldon*, and *Wymbledon*, is supposed to have been derived from one of its early proprietors.

The principal feature in the parish is Wimbledon Park, which comprises 922 acres, and contains a sheet of water covering a space of about 40 acres. This park is one of the finest in the county, and contains very fine timber, especially evergreen oaks and cedars; one of the latter measures, 2 feet from the ground, 19 feet in circumference, and one of its limbs 8 feet. In the pleasure-grounds is a curious sarcophagus, besides several blocks of marble taken from the French during the war, and presented to the late Earl Spencer, then first Lord of the Admiralty, said to have been brought from Pompeii and intended for Buonaparte. The common is surrounded by seats of the nobility and gentry, and exhibits at the south-west angle a circular encampment with a single ditch, including a surface of seven acres; the trench is very deep and perfect. It is said to mark the site of a battle, fought in 568, between Ceawlin, King of the West Saxons, and Ethelbert, King of Kent, in which the latter was defeated, and his two generals, Oslac and Cnebban, slain. At the north-east angle of the common is the village, consisting of one street, containing many respectable houses; and in detached situations are numerous handsome seats and pleasant villas. A little to the north of the encampment is a well, the water of which has never been known to freeze. The London and South-Western railway passes through the parish, and about half a mile from the church is a station. The mills of the English Copper Company are in this parish; and there are also works for the printing of calico. A pleasure fair is held on the first Monday after Lady-Day, and the two following days.

The living is a perpetual curacy; net income, £170; patrons, Dean and Chapter of Worcester, as appropriators of the rectory, which is valued in the king's books at £35. 2. 11. The church, situated about a quarter of a mile north of the village, was erected in 1787, on the site of a former, which had fallen to decay. It is a neat edifice, in the Grecian style of architecture: on the south side is Cecil chapel, an ancient building in which are several portions of mail armour and monuments, one an altar-tomb of black marble to the memory of Sir Edward Cecil: in the east window are some remains of painted glass representing the arms of the families of Leeds, Salisbury, Dorset, &c., and in the churchyard are several handsome mausoleums and monuments, one of the latter to the memory of G. S. Newton, R.A., a painter of considerable merit. There is a place of worship for Independents. Here is a National school: the school-house, with a garden and field annexed, were given by John Earl Spencer, in 1773. This school and an infants' school are supported by subscription, aided by a bequest, in 1650, from Dorothy Cecil, daughter of Edward Viscount Wimbledon, who gave a rent-charge of £22. 2. 6., for teaching poor children, and keeping her father's tomb in repair. Five almshouses were erected in 1839 by subscription, and are endowed with the interest of £1000, the profits of a fancy fair held in the grounds of Wimbledon House, belonging to Mrs. Marryat, which are laid out with excellent taste, especially the gardens, which comprise about three acres, and contain some of the rarest plants in the kingdom. There is an ancient oak in the grounds, measuring, three feet from the ground, 22 feet in circumference. On digging, in 1838, in the grounds of Belvidere House, to make an artificial piece of water, two fine figures of white marble, as large as life, were discovered several feet below the surface, one representing Summer, the other Winter, also a small group; they are in the possession of J. C. Peach, Esq., of Belvidere House. There are several bequests for the poor. It is said that Catherine Parr, after the death of Henry VIII., resided in a house in the village, now a large school for young gentlemen. The celebrated Lord Burleigh is said to have resided here, and planted the magnificent avenue of elms on the common. Judge Park resided many years in the parish; the parishioners have erected a monument in the church to his memory.

WIMBLINGTON, a hamlet, in the parish of DODDINGTON, union and hundred of NORTH WITCHFORD, Isle of ELY, county of CAMBRIDGE, 4 miles (S.) from March; containing 965 inhabitants. There is a place of worship for Wesleyan Methodists. A school was directed to be founded in 1714, by Mr. Thomas Eaton, for teaching 40 children: it was endowed with lands of considerable value, but in consequence of a suit in Chancery, arising from the misapplication of the premises, no school was erected till 1817: the rental of the estates is £144 per annum.

WIMBOLDSLEY, a township, in the parish of MIDDLEWICH, union and hundred of NORTHWICH, Southern Division of the county of CHESTER, $2\frac{1}{2}$ miles (S. W.) from Middlewich; containing 102 inhabitants. The Grand Junction railway passes through the township. Lea Hall was for a considerable period the residence of the celebrated physician, Dr. Fothergill, who died in 1780, having been considered for thirty years as standing at the head of the medical profession; his works are chiefly on medical subjects, and have been published in three octavo volumes, with an account of his life prefixed.

WIMBORNE (*ALL SAINTS*, or *ALLHALLOWS*), a parish, in the union of WIMBORNE and CRANBORNE, hundred of CRANBORNE, Wimborne Division of the county of DORSET, $\frac{1}{2}$ a mile (N.) from Wimborne-St. Giles, with which the population is returned. The living is a rectory, united in 1732 to that of Wimborne-St. Giles, and valued in the king's books at £9. 4. $4\frac{1}{2}$. The church, which formerly appears to have been the mother church of Wimborne-St. Giles, was pulled down in 1733.

WIMBORNE (*ST. GILES*), a parish, in the union of WIMBORNE and CRANBORNE, hundred of WIMBORNE-ST. GILES, Wimborne Division of the county of DORSET, $2\frac{1}{2}$ miles (S. W. by W.) from Cranborne; containing, with Wimborne-All Saints, 384 inhabitants. The living is a rectory, with that of Wimborne-All Saints united, valued in the king's books at £12. 13. 4.; patron, Earl of Shaftesbury. The tithes have been commuted for a rent-charge of £527. 8., subject to the payment of rates, which on the average have amounted to £57; the glebe comprises $107\frac{1}{2}$ acres, valued at £166. 5. per annum. The church was rebuilt in 1732, on the union of the two livings: it stands near the seat of the Earl of Shaftesbury, and is the burial-place of the family; its tower is crowned with balustrades, and at each angle is an urn surmounted by a steel vane. A National school has been established. Here are almshouses for eleven poor people, founded in 1624, by Sir Anthony Ashley, Bart., and endowed with a large farm in the parish of Gussage-All Saints.

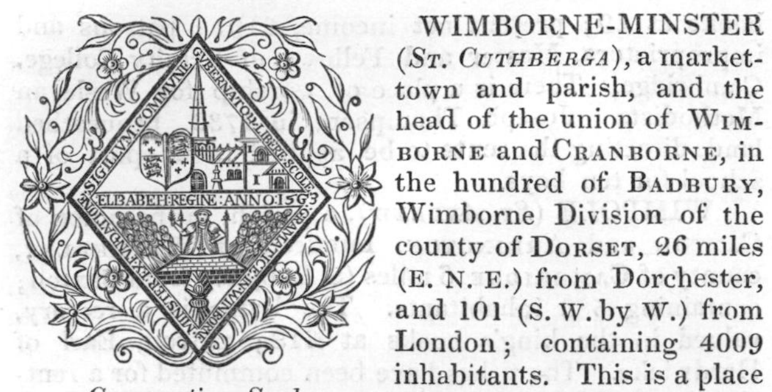

Corporation Seal.

WIMBORNE-MINSTER (*St. Cuthberga*), a market-town and parish, and the head of the union of Wimborne and Cranborne, in the hundred of Badbury, Wimborne Division of the county of Dorset, 26 miles (E. N. E.) from Dorchester, and 101 (S. W. by W.) from London; containing 4009 inhabitants. This is a place of very remote antiquity; in the time of the Romans it was of considerable importance as a station to their camp at Badbury, and by them was denominated *Vindogladia*, or *Ventageladia*, terms descriptive of its situation near to, or between, two rivers. The Saxon appellation of *Vinburnan*, whence the present name is obviously deduced, is of similar import; the epithet of Minster, from its ancient monastery, having been added as a term of distinction. Some have supposed this to have been the scene of the battle between Kearl, Earl of Devon, and the Danes, in 851, in which the latter were defeated; but Bishop Gibson states this to have occurred at Wenbury, in Devonshire, with which he endeavours to identify *Wicganbeorche*, the place where, in the Saxon Chronicle, it is stated to have taken place. About the commencement of the tenth century, Edward the Elder, in the beginning of his reign, being opposed by Ethelwald, son of his uncle Ethelbert, who aspired to the crown, encamped at Badbury, with a considerable army, and advanced upon Wimborne, where Ethelwald had fortified himself with a small force, which he captured, after an ineffectual resistance from the latter. But the principal cause of its celebrity was a nunnery, founded previously to 705, and dedicated to the Virgin Mary, by St. Cuthberga, daughter of Cenred, and sister of Ina, both kings of the West Saxons, which, about the year 900, being destroyed by the Danes, was subsequently converted into a house for Secular canons, the revenue of which, at the dissolution, was valued at £131. 14. The foundress became an inmate of the nunnery, where she died, and was buried in the church, of which, having been canonized, she was made the tutelar saint. The town is situated in a fertile vale, near the confluence of the rivers Stour and Allen, on the main road from London to Poole: the streets are irregular, and the houses in general of mean appearance. At its eastern extremity the Allen divides into two branches, over which are two bridges. Leland thus describes it:—"the town is yet meatly good, and reasonably well inhabited. It hath bene a very large thing, and was in price in the tyme of the West Saxon kinges. Ther be in and about it diverse chappelles, that in tymes paste were, as I have learnid, paroche chirchis of the very town of Wimburne." And in another place he says:—"the soile about Wimburn-Minstre self is very good for corne, grasse, and woodde." The town-hall, which formerly stood near the square, has long since fallen into decay: it occupied the site of St. Peter's chapel, sometimes styled the king's free chapel, which, having been neglected soon after the Reformation, was, with the cemetery, containing about one acre of ground, vested in the corporation, and their successors in fee, for the erection of a town-hall, the residue of the profits to be applied towards the maintenance of the choristers in the church. The market is on Friday; and fairs are held on the Friday before Good Friday and on September 14th, each for two days, for horses and cattle. Constables are appointed at the manorial court held annually at Michaelmas; and the town has been made a polling-place in the election of parliamentary representatives for the county.

On the establishment of the Secular canons, when the nunnery was destroyed by the Danes, the church became collegiate, and a royal free chapel, having been declared by letters of Edward II., in the eleventh year of his reign, to be exempt from all ordinary jurisdiction, imposition, &c. In Leland's time the society consisted of a dean, four prebendaries, five cantuarists, three vicars, and four secondaries. The dean and prebendaries maintained four priests and four clerks to serve the cure; *viz.*, in the collegiate church, and in St. Peter's, Kingston, and Holt chapels. On the dissolution of the college, its possessions lapsed to the crown; and Elizabeth, in the fifth year of her reign, on re-establishing the school, appointed twelve of the inhabitants governors, whom she incorporated, with a common seal, and granted in trust to them the tithes of the parish, and all other endowments of the college and school, with all ecclesiastical rights and spiritual jurisdiction previously belonging to the college and prebends. In the reign of Charles I., the governors having surrendered these possessions, the king re-granted them in full, on the condition of their providing the necessary officers for the service of the church and the school, with all ecclesiastical jurisdiction within the parish, and power to appoint the official and registrar of the peculiar court. Three incumbents are elected by the governors, who serve the church in rotation weekly; they also appoint three clerks, an organist, three singing men, and six singing boys; a visitation court is held annually. The chapels of Kingston and St. Peter have fallen into ruins; that at Holt remains. The church, commonly called the Minster, is a large cruciform structure, with a quadrangular tower rising from the intersection, and another at the west end, the former in the Norman style, the latter in the later English; the east window is in the early English style. A tempest destroyed the spire about 1600, and it has not since been replaced. The chancel and choir are approached from the nave by a flight of steps, and are supported by pillars: in the choir are sixteen stalls, with canopies of carved oak. Very extensive repairs and restorations have been going on for the last three or four years, which are now completed, and have greatly increased the splendour of this interesting edifice. Five stained glass windows have been put up at the altar end of the church, of which the three at the east end were presented by Mr. Bankes, the one on the north side of the altar by the Earl of Devon, and the opposite one, on the south, by the Duke of Beaufort. St. Cuthberga is supposed to have been entombed in the wall of the chancel: here also was King Ethelred's tomb, of which the brass plate fixed in the floor is all that remains. On the south side of the choir is an altar-tomb, with the effigies of the Duke and Duchess of Somerset, the parents of Margaret Countess of Richmond, mother of Henry VII.; on the opposite side is a similar tomb, but without figures, to the memory of Gertrude, Marchioness of Exeter, mother of the unfortunate Edward Courtenay, last Earl of Devonshire; and in the south aisle is a monument, with an armed recumbent figure, to Sir Edmund Uvedale, Knt., dated 1606. Under the area of the chancel is a small crypt, called the Dungeon, having pointed arches, with bold circular groins to support the roof; it was formerly used as a chantry, of which the church contained a great number. On the outside of the Bell tower is the figure of a soldier, with a hammer in each hand, with which it strikes the quarters of the hour. There are places of worship for Baptists, Independents, and Wesleyan Methodists; also a Roman Catholic chapel at Stapehill, in the vicinity.

The free grammar school, originally founded by Margaret Countess of Richmond, in 1497, was refounded by Queen Elizabeth, in 1563, for the gratuitous instruction of all applicants, without limitation, and designated Queen Elizabeth's free grammar school in Wimborne-Minster: the management is vested in the twelve governors. St. Margaret's hospital, for poor persons, situated at the west end of the town, is of ancient and obscure foundation; it consists of seven good tenements for five poor men and two women: the revenues are under the direction of the lord of the manor of Kingston-Lacy, and his steward, who conjointly nominate the almspeople: in a chapel attached to it divine service is occasionally performed. A second hospital, called Courtenay's, situated at the east end of the town, was built pursuant to the will of Gertrude Marchioness of Exeter, bearing date 1557; the governor and poor persons are incorporated and have a common seal: there are six almspeople on the foundation. At Pamphill, in this parish, are a school and almshouse, founded pursuant to the will of Roger Gillingham, dated July 2nd, 1695: the schoolmaster and almspeople are nominated and appointed by the governors of the free grammar school in this town; the former receives £20, and each of the latter £5 per annum. Another school is partly supported by subscription. The poor law union of Wimborne and Cranborne comprises 10 parishes or places, under the care of 13 guardians, and contains a population of 7933, according to the census of 1831. This is supposed to be the birthplace of Matthew Prior, the statesman and poet, who was educated at the free grammar school. The Duke of Monmouth, after his escape from the battle of Sedgmoor, is stated to have been arrested in a small enclosure, called Shagsheath, near this place; but this is doubted by some, who are of opinion that his capture took place near Ringwood. Badbury camp, in the vicinity, is a circular intrenchment, surrounded by three ramparts, enclosing an area of eighteen acres: Roman coins, urns, and a sword, were dug up in 1665.

WIMBOTSHAM (*St. Mary*), a parish, in the union of Downham, hundred of Clackclose, Western Division of the county of Norfolk, 1¼ mile (N. by E.) from Downham-Market; containing 476 inhabitants. The living is a discharged rectory, annexed to the vicarage of Stow-Bardolph, and valued in the king's books at £5. 6. 8.

WIMESWOULD (*St. Mary*), a parish, in the union of Loughborough, hundred of East Goscote, Northern Division of the county of Leicester, 5¼ miles (N. E. by E.) from Loughborough; containing 1276 inhabitants. The living is a discharged vicarage, valued in the king's books at £9; present net income, £191; patrons and impropriators, Master and Fellows of Trinity College, Cambridge. There is a place of worship for Wesleyan Methodists. Joseph Thompson, in 1733, bequeathed land, directing the rents to be applied in support of a school for ten boys.

WIMPOLE (*St. Andrew*), a parish, in the union of Caxton and Arrington, hundred of Wetherley, county of Cambridge, 6 miles (S. E. by S.) from Caxton; containing 583 inhabitants. The living is a rectory, valued in the king's books at £18; patron, Earl of Hardwicke. The tithes have been commuted for a rent-charge of £567, subject to the payment of rates, which on the average have amounted to £65; the glebe comprises one acre. The church has been enlarged by fitting up a private chapel with seats, and 100 free sittings have been provided, the Incorporated Society having granted £75 in aid of the expense: it contains various monuments to the Hardwicke family, among which is that to the memory of Lord Chancellor Hardwicke, who died in 1764, and was here interred. A school is supported by the Earl of Hardwicke.

WINCANTON (*St. Peter and St. Paul*), a market-town and parish, and the head of a union, in the hundred of Norton-Ferris, Eastern Division of the county of Somerset, 34 miles (E.) from Taunton, and 108 (W. by S.) from London; containing 2123 inhabitants. This place is of very great antiquity; it was anciently called *Wyndcaleton*, and derived its name from its situation on the windings of the river Cale, by which it is bounded on the west. At a very early period it was the scene of many sanguinary conflicts between the Britons and Saxons, and subsequently of numerous encounters between the latter and the Danes, who made frequent irruptions into this part of the country. During the parliamentary war, some of the earliest engagements between the contending parties took place in the immediate vicinity of this town, in which, according to Burnet's "History of his own Times," was shed the first blood in the Revolution of 1688; though some state this to have occurred at Cirencester. In 1747, it suffered material injury from an accidental fire, which destroyed a considerable portion of the town, to which calamity may be attributed the regular and uniform appearance it afterwards assumed.

The town is pleasantly situated on the western declivity of a hill rising gently from the river Cale, and consists principally of four regular streets, containing some well-built houses. The environs, which are pleasant, abound with interesting scenery, and on the south is an uninterrupted view of the fine Vale of Blackmore, extending for many miles: the land is extremely fertile; and within a short distance of the town are several gentlemen's seats. The manufacture of linen and bed-ticking was formerly carried on to a considerable extent: but within the last few years it has greatly declined: a branch of the silk manufacture has been introduced. The market is on Wednesday, and is well supplied with corn, cattle, cheese, and butter: the fairs are on Easter-Tuesday and September 29th. The town, which is within the jurisdiction of the county magistrates, is divided into two parts, called the Borough and the Tything; two constables for the former, and a tythingman for the latter, are annually appointed at the manorial court, and a court leet for the hundred is also held annually. It has been made a polling-place for the eastern division of the county. The living is a perpetual curacy; net income, £123; joint patrons, U. and G. Messiter, Esqrs., as owners of the impropriate rectory. The church is a spacious and neat edifice, with a square embattled tower; the interior consists of a nave, aisles, and chancel, and, as well as the exterior, was repaired in 1835, and enlarged by the addition of 350 sittings, 200 of which are free, the Incorporated Society having granted £150 in aid of the expense. There is a place of worship for Independents; also a National school. Various charitable bequests have been made for distribution among the poor. The poor law union of Wincanton comprises 39 parishes or places, and contains a population of 21,096, according to the census of 1831; the number of guardians is 43. At Stavordale, the north-eastern extremity of the parish, a small priory of Augustine canons, dedicated to St. James, is said to have been built by Sir William Zouch, which, in the 24th of Henry VIII., was annexed to the priory of Taunton: the remains, especially the roof and some portions of the chapel, are in good preservation. The Earl of Ilchester, among his other inferior titles, enjoys that of Lord Ilchester and Stavordale. At Horwood, about a mile south-east of the town, are two mineral springs, resembling in their properties those at Cheltenham. An urn, containing several Roman coins, was discovered in the parish, many years since.

WINCEBY (*St. Margaret*), a parish, in the union of Horncastle, hundred of Hill, parts of Lindsey, county of Lincoln, 4 miles (E. by S.) from Horncastle; containing 65 inhabitants. The living is a discharged rectory, valued in the king's books at £6. 0. 2½.; present net income, £276: it is in the patronage of the Crown. A school is supported by the rector and proprietor of the parish. A battle was fought here during the parliamentary war, in which the king's troops were defeated.

WINCH, EAST (*All Saints*), a parish, in the union and hundred of Freebridge-Lynn, Western Division of the county of Norfolk, 5½ miles (S. E. by E.) from Lynn-Regis; containing 466 inhabitants. The living is a discharged vicarage, valued in the king's books at £8; patron, E. Kent, Esq.; impropriators, the Landowners. The impropriate tithes have been commuted for a rent-charge of £3. 10., and the vicarial for £184, subject to the payment of rates, which on the latter have on the average amounted to £22; the glebe comprises 12½ acres, valued at £12. 10. per annum. In the east window of the chancel are the arms of Vere and Howard; and on the north side is the ancient chapel of St. Mary, the family burial-place of the latter.

WINCH, WEST (*St. Mary*), a parish, in the union and hundred of Freebridge-Lynn, Western Division of the county of Norfolk, 3 miles (S. by E.) from Lynn-Regis; containing 394 inhabitants. The living is a rectory, valued in the king's books at £9. 13. 4.; present net income, £353: it is in the patronage of the Crown. Here are two schools, in which 20 children are instructed at the expense of Lady Harriet Gurney: there is also a Sunday National school, supported by contributions.

WINCHAM, a township, in the parish of Great Budworth, union of Northwich, hundred of Bucklow, Northern Division of the county of Chester, 2 miles (N. E. by E.) from Northwich; containing 589 inhabitants.

WINCHCOMB (*St. Peter*), a market-town and parish, and the head of a union, in the Lower Division of the hundred of Kiftsgate, Eastern Division of the county of Gloucester, 15½ miles (N. E. by E.) from Gloucester, and 95 (W. N. W.) from London; containing 2514 inhabitants. This place, which is of equal antiquity and importance, was anciently called *Wincelcumb* (from the Saxon *Wincel*, a corner, and *comb*, a valley, descriptive of its situation in a nook or corner of the vale), of which its modern name is obviously a contraction. During the Heptarchy, if not the metropolis of the kingdom of Mercia, it was at least the residence of some of the Mercian kings, of whom Offa founded a nunnery here, in 787. Cenulph, who succeeded to the throne of that kingdom, after the death of Egferth, Offa's son, who survived his father only a few months, had a palace here, and, in 798, laid the foundation of the stately abbey, for 300 monks of the Benedictine order, which he endowed with an ample revenue, and dedicated with unusual splendour to the Blessed Virgin Mary. After the conclusion of the ceremony, which was conducted by Wulfred, Archbishop of Canterbury, assisted by twelve other prelates, in the presence of the king himself, Cuthred King of Kent, Sired King of the East Saxons, ten dukes, and the flower of the Mercian nobles, Cenulph, leading to the high altar his captive, Ethelbert Pren, the usurper of the kingdom of Kent, whom he had made prisoner, generously restored him to his liberty without fine or ransom. Cenulph, in the year 819, was buried in the abbey which he had founded, where also the remains of his son and successor, Kenelm, were deposited. The young king, after a reign of one year, having been, according to some, accidentally slain, or according to others, cruelly murdered, at the instigation of his unnatural sister Quendreda, in the hope thereby of securing to herself the throne, was first obscurely buried, and afterwards, on the discovery of the deed, removed with much funeral pomp, and interred near his father in the abbey church. He was at length canonized, and the numerous pilgrimages made to his shrine much augmented the revenue of the monastery, which was subsequently re-dedicated to the Virgin Mary and St. Kenelm. It was afterwards in the possession of Secular priests, and had almost fallen into decay, when Oswald, Bishop of Worcester, in the year 985, reformed its discipline, recovered the lands of which it had been deprived, and restored to it the Benedictine monks, who held it till the dissolution. This was a mitred abbey, the first summons of the abbot to parliament now on record being in 1265: its possessions were numerous, for, at the period of the Norman survey, no fewer than nineteen manors were annexed to it, independently of Winchcomb itself; but the monks, having opposed the Conqueror, were by him deprived of many of their lands. At the dissolution the revenue was £759. 11. 9. The building is reported to have been exceedingly magnificent, and the establishment so prosperous at one period, that it is said to have been "equal to a little university." Very few traces of it, however, remain, but the memorial is preserved in the name of part of a hamlet, which is still called the Abbey demesnes. Of the civil history of the place few particulars are recorded: the town appears to have been walled, and to the south of the church there was an ancient fortress, or castle, which, according to Leland, having fallen into decay, and the ruins being overspread with ivy, gave the name of Ivy castle to a spot which is now occupied only by a few cottages and gardens.

Winchcomb is situated in a beautiful vale, at the northern base of the Cotswold hills, by which it is sheltered nearly on every side, and is watered by the little river Isbourne, which flows close to it on the south-east; it consists principally of three streets, extending in a long line from east to west, with North-street and a few smaller ones branching from them. The houses are in general low and of indifferent appearance, and from its being but little of a thoroughfare, the place preserves an air of seclusion and tranquillity, and has that venerable character which denotes an old Anglo-Saxon town: it is abundantly supplied with excellent water from wells and springs. The cultivation of tobacco, which is said to have been first planted here after its introduction into the kingdom, in 1583, was for a considerable time a source of much profit to the inhabitants: but in the 12th of Charles I., that trade being restrained, the plantations were neglected. The principal branches of manufacture at present carried on are those of paper and silk, for the former of which there are in the neighbourhood two large mills, and one for the latter; there is also a tan-yard on a moderate scale. Other minor branches are cotton stockings and pins; and agricultural operations, with the spinning of linen and woollen, afford nearly constant employment to the parochial poor. The market is on Saturday: the fairs are on the last Saturday in March, May 6th, and July 28th, for horses, cattle, and sheep; and two fairs are held at Michaelmas for the hiring of servants. Previously to the time of Canute, Winchcomb, with a small surrounding district, was a county of itself; but in the reign of that monarch, according to an ancient manuscript in the cathedral church of Worcester, Edric, who governed under him as viceroy, "joined the sheriffdom of Winchelscomb, which was entire within itself, to the county of Gloucester." In the reign of Edward the Confessor the town was made a borough, and the government was vested in two bailiffs and twelve burgesses, of whom the former have till lately been annually elected by the lord of the manor, but exercise no jurisdiction, the charter having for many years ceased to be acted upon. It has lately been made a polling-place for the eastern division of the shire.

The living is a discharged vicarage, valued in the king's books at £3. 4.; present net income, £134; patron, Lord Sudeley; impropriators, several proprietors of land. The church, partly erected by Abbot William, in the reign of Henry VI., and completed at the expense of the parishioners, munificently assisted by Ralph Boteler, Lord of Sudeley, is a spacious and handsome structure, in the later style of English architecture, with a lofty square embattled tower, crowned with pinnacles; the walls are embattled, and strengthened with buttresses, also terminating with pinnacles; the south porch, of which the roof is elaborately groined and highly enriched, is a beautiful specimen of the style. The interior, consisting of a nave, aisles, and chancel, is of appropriate character; the nave is separated from the aisles by octagonal pillars and obtusely pointed arches, and from the chancel by an ancient carved oak screen. At Gretton, in this parish, is a chapel of ease. There are places of worship for Baptists and Wesleyan Methodists. A free grammar school was founded in 1522, by Henry VIII., who endowed it with £9. 4. 6. per annum, which was afterwards confirmed by Queen Elizabeth. The school, after being long continued in a house belonging to the corporation, was united to a grammar school subsequently founded by Lady Frances Chandos, for which she erected a school-house in St. Nicholas'-street, endowing it with certain lands and tenements, for the education of fourteen boys; the income, arising from nearly 20 acres of land, is £45 per annum, which is received by the master of the king's grammar school, who pays a sub-master to teach the boys reading, writing, and arithmetic. A school for teaching children to read was founded by George Townsend, Esq., who endowed it with £5 per annum, as a salary for the master (since increased to £20 by the trustees), and also left funds for apprenticing the children, with whom a premium of £15 is given. There are unendowed almshouses for six poor families, founded by Lady Dorothy Chandos, and various charitable bequests for the distribution of bread, clothing, and money to the poor. The poor law union of Winchcomb comprises 30 parishes or places, and contains a population of 9715, according to the census of 1831; it is under the care of 32 guardians. There are two mineral springs in the parish, one a strong saline, the other chalybeate, and nearly similar to those of Cheltenham. In addition to the abbey of St. Mary, previously noticed, were a church, dedicated to St. Nicholas, in the east part of the town, of which there are no remains, and an ancient hospital. About half a mile from the town are the beautiful remains of the magnificent castle of Sudeley, formerly belonging to the Botelers, lords of Sudeley, which is noticed in the article on SUDELEY-MANOR. Tidenham of Winchcomb, Bishop of Worcester, and physician to Richard II., is supposed to have been a native of this town; and Dr. Christopher Mercet, an eminent naturalist and philosopher, was born here in 1614.

WINCHELSEA (*St. Thomas*), a borough and parish, (formerly a market-town), having separate jurisdiction, in the union of RYE, locally in the hundred of Guestling, rape of HASTINGS, Eastern Division of the county of SUSSEX, 74 miles (E. by N.) from Chichester, and $66\frac{3}{4}$ (S. E.) from London; containing 772 inhabitants. The ancient town of this name, situated near the Camber Point, was a place of considerable importance in the time of the Romans, and was subsequently destroyed by an inundation of the sea, about the close of the thirteenth century. The present town, which is situated at the distance of a mile and a half from the sea, was built upon an eminence well adapted to prevent a similar accident, in the reign of Edward I., who gave land for that purpose, and contributed largely towards its erection. The site, originally called Higham, was, by the munificence of that monarch, surrounded with walls and defended by three strong gates, which formed the principal entrances, and are still in good preservation. In the reign

Arms.

of Henry III., Winchelsea and Rye were annexed to the cinque-ports, but more as appendages than equal ports, being members of the port of Hastings; in the different charters granted to these towns, they are invariably styled "ancient towns." The new town was invested with the same privileges as the old, and, enjoying all the benefits of the cinque-ports, it rapidly acquired a considerable degree of commercial importance. In the reign of Edward III. it sustained material injury from the French, who, having landed on this part of the coast, burnt a considerable portion of it; and it was subsequently plundered by the Spaniards, in the reign of Richard II. But it experienced the greatest injury from the retiring of the sea, about the close of Elizabeth's reign, by which its harbour was destroyed, and its commercial importance annihilated. The town occupies a space nearly two miles in circumference, divided into squares by streets intersecting each other at right angles, probably after the plan of the ancient town. Neither any trade nor manufacture is carried on at present: the market has fallen into disuse, but a fair is still held on May 14th for cattle. The Royal Military canal commences at Cliff-End, and passes by this town parallel with the shore, till it enters the sea at Shorncliff, near Hythe.

Corporation Seal of Winchelsea.

Obverse. *Reverse.*

According to the ancient charter, the government is vested in a mayor and twelve jurats, who are justices of the peace within the ancient town and its liberties. The borough received the elective franchise in the 42nd of Edward III., from which period till the 2nd of William IV it continued to return two members, but was disfranchised by the act then passed to amend the representation. The right of election was vested in the freemen, eleven in number; the mayor was the returning officer. Jointly with Hastings, it sends canopy bearers on the occasion of a coronation, these two places being entitled to every third turn, in common with the other cinque-ports. The living is a discharged rectory, valued in the king's books at £6. 13. 4.; present net income, £278; patrons, Trustees of Sir W. Ashburnham, Bart. The church is the only remaining portion of a very fine structure, the whole of which, except the chancel, is gone to decay: it presents an elegant specimen of the early and decorated styles of English architecture. On the north side are some stalls and a piscina of beautiful design, and there are several splendid monuments, including three supposed to be monuments of Knights Templars, cross-legged and in armour, of which one, in particular, is hardly excelled by any in the kingdom. There is a place of worship for Wesleyan Methodists; also a Sunday National school. In addition to the church of St. Thomas were anciently two other parochial churches, dedicated respectively to St. Leonard and St. Giles. The remains of antiquity still visible are, the ruins of Camber castle, erected by Henry VIII., a circular fortress with a round tower, which was the keep; the ancient gates of the town; and the interesting ruins of a monastery of Grey friars, founded by Edward II. Robert, Archbishop of Canterbury, who died in 1313, was a native of this town. Winchelsea gives the title of Earl to the family of Finch.

WINCHENDON, NETHER (*St. Nicholas*), a parish, in the union of Aylesbury, hundred of Ashendon, county of Buckingham, 7 miles (W. by S.) from Aylesbury; containing 294 inhabitants. The living is a perpetual curacy; patron, Sir W. B. Cave, Bart.

WINCHENDON, UPPER (*St. Mary Magdalene*), a parish, in the union of Aylesbury, hundred of Ashendon, county of Buckingham, 6 miles (W. by N.) from Aylesbury; containing 223 inhabitants. The living is a donative, valued in the king's books at £7. 17.; present net income, £60; patron and impropriator, Duke of Marlborough.

WINCHESTER, a city, having separate jurisdiction, and the head of a union, locally in the hundred of Buddlesgate, Winchester and Northern Divisions of the county of Southampton, of which it is the capital, 63 miles (S. W. by W.) from London; containing, with the population of the soke liberty, 8767 inhabitants.

Arms.

This place, called by the ancient Britons *Caer Gwent*, from the whiteness of its chalky soil, was the *Venta Belgarum* of Ptolemy and Antoninus; and, on its subsequent occupation by the Saxons, it obtained the appellation of *Wintan-Ceaster*, from which its present name is derived. It was probably first inhabited by the Celtic Britons, who emigrated from the coasts of Armorica in Gaul, and established themselves in this part of the Island, where, finding well-watered valleys, fertile plains, and shady forests, adapted to their support, and suited to the exercise of their religious rites, they fixed their chief residence, and continued in undisturbed possession till within a century prior to the Christian era, when they were expelled by a tribe of the Belgæ, who, after having established themselves on the southern coasts, concentrated their forces, and advancing into the country, made this one of their principal settlements. Among the several towns which were called *Ventæ*, this became the most important, and, prior to the Roman invasion, was the capital of the Belgian territory in Britain: it retained its pre-eminence till it fell under the power of the Romans, who, having achieved the conquest of this part of the island, under Vespasian, made it one of their principal stations. In the year 50, Ostorius Scapula fortified all the cities of the Belgæ between Anton, or the Southampton river, and the Severn, and placed garrisons in them, as a defence from the frequent assaults of the Britons, who were ever on the alert to surprise the enemy, and to recover the towns of which they have been deprived. The fortifications of this station may be still discerned in various places; and on Catherine hill, within a mile of the present city, are vestiges of a Roman camp, quadrangular in form, and defended by strong intrenchments: this, which was probably the *castra æstiva* of the station, communicated with the Roman road between Porchester and Winchester on one side, with the river on the other, and with the several roads leading to the neighbouring stations of *Vindonum*, or Silchester, and *Sorbiodunum*, or Old Sarum. Two Roman temples are said to have been erected near the site of the present cathedral, one consecrated to Apollo, and the other to Concord; and, among other evidences of Roman occupation, sepulchres have been recently discovered without the walls of the city to the north and east, nine of which, on being opened in 1789, were found to contain human bones, urns of black pottery elegantly formed, a coin of Augustus Cæsar, fibulæ, and other Roman relics. Carausius and Alectus, who assumed the imperial purple in Britain, are said to have fixed their residence in this place, where their coins have been discovered in greater profusion than in any other part of the kingdom. Soon after the establishment of Christianity in the island, a monastery was founded here, of which Constans, son of Constantine, was one of the brethren; but being allured by his father from his devotional retirement, to take the command of the forces in Spain, he was, by the revolt of his general, made prisoner, and afterwards put to death.

After the departure of the Romans from Britain, Vortigern, who had previously exercised authority over the western part of the island, being elected king, in order to defend it from the incursions of the Picts and Scots, who were making continual depredations, made Winchester the metropolis of the kingdom, and it was subsequently the residence of his successors. On the invasion of Britain by the Saxons, under Cerdic, and the defeat of the united Britons in the New Forest, it became the capital of the Saxon kingdom of Wessex, and the residence of the conqueror, who was crowned King of the West Saxons. Cerdic, having, in conjunction with his son Cenric, spent several years in extending his dominions and in giving security to his conquests, died, and was buried here in 534: during his government, the monastery was converted into a Pagan temple, and appropriated to the service of the Saxon deities. In 635, St. Birinus, whom Pope Honorius had sent into Britain, to propagate the Christian faith in those parts of the island which were still in Pagan darkness, met with a favourable reception from Cynegils, who, in conjunction with his son Cwichelm, was then king of the West Saxons. Cynegils, by the persuasion of Oswald, King of Northumbria, who afterwards espoused his daughter, Kineburga, was baptized at York; and, in the following year, his son and many of his subjects were converted to Christianity, which from that time began to flourish in this part of the island. This prince subsequently began to collect materials for building a cathedral in Winchester, intending to make it the seat of a Bishopric, which St. Birinus had, in the mean time, established at Dorchester, where his son Cwichelm held his court; but the design was frustrated by his death, which happened about six years after that of Cwichelm, who died the year after his conversion. Cenwahl, his second son, succeeding to the throne after the death of his father and elder brother, the people again relapsed into Paganism, under a prince who refused to acknowledge the new religion; but on his conversion to Christianity and baptism by St. Birinus, Cenwahl, in 648, completed the cathedral, which he dedicated to St. Birinus, St. Peter, and St. Paul, and founded, and amply endowed, a monastery near the site. About ten years after the death of St. Birinus, who was buried at Dorchester, Cenwahl divided the see into two portions, assigning the northern part of his kingdom to Dorchester, and the southern part to Winchester, to the cathedral of which latter place the remains of St. Birinus were removed, by Hedda, the fifth bishop. Egbert, who succeeded to the throne of Wessex, in 800, after many severe struggles for empire, obtained the sovereignty of all the other kingdoms of the Heptarchy, of which he was crowned sole monarch, in the cathedral church of Winchester, in 827, in the presence of a wittenagemote, or great assembly of the people. On this occasion he published an edict, abolishing all distinctions, and commanding all his subjects, in every part of his dominions, to be called English. This union of the kingdoms greatly promoted the importance of Winchester, which, from being the capital only of Wessex, became the metropolis of the kingdom. Ethelwolph, who succeeded Egbert, dated his charter from this city in 855, for the general establishment of tithes, which was signed in the cathedral church, by himself, by Burhred, King of Mercia, and Edmund, King of the East Angles (his tributary vassals), and by the chief nobility and prelates.

About this time the city seems to have been in a flourishing condition, and a commercial guild was established in it, under royal protection, at least a century earlier than in any other part of the kingdom. During the reigns of Ethelwolf and Ethelbald, St. Swithin, a native either of the city or of the suburbs, presided over the see: by his advice the latter monarch enclosed the cathedral and the cloisters with a wall and fortifications, to defend them from the predatory attacks of the Danes, who, at this period, were beginning to make frequent incursions upon this part of the coast, and who, in the succeeding reign, having landed in considerable numbers at Southampton, advanced to Winchester, where they committed the most barbarous outrages. On retiring to their ships, however, they were attacked, routed with great slaughter, and dispossessed of the immense quantity of plunder which they had taken in the city. On this occasion, the cathedral and monastic buildings, which had been previously fortified, escaped without injury. About the year 872, after repeated battles fought with varied success, in which Ethelbert was assisted by his younger brother Alfred, a band of those rapacious pirates assaulted the city, in which they made dreadful havoc; the cathedral was greatly damaged, and the ecclesiastics were inhumanly massacred. After the victory subsequently obtained over them by Alfred, Winchester was restored to its former importance, and again became the seat of government; and Alfred, who had fixed his chief residence here, ordered a general survey of the country to be made, and deposited in the archives of the city, which was thence called the *Codex Wintoniensis*. This monarch founded a monastery on the north side of the cathedral, for his chaplain, St. Grimbald, intending it also as a place of interment for himself and family; but dying before it was completed, he was buried in the cathedral, from which his remains were subsequently removed, and deposited in the new minster. In the reign of Athelstan six mints were established in the city, for coining as many different kinds of money: during this reign, the legendary battle between Guido, Earl of Warwick, and a Dane of gigantic stature, named Colbrand, is said to have taken place in a meadow near the city, on a spot of ground still called Danemark. In commemoration of this combat, which many historians regard as fabulous, there are numerous traditionary records; and, in the north wall of the city, the turret, called Athelstan's chair, from which that monarch is said to have viewed the battle, and a representation of it in stone, are stated to have existed formerly; and the battle-axe of Colbrand was kept in the cathedral till after the reign of James I.

In the reign of Edgar, a law was made to prevent frauds arising from the diversity of measures in different parts of the kingdom, and for the establishment of a legal standard measure to be used in every part of his dominions; the standard vessels made by order of that monarch were deposited in this city, from which circumstance originated the appellation "Winchester measure:" the original bushel is still preserved in the guildhall. In this reign, St. Ethelwold, a native of Winchester, who presided over the see, partly rebuilt the cathedral, which, on its completion in the following reign, he re-consecrated, in the presence of King Ethelred, Dunstan, Archbishop of Canterbury, and the principal nobility and prelates of the kingdom; and included in the dedication the name of St. Swithin, whose remains, buried at his own request in the churchyard, were removed and re-interred in the cathedral under a magnificent shrine, which had been prepared for that purpose by King Edgar. During the prelacy of St. Ethelwold, the Secular canons who officiated in the cathedral, being married men, had been, under the directions of Dunstan, Archbishop of Canterbury, replaced by Benedictine monks from the abbey of Abingdon; but on the accession of Edward the Martyr to the throne, Elfrida endeavoured to reverse that measure, which had been adopted generally throughout the kingdom, and by her influence caused the suppression of three Benedictine abbeys, which St. Ethelwold had founded, transferring their possessions to the married clergy. In consequence of these proceedings, a synod was held in the refectory of the old monastery in this city, in which the general dissolution of all monasteries was debated; but the measure was negatived by the intervention of one of those supposed miracles which were not uncommon upon such occasions. Ethelred, in 1002, having resolved upon the extermination of the Danes, by a general massacre throughout the kingdom, despatched secret letters to every part of his dominions for that purpose, which was carried into effect with the greatest inhumanity: such as were not actually put to death were mutilated and rendered incapable of any military service; and, in commemoration of that barbarous policy, the "Hoctide sports," so called from cutting the hamstrings of the victims, were instituted by that king, and continued, till within the last few ages, to be celebrated on the Monday in the third week after Easter. The retaliating vengeance of the Danes under Sweyn, King of Denmark, did not reach Winchester till some time after it had been inflicted on other parts of the kingdom; and on their approach in 1013, the inhabitants sued for peace, and gave hostages for the performance of any conditions.

After the partition of the kingdom between Edmund Ironside and Canute, the latter, having obtained the entire sovereignty, divided it into four parts, three of which he entrusted to the government of subordinate rulers; but reserving the fourth and most important under his own administration, he fixed his seat of government at Winchester, and greatly enriched the cathedral church, to which, after the memorable reproof of his courtiers at Southampton, for their flattery, he presented his regal crown, depositing it over the high altar, and making a vow never to wear it more. This monarch here held a general assembly of the nobility, in which he enacted laws for the government of the kingdom, and for the preservation of the royal forests and chaces. On the death of Hardicanute, in 1041, Edward the Confessor was crowned with great pomp and splendour in the cathedral, to which he granted an additional charter, and ordered a donation of half a mark to the master of the choir, and a cask of wine and 100 cakes of white bread to the convent, as often as a king of England should wear his crown in that city. During this reign, Queen Emma, his mother, by her own desire, to vindicate her innocence of the crime of incontinence, with which she had been aspersed, underwent the trial of the fiery ordeal in the cathedral church; and being conducted by two bishops, in the presence of the king and a crowded assembly of nobles and of the people, she is stated to have walked barefoot over nine heated ploughshares, without receiving the smallest injury. In gratitude for her deliverance, she enriched the possessions of the church with nine additional manors; the same number was also added by Bishop Alwyn, her kinsman and her asserted paramour, and the manors of Portland, Weymouth, and Wyke, were given on this occasion by the king: the first great seal of England was, in the course of this reign, made and kept in the city.

At the time of the Conquest, William fixed his principal residence at Winchester, which he made the seat of government, and built a strong castle at the south-west extremity of the city, in order to keep his new subjects in awe. Here he enacted most of his laws, and framed political measures for the security of his government, among which were the institution of the Curfew, and the general survey and estimate of the property of his subjects, called the Roll of Winchester, or Domesday-book, a probable imitation, or enlargement, of the *Codex Wintoniensis* of Alfred. Though he occasionally resided in London, which was growing into importance, and more especially during the latter part of his reign, yet he invariably celebrated the festival of Easter in this city. Soon after his establishment on the throne, Waltheof, who had been married to his niece Judith, and was created Earl of Huntingdon, being charged with entering into a conspiracy against him, was beheaded on St. Giles' hill, near Winchester. In 1079, Walkelyn, a relation of the Conqueror's, and bishop of the see, began to rebuild the cathedral and the adjoining monastery; for which purpose he obtained from the king a grant of timber from the woods in the vicinity. The building was completed in 1093, and dedicated, with great pomp, in the presence of all the bishops and abbots in the kingdom. On the death of Walkelyn, in 1098, William Rufus, who was crowned here, seized upon the bishoprick, and held it till the year 1100, when, being killed while hunting in the New Forest, his body was brought into the city on the following day, in a cart belonging to a charcoal maker, and interred in the centre of the choir of the cathedral: the lineal descendants of this man, whose name was Purkis, still pursue that occupation in the same place, which is within a few hundred yards of the spot where the monarch fell. On the death of Rufus, his elder brother Robert being then on a crusade, Henry, his younger brother, hastened to Winchester; and having made himself master of the royal treasure, he, in the presence of the reluctant nobles, drew his sword, and secured his pretensions to the crown by seizing and placing it upon his head. In this year he espoused Matilda, daughter of Malcolm III., King of Scotland, who had assumed the veil in the monastery of St. Mary, in this city, but had not taken the vows; by which marriage the royal Saxon and Norman lines were united; and on the birth of a son, in the following year, he conferred many additional privileges on the inhabitants. In the same year also a dreadful fire broke out, which destroyed the royal palace, the mints, the guildhall, a considerable portion of the city, and many of the public records. Henry, by the advice of Roger, Bishop of Sarum, ordered a general meeting of the masters of the several mints to assemble at Winchester, on Christmas-day in 1125, to investigate the state of the coin, which had been generally debased throughout the kingdom; and after due examination they were, with the exception of three of the Winchester mint-masters, found guilty of gross fraud, and each punished by the loss of his right hand. An entirely new coinage was ordered to be made, and the management of it was exclusively confined to those of the mint-masters of this city who had been declared innocent. About the same time, Henry, to prevent frauds in the measurement of cloth, ordered a standard yard, of the length of his own arm, to be made and deposited here with the standard measures of Edgar.

At this time Winchester appears to have attained its highest degree of prosperity: it was the seat of government, and the residence of the monarch; the royal mint, the treasury, and the public records were kept here; it had also a magnificent royal palace, a noble castle erected by the Conqueror, and another not less considerable, which was subsequently built as an episcopal palace for the bishops, with various stately public buildings, and numerous mansions for the residence of the nobility and gentry connected with the court: it had three royal monasteries, exclusively of inferior religious houses; a splendid cathedral, in which many of the monarchs of England had been crowned, and were interred; and a vast number of parish churches, of which Stowe relates that not less than 40 were destroyed in the war between Stephen and Matilda: its population was great, and its suburbs, in every direction, extended a mile further than they do at present; it was the general thoroughfare from the eastern to the western parts of the kingdom; it had a considerable manufactory for woollen caps, and enjoyed an extensive commerce with the continent, from which it imported wine, in exchange for its manufactures; and was a place of great resort for its numerous fairs, which were frequented by persons from various parts of the kingdom. On the death of Henry I., the city suffered greatly in the war which followed in the reign of Stephen, who having seized into his own hands the episcopal palaces throughout the kingdom, a synod was held here, to protest against the injustice of that measure, and to concert means of obtaining redress: at this meeting it was resolved that the assembled prelates should prepare an address, and send a deputation to the king, who then resided at the palace of Winchester, which was accordingly done, but the king, without paying the least attention to it, left the city and departed for London.

At this conjuncture the Empress Matilda landed on the coast of Sussex, to dispute Stephen's title to the throne, and the royal castle of Winchester was secured by a party in her interest; but, through the influence of Henry de Blois, brother of the king, who was then bishop of the see, the city was preserved in its allegiance to Stephen. On the subsequent captivity of the king, who was made prisoner in the war, and the acknowledgment of Matilda's claim to the crown by the greater part of the kingdom, the bishop abandoned the cause of his brother; and, having gone out with a solemn procession of his clergy, to meet the empress at Magdalene hill, conducted her and her partisans into the city with great ceremony. Her haughtiness, however, having excited disgust in the minds of the citizens, and the public opinion beginning to change in favour of the captive king, the bishop relaxed in his attention and deference to the empress, who, on that account, summoned him to wait upon her at the castle, where she then resided. Having returned an ambiguous answer to her summons, the bishop immediately began to put his castle of Wolvesey into a state of defence, and had scarcely completed its fortifications, when it was closely invested by the forces of Matilda, under the command of Robert Earl of Gloucester, her natural brother, and her uncle, David King of Scotland. A considerable body of Stephen's party having taken up arms, marched to the relief of the bishop: the armies on both sides were numerous and well appointed, and the city suffered dreadful havoc from their hostilities, which were carried on in the very centre of it, for several weeks, with the utmost acrimony. The party of the empress had possession of the royal castle and the northern part of the city; the king's party held the castle of Wolvesey, the cathedral, and the southern parts, and, discharging fireballs from Wolvesey castle, destroyed the abbey of St. Mary, the houses of the opposite party, and almost all the north part of the city, and ultimately succeeded in confining his opponents within the limits of the royal castle. The supply of water being cut off, and provisions beginning to fail, the garrison began to entertain thoughts of surrendering; but, having previously spread a report of Matilda's sickness and death, they obtained a truce for her interment, and placing her in a coffin, she was carried out through the army and escaped in safety to Gloucester. In the mean time, the Earl of Gloucester, with the King of Scots, taking advantage of the truce, made a sally from the castle; but being pursued, the earl was taken prisoner at Stockbridge, and subsequently exchanged for the captive monarch. Stephen, immediately on his liberation, repaired to Winchester, and began to strengthen the fortifications of the castle by the addition of new works; but, while engaged in that undertaking, an army, which had been newly raised in the adjoining counties, marched against him, and he was compelled to abandon his design, and save himself by flight. During this war, the bishop held a synod here, by an act of which it was decreed, that ploughs should have the same privileges of sanctuary as churches; and a sentence of excommunication was issued against all who should molest any person employed in agriculture. On the conclusion of this war, during which nearly one-half of the city was destroyed, the treaty between Stephen and Henry, the son of Matilda, the terms of which had been agreed upon at Wallingford Castle, was ratified at Winchester, by general consent.

Henry II., on his accession to the throne, was crowned here with his queen Margaret, and held a parliament in 1172; and here also, in 1184, his daughter, the Duchess of Saxony, gave birth to a son, named William, from whom the illustrious house of Hanover is supposed to have sprung. This monarch conferred many privileges upon the city, among which was that of being governed by a mayor and a subordinate bailiff: during his reign a calamitous fire, which began in the mint, destroyed the greater part of the town. On the death of Henry, his son, Richard I., surnamed Cœur de Leon, having secured the royal treasure in this city, was crowned in London; but, on his ransom from the captivity into which he had subsequently fallen, in returning from the Crusades, he had the ceremony of his coronation performed with great pomp in the cathedral of Winchester. In 1207, King John held a parliament here, in which he imposed a tax of one thirteenth part on all moveable property; and in the same year his queen gave birth to a son, who, from the place of his nativity, was surnamed Henry of Winchester. In the year following, in consideration of 200 marks paid down, and an annual payment of £100, that monarch granted the inhabitants a charter of incorporation, confirming all previous privileges; and, on his subsequent submission to the pope, he received absolution in the chapter-house of the monastery from sentence of excommunication, which had been pronounced against him by the legate of Pope Innocent III. Henry III., during his minority, kept his court here, under the guardianship of the Earl of Pembroke, and, after the death of that nobleman, under that of Peter de Rupibus, Bishop of Winchester. The residence of the king contributed materially to restore Winchester to the importance it had enjoyed previously to the war between Stephen and Matilda; but this advantage was greatly diminished by the numerous bands of lawless plunderers in the city and its vicinity, with which many of the inhabitants, and even members of the king's household, were connected. The depredations committed by these associations for the purpose of rapine were at length suppressed by the firmness and resolution of the king, but not till after 30 of the offenders had been brought to trial and publicly executed. During the war between this monarch and the barons, the city experienced considerable devastation, and suffered severely from the violence of both parties, who alternately had possession of it. After the battle of Evesham, the king held several parliaments here, in which all who had borne arms against him were attainted; but, with the exception of the Montfort family, none of these attainders were carried into execution, and the highest penalty inflicted did not exceed five years' rent of the forfeited estates. The celebrated trial of John Plantagenet, Earl of Surrey, took place here, for the murder of Alan de la Zouch, Chief Justice of Ireland, whom that nobleman killed on the bench in Westminster Hall, on being summoned before him to give evidence of the tenure by which he held his estates. On his oath, and on that of 24 compurgators, that he did not strike the judge from preconceived malice, the earl was acquitted, and fined 1200 marks. Edward I. also held several parliaments at Winchester, in one of which the celebrated ordinances, afterwards called the Statutes of Winchester, were passed; but the royal residence for the greater part was transferred to London, which, having risen into higher importance, had now become the metropolis of the kingdom; and Winchester, which hitherto had held the first rank among the cities of the empire, began to decline. Towards the end of his reign, this monarch, offended at the escape of a foreign hostage, who had been confined in the castle under the custody of the mayor, deprived the city of all its privileges, which were subsequently restored. Soon after the death of Edward II., a parliament was held here by Queen Isabel and Mortimer, in which Edmund of Woodstock, Earl of Kent, was arraigned, on a charge of high treason, and condemned to death. For this purpose he was led to a scaffold erected opposite to the castle gate; but so strong was the feeling of disgust which that iniquitous sentence excited, that no person could be found to carry it into execution, till, towards evening, a prisoner under sentence of death was prevailed upon, by the offer of pardon, to behead the earl.

Edward III. having made Winchester a staple for the sale of wool, the merchants erected large warehouses for conducting that lucrative trade, and the city began to recover its commercial importance; but its progress was interrupted by the destruction of Portsmouth and Southampton, in 1337, by the French, who were, however, repulsed from this city, and the following year by the plague, which ten years afterwards raged violently in the neighbourhood, and ultimately by the removal of the staple to Calais, in 1363. During this reign, Bishop Edynton, who was treasurer and chancellor to the king, commenced rebuilding the nave of the cathedral, which was completed by his successor, William of Wykeham, who, for his skill in architecture, was employed by Edward III. to superintend the erection of part of Windsor castle. Richard II. and his queen, visited Winchester in 1388; and, in 1392, that monarch removed to it his parliament from London, which was then suffering a suspension of its privileges under the king's displeasure. The marriage of Henry IV. with the Dowager Duchess of Bretagne was solemnized in the cathedral, by Bishop Wykeham, in 1401; and on the death of that prelate, Henry, afterwards Cardinal Beaufort, son of John of Gaunt, was by that monarch appointed to the see. Here Henry V. gave audience to the French ambassadors, whose insolence on that occasion led to the invasion of France, which soon followed. Henry VI. was a great benefactor to the city, which he frequently visited; and in 1449 he held a parliament here, which continued to sit for several weeks. In the course of this reign, however, its trade and population so greatly declined, that, in petitioning the king for the renewal of a grant conferred by his predecessor in 1440, the inhabitants represented that 997 houses were deserted, and 17 parish churches closed. Bishop Waynfleet having succeeded to the see, that monarch honoured the ceremony of his installation with his presence; and in the following reign, the queen of Henry VII. resided in the castle, where she gave birth to a son, whom, to conciliate the Welsh, the king named Arthur, in honour of the British hero of that name. In 1522, Henry VIII., in company with his royal guest, Charles V., came hither, where he spent several days: on this occasion the celebrated round table, at which the renowned king Arthur and his knights used to dine, and which was preserved in the castle, was fresh painted, and an inscription placed beneath it, in commemoration of the royal visit.

The dissolution of the monasteries, which took place during this reign, and the demolition of many of the religious establishments, completed the downfall of this once splendid and opulent city, and reduced it to a mere shadow of its former grandeur. On the accession of Mary, some transient gleams of returning prosperity revived, for a time, a hope of restoration: the marriage of that queen with Philip of Spain was solemnized in the cathedral, and several of the estates which had been alienated during the former reigns were restored to the see; but the real importance of Winchester had subsided, and, in a charter obtained for it from Elizabeth, through the solicitation of Sir Francis Walsingham, it is described as "having fallen into great ruin, decay, and poverty." On the death of Elizabeth, Sir Benjamin Tichborne, high sheriff of Hampshire, instantly on learning that event, and without waiting for any instructions from the lords of the privy council, who had been for many hours in close deliberation, proclaimed the accession of James I., in the city of Winchester. For this spirited conduct he was afterwards rewarded with the hereditary grant of the royal castle, and a pension of £100 per annum, during his own life and that of his eldest son, on whom the king also conferred the honour of knighthood. In the first year of this reign, the trial of Sir Walter Raleigh, Lords Cobham and Grey de Wilton, and others, on a charge of conspiracy, took place here, and Sir Walter Raleigh, though reprieved, was removed to the Tower of London, where he passed thirteen years in confinement.

At the commencement of the parliamentary war, Sir William Waller took possession of the castle for the parliament; but towards the close of the year 1643, it was retaken and garrisoned for the king, by Sir William, afterwards Lord Ogle, and the city was appointed the general rendezvous of the army then forming in the west for the re-establishment of the king's authority. Fortifications were at that time constructed round it, and more especially on the east and west sides, where vestiges of the intrenchments are still discernible; but the vigilance and activity of Sir William Waller disconcerted the enterprise, and on the subsequent defeat of Lord Hopton's party on Cheriton Down, Waller obtained possession of the city without difficulty. The castle, notwithstanding, held out for the king; and on the retreat of the parliamentarians to join the forces of the Earl of Essex, who was then laying siege to Oxford, the city also fell into the hands of the royalists. After the battle of Naseby, Cromwell was sent with an army to reduce Winchester, which, after being repeatedly summoned, refused to surrender, and the siege was immediately commenced. The garrison made a resolute defence, but after a week's resistance capitulated on honourable terms; the castle was immediately dismantled, and the works blown up; the fortifications were demolished, together with the Bishop's castle of Wolvesey, and several churches and public buildings. The wanton violence of the parliamentary troops was manifested in defacing the cathedral, destroying its monuments, violating the tombs, and in the indiscriminate insult offered to the relics of the illustrious dead, whose bones were scattered about the church; the statues of James and Charles, at the entrance of the choir, were thrown down, and the communion plate and other valuables belonging to the church were carried away. After the Restoration, the king chose Winchester for his residence during the intervals of his absence from London, and purchased the remains of the ancient castle, on the site and with the materials of which he began to erect an extensive and magnificent palace. The example of the king was followed by many of his nobility, who began to build splendid mansions, and Winchester once more exhibited signs of retrieving its distinction; but the death of the king, before the completion of these works, put an end to those flattering prospects. The palace, upon which considerable sums had been expended, was left unfinished; and after having been at various times used as a place of confinement for prisoners of war, and for various other purposes, was ultimately converted into barracks for the military. On the defeat of the Duke of Monmouth, in the reign of James II., Alice Lisle, widow of John Lisle, Esq., member for the city during the parliamentary war, and one of the judges who passed sentence of condemnation upon Charles I., was brought to trial in this city before the notorious Judge Jeffreys, on a charge of harbouring and concealing parties who were concerned in that rebellion: of this charge, though in opposition to the assertion of the jury that they were not satisfied with the evidence, she was pronounced guilty, and condemned to be burnt, but that sentence was changed into decapitation, which was accordingly carried into execution in 1685. Queen Anne, after her accession to the throne, paid a visit to the city, accompanied by Prince George of Denmark, on whom the royal palace of Charles II. had been settled at the time of his marriage, in the event of his surviving the queen, his consort.

The city is pleasantly situated on the eastern acclivity of an eminence rising gradually from the river Itchen, which is navigable to Southampton, and consists of one spacious regular street passing through the centre, and intersected at right angles by several smaller streets, extending in a parallel direction for about half a mile through the breadth of the city, which is nearly the same as its length: extensive hills, or downs, encircle it on the east and west. The principal parts of the city are within the limits of the ancient walls, which were of flint, strongly cemented with mortar, and defended by turrets at short intervals. The chief entrances from the suburbs were through four ancient gates, of which only the West gate is remaining, and, though it has undergone considerable alteration, still retains much of its ancient character: the other gates were removed by the commissioners appointed in 1770, by act of parliament, for the general improvement of the town. Over the Itchen, of which several branches intersect the town, is a handsome and substantial bridge of stone. At a small distance beyond the West gate an obelisk has been erected, on the spot where the people of the neighbouring country used to deposit their provisions, for the supply of the city during the time of the plague, the inhabitants leaving the stipulated sum for payment, to prevent any communication of the contagion. In the centre of the High-street is the city cross, 43 feet high, an elegant pyramidal structure, in the later style of English architecture, consisting of three successive stages, richly ornamented with open arches, canopied niches, and crocketed pinnacles, erected by the fraternity of the Holy Cross, instituted by Henry VI. In one of the niches of the second stage is a figure, supposed by some to be that of St. John the Evangelist, but, more probably, by others, to be that of St. Lawrence, to whom the church near the spot is dedicated: this beautiful relic owes its preservation to the spirited conduct of the inhabitants, who by force resisted an attempt, on the part of the commissioners of the pavements, to take it down, and drove away the workmen employed for that purpose. The houses are in general substantial and well built, and many of them possess an appearance of great antiquity: the city is paved, lighted with gas, and amply supplied with water of excellent quality. More houses are now in course of erection than for the last 50 years, and from its salubrity and situation it is recommended by the faculty to invalids. The London and South-Western railway passes a little to the west of the city, and will improve its commerce. A public subscription library has been established in High-street, within the last few years, which is supported by 100 proprietary members, whose shares are five guineas, and an annual subscription of one guinea and a half, and by annual subscribers of two guineas each. The theatre, in Gaol-street, a neat building handsomely fitted up, is occasionally opened by the Southampton company of comedians, but is very indifferently supported. Miscellaneous concerts and balls are held in St. John's rooms, in which also the general winter assemblies and subscription concerts usually take place. Hot, cold, vapour, and shower baths have been erected in High-street, for the use of the inhabitants. Races are held annually in July, on Worthy Down, about four miles from the city. On the site of the ancient castle is the unfinished palace of Charles II., now called the King's House, which, had it been completed according to the original design, would have been one of the most spacious and magnificent palaces in Europe: the front is 328 feet in length, and the principal story contained a splendid suite of state apartments: this building has been converted into an extensive and handsome range of barracks for the district, capable of containing 2000 men, with spacious grounds for exercise.

The trade of Winchester is very unimportant; it was formerly considerable for the manufacture of woollen caps, but at present there is only an extensive manufactory for sacking, and a very little business is carried on in wool-combing: the spinning of silk was introduced here a few years since, but the undertaking totally failed. A canal to Woodmill, about two miles above the Itchen Ferry, near Southampton, supplies the town with coal and the heavier articles of merchandise. The market days are Wednesday and Saturday, the latter for corn: the market-house, erected in 1772, is a handsome and commodious building, in every respect adapted to its use: the new corn-exchange affords excellent accommodation, and is a considerable ornament to the town. The fairs are on the first Monday in Lent, August 2nd, September 12th, and October 24th, for horses and pedlery; the first and last are held in the city, and the two others on the hills immediately adjoining; the September fair, which is held on St. Giles's hill, is a very large cheese fair.

Corporation Seal.

Winchester received its first regular charter of incorporation from Henry II., in 1184, 22 years before London was incorporated; and among other privileges conferred by that monarch, was the superintendence of the kitchen and laundry of the kings at the ceremony of their coronation. By this charter, confirmed and extended by succeeding sovereigns, and re-modelled by Queen Elizabeth, the government was vested in a mayor, high steward, recorder, two bailiffs, six aldermen, and 24 common-councilmen, assisted by a town-clerk, two coroners, four serjeants-at-mace, and subordinate officers. The mayor, recorder, and aldermen, were justices of the peace within the city and liberties. The corporation now consists of a mayor, six aldermen, and 18 councillors, under the act of the 5th and 6th of William IV., cap. 76, for an abstract of which see the Appendix, No. I. The borough is divided into three wards, and the municipal and parliamentary boundaries are co-extensive. The mayor, late mayor, and recorder, are justices of the peace, with six others appointed by commission. The police force consists of an inspector, sergeant, and seven men. The city first exercised the elective franchise in the 23rd of Edward I., since which time it has regularly returned two members to parliament. The right of election was formerly vested in the members of the corporation, who, by the privilege of electing freemen, might add to their number, which, having been augmented by the appointment of additional residents, then amounted to above 100, of whom the majority were non-resident; but the privilege has been extended to the £10 householders of an enlarged district, which has been constituted the new borough, comprising by estimation 715 acres, the limits of which are minutely described in the Appendix, No. II.: the old borough contained 692 acres. The number of electors at the last registration was 618, of whom 46 were freemen: the mayor is the returning officer. The recorder holds quarterly courts of session for all offences not capital, and a court of record is also held every quarter for the recovery of debts to any extent. Petty sessions are held twice a week. The Cheyney court, so called from its having been anciently held under an oak (*chêne*), which makes its origin revert to the time of the Druids, is an ancient court of the Bishops of Winchester, for the determining of actions, and the recovery of debts to any amount: its jurisdiction extends over all places which ever belonged to the see of Winchester, or the convent of St. Swithin, including 100 parishes, tythings, and hamlets, in the county of Southampton, some of which are 30 miles distant from the city: this court is held in the Cathedral Close weekly; the jury is selected from the liberty, or soke, of Winchester, and the judge is appointed by the bishop. The town-hall, a handsome structure, in the Grecian style of architecture, and of the Doric order, was rebuilt in 1713, on the site of a more ancient hall, which was erected on the foundation of a former, burnt down in 1112: the front is decorated with a well-executed statue in bronze of Queen Anne, which was presented to the corporation by George Brydges, Esq., who represented the city in seven successive parliaments. In the muniment-room, over the west gate of the city (formerly the armoury), are preserved the public records of the city, the original Winchester bushel, made by order of King Edgar, the standard yard of Henry, and the standard measures of succeeding sovereigns, with various other remains of antiquity. The city bridewell is now converted into a commodious police station, with a fire-engine house adjoining; and the city contracts with the magistrates of the county for the maintenance of prisoners.

The assizes and general quarter sessions for the county are held here, as the county town; and Winchester is, under the act of the 2nd and 3rd of Wm. IV., cap. 64, the place of election and a polling-place for the northern division of the county. The several courts are held in the chapel of the old castle, which has been converted into a county-hall, and appropriately fitted up for that purpose. This building is 110 feet in length: at the east end is suspended the celebrated round table, attributed to the renowned King Arthur, but which, with greater probability, is said to have been introduced by King Stephen, with a view to prevent disputes for precedence: it is made of oaken planks, and is eighteen feet in diameter, and ornamented with a figure of King Arthur, and the names of his knights, as collected from the romances of the times, in the costume and characters of the reign of Henry VIII.; in several parts it is perforated by bullets, probably by Cromwell's soldiers, while in possession of the city. An extensive common gaol for the county was erected in Gaol-street, in 1778, upon the principle recommended by the philanthropist, Howard: it comprises two yards for debtors, with two day-rooms on the poor side, and one day-room on the master's side; also five yards and five day-rooms for male felons, and two yards and two day-rooms for females: the prisoners not condemned to hard labour are employed in useful occupations and receive a portion of their earnings on their discharge: there are four separate infirmaries, a chapel, and other requisite offices. The county bridewell, in Hyde-street, was erected in 1786, and is a spacious structure, containing fourteen wards, six work-rooms, fourteen day-rooms, and eighteen airing-yards, in some of which are tread-wheels and capstans: the prisoners here are also employed in useful labour, and receive one fourth of their earnings on being discharged: the building, which is well adapted to the classification of the prisoners, comprises also an infirmary, and a chapel, in which divine service is performed twice on the Sunday, and morning prayer daily.

Arms of the Bishopric.

The origin of the diocese may be traced to the early part of the seventeenth century, when Cynegils, the first Christian king of the West Saxons, having been converted by St. Birinus, resolved to make his capital the seat of a bishopric, and began to collect materials for building a cathedral, which was afterwards accomplished by his son, Cenwahl, in 646. The establishment having been dispersed by the Danes, in 867, Secular priests were substituted the year following, who remained till 963, when Ethelwold, by command of King Edgar, expelled them, and supplied their place with monks of the Benedictine order from Abingdon: these kept possession of it without molestation, and it continued to flourish, enriched with royal donations and ample endowments, till the dissolution, at which time its revenue amounted to £1507. 17. 2. It was afterwards refounded by Henry VIII., for a bishop, dean, chancellor, twelve prebendaries, two archdeacons, six minor canons, ten lay clerks, eight choristers, and other officers. The cathedral, situated in an open space near the centre of the city, towards the south-east, and originally dedicated to St. Peter, St. Paul, and St. Swithin, was, upon the establishment of the present society by Henry VIII., dedicated to the Holy and Undivided Trinity. It is a spacious, massive, and splendid cruciform structure, chiefly in the Norman style of architecture, with a low tower rising from the centre, richly ornamented in its upper stages. The original building, as erected by Bishop Walkelyn, in 1079, was one of the most splendid and magnificent specimens of the Norman style in the kingdom: it was subsequently enlarged by Bishop Edington, and a considerable part was rebuilt by the celebrated William of Wykeham, who, adopting the later style of English architecture, which prevailed in his time, endeavoured to make the original style conform to that model. By this means the character of the architecture has been materially changed, though, from its extent and the loftiness of its proportions, it retains, notwithstanding the discrepancy of some parts, an air of stately grandeur, and displays many features of great beauty. The principal parts of the original structure are the transepts, in which the chief alteration is the insertion of windows in the later style; and the tower, which preserves its original character. The west front is an elegant composition, in the later style of English architecture, comprising three highly enriched porches of beautiful design. Some part of the eastern portion is in the finest character of the early English style, with occasional insertions of later date, particularly the clerestory windows of the choir; and in other parts of the building are various specimens of the early English at different periods, all remarkable for the excellence of their details. In some few instances are found small portions of the decorated, merging into the later English, of which, in various parts of the building, there are progressive series, from its commencement to the period of its utmost perfection. The interior, from the amplitude of its dimensions, and the loftiness of its elevation, is strikingly impressive : the nave is separated from the aisles by a long range of massive circular columns, twelve feet in diameter, and of proportionate height, which, in order to make them assimilate with the pointed arches that have been introduced within the circular Norman arches, have been cased with clustered pillars, and embellished with appropriate ornaments. In some of the intervals between the columns, which are two diameters in width, are various chantry and sepulchral chapels: the roof is elaborately groined, and richly ornamented with delicate tracery, embellished with the armorial bearings and devices of John of Gaunt, Cardinal Beaufort, and Bishops Waynfleet and Wykeham, which are continued along the fascia, under the arches of the triforium. The transepts, in which are several chapels and altars of exquisite beauty, are in the original Norman style of architecture; the central part is separated from the aisles by massive circular columns and arches, rising in successive series and with varied ornaments to the roof. The west aisle of the south transept has been partitioned off for a chapter-house, and at the extremity of the north transept is a beautiful Catherine-wheel window. At the eastern extremity of the nave, a flight of steps leads into the choir through a beautiful screen lately erected; and on each side of the entrance are niches containing ancient bronze statues of James I. and Charles I.

The choir, which includes the lower stage of the central tower, is in the early English style, with some insertions, and is lighted by a handsome range of clerestory windows in the later style: the original roof of the tower is concealed by an embellished ceiling, in the centre of which is an emblematical representation of the Trinity, with an inscription. The vaulting is supported by ribs springing from four busts of James I. and Charles I., in alternate succession, dressed in the costume of their times, above each of which is a motto, and, among various other ornaments, are the initials and devices of Charles I. and his queen, Henrietta Maria, with their profiles in medallions. The roof of the choir, from the tower to the east end, is richly groined, and embellished with a profusion of armorial bearings, devices, and other ornaments, exquisitely carved and richly painted and gilt; among them are the armorial bearings of the houses of Tudor and Lancaster, and those of the sees of Exeter, Bath and Wells, Durham, and Winchester, over which Bishop Fox, who superintended this work, successively presided. From the altar to the east window, the embellishments are emblematical of Scripture history; and among them are the instruments of the Crucifixion, and the faces of Pilate and his wife, and of the high priest and others; the whole of which embellishments have been judiciously renewed during the recent repairs of the edifice. The east window is of excellent proportion and design, and is embellished with remains of ancient stained glass of rich hue: the subjects are chiefly the Apostles and Prophets, and some of the bishops of the see, with appropriate symbols and legends: many of the figures were mutilated by the soldiery, when they defaced the cathedral, at which time also the painted glass generally was destroyed; the fragments that remain bear ample testimony to their original merit. The bishop's throne, prebendal stalls, and pulpit, are excellent specimens of tabernacle-work of appropriate character; the altar, in front of which is a beautiful tessellated pavement, is embellished with a painting, by West, of Christ raising Lazarus from the dead. Behind the altar, and separating it from the Lady chapel, is a finely-carved stone screen of beautiful design, elaborately enriched with canopied niches and other appropriate ornaments; the statues, which formerly filled the niches now vacant, were destroyed by Cromwell's soldiers. On each side of the altar, separating the presbytery from the aisles, are partitions of stone divided into compartments, and richly ornamented with arches, and with shields of armorial bearings and other devices: above the several compartments are placed six mortuary chests, richly carved and gilt, and surmounted by crowns, containing the bones of several of the Saxon kings and prelates, which were collected and deposited in them by Bishop Fox. In the south aisle of the choir is the sumptuous chapel, or chantry, of Bishop Fox, which, for its richness and minutely elaborate ornaments, is perhaps unequalled, either in the multiplicity of its parts, or in the fidelity of its details: in a niche under one of the arches is a recumbent figure of the founder, wrapped in a winding sheet, with the feet resting on a skull; the roof is finely groined, and embellished with the royal arms of the house of Tudor, richly emblazoned, and with the armorial bearings of the bishop, and the pelican, his favourite device. In the north aisle of the choir is the sepulchral chapel of Bishop Gardiner, an unsightly mixture of the later English and Grecian styles of architecture, and in a greatly dilapidated state. Behind the altar is a chapel, in which was kept the magnificent shrine of St. Swithin, the costly gift of King Edgar, said to have been of silver, richly gilt, and profusely ornamented with jewels. The Lady chapel, on each side of which is a smaller chapel, terminates the eastern extremity of the cathedral: it was built by Bishop de Lucy, and enlarged and beautified by Priors Hunton and Silkstede, whose initials and devices are worked into the groinings of the roof; the portrait of the latter, with his insigna of office, is still visible over the piscina, and on the walls are traces of paintings in fresco, representing subjects of scriptural, profane, and legendary history, now in a very imperfect state. The marriage of Queen Mary with Philip of Spain was solemnized in this chapel.

The magnificent chantry of Cardinal Beaufort, of Purbeck marble, is a highly-finished structure, in the later style of English architecture, and abounds with architectural beauty of the highest order, and with embellishments of the richest character; the roof, which is delicately groined, and enriched with fan tracery of elegant design, is supported on slender clustered columns of graceful proportion. It contains the tomb of the founder, on which is his effigy in a recumbent posture, in his robes as cardinal; around the cornice was an inscription in brass, which has been torn away by violence; and at the upper end of the chantry, enclosing the altar, are some beautiful canopied niches, crowned with crocketed pinnacles, from which the statues were taken by the parliamentarian soldiers. Bishop Waynfleet's chantry is in the same style, and of equal beauty with that of Cardinal Beaufort's, and, from the attention paid to it by the trustees of his foundation at Magdalene College, is kept in good repair; it contains the tomb of the bishop, on which is his effigy in his pontificals, in the attitude of prayer. There are various other chapels in this spacious and extensive pile, among which are, that of Bishop Langton, containing some fine carvings in oak, his tomb stripped of all its ornaments; and that of Bishop Orleton, of whom no memorial is preserved in the chapel; the roof is vaulted, and profusely ornamented with the figures of angels: on the north side is the tomb of Bishop Mews, a distinguished adherent to the cause of Charles I., who, after having served as an officer in the royal army, entered into holy orders, and was promoted to the see of Winchester: in this chapel is also a monument to the memory of Richard Weston, Earl of Portland, and lord high treasurer in the reign of Charles I. Underneath the high altar, and formerly accessible by a stone staircase leading from that part of the cathedral called the "Holy Hole," as being the depository of the remains of saints, are vestiges of the ancient Norman crypt built by Ethelwold; the walls, pillars, and groining are in their original state, and remarkable for the boldness and simplicity of their style; a new crypt, in the later style, has been built underneath the eastern end of the Lady chapel. Among the monuments, in addition to those in the sepulchral chapels, is the tomb of William Rufus, in the centre of the choir, of grey marble, raised about two feet above the surface of the pavement; his bones had been removed into one of the mortuary chests prior to the parliamentary war, during which the tomb was re-opened, and among the remaining ashes were found a large gold ring, a small silver chalice, and some pieces of cloth embroidered with gold: there are also the tombs of Hardicanute; Earl Beorn, son of Ertrith, sister of Canute; Richard, second son of William the Conqueror; Bishops Peter de Rupibus, Henry de Blois, Hoadly, Willis, and other distinguished prelates; Sir John Clobery, who assisted General Monk in planning the restoration of Charles II.; Sir Isaac Townsend, knight of the garter; the Earl of Banbury; Dr. Joseph Warton; Izaak Walton, and numerous other illustrious and distinguished persons. The ancient font, of black marble, supported on pillars of the same material, is of square form, and has the faces rudely sculptured with designs emblematical of the diffusion of the Holy Spirit, and with subjects from the legendary history of St. Nicholas; it is supposed to be of the time of Bishop Walkelyn. The whole length of this magnificent cathedral is 545 feet, from east to west, and the breadth along the transepts, 186; the mean breadth of the nave is 87, and that of the choir 40; the height of the tower is 140 feet, and its sides are 50 feet broad. The great cloisters, which enclosed a quadrangular area 180 feet in length, and 174 in breadth, were destroyed in the reign of Queen Elizabeth. On the east side of the quadrangle is a dark passage, which led to the infirmary and other offices belonging to the ancient monastery; and to the south of it is a doorway, that formerly led to the chapter-house, the site of which is now occupied by the Dean's garden, in the walls of which are some of the pillars and arches still remaining. The refectory is now divided into two stories; under it are two kitchens, the roofs of which are vaulted in the Norman style, and supported on a single central column, still remaining. The Prior's hall and some other apartments now form the deanery, and other remains of the conventual buildings may be traced in the gardens of the prebendal houses, which occupy what is termed the Cathedral Close, an extra-parochial district.

Winchester comprises the parishes of *St. Bartholomew*, which is partly in the Soke liberty; *St. Lawrence* the mother church; *St. Mary Kalendar; St. Maurice; St. Peter Colebrook;* and *St. Thomas*, within the city: and the parishes of *St. Faith, St. John, St. Michael, St. Peter Cheesehill, St. Martin Winnall*, and *St. Swithin*, within the Soke liberty. The living of *St. Bartholomew's* is a discharged vicarage, valued in the king's books at £ 0; present net income, £82: it is in the patronage of the Crown. The church, in Hyde-street, which is well adapted to the accommodation of all the parishioners, is not entitled to particular architectural notice. The living of *St. Lawrence's* is a discharged rectory, valued in the king's books at £6. 5.; present net income, £56: it is in the patronage of the Crown. The church, situated in the square, is an ancient structure with a lofty square tower, and consists only of one large aisle into which, on taking possession of his see, the bishop makes a solemn entry. The living of *St. Mary Kalendar's* is a rectory, united to that of St. Maurice, and valued in the king's books at £7. The church has been destroyed. The living of *St. Maurice's* is a rectory, to which those of St. Mary Kalendar, St. Peter Colebrook, St. George and St. Mary Wood, are united, valued in the king's books at £6. 7. 6.; present net income together, £145; patron, the Bishop. The church, in High-street, which was formerly the chapel of an ancient priory, is about to be rebuilt; the Incorporated Society having granted £400 in aid of the expense, for which 480 free sittings will be provided. The living of *St. Peter's Colebrook* is a rectory, united to that of St. Maurice, and valued in the king's books at £3. 4. 2. The church has been destroyed, as also have those of St. George and St. Mary Wood, the livings of which are valued in the king's books, the former at £3. 5. 8., and the latter at £2. The living of *St. Thomas'* is a discharged rectory, with that of St. Clement united, valued in the king's books at £13. 17. 8½.; net income, £145; patron, the Bishop. The church contains 300 free sittings, the Incorporated Society having granted £300 in aid of the expense: it is an ancient structure, in the Norman style, with a low tower: the interior consists of a nave and one aisle, separated by massive circular columns. The church of St. Clement's has been demolished. The living of *St. Faith's* is a sinecure rectory, annexed to the mastership of the hospital of St. Cross, which is extra-parochial, and in the chapel of which the parishioners attend divine service, the church of St. Faith having been demolished for more than two centuries. The living of *St. John's* is a perpetual curacy, with the rectory of St. Peter's Southgate united; net income, £82; patron, the Bishop. The church is an ancient structure, in the Norman style of architecture, with a massive tower and turret: the interior consists of a nave and two aisles, separated by massive circular columns. The church of St. Peter's Southgate has been destroyed. The living of *St. Michael's* is a discharged rectory, valued in the king's books at £5. 17. 11.; present net income, £104; patron, the Bishop. The church, which, with the exception of the ancient tower, has been rebuilt, is a handsome edifice, in the later style of English architecture, and consists of a spacious nave and chancel: it contains 620 free sittings, the Incorporated Society having granted

£350 in aid of the expense. The living of *St. Peter's Cheesehill* is a discharged rectory, valued in the king's books at £14. 9. $9\frac{1}{2}$.; present net income, £94: it is in the patronage of the Crown. The church is a neat plain structure, with a tower. The living of *St. Swithin's* is a discharged rectory, valued in the king's books at £6. 6. $10\frac{1}{2}$.; present net income, £65: it is in the patronage of the Crown. The church is over a postern called King's Gate, which is ascended by a staircase of stone, and formerly used as the church for the servants employed in the great priory of St. Swithin. The living of *St. Martin's Winnall* is a rectory, valued in the king's books at £5; present net income, £170; patron, the Bishop. The church was rebuilt in 1786, and consists of one aisle and a small tower. There are places of worship for Baptists, Independents, and Wesleyan Methodists. An elegant Roman Catholic chapel, in the later style of English architecture, dedicated to St. Peter, was erected in 1792, in St. Peter-street: the exterior is richly ornamented with canopies supported on corbels of antique heads of sovereigns and bishops, which crown the lofty windows of elegant tracery, and a frieze embellished with devices illustrative of the history of Winchester, surmounted by a parapet pierced in quatrefoils, and enriched with crocketed pinnacles rising from panelled buttresses, and terminating with crosses richly gilt: the interior is splendidly decorated; the windows of ground glass are painted in quatrefoil, and embellished with paintings of the principal saints and kings connected with the city; and on the north side, in which there are no windows, are corresponding panels painted in chiaro oscuro, with subjects from Scripture history: the altarpiece is ornamented with a good painting of the Transfiguration, by Mr. Cave, sen., from the original by Raphael. At the entrance of the walk leading to the chapel is an ancient Norman portal, which was removed from the church of St. Mary Magdalene's hospital. Nearly opposite is the convent, a large and handsome brick edifice, called the Bishop's House, consisting of Benedictine nuns removed from Brussels.

Winchester college holds a pre-eminent rank among the public literary institutions of the kingdom, and from a very early period has been distinguished as a seat of preparatory instruction. A grammar school had been established prior to the commencement of the twelfth century, on the site of which, in 1387, Bishop Wykeham, who received his early education in it, erected the present magnificent college, which he amply endowed for a warden, ten Secular priests, who are perpetual fellows, three priests' chaplains, three clerks, sixteen choristers, a first and second master, and 70 scholars, intending it as a preparatory seminary for his foundation of New College, Oxford, which he had completed the year before. For the government of the college, the bishop drew up a code of statutes, which, from their judicious adaptation to the purposes of the institution, were adopted by Henry VI. for the regulation of the establishments founded by that monarch at Eton and Cambridge. Under the influence of those salutary regulations, the college continued to flourish till the time of the dissolution, when its revenue amounted to £639. 8. 7.; but its reputation was held in such estimation, that it obtained a special exemption from the operation of that general measure. The buildings, which were completed in 1393, occupy two spacious quadrangles, the entrance into the outermost of which is through a noble turreted gateway, under a finely-pointed arch, surmounted by a canopy resting on the bust of a king on the one side, and of a bishop on the other, probably representing the founder and his royal patron. In the groining underneath the tower are the arms of the founder, and on the face of it, over the entrance, is a canopied niche, in which is a crowned statue of the Virgin, holding in the right hand a sceptre, and on the left arm a figure of the infant Jesus. On the opposite side of this quadrangle is a gateway leading into the second court, above which is a tower ornamented in front with three beautiful niches, enriched with canopies and crocketed pinnacles; in the central niche is a statue of the Virgin, with a book in the left hand, the right raised towards a figure of the angel Gabriel, which occupies the niche on that side; and in the niche on the left hand is the statue of the founder, in his episcopal robes, crowned with a mitre. The buildings surrounding the inner quadrangle are principally in the later style of English architecture, of which they exhibit an elegant specimen. The grand hall and the chapel occupy the south side; the former is lighted by a handsome range of well-proportioned windows enriched with elegant tracery; the roof, which is of timber, is finely arched, and the beams, which are handsomely ornamented, are supported by ribs springing from corbels decorated with coloured busts of kings and bishops: this noble room is ascended by a flight of steps at the south-west angle of the quadrangle, and at the western extremity, under an enriched canopy, is a figure of St. Michael piercing the dragon. In the centre of this side is the stately tower of the chapel, surmounted with turrets and crowned with pinnacles, the work of a later period than the building by Wykeham, and said to have been erected by the Warden Thurbern. The entrance into the chapel is by a vestibule, the ceiling of which is elaborately enriched, and in which are placed the ancient stalls, removed from the chapel, in 1681, by Dr. Nicholas, and some ancient brasses. The interior is beautifully arranged; the roof is finely groined; the windows are enriched with elegant tracery, ornamented with paintings of kings, saints, prelates, and nuns; and in the great east window is a representation of the Genealogy of Christ, the Crucifixion, and the Resurrection; the altar is embellished with a painting of the Salutation, by De Moine, presented to the College by the late head-master, Dr. Burton. Between the stairs leading to the hall and the entrance to the chapel is a passage conducting to the school and play-ground. The school-room is a plain brick building, erected in 1687, at an expense of £2600: over the entrance is a statue of Bishop Wykeham, finely executed in bronze, and presented to the College by Mr. C. G. Cibber, which has been injudiciously painted and gilt; and at the other extremity of the school-room are the statues for the government of the students, written in Latin. To the south of the chapel are the cloisters, enclosing a quadrangular area 132 feet square, and apparently of the date of the 15th century: they contain many ancient brasses, and in the centre of the enclosed area is a chantry chapel, erected by Mr. John Fromond, a liberal benefactor to the foundations of Bishop Wykeham. This building, the ceiling of which is strongly vaulted, is now appropriated as the college library, and contains a select and valuable collection of works, and a small museum of natural curiosities. The other sides of the quadrangle are composed of the houses and apartments of the warden, fellows, the head and second masters, and other members of the establishment; and contiguous to the college is a spacious quadrangular building, for the residence of gentlemen commoners not on the foundation, of whom the number is very considerable: the college, chapel, and school, were completely repaired in 1795. An annual visitation is held in July, by the warden and two of the fellows of New College, Oxford, at which there is an examination of the candidates for the vacant fellowships in that college, which, by the will of the founder, are to be supplied from this establishment, and of which there are, on an average, three in the year. At that time also is held, by the same persons, with the addition of the warden, subwarden, and head-master, an election of boys for admission on the foundation of Winchester college: the qualification for candidates by the statutes is, that they must be "*pauperes et indigentes scholares*;" boys are not eligible till above eight years of age. There are several scholarships and exhibitions for such as fail in obtaining fellowships in New College; and there is also a superannuated fund belonging to the establishment, founded by Dr. Cobden, Archdeacon of London, in 1784. In this noble institution many eminent prelates and literary characters have received their early education: among these were, Sir Thomas Brown, Sir Thomas Wooton, Sir Thomas Ryves; the poets Otway, Phillips, Young, Somerville, Pitt, Collins, Warton, and Hayley; and others distinguished for their genius and literary acquirements.

Christ's hospital was founded in 1586, by Peter Symonds, Esq., who endowed it with lands producing more than £420 per annum, for the support of six unmarried men above 50 years of age, who reside in the hospital, and are supplied with clothing and food; and for the clothing, maintenance, and education of four poor boys, from seven till fourteen years of age. There are two exhibitions, of £10 per annum each, tenable for four years, to Oxford and Cambridge, and with such as do not obtain them an apprentice-fee of £30 is given, on their leaving the hospital. Another school for 10 girls, who are educated and clothed, is endowed by Mrs. Imber. A charity school for clothing, instructing, and apprenticing 30 boys, is supported by subscription: there is a Central National school in Colebrook-street; and there are Sunday schools in connection with the established church and the several dissenting congregations. A diocesan board of education, in connection with the National Society, has been recently formed.

The hospital of St. Cross, about a mile south of the city, and beautifully situated on the bank of the river Itchen, was founded in 1132, by Bishop Henry de Blois, brother of King Stephen, who endowed it for the residence and maintenance of a master, steward, four chaplains, thirteen clerks, seven choristers, and thirteen poor brethren, and for the daily entertainment of 100 of the most indigent men of the city, who dined together in a common hall, called the "hundred menne's hall." Bishop Wykeham, on his appointment to the see of Winchester, in 1366, finding that the revenue of the hospital was misapplied, succeeded, after a tedious litigation, in re-establishing the institution according to the intention of the founder, and placed it on so secure a basis, that Henry de Beaufort, wishing to found some permanent charity, preferred an augmentation of the original endowment to the foundation of a new institution, and added two priests, increased the number of poor men to 35, appointed three sisters to attend upon them when ill, and greatly enlarged the buildings. At the suppression of monasteries its revenue was valued at £184. 4. 2.: it was exempted from dissolution, but suffered materially during the war in the reign of Charles I.: the present establishment consists of a master, chaplain, steward, and thirteen brethren. The buildings occupied two quadrangular areas, but the south side of the inner quadrangle has been taken down. The entrance gateway, erected by Cardinal Beaufort, is a good specimen of the later style of English architecture, surmounted by a lofty tower, the front of which is ornamented with three handsome niches, one of them containing the figure of the Cardinal in a kneeling posture, and on the cornice above the arch are busts of John of Gaunt, Henry IV. and V., and of his predecessor, Bishop Wykeham, with other devices; and in the spandrils, on each side of the arch, are the arms of the founder. In the inner court is the church, or chapel, of St. Cross, in which the parishioners of St. Faith's attend divine service: it is an ancient and interesting cruciform structure, comprising a series of styles, passing, by gradual and almost imperceptible transitions, from the Norman to the early and decorated styles of English architecture. The low tower rising from the centre is in the Norman style: the nave is separated from the aisles by ranges of pillars, of which some are of the massive circular character, and others clustered in the style of the early English, with pointed arches, which, towards the west end, merge into the decorated style: most of the arches of the chancel are pointed, and the windows generally towards the east end are Norman, with circular arches and zigzag ornaments: the groining of the roof towards the east is replete with ornaments of the Norman style; and that of the western part, which appears to have been the work of Cardinal Beaufort, is embellished with shields of the armorial bearings of the Cardinal, Bishop Wykeham, and of the College. The west front is an elegant composition in the early English style, with appropriate embellishments; and the west window, of five lights, is richly ornamented with painted glass, representing the figures of various saints, and emblazoned with armorial devices; over the stalls in the choir are sculptured figures of the most conspicuous subjects of scripture history. Among the funeral monuments are, an ancient brass in memory of John de Campden, the friend of Wykeham, and a modern mural tablet to Wolfran Cornwall, Esq., Speaker of the House of Commons. The living is a perpetual curacy, with the rectory of St. Faith's annexed, endowed with £200 private benefaction, and £200 royal bounty, and in the patronage of the Bishop of Winchester. The remaining buildings of the hospital comprise the apartments of the brethren, each of whom has three chambers for his own use, with a separate garden; the refectory; and the master's apartments, which are spacious and commodious. On the east side of the quadrangle, extending from the north transept of the church, is the ancient ambulatory, an open portico 135 feet long, above which are the infirmary and the nuns' chambers, so called from their having been appropriated to the use of the three sisters placed on the foundation, by Cardinal Beaufort, to attend the brethren when unwell.

The county hospital, or infirmary, in Parchment-street, the first institution of the kind established in the kingdom, was founded in 1736. The buildings, ascended by a fine flight of steps, comprise a centre and two wings, one having been recently added at the northern end, and are in every respect well adapted to the purposes of the institution. St. John's hospital, now called St. John's House, in High-street, is a very ancient establishment, said to have been founded in the year 933, by St. Brinstan, Bishop of Winchester, and to have become the property of the Knights Templars, upon the suppression of which order it was refounded, by permission of Edward II., for sick and lame soldiers, pilgrims, and necessitous wayfaring men, who had their lodging and other necessaries for one night, or longer, in proportion to their wants. After the dissolution, the site and remains were given to the corporation, who converted the great hall into a public room, in which meetings of the corporation, and public assemblies and concerts, are held. The hall is elegantly fitted up, chiefly by a donation of £800 from the late Colonel Bridges: it is embellished with a portrait of that gentleman, and a full length portrait of Charles II., in his robes of state, painted by Sir Peter Lely, and presented to the corporation by that monarch: in an adjoining room, called the council-chamber, are the city tables, recording a brief chronological account of its principal historical events. The ancient chapel of this hospital is now used as a school-room, in which 24 boys are instructed by a master, who is paid £22 annually by the trustees of Mr. William Over, who, in 1701, bequeathed an estate for that purpose. In an inner court of the northern part of the hospital are the almshouses, founded in 1558, by Ralph Lamb, Esq., who endowed them for the support of six poor widows of the city: the funds of this hospital were, a few years since, greatly augmented by the successful issue of a suit in Chancery directed to their investigation. Near the cathedral are almshouses, founded in 1672 by Bishop Morley, for the residence and support of ten clergymen's widows. There are various funds for charitable uses at the disposal of the corporation, among which is Sir Thomas White's charity, for loans without interest to young tradesmen; also divers sums for distribution among the poor, and numerous other charitable bequests. The poor law union of New Winchester comprises 33 parishes or places, under the care of 37 guardians, and contains a population of 17,062, according to the census of 1831.

Among the ancient monastic institutions, in addition to those already described, was Hyde abbey, originally the new minster founded by Alfred the Great, adjoining the site of the present cathedral, which, by way of distinction, was thence called the Old Minster. The foundation, after the death of Alfred, was completed by his son, Edward the Confessor, and placed under the superintendence of St. Grimbald, who established a fraternity of canons regular, who were afterwards expelled by Bishop Ethelwold, and replaced by monks of the order of St. Benedict. Alwyn, the eighth abbot in succession from St. Grimbald, was uncle of Harold, and, with twelve of his monks, assisted that monarch at the battle of Hastings, in which he was slain with his brethren. In resentment of this, William the Conqueror, treated the New Minster with the utmost rigour, seized upon its revenue, and would not allow a new abbot to be appointed. About three years after, he, however, permitted an abbot to be chosen, and restored some of the abbey lands, giving others in exchange for the remainder. The nuisances which had arisen, from the stagnation of the stream of water brought, in its immediate vicinity, to supply the fosse which had been dug round the castle erected by the Conqueror, and from its contiguity to the Old Minster, induced the fraternity to build a new abbey at a greater distance, and the present edifice was erected on a spot near the north wall of the city, called Hyde meadow, from which it took its name. Into this the remains of Alfred, his queen Alswitha, his sons Ethelred and Edward the Elder, Elfleda, Ethelhida, and king Edwy, were removed and re-interred. In the contest between Stephen and Matilda the abbey was burnt to the ground by the fire balls thrown from Wolvesey castle, but was rebuilt, with greater magnificence, in the reign of Henry II., and the abbot was invested with the privilege of a seat in parliament. It continued to flourish till the dissolution, at which time its revenue was £865. 1. 6.: it was soon after demolished, and very small portions of the monastic buildings are at present remaining, among which are, the tower of St. Bartholomew's church, some of the offices, and part of a large barn, with one gateway containing a regal head in the groining of the arch. On the site of the abbey a new bridewell has been erected, in digging the foundations of which many stone coffins, chalices, patins, rings, busts, capitals of ancient columns, and other fragments of sculpture, were found. Of these, the most interesting is a stone inscribed "Alfred Rex, 881," in Saxon characters, now in the possession of H. Howard, Esq., of Corby Castle, in the county of Cumberland. The abbey of St. Mary was founded by Alswitha, wife of Alfred, and, after the king's death, was the place of her retirement. Edburga, daughter of Edward the Elder, became abbess of this convent, which, in the reign of Edgar, was amply endowed by Bishop Ethelwold, who prescribed for the observance of the nuns the more severe rules of the order of St. Benedict. Many Saxon ladies of royal and noble lineage were sisters in this establishment, in which Matilda, wife of Henry I., received her education. The original buildings were destroyed in the war during the reign of Stephen, and restored in the following reign by Henry II., who was a liberal benefactor. At the time of the dissolution, its revenue was £179. 7. 2.; it remained for a few years after that period, when its abbess and eight of the nuns received small pensions, and the rest of the inmates were dispossessed: the only visible remains are in a large modern mansion, which has been partly built with the materials of the abbey. In the meadow of St. Stephen, near the Bishop's palace of Wolvesey, was a college, founded in 1300 by Bishop Pontoys, which he dedicated to St. Elizabeth, a daughter of the King of Hungary, and endowed for a provost, six chaplains, priests, six clerks, and six choristers: its revenue at the dissolution was £112. 17. 4. A monastery, dedicated to St. James, was founded in the abbey churchyard by John, or Roger, Inkpenne, who, in 1318, endowed it for a warden and several priests. In the church of St. Maurice was the fraternity of St. Peter; and in that of St. Mary Kalendar, a college, the revenue of which was granted to the corporation, in the reign of Philip and Mary. The hospital of St. Mary Magdalen was an ancient building, situated on Magdalen hill, and supposed to have been erected and endowed by one of the Bishops of Winchester about the close of the 12th century: but in 1665, the king ordered the inmates to be removed to the city of Winchester, and the ancient hospital buildings, being in a state of ruin, were taken down, and six tenements with three rooms each were built in St. John's parish, in the East Soke. The institution consists of a warden and four brothers and four sisters: the annual income of the charity amounts on an average to about £154. There were also convents of Augustine, Carmelite, Dominican, and Franciscan friars, the sites of which were, after the dissolution, granted to the college. Among the illustrious and eminent natives of this city were, Henry III.; Eleanor, youngest daughter of Edward I., who died in her infancy; and Prince Arthur, eldest son of Henry VIII., who died at Ludlow, and was buried in the cathedral church of Worcester. Winchester gives the title of Marquess to the distinguished family of Paulet.

WINCHFIELD (*St. Mary*), a parish, in the union of Hartley-Wintney, hundred of Odiham, Odiham and Northern Divisions of the county of Southampton, 2½ miles (N. E.) from Odiham; containing 277 inhabitants. The living is a rectory, valued in the king's books at £8. 16. 10½.; present net income, £247; patron, Rev. H. E. St. John. The Basingstoke canal passes through the parish; and the Winchfield and Hartley Row station of the London and South-Western railway is within its limits.

WINCHMORE-HILL, a chapelry, in the parish and hundred of Edmonton, county of Middlesex, 8 miles (N.) from London: the population is returned with the parish. The chapel, dedicated to St. Paul, and consecrated in 1828, was erected at an expense of about £5000, of which sum, £3843. 6. 3. was granted by the parliamentary commissioners, the remainder having been raised by subscription. There are places of worship for the Society of Friends, Independents, and Wesleyan Methodists; also a National school. Dr. Fothergill, an eminent physician, and a member of the Society of Friends, was buried here.

WINCLE, a chapelry, in the parish of Prestbury, union and hundred of Macclesfield, Northern Division of the county of Chester, 5½ miles (S. E. by S.) from Macclesfield; containing 453 inhabitants. The living is a perpetual curacy; net income, £116; patron, Vicar of Prestbury. The chapel was erected about 1642.

WINDER, a township, in the parish of Lamplugh, Allerdale ward above Derwent, Western Division of the county of Cumberland, 5¼ miles (E. by S.) from Whitehaven: the population is returned with the parish.

WINDER, LOW, a township, in the parish of Barton, West ward and union, county of Westmorland, 5¼ miles (S. by W.) from Penrith; containing 19 inhabitants.

WINDERMERE (*St. Martin*), a parish, in Kendal ward and union, county of Westmorland, 9 miles (W. N. W.) from Kendal; containing 1632 inhabitants. This parish derives its name from the beautiful lake, anciently called Wynandermere, which is twelve miles in length, by about one in breadth, and is in depth 40 fathoms. It is studded with many picturesque islands, the principal of which, Belle Isle, the property of Mr. Curwen, is richly wooded, and adorned with an elegant circular mansion in the Italian style. In the centre of it formerly stood Holm House, which was besieged for the parliament by Col. Briggs, who, on the siege of Carlisle being raised, was obliged to abandon it. On Lady Holm, a smaller island in this lake, of which the whole belongs to Applethwaite township, formerly stood a chapel, dedicated to the Virgin Mary. Great quantities of char, taken in this lake during the winter months, are potted and sent to different parts of the kingdom. On its margin are several good inns for the accommodation of visiters, at three of which, regattas are held annually about the beginning of September, and yacht races take place occasionally: these are attended by families of distinction, and terminate with balls, exhibitions of fireworks, &c. For a further account of this lake, see the article on the county of Westmoreland. The living is a rectory, valued in the king's books at £24. 6. 8.; present net income, £253: it is in the patronage of the family of Lady Le Fleming. The church stands in the village of Bowness: the east window, which is of stained glass, formerly belonged to Furness abbey, at the dissolution of which it was purchased by the parishioners and placed here.

WINDFORD (*St. Mary and St. Peter*), a parish, in the union of Bedminster, hundred of Hartcliffe with Bedminster, Eastern Division of the county of Somerset, 6½ miles (S. W. by S.) from Bristol; containing 865 inhabitants. The living is a rectory, valued in the king's books at £21. 12. 11.; present net income, £526; patrons, Provost and Fellows of Worcester College, Oxford. Here are two schools; one is endowed for teaching six children, and the other for four.

WINDLE, a township, in the parish and union of Prescot, hundred of West Derby, Southern Division of the county palatine of Lancaster, 1 mile (N.) from St. Helens; containing 5825 inhabitants. A school is endowed with £545 invested in turnpike-road trusts; two others are supported by the trustees of Sarah Cowley's charity; and another, for the children of Roman Catholics, is supported by an annual donation of £30.

WINDLESHAM (*St. John the Baptist*), a parish, in the union of Chertsey, First Division of the hundred of Wokeing, Western Division of the county of Surrey; containing, with the hamlet of Bagshot, 1912 inhabitants. The living is a rectory, valued in the king's books at £10. 9. 7.; present net income, £404: it is in the patronage of the Crown. The church is in the early English style, and has been recently enlarged; it contains 175 free sittings, the Incorporated Society having granted £100 in aid of the expense. There is a chapel of ease at Bagshot, also a small place of worship for Baptists. A National school is endowed with £175 three per cents., bequeathed by the Rev. Edward Cooper, late rector. Hool Mill, in this parish, erected by an abbot of Chertsey, in the reign of Edward III., is subject to a permanent rent-charge of £8 in support of the poor; and almshouses for six poor men and women were erected by James Butler, Esq. The parish also participates in Henry Smith's charity. There are numerous chalybeate springs.

WINDLESTON, a township, in the parish of St. Andrew, Auckland, union of Auckland, South-Eastern Division of Darlington ward, Southern Division of the county palatine of Durham, 4 miles (E. S. E.) from Bishop-Auckland; containing 201 inhabitants. Sir Robert Eden built a school and endowed it with £15 per annum, at Rushyforth, in this township; it is in connection with the National Society.

WINDLEY, a township, in the parish of Duffield, union of Belper, hundred of Appletree, Southern Division of the county of Derby, 6¾ miles (N. N. W.) from Derby; containing 204 inhabitants.

WINDRIDGE, a ward, in the parish of St. Stephen, hundred of Cashio, or liberty of St. Albans, county of Hertford, 1¾ mile (W. S. W.) from St. Albans: the population is returned with the parish.

WINDRUSH (*St. Peter*), a parish, in the union of Northleach, Lower Division of the hundred of Slaughter, Eastern Division of the county of Gloucester, 5¼ miles (E.) from Northleach; containing 291 inhabitants. The living is a discharged vicarage, united to that of Sherborne in 1776, and valued in the king's books at £5. A day and two Sunday schools are chiefly supported by Lord and Lady Sherborne.

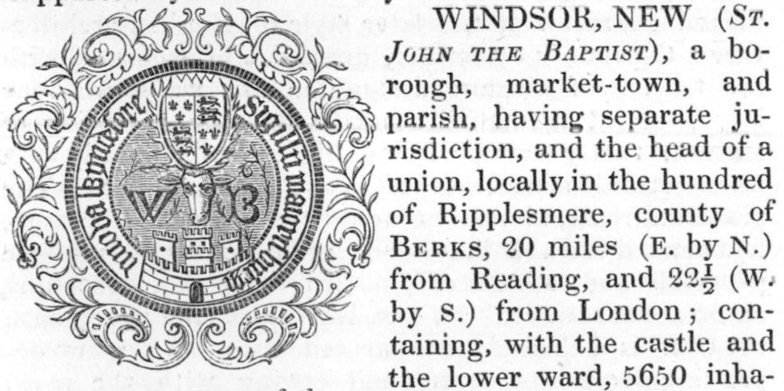

Seal and Arms of the Borough.

WINDSOR, NEW (*St. John the Baptist*), a borough, market-town, and parish, having separate jurisdiction, and the head of a union, locally in the hundred of Ripplesmere, county of Berks, 20 miles (E. by N.) from Reading, and 22½ (W. by S.) from London; containing, with the castle and the lower ward, 5650 inhabitants. This place owes its origin to a more ancient town, about two miles distant, called by the Saxons, from the winding course of the river Thames, *Windleshora*, of which its present name, Windsor, is an abbreviation. The first authentic notice of that town, which had been the residence of the Saxon kings, occurs in an ancient charter of Edward the Confessor, granting it, with all its appendages, to the monks of Westminster, in whose possession it remained till the Conquest. William, soon after his establishment on the throne, struck with the beauty of its situation on the bank of the Thames, and the peculiar adaptation of the surrounding country to the pleasures of the chace, procured it from the monastery of Westminster, in exchange for some lands in the county of Essex, and made it his occasional residence while pursuing the diversion of hunting. On a hill in the neighbourhood that monarch erected a fortress, where he held his court in 1070; and, two years afterwards, assembled there a synod of the nobility and prelates, in which the question of precedency between the sees of Canterbury and York was discussed, and decided in favour of the former. Around this fortress he laid out extensive parks, enlarged the boundaries of the neighbouring forest, and enacted severe laws for the preservation of the game. Old Windsor continued to be the royal residence of William and his successors till 1110, when Henry I., having partly rebuilt and considerably enlarged the fortress which his father had erected, by the addition of a suite of apartments, converted it into a palace, in which he occasionally resided and kept his court. From this time the importance of the ancient town began to decline, and subsequently a new town arose in the immediate vicinity of the castle, which was distinguished by the appellation of New Windsor. In the treaty of peace between Stephen and Matilda, the castle is referred to by the name of "Mota de Windsor;" and after the death of Stephen, Henry II. held a council here, in 1170. When Richard I. embarked on his expedition to the Holy Land, the castle became the residence of the Bishop of Durham, to whom, in conjunction with the Bishop of Ely, that monarch had entrusted the administration of the government in his absence. King John, during his contest with the barons, resided in the castle, which was at that time considered the next strongest fortress after the Tower of London: it was ineffectually besieged by the barons, to whom, in the succeeding reign, it was ceded by treaty; but in the following year it was surprised and taken by the king, who made Windsor the principal rendezvous of his forces. Henry III. erected a barbican, and strengthened the fortifications and outworks of the castle, which, during the baronial wars in that monarch's reign, was alternately taken and retaken by the contending parties, till Prince Edward finally obtained possession and held it for his father. On the succession of that prince to the throne, the castle was frequently the place of his residence, and four of his children were born at Windsor, which was likewise the favourite retreat of his queen Eleanor. Edward III., who was also born here, rebuilt the royal palace on a more extensive and magnificent scale, enlarged the castle with additional towers, erected the keep, and, near it, a tower of high elevation, named Winchester tower, after William of Wykeham, Bishop of Winchester, whom that monarch had made superintendent of his buildings. The same sovereign erected the collegiate chapel of St. George, in which he established a dean and twelve canons; and the magnificent hall of St. George, as a banqueting-house for the knights of the royal order of the garter, of which he was the founder; and surrounded the whole with a strong wall and rampart, faced with stone, and encompassed with a moat. While this monarch occupied the throne, two sovereigns were prisoners in the castle at the same time, *viz.*, John King of France, and David King of Scotland, the latter of whom he captured, after the reduction of that country. Edward IV. enlarged, and partly rebuilt, the collegiate chapel, the choir of which was vaulted by Henry VII., who also erected the lofty pile of building adjoining the state apartments in the upper ward. Henry VIII. added materially to the buildings by the erection of the prebendal houses and the gateway leading into the lower ward. Edward VI. and Queen Mary both made Windsor their residence; and, among other improvements, constructed a fountain in the centre of the upper quadrangle, from which the whole of the castle was supplied with water. Queen Elizabeth, after her accession to the throne, resided occasionally in the palace, to which she added some buildings adjoining the Norman gateway, and that part adjoining the buildings of Henry VII., which is called Queen Elizabeth's gallery, and raised the noble terrace on the north side of the castle, commanding a beautiful view of Eton College, and an extensive prospect over the vale of the Thames.

During the parliamentary war, the castle, which had received several additions in the reign of Charles I., was seized and garrisoned by the parliament, who, notwithstanding an attack of Prince Rupert, in 1642, to regain possession of it for the king, retained it in their hands till the conclusion of the war. After the Restoration, Charles II. repaired the injuries it had suffered, and greatly embellished the interior; and James II. and William III. ornamented the state apartments with a splendid collection of paintings. In almost every succeeding reign this interesting structure continued to receive additional embellishment; and, in the reign of George III., the alterations and additions were conducted on a larger scale, and with a stricter regard to the restoration and preservation of the original character of the building, than in that of any of his predecessors since the time of Edward III. In the reign of George IV., the varied attractions of Windsor induced that monarch to make it his principal residence: and, under the influence of a correct and refined taste, which duly appreciated the merits and the beauty of the ancient style of English architecture, a design was formed for the enlargement and decoration of the castle, of which a considerable part was accomplished under his immediate superintendence. For carrying this design into effect, divers sums, amounting to £771,000, were granted by parliament for the buildings alone, and, among the various plans which had been submitted for that purpose, the design of Mr. Jeffrey Wyatt was, on the approbation of His Majesty, adopted by government. Under this plan, several parts of the old building, which had been injudiciously engrafted on the main edifice (and which, not only from their want of harmony with the general character, diminished the unity of its design, but, from their projection into the main avenue, destroyed the perspective and obstructed the approach), were entirely removed; portions of freehold land within the park, belonging to private individuals, were purchased, and made to conform, in their appearance, with the varied beauty of the grounds; the height of the buildings generally throughout the castle was increased by an additional story; several new towers were erected; windows of lofty dimensions and of elegant and appropriate design generally inserted; some splendid gateway entrances from the principal approaches formed in a style of commensurate grandeur; which, with subsequent improvements, have rendered this interesting structure, with its appendant gardens, parks, and pleasure grounds, pre-eminently adapted to the purposes of a royal residence.

The castle occupies more than twelve acres of ground, and comprises the upper, lower, and middle wards. The principal approach is from the Little, or Home, park, through a lofty gateway, flanked on one side by the York, and on the other by the Lancaster, tower, both stately and massive structures, 100 feet high, crowned with projecting battlements supported on corbels. This gateway, which ranges in a line with the noble avenue of stately elms in the Great park, called the Long Walk, was erected by George IV., whose name it bears; the first stone was laid by that sovereign on the 12th of August, 1828, when His Majesty was pleased to change the name of the architect from Wyatt to Wyatville, upon whom he subsequently conferred the honour of knighthood. It is a noble and stately structure, and forms an entrance of correspondent grandeur into the upper ward, a spacious quadrangle, to which also are entrances through St. George's gate at the south-west, leading from the town, and the ancient Norman gateway at the west, from the middle and lower wards. On the north side of this quadrangle are the state apartments, which are open to the inspection of the public; on the east, Her Majesty's private apartments; on the south side are apartments for Her Majesty's visiters; and on the west the round tower, or keep, to the front of which has been removed, from the centre of the quadrangle, an equestrian statue in bronze of Charles II., in the Roman costume, on a marble pedestal ornamented with sculpture. The entrance to the state apartments is by a tower of very imposing character, projecting into the quadrangle, and in a line with King George the Fourth's gateway, and the long avenue in the Great park. This entrance leads into the grand hall and staircase, constructed in the reign of George III., under the superintendence of the late Mr. James Wyatt. The approach to the state apartments is by a superb vestibule, divided into three parts by ranges of finely-clustered columns and gracefully-pointed arches, in the most finished character of the later style of English architecture: the roof is elaborately groined, and decorated with fan tracery of elegant design: in the walls are four larger and three smaller niches for the reception of statues, richly canopied and highly embellished with architectural ornaments of beautiful character. The grand staircase, divided in the centre by a broad landing, is defended with a balustrade of bronze, with massive pedestals and capitals of polished brass, and lighted by an octagonal lantern 100 feet high from the pavement: the roof is delicately ornamented with fan tracery depending from the centre, and ending with the royal arms encircled by the garter. At the termination of the grand staircase is the queen's drawing-room; over the folding doors are the royal arms in artificial stone, and on each side are shields of arms of several of the British monarchs, supported by angels: the internal decorations of this splendid apartment are of the most superb character; the ceiling is beautifully painted in compartments, representing the Restoration of Charles II., the Labours of Hercules, and other subjects, and bordered with flowers and fruit, and ornaments richly gilt; the mirrors, chandeliers, and furniture, are in a corresponding style of elegance; a choice selection of paintings, by the first masters, is finely displayed, and the embellishments are disposed with the most refined taste, and the various arrangements are on a scale of the most splendid magnificence. The audience-chamber, of which the ceiling is painted with an allegorical representation of the re-establishment of the Church of England, is beautifully decorated with hangings of blue silk richly embroidered: the chair and canopy of state are superbly rich: the collection of paintings, chiefly historical, represent the victories of Edward III., painted by West; and the first installation of the knights of the order of the garter, in which more than 100 figures are finely grouped. The presence-chamber, and the whole suite of these magnificent state apartments, are in a style of correspondent grandeur, and equally remarkable for the stateliness of their proportions, and the elegance and splendour of their embellishments.

The new ball-room is finished in the most elaborate style of Louis XIV. : the walls and ceiling are panelled in compartments, highly ornamented and richly gilt; in the larger panels of the former are some superb specimens of rich tapestry, most exquisitely worked, representing the history of Jason and the Golden fleece; the colours are singularly vivid, and at the same time so softened by the skilful combination of light and shade, as to have all the force and delicacy of the finest painting; and in the intermediate panels are six mirrors, of large dimensions and great brilliancy. A pair of elegant folding doors, panelled and ornamented to correspond with the walls, open into St. George's Hall, a spacious and lofty apartment, appropriated as a banquet-room for the knights of the order of the garter: it is nearly 200 feet in length, and of proportionate width and elevation; the lofty arched ceiling is supported on beams, springing from corbels decorated with shields, on which are richly emblazoned the arms of the original knights, and divided into thirteen compartments, subdivided into panels of bold design, containing nearly 700 shields, emblazoned with the arms of the knights of the order up to the present time: at the east end, under a rich canopy, is the throne of Her Majesty, who is the sovereign of the order, at the back of which are Her Majesty's arms, and on each side are those of twelve preceding sovereigns, richly carved and emblazoned, and also those of Edward III. and the Black Prince: the mantel-piece is a massive and elegant piece of workmanship of Dove marble, richly sculptured in flowers and foliage, with the initials of George IV. Part of the plan for the general improvement of the castle, by Sir Jeffrey Wyatville, was the removal of the grand staircase, occupying the south end of the vestibule, leaving, by that means, an uninterrupted and magnificent hall, 140 feet in length and 40 feet wide, commanding, on the north, an extensive prospect over Eton, and on the south, of the long avenue, terminating with the pedestal on which has been placed a colossal statue of George III. The new grand staircase was to have an open communication with the grand hall below, and lead into an elegant vestibule above, communicating with the royal state apartments, St. George's Hall, the new guard chamber, and the Waterloo chamber. In the guard chamber have been deposited, on pedestals erected for the purpose, under richly canopied niches, suits of ancient armour, the coats of mail of John King of France, and David King of Scotland, with other military trophies; and on other pedestals, busts of the Duke of Wellington and the Duke of Marlborough, and on a pedestal formed of the frustum of the mast of the Victory, the bust of Admiral Lord Nelson. In the Waterloo chamber, a magnificent apartment, 100 feet in length, 46 feet wide, and 45 feet high, and lighted by a lantern, have been arranged the portraits of the various Sovereigns, Popes, Cardinals, Ministers of State, Ambassadors, Generals, and others connected with the prosecution of the war on the continent, and in the negociation of the late peace, painted for George IV. by Sir Thomas Lawrence, at an expense of more than £36,000, paid from the privy purse. The entrance to Her Majesty's private apartments is in the south-east angle of the quadrangle, through a handsome hall, from which is an ascent by a double staircase of great architectural beauty, lighted by a double lantern of elegant design, into a corridor 500 feet in length, communicating with Her Majesty's apartments on the east, and with the visiters' apartments on the south. The ceiling of this splendid gallery is panelled in compartments, with delicate tracery richly gilt; and the walls are decorated with paintings by the most eminent masters of the old and modern schools: the furniture is of the most sumptuous and elegant character, and the whole, enriched with every architectural ornament which the later style has combined, has an air of costly grandeur and stately magnificence. The private apartments consist of a dining-room, drawing-room, smaller drawing-room, and library, with bed-rooms, dressing-rooms, boudoir, and various other apartments. These rooms are of the most splendid and sumptuous elegance; they are decorated with every ornament that ingenuity can devise, or wealth purchase, and lighted with superb oriel windows of elegant design, enriched with tracery, which not only give an air of impressive beauty to their internal grandeur, but add greatly to the external embellishments of the castle: the rooms for Her Majesty's servants occupy the lower and higher stories of the palace. In front of the private apartments is a parterre, 400 feet in length and of equal breadth, surrounded by a broad terrace rampart wall with bastions; in the area are numerous statues finely sculptured, and under the terrace on the north side is an orangery, 250 feet in length, the front of which forms a long series of finely pointed arches with tracery.

The middle ward comprises the round tower, or keep, which was formerly the residence of the constable, whose office was both of a military and a civil nature. In his military character he was entrusted with the command of the castle, and with the custody of every thing contained in it, assisted by a lieutenant-governor, or deputy, who possessed equal authority during his absence: in his civil capacity, he was judge of a court of record having jurisdiction over the precincts of the forest, which extends seventy-seven miles and a half in circumference; but that office is now vested in a steward, assisted by a janitor, who is keeper of the prison, though no process has issued from it for many years. The round tower, which is of very spacious dimensions, has been raised 32 feet higher than its original elevation, and is crowned with a projecting machicolated battlement, supported on massive corbels and arches; and surmounted on the eastern part of the circumference by a newly erected turret, on the summit of which the royal standard is displayed during Her Majesty's presence at the castle. The lower part of the tower is surrounded by a rampart, in which are embrasures for seventeen pieces of cannon: the ascent to it is by a flight of 100 stone steps; the roof of the staircase is supported by corbels, consisting of busts of kings, knights, angels, and other figures, many of which are in good preservation; at the summit of the staircase is a large piece of cannon, pointed at the entrance through an aperture in the wall; and from the rampart a strong arched gateway, grooved for a portcullis, leads into the main tower, formerly appropriated to the reception of state prisoners of high rank.

The lower ward, or quadrangle, into which is an entrance leading from the town through Henry the Eighth's gateway, flanked with two lofty massive towers, comprises the collegiate chapel of St. George, beyond which, on the north side, are the houses of the dean, canons, minor canons, and other officers of the college, and various towers, among which are those of the Bishop of Winchester, who is prelate, and the Bishop of Salisbury, who is chancellor, of the order of the garter; a tower, formerly belonging to Garter King at Arms, of which a small portion only remains, and a store tower. Apartments have been also fitted up in this ward for the commanding officer and officer on guard, who, though subordinate to the constable, or governor of the castle, has the command of a company of the royal foot guards, always on duty here. On the south sides are the houses assigned to the thirteen poor knights on the royal foundation, each of whom has a pension of about £40 per annum, and wears a scarlet gown and a purple mantle, with the cross of St. George embroidered on the left shoulder; and a building appropriated to their governor: there are also houses for five additional knights, on the foundations of Sir Peter le Maire and Sir Francis Crane. By the will of Mr. Samuel Travers, in 1728, provision was made for seven disabled, or superannuated, lieutenants of the Royal Navy, to each of whom a pension of £60 per annum was assigned; they are chosen by the trustees, and their residence is at Travers College, Datchet-lane. In an apartment in the deanery, called the garter-room, the arms of the sovereign and knights companions of the order are emblazoned; and an ancient screen is decorated with the arms of Edward III., and of the several sovereigns and knights companions of the order from its original foundation: this apartment is at present used as a robing-room on days of installation.

The collegiate chapel of St. George (of which the establishment consists of a dean, twelve canons, seven minor canons, thirteen clerks, ten choristers, a steward, treasurer, and other officers) was, as before observed, originally built by Edward III., on the site of a smaller chapel erected by Henry I., and dedicated to Edward the Confessor. It was considerably enlarged by Edward IV., materially enriched by Henry VII., and repaired, restored, and greatly embellished by George III., who expended £20,000 in its improvement. It is a beautiful cruciform structure, in the purest character of the later style of English architecture, of which it displays one of the finest specimens in the kingdom: the transepts project in an octagonal form from the main building, and at the extremities of the aisles are lateral octangular projections, forming sepulchral chapels. Pierced parapets of elegant design are principally the external embellishments, and buttresses crowned with square embattled turrets. The interior is finely arranged; the walls are panelled throughout in one general design, of which the windows, enriched with tracery and divided by battlemented transoms, form an integral part. The nave is separated from the aisles by seven pointed arches and piers of peculiar beauty, adapted to the contrast of light and shade with singular effect; its roof and that of the choir are elaborately groined, richly embellished with fan tracery of beautiful design, and splendidly decorated with shields of armorial bearings and heraldic devices, highly emblazoned: it is lighted by an elegant range of clerestory windows, which are continued round the transepts; and the great west window, which occupies the whole of the western extremity above the entrance, of an elaborate and beautiful arrangement of panels, enriched with tracery, and embellished with ancient stained glass of unrivalled brilliancy. The choir, in which the installation of the knights takes place, and of which the general arrangement is, with the exception of the roof being more enriched with fan tracery, similar to that of the nave, is separated from it by a screen of artificial stone, from the manufactory of Coade, of appropriate character and beautiful design, ornamented with several devices illustrative of the order of the garter. On each side are the stalls of the sovereign and knights companions of the most noble order of the garter, enriched with historical and emblematical carvings, and with the names and heraldic honours of the knights richly emblasoned: the curtains and cushions are of blue velvet with gold fringe, and on the canopies of the several stalls are deposited the sword, helmet, mantle, and crest of the knights, above which are their banners of silk, emblazoned with their several armorial bearings and heraldic honours. The stall of the sovereign, whose banner is of velvet mantled with silk, and considerably larger than that of the knights companions, is on the right hand of the entrance. The other stalls, originally 25 in number, and increased to 31, occupy the north and south sides of the western part of the choir. The altar is embellished with a painting of the Last Supper, by West, which is considered to be one of the best productions of that artist; and the wainscot surrounding the presbytery is richly ornamented with the arms of Edward III., Edward the Black Prince, and those of the knights who originally composed the order, finely carved. In the east window is a beautiful painting, on glass, of the Resurrection, in three compartments, finely executed by Jarvis and Forrest, from a design by West, at an expense of £4000; and in the windows on the north and south sides of the altar are the arms of the sovereign, and of the several knights companions who subscribed to defray that expense. The east window of the south aisle is embellished with a painting, on glass, of the Angels appearing to the Shepherds, and in the west window is one of the Nativity; the west window of the north aisle is ornamented with a painting of the Adoration of the Magi; and at the eastern extremity is a chapter-room, forming an approach to the royal closet on the north side of the altar.

The various monumental chapels are separated from the aisles by screens of elegant and appropriate character, and in the south transept is a modern font of good design. At the east end of the north aisle are deposited the remains of Edward IV.; over the tomb is a black marble slab, on which is the inscription "Edward IV. and his Queen, Elizabeth Widville;" an elegant monument of iron, beautifully wrought, and representing a pair of gates between two antique towers, of elaborate design, which formerly covered the tomb, has been removed to the choir on the north side of the altar. In 1789, a small aperture was discovered in the side of this vault by some workmen, and, upon its enlargement by order of the canons, the skeleton of that monarch was found in a leaden coffin, enclosed in one of wood. In the opposite aisle, near the choir, were deposited the remains of Henry VI., which were removed from Chertsey in Surrey, by order of Henry VIII. Near the ascent to the altar is the royal vault, in which were interred the remains of Henry VIII., and his queen, Jane Seymour; and of Charles I., whose coffin being opened by order of George IV., when Prince Regent, the remains were found in a very perfect state, the countenance being as fresh as when they were interred. In a small chapel at the east end of the south aisle are the monuments of Edward Earl of Lincoln, and Richard Beauchamp, Bishop of Salisbury, first chancellor of the order of the garter. In the same aisle is a small chantry, erected in 1522, by John Oxenbridge, a canon, and a benefactor to the chapel; adjoining which is King's, or Aldworth, chapel, probably erected by Dr. Oliver King, Bishop of Bath and Wells, whose remains are interred in it. Opposite to this chapel are some panels of oak, on which are carved the arms and devices of Prince Edward (son of Henry VI.), Edward IV., and Henry VII., whose portraits, in full length, are painted on the panels. Near the centre of the aisle is the chapel of Sir Reginald Bray, in which he is interred; and at the west end is the Beaufort chapel, containing the monuments of Henry Somerset, Duke of Beaufort, of white marble, elegantly decorated with sculpture; and of Charles Somerset, Earl of Worcester, and his lady Elizabeth: on this tomb are the effigies of the earl, dressed in the habit of the order, and of his lady in her robes of state. In the centre of the north aisle is Rutland chapel, in which is an alabaster monument to the memory of Sir George Manners, Lord Roos, and Lady Anne, his wife, niece to Edward IV.: on the tomb are the figures of Sir George in armour, and his lady in her robes of state, and round it are the effigies of their children. In this chapel, in which Sir Thomas Syllinger and his wife, Anne Duchess of Exeter, and sister of Edward IV., were also interred, is a beautiful marble tablet to the memory of Major Packe, killed at the battle of Waterloo, in which he is represented as being raised from the field by a brother officer, finely sculptured in alto relievo. In the same aisle, near the choir, is the chapel of St. Stephen, decorated internally with paintings illustrative of the life and death of that martyr: this chapel was erected by Elizabeth, widow of Lord William Hastings, whose remains were deposited in it after his decapitation by Richard III.: in the south aisle of the choir is the chapel of St. John the Baptist, similarly decorated with paintings illustrative of his history. At the south-west corner of the church is Urswick's chapel, founded by Dr. Christopher Urswick, Dean of Windsor, who contributed greatly, with Sir Reginald Bray, to the completion of the church: it contains the cenotaph of the Princess Charlotte, beautifully executed in white marble, by Mr. Matthew Wyatt. In the lower compartment is the corpse of the Princess lying on a bier covered with drapery, under which the outline of the form is admirably traced, having the right arm hanging over the side of the bier, and at the corners are female figures kneeling, with their heads resting on it and veiled with drapery. In the upper compartment, the Princess appears with a countenance animated with hope, and, having drawn aside the curtains of her sepulchre, is rising from the tomb attended by angels, of whom one is bearing her infant in its arms. There are several other chapels; and, in various parts of this imposing and elegant structure, numerous interesting and highly admirable specimens of magnificent decoration and costly embellishment.

At the east end of the collegiate chapel is a chapel erected by Henry VII., as a place of interment for himself and his successors; but that monarch afterwards changing his purpose, it remained in a neglected state till the reign of Henry VIII., when Cardinal Wolsey, by permission of the king, began to erect a splendid tomb, the design of which exceeded in magnificence that of Henry VII., in Westminster abbey. The Cardinal died before it was completed, and was buried in Leicester abbey; and the unfinished sepulchre was destroyed in the parliamentary war. James II. converted this building into a chapel, and employed the artist Verrio to ornament the walls and ceiling with paintings; but the populace, excited by the public performance of the Romam Catholic rites, furiously assailed the building, destroying the windows and interior decorations; and in this ruined state it remained till George III. ordered it to be repaired, and subsequently constructed within it a royal mausoleum for himself and his successors. In clearing away the ground for this purpose, the workmen discovered two coffins in a stone recess, in one of which were the remains of Mary, daughter of Edward IV., and Elizabeth Widville, and in the other, those of their third son, George Duke of Bedford; the remains of both were re-interred in the same tomb with those of their parents. The mausoleum occupies the whole extent of the building, to which it may be regarded as a crypt: the roof is supported on massive octagonal columns and four pointed arches, in each of which are four shelves of stone, each capable of containing two coffins; at the east end are five niches for one coffin each, and in the centre is a range of twelve low altar-tombs, intended for the coffins of the sovereigns. The roof is strongly groined with ribs springing from the capitals of the columns: the entrance, through a pair of brazen gates, is by a subterraneous passage under the altar of St. George's chapel, into which is the descent by a platform lowered by machinery. The first coffin placed in the royal mausoleum was that of the Princess Amelia, daughter of George III.; after which, in succession, have been deposited the remains of the Duchess of Brunswick; the Princess Charlotte, daughter of George IV., and her infant son; Queen Charlotte; the Duke of Kent, fourth son of George III.; George III.; Princess Elizabeth of Clarence, infant daughter of William IV.; the Duke of York, second son of George III.; George IV.; the late Duke of Gloucester; and his late Majesty William IV.[a]: the two younger sons of George III., Prince Alfred and Prince Octavius, whose remains were removed from Westminster abbey, were also re-interred in the royal sepulchre. The chapel above the mausoleum is intended as a chapter-house for the order of the garter: it is lighted by a fine range of elegant windows with tracery, which surround the building, and form a beautiful group at the east end, which is hexagonal: the west end is ornamented with a large window of elegant design, in the compartments of which full-length paintings of the sovereigns and knights companions of the order will be placed.

The palace is surrounded on all sides, except the west, by a spacious and noble terrace above 2500 feet in extent, faced with a strong rampart of hewn stone, and having, at convenient intervals, easy slopes leading down to the park. The smaller park, which is also called the Home park, immediately surrounding the north-north-east and south sides of the castle, is about four miles in circumference, and was enclosed by William III., with a brick wall. Immediately under the terrace, on the east side of the castle, is a beautiful lawn, laid out in shrubberies and walks, called the Slopes, and extending, on the east side of the park, from the north terrace to the Adelaide Lodge: the grounds are beautifully diversified with forest trees and sylvan scenery. On the opposite side of the road is Frogmore Lodge, which was purchased by Queen Charlotte, and is now the favourite retreat of the Princess Augusta: the gardens and pleasure grounds are tastefully laid out, abounding with beautiful scenery, and containing many interesting objects, among which are a hermitage, designed by the Princess Elizabeth, a highly picturesque ruin, by Mr. James Wyatt, situated on the margin of a beautiful piece of water: in the interior is an elegant apartment, in which are, the effigy of an infant reposing on a cushion, and a monumental tablet to the memory of the Princess Charlotte, in which the countenance of the Princess, and the representation of her infant, are exquisitely sculptured. Many festive meetings were held here during the life of Queen Charlotte, the last of which was in honour of the fiftieth anniversary of the accession of George III. The Long Walk, extending from the upper quadrangle of the castle into the Great park, is continued in a direct line for three miles, forming a noble avenue of double rows of elms, 77 yards wide, and, at the opposite extremity, ascending a hill of considerable elevation, on which the first stone of a monument, in honour of his royal father, was laid by George IV., in 1829. The monument consists of a colossal statue in bronze, 25 feet high, by Westmacott, placed on a pedestal 40 feet high, forming a conspicuous object from the castle. Near this spot is Cumberland Lodge, the residence of the late Duke of Cumberland. This park, eighteen miles in circumference, abounds with forest scenery of great beauty, and is agreeably diversified with hill and valley, and with wood and water. Virginia water, issuing from a valley commencing near the back of Cumberland Lodge, after winding for several miles through the varied scenery of the Great park, expands towards the south-east into a beautiful lake, more than a mile in length and of considerable breadth, bounded by a verdant lawn surrounded with extensive plantations of various kinds of trees, and terminated by a fine cascade, a view of which is obtained from a bridge on the high road over the rivulet formed by the waste water of the lake, and running into the Thames near Chertsey. On the margin of the lake an elegant temple and a fishing gallery, of very light and beautiful design, have been erected. There is also a noble and magnificent ruin, consisting of numerous ancient columns of marble brought from the ruins of Corinth, and classically arranged and re-constructed by Sir Jeffrey Wyatville. The grounds are planted with shrubs and flowers, and laid out in pleasant walks. The surface of the lake is enlivened with pleasure boats and with several beautiful models of ships, among which is an elegant model of the Euryalus frigate, presented by Captain Inglis. George IV. took possession of the castle, after its partial restoration and improvement, on the 9th of December, 1828, and the first public act of that monarch was to confer the honour of knighthood on the architect by whom the additional buildings were erected and the improvements accomplished, and to add to his family arms a quartering of George the Fourth's gateway, with the motto "Windsor," both of His Majesty's suggestion. From the magnificence, extent, and grandeur of its buildings; the beauty and richness of the surrounding scenery, diversified with hills and vales, and enlivened with the frequent windings of the Thames, and the peaceful waters of its inland lake; the luxuriant woodlands within its enclosures; and the extensive and majestic forest in the vicinity, Windsor must unquestionably be regarded as one of the most spacious and magnificent palaces in Europe.

The town is pleasantly situated on the acclivities of the hill on which the castle is built, and consists of six principal streets, intersected by several smaller: it is well paved and lighted with gas, and amply supplied with water: the houses are in general of brick, and of respectable appearance, and several in the more modern part of the town are handsome and well built. The approach from Datchet is strikingly beautiful, and at the other extremity is an elegant iron bridge of three arches, resting on piers of granite, the first stone of which was laid in 1822, by the late Duke of York, connecting the town with Eton, on the opposite side of the Thames. The environs abound with varied scenery of every description, and the neighbourhood is enlivened with the windings of the river. Considerable improvements have lately been made, among which are, the removal of the ancient edifices of lath and plaster, and the erection of some handsome ranges of building fronted with stone, in which the materials of the lodges that were taken down for the improvement of the castle have been used: among the more recent erections are, York Place, Brunswick Terrace, and Augusta Place. On the west side of High-street is a meadow comprising more than two acres, called the Bachelors' Acre, which, by the award of the commissioners of the forest enclosure, was appropriated to the commonalty of the borough for their amusements. It is bounded on the east and south sides by a high bank; on the summit is a broad terrace, at the end of which is an obelisk, with inscriptions on the pedestal, commemorative of the fiftieth anniversary of the accession of George III., and of the visit of her Majesty and the Princesses, upon that occasion, to partake of the old English fare provided for the assembled populace. The barracks, for 1000 infantry, form a handsome and commodious range of building: they were erected in 1795, and enlarged to their present extent in 1803. The cavalry barracks, about half a mile from the town, on the road to Winkfield, are handsomely built, and occupy an open, healthy, and pleasant situation. The theatre, in Thames-street, a small commodious building, erected in 1815, at an expense of £6000, advanced on transferable shares, is open during the Ascot races and the vacations at Eton. The public library, in Castle-street, is well supported, and there is also a subscription circulating library. Windsor, though possessing all the advantages of a navigable river, and other favourable circumstances, among which may now be reckoned a station on the Great Western railway at Slough, about two miles distant, affording great facility of communication with London, has neither any particular branch of manufacture nor any trade, except what is necessary for the supply of the inhabitants: it has long been celebrated for the quality of the ale brewed here, of which considerable quantities are sent to London and other towns. It is indebted equally for its origin and its continued prosperity to the erection of the castle, and to its selection as a royal residence. The market days are Wednesday and Saturday, the latter chiefly for corn, which is pitched in the market-place; the fairs are on Easter-Tuesday, July 5th, and October 24th. A commodious market-place has been constructed for the sale of butchers' meat and other provisions: the area underneath the guildhall is appropriated to the use of the corn market.

The inhabitants were first incorporated in the fifth of Edward I., from which time this was the county town till 1314, when Edward II. transferred that distinction to Reading. The charter was extended and confirmed in various successive reigns till that of Charles II., by whom the government was vested in a body corporate, the members of which were styled "The Brethren of the Guildhall," and constituted the Common Council, 28 in number. Of this fraternity, 13 were called benchers, and 10 chief benchers or aldermen, of whom the mayor for the time being was one; the remainder were called "Younger Brethren of the Guild," or Common-Councilmen. The officers of the corporation were a mayor, high steward, town-clerk, two bailiffs, and others: the mayor and one of the aldermen were justices of the peace. The corporation at present consists of a mayor, 6 aldermen, and 18 councillors, under the act of the 5th and 6th of William IV., cap. 76, for an abstract of which see the Appendix, No. I. The borough is divided into two wards, and the municipal and parliamentary boundaries are made co-extensive. The mayor, late mayor, and recorder, are justices of the peace, and five others are appointed by the Crown. The police establishment consists of a superintendent, a sergeant, and 10 police constables. Quarterly courts of session are held for all offences not capital; and the corporation have power to hold a court of record, but it has been long in disuse. The borough first exercised the elective franchise in the 30th of Edward I., and sent members to parliament till the 14th of Edward III., from which time it discontinued till the 25th of Henry IV., but since that period it has regularly returned two members. The right of election was formerly vested in the inhabitants generally paying scot and lot, and resident within the borough (which originally comprised 2510 acres), at least six months previously to the election, in number about 600; but the privilege has been extended to the £10 householders of an enlarged district, which has been constituted the new borough, and comprises 2625 acres, the limits of which are minutely described in the Appendix, No. II. The number of voters now registered is 667; the mayor is the returning officer. The guildhall is a spacious and handsome building, in High-street, erected in 1686; it is supported on columns and arches of Portland stone, and ornamented at the north end with a statue of Queen Anne, and at the south with one of Prince George of Denmark; the chamber, in which the public business of the corporation is transacted, is decorated with portraits of all the sovereigns from James I. to Queen Anne, George III. and his queen, and George IV., and with those of Prince Rupert, Archbishop Laud, and some others. The common gaol and house of correction for the borough, which was rebuilt at the expense of George III., is in George-street, and contains seven wards, with an airing-yard, being capable of receiving eighteen prisoners.

Seal used by the Corporation for general purposes.

The living is a discharged vicarage, valued in the king's books at £15. 3. 4.; present net income, £400: it is in the patronage of the Crown; impropriator, — Kepple, Esq. The church is a handsome structure, in the later style of English architecture, with a lofty square embattled tower, crowned with pinnacles, erected in 1822, upon the site of a former church, which, having become greatly dilapidated, was taken down in 1820: the expense amounted to £14,040. 17. 3., towards defraying which, George III. contributed £1050, the Incorporated Society for building churches £750 (for which 600 free sittings were provided), £4000 was raised by subscription, and the remainder by a rate on the inhabitants. The interior is elegantly arranged; the nave is separated from the aisles by a range of six lofty clustered columns and pointed arches, which support the roof: the altar is embellished with an excellent painting of the Last Supper, found in one of the chantries in St. George's chapel, where it is supposed to have been secreted during the parliamentary war, and, after having been restored to its place over the altar of that chapel, presented to this church by George III., in 1788: the screen is of oak, richly carved, to correspond with two massive chairs presented by the Princess Augusta; and the rail, which surrounds the chancel, is elaborately carved with beautiful devices of pelicans feeding their young, and with fruit and foliage, supposed to be the work of the celebrated Gibbons, and formerly belonging to the chapel of St. George. Under small arches, at the east end of the church, are the royal closets, fitted up with crimson drapery; and the corporation seat is beautifully ornamented with tabernacle-work, and surmounted by an enriched canopy. There are several ancient monuments, among which may be noticed the sarcophagus of Chief Justice Reeve, with busts of himself and his lady, by Schemacker; that of Edward Jobson and Eleanor his wife, with their effigies, and those of their ten children, in the costume of the sixteenth century; and others, which have been carefully preserved on taking down the old church, and replaced in the new edifice. A subscription is now in progress for building another church. There are places of worship for Independents and Wesleyan Methodists.

On the north side of the church is the charity school, established in 1705, by subscription, for clothing and instructing children: the annual income, arising from several benefactions vested in the funds, and an annual payment of £24. 15. from the Exchequer, amounts to £167. 4., which sum is increased by annual subscriptions and collections after charity sermons. Thirty-six boys and the same number of girls are clothed and instructed: they are nominated by the dean and two senior canons, the mayor and two senior aldermen, and the vicar, of Windsor, who are the trustees. The school-house, with apartments for the master and mistress, was erected by means of a legacy of £500, bequeathed by Theodore

Randue, Esq. The ladies' charity school was established in 1784, by subscription, under the patronage of Queen Charlotte: the annual income, arising from endowments by several benefactors, is £56. 7.; 10 girls are clothed and instructed, and fitted for domestic service. Another school is endowed for 12 boys. The National school is supported by subscription; and there are funds, bequeathed by Mrs. Barker, for teaching children of this parish and the parishes of Egham, in Surrey, and Yately, in the county of Southampton; by Mr. Marrat, for teaching children of the parishes of Windsor and Clewer; and by Mr. Panton, for the endowment of the Sunday school. Archbishop Laud bequeathed £50 per annum to the parish, to be employed for two following years in apprenticing five poor boys; and, every third year, in giving marriage portions to three poor maidens of the town: this charity has been augmented, with a bequest of £1000, by Theodore Randue, Esq., with which, increased by £250 added by his executors, an estate has been purchased, yielding a rental of £128. 13., of which, £15 each are given for two successive years, to five such apprentices on Archbishop Laud's foundation as have faithfully served their terms; and, every third year, £26 each to poor maidens, who have lived for three years in the same family, with a good character.

An hospital for eight poor men and women was founded in 1503, and endowed in 1570 by Mr. Thomas Brotherton, and by Mr. Richard Gillis in 1666, with funds now producing £102. 16. per annum: the number of inmates has been augmented to twelve. A college for seven poor persons was founded in Datchet-lane by Mr. Travers, the income of which, augmented by a bequest of Lieutenant Robert Braithwaite, is £564. 3. 4. per annum, which, after deducting the necessary expenses, is divided among the inmates. Near the Pitfields, four almshouses were founded in 1688, by Mr. Richard Reeve, who endowed them with funds from which the inmates now receive £7. 10. 6. per annum: and there is an unendowed almshouse in Park-street, in which twelve poor men are supported by the parish. The royal general dispensary, in Church-street, was established in 1818, and is supported by subscription. The hospital for invalid soldiers was erected by George III., in 1784, on land which had been presented to his Majesty by the corporation; it is a neat and commodious building, well adapted to the purpose, and capable of receiving 40 patients. A charity for the relief of lying-in women was established in 1801, by a society of ladies, and is supported by subscription. Mrs. Phebe Thomas, in 1821, bequeathed funded property, from the proceeds of which twelve poor widows receive £10 per annum each; and the number will probably be augmented on the death of some annuitants named in the will of that lady. There are also numerous other bequests for distribution among the poor, for apprenticing children, and for other charitable purposes. The poor law union of Windsor comprises 6 parishes or places, and contains a population of 15,986, according to the census of 1831; it is under the care of 18 guardians. The parochial rates have been materially diminished by three several bequests, and also by three successive grants from the crown, as compensation for the loss of rates on houses purchased by the crown, and taken down for the improvements of the castle and approaches. Near the Long Walk, in the park, some labourers discovered a mineral spring, which was fast growing into repute; but the crowds of persons who frequented it proving a great annoyance, it was closed up, and a building of wood erected over the well. Among the illustrious natives of Windsor were, John, eldest son of Edward I., who died in his infancy, and was interred at Westminster, in 1273; Eleanor, eldest daughter of the same monarch, who was born in 1266, and married, by proxy, to Alphonso, King of Arragon, who died before the consummation of the marriage (she was afterwards married to Henry, Earl of Burg, in France, from whom the house of Anjou and the kings of Sicily are descended, and died in 1298); Margaret, third daughter of Edward I., born in 1275, and married to John, second Duke of Brabant, from whose son, John, the third duke, the Dukes of Burgundy were descended; Mary, the sixth daughter of the same monarch, born in 1279, who, when ten years of age, entered a nunnery at Amesbury, in the county of Wilts; Edward III., son of Edward II., in 1312, the first of the English sovereigns who issued gold coin, the pieces being called rose-nobles; William, the sixth son of Edward, who died in his infancy; and Henry VI., son of Henry V., who died by violence, in 1471. Windsor gives the title of Earl to the family of Stuart, Marquesses of Bute.

WINDSOR, OLD (*St. Peter*), a parish, in the union of Windsor, hundred of Ripplesmere, county of Berks, 2 miles (S. E. by S.) from New Windsor; containing 1453 inhabitants. Prior to the Conquest, this place is said to have been the residence of several Saxon kings, but after the improvements made by Henry I. in the fortress erected at New Windsor by William the Conqueror, it speedily lost its original importance. It is beautifully situated on the river Thames, which affords facility of communication with Oxford and London, and within four miles of the Slough station of the Great Western railroad. A pleasure fair is annually held in the village. The living is a discharged vicarage, valued in the king's books at £8. 6. 8.; present net income, £270: it is in the patronage of the Crown; impropriator, Rev. G. Isherwood. The church is a very ancient structure; and a district church is now being built by subscription in the western extremity of the parish. There is a chapel in the Great Park, a royal donative, at which there is morning service for parishioners. In the churchyard are several tombs of noble individuals, and other distinguished characters. A parochial school (now an almshouse), and four cottages with gardens attached, for poor persons, were erected in 1797, and endowed with land, which is now divided into allotments and let to 40 poor persons. The Onslow and Jubilee school of industry, founded by a bequest of £23 per annum by Lady Onslow for teaching gardening and agriculture to boys during one half of the day, and instructing them during the other half on the National plan, and also for preparing girls for creditable service, is further supported by subscription. The Roman road from Silchester passes through a part of the parish.

WINESTEAD (*St. German*), a parish, in the union of Patrington, Southern Division of the wapentake of Holderness, East Riding of the county of York, 2 miles (N. W. by N.) from Patrington; containing 145 inhabitants. The living is a rectory, valued in the king's books at £12; present net income, £247: it is in the patronage of Mrs. Hildyard. The church is an ancient building surrounded by stately trees: in front of the pulpit is a monument, with a recumbent statue in armour, to the memory of Sir Robert Hildyard. A school is supported by Mrs. Hildyard. The celebrated Andrew Marvel, M.P. for Hull in the time of Charles I., was born here, March 31st, 1621, during the incumbency of his father.

WINFARTHING (*St. Mary*), a parish, in the union of Guiltcross, hundred of Diss, Eastern Division of the county of Norfolk, $4\frac{1}{4}$ miles (N. by W.) from Diss; containing 703 inhabitants. The living is a discharged rectory, valued in the king's books at £12; patron, Earl of Albemarle.

WINFIELD, a township, in the parish and hundred of Wrotham, lathe of Aylesford, Western Division of the county of Kent, 5 miles (S.) from Wrotham: the population is returned with the parish.

WINFORD-EAGLE (*St. Lawrence*), a parish, in the union of Dorchester, hundred of Tollerford, Dorchester Division of the county of Dorset, $8\frac{1}{2}$ miles (W. N. W.) from Dorchester; containing 134 inhabitants. The living is annexed to the vicarage of Toller-Fratrum. Here is a Sunday National school. On Fernham down are several barrows, in one of which seventeen urns, containing bones and ashes, have been discovered. Dr. Thomas Sydenham, an eminent physician, was born here, in 1624; he died in 1689.

WINFORTON (*St. Mary*), a parish, in the union of Kington, hundred of Huntington, county of Hereford, 6 miles (N. E. by E.) from Hay; containing 158 inhabitants. The living is a rectory, valued in the king's books at £9. 6. 8.; present net income, £214: it is in the patronage of the Executors of the late John Freeman, Esq. A railroad from Hay to Kington passes through the parish. A school is endowed by the late Mr. Freeman with £12 per annum, and a house and garden.

WINFRITH-NEWBURGH (*St. Christopher*), a parish, in the union of Wareham and Purbeck, hundred of Winfrith, Wareham Division of the county of Dorset, 9 miles (W. by S.) from Wareham; containing 891 inhabitants. This is a very extensive and ancient parish, giving name to the hundred, and formerly belonged to the family of Newburgh, who had a seat here, of which there are no traces. Near the hamlet of Bromehill, a rivulet, tributary to the Frome, is crossed by three bridges, erected in 1769, at the joint expense of Edward Weald and James Frampton, Esqrs. The living is a rectory, with West Lullworth annexed, valued in the king's books at £23. 14. $4\frac{1}{2}$.; present net income, £440; patron, Bishop of Salisbury: two-thirds of the great tithes of corn, wool, and lambs, belong to the vicar of Chipping-Campden, in the county of Gloucester. The church has a fine Norman doorway and an embattled tower; the nave is covered with lead. Here is a Sunday National school.

WING (*All Saints*), a parish, in the union of Leighton-Buzzard, hundred of Cottesloe, county of Buckingham, 3 miles (S. W. by W.) from Leighton-Buzzard; containing 1152 inhabitants. The living is a vicarage, valued in the king's books at £18. 16. 3.; present net income, £338; patron and impropriator, Samuel Jones Loyd, Esq., by recent purchase of the Wing estate from the Earl of Chesterfield. The church is a remarkably fine structure. There is a place of worship for Wesleyan Methodists; also a National school for boys. Dormer's hospital was founded in 1596, by Lady Pelham, widow of Sir William Dormer, for eight poor persons: it was endowed by her with property, together with a bequest of Sir William Stanhope in 1772, now producing an annual income of £72. 6. 8., which is equally divided among the eight inmates. A fund of about £20 per annum, arising from bequests by Lady Carnarvon, William Hoare, William Dent and William Robinson, is, together with about £10. 15. per annum under the charity of Thomas Pratt of Wingrave, distributed in great coats and other clothing. Robert Shepherd, in 1685, bequeathed two rent-charges, amounting to £5. 16. 8., to be distributed in bread every Sunday, after deducting 20*s.* to be paid to the minister for catechising the children, and 10*s.* for a sermon. A Benedictine priory, a cell to the monastery of St. Nicholas at Angiers in France, was founded at Ascot, in this parish, by the Empress Maud, which, after the suppression, came into the possession of Cardinal Wolsey.

WING (*St. Peter and St. Paul*), a parish, in the union of Uppingham, hundred of Martinsley, county of Rutland, 3 miles (N. E.) from Uppingham; containing 307 inhabitants. The living is a rectory, valued in the king's books at £7. 5. 5.; present net income, £340: it is in the patronage of the Crown.

WINGATE, a township, in the parish of Kelloe, union of Easington, Southern Division of Easington ward, Northern Division of the county palatine of Durham, $6\frac{1}{2}$ miles (E. S. E.) from Durham; containing 115 inhabitants.

WINGATES, a township, in the parish of Long Horsley, union of Rothbury, Western Division of Morpeth ward, Northern Division of the county of Northumberland, 6 miles (S. E. by S.) from Rothbury; containing 163 inhabitants. Peter Silcock, in 1818, bequeathed £50, directing the interest to be applied for teaching five children in a school built by subscription.

WINGERWORTH, a parish, in the union of Chesterfield, hundred of Scarsdale, Northern Division of the county of Derby, $2\frac{3}{4}$ miles (S. by W.) from Chesterfield; containing 471 inhabitants. The living is a perpetual curacy; net income, £77; patron and appropriator, Dean of Lincoln. Here is a Roman Catholic chapel. A day and Sunday school is partly supported by an endowment of £19 per annum. The Ikeneld-street passes through the parish, in which large quantities of coal, iron-stone, and freestone, are obtained. The hall was taken possession of and garrisoned for the parliament, in 1643; the present large and elegant mansion was erected in 1728. On Stonedge cliff are several basins and two seats, excavated in the rock. The brass head of a catapulta was found a few years ago on the Roman road.

WINGFIELD (*St. Andrew*), a parish, in the union and hundred of Hoxne, Eastern Division of the county of Suffolk, $5\frac{3}{4}$ miles (E. by N.) from Eye; containing 668 inhabitants. The living is a perpetual curacy; net income, £100; patron and appropriator, Bishop of Norwich. The church, which is a spacious and handsome structure, was made collegiate in 1362. In the chancel, of which the architecture is highly enriched, are some superb monuments of the Wingfields and De la Poles: among those of the latter family is one to the memory of Michael, second Earl of Suffolk, and one to John, Duke of Suffolk. Michael, first Earl of Suffolk, in the eighth of Richard II., built the castle, of which the south front still remains, and the west side has been converted into a farm-house: these ruins, which are surrounded by a moat, are situated about a quarter of a mile north-west of the church, on a thickly-wooded plain, and are now the property of Lord Berners. A large school-house has been erected by subscription in the churchyard. Of the college, founded on the south side of the church by the will of Sir John Wingfield, in 1362, for a provost and nine priests, all that remains is the west side of the quadrangle, now used as a farm-house; it was valued at the surrender in 1534, at £50. 3. 5½. per annum.

WINGFIELD, NORTH (*St. Lawrence*), a parish, in the union of CHESTERFIELD, hundred of SCARSDALE, Northern Division of the county of DERBY, 4½ miles (S. S. E.) from Chesterfield; containing 1691 inhabitants. The living is a rectory, valued in the king's books at £21. 6. 3.; present net income, £772; patron, G. H. Barrow, Esq. The church is a large handsome structure, situated at a distance from the village. There is a place of worship for Wesleyan Methodists. A school is endowed with £2. 10. per annum for instructing five children. The Roman Ikeneld-street may be traced in this parish.

WINGFIELD, SOUTH (*All Saints*), a parish, in the union of BELPER, hundred of SCARSDALE, Northern Division of the county of DERBY, 2¼ miles (W.) from Alfreton; containing 1091 inhabitants. The manor-house, now an extensive and interesting ruin, was a splendid and spacious edifice, erected by Ralph Lord Cromwell, Lord Treasurer in the reign of Henry VI., it was afterwards, for several generations, one of the principal seats of the Earls of Shrewsbury; and Mary Queen of Scots, while in the custody of George, the sixth earl, passed some months here, in 1569; she was here also in November and December, 1584. At the commencement of the parliamentary war, Wingfield manor-house was garrisoned for the parliament; but being taken by the Earl of Newcastle, towards the close of the year 1643, it was then made a royal garrison: in 1644 it sustained a siege, but surrendered to the parliament in August: in 1646, it was dismantled by order of parliament. In 1774, a considerable part of the mansion was pulled down, to build a modern house near it with the materials. The village is large, and possesses a considerable and increasing trade in the weaving of stockings, for which there are about 200 frames in operation: coal is obtained in the parish. The living is a discharged vicarage, endowed with the rectorial tithes, and valued in the king's books at £6. 13. 4.; present net income, £324; patron, Duke of Devonshire. There is a place of worship for Wesleyan Methodists. Samuel Newton, in 1683, gave £200 for charitable uses, with which sum certain lands were purchased, now producing £33 per annum, £17 of which is applied for teaching twenty children. The estate called Strelley's charity, at Okerthorpe, in this parish, is now let for £55 a year, of which income £20 was directed by the donor to be applied for apprenticing two poor boys, and £10 for exhibitions for two poor scholars at the University. The ancient Ikeneld-street passes through the parish.

WINGFORD, county of CAMBRIDGE.—See WENTWORTH.

WINGHAM (*St. Mary*), a parish (formerly a market-town), in the union of EASTRY, hundred of WINGHAM, lathe of ST. AUGUSTINE, Eastern Division of the county of KENT, 9 miles (S. E.) from Canterbury; containing 1115 inhabitants. This place, situated on the high road from Canterbury to Sandwich, in a neighbourhood abounding with genteel residences, was formerly part of the ancient possessions of the see of Canterbury, to which it was granted in the early part of the Heptarchy; and in the 36th of Henry III., Archbishop Boniface obtained for the inhabitants the grant of a market. The archbishops had a palace here, in which they frequently resided and entertained several of the kings of England on their way to and from the continent. The manor was exchanged for other lands by Archbishop Cranmer in the reign of Henry VIII. The market has been disused, but considerable fairs for cattle are held on May 12th and Nov. 12th. The Wingham hops are considered the best grown in the county. An Horticultural Society, of which Lady Bridges is the patroness, was established in August 1835, and is well supported: there are three shows in the year. The petty sessions for the division are held here. The living is a perpetual curacy; net income, £114; patron, N. Bridges, Esq.; impropriator, W. Greville, Esq. There is a place of worship for Independents; and a National school has been established, and is partly supported by subscription. Sir James Oxenden, Bart., in 1686, founded a free school, and endowed it with £16 per annum, which is paid for the instruction of twenty boys. An infants' school is about to be formed by the Rev. C. Livingstone, for which Sir Brook W. Bridges has given a site for the school-room, and Sir Henry Oxenden the timber. A college for a provost and six canons in the church was founded in 1286, by John Peckham, Archbishop of Canterbury, which at the dissolution had a gross revenue of £208. 14. 3½., and was granted by Edward VI. to Sir Henry Palmer: on or near its site a stone coffin and some other relics of antiquity have been found. William de Wengham, Bishop of London, and Chancellor in the reign of Henry III., was a native of this parish.

WINGRAVE (*St. Peter and St. Paul*), a parish, in the union of AYLESBURY, hundred of COTTESLOE, county of BUCKINGHAM, 5½ miles (N. E.) from Aylesbury; containing 783 inhabitants. The living is a discharged vicarage, valued in the king's books at £9. 9. 7.; present net income, £98: it is in the patronage of the Countess of Bridgewater; impropriator, O. Oldham, Esq. There is a place of worship for Independents; also a school endowed with £2 per annum, for which five children are instructed. At Rowsham there was formerly a chapel. Thomas Pratt, in 1615, bequeathed some property, now producing a rental of £35. 6., to be distributed every Lady-day and Michaelmas among the poor of the several parishes and hamlets of Wingrave, Wing, Mentmore, Cheddington, Bettlow and Aldwick; and there is some property, known as Bailey's House Gift, and let for £17. 10. per annum, which is applied to repairing the church: there are also other trifling bequests, distributed among the poor in clothing, &c.

WINKBOURN, a parish, in the union of SOUTHWELL, Northern Division of the wapentake of THURGARTON, Southern Division of the county of NOTTINGHAM, 3 miles (N.) from Southwell; containing 134 inhabitants. The living is a donative, in the patronage of P. Pegge Burnell, Esq. The church, a large ancient edifice, formerly belonged, with the liberty, to the Knights Hospitallers of Newland, in the county of York; at the dissolution they were granted by Edward VI. to the family of Burnell. A school is endowed with £30 per annum, and a house and garden.

WINKFIELD (*St. Mary*), a parish, in the union of EASTHAMPSTEAD, hundred of RIPPLESMERE, county of BERKS, 5½ miles (S. W. by W.) from New Windsor; containing 2009 inhabitants. This place is pleasantly situated on the road from London through Windsor Forest to Reading. The living is a discharged vicarage, valued in the king's books at £8. 5. 10.; present net income, £343; patrons and appropriators, Dean and Chapter of Salisbury. There is a place of worship for Independents. The Earl of Ranelagh, in 1710, built a chapel on Winkfield plain, in which service is daily performed, and attached to it a free school for 22 boys, and another for 22 girls; he endowed them with property in Ireland, and directed that the schoolmaster should be in holy orders. In 1715, Thomas Maule, Esq., bequeathed £500; in 1783, Thomas Hatch, who had been educated here, £500; and in 1809, John Tow left £500 four per cent. stock, in augmentation of the income, which altogether amounts to upwards of £350 per annum. The children are clothed and educated, and afterwards apprenticed with a premium of £5 each. Here is also a National school.

WINKFIELD (*St. Andrew*), a parish, in the union and hundred of BRADFORD, Westbury and Northern Divisions, and Trowbridge and Bradford Subdivisions, of the county of WILTS, 2 miles (W. S. W.) from Trowbridge; containing 288 inhabitants. The living is a rectory, valued in the king's books at £5. 16. 5½.; present net income, £237; patrons, Rev. John Hall and others. There is a monument in the church, erected by his pupils to the memory of the Rev. Edward Spencer, who was for 43 years rector of this parish, and who died in 1819, in the 80th year of his age. A National school is supported by subscription.

WINKLEY, or WINCKLEIGH (*All Saints*), a parish in the union of TORRINGTON, hundred of WINKLEY, South Molton Division, (except the tything of Loosbear, which is in the Black Torrington and Shebbear Division) and Northern Division of the county of DEVON, 6½ miles (S. W.) from Chulmleigh; containing 1596 inhabitants. The living is a vicarage, valued in the king's books at £21. 8. 9.; present net income, £215; patrons and appropriators, Dean and Chapter of Exeter. This parish, with the tything of Loosbear, forms a distinct hundred, to which it gives name: the new road from Torrington to Exeter passes through it, and the scenery is agreeably enlivened with the grounds of Winkley Court, the property of J. T. Johnson, Esq. A fair for cattle is held on the Monday after July 7th; and courts leet and baron annually. Here is a National Sunday school; also an endowed almshouse, called Gidley's, for poor widows.

WINKSLEY, a chapelry, in the parish of RIPON, Lower Division of the wapentake of CLARO, West Riding of the county of YORK, 4¾ miles (W.) from Ripon; containing 259 inhabitants. The living is a perpetual curacy; net income, £68; patrons, Dean and Chapter of Ripon. The chapel, dedicated to St. Oswald, has been enlarged, and 195 free sittings provided, the Incorporated Society having granted £150 in aid of the expense.

WINKTON, a tything, in the parish and hundred of CHRISTCHURCH, Ringwood and Southern Divisions of the county of SOUTHAMPTON, 2½ miles (N. W. by N.) from Christchurch, with which the population is returned.

WINLATON, a parochial district, in the union of GATESHEAD, Western Division of CHESTER ward, Northern Division of the county palatine of DURHAM, 6¼ miles (W. by S.) from Gateshead; containing 3951 inhabitants. The village occupies an elevated site between the rivers Tyne and Derwent, and owes its rise to the extensive iron-works removed hither from Sunderland by Sir Ambrose Crawley, about 1690, for carrying on which the neighbourhood affords peculiar advantages, in the abundance of coal, the facility of water carriage, &c. Edge-tools, files, and nail-rods, are the principal articles manufactured; and those of a heavier kind, such as anchors, anvils, pumps, and cylinders for steam-engines, chain cables, spades, shovels, saws, and cast-iron utensils of every description, are made at Swalwell. A code of laws for the workmen was also established, which, in a great measure, has superseded the general law, and under which a court of arbitrators sits every ten weeks, by whom justice is promptly administered, at a very trifling expense. The principal manufacturers have large warehouses in Thames-street, London, and at Greenwich, and constantly employ two vessels, of about 300 tons' burden each, in transporting their goods thither and to Newcastle for sale: there are also works for refining lead on the banks of the Tyne. A chapel was built here in 1705, on the site of a more ancient one, which is said to have been demolished in 1569; but having been suffered to go to ruin, a spacious school-room was erected, in 1816, on the spot, in which divine service was occasionally performed by the rector of Ryton, to which parish Winlaton formerly belonged, till the new church was built, when it was made parochial. The church was finished in 1828, at an expense of £2500, of which sum, £2000 was defrayed by the parliamentary commissioners, and £400 by the Incorporated Society, for which 400 free sittings have been provided: it is in the later English style of architecture. The living is now a rectory; net income, £265; patron, Bishop of Durham. At the village of Blaydon there is a place of worship for Methodists; and National schools have been established in the district.

WINMARLEIGH, a township, in the parish and union of GARSTANG, hundred of AMOUNDERNESS, Northern Division of the county palatine of LANCASTER, 2 miles (N. W.) from Garstang; containing 275 inhabitants.

WINNALL, a parish, in the union of NEW WINCHESTER, hundred of FAWLEY, Winchester and Northern Divisions of the county of SOUTHAMPTON, ¾ of a mile (N. N. E.) from Winchester; containing 115 inhabitants. The living is a rectory, valued in the king's books at £5; patron, Bishop of Winchester. The tithes have been commuted for a rent-charge of £170, subject to the payment of rates, which on the average have amounted to £13.

WINNERSH, a liberty, in the parish of HURST, union of WOKINGHAM, hundred of SONNING, county of BERKS; containing 531 inhabitants.

WINNINGTON, a township, in the parish of GREAT BUDWORTH, union of NORTHWICH, Second Division of the hundred of EDDISBURY, Southern Division of the county of CHESTER, 1 mile (N. W.) from Northwich; containing 256 inhabitants. It is situated on the banks of the Weever, over which there is a stone bridge.

WINNINGTON, a township, in the parish of MUCKLESTON, Northern Division of the hundred of PIREHILL and of the county of STAFFORD, 4½ miles (N. E.) from Drayton-in-Hales; containing 249 inhabitants.

WINNOW, ST., a parish, in the union of BODMIN, hundred of WEST, Eastern Division of the county of CORNWALL, 2½ miles (S. E.) from Lostwithiel; containing 1048 inhabitants. The living is a vicarage, with the chapel of Nighton, in the patronage of the Dean and Chapter of Exeter (the appropriators), valued in the king's books at £5; present net income, £197. The navigable river Fowey runs on the west and south of this parish, and is crossed by a bridge, on the high road from Bodmin to Plymouth, at Resprin, where are the ruins of a chapel of ease. On Beacon hill a square battery was constructed by the royalists, a short time before the capitulation of the army of the parliament, in 1644.

WINSCALES, a township, in the parish of WORKINGTON, union of COCKERMOUTH, ALLERDALE ward above Derwent, Western Division of the county of CUMBERLAND, 2½ miles (S. E.) from Workington; containing 100 inhabitants.

WINSCOMBE (*St. James*), a parish, in the union of AXBRIDGE, hundred of WINTERSTOKE, Eastern Division of the county of SOMERSET, 2 miles (N. by W.) from Axbridge; containing 1526 inhabitants. The living is a discharged vicarage, valued in the king's books at £16. 2. 11.; present net income, £201; patrons and appropriators, Dean and Chapter of Wells. The church is a handsome structure, with a stately western tower crowned with pinnacles. Symons Cardinbrook, in 1761, gave the residue of his estate to be applied for teaching poor children: the school-room was erected by subscription, aided by about £60 from this bequest; the permanent annual income is £15. There is also a National school.

WINSFORD (*St. Mary Magdalene*), a parish, in the union of DULVERTON, hundred of WILLITON and FREEMANNERS, Western Division of the county of SOMERSET, 5 miles (N. by W.) from Dulverton; containing 524 inhabitants. The living is a vicarage, valued in the king's books at £14. 13. 9.; present net income, £330; patrons, Master and Fellows of Emanuel College, Cambridge; impropriators, Sir T. D. Acland, Bart., and others. The impropriate tithes have been commuted for a rent-charge of £130, and the vicarial for £360, subject to the payment of rates; the glebe comprises 92 acres, valued at £80 per annum. There is a place of worship for Wesleyan Methodists; also a day and Sunday school, partly supported by Sir T. D. Acland, Bart.

WINSHAM, a parish, in the union of CHARD, and forming one of the four detached portions which constitute the Eastern Division of the hundred of KINGSBURY, Western Division of the county of SOMERSET, 4 miles (E. by S.) from Chard; containing 932 inhabitants. The living is a vicarage, valued in the king's books at £14. 13. 4.; present net income, £287; patron, Dean of Wells; impropriator, H. H. Henley, Esq. The church is an ancient structure, with a tower rising from the centre. The manufacture of narrow woollen cloth is carried on at the village. Sir Matthew Holworthy, in 1680, gave certain premises, producing about £6 per annum, which is applied towards the instruction of 24 boys and 24 girls, in a school erected by subscription in 1818, in union with the National Society.

WINSHILL, a township, in the parish and union of BURTON-UPON-TRENT, hundred of REPTON and GRESLEY, Southern Division of the county of DERBY, 1½ mile (E. N. E.) from Burton-upon-Trent; containing 342 inhabitants.

WINSKILL, a township, in the parish of ADDINGHAM, union of PENRITH, LEATH ward, Eastern Division of the county of CUMBERLAND, 6¾ miles (N. E. by E.) from Penrith: the population is returned with the township of Hunsonby.

WINSLADE (*St. Mary*), a parish, in the union and hundred of BASINGSTOKE, Basingstoke and Northern Divisions of the county of SOUTHAMPTON, 3 miles (S. by E.) from Basingstoke; containing 134 inhabitants. The living is a discharged rectory, valued in the king's books at £6. 12. 1.; present net income, £164; patron, Lord Bolton. A school is supported by Lord and Lady Bolton.

WINSLEY, a joint hamlet with Snitterton, in the parish of DARLEY, union of BAKEWELL, hundred of WIRKSWORTH, Southern Division of the county of DERBY, 3½ miles (W. N. W.) from Matlock; containing, with Snitterton, 671 inhabitants.

WINSLEY, a joint chapelry with Limpley-Stoke, in the parish of GREAT BRADFORD, union and hundred of BRADFORD, Westbury and Northern Divisions, and Trowbridge and Bradford Subdivisions, of the county of WILTS, 1½ mile (W.) from Bradford; containing, with Limpley-Stoke, 2847 inhabitants. The chapel is dedicated to St. Nicholas. There is a place of worship for Wesleyan Methodists.

WINSLEY, a joint chapelry with Hartwith, in the parish of KIRKBY-MALZEARD, union of PATELEY-BRIDGE, Lower Division of the wapentake of CLARO, West Riding of the county of YORK, 3 miles (W. N. W.) from Ripley; containing, with Hartwith, 943 inhabitants.

WINSLOW (*St. Lawrence*), a market-town and parish, and the head of a union, in the hundred of COTTESLOE, county of BUCKINGHAM, 6½ miles (S. E.) from Buckingham, and 50 (N. W.) from London; containing 1290 inhabitants. This town is of considerable antiquity, having been given by King Offa to the abbey of St. Albans, so early as 794: it is situated on the brow of a hill, and consists of three principal streets regularly built and of neat appearance; the houses are chiefly of brick; water is amply supplied from wells. The land in the vicinity is extremely fertile, and in a high state of cultivation: lace-making is carried on to a small extent. The white poppy was so successfuly cultivated here, in 1821, as to produce 60 lb. of opium, worth at least £75, from four acres, and 143 lb., in the next year, from eleven acres; for which, on both occasions, the prize of thirty guineas was awarded by the Society for the Encouragement of Arts, Manufactures, and Commerce. The market, granted by charter of Henry III., is on Thursday; a small quantity of corn is pitched in the market-house. Fairs are held on February 18th, March 20th, Holy Thursday, August 21st, September 22nd, and November 26th, for cattle; on the Thursday before Old Michaelmas-day, and the first and second Thursdays following, are statute fairs. The living is a discharged vicarage, valued in the king's books at £11. 5. 10.; present net income, £185: it is in the patronage of the Crown; impropriator, W. S. Lowndes, Esq. The church is a spacious and venerable structure, in the later style of English architecture, with a square embattled tower at the west end; it has been repewed and 130 free sittings provided, the Incorporated Society having granted £100 in aid of the expense. There are places of worship for Baptists, Independents, and Wesleyan Methodists. A school in which 12 sons of poor labourers are taught free, and 8 others pay a small sum, was founded and endowed by Joseph Rogers, in 1724, with property now producing an annual income of £30. Here is also a Sunday National school. Coal and bread are annually distributed among the poor to the amount of about £35, derived from various bequests. The poor law union of Winslow comprises 17 parishes or places, containing a population of 7847, according to the census of 1831, and is under the superintendence of 18 guardians.

WINSLOW, a township, in the parish and union of BROMYARD, hundred of BROXASH, county of HEREFORD, 2½ miles (S. W. by W.) from Bromyard; containing 450 inhabitants.

WINSON, a chapelry, in the parish of BIBURY, union of NORTHLEACH, hundred of BRADLEY, Eastern Division of the county of GLOUCESTER, 5 miles (S. S. W.) from Northleach; containing 176 inhabitants. The chapel is dedicated to St. Michael.

WINSTANLEY, a township, in the parish and union of WIGAN, hundred of WEST DERBY, Southern Division of the county palatine of LANCASTER, 3¼ miles (S. W. by W.) from Wigan; containing 731 inhabitants.

WINSTER, a market-town and chapelry, in the parish of YOULGRAVE, union of BAKEWELL, hundred of HIGH PEAK, Northern Division of the county of DERBY, 19 miles (N. N. W.) from Derby, and 145 (N. N. W.) from London; containing 962 inhabitants. This small town is situated midway between the river Derwent and the Cromford and High Peak railway, about three miles from each, with the latter of which a branch communication is contemplated; it is badly supplied with water, which in dry seasons is only to be procured at the distance of a mile. The inhabitants are chiefly employed in mining, which was formerly much more extensively carried on; the cotton trade at one period was established, but has ceased for some years. The market, on Saturday, is very indifferently attended; and the four fairs formerly held annually have declined. Winster is within the jurisdiction of a court of pleas held at Tutbury every third Tuesday, for the recovery of debts under 40*s*. The living is a perpetual curacy; net income, £370; patrons, the Inhabitants; impropriator, Duke of Devonshire. In 1702, Mrs. Anne Phermey and Mrs. H. Fenshaw bestowed one-fourth of the tithes of corn and hay in the township on the minister. The chapel, dedicated to St. John the Baptist, is partly in the Norman style of architecture, and partly of a later date, with a tower nearly covered with ivy. Primitive and Wesleyan Methodists have each a place of worship. Thomas Eyre, Esq., in 1717, bequeathed £20 per annum for the instruction of 20 children; and an annuity of £5 was left in 1718, by Robert Moore, for teaching five more. In the neighbourhood are several barrows, in one of which, opened in 1768, two glass vessels were found, containing some clear, but green-coloured, water, also a silver bracelet, some glass beads, and other trinkets.

WINSTER, a chapelry, in the parish, union, and ward of KENDAL, county of WESTMORLAND, 7 miles (W.) from Kendal: the population is returned with the township of Undermilbeck, which is in the parish of Windermere. The living is a perpetual curacy; net income, £61; patron, Vicar of Kendal. This chapelry once formed part of the chapelry of Crook, and the inhabitants still contribute towards the repairs of that chapel.

WINSTON (*St. Andrew*), a parish, in the union of TEESDALE, South-Western Division of DARLINGTON ward, Southern Division of the county palatine of DURHAM, 6½ miles (E.) from Barnard-Castle; containing 327 inhabitants. The living is a rectory, valued in the king's books at £9. 18. 1½.; present net income, £557; patron, Bishop of Durham. The village is situated on an elevation rising from the northern bank of the river Tees, which is crossed by a handsome stone bridge of one arch, 111 feet in the span, built in 1764. A day and Sunday school is partly supported by an endowment of £3. 10. per annum, being the interest of £70 given by Lord Crewe's trustees.

WINSTON (*St. Andrew*), a parish, in the union of BOSMERE and CLAYDON, hundred of THREDLING, Eastern Division of the county of SUFFOLK, 1¼ mile (S. S. E.) from Debenham; containing 398 inhabitants. The living is a vicarage, valued in the king's books at £9. 3. 9.; present net income, £169; patrons and appropriators, Dean and Chapter of Ely.

WINSTONE (*St. Bartholomew*), a parish, in the union of CIRENCESTER, hundred of BISLEY, Eastern Division of the county of GLOUCESTER, 6¼ miles (N. W. by N.) from Cirencester; containing 164 inhabitants. The living is a discharged rectory, valued in the king's books at £7. 10.; present net income, £238; patron, Rev. E. Reed. There is a place of worship for Baptists; also a Sunday National school. The ancient Ermin-street passes through the parish.

WINTERBOURNE, a chapelry, in the parish of CHIEVELEY, union of NEWBURY, hundred of FAIRCROSS, county of BERKS, 3½ miles (N. N. W.) from Newbury; containing 326 inhabitants. The chapel is dedicated to St. James. A school is endowed with £10 per annum; and another is supported by the clergyman.

WINTERBOURNE (*St. Martin*), a parish, in the union of DORCHESTER, hundred of GEORGE, Dorchester Division of the county of DORSET, 3 miles (W. S. W.) from Dorchester; containing 369 inhabitants. The living is a discharged vicarage, valued in the king's books at £9. 15.; present net income, £73; patron, Bishop of Salisbury; impropriator, H. Sturt, Esq. The church has a neat embattled tower crowned with pinnacles. A market, granted by Henry III., was formerly held here; and a fair is still kept on St. Martin's day. In this parish is Maiden Castle, one of the strongest and most extensive Roman camps in the West of England, which, according to Ptolemy, was the *castra æstiva* of the garrison of *Dunium*, afterwards called *Durnovaria*, the capital of the Durotriges: it consists of a treble ditch and rampart, enclosing an irregular oval area of 44 acres, but the entire work covers an extent of 115½ acres. There are two very intricate entrances, that at the east end being defended by five, and that at the west end by six, ditches and ramparts. Near the former passes the vicinal road leading from Dorchester to Weymouth, and to the latter extends a branch from the *Via Iceniana*, which passes about a mile north of the camp. The summit commands an extensive prospect of barrows stretching for many miles along the tops of the hills to the southward.

WINTERBOURNE (*St. Mary*), formerly a parish (now divided into two), in the union of CLIFTON, Upper Division of the hundred of LANGLEY and SWINEHEAD, Western Division of the county of GLOUCESTER, 6½ miles (N. E. by N.) from Bristol; containing 2889 inhabitants. It has been divided into two distinct and separate parishes by Her Majesty's Commissioners, under the act of the 58th of George III., cap. 45. The living is a rectory, valued in the king's books at £27. 7. 6.; present net income of the old benefice, £1187; patrons, President and Fellows of St. John's College, Oxford. A chapel has been erected at Frenchay, in the chapelry of Hambrook. There is a place of worship for Wesleyan Methodists. The manufacture of hats is carried on extensively here. Fairs are held on June 29th and October 18th. A National school is supported by subscription. In the parish is Northwoods Park, situated on a fine eminence on the old turnpike-road to Gloucester, where a handsome freestone building, in the Grecian style of architecture, and so constructed with iron joists and composition floors as to be perfectly fire-proof, has been erected by Henry Hawes Fox, M.D., for the cure of a limited number of insane patients of the higher class of society. In the construction of the building, the panopticon plan has been adopted to the fullest extent compatible with the comfort of the invalids, and affording to the resident superintendent the fullest control over the proceedings of the patients and the servants of the establishment. The estate comprises 200 acres of freehold land, highly ornamented with extensive shrubberies, plantations, and full-grown timber, and singularly combining the most complete retirement with the greatest cheerfulness; the exercising grounds are very spacious, and command some beautiful and extensive prospects. The system of treatment is founded upon the principle of temporary separation from the immediate family and friends of the patients, association with invalids under similar circumstances, and constant mental and corporeal occupation and recreation, with the most perfect freedom from all personal restraint, and the benign influence which, in certain stages of the malady, religion is so well calculated to afford.

WINTERBOURNE-ABBAS (*St. Mary*), a parish, in the union of DORCHESTER, hundred of EGGERTON, Dorchester Division of the county of DORSET, 4¾ miles (W.) from Dorchester; containing 133 inhabitants. The living is a rectory, with that of Winterbourne-Steepleton united, valued in the king's books at £13. 17. 6.; present net income, £435; patrons, Rector and Fellows of Lincoln College, Oxford. The stream called the South Winterbourne, which runs through the parish, rises about a mile west of this place, in the vicinity of an ancient British temple, consisting of nine rude stones of unequal height, placed in a circular form, the diameter of which is 28 feet. Half a mile to the westward are the remains of a cromlech, and there are several other erect stones in the vicinity. This neighbourhood is supposed to have been the scene of some remarkable action, from the great number of tumuli scattered about in different directions.

WINTERBOURNE-ANDERSTON, county of DORSET.—See ANDERSTON.

WINTERBOURNE-BASSETT (*St. Catherine*), a parish, in the union of MARLBOROUGH, hundred of SELKLEY, Marlborough and Ramsbury, and Northern, Divisions of the county of WILTS, 7¾ miles (N. W.) from Marlborough; containing 288 inhabitants. The living is a rectory, valued in the king's books at £18. 9. 7.; present net income, £634; patrons, President and Fellows of Magdalene College, Oxford. Among various other Druidical remains are, a double circle of rude stones, a barrow surrounded with large stones, and the supposed site of houses once occupied by archdruids.

WINTERBOURNE-CAME (*St. Peter*), a parish, in the union of DORCHESTER, partly in the hundred of CULLIFORD-TREE, and partly within that portion of the liberty of FRAMPTON which is in the Dorchester Division of the county of DORSET, 2¾ miles (S. E. by S.) from Dorchester; containing 80 inhabitants. The living is a rectory, to which that of Winterbourne-Farringdon was united in 1751, valued jointly in the king's books at £15. 5.; present net income, £251; patron, Hon. G. L. D. Damer. Here was anciently a small Benedictine nunnery, supposed to have been a cell to the abbey of Caen in Normandy.

WINTERBOURNE-CLENSTONE (*St. Nicholas*), a parish, in the union of BLANDFORD, hundred of COOMBS-DITCH, Blandford Division of the county of DORSET, 4¼ miles (S. W.) from Blandford-Forum; containing 84 inhabitants. This parish was anciently more populous and of much greater importance than it is at present: it contained three churches, the livings of which were rectories; and the foundations of numerous houses are still discernible within its limits. A little to the south of the church, on the side of a hill, commences Coombs-ditch, which gives name to the hundred, and where the courts were formerly held: it is thought by Dr. Stukeley to have been a long rampart and ditch of the first colony of the Belgæ. The living is a rectory, valued in the king's books at £6. 18. 1½.; present net income, £160; patron, E. M. Pleydell, Esq.

WINTERBOURNE-DANTSEY, or DANNERY, (*St. Edward*), a parish, in the union of AMESBURY, hundred of ALDERBURY, Salisbury and Amesbury, and Southern, Divisions of the county of WILTS, 3¾ miles (N. E. by N.) from Salisbury; containing 161 inhabitants. The living is a perpetual curacy, in the patronage of the Prebendary of Chute and Chisenbury in the Cathedral Church of Salisbury; net income, £58: the impropriation belongs to Miss M. A. Skinner.

WINTERBOURNE, EARLS (*St. Michael*), a parish, in the union of AMESBURY, hundred of ALDERBURY, Salisbury and Amesbury, and Southern, Divisions of the county of WILTS, $3\frac{1}{2}$ miles (N. E. by N.) from Salisbury; containing 243 inhabitants. The living is a perpetual curacy, in the patronage of the Prebendary of Chute and Chisenbury in the Cathedral Church of Salisbury, the appropriator; net income, £34. There is a place of worship for Wesleyan Methodists; also a National school. Near the village is an ancient earthwork, called Chlorus Camp, or Figbury Ring, from its circular form, including an area of about fifteen acres.

WINTERBOURNE-FARRINGDON, or ST. GERMAN'S, formerly a parish, now claiming to be extra-parochial, in the hundred of CULLIFORD-TREE, Dorchester Division of the county of DORSET, $2\frac{1}{2}$ miles (S.) from Dorchester: the population is returned with Herringstone. The living was a discharged rectory, valued in the king's books at £7. 3. $6\frac{1}{2}$.; in 1751, it was united to the rectory of Winterbourne Came.

WINTERBOURNE-GUNNER, or CHERBOROUGH (*St. Mary*), a parish, in the union of AMESBURY, hundred of ALDERBURY, Salisbury and Amesbury, and Southern, Divisions of the county of WILTS, 4 miles (N. E. by N.) from Salisbury; containing 166 inhabitants. The living is a rectory, valued in the king's books at £12. 16. $10\frac{1}{2}$.; present net income, £173; patron and incumbent, Rev. C. J. Colman.

WINTERBOURNE-HERRINGSTONE, county of DORSET.—See HERRINGSTONE.

WINTERBOURNE-HOUGHTON (*St. Andrew*), a parish, in the union of BLANDFORD, hundred of PIMPERNE, Blandford Division of the county of DORSET, $4\frac{1}{2}$ miles (W. S. W.) from Blandford-Forum; containing 265 inhabitants. The living is a rectory, valued in the king's books at £13. 13. 4.; patron, E. M. Pleydell, Esq. The tithes have been commuted for a rent-charge of £180, subject to the payment of rates, which on the average have amounted to £18; the glebe comprises 118 acres, valued at £80 per annum. The church is an ancient structure, remarkable only for the remains of the original rood-loft, which are still in good preservation. A school is supported by the minister. In the reign of Edward II., this place belonged to Hugh le Despencer, on whose execution at Bristol it escheated to the crown. Considerable quantities of spar are found in some coppices within the parish.

WINTERBOURNE-KINGSTON (*St. Nicholas*), a parish, in the union of BLANDFORD, hundred of BEER-REGIS, Wareham Division of the county of DORSET, $6\frac{1}{2}$ miles (S. S. W.) from Blandford-Forum; containing 564 inhabitants. The living is annexed to the vicarage of Beer-Regis; impropriators, E. M. Pleydell, Esq., and others. Here is a Sunday National school.

WINTERBOURNE-MONKTON, a parish, in the union of DORCHESTER, hundred of CULLIFORD-TREE, Dorchester Division of the county of DORSET, 2 miles (S. W. by S.) from Dorchester; containing 101 inhabitants. The living is a rectory, valued in the king's books at £8; present net income, £157; patron, Earl of Ilchester. An Alien priory, subordinate to the priory of West, or de Vasto, of the order of Clugny, is said to have existed here before the 15th of John.

WINTERBOURNE-MONKTON (*St. Mary Magdalene*), a parish, in the union of MARLBOROUGH, hundred of SELKLEY, Marlborough and Ramsbury, and Northern, Divisions of the county of WILTS, 7 miles (W. N. W.) from Marlborough; containing 263 inhabitants. The living is a discharged vicarage, united in 1747 to that of Avebury, and valued in the king's books at £5.

WINTERBOURNE-STEEPLETON (*St. Michael*), a parish, in the union of DORCHESTER, hundred of UGGSCOMBE, Dorchester Division of the county of DORSET, 4 miles (W. by S.) from Dorchester; containing 176 inhabitants. The living is a rectory, united to that of Winterbourne-Abbas, and valued in the king's books at £10. 4. 7. The church, situated in the middle of the parish, is covered with lead, and has a tower surmounted by a stone spire, which, and that at Iwerne-Minster, are the only spires in the county. Here is a school in which six children are paid for by R. W. Andrews, Esq.

WINTERBOURNE-STOKE (*St. Peter*), a parish, in the union of AMESBURY, hundred of BRANCH and DOLE, Salisbury and Amesbury, and Southern, Divisions of the county of WILTS, 5 miles (W. by S.) from Amesbury; containing 272 inhabitants. The living is a discharged vicarage, valued in the king's books at £11. 2. 8.; present net income, £172; patron and impropriator, Lord Ashburton. The church has been recently repaired and repewed, and the chancel rebuilt. A school is partly supported by private charity.

WINTERBOURNE-STRICKLAND, a parish, in the union of BLANDFORD, hundred of PIMPERNE, Blandford Division of the county of DORSET, $3\frac{3}{4}$ miles (W. & W.) from Blandford-Forum; containing 401 inhabitants. The living is a rectory, valued in the king's books at £16. 6. 3.; present net income, £367; patron, Hon. H. D. Damer. The church, situated nearly in the centre of the parish, has an embattled tower crowned with pinnacles, and was repaired about 1716. A school is partly supported by private subscription. Quarrelston House, an ancient quadrangular building, formerly the seat of the Binghams, has been, for the greater part, pulled down at different times within the last half century.

WINTERBOURNE-WHITCHURCH (*St. Mary*), a parish, in the union of BLANDFORD, hundred of COOMBS-DITCH, Blandford Division of the county of DORSET, $5\frac{1}{2}$ miles (S. W.) from Blandford-Forum; containing 513 inhabitants. The living is a discharged vicarage, valued in the king's books at £7. 16. $10\frac{1}{2}$.; present net income, £97; patron, Bishop of Salisbury; impropriators, E. M. Pleydell and H. C. Compton, Esqrs. The church is a long narrow edifice, with a south transept, and a low embattled tower rising from the intersection; it contains a curious ancient font. Here is a National school. The Rev. Samuel Wesley, father of John and Charles Wesley, founders of the sect of Methodists, and author cf several poems on religious subjects, was born here during the incumbency of his father, who was ultimately ejected for non-conformity; he died in 1735.

WINTERBOURNE-ZELSTONE (*St. Mary*), a parish, in the union of BLANDFORD, hundred of RUSHMORE, Blandford Division of the county of DORSET, $6\frac{1}{4}$ miles (S. by E.) from Blandford-Forum; containing 233 inhabitants. The living is a rectory, valued in the king's books at £13. 11. 3.; present net income, £258; patron, J. J. Farquharson, Esq. The church has a lofty embattled tower. Here is a school in which five children are paid for by subscription.

WINTERBURN, a joint township with Flasby, in the parish of GARGRAVE, union of SKIPTON, Eastern Division of the wapentake of STAINCLIFFE and EWCROSS, West Riding of the county of YORK, 7 miles (N. W. by N.) from Skipton: the population is returned with Flasby. There is a place of worship for Independents.

WINTERINGHAM (*All Saints*), a parish, in the union of GLANDFORD-BRIDGE, Northern Division of the wapentake of MANLEY, parts of LINDSEY, county of LINCOLN, 7½ miles (W.) from Barton-upon-Humber; containing 726 inhabitants. The living is a rectory, valued in the king's books at £28; present net income, £657; patron, Earl of Scarborough. There is a place of worship for Wesleyan Methodists. According to Dr. Stukeley, the ancient name of this place was *Abontrus*.

WINTERSET, a township, in the parish of WRAGBY, wapentake of STAINCROSS, West Riding of the county of YORK, 5¾ miles (S. E. by E.) from Wakefield; containing 149 inhabitants.

WINTERSLOW (*All Saints*), a parish, comprising East and West Winterslow, in the union and partly in the hundred of ALDERBURY, and partly in the hundred of AMESBURY, Salisbury and Amesbury, and Southern, Divisions of the county of WILTS; containing 749 inhabitants. East Winterslow is 7¼ miles (N. E. by E.) and West Winterslow, 6½ miles (E. N. E.) from Salisbury. The living is a rectory, valued in the king's books at £18. 13. 4.; present net income, £784; patron, Rev. H. E. Fryer. There is a place of worship for Wesleyan Methodists. A day and a Sunday school, conducted upon the National plan, are endowed with a small portion of the poor's estate: the remainder, about £28 per annum, is distributed among the poor. One poor man and one poor woman, appointed by the rector, and not receiving relief from the parish, are entitled to £10 per annum each, chargeable upon the rectorial tithes of Pitton and Farley, under the will of John Thistlethwaite, who died in 1724. Near Winterslow Hut are three barrows, in one of which was discovered, a few years ago, an arched vault, constructed of rude flints, wedged together remarkably secure, enclosing two large sepulchral urns inverted, which were found to contain ashes enveloped in linen of a very fine texture, burnt bones, beads of red amber, a metal pin, a two-edged lance-head of brass, with hair of a beautiful brown colour, and other relics, supposed to have been those of some illustrious British female. The Roman road from Salisbury to Winchester passes through this parish.

WINTERTON (*All Saints*), a parish, in the union of GLANDFORD-BRIDGE, Northern Division of the wapentake of MANLEY, parts of LINDSEY, county of LINCOLN, 8¼ miles (W. S. W.) from Barton-upon-Humber; containing 1295 inhabitants. The living is a discharged vicarage, valued in the king's books at £8; present net income, £69: it is in the patronage of the Crown; impropriator, N. Graburn, Esq. There is a place of worship for Wesleyan Methodists. A meeting of farmers, for the sale of corn, &c., takes place here every Wednesday, but there is no established market. A fair for cattle is held on July 6th, and it is intended to apply for another to be held in the autumn. Ten poor children are instructed for a rent-charge of £3. 3. a year, the bequest of Richard Beck, in 1728. Here is also a Sunday National school. Three curious tesselated pavements were discovered in 1747: the Fosse-road terminates at this place.

WINTERTON (*All Saints*), a parish, in the hundred of WEST FLEGG, Eastern Division of the county of NORFOLK, 5½ miles (N. by W.) from Caistor; containing 631 inhabitants. This place had formerly a market and a fair, which have been long disused: the fishery affords employment to a considerable number of the inhabitants. On a promontory, called Winterton Ness, are two lighthouses, one supplied with coal, the other with oil. The living is a rectory, with that of East Somerton annexed, valued in the king's books at £20. 13. 4.; present net income, £478; patron, J. Hume, Esq. Here is a National school. Several large bones were found on the cliff, in 1665, one of which, supposed to be that of a man's leg, was three feet two inches in length, and weighed 57lb.

WINTHORPE (*St. Mary*), a parish, in the union of SPILSBY, Marsh Division of the wapentake of CANDLESHOE, parts of LINDSEY, county of LINCOLN, 11 miles (E. by N.) from Spilsby; containing 244 inhabitants. The living is a discharged vicarage, united in 1729 to that of Burgh-in-the-Marsh, and valued in the king's books at £8.

WINTHORPE (*All Saints*), a parish, in the union, and Northern Division of the wapentake, of NEWARK, Southern Division of the county of NOTTINGHAM, 1¾ mile (N. N. E.) from Newark; containing 228 inhabitants. The living is a rectory, valued in the king's books at £7. 11. 0½.; present net income, £170; patron and incumbent, Rev. R. Rastall. The village is situated on the banks of the Trent: the Fosse-road passes through the parish.

WINTNEY, HARTLEY, county of SOUTHAMPTON.—See HARTLEY-WINTNEY.

WINTON, a township, in the parish of KIRKBY-STEPHEN, East ward and union, county of WESTMORLAND, 1¼ mile (N. N. E.) from Kirkby-Stephen; containing 267 inhabitants. The village is large, and the houses well built and of handsome appearance. A free school, erected by subscription in 1659, is endowed with £14. 8. a year, £11 of which arises from land bequeathed in 1681, by Robert Waller, and the rest from £112 left by Richard Munkhouse, in 1722: it is open to all children of the township, who pay a small quarterage. John Langhorne, D.D., joint translator of Plutarch, author of "Fables of Flora," and other works; his brother, William Langhorne, M.A., his assistant in the above translation; and Richard Burn, LL.D., the eminent law-writer and historian, were natives of this place, and received the rudiments of their education at the school.

WINTON, a township, in the parish of KIRBY-SIGSTON, union of NORTH-ALLERTON, wapentake of ALLERTONSHIRE, North Riding of the county of YORK, 3¾ miles (N. E. by E.) from North-Allerton; containing 145 inhabitants.

WINTRINGHAM (*St. Peter*), a parish, in the union of MALTON, wapentake of BUCKROSE, East Riding of the county of YORK; containing 589 inhabitants, of which number, 347 are in the township of Wintringham, 6½ miles (E. by N.) from New Malton. The living is a donative, in the patronage of Sir W. Strickland, Bart.

WINWICK (*All Saints*), a parish, in the union of OUNDLE, partly in the hundred of LEIGHTONSTONE, county of HUNTINGDON, and partly in that of POLEBROOK, Northern Division of the county of NORTHAMPTON, 6¼ miles (S. E.) from Oundle; containing 326 inhabitants. The living is a discharged vicarage, valued in the king's books at £7. 16. 10.; present net income, £66; patron and impropriator, Lord Montagu. Sarah Ruff, in 1721, bequeathed an estate, the rent of which, amounting to £20 per annum, is distributed in money, coal, and clothing among the poor.

WINWICK (*St. Oswald*), a parish, in the union of WARRINGTON, hundred of WEST DERBY, Southern Division of the county palatine of LANCASTER; comprising the borough town of Newton-in-Mackerfield, with several chapelries and townships, and containing 17,961 inhabitants, of which number, 603 are in the township of Winwick, 3 miles (N.) from Warrington. The living is a rectory, valued in the king's books at £102. 9. 9½.; patron, Earl of Derby. The tithes of the several townships have been commuted separately, and on the whole amount to a rent-charge of £2086, subject to the payment of rates. The impropriate tithes of the township of Culcheth have been commuted for a rent-charge of £345. The church is an ancient edifice with a lofty spire, said to be coeval with the establishment of the Christian religion in this country. A new church, built at Croft, in this parish, was consecrated by the Bishop of Chester, on the 14th of November, 1833; and Her Majesty's commissioners have agreed to afford facilities for obtaining a site for a chapel at Newton-in-Mackerfield. There is a place of worship for Wesleyan Methodists. Between this village and the town of Newton is an elevated piece of ground, called Redbank, from its having been, in 1648, the scene of an obstinately contested battle between Oliver Cromwell and the Scots, when the latter were defeated with terrible slaughter. Considerable manufactures of cotton, fustian, locks, hinges, and various other articles, are carried on within the parish. Southworth Hall, in this parish, once belonged to the Roman Catholic college of Stonyhurst, and part of it is still used as a chapel by professors of that religion. A free grammar school was founded in 1618, by Gualter Legh, Esq., who endowed it with an annuity of £10, which by subsequent benefactions was augmented to £34: the school-room was built by Sir Peter Legh. One of the prescribed rules for the government of this charity is, that if the school be vacant for one whole year, the salary for that time shall devolve to the heirs of Francis, Thomas, and Peter Legh. Here is also a National school. In observance of a custom for some time established, the rector pays annually the rental of their cottages for six poor industrious tenants, placing over the door of each a board, on which is inscribed the date when its occupant enjoyed the benefaction.

WINWICK (*Holy Trinity*), a parish, in the union of DAVENTRY, hundred of GUILSBOROUGH, Southern Division of the county of NORTHAMPTON, 8¾ miles (N. N. E.) from Daventry; containing 159 inhabitants. The living is a rectory, valued in the king's books at £15. 6. 8.; present net income, £567; patron, Bishop of Lincoln. The Grand Union canal passes through the parish. Here are the remains of an old mansion, which have been converted into a farm-house; the gateway is a curious antique structure.

WIRKSWORTH (*St. Mary*), a market-town and parish, in the union of BELPER, chiefly in the hundred of WIRKSWORTH, but partly in the hundred of APPLETREE, Southern Division, and partly in that of HIGH PEAK, Northern Division, of the county of DERBY; containing 7754 inhabitants, of which number, 4082 are in the town of Wirksworth, 13 miles (N. N. W.) from Derby, and 139 (N. W. by N.) from London. This place, formerly written *Wiresworth*, *Werchestworde*, and *Wyrkysworth*, is of very great antiquity, and is supposed to derive its name from the valuable lead-works in the neighbourhood, which, by an inscription on a pig of lead, found in 1777, appear to have been worked so early as the time of the Emperor Adrian, at the commencement of the second century; and the Saxons subsequently carried on mining operations here on an extensive scale. In 1714, Eadburga, abbess of Repton, to which abbey Wirksworth then belonged, sent hence to St. Guthlac, patron saint of Croyland abbey, a leaden coffin; and in 835, Kenwara, another abbess of Repton, granted her estate at *Wercesvorde* to Humbert, on condition that he gave annually lead worth £15 to Archbishop Ceolnoth, for the use of Christ Church at Canterbury. In Domesday-book Wirksworth is described as the property of the Crown, having a church, a priest, and three lead mines; and it remained so until King John, in the fifth year of his reign, granted it to William de Ferrers, in whose family it continued until the attainder of his descendant, Robert, in the reign of Henry III., by which monarch it was, in 1265, given to his son, Edmund Earl of Lancaster, and the manor has since that period constituted a part of the possessions of the duchy of Lancaster.

The town is situated in a valley nearly surrounded with hills, at the southern extremity of the mining district; it is not regularly lighted or paved, and is supplied with water brought by pipes from the hills on its eastern side. The chief employment of the inhabitants arises from the lead mines, but some of them are also engaged in the cotton manufacture; and there are, in the town and its immediate neighbourhood, three establishments in which common ginghams are made, and others for the productions of hosiery, hats, tape, silk, and for woolcombing. The Cromford canal, and the Cromford and High Peak railway, commence in this parish; the former about a mile and a half north of the town, near where it crosses the river Derwent, by means of an aqueduct; and the latter about half a mile north. The mines and miners of the neighbourhood are governed by ancient customs, confirmed by a commission of inquiry in 1287; and all disputes and offences are determined at the Barmote courts, held twice a year before the steward, in the moot-hall, a handsome stone building, erected in 1814, by the Hon. Charles Bathurst, then Chancellor of the duchy. In this hall is deposited the ancient brass dish, the standard from which those used for measuring the ore are made, which must be brought to be corrected by it, at least twice a year, by all the miners. The code of laws and regulations by which these courts are governed is very similar to that in force in the mining districts of the duchy of Cornwall: one remarkable custom is, that each person has the privilege of digging and searching for lead-ore in any part of the king's field, which, with a few exceptions, comprehends the whole wapentake; and should he discover a vein of lead, he has a right to work it, and erect buildings necessary for that purpose, without making any compensation to the owner of the land. A market on Wednesday, and an annual fair for three days, were granted by Edward I., in 1305, to Thomas Earl of Lancaster: Tuesday is now the market day, for provisions generally; and there are fairs on Shrove-Tuesday, Easter-Tuesday, May 12th, July 8th, September 8th, and the third Tuesday in November, for cattle, the last being also a statute fair. The town is governed by a constable and headborough, and a petty session is held by the county magistrates, on Tuesday in each week. Two courts baron, at Easter and Michaelmas, and a court leet at Easter, are held annually for the king's manor, by the lessee of the crown; and a court is also held for the rectorial manor, under the Dean of Lincoln, as appropriator: there is also another manor within the parish, for which no courts are held, called the Holland, or Richmond, manor, granted in 1553, by Henry VIII., to Ralph Gell, Esq., which now belongs to his descendant, Philip Gell, Esq., of Hopton. This town has been made a polling-place for the southern division of the county.

The living is a vicarage, valued in the king's books at £42. 7. 8½.; present net income, £164; patron and appropriator, Dean of Lincoln. The vicar is entitled, by custom, to every fortieth dish (of fourteen pints) of lead-ore raised in the parish, but the quantity of late years has been very small. The church is a handsome structure, in the later English style, with a square tower, supported in the centre by four large pillars, and contains some ancient monuments. There are places of worship for Baptists, Independents, and Wesleyan Methodists. The free grammar school, adjoining the churchyard, was founded and endowed by Anthony Gell, Esq., of Hopton, in 1576, and has been rebuilt in the English style of architecture, at an expense of about £2000: the present income is upwards of £250 per annum. This school, in common with those of Ashbourn and Chesterfield, is entitled, next after the founder's relatives, to two fellowships and two scholarships at St. John's College, Cambridge, founded by the Rev. James Beresford, vicar of this parish, who died in 1520. Here is also a school for girls, supported by subscription. Almshouses for six poor men, near the school, were also founded and endowed by Anthony Gell, Esq., the inmates receive a small monthly allowance. Elizabeth Bagshaw, in 1797, bequeathed £2000 three per cent. consols. to trustees, for the benefit of the poor, the dividends of which, amounting to £56 per annum, are distributed in sums of £1 each: there are many other small donations and bequests, producing together a considerable sum, which is annually given to the poor of the town and parish; and a library for their use has been established, and is supported by subscription. In 1736, a quantity of Roman coins was discovered; and spars, fluors, &c., were found in great variety in the neighbourhood. Here were also some mineral springs, but they have been destroyed by draining the mines.

WIRSWALL, a township, in the parish of WHITCHURCH, union and hundred of NANTWICH, Southern Division of the county of CHESTER, 5½ miles (E. S. E.) from Malpas; containing 83 inhabitants. The Chester canal passes along its western boundary.

WISBECH (*St. Mary*), a parish, in the union and hundred of WISBECH, Isle of ELY, county of CAMBRIDGE, 2 miles (N. W.) from Wisbech-St. Peter; containing 1524 inhabitants. The living is annexed to the vicarage of Wisbech-St. Peter. The appropriate tithes have been commuted for a rent-charge of £1475, and the vicarial for £879, subject to the payment of rates, which on the average have respectively amounted to £175 and £104. The church is principally in the later style of English architecture, with a large square tower. In the hamlet of Guyhirn is a neat chapel, of which the living is a perpetual curacy, in the patronage of the Vicar of Wisbech-St. Peter. A school is endowed with 21 acres of land, for which 21 children are educated; two other schools are partly supported by the vicar.

Corporation Seal.

WISBECH (*St. Peter and St. Paul*), a sea-port, borough, market-town, and parish, and the head of a union, in the hundred of WISBECH, Isle of ELY, county of CAMBRIDGE, 43 miles (N.) from Cambridge, and 94 (N. by E.) from London; containing 7253 inhabitants. This place is of great antiquity, and is noticed in a charter by which, in 664 Wulphere' son of Peada, King of the Mercians, granted to the abbey of Medehamstead, now Peterborough, "the lands from Ragwell, 5 miles to the main river that goeth to Elm and to Wisbece." In the Norman survey it is mentioned under the same appellation, which it retained till the reign of Edward I., from which period till the time of Henry VI. it was invariably written "Wysebeche." The name is supposed to be derived from the river Ouse, then called the Wise, and from the Saxon "bec," signifying either a running stream or a tongue of land at the confluence of two rivers, which, previously to the diversion of their streams, was descriptive of its situation at the confluence of the Ouse with the river Nene. From the date of Wulphere's charter little is recorded of the history of this place till the year 1000, when the manor of Wisbech is said to have been given to the abbot and convent of Ely, by Oswi, and Leoflede, daughter of Brithnod, the first abbot, on the admission into that monastery of their son Ailwin, afterwards bishop of Elmham. William the Conqueror, in the last year of his reign, erected a strong castle here, which he placed under the command of a governor, styled a constable, with a strong garrison, to keep the refractory barons in submission, and to check the ravages of the outlaws, who made frequent incursions from the neighbouring fens into the upland parts of the county. In 1190, Richard I. granted to the tenants of Wisbech Barton Manor exemption from toll in all towns or markets throughout England, which grant was confirmed by King John, who, in 1216, visited the town, and is supposed to have taken up his residence in the castle, on leaving which that monarch, attempting to cross the wash at an improper time, lost all his carriages, treasure, and regalia. The greater part of the town, together with the castle, was destroyed in 1236 by an inundation of the sea, but was soon afterwards restored; and the castle subsequently falling into dilapidation, Bishop Morton, towards the close of the 15th century, erected on its site another of brick, which became an episcopal palace of the Bishops of Ely. In the reign of Elizabeth, the castle was appropriated to the confinement of state prisoners, and during the protectorate of Cromwell it was purchased by Thurloe, afterwards his secretary, who made it his occasional residence. After the Restoration, it again reverted to the Bishops of Ely, and was sold in 1793: all remains of it have disappeared in the recent improvements of the town, which is at present the most flourishing place in the Isle of Ely.

The town is situated on both sides of the river now called the Nene, over which is a handsome stone bridge of one elliptical arch, 72 feet in the span; the streets are regularly formed, the houses in general well built, and on the site of the ancient castle, which was purchased by an architect and taken down in 1816, a handsome crescent of more than 50 houses has been erected: the town is well paved, and lighted with gas. From the late improvement in the system of draining, a great portion of previously unproductive land in the vicinity of the town has been brought into a high state of cultivation, and on every side are seen fertile corn-fields and luxuriant pastures. A permanent literary society was established in 1781, who have a library containing more than 3000 volumes: and there is also a theological library, in which are many valuable works of the most eminent of the ancient divines. There is a reading-room; also a neat theatre; and assemblies are held in rooms appropriately fitted up. A commodious building has been erected, in which are hot, cold, and sea-water baths, furnished with dressing-rooms and every requisite appendage. About a century since, the principal articles of trade were, oil, for the preparation of which there were seven mills in the town; and butter, of which not less than 8000 firkins were sent annually to London. The importance of this place as a sea-port has much increased of late years, and the trade has greatly improved; the principal exports are corn, rape seed, long wool (of which great quantities are sent to the clothing districts in Yorkshire), and timber, which is brought to this place from the county of Northampton, and it is now one of the principal places of export for wheat in the kingdom: the chief imports are wine, deals, and coal. The navigation of the river above the town was, many years since, greatly improved by a straight cut from Peterborough, forming a communication with the upland country, and supplying Peterborough, Oundle, and Northampton with various commodities; and below the town, very extensive works have been executed by the commissioners of the Nene-Out-fall, which have greatly improved the drainage of large tracts of land in the neighbourhood, and made the navigation to the sea perfect: vessels of large burden now approach the town, and load and unload at the quay and granaries. In 1839, the tonnage duties were paid on 97,119 tons. In 1794, a canal was cut from the river at Wisbech to the Old Nene at Outwell, and thence to the Ouse at Salter's Lode Sluice, opening a communication with Norfolk, Suffolk, and the eastern counties. A packet arrives at this port from Peterborough every Tuesday and Friday, and departs every Wednesday and Sunday morning. No manufactures are at present carried on, but great numbers of sheep and oxen, which are fed in the neighbouring pastures, and grow to a large size, are sent twice in the week to the London market. The market is on Saturday; and there are fairs on the Saturday before Palm-Sunday, and the Saturday before Lady-day, for hemp and flax; also a considerable horse fair, formerly held on the Wednesday, but in future to be on the Thursday, before Whit-Sunday, which is numerously attended by the London dealers; and a large cattle fair on August 12th, at which 3000 head of cattle have been brought for sale: the market and fairs are held by the corporation on lease from the Bishop of Ely, who is lord of the manor. The market-place is a spacious open area.

The guild of the Holy Trinity, established in 1379, being found at the time of the dissolution to have supported a grammar school and maintained certain piers, jetties, and banks, "against the rage of the sea," was in 1549 restored by Edward VI., who also gave the inhabitants a charter of incorporation, which was renewed by James I. in 1611, and confirmed by Charles II. in 1669. By this charter ten of the principal inhabitants, being freeholders or householders to the amount of 40*s.* per annum, were annually elected by the burgesses at large on November 2nd, for the management of the estates and affairs of the corporation; they were styled capital burgesses, and one of them was appointed town-bailiff; a town-clerk and other officers were also appointed. The corporation then exercised no magisterial authority. At present it consists of a mayor, 6 aldermen, and 18 councillors, under the act of the 5th and 6th of William IV., cap. 76, for an abstract of which see the Appendix, No. I.: the borough is divided into two wards. The mayor and late mayor are justices of the peace, and three others are appointed by the Crown. The quarter sessions for the Isle of Ely are held here and at Ely alternately; petty sessions for the division are also held here, and a court of requests for debts under 40*s*. The police establishment consists of a superintendent, sergeant, and constables. The town has also been made a polling-place for the Isle of Ely. The town-hall is embellished with the town arms, a painting of Edward VI., and portraits of Dr. Jobson, the late vicar, (who was a considerable benefactor to the town), and Thomas Clarkson, Esq., the strenuous advocate of Negro Emancipation; a part of it is let to the commissioners of the customs for the port. The shire-hall is annexed to the gaol, which was rebuilt in 1807, and contains seventeen wards for prisoners of both sexes, and two for debtors, with airing-yards, a chapel, and a tread-mill, at which eleven men can work, and which was erected at an expense of £600.

The living is a vicarage, with Wisbech-St. Mary annexed, valued in the king's books at £26. 13. 4.; present net income, £779; patron, Bishop of Ely; appropriators, Dean and Chapter of Ely. The church is a spacious ancient structure, partly Norman, but chiefly in the decorated style of English architecture, with a lofty square embattled tower in the later style; it has two naves under one roof, divided in the centre by a beautiful range of light clustered pillars with pointed arches, and from their respective aisles by low massive pillars and circular Norman arches; the north aisle of the chancel is in the decorated style, and there is a fine window of the same character at the west end of the south aisle of the nave. A handsome chapel of ease, of an octagonal form, has been erected on the opposite side of the river, in the old market, at an expense of £9364, raised by subscription among the inhabitants, to meet the liberal offer of Dr. Jobson, the late vicar, who conveyed in fee a real estate of more than £5000 in value, as an endowment for the minister, to whom the rents and profits are given in perpetuity: the chapel was opened for divine service on January 13th, 1831; the preferment is in the gift of Trustees, consisting of the subscribers to the erection of the chapel, or shareholders in their places; net income, £200. There are places of worship for Baptists, the Society of Friends, Independents, Johnsonians, Wesleyan Methodists, Presbyterians, and Unitarians. The free grammar school is of very ancient foundation, and the appointment of a master, in 1446, by the guild of the Holy Trinity, is still on record; its original endowment has been augmented by bequests from Thomas Parke and John Crane, Esqrs., for increasing the master's stipend, which, including perquisites, amount to £200 per annum. Belonging to the school are four bye-fellowships, of £10 per annum each, founded at Peter-House, Cambridge, by Thomas Parke, Esq., in 1628; and two scholarships for youths of Wisbech, originally of £8, which are now worth £70, per annum each. The school buildings occupy the site of the old town-hall, in Ship-lane. Archbishop Herring, the present Bishop of Kildare, and Thomas Clarkson, Esq., were educated at this school. A National school is partly supported by subscription, and is endowed with lands which, in 1814, were let for £55 per annum. A fund for lending money to poor tradesmen, free of interest, was bequeathed by Mr. John Crane, of Cambridge, in 1652, which was increased by a gift of £300 from Mr. William Holmes. There are several almshouses for the poor, and many valuable charities belonging to the town. The poor law union of Wisbech comprises 20 parishes or places, nearly equally divided between the counties of Cambridge and Norfolk, and contains a population of 27,703, according to the census of 1831; it is under the superintendence of 46 guardians. Here was anciently an hospital, dedicated to St. John the Baptist, of which no traces are now discernible.

WISBOROUGH-GREEN (*St. Peter*), a parish, in the union of PETWORTH, partly in the hundreds of ROTHERBRIDGE and WEST EASWRITH, but chiefly in the hundred of BURY, rape of ARUNDEL, Western Division of the county of SUSSEX, $6\frac{1}{4}$ miles (N. E. by E.) from Petworth; containing 1782 inhabitants. This parish is bounded on the north by the county of Surrey, and the village is pleasantly situated on the road from Billinghurst to Petworth. Fairs are held annually on the 16th of July and 20th of November, for horses, cattle, sheep, and pigs. There are several feeders to the river Arun in this parish, which is intersected by the Arun and Wey canal. The living is a discharged vicarage, valued in the king's books at £9. 18. $0\frac{1}{2}$.; present net income, £346; patron and appropriator, Prebendary of Wisborough in the Cathedral Church of Chichester. The church is principally in the early style of English architecture, with a tower surmounted by a lofty shingled spire, which forms a conspicuous object for many miles round; it contains some monuments to the Napper and King families. At Loxwood End, in this parish, is a chapel. There is also a place of worship for Independents. The workhouse has been appropriated for the poor children of the union.

WISETON, or WYESTON, a township, in the parish of CLAYWORTH, union of EAST RETFORD, North-Clay Division of the wapentake of BASSETLAW, Northern Division of the county of NOTTINGHAM, 5 miles (E. S. E.) from Bawtry; containing 118 inhabitants. The Chesterfield canal passes in the vicinity. There is a school in which four children are instructed at the expense of Earl Spencer.

WISHAW (*St. Chad*), a parish, in the union of ASTON, Birmingham Division of the hundred of HEMLINGFORD, Northern Division of the county of WARWICK, $3\frac{3}{4}$ miles (E. S. E.) from Sutton-Coldfield; containing 216 inhabitants. The living is a rectory, valued in the king's books at £5. 5.; present net income, £370: it is in the patronage of the Folliet and Jesson families. The Birmingham and Fazeley canal passes through the parish. Lady Hackett, in 1710, gave £100, directing the interest to be applied for teaching six children: there is also a rent-charge of ten shillings, bequeathed in 1744, by Thomas Bayliss, for the education of one poor boy: the school is further supported by the Hon. Mrs. Berkeley Noel.

WISHFORD, GREAT (*St. Giles*), a parish, in the union of WILTON, hundred of BRANCH and DOLE, Salisbury and Amesbury, and Southern, Divisions of the county of WILTS, $2\frac{1}{2}$ miles (N. N. W.) from Wilton; containing 361 inhabitants. The living is a rectory, valued in the king's books at £17. 10. $7\frac{1}{2}$.; patron, Earl of Pembroke. The tithes have been commuted for a rent-charge of £430, subject to the payment of rates, which on the average have amounted to £66; the glebe comprises $17\frac{1}{2}$ acres, valued at £36. 17. per annum. Sir Richard Howe, Bart., in 1728, founded a free school, and endowed it with the tithes of Asserton, in the parish of Berwick-St. James, which now produce £63. 5. per annum. Sir Richard Grobham, in 1628, founded an almshouse for four aged men of this parish, and endowed it with property now producing a rental of about £80: the almshouse consists of four tenements under one roof, a common bakehouse, and a garden; each tenement contains a kitchen, a scullery, and a room above. A fund for apprenticing children, amounting to about £10. 10. per annum, was bequeathed by Daniel Oland, in 1735.

WISHFORD, LITTLE, a tything, in the parish of SOUTH NEWTON, hundred of BRANCH and DOLE, Salisbury and Amesbury, and Southern, Divisions of the county of WILTS, 3 miles (N. N. W.) from Wilton: the population is returned with the parish.

WISLEY, a parish, in the union of GUILDFORD, Second Division of the hundred of WOKEING, Western Division of the county of SURREY, $2\frac{1}{2}$ miles (N. by E.) from Ripley; containing 155 inhabitants. The living is a discharged rectory, with the vicarage of Pyrford annexed, valued in the king's books at £40. 19.; present net income, £210; patron, Earl of Onslow. The church is an ancient edifice, in the early English style. The parish participates in the benefits of Henry Smith's charity.

WISPINGTON (*St. Margaret*), a parish, in the union of HORNCASTLE, Southern Division of the wapentake of GARTREE, parts of LINDSEY, county of LINCOLN, $4\frac{1}{4}$ miles (W.N. W.) from Horncastle; containing

91 inhabitants. The living is a discharged vicarage; net income, £190; patron and impropriator, C. Turnor, Esq.

WISSETT (*St. Andrew*), a parish, in the union and hundred of Blything, Eastern Division of the county of Suffolk, 2 miles (N. W.) from Halesworth; containing 419 inhabitants. The living is a perpetual curacy; net income, £105; patron and impropriator, Sir E. C. Hartopp, Bart. The church has two Norman doors and a round tower.

WISTANSTOW (*Holy Trinity*), a parish, in the union of Church-Stretton, partly in the hundred of Purslow, but chiefly in that of Munslow, Southern Division of the county of Salop, 9¾ miles (N. W. by N.) from Ludlow; containing 989 inhabitants. The living is a rectory, valued in the king's books at £18; present net income, £764; patron, Earl of Craven. Here is a National school.

WISTASTON (*St. Mary*), a parish, in the union and hundred of Nantwich, Southern Division of the county of Chester, 2½ miles (N. E. by E.) from Nantwich; containing 350 inhabitants. The living is a discharged rectory, valued in the king's books at £4. 0. 3.; present net income, £139; patron, J. W. Hammond, Esq. The church was rebuilt in 1826, and 120 free sittings provided, the Incorporated Society having granted £100 in aid of the expense. A school is supported by a private individual. About £15 per annum, the rent of an estate purchased with bequests by Ann Ball, in 1604, and others, is distributed on St. Thomas' day among poor parishioners.

WISTESTON, a chapelry, in the parish of Marden, hundred of Broxash, county of Hereford, 7 miles (N. by E.) from Hereford: the population is returned with the parish. The living is a perpetual curacy; net income, £48; patron, W. Vale, Esq.; impropriator, James Beebee, Esq.

WISTON, otherwise WISSINGTON (*St. Mary*), a parish, in the union of Sudbury, hundred of Babergh, Western Division of the county of Suffolk, 1½ mile (W. by S.) from Nayland; containing 249 inhabitants. The living is a vicarage, endowed with the rectorial tithes, and valued in the king's books at £4. 19. 4½.: it is in the patronage of the Crown. The tithes have been commuted for a rent-charge of £440, subject to the payment of rates. The church has a rich and very curious Norman door, with a north entrance in the same style; and the chancel is separated from the nave by an enriched arch. The river Stour bounds the parish on the south, where it is crossed by a bridge.

WISTON (*St. Mary*), a parish, in the union of Thakeham, hundred of Steyning, rape of Bramber, Western Division of the county of Sussex, 1½ mile (N. N. W.) from Steyning; containing 296 inhabitants. The living is a rectory, valued in the king's books at £12. 13. 4.; present net income, £340; patron, C. Goring, Esq. The church is principally in the decorated style of English architecture. Here is a National school.

WISTOW (*St. John the Baptist*), a parish, in the union of St. Ives, hundred of Hurstingstone, county of Huntingdon, 3¾ miles (S. S. W.) from Ramsey; containing 404 inhabitants. The living is a rectory, valued in the king's books at £10. 17. 8½.; present net income, £354; patron, Rev. G. Mingaye.

WISTOW (*St. Winston*), a parish, in the union of Billesdon, hundred of Gartree, Southern Division of the county of Leicester, 7¼ miles (S. E. by S.) from Leicester; containing 298 inhabitants. The living is a discharged vicarage, valued in the king's books at £8. 18. 4.; present net income, £92; patron and impropriator, Sir Henry Halford, Bart., by whom the church, which contains a splendid marble monument to the first baronet, has been repaired and beautified at a considerable expense. There is a chapel of ease at Newton-Harcourt, in this parish. On the night previous to the battle of Naseby, King Charles I. slept at Wistow Hall, in this parish, which is now the seat of Sir Henry Halford, Bart., G.C.B. The house has been considerably enlarged, and the grounds laid out with much taste by the present proprietor, at an expense of £20,000. In the library is a splendid clock, ornamented with a bust of George IV., studded with diamonds, and valued at 600 guineas, which, shortly after the decease of that monarch, was presented by six members of the Royal family to Sir Henry Halford, as a tribute to his skill and assiduity as their physician. In the grounds are a male and female Emu, given by George IV., and with that sovereign's permission adopted by Sir Henry as his armorial bearings; also several kangaroos, presented by William IV., a Paul bull and cow, and several other foreign beasts and birds. The parish is intersected by the union canal.

WISTOW (*All Saints*), a parish, in the union of Selby, partly within the liberty of St. Peter's, East Riding, but chiefly in the Lower Division of the wapentake of Barkstone-Ash, West Riding, of the county of York, 3 miles (N. W. by N.) from Selby; containing 665 inhabitants. The living is a discharged vicarage, in the patronage of the Prebendary of Wistow in the Cathedral Church of York (the appropriator), valued in the king's books at £8; present net income, £221. The church is an ancient structure, with a tower. There are places of worship for Wesleyan and Primitive Methodists. A school is endowed with £5 per annum by Mr. Shaw, for the education of ten poor boys.

WISWELL, a township, in the parish of Whalley, union of Clitheroe, Higher Division of the hundred of Blackburn, Northern Division of the county palatine of Lancaster, 3 miles (S. by E.) from Clitheroe; containing 724 inhabitants. The extension of the manufacture in spinning cotton-thread, and the weaving and printing of calico, have caused a considerable increase in the population of this township within the last few years. Here is a Sunday National school.

WITCHAM (*St. Martin*), a parish, in the hundred of South Witchford, union and Isle of Ely, county of Cambridge, 5½ miles (W.) from Ely; containing 519 inhabitants. The living is a vicarage, valued in the king's books at £8. 11. 0½.; present net income, £100; patrons and appropriators, Dean and Chapter of Ely. There is a place of worship for Wesleyan Methodists; also a Sunday school supported by the curate.

WITCHAMPTON (*All Saints*), a parish, in the union of Wimborne and Cranborne, hundred of Cranborne, Wimborne Division of the county of Dorset, 4½ miles (N. by W.) from Wimborne-Minster; containing 481 inhabitants. The living is a rectory, valued in the king's books at £12. 12. 3½.; patron, H. C. Sturt, Esq. The tithes have been commuted for a rent-charge of £269. 10. 9., subject to the payment of rates; the glebe comprises 19½ acres. The church is a large handsome edifice, in the decorated style of English architecture, with a square tower. There is a place of worship for Wesleyan Methodists; also a school supported by C. Sturt, Esq. This parish is intersected by the rapid river Allen, on the banks of which is a paper-mill, affording employment to 20 persons. Four almshouses for aged persons were erected in 1831, at the expense of Mr. Sturt. Here are the remains of an old monastery, once subordinate to the abbots of Crawford, and now converted into a barn, the property of Sir George Bingham.

WITCHFORD (*St. Nicholas*), a parish, in the hundred of South Witchford, union and Isle of Ely, county of Cambridge, 3 miles (W. S. W.) from Ely; containing 461 inhabitants. The living is a vicarage, in the patronage of the Dean and Chapter of Ely (the appropriators), valued in the king's books at £9. 18. 9.; present net income, £142. A school is partly supported by the rector.

WITCHINGHAM, GREAT (*St. Mary*), a parish, in the union of St. Faith, hundred of Eynsford, Eastern Division of the county of Norfolk, 2 miles (S.) from Reepham; containing 582 inhabitants. The living is a discharged vicarage, with the rectory of Little Witchingham annexed, valued in the king's books at £4. 17. 11.; present net income, £658; patrons and impropriators, Warden and Fellows of New College, Oxford. Here is a National school, partly supported by an endowment of ten guineas per annum.

WITCHINGHAM, LITTLE (*St. Faith*), a parish, in the union of St. Faith, hundred of Eynsford, Eastern Division of the county of Norfolk, 2¼ miles (S. E. by S.) from Reepham; containing 62 inhabitants. The living is a rectory, annexed to the vicarage of Great Witchingham, and valued in the king's books at £5.

WITCHLING (*St. Margaret*), a parish, in the union of Hollingbourn, hundred of Eyhorne, lathe of Aylesford, Western Division of the county of Kent, 2½ miles (N. N. E.) from Lenham; containing 128 inhabitants. The living is a discharged rectory, valued in the king's books at £4. 1. 8.; present net income, £158; patron, Rev. Edwin Bosanquet.

WITCOMB-MAGNA (*St. Mary*), a parish, in the union of Cheltenham, Upper Division of the hundred of Dudstone and King's-Barton, Eastern Division of the county of Gloucester, 3½ miles (N. E. by N.) from Painswick; containing 174 inhabitants. The living is a discharged rectory, valued in the king's books at £4. 6. 8.; patron, Sir W. Hicks, Bart. The tithes have been commuted for a rent-charge of £132, subject to the payment of rates, which on the average have amounted to £10; the glebe comprises one acre, valued at £2 per annum. A small school is supported by subscription, and there is a Sunday National school. Near the foot of Cooper's Hill, on a delightful spot in this parish, the remains of a Roman villa, with a sacrarium, baths, &c., were discovered in 1818, the walls of which, to the height of nearly six feet, are remaining, some of them being covered with stucco painted in panels of different colours, elegantly ornamented with ivy leaves. Several of the apartments were paved with red sandstone, others with beautiful mosaic work, and in many of them have been found fragments of columns, and cornices of white marble, numerous coins of the Lower Empire, from Constantine to Valentinian and Valens, various domestic utensils, and other relics in copper and iron, an axe, a British hatchet, and the skulls of bullocks and goats, with fragments of stags' horns, &c.

WITHAM (*St. Nicholas*), a market-town and parish, and the head of a union, in the hundred of Witham, Northern Division of the county of Essex, 8 miles (N. E. by E.) from Chelmsford, and 37 (N. E. by E.) from London; containing 2735 inhabitants. The original erection of this town, or at least that part of it which is situated on Cheping Hill, is attributed to Edward the Elder, about the commencement of his reign, and was subsequently in the possession of the Knights Templars, who had a preceptory at Cressing, about three miles distant. Some consider this to have been the Roman station *Canonium* of Antoninus, which opinion receives confirmation from the quantity of Roman bricks in the walls of the church, and from Roman coins of different Emperors, which have been discovered in levelling the fortifications. There are some remains of a circular camp, defended by a double vallum, yet visible in the vicinity of the town. A mansion here, formerly the property of the Earl of Abercorn, and now in the possession of the Rev. Henry Du Cane, has been repeatedly honoured by the presence of royalty; George II. rested at it in his progress to and from his Hanoverian dominions, and Queen Charlotte, consort of George III., was received here on her first arrival in England. The town, which is pleasantly situated near the confluence of a small stream, called the Braine, with the river Blackwater, on the main road from London to Colchester, is of respectable appearance, and consists principally of one long street; it is lighted with gas, paved, and amply supplied with water from wells. There are no branches of manufacture; the principal trade is what arises from its situation on a great public thoroughfare, and the requisite supply of the inhabitants. The market, granted by Richard I., and kept originally at Cheping Hill, from which it was removed by Richard II., is on Tuesday; and fairs are held on the Monday before Whit-Sunday, June 4th, and Sept. 14th. The county magistrates hold here petty sessions for the division, every Tuesday; and manorial courts are held, as occasion requires, at which constables and other officers are appointed.

The living is a vicarage, valued in the king's books at £22. 0. 7½.; present net income, £473; patron and appropriator, Bishop of London. The church, situated at Cheping Hill, about half a mile north of the town, is a spacious and handsome edifice, with a tower of brick, in the later style of English architecture, and contains many ancient monuments, and a large tomb erected in the reign of Elizabeth, to the memory of Judge Southcote and his lady, by whose effigies it is surmounted. The vicarage-house is a handsome residence. There are places of worship for Baptists, the Society of Friends, Independents, and Roman Catholics. A National school is supported partly by the rent of a house conditionally bequeathed in 1630, by Dame Catherine Barnardiston, and partly by voluntary contributions. Another school for girls is partly supported by subscription. Two almshouses on Cheping Hill, for four poor widows, were endowed by Thomas Green, Esq., in 1491, with a farm in Springfield, let for £80 a year; the inmates have an allowance of money and fuel: this bequest, having been lost for 80 years, was recovered in Chancery, chiefly through the care and exertions of Dr. Warley, vicar of the parish. An almshouse for two poor widows was founded in the reign of Charles I., by means of a bequest from George Armond, Esq.: others, founded by Matthew Harvey, Esq., are now occupied by nine poor persons chosen by the pastor and deacons of the Independent chapel; and five, founded by an unknown donor, for ten poor widows, were endowed, in 1687, with a farm at Goldhanger, and another at Fairstead, and have an income of £165 per annum. Dr. Warley, amongst other benefactions, in 1719, bequeathed £100 in aid of an orthodox school for poor children, to be elected by the master and churchwardens for the time being. C. Barnardiston bequeathed £100 to be distributed in bread and fuel to the poor. The poor law union of Witham comprises 17 parishes or places, under the superintendence of 23 guardians, and contains, according to the census of 1831, a population of 14,432. There is a mineral spring in the neighbourhood, which was formerly in great repute.

WITHAM-FRIARY (*St. Mary*), a parish (formerly an extra-episcopal liberty), in the union and hundred of FROME, Eastern Division of the county of SOMERSET, 5½ miles (S. S. W.) from Frome; containing, exclusively of Charter-house-on-Mendip, which is in the hundred of Winterstoke, 574 inhabitants. This parish comprises about 5414 statute acres, of which about 4000 are at present the property of the Duke of Somerset; it was originally the property of the Wyndham family, and a splendid residence was built here by the Earl of Egremont, who died in 1763. The living is a perpetual curacy; net income, £106; patron, Duke of Somerset. In 1828, the church was repaired and considerably enlarged by subscription, aided by a grant of £100 from the Incorporated Society, in consideration of which 102 additional sittings are free; and in 1830 a neat parsonage-house was erected near the village. A capacious school-room was built near the church by the Duke of Somerset. Although the parish was known to be of peculiar and exempt jurisdiction, there are no records showing that any privileges were exercised prior to the year 1785, when an official was appointed: the manor having come into the possession of the Duke of Somerset, his Grace, in 1826, signified his intention of relinquishing his peculiar rights, provided the Governors of Queen Anne's bounty would contribute to the augmentation of the living; and having charged his estate in the parish with a permanent stipend for the minister, the usual licence to the incumbent was issued by the Bishop of Bath and Wells, in 1827, and no peculiar rights whatever are now exercised. Here was anciently a nunnery; and subsequently, in 1181, a monastery, said to be the first establishment of Carthusians in England, was founded by Henry II., in honour of the Blessed Virgin, St. John the Baptist, and All Saints, which at the dissolution had a revenue of £227. 1. 8.: the ruins were taken down in 1764, and a farm-house now stands upon its site.

WITHAM, NORTH (*St. Mary*), a parish, in the union of GRANTHAM, wapentake of BELTISLOE, parts of KESTEVEN, county of LINCOLN, 1½ mile (S. by W.) from Colsterworth; containing 273 inhabitants. The living is a rectory, valued in the king's books at £6. 19. 2.; present net income, £300; patron, Viscount Downe.

WITHAM-ON-THE-HILL (*St. Andrew*), a parish, in the union of BOURNE, wapentake of BELTISLOE, parts of KESTEVEN, county of LINCOLN, 4¼ miles (S. W.) from Bourne; containing 530 inhabitants. The living is a vicarage, valued in the king's books at £6. 1. 0½.; present net income, £107; patron and impropriator, General Johnson. There is a place of worship for Wesleyan Methodists. James Thompson, in 1719, bequeathed a rent-charge of £4 and the produce of certain land, for teaching twenty poor children.

WITHAM, SOUTH (*St. John the Baptist*), a parish, in the union of GRANTHAM, wapentake of BELTISLOE, parts of KESTEVEN, county of LINCOLN, 3¼ miles (S. by W.) from Colsterworth; containing 410 inhabitants. The living is a discharged rectory, valued in the king's books at £3. 12. 11.; present net income, £120; patron, Lord Huntingtower. The river Witham has its source in this parish. Three poor children are instructed for an annuity of £1. 10., the gift of the Rev. Mr. Troughton. A preceptory of Knights Templars existed here so early as 1164, which afterwards devolved upon the Hospitallers.

WITHCALL (*St. Martin*), a parish, in the union of LOUTH, Wold Division of the hundred of LOUTH-ESKE, parts of LINDSEY, county of LINCOLN, 3¾ miles (S. W. by W.) from Louth; containing 72 inhabitants. The living is a discharged rectory, valued in the king's books at £11. 16. 10.; present net income, £516: it is in the patronage of the Crown.

WITHCOTE, a parish, in the union of BILLESDON, and forming, with the parishes of Cold-Overton and Somerby, a detached portion of the hundred of FRAMLAND, locally in that of East Goscote, Northern Division of the county of LEICESTER, 9½ miles (S. S. E.) from Melton-Mowbray; containing 32 inhabitants. The living is a discharged rectory, valued in the king's books at £6. 9. 4½.; present net income, £133; patron, Rev. Henry Palmer, proprietor of the lordship, which comprises from 700 to 800 acres, who has elegantly rebuilt the church, and thoroughly repaired the family mansion, and enlarged it with an additional wing.

WITHERIDGE (*St. John the Baptist*), a parish, in the union of SOUTH MOLTON, hundred of WITHERIDGE, South Molton and Northern Divisions of the county of DEVON, 8¼ miles (E.) from Chulmleigh; containing 1263 inhabitants. The living is a vicarage, valued in the king's books at £23. 10. 5.; patron, incumbent, and impropriator, Rev. W. P. Thomas. The impropriate and vicarial tithes have each been commuted for rent-charges of £350, subject to the payment of rates, which on the average on each have amounted to £26. The church has a stone pulpit highly enriched. This is a decayed borough and market-town. A fair for cattle is held on June 24th; and there are still great markets on the Wednesday after September 21st, and the first Wednesday in November. Richard Melhuish, Esq., in 1799, gave £500 stock, the dividends arising from which are applied to the instruction of 40 poor children. William Chapple, the antiquary, was born here; he died in 1755.

WITHERLEY (*St. Peter*), a parish, in the hundred of SPARKENHOE, Southern Division of the county of LEICESTER, 1½ mile (E. by S.) from Atherstone; containing 492 inhabitants. The living is a rectory, valued in the king's books at £16. 2. 3½.; present net income, £500; patron and incumbent, Rev. J. Roberts. The church has one of the finest spires in the county; it is 156 feet high. The old Watling-street, which here separates Leicestershire from Warwickshire, crosses the river Anker at Witherley bridge. In this parish is Mancetter, the site of the great Roman station *Manduessedum*.

WITHERN (*St. Margaret*), a parish, in the union of Louth, Wold Division of the hundred of Calceworth, parts of Lindsey, county of Lincoln, 4½ miles (N. N. W.) from Alford; containing 390 inhabitants. The living is a rectory, valued in the king's books at £18. 10. 2½.; patron, Robert Vyner, Esq. The tithes have been commuted for a rent-charge of £484, subject to the payment of rates, which on the average have amounted to £67; the glebe comprises 49 acres, valued at £45 per annum. There is a place of worship for Wesleyan Methodists. George Stovins, in 1726, bequeathed £100, the interest of which, with that of £50 previously given by the Rev. William Jones, is applied for teaching and partly clothing poor children.

WITHERNSEA, a chapelry, in the parish of Hollym, union of Patrington, Southern Division of the wapentake of Holderness, East Riding of the county of York, 19 miles (E. by S.) from Kingston-upon-Hull; containing 130 inhabitants. The chapel, dedicated to St. Nicholas, and now in ruins, was apparently the remains of a magnificent building, probably the church of a priory which existed here in the reign of John, a cell to the abbey of Albemarle, in France.

WITHERNWICK (*St. Alban*), a parish, in the union of Skirlaugh, partly within the liberty of St. Peter's, but chiefly in the Northern Division of the wapentake of Holderness, East Riding of the county of York, 12 miles (E.) from Beverley; containing 443 inhabitants. The living is a discharged rectory, in the patronage of the Prebendary of Holme in the Cathedral Church of York, valued in the king's books at £6. 7. 1. The church is an ancient structure. There is a place of worship for Wesleyan Methodists. Matthias North gave land producing a trifling income for teaching and apprenticing children of this parish.

WITHERSDALE (*St. Mary Magdalene*), a parish, in the union and hundred of Hoxne, Eastern Division of the county of Suffolk, 3 miles (S. E. by E.) from Harleston; containing 194 inhabitants. The living is a discharged rectory, annexed to the vicarage of Fressingfield, and valued in the king's books at £6. 16. 8. The church is a small edifice without either tower or spire.

WITHERSFIELD (*St. Mary*), a parish, in the union and hundred of Risbridge, Western Division of the county of Suffolk, 2¼ miles (N. W. by N.) from Haverhill; containing 545 inhabitants. The living is a rectory, valued in the king's books at £9. 17. 1.; present net income, £465; patron, G. T. W. H. Duffield, Esq. Here is a Sunday National school.

WITHERSLACK, a chapelry, in the parish of Beetham, union and ward of Kendal, county of Westmorland, 7½ miles (W. N. W.) from Milnthorpe; containing 488 inhabitants. The living is a perpetual curacy; net income, £93; patrons, Trustees of Barwick's charity. The chapel, dedicated to St. Paul, was built and endowed in 1664, by Dr. John Barwick, a native of this place, and Dean of St. Paul's, London, who bequeathed the impropriate rectory of Lazonby, to which his brother, Peter Barwick, Esq., M.D., added an estate near Kirk-Oswald, to provide an annuity of £26 to the curate for teaching 40 children, one of £4 for repairing the chapel, and another of £10 for placing out apprentices and as a marriage portion to poor maidens within the chapelry. These allowances have been considerably augmented by the increased value of the lands, which now let for about £400 a year, and the treasurers were enabled, in 1824, to erect a girls' school on the same foundation, and have given several marriage portions, of £30 and £40 each, to deserving females. The fishery in the river Belo, which passes through the chapelry, belongs to the Earl of Derby, who holds his manorial court at the Derby Arms, on the second Tuesday after Trinity: the ancient hall has been converted into a farm-house. About a mile from the chapel a chalybeate spring was discovered, and named Holy Well, in 1656, but it has since disappeared.

WITHERSTONE, a parish, in the hundred of Eggerton, Bridport Division of the county of Dorset, 5 miles (E. N. E.) from Bridport: the population is returned with the parish of Poorstock. This ancient parish is now almost depopulated, being reduced to one farm-house, the occupier of which pays church and poor rates to Poorstock, the parish church having been suffered to go to decay soon after the Reformation. The living is a sinecure rectory, valued in the king's books at £2. 13. 4.; present net income, £94; patron, Earl of Dorchester.

WITHIEL (*St. Uvell*), a parish, in the union of Bodmin, Eastern Division of the hundred of Pyder and of the county of Cornwall, 5 miles (W. by S.) from Bodmin; containing 406 inhabitants. The living is a rectory, valued in the king's books at £10; present net income, £324; patron, Sir R. R. Vyvyan, Bart. The church contains an ancient font enriched with sculpture: it was repaired and repewed in 1820, by the late rector, and a gallery and organ were erected in 1831, by the Rev. V. F. Vyvyan, the present rector. This parish formerly belonged to the priory at Bodmin, of which establishment Prior Vyvyan, whose tomb is in the church of that place, founded the church of Withiel. At the entrance to the rectory-house is one of those ancient crosses with which this part of the country abounds. A National day and Sunday school are chiefly supported by the rector. Several sepulchral urns have been dug up at different times in various parts of the parish. Sir Beville Grenville, a distinguished royalist commander during the civil war of the 17th century, was born at Brynn, in this parish.

WITHIELL-FLOREY (*St. Mary Magdalene*), a parish, in the union of Williton, hundred of Taunton and Taunton-Dean, Western Division of the county of Somerset, 5¾ miles (N. E.) from Dulverton; containing 89 inhabitants. The living is a perpetual curacy; net income, £59; patron, Sir T. B. Lethbridge, Bart.

WITHINGTON (*St. Michael*), a parish, in the union of Northleach, hundred of Bradley, Eastern Division of the county of Gloucester, 6 miles (W.) from Northleach; containing 743 inhabitants. The living is a rectory, valued in the king's books at £30; present net income, £686; patron, Bishop of Worcester. The church is a cruciform structure, principally in the Norman style, and partly of later date: among the monuments is a handsome one to the memory of Sir John How, his wife, and nine children, in a small cross aisle on the south side of the church, the burial-place of that family. There are two schools, one for boys and one for girls, and one in the hamlet of Foxcote, supported by the interest of money left to the parish. The Rev. W. Osborn, D.D., who held the living, left £100 for apprenticing poor children, and John Rich, Esq., gave £100 also for the same purpose: there are other benefactions for the use of the poor. A Roman pavement was discovered in this parish, in 1811, a part of which was deposited in the British Museum.

WITHINGTON (*St. Peter*), a parish, in the hundred of Broxash, union and county of Hereford, 4½ miles (E. N. E.) from Hereford; containing 723 inhabitants. The living is a discharged vicarage, with the perpetual curacy of Preston-Wynne annexed, valued in the king's books at £5. 1.; present net income, £250; patron, Dean of Hereford, who, with the Chapter, is the appropriator. There is a place of worship for Baptists.

WITHINGTON, a township, in the parish of Manchester, union of Chorlton, hundred of Salford, Southern Division of the county palatine of Lancaster, 4 miles (S.) from Manchester; containing 1048 inhabitants.

WITHINGTON (*St. John the Baptist*), a parish, in the union of Atcham, Wellington Division of the hundred of South Bradford, Northern Division of the county of Salop, 6¼ miles (E.) from Shrewsbury; containing 193 inhabitants. The living is a perpetual curacy; net income, £80; patron and appropriator, Rector of Upton-Magna. The Shrewsbury canal passes through the parish.

WITHINGTON, LOWER, a township, in the parish of Prestbury, union and hundred of Macclesfield, Northern Division of the county of Chester, 7 miles (N. N. W.) from Congleton; containing 584 inhabitants. There is a place of worship for Wesleyan Methodists. Tunsted, a hill in this township, is supposed, from its Saxon etymology, *viz.*, the place of a town, to have been the site of an ancient ville of some consequence. A school is supported by Mrs. Parker, for the instruction of poor children.

WITHINGTON, a hamlet, in the parish of Leigh, Southern Division of the hundred of Totmonslow, Northern Division of the county of Stafford: the population is returned with the parish.

WITHINGTON, OLD, a township, in the parish of Prestbury, union and hundred of Macclesfield, Northern Division of the county of Chester, 7½ miles (N. N. W.) from Congleton; containing 191 inhabitants. A school containing eight girls is supported by a lady.

WITHNELL, a township, in the parish and hundred of Leyland, union of Chorley, Northern Division of the county palatine of Lancaster, 5 miles (N. E. by N.) from Chorley; containing 1251 inhabitants.

WITHYBROOK (*All Saints*), a parish, in the union of Foleshill, Kirby Division of the hundred of Knightlow, Northern Division of the county of Warwick, 8 miles (N. E. by E.) from Coventry; containing 318 inhabitants. The living is a discharged vicarage, annexed to that of Monk's-Kirby, and valued in the king's books at £8. 6. 8. The Oxford canal passes through the parish.

WITHYCOMBE (*St. Nicholas*), a parish, in the union of Williton, hundred of Carhampton, Western Division of the county of Somerset, 2½ miles (S. E.) from Dunster; containing 332 inhabitants. The living is a rectory, valued in the king's books at £10. 11. 5½.; present net income, £203; patron, T. Hutton, Esq. In this parish is a Druidical circle, formed of rude stones, not far from which are two cairns.

WITHYCOMBE-RAWLEIGH, a parish, in the union of St. Thomas, hundred of East Budleigh, Woodbury and Southern Divisions of the county of Devon; containing 1063 inhabitants. The living is a perpetual curacy, annexed to the vicarage of East Budleigh. The impropriate tithes have been commuted for a rent-charge of £200, and the vicarial for £210, subject to the payment of rates, which on the latter have on the average amounted to £10; the glebe comprises 4 acres, valued at £6 per annum. A portion of the ancient church was taken down about 1745, and a new church erected about half a mile from Exmouth. A day and Sunday school is supported by subscription; in another, 6 girls are clothed and instructed from the proceeds of an endowment. This parish is bounded on the west by the navigable river Exe. The manor was formerly held by the tenure of finding the king, whenever he should hunt in Dartmoor, two good arrows stuck in an oaten cake. Here is an almshouse, with a trifling endowment.

WITHYHAM (*St. Michael*), a parish, in the union of East Grinstead, hundred of Hartfield, rape of Pevensey, Eastern Division of the county of Sussex, 7½ miles (E. S. E.) from East Grinstead; containing 1610 inhabitants. The living is a rectory, valued in the king's books at £25. 5. 5.; present net income, £717; patron, Earl of Delawarr. The church is principally in the later style of English architecture; it was rebuilt in 1624, by Richard, Earl of Dorset, who was interred here. Her Majesty's Commissioners have agreed to afford facilities for obtaining a site for a church at Crowborough. A National school has been established.

WITHYPOOLE (*St. Andrew*), a parish, in the union of Dulverton, hundred of Williton and Freemanners, Western Division of the county of Somerset, 7 miles (N. W.) from Dulverton; containing 212 inhabitants. The living is a perpetual curacy, annexed to the rectory of Hawkridge.

WITLEY (*All Saints*), a parish, in the union of Hambledon, Second Division of the hundred of Godalming, Western Division of the county of Surrey, 4 miles (S. W. by S.) from Godalming; containing 1376 inhabitants. This parish is situated on the road from Godalming to Petworth; the land is elevated, and a telegraph has been erected here, which forms one of the stations on the Portsmouth line. A pleasure fair is held annually on the 23d of April. The living is a discharged vicarage, with that of Thursley annexed, valued in the king's books at £17. 15. 10.; present net income, £182; patron and incumbent, Rev. J. Chandler; impropriators of Witley, J. Leech, Esq., and the Rev. J. Chandler; impropriator of Thursley, J. Knowles, Esq. The church is a cruciform structure, principally in the early style of English architecture, with a central tower surmounted by a spire, and contains monuments to the Chandler and Webb families, and some ancient brasses with Norman details. A chapel of ease, dedicated to St. John the Evangelist, has been built at Milford by subscription. There is a place of worship for Calvinists. A parochial school is supported by subscription; and the parish participates in Henry Smith's charity. Many ancient coins of gold and silver have been discovered here.

WITLEY, GREAT (*St. Michael*), a parish, in the union of Martley, Lower Division of the hundred of Doddingtree, Hundred-House and Western Divisions of the county of Worcester, 4 miles (S.) from Stourport; containing 386 inhabitants. The living is a rectory, valued in the king's books at £7. 6. 3.; present net income, £391; patron, Lord Ward. The church, which adjoins the mansion of Witley Court, is an elegant structure, erected about 1760, by the first Lord Foley and his widow, to whom it contains a superb monument, by Rysbrach; the windows painted by Price, in 1719, and the ceiling by Verrio, were brought from the chapel at Canons, when that princely mansion of the late Duke of Chandos was pulled down, and the materials sold. This parish is intersected by the roads from Worcester to Ludlow, and from Staffordshire into Herefordshire. Witley Court, until recently the seat of the Foley family, and now, by purchase, the property of Lord Ward, is a spacious and elegant mansion, beautifully situated in a park abounding with picturesque scenery, and of which a part formerly belonged to a religious house, and pays no tithes. The female part of the population are employed in making gloves for the manufacturers of Worcester. Pure limestone is burnt for the purposes of building and agriculture, and limestone of an inferior sort is procured for the use of the roads. An abundant supply of water is obtained from the Abberley and Woodbury hills, which are composed of ferruginous or basaltic gravel. The interest of £126 consolidated annuities, purchased with various bequests, is annually distributed in money, tea, coal, &c. The poor children of the parish attend the Lancasterian school in the adjoining parish of Abberley. Fossils of various descriptions are found in the parish, and on the crown of Woodbury Hill are the remains of a British camp.

WITLEY, LITTLE, a chapelry, in the parish of Holt, union of Martley, Lower Division of the hundred of Oswaldslow, Worcester and Western Divisions of the county of Worcester, 6¾ miles (S. S. W.) from Stourport; containing 287 inhabitants. The chapel is dedicated to St. Michael.

WITNESHAM (*St. Mary*), a parish, in the union of Woodbridge, hundred of Carlford, Eastern Division of the county of Suffolk, 4½ miles (N. by E.) from Ipswich; containing 562 inhabitants. The living is a rectory, valued in the king's books at £18. 13. 4.; present net income, £463; patrons, Master and Fellows of St. Peter's College, Cambridge. Here is a National school, supported by the rector. A singular discovery took place in 1820, in this parish; on removing some earth, the skeleton of a man in armour, with his horse, supposed, from the steel cap by which his head was defended, to have been buried during some of the civil wars, was exposed to view. The family of Meadows, from a branch of which the present Earl Manvers is descended, have had a seat here since the time of Richard III. Bishop Latimer was instituted to this rectory in 1538, and held it till 1554.

WITNEY (*St. Mary*), a market-town and parish, and the head of a union, in the hundred of Bampton, county of Oxford; containing 5336 inhabitants, of which number, 3190 are in the town of Witney, 11 miles (W. by N.) from Oxford, and 65 (W. N. W.) from London. This place, anciently called *Whitteney*, was a town of some importance prior to the Conquest, and was one of the manors given to the monastery of St. Swithin, at Winchester, in the reign of Edward the Confessor, by Bishop Ailwyn, in gratitude for the deliverance of Queen Emma, mother of that monarch, from the reputed fiery ordeal which she underwent in the cathedral church of that city, in vindication of her innocence of a charge of incontinence. In the reign of Edward II., solemn tournaments were held here, between Henry Bohun, Earl of Hereford, and Aymer de Valence, Earl of Pembroke, who was a great benefactor to the town, which in the fifth year of that reign, was made a borough, and returned two members to parliament, from which it was released, on petition of the inhabitants, in the 33d of Edward III. The town is pleasantly situated on the river Windrush, a stream abounding with trout and crayfish, and much resorted to by the students from Oxford, over which a substantial stone bridge of three arches was erected in 1822, and on the high road from London to Cheltenham and Gloucester: it consists principally of two streets, containing neat well-built houses, and has a clean and respectable appearance: the environs are pleasant, and the ground in the vicinity is agreeably varied with hill and dale. Witney has long been celebrated for its staple manufacture of blankets, which have been invariably regarded as superior, both in texture and colour, to all others: the latter quality is attributable to the peculiar properties of the water of the Windrush. The blanket-weavers of the town were incorporated in the 10th year of the reign of Queen Anne, under the designation of "the Master, Assistants, Wardens, and Commonalty of Blanket-Weavers inhabiting in Witney, in the county of Oxford, or within 20 miles thereof." At that time the manufacturers had 150 looms in full operation, affording employment to more than 3000 persons, and consuming weekly about 200 packs of wool. The charter continued in force for some years, and under its provisions the company enacted laws for the regulation of the trade; but, in process of time, it was found to interfere with improvements in the manufacture, and having become incompatible with the interests of the trade, as at present conducted, it has nearly fallen into disuse. The number of persons now employed is about 1000, and the annual consumption of wool about 10,000 packs. Rough coatings, tilting for barges and waggons, and felting for paper-makers, are made also to a considerable extent: the glove trade affords employment to a small number of persons; and wool-stapling, as connected with the manufactures of the town, is carried on extensively: there is also a considerable trade in malt. The market is on Thursday, and a market recently established for cattle and sheep is held on the last Thursday in each month. The fairs are on the Tuesday in Easter week, Holy Thursday, July 10th, the Thursday after Sept. 8th, the Thursday before Oct. 10th, and December 4th. The town is within the jurisdiction of the county magistrates; and two bailiffs, assisted by two constables and other officers, are appointed by the jury at the court leet for it, held annually: a court baron is held twice in the year by the Duke of Marlborough, as lessee under the Bishop of Winchester. It is a polling-place for the county. A handsome blanket hall was erected in 1721: the town-hall is a neat stone building, with a piazza for the use of the market; and the market-cross, in the market-place near the town-hall, was erected in 1683, and repaired in 1811.

The living comprises a rectory and a vicarage, united, in the 9th of Charles I., into one benefice, by the designation of the rectory of Witney, with a reservation of the dues and fees of each, as if separate; the rectory is valued in the king's books at £47. 9. 4½., and the vicarage at £9. 12. 6.; present net income, £1290; patron, Bishop of Winchester. The church is a spacious and elegant cruciform structure, in the early, decorated, and later styles of English architecture, with a square central tower, having octagonal turrets at the angles, and surmounted by a lofty spire, panelled in compartments and richly ornamented: the nave is separated from the aisles by handsome piers and finely-pointed arches, and is lighted by a range of clerestory windows of the later style: the transepts are large, and the western, which is in the decorated style, is lighted by an elegant window of seven lights: the chancel, which is small, is in the early English style of architecture, and is lighted with windows of delicate tracery: there are several monumental effigies in the transepts, and many ancient tombs in various parts of the church, and in the chancel is a piscina of elegant design. There are places of worship for the Society of Friends, Independents, and Wesleyan Methodists: that for Independents was erected by W. Townsend, Esq., at an expense of £2000. The free grammar school, in Church Green, was founded under an act of parliament in 1664, by Henry Box, Esq., a native of this town, and citizen of London, who endowed it with a rent-charge of £63 on an estate at Longworth, in Berkshire. It is under the direction of the Grocers' Company, who are trustees, and the control of the Provost and four senior Fellows of Oriel College, Oxford, as visiters; and is conducted by a principal and a sub-master: the number of scholars is limited to twenty, who must be natives of Witney. The buildings comprise a spacious school-room, with a library, dwelling-house for the master, and a large play-ground in front. A free school was founded in 1723, by Mr. John Holloway, who endowed it with lands producing about £135 per annum, for instructing, clothing, and apprenticing the sons of journeymen weavers of Witney and Hailey: there are at present ten scholars from the former, and five from the latter place, who are clothed every year, and with each of whom an apprentice-fee of £15 is given, on leaving the school. The same benefactor erected almshouses for six widows of blanket-weavers, and endowed them with lands producing an income of about £85 per annum. Mr. William Blake, of the parish of Cogges, in 1693, endowed a school here with £26 per annum, which is paid to a mistress for teaching 40 children; and a National school is supported by subscription. Some ancient almshouses, on Church Green, were taken down, and six substantial houses erected, in 1795, by the feoffees of the charity estates: these houses are at present let to tenants, and the rents are distributed among the poor. Six neat almshouses for aged and unmarried women were erected in 1828, by Mr. Townsend, and are endowed with about 6*s.* per week each. There are also several charitable bequests for distribution among the poor of the parish. The poor law union of Witney comprises 42 parishes or places, and contains, according to the census of 1831, a population of 21,898; it is under the superintendence of 47 guardians. The Roman Akeman-street passes near the town.

WITSTON or WHITSON, a parish, in the union of NEWPORT, Lower Division of the hundred of CALDICOTT, county of MONMOUTH, 6½ miles (S. E. by E.) from Newport; containing 106 inhabitants. The living is a discharged vicarage, valued in the king's books at £6. 7. 8½.; present net income, £180; it is in the alternate patronage of the Chapter of Llandaff and the Provost of Eton College, the impropriators. The church, according to tradition, belonged to Portown, a town once situated in the neighbourhood, but swallowed up by the sea at some remote period.

WITTENHAM, LITTLE (*St. Peter*), a parish, in the union of WALLINGFORD, hundred of OCK, county of BERKS, 4¼ miles (N. W. by N.) from Wallingford; containing 113 inhabitants. The living is a rectory, valued in the king's books at £17. 10.; present net income, £400; patron, Rev. J. Hilliard. The church contains monuments to the Dunche family. Sinodun hill, in the neighbourhood, is surrounded by an ancient intrenchment, supposed to be of British origin, and to have been afterwards occupied by the Romans. Roman antiquities are found here occasionally.

WITTENHAM, LONG, or EARLS (*All Saints*), a parish, in the union of WALLINGFORD, hundred of OCK, county of BERKS, 4 miles (S. W.) from Abingdon; containing 547 inhabitants. The living is a discharged vicarage, valued in the king's books at £12. 12. 6.; present net income, £166; patrons and impropriators, Rector and Fellows of Exeter College, Oxford. The parish is bounded on the north by the Isis. A National and an infants' school are supported by voluntary contributions. Funeral urns and other Roman antiquities are found here.

WITTERING-EAST, a parish, in the union of WEST HAMPNETT, hundred of MANHOOD, rape of CHICHESTER, Western Division of the county of SUSSEX, 6¾ miles (S. W. by S.) from Chichester; containing 226 inhabitants. This parish is bounded on the south by the English Channel, and the village is situated near the sea. From the mouth of Chichester harbour to the extremity of Selsey Hill, a distance of nearly eight miles, the sea has absorbed a very considerable portion of the prebendal manor of Bracklesham; and the bay thus formed, called Bracklesham bay, affords at low water a delightful ride upon the sands, which are particularly firm and level, occasionally interspersed with patches of soft clay, in which beautiful fossil shells are found. The living is a discharged rectory, valued in the king's books at £6. 16. 8.; present net income, £190; patron, Bishop of Chichester. The church is an ancient structure with a Norman font, and a fine southern doorway in that style highly enriched. There was formerly at Bracklesham an ancient endowed chapel, annexed to the vicarage by Bishop Shirborne in 1518, of which nothing at present remains. There is a preventive station near the village.

WITTERING-WEST, a parish, in the union of WEST-HAMPNETT, hundred of MANHOOD, rape of CHICHESTER, Western Division of the county of SUSSEX, 7¾ miles (S. W.) from Chichester; containing 606 inhabitants. This parish, which is bounded on the south by the English Channel, and on the west by the mouth of Chichester harbour, was visited, in 477, by Ella, a Saxon adventurer, who, with his three sons, landed here and defeated the Britons, who were drawn up to oppose him; and a place on the shore is still called Ella-nor-point. It was the occasional residence of the Bishops of Chichester from the 13th to the 16th century, when the Episcopal palace, now called Cakeham Manor Place, became the property of the Ernley family. The ancient mansion has partly disappeared, and the remainder has been converted into a farm-house. Bishop Shirborne, induced by the beauty of the sea view, bounded by the Isle of Wight, built a lofty hexagonal tower of brick, which is still remaining, and from the summit of which the most extensive and magnificent prospects are obtained. Three preventive stations have been established along the coast. The living is a discharged vicarage, valued in the king's books at £10. 3. 4.; present net income, £165; patron and appropriator, Prebendary of Wittering in the Cathedral Church of Chichester. The church is an ancient edifice, combining various styles of English architecture, with a tower on the north side; it contains some fine specimens of carving in oak in the pews, and three stalls in the chancel, with a very fine Norman font of cylindrical form, and a monument of Caen stone, with effigies of William Ernley and family, richly sculptured in bass relief, and carved representations of the Resurrection and the Salutation of the Virgin. Here is a Sunday National school. The inhabitants have the privilege of sending eight poor boys to the school at Chichester, founded in 1702 by Oliver Whitby, a former prebendary of Wittering. A few years since, several Roman coins of the Emperors Constantine, Valentinius and others, were found in a field in this parish.

WITTERSHAM (*St. John the Baptist*), a parish, in the union of TENTERDEN, hundred of OXNEY, lathe of SHEPWAY, Eastern Division of the county of KENT, 5¼ miles (S. by E.) from Tenterden; containing 919 inhabitants. The living is a rectory, in the patronage of the Archbishop of Canterbury, valued in the king's books at £15. 8. 6½. The tithes have been commuted for a rent-charge of £730, subject to the payment of rates; the glebe comprises 17 acres, valued at £27 per annum. The church has portions in various styles of architecture. There is a place of worship for Wesleyan Methodists; and a National school has been established, which is supported from land purchased for it by the Rev. W. Cornwallis, in 1820. Another school is supported by subscription.

WITTON, a parochial chapelry, included in the parish of GREAT BUDWORTH, union and hundred of NORTHWICH, Southern Division of the county of CHESTER, ¼ of a mile (E.) from Northwich; containing 2912 inhabitants. The living is a perpetual curacy; net income, £168; patron, Edward Greenall, Esq. The chapel, dedicated to St. Helen, is a noble and spacious structure, in the later English style of architecture, with an embattled tower. The free grammar school, adjoining the cemetery, was founded in 1588, by Sir John Deane, who endowed it with a salt-work at Northwich, and certain houses and lands in other parts of the county, which property belonged to the college of St. John the Baptist, and its dissolved guild of St. Anne, in the city of Chester; to the priory of Norton, Cheshire; and to that of Basingwerk, in the county of Flint; and now produces an annual income of about £270. None, except the kinsfolk of the founder, have claims to admission on the foundation, unless their parents reside within the chapelry. Its statutes, in some respects, are similar to those of Harrow, but in the most essential points they are the same as those of St. Paul's school, London. The master is elected by the twelve feoffees appointed under the will of the founder, assisted by certain of the inhabitants, and approved by the bishop and the master of the King's school, Chester. The school-house, which was rebuilt about a century ago, is a substantial structure of brick and stone, with a porters' lodge attached, having also a commodious suite of apartments, occupied by the master, with a spacious class-room over the school. The Queen is visiter. A National school is supported by subscription.

WITTON, or WYTTON (*All Saints*), a parish, in the union of ST. IVES, hundred of HURSTINGSTONE, county of HUNTINGDON, 2½ miles (W. by N.) from St. Ives; containing 277 inhabitants. The living is annexed to the rectory of Houghton. Two schools are chiefly supported by private individuals. The river Ouse passes through the parish.

WITTON (*St. Margaret*), a parish, in the union and hundred of BLOFIELD, Eastern Division of the county of NORFOLK, 5½ miles (E.) from Norwich; containing 144 inhabitants. The living is a discharged rectory, consolidated, with that of Brundall, with the rectory of Little Plumstead, and valued in the king's books at £6. 13. 4. The tithes have been commuted for a rent-charge of £235, subject to the payment of rates, which on the average have amounted to £20; the glebe comprises 17 acres, valued at £38. 5. per annum.

WITTON (*St. Margaret*), a parish, in the hundred of TUNSTEAD, Eastern Division of the county of NORFOLK, 3½ miles (E. by N.) from North Walsham; containing 295 inhabitants. The living is a discharged vicarage, valued in the king's books at £4. 13. 1½.; present net income, £136; patron and appropriator, Bishop of Ely. John Norris, in 1777, bequeathed £855 three per cent. consols., directing, of the dividends, £10 to be appropriated for the education of 12 poor children, £10. 10. to the vicar for extra duty at Witton church, and the remainder in relief to the poor: the £10 is paid to a schoolmaster at Bacton. A fund of £21 per annum, the produce of bequests from Richard Drake, in 1649, and Robert Annison, is distributed among the poor.

WITTON, EAST (*St. Ella*), a parish, in the union of LEYBURN, wapentake of HANG-WEST, North Riding of the county of YORK, 2½ miles (S. E. by S.) from Middleham; containing 687 inhabitants. The living is a vicarage, valued in the king's books at £5. 3. 6½.; present net income, £93; patron and impropriator, Marquess of Ailesbury. The church is an elegant structure, in the later English style; the first stone was laid in 1809, and the building was completed in 1812, at the expense of the Earl of Ailesbury, in commemoration of the fiftieth anniversary of the reign of George III. In the neighbourhood are quarries of excellent freestone, well adapted for grindstones. A school, now in connection with the National Society, was erected in 1817, by the Marquess of Ailesbury, by whom it is chiefly supported; the master's salary is £60 per annum. About a mile east of the village are the ruins of Jervaulx abbey, founded about the middle of the twelfth century, by Akarius, in honour of the Virgin Mary, which at the dissolution had a revenue of £455. 10. 5. These interesting remains having been cleared from briars and rubbish, considerable portions of the abbey church, with its cross aisles, choir, and chapter-house, also several tombs and stone coffins, are now plainly visible: the tesselated pavement of the great aisle was also discovered, apparently in a perfect state, but, by exposure to the air, it soon crumbled to dust.

WITTON-GILBERT (*St. Michael*), a parish, in the union of Chester-le-Street, Western Division of Chester ward, Northern Division of the county palatine of Durham, $3\frac{1}{2}$ miles (N. W.) from Durham; containing, with the outside portion of the parish, 417 inhabitants. The living is a perpetual curacy, with the rectory of Kimbleworth united; patrons and appropriators, Dean and Chapter of Durham. The tithes have been commuted for a rent-charge of £301. 11. 5., subject to the payment of rates, which on the average have amounted to £24. The church is a small building, without a tower. An hospital for five lepers was founded near it, at an early period, by Gilbert De la Ley, the only fragment of which is a pointed window in a farm-house now occupying its site. Jane Finney, in 1728, gave certain land, now let for £12 a year, for teaching eight poor children. Here is a National school.

WITTON-LE-WEAR (*St. Philip and St. James*), a parish, in the union of Auckland, North-Western Division of Darlington ward, Southern Division of the county palatine of Durham, $4\frac{1}{4}$ miles (W. N. W.) from Bishop-Auckland; containing 502 inhabitants. The living is a perpetual curacy; net income, £94; patron and impropriator, Sir William Chaytor, Bart. The church is an ancient structure. The village is situated on the southern acclivity of an eminence rising from the north bank of the river Wear, which is here crossed by a bridge. On the south side of the river is Witton castle, built about 1410, formerly the baronial mansion of the Lords de Eure, many of whom signalized themselves in the border warfare: it is a large oblong edifice, with towers and turrets, and a handsome gateway entrance in the Norman style has been recently added by Sir W. Chaytor. In the great civil war it was held by Sir William D'Arcy for the king, and was besieged and taken by the parliamentarians, under Sir Arthur Haslerigg. The castle and the estate have been lately sold by Sir W. Chaytor to D. Maclean, Esq., for nearly £100,000. Coal abounds in the neighbourhood; and the Stockton and Darlington railway terminates at Witton Park colliery. There is a commodious grammar school, founded by John Cuthbert, Esq., and formerly endowed with the interest of £200; but the building has been long occupied as a private boarding school, and the endowment transferred to a National school.

WITTON, LONG, a township, in the parish of Hartburn, union, and Western Division of the ward, of Morpeth, Northern Division of the county of Northumberland, 8 miles (W. by N.) from Morpeth; containing 143 inhabitants.

WITTON, NETHER (*St. Giles*), a parish, in the union, and Western Division of the ward, of Morpeth; Northern Division of the county of Northumberland; containing 520 inhabitants, of which number, 329 are in the township of Nether Witton, $7\frac{3}{4}$ miles (W. N. W.) from Morpeth. The living is annexed to the vicarage of Hartburn. Here is a National school.

WITTON-SHIELS, a township, in the parish of Long Horsley, union, and Western Division of the ward, of Morpeth, Northern Division of the county of Northumberland, 7 miles (N. W by. W.) from Morpeth; containing 13 inhabitants. A strong tower, erected in 1608 by Sir Nicholas Thornton, has been converted into a Roman Catholic chapel.

WITTON, UPPER, a hamlet, in the parish of Aston, Birmingham Division of the hundred of Hemlingford, Northern Division of the county of Warwick, $3\frac{1}{2}$ miles (N. by E.) from Birmingham: the population is returned with the parish.

WITTON, WEST, a parish, in the union of Leyburn, wapentake of Hang-West, North Riding of the county of York, $4\frac{1}{2}$ miles (W.) from Middleham; containing 552 inhabitants. The living is a perpetual curacy; net income, £103; patron and impropriator, Lord Bolton. The church is a small building, supposed to have been erected in the reign of Henry I.: from the churchyard is a delightful view over Wensley dale. There is a place of worship for Roman Catholics. A National school has been established. On an eminence, called Penhill, are vestiges of an ancient castle, formerly belonging to Ralph Fitz-Randal.

WIVELISCOMBE (*St. Andrew*), a market-town and parish, in the union of Wellington, and forming, with the parishes of Ash-Priors, Bishops-Lydiard, and Fitzhead, one of the two unconnected portions which constitute the Western Division of the hundred of Kingsbury, Western Division of the county of Somerset, 28 miles (W.) from Somerton, and 155 (W. by S.) from London; containing 3047 inhabitants. This place is of considerable antiquity, but neither its origin nor the etymology of its name can be traced with certainty. Conjecture has deduced the latter from the Saxon *Willi* or *Vili*, signifying many, and *Combe*, a deep ravine or dell, of which there are several in the immediate environs. The town occupies a gentle eminence, in an extensive valley enclosed by lofty hills, which suddenly break into deep ravines: the houses are in general neat and well built, and, by the removal of several of the more ancient, the streets have been widened, and the general appearance of the town greatly improved. The inhabitants are supplied with water conveyed by pipes from a spring on Mawndown, a hill about a mile distant. A considerable woollen manufacture is carried on, but not to so great an extent as formerly: the articles consist chiefly of slave clothing for the West India markets, swan-skins for the Newfoundland fishery, and blankets for the home trade: the number of persons regularly employed varies from 800 to 1000. The diversion of the mail-road through this town has materially contributed to promote its various interests. The markets are on Tuesday and Saturday; the former is the principal, and a great deal of business is transacted in corn, &c. A great market for prime oxen, of the North Devon breed, considered to be the largest in the West of England, is held on the last Tuesday in February. Fairs are on May 12th for oxen and other cattle, and Sept. 25th for sheep. The town is under the superintendence of a bailiff and portreeve, with aletasters and other officers, all of whom are chosen at a court leet held annually. It is said to have been formerly a parliamentary borough, and that it was relieved from the elective franchise on petition. It has lately been made a polling-place for the western division of the county.

The living is a vicarage, in the patronage of the Prebendary of Wiveliscombe in the Cathedral Church of Wells (the appropriator), valued in the king's books at £27. 0. 10.; present net income, £300. The church was erected, a few years since, at an expense of £6000, raised on the security of the parish rates, to be paid off in twenty years, aided by a general subscription, and a grant of £500 from the Incorporated Society, for which 460 free sittings have been provided: it is a very handsome edifice, in the ancient style of English architecture. There is a place of worship for Independents; and also a Sunday National school. An infirmary was established in 1804, through the exertions of the medical men of the town and neighbourhood. In this parish are two ancient encampments; the one on an eminence at a place called Castle, of a circular form and very perfect; the other at Courtneys, square and evidently of Roman origin. Here are also some remains of an old episcopal palace, particularly an archway leading into the workhouse, and the kitchen, which is nearly entire. In digging for the foundation of the new church, it was discovered that the tower of the former had been erected upon the foundations of a more ancient building; a variety of Roman and Saxon coins was also found, together with some Nuremberg counters, used by the monks in their calculations on the Abacus.

WIVELSFIELD (*St. John the Baptist*), a parish, in the union of Chailey, partly in the hundred of Street, rape of Lewes, and partly in the hundred of Burley Arches, rape of Pevensey, Eastern Division of the county of Sussex, 4 miles (S. E.) from Cuckfield; containing 559 inhabitants. The living is a perpetual curacy; net income, £94; patron and impropriator, R. Tanner, Esq. The church is principally in the early style of English architecture. There is a place of worship for Independents. Six poor children are instructed for £5 per annum, arising from the sum of £100 bequeathed by Mrs. Frances More, in 1723. The line of the London and Brighton railway passes through the parish. The late Countess of Huntingdon resided here, and the late Rev. Mr. Romaine frequently visited this place.

WIVENHOE (*St. Mary*), a parish, in the union of Lexden and Winstree, Colchester Division of the hundred of Lexden, Northern Division of the county of Essex, 4½ miles (S. E. by E.) from Colchester; containing 1714 inhabitants. The village is situated on the Coln river, a little below Colchester, and much shipping belongs to the port: it has a regular custom-house establishment, with a commodious quay, whence the noted Colchester oysters are shipped for the London and other markets. The greater portion of the male population are employed in the oyster and other fisheries, and as pilots through the intricate navigation of the Eastern coast. The living is a discharged rectory, valued in the king's books at £10; patron, N. C. Corsellis, Esq. The tithes have been commuted for a rent-charge of £440, subject to the payment of rates, which on the average have amounted to £117; the glebe comprises 29 acres, valued at £29 per annum. The church has been enlarged, and 105 free sittings provided, the Incorporated Society having granted £40 in aid of the expense; it is an ancient structure, in the early English style, with a square embattled tower. There is a place of worship for Independents; and a National school has been established. In 1718, Mr. Feedham left £50 (which at the enclosure was exchanged for an allotment of land in the parish), directing the proceeds to be employed annually in clothing the poor widows of sailors; and Mr. Sanford, who died in 1829, bequeathed £2. 10. per annum, to be distributed among ten poor communicants residing within the parish.

WIVETON (*St. Mary*), a parish, in the union of Walsingham, hundred of Holt, Western Division of the county of Norfolk, ¼ of a mile (W. by S.) from Clay; containing 218 inhabitants. The living is a discharged rectory, valued in the king's books at £15; present net income, £207; patrons, Executors of the late G. Wyndham, Esq. Ralph Greenaway, in 1729, bequeathed property now consisting of the rectorial tithes of Briston, with a barn and an acre of land, and £1141. 11. 5. three per cent. consols., and producing an income of £264. 5., for the repairs of the church, and a weekly distribution of bread and money among the poor inhabitants.

WIX, or WEEKS (*St. Mary*), a parish, in the union and hundred of Tendring, Northern Division of the county of Essex, 4½ miles (E. S. E.) from Manningtree; containing 832 inhabitants. This parish is about eight miles in circumference; the soil is extremely fertile. The living is a perpetual curacy; net income, £120; patron, Rev. R. J. Scott. The church is a small edifice, built of the ruins of a former structure, which had gone to decay. Five pounds a year, bequeathed by a Mr. Clarke, is applied in support of a National school. A Benedictine nunnery, in honour of the Virgin Mary, was founded here in the time of Henry I., by Walter Mascherell and others, which, at its suppression, was valued at £92. 12. 3., and was granted to Cardinal Wolsey, towards erecting and endowing his intended colleges.

WIXFORD (*St. Milburg*), a parish, in the union of Alcester, Stratford Division of the hundred of Barlichway, Southern Division of the county of Warwick, 2 miles (S.) from Alcester; containing 108 inhabitants. The living is annexed to the rectory of Exhall.

WIXOE, a parish, in the union and hundred of Risbridge, Western Division of the county of Suffolk, 4 miles (W. S. W.) from Clare; containing 146 inhabitants. The living is a discharged rectory, valued in the king's books at £5. 13. 1½.; present net income, £180; patron, J. P. Elwes, Esq. The church has a handsome Norman doorway on the south side.

WOBURN (*St. Mary*), a market-town and parish, and the head of a union, in the hundred of Manshead, county of Bedford, 15 miles (S. W. by S.) from Bedford, and 42 (N. W. by N.) from London; containing 1827 inhabitants. This town, which, having suffered severely from fire in the year 1595, and again in 1724, has been almost entirely rebuilt, occupies a gentle eminence on the main road from London to Leeds, and consists of four broad and handsome streets, which intersect each other at right angles. The approaches to it, both from the north and the south, are kept in excellent repair, and have been recently embellished by two ornamental houses corresponding in architectural character with the new market-house, in the centre of the town, a handsome oblong edifice in the Tudor style of architecture, erected by the late Duke of Bedford in 1830, from designs by Mr. Blore: the sides of this building have each four cloister arches filled with iron work; at the east end is a neat arched doorway, over which is a handsome oriel window, and the north-east angle has a square tower, with a spiral roof of lead surmounted by a vane. The lower part of the building is principally appropriated to the use of the butchers of the town and neighbourhood; and the upper story comprises a splendid apartment for holding the manorial courts, and for the use of the county magistrates, who hold here a petty session for the hundred on the first Friday in every month. The market is on Friday; and fairs are held on January 1st, March 23rd, and October 6th: the spring fair is noted for an abundant supply of horses and cattle. The manufacture of thread-lace formerly constituted a principal branch of business, but of late has been entirely discontinued, and some attempts have been made to introduce that of plat from Tuscan straw, as a more healthy and advantageous occupation for the children of the poor. Assemblies, respectably attended, are occasionally held during the winter months. The town is singularly neat and improving; and the beauty of its site is greatly enhanced by the evergreen woods in its immediate vicinity, which were planted by John, fourth Duke of Bedford, and occupy 200 acres in extent. Near the market-house is a fountain, or reservoir, in the Tudor style of architecture, for supplying water in case of fire, erected at the sole expense of the late Duke of Bedford.

The living is a donative curacy; net income, £251; patron and impropriator, Duke of Bedford. The church was erected by Robert Hobbs, the last abbot of Woburn: it presents a singularly beautiful appearance, being nearly covered with ivy. The old quadrangular embattled tower, terminating in pinnacles and surmounted by a cupola, which was built by Abbot Hobbs, and formerly stood detached from the main building, was taken down and rebuilt in the later English style from the lower stage in 1830, by the Duke of Bedford, under the superintendence of Mr. Blore, and was then joined to the north aisle by a vestry-room and gallery above: it rises to the height of 90 feet, and is surmounted by an octagonal stone lantern: at each angle is a lofty pinnacle, panelled and crocketed, with a finial: the lantern has eight ornamented arches, supporting the roof, which rises spirally with crockets to a handsome finial. The building has two neat windows of similar character, in which the Tudor arch has been introduced, and a large octagonal turret at the junction; pinnacles have also been added to the ends of the aisles, and a cross patée to that of the nave. In the interior of the church is a curious alabaster monument of the Stanton family, consisting of twelve figures in the attitude of prayer, besides some other ancient sepulchral memorials. The fine altar-piece of the Nativity, by Carlo Maratti, was the gift of the late Duke of Bedford, who also adorned the building with a new window of five lights, with enriched and cinquefoil arched mullions, and the upper part embellished with stained glass, and with figures of the evangelists and four of the patriarchs. There are places of worship for Independents and Wesleyan Methodists. Adjacent to the church, and now made to correspond with it in architectural character, is the Free School, established in 1582, by Francis, the second Earl of Bedford, for the instruction of 35 boys. In 1808 the system of education was changed for the Lancasterian plan, the Duke of Bedford increasing the original endowment to £50 per annum, and the remaining sum necessary for its support being raised by the contributions of the inhabitants. In 1825, a similar school for girls was established under the patronage of the Duke and Duchess, in which the children are instructed in needlework and the manufacture of Tuscan plat, &c. Twelve almshouses were founded in 1672, and endowed by John, fourth Duke of Bedford, for the residence and maintenance of 24 poor widows; the buildings are situated at the north end of the town. The poor law union of Woburn comprises 16 parishes or places, under the care of 20 guardians, and contains, according to the census of 1831, a population of 10,633.

In the immediate vicinity of the town is Woburn Abbey, with its noble and extensive park, the seat of His Grace the Duke of Bedford, which occupies the site of a Cistercian abbey, founded in 1145 by Hugh de Bolebec, the revenue of which, at the dissolution, was valued at £430. 13. 11. The site, with a great part of the lands, was granted, in 1549, by Edward VI. to John, first Earl of Bedford. In the middle of the last century the abbey was almost entirely rebuilt, by Flitcroft, since which time considerable enlargements have been made, under the superintendence of the late Mr. Henry Holland, who erected also the principal entrance to the park from London, a handsome façade decorated with Ionic three-quarter columns, surmounted by the ducal arms and crest. The abbey, which is approached from this entrance through an extent of rich park scenery and by the margin of an artificial lake, occupies the four sides of a quadrangle, and comprises various suites of apartments magnificently furnished, and adorned with paintings by the most celebrated masters,—the print-room, by modern artists exclusively; the summer drawing-room, by Canaletti; and the winter drawing-room, by the late Henry Bone, Esq., the admirable painter in enamel; whilst a rich collection of upwards of 280 portraits of distinguished family and other characters adorns the gallery, or is interspersed through the corridors and chief apartments. The library, 56 feet in length by more than 23 in breadth, is stored with the most splendid illustrative and other works, of the highest class of reputation. The principal state rooms are in the west front, which is of the Ionic order: the private apartments adjoin the library on the south, having immediately in front a terrace arranged as an ornamental flower garden. In the hall is temporarily placed a beautiful tesselated pavement, formed out of the fragments of one much larger, discovered in 1823 on the site of a Roman villa near the *Porta Portese*. A covered arcade conducts from the private apartments to the Sculpture Gallery, formed by the munificent taste of the late Duke, 138 feet long by 25 wide, in which, amongst other valuable works of art by ancient sculptors, are deposited some of the finest productions of Chantrey, Westmacott, and Thorvaldsen;

the celebrated group of the Graces, by Canova; and the magnificent Lanti or Bedford Vase. At the east end is a temple dedicated to Liberty, on the model of the little Ionic temple on the Ilissus, given by Stuart, of which the interior is decorated with busts, by Nollekens, of the Rt. Hon. Charles James Fox and six of his most distinguished political friends. The interior of the gallery is formed into three compartments by two screens of columns, consisting of eight shafts of rare antique marbles; the floor is partially inlaid with Devonshire marble, and the exterior front of the building is embellished with groups of figures by Westmacott, representing a Dance of Cupids and the progress of society from rudeness to civilization and refinement. The pleasure grounds contain many objects of great attraction, among which are a conservatory, and a heathery designed by Sir Jeffrey Wyatville; extensive shrubberies of American and other evergreen plants; a rosarium; a salictum, or collection of indigenous and foreign willows; a pinetum; a Chinese dairy, erected by Holland in the most costly style; and a delightful rustic aviary. The eastern side of the quadrangle is occupied by the tennis-court and riding-house. The park abounds with fine timber, and is well stocked with red and fallow deer. The oak tree on which Robert, or Roger, Hobbs, the last abbot of Woburn, was hanged, pursuant to the mandate of Henry VIII., is still pointed out. In 1572 Queen Elizabeth made a journey to the mansion; and in 1645, when Charles I. visited the Earl of Bedford, the overtures of the parliamentary commissioners were privately submitted to him here, prior to being offered to him formally in public. The Park Farm, situated midway between the abbey and the town, and well worthy of the attention of strangers, was planned and in a great measure completed by that distinguished patron of agriculture, Francis, fifth Duke of Bedford, whose delightful rural festivals, or annual sheep-shearings, were held upon this spot, and were attended by the most eminent farmers and graziers in Great Britain, and by several from other parts of Europe, and even from America. Within a short distance, in a romantic part of the park, stands a very pretty summer retreat, called the Thornery, built in a rustic style of elegance, in the midst of a sheltered lawn and flower garden enclosed with forest trees; and the numerous keepers' lodges that occur at intervals in other parts of this beautiful domain, particularly that near Aspley wood, on the Northampton road, called Henry the Seventh's cottage, from the Tudor style of architecture adopted by the late Mr. Repton in its erection, bespeak the taste and spirit of its late noble proprietor.

WOKEFIELD, a tything in the parish of STRATFIELD-MORTIMER, union of BRADFIELD, hundred of THEALE, county of BERKS; containing 160 inhabitants.

WOKEING (*St. Peter*), a parish (formerly a market-town), in the union of GUILDFORD, First Division of the hundred of WOKEING, Western Division of the county of SURREY, 3 miles (W. by N.) from Ripley; containing 1975 inhabitants. This was one of the royal demesnes of Edward the Confessor, and was afforested in 1154, by Henry II., whose successor gave it to Alan, Lord Basset; but, in the reign of Edward II., it belonged to the Despencers, and on their attainder was given, by Edward III., to Edmund of Woodstock, from which time it had various distinguished owners till the time of Edward IV., who, it is recorded, kept his Christmas in 1480, at the royal palace of Wokeing. Henry VII. afterwards repaired and enlarged it, for the residence of his mother, Margaret Countess of Richmond, who died there. Henry VIII. used it as a summer retreat, where he sometimes entertained Wolsey, and on one of these occasions, in September 1551, that prelate was first informed, by a letter from the pope, of his elevation to the dignity of Cardinal. James I. granted Wokeing to Sir Edward Zouch; but, in the reign of Charles I., it again belonged to the crown, and was bestowed upon Barbara Duchess of Cleveland; it subsequently passed, by purchase, through various hands to Richard Lord Onslow, an ancestor of the Earl of Onslow, its present proprietor. There are now no remains of the palace, except its foundations and the guard-room; the Zouches having removed the greater part of the building, to erect another mansion at Hoe Place, in the neighbourhood. Sutton Place, a fine specimen of the style of building which prevailed in the sixteenth century, was erected in 1529, by Sir Richard Weston: it was of a quadrangular form, enclosing a square area 80 feet in dimensions, with a noble gateway, having lofty hexagonal turrets at the angles. A great part of this magnificent structure was burned down, during a visit of Queen Elizabeth, and the remainder, consisting of the south-west side and north-east front, continued in a ruinous state till 1721, when it was repaired and embellished by the late John Weston, Esq. The front of this building has been lately taken down. The Basingstoke canal, and the London and South-Western railway, pass through the parish, and the latter has one of its principal stations within a mile from the village, which is situated on the river Wey. There are a paper-manufactory and a brewery. The market, formerly held on Tuesday, is now disused. There is a fair on Whit-Tuesday; and courts leet and baron are held annually. The living is a vicarage, valued in the king's books at £11. 0. 5.; present net income, £234; patron, Earl of Onslow; impropriators, H. Halsey, Esq., and others. The church is partly in the early and partly in the decorated style of English architecture: it contains some brasses and a few monuments. There are places of worship for Baptists and Independents, and a Roman Catholic chapel. A parochial school is supported by subscription. In a field near the village is a lofty circular tower, supposed to have been a light-house to guide over the heath to the palace; and there was formerly a religious house at Homitage.

Corporation Seal.

WOKINGHAM (*All Saints*), a market-town and parish, having separate jurisdiction, and the head of a union, situated partly in, and forming a detached portion of, the hundred of AMESBURY, Wokingham and Southern Divisions of the county of WILTS, but chiefly in the hundred of SONNING, county of BERKS, 7 miles (E. S. E.) from Reading, and 32 (W. S. W.) from London; comprising the Berkshir and Town divisions, and containing 3139 inhabitants. This town, situated within the prescribed limits of Windsor Forest, is of triangular form, and consists of several streets irregularly built, meeting in a central area; water is supplied from wells in abundance: the atmosphere is considered particularly salubrious, and the inhabitants are remarkable for longevity. The manufacture of silk, gauze, and shoes, and the malting and flour trades, are the prevailing branches of business. The market, which is on Tuesday, is one of the most noted in the kingdom for its abundant supply of poultry: the fairs are on April 23rd, June 11th (both of little importance, and not regularly held), October 11th, and November 2nd, chiefly for cattle. The government of the town is vested in an alderman, seven capital burgesses, a high steward, recorder, and town-clerk: the charter has been possessed from time immemorial. The alderman is elected from among the capital burgesses, annually on the Wednesday in Easter week, and becomes the chief magistrate for the year ensuing: the alderman, high steward, and recorder, are justices of the peace, with exclusive jurisdiction. The corporation hold half-yearly courts of session for minor offences, and are authorised by charter to hold a court of requests, for the recovery of small debts; but this power has not been exercised for some years. This being the only town in the forest, all the forest courts are held here; and manorial courts are held as occasion requires. Petty sessions take place, on the first and third Tuesdays in the month, for the Wokingham, or Forest, division of the county; and the town has been made a polling-place for the shire. The town-hall, which is over the market-house, is an ancient building in the centre of the town; it was repaired about 15 years ago, at an expense of £1100, defrayed by subscription: balls are occasionally held here in the winter season.

The living is a perpetual curacy; net income, £126; patron, — Jacob, Esq.; appropriator, Dean of Salisbury. The church is an ancient structure. There are places of worship for Baptists and Wesleyan Methodists. The free school, in Down-street, for children of both sexes, is supported by the proceeds of various bequests left for the instruction of children, particularly a rent-charge of £20 on certain lands in this parish, by Dr. Charles Palmer, in 1711, and other gifts, amounting together to £31. 15., and by voluntary contributions; thirty-six boys and twelve girls receive instruction, and, with the master, are appointed by the corporation: there are likewise other scholars of both sexes in the establishment. The National school, in which 200 children of both sexes are taught, is supported partly by some small bequests, and partly by subscription, and is held in a substantial brick building, erected in 1825, at an expense of £700. Eight almshouses near the church, founded and endowed by Mr. John Westend, an inhabitant of this town, in 1451, are occupied by sixteen poor men and women, who receive a small allowance of fuel. At Luckley Green, in this parish, about a mile from the town, is an hospital, founded in 1665 by Henry Lucas, Esq., for sixteen poor pensioners and a master, under the superintendence of the Drapers' Company in London: the pensioners are chosen alternately from sixteen of the neighbouring parishes in the counties of Berks and Surrey. Attached to the hospital, which is a handsome brick building, erected at an expense of £2320, is a chapel, with a residence for the minister, who is the perpetual curate of the parish. Archbishop Laud bequeathed £50 per ann. to be expended every third year in portioning poor maidens, and for the two other years, in apprenticing boys of the parish. Mr. Staverton left a house in Staines, directing the rental, now £20, to be applied in purchasing a bull to be baited annually on St. Thomas' day, and afterwards given to the poor: the baiting of the bull has been long discontinued, and the amount is distributed, with the produce of some other bequests, among the poor in money, coal, and clothing. The poor law union of Wokingham extends into the counties of Berks, Oxford, and Wilts, and comprises 16 parishes or places, containing a population of 11,888, according to the census of 1831; it is under the superintendence of 20 guardians. Dr. Thomas Goodwin, who, after various promotions, was eventually raised to the see of Bath and Wells, was a native of this town, and received the elements of his education in the free school: in the chancel of the church is a monument to his memory, with an inscription written by his son, who was Bishop of Hereford.

WOLD, county of Northampton.—See OLD.

WOLDHAM, county of Kent. — See WOULDHAM.

WOLD-NEWTON, East Riding of the county of York.—See NEWTON, WOLD.

WOLDINGHAM, or WALDINGHAM, a parish, in the union of Godstone, Second Division of the hundred of Tandridge, Eastern Division of the county of Surrey, 5 miles (N. E. by N.) from Godstone; containing 48 inhabitants. The living is a donative curacy; net income, £14; patron, J. F. Jones, Esq. The church is a small neat edifice, rebuilt by the patron about ten years since: it occupies an elevated site. The parish participates in Henry Smith's charity. Two brass fibulæ, iron arrow heads, and celts, have been found on Upper Court Lodge farm.

WOLFERLOW (*St. Andrew*), a parish, in the union of Bromyard, hundred of Broxash, county of Hereford, 5½ miles (N. by E.) from Bromyard; containing 134 inhabitants. The living is a discharged vicarage, valued in the king's books at £4. 4. 9.; present net income, £177; patron, Sir T. E. Winnington, Bart.

WOLFHAMCOTE (*St. Peter*), a parish, in the union of Rugby, Southam Division of the hundred of Knightlow, Southern Division of the county of Warwick, 3¾ miles (N. W. by W.) from Daventry; containing 372 inhabitants. The living is a discharged vicarage, valued in the king's books at £12. 18. 2.; present net income, £73; the patronage and impropriation belong to Miss Tibbits. The Oxford canal passes through the parish. In sinking a well here some years ago, a vault, containing several urns and coins, was discovered.

WOLFORD, GREAT (*St. Michael*), a parish, in the union of Shipston-upon-Stour, Brails Division of the hundred of Kington, Southern Division of the county of Warwick, 4 miles (S. by W.) from Shipston-upon-Stour; containing 580 inhabitants. The living is a discharged vicarage, valued in the king's books at £8; present net income, £131; patrons and appropriators, Warden and Fellows of Merton College, Oxford. A church, capable of containing about 500 persons, has been erected on the site of the former, which, having become dilapidated, was taken down in 1833; it contains 180 free sittings, the Incorporated Society having granted £200 in aid of the expense. A school is partly supported by subscription.

WOLFORD, LITTLE, a hamlet, in the parish of GREAT WOLFORD, union of SHIPSTON-UPON-STOUR, Brails Division of the hundred of KINGTON, Southern Division of the county of WARWICK, 3 miles (S.) from Shipston-upon Stour; containing 280 inhabitants. Here is an old mansion, which is now and has long been in the possession of the Ingram family, one of the most ancient in this county, and part of which is known to have existed so early as the reign of King John. A school is chiefly supported by the landed proprietor.

WOLLASTON (*ST. ANDREW*), a parish, in the union of CHEPSTOW, and forming, with that of Tidenham, a detached portion of the hundred of WESTBURY, Western Division of the county of GLOUCESTER, 5¼ miles (N. E.) from Chepstow; containing 880 inhabitants. At the time of the Norman survey, William Count D'Eu, who, after a judicial combat at Salisbury, was executed for high treason, was lord of the principal part of this parish, which was afterwards granted to the family of the Clares, who gave to Tintern Abbey the manor and church, together with several granges stretching across the parish from the river Wye to the Severn. Their representatives, however, claimed the rights of chace and free warren, and other privileges, within the parish, as part of the liberty of the Earls Marshal, but the interests of both are now vested in the Beaufort family, in right of the lordship marcher of Strigul or Chepstow, and of the Crown grants made to the Somersets after the dissolution. The parish, towards the river Wye, is bounded by a range of limestone hills, and towards the Severn by a rich vale of red marl. The living is a discharged rectory, with Alvington and Lancaut consolidated, valued in the king's books at £13. 11. 5.; patron, Duke of Beaufort. The tithes have been commuted for a rent-charge of £327. 12., subject to the payment of rates, which on the average have amounted to £15; the glebe comprises 46½ acres, valued at £69 per annum. The church is a small cruciform edifice, partly in the Norman style of architecture. A Sunday National school is supported by subscription, aided by a bequest of £2 per annum by Richard Clayton, in 1605.

WOLLASTON (*ST. MARY*), a parish, in the union of WELLINGBOROUGH, hundred of HIGHAM-FERRERS, Northern Division of the county of NORTHAMPTON, 3 miles (S. S. E.) from Wellingborough; containing 973 inhabitants. The living is a discharged vicarage, with that of Irchester annexed, valued in the king's books at £13. 6. 8.; present net income, £440; patron, Rev. W. W. Dickens; the tithes have merged in the enclosure of waste lands. The church is a handsome cruciform structure, with a stately tower rising from the intersection, and surmounted by a spire. There are places of worship for Independents and Wesleyan Methodists. The sum of £5. 4. per annum, the produce of bequests from Thomas and Charles Neale, in 1674 and 1730, together with £5. 4. per annum by Jonathan Beetle, in 1800, is distributed weekly in bread among the poor.

WOLLASTON, a chapelry, in the parish of ALBERBURY, union of ATCHAM, hundred of FORD, Southern Division of the county of SALOP, 10½ miles (W.) from Shrewsbury; containing 383 inhabitants. The living is a perpetual curacy; net income, £95; patron, Vicar of Alberbury; impropriators, Warden and Fellows of All Souls' College, Oxford. The chapel is dedicated to St. Michael.

WOLLATON (*ST. LEONARD*), a parish, in the union of BASFORD, Southern Division of the wapentake of BROXTOW, Northern Division of the county of NOTTINGHAM, 3 miles (W.) from Nottingham; containing 537 inhabitants. The living is a discharged rectory, with the perpetual curacy of Cossal annexed, valued in the king's books at £14. 2. 6.; patron, Lord Middleton. The tithes have been commuted for a rent-charge of £79. 4. 10., subject to the payment of rates, which on the average have amounted to £130; the glebe comprises 51½ acres, valued at £77. 12. per annum. The church exhibits a mixture of the several styles of English architecture. The Nottingham canal passes through the middle of the parish, in various parts of which coal mines have been wrought from time immemorial. Wollaton Hall, the seat of Lord Middleton, is a large and lofty edifice, in the Elizabethan style, built by Sir Francis Willoughby, entirely of freestone, which was brought from Ancaster, in the county of Lincoln, in exchange for coal obtained upon the estate.

WOLLEY (*ALL SAINTS*), a parish, in the union of BATH, hundred of BATH-FORUM, Eastern Division of the county of SOMERSET, 3 miles (N.) from Bath; containing 104 inhabitants. The living is a rectory, annexed to that of Bathwick.

WOLNEY, a joint township with Hawcoat, in the parish of DALTON-IN-FURNESS, hundred of LONSDALE, north of the sands, Northern Division of the county palatine of LANCASTER, 7½ miles (S. W.) from Ulverstone; containing, with Hawcoat, 848 inhabitants. In November, 1839, C. D. Archbold, Esq., employed a number of men to dig for the remains of a vessel said to have been wrecked, several centuries ago, on the western shore of the isle of Wolney; but nothing more than a few decayed planks, timbers, and pieces of iron, were found, with the exception of several guns, onc of which measured 10 feet in length, formed of thick plates of iron hooped; a great number of stone balls, made chiefly of a close grained granite, about 8 or 12lbs. weight each; and 18lb. shot of hammered iron, and some small ones cast, which are enveloped in lead; a pair of compasses of bronze, of very antique fashion; some old swords, a buckle, and a number of other articles. These curious remains are placed in the repository at Woolwich.

WOLSELEY, a hamlet, in the parish of COLWICH, Southern Division of the hundred of PIREHILL, Northern Division of the county of STAFFORD: the population is returned with the parish. The village is situated at the junction of the London, Liverpool, and Chester roads, where there is a large inn and posting-house; the hall is the property of Sir Charles Wolseley, Bart.

WOLSINGHAM (*ST. MARY AND ST. STEPHEN*), a market-town and parish, in the union of WEARDALE, North-Western Division of DARLINGTON ward, Southern Division of the county palatine of DURHAM, 16 miles (W. S. W.) from Durham, and 259 (N. N. W.) from London; containing 2239 inhabitants. The town, which is irregularly built, is situated on the north bank of the Wear. There are manufactures of linen, woollen cloth, edge-tools, and implements of agriculture, in which, and in the neighbouring coal, lead, and limestone works, a great proportion of the population is employed. The market and fairs are held by grant from the Bishop of Durham; the former is on Tuesday, and the latter are on May 12th, and Oct. 2nd, for cattle and all sorts of merchandise. The county magistrates hold a petty session for the division here, every Wednesday; and a court leet and baron, under the Bishop of Durham, as lord of the manor, is held twice a year, at which debts under 40*s*. are recoverable: its jurisdiction extends to Stanhope, Bishopley, North and South Bedburn, Lynsack, and Softley. In 1824, a new town-hall was erected and covered with a roof, but it yet remains unfinished, from want of funds to complete the work. The living is a rectory, valued in the king's books at £31. 13. 4.; patron, Bishop of Durham. The tithes have been commuted for a rent-charge of £899. 4., subject to the payment of rates; the glebe comprises 12 acres, and, with the house, is valued at £42 per annum. The church, situated on rising ground to the north-west of the town, is a neat plain building with a low tower, and has a font of Weardale marble, beautifully variegated with petrifactions of shells, &c. There are places of worship for Baptists and Primitive and Wesleyan Methodists. The grammar school, founded in 1613, with a residence for the master, was rebuilt in 1786, upon a piece of waste land granted by the Bishop of Durham and the landholders of the parish, by whom it was endowed with sixteen acres of land. On the enclosure of the moor, seven acres and a quarter more were added: eighteen boys are appointed by the trustees, who are nine in number. Several bequests in money, particularly by Jonathan and George Wooler, have since been made, amounting to about £200, for the interest of which eight additional boys are taught on this foundation. Another school is partly supported by a donation of £5 from Lord Crewe's trustees; and 48 girls are also instructed in a school supported by the Misses Wilson of this place. The Rev. W. Wilson, the rector, supports two Sunday schools. Contiguous to a field called Chapel-Walls are the remains of an extensive building, surrounded by a moat, supposed to be those of the manor-house of the Bishop of Durham, attached to Wolsingham park. There are two chalybeate springs in the neighbourhood, and a sulphureous spring about two miles east of Wolsingham, on an estate called Bradley.

WOLSTAN (*St. Margaret*), a parish, in the union of Rugby, partly in the Kirby, but chiefly in the Rugby, Division of the hundred of Knightlow, Northern Division of the county of Warwick, 5½ miles (E. S. E.) from Coventry; containing 968 inhabitants. The London and Birmingham railway passes through the parish, in which is the Brandon station. The living is a vicarage, valued in the king's books at £15. 10.; it is in the patronage of Mrs. Scott. The church is a large cruciform structure. There is a place of worship for Baptists. An Alien priory, a cell to the abbey of St. Peter *super Divam* in Normandy, was founded here soon after the Conquest, and, at its suppression, granted by Richard II. to the Carthusian priory at Coventry. On the southern bank of the Avon are vestiges of a Roman encampment.

WOLSTANTON (*St. Margaret*), a parish, in the union of Wolstanton and Burslem, Northern Division of the hundred of Pirehill and of the county of Stafford; containing 10,853 inhabitants, of which number, 1083 are in the township of Wolstanton, 1½ mile (N. by E.) from Newcastle-under-Lyme. This parish is divided into the North and South sides or divisions, the former consisting of the chapelry of New-chapel, or Thursfield, and each containing several townships or hamlets, embracing in the whole upwards of 10,000 acres. There are numerous manufactories of china and earthenware, collieries, brick and tile works, &c. Brieryhurst, or Brerehurst, the most northerly hamlet, embraces the eastern side of Mow Cop, a lofty and rugged hill, which separates the counties of Staffordshire and Cheshire, and on the top of which a summer-house resembling a castle in ruins is conspicuous. Mines of coal and ironstone are extensively wrought at and near Kidsgrove, in this and the adjoining hamlet of Ranscliff; and several blast furnaces have been lately established here for smelting iron-ore, by Thomas Kinnersly, Esq. That township comprises 922 acres, and is little productive except in mineral treasures. The southern division of the parish is principally agricultural, but Tunstall and other northern hamlets contain extensive potteries, tileries, and collieries. The Grand Trunk canal passes through the parish on its summit level, and leaves it northwardly through two parallel tunnels which run under Harecastle Hill. Sir Nigel Gresley's canal, extending from the Apedale collieries and iron-furnaces to Newcastle, also runs through the west side of the parish. The living is a vicarage; patron and impropriator, Ralph Sneyd, Esq.: the impropriate rectory is valued in the king's books at £32. 3. 9. The tithes have been commuted for a rent-charge of £500, subject to the payment of rates. The church is an ancient structure, with a nave, side aisles, and a spacious chancel, in which is a curious monument to the memory of Sir William Sneyd, of Bradwell, with others to members of the same family; and being seated on an eminence, with its lofty spire, forms a conspicuous feature in this part of the country. The chapel at Thursfield, or New-chapel, is a neat brick building with a cupola and bell, founded by some of the proprietors of land in the neighbourhood, in the early part of Queen Elizabeth's reign, for the accommodation of the eight northern hamlets of Wolstanton: it is a perpetual curacy, with an endowment of about £70 per annum, and in the patronage of C. B. Lawton, Esq. A district church has been lately erected at Tunstall, at a nearly central point between the parish church and the ancient chapel of New-chapel, which are more than four miles apart; and at Brieryhurst, Mr. Kinnersly has erected and endowed an elegant church, having a tower furnished with six bells and a clock, besides a parsonage and school house, all situated near each other, in a secluded spot surrounded by woods. Her Majesty's commissioners have also approved of a plan for erecting a new church at Chesterton. There are in different parts of the parish several dissenting places of worship. At Thursfield is a Latin and English free school, founded by Dr. Robert Hulme, with an income now exceeding £100 per annum; 18 boys are instructed gratuitously. The township of

Wolstanton embraces 840 acres, and is within the copyhold manor of Newcastle and parcel of the Duchy of Lancaster: the other hamlets, except Knutton and Chesterton, are all within the manor of Tunstall. James Brindley, the celebrated engineer, was buried in the cemetery of New-chapel, where a plain altar-tomb is erected to his memory.

WOLSTONE, a chapelry, in the parish of UFFINGTON, union of FARRINGDON, hundred of SHRIVENHAM, county of BERKS, $5\frac{1}{4}$ miles (S. by E.) from Great Farringdon; containing 270 inhabitants. The chapel is dedicated to All Saints.

WOLTERTON (*ST. MARGARET*), a parish, in the union of AYLSHAM, hundred of SOUTH ERPINGHAM, Eastern Division of the county of NORFOLK, $4\frac{1}{4}$ miles (N. N. W.) from Aylsham; containing 41 inhabitants. The living is a discharged rectory, with that of Wickmere annexed, valued in the king's books at £8; present net income, £495; patron, Earl of Orford. The church is supposed to have been rebuilt by John de Wulterton, whose effigy, with that of his wife, still remains in one of the windows, and in others were representations of the twelve Apostles.

WOLVERHAMPTON (*ST. MARY AND ST. PETER*), a parish, and the head of a union, comprising the townships and market-towns of Bilston and Wolverhampton (of which the latter has been constituted a borough, the former also sharing in the franchise), in the Northern Division of the hundred of SEISDON, and partly in the Southern Division of the hundred of OFFLOW, and partly in the Eastern Division of the hundred of CUTTLESTONE, Southern Division of the county of STAFFORD; and containing 48,184 inhabitants, of which number, 24,732 are in the town of Wolverhampton, 16 miles (S.) from Stafford, and 123 (N. W.) from London. This place, which is of considerable antiquity, was called "Hanton" or "Hamton" prior to the year 996, when Wulfrana, sister of King Edgar, and widow of Aldhelm, Duke of Northampton, founded a college here, for a dean and several prebendaries, or Secular canons, and endowed it with so many privileges, that the town, in honour of Wulfrana, was called *Wulfranis Hamton,* of which its present name is a corruption. The college, under the same government, continued till the year 1200, when Petrus Blesensis, who was then dean, after fruitless attempts to reform the dissolute lives of the brethren, surrendered the establishment to Hubert, Archbishop of Canterbury; and it was subsequently annexed by Edward IV. to the deanery of Windsor. In 1258, the town obtained from Henry III. the grant of a market and a fair, from which time no circumstance of historical importance occurs till 1590, when a considerable part of it was destroyed by a fire, which continued burning for five days. In the parliamentary war, Charles I., accompanied by his sons, Charles Prince of Wales, and James Duke of York, visited Wolverhampton, where he was received with every demonstration of loyalty by the inhabitants, who, in aid of the royal cause, raised a liberal subscription, towards which Mr. Gough, ancestor of the learned antiquary of that name, contributed £1200. Prince Rupert, in 1645, fixed his headquarters in the town, while the king was encamped at Bushbury; and, immediately after the battle of Naseby, Charles marched into it, and quitted the day following.

The town is situated on an eminence, in a district abounding with mines of coal, iron, and limestone, and consists of several streets diverging from the market-place to the several roads from which they take their names. Among the improvements effected, under the provisions of an act obtained about 1814, is a new entrance on the east from Bilston, which, by means of a street crossing the town, nearly in a direct line, communicates on the west with Salop-street, leading towards Shrewsbury. The houses are in general substantial and neatly built of brick, many of them being modern and handsome; but in the smaller streets are several dwellings of more ancient appearance: the town is irregularly paved, well lighted with gas, and supplied with water by wells sunk to a great depth in the rock on which it is built. A public subscription library was established in 1794, which contains more than 5000 volumes, and for which a neat and commodious building was erected in the year 1816, when a news-room was added: over the library is a suite of rooms in which assemblies and concerts, under the superintendence of the Harmonic Society, take place. A neat theatre, well arranged for the purpose, is opened occasionally: prior to its erection, the celebrated Mrs. Siddons, and her brother, J. P. Kemble, performed in the town-hall, since taken down, where they first developed those talents which procured for them so distinguished a reputation. Races are held annually in August, and are well attended: the course is an extensive area near the town, on which an elegant stand has been erected. The manufacture of the finer steel ornaments, which was formerly carried on extensively, and brought to the highest perfection, in this town, has given place to the heavier articles of steel and iron, of which the principal are, smiths' and carpenters' tools of every description, files, nails, screws, gun-locks, hinges, steel-mills, and machinery; locks, for the making of which the town has long been celebrated; furnishing ironmongery and cabinet brasses, with every branch of the iron-manufacture; and brass, tin, Pont-y-pool and japanned wares in great variety. Here are also extensive chemical works for the manufacture of oil of vitriol, aqua-fortis, and other chemical preparations connected with medicine and manufactures; and there are two large mills for spinning worsted, and numerous other manufacturing establishments. The Rowley rag-stone is found in the coal mines in the parish, frequently in large masses, sometimes penetrating the thick stratum of coal at a depth of from 300 to 400 feet from the surface. The Birmingham canal, which forms a junction with the Staffordshire and Worcestershire canal, passes close to the town, on the west and north, where it is joined by the Essington and Wyrley canal, which terminates here, and affords facility of conveyance to every part of the kingdom: fly-boats proceed daily for London, Liverpool, and Chester, from the wharfs at Walsall-street and Horsleyfield, and twice a week for Derby, Nottingham, and Hull. In 1834, an act was obtained to enable the Grand Junction Railway Company to make a branch to this town, and a station on that line, which passes near the town, has been established here. The market is on Wednesday; and the fair, which continues for three days, the first being for cattle, commences on July 10th. The market-place is a large area, in the centre of which is a cast-iron column 45 feet high, supporting a large gas-lantern. The town is within the jurisdiction of the county magistrates, who hold petty sessions for the north and south divisions. By the act of the 2nd of William IV., cap. 45, it has been constituted a borough, returning two members to parliament, the right of election being vested in the £10 householders of a district comprising 18,604 acres, the limits of which are minutely described in the Appendix, No. II. The present number of registered electors is 2643, of whom 1315 are in Wolverhampton, 558 in Bilston, 337 in Willenhall, 75 in Wednesfield, and 358 in Sedgley: the constable of the manor of the deanery of Wolverhampton is the returning officer. A court of requests, for the recovery of debts not exceeding £5, is held in the public office, Prince's-street, every fourth Friday, under an act passed in the 48th of George III.; its jurisdiction extends over Wolverhampton and Wednesfield, and the parishes of Brewood, Pattingham, Busbury, and Penn. The lords of the manor hold a court leet, at which two constables (one of them chosen by the dean) and other officers, are annually appointed. The town has been made a polling-place for the southern division of the county.

The living is a perpetual curacy; net income, £193; patron and appropriator, Dean of Windsor, as incumbent of the ancient deanery of Wolverhampton. The church, formerly collegiate, and one of the king's free chapels, to which many immunities were granted, is a spacious cruciform structure, partly in the early decorated, but principally in the later, style of English architecture, with a handsome square embattled tower rising from the centre, the upper part of which is a very fine specimen of the later style. It has been lately repewed by subscription, and 560 free sittings provided, the Incorporated Society having granted £300 in aid of the expense. The piers and arches of the nave and transepts, if not of the early English, are of that style merging into the decorated; and the pulpit, of one entire stone, is richly embellished with sculpture. An octagonal font, of great antiquity, is supported on a shaft, the faces of which are embellished with the figures of St. Anthony, St. Paul, and St. Peter, in bass-relief, and is richly ornamented with bosses, flowers, and foliage. In the chancel are, a fine statue of brass, erected in honour of Admiral Sir Richard Leveson, who commanded under Sir Francis Drake against the Spanish Armada, and a monument to the memory of Col. John Lane, the protector of Charles II. after the battle of Worcester: in what was anciently the Lady chapel is an alabaster monument to John Lane and his wife, the former represented in armour. In the churchyard, which is enclosed with a handsome iron palisade, is a column twenty feet high, divided into compartments, highly enriched with sculpture of various designs, supposed to be either of Saxon or Danish origin. Near its south-western corner is a large vault, the roof of which is finely groined, and supported on one central pillar; the walls are three yards in thickness, and on both sides of the doorway are slight vestiges of sculpture; the interior is in good preservation. It appears to have been the basement story of some edifice, probably connected with the monastery of Wulfrana, the exact site of which has not been ascertained. The living of St. John's is a perpetual curacy; net income, £200; patron, Earl of Stamford and Warrington. The church is an elegant modern structure, in the Grecian style of architecture, with a handsome tower surmounted by a lofty and finely proportioned spire; the prevailing character is a mixture of the Ionic and Corinthian orders. A pleasing and appropriate effect is produced from the arrangement of the interior; the altar is ornamented with a good painting of the Descent from the Cross, by Barney, a native of the town. A new church of the Grecian Doric order, with a tower and spire, was erected in 1830, at an expense of £10,325. 3. 6., towards defraying which the inhabitants subscribed £3500, under the act of the 58th of George III.: it is dedicated to St. George, and, under the 21st section of the same act, has been made a district church. The living is a perpetual curacy; net income, £155; patron, Dean of Windsor. A church, dedicated to St. Paul, has been recently erected by the Rev. W. Dalton; and three additional churches are about to be built, at an expense of £10,000 each. There are places of worship for Baptists, Independents, Wesleyan Methodists, Methodists of the New Connection, Unitarians, and Roman Catholics: there is also a Roman Catholic chapel at Black-Ladies, and another at Long-Birch. A Roman Catholic bishop is resident here.

The free grammar school was founded, under letters patent of Henry VIII., in 1513, by Sir Stephen Jenyns, Knt., a native of this town, and lord mayor of London in 1508, who endowed it with estates in the parish of Rushoe, in the county of Worcester, producing an income, aided by other benefactions, of about £1170 per annum: the management, originally in the Master and Wardens of the Merchant Tailors' Company in London, is now, by a decree of the court of Chancery, vested in 40 trustees, including the Bishop of Lichfield and the two county members, for the time being. The building was erected, in 1713, by the Merchant Tailors' Company, who, on some disagreement with the inhabitants, subsequently petitioned the Lord Chancellor to be released from the governorship. Sir William Congreve; John Abernethy, Esq.; and John Pearson, Esq., Advocate General of India, were educated at this school. The Blue-coat charity school, for 100 boys and 50 girls, who are educated and clothed, is an ancient establishment of unknown origin: it has an endowment arising from a farm at Seisdon, tenements in the town, and funded property, purchased with accumulated benefactions, producing more than £240 per annum, and is further liberally supported by subscription: residences for the master and mistress are attached. National day and Sunday schools are supported by subscription; there are also a Diocesan and a British school. A dispensary was instituted in 1821, and a handsome and commodious building was erected by subscription, in 1826: its income, arising from bequests and annual subscriptions, is about £400 per annum. There are numerous charitable bequests for distribution among the poor; but one of the most praiseworthy institutions is the establishment of the "Union Mill," erected in 1813, at an expense of £14,000, raised in shares, for the purpose of grinding corn for the poor on easy terms, and supplying them with cheap flour and bread. The poor law union of Wolverhampton includes only a portion of the parish, comprising, with the town itself, the three chapelries of Bilston, Wednesfield and Willenhall, and containing a population of 46,937, according to the census of 1831; it is under the superintendence of 25 guardians.

WOLVERLEY, a township, in the parish of Wem,

Whitchurch Division of the hundred of NORTH BRADFORD, Northern Division of the county of SALOP; containing 62 inhabitants.

WOLVERLEY (*St. John the Baptist*), a parish, in the union of KIDDERMINSTER, partly in the Upper Division of the hundred of HALFSHIRE, but chiefly in, and forming a detached portion of, the hundred of OSWALDSLOW, Kidderminster and Western Divisions of the county of WORCESTER, 2 miles (N. by W.) from Kidderminster; containing 1840 inhabitants. The living is a vicarage, in the patronage of the Dean and Chapter of Worcester (the appropriators), valued in the king's books at £13. 6. 8.; present net income, £250. The church is a neat edifice of brick, on an elevated site. There is a place of worship at Cookley for Wesleyan Methodists. Here is a very extensive establishment for the manufacture of iron, tin plates, and wire, called the Cookley Iron Works. The Staffordshire and Worcestershire canal, and the river Stour, pass through the parish. William Seabright, in 1620, bequeathed property in and near London, now producing a rental of £633. 4., to establish a free grammar school for the children of Wolverley; a weekly supply of bread to the poor of Wolverley, Old Swinford, Kidderminster, Chaddesley-Corbett, and Bewdley, in the county of Worcester; Kinver, in the county of Stafford; and Alveley, in the county of Salop; and the surplus to be appropriated to the repair of the church and bridges in this parish. In 1829 it was determined, in consequence of the improved state of the school funds, to extend that charity; and in furtherance of this object the premises were re-erected, and now constitute a handsome range of buildings in the later style of English architecture, comprising a Latin school in the centre; a spacious school-room at each wing, one for boys and the other for girls; adjoining residences, with gardens attached, for the first and second masters; and a committee-room. The total number of pupils, exclusively of the head-master's boarders, is about 120 boys and 90 girls: the children in the lower schools are instructed upon the National system. The sum of £3. 0. 8. per ann. is paid to each of the seven parishes above named, for distribution in bread; and the trustees expend annually about £30 in clothing and £15 in coal for the poor of Wolverley. John Smith, in 1823, bequeathed £600 for founding an afternoon lectureship, about one-fourth of the interest to be applied to the relief of superannuated husbandmen and widows. John Baskerville, an eminent printer, was born here in 1706; he died in 1775.

WOLVERTON (*Holy Trinity*), a parish, in the union of POTTERSPURY, hundred of NEWPORT, county of BUCKINGHAM, 1 mile (E. N. E.) from Stony-Stratford; containing 417 inhabitants. The living is a discharged vicarage, valued in the king's books at £10. 3. 9.; present net income, £38; patrons and impropriators, Trustees of the late Dr. Radcliffe. Here is a National school. Catherine Featherstone, in 1711, bequeathed £100 for the benefit of the incumbent, and £50 for supplying coal annually to the poor. Rather more than one mile distant is a station of the London and Birmingham railway.

WOLVERTON (*St. Peter*), a parish, in the union and hundred of FREEBRIDGE-LYNN, Western Division of the county of NORFOLK, $2\frac{3}{4}$ miles (N. by W.) from Castle-Rising; containing 163 inhabitants. The living is a rectory, valued in the king's books at £12; patron, H. Henley, Esq. The tithes have been commuted for a rent-charge of £260, subject to the payment of rates, which on the average have amounted to £36. 7.; the glebe comprises 22 acres, and is valued at £13. 2. per annum.

WOLVERTON (*St. Catherine*), a parish, in the union and hundred of KINGSCLERE, Kingsclere and Northern Divisions of the county of SOUTHAMPTON, 8 miles (N. W.) from Basingstoke; containing 229 inhabitants. The living is a rectory, valued in the king's books at £13. 2. $8\frac{1}{2}$.; present net income, £296; patron, Duke of Wellington. A rent-charge of £16 was bequeathed by Sir John Browne for the poor not receiving parochial aid.

WOLVERTON (*St. Mary*), a parish, in the union of STRATFORD-UPON-AVON, Snitterfield Division of the hundred of BARLICHWAY, Southern Division of the county of WARWICK, 5 miles (W. S. W.) from Warwick; containing 166 inhabitants. The living is a rectory, valued in the king's books at £7. 10. $7\frac{1}{2}$.; present net income, £300; patron and incumbent, Rev. James Roberts.

WOLVES-NEWTON (*St. Thomas à Beckett*), a parish, in the union of CHEPSTOW, Upper Division of the hundred of RAGLAND, county of MONMOUTH, $5\frac{1}{4}$ miles (E. by S.) from Usk; containing 248 inhabitants. The living is a rectory, valued in the king's books at £8. 2. $8\frac{1}{2}$.; present net income, £190: it is in the patronage of the Crown.

WOLVEY (*St. John the Baptist*), a parish, in the Kirby Division of the hundred of KNIGHTLOW, Northern Division of the county of WARWICK, $5\frac{1}{4}$ miles (S. E. by E.) from Nuneaton; containing 935 inhabitants. The living is a discharged vicarage, valued in the king's books at £6. 6. $5\frac{1}{2}$.; present net income, £260; patrons and impropriators, Prebendary of Wolvey in the Cathedral Church of Lichfield, and others. The church contains 175 free sittings, the Incorporated Society having granted £150 in aid of the expense: there is a monument, 500 years old, of the Clinton family, formerly resident in the parish. Here is a Sunday National school. Little Copston, now called Smockington, in this parish, situated on the line of the Roman Watling-street, was anciently a considerable village, and had a chapel.

WOLVISTON, a chapelry, in the parish of BILLINGHAM, union of STOCKTON, North-Eastern Division of STOCKTON ward, Southern Division of the county palatine of DURHAM, $4\frac{1}{2}$ miles (N. by E.) from Stockton-upon-Tees; containing 582 inhabitants. The living is a perpetual curacy; net income, £97; patrons and appropriators, Dean and Chapter of Durham. The chapel, dedicated to St. Peter, has been enlarged by the addition of 190 sittings, 130 of which are free, the Incorporated Society having granted £100 in aid of the expense. Here is a National school.

WOMBLETON, a township, in the parish of KIRKDALE, union of HELMSLEY-BLACKMOOR, partly within the liberty of ST. PETER'S, East Riding, and partly in the wapentake of RYEDALE, North Riding, of the county of YORK, 4 miles (E. by S.) from Helmsley; containing 262 inhabitants, exclusively of that part which is in the liberty of St. Peter's, which, with part of Newton, contains 220. There is a place of worship for Wesleyan Methodists.

WOMBOURNE (*St. Benedict*), a parish, in the union, and Southern Division of the hundred, of SEISDON, Southern Division of the county of STAFFORD, 4 miles (S. W. by S.) from Wolverhampton; containing 1647 inhabitants. The living is a discharged vicarage, with that of Trysull annexed, valued in the king's books at £12. 12. 8½.; present net income, £608: it is in the patronage of certain Trustees. The church contains 190 free sittings, the Incorporated Society having granted £150 in aid of the expense: there is an elegant monument, by Chantrey, in memory of the late R. B. Marsh, Esq. The Staffordshire and Worcestershire canal passes through the parish.

WOMBRIDGE (*St. Mary and St. Leonard*), a parish, in the union of WELLINGTON, Wellington Division of the hundred of SOUTH BRADFORD, Northern Division of the county of SALOP, 2¼ miles (E.) from Wellington; containing 1855 inhabitants. The living is a perpetual curacy, in the patronage of William Charlton, Esq., as lord of the manor; net income, £82. The church has been enlarged, and 290 free sittings provided, the Incorporated Society having granted £270 in aid of the expense. The Shrewsbury, Shropshire, and Marquess of Stafford's, canals form a junction in this parish, which is intersected by the old Watling-street and the great Holyhead road, also by several railways communicating with the extensive coal and iron mines at Ketley and in the neighbourhood, which have been worked for centuries; but the most considerable iron-works were established here in 1818. At Oaken-Gates a small customary market is held. There are slight remains near the church of a priory of Black canons, founded in the reign of Henry I., by William Fitz-Alan, which at the dissolution had a revenue of £72. 15. 8.

WOMBWELL, a chapelry, in the parish of DARFIELD, Northern Division of the wapentake of STRAFFORTH and TICKHILL, West Riding of the county of YORK, 4½ miles (S. E. by E.) from Barnesley; containing 836 inhabitants. The chapel, a neat structure with a tower, has been enlarged, and 100 free sittings provided, the Incorporated Society having granted £100 in aid of the expense. A school is partly supported by a gratuity of £15 per annum from the Trustees of Mr. George Ellis's charity, who also allow £5 per annum for the instruction of six poor children in another school.

WOMENSWOULD (*St. Margaret*), a parish, in the union of BRIDGE, hundred of WINGHAM, lathe of ST. AUGUSTINE, Eastern Division of the county of KENT, 5 miles (S. by W.) from Wingham; containing 263 inhabitants. The living is a perpetual curacy, annexed to that of Nonington. There are several handsome monuments in the church, some of which, to the Montresor family, are by Chantrey. A school is supported by a private individual.

WOMERSLEY (*St. Martin*), a parish, in the Lower Division of the wapentake of OSGOLDCROSS, West Riding of the county of YORK; containing 843 inhabitants, of which number, 364 are in the township of Womersley, 5 miles (E. S. E.) from Pontefract. The living is a discharged vicarage, valued in the king's books at £6. 11. 5½.; present net income, £258; patron and impropriator, Lord Hawke. The church is a handsome structure with a lofty spire, situated on an eminence in the centre of the parish. On the southern bank of the river Went, which runs through this parish, are quarries of a fine freestone, whence a rail-road passes over the stream, runs through the township of Little Smeaton, and meets the new line of navigation made by the Aire and Calder Company: there are also extensive quarries of limestone. A neat building has been erected, comprising a dwelling-house for a schoolmaster, and two school-rooms for boys and girls, who are taught on the National system: the site was given by Lord Hawke, and the expense of the building was defrayed by liberal grants from the parent institution in London, and the York Diocesan Society, aided by voluntary contributions. At Walden-Stubbs is an ancient hall, once the seat of a family of the name of Shuttleworth, now occupied as a farm-house.

WONASTOW (*St. Wonnow*), a parish, in the Lower Division of the hundred of SKENFRETH, union and county of MONMOUTH, 2 miles (W. S. W.) from Monmouth; containing 149 inhabitants. The living is a discharged vicarage, valued in the king's books at £4. 15. 5.; present net income, £95; patron and impropriator, Sir W. Pilkington, Bart.

WONERSH (*St. John the Baptist*), a parish, in the union of HAMBLEDON, First Division of the hundred of BLACKHEATH, Western Division of the county of SURREY, 3½ miles (S. S. E.) from Guildford; containing, with the hamlet of Shimley Green, 1069 inhabitants. This place is situated on the road from Guildford to Brighton; and in the village is a mill for dressing leather, which affords employment to 40 persons. Charles II., after his restoration, granted a market and fair to be kept on Shimley Green; but the former has fallen into disuse, and the latter dwindled to a small pleasure fair held on the 11th of June. The manor-house of Tangley, originally a hunting box of King John's, was, in 1585, converted into a residence for the family of Sir Francis Duncombe. The living is a discharged vicarage, valued in the king's books at £15. 1. 3.; present net income, £137; patron and impropriator, Lord Grantley. The old church, with the exception of the north wall and tower, was rebuilt in 1795, and is picturesquely situated in Wonersh park, the seat of Lord Grantley: at the east end of the north aisle is the Grantley family vault, in which is interred Judge Chapple, whose daughter married Sir Fletcher Norton, Speaker of the House of Commons for 12 years, and raised to the peerage in 1782, by the title of Lord Grantley, Baron of Markenfield. There are two places of worship for Independents. The Wey and Arun canal passes through the parish, which abounds with iron-stone. Six poor boys are instructed for a rent-charge of £4, the bequest of Henry Chennell, in 1672; and four are taught for a like sum left by Richard Gwynne, in 1698. Here is a Sunday National school.

WONSTON, or WONSINGTON, (*Holy Trinity*), a parish, in the union of NEW WINCHESTER, hundred of BUDDLESGATE, Winchester and Northern Divisions of the county of SOUTHAMPTON, 5¾ miles (S.) from Whitchurch; containing 740 inhabitants. The living is a rectory, valued in the king's books at £46. 15. 7½.; patron, Bishop of Winchester. The tithes have been commuted for a rent-charge of £1150, subject to the payment of rates, which on the average have amounted to £213; the glebe comprises 20 acres, valued at £28 per annum. The church was burnt down in 1714, and rebuilt, for which purpose a benefaction of £10 is recorded as a gift of John Wallop, Esq.; it was repewed and beautified in 1829, at an expense of £750, defrayed by the Rev. Mr. Dallas, the present rector, by which 169 additional sittings were obtained; it has a fine window of painted glass, presented by the Hon. and Rev. Augustus Legge, 31 years rector of the parish. A National and an infants' school are supported by the rector, who has a printing press, from which are issued religious publications for the poor of the parish. Another school is endowed with £7. 16. per annum for the instruction of 13 children.

WOOBURN (*St. Paul*), a parish (formerly a market-town), in the union of WYCOMBE, hundred of DESBOROUGH, county of BUCKINGHAM, 3 miles (W. S. W.) from Beaconsfield; containing 1927 inhabitants. The living is a discharged vicarage, valued in the king's books at £12; present net income, £138; patron and impropriator, James Dupré, Esq. The church is a stately edifice, in the later style of English architecture, with a very handsome tower and a curiously carved font; it contains some monuments to the Bertie and Wharton families, of whom Philip, Lord Wharton, in 1694, gave a rent-charge of £22. 10., to be paid to the vicar for preaching an evening lecture every Sunday. There are places of worship for Independents and Wesleyan Methodists; also two schools, partly supported by subscription. A rivulet, rising at West Wycombe, flows through this parish, turning in its course several paper, mill-board, and flour mills. Many of the female inhabitants are employed in the manufacture of bone-lace. The market, which was held on Friday, and a fair on the festival of the translation of St. Edward, were granted by Henry VI., but they have been long disused, and fairs are now held, for horses, cattle, and sheep, on May 4th and November 12th. Wooburn House occupies the site of a noble palace, formerly the residence of the Bishops of Lincoln. A school on the Lancasterian system, and one for infants, are supported by subscription. The sum of £27. 10., the rental of an allotment of 28 acres of waste land under an enclosure act, and of a piece of meadow land called the church estate, is principally distributed among the poor. This parish is entitled to a distribution of bibles under Lord Wharton's charity; and there is a small apprenticing fund arising from a bequest by Esther Butterfield, in 1676.

WOOD, or WOODCHURCH, a ville, and member of the cinque-port liberty of DOVOR, in the union of the ISLE OF THANET, locally in the hundred of Ringslow, or the Isle of Thanet, lathe of ST. AUGUSTINE, Eastern Division of the county of KENT, 3 miles (S. W. by W.) from Margate; containing 292 inhabitants. Here are the ruins of a chapel of ease to the vicarage of Monkton: it was dedicated to St. Mary Magdalene.

WOOD-DALLING (*St. Andrew*), a parish, in the union of AYLSHAM, hundred of EYNSFORD, Eastern Division of the county of NORFOLK, 3 miles (N. by W.) from Reepham; containing 512 inhabitants. The living is a discharged vicarage, annexed to the rectory of Swannington, and valued in the king's books at £8. 8. 4.

WOOD-EATON (*Holy Rood*), a parish, in the union of HEADINGTON, hundred of BULLINGDON, county of OXFORD, 4 miles (N. N. E.) from Oxford; containing 86 inhabitants. The living is a rectory, valued in the king's books at £10. 0. 10.; patron, Richard Weyland, Esq. The tithes have been commuted for a rent-charge of £149. 4., subject to the payment of rates, which on the average have amounted to £3. 19. 10.; the glebe comprises 10 acres, valued at £10 per annum. A parochial school is supported by Mr. Weyland. There are good stone quarries in the parish. John Collins, a distinguished mathematician, was born here in 1624: he died in 1683.

WOOD-NORTON, a parish, in the union of AYLSHAM, hundred of EYNSFORD, Eastern Division of the county of NORFOLK, 7 miles (N. W.) from Reepham; containing 315 inhabitants. The living comprises the united rectories of *All Saints* and *St. Peter*, with the rectory of Swanton-Novers annexed, valued in the king's books at £7. 12. 3½.; present net income, £709; patrons, Dean and Canons of Christ-Church, Oxford. The church, dedicated to All Saints, has no steeple, the bells being hung on a frame in the churchyard. The church of St. Peter has been long demolished.

WOOD-RISING (*St. Nicholas*), a parish, in the union of MITFORD and LAUNDITCH, hundred of MITFORD, Western Division of the county of NORFOLK, 2½ miles (W. N. W.) from Hingham; containing 127 inhabitants. The living is a discharged rectory, valued in the king's books at £4. 18. 4.; patron, John Weyland, Esq. The tithes have been commuted for a rent-charge of £250, subject to the payment of rates, which on the average have amounted to £26; the glebe comprises 17 acres, valued at £26. 12. 6. per annum. A school, commenced in 1829, is chiefly supported by Mr. Weyland.

WOOD-THORPE, a hamlet, in the parish and union of LOUGHBOROUGH, hundred of WEST GOSCOTE, Northern Division of the county of LEICESTER, 1½ mile (S.) from Loughborough; containing 90 inhabitants.

WOODBANK, otherwise ROUGH-SHOTWICK, a township, in the parish of SHOTWICK, union of GREAT BOUGHTON, Higher Division of the hundred of WIRRALL, Southern Division of the county of CHESTER, 5½ miles (N. W.) from Chester; containing 51 inhabitants. Here is a National school.

WOODBASTWICK (*St. Fabian and St. Sebastian*), a parish, in the union of BLOFIELD, hundred of WALSHAM, Eastern Division of the county of NORFOLK, 5¼ miles (N. W.) from Acle; containing 288 inhabitants. The living is a discharged vicarage, with the rectory of Panxworth annexed, valued in the king's books at £6; patron and impropriator, J. Cator, Esq. The impropriate tithes have been commuted for a rent-charge of £292. 3. 10., and the vicarial for £143. 6. 11., subject to the payment of rates; the glebe comprises 26 acres, valued at £52 per annum. A school is partly supported by private charity; and a Sunday school is supported by Mr. Cator.

WOODBOROUGH (*St. Swithin*), a parish, in the union of BASFORD, Southern Division of the wapentake of THURGARTON and of the county of NOTTINGHAM, 7½ miles (N. E. by N.) from Nottingham; containing 774 inhabitants. The living is a perpetual curacy, in the patronage of the Collegiate Church of Southwell (the appropriators), valued in the king's books at £4; present net income, £93. The church has a fine Norman doorway, and the east window exhibits some remains of ancient stained glass. The Doverbeck, a considerable stream turning several mills, runs through the parish. The stocking frame was invented here by William Lew, in 1528: about 150 of them are usually at work in the village. The free school was built and endowed in 1739, by Mr. Wood, and enlarged by the Rev. Richard Oldanes, the master, in 1763: the income, about £90 a year, arises from certain land at Blidworth, and other premises at Stapleford.

WOODBOROUGH (*St. Mary Magdalene*), a parish, in the union of Pewsey, hundred of Swanborough, Everley and Pewsey, and Northern, Divisions of the county of Wilts, 3½ miles (W.) from Pewsey; containing 372 inhabitants. The living is a rectory, valued in the king's books at £10; patron, G. H. W. Heneage, Esq. The tithes have been commuted for a rent-charge of £294. 12., subject to the payment of rates, which on the average have amounted to £39; the glebe comprises 70 acres, valued at £104. 10. per annum. There is a place of worship for Wesleyan Methodists; also a Sunday National school. The Kennet and Avon canal passes through the parish, on the banks of which is Honey-street wharf: a considerable trade in timber and coal is carried on, and a great number of canal boats and barges are built.

WOODBRIDGE (*St. Mary*), a market-town and parish, and the head of a union, in the hundred of Loes, Eastern Division of the county of Suffolk, 7½ miles (E. N. E.) from Ipswich, and 76½ (N. E. by E.) from London; containing 4769 inhabitants. This town is of considerable antiquity, for, in the time of Edward the Confessor, the prior and convent of Ely had possessions here, and their successors still hold the manor of Kingston: the name is thought to be a corruption of *Wodenbryge*, from the Saxon god Woden. Towards the termination of the twelfth century a priory of Augustine canons was founded here, by Ernaldus Rufus and others, and dedicated to the Virgin Mary, the revenue of which, at the dissolution, was valued at £50. 3. 5.: a house built on the site by one of the Seckfords, now in the possession of the Carthew family, still retains the name of the Abbey. In 1666, upwards of 327 inhabitants died of the plague, and were buried, according to tradition, at Bearman's Hill, in the vicinity. The town is pleasantly situated on the north side of the river Deben, in the direct road from London to Yarmouth, and occupies the slope of a hill surrounded by beautiful walks: it consists of two principal streets, a spacious square called Market Hill, and several narrow streets and lanes, and is paved, lighted, and amply supplied with water: the atmosphere is highly salubrious, and the general appearance of the town is neat and respectable; from the summit of the hill is a commanding view of the river Deben to its influx into the sea. A small theatre was built in 1813; and concerts are held occasionally. During the war, barracks were erected on the high ground about half a mile north-west of the town, with accommodation for 750 cavalry and 4165 infantry; but they were pulled down on the restoration of peace. The trade principally consists in the exportation of corn, flour, and malt; and in the importation of coal, timber, foreign wine, spirits, porter, grocery, drapery, and ironmongery. The shipping of late years has greatly increased: the amount of duties paid at the custom-house, for the year ended January 5th, 1837, was about £2000. Vessels sail weekly from this port to London, and many others are employed in trading with Newcastle, Hull, and the Continent; one or two sail direct to Liverpool, from which place they bring back salt; and there is a small trade to the Baltic for timber. A manufacture of salt, of peculiarly fine quality, was formerly carried on here; and there was a brisk business in ship-building, but both have declined. The river Deben, near its mouth, forms the haven of Woodbridge, from which it is navigable for vessels of 120 tons' burden to the town; and on its bank are two excellent quays, the common quay, where the general exports and imports are shipped and landed, and the limekiln quay, where there are two docks, in which small ships of war and other vessels were formerly built. The custom-house is in Quay-lane. The market is on Wednesday, for corn, cattle, and provisions; and fairs are held on April 5th and October 23rd. The quarter sessions for the liberty of St. Ethelred, and the hundreds of Colneis, Carlford, Loes, Plomesgate, Wilford, and Thredling, are held here; and petty sessions take place every Wednesday. The town has been made a polling-place for the eastern division of the county. The sessions-hall, under which is the corn market, in the centre of the Market Hill, erected in 1587 by Thomas Seckford, Esq., has recently undergone some extensive repairs; it is a handsome and lofty edifice of brick. On an adjacent eminence is the bridewell, rebuilt in 1804, and containing two wards and three airing-yards, the prisoners are employed in pumping water for the use of the prison and house.

The living is a perpetual curacy, to which the impropriate rectory was annexed in 1667, by Mrs. Dorothy Seckford; net income, £439; patron and incumbent, Rev. T. W. Salmon. The church was built by John Lord Seagrave, in the reign of Edward III., and the tower and north portico in that of Henry VI.: on the north side of the chancel is an elegant private chapel, built in the reign of Elizabeth, by Thomas Seckford, Esq., in which, over the family vault, is a tomb without an inscription, probably erected to his memory: the north portico is adorned with sculpture, in relief, representing the conflict of St. Michael and the Dragon. The tower is stately and magnificent, and, like the church, is constructed of dark flint intermixed with freestone, and, towards the upper part, formed into elegant devices; the summit is crowned with battlements, having finials at the angles, which are surmounted by vanes, and decorated in the intervals with badges of the four Evangelists. There are places of worship for Baptists, the Society of Friends, Independents, and Wesleyan Methodists. The free grammar school, in Well-street, was founded in 1662, by Mrs. Dorothy Seckford and others, and is endowed with property producing about £37 per annum, with a house for the master, for which ten children are instructed. The master is elected by the heirs of the founder and the perpetual curate, or, in default of such, by the lord of the priory manor, the perpetual curate, the two churchwardens, and the three principal owners, and three principal occupiers, of land in the parish. A National and a Lancasterian school are partly supported by subscription. Almshouses were erected, in 1587, by Thomas Seckford, Esq., master of the court of requests, for the residence of thirteen poor unmarried men, with another house for three poor women, to attend them as nurses, and endowed with an estate in the parish of St. James Clerkenwell, London, which, in 1767, produced an income of £568 per annum; but more than £20,000 having been expended on it, such is the improving state of the property, that the rental is expected eventually to produce between £5000 and £6000 per annum: the present income is about £3000. This institution is placed under the government of the Master of the Rolls and the Chief Justice of the Common Pleas, by patent of Queen Elizabeth. New and handsome almshouses have been recently erected, and it is intended to add to them a chapel for the use of the inmates. There are, besides, different benefactions, amounting to about £150 per annum, for the benefit of the poor. The poor law union of Woodbridge comprises 46 parishes or places, under the care of 49 guardians, and contains, according to the census of 1831, a population of 22,163. Various relics of antiquity, especially fragments of warlike instruments, have been occasionally found in the vicinity. Christopher Saxton, the publisher of the first county maps, was a native of this place, and servant to Thomas Seckford, Esq., mentioned above, who resided in a mansion at Great Bealings, about a mile and a half distant: they were published under his patronage, in 1579, and dedicated to Queen Elizabeth.

WOODBURY (*St. Swithin*), a parish (formerly a market-town), in the union of St. Thomas, hundred of East Budleigh, Woodbury and Southern Divisions of the county of Devon, 3 miles (E. by S.) from Topsham; containing 1673 inhabitants. The living is a perpetual curacy; net income, £150; patrons and appropriators, Custos and College of Vicars Choral in the Cathedral Church of Exeter. The church contains some ancient monuments, among which is one to Chief Justice Sir Edmund Pollexfen. The navigable river Exe bounds the parish on the west. A school, in connection with the National Society, is supported by the several endowments of Thomas Weare, in 1691; William Holwell, M.D., in 1707; and Esaias Broadmead, in 1728; amounting in the whole to about £37 per annum. Here is an ancient earthwork, called Woodbury Castle, an enclosure of irregular form, deeply intrenched, on the edge of a lofty hill commanding a beautiful prospect of the river Exe up the vale to Exeter and Honiton, Hall down, and the sea.

WOODCHESTER (*St. Mary*), a parish, in the union of Stroud, hundred of Longtree, Eastern Division of the county of Gloucester, $2\frac{1}{2}$ miles (S.W.) from Stroud; containing 887 inhabitants. This place is supposed to have derived its name from having been the site of a Roman station, which, from the numerous and interesting relics of antiquity that have been found here, appears to have been the residence of the Roman proprætor, or perhaps of the Emperor Adrian. The village is situated on an eminence forming part of a range of hills which enclose a beautiful and fertile vale; the surrounding scenery is richly diversified, and the neighbourhood abounds with interesting features. Spring Park, the seat of Lord Ducie, is a splendid residence beautifully situated, and combining a rich variety of picturesque scenery; and the Priory, the seat of Sir Samuel Wathen, a fine old mansion near the church, forms a pleasing object in the landscape. The manufacture of woollen cloths is carried on extensively, and in the neighbourhood of the village are not less than eight mills in constant operation. The living is a rectory, valued in the king's books at £10; patron, Lord Ducie. The tithes have been commuted for a rent-charge of £265, subject to the payment of rates, which on the average have amounted to £58; the glebe comprises 30 acres, valued at £52. 10. per annum. The church contains a fine monument to the memory of Sir George Huntley. There is a place of worship for Baptists. Robert Bridges, in 1722, bequeathed £500 for teaching, clothing, and apprenticing poor boys: but the school of St. Chloe, at Minchinhampton, being open to boys of this parish, the produce, now £50 per annum, is appropriated solely to the clothing and apprenticing of boys, with a premium of £15 each. Six schools are endowed with £10 per annum each, and two with £12 per annum. Here is also a National Sunday school. Among the antiquities that have been found at this place are foundations and ruins of buildings, fragments of statues, stags' horns, glass, pottery, coins of the Lower Empire, a coin of Adrian, one of Lucilla, and a noble tesselated pavement, of which an engraving was exhibited to the Society of Antiquaries by Samuel Lysons, Esq., F.A.S., who published an elaborate account of these relics, in 1797. The design of this pavement, which is superior to any thing of the kind yet found in the kingdom, is a circle, 25 feet in diameter, within a border 48 feet 10 inches square, divided into 24 compartments, and enriched with figures of animals and various architectural ornaments.

WOODCHURCH (*Holy Cross*), a parish, in the union, and Lower Division of the hundred, of Wirrall, Southern Division of the county of Chester; containing 929 inhabitants, of which number, 78 are in the township of Woodchurch, $6\frac{3}{4}$ miles (N. by W.) from Great Neston. This parish, like most others in this part of the hundred of Wirrall, probably abounded with oak trees, though at present there is but little of that species of timber in the neighbourhood: the pews of the church and the wood-work in most of the older buildings are of split oak; and the name of the parish appears to be descriptive of the situation of the church either in, or contiguous to, a wood. The living is a rectory, valued in the king's books at £25. 9. 2.; present net income, £827: it is in the patronage of Mrs. Ellen King, in whose ancestors the advowson has remained for some centuries. The church, though internally modernized, is a very ancient structure, in the Norman style of architecture, and contains an ancient font embellished with emblematical sculpture. William Gleave, Esq., alderman of London, in 1665, left £500 for the erection and endowment of a free school, of which sum, £100 was to be laid out in the erection of the school-house, and £400 in the purchase of land; with the latter sum an estate was purchased at Newton *cum* Larton, in the adjoining parish of West Kirby, consisting of a farmhouse and 21 Cheshire acres of land, producing to the master an income of £57. 15. per annum. James Goodaker, of Barnston, in this parish, in 1525, left 20 marks to buy 20 yoke of bullocks, for which subsequently cows were substituted; and the Rev. Richard Sherlock, in 1677, left £50 to the township of Oxton, to be laid out in the purchase of cows, which are let to the poor, who pay 5 per cent. annually on the purchase money for the hire of the cows, and gain a livelihood by selling milk; there are also several charitable bequests, the interest of which is distributed in bread to the poor.

WOODCHURCH (*All Saints*), a parish, in the union of TENTERDEN, hundred of BLACKBOURNE, Lower Division of the lathe of SCRAY, Western Division of the county of KENT, 4 miles (E. by N.) from Tenterden; containing 1187 inhabitants. The living is a rectory, in the patronage of the Archbishop of Canterbury, valued in the king's books at £26. 13. 4. The tithes have been commuted for a rent-charge of £682, subject to the payment of rates, which on the average have amounted to £94; the glebe comprises 15 acres, valued at £28 per annum. The church is partly in the early, and partly in the later, style of English architecture, with a tower surmounted by a spire; it contains numerous ancient monuments. There is a place of worship for Wesleyan Methodists.

WOODCOT, a township, in the parish of WRENBURY, union and hundred of NANTWICH, Southern Division of the county of CHESTER, $3\frac{3}{4}$ miles (S. W. by W.) from Nantwich; containing 30 inhabitants.

WOODCOTE, a liberty, in the parish of SOUTH STOKE, hundred of DORCHESTER, county of OXFORD, $5\frac{1}{2}$ miles (S. S. E.) from Wallingford: the population is returned with the parish. The chapel is dedicated to St. Leonard. Here is a National school.

WOODCOTE, a chapelry, in the parish of SHERIFF-HALES, union of NEWPORT, Newport Division of the hundred of SOUTH BRADFORD, Northern Division of the county of SALOP, 3 miles (S. E. by S.) from Newport; containing 195 inhabitants.

WOODCOTS, a tything, in the parish of HANDLEY, hundred of SIXPENNY-HANDLEY, Wimborne Division of the county of DORSET: the population is returned with the parish.

WOODCUTT, a parish, in the union of KINGSCLERE, hundred of PASTROW, Kingsclere and Northern Divisions of the county of SOUTHAMPTON, 5 miles (N. N. W.) from Whitchurch; containing 90 inhabitants. The living is a donative; net income, £20; patrons and impropriators, Trustees of the late E. Temple, Esq. A school is partly supported by subscription.

WOOD-DITTON, county of CAMBRIDGE.—See DITTON, WOOD.—*And other places having a similar distinguishing prefix will be found under the proper name.*

WOODEN, a township, in the parish of LESBURY, Eastern Division of COQUETDALE ward, Northern Division of the county of NORTHUMBERLAND, $4\frac{1}{2}$ miles (S. E. by E.) from Alnwick: the population is returned with the parish.

WOODEND, a hamlet, in the parish of BLAKESLEY, union of TOWCESTER, hundred of GREEN'S-NORTON, Southern Division of the county of NORTHAMPTON, $5\frac{1}{2}$ miles (W. by N.) from Towcester; containing 302 inhabitants.

WOODFORD, a township, in the parish of PRESTBURY, union and hundred of MACCLESFIELD, Northern Division of the county of CHESTER, 6 miles (S. by W.) from Stockport; containing 403 inhabitants.

WOODFORD (*St. Mary*), a parish, in the union of WEST HAM, hundred of BEACONTREE, Southern Division of the county of ESSEX, 8 miles (N. E. by N.) from London; containing 2548 inhabitants. This parish, so called from the ford in the wood, or forest, where is now Woodford-bridge, is about three miles in length and two miles in breadth, comprising about 2000 acres of fertile land, of which the chief portion is meadow and pasture. Woodford-bridge is a beautiful posting village, situated on the confines of Epping Forest, on the main road from London to Newmarket: the houses are in general detached, and irregularly arranged on the undulating declivities of a rising ground, skirted at the bottom by the river Roden, beautifully interspersed with trees, and disclosing at intervals mansions of a superior character, which are principally occupied by wealthy merchants of the metropolis. In different parts of the parish are some fine and extensive views into the counties of Essex and Kent. A nearer communication with the metropolis has been recently opened, by the construction of a new road from the highest part of the village, near the Castle Inn, which passes through the forest into the Lea Bridge road. The custom of Borough English, by which the younger son inherits, prevails in this manor. The living is a rectory, valued in the king's books at £11. 12. 1.; patron, Hon. W. P. T. L. Wellesley. The tithes have been commuted for a rent-charge of £670, subject to the payment of rates, which on the average have amounted to £43. 15. 10.; the glebe comprises 15 acres, valued at £31. 17. 6. per annum. The church was erected on the site of a former edifice, in 1817, at an expense of nearly £9000, defrayed partly by subscription, and partly by a rate: it is situated in the lowest part of the village, on the west side of the London road, and is an elegant edifice, in the ancient style of English architecture, with a square embattled tower; the aisles are separated from the nave by six pointed arches carried up to the roof, which is of open wood work, and surmounted in the centre by an octangular lantern tower; the east window is of stained glass, and divided into three compartments, containing figures of our Saviour, the four Evangelists, St. Peter, and St. Paul; there are some good monuments. In the churchyard is a splendid Corinthian column of marble, about 40 feet in height, erected to the memory of the family of Godfrey, which flourished many years in Kent; also a tomb with a column entirely covered with ivy, of picturesque appearance, and a remarkably fine old yew tree. An episcopal chapel has been built; and there are places of worship for Independents and Wesleyan Methodists. A National school, in which 75 children of both sexes are educated, of whom 50 are clothed, is supported by voluntary contributions. This parish is entitled to send two boys for gratuitous instruction to each of the schools founded at Chigwell, in 1629, by Dr. Samuel Harsnett, Archbishop of York; it has also the perpetual right of presenting two boys to Christ's Hospital, London, granted by Thomas Foulkes, in 1686. In 1828, a parochial library was established in the village. At Woodford Wells is a mineral spring, formerly in high estimation, but now little resorted to.

WOODFORD (*All Saints*), a parish, in the union of DAVENTRY, hundred of CHIPPING-WARDEN, Southern Division of the county of NORTHAMPTON, $7\frac{1}{2}$ miles (S. S. W.) from Daventry; containing 827 inhabitants. The living is a discharged vicarage, valued in the king's books at £6. 10.; present net income, £275: it is in the patronage of the Crown; impropriators, Rev. Sir H. Dryden, Bart., and G. Hitchcock, Esq. There is a place of worship for Moravians. A Sunday school is endowed with £6 per annum. Peter Grey, in 1577, bequeathed an estate now producing an annual income of £55, which sum is expended in coal for distribution among the poor. At Hinton, in this parish, is a mineral spring.

WOODFORD (*St. Mary*), a parish, in the union of THRAPSTONE, hundred of HUXLOE, Northern Division of the county of NORTHAMPTON, 2½ miles (S. W. by W.) from Thrapstone; containing 639 inhabitants. The living is a rectory in united medieties, valued jointly in the king's books at £22. 9. 7.; present net income, £497; patron, Lord St. John. A school for girls is supported by a lady. In the neighbourhood are three tumuli, near which have been found Roman tiles, fragments of tesselated pavement, an urn, and two small coins of the Lower Empire, inscribed *Constantinopolis*.

WOODFORD (*All Saints*), a parish, in the union of AMESBURY, hundred of UNDERDITCH, Salisbury and Amesbury, and Southern, Divisions of the county of WILTS, 4¼ miles (N. N. W.) from Salisbury; containing 397 inhabitants. The parish is bounded on the east by the river Avon, and here was formerly a palace of the Bishops of Salisbury, but no traces of it are now visible. Charles II., after the battle of Worcester, was concealed in this neighbourhood. The living is a vicarage, consolidated with that of Wilsford, and valued in the king's books at £13. 10. The church contains 118 free sittings, the Incorporated Society having granted £15 in aid of the expense. Here is a National school.

WOODFORD-GRANGE, an extra-parochial liberty, in the Southern Division of the hundred of SEISDON and of the county of STAFFORD; containing 18 inhabitants.

WOODGARSTON, a tything, in the parish of MONK'S-SHERBORNE, hundred of CHUTELY, Basingstoke and Northern Divisions of the county of SOUTHAMPTON, 4½ miles (N. W. by W.) from Basingstoke: the population is returned with the parish.

WOODGREEN, an extra-parochial liberty, in the Northern Division of the hundred of NEW FOREST, Ringwood and Southern Divisions of the county of SOUTHAMPTON, 3 miles (N. E. by E.) from Fordingbridge; containing 363 inhabitants.

WOODHALL (*St. Margaret*), a parish, in the union of HORNCASTLE, Southern Division of the wapentake of GARTREE, parts of LINDSEY, county of LINCOLN, 2¾ miles (W. S. W.) from Horncastle; containing 196 inhabitants. The living is a discharged vicarage, valued in the king's books at £13; present net income, £70; patron and appropriator, Bishop of Lincoln.

WOODHALL, a joint township with Brackenholme, in the parish of HEMINGBROUGH, wapentake of OUZE and DERWENT, East Riding of the county of YORK, 5¼ miles (N. W. by W.) from Howden: the population is returned with Brackenholme.

WOODHALL, a township, in the parish of HARTHILL, union of WORKSOP, Southern Division of the wapentake of STRAFFORTH and TICKHILL, West Riding of the county of YORK, 9½ miles (S. S. E.) from Rotherham: the population is returned with the parish.

WOODHAM, a hamlet, in the parish of WADDESDON, union of AYLESBURY, hundred of ASHENDON, county of BUCKINGHAM, 8½ miles (W. N. W.) from Aylesbury; containing 38 inhabitants.

WOODHAM, a township, in the parish of AYCLIFFE, union of SEDGFIELD, South-Eastern Division of DARLINGTON ward, Southern Division of the county palatine of DURHAM, 7 miles (E. S. E.) from Bishop-Auckland; containing 204 inhabitants. James I., on his way to take possession of the Crown of England, halted at this place, where he attended a horse-race.

WOODHAM-FERRIS (*St. Mary*), a parish, in the hundred of CHELMSFORD, Southern Division of the county of ESSEX, 4½ miles (S. S. E.) from Danbury; containing 826 inhabitants. This parish, which is bounded on the south by the river Crouch, derived its name from its situation in a thickly-wooded district, and its adjunct from the noble family of Ferrers, to whom the lands chiefly belonged at the time of the Norman survey. About a mile from the church is Edwin Hall, a handsome mansion, erected by Edwin Sandys, Archbishop of York. The living is a rectory, valued in the king's books at £28. 13. 4.; present net income, £661; patron, Sir B. W. Bridges, Bart. The church is an ancient edifice, with a tower of brick, and contains an elegant monument to the memory of Cecilia, wife of the above-named archbishop. Here is a National school. At Bikinacre, in this parish, there was anciently a hermitage, which was superseded by a priory of Black canons, founded and endowed by Maurice Fitz-Jeffrey, in consideration of certain sums of money due from him to Henry II.; it was dedicated to St. John the Baptist, and being almost deserted in the time of Henry VII., was then annexed to St. Mary's Spittal, London.

WOODHAM-MORTIMER (*St. Margaret*), a parish, in the union of MALDON, hundred of DENGIE, Southern Division of the county of ESSEX, 2½ miles (S. W. by W.) from Maldon; containing 339 inhabitants. This parish, called in some documents Little Woodham, derives its present adjunct from the family of Mortimer, to whom it anciently belonged. The living is a rectory, valued in the king's books at £6. 13. 4.: it is in the patronage of Eliza Wigg. The tithes have been commuted for a rent-charge of £340, subject to the payment of rates, which on the average have amounted to £57. 8. 9.; the glebe comprises 45 acres, valued at £56. 13. 6. per annum. The church has a richly-carved altar-piece. A National school has been established. In the marshes near Crouch river are several barrows.

WOODHAM-WALTER (*St. Michael*), a parish, in the union of MALDON, hundred of DENGIE, Southern Division of the county of ESSEX, 2½ miles (E. N. E.) from Danbury; containing 538 inhabitants. This parish is separated from the hundred of Witham by the river Chelmer, and is amply supplied with water from numerous springs; the lands are well cultivated, and the scenery beautifully diversified. An ancient mansion, called the Fort, is said to have been for some time the residence of Queen Elizabeth, during the reign of Mary. The village called Brook-Street, from a stream which flows through that part of the parish, consists of a few good houses. The living is a rectory, valued in the king's books at £12. 13. 1½.; present net income, £437; patron, Rev. L. Way. The church is a neat edifice in good repair, and in the chancel are some remains of ancient stained glass. Here is a National school.

WOODHAY, EAST (*St. Martin*), a parish, in the union of KINGSCLERE, hundred of EVINGAR, Kingsclere and Northern Divisions of the county of SOUTHAMPTON, 10½ miles (N. N. W.) from Whitchurch; containing 1269 inhabitants. The living is a rectory, with the perpetual curacy of Ashmansworth annexed, valued in the king's books at £21. 6. 0½.; patron, Bishop of Winchester. The tithes have been commuted for a rent-charge of £1008. 10., subject to the payment of rates, which on the average have amounted to £205; the glebe comprises 43 acres, valued at £40 per annum. The church was rebuilt at the expense of the parishioners, in 1823, and contains monuments to Bishops Kenn and Louth, who held the living. Fifteen children are taught in an infants' school for £9 a year, the interest of a bequest from the Rev. Joshua Wakefield, in 1753. A National school is supported by voluntary contributions. Here was formerly a palace belonging to the Bishops of Winchester.

WOODHAY, WEST (*St. Lawrence*), a parish, (formerly a market-town), in the union of HUNGERFORD, hundred of KINTBURY-EAGLE, county of BERKS, 6 miles (S. E.) from Hungerford; containing 127 inhabitants. The living is a rectory, valued in the king's books at £4. 4. 3.; present net income, £230; patron, Rev. John Sloper. The market, which was on Tuesday, was granted to one of the Barons St. Amand, in 1318, but it has for many years been disused. Here is a mansion, built in 1636, by Inigo Jones, from the drawing-room of which is a view of Windsor Castle, 36 miles distant.

WOODHEAD, a chapelry, in the parish of MOTTRAM-IN-LONGDENDALE, hundred of MACCLESFIELD, Northern Division of the county of CHESTER, 13¾ miles (E. N. E.) from Stockport: the population is returned with the parish. The living is a perpetual curacy; net income, £85; patron, Bishop of Chester. The chapel is a neat structure. There is a place of worship for Calvinistic Methodists.

WOODHORN (*St. Mary*), a parish, in the union, and Eastern Division of the ward, of MORPETH, Northern Division of the county of NORTHUMBERLAND; containing 1416 inhabitants, of which number, 155 are in the township of Woodhorn, 8 miles (E. N. E.) from Morpeth. The living is a vicarage, valued in the king's books at £21. 15. 7½.; present net income, £518; patron and appropriator, Bishop of Durham. The church is an ancient edifice, with the exception of the chancel, which has been added by the Master and Wardens of the Mercers' Company, London. At Newbiggin, in this parish, is a chapel of ease. The Rev. Dr. Triplett, in 1640, bequeathed a rent-charge of £5 for clothing and apprenticing poor children; and the late Viscountess Bulkeley, in 1826, bequeathed £500, which was invested in the purchase of £642. 1. 1. three per cent. consols., and the dividends are distributed among the poor at Christmas.

WOODHORN, a demesne, in the parish of WOODHORN, union, and Eastern Division of the ward, of MORPETH, Northern Division of the county of NORTHUMBERLAND, 8 miles (E. N. E.) from Morpeth; containing 9 inhabitants.

WOODHOUSE, a chapelry, in the parish and union of BARROW-UPON-SOAR, hundred of WEST GOSCOTE, Northern Division of the county of LEICESTER, 3 miles (S. W.) from Loughborough; containing 1262 inhabitants. The living is a perpetual curacy; net income, £73; patron, Vicar of Barrow. The chapel is dedicated to St. Mary. A church has been erected at Woodhouse Eaves, in Charnwood forest, and was consecrated on the 5th of September, 1837: it contains 400 free sittings, the Incorporated Society having granted £350 in aid of the expense. There is a place of worship for Wesleyan Methodists. Thomas Rawlins, in 1691, left an estate now producing £88. 10. per annum, of which £24 is paid to a master for teaching 34 poor children of Woodhouse, Quorndon, and Barrow, and £2 for books; £2 each to Quorndon and Woodhouse, for apprenticing two boys; £2 each to the two oldest trustees, and the remainder to the poor: the school is now conducted on the National system.

WOODHOUSE, a township, in the parish of SHILBOTTLE, union of ALNWICK, Eastern Division of COQUETDALE ward, Northern Division of the county of NORTHUMBERLAND, 5¼ miles (S. E. by S.) from Alnwick; containing 31 inhabitants.

WOODHOUSE, a joint township with Burntwood and Edgehill, in the parish of ST. MICHAEL, LICHFIELD, union of LICHFIELD, Southern Division of the hundred of OFFLOW and of the county of STAFFORD, 2¾ miles (W. by S.) from Lichfield; containing 206 inhabitants.

WOODHOUSE-HALL, an extra-parochial liberty, in the Hatfield Division of the wapentake of BASSETLAW, Northern Division of the county of NOTTINGHAM, 6½ miles (S. W. by S.) from Worksop; containing 11 inhabitants.

WOODHOUSES, a township, in the parish of MAYFIELD, Southern Division of the hundred of TOTMONSLOW, Northern Division of the county of STAFFORD, 4 miles (N. W. by W.) from Ashbourn; containing 28 inhabitants.

WOODHURST (*All Saints*), a parish, in the union of ST. IVES, hundred of HURSTINGSTONE, county of HUNTINGDON, 4 miles (N.) from St. Ives; containing 408 inhabitants. The living is united, with that of Old Hurst, to the vicarage of St. Ives. A fund of £10. 5. per annum, the rent of five acres of land, is distributed in money and coal among the poor.

WOODKIRK, West Riding of the county of YORK. —See ARDSLEY, WEST.

WOODLAND, a tything, in the parish and hundred of CREDITON, Crediton and Northern Divisions of the county of DEVON; containing 268 inhabitants.

WOODLAND, a chapelry, in the parish of IPPLEPEN, union of NEWTON-ABBOTT, hundred of HAYTOR, Teignbridge and Southern Divisions of the county of DEVON, 2 miles (E. by S.) from Ashburton; containing 237 inhabitants. The living is a perpetual curacy; net income, £56; patrons, the Parishioners; appropriators, Dean and Canons of Windsor. A school is supported by subscription.

WOODLAND, a township, in the parish of COCKFIELD, union of TEESDALE, South-Western Division of DARLINGTON ward, Southern Division of the county palatine of DURHAM, 7¼ miles (N. by E.) from Barnard-Castle; containing 223 inhabitants. Coal is obtained in the neighbourhood. There is a place of worship for Wesleyan Methodists; also a Sunday National school.

WOODLAND, a chapelry, in the parish of KIRKBY-IRELETH, hundred of LONSDALE, north of the sands, Northern Division of the county palatine of LANCASTER, 8¼ miles (N. N. W.) from Ulverstone; containing 302 inhabitants. The living is a perpetual curacy; net income, £68; patrons, the Landowners, who, in 1822, repaired the chapel.

WOODLAND-EYAM, a township, in the parish of EYAM, union of BAKEWELL, hundred of HIGH PEAK, Northern Division of the county of DERBY; containing 213 inhabitants.

WOODLAND-HOPE, a hamlet, in the parish of HOPE, union of CHAPEL-EN-LE-FRITH, hundred of HIGH PEAK, Northern Division of the county of DERBY; containing 273 inhabitants.

WOODLANDS, or LAMBORNE-WOODLANDS, a district chapelry, in the parish of LAMBORNE, union of HUNGERFORD, county of BERKS, 4½ miles (N.) from Hungerford; containing 400 inhabitants. The chapel, dedicated to St. Mary, with a house for the minister, was erected and endowed at the sole expense of the Misses Seymour, of Speen; it is a neat edifice, in the early English style, containing 270 sittings, half of which are free, and was consecrated in Sept., 1837. The living is a perpetual curacy, in the patronage of the founders, who have appointed the Rev. John Bacon, grandson of the celebrated sculptor. A National school, in which are 40 boys and 40 girls, was established and is chiefly supported by the Misses Seymour.

WOODLANDS, a tything, in the parish of HORTON, union of WIMBORNE and CRANBORNE, hundred of KNOWLTON, Wimborne Division of the county of DORSET, 4¼ miles (S. S. W.) from Cranborne; containing 423 inhabitants. The unfortunate Duke of Monmouth, after his flight from the battle of Sedgmore in Somersetshire, is stated to have been found here by his enemies, in a ditch under an ash tree, which is inscribed with the various names of those who have since visited the spot. A fair, which was transferred hither from Knolton, is held on July 5th. There is an old episcopal chapel in ruins; also a place of worship for Wesleyan Methodists.

WOODLANDS, a joint tything with Chaddenwicke, in the parish and hundred of MERE, Hindon and Southern Divisions of the county of WILTS; containing, with Chaddenwicke, 716 inhabitants.

WOODLEIGH (*ST. MARY*), a parish, in the union of KINGSBRIDGE, hundred of STANBOROUGH, Stanborough and Coleridge, and Southern, Divisions of the county of DEVON, 3 miles (N.) from Kingsbridge; containing 297 inhabitants. The living is a rectory, valued in the king's books at £22. 8. 4.; present net income, £392; patron, Rev. R. Edmonds. The church contains an altar-tomb, representing the Resurrection of our Saviour.

WOODLESFORD, a township, in the parish of ROTHWELL, union of WAKEFIELD, Lower Division of the wapentake of AGBRIGG, West Riding of the county of YORK, 6 miles (N. N. E.) from Wakefield: the population is returned with Outton. Here are manufactures of paper and earthenware.

WOODLEY, a joint township with Sandford, in the parish and hundred of SONNING, union of WOKINGHAM, county of BERKS, 3½ miles (E. by N.) from Reading; containing 796 inhabitants. Here is a Roman Catholic chapel. A school for the education of poor children is supported by subscription.

WOODMANCOTE, a tything, in the parish of NORTH CERNEY, hundred of RAPSGATE, Eastern Division of the county of GLOUCESTER, 5 miles (N. by W.) from Cirencester: the population is returned with the parish.

WOODMANCOTE, a parish, in the union of STEYNING, hundred of TIPNOAK, rape of BRAMBER, Western Division of the county of SUSSEX, 5 miles (N. E. by E.) from Steyning; containing 342 inhabitants. The living is a rectory, valued in the king's books at £13. 1. 10½.; present net income, £369: it is in the patronage of the Crown. The church is principally in the early style of English architecture.

WOODMANCOTT, a hamlet, in the parish of BISHOP'S-CLEEVE, union of WINCHCOMB, hundred of CLEEVE, or BISHOP'S-CLEEVE, Eastern Division of the county of GLOUCESTER, 3½ miles (W. by S.) from Winchcomb; containing 267 inhabitants.

WOODMANCOTT (*ST. JAMES*), a parish, in the union of BASINGSTOKE, hundred of MAINSBOROUGH, Winchester and Northern Divisions of the county of SOUTHAMPTON, 8 miles (S. W.) from Basingstoke; containing 92 inhabitants. The living is annexed to the rectory of Brown-Candover, where the parishioners formerly interred; but the burial-ground here was consecrated in 1838.

WOODMANSEY, a joint township with Beverley Park, in the parish of ST. JOHN, BEVERLEY, union, and liberties of the borough, of BEVERLEY, East Riding of the county of YORK, 2¼ miles (S. E. by E.) from Beverley; containing, with Beverley Park, 360 inhabitants.

WOODMANSTERNE (*ST. PETER*), a parish, in the union of CROYDON, First Division of the hundred of WALLINGTON, Eastern Division of the county of SURREY, 3 miles (S.) from Carshalton; containing 184 inhabitants. This parish, which is also called Woodmanstone, or Woodmansthorne, occupies an elevated site, is well wooded, and contains several ponds of water, one of which, near the church, is named Mere Pond. In the grounds of the Oaks, formerly an inn, but converted into a hunting seat by the late General Burgoyne, is an old beech tree, remarkable for its boughs having grown fast to one another. Shortes House, in this parish, is a very ancient building, with curiously carved wainscoting. The living is a rectory, valued in the king's books at £11. 7. 6.; present net income, £301: it is in the patronage of the Crown. The church is a small neat edifice. A National school for 50 children was erected in 1839, by subscription.

WOODNESBOROUGH (*ST. MARY*), a parish, in the union and hundred of EASTRY, lathe of ST. AUGUSTINE, Eastern Division of the county of KENT, 1¾ mile (W. S. W.) from Sandwich; containing 822 inhabitants. The living is a vicarage, valued in the king's books at £10. 0. 7½.; patrons and appropriators, Dean and Chapter of Rochester. The tithes have been commuted for a rent-charge of £350, subject to the payment of rates. The church is principally in the decorated style of English architecture. On Woodnesborough hill is a lofty artificial mount, supposed by some to have been either the place where the Saxon idol Woden was worshipped, or the burial-place of Vortimer; whilst others state it to be the *Woodnesbeorth* of the Saxon Chronicle, and the scene of the battle between Celred and Ina, kings of Mercia and the West Saxons, in 715. A fine gold coin, bearing on one side the figure of an armed warrior, and on the other that of Victory, was found here in 1514.

WOODSETTS, a township, in the parish of LAUGHTON-EN-LE-MORTHEN, union of WORKSOP, partly within the liberty of ST. PETER'S, and partly in the Southern Division of the wapentake of STRAFFORTH and TICKHILL, West Riding of the county of YORK, 4½ miles (N. W. by W.) from Worksop; containing 146 inhabitants.

WOODSFIELD, a hamlet, in the parish of POWICK, Lower Division of the hundred of PERSHORE, Upton and Western Divisions of the county of WORCESTER, 5½ miles (S. S. W.) from Worcester; containing 42 inhabitants. Here are the remains of a chapel, formerly dependent upon the church of Great Malvern.

WOODSFORD, a parish, in the union of DORCHESTER, hundred of WINFRITH, Dorchester Division of the county of DORSET, 5½ miles (E.) from Dorchester; containing 182 inhabitants. The living is a rectory, valued in the king's books at £4. 9. 9½.; present net income, £234; patron, H. C. Sturt, Esq. The church is a small ancient structure with a low tower. The parish is bounded on the north by the river Frome, upon which are the remains of an ancient castle, built by Guido de Brient, of a quadrangular form, one side of which has been converted into a farm-house, a very lofty building: the principal entrance was on the west, where there is still an ancient staircase, and in the south-east corner is another, both pierced with narrow apertures for arrows, or small arms. The offices on the basement story are all vaulted with stone; above is an apartment called the Queen's room, with the vestiges of a chapel, and around the whole are traces of a moat.

WOODSIDE, a township, in the parish of WESTWARD, ALLERDALE ward below Derwent, Western Division of the county of CUMBERLAND, 2 miles (N. W. by N.) from Temple-Sowerby; containing, with Rosley, 650 inhabitants. The rivers Eden and Eamont unite their streams here.

WOODSIDE, a township, in the parish of SHIFFNALL, Shiffnall Division of the hundred of BRIMSTREE, Southern Division of the county of SALOP, 3 miles (S. S. E.) from Shiffnall; containing 379 inhabitants.

WOODSIDE-QUARTER, a township, in the parish and union of WIGTON, ward, and Eastern Division of the county, of CUMBERLAND, 3 miles (E. by N.) from Wigton; containing 750 inhabitants. A school is partly supported by the interest of £3000 raised by subscription.

WOODSIDE-WARD, a township, in the parish of ELSDON, union of ROTHBURY, Southern Division of COQUETDALE ward, Northern Division of the county of NORTHUMBERLAND, 1½ mile (N.) from Elsdon; containing 131 inhabitants.

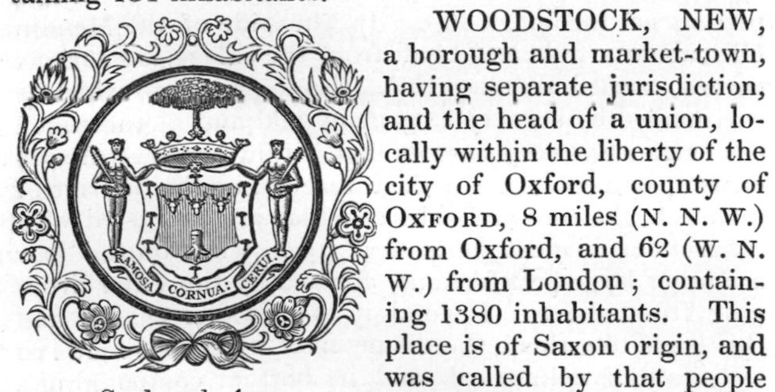

Seal and Arms.

WOODSTOCK, NEW, a borough and market-town, having separate jurisdiction, and the head of a union, locally within the liberty of the city of Oxford, county of OXFORD, 8 miles (N. N. W.) from Oxford, and 62 (W. N. W.) from London; containing 1380 inhabitants. This place is of Saxon origin, and was called by that people *Vudestoc*, signifying a woody place. It appears to have been chosen, at an early period, as an abode of royalty, and the manor-house, as it was called, is supposed to have been built upon the site of a Roman villa. Alfred the Great resided here whilst translating Boëthius: Ethelred held a council, or parliament here, and made several of the statutes enumerated by Lambard, in his collection of Anglo-Saxon laws; and it subsequently became a favourite residence of Henry I., who in a great measure rebuilt the place, surrounded the park with a wall, and stocked it with wild animals. In the reign of Henry II., Woodstock is celebrated as the residence of the fair Rosamond, whose romantic adventures are so interwoven with the history of that monarch: Henry here received Rhŷs, Prince of Wales, when he, in 1163, came to do homage. Edmund of Woodstock, the second son of Edward I., was born here in 1330; as were also Edward the Black Prince, and Thomas of Woodstock, sons of Edward III. Henry VII. added considerably to the buildings of the palace, particularly the front and the principal gate-house, which was for some time the place of confinement of Queen Elizabeth, during the reign of her sister Mary: on her accession to the throne, the palace was occasionally selected as her residence, and the town was distinguished by her favours. During the civil war of the 17th century, the palace was besieged and much damaged, the furniture was sold, and the building allotted by Cromwell to three of his partisans, two of whom sold their portions in 1652; the third portion, consisting of the gate-house and some adjoining buildings, was suffered to remain, and was, after the Restoration, converted into a dwelling-house by Lord Lovelace, who resided in it for several years; it was finally taken down by Sarah Duchess of Marlborough, and the only indications of its site are two fine sycamore trees in Blenheim Park. The manor and honour continued in the crown until the reign of Queen Anne, when it was granted to the celebrated Duke of Marlborough, for whom a splendid mansion was erected, at the expense of the nation, and called Blenheim, after the scene of one of his victories, as a recompense for his great military and diplomatic services. The town is very pleasantly situated on an eminence, on the eastern bank of the Glyme, an expansion of which forms the lake in Blenheim park, and afterwards joining the Evenload in the vicinity, both fall into the Isis: the streets are clean and spacious; and the houses, which are mostly built of stone, are generally large and handsome. The manufacture of gloves is the principal branch of trade, and, although fluctuating, is carried on to a considerable extent. The manufacture of various articles of fine steel has very much decayed since the rise of Birmingham and Sheffield: those made here formerly brought very high prices, from the beauty of the workmanship and the brightness of the polish, and are remarkable from having been generally made of old nails of horse-shoes, formed into bars. Queen Elizabeth, among other privileges, granted a wool staple, of which there are no remains, and a market to be held on Friday, but the principal market day is Tuesday. There are fairs on Tuesday after Feb. 2nd, April 5th, Tuesday at Whitsuntide, Aug. 2nd, Oct. 2nd (a great mart for cheese), Tuesday after Nov. 1st, and Dec. 17th.

This place, which has long been a borough by prescription, was incorporated in 1453, by Henry VI., whose charter was confirmed and enlarged by succeeding monarchs, the last of whom, Charles II., granted that under which the corporation now acts; a restrictive charter was forced on it by James II., but soon after set aside by proclamation. The members of the corporation are in number twenty-five, *viz.*, five aldermen, from among whom a mayor is annually chosen, a high steward, a recorder, assisted by seventeen common-councilmen, and a town-clerk; two of the common-councilmen act as chamberlains in rotation. This borough, which originally comprised only 45 acres, was always privileged to send two members to parliament, though the right was only exercised without interruption from the 13th of Elizabeth to the 2nd of William IV., when, by the act then passed to amend the representation, it was deprived of one. The right of election was formerly vested in the mayor, aldermen, and freemen, in number about 150, nearly one-half being non-resident; but the privilege has been extended to the £10 householders of an enlarged district, which has been constituted the new elective borough, and comprises 21,712 acres, the limits of which are minutely described in the Appendix, No. II.: the mayor is the returning officer. The corporation, as lords of the manor, hold a court leet annually; and a court for the recovery of debts under £15 within the borough is held every month, but only *pro forma* for many years past. The town-hall is a handsome stone building, erected about the year 1766, by the Duke of Marlborough, after a design by Sir William Chambers.

Woodstock, though for all civil purposes a parish, and by far the more populous, is only a chapelry to the rectory of Bladon: a parsonage-house was erected here by Bishop Fell, in which the rector may optionally reside. The chapel, dedicated to St. Mary, was principally rebuilt in 1785, on the site of a chantry founded by King John: in the original part of the building, which forms the south side of the chapel, is a round-headed doorway of early Norman architecture, composed of red stone, ornamented with chevron work; in the interior are three massy columns supporting pointed arches, with capitals having various grotesque sculptures of the human countenance: the more modern part of the structure has been erected in a style no way corresponding with this ancient part, but it has a tower of good proportions. Particular Baptists and Wesleyan Methodists have each a place of worship. The free grammar school was founded and endowed in 1585, by Richard Cornwall, a native of the town, and further endowed in 1616, by Thomas Fletcher, with rent-charges of £12. Eight boys and eight girls are instructed and clothed, under the benefaction of the Rev. Sir Robert Cocks, Bart., formerly rector of Bladon with Woodstock, now producing upwards of £60 per annum; the remainder of the endowment being given as apprentice fees with the boys when leaving the school. Here is a Sunday National school. Almshouses for six poor widows, standing near the southern entrance of the town, were built in 1793, by the late Duchess of Marlborough. There are various bequests for the benefit of the poor, which are distributed amongst them in money, clothing, &c. In the Rolls of the reign of Henry III., mention is made of an almshouse, built near the king's manor, which Tanner thinks may be the same with the hospital of St. Mary the Virgin and St. Mary Magdalene, for which a protection was granted to beg, by patent of the 1st of Edward III. The poor law union of Woodstock comprises 33 parishes or places, under the care of 37 guardians, and contains, according to the census of 1831, a population of 13,219. Roman coins, especially of Constantine, are occasionally dug up within the limits of the borough; and the Akeman-street-way, an ancient Roman road, passes through the northern part of Blenheim park. Chaucer, the father of English poets, resided, and is said by some to have been born, here. Woodstock gives the title of Viscount to the Duke of Portland.

WOODSTONE (*St. Augustine*), a parish, in the union of Peterborough, hundred of Norman-Cross, county of Huntingdon, ¾ of a mile (S. W. by W.) from Peterborough; containing 243 inhabitants. The living is a rectory, valued in the king's books at £7. 11. 3.; present net income, £342; patron, R. J. Tompson, Esq. John and Mary Walsham, in 1728, gave property now producing an annual income of £47, for the establishment of a school and other charitable purposes: £8 per annum is paid for instructing sixteen children, and the residue is distributed among the poor on St. Thomas's day.

WOODTHORPE, a township, in the parish of North Winkfield, union of Chesterfield, hundred of Scarsdale, Northern Division of the county of Derby, 4 miles (S.) from Chesterfield; containing 231 inhabitants.

WOODTON (*All Saints*), a parish, in the union of Loddon and Clavering, hundred of Loddon, Eastern Division of the county of Norfolk, 5 miles (N. W.) from Bungay; containing 539 inhabitants. The living is a rectory, valued in the king's books at £6. 13. 4.; present net income, £505; patron, A. I. Suckling, Esq.

WOOD-WALTON, county of Huntingdon.—See WALTON, WOOD.

WOODYATES, WEST, an extra-parochial liberty, in the hundred of Wimborne-St.-Giles, Wimborne Division of the county of Dorset, 5½ miles (N. N. W.) from Cranborne; containing 18 inhabitants. On the neighbouring downs are numerous barrows, also a vast rampart and fosse, termed Grimesditch, crossed by the Roman road from Dorchester to Old Sarum.

WOOKEY (*St. Matthew*), a parish, in the union of Wells, hundred of Wells-Forum, Eastern Division of the county of Somerset, 1¾ mile (W.) from Wells; containing 1100 inhabitants. The living is a discharged vicarage, in the patronage of the Subdean of Wells (the appropriator), valued in the king's books at £12. 15. 10.; present net income, £309. In the side of the Mendip hills, about a mile and a half from the village, is a curious cavern, termed Wookey Hole, the approach to which is surrounded by scenery extremely wild and picturesque: the entrance is very narrow, but within are several spacious apartments, one of them resembling the interior of a church, the roof and sides of which are encrusted with concretions of most fantastical form, while on the floor are other large petrifactions, formed by the water dropping from above. Beyond it is a smaller cavity, and this leads to a third, the diameter of which is about 120 feet, its roof cylindrical, and its bottom composed of a fine sand, on one side of which runs a very cold and pure stream of water, the primary source of the river Ax.

WOOL (*Holy Rood*), a parish, in the union of Wareham and Purbeck, liberty of Bindon, Wareham Division of the county of Dorset, 6 miles (W. by S.) from Wareham; containing 467 inhabitants. The living is annexed to the vicarage of Coombe-Keynes. A fair for cattle is held on May 3rd. Nearly half a mile to the south are the remains of Bindon abbey, founded in 1172 by Robert de Newburgh and Matilda his wife, in honour of the Virgin Mary, for monks of the Cistercian order, whose revenue, at the dissolution, was £229. 2. 1. A school is partly supported by subscription.

WOOLASCOTT, a township, in that part of the parish of St. Mary, Shrewsbury, which is in the hundred of Pimhill, Northern Division of the county of Salop: the population is returned with the parish.

WOOLAVINGTON (*St. Mary*), a parish, in the union of Bridg-water, hundred of Whitley, Western Division of the county of Somerset, 4 miles (N. E.) from Bridg-water; containing 412 inhabitants. The living is a vicarage, endowed with a portion of the rectorial tithes, with that of Puriton annexed, and valued in the king's books at £11. 7. 11.; present net income, £352; patrons, Dean and Canons of Windsor, the appropriators of the remainder of the rectorial tithes both of Woolavington and Puriton. The church has a small sepulchral chapel attached. A fair for cattle and sheep is held on October 18th. Here is a National school.

WOOLBEDING, a parish, in the union of Midhurst, hundred of Easebourne, rape of Chichester, Western Division of the county of Sussex, 1½ mile (N. W.) from Midhurst; containing 307 inhabitants. This parish is intersected by the river Rother, and forms part of the new borough of Midhurst under the late Municipal Reform Act. Near the conservatory of Woolbeding House is a bronze fountain, removed from the quadrangle at Cowdray; and near it is a remarkable tulip tree, the trunk of which is eight feet in girth, at a height of three feet from the ground. The living is a rectory, valued in the king's books at £7. 0. 10.; present net income, £227: it is in the patronage of the Hon. Mrs. George Ponsonby. The church is situated in a very sequestered spot, and in the chancel window is some ancient stained glass, removed from the priory of Mottesfont, in Hampshire, by the Rev. Sir Henry Mill, when rector of the parish.

WOOLBOROUGH, or WOLBOROUGH (*St. James*), a parish, in the union of Newton-Abbott, hundred of Haytor, Teignbridge and Southern Divisions of the county of Devon, 1 mile (S.) from Newton-Abbott; containing, with the town of Newton-Abbott, 2194 inhabitants. The living is a donative; net income, £235; patron and impropriator, Earl of Devon. The church has some fine screen-work across the nave and aisles: the chancel underwent great improvement a few years since, and a new and handsome altar-piece of stone has been erected. A school is endowed with about £90 a year, arising from certain property bequeathed, in 1788, by Hannah Maria Bearne, for charitable uses; there is also a National school, supported by subscription. Almshouses for four clergymen's widows were founded by Lady Lucy Reynell.

WOOLDALE, a township, in the parish of Kirk-Burton, union of Huddersfield, Upper Division of the wapentake of Agbrigg, West Riding of the county of York, 6¼ miles (S.) from Huddersfield; containing 3993 inhabitants. This place was anciently called *Wolves-dale*, from its abounding with those animals. Here are fulling and scribbling mills, and an extensive manufacture of woollen cloth. There are places of worship for the Society of Friends and Unitarians. A school-room was built by means of a legacy, aided by subscription, on part of the waste given by the Duke of Leeds, and a house for the master was erected, also by subscription, in 1821.

WOOLER (*St. Mary*), a market-town and parish, in the union, and Eastern Division of the ward, of Glendale, Northern Division of the county of Northumberland, 46 miles (N. N. W.) from Newcastle-upon-Tyne, and 318 (N. N. W.) from London; containing 1926 inhabitants. This place occupies the eastern declivity of the Cheviot hills, and near it is the village of Humbledon, celebrated for the memorable victory gained by Percy, Earl of Northumberland, in the reign of Henry IV., over a Scottish army of 10,000 men, under the command of Earl Douglas: the engagement took place on a plain within a mile north-west of the town, where a stone pillar has been erected, commemorative of the event. A great part of the town was destroyed by fire in 1722, since which period it has not made any considerable advances towards improvement. It consists of several streets diverging from the market-place, which is in the centre, and is indifferently paved, and supplied with water from a fountain erected at the public expense: a good trout stream flows through the lower part of it, and falls into the river Till: the houses are mostly old, and the general appearance of the place is unfavourable. The situation, though mountainous, is extremely salubrious, the town having formerly been much resorted to by invalids, who have ceased to frequent it, probably from a want of accommodation. A public subscription library is well supported; and a mechanics' institute was established in 1827. The market is on Thursday; and fairs are held on May 4th and October 27th, for horses, cattle, and sheep; and on the Tuesday in Whitsun-week a general fair is held, on a hill at a short distance from the town, which, from that circumstance, is called Whitsun Bank fair. The lord of the manor holds a court leet and baron annually, within three weeks after Easter, at which constables and other officers for the ensuing year are appointed. The town has been made a polling-place for the northern division of the county.

The living is a vicarage, endowed with a portion of the rectorial tithes, and valued in the king's books at £5. 8. 1½.; present net income, £478; patron, Bishop of Durham; impropriator of the remainder of the rectorial tithes, Earl of Tankerville. The church, rebuilt in 1765, on the site of the ancient structure which was destroyed by fire, is a neat edifice, and has been enlarged by the addition of 500 sittings, 230 of which are free, the Incorporated Society having granted £150 in aid of the expense; it occupies an eminence commanding an extensive and richly-varied prospect. There are places of worship for Baptists, Burghers, and Presbyterians, also a Scotch Relief Church and a Roman Catholic chapel. The free grammar school, founded and endowed by the Earl of Tankerville, for six boys of the town, has received an additional endowment, for four more, in compensation for a lapsed gift of £100, bequeathed for that purpose by Mrs. Chisholme. There are National and Sunday schools in connection with the established church and the several dissenting congregations; also a dispensary. On a circular mount near the town are the remains of an ancient tower, apparently of Norman origin. There are many intrenchments in the vicinity, of which the most remarkable is "Humbledon Hugh," about a mile north-west of the town; it is circular in form, with a large cairn on the summit; the sides of the hill are formed into terraces, about twenty feet broad, of which there are three successive tiers, which, being filled with soldiers, presented a formidable resistance to any assailing force: in the plain beneath is the pillar commemorating the victory of Earl Percy.

WOOLFARDISWORTHY, a parish, in the union of BIDEFORD, hundred of HARTLAND, Great Torrington and Northern Divisions of the county of DEVON, 9½ miles (S. W. by W.) from Bideford; containing 840 inhabitants. The living is a perpetual curacy; net income, £91; patron and impropriator, Rev. W. C. Loggin. The impropriate tithes have been commuted for a rent-charge of £470, subject to the payment of rates, which on the average have amounted to £20. The church has an enriched Norman doorway and font.

WOOLFARDISWORTHY (*Holy Trinity*), a parish, in the union of CREDITON, hundred of WITHERIDGE, South Molton and Northern Divisions of the county of DEVON, 6 miles (N. by W.) from Crediton; containing 226 inhabitants. The living is a rectory, valued in the king's books at £9. 19. 4½.; present net income, £258; patron and incumbent, Rev. John Hole, who supports a day and Sunday school. Berry Castle, an ancient Roman encampment, is in this parish.

WOOLFERTON, a township, in the parish of RICHARD'S-CASTLE, hundred of MUNSLOW, Southern Division of the county of SALOP, 3 miles (S. E.) from Ludlow: the population is returned with the parish. The Stourport canal passes through the township.

WOOLHAMPTON (*St. Peter*), a parish, in the union of NEWBURY, hundred of THEALE, county of BERKS, 7¼ miles (E.) from Newbury; containing 364 inhabitants. The living is a rectory, valued in the king's books at £7. 17. 6.; present net income, £202; patron and incumbent, Rev. Miles L. Halton. Here is a Roman Catholic chapel. A school is supported by the Countess of Falmouth. The navigable river Kennet runs through the parish.

WOOLHOPE (*St. George*), a parish, in the union of LEDBURY, hundred of GREYTREE, county of HEREFORD, 7¾ miles (W. by S.) from Ledbury; containing 880 inhabitants. The living is a discharged vicarage, with that of Fownhope annexed, valued in the king's books at £7. 12. 8½.; present net income, £626; patrons and appropriators, Dean and Chapter of Hereford. A school is endowed with £6 per annum, for which 12 children are instructed.

WOOLLAND, a parish, in the union of STURMINSTER, hundred of WHITEWAY, Sturminster Division of the county of DORSET, 8 miles (W. by N.) from Blandford-Forum; containing 119 inhabitants. The living is a donative; net income, £35; it is in the patronage of G. C. Loftus, Esq., and Mrs. Loftus. The church was rebuilt in 1745, a little to the westward of the ancient site.

WOOL-LAVINGTON, EAST and WEST, a parish, in the union of MIDHURST, hundred of ROTHERBRIDGE, rape of ARUNDEL, Western Division of the county of SUSSEX, 4½ miles (S. W. by S.) from Petworth; containing 338 inhabitants. This parish is pleasantly situated at the base of the northern acclivity of the downs, which are here extremely beautiful and picturesque, including the hanging woods in Wool-Lavington Park, and most extensive views over the Weald: the cultivated lands are widely detached, some portions of the parish being nine miles distant from each other. The present mansion in the park was built by the late John Sargent, Esq., the intimate friend of Hayley the poet, and himself author of several elegant poems, the chief of which are The Mine, The Vision of Stonehenge, and Mary Queen of Scots. The parish is included within the new borough of Midhurst, under the Reform Act. The living is a rectory, valued in the king's books at £9; present net income, £164; it is in the patronage of Mrs. Sargent. The church, pleasantly situated near Wool-Lavington House, is a neat edifice, in the early and later English styles; and near it is a neat school-house, with a residence for the mistress, erected at the sole expense of the Rev. H. E. Manning, the rector, for girls of this and the adjoining parish of Graffham, and chiefly supported by the rector; the school is on the National system. John Dyson, in 1836, bequeathed £8 per annum to be divided between two aged people attending the church.

WOOLLEY, a tything, in the parish of CHADDLEWORTH, hundred of KINTBURY-EAGLE, county of BERKS, 6 miles (W.) from East Ilsley: the population is returned with the parish.

WOOLLEY (*St. Mary*), a parish, in the hundred of LEIGHTONSTONE, union and county of HUNTINGDON, 5 miles (N. E. by N.) from Kimbolton; containing 58 inhabitants. The living is a rectory, valued in the king's books at £9. 9. 2.; present net income, £111; patron, J. Cockerell, Esq. The church has a western tower crowned with a handsome cupola.

WOOLLEY, a chapelry, in the parish of ROYSTON, wapentake of STAINCROSS, West Riding of the county of YORK, 5¾ miles (N. by W.) from Barnesley; containing 553 inhabitants. The living is a perpetual curacy; net income, £200; patron, G. Wentworth, Esq. The chapel, dedicated to St. Peter, is an ancient structure, with a tower; the windows are decorated with stained glass. Sixteen poor children are instructed for £16. 7. per annum, arising from the rent of certain land bequeathed by Nicholas Burley, and a trifling annuity from the parish fund; five others are paid for by Mrs. Wentworth.

WOOLLOS, ST., a parish, in the union of NEWPORT, Upper Division of the hundred of WENTLLOOGE, county of MONMOUTH, adjacent to the western side of the town of Newport, and containing, with that town, 7062 inhabitants. The living is a discharged vicarage, with the perpetual curacy of Bettws annexed, valued in the king's books at £7. 3. 11½.; present net income, £278; patron and appropriator, Bishop of Gloucester and Bristol. A church has been erected at Newport. There is a place of worship for Independents; also a National school for girls, and a Lancasterian school, partly supported by subscription.

WOOLPIT (*St. Mary*), a parish, in the union of STOW, hundred of THEDWASTRY, Western Division of the county of SUFFOLK, 5¾ miles (N. W. by W.) from Stow-Market; containing 880 inhabitants. This place is situated on the high road from Ipswich to Bury-St. Edmunds, and was formerly a market-town: it is celebrated for a remarkably fine vein of brick earth, and the white bricks made here are in great estimation. One of the largest horse fairs in England is held here, on Sept. 16th, and a large fair for bullocks on the 18th and 19th. The living is a rectory, valued in the king's books at £6. 18. 9.; present net income, £350; it is in the patronage of Mrs. Luke Flood Page. The church is partly in the decorated, and partly in the later, style of English architecture, and has a very beautiful porch. Here are a National and an infants' school. An image of the Virgin Mary was much visited here before the Reformation, and a well, called "Our Lady's Well," is still held in great repute for its medicinal properties.

WOOLSINGTON, a township, in the parish of NEWBURN, union and Western Division of CASTLE ward, Southern Division of the county of NORTHUMBERLAND, $5\frac{1}{4}$ miles (N. W. by N.) from Newcastle-upon-Tyne; containing 57 inhabitants.

WOOLSTANWOOD, a township, in the parish, union, and hundred of NANTWICH, Southern Division of the county of CHESTER, $3\frac{1}{4}$ miles (N. N. E.) from Nantwich; containing 70 inhabitants.

WOOLSTASTON, a parish, in the union of CHURCH-STRETTON, hundred of CONDOVER, Southern Division of the county of SALOP, $3\frac{1}{2}$ miles (N.) from Church-Stretton; containing 89 inhabitants. The living is a rectory; net income, £162; patron, W. W. Whitmore, Esq,

WOOLSTHORPE (*St. James*), a parish, in the union of GRANTHAM, wapentake of WINNIBRIGGS and THREO, parts of KESTEVEN, county of LINCOLN, $6\frac{1}{4}$ miles (W. by S.) from Grantham; containing 650 inhabitants. The living is a rectory, valued in the king's books at £13. 3. 8½.; present net income, £191; patron, Duke of Rutland, who supports a school for 25 girls, who are also clothed.

WOOLSTON, a hamlet, in the parish of NORTH CADBURY, hundred of CATSASH, Eastern Division of the county of SOMERSET, $2\frac{3}{4}$ miles (S.) from Castle-Cary: the population is returned with the hamlet of Yarlington.

WOOLSTONE (*St. Martin*), a parish, in the union of TEWKESBURY, and forming a detached portion of the Lower Division of the hundred of DEERHURST, Eastern Division of the county of GLOUCESTER, 5 miles (W. N. W.) from Winchcomb; containing 92 inhabitants. The living is a rectory, valued in the king's books at £13. 6. 0½.; patron, Earl of Coventry. The tithes have been commuted for a rent-charge of £153, subject to the payment of rates, which on the average have amounted to £9; the glebe comprises $32\frac{1}{2}$ acres, valued at £28. 19. per annum. The church was rebuilt in 1499. Here is a National school.

WOOLSTONE, a joint township with Martinscroft, in the parish and union of WARRINGTON, hundred of WEST DERBY, Southern Division of the county palatine of LANCASTER, $2\frac{1}{2}$ miles (E. by N.) from Warrington; containing, with Martinscroft, 578 inhabitants. Here is a Roman Catholic chapel.

WOOLSTONE, a tything, in the parish of HOUND, hundred of MANSBRIDGE, Southampton and Southern Divisions of the county of SOUTHAMPTON, $1\frac{1}{2}$ mile (S. E. by E.) from Southampton: the population is returned with the parish.

WOOLSTONE, GREAT (*Holy Trinity*), a parish, in the union of NEWPORT-PAGNELL, hundred of NEWPORT, county of BUCKINGHAM, 3 miles (N. E.) from Fenny-Stratford; containing 120 inhabitants. This parish is bounded on the east by a branch of the river Ouse, and is intersected by the Grand Junction canal. The living is a discharged rectory, valued in the king's books at £8. 16. 1.; present net income, £157; patron, J. C. Neild, Esq. The church, a handsome edifice in the later style of English architecture, was built in 1832, at the sole expense of T. S. I. Baily, Esq., of Shenley House, who is lord of the manor, the former structure having fallen into decay, and forms an interesting feature in the view from the surrounding neighbourhood.

WOOLSTONE, LITTLE (*Holy Trinity*), a parish, in the union of NEWPORT-PAGNELL, hundred of NEWPORT, county of BUCKINGHAM, $3\frac{1}{4}$ miles (N. E.) from Fenny-Stratford; containing 124 inhabitants. The living is a discharged rectory, valued in the king's books at £8. 6. 1.; present net income, £102: it is in the patronage of the Crown. The church is in a ruinous condition. This parish is bounded on the east by a branch of the river Ouse, and is intersected by the Grand Junction canal. There is a small portion of land, bequeathed by a former rector of the parish, the rent of which is appropriated to the apprenticing of one boy yearly.

WOOLSTROP, a hamlet, in the parish of QUEDGLEY, Middle Division of the hundred of DUDSTONE and KING'S-BARTON, Eastern Division of the county of GLOUCESTER, 5 miles (S. W. by W.) from Gloucester; containing 39 inhabitants.

WOOLTON, LITTLE, a township, in the parish of CHILDWALL, union of PRESCOT, hundred of WEST DERBY, Southern Division of the county palatine of LANCASTER, $4\frac{1}{2}$ miles (S. W. by S.) from Prescot; containing 734 inhabitants.

WOOLTON, MUCH, a chapelry, in the parish of CHILDWALL, union of PRESCOT, hundred of WEST DERBY, Southern Division of the county palatine of LANCASTER, $5\frac{1}{4}$ miles (S. W. by S.) from Prescot; containing 1344 inhabitants. The living is a perpetual curacy; net income, £170; patron, Vicar of Childwall. Here is a Roman Catholic chapel; also a National school, and a school for the children of Roman Catholics.

WOOLVERCOTT (*St. Peter*), a parish, in the union of WOODSTOCK, hundred of WOOTTON, county of OXFORD, $2\frac{3}{4}$ miles (N. N. W.) from Oxford; containing 524 inhabitants. The living is a perpetual curacy; net income, £80; patrons and impropriators, Warden and Fellows of Merton College, Oxford. The church is situated on the banks of the Isis, and contains, in a sepulchral chapel on the north side, a stately monument to the family of Walter, of whom David was high sheriff of the county, and commanded a regiment of horse under Charles I., in the parliamentary war. A National school is partly supported by subscription. A Benedictine abbey, in honour of the Virgin Mary and St. John the Baptist, was founded in the hamlet of Godstow, in 1138, by a pious lady, called Editha, which at the dissolution had a revenue of £319. 18. 8. Henry II. was a great benefactor to it; and within its walls fair Rosamond was interred.

WOOLVERSTONE (*St. Michael*), a parish, in the union and hundred of SAMFORD, Eastern Division of the county of SUFFOLK, $4\frac{1}{2}$ miles (S. by E.) from Ipswich; containing 235 inhabitants. This parish is bounded on the east by the navigable river Orwell, on the western bank of which is the hall, a spacious and elegant mansion, the seat of Archdeacon Berners, delightfully situated. The living is a discharged rectory, consolidated with the rectory of Erwarton, and valued in the king's books at £5. 8. 9. The tithes have been commuted for a rent-charge of £230, subject to the payment of rates, which on the average have amounted to £34; the glebe consists of 30 acres, valued at £45. 10. per annum. The church is situated within the park, and has been recently restored and beautified. A school is partly supported by the rector.

WOOLVERTON (*St. Lawrence*), a parish, in the union and hundred of Frome, Eastern Division of the county of Somerset, 4½ miles (N. by E.) from Frome; containing 207 inhabitants. The living is a discharged rectory, consolidated with that of Road, and valued in the king's books at £7. 1. 3. This parish is bounded on the east by the river Frome, by which it is separated from that of Road.

WOOLWICH (*St. Mary Magdalene*), a market-town and parish, in the union of Greenwich, hundred of Blackheath, lathe of Sutton-at-Hone, Western Division of the county of Kent, 8 miles (E. by S.) from London; containing 17,661 inhabitants. This place, originally a small fishing town, unnoticed by any of the earlier Kentish historians, owes its present importance, among other causes, to its situation on the river Thames, which in this part is nearly three quarters of a mile broad, and of sufficient depth, at the lowest state of the tide, for ships of the largest burden. In the reign of Henry VII., a ship of war of 1000 tons' burden was built here, which that monarch named the "Harry Grace de Dieu;" but it does not appear that any regular establishment for ship-building had been formed previously to the reign of Henry VIII., who constructed a royal dock-yard here, which was enlarged by Queen Elizabeth, and has continued progressively to increase in every succeeding reign. The "Sovereign of the Seas," the largest ship that had ever been built in England, was launched from this dock-yard in the reign of Charles I.: this ship, which was of 1637 tons' burden, and carried 176 guns, was richly ornamented with carving and gilding; from which circumstance, combined with the destructive efficacy of its heavy ordnance in the war with the Dutch, it obtained from that people the appellation of the "Golden Devil." In the reign of George I., the cannon for the Board of Ordnance was cast in a foundry situated in Moorfields, which having been destroyed by an explosion, occasioned by dampness in the moulds at the time of pouring in the liquid metal, the establishment was removed to Woolwich, and placed under the superintendence of Mr. Andrew Schalch, a native of Schaffhausen in Switzerland, who, travelling for improvement, visited the foundry in Moorfields at the time when preparations were in progress for casting several pieces of ordnance on the day following, in the presence of many of the nobility, general officers, and a large concourse of people. Mr. Schalch, who had obtained permission to inspect the process, minutely examined the preparations, and instantly perceiving the improper state of the moulds, warned the surveyor-general of the ordnance, and the superintendent of the foundries, of the lurking danger; and they, sensible of the justness and importance of his apprehensions, retired with their friends and all whom they could persuade to accompany them, in time to escape the effect of the explosion, by which several lives were lost, and many of the workmen dreadfully burnt and mangled. The Board of Ordnance subsequently finding this gentleman duly qualified, authorised him to choose a commodious situation within twelve miles of the metropolis, for the erection of a new foundry; and, after having visited several places, he selected the Warren at Woolwich for that purpose. The first specimens of ordnance cast under his superintendence being highly approved of, he was appointed master founder, which office he held for nearly 60 years, with so much skill and attention that, during this long period, not a single accident occurred. This circumstance may be considered the origin of the present arsenal, the subsequent extension and establishment of which, with the augmentation of the artillery, whose head-quarters were fixed here, the institution of the Royal Military Academy, and various other foundations, has raised the town to a degree of importance, as a grand naval and military depôt, without a parallel in any empire of the world.

The town is situated on elevated ground, rising gradually from the south bank of the river Thames, on the opposite side of which, in the county of Essex, is a detached part of the parish: it consists of one main street, extending nearly a mile parallel with the river, from which numerous other streets branch off in various directions, and is partly included in the parish of Plumstead. The houses in that part of it which may be considered the principal street are of ancient appearance, occasionally interspersed with substantial and well-built dwellings; but the other streets consist of modern houses, principally erected for the accommodation of the artificers and labourers employed in the dock-yard, arsenal, and other public works. The upper part of the town, towards the common and the Charlton road, is elevated and pleasant, and contains several ranges of well-built and handsome modern houses: the environs abound with rich woodland scenery, agreeably diversified with the windings of the Thames, sometimes seen in pleasing combination, and at others in striknig contrast. The town is partially paved under the superintendence of commissioners annually chosen under the provisions of an act of parliament passed in the 47th of George III.; lighted with gas by a company established by act of parliament; and amply supplied with water from the works of the Kent Water Company.

The public buildings are all on a scale of vast extent, and most of them in a style of magnificence corresponding with the importance of the purposes to which they are applied. The dock-yard commences near the village of New Charlton on the west, and extends nearly a mile along the bank of the river to the east; the breadth varies from one to two furlongs; the principal entrance is through a stone portal, of which the piers are ornamented with anchors sculptured in stone. On the left hand, within the walls, is a house for a commissioner, and on the right are the houses belonging to the principal officers of the yard. Beyond these is the smithery, a spacious and lofty building, in which are, a steam-engine of 20-horse power, which works two large lift-hammers, weighing nearly four tons each, and one of 14-horse power, working three tilt-hammers, of less weight: there is another steam-engine, of 14-horse power, for blowing the fires throughout the smithery, and there are several blast furnaces for converting scrap iron into pigs, and a machine for rolling iron. Knees, keelsons, breast-pieces, and all other iron work connected with ship-building, are manufactured here, and also anchors of the largest size, great numbers of which are kept in readiness for supplying the Royal Navy. There are two dry docks, one of which is double, for the repairing of vessels, and several slips in which ships of war of the largest dimensions are built, under lofty sheds lighted from the roof. An extensive building has been recently erected for an engineering foundry, and the manufacture of steam-engine boilers and the requisite machinery for the steam-vessels, which are now built here, some of which are of very great tonnage. A capacious basin, 400 feet long, and 290 feet in mean breadth, has been excavated, capable of receiving ships of the largest class, the entrance into which from the river is by a caisson of large dimensions; the embankment is secured by strong sloping walls of brick, coped with massive blocks of stone. The line of wharfage is very extensive and of proportionate breadth: there are a mast-pond, a boat-pond, and several mast-houses and boat-houses, also extensive ranges of timber-sheds, storehouses of every kind upon the largest scale, a mould loft, and every requisite arrangement for the purposes of the establishment. Many fine first and second rate ships, and others of smaller dimensions, of which there are always several upon the stocks, have been built here. In the eastern part of the town was, until recently, the rope-yard, an extensive range of building, three stories high, and about 1080 feet in length, in which ropes of various sizes, cordage for rigging the ships, and cables of the largest size, were made; but this establishment has been removed from Woolwich, and its site is now covered with houses forming Beresford-street.

To the east of the site of the rope-yard is the royal arsenal, under the control of the Master-General and the Honourable Board of Ordnance: this magnificent establishment comprises within the boundary walls more than 100 acres, and, including the canal, it is 142, the greater part of which is within the adjoining parish of Plumstead. The principal entrance is through a spacious and lofty central gateway for carriages, with smaller entrances on each side; the inner piers are ornamented with small piles of shot, and the outer piers, which are loftier, are surmounted by mortars and piles of shells. Nearly opposite the entrance is a range of handsome houses, appropriated to the commandant of the garrison, the field-officers of the royal artillery, and the principal officers attached to some of the departments; the chief of which are, the inspector of artillery's department, the royal carriage department, the royal engineer's department, the storekeeper's department, superintendent of shipping, and the royal laboratory. In addition to these are immense ranges of store-houses, forming a grand national depôt of warlike stores, of every description, for the naval and military departments of the service. On the right hand of the entrance is a range of buildings used, till within the last few years, as an academy for part of the gentlemen cadet company, in connection with the Royal Military Academy, and now occupied partly as store-rooms and partly as dwelling-houses. On the left hand is a handsome guard-house, with a portico of four columns of Portland stone, beyond which is the royal brass-foundry, erected by Vanbrugh, a spacious and lofty building of red brick, ornamented with stone and roofed with slate, which is perforated for ventilation. Over the entrance are the royal arms, handsomely carved in stone, above which is a neat cupola: it contains three large furnaces for casting brass ordnance only, the largest of which will melt eighteen tons of metal at one time: to avoid all danger of explosion, the moulds are heated to a considerable degree before the metal is allowed to run into them. To the east of the foundry are appropriate workshops for boring and engraving the cannon: the machine for boring is well adapted to the purpose; the gun is fixed to a shaft, which is turned round by horse power, applied to a horizontal wheel working into a wheel at the end of the shaft, which gives it a vertical motion; the cutter, or borer, is placed in a frame, and adjusted by iron clamps and screws, and is moved gradually forward, as the cavity deepens, by machinery adapted to that purpose: the time requisite for boring a nine-pounder is usually three days, and longer in proportion for pieces of a larger calibre. To the east of the buildings appropriated to the boring and engraving of cannon are the workshops of the royal carriage department, for the construction and manufacture of gun carriages for naval and land service, and of all carts, ammunition waggons, and other carriages used in the ordnance department; in these workshops are steam-engines applied to the working of circular and other saws for converting timber, and machinery of ingenious construction for planing wood, and for turning wood and metal. In a line with this range of buildings is the royal engineer department, under the direction of which are the erection and repair of all buildings belonging to the Board of Ordnance within a limited distance of Woolwich. To the north-west of the foundry is the royal laboratory, in which are made up blank and ball cartridges for small arms, cartridges for cannon of all descriptions, grape and case shot, and all combustible articles, and a variety of other important duties relating to the naval and military service is performed; a powerful hydraulic press has been recently introduced for making leaden bullets by pressure, instead of casting them as formerly. In addition to the work-shops for these purposes, are the buildings by Vanbrugh, originally used as the academy for cadets, but in which are now deposited some ingenious models of fire-ships, completely prepared for explosion, models of various fireworks exhibited on public occasions, various machines for ascertaining the strength of gunpowder, with samples of the different ingredients for making it, in their several stages of purification and refinement; specimens of the various kinds used by different nations, and of the several degrees of fineness, from the coarsest used for the heaviest ordnance to the finest for small arms; numerous models of muskets, and arms of different construction; and a variety of interesting and curious objects.

A line of wharfage, with a commodious quay, accessible to ships of large burden, extends for many hundred yards along the bank of the Thames, on which is a spacious and magnificent range of store houses, occupying three sides of a quadrangle, the area of which is filled with vast quantities of shot and shell of every size, ranged in regular quadrangular and pyramidal piles, and duly numbered. The buildings are of light brick, with quoins, cornices, pilasters, and pediments of stone, and handsomely embellished with appropriate ornaments; the central range, comprising three stories, is connected with the wings, which are two stories high, by handsome arched portals of stone forming the entrances into the quadrangle, and surmounted with balustraded corridors, communicating with the principal stories of each range. In the basement story of the principal range are deposited general stores for the naval service; in the second are the harness and other equipments for the royal horse artillery: and in the upper story, stores of various descriptions. The east wing is appropriated to the reception of stores for garrison and field services, with a large assortment of nails and other necessaries; and the west contains the stores and various implements used by the sappers and miners, and those for making intrenchments and constructing fortifications, among which are sandbags, axes, shovels, spades, barrows, grates for heating shot, and numerous other articles, and an extensive collection of samples of different materials, and patterns of various implements, with which the several articles furnished to the Board of Ordnance are compared, before they are received into the depôt. On the ground-floors of these store-rooms are iron tram-roads, upon which carriages constructed for the purpose, once put in motion, will run, when heavily loaded, from one extremity to the other, for the conveyance of stores to the wharf. To the east and west of the principal buildings are smaller quadrangular ranges of store-houses, of one and two stories in height: in both of these, the ranges parallel with the river are of one story, and are appropriated as repositories for carriages; in the lower story of the eastern range are stores of oil and cement; and in the upper, a general repository of stores of various kinds; the lower story of the western range is for the reception of carriages, and the upper is the depôt of clothing, for the royal artillery and for the sappers and miners; and in the centre of each of these smaller quadrangles are painters' shops. There are also various ranges of building for warehouses in different parts of the enclosure. To the south of the principal quadrangle are immense quantities of iron ordnance of various calibres, placed on iron skidding, and ranged in double files, extending many hundred yards in length, and, with small intervals between the rows, spreading over several acres of ground: large quantities of iron carriages for guns, and beds for mortars, are ranged at the extremity and around the space occupied by the ordnance; and numerous mortars of the largest calibre are disposed in various parts of the ground.

The arsenal is bounded on the south-east by a canal, 35 feet broad, on the banks of which are wooden buildings for the manufacture of the Congreve rockets, under the superintendence of the officers of the royal laboratory; and towards the south-eastern extremity of the boundary wall is pleasantly situated, on the road to Plumstead, the house appropriated to the residence of the storekeeper and paymaster. A little to the west is a saw-mill, worked by a steam-engine of 20 horse power, for sawing trees and rough timber into planks of any required thickness, to which the saws, fixed in frames and worked perpendicularly with great efficacy, can be adjusted at pleasure; there are also circular and other saws, with machinery of a very ingenious description, for turning and other purposes, all under the direction of the officers belonging to the royal carriage department. At a short distance from the royal arsenal, on the road to Woolwich common, are the barracks for the sappers and miners, a neat substantial and commodious range of building, capable of receiving from 250 to 270 men. Adjoining these is the grand depôt of field train artillery, consisting of a central building appropriated as offices for the director-general of the field train, and other officers of the department, and five spacious sheds, averaging each 300 feet in length, in which are deposited, in double files, an immense number of guns, mounted on field carriages, and supplied with a due proportion of stores and ammunition, in readiness at a minute's notice for immediate service. To the south of the depôt is the ordnance hospital, a handsome range of building, containing apartments for a resident surgeon and apothecary, and other officers, and for the servants of the establishment, with wards for the reception of 700 patients, a medical library, and other requisite offices: it is under the superintendence of the director-general and medical staff of the garrison, from which all the ordnance medical establishments abroad are supplied.

The barracks for the royal foot and horse artillery form a spacious and splendid pile of building, of which the principal front, facing the common, is 340 yards in length. The principal entrance is through a central portal of three arches, divided by lofty columns of the Doric order, supporting pedestals surmounted with military trophies, and above the central arch are the royal arms, finely sculptured in stone. The building, which is of light brick, ornamented with Portland stone, consists of six principal ranges, connected by four lower buildings, in front of which are colonnades of the Doric order, surmounted by balustrades: on the second range, on the east side of the entrance, is a handsome cupola, in which is a clock; and on the corresponding building on the west side is a similar cupola, with a wind-dial. The chapel, which is neatly fitted up, contains 1000 sittings, and is regularly opened for divine service: the library and reading-room are well supplied with works of general literature and periodical publications. The mess-room is a splendid apartment, 60 feet in length, 50 feet wide, and of proportionate height; at one end is a circular recess, in which is a music gallery, and at the other, a handsome range of windows looking upon the common: from the ceiling, which is ornamented with groining above the cornice, three elegant cut-glass chandeliers are suspended, and the whole arrangement is in the style of an elegant assembly-room. Attached to it is a suite of apartments, comprising a drawing-room of appropriate character, with retiring and ante-rooms: in this elegant suite of rooms the officers of the garrison give frequent balls to the gentry of the vicinity; and, in 1830, they had the honour of entertaining King William IV. and Queen Adelaide, who accompanied His Majesty on his visit to review the royal artillery. The principal entrance forms an avenue, 220 yards in length, and terminates with a handsome arched portal, dividing the buildings into two spacious quadrangles, round which are the stabling and barracks for the horse artillery; and at the extremity of the east quadrangle is a spacious riding-school of elegant design, near which is a large brick building used as a racket-court by the officers of the garrison: the whole establishment is arranged for the accommodation of from 3000 to 4000 men.

The Parade, in front of the barracks, is about 60 yards in breadth, adjoining the common, which, in this part, is a fine level lawn, appropriated for the exercise of the foot artillery. In the centre of the Parade are ranged several beautiful pieces of artillery, mounted on carriages of bronze, richly chased and ornamented. Among these is a very large piece of ordnance taken at the siege of Bhurtpoor, in the East Indies, and presented by the captors to the King of England: it is mounted on a splendid carriage of bronze: the breech, which is of unusually large proportion, rests upon the shoulders of a lion couchant, beautifully executed: one side of the carriage is ornamented with a view of the citadel of Bhurtpoor in a medallion, and the other bears an inscription commemorative of its capture: the wheels are solid, with a face of Apollo, or the sun, forming the nave, and the beams of the sun the radii. The more distant part of the common is appropriated to the exercise of the horse artillery. At a short distance to the north of the artillery barracks are those for the Woolwich division of marines, a plain and irregular building, with accommodation for about 450 men, adjoining which is an hospital for seamen and marines.

Adjoining the field to the west of the royal artillery barracks is the repository, for the exercise and general instruction of all persons belonging to the artillery, occupying an extensive piece of ground, tastefully laid out in parterres and walks leading to the several buildings; nearly opposite to the entrance are the modelling rooms for the use of the officers and men, in which are models, and drawings of projected improvements in the construction of gun-carriages and implements of war, and in which various mechanical experiments are performed for that purpose. In a shed adjoining them are preserved the funeral car of Napoleon, brought from St. Helena; a travelling oven used by the French army in their campaigns under Buonaparte, and some other curiosities; and in various parts of the ground are numerous pieces of brass ordnance, of different kinds, taken from the enemy, among which are, two captured at the battle of Malplaquet, with three barrels each, and several others of very singular construction. The ground, which is in many places unequal and precipitous, rising abruptly from several pieces of water, by which it is intersected, is made available for the practice of the artillery corps in the construction of bridges of pontoons, for transporting artillery across rivers, in the managing of gun-boats, and in the more difficult and arduous exercises of their duty; heavy pieces of artillery are manœuvred under every possible disadvantage of situation, lowered down deep declivities, and raised up precipitous heights, by a variety of contrivances: in various parts of the ground are intrenchments of earth and batteries of turf, which are thrown up by the students for their improvement in the art of fortification. On the north of the entrance is the rotunda, or model-room, a spacious circular apartment, 115 feet in diameter, originally erected in the gardens of Carlton palace by George IV., when Prince Regent, for the entertainment of the allied sovereigns, on their visit to this country after the peace of 1814, and presented by that monarch to the garrison. The roof, in the form of a tent, is supported on a lofty central Doric column, on the pedestal of which are various kinds of ancient armour, coats of mail, helmets and other military trophies, with specimens of the small arms of various nations. The room is lighted by a range of windows in the several compartments into which it is divided, and in which a vast number of beautiful and well-finished models of machinery are arranged, with apparatus for military and naval warfare: among these are guns and weapons of various descriptions, boats, pontoons, carriages, and implements, a variety of missiles, and Congreve and other rockets, with machines for discharging them singly or in vollies; a block of wood, fifteen inches square, pierced through and shattered by a Congreve rocket, which is wedged within the fissure, is preserved as a specimen of the destructive efficacy of this invention. Around the inner circle is arranged a most interesting variety of larger models, finished with the most scrupulous and minute exactness: among these are a bomb ship, with the whole apparatus for throwing the shells; a ship for the transport of horses, with the apparatus for slinging them, and the several arrangements for their management on board; models of all the royal dock-yards; the lines and fortifications of Portsmouth; the breakwater at Plymouth; the island of St. Kits; Cumberland fort; the citadel of Messina; the floating battery at South Carolina; the town of Quebec; the rock of Gibraltar, with the fortifications and batteries formed by excavated passages in the solid rock, and fine specimens of the strata highly polished; Fort William, in Bengal; Rio Janeiro; with a beautiful model of St. James's Park, and the several buildings erected in it on the occasion of the celebration of peace; a pair of kettle drums, of which the larger weighs more than 4 cwt. and 3 qrs., taken from the cathedral of Strasbourg; a lever target upon a new construction, and an infinite number of interesting and ingenious specimens of the adaptation of science to the invention, or improvement, of machinery connected with the art of war. On the south-west part of the common is the veterinary hospital for the horse artillery, under the control of the commandant and the superintendence of a veterinary surgeon and assistants; this building, which is well adapted to its use, is situated in the parish of Charlton. Between the repository and the veterinary hospital are 50 cottages, neatly built of brick, containing two apartments each, for the accommodation of 100 married soldiers.

At the south-eastern extremity of the common, opposite to the artillery barracks, is the Royal Military Academy, established in 1741, originally for the instruction of officers and men belonging to the military department of the ordnance, but now appropriated exclusively to gentlemen cadets, the number of whom varies from 100 to 140. The establishment is under the superintendence of a governor, who is always master-general of the ordnance for the time being; a lieutenant-governor and an inspector, who are officers of high rank in the artillery and engineer departments; a professor of mathematics, and four mathematical masters; a professor of fortification, and one master; a drawing-master of military plans, surveying, and perspective; a drawing-master for landscape; a French master; and a lecturer on chymistry. The buildings form a spacious pile in the early English, and partly in the Elizabethan, style of architecture, and comprise a central range with angular octagonal towers crowned with domes, containing on the basement story the entrance-hall and school-rooms, and, in a central situation between them, an apartment originally intended for the inspector, but used only as a receptacle for stores, and as a place from which hot air is distributed for warming the building: above these is the grand hall, in which the public examinations are held. The centre is connected, by corridors, with two wings in the Elizabethan style, with turrets at the angles, and containing the apartments for the residence of the cadets: behind the central range is the refectory, a spacious hall with a lofty timber-framed roof, lighted by windows of appropriate character, adjoining which are the kitchen and domestic offices. On the east side of the common are the houses of the professors, and some handsome ranges of building, including the quarters of the field-officers of the garrison, and several private residences.

There is no trade except what is requisite for the supply of the inhabitants, nor any particular branch of manufacture carried on here. The intercourse with the metropolis is great, and is facilitated by passage-boats on the river, by carriages direct, and by vans, which run every half hour from the Ship Tavern to Greenwich, whence the distance is traversed in the course of about ten minutes by the railway. There was an ancient ferry to the Devil's House, originally Du Val's house, on the opposite side of the river, but the present ferry is nearly a mile further up the river. Hulks are moored off Woolwich, for the reception of convicts whose sentence of transportation is commuted for hard labour at home, and who are employed in the dock-yard, arsenal, and public works. The market is on Friday; and, under the provisions of the local act before mentioned, markets are also held on Wednesday and Saturday: a market-house was erected a few years since, but it has never been used for that purpose, and the entrances are now closed up, the building having been appropriated to the reception of stores. By the act of the 2nd of William IV., cap. 45, Woolwich has been incorporated within the limits of the new borough of Greenwich; the number of tenements rated at £10 and upwards, within the parish, is upwards of 1300. The town is within the jurisdiction of the county magistrates, who hold their sittings every Monday and Friday at the King's Arms hotel; and a petty session for the division is held at the Green Man, at Blackheath, on the first Thursday in every month. The court of requests for the hundred of Blackheath, and other places in the county of Kent, is held under commissioners annually appointed by the act of the 47th of George III., at the Crown and Anchor tavern, every alternate Friday, for the recovery of debts not exceeding £5.

The living is a rectory, valued in the king's books at £7. 12. 6.; present net income, £740; patron, Bishop of Rochester. The church was rebuilt by act of parliament passed in the 5th of George II., at an expense of £6500, towards defraying which £3000 was appropriated from the grant of Queen Anne, for building fifty new churches, and the remainder raised by contributions of the inhabitants: it is situated on an eminence overlooking the dock-yard and the river, and is a neat building of brick with a square tower, ornamented with copings and cornices of stone; the interior, in which several standards taken from the enemy are deposited, is lofty and well arranged; the galleries are supported on Ionic columns of good proportion: in the churchyard are numerous monuments to officers of the royal artillery, among which is one to the memory of Lieutenant-General Williamson, whose wife was lineally descended from Robert, second King of Scotland. The ordnance chapel, on the road to Plumstead, a plain commodious building, and the chapel in the barracks, are the only additional episcopal churches, to both of which chaplains are appointed by the Honourable the Board of Ordnance. A chapel of ease has been erected on the site of the late rope-yard; and near the entrance of the Arsenal is a handsome proprietary chapel, erected by subscription in 1838; it is in the Grecian style, with a handsome Ionic portico of six columns supporting a triangular pediment. It is also in contemplation to erect another new church. There are places of worship for Baptists, Independents, Wesleyan and Welsh Methodists, and for a society calling themselves Arminian Bible Christians, also a Scottish church, and a Roman Catholic chapel. Mrs. Mary Wiseman, in 1758, bequeathed £1000 South Sea annuities for educating and clothing six orphan sons of shipwrights of Her Majesty's dock-yard, and for apprenticing them to the same business in the yard: this property, by accumulated savings, now produces £86. 5. per annum, for which ten boys are clothed, educated and apprenticed, according to the intention of the testator, when practicable, in the dock-yard, or, when not so, to shipwrights out of the yard, or to other trades when such cannot be found. Mrs. Withers, in 1750, bequeathed £600 Old South Sea annuities, of which £100 was to be laid out in building a school-room, with an apartment for a mistress, who was to receive the dividends on the remainder, for instructing 30 girls maintained in the workhouse; and the further sum of £600, in the same funds, to augment the salary of the schoolmistress, on condition of her teaching as many children, nominated by the rector, as would make up the number to 30, when so many might not be at any time in the workhouse: the building adjoins the old parish workhouse, and the salary arising from the endowment is £33 per annum. The National school, in Powis-street, in which 200 boys, and a similar establishment in the churchyard, in which 130 girls, are instructed, are supported by subscription; the lease of the present buildings is about to expire, and the subscribers contemplate the purchase of a freehold site for an appropriate building. There are also a British and Foreign school, and an infants' school, supported by the same means. Attached to Enon chapel is an endowed school, in which 120 boys and 70 girls are taught free of expense. An almshouse for five aged widows was founded about the year 1560, by Sir Martin Bowes, who endowed it with a portion of the produce of lands and tenements vested for charitable uses in the Company of Goldsmiths, London, by whom the almshouses were rebuilt in 1771: the buildings consist of five houses of four apartments each, and are inhabited by five widows, parishioners of Woolwich, above 50 years of age. There are also several other bequests for charitable purposes, and for distribution among the poor.

WOOPERTON, a township, in the parish of Eg-

LINGHAM, union of GLENDALE, Northern Division of COQUETDALE ward and of the county of NORTHUMBERLAND, 6¼ miles (S. E. by S.) from Wooler; containing 107 inhabitants.

WOORE, a chapelry, in the parish of MUCKLESTON, union of DRAYTON, Drayton Division of the hundred of NORTH BRADFORD, Northern Division of the county of SALOP, 6¾ miles (N. N. E.) from Drayton-in-Hales; containing 400 inhabitants. The living is a perpetual curacy; net income, £50: the right of presentation is in dispute. The chapel has been enlarged, and 160 free sittings provided, the Incorporated Society having granted £145 in aid of the expense. Here is a National school; and about 30 children are taught in a Sunday school, for an annuity of £10, the bequest of William Elkins, in 1593, to which has been added one of £5, bequeathed by Randolph Woolley, in 1615.

WOOTHORPE, a hamlet, in the parish of ST. MARTIN, STAMFORD-BARON, union of STAMFORD, soke of PETERBOROUGH, Northern Division of the county of NORTHAMPTON, 6 miles (N. W. by N.) from Wansford; containing 49 inhabitants. A small Benedictine nunnery, dedicated to St. Mary, existed here in the time of Henry I., and was united, in the reign of Edward III., to the convent of our Lady St. Mary and St. Michael, at Stamford-Baron.

WOOTTON (*ST. MARY*), a parish, in the hundred of REDBORNESTOKE, union and county of BEDFORD, 4½ miles (S. W.) from Bedford; containing 1051 inhabitants. The living is a vicarage, valued in the king's books at £13. 6. 8.; present net income, £236; the patronage and impropriation belong to Lady Payne. The church contains numerous monuments to the Monoux family. There is a place of worship for Wesleyan Methodists.

WOOTTON (*ST. PETER*), a parish, in the union of ABINGDON, hundred of HORMER, county of BERKS, 3½ miles (N. W. by N.) from Abingdon; containing 340 inhabitants. This place was formerly a chapelry in the parish of Cumner, and was made a separate parish early in the last century. The living is a perpetual curacy, annexed to that of South Hinksey. Here is a Sunday National school.

WOOTTON, a hamlet, in the parish of ST.-MARY-DE-LODE, GLOUCESTER, Upper Division of the hundred of DUDSTONE and KING'S-BARTON, union, and Eastern Division of the county, of GLOUCESTER, ¾ of a mile (E. by S.) from Gloucester; containing 804 inhabitants.

WOOTTON (*ST. MARTIN*), a parish, in the union of DOVOR, hundred of KINGHAMFORD, lathe of ST. AUGUSTINE, Eastern Division of the county of KENT, 9½ miles (S. E. by S.) from Canterbury; containing 128 inhabitants. The living is a rectory, valued in the king's books at £8. 10. 2½.; present net income, £239; patrons, Sir J. W. E. Brydges, Bart., and Sir J. W. H. Brydges, Knt. The church is in the early style of English architecture. Sir Samuel Egerton Brydges, Bart., who died in 1837, was a native of this place; he was born in 1762.

WOOTTON (*ST. ANDREW*), a parish, in the union of GLANDFORD-BRIDGE, Northern Division of the wapentake of YARBOROUGH, parts of LINDSEY, county of LINCOLN, 5¾ miles (S. E.) from Barton-upon-Humber; containing 459 inhabitants. The living is a discharged vicarage, valued in the king's books at £4. 18. 4.; present net income, £203; patron, J. Giffard, Esq.; impropriator, Lord Yarborough. There is a place of worship for Wesleyan Methodists. A school is endowed with £7 per annum, for which three children are instructed.

WOOTTON (*ST. GEORGE THE MARTYR*), a parish, in the union of HARDINGSTONE, hundred of WYMMERSLEY, Southern Division of the county of NORTHAMPTON, 2½ miles (S. by E.) from Northampton; containing 643 inhabitants. The living is a rectory, valued in the king's books at £21. 15.; present net income, £300; patrons, Rector and Fellows of Exeter College, Oxford. The rent of an allotment of four acres of land, awarded under an enclosure act, and now let for about £7 per annum, is appropriated to the support of a Sunday National school: there are some trifling bequests for the poor.

WOOTTON (*ST. MARY*), a parish, in the union of WOODSTOCK, hundred of WOOTTON, county of OXFORD, 2¼ miles (N. by W.) from Woodstock; containing 1060 inhabitants. The living is a rectory, valued in the king's books at £15. 2. 8½.; present net income, £783; patrons, Warden and Fellows of New College, Oxford. The church is partly in the Norman style, but principally of later date. At Old Woodstock, and in other parts of the parish, the manufacture of gloves is carried on. Charles Parrott, in 1785, bequeathed £2300 India annuities, now producing about £90 per annum, for the maintenance, education, and apprenticing of twelve boys; and in 1835 a school-house was built. Another school was endowed by the Rev. Lancelot Charles Lee, for clothing and teaching 6 girls. Numerous vestiges of Roman occupation have been discovered at various times; and on Chaldon hill are the remains of an exploratory camp, near which passes the old Roman road, Akeman-street.

WOOTTON (*ST. LAWRENCE*), a parish, in the union of BASINGSTOKE, hundred of CHUTELY, Basingstoke and Northern Divisions of the county of SOUTHAMPTON, 4¼ miles (W. by N.) from Basingstoke; containing 847 inhabitants. The living is a vicarage, valued in the king's books at £10. 2. 3½.; present net income, £211; patrons and appropriators, Dean and Chapter of Winchester. The church is very ancient, and has a Norman doorway, with pillars and arches of the same character, and several windows in the early English style: it contains a handsome marble monument to Sir Thomas Hooke, Bart., and several to the family of Wither. A parochial school for 60 children is supported by subscription. The London and South-Western railway passes through the parish.

WOOTTON (*ST. EDMUND*), a parish, in the liberty of EAST MEDINA, Isle of Wight Division of the county of SOUTHAMPTON, 4 miles (N. E.) from Newport; containing 55 inhabitants. The parish is bounded on the north by the Motherbank, and on the east by an inlet of the sea, across which there is a narrow causeway, called Wootton bridge, upwards of 900 feet in length, and on the high road from Ryde to Newport. On an eminence south of the bridge is Fern Hill, the seat of Samuel Sanders, Esq., a curious edifice with a lofty handsome tower, having somewhat the appearance of a church: it was erected by the late Duke of Bolton, when governor of the island, and commands a noble prospect of Spithead and the adjacent parts of Hampshire. The living is a rectory, valued in the king's books at £7. 16. 0½.; present net income, £240; patron and incumbent, Rev. R. W. White. There is a place of worship for Wesleyan Methodists. At Wootton farm is an ancient oak of remarkably large dimensions, being 47 feet in girth.

WOOTTON, a township, in the parish of ECCLESHALL, Northern Division of the hundred of PIREHILL and of the county of STAFFORD; containing 150 inhabitants.

WOOTTON, a township, in the parish of ELLASTONE, Southern Division of the hundred of TOTMONSLOW, Northern Division of the county of STAFFORD, 4½ miles (W. by S.) from Ashbourn; containing 269 inhabitants. Wootton Lodge is a handsome mansion, built by Inigo Jones, and occupies an elevated site on Weaver Hill, one of the loftiest in the neighbourhood, and abounding with a variety of minerals, of which limestone has only been worked hitherto. Jean Jacques Rousseau spent about 18 months at the lodge of this place.

Seal and Arms.

WOOTTON - BASSETT (*ALL SAINTS*), a market-town and parish (formerly a representative borough), in the union of CRICKLADE and WOOTTON-BASSETT, hundred of KINGSBRIDGE, Swindon and Northern Divisions of the county of WILTS, 36 miles (N. by W.) from Salisbury, and 87 (W.) from London; containing 1896 inhabitants. This place, which appears to have been originally of greater importance than it is at present, was, at the time of the Norman Conquest, called *Wodeton*, from *wode*, a wood, and *tun*, a town: about a century after that period it became the property of the noble family of Bassett, from whom it derived the adjunct to its name. It is pleasantly situated on elevated ground, commanding extensive and pleasingly diversified prospects of the surrounding country, which is extremely fertile and in a high state of cultivation, and consists principally of one street, nearly half a mile in length, the houses in which are in general indifferently built and of mean appearance. The manufacture of broad cloth, which was formerly carried on here, has entirely ceased, and there is now neither any branch of manufacture nor trade, beyond what is requisite for the supply of the inhabitants. The Wilts and Berks canal passes within half a mile to the south of the town; and the line of the Great Western railway runs through the parish. The market is on Tuesday; and the fairs, formerly six in number, have been reduced to two, which are held on the Mondays next after the feasts of Pentecost and St. Bartholomew. The town received its first charter of incorporation in the reign of Henry VI., under which, renewed by Charles II., in the 31st year of his reign, the government is vested in a mayor, two aldermen, and twelve capital burgesses, assisted by a town-clerk and subordinate officers. The mayor is annually chosen by the corporation, who also elect the town-clerk; and the inferior officers are appointed by the mayor, who, with the free tenants in the borough, anciently enjoyed the privilege of free common in Fasterne great park, which contained nearly 2000 acres. The borough first exercised the elective franchise in the 25th of Henry VI., from which time it regularly returned two members to parliament, but was disfranchised by the act of the 2nd of William IV., cap. 45. The right of election was vested in the inhabitant householders paying scot and lot, in number exceeding 200.

The living is a vicarage, valued in the king's books at £12; present net income, £461; patron and impropriator, Earl of Clarendon. The church is an ancient structure: in cleaning the south wall, some years since, a curious painting was discovered, the subject of which was the murder of Thomas à Becket, executed in a rude style. There is a place of worship for Independents. The free school was founded in 1688, by Richard Jones, and endowed with lands vested in trustees, now producing about £25 per annum. In another, 18 girls are paid for by the Earl of Clarendon. An ancient hospital, dedicated to St. John, which formerly existed here, was, during the reign of Henry IV., granted and united to the priory of Bradenstoke, in this county. The old manor-house has been converted into a farm-house. At a short distance below the town is a mineral spring, possessing the same properties as that of Cheltenham waters, and much used by those residing in the neighbourhood, though not generally known.

WOOTTON-COURTNEY (*ALL SAINTS*), a parish, in the union of WILLITON, hundred of CARHAMPTON, Western Division of the county of SOMERSET, 4 miles (W.) from Dunster; containing 426 inhabitants. The living is a rectory, valued in the king's books at £16. 8. 9.; present net income, £386; patrons, Provost and Fellows of Eton College. The church is a handsome structure. A school is partly supported by the clergyman, and there is a Sunday National school. Iron ore is found at the foot of a hill called Dunkery, which has an elevation of 1650 feet above the level of the sea.

WOOTTON-FITZPAIN, a parish, in the union of BRIDPORT, hundred of WHITCHURH-CANONICORUM, Bridport Division of the county of DORSET, 4 miles (N. E. by N.) from Lyme-Regis; containing 455 inhabitants. The living is a rectory, valued in the king's books at £8. 15.; present net income, £146; patron, James Drew, Esq. A day and Sunday school is supported by Mrs. Drew.

WOOTTON-GLANVILLE (*ST. MARY*), a parish, in the union of CERNE, hundred of BUCKLAND-NEWTON, Sherborne Division of the county of DORSET, 7½ miles (S. S. E.) from Sherborne; containing 331 inhabitants. The living is a rectory, valued in the king's books at £12; patron, Rev. J. Wickens. The tithes have been commuted for a rent-charge of £315, subject to the payment of rates; the glebe comprises 22 acres, valued at £25 per annum. The church is principally in the decorated style, with a low embattled tower of later date; in the windows are some fragments of ancient stained glass: it was repaired and newly pewed in 1741, and contains an altar-tomb with a recumbent figure, also several monuments and inscriptions.

WOOTTON, LEEK (*ALL SAINTS*), a parish, in the union of WARWICK, Kenilworth Division of the hundred of KNIGHTLOW, Southern Division of the county of WARWICK, 2¼ miles (S.) from Kenilworth; containing 433 inhabitants. The living is a vicarage, valued in the king's books at £5. 12. 1.; present net income, £300; patron and impropriator, Lord Leigh. A day and Sunday National school is endowed with the interest of £240.

WOOTTON-NEWLAND, a tything, in the parish of WOOTTON-GLANVILLE, hundred of BUCKLAND-NEWTON, Sherborne Division of the county of DORSET: the population is returned with the parish.

WOOTTON, NORTH, a parish, in the union and hundred of SHERBORNE, Sherborne Division of the county of DORSET, 2 miles (S. E. by S.) from Sherborne; containing 78 inhabitants. The living is a perpetual curacy; net income, £51; patron, Earl Digby; impropriator, Robert Gordon, Esq. The church was anciently a chapel of ease to the vicarage of Sherborne.

WOOTTON, NORTH (*ALL SAINTS*), a parish, in the union and hundred of FREEBRIDGE-LYNN, Western Division of the county of NORFOLK, 2 miles (W. by S.) from Castle-Rising; containing 179 inhabitants. The living is a discharged vicarage, valued in the king's books at £10; patron and impropriator, Col. Howard. The impropriate tithes have been commuted for a rent-charge of £62, and the vicarial for one of £200, subject to the payment of rates, which on the latter have, on the average, amounted to £14: the glebe contains only half an acre, valued at £1 per annum.

WOOTTON, NORTH, a parish, in the union of WELLS, hundred of GLASTON-TWELVE-HIDES, Eastern Division of the county of SOMERSET, 4 miles (W. S. W.) from Shepton-Mallet; containing 307 inhabitants. The living is annexed to the vicarage of Pilton. The church is a neat plain building. The interest of £104, bequeathed by John Humphreys, is applied in support of a Sunday school.

WOOTTON-RIVERS (*ST. ANDREW*), a parish, in the union of PEWSEY, hundred of KINWARDSTONE, Everley and Pewsey, and Southern, Divisions of the county of WILTS, 3 miles (N. E.) from Pewsey; containing 405 inhabitants. The living is a rectory, valued in the king's books at £7. 10. 5.; it is in the alternate patronage of the Master and Fellows of St. John's College, Cambridge, and the Principal and Fellows of Brasenose College, Oxford; but it must be given to one who has been a scholar at either from Somersetshire. The tithes have been commuted for a rent-charge of £390, subject to the payment of rates, which on the average have amounted to £75; the glebe comprises 49½ acres, valued at £73. 15. per annum.

WOOTTON, SOUTH (*ST. MARY*), a parish, in the union and hundred of FREEBRIDGE-LYNN, Western Division of the county of NORFOLK, 2¼ miles (S. W. by W.) from Castle-Rising; containing 177 inhabitants. The living is a discharged rectory, valued in the king's books at £8. 6. 8.; present net income, £226: it is in the patronage of the Crown.

WOOTTON-UNDER-WOOD (*ALL SAINTS*), a parish, in the union of AYLESBURY, hundred of ASHENDON, county of BUCKINGHAM, 7 miles (N. by W.) from Thame; containing 312 inhabitants. This parish is pleasantly situated within a mile of Dorton Spa, where a handsome pump-room and baths have been erected for the use of the rapidly increasing numbers whom the powerful efficacy of these waters attracts to the spot. Wootton House, an elegant mansion belonging to the Duke of Buckingham, built after the model of the old Buckingham Palace at Pimlico, is finely situated in a park richly embellished with wood, and diversified with a beautiful lake studded with picturesque islands. The living is a perpetual curacy; net income, £69; patron, Duke of Buckingham; appropriator, Archbishop of Canterbury. The church was repaired a few years since, when a spire of wood covered with copper, was added to the tower; and in the Grenville chapel, or south aisle, which was built in 1343, a *columbarium* has been constructed by the Duke of Buckingham, for the interment of his family. Two day and Sunday schools are supported by the Duchess of Buckingham, who also clothes 40 of the children.

WOOTTON-WAWEN (*ST. PETER*), a parish, in the union of STRATFORD-UPON-AVON, Henley Division of the hundred of BARLICHWAY, Southern Division of the county of WARWICK, 1½ mile (S.) from Henley-in-Arden; containing 2271 inhabitants. The living is a discharged vicarage, valued in the king's books at £11. 9. 7.; present net income, £379; patrons, Provost and Fellows of King's College, Cambridge; impropriators, Charles Mills and John Phillips, Esqrs. The church is principally in the later style of English architecture, with a handsome tower between the nave and the chancel; the south door is in the early English, and part of the south aisle is in the decorated style. At the east end of the north aisle is a desk in which are chained some old books, chiefly expositions, by various authors, on the four gospels, presented to the church at an early period after the Reformation, and intended for the use of the parishioners, with the permission of the Vicar. There is a chapel of ease at Ullenhall, in this parish, and there are endowed chapels at Henley and Bearley: there is also a place of worship for Baptists at Henley. In this parish, which is situated on the river Alne, and is intersected by the Stratford and Avon canal, is Wootton Hall, formerly the seat of Lord Carrington, who at the battle of Edge-Hill bravely redeemed the royal standard, as is recorded on his monument in Christ Church, Oxford. The title, originally conferred by Charles I. on the family of Smythe, after lying dormant for many years, was, during the administration of Mr. Pitt, conferred on a different family. The Dowager Lady Smythe, the lineal descendant and heiress of the first Carrington family, resided in this mansion till her decease in 1831, when the estate, manor, court leet, court baron, and view of frankpledge for the hundred and liberty of Partlow, descended to her only son, Sir Edward Joseph Smythe, Bart., of Acton-Burnell, in the county of Salop, the present proprietor. Over the front entrance of the hall are finely executed, in relief, the arms of Lord Carrington; and adjoining it is an elegant Roman Catholic chapel, of the Grecian Doric order, erected by the late Dowager Lady Smythe, in 1814. A school is endowed for the education of 30 boys. A Benedictine priory was founded here as a cell to the abbey of Conches in Normandy, to which abbey the church of St. Peter, with some lands in this place, had been given by Robert de Tonei, otherwise Stafford, son of Robert de Tonei, standard-bearer of Normandy, and confirmed by his descendants: at the dissolution of alien priories, the revenue was first granted by Richard II. to the priory of St. Anne near Coventry, and afterwards by Henry VI. to King's College, Cambridge.

Arms.

WORCESTER, a city, and county of itself, having exclusive jurisdiction, and the head of a union, locally in the county of Worcester, of which it is the capital, Worcester and Western Divisions of the county, 111 miles (N. W. by W.) from London; containing 18,610 inhabitants. This place, which is unquestionably of great antiquity, is, under the name of *Caer Guorangon*, enumerated by Nennius in his catalogue of cities belonging to the Britons, by whom, from the advantages of its situation near a fordable part of the river Severn, and on the confines of a thick forest, it was selected as a place of strength and security. On the expulsion of that people by the Romans, it was, with other British towns, retained by the conquerors; and, if not one of their principal stations, as some (judging from the Roman roads in the vicinity appearing to concentrate here) have supposed, it was very probably one of those fortresses which the prætor Ostorius erected on the banks of the Severn, to secure his conquests on that side of the river. It came again, on the departure of the Romans from Britain, into the possession of its ancient inhabitants, from whom it was taken, in 628, by Penda, King of Mercia, whose son Wolfhere, on his accession to the throne of that kingdom, appointed Osric his viceroy over the province of *Huiccia*, including the counties of Worcester and Gloucester, with part of Warwickshire, and forming a portion of the kingdom of Mercia. Osric, either repairing the Roman fortress, or erecting another in this city, which by the Saxons was called *Wigornaceastre*, made it his residence, and fortified it as a frontier against the Britons, who had retreated into the territories on the other side of the Severn. Sexulf, Bishop of Mercia, founded here the first Christian church within his diocese, which he dedicated to St. Peter; and in the reign of Ethelred, that monarch having resolved to divide the kingdom of Mercia into five separate dioceses, Osric prevailed upon him to establish one of them at *Wigornaceastre*, the metropolis of his province: and, in 679, Bosel was consecrated first bishop, by the style of *Episcopus Huicciorum*, and invested with full authority to preside over the ecclesiastical affairs of the province of Huiccia, or Wiccia. From the death of Osric nothing is recorded, either of the province or of the city, till the time of Offa, in one of whose charters Uhtred, a Wiccian prince, is styled *Regulus et Dux propriæ gentis Huicciorum* (ruler and duke of his own people the Huiccii), and his brother Aldred is described as *Subregulus Wigorniæ civitatis*, lieutenant of the city of Worcester, by licence of King Offa. After the union of the kingdoms of the Heptarchy, Alfred the Great appointed Duke Ethelred, a Mercian prince, to whom he gave his daughter Elfleda in marriage, to the government of Mercia; and in 894, Ethelred and Elfleda rebuilt the city, which had been destroyed by the Danes. Soon after this, Wærfred, Bishop of Worcester, desirous of defending the city and the cathedral from the future attacks of these rapacious invaders, obtained from Ethelred a grant of one moiety of the royal dues, with which he repaired the ancient seat of the Huiccian viceroys, and erected several fortresses around the cathedral, of which the only one now remaining is Edgar's tower. In 1041, a tax imposed by Hardicanute excited an insurrection of the citizens, who having seized the collectors, after endeavouring to shelter themselves in Edgar's tower, and put them to death, the king, to punish this outrage, sent an army to Worcester, and the inhabitants, abandoning the city, retired to the river island Bevere, in which they fortified themselves, determined to hold out to the last extremity. The forces of Hardicanute, having plundered and set fire to the city, attacked the inhabitants in their place of refuge, but were so vigorously repulsed that, after repeated fruitless attempts to dislodge them, the general was compelled to grant honourable terms of capitulation, and the inhabitants returned to their city, and repaired it.

Soon after the Conquest, a royal castle was erected here, of which Urso d'Abitot, who accompanied the Conqueror into England, was appointed constable, and made sheriff of the county. He extended the buildings of the castle, and, to the great annoyance of the monks, infringed upon the site of the cathedral, the outer ward having occupied what is now the College Green. In 1074, Roger Earl of Hereford, Ralph de Guader, Earl of East Anglia, and other powerful barons, entered into a conspiracy against William, and invited aid from Denmark; but their design having been discovered, they were obliged to enter the field before the expected succour arrived; and Bishop Wulstan, Urso d'Abitot, and Agelwy, abbot of Evesham, assisted by Walter de Lacey, assembled a body of troops to guard the passes of the Severn, intercepted their progress, and terminated the rebellion. The inhabitants, in 1088, having embraced the cause of William Rufus, the reigning monarch, Bernard de Neumarché, Lord of Brecknock, Osborn Fitz-Richard, Roger de Lacey, Ralph de Mortimer, and other partisans of his elder brother Robert, assembled a large force, and assaulted the city. On this occasion Bishop Wulstan armed his tenants, and retiring into the castle, with the citizens and their wives and children, animated the garrison to a resolute defence. The assailants set fire to the suburbs; but more intent upon plunder than prudent in securing their ground, they spread themselves over the open country, for the sake of pillage; and the garrison taking advantage of the opportunity, sallied from the castle, and advancing upon them suddenly, while in the act of ravaging the bishop's lands at Wick, captured or killed 500 men, and put the rest to flight. In 1113, the greater part of the city was destroyed by fire, which nearly consumed the cathedral and the castle: this calamity is supposed to have been inflicted by the Welsh, who at that time had resolved on the entire devastation of the English marches.

In the reign of Stephen, William de Beauchamp, constable of the castle, having embraced the cause of Matilda, drew upon him the resentment of that monarch, who deposed him from his government, and appointed in his place Waleran, Count of Meulant, whom he created Earl of Worcester. Matilda, in 1139, having gained several advantages in various parts of the kingdom, and greatly increased the number of her partisans, marched from Gloucester with a considerable force, and arriving before Worcester, laid siege to it; but before her arrival, the inhabitants had deposited every thing valuable in the cathedral, and made the necessary preparations for defending their city: the assailants attacked it on the south side, but being repulsed, they renewed the attack on the north side, and gaining an entrance, set fire to it in several places. Having succeeded in obtaining possession of the castle, William de Beauchamp was reinstated in his government by Matilda, and his appointment was subsequently confirmed by her son, Henry II. In 1149, Stephen, to punish the inhabitants for the assistance which they had given to his opponent, took the city and burnt it: but the castle having been strengthened with additional fortifications, resisted all his attempts, and Eustace, his son, having subsequently invested it without success, again set fire to the city in revenge. Worcester, which was so frequently the victim of intestine war and of accidental calamity, was fortified by Hugh de Mortimer against Henry II.; but on the approach of that monarch to invest it, Mortimer, on his submission, received pardon, and the city escaped damage. In 1189, it was almost totally destroyed by an accidental conflagration; and in 1202 it again suffered a similar calamity, when the cathedral and adjacent buildings were consumed, but the walls not being demolished, the injury was speedily repaired.

In the contest between King John and the barons, the latter having obtained the aid of Louis, Dauphin of France, the inhabitants adhered to their cause, and opening the gates of the city, received William Mareschall, son of the Earl of Pembroke, as governor of the castle for the Dauphin, in 1216; but Ranulph Earl of Chester, with a body of the royal forces, took that fortress by surprise, and afterwards obtained possession of the city. The inhabitants were made prisoners, and compelled by torture to discover their treasures; the soldiers of the garrison, who had taken sanctuary in the cathedral, were forcibly dragged out; the church and convent were plundered; and a fine of 300 marks was imposed upon the inhabitants, for the payment of which they were obliged to melt down the precious metals with which the shrine of St. Wulstan was enriched. In the course of the same year, that king was buried in the cathedral of this city. During this reign, Walter de Beauchamp, who had been appointed governor of the castle, having taken part with the barons, was deposed, and his lands confiscated. In 1217, the outer ward of the castle, which was contiguous to the cathedral, was granted to the monks for the enlargement of their close, by the Earl of Pembroke, guardian to the young king, since which time the Earls of Worcester have ceased to reside in it; the inner ward, comprising the citadel and keep, was alone kept up as a fortress for the protection of the city. In 1218, Bishop Sylvester obtained from Henry III. the grant of an annual fair for four days, in honour of St. Wulstan, to commence on the festival of St. Barnabas. During the reign of this monarch, a great tournament was celebrated here, in the year 1225, in which all who took part were subsequently excommunicated by Bishop Blois. A great part of the city, in 1233, was destroyed by an accidental fire, which greatly damaged the buildings of the cathedral. In 1263, Robert Ferrers, Earl of Derby, Peter de Montfort, son of Simon de Montfort, Robert Earl of Leicester, and others of the confederate barons, laid siege to the city, which they took after several assaults; they spared the church, but plundered the houses of the inhabitants, and put several Jews to death. After the battle of Lewes, in which Henry III. was made prisoner, that monarch was brought by the Earl of Leicester to Worcester, whence, together with his son, Prince Edward, he was removed to Hereford castle: the latter, having made his escape, repaired hither, where he assembled an army with which he defeated the earl and the confederate barons in the celebrated battle of Evesham. In 1299, the street leading to the suburb of St. John's was destroyed by an accidental fire, which also burnt down the wooden bridge over the Severn, which was afterwards replaced with one of stone. The city, in 1401, was plundered and partly burnt by the forces of Owain Glyndwr, in his repeated attacks upon the English frontiers in the reign of Henry IV., against whom he maintained a desultory warfare for a considerable time; but that monarch advancing against him, drove him back into Wales, and retiring after his victory to Worcester, took up his residence in that city, whence, after disbanding his army, he withdrew privately to London. In the reign of Edward IV., Queen Margaret, after the defeat of her party at the battle of Tewkesbury, and the subsequent murder of her son, was taken from a convent near that town, into which she had retired the day after the battle, by Lord Stanley, and brought before the king, who was then at Worcester. The Duke of Buckingham, in 1484, having raised an army of Welshmen to oppose the claim of Richard III. to the throne, a sudden inundation of the Severn impeded their progress, and disconcerted the enterprise; and after the battle of Bosworth Field, in which that monarch was slain, Worcester was seized for Henry VII.: several of the partisans of Richard were made prisoners here, and beheaded at the high cross, and a fine of 500 marks was paid to the king for the redemption of the city. In 1486, Sir Humphrey Stafford and his brother, Lord Lovell, having escaped from their sanctuary, at Colchester, levied a force of 3000 or 4000 men, and laid siege to this city; but on the approach of an army sent against them by the king, under the command of the Duke of Bedford, they raised the siege and dispersed. During the prelacy of Archbishop Whitgift, Sir John Russel and Sir Henry Berkeley came to the sessions here, with a large band of armed followers, to decide by force a quarrel which had arisen between them; but by the vigilance and activity of the bishop, who placed strong guards at the city gates, they were arrested and brought to his palace, when he prevailed upon them to deliver up their arms to his servants, and appeased their animosity. During the destructive pestilence that raged here in 1637, the inhabitants again abandoned the city, and shut themselves up in the island of Bevere.

In the parliamentary war, Worcester was the first city that openly declared in favour of the king, and the inhabitants opened their gates to admit Sir John Byron, at the head of 300 cavaliers, whom they assisted to fortify the city against the parliament. These, being afterwards joined by Lord Coventry with some troops of horse, and expecting further aid from the king, began to act on the defensive; but before the promised succours arrived, Colonel Fiennes, at the head of 1000 dragoons, and accompanied by the train bands from Oxford, and a detachment of the troops under Lord Say, arrived before the city, and summoned it to surrender to the parliament. The inhabitants indignantly refusing, he immediately commenced the assault; and a shot having been fired into the city, through a hole made in the gate, by one of the parliamentarians, the cavaliers sallied out on the assailants, and having killed several of Col. Fiennes' troops, returned without being pursued. Prince Rupert, with his brother Prince Maurice, arriving soon after with a considerable body of troops, joined Sir John Byron, and the royalists drew out their forces into Pitchcroft meadow, adjoining the town, to give the enemy battle. A spirited encounter took place, and was kept up for some time, but Prince Rupert perceiving a considerable reinforcement, under the Earl of Essex, advancing to the assistance of the parliamentarians, withdrew his forces into the city, where the engagement was continued till night, to the great disadvantage of the Prince, who, with a party of his troops, retreated to Hereford in disorder. On the same evening the Earl of Essex arrived, but, for fear of surprise, did not enter the city till the following morning, when the parliamentarian troops were quartered in the cathedral, which they stripped of its ornaments, destroyed the altar, and committed every kind of depredation: having explored the vaults, they found a large store of provisions and supplies which had been sent from Oxford for the king's use, and a considerable quantity of plate. The mayor and aldermen were taken into custody for having surrendered the city to the cavaliers, and sent under a strong guard to London; and 22,000 pounds' weight of plate was sent off under the same escort. A gallows was erected in the market-place, for the execution of such of the citizens as should be found guilty of having betrayed Col. Fiennes' soldiers to Prince Rupert; and a commission was appointed by authority of the parliament, under which Sir Robert Harlow and Sergeant Wilde were sent down, to secure the city and try the delinquents; and these officers, as a preliminary step, imposed a fine of £5000 on the inhabitants. After having repaired the fortifications, and obtained from the citizens a loan of £3000 for the parliament, the Earl of Essex divided his army, consisting of 24,000 men, into three brigades; two of them he detached in different directions, to intercept the king's forces on their march towards London; and leaving a garrison in the city, he advanced at the head of the third brigade to Shrewsbury, in pursuit of that part of the royal army which was headed by the king in person.

The citizens, after the departure of the earl and his army, still maintained their loyalty, and the corporation passed several resolutions in favour of the royal cause: they elected for mayor and sheriff two ardent royalists, provided additional ordnance and ammunition, strengthened the fortifications, and raised levies of money, which they transmitted for the king's use. These measures again drew upon them the vengeance of the parliament, and in March 1646, Sir William Brereton and Colonels Morgan and Birch appeared before the city, with a force of 2500 foot and horse, and demanded its surrender; this being peremptorily refused, they drew off their forces at night towards Droitwich, and advanced to assist in the siege of Lichfield. The citizens sent messengers for instructions to the king; who had escaped from Oxford, and was then at Newark; and in the meantime Gen. Fairfax, who was then at Headington, near Oxford, sent a letter to the governor of Worcester, requiring him to deliver up the city to the parliament, and on his refusal despatched Col. Whalley, with 5000 men, to reduce it. The garrison, which consisted of 1500 men, made a resolute defence, but after having sent various messengers to the king for instructions, and receiving no reply, their ammunition and provisions beginning to fail, and being in hourly expectation of the arrival of Fairfax, with an army of 10,000 foot and 5000 horse, they capitulated on honourable terms, on July 23rd. After a respite of five years, Worcester again became the seat of war; the citizens, still firm in their loyalty to the king, notwithstanding the opposition of the garrison, opened their gates to Charles II., who arrived at the head of a Scottish army of 12,000 men, attended by the Dukes of Hamilton and Buckingham, and other officers of distinction, on the 22nd of August, 1651; and, after some slight opposition from the garrison, entered in triumph, preceded by the mayor and corporation, by whom, on the following day, he was solemnly proclaimed. On the 28th, Cromwell, at the head of 17,000 men, arrived at Red Hill, within one mile of the city, where he fixed his head-quarters; and being soon after joined by the forces under Generals Fleetwood, Lambert, and Harrison, his army amounted to 30,000 men. General Lambert having surprised a detachment of the king's forces ordered to guard the pass of the Severn, approached to besiege the city. A general engagement now took place, and the parliamentarians were beginning to give way, when a reinforcement arriving from the other side of the Severn, the royal forces were overwhelmed, and compelled to retire into the city in disorder. A part of the Scottish troops laying down their arms, and the enemy advancing on all sides, every hope of victory was dispelled; Cromwell carried the royal fort by storm, putting all the garrison to the sword, and gained possession of the city: the king, attended only by Lord Wilmot, narrowly escaped by the back entrance of the house in which he was quartered, at the moment Col. Cobbet was entering at the front, to make him prisoner; and mounting a horse which had been prepared for him, rode to Boscobel, where he was hospitably entertained and concealed till he found means of escaping into France. The battle was sustained for some time with desperate valour; the citizens made their last stand at the town-hall, but without success, and the city was eventually given up to plunder. Cromwell describes his success upon this occasion as a "crowning mercy;" and, in token of his joy for the victory, he ordered a 60-gun ship, which was soon after launched at Woolwich, to be named the "Worcester."

The city is pleasantly situated at the base and on the acclivity of elevated ground, rising gently from the east bank of the river Severn, over which is a handsome stone bridge of five elliptical arches, connecting it with the suburb of St. John's, and built in 1780, at an expense of £29,843, towards defraying which Henry Crabb Boulton and John Walsh, Esqrs., members for the city, contributed £3000. It consists of several spacious and regular streets, of which the Foregate is a stately and lengthened avenue of handsome well-built houses, terminating with a fine view of St. Nicholas' church. The approaches exhibit rich and beautiful scenery, which, in many parts, is pleasingly diversified and strikingly picturesque. Bromsgrove Lickey to the north-east, the Malvern hills to the south-west, and the Shropshire hills and the Welsh mountains in the distance, are forcibly contrasted with the windings of the Severn, and the luxuriant vales, orchards, hop-grounds, and fertile meadows, for which the surrounding country is distinguished. The streets are well paved, lighted with gas, and amply supplied with river water by means of a steam-engine, erected on the eastern bank of the Severn at a place called Little Pitchcroft, in 1810. An act of parliament was obtained in 1823, for more effectually paving, lighting, and watching the city, under the authority of which several improvements have been effected. A public subscription library was established in Angel-street, in 1790, containing upwards of 5000 volumes: for this institution a building was a few years since erected by subscription, occupying a more eligible situation on the eastern side of the Foregate, near Sansom fields; on the basement story is a large and elegant reading and news-room, over which is an apartment for the library, appropriately fitted up. Two medical societies have been formed, the first in 1796, and the other, to which an extensive and well-assorted library is attached, in 1815; and there is a society for the encouragement and improvement of native artists, whose first exhibition of paintings took place in the town-hall, in Sept. 1818. The theatre, a neat and appropriate building, erected in 1780, by a tontine subscription in shares of £50 each, and handsomely fitted up, is opened occasionally; and assemblies and concerts are held in the large room at the town-hall. The musical festivals of the choirs of Worcester, Hereford, and Gloucester, take place here in the cathedral, every third year, and are attended by numerous and fashionable audiences: the surplus amount of receipts is appropriated to the benefit of the widows and orphans of the poorer clergy of the associated dioceses. A society for the promotion of this object was formed in 1778, under the patronage of Bishop North, whose successors are perpetual presidents; for this purpose the diocese is divided into four districts, to each of which two stewards are appointed. Races take place in Aug. and Nov.; at the former time they continue for three days, and are numerously attended: the course is on Pitchcroft meadow, where a grand stand has been erected, near the margin of the Severn, by which the course is bounded on one side.

The manufacture of broad cloth prevailed here to a very great extent in the reign of Henry VIII., at which time there were 380 looms, employing 8000 persons: on its decline the carpet manufacture was introduced, which, after flourishing for a short time, was transferred to Kidderminster. The present manufactures are those of porcelain and gloves, for the former of which this city has obtained a degree of reputation unequalled at home, and not surpassed abroad: the Worcester china is equally valued for its fineness and transparency, the elegance of its patterns, and the beauty of its embellishments. This branch of manufacture was established in 1751, by Dr. Wall and some other proprietors, and its progress has been rapid and successful: there are at present three manufactories, which have splendid shew-rooms, visited by persons travelling through Worcester with infinite gratification; from these the principal shops in the metropolis and other great towns are supplied with the most costly of their wares. The glove manufacture is conducted upon a very extensive scale, affording employment to not less than 8000 persons in the city, exclusively of many thousands in the neighbouring villages: the gloves made are in high estimation, not only in the several parts of England, but in the foreign markets, to which they are exported in great numbers. A distillery upon a large scale, a rectifying establishment, and a British wine manufactory, are successfully conducted; and extensive iron-foundries have been erected on the banks of the canal and the Severn: a considerable trade is carried on in hops, of which there are extensive plantations in the vicinity. The Worcester and Birmingham canal affords great facility of communication between the latter town and the Severn, and for the conveyance of goods from Manchester and the north of England, through Worcester; and the Severn, which is navigable for barges of considerable tonnage, and on the banks of which are commodious quays and spacious warehouses, contributes greatly to promote the trade and commercial prosperity of the city. The market days are Wednesday, Friday, and Saturday: the fairs are on the Saturday before Palm-Sunday, Saturday in Easter week, August 15th, and September 19th, which is a great fair for hops; a cattle fair is also held on the first Monday in Dec.; and there are markets free of toll on the second Monday in Feb., and on the first Mondays in May, June, July, and Nov. The market-place, nearly opposite the town-hall, in High-street, is a spacious and commodious area, erected in 1804, at an expense of £5050: the entrance is through a handsome arched portal of stone, ornamented with pillars of the Tuscan order, supporting a panelled entablature, on each side of which are smaller entrances; the interior is commodiously arranged for the sale of butchers' meat, fish, poultry, and various other articles, and behind it is the vegetable market. The corn market is a spacious area at the east end of Silver-street. The hop market is held in a spacious area opposite Berkeley chapel, at the south end of the Foregate: the buildings surrounding this area, formerly used as the city workhouse, have been converted into warehouses and offices, of which the rents are applied in aid of the poor rates of the several parishes which contributed to their erection: the sale of hops is very considerable, averaging annually about 25,000 pockets.

Corporation Seal.

Worcester was first constituted a city by Wulfhere, the sixth king of Mercia, and additional privileges were granted by Offa and Edgar; the inhabitants were incorporated by Henry I., whose charter was confirmed by Henry II., Richard I., and King John, and afterwards renewed, in 1227, by Henry III., who vested the government in two bailiffs, two aldermen, two chamberlains, twenty-four common-councilmen, and forty-eight assistants. Edward III., in the 12th of his reign, confirmed several former charters, and granted additional privileges, which were subsequently confirmed by Richard II., Henry IV., V., and VI., Edward IV., and Henry VII.; and in 1554 a charter was granted by Philip and Mary, declaring Worcester to be a city of itself, and incorporating the citizens by the designation of the "bailiffs, aldermen, chamberlains, and citizens of the city of Worcester." This and all preceding charters were confirmed by charter of the 19th of James I., by which the city and liberties of Worcester were constituted a county, and the corporation styled the "mayor, aldermen, and citizens." Under this charter the government was vested in a mayor, recorder, sheriff, six aldermen, twenty-four "capital citizens and councillors," and forty-eight capital citizens, with a town-clerk, two chamberlains, two coroners, and subordinate officers. The mayor, aldermen, and recorder, were justices of the peace within the city and the county of the city. The present corporation consists of a mayor, 12 aldermen, and 36 councillors, under the act of the 5th and 6th of William IV., cap. 76, for an abstract of which see the Appendix, No. I. The borough is divided into five wards: a sheriff is appointed by the council. The mayor, late mayor, and the recorder, are justices of the peace, *ex officio*, and nine others are appointed by the crown. The police establishment consists of an inspector, sergeant, and 22 police constables. The freedom is inherited by the eldest sons of freemen, and acquired by servitude. The city first exercised the elective franchise in the 23rd of Edward I., since which time it has regularly returned two members to parliament. The right of election was formerly vested in the freemen not receiving parochial relief, in number, including those non-resident, about 3000; but the non-resident freemen, except within seven miles of the borough (which originally comprised 320 acres), have been disfranchised, and the privilege has been extended to the £10 householders of an enlarged district, which has been constituted the new elective borough, and comprises 1253 acres, the limits of which are minutely described in the Appendix, No. II.: the sheriff is the returning officer. The recorder holds quarterly courts of session for all offences within the city and county of the city, not capital; and a court of record is held every Monday, for the recovery of debts to any amount. A sheriff's court is also held monthly. The town-hall is a handsome building of brick, with quoins, cornices, and ornaments of stone, consisting of a centre and two slightly-projecting wings, surmounted by a close-panelled parapet, decorated with urns and statues; in the centre is a statue of Justice, on each side of which are those of Peace and Plenty: the entrance is ornamented with two engaged columns of the composite order, on one side of which is a niche containing a statue of Charles I., and on the other a statue of Charles II.; the pediment over the entrance is ornamented with the city arms. In a niche occupying the central window of the principal story is a fine statue of Queen Anne; above is a circular pediment, in the tympanum of which are the arms of England, supported by angels. The lower room is divided into two parts, by the Crown bar on the north, and the Nisi Prius court on the south, and is decorated with portraits and ancient armour. On the upper story is the grand council-chamber, of the same dimensions as the lower room, with circular terminations, and divided into three compartments by two screens of columns crossing the room near the ends; it is lighted by numerous lustres, and is appropriately decorated for civic entertainments and for assemblies, which occasionally take place in it; opposite the principal entrance is a full-length portrait of George III., presented by that monarch when he visited the city in 1788. The new city gaol and bridewell was built in 1824, at an expense of £12,578. 12. 11.; it comprises eight distinct wards, eight day-rooms, eight airing-yards, two work-rooms, apartments for the sick, with separate rooms for male and female debtors, and a chapel, in which divine service is regularly performed; in one of the yards is a tread-wheel, which is exclusively used for the pumping of water to supply the prison. The county gaol and house of correction was erected in 1809, at an expense of £19,000: it is situated on the north-west side of the town, and comprises eleven wards, with day-rooms, work-rooms, airing-yards, and other requisites: the several departments are connected by small bridges with the keeper's house in the centre, in which is also the chapel: there is a tread-mill with three wheels for grinding corn for the use of the prison. The assizes and general quarter sessions, and the election of representatives for the western division of the shire, are held here.

Arms of the Bishopric.

Worcester was first erected into a see in the reign of Ethelred, and, in 679, Bosel was consecrated first bishop. The establishment, which was amply endowed by successive Saxon monarchs, consisted of Secular canons till the eighth century, when a convent, dedicated to St. Mary, was founded near the cathedral church of St. Peter, of which Ethelburga was abbess; on her death it was converted into a monastery for monks of the Benedictine order. The disputes which subsequently arose between the Secular clergy and the monks terminated in 960, by the surrender of the church of St. Peter to the latter, and the church of St. Mary became the cathedral of the diocese. After the Conquest the establishment continued to increase, and flourished till the dissolution, at which time its revenue was valued at £1386. 12. 10. It was refounded by Henry VIII., for a bishop, dean, archdeacon, ten prebendaries, ten minor canons, ten lay clerks, ten choristers, two schoolmasters, forty king's scholars, and other members. Prior to the passing of the act of the 6th and 7th of William IV., cap. 77, the jurisdiction of the see, with the exception of fifteen parishes and eight chapelries, extended over the whole of the county of Worcester, nearly one-third of the county of Warwick, the parishes of Brome and Clent in the county of Stafford, and the parish of Hales-Owen, in the county of Salop: by that act it is declared that the diocese shall consist of the whole counties of Worcester and Warwick. The ancient cathedral church of St. Peter, after its surrender to the monastery of St. Mary, was rebuilt by St. Oswald, in 983, but being destroyed by Hardicanute in 1041, Bishop Wulstan, in 1084, founded the present cathedral, which was subsequently enlarged and improved by several of his successors. It is a spacious and venerable pile, in the form of a double cross, with a noble and lofty square tower, rising from the centre to the height of 167 feet: the prevailing style of architecture is the early English, intermixed with portions in the Norman, the decorated, and the later English styles. The tower is a fine composition, enriched with series of canopied niches, in which are statues of kings and bishops, and embellished with sculpture of elegant design. The exterior possesses a simplicity of elegance, arising from the loftiness of its elevation and the justness of its proportions; the interior is remarkable for the airiness and lightness of its appearance, and, in many parts, for the correctness of its details and the appropriate character of its embellishments. Part of the nave contains specimens of the Norman style, and, in some places, portions in the decorated: it is separated from the aisles by lofty ranges of finely-clustered columns and pointed arches, and lighted by a fine range of clerestory windows, the tracery of which is in the later style: the roof is finely groined, and ornamented with bosses of flowers, antique heads, and other devices. The choir, to which is an ascent of several steps, is in the early English style; the groining of the roof and the details are in general of very elegant character, and in high preservation; the altar-screen is of carved stone, and the pulpit, also of stone and of octagonal form, is richly sculptured with symbols of the Evangelists, and devices illustrative of scripture history: the east window, as well as the great west window of the nave, are modern compositions in the later English style; and the bishop's throne and prebendal stalls are richly embellished with tabernacle-work. The Lady chapel, also in the same style, consisting of a nave and aisles, is among the earlier parts of the cathedral, being equally remarkable for the symmetry of its parts and the goodness of its preservation. In the south-eastern transept is the monumental chapel of Prince Arthur, son of Henry VII., in the later style of English architecture, of which it is an elegant specimen, containing his tomb highly enriched with sculpture, emblematical of the union of the houses of York and Lancaster, and other embellishments; adjoining it is the dean's chapel, and to the north the bishop's chapel, with others in various parts of the building. In the centre of the choir is the tomb of King John; the slab bearing the effigy of that monarch is of a date soon after his decease, but the tomb, which is in the later style, was probably erected at the same time as Prince Arthur's chapel. From a supposition, at that time generally prevailing, that this was only a cenotaph of the monarch, whose remains were interred in the Lady chapel, the Dean and Chapter, in 1797, resolved upon its removal to that spot; on opening it, however, a stone coffin was found, in which were the remains of the king, in good preservation, but on exposure to the air they mouldered to dust. There are several interesting monuments, among which those of Bishops Hough, Maddox, and Johnson, and of Mrs. Rae, are elegant specimens of sculpture. To the south of the cathedral are the cloisters, in the later style of English architecture, enclosing a spacious quadrangular area, on the south side of which is the ancient refectory of the monastery, in the decorated style of architecture, with some elegant windows, and a doorway highly enriched, now appropriated to the use of the king's school. On the eastern side is the chapter-house, in which is the library, an ancient building in the form of a decagon, the roof of which, finely groined, is supported on a central column: the windows are of modern insertion, and the walls are ornamented with a series of Norman intersecting arches. The episcopal palace is a modern embattled edifice of brick, decorated with stone, pleasantly situated on the margin of the Severn, and containing several spacious apartments: the drawing-room is ornamented with portraits of George III. and Queen Charlotte, between which is a marble tablet, recording their presentation to the bishop by their Majesties, who, when on a visit to Worcester, took up their abode in the palace.

The city comprises the parishes of *St. Alban, All Saints, St. Andrew, St. Clement, St. Helen, St. Martin, St. Nicholas, St. Peter,* and *St. Swithin:* those of St. Clement, St. Martin, and St. Peter, are partly in the Lower Division of the hundred of Oswaldslow. The living of St. Alban's is a discharged rectory, valued in the king's books at £5; present net income, £74; patron, the Bishop. The living of All Saints is a discharged rectory, valued in the king's books at £13. 12. 4½.; present net income, £138: it is in the patronage of the Crown. The living of St. Andrew's is a discharged rectory, valued in the king's books at £10. 5. 10.; present net income, £165; patrons, the Dean and Chapter. The church has undergone extensive reparation; the tower, in 1814, was cased with freestone; it is 90 feet in height, and is surmounted by an octagonal spire, 150 feet 6 inches high, regularly and symmetrically diminishing from 20 feet at the base to only six inches and five-eighths at the top, the whole terminated by a Corinthian capital, and surmounted by a gilt weathercock, and forming one of the most striking ornaments of the city: the spire was erected in 1751, by Nathaniel Wilkinson, a stone mason of the city. The living of St. Clement's is a discharged rectory, valued in the king's books at £5. 5.; present net income, £101; patrons, the Dean and Chapter. The church, a small old structure of stone, stood on the eastern bank of the Severn, although the principal part of the parish was on the western side of that river; but being much decayed, and liable to be flooded by the overflowing of the river, a new church, on an enlarged scale, was built, which was opened in 1823 It is in the Norman style, and is situated on the upper road to Henwick, &c.; the expense of its erection was nearly £6000, and was defrayed by a subscription among the parishioners, aided by the appropriation of several small benefactions, and a grant from the Society for Building Churches. The living of St. Helen's is a discharged rectory, valued in the king's books at £11; present net income, £136; patron, the Bishop. The church has been repewed, and 200 free sittings provided, the Incorporated Society having granted £70 in aid of the expense. The living of St. Martin's is a rectory, valued in the king's books at £15. 3. 4.; present net income, £378; patrons, the Dean and Chapter. A gallery has been erected in the church, and 140 free sittings provided, the Incorporated Society having granted £20 in aid of the expense. The living of St. Nicholas is a discharged rectory, valued in the king's books at £16. 10. 7½.; present net income, £260; patron, the Bishop. The church is a uniform modern structure, with a handsome steeple, and, from its situation in the more open part of the town, forms a conspicuous and interesting object in the perspective of the

Foregate and Broad-street. The living of St. Peter's is a vicarage, valued in the king's books at £12. 4. 2.; present net income, £233; patrons, the Dean and Chapter; appropriators, Dean and Canons of Christ-Church, Oxford. The church has been rebuilt, and contains 620 free sittings, the Incorporated Society having granted £600 in aid of the expense. Her Majesty's Commissioners have agreed to afford facilities for obtaining a site for a new church in this parish. The living of St. Swithin's is a discharged rectory, valued in the king's books at £15. 15.; present net income, £170: it is in the patronage of the Dean and Chapter. A district church was erected in 1834, at Blockhouse; it contains 500 sittings, 330 of which are free, the Incorporated Society having granted £300 in aid of the expense. There are places of worship for Baptists, the Society of Friends, those of the late Countess of Huntingdon's Connection, Independents, Wesleyan Methodists, and Roman Catholics.

The royal grammar school connected with the cathedral was founded at the time of that establishment by Henry VIII., for forty boys, of which number, ten are appointed by the dean, and three by each of the ten prebendaries: the scholars are admitted only for four years, and are required to undergo an examination in the rudiments of Latin, in which, if found deficient, they pay to the head-master £10 for the first year's instruction; they receive annually £2. 6. 8., out of which they have to find a surplice, and to pay £1. 10. per annum to the writing-master: there are two exhibitions to Balliol College, Oxford, founded by Dr. Bell, Bishop of Worcester, which are restricted to this diocese. The free grammar school was founded by Queen Elizabeth, in 1561, for twelve boys, to the three senior of whom she assigned thirteen shillings and fourpence per annum, for purchasing books: this school stands the third in claim to six scholarships founded by Sir Thomas Cookes, Bart., founder of Worcester College, Oxford, which lead to the six fellowships in that college by the same founder, as vacancies occur. The Rev. John Meek, in 1665, bequeathed to Magdalene Hall, Oxford, estates then producing £100 per annum for ten scholars from this school. Mr. Joseph Worfield, in 1642, assigned to the corporation certain lands and tenements in the parishes of Powick, Leigh, Wicke, and Bransford, in the county of Worcester, in trust for the maintenance and education of fourteen poor boys of the city, or of those parishes, to be elected from any schools whatever, and to be sent to either of the Universities for seven years: the income is about £240 per annum, which is appropriated to the payment of £30 each per annum to seven students in the University. The free school and Trinity almshouses, under the management of the six masters appointed by Queen Elizabeth, on establishing the free grammar school, were founded in 1558, by Mr. Thomas Wilde, who endowed them with land called Little Pitchcroft, and a part of the meadow of Great Pitchcroft, producing, with subsequent donations, an annual income of nearly £300: the school was intended as preparatory for the free grammar school, and the almshouses for the residence and support of aged inhabitants of the city: the buildings, situated partly in the parish of St. Nicholas, and partly in that of St. Swithin, consist of a school-room, with a dwelling-house for the master, and 29 apartments for the almspeople: there are at present six boys, who are maintained, clothed and instructed. Schools for the education of sixteen boys and eight girls were founded in 1713, by Bishop Lloyd, who endowed them with a small estate in the parish of Aston, in this county, producing at present about £80 per annum: in 1782, a house was purchased by subscription, and fitted up as a school-house, with dwellings for the master and mistress. Here is a National school for girls. A British and Foreign school is supported by subscription; and there are various Sunday and other schools.

St. Oswald's hospital was founded prior to 1268, and originally endowed for a master, chaplain, and four brethren; at the time of the dissolution it was given to the Dean and Chapter, but had been dispossessed of a considerable portion of the lands with which it was endowed. In 1660, Dr. John Fell, Bishop of Oxford, having been appointed to the mastership, successfully exerted himself for the recovery of its alienated property: a new charter of foundation was obtained in the fifteenth of Charles II., and almshouses for ten men and a chapel were erected. Thomas Haynes, Esq., in 1681, built rooms for six additional brethren, and added £50 per annum to its endowment: its present revenue is about £350, which is appropriated to the support of sixteen aged men and twelve women. The almshouses founded by Richard Inglethorpe, for six aged men and a woman to attend upon them, have an endowment of £53 per annum, exclusively of fines on the renewal of leases, which amount to a considerable sum, and have been rebuilt and enlarged for nine inmates. Mr. John Nash, alderman of the city, in 1661, founded ten almshouses, which he endowed with lands and tenements in Powick and the parish of St. Martin, for eight aged men, and two aged and unmarried women to wait upon them: the endowment produces at present an income of £367. 1. 8. per annum, which is appropriated to seventeen almspeople; he likewise left an apprenticing fund of £4 per annum to each of the nine parishes. Mr. Michael Wyatt, in 1725, left property in trust to the mayor and corporation, for the erection and endowment of almshouses for six freemen of the city: the premises are neatly built of brick, and comprise six tenements for aged men: the annual produce of the endowment is £49. Berkeley's hospital was founded in 1692, by Robert Berkeley, Esq., of Spetchley, in the county of Worcester, who endowed it with £6000 from the rents of his lands, in annual sums of £400, for twelve aged men and one aged woman of the city, and for the payment of £20 per annum to a chaplain for performing service in the chapel. Geary's almshouses, for four aged women, are endowed with about £30 per annum. Shewring's hospital was founded in 1702, by Mr. Thomas Shewring, alderman of the city, who endowed it with messuages, lands, and tenements, in and near Worcester, producing at present an income of nearly £150 per annum, for six aged women of the parishes of St. Swithin, All Saints, St. Andrew, St. Helen, and St. Clement, and one of the tything of Whistons, with preference to the kindred of the founder: the premises are neatly built of brick, and under the window in each apartment is a stone inscribed with the name of the parish from which, in case of vacancy, the tenant is to be chosen. Mr. William Jarviss, in 1772, bequeathed property, now producing £122. 13. 8. per annum, for the support of three aged freemen and one widow, and for apprenticing four boys, of the parish of St. Andrew annually; and in 1567, Mr. John Walsgrove bequeathed eight almshouses to the poor of that parish, which were subsequently endowed with premises for keeping them in repair, by his son, and £4 per annum by his grandson; the houses were rebuilt in 1825 at a considerable expense, and contain two apartments for each tenant. Some almshouses founded by Mr. Steynor, as residences for the poor of the parish, have been taken down, and the rents of some other houses, with which they were endowed, are now divided among eight poor people. There are numerous charitable bequests and donations, amounting in the aggregate to a very considerable sum per annum, for apprenticing poor children, lending money without interest to young tradesmen setting up in business, for distribution among the poor, and for various other charitable uses; in addition to which, Worcester is one of the cities partaking of Sir Thomas White's charity. The parish of St. Swithin is in possession of lands and houses derived from some unknown sources, the annual value of which is computed at £763: the income is appropriated to the repair of the church and relief of needy parishioners.

The city and county infirmary, erected in 1770, occupies an airy and appropriate situation, adjoining Pitchcroft meadow, and was completed at an expense of £6085. 9. 9., raised by subscription: it has two handsome fronts; the internal arrangements are well adapted, and a considerable quantity of garden and pleasure ground is attached to it. An institution for the relief of lying-in women has been established, and is supported by subscription, under the patronage and management of a committee of ladies. The house of industry, an extensive brick building, occupying an elevated situation to the east of the town, was erected by act of parliament obtained in 1792, for the accommodation of the incorporated parishes of the city: the buildings were erected at an expense of £7318, and the purchase of the land belonging to it was £2273. It consists chiefly of a central elevation and two wings, the first 116 feet in length, 44 in breadth, and 40 in height: on the roof of the southern wing is a capacious reservoir, filled with water by a pump in the brewhouse, whence it is distributed by pipes to the baths, and every other part of this extensive and well-arranged establishment. Behind the house are workshops for the men, and an hospital: further backward is a burial-ground, for those who die in the house; and in front is an extensive plot of ground, with a small building used for reading the burial service, in which the general poor of the united parishes are interred. The poor law union of Worcester comprises the seven parishes within the city, together with the parishes of St. Martin, St. Clement, St. Peter the Great, St. John Bedwardine, and the tything of Whistons, in the parish of Claines, partly within the city and liberties, and partly in the county of Worcester; and contains a population of 26,542, according to the census of 1831: the union is under the care of 22 guardians. A female penitentiary is supported by subscription; and there is also a dispensary.

Among the ancient monastic establishments were, an hospital, founded in the south-east part of the city, in honour of St. Wulstan, bishop of the see, in 1088, the revenue of which at the dissolution was £79. 12. 6.; the remains of this establishment, which was subsequently denominated the Commandery, and still retains that name, are considerable: a convent of Grey friars, without St. Martin's gate, founded about the year 1268, by the family of the Beauchamps, Earls of Warwick, the remains of which were for several years used as the city gaol: a convent of Dominican friars, in the west part of the city, the site of which is now covered with buildings: and a convent of White nuns of the Benedictine order, which existed at the time of the Conquest, and at the dissolution had a revenue of £53. 13. 7.; the site still bears the name of the White Ladies; a small portion of the buildings is visible, and a farm, about a mile from the city, called the Nunnery, is probably a part of its ancient demesne. The guild of the Holy Trinity was instituted by Henry IV., and, on its dissolution, was converted into an hospital by Queen Elizabeth. Among the distinguished prelates of the see were, the venerable Dr. Latimer, and Drs. Prideaux, Stillingfleet, and Hurd. Florence and William of Worcester were brethren in the monastery; Nicholas Facio de Duillier, a native of Switzerland, and author of several mathematical and philosophical works, resided here for 33 years, and was buried in St. Nicholas' church, in 1753; Dr. Thomas, son of Bishop Thomas, and author of a Survey of the Cathedral Church of Worcester; and Drs. Mackenzie, Johnstone, and Wall, eminent medical practioners, were also residents; the last introduced the manufacture of porcelain, and contributed, by an analysis of its medicinal springs, to bring Malvern into repute as a watering-place. Among the eminent natives was Edward Kelly, noted for his knowledge of chymistry and astrology, born in 1555; John Lord Somers, a celebrated lawyer; and Mr. Thomas White, a distinguished sculptor and architect. Worcester gives the inferior title of Marquess to the Duke of Beaufort.

WORCESTERSHIRE, an inland county, bounded on the west by Herefordshire, on the south and south-east by Gloucestershire, on the east and north-east by Warwickshire, on the north by Staffordshire and a detached portion of Shropshire, and on the north-west by Shropshire. It extends from 52° 0′ to 52° 30′ (N. Lat.), and from 2° 14′ to 3° 0′ (W. Lon.), and, including the detached portions, comprises an area of upwards of 780 square miles, or about 500,000 statute acres. The population amounts to 211,365. At the period of the Roman invasion of Britain, the district now included within the confines of Worcestershire is supposed to have been partly occupied by the ancient British tribe of the Cornavii, and partly by that of the Dobuni. Under the Roman dominion it was included in the division called *Flavia Cæsariensis*, but being then for the most part low and woody, it received but little attention from these conquerors. On the complete establishment of the Saxon Heptarchy, it was included in the kingdom of Mercia; and in the predatory invasions of the Danes it suffered, at a later period, in common with most other parts of the kingdom.

This county is in the diocese of Worcester (excepting fifteen parishes and eight chapelries, which are in that of Hereford, but, by the act of the 6th and 7th of William IV., cap. 77, are to be transferred to the diocese of Worcester), and in the province of Canterbury: it forms an archdeaconry, including the deaneries of Blockley, Droitwich, Evesham, Kidderminster, Pershore, Powick, Kington, Warwick, Wich, and Worcester: the total number of parishes is 171. For purposes of civil government it is divided into the five hundreds of Blackenhurst, Doddingtree, Halfshire, Oswaldslow, and Pershore, each of which is divided into Upper and Lower, excepting Oswaldslow, which has also a middle division. It contains the city of Worcester; the borough and market-towns of Bewdley, Droitwich, Dudley, Kidderminster, and Evesham; and the market-towns of Bromsgrove, Pershore, Shipston-upon-Stour, Stourbridge, Stourport, Tenbury, and Upton-upon-Severn. By the act of the 2nd of William IV., cap. 45, the county has been divided into two portions, called the Eastern and Western divisions, each empowered to send two members to parliament. The place of election for the former is Droitwich, and the polling-places are Droitwich, Pershore, Shipston, Stourbridge, Evesham, Dudley, Hales-Owen, Bromsgrove, and King's-Norton; that for the latter is Worcester, and the polling-places, Worcester, Upton, Stourport, and Tenbury. Two citizens are returned for the city of Worcester; and two burgesses for the borough of Evesham, and one each for those of Bewdley, Droitwich, Dudley, and Kidderminster, of which the last two were enfranchised by the act above named, and Droitwich was then deprived of one member, having formerly returned two. This county is included in the Oxford circuit: the assizes and quarter sessions are held at Worcester, where stands the county gaol and house of correction.

The form of the county nearly approaches a parallelogram, two-thirds of the area of which lie to the east of the Severn; but its boundaries are extremely irregular, and its detached portions numerous. Its general appearance, when viewed from the heights bordering it in different parts, is that of a rich plain, the more gentle elevations being hardly discernible. The Vale of the Severn, extending through it from north to south, a distance of about 30 miles, varies in breadth from a quarter of a mile to a mile, and contains about 10,000 acres. The Vale of Evesham is an indefinite tract in the south-eastern part of the county, including the Valley of the Avon, the adjoining uplands to the north of that river, and the whole of the vale land in the southern part of the county and the adjoining parts of Gloucestershire. To the north-east of Bromsgrove is a ridge of hills, called the Lickey, which extends to Hagley, and has various branches eastward: some of its highest peaks rise to the height of nearly 900 feet. The Abberley hills, in the north-western part of the county, extend over the parish of Abberley, and are seen to a great distance, rising to about the same height as the last-mentioned: Witley hill is a little to the southward of these. Bredon hill is another remarkable elevation, to the south of Pershore, and on the south-eastern side of the Avon, rising to the height of nearly 900 feet. But by far the loftiest tract is the Malvern hills, a chain extending from north to south, upon a base about six miles in length and from one to two in breadth: a line passing along the summit of this ridge separates Worcestershire from Herefordshire: the most elevated point attains the height of 1313 feet above the Severn. The views obtained from most of these eminences are remarkable for their beauty and extent, particularly those from the Malvern hills; and their rocky summits give a picturesque diversity to much of the scenery.

The soils are remarkable for their general fertility, and add a peculiarly rich verdure to a district presenting great beauty of outline, and enjoying an eminently fine climate. Those of the valleys traversed by the principal rivers consist of a deep rich sediment, which has been deposited by floods during a long series of ages: this sediment is in some places a pure clay, adapted to the making of bricks, but generally consists of a rich mould. The valleys of the Severn, the Avon, the Stour, the Selwarpe, and nearly all the smaller streams, consist of rich natural meadows and pastures; while that of the Teme abounds also with hop-plantations and orchards. In the valley of the Stour some small tracts of peat bog are found: the extent of this kind of soil, bordering on the rivers, is estimated at about 50,000 acres. Rich clay and loamy soils occupy nearly half the county in its middle, southern, and western districts, and, besides the ordinary crops of other counties, produce great quantities of hops and fruit. The substrata of the Vale of Evesham are various, being sometimes a yellowish gravel, and at others a clay, which is unfit for making bricks, as containing calcareous particles. About Kidderminster and Stourbridge light sandy soils prevail to a very great extent; some of them are poor and barren, as at Mitton and Wolverley; others rich and fertile: the sands of Wolverley are in many places of considerable depth, or terminate in a sandy rock of various depths. To the north-east of Bromsgrove, including the hilly cultivated tracts, a mixed gravel abounding in springs, and a gravelly loam, are found: in this north-eastern part of the county some of the hilly grounds have also a moist clay loam on a broken rocky substratum, and a lighter loam on clay, with a similar substratum: the lands on Bromsgrove Lickey have often a deep substratum of sand; but, on the higher parts of it, they frequently rest immediately upon an irregular granitic rock, or upon a soft pudding-stone. The waste lands in the eastern part of the county, which are but of small extent, have generally a deep black peaty soil. The Abberley hills and Witley hills have a strong wet clay, resting on limestone.

The soil and climate being well adapted to the production of every kind of grain in abundance, the agriculture of the county is less subject to any characteristic system than that of almost any other: the drill husbandry is practised chiefly on the hills and lighter soils. The amount of arable land is estimated at 360,000 acres: the crops most generally cultivated are, wheat, barley, oats, beans, peas, vetches, turnips, and hops. Four varieties of wheat are commonly sown, and annually occupy about 43,500 acres. Nearly 1000 acres in the sandy parts of the county are annually sown with rye, much of the produce of which is sold to be sown as early spring food for sheep. Barley is sown after turnips on all lands where the latter crop is grown, in conjunction with clover and grass seeds, and is calculated to occupy about 33,000 acres annually. Oats are grown on a much smaller scale, being seldom cultivated on the richer soils: beans are grown to a considerable extent on the strong soils, but peas only to a limited degree; vetches are in common cultivation in almost every part of the county, and are chiefly employed as green food for horses. Turnips are extensively and successfully grown on the more friable soils; the

Swedish turnip is also cultivated, but only to a small extent. The sands of Wolverley are famous for their produce of carrots and carrot seed, for the most part sold to persons who carry them to the markets of Birmingham, Stourbridge, or the populous parts of Staffordshire; much of the seed is frequently sold to the London dealers. Worcestershire has long been famous for the culture of hops, in all cases upon a deep rich loam, or a peaty soil, plentifully manured: many hop-grounds on the banks of the Teme receive occasional irrigation from the overflow of that river, whose waters are of a peculiarly fertilizing quality. Three sorts of hops are here cultivated, distinguished as red, green, and white; the produce, though varying extremely, is estimated to average about five cwt. per acre. The principal artificial grasses are red and white clover, trefoil, and rye-grass.

The extensive vales, particularly that of the Severn, consist of meadows and pastures of a particularly rich quality, occupying an extent of about 50,000 acres: almost any proportion of this land may be mown at pleasure, and a great quantity of hay is sent to the mining districts of Shropshire and Staffordshire. There are, besides, nearly 50,000 acres of permanent upland pasture, including parks and pleasure grounds. A part of the pastures is grazed by cows belonging to the dairies, the produce of which is chiefly butter, for home consumption and the supply of Birmingham, and cheese made of skimmed milk: in some dairies, however, cheese only, of a good quality, is made. Great numbers of sheep and cattle are fattened in the rich meadows, chiefly in the southern and western parts of the county; many cattle are also fattened in stalls during the winter, a very great proportion of them being driven to the London market. Lime is extensively used as a manure on the gravelly and sandy soils of the north-eastern part of the county, where marl is also occasionally employed; horn shavings, leather shreds, ashes, soot, and offal salt from the works at Droitwich, are in some places used for the same purpose to a small extent.

The cattle are of various sorts, few being bred in the county: those most esteemed are the Hereford and long-horned breeds, the latter being chiefly bought at the fairs of Staffordshire and Shropshire: besides these, almost all the other surrounding counties furnish Worcestershire with various kinds of cattle, to be fattened in its rich vales; for which purpose also great numbers of Welsh, Yorkshire, and even Scotch, cattle are imported. The only peculiar kind of sheep is the common, or waste land, breed, occupying all the wastes, except those in the southern parts of the county: they are without horns, and are supposed to have sprung from the same stock as the South Down sheep, and the Cannock-heath sheep of Staffordshire. In the enclosures are found the Cotswold (which also occupy the southern wastes), the Ryeland, the Leicester, the South Down, and various other breeds and mixtures; besides which, many of the Somersetshire, Wiltshire, and Dorsetshire breeds are annually brought in, chiefly for the purpose of being fattened. The hogs are chiefly of a large white slouch-eared kind: much bacon is consumed in the county; the surplus assists in the supply of the adjoining manufacturing districts of Warwickshire, Staffordshire, &c. The extent of land applied to the raising of vegetables for human food is estimated at about 5000 acres: there are very considerable horticultural tracts near the principal towns, more particularly on the north-eastern side of Worcester, and on the northern side of the town of Evesham, in the vicinity of which latter place there are about 300 acres of garden ground, which, besides producing all the other ordinary vegetables, supply the cities of Bath and Bristol, and the town of Birmingham, with considerable quantities of early peas and asparagus: great quantities of cucumbers and onions are exported from the same district, chiefly to the last-mentioned town; much onion seed is also produced there. This county has for many centuries been famous for its orchards, which flourish in a degree unknown to most other parts of the kingdom; they are situated chiefly around the towns, villages, and farm-houses, chiefly of the middle, southern, and western parts of the county, where the various kinds of fruit-trees are also frequently dispersed in the hedge-rows, and form an important source of profit. The average quantity of cider and perry made is remarkably great, for, besides supplying the consumption of the county, which is very considerable, a large surplus, together with great quantities of raw fruit, is exported to other parts of the kingdom.

Worcestershire is adorned with a plentiful store of timber: in many parts are oak coppices of different degrees of growth, and in some are small tracts of the finest oak and ash timber, particularly in the neighbourhood of the different seats: the most important produce of the underwoods is, poles for the hop-yards, and charcoal for the iron-works. The Forest of Wyre, on the north-western border of the county, near Bewdley, partly in this county and partly in that of Salop, is a great nursery for oak-poles and underwood. Some parts possess beech timber of excellent quality, and many of the precipitous heights bordering on the Severn, and the hills in some other places, are ornamented with large plantations of fir. The hedge-rows, too, throughout a large portion of the most fertile districts, are well stocked with some of the most valuable elm timber in the kingdom, more particularly in the parishes of Hartlebury, Elmley-Lovett, Ombersley, &c., great quantities of which are regularly cut down and sent to Birmingham, or exported by the Severn and the canals. On the borders of the rivers are many poplar and willow plantations, more particularly along the course of the Teme. The waste lands do not, at most, exceed 20,000 acres, and consist of high hilly tracts, or of small commons and wastes, dispersed in various quarters. Of the hilly wastes, the principal are the upper parts of the Malvern hills, which are very rocky; of Bredon hill, near Pershore; and of the Abberley and Witley hills, together with some of the unenclosed parts of Bromsgrove Lickey. Wyre Forest, to the left of Bewdley, besides its woodlands, comprises also a considerable portion of open land.

The mineral productions are of minor importance. Coal is obtained in the north-western part of the county, particularly at Mamble, which place communicates, by means of an iron railway, with the Leominster canal; and again at Pensax, where the small refuse is partly converted into coke, highly esteemed for the drying of hops, and is partly used for burning the limestone obtained at Witley hill, but the seam is only from two feet to two feet six inches thick, and lies at the depth of about twenty yards, from which the water is raised in buckets. Common rock salt and a species of gypsum are found at Droitwich. Limestone of the lias formation forms the substratum of nearly the whole south-eastern portion of the county, and is worked at South Littleton and other places; the kind called by geologists "mountain limestone" is found in the hills of the north-western part, and is burned in several places, particularly at Witley and Huddington. The town of Dudley is situated at the southern extremity of a range of limestone hills, which extends northward into Staffordshire; and this, upon which stands the castle and part of the town, is completely undermined by stupendous quarries. Freestone for building is obtained in several places. The Malvern hills are formed chiefly of a kind of decomposed granite, with which, on their northern side, gneiss is connected, and on their eastern, sienite. The precipitous swells of Bromsgrove Lickey are composed chiefly of quartz, a silicious stone, much resembling the granite of the Malvern hills: in the Broadway hills a reddish stone is quarried. In the Vale of Evesham, in the parishes of Badsey, the three Littletons, and Prior's-Cleeve, are quarries of a calcareous flag-stone, about three inches thick, and of a very durable quality, some of it bearing a fine polish: considerable quantities are raised for gravestones, kitchen floors, barn floors, &c., and much of it is exported by means of the Avon navigation. Brick clay, gravel, sand, and marl, exist in numerous places. The most remarkable fossil production is that found in the limestone at Dudley, thence called the "Dudley locust."

The manufactures are various, extensive, and important. Those of gloves and porcelain are carried on at Worcester. Stourbridge has an extensive manufacture of glass, which has long flourished both there and at Dudley: and at both places the iron manufacture is carried on to a very considerable extent. Nails, needles, and fish-hooks, are made at Bromsgrove, also at Redditch, on the border of Warwickshire. Kidderminster is famous for its carpets; and the manufacture of bombazines is still carried on, but not so extensively as formerly. On the river Stour and its tributary streams are several very considerable iron-works, in which the pig-iron from the foundries of Shropshire, Staffordshire, and other mining districts, is rendered malleable, and worked into bars, rods, sheet-iron, &c. The manufacture of salt, at Droitwich, is known to have been practised so early as the year 816, when this county formed part of the Saxon kingdom of Mercia: it is here made from inexhaustible brine springs, which lie at the depth of about 80 feet, and, when bored into, immediately rise and fill the pit dug to receive their waters. Worcester is the great mart for the hops, fruit, cider, and perry produced in this county and that of Hereford. The quantity of hops brought to market varies according to the plentifulness of the crop, but is supposed to average about 36,000 cwt. An idea of the quantity of fruit exported may be formed from the fact, that the cargoes sent northward, consisting chiefly of apples and pears, have, in some years, amounted to upwards of 2000 tons; besides which, considerable quantities are carried out of the county from the markets of Bewdley, Kidderminster, Bromsgrove, &c. The cider annually exported amounts to about 10,000 hogsheads, of 110 gallons each; the perry to about a tenth part of that quantity. Of the wheat, barley, and beans, grown in the county, the surplus is very great, and finds ready carriage to Birmingham and the populous parts of Staffordshire and Shropshire, or down the Severn, to be conveyed coastwise. Fat cattle, sheep, and hogs, are supplied for the London market, and the manufacturing districts of Warwickshire and Staffordshire: about 2000 packs of wool, of 240lb. each, are annually exported; as also are clover and grass seeds, hay, and timber.

The principal rivers are the Severn, the Upper Avon, the Teme, and the Stour. The Severn is navigable for vessels of 80 tons' burden as high as Worcester bridge, and for those of 60 in the higher part of its course through the county; but the navigation, though of great benefit and importance, is frequently impeded in the summer by sands and shoals. By the statute of the 30th of Charles II., cap. ix., the conservancy of the Severn, within the limits of the county, is granted to the magistrates of Worcestershire. The Upper Avon, so early as the year 1637, was made navigable, with the aid of locks, in the whole of its course through Worcestershire, a distance of about 20 miles. The Teme has too great a declivity, and its waters are too shallow, to admit of its being navigated higher than a small distance above Powick bridge: the scenery on its banks is particularly beautiful. The Stour is navigable for a short distance to some of the iron-works on its banks. The Trent and Severn, or, as it is more commonly called, the Staffordshire and Worcestershire, canal enters the county from Staffordshire, near Wolverley, and thence proceeds down the valley of the Stour, and by the town of Kidderminster, to the navigable channel of the Severn, at Stourport, into which it opens through a spacious basin: the length of that part of its course included in Worcestershire is about nine miles, in which it has nine locks, and a fall of 90 feet: this canal, one of the works of the celebrated Brindley, is that branch of the Grand Trunk canal which unites the navigation of the Severn with the water communication between the rivers Trent and Mersey: the act for its formation was obtained in 1766, and it was completed about the year 1770. The Droitwich canal, from that town to the Severn, down the valley of the Salwarpe, was constructed soon after the above, and by the same engineer: it is five miles and a half long, with five locks and a fall of about 60 feet: the cost of its formation was £25,000. The noble canal from Birmingham to the Severn, immediately below Worcester, called the Birmingham and Worcester canal, for vessels of 60 tons' burden, commences with a short tunnel in the vicinity of the first-mentioned town, where it communicates with the Birmingham, Birmingham and Fazeley, and Birmingham and Warwick, canals, and proceeds nearly southward, across two valleys, over which it is conveyed by extensive embankments to a little beyond King's-Norton, where it passes through another tunnel, upwards of a mile in length, and then, after completing its summit level of sixteen miles and three quarters from the wharfs at Birmingham, descends south-westward from the towns of Bromsgrove and Droitwich, by a lockage of 450 feet fall, to the Severn: it has also other tunnels, of smaller extent than the last; the act of parliament for its formation was obtained in 1791: its total length is 29 miles. The Dudley Extension canal branches from it near Selly Oak, and thence proceeds westward, through a long tunnel, to Hales-Owen, a short distance beyond which it is carried through another tunnel, and, on emerging, pursues a winding northerly course to Dudley, and there passes through a tunnel under the limestone hills, nearly two miles in length, into the county of Stafford, where it forms a junction with the Birmingham canal from that town to Wolverhampton: its total length is thirteen miles. The Stratford-upon-Avon canal branches from the Birmingham and Worcester canal near King's-Norton, and thence proceeds eastward, through a small tunnel, into Warwickshire. The Kington, Leominster, and Stourport canal was projected towards the close of the last century, the first act of parliament for the execution of the design having been obtained in 1791; but the expense was found much to exceed the sum at first computed, and only the part of its course between Leominster and Stourport has been completed. The line of the Birmingham and Gloucester railway enters this county from Birmingham, and passes a little to the east of Bromsgrove, Droitwich, and Worcester, to the west of Pershore, and enters the county of Gloucester a little to the north-east of Tewkesbury.

The Roman roads which crossed the county were, the *Ikeneld-street*, which passed northward, from Alcester in Warwickshire, through its north-western extremity, into Staffordshire; another which passed from Worcester into Shropshire; a third, from Worcester, southward by Upton, to Tewkesbury, where it joined the *Ikeneld-street;* and the *Ridge-way*, which bounds the county for several miles, on the east. Numerous vestiges of them are still visible; as also of a Fosse-way, which passes through the detached parish of Blockley, and an ancient road which crossed Hagley common, now called the King's Headland. Stukeley supposes Upton, on the banks of the Severn, to have been the *Ypocessa* of the Romans; and Worcester, from the termination of its name and other circumstances, appears to have been either a Roman station, or a fort. The remains of antiquity include few very remarkable objects. Near the Four-shire Stone, at a point where the counties of Worcester, Gloucester, Warwick, and Oxford meet, there is a small earthwork, supposed by Gough to be of British construction; and there are traces of other ancient encampments in the vicinities of Bredon, Kempsey, and Malvern; as also on Witchbury hill, Woodbury hill, and Conderton hill, in the parish of Overbury. Various coins of the Lower Empire have been found in the vicinity of Hagley, particularly near the large camp on Witchbury hill; and on Clent heath, about half a mile from Witchbury, are five barrows, assigned by popular tradition to the Romans, which, on being opened, were found to contain burnt wood, ashes, and bones. The number of religious houses, including colleges and hospitals, was about 28. There yet exist remains of the abbeys of Bordesley, Evesham, and Pershore; of the Commandery of St. Wulstan at Worcester; of the priories of Dodford and Great Malvern; and of the nunnery of Cokehill, in the parish of Inkberrow. There are also relics of the ancient castles of Dudley; Ham, near Clifton-upon-Teme; Hartlebury; and Holt. Worcestershire contains a considerable number of elegant mansions: among the principal are Croome Park, Hartlebury Castle, Hewell Park, Madresfield, Northwick Park, Ombersley Court, Witley Court, Hagley Park, Hanbury Hall, and Stanford Court. The mineral springs are very numerous; the most noted are the chalybeate waters of Bredon, Bromsgrove (which are also petrifying), Hallow Park near Worcester, Kidderminster, and Worcester; and those of other qualities at Abberton, near Naunton-Beauchamp, and at Churchill. But the Malvern wells, which possess various properties, are by far the most celebrated, and, in conjunction with the fine climate and scenery of the surrounding country, have rendered the town of Great Malvern a place of fashionable resort. The salt springs of Droitwich have been noticed above.

WORDWELL (*All Saints*), a parish, in the union of Thingoe, hundred of Blackbourn, Western Division of the county of Suffolk, 6 miles (N. by W.) from Bury-St. Edmund's; containing 69 inhabitants. The living is a discharged rectory, united to that of West Stow, and valued in the king's books at £7. 7. 3½. The church is a small ancient edifice, in the Norman style.

WORFIELD (*St. Peter*), a parish, in the union of Bridgenorth, Hales-Owen Division of the hundred of Brimstree, Southern Division of the county of Salop, 3¾ miles (N. E. by E.) from Bridgenorth; containing 1676 inhabitants. The living is a discharged vicarage, valued in the king's books at £16. 15.; patron and impropriator, W. Y. Davenport, Esq. The impropriate tithes have been commuted for a rent-charge of £1745, and the vicarial for one of £239, both subject to the payment of rates, which on the former have on the average amounted to £81, and on the latter to £10; the glebe comprises about two acres. The church contains 150 free sittings, the Incorporated Society having granted £60 in aid of the expense. Here is a National school. William Lloyd and Thomas Parker, in the year 1613, conveyed estates, now producing an annual income of about £46, in trust for the maintenance of a school, in which 24 boys are educated; and a school for girls is supported by subscription.

WORGRET, a tything, in the parish of East Stoke, hundred of Hundredsbarrow, Wareham Division of the county of Dorset, 1 mile (W.) from Wareham: the population is returned with the parish.

WORKINGTON (*St. Michael*), a market-town, sea-port, and parish, in the union of Cockermouth, Allerdale ward above Derwent, Western Division of the county of Cumberland; containing, exclusively of seamen, 7196 inhabitants, of which number, 6415 are in the town of Workington, 34 miles (S. W. by W.) from Carlisle, and 310 (N. W. by N.) from London. The only historical circumstance of interest connected with this place is the landing here, in 1568, of Mary Queen of Scots, when she sought an asylum in England, after her escape from the field of Langside: she was hospitably entertained at Workington Hall (the apartment she occupied being still called the Queen's chamber), until Queen Elizabeth gave directions for her removal to Carlisle castle. The town is situated on the southern bank of the Derwent, near its influx into the sea; and, in addition to the older part, which is narrow and irregular, contains some modern streets, in which are many handsome and well-built houses: it is not lighted, and is badly paved with pebbles, but well supplied with water from the Derwent. There is a small theatre in Christian-street, and an assembly and news room in the Square. The Hall, the ancient seat of the Curwens, occupies an eminence on the south side of the river, commanding beautiful views of the surrounding country, the sea, and Scotland. Upon the Cloffocks, an extra-parochial meadow, or island, situated north-east of the town, on the banks of the Derwent, races are annually held in August. A handsome stone bridge of three arches crosses the river, at the entrance into the town from Maryport, which was erected in 1763, at the expense of the county. The trade principally arises from the exportation of coal to Ireland, in which more than 100 vessels are employed. The harbour has been secured by the erection of a breakwater within these few years, and is now one of the safest on the coast: the entrance is lighted with gas. Great improvement has been also effected by enlarging the quays, owing to the indefatigable exertions of the late Mr. Curwen. The collieries give employment to about 500 persons; the mines are drained by an engine of more than 150-horse power. There are three ship-builders' yards, in which vessels of from 300 to 400 tons' burden are constructed, besides two patent slips: the manufacture of cordage and other articles connected with the shipping is carried on, though not so extensively as formerly; there is also a manufactory for imitation Leghorn hats, giving employment to upwards of 400 men, women, and children, during the summer months, in the preparation of the straw, which is grown in the neighbourhood: the manufacturer has received a patent for the invention. The salmon fishery, for which Camden mentions this place to have been famous, although not so productive as in his time, is still pursued in the Derwent and along the coast. The markets are on Wednesday and Saturday, that on the former day being the principal: it is a large corn market, and has recently been removed to Washington-street: there is another market-place, for butter, poultry, &c., which is connected with convenient shambles for butchers' meat. The fairs, on the 18th of May and Oct., have nearly fallen into disuse. Manor courts are held occasionally, and the county magistrates hold petty sessions, every Wednesday, at the public office in Udale-street.

The living is a rectory, valued in the king's books at £33. 5.; present net income, £966; patron, Henry Curwen, Esq. The church, situated at the west end of the town, and rebuilt in 1770, is a handsome structure, in the later style of English architecture, with a square tower. A district chapel, dedicated to St. John, has been erected, under the auspices of Her Majesty's commissioners for building churches, the first stone of which was laid on April 15th, 1822: it is a handsome building of the Tuscan order, with a portico and cupola, and the expense of its erection was upwards of £10,000. There are places of worship for Independents, Primitive and Wesleyan Methodists, Presbyterians, and Roman Catholics. A school in the square was founded in 1808, by the late Mr. Curwen, when the free grammar school was broken up, and is partly supported by H. Curwen, Esq. There is also a school of industry, established in 1816, for 30 girls. A dispensary, and several benevolent institutions for clothing the poor, are supported by voluntary contributions. On an eminence near the sea, at a short distance hence, are the remains of an ancient dilapidated building, called the Old Chapel, which, as it commanded an extensive view of Solway Firth and the Scottish coast, was probably used as a watch tower, to guard against the incursions of the Scots.

WORKSOP (*St. Mary and St. Cuthbert*), a market-town and parish, and the head of a union, in the Hatfield Division of the wapentake of Bassetlaw, Northern Division of the county of Nottingham, 26 miles (N.) from Nottingham, and 146 (N. N. W.) from London; containing 5566 inhabitants. This place, which in Domesday-book is written *Wirchesope*, and in other records of that period *Wyrksoppe* and *Wirkensop*, appears to have belonged, prior to the Conquest, to Elsi, a Saxon nobleman. It was afterwards granted by the Conqueror to Roger de Busli, and subsequently became the property of William de Lovetot, who, in the reign of Henry I., founded here a priory for canons Regular of the order of St. Augustine, the prior of which was, in the reign of Henry III., summoned to parliament. It passed, after a considerable period, by the marriage of the heiress of the Lovetots, to the family of Furnival; then to that of Nevill; and from that family to the Talbots, afterwards Earls of Shrewsbury, to whom, on the dissolution of monastic establishments, the revenue of the priory, then valued at £239, was granted by Henry VIII. From this family the manor descended by marriage to the Earls of Arundel, now Dukes of Norfolk, who held it as tenants in chief of the Crown, by the service of a knight's fee, and of procuring a glove for the king's right hand at his coronation, and supporting that hand while holding the sceptre; it has lately been sold to the Duke of Newcastle. In Dec. 1460, an engagement took place at Worksop between the forces of the Duke of York and those of the Duke of Somerset, when the latter were defeated. Gilbert, first Earl of Shrewsbury, who so much distinguished himself in the French wars under Henry V., built the magnificent mansion-house, afterwards the ducal residence, and the place of confinement of Mary Queen of Scots, in the sixteenth year of her captivity, she being at that time in the custody of the earl; and her son, James I., on the 20th of April, 1603, rested here, on his way to London to assume the English crown. In 1761, it was accidentally destroyed by fire, but was soon afterwards splendidly rebuilt by His Grace the Duke of Norfolk.

The town is situated in a pleasant valley, near the northern extremity of the forest of Sherwood, in the midst of a well-wooded and picturesque country; and the vicinity is ornamented by the magnificent seats of several noblemen, amongst which are Worksop Manor, now the noble mansion of the Duke of Newcastle, alluded to above, standing in a park eight miles in circumference; Welbeck Abbey, the seat of the Duke of Portland; Clumber, also the mansion of the Duke of Newcastle; and Thoresby, the seat of Earl Manvers. It is neat in its general appearance, and consists, in the higher and principal part, of one long street, with a second running into it at right angles; the houses are well built, and the town is paved, lighted by subscription with gas, and well supplied with water. Camden describes Worksop as famous for the production of liquorice, which has long since ceased to be cultivated. Malt, which is made in considerable quantities, barley being much grown in the surrounding country, is the principal article of trade; and the Chesterfield canal, passing on the northern side of the town, affords every facility for its conveyance to Manchester and the other markets to which it is chiefly sent: on this canal are wharfs communicating with the town, and to the east it crosses the river Ryton by an aqueduct. The market is on Wednesday; and there are fairs on March 31st and Oct. 14th, for horses and cattle; and a statute fair about three weeks after. Constables are chosen at the annual court leet of the lord of the manor. It is in contemplation to take down the old moot-hall (which has been many years in a dilapidated state), and some of the adjoining buildings, and to erect on their site a handsome structure, comprising a town-hall, assembly-room, prison, market-house, &c. Worksop is a polling-place for the borough of East Retford.

The living is a vicarage, valued in the king's books at £12. 4. 2.; present net income, £388; patron, Duke of Norfolk; impropriator, Duke of Portland. The church, standing on the eastern side of the town, comprises the western portion of the priory church, and its cathedral-like towers form an interesting object in the view of the town: it is one of the principal remaining specimens of Norman architecture, but in the exterior much of the English style has been mixed with it. The western entrance is under a beautiful receding Norman arch, with zigzag ornaments, and the towers which surmount it have circular and pointed arched windows, in different gradations. The nave is separated from the aisles by eight pillars, alternately cylindrical and octangular, supporting circular arches ornamented with quatrefoils, above which are two tiers of windows; the old pulpit and reading-desk have been lately taken down and replaced by new ones: the church contains some few ancient monuments. At the south-eastern extremity of the church are the remains of the beautiful chapel of St. Mary, forming an interesting ruin, the ornamental parts of which are most richly executed, and the windows considered some of the most perfect models of the lancet shape in the kingdom. On the northern side, and contiguous to the church, are some fragments of the walls of the priory, and in the meadows below are extensive traces of the foundation. The priory well is still in high estimation, for the purity and softness of the water. The principal gateway to the priory still exists, forming the entrance towards the church: it is in the later English style, and is 20 yards in front, with a pediment, in the tympanum of which is a niche with much tabernacle work, and in it a figure in a sitting posture. Above is a window of twelve lights, also two canopied niches of great beauty, which contain figures described by Dodsworth (who wrote when they were in a much better state of preservation than at present,) as those of armed knights, each bearing a shield, that on the west charged with a lion rampant for Talbot, and that on the east bearing a bend between six mantletts for Furnival. The room over the gateway is used as the National school for boys: the stone staircase leading to it is entered by an elegant porch rising about two-thirds of the height of the whole front, the doors, windows, niches, roof, &c., of which are of the most beautiful proportions and elaborate workmanship. At Shireoaks there is a neat chapel, built and endowed in 1809 by the Rev. John Hewitt, then lord of the manor; and at Scofton, close to the hamlet of Osberton, a handsome chapel capable of accommodating upwards of 200 persons has been erected and endowed by Geo. Savile Foljambe, Esq., to whom the right of presentation belongs: it was consecrated Dec. 30th, 1834. There are places of worship for Independents and Wesleyan Methodists; and near the manor-house is a chapel for Roman Catholics, who are numerous in the neighbourhood. In the National school, which is supported chiefly by subscription, upwards of 200 children of both sexes are educated and partially clothed. There are some small endowments for the benefit of the poor. The poor law union of Worksop embraces portions of the counties of Nottingham, Derby, and York, and comprises 26 parishes or places, under the care of 29 guardians; the population, according to the census of 1831, amounts to 16,111. On a hill, at the western side of the town, the site of the ancient castle of the Lovetots may still be traced; and in the park of the manor are some tumuli, which, from fragments discovered in them, appear to be ancient British. The hamlet of Shireoaks is so named from an oak whose branches are said to have overshadowed a portion of the three counties of Nottingham, Derby, and York. At Osberton, human bones, stone coffins, an antique font, some stained glass, &c., have been found at various times, the supposed remains of a church: they are preserved at Osberton Hall. The ruins of the ancient manor-house of Gateford, another hamlet in the parish, with its gables, moats, &c., are still visible; and near them, in 1826, several Roman coins of Nero and Domitian were found.

WORLABY, an extra-parochial liberty, in the union of LOUTH, hundred of HILL, parts of LINDSEY, county of LINCOLN, 7 miles (S.) from Louth; containing 34 inhabitants.

WORLABY (*ST. CLEMENT*), a parish, in the union of GLANDFORD-BRIDGE, Northern Division of the wapentake of YARBOROUGH, parts of LINDSEY, county of LINCOLN, 5½ miles (N. by E.) from Glandford-Bridge; containing 309 inhabitants. The living is a discharged vicarage, valued in the king's books at £6. 8. 4.; present net income, £278; patron and impropriator, J. Webb, Esq.

WORLDHAM, EAST, a parish, in the union and hundred of ALTON, Alton and Northern Divisions of the county of SOUTHAMPTON, 2½ miles (E. by S.) from Alton; containing 212 inhabitants. The living is a discharged vicarage, valued in the king's books at £5. 18. 1½.; present net income, £142; patrons and impropriators, President and Fellows of Magdalene College, Oxford. Here is a Sunday National school.

WORLDHAM, WEST (*ST. NICHOLAS*), a parish, in the union and hundred of ALTON, Alton and Northern Divisions of the county of SOUTHAMPTON, 2½ miles (S. E. by E.) from Alton; containing 96 inhabitants. The living is a perpetual curacy; net income, formerly £38, which has been augmented with £200 from Winchester College and £200 Queen Anne's Bounty; patrons and impropriators, Warden and Fellows of Winchester College. A school is partly supported by the clergyman.

WORLE (*ST. MARTIN*), a parish, in the union of AXBRIDGE, hundred of WINTERSTOKE, Eastern Division of the county of SORMERSET, 8 miles (N. W.) from Axbridge; containing 770 inhabitants. The living is a discharged vicarage, valued in the king's books at £12. 15.; present net income, £277: it is in the patronage of the Crown; impropriators, Trustees of a charity. The church is a neat structure, with a tower surmounted by a small spire. There is a place of worship for Wesleyan Methodists; also a National school. This parish produces very fine limestone and *lapis calaminaris:* the inhabitants feed great numbers of poultry, and dispose of them to the visiters at Weston-super-Mare, which has become a thriving watering-place. In the vicinity are vestiges of a Roman camp.

WORLESTON, a township, in the parish of ACTON, union and hundred of NANTWICH, Southern Division of the county of CHESTER, $1\frac{3}{4}$ mile (N.) from Nantwich; containing 367 inhabitants. A school is partly supported by subscription.

WORLINGHAM (*All Saints*), a parish, in the union and hundred of WANGFORD, Eastern Division of the county of SUFFOLK, $1\frac{1}{4}$ mile (S. E. by S.) from Beccles; containing 202 inhabitants. The living is a rectory, with that of Worlingham Parva annexed, valued in the king's books at £12; present net income, £260: it is in the patronage of the Crown. The church of Worlingham Parva, which was dedicated to *St. Peter*, has been demolished. A school is partly supported by the Countess of Gosford and the clergyman. The navigable river Waveney runs on the northern side of the parish. Part of the rents of the town-estate is appropriated to teaching poor children. This place give the title of Baron to the Earl of Gosford, who has a seat here.

WORLINGTON (*All Saints*), a parish, in the union of MIDDENHALL, hundred of LACKFORD, Western Division of the county of SUFFOLK, $1\frac{1}{4}$ mile (W. S. W.) from Mildenhall; containing 368 inhabitants. The living is a rectory, valued in the king's books at £19. 6. 8.; present net income, £197: the late Hon. Thomas Windsor presented in 1818. Worlington Hall and manor are the property of Sir F. G. Cooper, Bart. A school for girls is supported by subscription. The parish is bounded on the north by the navigable river Lark, over which is a ferry.

WORLINGTON, EAST (*St. Mary*), a parish, in the union of SOUTH MOLTON, hundred of WITHERIDGE, South Molton and Northern Divisions of the county of DEVON, 6 miles (E.) from Chulmleigh; containing 292 inhabitants. The living is a rectory, valued in the king's books at £7. 15. 10.; present net income, £208; patron, Hon. N. Fellowes. In the neighbourhood are the remains of an ancient cross. Roman coins have been found here.

WORLINGTON, WEST (*St. Mary*), a parish, in the union of SOUTH MOLTON, hundred of WITHERIDGE, South Molton and Northern Divisions of the county of DEVON, $5\frac{1}{2}$ miles (E.) from Chulmleigh; containing 187 inhabitants. The living is a rectory, valued in the king's books at £8. 15. 10.; present net income, £155; patron, Lewis Buck, Esq. Within the parish are the ruins of a castellated mansion, the ancient seat of the Affetons.

WORLINGWORTH (*St. Mary*), a parish, in the union and hundred of HOXNE, Eastern Division of the county of SUFFOLK, 5 miles (N. W.) from Framlingham; containing 729 inhabitants. The living is a rectory, with that of Southolt annexed, valued in the king's books at £19. 12. $3\frac{1}{2}$.; patron, Lord Henniker. The tithes have been commuted for a rent-charge of £680, subject to the payment of rates, which on the average have amounted to £170; the glebe comprises 52 acres, valued at £78 per annum. The church is principally in the later style of English architecture, and has an enriched font with a lofty and elegant cover. John Baldry, in 1689, bequeathed a house and land; and William Godbold, in 1698, left other land, for teaching poor children, for which 40 are instructed. A school-house was built in 1825 on land belonging to the parish, at the expense of Mr. John Cordy, who thus obtained the privilege of sending two children from the parish of Southolt to the school.

WORMBRIDGE (*St. Thomas the Apostle*), a parish, in the union of DORE, hundred of WEBTREE, county of HEREFORD, 9 miles (S. W.) from Hereford; containing 121 inhabitants. The living is a donative curacy, in the patronage of Edward Bolton Clive, Esq., the impropriator; net income, £51.

WORMEGAY (*Holy Cross*), a parish, in the union of DOWNHAM, hundred of CLACKCLOSE, Western Division of the county of NORFOLK, $7\frac{1}{2}$ miles (N. N. E.) from Downham-Market; containing 323 inhabitants. The living is a perpetual curacy; patron, Bishop of Norwich. The impropriate tithes have been commuted for a rent-charge of £349. 5.; subject to the payment of rates; the glebe comprises 3 acres, valued at £2. 10. per annum. Her Majesty's Commissioners have agreed to afford facilities for obtaining a site for a new church here. A priory of Black canons, in honour of the Virgin Mary, the Holy Cross, and St. John the Evangelist, was founded here in the reign of Richard I., or John, and, in 1468, was united to the priory of Pentney, to which it became a cell.

WORMHILL, a chapelry, in the parish of TIDESWELL, hundred of HIGH PEAK, Northern Division of the county of DERBY, $2\frac{1}{4}$ miles (W. S. W.) from Tideswell; containing 313 inhabitants. The living is a perpetual curacy; net income, £270; patrons, certain Trustees. The chapel is dedicated to St. Margaret. A school is endowed with £2 per annum. The river Wye runs in the vicinity through the most picturesque scenery, particularly that of Chee dale, in this chapelry.

WORMINGFORD (*St. Andrew*), a parish, in the union of LEXDEN and WINSTREE, Colchester Division of the hundred of LEXDEN, Northern Division of the county of ESSEX, $3\frac{3}{4}$ miles (W. S. W.) from Nayland; containing 543 inhabitants. This parish is situated on the navigable river Stour, from a ford across which, and a former proprietor of the manor, it derives its name: the surface rises gradually from the bank of the river to a considerable elevation: the soil is sandy, with a large intermixture of clay. The living is a vicarage, endowed with a portion of the rectorial tithes, and valued in the king's books at £7. 13. 4.; patron, and impropriator of the remainder of the rectorial tithes, John J. Tufnel, Esq. The impropriate tithes have been commuted for a rent-charge of £496. 17. 6., and the vicarial for £369, subject to the payment of rates, which on the average have amounted respectively to £82 and £61, the glebe comprises 4 acres, valued at £5. 19. 4. per annum. The church is a small ancient edifice, with a low square tower. Here is a National school endowed with £10 per annum.

WORMINGHALL (*St. Peter*), a parish, in the union of THAME, hundred of ASHENDON, county of BUCKINGHAM, $4\frac{3}{4}$ miles (W. N. W.) from Thame; containing 297 inhabitants. The living is a discharged vicarage, valued in the king's books at £6. 18. 10.; present net income, £58; patron and impropriator, Viscount Clifden. A market, formerly held here on

Thursday, was granted to John de Rivers, in 1304, with a fair on the festival of St. Peter and St. Paul. An almshouse for four poor women and six poor men, was founded in 1670, by John King, and endowed by him with property now producing a rental of about £80. There is also a fund of about £20 per annum, arising from various bequests, distributed in bread among the poor.

WORMINGTON (*Holy Trinity*), a parish, in the union of Winchcomb, Lower Division of the hundred of Kiftsgate, Eastern Division of the county of Gloucester, 5 miles (N. by E.) from Winchcomb; containing 96 inhabitants. The living is a discharged rectory, valued in the king's books at £7. 15. 5.; present net income, £143; patron, Josiah Gist, Esq., who supports a Sunday school.

WORMLEIGHTON (*St. Peter*), a parish, in the union of Southam, Burton-Dassett Division of the hundred of Kington, Southern Division of the county of Warwick, 5¾ miles (S. S. E.) from Southam; containing 161 inhabitants. The living is a discharged vicarage, valued in the king's books at £6. 13. 4.; present net income, £80; patron and impropriator, Earl Spencer. The Oxford canal passes through the parish.

WORMLEY (*St. Lawrence*), a parish, in the union of Ware, hundred and county of Hertford, 2¼ miles (N. by E.) from Cheshunt; containing 471 inhabitants. The living is a rectory, valued in the king's books at £10. 12. 3½.; present net income, £235; patron, Sir Abraham Hume, Bart. The church has a Norman doorway, and, at the west end, a square wooden tower; it contains several tablets, altar-tombs, and other sepulchral memorials. A school for boys is supported by Sir A. Hume. The New River runs through the parish, and the river Lea bounds it on the east.

WORMSHILL (*St. Giles*), a parish, in the union of Hollingbourn, hundred of Eyhorne, lathe of Aylesford, Western Division of the county of Kent, 5 miles (S. S. W.) from Sittingbourne; containing 186 inhabitants. This parish comprises 1440 acres of land, of which 650 are arable, and is intersected by the boundary line between East and West Kent. The living is a rectory, valued in the king's books at £10; present net income, £243; patrons, Governors of Christ's Hospital, London. The church, a plain building of flint with a low square tower, contains a few fragments of stained glass. The parsonage-house, with a glebe of 30 acres, has been lately much improved. Ten acres of land, by an unknown benefactor, producing £6 per annum, are applied to teaching poor children.

WORMSLEY (*St. Mary*), a parish, in the union of Weobley, hundred of Grimsworth, county of Hereford, 3½ miles (S. E. by S.) from Weobley; containing 102 inhabitants. The living is a perpetual curacy, in the patronage of T. A. Knight, Esq.

WORPLESDON (*St. Mary*), a parish, in the union of Guildford, First Division of the hundred of Woking, Western Division of the county of Surrey, 3½ miles (N. N. W.) from Guildford; containing 1360 inhabitants. This parish comprises about 6700 acres, of which about 300 are woodland, and nearly 2200 common or waste. The Wey and Arun navigation passes through it, and a telegraph, which is one on the projected line to Plymouth, has been erected near the church, at an expense of £2000, and is inhabited by a lieutenant of the royal navy, who has the management of it. The living is a rectory, valued in the king's books at £24. 13. 9.; patrons, Provost and Fellows of Eton College. The tithes have been commuted for a rent-charge of £1068, subject to the payment of rates, which on the average have amounted to £240; the glebe comprises 76 acres, valued at £76. 10. per annum. The church is an ancient structure, in the early English style: it contains some interesting monuments, and the east window is embellished with stained glass, collected and arranged at the expense of the Rev. W. Roberts, rector, in 1802. The Rev. Dr. Moore, in 1706, bequeathed £200, directing the interest to be applied in teaching poor children. In 1829, the remains of a Roman tesselated pavement were discovered on Broad-Street Common; the building, of which it formed the floor, was 62 feet long and 23 feet wide within the walls, and divided into five separate apartments, with a passage on the western side extending through the whole length; the tesseræ were of iron-stone about one inch square, and beneath it were found three Roman coins in a very corroded state.

WORSALL, HIGH, a chapelry, in the parish of North Allerton, union of Stockton, wapentake of Allertonshire, North Riding of the county of York, 4 miles (S. S. W.) from Yarm; containing 133 inhabitants. The living is a perpetual curacy; net income, £60; patron, Vicar of North Allerton.

WORSALL, LOW, a township, in the parish of Kirk-Leavington, union of Stockton, Western Division of the liberty of Langbaurgh, North Riding of the county of York, 3 miles (S. W.) from Yarm; containing 164 inhabitants.

WORSBROUGH, a chapelry, in the parish of Darfield, wapentake of Staincross, West Riding of the county of York, 2½ miles (S. by E.) from Barnesley; containing 2677 inhabitants. The living is a perpetual curacy; net income, £140; patron and appropriator, Rector of Darfield. The chapel, dedicated to St. Mary, is an elegant building, principally in the later style of English architecture, with a low tower surmounted by a spire: it has been repewed and galleries erected, by which means 334 additional sittings have been provided, 274 of which are free, the Incorporated Society having granted £200 in aid of the expense. A free school here is endowed with an annual pension of £4. 15. from the Crown, an annuity of £13. 6. 8. bequeathed by John Rayney, in 1631, and a house and land let for £13 per annum. A school for girls is endowed with £15 per annum; and another school is partly supported by subscription.

WORSLEY, a township, in the parish of Eccles, hundred of Salford, Southern Division of the county palatine of Lancaster, 6 miles (W. by N.) from Manchester; containing 7839 inhabitants. In the 10th of George II., an act was obtained for making navigable the river called Worsley brook, but the design was not carried into effect; and, in the 32nd of the same reign, the celebrated Duke of Bridgewater obtained an act, and subsequently other acts, enabling him to construct a series of canals from his extensive collieries here to different places, affording the means of conveying coal, &c., through a populous manufacturing district: the underground canals and tunnels at Worsley are said to be eighteen miles in length, and their construction to have cost £168,960. Worsley Archers' Society, formed in Aug. 1826, consists of 24 members, who hold their meetings every Wednesday at the Grapes' Inn, from the first Wednesday in April to the first in Oct. There is a place of worship for Wesleyan Methodists.

WORSTEAD (*St. Mary*), a parish, (formerly a market-town), in the hundred of Tunstead, Eastern Division of the county of Norfolk, $2\frac{3}{4}$ miles (S. S. E.) from North Walsham, and 121 (N. E. by N.) from London; containing 830 inhabitants. This place was formerly celebrated for the invention and manufacture of woollen twists and stuffs, thence called *worsted* goods; but this branch of trade, soon after its introduction by the Flemings, in the reign of Henry I., was, on the petition of the inhabitants of Norwich, removed to that city in the reign of Richard II., where it was finally established in the reign of Henry IV. The town at present has neither any manufacture nor trade: a navigable canal, which joins the sea at Yarmouth, passes through it. The market is entirely disused; but a fair for cattle is held on May 12th, and another at Scotto, an adjoining parish, on Easter Tuesday. A manorial court is held annually, at which constables and other officers are appointed. The living is a discharged vicarage, valued in the king's books at £10; present net income, £251; patrons and appropriators, Dean and Chapter of Norwich. The church is a spacious and elegant structure, partly in the decorated, and partly in the later, style of English architecture, with a lofty square embattled tower, strengthened with enriched buttresses, and crowned with pinnacles, forming, both in its combinations and details, a beautiful specimen of the decorated style: the chancel and the nave are principally in the later style, and are ornamented with screen-work of wood richly carved: the font is peculiarly rich, the sides being highly ornamented, and the pedestal on which it is supported is relieved with buttresses and canopied niches, and the risers of the steps are pannelled in compartments; the cover is of tabernacle work elegantly designed. There is a place of worship for Baptists, connected with which is a Sunday school, conducted on the National plan, and supported by subscription. An almshouse was founded in 1821, by Samuel Chapman, for 12 poor aged persons, either married couples or single persons, and endowed with a portion of the rent of an estate at Hellesdon, amounting to about £10 per annum, which sum is expended in coal for the inmates. The Rev. Henry Wharton, in 1694, bequeathed a rental of £30 to be applied in beautifying the church; and about £12 per annum, arising from bequests by Charles Themylthorpe, in 1721, and others, is distributed among the poor in bread and money.

WORSTHORN, a township, in the parish of Whalley, union of Burnley, Higher Division of the hundred of Blackburn, Northern Division of the county palatine of Lancaster, $2\frac{1}{4}$ miles (E.) from Burnley; containing 798 inhabitants. A chapel was built here in 1833, containing 650 sittings, 450 of which are free, the Incorporated Society having granted £350 in aid of the expense. A day and Sunday National school is partly supported by subscription; it commenced for daily instruction in 1831.

WORSTON, a township, in the parish of Whalley, union of Clitheroe, Higher Division of the hundred of Blackburn, Northern Division of the county palatine of Lancaster, $2\frac{1}{4}$ miles (E. N. E.) from Clitheroe; containing 129 inhabitants.

WORSTON, a township, in that part of the parish of St. Mary and St. Chad, Stafford, which is in the Southern Division of the hundred of Pirehill, Northern Division of the county of Stafford; containing 25 inhabitants. Here are a large corn and a silk mill.

WORTH, a township, in the parish of Prestbury, union and hundred of Macclesfield, Northern Division of the county of Chester, 6 miles (S. S. E.) from Stockport; containing 490 inhabitants, who are chiefly employed in the neighbouring collieries.

WORTH, or WORD (*St. Peter and St. Paul*), a parish, in the union and hundred of Eastry, lathe of St. Augustine, Eastern Division of the county of Kent, $1\frac{1}{2}$ mile (S.) from Sandwich; containing 411 inhabitants. The living is annexed to the vicarage of Eastry.

WORTH, a parish, in the union of East Grinstead, hundred of Buttinghill, rape of Lewes, Eastern Division of the county of Sussex, 2 miles (E. S. E.) from Crawley; containing 1859 inhabitants. The living is a rectory, valued in the king's books at £13. 13. 4.; patron and impropriator, Rev. J. M. Bethune, LL.D. The tithes have been commuted for a rent-charge of £1100, subject to the payment of rates. There are places of worship for dissenters. Sixteen children are instructed for a rent-charge of £8, the gift of Timothy Shelley, in 1767. The line of the London and Brighton railway passes through the parish.

WORTH-MATRAVERS (*St. Nicholas*), a parish, in the union of Wareham and Purbeck, hundred of Rowbarrow, Wareham Division of the county of Dorset, $3\frac{1}{2}$ miles (S. S. E.) from Corfe-Castle; containing 356 inhabitants. The living is a discharged vicarage, valued in the king's books at £8. 8. 4.; present net income, £133; patron and impropriator, Rev. T. O. Bartlett. The church is a very ancient structure. A school is supported by John Calcraft, Esq.; and there is a Sunday National school. This parish has the English Channel on the south, where is the noted cliff called St. Alban's Head, with a signal-house on its summit; also the remains of a very ancient chapel, dedicated to St. Aldhelms, built and vaulted with stone, and supported by a single massive pillar with four arches, meeting in a point at the crown: it is entered through a semicircular doorway in the north side, but has no window, only a hole on the south side. Near Quarr, which anciently belonged to the Cullifords, marble was once quarried.

WORTHAM (*St. Mary*), a parish, in the union and hundred of Hartismere, Western Division of the county of Suffolk, $5\frac{1}{2}$ miles (N. W.) from Eye; containing 1016 inhabitants. This parish is bounded on the north by the river Waveney, which separates the counties of Norfolk and Suffolk. The living is a rectory, formerly in medieties, respectively called Everard and Jervis, now consolidated; the former is valued in the king's books at £13. 2. $8\frac{1}{2}$., and the latter at £13. 1. $0\frac{1}{2}$.; present net income, £521; patron and incumbent, Rev. R. Cobbold. The church had a spacious round tower, now in ruins.

WORTHEN (*All Saints*), a parish, partly in the hundred of Cawrse, county of Montgomery, North Wales, but chiefly in the hundred of Chirbury, Southern Division of the county of Salop; containing 2668 inhabitants, of which number, 2290 are in that part of

the parish which is in the county of Salop, and which is divided into the quarters of Bing-Weston, Bromblow, Upper Heath, and Worthen, 9 miles (N. E.) from Montgomery. The living is a rectory, with the chapelry of Wolston annexed, valued in the king's books at £28. 14. 7.; present net income, £1279; patrons, Warden and Fellows of New College, Oxford. This place had formerly a market on Wednesday, and two annual fairs, granted by Henry III. In this and the neighbouring parishes is a very singular ridge of stones, termed Stiperstones, extending several miles towards Shrewsbury, and said to be the ancient boundary between England and Wales; there are still the remains of two old castles upon the same line. In the neighbourhood are considerable lead mines, some of which were worked by the Romans, in the time of Adrian; and the manufacture of flannel is carried on to a limited extent. A school is supported by the rector and the Rev. Sir Edward Kynaston, Bart. A fund of £12. 7. 6. per annum, arising from the bequests of John Powel, Robert Nicholess, and Martha Scarlet, is distributed among the poor on St. Thomas' day.

WORTHEN, a quarter, in the parish of WORTHEN, hundred of CHIRBURY, Southern Division of the county of SALOP: the population is returned with the parish.

WORTHING (*ST. MARGARET*), a parish, in the union of MITFORD and LAUNDITCH, hundred of LAUNDITCH, Western Division of the county of NORFOLK, 4 miles (N. by E.) from East Dereham; containing 138 inhabitants. The living is a rectory, annexed to that of Swanton-Morley. The church is an ancient structure, with a round tower.

WORTHING, a market-town and bathing-place, in the parish of BROADWATER, hundred of BRIGHTFORD, rape of BRAMBER, Western Division of the county of SUSSEX, 20 miles (E. by S.) from Chichester, and 56 (S. by W.) from London: the population is returned with the parish. This fashionable and attractive watering-place, a few years before the French Revolution, was only a small village. It is indebted in a great degree for the foundation of its celebrity to the late Princess Amelia, who was advised by her physicians to reside here during the summer of 1797; and was subsequently honoured by the visits of the Princess Charlotte, the present King of Hanover, the Duke of Gloucester, Princess Augusta and Princess Sophia. Its situation forms a strong recommendation to visitors, more especially to invalids, as the South Down hills, which approach to within two miles of the town, completely shelter it from the north and east winds, and protect it from the cold to which other bathing-places on this coast are in the winter subject. The town is lighted with gas, paved, and abundantly supplied with water by commissioners appointed for its improvement, who have erected an elegant town-hall, in which the magistrates for the division hold petty sessions every alternate week. It contains some good streets, handsome terraces, crescents, and detached villas; and in front of the esplanade are several handsome houses and hotels, baths, and every requisite to render it not only a place of fashionable resort during the summer, but also a favourite winter residence. The esplanade is nearly ¾ of a mile in length, and is 20 feet wide, forming a neat gravelled terrace, the waves flowing up close to its base; and running in a line parallel with the carriage road; the views from it are most extensive, commanding the English Channel, the Isle of Wight, Brighton, and the whole range of coast as far as Beachy Head. The sands, which are level, and extend for several miles, afford excellent carriage drives, and facilities of equestrian exercise or promenades. The royal baths, erected in 1823, form a costly and elegant building, comprising India, medicated, vapour, champooing, hot, cold, shower, and Douce baths, with every requisite, including reading-rooms. The Parisian baths, a similar establishment, are also elegantly fitted up. The theatre, a small but neat building, is opened in the season; the libraries and reading-rooms are well supplied, and every description of attraction and accommodation has been provided for the many respectable families who resort hither. A literary society and a mechanics' institution have been lately established. The principal market is on Saturday, and there is a corn market on alternate Wednesdays, and one for vegetables daily: the market-place is a neat quadrangular erection. A fishery, for mackarel in the spring and herrings in the autumn, has been established here, and great quantities of the former are sent to the London market; and soles, cod, shrimps, and prawns, are caught in abundance. The town has lately been made a polling-place for the western division of the county. The chapel is a handsome building: it was erected in 1812, at an expense of £14,000, raised by the inhabitants, and is a neat edifice of pale brick, with a portico of mixed Doric and Tuscan character, and a bold though low turret; the interior is neat and unadorned, and there is a good organ: it contains 1100 sittings, 150 of which are free. The living is a perpetual curacy; net income, £150; patron, Rector of Broadwater. It is in contemplation to erect a new church here. There are places of worship for Independents and Wesleyan Methodists. National schools are partly supported by subscription; and there is an infants' school. Miss Hawes, in 1828, bequeathed £1000 four per cents., one-fourth part of which she appropriated to the schools, and the remainder to be distributed in clothes and food among the poor. A savings bank was established in 1817; and a dispensary, and several other institutions, have been formed for the benefit of the poor, which are supported by voluntary contributions.

WORTHINGTON, a township, in the parish of STANDISH, union of WIGAN, hundred of LEYLAND, Northern Division of the county palatine of LANCASTER, 3½ miles (N. by W.) from Wigan; containing 124 inhabitants.

WORTHINGTON, a chapelry, in the parish of BREEDON, hundred of WEST GOSCOTE, Northern Division of the county of LEICESTER, 4¼ miles (N. E.) from Ashby-de-la-Zouch; containing 1211 inhabitants. The living is a perpetual curacy; net income, £100; patron, Lord Scarsdale. The chapel is dedicated to St. Matthew.

WORTHY, ABBOT'S, a tything, in the parish of KING'S-WORTHY, hundred of MITCHELDEVER, Winchester and Northern Divisions of the county of SOUTHAMPTON, 2 miles (N. N. E.) from Winchester: the population is returned with the parish.

WORTHY, HEADBOURN (*ST. MARTIN*), a parish, in the union of NEW WINCHESTER, hundred of BARTON-STACEY, Winchester and Northern Divisions of the county of SOUTHAMPTON, 2 miles (N. by E.) from Winchester; containing 190 inhabitants. The living is a rectory, valued in the king's books at £15. 12. 1.; patrons, Trustees of Dr. Radcliffe, for a member of University College, Oxford. The tithes have been commuted for a rent-charge of £385, subject to the payment of rates, which on the average have amounted to £35; the glebe comprises $44\frac{1}{2}$ acres, valued at £35 per annum. Here is a National school. The London and South-Western railway passes through the parish. Joseph Bingham, the ecclesiastical historian, was rector of this parish, and was interred here in 1723.

WORTHY, KING'S (*St. Mary*), a parish, in the union of New Winchester, partly in the hundred of Mitcheldever, but chiefly in the hundred of Barton-Stacey, Winchester and Northern Divisions of the county of Southampton, $2\frac{1}{4}$ miles (N. N. E.) from Winchester; containing 345 inhabitants. The living is a rectory, valued in the king's books at £22. 12. 6.; patron, Sir Thomas Baring, Bart. The tithes have been commuted for a rent-charge of £450, subject to the payment of rates, which on the average have amounted to £74; the glebe comprises 10 acres, valued at £22. 10. per annum. The church has been enlarged, and 150 free sittings provided, the Incorporated Society having granted £85 in aid of the expense. The London and South-Western railway passes through the parish. A National school has been established.

WORTHY, MARTYR (*St. Swithin*), a parish, in the union of New Winchester, partly in the hundred of Bountisborough, but chiefly in that of Fawley, Winchester and Northern Divisions of the county of Southampton, 3 miles (N. E. by N.) from Winchester; containing 219 inhabitants. The living is a rectory, valued in the king's books at £15. 10. $2\frac{1}{2}$.; present net income, £343; patron, Bishop of Winchester. Twelve boys and ten girls are instructed in a National school, for a rent-charge of £6. 13. 4., the bequest of Agnes Parnell, in 1589.

WORTING (*St. Thomas à Becket*), a parish, in the union of Basingstoke, hundred of Chutely, Basingstoke and Northern Divisions of the county of Southampton, $2\frac{1}{4}$ miles (W.) from Basingstoke; containing 120 inhabitants. The living is a rectory, valued in the king's books at £8. 17. $8\frac{1}{2}$.; patron, Rev. Lovelace Bigg Wither. The tithes have been commuted for a rent-charge of £277, subject to the payment of rates; the glebe comprises $5\frac{1}{2}$ acres, valued at £9 per annum. A parochial school for 40 children is supported by subscription. Dr. Pelham Warren, an eminent physician, lived and was interred here. The London and South-Western railway passes through the parish.

WORTLEY, a tything, in the parish of Wotton-under-Edge, Upper Division of the hundred of Berkeley, Western Division of the county of Gloucester: the population is returned with the parish.

WORTLEY, a chapelry, in the parish of St. Peter, town and liberty of Leeds, West Riding of the county of York, $2\frac{1}{2}$ miles (W. S. W.) from Leeds; containing 5944 inhabitants. The manufacture of woollen cloth is extensively carried on here. In the neighbourhood coarse earthenware and tobacco-pipes are made from clay obtained upon the spot. The living is a perpetual curacy; net income, £147; patrons, Trustees; impropriators, Master and Fellows of Trinity College, Cambridge. The chapel is a neat modern structure. There are places of worship for Independents and Wesleyan and Primitive Methodists; also a school endowed for the instruction of ten children, and a Sunday National school.

WORTLEY, a chapelry, and the head of a union, in the parish of Tankersley, wapentake of Staincross, West Riding of the county of York, $5\frac{1}{2}$ miles (S. S. W.) from Barnesley; containing 918 inhabitants. The living is a perpetual curacy; net income, £105; patron, Lord Wharncliffe; impropriator, Rector of Tankersley. Here is a school, in which 20 children are paid for by Lady Wharncliffe. A farm, known as the poor's estate, is supposed to have been given by a Countess of Devonshire, in the reign of Charles I.: the rental, amounting to £29, is distributed among the poor. The poor law union comprises 12 chapelries and townships, containing a population of 23,713, according to the census of 1831, and is under the care of 20 guardians. This place, which is delightfully situated on the river Don, and embosomed in fine woods, is celebrated in the ancient poem of "The Dragon of Wantley."

WORTON, a hamlet, in the parish of Cassington, hundred of Wootton, county of Oxford; containing 75 inhabitants.

WORTON, a tything, in the parish of Potterne, union of Devizes, hundred of Potterne and Cannings, Devizes and Northern Divisions of the county of Wilts, $3\frac{1}{2}$ miles (S. W.) from Devizes; containing 302 inhabitants. There is a place of worship for Wesleyan Methodists.

WORTON, NETHER (*St. James*), a parish, in the union of Woodstock, hundred of Wootton, county of Oxford, $3\frac{3}{4}$ miles (W. S. W.) from Deddington; containing 94 inhabitants. This parish is watered by the river Swere, which adds much to the interesting scenery of the neighbourhood, and on its banks are some rich dairy farms. The living is a perpetual curacy; net income, £40; patron, Joseph Wilson, Esq. The east and north sides of the church were rebuilt by the patron at his sole cost; he also supports a school for children of this and the adjoining parish.

WORTON, OVER (*Holy Trinity*), a parish, in the union of Woodstock, hundred of Wootton, county of Oxford, 4 miles (S. W. by W.) from Deddington; containing 56 inhabitants. This parish abounds with building stone of good quality; the hall, the rectory-house, the grange, and all the houses in the parish, have been rebuilt by the proprietor within the last 20 years. The living is a rectory, valued in the king's books at £6. 2. $8\frac{1}{2}$.; present net income, £208; patron, Rev. William Wilson. The church is an ancient structure, romantically situated on a rocky eminence, richly wooded, and being almost covered with ivy, has a very picturesque appearance. In front of the rectory-house is the pedestal of an ancient cross.

WORTWELL, a hamlet, in the parish of Reddenhall, union of Depwade, hundred of Earsham, Eastern Division of the county of Norfolk, $2\frac{3}{4}$ miles (N. E. by E.) from Harleston; containing 537 inhabitants.

WOTHERSOME, a township, in the parish of Bardsey, Lower Division of the wapentake of Skyrack, West Riding of the county of York, $3\frac{1}{2}$ miles (S.) from Wetherby; containing 21 inhabitants.

WOTTON-ABBAS, a liberty, in the parish and hundred of Whitchurch-Canonicorum, Bridport Division of the county of Dorset, $4\frac{3}{4}$ miles (N. E. by N.)

from Lyme-Regis: the population is returned with the parish. The liberty is of great extent, stretching from the river Char to the Ax, which bounds the counties of Devon and Dorset. Courts leet and baron are held here; and an annual fair, granted in the 7th of Queen Anne, on the Wednesday before the festival of St. John the Baptist, is kept on a lofty hill, called Lambert's Castle, the summit of which, in the form of the letter D, is fortified with triple trenches and ramparts, enclosing twelve acres, and having several entrances.

WOTTON (*St. John the Evangelist*), a parish, in the union of Dorking, First Division of the hundred of Wotton, Western Division of the county of Surrey, 3 miles (W. S. W.) from Dorking; containing 651 inhabitants. This parish, which gives name to the hundred, is about nine miles in length and scarcely more than one mile in average breadth. The soil is various, and the lands are watered by two streams rising in the northern declivity of Leeth hill, which uniting, fall into the river Wey near Shalford. Another stream rises under this hill and falls into the river Arun. There are considerable woods of oak, ash, beech, hazel and birch. On the summit of Leeth hill, which is the highest in the county, Richard Hall, Esq., in 1766, erected a tower, which is 999 feet above the level of the sea, commanding an extreme view of the Woulds of Surrey and Sussex, with the English Channel in the distance, and northward a fine view of Reigate and the valley of the Thames, with the hills of Harrow, Hampstead, and Highgate. Mr. Hall was, at his own request, interred underneath the tower. The living is a rectory, valued in the king's books at £12. 18. 9.; patron, William J. Evelyn, Esq. The tithes have been commuted for a rent-charge of £548, subject to the payment of rates. The church, a handsome structure with a tower at the west end, contains several monuments to the Evelyn family. There is an ancient chapel, dedicated to St. John the Baptist, at Oakwood: it is endowed, and in the patronage of Mr. Evelyn. The sum of £30 per annum was bequeathed in 1817, by William Glanville, Esq., of which £2 each is given to 5 poor boys, who, on the anniversary of his death, have to repeat the Lord's prayer, the creed, and ten commandments, with other portions of Scripture, at his tomb in the churchyard; and the remaining £20, in apprentice premiums of £10 to two boys. Eight girls are instructed at the expense of Mrs. Evelyn; and another school is partly supported by a bequest of £7. 10. per annum by the family of Haynes. Lady Evelyn left £500 to this parish and that of Abinger; and the parish also participates in the charity of Henry Smith. John Evelyn, a great benefactor to the Royal Society, of which he was a member, and the author of Sylva, and several other works, was born and buried here; and Sir Samuel Romilly resided many years in the parish.

WOTTON-UNDER-EDGE (*St. Mary*), a market-town and parish, in the union of Dursley, Upper Division of the hundred of Berkeley, Western Division of the county of Gloucester, 19 miles (S. S. W.) from Gloucester, and 108 (W. by N.) from London; containing 5482 inhabitants. The name of this place, formerly *Wotton under Ridge*, is descriptive of its situation beneath the western ridge of the Cotswold hills. The old town, which stood in the rear of the present town, was destroyed by fire in the reign of John; the site is still called the Old Town, and a spot there called "the Brands" is commemorative of the fire. On the restoration of the town, a market and fair, with various municipal privileges, were granted by Henry III. to Maurice Lord Berkeley, in 1254, which laid the foundation of its subsequent importance. During the civil war of the 17th century, a garrison was maintained here in the interest of the king. The present town is situated on a gentle eminence, and consists of five streets, besides the site of the old town; the houses are in general well built and of neat appearance. It has long been celebrated for the manufacture of fine broad cloth, which affords employment to the inhabitants of the town and vicinity; on a small stream, which flows to the west of the town, are several water-mills connected with the manufacture. The market is on Friday; and there is a fair annually on Sept. 25th, for cattle and cheese: a fair for cattle on the Tuesday preceding March 25th has also been recently established. A mayor is chosen annually in Oct., at the manorial court leet, but he has no magisterial authority; at the termination of his mayoralty he becomes an alderman. Petty sessions for the division are held here once a fortnight, on Friday; and the town has been made a polling-place for the western division of the shire.

The living is a vicarage, valued in the king's books at £13. 10.; present net income, £112; patrons and appropriators, Dean and Canons of Christ-church, Oxford. The church, which has recently undergone considerable repairs, is a spacious and handsome structure, having a tower with battlements and pinnacles, and containing some curious sepulchral memorials: it includes 750 free sittings, the Incorporated Society having granted £250 in aid of the expense. There are places of worship for Baptists, Independents, and Wesleyan Methodists. The free grammar school was founded and endowed by Lady Catherine Berkeley, under letters patent from Richard II., in 1385; and being supposed to have become forfeited in the reign of Edward I., by the act for the dissolution of chantries, James I., in 1622, on the petition of the inhabitants, confirmed and established it, under the title of "the Free Grammar School of Lord Berkeley," to consist of one master and five or more poor scholars, who should be a body corporate, the election of the master and scholars being vested in his lordship and his heirs, or, in default of issue, in the lord of the manor. The total annual income is £376. 12. 6. There are ten boys on the foundation, who are allowed £6 per annum for books and other purposes, and have the privilege of an exhibition at the University, with an allowance of £60 for that purpose. The Blue-coat school is endowed with £60 per annum from the funds of the general hospital trust, and with the produce of sundry bequests; the annual income is £136. 13.; 30 boys on the foundation are clothed, educated, and apprenticed, and about ten others receive gratuitous instruction: the school-house was erected about 1714, partly from the funds of Hugh Perry's estate, and partly by subscription. Here is also a National day and Sunday school. An hospital for twelve persons of both sexes, founded in 1630, by Hugh Perry, Esq., alderman of London; another for six aged persons of both sexes, founded by Thomas Dawes, in 1712; and the general hospital, are situated in Church-lane, and form three sides of a square, with an open court in the middle, and a chapel at the north end. Hugh Perry, Esq., also procured a supply of water for the town at his own expense. Sir Jonathan Dawes, sheriff of London, gave £1000 for the relief of the poor. On Westridge, in this parish, are the remains of a square camp, with double intrenchments, partly covered with wood, which is called Becketsbury.

WOUGHTON-ON-THE-GREEN (*St. Mary*), a parish, in the union of Newport-Pagnell, hundred of Newport, county of Buckingham, 2½ miles (N. by W.) from Fenny-Stratford; containing 303 inhabitants. This parish, which is intersected on the east by the turnpike road from Newport-Pagnell to Aylesbury, and on the west by the Grand Junction canal, derives its name from the situation of its village round a pleasant green of an oblong form, at the east end of which is the church. The living is a rectory, valued in the king's books at £16. 9. 7.; present net income, £107; patron and incumbent, Rev. Francis Rose, who is also lord of the manor. The church is in the later style of English architecture; and in the north wall of the chancel is a full length statue in memory of one of the Muxon family, who are said to have been formerly owners of a great part of the parish. There is a bequest in houses and lands from an unknown benefactor, the rents of which, together with that of some land awarded under an enclosure act in 1769, amounting in the aggregate to about £20 per annum, are applied partly to the repairs of the church, and partly to the support of the poor.

WOULDHAM (*All Saints*), a parish, in the union of Malling, hundred of Larkfield, lathe of Aylesford, Western Division of the county of Kent, 3¼ miles (S. S. W.) from Rochester; containing 247 inhabitants. This parish comprises about 1530 acres of land, of which about 1350 are in good cultivation. The village is situated on the eastern bank of the river Medway, across which is an ancient dam supposed to have stood for nearly 1000 years. Star Castle, an ancient manor in this parish, is tithe-free; and another is partly so; both belonged to the abbey of West Malling. The living is a discharged rectory, valued in the king's books at £14. 6. 5½.; present net income, £198; patron, Bishop of Rochester. The church is an ancient edifice, in the early English style, and contains a Norman font. The parsonage-house has been enlarged and improved by the present incumbent.

WRABNESS (*All Saints*), a parish, in the union and hundred of Tendring, Northern Division of the county of Essex, 4¾ miles (E.) from Manningtree; containing 248 inhabitants. This parish is bounded on the north by the river Stour, which is here navigable. The living is a discharged rectory, valued in the king's books at £8; present net income, £317: it is in the patronage of the Crown. The church is a small ancient edifice, and had originally a tower of stone, which has been replaced by a belfry turret of wood. There is a place of worship for Wesleyan Methodists; also a National school.

WRAGBY (*All Saints*), a market-town and parish, in the union of Horncastle, Western Division of the wapentake of Wraggoe, parts of Lindsey, county of Lincoln, 10½ miles (E. N. E.) from Lincoln, and 139½ (N. by W.) from London; containing 601 inhabitants. This place, noticed by Leland as a village giving name to a small beck, or stream, which flowed by it, in its course from Panton to Bardney abbey, is of some antiquity, but is not distinguished by any event of historical importance. From an inconsiderable village it was raised to a market-town by George Duke of Buckingham, who, in 1671, obtained for it the grant of a market and three annual fairs, two of which are still held. The town is pleasantly situated on the high road from Lincoln to Horncastle, at the point where it meets the road to Louth, and consists of neatly-built houses. The environs comprise an extensive tract of fertile land, in the cultivation of which the inhabitants are principally employed. There is very little trade except what arises from its situation on a public road, and is requisite for the supply of the inhabitants. The market is on Thursday; and fairs are annually held on Holy Thursday and September 29th, for sheep and cattle. The town is under the jurisdiction of the county magistrates, and within that of a court of requests for the wapentake of Wraggoe, established by an act passed in the 19th of George III., for the recovery of debts under 40*s*. It has also lately been made a polling-place for the parts of Lindsey. The living is a vicarage, united in 1735 to the rectory of East Torrington, and valued in the king's books at £8. 4. 2.; impropriator, C. Turnor, Esq. The church is an ancient structure, principally in the later style of English architecture, and contains several sepulchral memorials. The first stone of a new church was laid on the 28th of April, 1837. There is a place of worship for Wesleyan Methodists. William Hansard, in 1632, bequeathed a rent-charge of £30 for teaching 20 poor boys of the parish. Sir Edmund Turnor, Knt., founded an almshouse, with a chapel, for six clergymen's widows, and for six aged widowers, or widows, of Wragby, which, in 1707, he endowed with a rent-charge of £100, out of which is paid a stipend to the vicar for officiating in the chapel.

WRAGBY (*St. Michael*), a parish, chiefly in the Upper Division of the wapentake of Osgoldcross, but partly in the wapentake of Staincross, West Riding of the county of York, 5 miles (S. W.) from Pontefract; containing 756 inhabitants. The living is a donative, in the patronage of Charles Winn, Esq. The church is principally in the later style of English architecture. An annuity of £6. 5., paid out of the revenue of the duchy of Lancaster, is applied in support of a school for the poor children of the parish.

WRAGHOLME, a hamlet, in the parish of Grainthorpe, union of Louth, Marsh Division of the hundred of Louth-Eske, parts of Lindsey, county of Lincoln, 9 miles (N. N. E.) from Louth: the population is returned with the parish.

WRAMPLINGHAM (*St. Peter and St. Paul*), a parish, in the hundred of Forehoe, Eastern Division of the county of Norfolk, 3 miles (N. by E.) from Wymondham; containing 247 inhabitants. The living is a discharged rectory, valued in the king's books at £5. 4. 9½.; patron, R. Marsham, Esq. The tithes have been commuted for a rent-charge of £260, subject to the payment of rates, which on the average have amounted to £34; the glebe comprises 34 acres, valued at £45. 18. per annum. The church has a round tower. Here is a National Sunday school.

WRANGLE (*St. Peter and St. Paul*), a parish, in the union of Boston, wapentake of Skirbeck, parts of Holland, county of Lincoln, 9 miles (N. N. E.) from Boston; containing 1030 inhabitants. The living is a vicarage, valued in the king's books at £9. 18. 6½.;

present net income, £868; patron and incumbent, Rev. Thomas Bailey Wright; impropriators, J. Linton and J. Roper, Esqrs. The church contains a curious monument to Sir John Reade, Knt. The Rev. Thomas Alenson bequeathed land now producing £120 per annum, one moiety of which is applied for teaching poor children, and the other for the maintenance of five poor people.

WRANTAGE, a tything, in the parish and hundred of NORTH CURRY, Western Division of the county of SOMERSET, $5\frac{1}{2}$ miles (E. by S.) from Taunton: the population is returned with the parish.

WRATTING, GREAT (*St. Mary*), a parish, in the union and hundred of RISBRIDGE, Western Division of the county of SUFFOLK, $2\frac{3}{4}$ miles (N. E. by N.) from Haverhill; containing 344 inhabitants. The living is a rectory, with that of Little Wratting annexed, valued in the king's books at £8; present net income, £450; patron and incumbent, Rev. T. B. Syer. Salmon supposes this place to have been the Roman station *Ad Ansam*: numerous remains of Roman antiquity have been dug up in the parish.

WRATTING, LITTLE, a parish, in the union and hundred of RISBRIDGE, Western Division of the county of SUFFOLK, $5\frac{3}{4}$ miles (W. by N.) from Clare; containing 212 inhabitants. The living is a rectory, annexed to that of Great Wratting, and valued in the king's books at £4. 19. $9\frac{1}{2}$. The Turnor family had formerly a seat in this parish, at Blunt's Hall.

WRATTING, WEST (*St. Andrew*), a parish, in the union of LINTON, hundred of RADFIELD, county of CAMBRIDGE, $5\frac{1}{4}$ miles (N. E. by N.) from Linton; containing 763 inhabitants. This parish, so called from its position with respect to Great and Little Wratting, has facility of communication with Newmarket, Cambridge, Saffron-Walden, and London; the air is very salubrious. A pleasure fair is held on Whit Monday. The living is a vicarage, valued in the king's books at £7. 17. $3\frac{1}{2}$.; present net income, £215; patrons and appropriators, Dean and Chapter of Ely. The church and the vicarage-house were repaired and improved, at an expense of £767, by Sir John Jacob, who died in 1740. Parochial schools are supported by the Rev. J. H. Watson, vicar, and Sir Charles Watson, Bart.

WRATTON, or WRAYTON, a joint township with Melling, in the parish of MELLING, hundred of LONSDALE, south of the sands, Northern Division of the county palatine of LANCASTER, 5 miles (S.) from Kirkby-Lonsdale: the population is returned with Melling.

WRAWBY (*St. Mary*), a parish, in the union of GLANDFORD-BRIDGE, Southern Division of the wapentake of YARBOROUGH, parts of LINDSEY, county of LINCOLN, 2 miles (N. E. by E.) from Glandford-Bridge; containing 2418 inhabitants. The living is a discharged vicarage, valued in the king's books at £9. 14. 7.; present net income, £220; patrons and appropriators, Master and Fellows of Clare Hall, Cambridge. There is a chapel of ease at Glandford-Bridge, in this parish.

WRAXALL (*St. Mary*), a parish, in the union of BEAMINSTER, hundred of EGGERTON, Bridport Division of the county of DORSET, 8 miles (E. by S.) from Beaminster; containing 70 inhabitants. The living is a rectory, united in 1758 to that of Rampisham, and valued in the king's books at £5.

WRAXALL (*All Saints*), a parish, in the union of BEDMINSTER, hundred of PORTBURY, Eastern Division of the county of SOMERSET, $6\frac{1}{2}$ miles (W. by S.) from Bristol; containing 802 inhabitants. The living is a rectory, valued in the king's books at £49. 11. 8.; patron, Rev. James Vaughan. The tithes have been commuted for a rent-charge of £520, subject to the payment of rates; the glebe comprises 30 acres, valued at £37. 10. per annum. A fair is held here at Allhallow-tide, which continues six days. A school adjoining the churchyard was erected by Richard Vaughan, who endowed it with £300, the interest of which, together with about £6 a year bequeathed by Elizabeth Martindale, is applied for teaching poor children. Another school containing 20 girls is supported by Mrs. Kingston. On Leigh down, about a mile from Fayland's Inn, in this parish, is an irregular intrenchment, and near it another of a circular form, called the Old Fort. On the same down, upon opening a tumulus in 1815, several hundred Roman coins of the Lower Empire were discovered, with fragments of ancient urns; and many other indications of the residence of the Romans have been observed in the neighbourhood.

WRAXALL, NORTH (*St. James*), a parish, in the union and hundred of CHIPPENHAM, Chippenham and Calne, and Northern, Divisions of the county of WILTS, 7 miles (W. by N.) from Chippenham; containing 415 inhabitants. The living is a rectory, valued in the king's books at £15. 9. 2.; patron, W. Heneage, Esq. The tithes have been commuted for a rent-charge of £391. 8., including the glebe, which comprises 87 acres, valued at £174 per annum: the rates on the average have amounted to £41. Here is a National school.

WRAXALL, SOUTH, a chapelry, in the parish of GREAT BRADFORD, union and hundred of BRADFORD, Westbury and Northern Divisions, and Trowbridge and Bradford Subdivisions, of the county of WILTS, 5 miles (W. by N.) from Melksham; containing 389 inhabitants. The chapel, dedicated to St. James, contains 100 free sittings, the Incorporated Society having granted £200 in aid of the expense.

WRAY, a township, in the parish of MELLING, hundred of LONSDALE, south of the sands, Northern Division of the county palatine of LANCASTER, 10 miles (N. E. by E.) from Lancaster; containing 586 inhabitants, several of whom are employed in making nails. Her Majesty's Commissioners have agreed to afford facilities for obtaining a site for a church here. Richard Pooley, in 1685, bequeathed £20 for the erection of a school, and £200 to purchase land for its support; the annual income is about £35, which sum, with about £4 per annum, arising from a bequest by Mary Thompson, in 1803, is applied in aid of a National school.

WREA, a township, in the parish of KIRKHAM, union of the FYLDE, hundred of AMOUNDERNESS, Northern Division of the county palatine of LANCASTER, $1\frac{3}{4}$ mile (W. by S.) from Kirkham: the population is returned with the chapelry of Ribby.

WREAY, a chapelry, in the parish of ST. MARY, CARLISLE, union of CARLISLE, CUMBERLAND ward, Eastern Division of the county of CUMBERLAND, $5\frac{3}{4}$ miles (S. E. by S.) from Carlisle; containing 166 inhabitants. The living is a perpetual curacy; net income, £86; patrons and appropriators, Dean and Chapter of Carlisle. The chapel, dedicated to St. Mary, was consecrated in 1739. A school, erected by subscription in 1760, was endowed by John Brown, in 1763, with £200, which was laid out in land now producing £15 a year, for the education of poor children.

WRECKLESHAM, a joint township with Bourn, in the parish and hundred of FARNHAM, Western Division of the county of SURREY, 1¾ mile (S. W. by S.) from Farnham; containing 770 inhabitants. A chapel has been erected here, and 200 free sittings provided, the Incorporated Society having granted £300 in aid of the expense.

WREIGH-HILL, a township, in the parish and union of ROTHBURY, Western Division of COQUETDALE ward, Northern Division of the county of NORTHUMBERLAND, 5¼ miles (W.) from Rothbury; containing 27 inhabitants. It is bounded on the south by the river Coquet, and was anciently called Wreck Hill, probably from having been destroyed in 1412 by a band of Scottish freebooters, who killed most of the inhabitants. In 1665, almost the entire population was swept off by the plague, since which event great quantities of human bones have been discovered on the spot where the victims were interred. There are strata of limestone and freestone in this township. George Coughran, the celebrated youthful mathematician, was born here.

WRELTON, a township, in the parish of MIDDLETON, PICKERING lythe and union, North Riding of the county of YORK, 2¾ miles (W. N. W.) from Pickering; containing 172 inhabitants.

WRENBURY (*St. Margaret*), a parish, in the union and hundred of NANTWICH, Southern Division of the county of CHESTER; containing 903 inhabitants, of which number, 524 are in the township of Wrenbury with Frith, 4¾ miles (S. W. by W.) from Nantwich. The living is a perpetual curacy; net income, £132; patron, Vicar of Acton. The church has a fine carved oak ceiling and an elegant tower. A school is endowed with the interest of £230, for which 10 boys are instructed; and a day and Sunday school, in which 40 children are taught, is supported by Mrs. Starkey and Mrs. Hewitt. A branch of the Chester canal passes through the parish.

WRENINGHAM, GREAT and LITTLE (*All Saints*), a parish, in the union of HENSTEAD, hundred of HUMBLEYARD, Eastern Division of the county of NORFOLK, 4¼ miles (E. S. E.) from Wymondham; containing 409 inhabitants. The living is a rectory, annexed to that of Ashwellthorpe, and valued in the king's books at £10. The church is dedicated to *All Saints*: that of Little Wreningham, which was dedicated to *St. Mary*, has been long demolished.

WRENTHAM (*St. Nicholas*), a parish, in the union and hundred of BLYTHING, Eastern Division of the county of SUFFOLK, 4½ miles (N. by W.) from Southwold; containing 1022 inhabitants. The living is a rectory, valued in the king's books at £21. 6. 8.; present net income, £483; patron, Sir T. S. Gooch, Bart. The church has been enlarged, and 312 free sittings provided, the Incorporated Society having granted £200 in aid of the expense. There is a place of worship for Independents; also a National school. Wrentham Hall, a mansion in the Elizabethan style, for many ages the seat of the Brewster family, was taken down by the late Sir Thomas Gooch, Bart. William Wotton, a learned divine, was born here in 1666; he died in 1726; and William Ames, another learned divine, was rector of this parish.

WRENTHORP, a joint township with Stanley, in the parish and union of WAKEFIELD, Lower Division of the wapentake of AGBRIGG, West Riding of the county of YORK, 1½ mile (N. N. E.) from Wakefield: the population is returned with Stanley. The manufacture of woollen cloth, &c., is carried on here.

WRESSEL (*St. John of Beverley*), a parish, in the union of HOWDEN, Holme-Beacon Division of the wapentake of HARTHILL, East Riding of the county of YORK; containing 386 inhabitants, of which number, 183 are in the township of Wressel with Loftsome, 3¾ miles (N. W.) from Howden. The living is a discharged vicarage, valued in the king's books at £5. 13. 9.; present net income, £157; patron and impropriator, Colonel Wyndham. The church is a very ancient building. Here are some remains of Wressel castle, built by Thomas Percy, Earl of Worcester, who was made prisoner at the battle of Shrewsbury, and afterwards beheaded; this once princely mansion continued to be a seat of the Northumberland family till the civil war in the reign of Charles I., when it was demolished by order of the parliament. The Hull and Selby railway passes near this place.

WRESTLINGWORTH, (*St. Peter*), a parish, in the union and hundred of BIGGLESWADE, county of BEDFORD, 6 miles (E. N. E.) from Biggleswade; containing 448 inhabitants. The living is a rectory, valued in the king's books at £7. 6. 8.; present net income, £135: it is in the patronage of the Crown.

WRETHAM, EAST (*St. Ethelbert*), a parish, in the union of THETFORD, hundred of SHROPHAM, Western Division of the county of NORFOLK, 5¾ miles (N. E. by N.) from Thetford; containing 325 inhabitants. The living is a rectory, with that of West Wretham annexed, valued in the king's books at £11. 12. 3½.; present net income, £547; patron, W. Birch, Esq., who supports a Sunday school.

WRETHAM, WEST (*St. Lawrence*), a parish, in the union of THETFORD, hundred of SHROPHAM, Western Division of the county of NORFOLK, 5½ miles (N. N. E.) from Thetford: the population is returned with the parish of East Wretham. The living is a rectory, annexed to that of East Wretham, and valued in the king's books at £12. 11. 3.

WRETTON (*All Saints*), a parish, in the union of DOWNHAM, hundred of CLACKCLOSE, Western Division of the county of NORFOLK, 1 mile (W.) from Stoke-Ferry; containing 523 inhabitants. The living is a perpetual curacy, annexed to that of Wereham; impropriator, E. R. Pratt, Esq. There is a place of worship for Wesleyan Methodists; also a Sunday National school.

WRIBBENHALL, a hamlet, in the parish of KIDDERMINSTER, Lower Division of the hundred of HALFSHIRE, Kidderminster and Western Divisions of the county of WORCESTER, situated on the left bank of the Severn, immediately opposite Bewdley, and connected with that town by a noble bridge over the river: the population is returned with the parish. Here is a chapel for the service of the church of England, but not consecrated: it was erected in the year 1701, at the expense of the inhabitants of the hamlet, on a plot of waste land belonging to Lord Foley, and was subsequently claimed by his lordship, as lord of the manor: this gave rise to litigation, and after various decisions, it was given in his favour. Since that period his lordship has continued to appoint the minister, who held his situation solely by virtue of such presentation, until its existence was legalised by a clause in an act of parliament which passed in the early part of the reign of George IV., relating to dissenting places of worship, which excepts from its provisions all chapels wherein the service of the church of England had previously been performed: it is exempt from all ecclesiastical jurisdiction. Here is a National school.

WRIGHTINGTON, a township, in the parish of ECCLESTON, union of WIGAN, hundred of LEYLAND, Northern Division of the county palatine of LANCASTER, 4 miles (N. W.) from Wigan; containing 1601 inhabitants.

WRINEHILL, a joint township with Checkley, in the parish of WYBUNBURY, union and hundred of NANTWICH, Southern Division of the county of CHESTER, 7 miles (E. S. E.) from Nantwich: the population is returned with Checkley.

WRINGTON (*ALL SAINTS*), a parish (formerly a market-town), in the union of AXBRIDGE, and in a detached portion of the hundred of BRENT with WRINGTON, Eastern Division of the county of SOMERSET, 7 miles (N. N. E.) from Axbridge; containing 1540 inhabitants. This place, which is situated near the Mendip hills, is not distinguished either for trade or for any other branch of manufacture: the inhabitants are principally employed in agricultural pursuits, especially in the cultivation of teasel, of which great quantities are produced in the neighbourhood, for the supply of the clothiers in the adjoining districts, who use it in dressing the cloth. The town consists principally of two streets, intersecting each other obliquely, with other houses irregularly built, and in detached situations. The market, originally granted in the reign of Edward II., was held on Tuesday; and a fair was held annually on September 9th; but both have fallen into disuse. The county magistrates hold petty sessions here.

The living is a rectory, valued in the king's books at £39. 9. 4½.; patron, Duke of Cleveland. The tithes have been commuted for a rent-charge of £600, subject to the payment of rates; the glebe comprises 54 acres, valued at £60 per annum. The church, situated at the south-west extremity of the town, is a spacious and handsome structure, in the later style of English architecture, with a square embattled tower, surmounted by angular turrets crowned with pinnacles. There are places of worship for Independents and Wesleyan Methodists. Mr. George Legg, in 1704, bequeathed nine acres of land, now producing £20 per annum, for the instruction of twelve children of both sexes in reading the bible: and there was a school of industry, in which from 30 to 40 children were instructed, and which was partly supported by a dividend on £50 stock, the bequest of Mrs. Webb: these schools have been united to the National schools which have been established. The late Mrs. Hannah More resided for 25 years in a cottage, built by herself and her sisters, at Barley Wood, in this parish; and John Locke, the eminent philosopher, was born in an old thatched house on the north side of the churchyard, in 1632; after his decease, in 1704, an urn with an appropriate inscription was presented by Mrs. Montague to Mrs. Hannah More, and erected in the pleasure grounds at Barley Wood. Dr. John Rogers, an eminent divine, held the rectory of this parish.

WRITHLINGTON (*ST. MARY MAGDALENE*), a parish, in the union of FROME, hundred of KILMERSDON, Eastern Division of the county of SOMERSET, 7 miles (N. W. by N.) from Frome; containing 245 inhabitants. The living is a discharged rectory, valued in the king's books at £5. 7. 8½.; present net income, £166; patron, Prebendary of Writhlington in the Cathedral Church of Salisbury. In this parish are extensive coal mines and quarries of freestone: white lias and fullers' earth are also obtained here.

WRITTLE (*ALL SAINTS*), a parish (formely a market-town), in the union and hundred of CHELMSFORD, Southern Division of the county of ESSEX, 2½ miles (W. by S.) from Chelmsford; containing 2348 inhabitants. Morant and other writers have placed here the *Cæsaromagus* of Antoninus; and the remains of a royal palace, built by King John in 1211, which occupied an acre of ground surrounded by a deep moat, are still visible. This place has been long divested of the greater part of its trade by the rising importance of the neighbouring town of Chelmsford, but malting and brewing are still carried on, and there is an oil-mill in the vicinity. Courts leet and baron are held here, and the inhabitants have the privilege of appointing their own coroner. The parish, which is the most extensive in the county, is about 52 miles in circumference, and abounds with every variety of surface and scenery. The soil is generally fertile, and much of it adapted for wheat; hops also of good quality are grown in several parts. The living is a vicarage, with the donative of Roxwell annexed; net income, £718; patrons and impropriators, Warden and Fellows of New College, Oxford. The church is an ancient and spacious structure, with a massive square tower surmounted by a lantern turret, and contains numerous elegant and interesting monuments. There is a place of worship for Independents; also a National school. Almshouses for six poor people, which were rebuilt in 1820, were endowed with land now producing £55 per annum by Thomas Hawkins, in 1607; and John Blencowe, in 1774, founded a free school for the education of the poor children of Writtle and Roxwell. The income at present is £82 per annum, of which, two-thirds are given to the parish of Writtle, and the remainder to that of Roxwell, and paid to the master and mistress of the National school here, for teaching 33 boys and 22 girls. About four miles north-east of the church, in the middle of a wood, called Highwood Quarter, a hermitage was founded, in the reign of Stephen, which, in that of Henry II., was attached to St. John's abbey, Colchester.

WROCKWARDINE (*ST. PETER*), in the union of WELLINGTON, Wellington Division of the hundred of SOUTH BRADFORD, Northern Division of the county of SALOP, 2 miles (W. by N.) from Wellington; containing 2528 inhabitants. It has been lately divided into two separate parishes. The living is a discharged vicarage, valued in the king's books at £7. 8. 6., and in the patronage of the Crown; impropriators, Duke of Cleveland and others. The impropriate tithes have been commuted for a rent-charge of £208. 6. 8., and the vicarial for £310. 19. 4., subject to the payment of rates; the glebe comprises one acre, valued at £3. 10. per annum. The impropriate tithes of Charlton have been commuted for £133, and there is a rent-charge of £32. 14., payable to the vicar. The church is a venerable edifice of red stone, substantially built and in good repair. A church in the Grecian style of architecture, with a tower, was erected at Wrockwardine Wood in 1833, at an expense of £1550, under an act of the 58th of George III: it contains 430 free sittings, the Incorporated Society having granted £300 in aid of the expense. The living is vicarial; net income, £81; patron, Vicar of Wrockwardine, with reversion to the Crown; impropriator, E. Cludde, Esq. The Shrewsbury canal passes through the parish, which is bounded on the north by the river Tern. In the neighbourhood are extensive mines of coal and iron-stone, also a mineral spring of considerable celebrity, called Admaston Spa: there is also a manufacture of glass. Here are schools in connection with the National Society. About £16 per annum is distributed among the poor at Christmas and on Good Friday.

WROOT (*St. Pancras*), a parish, in the union of Thorne, Western Division of the wapentake of Manley, parts of Lindsey, county of Lincoln, 8 miles (N. E. by N.) from Bawtry; containing 289 inhabitants. The living is a rectory, valued in the king's books at £3. 7. 8½.; present net income, £260: it is in the patronage of the Crown. There is a place of worship for Wesleyan Methodists; and two schools are supported by endowment.

WROTHAM (*St. George*), a parish (formerly a market-town), in the union of Malling, hundred of Wrotham, lathe of Aylesford, Western Division of the county of Kent, 11 miles (W. N. W.) from Maidstone, and 24 (S. E. by E.) from London; containing 2601 inhabitants. This is a place of very remote antiquity: that it was a town of the ancient Britons is probable from various discoveries of British coins, and fragments of brass armour and military weapons; other circumstances lead to the conclusion that it was afterwards a Roman station, and the ancient military way from Oldborough to Stane-street passed through it. Woodland, or Week, now only a hamlet in this parish, was formerly a parish of itself. The town is situated near the foot of the chalk hills, and consists principally of two streets crossing each other on the high road from London to Maidstone; in the centre is the market-place, where was formerly a public well, now filled up. Some paper is manufactured at Basted. The market has been discontinued for many years, but whenever there is a fifth Tuesday in the month, a cattle market is held; there is a fair on May 4th. A palace for the Archbishops of Canterbury formerly stood here, of which the terrace and a few offices alone remain. Wrotham-hill, immediately above the town, affords one of the finest prospects in England. The living comprises a sinecure rectory and a vicarage, in the patronage of the Archbishop of Canterbury: the former, with the vicarage of Stanstead annexed, is valued in the king's books at £50. 8. 1½.; and the latter, with Woodland annexed, at £22. 5. 10.; present net income, £2061. The church is an ancient and spacious structure, with a mixture of the various styles, from the Norman to the later English; it contains sixteen stalls.

WROTTESLEY, a hamlet, in the parish of Tettenhall, Southern Division of the hundred of Seisdon and of the county of Stafford; containing 246 inhabitants. Here are vestiges of an ancient city, from three to four miles in circuit, with streets running in different directions. Within its limits huge square stones have been dug up, with hinges, a dagger, and other relics, supposed to be Roman.

WROUGHTON (*St. John the Baptist and St. Helen*), a parish, in the union of Highworth and Swindon, and in a detached portion of the hundred of Elstub and Everley, Swindon and Northern Divisions of the county of Wilts, 3 miles (S. W. by S.) from Swindon; containing 1545 inhabitants. The living is a vicarage, valued in the king's books at £12; present net income, £160: the rectory (an impropriation) is valued at £31. 4. 4½.; patron of both, Bishop of Winchester; impropriator, Rev. R. Pretyman. There is a place of worship for Wesleyan Methodists. Thomas Benit, in 1743, gave land now producing more than £20 a year, for the endowment of a school, which is now in connection with the National Society.

WROXETER (*St. Andrew*), a parish, in the union of Atcham, Wellington Division of the hundred of South Bradford, Northern Division of the county of Salop, 5¾ miles (S. E. by E.) from Shrewsbury; containing 636 inhabitants. This place, which is noticed by Ninnius, in his catalogue of British cities, under the appellation of "Caer Vrauch," is supposed to have obtained that appellation from its situation near the Wrekin mountain; and most probably from the same source also it was called by the Saxons "Wrekinceastre," from which its modern name is obviously derived. By most antiquaries it is identified with the *Uriconium* of Antoninus, and the *Viriconium* of Ptolemy, an important Roman station on the north-east bank of the Severn, in the bed of which, at low water, may be traced some foundations of an ancient stone building, supposed to have been a bridge. The Roman Watling-street passed through the centre of the station, and crossed the river at Wroxeter ford, from which point it branched off towards Church-Stretton. The city was enclosed with walls three yards in thickness, and extending for three miles in circumference, surrounded by a rampart and fosse; it flourished for a considerable time as the metropolis of the Cornavii, but suffered greatly during the Saxon wars, and is said to have been destroyed by the Danes. The parish is bounded on the west by the Severn, and contains some coal. The living is a discharged vicarage, valued in the king's books at £11. 8.; present net income, £330; patron, Duke of Cleveland. A free grammar school, situated at Donnington, in this parish, was founded in 1627, by Thomas Alcock, who endowed it with 20 marks per annum for the gratuitous instruction of boys of this parish and of Uppington, which endowment was augmented, in 1652, with a bequest of the same amount by Richard Stevinton: it is entitled to two exhibitions to Christ-Church College, Oxford, founded by Mr. Careswell, who also founded others in that college for scholars of Bridgenorth, Newport, Shiffnall, Shrewsbury, and Wem. The sums allowed to the exhibitioners are £60 to each under graduate, and £70 to each under-graduate being a commoner; £21 to each bachelor of arts, if not resident, and £60 if resident; and £27 to each master of arts. The master is appointed by the Duke of Cleveland, who is lord of the manor. Two other schools are partly supported by subscription. There are some small benefactions to the poor. Of the ancient city of *Uriconium,* from the ruins of which has arisen the present town of Shrewsbury, there are still some portions remaining; and within the area have been found numerous coins and vestiges of Roman antiquity; the former, which are principally of the Lower Empire, and which the tenants are by their leases bound to deliver to the lord of the manor, are here called "Dynders," probably a corruption of *Denarii,* for the coining of which several moulds of fine compact clay have been found here: among the latter are part of a square sudatory, now destroyed, but of which a beautiful model in wood is preserved in the library of Shrewsbury school, in which also are three original sepulchral inscriptions dug up here; and another subsequently discovered has been placed against the wall of the vicarage-house; many urns and numerous skeletons have been found imbedded in red clay, and, in the hollow way towards the river Severn, several skulls have been discovered.

WROXHALL (*St. Leonard*), a parish, in the union of Warwick, Snitterfield Division of the hundred of Barlichway, Southern Division of the county of Warwick, 6 miles (N. W. by N.) from Warwick; containing 181 inhabitants. The living is a donative, in the patronage of the Representative of the late Christopher Roberts Wren, Esq. The church forms the north side of the quadrangular edifice called Wroxhall Abbey, founded by Hugh de Hatton, about the close of the reign of Henry I., for Benedictine nuns, whose revenue at the dissolution was valued at £78. 10. 1.: the mansion is occupied by the widow of the late C. R. Wren, Esq., fourth in descent from Sir Christopher Wren, who purchased the estate from the family of Burgoyne, about the year 1713.

WROXHAM (*St. Mary*), a parish, in the union of St. Faith, hundred of Taverham, Eastern Division of the county of Norfolk, $2\frac{1}{2}$ miles (S. E.) from Coltishall; containing 368 inhabitants. The living is a discharged vicarage, with that of Salhouse united, valued in the king's books at £7. 17. 1.; present net income, £335; patron, S. Trafford, Esq.: the impropriation belongs to Lord Suffield and Mrs. Burroughes.

WROXTON (*All Saints*), a parish, in the union of Banbury, hundred of Bloxham, county of Oxford, 3 miles (W. N. W.) from Banbury; containing 780 inhabitants. This place was principally distinguished for an extensive monastery founded in the reign of Henry III., for a prior and brethren of the Augustine order, and dedicated to the Blessed Virgin, by Michael Belet, who endowed it with the lordships of Wroxton and Balscott; at the dissolution its revenue was £78. 14. $8\frac{1}{2}$.; and it was granted to Sir Thomas Pope, who bestowed it on the Society of Trinity College, Oxford. On its suppression, part of the buildings was demolished, and of the remainder, some portions are incorporated with the venerable mansion subsequently erected on its site by William Pope, first Earl of Downe, in 1618, which still retains the name of Wroxton Abbey, and is now one of the seats of the noble family of Guilford. The present mansion is very extensive and beautifully situated: the entrance, through a porch in the west front, leads into a spacious hall, from which, under a projecting screen of elaborate design, is the entrance to the principal apartments, and from the centre of the ceiling is a fine pendant of elegant fan tracery. The dining-room has a beautifully enriched ceiling, and the walls are hung with some family portraits by eminent masters, of which, in various parts of the house, there is an extensive collection; the library, an elegant room, in the later English style, contains a rare and valuable collection of works, among which are numerous volumes on the politics of modern Europe, and the secret history of courts; and the chapel is embellished with a handsome window of ancient stained glass. The demesne is highly embellished, and the pleasure grounds adjoining the house are laid out with great taste. The living is a vicarage not in charge; net income, £137: the patronage and impropriation belonged to the Ladies North. The church, situated on elevated ground near the abbey, is a handsome structure, in the later style of English architecture; the oak roof is still preserved in its original character, and at the west end of the nave is an ancient stone font ornamented with sculpture and figures of six of the apostles: at the north-east angle of the chancel is a splendid altar-tomb, with the recumbent effigies of Sir William Pope, first Earl of Downe, and Lady Anne, his wife, richly habited in the costume of the seventeenth century; there are also tablets to Francis Lord Guilford, and Lady Elizabeth, his wife, to Francis Earl of Guilford, and his three wives, to Lord North (prime minister), and to the Lady of Lord Keeper Guilford. There are places of worship for Independents and Wesleyan Methodists; also a day and Sunday school.

WUERDALE, a joint township with Wardle, in the parish and union of Rochdale, hundred of Salford, Southern Division of the county palatine of Lancaster, 2 miles (N. E.) from Rochdale; containing, with Wardle, 6754 inhabitants. A church in the later English style, with a campanile turret, was erected in 1833, at an expense of £3071. 10. 7., under the act of the 58th of George III.

WYASTON, a township, in the parish of Edlaston, hundred of Appletree, Southern Division of the county of Derby, $3\frac{1}{4}$ miles (S. by E.) from Ashbourn: the population is returned with the parish.

WYBERTON (*St. Leodegar*), a parish, in the union of Boston, hundred of Kirton, parts of Holland, county of Lincoln, $2\frac{1}{4}$ miles (S.) from Boston; containing 530 inhabitants. The living is a rectory, valued in the king's books at £33. 6. 8.; present net income, £597; patron and incumbent, Rev. Martin Sheath. The parsonage-house commands a fine view of Boston church.

WYBUNBURY (*St. Chad*), a parish, in the union and hundred of Nantwich, Southern Division of the county of Chester; containing 4193 inhabitants, of which number, 445 are in the township of Wybunbury, $3\frac{1}{2}$ miles (E. S. E.) from Nantwich. The living is a vicarage, valued in the king's books at £13. 12. 1.; present net income, £295; patron and appropriator, Bishop of Lichfield. The church was rebuilt in 1595, and again in 1832; it contains 520 free sittings, the Incorporated Society having granted £300 in aid of the expense. Her Majesty's Commissioners have agreed to afford facilities for obtaining a site for a church at Doddington, in this parish. There is a place of worship for Wesleyan Methodists. A school, founded by the late Sir Thomas Delves, Bart., is conducted on the National system, and attended by 140 boys, of whom 20 receive annually a blue coat and cap each; the master has a salary of £90, and a house. The same individual endowed a school for ten girls, each of whom has a blue gown and bonnet annually; also four others in different parts of the parish, which afford instruction to about 70 girls. In the churchyard there is a school called the Wybunbury Charity, built by subscription about 80 years since, and endowed by several persons for the instruction of 20 boys. An hospital, dedicated to the Holy Cross and St. George, for a master and brethren, existed here before 1464.

WYCLIFFE, a parish, in the wapentake of GILLING-WEST, North Riding of the county of YORK, 2½ miles (E. N. E.) from Greta-Bridge; containing 156 inhabitants. The living is a rectory, valued in the king's books at £14. 12. 1.; present net income, £456; patron, Sir C. Constable, Bart. The church was rebuilt in the reign of Edward III. Here is a Roman Catholic chapel. This is said to be the birthplace of Wickliffe, the Reformer.

WYCOMBE, a hamlet, in the parish of ROTHLEY, union of MELTON-MOWBRAY, hundred of EAST GOSCOTE, Northern Division of the county of LEICESTER, 4½ miles (N. N. E.) from Melton-Mowbray: the population is returned with the chapelry of Chadwell.

Seal and Arms.

WYCOMBE, HIGH, or CHIPPING (*ALL SAINTS*), a borough, market-town, and parish, and the head of a union, in the hundred of DESBOROUGH, county of BUCKINGHAM, 31 miles (S. S. E.) from Buckingham, and 29 (W. by N.) from London; containing 6299 inhabitants, of which number, 3198 are in the borough. This place, which is evidently of great antiquity, is by some supposed to have been occupied by the Romans. In the vicinity a tesselated pavement, nine feet square, was discovered in 1774; and among the numerous Roman coins that have been found were some of Antoninus Pius, Marcus Aurelius, and other Roman emperors. Of its occupation by the Saxons, the prefix to its name, "Cheaping," signifying a market, is an evident proof; and in the immediate neighbourhood of the town are the remains of a strong double intrenchment, called Desborough Castle, which was probably thrown up by that people to check the progress of the Danes. The only historical event connected with the place is a successful attack on the parliamentary troops quartered here, by Prince Rupert, after the battle of Reading. The town is pleasantly situated on a fine rivulet, called the Wycombe stream, which, after winding through the adjoining meadows, flows into the Thames below Marlow: it consists of one principal street, on the high road from London to Oxford, Cheltenham, and Worcester, from which some smaller streets branch off in various directions. The houses are in general well built; many of them are spacious and handsome, and the town has a prepossessing appearance of cheerfulness and great respectability. On each side are some hills, richly wooded; from that on the south are seen the park, and part of Wycombe Abbey, the seat of Lord Carrington, with its rich plantations. The environs abound with pleasingly varied scenery, and the surrounding district is luxuriantly fertile, and in the highest state of cultivation. The manufacture of paper is carried on to a very considerable extent, for which there are more than 30 mills on the banks of the Wycombe stream, besides six flour-mills. The making of lace affords employment to more than 1000 of the inhabitants, and chairs are made in great numbers: there is a considerable trade in malt, and the town derives a great degree of traffic from its situation on a public thoroughfare. The market, which is extensively supplied with corn, is on Friday. Cattle fairs are held on the 2nd Wednesday in April and on the 28th of October, a wool fair on the last Wednesday in June, and a statute and pleasure fair on the Monday next before Michaelmas-day.

Wycombe, though governed by a mayor in the reign of Edward III., received its first regular charter of incorporation from Henry VI., which was confirmed and extended in the reigns of Elizabeth, James I., and Charles II., by which the government was vested in a mayor, recorder, two bailiffs, twelve aldermen, and an indefinite number of burgesses, assisted by a town-clerk and other officers. The corporation at present consists of a mayor, 4 aldermen, and 12 councillors, under the act of the 5th & 6th of William IV., cap. 76, for an abstract of which see the Appendix, No. I.: the mayor and late mayor are justices of the peace. The borough, which for municipal purposes comprises only 134 acres, first exercised the elective franchise in the 28th of Edward I., since which time it has continued to return two members to parliament. The right of election was formerly vested in the mayor, bailiffs, and burgesses not receiving alms; but the privilege has been extended to the £10 householders of the entire parish, which has been constituted the new elective borough, and comprises 6310 acres. The present number of registered voters is 339: the mayor is the returning officer. The police establishment consists of one day and two night constables. The corporation hold occasional sessions for all offences not capital; and under their charter they have power to hold a court of record, for the recovery of debts under £40, but this court has been disused for 150 years. The town-hall, erected in 1757, at the expense of the Earl of Shelburne, is a commodious and neat structure of brick, supported on stone pillars.

The living is a discharged vicarage, valued in the king's books at £23. 17. 1.; present net income, £140; patron, Marquess of Lansdowne; impropriator, William Terry, Esq. The church is an ancient and venerable structure, in the early style of English architecture, with a square embattled tower, which has been subsequently ornamented and crowned with pinnacles. The interior consists of a nave, aisles, and chancel, the last of which is separated from the nave by an ancient oak screen: it contains several ancient and interesting monuments, among which are, one to the memory of Henry Petty, Earl of Shelburne, who died in 1751; one to the lady of the late earl, and several to the families of Archdale, Llewellyn, and Bradshaw. There are two places of worship for Independents, and one each for Baptists, Wesleyan Methodists, and the Society of

Friends. An ancient hospital for lepers, dedicated to St. Margaret and St. Giles, and another dedicated to St. John the Baptist, for a master, brethren, and sisters, were founded here in the reign of Henry III.; the latter was granted by Elizabeth to the mayor and the corporation, and the endowment, which was augmented by a bequest of £1000 from Mrs. Mary Bowden in 1790, producing altogether an annual income of £290. 16. 1. per annum, is appropriated to the maintenance of a grammar school, and almshouses for four aged persons. The school premises are situated in High-street, and comprise under one roof a residence for the master and a school-room, with a garden and orchard attached: the master has the free use of the premises, and receives a salary of £70, of which £30 is from Mrs. Bowden's bequest, and is paid in consideration of the boys being taught writing and arithmetic in addition to the classics. Adjoining the school-room, on the western side, are the almshouses. A school on the British system has been erected, and is partly supported by subscription. A Sunday school, now on the National system, was established in this town by Miss Hannah Ball, in 1769, 14 years prior to their introduction by Mr. Raikes, of Gloucester, in 1783, to whom some attribute their origin, others to the Rev. Thomas Stock of the same place. The almshouses in Crendon-lane were founded in 1677, by John Lane, who endowed them with property now producing £23 per annum: they are occupied by two poor widows. There are several other almshouses, which, with various benefactions, amounting to a considerable sum annually, are under the control of trustees appointed by the Lord Chancellor, who appropriate these funds to general purposes of relief. Wycombe is also entitled to a share of Lord Warton's bible charity. Previously to the completion of the Royal Military College at Sandhurst, a department of that institution was established at Wycombe, which was, in 1802, removed to Great Marlow. The poor law union of Wycombe comprises 33 parishes or places, containing a population of 33,947, according to the census of 1831, and is under the care of 41 guardians. Dr. Gumble, who wrote the Life of General Monk, and is supposed to have assisted him in effecting the restoration of Charles II., was vicar of this parish. The learned William Alley, Bishop of Exeter, and one of the translators of the Bible, in the reign of Queen Elizabeth; and Charles Butler, author of a Treatise on Rhetoric, and other works, were natives of this town. Wycombe gives the titles of Earl and Baron to the Marquess of Lansdowne.

WYCOMBE, WEST (*St. Lawrence*), a parish, in the union of Wycombe, hundred of Desborough, county of Buckingham, 2½ miles (N. W. by W.) from High Wycombe; containing 1901 inhabitants, many of whom are employed in the making of lace and chairs. The living is a discharged vicarage, valued in the king's books at £11. 9. 7.; present net income, £250; patron and impropriator, Sir. J. Dashwood King, Bart. The church, which is surrounded by an ancient intrenchment, was erected in 1763, at the expense of Lord le Despenser; it is an elegant structure, in the Grecian style, with a profusion of Mosaic ornaments, and containing some handsome monuments; in the adjoining mausoleum is one of considerable beauty to the memory of Sarah, Baroness le Despenser, with many memorials of the Dashwood family and others; and in one of its recesses was deposited, in 1775, an urn enclosing the heart of Paul Whitehead, the poet, which he had bequeathed to Lord le Despenser. The church occupies an eminence finely clothed with woods, emerging from which the tower and the mausoleum form objects strikingly picturesque. There are places of worship for Independents and Wesleyan Methodists; also a Sunday National school. This parish has an interest in the charity of Katherine Pye, described in the article on Risborough (Prince's). In the neighbourhood is an ancient camp, doubly intrenched, called Desborough Castle, which gives name to the hundred; vestiges of buildings, together with window-frames of stone, similar to those of a church, have been discovered on its site. Under the hill on which the church stands is a cave, dug by Lord le Despenser at the time of building the church.

WYDDIALL (*St. Giles*), a parish, in the union of Buntingford, hundred of Edwinstree, county of Hertford, 1½ mile (N. E.) from Buntingford; containing 243 inhabitants. The living is a rectory, valued in the king's books at £16; present net income, £290; patron, C. Ellis Heaton, Esq. The church has an embattled tower at the west end, and contains several monuments: on the north side of the chancel is a chapel, in which are some remains of fine stained glass, representing the Crucifixion.

WYE (*St. Martin and St. Gregory*), a parish, (formerly a market-town), in the union of East Ashford, hundred of Wye, Upper Division of the lathe of Scray, Eastern Division of the county of Kent, 4 miles (N. E.) from Ashford, and 56 (E. S. E.) from London; containing 1639 inhabitants. This place is of great antiquity, and was formerly of considerable importance; it was the head of a royal manor having extensive jurisdiction, and formed part of the demesne lands of the Saxon kings prior to the Conquest, when, with all its appendages, liberties, and royal customs, it was granted to the abbey of Battle, in Sussex, and continued to form part of its possessions till the dissolution. Leland calls it "a pratie market townelet," but the market has been long discontinued. The town, which at present is little more than a considerable village, is pleasantly situated near the right or eastern bank of the river Stour, over which is a stone bridge of five arches, built in 1638. The houses, which are neatly built, are principally ranged round a green, and in two parallel and two cross streets. On the river, a little above the bridge, is a corn-mill. Fairs are held on May 29th and Oct. 11th. The living is a perpetual curacy; net income, £101; patron, Earl of Winchilsea, to whom, and the Archbishop of Canterbury, the impropriation belongs. The church was rebuilt and made collegiate by John Kemp, a native of this town, who was first preferred to the bishopric of Rochester, and having successively presided over several other sees, was lastly translated to the archbishopric of Canterbury and made Cardinal. In 1447, he here founded a college for a Master or Provost and Secular canons, the revenue of which, at the dissolution, was valued at £93. 2. The church was a beautiful cruciform structure, with a central tower surmounted by a spire, and had all the usual parts of a large collegiate church; in 1572, the spire was injured by lightning, and, having been restored, fell, in 1686, and destroyed a portion of the east end of the church, together with all the monuments in the chancels, among which was the tomb of the father and mother of the founder; the east end was partly rebuilt in 1701, but on a much smaller scale. There is a place of worship for Wesleyan Methodists. The free grammar school was founded by grant from Charles I. of the rectories of Boughton-Aluph, Beuset, and Newington, and other premises, to Robert Maxwell and his heirs, for affording classical instruction to the poor children of parishioners. This school having fallen into decay, was revived in 1832; it is endowed with £10 per annum, and a house, and the master is also paid £5. 5. annually by each of the pupils, of whom there are 20 attending the school. An exhibition, originally of £10 per annum, to Lincoln College, Oxford, was attached to this school by Sir George Wheeler, in 1723, which was augmented to £20, in 1759, by his son, the Rev. Granville Wheeler: in default of candidates from this establishment, it is open to any other grammar school in the kingdom. A free school for children of both sexes was founded and endowed in 1708, by Lady Joanna Thornhill; the present annual income is £193. 10. 6., for which about 30 boys and 35 girls are now instructed, the school being open to all children of the poor, who are nominated by the trustees, consisting of the ministers of Wye and the four adjoining parishes, and the heirs of three other persons: the salary of the master is £40, and that of the mistress is £25. In 1723, Sir George Wheeler devised the ancient collegiate buildings and lands for the respective residences and schools of the master of the grammar school and the master and mistress under Lady Thornhill's charity: these establishments, therefore, now occupy the college green, the former the south, and the latter the north, side. An almshouse for the residence of six poor persons was founded by Sir Thomas Kemp. Olanteigh, in this parish, was formerly the seat of the families of Kemp and Thornhill, and is supposed to have been the birthplace of Archbishop Kemp, and also of his nephew, Thomas Kemp, Bishop of London; it passed from the Kemps to the Thornhills, and from them to the family of Sawbridge, of whom Alderman Sawbridge, who was buried in the church, and his sister, Mrs. Catherine Macauley Graham, author of a History of England, were born there. Several years since, in making a sunk fence on the grounds of Olanteigh, two human skeletons were found on one side of a large tumulus, together with several small pieces of iron, two of which appeared to have been spear-heads. Withersdane, a hamlet in this parish, was anciently celebrated on account of a holy well, consecrated by St. Eustace. Dr. Plott, the celebrated antiquary and naturalist, received his early education at Wye College.

WYERSDALE, NETHER, a township, in the parish and union of GARSTANG, hundred of AMOUNDERNESS, Northern Division of the county palatine of LANCASTER, 4 miles (N. N. E.) from Garstang; containing 770 inhabitants. Here is a large establishment for spinning and manufacturing cotton. A school is endowed with £6 per annum, and another is partly supported by the proprietor of Scorton works.

WYERSDALE, OVER, a chapelry, in the parish of LANCASTER, hundred of LONSDALE, south of the sands, Northern Division of the county palatine of LANCASTER, 6 miles (N. N. E.) from Garstang; containing 872 inhabitants. The living is a perpetual curacy; net income, £135; patron, Vicar of Lancaster. William Cawthorne, in 1683, gave a school-house, with a messuage and land, for the use of a schoolmaster, also a rent-charge of £15, for which 30 boys are instructed; and another school has an allowance of £20 per annum from the Society of Friends. Some monks from the abbey of Furness settled here, but, in 1188, they removed to Ireland, and founded Wythney abbey.

WYFORDBY, or WYVERBY (*St. Mary*), a parish, in the union of MELTON-MOWBRAY, hundred of FRAMLAND, Northern Division of the county of LEICESTER, 3 miles (E.) from Melton-Mowbray; containing, with the chapelry of Brantingby (with which it is assessed to the poor, constables', and county rates), 98 inhabitants. The living is a rectory, valued in the king's books at £6; present net income, £137; patron, Sir E. C. Hartopp, Bart. A Sunday school is supported by Lady Hartopp. This parish is intersected by the river Eye and the Oakham canal.

WYHAM (*All Saints*), a parish, in the union of LOUTH, wapentake of LUDBOROUGH, parts of LINDSEY, county of LINCOLN, 7¼ miles (N. W. by N.) from Louth; containing 94 inhabitants. The living is a discharged rectory, valued in the king's books at £8; present net income, £195; patron, J. F. Heneage, Esq.

WYKE-CHAMPFLOWER, a chapelry, in the parish and hundred of BRUTON, Eastern Division of the county of SOMERSET, 1½ mile (W.) from Bruton; containg 93 inhabitants. The living is a perpetual curacy; net income, £30; patron and impropriator, Sir H. H. Hoare, Bart. The chapel is dedicated to St. Mary.

WYKE-HAMON, formerly a parish, now a hamlet, in the parish of WICKEN, hundred of CLELEY, Southern Division of the county of NORTHAMPTON: the population is returned with Wicken, with which the rectory has long since been consolidated.

WYKE-REGIS (*All Saints*), a parish, in the union of WEYMOUTH, and liberty of WYKE-REGIS and ELWALL, Dorchester Division of the county of DORSET, 1¼ mile (W. S. W.) from Weymouth; containing 1197 inhabitants. The living is a rectory, with Weymouth annexed, valued in the king's books at £19. 7. 1.; present net income, £623; patron, Bishop of Winchester. The church is a large and ancient pile, with a lofty embattled tower, and contains 170 free sittings: it is the mother church of Weymouth, and the usual burial-place of its inhabitants. At Smallmouth, in this parish, is a ferry to the Isle of Portland.

WYKEHAM (*All Saints*), a parish, in the union of SCARBOROUGH, PICKERING lythe, North Riding of the county of YORK, 6½ miles (S. W. by W.) from Scarborough; containing 605 inhabitants. The living is a perpetual curacy, in the patronage of the Hon. M. Langley. The church was repaired and beautified at the expense of the late Richard Langley, Esq. Three schools are partly supported by the Hon. M. Langley. A priory of Cistercian nuns, in honour of the Blessed Virgin Mary, was founded here about 1153, by Pain Fitz-Osbert, which at the dissolution had a revenue of £25. 17. 6.: there are still some remains of the church belonging to it.

WYKEHAM, EAST, a parish, in the Wold Division of the hundred of LOUTH-ESKE, parts of LINDSEY, county of LINCOLN, 7 miles (W. by N.) from Louth;

containing 31 inhabitants. The living is a discharged vicarage, in the patronage of — Ferrand, Esq. The church is in ruins.

WYKEHAM, WEST, a parish, in the Eastern Division of the wapentake of WRAGGOE, parts of LINDSEY, county of LINCOLN, 7½ miles (W. by N.) from Louth. The living is a vicarage, valued in the king's books at £3. 6. 8., and in the patronage of the Crown. The church is in ruins.

WYKEN, a parish, in the union of FOLESHILL, county of the city of COVENTRY, 3 miles (N. E. by E.) from Coventry; containing 104 inhabitants. The living is a perpetual curacy; net income, £115; patron and impropriator, Earl of Craven.

WYLAM, a township, in the parish of OVINGHAM, union of HEXHAM, Eastern Division of TINDALE ward, Southern Division of the county of NORTHUMBERLAND, 9 miles (W.) from Newcastle-upon-Tyne; containing 887 inhabitants. The whole township was an appurtenance to the monastery of Tynemouth, and was granted by the Crown to a branch of the Fenwick family, of Fenwick Tower, from whom it passed to the Blackett family in the reign of Charles II., and is now the property of Christopher Blackett, Esq. The river Tyne, over which here is a bridge, erected by subscription in 1835, separates it from the line of the Newcastle and Carlisle railway. On both sides of the river are extensive collieries, belonging to Mr. Blackett; and an iron-foundry has been established by the Messrs. Thompson, a rich vein of iron-stone running through the township, chiefly on the south side of the river. There are also quarries of excellent stone, applicable for building and other purposes. The village contains a place of worship for Wesleyan Methodists, and there is a National school. At the west end of it is Wylam Hall, an ancient building (formerly a *peel*, or strong house), containing an arch of stone, sixty feet in length, which was used as a place of safety for cattle: the hall is now held on lease from Mr. Blackett, and inhabited by Mr. Thompson, proprietor of the iron-works.

WYLDECOURT, a tything, in the parish of HAWKCHURCH, hundred of CERNE, TOTCOMBE, and MODBURY, Bridport Division of the county of DORSET; containing 316 inhabitants.

WYLY, county of WILTS.—See WILY.

WYMERING (*St. Peter and St. Paul*), a parish, in the union of FAREHAM, hundred of PORTSDOWN, Fareham and Southern Divisions of the county of SOUTHAMPTON, 4¼ miles (W.) from Havant; containing 578 inhabitants. The living is a vicarage, annexed to the rectory of Widley. The northern end of Portsea island, where Hilsea barracks stand, and across which are strong lines of defence, forms a part of this parish, and is connected with the main portion of it by Pos bridge, which crosses the narrow channel between Portsmouth and Langston harbours, and terminates in a *tête du pont*. Great and Little Horsea islands, at the upper end of the former harbour, are also in this parish.

WYMINGTON (*St. Lawrence*), a parish, in the union of WELLINGBOROUGH, hundred of WILLEY, county of BEDFORD, 6¼ miles (N.) from Harrold; containing 257 inhabitants. The living is a rectory, valued in the king's books at £10; present net income, £90; patron, Rev. R. Midgley. The church, a handsome structure, in the later style of English architecture, is said to have been built some time in the fourteenth century, by John Curteys, the lord of the manor, and mayor of the staple at Calais.

WYMONDHAM (*St. Peter*), a parish, in the union of MELTON-MOWBRAY, hundred of FRAMLAND, Northern Division of the county of LEICESTER, 6½ miles (E.) from Melton-Mowbray; containing 746 inhabitants. The living is a rectory, valued in the king's books at £12; present net income, £391; it is in the patronage of the Crown. This is a place of great antiquity, being still surrounded by the remains of its ancient walls: the Oakham canal runs through the parish. Sir John Sedley, in 1637, endowed with land a school for all the poor boys of the parish.

WYMONDHAM, or WINDHAM (*St. Mary the Virgin*), a parish, in the hundred of FOREHOE, Eastern Division of the county of NORFOLK, 9 miles (W. S. W.) from Norwich, and 100 (N. E. by N.) from London; comprising the market-town of Wymondham, which forms the in-soken, and the divisions of Downham, Market-street, Silfield, Suton, Towngreen, and Wattlefield, which form the out-soken; and containing 5485 inhabitants. This town derives its name from the Saxon words *Win Munde Ham*, which signify "a pleasant village on a mount," and is indebted for its importance to the foundation of a priory of Black monks, at first a cell to the abbey of St. Alban's, by William d'Albini, or Daubeny, in the reign of Henry I. This monarch endowed the monastery with lands and with the privilege of appropriating all wrecks between Eccles, Happisburgh, and Tunstead, and with an annual rent, in kind, of 2000 eels from the village of Helgay. About 1448, it was elevated to the rank of an abbey, and at the dissolution its revenue was valued at £72. 5. 4. The two Ketts, who disturbed this county in the reign of Edward VI., were accustomed to assemble their followers under an oak, of which part yet remains in the vicinity of the town; after their defeat by the Earl of Warwick, the elder was hanged in chains on the castle of Norwich, and the younger upon the lofty steeple of the church of Wymondham, of which town they were both natives. In 1615, 300 houses were destroyed by fire: and, in 1631, the plague raged with great fury among the inhabitants. The town, which is situated on the main road from Norwich to London, is of considerable extent, and is well supplied with water from springs: it has much improved of late years. Different branches of weaving are carried on in private houses, from 1000 to 1200 persons being employed by Messrs. Tipple and Son, the only firm in the town; the chief articles are bombazines and crapes. The market, granted by charter of King John in 1203, is held on Friday: the fairs are on Feb. 14th and May 17th, principally for cattle, horses, and pedlery; statute fairs for hiring servants are held occasionally: in the market-place is an ancient cross. A court leet is held annually for the appointment of constables, and manorial courts are held as occasion requires: the inhabitants enjoy the privilege of exemption from serving on juries at assizes and sessions. The house of correction contains three wards, with day-rooms and two airing-yards.

The living is a discharged vicarage, valued in the king's books at £10. 14. 4½.; present net income, £515; patron and appropriator, Bishop of Ely. The church, which comprises the eastern part of the abbey church, is a fine structure, in various styles of architecture, with the remains of the central tower, and another at the west end. Amid the ruins of the ancient conventual edifice, are some Norman arches with low massive columns, and fragments of old walls: the more modern parts of the building are, the north aisle, porch, and towers. In the interior is a large font, adorned with carved work, and elevated on steps; in the chancel is the tomb of the founder, who died in 1156, and several sepulchral memorials to the d'Albini family and other noble persons. There are places of worship for Baptists, the Society of Friends, Independents, and Wesleyan Methodists. A free grammar school was founded in the reign of Elizabeth, and endowed with part of the property of the guilds belonging to the collegiate church: a scholarship in Corpus Christi College, Cambridge, was attached to it, in 1574, by Archbishop Parker; another, in 1580, by John Parker, Esq.; and in 1659, a share in an exhibition for scholarships, to the same college, given by Edward Coleman, Esq.: the scholars must be natives, and have continued at the school for two years without intermission, and be fifteen years of age: the master's salary is about £60 per annum. There is also a bequest of a house and land, by a person unknown, producing about £100 per annum, for this school, part being distributed among the poor. The school is kept in an ancient chapel, dedicated to St. Mary and St. Thomas à Becket, at the bridge. A school for girls is partly supported by endowment. The Rev. John Hendry, in 1722, bequeathed £400 to be vested in the purchase of land, and the rental to be given to the vicar, on condition that two sermons be preached in the church every Sunday throughout the year; also a rent-charge of £13. 10. for preaching a sermon every Friday in Lent: the same benefactor bequeathed a small estate for the use of the charity school, chargeable with the payment of 50*s.* annually to indigent old maids of Wymondham, and ten shillings to the poor of Crownthorpe.

WYMONDLEY, GREAT, a parish, in the union of HITCHIN, hundred of BROADWATER, county of HERTFORD, 2 miles (E. by S.) from Hitchin; containing 321 inhabitants. The living is a vicarage, with which that of Ippolitts was united in 1685; net income, £301; patrons and impropriators, Master and Fellows of Trinity College, Cambridge. The church is ancient, having a Norman arch between the nave and the chancel, with an embattled tower. The manor is held by the service of cupbearer to the kings of England, at their coronation. James Lucas, in 1821, bequeathed £150 three per cent. consols., directing the dividends to be distributed among the poor of Great and Little Wymondley, in the proportion of two-thirds to the former, and one-third to the latter place: there are also some other trifling bequests.

WYMONDLEY, LITTLE (*St. Mary*), a parish, in the union of HITCHIN, hundred of BROADWATER, county of HERTFORD, $2\frac{1}{4}$ miles (S. E. by E.) from Hitchin; containing 226 inhabitants. The living is a donative curacy; net income, £20; patron and impropriator, S. H. U. Heathcote, Esq. The church contains, among other sepulchral memorials, some very ancient gravestones. A priory of Black canons, in honour of St. Lawrence, was founded here, in the time of Henry III., by Richard Argentein, which at the dissolution had a revenue of £37. 10. 6. There are no remains of the building; its site is marked by some avenues of stately box trees, and there is an ancient well, to the water of which tradition ascribes considerable efficacy. In the village is a college for educating Protestant dissenting ministers, founded in 1729, by W. Coward, Esq., with a chapel attached. This establishment originated at Northampton, and the celebrated Dr. Doddridge was its first theological professor. It possesses a valuable library of about 10,000 volumes, and an extensive and complete philosophical apparatus. There are two professorships, one including the theological, philosophical, and mathematical departments; the other, every branch of classical literature.

WYRARDISBURY, or WRAYSBURY (*St. Andrew*), a parish, in the union of ETON, hundred of STOKE, county of BUCKINGHAM, 3 miles (S. W. by S.) from Colnbrook; containing 682 inhabitants. The living is a vicarage, with that of Langley-Marish annexed, valued in the king's books at £14. 10. 5.; present net income, £445; patrons and appropriators, Dean and Canons of Windsor. In this parish is Magna Charta island, a small islet in the Thames, on which King John, at the instance of the barons, signed that celebrated charter of English liberty: it is the property of George Simon Harcourt, Esq., of Ankerwycke House, who has laid out the ground in beautiful designs, and by whose permission this interesting spot is open to visitors on Monday, Wednesday, and Saturday, in each week during summer, on the condition of complying with certain printed rules drawn up to preserve the grounds, &c., from damage. Here are some copper-mills. A Benedictine nunnery, in honour of St. Mary Magdalene, was founded at Ankerwycke, in this parish, in the time of Henry II., by Sir Gilbert de Montfichet, which at the dissolution was valued at £45. 14. 4. A day and Sunday school is supported by subscription. There are six old cottages known as the almshouses, but without any endowment or any document to denote by whom they were given: they are occupied rent-free by aged poor persons. William Gill, in 1798, bequeathed to the poor £300 four per cent. consols., which was subsequently augmented by a further bequest of £100 from Thomas Wright; and the interest, amounting to £13. 8., is distributed on Christmas-day. John Lee, in 1807, gave two annuities to the Corporation of the Sons of the Clergy, in trust to pay £26 per annum to a Sunday afternoon lecturer. The parish is also in possession, from some unknown source, of property called the Church and the Bridge lands, now let for about £46 per annum.

WYRE-PIDDLE, a chapelry, in the parish of FLADBURY, union of PERSHORE, Middle Division of the hundred of OSWALDSLOW, Pershore and Eastern Divisions of the county of WORCESTER, $1\frac{3}{4}$ mile (N. E. by E.) from Pershore; containing 175 inhabitants.

WYRLEY, GREAT, a township, in the parish of CANNOCK, union of PENKRIDGE, Eastern Division of the hundred of CUTTLESTONE, Southern Division of the county of STAFFORD, $6\frac{1}{2}$ miles (N. N. W.) from Walsall; containing 531 inhabitants.

WYRLEY, LITTLE, a township, in the parish of NORTON-UNDER-CANNOCK, Southern Division of the hundred of OFFLOW and of the county of STAFFORD, $7\frac{1}{4}$ miles (W. S. W.) from Lichfield: the population is returned with the parish. Several of the inhabitants are employed in the Brownhill coal-mine, the shaft of which is 90 yards in depth, and the strata three yards thick.

WYSALL (*Holy Trinity*), a parish, in the union of LOUGHBOROUGH, Southern Division of the wapentake of RUSHCLIFFE and of the county of NOTTINGHAM, 8¾ miles (S. by E.) from Nottingham; containing 271 inhabitants. The living is a discharged vicarage, valued in the king's books at £4. 11. 0½.; present net income, £123; patron and impropriator, Earl of Gosford.

WYTHALL, a chapelry, in the parish of KING'S-NORTON, Upper Division of the hundred of HALF-SHIRE, Northfield and Eastern Divisions of the county of WORCESTER, 8 miles (N. E. by E.) from Bromsgrove: the population is returned with the parish. The living is a perpetual curacy; net income, £80; patrons, Dean and Chapter of Worcester; impropriator, Vicar of Bromsgrove. The chapel is dedicated to St. Mary.

WYTHAM (*All Saints*), a parish, in the union of ABINGDON, hundred of HORMER, county of BERKS, 3 miles (N. W.) from Oxford; containing 218 inhabitants. The living is a rectory, valued in the king's books at £7. 5. 2½.; present net income, £306; patron, Earl of Abingdon. A day and Sunday school, consisting of about thirty children of both sexes, is chiefly supported by the Countess of Abingdon. Here was anciently a nunnery, which was originally founded at Abingdon, by the sister of King Ceadwalla, and afterwards removed hither; but, during the war between Offa and Cynewulf, it was demolished by the nuns, who had suffered great annoyance from a castle having been erected in the neighbourhood.

WYTHBURN, a chapelry, in the parish of CROSTHWAITE, union of COCKERMOUTH, ALLERDALE ward below Derwent, Western Division of the county of CUMBERLAND, 8¼ miles (S. E. by S.) from Keswick: the population is returned with St. John's. The living is a perpetual curacy; net income, £82; patron, Vicar of Crosthwaite. The boundaries of the counties of Cumberland and Westmorland are here marked by "Dunmaile-Raise Stones," which are said to commemorate the defeat of the last king of Cumberland, by Edmund, the Saxon monarch, of whom Malcolm, King of Scotland, held Cumberland in fee. Thirlmere lake is within this chapelry.

WYTHORP, a chapelry, in the parish of LORTON, union of COCKERMOUTH, ALLERDALE ward above Derwent, Western Division of the county of CUMBERLAND, 5 miles (E. by S.) from Cockermouth; containing 121 inhabitants. The living is a perpetual curacy; net income, £51; patrons, the Proprietors. The chapel is situated on an eminence above the western bank of Bassenthwaite lake. The ancient hall has been converted into a farm-house. The Rev. John Hudson, a learned divine and critic, was born here, in 1662; he died in 1719.

WYTON, a township, in the parish of SWINE, union of SKIRLAUGH, Middle Division of the wapentake of HOLDERNESS, East Riding of the county of YORK, 5½ miles (N. E. by E.) from Kingston-upon-Hull; containing 93 inhabitants.

WYVERSTONE (*St. George*), a parish, in the union and hundred of HARTISMERE, Western Division of the county of SUFFOLK, 6¾ miles (N.) from Stow-Market; containing 316 inhabitants. The living is a discharged rectory, valued in the king's books at £8. 14. 9½.; present net income, £273; patron, John Moseley, Esq.

WYVILL, a parish, in the union of GRANTHAM, wapentake of WINNIBRIGGS and THREO, parts of KESTEVEN, county of LINCOLN, 6 miles (N. W.) from Colsterworth: the population is returned with the parish of Hungerton. The living is a discharged rectory, with that of Hungerton united; net income, £35; patron, Bishop of London. The church is in ruins: the inhabitants attend that at Harlaxton.

Y

YADDLETHORPE, a hamlet, in the parish of BOTTESFORD, union of GLANDFORD-BRIDGE, Eastern Division of the wapentake of MANLEY, parts of LINDSEY, county of LINCOLN, 8 miles (W.) from Glandford-Bridge; containing 106 inhabitants.

YAFFORTH, a chapelry, in the parish of DANBY-WISK, union of NORTH ALLERTON, wapentake of GILLING-EAST, North Riding of the county of YORK, 1½ mile (W. by N.) from North Allerton; containing 165 inhabitants.

YALDING (*St. Peter and St. Paul*), a parish, (formerly a market-town), in the union of MAIDSTONE, hundred of TWYFORD, lathe of AYLESFORD, Western Division of the county of KENT, 6 miles (S. W.) from Maidstone; containing 2460 inhabitants. This parish is intersected by different branches of the Medway, and upon two of the larger streams stands the village, approached by a long narrow stone bridge, besides which there is another in the parish, called Twyford bridge. The river is navigable to this place for barges, by which a considerable traffic in timber, corn, and coal, is carried on. The market has been long disused; a fair for cattle is held on October 15th. The living is a vicarage, valued in the king's books at £20. 18. 9.; present net income, £1184; patrons and impropriators, — Warde, and — Holmes, Esqrs. The church is principally in the decorated style of English architecture. William Cleave, Esq., in 1665, founded a free school, and endowed it with a farm now let for £50 a year. A charity school, founded in 1711, for girls and young children, has been endowed by Mrs. Alchorn and Mrs. Warde, sisters, with a school-house, besides certain lands and other premises, the rents of which are paid half-yearly to a schoolmistress for teaching from 20 to 24 children, under the superintendence of the vicar.

YANWATH, a joint-township with Eamont-Bridge, in the parish of BARTON, WEST ward, and union, county of WESTMORLAND, 2 miles (S. by W.) from Penrith; containing, with Eamont-Bridge, 327 inhabitants. The ancient hall, a quadrangular castellated building, is now occupied as a farm-house, and about a mile from it are vestiges of a circular camp, called Castle Steads.

YANWORTH, a chapelry, in the parish of HAZLETON, union of NORTHLEACH, hundred of BRADLEY, Eastern Division of the county of GLOUCESTER, 3½ miles (W. by S.) from Northleach; containing 123 inhabitants. The chapel, dedicated to St. Michael, is a chapel of ease to Hazleton, where the inhabitants of Yanworth anciently buried their dead; but since the latter part of the last century this has been their usual place of sepulture.

YAPHAM, a chapelry, in the parish and union of POCKLINGTON, Wilton-Beacon Division of the wapentake of HARTHILL, East Riding of the county of YORK, 2½ miles (N. N. W.) from Pocklington; containing 137 inhabitants. The living is a perpetual curacy, annexed to the vicarage of Pocklington. Twelve children are educated for an annuity of £12, paid out of the produce of the chapel lands. There is no burial-ground at Yapham; the inhabitants inter their dead at Pocklington.

YAPTON, a parish, in the union of WEST HAMPNETT, hundred of AVISFORD, rape of ARUNDEL, Western Division of the county of SUSSEX, 5 miles (S. W.) from Arundel; containing 578 inhabitants. This parish is situated on the road from Arundel to Bognor, and is intersected by the Arundel and Portsmouth canal: the soil is a rich loam, producing excellent crops of grain. The living is a discharged vicarage, united to that of Walberton, and valued in the king's books at £7. 10. 11½.; impropriators, Ignio Thomas, Esq., and others. The church is principally in the early style of English architecture, with a tower at the west end, and contains an ancient font of curious design, and several neat monuments. There was formerly an ancient chapel at Bilsom, now converted into cottages. Stephen Roe, in 1766, bequeathed £1200 three per cent. South Sea Annuities, producing £36 a year, of which sum, £20 is applied for teaching 20 children in the National school, and the residue to other charitable purposes.

YARBOROUGH, or YARBURGH (*St. John the Baptist*), a parish, in the union of LOUTH, Marsh Division of the hundred of LOUTH-ESKE, parts of LINDSEY, county of LINCOLN, 4¾ miles (N. E. by N.) from Louth; containing 175 inhabitants. The living is a discharged rectory, valued in the king's books at £9. 13. 6.; present net income, £226; patron, N. E. Yarburgh, Esq. There is a place of worship for Wesleyan Methodists. Yarborough gives the title of Baron to the family of Pelham.

YARBOROUGH, a hamlet, in the parish of CROXTON, Eastern Division of the wapentake of YARBOROUGH, parts of LINDSEY, county of LINCOLN, 8 miles (N. E.) from Glandford-Bridge: the population is returned with the parish. Here are the remains of a very extensive camp, upon the site of which vast numbers of Roman coins have been discovered.

YARCOMBE (*St. John the Baptist*), a parish, in the union of CHARD, hundred of AXMINSTER, Honiton and Southern Divisions of the county of DEVON, 5½ miles (W.) from Chard; containing 804 inhabitants. The living is a discharged vicarage, valued in the king's books at £28; present net income, £607: it is in the patronage of the Crown; impropriator, Sir H. F. T. S. Drake. There is a place of worship for Baptists. A National school is partly supported by subscription.

YARDLEY (*St. Lawrence*), a parish, in the union of BUNTINGFORD, hundred of ODSEY, county of HERTFORD, 4½ miles (W. S. W.) from Buntingford; containing 599 inhabitants. The living is a discharged vicarage, valued in the king's books at £12; present net income, £242; patrons, Dean and Chapter of St. Paul's, London; impropriator, J. Murray, Esq. The church has an embattled tower surmounted by a spire covered with lead. Here is a National school. Chauncey, the historian of Hertfordshire, was interred here.

YARDLEY (*St. Edburgh*), a parish, in the union of SOLIHULL, Upper Division of the hundred of HALFSHIRE, Northfield and Eastern Divisions of the county of WORCESTER, 4½ miles (E.) from Birmingham; containing 2488 inhabitants. This parish is situated on the high road from Birmingham to Coventry, and is separated from the county of Warwick by a small rivulet. The London and Birmingham railway passes through it. Great quantities of tiles are made in the neighbourhood, and conveyed to Birmingham, whence they are sent by the canals to various parts of the kingdom. The living is a discharged vicarage, valued in the king's books at £9. 19. 4½.; present net income, £463; patron and impropriator, Col. Greswolde. The church exhibits various specimens of the English style of architecture, with a fine tower and spire of the later date. The inhabitants have, from a very early period, enjoyed the benefit of certain lands and rent-charges granted to trustees for their use by various benefactors: the present annual revenue of this charity amounts to £833. 19.; and it is appropriated to the maintenance of two schools, in paying house-rent for poor parishioners, repairs of the church and bridges, a distribution of bread and money twice a year, and apprenticing poor children. In Yardley free school, which is near the church, and is an ancient building of two rooms, with a house and garden for the free use of the master adjoining, from 60 to 80 boys are instructed and supplied with stationery gratuitously: the master receives a salary of £100 per annum. The other school is situated at Hall Green, and in it about 50 children are gratuitously instructed by a master who has a salary of £40, with the free use of the school-house, and permission to take four pay scholars. Fifty-one small cottages, with gardens attached, belonging to the charity, are occupied rent-free by poor persons; besides which the trustees annually pay a considerable sum as rent for non-resident parishioners, and £26 to a surgeon for attendance and medicine to all the resident poor. About five children are apprenticed annually, and £61. 14. is distributed in bread and money at Midsummer and Christmas. Here is a National school; also a school called the Vicarage-school, chiefly supported by the family of the vicar. Job Marston, in 1701, bequeathed property for building and endowing a chapel at Hall Green: the present annual rental amounts to £140, which sum, after defraying the charge of repairs, is paid by the trustees to the minister: this chapel, which was consecrated by Bishop Lloyd on the 25th of May, 1704, is a free chapel and donative, and the minister is appointed by the trustees. The same testator also bequeathed property now producing £111 per annum, which is appropriated to a distribution of clothing, bread, &c., and in apprenticing one or two children annually. Henry Greswold Lewis, in 1829, gave £1500, directing the dividends to be expended in clothing, bread, and meat for three poor men and three poor women, of each of the parishes of Yardley, Radford-Semele, and Solihull.

YARDLEY-GOBION, a hamlet, in the parish and union of POTTERS-PURY, hundred of CLELEY, Southern Division of the county of NORTHAMPTON, 3½ miles (N. N. W.) from Stony-Stratford; containing 594 inhabitants. There is a place of worship for Independents.

YARDLEY-HASTINGS (*St. Andrew*), a parish, in the union of HARDINGSTONE, hundred of WYMMERSLEY, Southern Division of the county of NORTHAMPTON,

8½ miles (E. S. E.) from Northampton; containing 1051 inhabitants. The living is a rectory, to which a portion of the rectory of Denton is annexed, valued in the king's books at £13. 16. 0½.; present net income, £355; patron, Marquess of Northampton. There is a place of worship for Independents. North of the church are the ruins of an ancient mansion, once the seat of the family of Hastings, Earls of Pembroke. A fair is held on Whit-Monday. The Rev. Edward Lye, author of the Anglo-Saxon Dictionary, died rector of this parish, in 1769. A day and Sunday National school is partly supported by subscription: there are two Sunday schools, one in connection with the Established Church, the other appertaining to Independents, and both supported by like means; in addition to which a sewing school is chiefly supported by the Marquess of Northampton.

YARKHILL (*St. John the Baptist*), a parish, in the union of LEDBURY, hundred of RADLOW, county of HEREFORD, 7¼ miles (E. by N.) from Hereford; containing 409 inhabitants. The living is a discharged vicarage, valued in the king's books at £3. 19. 3.; present net income, £125; patrons, Dean and Chapter of Hereford; impropriator, Master of Ledbury Hospital.

YARLESIDE, a division, in the parish of DALTON-IN-FURNESS, hundred of LONSDALE, north of the sands, Northern Division of the county palatine of LANCASTER, 2 miles (S.) from Dalton; containing 499 inhabitants. It comprises several small villages and hamlets.

YARLETT, a liberty, in the parish of WESTON-UPON-TRENT, Southern Division of the hundred of PIREHILL, Northern Division of the county of STAFFORD; containing 21 inhabitants. This liberty comprises about 400 acres of land, the property of — Tunnicliff, Esq., who resides at the hall, a neat mansion situated on a gentle declivity.

YARLINGTON (*St. Mary*), a parish, in the union of WINCANTON, hundred of BRUTON, Eastern Division of the county of SOMERSET, 3½ miles (W.) from Wincanton; containing 283 inhabitants. The living is a rectory, valued in the king's books at £16. 1. 3.; patron, Rev. Robert G. Rogers. The tithes have been commuted for a rent-charge of £244, subject to the payment of rates; the glebe comprises 38 acres, valued at £72 per annum. The church has an embattled tower on the south side. Here is a Sunday school, in which about 50 children are instructed at the expense of the clergyman. On the south-west declivity of Godshill, in this parish, is a double-intrenched camp, from which there is an extensive prospect.

YARLINGTON, a hamlet, in the parish of NORTH CADBURY, hundred of CATSASH, Eastern Division of the county of SOMERSET: the population is returned with the parish.

YARM (*St. Mary Magdalene*), a market-town and parish, in the union of STOCKTON, Western Division of the liberty of LANGBAURGH, North Riding of the county of YORK, 44 miles (N. N. W.) from York, and 238 (N. by W.) from London; containing 1636 inhabitants. This town, which is situated on a low peninsula formed by the river Tees, appears to have been formerly of more importance than it is at present; its decline may be attributed partly to its vicinity to the rising town of Stockton, distant only four miles, and partly to its having been exposed, from its low situation, to floods. On the 17th of February, 1758, from a sudden thaw on the western hills, the water rushed down upon the town, seven feet in depth, destroying and carrying off live stock and household goods to a considerable extent; but the most destructive of these inundations occurred in November 1771, when the Tees rose so high as to occasion, in some parts of the town, a depth of 20 feet, which, in addition to the destruction of a great quantity of valuable property, occasioned also the loss of some lives: it has been since occasionally subject to similar, though much less violent, calamities. The principal street, which extends north and south, is spacious, and contains some good houses. A bridge of five arches across the Tees, built in 1400 by Walter Skirlaw, Bishop of Durham, has been much improved. A beautiful iron bridge of one arch, of 180 feet span, was constructed in 1805; but, owing to some defect in the foundation, it fell down on January 12th, 1806, a short time before it was to have been opened to the public, and has not been replaced. The trade principally consists in the exportation of agricultural produce, and the town also participates with Stockton in the salmon fishery of the Tees, the tide flowing a short distance above it: the corn trade, which at one period was carried on to a considerable extent, has much declined. In addition to the advantages derived from the navigation of the river Tees, a branch from the Stockton and Darlington railway facilitates the transit of goods. The market is on Thursday; and there are fairs on the Thursday before April 5th, on Ascension day, August 2nd, and October 19th and 20th, for horses, cattle, and cheese, of which last article great quantities are sold on the latter day. A court for the recovery of small debts is held here, twice in the year, by the lord of the manor.

The living is a perpetual curacy; net income, £151; patron and appropriator, Archbishop of York. The appropriate tithes have been commuted for a rent-charge of £265, subject to the payment of rates; the glebe consists of about 2 acres. The church, situated at the west side of the town, is a plain neat structure, rebuilt in 1730: it has a handsome painted window, containing a full-length figure of Moses delivering the law on Mount Sinai. The Society of Friends, Independents, Primitive and Wesleyan Methodists, and Roman Catholics, have each a place of worship. The free grammar school was founded in the reign of Queen Elizabeth, by Thomas Conyers; the endowment was increased in 1700, by a bequest from William Chaloner, of £400 three per cent. stock, which, together with the preceding, produces at present an income of £21, as a salary to the master, who has also rent-free a small house of two apartments adjoining the school-room. The same Mr. Chaloner also bequeathed £100 four per cent. stock to the minister of Yarm, for four annual Sunday evening lectures in the parish church. A National school was erected in 1816, by subscription, and is supported by voluntary contributions: some small bequests are annually distributed among the poor. An hospital, dedicated to St. Nicholas, founded in the year 1185, and valued at its dissolution at £5 per annum; and a house of Black friars, founded in the thirteenth century, both by the family of De Brus, existed here; but no trace of either is discernible. The site of the latter is now occupied by a commodious mansion, called the Friarage, the grounds of which, extending about a mile along the banks of the Tees, are beautifully laid out.

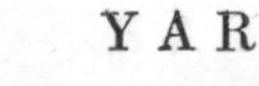
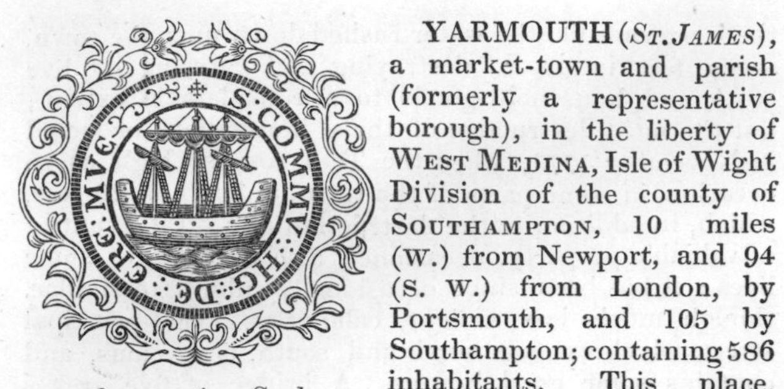

Corporation Seal.

YARMOUTH (*St. James*), a market-town and parish (formerly a representative borough), in the liberty of West Medina, Isle of Wight Division of the county of Southampton, 10 miles (W.) from Newport, and 94 (S. W.) from London, by Portsmouth, and 105 by Southampton; containing 586 inhabitants. This place, which derives its name from its situation on the river Yar, was formerly of much greater extent and importance than it is at present; but it had suffered severely from attacks of the French, by whom it was, in the reign of Richard II., pillaged and entirely burned, and on two subsequent occasions it was nearly destroyed by them. The town field, laid out regularly in right angles, though now destitute of buildings, clearly appears to have been originally the site of a part of the town. Yarmouth is situated on a bank sloping to the sea, on the eastern point of land at the mouth of the Yar, and consists of several neat streets, for the most part running east and west: the houses, which are of freestone, are in general well built and of neat appearance, and public baths have been recently established. At its western extremity are a castle and small fort, erected by Henry VIII.: the latter, which occupies the site of a church, or ancient religious house, and consists of a platform with eight guns, and houses for the garrison, has recently been granted by the Board of Ordnance to the corporation, with a view to its removal, for the purpose of improving the quay and the landing-places. A large house near the former, which has been converted into an inn, was erected by Sir Robert Holmes, for the reception of Charles II., a portrait of whom, during his stay here, was painted by Sir P. Lely, and is in the possession of the Holmes family. The trade is now very limited: a considerable quantity of fine white sand, used in the manufacture of flint glass and the finer sorts of British china, is obtained from some pits on the shore of Alum Bay, near the Needles: the principal imports are, coal from Sunderland, and timber from the New Forest. A constant intercourse by boats is kept up with the opposite town of Lymington, and before the general use of steam-boats, this was considered the safest and most expeditious passage to the island. The market is on Friday; and a fair is held on July 25th. The market-house is a neat building, with a hall over it, in which the several courts are held, and the public business of the corporation is transacted. The original charter of incorporation was granted by Baldwyn de Redvers, Earl of Devon, and confirmed by Edward I.; but that under which the corporation now acts was granted in the 7th of James I., which ordains the appointment of a mayor and twelve capital burgesses, with power to choose a steward, a town-clerk, and a sergeant-at-mace, and to create freemen, but this last privilege is not now exercised. The borough courts are held by the mayor and steward, and the corporation is entitled to all the fines, forfeitures, and profits of the courts, with many other privileges. The borough first sent members to parliament in the 23rd of Edward I., but made no other return until the 27th of Elizabeth, from which period it exercised the privilege without intermission until the 2nd of William IV., when it was disfranchised by the act then passed to amend the representation. The right of election was vested in the mayor and burgesses; and the mayor was the returning officer.

The living is a discharged rectory, net income, £43: it is in the patronage of the Crown. The church, situated in the centre of the town, is a neat structure, consisting of a nave and chancel, on the south side of which is a sepulchral chapel, containing a handsome statue of the full size, in Parian marble, of Sir Robert Holmes, formerly Governor of the Isle of Wight. There are places of worship for Baptists and Wesleyan Methodists. A National and a Lancasterian school are partly supported by subscription. The sum of £30 was bequeathed by Thomas Lord Holmes, of which £10 is distributed annually to the poor not receiving parochial relief, £10 is paid towards apprenticing a boy out of the parish, and the remaining £10 is given to the minister. There are some vestiges of a Roman station, on the site of which a house has been built, occupied as a private residence. Many Saxon customs not generally observed are still retained at Yarmouth, such as decorating the coffins of the dead with flowers and evergreens, the mourners carrying bunches of rosemary, &c.

Arms.

YARMOUTH, GREAT (*St. Nicholas*), a sea-port, borough, market-town, and parish, having separate jurisdiction, locally in the Eastern Division of the hundred of Flegg, Eastern Division of the county of Norfolk, 23 miles (E. by S.) from Norwich, and 123 (N. E.) from London; containing 21,115 inhabitants. This place, which, from its extensive and prosperous trade and many other advantages and privileges, may be considered the most flourishing port on this part of the coast, derives its name from its situation at the mouth of the river Yare, which here falls into the ocean. It occupies ground originally covered by the sea, which, on its receding, left a bank of sand whereon a few fishermen settled, the first of whom, denominated *Fuller*, imparted his name to the higher portion, still called *Fuller's hill*. As the bank increased in extent and density, the population augmented; but the channel of the northern branch of the Yare, on which the first settlers fixed their habitations, becoming choked up with sand, they, in 1040, removed to the southern branch. The earliest authentic record of Yarmouth is in Domesday-book, in which it is described as "the king's demesne, and having 70 burgesses." Its fishery, at an early period, attracting many residents, a charter was granted, at the request of the inhabitants, by Henry III., allowing them to enclose the burgh, on the land side, with a wall and moat; the former was 2240 yards in length, and had sixteen towers and ten gates. A castle, having four watch towers, and upon which a fire beacon was placed in 1588, was also built about this time in the centre of the town: in the last-named year a mound, called South Mount, was thrown up and crowned with heavy ordnance, and the place was then considered impregnable. The castle having been demolished in 1621, and the changes introduced into the system of warfare rendering further defences necessary, strong parapets were constructed in front of the town, and cannon planted on them, facing the sea: the circuit of the fortifications thus completed was nearly two miles and a half. The only military operation in which the inhabitants have been ever actually engaged was their gallantly repulsing Kett, when, in his rebellion, he attempted, at the head of 20,000 men, to take the town by assault. But though Yarmouth has been only slightly visited by the scourge of warfare, it has suffered severely from the plague, to which, in 1348, upwards of 7000 persons fell victims; in 1579, upwards of 2000; and more than 2500 in 1664.

The town occupies an extent of 153 acres, on the western bank of a peninsula formed by the river Yare and the sea, and is connected with South Town, or Little Yarmouth, which is on the opposite bank, by a drawbridge. It is of a quadrangular form, about a mile long and half a mile broad, and consists of four good streets parallel with each other, a handsome new street leading to the quay, on which is a noble range of buildings, and a great number of narrow rows intersecting the principal streets at right angles: it is lighted with gas, and well supplied with fresh water, and the streets are kept remarkably clean. There are several very ancient houses, one of which, built in 1596, was the residence of a granddaughter of Oliver Cromwell, and is now the property and residence of Charles Palmer, Esq., F.S.A. In the drawing-room, a meeting of the principal officers of the Parliamentarian army is said to have been held for the purpose of deciding the fate of Charles I.; this apartment, which is elaborately ornamented with rich carved work, has been restored to its pristine state by the present proprietor. The theatre, a neat and commodious edifice, erected in 1778, near the market-place, is open during the summer months; and races take place, annually in August, on the Denes, a fine down south of the town. The bathing-houses on the beach, near the jetty, possess every accommodation for visiters; and adjoining is a public-room, built in 1788, where balls and concerts occasionally take place. There are very pleasant walks on the quay and beach; and the extensive sea view, enlivened by the number of vessels in the roads, is a source of considerable gratification to the frequenters of this sea-port, which is resorted to as a watering-place. The barracks on the South Denes, near the beach, form a magnificent quadrangular range of buildings, designed by Mr. Pilkington, and erected at a cost of £120,000; the building was used as a naval hospital during the war. The armoury in South Town will contain, exclusively of other military and naval stores, 10,000 stand of arms. Between the barracks and the entrance to the harbour, a grand fluted column, 130 feet high, and surmounted by a statue of Britannia, has been erected to the memory of Admiral Lord Nelson, and, as a landmark, well supplies to seamen the loss of Gorleston steeple, which was blown down in the year 1813. On the quay is the custom-house; and within a short distance is a public library, with a good collection of books: there are also subscription reading-rooms adjoining the library. A handsome suspension chain bridge, of 86 feet span, has been constructed at the northern part of the quay over the river Bure, by Robert Cory, Esq., in whom the property is vested; and a new road, in communication with this bridge, completed under an act of parliament, has shortened the distance between this town and Norwich about $3\frac{3}{4}$ miles.

Yarmouth is not a manufacturing town, though a considerable establishment for winding and throwing silk has been formed in connection with a larger concern at Norwich, and for which buildings have been erected on the site formerly occupied by the barracks, at the north of the town. There are extensive yards for shipbuilding, with corresponding rope-walks, and several large breweries. A considerable trade is carried on coastwise in malt, corn, flour, coal, timber, and other articles. A direct intercourse prevails with the Baltic, the Mediterranean, Portugal, and other parts of the continent; and a regular communication by steam-vessels is kept up internally with Norwich, and coastwise with London and the north of England. But the principal source of trade by which this town is supported is the herring fishery, which is usually productive to a remarkable extent. The fish, when cured, or dried, for both of which processes there are very extensive establishments, are not only sent to all parts of the kingdom, but exported in considerable quantities to other parts of the world, particularly to the West Indies. Many vessels from other parts of the coast fish here, and some, at a defined distance, from other countries, especially France and the Netherlands. The situation of Yarmouth, in a commercial point of view, affords unusual advantages. The Yare is here navigable for vessels of 250 tons' burden; and to Norwich, a distance of 32 miles, for smaller vessels, without the intervention of locks. The Waveney, which falls into the Yare, is navigable by Beccles to Bungay, a distance of 20 miles; and the Bure, which also joins the Yare, by Horstead to Aylsham, 30 miles, and another branch to North Walsham, 25 miles hence, thus opening an extensive and valuable channel of inland communication. In 1836, an act was passed for making a railway from London to Norwich and Yarmouth, to be called the Eastern Counties railway, which is now in progress. Many attempts have been unsuccessfully made to form a safe harbour, at the enormous expense of above £240,000; but the present one, which is the seventh that has been constructed, was projected and executed, at an expense of about £4200 only, by Jans Johnson, a native of Holland, and affords secure anchorage at all times. In 1835, an act was passed for improving the haven and the several rivers connected therewith, also for repairing or rebuilding the bridge over the haven, and St. Olave's bridge over the river Waveney. At the entrance of the Yare are two piers; that on the south is 1230 feet long, and forms an agreeable promenade; and that on the north is 400 feet in length, erected on wooden piles, and secured by iron railing. The quay, which in length and beauty of construction ranks the first in England, is a very great ornament to the town, and its centre is formed into a beautiful walk, planted on each side with trees. A duty of fifteen pence per chaldron, producing about £8000 per annum, is levied on all coal brought to the port, and applied, under the direction of twelve commissioners, to keeping the jetties and piers in repair, and deepening and clearing the river: the commissioners are chosen, three each, by the corporations of Yarmouth and Norwich, and by the magistrates of Norfolk and Suffolk. The number of vessels belonging to this port is upwards of 500, exclusively of small craft; and the amount of duties paid at the custom-house, for the year ended January 5th, 1837, was £63,783. The navigation of the coast is very dangerous; but the Roads, in which are two floating light-houses, are frequently resorted to by the North Sea fleet, and merchant vessels are constantly repairing to them for shelter. The market is on Wednesday and Saturday; and fairs are held on the Monday and Tuesday at Shrovetide, and on the Friday and Saturday in Easter week.

Corporation Seal.

Obverse. *Reverse.*

Prior to the reign of King John, the town was governed by a provost appointed by the Crown; but a charter of incorporation granted by that monarch in the ninth year of his reign, empowered the burgesses to choose their own magistrates, called bailiffs, of whom four were elected, who were authorised to hold a court of hustings, now called the Burgh court. These privileges were extended by charters granted by succeeding sovereigns, of which that of Edward II. granted tronage to the burgesses, and exemption from serving on any assizes, juries, or inquisitions, out of the borough. Richard II. united Kirtley-road to Yarmouth. By charter of Henry VII., aldermen and coroners were appointed, and power was given to elect justices of the peace. Elizabeth granted a charter to hold an admiralty court weekly, with power to try all maritime causes, except piracy; and James I. confirmed the foregoing, adding the power to try pirates, and defined the admiralty jurisdiction to be from Winterton Ness, in Norfolk, to Easton Ness, in Suffolk, and seven leagues eastward from all sea banks and shores; ship-owners are also exempted from paying harbour dues at Dovor, Ramsgate, Rye, and other harbours on the coast. The charter granted by Queen Anne, in 1702, vested the government in a mayor, high steward, recorder, sub-steward, eighteen aldermen (including the mayor), and thirty-six common councilmen, assisted by a town-clerk, registrar of the admiralty court, four proctors, who were also attorneys of the Burgh court, a water-bailiff, marshal and gaoler, six sergeants-at-mace, and pier-master. The mayor, high-steward, recorder, and deputy-mayor, and such aldermen as had passed the chair, were justices of the peace. The corporation at present consists of a mayor, high-steward, recorder, twelve aldermen, and thirty-six councillors, assisted by a town-clerk, water-bailiff, gaoler, three sergeants-at-mace, and other officers, appointed under the act of the 5th and 6th of William IV., cap. 76, for an abstract of which see the Appendix, No. I.; and the borough, formerly consisting of eight wards, is now divided into six. The mayor and the recorder are *ex officio* justices of the peace, and ten others are appointed by the Crown. Courts of session are held quarterly before the recorder; but the borough court has been discontinued, and the four attornies who formerly practised there have each a pension. The freedom is obtained by birth and servitude: all sons of freemen become free on attaining 20 years of age, but cannot vote until they are 21. The borough, which originally comprised 1460 acres, first sent members to parliament in the reign of Edward I.: the elective franchise was formerly vested in the freemen, in number about 2000: but the privilege has been extended to the £10 householders of an enlarged district, which has been constituted the new borough, and comprises 2823 acres, of which the limits are minutely described in the Appendix, No. II.: the mayor is the returning officer.

Admiralty Seal, now disused.

The admiralty jurisdiction was totally abolished by the Municipal Reform Act; the last court of admiralty was held on the 7th of September, 1835. The registrar has a pension of £150 per annum for life, and the four proctors and the marshal have also pensions. A court of requests, for the recovery of debts under 40*s*., is held every Monday: the jurisdiction extends throughout the borough and liberties. A court leet and a court of pie-poudre are also held. The jurisdiction of the corporation, by the charter of the 20th of Charles II., extends to South Town, or Little Yarmouth, in the county of Suffolk, and, as conservators of the Yare, Waveney, and Bure, for ten miles upon each of those rivers. The inhabitants are not liable to serve on juries for the county, nor to the payment of county rates, as the corporation supports the gaol, and maintains the prisoners; and writs, unless accompanied with a *non omittas*, can only be executed under the warrant of the mayor, and by one of his officers. The town has been made a polling-place for the eastern division of the county. The town-hall, near the centre of the quay, is an elegant building of the Tuscan order, with a handsome portico in front; it is also the mansion-house, and under the control of the mayor for the time being. The council-chamber, in which public meetings and assemblies are held, is a splendid room, ornamented with a fine portrait of George I., in his robes; the card-room is spacious, and contains paintings, by Butcher, of the quay, the Roads, and the market-place, and a portrait of Sir Robert Walpole, who was high steward.

The living is a perpetual curacy; net income, £430; patrons, Dean and Chapter of Norwich. In Domesday-book mention is made of a church dedicated to St. Benedict, probably founded by the barons of the cinque-ports, of which the foundations are still visible, about a mile from the entrance of the town. The present church, situated in the north-east part of the town, was founded by Herbert de Lozinga, Bishop of Norwich, about 1101, and appropriated to the prior and monks of the Holy Trinity at Norwich, who had a cell here: he erected only the cross, which constitutes the present nave and transepts; the aisles were added in 1250, and in the following year it was dedicated to St. Nicholas. It is a handsome cruciform structure, in the early, decorated, and later, styles of English architecture, with a central tower and spire, four turrets at the west end surmounted with pinnacles, and an elegant south porch. Seventeen oratories, each with an image, altar, lights, &c., and supported by a guild, were founded in it: on the tower was formerly a wooden spire, which appeared crooked from whatever side viewed, but it was replaced by the present one in 1804. The chapel of ease, dedicated to St. George, a handsome edifice, built in 1716, is supported by a duty of one shilling per chaldron on all coal consumed in the parish: two ministers were formerly appointed to serve in it by the mayor and corporation, but the patronage has been taken from that body, and must be sold; net income, £200. An additional church, dedicated to St. Peter, and in the later English style, with a lofty square tower, was erected near the White Lion Gates, on the north side of the road to the jetty, in 1833, at an expense of £7596. 4., raised by subscription, aided by a grant from the Commissioners for Building and Enlarging Churches. The living is a perpetual curacy; net income, £160; patron, Incumbent of St. Nicholas'. There are places of worship for Particular Baptists, the Society of Friends, Independents, Primitive and Wesleyan Methodists, and Unitarians, also a Roman Catholic chapel. The free grammar school, in the market-place, commonly called the children's hospital school, was founded by the corporation, in 1651, and was part of St. Mary's hospital: it is now a free school for reading, writing, and arithmetic only; thirty boys and twenty girls are clothed, maintained, instructed, and apprenticed. The present revenue of the charity, independently of fines upon the renewal of leases, is £856. 19., of which £100 per annum, with a septennial fine of £100, is derived from an estate in Ireland, now worth £6000 per annum, but of which a lease for 1000 years at the above rental was granted in 1714. The master, who, with the assistance of his daughter, instructs all the children, receives a salary of £120, with the use of a free residence: there is also a resident nurse, with a salary of £15 and her board: and a surgeon is paid £10 per annum for attendance and medicine. There is also a Lancasterian school on the Chapel Denes, supported by subscription; and a school of industry for girls is partly supported by like means. There is a charity school connected with the dissenters of the Old Meeting House, which is endowed with funded property purchased with the amount of several legacies, and producing a dividend of £17, which is paid to the master, who instructs ten boys and six girls gratuitously. Two other schools, one called the "North End school," and the other the "South End school," are partly supported by subscription. There is also a day and Sunday school, supported by subscription. The Rev. Edward Warnes, in 1694, bequeathed an estate now let for £375 per annum, which is distributed at Easter and Christmas among orphans and widows, those of clergymen having the preference. There is a lending fund of £130, bequeathed by Catherine Rogers and William Southwell. The Fishermen's hospital, of a quadrangular form, comprising twenty houses of two rooms each, for the accommodation of that number of fishermen and their wives, had an annual income of £160, paid by the Treasury, and originally granted by the Crown as a reduction of the duty then levied upon all beer carried to sea, and £56. 10. derived from various private benefactions; but the allowance from the Treasury is now reduced to 2*s.* 6*d.* during the winter and 2*s.* in summer, paid weekly to such of the inmates as were there in 1830, and when these die, the allowance is to cease. Seventy-eight houses in different parts of the town are occupied rent-free by paupers; and an annual sum of £62. 10. is distributed by the charity trustees in money, bread, and coal, among the inmates. In the arrangements under the Poor Law Amendment Act, Great Yarmouth supports its own poor, who are under the care of 16 guardians. Besides the cell belonging to the Holy Trinity at Norwich, and the hospital of St. Mary, here were also a cell of Austin friars belonging to the priory of Gorleston, two lazar-houses, and houses of Black, Grey, and White friars, many fragments of which remain, as well as of the ancient walls of the town. Yarmouth gives the inferior title of Earl to the Marquess of Hertford.

YARNFIELD, a hamlet, in the parish of MAIDEN-BRADLEY, union of MERE, hundred of NORTON-FERRIS, Eastern Division of the county of SOMERSET, 5½ miles (E. N. E.) from Bruton; containing 91 inhabitants.

YARNSCOMBE (*St. Andrew*), a parish, in the union of TORRINGTON, hundred of HARTLAND, Great Torrington and Northern Divisions of the county of DEVON, 6 miles (N. E. by E.) from Great Torrington; containing 498 inhabitants. The living is a discharged vicarage, valued in the king's books at £7. 11. 11.; present net income, £129; it is in the patronage of the Crown; impropriator, Lord Rolle.

YARNTON, or YARINGTON, a parish, in the union of WOODSTOCK, hundred of WOOTTON, county of OXFORD, 4¼ miles (N. W. by N.) from Oxford; containing 299 inhabitants. The living is a discharged vicarage, valued in the king's books at £8. 5. 5.; present net income, £217; it is in the patronage of Sir George Dashwood, Bart., for three turns, and of the Warden and Fellows of All Souls' College, Oxford, for one; impropriators, Rector and Fellows of Exeter College, Oxford. The church is an ancient structure, with a tower built in 1612, by Sir Thomas Spencer, who also built the aisle in which he is interred, as a sepulchral chapel for his family, who resided in the ancient manor-house near the church, the remains of which are now occupied as a farm-house. In a recess in the aisle is an altar-tomb, with recumbent effigies of Sir William Spencer and his lady. The churchyard contains an ancient cross embellished with figures in full length, now much mutilated. A National school is partly supported by endowment.

YARPOLE (*St. Leonard*), a parish, in the union of LEOMINSTER, hundred of WOLPHY, county of HEREFORD, 5 miles (N. N. W.) from Leominster; containing 651 inhabitants. The living is a vicarage, annexed to the rectory of Croft; impropriators, Trustees of Lucton School. A school is supported by subscription, aided by a small endowment.

YARWELL (*St. Mary Magdalene*), a parish, in the union of OUNDLE, hundred of WILLYBROOK, Northern Division of the county of NORTHAMPTON, 1¼ mile (S. by W.) from Wansford; containing 369 inhabitants. The living is annexed to the vicarage of Nassington. There are about 18 acres of land, derived from some unknown source, and now let for £29. 10. per annum, half of which sum is distributed at Christmas among poor widows and others not receiving parochial relief.

YATE (*St. Mary*), a parish, in the union of Chipping-Sodbury, Upper Division of the hundred of Henbury, Western Division of the county of Gloucester, 1 mile (W.) from Chipping-Sodbury; containing 824 inhabitants. The living is a rectory, valued in the king's books at £30. 18. 11½.; present net income, £851; patron, W. S. Goodenough, Esq. Coal is procured here in considerable quantities; and this has lately been made a polling-place for the western division of the shire.

YATE, a joint township with Pick-up-Bank, in the parish of Whalley, union, and Higher Division of the hundred, of Blackburn, Northern Division of the county palatine of Lancaster, 4 miles (S. E.) from Blackburn; containing, with Pick-up-Bank, 1209 inhabitants.

YATE, GREAT, a township, in the parish of Croxden, Southern Division of the hundred of Totmonslow, Northern Division of the county of Stafford, 5½ miles (N. N. W.) from Uttoxeter: the population is returned with the parish.

YATEHOUSE, a joint township with Byley, in the parish of Middlewich, union and hundred of Northwich, Southern Division of the county of Chester, 1¾ mile (N. by E.) from Middlewich: the population is returned with Byley.

YATELY (*St. Peter*), a parish, in the hundred of Crondall, Odiham and Northern Divisions of the county of Southampton, 4½ miles (N. W.) from Farnborough; containing 1874 inhabitants. This parish is situated on the high road from London to Southampton, and comprises the tythings of Yately, Cove, and Hawley, which last was constituted a district parish in 1839. A cattle fair is held annually on the 8th of November. The London and South-Western railway passes through the parish, and here is a station on the line. The living is a perpetual curacy; net income, £72; patron and impropriator, Master of the Hospital of St. Cross. There is a place of worship for Baptists. A National school is endowed with £9. 6. 8. a year, being one third of the income arising from land bequeathed for charitable purposes by Mary Barker, in 1706. The parish is, for the relief of the poor, under Gilbert's act, in the union of Farnborough.

YATESBURY (*All Saints*), a parish, in the union and hundred of Calne, Marlborough and Ramsbury, and Northern, Divisions of the county of Wilts, 4½ miles (E. by N.) from Calne; containing 274 inhabitants. The living is a rectory, valued in the king's books at £17. 3. 4.; present net income, £438; patron, Colonel James Kyrle Money.

YATTENDON (*St. Peter and St. Paul*), a parish, in the union of Bradfield, hundred of Faircross, county of Berks, 6½ miles (N. E.) from Newbury; containing 241 inhabitants. The living is a rectory, valued in the king's books at £14. 6. 8.; present net income, £384; patron, Rev. J. F. Howard. Two schools are partly supported by subscription. Here was formerly a weekly market, on Tuesday, granted in 1258, with a fair on the festival of St. Nicholas, to Peter de Etyndon, and confirmed in 1319, to John de la Beche, with another fair on the festival of St. Peter and St. Paul; these have long been disused, but an annual fair is held on the 10th of July. Thomas Carte, the historian, wrote the greater part of his history of England at this place; he died in 1754, and was buried in the church, without any monument to his memory. A castle, said to have been inhabited by King Alfred, once occupied the site of the present manor-house; and a large field in the parish, where Alfred gained a decisive victory over the Danes, is still called England's Field.

YATTON, a chapelry, in the parish of Much Marcle, union of Ross, hundred of Greytree, county of Hereford, 5½ miles (N. E. by N.) from Ross; containing 211 inhabitants.

YATTON (*St. Mary*), a parish, in the union of Bedminster, hundred of Winterstoke, Eastern Division of the county of Somerset, 8 miles (N.) from Axbridge; containing 1865 inhabitants. The living is a vicarage, with that of Kenn annexed, in the patronage of the Prebendary of Yatton in the Cathedral Church of Wells (the appropriator), valued in the king's books at £30; present net income, £422. The church is a stately cruciform structure, with a tower in the centre, formerly surmounted by a spire. Nine poor children are instructed for £10. 10. a year, arising from a donation of £120 and certain land, by John Lane. Here is a National school. On Cadbury hill, in the vicinity, are vestiges of an ancient fortification. In 1782, thirteen human bodies, some of them fresh and of unusual size, and a stone coffin, were discovered in a limestone quarry, about two feet and a half below the surface of the earth.

YATTON-KEYNALL (*St. Margaret*), a parish, in the union and hundred of Chippenham, Chippenham and Calne, and Northern, Divisions of the county of Wilts, 4¼ miles (N. W. by W.) from Chippenham; containing 419 inhabitants. The living is a rectory, valued in the king's books at £8. 7. 1.; present net income, £500; patron, Rev. — Daubeny.

YAVERLAND, a parish, in the liberty of East Medina, Isle of Wight Division of the county of Southampton, 8 miles (E. S. E.) from Newport; containing 96 inhabitants. The living is a rectory, valued in the king's books at £6. 6. 10½.; it is in the patronage of Mrs. A. Wright. The tithes have been commuted for a rent-charge of £240, subject to the payment of rates, which on the average have amounted to £34; the glebe contains 12 acres, valued at £20 per annum. The church has a Norman door, but is principally in the later English style.

YAXHAM (*St. Peter*), a parish, in the union of Mitford and Launditch, hundred of Mitford, Western Division of the county of Norfolk, 2½ miles (S. E. by S.) from East Dereham; containing 505 inhabitants. The living is a rectory, with that of Welborne annexed, valued in the king's books at £10. 0. 10.; patron and incumbent, Rev. J. Johnson. The tithes have been commuted for a rent-charge of £500, subject to the payment of rates, which on the average have amounted to £126; the glebe comprises 46½ acres, valued at £69. 10. per annum.

YAXLEY (*St. Peter*), a parish (formerly a market-town), in the union of Peterborough, hundred of Norman-Cross, county of Huntingdon, 14 miles (N. N. W.) from Huntingdon, and 73 (N. by W.) from London; containing 1140 inhabitants. The village is irregularly, but neatly, built, extending for a considerable distance along the high road from Stilton to Farcet, and is amply supplied with water. At a short distance to the east is Whittlesea mere, one of the most extensive sheets of water in the kingdom; it is six miles in length, and three miles broad, and abounds with fish. The barracks of Norman-Cross, in this parish, were used, during the late war, as a place of confinement for French prisoners, but are now partly dismantled. The neighbourhood is extremely productive of sedges and reeds, the preparation of which affords employment to a considerable portion of the inhabitants: a fair is held annually on Holy Thursday, for cattle. The living is a discharged vicarage, valued in the king's books at £11; present net income, £177: it is in the patronage of the Crown; impropriator, Earl of Carysfort. The church, situated on an eminence at the western extremity of the town, is a handsome structure, principally in the later style of English architecture, with some portions of earlier date; the tower is surmounted by a finely-proportioned crocketed spire, supported by flying buttresses. There is a place of worship for Independents. The workhouse and school were established under the wills of Frances and Jane Proby, who, in 1712, bequeathed property for that purpose to the parishes of Yaxley, Elton, and Flitton: the share of the bequest appropriated to Yaxley amounts to about £70 per annum, out of which a schoolmaster, who has the free use of the school premises, receives about £50 for gratuitously instructing 20 boys of the parish. A school for girls is chiefly supported by the clergyman.

YAXLEY (*St. Mary*), a parish, in the union and hundred of Hartismere, Western Division of the county of Suffolk, 1½ mile (W.) from Eye; containing 478 inhabitants. The living is a vicarage, valued in the king's books at £6. 6. 5½.; present net income, £150; patron, J. T. Mott, Esq.; impropriators, Sir E. Kerrison, Bart., and others. Yaxley Hall was the seat of a family who took their name from the parish.

YAZOR (*St. John the Baptist*), a parish, in the union of Weobley, hundred of Grimsworth, county of Hereford, 4½ miles (S.) from Weobley; containing 196 inhabitants. The living is a discharged vicarage, endowed with a portion of the rectorial tithes, annexed to the rectory of Bishopstone, and valued in the king's books at £5. 12. 6.; impropriators of the remainder of the rectorial tithes, Sir R. Price, Bart., and — Arkwright, Esq.

YEADEN, a township, in the parish of Guisley, Upper Division of the wapentake of Skyrack, West Riding of the county of York, 7 miles (N. N. E.) from Bradford; containing 2761 inhabitants, many of whom are employed in the manufacture of woollen cloth and worsted goods, and in a scribbling-mill and extensive bleaching-grounds in the neighbourhood. This is a large clothing village, situated on a lofty moorland hill, on the north side of Airedale, and consisting of three divisions, called Upper and Lower Yeaden, and Henshaw, the first of which is the largest. The township contains 1730 acres of land, mostly the property of W. R. C. Stansfeld, Esq., the lord of the manor. There are two places of worship for Wesleyan Methodists, one of which was erected in 1835, at the cost of about £2000; and one for Association Methodists, built in 1831. A school-house was built in 1823, for children of Rawden and Yeaden.

YEALAND-CONYERS, a township, in the parish of Warton, hundred of Lonsdale, south of the sands, Northern Division of the county palatine of Lancaster, 2¼ miles (W. S. W.) from Burton-in-Kendal; containing 294 inhabitants, who are chiefly employed in the spinning of flax and the manufacture of linen. Here is a Roman Catholic chapel.

YEALAND-REDMAYNE, a township, in the parish of Warton, hundred of Lonsdale, south of the sands, Northern Division of the county palatine of Lancaster, 3 miles (S. W.) from Burton-in-Kendal; containing 227 inhabitants.

YEALMPTON (*St. Bartholomew*), a parish, in the union of Plympton-St. Mary, hundred of Plympton, Ermington and Plympton, and Southern, Divisions of the county of Devon, 3¾ miles (S. E. by S.) from Earl's-Plympton; containing 1262 inhabitants. The navigable river Yealm, which gives name to the town, here flows through much pleasing scenery, and is crossed by a bridge at the village. Yealmpton was anciently denominated a borough; it is now much decayed, though still a very genteel place, the neighbourhood being adorned by some elegant seats. A great cattle market is held on the fourth Wednesday in every month. Near the church are the ruins of a building, once, probably, the residence of the prebendary: according to tradition, it was a palace of the Saxon kings, having been occupied by Ethelwold, whose lieutenant, Lipsius, was buried here. At Kitley, the fine mansion of the family of Bastard, is a collection of the most valuable productions of Sir Joshua Reynolds. The living is a vicarage, with that of Revelstoke annexed, valued in the king's books at £35. 19. 4½.; present net income, £392; patron and appropriator, Prebendary of King's-Teignton in the Cathedral Church of Salisbury. The church is partly in the early, and partly in the later, style of English architecture, with two stone stalls, enriched with trefoil arches. Two schools are supported by E. P. Bastard, Esq.

YEARDSLEY, a joint township with Whaley, in the parish of Taxall, union and hundred of Macclesfield, Northern Division of the county palatine of Chester, 10 miles (S. E. by E.) from Stockport: the population is returned with Whaley. This place is situated on the west bank of the river Goyt, and on the road from Manchester to Buxton. It appears to have been the property of the Jodrells since the time of Henry VI., and Sir Francis Jodrell, of Henbury, is the present proprietor. There are some very productive collieries, which are extensively worked; and one of the seams of coal is crossed by a vein of lead-ore, which is of very rare occurrence. In the village, which is of considerable antiquity, a manufactory of tape is carried on upon a limited scale. The Peak Forest canal commences at this place; and a railway from it to the Cromford canal, which will open a direct line from Manchester to London, and to all parts of the kingdom, is in progress. There is a place of worship for Wesleyan Methodists, with a Sunday school attached.

YEARSLEY, a township, in the parish of Coxwold, union of Easingwould, wapentake of Birdforth, North Riding of the county of York, 5½ miles (N. E. by E.) from Easingwould; containing 164 inhabitants. A school is partly supported by George Womble, Esq.

YEAVELEY, a chapelry, in the parish of SHIRLEY, hundred of APPLETREE, Southern Division of the county of DERBY, 4½ miles (S.) from Ashbourn; containing 271 inhabitants. There is a place of worship for Independents. Here was a preceptory of the Knights Hospitallers, dedicated to St. Mary and St. John the Baptist, to which Sir William Meynell was a great benefactor in 1268, and which, at the dissolution, had a revenue of £107. 3. 8. The chapel, now called Stydd chapel, has fallen to ruins, but it is in contemplation to rebuild it.

YEAVERING, a township, in the parish of KIRK-NEWTON, union and Western Division of the ward of GLENDALE, Northern Division of the county of NORTHUMBERLAND, 4½ miles (W. N. W.) from Wooler; containing 68 inhabitants. In this township is Yeavering Bell, a lofty conical mountain rising to the height of more than 2000 feet from the vale. Its summit, which is level, and 1000 yards in circuit, is encompassed by the remains of an ancient wall, eight yards in breadth, built on the very edge of the hill, with an entrance on the south; within this is another wall, defended by a ditch, and in the centre of the area is a large cairn hollowed like a bowl. There are several smaller circles on other parts of the hill, with vestiges of a grove of oaks, strongly indicating that these works were constructed by the Druids. In the neighbourhood are an immense cairn and a cluster of rocks, respectively called Tom Tallan's grave and crag. Yeavering was the residence of some of the Saxon kings of Northumbria, particularly Edwin, after his conversion: here Paulinus was employed in baptizing other converts in the river Glen, close by. Near the village is a rude column of stone commemorating the victory gained in 1415, by the Earl of Westmorland, with an English force of 440 men, over Sir Robert Umfranville, at the head of a Scottish army of 4000.

YEDDINGHAM (*St. John the Baptist*), a parish, in the union of MALTON, wapentake of BUCKROSE, East Riding of the county of YORK, 8¼ miles (N. E.) from New Malton; containing 109 inhabitants. The living is a vicarage, valued in the king's books at £5. 4. 2.; present net income, £205; patron, Earl Fitzwilliam; impropriator, Mark Foules, Esq. The village is situated on the navigable river Derwent. Here is a National school. A Benedictine nunnery, in honour of the Blessed Virgin Mary, was founded before 1163, by Roger de Clere, which at the dissolution had a revenue of £26. 6. 8.

YELDERSLEY, a hamlet, in the parish of ASHBOURN, hundred of APPLETREE, Southern Division of the county of DERBY, 3½ miles (E. S. E.) from Ashbourn; containing 226 inhabitants. A school is partly supported by endowment.

YELDHAM, GREAT (*St. Andrew*), a parish, in the union of HALSTED, hundred of HINCKFORD, Northern Division of the county of ESSEX, 7 miles (N. W. by N.) from Halsted; containing 673 inhabitants. The village is pleasantly situated in a retired part of the parish. Many women and children are employed in the straw-plat manufacture. The living is a rectory, valued in the king's books at £20; present net income, £395; patron, John Martin Cripps, Esq. The church is a small ancient edifice with a very handsome tower; by a new arrangement of the pews 124 additional sittings have been gained, of which 84 are free, in consideration of a grant of £40 from the Incorporated Society: the churchyard is planted with avenues of fir trees. A National school is partly supported by an endowment bequeathed by John Symonds, in 1691. The gravelly soils are replete with fossils.

YELDHAM, LITTLE, a parish, in the union of HALSTED, hundred of HINCKFORD, Northern Division of the county of ESSEX, 9 miles (N. N. W.) from Halsted; containing 371 inhabitants. This parish is situated in a pleasant and healthy district: the soil is strong and rather wet, but the lands are generally in profitable cultivation. The village is pleasantly situated and contains several well-built and handsome houses. The living is a rectory, valued in the king's books at £8; present net income, £222: it is in the patronage of the Crown. The church is a spacious and venerable structure, with a handsome square embattled tower: on the south side of the chancel is a small chapel belonging to the family De La Pole; the altar-piece is richly embellished, and has a window of stained glass. Here is a Sunday National school.

YELFORD-HASTINGS, a parish, in the union of WITNEY, hundred of BAMPTON, county of OXFORD, 3½ miles (S.) from Witney; containing 17 inhabitants. This parish originally belonged to the family of Hastings; the estate was afterwards purchased by Mr. Speaker Lenthall, whose descendant, K. J. W. Lenthall, Esq., is the lord of the manor and patron of the living, which is a discharged rectory, valued in the king's books at £4. 3. 6½.; present net income, £108.

YELLING (*Holy Cross*), a parish, in the union of CAXTON and ARRINGTON, hundred of TOSELAND, county of HUNTINGDON, 6 miles (E. N. E.) from St. Neots; containing 326 inhabitants. The living is a rectory, valued in the king's books at £14. 10. 5.; present net income, £380: it is in the patronage of the Crown. The rent of an allotment of about 14 acres of land, amounting to £14. 14. 7., is chiefly distributed among the poor.

YELVERTOFT (*All Saints*), a parish, in the union of RUGBY, hundred of GUILSBOROUGH, Southern Division of the county of NORTHAMPTON, 9¾ miles (N. by E.) from Daventry; containing 596 inhabitants. The living is a rectory, valued in the king's books at £25. 0. 10.; present net income, £487: it is in the patronage of the Earl of Craven. There is a place of worship for Independents. The Union canal passes through the parish. Thirty poor children are instructed for £35 a year, arising from land bequeathed by Mrs. Ashby, in the year 1719.

YELVERTON (*St. Mary*), a parish, in the union of LODDON and CLAVERING, hundred of HENSTEAD, Eastern Division of the county of NORFOLK, 5¾ miles (S. E. by S.) from Norwich; containing 80 inhabitants. The living is a discharged rectory, with that of Alpington annexed, valued in the king's books at £10; it is in the patronage of the Crown. The tithes have been commuted for a rent-charge of £400, subject to the payment of rates, which on the average have amounted to £64; the glebe comprises 21 acres, valued at £42 per annum. The church contains a Norman font. Here is a school on the National plan, which was commenced in 1823, and is aided by contributions from the National Society and the rector.

Corporation Seal.

YEOVIL (*St. John the Baptist*), a market-town and parish, and the head of a union, in the hundred of Stone, Western Division of the county of Somerset, 9½ miles (S. S. E.) from Somerton, and 122 (W. S. W.) from London; containing 5921 inhabitants. This place, which, from the discovery of tesselated pavements and other relics of antiquity, is supposed to have been known to the Romans, derives its name from the river Yeo, or Ivel, the *Velox* of Ravennas, which, having its source in seven springs near Sherborne, separates the counties of Somerset and Dorset, and passes this place at a short distance to the east, beneath a stone bridge of three arches, near which it receives a small stream, turning three mills, which bounds the town on the south. It was anciently called the town, borough, lordship, and hundred of Yeovil, and included a district which soon after the Conquest fell into the possession of the crown; part of these lands was assigned by the name of the manor of Yeovil to the rector of the church of St. John the Baptist in this town, by one of the kings of England, who also granted him a weekly market on Friday, view of frankpledge, and several other rights and privileges. The inhabitants were also incorporated under the designation of the Provost and Commonalty of Yeovil; and a daily court of pie-poudre was held by the provost on behalf of the rector, but has been long discontinued. The manor was held by the successive rectors till the year 1418, when the then rector resigned the church, together with the town and lordship, to Henry V., who granted the manor, with all its rights and privileges, and the rectory, to the abbot and convent of the Virgin Mary and St. Bridget, which that monarch had founded at Sion, in the county of Middlesex. This grant was confirmed by Edward IV., and after the dissolution of the monasteries the manor was settled by Henry VIII. on his Queen, Catherine, who held it till her death, when it reverted to the crown, and was subsequently granted by James I., in the 8th of his reign, to George and Thomas Whitemore, who in the year following sold it to Sir Edward and Sir Robert Phelips, whose descendant, John Phelips, of Montacute, Esq., is the present lord. In 1449, an accidental fire consumed 117 houses in the town, of which 45 belonged to different chantries, and on this occasion an indulgence of 40 days was granted to all who contributed to repair the loss. The town consists of numerous streets, many of them spacious; and the houses, of which several are of stone, are in general well built. It is well supplied with water from springs, which rise at a short distance, and is sheltered on the north by a range of hills, which, as well as the adjacent country, are in a high state of cultivation; the metropolis is chiefly supplied with what is called Dorset butter from the numerous dairy farms in this neighbourhood. On the south-east are three remarkable hills, from the summit of one of which, Newton hill, the English and Bristol Channels can be discerned. The inhabitants were formerly engaged in the woollen manufacture to a great extent, but this has been superseded by that of leather gloves, which are made here to the extent of 4000 dozen per week, affording employment to many hundred persons in the neighbouring villages. The market is on Friday, and every alternate Friday is the great market: corn, cattle, pigs, bacon, butter, cheese, hemp, and flax, are sold in considerable quantities; in the purchase and sale of the two last articles upwards of £1000 is frequently returned in one day. Fairs are held on June 28th and November 17th, for horses, cattle, and pedlery, and continue for two days each. The market-house is supported on stone pillars. The government of the town, which is a corporation by prescription, is vested in a portreeve and eleven burgesses; a macebearer and two constables are chosen for the town, and two constables for the parish, which has a distinct jurisdiction: the portreeve exercises magisterial authority while in office. A court of record is held every three weeks, and a court leet for the borough annually, by the lord of the manor.

The living is a vicarage, with that of Preston annexed, valued in the king's books at £18; present net income, £391; patron, John Phelips, Esq.; impropriators, H. Halsey and J. Newman, Esqrs. The church is a fine old cruciform structure, near the centre of the town, in the ancient style of English architecture, with a tower at the west end surmounted by a balustrade and according to Leland contained the chantries of St. John the Baptist, the Holy Cross, the Holy Trinity, and the Virgin Mary; a gallery has been erected, containing 200 free sittings, the Incorporated Society having granted £62 in aid of the expense. At its western end stands an ancient building, now used as a school-room, of much older date than the church itself, which was probably a chapel. There are places of worship for Particular Baptists, Independents, Wesleyan Methodists, and Unitarians. A free school, originally founded in 1707, by subscription, has been endowed with sundry bequests, especially those of Francis Cheeseman, in 1711; of John Noyes, in 1718, who bequeathed the manor of Lea, and estates in Romsey Extra, producing about £150 per annum, partly extended to Romsey and Fisherton-Anger, for instructing, clothing, and apprenticing poor children, and afterwards for setting some of them up in business; and of Elizabeth Boucher, in 1725. About 30 boys are educated on the various foundations, and some of them are also clothed and apprenticed. An almshouse for a master, two wardens, and twelve poor persons, of either sex, was founded in 1476, by John Woburne, minor canon of St. Paul's Cathedral, and endowed to a considerable extent with landed property; a chapel is annexed to the institution. The portreeve's almshouses, in Back-street, for four poor women, each of whom receives a small allowance, are of unknown origin. The poor law union of Yeovil comprises 35 parishes or places, containing a population of 25,581, according to the census of 1831, and is under the care of 47 guardians. In the hamlets of Kingston, Marsh, and Hensford were ancient chapels, formerly used as oratories, but afterwards as places of public worship, dependent on the mother church, in which the inhabitants of those places had a right of sepulture; the places appropriated for that purpose are still pointed out in the church.

YEOVILTON (*St. Bartholomew*), a parish, in the union of Yeovil, hundred of Somerton, Western Division of the county of SOMERSET, 1½ mile (E.) from Ilchester; containing 275 inhabitants. The living is a rectory, valued in the king's books at £26. 9. 2.; present net income, £445; patron, Bishop of Bath and Wells. The river Yeo bounds the parish on the south. A National school is supported by subscription.

YETLINGTON, a joint township with Callaley, in the parish of WHITTINGHAM, union of ROTHBURY, Northern Division of COQUETDALE ward and of the county of NORTHUMBERLAND, 12 miles (W. S. W.) from Alnwick: the population is returned with Callaley.

YETMINSTER (*St. Andrew*), a parish (formerly a market-town), in the union of SHERBORNE, hundred of YETMINSTER, Sherborne Division of the county of DORSET, 5¼ miles (S. W.) from Sherborne; containing 1199 inhabitants. This extensive parish lies on the western border of the county, and gives name to the hundred. The village, which is situated near the river Ivel, consists of a long well-built street, having still the appearance of a town. In the year 1300, the Bishop of Sarum obtained a grant from Edward I. for a market and fair, which was confirmed by Richard II., but the market has been long disused, and fairs are now held on April 23rd and October 1st. The living is a discharged vicarage, in the patronage of the Prebendary of Yetminster in the Cathedral Church of Salisbury (the appropriator), valued in the king's books at £20. 14. 7.; present net income, £277. The church is a large ancient structure with a lofty tower, crowned with battlements and pinnacles. There are chapels of ease at Leigh and Chetnole. The Hon. Robert Boyle, in 1699, bequeathed an estate, now producing more than £70 per annum, for teaching 20 poor boys: here is a National school.

YIELDING, or YELDEN (*St. Mary*), a parish, in the hundred of STODDEN, union and county of BEDFORD, 3¾ miles (E.) from Higham-Ferrers; containing 276 inhabitants. The living is a rectory, valued in the king's books at £13. 13. 4.; present net income, £317; patron and incumbent, Rev. E. S. Bunting.

YOCKLETON, a township, in the parish of WESTBURY, hundred of FORD, Southern Division of the county of SALOP, 6 miles (W. by S.) from Shrewsbury: the population is returned with the township of Westbury.

YOKEFLEET, East Riding of the county of YORK. —See YORKFLEET.

Arms.

YORK, a city and county of itself, having exclusive jurisdiction, and the head of a union, locally in the East Riding of the county of YORK, of which it is the capital, 198 miles (N. N. W.) from London; containing 25,359 inhabitants. The origin of this ancient city, in Nennius' catalogue called *Caer Ebrauc*, is involved in obscurity. According to Llwyd, the learned Welsh antiquary, it is identified with the city called by the Britons *Caer Effioc*, and, among the towns of the Brigantes mentioned by Ptolemy, with the *Eboracum* of the Romans. The latter name is probably a modification of the former, on its becoming the station of the sixth legion, sent into Britain by Adrian. The early importance of the city must unquestionably be attributed to the Romans, who fixed a colony here, and made this the metropolis of their empire in Britain. The Emperor Adrian fixed his principal station in this city, in 124, while engaged in restraining the incursions of the northern hordes. In the reign of Commodus, the Caledonians having made a successful irruption into Britain, attacked and routed the Roman army, and laid waste the open country as far as the city of York; but Marcellus Ulpius, who had been sent over from Rome, aided by the ninth legion, at that time stationed in the city, quickly routed them with great slaughter, and drove them back within their own territory. The Emperor Severus, in the fourteenth year of his reign, finding that the city of York was besieged by the northern Britons, came over into Britain, with his sons Caracalla and Geta and a numerous army, and attended by his whole court; the besiegers, on his approach, retired towards the north, and intrenched themselves behind the rampart which his predecessor Adrian had constructed, to defend the inhabitants from their assaults. The emperor, leaving his son Geta in the city, to administer justice during his absence, advanced with Caracalla to give the Britons battle, and, though from age and infirmity obliged to be carried in a litter, routed them with great slaughter; and leaving Caracalla to complete his victory, and to superintend the erection of a strong wall of stone nearly 80 miles in length, which he ordered to be built near the rampart of earth raised by Adrian, as a more effectual barrier against their future incursions, returned to York, where he spent the remainder of his days. The Caledonians again taking up arms, Severus sent out his legions with positive instructions to give no quarter, but to put men, women, and children indiscriminately to the sword. During this period the city was in its highest degree of splendour; the residence of the court, and the resort of numerous tributary kings and foreign ambassadors, conferred upon it a distinction almost unsurpassed among the cities of the world, and obtained for it the appellation of "*Altera Roma*," to which city, in these respects more than in any fancied resemblance of design, it might not unaptly have been compared. Severus died in his palace here, in 212, and his funeral obsequies were performed with great solemnity on the west side of the city, near Ackham; in the immediate vicinity of the spot are three natural sandhills, called Severus' hills, upon which the ceremony is supposed to have been performed: his remains were deposited in a costly urn, and sent to Rome, where they were placed in the sepulchre of his ancestors. Constantius Chlorus, another of the Roman emperors who resided for some time in Britain, died also in this city, in 307. His son, Constantine the Great, who at the time of his father's death was at York, was proclaimed emperor by the army. Of the grandeur of the city during its occupation by the Romans, numerous vestiges have been discovered, and various remains of Roman architecture have been found. Of these, the principal are, a polygonal tower, with the south wall of the Mint yard; a votive altar to the tutelar genius of the place; an altar dedicated to the household and other gods by Ælius Marcianus; a cemetery without Micklegate Bar, in which many urns, containing ashes and burnt bones, have been recently dug up; also a small coffin of red clay, and a leaden coffin, of large dimensions, enclosed with oak ; besides numerous coins and various other relics.

After the departure of the Romans from Britain, the city suffered greatly from the depredations of the Scots and Picts, by whom it was frequently assailed; and after the arrival of the Saxons it experienced considerable devastation in the wars which arose between the Britons and their new allies, in the many contests for empire during the establishment of the several kingdoms of the Heptarchy, and in the mutual wars of their several monarchs for the extension of their territories. By the Saxons the city was called *Euro wic*, *Euore wic*, and *Eofor wic*, all descriptive of its situation on the river Ouse, which, according to Leland, was at that time called the *Eure*; and from these Saxon appellations its present name is most probably contracted. Edwin, King of Northumbria, made this place the metropolis of his kingdom, and upon his conversion to Christianity, soon after his marriage with Ethelburga, daughter of Ethelbert, King of Kent, in 624, erected it into an archiepiscopal see, of which he made Paulinus, Ethelburga's confessor, primate. This monarch founded a church, which he dedicated to St. Peter, and his example in embracing the Christian faith was followed by vast numbers of his subjects, who, under the influence of Paulinus' ministry, were converted to Christianity. On the death of Edwin, who was killed in battle in 633, while resisting an attack of the Britons under Cadwallo, assisted by Penda, King of Mercia, the city suffered severely from the ravages of the confederated armies, who devastated it with fire and sword, and massacred the inhabitants. Ethelburga fled into Kent, accompanied by Paulinus; and the newly-erected church, which was scarcely finished, lay neglected for some time, till it was restored by Oswald, Edwin's successor, who, collecting a small army, after a fierce and sanguinary conflict, slew Cadwallo and the chief of his officers, and regained possession of his kingdom.

After the union of the several kingdoms of the Heptarchy, York again became a place of importance, and in the ninth century was the seat of commerce and of literature, as far as they then prevailed in the kingdom. During the Danish incursions it was reduced to ashes, and having been rebuilt, it finally became one of the principal settlements of those rapacious invaders, who kept possession of it till Athelstan attacked and expelled them from the city, and demolished the castle which they had erected for their defence. In the peaceful times which followed, the city gradually recovered, and continued to flourish till the Conquest, at which time, according to the Norman survey, it contained six shires, exclusively of the archbishop's; one of these lay waste in consequence of the demolition of the castles, in the other five were 1428 houses, and in the archbishop's 200 houses. William the Conqueror placed strong garrisons in the two castles which remained, both to overawe the inhabitants, and to protect the city from the attempts of the Saxon nobility, who, refusing to submit to his government, had gone over into Denmark, to incite Sweyn, king of that country, to invade Britain for the recovery of a throne which had descended to him from his ancestors. In 1069, Sweyn sent his two sons, Harold and Canute, with 240 ships and a numerous army, who, having arrived in the Humber, disembarked their forces and advanced to York, laying waste the country through which they marched: on their arrival before the city they were joined by Edgar Atheling, who, with a large number of the English exiles, had arrived from Scotland for the same purpose. The garrison, to prevent them from fortifying themselves in the suburbs, set fire to the houses; but the wind being high, the flames communicated to the city, and, during the consternation of the inhabitants, the enemy entered and made themselves masters of it. The successful Danes then proceeded northward, and, after subduing the greater part of Northumberland, finding their further progress arrested by the severity of the winter, returned to York, where they took up their winter quarters. William was unable, from the severity of the weather, to bring an army against them till the spring, when he advanced with his forces and encamped near the confluence of the rivers Humber and Trent, and, after a severe and obstinate battle, obtained a triumphant victory; Harold and Canute escaped, with a few of their principal officers, to their ships, and Edgar Atheling, with great difficulty, effected his retreat into Scotland. William, attributing the first success of the Danes to the treachery of the citizens, took signal vengeance on them, burnt the city, and laid waste the neighbouring country, which, from the Humber to the Tyne, remained for several years in a state of desolation. From this signal calamity York gradually recovered in the two succeeding reigns. Archbishop Thomas repaired the cathedral, for temporary use, by covering the remaining walls with a roof, and afterwards, finding that they had been essentially injured by the fire, he pulled them down and rebuilt the church. Though continually exposed to the assaults of the Scots, it continued progressively to advance in importance; and, in 1088, a splendid monastery for monks of the Benedictine order was erected, and dedicated to St. Mary, of which William Rufus laid the first stone. In the reign of Stephen the city was almost entirely consumed by an accidental fire, which is stated to have destroyed the cathedral, the monastery, with some other religious houses, and 39 parish churches.

In 1138, David King of Scotland, whom Matilda had engaged in her interest, by a promise of ceding to him the county of Northumberland, laid siege to York; but Archbishop Thurstan, though at that time confined to his bed by illness, assembled the nobility and gentry, who, under the conduct of Ralph, Bishop of Durham, his deputy, advanced against him, and put him to flight with considerable loss. In the reign of Henry II., one of the first meetings distinguished in history by the name of Parliament was held here in 1169, at which William King of Scotland, accompanied by all his barons, abbots, and prelates, attended, and did homage to Henry in the cathedral, acknowledging him and his successors his superior lords. In the reign of Richard I., a general massacre of the resident Jews took place, under circumstances of peculiar atrocity; the fury of the populace had first been excited against them for mingling with the crowd at the king's coronation in London, and, in spite of a proclamation in their favour by the king, the same spirit of persecution manifested itself in many of the large towns, especially in York, where many of the victims, having taken refuge in the castle, after defending it for some time against their assailants, perished by their own hands, after putting their wives and children to death. In 1221, Alexander King of Scotland, who the year before had met Henry III. at York, had another interview with that monarch here, when he espoused the Lady Joan, sister of the king, and at the same time Hubert de Berg married the Lady Margaret, sister of Alexander; these marriages were both solemnised in the city, in presence of the king. In 1237, Cardinal Otto, the pope's legate, negociated a peace between the kings of England and Scotland, who met here for that purpose; and, in 1252, Alexander III., King of Scotland, came to York, attended by a large retinue of his nobility, and celebrated his marriage with Margaret, daughter of Henry III. Upon this occasion considerable festivities took place; the Scottish king, with his retinue, was lodged in a separate part of the city appropriated to their use, and he and 20 of his principal attendants received the honour of knighthood. In the reign of Edward I., a parliament was held here, which was attended by most of the barons and principal nobility; the great charter, with the charter of forests, was renewed with great solemnity, and the Bishop of Carlisle pronounced a curse upon all who should attempt to violate it. The Scottish lords, who were summoned to attend this parliament, not making their appearance, the English lords decreed that an army should be sent, under the command of the Earl of Surrey, to relieve Roxburgh, which the Scots were at that time besieging. After the battle of Bannockburn, in 1315, Edward II. came to York, and held a council, in which it was decreed to send a force for the defence of Berwick, then threatened with siege by Robert Bruce; and, in 1322, the Earl of Hereford, who, with the Earl of Lancaster, had rebelled against the king, having been killed at Boroughbridge by Adam de Hercla, who had been sent against him, his body was conveyed hither, where also many of his partisans were hanged, drawn, and quartered. After the suppression of this rebellion, which had been excited to free the kingdom from the influence of the De Spencers, the king held a parliament in this city, in which the decree made in the preceding year at London, for alienating their estates, was reversed, and the elder Spencer created Earl of Winchester. At this parliament the several ordinances made at different times were examined, and such of them as were confirmed were, by the king's order, directed to be called statutes; the clergy of the province of York granted the king a subsidy of fourpence in each mark, and Edward, the king's son, was created Prince of Wales and Duke of Aquitaine. After the breaking up of the parliament, Aymer de Valence was arrested, on his return, by order of the king, and brought back into this city, on a charge of having secretly abetted the barons in their rebellion, and of having contributed to excite the late disturbances; but, upon the intercession of several noblemen, he was released, on payment of a fine, and taking an oath of fidelity and allegiance to the king. This monarch, having collected an army to oppose Robert Bruce, who was then desolating the English border, was surprised by the enemy, and with difficulty escaped into the city.

In the beginning of the reign of Edward III., the Scots having sent three armies to lay waste the English border, and take possession of the adjoining counties, the king collected an army, with which he marched to York, where he was soon after joined by Lord John Beaumont, of Hainault, with a considerable body of forces. The Scots, being informed of these preparations, sent ambassadors to York, to negociate a treaty of peace; upon the failure of which, Edward advanced against them with his army, and, enclosing them in Stanhope Park, had nearly made them prisoners; but by the treachery of Roger Mortimer, who opened a road for their escape, they withdrew their forces, and Sir William Douglas assaulting Edward's camp by night, nearly succeeded in killing the king: on the failure of his attempt, the Scots, after doing what mischief they could, retreated within their own territories. Beaumont, after receiving an ample reward for his services, returned to his own dominions, and a marriage was soon after negociated between his niece and the king, which was solemnized at York, in 1327. After the battle of Hallidown Hill, in 1333, Edward retired to York, where he held a parliament, in which Edward Balliol, whose cause he had embraced, in opposition to David Bruce, was summoned to attend him; but Balliol, having sent messengers to excuse his attendance, afterwards met the king at Newcastle. In 1335, Edward took up his residence in the monastery of the Holy Trinity in this city, and held a council, in which the Bishop of Durham, then Chancellor, resigned the great seal into his hands, and he immediately delivered it to the Archbishop of Canterbury, who took the usual oaths of office in the presence of the council, and on the same day proceeded to the church of the monastery of the Blessed Mary, where he affixed it to several deeds. Richard II., while on his expedition against the Scots, in 1385, passed some time in this city, which he also visited in 1389, in order to adjust some differences that had arisen between the ecclesiastical and civil authorities. On this occasion the monarch took his own sword from his side, and presented it to William de Selby, the mayor, to be borne in all public processions before him and his successors, whom he dignified with the title of Lord Mayor, which honour has been ever since retained, and is possessed by no other city, except those of London and Dublin. This monarch, in the nineteenth year of his reign, erected the city into a county of itself, and appointed two sheriffs, in lieu of the three bailiffs that previously formed a part of the corporation, and presented the first mace to the city, and a cap of maintenance to the sword-bearer: during this reign, Edmund Langley, fifth son of Edward III., was created the first Duke of York. In the reign of Henry IV., the Earl of Northumberland and Lord Bardolph, who, after the defeat of an insurrection against that monarch, headed by the Earl of Nottingham and the Archbishop of York, had retired into Scotland, raised some forces in that country, and made an irruption into the northern part of the kingdom; but Sir Thomas Rokesby, sheriff of Yorkshire, having levied some forces, defeated them in a battle in which both those noblemen were slain; and the king, marching into York, found several of the earl's adherents in the city, of whom some were ransomed and others punished; the earl's head was severed from his body, and, being sent to London, was fixed upon the bridge.

During the war between the houses of York and Lancaster, this city was occasionally connected with the contending parties, and though not actually the seat of war, several of the battles took place in the neighbourhood. In the reign of Henry VI., Edward, Duke of York, who had raised an army in support of his claim to the crown, was killed in the battle of Wakefield, and his body being afterwards found among the slain, the head was struck off by order of Queen Margaret, and fixed upon the gate of York, with a paper crown upon it, in derision of his pretended title. In 1461, soon after the assumption of the crown by Edward IV., Queen Margaret, having levied an army of 60,000 men, made another effort to regain the crown, and advancing towards York, was met by Edward and the Earl of Warwick with 40,000 men; the armies met at Towton, and a sanguinary battle ensued, in which 36,776 men are said to have been slain. During the engagement, Henry and Margaret remained in the city of York; but, on hearing of the total defeat of their army, fled with great precipitation into Scotland. After the restoration of Henry VI., Edward IV. landed at Ravenspur, in Yorkshire, in 1471, and advanced to York without opposition. On his arrival he hesitated to enter the gates, for fear of treachery; but being informed by the mayor and citizens that, provided he sought only to recover his dukedom of York, and not to lay his hand upon the crown, he might enter with safety, he took up his abode there, after swearing to a priest, who met him on his entrance, to treat the citizens with courtesy, and to be faithful and obedient to the king. Having remained at York for some time, he left a garrison in the city and marched towards London; and meeting with the army of the Earl of Warwick, near Barnet, a sanguinary battle took place, in which the earl, his brother, and several of his principal officers, were slain, and Edward, after this victory, was peaceably established on the throne. Richard III. arrived at York in the year 1483, and was crowned with great solemnity and pomp in the cathedral church, by Archbishop Rotherane. In the year 1503, Margaret, eldest daughter of Henry VII., visited the city, in which she remained for some days.

In the reign of Henry VIII. the art of printing was first established in York, by Hugo Goes, the son of an ingenious printer at Antwerp. At the time of the dissolution of monasteries, during this reign, there were in the city of York, besides the cathedral, forty-one parish churches, seventeen chapels, sixteen hospitals, and nine religious houses, including the monastery of St. Mary: with the suppression of the monasteries, ten parish churches were demolished, and their revenues and materials appropriated to secular uses. In consequence of these proceedings, the insurrection called the Pilgrimage of Grace originated in Yorkshire, and in a short time 40,000 men, headed by Robert Aske, and attended by priests with sacred banners, took possession of this city and of Hull. The Duke of Norfolk being sent against them, they were ultimately dispersed, their principal leaders were taken and executed, and Aske was brought to York, where he was hanged upon Clifford's tower. After the suppression of this insurrection, Henry made a tour through the county, on the border of which he was met by 200 of the principal gentry, with 4000 of the yeomanry on horseback, who made their submission to the king, by Sir Matthew Bowes, their speaker, and presented him with £900: on his advance towards the city from Barnsdale, the abbot of York, attended by 300 priests, went out to meet him, and presented him with £600; and on his entering it, the lord mayor, with the mayors of Newcastle and Hull, who had repaired to York to meet him, received him with great pomp and ceremony, and in token of their submission presented him with £100 each. Henry remained at York for twelve days, and established there a president and council, under the great seal of oyer and terminer, and after making several other arrangements, departed for Hull, where he threw up some additional fortifications. During the reign of Elizabeth, an insurrection to restore the Roman Catholic religion was headed by Thomas Percy, Earl of Northumberland, and Charles Neville, Earl of Westmorland, on the failure of which, Simon Digby, of Askew, and John Fulthorpe, of Iselbeck, Esqrs., who had been made prisoners, were taken from York Castle to Knavesmire, where they were executed. The Earl of Westmorland escaped out of the country, but the Earl of Northumberland, being taken prisoner, and attainted by parliament, was beheaded at York, and his head placed on the Micklegate bar. James I. resided for some time at the manor palace in this city; and, in 1633, Charles I. visited York, where, in 1639, he held a council at the palace, and made the city the chief rendezvous of the troops destined to march against the Scottish rebels. During his visit, the king, who was then 39 years of age, ordered the Bishops of Ely and Winchester to wash the feet of 39 beggars, first in warm water, and afterwards with wine, which ceremony was performed in the south aisle of the cathedral. The king afterwards gave to each of them a purse containing 39 silver pence, several articles of wearing apparel, and a quantity of wine and provisions. Before leaving the city, he dined with the lord mayor and corporation, and expressed his satisfaction at the hospitality with which he had been entertained, by conferring the honour of knighthood on the lord mayor and the recorder. While Charles remained here, the Scots demanded an audience to express their grievances, and ultimately succeeded in obtaining a treaty of peace, after which the king disbanded his army, and returned to London.

Previously to the commencement of the parliamentary war, the king, to avoid the importunity of the parliament, who petitioned for the exclusive control of the militia, and for other privileges subversive of the royal authority, removed to this city, and was received by the inhabitants with every demonstration of loyalty and affection. He sent a message to both houses of parliament, and afterwards advanced to Hull to secure the magazine which had been left in that town, upon disbanding the army raised to oppose the Scots; but, on being denied admission by Sir John Hotham, the parliamentary governor, he returned to York. The parliament soon after appointed a commission to reside in the city, to strengthen their party, and to watch the movements of the king; and on their passing an ordinance for embodying the militia, the king ordered his friends to meet him in this city, whither he directed the several courts to be in future adjourned. The Lord-Keeper Littleton, being ordered by the parliament not to issue the writs, apparently obeyed; but on the first opportunity made his escape to York, and bringing with him the great seal, joined the royal party, for which he was afterwards proclaimed by the parliament a traitor and a felon. On May 27th, 1642, the king issued a proclamation, dated from his court at York, appointing a public meeting of the nobility and gentry of the neighbourhood to be held at Heworth moor, on the 3rd of June. This meeting was attended by more than 70,000 persons, who, on his Majesty's approach, accompanied by his son, Prince Charles, and 150 knights in complete armour, and attended with a guard of 800 infantry, greeted him with the loudest acclamations of loyalty and respect. The king, in a short address, explained the particulars of the situation in which he was placed, and thanking them for their assurances of loyalty and attachment, returned to the city, where, after keeping his court for more than five months, during which time every attempt at negociation had failed, he advanced to Nottingham, and there erected his standard. In 1644, the parliamentary army, under Sir Thomas Fairfax, the Earl of Leven, and the Earl of Manchester, besieged the city, which was defended by the Marquess of Newcastle, and in a state of great distress; but hearing that Prince Rupert was approaching with an army to its relief, they raised the siege, and encamped on Marston moor, about six miles from York, where they awaited the arrival of the royalists. The armies, which were nearly equal in number, each consisting of about 25,000 men, met on July 2nd, when, after a long and sanguinary engagement, the royalists were defeated: the parliamentarians, after this signal victory, returned to the siege of York, which, having held out nearly four months, surrendered upon honourable terms. On January 1st, 1645, the great convoy, under the conduct of General Skippon, arrived at York with the sum of £200,000, which, according to treaty, was paid to the Scots for surrendering to the parliament the person of the unfortunate monarch, who, relying upon their fidelity, had entrusted himself to their protection. After the Restoration, Charles II. was proclaimed here with triumphant rejoicings.

York was connected with several of the proceedings which led to the Revolution of 1688: James II. had attempted to introduce the Roman Catholic religion into the city, and for this purpose had converted one of the large rooms in the manor palace into a chapel, in which the service was performed according to the Romish ritual. This attempt, together with some arbitrary proceedings on the part of the court, gave great offence to the citizens; and in a general meeting appointed to vote a loyal address to the king, on the rumoured landing of the Prince of Orange, they resolved to add to their address a petition for a free parliament and redress of grievances. On November 19th, the Duke of Newcastle, lord-lieutenant of the county, arrived in the city to preside at a county meeting for the same purpose; but finding that several of the deputy-lieutenants had joined with the citizens in their petition, retired the next day in disgust. The meeting took place in the guildhall, where a petition was framed in addition to the address; but during the proceedings, a rumour being raised of an insurrection of the papists, the party rushed from the hall, and, headed by some gentlemen on horseback, advanced towards the troops of militia, at that time on parade, crying out "A free parliament, the Protestant religion, and No Popery." The militia immediately joined them, and having secured the governor and the few regular troops then in the city, they placed guards at the several entrances leading into the town. On the following day they summoned a public meeting, passed resolutions, and issued a declaration explanatory of their proceedings. On the 24th they attacked, plundered, and destroyed the houses belonging to the principal Roman Catholics in the city, together with their chapels; and on December 14th, a congratulatory address was voted by the lord mayor and corporation, to the Prince of Orange, who, and also his consort, were proclaimed on February 17th, by the title of King William and Queen Mary, amidst general acclamation. During the rebellion in 1745, the inhabitants raised four companies of infantry, called the York Blues, for the protection of the city against the attempts of the insurgents. In 1789, their Royal Highnesses the Prince of Wales and the Duke of York visited the races, on the conclusion of which they entered Earl Fitzwilliam's carriage, and were drawn into the city by the populace, who took the horses from the carriage, amidst the loud congratulations of the assembled inhabitants. On February 2nd, 1829, the inhabitants were greatly alarmed by the appearance of smoke issuing from the roof of the cathedral, and, on inspection, had the mortification to find that the choir of that beautiful structure was in flames. Every possible assistance was immediately obtained, but the beautiful tabernacle-work, the roof, and every thing combustible in that part of the church, were destroyed, and several of the piers and the finer masonry materially injured. This lamentable destruction, which was regarded as a national calamity, was the work of a lunatic, who had secreted himself for that purpose in the cathedral, after the performance of the evening service, and, under the influence of a fanatical delusion, set fire to this magnificent pile. Within a very short time after, a sum of £50,000 was subscribed, principally within the county, and a large quantity of well-seasoned timber, of the value of £5000, was contributed by government from the royal dock-yards, for the restoration of the building.

The city is pleasantly situated on the bank of the river Ouse, near its confluence with the Foss, and is nearly three miles in circumference: it is almost surrounded with walls, generally supposed to have been raised by the Romans, and restored in the reign of Edward I., but they were much damaged during the parliamentary war, and remained in a very dilapidated state till 1831, when they were repaired by subscription, and the walk round the top restored, forming at present a beautiful promenade. They are defended by four ancient gates, forming the principal entrances, namely, Micklegate Bar, to the south-west; Bootham Bar, to the north-west; Monk Bar, to the north-east; and Walmgate Bar, to the south-east. Terminating that part of the wall which extends from Walmgate Bar, on the north-west, to the edge of the marsh formed by the waters of the Foss and other smaller streams, is the Red Tower, built of brick: the inner face of this part of the wall presents a series of arches, and the same is seen in other parts. Besides these principal gates, there were five posterns, or smaller entrances, which took their names from the streets and parts of the city to which they led, and were severally called North-street, Skeldergate, Castlegate, Fishergate, and Layerthorpe posterns; but Skeldergate and Castlegate posterns have been removed. There are six bridges, of which the principal, over the river Ouse, was begun in 1810, and completed in 1820, at an expense of £80,000: it is a handsome and substantial structure of three arches of freestone, forming a communication between the parts of the city which are on opposite sides of the river. A handsome stone bridge has been erected over the Foss, of which the first stone was laid in 1811; and over the same river are four other bridges, affording communication with the suburbs. The city has of late years undergone considerable improvement under a body of local commissioners: it is well paved, lighted with gas by a company whose extensive works were erected in 1824, and by another company called the York Union Gas Company, established in 1837; and amply supplied with water by the York Company's water-works. It is progressively increasing in size: in the adjacent township of Fulford, a row of new and very superior buildings, called New Walk Terrace, has been erected, separated by a drain only from the city liberty, and in all probability it will ere long extend itself at many points into this township. On the north-west, the continuous buildings stretch out of the borough a considerable distance into the township of Clifton, and on the north-east they nearly extend into the township of Heworth. Heworth Moor was enclosed in 1817, since which period a great number of substantial and excellent houses have been built in that neighbourhood, extending along the side of the Malton road; and many market gardens are cultivated in this thriving and populous district. Interspersed throughout the neighbourhood are numerous mansions of persons in affluent circumstances, which, with their gardens and pleasure grounds, contribute materially to enrich the scenery. Of the ancient castle, erected by William the Conqueror, there remains only the mount, thrown up with prodigious labour, on which is an ancient circular building, called Clifford's Tower, appearing to have been the keep, which was reduced to its present ruinous condition by an accidental fire in the year 1685. The ancient fortress, after it was dismantled by Cromwell, remained in a dilapidated state for several years; its site is now occupied by the county prison.

The subscription library was established in 1794, and contains a well-assorted collection in every department of literature, at present exceeding 16,000 volumes: a handsome building was erected for the purpose in 1811, of which the ground-floor is occupied by a subscription news-room, well furnished with periodical publications. There are also three other subscription news-rooms, all of which are well supported. The Philosophical Society was instituted in 1822, and in 1826 obtained from the crown a grant of three acres of land, for the erection of a suitable building and the establishment of a botanic garden: among other subjects it embraces the geology, natural history, and antiquities of the county. Its meetings were held, and the museum deposited, in a house at the extremity of Ousebridge; but a handsome and commodious building was erected a few years since, for the use of the society, on part of the site of the venerable abbey of St. Mary, by voluntary subscription of the members, assisted by the noblemen and gentlemen of the county: it is in the Grecian style of architecture, and of the Doric order, and is surrounded by about three acres of land laid out as a botanic garden, and ornamented with shrubberies, pleasure grounds, and plantations. The meetings of the society, which are in general well attended, are held on the first Tuesdays in January, February, March, April, July, October, November, and December. The Yorkshire Central Agricultural Association was formed in 1832, under the auspices of the Earl of Harewood, who is its patron. The theatre was erected in 1769, and in 1822 was considerably enlarged, greatly improved, and elegantly fitted up: it is brilliantly lighted with gas, and is opened by the York company of comedians, in the first week in March, and continues open till the first week in May; the company also perform during the assizes and the race-week. Concerts and assemblies are held periodically during the winter season, in a splendid suite of rooms in Blake-street, erected after a design by Lord Berlington, in 1730, upon a scale of sumptuous magnificence, unparalleled in any town in the kingdom: the entrance is by an elegant vestibule into the principal room, which is 112 feet in length, 40 feet wide, and 40 feet in height, ornamented in the lower part with a range of Corinthian columns and an enriched cornice, from which rises a series of the Composite order, surmounted by an appropriate cornice, and decorated with wreaths of fruit and foliage: this room is lighted by thirteen brilliant chandeliers suspended from the ceiling, each of which consists of eighteen branches. On the right hand of the large room is a smaller, in which the subscription assemblies are held, of which there are generally six or seven, and the subscription concerts, of which there are generally four, during the season, exclusively of benefit concerts, and the assize and race balls, which are held in the larger room: the smaller room, which is elegantly fitted up, is 66 feet in length and 22 feet wide, and the ceiling is richly ornamented: there are also other apartments and ante-rooms, forming altogether a splendid suite. The new concert-rooms, adjoining the old assembly-rooms, were erected in 1824, at an expense of £9400, raised from the profits of the musical festivals, and were opened to the public in 1828; the principal room is 92 feet long, 60 feet wide, and 45 feet high, and will afford accommodation for 1800 persons.

The York musical festival was instituted in 1823, and has been liberally patronised, not only by the nobility and gentry resident in the county, but also by families of the highest distinction in every part of the kingdom. The nave of the spacious cathedral is fitted up on these occasions for the performance of sacred music: the orchestra combines the united talents of the metropolis with the professional skill of every other part of the kingdom, and the performances rank among the most profitable and attractive of these periodical festivals. Miscellaneous concerts are held also in the large concert-rooms during the period of the festival; and the proceeds, after deducting the expenses, are appropriated to the York County Hospital, and the general infirmaries of Hull, Leeds, and Sheffield. The races are held in May and August, and are in general numerously attended; the course, on Knavesmire, about a mile from the town, on the road to Tadcaster, is well adapted to the purpose: the grand stand, erected by subscription in 1754, is nearly 300 feet in length, with a balustrade projecting in front, supported on a rustic arcade. The cold baths, near the New Walk, are provided with dressing-rooms and every requisite accommodation for ladies and gentlemen: and at Lendal tower, adjoining the water-works, is an establishment of hot, cold, tepid, and vapour baths. The cavalry barracks, about a mile to the south-west of the city, were erected in 1796, at an expense of £30,000, including the purchase of twelve acres of ground, which are attached to them, for parade, and for performing the different evolutions: the buildings are handsome and commodious, and include arrangements for three field-officers, five captains, nine subalterns, and 240 non-commissioned officers and privates, with stabling for the requisite number of horses.

The city is not much distinguished either for its commerce or manufactures; the trade principally arises from the supply of the inhabitants and the numerous opulent families in the neighbourhood. Several linen factories have been recently established, but are not carried on to any great extent; the manufacture of glass was introduced in 1797; and some works for white and red lead are conducted upon a moderate scale. Carpets, worsted lace for liveries, gloves, and combs, are made in moderate quantities; and there are some chymical laboratories and iron-foundries. The river Ouse is navigable as far as the bridge, for vessels of 80 tons' burden; and ships of 150 tons' burden trade with London. The trustees for the river Ouse have expended large sums in improving the navigation, and steamers now ply between this place and Hull at any time of the tide. Great quantities of coal are brought hither in barges of 30 and 40 tons' burden; and from the junction of the Foss with the Ouse is a navigable communication to the parish of Sheriff-Hutton, in the North Riding. The York and North Midland railway, lately opened, is carried over the river Ouse at Poppleton by a viaduct 300 feet long: it crosses the Leeds and Selby railway and unites the city with the North Midland railway. The Great North of England railway commences at York, and is in a state of great forwardness. These lines are intended to form a better communication with the west of Yorkshire and the north and south of England, and will probably add greatly to the commerce of the city. The market days are Tuesday, Thursday, and Saturday; the last, which is the principal, is for corn. Fairs for cattle and horses, at which very large quantities of live stock are disposed of, are held every fortnight, and on Whit-Monday, St. Peter's day, Lammas-day, and some other festivals during the year, in a spacious market-place without the city walls, near Walmgate Bar, in the construction of which, and in the erection of a handsome inn contiguous to it, the corporation have expended upwards of £10,000. A fair for leather is held every month; a fair for wool is held on Peaseholm Green every Thursday, from Lady-day to Michaelmas, which is well attended; a fair for flax on the Saturdays before Michaelmas, Martinmas, Christmas, Lady-day, St. Peter's day, Lammas-day, and Whit-Monday; and a large horse fair, without Micklegate Bar, in the week next before Christmas. In the session of parliament of 1833, an act was obtained for improving and enlarging the market-places within the city of York, and rendering the approaches thereto more commodious; and for regulating and maintaining the several markets and fairs held within the city and its suburbs; also for amending a previous act for paving, lighting, watching, and improving the city, and for other purposes.

Corporation Seal.

Obverse. *Reverse.*

The earliest charter granted to the inhabitants was by Henry II., confirming to them all the liberties they held in the time of Henry I. Richard I., in the 1st of his reign, granted them an exemption from toll and all customs in England and Normandy; and John, in the first of his reign, granted them a charter confirming all former privileges, and bestowing on the inhabitants the city, subject to a fee-farm rent of £160, payable half-yearly into the treasury. Confirmatory charters were subsequently granted by Henry III., Edward II. and III., and Richard II., which last monarch, by a second charter, erected the city, with the district adjoining it, into a county of itself, dignified the mayor with the title of Lord, and in lieu of the three bailiffs, appointed two sheriffs. Charters were subsequently granted by Henry VI., Edward IV., Henry VII., Elizabeth, and Charles I. and II.; the charter of the last-named monarch, as attested by that of the 10th of George IV., vested the government in a lord mayor, 12 aldermen, two sheriffs, and certain of the ex-sheriffs (who together were called the twenty-four, though generally exceeding that number), and 72 common-councilmen, assisted by a recorder, two city counsel and a town-clerk, with chamberlains, sword-bearer, mace-bearer, four sergeants-at-mace, and other officers. The lord mayor, aldermen, sheriffs, and ex-sheriffs, constituted what was called the Upper House; and the common-councilmen and other officers, the Lower House. The lord mayor, recorder, two city counsel, and the aldermen, were justices of the peace, with exclusive jurisdiction within the city and the county of the city. The present corporation consists of a lord mayor, 12 aldermen, and 36 councillors, appointed under the act of the 5th and 6th of William IV., cap. 76, for an abstract of which see the Appendix, No. I.; and the city, which formerly consisted of four, is now divided into six wards. The lord mayor, the ex-lord mayor, and recorder, are justices of the peace *ex officio*, and others are appointed by the crown; a sheriff is also chosen by the town council. The freedom is inherited by all the sons of freemen on their coming of age, and acquired by apprenticeship to a resident freeman. The city, which comprises 1938 acres, first exercised the elective franchise in the 49th of Henry III., since which time it has regularly returned two members to parliament. The right of election was formerly vested in the corporation and freemen generally, in number nearly 4000; but the privilege has been extended to the £10 householders of an enlarged district, which has been constituted the new elective borough, and comprises 2805 acres, the limits of which are minutely described in the Appendix, No. II. The number of voters at the last registration was 3326: the sheriff is the returning officer. The county of the city formerly included the whole of the city and the ainsty, which was previously a wapentake of the West Riding, and was separated therefrom by charter of the 27th of Henry VI., and placed under the exclusive jurisdiction of the city magistrates; but by the act of the 5th and 6th of William IV., cap. 76, it has been separated from the city, and again annexed to the West Riding. It comprises the parishes of Acomb, Askham-Bryan, Askham-Richard, Bilbrough, Bilton, Bishopsthorpe, Bolton-Percy, Healaugh, Holy Trinity, Long Marston, Moor-Monkton, Rufforth, Thorp-Arch, Walton, and Wighill, and parts of those of Acaster-Malbis, Kirk-Hammerton, St. Mary Bishopshill Senior, St. Mary Bishopshill Junior, Stillingfleet, and Tadcaster. Courts of assize for the city and county of the city are opened by the judges on the northern circuit, under a separate commission, on the same day as the assizes for the county: at these courts, which are held in the guildhall, the lord mayor takes the chair in presence of the judge, who sits on his right hand. Courts of quarter session are held before the recorder, for all offences not capital; the lord mayor and one of the justices hold a petty session twice in the week; and a court of record, held weekly by prescription, for the recovery of debts to any amount, in which the sheriffs presided, is now, under the provisions of the Municipal Reform Act, held before the recorder. To the corporation belongs the conservancy of the rivers Aire, Derwent, Don, Ouse, Wharfe, and some parts of the Humber. The mansion-house, erected in the year 1726, for the residence of the chief magistrate, is a stately and handsome edifice, containing a splendid suite of apartments; the banquet hall is lighted by a double range of windows, and ornamented with a large collection of well-painted portraits, among which are those of William III.; George II.; George IV., when Prince of Wales, presented by His Royal Highness to the corporation in 1811; Lord Dundas, painted by Jackson, in 1822; Lord Bingley; Sir William Mordaunt Milner, Bart.; Sir John Lister Kaye, Bart.; and other eminent persons. The guildhall is a handsome structure, in the later style of English architecture, erected in 1446: the hall is appropriated to the use of the courts, and for the transaction of corporate affairs and the election of members and officers of the corporation. The council-chamber, adjoining the guildhall, was erected in 1819, when the buildings anciently used for that purpose, and situated on the old bridge over the Ouse, were taken down: the upper room is assigned to the meetings of the lord mayor, aldermen, and councillors; and the lower apartment is appropriated to various public purposes. The Merchants' Hall is situated in Fossgate street, and is almost the only remains of the numerous ancient guilds formerly incorporated for the regulation of the trade of the city. The common gaol for the city and county of the city was erected in 1807, at the joint expense of the city and the ainsty, towards which the former contributed three-fifths, and the latter two-fifths: it is a substantial stone building, consisting of three stories, surmounted by a cupola and vane, and is now used as the house of correction for the city, the county gaol in the castle being now used as the common gaol. The house of correction for the city and county of the city, erected in 1814, at the expense of the city and ainsty, has been taken down; it was sold to the York and North Midland railway company.

The liberty of St. Peter has jurisdiction over fifty-two townships or parts of townships in the East Riding, forty-two in the North Riding, and twenty-six in the West Riding: it has a separate clerk of the peace, and other officers, and is entirely exempt from the jurisdiction of the county magistrates; the petty sessions for this liberty are always held at the hall of pleas within the close of the Cathedral Church; for an account of its other courts, see the article on the county. The general assizes for the county, and the election of knights of the shire for the North Riding, take place at York. The site of the ancient castle, which, on its being dismantled after the parliamentary war, was converted into a prison, is at present occupied by the county hall and common gaol, erected in 1701, and forming three sides of a quadrangle, near the confluence of the Ouse and the Foss: now approached by a handsome gateway and porter's lodge in the new wall, fronting Tower-street, and near the north-western side of Clifford's Tower. The county hall, which occupies the western range, is a handsome structure, in the Grecian style of architecture, erected in 1777, with a noble portico of six lofty columns of the Ionic order, above which are the king's arms, a figure of Justice, and other emblematical ornaments: the hall is 150 feet long, and 45 feet wide; at one end is the crown bar, and at the other the court of nisi prius, each lighted by an elegant dome, supported on twelve pillars of the Corinthian order. On the east side of the quadrangle are the apartments of the clerk of assize, the office of the court of record, the indictment office, hospital rooms, and cells for female prisoners: this range, which is 150 feet in length, is fronted with a handsome colonnade of the Ionic order. The old county gaol occupies the south side of the quadrangle; and in 1836 a very large addition, called the New Works, was completed, at an expense of £203,530, including a massive boundary wall, 32 feet high, with pierced battlements, recessed gateway, and projecting towers: the new prison, which stands on the north-east side of Clifford's Tower, comprises four radiating double wings, with eight airing-yards, and in the centre is the governor's house, which commands inspection over the whole: the buildings are fire-proof, being constructed entirely of stone and wrought-iron.

Arms of the Archbishopric.

The city was constituted an archiepiscopal see by Edwin, King of Northumberland, who, after his conversion to Christianity, in 627, erected a church here, which he dedicated to St. Peter, and made Paulinus, the confessor of his queen, Ethelburga, first archbishop. After the death of Edwin, who was killed in battle, Paulinus was compelled to abandon the province to the fury of the Britons, who, under Cadwallo, assisted by the King of Mercia, took possession of the city, and, accompanied by Ethelburga, found an asylum in the kingdom of Kent. During his absence the newly-founded establishment fell into decay, but was restored by Oswald, the successor of Edwin, who, after a successful battle with the Britons, expelled them from the city, and recovered possession of his capital. Paulinus, dying in Kent, was succeeded in the government of the see and province by Cedda, who held it till the return of Wilfrid from France, whither he had been sent for consecration, and where he remained for three years. The establishment, under Wilfrid and his successors, remained upon its original foundation till after the Conquest, when Thomas, chaplain to William the Conqueror, being made archbishop, constituted the several dignitaries and prebendaries, and established the first regular chapter. After frequent disputes for precedency with the Archbishop of Canterbury, which were carried on for many years with the greatest animosity, it was ultimately decided in favour of Canterbury, the archbishop of that see being styled Primate of all England, as a superior designation to that of the archbishop of York, who is styled Primate of England. The archbishop of York, who is also lord high almoner to the king, takes precedency of all dukes who are not of the blood royal, and of all the chief officers of state, with the exception of the lord high chancellor; he places the crown on the head of the queen at coronations; and, in the county of Northumberland, has the power and privileges of a prince palatine: he was formerly styled metropolitan of Scotland. The province of York comprises the sees of York, Carlisle, Chester, Durham, Sodor and Man, and Ripon; the county of Nottingham, now in the diocese and province of York, will, under the act of the 6th and 7th of William IV., c. 77, be transferred to the diocese of Lincoln and province of Canterbury, thus leaving the diocese of York to consist of the county of York, except such parts thereof as have been included in the new diocese of Ripon. The ecclesiastical establishment consists of an archbishop, dean, chancellor, precentor, subdean, succentor, four archdeacons, four canons residentiary, twenty-four prebendaries, chancellor of the diocese, a subchanter, and four vicars choral, seven lay clerks, six choristers, organist, and other officers. The canons residentiary are appointed by the dean, who must choose them out of the prebendaries: the dean and the four residentiaries constitute the chapter. The treasurership, erected in the year 1090, was dissolved and made a lay fee by King Edward VI., as were also the prebends of Wilton and Newthorpe, annexed thereto. In the 26th of Henry VIII., the revenue of the archbishop was valued at £2035. 3. 7., that of the canons residentiary at £439. 2. 6., and that of the dean at £308. 10. 7.

The cathedral, originally founded by Edwin, after having been frequently demolished and restored, was destroyed by an accidental fire in 1137. It remained in a desolate state for some time, till Archbishop Roger, in 1171, rebuilt the choir, and in the reign of Henry III., Walter de Gray built the south transept. In the beginning of the reign of Edward I., John le Romaine, treasurer of the church, built the north transept and a central tower; and, in 1291, his son of the same name, who was made archbishop, laid the foundation of the nave, which was, 40 years afterwards, completed by Archbisop William de Melton, who also built the west front and the two western towers. Archbishop Thoresby, in 1361, rebuilt the choir in a style better adapted to the character of the nave, to which it was before greatly inferior; and, in 1370, the central tower was taken down, and in the course of eight years completely rebuilt in a more appropriate manner; and the whole edifice at present displays a regular series of the richest and purest specimens of the various styles of English architecture, with some remains of the Norman, of which the only portion now entire is the crypt, under the eastern part of the church. The distant view of this extensive and magnificent pile, towering above the churches and other buildings of the city, and equally unrivalled in the magnitude of its dimensions and the richness of its embellishment, is strikingly impressive. The cathedral is a cruciform structure, with the addition of two lateral projections between the central tower and the east end, which are called the light transepts, and is 524½ feet in length from east to west, and 222 in breadth along the principal transepts. The west front, which is divided into three compartments by richly panelled buttresses of four stages, terminating with boldly crocketed finials, is almost covered with a profusion of the most richly varied sculpture, comprising several canopied niches, in which are statues. The central compartment contains the principal entrance, a beautiful pointed and richly moulded arch, supported on a series of slender clustered columns, surmounted by a straight angular canopy with crocketed pinnacles, and ornamented with canopied niches, in which are statues of the Archbishops Melton, Percy, and Vavasour. The principal arch is divided, by a slender clustered pillar in the centre, into two smaller cinquefoiled arches, forming a double doorway, and having the spandril decorated with a circular window of elegant tracery. On each side of the principal entrance are two series of trefoiled arches, with delicately feathered canopies, terminating in crocketed finials; and above it is the beautiful west window of eight lights, enriched with elegant tracery, and surmounted by an acutely angular canopy and parapet pierced in beautiful design, behind which is seen the gable of the roof of the nave, with an angular pediment, perforated with great delicacy, and having on the apex a richly crocketed pinnacle. The entrances to the aisles are through plainer arches, above which are elegant windows of three lights, with tracery surmounted by canopies similar to that over the west window. The western towers, which are uniform and of graceful elevation, are strengthened with double buttresses at the angles, highly enriched with canopies and pinnacles at the offsets, and which, after diminishing in four successive stages, die away under the cornice, which is carried round the upper part of the towers. Above the windows previously noticed are others of five lights with tracery, and crowned by the embattled parapet, which is carried round the nave; the upper portion of the towers is ornamented with large belfry windows of three lights with tracery, surmounted by a delicately feathered ogee canopy, terminating in a lofty crocketed finial; the summits are wreathed with pierced embattled parapets, and crowned with eight boldly crocketed pinnacles.

The north and south sides of the cathedral are strengthened with buttresses terminating with pinnacles; and a delicately pierced parapet is continued round the walls of the nave. The transepts, which are in the early style of English architecture, are nearly similar in design, though differing in the minuter details: the entrance to the south transept is through an elegant porch, ascended by a double flight of steps, above which are three lofty lancet-shaped windows, divided only by panelled and enriched buttresses; and a large circular window, surrounded with rich and varied mouldings, occupies the centre of a triangular pediment, at each end of which is an octagonal turret, and on the apex a crocketed finial. The front of the north transept is ornamented with five lofty narrow windows, which occupy the principal part, above which are five others, of unequal height, ranged in the triangular pediment. The central tower, which rises to the height of 213 feet, is a massive square structure relieved on each of its faces by two large windows of three lights, separated and bounded at each side by enriched buttresses, terminating in crocketed finials; the crown of the arch of the windows is surmounted by an enriched canopy, and the summit of the tower is wreathed with a delicately pierced and embattled parapet. The east front, which is one of the finest compositions extant, is divided into three compartments by four octangular buttresses, terminating in crocketed pinnacles, and profusely ornamented with canopied niches, in which are, a figure of an archbishop seated, holding in his left hand the model of a church, and having the right hand raised; a statue of Vavasour, in tolerable preservation; and one, much mutilated, said to be that of Lord Percy. The magnificent window, of nine lights, filled with rich and intricate tracery, occupies the whole of the central compartment, and is surmounted by an enriched ogee canopy, above which is some highly elaborate and beautiful tabernacle-work, and in the centre, a square turret, with a crocketed finial.

On entering the cathedral from the west end, the vastness of its dimensions, the justness of its proportions, and the simplicity and beauty of the arrangement, produce an intense impression of grandeur and magnificence. The nave is separated from the aisles by long ranges of finely clustered columns, of which the central shafts rise to the roof, which is plainly groined, and the others support a series of gracefully pointed arches, in the decorated style, chastely and appropriately enriched. The triforium consists of openings of five lofty narrow trefoiled arches, with acute angular canopies, richly ornamented. The clerestory is a noble range of windows, divided by slender mullions into five lights, having in the crown of the arch a circular light, with geometrical tracery of beautiful design: the aisles are lighted by an elegant range of windows of three lights, with quatrefoiled circles and tracery, and the walls below them are decorated with panels and tracery, and with canopied niches with crocketed pinnacles. At the eastern extremity of the nave is the lantern tower, supported on four lofty clustered columns and finely pointed arches, the windows of which diffuse a pleasing light over the transepts and eastern portion of the nave, which, when viewed from this point, derives increased effect from the great west window, which is filled with flowing tracery of the most delicate and beautiful character. The transepts in the early style of English architecture, are dissimilar only in the minuter details and the arrangement of the ends. The central part is separated from the aisles by clustered columns and sharply pointed arches; the triforium consists of four arches separated by small pillars resembling the Norman, and included in a large circular arch, having in the spandril a cinquefoiled, and on each side of it a quatrefoiled, circle; the clerestory consists of ranges of five sharp-pointed arches, of which the three central only admit light; the roof, which is of wood, is groined like that of the nave: the aisles of the transepts are lighted with double lancet-shaped windows, beneath which is a series of blank trefoiled arches. The choir is separated from the nave by a splendid stone screen supporting the organ, and divided into fifteen compartments, containing a series of richly canopied niches, in which are placed, on elegant pedestals, the statues of the kings of England, from William the Conqueror to Henry VI.: the statue of the last monarch was removed from its niche in the reign of James I., whose statue was substituted in its place; but a statue of Henry VI., from the chisel of Michael Taylor, sculptor, of York, now occupies the niche from which that of James I. has been removed. Nearly in the centre of the screen is the doorway leading into the choir, an obtuse arch supported on slender clustered columns, with an ogee canopy, terminating with a crocketed finial. Above the niches in which are the statues of the kings are series of narrow shrines, richly canopied, and containing smaller statues, and above them a series of angels, beautifully sculptured; the whole is surmounted with bands of delicate tracery, and enriched with the most elaborate sculpture.

The choir, of which the roof is loftier and more intricately groined than that of the nave, is a beautiful specimen of the later style of English architecture. The piers and arches are similar to those of the nave, and the intervals between the arches are embellished with shields of armorial bearings: the openings in the triforium consist of a series of five cinquefoiled arches with canopies and crocketed finials, divided in the centre by horizontal transoms; and the clerestory is a beautiful range of windows of five lights, with cinquefoiled heads, having the crown of the arch enriched with elegant tracery. The walls of the aisles of the choir are panelled and enriched with tracery corresponding with the character of the windows, which, as well as the groining of the roof, is similar to those of the nave. The magnificent east window, of nine lights, occupies nearly the whole of the east end of the choir, and is embellished with nearly 200 subjects from sacred history, painted in glass; the upper section of the window is occupied with intricate tracery, elaborately wrought into a series of canopies, running up to the crown of the arch, and containing projecting busts, and the outer border is enriched with small tabernacles, containing half-length figures; the window is divided, nearly in the centre, by an embattled transom, in which a light gallery is wrought, affording an unobstructed view of the whole cathedral. Behind the altar, to which is an ascent of fifteen steps, and separating it from the Lady chapel, is an elaborately enriched and beautiful stone screen, divided into compartments by slender panelled buttresses terminating with crocketed pinnacles; each compartment contains, in the lower division, a triple shrine of niches richly canopied, and in the upper, a beautiful open arch, separated by slender mullions into three divisions, enriched with elegant tracery, and surmounted by a square head, of which the spandrils are pierced in quatrefoil circles; above these is a delicate open embattled parapet, pierced alternately into triple cinquefoiled arches and circles of double quatrefoil, with shields of armorial bearings. The intervals of this exquisitely wrought and highly enriched screen have been filled with plate-glass, affording a view of the eastern portion of the choir and of the magnificent east window. On each side of the choir, and on each side of the entrance under the organ, are the prebendal stalls, of oak richly carved, and surmounted with canopies of tabernacle-work: at the east end are the bishop's throne and pulpit, opposite to each other, both elaborately ornamented; and in the centre is the desk for the vicars choral, enclosed with tabernacle-work, on the north side of which is an eagle of brass on a pedestal. The pavement of the choir and nave has been beautifully relaid in mosaic work. The Lady chapel is perfectly similar to the choir, of which it is only a continuation, and contains some beautiful monuments. Beneath the altar is an ancient crypt of Norman architecture, with low massive circular columns with varied capitals, supporting a plainly groined roof; it was built with the materials of Archbishop Thomas' church, by Archbishop Thoresby. On the south side of the choir are three chapels, or rather vestries, in which are several ancient chests. In the inner vestry, or council-chamber, is a large press containing many of the ancient records of the church, and a large horn of ivory, presented by Ulphus, Prince of West Deira, with all his lands and revenues, to the cathedral, which, after having been lost and stripped of its gold ornaments, was restored to the church by Henry Lord Fairfax. The lands which are held by this horn are situated a little to the eastward of the city, and are of great value. Here is also preserved a large and elegant bowl, edged with silver doubly gilt, and standing upon three silver feet, originally presented by Archbishop Scroope, in 1398, to the company of Cordwainers of this city: in the middle of it are the armorial bearings of that company, richly embossed; and round the rim is an inscription in old English characters. Among the other curiosities are, a state canopy of gold tissue, given by the citizens in honour of James I., on his first visit to York; a superb pastoral staff of silver, about seven feet long, with the figure of the Virgin and the Infant placed under the crook, given by Catherine of Portugal, Queen Dowager of England, to her confessor, on his being appointed to the archbishopric by James II., in the year 1689, and said to have been wrested from him by the Earl of Darnley, when he went in procession to the minster, and deposited in the care of the Dean and Chapter, in whose possession it has remained ever since; and an antique chair, said to be coeval with the cathedral, in which several of the kings of England have been crowned, and which is placed within the rails of the altar, when the archbishop officiates, for his use.

The monument of Archbishop Walter de Grey consists of two tiers of trefoiled arches, supported on slender columns, sustaining a canopy of niches with angular pediments and finials, under which, on an altar-tomb, is the recumbent effigy of the prelate in his pontifical robes. The tomb of Archbishop Godfrey is in the shape of a coffin, under a canopy of trefoiled arches, having the sides decorated with plain shields in quatrefoiled circles. Beneath an arch at the east end is the monument of Archbishop Henry Bowett, a beautiful composition in the later style: an obtusely pointed arch supports a highly enriched canopy of elaborate and delicate tabernacle-work; above the arch are three lofty shrines, in each of which is a statue on a pedestal, and beneath the canopy is a slab of marble, enclosed with a parapet pierced in quatrefoil. The monument of Archbishop Thomas Savage has a recumbent figure of the prelate on an altar-tomb, under a square-headed straight-lined arch, in the spandrils of which are shields of arms supported by unicorns, and angels lifting up their censers, and the cornice is ornamented with five projecting angels, bearing shields of the same arms. There are also several large stone coffins, some recumbent figures of knights, and numerous tombs of archbishops, of which that of Archbishop Roger is the most ancient. In the north aisle of the choir is a recumbent statue in alabaster, commonly, but erroneously, said to be that of Prince William de Hatfield, second son of Edward III., under a rich and beautiful canopy; and in the north transept is the tomb of John Haxby, treasurer of the church, on which, according to ancient usage, payments of money for the church estates are still occasionally made. There are numerous other monuments and tombs in various parts of the church; among which are those of Sir William Ingram, Knt., commissary of the prerogative court; Charles Howard, Earl of Carlisle; Frances Cecil, Countess of Cumberland; a statue of William Wentworth, Earl of Strafford, son of the minister of Charles I.; and a monument to William Burgh, LL.D., on which is an emblematical figure of Faith, finely sculptured by Westmacott. From the north transept a passage leads to the chapter-house, an elegant and highly enriched octagonal structure, in the decorated style of English architecture, with a lofty and elaborately groined roof of wood, without a central pier, profusely ornamented with sculpture in various devices: seven sides of the octagon are occupied by large windows of elegant tracery, embellished with shields of armorial bearings painted on glass; below the windows are forty-four stalls of rich tabernacle-work of Petworth marble: the finely clustered pillars between the windows are perforated for a narrow gallery, which is carried round the whole building above the cornice of the stalls; the eighth side is solid, and enriched with tracery corresponding with the windows, and the arch forming the doorway is divided into two trefoiled arches by a clustered column in the centre, above which is a statue of the Virgin with the Infant in her arms, enshrined in a canopied niche; above the entrance is a series of niches, in which were formerly statues in silver of Our Saviour and the twelve Apostles. The vestibule is of beautiful design; the windows are large and enriched with tracery of exquisite delicacy, and the walls beneath them are ornamented with tracery of corresponding character. The building now used for the library, anciently a chapel belonging to the archiepiscopal palace, is situated a short distance to the north-west of the cathedral, and having undergone complete repair, exhibits a fine specimen of early Anglo-Norman architecture; it contains a valuable collection of works on theology and general literature. The recent removal of ancient buildings to the north of the cathedral has disclosed a series of very beautiful Norman arches, which formed part of the archiepiscopal palace, and which, though greatly mutilated, are peculiarly fine in their details.

PARISHES IN THE CITY OF YORK.

PARISHES.	LIVINGS.	Value in the King's Books. £. s. d.	Present Net Income. £.	PATRONS.	Population.
All Saints, North-street	Discharged Rectory	4 7 11	107	The Crown	1216
All Saints, Pavement, with } united	Discharged Rectory	5 16 10½	100	The Crown	508
St. Peter the Little	Discharged Vicarage				632
St. Crux	Discharged Rectory	6 16 6	94	The Crown	874
St. Cuthbert, with } united	Discharged Rectory	5 10 10	233	The Crown	976
St. Helen on the Walls and	Discharged Rectory	....			422
All Saints in Peaseholm	Discharged Rectory	....			407
St. Denis in Walmgate, with } united	Discharged Rectory	4 0 10	90	The Crown and G. Palmer, Esq., alternately	1718
St. George and Naburn	Discharged Rectory				
St. Olave, with } united	Perpetual Curacy	....	138	Earl de Grey	1052
St. Giles					
St. Helen Stonegate	Discharged Vicarage	4 5 5	103	The Crown	707
St. John at Ousebridge-end	Perpetual Curacy	....	209	Dean and Chapter	926
St. Lawrence, with } united	Discharged Vicarage	5 10 0	83	Dean and Chapter, the appropriators	830
St. Nicholas	....	....			103
St. Margaret Walmgate, with } united	Discharged Rectory	4 9 9½	124	The Crown	1034
St. Peter-le-Willows	Discharged Rectory				413
St. Martin, Coney-street	Discharged Vicarage	4 0 0	97	Dean and Chapter	586
St. Martin Micklegate, with } united	Discharged Rectory	5 16 3	243	Trustees of H. Willoughby, Esq., and others	547
St. Gregory	Discharged Vicarage				
St. Mary Bishopshill Senior	Discharged Rectory	5 0 10	226	The Crown	1096
St. Mary Bishopshill Junior with } united	Discharged Vicarage	10 0 0	144	Dean and Chapter, the appropriators	2171
Upper Poppleton, and	Perpetual Curacy	....			
Copmanthorpe	Chapel of Ease	....			
St. Mary Castlegate	Discharged Rectory	2 8 6½	76	The Crown	946
St. Michael-le-Belfry, with } united	Perpetual Curacy	....	140	Dean and Chapter, the appropriators	1692
St. Wilfrid	Discharged Rectory	2 0 10			443
St. Michael Spurrier-gate, or Ouse-bridge	Discharged Rectory	8 12 1	91	The Crown	645
St. Sampson	Perpetual Curacy	....	109	Subchanter and Vicars Choral	995
St. Saviour, with } united	Discharged Rectory	5 6 8	173	The Crown	1455
St. Andrew					238
Holy Trinity, King's-ct., or Christ-church	Discharged Vicarage	8 0 0	87	Master of Wells Hospital, the impropriator	706
Holy Trinity, Micklegate	Discharged Vicarage	....	93	The Crown	1108
Holy Trinity, Goodramgate, with } united	Discharged Rectory	....	138	Archbishop of York	540
St. John Delpike, and	Discharged Rectory	....			350
St. Maurice without Monk-bar	Discharged Vicarage	12 4 9½			1114

The churches are in general in the later style of English architecture, but several of them contain portions in the Norman and early English styles. The church of *All Saints on the Pavement* is a very ancient structure, said to have been built on the site and with the ruins of the Roman *Eboracum*; it has an octagonal lantern tower with large windows of elegant tracery, in which was formerly a lamp to guide travellers across the forest of Galtres: the chancel was taken down, in 1782, for the enlargement of the market-place, but since the removal of the market the site has been added to the cemetery. The church of *All Saints in North-Street* has some ancient stained glass in the windows, and in the south wall the mutilated remains of a Roman sepulchral monument. The church of *St. Crux* has a neat square tower of brick, surmounted by a dome, and declining considerably from a perpendicular line; in the chancel is a monument to the memory of Sir Robert Walton, twice Lord Mayor of the city, with the effigies of himself, his wife, and three children. The church of *St. Cuthbert* is a neat edifice in the later style, with some ancient portions: the windows were formerly embellished with stained glass, of which some portions are remaining. Near the site many Roman antiquities have been found, consisting of urns, pateræ, and part of the foundation of an apparently Roman building. The church of *St. Denis in Walmgate*, originally a spacious structure, has been much reduced by taking down the western part, which, from the insecurity of the foundation, was giving way; and the spire, which was perforated by a ball during the parliamentary war, has been replaced with a square tower of indifferent character: little remains of the original architecture, except the entrance door, which belonged to an ancient porch that has been removed. In the interior are, a mural tablet with a female figure in the attitude of prayer, erected to Mrs. Dorothy Hughes; and an elegant marble monument to Robert Welbourne Hotham, Esq., sheriff of York in 1801: in the north aisle is a sepulchral chapel of the Earls of Northumberland, in which Earl Henry, who fell at the battle of Towton Field, was interred. The church of *St. Helen*, supposed to have been originally a temple of Diana, was rebuilt in the reign of Mary, and the ground of the churchyard, which had risen to an enormous height, was levelled and marked out as the site of St. Helen's square: the present structure, which has an elegant octagonal tower, has been much modernized, and most of the painted glass has been removed. Near the entrance is a Norman font lined with lead, and ornamented with antique sculpture: there are several monuments, and two mural tablets to the memory of Barbara and Elizabeth Davyes, two maiden sisters, who died in 1765 and 1767, each 98 years of age. The steeple of the church of *St. John* was blown down in 1551, and has not been rebuilt: the interior contains a monument to Sir Richard York, Knt., lord mayor of the city in 1469: the churchyard has been much curtailed by the improvement near Ouse bridge. The church of *St. Lawrence* was nearly destroyed, during the siege of York, by the parliamentarian forces, and lay in ruins till 1669, when it was repaired: it consists only of a nave, with a square embattled tower. Over the altar is a large handsome window, with some remains of ancient stained glass; and there are some neat marble tablets to deceased members of the family of Yarburgh. The porch has been removed, but at the entrance is a fine Norman arch, with three mouldings ornamented with flowers; in the north wall of the church is a large grit-stone, supposed to have been a Roman altar; and in the churchyard wall are two antique statues. The church of *St. Margaret in Walmgate* is an ancient building of brick, with a steeple of the same material: the only interesting feature is a Norman porch, removed from the dissolved hospital of St. Nicholas: at the entrance is a semicircular arch, resting on single columns, and having four mouldings ornamented alternately with the signs of the zodiac, emblematical representations of the seasons, and grotesque figures. It has been repaired and a gallery erected, by which 150 free sittings have been provided, the Incorporated Society having granted £80 in aid of the expense. The church of *St. Martin in Micklegate* is a neat ancient structure, with a more modern steeple, built in 1677; the windows contain some portions of beautiful stained glass, and in the exterior of the walls of the church, and in the walls of the churchyard, are some remains of mutilated Roman sculpture. The church of *St. Martin the Bishop*, in Coney-street, is an elegant structure, in the later style of English architecture, with a square embattled tower: the interior is spacious and appropriately arranged. Among the monuments are, one to Sir William Sheffield and his lady, with busts and the family arms; and a plain marble tablet to Elizabeth, wife of Robert Porteus, and another of Beilby Porteus, Bishop of London. The church of *St. Mary Bishopshill Sen.* has portions in the early and decorated styles of English architecture, of which the details are very good: and that of *St. Mary Bishopshill Jun.* has a Norman tower; some of the piers and arches are in the early English style, with portions of a later date. The church of *St. Mary in Castlegate* has a very handsome and lofty spire, and contains several ancient monumental inscriptions. In digging a grave in this church, a copper plate was found, which had been fastened on the inside of the lid of the coffin of a priest, who was executed for the plot of 1680. The church of *St. Maurice* is a very ancient structure: the interior has been recently repaired and modernized. The church of *St. Michael le Belfry* is a spacious and elegant edifice, in the later style of English architecture, erected on the site of a more ancient church, which was taken down in 1535: the nave is separated from the aisles by slender clustered columns and finely pointed arches; and the interior is handsomely arranged, with the exception of the altar, which is of the Corinthian order, and consequently inappropriate to the general character of the building. *St. Michael's in Spurriergate* is a very ancient structure; the west end is built of gritstone, in large masses. The church of *St. Olave*, adjoining the ruins of St. Mary's abbey, and a very ancient edifice, was destroyed, during the siege of York, by the parliamentarian forces, who used the roof as a platform for their cannon: it was rebuilt in 1722, with stone taken from the ruins of the abbey. The interior is modern and neatly arranged, the east window contains some excellent stained glass, and there are some neat mural tablets. The church of *St. Sampson* is an ancient edifice, in the later style of English architecture, with a square embattled tower, on the west side of which is a sculptured figure of the tutelar saint, and on which may be perceived its perforation by a cannon ball during the siege of the city. There were formerly three chantry chapels in this church; most of the painted glass has been removed from the windows, and the monumental inscriptions have been greatly defaced. The church of *St. Saviour* is an ancient structure, with a handsome tower surmounted by a wooden cross: it contains 200 free sittings, the Incorporated Society having granted £100 in aid of the expense: the interior is very neatly arranged; the windows contain considerable portions of ancient stained glass, and there are several ancient monuments. The church of *St. Trinity in Micklegate* is an ancient structure, principally in the Norman style of architecture, with portions of a later date; the tower preserves its original Norman character, but the church has been greatly mutilated; it formerly belonged to the priory of the Holy Trinity, of which some ruined arches may be traced, and a gateway is still remaining in good preservation. The church of *St. Trinity in Goodramgate* is an ancient edifice, in which were formerly three chantry chapels; over the altar is a fine window, containing some beautiful specimens of stained glass; there are also some very ancient monumental inscriptions. The church of *St. Trinity in King's-court*, usually called *Christchurch*, is an ancient edifice, to which there is a descent of several steps; it was considerably reduced at the east end in 1830, in order to widen Collier-gate, and was then repewed. The Roman palace was situated near this church, on the side of which is a ditch, still called King's ditch, which is supposed to have bounded the demesne. In addition to the several churchyards, a public cemetery has been established, on the Fulford road, comprising 8½ acres, beautifully laid out and inclosed at a cost of about £6000.

There are places of worship for the Society of Friends, Independents, Primitive, Wesleyan, and Association Methodists, and Unitarians; also two Roman Catholic chapels.

The free grammar school, in the Cathedral Close, was erected in 1546, and endowed with £12 per annum by Robert Holgate, Archbishop of York; it is under the inspection of the archbishop for the time being, who appoints the master. Another free grammar school, in the "Horseayer," was founded by charter of Philip and Mary, and endowed by the Dean and Chapter with the lands of the hospital of St. Mary, originally founded in 1330, by Robert de Pykering, Dean of York, the site and revenue of which, on its suppression, were granted to that body: the endowment was subsequently augmented, in the reign of Elizabeth, with £4 per annum, charged on the manor of Hartesholm, in the county of Lincoln, by Robert Dallison, chanter in the cathedral church of that city: the master is appointed by the Dean and Chapter of York, and has a residence rent-free; the number of scholars, which is regulated by the Dean and Chapter, seldom exceeds 23: the school was formerly held in part of the old church of St. Andrew, but has been removed into the cathedral; and the old school-room is now occupied by an infants' school, supported by subscription. Three schools were erected in Walmgate, Friar Wells, and Bishopshill respectively, by Mr. John Dodsworth, ironmonger of the city, who endowed them with £10 per annum each, for the gratuitous instruction of children of the parishes near which they are situated. The Blue-coat school for boys, held in an ancient building on Peaseholm Green, called St. Anthony's Hall; and the Grey-coat school for girls, for which an appropriate building was erected near Monkgate bar, were established by the lord mayor and corporation, in 1705; they are liberally supported by subscription, and with the interest of donations vested in the funds, among which was a legacy of £4000 by Thomas Wilkinson, Esq., of Highthorne, alderman of York, in 1820, which, though void in law, was given to the charity by the testator's relative and executor, William Hotham, Esq., also alderman of the city; 64 boys and 43 girls are clothed, maintained, and instructed in reading, writing, and arithmetic, in this establishment; the boys are also taught to weave, and are apprenticed on leaving the school; and the girls are qualified to become useful servants, and placed out in respectable families. A charity school was founded in 1773, by Mr. William Haughton, who bequeathed £1300 for its erection and endowment, and £290 more after the demise of certain annuitants, for the instruction of 20 poor children of the parish of St. Crux, near the church of which a commodious school-house has been erected: the income is about £180 per annum. The same benefactor left £500, directing the interest to be appropriated to the payment of the rents of poor widows of that parish; and £1000 to be lent without interest to 40 poor tradesmen, but this sum has been reduced to £232. 6. by litigation, to establish the will of the testator. A school is endowed by Lady Hewley with £12 per annum for the instruction of 15 children. At Monkgate is the institution called "Manchester College," removed hither from that town in 1803, for the maintenance and education of young men for the ministry among the Independents, supported by donations and subscriptions. A spinning school was established in 1782, by Mrs. Cappe and Mrs. Gray, to instruct children in spinning worsted; but the plan was soon changed, and 40 girls are now taught to read, sew, and knit, and are principally clothed from funds raised by subscription. The school for the blind was established in 1836, in memory of the late William Wilberforce, Esq., by subscription and donations amounting to £9000; it is held in the manor-house. Two National schools were formed in 1812. A Lancasterian school for girls was originally established in 1813, and removed in 1816 to St. Saviour's gate; it is partly supported by subscription. A Roman Catholic school is partly supported by endowment; and there are Sunday schools in connection with the established church and the several dissenting congregations.

An hospital was founded by Alderman Agar, who endowed it with lands forming part of the estate of Lord Middleton, for six aged widows. The hospital of St. Catherine, formerly a house for the reception of poor pilgrims, has been converted into an almshouse for the residence of four aged widows, and the original endowment has been augmented by benefactions from Mrs. Nicholson, James Luntley, Henry Myers, and others. An Hospital was founded in 1717, by Dr. Colton and Mary his wife, who endowed it with lands at Cawood and Thorpe-Willoughby, for eight aged women. An hospital was founded at Bootham, in 1640, by Sir Arthur Ingram, alderman of York, who endowed it with £5 per annum each for ten aged women, and 20 nobles to a chaplain, to read prayers: the buildings consist of ten neat cottages, containing two rooms each, with a chapel in the centre; the inmates are nominated by the Marquess of Hertford, a lineal descendant of the founder. Mason's hospital was founded in 1732, by Mrs. Mason, who endowed it for six aged widows. Mrs. Anne Middleton, in 1655, bequeathed £2000 for the erection and endowment of an hospital for 20 widows of poor freemen of York, which bequest was augmented by a legacy of £200 from Thomas Norfolk, in 1780, and a donation of £100 from Jonathan Gray, in 1830: this hospital is situated in Skeldergate, and was entirely rebuilt by the corporation, in 1829, at an expense of nearly £2000. Near Marygate is the Old Maids' hospital, founded in 1725, by Miss Mary Wandesford, who endowed it with an estate at Brompton-upon-Swale, near Richmond, a mortgage of £1200 and £1200 South Sea Stock, for ten maiden gentlewomen, members of the church of England, and a reader or preacher: the inmates receive an annual payment of £16. 17. 4., and the reader a stipend of £15 per annum, for reading morning prayer every Wednesday and Friday in the chapel of the hospital. St. Thomas' hospital, without Micklegate Bar, was originally founded for the fraternity of Corpus Christi: after its dissolution it was repaired, in 1787, and endowed with a portion of £2137. 8. 2. stock, by William Luntley, glover; with £25 per annum by Lady Conyngham; and with £100 by John Hartley; from the produce of these sums twelve aged widows receive each £6 per annum. Trinity hospital was founded in 1373, by John de Rawcliffe, for a priest, five brethren, and five sisters: the Merchants' Company, upon its dissolution, in the reign of Edward VI., having obtained possession of the building, re-endowed it for ten aged persons of both sexes. The hospital founded by Sir Thomas Walter, in 1612, and endowed by him with £3 per annum for a reader, and £2 per annum each to ten aged persons, payable out of the lordship of Cundall, has, from some unknown cause, been reduced; there are at present only seven inmates. An almshouse in St. Denis-lane, originally founded by the Company of Cordwainers, after having fallen into a state of dilapidation and decay, was rebuilt by Mr. Hornby, at his own cost, and affords a comfortable residence for four decayed members of that fraternity, but has no endowment. An hospital was founded early in the last century, by Percival Winterskelf, Gent., who endowed it for six aged persons. Lady Hewley's hospital, founded in 1708, is a neat brick building, comprising ten houses, for ten aged women, who receive each £15 per annum from that lady's endowment; the same person also bequeathed large sums of money for other charitable uses. An hospital near Foss bridge was founded by Mrs. Dorothy Wilson, who, in 1717, endowed it with lands at Skipwith and Nun-Monkton, for ten aged women, who receive each £15 per annum from the same funds: a salary of £20 per annum is paid to a schoolmaster for teaching 20 boys, who have each a new suit of clothes annually; £2 per annum is given to a schoolmistress, for teaching six children to read, and the same sum to three blind people: the hospital is a neat brick building, rebuilt a second time in 1812. The hospital in Castlegate was founded and endowed in 1692, by Sir Henry Thomson, for the support of six poor men: this endowment, which produces an annual rental of £81, has been subsequently augmented by bequests from Thomas Norfolk and John Girdler.

The county hospital originated in 1740, by the benevolence of Lady Hastings, who bequeathed £500 for the relief of the diseased poor of the county; other donations and subscriptions being subsequently obtained, the present edifice, in Monkgate, was soon afterwards erected. The city dispensary, in New-street, for which a commodious building was erected in 1828, administers extensive relief, and is liberally supported by subscription. The lunatic asylum, without Bootham Bar, was established in 1774, and has undergone considerable alteration, and received great additions: in 1817, a large building was erected behind the former, intended solely for the reception of female patients; it is a handsome and commodious edifice, and the whole is surrounded with gardens and pleasure grounds; it is supported partly by subscription, and by the moderate weekly payment of patients for their board, who are admitted on producing a proper certificate. About a mile from York, in the village of Heslington, is a similar institution, called The Retreat, established by the Society of Friends, in 1796: the building, which was erected at an expense of £12,000, forms a handsome quadrangular range with an entrance lodge, and is well adapted to the reception of patients, who pay a moderate sum in proportion to their circumstances. There are numerous institutions for relieving the distress and alleviating the sufferings of the poor; among which are the Charitable Society, for the relief of distressed objects resident in the city; the Benevolent Society, for the casual relief of strangers; the Lying-in Society, the Clothing Society, a society for the encouragement of female servants, and various others. The corporation have at their disposal for distributing in coal, bread, and other relief among the poor, and for apprenticing children, a fund of nearly £200 per annum, arising from the bequests of various individuals; and a lending fund, a portion of which is derived from the charity of Sir Thomas White. Among the most munificent benefactors to the poor were, the Countess Dowager of Conyngham, who bequeathed £20,000 for charitable purposes; Mr. John Allen, who, with several other sums, bequeathed £140 per annum for the erection and endowment of an hospital for twelve aged men, who receive each £12 per annum; and numerous others. The poor law union of York embraces also a portion of the North Riding, and comprises 79 parishes or places, containing, according to the census of 1831, a population of 39,645; it is under the care of 79 guardians.

Near the city are the beautiful ruins of the venerable abbey of St. Mary, founded in 1088, by William Rufus, who laid the first stone of the building, and amply endowed it for monks of the Benedictine order: it flourished till the dissolution, at which time its revenue was £2085. 1. 5. Among other ancient remains are the crypt of the hospital of St. Leonard, originally founded in the reign of William the Conqueror, and dedicated to St. Peter, previously to erecting a church in it by King Stephen, dedicated to St. Leonard, by which name it was afterwards distinguished; at the dissolution its revenue was estimated at £500. 11. 1.: the crypt has been for many years used as wine vaults. Considerable portions of the ancient city walls are remaining; they were in a greatly dilapidated condition, but have been almost entirely repaired by public subscription. Among the eminent natives were, Constantine the Great, the first Roman emperor that embraced Christianity; Flaccus Albannus, the pupil of Bede; Waltheof, Earl of Northumberland, son of the gallant Siward; Thomas Morton, successively Bishop of Chester, Lichfield and Coventry, and Durham; and among those of more recent date may be noticed Gent, an eminent printer and historian; Swinburn, a distinguished lawyer and civilian; Flaxman, the celebrated sculptor; and several other eminent characters. York gave the title of Duke to Prince Frederick, second son of King George III., who died January 5th, 1827.

YORKFLEET, a township, in the parish and union of HOWDEN, wapentake of HOWDENSHIRE, East Riding of the county of YORK, 6¼ miles (S. E. by E.) from Howden; containing 190 inhabitants. The village is situated on the river Ouse.

YORKSHIRE, a maritime county, and by far the largest in England, bounded on the south by the Humber and the counties of Lincoln, Nottingham, and Derby; on the south-west, for a short distance, by that of Chester; on the west by Lancashire; on the north-west by Westmorland; on the north by Durham; and on the north-east and east by the North Sea. It extends from 53° 19° to 54° 40′ (N. Lat.), and from 10′ (E. Lon.) to 2° 40′ (W. Lon.); and includes an area of 3,815,040 statute acres, or nearly 5961 square miles. The population amounts to 1,371,359.

The ancient British inhabitants of this territory were the Brigantes, the most numerous and powerful of all the tribes that shared in the possession of Britain before its Conquest by the Romans. The latter succeeded in subjugating them about the year 71, after defeating them in several sanguinary battles, and ravaging the whole of their country. The Romans then fixed their principal station in the north at *Eboracum*, now York, which held the rank of a *municipium*, or free city, and from which central point their cohorts, dispersed in every direction, retained the surrounding country in obedience, though the territory at present included within the limits of this county suffered repeatedly during this period from the incursions of the northern barbarians. The Caledonians having overrun a great part of the country to the north of the Humber, the Emperor Adrian arrived in Britain, in the year 120, to oppose them in person, and fixed his residence at *Eboracum:* on his approach the invaders retired, and the emperor, having made provisions for the future security of the province, soon returned to Rome. But no sooner had he departed than the Caledonians renewed their predatory inroads, which became more frequent and extensive, until, in the reign of Antoninus Pius, the Brigantes having at the same time attempted to throw off the Roman yoke, that emperor sent Lollius Urbicus with strong reinforcements to suppress these commotions: this commander, having first reduced the revolted Brigantes, drove the Caledonians northward into the highlands of Scotland, and thus restored tranquillity. This people, however, having renewed their irruptions, in the year 207, the Emperor Severus came over with a numerous army, and immediately advanced to York, whence, having rejected all overtures for peace, he marched northward and expelled them, leaving to his son Caracalla the command of the army, and the care of repairing Adrian's rampart. Severus, labouring under indisposition, retired, in the year 211, to York, where he expired, and his obsequies and apotheosis were solemnised with great magnificence. Constantius Chlorus, Emperor of the West, resided for a long time at York, where he also died, in 307: he was succeeded by his son Constantius, who was saluted emperor by the Roman soldiery in that city, and who soon after collected a powerful army, composed chiefly of native Britons, and departed for the continent. The barbarians of the north again renewed their incursions, about the year 364, but were at length repelled by the Roman General Theodosius, in 368. In the latter period of the Roman empire in Britain, the territory at present contained in Yorkshire was included in the division called *Maxima Cæsariensis*. After the accession of Honorius, one of the sons of Theodosius, to the empire of the West, in 393, the invasions of the Picts and Scots became incessant, and their progress was every where marked with desolation; and when the Romans, about the year 410, abandoned Britain, in order to defend their continental dominions, the Romanized Britons fell into a state of anarchy, amidst which it is only known of Yorkshire, that it formed the greater part of a British kingdom, named Diefyr, or Deira, the conquest of which by the Saxon chieftains was not achieved until after a lapse of 111 years from the first arrival of Hengist in Kent. Bernicia, situated to the north of the Roman wall, having been subjugated by Ida, about the year 547, Ella, another Saxon leader, about the year 560, penetrated southward from that territory, and effected the conquest of Deira: these two kingdoms, afterwards united into one sovereignty by Ethelfrith of Bernicia, derived, from their situation to the north of the Humber, the name of Northumberland, or Northumbria. It was in the year 628, during the reign of Edwin, the next Northumbrian monarch, who had married a Christian Princess, named Ethelburga, sister of Ethelbald, King of Kent, that Christianity was first introduced into this part of Britain. In 633, Penda, King of Mercia, having entered into a league with Cadwallo, King of North Wales, against Edwin of Northumbria, the united forces of these confederate princes invaded the dominions of the latter, who opposed them at Hatfield, in the west Riding, about seven miles to the north-east of Doncaster, where a desperate battle ensued, in which the Northumbrian monarch, together with one of his sons and the greater part of his army, perished: the victors then ravaged Northumbria with merciless cruelty, and this powerful kingdom became once more divided into two separate sovereignties; Osric, nephew of Edwin, succeeding to the precarious throne of Deira, and restoring paganism in his dominions, while Eanfrid, son of Ethelfrith, ascended that of Bernicia. Osric, having besieged Cadwallo in York, was killed, and his army totally routed, in attempting to repulse the Welsh prince, who had made a vigorous sortie: and, during the space of a year, Cadwallo remained master of York, desolating the whole country of Deira: he also put to death Eanfrid, King of Bernicia, but, in 634, was defeated and slain, with the flower of his army, by Oswald, brother of Eanfrid, who thereupon succeeding without opposition to the throne of Northumbria, fixed his residence at York, restored Christianity, and completed the building of the church, which Edwin had left unfinished.

Penda, King of Mercia, preparing to invade Northumbria, Oswald hastily entered his dominions, but was defeated and slain in Shropshire, in 642, and Penda ravaged his territory: the Bernicians placed Oswy, the brother of Oswald, on the throne of their kingdom; and in the following year Oswin, the grandson of Edwin, was elected and crowned king of Deira. Oswy soon asserted his claim to the throne of York; and Oswin, being of a religious and unsuspecting, rather than of a martial disposition, was betrayed into the hands of Oswy, who inhumanly put him to death. The people of Deira immediately elected Adelwald, nephew of Oswin, for their king; and this monarch, having been induced to enter into a league with the kings of Mercia and East Anglia, against the sovereign of Bernicia, the confederated forces encountered those of Oswy, on the northern bank of the Aire, near Leeds, in 655. But Adelwald, seeing that the victory of either party would be equally dangerous to him, took no part in the action which ensued; and though the Mercian king Penda attacked the Bernicians with great impetuosity, not doubting of success, yet his soldiers, as soon as they perceived Adelwald withdrawing his forces, suspecting treachery, began to give way; and though the kings of Mercia and East Anglia made great efforts to rally their troops, both of them were slain, and their army was routed with terrible slaughter. On the peaceful death of Adelwald, Oswy succeeded to the entire dominion of Northumbria, but his affection for his natural son Alfred induced him to make him king of Deira; and on the death of Oswy, in 670, his son Egfrid succeeded him in the kingdom of Bernicia; the Deirians, however, revolted against Alfred, and put themselves under the dominion of Egfrid, on whose death, after an active reign of fifteen years, Alfred was recalled to assume the sway over Northumbria, the dominion of which was never again divided. A few of the succeeding reigns, though short, were marked by no act of peculiar violence; but the instances of ferocity, treason, and rebellion, which disfigure the annals of this northern kingdom, from the close of the reign of Eadbert to the commencement of the ninth century, present one of the most appalling pictures to be found in the history of any age or country: within the short space of 50 years eight kings were successively hurled from this blood-stained throne by expulsion or assassination. A region of civil discord was ill prepared to resist the victorious arms of King Egbert, who, from the conquest of Mercia, advanced to that of Northumbria: the reigning prince, Eanred, submitted without an appeal to arms, and accepted the same terms that had been granted to East Anglia and Mercia, according to which Northumbria was to remain a distinct, but tributary, kingdom.

About the middle of the ninth century, Ragnar Lodbrog, a celebrated Danish pirate, was wrecked, with two vessels of a size unusually large at that period, on the coast of Northumbria, in which country fresh disputes for the throne had arisen: and having succeeded in landing, he moved forward to plunder and ravage, regardless of his fate, but was soon opposed by Ella, one of the rival kings, with the whole of his forces, and a fierce, though unequal, conflict ensued, in which, after seeing most of his followers fall around him, Ragnar was at last overpowered and made prisoner, and soon after cruelly put to death. A more powerful force than had ever before sailed from Denmark soon after approached the English coasts, under the command of Inguar and Ubba, two sons of Ragnar, and, in 867, after having wintered on the coast of East Anglia, entered the Humber and ravaged Holderness, slaughtering such of the inhabitants as were unable to save themselves by flight. Advancing with insatiable avidity and ruthless vengeance, they destroyed with fire and sword all the country near the northern shores of the Humber, and near York defeated Osbert, the rival of Ella in the sovereignty of Northumbria, who was slain in the action, together with great numbers of his men. The Danes then entered York, to which city Ella was advancing in aid of his rival, and near which he was met by the North-men, who slew him, and routed his army with great slaughter. Northumbria now, from an Anglo-Saxon, became a Danish, kingdom, of which this county formed by far the largest and most important part. Inguar established his throne at York, which city was colonised by his followers, and extended his sway over the whole country from the Humber to the Tyne. The Danes, no longer fighting only for plunder, but for dominion, in 868, moved southward into Mercia, and returned the following year with a rich booty. In the spring of 870, several large bodies of their army again marched into the more southern provinces; and the storm which had been gathering at York, and in its vicinity, extended its direful effects over the whole of them. In the year 878, the Northumbrian Danes acknowledged the paramount sovereignty of the Saxon king, Alfred, but were, notwithstanding, governed by their own chieftains, one of whom bore the title of king, and had his principal residence at York. In 910, hostilities having arisen between the Danes and the Saxons, Edward the Elder ravaged a great part of Northumbria, and totally routed the Danes, slaying two of their kings, Halfden and Eowils, together with many of their great officers, and several thousand of their soldiers. Athelstan, who ascended the Anglo-Saxon throne in 924, with a large army expelled the Danish chieftains, and made himself master of all Northumbria. Anlaf, one of the expelled princes, having entered into alliance with different chieftains of Ireland and Wales, and with Constantine King of Scotland, soon after entered the Humber with a fleet of 615 ships, filled with warriors: these troops having disembarked, the Saxons abandoned the stations that were weakly fortified; but the stronger fortresses, being well garrisoned, resisted the attacks of the invaders, and gave time for Athelstan to prepare for the contest. Both parties having concentrated their forces, a sanguinary and decisive conflict took place, in which the confederates were totally defeated; and the king of Scotland, and six Welsh and Irish kings, with twelve of their earls and general officers, and a vast number of their followers, were slain. The issue left the Anglo-Saxon monarch master of all Northumbria, which (its population being chiefly Danish) he held in subjection by numerous garrisons, and totally destroyed the castle of York. Some time afterwards, Eric, King of Norway, being expelled by his subjects, was kindly received by Athelstan, who placed him on the throne of Northumbria, as a vassal of the Anglo-Saxon crown: Eric fixed his habitation at York, which thus again became a royal residence. On the death of Athelstan, Anlaf, having obtained assistance from Olaus, King of Norway, once more entered this principality, and, appearing before the gates of York, was admitted by the citizens, whose example was followed in most of the other towns, the English garrisons being either expelled or slaughtered by the inhabitants, who were for the most part Danish. Anlaf then extended his conquests into Mercia, and, by a treaty with Edmund, the successor of Athelstan on the Anglo-Saxon throne, was confirmed in his title to the kingdom of Northumbria. From this period to the final subjugation of the Northumbrian kingdom by Edred in 951, its imperfect history is very confused: after that event it was governed by a succession of earls or viceroys, who, like the ancient kings, had their residence at York.

This county is in the dioceses of York and Ripon, in the province of York, and forms the three archdeaconries of York (or of the West Riding), the East Riding, and Cleveland, and part of that of Richmond: the archdeaconry of York is subdived into the deaneries of the city and ainsty of York, Craven, Doncaster, and Pontefract; that of the East Riding into those of Buckrose, Dickering, Harthill and Hull, and Holderness; and that of Cleveland into those of Bulmer, Cleveland, Ryedale, and Ripon; while that of Richmond comprises, in this county, those of Boroughbridge, Catterick, Richmond, and part of Lonsdale: the total number of parishes is 604.

The grand civil and military division of Yorkshire is into three *ridings*,—West, North, and East, the term *riding* being corrupted from *trithing*, a third part: the West Riding is subdivided into the wapentakes of Agbrigg (Upper and Lower), Barkstone-Ash (Upper and Lower), Claro (Upper and Lower), Morley, Osgoldcross (Upper and Lower), Skyrack (Upper and Lower), Staincliffe and Ewcross (East and West), Staincross, and Strafforth and Tickhill (North and South), with the liberty of Ripon and soke of Doncaster; and by the act of the 5th and 6th of William IV., cap. 76, the ainsty has been severed from the county of the city of York, and annexed to the West Riding. The North Riding is divided into the wapentakes of Allertonshire, Birdforth, Bulmer, Gilling-East, Gilling-West, Hallikeld, Hang-East, Hang-West, and Ryedale, also Pickering Lythe, and the liberties of Langbaurgh and Whitby-Strand; and the East Riding into the wapentakes of Buckrose, Dickering, Harthill (Bainton-Beacon, Holme-Beacon, Hunsley-Beacon, and Wilton-Beacon, divisions), Holderness (Middle, North, and South), Howdenshire, and Ouze and Derwent, besides which it comprehends within its limits the liberty of St. Peter of York, the borough and liberties of Beverley, and the county of the town of Kingston-upon-Hull, which comprises a few parishes in the neighbourhood of that place. Yorkshire contains the city of York; the borough, market, and sea-port towns of Hull, Scarborough, and Whitby; the borough and market towns of Beverley, Bradford, Doncaster, Halifax, Huddersfield, Knaresbrough, Leeds, Malton, North-Allerton, Pontefract, Richmond, Ripon, Sheffield, Thirsk, and Wakefield; the market and seaport-town of Bridlington; and the market-towns of Askrigg, Barnesley, Bawtry, Bedale, Bingley, Boroughbridge, Dewsbury, Guisborough, Hawes, Hedon, Helmsley, Howden, Keighley, Kirkby-Moorside, Leyburn, Market-Weighton, Masham, Otley, Patrington, Penistone, Pickering, Pocklington, Reeth, Rotherham, Sedbergh, Selby, Settle, Sherburn, Skipton, South Cave, Stokesley, Tadcaster, Thorne, Wetherby, and Yarm. On the disfranchisement of the Cornish borough of Grampound, the privilege of returning to parliament two additional members was granted to this large and populous county, which accordingly then sent four; and under the act passed to amend the representation in the 2nd of William IV., two more were added, making two for each Riding. The place of election for the North Riding is York, and the polling-places are York, Malton, Scarborough, Whitby, Stokesley, Guisborough, Romaldkirk, Richmond, Reeth, Askrigg, Thirsk, North-Allerton, and Kirkby-Moorside: that for the East Riding is Beverley, and the polling-places are Beverley, Hull, Driffield, Pocklington, Bridlington, Howden, Hedon, and Settrington; and that for the West Riding is Wakefield, and the polling-places, Wakefield, Sheffield, Doncaster, Snaith, Huddersfield, Halifax, Bradford, Barnesley, Leeds, Keighley, Settle, Knaresbrough, Skipton, Pately-Bridge, Dent, Ripon, New Delph, Pontefract, Rotherham, Hebden-Bridge, Holmfirth, Otley, Aberford, and Birstal. Two citizens are returned for the city of York; two burgesses for each of the boroughs, except North-Allerton and Thirsk, which, under the act of the 2nd of William IV., cap. 45, have each been deprived of one; and except also the newly-enfranchised boroughs of Huddersfield, Wakefield, and Whitby, which are empowered to send only one each: the remaining towns to which the privilege of exercising the elective franchise was granted by the above-named act are Bradford, Halifax, Leeds, and Sheffield; and those which were entirely disfranchised are Aldborough, Boroughbridge, and Hedon. This shire is included in the Northern circuit: the assizes are held at York, where is the county gaol. The quarter sessions for the West Riding are held as follows: the Easter sessions at Pontefract; the Midsummer quarter sessions at Skipton, whence they are adjourned to Bradford, and thence to Rotherham; the Michaelmas quarter sessions begin at Knaresbrough, whence they are adjourned to Leeds, and thence to Sheffield; the Christmas quarter sessions commence at Wetherby, and are adjourned to Wakefield, and thence to Doncaster: on the termination of each session there is an adjournment to Wakefield for the purpose of inspecting the prison, which generally takes place within a month or six weeks after that time. In pursuance of an act passed in the year 1704, the office for the registration of deeds, conveyances, and wills, relating to property within the West Riding, was established at Wakefield, where also are kept the records of the sessions. The quarter sessions for the North and East Ridings are held respectively at North-Allerton and Beverley, in each of which towns are also offices for the registration of all deeds relating to landed property within those Ridings.

One of the most remarkable peculiarities in the civil and military jurisdiction of Yorkshire is, that each of its Ridings has a distinct lord-lieutenant. The ainsty of York was formerly a wapentake of the West Riding; but, in the 27th of Henry VI., it was annexed to the city, and placed under its immediate jurisdiction; it now again forms part of the West Riding, as before stated. The liberty of St. Peter's comprehends all those parts of the county that belong to the cathedral church of St. Peter at York: the jurisdiction is separate and exclusive, and it has its own magistrates, steward, bailiff, coroners, and constables: amongst its privileges, the inhabitants are exempt from the payment of all manner of tolls throughout England, Wales, and Ireland, on the production of a certificate from the under-steward. Quarter sessions are held for this liberty at the sessions-house in the Minster-yard at York; and a court is held in the hall every three weeks, where pleas in actions of debt, trespass, replevin, &c., to any amount whatever, arising within the liberty, are heard. There is also a court leet and view of frankpledge for the whole liberty, held twice a year, *viz.*, on the Wednesday in Easter week, and on the first Wednesday after New Michaelmas-day. Sessions for the Archbishop of York's liberty of Cawood, Wistow, and Otley, are held at Otley, in Jan. and April, and at Cawood, in April and Oct.

The West Riding, which, whether considered with regard to its extent and population, or to its trade and manufactures, is by far the most important, is bounded on the north by the North Riding; on the east by the ainsty, and the river Ouse, to its junction with the Trent; and on the south and west, by the arbitrary limits of the county: its greatest length, from east to west, is 95 miles; its extreme breadth, from north to south, 48 miles; and its circumference, about 320 miles, including an area of 2450 square miles, or 1,568,000 statute acres: its population is 976,350. The surface of this portion of Yorkshire is much diversified, but may be divided into three large districts, gradually varying from a level and marshy to a rocky and mountainous region. The flat and marshy district, forming part of the extensive Vale of York, lies along the borders of the Ouse, and in most places extends westward as far as within three or four miles of an imaginary line drawn from Doncaster to Sherburn: the general level is broken only by low sandy hills, which occur in the vicinities of Snaith, Thorne, and Doncaster, and the altitude of which is seldom more than 50 feet above the level of the sea; so that the great rivers Ouse, Aire, and Don, which traverse this extensive tract, have often changed their channels. The middle parts of the Riding, as far westward as Sheffield, Bradford, and Otley, contain a variety of beautiful scenery, formed chiefly by noble hills of gentle ascent; but further westward the county becomes rugged and mountainous, scarcely any thing being seen beyond Sheffield, in that direction, but high black moors, which, running north-westward, join the lofty hills of Blackstone Edge, on the border of Lancashire. The north-western part of the Riding, forming the western part of the district of Craven, presents a confused heap of rocks and mountains, among which Pennygant, Wharnside, and Ingleborough, are particularly conspicuous.

The North Riding, the next most extensive division, is bounded on the north by the river Tees; on the north-east and east by the ocean; on the south-east by the rivers Hertford and Derwent, which separate it from the East Riding; on the south by the river Ouse and the West Riding; and on the west by the county of Westmorland: its greatest length is 83 miles, from east to west; its extreme breadth 47 miles, from north to south; and it comprises an area of 1,311,187 acres, or about 2048 square miles: its population is 190,756. The face of the country along the coast, from Scarborough nearly to the mouth of the Tees, is hilly and bold, the cliffs overhanging the beach being generally from 60 or 70 to 150 feet high; while Stoupe Brow, vulgarly "Stow Brow," about seven miles to the south of Whitby, rises to the stupendous height of 893 feet. From the ordinary elevation of the cliff the ground rises, in most places very rapidly, to the height of 300 or 400 feet; and the maritime tract thus formed, comprising about 64,920 acres, is tolerably productive. A little further inland, successive hills, rising one above another, form the elevated tract of the Eastern Moorlands: this wild and mountainous district, which occupies a space of about 30 miles in length from east to west, and 15 in breadth from north to south, is intersected by numerous beautiful and fertile dales, some of which are rather extensive; but, rising to the height of upwards of 1000 feet, the general aspect of this extensive tract is bleak and dreary, and the whole is destitute of wood, excepting only a few dwarfish trees among the scattered habitations in the valleys. On the roads leading from Whitby to Guisborough, Stokesley, and Pickering, at the distance of a few miles, commence dreary and extensive wastes, bounded only by the horizon. Some of the hills, however, near the edges of this rugged and mountainous region command picturesque and magnificent prospects. But the most remarkable object in the topography of these wilds is the singular peaked mountain called Rosebury - Topping, situated near the village of Newton, about a mile to the eastward of the road from Guisborough to Stokesley, which rises to the height of 1488 feet above the level of the sea, and is a noted landmark: the view from its summit is celebrated for its great extent and variety. The total extent of the Eastern Moorland district is 298,625 acres. The Vale of Cleveland, situated to the north-west of these mountains, is the fruitful tract bordering on the river Tees, in the lower part of its course; in this county it comprises an area of 70,444 acres, the whole under cultivation, and is lightly marked with gentle eminences. The extensive Vale of York is considered by Mr. Tuke, author of the "General View of the Agriculture of the North Riding of Yorkshire, drawn up for the consideration of the Board of Agriculture," to reach from the border of the Tees to the southern confines of the county, the northern part of it only being included in this Riding: this part, bounded on each side by the Eastern and the Western Moorlands, has a gentle slope, from the border of the river Tees, southward as far as York, where it sinks into a perfect flat; between the Tees and York, however, its ordinarily level surface is broken by several bold swells; and on the east it is separated from Ryedale by a range of hills, called by Mr. Marshall, in his Rural Economy of Yorkshire, the "Howardian Hills." This part of the vale, together with these hills, comprises an extent of 456,386 acres, of which about 15,000 are uncultivated. Ryedale (so called from its being traversed by the river Rye) and the East and West Marishes, form one extensive level, situated between the Eastern Moorlands and the river Derwent, and contain 103,872 acres, of which about 3000 are waste: the surface of its lower parts is flat, but towards the north it rises with a gentle ascent for three or four miles towards the foot of the moors; its lower levels are also broken by several isolated swells of considerable extent and elevation: the Marishes are separated from Ryedale by the Pickering-beck. The Western Moorlands, occupying the rest of the North Riding, to the west of the Vale of York, and of far greater elevation than the Eastern Moorlands, resemble in general character the mountainous parts of Craven, and are, like them, intersected by numerous fertile dales: their total extent is 316,940 acres.

The East Riding is bounded on the north and north-west by the little river Hertford, and the Derwent as far down as Stamford-bridge, about a mile above which place an irregular boundary line commences, which joins the Ouse, about a mile below York: from this point it is bounded, on the west and south-west, by the Ouse, on the south by the Humber, and on the east by the North Sea: its greatest length is 52 miles, from south-east to north-west; its extreme breadth 42 miles, from south-west to north-east; and it includes an area of 819,193 statute acres, or nearly 1280 square miles: its population, including the city of York and the ainsty, is 204,253. This division of Yorkshire is far less conspicuously marked with the bolder features of nature than the other parts of the county. It may be distinguished into three districts, *viz.*, the Wolds and the two level tracts, one of which lies to the east, the other to the west and north of that elevated region. The Wolds are a magnificent assemblage of lofty chalk hills, extending from the banks of the Humber in the vicinity of Hessle, in a northerly direction, to the neighbourhood of New Malton on the Derwent, whence they range eastward, within a few miles of the course of that river, to the coast, where they form the lofty promontory of Flamborough Head, and, in the vicinities of the villages of Flamborough, Bempton, and Specton, rise in cliffs to the height of 100, and in some places of 150, feet. Their surface is for the most part divided into numerous extensive swells, by deep, narrow, and winding valleys, and occupies an extent of about 400,000 acres. Their eastern side, at Bridlington, sinks into a perfect flat, which continues for eight or nine miles southward. At the distance of about seven miles southward of Bridlington, however, the wapentake of Holderness begins, the eastern-part of which, towards the sea-coast, is a finely varied country, in which is situated Hornsea mere, the largest lake in the county, being about a mile and three quarters long, and three quarters of a mile across in the broadest part; but the western edge is a fenny tract of about four miles in breadth, and extending nearly 20 miles in length, southward, to the banks of the Humber: these fenny lands are provincially called "Cars." The southern part of Holderness also falls into marshes, bordering on the Humber; and the county terminates south-eastward in the long low promontory of Spurnhead, the *Ocellum Promontorium* of Ptolemy. The Humber is known to have made, in former ages, considerable encroachment on the shores of Holderness; but in later times it has gradually receded from very extensive tracts. About the commencement of the reign of Charles I., an island, since called Sunk Island, began to appear in the Humber, nearly opposite Patrington; at first a few acres only were left dry at low water; but, as it increased in extent every year, it was at last embanked, and converted into pasture ground: successive embankments were made, large tracts being at each time secured, until at the present period it comprises about 4700 acres of fertile land, and towards the west end is separated from the Holderness marshes only by a ditch a few feet broad: it is held on lease from the crown. The Holderness marshes have also been increased by the retiring of the waters of the Humber; and a large tract of land, called "Cherry-cob Sands," which was left dry, and embanked in the same manner as Sunk Island, is more particularly worthy of notice. The third natural division of the East Riding, which extends from the western foot of the Wolds to the boundary of the West Riding, is commonly called "The Levels," and, though generally fertile, and interspersed with villages and hamlets, is every where flat and uninteresting. One of the most important agricultural improvements in the county is the drainage of the cars and marshes of this division of it, together with those in the North Riding, bordering on the course of the Derwent.

The "Holderness Drainage" lies chiefly adjoining to and on the eastern side of the river Hull, extends from north to south about eleven miles, and contains 11,211 acres: in 1762, an act of parliament was obtained for draining this level, much of which before that period was of small value, being usually covered with water for above half the year: the execution of this drainage was vested in trustees, appointed by the owners of land within the limits of its operation. The "Beverley and Barmston Drainage," executed under the provisions of an act passed about the year 1792, lies parallel to the last, but on the opposite side of the river Hull, and extends from the sea-shore at Barmston, a few miles south of Bridlington, along the course of that river nearly to Kingston-upon-Hull, a distance of about 24 miles: its northern part contains more than 2000 acres, and has an outfall into the sea at Barmston; while the southern division, extending southward from Foston, contains upwards of 10,000 acres, and has its outlet into the river Hull at a place called Wincolmlee. The "Keyingham Drainage," lying between Sunk Island and the main land, was originally completed under an act passed in the year 1722; a new act was obtained in 1802, under which the course of the drainage in some parts was altered, and an additional quantity of land included, making a total of 5500 acres: the execution of this was vested in three commissioners, and on a vacancy occurring by death or resignation, another commissioner is elected by the proprietors. The "Hertford and Derwent Drainage" contains upwards of 10,500 acres, of which 4500 are in the East, and the remainder in the North, Riding: the act for this was obtained in the year 1800, and its execution was vested in three directors and three commissioners: the directors have a power to levy an annual assessment, not exceeding an average of three shillings per acre, for the purpose of maintaining and repairing the existing works and drains, and also of further making such new works as may, from time to time, become necessary. Spalding Moor and Walling Fen, lying to the westward of the southern part of the Wolds, were drained, allotted, and enclosed, about 50 years since, under the provisions of the same act of parliament.

The ainsty of York is situated to the west, and on the south-western side, of the Ouse, which borders it from the mouth of the Nid to that of the Wharfe, separating it first from the North, and afterwards from the East, Riding: from the West Riding it is separated for some distance by the Nid, and afterwards by a line including Wilstrop, Cattle-Bridge, Bickerton, and Thorp-Arch, and terminating at the junction of the Wharfe with the Ouse: its circumference is 32 miles, and its population amounts to 9102. The surface and scenery of this tract have the same general character as the rest of the Vale of York, of which it forms a part: the western portion of it is diversified by various gentle swells; while the eastern, adjoining the Ouse, is an entire flat, abounding with excellent meadows and pastures. The whole district of the ainsty was anciently a forest, but it was disafforested by the charters of Richard I. and John.

The climate of Yorkshire is as various as its surface. The soils comprise all the varieties common in the kingdom, from the deep strong clay and rich loam to the worst kind of peat earth. In the West Riding the prevailing quality is loam, the value of which depends, in a great measure, upon the nature of the substratum. Much of the low ground in this division of the county, bordering on the Ouse, has a clayey soil of a very tenacious quality, the cultivation of which is subject to many difficulties, although it is capable of producing the most abundant crops; this and a rich loamy soil are the predominating kinds, but are intermixed with some sandy and moorish tracts. The middle of the Riding is occupied chiefly by loam resting on limestone; and the same kind of soil, with a similar basis, although intermixed in many places with tracts of moor, of different qualities, prevails even to its western limits. In the North Riding the soils of the coast district are, a brownish clay, a clayey loam, a lightish soil upon alum shale; a loam upon freestone, or, as it is here called, "gritstone;" and, in some valleys to the west of Whitby, a deep rich loam. The soil of Cleveland is generally a fertile clay, with occasionally some clayey loam, and a fine red sandy soil, the latter of which is found chiefly between Marsh and Worsall, and about Crathorne, near the moors. In the Vale of York, the soil, though varying greatly, is for the most part fertile, and comprises rich gravelly loams; rich strong loams; rich hazel loams; strong and fertile, gravelly, and cold clays; sandy loams of various qualities, sometimes intermixed with cobblestones and coarse gravel; loamy soils upon limestone; cold and springy soils; and some small tracts of swampy and peaty land: there is also a cold thin clay upon what is called "moorband," a stratum from six inches to a foot in thickness, of a ferruginous and ochreous appearance, which, wherever found, is attended with great sterility. On the southern side of the Howardian hills good clayey and loamy soils prevail, as also at the western extremity: in other places the soil is frequently thin and poor, resting immediately upon a gritstone, and limestone, substratum. The soils of Ryedale are for the most part of extraordinary fertility, comprising a hazel-loam upon a clay substratum, and a deep warp, or silt, washed down from the higher country by the floods of many former ages, and deposited upon a gravel or clay; but some old clayey and yellow loamy soils, mixed with sandy pebbles, of less fertility than the above, are occasionally met with; the detached swells have a rich strong clay, one only excepted in the vicinity of Normanby, which is sandy: the northern margin of this vale has for the most part a deep loamy soil, resting on an imperfect reddish sandstone; but approaching the moors, the soil gradually becomes stiffer and less fertile, though in some places it is a yellow sand. The soil of the Marishes is chiefly clay, with some sandy loam, gravel, and peat, the whole very low and wet. Most of the narrow valleys of the Eastern Moorlands contain more or less of a black moory soil, resting upon clay; of a sandy soil, in some places intermixed with large gritstones, upon a shaly rock; and of a light loam lying on gritstone; on their eastern side is also found a stiffish loam upon limestone, and a deep sandy loam upon whinstone; and in the lowest situation a light loam upon gravel, or freestone. In the Western Moorlands, the soil of the bottom of Wensleydale is generally a rich loamy gravel; that on the sides of the hills, by which it is enclosed, a good loam, in some places rather stiff, and sometimes resting on gritstone, but generally on limestone: there are also some small tracts of clay and peat. Swaledale and the smaller dales are very similar to this in their soils, and several of those which discharge their waters into the Tees are peculiarly fertile.

In the East Riding, the whole country to the eastward of the Wolds is occupied by clays and loams, of occasionally varying quality; excepting only a narrow tract of gravelly land, which extends for two or three miles both to the north and south of Rise, near Hornsea, and is excellent turnip land. The Holderness marshes, bordering on the Humber, below Hull, have a strong clayey loam, the fertility of which is almost unequalled. The Wolds are composed almost entirely of chalk; their soils are, therefore, warm and calcareous: the most prevailing kind is a friable and rather light loam, having a mixture of chalky gravel, in some parts very shallow; a deeper sandy loam, resting immediately upon the chalk, is also found in some places. The soil of the "Levels" is, in some parts, clayey; in others, sandy. An extensive sandy, and in some places moory, tract stretches across the middle of them from South Cave, north-westward, to York; but near the banks of the Derwent and the Ouse the predominant soils are a clayey loam and a very strong clay; the latter chiefly prevails from Gilberdike to Howden, and thence extends quite to the Ouse. The tract of ground called "Marshland," situated below the junction of the Aire with the Ouse, is supposed to have been, at some former period, wholly covered with water, and the soil is for the most part of that sort which, in many places, is known by the name of *water-fat*. The Vale of the Derwent, extending from the vicinity of the coast westward towards York, is remarkable for the great variety of its soils, which, however, are generally fertile; they include a very light fine sand, loams of various qualities, a strong clay of divers colours, lying in some places upon a coarse hard limestone abounding with shells, and a black peat, which extends along the course of the Hertford to its junction with the Derwent, and thence to Yeddingham bridge. The soils of the ainsty bear a character of general fertility: they are chiefly loams of rather various kinds, some on a calcareous, and others on a gravelly, substratum. The low grounds adjoining the rivers have a soil formed chiefly of the alluvial matter washed from the surrounding higher grounds: those on the banks of the Ouse are most remarkable for their fertility. All that portion of the West Riding included between the river Ouse and an imaginary line drawn from Ripley southward by Leeds, Wakefield, and Barnesley, to Rotherham, is principally employed in the production of corn; while the land in the vicinity of the manufacturing towns is under no peculiar system of husbandry: the amount of arable land in the western parts of the West Riding is extremely small. In the North Riding, in the Vale of York, one third of the land is in tillage; on the western end of the Howardian hills, and thence to Thirsk, only one-fourth; on the rest of the Howardian hills, nearly one-half; in Ryedale, the Marishes, and the northern part of the coast, about one-third; the southern part of the coast one-fourth; in Cleveland, one-half; in the dales of the eastern moors, one-fifth; and in those of the western moors, hardly any. In the East Riding the proportion of land under tillage, on the Wolds, is two-thirds; in Holderness rather more than one-third, and towards the south-eastern extremity of the county considerably more; and in Howdenshire, and to the west of the Wolds, somewhat less than one-third.

Every kind of agricultural crop is cultivated in this county: and the systems of tillage, on account of the great diversity of soils and situations, are extremely various. Wheat is grown to a great extent on all the lower and more fertile lands; and no other district in the north of England, in proportion to its size, is considered to produce so much of it, or of so good a quality, as Cleveland, whence large quantities are shipped to the southern coast of England, and much is conveyed to Thirsk and Leybourn, where it is bought up for the manufacturing districts. Rye is sometimes sown on the lighter soils, more particularly of the North

Riding, where wheat is not unfrequently mixed with it: of this mixture, provincially called "meslin," the common household bread of that portion of the county is chiefly made. The quantity of land annually sown with barley is no where remarkably great, except on the Wolds, the soil of which is peculiarly adapted to its culture: in the North Riding, in Ryedale and the dales of the Eastern Moorlands, are occasionally seen plots of the species provincially called *big*, which is six-rowed barley; and of *bear*, four-rowed. Besides being occasionally grown in other places, oats are very much cultivated in all the arable parts of the North Riding, more particularly in Ryedale, which district is as remarkable for the quantity and excellent quality of its oats, as Cleveland is for those of its wheat: two crops are here always taken in succession, and frequently three: in the western parts of the West Riding, too, this corn is the prevailing crop: oaten bread is in common use in the manufacturing districts of the West Riding. Considerable quantities of flax are grown in the West Riding, in the neighbourhood of Selby; in the East Riding, about Howden and on the eastern bank of the Derwent; and in the North Riding, a small quantity in Ryedale, and a few other situations. Woad, for dyeing, is cultivated in the neighbourhood of Selby, among red clover. In the vicinity of York mustard is a valuable article of cultivation, and fields of it are occasionally seen in different places in the northern and eastern parts of the county: that which is grown near York is prepared for use in mills at that city, and is afterwards sold as Durham mustard. The wapentake of Barkstone-Ash, in the eastern part of the West Riding, is distinguished for an extensive growth of teasel, which is also occasionally cultivated to a small extent in different other places having a strong soil: it is purchased by the cloth-dressers, for the purpose of raising the nap on cloth, before it undergoes the operation of shearing. Sainfoin is grown in different situations. On the richer soils the principal artificial grasses are, however, red clover, when the next crop is to be wheat; and white clover and hay seeds, when the land is to remain in pasture; sometimes only the hay-seeds are sown, or trefoil, or rye-grass added. In the North Riding the cultivation of grasses is little attended to, except in the country lying between Boroughbridge and Catterick; they are here chiefly sown where the land is intended to remain permanently under grass, and consist generally of white clover, trefoil, rib-grass, and hay-seeds, with which some mix red clover, while others sow rye-grass, instead of the hay-seeds.

The grass lands are very extensive, for, besides the tracts included with the arable districts in the large proportion above stated, the productive parts of the western side of the county are kept almost exclusively in grass, and from Ripley to its western extremity the whole country is employed in grazing; while corn, and that almost entirely oats, is raised only in very small quantities on the inferior moorish soils. The old pasture lands, forming by far the greater portion of the grass lands, have remained in that state from time immemorial, and in the West Riding are frequently mown, producing hay held in great esteem. Some of them are, nevertheless, of a very mean quality, and, especially in the North Riding, are often covered with thistles, ant-hills, and occasionally furze: in the dales of the Western Moorlands, however, remarkably great attention is paid to the meadows. The extent of natural meadow, namely, such as derives the whole, or the greater part, of its fertility from the overflow of rivers, is not very great: many of the old fields of this kind in the Vale of York and Ryedale have been constantly mown for ages, and are still highly productive. The East Riding contains the smallest quantity of grass land, its sheep pastures on the Wolds, for which it was formerly so distinguished, having been mostly brought under various courses of tillage; but it contains, on the banks of the Derwent, above Malton, and again at Cottingwith, low tracts of marshy meadows, occasionally overflowed by that river, which produce abundant crops of coarse flaggy hay, of which that obtained from the last-mentioned district is of a peculiarly nutritive quality. The whole of the West Riding is an eminent grazing district, where cattle and sheep of all kinds are fattened to great perfection, chiefly to supply the manufacturing parts of Yorkshire and Lancashire: for this purpose, great numbers of lean cattle and sheep are annually brought from Scotland and the northern counties contiguous to Yorkshire. It has also numerous small dairies, for the supply of its own manufacturing towns and those of Lancashire with butter; and some large ones in the vicinity of the large towns, to which the milk is chiefly sold. In the North Riding, the pastures are for the most part appropriated to the dairy; though grazing is also practised in some parts of it, more particularly in the Vale of York: the butter produced in this Riding is chiefly packed in firkins, and sold to factors, who ship it for the London and other markets. In the East Riding, grazing and fattening, also stall-feeding, are practised to a very considerable extent.

The manures are, lime, which is used in almost every part of the county; rape-dust; bones, great quantities of which are imported from abroad; horn-shavings, and several other articles of refuse from the manufacturing towns; kelp-ashes and peat-ashes, in the North Riding; sea-*wreck*, or sea-weed, which is frequently thrown upon the coast by the tide; sea-sand; whale blubber, and the refuse of the oil; and, in the East Riding, chalk: in this division, also, straw is frequently spread upon the land and burned. Extensive tracts, bordering on the Ouse and the Humber, in the East Riding, and the eastern parts of the West Riding, are rendered of extraordinary fertility by the practice of warping; this is the admitting of the tide, which rises higher than their level, to overflow them, and afterwards allowing it to retire from them at its ebb, when it deposits a thin bed of mud and salts, provincially called *warp:* this operation is performed chiefly by means of a clough, or inlet in the bank of the river, walled strongly on each side, and a floodgate fixed in the middle, which, as the tide falls, permits the gentle egress of the waters which had been admitted upon the land previously banked and prepared, through a smaller opening on a higher level. The West Riding is the only division containing any considerable extent of irrigated meadows, which are most common in the manufacturing districts. A dry limestone ridge, about four miles broad, extending from east to west, northward of Ryedale, and in the vicinity of Kirkby-Moorside, was ingeniously supplied with water, in the latter part of the last century, by means of artificial rills, brought down from the much loftier tract of the Eastern Moorlands. Artificial ponds, of a peculiar construction, for catching and preserving the rain water, are very common on the Wolds of the East Riding, and in different parts of the North Riding that require such accommodation. With regard to implements of husbandry, the common Rotherham plough, sometimes called the Dutch plough, is in general use, except upon the Wolds, where the clumsy, heavy, old-fashioned foot-plough, having a short straight wooden mould-board, is chiefly employed.

The cattle for which the West Riding is most noted are the hardy, long-horned, or Craven breed: the cattle and sheep brought into this division of the county, for the purpose of being fattened, include almost all the different varieties reared in Britain, though the greater number are Scotch. Short-horned cattle are the prevailing kind in the eastern parts of this Riding, in the East Riding, and in the North Riding, excepting only its westernmost districts. The short-horned cattle of the northern part of the Vale of York, and of Cleveland, where considerable numbers are bred, are generally known by the name of the Teeswater breed; and in the south of England by that of the Holderness cattle, from the district of that name in the East Riding, where this breed was either originally established, or first so improved as to bring it into notice, and where, in common with the tracts before mentioned, the best of the sort are still to be met with: these are also occasionally called Durham, or Dutch cattle, and the cows are in great demand in some of the southern counties of England, more particularly near London, as their produce of milk is remarkably great. Many excellent cattle of this kind are also bred in Ryedale and the Marishes, and on the Howardian hills; as also in the Eastern Moorlands, and along the coast of the North Riding, in which districts, however, they are not quite so large as those bred near the Tees. In the Western Moorlands are found some small long-horned cattle, and a mixed breed, between the long-horned and the short-horned species, which also occupies a considerable portion of the West Riding, including Nidderdale and the adjacent country, and is held in great esteem: in the lower parts of the dales of the Western Moorlands, many of the short-horned breed are also kept. The working of oxen is most common in the eastern part of the North Riding: the cattle of these districts being, from their natural strength and hardihood, well adapted to the purpose, are trained to labour at two or two years and a half old, and are worked until five or six years old.

The kinds of sheep are very numerous and much intermixed. Those bred upon the moors of the mountainous parts of the West Riding, which are supposed to be native, are generally called the Penistone breed, from the name of the market-town at which they are chiefly sold. The Dishley breed is common in the southern and eastern parts of this Riding. The sheep of the old stock of the northern part of the Vale of York, and of Cleveland, generally called Teeswater sheep, are very large, with long, dry, and harsh wool; but within the last 40 years these sheep have been greatly intermixed with the Dishley, Northumberland, and some other breeds, and the varieties thus produced occupy all the low lands and rich cultivated tracts of the North Riding. The next in point of number in this division of the county is the hardy unmixed breed which occupies the summits of both the Moorlands; but the greater number of the sheep on the Moorlands are temporary flocks, of a kind called "Short Scots," to distinguish them from a larger breed of Scotch sheep, called "Long Scots." A peculiar, but far less numerous, race occupies a middle region in the western part of the county, the grassy summits of the calcareous hills, and the higher enclosed lands of the Western Moorlands, being a pure, unmixed, and hardy race, much resembling the Old Wiltshire breed: their fleeces are generally thick, dry, and harsh, but some of them produce a very fine wool, used in the hosiery manufacture, for which the dales of these Moorlands are so celebrated. The native sheep of the East Riding are the Holderness and the Wolds breeds, which have of late years been much intermixed with the Leicester, the former of these resembling the Lincolnshire sheep; the Wolds breed is small, hardy, and active, with a short thick fleece of fine clothing wool. The South Down breed has been introduced upon the Wolds, and is gradually extending itself. The hogs are of various kinds: in the western part of the West Riding many hogs are fattened upon oatmeal, and their flesh sold for the manufacturing districts of Lancashire, excepting only the hams, which are usually sent to the London market: from the East Riding many are sold into Lincolnshire, and are thence, in many instances, forwarded to the metropolis. In the West Riding few horses are bred, except in the eastern parts of it; but the North and East Ridings have long been famous for the breeding and rearing of horses, chiefly adapted for the coach and the saddle, for which purposes they are not excelled by any in the kingdom: the strongest are chiefly employed as coach-horses; and the lighter for the field, the road, and the army. The principal fairs, for horses of every description, are at Beverley, Howden, Malton, and York, and are resorted to by numerous dealers from London and all parts of the kingdom. At an annual fair held at Hull, in October, great numbers of colts are also purchased by the Lincolnshire farmers and graziers, who keep them until they are four years old, when they are sold to the London dealers at Horncastle fair.

A great deal of oak and ash timber is produced in the West Riding, and great attention is paid to the management of the woods by their proprietors: the timber meets with a ready sale to the ship-building and manufacturing towns; and much is also used in the mines and collieries. The extent of the woodlands in the North Riding is estimated at about 25,000 acres, dispersed in all quarters, the Moorlands and Cleveland having the smallest proportion: exclusively of the above, this division also produces a considerable quantity of timber in its hedge-rows, more particularly in the Vale of York, on the Howardian hills, and in Ryedale. The spontaneous produce of the best woodlands is oak, ash, and broad-leaved, or wych elm; of those in mountainous situations, chiefly birch and alder; and of the hedge-rows, various kinds of trees, for the most part of artificial plantation. In this Riding it is the custom to sell the falls of wood to professed wood-buyers, who cut up the trees on the spot, according to the purposes for which the different parts of them are best calculated: the ports of Scarborough and Whitby consume most of the ship timber, excepting only such as grows towards its western extremity: the oak timber grown in the greater part of this Riding, though not large, is extremely hard and durable: the only peculiar application of the ash timber, which grows abundantly and in great perfection, is in the manufacture of butter-firkins, in which it is chiefly consumed. Plantations have been made on the sides and summits of several of the Moorland and other barren hills, chiefly of Scotch fir, larch, and spruce, a few oaks, &c. The East Riding is little remarkable for its timber; the natural woods are confined chiefly to the levels lying between the rivers Ouse and Derwent and the Wolds, where there are also abundance of timber-trees in the hedge-rows of old enclosures: the only woods to the east of the Wolds are those of Rise and Burton-Constable. The fine elevations of the Wolds have been ornamented in different parts by extensive plantations of Scotch and spruce firs, larch, beech, ash, &c., to the amount of several thousand acres; and various other plantations have been made in the low country to the west of them.

The wastes are very extensive, and about the end of the last century were calculated to amount in the whole to 849,272 acres; but the amount has, since that period, been considerably lessened by numerous enclosure acts, obtained both for the detached wastes and for parts of the Moorlands. The surface of some of the higher hills of the Eastern Moorlands is entirely covered with large freestones; while upon others are extensive beds of peat bog, in many places very deep, frequently not passable, and never without danger: these are invariably overgrown with ling, in some places mixed with bent and rushes. Near the old enclosures are some considerable tracts of loamy and sandy soils, producing furze, fern (here called "brackens"), thistles, and coarse grass, with but little ling; but wherever ling is the chief produce, the soil is invariably black moor, or peat. The subsoils of these extensive wastes are various: in some places a yellowish, in others a reddish, clay occurs; a loose freestone rubble, resting either upon a freestone rock or upon clay, is also very common; and in different other places is found a rotten earth of a peaty quality (which produces very luxuriant ling, bent, and rushes), a hard cemented reddish sand, or a grey sand; the basis of the whole is freestone. The Hamilton hills, forming the western end of these wastes, are, however, very different, having generally a fine loamy soil on a limestone rock, which produces great quantities of coarse grass and bent, in some places intermixed with ling, more particularly towards the south-western parts of them. The mountains of the western side of the county differ materially in their produce from the Eastern Moorlands: some, instead of black ling, are covered with a fine sweet grass; others with extensive tracts of bent; and though the higher parts produce ling, it is generally mixed with a large proportion of grass, bent, or rushes: the soil on the lower parts is a fine loam, in many places rather stiff, resting upon a hard blue limestone: the bent generally covers a strong soil lying upon a gritstone or freestone rock; the black ling, a reddish peat upon a red subsoil, or, in many places, a loose grit rubble, beneath which is a gritstone rock. Some of the lower tracts of the eastern moors, the lower parts of the western moors in general, and in some instances the higher parts of the latter, are stinted pastures during the summer, and those who have that limited right in summer have a right in winter of turning upon them whatever quantity of stock they choose: these pastures are chiefly stocked with young cattle, horses, and such sheep as are intended to be sold off the same year. The remainder of the moors is common without stint, and is stocked for the most part with sheep, though a small, hardy, and very strong kind of horses is also bred and reared upon the Western Moorlands, and chiefly sold to the manufacturing parts of the West Riding and of Lancashire. The Moorland sheep are remarkable for their wretched appearance and great activity: they are wholly supported on these mountain wastes, and their mutton is of a particularly fine quality. The wastes of the East Riding consist chiefly of low, sandy, barren, and moory tracts lying between the Wolds and the rivers Ouse and Derwent, and the chief natural produce of which is short heath. The common fuel throughout the county is coal, with which the North Riding, as far south as Thirsk, is for the most part supplied by the collieries in the county of Durham, from which it is brought in one-horse carts to the coast of the East Riding from Newcastle and Sunderland, and the rest of the same division from the West Riding: in the Moorlands of the North Riding much peat is used.

To the geologist Yorkshire affords interesting fields of study: all its strata, with slight variations, dip eastward, those which appear at its western extremities being of the oldest formation. The mineral productions are various and important, and have given rise, and afford support, to some of its principal manufactures: they consist chiefly of coal, iron, lead, stone of various qualities, and alum. The best coal is obtained in the West Riding, which comprises one of the most valuable and extensive coal fields in the kingdom. This coal district is bounded on the east by a narrow range of magnesian limestone, extending from Tickhill northward by Doncaster, Ferrybridge, Wetherby, Knaresbrough, and Ripon, and consists of a great number of alternations of sand-stone, clay, shale, coal, and iron-stone, which form the substrata of the most populous parts of the Riding. Its surface is characterized by successive parallel ranges of high ground, extending in length from north to south: the ascent to these hills on their western sides is abrupt, while on the east they decline more gradually, each one to the foot of the next range, under which its strata dip. Next to the magnesian limestone and its subjacent sand, proceeding westward, appear, first, the blue shale and thin coal of the Vale of Went, and then the grit freestone of Ackworth and Kirby, beneath which is found the swift-burning coal of Wragby, Shafton, Crofton, and other places in the great clay district of the Dearn below Barnesley, and of the Calder below Wakefield. These various measures rest upon the grit freestone of Rotherham, Barnesley, Newmiller Dam, and East Ardsley, through which pits are sunk near Barnesley to several thick seams of hard furnace coal, one of them as much as ten feet thick. The next great sand-stone stratum forms high grounds, and frequently projects beyond the general range into detached hills: it occurs near Sheffield, Wentworth Park, and Bretton Park, and forms the high ground of Horbury and Dewsbury, and of Middleton, near Leeds; beneath it are found valuable beds of iron-stone, which are worked at Rotherham, Haigh-bridge, Low Moor, and several other places, where an abundance of muscle shells is found in contact with them: contiguous to this iron-stone are several strata of excellent coal. Next in the series lies the sand-stone of Wortley-Chapel, Silkstone, Elmley, and Whitley-hall, with the valuable bituminous coal of Silkstone and Flockton, the best seams of the whole formation: this rock, entering the West Riding from Derbyshire, and passing by Sheffield, Penistone, Huddersfield, Elland Edge, and the Clayton heights, afterwards takes its course parallel with the river Aire, by Idle and Chapel-Allerton, towards the magnesian limestone: in this part of the coal district, near Sheffield, Bradford, and Leeds, is dug the *galliard* stone, so much in request for making and mending the roads.

The coal mines are most numerous in the tract between Leeds and Wakefield, and in the neighbourhoods of Bradford, Barnesley, and Sheffield. Characterised by its irregular texture, its numerous quartz pebbles, and its frequently craggy surface, the millstone grit, with soft alternations both above and below it, occupies the wide and barren moors to the west of Sheffield, Penistone, Huddersfield, Bradford, Otley, Harrogate, Ripley, and Masham: in the numerous alternations of this stone, thin seams of coal frequently occur, and in certain situations are worked with advantage. Of the millstone-grit, an excellent and almost imperishable building stone, great quantities are annually sent down the rivers Don and Aire. The summits of Wharnside, Ingleborough, Pennygant, and other lofty mountains on the western boundary of the county, are crowned with coal measures, but their base consists wholly of limestone. The principal lead mines in the West Riding are at Grassington, about ten miles west of Pateley-Bridge, and are found in a limestone tract which occupies also a great part of Craven; but here the ores are far less abundant than in the vales of the Nid and the Wharfe. Hongill Fells, on the western boundary of the county, consist of the kind of slate called by geologists *grey wacké*. In the North Riding seams of an inferior kind of coal, which is heavy, sulphureous, and burns entirely away to white ashes, are wrought in different parts of both the Eastern and Western Moorlands, at Gilling Moor on the Howardian hills, and in the Vale of York, between Easingwould and Thirsk. Cleveland and the coast of this Riding abound in all their hills with inexhaustible beds of aluminous strata; and extensive works for the manufacture of alum have been established in the vicinity of Whitby, where the art is stated to have been first introduced from Italy, in the year 1595. Alum is also found, but not worked, in the Eastern Moorlands and in the vicinity of Bradford. In the Western Moorlands are many lead mines, some of which have been, and others still are, very valuable: these are situated in Swaledale, Arkendale, and the neighbouring valleys: their annual produce is estimated at 6000 tons, of which one half is yielded by the mines of Swaledale.

Veins of copper have been discovered at Richmond and at Middleton-Tyas, at which latter place that metal was worked about the middle of the last century: copper pyrites is also found in considerable quantities in all the alum mines, and copperas was formerly extracted from it. Great quantities of iron-stone are found in Bilsdale, Bransdale, and Rosedale, in the Eastern Moorlands, where iron seems to have been extensively manufactured in ancient times; but Ayton is the only place where forges have been erected at a modern period, and these are now abandoned. The iron-ore found in the northern parts of the Eastern Moorlands is sometimes in detached pieces, but more frequently in regular strata, of from six to fourteen inches thick, dipping towards the south: in the neighbourhood of Whitby, some of these beds are wrought, and their produce carried to the works in the north, where this ore is of great use in fluxing the more obdurate ores there obtained. Freestone, or gritstone, of an excellent quality for building, is found in many parts of this Riding, particularly on Gatherly Moor, near Richmond, at Renton, near Boroughbridge, in the neighbourhood of Whitby, in all parts of the Eastern Moorlands, of which it forms the chief basis, and in many parts of the Western. Nor is limestone less abundant: the Western Moorlands in a great measure consist of it; the Hamilton and Howardian hills, almost entirely; and a narrow ridge, producing lime of a peculiarly excellent quality for agricultural purposes, extends for at least 30 miles along the southern edge of the Eastern Moorlands: various isolated masses are also found in different situations. In Coverdale, one of the smaller valleys of the Western Moorlands, and at Pen-hill, between this and Wensleydale, a kind of flag-stone, used for covering roofs, is dug; and in Swaledale a kind of purple slate, resembling that of Westmorland, but thicker and coarser, the use of which extends little beyond the spot where it is produced. Marble of various kinds, some much resembling that worked in Derbyshire, and some, in closeness of texture and distinctness of colours, superior to it, is found in many parts of the calcareous hills of the Western Moorlands, but is only used for burning into lime, or mending the roads: some of the limestone on the northern margin of Ryedale also greatly resembles the marble of Derbyshire, and is susceptible of nearly an equal polish. In the vicinity of the small river Greta, and in other places in the north-western extremity of the county, large blocks of a light red granite are found scattered over the surface, and in some places a light grey kind of the same stone. Gypsum, or alabaster, is found in the Vale of York, in the North Riding, and in some parts of the levels of the East and West Ridings: near Thornton-bridge, on the Swale, where it is worked for the use of the plasterers of the neighbourhood, it lies in strata several feet thick, and in some places not more than four feet from the surface. The principal mineral productions of the East Riding are, the chalk of the Wolds, which is occasionally used in building, and frequently for burning into lime; and the coarse hard limestone of the vale of the Derwent, which is of little value either for building or burning: the springs in the chalk are remarkably powerful, and many of them breaking out through the gravel at the eastern foot of the Wolds, combine to form the river Hull. In the gravel beds resting on the chalk, to the east of where this substance appears next the surface, very perfect remains of large animals are found: vertebræ, eighteen feet in length, and from eight to ten inches in diameter, have here been exhumed, as are frequently teeth, measuring from eight to ten inches in circumference. The strata of the West Riding contain few fossil remains except at Bradford, where, in a stratum of sandstone, are found beautiful impressions of euphorbium, bamboo cane, and other tropical productions; at a little distance from Knaresbrough a bed of strontian earth exists, which is very rare in this kingdom. Various remarkable petrifactions of animals have been discovered in the alum rocks in the vicinity of Whitby, as also *cornua ammonis*, or snake stones: some of the strata in the same neighbourhood also contain petrified cockle, oyster, and scallop shells; jet and petrified wood; and *trochitæ*, or "thunderbolts," as they are vulgarly called, which are singular conical stones of from half an inch to an inch and a half in diameter at the base, and from two to five or six inches long. Great quantities of remarkable chrystals of *gypsum selenites* and *prismaticum* are discovered in a bed of clay at Knapton, in the East Riding.

The manufactures, the most valuable and extensive of which are confined to the West Riding, are of the highest degree of importance to the kingdom, as well as to the multitudes to whom they afford subsistence, and, in numerous instances, wealth. The two distinguishing manufactures are those of woollen goods and cutlery; the seat of the former is the district including the towns of Leeds, Halifax, Huddersfield, Bradford, and Wakefield; and that of the latter, Sheffield and its vicinity. The principal inducement for the establishment of these great works in the situations which they now occupy, was the plentiful supply of water and fuel for giving motion to machinery, and for the various other purposes of their several departments. The river Aire is the eastern boundary of the clothing district, which extends over the county thence to the mountain ridge separating this county from that of Lancaster. The great bulk of the woollen manufactures consisted formerly of the coarser kinds of cloth; but at present "Yorkshire cloth" no longer conveys the exclusive idea of inferiority, as the manufacturers now produce also great quantities of black and blue superfine cloths of distinguished merit. Until of late years, when numerous extensive factories have been erected (in which the whole process of making cloth, from the first breaking of the wool to the finishing of the piece ready for the consumer is completed), the first stages of the manufacture were carried on in villages and hamlets, where the wool underwent the respective operations of spinning, weaving, and fulling: this, however, is now only partially the case; the cloth from these scattered establishments is sent in its unfinished state to the cloth halls in the respective towns, where it is sold to the merchants, who have it dressed under their own direction. Besides broad and narrow cloths of various qualities, serges, and kerseymeres, the woollen manufactures of the West Riding include also great quantities of ladies' cloths, such as pelisse cloths and shawls; stuff goods of various kinds; camblets, shalloons, tammies, duroys, everlastings, calimancoes, moreens, shags, serges, baize, &c. Carpets much resembling those of Scotland are manufactured on a very extensive scale at Dewsbury, where is one of the largest factories for this article, and for woollen cloths and blankets, in the kingdom. Several very large factories have been established for spinning flax for canvass, linen, sacking thread, &c.: an extensive branch of the Manchester cotton trade is also carried on; and at Barnesley the manufacture of linen prevails. There is a considerable trade in the spinning of worsted yarn, and in the manufacture of wool cards and combs. The Leeds pottery enjoys a very considerable reputation both in the British dominions and in foreign countries: the wholesale tobacco trade is also carried on to a great extent in that town, where there are mills for preparing the raw material.

Sheffield has, from a very remote period, been famous for its manufacture of cutlery, which, however, was of very small extent until the early part of the seventeenth century, when it began gradually to increase; and by an act of parliament, passed in 1624, the cutlers of the liberty of Hallamshire, comprising the town of Sheffield and the adjacent country, were erected into a corporate body, which at present consists of between 3000 and 4000 members. Even until the middle of the last century the trade of Sheffield was still limited and precarious, but at that period various new branches of manufacture were introduced, more particularly that of plated goods. Its present manufactures, branches of which are also carried on in the numerous villages and hamlets in the surrounding country, to the distance of about seven miles from the town, include all kinds of cutlery and plated goods, edge-tools, combs, cases, buttons, fenders, files, anvils, joiners'-tools, lancets, ink-stands, nails, snuffers, saws, scythes, hay and straw knives, sickles, shears, awls, bellows, and an endless variety of other articles of hardware. There are also several foundries for iron, brass, and Britannia metal, and extensive works for the refining of steel: the iron-works at Rotherham are particularly celebrated, and produce all kinds of articles in cast-iron, and much wrought iron, in bars, sheets, and rods, together with tinned plates and steel. At Sheffield is also a minor manufacture of hair seating, besides a more considerable one of carpets. In the dales of the Eastern Moorlands and in Cleveland some coarse linens are manufactured by the small farmers; and at Crathorne in Cleveland, and various places near the Hamilton hills, are considerable bleaching establishments. The dales of the Western Moorlands have long been famous for their manufacture of knit worsted and yarn stockings, but this has been, in a great measure, superseded by the spinning of worsted for the manufactures of the West Riding. Cotton-mills have been erected in Wensleydale, at Easingwould, and at Masham, at which latter place is also a worsted-mill, and in its vicinity shalloons and shags are manufactured to a small extent. York and the East Riding have various isolated manufactures, the whole of which are mentioned under the heads of the places where they are respectively carried on. In the vicinities of York and Hull a kind of coarse earthenware is made, as well as bricks and tiles; and on Walling Fen, near Howden, great quantities of white bricks are made from a blue clay found there, which are exported in various directions, being in great demand for superior buildings, on account of their beauty of colour, accuracy of form, and durability. Almost every town in the North Riding, and many in the other parts of the county, have tanners and tawers, who manufacture the hides and skins produced in their respective neighbourhoods. To this enumeration of manufactures may also be added the building and rigging of ships, which is carried on to a considerable extent at Hull and Whitby, and in a minor degree at Scarborough and Thorne: at the three first-mentioned places are considerable manufactures of sail-cloth and cordage. The chief port of the county is Hull, which may be considered the fourth in England; besides this it possesses, of a smaller class, those of York, Selby, Goole, Thorne, Bridlington, Scarborough, and Whitby. The commerce is of a very extensive and diversified character: the foreign and coasting trade is wholly centred in the above-mentioned ports, more particularly in that of Hull, through which is poured an immense quantity of manufactured goods, coal, stone, &c., from the West Riding, and of cotton-twist and manufactured cottons from Lancashire, the latter of which articles are chiefly forwarded to Hamburgh. Hull and Whitby share largely in the Greenland fishery; and their imports of timber, deals, hemp, flax, &c., from the Baltic, are very considerable. The internal commerce of the West Riding is very extensive, and is greatly facilitated by an excellent system of artificial navigation. A considerable quantity of corn is exported from Hull, Bridlington, and Scarborough to London, and the collieries of the north; and from the various principal markets of the East and North Ridings great quantities of grain are sent by water carriage into the western division of the county, from which the first-mentioned division receives in return coal, lime, flag stones, bricks and tiles, and sundry other articles. A large quantity of hams and bacon is annually sent from the eastern parts of Yorkshire to the metropolis and other populous districts of the kingdom.

The principal rivers are, the Northern Ouse (so called to distinguish it from the Ouse of Buckinghamshire), the Swale, the Ure, the Wharfe, the Derwent, the Aire, the Calder, the Don, the Hull, the Tees, and the Esk, all of which, except the two latter, pour their waters through the great estuary of the Humber. The Humber is navigable up to Hull for ships of the largest burden; the Ouse up to the newly-formed port of Goole, for vessels drawing not more than sixteen feet of water, and to York, for those of 140 tons' burden: above that city it is navigable for barges of 30 tons' burden, as also is the Ure past Boroughbridge to Ripon, and the Swale, only for a very few miles: the spring tides would turn the current of the Ouse to a little above York, were it not that they are obstructed by locks about four miles below that city. The Wharfe is navigable as far as Tadcaster; the Derwent is navigable for vessels of 25 tons' burden to New Malton, above which town the navigation has been continued to Yeddingham bridge, a further distance of about nine miles. The Aire becomes navigable at Leeds, and a few miles lower, near Castleford, is joined by the Calder, which is navigable to Salter-Hebble, near Halifax. The Don having been joined by the powerful stream of the Rother, unites with the Ouse at Goole; the lower part of its channel, from the vicinity of Snaith, being artificial, and usually called the Dutch river: in 1751, this river was made navigable to Tinsley, three miles below Sheffield, and, under the provisions of an act of parliament passed in 1815, this navigation has been continued by a cut, called the Tinsley canal, to Sheffield. The Hull falls into the Humber at Kingston-upon-Hull, where its mouth forms a secure but narrow haven: this river is navigable to Frodingham bridge, several miles above Beverley (with which town it communicates by means of a short cut), whence the navigation is continued by a canal to Great Driffield. Another canal extends eastward from the river Hull to Leven, a distance of about three miles. The Tees is navigable for vessels of 60 tons' burden to a short distance above Yarm, where the spring tides rise about seven feet: below Stockton it spreads into the fine estuary of Redcar, three miles broad.

The canals are nearly all within the limits of the West Riding. Under this head, however, may be classed the small navigable river Foss, the channel of which is believed to have been originally formed by the Romans, to effect the drainage of an extensive level tract lying between the Ouse and the Howardine hills, near the western extremity of which it rises, and thence takes first a south-easterly, and then a southerly, course to the Ouse, at York: at the end of the last century the navigation was made perfect from York to Sheriff-Hutton, a distance of about fourteen miles, under the provisions of an act of parliament passed in the year 1793. Market-Weighton and Hedon, which are both situated in the East Riding, and are considerable markets for corn, have each the advantage of a navigable canal to the Humber. The canals of the West Riding, in alphabetical order, are as follows; the Barnesley canal, which commences in the navigable channel of the river Calder, a little below Wakefield, and, taking a southerly direction, unites with the Dearn and Dove canal near Barnesley: its length is only fifteen miles, but it is of great importance, as forming part of the line of navigation from Sheffield to Barnesley, Wakefield, Leeds, Huddersfield, Manchester, and Liverpool. The Bradford canal, which is only three miles in length, commences in the Leeds and Liverpool canal at Windhill, in the parish of Idle, and terminates at Bradford, where extensive railways connect it with the collieries and iron-works of Low Moor and Bowling. The Dearn and Dove canal commences in a side cut from the river Don, between Swinton and Mexborough, and, passing north-westward, terminates in the Barnesley canal, at Eyming's Wood, after a course of nine miles: together with the Barnesley canal it forms a line connecting the navigable channel of the Don with that of the Calder. From the newly-formed commercial docks at Goole a canal passes westward to the river Aire, at Ferrybridge, and thus completes the water communication between that rising port and the manufacturing districts of the West Riding, together with the counties of Lancaster, Chester, and Stafford. The Huddersfield canal, nineteen miles and a half long, commences in Sir John Ramsden's canal, on the southern side of that town, and, proceeding westward, passes near Saddleworth, through the range of mountains on the borders of Yorkshire and Lancashire, by one of the largest tunnels in the kingdom, being nearly three miles and a half in length, and terminates in the latter county in the Manchester, Ashton, and Oldham canal. The Leeds and Liverpool canal enters this county from Colne in Lancashire, whence it proceeds by Skipton, Keighley, and Bingley, and across the river Aire, near Shipley, to Leeds, where it terminates in the Aire navigation: this extensive and important canal connects, by a direct water communication, the ports of Liverpool and Hull with the large manufacturing town of Leeds. The Ramsden canal, four miles in length, commences in the Calder and Hebble navigation at Cooper's bridge, and terminates in the Huddersfield canal at the King's Mills, near Huddersfield; thus completing, in conjunction with the Huddersfield canal, the important line of water communication between Manchester and the great manufacturing towns of Yorkshire. The Rochdale canal, entering from Rochdale in Lancashire, terminates in the Calder and Hebble Navigation at Sowerby-bridge, two miles from Halifax. The Stainforth and Keadby canal, partly in this county, and partly in the Isle of Axholme, in Lincolnshire, branches from the navigation of the Don at Fishlake, near Stainforth, and, passing by Thorne, terminates in the Trent, at Keadby, after a course of fifteen miles. The Leeds and Selby railway was one of the first commenced; the line is continued from Selby to Hull by the Hull and Selby railway. The line of the Manchester and Leeds railway enters the county at Longfield, and passing near Halifax, Dewsbury, and Wakefield, is joined by the York and North Midland railway in the township of Altofts, and then proceeds to Leeds. A short railway has been formed between Sheffield and Rotherham, which is also connected with the York and North Midland railway close to the latter town. The Manchester and Sheffield line of railway enters the county between two branches of the river Don, to the west of Penistone, and passing close to the north of that town, proceeds to Sheffield. The Whitby and Pickering railway connects these towns, thus establishing a communication for the transportation of the produce of the latter to the former, as a port for its export. The last great undertaking in this branch of communication is the formation of the Great North of England railway, which proceeds nearly in a straight line north-west-by-west from York to the vicinity of Darlington, in Durham, near which it crosses the river Tees, and in its course passes close to the towns of Thirsk and North-Allerton.

Besides the great station of *Eboracum*, at York, the chief seat of the Roman power in Britain, this county contained also, in the West Riding, the stations of *Isurium*, at Aldborough; *Legiolum*, a little below the junction of the rivers Aire and Calder; *Danum*, at Doncaster; *Olicana*, at Ilkley; *Cambodunum*, at Slack, near Halifax; and *Calcaria*, at Tadcaster: in the North Riding, those of *Cataractonium*, at Catterick; and *Derventio*, at Stamford-Bridge, or at Alby, a mile further northward: and in the East Riding, *Delgovitia*, at Londesborough; and *Prætorium*, at Patrington. The most durable of the works of this people were the roads which they constructed, in order to facilitate the communication between their military stations, several of which traversed Yorkshire in different directions, and remains of some of them may yet be traced in various parts of it: the common centre from which they diverged was *Eboracum*, or York. The line of the great road, since called the Watling-street, which ran the whole length of England, from the coast of Kent to the wall of Severus, enters from Nottinghamshire in the vicinity of Bawtry, and passes through Doncaster, Barnsdale, Pontefract Park, Castleford, Tadcaster, York, Aldborough, and Catterick, into the county of Durham at Pierse-Bridge. Another military road entered from Manchester, and passed through the vicinity of Halifax, and by Wakefield, to the Watling-street. Another similar road, from Chesterfield, on the north-western confines of Derbyshire, passed by Sheffield, Barnesley, Hemsworth, and Ackworth, to the Watling-street, at or near Pontefract: a vicinal way also appears to have passed through Pontefract, in a southerly direction, to the villages of Darrington, Wentbridge, Smeaton, Campsall, and Hatfield. From York a Roman road ran to Malton, and appears to have there divided into two branches, one, now commonly called Wade's Causeway, leading to Dunsley bay, in the neighbourhood of Whitby; the other to Scarborough and Filey: another road passed from York, by Stamford-Bridge, Fridaythorpe, and Sledmere, and across the Wolds, to Bridlington bay, called by Ptolemy *Gabrantovicoum Sinus Portuosus*, or *Salutaris*. Further to the south was a Roman road from York, by Stamford-Bridge and Londesborough, to Patrington; from Londesborough, a branch of this, formerly called Humber-Street, passed in a straight line southward to the village of Brough on the Humber. The most remarkable antiquities exist in the remains of ancient castles and religious edifices; but there are also several specimens of military and other works of a more remote period. The three gigantic obelisks of single stones, vulgarly called the Devil's Arrows, situated near Boroughbridge, are by some thought to be Druidical, and by others of Roman origin. Traces of Roman encampments are found in several places, and the remains of their roads are more particularly conspicuous on the Eastern Moorlands, where the ancient road from Malton to Dunsley bay, now called Wade's Causeway, is in excellent preservation, being twelve feet broad, in some places raised more than three feet above the surface, and paved with flint pebbles; and on the Wolds, where the Roman road from York to Bridlington bay may be traced for many miles. The only remains of Roman structures now to be seen in York, the site of the ancient *Eboracum*, are the polygonal tower and the south wall of the mint yard. A vast variety of Roman antiquities has at different times been found in York and its vicinity, in digging the cellars, drains, and foundations of houses, such as altars, sepulchral and other urns, sarcophagi, coins, signets (both cameos and intaglios), fibulæ, &c. Roman urns, coins, &c., have been discovered in several other situations near the stations and roads of that people. Many ancient tumuli are discernible in various parts of the county, particularly on the Wolds. Besides the Roman encampments, others of the Saxons and the Danes may be traced in several places in the North and West Ridings. The remarkable assemblage of rocks, called Bramham Crags, about nine miles north-west of Ripon, are supposed, from the peculiar marks of rude sculpture which some of them exhibit, to have been a celebrated Druidical temple.

The number of religious houses was about 106, including seven Alien priories: the ruins of several of them are amongst the most beautiful and picturesque in the kingdom. The principal ruins of abbeys are those of St. Mary's at York; of Fountains, Kirkstall, Roche, and Selby, in the West Riding; and of Byland, Reivaulx, Easby, Eggleston, and Whitby, in the North Riding; and of priories, those of Bolton and Knaresbrough, in the

West Riding; of Guisborough, Mountgrace, and Wikeham, in the North Riding; and of Bridlington, Kirkham, and Watton, in the East Riding. The most distinguishing remains of ancient fortresses, besides Clifford's Tower at York, are those of the castles of Cawood, Conisbrough, Harewood, Knaresbrough, Pontefract, Great Sandall, Skipton, and Tickhill, in the West Riding; of Helmsley, Malton, Mulgrave, Pickering, Richmond, Scarborough, Sheriff-Hutton, and Skelton, in the North Riding; and of Wressle, in the East Riding. The most remarkable ancient mansions are, Temple-Newsome, near Leeds; and Gilling Castle, near Helmsley, formerly the seat of the ancient family of Fairfax; besides which, several in different parts of the county are now occupied as farm-houses. Yorkshire contains a great number of elegant seats of more modern erection, belonging to the nobility and gentry who possess estates within its limits: some of those more particularly worthy of mention in the West Riding are, Wentworth House, Wentworth Castle or Stambrough Hall, Methley Park, Thundercliffe Grange, Sandbeck Park, Newby Hall, Harewood House, Scarthingwell Hall, and Allerton-Mauleverer; in the North Riding, Hornby Castle, Stanwick, Castle-Howard, and Mulgrave Castle; in the East Riding, Londesborough; and in the ainsty, Bishopthorpe; near York, the archiepiscopal palace.

The chalybeate and sulphureous springs of Harrogate, discovered in 1571, are of great celebrity, and have rendered that once obscure hamlet one of the principal watering-places in the North of England. Askerne, about eight miles north of Doncaster, has of late years become much noted for its medicinal waters, which much resemble those of Harrogate, both in smell and taste, but differ from them in their operation. The chalybeate and saline springs of Scarborough, discovered early in the seventeenth century, have long been celebrated and greatly resorted to. In May, 1822, a mineral spring was discovered a mile to the south-east of Guisborough, which is greatly resorted to by persons labouring under different complaints: the waters are diuretic. There are, besides, mineral springs of various qualities at Aldfield, Boston, Gilthwaite, Horley Green, Ilkley, and Knaresborough, in the West Riding; a chalybeate spring at Bridlington Quay, on the coast of the East Riding; and a noted mineral spring at Thorp Arch, in the ainsty. At Knaresbrough is the celebrated dropping and petrifying well; and at the bottom of Giggleswick Scar, near the village of Giggleswick, is a spring which ebbs and flows at irregular periods. On the Wolds, and near Cottingham on their eastern side, are periodical springs, which sometimes emit very powerful streams of water for a few months successively, and then become dry for years. Some of the most remarkable waterfalls are, Thornton Force, formed by a small stream which is driven down a precipice of about 30 yards in height, and is situated near the village of Ingleton, in the West Riding, and in the vicinity of Thornton Scar, a tremendous cliff of about 300 feet in height; the cataract of Malham Cove, which is 300 feet high; Aysgarth Force; Hardrow Fall; High Force, or Fall, on the Tees; Mallin Spout; Egton; and Mossdale Fall, all in the North Riding. Among the natural curiosities of this county must also be enumerated its caves, the principal of which, situated among the Craven mountains, are, Yordas Cave, in a mountain called Greg-roof, and Weathercote Cave, both of them in the vicinity of Ingleton, and in the latter of which is a stupendous cataract of twenty yards fall; Hurtlepot and Ginglepot, near the head of the subterranean river Wease, or Greta; and Donk Cave, near the foot of Ingleborough. At the foot of the mountain called Pennigant, in the same neighbourhood, are two frightful orifices, called Hulpit and Huntpit Holes, through each of which runs a subterraneous brook, passing under ground for about a mile, and then emerging, one at Dowgill Scar, and the other at Bransil-head.

YOULGRAVE (*All Saints*), a parish, partly in the hundred of Wirksworth, and partly in that of High Peak, Northern Division of the county of Derby; containing 3681 inhabitants, of which number, 951 are in the township of Youlgrave, 3 miles (S. by W.) from Bakewell. The living is a discharged vicarage, valued in the king's books at £9. 4. 7.; present net income, £214; patron and impropriator, Duke of Devonshire. The church is partly Norman, and partly of later date. There is a place of worship for Wesleyan Methodists. A school was erected by subscription, about 1765, and in 1824 a residence for the master was added by the Duke of Rutland, who pays for the instruction of 12 children.

YOULTHORPE, a joint township with Gowthorpe, in the parish of Bishop-Wilton, union of Pocklington, partly within the liberty of St. Peter's, and partly in the Wilton-Beacon Division of the wapentake of Harthill, East Riding of the county of York, $5\frac{1}{4}$ miles (N. W. by N.) from Pocklington; containing 106 inhabitants.

YOULTON, a township, in the parish of Alne, wapentake of Bulmer, North Riding of the county of York, $6\frac{1}{2}$ miles (S. S. W.) from Easingwould; containing 59 inhabitants.

YOXFORD (*St. Peter*), a parish, in the union and hundred of Blything, Eastern Division of the county of Suffolk, $23\frac{1}{2}$ miles (N. E.) from Ipswich; containing 1149 inhabitants. The village is situated in a remarkably pleasant and genteel neighbourhood, on the high road from Ipswich to Yarmouth, and consists principally of one well-built street of modern houses, with two commodious inns. Cockfield Hall, the seat of Sir Charles Blois, Bart., is a handsome mansion of the time of James I. The living is a vicarage, valued in the king's books at £5. 14. 2.; present net income, £161; patron and incumbent, Rev. Dr. Roberts; impropriators, Earl of Stradbroke and Sir Charles Blois, Bart. The church has been lately enlarged by subscription, aided by a grant of £140 from the Incorporated Society, for which 280 free sittings have been provided; it contains some ancient brasses and handsome monuments. A National school has been recently built by subscription.

YOXHALL (*St. Peter*), a parish, in the union of Lichfield, Northern Division of the hundred of Offlow and of the county of Stafford, $7\frac{1}{2}$ miles (N. N. E.) from Lichfield; containing 1582 inhabitants. The living is a rectory, valued in the king's books at £17. 6. 8.; present net income, £508; patron, Lord Leigh. The church exhibits various styles of architecture, from the Norman to the later English. Here is a Roman Catholic chapel. A free school, founded in 1695 by Thomas Taylor, is endowed with various bequests, producing about £20 per annum, and is now in connection with the National Society. In levelling a piece of ground, about 40 vessels of a soft brown earthenware, containing ashes and human bones, were taken up some years ago.

Z

ZEAL-MONACHORUM (*St. Peter*), a parish, in the union of Crediton, hundred of North Tawton, South Molton and Northern Divisions of the county of Devon, $1\frac{1}{4}$ mile (N.) from Bow; containing 747 inhabitants. The living is a rectory, valued in the king's books at £17. 8. 9.; present net income, £388; patron and incumbent, Rev. John Comyns. Seven poor children are educated for an annuity of £5, the bequest of Weekes Hole, Esq., in 1768; the school is further supported by subscription.

ZEAL, SOUTH, a chapelry, in the parish of South Tawton, hundred of Wonford, Crockernwell and Southern Divisions of the county of Devon, $4\frac{1}{2}$ miles (E. S. E.) from Oakhampton: the population is returned with the parish. This is a decayed borough and market-town: the market has been long disused, but there is a fair for cattle on the Tuesday following the martyrdom of Thomas à Becket. The chapel, dedicated to St. Mary, is now used as a school-house.

ZEALS, a tything, in the parish and hundred of Mere, Hindon and Southern Divisions of the county of Wilts, 2 miles (W. by S.) from Mere; containing 510 inhabitants.

ZENNOR (*St. Sennar*), a parish, in the union of Penzance, Western Division of the hundred of Penwith and of the county of Cornwall, 5 miles (W. S. W.) from St. Ives; containing 811 inhabitants. This parish is situated near the western extremity of the Bristol Channel, by which it is bounded on the north; the line of the coast is in some parts alternated with small bays and with projecting headlands, one of which is named Gurnard's Head. There are some tin mines in the parish, but the substratum of the greater part of the soil is a species of moorstone. The living is a discharged vicarage, valued in the king's books at £5. 5. $0\frac{1}{2}$.; present net income, £179; patron, Bishop of Exeter; impropriator, George John, Esq. There was formerly a chapel at Kerrow, of which there are still some remains. Here is a place of worship for Wesleyan Methodists. The remains of a Druidical circle, and some barrows, are the principal relics of antiquity; at Foage is a stone barrow with a kistvaen, about a furlong from which there is an ancient cromlech.

APPENDIX, N°. I.

ABSTRACT of certain Sections of the Act of the 5th and 6th of William IV., cap. 76, intituled "An Act to provide for the Regulation of Municipal Corporations in *England* and *Wales;*" with the Amendments made by the Act of the 1st of Victoria, cap. 78.

1. After reciting that divers Bodies Corporate have been constituted within the Cities, Towns, and Boroughs of England and Wales, for the good government of the same; and that it is expedient that the Charters by which the said Bodies Corporate are constituted be altered; accordingly repeals all Acts, Charters, and Customs inconsistent with this Act, relating to the Boroughs named in the Schedules (A) and (B).

3. No Freedom to be hereafter acquired by Gift or Purchase.

6. Corporations to be styled, Mayor, Aldermen, and Burgesses.

7. Boundaries of certain Boroughs to be those settled by the 2nd and 3rd of William IV., cap. 64; Boundaries of other Boroughs to remain, until altered by Parliament; but no Place detached from the main part of such Borough or Town Corporate (except the Liberties) shall be included therein: these Detached Parts, by the 8th Section, are declared to be Part of the adjoining County.

9. Every Male Occupant, of full Age, of any House, Warehouse, Counting-house, or Shop, within any Borough, for the Year preceding the last day of August in any Year, and the whole of each of the Two preceding Years, and who shall have been an inhabitant Householder within the said Borough, or within Seven Miles, shall, if duly Enrolled, be a Burgess of such Borough, and Member of the Body Corporate; but he must have been Rated for the Relief of the Poor of the Parish wherein such Premises are situated in respect thereof during his Occupation, and have paid on or before the last day of August as aforesaid all such Rates, including therein all Borough Rates, if any, under this Act, except such as shall become payable within Six Calendar Months next before the said last day of August; but such Rating and Occupation need not be of the same Premises or in the same Parish.

11. Occupiers may Claim to be Rated, and on paying the last Rate due, the Overseer must Rate them; if he shall refuse, such Occupiers shall nevertheless be deemed Rated.

12. In case of Titles by Descent, Marriage, Marriage Settlement, Devise, or Promotion to any Benefice or Office, the Occupation and Rating of the Parties from and to whom the Title is derived may be reckoned conjointly. [The Act of Victoria, in this case, declares that it shall not be necessary, in support of the Title of such Person to be Enrolled, to prove that he was an inhabitant Householder within the said Borough, or within Seven Miles, or that he was an Occupant or Rated within the same, before the Title to the Premises devolved upon him; and that the Rating in the name of the former Occupier shall be sufficient.]

14. Declares that it shall not be necessary for any Person to be free of any City, Town, or Borough, or of any Guild, &c., to keep a Shop, or exercise any Trade or Handicraft therein.

25. Mayor, Aldermen, and Councillors to be chosen in every Borough, who together shall constitute the Town Council; the number of Councillors to be that mentioned in conjunction with the name of the Borough in Schedules (A) and (B), and the number of Aldermen to be One-third of the Councillors; the day of Election, November 9th, every Third Year; the Council to elect from the Councillors, or from Persons qualified to be Councillors, the Aldermen, or a sufficient number to fill the Places of those who then go out of Office, namely, One-half, those to go out who have been Aldermen for the longest time without Re-election; but any Aldermen may be forthwith re-elected, but may not Vote in the Election of a new Alderman.

26. Mayor and Aldermen to be Members of the Council during their Offices.

27. Extraordinary Vacancies in the Office of Aldermen to be filled up by the Council; and the Person so Elected is to remain in Office as long as the Alderman in whose room he was elected would have done.

28. No Minister shall be elected a Councillor, nor any Person not entitled to be on the Burgess List, nor unless seised or possessed of Real or Personal Estate, or both; viz., in all Boroughs divided into Four or more Wards, £1000, or Rated to the Poor upon the annual value of not less than £30; and in all Boroughs divided into less than Four Wards, or not divided into Wards, of £500, or Rated at not less than £15 per annum, or while he shall hold any Office or place of Profit, other than Mayor, in the gift or disposal of the Council, shall have, directly or indirectly, any share or interest in any Contract or Employment with, by, or on behalf of, such Council; but not if a Shareholder of any Company which shall Contract with the Council for Lighting or supplying with Water or insuring against Fire any part of the Borough.

29. Councillors, Auditors, and Assessors are to be Elected by the Burgesses on the Roll for the time being. [By the Act of the 1st of Victoria, cap. 78, sec. 15, Auditors and Assessors are disqualified to be of the Council.]

30. Councillors to be chosen on the 1st of November every Year.

31. One-third part of the Council to go out of Office annually, viz., those who have been longest in Office without Re-election; but any one may be forthwith Re-elected.

32. Such Elections are to be held before the Mayor and Assessors.

36. An Alderman to be chosen to preside at Election in case of the death or inability of the Mayor.

37. Auditors and Assessors to be elected each 1st of March; but no Burgess may Vote for more than one; and no Burgess shall be eligible who shall be of the Council, or Town-clerk, or Treasurer.

39. Enacts that every Borough in the Schedule (A) shall be divided into the number of Wards mentioned in conjunction therewith.

48. Penalties on Mayor, Alderman, or Assessor neglecting to comply with the Provisions of this Act, £100; and on Overseers, £50; to be recovered by any Person within Three Months; one half for the Plaintiff and the other for the Treasurer of the Borough Fund.

49. Council to elect the Mayor every Year from the Aldermen or Councillors.

50. Mayor, Aldermen, and Councillors, Auditors, and Assessors, not to act until they have made a Declaration of acceptance of Office; and the Aldermen are to make a Declaration of Qualification once in Three Years, if required so to do by any Two Members of the Council; but the Declaration also required by the 9th of George IV., c. 17, not to be dispensed with.

51. Every Burgess elected to the office of Alderman, Councillor, Auditor, or Assessor, and every Councillor elected to the Office of Mayor, shall accept the office in Five Days, or pay such a Fine to the Borough Fund as the Council shall declare, not exceeding £50 for Burgesses, and £100 for Mayor, which Fine may be levied by Distress; but exempts Persons Disabled by Infirmities, and 65 Years Old, and who have Served or Fined within Five Years, and Officers in Service.

52. If any Mayor, Alderman, or Councillor, shall be declared Bankrupt or become Insolvent, or absent himself from the Borough for more than Six Months at the same time, unless in case of illness, he shall lose his Office, and be liable to the Fine in the last Section; but becomes re-eligible on obtaining his Certificate or paying his Debts in full.

53. Penalty on Persons not Qualified, &c., acting as Mayor, Alderman, or Councillor, £50, to be applied as directed in Section 48.

54. Disqualifies Persons convicted of offering or receiving Bribes from voting at any election within the Borough, Municipal or Parliamentary, and imposes a Fine of £50.

57. The Mayor to be a Justice of the Peace for the Borough, and Returning Officer at Parliamentary Elections.

58. Empowers the Council to appoint a Town-Clerk, Treasurer, and other Officers.

61, 62, and 63. The Councils of certain places therein mentioned to appoint their Sheriff; and in Boroughs having a separate Court of Quarter Session, a Coroner, not being an Alderman or Councillor; in other Boroughs the County Coroners are to act.

69. All acts of the Council to be decided by a Majority of the Councillors present; a third Part of the whole Number to be a Quorum; quarterly Meetings are to be held for general business.

71. Trustees of Charity Estates to be appointed by the Lord Chancellor, in order that the Administration thereof be kept distinct from that of the Public Stock and Borough Fund.

72 and 73. Council to act as Trustees where Corporators were *ex-officio* sole Trustees, except of Charities; and to appoint a Number to be joint Trustees.

75. Transfers to the Councils the Powers of Trustees, for paving, lighting, cleansing, watching, regulating, supplying with water, and improving Boroughs.

76. The Council to appoint a Watch Committee, which may appoint Constables, and the Constables may act for the County as well as Borough.

87. Empowers the Council to extend the Provisions of any Local Act to such parts of Boroughs as were not previously within its operation.

88. Council may assume the Powers of Inspectors under the Act of the 3rd and 4th of William IV., c. 90, for lighting any Part of the Borough not within a Local Act.

90. Council may make Bye-Laws, but only when at least Two-thirds of them are present; and may impose Fines not exceeding £5.

92. The Surplus, if any, of the Borough Fund to be applied, under the direction of the Council, for the public benefit of the Inhabitants and improvement of the Borough; and if insufficient for the above purposes, the Council shall order a Rate to make up the deficiency, in the nature of a County Rate, with power to appeal.

98. The Royal Commission may be issued for certain Persons to act as Justices in any such Boroughs, on Petition of the Council; but they must reside within Seven Miles.

99. Councils may make Bye-laws on which the Crown may appoint Stipendiary Magistrates.

101. Makes it unnecessary for Justices to be qualified by Estate, and precludes them from sitting at Quarter Sessions.

103. On Petition to the Privy Council, Her Majesty may grt ana separate Court of Quarter Sessions, and appoint a Recorder, who is to be a Justice of the Peace for the Borough, and not a Member of Parliament, Alderman, Councillor, or Police Magistrate.

105. Sessions of the Peace to be held for the Borough, the Recorder to be the sole Judge, to take cognizance of Crimes as any Court of Quarter Sessions.

107. Abolishes capital Jurisdictions, and all other criminal Jurisdictions in the Boroughs, other than are specified in this Act.

108. Abolishes chartered Admiralty Jurisdictions.

111 and 112. County Magistrates to have jurisdiction in all Boroughs which have not a separate Court of Quarter Sessions of the Peace under this Act; but Boroughs which have such Court are not to be assessed to County Rates, except as thereafter.

117. Such Boroughs to pay such proportion of the other County Expenditure as they would have done if this Act had not passed.

118. Borough Courts of Record to be holden as heretofore; and where a Barrister of Five Years' standing shall be Judge, Actions may be tried to the amount of £20; but not where the Title to Land, &c., is in question.

119. Registrar and other Officers of the Court to be appointed by the Council.

121. Every Burgess of any Borough wherein there shall be separate Sessions of the Peace, or a Court of Record for the trial of Civil Actions (unless exempt or disqualified otherwise than in respect of Property from serving on juries by the 6th of George IV., cap. 50,) shall be Qualified and Liable to Serve on Grand Juries in such Borough, and also upon Juries for the

Trial of all Issues joined in any Court of Quarter Sessions, and in any Court of Record for the trial of Civil Actions within the Borough; but no one is to be summoned more than once a Year.

122. Members of the Council, &c., exempt from Serving on Juries; and Burgesses of Boroughs which have Quarter Sessions, exempt from Juries of County Quarter Sessions.

123. Abolishes and Repeals all Chartered Exemptions from Serving on Juries in the Queen's Courts.

134. and 135. Jurisdiction of the Cinque Ports preserved, as therein mentioned.

139. Advowsons belonging to Bodies Corporate to be Sold, as the Ecclesiastical Commissioners may direct; the vacancy, before the completion of such Sale, to be supplied by the Bishop of the Diocese.

141. Her Majesty to grant Charters to such Towns or Boroughs as are not Bodies Corporate, on Petition of the Inhabitants, and by Advice of Her Privy Council, and to extend thereto the Powers and Provisions of this Act.

SCHEDULES TO WHICH THIS ACT REFERS.

SCHEDULE (A).

ENGLAND AND WALES.

BOROUGHS WHICH ARE TO HAVE A COMMISSION OF THE PEACE.

SECTION 1.—PARLIAMENTARY BOUNDARIES TO BE TAKEN UNTIL ALTERED BY PARLIAMENT.

Borough.	Wards.	Aldermen.	Councillors.
Aberystwith	0	4	12
Abingdon	0	4	12
Barnstaple	2	6	18
Bath	7	14	42
Bedford	2	6	18
Berwick-upon-Tweed	3	6	18
Bridg-water	2	6	18
Bridport	2	6	18
Bristol	10	16	48
Bury St. Edmunds	3	6	18
Cambridge	5	10	30
Canterbury	3	6	18
Cardiff	2	6	18
Carlisle	5	10	30
Carmarthen	3	6	18
Carnarvon	2	6	18
Chester	5	10	30
Chichester	2	6	18
Colchester	3	6	18
Dartmouth	0	4	12
Denbigh	0	4	12
Derby	6	12	36
Devizes	2	6	18
Dorchester	0	4	12
Dover	3	6	18
Durham	3	6	18
Evesham	0	4	12
Gateshead	3	6	18
Gloucester	3	6	18

Borough.	Wards.	Aldermen.	Councillors.
Guilford	0	4	12
Harwich	0	4	12
Haverfordwest	0	4	12
Hereford	3	6	18
Hertford	0	4	12
Ipswich	5	10	30
Kendal	3	6	18
Kidderminster	3	6	18
Kingston-upon-Hull	7	14	42
King's Lynn	3	6	18
Leeds	12	16	48
Leicester	7	14	42
Leominster	0	4	12
Lichfield	2	6	18
Liverpool	16	16	48
Macclesfield	6	12	36
Monmouth	0	4	12
Neath	0	4	12
Newark	3	6	18
Newcastle-under-Lyme	2	6	18
Newcastle-upon-Tyne	7	14	42
Newport, Monmouth	2	6	18
Newport, (Isle of Wight)	2	6	18
Northampton	3	6	18
Norwich	8	16	48
Nottingham	7	14	42
Oxford	5	10	30
Pembroke	2	6	18

Borough.	Wards.	Aldermen.	Councillors.
Poole	2	6	18
Portsmouth	7	14	42
Preston	6	12	36
Reading	3	6	18
Ripon	0	4	12
Rochester	3	6	18
St. Alban's	0	4	12
Sarum, New	3	6	18
Scarborough	2	6	18
Shrewsbury	5	10	30
Southampton	5	10	30
Stafford	2	6	18
Stamford	2	6	18
Stockport	7	14	42
Sudbury	0	4	12
Sunderland	7	14	42
Swansea	3	6	18
Tiverton	3	6	18
Truro	2	6	18
Warwick	2	6	18
Wells	0	4	12
Weymouth and Melcombe Regis	2	6	18
Wigan	5	10	30
Winchester	3	6	18
Windsor	2	6	18
Worcester	6	12	36
Yarmouth, Great	6	12	36

SECTION 2.—MUNICIPAL BOUNDARIES TO BE TAKEN UNTIL ALTERED BY PARLIAMENT.

Borough.	Wards.	Aldermen.	Councillors.
Andover	0	4	12
Banbury	0	4	12
Beverley	2	6	18
Bewdley	0	4	12
Bideford	0	4	12
Boston	3	6	18
Brecon	0	4	12
Bridgenorth	0	4	12
Clitheroe	0	4	12
Chesterfield	0	4	12
Congleton	3	6	18
Coventry	6	12	36
Deal	2	6	18
Doncaster	3	6	18
Exeter	6	12	36
Falmouth	0	4	12
Grantham	0	4	12
Gravesend	2	6	18
Grimsby	0	4	12
Hastings	3	6	18
Kingston-upon-Thames	3	6	18
Lancaster	3	6	18
Lincoln	3	6	18
Liskeard	0	4	12
Louth	2	6	18
Ludlow	0	4	12
Maidstone	3	6	18
Maldon	0	4	12
Newbury	0	4	12
Oswestry	2	6	18
Penzance	2	6	18
Plymouth	6	12	36
Pontefract	0	4	12
Richmond	0	4	12
Romsey	0	4	12
St. Ives	0	4	12
Saffron Walden	0	4	12
Stockton	2	6	18
Tewkesbury	0	4	12
Walsall	3	6	18
Welchpoole	0	4	12
Wenlock	3	6	18
Wisbech	2	6	18
York	6	12	36

SCHEDULE (B).

ENGLAND AND WALES.

BOROUGHS WHICH ARE NOT TO HAVE A COMMISSION OF THE PEACE, UNLESS ON PETITION AND GRANT.

SECTION 1.—PARLIAMENTARY BOUNDARIES TO BE TAKEN UNTIL ALTERED BY PARLIAMENT.

Borough.	Wards.	Aldermen.	Councillors.
Arundel	0	4	12
Beaumaris	0	4	12
Cardigan	0	4	12
Llanidloes	0	4	12
Pwllheli	0	4	12
Ruthin	0	4	12
Tenby	0	4	12
Thetford	0	4	12
Totnes	0	4	12

SECTION 2.—MUNICIPAL BOUNDARIES TO BE TAKEN UNTIL ALTERED BY PARLIAMENT.

Borough.	Wards.	Aldermen.	Councillors.
Basingstoke	0	4	12
Beccles	0	4	12
Blandford-Forum	0	4	12
Bodmin	0	4	12
Buckingham	0	4	12
Calne	0	4	12
Chard	0	4	12
Chippenham	0	4	12
Chipping Norton	0	4	12
Daventry	0	4	12
Droitwich	0	4	12
Eye	0	4	12
Faversham	0	4	12
Folkestone	0	4	12
Flint	0	4	12
Glastonbury	0	4	12
Godalming	0	4	12
Godmanchester	0	4	12
Helstone	0	4	12
Huntingdon	0	4	12
Hythe	0	4	12
Launceston	0	4	12
Llandovery	0	4	12
Lyme-Regis	0	4	12
Lymington	0	4	12
Maidenhead	0	4	12
Marlborough	0	4	12
Morpeth	0	4	12
Penryn	0	4	12
Retford, East	0	4	12
Rye	0	4	12
Sandwich	0	4	12
Shaftesbury	0	4	12
South Wold	0	4	12
South Molton	0	4	12
Stratford-on-Avon	0	4	12
Tamworth	0	4	12
Tenterden	0	4	12
Torrington	0	4	12
Wallingford	0	4	12
Wycombe, Chipping	0	4	12

SCHEDULE (C.)

Berwick-upon-Tweed.	Northumberland.	Exeter.	Devonshire.
Bristol.	Gloucestershire.	Kingston-upon-Hull.	Yorkshire.
Chester.	Cheshire.	Newcastle-upon-Tyne.	Northumberland.

APPENDIX, Nº II.;

SHEWING the Boundaries of the old and new Boroughs in England, as adopted and defined by the Act passed in the 2nd and 3rd of William IV., cap. 64, intituled "An Act to settle and describe the Divisions of Counties, and the Limits of Cities and Boroughs, in *England* and *Wales*, in so far as respects the Election of Members to serve in Parliament."

1.—COUNTY OF BEDFORD.

BEDFORD.—The old Borough of Bedford.

2.—COUNTY OF BERKS.

ABINGDON.—The old Borough of Abingdon.

READING.—The old Borough of Reading.

WALLINGFORD.—The old Borough of Wallingford; the several parishes of Brightwell, Sotwell, North Moreton, South Moreton, Bensington, Crowmarsh, and Newnham-Murren; the Liberty of Clapcot, and the Extra-parochial Precinct of the Castle; and also all such parts of the several Parishes of Cholsey, Aston-Tirrold, and Aston-Upthorp as are situate on that Side of the Line next herein after described, on which the Town of Wallingford lies; (that is to say,)

From Blewberry, along the Road called "The Ikeneld Way," to the Point on King's Standing Hill at which the same meets the Boundary of the Parish of Cholsey; thence, Eastward, along the Boundary of the Parish of Cholsey to the Point at which the same reaches the River Thames.

NEW WINDSOR.—The old Borough of New Windsor, the Lower Ward of the Castle, and so much of the Parish of Clewer as is situated to the East of the following Boundary; (that is to say,)

From the Point at which the Goswell Ditch joins the River Thames, along the Goswell Ditch to the Point at which the same meets Clewer Lane; thence, Westward, along Clewer Lane to a Point twenty-five yards distant from the Point last described; thence in a straight line to the North-western Corner of the Enclosure Wall of the Cavalry Barracks; thence along the Western Enclosure Wall of the Cavalry Barracks to the Point at which the same cuts the Boundary of the Parish of New Windsor.

3.—COUNTY OF BUCKINGHAM.

AYLESBURY.—The hundreds of Aylesbury, Ashendon, and Cottesloe, now commonly called the three hundreds of Aylesbury.

BUCKINGHAM.—The several Parishes of Buckingham, Maids-Moreton, Thornborough, Padbury, Hillersden, Preston-Bissett, Tingewick, and Radclive-cum-Chackmore.

GREAT MARLOW.—The several Parishes of Great Marlow, Little Marlow, Medmenham, and Bisham.

CHIPPING-WYCOMBE.—The Parish of Chipping-Wycombe.

4.—COUNTY OF CAMBRIDGE.

CAMBRIDGE.—The old Borough of Cambridge.

5.—COUNTY OF CHESTER.

NORTHERN DIVISION.

MACCLESFIELD.—From the Point at which the Boundary of the Borough of Macclesfield meets the Leek Road near Moss Pool, Southward, along the Leek Road to the Bridge over the Macclesfield Canal; thence, Eastward, along the Macclesfield Canal to the Point at which the same meets the Boundary of the Borough; thence, Eastward, along the Boundary of the Borough to the Point at which the same is again met by the Macclesfield Canal; thence, Northward, along the Macclesfield Canal to the Point at which the same crosses Shore's Clough Brook; thence, Westward, along Shore's Clough Brook to the Point at which the same meets the Boundary of the Township of Hurdsfield; thence, Southward, along the Boundary of the Township of Hurdsfield to the Point at which the same meets the Boundary of the Borough of Macclesfield; thence, Westward, along the Boundary of the Borough of Macclesfield to the Point first described.

STOCKPORT.—The Township of Stockport, and the respective Hamlets of Brinksway and Edgeley, together with those parts of the respective Townships of Brinnington and Heaton-Norris which are included within the following Boundaries respectively; (that is to say,)

BRINNINGTON.—From the Point at which the Boundary of the Township of Stockport would be cut by a straight line to be drawn from the Bridge over the River Mersey on the Bredbury and Hyde Road to the Corn Mill in the Township of Heaton-Norris, between the Manchester and Stockport Canal and the Reddish Road, and now in the occupation of Mr. Walmsley, along such straight line to the Point at which the same cuts the River Tame; thence along the River Tame to the Point at which the same meets the Boundary of the Township of Stockport; thence, Eastward, along the Boundary of the Township of Stockport to the Point first described.

HEATON-NORRIS.—From the Point at which the Boundary of the Township of Heaton-Norris meets the Manchester Road, between a Public-house called the Ash, and Danby Lane, along the Manchester Road to the Point at which the same meets Danby Lane; thence along Danby Lane to the Point at which the same is cut by a straight line drawn thereto from the First Mile Stone on the Altrincham Road through the Western Angle of the Public-house called the Heaton-Norris Club House; thence along the said straight line to the Point at which the same meets the Southern Boundary of the Township of Heaton-Norris; thence, Eastward, along the Boundary of the Township of Heaton-Norris to the Point first described.

SOUTHERN DIVISION.

CHESTER.—The old City of Chester, and also the Space included within the following Boundary; (that is to say,)

From the Second City Boundary Stone in Boughton Ford Mead, and on the Eastern Bank of the River Dee, in a straight line to the Western Extremity of a Lane which leads from Stock Lane to Boughton Heath; thence in a straight line to the Southern Extremity of Heath Lane; thence along Heath Lane to the Point at which the same joins the Christleton Road; thence along the Christleton Road to the Point at which the same is joined by New Lane; thence along New Lane to the Point at which the same meets Filkin Lane; thence along Filkin Lane to the Point at which the same joins, at Asp-Tree Turnpike Gate, the Tarvin Road; thence along the Tarvin Road to Tarvin Bridge; thence along the Nantwich Canal to the Point at which the same meets the old City Boundary; thence, Southward, along the old City Boundary to the Second City Boundary Stone aforesaid.

6.—COUNTY OF CORNWALL.

EASTERN DIVISION.

BODMIN.—The several Parishes of Bodmin, Lanivet, Lanhydrock, and Helland.

LAUNCESTON.—The old Borough of Launceston and the Parish of St. Stephen, and all such Parts of the several Parishes of Lawhitton, St. Thomas the Apostle, and South Petherwin as are without the old Borough of Launceston.

LISKEARD.—The Parish of Liskeard, and also all such Parts of the old Borough of Liskeard as are without the Parish of Liskeard.

WESTERN DIVISION.

HELSTON.—The old Borough of Helston, the Parish of Sithney, and also the Space included within the following Boundary; (that is to say,)

From Coverack Bridge, over the River Loo, in a straight line across the Wendron Road to the Western Extremity of a Lane leading by Wheal Ann to Graham Mine; thence along the said Lane to the Point at which the same meets a small Stream; thence, Southward, along the said Stream to the Point at which the same meets a Lane leading from Wendron to Trecoose and Constantine; thence, Eastward, along the said Lane to Treecoose and Constantine, to the Point at which the same meets the Boundary of the Parish of Wendron; thence, Southward, along the Boundary of the Parish of Wendron to Coverack Bridge.

ST. IVES.—The Old Borough of St. Ives, and the respective parishes of Lelant and Towednack.

PENRYN and FALMOUTH.—From the Point, on the North of Penryn, at which the Boundary of the old Borough leaves the Boundary of the Parish of Mylor, Westward, along the Boundary of the old Borough to the Point at which the same meets the Road from Penryn to Helston; thence, in a straight line to the Point, called Hill Head, at which the Road to Penryn from

Budock joins the Road to Penryn from Constantine; thence in a straight line to the nearest Point of the Boundary of the Parish of Falmouth; thence, Southward, along the Boundary of the Parish of Falmouth to the Point at which the same meets the Boundary of the detached Portion of the Parish of Budock; thence in a straight line to the Northern Point at which the Boundary of the detached Portion of the Parish of Budock leaves the Boundary of the Parish of Falmouth; thence, Westward, along the Sea Coast to the Point at which the same is met by the Boundary of the Parish of Gluvias; thence, Eastward, along the Boundary of the Parish of Gluvias to the Point first described.

Truro.—From Bosvigo Bridge over the Kenwyn River, and on the Boundary of the old Borough, along Bosvigo Lane, to the Point at which the same joins the Redruth Road; thence along the Redruth Road to the Point at which the same is joined, near Chapel-Hill Gate, by Green Lane; thence along Green Lane to the Point at which the same joins the Falmouth Road; thence along an Occupation Road leading through Newham-Farm Land to the Point at which such Occupation Road meets Newham-Farm Lane; thence along a Fence which proceeds from Newham-Farm Lane, and is the South-western Boundary of Two Fields respectively called Great Beef Close and Little Beef Close, to the Point at which such Fence meets the North-western Fence of a Field called Bramble Close; thence, Eastward, along the Fence of Bramble close to the Point at which the same reaches the Shore of Calenick Creek; thence along the Shore of Calenick Creek to Lower Newham Wharf; thence, in a straight line across the Truro and Falmouth River to the South-eastern Extremity of Sunny-Corner Wharf; thence in a straight line to Sunny-Corner; thence in a straight line to the Point at which Trenack Lane would be cut by a straight line to be drawn from the Eastern Extremity of Newham-Farm Lane to the Point called Hill Head, at which St. Clement's Lane meets the St. Austell old Turnpike Road; thence in a straight line to Mitchell-Hill Gate, on the old London Road; thence in a straight line to the Point at which the Boundary of the old Borough would be cut by a straight line to be drawn from Mitchell-Hill Gate to Kenwyn Church; thence, Northward, along the Boundary of the old Borough to Bosvigo Bridge.

7.—COUNTY OF CUMBERLAND.

EASTERN DIVISION.

Carlisle.—The ancient City of Carlisle, and the respective Townships of Botchergate and Rickergate, and also all such Part of the Township of Caldewgate as is comprised within the Boundary hereinafter described; (that is to say,)

From the Bridge over the River Caldew uniting the Township of Caldewgate with the old City of Carlisle, Southward, along the River Caldew to the Point at which the same leaves the Boundary of the Township of Caldewgate; thence, Westward, along the Boundary of the Township of Caldewgate to the Point at which the Road from the Kell Houses to Carlisle joins the Wigton Road; thence in a straight line to the Point at which the Bye-Road from Stainton, over the Summer House Ford in the River Eden, and across the Canal from the Solway to Carlisle, meets the Road from Great and Little Orton to Carlisle at a place called New Town; thence along the said Road from Stainton to the Point at which the same reaches the Summer House Ford; thence along the Boundary of the Township of Caldewgate to the Bridge first described.

WESTERN DIVISION.

Cockermouth.—The several Townships of Cockermouth, Eaglesfield, Brigham, Papcastle, and Bridekirk; and also that detached Portion of the Township of Dovenby which lies between the respective Townships of Papcastle, Bridekirk, and Cockermouth.

Whitehaven.—From the Point on the Sea Coast, North of Whitehaven, at which the Boundary of the Township of Preston Quarter meets the Boundary of the Township of Moresby, Eastward, along the Boundary of the Township of Preston Quarter, to the Point at which the stream which flows through the Village of Hensingham falls into the Poe Beck; thence in a straight line to the Point on the Sea Coast at which the Boundary of the Township of Preston meets the Boundary of the Township of Sandwith; thence along the Sea Coast to the Point first described.

8.—COUNTY OF DERBY.

SOUTHERN DIVISION.

Derby.—The old Borough of Derby.

9.—COUNTY OF DEVON.

NORTHERN DIVISION.

Barnstaple.—From the new Bridge over Braddiford Water, on the New Braunton Road, along the Hedge which is the Eastern Boundary of the East Pillow Marsh Field, to the Point at which the same cuts Poleshill Lane; thence along Poleshill Lane to the Point at which the same meets Hall's Mill Lane; thence along Hall's Mill Lane to the Point at which the same meets the Mill Leat; thence along the Mill Leat to the Point at which the same meets Shearford Lane; thence along Shearford Lane to the Point at which the same joins the Roborough Road; thence along the Roborough Road to the Point at which the same is met by Smoky House Lane; thence along Smoky House Lane to the Point at which the same is cut by a Hedge which divides the Field called "Great Mill Close" from the Field called "Little Mill Close;" thence along the last-mentioned Hedge, and in a Line in continuation of the Direction thereof, to the Point at which such Line cuts the River Yeo; thence, Eastward, along the Boundary of the old Borough of Barnstaple to the Point at which the same meets, in Cooney Cut, the South-eastern Fence of a Field called "Ham;" thence, along the last-mentioned Fence to the Point at which the same cuts Land Key Road; thence in a straight line to the Point on Rumson Hill at which Windy Ash Lane meets the Brindon Cross Road; thence along Windy Ash Lane to the Point at which Wood Street Water crosses the same; thence along Wood Street Water to the Point at which the same joins the River Taw; thence along the River Taw to the Point at which the same is joined by the River Yeo; thence along the River Yeo to the Swing Bridge on the New Braunton Road; thence along the New Braunton Road to the new Bridge first described.

TIVERTON.—The Parish of Tiverton.

SOUTHERN DIVISION.

ASHBURTON.—The Parish of Ashburton.

DARTMOUTH.—From the Point on the Sea Coast at which the Boundary of the Parish of Townstall meets the Boundary of the Parish of Stoke-Fleming, Northward, along the Boundary of the Parish of Townstall, to the Point at which the same meets the Stoke Road; thence along the Stoke Road, passing Swallaton Cross and Swallaton Gate, to the Point at which the Stoke Road meets the Milton Road; thence along the Milton Road to the Point at which the same is met by the Boundary of the Parish of Townstall; thence, Westward, along the Boundary of the Parish of Townstall to the Point at which the same reaches Old Mill Creek; thence along the Low-water Mark to the Point first described.

DEVONPORT.—The parish of Stoke-Damerall, and the Township of Stonehouse.

EXETER.—From the Turnpike Gate on the Morton Road, Southward, along Cowick Lane to the Point at which the same meets Stone Lane; thence along Stone Lane to the Point at which the same meets the Road from Exeter to Alphington; thence, Southward, along the Road from Exeter to Alphington to the Point at which the same is joined by Marsh Barton Lane; thence along Marsh Barton Lane to the Point at which the same reaches the Western Branch of the River Exe; thence in a straight line to the Point at which Abbey Lane meets the Eastern Branch of the River Exe; thence, Southward, along the Leat to the Point at which the same is joined by the Brook which runs down through East Wonford; thence along the said Brook to the Point at which the same crosses the Old Stoke and Tiverton Road near the Road to Mincing Lake Farm; thence along the Old Stoke and Tiverton Road to the Point at which the same meets the Boundary of the County of the City; thence, Northward, along the Boundary of the County of the City to the Point near Foxhays at which a Branch of the River Exe, flowing through Exwick, joins the main Stream thereof; thence in a straight line to the Point at which the Road from Exwick to the Turnpike Gate on the Morton Road is joined by a Road leading from Foxhays to Cleave; thence along the said Road from Exwick to the Turnpike Gate on the Morton Road to the Point at which the same reaches such Turnpike Gate.

HONITON.—The Parish of Honiton.

PLYMOUTH.—From the North-eastern Boundary Stone in a straight line to the nearest Point of the Line of the Embankment; thence, Southward, along the Line of the Embankment to the Point at which the same meets the Boundary of the old Borough; thence, Southward, along the Boundary of the old Borough to the Point first described.

TAVISTOCK.—The Parish of Tavistock, except the Manor of Cudliptown.

TOTNES.—The Parish of Totnes, and the Manor of Bridgetown.

10.—COUNTY OF DORSET.

BRIDPORT.—From the Toll Bar on the Exeter Road in a straight line to the Northern Extremity of the Fence which separates the Field called "Marland Five Acres" from the Field called "Higher Girtups and Dogholes;" thence along the Western Fence of the Field Higher Girtups and Dogholes to the Point at which the same reaches a Lane leading into Mead Lane; thence along the said Lane leading into Mead Lane to the Point at which the same reaches Mead Lane; thence along Mead Lane to the Point at which the same joins the Chard Road; thence, Northward, along the Chard Road to the Point at which the same is joined by the first Lane on the Right, called "Green Lane;" thence in a straight line to Allington Mill; thence in a straight line to the Point at which Coneygere Lane joins the Pymore Road; thence along Coneygere Lane to the Point at which the same joins the Beaminster Road; thence in a straight line to the Bridge over the River Asher close by the Flood Houses; thence along the River Asher to the Point at which the same would be cut by a straight line to be drawn from the Eastern Extremity of Coneygere Lane to the Turnpike Gate on the Dorchester Road; thence along the said straight line to the Turnpike gate on the Dorchester Road; thence, Southward, along the Dorchester Road to the Point at which the same is joined by Bothenhampton Lane; thence along Bothenhampton Lane to the Point at which the same is met by the Stream which forms the Boundary between the respective Parishes of Wallditch and Bothenhampton; thence along the said Stream to the Point at which the same falls into the River Asher; thence down the River Asher (following the Easternmost Branch thereof at the Points at which the same divides into Two Branches) to Squibs Bridge; thence in a straight line to the South-eastern Corner of Keemy Cottage on the Bothenhampton Road; thence in a straight line to the Eastern Extremity of Wonderwell Lane; thence, Westward, along Wonderwell Lane to the Point at which the same joins the Burton-Bradstock Road; thence, Southward, along the Burton-Bradstock Road to Wich Gate; thence in a straight line through the Bombardier's House to the Sea Coast; thence along the Sea Coast to the Eastern Extremity of West Cliff; thence, Northward, along West Cliff, and along the Western Boundary of the Ship Yard of Messieurs Matthews and Company, to the Point at which the same meets the Boundary of the Field called "Pitfield Marsh;" thence, Northward, along the Boundary of Pitfield Marsh to the Point at which the same meets the River Brit at Ire Pool; thence up the River Brit to the Point at which the same is joined by the Stream which forms the Boundary between the respective Parishes of

Symondsbury and Allington; thence along the last-mentioned Stream to the Point at which the same meets the Fence which runs down thereto from the Toll Bar at the Exeter Road; thence along the last-mentioned Fence to the Toll Bar on the Exeter Road.

Dorchester.—From the second or middle Bridge on the Sherborne Road, along the Northern Branch of the River Frome, passing under Grey's Bridge, to the Point at which such Northern Branch is met, near Stanton's Cloth Factory, by the Boundary of the Parish of Fordington; thence, Southward, along the Boundary of the Parish of Fordington to the Point at which the same meets the Wareham Road; thence, Westward, along the Wareham Road to the Turnpike Gate; thence in a straight line to the Centre of the Barrow called "Two Barrows;" thence in a straight Line to the Centre of the Amphitheatre called Maumbury Ring; thence in a straight line to the Centre of the Barrow called Lawrence Barrow, near the Exeter Road; thence in a straight line to the South-western Corner of the Barrack Wall; thence, Northward, along the Barrack Wall and Palisade to the Point at which such Palisade meets the Southern Branch of the River Frome; thence in a straight line to the second or middle Bridge on the Sherborne Road.

Lyme-Regis.—The respective Parishes of Lyme-Regis and Charmouth.

Poole.—The County of the Town of Poole, the Parish of Hamworthy, and the respective Tythings of Parkston and Longfleet.

Shaftesbury.—The old Borough of Shaftesbury; the several Out-Parishes of Holy Trinity, St. James, and St. Peter; the several Parishes of Cann-St. Rumbold, Motcomb, East Stower, Stower-Provost, Todbere, Melbury-Abbas, Compton-Abbas, Donhead-St. Mary, and St. Margarets Marsh; and the Chapelry of Hartgrove.

Wareham.—The old Borough of Wareham; the Parishes of Corfe-Castle and Beer-Regis; the several Out-Parishes of Lady Saint Mary, Holy Trinity, and Saint Martin; and the Chapelry of Arne; that part of the Parish of East Stoke which adjoins the Eastern Boundary of the old Borough of Wareham; and also such part of the Parish of East Morden as is comprised within the following Boundary; (that is to say,)

From the Point at which the Boundary of the Parish of East Morden meets the Southern Boundary of Morden Park Wood, Southward, along the Boundary of Morden Park Wood, to the point at which the same meets the Sherford Lake; thence, Eastward, along the Sherford Lake to the Point at which the same meets the Boundary of the Parish of East Morden; thence, Southward, along the Boundary of the Parish of East Morden to the Point first described.

Weymouth and Melcombe-Regis.—From the old Sluice on the Wareham Road in a straight line to the Point at which the Northern Wall of the old Barrack Field meets the Dorchester Road; thence along the said Northern Wall, and in a Line in the Direction thereof, to the Point at which such Line meets the Boundary of the old Borough; thence, Northward, along the Boundary of the old Borough to the Point at which the same meets the Upper Wyke Road; thence, Westward, along the Upper Wyke Road to the Point at which the same is joined by a Cross Road leading to the Lower Wyke Road, otherwise called Buxton's Lane; thence along the said Cross Road to the Point at which the same joins the said Lower Wyke Road; thence along the said Lower Wyke Road to the Point at which the same joins the Sandsfoot Castle Road; thence, Northward, along the Sandsfoot Castle Road to the Ponit at which the same is met by the Footpath leading by Lovel's Farm to Bincleves; thence along the said Footpath to the Point at which the same reaches the Edge of the Cliff at Bincleves; thence along the Sea Coast to the old Sluice aforesaid.

11.—COUNTY OF DURHAM.

NORTHERN DIVISION.

Durham.—From Shincliffe Bridge over the River Wear, on the Stockton Road, along the Stockton Road, to the Point at which the same is met by a Lane leading into the Darlington Road; thence along the said Lane to the Point at which the same joins the Darlington Road; thence along the Darlington Road to the Point at which the same is met by Potter's Lane; thence along Potter's Lane to the Point at which the same meets Quarry Head Lane; thence along Quarry Head Lane to the Point at which the same meets Margery Lane; thence along Margery Lane to the Point at which the same meets Flass Lane; thence along Flass Lane to the Point at which the same meets a lane leading into the newly-cut Turnpike Road which forms the Commencement of the Newcastle Road; thence along the last-mentioned Lane to the Point at which the same joins the said newly-cut road; thence, Northward, along the said newly-cut Road to the Point at which the same joins the old Line of the Newcastle Road; thence in a straight line through the Northernmost of the Two Outbuildings attached to Kepier's Hospital to the River Wear; thence along the River Wear to the Point at which the same meets Kepier Lane; thence along Kepier Lane, passing under the old Arches of the Hospital, to the Point at which the same Lane is joined, on the South-west of High Grange Farm, by a Lane leading into the Loaning Head Road; thence along the last-mentioned Lane, crossing the Sunderland Road, to the Point at which the same Lane joins the Loaning Head Road; thence along the Loaning Head Road to the Point at which the same is met by a Beck running close to the North of Pellaw Wood and to the South of Gilesgate Church; thence along the said Beck to the Point at which the same falls into the River Wear; thence along the River Wear to Shincliffe Bridge.

Gateshead.—The Parish of Gateshead, and also all such Part of the Chapelry of Heworth in the Parish of Jarrow as is situated to the West of a straight line to be drawn from Kirton Toll Gate House to Blue Quarry Mill, and prolonged each way to the Boundary of the Parish of Gateshead.

South Shields.—The respective Townships of South Shields and Westoe.

Sunderland.—The Parish of Sunderland, and the several Townships of Bishop-Wearmouth, Bishop-Wearmouth-Panns, Monk-Wearmouth, Monk-Wearmouth-Shore, and Southwick.

12.—COUNTY OF ESSEX.

NORTHERN DIVISION.

COLCHESTER.—The old Borough of Colchester.

HARWICH.—The old Borough of Harwich.

SOUTHERN DIVISION.

MALDON.—The old Borough of Maldon, and the Parish of Heybridge.

13.—COUNTY OF GLOUCESTER.

EASTERN DIVISION.

CHELTENHAM.—The Parish of Cheltenham.

CIRENCESTER.—The Parish of Cirencester.

GLOUCESTER.—From the old City Boundary Stone on the Western side of the Lane called Castle Lane, leading from Westgate Street to the County Gaol, Northward along the old City Boundary to the Boundary Stone, South of the London Road, which marks the Easternmost Point of the old City Boundary; thence in a straight line through the Eastern Corner of the Mill upon the River Twiver, between the old City Boundary and the Tram-road from the Gloucester and Berkeley Canal to Cheltenham, to the said Tram-road: thence along the said Tram-road to the Point at which the same is met by Barton Lane; thence along Barton Lane to the Point at which the same crosses the Sud Brook; thence along the Sud Brook to the Point at which the same falls into the Gloucester and Berkeley Canal; thence along the Gloucester and Berkeley Canal to the Point at which the same is met by the old City Boundary; thence, Westward, along the old City Boundary to the Point first described.

STROUD.—The several Parishes of Stroud, Bisley, Painswick, Pitchcomb, Randwick, Stonehouse, Leonard-Stanley, King's-Stanley, Rodborough, Minchinhampton, Woodchester, Avening, and Horsley, except that Part of the Parish of Leonard-Stanley which is called Lorridge's Farm, and is surrounded by the Parish of Berkeley.

TEWKESBURY.—The Parish of Tewkesbury.

14.—COUNTY OF HANTS.

NORTHERN DIVISION.

ANDOVER.—The respective Parishes of Andover and Knights-Enham, and the Tything of Foxcot.

PETERSFIELD.—The old Borough of Petersfield, and the Tything of Sheet; the several Parishes of Buriton, Lyss-Turney, and Froxfield; the several Tythings of Ramsden, Langrish, and Oxenbourn, in the Parish of East-Meon; and also the Parish of Steep, except the respective Tythings of North and South Ambersham.

WINCHESTER.—From St. Winnall's Church in a straight line to the Cottage on the New Alresford Road which is North-west of the White House on St. Giles's Hill; thence in a straight line to the Turnpike Gate at Barr End; thence in a straight line to the Point at which the Gosport Road joins the Southampton Road; thence in a straight line to the Point at which an Angle is made in the Northern Bank of the Lane leading from St. Cross to Compton Down, perpendicularly above the deep Hollow in the said Lane; thence in a straight line to the Cock Lane Turnpike Gate: thence in a straight line to the Three Horse Shoes Public-house on the Week Road; thence in a straight line to the House on the Andover Road which is immediately North-west of the Point at which the Boundary of the City of Winchester crosses the same Road; thence in a straight line to the South-eastern Corner of the Fir Plantation on the Western Side of the Basingstoke Road; thence in a straight line to St. Winnall's Church.

SOUTHERN DIVISION.

CHRISTCHURCH.—The Parish of Christchurch, and the Chapelry of Holdenhurst, except such Part of the Tything of Hurn in the Parish of Christchurch as is situated to the North of the following Boundary; (that is to say,)

From the Point at which the Western Boundary of the Parish of Christchurch crosses the Road from Dudsbury to Hurn Bridge, in a straight line to the South-western Corner of Merritown Common; thence along the Southern Boundary of Merritown Common and of Hurn Common to the Point at which the Southern Boundary of Hurn Common reaches the Moor's River; thence in a straight line to the Southern Boundary Post of the Parish of Christchurch on the Ringwood Road, close by Fillybrook Plantation.

LYMINGTON.—The Parish of Lymington, and also such Part of the Parish of Boldre as is comprised within the following Boundary (that is to say,)

From East-end Bridge, on the Eastern Boundary of the Parish of Boldre, in a straight line through Boldre Church to the Western Bank of Lymington River; thence, Southward, along the Western Bank of Lymington River to the Point at which the same meets the Boundary of the Parish of Boldre; thence, Southward, along the Boundary of the Parish of Boldre to East-end Bridge aforesaid.

PORTSMOUTH.—The old Borough of Portsmouth, and the Parish of Portsea.

SOUTHAMPTON.—The Town and County of the Town of Southampton.

15.—COUNTY OF HEREFORD.

HEREFORD.—The whole Space contained within the Boundary of the Liberties of the City of Hereford, including Castle Green.

LEOMINSTER.—The Parish of Leominster.

16.—COUNTY OF HERTFORD.

ST. ALBANS.—From the Turnpike Gate on the London Road East of St. Albans, called St. Albans Gate, in a straight line to the Point at which the Boundary of the old Borough crosses the River at the bottom of the Cotton Mill Lane; thence, Southward, along the Boundary of the old Borough to the Point at which the Western Boundary of the Parish of St. Albans leaves the River; thence in a straight line through the South-eastern Corner of St. Michael's Churchyard, to the Hempstead Road; thence, Northward, along the Hempstead Road to the Point at which the same meets the Road leading to Gorehambury, formerly the Redbourn Road; thence in a straight line to the Western Extremity of the Tongue of Land in the River just above Kingsbury Fish-pond; thence in a straight line to the Side Bar belonging to Kingsbury Turnpike Gate, by the Side of the new Redbourn Road; thence, Eastward, in a straight line to the Point at which the Boundary of the old Borough meets Luton Lane; thence, Eastward, along the Boundary of the old Borough to the Point at which the same crosses Sweetbriar Lane; thence in a straight line to St. Albans Turnpike Gate aforesaid.

HERTFORD.—From the Corporation Post at the Bottom of Port Hill, along the Bengeo Road to the Point at which the same is cut by the Northern Fence of Port Hill Field; thence along the Northern and Western Fences of Port Hill Field to the Point at which such Western Fence cuts the Mole Wood Hill Road; thence in a straight line through Sele Farm Bridge to the Stevenage Road; thence in a straight line to the Point at which the Hertingfordbury Road is crossed by the Boundary of the Out-Borough of Hertford; thence, Southward, along the Boundary of the Out-Borough of Hertford to the Corporation Post at the Bottom of Port Hill.

17.—COUNTY OF HUNTINGDON.

HUNTINGDON.—The old Borough of Huntingdon, and the Parish of Godmanchester.

18.—COUNTY OF KENT.

EASTERN DIVISION.

CANTERBURY.—From the Westernmost Point, near St. Jacob's, at which the Boundary of the City Liberties meets the Ashford Road, in a straight line to the Point at which the respective Boundaries of the Parishes of Harbledown, St. Dunstan, and Holy Cross Westgate meet; thence, Northward, along the Eastern Boundary of the Parish of Harbledown to the Point at which the same turns North-westward near the Whitstable Road; thence in a straight line, in the Direction of St. Stephen's Church, to the Point at which such straight line cuts the Boundary of the Parish of St. Stephen; thence, Eastward, along the Boundary of the Parish of St. Stephen to the Point at which the same meets the Boundary of the Parish of Holy Cross Westgate; thence, in a straight line, through the Point at which the Road to St. Stephen's Church meets the Road to Sturry, to the nearest Branch of the River Stour; thence along the said Branch of the River Stour to the Corporation Stone, Number 5; thence, Eastward, along the Boundary of the City Liberties, including the whole of the Borough of Longport, to the Point first described.

DOVOR.—From the Jetty, along the Boundary of the Liberties of the Town and Port of Dovor, on the Eastern Side of the Castle, and through the Parish of Charlton, to the Boundary Stone at which the Boundary of the said Liberties meets the Boundary of the Parish of Buckland in Back Lane; thence along Back Lane to the Point at which the same meets the Road leading down to Crabbe Turnpike Gate on the London Road; thence in a straight line, in a Westerly Direction, to the Point at which the Boundary of the Parish of Buckland crosses the London Road; thence along the Boundary of the Parish of Buckland to the Point at which the same crosses the River; thence in a straight line to the Point at which the Boundary of the Parish of Buckland meets the Road leading to Combe Farm; thence along the Boundary of the Parish of Buckland to the Point at which the Boundary of the Parish of Hougham is intersected by the Boundary of the Liberties aforesaid; thence along the Boundary of the said Liberties to the Sea Coast; thence along the Sea Coast to the Jetty.

Hythe.—The old Borough of Hythe; the Liberties of the Town of Folkestone; and the several Parishes of West Hythe, Saltwood, Cheriton, Folkestone, and Newington, except that detached Part of the Parish of Newington, called Marwood Land.

Sandwich.—The several Parishes of St. Mary, St. Peter, and St. Clement; and the Extra-parochial Precinct of St. Bartholomew, Sandwich; the Parish of Deal; and the Parish of Walmer.

WESTERN DIVISION.

Chatham.—From the Easternmost Point at which the Boundary of the City of Rochester meets the Right Bank of the River Medway, Southward, along the Boundary of the City of Rochester to the Boundary Stone of the said City marked 5; thence in a straight line to the Windmill in the Parish of Chatham on the top of Chatham Hill; thence in a straight line to the Oil Windmill in the Parish of Gillingham, between the Village of Gillingham and the Fortifications; thence in a straight line through Gillingham Fort to the Right Bank of the River Medway; thence along the Right Bank of the River Medway to the Point first described.

Greenwich.—From the Point at which the Royal Arsenal Canal at Woolwich joins the River Thames, along the said Canal to the Southern Extremity thereof; thence in a straight line to the South-western Corner of the Ordnance Storekeeper's House; thence in a straight line, in the Direction of a Stile in the Footpath from Woolwich to Plumstead Common, over Sand Hill, to the Boundary of the Parish of Woolwich; thence, Southward, along the Boundary of the Parish of Woolwich to the Point at which the same meets the Boundary of the Parish of Charlton; thence, Westward, along the Boundary of the Parish of Charlton to the Point at which the same turns Southward near the Dovor Road; thence along the Dovor Road to the nearest Point of the Boundary of the Parish of Greenwich; thence, Westward, along the Boundary of the Parish of Greenwich to the Point at which the same turns abruptly to the South, close by the Dovor Road; thence in a straight line, in a Westerly Direction, to the nearest Point of the Boundary of the Parish of Greenwich; thence, Westward, along the Boundary of the Parish of Greenwich to the Point at which the same meets the Boundary of the Parish of St. Paul Deptford; thence, Southward, along the Boundary of the Parish of St. Paul Deptford to the Point at which the same meets the River Thames; thence along the River Thames to the Point first described.

Maidstone.—The old Borough of Maidstone.

Rochester.—The whole Space comprised within the Boundaries of the Liberties of the old City of Rochester, and also such Parts of the respective Parishes of Stroud and Frindsbury as are situated between the Left Bank of the River Medway, and the Boundary hereafter described; (that is to say,)

From the Entrance from the River Medway of the Thames and Medway Canal, along a Footpath which leads up the Hill towards Upnor, to the Point (on the Top of the Hill) at which the same is met by a Road or Path leading towards Frindsbury Church; thence along such Road or Path to the Point at which the same joins Parsonage Lane; thence along Parsonage Lane to the Point at which the same joins the Road from Frindsbury to Hoo; thence in a straight line to the Northernmost Angle of the Boundary of the Parish of Strood; thence, Westward, along the Boundary of the Parish of Strood to the Point at which the same meets the London Road; thence towards Rochester along the London Road to the Point at which the same is joined by the Road from the Three Crouches; thence in a straight line to the Point at which the left Bank of the River Medway would be cut by a straight line to be drawn from the Point last described to Fort Clarence.

19.—COUNTY OF LANCASTER.

NORTHERN DIVISION.

Blackburn.—The Township of Blackburn.

Clitheroe.—The respective Chapelries of Downham and Clitheroe; and the Four Townships of Whalley, Wiswell, Pendleton, and Henshorn, and Little Mitton and Colcoats.

Lancaster.—From the Point on the River Lune at which the respective Boundaries of the Townships of Lancaster, Skerton, and Heaton-with-Oxcliffe meet, Westward, along the Boundary of the Township of Lancaster to the Point at which the respective Boundaries of the Townships of Lancaster, Bulk, and Quernmoor meet; thence in a straight line to the Aqueduct Bridge over the Caton Road; thence, Northward, along the Canal from Preston to Kendal to the Fourth Bridge over the same from the Aqueduct; thence in a straight line to the Point at which Bracken Lane meets Scale Lane; thence along Scale Lane to the Point at which the same reaches the River Lune; thence along the River Lune to the Point first described.

Preston.—The old Borough of Preston, and the Township of Fishwick.

SOUTHERN DIVISION.

Ashton-under-Line.—The whole Space over which the Provisions of an Act passed in the Seventh and Eighth Years of the

Reign of His late Majesty King George the Fourth, and intituled "An Act for lighting, cleansing, watching, and otherwise "improving the Town of Ashton-under-Line, in the county Palatine of Lancaster, and for regulating the Police thereof," at present extend.

BOLTON-LE-MOORS.—The several Townships of Great Bolton, Little Bolton, and Haulgh, except that detached part of the Township of Little Bolton which is situate to the North of the Town of Bolton.

BURY.—From the Point in the Hamlet of Starling at which a Boundary Stone marks the Boundary of the respective Townships of Elton and Ainsworth, along the Lane from Starling to Walshaw Lane, to the Point in the Hamlet of Walshaw Lane at which a Boundary Stone marks the Boundary of the respective Townships of Elton and Tottington Lower End; thence, Eastward, along the Boundary of the Township of Elton to the Point at which the same meets the Woodill Brook; thence in a straight line to the Point at which the Pigs Lea Brook falls into the River Irwell; thence, Eastward, along the Boundary of the Township of Bury to the Point at which the same meets the Boundary of the Township of Elton; thence, Westward, along the Boundary of the Township of Elton to the Point first described.

LIVERPOOL.—From the Western Extremity of Dingle Lane, on the South of the Town, along Dingle Lane, to the Point at which the same meets Ullet Lane; thence along Ullet Lane to the Point at which the same meets Lodge Lane; thence along Lodge Lane to the Point at which the same meets Smithdown Lane; thence along Smithdown Lane to the Point at which the same is met by the Boundary of the Township of Wavertree; thence, Northward, along the Boundary of the Township of Wavertree to that Point thereof which is nearest to the Southern-eastern Corner of the Wall of the new Botanic Gardens; thence in a straight line to the said South-eastern Corner; thence along the Eastern Wall of the new Botanic Gardens to the Point at which such Wall reaches Edge Lane; thence, Eastward, along Edge Lane to a Point Seventy-four Yards distant from the Point last described; thence in a Line Parallel to the new Street called Grove Street to the Point at which such parallel line reaches the London Road; thence along the London Road to the Point at which the same is joined by Deane Street; thence in a straight line to the Boundary Stone in Rake Lane, near the Southern Extremity of Whitefield Lane; thence, Northward, along the Boundary of the township of Everton to the Point at which the same joins the Boundary of the Township of Kirkdale: thence, Northward, along the Boundary of the Township of Kirkdale to the Point at which the same reaches the High-water Mark of the River Mersey; thence along the High Water Mark of the River Mersey to that Point thereof which is nearest to the Point first described; thence in a straight line to the Point first described.

MANCHESTER.—The several Townships of Manchester, Chorlton Row otherwise Chorlton-upon-Medlock, Ardwick, Beswick, Hulme, Cheetham, Bradford, Newton, and Harpurhey.

OLDHAM.—The several Townships of Oldham, Chadderton, Crompton, and Royton.

ROCHDALE.—The Space defined in the 101st Section of an Act passed in the Sixth Year of the reign of His late Majesty King George the Fourth, and intituled "An Act for lighting, cleansing, watching, and regulating the Town of Rochdale, in the "County Palatine of Lancaster."

SALFORD.—From the Northernmost Point at which the Boundary of the Township of Salford meets the Boundary of the Township of Broughton, Northward, along the Boundary of the Township of Broughton, to the Point at which the same meets the Boundary of the Township of Pendleton; thence, Westward, along the Boundary of the Township of Pendleton to the Point at which the same meets the Boundary of the detached Portion of the Township of Pendlebury; thence, Southward, along the Boundary of the detached Portion of the Township of Pendlebury to the Point at which the same meets the Boundary of the Township of Salford; thence, Westward, along the Boundary of the Township of Salford to the Point first described.

WARRINGTON.—The respective Townships of Warrington and Latchford; and also those Two detached Portions of the Township of Thelwall which lie between the Boundary of the Township of Latchford and the River Mersey.

WIGAN.—The Township of Wigan.

20.—COUNTY OF LEICESTER.

SOUTHERN DIVISION.

LEICESTER.—The old Borough of Leicester, and the Space over which the Magistrates of the old Borough of Leicester at present exercise a Jurisdiction concurrently with the Magistrates of the County of Leicester, including the Castle View.

21.—COUNTY OF LINCOLN.

PARTS OF LINDSEY.

LINCOLN.—The old City of Lincoln, the Bail and Close, and a certain Common, belonging to the Freemen of Lincoln, called Canwick Common, together with all Extra-parochial Places, if any, which are surrounded by the old City of Lincoln, the Bail and Close, and the said Common, or any or either of them, or by the Boundaries or Boundary of any or either of them.

GREAT GRIMSBY.—The several Parishes of Great Grimsby, Great Coates, Little Coates, Bradley, Laceby, Waltham, Scartho, Clee, Weelsby, and Cleethorpe.

PARTS OF KESTEVEN AND HOLLAND.

BOSTON.—The old Borough of Boston, the Parish of Skirbeck, and the Hamlet of Skirbeck Quarter, including the Fen Allotment of the Hamlet of Skirbeck Quarter, but not the Fen Allotment of the Parish of Skirbeck.

GRANTHAM.—The Parish of Grantham, (including the several Townships of Spittlegate, Manthorp with Little Gonerby, and Harrowby,) and that Part of the Parish of Somerby which is contained between the Boundary of the Parish of Grantham and High Dyke.

STAMFORD.—The old Borough of Stamford, and such Part of the Parish of Saint Martin Stamford Baron as lies between the Boundary of the old Borough and the following Boundary; (that is to say)

From the Westernmost Point at which the Boundary of the Parish of Saint Martin meets the Boundary of the old Borough, Southward, along the Boundary of the Parish of Saint Martin, to the Northernmost Point at which the same meets the Woothorpe Road; thence in a straight line to the Southern Tower, on the London Road, of the Gateway to Burghley House; thence, Northward, along the Wall of Burghley Park to the Point at which the same meets an Occupation Road called the "New Road," which runs from the Barnack and Pilsgate Road to the River Welland; thence along the said Occupation Road, and in a Line in continuation of the Direction thereof, to the Point at which such Line cuts the Boundary of the old Borough.

22.—COUNTY OF MIDDLESEX.

FINSBURY.—The several Parishes of Saint Luke, Saint George the Martyr, Saint Giles-in-the-Fields, Saint George Bloomsbury, Saint Mary Stoke Newington, and Saint Mary Islington; the several Liberties or Places of Saffron Hill, Hatton-Garden, Ely Rents, Ely Place, the Rolls, Glass House Yard, and the Charter House; Lincolns Inn and Greys Inn; the Parish of Saint James and Saint John Clerkenwell, except that Part thereof which is situated to the North of the Parish of Islington; those Parts of the respective Parishes of Saint Sepulchre and Saint Andrew Holborn, and of Furnivals Inn and Staple Inn respectively, which are situated without the Liberty of the City of London.

LONDON.—The whole Space contained within the exterior Boundaries of the Liberties of the City of London, including the Inner Temple and the Middle Temple.

MARY-LE-BONE.—The several Parishes of Saint Mary-le-bone, Saint Pancras, and Paddington.

TOWER HAMLETS.—The several Divisions of the Liberty of the Tower, and the Tower Division of Ossulstone Hundred.

WESTMINSTER.—The old City and Liberties of Westminster, and the Duchy Liberty.

23.—COUNTY OF MONMOUTH.

MONMOUTH DISTRICT.

MONMOUTH.—The Parish of Monmouth, and all such Parts of the old Borough of Monmouth as lie without the Parish of Monmouth.

NEWPORT.—From the Point, on the South of the Town, at which the Mendle Gief Road is joined by a Husbandry Road leading to Hundred Acres Gout, along the Mendle Gief Road, to the Point at which the same meets the Cardiff Road; thence, Westward, along the Cardiff Road to the Point at which the same meets the Streamlet from Cwrty-bella Well; thence along the said Streamlet to the Pool on the Western Side of Friar's Garden Wall; thence along the Watercourse up from the said Pool to another Pool on the Western Side of Bull Field; thence along the Western Fence of Bull Field to the Point at which the same Fence cuts the Road from Stow to Risca; thence, Westward, along the Road from Stow to Risca to the Point at which the same is cut by the Fence which runs Northward from the East End of the Cottages belonging to John Ricketts; thence along the last-mentioned Fence to the North-western Corner of the Field of which it is the Western Boundary; thence, Eastward, along the Northern Fence of the last-mentioned Field to the Point at which the same is intersected by the Fence of the adjoining Field; thence, Northward, along the last-mentioned Fence to a Well Head; thence along the Stream leading therefrom to the Point at which the same meets the Boundary of the old Borough; thence, Northward, along the Boundary of the old Borough to the Point at which the same meets the River Usk at the Mouth of Cridan Pill; thence along the River Usk to the Point at which the same is joined by a Pill opposite the Castle; thence along the said Pill to the Gout; thence along the Watercourse, in a Direction nearly due East, to the Point at which the same meets the new Road to Caerleon; thence along the new Road to Caerleon to the Point at which the same joins the old Road to Christ Church; thence along the New Reen to the Point at which the same meets Liswerry Pill; thence along Liswerry Pill to the Point at which the same joins the River Usk; thence along the River Usk to the Point at which the same is joined by Hundred Acres Gout; thence along Hundred Acres Gout to the Point at which the same is met by the said Husbandry Road leading thereto from the Mendle Gief Road; thence along the said Husbandry Road to the Point first described.

USK.—From the Bridge on the North of the Town, called "Cwm-cayo Bridge," along the Brook over which the said Bridge is built, to the Point at which the same falls into the River Usk; thence down the River Usk, and along the Boundary of the old Borough, to the Point at which the same cuts the Mill-Stream; thence in a straight line to the Farm House of Little Castle Farm; thence along the Eastern side of the Fence of the Farm-yard of Little Castle Farm to the North-eastern Corner of such Farm-yard; thence in a straight line to the Oak Tree in the Wood Hedge on the summit of Lady Hill; thence in a straight line to the Point at which Cwm-cayo Brook would be cut by a straight line to be drawn from the Tree last described to Cwm-cayo Bridge; thence along Cwm-cayo Brook to Cwm-cayo Bridge.

24.—COUNTY OF NORFOLK.

EASTERN DIVISION.

NORWICH.—The City and County of the City of Norwich, together with all such Extra-parochial Places as are contained within the outer Boundary of the City and County of the City of Norwich.

GREAT YARMOUTH.—The old Borough of Great Yarmouth, and the Parish of Gorleston.

WESTERN DIVISION.

KING'S-LYNN.—The old Borough of King's-Lynn.

THETFORD.—The old Borough of Thetford, under the Act, but the Parliamentary Boundary has since been altered by Government.

25.—COUNTY OF NORTHAMPTON.

NORTHERN DIVISION.

PETERBOROUGH.—The Parish of Saint John the Baptist, Peterborough, together with the Extra-parochial District known by the Name of "The Minster Precincts."

SOUTHERN DIVISION.

NORTHAMPTON.—The old Borough of Northampton.

26.—COUNTY OF NORTHUMBERLAND.

NORTHERN DIVISION.

BERWICK-UPON-TWEED.—The Parish of Berwick, and the respective Townships of Tweedmouth and Spittal.

MORPETH.—The several Townships of Morpeth, Buller's Green, Newminster Abbey, Catchburn with Morpeth Castle, and Stobhill, Hepscot, and Tranwell with High Church, and the Parish of Bedlington.

SOUTHERN DIVISION.

NEWCASTLE-UPON-TYNE.—The Town and County of the Town of Newcastle, and the several Townships of Byker, Heaton, Jesmond, Westgate, and Elswick.

TYNEMOUTH AND NORTH SHIELDS.—The several Townships of Tynemouth, North Shields, Chirton, Preston, and Cullercoats.

27.—COUNTY OF NOTTINGHAM.

NORTHERN DIVISION.

NOTTINGHAM.—The County of the Town of Nottingham.

RETFORD (EAST).—The Borough of East Retford, and the Hundred of Bassetlaw.

SOUTHERN DIVISION.

NEWARK-UPON-TRENT.—The old Borough of Newark.

28.—COUNTY OF OXFORD.

BANBURY.—The Parish of Banbury.

OXFORD.—From the Tree on the East of the City called "Joe Pullen's Tree," in a straight line to the Boundary Stone in the Lane called "Mrs. Knapp's Free Board;" thence along the said Lane to the Western Extremity thereof; thence in a straight line to the Centre of the Island situate at the Junction of the Stream called "Harson's Heat" with the River Cherwell; thence, Westward, along the River Cherwell to the Point at which the same joins the old City Boundary; thence, Westward, along the old City Boundary to the Point at which the River Cherwell divides into Two Streams; thence along the Easternmost of such Two Streams to King's Mill; thence in a straight line to the Easternmost Part of King's Mill; thence in a straight line to "Joe Pullen's Tree."

NEW WOODSTOCK.—The old Borough of New Woodstock; the several Parishes of Bladon, Begbrooke, Shipton-on-Cherwell, Hampton-Gay, Tackley, Wootton, Stonesfield, Combe, and Handborough; the Parish of Kidlington, except the respective Hamlets of Gosford and Water-Eaton; the Hamlet of Old Woodstock, and Blenheim Park.

29.—COUNTY OF SALOP.

NORTHERN DIVISION.

SHREWSBURY.—From the Point at which the River Severn is joined by a Stream or Watercourse which flows by the Dog Kennel, and under Bow Bridge, along the said Stream or Watercourse to the Point at which the same reaches the Road leading from Old Heath into the Chester Road; thence along the said Road from Old Heath to the Point at which the same joins the Chester Road; thence along the Chester Road to the Point at which the same is met by a Watercourse which runs round the Corporation Gardens and Round Hill, and joins the River Severn near the House called "The Flash:" thence along the last-mentioned Watercourse to the Point at which the same reaches the old Baschurch Road; thence along the old Baschurch Road to the Point at which the same is met by a Footpath leading along the Wall of Flash House towards the River Severn; thence along the said Footpath to the Point at which the same meets again the last-mentioned Watercourse; thence along the last-mentioned Watercourse to the Point at which the same joins the River Severn; thence along the River Severn to the Point at which the same is met by the common Boundary of the respective Parishes of Saint Chad and Saint Julian; thence, Eastward, along the Boundary of the Parish of Saint Chad to the Point at which the same reaches a Lane or Road which leads from the Montgomery Road into Lands belonging to Mrs. Cartwright; thence along such Lane or Road to the Point at which the same joins the Montgomery Road; thence in a straight line to the Point at which the Stream from the Conduit Head joins the Radbrook Stream; thence along the Radbrook Stream to the Point at which the same reaches Kingsland Lane; thence along Kingsland Lane to the Point at which the same joins the Bishop's-Castle Road; thence along the Bishop's-Castle Road to the Point at which the same is met by the Boundary of the Parish of Saint Julian; thence, Eastward, along the Boundary of the Parish of Saint Julian to the Point at which the same meets the Boundary of the Parish of Holy Cross; thence, Eastward, along the Boundary of the Parish of Holy Cross to the Point first described.

SOUTHERN DIVISION.

BRIDGENORTH.—The old Borough of Bridgenorth, and the several Parishes of Quatford, Oldbury, Tasley, and Astley Abbot's.

LUDLOW.—From the Point on the South of the Town at which Dirty Brook joins the River Teme, North-eastward, along the Boundary of the Township of Ludford to that Point thereof which is nearest to the South-western Corner of the Piece of Land called Rock Close; thence in a straight line to the South-western Corner; thence along the Western Fence of Rock Close to the Point at which the same cuts the Road to the Sheet; thence towards Ludlow along the Road to the Sheet to the Point at which the same is joined by a Road leading by Gallows Bank into Rock Lane; thence along the last-mentioned Road to the Point at which the same reaches Rock Lane; thence along Rock Lane to the Point at which the same is joined by a Road to the Sandpits Turnpike; thence along the said Road to the Sandpits Turnpike to the Point at which the same is met by the Eastern Fence of the Garden of the Public-house called the Cross Keys; thence in a straight line to the Point at which Fishmoor Brook would be cut by a straight line to be drawn from the Point last described to Stanton Lacy House; thence along the Fishmore Brook to the Point at which the same joins the River Corve; thence up the River Corve to the Point at which the same meets the Fence which separates the Lands occupied by Mr. William Russell from the Lands occupied by Mr. Henry Lloyd; thence along the last-mentioned Fence to the Point at which the same meets the Shrewsbury Road; thence along the Fence which separates the Two Fields respectively called the "Lease Piece" and "Pike Field" to the Point at which such Fence meets the Burway Road; thence, Northward, along the Burway Road to the Point at which the same is met by the Fence which separates the Two Fields respectively called "The Marshes" and "The Ox Pasture;" thence along the last-mentioned Fence to the Point at which the same meets the River Teme; thence in a straight line to the Point at which the Fence which divides the Lands of the Honourable Robert Henry Clive from Lands of the Corporation of Ludlow, in the occupation of Mr. William Smith, meets the Prior Halton Road; thence towards Ludlow along the Prior Halton Road to the Point at which the same is met by the Fence which divides the Lands of the Corporation of Ludlow, occupied by the late Mr. Johnnes and Mr. George Anderson, from Lands of the said Corporation occupied by the late Mr. Anthony Jones and Mr. Robert Meyrick; thence along the last-mentioned Fence to the Point at which the same meets the Brick House Road; thence in a straight line to the Eastern Corner of Whitecliff Coppice; thence, Southward, along the North-Eastern Fence of Whitecliff Coppice to the Point at which the same meets the Boundary of the Township of Ludford; thence, Southward, along the Boundary of the Township of Ludford to the Point first described.

WENLOCK.—The old Borough of Wenlock.

30.—COUNTY OF SOMERSET.

EASTERN DIVISION.

BATH.—The old City of Bath, the respective Parishes of Bathwick and Lyncombe and Widcombe, and also that Part of the Parish of Walcott which lies without the old City of Bath and adjoins the Boundary of the old City of Bath.

BRISTOL.—From the Point on the North-east of the City at which the Eastern Boundary of the Out-Parish of Saint Paul meets the North-western Boundary of the Out-Parish of St. Philip and Jacob, Eastward, along the Boundary of the Parish of St. Philip and Jacob to that Point thereof which is nearest to the Point at which the Wells Road leaves the Bath Road; thence in a straight line to the said Point at which the Wells Road leaves the Bath Road; thence along the Wells Road to the Knowle Turnpike Gate; thence along the Road which leads from the Knowle Turnpike Gate to Bedminster Church to the Point at which the same is crossed by Bedminster Brook; thence along Bedminster Brook to the Point at which the same crosses the Road from Locks Mill to Bedminster; thence along the last-mentioned Road, passing the Southern Extremity of the Village of Bedminster, to the Point at which the same meets the Brook at Marsh Pit; thence along the last-mentioned Brook to the Point at which the same meets the Boundary of the Parish of Clifton; thence, Northward, along the Boundary of the Parish of Clifton to the Boundary Stone marked (C P) and (W P 12), marking the North-eastern Angle of the Boundary of the Parish of Clifton, and situate on Durdham Down, East of the Shirehampton Road; thence in a straight line to the Southernmost Point at which the Boundary of the Tything of Bishop's-Stoke meets Parry's Lane; thence, Eastward, along the Boundary of the Tything of Bishop's-Stoke to the Point at which the same joins the Boundary of the Out-Parish of St. Paul; thence, Northward, along the Boundary of the Out-Parish of St. Paul to the Point first described.

FROME.—From Cottle's Oak Turnpike Gate, along Barton Lane, to the Point at which the same meets Green Lane; thence along Green Lane to the Point at which the same meets the Lane to Hellicar's Grave; thence along the Lane to Hellicar's Grave to the Southern Extremity thereof; thence in a straight line through Plaguy House into Grove Lane; thence in a straight line to the Point at which the Road from Tytherington is met by the Lane to Adderwell at a place called the Mount; thence along the Lane to Adderwell to the Eastern extremity thereof near Bellows Hole; thence in a straight line to the Point at which Frome River would be cut by a straight line to be drawn from the Point last described to the House called "Mrs. White's" or "Southfield Farm House;" thence, Northward, along Frome River to the Point at which the same is joined by Rodden Lake Streamlet; thence along Rodden Lake Streamlet to Rodden Bridge at the end of Rodden Lane; thence along Rodden Lane to the Point called Clink Crossways; thence in a straight line to the Twelfth Mile Stone on the Bath Road; thence in a straight line to the North-eastern Corner of Mr. Shepherd's Garden Wall; thence in a straight line, through the House of Thomas Ball and Mrs. Slade, to Frome River; thence along Frome River to the Northernmost Part of the Buildings of the Dye House, late the property of Samuel Button; thence, in a straight line to the Centre of Kissing Batch Pound; thence in a straight line to Cottle's Oak Turnpike Gate.

WELLS.—From the Point on the North-east of the City at which the old City Boundary meets Back Lane, along Back Lane to the Point at which the same joins the Bath Road; thence in a straight line across the Bath Road to the Northern Extremity of Drang Lane; thence along Drang Lane, and along the Footpath across Drang Meadow, to the Point at which such Footpath joins the Road which leads to the Turnpike on the Shepton-Mallet Road; thence, Westward, along the Road so joined to the next City Boundary Stone; thence, Southward, along the old City Boundary to the Point first described.

WESTERN DIVISION.

BRIDG-WATER.—From the Easternmost Point at which the Boundary of Three Elm Field meets the River Parret, Westward, along the Boundary of Three Elm Field to the Point at which the same meets Reed Moor Pill; thence, Westward, along Reed Moor Pill to the Point at which the same reaches the Southern Boundary of the Two Fields respectively called the Pasture Ground; thence in a straight line to the Point at which the Boundary of the Parish of Wembdon would be cut by a straight line to be drawn from the Point last described to the Spire of Bridg-water Church; thence, Southward, along the Boundary of the Parish of Wembdon to the Point at which the same meets the Cannington Road; thence, Westward, along the Cannington Road to the Point at which the same is met by the Boundary of the Field called Six Acres; thence, Westward, along the Boundary of the Field called Six Acres to the Point at which the same meets, near the Horse and Jockey Inn, the Road from West Street; thence, Westward, along the said Road from West Street to the Point at which the same is met by the Western Boundary of Matthew's Field; thence along the Western Boundary of Matthew's Field to the Point at which the same meets the Town Mill Leat: thence along the Town Mill Leat to the Point at which the same reaches the South-eastern Corner of Matthew's Field; thence in a straight line to the Point at which Hamp Brook meets Hamp Lane; thence along Hamp Lane to the Point at which the same joins West Road; thence along West Road to the Point at which the same is joined by Row's Lane; thence along Row's Lane to the Point at which the same meets the Fence which encloses the Grounds of the House called "Hamp," belonging to John Chapman, Esquire; thence, Southward, along the last mentioned Fence to the Point at which the same meets a Stream at Barland Lane Bridge; thence along the said Stream to the Point at which the same falls into the River Parret at Barland Clize; thence, Westward, along the River Parret to the Point at which the same is joined by the Boundary of the Northernmost of the Two contiguous Fields respectively called Five Acres; thence, Eastward, along the Boundary of the last-mentioned Field to the Point at which the same meets the Boundary of the Field called Four Acres; thence, Northward, along the Boundary of the Field called Four Acres to the Point at which the same meets the Boundary of a Field called Five Acres; thence, Eastward, along the Boundary of the last-mentioned Field called Five Acres to the Point at which the same meets the Weston-Zoyland Road; thence, Eastward, along the Weston-Zoyland Road to the Point at which the same is met by an Occupation Road leading towards the North; thence along the said Occupation Road to the Northern Extremity thereof; thence along the Fence which is the Western Boundary of the Fields respectively called Ten Acres, Seven Acres, and Five Acres, formerly belonging to Alexander Popham, Esquire, to the Point at which such

Fence meets the Fence of a Field called The Hundred Acres; thence in a straight line to the Southern Extremity, close by a Penfold, of the Fence which divides the Two Fields respectively called "Part of the Hundred Acres;" thence, Eastward, along the boundary of the Easternmost of the Two last-mentioned Fields to the Point at which such Boundary meets the Bath Road; thence, Northward, along the Boundary of the Field called Small Croft to the Point at which the same meets the Bristol Road; thence, Westward, along the Boundary of Great Castle Field to the Point at which the same meets the River Parret; thence along the River Parret to the Point first described.

TAUNTON.—From the Point on the North-west of the Town at which Mill Lease Stream crosses Greenway Lane, along Greenway Lane to the Point at which the same joins the Kingston Road; thence along the Kingston Road to the Point at which the same is joined by the Cheddon Road; thence along the Cheddon Road to the Point at which the same is joined by Priors Wood Lane; thence along Priors Wood Lane to the Point at which the same is met by the Obridge Stream; thence along the Obridge Stream to the Point at which the same falls into the River Tone; thence, Southward, along the River Tone to the Point at which the same is met by Mill Lane; thence along Mill Lane to the Point at which the same joins the Bridg-water Road; thence along the Bridg-water Road to the Point at which the same is joined by Bath Pool Lane; thence in a straight line to Stream Plat Bridge; thence along the Stream over which Stream Plat Bridge is built, through Holway Bridge, to the Point at which the same Stream meets the Boundary of the Parish of Wilton at Cuckoo Corner; thence, Westward, along the Boundary of the Parish of Wilton to the Point at which the same meets Sherford Stream; thence along Sherford Stream to the Point at which the same meets Sherford Lane; thence along Sherford Lane to the Point at which the same joins the Honiton Road; thence along the Honiton Road to the Point at which the same is joined by Hoverland Lane; thence along Hoverland Lane to the Point at which the same meets Ganton Stream; thence along Ganton Stream to the Point at which the same meets the Boundary of the Parish of Wilton; thence, Northward, along the Boundary of the Parish of Wilton to the Point at which the same meets the Bishops Hull Road; thence, Northward, along the Bishops Hull Road to the Point at which the same is joined by Long Run Lane; thence in a straight line to the Turnpike House on the Staplegrove Road; thence along the Staplegrove Road to the Point at which the same is crossed by Mill Lease Stream; thence along Mill Lease Stream to the Point first described.

31.—COUNTY OF STAFFORD.

NORTHERN DIVISION.

NEWCASTLE-UNDER-LYME.—The old Borough of Newcastle-under-Lyme, and the Portion of the Parish of Stoke-upon-Trent which is surrounded partly by the Boundary of the old Borough of Newcastle-under-Lyme and partly by the Boundary of the Township of Knutton.

STAFFORD.—From the Point at which the Boundary of the old Borough is cut by a straight line drawn from the Windmill near the Bridge on the Doxey Road to the Stile at the Southern End of the Footpath from the Newport Road into the Penkridge Road, along the said straight line to the Point at which the same meets the Penkridge Road; thence, Southward, along the Penkridge Road to the Point at which a Stream of Water running along the Eastern Side of that Road turns Eastward therefrom; thence along the said Stream to the Point at which the same meets Spittal Brook; thence along Spittal Brook to the Point at which the same meets the River Sow; thence along the River Sow to the Point at which the same meets the Boundary of the old Borough; thence, Northward, along the Boundary of the old Borough to the Point first described.

STOKE-UPON-TRENT.—The several Townships of Penkhull with Boothen, Tunstall, Burslem, Hanley, Shelton, Fenton-Vivian, Lane-End, Fenton-Calvert, and Longton, the Vill of Rushton Grange, and the Hamlet of Sneyd.

SOUTHERN DIVISION.

LICHFIELD.—The County of the City of Lichfield, and the Place called The Close, which is encompassed by the said County.

TAMWORTH.—The Parish of Tamworth.

WALSALL.—The Parish of Walsall, except that detached Part thereof which is surrounded by the respective Parishes of Aldridge and Rushall, and the Chapelry of Pelsall.

WOLVERHAMPTON.—The several Townships of Wolverhampton, Bilston, Willenhall, and Wednesfield, and the Parish of Sedgley.

32.—COUNTY OF SUFFOLK.

EASTERN DIVISION.

IPSWICH.—The old Borough of Ipswich.

WESTERN DIVISION.

BURY-ST.-EDMUNDS.—The old Borough of Bury-St.-Edmunds.

EYE.—The several Parishes of Eye, Hoxne, Denham, Redlingfield, Occold, Thorndon, Braisworth, Yaxley, Thrandeston, Broome, and Oakley.

SUDBURY.—The old Borough of Sudbury, and the Township or Hamlet of Ballingdon cum Brundon; together with all or any Extra-parochial Places or Place surrounded by the Boundaries either of the old Borough of Sudbury or of the Township or Hamlet of Ballingdon cum Brundon.

33.—COUNTY OF SURREY.

EASTERN DIVISION.

LAMBETH.—The Parish of St. Mary Newington, the Parish of St. Giles Camberwell, except the Manor and Hamlet of Dulwich, and also such Part of the Parish of Lambeth as is situate to the North of the Line hereinafter described, including the Extra-parochial Space encompassed by such Part:

From the Point at which the Road from London to Dulwich by Red Post Hill leaves the Road from London over Herne Hill in a straight line to St. Matthew's Church at Brixton; thence in a straight line to a Point in the Boundary between the respective Parishes of Lambeth and Clapham One Hundred and Fifty Yards South of the Middle of the Carriage-way along Acre Lane.

REIGATE.—The Parish of Reigate.

SOUTHWARK.—The old Borough of Southwark, including the Mint and Manor of Suffolk; the several Parishes of Rotherhithe, Bermondsey, and Christ Church; and the Clink Liberty of the Parish of St. Saviour.

WESTERN DIVISION.

GUILDFORD.—From the Point on the North of the Town at which a Creek leading from Dapdune House joins the River Wey, in a straight line to the Point at which the Road called the New Road joins the Stoke Road; thence along the New Road to the Point at which the same joins the Kingston Road; thence along the Kingston Road to the Point at which the same joins Cross Lane; thence along Cross Lane to the Point at which the same joins the Epsom Road; thence in a straight line to the Point in Chalky Lane at which the Boundary of Trinity Parish leaves the same; thence along the Southern Boundary of Trinity Parish to the Point at which such Boundary enters Gaol Lane; thence in a straight line to the Point at which the River Wey turns abruptly to the North at a Wharf close by the Horsham Road; thence in a straight line to the Point at which the Path from Guildford across Bury Fields abuts on the Portsmouth Road; thence in a straight line to the South-western Corner of Cradle Field; thence along the Western Hedge of Cradle Field to the Point at which the same cuts the old Farnham Road; thence in a straight line towards Worplesdon Semaphore to the Point at which such Line cuts the new Farnham Road; thence in a straight line to the Point first described.

34.—COUNTY OF SUSSEX.

EASTERN DIVISION.

BRIGHTHELMSTONE.—The respective Parishes of Brighthelmstone and Hove.

HASTINGS.—The Town and Port of Hastings and its Liberties, including that detached Part of the Parish of St. Leonard which lies near the Town of Winchelsea, and including also the Liberty of the Sluice, but excluding all such other Parts of the old Borough of Hastings as are detached from the main Body thereof.

LEWES.—From the Town Mill on the North-western Side of the Town in a straight line to the Smock Windmill, which is the most southerly of the Two Windmills called "the Kingstone Mills;" thence in a straight line to the Point at which the Boundary of the Parish of Southover crosses the Cockshut Stream; thence along the Cockshut Stream to the Point at which the same joins the River Ouse; thence along the River Ouse to the Point at which the same would be cut by a straight line to be drawn from the Point last described to the Point on the Eastern Cliff, known as the site of an old Windmill; thence in a straight line to the said Point on the Eastern Cliff; thence in a straight line to the Windmill called "Malling Mill;" thence in a straight line to the Point at which the Stream which turns the Paper Mill falls into the River Ouse; thence in a straight line to the Town Mill.

RYE.—The ancient Towns of Rye and Winchelsea, the several Parishes of Rye, Peasmarsh, Iden, Playden, Winchelsea, East Guildford, Icklesham, and Udimore, and also that part of the Parish of Brede which lies between the Parishes of Udimore and Icklesham.

WESTERN DIVISION.

ARUNDEL.—The Parish of Arundel.

CHICHESTER.—From the Eastern Extremity of the Boundary of the old City Liberty at St. James' Post, Northward, along the said Boundary to the Point at which the same meets the old Broill Road; thence in a straight line to the Westernmost Point at which the Boundary of the Parish of St. Peter the Great meets the Boundary of the Parish of St. Bartholomew; thence, Southward, along the Boundary of the Parish of St. Bartholomew to the Point at which the same crosses the new Road to Fishbourn; thence in a straight line to the Turnpike Gate on the Stockbridge Road; thence in a straight line to the Canal Bridge adjoining the Basin; thence in a straight line to the Southern Extremity of Snag Lane; thence in a straight line to the Southern Extremity of Cherry Orchard Lane; thence in a straight line to the Point at which the Rumboldsweek Road meets the Oving Road; thence in a straight line to the Point first described.

HORSHAM.—The Parish of Horsham.

MIDHURST.—The several Parishes of Midhurst, Easebourne, Heyshot, Chithurst, Graffham, Didling, and Cocking; and the Tything of South Ambersham in the Parish of Steep; that part of the Parish of Bignor which is surrounded by the Parish of Easebourne; those Parts of the several Parishes of Woolavington, Bepton, and Woolbeding which adjoin the Parish of Midhurst; that Part of the Parish of Linch which adjoins the said Part of the Parish of Bepton; and also that Part of the Parish of Linch in which Woodmans Green is situate; all such Parts of the respective Parishes of Stedham and Iping as are not situated to the North of the Cross Road which runs from Woodmans Green, between the North End Farm and Hobberts Farm, to Milland Marsh; the Parish of Trotton, except that Part thereof which lies to the North of the Cross Road from Vining Common to Home Hill and Cobed Hall called Lonebeech Lane; and all such Parts of the respective Parishes of Selham and Lodsworth, and of the Tything of North Ambersham, as are not situated to the North of the Brook which runs from Cooks Bridge on the London Road to Lickfold Bridge.

SHOREHAM (NEW).—The Rape of Bramber.

35.—COUNTY OF WARWICK.

NORTHERN DIVISION.

BIRMINGHAM.—The respective Parishes of Birmingham and Edgbaston, and the several Townships of Bordesley, Duddeston and Nechels, and Deritend.

COVENTRY.—The City of Coventry and the Suburbs thereof.

SOUTHERN DIVISION.

WARWICK.—The old Borough of Warwick.

36.—COUNTY OF WESTMORLAND.

KENDAL.—The respective Townships of Kendal and Kirkland, and all such Parts of the Township of Nether Graveship as adjoin the Township of Kendal.

37.—ISLE OF WIGHT.

NEWPORT.—From the Point on the South of the Town at which the Footpath to Shide joins the Niton Road at Trattles Butt, in a straight line to the House in the Parish of Carisbrooke which belongs to Joshua Spickernell, and is now in the occupation of Mrs. Stanborough; thence in a straight line across the Gatcomb Road to the House which belongs to James Barlow Hoy, Esquire, and is now in the occupation of James Dennett; thence in a straight line in the Direction of West Mill to the Point at which such straight line cuts the Lukeley or Carisbrooke Stream; thence, Northward, along the Lukeley or Carisbrooke Stream to the Point at which the same meets the Boundary of the old Borough; thence, Northward, along the Boundary of the old Borough to Pan Bridge; thence in a straight line to the Point at which the Footpath to Shide meets Church Litton Lane; thence along the said Footpath to the Point first described.

38.—COUNTY OF WILTS.

NORTHERN DIVISION.

CALNE.—The Parish of Calne, and also those Parts of the respective Parishes of Blackland and Calstone-Willington which are surrounded by the Parish of Calne, including all such parts, if any, of the old Borough of Calne as are without the Parish of Calne.

CHIPPENHAM.—The several Parishes of Chippenham, Hardenhuish, and Langley-Burrel, and the Extra-parochial Space called Pewisham.

CRICKLADE.—The Hundred of Highworth, Cricklade, and Staple, and the Hundreds of Kingsbridge and Malmesbury.

DEVIZES.—The old Borough of Devizes, including the respective Parishes of St. John the Baptist and the Blessed Virgin Mary, and also so much of the Chapelry of St. James and of the Parish of Rowde as lies between the Boundary of the old Borough and the following Boundary; (that is to say,)

From the Point at which the Boundary of the Parish of St. John the Baptist would be cut by a straight line to be drawn from the Dairy Farm House on the Chippenham Road, called Ox House, to the Round Tower of the new County Bridewell, in a straight line to Ox House; thence in a straight line to a House occupied by Mr. Mayo, called Brow Cottage; thence in a straight line to the Point at which the Towing Path of the Kennet Canal meets Dye House Lane; thence, Eastward, along the Kennet Canal to the Point at which the same turns Northward near London Bridge; thence in a straight line drawn due East to a Point One Hundred Yards distant; thence in a straight line to Mr. Gundry's House on the Salisbury Road; thence in a straight line to a House called Southgate, occupied by Mr. Slade; thence in a straight line to the Southernmost Point at which Gallows Acre Lane is met by the Boundary of the Parish of St. John the Baptist.

MALMESBURY.—The old Borough of Malmesbury, the respective Out Parishes of St. Paul, Malmesbury, and St. Mary, Westport, and the several Parishes of Brokenborough, Charlton, Garsdon, Lea, Great Somerford, Little Somerford, Foxley, and Bremilham.

MARLBOROUGH.—The old Borough of Marlborough and the Parish of Preshute.

SOUTHERN DIVISION.

SALISBURY.—From the South-western Extremity of the Wall of the Poorhouse at Fisherton Anger, in a straight line to a Point in the Wilton Road, which is Three hundred and Thirty Yards distant from the Point at which the Wilton Road joins the Devizes Road; thence in a straight line to a Point in the Devizes Road which is Six hundred and Forty Yards distant from the Point at which the Wilton Road joins the Devizes Road; thence in a straight line to the Point at which the Stratford Road joins the Marlborough Road; thence in a straight line to the Point called Whipping Cross Tree; thence in a straight line to the Point at which the Road from Salisbury to Laverstock joins the Road from Salisbury to Clarendon; thence in a straight line to the Point at which the Eastern Boundary of the City meets the River Avon; thence along the River Avon to the Point at which the same joins the River Nadder; thence along the River Nadder to the Point first described.

WESTBURY.—The Parish of Westbury.

WILTON.—The several Parishes of Wilton, Fugglestone, Stratford-under-the-Castle, Great Durnford, Woodford, South Newton, Wishford, Barford, Burcombe, Netherhampton, West Harnham, and Britford; such Part of the Parish of Fisherton-Anger as will not by the Provisions of this Act be included within the Boundary of the City of Salisbury; and also all such Parts of the several Parishes of Bishopston, Stratford-St.-Anthony, Coombe-Bisset, and Homington, as are situated to the North of a straight line to be drawn from Odstock Church to the Point on Coombe Hill at which a Fence dividing the Down from the cultivated Land meets the old Road from Salisbury to Blandford, and thence through the Centre of the Clump of Trees called Fallstone Middle Nursery to the Western Boundary of the Parish of Bishopston; together with all such Part of the Extra-parochial Place called Grovely Wood as is situate to the East of a straight line to be drawn from the Point at which the Western Boundary of the Parish of Wishford meets the Northern Boundary of Grovely Wood, to the Point at which the Western Boundary of the Parish of Barford meets the Southern Boundary of Grovely Wood.

39.—COUNTY OF WORCESTER.

EASTERN DIVISION.

DROITWICH.—The old Borough of Droitwich; the several Parishes of Dodderhill, Hampton-Lovett, Doverdale, Salwarp, Martin-Hussingtree, Oddingley, Hadsor, Hindlip, Himbleton, and Elmbridge; the Moreway-end Division and the Broughton Division of the Parish of Hanbury; the Extra-parochial Places called Crutch and Westwood Park; together with the Two Parts of the respective Parishes of Claines and Warndon which are surrounded by the respective Parishes of Hindlip and Martin-Hussingtree; and also the Extra-parochial Place called Shell, and the detached Part of the Parish of Inkberrow, which are respectively contained between the Parish of Himbleton and the Broughton Division of the Parish of Hanbury.

DUDLEY.—The Parish of Dudley.

EVESHAM.—The old Borough o Evesham.

WESTERN DIVISION.

BEWDLEY.—The Parish of Ribbesford, and the several Hamlets of Wribbenhall, Hoarstone, Blackstone, Netherton, and Lower Mitton with Lickhill.

KIDDERMINSTER.—From the Point at or near Proud Cross at which the Boundary of the old Borough meets the Broomfield Road, along the Boundary of the old Borough, to the Point at which the Abberley Road meets the Black Brook; thence, Westward, along the Abberley Road to the first Point at which the same is met by a Hedge running due South therefrom; thence along the said Hedge to its Southern Extremity near a Stone Quarry; thence in a straight line to the said Stone Quarry; thence in a straight line to the First Mile Stone on the Bewdley Road; thence, Westward, along the Bewdley Road to the Point at which the same is joined by a Footpath leading to the Stourport Road; thence along the said Footpath to the Point at which the same meets the Boundary of the old Borough; thence, Southward, along the Boundary of the old Borough to the Point at which the same meets the South-eastern Fence of a Wood called "The Copse," situated on the Eastern Bank of the River Stour; thence along the said Fence to the Point at which the same meets Hoo Lane; thence across Hoo Lane, over a Stile called "Gallows Stile," along a Footpath leading from the said Stile to the Lane from Hoo-Brook to Comberton Hill, to the Point at which the last-mentioned Footpath meets the Lane from Hoo-Brook to Comberton Hill; thence, Northward, along the Lane from Hoo Brook to Comberton Hill to the Point at which the same meets the Boundary of the old Borough; thence, Northward, along the Boundary of the old Borough to the Point first described.

WORCESTER.—From the Liberty Post on the Tewkesbury Road, Southward, along the Tewkesbury Road, to the Point beyond the Turnpike at which the same Road is met by Duck Brook; thence along Duck Brook to the Point at which the same crosses the London Road; thence in a straight line to the Western Extremity of the Road which leads out of the London Road to Lark Hill; thence along the said Road to Lark Hill to the Eastern Extremity thereof; thence along a Footpath leading to the New Town Road to the Point at which the same reaches the New Town Road; thence, Westward, along the New Town Road to the Point at which the same is crossed by a Footpath leading from the House of Industry to the Porte Fields Road; thence along the last-mentioned Footpath to the Point at which the same joins the Porte Fields Road; thence along a Footpath which leads from the Porte Fields Road, past Rainbow Villa, into the Astwood Road, to the Point at which such Footpath joins the Astwood Road; thence along a Road which leads from the Astwood Road to the Whey Tavern to the Point at which such Road crosses the Worcester and Birmingham Canal; thence along the Worcester and Birmingham Canal to the Bridge which is nearest to Gregory's Mill; thence along the Road leading from the said Bridge to the Birmingham Road to the Point at which the same is crossed by the Barborne Brook; thence along the Barborne Brook to the Point at which the same falls into the River Severn; thence along the River Severn to the Point at which the same is met by the Boundary of the Parish of St. Clement; thence, Westward, along the Boundary of the Parish of St. Clement to the Point at which the same meets the Boundary of the Township of St. John; thence, Westward, along the Boundary of the Township of St. John to the Point at which the same meets the Hereford Road; thence along the Hereford Road to the Point at which the same is met by Powick Lane, leading to Powick Bridge; thence, Southward, along Powick Lane to the Point at which the same terminates in a Footpath; thence in a straight line to the Point at which Cut Throat Lane is met by a Footpath leading from Broughton Fields to the Malvern Road; thence along the last-mentioned Footpath to the Point at which the same joins the Malvern Road; thence, Northward, along the Malvern Road to the Point at which the same meets the Boundary of the Township of St. John; thence, Eastward, along the Boundary of the Township of St. John to the Point at which the same meets the Boundary of the Parish of St. Clement; thence, Eastward, along the Boundary of the Parish of St. Clement to the Point at which the same meets the River Severn; thence, Southward, along the River Severn to the Point at which the same is met by the old City Boundary; thence, Southward, along the old City Boundary to the Liberty Post aforesaid.

40.—COUNTY OF YORK.

NORTH RIDING.

MALTON.—The respective parishes of St. Leonard and St. Michael New Malton, the Parish of Old Malton, and the Parish of Norton.

NORTHALLERTON.—The respective Townships of Northallerton and Romanby, and the Chapelry of Brompton.

RICHMOND.—The respective Parishes of Richmond and Easby.

SCARBOROUGH.—The Parish of Scarborough, together with the Extra-Parochial Precinct of Scarborough Castle.

THIRSK.—The several Townships of Thirsk, Sowerby, Carlton-Minniott, Sand-Hutton, Bagby, and South Kilvington.

WHITBY.—The several Townships of Whitby, Ruswarp, and Hawsker-cum-Stainsacre.

YORK.—From the ancient Barn on the Easingwould Road, Two Hundred Yards beyond the First Mile Stone on that Road, in a straight line to the Lady or Clifton Mill; thence in a straight line to the Pepper or Stray Mill; thence in a straight line to the Point at which the Stockton Road would be cut by a straight line to be drawn thereto from the Pepper or Stray Mill through the New Manor House; thence along the Stockton Road to the Point at which the same is joined by a Lane leading from the Eastern Extremity of the Village of Heworth towards the North; thence in a straight line to the Point at which the Tang Hall Beck would be cut by a straight line to be drawn from the Point last described to Heslington Mill; thence along Tang Hall Beck to the Point at which the same crosses the Boundary of the County of the City of York; thence, Southward, along the Boundary of the County of the City of York to the Point at which the same would be cut by a straight line to be drawn thereto from the South-eastern Corner of the Barracks through Lamel Mill; thence in a straight line to the South-eastern Corner of the Barracks; thence along the Southern Wall of the Barracks to the Point at which the same cuts the Selby Road;

thence along the Selby Road to the Point at which the same is joined by Fulford Church Lane; thence along the Northern Hedge of Fulford Church Lane to the Point at which the same ceases to be continuous, close by a Farm Building belonging to Mr. Ellis; thence in a straight line, in the Direction of the said Hedge, to the River Ouse; thence along the River Ouse to the Southernmost Point at which the same is met by the Boundary of the City Liberty; thence, Westward, along the Boundary of the City Liberty to the Point at which the same again meets the River Ouse; thence along the River Ouse to the Point at which the same would be cut by a straight line to be drawn from the Barn first described to Acomb Church; thence in a straight line to the Barn first described.

EAST RIDING.

BEVERLEY.—The several Parishes of St. Mary, St. Martin, and St. Nicholas, and also such Part of the Parish of St. John as is comprised within the Liberties of Beverley.

KINGSTON-UPON-HULL.—The several Parishes of St. Mary, the Holy Trinity, Sculcoates, and Drypool; together with the Extra-parochial Space called Garrison-side, and all other Extra-parochial Places, if any, which are surrounded by the Boundaries of the said Parishes of St. Mary, the Holy Trinity, Sculcoates, and Drypool; or any or either of them; and also all such Part of the Parish of Sutton as is situated to the South of a straight line to be drawn from Sculcoates Church to the Point at which the Sutton Drain meets the Summergangs Drain.

WEST RIDING.

BRADFORD.—The several Parishes of Bradford and Manningham and Bowling, and the Township of Horton, including the Hamlets of Great and Little Horton.

HALIFAX.—From the Point on the North of the Town at which the respective Boundaries of the several Townships of Halifax, North Owram, and Ovenden meet, Westward, along the Boundary of the Township of Halifax, to the Point at which the same meets the Road leading from a House called Shay to Bank Top; thence along the said Road from Shay to Bank Top to the Point at which the same meets the Road leading from South Owram to North Owram; thence along the said Road from South Owram to North Owram to God Lane Bridge; thence in a straight line to the South-Eastern Corner of New Town on the Bradford Road; thence in a straight line to the Point first described.

HUDDERSFIELD.—The Township of Huddersfield.

KNARESBOROUGH.—The Boundary described in the Second Section of an Act passed in the Fourth Year of the Reign of His late Majesty King George the Fourth, and intituled "*An Act for paving, lighting, watching, cleansing, and improving the Town " of Knaresborough, in the West Riding of the county of York,* and that Part of the Township of Scriven-with-Tentergate " which adjoins the said Town and is called Tentergate."

LEEDS.—The Parish of Leeds.

PONTEFRACT.—The old Borough and Township of Pontefract, and the Extra-parochial Space called the Pontefract Park District: and also the several Townships of Tanshelf, Monkhill, Nottingley, Ferrybridge, and Carleton.

RIPON.—The Township of Ripon; and also such Part of the Township of Aismunderby-cum-Bondgate as is situate to the North of the Point on the South of the Town of Ripon at which the Ripley Road meets the Littlethorpe Road, and which is the Southern Extremity of the nearly disjointed Portion of the Township of Aismunderby-cum-Bondgate.

SHEFFIELD.—The Parish of Sheffield.

WAKEFIELD.—From the Southernmost Point at which the Boundary of the Township of Wakefield leaves the River Calder, along the Boundary of the Township of Wakefield, to the Point at which the same is intersected by a Hedge running nearly North close by the Western Side of Park Gate Farm; thence in a straight line to the Point at which the Footpath leading to St. Swithin's Well joins the Footpath from East Moor to Old Park; thence in a straight line to the Point at which the Stanley Road would be cut by a straight line to be drawn from the Point last described to the Cupola of the Lunatic Asylum; thence along the Stanley Road to the Point at which the same is met by the East Moor Road; thence along the East Moor Road to the Point at which the same meets the Boundary of the Township of Wakefield; thence, Westward, along the Boundary of the Township of Wakefield to the Point at which the same meets the Boundary of the detached Portion of the Township of Alverthorp which lies North of the Township of Wakefield; thence, Westward, along the Boundary of the said detached Portion of the Township of Alverthorp to the Point at which the same joins again the Boundary of the Township of Wakefield; thence, Southward, along the Boundary of the Township of Wakefield to the Point at which the same meets Balne Lane; thence along Balne Lane to the Point at which the same is met by Humble Jumble Lane; thence along Humble Jumble Lane to the Point at which the same meets the Footpath to Flanshaw Lane; thence along the Footpath to Flanshaw Lane to the Point at which the same meets Smithson's Rail-road; thence along Smithson's Rail-road to the Point at which the same meets the Dewsbury Road; thence along the Dewsbury Road to the Point at which the same meets the New or Occupation Road which unites the Dewsbury and Horbury Roads; thence along the said New Road to the Point at which the same meets the Park Wall of Thornes House; thence, Northward, along the said Wall to the Point at which the same meets the Road from Thornes to Horbury; thence along the Road from Thornes to Horbury to the Point at which the same meets the Stream called "The Gilsike;" thence along the said Stream to the Point at which the same falls into the River Calder; thence along the River Calder to the Point first described.

APPENDIX, N°. III.

As the following Report was not laid before the House of Commons, until the greater portion of the present edition was in print, an Abstract of it is here inserted.

ENGLAND.

A Return of all Agreements for the Commutation of Tithes, which have been Confirmed by the Tithe Commissioners, from the 1st day of January, 1839, to the 1st day of July, 1839;—(being in continuation of the Return presented on the 5th day of February, 1839, No. 68, ordered to be printed 28th February, 1839.)

COUNTY, PARISH, AND CHAPELRY.	Average Amount of Rates during seven years preceding Christmas, 1835.	RENT-CHARGE.	GLEBE. Acreage.	GLEBE. Value.	TITHE-OWNER.
	£. s. d.	£. s. d.	A. R. P.	£. s. d.	
BEDFORD:					
Bletsoe	88 17 1	333 18 9	34 0 0	59 10 0	rector.
Eyeworth	....	275 0 0	....	....	impropriator.
—	5 0 0	115 0 0	1 2 0	1 10 0	vicar.
Hatley, Cockayne	not stated	191 13 4	14 2 4	14 0 0	rector.
Knotting	77 4 0	319 18 0	0 3 16	2 0 0	rector.
BERKS:					
Aldermaston	116 7 6	535 0 0	....	....	impropriators.
Avington	15 10 0	299 10 0	6 0 0	9 0 0	rector.
Blewberry	135 19 0	1100 0 0	217 2 1	200 0 0	prebendary of Blewberry.
—	32 14 0	232 13 0	1 3 1	....	vicar.
Brimpton	29 10 5	320 0 0	15 0 16	30 0 0	vicar.
Buscot	60 16 3	535 0 0	64 0 28	103 0 0	rector.
Frilsham	29 10 0	188 0 0	29 2 0	22 2 6	rector.
Hampstead-Norris	217 12 0	916 13 6	152 2 31	190 18 0	impropriator.
—	66 19 0	313 14 3	135 0 11	216 0 0	vicar.
Hatford	28 3 3	259 0 0	53 0 0	79 10 0	rector.
Ilsley, East	126 9 5	700 0 0	63 1 22	78 15 0	rector.
Padworth	45 12 5	250 0 0	28 0 0	25 0 0	rector.
Pangbourne	173 4 7	609 0 0	3 0 0	6 0 0	rector.
Reading, St. Giles	89 18 6	512 0 0	....	....	vicar.
Tidmarsh	18 0 0	220 0 0	28 3 3	not stated	rector.
Waltham, Bright	141 5 2	700 0 0	86 1 0	135 0 0	rector.
Waltham, St. Lawrence	....	*350 0 0	....	....	impropriator. *Of this sum £40 is payable to the vicar, value of lands allotted to him.
BUCKS:					
Aston-Sandford	10 10 2½	107 0 0	53 1 35	51 12 0	rector.
Astwood	55 0 0	230 4 0	1 3 0	3 10 0	vicar.
Burnham (exclusive of Lower Boveny).	....	750 0 0	....	....	impropriator.
—	78 4 4	634 16 6	24 1 3	48 0 0	vicar.
Fingest	35 0 0	186 0 0	26 0 13	20 16 0	rector.
Hampden, Great	59 9 10	295 0 0	37 0 0	35 0 0	rector.
Horsendon	18 18 2	148 13 0	20 0 0	35 0 0	rector.
Leckhampstead	....	500 0 0	79 2 0	111 16 0	rector.
Marsh-Gibbon	152 7 0	500 0 0	127 2 20	114 6 0	rector.
Medmenham	....	10 0 0	....	....	impropriator.
—	20 0 0	200 0 0	3 0 0	5 0 0	vicar.
Shabbington	79 7 9	380 0 0	97 0 37	170 0 0	vicar.
Stratford, Water	36 10 11	300 0 0	38 0 26	56 0 0	rector.

A RETURN OF ALL AGREEMENTS FOR THE COMMUTATION OF TITHES.

COUNTY, PARISH, AND CHAPELRY.	Average Amount of Rates during seven years preceding Christmas, 1835.	RENT-CHARGE.	GLEBE. Acreage.	GLEBE. Value.	TITHE-OWNER.
BUCKS—*continued.*	£. s. d.	£. s. d.	A. R. P.	£. s. d.	
Walton	27 0 0	195 0 0	48 1 5	96 10 0	rector.
Wyrardisbury, otherwise Wraysbury.	10 16 0	377 0 0	49 3 35	90 0 0	- - dean and canons of St. George's Chapel, Windsor.
—	2 14 0	154 0 0	18 2 36	35 0 0	vicar.
CAMBRIDGE:					
Chesterton	90 0 0	500 0 0	90 2 0	99 11 0	Trinity College, Cambridge.
—	26 2 10	180 0 0	27 2 1	30 5 0	vicar.
Cheveley	90 0 0	695 0 0	37 0 0	59 4 0	rector.
Chippenham	56 11 0	325 0 0	18 0 0	18 0 0	vicar.
Croydon-cum-Clapton	86 13 0	528 0 0	10 0 0	14 0 0	rector.
Doddington	670 0 0	9956 4 4	59 1 7	110 18 0	rector.
Littleport	35 0 0	247 19 0	81 0 0	121 10 0	Clare Hall, Cambridge.
—	300 0 0	1931 0 0	76 0 0	95 0 0	vicar.
Melbourne	- not stated -	860 0 0	186 0 34	138 10 0	dean and chapter of Ely.
—	....	220 0 0	43 2 11	32 5 0	vicar.
Morden, Steeple	86 10 0	704 10 0	201 2 0	220 0 0	New College, Oxford.
—	21 10 0	235 0 0	21 2 0	30 0 0	vicar.
Papworth St. Agnes	49 15 8	308 0 0	62 3 7	45 0 0	rector.
CHESTER:					
Bromborrow	....	249 2 0	5 3 6	11 11 6	impropriator.
Budworth, Little	....	163 14 0	5 2 16	14 14 0	bishop of Chester.
Churton-Heath	....	17 0 0	....	....	vicar of Saint Oswald.
Dodleston	- not stated -	625 0 0	32 2 0	48 15 0	rector.
Lawton, Church	- not stated -	260 0 0	37 0 0	74 0 0	rector.
Shocklach	....	240 0 0	....	....	impropriators.
—	4 5 6	100 0 0	1 3 0	4 10 0	perpetual curate.
Swettenham	....	273 10 0	16 1 8	38 0 0	rector.
CORNWALL:					
Anthony, St., in Roseland	- not stated -	118 0 0	....	....	impropriator.
Cardinham	10 0 0	450 0 0	197 0 0	100 0 0	rector.
—	....	50 0 0	....	....	
Ewe, St.	....	10 0 0	....	....	impropriator.
—	....	640 0 0	88 1 4	90 0 0	rector.
Genny's, St.	18 5 0	160 0 0	20 0 0	....	vicar.
—	33 10 0	220 0 0	....	....	impropriator.
Ludgvan	....	800 0 0	38 2 1	not stated	rector.
Luxulian	....	225 0 0	....	....	impropriator.
—	....	230 0 0	10 3 0	not stated	vicar.
Mawnan	....	304 10 0	38 1 19	42 0 0	rector.
CUMBERLAND:					
Kirklington	9 8 10	52 9 3¾	15 12 11	7 0 0	rector.
DERBY:					
Barlborough	....	583 0 0	73 0 36	95 0 0	rector.
Blackwell	....	170 0 0	....	....	impropriator.
—	21 12 0	101 0 0	1 2 3	1 4 4	vicar.
Brampton	34 4 0	410 0 0	13 2 7	16 17 6	dean of Lincoln.
—	....	90 0 0	12 1 0	15 12 0	perpetual curate.
Clown	....	330 0 0	67 0 0	72 0 0	rector.
Darley with Little Rowsley	....	253 0 0	126 1 36	190 0 0	rector.
Duffield	....	458 0 0	120 0 16	240 0 0	impropriator.
—	....	10 0 0	12 0 0	24 0 0	vicar.
Staveley	....	605 9 7	....	....	impropriator.
—	....	605 9 7	92 1 31	147 0 0	rector.
DEVON:					
Ashton	24 10 0	256 0 0	55 0 0	27 0 0	rector.
Atherington	....	400 0 0	200 0 0	not stated	rector.

A RETURN OF ALL AGREEMENTS FOR THE COMMUTATION OF TITHES.

COUNTY, PARISH, AND CHAPELRY.	Average Amount of Rates during seven years preceding Christmas, 1835.	RENT-CHARGE.	GLEBE. Acreage.	GLEBE. Value.	TITHE-OWNER.
	£. s. d.	£. s. d.	A. R. P.	£. s. d.	
DEVON—*continued.*					
Bittadon	...	73 0 0	23 0 0	27 0 0	rector.
Buckland, Egg	...	200 0 0	...	...	impropriator.
—	22 0 0	506 0 0	32 0 0	...	vicar.
Challacombe	1 4 0	186 0 0	59 0 0	30 0 0	rector.
Churchstow	...	325 0 0	14 0 0	25 0 0	- - corporation of Exeter as charity trustees.
Countisbury	- not stated -	105 0 0	14 0 3	20 0 0	archdeacon of Barnstaple.
Crediton	...	1770 0 0	...	...	- - twelve governors of the hereditaments and goods of the church of Crediton, otherwise Kryton.
Denbury	...	180 0 0	16 0 0	45 0 0	rector.
Dowland	13 16 10	125 0 0	...	...	impropriator.
Feniton	3 0 0	280 0 0	70 0 0	140 0 0	rector.
Harpford	19 16 6	130 0 0	...	...	impropriator.
—	11 0 0	146 0 0	8 0 0	16 0 0	vicar.
Ideford	17 13 4	255 0 0	59 0 0	100 0 0	rector.
Kerswell, Abbots	...	110 0 0	...	...	vicar of Cornworthy.
—	16 10 0	200 0 0	48 2 10	55 0 0	vicar.
Leigh, North	...	169 10 0	42 0 0	75 0 0	rector.
Lympston	...	263 0 0	11 3 0	28 0 0	rector.
Meeth	4 0 0	221 10 0	25 0 0	25 0 0	rector.
Sandford	...	1150 0 0	...	...	- - twelve governors of the hereditaments and goods of the church of Crediton, otherwise Kryton.
Satterleigh	2 0 0	67 0 0	28 0 0	35 0 0	rector.
Sheldon	...	140 0 0	2 0 0	not stated -	perpetual curate.
Tamerton-Foliott	...	134 1 0	...	...	impropriators.
—	...	338 10 0	5 0 0	5 0 0	vicar.
Tavy, St. Peter	9 0 0	235 0 0	64 0 11	64 0 0	rector.
Teignton, Drews	...	614 17 0	...	...	impropriator.
Ven-Ottery	6 0 0	125 0 0	16 0 0	25 0 0	rector.
Virginstow	11 10 0	115 0 0	40 0 0	35 0 0	rector.
Werrington	16 10 0	290 0 0	...	...	impropriators.
Woodland	3 10 0	200 0 0	...	...	- - dean and canons of St. George's Chapel, Windsor.
DORSET:					
Allington	...	190 0 0	3 1 0	7 0 0	impropriator.
Batcombe	6 8 0	130 0 0	45 0 0	50 0 0	rector.
Bettiscombe	8 5 0	128 0 0	54 1 32	75 0 0	rector.
Broadwinsor	...	530 0 0	85 0 0	106 0 0	- - vicars choral of the cathedral church of Sarum.
—	...	2 10 0	...	...	impropriator.
—	...	750 0 0	9 0 0	17 0 0	vicar.
Cattistock	...	500 0 0	28 0 38	56 0 0	rector.
Cerne, Up	3 2 3	165 0 0	15 0 0	20 0 0	rector.
Chelborough, East	...	160 0 0	110 0 0	130 0 0	rector.
Chelborough, West	...	82 0 0	29 1 5	40 0 0	rector.
Chickerill	23 10 0	280 0 0	45 2 8	80 0 0	rector.
Compton-Vallence	53 15 0	235 0 0	107 1 12	116 0 0	rector.
Fifehead-Magdalen	15 0 0	245 0 0	24 2 0	54 5 0	vicar.
Haydon	...	120 0 0	33 3 8	29 0 0	vicar.
Hook	...	41 0 0	36 0 0	36 0 0	rector.
Langton, Long Blandford	25 0 0	352 0 0	78 3 8	62 0 0	rector.
Litton-Cheney	...	650 0 0	116 3 18	150 0 0	rector.
Melbury-Bubb	7 0 0	170 0 0	48 0 25	75 0 0	rector.
Monckton, Tarrant	- not stated -	500 0 0	55 3 37	...	impropriator.
Okeford, Child	...	250 0 0	70 3 32	140 0 0	rector.
Poorstock	20 0 0	303 0 0	70 0 0	100 0 0	dean and chapter of Salisbury.
—	...	230 0 0	...	...	- - vicar of the parish, with West Milton annexed.
Stalbridge	175 18 10½	1200 0 0	53 0 0	100 0 0	rector.
Steepleton-Preston	...	95 0 0	...	...	rector.
Stockwood	6 18 10	125 0 0	42 2 22	55 0 0	rector.
Sturminster-Marshall	22 10 7	469 0 0	7 3 4	10 5 0	Eton College.
—	6 3 8	120 0 0	122 3 9	180 0 0	vicar.

A RETURN OF ALL AGREEMENTS FOR THE COMMUTATION OF TITHES.

COUNTY, PARISH, AND CHAPELRY.	Average Amount of Rates during seven years preceding Christmas, 1835.	RENT-CHARGE.	GLEBE.		TITHE-OWNER.
			Acreage.	Value.	
DORSET—*continued.*	£. s. d.	£. s. d.	A. R. P.	£. s. d.	
Sturminster-Newton-Castle, with the Tithing of Bagber	- not stated -	185 0 0			impropriators.
		775 0 0	81 0 0	120 0 0	vicar.
Swanwich	69 11 8	400 0 0	19 2 13	16 10 0	rector.
Winford-Eagle		6 10 0			impropriator.
—		150 0 0			vicar.
Winterbourne, Clenstone	17 7 6	199 10 0	2 2 21	5 0 0	rector.
Winterbourne, Whitchurch	5 7 6	95 0 0			vicar.
Winterbourne, Zelstone		230 0 0	32 1 33	67 0 0	rector.
Wyke-Regis		550 0 0	26 3 16	100 0 0	rector.
DURHAM:					
Bishopton		110 10 0	55 0 0	55 0 0	vicar of Bishopton.
—		528 1 0			- - master and brethren of Christ's Hospital, in Sherburn.
Craike	48 1 1	678 0 0	52 0 0	52 0 0	rector.
Shincliffe		191 18 9			- - prebendary of the 5th canonry of Durham.
—		4 9 7			perpetual curate.
Stainton, Great		200 0 0	55 0 0	68 15 0	rector of Stainton.
Stranton		220 0 0	house&garden	20 0 0	vicar.
—	3 6 0	103 6 0			impropriator.
ESSEX:					
Aveley	59 0 0	453 15 0	30 0 0	30 0 0	- - dean and chapter of St. Paul's Cathedral.
—	40 0 0	327 0 0	1 0 0	1 0 0	vicar.
Baddow, Little	60 0 0	358 0 0	40 0 0	40 0 0	appropriate rector.
—	35 0 0	197 5 6	5 0 0	2 10 0	vicar.
—	- not stated -	126 17 0			impropriators.
Bardfield, Little	107 16 0	465 0 0	63 1 20	82 10 0	rector.
Birdbrook	154 0 0	600 0 0	99 0 1	148 10 0	rector.
Borley	65 1 2	276 10 0	10 2 13	16 0 0	rector.
Brightlingsea	115 17 6	240 0 0			impropriator.
—	11 0 0	150 0 0			vicar.
Bromley, Little	81 9 3	560 0 0	11 0 0	16 10 0	rector.
Broxted	153 14 10	660 0 0			impropriator.
—		200 0 0			vicar.
Burstead, Great	62 11 6	268 10 0			impropriator.
—	41 14 6	177 10 0			vicar.
Buttsbury	71 16 0	323 6 8			impropriator.
—			8 0 36	10 0 0	perpetual curate.
—		6 13 4			priest and paupers of Gyng-Petre.
Chadwell	48 8 0	480 0 0	40 0 0	40 0 0	rector.
Childerditch	27 19 4	172 0 0	17 3 10	22 0 0	vicar.
Colne, Earl's		242 14 9			impropriators.
—	126 1 6	670 0 0	1 0 35	1 18 6	vicar.
Corringham	45 14 8	830 0 0	29 0 0	22 0 0	rector.
Danbury	101 13 1	569 0 0	22 2 0	22 10 0	rector.
Denzie	82 16 0	732 0 0	13 0 0	24 0 0	rector.
Easton, Great	232 5 11	740 0 0	84 0 0	75 12 0	rector.
Foxearth	18 16 0	68 16 0			- - trustees of the Guildhall Feoffment Charity, in Bury Saint Edmund's.
—	83 17 0	435 0 0	23 0 0	40 0 0	rector.
Goldhanger	92 4 8	580 0 0	27 0 0		rector.
Greenstead	55 0 0	292 0 0	3 0 0	5 0 0	rector.
Henny, Little	15 5 5	88 0 0	11 0 0	11 0 0	rector.
Holland, Little	18 13 0	164 0 0			impropriator.
—	9 7 0	57 10 0	1 1 17	2 0 0	vicar.
Ilford, Little	8 4 3	310 0 0	36 0 13	100 0 0	rector.
Ingatestone	73 7 3	560 0 0	1 0 0		rector.
Ingrave	62 6 10	290 0 0	70 0 0	70 0 0	rector.
Latchingdon-cum-Lawling	130 0 0	900 0 0	44 0 0	45 0 0	rector.
Lawford	137 2 2	720 0 0	35 0 0	40 0 0	rector.
Liston	54 10 11	200 0 0	18 0 0	36 0 0	rector.

LONDON :
GILBERT AND RIVINGTON, PRINTERS,
ST. JOHN'S SQUARE.

5R